DIRECTORY OF RESEARCH GRANTS 2010

33rd Edition
Volume 1

WITHDRAWN

Schoolhouse Partners LLC
Nashville, Indiana

Why 'Partners'? Partners work together as a team, with each person contributing different skills and expressing his or her individual interests and opinions to the unity and efficiency of the group in order to achieve common goals. Our aim is to make worthwhile resources available to you so that you and your institution or organization can and will achieve those goals.

Copyright © 2010 by Schoolhouse Partners LLC
Schoolhouse Partners, Post Office Box 2059, Nashville, IN 47448
www.schoolhousepartners.net
Published 1975, 33rd edition 2010

Printed and bound in the United States of America

ISBN 978-0-9841725-2-8
0-9841725-2-1

Table of Contents

Introduction

For over three decades the GRANTS Database and its print complements, including the annual Directory of Research Grants, have provided the research community with current, accurate information regarding funding for research-related programs and projects, scholarships, fellowships, conferences, and internships. To meet the increased information needs of professionals, the GRANTS Database, researched and edited by members of the Schoolhouse Partners editorial staff, has grown significantly. This 33rd edition of the annual Directory of Research Grants offers factual and concise descriptions of more than 5,800 research funding programs. With this new edition both experienced and novice grantseekers will be better able to target only the most applicable programs for their research needs.

Directory of Research Grants 2010

The *Directory of Research Grants 2010* features listings of programs that offer non-repayable research funding for projects in medicine, the physical and social sciences, the arts and humanities, education, community and economic development, and children and youth programs. Listings in the main section of the *Directory* contain annotations describing each program's focuses and goals, program requirements explaining eligibility, funding amounts, deadlines, *Catalog of Federal Domestic Assistance* program number (for U.S. government programs only), sponsor name and address, contact information, and the sponsor's Internet Web address. Grantseekers with access to the Internet can use the addresses to locate further information about the organizations and their application procedures. Internet addresses are also provided, when available, within listings. Some of the grant programs listed in the main section have geographic restrictions for applicants.

Indexes. The Subject Index of this *Directory* lists all program titles—with accession numbers—under their applicable subject terms. Other indexes follow, including the Sponsoring Organizations Index, which lists program sponsors alphabetically along with their programs; the Grants by Program Type Index, which lists 38 program categories, such as Basic Research, Fellowships, Travel Grants, etc., with the grants that fall within their scope; and the Geographic Index, which lists programs that have state, regional, or international focus. See How to Use This Directory on pages v-vii for sample index entries.

Using the *Directory* for Grantseeking

By using the *Directory of Research Grants*, grantseekers can match the needs of their particular programs with those sponsors offering funding in the researchers' area of interest. The information listed here is meant to eliminate the costs incurred by both grantseekers and grantmakers when inappropriate proposals are submitted for a sponsor's funding program. However, because the GRANTS Database is updated daily basis, with program listings continually added, deleted, and revised, grantseekers using this *Directory* may also search the GRANTS database online through *GrantSelect* (www.grantselect.com).

All new and revised information has been taken from (1) the sponsor's updated of previously published program statements included in earlier editions of GRANTS publications, (2) questionnaires sent to sponsors whose programs were not listed in previous editions, or (3) other materials published by the sponsor and furnished to Schoolhouse Partners. Updated information for U.S. government programs includes new and revised program information published in the *Federal Register*; the latest edition of the *Catalog of Federal Domestic Assistance*; the *NIH Guide*, published weekly by the National Institutes of Health; and the *NSF E-Bulletin*, a monthly publication of the National Science Foundation. Included in this edition are identifying document numbers from the NIH and NSF publications. Located at the ends of the program descriptions in certain entries, the numbers indicated the ongoing NIH program number (PA) or the request for applications number (RFA). For programs of the National Science Foundation, the *NSF Bulletin* number appears. This information will help users identify the programs when seeking additional information from program staff.

While Schoolhouse Partners has made every effort to verify that all information is both accurate and current within the confines of format and scope, the publisher does not assume and hereby disclaims any liability to any party for loss or damages caused by errors or omissions in this *Directory*, whether such errors or missions result from accident, negligence, or any other cause. Anyone having questions regarding the content, format, or any other aspect of the *Directory of Research Grants 2010*, GrantSelect, or the GRANTS Database should contact the Editors, GRANTS, Schoolhouse Partners or editor@schoolhousepartners.net.

How to Use This Directory

The *Directory of Research Grants 2010* is designed to allow the user quick and easy access to information regarding funding programs in a researcher's specific area of interest. This *Directory* is composed of a main section, Grant Programs, which lists grant programs in alphabetical order, and three indexes: the Subject Index, the Grants by Program Type Index, and the Geographic Index.

GRANT PROGRAMS

Each listing in this section consists of the following elements: an annotation describing each program's focus and goals, requirements explaining eligibility, funding amounts, application and renewal dates, the *Catalog of Federal Domestic Assistance* program number (for U.S. government programs only), sponsor information, contact information, and Internet address.

GRANT TITLE ———

ACCESSION NUMBER

Paul G. Allen Family Foundation Grants　　　　4298

The single foundation, created through the consolidation of Allen's six previous foundations (The Allen Foundation for the Arts, The Paul G. Allen Charitable Foundation, The Paul G. Allen Foundation for Medical Research, The Paul G. Allen Forest Protection Foundation, The Allen Foundation for Music, and The Paul G. Allen Virtual Education Foundation), will continue to focus on the Allen family's philanthropic interests in the areas of arts and culture, youth engagement, community development and social change, and scientific and technological innovation. The Arts and Culture Program fosters creativity and promotes critical thinking by helping strong arts organizations become sustainable and supporting projects that feature innovative and diverse artistic forms.

GRANT DESCRIPTION

The Youth Engagement Program improves the way young people learn by supporting organizations that use innovative teaching strategies and provide opportunities for children to address issues relevant to their lives. The Community Development and Social Change Program promotes individual and community development by supporting initiatives and organizations that provide access to resources and opportunities. The Scientific and Technological Innovation Program advances promising scientific and technology research that has the potential to enhance understanding and stewardship of the world in which we live. Organizations may only receive one grant per year. Organizations must not have any delinquent final reports due to any of the Paul G. Allen Foundations for previous grants. Grantseekers are encouraged to apply through the online application process, where basic organizational and project information will be requested. Guidelines are available online.

REQUIREMENTS ———

Requirements 501(c)3 tax-exempt organizations, status from the Internal Revenue government entities, and IRS-recognized tribes are eligible. Eligible organizations must be located in, or serving populations of, the Pacific Northwest, which includes Alaska, Idaho, Montana, Oregon, and Washington.

RESTRICTIONS ———

Restrictions In general, the foundation will not consider requests for general fund drives, annual appeals, or federated campaigns; special events or sponsorships; direct grants, scholarships, or loans for the benefit of specific individuals; projects of organizations whose policies or practices discriminate on the basis of race, ethnic origin, sex, creed, or sexual orientation; contributions to sectarian or religious organizations whose principle activity is for the benefit of their own members or adherents; loans or debt retirement; projects that will benefit the students of a single school; general operating support for ongoing activities; or projects not aligned with the foundation's specified program areas. 509(a) private foundations are ineligible.

APPLICATION/
DUE DATE ———

Date(s) Application Is Due Mar 31; Sep 30.

Contact Grants Administrator, (206) 342-2030; fax: (206) 342-3030; ——— CONTACT
email: info@pgafamilyfoundation.org

INTERNET ADDRESS ———

Internet http://www.pgafamilyfoundation.org

Sponsor Paul G. Allen Family Foundation ——— SPONSOR INFORMATION
505 Fifth Ave S, Ste 900
Seattle, WA 98104

SUBJECT INDEX

The most effective way to access specific funding programs is through the Subject Index. This index lists the subject terms with applicable grants program titles – and their accession numbers – alphabetically under each term. Terms were assigned to target the specific area of research designated in the description of each program. Cross-references are used to link subjects and assist the user in finding specific grant information.

Following are general guidelines that can make your search of this index more successful. First, check under the specific topic of interest rather than a more general term. For instance, if you are interested in chemical engineering, look under "Chemical Engineering" rather than "Engineering." Items indexed under "Engineering" indicate funding in broad areas of engineering.

Use general headings when you want grants covering broader areas or if you can't find a specific topic. For example, many grants list funding for humanities research, health programs, or science and technology. To find these grants use such headings as "Humanities," "Medical

Programs," "Science," or "Technology." For additional grant information on more specific humanities research opportunities, such as in American History or Cultural Anthropology, also check under the topics "United States History" and "Anthropology, Cultural."

Many of the grant programs provide funding for research-related scholarships, faculty fellowships, dissertations, undergraduate education, conferences, or internships. If grant funds are designated for specific disciplines, you will find the items under the specific subject. Scholarships and fellowships are also listed under the terms "Native American Education," "African Americans (Student Support)," "Hispanic Education," "Minority Education," and "Women's Education."

Grants concerning study of a particular country are listed under the name of the country. Grants concerning the history, literature, art and language of a country are listed under the name of the country also, e.g., "Chinese Art" and "Chinese Language/Literature."

SUBJECT TERM ———————— **Education**
A.L. Mailman Family Foundation Grants, 5
AAAS Science & Technology Policy Fellowships - Health, Education & Human Services, 22
AARP Andrus Foundation Grants, 104
Abbott Laboratories Fund Grants, 125
ACIE Host University Edmund S. Muskie/Freedom Support Act Graduate Fellowships, 156
ACT Awards, 189
Akonadi Foundation Anti-Racism Grants, 351 ——————————PROGRAM TITLE
Akron Community Foundation Grants, 352
Albert and Margaret Alkek Foundation Grants, 370
Albuquerque Community Foundation Grants, 379
Alcoa Foundation Grants, 380
Alcon Foundation Grants Program, 382

GEOGRAPHIC INDEX

This index lists programs that have state, regional, or international geographic focus. The Geographic Index is arranged by state, followed by Canadian programs, then by international programs by country, and lists grant program titles and their corresponding accession numbers.

COUNTRY———————— **United States**

Alabama ———————————————————————— STATE
3M Fndn Grants, 2
Alabama Humanities Fndn Grants Program, 368
Arkema Inc. Fndn Science Teachers Program, 705
CDC Injury Control Research Centers Grants, 1198
DOE Experimental Program to Stimulate
PROGRAM TITLE ———————— Competitive Research (EPSCoR), 1653
Hill Crest Fndn Grants, 2293
Linn-Henley Charitable Trust Grants, 2777
NOAA Community-Based Restoration Program (CRP) Grants, 3843
Southern Company's Longleaf Pine Reforestation Fund, 4726

PROGRAM TYPE INDEX

This index is broken into 38 categories according to the type of program funded:

- Adult Basic Education
- Adult/Family Literacy Training
- Awards/Prizes
- Basic Research
- Building Construction and/or Renovation
- Capital Campaigns
- Centers: Research/Demonstration/Service
- Citizenship Instruction
- Community Development
- Consulting/Visiting Personnel
- Cultural Outreach
- Curriculum Development/Teacher Training
- Demonstration Grants
- Development (Institutional/Departmental)
- Dissertation/Thesis Research Support
- Educational Programs
- Environmental Programs
- Exhibitions, Collections, Performances, Video/Film Production
- Faculty/Professional Development
- Fellowships
- General Operating Support
- Graduate Assistantships
- International Exchange Programs
- International Grants
- Job Training/Adult Vocational Programs
- Matching/Challenge Funds
- Materials/Equipment Acquisition (Computers, Books, Tapes, etc.)
- Preservation/Restoration
- Publishing/Editing/Translating
- Religious Programs
- Scholarships
- Seed Grants
- Service Delivery Programs
- Symposia, Conferences, Workshops, Seminars
- Technical Assistance
- Training Programs/Internships
- Travel Grants
- Vocational Education

Government and Organization Acronyms

AAAAI	American Academy of Allergy Asthma and Immunology
AAAS	American Association for the Advancement of Science
AACAP	American Academy of Child and Adolescent Psychiatry
AACN	American Association of Critical Care Nurses
AACR	American Association of Cancer Research
AAFCS	American Association of Family and Consumer Sciences
AAFP	American Academy of Family Physicians Foundation
AAFPRS	American Academy of Facial Plastic and Reconstructive Surgery
AAR	American Academy in Rome
AAS	American Antiquarian Society
AASL	American Association of School Libraries
AAUW	American Association of University Women
ACC	Asian Cultural Council
ACLS	American Council of Learned Societies
ACM	Association for Computing Machinery
ACS	American Cancer Society
ADA	American Diabetes Association
ADHF	American Digestive Health Foundation
AF	Arthritis Foundation
AFAR	American Federation for Aging Research
AFOSR	Air Force Office of Scientific Research
AFUD	American Foundation for Urologic Disease
AFUW	Australian Federation of University Women
AGS	American Geriatrics Society
AHA	American Heart Association
AHAF	American Health Assistance Foundation
AFHMR	Alberta Heritage Foundation for Medical Research
AHRQ	Agency for Healthcare Research and Quality
AICR	American Institute for Cancer Research
AIIS	American Institute for Indian Studies
AJA	American Jewish Archives
ALA	American Library Association
ALISE	Association for Library and Information Science Education
AMNH	American Museum of Natural History
AMS	American Musicological Society
ANL	Argonne National Library
ANS	American Numismatic Society
AOA	American Osteopathic Association
APA	American Psychological Association
APSA	American Political Science Association
ARIT	American Research Institute in Turkey
ARO	Army Research Office
ASA	American Statistical Association
ASCSA	American School of Classical Studies at Athens
ASECS	American Society for Eighteenth-Century Studies
ASF	American-Scandinavian Foundation
ASHRAE	American Society of Heating, Refrigerating, and Air Conditioning Engineers
ASME	American Society of Mechanical Engineers
ASNS	American Society for Nutritional Sciences
ASPRS	American Society of Photogrammetry and Remote Sensing
AWU	Associated Western Universities
AWWA	American Water Works Association
BA	British Academy
BWF	Burroughs Wellcome Fund
CBIE	Canadian Bureau for International Education
CCFF	Canadian Cystic Fibrosis Foundation
CDC	Centers for Disease Control and Prevention
CES	Council for European Studies
CF	The Commonwealth Fund
CFF	Cystic Fibrosis Foundation
CFPC	College of Family Physicians of Canada
CFUW	Canadian Federation of University Women
CHEA	Canadian Home Economics Association
CIES	Council for International Exchange of Scholars
CIUS	Canadian Institute of Ukrainian Studies
CLA	Canadian Lung Association
CLF	Canadian Liver Foundation
CRI	Cancer Research Institute
DAAD	Deutscher Akademische Austauschdienst (German Academic Exchange Service)
DHHS	Department of Health and Human Services
DOA	Department of Agriculture
DOC	Department of Commerce
DOD	Department of Defense
DOE	Department of Energy
DOI	Department of the Interior
DOJ	Department of Justice
DOL	Department of Labor
DOS	Department of State
DOT	Department of Transportation
EFA	Epilepsy Foundation of America
EPA	Environmental Protection Agency
ESF	European Science Foundation
ETS	Educational Testing Service
FCAR	Formation de Chercheurs et L'Aide a la Recherche
FDA	Food and Drug Administration
FIC	Fogarty International Center
GAAC	German American Academic Council

HHMI	Howard Hughes Medical Institute
HRSA	Health Resources and Services Administration
HUD	Department of Housing and Urban Development
IRA	International Reading Association
IREX	International Research and Exchanges Board
IUCP	Indiana University Center on Philanthropy
JDF	Juvenile Diabetes Foundation International
JMO	John M. Olin Foundation
JSPS	Japan Society for the Promotion of Science
KFC	Kidney Foundation of Canada
LSA	Leukemia Society of America
MHRC	Manitoba Health Research Council
MLA	Medical Library Association
MMA	Metropolitan Museum of Art
MSSC	Multiple Sclerosis Society of Canada
NAACP	National Association for the Advancement of Colored People
NARSAD	National Alliance for Research on Schizophrenia and Depression
NASM	National Air and Space Museum
NATO	North Atlantic Treaty Organization
NCCAM	National Center for Complementary and Alternative Medicine
NCI	National Cancer Institute
NCIC	National Cancer Institute of Canada
NCRR	National Center for Research Resources
NEH	National Endowment for the Humanities
NEI	National Eye Institute
NFID	National Foundation for Infectious Diseases
NFWF	National Fish and Wildlife Foundation
NHGRI	National Human Genome Research Institute
NHLBI	National Heart, Lung and Blood Institute
NIA	National Institute on Aging
NIAAA	National Institute on Alcohol Abuse and Alcoholism
NIAF	National Italian American Foundation
NIAID	National Institute of Allergy and Infectious Diseases
NIAMS	National Institute of Arthritis and Musculoskeletal Skin Diseases
NICHD	National Institute of Child Health and Human Development
NIDA	National Institute on Drug Abuse
NIDCD	National Institute on Deafness and Other Communication Disorders
NIDCR	National Institute of Dental and Craniofacial Research
NIDDK	National Institute of Diabetes, and Digestive and Kidney Diseases
NIDRR	National Institute on Disability and Rehabilitation Research
NIEHS	National Institute of Environmental Health Sciences

NIGMS	National Institute of General Medical Sciences
NIH	National Institutes of Health
NIJ	National Institute of Justice
NIMH	National Institute of Mental Health
NINDS	National Institute of Neurological Disorders and Stroke
NINR	National Institute of Nursing Research
NIOSH	National Institute for Occupational Safety and Health
NIST	National Institute of Standards and Technology
NKF	National Kidney Foundation
NL	Newberry Library
NLM	National Library of Medicine
NMF	National Medical Fellowships, Inc.
NMSS	National Multiple Sclerosis Society
NOAA	National Oceanic and Atmospheric Administration
NRC	National Research Council
NSERC	Natural Sciences and Engineering Research Council of Canada
NSF	National Science Foundation
OAH	Organization of American Historians
ONF	Oncology Nursing Foundation
ONR	Office of Naval Research
OREF	Orthopaedic Research and Education Foundation
ORISE	Oak Ridge Institute for Science and Education
OSI	Open Society Institute
PDF	Parkinson's Disease Foundation
PhRMA	Pharmaceutical Research and Manufacturers of American Foundation
PHSC	The Photographic Historical Society of Canada
RCPSC	Royal College of Physicians and Surgeons of Canada
RSC	Royal Society of Canada
RWJF	Robert Wood Johnson Foundation
SLA	Special Libraries Association
SSHRC	Social Sciences and Humanities Research Council of Canada
SSRC	Social Science Research Council
STTI	Sigma Theta Tau International
USHMM	United States Holocaust Memorial Museum Research Institute
USIA	United States Information Agency
USIP	United States Institute of Peace
WAWH	Western Association of Woman Historians

1 in 9: Long Island Breast Cancer Action Coalition Grants   1

The Coalition's mission is to promote awareness of the breast cancer epidemic through education, outreach, advocacy, and direct support of research which is being done to find the causes of and cures for breast cancer and other related cancers. The Coalition's major goals are to: raise awareness of the epidemic of breast cancer and keep the disease in the forefront; obtain funding for research; investigate the link between pesticides and breast cancer since Long Island is a farm area; increase detection of breast cancer; and find the causes, a cure and how to prevent breast cancer.

Geographic Focus: New Jersey; New York; Pennsylvania
Contact: Geri Barish, Pres.; (516) 374-3190; fax: (516) 569-1894; email: info@1in9.org
Internet: http://www.1in9.org/
Sponsor: Long Island Breast Cancer Action Coalition
P.O. Box 729
Baldwin, NY 11510-0374

1772 Foundation Fellowships   2

The 1772 Foundation is a national, non-profit organization whose mission is to fund those entities that preserve and enhance America's built and natural environment. The Foundation offers paid internship and fellowship opportunities for highly qualified undergraduate seniors and graduate students pursuing a degree in historic preservation or a closely allied field such as museum studies, architectural history, history, landscape architecture and planning at an accredited college or university. Examples of topics of current interest for the 1772 Foundation include: the sustainability of historic sites, an inventory of African American historic sites/museums, revolving funds as a preservation tool, and agricultural education and programming at historic sites.

Requirements: No grants to individuals, or for operating support, scholarships, professional fees, studies and reports, books, strategic planning, endowments or sabbaticals.
Geographic Focus: Connecticut; Maine; Massachusetts; New Jersey; Rhode Island
Date(s) Application is Due: Mar 14
Contact: Mary Anthony, Executive Director; (860) 928-1772 or (860) 928-2125; email: maryanthony@1772Foundation.org
Internet: http://www.1772foundation.org/
Sponsor: 1772 Foundation
P.O. Box 112
Pomfret Center, CT 06259

1976 Foundation Grants   3

Established in Pennsylvania in 1976, the 1976 Foundation gives primarily to organizations that support services for the blind, pediatric medicine, health care and research. The average grant ranges up to $10,000, though larger grants will be considered. Though there are no specific guidelines or forms, the annual deadline is June 30. Applicants should contact the office in writing, explaining the project and amount needed.

Geographic Focus: Pennsylvania
Samples: Children's Hospital of Philadelphia, Philadelphia, Pennsylvania, $10,000 (2007); Purnell School 51, Pottersville, New Jersey, $10,000 (2007).
Amount of Grant: $10,000 maximum
Date(s) Application is Due: Jun 30
Contact: Nathaniel Peter Hamilton, President; (610) 254-9401; fax: (610) 254-9404
Sponsor: 1976 Foundation
200 Eagle Road, Suite 308
Wayne, PA 19087-3115

1st Touch Foundation Grants   4

Established by Derek Lee in 2006 in an effort to help his daughter Jada, giving is centered around the testing and researching of Leber's Congenital Amaurosis (LCA), an extremely rare disease which causes severe vision loss and blindness. The Foundation's primary long-range mission is to help researchers find a cure and eradicate this disease. It will also continue to provide grants for higher education, and help those in its local communities. There are no specific guidelines, application forms, or deadlines with which to adhere, so applicants should contact the Foundation office directly.

Geographic Focus: California
Contact: Justin Bass, Secretary; (415) 421-0535; Heather Scherber, Director; email: hscherber@1sttouch.org
Internet: http://1sttouch.org/index2.html
Sponsor: 1st Touch Foundation
PMB 10, 5480 Dewey Drive, #150
Fair Oaks, CA 95628

25th Anniversary Foundation Grants   5

Established by Arista Records, which was founded by Clive Jay Davis in the late 1970s, the 25th Anniversary Foundation was named to honor a quarter-century of company number one hits. The Foundation's primary field of interest is supporting AIDS research at nationally recognized organizations. There are no particular application formats or deadlines with which to adhere, and potential applicants should first approach the Foundation in writing, outlining their overall need.

Samples: City of Hope, Los Angeles, California, $10,000—for AIDS research (2008); T.J. Martel Foundation, New York, New York, $10,000—for AIDS research (2008).
Amount of Grant: $10,000 average
Contact: Robert J. Sorrentino, President, c/o Bertelsmann Music Group; (212) 782-1137 or (212) 782-1000; fax: (212) 782-1010
Sponsor: 25th Anniversary Foundation (formerly Arista Records Foundation, Inc.)
1745 Broadway
New York, NY 10019

3M Canada National Teaching Fellowships   6

Nominations of faculty members teaching at Canadian universities for one-year teaching fellowships are invited. The program embodies the highest ideals of teaching excellence and scholarship with a commitment to enhance the educational experience of every learner. Up to 10 awards, presented annually at the STLHE conference in June, recognize exemplary contributions to educational and teaching excellence in Canadian universities. The award includes lifetime membership in the Society. 3M Fellowship winners will attend an all expense paid 3-day retreat at the Fairmont Le Chateau Montebello, to celebrate and share exceptional achievements in teaching and provide an opportunity for outstanding teachers to share experiences and ideas.

Requirements: Awards are open to all individuals currently teaching at a Canadian university, regardless of discipline or level of appointment.
Date(s) Application is Due: Aug 31
Contact: Dr. Arshad Ahmad, Program Coordinator; (514) 848-2424, ext. 2928; email: arshad@jmsb.concordia.ca; Sylvia Riselay, STLHE Administrator; (905) 525-9140, ext. 20130; email: riselays@mcmaster.ca
Internet: http://www.mcmaster.ca/3Mteachingfellowships/nom_info.html
Sponsor: 3M Canada / Society for Teaching and Learning in Higher Education (STLHE)
1280 Main Street West
Hamilton, ON L8S 4K1 Canada

3M Company Higher Education Grants   7

Parallel to its Foundation, the 3M Company makes charitable contributions to nonprofit organizations directly. Support is given on a national basis. Most giving is initiated through a Request for Proposal process that allows the company to focus our giving and maximize results. Areas of interest for Higher Education include: advancing academic excellence in science, engineering and business; and increased participation and retention of underserved people in these disciplines. Types of support include: building and renovation; capital campaigns; donated equipment; donated land; donated products; employee volunteer services; general operating support; in-kind gifts; internship funds; program development; seed money; technical assistance; and use of facilities.

Restrictions: No support is offered for: political, fraternal, social, veterans, or military organizations; propaganda or lobbying organizations; religious organizations not of direct benefit to the entire community; animal-related organizations; or disease-specific organizations. No grants are given to individuals, or for electronic media promotion or sponsorships, athletic events, non-3M equipment, endowments, emergency needs, conferences, seminars, workshops, symposia, fund raising or testimonial events, travel, or film or video production; no cause-related marketing.
Geographic Focus: Alabama; Alaska; Arkansas; California; Connecticut; Georgia; Hawaii; Illinois; Indiana; Iowa; Kentucky; Massachusetts; Michigan; Minnesota; Missouri; Nebraska; New Jersey; New York; Ohio; South Carolina; South Dakota; Texas; Utah; Wisconsin
Contact: Cynthia F. Kleven, Secretary; (651) 733-0144; fax: (651) 737-3061; email: cfkleven@mmm.com
Internet: http://solutions.3m.com/wps/portal/3M/en_U.S./CommunityAffairs/CommunityGiving/U.S./Apply/
Sponsor: 3M Company Contributions
3M Center, Building 225-01-S-23
Saint Paul, MN 55144-1000

3M Foundation Grants   8

The foundation supports organizations involved with arts and culture, K-12 education, higher education, the environment, and health and human services. Special emphasis is directed toward programs designed to help prepare individuals and families for success. Fields of interest are: arts; arts education; business education; disaster relief, preparedness, and services; economics; elementary and secondary education; employment training; engineering; environmental causes; family services; federated giving programs; health care; higher education; human services; mathematics; minorities; science programs; youth development; and youth, services. Types of support include: building construction and renovation, capital campaigns, curriculum development, employee matching gifts, general operating support, in-kind gifts, program development, and scholarship funds. The foundation utilizes an invitational Request For Proposal (RFP) process for organizations located in Minneapolis and St. Paul, Minnesota, and Austin, Texas. Application forms are not required.

Requirements: Established 501(c)3 tax-exempt organizations in all 3M communities are eligible.

Restrictions: No support for religious organizations, conduit agencies, political groups, fraternal organizations, social groups, or veteran organizations. No funding for hospitals, K–12 schools, military organizations, animal –related organizations, or disease– specific organizations. No grants to individuals, or for endowments, emergency operating support, advocacy and lobbying efforts, fundraising events and associated advertising, travel, publications, start–up needs, non –3M equipment, debt reduction, conferences, athletic events, or film or video production; no loans or investments.

Geographic Focus: Alabama; Alaska; Arkansas; California; Connecticut; Georgia; Hawaii; Illinois; Indiana; Iowa; Kentucky; Massachusetts; Michigan; Minnesota; Missouri; Nebraska; New York; Ohio; South Carolina; South Dakota; Texas; Utah; Wisconsin

Samples: Nature Conservancy (Arlington, VA)—to conserve and restore grasslands in northwestern and west-central Minnesota, and for conservation work along the lower Rio Grande River in Texas, $5.15 million.; U of Minnesota-Twin Cities (Minneapolis, MN)—to endow the 3M Science and Technology Fellowships; for the Carlson School of Management's new center, which will evaluate the business potential of new technologies; and for scholarships for undergraduates, $8.6 million and $1 million matching grant.

Amount of Grant: $47.1 million total

Contact: Cynthia F. Kleven, Secretary; (651) 733-0144; fax: (651) 737-3061; email: cfkleven@mmm.com

Internet: http://solutions.3m.com/wps/portal/3M/en_U.S./CommunityAffairs/CommunityGiving/U.S./K12/

Sponsor: 3M Foundation
3M Center, Building 225-01-S-23
Saint Paul, MN 55144-1000

A Rocha U.S.A Grants 9

A Rocha U.S.A is part of an international family of conservation organizations working to show God's love for all creation through community-based efforts focused on science and research, practical conservation, environmental education, and environmental health. The organization works toward the conservation of habitats and species world-wide through environmental education, conservation activities, and scientific research; and seeks to build a sense of community among individuals, communities, and their environments through biblical faith. Fields of interest include: community and economic development, environmental health, environmental education, research, and Jewish agencies and synagogues. Giving is both national and international in scope. There are no specific application formats or deadlines with which to adhere, and potential applicants should contact the office directly.

Contact: Thomas Rowley, Executive Director; (830) 992-7940; email: usa@arocha.org

Internet: http://www.arocha.org/int-en/index.html

Sponsor: A Rocha U.S.A, Inc.
P.O. Box 1338
Fredericksburg, TX 78624-1338

A Social Corporation Grants 10

The Fund, established in Wisconsin in 1987, has as its primary purpose to support research in medical physics within the Madison, Wisconsin, region. Specific application guidelines are available upon request. There are no particular deadlines with which to adhere, and applicant organizations should contact the fund administrator directly for further details.

Geographic Focus: Wisconsin

Contact: Charles Lescrenier, President; (608) 831-1188 or (800) 426-6391; fax: (608) 836-9201

Sponsor: A Social Corporation, Inc.
P.O. Box 620327, 2500 West Beitline Highway at University Avenue
Middleton, WI 53562-0327

A-T Medical Research Foundation Grants 11

The Foundation is a national 501(c) tax-exempt organizations offering a comprehensive program of information and support services for patients and their families on ataxia-telangiectasia, a lethal genetic disease that attacks children, causing progressive loss of muscle control, immune system problems, and a strikingly high rate of cancer, especially leukemia and lymphoma. The Foundation also offers patient and professional information and education materials, sponsors meetings and scientific workshops, funds research, and provides referrals to chapters and support groups. Seed grants and research grants are available.

Amount of Grant: $35,000-$255,000

Contact: Pamela Smith, President; (818) 906-2861; fax: (818) 906-2870; email: becca4435@aol.com or atmrf@aol.com

Sponsor: Ataxia Telangiectasia Medical Research Foundation
16224 Elisa Place
Encino, CA 91436-3320

A. V. Stout Fund Research Grants 12

The fund is dedicated to supporting research grants for New England wildlife and its habitat as well as improving management techniques through innovative ideas and programs. The research, ideas and programs supported by the fund will give resource administrators and policy makers better understanding in managing the political and environmental issues involved. The fund is open to qualified students, interns, resource managers and organizations. Grants average approximately $1,000 and usually will not exceed $3,000. In certain cases, as determined by the selection committee, grant amounts could exceed $3,000.

Requirements: All students, individuals and organizations whose projects are within the New England area and who are interested in the fund's objectives are eligible. All applications must be completed on the application form (see website) and attachments will not be accepted. Applicants are encouraged to seek other sources of funding to share project costs.

Geographic Focus: Connecticut

Amount of Grant: up to $3,000

Contact: Karen Outlaw, Executive Director; (212) 362-4831

Internet: http://www.norcrossws.org/Grants/Andystout.htm

Sponsor: Norcross Wildlife Foundation
250 West 88th Street, #806
New York, NY 10024

A.L. Mailman Family Foundation Grants 13

The foundation awards grants on the state, regional, and national levels with a focus on childcare, especially on issues of quality, caregiver training, teacher/student ratios, and connecting quality childcare to later achievements. The foundation also supports efforts to stimulate moral growth and the development of social responsibility and for research in and refinement of developmental, individualized education of young children. Types of support include curriculum development, matching/challenge grants, program development, publication, research grants, seed grants, and technical assistance. Prospective applicants should write a letter describing the proposed project. The letter should include a project summary, contact information, fit with Foundation objectives, and an estimated budget and timeframe. The Foundation board meets in April and October to authorize grants. Proposals should be sent to the foundation by June 15 for fall review and by January 15 for spring review. Letters of inquiry should be submitted by May 1 for fall proposals and by December 1 for spring proposals.

Restrictions: Grants do not support direct service organizations, childcare centers and schools, locally focused organizations, individuals, capital campaigns and endowments, scholarships, general operating expenses or deficit reduction, or organizations or projects outside of the United States.

Samples: Agenda for Children (Alliance Foundation for Comunity Health)—to develop a nationally replicable language development program model, $45,000 (2006); Enterprise Foundation, Enterprise Child Care (Columbia, MD)—to produce a videotape, particularly designed for providers who offer childcare in their homes, on ways to create safe and stimulating home-based childcare environments, $40,000.

Amount of Grant: $5,000-$50,000

Date(s) Application is Due: Dec 1; Jan 15; May 1; Jun 15

Contact: Luba H. Lynch, Executive Director; (914) 683-8089; fax: (914) 686-5519; email: info@mailman.org

Internet: http://www.mailman.org/guidelines/index.html

Sponsor: A.L. Mailman Family Foundation
707 Westchester Avenue
White Plains, NY 10604

A.O. Smith Community Grants 14

The foundation supports nonprofit organizations in communities where A.O. Smith Corporation has facilities. The foundation supports elementary and secondary school projects that focus on the quality of educational programs and curriculum development; civic, cultural, and social welfare of communities; and medical research and improved local health services. Types of support include continuing support, annual campaigns, building/renovation, scholarship funds, and employee matching gifts. Proposals should describe the project's benefits and constituency, budget information including other sources of funding, and how results will be reported. There is no deadline for applications, although proposals to be considered for the following year's foundation budget must be received by October 30.

Requirements: Nonprofit organizations in company areas may apply.

Restrictions: Grants do not support political organizations or organizations whose chief purpose is to influence legislation.

Amount of Grant: $500-$20,000 typically; $2,000 average

Date(s) Application is Due: Oct 30

Contact: Edward J. O'Connor; (414) 359-4000; fax: (414) 359-4064

Internet: http://www.aosmith.com/About/Detail.aspx?id=132

Sponsor: A.O. Smith Foundation
P. O. Box 245008
Milwaukee, WI 53224-9510

AAA Foundation for Traffic Safety Grants 15

The foundation sponsors research that not only identifies critical traffic safety problems, but also searches for underlying causes and possible solutions, as well as identifying early trends and offering solutions before the general public is aware of the problem. The foundation continues research into ways to help the growing numbers of teen drivers and older drivers evaluate and improve their driving performance. A sampling of recent research includes drunk driving, seeking additional solutions; seated for safety; headlight glare countermeasures; supplemental transportation programs for seniors; distracted driving, phase I; and longer combination vehicle safety study. Prior to submitting unsolicited proposals, applicants may email a one-page preproposal to gauge the foundation's general interest in the proposed topic.

Requirements: Research institutions and nonprofit organizations with 501(c)3 status are eligible.

Restrictions: The foundation does not fund research to develop new devices. No grants are made for community action initiatives or other purely local traffic safety programs.
Samples: Guidebook for Observing Occupant Restraint System Use; Older Driver Involvement in Injury Crashes; Drunk Driving—Seeking Additional Solutions; Children Seated for Safety
Contact: Betty Barksdale, Foundation Office Manager, (202) 638-5944; fax: (202) 638-5943; email: info@aaafoundation.org
Internet: http://www.aaafoundation.org/about_us/index.cfm
Sponsor: AAA Foundation for Traffic Safety
607 14th Street NW, Suite 201
Washington, DC 20005

AAA Minority Dissertation Fellowship Program 16

The association invites minority doctoral candidates in anthropology to apply for a nonrenewable dissertation writing fellowship. Students of any subfield or specialty in anthropology will receive equal consideration. Dissertation topics in all areas of the discipline are welcome. Doctoral students who require financial assistance to complete the write-up phase of the dissertation are urged to apply. The dissertation proposal must be approved by the applicant's committee prior to application. Application and guidelines are available online.
Requirements: An applicant must be a U.S. citizen; a member of an historically underrepresented ethnic minority group; enrolled in a full-time academic program leading to a doctoral degree in anthropology at the time of application; admitted to degree candidacy before the dissertation fellowship is awarded; and a member of the American Anthropological Association.
Samples: Rocio Magana, University of Chicago, $10,000, (2007-08); Shedra Amy Snipes, University of Washington, $10,000, (2006-07); Russell Rodriguez, University of California, $10,000, (2005-06); Lisa Anderson-Levy, University of Minnesota, $10,000, (2004-05).
Amount of Grant: $10,000
Date(s) Application is Due: Feb 15
Contact: Kathleen Terry-Sharp, Director; (703) 528-1902, ext. 3010; fax: (703) 528-3546; email: ksharp@aaanet.org
Internet: http://www.aaanet.org/profdev/awards/
Sponsor: American Anthropological Association
2200 Wilson Boulevard, Suite 600
Arlington, VA 22201-3357

AAAAI Allergy, Asthma and Immunology Education and Research Trust Clinical Research Grants 17

Formerly the Education and Research Trust (ERT), training program directors, deans, department chairpersons, and sponsors are invited to nominate promising investigators involved in research dealing with asthma for this grant. The award is a grant in aid provided to the program sponsoring the applicant and is intended to provide salary support. Application should include a biographical sketch, including a curriculum vita, an outline of past accomplishments, a summary of a current research project, and plans for future research.
Requirements: Applicants must be MDs or PhDs who are fellows in training or residents associated with an approved allergy and immunology program and are AAAAI members (or individuals who have submitted applications for membership by September 1).
Restrictions: Individuals with a faculty appointment and past AAAAI research award recipients will not be considered.
Samples: Seema Sharma Aceves, MD, PhD, Children's Hospital San Diego—for Evaluating Disease Severity and Tissue Remodeling in Pediatric Patients with Eosinophilic Esophagitis, $10,000 (2006); Kevin P. O'Brien, MD, Parker, Colorado—for Prevention of Atopy by Combination of Lactobacillus Supplements (pro-TH1) and Omega-3 Supplements (anti-TH2) During Pregnancy and Breastfeeding: A Randomized, Placebo-controlled Trial, $10,000 (2005).
Contact: Angela Gade, Development Senior Project Manager; (414) 272-6071; fax: (414) 276-6070; email: agade@aaaai.org or info@aaaai.org
Internet: http://www.aaaai.org/members/grants_awards
Sponsor: American Academy of Allergy, Asthma, and Immunology
555 East Wells Street, Suite 1100
Milwaukee, WI 53202-3823

AAAAI Allergy, Asthma and Immunology Education and Research Trust Faculty Development Award 18

The award offers an individual three years of financial support to concentrate his/her research efforts in either basic or clinical allergy and immunology. The objectives of this award are to recognize and support young researchers and promote the specialty of allergy/immunology. The recipient is expected to participate in a seminar at the AAAAI annual meeting, where awardees present their research.
Requirements: The award is open to researchers of junior faculty status—either instructor, assistant professor, or the equivalent grade. This award is directed toward MDs, but in unusual circumstances PhDs will be considered. Applicants must be U.S. or Canadian citizens or permanent residents who are AAAAI members (or have applied for membership by September 1).
Restrictions: Past recipients are ineligible.
Samples: Christine Seroogy, MD, University of Wisconsin School of Medicine and Public Health—for The role of toll-like receptors (TLRs) in CD25+ T regulatory cell development and function, $300,000 over three years (2007); Jordan S. Orange, MD, PhD, Children's Hospital of Philadelphia—for Directed Secretion Through the Immunological Synapse, $300,000 over three years (2005); Sanjeev Jain, MD, PhD,

University of Washington—for Development of Structure Based Inhibitors of Nitric Oxide Synthase for the Treatment of Asthma, $300,000 over three years (2003).
Amount of Grant: $100,000 per year for three years
Date(s) Application is Due: Nov 15
Contact: Angela Gade, Development Senior Project Manager; (414) 272-6071; fax: (414) 276-6070; email: agade@aaaai.org or info@aaaai.org
Internet: http://www.aaaai.org/art/
Sponsor: American Academy of Allergy, Asthma, and Immunology
555 East Wells Street, Suite 1100
Milwaukee, WI 53202-3823

AAAAI Phoenix Program Award 19

Funded by the ERT, the AAAAI Allergy Educational Forum and the AAAAI, the grant is a critical element in the development of new allergy/immunology training programs. This grant award can be utilized to effectively increase the size of programs, increase research staff, increase the number of fellows being trained and to develop future faculty members. It is hoped that such additions will increase research quantity and quality, teaching and training mission, and the delivery of health care. Awards average $50,000 to $100,000 per year for three successive years.
Requirements: U.S. college and university medical centers are eligible to apply.
Samples: Penn State University College of Medicine, $150,000 over three years (2006); University of Mississippi Medical Center, $250,000 over three years (2005); University of North Carolina-Chapel Hill, $250,000 over three years (2004).
Amount of Grant: $250,000 over three years maximum
Contact: Angela Gade, Development Senior Project Manager; (414) 272-6071; fax: (414) 276-6070; email: agade@aaaai.org or info@aaaai.org
Internet: http://www.aaaai.org/art/
Sponsor: American Academy of Allergy, Asthma, and Immunology
555 East Wells Street, Suite 1100
Milwaukee, WI 53202-3823

AAAAI Research Award Grants 20

The organization promotes and stimulates research and study in allergy, asthma and immunology. The goal of this program is to support the research by a fellow-in-training, resident, and post-doctoral scholar at an allergy and immunology or a clinical and laboratory immunology training program.
Requirements: Fellow-in-training, resident, or post-doctoral scholar affiliated with an allergy and immunology or clinical and laboratory immunology training program that is in good standing with Residency Review Committee and is fully-accredited by the Accreditation Council of Graduate Medical Education may apply. Applicants must be an AAAAI member or must submit a membership application with the award application.
Restrictions: Applicants with faculty status will not be considered. Past AAAAI research award recipients are not eligible.
Amount of Grant: $25,000 for one year
Contact: Angela Gade, Development Senior Project Manager; (414) 272-6071; fax: (414) 276-6070; email: agade@aaaai.org or info@aaaai.org
Internet: http://www.aaaai.org/members/grants_awards
Sponsor: American Academy of Allergy, Asthma, and Immunology
555 East Wells Street, Suite 1100
Milwaukee, WI 53202-3823

AAAAI Summer Clinical Research Fellowship Medical Student Grants 21

The Program goal is to award grants to medical students wishing to conduct clinical research in allergy/immunology during their summer recess in order to encourage their consideration of careers as a clinician in allergy/immunology. Grants will be awarded to outstanding medical students who wish to pursue clinical research. Clinical topics for research examples include optimizing health care of asthma, collecting a series of cases with an interesting diagnosis and preparing a literature review of the topic, a case report with a literature review, reviewing practice parameters related to management of a problem in allergy/immunology and applying to care in a clinic or office, as well as other topics of interest in the practice of allergy/immunology.
Requirements: Applicant must be a full-time medical student residing in the U.S. or Canada and will have successfully completed at least 8 months of medical school by May 15.
Restrictions: Past Summer Fellowship Grant recipients are not eligible.
Amount of Grant: $2,000
Date(s) Application is Due: Mar 3
Contact: Program Administrator; (414) 272-6071; fax: (414) 276-6070; email: info@aaaai.org
Internet: http://www.aaaai.org/members/grants%5Fawards/
Sponsor: American Academy of Allergy, Asthma, and Immunology (AAAAI)
555 East Wells Street, Suite 1100
Milwaukee, WI 53202-3823

AAAAI Summer Research Fellowship Medical Student Grants 22

Fellowships are awarded on an annual basis to assist medical students interested in pursuing research in allergy and immunology during their summer recess. The fellowships will be awarded to outstanding individuals who wish to pursue research in the following areas: physiology of allergic diseases, pharmacology of allergy and inflammation, basic cellular and molecular immunology, AIDS, and other topics pertinent to the understanding of allergic and immune mechanisms of disease.

Requirements: Applicant must be a full time medical student residing in the U.S. or Canada and must have successfully completed at least eight months of medical school by May 15.

Restrictions: Past recipients are ineligible.

Samples: Jason Konner, Aaron Diamond AIDS Research Ctr (New York, NY)—summer fellowship recipient, $1500.; Goron Lai, Johns Hopkins Asthma and Allergy Ctr (Baltimore, MD)—summer fellowship recipient, $1500.

Amount of Grant: $2,000

Date(s) Application is Due: Mar 3

Contact: Administrator; (414) 272-6071; fax: (414) 276-6070; email: info@aaaai.org

Internet: http://www.aaaai.org/members/grants%5Fawards/

Sponsor: American Academy of Allergy, Asthma, and Immunology (AAAAI)
555 East Wells Street, Suite 1100
Milwaukee, WI 53202-3823

AAAS Award for Public Understanding of Science and Technology 23

AAAS invites nominations for this annual award, which honors working scientists and engineers who make outstanding contributions to the 'popularization of science'. It is intended to encourage technical professionals to communicate with the public by popularizing their own work. Such efforts can include publications or broadcasts, lectures, museum presentations or exhibit designs, or any other outreach project. The Award shall be given annually to scientists or engineers who, while working in their fields, have also contributed substantially to public understanding of science and technology. A monetary prize of $5,000, a commemorative plaque, complimentary registration, and reimbursement for reasonable travel and hotel expenses to attend the AAAS Annual Meeting to receive the prize are given to the recipient. Nominators are encouraged to identify candidates whose contributions reach broad audiences that include women, minorities, disabled persons, and senior citizens.

Requirements: Eligible individuals include scientists and engineers (individual or a small group) from all disciplines (including social sciences and medicine) engaged in research, teaching, and related activities that have contributed substantially to the public's understanding of science or technology. Nominations should include name, position, institution, professional address and phone, and home address and phone of the candidate; name, position, institution, and professional address and phone of the nominator; a summary of the action or actions that form the basis for the nomination (about 250 words); a longer statement (no more than three pages) providing additional details of the actions for which the candidate is nominated, including one paragraph that describes the nominee's impact on the general public; the candidate's vita (three-page maximum); two letters of support; and at least one representative sample that illustrates the nominee's contribution. Books, videotapes, brochures, magazine articles, or other materials are welcome.

Restrictions: Only materials produced for general audiences, as opposed to professional or trade audiences, will be considered. Employees of the AAAS are not eligible. All materials become the property of AAAS.

Samples: 2006: S. James Gates, Jr., honored for sustained and career-long contributions to the public understanding of physics. 2005: Jane Lubchenco, honored for her exemplary commitment to, and leadership of, public understanding of science initiatives in public policy and professional arenas as a core aspect of her scientific practice. 2004: Eric S. Lander, honored for his excellence in communicating complex scientific ideas, and their implications for society, to the general public and policy-makers while actively engaged in a demanding research program.

Amount of Grant: $5,000

Date(s) Application is Due: Aug 15

Contact: Stacey Pasco; (202) 326-6441 or (202) 326-6670; fax: (202) 371-9849; email: jkass@aaas.org

Internet: http://www.aaas.org/aboutaaas/awards/public/index.shtml

Sponsor: American Association for the Advancement of Science
1200 New York Avenue, NW, Room 608
Washington, DC 20005

AAAS Entry Point Internships 24

Entry Point is a program of the American Association for the Advancement of Science (AAAS) offering Outstanding Internship Opportunities for Students with Disabilities in Science, Engineering, Mathematics, Computer Science, and some fields of Business. To meet the challenge of the competitive global economy in the new millennium, private industry and government research agencies must expand the pool of technical talent. The American Association for the Advancement of Science (AAAS) has developed unique partnerships with IBM, NASA, Merck, NOAA, Google, Lockheed Martin, CVS, NAVAIR, Pfizer, Infosys, and university science laboratories to meet their human resources needs. Working with its partners, AAAS identifies and screens undergraduate and graduate students with disabilities who are pursuing degrees in science, engineering, mathematics, computer science, and some fields of business, and places them in paid summer internships.

Requirements: Students must: be full-time undergraduate or graduate student, with a disability, majoring in a science or engineering field. Some fields of business are also considered; maintain a B average or better; and be a U.S. citizen or have a right-to-work permit.

Contact: Laureen Summers; (202) 326-6649; fax: (202) 371-9849; email: LSummers@aaas.org

Internet: http://ehrweb.aaas.org/entrypoint/

Sponsor: American Association for the Advancement of Science
1200 New York Avenue, NW
Washington, DC 20005-3920

AAAS International Scientific Cooperation Award 25

AAAS, in collaboration with its affiliated organizations, seeks to recognize an individual or a limited number of individuals working together in the scientific or engineering community for making an outstanding contribution to furthering international cooperation in science and engineering. Nominations for the award may be originated by anyone. CAIP administers the screening and selection process for the awarding of the AAAS Award for International Scientific Cooperation (ISC). The ISC award is presented each February at the AAAS Annual Meeting. A monetary prize of $5,000, a commemorative plaque, complimentary registration, and reimbursement for reasonable travel and hotel expenses to attend the AAAS annual meeting are given to the recipient.

Requirements: Nominations should be typed and should include the following information: the nominator's name, address, and phone number; the nominee's name and title, institutional affiliation, and address; a summary of the actions that form the basis for the nomination (about 250 words); a longer statement (no more than three pages) providing additional details of the actions for which the candidate is nominated; at least two letters of support with addresses and phone numbers; and the candidate's vita (no more than three pages) containing professional positions held. Any documentation—books, articles, etc.—that illuminates the significance of the nominee's achievements may also be submitted.

Restrictions: The award is open to all regardless of nationality or citizenship. Nominees must be living at the time of their nomination. Any individual or small group in the scientific and engineering community that has contributed substantially to the understanding or development of science or engineering across national boundaries is eligible for this award.

Samples: Michael Balick, Philecology Curator and Director of the Garden's Institute of Economic Botany and Vice President for Research and Training, The New York Botanical Garden, New York (2004); Mahabir P. Gupta, Research Professor of Pharmacognosy, School of Pharmacy, University of Panama, Panama, Republic of Panama; Executive Director, INTERCIENCIA Association, (2003); L.S. 'Skip' Fletcher, Regents Professor and Thomas Dietz Professor, Dept. of Mechanical Engineering, Texas A&M University, College Station, Texas (2002).

Amount of Grant: $5,000

Date(s) Application is Due: Aug 1

Contact: Linda Stroud, CAIP Coordinator; (202) 326-6659; fax: (202) 289-4958; email: lstroud@aaas.org

Internet: http://www.aaas.org/programs/international/caip/isc

Sponsor: American Association for the Advancement of Science
1200 New York Avenue, NW
Washington, DC 20005

AAAS Mentor Award 26

The two categories of the AAAS Mentor Awards (Lifetime Mentor Award and Mentor Award) both honor individuals who during their careers demonstrate extraordinary leadership to increase the participation of underrepresented groups in science and engineering fields and careers. These groups include: women of all racial or ethnic groups; African American, Native American, and Hispanic men; and people with disabilities. Both awards recognize an individual who has mentored and guided significant numbers of students from underrepresented groups to the completion of doctoral studies or who has impacted the climate of a department, college, or institution to significantly increase the diversity of students pursuing and completing doctoral studies. Nominations should include name, position, institution, professional address and phone, and home address and phone of the candidate; name, position, institution, and professional address and phone of the nominator; a summary of the actions that form the basis for the nomination (about 250 words); a letter of nomination that enumerates the ways in which the person reflects the purpose of this award; the candidate's vita (three-page maximum); the total number of students the candidate mentored at the bachelor's or master's level who went on to the doctoral level at other institutions, and the total number of underrepresented students the candidate mentored at the doctoral level; a maximum of five supporting letters from students and three supporting letters from colleagues representative of the different spheres in which the candidate has demonstrated effort, results, and commitment.

Requirements: The award is open to all regardless of nationality or citizenship. Nominees must be living at the time of their nomination. Commitment and extraordinary effort may be demonstrated by the number and diversity of students mentored; assisting students to present and publish their work, to find financial aid, and to provide career guidance; providing psychological support, encouragement, and essential strategies for life in the scholarly community; and continued interest in the individual's professional advancement. The Lifetime Mentor award winner will have served in the role of mentor for 25 or more years. The Mentor Award winner will have served in the role of mentor for less than 25 years.

Amount of Grant: $5,000, a commemorative plaque, complimentary registration, and reimbursement for reasonable travel and hotel expenses to attend the AAAS Annual Meeting.

Date(s) Application is Due: Jul 31

Contact: Yolanda George, Mentor Awards Coordinator; (202) 326-6670; fax: (202) 371-9849; email: ygeorge@aaas.org

Internet: http://www.aaas.org/aboutaaas/awards/mentor/index.shtml

Sponsor: American Association for the Advancement of Science
1200 New York Avenue, NW
Washington, DC 20005

AAAS Newcomb Cleveland Prize 27

The association's oldest award, funded by Newcomb Cleveland of New York City, is given to the author or authors of an outstanding paper published in the Research Articles or Reports sections of Science. Each annual contest starts with the first issue of June and ends with the last issue of the following May.

Requirements: An eligible paper is one that includes original research data, theory, or synthesis; is a fundamental contribution to basic knowledge or is a technical achievement of far-reaching consequence; and is a first-time publication of the author's own work. Reference to pertinent earlier work by the author may be included to give perspective. Readers of 'Science' are invited to nominate papers appearing in the Articles, Research Articles, or Reports sections. Nominations must be typed and provide the following information: title of paper, author's name, date of issue in which it was published, page number, and a brief statement of justification for nomination.

Restrictions: Self-nominations will not be accepted.

Samples: Past recipients: 2006, Jason R. Petta, Alexander C. Johnson, Jacob M. Taylor, Edward A. Laird, Amir Yacoby, Mikhail D. Lukin, Charles M. Marcus, Micah P. Hanson, Arthur C. Gossard for the research article 'Coherent Manipulation of Coupled Electron Spins in Semiconductor Quantum Dots', published online in 'Science' Express and in 'Science'; 2005, David D. Awschalom, Yuichiro K. Kato, Roberto C. Myers, and Arthur C. Gossard for the research article 'Observation of the Spin Hall Effect in Semiconductors', published in 'Science'; 2004, Brian Kuhlman, Gautam Dantas, Gregory C. Ireton, Gabriele Varani, Barry L. Stoddard and David Baker for 'Design of a Novel Globular Protein Fold with Atomic-Level Accuracy', published in 'Science'.

Amount of Grant: $25,000

Date(s) Application is Due: Jun 30

Contact: Sylvia Kihara, Newcomb Cleveland Prize Coordinator: (202) 326-6507; fax: (202) 289-7562; email: skihara@aaas.org

Internet: http://www.aaas.org/aboutaaas/awards/newcomb/index.shtml

Sponsor: American Association for the Advancement of Science
1200 New York Avenue, NW, Room 1044
Washington, DC 20005

AAAS Philip Hauge Abelson Prize 28

The prize was established by the AAAS Board of Directors and is presented at the AAAS Annual Meeting. It is awarded annually to either a public servant, in recognition of sustained exceptional contributions to advancing science, or to a scientist whose career has been distinguished both for scientific achievement and for other notable services to the scientific community.

Requirements: The award is open to all regardless of nationality or citizenship. Nominees must be living at the time of their nomination. Any individual or small group in the scientific and engineering community that has contributed substantially to the understanding or development of science or engineering across national boundaries is eligible for this award.

Restrictions: Each nomination must be seconded by at least two other AAAS members. The winner will be selected by a five-member judging panel.

Samples: Past recipients: 2006: Charles M. Vest, recognized for his effective leadership and outstanding contributions to the areas of public policy, university research, and education; 2005: Norman R. Augustine, honored for his outstanding contributions to U.S. science and technology policy, his unrelenting work to maintain U.S. scientific and technological preeminence, and his initiatives to strengthen the scientific partnerships between academia, industry, and government; 2004: Dr. Maxine Frank Singer, honored for her scientific accomplishments, leadership in the establishment of scientific policy, substantial contributions to the improvement of math and science education, and efforts to raise awareness and understanding in matters of science globally and to increase the presence of women and minorities in the scientific community.

Amount of Grant: A monetary prize of $5,000, a commemorative plaque, complimentary registration, and reimbursement for reasonable travel and hotel expenses to attend the AAAS Annual Meeting are given to the recipient.

Date(s) Application is Due: Sep 1

Contact: Stephen Nelson, Award Coordinator; (202) 326-6600; fax: (202) 289-4950; email: snelsond@aaas.org

Internet: http://www.aaas.org/aboutaaas/awards/abelson/index.shtml

Sponsor: American Association for the Advancement of Science
1200 New York Avenue, NW
Washington, DC 20005

AAAS Roger Revelle Fellowship in Global Stewardship 29

The Fellowship is designed to provide a unique opportunity for an accomplished scientist to address global stewardship issues by applying his or her broad, multidisciplinary background toward solutions to these important societal problems. The Revelle Fellow will engage with professionals in the public policy arena to strengthen his or her understanding of the intricacies of environmental policymaking and make practical contributions to the more effective use of scientific and technical knowledge in federal decision-making. The fellowship provides the unique opportunity to address sustainability challenges across multiple disciplines. The focus will be on human interaction with ecosystems, which includes, but is not limited to, population, sustainable development, food, oceans, global climate change, and related environmental concerns.

Requirements: Applications are invited from candidates in any physical, biological, or social science; any field of engineering; or any relevant interdisciplinary field. All applicants must have a PhD or an equivalent doctoral-level degree and at least three years of post-degree professional experience by the application deadline. Individuals with a master's degree in engineering and six years of post-degree, professional experience may apply. All applicants must be U.S. citizens.

Restrictions: Federal employees are not eligible.

Amount of Grant: $64,000 stipend

Date(s) Application is Due: Dec 15

Contact: Emily MacGillivray, Fellowships Program Coordinator; (202) 326-8979 or (202) 326-6700; fax: (202) 289-4950; email: emacgill@aaas.org or fellowships@aaas.org

Internet: http://fellowships.aaas.org/02_Areas/02_Global_Stewardship.shtml

Sponsor: American Association for the Advancement of Science
1200 New York Avenue, NW
Washington, DC 20005

AAAS Science & Technology Policy Fellowships - Congressional 30

Fellows spend one year working as special legislative assistants on the staffs of members of Congress or congressional committees, beginning in September. The program includes an orientation on congressional and executive branch operations, and a year-long seminar program on issues involving science and public policy. The fellowships are designed to provide a unique public policy learning experience, to demonstrate the value of science-government interaction, and to bring technical backgrounds and external perspectives to the decision-making process in government. AAAS will select and sponsor two Congressional Fellows.

Requirements: Prospective Fellows must demonstrate exceptional competence in some area of science or engineering; have a good scientific and technical background; be cognizant of and demonstrate sensitivity toward political and social issues; and, perhaps most importantly, have a strong interest and some experience in applying personal knowledge toward the solution of societal problems. Applications are invited from individuals in any physical, biological, or social science, any field of engineering, or any relevant interdisciplinary field. All applicants must have a Ph.D. or an equivalent doctoral-level degree by the application deadline (December 10). Individuals with a master's degree in engineering and at least three years of post-degree professional experience may apply. Persons from underrepresented minority groups and persons with disabilities are encouraged to apply.

Restrictions: Applicants must be U.S. citizens. Federal employees are not eligible. Will accept only online application submissions. Applications with fewer than three recommendation letters will not be considered.

Amount of Grant: Annual stipends are $70,000

Date(s) Application is Due: Dec 15

Contact: Paget Graham, Senior Program Manager; (202) 326-6700; fax: (202) 289-4950; email: pgraham@aaas.org or fellowships@aaas.org

Internet: http://fellowships.aaas.org/02_Areas/02_Congressional.shtml

Sponsor: American Association for the Advancement of Science
1200 New York Avenue, NW
Washington, DC 20005

AAAS Science & Technology Policy Fellowships - Diplomacy 31

Selected fellows apply their scientific and technical background to foreign policy, international affairs and development, and advancing research, programs, and treaties through international cooperation. Placement opportunities are available at Fogerty International Center, National Institutes of Health; Foreign Agriculture Service, U.S. Department of Agriculture; U.S. Agency for International Development; or U.S. Department of State

Requirements: Applicants must have a PhD or equivalent doctoral level degree at the time of application. Persons with a master's degree in engineering and at least three years of postdegree professional experience will also be considered. All applicants must be U.S. citizens. Prospective fellows must demonstrate exceptional competence in some area of science and engineering (multidisciplinary expertise is especially valued); be cognizant of the ways in which science and technology affect a broad range of international development and foreign policy issues; communicate and work effectively with decision makers and others outside of the scientific and engineering communities; exhibit a willingness and flexibility to tackle problems in a number of nonscientific areas; demonstrate sensitivity toward political, economic, and social issues; and have some experience and/or strong interest in applying knowledge toward the solution of problems in the area of foreign affairs or international development. Applications are invited from individuals in any physical, biological, or social science; any field of engineering; or any relevant interdisciplinary field. Some program areas and agencies seek additional qualifications. See the website for more specific guidelines.

Restrictions: Federal employees are not eligible.

Amount of Grant: $67,000-$87,000

Date(s) Application is Due: Dec 15

Contact: Rick Kempinski, Senior Program Manager; (202) 326-6700; fax: (202) 289-4950; email: rkempinski@aaas.org or fellowships@aaas.org

Internet: http://fellowships.aaas.org/02_Areas/02_index.shtml

Sponsor: American Association for the Advancement of Science
1200 New York Avenue, NW
Washington, DC 20005

AAAS Science & Technology Policy Fellowships - Energy, Environment, Agriculture & Natural Resources 32

Fellows chosen for this program engage in projects, programs, policies and outreach initiatives to protect environmental and human health, tackle energy challenges and opportunities, and to safeguard our air, water, land, and natural resources. Placement opportunities are available with the national Oceanic Atmospheric Association, National Science Foundation, U.S. Department of Agriculture, U.S.DA Forest Service, U.S. Department of Energy, and U.S. Environmental Protection Agency. Fellowships are awarded to highly qualified individuals interested in learning about the science-policy interface while applying their scientific and technical knowledge and analytical skills to the federal policy realm.

Requirements: Prospective fellows must have a PhD or equivalent doctoral level degree. Persons with a master's degree in engineering and at least three years of post-degree professional experience may apply. Candidates must demonstrate exceptional competence in some area of science or engineering, and an interest in applying their expertise to the economic and technical assessment of problems relating to human health, agriculture, and the environment. Applications are invited from individuals in any physical, biological, or social science; any field of engineering; or any relevant interdisciplinary field. Persons with a DVM, MD, or a PhD in the natural sciences or economics are especially encouraged to apply. Applicants must be U.S. citizens.

Restrictions: Federal employees are ineligible.

Amount of Grant: $67,000-$87,000

Date(s) Application is Due: Dec 15

Contact: Claire Hayes, Senior Program Manager; (202) 326-6700; fax: (202) 289-4950; email: chayes@aaas.org or fellowships@aaas.org

Internet: http://fellowships.aaas.org/02_Areas/02_index.shtml

Sponsor: American Association for the Advancement of Science

1200 New York Avenue, NW

Washington, DC 20005

AAAS Science & Technology Policy Fellowships - Executive Branch   33

The AAAS Science & Technology Policy Fellowships are extremely competitive, involving a formal, three-tier merit review and selection process. The reviewers and Selection Committee members identify the best scientists and engineers from the applicant pool who they believe will benefit most from the opportunities that a AAAS Fellowship provides, and who will offer significant expertise, skills, effort and new perspectives to hosting offices. The placement process for executive branch Fellows who will be working in federal agencies begins with a week of interviews in Washington, DC in early to mid-April. AAAS provides information about hosting offices and projects that match Fellows' expertise and offers guidance and support. Detailed information about the placement process is provided following selection as a AAAS Executive Branch Fellow.

Requirements: Applicants must have a PhD or equivalent doctoral level degree at the time of application. Persons with a master's degree in engineering and at least three years of postdegree professional experience will also be considered. All applicants must be U.S. citizens. Prospective fellows must demonstrate exceptional competence in some area of science and engineering (multidisciplinary expertise is especially valued); be cognizant of the ways in which science and technology affect a broad range of international development and foreign policy issues; communicate and work effectively with decision makers and others outside of the scientific and engineering communities; exhibit a willingness and flexibility to tackle problems in a number of nonscientific areas; demonstrate sensitivity toward political, economic, and social issues; and have some experience and/or strong interest in applying knowledge toward the solution of problems in the area of foreign affairs or international development. Applications are invited from individuals in any physical, biological, or social science; any field of engineering; or any relevant interdisciplinary field. Some program areas and agencies seek additional qualifications. See the website for more specific guidelines.

Restrictions: Federal employees are not eligible.

Amount of Grant: $67,000-$87,000

Date(s) Application is Due: Dec 15

Contact: Sage Russell, Associate Director; (202) 326-6700; fax: (202) 289-4950; email: srussell@aaas.org or fellowships@aaas.org

Internet: http://fellowships.aaas.org/02_Areas/02_index.shtml

Sponsor: American Association for the Advancement of Science

1200 New York Avenue, NW

Washington, DC 20005

**AAAS Science & Technology Policy Fellowships - Health, Education &   34
Human Services**

Fellows chosen for this program focus their experience to support improved programs, policies, planning, and risk analysis for initiatives in human and environmental health, biological threats, food safety, education and research, and to aid in developing and implementing regulations and providing oversight. Placement opportunities are available with the National Institutes of Health, National Science Foundation, U.S.DA Food Safety Inspection Services and U.S. Department of Health & Human Services. Fellowships are awarded to highly qualified individuals interested in learning about the science-policy interface while applying their scientific and technical knowledge and analytical skills to the federal policy realm.

Requirements: Prospective fellows must have a PhD or equivalent doctoral level degree. Persons with a master's degree in engineering and at least three years of post-degree professional experience may apply. Candidates must demonstrate exceptional competence in some area of science or engineering, and an interest in applying their expertise to the economic and technical assessment of problems relating to human health, agriculture, and the environment. Applications are invited from individuals in any physical, biological, or social science; any field of engineering; or any relevant interdisciplinary field. Persons with a DVM, MD, or a PhD in the natural sciences or economics are especially encouraged to apply. Applicants must be U.S. citizens.

Restrictions: Federal employees are not eligible.

Amount of Grant: $67,000-$87,000

Date(s) Application is Due: Dec 15

Contact: Angelique Dorazio-Sanders, Senior Program Manager; (202) 326-6700; fax: (202) 289-4950; email: fellowships@aaas.org

Internet: http://fellowships.aaas.org/02_Areas/02_index.shtml

Sponsor: American Association for the Advancement of Science

1200 New York Avenue, NW

Washington, DC 20005

**AAAS Science & Technology Policy Fellowships - National Defense &   35
Global Security**

Fellows spend one year contributing external perspective and expertise on technology in defense policy, national and international security and systems analysis, and assisting with project and program oversight. Fellows will be placed in one of the following: Federal Bureau of Investigation (FBI), Nuclear Regulatory Commission, U.S. Department of Defense or U.S. Department of Homeland Security. Fellows will work on a variety of issues related to defense policy, national security, weapons of mass destruction, technology applications, defense systems analysis and support, and program oversight and management. Assignments may involve significant interagency, congressional or international activity. The program includes an orientation on international affairs and executive branch and congressional operations, and a year-long seminar series on issues involving science, technology and public policy. Fellows should not expect to work specifically on technical issues related to their dissertations or previous post-doctoral appointments, but rather to apply their technical, organizational, and communication skills to technical and policy issues in the placement agency. Additionally, Fellows help increase the awareness of the placement agency and its departments as a challenging and rewarding career environment for scientists and engineers.

Requirements: All applicants must have a doctoral-level degree (PhD, MD, DVM, DSc, PharmD, and other terminal degrees), in any physical, biological, health/medical or social science, any field of engineering, or any relevant interdisciplinary field (individuals with a master, degree in engineering and at least three years of post-degree professional experience also may apply). Note: All requirements for the degree must be completed by the application deadline. A prospective fellow must demonstrate exceptional competence in their specialty appropriate to their career stage, and have the strong endorsement of three references. Show an understanding of the opportunities for science and engineering to support a broad range of non-scientific issues, and display a commitment to apply their scientific or technical expertise to serve society; Exhibit awareness and sensitivity to the political, economic and social issues that influence policy; Be articulate communicators, both verbally and in writing, to decision-makers and non-scientific audiences, and have the ability to work effectively with individuals and groups outside the scientific community; Demonstrate integrity and good judgment, and the flexibility and willingness to address policy issues outside their scientific realm.

Restrictions: Federal employees are ineligible. All applicants must be U.S. citizens (dual citizenship from the United States and another country is acceptable). Some program areas and host agencies seek additional qualifications. Holding a Title 42 position is considered federal employment and therefore renders an applicant ineligible for a AAAS Science & Technology Policy Fellowship. Individuals working in postdoctoral positions at government labs or agencies, for private contractors, and in state government posts are not federal employees and therefore are eligible for the AAAS Fellowships.

Amount of Grant: Stipend: (depending on degree level and post-degree experience) $67,000-$87,000. Plus professional development, travel and relocation reimbursement and health insurance

Date(s) Application is Due: Dec 15

Contact: Rick Kempinski, Senior Program Manager; (202) 326-6700; fax: (202) 289-4950; email: rkempinski@aaas.org or fellowships@aaas.org

Internet: http://fellowships.aaas.org/02_Areas/02_index.shtml

Sponsor: American Association for the Advancement of Science

1200 New York Avenue, NW

Washington, DC 20005

AAAS Science Journalism Awards   36

The awards, which are cosponsored by AAAS and Johnson & Johnson Pharmaceutical Research & Development, recognize outstanding reporting for a general audience and honor individuals (rather than institutions, publishers or employers) for their coverage of the sciences, engineering and mathematics. The winning journalists have helped to foster the public's understanding and appreciation of science. Independent screening and judging committees select the winning journalists and their entries based on scientific accuracy, initiative, originality, clarity of interpretation and value in fostering a better understanding of science by the public. One award in each of six categories is given annually: newspapers with daily circulation more than 100,000, newspapers with circulation of less than 100,000, general circulation magazines, radio, and television, and online. Additional International category - Children's Science News (Provide English translations of foreign-language entries whenever possible).

Requirements: Persons other than the author may submit entries in accordance with the rules of the competition. The contest is open to journalists who have published or broadcast in media outlets accessible to the general public. Except for the Children's Science News award, which is open to international media outlets, the submitted materials must have been published, broadcast or posted online within the United States by a U.S.-based news organization. Print articles must be available to the public by subscription or newsstand sales.

Restrictions: Items exclusively concerning health or medical treatment; Items published originally in AAAS publicaitons or produced by AAAS; Items by employees of AAAS or Johnson & Johnson Pharmaceutical Research Development, LLC. Winners from the previous year are not eligible. Individuals winning three times are no longer eligible.

Amount of Grant: $3,000 each, plus reimbursement for reasonable travel and hotel expenses. In cases of multiple authors or producers, only one person's travel expenses will be covered.

Date(s) Application is Due: Aug 1

Contact: Awards Coordinator; (202) 326-6781; fax: (202) 789-0455; email: media@aaas.org

Internet: http://www.aaas.org/aboutaaas/awards/sja/index.shtml

Sponsor: American Association for the Advancement of Science

1200 New York Avenue, NW

Washington, DC 20005

AAAS Scientific Freedom and Responsibility Award   37

The award is presented to honor scientists and engineers whose exemplary actions have served to foster scientific freedom and responsibility and recognizes scientists and engineers who have: acted to protect the public's health, safety, or welfare; or focused public attention on important potential impacts of science and technology on society by their responsible participation in public policy debates; or established important new precedents in carrying out the social responsibilities or in defending the professional freedom of scientists and engineers. Nominations should include nominator's name, address, and phone number; and the name and address of the nominee; a summary of the action or actions that form the basis for the nomination (about 250 words); a longer statement (no more than three pages) providing additional details of the actions for which the candidate is nominated; at least two letters of support with addresses and phone numbers; and the candidate's vita (no more than three pages). Any documentation—books, articles, etc.—that illuminates the significance of the nominee's achievements may also be submitted.

Requirements: The award is open to all regardless of nationality or citizenship. Eligible nominees are scientists and engineers who have acted to protect the public's health, safety, or welfare; or to focus public attention on potentially serious impacts of science and technology on society by their responsible participation in public policy debates; or to establish important new precedents in social responsibility or in defending the professional freedom of scientists and engineers.

Restrictions: All materials become the property of AAAS.

Samples: Past recipients: 2006, Eugenie Scott, the Dover High School Science Department, and R. Wesley McCoy - These dedicated individuals are honored for their determination to defend sound education in U.S. public schools by vigorously challenging attempts to introduce intelligent design into science classes; 2005, Dr. David Michaels, recognized for his commitment to obtain justice for workers whose health suffered from working in nuclear weapons programs, and for advocating scientific integrity in public policy making; 2004, The Recombinant DNA Advisory Committee (RAC) of the National Institutes of Health, for steadfast commitment to academic freedom in the face of mounting social and political pressure.

Amount of Grant: $5,000, a commemorative plaque, complimentary registration, and reimbursement for travel and hotel expenses to attend the AAAS Annual Meeting.

Date(s) Application is Due: Sep 1

Contact: Deborah Runkle; (202) 326-6794; fax: (202) 289-4950; email: drunkle@aaas.org

Internet: http://www.aaas.org/aboutaaas/awards/freedom/index.shtml

Sponsor: American Association for the Advancement of Science

1200 New York Avenue NW

Washington, DC 20005

AAB Berlin Prize Fellowships   38

The academy welcomes younger as well as established scholars, artists, and professionals who wish to engage in independent study in Berlin for an academic semester or, in special cases, for an entire academic year. Fellowships have been awarded to scholars working in a variety of disciplines, including history, political science, economics, German studies, art history, musicology, anthropology, law, and linguistics, as well as to writers and poets, public policy experts, journalists, composers, and artists. Fellows are encouraged to work in association with a Berlin institution such as a museum, library, archive, university, government agency, studio, or media organization. Guidelines and application are available online.

Requirements: All candidates should be permanently based in the United States and must be either U.S. citizens or permanent residents. They must have completed their doctorate or equivalent professional degree. Candidates need not work on German topics, but their project descriptions should explain how a residency in Berlin will contribute to further professional development.

Restrictions: Applications are accepted from all fields except for the visual arts, music composition, and music performance.

Samples: Henry Cole, Harvard U—for work on a fifth volume of poetry.; Jeremy King, Mount Holyoke College—for the project The Nationalism of East Central Europe, 1848-Present.; Donald Shriver, Union Theological Seminary—for the project Repentance in Politics.

Amount of Grant: $3,500-$5,000 per month

Date(s) Application is Due: Oct 25

Contact: Fellowship Administrator; (212) 588-1755; fax: (212) 588-1758; email: nyoffice@americanacademy.de

Internet: http://www.americanacademy.de/home/fellows/applications/

Sponsor: American Academy in Berlin

14 East 60th Street, Suite 604

New York, NY 10022

AAB Public Policy Fellowships   39

Short term fellowships of six to eight weeks are available for the Bosch Fellows in Public Policy. Fellows are sought from the fine arts including painting, sculpture, music, film, and drama; scholarly disciplines such as art history, history, philosophy, sociology, political science, and public policy; as well as from professional fields including architecture, law, business, economics, and journalism. Fellows are encouraged to take up an association with a Berlin institution such as a museum, library, archive, university, government agency, film studio, or media organization. Specialists on German topics as well as other persons for whom the cultural opportunities or political setting of Berlin offers an advantageous professional venue are encouraged to apply. Completed applications, with the exception of visual art and music applications, should be received by October 25 by the Berlin office only (ATTN: Application for Fellowship, Am Sandwerder 17-19, D-14109 Berlin, Germany; +49 30 804 83 0; fax: +49 30 804 83 111; email: applications@americanacademy.de).

Requirements: Candidates must be either American citizens or permanent residents and in both cases should be permanently living and working in the United States.

Amount of Grant: $3,000-$5,000 monthly

Date(s) Application is Due: Oct 25

Contact: Fellowship Administrator; (212) 588-1755; fax: (212) 588-1758; email: nyoffice@americanacademy.de

Internet: http://www.americanacademy.de/index.php?id=6

Sponsor: American Academy in Berlin

14 E. 60th Street, Suite 604

New York, NY 10022

AACAP Beatrix A. Hamburg Award for the Best New Research Poster   40
by a Child and Adolescent Psychiatry Resident

This award recognizes the author of the best new research poster presented at the AACAP Annual Meeting. The award provides a $1,000 honorarium. The recipient will receive a plaque at the Young Leader Awards Ceremony during the AACAP Annual Meeting. There is no application process for this award. The AACAP Program Committee will review all new research poster submissions by child and adolescent psychiatry residents to determine the recipient of this award. The recipient will be notified through an acceptance letter. Non-recipients will not be notified.

Requirements: The author must be a child and adolescent psychiatry resident at an accredited institution and AACAP member. Applicants must: be a child and adolescent psychiatry resident; attend an accredited institution; be an AACAP member; and attend and present a new research poster at the AACAP Annual Meeting.

Amount of Grant: $1,000 honorarium

Date(s) Application is Due: Jun 14

Contact: Stacie Hall, Director; (202) 966-7300, ext. 113; fax: (202) 966-2891; email: shall@aacap.org

Internet: http://www.aacap.org/cs/root/research_and_training_awards/aacap_beatrix_a_hamburg_award_for_the_best_new_research_poster_by_a_child_and_adolescent_psychiatry_resident_at_the_aacap_annual_meeting

Sponsor: American Academy of Child and Adolescent Psychiatry

3615 Wisconsin Avenue, NW

Washington, DC 20016-3007

AACAP Congressional Fellowships   41

The Program is designed to educate policy makers and Congressional staff about child and adolescent psychiatry and to foster awareness of children's mental health issues. This experience will allow the Fellow to develop a keen understanding of how public policy affects patient care, education, and health insurance issues. The AACAP Congressional Fellow will be placed with a Congressional office or Committee where they will gain invaluable experience as they assist in the development of legislative and public policy initiatives. The award includes an $85,000 stipend, up to $3,000 in relocation or moving expenses, and health insurance for the length of the fellowship program.

Requirements: Successful applicants must: be an AACAP member; be a PGY-4 psychiatry resident or beyond; have completed training in or be scheduled to complete training in one of the child and adolescent psychiatry programs (CAP training program, Combined General Psychiatry/CAP Program, or Triple Board Program); have a demonstrated interest in public policy issues; exhibit an interest in applying medical and scientific knowledge to the resolution of public policy issues; and be a U.S. citizen or permanent resident.

Amount of Grant: $85,000 stipend; up to $3,000 in relocation or moving expenses

Date(s) Application is Due: Feb 1

Contact: Gabe Robbins; (202) 966-7300, ext. 117; fax: (202) 966-2891; email: grobbins@aacap.org

Internet: http://www.aacap.org/cs/root/legislative_action/2008_congressional_fellowship_program

Sponsor: American Academy of Child and Adolescent Psychiatry

3615 Wisconsin Avenue, NW

Washington, DC 20016-3007

AACAP Educational Outreach Program for Child and Adolescent   42
Psychiatry Residents

The Educational Outreach Program (EOP) funding provides the opportunity for up to 50 child and adolescent psychiatry residents to receive a formal overview to the field of child and adolescent psychiatry, establish child and adolescent psychiatrists as mentors and experience the AACAP Annual Meeting. Participants will be exposed to the breadth and depth of the field of child and adolescent psychiatry, including research opportunities, access to mentors, and various networking opportunities. Participation in this program provides up to $750 for travel expenses.

Requirements: Applicants must: be child and adolescent psychiatry residents at the time of the AACAP Annual Meeting; be currently enrolled in a residency program in the United States; be residents in their first or second year of child fellowship training are eligible (Triple Boarders in their fourth or fifth year of training in their triple board programs are eligible); either be members of the AACAP or have a membership application pending at the time of application; and attend all AACAP Annual Meeting events specified by AACAP.

Amount of Grant: $750 maximum
Date(s) Application is Due: Jul 12
Contact: Gabe Robbins; (202) 966-7300, ext. 117; fax: (202) 966-2891; email: grobbins@aacap.org
Internet: http://www.aacap.org/cs/residents/eop-capfellows
Sponsor: American Academy of Child and Adolescent Psychiatry
3615 Wisconsin Avenue, NW
Washington, DC 20016-3007

AACAP Educational Outreach Program for General Psychiatry Residents   43

The Educational Outreach Program (EOP) for General Psychiatry residents provides the opportunity for up to 20 general psychiatry residents to receive a formal overview to the field of child and adolescent psychiatry, establish child and adolescent psychiatrists as mentors and experience the AACAP Annual Meeting. Participants will be exposed to the breadth and depth of the field of child and adolescent psychiatry, including research opportunities, access to mentors, and various networking opportunities. Participation in this program provides up to $1,500 for travel expenses to the AACAP Annual Meeting.

Requirements: Applicants must: be general psychiatry residents at the time of the AACAP Annual Meeting; be currently enrolled in a residency program in the United States; be residents in their first, second or third year of general psychiatry training (Triple Boarders in their first, second or third year of training in their triple board programs are eligible); either be members of the AACAP or have a membership application pending at the time of application; and attend all AACAP Annual Meeting events specified by AACAP.

Amount of Grant: $1,500 maximum
Date(s) Application is Due: Jul 12
Contact: Gabe Robbins; (202) 966-7300, ext. 117; fax: (202) 966-2891; email: grobbins@aacap.org
Internet: http://www.aacap.org/cs/residents/eop-generalresidents
Sponsor: American Academy of Child and Adolescent Psychiatry
3615 Wisconsin Avenue, NW
Washington, DC 20016-3007

AACAP Elaine Schlosser Lewis Award for Research on Attention-Deficit Disorder   44

This annual award acknowledges outstanding leadership and continuous contributions in the field of research by giving $5,000 for the best paper published in the Journal on attention-deficit disorder written by a child and adolescent psychiatrist. The award winner will be recognized at a Distinguished Awards Luncheon and make an Honors Presentation about his or her work during the AACAP Annual Meeting.

Requirements: Nomination letters must be accompanied by a CV for the individual nominated.
Samples: Margaret Weiss, M.D., Children's Hospital Vancouver, BC—Sleep Hygiene and Melatonin Treatment for Children and Adolescents With ADHD and Initial Insomnia (2006).
Amount of Grant: $5,000
Date(s) Application is Due: May 7
Contact: Alyssa Sommer, Research Coordinator; (202) 966-7300, ext. 157; fax: (202) 364-5925; email: asommer@aacap.org
Internet: http://www.aacap.org/cs/root/research_and_training_awards/aacap_journal_awards
Sponsor: American Academy of Child and Adolescent Psychiatry
3615 Wisconsin Avenue, NW
Washington, DC 20016-3007

AACAP George Tarjan Award for Contributions in Developmental Disabilities   45

This award recognizes a child and adolescent psychiatrist and Academy member who has made significant contributions in a lifetime career or single seminal work to the understanding or care of those with mental retardation and developmental disabilities. These contributions must have national and/or international stature and clearly demonstrate lasting effects. The contributions may be in areas of teaching, research, program development, direct clinical service, advocacy or administrative commitment. A cash prize of up to $1,000 will be awarded. The award winner will be recognized at a Distinguished Awards Luncheon and make an Honors Presentation about his or her work during the AACAP Annual Meeting.

Requirements: Nomination letters must be accompanied by a CV for the individual nominated.
Samples: Peter Tanguay, M.D., University of Louisville, Louisville, KY—Autism: Fixing the DSM-IV (2006).
Amount of Grant: $1,000
Date(s) Application is Due: Apr 30
Contact: Edwin Cook, Jr., M.D. or Byron King, M.D., Co-Chairs; (202) 966-7300; fax: (202) 966-2891; email: training@aacap.org

Internet: http://www.aacap.org/cs/distinguishedmembers
Sponsor: American Academy of Child and Adolescent Psychiatry
3615 Wisconsin Avenue, NW
Washington, DC 20016

AACAP Irving Philips Award for Prevention   46

This award recognizes a child and adolescent psychiatrist and Academy member who has made significant contributions in a lifetime career or single seminal work to the prevention of mental illness in children and adolescents. These contributions must have national and/or international stature and clearly demonstrate lasting effects. The contributions may be in the areas of teaching, research, program development, direct clinical service, advocacy or administrative commitment. The award pays $2,500 to the winner and a $2,000 donation to a prevention program or center of the awardee's choice. The award winner will be recognized at a Distinguished Awards Luncheon and make an Honors Presentation about his or her work during the AACAP Annual Meeting.

Requirements: Nomination letters must be accompanied by a CV for the individual nominated.
Samples: Charles Zeanah, M.D., Tulane University New Orleans, LA—Preventing Adverse Outcomes for Abandoned Children: The Bucharest Early Intervention Project (2006).
Amount of Grant: $2,500 to winner, plus $2,000 donation
Date(s) Application is Due: Apr 30
Contact: Harry H. Wright, M.D. Chair, Prevention Committee; (202) 966-7300; fax: (202) 966-2891; email: training@aacap.org
Internet: http://www.aacap.org/cs/distinguishedmembers
Sponsor: American Academy of Child and Adolescent Psychiatry
3615 Wisconsin Avenue, NW
Washington, DC 20016

AACAP Jeanne Spurlock Lecture and Award on Diversity and Culture   47

The lecture and award reflects the spirit of the nomination for outstanding contributions to diversity and culture and in some way encourages individuals from diverse cultural backgrounds to become child and adolescent psychiatrists. Nominees should be individuals who have made contributions in the areas of social awareness including: civil rights; spirituality and/or religion; social welfare; public information; scientific research; education and mentoring; and the arts (literature, theatre, music, painting, sculpture or photography). The award includes an honorarium of $2,500, with the award winner being recognized at a Distinguished Awards Luncheon where he or she will make an honors award presentation about his or her work during the AACAP Annual Meeting.

Amount of Grant: $2,500
Date(s) Application is Due: May 1
Contact: Jennifer Mecidus, Clinical Practice Manager; (202) 966-7300, ext. 137; fax: (202) 966-9518; email: jmedicus@aacap.org
Internet: http://www.aacap.org/cs/awards
Sponsor: American Academy of Child and Adolescent Psychiatry
3615 Wisconsin Avenue, NW
Washington, DC 20016-3007

AACAP Jeanne Spurlock Minority Medical Student Clinical Fellowship in Child and Adolescent Psychiatry   48

Up to fourteen (14) fellowships are available each year for minority medical students to explore a career in child and adolescent psychiatry, gain valuable work experience, and meet leaders in the child and adolescent psychiatry field. The fellowship opportunity provides up to $3,500 for 12 weeks of clinical training under a child and adolescent psychiatrist mentor. Research assignments may include responsibility for part of the observation or evaluation, developing specific aspects of the research mechanisms, conducting interviews or tests, use of rating scales, and psychological or cognitive testing of subjects. Contact AACAP for applications.

Requirements: African American, Asian American, Native American, Alaskan Native, Mexican American, Hispanic, and Pacific Islander students in accredited U.S. medical schools are eligible.
Amount of Grant: $3,500 each, plus expenses to attend the annual meeting
Date(s) Application is Due: Mar 1
Contact: Gabe Robbins; (202) 966-7300, ext. 117; fax: (202) 966-2891; email: grobbins@aacap.org
Internet: http://www.aacap.org/cs/students/opportunities/SpurlockClinical
Sponsor: American Academy of Child and Adolescent Psychiatry
3615 Wisconsin Avenue, NW
Washington, DC 20016

AACAP Jeanne Spurlock Research Fellowship in Substance Abuse and Addiction for Minority Medical Students   49

The Fellowship offers a unique opportunity for minority medical students to explore a research career in substance abuse in relation to child and adolescent psychiatry, gain valuable work experience, and meet leaders in the child and adolescent psychiatry field. The fellowship opportunity provides up to $3,500 for 12 weeks of summer research under a child and adolescent psychiatrist researcher/mentor. The research training plan must provide for significant contact between the student and the mentor and for exposure to state-of-the-art drug abuse and addiction research. The plan should include program planning discussions, instruction in research planning and implementation, regular meetings with the mentor, laboratory director, and the research group, and assigned readings. Research assignments may include responsibility for part of the observation or evaluation, developing specific aspects of the research mechanisms, conducting interviews or tests, use of rating scales, and psychological or cognitive testing of subjects.

Requirements: All fellowship participants must attend the AACAP Annual Meeting. Applications are considered from African-American, Native American, Alaskan Native, Mexican American, Hispanic, Asian, and Pacific Islander students in accredited U.S. medical schools.

Amount of Grant: $3,500 maximum

Date(s) Application is Due: Mar 1

Contact: Gabe Robbins; (202) 966-7300, ext. 117; fax: (202) 966-2891; email: grobbins@aacap.org or apartner@aacap.org

Internet: http://www.aacap.org/cs/students/opportunities/SpurlockResearch

Sponsor: American Academy of Child and Adolescent Psychiatry
3615 Wisconsin Avenue, NW
Washington, DC 20016

AACAP Junior Investigator Awards   50

The AACAP Junior Investigator Award offers two awards of up to $30,000 a year for two years for child and adolescent psychiatry junior faculty (assistant professor level or equivalent). The program is intended to facilitate innovative research. The research may be basic or clinical in nature but must be relevant to our understanding, treatment and prevention of child and adolescent mental health disorders. The award also includes the cost of attending the AACAP Annual Meeting for five days.

Requirements: Recipients are required to submit a poster or oral presentation on his or her research for the AACAP's Annual Meeting. Applicants must: be board eligible or certified in child and adolescent psychiatry; have a doctoral level degree and be in a faculty or independent research position; and have an on-site mentor who has had experience in the type of research that is being proposed that will normally include work with children and adolescents. Candidates must either be AACAP members or have a membership application pending (not paid by the award).

Restrictions: Applicants who have served as Principal Investigator on an NIH R01 grant are not eligible.

Amount of Grant: $30,000 maximum per year for two years

Date(s) Application is Due: Mar 1

Contact: Alyssa Sommer, Research Coordinator; (202) 966-7300, ext. 157; fax: (202) 364-5925; email: asommer@aacap.org

Internet: http://www.aacap.org/cs/root/research_and_training_awards/aacap_junior_investigator_awards

Sponsor: American Academy of Child and Adolescent Psychiatry
3615 Wisconsin Avenue, NW
Washington, DC 20016-3007

AACAP Klingenstein Third Generation Foundation Award for Research in Depression or Suicide   51

This annual award acknowledges outstanding leadership and continuous contributions in the field of research by giving $5,000 for the best paper on suicide and/or depression published in the Journal during the past year. The award winner will be recognized at a Distinguished Awards Luncheon and make an Honors Presentation about his or her work during the AACAP Annual Meeting.

Requirements: Nomination letters must be accompanied by a CV for the individual nominated.

Samples: David DeMaso, M.D., Children, Hospital Boston, MA—Depression Experience Journal: A Computer-Based Intervention for Families Facing Childhood Depression (2006).

Amount of Grant: $5,000

Date(s) Application is Due: May 7

Contact: Gabe Robbins; (202) 966-7300, ext. 117; fax: (202) 966-2891; email: grobbins@aacap.org

Internet: http://www.aacap.org/cs/root/research_and_training_awards/aacap_journal_awards

Sponsor: American Academy of Child and Adolescent Psychiatry
3615 Wisconsin Avenue, NW
Washington, DC 20016-3007

AACAP Norbert and Charlotte Rieger Award for Scientific Achievement   52

This award is given annually for the most significant article published in the Journal of the American Academy of Child and Adolescent Psychiatry (JAACAP) in the past year (July-May). The article must be written by a psychiatrist specializing in child and/or adolescent psychiatry and cannot be a review.

Requirements: The only submissions accepted are papers from child and adolescent psychiatrists that have been published in the Journal during the past year.

Samples: Gail Bernstein, M.D., University of Minnesota, Minneapolis, MN—School-Based Interventions for Anxious Children (2006).

Amount of Grant: $4,500 maximum

Date(s) Application is Due: May 7

Contact: Gabe Robbins; (202) 966-7300, ext. 117; fax: (202) 966-2891; email: grobbins@aacap.org

Internet: http://www.aacap.org/cs/root/research_and_training_awards/aacap_journal_awards

Sponsor: American Academy of Child and Adolescent Psychiatry
3615 Wisconsin Avenue, NW
Washington, DC 20016-3007

AACAP Norbert and Charlotte Rieger Psychodynamic Psychotherapy Award   53

The Award recognizes the best published or unpublished paper, written by an AACAP member that addresses the use of psychodynamic psychotherapy in clinical practice and fosters development, teaching, and practice of psychodynamic psychotherapy within child and adolescent psychiatry. Papers that express a novel hypothesis, raise questions about existing theory, or integrate new neuroscience and developmental psychotherapy research with psychodynamic principles may be nominated. Unpublished, new papers and papers published within the last three years may be submitted by their authors. Published papers may be nominated by any member of the AACAP. Authors may be senior or junior faculty members or residents. Delivery of the winning paper will be made at the AACAP Annual Meeting Honors Presentation.

Samples: Theodore Gaensbauer, M.D., University of Colorado Health Sciences Center, Denver, CO—Traumatized Young Children: The Treatment Process (2006).

Amount of Grant: $4,500

Date(s) Application is Due: May 7

Contact: Gabe Robbins; (202) 966-7300, ext. 117; fax: (202) 966-2891; email: grobbins@aacap.org

Internet: http://www.aacap.org/cs/2007_press_releases/rachel_seidel_md_wins_aacap_2007_norbert_and_charlotte_rieger_psychodynamic_psychotherapy_award

Sponsor: American Academy of Child and Adolescent Psychiatry
3615 Wisconsin Avenue, NW
Washington, DC 20016-3007

AACAP Pilot Research Award for Attention-Deficit Disorder, Supported by the Elaine Schlosser Lewis Fund   54

The Award for Attention-Deficit Disorder is available for a junior faculty or child psychiatry resident, and is supported by the Elaine Schlosser Lewis Fund. The award recipient will be encouraged to work with a child and adolescent psychiatric investigator with expertise in his or her particular area of interest. Work must be completed within one year of receipt of the award.

Requirements: Candidates must be board eligible, certified in child and adolescent psychiatry, or enrolled in a child psychiatry residency or fellowship program. Candidates must also have a faculty appointment in an accredited medical school or be in a fully accredited child and adolescent psychiatry clinical research or training program.

Restrictions: At the time of application, candidates may not have more than two years experience following graduation from residency/fellowship training. Candidates must not have any previous significant, individual research funding in the field of child and adolescent mental health. These include the following: NIMH/NIH Funding (Small Grants, T Award or R-01) or similar foundation or industry research funding.

Amount of Grant: $15,000

Date(s) Application is Due: May 7

Contact: Alyssa Sommer, Research Coordinator; (202) 966-7300, ext. 157; fax: (202) 364-5925; email: asommer@aacap.org

Internet: http://www.aacap.org/cs/root/research_and_training_awards/pilot_research_award_for_attention_disorder_for_a_junior_faculty_or_child_psychiatry_resident_supported_by_the_elaine_schlosser_lewis_fund

Sponsor: American Academy of Child and Adolescent Psychiatry
3615 Wisconsin Avenue, NW
Washington, DC 20016-3007

AACAP Pilot Research Award for Learning Disabilities, Supported by the Elaine Schlosser Lewis Fund   55

The Award for Learning Disabilities and Psychiatric Disorders is available for a junior faculty or child psychiatry resident, and is supported by the Elaine Schlosser Lewis Fund. The award encourages work with a child and adolescent psychiatric investigator with expertise in his or her particular area of interest. Work must be completed within one year of receipt of the award. Please note that travel to both the Annual Meeting and the Research Update Luncheon is not included in the award and is the responsibility of the recipient.

Requirements: Candidates must be board eligible, certified in child and adolescent psychiatry, or enrolled in a child psychiatry residency or fellowship program. Candidates must also have a faculty appointment in an accredited medical school or be in a fully accredited child and adolescent psychiatry clinical research or training program.

Restrictions: Candidates must not have any previous significant, individual research funding in the field of child and adolescent mental health. These include the following: NIMH/NIH Funding (Small Grants, T Award or R-01) or similar foundation or industry research funding. The recipient must be an AACAP member at the time of application and agree to present his or her research.

Amount of Grant: $15,000

Date(s) Application is Due: May 7

Contact: Alyssa Sommer, Research Coordinator; (202) 966-7300, ext. 157; fax: (202) 364-5925; email: asommer@aacap.org

Internet: http://www.aacap.org/cs/root/research_and_training_awards/pilot_research_award_for_learning_disabilities_for_a_junior_faculty_or_child_psychiatry_resident_supported_by_the_elaine_schlosser_lewis_fund

Sponsor: American Academy of Child and Adolescent Psychiatry
3615 Wisconsin Avenue, NW
Washington, DC 20016-3007

AACAP Pilot Research Awards, Supported by Eli Lilly and Company   56

Eight awards are available each year to junior faculty and child psychiatry fellows for pilot research with child and adolescent psychiatric researchers. Award recipients will be matched with child and adolescent psychiatric investigators working in their particular area of interest to act as consultants or mentors during the course of the award. Work must be completed within one year of receipt of the award.

Requirements: Candidates must be board eligible, certified in child and adolescent psychiatry, or enrolled in a child psychiatry residency or fellowship program. Candidates must have a faculty appointment in an accredited medical school or be in a fully accredited child and adolescent psychiatry clinical research or training program.

Restrictions: At the time of application, candidates may not have more than two years experience following graduation from residency/fellowship training. Candidates must not have any previous significant, individual research funding in the field of child and adolescent mental health. These include the following: NIMH/NIH Funding (Small Grants, T Award or R-01) or similar foundation or industry research funding.

Amount of Grant: $15,000
Date(s) Application is Due: May 7
Contact: Alyssa Sommer, Research Coordinator; (202) 966-7300, ext. 157; fax: (202) 364-5925; email: asommer@aacap.org
Internet: http://www.aacap.org/cs/root/research_and_training_awards/aacap_pilot_research_award_for_junior_faculty_and_child_psychiatry_fellows_supported_by_lilly_usa_llc
Sponsor: American Academy of Child and Adolescent Psychiatry
3615 Wisconsin Avenue, NW
Washington, DC 20016-3007

AACAP Quest for the Test Bipolar Disorder Pilot Research Award 57

The AACAP Quest for the Test Bipolar Disorder Pilot Research Award, supported by The Ryan Licht Sang Bipolar Foundation, offers $15,000 for child and adolescent psychiatry residents and junior faculty who have an interest in beginning a career in child and adolescent mental health research. Proposed research projects should address questions concerning the pathology or diagnosis of children and adolescents with bipolar disorder. The award also includes the cost of attending the AACAP Annual Meeting for five days.

Requirements: The recipient is required to submit a poster presentation on his or her research for the AACAP's Annual Meeting. Candidates must be board eligible, certified in child and adolescent psychiatry, or enrolled in a child psychiatry residency or fellowship program. Candidates must also have a faculty appointment in an accredited medical school or be in a fully accredited child and adolescent psychiatry clinical research or training program.

Restrictions: At the time of application, candidates may not have more than two years experience following graduation from residency/fellowship training. Candidates must not have any previous significant, individual research funding in the field of child and adolescent mental health. These include the following: NIH Funding (K or R level grants) or similar foundation or industry research funding.

Amount of Grant: $15,000
Date(s) Application is Due: May 7
Contact: Alyssa Sommer, Research Coordinator; (202) 966-7300, ext. 157; fax: (202) 364-5925; email: asommer@aacap.org
Internet: http://www.aacap.org/cs/root/research_and_training_awards/aacap_quest_for_the_test_bipolar_disorder_pilot_research_award
Sponsor: American Academy of Child and Adolescent Psychiatry
3615 Wisconsin Avenue, NW
Washington, DC 20016-3007

AACAP Rieger Service Program Award for Excellence 58

This award recognizes innovative programs that address prevention, diagnosis, or treatment of mental illnesses in children and adolescents, and serve as model programs to the community. This award of $4,500 is shared among the awardee and his or her service program. The award winner will be recognized at a Distinguished Awards Luncheon and make an Honors Presentation about his or her work during the AACAP Annual Meeting.

Requirements: Nomination letters must be accompanied by a CV and any support materials for the individual or organization nominated.
Samples: Bradley Stein, M.D., Ph.D., Rand Corporation, Pittsburgh, PA—The Development, Implementation, and Dissemination of an Effective School-based Program for Children Exposed to Violence (2006).
Amount of Grant: $4,500
Date(s) Application is Due: Apr 30
Contact: Kaye McGinty, M.D., and Mark Chenven, M.D., Co-Chairs; (202) 966-7300; fax: (202) 966-2891; email: aboccanelli@aacap.org
Internet: http://www.aacap.org/cs/distinguishedmembers
Sponsor: American Academy of Child and Adolescent Psychiatry
3615 Wisconsin Avenue, NW
Washington, DC 20016

AACAP Robert Cancro Academic Leadership Award 59

This award recognizes a currently serving General Psychiatry Training Director, Medical School Dean, CEO of a Training Institution, Chair of a Department of Pediatrics or Chair of a Department of Psychiatry for his or her contributions to the promotion of child and adolescent psychiatry. Named in honor of Robert Cancro, M.D., Chairman at New York University, this award offers $500 to the awardee. Nominations for the award may be made by Child and Adolescent Training Directors or Division Directors. The recipient of this award will receive a plaque and be recognized at the AACAP Annual Meeting. The award recipient will be honored at the Distinguished Awards Luncheon, Training Directors Dinner, and will provide an Honors Presentation on his or her work during the Annual Meeting.

Requirements: Nominations must include a CV for the individual nominated
Samples: Susan McLeer, M.D., Drexel University Philadelphia, PA—Money, Mergers and Hospital Closures: Building and Rebuilding Programs in the 21st Century (2006).
Amount of Grant: $500
Date(s) Application is Due: Apr 30
Contact: Jeffrey Hunt, M.D. and Dorothy Stubbe, M.D., Co-Chairs; (202) 966-7300; fax: (202) 966-2891; email:
Internet: http://www.aacap.org/cs/distinguishedmembers
Sponsor: American Academy of Child and Adolescent Psychiatry
3615 Wisconsin Avenue, NW
Washington, DC 20016

AACAP Robinson-Cunningham Award 60

The Award is given for the best manuscript written by a child and adolescent psychiatrist during residency training. The paper must involve children, adolescents, or their families and be published in a professional, peer-reviewed journal within 3-5 years of graduation from a residency training program. The recipient will receive a $100 honorarium and a plaque at the Young Leaders Awards Ceremony during the AACAP Annual Meeting.

Amount of Grant: $100
Date(s) Application is Due: May 7
Contact: Gabe Robbins; (202) 966-7300, ext. 117; fax: (202) 966-2891; email: grobbins@aacap.org
Internet: http://www.aacap.org/cs/root/research_and_training_awards/the_robinsoncunningham_award
Sponsor: American Academy of Child and Adolescent Psychiatry
3615 Wisconsin Avenue, NW
Washington, DC 20016-3007

AACAP Sidney Berman Award for the School-Based Study and Intervention for Learning Disorders and Mental Ilness 61

This award recognizes an individual or program that has shown outstanding achievement in the school-based study or delivery of intervention for learning disorders and mental illness. A cash prize of $4,500 will be awarded. The award winner will be recognized at a Distinguished Awards Luncheon and make an Honors Presentation about his or her work during the AACAP Annual Meeting.

Requirements: Nomination letters must be accompanied by a CV for the individual nominated or program information.
Amount of Grant: $4,500
Date(s) Application is Due: Apr 30
Contact: Steven Adelsheim, M.D. and Richard Mattison, M.D., Co-Chairs; (202) 966-7300; fax: (202) 966-2891; email: aboccanelli@aacap.org
Internet: http://www.aacap.org/cs/distinguishedmembers
Sponsor: American Academy of Child and Adolescent Psychiatry
3615 Wisconsin Avenue, NW
Washington, DC 20016

AACAP Simon Wile Leadership in Consultation Award 62

This annual award acknowledges outstanding leadership and continuous contributions in the field of liaison child and adolescent psychiatry. Nomination letters must be accompanied by a CV for the individual nominated. The award winner will be recognized at a Distinguished Awards Luncheon and make an Honors Presentation about his or her work during the AACAP Annual Meeting.

Requirements: Practicing child and adolescent psychiatrists are eligible for nomination.
Amount of Grant: $500 honorarium and duty of presentation at annual meeting
Date(s) Application is Due: Apr 30
Contact: Richard Martini, M.D., David DeMaso, M.D., and Read Sulik, M.D., Co-Chairs; (202) 966-7300; fax: (202) 966-2891; email: aboccanelli@aacap.org
Internet: http://www.aacap.org/cs/distinguishedmembers
Sponsor: American Academy of Child and Adolescent Psychiatry
3615 Wisconsin Avenue, NW
Washington, DC 20016

AACAP Summer Medical Student Fellowships 63

The American Academy of Child and Adolescent Psychiatry (AACAP) is offering summer fellowships for work with a child and adolescent psychiatrist mentor in a clinical or research setting. Sponsored by the AACAP Campaign for America's Kids, the Fellowships offer a chance for up to 6 medical students to explore a career in child and adolescent psychiatry, gain valuable work experience, and meet leaders in the child and adolescent psychiatry field. The fellowship opportunity provides up to $3,500 for 12 weeks of clinical or research training under a child and adolescent psychiatrist mentor. The anticipated application deadline for this fellowship is April.

Requirements: Applicants must be students enrolled in accredited U.S. medical schools.
Amount of Grant: $3,500
Contact: Gabe Robbins; (202) 966-7300, ext. 117; fax: (202) 966-2891; email: grobbins@aacap.org or apartner@aacap.org; Efrain Bleiberg, M.D.; (713) 275-5213; email: ebleiberg@menninger.edu
Internet: http://www.aacap.org/cs/students/summerfellowship
Sponsor: American Academy of Child and Adolescent Psychiatry
3615 Wisconsin Avenue, NW
Washington, DC 20016

AACAP Systems of Care Special Program Scholarships   **64**

The scholarship includes up to $750 for travel expenses to the AACAP's Annual Meeting. Residents may apply to both the Special Program Scholarship and the Educational Outreach Program (EOP); however, individuals cannot receive both awards at the same time. This includes airfare, hotel, and meals, maximum $75 per day.

Amount of Grant: $750

Date(s) Application is Due: Jul 13

Contact: Adriano Boccanelli, Clinical Practice Assistant; (202) 966-7300, ext. 133; fax: (202) 966-2891; email: aboccanelli@aacap.org

Internet: http://www.aacap.org/cs/root/medical_students_and_residents/residents/2009_systems_of_care_special_program_scholarship

Sponsor: American Academy of Child and Adolescent Psychiatry

3615 Wisconsin Avenue, NW

Washington, DC 20016

AACAP-NIDA K12 Career Development Awards   **65**

The grant provides up to five years of salary support and mentored addiction research training for qualified child and adolescent psychiatrists (CAPs) who intend to establish careers as independent investigators in mental health and addiction research. Letters of Intent are required and due on or before January 5. Full applications are due March 30, with new grants awarded on June 1. Applicants are strongly encouraged to contact program staff regarding the program and candidate qualifications before submitting Letters of Intent.

Requirements: Eligible applicants must be a child and adolescent psychiatrist with a commitment and goal of becoming an independent investigator in addiction-related research focused on children and adolescents. Applicants must also be U.S. citizens, non-citizen nationals, and/or individuals with permanent residence status. Women and minority candidates are especially encouraged to apply.

Amount of Grant: $140,000 maximum

Date(s) Application is Due: Jan 5

Contact: Stacia Fleisher; (202) 966-7300; fax: (202) 966-2891; email: sfleisher@aacap.org or research@aacap.org

Internet: http://www.aacap.org/cs/root/research_and_training_awards/aacap_nida12

Sponsor: American Academy of Child and Adolescent Psychiatry

3615 Wisconsin Avenue, NW

Washington, DC 20016

AACC International Alsberg-French-Schoch Memorial Lectureship Awards   **66**

This award, sponsored by the Corn Refiners Association, is the organization's highest award given every two to four years in recognition of scientists who have made significant and superior contributions to fundamental starch science. The award is given every two to four years and includes a $2,000 honorarium and a plaque, as well as the opportunity to present a lecture on some phase of starch science before an annual meeting of the Association. To be nominated for an award, contact information for both the person suggesting the nominee and the individual being nominated, as well as a summary of the nominee's career, accomplishments, and contributions should be forwarded directly to AACC International headquarters. Nominations should address the activities most relevant to the award, with a clear statement of the impact the nominee has had in his/her specific area of expertise.

Requirements: Nominee must be an AACC International member. All members are strongly encouraged to submit nominations for these awards at any time.

Samples: Yasuhito Takeda, Vice-President, Kagoshima University, Kagoshima, Kagoshima Prefecture, Japan (2007); Jay-lin Jane, Professor, Iowa State University, Ames, IA (2005).

Amount of Grant: $2,000

Date(s) Application is Due: Jan 1

Contact: Khalil Khan, AACC International Award Chair; (651) 454-7250; fax: (651) 454-0766; email: khalil.khan@ndsu.edu or aacc@scisoc.org

Internet: http://www.aaccnet.org/membership/awards.asp#ALSBERG

Sponsor: American Association of Cereal Chemists International

3340 Pilot Knob Road

St. Paul, MN 55121

AACC International Applied Research Award   **67**

The Award is presented for a significant body of distinguished contributions to the application of science in the cereals area. Of primary interest is the development of novel machinery, processes, products, patents, tests, methodologies, or procedures taking advantage of the wealth of accumulated knowledge from basic research as available in the literature together with basic/applied research performed by the applicant in developing the product/patent. The methodologies, applications, products and processes must have a clear and sustained record of adoption and/or commercialization and must have contributed positively to the development of science and knowledge of relevance to the cereal/food industry. The award will consist of a $2,000 honorarium and a plaque and will be given to an individual scientist or a team of scientists. Recipients of the Award are also accorded the status of AACC International Fellow for their contributions leading to this award. Awardees will have the opportunity to present a lecture during an annual meeting of the Association.

Requirements: Nominee must be an AACC International member. All members are strongly encouraged to submit nominations for these awards at any time.

Samples: Philip C. Williams, PDK Projects, Nanaimo, B.C. (2007).

Amount of Grant: $2,000

Date(s) Application is Due: Jan 1

Contact: Khalil Khan, AACC International Award Chair; (651) 454-7250; fax: (651) 454-0766; email: khalil.khan@ndsu.edu or aacc@scisoc.org

Internet: http://www.aaccnet.org/membership/awards.asp#Applied_Research_Award

Sponsor: American Association of Cereal Chemists International

3340 Pilot Knob Road

St. Paul, MN 55121

AACC International Biotechnology Division's Bruce Wasserman Young Investigator Award   **68**

The purpose of the Award is to recognize young scientists who have made outstanding contributions to the field of cereal biotechnology. The research recognized by this award should be relevant to the broad aims of the AACC International. The work can be either basic (e.g., improved understanding of structure-function relationships) or applied (e.g., new approaches for cereal production, breeding or utilization). For the purposes of this award, 'Cereal Biotechnology' is broadly defined, and encompasses any significant body of research using plants, microbes, genes, proteins or other biomolecules. Contributions in the disciplines of genetics, molecular biology, biochemistry, microbiology and fermentation engineering are all included. The research will be evaluated by its impact as measured as either enhanced knowledge of cereal functionality or improved cereal production or utilization. Achievements will be considered in relation to the age and experience of the nominee. AACC International membership is not required.

Requirements: Nominees must be no older than 40 by 1st of July, but nominations of younger scientists are particularly encouraged.

Amount of Grant: $1,000

Date(s) Application is Due: Jan 15

Contact: Michael J. Giroux; (406) 994-7877; fax: (406) 994-7600; email: mgiroux@montana.edu

Internet: http://www.aaccnet.org/divisions/divisionsdetail.cfm?CODE=BIOTECH

Sponsor: American Association of Cereal Chemists International

3340 Pilot Knob Road

St. Paul, MN 55121

AACC International Carl Wilhelm Brabender Award   **69**

The Award provides a plaque and a travel grant to enable the recipient to visit scientists and laboratories in other parts of the world to increase the awardee's knowledge and experience in the field. The only qualification is to be professionally active in rheology, specifically those engaged in the milling and baking industry. The Awardee shall present an award address at the AACC International Annual Meeting or at one of the three annual meetings of the AG at which the award is given.

Samples: Jozef Kokini, Director of the Center for Advanced Food Technology, Rutgers University (2002); Bert D'Appolonia, Professor Emeritus of cereal science, North Dakota State University (1998).

Date(s) Application is Due: Jan 1

Contact: Khalil Khan, AACC International Award Chair; (651) 454-7250; fax: (651) 454-0766; email: khalil.khan@ndsu.edu or aacc@scisoc.org

Internet: http://www.aaccnet.org/membership/awards.asp#BRABENDER

Sponsor: American Association of Cereal Chemists International

3340 Pilot Knob Road

St. Paul, MN 55121

AACC International Edith A. Christensen Award   **70**

In recognition of the central role that analytical methodology and approved methods have played in the history of AACC International and their future importance, the Edith A. Christensen Award for outstanding contributions in analytical methodology was created to recognize scientific contributions in analytical methodology, and service to the society through approved methods leadership. The award consists of a suitably inscribed plaque and a $1,500 honorarium.

Requirements: Nomination must be made by letter of nomination detailing the nominee's accomplishments in analytical methodology and the approved methods process, and two additional letters of support for the nominee.

Samples: Charles Gaines, $1,500—award recipient (2007); Philip C. Williams, $1,500—award recipient (2006); Edith A. Christensen, $1,500—award recipient (2005).

Amount of Grant: $1,500 honorarium

Date(s) Application is Due: Jan 1

Contact: Khalil Khan, AACC International Award Chair; (651) 454-7250; fax: (651) 454-0766; email: khalil.khan@ndsu.edu or aacc@scisoc.org

Internet: http://www.aaccnet.org/membership/awards.asp#Christensen

Sponsor: American Association of Cereal Chemists International

3340 Pilot Knob Road

St. Paul, MN 55121

AACC International Engineering and Processing Division's Stanley Watson Award   **71**

The Stanley Watson Award, established in 2001, is named in honor of an early corn wet-milling pioneer who started his research at the U.S.DA during the early 1940s. Watson made substantial contributions in the improvement of the wet-milling process. The award recognizes outstanding AACC International members who, through the application of engineering principles, have made significant contributions in the area of cereal/grain processing. The award is presented biannually. Nominations are accepted through June 12.

Requirements: Nominations must include: name of nominee; position; complete address and contact information; area of research; and a nominating statement detailing the

nominee's significant contributions in the area of cereal/grain processing via application of engineering principles.
Samples: Do Sup Chung—recipient (2003); Stanley Watson—recipient (2001).
Date(s) Application is Due: Jun 12
Contact: Girish Ganjyal, Division Chair; (406) 994-7877; fax: (406) 994-7600; email: girish.ganjyal@mgpingredients.com
Internet: http://www.aaccnet.org/divisions/divisionsdetail.cfm?CODE=ENG
Sponsor: American Association of Cereal Chemists International
3340 Pilot Knob Road
St. Paul, MN 55121

AACC International Excellence in Teaching Award 72
The Award is to be presented to an AACC International member and current teacher who has made significant contributions through teaching in the broad field of Cereal Science and Technology. The award consists of an honorarium of $1,500 and a suitably inscribed plaque. Nominations for the Award are due January 1. To be nominated for an award, contact information for both the person suggesting the nominee and the individual being nominated, as well as a summary of the nominee's career, accomplishments, and contributions should be forwarded directly to AACC International headquarters. Nominations should address the activities most relevant to the award, with a clear statement of the impact the nominee has had in his/her specific area of expertise.
Requirements: Nominee must be an AACC International member. All members are strongly encouraged to submit nominations for these awards at any time.
Samples: Ralph Waniska, Professor of Food Science and Technology, Texas A&M University (2007); Jan Delcour, Department of Food and Microbial Technology, Katholieke Universiteit Leuven, the Netherlands (2006); John R. N. Taylor, Department of Food Science, University of Pretoria, Pretoria, South Africa (2005).
Amount of Grant: $1,500
Date(s) Application is Due: Jan 1
Contact: Khalil Khan, AACC International Award Chair; (651) 454-7250; fax: (651) 454-0766; email: khalil.khan@ndsu.edu or aacc@scisoc.org
Internet: http://www.aaccnet.org/membership/awards.asp#EXCELLENCE%20IN%20TEACHING
Sponsor: American Association of Cereal Chemists International
3340 Pilot Knob Road
St. Paul, MN 55121

AACC International Fellowships 73
The AACC International Board of Directors established a Fellows program in 1985 to honor association members who have made distinguished contributions to the field of cereal science and technology in research, industrial achievement, leadership, education, administration, communication, or regulatory affairs. Any member who has been a member for 10 years and made such a contribution is eligible.
Samples: Tony Blakeney—2007 Fellow.; Jim Dexter—2007 Fellow.; Ravindra Chibbar—2006 Fellow.; Domenico Lafiandra—2006 Fellow.
Date(s) Application is Due: Jan 1
Contact: Khalil Khan, AACC International Award Chair; (651) 454-7250; fax: (651) 454-0766; email: khalil.khan@ndsu.edu or aacc@scisoc.org
Internet: http://www.aaccnet.org/membership/awards.asp#FELLOWS
Sponsor: American Association of Cereal Chemists International
3340 Pilot Knob Road
St. Paul, MN 55121

AACC International Manhattan Section Student Travel Awards 74
The Division strives to provide the latest information on technology to its membership. The Division supports professional development of students through Travel Awards and Scholarships. The Milling and Baking Division sponsors three student travel awards through the program administered by the AACC International Manhattan Section. The awards help students attend the AACC International Annual Meeting. Complete information and application forms are available online. Applications are due August 1.
Date(s) Application is Due: Aug 1
Contact: Robert A. Sombke, Division Chair; (701) 795-7274; fax: (701) 795-7251; email: rsombke@ndmill.com
Internet: http://www.aaccnet.org/divisions/divisionsdetail.cfm?CODE=MILLING
Sponsor: American Association of Cereal Chemists International
3340 Pilot Knob Road
St. Paul, MN 55121

AACC International Milling & Baking Division's John C. Halverson Memorial Lectureship 75
The John C. Halverson Memorial Lectureship is an award presented by the Milling and Baking Division to honor an individual for their contributions to the technology of milling and baking. The award consists of an honorarium and suitably inscribed certificate. The recipient is given the opportunity to present a technical paper detailing their findings and accomplishments.
Samples: Louise Slade—award recipient (2007); Harry Levine—award recipient (2007); Paul Seib—award recipient (2005).
Contact: Robert A. Sombke, Division Chair; (701) 795-7274; fax: (701) 795-7251; email: rsombke@ndmill.com
Internet: http://www.aaccnet.org/divisions/divisionsdetail.cfm?CODE=MILLING
Sponsor: American Association of Cereal Chemists International
3340 Pilot Knob Road
St. Paul, MN 55121

AACC International Milling & Baking Division's M. Rella Dwyer Graduate Fellowship 76
The Division strives to provide the latest information on technology to its membership. The Division supports professional development of students through Travel Awards and Scholarships. The M. Rella Dwyer Graduate Fellowship honors M. Rella Dwyer. A graduate fellowship award totaling $3,000 is made annually to a deserving student preparing a career in milling and baking.
Amount of Grant: $3,000
Contact: Robert A. Sombke, Division Chair; (701) 795-7274; fax: (701) 795-7251; email: rsombke@ndmill.com
Internet: http://www.aaccnet.org/divisions/divisionsdetail.cfm?CODE=MILLING
Sponsor: American Association of Cereal Chemists International
3340 Pilot Knob Road
St. Paul, MN 55121

AACC International Protein Division's Best Student Paper and Best Student Poster Award 77
The Division selects the best protein chemistry papers presented by a student at the AACC International Annual Meeting each year. The Protein Division Best Student Paper Awards consists of an engraved plaque, an honorarium, and recognition at the annual meeting. The Best Student Poster Award consists of an engraved plaque, an honorarium, and recognition at the annual meeting.
Samples: Shuping Yan, Kansas State University, $100—Best Student Poster Award (2007); O'Rouzaud-Sandez, Universidad de Sonora, Mexico, $75—Best Student Poster Award (2007).
Amount of Grant: $75 for best poster; $100 for best paper
Date(s) Application is Due: Jun 12
Contact: Scott R. Bean, Division Chair; (785) 776-2725; fax: (785) 537-5534; email: scott.bean@ars.usda.gov
Internet: http://www.aaccnet.org/divisions/divisionsdetail.cfm?CODE=PROTEIN#ProteinDivisionAwards
Sponsor: American Association of Cereal Chemists International
3340 Pilot Knob Road
St. Paul, MN 55121

AACC International Protein Division's Student Travel Grant 78
This organization within AACC International was formed in 1994 for members who have an interest in cereal proteins. A student travel grant of $250 is awarded to a student attending the annual meeting. Students presenting a poster or paper in the area of protein chemistry are eligible to apply. This award money will be $500 for the AACC International annual meeting being held in Honolulu, Hawaii.
Samples: S. Ng, Manchester Metropolitan University, United Kingdom, $250—Travel Grant Recipient (2007).
Amount of Grant: $250-$500
Date(s) Application is Due: Jun 12
Contact: Scott R. Bean, Division Chair; (785) 776-2725; fax: (785) 537-5534; email: scott.bean@ars.usda.gov
Internet: http://www.aaccnet.org/divisions/divisionsdetail.cfm?CODE=PROTEIN#ProteinDivisionAwards
Sponsor: American Association of Cereal Chemists International
3340 Pilot Knob Road
St. Paul, MN 55121

AACC International Protein Division's Walter Bushuk Graduate Award 79
The AACC International Walter Bushuk Graduate Research Award in Cereal Protein Chemistry for Ph.D. Students is presented to an individual for outstanding contributions in basic and/or applied research in cereal protein chemistry. This award recognizes research in relevant to the broad aims of AACC International. The research can be either fundamental/basic (such as improved understanding of structure-function relationships) or applied (such as development of new products, processes, techniques). The awardee will receive an honorarium of $750 and a plaque, and will be required to present a lecture at the AACC International annual meeting of the year in which the award is given.
Amount of Grant: $750 honorarium
Date(s) Application is Due: Apr 1
Contact: Scott R. Bean, Division Chair; (785) 776-2725; fax: (785) 537-5534; email: scott.bean@ars.usda.gov
Internet: http://www.aaccnet.org/divisions/divisionsdetail.cfm?CODE=PROTEIN#ProteinDivisionAwards
Sponsor: American Association of Cereal Chemists International
3340 Pilot Knob Road
St. Paul, MN 55121

AACC International Rheology Division's George W. Scott Blair Lecture Award 80
The Rheology Division od AACC International sponsors several awards to scientists and students working in the area of rheology, in order to recognize their accomplishments, subsidize their efforts and publicize the field of rheology in general. The purpose of the Award is to recognize scientists who have demonstrated outstanding ability in research in the area of rheology and texture as related to cereal based products. The Award consists of an engraved plaque, $500 honorarium and travel expenses (if necessary) to attend the AACC International Annual Meeting.
Amount of Grant: $500 plus travel expenses
Date(s) Application is Due: Jul 12

Contact: Alan A. Oppenheimer, Division Chair; (763) 764-6616; fax: (763) 764-8211; email: alan.oppenheimer@genmills.com
Internet: http://www.aaccnet.org/divisions/scottblairaward.asp
Sponsor: American Association of Cereal Chemists International
3340 Pilot Knob Road
St. Paul, MN 55121

AACC International Rheology Division's Isydore Hlynka Best Student Paper Award   81

The Rheology Division sponsors several awards to scientists and students working in the area of rheology, in order to recognize their accomplishments, subsidize their efforts and publicize the field of rheology in general. The Division selects the best rheology paper presented by a student at the annual meeting each year. The following year, the Isydore Hlynka Best Student Paper Award is presented to the student at the Rheology Division Luncheon. The award consists of an engraved plaque, $300 honorarium, and recognition at the annual meeting.
Amount of Grant: $300 honorarium
Date(s) Application is Due: Jul 12
Contact: Alan A. Oppenheimer, Division Chair; (763) 764-6616; fax: (763) 764-8211; email: alan.oppenheimer@genmills.com
Internet: http://www.aaccnet.org/divisions/divisionsdetail.cfm?CODE=RHEO#RheologyDivisionAwards
Sponsor: American Association of Cereal Chemists International
3340 Pilot Knob Road
St. Paul, MN 55121

AACC International Rheology Division's Travel Grants   82

The Rheology Division sponsors several awards to scientists and students working in the area of rheology, in order to recognize their accomplishments, subsidize their efforts and publicize the field of rheology in general. Student travel grants ($150 each) are awarded to two students attending the annual meeting. Students presenting a poster or paper in the area of rheology are eligible.
Amount of Grant: $150
Date(s) Application is Due: Jul 12
Contact: Alan A. Oppenheimer, Division Chair; (763) 764-6616; fax: (763) 764-8211; email: alan.oppenheimer@genmills.com
Internet: http://www.aaccnet.org/divisions/divisionsdetail.cfm?CODE=RHEO#RheologyDivisionAwards
Sponsor: American Association of Cereal Chemists International
3340 Pilot Knob Road
St. Paul, MN 55121

AACC International Rheology Division's Young Scientist Award   83

Sponsored by the Rheology Division of AACC International, the purpose of the Award is to recognize scientists who have completed their academic studies within the past five years and have demonstrated outstanding ability in research in the area of rheology and texture as related to cereal-based products. The Award is normally presented in odd years, and consists of an engraved plaque, $500 honorarium, and travel expenses (if necessary) to attend the AACC International Annual Meeting.
Amount of Grant: $500 plus travel expenses
Date(s) Application is Due: Jul 12
Contact: Alan A. Oppenheimer, Division Chair; (763) 764-6616; fax: (763) 764-8211; email: alan.oppenheimer@genmills.com
Internet: http://www.aaccnet.org/divisions/youngscientistaward.asp
Sponsor: American Association of Cereal Chemists International
3340 Pilot Knob Road
St. Paul, MN 55121

AACC International Thomas Burr Osborne Medal   84

The Medal is awarded to an individual whose research in the field of cereal chemistry has contributed significantly to the progress of the science. The award consists of an honorarium of $2,000 and a suitably inscribed medal in the form of a plaque. The recipient shall present an address at the AACC International Annual Meeting at which the medal is presented. Recipients of the Medal are also accorded the status of AACC International Fellow.
Samples: Herbert Wieser, DFA Lebensmittelchemie, Garching, Germany (2007); Bruce Stone, Emeritus Professor of Biochemistry, La Trobe University, Victoria, Canada (2006); Olin Anderson, Research Leader, Western Regional Research Center, Albany, California (2005).
Amount of Grant: $2,000
Date(s) Application is Due: Jan 1
Contact: Khalil Khan, AACC International Award Chair; (651) 454-7250; fax: (651) 454-0766; email: khalil.khan@ndsu.edu or aacc@scisoc.org
Internet: http://www.aaccnet.org/membership/awards.asp#BURR
Sponsor: American Association of Cereal Chemists International
3340 Pilot Knob Road
St. Paul, MN 55121

AACC International Young Scientist Research Award   85

The Award is presented to an individual for outstanding contributions in basic and applied research to cereal science with the expectation that contributions will continue. This award recognizes research relevant to the broad aims and interests of AACC International. The research can be either basic (such as improved understanding of structure-function relationships) or applied (such as development of new products, processes, or techniques) and nominees may be from government, academia or industry. Evidence of accomplishments can include awards, publications, invitations to lecture, patents and commercialized foods, processes and/or testing procedures. The research will be evaluated based on impact on advancing cereal science knowledge or improved production or utilization. Achievements will be considered in relation to the age and experience of the nominee. The recipient will receive a $1000 honorarium, a plaque, and will be encouraged to present a lecture at the AACC International meeting of the year in which the award is given.
Requirements: Nominees must be no older than 40 years by June 1 of the year the award is sought, but nominations of younger scientists are particularly encouraged. AACC International membership is required.
Samples: Vijay Singh, University of Illinois at Urbana-Champaign, Urbana, IL (2007).
Amount of Grant: $1,000
Date(s) Application is Due: Jan 1
Contact: Khalil Khan, AACC International Award Chair; (651) 454-7250; fax: (651) 454-0766; email: khalil.khan@ndsu.edu or aacc@scisoc.org
Internet: http://www.aaccnet.org/membership/awards.asp#YoungScientist
Sponsor: American Association of Cereal Chemists International
3340 Pilot Knob Road
St. Paul, MN 55121

AACN Clinical Inquiry Fund Grants   86

The fund provides small awards to qualified individuals carrying out clinical research projects that directly benefit patients and/or families. Interdisciplinary projects are encouraged. Funds may be awarded for new projects, projects in progress, and projects required for an academic degree as long as all other project criteria are met. The funds may be used to cover direct project expenses—i.e., printed materials, small equipment, and supplies including computer software.
Requirements: Principal Investigators must be nurses holding current AACN membership and must remain AACN members throughout the life of the grant funding.
Restrictions: Funds may not be used to pay salaries or institutional overhead or augment funding from formal grants.
Amount of Grant: $500
Date(s) Application is Due: Oct 1
Contact: Pamela Shellner, Clinical Practice Specialist; (949) 362-2050, ext. 321 or (800) 394-5995, ext. 321; fax: (949) 362-2020; email: research@aacn.org or info@aacn.org
Internet: http://www.aacn.org/WD/Practice/Docs/Grant_Table09-10.pdf
Sponsor: American Association of Critical-Care Nurses
101 Columbia
Aliso Viejo, CA 92656-4109

AACN Clinical Practice Grants   87

This $6,000 grant supports research focused on one or more AACN Research priorities. Research conducted in fulfillment of an academic degree is acceptable. Funds may be awarded for new projects, projects in progress and projects required for an academic degree as long as all other project criteria are met. Eligible projects may include research utilization studies, CQI projects or outcomes evaluation projects. Interdisciplinary and collaborative projects are encouraged and may involve interdisciplinary teams, multiple nursing units, home health, subacute and transitional care, other institutions or community agencies. Funds may be used to cover direct project expenses, such as printed materials, small equipment or supplies including computer software. Six awards are available each year. One award is funded up to $6,000 each year.
Requirements: Principal Investigators must be nurses holding current AACN membership and must remain AACN members throughout the life of the grant funding.
Restrictions: Funds may not be used to pay salaries or institutional overhead or augment funding from formal grants. Principal investigators who have received funding from AACN are ineligible to receive additional funding from AACN during the lifetime of the original award.
Amount of Grant: $6,000 maximum
Date(s) Application is Due: Oct 1
Contact: Pamela Shellner, Clinical Practice Specialist; (949) 362-2050, ext. 321 or (800) 394-5995, ext. 321; fax: (949) 362-2020; email: research@aacn.org or info@aacn.org
Internet: http://www.aacn.org/WD/Practice/Content/grant-CP.pcms?pid=1@menu=practice
Sponsor: American Association of Critical-Care Nurses
101 Columbia
Aliso Viejo, CA 92656-4109

AACN Critical Care Grants   88

AACN's research agenda promotes the creation of cultures of inquiry, broad sharing and evidence-based practice. This $15,000 grant supports research focused on one or more of AACN Research priorities. Research priority areas include: effective and appropriate use of technology to achieve optimal patient assessment, management, and/or outcomes; creating a healing, humane environment; processes and systems that foster the optimal contribution of critical care nurses; effective approaches to symptom management; and prevention and management of complication.
Requirements: Principal Investigators must be nurses holding current AACN membership and must remain AACN members throughout the life of the grant funding.
Restrictions: The proposed research may not be used to meet the requirements of an academic degree. Funds may not be used to pay salaries or institutional overhead or augment funding from formal grants. Principal investigators who have received funding

from AACN are ineligible to receive additional funding from AACN during the lifetime of the original award.
Amount of Grant: $15,000
Date(s) Application is Due: Oct 1
Contact: Pamela Shellner, Clinical Practice Specialist; (949) 362-2050, ext. 321 or (800) 394-5995, ext. 321; fax: (949) 362-2020; email: research@aacn.org or info@aacn.org
Internet: http://www.aacn.org/WD/Practice/Content/grant-CC.pcms?pid=1@menu=practice
Sponsor: American Association of Critical-Care Nurses
101 Columbia
Aliso Viejo, CA 92656-4109

AACN End of Life/Palliative Care Small Projects Grants   89
The end of life/palliative care grant provides an award to a qualified individual(s) carrying out a project focusing on end of life and/or palliative care outcomes in acute and/or critical care settings. Eligible projects may focus on any age group (neonatal to elderly) in the acute or critical care arena. They may include patient education programs, staff development programs, competency-based educational programs, CQI/outcomes evaluation projects, or small clinical research studies. Funds may be awarded for new projects, projects in progress and projects required for an academic degree as long as all other project criteria are met. Collaborative projects are encouraged and may involve interdisciplinary teams, multiple nursing units, home health, subacute and transitional care, other institutions or community agencies. Funds may be used to cover direct project expenses, such as printed materials, small equipment, supplies including computer software. Two awards are available each year. Each proposal can be funded for up to $500.
Requirements: Principal Investigators must be nurses holding current AACN membership and must remain AACN members throughout the life of the grant funding.
Restrictions: Funds may not be used to pay for salaries, speaker honorarium or costs, institutional overhead nor augment funding from formal large grants.
Amount of Grant: $500 maximum
Date(s) Application is Due: Oct 1
Contact: Pamela Shellner, Clinical Practice Specialist; (949) 362-2050, ext. 321 or (800) 394-5995, ext. 321; fax: (949) 362-2020; email: research@aacn.org or info@aacn.org
Internet: http://www.aacn.org/WD/Practice/Content/grant-EOL.pcms?pid=1@menu=Practice
Sponsor: American Association of Critical-Care Nurses
101 Columbia
Aliso Viejo, CA 92656-4109

AACN Evidence-Based Clinical Practice Grants   90
The Evidence-Based Clinical Practice Grant provides awards to grants that stimulate the use of patient-focused data and/or previously generated research findings to develop, implement and evaluate changes critical and acute care nursing practice. Funds may be awarded for new projects, projects in progress and projects required for an academic degree as long as all other project criteria are met. Eligible projects may include research utilization studies, CQI projects or outcomes evaluation projects. Interdisciplinary and collaborative projects are encouraged and may involve interdisciplinary teams, multiple nursing units, home health, subacute and transitional care, other institutions or community agencies. Funds may be used to cover direct project expenses, such as printed materials, small equipment or supplies including computer software. Six awards are available each year. Each proposal can be funded for up to $1,000.
Requirements: Principal Investigators must be nurses holding current AACN membership and must remain AACN members throughout the life of the grant funding.
Restrictions: Funds may not be used to pay salaries or institutional overhead or augment funding from formal grants. Principal investigators who have received funding from AACN are ineligible to receive additional funding from AACN during the lifetime of the original award.
Amount of Grant: $1,000 maximum
Date(s) Application is Due: Oct 1
Contact: Pamela Shellner, Clinical Practice Specialist; (949) 362-2050, ext. 321 or (800) 394-5995, ext. 321; fax: (949) 362-2020; email: research@aacn.org or info@aacn.org
Internet: http://www.aacn.org/WD/Practice/Content/grant-EBP.pcms?pid=1@menu=Practice
Sponsor: American Association of Critical-Care Nurses
101 Columbia
Aliso Viejo, CA 92656-4109

AACN Mentorship Grant   91
This award, cosponsored by Mallinckrodt Inc. and AACN, facilitates critical care nursing practice research between a novice and an experienced researcher. The novice researcher is a beginning researcher with limited or no research experience in the area of the proposed investigation. The novice researcher is also a registered nurse with current AACN membership. The grant may be used to fund research for an academic degree. The mentor must show strong evidence of research expertise in the proposed area of research to be pursued by the novice investigator.
Requirements: Principal Investigators must be nurses holding current AACN membership and must remain AACN members throughout the life of the grant funding.
Restrictions: The mentor may not be designated mentor in two consecutive years and may not be conducting research as part of an academic degree.
Amount of Grant: $10,000 maximum
Date(s) Application is Due: Oct 1

Contact: Pamela Shellner, Clinical Practice Specialist; (949) 362-2050, ext. 321 or (800) 394-5995, ext. 321; fax: (949) 362-2020; email: research@aacn.org or info@aacn.org
Internet: http://www.aacn.org/WD/Practice/Content/grant-Mentor.pcms?pid=1@menu=practice
Sponsor: American Association of Critical-Care Nurses
101 Columbia
Aliso Viejo, CA 92656-4109

AACN Physio-Control Small Projects Grants   92
The program provides awards to qualified individuals carrying out projects focusing on aspects of acute myocardial infarction, resuscitation, or sudden cardiac death. Such projects may include the use of defibrillation, synchronized cardioversion, noninvasive pacing, or interpretive 12-lead electrocardiogram. Eligible projects may include patient education programs, staff development programs, competency-based educational programs, continuous quality improvement projects, outcomes evaluation projects, or small clinical research studies. Funds may be awarded for new projects, projects in progress, and projects required for an academic degree as long as all other project criteria are met. Collaborative projects are encouraged and may involve interdisciplinary teams, multiple nursing units, home health, subacute and transitional care, other institutions, or community agencies. Funds may be used to cover direct project expenses, such as printed materials, small equipment, or supplies including computer software.
Requirements: Principal Investigators must be nurses holding current AACN membership and must remain AACN members throughout the life of the grant funding.
Restrictions: Funds may not be used to pay for salaries, speaker honorarium or costs, institutional overhead nor augment funding from formal large grants.
Amount of Grant: $1,500 maximum
Date(s) Application is Due: Oct 1
Contact: Pamela Shellner, Clinical Practice Specialist; (949) 362-2050, ext. 321 or (800) 394-5995, ext. 321; fax: (949) 362-2020; email: research@aacn.org or info@aacn.org
Internet: http://www.aacn.org/WD/Practice/Docs/Grant_Table09-10.pdf
Sponsor: American Association of Critical-Care Nurses
101 Columbia
Aliso Viejo, CA 92656-1491

AACN-Edwards Lifesciences Nurse-Driven Clinical Practice Outcomes Grants   93
Co-sponsored by Edwards Lifesciences and the American Association of Critical-Care Nurses, each grant supports a nurse experienced in research who is conducting a clearly articulated research study that relates to use of protocol-based care driven by nurses. A commitment is required to produce results that will be available and disseminated in a short timeframe. Proposals submitted by an interdisciplinary team or an experienced researcher mentoring a novice will also be considered. Funds may be used for original research or replication of existing research. Especially desirable are proposals that focus on any of the following at any point in the continuum of acute or critical care: implementation of nurse-driven protocols using technology, including minimally invasive devices; with evaluation of patient outcomes and nurses‚Äô role in implementing the protocols; or development and clinical evaluation of nurse-driven protocols. Two awards are available each year. Each proposal can be funded up to of $5,000.
Requirements: Principal Investigators must be nurses holding current AACN membership and must remain AACN members throughout the life of the grant funding.
Restrictions: Funds may not be used to pay salaries or institutional overhead or augment funding from formal grants. Principal investigators who have received funding from AACN are ineligible to receive additional funding from AACN during the lifetime of the original award.
Amount of Grant: $5,000 maximum
Date(s) Application is Due: Oct 1
Contact: Pamela Shellner, Clinical Practice Specialist; (949) 362-2050, ext. 321 or (800) 394-5995, ext. 321; fax: (949) 362-2020; email: research@aacn.org or info@aacn.org
Internet: http://www.aacn.org/WD/Practice/Content/grant-Edwards.pcms?pid=1@menu=practice
Sponsor: American Association of Critical-Care Nurses
101 Columbia
Aliso Viejo, CA 92656-4109

AACN-Philips Medical Systems Clinical Outcomes Grants   94
Co-sponsored by Philips Medical Systems and the American Association of Critical-Care Nurses, each grant supports experienced nurses in conducting clearly articulated research studies. The research must be relevant to clinical nursing practice in acute or critical care. Proposals submitted by an interdisciplinary team or an experienced researcher mentoring a novice will also be considered. Funds may be used for original research or replication of existing research. Qualified proposals should seek to achieve improved outcomes and/or system efficiencies in care of acute or critically ill individuals of any age. Especially desirable are proposals for hospital-based inquiry that focuses on any of the following: intervention strategies addressing technology integration into patient care; use of computerized medical record systems in assessing patient outcomes and managing care; improving specific patient care outcomes (e.g., clinical, safety, financial); use of simulation in clinical education of nurses; and one or more of the 2006 National Patient Safety Goals issued by the Joint Commission on Accreditation of Healthcare Organizations.
Requirements: Principal Investigators must be nurses holding current AACN membership and must remain AACN members throughout the life of the grant funding.
Restrictions: Funds may not be used to pay salaries or institutional overhead or augment funding from formal grants. Principal investigators who have received funding from

AACN are ineligible to receive additional funding from AACN during the lifetime of the original award.
Amount of Grant: $10,000 maximum
Date(s) Application is Due: Oct 1
Contact: Pamela Shellner, Clinical Practice Specialist; (949) 362-2050, ext. 321 or (800) 394-5995, ext. 321; fax: (949) 362-2020; email: research@aacn.org or info@aacn.org
Internet: http://www.aacn.org/WD/Practice/Content/grant-Philips.pcms?pid=1@menu=practice
Sponsor: American Association of Critical-Care Nurses / Philips Medical Systems
101 Columbia
Aliso Viejo, CA 92656-4109

AACN-Sigma Theta Tau Critical Care Grant   95
The grant, cosponsored by AACN and Sigma Theta Tau International, supports critical care nursing research and is awarded annually for a study relevant to critical care nursing practice. The proposed study may be used to meet requirements of an academic degree.
Requirements: Principal investigators must be a member of AACN and/or of Sigma Theta Tau International. The PI must have a minimum Masters Degree.
Restrictions: Funds may not be used to pay salaries or institutional overhead or augment funding from formal grants. Principal investigators who have received funding from AACN are ineligible to receive additional funding from AACN during the lifetime of the original award.
Amount of Grant: $10,000 maximum
Date(s) Application is Due: Oct 1
Contact: Pamela Shellner, Clinical Practice Specialist; (949) 362-2050, ext. 321 or (800) 394-5995, ext. 321; fax: (949) 362-2020; email: research@aacn.org or info@aacn.org
Internet: http://www.aacn.org/WD/Practice/Content/grant-STT.pcms?pid=1@menu=practice
Sponsor: American Association of Critical-Care Nurses
101 Columbia
Aliso Viejo, CA 92656-4109

AACR Award for Lifetime Achievement in Cancer Research   96
The AACR Award for Lifetime Achievement in Cancer Research was established and first given in 2004 to honor an individual who has made significant fundamental contributions to cancer research, either through a single scientific discovery or a body of work. These contributions, whether they have been in research, leadership, or mentorship, must have had a lasting impact on the cancer field and must have demonstrated a lifetime commitment to progress against cancer. Nominations may be made by any scientist, whether an AACR member or nonmember, who is now or has been affiliated with any institution involved in cancer research, cancer medicine, or cancer-related biomedical science. Nominations for the Award must be submitted electronically to the AACR no later than 4:00 p.m. United States Eastern Time on November 16.
Requirements: Nominations may be made on behalf of individuals who are living at the time of the nomination. Candidates need not be members of the AACR. Candidates need not be currently engaged in cancer research.
Restrictions: Institutions or organizations are not eligible for the Award. Paper nominations will not be accepted.
Date(s) Application is Due: Nov 16
Contact: Monique P. Eversley, Staff Associate; (267) 646-0576 or (215) 440-9300, ext. 1400; fax: (215) 440-9372; email: monique.eversley@aacr.org or awards@aacr.org
Internet: http://www.aacr.org/home/scientists/scientific-achievement-awards/lifetime-achievement-award.aspx
Sponsor: American Association for Cancer Research
615 Chestnut Street, 17th Floor
Philadelphia, PA 19106-4404

AACR Award for Outstanding Achievement in Cancer Research   97
This award is given in recognition of a young investigator on the basis of meritorious achievement in cancer research. The winner receives an honorarium of $5,000, presents a 50-minute lecture, and is given full support for the winner and a guest to attend the AACR Annual Meeting.
Requirements: Candidacy is open to all cancer researchers who are affiliated with any institution involved in cancer research, cancer medicine, or cancer-related biomedical science anywhere in the world. Such institutions include those in academia, industry, or government. In accordance with the wishes of the donor, the recipient must be no more than 40 years of age by the time the award is received.
Restrictions: Institutions or organizations are not eligible for the Award.
Amount of Grant: $5,000 honorarium
Date(s) Application is Due: Oct 15
Contact: Monique P. Eversley, Program Associate; (267) 646-0576; email: monique.eversley@aacr.org
Internet: http://www.aacr.org/home/scientists/scientific-achievement-awards/outstanding-achievement-award.aspx
Sponsor: American Association for Cancer Research
615 Chestnut Street, 17th Floor
Philadelphia, PA 19106-4404

AACR Award for Outstanding Achievement in Chemistry in Cancer Research   98
The Award will be given for outstanding, novel, and significant chemistry research, which has led to important contributions to the fields of basic cancer research; translational cancer research; cancer diagnosis; the prevention of cancer; or the treatment of patients with cancer. Such research may include, but is not limited to, drug discovery and design; structural biology; proteomics, metabolomics and biological mass spectrometry; chemical aspects of carcinogenesis; imaging agents and radiotherapeutics; and chemical biology. The winner of the award will give a 50-minute lecture during the AACR Annual Meeting, will receive a commemorative plaque and a $10,000 honorarium, and receive support for the winner and a guest to attend the Annual Meeting. Nominations for the Award must be submitted electronically no later than 4:00 p.m. United States Eastern Time on October 15.
Requirements: Candidacy is open to all researchers who are affiliated with any institution involved in cancer research, cancer medicine, or cancer-related biomedical science anywhere in the world. Such institutions include those in academia, industry, or government.
Restrictions: Institutions or organizations are not eligible for the Award. Paper nominations will not be accepted.
Amount of Grant: $10,000 honorarium
Date(s) Application is Due: Oct 15
Contact: Lauren Medvetz, Program Coordinator; (267) 646-0689; fax: (215) 440-9372; email: lauren.medvets@aacr.org
Internet: http://www.aacr.org/home/scientists/scientific-achievement-awards/chemistry-in-cancer-research-award.aspx
Sponsor: American Association for Cancer Research
615 Chestnut Street, 17th Floor
Philadelphia, PA 19106-4404

AACR Brigid G. Leventhal Scholar in Cancer Research Awards   99
The purpose of this award program is to enhance the education and training of early career scientists by providing financial support for their participation in AACR Annual Meetings and Special Conferences. The AACR-WICR Brigid G. Leventhal Scholar Award Committee makes selection for these competitive awards after careful consideration of the candidate, application and accompanying materials.
Requirements: Candidates must be: Women in Cancer Research members; full time scientists in training who are graduate students, medical students, residents, clinical fellows or equivalent, or postdoctoral fellows; and first authors on abstracts submitted for consideration for presentation at the AACR Annual Meeting or Special Conference which the applicant wants to attend.
Date(s) Application is Due: Sep 10
Contact: Mona E. Shater, Staff Assistant; (215) 440-9300 or (267) 646-0654; fax: (215) 440-9372; email: mona.shater@aacr.org or wicr@aacr.org
Internet: http://www.aacr.org/home/membership-/association-groups/women-in-cancer-research.aspx
Sponsor: American Association for Cancer Research
615 Chestnut Street, 17th Floor
Philadelphia, PA 19106-4404

AACR Career Development Awards for Pediatric Cancer Research   100
Two Career Development Awards will be presented, the AACR-Aflac Incorporated Career Development Award for Pediatric Cancer Research and the AACR Centennial Career Development Award for Childhood Cancer Research. These Awards are open to junior faculty who are in their first full-year, faculty appointment. The Awards provide two-year grants of $100,000 ($50,000 per year) for direct research expenses and/or salary support. Research projects must have direct applicability to pediatric cancer.
Requirements: Candidates must have acquired a doctoral degree and have completed postdoctoral studies or clinical fellowships on or after July 2 (two years prior to the start of the grant term). Candidates must be in their first full-time, faculty appointment and hold the title of Instructor, Research Assistant Professor, Assistant Professor, or an equivalent full-time faculty appointment at the start of the grant term (July 1).
Restrictions: Non AACR members must submit a satisfactory application for Active Membership with the grant application by December 10th. Employees or subcontractors of a government or a for-profit private industry are not eligible.
Amount of Grant: two-year grant $100,000 ($50,000 per year)
Date(s) Application is Due: Dec 18
Contact: Julia Laurence, Program Assistant; (267) 646-0655; fax: (215) 440-9372; email: julia.laurence@aacr.org or awards@aacr.org
Internet: http://www.aacr.org/home/scientists/research-funding—fellowships/career-development-awards.aspx
Sponsor: American Association for Cancer Research
615 Chestnut Street, 17th Floor
Philadelphia, PA 19106-4404

AACR Centennial Postdoctoral Fellowships in Cancer Research   101
AACR Centennial Postdoctoral Research Fellowships in Cancer Research will provide $60,000 per year for up to three years to provide salary support and research funding for clinical and postdoctoral fellows in the first, second, or third year of their fellowship status. In addition to financial flexibility these fellowships also provide career flexibility. Fellowship funds may be transferred to move with the grantee if the grantee's fellowship is completed during the grant term - a valuable commodity for a newly-independent investigator. Under this award mechanism, the postdoctoral trainee is considered the Lead Investigator and should write the proposal independently with appropriate direction from the mentor. Proposals may be submitted in any area of cancer research whether laboratory-based, translational, clinical, epidemiological, behavioral, or other. Applications must be completed online using the proposalCENTRAL website, with one paper copy submitted to the AACR office. Application instructions and program guidelines are available online.

Requirements: Applicants must: be in the 1st, 2nd, or 3rd year of their postdoctoral training at the start of the fellowship term; be members of the AACR or must submit an application for membership prior to the fellowship application deadline; have acquired a doctoral degree in a biological sciences-related field; and hold the title of Postdoctoral Fellow or Clinical Research Fellow at an academic facility, teaching hospital, or research institution.

Restrictions: Employees, trainees, or subcontractors of a national government or private industry are not eligible.

Amount of Grant: $60,000 per year for up to 3 years

Date(s) Application is Due: Feb 1

Contact: Hanna Hopfinger, Program Assistant; (267) 646-0665; fax: (215) 440-9372; email: hanna.hopfinger@aacr.org

Internet: http://www.aacr.org/home/scientists/research-funding—fellowships/research-funding.aspx

Sponsor: American Association for Cancer Research
615 Chestnut Street, 17th Floor
Philadelphia, PA 19106-4404

AACR Centennial Pre-doctoral Fellowships in Cancer Research   102

The AACR Centennial Pre-doctoral Research Fellowships will provide full-time graduate students a rare opportunity to invest in their future cancer research careers. The grant terms are flexible, allowing applicants to tailor their grant funds to fit their individual research funding and salary needs. Grantees may also transfer their funds to a postdoctoral fellowship position, should they advance in their careers during the grant term. Under this award mechanism, the pre-doctoral trainee is considered the Lead Investigator and should write the proposal independently with appropriate direction from the mentor. Proposals may be submitted in any area of cancer research whether laboratory-based, translational, clinical, epidemiological, behavioral, or other. A grant of $30,000 per year over a maximum period of three years will be awarded for salary support and research funding.

Requirements: Applicants must: be full-time graduate students who are pursuing a doctoral degree and engaged in a program of cancer research; be members of the AACR or must submit an application for membership prior to the fellowship application deadline; have completed at least one year of research training at the graduate level or have equivalent research experience; and be engaged in basic, clinical, translational or epidemiological cancer research. At the time of award acceptance the awardee must have advanced to degree candidacy.

Restrictions: Summer student positions and undergraduate summer studentships do not qualify. Employees, trainees, or subcontractors of a national government or private industry are not eligible.

Amount of Grant: $30,000 per year over a maximum period of three years

Date(s) Application is Due: Feb 1

Contact: Hanna Hopfinger, Program Assistant; (267) 646-0665; fax: (215) 440-9372; email: hanna.hopfinger@aacr.org

Internet: http://www.aacr.org/home/scientists/research-funding—fellowships/research-funding.aspx

Sponsor: American Association for Cancer Research
615 Chestnut Street, 17th Floor
Philadelphia, PA 19106-4404

AACR Clinical and Translational Research Fellowships   103

The AACR Clinical and Translational Research Fellowships consist of three different Fellowships: AACR-Astellas U.S.A Foundation Fellowship in Clinical/Translational Cancer Research, which is a one-year grant of $30,000; AACR-AstraZeneca Fellowship for Translational Lung Cancer Research, which is a two-year grant of $90,000; and, AACR-Bristol-Myers Squibb Oncology Fellowship in Clinical Cancer Research, which is a one-year grant of $40,000. These fellowships support the salary and benefits of the Fellow. A partial amount of funds may be designated for direct research support. The AACR requires applicants to submit both an online and a paper application. The online application deadline is December 10th, the application must be submitted using the proposalCENTRAL website at https://proposalcentral.altum.com. The paper application with original signatures and all supporting documents must be postmarked and sent no later then December 15.

Requirements: Applications are open to Postdoctoral Fellows and Clinical Research Fellows at an academic facility, teaching hospital, or research institution who will be in the 1st, 2nd, or 3rd year of their postdoctoral training at the start of the fellowship term.

Restrictions: Academic faculty holding the rank of assistant professor or higher, graduate or medical students, government employees, and employees of private industry are not eligible.

Amount of Grant: $30,000-$90,000 range

Date(s) Application is Due: Dec 10; Dec 15

Contact: Hanna Hopfinger; (267) 646-0665 or (215) 440-9300; fax: (215) 440-9372; email: hanna.hopfinger@aacr.org or awards@aacr.org

Internet: http://www.aacr.org/home/scientists/research-funding—fellowships/postdoctoral-fellowships.aspx

Sponsor: American Association for Cancer Research
615 Chestnut Street, 17th Floor
Philadelphia, PA 19106-4404

AACR Fellows Grants   104

The Fellowship program supports innovative research by a meritorious young investigator by presenting the Fellow with research funds to pursue an independent line of investigation within the context of his/her current Fellowship placement. By allowing a Fellow to acquire the equipment and supplies needed to pursue a new direction in his/her research program, the Fellows Grant assists the Fellow in developing preliminary data to support a future project or investigating a new technique that otherwise would not be possible in the absence of this funding. The Selection Committee favors proposals that provided evidence of an applicant's research initiative and creativity, and the support of a scientific mentor. The Committee also weighed the potential of the candidate to make meaningful contributions to the field, and promise of success as a cancer researcher in future years.

Requirements: Candidates must be, at the start of the grant term, in their third, fourth, or fifth year of their fellowship status.

Amount of Grant: $30,000-$40,000 per year

Date(s) Application is Due: Dec 18

Contact: Julia Laurence, Staff Assistant; (267) 646-0655; fax: (215) 440-9372; email: julia.laurence@aacr.org or awards@aacr.org

Internet: http://www.aacr.org/home/scientists/research-funding—fellowships/fellows-grants.aspx

Sponsor: American Association for Cancer Research
615 Chestnut Street, 17th Floor
Philadelphia, PA 19106-4404

AACR Gertrude Elion Cancer Research Award   105

This prestigious award provides a one-year research grant to an assistant professor at an academic or nonprofit research institute worldwide for salary and benefits, laboratory supplies, and limited domestic travel to support research in cancer etiology, diagnosis, treatment, or prevention (basic, translational, or clinical cancer research). The AACR requires applicants to submit both an online and a paper application. The online application must be submitted by 12:00 noon (United States Eastern Daylight Time) on February 27, using the proposalCENTRAL website at https://proposalcentral.altum.com. The paper application with original signatures and all supporting documents must be postmarked and sent no later then March 3.

Requirements: Candidates must be nominated by an AACR member and submit a detailed application. Candidates must have completed postdoctoral studies or clinical fellowships no later than July 1 of the award year and ordinarily not more than five years earlier. Candidates must be tenure-tracked scientists at the level of assistant professor at an academic institution anywhere in the world.

Restrictions: Tenured faculty in academia, employees of a national government, and employees of private industry are not eligible.

Amount of Grant: $50,000

Date(s) Application is Due: Feb 27; Mar 3

Contact: Julia Laurence, Staff Assistant; (215) 440-9300, ext. 102 or (267) 646-0655; fax: (215) 440-9372; email: laurence@aacr.org

Internet: http://www.aacr.org/home/scientists/research-funding—fellowships/elion-award.aspx

Sponsor: American Association for Cancer Research
615 Chestnut Street, 17th Floor
Philadelphia, PA 19106-4404

AACR Henry Shepard Bladder Cancer Research Grants   106

These two-year Grants will provide up to $250,000 in total support for innovative cancer research projects designed to accelerate the discovery, development, and application of new agents to treat bladder cancer and/or for pre-clinical research with direct therapeutic intent. Laboratory-based projects must present plans with clinical collaborators indicating how the work will be translated into the clinic. Similarly, clinical studies must show how the work was derived from basic preclinical work and how results will be channeled back to laboratory-based collaborators. Four grants will be awarded. The submission deadline is July 25, at 12:00 noon, U.S. Eastern Time.

Requirements: Independent investigators who are affiliated with any institution involved in cancer research, cancer medicine, or cancer-related biomedical science anywhere in the world may apply. There are no geographic, national, or residency status restrictions. Applicants must have acquired a doctoral degree in a related field.

Restrictions: Pre- and Postdoctoral fellows are not eligible. Employees or subcontractors of a national government or the for-profit private industry are not eligible to serve as the Principal Investigator for the purpose of these Grants. However, collaborations with such individuals are encouraged. Neither the members of the Scientific Review Committee nor the members of their individual laboratories are eligible for these Grants.

Amount of Grant: $250,000 maximum over two years

Date(s) Application is Due: Jul 25

Contact: Hanna Hopfinger, Program Assistant; (267) 646-0665; fax: (215) 440-9372; email: hanna.hopfinger@aacr.org

Internet: http://www.aacr.org/home/scientists/research-funding—training-grants/research-funding/research-funding—fellowships/bladder-cancer-research-grants.aspx

Sponsor: American Association for Cancer Research
615 Chestnut Street, 17th Floor
Philadelphia, PA 19106-4404

AACR Joseph H. Burchenal Memorial Award   107

This Award was established in 1996 to recognize outstanding achievements in clinical cancer research. It is named for the late Dr. Joseph H. Burchenal, Honorary Member and Past President of the AACR, and a major figure in clinical cancer research and chemotherapy. The winner of the Award for Outstanding Achievement in Clinical Cancer Research will give a 50-minute lecture during the AACR Annual Meeting, and receive an honorarium and support for attendance at the meeting.

Requirements: Candidacy is open to all cancer researchers who are affiliated with any institution involved in cancer research, cancer medicine, or cancer-related biomedical science anywhere in the world. Such institutions include those in academia, industry, or government.

Restrictions: Institutions or organizations are not eligible for the Award. Candidates may not nominate themselves.

Samples: Joseph R. Bertino, M.D., Interim Director & Chief Scientific Officer, The Cancer Institute of New Jersey (2008).

Date(s) Application is Due: Oct 15

Contact: Monique P. Eversley, Program Associate; (267) 646-0576; email: monique. eversley@aacr.org

Internet: http://www.aacr.org/home/scientists/scientific-achievement-awards/ burchenal-award.aspx

Sponsor: American Association for Cancer Research
615 Chestnut Street, 17th Floor
Philadelphia, PA 19106-4404

AACR Judah Folkman Career Development Award for Anti-Angiogenesis Research   108

The AACR Judah Folkman Career Development Award for Angiogenesis Research is open to junior faculty who completed postdoctoral studies or clinical fellowships no more then three years prior to the start of the grant term (Dec 1); who are in their first full-time, faculty appointment and hold the title of Instructor, Research Assistant Professor, Assistant Professor, or an equivalent full-time faculty appointment; and who are at an academic, medical, or research Institution. The Career Development Award provides a two-year grant of $100,000 ($50,000 per year) for direct research expenses and/or salary support. One Award will be presented.

Requirements: Applicants must have acquired a medical degree or hold a combined M.D./Ph.D and may not currently be a candidate for a further doctoral or professional degree. The Award is open to AACR members. Non-members must submit a satisfactory application for Active Membership and additional Curriculum Vitae with the grant application.

Restrictions: Employees, trainees, or subcontractors of a national government or private industry are not eligible. Exceptions may apply if an applicant holds a full-time position at a Veterans' Hospital or national laboratory in the U.S. Current recipients of any AACR grant are also not eligible.

Amount of Grant: two-year grant $100,000 ($50,000 per year)

Date(s) Application is Due: Aug 10

Contact: Julia Laurence, Program Assistant; (267) 646-0655; fax: (215) 440-9372; email: julia.laurence@aacr.org or awards@aacr.org

Internet: http://www.aacr.org/home/scientists/research-funding—fellowships/career-development-awards.aspx

Sponsor: American Association for Cancer Research
615 Chestnut Street, 17th Floor
Philadelphia, PA 19106-4404

AACR Kirk A. Landon and Dorothy P. Landon Foundation Prizes   109

These two major international awards recognize outstanding scientists who have made seminal basic and translational cancer research discoveries at the cutting edge of scientific novelty and significance, which have accelerated progress against cancer and have implications for future discoveries and contributions to cancer research. The Kirk A. Landon-AACR Prize for Basic Cancer Research recognizes significant, fundamental contributions to laboratory research. The Dorothy P. Landon-AACR Prize for Translational Cancer Research recognizes extraordinary achievement in translational cancer research—the interface between basic research and its application to the clinic in the areas of diagnosis, treatment, or the prevention of cancer. Both prizes bring heightened public attention to landmark achievements in the continuing effort to prevent and cure cancer through the presentation of dynamic lectures during the AACR annual meeting; and promote and reward continued, productive cancer research.

Requirements: The prizes are open to all cancer researchers who are affiliated with any institution involved in cancer research, cancer medicine, or cancer-related biomedical science anywhere in the world. Such institutions include those in academia, industry, or government. Candidates must be active researchers and have a record of recent publications.

Restrictions: Institutions or organizations are ineligible.

Samples: Arnold J. Levine, Ph.D., Professor, The Cancer Institute of New Jersey—Kirk A. Landon-AACR Prize for Basic Cancer Research (2008); John Mendelsohn, M.D., President, UT M.D. Anderson Cancer Center, Houston, Texas—Dorothy P. Landon-AACR Prize for Translational Cancer Research (2008).

Amount of Grant: $100,000 each

Date(s) Application is Due: Aug 25

Contact: Monique P. Eversley, Staff Associate; (267) 646-0576 or (215) 440-9300, ext. 1400; fax: (215) 440-9372; email: monique.eversley@aacr.org or awards@aacr.org

Internet: http://www.aacr.org/home/scientists/scientific-achievement-awards/landon-prizes.aspx

Sponsor: American Association for Cancer Research
615 Chestnut Street, 17th Floor
Philadelphia, PA 19106-4404

AACR Margaret Foti Award for Leadership and Extraordinary Achievements in Cancer Research   110

This Award will recognize a true champion of cancer research, who embodies the sustained commitment of Margaret Foti to the prevention and cure of cancer. The Award will be given to an individual whose leadership and extraordinary achievements in cancer research have made a major impact in the field. Such achievements may include contributions to the acceleration of progress in cancer research; raising national or international awareness of cancer research; or in other ways demonstrating a sustained commitment to cancer research. Nominations for the Award must be submitted electronically no later than 4:00 p.m. U.S. Eastern Time on November 16. The recipient will receive an honorarium, a commemorative plaque, and support for the winner to attend the AACR Annual Meeting.

Requirements: Candidacy is open to all individuals who are affiliated with any organization whose mission supports cancer research or any institution involved in cancer research, cancer medicine, or cancer-related biomedical science anywhere in the world. Such institutions include those in academia, industry, or government.

Restrictions: Institutions or organizations are not eligible for the Award. Paper nominations will not be accepted.

Date(s) Application is Due: Nov 16

Contact: Monique P. Eversley, Program Associate; (267) 646-0576 or (215) 440-9300, ext. 1400; fax: (215) 440-9372; email: monique.eversley@aacr.org or awards@aacr.org

Internet: http://www.aacr.org/home/scientists/scientific-achievement-awards/margaret-foti-award.aspx

Sponsor: American Association for Cancer Research
615 Chestnut Street, 17th Floor
Philadelphia, PA 19106-4404

AACR Minority Scholar in Cancer Research Awards   111

The AACR administers this award program that is supported by a grant from the Comprehensive Minority Biomedical Branch of the National Cancer Institute to provide funds for participation of meritorious minority scientists in the AACR Special Conferences Series (U.S. domestic conferences only). These awards are intended to enhance the education and training of minority researchers and to increase the visibility and recognition of minorities involved in cancer research.

Requirements: Candidates must be full-time graduate students, medical students, residents, clinical or postdoctoral fellows, or junior faculty members who are either engaged in cancer research or who have the training and potential to make contributions to this field. Candidates must also be citizens or permanent residents of the United States or Canada.

Restrictions: This program applies only to racial/ethnic minority groups identified by the National Cancer Institute as being traditionally underrepresented in cancer and biomedical research. These groups include African Americans/Blacks, Alaskan Natives, Hispanic Americans, Native Americans, and Native Pacific Islanders. Only citizens of the United States or Canada or scientists who are permanent residents of these countries may receive one of these awards.

Samples: Irene Alvarez, B.S., University of Arizona, Tucson, Arizona (2008); Cristina M. Contreras, B.S., UT Southwestern Medical Center, Dallas, Texas (2008); Vivian A. Galacia, B.S., University of Southern California, Los Angeles, California (2008).

Date(s) Application is Due: Jul 20

Contact: Mona E. Shater, Staff Assistant; (215) 440-9300 or (267) 646-0654; fax: (215) 440-9372 or (215) 440-9412; email: mona.shater@aacr.org or micr@aacr.org

Internet: http://www.aacr.org/home/scientists/meetings—workshops/special-conferences/metabolism-and-cancer/financial-support-for-attendance.aspx

Sponsor: American Association for Cancer Research
615 Chestnut Street, 17th Floor
Philadelphia, PA 19106-4404

AACR Minority-Serving Institution Faculty Scholar in Cancer Research Awards   112

The AACR is pleased to announce the availability of Scholar Awards in Cancer Research for full-time faculty members of Minority-Serving Institutions [Historically Black Colleges and Universities (HBCUs), Hispanic Serving Institutions (HSIs), and Tribal Colleges and Universities and other post-secondary institutions as defined by the U.S. Department of Education]. The purposes of this Award program is to increase the scientific knowledge base of faculty members at Minority-Serving Institutions, and to encourage them and their students to pursue careers in cancer research. AACR Minority-Serving Institution Faculty Scholar in Cancer Research Awards are presented by the American Association for Cancer Research to scientists at the level of Assistant Professor or above at a Minority-Serving Institution, who are engaged in meritorious basic, clinical, or translational cancer research.

Requirements: Candidates must have completed doctoral studies or clinical fellowships and hold full-time faculty status at an institution designated as a Minority-Serving Institution. Candidates must have acquired doctoral degrees in fields relevant to cancer research. Candidates must be citizens or permanent residents of the United States or Canada.

Samples: Joseph Agyin, Ph.D., UT Health Science Center, San Antonio, Texas (2008); Wellington Ayensu, M.D., Jackson State University, Jackson, Mississippi (2008); Rebecca Hartley, Ph.D., University of New Mexico, Albuquerque, New Mexico (2008).

Date(s) Application is Due: Jul 20

Contact: Mona E. Shater, Staff Assistant; (215) 440-9300 or (267) 646-0654; fax: (215) 440-9372 or (215) 440-9412; email: mona.shater@aacr.org or micr@aacr.org

Internet: http://www.aacr.org/home/scientists/meetings—workshops/special-conferences/metabolism-and-cancer/financial-support-for-attendance.aspx

Sponsor: American Association for Cancer Research
615 Chestnut Street, 17th Floor
Philadelphia, PA 19106-4404

AACR Outstanding Investigator Award for Breast Cancer Research   113

The AACR Outstanding Investigator Award for Breast Cancer Research, funded by Susan G. Komen for the Cure, will recognize an investigator of no more than 50 years of age whose novel and significant work has had or may have a far-reaching impact on the etiology, detection, diagnosis, treatment, or prevention of breast cancer. Such work may involve any discipline across the continuum of biomedical research, including basic, translational, clinical, and epidemiological studies. The recipient of the Award will receive a $10,000 honorarium and present a 25-minute lecture at the Annual San Antonio Breast Cancer Symposium. Nominations may be made by any scientist, whether an AACR member or nonmember, who is now or has been affiliated with any institution involved in cancer research, cancer medicine, or cancer-related biomedical science. Candidates may not nominate themselves. Nominations must be submitted online at https://proposalcentral.altum.com, no later than 4:00 p.m. United States Eastern Time on May 15th.

Requirements: All cancer researchers who are affiliated with any institution involved in cancer research, cancer medicine, or cancer-related biomedical science anywhere in the world may be nominated. Such institutions include those in academia, industry, or government. Candidates must be no more than 50 years of age at the time the Award is received.

Restrictions: Institutions or organizations are not eligible for the Award. Paper nominations will not be accepted.

Amount of Grant: $10,000 honorarium

Date(s) Application is Due: May 15

Contact: Monique P. Eversley, Program Associate; (267) 646-0576 or (215) 440-9300, ext. 1400; fax: (215) 440-9372; email: monique.eversley@aacr.org or awards@aacr.org

Internet: http://www.aacr.org/home/scientists/scientific-achievement-awards/outstanding-investigator-award.aspx

Sponsor: American Association for Cancer Research
615 Chestnut Street, 17th Floor
Philadelphia, PA 19106-4404

AACR Scholar-in-Training Awards   114

Scholar-in-Training Awards provide financial support for early-career scientists to attend the AACR Annual Meeting. Since its inception in 1986, the program has provided over 3,200 grants and has received support from more than 30 cancer research foundations, corporations, individuals, and other organizations dedicated to the fight against cancer. A stipend (from $400-$2,000 depending on the geographic location of the recipient or the type of award) will be presented to recipients on-site. Recipients may pick up their a check and welcome package at the Associate Member Resource and Career Center.

Requirements: Eligible candidates are graduate students, medical students and residents, clinical fellows or equivalent, and postdoctoral fellows. Eligible candidates must be the presenter of a proffered paper. Eligible candidates may be traveling within the U.S. or from abroad. Some awards are specifically designated for those traveling from Asia, Europe, and countries with emerging economies.

Restrictions: Employees or subcontractors of private industry are not eligible.

Amount of Grant: $400-$2,000

Date(s) Application is Due: Jul 30

Contact: Mona E. Shater, Staff Assistant; (215) 440-9300 or (267) 646-0654; fax: (215) 440-9372; email: mona.shater@aacr.org

Internet: http://www.aacr.org/home/scientists/meetings—workshops/special-conferences/metabolism-and-cancer/financial-support-for-attendance.aspx

Sponsor: American Association for Cancer Research
615 Chestnut Street, 17th Floor
Philadelphia, PA 19106-4404

AACR Scholar-in-Training Awards - Special Conferences   115

The AACR offers Scholar-in-Training Awards to enhance the education and training of early career scientists by providing financial support for their attendance at AACR Special Conferences. These awards are provided to offset a portion of the registration, travel, and subsistence expenses incurred in attending these conferences. A stipend (typically $400,Äì$1,400) will be presented to award recipients at the conference.

Requirements: All graduate and medical students, postdoctoral fellows, and physicians-in-training who submit abstracts may be considered for a Scholar-in-Training Award.

Restrictions: Employees or subcontractors of private industry are not eligible.

Amount of Grant: $400-$1,400

Contact: Mona E. Shater, Staff Assistant; (215) 440-9300 or (267) 646-0654; fax: (215) 440-9372 or (215) 440-9412; email: mona.shater@aacr.org or programs@aacr.org

Internet: http://www.aacr.org/home/scientists/meetings—workshops/travel-grants/scholar-in-training-awards—-special-conferences.aspx

Sponsor: American Association for Cancer Research
615 Chestnut Street, 17th Floor
Philadelphia, PA 19106-4404

AACR Team Science Award   116

The Award has been established by the AACR and Eli Lilly and Company to acknowledge and catalyze the growing importance of interdisciplinary teams to the understanding of cancer and/or the translation of research discoveries into clinical cancer applications. In addition, through the presentation of this Award, the AACR and Eli Lilly seek to effect change within the traditional cancer research culture by recognizing those institutions that value and foster interdisciplinary team science. These institutions will have demonstrated their support of a team science environment by creating mechanisms to enhance the required infrastructure, such as through pilot funding, technology transfer offices, shared resources, etc., and by presenting awards, honors, appointments,

and promotions to those who participate in interdisciplinary teams. Nominations for the 2009 Award will be accepted until November 16.

Requirements: For the purpose of this Award, a team is defined as a group of individuals representing interdisciplinary expertise, each of whom have made substantive and quantifiable contributions to the research being recognized. Team members may be working within the same institution or at several institutions. Candidacy is open to all cancer researchers who are affiliated with any institution involved in cancer research, cancer medicine, or cancer-related biomedical science anywhere in the world. Such institutions include those in academia, industry, or government.

Amount of Grant: $50,000

Date(s) Application is Due: Nov 16

Contact: Monique P. Eversley, Program Associate; (267) 646-0576 or (215) 440-9300, ext. 1400; fax: (215) 440-9372; email: monique.eversley@aacr.org or awards@aacr.org

Internet: http://www.aacr.org/home/scientists/scientific-achievement-awards/team-science-award.aspx

Sponsor: American Association of Cancer Research
615 Chestnut Street, 17th Floor
Philadelphia, PA 19106-4404

AACR Thomas J. Bardos Science Education Awards   117

AACR-Thomas J. Bardos Science Education Awards for Undergraduate Students are available to full-time third-year undergraduate students majoring in science. The purpose of these awards is to inspire young science students to enter the field of cancer research. The award consists of a waiver of registration fees for two annual AACR meetings and $1500/year for research and travel expenses. The award is for a two-year period.

Requirements: Candidates must be full-time, third year undergraduate students majoring in science. Applications from students who are not yet committed to cancer research are welcome.

Amount of Grant: $1,500 plus a wave of registration fee

Date(s) Application is Due: Dec 7

Contact: Mona E. Shater, Staff Assistant; (215) 440-9300 or (267) 646-0654; fax: (215) 440-9412; email: mona.shater@aacr.org or scienceeducation@aacr.org

Internet: http://www.aacr.org/home/scientists/meetings—workshops/travel-grants/aacr-thomas-j-bardos-science-education-awards-for-undergraduate-students-.aspx

Sponsor: American Association for Cancer Research
615 Chestnut Street, 17th Floor
Philadelphia, PA 19106-4404

**AACR-American Cancer Society Award for Research Excellence
in Cancer Epidemiology and Prevention**   118

The AACR and the American Cancer Society established this Award in 1992 to honor outstanding research accomplishments in the fields of cancer epidemiology, biomarkers, and prevention. Nominations may be made via letter from any scientist, whether an AACR member or nonmember, who is now or has been affiliated with any institution involved in cancer research, cancer medicine, or cancer-related biomedical science. The winner will receive an honorarium of $5,000 and give a 50-minute lecture during the AACR Annual Meeting.

Requirements: Candidacy is open to all cancer researchers who are affiliated with any institution involved in cancer research, cancer medicine, or cancer-related biomedical science anywhere in the world. Such institutions include those in academia, industry, or government.

Restrictions: Institutions or organizations are not eligible for the Award. Candidates may not nominate themselves.

Amount of Grant: $5,000 honorarium

Date(s) Application is Due: Oct 15

Contact: Monique P. Eversley, Program Associate; (267) 646-0576; email: monique.eversley@aacr.org

Internet: http://www.aacr.org/home/scientists/scientific-achievement-awards/acs-award.aspx

Sponsor: American Association for Cancer Research
615 Chestnut Street, 17th Floor
Philadelphia, PA 19106-4404

AACR-Colorectal Cancer Coalition Fellows Grant   119

The Fellowship program supports innovative research by a meritorious young investigator by presenting the Fellow with research funds to pursue an independent line of investigation within the context of his/her current placement. This is a one-year grant of $30,000. Research projects are restricted to translational or clinical cancer research that has an ultimate goal of developing or improving therapeutic interventions for patients with metastatic colorectal cancer. One grant will be awarded this cycle.

Requirements: Candidates must be, at the start of the grant term, in their third, fourth, or fifth year of their fellowship status.

Amount of Grant: $30,000

Date(s) Application is Due: Dec 18

Contact: Hanna Hopfinger, Grants Program Assistant; (267) 646-0665; fax: (215) 440-9372; email: hanna.hopfinger@aacr.org

Internet: http://www.aacr.org/home/scientists/research-funding—fellowships/fellows-grants.aspx

Sponsor: American Association for Cancer Research
615 Chestnut Street, 17th Floor
Philadelphia, PA 19106-4404

AACR-FNAB Fellows Grant for Translational Pancreatic Cancer Research   120

The AACR-FNAB Fellows Grant for Translational Pancreatic Cancer Research provides a one-year grant of $35,000. Research projects are restricted to translational pancreatic cancer research that includes the use of human tissue and has implications for therapeutic application and individualized medicine. Proposed research must have implications for individualized cancer treatment and must make use of human biopsies or samples, such as needle biopsies or circulating cancer cells. In vivo primary tumor explants meet these criteria, but xenografts from established cell lines do not. One grant will be awarded this cycle.

Requirements: Candidates must hold the title of Postdoctoral Fellow or Clinical Research Fellow at an academic facility, teaching hospital, or research institution. Candidates must hold this title at the start of the fellowship term on July 1, and continue to hold it during the entire fellowship term.

Samples: Sarat Chandarlapaty, M.D., Ph.D., Memorial Sloan-Kettering Cancer Center, New York, NY (2007); Sibele I. Meireles, Ph.D., Fox Chase Cancer Center, Philadelphia, PA (2006); Daniel C. Cho, M.D., Beth Israel Deaconess Medical Center, Boston, MA (2006).

Amount of Grant: $35,000

Date(s) Application is Due: Dec 18; Dec 23

Contact: Hanna Hopfinger, Program Assistant; (267) 646-0665 or (215) 440-9300; fax: (215) 440-9372; email: hanna.hopfinger@aacr.org or awards@aacr.org

Internet: http://www.aacr.org/home/scientists/research-funding—fellowships/fellows-grants.aspx

Sponsor: American Association for Cancer Research
615 Chestnut Street, 17th Floor
Philadelphia, PA 19106-4404

AACR-FNAB Fellows Grant for Translational Pancreatic Cancer Research   121

The Fellows Grants support innovative research by a meritorious young investigator by presenting the Fellow with research funds to pursue an independent line of investigation within the context of his/her current placement. This is a one-year grant of $35,000. Research projects are restricted to translational pancreatic cancer research that includes the use of human tissue and has implications for therapeutic application and individualized medicine.

Requirements: Candidates must be, at the start of the grant term, in their third, fourth, or fifth year of their fellowship status.

Amount of Grant: $35,000

Date(s) Application is Due: Dec 18

Contact: Julia Laurence, Staff Assistant; (215) 440-9300, ext. 102 or (267) 646-0655; fax: (215) 440-9372; email: laurence@aacr.org

Internet: http://www.aacr.org/home/scientists/research-funding—fellowships/fellows-grants.aspx

Sponsor: American Association for Cancer Research
615 Chestnut Street, 17th Floor
Philadelphia, PA 19106-4404

AACR-FNAB Foundation Career Development Award for Translational Cancer Research   122

Career development awards are two-year awards that support research by scientists who are engaged in meritorious cancer research at academic institutions. These awards provide important transitional support for direct research expenses as researchers move from the ranks of early career scientists to faculty status. The Awards are open to junior faculty who completed postdoctoral studies or clinical fellowships between specified dates and who are at an academic or medical institution. Research projects in this area should have direct applicability to translational cancer and be focused on any individualized therapeutic area. Projects should have implications for individualized cancer treatment and make use of human biopsies or samples, such as needle biopsies or circulating cancer cells. One grant will be presented annually. The submission deadline is December 3, with the decision date March 31.

Requirements: Candidates must have acquired a doctoral degree in a related field and may not currently be a candidate for a further doctoral or professional degree. If a candidate has obtained an equivalent degree at a foreign institution, information on the nature of the degree must be provided at the time of application.

Restrictions: Awards are restricted to investigators or institutions within the U.S. Employees of a national government and employees of private industry are ineligible.

Samples: Tara A. Young, M.D., Ph.D., University of California, Los Angeles, Los Angeles, CA, $50,000 - Project: High Resolution Cytogenetic Study of Archival Metastatic Choroidal Melanoma (2008-2010).

Amount of Grant: $50,000 per year for two years

Date(s) Application is Due: Dec 3

Contact: Julia Laurence, Staff Assistant; (267) 646-0655; fax: (215) 440-9372; email: julia.laurence@aacr.org or awards@aacr.org

Internet: http://www.aacr.org/home/scientists/research-funding—fellowships/research-funding/research-funding—fellowships/career-development-award-recipients.aspx

Sponsor: American Association for Cancer Research
615 Chestnut Street, 17th Floor
Philadelphia, PA 19106-4404

AACR-Genentech BioOncology Career Development Award for Cancer Research on the HER Family Pathway   123

Career development awards are two-year awards that support research by scientists who are engaged in meritorious cancer research at academic institutions. These awards provide important transitional support for direct research expenses as researchers move from the ranks of early career scientists to faculty status. The Awards are open to junior faculty who completed postdoctoral studies or clinical fellowships between specified dates and who are at an academic or medical institution. Research projects in this area should have direct applicability to basic, translational, or clinical cancer research focused on the HER Family Pathway. One grant will be presented annually. The submission deadline is December 3, with the decision date March 31.

Requirements: Candidates must be affiliated with an institution in the U.S. and must hold an M.D. or combined M.D./Ph.D. If a candidate has obtained an equivalent degree at a foreign institution, information on the nature of the degree must be provided at the time of application.

Restrictions: Employees of a national government and employees of private industry are ineligible.

Amount of Grant: $50,000 per year for two years

Date(s) Application is Due: Dec 3

Contact: Julia Laurence, Staff Assistant; (267) 646-0655; fax: (215) 440-9372; email: julia.laurence@aacr.org or awards@aacr.org

Internet: http://www.aacr.org/

Sponsor: American Association for Cancer Research
615 Chestnut Street, 17th Floor
Philadelphia, PA 19106-4404

AACR-Genentech BioOncology Fellow for Cancer Research in Angiogenesis   124

AACR Research Fellowships are open to Postdoctoral Fellows and Clinical Research Fellows at an academic facility, teaching hospital, or research institution who will be in the 1st, 2nd, or 3rd year of their postdoctoral training at the start of the fellowship term on July 1. This is a two-year term with a grant of $35,000 per year. One Fellowship will be awarded.

Requirements: The recipients of the Fellowships must attend the AACR Annual Meeting to accept the award.

Restrictions: Research projects are restricted to cancer research with direct applicability to angiogenesis.

Amount of Grant: $35,000 per year for two years

Date(s) Application is Due: Dec 3

Contact: Hanna Hopfinger, Program Assistant; (267) 646-0665; fax: (215) 440-9372; email: hanna.hopfinger@aacr.org

Internet: http://www.aacr.org/

Sponsor: American Association for Cancer Research
615 Chestnut Street, 17th Floor
Philadelphia, PA 19106-4404

AACR-GlaxoSmithKline Clinical Cancer Research Scholar Awards   125

These awards provide support of $4,000 which may be used by the recipient to support attendance at the Annual Meeting and any other AACR meeting within a two-year period. Up to $2,000 per meeting may be used, and reimbursement checks are provided after the meeting upon submission of proper documentation of travel, housing, registration, and/or subsistence expenses. Award recipients are still required to pay the Annual Meeting registration fee and make their own travel and housing arrangements. Selection is made by the Program Committee based upon the novelty, quality, and significance of the abstract submitted. Award recipients will receive notification separate from abstract acceptance and scheduling information.

Requirements: Graduate students, medical students and residents, clinical fellows or equivalent, and postdoctoral fellows (AACR members and nonmembers) who are presenters of a meritorious abstract in clinical cancer research are eligible.

Restrictions: Employees or subcontractors of private industry are not eligible.

Samples: Manisha Bhutani, M.D., UT M.D. Anderson Cancer Center (2008); David W. Cescon, M.D., Princess Margaret Hospital, University of Toronto (2008); Zeshaan A. Rasheed, M.D., Ph.D., Sidney Kimmel Comprehensive Cancer Center, Johns Hopkins University School of Medicine (2008).

Amount of Grant: $2,000-$4,000

Date(s) Application is Due: Nov 28

Contact: Mona E. Shater, Staff Assistant; (215) 440-9300 or (267) 646-0654; fax: (215) 440-9372; email: mona.shater@aacr.org

Internet: http://www.aacr.org/home/scientists/meetings—workshops/travel-grants/aacr-glaxosmithkline-clinical-cancer-research-scholar-awards.aspx

Sponsor: American Association for Cancer Research
615 Chestnut Street, 17th Floor
Philadelphia, PA 19106-4404

AACR-Minorities in Cancer Research Jane Cooke Wright Lectureship Awards   126

The AACR-Minorities in Cancer Research Jane Cooke Wright Lectureship was first presented in 2006. The Lectureship is intended to give recognition to an outstanding scientist who has made meritorious contributions to the field of cancer research and who has, through leadership or by example, furthered the advancement of minority investigators in cancer research. The recipient will present the AACR-Minorities in Cancer Research Jane Cooke Wright Lectureship during the AACR Annual Meeting, receive an honorarium and commemorative plaque, and receive support for the winner and a guest to attend the Annual Meeting.

Requirements: Candidacy is open to all cancer researchers who are affiliated with any institution involved in cancer research, cancer medicine, or cancer-related biomedical science anywhere in the world. Such institutions include those in academia, industry, or government. Nominations may be made by any scientist, whether an AACR member or nonmember, who is now or has been affiliated with any institution involved in cancer research, cancer medicine, or cancer-related biomedical science. Candidates may not nominate themselves. Nominations must be submitted online at https://proposalcentral. altum.com, no later than 4:00 p.m. United States Eastern Time on October 15th. Paper nominations will not be accepted.

Restrictions: Institutions or organizations are not eligible for the Lectureship.
Date(s) Application is Due: Oct 15
Contact: Monique P. Eversley, Staff Associate; (267) 646-0576 or (215) 440-9300, ext. 1400; fax: (215) 440-9372; email: monique.eversley@aacr.org or awards@aacr.org
Internet: http://www.aacr.org/home/scientists/scientific-achievement-awards/micr-wright-lectureship.aspx
Sponsor: American Association for Cancer Research
615 Chestnut Street, 17th Floor
Philadelphia, PA 19106-4404

AACR-National Brain Tumor Foundation Fellows Grant   127

The AACR-National Brain Tumor Society Fellowship, in memory of Bonnie Brooks, is open to Postdoctoral Fellows and Clinical Research Fellows at an academic facility, teaching hospital, or research institution. This fellowship provides a one-year grant of $40,000 to support the salary and benefits of Fellows engaged in Glioblastoma multiforme (GBM) research. A partial amount of funds may be designated for research support. It is anticipated that two fellowships will be awarded. The AACR requires applicants to submit both an online and a paper application. The online application must be submitted by 12:00 noon (United States Eastern Time) on August 10, the application must be submitted using the proposalCENTRAL website at https://proposalcentral. altum.com. The paper application with original signatures and all supporting documents must be postmarked and sent no later than August 12.

Requirements: Candidates must be, at the start of the grant term, in their third, fourth, or fifth year of their fellowship status.
Restrictions: Investigators holding the rank of Instructor, Adjunct Professor, Assistant Professor, Research Assistant Professor, the equivalent or higher are not eligible.
Amount of Grant: $40,000
Date(s) Application is Due: Aug 10; Aug 12
Contact: Hanna Hopfinger, Grants Program Assistant; (267) 646-0665; fax: (215) 440-9372; email: hanna.hopfinger@aacr.org
Internet: http://www.aacr.org/home/scientists/research-funding—fellowships/postdoctoral-fellowships.aspx
Sponsor: American Association for Cancer Research
615 Chestnut Street, 17th Floor
Philadelphia, PA 19106-4404

AACR-NCI International Investigator Opportunity Grants   128

Recognizing that there is no substitute for the collegial interaction and scholarly discussion that is a cornerstone of scientific advancement, these grants provide significant financial support for meeting registration, housing, travel, and incidentals to attend the annual American Association for Cancer Research Meeting. Grants are available for ten cancer researchers from countries where such opportunities are limited.

Requirements: Applicants must be nationals or permanent residents of, and must reside and conduct cancer research within, low and middle-income countries as defined by the World Bank. Applicants should also hold an doctorate or equivalent degree and be actively engaged in a program of cancer research with a record of cancer-focused publications in peer-reviewed scientific journals.
Restrictions: Funds may not be used for entertainment expenses (such as for tours or in-room movies), meals for other individuals, souvenirs, and other non-meeting expenses.
Samples: Hakan Akbulut, M.D., Professor, Medical Oncology, Ankara University, Ankara, Turkey (2008); Soo H. Teo, D.Phil., Chief Executive, Cancer Research Initiatives Foundation, Subang Jaya, Malaysia (2008); Jozsef Tovari, Ph.D., Lab Head, Department of Tumor Progression, National Institute of Oncology, Budapest, Hungary (2008).
Date(s) Application is Due: Sep 28;
Contact: Susan Waskey, Program Assistant; (215) 440-9300 or (866) 423-3965; fax: (215) 440-9313; email: susan.waskey@aacr.org
Internet: http://www.aacr.org/home/scientists/meetings—workshops/travel-grants.aspx
Sponsor: American Association for Cancer Research and the National Cancer Institute
615 Chestnut Street, 17th Floor
Philadelphia, PA 19106-4404

AACR-Pancreatic Cancer Action Network Career Development Award for Pancreatic Cancer Research   129

The Pancreatic Cancer Action Network-AACR Career Development Award represents a joint effort to encourage and support early career scientist who are in the first four years of a faculty appointment to conduct pancreatic cancer research and establish successful career paths in this field. The research proposed for funding may be basic, translational, clinical or epidemiological in nature and must have direct applicability and relevance to pancreatic cancer. The Award provides a two year grant of $200,000 ($100,000 per year) for direct research expenses, which may include salary and benefits of the grant recipient, postdoctoral or clinical research fellows, and/or research assistants, research/laboratory supplies and equipment. It is anticipated that three Career Development Awards will be funded.

Requirements: Candidates must have acquired a doctoral degree in a related field and may not currently be a candidate for a further doctoral or professional degree. If a candidate has obtained an equivalent degree at a foreign institution, information on the nature of the degree must be provided at the time of application.

Restrictions: Awards are restricted to investigators or institutions within the U.S. Employees of a national government and employees of private industry are ineligible.
Amount of Grant: $200,000 per year for two years
Date(s) Application is Due: Oct 28
Contact: Julia Laurence, Program Assistant; (267) 646-0655; fax: (215) 440-9372; email: julia.laurence@aacr.org or awards@aacr.org
Internet: http://www.aacr.org/home/scientists/research-funding—fellowships/career-development-awards.aspx
Sponsor: American Association for Cancer Research
615 Chestnut Street, 17th Floor
Philadelphia, PA 19106-4404

AACR-Pancreatic Cancer Action Network Innovative Grants   130

The AACR-PanCAN Pilot Grants represent a joint effort of the AACR and the Pancreatic Cancer Action Network to support innovative research focusing on the early detection or treatment of pancreatic cancer. Research projects may have direct application to pancreatic cancer or may demonstrate relevance to pancreatic cancer in order to encourage investigators with experience in other areas of research to apply their ideas to this challenging field. Proposals will be accepted for innovative research projects that have a direct application or demonstrate relevance to the early detection or treatment of pancreatic cancer; projects may be in any discipline of basic, translational, clinical, or epidemiological cancer research. Proposals that will develop preliminary data necessary to prepare and submit a competitive research grant application to a major federal funding agency will also be accepted. Special emphasis will be placed on research that is not duplicative of other efforts and has the potential for national application. Researchers who are affiliated with any institution involved in cancer research, cancer medicine, or cancer-related biomedical science within the United States may apply. There are no residency status restrictions. Each grantee will be awarded a total of $200,000, paid over the two-year term.

Requirements: Applicants must have acquired a doctoral degree in a related field. If an applicant has obtained an equivalent degree at a foreign institution, information on the nature of the degree must be provided at the time of application.
Restrictions: Employees or subcontractors of a national government or the for-profit private industry are not eligible to apply. Members of the Scientific Review Committee are not eligible for these grants. Current or past recipients of a PanCAN Pilot Grant are not eligible.
Amount of Grant: $200,000 over two years
Date(s) Application is Due: Oct 5; Nov 30
Contact: Julia Laurence, Program Assistant; (215) 440-9300, ext. 102 or (267) 646-0655; fax: (215) 440-9372; email: julia.laurence@aacr.org
Internet: http://www.aacr.org/home/scientists/research-funding—fellowships/pancreatic-cancer-action-network-aacr-innovative-grants.aspx
Sponsor: American Association for Cancer Research
615 Chestnut Street, 17th Floor
Philadelphia, PA 19106-4404

AACR-Prevent Cancer Foundation Award for Excellence in Cancer Prevention Research   131

he Award for Excellence in Cancer Prevention Research will be given to a scientist residing in any country in the world for his or her seminal contributions to the field of cancer prevention. Such investigations must have been conducted in basic, translational, clinical, epidemiological, or behavioral science in cancer prevention research. Further, these studies must have had not only a major impact on the field, but must also have stimulated new directions in this important area. The recipient of the Award will receive a $5,000 honorarium and will present a lecture at the Annual AACR International Conference. Nominations for the Award must be submitted electronically no later than 4:00 p.m. United States Eastern Time on Monday, July 31.

Requirements: All cancer researchers who are affiliated with any institution involved in cancer research, cancer medicine, or cancer-related biomedical science anywhere in the world. Such institutions include those in academia, industry, or government may be nominated for the Award. Candidates must currently maintain an active research program, have a record of recent publications, and be able to present the Award lecture at the Conference.
Restrictions: Institutions or organizations are not eligible for the Award. Paper nominations will not be accepted.
Amount of Grant: $5,000 honorarium
Date(s) Application is Due: Jul 31
Contact: Monique P. Eversley, Staff Associate; (267) 646-0576 or (215) 440-9300, ext. 1400; fax: (215) 440-9372; email: monique.eversley@aacr.org or awards@aacr.org
Internet: http://www.aacr.org/home/scientists/scientific-achievement-awards/prevent-cancer-award.aspx
Sponsor: American Association for Cancer Research / Prevent Cancer Foundation
615 Chestnut Street, 17th Floor
Philadelphia, PA 19106-4404

AACR-WICR Scholar Awards   132

AACR-WICR Scholar Awards are awarded to ten members of Women in Cancer Research who are scientists-in-training and presenters of meritorious scientific papers at AACR Annual Meetings.

Requirements: Eligible candidates for these awards must fit the following criteria, candidates must be: members of Women in Cancer Research (WICR); full time scientists-in-training who are graduate students, medical students, residents, clinical fellows or equivalent, or postdoctoral fellows; travelling from within the United States or abroad; first authors on abstracts submitted for consideration for presentation at the AACR Annual Meeting.

Date(s) Application is Due: Dec 1

Contact: Lauren Medvetz, Program Coordinator; (267) 646-0689; fax: (215) 440-9372; email: lauren.medvets@aacr.org

Internet: http://www.aacr.org/home/scientists/meetings—workshops/travel-grants/wicr-scholar-awards.aspx

Sponsor: American Association for Cancer Research and Women in Cancer Research (WICR)

615 Chestnut Street, 17th Floor
Philadelphia, PA 19106-4404

AACR-Women in Cancer Research Charlotte Friend Memorial Lectureship Awards   133

The AACR-Women in Cancer Research Charlotte Friend Memorial Lectureship was established in 1998 in honor of renowned virologist and discoverer of the Friend virus, Dr. Charlotte Friend, for her pioneering research on viruses, cell differentiation, and cancer. The lecture is intended to give recognition to an outstanding female or male scientist who has made meritorious contributions to the field of cancer research and who has, through leadership or by example, furthered the advancement of women in science. The winner of the Lectureship will present a 50-minute lecture during the AACR Annual Meeting, receive an honorarium and commemorative plaque, and receive support for the winner and a guest to attend the Annual Meeting.

Requirements: Candidacy is open to all cancer researchers who are affiliated with any institution involved in cancer research, cancer medicine, or cancer-related biomedical science anywhere in the world. Such institutions include those in academia, industry, or government.

Restrictions: Nominations may be made by any scientist, whether an AACR member or nonmember, who is now or has been affiliated with any institution involved in cancer research, cancer medicine, or cancer-related biomedical science. Candidates may not nominate themselves. Nominations must be submitted online at https://proposalcentral.altum.com, no later than 4:00 p.m. United States Eastern Time on October 15th. Paper nominations will not be accepted. Institutions or organizations are not eligible for the Lectureship.

Date(s) Application is Due: Oct 15

Contact: Lauren Medvetz, Program Coordinator; (267) 646-0689; fax: (215) 440-9372; email: lauren.medvets@aacr.org

Internet: http://www.aacr.org/home/scientists/scientific-achievement-awards/wicr-friend-lectureship.aspx

Sponsor: American Association for Cancer Research

615 Chestnut Street, 17th Floor
Philadelphia, PA 19106-4404

AACR Basic Cancer Research Fellowships   134

The AACR Basic Cancer Research Fellowships consist of two Fellowships, each one-year grants. The AACR Anna D, Barker Fellowship in Basic Cancer Research, funds $40,000, and the AACR-Astellas U.S.A Foundation Fellowships in Basic Cancer Research provides funding in the amount of $30,000. These grants are for salary and benefits to the Fellows. A partial amount of funds may be designated for direct research support. The recipients must attend the AACR Annual Meeting to accept the grants. The AACR requires applicants to submit both an online and a paper application. The online application must be completed by 12:00 noon (United States Eastern Time) on December 10, using the proposalCENTRAL website at http://propasalcentral.altum.com. The paper application with original signatures and all supporting documents must be postmarked and sent no later than December 15.

Requirements: Applications are open to Postdoctoral Fellows and Clinical Research Fellows at an academic facility, teaching hospital, or research institution who will be in the 1st, 2nd, or 3rd year of their postdoctoral training at the start of the fellowship term.

Restrictions: Academic faculty holding the rank of assistant professor or higher, graduate or medical students, government employees, and employees of private industry are not eligible.

Amount of Grant: $30,000 - $40,000 range

Date(s) Application is Due: Dec 10; Dec 15

Contact: Hanna Hopfinger, Grants Program Assistant; (267) 646-0665; fax: (215) 440-9372; email: hanna.hopfinger@aacr.org

Internet: http://www.aacr.org/home/scientists/research-funding—fellowships/postdoctoral-fellowships.aspx

Sponsor: American Association for Cancer Research

615 Chestnut Street, 17th Floor
Philadelphia, PA 19106-4404

AAES Engineering Journalism Award   135

The purpose of this American Association of Engineering Societies (AAES) award is to recognize outstanding reporting of an event or issue that furthers public understanding of engineering and is given in one of these three categories: daily newspapers, general circulation print media, and broadcast radio or television. Nominations must be articles published or broadcast in English between January 1 and December 31. Nominations are welcome from reporters, editors, publishers, and members of the engineering community. Print entries must be easily reproducible; electronic entries must include both a VHS or cassette and a script. The nomination cycle runs from October 1 to December 31.

Samples: Daniel Polin (2006); Kenneth Mandel (2006); Jon Palfreman (2005).

Amount of Grant: $5,000 maximum

Date(s) Application is Due: Dec 31

Contact: William Koffel, Executive Director; (202) 296-2237, ext. 201; fax: (202) 296-1151; email: wkoffel@aaes.org or info@aaes.org

Internet: http://www.aaes.org/communications/awards.asp

Sponsor: American Association of Engineering Societies

1620 I (Eye) Street, NW, Suite 210
Washington, DC 20006-4028

AAES National Engineering Award   136

Established in 1979, the National Engineering Award is presented on behalf of the engineering profession to those engineers whose leadership and accomplishments have particularly benefited humanity. The Award is presented for inspirational leadership and tireless devotion to the improvement of engineering education and to the advancement of the engineering profession, as well as to the development of sound public policies as an engineer-statesman. The award may be jointly presented to two or more engineers if a team effort is worthy of recognition.

Samples: Harold (Bud) E. Nelson (2006); E. Walter LeFevre (2005); Daniel R. Benigni (2004).

Date(s) Application is Due: Dec 31

Contact: William Koffel, Executive Director; (202) 296-2237, ext. 201; fax: (202) 296-1151; email: wkoffel@aaes.org or info@aaes.org

Internet: http://www.aaes.org/communications/awards.asp

Sponsor: American Association of Engineering Societies

1620 I (Eye) Street, NW, Suite 210
Washington, DC 20006-4028

AAF Richard Morris Hunt Architecture Fellowship   137

The Richard Morris Hunt Fellowship, sponsored by the American Architectural Foundation and The French Heritage Society as part of a commitment to stewardship of American and French heritage, is awarded to architects pursuing a career in historic preservation. The fellowship permits young architects from France and the United States to observe and practice the latest preservation technologies and techniques in each other, countries. Inaugurated in 1990 as part of the organizations' commitment to foster stewardship of architectural heritage, the recipients represent the best their countries have to offer in the field of historic preservation. The six-month fellowship alternates each year between a French and an American architect and carries a stipend of $25,000.

Amount of Grant: $25,000

Contact: Mary Felber, Director; (202) 626-7511; fax: (202) 626-7420; email: mfelber@archfoundation.org

Internet: http://www.archfoundation.org/aaf/aaf/Programs.Fellowships.htm

Sponsor: American Architectural Foundation

1799 New York Avenue, NW
Washington, DC 20006

AAF RTKL Traveling Fellowships   138

The RTKL Traveling Fellowship is intended to encourage and support foreign travel undertaken to further education toward a professional degree. The fellowship was established by a $25,000 contribution to the American Architectural Foundation by the architecture firm RTKL in honor of one of its founders, Francis T. Taliaferro. Each year, one $2,500 fellowship is awarded to a student submitting the winning proposal outlining a foreign itinerary which is directly relevant to his or her educational goals. Students must complete travel prior to graduation. Application forms are available at NAAB-accredited programs, or may be downloaded. Applications are accepted from students planning independent foreign travel outside the U.S. and those who have been selected to participate in established school travel programs. Applications must be postmarked by January 15.

Requirements: Applicants must be in the second-to-last year of a BArch or MArch program when applying and complete the travel prior to graduation. complete the travel prior to graduation. Those undertaking independent travel must be accepted in an advanced degree program.

Amount of Grant: $2,500

Date(s) Application is Due: Jan 15

Contact: Mary Felber, Director; (202) 626-7511; fax: (202) 626-7420; email: mfelber@archfoundation.org

Internet: http://www.archfoundation.org/aaf/aaf/Programs.Fellowships.htm

Sponsor: American Architectural Foundation

1799 New York Avenue NW
Washington, DC 20006

AAF Traveling Fellowship at the Sir John Soane, Museum   139

The American Architectural Foundation, Fellowship at the Sir John Soane, Museum provides a $5,000 grant annually to enable a student in graduate degree programs in the history of art, architecture, interior design or related fields to travel to England to pursue research on any aspect of the work of Sir John Soane or the Sir John Soane, Museum and collections. These include Egyptian antiquities, classical antiquities, casts, sculpture galleries, the history of museums, authentic restoration, architecture and architectural drawings c. 1650-1850, architectural models, architectural theory c. 1650-1850, neoclassical sculpture, the Grand Tour, Hogarth, George Dance Junior, Joseph Michael Gandy, Sir William Chambers, Sir Christopher Wren and Hawksmoor, architecture and decoration, Robert and James Adam and English Regency painting. In order to be eligible, an applicant must be enrolled in a graduate degree program in the history of art, architecture, the decorative arts, interior design or in a field appropriate to the fellowship, purpose. Applications for the fellowship will be accepted until March 1.

Amount of Grant: $5,000
Date(s) Application is Due: Mar 1
Contact: Mary Felber, Director; (202) 626-7511; fax: (202) 626-7420; email: mfelber@archfoundation.org
Internet: http://www.archfoundation.org/aaf/aaf/Programs.Fellowships.htm
Sponsor: American Architectural Foundation
1799 New York Avenue NW
Washington, DC 20006

AAFA Investigator Research Grants 140

AAFA is dedicated to finding the causes, new treatments and cures for asthma and allergic diseases. AAFA offers a program that sponsors seed grants for investigators wishing to explore new areas of scientific merit related to asthma and allergic diseases. Support has enabled 80% of the researchers who have received the AAFA research grant to win additional research funding from other sources totaling more than $18 million. The Asthma and Allergy of Foundation of America is the only patient organization in the United States that funds research grants for both asthma and allergies. At this time, the application process for grants is through an NIH referral system.
Samples: Stephen C. Dreskin, MD, PhD, University of Colorado, Boulder, Colorado (2006); Anthony A. Horner, MD, University of California, San Diego, CA—Analysis of airway (2005); David Lewis, MD, Stanford University School of Medicine—Mechanisms of viral induced asthma (2004).
Contact: Mary Brasler, Director of Programs and Services; (202) 466-7643, ext. 238; email: mbrasler@aafa.org
Internet: http://www.aafa.org/display.cfm?id=6,äÇ=64
Sponsor: Asthma and Allergy Foundation of America
1233 20th Street, NW, Suite 402
Washington, DC 20036

AAFCS International Graduate Fellowships 141

The Association encourages graduate study in family and consumer sciences and its subspecialties, such as textiles and clothing, nutrition, institutional management/food service systems administration, consumer studies, Cooperative Extension, and family and consumer sciences communication, education, and administration. International graduate fellowships are awarded to individuals who have exhibited the potential to make contributions to the family and consumer sciences profession.
Requirements: Applicants must be citizens or permanent residents of the United States. Applicant must show clearly defined plans for full-time graduate or undergraduate study during the time for which the fellowship is awarded. Each request for appropriate application form must be accompanied by an application fee of $40.
Contact: Fellowships and Awards Committee; (703) 706-4600; fax: (703) 706-4663; email: staff@aafcs.org
Internet: http://www.aafcs.org/programs/scholarships.htm
Sponsor: American Association of Family and Consumer Sciences
400 N. Columbus Street, Suite 202
Alexandria, VA 22314

AAFCS National Graduate Fellowships 142

AAFCS annually awards national graduate fellowships to support graduate study in areas such as family and consumer sciences, nutrition, textiles, and home economics. The fellowships are awarded to individuals who have exhibited the potential for making contributions to these professions. It is recommended that applicants first complete at least one year of professional family and consumer science experience by the time of application.
Requirements: Applicants must be citizens or permanent residents of the United States. Applicant must show clearly defined plans for full-time graduate or undergraduate study during the time for which the fellowship is awarded. Each request for appropriate application form must be accompanied by an application fee of $40.
Amount of Grant: $5,000 maximum
Date(s) Application is Due: Jan 7
Contact: Fellowships and Awards Committee; (703) 706-4600; fax: (703) 706-4663; email: staff@aafcs.org
Internet: http://www.aafcs.org/programs/fellowshipseven.html
Sponsor: American Association of Family and Consumer Sciences
400 N. Columbus Street, Suite 202
Alexandria, VA 22314

AAFP Foundation Joint Grants 143

The foundation supports clinical research projects within family practice that ultimately will result in improved patient care. This program includes a research grant stimulation program that is designed to stimulate research in family practice by supporting small and first-time research projects. Longer research proposals must be prepared in accordance with standard guidelines for grant proposals.
Requirements: Eligible to apply are individual family physicians, residents in family practice, and medical students planning to pursue the specialty of family practice; educational and health care institutions or organizations that will use the grant exclusively for programs or projects directly involved with family practice or family medicine; departments of family practice; family practice residency programs; and family practice organizations or associations. Research stimulation proposals may be reviewed at any time during the year depending on availability of funds.
Restrictions: Proposals seeking support for the cost of instituting programs or support for such activities as videotape production, designing curriculum, or implementing a project, etc., will be denied.

Amount of Grant: $30,000 maximum
Contact: Susie Morantz, Senior Program Manager; (913) 906-6000, ext. 4470 or (800) 274-2237; fax: (913) 906-6095; email: smorantz@aafp.org
Internet: http://www.aafpfoundation.org/x270.xml
Sponsor: American Academy of Family Physicians Foundation
11400 Tomahawk Creek Parkway, Suite 440
Leawood, KS 66211-2672

AAFP Foundation Pfizer Teacher Development Awards 144

The purpose of the awards is to honor community-based physicians who give up time from their practice to teach family medicine students and/or residents part time. Eligible candidates may serve as preceptors or as volunteer teachers at other sites, including family medicine teaching centers. Guidelines are available online.
Requirements: Applicant must have graduated from an ACGME-approved family practice residency program within the last seven years; have entered family medicine; be a member of the American Academy of Family Physicians; have entered or plan to enter part-time family medicine teaching (not less than four hours per month and no more than 32 hours per month—averaged over a year); and teach voluntarily or receive no more than $18,000 compensation for the educational time devoted to residents and/or students.
Amount of Grant: $2,000
Contact: Sondra Goodman, Grants Program Manager; (913) 906-6000, ext. 4457 or (800) 274-2237; fax: (913) 906-6095; email: sgoodman@aafp.org
Internet: http://www.aafpfoundation.org/x269.xml
Sponsor: American Academy of Family Physicians Foundation
11400 Tomahawk Creek Parkway, Suite 440
Leawood, KS 66211-2672

AAFP Foundation Wyeth Immunization Awards 145

This program offers family medicine residency programs the opportunity to receive recognition for identifying and developing creative solutions to overcoming barriers to childhood immunizations, thus increasing immunization rates, and to share the best practices that have benefited the communities in which they live. Applications are accepted annually for one of the following award track categories: Best Practices Award—overall achievement with systems already in place to overcome immunization barriers and achieve high rates within a certain time frame; Most Improved Award—overcame barriers and other challenges to greatly enhance immunization rates; and Implement New System Grant—residency programs seeking to implement a new system to help increase immunization rates in under-served children. Grant winners will receive: one of 10 monetary awards (four at $5,000 and six at $10,000) determined by rank of application as scored by review panel; an education scholarship for $1,000 to send a resident to the annual AAFP National Conference of Family Medicine Residents and Medical Students; an additional travel scholarship to send a resident to the AAFP Scientific Assembly (top six); and a plaque of recognition.
Requirements: Family medicine residency programs that have increased immunization rates within a defined time frame and can demonstrate the best practices implemented to achieve the higher compliance are eligible.
Amount of Grant: $1,000-$10,000
Date(s) Application is Due: Apr 21
Contact: Sondra Goodman, Grants Program Manager; (913) 906-6000, ext. 4457 or (800) 274-2237; fax: (913) 906-6095; email: sgoodman@aafp.org
Internet: http://www.aafpfoundation.org/x655.xml
Sponsor: American Academy of Family Physicians Foundation
11400 Tomahawk Creek Parkway, Suite 440
Leawood, KS 66211-2672

AAFP Research Stimulation Grants 146

This Program provides relatively small grants to stimulate research in family practice in the areas of preliminary efforts leading to a larger research project, pilot projects, and data collection for a larger research project. The applicant must include a discussion of how the proposed project is anticipated to lead to a larger project. Grant applications will be accepted twice a year: by March 1 and September 1.
Requirements: Eligible to apply are individual family physicians, family practice/medicine organizations or associations, residents in family practice, departments of family medicine, and educational and health care institutions directly involved in family practice/medicine.
Amount of Grant: $7,500
Date(s) Application is Due: Mar 1; Sep 1
Contact: Susie Morantz, Senior Program Manager; (913) 906-6000, ext. 4470 or (800) 274-2237; fax: (913) 906-6095; email: smorantz@aafp.org
Internet: http://www.aafpfoundation.org/x446.xml
Sponsor: American Academy of Family Physicians Foundation
11400 Tomahawk Creek Parkway, Suite 440
Leawood, KS 66211-2672

AAFPRS Ben Shuster Memorial Award 147

This award is given annually for the best paper based on clinical work or research in the field of facial plastic surgery by a resident or fellow in training. The paper must have been delivered at a national meeting (or its equivalent) within the preceding two years prior to the February 1 date of submission. Studies prepared during the first year after completion of residency training will be considered provided that the research was conducted during the author's residency training program or fellowship. A certificate and an award of $1,000 are presented.

Requirements: The competition is open to U.S. or Canadian residents or fellows in otolaryngology who are members of AAFPRS. Each entrant must be the sole or senior author and an AAFPRS member.
Amount of Grant: $1,000
Date(s) Application is Due: Feb 1
Contact: Awards Coordinator; (703) 299-9291; fax: (703) 299-8898; email: info@aafprs.org
Internet: http://www.aafprs.org/physician/awards_grants/awards.html
Sponsor: American Academy of Facial, Plastic, and Reconstructive Surgery
310 S Henry Street
Alexandria, VA 22314

AAFPRS Bernstein Grant 148

The purpose of the grant is to encourage original research projects that will advance facial plastic and reconstructive surgery. The grant may be used as seed money for research projects. Applicants should submit an application, a detailed proposal, the investigator's curriculum vita, and letters from the appropriate institutional review committees certifying institutional conformity to the U.S. government guidelines for human and animal experiments. The recipient should complete his or her work within three (3) years with interim reports due each year. All research grant monies awarded are intended to fund direct costs related to the research as described in the grant proposal.
Requirements: Applicants should be U.S. or Canadian AAFPRS fellow members. The primary criteria are that the research be original and have direct application to facial plastic and reconstructive surgery.
Amount of Grant: $25,000
Date(s) Application is Due: Jan 15
Contact: Research Grant Coordinator; (703) 299-9291; fax: (703) 299-8898; email: info@aafprs.org
Internet: http://www.aafprs.org/physician/awards_grants/bernstein_grant.html
Sponsor: American Academy of Facial, Plastic, and Reconstructive Surgery
310 South Henry Street
Alexandria, VA 22314

AAFPRS Fellowships 149

The Educational and Research Foundation for the American Academy of Facial Plastic and Reconstructive Surgery fellowship program provides postgraduate training in facial plastic surgery. The objectives of the fellowship program are to: provide an outstanding academic opportunity for the acquisition of specialized knowledge and skills in facial plastic surgery; develop trained specialists who will contribute to the ongoing development of facial plastic and reconstructive surgery; foster development of facial plastic and reconstructive surgery educators, especially in residency programs; and encourage the development of new skills and knowledge in facial plastic and reconstructive surgery through basic research and clinical trials. Eligible physicians may apply for 40 positions available each year.
Requirements: Applicants to the Fellowship Program must be physicians who are in or have completed an otolaryngology or plastic surgery residency program accredited by the Accreditation Council for Graduate Medical Education (ACGME) or Royal College of Physicians and Surgeons of Canada (RCPSC) or board-certified in otolaryngology-head and neck surgery or plastic surgery. Applicants must be members of the AAFPRS before submitting an application.
Restrictions: Applicants should not be full-time faculty members holding the rank of assistant professor or higher at the institution where the fellowship will take place.
Date(s) Application is Due: Feb 1
Contact: Fellowship Coordinator; (703) 299-9291; fax: (703) 299-8898; email: info@aafprs.org or fsanders@aafprs.org
Internet: http://www.aafprs.org/physician/benefits/ben_fellow.html
Sponsor: American Academy of Facial, Plastic, and Reconstructive Surgery
310 South Henry Street
Alexandria, VA 22314

AAFPRS Investigator Development Grant 150

The purpose of the Investigator Development Grant is to support the work of a young faculty member in facial plastic surgery conducting significant clinical or laboratory research and involved in the training of resident surgeons in research. One $15,000 grant may be awarded each year upon receipt of the signed agreement. Recipients should complete their work within three (3) years with interim reports due each year. An extension is possible upon written request to the Research Committee. The original proposal and all other requirements should be submitted to C.O.R.E. no later than January 15 or the next business day thereafter.
Requirements: An applicant must be an AAFPRS member at any level involved in the training of resident surgeons.
Amount of Grant: $15,000 maximum
Date(s) Application is Due: Jan 15
Contact: Research Grant Coordinator; (703) 299-9291; fax: (703) 299-8898; email: info@aafprs.org
Internet: http://www.aafprs.org/physician/awards_grants/investigator_grant.html
Sponsor: American Academy of Facial, Plastic, and Reconstructive Surgery
310 South Henry Street
Alexandria, VA 22314

AAFPRS Ira J. Tresley Research Award 151

This annual award is given to a U.S. or Canadian academy member for the best paper based on any research study in the field of facial plastic surgery. The competition is open to any U.S. or Canadian physician who is an AAFPRS member and who has been board certified for at least three years. The paper must have been presented at a national meeting or its equivalent within the preceding two years prior to the February 1 date of submission. A certificate and an award of $1,000 are presented.
Amount of Grant: $1,000
Date(s) Application is Due: Feb 1
Contact: Awards Coordinator; (703) 299-9291; fax: (703) 299-8898; email: info@aafprs.org
Internet: http://www.aafprs.org/physician/awards_grants/awards.html
Sponsor: American Academy of Facial, Plastic, and Reconstructive Surgery
310 South Henry Street
Alexandria, VA 22314

AAFPRS John Orlando Roe Award 152

This award, named for John Orlando Roe, the surgeon who accomplished the first rhinoplasty in 1887, includes a certificate and an award of $1,000 to be presented each year to an AAFPRS Fellow in an AAFPRS Foundation Fellowship Program who submits the best clinical research paper written during the current fellowship year.
Amount of Grant: $1,000
Date(s) Application is Due: Feb 1
Contact: Awards Coordinator; (703) 299-9291; fax: (703) 299-8898; email: info@aafprs.org
Internet: http://www.aafprs.org/physician/awards_grants/awards.html
Sponsor: American Academy of Facial, Plastic, and Reconstructive Surgery
310 S. Henry Street
Alexandria, VA 22314

AAFPRS Leslie Bernstein Resident Research Grants 153

Two resident research grants, tenable in the United States or Canada, may be awarded each year to stimulate resident research in facial plastic surgery on projects that are well conceived and scientifically valid. Residents are encouraged to enter early in their training so that their applications may be revised and resubmitted if not accepted the first time. This award may be integrated with other funding to complete a project. Interested residents should write for entry guidelines.
Requirements: U.S. and Canadian residents at any level who are AAFPRS members are eligible to apply, even if the research work will be done during their fellowship year. Grant recipients must complete their work in one (1) to two (2) years with interim reports due each year.
Amount of Grant: $5,000
Date(s) Application is Due: Dec 15; Jan 15
Contact: Research Grant Coordinator; (703) 299-9291; fax: (703) 299-8898; email: info@aafprs.org
Internet: http://www.aafprs.org/physician/awards_grants/grants.html
Sponsor: American Academy of Facial, Plastic, and Reconstructive Surgery
310 South Henry Street
Alexandria, VA 22314

AAFPRS Residency Travel Awards 154

Two Residency Travel Awards may be given each year for two outstanding papers in facial plastic and reconstructive surgery primarily authored by a resident or medical student in training. The paper must be submitted by February 1 for consideration, which will be presented at the Annual Fall Meeting. Entries should conform to the guidelines of the Archives of Facial Plastic Surgery, the official journal of the AAFPRS. Papers submitted will be judged anonymously. Only the cover page should contain the applicant's name or institutional identification.
Date(s) Application is Due: Feb 1
Contact: Awards Coordinator; (703) 299-9291; fax: (703) 299-8898; email: info@aafprs.org
Internet: http://www.aafprs.org/physician/awards_grants/awards.html
Sponsor: American Academy of Facial, Plastic, and Reconstructive Surgery
310 South Henry Street
Alexandria, VA 22314

AAFPRS Sir Harold Delf Gillies Award 155

This award is named for Sir Harold Delf Gillies, a British otolaryngologist who in September 1917 described the tubed pedicle flap. Dr. Gillies frequently visited the U.S. and lectured widely to surgeons of various specialties and was given the title Father of Plastic Surgery. A certificate and an award of $1,000 will be presented each year to an AAFPRS Fellow in an AAFPRS Foundation Fellowship Program who submits the best basic science research paper written during the current fellowship year.
Amount of Grant: $1,000
Date(s) Application is Due: Feb 1
Contact: Awards Coordinator; (703) 299-9291; fax: (703) 299-8898; email: info@aafprs.org
Internet: http://www.aafprs.org/physician/awards_grants/awards.html
Sponsor: American Academy of Facial, Plastic, and Reconstructive Surgery
310 South Henry Street
Alexandria, VA 22314

AAG Anne U. White Fund Grants 156

The Fund's income is to be used to encourage field research conducted by a member of the Association jointly with her or his partner. The Fund will serve the purposes Anne White held dear if it enables people, regardless of any formal training in geography, to

engage in useful field studies and to have the joy of working along side their partners. In the past, two awards have been made each year of approximately $1,500 each. Digital submissions are required. All applicants will be notified of the status of their submission within approximately 3 months after the deadline. Awardees may anticipate receiving their grants by late Spring.
Amount of Grant: $1,500
Date(s) Application is Due: Dec 31
Contact: Ehsan Khater, Coordinator; (202) 234-1450; fax: (202) 234-2744; email: ekhater@aag.org or grantsawards@aag.org
Internet: http://www.aag.org/Grantsawards/Annewhitefund.html
Sponsor: Association of American Geographers
1710 Sixteenth Street NW
Washington, DC 20009-3198

AAG Dissertation Research Grants   157
The following funds provide financial assistance to PhD candidates preparing dissertations in geography. The Robert D. Hodgson Fund gives preference to those topics that best reflect Dr. Hodgson's belief that the pursuit and understanding of geographic knowledge may lead to international cooperation. The Paul Vouras Fund gives preference to minority applicants for support of dissertation work. The Otis Paul Starkey Fund gives preference to regional study or significant problem areas in the United States or its possessions. Funds are to be used for direct research expenses and may not be used to cover overhead costs. Applicants may obtain application forms from the association.
Requirements: Any person is eligible to apply for a grant who has been a member of AAG for at least a year at the time of submitting the application, does not have a PhD degree at the time of the award, and has completed all PhD requirements except for the dissertation research.
Amount of Grant: $500 maximum
Date(s) Application is Due: Dec 31
Contact: Ehsan Khater, Coordinator; (202) 234-1450; fax: (202) 234-2744; email: ekhater@aag.org or grantsawards@aag.org
Internet: http://www.aag.org/Grantsawards/Dissertationresearch.html
Sponsor: Association of American Geographers
1710 Sixteenth Street NW
Washington, DC 20009-3198

AAG E. Willard and Ruby S. Miller Award in Geography   158
This annual award recognizes members of the Association who have made truly outstanding contributions to the geographic field due to their special competence in teaching or research. Funding for the award comes from the estate of Ruby S. Miller. More than one award may be awarded each year. Each award includes $1,000 and a commemorative plaque. Electronic submission is encouraged.
Requirements: Members from all fields of geography are eligible for the award. Nominations (including self-nominations) are welcome.
Amount of Grant: $1,000
Date(s) Application is Due: Dec 31
Contact: Ehsan Khater, Coordinator; (202) 234-1450; fax: (202) 234-2744; email: ekhater@aag.org or grantsawards@aag.org
Internet: http://www.aag.org/Grantsawards/miller.html
Sponsor: Association of American Geographers
1710 Sixteenth Street NW
Washington, DC 20009-3198

AAG International Geographic Information Fund Graduate Research Grants   159
The Association of American Geographers will award one or more small grants of up to $500 to support graduate research proposed by a student in any area of spatial analysis or geographic information science or systems. The award(s) will be presented at the AAG Annual Meeting. The program is designed to support college and university student career development in the academic areas of applied spatial data analysis or geographic information systems (GIS). Digital submissions are encouraged.
Requirements: AAG IGIF Graduate Research Awards are available to full-time students who are currently registered in an undergraduate or graduate degree program providing either a degree or explicit specialization in some area of applied spatial data analysis or GIS study at a duly accredited and recognized college, university or other educational institution located within the United States.
Amount of Grant: $500
Date(s) Application is Due: Dec 31
Contact: Ehsan Khater, Coordinator; (202) 234-1450; fax: (202) 234-2744; email: ekhater@aag.org or grantsawards@aag.org
Internet: http://www.aag.org/Grantsawards/IGIF.cfm
Sponsor: Association of American Geographers
1710 Sixteenth Street NW
Washington, DC 20009-3198

AAG International Geographic Information Fund Student Travel Grants   160
The Association of American Geographers will award one or more small grants of approximately $500 for the support of student travel to national and international symposia or specialized meetings sponsored by recognized professional organizations supporting activities associated with specific aspects of geographic data analysis and handling and geographic information systems. Support of student travel means reimbursement for the cost of transportation, accommodations, meals, and any other cost or expense reasonable related to travel for attendance at such meetings. The program is

designed to support college and university student career development in the academic areas of applied spatial data analysis or geographic information systems (GIS).
Requirements: Grants are available to full-time students who are currently registered in an undergraduate or graduate degree program providing either a degree or explicit specialization in some area of applied spatial data analysis or GIS study at a duly accredited and recognized college, university or other educational institution located within the United States.
Amount of Grant: $500
Date(s) Application is Due: Dec 31
Contact: Ehsan Khater, Coordinator; (202) 234-1450; fax: (202) 234-2744; email: ekhater@aag.org or grantsawards@aag.org
Internet: http://www.aag.org/Grantsawards/IGIF.cfm
Sponsor: Association of American Geographers
1710 Sixteenth Street NW
Washington, DC 20009-3198

AAG John Brinkerhoff Jackson Prize   161
The association offers this award for the best nonfiction book about the human geography of the United States. The prize will go to the writer who can best interpret geographical research in language that lay readers can understand. Publishers are invited to submit entries for the Jackson Prize competition, and should forward one copy of the published book to each of the four committee members listed at the web site. The prize, which carries an award of $1,000 to the author, is administered by the Association of American Geographers.
Requirements: The prize is restricted to books written by geographers, with preference given to those by U.S. citizens or permanent residents.
Restrictions: Textbooks, dissertations, and scholarly articles will not be considered.
Amount of Grant: $1,000
Date(s) Application is Due: Dec 31
Contact: Michael P. Conzen, Chair; (312) 702-8308; email: m-conzen@uchicago.edu
Internet: http://www.aag.org/Grantsawards/jackson_prize.cfm
Sponsor: Association of American Geographers
1710 Sixteenth Street NW
Washington, DC 20009-3198

AAG Mel Marcus Fund Grants for Physical Geography   162
The objective of The Mel Marcus Fund for Physical Geography is to carry on this tradition of excellence and humanity in field work. Grants from the Mel Marcus Fund for Physical Geography will foster personally formative participation by students in field-based physical geography research in challenging outdoor environments. Marcus Fund grants of up to $2000 will be awarded to faculty applicants to support travel and logistical costs of including students in field research in physical geography. Research sites may include international as well as national locations. Marcus grant funds may be used to cover travel to the research site, as well as expenses related to the logistical support of the student, such as food, fuel and lodging. Digital submissions are encouraged.
Restrictions: Student stipends are not eligible expenses under this grant.
Amount of Grant: $2,000 maximum
Date(s) Application is Due: Dec 31
Contact: Ehsan Khater, Coordinator; (202) 234-1450; fax: (202) 234-2744; email: ekhater@aag.org or grantsawards@aag.org
Internet: http://www.aag.org/Grantsawards/marcus.cfm
Sponsor: Association of American Geographers
1710 Sixteenth Street NW
Washington, DC 20009-3198

AAG Research Grants   163
Grants are given to support research and field work done by AAG members. Preference will be given to projects that may generate substantial support from private or federal granting agencies; may be pertinent to questions affecting important gaps in the spectrum of geographic knowledge and research; or will open up new areas, topics, or techniques for geographic research. Applicants should obtain forms from AAG.
Requirements: Any person is eligible to apply for a grant who has been a member of the association for at least two years at the time of submitting an application.
Restrictions: This grant will not be made to support doctoral dissertations.
Amount of Grant: $1,000 maximum
Date(s) Application is Due: Dec 31
Contact: Ehsan Khater, Coordinator; (202) 234-1450; fax: (202) 234-2744; email: ekhater@aag.org or grantsawards@aag.org
Internet: http://www.aag.org/Grantsawards/Generalresearch.html
Sponsor: Association of American Geographers
1710 Sixteenth Street NW
Washington, DC 20009-3198

AAHN Grant for Historical Research   164
The American Association for the History of Nursing is offering a research grant of $3,000 for new researchers. (Indirect costs of 8% are also available). It is expected that the research and new materials produced by the award winner will help ensure the growth of scholarly work focused on the history of nursing. The deadline for submission of applications is April 1. The AAHN Panel will make its decision about the award by May 15 and the recipient will be notified by June 1 of each year. Funding will start July 1 of the grant year and last for one year. A no-cost extension may be granted on request.
Requirements: Applicants must be members of AAHN and hold the doctorate. They may be faculty members or independent researchers.

Amount of Grant: $3,000
Date(s) Application is Due: Apr 1
Contact: David Stumph, Executive Director; (303) 422-2685; fax: (303) 422-8894; email: aahn@aahn.org
Internet: http://www.aahn.org/grants.html
Sponsor: American Association of the History of Nursing
10200 W. 44th Avenue, Suite 304
Wheat Ridge, CO 80033

AAHN Pre-Doctoral Research Grant 165

This grant is designed to encourage and support graduate training and historical research at the Masters and Doctoral levels. The grant will be in the amount of $2,000. Proposals will focus on a significant question in the history of nursing. Selection criteria include the scholarly merit of the proposal, consideration of the student's preparation for this study, the advisor's qualifications for guiding the study and the project's potential for contributing to scholarship in the field of nursing history.
Requirements: The student must be: enrolled in an accredited masters program or doctoral program; and a member of AAHN. The research advisor must be doctorally prepared with scholarly activity in the field of nursing history and prior experience in guidance of research training.
Amount of Grant: $2,000
Date(s) Application is Due: Apr 1
Contact: David Stumph, Executive Director; (303) 422-2685; fax: (303) 422-8894; email: aahn@aahn.org
Internet: http://www.aahn.org/grants.html
Sponsor: American Association of the History of Nursing
10200 W. 44th Avenue, Suite 304
Wheat Ridge, CO 80033

AAHPM Hospice and Palliative Medicine Fellowships 166

The program offers training programs to prepare physicians for careers in academic or community-based palliative medicine. Fellowships provide clinical experience and supervision for palliative medicine in hospitals, homes, hospices, chronic care institutions, and outpatient settings, and include consultational and longitudinal care exposure, as well as training in bereavement care. Some programs also provide opportunities for mentored research or teaching. Most programs begin in July, and clinical training generally lasts one year, allowing for board eligibility upon successful completion. For a complete list of U.S. palliative medicine fellowship training programs, members should visit the web site.
Requirements: Trainees must have already completed residency training in a primary specialty, such as internal or family medicine, or have similar clinical training and experience.
Contact: Fellowships Director; (847) 375-4712; fax: (847) 375-6475; email: info@aahpm.org
Internet: http://www.aahpm.org/fellowship/index.html
Sponsor: American Academy of Hospice and Palliative Medicine
4700 W. Lake Avenue
Glenview, IL 60025-1485

AAJA Fellowships 167

The fellowships are designed to enable members of the association to attend short-term professional training and skills development programs. These fellowships can be used to defray tuition, travel, food, lodging, and other related costs. News organizations of the recipients must grant them a leave of absence to attend the program. Fellowships must be used within 12 months of the award date. Applications are accepted on a case-by-case basis year-round. Grants will be awarded contingent upon proof of registration at, or acceptance to, the training program.
Requirements: Applicants must be AAJA members with at least three years of professional journalism experience.
Amount of Grant: $1,000 maximum
Contact: Albert Lee, Professional Programs Coordinator; (415) 346-2051, ext. 107; fax: (415) 346-6343; email: albertl@aaja.org
Internet: http://www.aaja.org/programs/professional/fellowships
Sponsor: Asian American Journalists Association
1182 Market Street, Suite 320
San Francisco, CA 94102

AALAS Foundation Grants 168

The foundation educational grants have a twofold purpose: to attract talented people to the field of laboratory animal science and to develop new training materials that promote the continuing education of AALAS members. The foundation awards grants for student clerkships to enhance the training of future laboratory animal personnel. The head of the training and/or research program sponsoring the clerkship should submit the application, not the student who will receive the education. Applications are judged on the strength of the training program, not the credentials of the students. Three aspects are considered—clinical experience, teaching experience, and research experience—in evaluating student training programs. The foundation also supports the development of training materials that address the continuing educational needs of AALAS members. After completion, AALAS will have unrestricted use of these materials and will make them available to the members. Interested persons should write for the brochure containing information required to apply for either type of grant.

Requirements: Persons currently in the field of laboratory animal science and those persons who are considering the field as a career are eligible. Applications for student clerkship programs must be submitted by the program sponsor.
Amount of Grant: $5,000 maximum
Date(s) Application is Due: Apr 15; Sep 15
Contact: John McCutchen; (901) 754-8620; fax: (901) 753-0046; email: john.mccutchen@aalas.org
Internet: http://aalasfoundation.org/about_grants_and_funding.html
Sponsor: American Association for Laboratory Animal Science
9190 Crestwyn Hills Drive
Memphis, TN 38125

AAMC Award for Distinguished Research 169

The Award for Distinguished Research in the Biomedical Sciences honors outstanding biomedical research related to health and disease. The research recognized should have contributed to the substance of medicine. Nominations may be made by anyone on the faculty or staff of a medical school or teaching hospital or by a member of an academic society.
Requirements: The nominee must be an individual who serves on the faculty of an AAMC member medical school or teaching hospital.
Amount of Grant: $5,000
Date(s) Application is Due: May 1
Contact: Sandra D. Gordon; (202) 828-0472; fax: (202) 828-1125; email: sgordon@aamc.org
Internet: http://www.aamc.org/about/awards/research.htm
Sponsor: Association of American Medical Colleges
2450 N Street NW
Washington DC 20037-1127

AANS Neurosurgery Research and Education Foundation Young 170
Clinician Investigator Award

This one-year, renewable award is available to North American neurosurgeons who are full-time faculty in teaching institutions and are in the early years of their careers. The purpose of the program is to fund pilot studies that could provide preliminary data to be used for strengthening applications for more permanent funding from other sources. Applications related to any field of neurosurgery are encouraged. The NREF encourages both patient-oriented clinical research and basic science research projects. The one-year award totals $40,000.
Restrictions: Applicant may not accept another award for the same project during the same time period. No more than one award per year will be awarded to the same institution.
Amount of Grant: $40,000
Date(s) Application is Due: Oct 31
Contact: Michele S. Gregory, Director of Development; (847) 378-0540 or (847) 378-0500; fax: (847) 378-0600; email: msg@aans.org or info@aans.org
Internet: http://www.aans.org/research/fellowship/nref_y.asp
Sponsor: Research Foundation of the American Association of Neurological Surgeons
5550 Meadowbrook Drive
Rolling Meadows, IL 60008-3852

AANS Neurosurgery Research and Education Foundation/Spine 171
Section Young Clinician Investigator Award

Established by the Neurosurgery Research and Education Foundation (NREF) of the American Association of Neurological Surgeons (AANS), the Joint Section on Disorders of the Spine and Peripheral Nerves Award grants support to young faculty who are pursuing careers in the area of spinal cord, vertebral column or peripheral nerve disorders. The purpose of the award is to fund pilot studies that provide preliminary data used to strengthen applications for more permanent funding from other sources. The one-year award totals $40,000.
Requirements: Applicants must be physicians or neurosurgeons who are full-time faculty in North American teaching institutions and in the early years of their careers.
Restrictions: Applicant may not accept another award for the same project during the same time period. No more than one award per year will be awarded to the same institution.
Date(s) Application is Due: Oct 31
Contact: Michele S. Gregory, Director of Development; (847) 378-0540 or (847) 378-0500; fax: (847) 378-0600; email: msg@aans.org or info@aans.org
Internet: http://www.aans.org/research/fellowship/nref_y.asp
Sponsor: Research Foundation of the American Association of Neurological Surgeons
5550 Meadowbrook Drive
Rolling Meadows, IL 60008-3852

AAO-HNS Medical Student Research Prize 172

The Medical Student Research Prize is given annually for a manuscript submitted to the Editor of Otolaryngology-Head and Neck Surgery for review and consideration for publication in the journal. An award winning manuscript will be automatically accepted as an oral or poster presentation at the Translational and Basic Research Presentations Program (previously called the Research Forum). First prize is $1,000, and the award is available in both clinical and basic science categories through the generosity of GYRU. S. ACMI ENT Division.
Requirements: All manuscripts submitted for this competition must have a medical student as senior author and must describe work that has not been placed under active consideration for publication other than in Otolaryngology-Head and Neck Surgery.

Amount of Grant: $1,000
Date(s) Application is Due: Mar 9
Contact: Stephanie L. Jones, Assistant Director; (703) 519-1586 or (703) 836-4444; email: sljones@entnet.org
Internet: http://www.entnet.org/educationandresearch/grantsandfellowships.cfm
Sponsor: American Academy of Otolaryngology-Head and Neck Surgery
1650 Diagonal Road
Alexandria, VA 22314-2857

AAO-HNS Resident Research Prizes 173

Resident Research Prizes are given annually for manuscripts submitted to the Editor of Otolaryngology-Head and Neck Surgery for review and consideration for publication in the journal. All award winning manuscripts will be automatically accepted as oral or poster presentations at the Translational and Basic Research Presentations Program (previously called the Research Forum). First prize is $2,000, second prize is $1,500, and third prize is $1,000. All awards are available in both clinical and basic science categories through the generosity of GYRU.S. ACMI ENT Division.
Requirements: All manuscripts submitted for this competition must have a resident as senior author and must describe work that has not been placed under active consideration for publication other than in Otolaryngology-Head and Neck Surgery.
Amount of Grant: $1,000-$2,000
Date(s) Application is Due: Mar 9
Contact: Stephanie L. Jones, Assistant Director; (703) 519-1586 or (703) 836-4444; email: sljones@entnet.org
Internet: http://www.entnet.org/educationandresearch/grantsandfellowships.cfm
Sponsor: American Academy of Otolaryngology-Head and Neck Surgery
1650 Diagonal Road
Alexandria, VA 22314-2857

AAO-HNSF CORE Research Grants 174

CORE Research Grants are available in three different categories: Research Project Grants—an independent or assisted project using existing skills and resources, and is hypothesis-based, with specific aims and timetable; Research Training Grants—a research project that is designed to help you refine existing research skill set or acquire new skills, and requires mentor and evidence of training process; and Career Development Grants—a formal, mentored period of research combined with coursework, designed to help researcher become independent researcher. Applicants should complete a Letter of Intent by December 15th, using the online application system, and should then submit the completed application before 12 o'clock midnight EST on January 15.
Date(s) Application is Due: Dec 15; Jan 15
Contact: Stephanie L. Jones, Assistant Director; (703) 519-1586 or (703) 836-4444; email: sljones@entnet.org
Internet: http://www.entnet.org/EducationAndResearch/coreGrants.cfm
Sponsor: American Academy of Otolaryngology-Head and Neck Surgery
1650 Diagonal Road
Alexandria, VA 22314-2857

AAO-HNSF Health Services Research Grants 175

The purpose of the Health Services Research Grant is to foster research that will improve the effectiveness and appropriateness of medical practice. Projects supported under this program will develop and disseminate scientific information on the effects of otolaryngology services and procedures on patients' survival, health status, functional capacity, and quality of life. The award is intended to promote increased participation by otolaryngologists in the rapidly expanding area of health services research. Proposed projects may be related to any area of otolaryngology-head and neck surgery, but must have direct or potential clinical significance for patients seen by otolaryngologist-head and neck surgeons. Grants range up to $10,000 maximum, are for one year, and are not renewable. Projects must be completed within two years of the award date; no-cost extensions are available upon written request. Letters of intent must be received by December 15, and completed applications must be submitted by January 15.
Requirements: Any otolaryngologist in the United States or Canada is eligible to apply for the Health Services Research Grant. Applicants may be independent practitioners, residents or fellows in an approved training program, or practitioners affiliated with academic or similar institutions. All applicants must be members in good standing of AAO-HNSF; Associate and Corresponding Members are not eligible to apply as Principal Investigator, but may participate actively in the proposed project.
Amount of Grant: $10,000 maximum total (direct and indirect) costs
Date(s) Application is Due: Dec 15; Jan 15
Contact: Stephanie L. Jones, Assistant Director; (703) 519-1586 or (703) 836-4444; email: sljones@entnet.org
Internet: http://www.entnet.org/EducationAndResearch/coreGrants.cfm
Sponsor: American Academy of Otolaryngology-Head and Neck Surgery Foundation
One Prince Street
Alexandria, VA 22314-3357

AAO-HNSF Maureen Hannley Research Training Awards 176

The purpose of the Maureen Hannley Research Training Award program is to foster the acquisition of contemporary basic or clinical research skills among new full-time academic otolaryngologist-head and neck surgeons. The award is intended as a preliminary step in clinical investigator career development and is expected to facilitate the recipient's preparation of a more comprehensive individualized research training plan suitable for submission to one of the National Institutes of Health Clinical Investigator Career Development Award (CIA) programs (K series). Grants range up to $15,000

maximum. The award is one year, and is non-renewable. Projects must be completed within two years of the award date; no-cost extensions are available upon written request. Letters of intent must be received by December 15, and completed applications must be submitted by January 15.
Requirements: Applicants in the United States or Canada and members in good standing of the AAO-HNSF are welcome. Applicants must have demonstrated potential for excellence in research and teaching and a serious commitment to an academic research career in otolaryngology-head and neck surgery. Priority will be given to senior residents, fellows or faculty who have completed residencies or fellowships within two years of the application receipt date. All candidates must be sponsored by the Chairperson of his/her Division or Department and by an official representative of the institution which would administer the Award and in whose name the application is formally submitted.
Amount of Grant: $15,000 maximum total (direct and indirect) costs
Date(s) Application is Due: Dec 15; Jan 15
Contact: Stephanie L. Jones, Assistant Director; (703) 519-1586 or (703) 836-4444; email: sljones@entnet.org
Internet: http://www.entnet.org/EducationAndResearch/coreGrants.cfm
Sponsor: American Academy of Otolaryngology-Head and Neck Surgery Foundation
One Prince Street
Alexandria, VA 22314-3357

AAO-HNSF Percy Memorial Research Award 177

The Percy Memorial Research Award is an annual grant-in-aid of a worthy research project proposed in any area within the scope of otolaryngology-head and neck surgery. The Award was established in memory of A. Edward Percy, Jr., M.D. and his parents. Projects must have direct or potential clinical significance for patients seen by otolaryngologists-head and neck surgeons. They must be designed so as to yield useful information within the period of award, but priority will be given to projects which are also innovative and promise to develop into new long range research programs which will attract funding from other sources. Grants range up to $10,000 maximum, are for one year, and are not renewable. Letters of intent must be received by December 15, and completed applications must be submitted by January 15.
Requirements: Applicants in the United States or Canada and members in good standing of AAO-HNSF are welcome to apply. Applicants should be experienced independent otolaryngologist investigators affiliated with academic or similar institutions eligible to apply for and administer Federal research awards. Fellows with substantial research experience are eligible to apply, as are individuals who have already competed successfully for independent research grant support from a private or Federal funding agency.
Restrictions: Funds may not be requested to pay any portions of the salaries of the principal investigator or of any support personnel with strictly secretarial or clerical responsibilities. Funds may also not be used for the purchase of any item of equipment costing more than $500.
Amount of Grant: $25,000 maximum total (direct and indirect) costs
Date(s) Application is Due: Dec 15; Jan 15
Contact: Stephanie L. Jones, Assistant Director; (703) 519-1586 or (703) 836-4444; email: sljones@entnet.org
Internet: http://www.entnet.org/EducationAndResearch/coreGrants.cfm
Sponsor: American Academy of Otolaryngology-Head and Neck Surgery Foundation
One Prince Street
Alexandria, VA 22314-3357

AAO-HNSF Rande H. Lazar Health Services Research Grant 178

The purpose of the Rande H. Lazar Health Services Research Grant is to support the gathering of socioeconomic data for otolaryngology. The award is intended to promote increased participation by otolaryngologists in the rapidly expanding area of health services research. Proposed projects may be related to any area of otolaryngology-head and neck surgery, but must have direct or potential clinical significance for patients seen by otolaryngologist-head and neck surgeons. Grants range up to $10,000 maximum, are for one year, and are not renewable. Letters of intent must be received by December 15, and completed applications must be submitted by January 15.
Requirements: Any otolaryngologist in the United States or Canada is eligible to apply for the Rande H. Lazar Health Services Research Grant. Applicants may be independent practitioners, residents or fellows in an approved training program, or practitioners affiliated with academic or similar institutions. All applicants must be members in good standing of AAO-HNSF; Associate and Corresponding Members are not eligible to apply as Principal Investigator, but may participate actively in the proposed project.
Amount of Grant: $10,000 maximum total (direct and indirect) costs
Date(s) Application is Due: Dec 15; Jan 15
Contact: Stephanie L. Jones, Assistant Director; (703) 519-1586 or (703) 836-4444; email: sljones@entnet.org
Internet: http://www.entnet.org/EducationAndResearch/coreGrants.cfm
Sponsor: American Academy of Otolaryngology-Head and Neck Surgery Foundation
One Prince Street
Alexandria, VA 22314-3357

AAO-HNSF Resident Research Awards 179

The purpose of this grant is to stimulate original resident research in otolaryngology projects that are well-conceived and scientifically valid, with the potential to advance otolaryngology. Proposed projects may be related to any area of otolaryngology-head and neck surgery, and should be designed in collaboration with a preceptor investigator and approved by the candidate's department chairperson and institution. Grants range up to $10,000 maximum. Projects must be completed within one year of the award date; no-cost extensions are available upon written request. Letters of intent must be received by December 15, and completed applications must be submitted by January 15. Up to eight awards are available annually.

Requirements: The award is open to any AAO-HNS member in good standing who is a resident in an accredited otolaryngology - head and neck surgery training program in the U.S. and Canada.

Amount of Grant: $10,000 maximum

Date(s) Application is Due: Dec 15; Jan 15

Contact: Stephanie L. Jones, Assistant Director; (703) 519-1586 or (703) 836-4444; email: sljones@entnet.org

Internet: http://www.entnet.org/EducationAndResearch/coreGrants.cfm

Sponsor: American Academy of Otolaryngology-Head and Neck Surgery Foundation
One Prince Street
Alexandria, VA 22314-3357

AAOA Foundation/AAO-HNSF Combined Research Grants 180

The purpose of this award is to support a collaborative AAOA Foundation research project by fostering basic or clinical research related to otolaryngic allergy, rhinology, or related immunology by otolaryngologists-head and neck surgeons. Grants range up to $10,000 maximum. Projects must be completed within two years of the award date; no-cost extensions are available upon written request. Letters of intent must be received by December 15, and completed applications must be submitted by January 15.

Requirements: Applicants must have demonstrated potential for excellence in research. There are no restrictions on career stage, education, or country of residence. AAOA members are preferred but not required. All candidates must be sponsored by the Chairperson of his/her Division or Department and by an official representative of the institution that would administer the Award and in whose name the application is formally submitted.

Amount of Grant: $10,000 maximum

Date(s) Application is Due: Dec 15; Jan 15

Contact: Stephanie L. Jones, Assistant Director; (703) 519-1586 or (703) 836-4444; email: sljones@entnet.org

Internet: http://www.entnet.org/EducationAndResearch/coreGrants.cfm

Sponsor: American Academy of Otolaryngic Allergy (AAOA) Foundation
One Prince Street
Alexandria, VA 22314-3357

AAOHN Foundation Experienced Researcher Grants 181

The foundation funds research focused on improving the health and safety of the nation's workforce. Research grants are awarded to support research in the field of occupational and environmental health by an experienced occupational and environmental health registered nurse. Proposals may be submitted that are in the early stages of development (prior to data collection). Two $10,000 grants are awarded annually.

Requirements: Experienced occupational and environmental health registered nurse researchers are eligible to apply.

Restrictions: Completed research projects will not be accepted for consideration.

Amount of Grant: $10,000

Date(s) Application is Due: Dec 1

Contact: Ann Cox, Executive Director; (770) 455-7757; fax: (770) 455-7271; email: ann@aaohn.org

Internet: http://www.aaohn.org/foundation/grants/index.cfm

Sponsor: American Association of Occupational Health Nurses Foundation
2920 Brandywine Road, Suite 100
Atlanta, GA 30341

AAOHN Foundation New Investigator Researcher Grants 182

The foundation funds research, scholarship, and leadership development grants focused on improving the health and safety of the nation's workforce. Grants are awarded to encourage research in the field of occupational and environmental health by a new or novice registered nurse principal investigator. Proposals may be submitted that are in the early stages of development (prior to data collection).

Restrictions: Completed research projects are ineligible.

Amount of Grant: $3,000

Date(s) Application is Due: Dec 1

Contact: Ann Cox, Executive Director; (770) 455-7757; fax: (770) 455-7271; email: ann@aaohn.org

Internet: http://www.aaohn.org/foundation/grants/index.cfm

Sponsor: American Association of Occupational Health Nurses Foundation
2920 Brandywine Road, Suite 100
Atlanta, GA 30341

AAP International Travel Grants 183

The AAP has set aside a minimum of twelve (12) five-hundred dollar ($500.00) grants to be awarded to categorical pediatric or combined-training pediatric residents who wish to complete a clinical pediatric elective in the developing world during residency. The selection committee is composed of members from the AAP Section on International Child Health and the Section on Residents. After the selection committee reviews the applications the recipients will be notified mid-May and mid-November.

Requirements: Must be a categorical pediatric or combined-training pediatric resident at a U.S. or Canadian university.

Samples: Gunjan Kamdar, New York Presbyterian (2006); Bill Finn, University of Rochester (2006); Raina Paul, Boston Combined Residency Program (2006).

Amount of Grant: $500

Date(s) Application is Due: Mar 15; Sep 15

Contact: Kimberley VandenBrook, Grants Administrator; (800) 433-9016, ext. 7134; email: kvandenbrook@core.com

Internet: http://www.aap.org/sections/ypn/r/funding_awards/international_travel.html

Sponsor: American Academy of Pediatrics
P.O. Box 927, 141 Northwest Point Boulevard
Elk Grove Village, IL 60007-1098

AAP Nutrition Award 184

One award is given annually for outstanding achievements in research relating to nutrition of infants and children. The award is made to an individual or for one project. There is no age restriction for the award; however, it is hoped that younger persons will be considered. The award is made possible by a grant from the Infant Formula Council. In addition to the honorarium, the award includes round-trip airfare and two days' lodging for the recipient and a guest to attend the annual meeting and receive the award.

Requirements: The competition is open to residents of the United States or Canada whose research has been completed and publicly reported. Separate letters should be written for each individual nominated; a letter should contain a description of the nominee's achievements, stating clearly the basis for the recommendation, including references to the literature that describes her/his work. Nominee's bibliography is to be submitted with the nominating letter together with copies of available reprints. Letters supporting the nomination are to be screened by the nominator and forwarded with the nomination.

Restrictions: An individual may not submit more than five letters supporting the nomination. Current members of the AAP Committee on Nutrition are not eligible for this award.

Samples: Virginia A Stallings (2008); Dennis M Bier (2007); William Dietz (2006); William W. Hay, Jr. (2005).

Amount of Grant: $3,000 plus round-trip tourist class airfare and lodging for two days

Date(s) Application is Due: Mar 1

Contact: Debra Burrows, Manager; (847) 434-4000; fax: (847) 434-8000; email: research@aap.org

Internet: http://www.aap.org/visit/nutrannouncemts.htm

Sponsor: American Academy of Pediatrics
P.O. Box 927, 141 Northwest Point Boulevard
Elk Grove Village, IL 60009-0927

AAP Resident Research Grants 185

The Program is designed to give pediatric residents with limited research experience an opportunity to initiate and complete research projects related to their professional interests. Projects may be related to the full spectrum of child health research, such as behavioral sciences, biomedical sciences, epidemiology, health services, perinatal/neonatal health, prevention, public health, quality improvement, quality measurement, or basic laboratory-based science. Research projects can be conducted for a maximum of 2 years and should be completed during the residency program.

Requirements: Applicants must be legal residents of the United States or Canada. Although awards are primarily intended for support of first- and second-year pediatric residents, individuals who have secured a position for a third or fourth year of residency beginning in July may also apply.

Amount of Grant: $2,000 maximum

Date(s) Application is Due: Feb 27;

Contact: Jeannine Hess, Grants Administrator; (800) 433-9016, ext. 7876; fax: (847) 434-8000; email: jhess@aap.org

Internet: http://www.aap.org/sections/ypn/r/funding_awards/research_grants.html

Sponsor: American Academy of Pediatrics
P.O. Box 927, 141 Northwest Point Boulevard
Elk Grove Village, IL 60007-1098

AAR Classical Society of the American Academy in Rome Summer Scholarships 186

The Classical Society of the American Academy in Rome offers two scholarships, one to a graduate student and one to a secondary school teacher of classical languages and/or classical civilization. The scholarships are awarded on a competitive basis and are to be used to enable the recipients to attend the summer session of the American Academy in Rome. Contact the director for applications.

Requirements: High school teachers and graduate students of Latin, ancient history, and the classics are eligible to apply.

Restrictions: High school students and college undergraduates are not eligible.

Date(s) Application is Due: Mar 1

Contact: Myles McDonnell, Director, (212) 751-7200 ext 41; email: catulussr@aol.com

Internet: http://www.aarome.org/summer/css

Sponsor: American Academy in Rome
7 E 60th Street
New York, NY 10022-1001

AAR Postdoctoral Fellowships 187

The Academy offers 11-month postdoctoral fellowships in the following areas: architecture, design, historic preservation and conservation, landscape architecture, literature, musical composition, visual arts, ancient studies, medieval studies, Renaissance and early modern studies, and modern Italian studies. Applications may be downloaded from the Web site or obtained by contacting the office.

Requirements: All applicants for postdoctoral fellowships must have received the PhD at the time of application. U.S. citizens and those foreign nationals who have lived in the U.S. for the three years immediately preceding the application deadline are eligible

to apply for the NEH postdoctoral fellowships and are not required to submit the application fee. Applicants for academy postdoctoral awards must be assistant professors, associate professors appointed within the prior two years, or independent scholars who have received the PhD within the prior seven years. Applicants for pre- and postdoctoral Rome Prize fellowships in the humanities must include the following materials in support of their applications: Six copies of the completed and signed application form; six copies of a current curriculum vitae, which includes languages read or spoken and level of fluency; six copies of a project proposal (on 8.5 x 11 paper), not to exceed four pages, describing the nature of the work you plan to undertake in Rome and identifying the special resources in Italy that will be important to your work (please include bibliographical notes for any sources cited, [not counted towards the four-page proposal statement limit]); and six copies of a supporting paper in the proposed field of study, not to exceed 20 pages, excluding bibliography and footnotes.
Amount of Grant: $21,000 maximum
Date(s) Application is Due: Nov 1
Contact: Shawn Miller, Programs Director, (212) 751-7200; fax: (212) 751-7220; email: info@aarome.org
Internet: http://www.aarome.org/rome_prize/index.htm
Sponsor: American Academy in Rome
7 E 60th Street
New York, NY 10022-1001

AAR Predoctoral Fellowships   188
The academy offers both 11-month and two-year predoctoral fellowships. Eleven-month awards are available in the following fields: architecture, design, historic preservation and conservation, landscape architecture, literature, musical composition, visual arts, ancient studies, medieval studies, Renaissance and early modern studies, and modern Italian studies. Two-year awards are limited to projects in the history of art and architecture of any period and to projects whose disciplinary focus is in ancient studies. Predoctoral fellowships are meant to provide scholars with the necessary time to research and complete their doctoral dissertations.
Requirements: Applicants must have fulfilled all predissertation requirements by the application deadline. A nonrefundable fee of $40 must accompany each application. Applicants must include the following materials in support of their applications: six copies of the completed and signed application form; six copies of a current curriculum vitae, which includes languages read or spoken and level of fluency; Six copies of a project proposal (on 8.5 x 11 paper), not to exceed four pages, describing the nature of the work you plan to undertake in Rome and identifying the special resources in Italy that will be important to your work (please include bibliographical notes for any sources cited [not counted towards the four-page proposal statement limit]); and six copies of a supporting paper in the proposed field of study, not to exceed 20 pages, excluding bibliography and footnotes.
Amount of Grant: $15,750
Date(s) Application is Due: Nov 1
Contact: Programs Department, (212) 751-7200; fax: (212) 751-7220; email: info@aarome.org
Internet: http://www.aarome.org/rome_prize/index.htm
Sponsor: American Academy in Rome
7 E 60th Street
New York, NY 10022-1001

AAR Rome Prize Fellowships   189
At its principal site in Rome, the academy operates a program of fellowships and residencies that has at its core the development of gifted American artists and scholars. Each year, through a national competition, the academy awards fellowships in the fine arts and pre- and postdoctoral fellowships in the humanities. Rome Prize winners pursue independent projects, which vary in content and scope, for periods ranging from six months to two years at the academy. To complement their work, a series of walks, talks and tours in and around Rome, Italy, and the Mediterranean is offered during the year. The academy's Rome Prize winners are part of a residential community of 65 to 70 people each year. The artists and scholars who make up this multidisciplinary community also have the opportunity to foster their work through exchanges with members of the Italian and newly-united European artistic and scholarly communities.
Requirements: Applicants for the Rome Prize fellowships must be U.S. citizens at the time of application, with the exception of those competing for postdoctoral fellowships sponsored by the National Endowment for the Humanities. U.S. citizens and those foreign nationals who have lived in the United States for the three years immediately preceding the application deadline may apply for the NEH postdoctoral fellowships. Graduate students in the humanities may apply only for predoctoral fellowships.
Restrictions: Previous winners of the predoctoral Rome Prize are not eligible to apply for the Academy's postdoctoral fellowships. Undergraduate students are not eligible for Rome Prize fellowships. There is an application fee based upon the postmark date and the number of disciplines in which you are applying. Those applicants applying solely for NEH post-doctoral fellowships are not required to submit application fees.
Amount of Grant: $11,500-$23,000
Date(s) Application is Due: Nov 1
Contact: Programs Department, (212) 751-7200; fax: (212) 751-7220; email: info@aarome.org
Internet: http://www.aarome.org/rome_prize/index.htm
Sponsor: American Academy in Rome
7 E. 60th Street
New York, NY 10022-1001

Aaron & Cecile Goldman Family Foundation Grants   190
The Aaron and Cecile Goldman Family Foundation, established in 1962 in the District of Columbia, offers its primary support to Jewish and local agencies, including education and human services, and to local children's programs and camps. Fields of interest include: arts and humanities; children and youth services; education; human services; Jewish agencies and synagogues; Jewish federated giving programs; recreational programs; and camps. Types of support include building and renovation, capital campaigns, emergency funding, endowments, equipment purchase, fellowships, matching/challenge grants, support for professions, program development, and scholarships. Applicants should forward a letter of proposal to the Foundation office, whose board members meet in the spring and fall of each year.
Requirements: One copy of the proposal letter is required. 501(c)3 organizations supporting the residents of Washington, DC, and its surrounding area are eligible to apply.
Restrictions: No grants are offered to individuals.
Geographic Focus: District of Columbia
Samples: The Institute for Jewish Leadership and Values, Rockville, Maryland, $18,000 (2008); Ethiopian Community Services and Development Council, Washington, DC, $15,000 (2008); Save a Childs Heart Foundation, Potomac, Maryland, $10,000 (2008).
Amount of Grant: $20,000 maximum
Contact: Jennifer Margolius Fisher; (202) 332-6600; fax: (202) 332-1800; email: jennifer@themargoliusfirm.com
Sponsor: Aaron & Cecile Goldman Family Foundation
3000 Connecticut Avenue NW, Suite 100
Washington, DC 20008-2509

AARP Andrus Foundation Grants   191
The foundation actively seeks new approaches to maximizing the independence of older persons. Grants are awarded under two initiatives: living with chronic health conditions, and aging and living environments. Three types of grants are awarded. Research grants support cutting-edge, applied psychosocial, behavioral, health, and policy research and the dissemination of research findings. Dissemination grants support translating research information into language and materials useful to older adults and the practitioners, service providers, and others who work on their behalf. Grants generally are awarded for projects with the potential for national impact and high visibility, and may include public education programs, major media events, and the production of consumer or practitioner-oriented materials. Funding will also be allocated for the packaging and dissemination of research results and products related to aging and financial security, but not for conferences or the publication of conference proceedings. Education and training grants support graduate students engaged in aging-related research. National programs that provide leadership development, professional development, and research training to students and young professionals receive preference. Grants are awarded only to national organizations, not to individuals or individual educational institutions. Submit a letter of inquiry by the listed deadline dates; full proposals are by invitation.
Requirements: Grants are awarded only to colleges, universities, and 501(c)3 non-profit organizations.
Restrictions: Grants do not fund capital campaigns, organizational infrastructure (bricks and mortar), or vehicle purchases; individuals; biomedical or basic (bench) research; programs and services for older adults; education and training programs for professionals; pilot or preliminary research studies; or organizational overhead expenses.
Samples: Lighthouse International (New York, NY)—to develop, implement, and evaluate a public-awareness campaign about key research findings on vision, vision impairment, and vision rehabilitation among elderly people, $150,000.; American Society on Aging (San Francisco, CA)—to sponsor four minority practitioners who are conducting research projects as part of their participation in a leadership and professional-development program, $33,400.; International Longevity Ctr (New York, NY)—to produce five consumer publications that address various trends in aging, $150,700.; National Hospice and Palliative Care Organization (Alexandria, VA)—for research needed to develop a more comprehensive set of performance measures for end-of-life professionals, $100,000.
Date(s) Application is Due: Mar 1; Jul 1; Nov 1
Contact: Grants Administrator, (800) 775-6776 or (202) 434-6190; fax: (202) 434-6458; email: andrus@aarp.org
Internet: http://www.aarp.org/about_aarp/nrta/nrta_past_present_and_future.html#top
Sponsor: American Association of Retired Persons
601 E Street NW
Washington, DC 20049

AARP Foundation Gerontology Graduate Scholarships and Fellowships   192
The association awards graduate scholarships and fellowships for research on aging. Gerontology and aging studies faculty are invited to nominate qualified students. This program is funded by the AARP Andrus Foundation. Faculty nomination and student application forms are available online. Annual deadline dates may vary; contact program staff for exact dates.
Requirements: Gerontology and aging studies faculty are invited to nominate qualified students for these awards.
Amount of Grant: $15,000; $7000 graduate scholarships
Contact: Derek Stepp, Director, (202) 289-9806 ext 122; fax: (202) 289-9824; email: dstepp@aghe.org
Internet: http://www.aghe.org/site/aghewebsite/section.php?id=8183
Sponsor: Association for Gerontology in Higher Education

1030 15th Street NW, Suite 240
Washington, DC 20005

AAS American Historical Print Collectors Society Fellowship 193

This fellowship for research on American prints of the 18th and 19th centuries or for projects using prints as primary documentation is funded by the American Historical Print Collectors Society and AAS. Fellows are selected on the basis of the applicant's scholarly qualifications, the scholarly significance or importance of the project, and the appropriateness of the proposed study to the society's collections.
Requirements: Doctoral candidates may apply.
Amount of Grant: $1000 per month
Date(s) Application is Due: Jan 15
Contact: Grants Administrator, (508) 755-5221; fax: (508) 754-9069; email: academicfellowships@mwa.org
Internet: http://www.americanantiquarian.org/ahpcs.htm
Sponsor: American Antiquarian Society
185 Salisbury Street
Worcester, MA 01609-1634

AAS Fellowships for Creative and Performing Artists and Writers 194

The program supports visiting fellowships for historical research by creative and performing artists, writers, film makers, journalists, and other persons whose goals are to produce imaginative, non-formulaic works dealing with pre-20th-century American history. The fellowships provide the time for uninterrupted research, reading, and collegial discussion at the society, between January 1 and December 31. Fellows receive a stipend plus an allowance for travel expenses. Application and guidelines are available online.
Requirements: Successful applicants are those whose work is for the general public rather than for academic or educational audiences.
Amount of Grant: $1200 stipend
Date(s) Application is Due: Oct 5
Contact: Fellowships Administrator, (508) 755-5221; fax: (508) 753-3311
Internet: http://www.americanantiquarian.org/artistfellowship.htm
Sponsor: American Antiquarian Society
185 Salisbury Street
Worcester, MA 01609-1634

AAS Historic Research Fellowships 195

AAS accepts applications for fellowships for historical research by creative and performing artists, writers, filmmakers, and journalists. Fellowships are for individuals whose research objectives are to produce works dealing with pre-20th century American history designed for the public rather than for the academic/educational community. Fellowship projects may include (but are not limited to): historical novels, performance of historical music or drama, poetry, documentary films, television programs, radio broadcasts, plays, libretti, screenplays, magazine or newspaper articles, costume designs, set designs, illustrations and other graphic arts, book designs, sculpture, paintings, other works of fine and applied art, and nonfiction works of history designed for general audiences of adults or children. Fellowships will allow recipients to conduct uninterrupted research, reading, and collegial discussion at AAS.
Amount of Grant: $1200 per month, plus travel expense allowance
Date(s) Application is Due: Oct 5
Contact: James David Moran, Director of Outreach, (508) 471-2131; email: JMoran@mwa.org
Internet: http://www.americanantiquarian.org/artistfellowship.htm
Sponsor: American Antiquarian Society
185 Salisbury Street
Worcester, MA 01609-1634

AAS Kate B. and Hall J. Peterson Fellowships 196

The society will award a number of fellowships for one to three months to scholars for research and writing - - including doctoral dissertations - - in any field of American history and culture through 1876. All awards will be made not only on the basis of the applicant's scholarly qualifications and the general interest of the project, but also on the appropriateness of the inquiry to the society's holdings.
Requirements: The fellowships are open to individuals, including foreign nationals, and students working on dissertations.
Restrictions: All recipients are expected to be in regular and continuous residence at the society's library during the period of the grant.
Amount of Grant: $1,700 per month
Date(s) Application is Due: Jan 15
Contact: Caroline Sloat, Director of Scholarly Programs; (508) 471-2130 or 508-755-5221; fax: (508) 753-3311; email: csloat@mwa.org or academicfellowships@mwa.org
Internet: http://www.americanantiquarian.org/peterson.htm
Sponsor: American Antiquarian Society
185 Salisbury Street
Worcester, MA 01609-1634

AAS Legacy Fellowships 197

The Legacy Fellowship is awarded for research on any topic supported by the collections. It is funded by the gifts of former fellows and research associates. This fellowship is awarded to an individual engaged in scholarly research and writing—including doctoral dissertations—in any field of American history and culture through 1876. The Fellow is selected on the basis of the applicant's scholarly qualifications, the scholarly significance or importance of the project, and the appropriateness of the proposed study to the Society's collections.
Requirements: The fellowships are open to individuals and students working on dissertations.
Restrictions: All recipients are expected to be in regular and continuous residence at the society's library during the period of the grant.
Amount of Grant: $1,700 per month
Date(s) Application is Due: Jan 15
Contact: Caroline Sloat, Director of Scholarly Programs; (508) 471-2130 or 508-755-5221; fax: (508) 753-3311; email: csloat@mwa.org or academicfellowships@mwa.org
Internet: http://www.americanantiquarian.org/legacy.htm
Sponsor: American Antiquarian Society
185 Salisbury Street
Worcester, MA 01609-1634

AAS Long-Term Visiting Academic Research Fellowships 198

Society offers long-term visiting academic research fellowships tenable for four to twelve months each year. All awards are for a period of residence to use the AAS library's resources for research and writing. Fellowships are awarded not only on the basis of the applicant's scholarly qualifications and the general interest of his/her project, but also on the appropriateness of the inquiry to the society's holdings—early U.S. history and culture through 1876. Recipients are expected to be in regular residence at the society's library during the period of the grant. NEH fellows must devote full time to their study and may not accept teaching assignments nor undertake any other major activities during the tenure of the award. Recipients may not hold any other major fellowships, except sabbaticals or other grants from their own institutions.
Requirements: These Fellowships are available for scholars beyond the doctorate.
Restrictions: The fellowships may not be awarded to degree candidates or for study leading to advanced degrees, nor may they be granted to foreign nationals unless they have been residents of the United States for at least three years preceding their award.
Amount of Grant: $35,000-$40,000 annually
Date(s) Application is Due: Jan 15; Oct 3; Oct 15
Contact: Caroline Sloat, Director of Scholarly Programs; (508) 471-2130 or 508-755-5221; fax: (508) 753-3311; email: csloat@mwa.org
Internet: http://www.americanantiquarian.org/longterm.htm
Sponsor: American Antiquarian Society
185 Salisbury Street
Worcester, MA 01609-1634

AAS Short-Term Visiting Academic Research Fellowships 199

The society awards short-term fellowships for one to three months. Fellowships are awarded not only on the basis of the applicant's scholarly qualifications and the general interest of his/her project, but also on the appropriateness of the inquiry to the society's holdings—early U.S. history and culture through 1876. Recipients are expected to be in regular residence at the society's library during the period of the grant. NEH fellows must devote full time to their study and may not accept teaching assignments nor undertake any other major activities during the tenure of the award. Recipients may not hold any other major fellowships, except sabbaticals or other grants from their own institutions.
Requirements: Short-term fellowships are available for scholars holding the PhD and for doctoral candidates engaged in dissertation research.
Restrictions: The fellowships may not be awarded to degree candidates or for study leading to advanced degrees, nor may they be granted to foreign nationals unless they have been residents of the United States for at least three years preceding their award.
Amount of Grant: $1,000 per month
Date(s) Application is Due: Jan 15
Contact: Caroline Sloat, Director of Scholarly Programs; (508) 471-2130 or 508-755-5221; fax: (508) 753-3311; email: csloat@mwa.org
Internet: http://www.americanantiquarian.org/acafellowship.htm
Sponsor: American Antiquarian Society
185 Salisbury Street
Worcester, MA 01609-1634

AAS/Mellon Post-Dissertation Fellowships 200

The purpose of the 12-month fellowship is to provide the recipient with time and resources to extend research and/or to revise the dissertation for publication. Any topic relevant to the society's library collections and programmatic scope—American history and culture through 1876—is eligible. Applicants may come from such fields as history, literature, American studies, political science, art history, music history, and others relating to America in the period of the society's coverage. The society welcomes applicants who have already been in contact with potential publishers and will assist the successful candidate in finding a publisher.
Requirements: Scholars who are no more than three years beyond receipt of the doctorate are eligible. Graduate students may apply if they expect to have the doctorate in hand by June 15.
Amount of Grant: $30,000 stipend
Date(s) Application is Due: Oct 15
Contact: Fellowship Administrator, (508) 755-5221; fax: (508) 754-9069; email: academicfellowships@mwa.org
Internet: http://www.americanantiquarian.org/mellon.htm
Sponsor: American Antiquarian Society
185 Salisbury Street
Worcester, MA 01609

AASA Educational Administration Scholarships 201
Scholarships are available for outstanding graduate students who intend to pursue the public school superintendency as a career. Applications are available online in the spring or from chairs of educational administration departments at universities that have such a program.
Requirements: Only one candidate may be submitted by each university, and they must be recommended by a dean and at least one faculty member, with supporting statements.
Amount of Grant: $2000
Date(s) Application is Due: Sep 30
Contact: Scholarships Administrator, (703) 875-0729; email: awards@aasa.org
Internet: http://www.aasa.org/awards_and_scholarships/scholarships/Graduate_Student.htm
Sponsor: American Association of School Administrators
801 N Quincy Street, Suite 700
Arlington, VA 22203-1730

AASF Lydia Donaldson Tutt-Jones Memorial Research Grant 202
The grant supports research that identifies attitudinal and behavioral contributors to African American academic success. The interest is to increase the body of knowledge about African American students who are excelling in school to aid in the replication of that success for others. Focus may be upon student or parental variables, or both. Recipients must prepare a publishable caliber paper upon completion of the project and provide a copy to the foundation. There is no special application form, recommended format, or length of material required. Interested individuals should submit a letter of interest, curriculum vita, a description of the proposed research project (including a timeline, plus a letter of recommendation from their faculty mentor, if students, or from their department chairperson, agency head, or officer of their professional association, if professionals). An original and five copies of the application materials should be submitted, along with a stamped, self-addressed postcard that will be used to acknowledge receipt of the application. Deadlines may vary. See website for more information.
Requirements: Graduate students and professionals are eligible. Students must be recommended by a faculty mentor who agrees to oversee the project and the submission of a publishable paper upon its completion.
Amount of Grant: $2000
Contact: Award Administrator, (954) 792-1117; fax: (954) 792-9191; email: info@blacksuccessfoundation.org
Internet: http://www.blacksuccessfoundation.org/lcdtj%20research%20grant.htm
Sponsor: African American Success Foundation
4330 W Broward Boulevard, Suite H
Fort Lauderdale, FL 33317-3753

AASL Collaborative School Library Media Award 203
The award recognizes and encourages collaboration and partnerships between school library media specialists and teachers through joint planning of a program, unit, or event in support of the curriculum and that uses media center resources. The award is given to the recipient's school library media center. The application must address the degree to which the project meets the standards outlined in Information Power: Building Partnerships for Learning. Application forms are available on the Web site. Annual deadline dates may vary.
Requirements: School library media specialist and teacher(s) who have worked together to execute a project, event, or program to further information literacy, independent learning, and social responsibility using resources of the school library media center may apply.
Samples: Melissa Bergin, Niskayuna (N.Y.) High School—award recipient, (2006); Valerie Edwards, Monona Grove High School (Monona, WI)—award recipient, $2500 (2005); Margaret Lincoln, Lakeview High School (Battle Creek, MI)—award recipient (2004); Leslie Preddy, Perry Meridian Middle School (Indianapolis, IN)—award recipient (2003).
Amount of Grant: $2,500
Date(s) Application is Due: Feb 1
Contact: Beverley Becker, AASL Deputy Executive Director; (800) 545-2433, ext. 4385; fax: (312) 664-7459; email: aasl@ala.org or bbecker@ala.org
Internet: http://www.ala.org/Template.cfm?Section=awards&template=/ContentManagement/ContentDisplay.cfm&ContentID=149331
Sponsor: American Association of School Librarians
50 E. Huron Street
Chicago, IL 60611-2795

AASL Distinguished School Administrator Award 204
This award, donated by Social Issues Resources Series, Inc., is given annually to an administrator of a school or group of schools for developing an exemplary school library media program and for having made an outstanding and sustained contribution toward expanding the role of the school library media center as an agency for the improvement of education. Nomination forms are available on the Web site. Annual deadline dates may vary.
Requirements: State, county, or district school superintendents and building principals, currently in administrative office and directly responsible for a school or group of schools at any level, are eligible for nomination. District administrators responsible for broad instructional leadership, such as assistant superintendents, directors of curriculum and instruction, and directors of elementary and/or secondary education, are also eligible.
Samples: Kay McBride, Library Specialist, Pasadena (Texas) ISD—award recipient, (2006); Michael Hart, Principal-Holy Rosary School (Tacoma, WA)—award recipient,

$200 (2005); Dr. William Harner, Principal-Greenville County School District (Greenville, SC)—award recipient (2004); Barbara De Spain, Lois Lenski Elementary School (Centennial, CO)—award recipient (2003).
Amount of Grant: $2,000
Date(s) Application is Due: Feb 1
Contact: Beverley Becker, AASL Deputy Executive Director; (800) 545-2433, ext. 4385; fax: (312) 664-7459; email: aasl@ala.org or bbecker@ala.org
Internet: http://www.ala.org/Template.cfm?Section=awards&template=/ContentManagement/ContentDisplay.cfm&ContentID=149334
Sponsor: American Association of School Librarians
50 E. Huron Street
Chicago, IL 60611-2795

AASL/Highsmith Research Grant 205
Cosponsored by the AASL and the Highsmith Company, Inc., this annual grant is presented to one or more AASL members who are school library media specialists, library educators, or library information science/education professors to conduct innovative research aimed at measuring and evaluating the impact of school library media programs on learning and education. Application forms are available on the Web site. Annual deadline dates may vary.
Restrictions: Applicants must be AASL personal members.
Samples: Judith Dzikowski (Syracuse, NY)—for Partners in Achievement: Libraries and Students (PALS) Improving Student Achievement through Data Use for Library Media Specialists, $5,000 (2007); Joyce Valenza (Rydal, PA)—for School Virtual Libraries: The Influence of Best Practices on High School Students' Information Seeking, $5,000 (2005); Robyn Young (Avon, IN)—for More than just Comics: Graphic Novels and their Effect on Student Achievement, $5000 (2004); Kathy Latrobe and Rhonda Taylor (Norman, OK)—for Study of LMS's Attributes for Ongoing Program Assessment, $5000 (2004).
Amount of Grant: $5,000 maximum
Date(s) Application is Due: Feb 1
Contact: Beverley Becker, AASL Deputy Executive Director; (800) 545-2433, ext. 4385; fax: (312) 664-7459; email: aasl@ala.org or bbecker@ala.org
Internet: http://www.ala.org/Template.cfm?Section=awards&template=/ContentManagement/ContentDisplay.cfm&ContentID=149332
Sponsor: American Association of School Librarians
50 E. Huron Street
Chicago, IL 60611-2795

AAUW American Dissertation Fellowships 206
American Fellowships support women doctoral candidates completing dissertations or scholars seeking funds for postdoctoral research leave from accredited institutions. Dissertation Fellowships are available to women who will complete their dissertation writing between April 1 and June 30. To qualify, applicants must have completed all course work, passed all required preliminary examinations, and received approval for their research proposal or plan by November 15 of the previous year. Open to applicants in all fields of study, except engineering. Scholars engaged in researching gender issues are encouraged to apply.
Requirements: Applicants must be U.S. citizens or permanent residents by the application postmark deadline. When comparing proposals of equal merit, special consideration will be given to women holding junior academic appointments who are seeking research leave, women who have held the doctorate for at least three years, and women whose educational careers have been interrupted.
Restrictions: Candidates may apply for only one of the awards described. Former recipients of these awards are not eligible to apply for additional American Fellowships or publication grants. Students holding any fellowship for writing a dissertation in the year prior to the AAUW Educational Foundation fellowship year are not eligible.
Amount of Grant: $20,000
Date(s) Application is Due: Nov 15
Contact: Todd Beckman, Fellowships and Grants; (319) 337-1716, ext. 60; email: aauw@act.org or foundation@aauw.org
Internet: http://www.aauw.org/education/fga/fellowships_grants/american.cfm
Sponsor: American Association of University Women Educational Foundation
Deptartment 60, 301 ACT Drive, P.O. Box 4030
Iowa City, IA 52243-4030

AAUW American Postdoctoral Research Leave Fellowships 207
American Fellowships support women doctoral candidates completing dissertations or scholars seeking funds for postdoctoral research leave from accredited institutions. Postdoctoral Research Leave Fellowships offer one-year support for women who will have earned a doctoral degree by Nov. 15. Postdoctoral fellowships are available in the arts and humanities, social sciences, and natural sciences; one is designated for a woman from an underrepresented group in any field. Limited additional funds may be available when matched by the fellow's institution.
Requirements: Applicants must be U.S. citizens or permanent residents by the application postmark deadline. When comparing proposals of equal merit, special consideration will be given to women holding junior academic appointments who are seeking research leave, women who have held the doctorate for at least three years, and women whose educational careers have been interrupted.
Restrictions: Candidates may apply for only one of the awards described. Former recipients of these awards are not eligible to apply for additional American Fellowships or publication grants.
Amount of Grant: $30,000

Date(s) Application is Due: Nov 15
Contact: Todd Beckman, Fellowships and Grants; (319) 337-1716, ext. 60; email: aauw@
act.org or foundation@aauw.org
Internet: http://www.aauw.org/education/fga/fellowships_grants/american.cfm
Sponsor: American Association of University Women Educational Foundation
Deptartment 60, 301 ACT Drive, P.O. Box 4030
Iowa City, IA 52243-4030

AAUW American Summer/Short-Term Research Publication Grants   208
The Grants fund women college and university faculty and independent researchers
to prepare research data for publication. Applicants may be tenure track, part-time, or
temporary faculty or new or established scholars and researchers at universities. Time
must be available for eight consecutive weeks of final writing, editing, and responding
to issues raised in critical reviews. Scholars with strong publishing records should seek
other funding.
Requirements: Applicants must be U.S. citizens or permanent residents by the application
postmark deadline. Applicants must have received their doctorates by the application
deadline.
Restrictions: Funds cannot be used for undertaking research. Candidates may apply for
only one of the awards described. Former recipients of these awards are not eligible to
apply for additional American Fellowships or publication grants.
Amount of Grant: $6,000
Date(s) Application is Due: Nov 15
Contact: Todd Beckman, Fellowships and Grants; (319) 337-1716, ext. 60; email: aauw@
act.org or foundation@aauw.org
Internet: http://www.aauw.org/education/fga/fellowships_grants/american.cfm
Sponsor: American Association of University Women Educational Foundation
Deptartment 60, 301 ACT Drive, P.O. Box 4030
Iowa City, IA 52243-4030

AAUW Career Development Grants   209
Grants support women currently holding a bachelor's degree who are preparing to
advance their careers, change careers, or re-enter the work force. Special consideration
is given to AAUW members, women of color, and women pursuing their first advanced
degree or credentials in nontraditional fields. Grants provide support for course work
beyond a bachelor's degree, including a master's degree, second bachelor's degree, or
specialized training in technical or professional fields. Funds are available for distance
learning. Course work must be taken at an accredited two- or four-year college or
university, or at a technical school that is fully licensed or accredited by an agency
recognized by the U.S. Department of Education.
Requirements: Applicants must be women who are U.S. citizens or permanent residents
of the United States who hold at least a baccalaureate degree and who have completed
their most recent degree at least five years before the July 1 effective date of the award.
Preference is given to AAUW members. Applicants must have applied or been accepted
at a two- or four-year accredited college or university; technical schools must be licensed
or accredited by an agency recognized by the U.S. Department of Education. Coursework
must be a prerequisite for a professional employment plan.
Restrictions: Academic grants are not available for doctoral level work. Institute grants
may not be used for internships, research, or independent study.
Amount of Grant: $2,000-$12,000
Date(s) Application is Due: Dec 15
Contact: Todd Beckman, Fellowships and Grants; (319) 337-1716, ext. 60; email: aauw@
act.org or foundation@aauw.org
Internet: http://www.aauw.org/education/fga/fellowships_grants/career_development.
cfm
Sponsor: American Association of University Women Educational Foundation
Deptartment 60, 301 ACT Drive, P.O. Box 4030
Iowa City, IA 52243-4030

AAUW Community Action Grants   210
One-year grants provide seed money for new projects. Two-year Community Action
Grants provide seed money to individual women, AAUW branches, AAUW state
organizations, and community-based nonprofit organizations. These grants provide start-
up funds for longer-term programs that address the particular needs of the community
and develop girls sense of efficacy through leadership or advocacy opportunities. Topic
areas for both one- and two-year grants are unrestricted, but should include a clearly
defined activity that promotes education and equity for women and girls. Funds support
planning activities, coalition building, program implementation, and evaluation.
Guidelines are available online.
Requirements: Applicants must be women who are U.S. citizens or permanent residents.
Grant projects must have direct public impact, be nonpartisan, and take place within
the U.S. or its territories.
Amount of Grant: $2,000-$7,000 one-year grants; $5,000-$10,000 two-year projects
Date(s) Application is Due: Jan 15
Contact: Todd Beckman, Fellowships and Grants; (319) 337-1716, ext. 60; email: aauw@
act.org or foundation@aauw.org
Internet: http://www.aauw.org/education/fga/fellowships_grants/community_action.
cfm
Sponsor: American Association of University Women Educational Foundation
Deptartment 60, 301 ACT Drive, P.O. Box 4030
Iowa City, IA 52243-4030

AAUW International Fellowships   211
Funds are for one year of graduate and postgraduate study or research at an approved
institution in the United States and are given to women of outstanding ability who are
citizens of countries other than the United States. Funding varies according to level
of study, which includes: a Master's / Professional Fellowship of $18,000; a Doctorate
Fellowship of $20,000; and a Postdoctoral Fellowship of $30,000. Strong preference will
be given to women whose credentials prove prior commitment to the amelioration of the
lives of other women and girls through civic, community, or professional work and whose
research proposals show a continued interest in the advancement of women. Applicants
are judged on their professional potential and on the importance of their projects to their
country of origin. Applications are available from August 1 to December 1.
Requirements: Candidates must have an academic degree equivalent to the U.S. bachelor's
degree; must have a specific plan of study or research that will advance the applicant's
professional competence; must intend to return home to pursue a professional career;
must devote full time to graduate work; and must have satisfactory English proficiency.
Applicants whose native language is not English must submit up-to-date scores on one
of four language proficiency examinations. Preference will be given to applicants who
have specific positions to return to in their own countries.
Restrictions: Previous and current recipients of AAUW fellowships are not eligible to
apply.
Amount of Grant: $18,000-$30,000
Date(s) Application is Due: Dec 1
Contact: Todd Beckman, Fellowships and Grants; (319) 337-1716, ext. 60; email: aauw@
act.org or foundation@aauw.org
Internet: http://www.aauw.org/education/fga/fellowships_grants/international.cfm
Sponsor: American Association of University Women Education Foundation
Deptartment 60, 301 ACT Drive, P.O. Box 4030
Iowa City, IA 52243-4030

AAUW Selected Professions Fellowships—Engineering Dissertation Awards   212
Selected Professions Fellowships are awarded to women who intend to pursue a full-
time course of study at accredited institutions during the fellowship year in one of the
designated degree programs where women's participation traditionally has been low.
Requirements: Applicants must be women who are citizens or permanent residents of
the United States and who expect to receive their doctoral degree in engineering at
the end of the fellowship year. Selected Fellowships are awarded for the following
programs: Architecture (M.Arch, M.S.Arch); Computer/Information Sciences (M.S.);
Engineering (M.E., M.S., Ph.D.); Engineering Dissertation Award; and Mathematics/
Statistics (M.S.).
Restrictions: Students holding any fellowship for the purpose of writing the dissertation
in the year prior to the AAUW fellowship year are not eligible to apply.
Amount of Grant: $20,000
Date(s) Application is Due: Dec 15
Contact: Todd Beckman, Fellowships and Grants; (319) 337-1716, ext. 60; email: aauw@
act.org or foundation@aauw.org
Internet: http://www.aauw.org/education/fga/fellowships_grants/selected.cfm
Sponsor: American Association of University Women Educational Foundation
Deptartment 60, 301 ACT Drive, P.O. Box 4030
Iowa City, IA 52243-4030

AAUW Selected Professions Fellows—Master's and First Professional Awards   213
Selected Professions Fellowships are awarded to women who intend to pursue a full-
time course of study at accredited institutions during the fellowship year in one of the
designated degree programs where women's participation traditionally has been low.
Selected Fellowships are awarded for the following programs: Architecture (M.Arch,
M.S.Arch); Computer/Information Sciences (M.S.); Engineering (M.E., M.S., Ph.D.);
Engineering Dissertation Award; and Mathematics/Statistics (M.S.). Fellowships
in the following degree programs are restricted to women of color, who have been
underrepresented in these fields: Business Administration (M.B.A., E.M.B.A.); Law
(J.D.); and Medicine (M.D., D.O.).
Requirements: Applicants must be U.S. citizens or permanent residents of the United
States. Special consideration is given to applicants who show professional promise in
innovative or neglected areas of research and/or practice in public interest concerns.
Amount of Grant: $5,000-$12,000
Date(s) Application is Due: Jan 10
Contact: Todd Beckman, Fellowships and Grants; (319) 337-1716, ext. 60; email: aauw@
act.org or foundation@aauw.org
Internet: http://www.aauw.org/education/fga/fellowships_grants/selected.cfm
Sponsor: American Association of University Women Educational Foundation
Deptartment 60, 301 ACT Drive, P.O. Box 4030
Iowa City, IA 52243-4030

ABA Minority Legal Opportunity Scholarships   214
Each year, the program awards scholarships, which are renewable for two additional
years, to minority first-year law students attending an ABA-accredited law school.
Selected applicants are required to demonstrate admission to and plans to enroll at an
ABA-accredited law school before receiving scholarship funds.
Requirements: U.S. students who are members of racial/ethnic minorities that have been
underrepresented in the legal profession, who have achieved a minimum cumulative 2.5
grade-point average, are eligible.
Restrictions: Law students who have completed one or more semesters or years of law
school are not eligible.
Amount of Grant: $5000 annually; $15,000 total

Date(s) Application is Due: Mar 1
Contact: Scholarship Administrator, (312) 988-5415; email: mastronardi@staff.abanet.org
Internet: http://www.abanet.org/fje/losfpage.html
Sponsor: American Bar Association
750 N Lake Shore Drive
Chicago, IL 60611

Abbott Fund Global AIDS Care Grants 215

Abbott Global AIDS Care programs focus on four key areas: strengthening health care systems, helping children affected by HIV/AIDS, preventing mother-to-child transmission of HIV, and expanding access to testing and treatment. Together with our implementing partner organizations, Abbott and Abbott Fund are pioneering innovative model programs, and sharing progress and lessons learned to encourage the adaptation of successful programs.
Requirements: Grants are made to tax-exempt organizations supporting global HIV/AIDS programs.
Restrictions: Contributions will not be made to individuals; for-profit entities; purely social organizations; political parties or candidates; sectarian religious organizations; advertising; symposia, conferences, and meetings; ticket purchases; memberships; business-related purposes; volunteer efforts of non-Abbott employees; or marketing sponsorships.
Geographic Focus: Arizona; California; Illinois; Kansas; Massachusetts; Michigan; New Jersey; New York; North Carolina; Ohio; Puerto Rico; Texas; Utah; Virginia
Amount of Grant: $10,000-$100,000
Contact: Cindy Schwab, Vice President; (847) 937-7075; fax: (847) 935-5051; email: cindy.schwab@abbott.com
Internet: http://www.abbottfund.org/sections/what.html
Sponsor: Abbott Fund
Department 379, Building 6D, 100 Abbott Park Road
Abbott Park, IL 60064-3500

Abbott Fund Science and Medical Innovation Grants 216

The Abbott Fund Science and Medical Innovation program supports new approaches to learning that are designed to foster a better understanding of science and medical innovation and the value they bring to improving human health. Partnerships support K-12 education through postgraduate level research including funding and volunteer assistance for student and educator workshops and programs; career development; and educational grants for college and university science facilities, scholarships and fellowships.
Requirements: Grants are made to tax-exempt organizations supporting company operating areas in Arizona, California, Illinois, Kansas, Massachusetts, Michigan, New Jersey, New York, North Carolina, Ohio, Puerto Rico, Texas, and Virginia, and Utah. It also supports the communities of: Abingdon, England; Brockville, Canada; Campoverde, Italy; Clonmel, Ireland; Cootehill, Ireland; Delkenheim, Germany; Kanata, Canada; Katsuyama, Japan; Ludwigshafen, Germany; Queenborough, England; Rio de Janeiro, Brazil; Sligo, Ireland; and Zwolle, the Netherlands.
Restrictions: Contributions will not be made to individuals; for-profit entities; purely social organizations; political parties or candidates; sectarian religious organizations; advertising; symposia, conferences, and meetings; ticket purchases; memberships; business-related purposes; volunteer efforts of non-Abbott employees; or marketing sponsorships.
Geographic Focus: Arizona; California; Illinois; Kansas; Massachusetts; Michigan; New Jersey; New York; North Carolina; Ohio; Puerto Rico; Texas; Utah; Virginia
Amount of Grant: $20,000-$100,00
Contact: Cindy Schwab, Vice President; (847) 937-7075; fax: (847) 935-5051; email: cindy.schwab@abbott.com
Internet: http://www.abbottfund.org/sections/what.html
Sponsor: Abbott Fund
Department 379, Building 6D, 100 Abbott Park Road
Abbott Park, IL 60064-3500

Abbott-ASM Lifetime Achievement Award 217

The award, sponsored by Abbott Laboratories, is given to honor a distinguished scientist for a lifetime of outstanding contributions in fundamental biomedical research in any of the microbiological sciences. The award consists of a cash prize, framed certificate, medallion, and national or international travel expenses incidental to receiving the award at the ASM general meeting. Guidelines are available online. All nomination materials must be submitted electronically.
Requirements: Mature scientists, both active and retired, from all areas of microbiology are eligible.
Restrictions: Self-nominations are not be accepted.
Samples: Norman R. Pace, $20,000—award recipient (2007); R. John Collier, $20,000—award recipient (2006); Jonathan Beckwith, $20,000—award recipient (2005).
Amount of Grant: $20,000, a commemorative piece, and travel to the ASM General Meeting
Date(s) Application is Due: Oct 1
Contact: Awards Committee, (202) 942-9226; fax: (202) 942-9353; email: awards@asmusa.org
Internet: http://www.asm.org/Academy/index.asp?bid=2587
Sponsor: American Society for Microbiology
1752 N Street, NW
Washington, DC 20036-2804

Abe Schecter Graduate Scholarship in Electronic News 218

Any continuing or incoming graduate student whose career objective is electronic news or electronic news teaching or research may apply for this scholarship. The winner also receives an expense-paid trip to the RTNDF annual conference. Scholarships will be awarded on an international basis. Applications are available on the Web site.
Requirements: Applicants must be currently enrolled in school, be in good standing, and have at least one full year of school remaining.
Restrictions: Applicants may apply in more than one category but are restricted to one award. Previous winners are not eligible.
Amount of Grant: $2000
Date(s) Application is Due: May 9
Contact: Irving Washington, (202) 467-5218; fax: (202) 223-4007; email: irvingw@rtndf.org
Internet: http://www.rtnda.org/asfi/scholarships/graduate.shtml
Sponsor: Radio and Television News Directors Foundation
1000 Connecticut Avenue NW, Suite 615
Washington, DC 20036

Abraham Lincoln Fellowships in Constitutional Government 219

The fellowship is awarded to bright, ambitious Americans who want to reclaim limited constitutional government. Each fellow will attend a one-week seminar in constitutional theory and practice led by eminent scholars. Fellows also will be invited to attend occasional weekend seminars in and around the nation's capital and will be honored at the institute's annual Lincoln Day Colloquium and Dinner. Fellowships include a stipend, all travel and lodging expenses, and most meals. Up to 10 fellowships will be awarded this year. Guidelines and application are available online.
Requirements: Participation is open to young patriots working on matters of national public policy and politics. This includes, but is not limited to, U.S. government employees, public policy research institutions, and print and broadcast media.
Amount of Grant: $1500 stipend
Date(s) Application is Due: May 6
Contact: Thomas Krannawitter, Vice President, (909) 621-6825; fax: (909) 626-8724; email: tkrannawitter@claremont.org
Internet: http://www.claremont.org/projects/lincoln/index.html
Sponsor: Claremont Institute for the Study of Statesmanship and Political Philosophy
937 W Foothill Boulevard, Suite E
Claremont, CA 91711

ACAAI Foundation Research Grants 220

The Foundation develops, promotes and funds clinical research and educational programs related to allergy, asthma & immunology. The Foundation is dedicated to: strengthening, supporting and funding allergy/immunology training programs; heightening awareness of the critical role of the allergist in cost-effective treatment of asthma and other allergic and immunologic diseases; and supporting asthma camp programs to teach children and their parents how to live with asthma. Programs include clinical fellowship stipends, fellow-in-training research grants, young faculty support awards, scholar, return awards, and support for asthma camps.
Amount of Grant: $250,000 maximum
Contact: Program Administrator, (847) 427-1200; fax: (847) 427-1294; email: mail@acaai.org
Internet: http://www.acaai.org/Member/ACAAI_Foundation/ACAAI_Foundation.htm
Sponsor: American College of Allergy, Asthma & Immunology Foundation
85 West Algonquin Road, Suite 550
Arlington Heights, IL 60005

Academic and Research Librarian of the Year Award 221

This award is presented by the Association of College and Research Libraries and funded by the Baker & Taylor Company to recognize an individual member of the library profession who is making an outstanding national or international contribution to academic or research librarianship and library development. Individuals nominated should have demonstrated achievements in such areas as service to the organized profession through ACRL and related organizations, significant and influential research on academic or research library service, publication of a body of scholarly and/or theoretical writing contributing to academic or research library development, or planning and implementing a library program of such exemplary quality that it has served as a model for others.
Requirements: Individuals may nominate themselves or others.
Amount of Grant: $3000
Date(s) Application is Due: Dec 1
Contact: Megan Bielefeld, ACRL Program Coordinator, (800) 545-2433 ext 2514 or (312) 280-2514; email: mbielefeld@ala.org
Internet: http://www.ala.org/Template.cfm?Section=Awards17&template=/ContentManagement/ContentDisplay.cfm&ContentID=93938
Sponsor: Association of College and Research Libraries
50 E Huron Street
Chicago, IL 60611-2795

Academy for International and Area Studies Fellowships Program 222

The Academy Scholars Program has been established by the Harvard Academy for International and Area Studies to assist young scholars who are preparing for an academic career involving both a social science discipline and a particular area of the world. Those selected as academy scholars are given time, guidance, access to Harvard facilities, health insurance, and substantial financial assistance as they work for two years conducting either dissertation or postdoctoral research in their chosen fields and areas. The senior scholars, a distinguished group of senior Harvard faculty members, act as mentors to the academy scholars to help them achieve their intellectual potential. The fellowships are offered for research conducted while in residence at the Center for International Affairs, although travel for research is allowed.

Requirements: Predoctoral applicants must have completed all coursework and general examinations by the beginning of the first year for which they seek support. Competitive candidates must have already made some significant progress on their dissertations.

Amount of Grant: $25,000 for predoctoral students; $40,000 for postdoctoral scholars

Contact: Beth Baiter, Fellowship Coordinator, (617) 495-4432; fax: (617) 384-9259; email: bbaiter@wcfia.harvard.edu

Internet: http://www.wcfia.harvard.edu/progdetail.asp?ID=18

Sponsor: Harvard University Weatherhead Center for International Affairs

1737 Cambridge Street, Coolidge Hall

Cambridge, MA 02138

Academy of American Franciscan History Dissertation Fellowships 223

The Academy of American Franciscan History is accepting applications for four dissertation fellowships, each worth $10,000. As many as two of these fellowships will be awarded for projects dealing with some aspect of the history of the Franciscan Family in Latin America, including the United States Borderlands, Mexico, Central and South America. Up to another two fellowships will be awarded to support projects dealing with some aspect of the history of the Franciscan Family in the rest of the United States and Canada. Projects may deal with any aspect of the history of the Franciscan Family, including any of the branches of the Family, male, female, tertiary, Capuchin, etc. The deadline for applications will be February 1.

Requirements: The recipient must be engaged in full-time research during the period of the fellowship. The applicant must be a doctoral candidate at a University in the Americas, and the bulk of the research should be conducted in the Americas.

Amount of Grant: $10,000

Date(s) Application is Due: Feb 1

Contact: Dr. Jeffrey M. Burns, Director; (510) 548-1755; email: acadafh@aol.com or acadafh@fst.edu

Internet: http://aafh.org/Scholarships.html

Sponsor: Academy of American Franciscan History

1712 Euclid Avenue

Berkley, CA 94709-1208

AcademyHealth Alice S. Hersh New Investigator Awards 224

The Alice S. Hersh New Investigator Award recognizes the contribution of new scholars to the field of health services research. Eligible nominees are those whose first research appointments were no earlier than January 1, 2002. Post-doctoral positions will be considered research appointments unless these years provided training necessary for entry into the field. In addition, the eligibility period may be extended for a nominee who has taken a formal leave of absence from their employer. The awards will be presented at the Annual Research Meeting in June. All awardees receive complimentary registration and lodging to attend the meeting and a $2,500 prize from the Alice S. Hersh Memorial Fund.

Requirements: Nominators must be members of AcademyHealth although nominees need not be. Nominees from academic and non-academic institutions are encouraged. Nominations must include one letter of nomination (no more than two pages) that reviews the nominee's contribution to the field and specifies the date of the nominee's first research appointment, and must include curriculum vitae.

Amount of Grant: $2,500

Date(s) Application is Due: Mar 2

Contact: Jennifer Muldoon, Director; (202) 292-6770; fax: (202) 292-6870; email: jennifer.muldoon@academyhealth.org

Internet: http://www.academyhealth.org/content.cfm?ItemNumber=954@navItemNumber=2342

Sponsor: AcademyHealth

1150 17th Street NW, Suite 600

Washington, DC 20036

AcademyHealth HSR Impact Awards 225

The program requests nominations of health services research that has made a positive impact on health policy or practice. The winning impacts will be disseminated widely as part of AcademyHealth's ongoing efforts to promote the field of health services research and communicate its value for health care decision-making. Nominated research may be published or unpublished, a single study or a body of work, the work of an individual or a team. The time frame for when the research was conducted is open, but the impact should have occurred recently. Nominators and nominees are not required to be AcademyHealth members. Self-nominations are accepted. Award winners are announced at the National Health Policy Conference, February 6 and 7, and the winners receive complimentary registration, travel, and lodging to the conference.

Requirements: Each nomination must include a summary describing the research used, the method of translation, and the impact on policy and/or practice (no more than 500 words); a letter of recommendation from the user of this research discussing the impact the research has had; contact information for the lead researcher; and contact information for the nominator (may be the lead researcher).

Amount of Grant: $2000

Date(s) Application is Due: Jul 29

Contact: Jennifer Muldoon, Director; (202) 292-6770; fax: (202) 292-6870; email: jennifer.muldoon@academyhealth.org

Internet: http://www.academyhealth.org/awards/hsrimpacts.htm

Sponsor: AcademyHealth

1150 17th Street NW, Suite 600

Washington, DC 20036

AcademyHealth Nemours Child Health Services Research Awards 226

The Nemours Child Health Services Research Award recognizes the scientific work of emerging scholars in the field of child health services research, particularly research on quality improvement of pediatric health services. Recipients must be within seven years of entry into the field of child health services research as of December 31, 2007. Year of entry will be judged by the nominee's first publication in child health services research. The winner will receive $1,000 in recognition of his/her contribution to child health services research, and the award will be presented at the Child Health Services Research Meeting, held in in Washington, D.C., in conjunction with AcademyHealth's Annual Research Meeting in June.

Requirements: Nominations should include: Letter of nomination that identifies when the candidate first entered the field (i.e., title and year of first publication in child health services research); Curriculum vitae that includes a detailed list of accomplishments that merit this recognition. Self-nominations will be accepted.

Amount of Grant: $1,000

Date(s) Application is Due: Apr 15

Contact: Jennifer Muldoon, Director; (202) 292-6770; fax: (202) 292-6870; email: jennifer.muldoon@academyhealth.org

Internet: http://www.academyhealth.org/content.cfm?ItemNumber=982&navItemNumber=2342

Sponsor: AcademyHealth

1150 17th Street NW, Suite 600

Washington, DC 20036

AcademyHealth PHSR Research Article of the Year Awards 227

Public Health Systems Research (PHSR) is a field of inquiry examining the organization, financing, performance, and impact of health systems - defined as the constellation of governmental and non-governmental actors that influence population health, including health care providers, insurers, purchasers, public health agencies, community-based organizations and entities that operate outside the traditional sphere of health care. The Article of the Year Award recognizes the best scientific work that the field of PHSR has produced and published in the previous year. The awardee will receive complimentary registration for both the PHSR Interest Group Annual Meeting and AcademyHealth, Annual Research Meeting, travel assistance, and a $1,000 cash award. Nominated articles must have been published in a peer-reviewed journal between January 1 and December 31 of the previous year, using either qualitative or quantitative techniques, articles must present, analyze, and comment on new data or synthesize and analyze data that have already been collected. The selection will be based on the article, contribution to the understanding of public health systems; provision of new insights to the field of PHSR; and potential to advance the field and/or challenge current thinking. Nominators must be members of AcademyHealth, although nominees need not be. Self-nominations are acceptable.

Requirements: Nominations must: be received at AcademyHealth by April 15; include the name and contact information for both the nominator and nominee and a copy of the article(s). Nominations may be submitted by email or postal mail. email to: PHSR@academyhealth.org. Mail to: AcademyHealth, Attention: PHSR Article of the Year Committee, 1150 17th Street, NW, Suite 600, Washington, DC 20036.

Amount of Grant: $1,000

Date(s) Application is Due: Apr 15

Contact: Jennifer Muldoon, Director; (202) 292-6770; fax: (202) 292-6870; email: jennifer.muldoon@academyhealth.org

Internet: http://www.academyhealth.org/Communities/content.cfm?ItemNumber=2536

Sponsor: AcademyHealth

1150 17th Street NW, Suite 600

Washington, DC 20036

ACC American Artists and Museum Professionals in Asia Fellowships 228

American Artists and Museum Professionals in Asia is a grant program established at the ACC in 2007 through the support of a three-year donation from The Henry Luce Foundation. The program provides individual fellowships to American artists and American museum professionals for research, study, and creative and collaborative projects in those countries of Asia extending from Burma eastward through Korea and Indonesia.

Requirements: Americans seeking aid to undertake activities in Asia are eligible to apply. Completed applications must be submitted at least six months prior to the planned date of project implementation. Applicants should send a brief description of the activity for which assistance is sought. If the proposed activity falls within the council's guidelines, additional application materials will be provided.

Restrictions: The Council is unable to consider proposals for the following: personal exhibitions; individual performance tours; undergraduate study; or activities conducted by individuals in their home countries.

Date(s) Application is Due: Feb 1

Contact: Program Director; (212) 843-0403; fax: (212) 843-0343; email: acc@accny.org

Internet: http://www.asianculturalcouncil.org/programs.html

Sponsor: Asian Cultural Council
6 West 48th Street, 12th Floor
New York, NY 10036-1802

ACC Asian Art and Religion Fellowships   229
This Fellowship program focusing on the relationship between the artistic and religious traditions of Asia enables American scholars, specialists, and artists to conduct research and undertake projects in Asia involving the interdisciplinary analysis of religion and the arts. The Council awards up to five research fellowships or travel grants each year through this program. Asian Art and Religion Fellowships have supported research in India on the iconography of Hindu deities, in Indonesia on the relationship between tantric religious thought and masked performance, and in Thailand on Buddhist architecture of northern Thailand.
Requirements: Asian individuals in the visual and performing arts seeking grant assistance to conduct research, study, receive specialized training, undertake observation tours, or pursue creative activity in the United States are eligible to apply for fellowship support from the Council. Americans seeking aid to undertake activities in Asia are also eligible to apply. Completed applications must be submitted at least six months prior to the planned date of project implementation. Applicants should send a brief description of the activity for which assistance is sought. If the proposed activity falls within the council's guidelines, additional application materials will be provided.
Restrictions: The Council is unable to consider proposals for the following: personal exhibitions; individual performance tours; undergraduate study; or activities conducted by individuals in their home countries.
Date(s) Application is Due: Feb 1
Contact: Program Director; (212) 843-0403; fax: (212) 843-0343; email: acc@accny.org
Internet: http://www.asianculturalcouncil.org/programs.html
Sponsor: Asian Cultural Council
6 West 48th Street, 12th Floor
New York, NY 10036-1802

ACC China Onsite Seminar Program in Art History   230
The program awards one grant each year to an American university to provide special opportunities for American graduate students of Chinese art history to enlarge their understanding of Chinese art in its cultural context. At the same time, the program introduces art history students in China to American theoretical approaches to studies in this field. There are two deadline dates for this award.
Requirements: Arts organizations and educational and cultural institutions are eligible to apply.
Date(s) Application is Due: Feb 1; Aug 1
Contact: Director, (212) 812-4300; fax: (212) 812-4299; email: acc@accny.org
Internet: http://www.asianculturalcouncil.org/programs.html
Sponsor: Asian Cultural Council
437 Madison Avenue, 37th Fl
New York, NY 10022

ACC Fellowships and Projects   231
Many of the Asian Cultural Council's fellowships are awarded with unrestricted grant funds, primarily to support artists and specialists from Asia pursuing research, training, and creative programs in the United States. On average each year, unrestricted funds provide support for sixteen individual fellowships and seventeen project grants.
Requirements: Asian individuals in the visual and performing arts seeking grant assistance to conduct research, study, receive specialized training, undertake observation tours, or pursue creative activity in the United States are eligible to apply for project or fellowship support from the Council. Americans seeking aid to undertake activities in Asia are also eligible to apply. Completed applications must be submitted at least six months prior to the planned date of project implementation. Applicants should send a brief description of the activity for which assistance is sought. If the proposed activity falls within the council's guidelines, additional application materials will be provided.
Restrictions: The Council is unable to consider proposals for the following: personal exhibitions; individual performance tours; undergraduate study; or activities conducted by individuals in their home countries.
Date(s) Application is Due: Feb 1
Contact: Program Director; (212) 843-0403; fax: (212) 843-0343; email: acc@accny.org
Internet: http://www.asianculturalcouncil.org/programs.html
Sponsor: Asian Cultural Council
6 West 48th Street, 12th Floor
New York, NY 10036-1802

ACC Ford Foundation Fellowships   232
This fellowship program provides support for training, travel, and research in the United States by individuals from Asia engaged in the study, documentation, and preservation of Asian traditional arts. Funded by an endowment grant from the Ford Foundation, this program offers awards in archaeology, art history, conservation, dance ethnology, ethnomusicology, museology, and other disciplines involving traditional Asian culture. Fellowships are generally awarded for periods ranging from three to 12 months. Completed applications must be submitted at least six months prior to the planned date of project implementation. Applicants should send a short description of the activity for which assistance is being requested. If the proposed activity falls within the council's guidelines, application materials will be provided by ACC.

Requirements: Asian individuals in the visual and performing arts seeking grant assistance to conduct research, study, receive specialized training, undertake observation tours, or pursue creative activity in the United States are eligible to apply for project or fellowship support from the Council. Americans seeking aid to undertake activities in Asia are also eligible to apply. Completed applications must be submitted at least six months prior to the planned date of project implementation. Applicants should send a brief description of the activity for which assistance is sought. If the proposed activity falls within the council's guidelines, additional application materials will be provided.
Restrictions: The Council is unable to consider proposals for the following: personal exhibitions; individual performance tours; undergraduate study; or activities conducted by individuals in their home countries.
Date(s) Application is Due: Feb 1
Contact: Program Director; (212) 843-0403; fax: (212) 843-0343; email: acc@accny.org
Internet: http://www.asianculturalcouncil.org/programs.html#4
Sponsor: Asian Cultural Council
6 West 48th Street, 12th Floor
New York, NY 10036-1802

ACC Hong Kong Arts Program Fellowships   233
The Program, made possible by donations from a broad group of donors in Hong Kong, provides fellowships to especially promising artists, students, and scholars from Hong Kong and other parts of China for research, study, and creative work in the United States. Limited support is also available for projects sponsored by Hong Kong institutions that involve Hong Kong-United States or Hong Kong-Asia cultural exchange. The ACC is able to award grant support to approximately ten individual and institutional applicants each year. Applicants should send a short description of the activity for which assistance is being requested. If the proposed activity falls within the council's guidelines, further application materials will be provided.
Requirements: Asian individuals in the visual and performing arts seeking grant assistance to conduct research, study, receive specialized training, undertake observation tours, or pursue creative activity in the United States are eligible to apply for project or fellowship support from the Council. Americans seeking aid to undertake activities in Asia are also eligible to apply. Completed applications must be submitted at least six months prior to the planned date of project implementation. Applicants should send a brief description of the activity for which assistance is sought. If the proposed activity falls within the council's guidelines, additional application materials will be provided.
Restrictions: The Council is unable to consider proposals for the following: personal exhibitions; individual performance tours; undergraduate study; or activities conducted by individuals in their home countries.
Date(s) Application is Due: Feb 1
Contact: Program Director; (212) 843-0403; fax: (212) 843-0343; email: acc@accny.org
Internet: http://www.asianculturalcouncil.org/programs.html#5
Sponsor: Asian Cultural Council
6 West 48th Street, 12th Floor
New York, NY 10036-1802

ACC Humanities Fellowship Program   234
The program assists American scholars, doctoral students, and specialists in the humanities to undertake research, training, and study in Asia in the following fields: archaeology; conservation; museology; and the theory, history, and criticism of architecture, art, dance, film, music, photography, and theater. Fellowship grants range in duration from one to nine months. The program also supports American and Asian scholars participating in international conferences, exhibitions, visiting professorships, and similar projects. Funding includes international travel, maintenance, per diem, and related research expenses. Completed applications must be submitted at least six months prior to the planned date of implementation. Applicants should send a short description of the activity for which assistance is being sought. If the proposed activity falls within the council's guidelines, application materials will be provided.
Requirements: American scholars, doctoral students, and specialists in the humanities are eligible to apply.
Restrictions: The Council is unable to consider proposals for the following: personal exhibitions; individual performance tours; undergraduate study; or activities conducted by individuals in their home countries.
Date(s) Application is Due: Feb 1
Contact: Program Director; (212) 843-0403; fax: (212) 843-0343; email: acc@accny.org
Internet: http://www.asianculturalcouncil.org/programs.html#6
Sponsor: Asian Cultural Council
6 West 48th Street, 12th Floor
New York, NY 10036-1802

ACC Japan-U.S. Arts Program Fellowships   235
The Program provides grant fellowships to individuals and institutions in Japan and the United States for exchange activities that encourage the study and understanding of Japanese art and culture. The individual fellowships enable Japanese artists, scholars, and specialists to visit the United States for research, observation, and creative work and allow their American counterparts to visit Japan for similar purposes. The council also provides assistance for performances, exhibitions, and other projects of unusual importance for the development of Japanese-American cultural exchange. The program is administered with the assistance of the ACC's office in Tokyo. Annually, the program supports approximately thirteen individuals and seven institutional projects. Applicants

should send a short description of the activity for which assistance is being requested. If the proposed activity falls within the council's guidelines, application materials will be provided by the ACC.

Requirements: Asian individuals in the visual and performing arts seeking grant assistance to conduct research, study, receive specialized training, undertake observation tours, or pursue creative activity in the United States are eligible to apply for project or fellowship support from the Council. Americans seeking aid to undertake activities in Asia are also eligible to apply. Completed applications must be submitted at least six months prior to the planned date of project implementation. Applicants should send a brief description of the activity for which assistance is sought. If the proposed activity falls within the council's guidelines, additional application materials will be provided.

Restrictions: The Council is unable to consider proposals for the following: personal exhibitions; individual performance tours; undergraduate study; or activities conducted by individuals in their home countries.

Date(s) Application is Due: Feb 1

Contact: Program Director; (212) 843-0403; fax: (212) 843-0343; email: acc@accny. org

Internet: http://www.asianculturalcouncil.org/programs.html#7

Sponsor: Asian Cultural Council
6 West 48th Street, 12th Floor
New York, NY 10036-1802

ACC John D. Rockefeller III Award   236

This annual award is presented to an individual from Asia or the United States who has made an especially significant contribution to the understanding, practice, or study of the visual or performing arts of Asia. The award will enable recipients to pursue work in some aspect of the arts of Asia through international travel and research. Applications are accepted anytime.

Requirements: Individuals from Asia and the United States who are active in any field of the visual or performing arts of Asia, whether affiliated with an institution or working independently, are eligible for award consideration. Candidates are nominated by artists, scholars, and others professionally involved in Asian art and culture.

Samples: Lin Hwai Min, dancer/choreographer and director of the Cloudgate Dance Theater, Taiwan (2006).

Amount of Grant: $30,000

Contact: Program Director; (212) 843-0403; fax: (212) 843-0343; email: acc@accny. org

Internet: http://www.asianculturalcouncil.org/programs.html#12

Sponsor: Asian Cultural Council
6 West 48th Street, 12th Floor
New York, NY 10036-1802

ACC Mekong Region Fellowship   237

The Mekong Region Fellowship program is funded by the Rockefeller Foundation. The program assists individual artists and specialists from Burma, Cambodia, Laos, Thailand, Vietnam, and China, Yunnan Province in undertaking research, study, and creative projects in the United States and Asia. The program also provides funds for special projects that encourage regional collaboration and partnership.

Requirements: Asian individuals in the visual and performing arts seeking grant assistance to conduct research, study, receive specialized training, undertake observation tours, or pursue creative activity in the United States are eligible to apply for project or fellowship support from the Council. Americans seeking aid to undertake activities in Asia are also eligible to apply. Completed applications must be submitted at least six months prior to the planned date of project implementation. Applicants should send a brief description of the activity for which assistance is sought. If the proposed activity falls within the council's guidelines, additional application materials will be provided.

Restrictions: The Council is unable to consider proposals for the following: personal exhibitions; individual performance tours; undergraduate study; or activities conducted by individuals in their home countries.

Date(s) Application is Due: Feb 1

Contact: Program Director; (212) 843-0403; fax: (212) 843-0343; email: acc@accny. org

Internet: http://www.asianculturalcouncil.org/programs.html#8

Sponsor: Asian Cultural Council
6 West 48th Street, 12th Floor
New York, NY 10036-1802

ACC Ock Rang Cultural Foundation Fellowship Program   238

The Ock Rang Cultural Foundation in Seoul, Korea, which operates the Dongsung Art Center, provides funding to the ACC that supports one to four fellowships each year. The program funds artists and specialists from Korea pursuing research, study, and creative work in the United States, primarily in the technical theater arts. The program also funds American specialists traveling to Korea and Asian artists presenting workshops and surveying arts activities in Korea. There are two deadline dates for this award.

Date(s) Application is Due: Feb 1; Aug 1

Contact: Director, (212) 812-4300; fax: (212) 812-4299; email: acc@accny.org

Internet: http://www.asianculturalcouncil.org/programs.html#1

Sponsor: Asian Cultural Council
437 Madison Avenue, 37th Fl
New York, NY 10022

ACC Philippines Fellowship Program   239

The Program was formally inaugurated in 2000 in association with the ACC Philippines Foundation, a foundation established in Manila to collaborate with the ACC in raising funds and making grants for Philippines-U.S. cultural exchange. Through the ACC Philippines Foundation a range of donors in both countries are contributing to the Program, which emphasizes support for artists, scholars, and specialists from the Philippines undertaking research, study, and creative work in the United States. Some grants are also made to Americans pursuing similar activities in the Philippines and to institutions engaged in Philippines-U.S. or Philippines-Asia exchange projects. Applicants should send a short description of the activity for which assistance is being requested. If the proposed activity falls within the council's guidelines, application materials will be provided by the ACC.

Requirements: Asian individuals in the visual and performing arts seeking grant assistance to conduct research, study, receive specialized training, undertake observation tours, or pursue creative activity in the United States are eligible to apply for project or fellowship support from the Council. Americans seeking aid to undertake activities in Asia are also eligible to apply. Completed applications must be submitted at least six months prior to the planned date of project implementation. Applicants should send a brief description of the activity for which assistance is sought. If the proposed activity falls within the council's guidelines, additional application materials will be provided.

Restrictions: The Council is unable to consider proposals for the following: personal exhibitions; individual performance tours; undergraduate study; or activities conducted by individuals in their home countries.

Date(s) Application is Due: Feb 1

Contact: Program Director; (212) 843-0403; fax: (212) 843-0343; email: acc@accny. org

Internet: http://www.asianculturalcouncil.org/programs.html#9

Sponsor: Asian Cultural Council
6 West 48th Street, 12th Floor
New York, NY 10036-1802

ACC Residency Program in Asia   240

The program in Asia assists individual American artists, scholars, and professionals undertaking research, teaching, and creative residencies at cultural and educational institutions in East and Southeast Asia. Projects supported in the Program demonstrate close collaboration in design and execution between the visiting American specialist and the host organization and produce tangible results such as publication, course development, or the creation of new artistic work. The program helps to foster the growth of Asian cultural studies in the U.S. and of American cultural studies in Asia, as well as to stimulate long-term relationships in the arts and humanities between American individuals and institutions and their colleagues in Asia. Eligible fields for consideration include archaeology, architecture (design, history, and theory), art history, conservation, crafts, dance, film, museology, music, painting, photography, printmaking, sculpture, theater, and video.

Requirements: Applications may be submitted by the participating American specialist or by the host institution in Asia.

Restrictions: The Council is unable to consider proposals for the following: personal exhibitions; individual performance tours; undergraduate study; or activities conducted by individuals in their home countries.

Date(s) Application is Due: Jan 15; Aug 1

Contact: Program Director, (212) 812-4300; fax: (212) 812-4299; email: acc@accny. org

Internet: http://www.asianculturalcouncil.org/programs.html#1

Sponsor: Asian Cultural Council
437 Madison Avenue, 37th Floor
New York, NY 10022

ACC Starr Foundation Fellowship Program   241

The contemporary visual arts of Asia are the focus of this fellowship program, established with an endowment grant from the Starr Foundation. Assistance is provided to artists and arts specialists from Asia for creative activity, research, training, and observation in the United States in the fields of art criticism, crafts, design, film, museology, painting, photography, printmaking, and sculpture. Assistance is provided for periods ranging from one month to one year. Approximately seventeen Fellowship grants are given annually supporting artists, curators, and critics from Asia traveling to the United States. Completed applications must be submitted at least six months prior to the planned date of project implementation.

Requirements: Asian individuals in the visual and performing arts seeking grant assistance to conduct research, study, receive specialized training, undertake observation tours, or pursue creative activity in the United States are eligible to apply for project or fellowship support from the Council. Americans seeking aid to undertake activities in Asia are also eligible to apply. Completed applications must be submitted at least six months prior to the planned date of project implementation. Applicants should send a brief description of the activity for which assistance is sought. If the proposed activity falls within the council's guidelines, additional application materials will be provided.

Restrictions: The Council is unable to consider proposals for the following: personal exhibitions; individual performance tours; undergraduate study; or activities conducted by individuals in their home countries.

Date(s) Application is Due: Feb 1

Contact: Program Director; (212) 843-0403; fax: (212) 843-0343; email: acc@accny. org

Internet: http://www.asianculturalcouncil.org/programs.html#10

Sponsor: Asian Cultural Council

6 West 48th Street, 12th Floor
New York, NY 10036-1802

ACC Taiwan Fellowship Program   242

The program makes grants to artists, scholars, and specialists from Taiwan for research, study, and creative work in the United States. The program also supports American and Asian arts specialists visiting Taiwan to participate in cultural exchange projects organized by local arts institutions. Application procedures and grantee programming are administered with the assistance of the ACC-SAACF office in Taipei. The program supports up to eight grant recipients each year. Applicants should send a short description of the activity for which assistance is being sought. If the proposed activity falls within the council's guidelines, application materials will be provided.

Requirements: Asian individuals in the visual and performing arts seeking grant assistance to conduct research, study, receive specialized training, undertake observation tours, or pursue creative activity in the United States are eligible to apply for project or fellowship support from the Council. Americans seeking aid to undertake activities in Asia are also eligible to apply. Completed applications must be submitted at least six months prior to the planned date of project implementation. Applicants should send a brief description of the activity for which assistance is sought. If the proposed activity falls within the council's guidelines, additional application materials will be provided.

Restrictions: The Council is unable to consider proposals for the following: personal exhibitions; individual performance tours; undergraduate study; or activities conducted by individuals in their home countries.

Date(s) Application is Due: Feb 1

Contact: Director; (212) 843-0403; fax: (212) 843-0343; email: acc@accny.org

Internet: http://www.asianculturalcouncil.org/programs.html#11

Sponsor: Asian Cultural Council
6 West 48th Street, 12th Floor
New York, NY 10036-1802

Access Fund Climbing Preservation Grants   243

The program awards grants to projects that preserve and enhance climbing opportunities and conserve the climbing environment throughout the United States. Grants support projects that encourage access to or enhance opportunities for climbing; be supported by the climbing community; raise awareness about climber responsibilities through stewardship projects; develop or support partnerships with resource management agencies, conservation organizations, land trusts, and local climbing groups; reduce climber impacts on natural and cultural resources; develop knowledge about natural and cultural resource values where the information is used to open climbing areas or mitigate climbing impacts; include volunteer labor and/or pro bono services; use some matching funds; and be located within the United States. Requests over $10,000 are considered if the project has national significance and offers a high percentage of matching funds. Contact the office before submitting an application for a large grant.

Requirements: Applications may be submitted by local climbing groups or organizations; governmental agencies that wish to sponsor or organize a local project; and research groups, conservation organizations, and land trusts. 501(c)3 tax-exempt status is not necessary.

Samples: Indian Creek, Utah, $4,000—for trail design and construction in the Broken Tooth and Scarface areas (2007); Colorado Division of Wildlife, Boulder Canyon, Colorado, $500—for cliff nesting raptor signage (2007); Illinois Climbers Association, Illinois, $1,800—for organization start up and incorporation costs (2007).

Amount of Grant: $500-$10,000

Date(s) Application is Due: Mar 1; Jun 15; Sep 15

Contact: Ellen Jardine, Business & Finance Manager; (303) 545-6772, ext. 107; fax: (303) 545-6774; email: ellen@accessfund.org

Internet: http://www.accessfund.org/cons/guidelines.php

Sponsor: Access Fund
P.O. Box 17010, 207 Canyon, Suite 201
Boulder, CO 80306

ACCP Frontiers Career Development Research Awards   244

Supports previously unmet or under-served areas of Pharmacy-Based Health Services Research, Clinical Research, or Translational Research. Through its awards, the Institute is especially interested in supporting: research that helps to pursue one or more of the priority areas outlined in the Research Agenda of the American College of Clinical Pharmacy (see below); the career development of clinical pharmacy investigators; health services research that assesses the impact of clinical pharmacy services on the use, costs, quality, accessibility, delivery, organization, financing, and outcomes of health care, including the development of payment models for clinical pharmacists‚Äô services; and clinical and translational research, including areas that extend beyond those funded by other ACCP awards and grants.

Requirements: Any ACCP Full Member, Associate Member, or Affiliate Member is eligible to apply.

Restrictions: Proposals submitted concurrently with an application for membership will not be considered.

Amount of Grant: $10,000-$30,000

Date(s) Application is Due: Jun 1

Contact: Cathy Englund, Executive Secretary; (913) 492-3311; fax: (913) 492-0088; email: cenglund@accp.com

Internet: http://www.accp.com/frontiers/research.php#resfel

Sponsor: American College of Clinical Pharmacy
13000 W. 87th Street Parkway
Lenexa, KS 66215-4530

ACCP Hematology/Oncology PRN Minisabbatical Program   245

The Minisabbatical Program is an opportunity for members of the Hematology/Oncology PRN exclusively to gain or expand their skills in practice or research under the guidance of experts in oncology pharmacotherapy. The major purposes are: to learn new clinical skills or approaches to patient care that will enhance the applicant's practice in oncology pharmacotherapy (e.g., patient management, clinic development, payment for services); and to learn new research skills that will enhance or expand the applicant's research program in oncology pharmacotherapy (e.g., laboratory techniques, pharmacokinetic modeling).

Requirements: The applicant for the program must be a current member of the Hematology/Oncology PRN.

Restrictions: The following individuals will not be considered as candidates: current residents or fellows; members of the Minisabbatical Awards Committee; or persons in receipt of a minisabbatical award in the past 5 years.

Amount of Grant: $2,750 for recipient; $750 for mentor

Date(s) Application is Due: Jun 2

Contact: Cathy Englund, Executive Secretary; (913) 492-3311; fax: (913) 492-0088; email: cenglund@accp.com

Internet: http://www.accp.com/frontiers/rihmonmini.php

Sponsor: American College of Clinical Pharmacy
13000 W. 87th Street Parkway
Lenexa, KS 66215-4530

ACCP Infectious Diseases PRN Minisabbatical Program   246

The minisabbatical program is an opportunity for members of the ID PRN exclusively to gain or expand their skills in practice or research under the guidance of experts in infectious diseases pharmacotherapy. The major purposes are: to learn new clinical skills or approaches to patient care that will enhance the applicant's practice in infectious diseases (e.g., patient evaluation techniques, practice management, antimicrobial surveillance, etc.); and to learn new research skills that will enhance or expand the applicant's research program in infectious diseases (e.g., laboratory techniques, pharmacokinetic and pharmacodynamic modeling, epidemiology, ID outcomes research).

Requirements: The applicant for the ID PRN Minisabbatical Program must be a current member and maintain membership with the ID PRN for the duration of project.

Restrictions: The following individuals will not be considered as candidates: current students, residents or fellows at time of application; members of the Minisabbatical Awards Committee; if currently funded by the NIH or have other multi-year peer reviewed funding; or persons in receipt of a minisabbatical award in the past 5 years.

Amount of Grant: $,3000 for recipient; $1,000 for mentor

Date(s) Application is Due: Jun 2

Contact: Cathy Englund, Executive Secretary; (913) 492-3311; fax: (913) 492-0088; email: cenglund@accp.com

Internet: http://www.accp.com/frontiers/riinfdmini.php

Sponsor: American College of Clinical Pharmacy
13000 W. 87th Street Parkway
Lenexa, KS 66215-4530

ACCP Infectious Diseases Research Fellowships   247

The purpose of ACCP fellowships is to support the development of Clinical Scientists through postgraduate fellowship experiences, or through the support of graduate students who have at least completed their qualifying examinations. Research activities must relate directly to the stated areas of emphasis. Fellowships support a research training experience and not a clinical pharmacy residency program. A pharmacy fellowship is a directed, highly individualized postgraduate program designed to prepare the participant to become an independent researcher. Fellowships exist primarily to develop competency in the scientific research process. The ACCP Research Institute believes that clinical pharmacy fellowships should be at least two years in length, and will give strong preference to funding applications that describe a multi-year training program. This grant will provide the fellow's stipend for either the first, second, or third year. Preceptors will have the option to receive the fellowship's entire funding over one year, or half funding for each of two years.

Requirements: Preceptor: An ACCP Full Member at the time application is made. Fellowship applications submitted concurrently with an application for Full Membership will not be considered. Fellow: An ACCP Member, Associate Member, or applicant for member. If a graduate student, (a) preference will be given to supporting students who have received or are concurrently enrolled in a Pharm.D. degree program; and (b) the Fellow must at least have completed his/her qualifying examinations.

Restrictions: Ineligible applicants include: ACCP Board of Regent members; ACCP Research Institute Board of Trustee members; Grants and Fellowship Selection Committee members; preceptor of an ACCP Research Institute fellowship within the previous year.

Amount of Grant: $40,500

Date(s) Application is Due: Mar 23

Contact: Cathy Englund, Executive Secretary; (913) 492-3311; fax: (913) 492-0088; email: cenglund@accp.com

Internet: http://www.accp.com/frontiers/research.php#resfel

Sponsor: American College of Clinical Pharmacy
13000 W. 87th Street Parkway
Lenexa, KS 66215-4530

ACCP Oncology Research Fellowships   248

This fellowship, supported by a grant from sanofi-aventis, supports the development of Clinical Scientists in oncology through postgraduate fellowship experiences, or through the support of graduate students who have at least completed their qualifying examinations. Research activities must related to the stated area of emphasis. This grant provides the fellow's stipend of $38,000, travel support of $1,000, and personnel fringe benefits of $1,500.

Requirements: Preceptor: An ACCP Full Member at the time application is made. Fellowship applications submitted concurrently with an application for Full Membership will not be considered. Fellow: An ACCP Member, Associate Member, or applicant for member. If a graduate student, (a) preference will be given to supporting students who have received or are concurrently enrolled in a Pharm.D. degree program; and (b) the Fellow must at least have completed his/her qualifying examinations.

Restrictions: Ineligible applicants include: ACCP Board of Regent members; ACCP Research Institute Board of Trustee members; Grants and Fellowship Selection Committee members; preceptor of an ACCP Research Institute fellowship within the previous year.

Amount of Grant: $40,500
Date(s) Application is Due: Mar 3
Contact: Cathy Englund, Executive Secretary; (913) 492-3311; fax: (913) 492-0088; email: cenglund@accp.com
Internet: http://www.accp.com/frontiers/research.php#resfel
Sponsor: American College of Clinical Pharmacy
13000 W. 87th Street Parkway
Lenexa, KS 66215-4530

ACCP Pain & Palliative Care PRN Minisabbatical Program 249

The Minisabbatical Program is an opportunity for members of the Pain & Palliative Care PRN exclusively to gain or expand their skills in practice or research under the guidance of experts in pain & palliative care pharmacotherapy. The Pain & Palliative Care minisabbatical program is made possible by an unrestricted educational grant from Amgen. The major purposes are: to learn new clinical skills or approaches to patient care that will enhance the applicant's practice in pain and palliative care pharmacotherapy; and to learn new research skills that will enhance or expand the applicant's research program in pain and palliative care pharmacotherapy.

Requirements: The applicant for the minisabbatical program must be a current member of the Pain & Palliative Care PRN.

Restrictions: The following individuals will not be considered as candidates: current residents or fellows; members of the Minisabbatical Awards Committee; or persons in receipt of a minisabbatical award in the past 5 years.

Amount of Grant: $2,750 for recipient; $750 for mentor
Date(s) Application is Due: Jun 1
Contact: Cathy Englund, Executive Secretary; (913) 492-3311; fax: (913) 492-0088; email: cenglund@accp.com
Internet: http://www.accp.com/frontiers/rinephmini.php
Sponsor: American College of Clinical Pharmacy
13000 W. 87th Street Parkway
Lenexa, KS 66215-4530

ACCP Pharmacotherapy Investigator Development Research Awards 250

The purpose of these awards is to provide funding for research projects that will: help pursue one or more of the priority areas outlined in the Research Agenda of the American College of Clinical Pharmacy; contribute to the development of the principal investigator, research career; promote the safe, effective, and cost-effective use of medications; and advance the practice of clinical pharmacy. This includes, but is not limited to, investigations aimed at generating preliminary data or seed money for pilot projects. The award may serve as the total support for a project or supplement an existing research effort as long as a specific portion of the research is identified as being made possible by this award, and provided that the investigator states specifically how the balance will be funded and provides evidence of its guaranteed availability. The overall goal of the awards is to assist developing researchers in improving their competitiveness for extramural funding and to support their evolving research careers. In this context, projects that involve the use of animals or in vitro models will be considered as long as the applicant adequately justifies the request.

Requirements: The principal investigator of the award application must: be an ACCP member; be ten or fewer years since completion of his/her formal training or first academic appointment; not be currently funded by the NIH or have other multi-year peer reviewed funding; or not be a student, resident, or fellow.

Restrictions: The award must not duplicate funding for another research project.

Amount of Grant: $20,000
Date(s) Application is Due: Jul 16
Contact: Cathy Englund, Executive Secretary; (913) 492-3311; fax: (913) 492-0088; email: cenglund@accp.com
Internet: http://www.accp.com/frontiers/research.php#resfel
Sponsor: American College of Clinical Pharmacy
13000 W. 87th Street Parkway
Lenexa, KS 66215-4530

ACE Fellows Program 251

The program supports leadership development program in higher education. Fellows spend an extended period of time on another campus, working directly with presidents and other senior leaders, observing firsthand how the institution and its leaders address strategic planning, resource allocation, development, policy, and other issues and challenges. Fellows observe and participate in key meetings and events, and take on special projects and/or assignments while under the mentorship of a team of experienced administrators. The program also includes participation in three week-long national seminars, visits to other campuses, attendance at national meetings, and contact with a national network of higher education leaders.

Requirements: Candidates must have demonstrated records of leadership in institutionwide contexts and must be nominated by the institution's president or other senior officer who agrees to pay the candidate's salary and benefits.

Date(s) Application is Due: Nov 1
Contact: Marlene Ross, Director, (202) 939-9420; fax: (202) 785-8056; email: fellows@ace.nche.edu
Internet: http://www.acenet.edu/programs/fellows
Sponsor: American Council on Education
1 Dupont Circle
Washington, DC 20036-1193

ACF Adoption Opportunities Project Grants 252

The Adoption Opportunities program provides discretionary funds for projects designed to eliminate barriers to adoption and help find permanent families for children who would benefit from adoption, particularly children with special needs. Some of the major program areas include: the development and implementation of a national adoption information exchange system; increasing services in support of the placement in adoptive families of minority children who are in foster care and have the goal of adoption, with a special emphasis on the recruitment of minority families; increasing post-legal adoption services for families who have adopted children with special needs; and, supporting the placement of children in kinship care arrangements, pre-adoptive, or adoptive homes.

Requirements: Eligible entities include States, local government entities, federally recognized Indian Tribes and tribal organizations, faith-based and community organizations, colleges and universities, public or private non-profit licensed child welfare or adoption agencies, and adoption exchanges.

Geographic Focus: Alabama; Alaska; Arizona; Arkansas; California; Colorado; Connecticut; Delaware; District of Columbia; Florida; Georgia; Guam; Hawaii; Idaho; Illinois; Indiana; Iowa; Kansas; Kentucky; Louisiana; Maine; Maryland; Massachusetts; Michigan; Minnesota; Mississippi; Missouri; Montana; Nebraska; Nevada; New Hampshire; New Jersey; New Mexico; New York; North Carolina; North Dakota; Ohio; Oklahoma; Oregon; Pennsylvania; Puerto Rico; Rhode Island; South Carolina; South Dakota; Tennessee; Texas; U.S. Virgin Islands; Utah; Vermont; Virginia; Washington; West Virginia; Wisconsin; Wyoming

Amount of Grant: $300,000
Contact: Geneva Ware-Rice, (202) 205-8354
Internet: http://www.acf.hhs.gov/programs/fbci/progs/fbci_adoption.html
Sponsor: Administration on Children, Youth and Families (ACYF) - Family and Youth Services Bureau
P.O. Box 13505
Silver Spring, MD 20911-3505

ACGT Investigators Grants 253

The Investigators Award in Clinical Translation of Gene Therapy for Cancer distributes funds over 3-5 years, inclusive of a maximum of 10% indirect costs. Funds may be used at the recipient's discretion for salary, technical assistance, supplies, animals or capital equipment, but may not support staff not directly related to the project, e.g. secretaries or administrative assistants. Purchase of equipment is not allowed in the final year of the grant. ACGT is accepting grant applications to produce and release-testing of the clinical trial agents under cGMP, conduct the necessary pre-clinical pharmacological and toxicological studies in appropriate animal models, and/or conducting the clinical translational trials in patients in support of an Investigative New Drug application to the FDA.

Requirements: The candidate must hold an MD, PhD, or equivalent degree and be a tenure-track or tenured faculty. The investigator must be conducting original research as an independent faculty member. ACGT has no citizenship restrictions, and research supported by the award must be conducted at medical schools and research centers only in the United States. While the unambiguous demonstration of preclinical efficacy in cancer treatment by gene therapy is a pre-requisite, entering into the clinical trial during the funding period is also a requirement.

Amount of Grant: $1 million maximum over three to five years
Contact: Grace Pedersen, Foundation Administrator; (203) 358-8000, ext. 495; email: aneslage@acgtfoundation.org
Internet: http://www.acgtfoundation.org/grants.html
Sponsor: Alliance for Cancer Gene Therapy
96 Cummings Point Road
Stamford, CT 06902

ACGT Young Investigator Grants 254

The alliance funds research aimed at furthering the development of gene therapy approaches to the treatment of cancer. The overall objectives of this grant are to advance gene therapy into the causes, treatment, and prevention of all types of cancer by promoting development of novel and innovative studies by young investigators. The emphasis of this initiative is to promote basic, and preclinical research approaches utilizing cells and genes as medicine. The six main areas of research ACGT will support are: tumor-specific replicating viruses and bacteria, anti-angiogenesis, immune-mediated gene therapy and cancer vaccines, oncogenes/suppressor oncogenes/apoptosis, tumor targeting and vector development, and other cancer gene therapy research. The three-year grant may be used at the recipient's discretion for salary, technical assistance, supplies, animals or capital equipment, but may not support staff not directly related to the project, e.g. secretaries or administrative assistants. Purchase of equipment is not allowed in the third year of the grant. Continued support is contingent upon submission and approval of a noncompetitive renewal application each year.

Requirements: Candidates must hold an MD, MPH, PhD, or equivalent degree and be a tenure-track assistant professor within five years of their initial appointment to this rank, at the time of award activation. The investigator must be conducting original research as an independent faculty member. ACGT has no citizenship restrictions, and research supported by the award must be conducted at medical schools and research centers only in the United States.

Amount of Grant: $500,000 maximum over three years
Date(s) Application is Due: Sep 21
Contact: Grace Pedersen, Foundation Administrator; (203) 358-8000, ext. 495; email: aneslage@acgtfoundation.org
Internet: http://www.acgtfoundation.org/grants.html
Sponsor: Alliance for Cancer Gene Therapy
96 Cummings Point Road
Stamford, CT 06902

ACHE Foster G. McGaw Student Scholarship 255

The scholarship is designated for students enrolled in their final year of a healthcare management graduate program. One scholarship is awarded per candidate. Application is available online in the Fall.

Requirements: Applicant must be a student associate in good standing of the American College of Healthcare Executives; enrolled in full-time study for the upcoming fall term, which is the student's final year of didactic work in a healthcare management graduate program; demonstrate financial need; and a U.S. or Canadian citizen.

Restrictions: Previous recipients of this scholarship and the Albert W. Dent Graduate Student Scholarship are ineligible.

Amount of Grant: $3500
Date(s) Application is Due: Mar 31
Contact: Thomas Killam, (312) 424-9400; fax: (312) 424-0023; email: tkillam@ache.org
Internet: http://www.ache.org/Faculty_Students/mcgaw_scholarship.cfm
Sponsor: American College of Healthcare Executives
1 N Franklin Street, Suite 1700
Chicago, IL 60606-3491

Achelis and Bodman Foundations Grants 256

Both Foundations concentrate their grant programs in New York City, although The Bodman Foundation also makes grants in northern New Jersey. Funding is concentrated in six program areas: Arts and Culture, Education, Employment, Health, Public Policy, and Youth and Families.

Requirements: Nonprofit organizations based in New York City and northern New Jersey that are tax-exempt under Section 501(c)3 of the Internal Revenue Code and fall within the program areas of the foundations are welcome to submit an inquiry or proposal letter by regular mail. Do not send initial inquiries by email or fax; also do not send CDs, DVDs, computer discs, video tapes, or proposals through the Internet or by facsimile. An initial inquiry to the foundations should include only the following items: (1) A proposal letter that briefly summarizes the history of the project, need, research, objectives, time period, key staff, project budget, and evaluation plan. (2) Latest annual report. (3) Latest complete audited financial statements. (4) IRS 501(c)3 tax-exemption letter. There are no deadlines and applications may be submitted at any time during the year. Each request is reviewed by staff and will usually receive a written response within thirty days. Those requests deemed consistent with the interests and resources of the Foundations will be evaluated further and more information will be requested. Staff may request a site visit, conference call, or meeting. All grants are reviewed and approved by the Trustees at one of their three board meetings in May, September, or December.

Restrictions: The foundations generally do not make grants for the following purposes or program areas: Nonprofit organizations outside of New York and New Jersey; Annual appeals, dinner functions, and fundraising events; Endowments and capital campaigns; Loans and deficit financing; Direct grants to individuals; Individual day-care and after-school programs; Housing; Organizations or projects based outside the U.S.; Films or video projects; Small art, dance, music, and theater groups; Individual K-12 schools (except charter schools); National health and mental health organizations; Government agencies or nonprofit organizations significantly funded or reimbursed by government agencies. Limited resources prevent the foundations from funding the same organization on an ongoing annual basis.

Geographic Focus: New Jersey; New York
Samples: Foundation for Cultural Review - (Achelis, 2007) $35,000 toward the redesign of the website of the New Criterion magazine.; High 5 Tickets to the Arts - (Bodman, 2006) $15,000 for the Take 5 Program for disadvantaged youth.; Coalition for Educational Freedom - (Bodman, 2007) $50,000 toward the creation of a state-wide school choice organization in Connecticut.; Center for Alternative Sentencing and Employment Services - (Achelis, 2007) $35,000 toward the Career Exploration Program for young ex-offenders in New York City.; Jericho Project - (Achelis, 2007) $50,000 toward the Vocational and Educational Program for those with substance abuse or mental health problems.; Brain Trauma Foundation - (Achelis, 2007) $50,000 for the Veterans Information Initiative.; Children of Bellevue - (Bodman, 2007) $25,000 for the Frances L. Loeb Child Protection and Development Center at Bellevue Hospital Center

Contact: John B. Krieger, Executive Director; (212) 644-0322; fax: (212) 759-6510; email: main@achelis-bodman-fnds.org
Internet: http://foundationcenter.org/grantmaker/achelis-bodman/guidelines.html
Sponsor: Achelis and Bodman Foundations
767 Third Avenue, 4th Floor
New York, NY 10017

ACICS Dissertation Research Fellowship 257

ACICS supports an annual competitive fellowship to support dissertation research in the realm of private career school accreditation. The fellowship will be awarded to the winner of a competition for the most promising 400-word dissertation proposal abstract. The applicant with the winning proposal will receive $1000, $3000 upon approval of the research proposal by the student's graduate committee; and $4000 after the fellow passes the final oral dissertation defense and submits a copy of the dissertation. Applicants must submit a 400-word research proposal abstract, evidence that the applicant has reached the dissertation stage of a doctoral program in an accredited doctoral degree-granting institution, curriculum vita, and three letters of reference (two from faculty at the individual's current institution and one from a professional, work-related contact).

Amount of Grant: $8000
Date(s) Application is Due: May 1
Contact: Dr. Kathleen Prince, Director of Quality Enhancement, (202) 336-6775; fax: (202) 842-2593; email: kprince@acics.org
Internet: http://www.acics.org/applications/fellowship.asp
Sponsor: Accrediting Council for Independent Colleges and Schools
750 First Street NE, Suite 980
Washington, DC 20002-4241

ACIE Host University Edmund S. Muskie/Freedom Support Act Graduate Fellowships 258

The program is accepting applications from universities to host scholars from the New Independent States for the academic year. The Open Society Institute sponsors professors in the fields of education, environmental management, law, and public health. The American Councils for International Education sponsors professors in the fields of business administration, economics, international affairs, journalism and mass communications, library and information science, public administration, and public policy. Host institutions are selected on the basis of their academic strength, experience working with international students, ability to facilitate internships, ability to assign faculty advisors as well as a program coordinator, ability to contribute financially toward tuition in the form of cost-share, and ability to mentor and evaluate the academic progress and cultural adaptation of the fellow(s).

Requirements: Applicants must be a citizen of one of the 12 participating countries, under the age of 40, with an undergraduate degree. Applicants must also currently reside in one of the 12 participating countries, have a professional aptitude and leadership potential, and must be committed to returning to their home country after fellowship. The participating countries are Armenia, Azerbaijan, Belarus, Georgia, Kazakhstan, Kyrgyzstan, Moldova, Russia, Tajikistan, Turkmenistan, Ukraine, and Uzbekistan.

Contact: Susan Frarie, (202) 833-7522; fax: (202) 833-7523; email: frarie@actr.org or general@americancouncils.org
Internet: http://www.americancouncils.org/program.asp?PageID=121&ProgramID=22
Sponsor: American Councils for International Education
1776 Massachusetts Avenue NW, Suite 700
Washington, DC 20036

ACIE Title VIII Research Scholar Program 259

Funded by the U.S. Department of State, Program for the Study of Eastern Europe and The Independent States of the Former Soviet Union (Title VIII), the American Councils Research Scholar program provides full support for graduate students, faculty, and independent scholars seeking to conduct research for three months to nine months in Belarus, Central Asia, Russia, the South Caucasus, Ukraine, and Moldova. Scholars may apply for support for research in more than one country during a single trip, provided they intend to work in the field for a total of three to nine months. Participants in the program will receive international airfare from the scholar's home to his/her host city overseas; academic affiliation at a leading local university; visa(s) arranged by American Councils in direct collaboration with academic host institutions in order to facilitate archive access and guarantee timely visa registration; housing in a university dormitory or with a local host family (American Councils also provides informal assistance in locating apartments in some cities); a monthly living stipend; financial and logistical support for travel within the region as required by research. Health insurance of up to $50,000 per accident or illness; ongoing logistical support from American Councils offices throughout the region, including in-country orientation programs and 24-hour emergency aid. Proposals will be reviewed by an independent selection committee of leading U.S. scholars in the social sciences and the humanities. Award deadlines are twice per year. Applications for summer, fall, and academic year programs are due on January 15; applications for spring programs are due October 1.

Requirements: U.S. graduate students, faculty, and post-doctoral researchers are eligible to apply.

Amount of Grant: $5000-$25,000
Date(s) Application is Due: Jan 15; Oct 1
Contact: Research Scholar Program Manager, (202) 833-7522; email: outbound@americancouncils.org
Internet: http://www.americancouncils.org/program.asp?PageID=121&ProgramID=15
Sponsor: American Councils for International Education
1776 Massachusetts Avenue NW, Suite 700
Washington, DC 20036

ACIE U.S.-Eurasia Awards for Excellence in Teaching (TEA) 260

The program provides a professional development opportunity for award-winning U.S. teachers to utilize their talents and expertise to improve the quality of secondary education in Russia and to create linkages and learning partnerships between U.S. and Russian schools. Participants will take part in a three-day cross-cultural symposium,

Celebrating Teaching Excellence Across Cultures, and a three-week exchange program with teachers from Armenia, Azerbaijan, Georgia, Kazakhstan, Kyrgyzstan, Russia, Tajikistan, Turkmenistan, Ukraine and Uzbekistan who have won the equivalent program in their country. Application and guidelines are available online.

Requirements: U.S. middle school and high school teachers of the English, English as a Foreign Language (EFL), social studies, history, math, science, or information technology who have been recognized for excellence in teaching at the national, state, or local level.

Restrictions: Applicant must be a U.S. citizen or permanent resident.

Date(s) Application is Due: Feb 28

Contact: Ben Dunbar, (202) 833-7522; email: teachers@americancouncils.org

Internet: http://www.americancouncils.org/newsDetail.php?news_id=NjY=

Sponsor: American Councils for International Education

1776 Massachusetts Avenue NW, Suite 700

Washington, DC 20036

ACLS American Research in the Humanities in China   261

The program is funded by the National Endowment for the Humanities. Stipends are calculated based upon the awardee, academic rank and the length of his/her stay in China. Applicants must submit a carefully formulated research proposal that reflects an understanding of the present Chinese academic and research environment. The proposal should include a persuasive statement of the need to conduct the research in China. Stipends for four months of research cannot exceed $20,000. The maximum award is $40,000.

Requirements: This program is open to scholars in the humanities and humanities-related social sciences who have received a Ph.D. or its equivalent by the time of application. An applicant must: be a citizenship or permanent resident status of the United States who has lived in the United States continuously for at least three years as of the application deadline date; and hold a Ph.D. degree conferred prior to the application deadline.

Amount of Grant: $40,000 maximum

Date(s) Application is Due: Nov 12

Contact: Nicole Stahlmann, Director of Fellowship Programs; (212) 697-1505, ext. 134; fax: (212) 949-8058; email: nstahlmann@acls.org

Internet: http://www.acls.org/grants/Default.aspx?id=524

Sponsor: American Council of Learned Societies

633 Third Avenue

New York, NY 10017-6795

ACLS Charles A. Ryskamp Research Fellowships   262

These fellowships support advanced assistant professors and non-tenured associate professors in the humanities and related social sciences whose scholarly contributions have advanced their fields and who have well designed and carefully developed plans for new research. Appropriate fields of specialization include but are not limited to: anthropology, archeology, art history, economics, geography, history, languages and literatures, law, linguistics, musicology, philosophy, political science, psychology, religion, and sociology. Proposals in the social science fields listed are eligible only if they employ predominantly humanistic approaches (e.g., economic history, law and literature, political philosophy). Proposals in interdisciplinary and cross-disciplinary studies are welcome, as are proposals focused on any geographic region or on any cultural or linguistic group. The ultimate goal of the project should be a major piece of scholarly work by the applicant that will take the form of a monograph or other equally substantial form of scholarship. The fellowships support an academic year of research (nine months), plus an additional summer's research (two months) if justified. Fellows have three years to use the funds awarded them. Completed applications must be submitted through the ACLS Online Fellowship Application system.

Requirements: The program is open to tenure-track assistant professors and non-tenured associate professors who by September 30 will have successfully completed their institution's last reappointment review before tenure review (if the institution does not have multi-year contracts, the guideline will mean having passed three annual reappointment reviews), and whose tenure review will not be complete before February 1. Applicants must hold the PhD or equivalent and be employed at academic institutions in the United States, remaining so for the duration of the fellowship.

Restrictions: ACLS does not fund creative work (e.g., novels or films), textbooks, straightforward translation, or pedagogical projects.

Amount of Grant: $64,000, plus $2,500 for research and travel

Date(s) Application is Due: Oct 2

Contact: Nicole Stahlmann, Director of Fellowship Programs; (212) 697-1505, ext. 134; fax: (212) 949-8058; email: nstahlmann@acls.org

Internet: http://www.acls.org/grants/Default.aspx?id=408

Sponsor: American Council of Learned Societies

633 Third Avenue

New York, NY 10017-6795

ACLS Chinese Fellowships for Scholarly Development   263

This program is for scholars in the humanities to do research in the People's Republic of China on China or the Chinese portion of a comparative study. Grants are offered for four to twelve months of research in China. Applicants should demonstrate that they have fully utilized the available resources in the United States and are prepared by virtue of study, training, and planning to take full advantage of an opportunity to do research in China. The fellowship includes a monthly stipend and travel allowance.

Requirements: This program supports individuals with the MA, PhD, or equivalent. Scholars must be U.S. citizens and permanent residents who have lived in the United States continuously for at least three years by the application deadline.

Restrictions: Chinese scholars may not apply directly. Scholars who have previously visited the United States for five months or more, or who are enrolled in degree programs, are not eligible.

Date(s) Application is Due: Nov 14

Contact: Nicole Stahlmann, Director of Fellowship Programs; (212) 697-1505, ext. 134; fax: (212) 949-8058; email: nstahlmann@acls.org

Internet: http://www.acls.org/grants/Default.aspx?id=526

Sponsor: American Council of Learned Societies

633 Third Avenue

New York, NY 10017-6795

ACLS Collaborative Research Awards   264

ACLS invites applications for the inaugural competition for the ACLS Collaborative Research Awards. These awards support collaborative research in the humanities and related social sciences. A grant from The Andrew W. Mellon Foundation supports this program. Collaborations need not be interdisciplinary or inter-institutional, but must involve at least two scholars; applicants at the same institution must demonstrate why local funding is insufficient to support the project. It is hoped that projects of successful applicants will help demonstrate the range and value of both collaborative research and inquiry in the humanities, and model how such collaboration may be carried out successfully. Collaborations that involve the participation of assistant and associate faculty members, or that of scholars at different kinds of institutions, are particularly encouraged. Awards amounts will range from $60,000 to $140,000 in total, depending on the nature and duration of the collaboration, the kinds of expenses projected to carry out the research, and the number of collaborators. Salary-replacement stipends are based on academic rank: up to $35,000 for Assistant Professor and career equivalent; up to $40,000 for Associate Professor and career equivalent; and up to $60,000 for full Professor and career equivalent.

Requirements: The project coordinator must have an appointment at an institution of higher education; other project members may be independent scholars. The project coordinator must be at a U.S.-based institution; other project members may be at institutions outside the United States or may be independent scholars. All project collaborators must hold a Ph.D. degree or its equivalent in publications and professional experience.

Restrictions: ACLS does not fund creative work (e.g., novels or films), textbooks, straightforward translation, or pedagogical projects.

Amount of Grant: $140,000 maximum per project

Date(s) Application is Due: Nov 12

Contact: Nicole Stahlmann, Director of Fellowship Programs; (212) 697-1505, ext. 134; fax: (212) 949-8058; email: nstahlmann@acls.org

Internet: http://www.acls.org/grants/Default.aspx?id=3154

Sponsor: American Council of Learned Societies

633 Third Avenue

New York, NY 10017-6795

ACLS Digital Innovation Fellowships   265

This program invites applications to pursue digitally based research projects in all disciplines of the humanities and humanities-related social sciences. It is hoped that projects of successful applicants will help advance digital humanistic scholarship by broadening understanding of its nature and exemplifying the robust infrastructure necessary for creating further such works. The fellowships are intended to support an academic year dedicated to work on a major scholarly project that takes a digital form. Projects might include but are not limited to digital research archives, new media representations of extant data, innovative databases, and digital tools that further humanistic research. The fellowships are intended as salary replacement and may be held concurrently with other fellowships and grants and any sabbatical pay up to an amount equal to the candidate's current academic year salary. All applications must include the endorsement of a senior administrator of the applicant's institution. This endorsement should include discussion of how the institution's existing cyber infrastructure complements and supports the technologies to be developed for the specified project. Completed applications must be submitted through the ACLS online application system. Guidelines are available online.

Requirements: This program is open to scholars in all fields of the humanities and the humanistic social sciences. Applicants must have a PhD degree conferred prior to the application deadline. (An established scholar who can demonstrate the equivalent of the Ph.D. in publications and professional experience may also qualify.) U.S. citizenship or permanent resident status is required as of application deadline.

Restrictions: ACLS does not support creative works (e.g., novels or films), textbooks, straightforward translations, or purely pedagogical projects.

Amount of Grant: $60,000 maximum stipend; $25,000 maximum project costs

Date(s) Application is Due: Oct 2

Contact: Nicole Stahlmann, Director of Fellowship Programs; (212) 697-1505, ext. 134; fax: (212) 949-8058; email: nstahlmann@acls.org

Internet: http://www.acls.org/grants/Default.aspx?id=508

Sponsor: American Council of Learned Societies

633 Third Avenue

New York, NY 10017-6795

ACLS Dissertation Fellowships in East European Studies   266

Dissertation fellowships are offered in the social sciences and humanities relating to Albania, Bulgaria, Czech Republic, Estonia, Slovakia, Hungary, Latvia, Lithuania, Poland, Romania, and the former Yugoslavia. Doctoral candidates may apply for support of dissertation research or writing to be undertaken at any university or institution outside of East Europe, although short visits to the area may be proposed as part of a coherent program primarily based elsewhere. Fellowships require an academic year of nine to 12 months devoted solely to work on the dissertation.

Requirements: Currently enrolled graduate students who will have completed all requirements for the doctorate except the dissertation by June may apply for one-year, non-renewable support to complete the dissertation.

Amount of Grant: $18,000 maximum
Date(s) Application is Due: Nov 12
Contact: Nicole Stahlmann, Director of Fellowship Programs; (212) 697-1505, ext. 134; fax: (212) 949-8058; email: nstahlmann@acls.org
Internet: http://www.acls.org/grants/Default.aspx?id=532
Sponsor: American Council of Learned Societies
633 Third Avenue
New York, NY 10017-6795

ACLS Early-Career Postdoctoral Fellowships in East European Studies   267

Support is provided for six to 12 months of postdoctoral research and training in the social sciences and humanities relating to Albania, Bulgaria, Czech Republic, Estonia, Hungary, Latvia, Lithuania, Poland, Romania, Slovakia, and the successor states of Yugoslavia. Scholars may apply for fellowships for periods of six to 12 consecutive months of full-time research and writing between July 1 and September 1 of the following year. The fellowships are intended primarily as salary replacement to provide free time for research; the funds may be used to supplement sabbatical salaries or awards from other sources, provided they intensify or extend the contemplated research. Five to seven fellowships are awarded each year.

Requirements: An applicant must: be a citizen or permanent resident of the United States as of the application deadline date; hold a Ph.D. degree conferred prior to the application deadline (however, an established scholar who can demonstrate the equivalent of the Ph.D. in publications and professional experience may also qualify); and be at an early career stage.

Restrictions: These fellowships are to be used for work outside East Europe, although short visits to the area may be proposed as part of a coherent program primarily based elsewhere. Tenured faculty are not eligible.

Amount of Grant: $25,000 maximum
Date(s) Application is Due: Nov 12
Contact: Nicole Stahlmann, Director of Fellowship Programs; (212) 697-1505, ext. 134; fax: (212) 949-8058; email: nstahlmann@acls.org
Internet: http://www.acls.org/grants/Default.aspx?id=534
Sponsor: American Council of Learned Societies
633 Third Avenue
New York, NY 10017-6795

ACLS East European East Heritage Speakers Research Grant   268

A grant of $20,000 will be made to an individual or collaborative team for a research project on heritage speakers of an East European language in the United States. Heritage speakers constitute an untapped national resource for academic and policy research, but there has been little study of their special needs for bringing their linguistic skills up to a professional level. The project should culminate in: an analytical paper; and a proposal for an advanced-mastery course for community of heritage speakers that was studied. There is no formal application form.

Amount of Grant: $20,000
Date(s) Application is Due: Jan 16
Contact: Nicole Stahlmann, Director of Fellowship Programs; (212) 697-1505, ext. 134; fax: (212) 949-8058; email: nstahlmann@acls.org; Olga Bukhina, Coordinator of International Programs; (212) 697-1505, ext. 130; fax: (212) 949-8058; email: obukhina@acls.org
Internet: http://www.acls.org/grants/Default.aspx?id=3152
Sponsor: American Council of Learned Societies
633 Third Avenue
New York, NY 10017-6795

ACLS East European Language Training Grants   269

Individuals may apply for grants for summer study of Albanian, Bosnian-Croatian-Serbian, Bulgarian, Czech, Hungarian, Macedonian, Polish, Romanian, Slovak, or Slovene. Applicants should present proposals for attendance at intensive courses offered by institutions of higher education in the U.S., or, in exceptional cases, for study at the advanced level in courses in Eastern Europe. These grants are intended for people who will use the East European languages in academic research or teaching. In addition, grants will be available to U.S. institutions to support intensive summer course instruction in Albanian, Bosnian-Croatian-Serbian, Bulgarian, Czech, Estonian, Hungarian, Latvian, Lithuanian, Macedonian, Polish, Romanian, Slovak, or Slovene.

Requirements: Applicants must be citizens or permanent residents of the United States, and must have completed, at minimum, a four-year college degree. They may request support for beginning, intermediate, or advanced language study, and the application will request the name of the institution they wish to attend, along with a statement of the significance of this training for their career plans.

Amount of Grant: $2,000 maximum, individuals; $10,000 maximum, organizations
Date(s) Application is Due: Jan 16
Contact: Nicole Stahlmann, Director of Fellowship Programs; (212) 697-1505, ext. 134; fax: (212) 949-8058; email: nstahlmann@acls.org
Internet: http://www.acls.org/grants/Default.aspx?id=542
Sponsor: American Council of Learned Societies
633 Third Avenue
New York, NY 10017-6795

ACLS East European Studies Program Travel Grants   270

Pending confirmation of funding, grants will be made to support travel for presentation of papers at scholarly conferences taking place before December 31, ($1,000 for travel within North America; $2,500 for intercontinental travel). The purpose of these grants is to build the East European (EE) field and to encourage comparative study that includes East European studies. Topics of papers must be related to Eastern Europe, but applications are welcome from EE specialists proposing to attend conferences whose general theme lies outside Eastern Europe as well as from scholars whose area expertise lies outside EE who propose to attend an EE-focused event.

Requirements: Applicants must be a citizens or permanent residents of the United States.
Restrictions: Funds may not be used in Western Europe.
Amount of Grant: $1,000-$2,500
Date(s) Application is Due: Jan 30
Contact: Nicole Stahlmann, Director of Fellowship Programs; (212) 697-1505, ext. 134; fax: (212) 949-8058; email: nstahlmann@acls.org
Internet: http://www.acls.org/grants/Default.aspx?id=538
Sponsor: American Council of Learned Societies
633 Third Avenue
New York, NY 10017-6795

ACLS Fellowships   271

The program invites research applications in all disciplines of the humanities and humanities-related social sciences. Appropriate fields of specialization include but are not limited to: anthropology, archeology, art history, economics, geography, history, languages and literatures, law, linguistics, musicology, philosophy, political science, psychology, religion, and sociology. Proposals in the social science fields listed are eligible only if they employ predominantly humanistic approaches (e.g., economic history, law and literature, political philosophy). Proposals in interdisciplinary and cross-disciplinary studies are welcome, as are proposals focused on any geographic region or on any cultural or linguistic group. The fellowships are intended as salary replacement and may be held concurrently with other fellowships, grants and any sabbatical pay to reach that goal.

Requirements: The applicant must have had a PhD degree conferred two years prior to the listed application deadline. An established scholar who can demonstrate the equivalent of the PhD in publications and professional experience may also apply. The applicant must have U.S. citizenship or permanent resident status as of the application deadline. The most recent supported research leave must have concluded two years prior to July 1. (Supported research leave is defined as the equivalent of one semester or more of time free from teaching or other employment to pursue scholarly research or writing supported by sabbatical pay or other institutional funding, fellowships and grants, or a combination of these.) This definition will apply to independent scholars as well as those with institutional affiliations.

Restrictions: ACLS does not fund creative work (e.g., novels or films), textbooks, straightforward translation, or pedagogical projects.
Amount of Grant: $60,000 maximum, full professor and career equivalent; $40,000 maximum, associate professor and career equivalent; $30,000 maximum, assistant professor and equivalent
Date(s) Application is Due: Oct 2
Contact: Nicole Stahlmann, Director of Fellowship Programs; (212) 697-1505, ext. 134; fax: (212) 949-8058; email: nstahlmann@acls.org
Internet: http://www.acls.org/grants/Default.aspx?id=380
Sponsor: American Council of Learned Societies
633 Third Avenue
New York, NY 10017-6795

ACLS Frederick Burkhardt Residential Fellowships for Recently   272
Tenured Scholars

The objectives of this program are to encourage more adventurous, more wide-ranging, and longer term patterns of research than are current in these disciplines; to link a small number of outstanding scholars and their projects to one of a limited number of residential study centers with an established record of advancing multi-disciplinary scholarship; and to sustain the scholarly momentum of the emerging intellectual leaders in fields of the humanities and related social sciences. Appropriate fields of specialization include but are not limited to: anthropology, archaeology, art history, economics, geography, history, languages and literatures, law, linguistics, musicology, philosophy, political science, psychology, religion, and sociology. Proposals in the social science fields listed above are eligible only if they employ predominantly humanistic approaches (e.g., economic history, law and literature, political philosophy). Proposals in interdisciplinary and cross-disciplinary studies are welcome, as are proposals focused on any geographic region or on any cultural or linguistic group. Fellowships are intended to support an academic year (normally nine months) of residence at any one of nine national residential research centers. Applications must be submitted online; guidelines also are available online.

Requirements: The Burkhardt Fellowship Program is open to recently tenured humanists‚Äîscholars who will have begun their first tenured contracts by the application deadline but began their first tenured contracts no earlier than the fall 2002 semester or quarter. An applicant must be employed in a tenured position at a degree-granting academic institution in the U.S., remaining so for the duration of the fellowship. U.S. citizenship or permanent residency is not required, and previous supported research leaves do not affect eligibility for the Burkhardt Fellowship.

Restrictions: ACLS does not fund creative work (e.g., novels or films), textbooks, straightforward translation, or pedagogical projects.
Amount of Grant: $75,000
Date(s) Application is Due: Oct 2

Contact: Nicole Stahlmann, Director of Fellowship Programs; (212) 697-1505, ext. 134; fax: (212) 949-8058; email: nstahlmann@acls.org
Internet: http://www.acls.org/grants/Default.aspx?id=480
Sponsor: American Council of Learned Societies
633 Third Avenue
New York, NY 10017-6795

ACLS New Perspectives on Chinese Culture and Society Grants   273
This program is intended to support projects that bridge disciplinary or geographic boundaries, engage new kinds of information, develop fresh approaches to traditional materials and issues, or otherwise bring innovative perspectives to the study of Chinese culture and society. The program awards funds in support of planning meetings, workshops, and/or conferences leading to publication of scholarly volumes. Proposals are expected to be empirically grounded, theoretically informed, and methodologically explicit. The program will support collaborative work of three types: formal research conferences intended to produce significant new research published in a conference volume; workshops or seminars, designed to less formally facilitate new research on newly available or inadequately researched problems, data, or texts; and planning meetings to organizers of the above-described types of projects. There are no application forms. Guidelines are available online.
Requirements: The program aims to provide opportunities for interchange among scholars who may not otherwise have chances to work with one another.
Restrictions: Proposals for activities that involve scholars primarily from one institution and that fall within an institution's normal range of colloquia, symposia, or seminar series will not be supported. In addition, the program does not normally support regular scheduled meetings, conventions, or parts thereof. Funds cannot be used to cover the following expenses: direct research expenses, such as those of field work, obtaining research materials, or analysis of data; salaries or released time for organizers or participants; honoraria or speaker's fees for organizers or U.S. participants; purchase of equipment; or institutional overhead (direct administrative costs are allowed).
Amount of Grant: $25,000 maximum for formal research conferences; $15,000 maximum for workshops or seminars; $6000 for planning meetings
Date(s) Application is Due: Nov 12
Contact: Nicole Stahlmann, Director of Fellowship Programs; (212) 697-1505, ext. 134; fax: (212) 949-8058; email: nstahlmann@acls.org
Internet: http://www.acls.org/grants/Default.aspx?id=528
Sponsor: American Council of Learned Societies
633 Third Avenue
New York, NY 10017-6795

ACM A.M. Turing Award   274
This award is the association's most prestigious technical award and is given to an individual selected for contributions of a technical nature made to the computing community. The contributions should be of lasting importance to the computer field. Nomination forms should be obtained from ACM headquarters.
Requirements: Each nomination should consist of name, address, and phone number of person making the suggestion; name and address of candidate; statement of 200 to 500 words explaining why the candidate deserves the award; and if possible, names and addresses or telephone numbers of others who agree with the recommendation.
Samples: Francis E. Allen—award recipient, $100,000 (2006); Vinton Cerf—award recipient, $100,000 (2004); Robert Kahn—award recipient, $100,000 (2004).
Amount of Grant: $100,000
Contact: Rosemary McGuinness, Headquarters Liaison, (212) 626-0561; email: mcguinness@acm.org
Internet: http://www.acm.org/awards/taward.html
Sponsor: Association for Computing Machinery
2 Penn Plaza, Suite 701
New York, NY 10121-0701

ACM Doctoral Dissertation Award   275
This award is given to the individual who has been judged to have prepared the most outstanding doctoral dissertation on computer science and engineering during the previous year. Dissertations will be reviewed for technical depth and significance of the research contribution, potential impact on theory and practice, and quality of presentation. A committee of five individuals serving staggered five-year terms performs an initial screening to generate a short list, followed by an in-depth evaluation to determine the winning dissertation. Nomination forms are available from ACM headquarters. The winning dissertation is published by Springer.
Requirements: Each nominated dissertation must have been accepted by the department within a 12-month period prior to August 31. Exceptional dissertations completed in August, but too late for submission last year will be considered. Only English language versions will be accepted.
Restrictions: Nominations are limited to one per university or college, from any country, unless more than 10 Ph.D.s are granted in one year, in which case two may be nominated.
Samples: Yi-Ren Ng—award recipient, $5000 (2006); Ben Liblit—award recipient, $5000 (2005) Boaz Barak—award recipient, $5000 (2004); AnHai Doan—award recipient, $5000 (2003).
Amount of Grant: $5000
Date(s) Application is Due: Aug 31
Contact: Priya Narasimhan, (412) 268-8801; fax: (412) 268-1194; email: Priya@cs.cmu.edu
Internet: http://awards.acm.org/homepage.cfm?awd=146

Sponsor: Association for Computing Machinery
2 Penn Plaza, Suite 701
New York, NY 10121-0701

ACM Eckert-Mauchly Award   276
This award is given for contributions to computer and digital systems architecture where the field of computer architecture is considered at present to encompass the combined hardware-software design and analysis of computing and of digital systems. The award is administered jointly by ACM and the IEEE Computer Society.
Requirements: Nomination should consist of name, address, and phone number of person making the suggestion; name and address of candidate; a statement of 200-500 words explaining why the candidate deserves the award; and, if possible, the names and addresses or telephone numbers of others who agree with the recommendation.
Samples: James H. Pomerene—award recipient, $5000 (2006); Robert P. Colwell—award recipient, $5000 (2005); Frederick Brooks, Jr.—award recipient, $5000 (2004); Joseph Fisher—award recipient, $5000 (2003).
Amount of Grant: $5000
Date(s) Application is Due: Oct 31
Contact: Alan Berenbaum, (631) 435-6480; email: aberenbaum@acm.org
Internet: http://awards.acm.org/homepage.cfm?awd=148
Sponsor: Association for Computing Machinery
2 Penn Plaza, Suite 701
New York, NY 10121-0701

ACM Grace Murray Hopper Award   277
This award is given to the outstanding young computer professional of the year, selected on the basis of a single recent major technical or service contribution. This candidate must have been 35 years of age or less at the time the qualifying contribution was made. Nominating forms should be obtained from ACM headquarters. This award is provided by Google.
Requirements: Each nomination submitted should consist of name, address, and phone number of person making the nomination; name and address of candidate; a statement of 200-500 words explaining why the candidate deserves the award; and if possible, the names and addresses or telephone numbers of others who agree with the recommendation.
Samples: Daniel Klein—award recipient, $15,000 (2006); Omer Reingold—award recipient, $15,000 (2005); Jennifer Rexford—award recipient, $15,000 (2004); Stephen Keckler—award recipient, $15,000 (2003).
Amount of Grant: $15,000
Date(s) Application is Due: Oct 31
Contact: Gabriel (Gabby) Silberman, (212) 415-6908; fax: (631) 342-5013; email: Gabby.Silberman@ca.com
Internet: http://awards.acm.org/homepage.cfm?awd=145
Sponsor: Association for Computing Machinery
2 Penn Plaza, Suite 701
New York, NY 10121-0701

ACM Karl V. Karlstrom Outstanding Educator Award   278
This award is presented annually to an outstanding educator who is appointed to a recognized educational baccalaureate institution, recognized for advancing new teaching methodologies or effecting new curriculum development or expansion in computer science and engineering, or making a significant contribution to the educational mission of the ACM. Those who have been teaching for 10 years or less will be given special consideration. The prize is supplied by Prentice-Hall Publishing Company.
Requirements: Each nomination should consist of name, address, and phone number of person making the suggestion; name and address of candidate; statement of 200 to 500 words explaining why the candidate deserves the award; and if possible, names and addresses or telephone numbers of others who agree with the recommendation.
Samples: Stuart J. Russell—award winner, $5000, (2005); Sartaj Sahni—award winner, $5000, (2003); John Gorgone—award winner, $5000, (2002).
Amount of Grant: $5,000
Date(s) Application is Due: Oct 31
Contact: Jeffrey Ullman, (650) 494-8016; email: ullman@gmail.com
Internet: http://awards.acm.org/homepage.cfm?awd=147
Sponsor: Association for Computing Machinery
2 Penn Plaza, Suite 701
New York, NY 10121-0701

ACM Paris Kanellakis Theory and Practice Award   279
The award honors specific theoretical accomplishments that have had a significant and demonstrable effect on the practice of computing. This award is endowed by contributions from the Kanellakis family, with additional financial support provided by ACM's Special Interest Groups on Algorithms and Computational Theory (SIGACT), Design Automaton (SIGDA), Management of Data (SIGMOD), and Programming Languages (SIGPLAN), the ACM SIG Projects Fund, and individual contributions.
Requirements: Each nomination should consist of name, address, and phone number of person making the suggestion; name and address of candidate; statement of 200 to 500 words explaining why the candidate deserves the award; and if possible, names and addresses or telephone numbers of others who agree with the recommendation.
Samples: Robert Brayton—award recipent, $5000 (2006); Yoav Freund, Columbia University, and Robert Schapire, Princeton University—award recipients, $5000 (2004); Gary Miller, Michael Rabin, Robert Solovay, Volker Strassen—award recipents, $5000 (2003); Peter Franaszek—award recipients, $5000.

Amount of Grant: $5,000
Date(s) Application is Due: Oct 31
Contact: Anna Karlin, (206) 543-9344; email: karlin@cs.washington.edu
Internet: http://awards.acm.org/homepage.cfm?awd=147
Sponsor: Association for Computing Machinery
2 Penn Plaza, Suite 701
New York, NY 10121-0701

ACM Software System Award 280

Provided by IBM, this award goes to an institution or individual(s) recognized for developing a software system that has had a lasting influence, reflected in contributions to concepts, in commercial acceptance, or both.
Requirements: Each nomination should consist of name, address, and phone number of person making the suggestion; name and address of candidate; statement of 200 to 500 words explaining why the candidate deserves the award; and if possible, names and addresses or telephone numbers of others who agree with the recommendation.
Samples: Bertrand Meyer—award recipient, $10,000 (2006); Robert S. Boyer, Matt Kaufmann, and J. Strother Moore—award recipients, $10,000 (2005); Raghuram Bindignavle, Simon Lam, Shaowen Su, and Thomas Y. C. Woo—award recipients, $10,000 (2004); Stuart Feldman—award recipient, $10,000 (2003); James Gosling—award recipient, $10,000.
Amount of Grant: $10,000
Date(s) Application is Due: Oct 31
Contact: Frank Tompa, (519) 888-4567; email: fwtompa@uwaterloo.ca
Internet: http://awards.acm.org/homepage.cfm?awd=149
Sponsor: Association for Computing Machinery
2 Penn Plaza, Suite 701
New York, NY 10121-0701

ACOG History of American Obstetrics and Gynecology Fellowship 281

ACOG and Ortho-McNeil Pharmaceutical Corporation jointly sponsor the fellowship in the History of American Obstetrics and Gynecology each year. The recipient of the fellowship spends one month in the Washington, DC, area working full-time to complete their specific historical research project. Although the fellowship will be based in the ACOG History Library, the fellow is encouraged to use other national, historical, and medical collections in the Washington, DC, area. The results of this research must be disseminated through either publication or presentation at a professional meeting. Application and guidelines are available online.
Requirements: ACOG junior fellows and fellows are encouraged to apply.
Amount of Grant: $5000
Date(s) Application is Due: Oct 1
Contact: Debra Scarborough, (202) 863-2578; fax: (202) 484-1595; email: history@acog.org
Internet: http://www.acog.org/from_home/departments/dept_notice.cfm?recno=21&bulletin=254
Sponsor: American College of Obstetricians and Gynecologists
409 12th Street SW
Washington DC 20024-2588

ACRL Doctoral Dissertation Fellowship 282

This fellowship, sponsored by the Thomsn Scientific, is intended to foster research in academic librarianship by encouraging and assisting doctoral students in the field with their dissertation research. Qualified students who wish to be considered for the fellowship may apply by submitting a brief proposal that includes the following: a description of the research, including significance and methodology; a schedule; a budget; the name of dissertation advisor and committee members; a cover letter from dissertation advisor endorsing the proposal; and an up-to-date curriculum vita.
Requirements: An applicant must be an active doctoral student in the academic librarianship area in a degree-granting institution, have completed all coursework, and have had a dissertation proposal accepted by the institution. The applicant need not be an ACRL member.
Restrictions: Recipients of the fellowship may not receive it a second time.
Samples: Jean E. Dryden—award recipient (2007); Susan Ward Aber—award recipient (2005); Judy Jeng—award recipient (2004); Yung-Rang Laura Cheng—award recipient (2003); Charlotte Ford—award recipient, $1000 (2002); Laurie Bonnici—award recipient, $1000.; Alenka Sauperl—award recipient, $1500.
Amount of Grant: $1500
Date(s) Application is Due: Dec 1
Contact: Megan Bielefeld, (312) 280-2514; email: mbielefeld@ala.org
Internet: http://www.ala.org/Template.cfm?Section=Grants_and_Fellowships&template=/ContentManagement/ContentDisplay.cfm&ContentID=29778
Sponsor: Association of College and Research Libraries
50 E. Huron Street
Chicago, IL 60611

ACRL Miriam Dudley Librarian Instruction Award 283

This grant recognizes a significant contribution by a librarian to the advancement of instruction in a college or research library environment. Nominees should have achieved distinction in one or more of the following areas: planning and implementation of an academic instruction program in a library environment that has served as a model for other programs nationally or regionally; production of a body of research and publication that has a demonstrable impact on the concepts and methods of teaching

and information-seeking strategies in a college or research institution; sustained participation in organizations, at the national or regional level, devoted to the promotion and enhancement of academic instruction in a library environment; or promotion, development, and integration of education for instruction in ALA-accredited library schools or professional continuing education programs that have served as models for other courses and programs. Nominations must include a letter including the name, address, and phone number of the nominator and the nominee; a narrative statement supporting the nomination and addressing how the nominee meets the criteria for the award; and nominee's current vita. Send nominating package to: IS Miriam Dudley Award chair, Lisa J. Hinchliffe, Coordinator for Information Literacy, University of Illinois, 1408 W. Gregory Drive, Library 434, Urbana, IL 61801.
Restrictions: Nominees will be judged on an individual basis; this award cannot be given to a pair or group of persons.
Samples: Mary Jane Petrowski—award recipient, (2006); Ilene F. Rockman—award recipient, $1000 (2005); William Miller—award recipient, $1000 (2004); Loanne Snavely—award recipient, $1000 (2003); Randy Hensley—award recipient, $1000.; Patricia Iannuzzi—award recipient, $1000.; Carol Collier Kuhlthau—award recipient, $1000.
Amount of Grant: $1000
Date(s) Application is Due: Dec 1
Contact: Megan Bielefeld, Program Officer, (800) 545-2433 ext. 2514; email: mbielefeld@ala.org
Internet: http://www.ala.org/ala/acrl/acrlawards/miriamdudley.htm
Sponsor: Association of College and Research Libraries
50 E Huron Street
Chicago, IL 60611-2795

ACS Master's Training Grants in Clinical Oncology Social Work 284

These grants are awarded annually to qualifying institutions that train clinical oncology social workers to provide cancer patients and their families with psychosocial services. Grants are available to second-year students in a master's program and post-master's social workers within five years of graduation. The master's training must introduce social workers to the special needs of cancer patients and their families. Contact the society for application forms.
Requirements: Applications will be considered from institutions that identify clinical oncology social work training activities, cancer programs with defined psychosocial support services, and relationships with schools accredited by the Council on Social Work Education.
Amount of Grant: $12,000 maximum
Date(s) Application is Due: Oct 15
Contact: Extramural Grants Department, (404) 329-7558; fax: (404) 321-4669; email: grants@cancer.org
Internet: http://www.cancer.org/docroot/RES/content/RES_5_2x_Masters_Training_Grants_in_Clinical_Oncology_Social_Work.asp?sitearea=RES
Sponsor: American Cancer Society
1599 Clifton Rd NE
Atlanta, GA 30329-4251

ACS Audrey Meyer Mars International Fellowships in Clinical Oncology 285

The purpose of the fellowships is to provide one year of advanced training in clinical oncology at participating cancer centers in the United States to qualified physicians and surgeons from other countries, particularly countries where advanced training is not readily available. Training will be conducted at one of the United States cancer centers participating in the program; a list of these centers is available on the Web site.
Requirements: Eligible applicants must be qualified physicians or surgeons who have demonstrated an interest in clinical cancer management and who desire advanced training in clinical oncology; be accepted for training by one of the participating institutions and must have fulfilled all requirements of the institution and of the state in which the institution is located; and must have passed the ECFMG or the Test of English as a Foreign Language (TOEFL). Institutions should review applications to determine whether the applicant meets all institutional requirements and would be accepted for training; forward the completed nomination form to ACS by January 15 of the year in which the fellowship is to begin.
Restrictions: Applications for training in basic cancer research will not be accepted for the award. Institutions should not submit applications for candidates they would not accept for training and must agree not to recruit fellows sponsored by the program for permanent positions.
Amount of Grant: $45,000 maximum
Date(s) Application is Due: Dec 1
Contact: Extramural Grants Program, (404) 329-7558; fax: (404) 321-4669; email: grants@cancer.org
Internet: http://www.cancer.org/docroot/res/content/res_5_2x_audrey_meyer_mars_international_fellowships_in_clinical_oncology.asp?sitearea=res
Sponsor: American Cancer Society
1599 Clifton Rd NE
Atlanta, GA 30329-4251

ACS Cancer Control Career Development Awards for Primary Care Physicians 286

ACS annually awards three three-year career development awards to physicians specializing in primary care (e.g., family practice, internal medicine, pediatrics, and obstetrics and gynecology). These awards are intended to encourage and assist in the development of promising candidates who will pursue academic careers in primary care specialties. It is anticipated that physicians trained under these awards will improve cancer control through involvement in primary care practice, education, and research activities related to cancer control. Awards are made for three years with progressive stipends.

Requirements: Candidates must be U.S. citizens or permanent residents; must hold an MD, DO, or equivalent degree; and must have completed residency requirements of the appropriate primary care specialty board. Candidate must hold academic rank from instructor to assistant professor and must not be more than 10 years out of training.

Restrictions: Applicants may not have academic rank above that of assistant professor and must not be tenured or be the section head (or equivalent) in his/her discipline. Additionally, the applicant may not be training as a fellow at the time of the award.

Amount of Grant: $50,000 first year, $55,000 second year, $60,000 third year; stipends paid in monthly installments

Date(s) Application is Due: Oct 15

Contact: Extramural Grants Department, (404) 329-7558 or (800) 875-2562; fax: (404) 321-4669; email: grants@cancer.org

Internet: http://www.cancer.org/docroot/RES/content/RES_5_2x_Cancer_Control_Career_Development_Awards_for_Primary_Care_Physicians.asp?sitearea=RES

Sponsor: American Cancer Society
1599 Clifton Rd NE
Atlanta, GA 30329-4251

ACS Doctoral Scholarships in Cancer Nursing 287

Doctoral degree scholarships in cancer nursing are awarded to graduate students pursuing doctoral study in the following cancer nursing fields: research, education, administration, or clinical practice. Awards are made for up to four years.

Requirements: Applicants must be currently enrolled in or applying to a doctoral degree program and have current licensure to practice as a registered nurse. American citizens or permanent residents are eligible.

Amount of Grant: $15,000 per year

Date(s) Application is Due: Dec 1

Contact: Extramural Grants Department, (404) 329-7558; fax: (404) 321-4669; email: grants@cancer.org

Internet: http://www.cancer.org/docroot/RES/content/RES_5_2x_Doctoral_Degree_Scholarships_in_Cancer_Nursing.asp?sitearea=RES

Sponsor: American Cancer Society
1599 Clifton Rd NE
Atlanta, GA 30329-4251

ACS Institutional Research Grants 288

The IRG program provides seed money for the initiation of promising new projects by independent junior investigators so they may obtain preliminary results that will enable them to compete successfully for national research grants.

Requirements: Only full-time, tenure-track JHU faculty members at the ranks of Instructor or Assistant Professor or equivalent who are within the first six years of their first independent research or faculty appointment may apply. Relevance to the cancer problem is important and must be well documented. Within this research scope, a wide breadth of research approaches will be considered; these include basic laboratory studies and clinical research.

Restrictions: Senior investigators, postdoctoral fellows and junior investigators who have competitive national research grants or who have received prior support form the IRG are not eligible.

Amount of Grant: $30,000 per year

Date(s) Application is Due: Mar 1

Contact: Grace Bigelow, Research Projects Administration, (410) 516-5256; fax: (410) 516-7775; email: graceb@jhu.edu

Internet: http://www.cancer.org/docroot/RES/content/RES_5_2x_Institutional_Research_Grants.asp?sitearea=RES

Sponsor: American Cancer Society
3100 Wyman Park Drive, Suite W400
Baltimore, MD 21211

ACS Postdoctoral Fellowships 289

Fellowships are designed to enable a new investigator to qualify for an independent career in cancer research. Postdoctoral fellowships may be made for one, two, or three years. Stipend payments are made directly to the individual, or the institution if requested, at the beginning of each month. Travel funds are paid to the individual. Institutional allowances are paid annually at the start of the grant and on the anniversary date thereafter. Awards are made for one to three years with progressive stipends. Application forms are available upon request; call (404) 329-7558.

Requirements: Applicants must be citizens or permanent residents of the United States (the latter must provide notarized evidence of their legal resident alien status) and shall have been awarded a doctoral degree prior to the activation date of the grant. Application must be endorsed by applicant's mentor and the head of the department in which the training will be received. A plan of training must be formulated and agreed upon by the mentor and the applicant and described in detail in the application.

Restrictions: Awards will not be made to applicants who have completed five or more years of postdoctoral training prior to the start date of the fellowship.

Amount of Grant: $40,000, $42,000, and $44,000 for the first, second, and third years, respectively, and $4000 maximum per year institutional allowance

Date(s) Application is Due: Apr 1; Oct 15

Contact: Extramural Grants Department, (404) 329-7558; fax: (404) 321-4669; email: grants@cancer.org

Internet: http://www.cancer.org/docroot/RES/content/RES_5_2x_Postdoctoral_Fellowships.asp?sitearea=RES

Sponsor: American Cancer Society
1599 Clifton Rd NE
Atlanta, GA 30329-4251

ACS Research Scholar Grants for Beginning Investigators 290

The grants support basic, preclinical, or epidemiologic research projects initiated by investigators in the first eight years of independent cancer research careers. There is no deadline; however, a letter of intent is required to obtain applications and instructions.

Amount of Grant: $72,000 maximum

Contact: Research Scholar Grants for Beginning Investigators, (404) 329-7558; fax: (404) 321-4669; email: grants@cancer.org

Internet: http://www.cancer.org/docroot/RES/RES_0.asp

Sponsor: American Cancer Society
1599 Clifton Rd NE
Atlanta, GA 30329

ACS Research Scholar Grants for Health Services and Health Policy Research 291

The grants support research projects centered on health services and health policy initiated by investigators at any stage in their careers. The initial award includes 20 percent for indirect costs for up to four years, and may be renewed once for up to four years.

Amount of Grant: $200,000 maximum per year

Date(s) Application is Due: Apr 1; Oct 15

Contact: Extramural Grants Department, (404) 329-7558; fax: (404) 321-4669; email: grants@cancer.org

Internet: http://www.cancer.org/docroot/RES/content/RES_5_2x_Research_Scholar_Grants_For_Health_Services_and_Health_Policy_and_Outcomes_Research.asp?sitearea=RES

Sponsor: American Cancer Society
1599 Clifton Rd NE
Atlanta, GA 30329-4251

ACS Research Scholar Grants in Psychosocial and Behavioral and Cancer Control Research 292

The grants support research projects focusing on the psychosocial and behavioral aspects of cancer by independent investigators at any stage in their careers. Applications are encouraged in which an individual at an early career stage is coprincipal investigator with an established researcher. The initial award supports four years of research and may be renewed once for four additional years.

Requirements: Individuals must not have held an independent position for more than six years at the time the application is submitted.

Amount of Grant: $400,000 maximum

Date(s) Application is Due: Apr 1; Oct 15

Contact: Extramural Grants Department, (404) 329-7558; fax: (404) 321-4669; email: grants@cancer.org

Internet: http://www.cancer.org/docroot/RES/content/RES_5_2x_Research_Scholar_Grants_in_Psychological_and_Behavioral_Research_for_Beginning_and_Senior_Investigators.asp?sitearea=RES

Sponsor: American Cancer Society
1599 Clifton Rd NE
Atlanta, GA 30329-4251

ACSV Artist Fellowships 293

Artist fellowships are available in the categories of visual arts and literary arts. The deadline for visual arts is November 8 and the deadline for literary arts is November 15. Annual deadline dates may vary; contact program staff for exact dates.

Requirements: Applicants must be a resident of Santa Clara County for at least one year prior to the application deadline; OR be a resident of an immediately bordering county for at least one year who can demonstrate through professional history that artistic activity has had significant impact on residents of Santa Clara County.

Geographic Focus: California

Amount of Grant: $4000

Date(s) Application is Due: Nov 8; Nov 15

Contact: Diem Jones, Director of Grants Program, (408) 998-2787 ext 207; fax: (408) 971-9458; email: djones@artscouncil.org

Internet: http://www.artscouncil.org/af.htm

Sponsor: Arts Council of Silicon Valley
4 N Second Street, Suite 210
San Jose, CA 95113-1305

ACT Awards 294

The program is designed to benefit programs that aim to help middle and high school students reach their career and educational goals. The awards target programs that work with students whose circumstances and environments could keep them from continuing their education or career training. Winners receive awards in the form of ACT programs, services, consultation and resources. Awards are provided based on the benefit a program is likely to offer its defined at-risk population; overall program design; how readily it could be replicated elsewhere; and how much support a program has.

Requirements: Proposed projects may be programs or studies undertaken with at-risk students in grades eight through 12 or in two- and four-year colleges and universities.

Restrictions: Awards do not include salaries or stipends for award recipients or other personnel, travel expenses, conference fees, tuition costs, or any similar types of direct expenses.
Amount of Grant: $6000-$8000
Date(s) Application is Due: May 15
Contact: Director of Policy Research, (319) 337-1293; email: noethr@act.org
Internet: http://www.act.org
Sponsor: ACT Inc
P.O. Box 168, 500 ACT Drive
Iowa City, IA 52243-0168

ACT Summer Internship Program 295
The summer internship program is offered annually to outstanding graduate students interested in careers in testing and measurement. The eight-week program, which runs from early June through the end of July at ACT's national headquarters in Iowa City, IA, will provide interns with practical experience through direct interaction with professional and administrative staff responsible for research and development of ACT's testing programs. Internships will be offered in five categories including test development, I/O psychology, sychometric and statistical analysis, vocational psychology, and market research. Candidates may apply for more than one category, if interested and qualified. Interns selected to participate will be paid a stipend plus round-trip transportation costs. A supplemental living allowance for accompanying spouse and/or dependents is also available. Application forms are available on the Web site.
Requirements: Graduate students enrolled in master's or doctoral programs in educational measurement, evaluation, mathematical and applied statistics, research, industrial/organizational psychology, and related fields are eligible. To apply, applicants must submit a completed application form, resume, a clear copy of graduate and undergraduate transcripts, and two letters of recommendation from persons who are familiar with the applicant's academic work and professional background.
Amount of Grant: $5000 stipend
Date(s) Application is Due: Feb 13
Contact: Human Resources Dept, (319) 337-1763; email: working@act.org
Internet: http://www.act.org/humanresources/jobs/intern.html
Sponsor: American College Testing
P.O. Box 168, 500 ACT Drive
Iowa City, IA 52243-0168

Acton Institute for the Study of Religion and Liberty Calihan Academic Fellowships 296
The Calihan Academic Fellowships provide scholarships and research grants to future scholars and religious leaders whose academic work shows outstanding potential. Applicants studying theology, philosophy, economics, or related fields must demonstrate the potential to advance understanding in the relationship between theology and the principles of the free and virtuous society. Such principles include recognition of human dignity, the importance of the rule of law, limited government, religious liberty, and freedom in economic life. Awards are open to seminarians or graduate students in theology, philosophy, religion, economics, or in related fields.
Requirements: Strong academic performance is essential. Candidates must: have a demonstrated interest in the themes of the Acton Institute; and display the potential to advance the themes of the Acton Institute.
Amount of Grant: $3,000 maximum
Date(s) Application is Due: Jul 15; Oct 31
Contact: Kara Eagle, Programs Officer; (616) 454-3080; fax: (616) 454-9454; email: keagle@acton.org
Internet: http://www.acton.org/programs/students/calihan_academic_scholarship.php
Sponsor: Acton Institute for the Study of Religion and Liberty
161 Ottawa NW, Suite 301
Grand Rapids, MI 49503

Acton Institute for the Study of Religion and Liberty Novak Award 297
Named after distinguished American theologian Michael Novak, this $10,000 award rewards new outstanding scholarly research into the relationship between religion, economic freedom, and the free and virtuous society. This award recognizes those scholars early in their academic career who demonstrate outstanding intellectual merit in advancing the understanding of theology's connection to human dignity, the importance of the rule of law, limited government, religious liberty, and freedom in economic life. The nomination deadline is November 1, and the application deadline is December 1.
Requirements: Professors, university faculty members, and other scholars may nominate qualified individuals for the Novak Award by completing the online nomination form.
Samples: Professor Carlos Hoevel—for A Rosminian Vision for the Post-Crisis Global Economy, $10,000 (2008); Dr. Andrea Schneider—for Centesimus Annus: Law, Human Rights, and the Free Society, $10,000 (2007); Dr. Jan Klos—for Spiritualizing the World: Secularism as a Religion - Challenges for Faith and Liberty in Europe, $10,000 (2006).
Amount of Grant: $10,000 stipend per year
Date(s) Application is Due: Nov 30
Contact: Kara Eagle, Programs Officer; (616) 454-3080; fax: (616) 454-9454; email: keagle@acton.org
Internet: http://www.acton.org/programs/students/novak.php
Sponsor: Acton Institute for the Study of Religion and Liberty
161 Ottawa NW, Suite 301
Grand Rapids, MI 49503

ACTR/ACCELS Combined Research and Training Program 298
The program allows faculty and graduate students to conduct research in the New Independent States (NIS) in the humanities or social sciences. Awards are available for research projects lasting three to nine months. Combined research and language training programs also are available. The January deadline is for the summer, fall and academic year; the October deadline is for spring programs. Applications are available on the Web site as soon as they are available or may be obtained by contacting the program directly.
Requirements: U.S. citizens (graduate students, postdoctoral scholars, and faculty) who have the necessary background to conduct advanced research in the NIS are eligible.
Amount of Grant: $5000-$25,000
Date(s) Application is Due: Jan 15; Oct 1
Contact: Research Grants Administrator, (202) 833-7522; fax: (202) 833-7523; email: general@actr.org
Internet: http://www.americancouncils.org/program.asp?PageID=121&ProgramID=13
Sponsor: American Council of Teachers of Russian/American Council for Collaboration in Education and Language Study
1776 Massachusetts Avenue NW, Suite 700
Washington, DC 20036

Actuarial Foundation Research Grants 299
The foundation supports research that is of significant value to future endeavors of the actuarial profession. New knowledge and techniques developed in other fields, as well as enhanced applications of existing actuarial principles and techniques, that can be applied by actuaries are given priority, including those of a multidisciplinary nature. Jointly sponsored seminars and call for paper projects with or without linked symposia that promote research, as well as individual research papers, books and other projects will be considered. The Foundation is particularly interested in partnering with other organizations to promote research. The foundation does not require use of a standard proposal. Proposals are accepted at any time. Interested individuals or organizations should begin the process with a brief letter of interest.
Date(s) Application is Due: Oct 1
Contact: Grants Administrator, (847) 706-3535; fax: (847) 706-3599
Internet: http://www.actuarialfoundation.org/research_edu/research.htm
Sponsor: Actuarial Foundation
475 N Martingale Road, Suite 600
Schaumburg, IL 60173-2226

ACVIM Foundation Clinical Investigation Grants 300
The Foundation improves the health and well-being of animals by funding humane studies and communicating information about this work to the veterinary community and the general public. It supports leading veterinary scientists as they take on a range of critical animal health issues within the specialties of small and large animal internal medicine, cardiology, neurology, and oncology. This work, in turn, holds promise for better preventive care, better nutrition, and more sensitive screening and improved treatment for such devastating diseases as cancer, epilepsy, cardiomyopathy, and kidney failure. The Foundation's main goal is to support clinical investigations that lead to new diagnostic, treatment, and prevention techniques. Resident applicants may request up to $12,500 per year, while diplomate applicants may request up to $15,000. One $20,000 grant will be awarded to a collaborative/multi-center proposal, with preference toward private practice/academia partnerships. Funding for multi-year projects up to 3 years will be considered, subject to annual non-competitive renewal based on availability of funds and demonstration of adequate progress on the project. The foundation will consider funding for multi-year projects up to three years.
Samples: Dr. Jill Beech, University of Pennsylvania School of Veterinary Medicine—Evaluation of Endogenous Alpha MSH Concentrations at Different Seasons and in Response to TRH in Normal Horses and Those with Pituitary Hyperplasia (2006); Dr. Stephanie Kottler, University of Missouri-Columbia College of Veterinary Medicine—Prevalence of Staphylococcus aureus and MRSA carriage in three populations (2006).
Amount of Grant: $20,000 maximum
Date(s) Application is Due: Sep 15
Contact: Angela E. Frimberger, Executive Director; (303) 231-9933 or (800) 245-9081; email: info@acvimfoundation.org
Internet: http://www.acvimfoundation.org/grants/info.html
Sponsor: American College of Veterinary Internal Medicine (ACVIM)
997 Wadsworth Boulevard, Suite A
Lakewood, CO 80214

ADA Career Development Awards 301
The awards assist exceptionally promising new investigators in their transition to the level of established investigators. The awards support such individuals' salary and research program for five years to enable the investigators to initiate an independent research effort that would result in sufficient accomplishments that in turn would qualify the investigators for long-term research funding. Application request forms are available on the Web site.
Requirements: Applicants must hold full-time or clinical faculty positions or the equivalent at university-affiliated institutions within the United States or one of its possessions; have an MD or PhD degree, or in the case of other health professions, an appropriate advanced degree; and have only two to five years postdoctoral/postfellowship research experience. Applicants must be U.S. citizens or permanent residents or have applied for permanent resident status and must be members of ADA's professional section.

Restrictions: Time spent in residency training is not considered equivalent to research experience.
Amount of Grant: $200,000 maximum per year for salary and expenses
Date(s) Application is Due: Jan 15
Contact: Research Department, (703) 299-2071; fax: (703) 549-1715; email: research@diabetes.org
Internet: http://www.diabetes.org/diabetes-research/research-grant-application-forms/development-awards.jsp
Sponsor: American Diabetes Association
1701 N Beauregard Street
Alexandria, VA 22311

ADA Clinical Research Awards 302
The grants provide a stipend for three years for studies undertaken that involve humans directly. In vitro research on human blood or tissue samples does not qualify under this program unless there has been a major in vivo intervention (i.e., diet, drugs, exercise, etc.) and the study protocol is designed specifically to quantitate the effect of this manipulation on the tissue being examined in vitro. The funds may be used for equipment, supplies, and/or technician support. Up to $20,000 of the funds may be used as salary support for the principal investigator. Awards are made for up to three years. Application request forms are available on the Web site.
Requirements: Applicants must be U.S. citizens or permanent residents or have applied for permanent resident status, hold an MD or PhD degree (for other health professionals, an appropriate health- or science-related degree), and hold a faculty-level appointment at a university-affiliated institution within the United States or U.S. possessions. Award recipients must be members of the ADA's professional section.
Amount of Grant: $20,000-$100,000
Date(s) Application is Due: Jan 15; Jul 15
Contact: Research Department, (703) 299-2071 or (703) 549-1500; fax: (703) 549-1715; email: research@diabetes.org
Internet: http://www.diabetes.org/diabetes-research/research-grant-application-forms/nationwide-research-awards.jsp
Sponsor: American Diabetes Association
1701 N Beauregard Street
Alexandria, VA 22311

ADA Foundation Dental Literature Grants 303
The foundation will award grants to applicants who submit competitive proposals to complete a systematic literature review on a high-priority research topic supporting evidence-based dentistry. The program focuses on available dental literature addressing four specific clinical questions. Applicants are asked to choose one of the following questions for a systematic literature review. At what frequency is dental prophylaxis effective in preventing periodontitis in individuals with and without known risk factors? Does correcting malocclusion in children and adults reduce the risk of periodontal disease? What are the clinical, biological, psychosocial, and/or economic outcomes of treating a pulpally involved (periodontally sound) single tooth through: endodontic care, extraction and implant placement, fixed partial denture, and/or extraction without implant placement? What are the longitudinal beneficial and harmful effects of endodontic services compared to extraction and implant placement? Proposals must be submitted in hard copy. Guidelines are available online.
Requirements: Applicants must be affiliated with a dental school or advanced education program accredited by the Commission on Dental Accreditation, a dental specialty organization or a national dental related organization; should be knowledgeable and experienced in systematic review; have a demonstrated track record in research; and have published studies in clinical dental research.
Amount of Grant: $50,000 maximum
Date(s) Application is Due: Jan 30
Contact: Lisa Barron, at (312) 440-4639; email: barronl@ada.org
Internet: http://www.adafoundation.org/prof/resources/pubs/adanews/adanewsarticle.asp?articleid=1640
Sponsor: American Dental Association Foundation
211 E Chicago Avenue
Chicago, IL 60611-2616

ADA Medical Scholars Awards and Physician-Scientist Training Awards 304
The goal of the program is to produce new leaders in diabetes research. The medical scholars program provides one year of research support to students in medical school; the physician-scientist award provides three years of support for the doctoral portion of an MD/PhD degree. Applications may be requested from the Web site.
Amount of Grant: $30,000 Medical Scholars; $30,000 per year Physician-Scientist
Date(s) Application is Due: Jan 15
Contact: Research Department, (703) 299-2071 or (703) 549-1500; fax: (703) 549-1715; email: research@diabetes.org
Internet: http://www.diabetes.org/diabetes-research/research-grant-application-forms/ADA-grant-opportunities/ADA-current-grant-opportunities.jsp
Sponsor: American Diabetes Association
1701 N Beauregard Street
Alexandria, VA 22311

ADA Mentor-Based Postdoctoral Fellowships 305
The purpose of this program is to support the training of scientists in an environment most conducive to beginning a career in diabetes research. An award is given to an established and active investigator in diabetes re-search for the annual stipend support of a postdoctoral fellow to work closely with the mentor. The applicant investigator will be responsible for the selection of the qualified fellow. The term of the award is 4 years. Funding for each year will be contingent upon submission of the name and CV of a fellow by June 1 each year.
Requirements: The postdoctoral fellow must have an MD or PhD degree, no more than three years of postdoctoral research experience, and cannot serve an internship or residency during the award period. There are no citizenship requirements for the fellow. The investigator must meet citizenship and employment eligibility requirements. Award recipients must be members of ADA's professional section.
Amount of Grant: $45,000 maximum per year; $3000 maximum per year for laboratory supply costs; $1000 maximum for travel
Date(s) Application is Due: Jul 15
Contact: Research Department, (703) 299-2071 or (703) 549-1500; fax: (703) 549-1715; email: research@diabetes.org
Internet: http://www.diabetes.org/diabetes-research/research-grant-application-forms/training-awards.jsp#medical
Sponsor: American Diabetes Association
1701 N Beauregard Street
Alexandria, VA 22311

ADA Minority Dental Student Scholarships 306
Scholarships are awarded to minority students who are entering their second year of dental school. The number of scholarships awarded is dependent on funds available. Application forms are available at accredited dental schools.
Requirements: Applicants must be citizens of the United States from one of the following minority groups currently underrepresented in the dental profession: Native American, African American, or Hispanic. Students selected must have been accepted by a dental school in the United States that is accredited by the Commission on Dental Accreditation. Applicants must demonstrate a minimum need of $2500 and maintain a 3.0 grade point average on a 4.0 scale.
Amount of Grant: $2500 maximum per year
Date(s) Application is Due: Jul 31
Contact: Health Foundation, (312) 440-2547; fax: (312) 440-3526; email: adaf@ada.org
Internet: http://www.ada.org/ada/prod/adaf/prog_scholarship_prog.asp
Sponsor: ADA Endowment and Assistance Fund Inc
211 E Chicago Avenue
Chicago, IL 60611

ADA Research Awards 307
This program provides grant support to both new and established investigators. Applications will be considered in any area that is relevant to the etiology or pathophysiology of diabetes and its complications. Awards are made for up to three years for the purchase of equipment, supplies, and salary for technical assistance. Up to $20,000 may be used for the investigator's salary. Up to 15 percent of the award may be used for indirect costs.
Requirements: Applicants must hold MD or PhD degrees or, in the case of other health professions, an appropriate advanced degree; must be U.S. citizens or permanent residents or have applied for permanent resident status; and must hold full-time or clinical faculty positions or the equivalent at university-affiliated institutions within the United States or U.S. possessions. Award recipients must be members of the ADA's professional section.
Amount of Grant: $20,000-$100,000 per year for two years
Date(s) Application is Due: Jan 15; Jul 15
Contact: Magda Galindo, Research Program Manager, (703) 299-2071 or (703) 549-1500; fax: (703) 549-1715; email: mgalindo@diabetes.org. or research@diabetes.org
Internet: http://www.diabetes.org/diabetes-research/research-grant-application-forms/ADA-grant-opportunities/ADA-current-grant-opportunities.jsp
Sponsor: American Diabetes Association
1701 N Beauregard Street
Alexandria, VA 22311

Adam Richter Charitable Trust Grants 308
The Trust, which is administered by the Bank of America in Dallas, Texas, was established in California in 1994. Offering grants on a national basis, its primary fields of interest include: the arts; animal welfare; Christian agencies and churches; the environment; health organizations; higher education; hospitals; and human services. There are no formal applications or deadlines, and interested applicants should begin by
Samples: Habitat for Humanity International, San Francisco, CA, $13,082 (2007); Greenpeace, Washington, DC, $2,000 (2007); Sonoma Land Trust, Santa Rosa, CA, $12,000 (2007).
Contact: c/o Michael Schlebach, Specialty Asset Manager; (214) 922-9290
Sponsor: Adam Richter Charitable Trust
411 N. Akard Street
Dallas, TX 75201-3307

Adaptec Foundation Grants 309
The corporate contributions program is primarily directed, although not exclusively, to organizations and programs serving the communities where its employees work and live, with special emphasis in the Silicon Valley area. Funding priorities include education and research—programs that lower the drop-out rate of youth in accredited schools and colleges, address career development and preparation, job training and placement, educational scholarship assistance, and investment in higher education (primarily in

engineering-related university programs); health and human services—research cures for devastating diseases, emergency shelter and subsistence for the homeless, treatment and assistance for abused spouses and children, drug and alcohol rehabilitation, and supporting organizations that strengthen youths and families; helping people enjoy life more fully through music, dance, art museums, opera, theater, and performing arts groups; and $25-$500 employee matching gifts. The Corporate Contributions Committee meets quarterly to consider requests.

Restrictions: In general, Adaptec does not make contributions to any organization that does not have 501(c)3 or comparable status; individuals, except through the Adaptec Scholarship program; churches, synagogues, and other religious groups. (Requests are considered from religious-sponsored and other groups whose activities do not support any specific religious doctrine or ethnic/cultural group.); political or fraternal organizations; underwrite or assist with the development of films, television, or video/radio productions, whether for commercial, independent, or nonprofit ventures, unless such activity is in support of the computer I/O software/hardware industry and is directly related to the business of the company; tickets for raffles, contests, or other fundraising activities (except requests from major fundraising organizations that meet the defined objectives of the contribution program, such as United Way, the Silicon Valley Charity Ball and the San Jose Symphony); organizations whose primary objective is to assist, benefit, and address animal welfare or environmental issues, such as the preservation of endangered species and plants or conservation efforts to preserve the quality of wildlife, water, soil, or air; reduce debts, fund general operating expenses or capital improvement projects, or retroactively fund activities or programs that are already completed; and school sports leagues and events, and other non-educational related activities such as band, debate teams, and goodwill trips.

Amount of Grant: $250-$2,500
Contact: Corporate Contributions; (408) 945-8600; fax: (408) 262-2533
Internet: http://cms.adaptec.com/ko-KR/company/about/_corporate/adaptec_corpcontributions.htm?nc=/ko-KR/company/about/_corporate/adaptec_corpcontributions.htm
Sponsor: Adaptec Foundation
691 South Milpitas Boulevard, MS-15
Milpitas, CA 95035-5473

ADC Telecommunications STEM Education Grants 310

The corporation supports science, math, technology, economic and business education, with a focus on K-12 public, charter, alternative, and culturally specific schools. Also of interest are industrial arts programs in public and private schools, public colleges, and graduate and postgraduate programs at public and private institutions.

Requirements: Grants are awarded to Minnesota nonprofits in Twin Cities Metro Area. Grants are also awarded in Marietta, Georgia, Santa Teresa, New Mexico, Raleigh, North Carolina, and Sidney, Nebraska. There is also grantmaking outside of the U.S.
Restrictions: Requests for in-kind giving, general operating grants, capital grants or fundraisers including customer-sponsored charity benefit events are not considered.
Samples: Education Cooperative Service Unit of Metro Twin Cities Area —for a young inventors program, $2000.; Minnesota Institute for Talented Youth (MN)—for summer math and science classes, $3000.; Saint Louis Park Junior High School (MN)—for the Odyssey of the Mind state champion team's trip to the world competition, $1500.
Contact: Bill Linder-Scholer, Executive Director ADC Foundation; (952)917-0580; email: bill.linder-scholer@adc.com
Internet: http://www.adc.com/aboutadc/adcfoundation
Sponsor: ADC Telecommunications
P.O. Box 1101
Minneapolis, MN 55440-1101

ADEC Agricultural Telecommunications Grants 311

Grants will be awarded competitively to encourage the development and utilization of an agricultural communications network to facilitate and to strengthen agricultural extension, resident education, and research, and domestic and international marketing of United States agricultural commodities and products through a partnership between eligible institutions and the Department of Agriculture. The network will employ satellite and other telecommunications technology to disseminate and to share academic instruction, cooperative extension programming, agricultural research, and marketing information. Proposals will be accepted from new applicants (lead institutions) and repeat applicants (lead institutions). Proposals may be submitted electronically via the Web site, or applicants may contact the office for application materials.

Requirements: Proposals are invited from accredited institutions of higher education.
Amount of Grant: $50,000 maximum
Date(s) Application is Due: July 15
Contact: Dr. Janet Poley, President, (402) 472-7000; fax: (402) 472-9060; email: jpoley@unl.edu
Internet: http://www.adec.edu/fed-pgms.html
Sponsor: American Distance Education Consortium
P.O. Box 830952
Lincoln, NE 68583-0952

Adelle and Erwin Tomash Fellowship in the History of Information Processing 312

The Charles Babbage Institute at the University of Minnesota is accepting applications for this fellowship to be awarded for the academic year to a graduate student whose dissertation will address some aspect of the history of computers and information processing. Topics may be chosen from the technical history of hardware or software; economic or business aspects of the information processing industry; or other topics in the social, institutional, or legal history of computing. Theses that consider technical issues in their socioeconomic context are especially encouraged. The fellowship may be held at the home academic institution, the Babbage Institute, or any other location where there are appropriate research facilities.

Requirements: Priority will be given to students who have completed all requirements for the doctoral degree except the research and writing of the dissertation. Applicants should send biographical data and a research plan containing a statement and justification of the research problem, a discussion of procedure for research and writing, information on availability of research materials, and evidence of faculty support for the project. Applicants should also arrange for three letters of reference and certified transcripts of college credits to be sent directly to the institute. This fellowship has no geographical restrictions, and students from any country are eligible to apply.
Amount of Grant: $10,000; $2000 maximum for tuition, fees, travel to Babbage Institute and relevant archives, and other approved research expenses
Date(s) Application is Due: Jan 15
Contact: Jeffrey Yost, Associate Director, (612) 642-5050; fax: (612) 625-8054; email: yostx003@tc.umn.edu
Internet: http://www.cbi.umn.edu/research/tomash.html
Sponsor: Charles Babbage Institute, University of Minnesota
211 Anderson Library
Minneapolis, MN 55455

ADHF Miles and Shirley Fiterman Foundation Basic Research Awards 313

Two awards are offered: Hugh R. Butt Award in Hepatology or Nutrition, and the Joseph B. Kirsner Award in Gastroenterology. The awards are funded by the Miles and Shirley Fiterman Foundation and recognize excellence in clinical research in hepatology or nutrition and enterology and help support the clinical research efforts of the recipients. Each award, given annually, is provided to the awardee's institution to support his or her research. There are no application forms. A nominating letter, curriculum vita, bibliography, and five reprints of the candidate's most important work should be received by the listed deadline date for applications.

Requirements: Young to midlevel investigators will be ranked above senior people whose contributions, although substantive, may have been accomplished in the distant past and whose major current activities do not include active, cutting-edge research activities.
Amount of Grant: $35,000
Date(s) Application is Due: Jan 14
Contact: Grants Administrator, (301) 222-4005; fax: (301) 222-4010; email: info@fdhn.org
Internet: http://www.fdhn.org/html/awards/elect_app.html
Sponsor: American Digestive Health Foundation
4930 Del Ray Avenue
Bethesda, MD 20814

ADHF/Elsevier Research Initiative Grant 314

This research initiative grant is offered to investigators for support of pilot research projects in gastroenterology-related areas. The objectives of the program are to provide funds for new investigators to help them establish their research careers and to support pilot projects that represent new research directions for established investigators. The intent is to stimulate research in gastroenterology-related areas by permitting investigators to obtain new data that can ultimately provide the basis for subsequent grant applications of more substantial funding and duration. The awards may be used for salary, supplies, or equipment.

Requirements: Investigators must possess an MD or PhD degree (or equivalent) and must hold faculty positions at an accredited North American institution.
Restrictions: Applicants may not hold awards on a similar topic from other agencies.
Amount of Grant: $25,000 per year
Date(s) Application is Due: Jan 14
Contact: Administrator, (301) 222-4005; fax: (301) 222-4010; email: awards@fdhn.org
Internet: http://www.fdhn.org/html/awards/elect_app.html
Sponsor: American Digestive Health Foundation
4930 Del Ray Avenue
Bethesda, MD 20814

Administration on Aging Alzheimer's Disease Demonstration Grants to States 315

The purpose of this program is to develop models of assistance for persons with Alzheimer's disease and their families, and improve the responsiveness of existing home- and community-based care systems for persons with Alzheimer's disease and related disorders and their families. HHS-2005-AoA-AZ-0502

Requirements: Eligibility for grant awards is limited to state agencies. Only one application per state will be accepted.
Amount of Grant: $250,000-$325,000
Date(s) Application is Due: May 27
Contact: Margaret Tolson, Program Officer, (202) 357-3440; fax (202) 401-7620; email: Margaret.Tolson@aoa.hhs.gov
Internet: http://www.aoa.dhhs.gov/doingbus/fundopp/fundopp.asp
Sponsor: Administration on Aging
200 Independence Avenue SW
Washington, DC 20201

Advancing Newborn Medicine Fellowship Grant Program 316

The program's objectives are to promote research in the area of neonatal cardiorespiratory medicine, support the professional growth of neonatology fellows, and contribute to improvements in the medical care of hospitalized newborns. Fellows are invited to submit proposals stating a specific question to be answered, methods to be used, and type of analysis to be performed.

Requirements: Proposals are accepted from individuals in accredited neonatal-perinatal medicine training programs in the United States.

Amount of Grant: $7500 each

Date(s) Application is Due: Nov 30

Contact: Glen Santiago, Phase Five Communications, (888) 706-3899

Internet: http://www.forestpharm.com

Sponsor: Forest Pharmaceuticals Inc
909 Third Avenue
New York, NY 10022

AED New Voices Fellowship Program 317

The program seeks to strengthen small nonprofit organizations by supporting resident fellows who can bring a new voice to an organization and its field of work. Applications are reviewed under one of the following categories: foreign policy, peace and security; HIV/AIDS; international economic policy international human rights; migrant and refugee rights; racial justice and civil rights; and women's and reproductive rights. Awards include support for fellow's salary, fringe benefits, and to purchase a computer. Preference is given to nonprofits with budgets between $100,000 and $2 million. Applications are prepared jointly by an organization and its proposed fellow.

Requirements: 501(c)3 nonprofit organizations that reflect diverse educational, cultural, and experiential backgrounds are eligible. Potential fellows should have completed an undergraduate or graduate degree or have comparable education, skills, and relevant experience.

Restrictions: Groups with budgets of more than $5 million are not eligible.

Amount of Grant: $100,000 average salary over two years

Date(s) Application is Due: Jan 10

Contact: Program Officer, (202) 884-8607; fax: (202) 884-8400; email: newvoice@aed.org

Internet: http://www.aed.org/newvoices

Sponsor: Academy for Educational Development
1825 Connecticut Avenue NW, Suite 744
Washington, DC 20009-5721

AEG Foundation Legget Fund Grants 318

The Fund was established by the Foundation (then the Engineering Geology Foundation) in 1993 to support publications and public outreach in engineering geology and environmental geology that serve as information resources for the professional practitioner, students, faculty, and the public. The fund also supports public education about the interactions between the works of mankind and the geologic environment. Public outreach activities that the fund can support include K-12 teacher resources, publications to inform the public about the interactions between the works of humankind and the geologic environment, and signage or explanatory materials at public locations of interest for their engineering geology or environmental geology significance. The grants are intended to support technical and production editing of eligible publications, production costs, page charges, and conversion of print media to electronic form. Public education project grants from the Legget Fund are made to support technical and production editing, and production costs, of eligible works.

Restrictions: The Fund does not support research costs of individuals or organizations leading to publication or production, nor does it support general and administrative (overhead) costs, technical or scientific consulting fees, or equipment purchase or maintenance costs of organizations or individuals receiving grants from it.

Amount of Grant: $1,000-$50,000

Contact: Becky Roland, Chief Operating Officer; (303) 757-2926; fax: (303) 757-2969; email: broland@aegweb.org

Internet: http://www.aegfoundation.org/funds/legget

Sponsor: Association of Engineering Geologists (AEG) Foundation
3773 Cherry Creek N. Drive, Suite 575
Denver, CO 80209

AEGON Transamerica Foundation Grants 319

The foundation supports organizations involved with arts and culture, education, cancer, ALS research, hunger, human services, and the disabled. Giving on a national basis in areas of company operations, with some emphasis on Little Rock, Arkansas, California, Atlanta, Georgia, Cedar Rapids, Iowa, Louisville, Kentucky, Baltimore, Maryland, Charlotte, North Carolina, New York, Pennsylvania, and Fort Worth, Texas.

Requirements: Grant recipient must be a 501(c)3 organization.

Restrictions: Grants are not awarded to individuals. Unsolicited proposals not accepted.

Geographic Focus: Arkansas; California; Georgia; Iowa; Kentucky; Maryland; New York; North Carolina; Pennsylvania; Texas

Samples: Johns Hopkins University, Baltimore, MD, $2,500,000 (2008); United Way of Central Carolinas, Charlotte, NC, $46,676 (2008); Theater Cedar Rapids, Cedar Rapids, IA, $150,000 (2008).

Amount of Grant: $50 - $2,500,000 range

Contact: Grants Administrator, (319) 398-8935

Sponsor: AEGON Transamerica Foundation
4333 Edgewood Road, NE
Cedar Rapids, IA 52499-3210

AERA Dissertation Grants 320

AERA invites education policy- and practice-related dissertation proposals using NCES, NSF, and other national data bases. Dissertation grants are available for advanced doctoral students and are intended to support the student while writing the doctoral dissertation. Dissertation topics may cover a wide range of policy- or practice-related issues. Researchers must include the analysis of data from at least one NSF or NCES data set in the dissertation. Additional large-scale nationally representative data sets may be used in conjunction with the obligatory NSF or NCES data set. If international data sets are used, the study must include U.S. education. Minority researchers are strongly encouraged to apply. A total of two hard copies of all required materials must be submitted and received by the deadline above. Guidelines are available online.

Requirements: Applicants may be U.S. citizens, U.S. permanent residents, or non-U.S. citizens.

Amount of Grant: $15,000 maximum

Date(s) Application is Due: Mar 1; Sep 1

Contact: Jeanie Murdock, (805) 964-5264; email: jmurdock@aera.net

Internet: http://www.aera.net/grantsprogram/res_training/diss_grants/DGFly.html

Sponsor: American Educational Research Association
5662 Calle Real, No 254
Goleta, CA 93117-2317

AERA Research Grants 321

AERA invites education policy- and practice-related research proposals using NCES, NSF, and other national databases. One- and two-year research grants are available for faculty at institutions of higher education, postdoctoral researchers, and other doctoral-level researchers. Research topics may cover a wide range of policy- or practice-related issues. Researchers must include the analysis of data from at least one NSF or NCES data set in the proposed study. Successful principal investigators will consult with NCES or NSF staff regarding their proposed research projects and the handling of NCES, NSF, and other federal agency data sets pertinent to their projects. Guidelines are available online.

Requirements: Applicants for may be U.S. citizens, U.S. permanent residents, or non-U.S. citizens. Applicants must have received the doctoral degree by the start date of the grant.

Restrictions: Institutions may not charge indirect costs on these awards.

Amount of Grant: $20,000 maximum one-year projects; $35,000 maximum two-year projects

Date(s) Application is Due: Mar 1; Sep 1; Jan 8

Contact: Jeanie Murdock, (805) 964-5264; email: jmurdock@aera.net

Internet: http://www.aera.net/grantsprogram/res_training/res_grants/RGFly.html

Sponsor: American Educational Research Association
5662 Calle Real, No 254
Goleta, CA 93117-2317

Aerospace Postdoctoral Fellowships 322

Under this program, fellows conduct research at the NASA-Johnson Space Center and then work at the University of Houston and the University of Houston Clear Lake, where they transfer their research expertise and experiences to the academic community. Each research project is directed by a team consisting of at least one UH and/or UHCL faculty member and at least one senior research staff member of JSC. The fellowship program will enable UH/UHCL faculty to organize far stronger proposals for external support through stand alone proposals and for proposals in association with NASA-JSC area organizations. In addition, the program will stimulate publications, new intellectual property, new graduate courses, and several graduate fellowships.

Requirements: Applicants must have PhD degrees and up to three years relevant experience. The program is open to U.S. citizens and foreign nationals.

Amount of Grant: $50,000 per year

Contact: Aerospace Postdoctoral Fellowships, (713) 743-3524; fax: (713) 743-3589; email: dcriswell@uh.edu

Internet: http://www.isso.uh.edu

Sponsor: University of Houston
617 Science and Research 1, MC 5005
Houston, TX 77204

AF Doctoral Dissertation Award for Arthritis Health Professionals 323

The award provides one or two years of salary and/or research support to projects related to arthritis management and/or comprehensive patient care in rheumatology practice, research, or education. A dissertation project is preferred. Suitable studies include, but are not limited to, functional, behavioral, nutritional, occupational, or epidemiological aspects of patient care and management. Drug studies and laboratory in vitro studies are not appropriate. Application and guidelines are available online. Annual deadline dates may vary; contact program staff for exact dates.

Requirements: The awards are designed for predoctoral students entering the research phase of their programs. The doctoral chairperson must approve the project. A candidate must have membership or eligibility for membership in his/her professional organization. Applicants must be U.S. citizens or have permanent residence.

Restrictions: The award does not support laboratory research.

Amount of Grant: $30,000 per year

Date(s) Application is Due: Sep 1

Contact: Research Department, (404) 965-7537

Internet: http://www.arthritis.org/research/ProposalCentral.asp

Sponsor: Arthritis Foundation
1330 W Peachtree Street, Suite 100
Atlanta, GA 30309

AF Investigator Awards   **324**

The purpose of the award is to provide support to physicians and scientists in research fields broadly related to arthritis for the period between completion of postdoctoral fellowship training and establishment as an independent investigator. A senior scientist familiar with the applicant's area of research should be designated as the sponsor. The sponsor and the chairperson of the relevant academic department are responsible for stating the role of the applicant within the department, promising protection of time for research activities related to the award, guaranteeing space for the investigative work, and outlining future opportunities for the applicant. Each investigator is expected to devote at least 80 percent of his/her professional time to laboratory research. Permission for U.S. applicants to work abroad may be granted but the award may not be used for sabbatical leave. Awards will begin July 1 and will be renewed each year for a period of three years pending receipt of satisfactory evidence of progress. The awards will be paid to the sponsoring institution for stipend only, stipend plus research expenses, or research expenses only; the budget should indicate how the grant will be allocated. Application and guidelines are available online.

Requirements: Applicants must be U.S. citizens or have permanent residence. Applicants must have completed a minimum of three and a maximum of six years postdoctoral research experience, and must hold an MD, PhD, or equivalent degree, and have demonstrated distinction and productivity in research.

Restrictions: The applicant may not hold an NIH R01, NSF Grant, FIRST, Howard Hughes, Pew, VA Merit, Wellcome, Searle, Arthritis Foundation Biomedical Science Grant or Clinical Science Grant or equivalent award at the time of application. If such an award is made for a different project after the start date of the Arthritis Investigator Award, the individual may retain his/her Arthritis Foundation award. Individuals at the NIH and CDC are not eligible to apply.

Amount of Grant: $75,000 annually plus $1000 institutional grant

Date(s) Application is Due: Sep 1

Contact: Research Department, (404) 965-7537; email: grantsupport@arthritis.org

Internet: http://www.arthritis.org/research/ProposalCentral.asp#career

Sponsor: Arthritis Foundation
1330 W Peachtree Street
Atlanta, GA 30309

AF New Investigator Research Grants for Arthritis Health Professionals   **325**

These research grants are to encourage PhD level health professionals who have research expertise to design and carry out innovative research projects related to the rheumatic diseases. The grant is intended to provide support for the period between completion of doctorate work and establishment as an independent investigator. The research project must be related to arthritis management and/or comprehensive patient care in rheumatology practice, research, or education. Suitable studies include, but are not limited to, functional, behavioral, nutritional, occupational, or epidemiological aspects of patient care and management. Individuals with limited research experience must apply in conjunction with a supervisor or co-investigator with demonstrated research expertise in the applicant's area of study. Awards are made for three years. Application and guidelines are available online.

Requirements: Grants are designed for applicants with a PhD or equivalent doctoral degree and demonstrated research experience. These awards are meant to encourage investigators who have received a doctoral degree within the last five years. A candidate must have membership or eligibility for membership in her/his professional organization.

Restrictions: Drug studies and laboratory in vitro studies are not appropriate. Individuals at the NIH and CDC are not eligible to apply. Not for laboratory research. MDs are not eligible.

Amount of Grant: $50,000 annually

Date(s) Application is Due: Sep 1

Contact: Research Department, (404) 965-7636; email: grantsupport@arthritis.org

Internet: http://www.arthritis.org/research/ProposalCentral.asp#grants

Sponsor: Arthritis Foundation
1330 W Peachtree Street
Atlanta, GA 30309

AF Physician Scientist Development Award   **326**

The award, which is cosponsored by the Scleroderma Foundation, is designed to enable MDs and those with an equivalent medical degree with no more than 1-1/4 years of research experience at the time of application to embark on careers in biomedical and/or clinical research related to the understanding of arthritis and the rheumatic diseases. Candidates with more than 1-1/2 years of postdegree laboratory experience must apply to the foundation's postdoctoral fellowship program. This is a two-year award that may be renewed for a third year. The stipend includes salary and fringe benefits. Application and guidelines are available online.

Requirements: Only physicians with no more than 1-1/4 years of research training at the time of application are eligible. Individuals must have received an MD, DO, or equivalent medical degree from an accredited institution within the past seven years; completed training in internal medicine or pediatrics; and will have completed at least one year of specialty training as of the start date of the award. For those awards funded by the Scleroderma Foundation, applicants will have completed at least one year of specialty training in rheumatology. Applicants must be U.S. citizens or have permanent residence.

Restrictions: MD/PhDs, DO/PhDs, and PhDs are not eligible.

Amount of Grant: $50,000 per year

Date(s) Application is Due: Sep 1

Contact: Research Department, (404) 965-7537; email: grantsupport@arthritis.org

Internet: http://www.arthritisfoundation.org/research/ProposalCentral.asp

Sponsor: Arthritis Foundation
1330 W Peachtree Street
Atlanta, GA 30309

AF Postdoctoral Research Fellowships   **327**

The grant provides a salary stipend for MDs, DOs, PhDs, or equivalent for a three-year period. Ninety percent time must be devoted to arthritis-related research. A stipend and institutional grant are awarded. There are no citizenship requirements for this award. Application and guidelines are available online.

Requirements: MDs are not eligible after six years of postdoctoral research training (or seven years in the case of a clinical training program, which includes one year in the laboratory).

Restrictions: MDs are not eligible after six years of laboratory training or seven years in the case of a clinical training program that includes one year in the laboratory. PhDs are not eligible after four years of postdegree laboratory experience. Individuals at or above the assistant professor level, or those who have tenured positions, are ineligible; however, if one is promoted or establishes tenure after receiving an award, he/she may continue to receive the award for the remainder of the fellowship period.

Amount of Grant: $50,000 stipend plus $500 institutional grant

Date(s) Application is Due: Sep 1

Contact: Research Department, (404) 965-7537; email: grantsupport@arthritis.org

Internet: http://www.arthritis.org/research/proposalcentral.asp

Sponsor: Arthritis Foundation
1330 W Peachtree Street
Atlanta, GA 30309

AF Target Identifications in Lupus Grants   **328**

The purpose of the grants program is to fund highly meritorious and innovative research directed to the identification of targets for new therapies in the treatment of lupus. This award is open to nonprofit and industry applications from anywhere in the world. Grants are made for two years with a possibility of renewal. Application and guidelines are available online.

Requirements: Individuals with doctoral degrees (MD, PhD, DO, or equivalent) are eligible to apply. Scientific independence as evidenced by direction of a research program or a publication record or other experience that establishes scientific leadership is necessary to apply.

Amount of Grant: $250,000 per year

Date(s) Application is Due: Sep 1

Contact: Research Department, (404) 965-7537; email: grantsupport@arthritis.org

Internet: http://www.arthritis.org/research/ProposalCentral.asp

Sponsor: Arthritis Foundation
1330 W Peachtree Street
Atlanta, GA 30309

AFAR CART Fund Grants   **329**

The goal of the CART Fund is to encourage exploratory and developmental AD research projects within the United States by providing support for the early and conceptual plans of those projects that may not yet be supported by extensive preliminary data but have the potential to substantially advance biomedical research. This proposal should be distinct from those projects designed to increase knowledge in a well established area unless it is intended to extend previous discoveries toward new directions or applications. Applications may encompass a project period of up to two years with a combined budget for direct cost up to $250,000. No indirect costs are allowed. Domestic public and private institutions are eligible, such as universities, colleges, hospitals and laboratories. This is for NEW projects only. Up to two $250,000 awards will be available this year. Dr. John Trojanowski will chair a scientific review group that will triage the Letters-of-Intent and select a maximum of fifteen deemed to have the highest merit. Those selected will be invited to submit a subsequent standard grant application from which the final recommendation will be made by the review group. The final selection will be made by the CART Fund selection committee. Notifications of the finalists (maximum of 15) will be mailed by January 7th. Investigators whose proposals are accepted for further consideration will be sent a full application form which must be completed and returned by February 16th.

Requirements: The CART Fund is inviting interested applicants from within the United States to submit a Letter-of-Intent that includes sufficient detail to communicate the importance of your study as well as information on its feasibility. Submit email questions and Letters-of-Intent to Dr. James B. Puryear, Chairman, CART Grants Program: jimpuryear@comcast.net.

Restrictions: Applications will be deemed ineligible from for-profit organizations as well as those already supported by regular or program grants.

Samples: Sanjay W. Pimplikar, Ph.D. - The Cleveland Clinic, OH (2008); Todd E. Golde, M.D., Ph.D. - Mayo Clinic, Jacksonville, FL (2007); Claudio Soto, Ph.D. - University of Texas Medical Branch, TX (2006);

Amount of Grant: $250,000

Date(s) Application is Due: Feb 16

Contact: Grants Manager; (212) 703-9977 or (888) 582-2327; fax: (212) 997-0330; email: grants@afar.org or info@afar.org

Internet: http://www.afar.org/CART.html

Sponsor: American Federation for Aging Research
55 West 39th Street, 16th Floor
New York, NY 10018

AFAR Medical Student Summer Research Training in Aging Program   330

The program provides eight- to 12-week scholarships for clinical geriatrics and aging research. Scholars receive training from top experts in geriatrics and/or gerontology and other disciplines such as physiology, molecular biology, neurology, and epidemiology at one of four national training centers or, for a limited number of medical schools, at their own institutions. Scholars also participate in research initiatives in basic, clinical, or health services research relevant to care of the elderly. Following the scholarship, scholars participate in aging research and/or related activities under the supervision of their home-site sponsor. Most scholars will conduct training and research during the summer months. Application and guidelines are available online.

Requirements: Any allopathic or osteopathic medical student in good standing who has completed at least one year of medical school and is a U.S. citizen or permanent resident may apply. Each student submits an application, which must be accompanied by a letter of support from an individual who knows the candidate and by a biographical sketch from the home school faculty member who has agreed to nurture the work of the scholar's geriatric career development.

Amount of Grant: $1731 monthly stipend

Date(s) Application is Due: Feb 7

Contact: Grants Manager, (212) 703-9977; fax: (212) 997-0330; email: amfedaging@aol.com

Internet: http://www.afar.org/grants.html

Sponsor: American Federation for Aging Research
70 W 40th Street
New York, NY 10018

AFAR Medical Student Training in Aging Research Program   331

The Medical Student Training in Aging Research Program (MSTAR) provides medical students, early in their training, with an enriching experience in aging related research and geriatrics, under the mentorship of top experts in the field. This program introduces students to research and academic experiences that they might not otherwise have during medical school. Students participate in an eight to twelve week structured research, clinical, and didactic program in geriatrics, appropriate to their level of training and interests. Students may train at a National Training Center supported by the National Institute on Aging or, for a limited number of medical schools, at their own institution. Refer to the AFAR website for a complete listing of the participating institutions.

Requirements: Any allopathic or osteopathic medical student in good standing, who will have successfully completed one year of medical school at a U.S. institution by June, may apply. Applicants must be citizens or non-citizen nationals of the United States, or must have been lawfully admitted for permanent residence (i.e., in possession of a currently valid Alien Registration Receipt Card I-551, or some other legal verification of such status.)

Restrictions: Individuals on temporary or student visas and individuals holding PhD, MD, DVM, or equivalent doctoral degrees in the health sciences are not eligible.

Date(s) Application is Due: Feb 6

Contact: Grants Manager; (212) 703-9977 or (888) 582-2327; fax: (212) 997-0330; email: grants@afar.org or info@afar.org

Internet: http://www.afar.org/medstu.html

Sponsor: American Federation for Aging Research
55 West 39th Street, 16th Floor
New York, NY 10018

AFAR Paul Beeson Career Development Awards in Aging Research   332
for the Island of Ireland

The Paul Beeson Career Development Awards In Aging Research Program offers faculty development awards to outstanding junior physician faculty committed to academic careers in aging-related research, teaching, and practice. The goals of the program are: to encourage and assist the development of future leaders in the field of aging by supporting faculty members early in their careers to gain additional research training as needed and to establish independent programs in aging research; to deepen the commitment of research institutions to academic research in aging and to translating research outcomes to geriatric medicine by involving mentor and recipient in establishing and advancing the recipient's career in aging research; to expand medical research on aging broadly defined as including the biology of aging, maintenance of health and independence in old age, diseases and disabilities of old age and issues in their clinical management, and systems of care for the elderly. To maximize the educational/training opportunities of this fellowship program, Beeson Ireland Scholars are encouraged to spend three to six months abroad with a research team in the U.S. in the second or the third year of the award. The goals of this rotation are varied and can be for purposes of training in techniques, data collection, data analysis, etc. to enhance the Scholars' transition to independence. This is unique to the Beeson Ireland Program. To date, the program has provided awards to 149 very promising junior faculty at institutions in the United States as well as 3 junior faculty on the Island of Ireland. The program was established in 1994 by The Atlantic Philanthropies, The John A. Hartford Foundation, The Commonwealth Fund and The Starr Foundation, and administered by the American Federation for Aging Research (AFAR.) In 2004, The Atlantic Philanthropies, The John A. Hartford Foundation, The Starr Foundation and AFAR entered a partnership with The National Institute on Aging (NIA.) Until 2006, the program soley funded Scholars at U.S.-based institutions with support for beginning in 2007, with support from The Atlantic Philanthropies, Scholars are also funded in Ireland (The Republic of Ireland and Northern Ireland.) The scholar will receive a grant for three years. The salary will be at specialty registrar/clinical lecturer level to protect a minimum of 75% of the scholar's time for research, with the remainder available annually for research support, including an institutional overhead on the research support component.

Requirements: Nominations for the Paul Beeson Career Development Awards In Aging Research Program are to be made by the deans of medical schools (or equivalent) in Ireland (the Republic of Ireland and Northern Ireland.) Institutions may submit as many applications as they wish. To be eligible for nomination, a candidate must: be a U.S. physician or an Irish citizen physician of the Republic of Ireland, a UK citizen of Northern Ireland, an EU citizen living and working in the Island of Ireland or a non-EU citizen resident in the island of Ireland with a valid work permit; commit at least 75% of his/her full-time professional effort to the goals of this award; be a medical graduate who has recently completed or is about to complete a doctoral degree (PhD/MD) and be undertaking or have recently completed higher clinical training; have at least one research publication in a high-impact journal. If the candidate is not the first author, he/she must provide a short description of his/her contribution to the research and the writing of the paper; this should be substantial. For each scholar, a senior faculty member at the scholar's institution must be selected to serve as a mentor to help guide the scholar's research and career planning and provide access to organizations, programs, and colleagues helpful to the scholar's efforts. More than one mentor may be selected. Letters of endorsement including specific information on institutional support for the scholar should be provided by the dean, the relevant department chairperson (or equivalent - may be one person), and the mentor. In addition, three letters of reference should be provided by other faculty members and/or senior professionals with whom the scholar has worked and who are well acquainted with his/her capabilities, potential, accomplishments, and commitment. All candidates must submit applications endorsed by the Dean of School of Medicine (or equivalent).

Date(s) Application is Due: Jan 22

Contact: Grants Manager; (212) 703-9977 or (888) 582-2327; fax: (212) 997-0330; email: beeson@afar.org

Internet: http://afar.org/BeesonIreland.html

Sponsor: American Federation for Aging Research
55 West 39th Street, 16th Floor
New York, NY 10018

AFAR Research Grants   333

The major goal of this program is to assist in the development of the careers of junior investigators committed to pursuing careers in the field of aging research. Of particular interest are research projects concerned with understanding the basic mechanisms of aging; projects investigating age-related diseases, especially if approached from the point of view of how basic aging processes may lead to these outcomes; and projects concerning mechanisms underlying common geriatric functional disorders, as long as these include connections to fundamental problems in the biology of aging. Examples of promising areas of research include: aging and immune function; genetic control of longevity; neurobiology and neuropathology of aging; invertebrate or vertebrate animal models; cardiovascular aging; aging and cellular stress resistance; metabolic and endocrine changes; age-related changes in cell proliferation; caloric restriction and aging; DNA repair and control of gene expression; biology of the menopause; and aging and apoptosis. Application and instruction sheet are available online.

Requirements: The applicant must be an independent investigator with assigned independent space and must be within the first four years of a junior faculty appointment (instructor, assistant professor or equivalent) by July 1st.

Restrictions: The program does not provide support for postdoctoral fellows in the laboratory of a senior investigator; investigators who have already received major independent funding for research on aging, such as an R01 grant or a grant of equal to or greater than $100,000 from another private funding source; senior faculty, i.e., at the rank of associate professor level or higher; and projects that deal strictly with clinical problems, such as the diagnosis and treatment of disease, health outcomes, or the social context of aging. Former AFAR grant recipients are not eligible to reapply. Applicants for the Glenn/AFAR Breakthroughs in Gerontology (BIG) program cannot also submit an application for this research grant.

Amount of Grant: $75,000

Date(s) Application is Due: Dec 16

Contact: Grants Manager; (212) 703-9977 or (888) 582-2327; fax: (212) 997-0330; email: grants@afar.org or info@afar.org

Internet: http://afar.org/afar99.html

Sponsor: American Federation for Aging Research
55 West 39th Street, 16th Floor
New York, NY 10018

AFAR/Merck Junior Investigators In Geriatric Clinical Pharmacology   334

The training program provides the opportunity for two fellows to acquire competency in clinical pharmacology, geriatrics/gerontology, and research in geriatric clinical pharmacology. Applications must be submitted by an institution on behalf of an individual candidate for the fellowship.

Requirements: The candidate must be board certified or eligible in a primary specialty by July 1. Proof of board eligibility must be submitted. At the time of application, the candidate must be within three years of having completed postdoctoral or fellowship training. Previous training in geriatrics or clinical pharmacology is not required, but one or the other is highly desirable. Candidate must be a citizen or a permanent resident of the United States.

Samples: Dr. Daniel Kaufer, Assistant Professor of Psychiatry and Neurology, U of Pittsburgh (Pittsburgh, PA)—for research to identify risk factors for anticoagulant-related bleeding, $100,000.

Amount of Grant: $60,000 annually for two years, average

Date(s) Application is Due: Nov 3

Contact: Grants Manager, (212) 703-9977; fax: (212) 997-0330; email: grants@afar.org

Internet: http://www.afar.org/grants.html
Sponsor: American Federation for Aging Research
70 W 40th Street
New York, NY 10018

AFB Ferdinand Torres Scholarship 335

The scholarship is awarded to a legally blind full-time undergraduate or graduate student who presents evidence of economic need. U.S. citizens are eligible. Preference will be given to applicants residing in the New York City metropolitan area and new immigrants to the United States. Guidelines are available online.
Requirements: Applicants must submit proof of legal blindness; proof of acceptance in an accredited undergraduate or graduate program; evidence of economic need; proof of U.S. residence (e.g., telephone bill, lease, utility bill); and, for new immigrants, a description of country of origin and reason for coming to the United States.
Amount of Grant: $1500
Date(s) Application is Due: Apr 30
Contact: Julie Tucker, (212) 502-7661; fax: (212) 502-7771; email: juliet@afb.net
Internet: http://www.afb.org/section.asp?Documentid=1842
Sponsor: American Foundation for the Blind
11 Penn Plz, Suite 300
New York, NY 10001

AFB Karen D. Carsel Memorial Scholarship 336

This scholarship is given annually to a legally blind full-time graduate student who presents evidence of economic need. Contact the foundation for an application packet. Annual deadline dates may vary; contact program staff for exact dates.
Requirements: Applicant must submit application and the following information: proof of U.S. citizenship; evidence of legal blindness; certified transcripts of grades; proof of acceptance into a full-time graduate program; evidence of economic need; three letters of reference; and a typewritten statement of no more than two double-spaced pages describing educational and personal goals, strengths, weaknesses, hobbies, achievements, information on any college financial assistance received, and how scholarship monies will be used.
Amount of Grant: $500
Date(s) Application is Due: Apr 30
Contact: Julie Tucker, Executive Secretary, Information Center, (212) 502-7661; fax: (212) 502-7771; email: juliet@afb.net
Internet: http://www.afb.org/info_document_view.asp?documentid=1842
Sponsor: American Foundation for the Blind
11 Penn Plz, Suite 300
New York, NY 10011

AFB Rudolph Dillman Memorial Scholarship 337

Scholarships are offered to undergraduate or graduate students who are studying in the field of rehabilitation and/or education of persons who are blind or visually impaired. One of these grants is specifically for a student who meets all requirements and submits evidence of economic need.
Requirements: Legally blind students are eligible. Applicants must submit proof of acceptance in an accredited undergraduate or graduate program studying in the field of rehabilitation and/or education of persons who are blind or visually impaired.
Restrictions: Previous recipients are ineligible.
Amount of Grant: $2500
Date(s) Application is Due: Apr 30
Contact: Brent Hopkins, Program Contact, (212) 502-7676; fax: (212) 502-7777; email: bhopkins@afb.net
Internet: http://www.afb.org/info_document_view.asp?documentid=1907
Sponsor: American Foundation for the Blind
11 Penn Plz, Suite 300
New York, NY 10011

AFCEA Educational Foundation Postgraduate Fellowships 338

AFCEA Fellowships of $15,000 each will be awarded to full-time postgraduate students who have earned Master's degrees or the equivalent and who are currently enrolled in a doctoral degree program in electrical, electronic, chemical or communications engineering, physics, math, or computer science. The dissertation title or abstract of the specific area of research is required. The intent of the AFCEA Fellowship is to reward excellence of demonstrated effort at the doctoral level of study, rather than the potential of such excellence. Please read the requirements carefully to determine specific eligibility before submitting an application. Applications are available November through January 1.
Requirements: Candidates must be U.S. citizens and nominated by the dean of the college of engineering at any accredited university in the United States.
Amount of Grant: $15,000
Date(s) Application is Due: Feb 1
Contact: Norma Corrales, Scholarships and Awards, (703) 631-6149 or (800) 336-4583 ext 6149; email: scholarship@afcea.org.
Internet: http://www.afcea.org/scholarships/scholarships_details.asp?ID=7
Sponsor: Armed Forces Communications and Electronics Association Educational Foundation
4400 Fair Lakes Ct
Fairfax, VA 22033

Affymetrix Corporate Contributions Grants 339

Affymetrix makes charitable donations to nonprofit organizations through its corporate philanthropy program. The company focuses its giving in three main areas: education—programs that support science and math education, focusing on K-12 students and their teachers; ethics—organizations that help foster an ongoing public dialog about genetic-related ethics; and cancer research and advocacy—nonprofit organizations working in the areas of disease research and advocacy, with a specific emphasis on cancer. Requests are reviewed four times per year. The average grant size is approximately $2500.
Requirements: 501(c)3 nonprofit organizations are eligible.
Restrictions: The corporation does not provide funding for advertising journals or booklets; fundraising events such as telethons, walkathons, and races; specific performances or concerts; sporting events; endowment campaigns; film, video, television, or radio projects; grants to individuals; political causes or candidates; or organizations that practice discrimination or limit membership on the basis of race, creed, gender, age, sexual orientation, or national origin.
Contact: Contributions Manager, email: outreach@affymetrix.com
Internet: http://www.affymetrix.com/corporate/outreach/corporate.affx
Sponsor: Affymetrix
3420 Central Expwy
Santa Clara, CA 95051

AFLAC Inc Grants 340

The company awards grants to nonprofit organizations for cancer research, health-related projects, and community services. International contributions are determined by the company's senior officers. Submit a letter of inquiry that includes an organization description, amount requested, purpose of request, recently audited financial statement, and proof of tax-exempt status.
Requirements: Nonprofit organizations are eligible.
Restrictions: Grants are not made to individuals.
Samples: National Museum of African American History and Culture (Washington, DC)—to develop this new museum, whose collections will cover such topics as slavery, the Harlem Renaissance, and the civil-rights movement, $1 million (2005).
Contact: Patsy Thomas, (706) 317-6109; fax: (706) 320-2288; email: pthomas3@aflac.com
Internet: http://www.aflac.com
Sponsor: AFLAC Inc
1932 Wynnton Road
Columbus, GA 31999

AFOSR Artificial Intelligence Research Grants 341

This program sponsors research into ways to make the best use of uncertain information; share and disseminate information; increase the accuracy, speed, and economy of the recognition and identification process; and aid the intelligence analyst. The program concentrates on research needed to develop large-scale intelligent systems that can address practical Air Force needs. To that end, means are sought to scale up those methods that work for small knowledge-based systems. To aid the information analyst in fusing information from diverse modalities, the program seeks means to combine numeric and symbolic inference methods. Research could also focus on integrating probabilistic reasoning methods with traditional formal logic methods, and perhaps with other forms of computation. Qualitative methods that will drastically simplify computation and increase performance robustness are also of interest. The program seeks to develop technology that will support decision making. To that end, research is needed to develop intelligent agents capable of gathering information, reducing data to a manageable amount of essential information, and cooperating with other agents to solve problems. Research is also needed to combine artificial intelligence methods with operations research tools to overcome inefficiencies in solving some mission-critical Air Force problems (e.g., scheduling in a distributed, dynamic environment). Intelligent tutoring is an area of increased interest to the Air Force. The focus of this effort is to develop efficient computer-mediated tools for instructional delivery both for training and tutoring, with the objective of reducing personnel needs and optimizing tutoring and training. Adaptive teaching systems that model the trainee and attempt to understand his or her responses by simulating these models is one area supported within this program. Research tasks in intelligent tutoring are linked to the Human Resource Laboratory of the Air Force Armstrong Laboratory, where the evaluation and experimentation with actual trainees occurs. Proposals are accepted at any time.
Requirements: Prior to submitting an informal or formal proposal, investigators are encouraged to explore, with the program manager AFOSR's possible interest in their research ideas.
Contact: Dr. Robert L. Herklotz, AFOSR/NM, (703) 696-6565; fax: (703) 696-8450; email: robert.herklotz@afosr.af.mil
Internet: http://www.afosr.af.mil/pdfs/BAA2005-1.pdf
Sponsor: Air Force Office of Scientific Research
4015 Wilson Boulevard, Room 713
Arlington, VA 22203-1954

AFOSR Atomic and Molecular Physics Research Grants 342

This program involves experimental and theoretical research on the properties and interactions of atoms and molecules and forms the basic underpinning of a large range of technological applications in navigation, guidance, communications, low- and high-altitude nuclear weapons effects phenomenology, directed energy weaponry, and lasing mechanisms. Among the topics of interest are the following: apply cooling and trapping techniques to a broad range of problems, including high-resolution spectroscopy and cold atom collisions particularly between atoms in excited quantum states; develop high-precision techniques for navigation, guidance, and remote sensing'particularly those suited to use in an orbital environment; study and understand the evolution of cold atomic systems into ultralow-density condensed matter systems; study the formation and evolution of cold (<1 K) plasmas; study the dynamics of single, large molecules in complex systems; Study antiproton capture, confinement, transport, injection, and annihilation processes'particularly those leading to the formation and storage of anti-hydrogen; develop novel techniques for production of high-power microwaves, X-rays, and gamma rays; understand the interaction of atoms and molecules with strong fields; and study cross-sections of atmospheric species. Proposals are accepted at any time.

Requirements: Scientists are encouraged to contact the program manager before submitting a formal proposal to determine whether their research interests would match existing programs.

Contact: Dr. Anne Matsuura, Physics and Electronics, AFOSR/NE, (703) 696-6204; fax: (703) 696-8481; email: anne.matsuura@afosr.af.mil

Internet: http://www.afosr.af.mil/pdfs/BAA2005-1.pdf

Sponsor: Air Force Office of Scientific Research
4015 Wilson Boulevard, Room 713
Arlington, VA 22203-1954

AFOSR Combustion and Diagnostic Research Grants   343

Fundamental understanding of the physics and chemistry of multiphase turbulent reacting flows is essential for improving performance of air-breathing propulsion systems. The program is interested in innovative research proposals that use simplified configurations for experimental and theoretical investigations. Highest priority is assigned to studies of supersonic combustion, atomization and spray behavior of slurries and liquids, fuel combustion chemistry, supercritical fuel behavior in precombustion and combustion environments, and novel diagnostic methods for experimental measurements. Other topics of interest include turbulent combustion, soot formation, and interactive control. The program also seeks innovative approaches to produce reduced models of turbulent combustion. These models would improve upon current capability by producing prediction methods that are both quantitatively accurate and computationally tractable. They would address all aspects of multiphase turbulent reacting flow, including such challenging objectives as predicting the concentrations of trace pollutant and signature producing species as products of combustion. Approaches such as novel subgrid-scale models for application to large eddy simulations of subsonic and supersonic combustion are of interest. Proposals are accepted at any time.

Requirements: Prior to submitting an informal or formal proposal, investigators are encouraged to explore, with the program manager, AFOSR's possible interest in their research ideas.

Contact: Dr. Julian M. Tishkoff, AFOSR/NA, (703) 696-8478; fax: (703) 696-8451; email: julian.tishkoff@afosr.af. mil

Internet: http://www.afosr.af.mil/pdfs/BAA2005-1.pdf

Sponsor: Air Force Office of Scientific Research
4015 Wilson Boulevard, Room 713
Arlington, VA 22203-1954

AFOSR Defense University Research Instrumentation Program   344

This program is designed to improve the capabilities of U.S. universities to conduct research and to educate scientists and engineers in areas important to national defense by providing funds for the acquisition of research equipment. AFOSR BAA 05-4

Requirements: U.S. higher education institutions with degree-granting programs in science, math, and engineering are eligible to apply.

Amount of Grant: $60,000-$990,000 range; $207,000 average

Date(s) Application is Due: Aug 25

Contact: Douglas Glenn Holloway, (703) 696-5944; fax: (703) 696-7320; email: glenn. holloway@afosr.af.mil

Internet: http://www.afosr.af.mil/docs/DURIP06BAA19May05.pdf

Sponsor: Air Force Office of Scientific Research
PIE/DURIP, 875 N Randolph Street, Suite 325, Room 3112
Arlington, VA 22203

AFOSR Electromagnetics Research Grants   345

Conduct research in electromagnetics to produce conceptual descriptions of electromagnetic properties of novel materials/composites (such as photonic band gap media) and simulate their uses in various operational settings. Evaluate methods to recognize and track targets and to penetrate tree cover or other dispersive media with wide band radar (propagation of precursors for example) and design switches/transmitters to produce such pulses. Develop computational electromagnetic simulation codes that are rapid and accompanied by rigorous error estimates/controls. Proposals are accepted at any time.

Requirements: Investigators are encouraged to discuss their projects with the program manager prior to submission of a proposal.

Contact: Dr. Arje Nachman, Mathematics and Space Sciences, AFOSR/NM, (703) 696-8427; fax: (703) 696-8450; email: arje.nachman@afosr.af.mil

Internet: http://www.afosr.af.mil/pdfs/BAA2005-1.pdf

Sponsor: Air Force Office of Scientific Research
4015 Wilson Boulevard, Room 713
Arlington, VA 22203-1954

AFOSR Metallic Materials Research Grants   346

The goal of research in metallic materials is to provide the fundamental knowledge required to develop metallic alloys for aerospace applications. Potential applications of these materials include turbine airfoils and disks, engine casings and nozzle components,

rocket propulsion components, airframe and spacecraft structural components, and hypersonic vehicle skins. Research on improved metallic structural materials capable of operating with higher specific mechanical properties is encouraged. Research on new actuation/sensing alloys are also of interest. This goal will be accomplished by understanding the relationships between processing, chemistry, and structure on the one hand and properties of metallic and metallic composite materials on the other. Specific scientific topics include the development and experimental verification of theoretical and computational (atomistic) models, processing science, phase transformations, interfacial phenomena, strengthening mechanisms, plasticity, creep, fatigue, environmental effects, and fracture of structural metallic materials. Materials currently under research include lightweight structural metals, refractory metals, intermetallic alloys, amorphous alloys and their composites, and microlaminated materials. Proposals are accepted at any time.

Requirements: Prior to submitting an informal or formal proposal, investigators are encouraged to explore, with the program manager, AFOSR's possible interest in their research ideas.

Contact: Dr. Craig S. Hartley, AFOSR/NA, (703) 696-8523; fax: (703) 696-8451; email: craig.hartley@afosr.af.mil

Internet: http://www.afosr.af.mil/pdfs/BAA2005-1.pdf

Sponsor: Air Force Office of Scientific Research
4015 Wilson Boulevard, Room 713
Arlington, VA 22203-1954

AFOSR Molecular Dynamics Research Grants   347

The objectives of this program are to understand, predict, and control the reactivity and flow of energy in molecules. This knowledge will be used in atmospheric chemistry, high-energy density material research, chemical laser research, and other chemical systems where predictive capabilities and control of chemical reactivity and energy flow at a detailed molecular level will be of importance. Areas of interest in atmospheric chemistry include the dynamics of ion-molecule reactions relevant to processes in weakly ionized plasmas in the ionosphere, gas-surface interactions in space, and reactive energy transfer processes that produce and affect radiant emissions in the upper atmosphere. Research on high-energy density matter for propulsion applications investigates novel concepts for storing chemical energy in low-molecular-weight systems, the stability of energetic molecular systems, and the storage of energetic species in cryogenic solids. Research in energy transfer and energy storage in metastable states of molecules supports an interest in new concepts for chemical lasers. Also of interest is the study of the structure, stability, and growth of metal/ceramic interfaces. Fundamental studies aimed at developing predictive capabilities for and control of chemical reactivity, bonding, and energy transfer processes are encouraged. Unsolicited proposals are welcomed at any time.

Requirements: Prior to submitting an informal or formal proposal, investigators are encouraged to explore, with the program manager, AFOSR's possible interest in their research ideas.

Contact: Dr. Michael R. Berman, AFOSR/NL, (703) 696-7781; fax: (703) 696-8449; email: michael.berman@afosr.af.mil

Internet: http://www.afosr.af.mil/pdfs/BAA2005-1.pdf

Sponsor: Air Force Office of Scientific Research
4015 Wilson Boulevard, Room 713
Arlington, VA 22203-1954

AFOSR Optimization and Discrete Mathematics Research Grants   348

The goal of this program is to develop mathematical methods for solving large or complex problems, such as those occurring in logistics, engineering design, or strategic planning. These problems can often be formulated as mathematical programs. Therefore, research is directed at linear and nonlinear programming methods, especially when formulated for the solution of selected Air Force problems. In addition to the evolution of traditional solution methods, the program supports new algorithmic paradigms (e.g., simulated annealing, genetic algorithms). It supports research in discrete event systems, especially as it relates to Air Force transportation, manufacturing, command and control systems, and battlefield management. The program is particularly interested in the control of discrete event systems through models that combine simulation and optimization. Proposals are accepted at any time.

Requirements: Prior to submitting an informal or formal proposal, investigators are encouraged to explore, with the program manager, AFOSR's possible interest in their research ideas.

Contact: Dr. Neal Glassman, Mathematics and Space Sciences, AFOSR/NM, (703) 696-9548; fax: (703) 696-8450; email: neal.glassman@afosr.af.mil

Internet: http://www.afosr.af.mil/pdfs/BAA2005-1.pdf

Sponsor: Air Force Office of Scientific Research
4015 Wilson Boulevard, Room 713
Arlington, VA 22203-1954

AFOSR Organic Matrix Composites Research Grants   349

This program addresses the materials science issues relating to the use of polymers in aerospace and space structures such as airframes, engine components, rocket, launch vehicles and satellites. The goal is to provide the scientific base that will lead to higher performance, more durable, and more affordable structures for Air Force applications. The approach is to address issues relating to the development of improved performance or lower cost polymer-matrix composite (PMC) systems and the processing and the utilization of these structures during deployment. Chemistry and processing of structural adhesives and polymeric precursors for ceramic and carbon-carbon structures are also within the scope of this program. Materials issues relating to all material preforms and processing leading to the end components are of interest. Examples of these include

resin chemistry and formulations, prepregs processing, dry preforms, layups, and cure processes. The emphasis of the current program is to study the environmental effects on the long-term properties of polymer matrix composites. These environmental effects include a harsh processing environment (e.g., high-temperature processing), application environments (e.g., high- temperature exposure under pressurized conditions), and service environments (e.g., moisture, solvents). Research will address the chemistry and physics of the degradation mechanisms that lead to deterioration of the performance of the PMC structures. The scope will cover the matrix, reinforcement, interphase, and composite as a whole. The results of this research will lead to accurate prediction of PMC structures' service life and to alternative material systems, processing procedures, and service practices that can increase the service life of these structures. Proposals are accepted at any time.
Requirements: Investigators are encouraged to discuss their projects with the program manager prior to submission of a proposal.
Contact: Dr. Charles Y-C Lee, AFOSR/NL, (703) 696-7779; fax: (703) 696-8449; email: charles.lee@afosr.af.mil
Internet: http://www.afosr.af.mil/pdfs/BAA2005-1.pdf
Sponsor: Air Force Office of Scientific Research
4015 Wilson Boulevard, Room 713
Arlington, VA 22203-1954

AFOSR Physical Mathematics and Applied Analysis Research Grants   350
Conduct research in physical mathematics and applied analysis to develop accurate models of physical phenomena to enhance the fidelity of simulation. Investigate the properties of coherently propagating ultrashort laser pulses (both the currently examined IR frequencies and possible extension to UV frequencies) through the air and their exploitation in areas such as electronic warfare (ancillary production of HPM), laser-guided munitions (possible propagation through obscurants), and irradiation of chem/bio clouds. Develop algorithms to simulate nonlinear optical effects within fiber lasers (with weaponization in mind) and nonlinear optical media. Study feasibility of designing reconfigureable warheads by suitable placement/timing of microdetonators as well as the operation of pulsed detonation engines. Pursue description of dynamics of internal stores released from transonic/supersonic platforms as well as the fluid dynamics accompanying curved, rotating jet turbine blades. Also the dynamics of the atmosphere near and above the tropopause with an emphasis on the understanding of turbulence and its production by topography and storms is of interest. Other areas of interest include the understanding of chaos in circuitry such as missile guidance systems, prediction of effective properties of various composite media, advanced fracture mechanics theories, which also include thermal loading such as might be produced by exposure to a strong laser.
Requirements: Prior to submitting an informal or formal proposal, investigators are encouraged to explore, with the program manager, AFOSR's possible interest in their research ideas.
Contact: Dr. Arje Nachman, Mathematics and Space Sciences, AFOSR/NM, (703) 696-8427; fax: (703) 696-8450; email: arje.nachman@afosr.af.mil
Internet: http://www.afosr.af.mil/pdfs/BAA2005-1.pdf
Sponsor: Air Force Office of Scientific Research
4015 Wilson Boulevard, Room 713
Arlington, VA 22203-1954

AFOSR Polymer Chemistry Research Grants   351
The goal of this research area is to gain a better understanding of the influence of chemical structures and processing conditions on the properties and behaviors of polymeric and organic materials. Research interests include photonic polymers, polymers with interesting electronic properties, polymer blends, liquid crystals and liquid crystalline polymers, and durable coatings for aircraft and nanostructures. Proposals with innovative material concepts that will extend our understanding of the structure-property relationship of these materials and achieve significant property improvement over current state-of-the-art knowledge are sought. Research interests include creating new and improved properties, achieving reproducible properties, and addressing durability of properties during the life cycle of Air Force systems. Material concepts that can improve on the above-mentioned optical, electronic, and mechanical properties of polymers are sought. These concepts include, but are not limited to, polymer blends, liquid crystals and liquid crystalline polymers, and nanostructures. Proposals are accepted at any time.
Requirements: Prior to submitting an informal or formal proposal, investigators are encouraged to explore, with the program manager, AFOSR's possible interest in their research ideas.
Contact: Dr. Charles Lee, Chemistry and Life Sciences, AFOSR/NL, (703) 696-7779; fax: (703) 696-8449; email: charles.lee@afosr.af.mil
Internet: http://www.afosr.af.mil/pdfs/BAA2005-1.pdf
Sponsor: Air Force Office of Scientific Research
4015 Wilson Boulevard, Room 713
Arlington, VA 22203-1954

AFOSR Quantum Electronic Solids Research Grants   352
This program focuses on materials that exhibit cooperative quantum electronic behavior, with the primary emphasis on superconductors—and any conducting materials with surfaces that can be modified and observed through the use of scanning tunneling—and related atomic-force microscopies. The program also focuses on device concepts using these materials for electromagnetic detection and signal processing in Air Force systems. The long-standing materials aspects of the program are based on the fabrication, characterization, and electronic behavior of superconducting thin films, which ultimately can lead to the discovery of new and improved electronic circuit elements. Two main objectives are to understand the mechanisms that give rise to superconductivity in

selected ceramics and to produce high-quality Josephson tunneling structures. Recently the program has been expanded to include bulk superconducting materials that can be useful in producing current-carrying wires in power applications. A continuing interest in this program is the search for new electronic device concepts that involve superconductive elements, either alone or in concert with semiconductors and normal metals; there is also interest in understanding high-power absorption in high-temperature superconducting materials at microwave frequencies. A minor aspect of this program is the inclusion of scanning probe techniques to fabricate, characterize, and manipulate atomic-, molecular-, and nanometer-scale structures, with the goal of producing a new generation of improved sensors, resulting in the ultimate miniaturization of analog and digital circuitry. Proposals are accepted at any time.
Requirements: Investigators are encouraged to discuss their projects with the program manager prior to submission of a proposal.
Contact: Dr. Harold Weinstock, Physics and Electronics, AFOSR/NE, (703) 696-8572; fax: (703) 696-8481; email: harold.weinstock@afosr.af.mil
Internet: http://www.afosr.af.mil/pdfs/BAA2005-1.pdf
Sponsor: Air Force Office of Scientific Research
4015 Wilson Boulevard, Room 713
Arlington, VA 22203-1954

AFOSR Semiconductor Materials Research Grants   353
This research area is directed toward developing advanced optoelectronic and electronic materials and structures to provide improvements required for future Air Force systems. The focus is currently on growth and use of semiconductors in bulk structures, single heterostructures, quantum wells, superlattices, quantum wires, and quantum dots. Proposals are sought for significant advances in these areas, or expansion to novel application of materials such as organic polymers, amorphous, and polycrystalline materials, with estimates comparing potential improvements to present capabilities and the impact on Air Force capabilities. Wavelength ranges of interest span the spectrum from UV, visible, NIR, MWIR, FIR, and extending into the terahertz range. Also of interest are the semiconductor systems that exhibit ferromagnetism, which may lead to semiconductor spintronic devices. An understanding of these materials is important to device development. Proposals area accepted at any time.
Contact: Dr. Anne Matsuura, AFOSR/NE, (703) 696-6204; fax: (703) 696-8481; email: anne.matsuura@afosr.af.mil
Internet: http://www.afosr.af.mil/pdfs/BAA2005-1.pdf
Sponsor: Air Force Office of Scientific Research
4015 Wilson Boulevard, Room 713
Arlington, VA 22203-1954

AFOSR Sensory Systems Research Grants   354
The sensory systems program pursues an understanding of biological sensing mechanisms and investigates the integration of multiple sensory systems in human perception. Emphasis is on studies that can contribute the basic science foundation needed to inform new approaches to enhance human performance. This program supports research that coordinates empirical studies with mathematical or computational modeling. The development of theoretical models is desired, in part, for their eventual application to human factors problems, such as those that arise in the design of human systems to, for example, assist spatial orientation or navigation, find, track, and manipulate objects, or respond to acoustic information from multiple, simultaneous sources. The current emphasis of this program is on the dynamic integration of multiple sensory inputs in human performance. One ongoing effort deals with the integration of auditory, visual, vestibular, and somatosensory inputs in response to non-standard gravito-inertial forces. Another deals with the coordination of head and eyes in tracking moving targets. A third effort studies several aspects of spatial audition, including sound localization, distance perception, and auditory cueing of visual search. The program is multi-disciplinary, drawing upon expertise in areas such as neurophysiology, computer and electrical engineering, biology, mathematics, and experimental psychology. Applicants are encouraged to develop collaborative relationships with scientists in the Air Force Research Laboratory. Proposals are accepted at any time.
Requirements: Prior to submitting an informal or formal proposal, investigators are encouraged to explore, with the program manager, AFOSR's possible interest in their research ideas.
Contact: Dr. Willard Larkin, Chemistry and Life Sciences, AFOSR/NL (703) 696-7793; fax: (703) 696-8449; email: willard.larkin@afosr.af.mil
Internet: http://www.afosr.af.mil/pdfs/BAA2005-1.pdf
Sponsor: Air Force Office of Scientific Research
4015 Wilson Boulevard, Room 713
Arlington, VA 22203-1954

AFOSR Signals Communication and Surveillance Research Grants   355
This research activity is concerned with the systematic analysis and interpretation of variable quantities in media that are intended to convey information. Communications signals and surveillance images are of special importance. Signals are physically generated, propagated through electromagnetic or other media, and recaptured for use at a receiving mechanism. Modern radar, infrared, and electro-optical sensing systems produce large quantities of raw signaling that exhibit hidden correlations, are vulnerable to distortion by noise, and retain features tied to a particular physical origin. Statistical research that treats spatial and temporal dependencies in such data is necessary to exploit the usable information within. An outstanding need in the treatment of signals is to develop resilient algorithms for data representation in fewer bits (compression), image reconstruction/enhancement, and spectral/frequency estimation in the presence

of external corrupting factors. These factors can involve deliberate interference, noise, ground clutter, and multipath effects. Proposals are accepted at any time.

Requirements: Prior to submitting an informal or formal proposal, investigators are encouraged to explore, with the program manager, AFOSR's possible interest in their research ideas.

Contact: Dr. Jon Sjogren, AFOSR/NM, (703) 696-6564; fax: (703) 696-8540; email: jon.sjogren@afosr.af.mil

Internet: http://www.afosr.af.mil/pdfs/BAA2005-1.pdf

Sponsor: Air Force Office of Scientific Research
4015 Wilson Boulevard, Room 713
Arlington, VA 22203-1954

AFOSR Software and Systems Research Grants   356

The goal of this research program is to develop advanced computing technology to support future Air Force needs in battlespace information management. Computing research is sought to meet several challenges: control and integration of the vast amounts of information flowing through battlespace computer networks, protection of friendly information resources, and complexities in software and algorithm development in support of dynamic planning and execution control. Basic research is needed in a number of areas to build battlespace information systems of the future. For example, mathematical foundations of information fusion must be established—robust, integrated fusion architectures for handling increasing diversity of input sources are especially important. The program also is interested in foundational approaches to the specification and design of agents for network management, for information retrieval, filtering, summarizing, and for planning. For network protection, researchers will focus on determining and analyzing network security properties at all network layers and examining how to ensure that a network possesses these properties. New approaches to intrusion detection and attack recovery are also needed. Basic research that anticipates the nature of future information system attacks is critical to the survivability of these systems. In the area of software and algorithm development, we seek mathematical approaches for the specification, design, and analysis of distributed software systems. Rigorous mathematical methods, especially those that involve aspects of timing, control, dependability, and security, will be crucial to development of future battlespace information systems. New approaches for overcoming the increasing computational complexity of these systems are essential. Proposals are accepted at any time.

Requirements: Prior to submitting an informal or formal proposal, investigators are encouraged to explore, with the program manager, AFOSR's possible interest in their research ideas.

Contact: Dr. Robert L. Herklotz, Mathematics and Space Sciences, AFOSR/NM, (703) 696-6565; fax: (703) 696-8450; email: robert.herklotz@afosr.af.mil

Internet: http://www.afosr.af.mil/pdfs/BAA2005-1.pdf

Sponsor: Air Force Office of Scientific Research
4015 Wilson Boulevard, Room 713
Arlington, VA 22203-1954

AFOSR Space Electronics Research Grants   357

This research program addresses Air Force requirements for advanced high-performance electronic devices. Depending upon the specific requirement, this calls for high-efficiency, greater-speed, higher-power, lower-noise, low-voltage/low-power performance, etc. Greater emphasis is given to analog devices than to digital and optoelectronic structures. Emphasis is shifting away from more traditional compound semiconductor materials, such as GaAs and InP, to emerging materials such as the wide bandgap GaN family. There is also interest in the understanding and electronic applications of so-called 'wet Al-oxides' formed by the oxidation of AlAs and related materials. The effects of radiation (natural and man made) on these and other electronic devices are important concerns. Proposals are accepted at any time.

Requirements: Prior to submitting an informal or formal proposal, investigators are encouraged to explore, with the program manager, AFOSR's possible interest in their research ideas.

Contact: Dr. Gerald Witt, Physics and Electronics, AFOSR/NE, (703) 696-8571; fax: (703) 696-8481; email: gerald.witt@afosr.af.mil

Internet: http://www.afosr.af.mil/pdfs/BAA2005-1.pdf

Sponsor: Air Force Office of Scientific Research
4015 Wilson Boulevard, Room 713
Arlington, VA 22203-1954

AFOSR Space Power and Propulsion Research Grants   358

Research activities fall into three areas: nonchemical orbit-raising propulsion, chemical propulsion, and plume signatures/contamination resulting from both chemical and nonchemical propulsion. Research in the first area is directed primarily at advanced space propulsion and is stimulated by the need to transfer payloads between orbits, station-keeping, and pointing. It includes studies of the sources of physical (nonchemical) energy and the mechanisms of release. Emphasis is on understanding electrically conductive flowing gases (plasmas) that serve to convert beamed or electrical energy into kinetic form. Theoretical and experimental investigations are being conducted on the phenomenon of energy coupling and the transfer of plasma flows in electrode and electrodeless systems under plasma dynamic environments. Topics of interest include characteristics of pulsed and steady-state plasmas; scaling physics; characteristics of equilibrium and non-equilibrium flowing plasma; characteristics of electrical and hydrodynamic flows; instabilities of plasma bulk and wall layers; interactions of plasma-surface, plasma-electrode, plasma-magnetic, and plasma-electric fields; losses to inert parts; characteristics of plasmas in high-magnetic fields and pressures; and plasma diagnostics (new and unique noninterfering measuring techniques). Research is sought

on chemical propulsion to predict and suppress combustion instabilities in liquid rocket systems and pulsed detonation rocket engines. Topics of interest include the modeling of the coupling among unsteady flows, combustion, acoustic fields, and chemical kinetics, detonation phenomenon, modeling using novel tools such as molecular dynamics, direct simulation Monte Carlo, and hybrid approach. Proposals are accepted at any time.

Requirements: Prior to submitting an informal or formal proposal, investigators are encouraged to explore, with the program manager, AFOSR's possible interest in their research ideas.

Contact: Dr. Mitat Birkan, Aerospace and Materials Sciences, AFOSR/NA, (703) 696-7234; fax: (703) 696-8451; email: mitat.birkan@afosr.af.mil

Internet: Space Electronics

Sponsor: Air Force Office of Scientific Research
4015 Wilson Boulevard, Room 713
Arlington, VA 22203-1954

AFOSR Space Sciences Research Grants   359

The program's research interests include, but are not limited to, the hazards to spacecraft caused by space debris, interplanetary dust, asteroids, and comets; the structure and dynamics of the solar interior and their role in driving solar activity; the mechanism(s) responsible for heating the solar corona and accelerating it outward as the solar wind; coronal mass ejections (CMEs) and solar flares; the coupling between the solar wind, the magnetosphere, and the ionosphere; the origin and energization of magnetospheric plasma; and the triggering and temporal evolution of geomagnetic storms. By specifying the flow of mass, momentum, and energy from the sun to the earth, and by forecasting the plasma phenomena that mediate the flow of energy through space, the program's goal is to develop a global, coupled solar-terrestrial model that connects solar activity with the deposition of energy in the earth's upper atmosphere. The program also is strongly interested in advanced deep space surveillance techniques to observe and track Near Earth Objects and other physical threats to Air Force systems. In this regard, we are looking for innovative astronomical observation techniques that involve advanced image processing and/or sensor technology. Astrophysical or astronomical research and observations that investigate stellar-planetary interactions in general and physical processes occurring in the sun in particular are also of interest. Proposals are accepted at any time.

Requirements: Investigators are encouraged to explore their projects with the program manager prior to submitting the proposal.

Contact: Major David Byers, AFOSR/NM, (703) 696-8411; fax: (703) 696-8450; email: david.byers@afosr.af.mil

Internet: http://www.afosr.af.mil/pdfs/BAA2005-1.pdf

Sponsor: Air Force Office of Scientific Research
4015 Wilson Boulevard, Room 713
Arlington, VA 22203-1954

AFOSR Structural Mechanics Research Grants   360

The objective of this research program is to study solid mechanics fundamentals and structural principles that are necessary to ensure the integrity of current and future aerospace structures, including aircraft, missiles, and spacecraft. Proposals are sought that will lead to a fundamental understanding of the behavior of structures that are composed of current metallic materials as well as advanced composite materials. Proposals are also sought that will develop principles to predict nonlinear aerospace structural characteristics under coupled fluid, thermal, and mechanical loads. We are interested in solid mechanics principles that govern nonlinear coupled deformation and damage mechanisms that dictate anisotropic and heterogeneous medium response and structural performance. Topics such as damage localization, instability formation, homogenization, energy dissipation, and local and global response correlation are of interest. Structural nonlinear behavior and control owing to coupled mechanical, fluid, acoustic, and thermal loads are important to the design and performance prediction of aerospace systems. Fluid-structure interaction, aerothermoelasticity, and the development of intelligent materials and structures are of interest to this program. The degradation of materials and structures over long periods of service is also of interest, since current Air Force weapon systems will remain in service much longer than originally anticipated. This research includes the prediction of material degradation under combined mechanical and environmental loads, as well as the nondestructive detection and quantification of internal damage (e.g., corrosion, fatigue cracking). Proposals are accepted at any time.

Requirements: Prior to submitting an informal or formal proposal, investigators are encouraged to explore, with the program manager, AFOSR's possible interest in their research ideas.

Contact: Captain Clark Allred, Aerospace and Materials Sciences, AFOSR/NA (703) 696-7259; fax: (703) 696-8451; email: clark.allred@afosr.af.mil

Internet: http://www.afosr.af.mil/pdfs/BAA2005-1.pdf

Sponsor: Air Force Office of Scientific Research
4015 Wilson Boulevard, Room 713
Arlington, VA 22203-1954

AFOSR Theoretical Chemistry Research Grants   361

The major objective of the theoretical chemistry program is to develop new methods that can be utilized as predictive tools for designing new materials and improving processes important to the Air Force. These new methods can be applied to areas of interest to the Air Force including the structure and stability of molecular systems that can be used as advanced propellants; molecular reaction dynamics; and the structure and properties nanostructures and interfaces. Interest in advanced propellants is concentrated in the High Energy Density Matter (HEDM) Program, which aims to develop new propellant systems that can double the current payload capacity that can be put into orbit. Theoretical chemistry is used to predict promising energetic systems, to assess their stability, and to guide the efficient synthesis of selected candidates. These tools will help identify the most promising synthetic reaction pathways and predict the effects of condensed media effects on synthesis. This program is also seeking to identify novel energetic molecules and investigating the interactions that control or limit the stability of these systems. Particular interests in reaction dynamics include developing methods to seamlessly link electronic structure calculations with reaction dynamics, and using theory to describe and predict the details of ion-molecule reactions and electron-ion dissociative recombination processes relevant to ionospheric and space effects on Air Force systems. Interest in nanostructures and materials includes work on catalysis, surface-enhanced processes mediated by plasmon resonances, and bonding at metal-ceramic interfaces. This program also encourages the development of new methods and algorithms that take advantage of parallel computing architectures to predict properties with chemical accuracy for systems having a very large number of atoms that span multiple time and length scales. Proposals are accepted throughout the year.

Requirements: Investigators are encouraged to discuss their projects with the program manager prior to submitting a proposal.

Contact: Dr. Michael Berman, Chemistry and Life Sciences, AFOSR/NL, (703) 696-7781; fax: (703) 696-8449; email: michael.berman@afosr.af.mil

Internet: http://www.afosr.af.mil/pdfs/BAA2005-1.pdf

Sponsor: Air Force Office of Scientific Research
4015 Wilson Boulevard, Room 713
Arlington, VA 22203-1954

AFOSR Turbulence and Rotating Flows Research Grants 362

This program seeks to advance fundamental understanding of the complex, unsteady flows occurring in gas turbine engines and to apply that understanding to the development of physically based predictive models and innovative concepts for controlling complex internal flows. The program also addresses a broader class of flow-control problems related to generic technologies such as fluid thrust vectoring, internal flow tailoring, high lift, enhanced mixing, noise and signature reduction, aero-optics, aeroacoustics, drag reduction, electronic cooling, compressor stability, heat transfer, and thermal management. Primary emphasis is placed on understanding and controlling fundamental flow processes using active flow control approaches, including the exploration of emerging microelectromechanical systems (MEMS) technology for aerodynamic measurement and control. A particular challenge is the exploration of innovative actuator concepts for fluids-based flow and flight control strategies. The program is also interested in ideas exploring frontiers in fluid mechanics relative to fundamental flow processes occurring in microscale devices and systems, and in the potential for MEMS-based approaches to the control of microscale flows relative to electronic cooling and materials-processing technology. Research contributing to the understanding of flow instabilities and the mechanisms of transition from laminar to turbulent flow in both bounded and free-shear flows is of interest—especially the receptivity of linear and nonlinear transition processes to background and imposed flow disturbances—as is the impact on flow controllability. Improved turbulence modeling approaches are sought for the prediction of flow and heat transfer in highly strained and unsteady turbulent environments (e.g., gas turbine engines). In this context, original ideas are sought for modeling turbulent transport, especially ideas for incorporating the physics of turbulence into predictive models. The program is also interested in improved subgrid models for LES methods, especially in the near-wall region. High-quality turbulent flow data relevant to the advancement of transport and subgrid models for high-Reynolds number turbulent flows are also of interest. Research that addresses fundamental flow phenomena occurring in gas turbine engines, emphasizing the roles of unsteadiness and three-dimensionality in determining the performance, stability, and heat-transfer characteristics of these internal flows, is encouraged. Active control strategies for rotating stall and surge instabilities in gas turbine engine compressors are of interest. Of particular concern is the phenomenon of unsteady flow-induced forced blade response and its impact on high-cycle fatigue of turbine engine components. Another principal concern is the prediction and control of heat transfer in gas turbines, including the effectiveness of both film-cooling and internal-cooling flows. The principal areas of interest include blade wake effects, shock impingement effects, high free-stream turbulence, stagnation-point heating, blade tip clearance flows, blade hub juncture flows, and transition heat transfer phenomena. Proposals are accepted at any time.

Requirements: Prior to submitting an informal or formal proposal, investigators are encouraged to explore, with the program manager, AFOSR's possible interest in their research ideas.

Contact: Dr. Thomas Beutner, AFOSR/NA, (703) 696-6961; fax: (703) 696-8451; email: tom.beutner@afosr.af.mil

Internet: http://www.afosr.af.mil/pdfs/BAA2005-1.pdf

Sponsor: Air Force Office of Scientific Research
4015 Wilson Boulevard, Room 713
Arlington, VA 22203-1954

AFOSR Unsteady Aerodynamics and Hypersonics Research Grants 363

This program is focused on providing the fundamental fluid mechanics research base for future systems. Through a balance of experiments, analytical modeling approaches, and numerical simulations of the relevant flow physics, a fundamental understanding of the basic fluid flow fields associated with future complex configurations is achieved. This increased knowledge base will provide flow field prediction methods and flow control approaches that, in the short term, will reduce the weight and cost of future systems, and in the long term, will enable completely new, revolutionary vehicle designs that are unacceptable today due to aerodynamic performance constraints. Research areas

of interest include understanding the basic mechanisms present in time-dependent aerodynamic flows of all types, separated flows, separation control, circulation control, and vortical flows. Low-order flow modeling approaches that lead to adaptive control methods are desired. Internal and external flow tailoring for aerodynamic shape change is of interest. Nonlinear aerostructure interaction research, including flow-control approaches for suppression of destructive flow-structure interactions, is also of interest. Aeroacoustics research, especially as it applies to airframe noise or sonic fatigue, would also be considered a part of the aerostructure interaction subthrust. A major research area of interest is high-speed boundary layers. Quiet wind tunnel research, cross-flow instability mechanisms, and the receptivity of high-speed boundary layers to external disturbances are all areas of interest. Weakly ionized flows used for high-speed vehicle drag reduction are of interest as well as secondary jet injection for high-speed flight control. Finally, the fundamental flow physics associated with the airframe integration of combined cycle propulsion systems of all types (ramjet, scramjet, PDEs) is of interest, particularly the time-dependent characteristics of the inlet and nozzle flow fields. Proposals are accepted at any time.

Requirements: Prior to submitting an informal or formal proposal, investigators are encouraged to explore, with the program manager, AFOSR's possible interest in their research ideas.

Contact: Dr. John Schmisseur, AFOSR/NA, (703) 696-6962; fax: (703) 696-8451; email: john.schmisseur@afosr.af.mil

Internet: http://www.afosr.af.mil/pdfs/BAA2005-1.pdf

Sponsor: Air Force Office of Scientific Research
4015 Wilson Boulevard, Room 713
Arlington, VA 22203-1954

AFP Research Grants 364

The council awards annual research grants to individuals who wish to undertake research projects that enrich our knowledge of philanthropic fundraising and improve its practice. Preference will be given to research focused on Canada and the United States, regardless of where the research is based. Guidelines are available online.

Amount of Grant: $10,000 maximum

Date(s) Application is Due: Mar 1; Sep 1

Contact: Awards Administrator, (703) 519-8469; fax: (703) 684-0540; email: cwilliams@afpnet.org

Internet: http://www.afpnet.org/tier3_cd.cfm?folder_id=930&content_item_id=1531

Sponsor: Association of Fundraising Professionals
1101 King Street, Suite 700
Alexandria, VA 22314

AFP Skystone Ryan Prize for Research on Fundraising and Philanthropy 365

The award is made to the author of a published book or monograph of at least 50 pages that is based on applied or basic research in fundraising or philanthropy, and that was published by a commercial publishing house or a professional organizations during the 23 months preceding the November 1 deadline. Nominations are required by the listed application deadline.

Restrictions: Articles, directories, op-ed pieces, and self-published or unpublished works will not be considered.

Amount of Grant: $3000

Date(s) Application is Due: Nov 1

Contact: Prize Administrator, (800) 666-3863 or (703) 684-0410; fax: (703) 684-0540

Internet: http://www.afpnet.org/tier3_cd.cfm?folder_id=891&content_item_id=1544

Sponsor: Association of Fundraising Professionals
1101 King Street, Suite 700
Alexandria, VA 22314

African American Heritage Grants 366

Grants are awarded to assist organizations in the preservation and promotion of historic African American properties and sites in Indiana. Awards are made on a four-to-one matching basis, funding 80% of the total project cost up to $2,500, whichever is less.

Requirements: Civic groups, schools, libraries, historical societies, and other nonprofit agencies are eligible to apply for grants for organizational assistance, studies assisting in or leading to the preservation of a historic African American place, and programs promoting the preservation, interpretation, and/or visitation of a historic African American place. Contact the regional community preservation specialist that serves your community (see website for list of regional offices or contact the state headquarters office) for guidelines and forms.

Restrictions: Properties must be located in Indiana.

Geographic Focus: Indiana

Samples: Historic Eleutherian College Inc. - $2,500 to pay for a comprehensive master restoration and conservation plan for the c.1850 former African American college, which is now a National Historic Landmark.

Amount of Grant: $500 - $2,500

Contact: Carla Jones, Receptionist, State Headquarters; (317) 639-4534; fax: (317) 639-6734; email: info@historiclandmarks.org

Internet: http://www.historiclandmarks.org/help/grants.html

Sponsor: Historic Landmarks Foundation of Indiana
340 W. Michigan Street
Indianapolis, IN 46202

AFRL/NRC Resident Research Associateship (RRA) Program 367

The council administers postdoctoral and senior research awards through its associateship programs office, part of the Policy and Global Affairs Division. The programs are sponsored by 30 federal laboratories and NASA Research Centers at over 100 locations in the United States and overseas. Awards are made to doctoral level scientists and engineers who can apply their special knowledge and research talents to research areas that are of interest to them and to the host laboratories and centers. Each awardee works in collaboration with a research adviser, who is a staff member of the federal laboratory. In addition to traditional postdoctoral and senior awards, the associateship programs offer summer faculty awards, combined teaching and research awards, research management awards, opportunities in the NASA Astrobiology Institute, and international opportunities including a joint program with the Alexander von Humboldt Foundation.

Requirements: Postdoctoral research associateships are awarded to U.S. citizens and permanent residents who have held doctorates for less than five years at the time of application. Senior research associateships are awarded to individuals who have held doctorates for more than five years, have significant research experience, and are recognized internationally as experts in their specialized fields as evidenced by numerous publications in reviewed journals, invited presentations, authorship of books or book chapters, and professional society awards of international stature. U.S. citizenship is not a requirement for senior research associateships.

Amount of Grant: $36,000-$65,000 (higher for senior researchers)
Date(s) Application is Due: Feb 1; May 1; Aug 1; Nov 1
Contact: Associateship Programs, (202) 334-2760; fax: (202) 334-2759; email: rap@nas.edu
Internet: http://www4.nationalacademies.org/pga/rap.nsf
Sponsor: National Research Council
500 Fifth Street NW
Washington, DC 20001

AFUD Student Summer Fellowships 368

Each year 10 fellowships are awarded to attract the best medical students to work in urology research laboratories. These fellowships are to be held in the summer months. A pre-approved study program and sponsor recommendation are required. Application forms may be obtained from the AFUD. Applications are available on the Web site.

Requirements: Applicant must be a student at an accredited medical school in the United States.
Amount of Grant: $1000
Date(s) Application is Due: Sep 1
Contact: Anthony Caputi, Manager, Research Scholar Program, (410) 689-3990; fax: (410) 468-1808; email: anthony@afud.org
Internet: http://www.afud.org/research/application/summerprogram.asp
Sponsor: American Foundation for Urologic Disease
1128 N Charles Street
Baltimore, MD 21201

AFUW Diamond Jubilee Bursary 369

The program offers a bursary to assist men and women to undertake and complete higher degrees at Australian universities and to encourage advanced scholarship and original research by university men and women. The bursary may be used for the purchase or hire of equipment, field trip or research expenses, thesis publication costs, fees, dependent care expenses incurred because of study commitments, or short-term assistance with living expenses. The Bursary must be used within twelve months of the date of the award. There is no restriction on field of study or citizenship. The bursary may be held concurrently with awards offered by organizations other than the AFUW-SA Inc. Trust Fund. Selection of winners is based primarily on academic merit, but also on financial need, the importance of the purpose for which the bursary will be used to the progress or completion of the degree, community activities and other interests.

Requirements: Applications are invited from men and women who are enrolled in MA or PhD programs at Australian universities and who are Australian or overseas citizens.
Restrictions: Applicants may not be in full-time paid employment or on fully paid study leave during the tenure of the bursary.
Amount of Grant: $A4000 maximum each
Date(s) Application is Due: Mar 2
Contact: Gaynor Reeves, Fellowships Convener; AFUW-SA Trust Fund; (02) 4952 3174; fax: (02) 4952 3070; email: gay.reeves@bigpond.com or gaynor.reeves@gmail.com
Internet: http://www.afuwsa-bursaries.com.au/
Sponsor: Australian Federation of University Women—South Australia Trust Fund
G.P.O. Box 634
Adelaide, South Australia 5001 Australia

AFUW Doreen McCarthy, Barbara Crase, Cathy Candler & Brenda Nettle Bursaries 370

The program offers a bursary to assist women to undertake and complete higher degrees at Australian universities and to encourage advanced scholarship and original research by university women. The bursary may be used for the purchase or hire of equipment, field trip or research expenses, thesis publication costs, fees, dependent care expenses incurred because of study commitments, or short-term assistance with living expenses. The Bursary must be used within twelve months of the date of the award. Selection of winners is based primarily on academic merit, but also on financial need, the importance of the purpose for which the bursary will be used to the progress or completion of the degree, community activities and other interests.

Requirements: Applications are invited from women who are enrolled in MA or PhD programs at Australian universities and who are Australian or overseas citizens.
Restrictions: Applicants may not be in full-time paid employment or on fully paid study leave during the tenure of the bursary.

Amount of Grant: $A5000 maximum each
Date(s) Application is Due: Mar 2
Contact: Gaynor Reeves, Fellowships Convener; AFUW-SA Trust Fund; (02) 4952 3174; fax: (02) 4952 3070; email: gay.reeves@bigpond.com or gaynor.reeves@gmail.com
Internet: http://www.afuwsa-bursaries.com.au/
Sponsor: Australian Federation of University Women—South Australia Trust Fund
G.P.O. Box 634
Adelaide, South Australia 5001 Australia

AFUW Fellowships 371

Regular AFUW Fellowships are offered annually through special donations from AFUW-Qld. The Fellowships are available to women who are Australian citizens or permanent residents and enrolled for a PhD degree by research in an Australian university. Funding is to be used for a specific project or purpose for PhD research in any field. Awards are made both by AFUW Inc and by the State and Territory Associations that comprise the Federation.

Requirements: Applications must be made electronically on the form supplied, either as a Word document or in rich text format.
Amount of Grant: $A4000 maximum each
Date(s) Application is Due: May 31
Contact: Gaynor Reeves, Fellowships Convener; AFUW-SA Trust Fund; (02) 4952 3174; fax: (02) 4952 3070; email: gay.reeves@bigpond.com or gaynor.reeves@gmail.com
Internet: http://www.afuw.org.au/ScholarshipsFS.htm
Sponsor: Australian Federation of University Women—South Australia Trust Fund
53 Verulam Road
Lambton NSW, Australia 2299 Australia

AFUW Georgina Sweet Fellowships 372

Georgina Sweet Fellowships are available to women postgraduates enrolled in or accepted by an Australian university for advanced post-first degree study or research in any field. Members of the Australian Federation of University Women (AFUW Inc.) should forward their application direct to the Fellowships Officer. Members of other International Federation of University Women (IFUW) Federations or Associations must forward their applications through their National Federation or Association.

Requirements: Applicants must be a financial member of a National Federation or Association of the International Federation of University Women or an independent member of IFUW.
Amount of Grant: $A6000 maximum each
Date(s) Application is Due: May 31
Contact: Gaynor Reeves, Fellowships Convener; AFUW-SA Trust Fund; (02) 4952 3174; fax: (02) 4952 3070; email: gay.reeves@bigpond.com or gaynor.reeves@gmail.com
Internet: http://www.afuw.org.au/ScholarshipsFS.htm
Sponsor: Australian Federation of University Women—South Australia Trust Fund
53 Verulam Road
Lambton NSW, Australia 2299 Australia

AFUW Heather Latz Bursary 373

The AFUW Heather Latz Bursary is intended to assist women to undertake and complete higher degrees at Australian universities and to encourage advanced scholarship and original research by university women. The bursary may be used for the purchase or hire of equipment, field trip or research expenses, thesis publication costs, fees, dependent care expenses incurred because of study commitments, or short-term assistance with living expenses. The Bursary must be used within twelve months of the date of the award. There is no restriction on field of study or citizenship. The bursary may be held concurrently with awards offered by organizations other than the AFUW-SA Inc. Trust Fund. Selection of winners is based primarily on academic merit, but also on financial need, the importance of the purpose for which the bursary will be used to the progress or completion of the degree, community activities and other interests.

Requirements: Applications are invited from women who are enrolled in MA or PhD programs at Australian universities and who are Australian or overseas citizens.
Restrictions: Applicants may not be in full-time paid employment or on fully paid study leave during the tenure of the bursary.
Amount of Grant: $A4000 maximum each
Date(s) Application is Due: Mar 2
Contact: Gaynor Reeves, Fellowships Convener; AFUW-SA Trust Fund; (02) 4952 3174; fax: (02) 4952 3070; email: gay.reeves@bigpond.com or gaynor.reeves@gmail.com
Internet: http://www.afuwsa-bursaries.com.au/
Sponsor: Australian Federation of University Women—South Australia Trust Fund
G.P.O. Box 634
Adelaide, South Australia 5001 Australia

AFUW Padnendadlu Graduate Bursary 374

The aim of this bursary is to assist Aboriginal and/or Torres Strait Islander Australian women towards the completion of either a Graduate Diploma or Graduate Certificate at a South Australian university. A Bursary may be used for the purchase or hire of equipment, field trip or research expenses, thesis publication costs, fees incurred because of study commitments, conference attendance or short-term assistance with living expenses. The Bursary must be used within twelve months of the date of the award. The amount of the bursary will not exceed $2,000 but may be less depending on the requirements of the successful applicant.

Requirements: This application form should be used by Aboriginal and/or Torres Strait Islander Australian women undertaking a postgraduate degree at a South Australian university either via research or coursework.

Restrictions: Applicants may not be in full-time paid employment or on fully paid study leave during the tenure of the bursary.
Amount of Grant: $A2000 maximum each
Date(s) Application is Due: Mar 2
Contact: Gaynor Reeves, Fellowships Convener; AFUW-SA Trust Fund; (02) 4952 3174; fax: (02) 4952 3070; email: gay.reeves@bigpond.com or gaynor.reeves@gmail.com
Internet: http://www.afuwsa-bursaries.com.au/
Sponsor: Australian Federation of University Women—South Australia Trust Fund
G.P.O. Box 634
Adelaide, South Australia 5001 Australia

AFUW Padnendadlu Postgraduate Bursary   **375**
The aim of the Padnendadlu Postgraduate Bursary is to assist Aboriginal and/or Torres Strait Islander Australian women to undertake and complete higher degrees at South Australian universities and to encourage advanced scholarship and original research. A Bursary may be used for the purchase or hire of equipment, field trip or research expenses, thesis publication costs, fees incurred because of study commitments, conference attendance or short-term assistance with living expenses. The Bursary must be used within twelve months of the date of the award.
Requirements: This application form should be used by Aboriginal and/or Torres Strait Islander Australian women undertaking a postgraduate degree at a South Australian university either via research or coursework.
Restrictions: Applicants may not be in full-time paid employment or on fully paid study leave during the tenure of the bursary.
Amount of Grant: $A5000 maximum each
Date(s) Application is Due: Mar 2
Contact: Gaynor Reeves, Fellowships Convener; AFUW-SA Trust Fund; (02) 4952 3174; fax: (02) 4952 3070; email: gay.reeves@bigpond.com or gaynor.reeves@gmail.com
Internet: http://www.afuwsa-bursaries.com.au/
Sponsor: Australian Federation of University Women—South Australia Trust Fund
G.P.O. Box 634
Adelaide, South Australia 5001 Australia

AFUW Padnendadlu Undergraduate Bursary   **376**
The bursaries are envisaged as a short-term aid to assist with the completion of the academic year. Examples of purposes for which the Bursary may be used are equipment or book costs, fees incurred because of study commitments, short-term assistance with living expenses or any other purpose that will assist with the completion of the academic qualification for which the applicant is enrolled. The Bursary may be held concurrently with awards offered by organizations other than The AFUW-SA Inc. Trust Fund.
Requirements: This Bursary is open to Aboriginal and/or Torres Strait Islander Australian women undergraduates at any of the South Australian universities. Applicants must be undertaking subjects for the final year of their Bachelor degree, or undertaking an Honors year.
Restrictions: Women in full-time paid employment or on fully-paid study leave during the tenure of the Bursary are not eligible to apply, nor are women who won this bursary in the previous year.
Amount of Grant: $A700 maximum each
Date(s) Application is Due: Mar 2
Contact: Gaynor Reeves, Fellowships Convener; AFUW-SA Trust Fund; (02) 4952 3174; fax: (02) 4952 3070; email: gay.reeves@bigpond.com or gaynor.reeves@gmail.com
Internet: http://www.afuwsa-bursaries.com.au/
Sponsor: Australian Federation of University Women—South Australia Trust Fund
G.P.O. Box 634
Adelaide, South Australia 5001 Australia

AFUW South Australia Jean Gilmore, Thenie Baddams & Daphne   **377**
 Elliott Bursaries
The program offers a bursary to assist women to undertake and complete higher degrees at Australian universities and to encourage advanced scholarship and original research by university women. The bursary may be used for the purchase or hire of equipment, field trip or research expenses, thesis publication costs, fees, dependent care expenses incurred because of study commitments, or short-term assistance with living expenses. The bursary must be used within 12 months of the date of the award. Bursaries may be shared by two successful applicants.
Requirements: Applications are invited from women who are enrolled in MA or PhD programs at Australian universities and who are Australian or overseas citizens.
Restrictions: Applicants may not be in full-time paid employment or on fully paid study leave during the tenure of the bursary.
Amount of Grant: $A6000 maximum each
Date(s) Application is Due: Mar 2
Contact: Gaynor Reeves, Fellowships Convener; AFUW-SA Trust Fund; (02) 4952 3174; fax: (02) 4952 3070; email: gay.reeves@bigpond.com or gaynor.reeves@gmail.com
Internet: http://www.afuwsa-bursaries.com.au/
Sponsor: Australian Federation of University Women—South Australia Trust Fund
G.P.O. Box 634
Adelaide, South Australia 5001 Australia

AFUW Winifred E. Preedy Postgraduate Bursary   **378**
The program offers a bursary to assist women to undertake and complete higher degrees in dentistry or allied fields at Australian universities. The bursary may be used for the purchase or hire of equipment, field trip or research expenses, thesis publication costs, fees, dependent care expenses incurred because of study commitments, or short-term

assistance with living expenses. The bursary must be used within 12 months of the date of the award. Bursaries may be shared by two successful applicants.
Requirements: Applications are invited from women who are enrolled in MA or PhD programs in dentistry or a related field at Australian universities and have completed at least one year of a postgraduate degree.
Restrictions: Applicants may not be in full-time paid employment or on fully paid study leave during the tenure of the bursary, nor be previous winners of the Winifred E. Preedy Bursary.
Amount of Grant: $A4000 maximum each
Date(s) Application is Due: Mar 2
Contact: Gaynor Reeves, Fellowships Convener; AFUW-SA Trust Fund; (02) 4952 3174; fax: (02) 4952 3070; email: gay.reeves@bigpond.com or gaynor.reeves@gmail.com
Internet: http://www.afuwsa-bursaries.com.au/
Sponsor: Australian Federation of University Women—South Australia Trust Fund
G.P.O. Box 634
Adelaide, South Australia 5001 Australia

AGI Minority Participation Program   **379**
Scholarships are available for geoscience and geoscience education majors to assist with their graduate degrees. The term geoscience is used broadly to include major study in the fields of geology, geophysics, geochemistry, hydrology, meteorology, physical oceanography, planetary geology, and earth science education. Applications are available on the Web site.
Requirements: Applicants must be U.S. citizens and members of the following ethnic minority groups: African American, Hispanic, and Native American (American Indian, Eskimo, Samoan, or Hawaiian). Applicants must be enrolled in an accredited institution and must meet financial need criteria.
Amount of Grant: $500-$3000
Date(s) Application is Due: Mar 1
Contact: Cindy Martinez, Project Manager, (703) 379-2480; fax: (703) 379-7563; email: cmm@agiweb.org
Internet: http://www.agiweb.org/mpp/index.html
Sponsor: American Geological Institute
4220 King Street
Alexandria, VA 22302-1507

AGS Edward Henderson Student Award   **380**
The award is presented to a medical student interested in pursuing a career in geriatrics, who has demonstrated excellence in the field. The student must be nominated by one faculty member and have at least two supporting letters of nomination from other faculty. Annual deadline dates may vary; contact program staff for exact dates.
Requirements: A student must have demonstrated a commitment to the field of geriatrics through leadership in areas pertinent to geriatrics; initiation of new information or programs in geriatrics; or scholarship in geriatrics through original research or reviews.
Amount of Grant: $500 travel stipend to attend the annual meeting
Contact: Dennise McAlpin, (212) 308-1414 ext 321; fax: (212) 832-8646; email: dmcalpin@americangeriatrics.org or info.amger@americangeriatrics.org
Internet: http://www.americangeriatrics.org/news/meeting/awrdbroc.shtml#Henderson
Sponsor: American Geriatrics Society
350 5th Avenue, Suite 801
New York, NY 10018

AGS Foundation T. Franklin Williams Scholars Award   **381**
This program supports physician-scientists committed to improving the health care of older adults. The award is intended to allow individuals to initiate and ultimately sustain a career in research and education. Junior faculty members devoting 75 percent time to research are eligible to receive project support. The award must be matched. Guidelines and request for application are available online.
Requirements: To be eligible, the applicant must have an MD or DO degree; and hold a full-time faculty appointment at a U.S. academic medical institution at the level of assistant professor for no longer than four years at the time the grant becomes effective. The applicant must have at least two sponsors who are committed to providing guidance and collaboration throughout the course of the proposed project.
Amount of Grant: $37,500 per year for two years
Date(s) Application is Due: Feb 28
Contact: Award Administrator, (800) 247-8779 or (212) 308-1414; fax: (212) 832-8646; email: sreinthaler@americangeriatrics.org
Internet: http://www.frycomm.com/ags/Franklin_Williams/index.asp
Sponsor: AGS Foundation for Health in Aging
350 Fifth Avenue, Suite 801
New York, NY 10118

AGS John A. Hartford Foundation Geriatrics for Specialists Initiative   **382**
AGS, through a program funded by the John A. Hartford Foundation, seeks proposals for specialty-specific initiatives from academic training centers to develop, initiate, and evaluate programs designed to increase education for residents in the geriatrics aspect of their disciplines. The overall project addresses the urgent need to create a structure for developing leaders in geriatrics in academic surgery and related medical specialties. The program was created to assure that specialty residency programs adopt specific learning objectives and curricula in geriatric care as part of a targeted effort to enhance residents' knowledge, skills, and attitudes relevant to care of the older patient, in order to

improve dissemination of new clinical research findings in geriatrics into the surgical and medical specialties. Eight two-year grants will be awarded. Application and guidelines are available online.

Requirements: Proposals must be from institutions that have well-developed academic programs in the applicant disciplines and in geriatrics. The proposal should be submitted by a member of the specialty department or division.

Amount of Grant: $16,000 per year

Date(s) Application is Due: Apr 8

Contact: Grants Administrator, (212) 308-1414; fax: (212) 832-8646

Internet: http://www.americangeriatrics.org/2005_GESR.shtml

Sponsor: American Geriatrics Society

350 Fifth Avenue, Suite 801

New York, NY 10118

AGS Student Research Award   383

An award will be given to the student presenting the most outstanding paper or poster at the AGS annual meeting. The awardee will be chosen based on originality, scientific merit, and relevance of the research. The awardee will present his/her research at the annual meeting. Annual deadline dates may vary; contact program staff for exact dates.

Requirements: Students must submit an abstract of research or research in progress on the official AGS abstract form. In addition, the student's curriculum vita and a letter from the student's advisor verifying the student's contribution to the work must be provided.

Amount of Grant: $1500 travel stipend to attend the annual meeting

Contact: Dennise McAlpin, Program Contact, (212) 308-1414 ext 321; fax: (212) 832-8646; email: dmcalpin@americangeriatrics.org or info@americangeriatrics.org

Internet: http://www.americangeriatrics.org/education/geristudents/Scholarships_Awards

Sponsor: American Geriatrics Society

350 Fifth Avenue, Suite 801

New York, NY 10118

AGS/Merck New Investigator Awards   384

The awards, funded through an educational grant from Merck U.S. Human Health, are given to individuals whose original research, as presented in a submitted abstract, reflects new and relevant research in geriatrics. Topics are invited in basic research, clinical investigation, clinical medicine, public health, and research in the fundamental neurosciences. Awards will be chosen based on originality, scientific merit, relevance of the research, and the applicant's overall academic accomplishments. The awards are intended, in part, to cover awardees' travel expenses to attend the AGS annual meeting. Annual deadline dates may vary; contact program staff for exact dates.

Requirements: The awards are restricted to fellows in training and new and junior investigators holding an academic appointment not longer than five years postfellowship. The work reported must not have been published prior to the date of application.

Restrictions: Materials presented at other national meetings will not be accepted.

Amount of Grant: $500

Contact: Dennise McAlpin, (212) 308-1414; fax: (212) 832-8646; email: dmcalpin@americangeriatrics.org or info.amger@americangeriatrics.org

Internet: http://www.americangeriatrics.org/news/meeting/am2004_newInvestAward.shtml

Sponsor: American Geriatrics Society

350 Fifth Avenue, Suite 801

New York, NY 10118

AHA Beginning Grants-in-Aid   385

Grants-in-aid promote the independent status of promising beginning scientists. Research areas broadly related to cardiovascular function and disease or stroke, or to related clinical, basic science, and public health problems are funded. All basic disciplines as well as epidemiological, community, and clinical investigations that bear on cardiovascular and stroke problems are eligible for funding. Awards are made for two years.

Requirements: Applicants must have an MD, PhD, DO, or equivalent and be initiating an independent research career. Faculty up to and including assistant professor may apply. Applicants must have a full-time faculty/staff appointment at activation and must be U.S. citizens, permanent residents, exchange visitors, temporary workers in a specialty occupation, Canadian or Mexican citizens engaging in professional activities, temporary workers with extraordinary abilities in the sciences, or students with student visas.

Geographic Focus: Arizona

Amount of Grant: $62,000 per year maximum

Date(s) Application is Due: Jan 5

Contact: Affiliate Research Services, (214) 706-1158; fax: (214) 706-1341; email: affil@heart.org

Internet: http://www.americanheart.org/presenter.jhtml?identifier=2483

Sponsor: American Heart Association—Desert Mountain Affiliate

7272 Greenville Avenue

Dallas, TX 75231-4596

AHA Established Investigator Grants   386

The program supports the career development of highly promising clinician-scientists and PhDs who have recently acquired independent status by funding high quality, innovative research projects for which no previous financial support has been obtained from other granting agencies. Grants provide four-year project and salary support. At the time of award activation, the time period since applicant's first faculty appointment must be four to nine years. Approximately 50 grants will be awarded annually. Award amount includes salary/fringe and 10 percent indirect costs.

Requirements: U.S. citizens or permanent residents with the MD, PhD, DO, or equivalent doctoral degrees are eligible. Eligible applicants must have full-time faculty/staff appointment at application, up to and including Associate Professor (or equivalant). Applicant must have at least four years but no more than nine years since first full-time faculty/staff appointment.

Restrictions: Current and past principal investigators are ineligible.

Amount of Grant: $100,000 annually, including 10 percent indirect costs

Contact: Division of Research Administration, National Center, (214) 706-1158; fax: (214) 706-1341; email: ncrp@heart.org

Internet: http://www.americanheart.org/presenter.jhtml?identifier=9713

Sponsor: American Heart Association

7272 Greenville Avenue

Dallas, TX 75231-4596

AHA New York State Affiliate Postdoctoral Fellowships   387

Research fellowships are awarded for a two- to three-year period to help a trainee initiate a career in cardiovascular research while obtaining significant research results.

Requirements: Applicants must be MDs, CVMs, PDs, or PhDs, no more than three years postdoctoral, who agree to commit full-time effort to research. U.S. citizens, permanent residents, exchange visitors, temporary workers in specialty occupations, Canadian or Mexican citizens engaging in professional activities, temporary workers with extraordinary ability in the sciences, and those with student visas are eligible. Postdoctoral applicants who are outside the United States at time of application must provide visa documentation prior to award activation.

Restrictions: The program is not intended for individuals of faculty rank.

Amount of Grant: $38,000 maximum

Date(s) Application is Due: Jan 13; Nov 15

Contact: Grants Administrator, (214) 706-1158 or (214) 706-1744; fax: (214) 706-1341; email: ncrp@heart.org

Internet: http://www.americanheart.org/presenter.jhtml?identifier=2402

Sponsor: American Heart Association—New York State Affiliate

7272 Greenville Avenue

Dallas, TX 75231-4596

AHA New York State Affiliate Scientist Development Grants   388

The purpose of the program is to help promising beginning scientists to move from completion of research training to the status of independent investigators. The focus of the research should be broadly related to CV function and disease, stroke, or to related clinical, basic science, and public health problems. The duration of the award is three years.

Requirements: Eligible to apply are U.S. citizens or permanent residents; individuals of up to and including assistant professor (or equivalent) rank at application; individuals holding MD, PhD, DO, or equivalent at application. At activation, no more than four years should have elapsed since first full-time faculty/staff appointment. At activation, applicant must have full-time faculty/staff appointment.

Restrictions: At activation, applicants shall have received no prior national level grant. Awards are non-renewable.

Amount of Grant: $66,000 maximum per year

Date(s) Application is Due: Jan 13

Contact: Grant Administrator, (214) 706-1158 or (214) 706-1744; fax: (214) 706-1341; email: ncrp@heart.org

Internet: http://www.americanheart.org/presenter.jhtml?identifier=2402#sdg

Sponsor: American Heart Association—New York State Affiliate

7272 Greenville Avenue

Dallas, TX 75231-4596

AHA Postdoctoral Fellowship   389

The Pacific Mountain Affiliate focuses on services in its region, including Alaska, Arizona, Colorado, Hawaii, Idaho, Montana, New Mexico, Oregon, Washington and Wyoming. Fellowships assist postdoctoral trainees to initiate a career in cardiovascular research. Research funded is broadly related to cardiovascular function and disease, stroke, or to related clinical, basic science, and public health problems. All basic disciplines as well as epidemiological, community, and clinical investigations that bear on cardiovascular and stroke problems are funded. Awards are made for two to three years.

Requirements: Applicants must be U.S. citizens, permanent residents, exchange visitors, temporary workers in a specialty occupation, Canadian or Mexican citizens engaging in professional activities, temporary workers with extraordinary abilities in the sciences, or students with student visas. Applicants must have an MD, PhD, DO, or equivalent at the time the fellowship is activated.

Amount of Grant: $40,568-$48,428

Date(s) Application is Due: Jan 10

Contact: Affiliate Research Services, (214) 706-1158, (214) 706-1744, (214) 706-1457; fax: (214) 706-1341; email: ncrp@heart.org

Internet: http://www.americanheart.org/presenter.jhtml?identifier=2418

Sponsor: American Heart Association—Pacific Mountain Affiliate

7272 Greenville Avenue

Dallas, TX 75231-4596

AHA Predoctoral Fellowship   390

Fellowships help students initiate careers in cardiovascular research by providing research assistance and training. Research broadly related to cardiovascular function and disease, stroke or to related clinical, basic science, and public health problems will be funded. The fellowships are awarded in basic disciplines as well as epidemiological, community and clinical investigations that bear on cardiovascular and stroke problems. Awards are made for one or two years.

Requirements: Predoctoral PhD, MD, DO (or equivalent) students seeking research training with a sponsor/mentor prior to embarking on a research career may apply. Applicants must be U.S. citizens, permanent residents, exchange visitors, temporary workers in a specialty occupation, Canadian or Mexican citizens engaging in professional activities, temporary workers with extraordinary abilities in the sciences, or students with student visas.

Restrictions: An awardee may not hold another association award concurrently. No more than one fellow per sponsor will be funded. Funds may not be used to pay tuition.

Geographic Focus: Arizona; Colorado; New Mexico; Wyoming

Amount of Grant: $24,000 maximum

Date(s) Application is Due: Jan 10

Contact: Affiliate Research Services, (214) 706-1457, (214) 706-1158, (214) 706-1744; fax: (214) 706-1341; email: ncrp@heart.org

Internet: http://www.americanheart.org/presenter.jhtml?identifier=2418

Sponsor: American Heart Association—Pacific Mountain Affiliate

7272 Greenville Avenue

Dallas, TX 75231-4596

AHA Scientist Development Grants 391

The program supports highly promising beginning scientists in their progress toward independence and bridges the gap between completion of research training and readiness for competition as independent investigators. Grants provide four-year project and salary support. Applications may be submitted for review in the final year of a postdoctoral research fellowship or in the first four years of a faculty appointment. Peer review criteria will include originality and scientific merit of the proposed project (which cannot overlap with other funded work), prior productivity of the applicant, and evidence that the award will promote independent status for the applicant. Approximately 70 grants will be made annually. The award includes salary/fringe, project support, and 10 percent for indirect costs. The award is not renewable.

Requirements: U.S. citizens or permanent residents with the MD, PhD, DO, or equivalent doctoral degrees are eligible. Eligible applicants must have full-time faculty/staff appointment at activation, up to and including associate professor or equivalent.

Restrictions: Applicants cannot hold or have held any other national award. Applicant must have no more than four years since first full-time faculty/staff appointment at activation.

Amount of Grant: $65,000 maximumannually

Date(s) Application is Due: Jan 9

Contact: Division of Research Administration, National Center, (214) 706-1158, (214) 706-1744; fax: (214) 706-1341; email: ncrp@heart.org

Internet: http://www.americanheart.org/presenter.jhtml?identifier=3004142

Sponsor: American Heart Association

7272 Greenville Avenue

Dallas, TX 75231-4596

AHAF Alzheimer's Disease Research Grants 392

AHAF funds outstanding scientists and physicians in neurobiology, physiology, pathology, molecular and developmental biology, chemistry, pharmacology, epidemiology, and surgery who are conducting research to better understand and/or treat Alzheimer's disease. Junior or senior investigators will be considered. Grants are awarded for up to two years and are renewable. Grant applications are reviewed by a scientific review committee on a competitive peer-review system.

Requirements: Grants are awarded to universities, medical centers, and independent research institutions. AHAF funds grants for research at nonprofit organizations only. The principal investigator must hold the academic rank of assistant professor (or equivalent) or higher.

Restrictions: Funding is not provided for overhead, construction, or building expenses.

Amount of Grant: $300,000 maximum per year

Date(s) Application is Due: Oct 19

Contact: Kara Hurst, Grants Coordinator; (800) 437-2423 or (301) 948-3244; fax: (301) 258-9454; email: khurst@ahaf.org

Internet: http://www.ahaf.org/alzdis/research/grants.htm

Sponsor: American Health Assistance Foundation

22512 Gateway Center Drive

Clarksburg, MD 20871

AHAF Macular Degeneration Research Grants 393

The foundation awards research grants to advance study of macular degeneration. Applications are evaluated based on the scientific merit of the proposal, the feasibility of the proposed research, the potential of the research to lead to better understanding and treatment of eye diseases, and the demonstrated ability of the investigator to complete the research. Application materials are available on the Web site.

Amount of Grant: $50,000 maximum per year

Date(s) Application is Due: July 10; Oct 29

Contact: Kara Hurst, Grants Coordinator; (800) 437-2423 or (301) 948-3244; fax: (301) 258-9454; email: khurst@ahaf.org

Internet: http://www.ahaf.org/macular/research/grants.htm

Sponsor: American Health Assistance Foundation

22512 Gateway Center Drive

Clarksburg, MD 20871

AHAF National Glaucoma Research Grants 394

AHAF funds outstanding scientists and physicians with expertise in cell and molecular biology, physiology, biochemistry, endocrinology, and pharmacology. Grants are awarded on the basis of scientific merit of the proposal, the relevance of the research, and the potential impact of the proposed study on better understanding and/or treatment of glaucoma. Applications are reviewed by a scientific review committee on a competitive peer-review system. AHAF is interested in receiving focused research grant applications from investigators at all stages of their careers. AHAF is particularly interested in new investigators with little or no previous grant support and established investigators with new ideas or directions for their research.

Requirements: AHAF grants are awarded to universities, medical centers, and independent research institutions. AHAF provides grants for research at nonprofit organizations only.

Restrictions: Grants are not made to individuals. Funding is not provided for overhead costs, construction, or building expenses.

Samples: Teri Belecky-Adams, Ph.D., Trustees of Indiana University, $100,000.; David Calkins, Ph.D., Vanderbilt University Medical Center, Nashville, TN, $100,000.; Mortimer Civan, M.D., University of Pennsylvania, Philadelphia, PA, $100,000.

Amount of Grant: $100,000 maximum

Date(s) Application is Due: Oct 27

Contact: Kara Hurst, Grants Coordinator; (800) 437-2423 or (301) 948-3244; fax: (301) 258-9454; email: khurst@ahaf.org

Internet: http://www.ahaf.org/glaucoma/research/glresrch.htm

Sponsor: American Health Assistance Foundation

22512 Gateway Center Drive

Clarksburg, MD 20871

AHEPA Educational Foundation Scholarships 395

The AHEPA Educational Foundation's scholarship programs promote, encourage, induce, and advance education at the college, university, and graduate school level. A number of named undergraduate and graduate scholarships are available. Application and guidelines are available online. Annual deadline dates may vary; contact program staff for exact dates.

Amount of Grant: $3000 - $5000

Date(s) Application is Due: Jun 1

Contact: AHEPA Headquarters, (202) 232-6300; fax: (202) 232-2140; email: ahepa@ahepa.org

Internet: http://www.ahepa.org/educ_foundation/index.html

Sponsor: American Hellenic Educational Progressive Association

1909 Q Street NW, Suite 500

Washington, DC 20009

AHFMR Clinical Fellowships 396

The program is designed for highly qualified individuals who hold an M.D. or D.D.S. degree, and who anticipate undertaking a career in health related or clinical research in Alberta. The awards consist of a stipend and a research allowance. A Clinical Fellowship award is normally tenable for a maximum of 3 years. However, if the trainee is registered in a graduate program, the term may be extended.

Requirements: Normally, an award will be held within the Province of Alberta; however, candidates who are Canadian citizens or permanent residents with records of outstanding performance in postgraduate training may seek research training elsewhere, if sponsored by an Alberta faculty with an expressed interest in future recruitment of the candidate. Candidates must hold an MD or DDS degree and have received a significant portion of their postgraduate training in Alberta.

Amount of Grant: $20,000-$50,000

Date(s) Application is Due: Mar 1; Oct 1

Contact: Pamela Valentine, Director of Grants and Awards; (780) 423-5727; fax: (780) 429-3509; email: pamela.valentine@ahfmr.ab.ca

Internet: http://www.ahfmr.ab.ca/grants/Clin-fellow.php

Sponsor: Alberta Heritage Foundation for Medical Research

10104-103 Avenue, Suite 1500

Edmonton, Alberta T5J 4A7 Canada

AHFMR Clinical Investigator Awards 397

The Award allows highly qualified clinicians in the early stages of their careers in health research to commit the majority of their time (greater than 75%) to research. This award includes a $110,000 annual contribution toward salary and benefits, a $10,000 annual research prize, a $35,000 annual research allowance, and additional funds for renovations, operations, relocation, and equipment. In all cases, the relevance of the applicant's proposed research to human health should be apparent. All applicants must name a mentor or group of mentors who will foster their early career development.

Requirements: Canadian citizens or permanent residents of Canada who hold an MD or DDS degree, have completed all requirements for clinical specialty recognition, and qualify to hold a full-time position in a clinical department of the sponsoring institution are eligible to apply.

Amount of Grant: $155,000

Date(s) Application is Due: Sep 15

Contact: Pamela Valentine, Director of Grants and Awards; (780) 423-5727; fax: (780) 429-3509; email: pamela.valentine@ahfmr.ab.ca

Internet: http://www.ahfmr.ab.ca/grants/ClinInvest.php

Sponsor: Alberta Heritage Foundation for Medical Research

10104-103 Avenue, Suite 1500

Edmonton, Alberta T5J 4A7 Canada

AHFMR Community Support Program Grants 398
Community support program grants will provide two types of awards: Community Health Research Visiting Lecturer Awards and Community Health Research Travel Grants. The former (Community Health Research Visiting Lecturer Awards) are intended to assist Alberta's non-profit organizations in attracting outstanding health researchers to the province. The latter (Community Health Research Travel Grants) facilitate contact among health researchers in Alberta, as well as with national and international colleagues. Candidates from a wide variety of disciplines may apply as long as the purpose of the travel is to promote community health research.
Requirements: The Community Health Research Visiting Lecturer Awards program is open to community organizations that are directly involved in health research. The Universities of Alberta, Calgary and Lethbridge are not eligible to apply to this program but must submit requests to the AHFMR Visiting Lecturer Program. The Community Health Research Travel Grants program is open to individuals (independent investigators) actively engaged in health research in a non-profit organization in Alberta are eligible to apply.
Restrictions: University faculty are not eligible to apply for these awards.
Date(s) Application is Due: Jan 9; Apr 14; Jul 10; Oct 9
Contact: Pamela Valentine, Director of Grants and Awards; (780) 423-5727; fax: (780) 429-3509; email: pamela.valentine@ahfmr.ab.ca
Internet: http://www.ahfmr.ab.ca/grants/commsupp.php
Sponsor: Alberta Heritage Foundation for Medical Research
10104-103 Avenue, Suite 1500
Edmonton, Alberta T5J 4A7 Canada

AHFMR Dr. Lionel E. McLeod Health Research Scholarship 399
The award enables academically superior graduate students to undertake full-time training in medical or health research. The Scholarship is tenable through the Faculty of Medicine in any area related to health including: medicine; health services research; social and behavioral sciences (sociology, anthropology, psychology, economics, or political science); quantitative and population sciences (epidemiology and biostatistics, operations research, decision sciences, or computer sciences); and the humanities (philosophy, history, or theology).
Requirements: Candidates must be students at the University of Alberta, the University of Calgary, or the University of British Columbia and sponsored by the Faculty of Medicine.
Amount of Grant: $21,500
Contact: Pamela Valentine, Director of Grants and Awards; (780) 423-5727; fax: (780) 429-3509; email: pamela.valentine@ahfmr.ab.ca
Internet: http://www.ahfmr.ab.ca/grants/mcleod.php
Sponsor: Alberta Heritage Foundation for Medical Research
10104-103 Avenue, Suite 1500
Edmonton, Alberta T5J 4A7 Canada

AHFMR Establishment and Independent Establishment Grants 400
The two types of grants available include: Establishment Grants—AHFMR provides funds to assist reseachers (newly recruited to Alberta) in the start-up of their laboratories and/or research projects; and Independent Establishment Grants—open to new Alberta Investigators who are not candidates for Heritage personnel awards (it is understood that these investigators will commit much of their time to medical or health research). Those sponsoring investigators for these awards must notify the Foundation of their intent to do so at least four months in advance of the application deadline, and request a determination of the eligibility of the candidate. The Foundation will not accept applications for which advanced notice has not been received. There are two categories of grant funding: Operating Funds and Major Equipment Funds.
Requirements: Eligibility to apply for these grants is restricted to applicants recruited to leadership positions (such as Department Head or Director of a centre, institute or division), where the recruitment is expected to stimulate the development of research in a field or discipline of interest to the Foundation.
Date(s) Application is Due: Sep 15
Contact: Pamela Valentine, Director of Grants and Awards; (780) 423-5727; fax: (780) 429-3509; email: pamela.valentine@ahfmr.ab.ca
Internet: http://www.ahfmr.ab.ca/grants/EstabIndep.php
Sponsor: Alberta Heritage Foundation for Medical Research
10104-103 Avenue, Suite 1500
Edmonton, Alberta T5J 4A7 Canada

AHFMR Fast-Track Studentships and Fellowships 401
AHFMR will accept, at any time, applications from Alberta-based institutions for full-time Studentship or Fellowship awards for outstanding candidates currently training outside of Alberta. The fast-track program is intended to assist these institutions in recruiting highly qualified trainees to the province. Up to a maximum of 10 awards per calendar year will be awarded through the fast-track review process in each category. Applications to the program will be reviewed on a first come-first serve basis.
Requirements: Applications must include a letter from the Dean/designate of the faculty indicating that the proposal for a fast-track award has been reviewed and stating why a fast-track application should be considered for the candidate in question. Please refer to the sections on Full-Time Fellowships or Full-Time Studentships for further details regarding eligibility for training awards.
Amount of Grant: $20,000-$23,000 for studentships; $35,000-$50,000 for fellowships
Contact: Pamela Valentine, Director of Grants and Awards; (780) 423-5727; fax: (780) 429-3509; email: pamela.valentine@ahfmr.ab.ca
Internet: http://www.ahfmr.ab.ca/grants/fasttrack.php

Sponsor: Alberta Heritage Foundation for Medical Research
10104-103 Avenue, Suite 1500
Edmonton, Alberta T5J 4A7 Canada

AHFMR Forefront Executive-in-Residence Award 402
ForeFront supports career, industry and innovation development. Alberta's universities are integral to the innovation system and the creation of the medical and health industries through research, technology transfer, and spin-off company creation. The Executive-in-Residence Award enables the technology transfer offices to recruit senior-level management who have relevant industry experience and appropriate networks to assist university spin-off companies with the commercialization of technologies. The goal is to advance technologies and build better companies that will both improve health care and diversify and contribute to the economy.
Samples: David Rafter, University Technologies International, Calgary, Alberta, Canada (2007); Ronald Matheson, TEC Edmonton, Edmonton, Alberta, Canada (2006).
Contact: Tina Blake; (780) 423-5727; fax: (780) 429-3509; email: tina.blake@ahfmr.ab.ca
Internet: http://www.ahfmr.ab.ca/forefront/EIR.php
Sponsor: Alberta Heritage Foundation for Medical Research
10104-103 Avenue, Suite 1500
Edmonton, Alberta T5J 4A7 Canada

AHFMR Forefront Internship Program Grants 403
The Internship Program addresses the need for highly trained staff and management of the Alberta-based health, medical products and biotechnology industries. This Program is intended to support technology commercialization training and experience for individuals with an appropriate background in science and/or business. Although the work plan should identify the specific area of focus, the intern should receive a broad exposure to all of the following areas: product development; intellectual property strategy; regulatory requirements; market research and strategy; and business planning and development. Interns are awarded a $5,000 training allowance, which may be used for expenses related to educational workshops/seminars/courses necessary for relevant commercialization training, plus a $45,000 to $50,000 stipend.
Requirements: The applicant must hold, or be in the final year of a degree in business, management, or science. The applicants should have some relevant experience in the business or private sector. Applicants must be sponsored by an Alberta organization actively engaged in the commercialization of medical or health related technology. The organization must have the experience and resources to provide an appropriate training environment, and designate one person within the organization who will be the primary supervisor for the intern. The organization must agree to provide direct and continuous supervision.
Amount of Grant: $45,000-$50,000 per year
Date(s) Application is Due: Feb 15
Contact: Carla Weyland, ForeFront Officer; (780) 423-5727; fax: (780) 429-3509; email: carla.weyland@ahfmr.ab.ca
Internet: http://www.ahfmr.ab.ca/forefront/internship.php
Sponsor: Alberta Heritage Foundation for Medical Research
10104-103 Avenue, Suite 1500
Edmonton, Alberta T5J 4A7 Canada

AHFMR Forefront MBA Studentship Award 404
This award will enable students with a background in the medical/health/life sciences to apply their knowledge and pursue a career in management in the medical/health industry by supporting their education and hands-on training in the University of Alberta Master of Business Administration program, specializing in Technology Commercialization. The Tech Com MBA builds and delivers a tool kit that emphasizes the importance of capital, management practices, infrastructure, information, and networks. The goal of the Tech Com MBA is to position graduates for a leadership role in today's leading edge, high technology industries. The main objectives are to: encourage Albertans with a background in science to explore careers that will give them the opportunity to apply their knowledge to commercializing medical discoveries that will lead to improved health; and build capacity of knowledgeable, highly-skilled technology managers to meet the needs of the Alberta medical/health industry.
Samples: Dr. Stephanie Minnema, Calgary, Alberta, Canada (2007).
Contact: Tina Blake; (780) 423-5727; fax: (780) 429-3509; email: tina.blake@ahfmr.ab.ca
Internet: http://www.ahfmr.ab.ca/forefront/mba.php
Sponsor: Alberta Heritage Foundation for Medical Research
10104-103 Avenue, Suite 1500
Edmonton, Alberta T5J 4A7 Canada

AHFMR Forefront MBT Studentship Awards 405
The award will enable students to pursue a career in the medical, health, and biotechnology industries by supporting their education and hands-on training in the Master of Biomedical Technology (MBT) Program at the University of Calgary. The MBT Program takes a multi-disciplinary approach and exposes students to a broad range of courses and cutting-edge technologies that will ensure the students acquire practical competencies and extensive knowledge of concepts in biomedical and bioinformatics disciplines, including the application of this knowledge through commercialization of medical and health-related innovations. The main objectives are to: encourage individuals with a background in science to explore careers that will give them the opportunity to apply their knowledge to commercializing medical discoveries that will lead to improved

health; and build capacity of knowledgeable, highly-skilled technology managers to meet the needs of the Alberta medical/health industry.
Samples: Paul Lam, BS Degree (2007); Jenna Latanville, BS Degree (2007); Jennifer Lukus, BS Degree (2007); Katarzyna Okoniewska, BS Degree (2007).
Contact: Tina Blake; (780) 423-5727; fax: (780) 429-3509; email: tina.blake@ahfmr.ab.ca
Internet: http://www.ahfmr.ab.ca/forefront/mbt.php
Sponsor: Alberta Heritage Foundation for Medical Research
10104-103 Avenue, Suite 1500
Edmonton, Alberta T5J 4A7 Canada

AHFMR Full-Time Fellowships 406
The program is designed to enable highly qualified doctoral graduates to prepare for careers in medical or health research as independent investigators. A fellowship award will provide to the host institution funding for one year's stipend, its associated benefits, and a research allowance. Fellows may engage in teaching activities related to their research discipline a maximum of 20 percent of their time.
Requirements: Candidates must have a Ph.D., M.D., D.D.S., D.V.M. or D.Pharm. degree. Normally, support will not be provided beyond 6 years after receipt of the Ph.D. degree, or beyond 8 years after receipt of the M.D., D.D.S., D.V.M. or D.Pharm. degrees.
Amount of Grant: $35,000-$50,000
Date(s) Application is Due: Mar 1; Oct 1
Contact: Pamela Valentine, Director of Grants and Awards; (780) 423-5727; fax: (780) 429-3509; email: pamela.valentine@ahfmr.ab.ca
Internet: http://www.ahfmr.ab.ca/grants/FT-fellow.php
Sponsor: Alberta Heritage Foundation for Medical Research
10104-103 Avenue, Suite 1500
Edmonton, Alberta T5J 4A7 Canada

AHFMR Full-Time M.D./Ph.D. Studentships 407
Full-Time M.D./Ph.D. Studentships are intended to provide an opportunity for exceptional candidates, who wish to pursue careers as Clinical Investigators, to study for the M.D. and the Ph.D. degrees simultaneously. AHFMR support is complementary to the formal M.D./Ph.D. programs at the University of Alberta and the University of Calgary. M.D./Ph.D. studentship awards are tenable only at an Alberta-based university.
Requirements: Applicants are limited to those students that formally enter the MD/PhD program not later than the start of their second year of PhD studies at either the University of Alberta or the University of Calgary. AHFMR's offer of support is contingent on the granting of complementary stipend support from the Faculty of not less than 15% of the value of the AHFMR award, for the duration of the award. Furthermore, it is expected that the universities will undertake to provide administrative support to the offices of the coordinators of the M.D./Ph.D. programs.
Amount of Grant: $20,000
Date(s) Application is Due: Mar 1; Oct 1
Contact: Pamela Valentine, Director of Grants and Awards; (780) 423-5727; fax: (780) 429-3509; email: pamela.valentine@ahfmr.ab.ca
Internet: http://www.ahfmr.ab.ca/grants/FT-student.php#MPS
Sponsor: Alberta Heritage Foundation for Medical Research
10104-103 Avenue, Suite 1500
Edmonton, Alberta T5J 4A7 Canada

AHFMR Full-Time Studentships 408
AHFMR Full-time Studentships enable academically superior students to undertake full-time research training in the basic biomedical sciences or in clinical research. The award consists of a stipend and a research allowance. Approved uses of the research allowance include: the purchase of scientific materials, supplies and expendables; the purchase of minor equipment; computer software programs; costs for the use of libraries, or computers; costs associated with the publication of research results; travel expenses to attend scientific meetings; and purchase of books, periodicals and journals.
Requirements: Candidates must have been accepted into, or be currently engaged in, a full-time, thesis-based, graduate program at an Alberta-based university in a health-related discipline leading to a Master's or doctoral degree.
Restrictions: This award is not available to students registered in a course-based program. Normally, support will not be provided beyond 6 years of enrollment in graduate school. Candidates who have interrupted their training for parenting or other reasons and who have consequently exceeded this time limit, may apply for studentship support. In such cases, however, the candidate is advised to clearly explain the nature of his/her particular circumstances at the time of application.
Amount of Grant: $20,000 plus research allowance
Date(s) Application is Due: Mar 1; Oct 1
Contact: Pamela Valentine, Director of Grants and Awards; (780) 423-5727; fax: (780) 429-3509; email: pamela.valentine@ahfmr.ab.ca
Internet: http://www.ahfmr.ab.ca/grants/FT-student.php
Sponsor: Alberta Heritage Foundation for Medical Research
10104-103 Avenue, Suite 1500
Edmonton, Alberta T5J 4A7 Canada

AHFMR Health Research Career Renewal Awards 409
Research Career Renewal Awards allow qualified Alberta-based faculty members to obtain rigorous training in any of the following disciplines: epidemiology; biostatistics; psychosocial sciences; and clinical experimental method and design. Training must

be undertaken in a center that is renowned for the quality of research and training in the particular area of study. The Career Renewal award will provide a contribution to stipend and/or a research allowance (which can be applied toward offsetting tuition costs, for example).
Requirements: An Alberta-based institution must propose and sponsor the candidate (who must be a full-time faculty member) in addition to committing to provide an adequate research environment upon the trainee's return.
Date(s) Application is Due: Mar 1; Oct 1
Contact: Pamela Valentine, Director of Grants and Awards; (780) 423-5727; fax: (780) 429-3509; email: pamela.valentine@ahfmr.ab.ca
Internet: http://www.ahfmr.ab.ca/grants/CareerAward.php
Sponsor: Alberta Heritage Foundation for Medical Research
10104-103 Avenue, Suite 1500
Edmonton, Alberta T5J 4A7 Canada

AHFMR Health Research Full-Time Studentships 410
The Full-Time Health Research Studentship program enables academically superior students to undertake full-time training in health research. The award supports training in: research on the organization and delivery of health care; technology assessment; community health; health promotion; disease prevention; and related disciplines. The Health Research Studentship consists of a stipend and a research allowance.
Requirements: Candidates must have been accepted into, or be currently studying in, a full-time, thesis-based, graduate program at an Alberta-based university in a health-related discipline leading to a Master's or doctoral degree. This award is not available to students registered in a course-based program. Normally, support will not be provided beyond 6 years of enrollment in graduate school. Candidates who have interrupted their training for parenting or other reasons and who have consequently exceeded this time limit, may apply for studentship support. In such cases, however, the candidate is advised to clearly explain the nature of his/her particular circumstances at the time of application.
Amount of Grant: $20,000 plus research allowance
Date(s) Application is Due: Mar 1; Oct 1
Contact: Pamela Valentine, Director of Grants and Awards; (780) 423-5727; fax: (780) 429-3509; email: pamela.valentine@ahfmr.ab.ca
Internet: http://www.ahfmr.ab.ca/grants/FT-student.php#HRS
Sponsor: Alberta Heritage Foundation for Medical Research
10104-103 Avenue, Suite 1500
Edmonton, Alberta T5J 4A7 Canada

AHFMR Health Research Fund Grants 411
The Health Research Fund (HRF), an operating grant program that has been in existence for over ten years, is a collaborative program with the funding originating from Alberta Health and Wellness and with AHFMR doing the program administration. The Fund supports relevant, high quality health research studies in the areas of: health services; population health; health technology assessment. There are no specific deadlines.
Requirements: The HRF competition is aimed at investigative teams which include both researchers and decision-makers or users of health research information working together on issues relating to the appropriate research areas. Proposals that do not include a relevant decision-maker and/or user of health research information as part of the investigative team are not eligible for funding. Applications may be sponsored by either non-profit decision-making organizations (health agencies, associations, service provider organizations, regional or provincial health authorities, Ministry of Health and Wellness) or non-profit research organizations (universities, hospitals, research institutions or centers). The sponsoring organization must be Alberta-based and have been in existence for at least one year.
Restrictions: Funding will not be provided for the following activities/items: projects consisting strictly of basic biomedical or clinical research (such as clinical trials); projects that are limited to program development or evaluation; projects that involve issues not related to human health; projects that duplicate services already provided; staff training, unless such training is required for the proposed project; purchase of furniture, or renting/leasing space and office equipment, (unless such items are not provided for by existing organizations).
Amount of Grant: $50,000 - $75,000 for two years
Contact: Pamela Valentine, Director of Grants and Awards; (780) 423-5727; fax: (780) 429-3509; email: pamela.valentine@ahfmr.ab.ca
Internet: http://www.ahfmr.ab.ca/grants/HRF/hrfguide.php
Sponsor: Alberta Heritage Foundation for Medical Research
10104-103 Avenue, Suite 1500
Edmonton, Alberta T5J 4A7 Canada

AHFMR Heritage Summer Studentships 412
AHFMR Summer Studentships offer motivated students with exceptional academic records an opportunity to participate in medical or health research in Alberta during the summer months. The award is meant to encourage students to consider pursuing formal training and a career in health research. The award is tenable for a minimum of 2 months, and a maximum of 4 months during the period of May to August. The award consists of a $1300 per month stipend only.
Requirements: Those candidates who meet one of the following criteria are eligible to apply: registered in an Alberta-based undergraduate degree program in a medical or health-related field; registered in an undergraduate degree program outside of Alberta, and desiring to engage in research during the summer at an Alberta institution; registered in an M.D. program, and who may also hold an undergraduate or graduate degree; exceptional high school students with records of participation in the health care

system, and a clear interest in pursuing a health research career; or currently in the last term of their undergraduate degrees and who have applied either to Medical School or to a graduate program that would start in the coming Fall.
Amount of Grant: $2,600-$5,200 stipend
Date(s) Application is Due: Feb 1
Contact: Christina Van Gelderen; (780) 423-5727; fax: (780) 429-3509; email: christina.van.gelderen@ahfmr.ab.ca
Internet: http://www.ahfmr.ab.ca/grants/Sum-student.php
Sponsor: Alberta Heritage Foundation for Medical Research
10104-103 Avenue, Suite 1500
Edmonton, Alberta T5J 4A7 Canada

AHFMR Infrastructure Funding Grants   413
AHFMR has developed a program to support infrastructure costs associated with the Canada Research Chairs (CRC) program. Infrastructure support for the Chairs comes from the Canada Foundation for Innovation (CFI), which covers 40% of total eligible costs. The remaining costs come from other funding partners. AHFMR has developed a program that will provide a contribution to the 60% of the cost of the CFI-eligible projects that normally comes from the universities and their funding partners.
Requirements: The Program provides support for successful CRC holders who are: faculty members at an Alberta institution, who are ineligible to apply for AHFMR Establishment awards and are principal investigators on active Canadian Institute for Health Research (CIHR) Operating Grants: or are currently recipients of unconditional AHFMR Personnel Awards and are principal investigators on non-AHFMR peer-reviewed operating grants.
Restrictions: Investigators being recruited to the province, and who are eligible to apply for, or have received AHFMR Establishment grants are not eligible for consideration under this program.
Amount of Grant: $62,500 average
Contact: Pamela Valentine, Director of Grants and Awards; (780) 423-5727; fax: (780) 429-3509; email: pamela.valentine@ahfmr.ab.ca
Internet: http://www.ahfmr.ab.ca/grants/crcinfo.php#1
Sponsor: Alberta Heritage Foundation for Medical Research
10104-103 Avenue, Suite 1500
Edmonton, Alberta T5J 4A7 Canada

AHFMR Institutional Support Program Grants   414
AHFMR offers a variety of programs to help the members of the research community in Alberta access the most up-to-date information and technical developments in medical and health-related research. Funding for visiting scientists includes: Heritage visiting professorships; and Heritage visiting scientists (to or from Alberta). Funding for conferences and meetings includes: conferences and symposia; and local workshops, retreats or planning sessions.
Requirements: Faculty members working in health research at the Alberta-based universities may apply. In the case of a visiting professorship or scientist, the sponsoring institution must identify the candidate's special expertise, as well as the expected long-term research benefits of the visit.
Contact: Pamela Valentine, Director of Grants and Awards; (780) 423-5727; fax: (780) 429-3509; email: pamela.valentine@ahfmr.ab.ca
Internet: http://www.ahfmr.ab.ca/grants/institutional.php
Sponsor: Alberta Heritage Foundation for Medical Research
10104-103 Avenue, Suite 1500
Edmonton, Alberta T5J 4A7 Canada

AHFMR Interdisciplinary Team Grants   415
The Alberta Heritage Foundation for Medical Research (AHFMR) Interdisciplinary Team Grants Program provides opportunities for high-quality, internationally recognized teams of investigators to complete research initiatives with defined health outcomes. Funds available in this competition are to support collaborative, interdisciplinary and multi-institutional teams that address important research questions, health problems or issues in defined areas of research that are aligned with strategic research priorities of the Foundation and Alberta. The Program will provide up to $1 million per year per team for up to five years. Funding can be used for a broad range of research-related costs including research infrastructure, research operating costs, core administrative/management costs, scientific support for the team (including the costs associated with the recruitment of new investigators to the team), collaborative/linkage activity and knowledge exchange/translation activity.
Requirements: Each eligible AHFMR Interdisciplinary Team Grant application will include: A Team Leader—the Team Leader must be an established researcher with proven leadership skills and experience who will act as research program director and who will assume administrative responsibility for the grant. It is expected that the Team Leader will devote a significant and appropriate portion of her/his time to these tasks. The Team Leader will have their primary academic appointment at an Alberta-based university; at least two additional independent investigators with established research track records. Teams with a nucleus of experienced investigators are encouraged to include promising new investigators as part of their group; team members who collectively have an extensive record of success, are creative and original in their approach to research and its translation, and who have experience working in research teams. The specific contribution of each team member and end-user partner, where applicable, must be described; representation from more than one research discipline and from more than one "research pillar" (i.e. biomedical; clinical science; health systems and services; and the social, cultural and other factors that affect the health of populations.);

and representation from more than one Alberta-based university. Multi-institutional collaboration is strongly encouraged in this Program.
Amount of Grant: $1,000,000 maximum
Date(s) Application is Due: Nov 14
Contact: Pamela Valentine, Director of Grants and Awards; (780) 423-5727; fax: (780) 429-3509; email: pamela.valentine@ahfmr.ab.ca
Internet: http://www.ahfmr.ab.ca/grants/team_guidelines.php
Sponsor: Alberta Heritage Foundation for Medical Research
10104-103 Avenue, Suite 1500
Edmonton, Alberta T5J 4A7 Canada

AHFMR Major Equipment Grants   416
AHFMR holds an annual competition to assist Alberta's medical or health researchers in obtaining major scientific equipment (defined as equipment costing more than $10,000 per unit or setup). It is AHFMR's expectation that large requests (e.g. greater than $100,000) will include contribution from other sources. Major equipment funds may be applied toward the purchase of the equipment, as well as toward the following expenses: installation; freight; transportation; duty; sales tax; and GST.
Requirements: An Alberta-based institution must sponsor applications on behalf of full-time, competitively-funded, productive faculty members.
Amount of Grant: $10,000-$100,000 per unit
Date(s) Application is Due: Mar 1
Contact: Pamela Valentine, Director of Grants and Awards; (780) 423-5727; fax: (780) 429-3509; email: pamela.valentine@ahfmr.ab.ca
Internet: http://www.ahfmr.ab.ca/grants/equipment.php
Sponsor: Alberta Heritage Foundation for Medical Research
10104-103 Avenue, Suite 1500
Edmonton, Alberta T5J 4A7 Canada

AHFMR Medical or Health Scholars Investigator Grants   417
Heritage Medical or Health Scholars are investigators who have recently completed their postdoctoral research training, and are currently seeking their first faculty-level appointments, or who are at the Heritage Population Heath Investigator or Heritage Clinical Investigator levels. Candidates must demonstrate an ability to initiate and conduct independent and collaborative research, as well as an interest and ability, or potential ability, to train future investigators. The Scholarship award includes a contribution toward salary and benefits, and may also include a research allowance and funds for renovations, and relocation.
Requirements: Candidates for the Heritage Scholar award must possess a M.D., D.D.S., D.V.M., Ph.D. or equivalent qualification. Furthermore, they must be eligible to hold a full-time appointment with the sponsoring institution.
Amount of Grant: $20,000
Date(s) Application is Due: Sep 15
Contact: Pamela Valentine, Director of Grants and Awards; (780) 423-5727; fax: (780) 429-3509; email: pamela.valentine@ahfmr.ab.ca
Internet: http://www.ahfmr.ab.ca/grants/scholars.php
Sponsor: Alberta Heritage Foundation for Medical Research
10104-103 Avenue, Suite 1500
Edmonton, Alberta T5J 4A7 Canada

AHFMR Medical or Health Scientist Awards   418
The Heritage Medical or Health Scientist Award is the most senior personnel award offered by AHFMR. Heritage Scientists must be established, internationally recognized scientific leaders in their fields of study. Their stature should be apparent by: the quality of their publication records; the originality and vision of their research proposals; the quality and number of their trainees; their ability to attract and retain research grants; and the leadership qualities they exhibit locally, nationally, and internationally. The Scientist award includes a contribution toward salary and benefits, and may also include a research allowance and funds for renovations, and relocation.
Requirements: Candidates for the Heritage Scientist award must possess a M.D., D.D.S., D.V.M., Ph.D. or equivalent qualification. Furthermore, they must be eligible to hold a full-time appointment with the sponsoring institution. AHFMR expects a time commitment of not less than 75% to health research activities. The department chair and the dean will designate the time expected for teaching, clinical duties, examining, and other activities.
Amount of Grant: $10,000-$40,000
Date(s) Application is Due: Sep 15
Contact: Pamela Valentine, Director of Grants and Awards; (780) 423-5727; fax: (780) 429-3509; email: pamela.valentine@ahfmr.ab.ca
Internet: http://www.ahfmr.ab.ca/grants/scientists.php
Sponsor: Alberta Heritage Foundation for Medical Research
10104-103 Avenue, Suite 1500
Edmonton, Alberta T5J 4A7 Canada

AHFMR Part-Time Fellowships   419
AHFMR Part-Time Fellowships enable highly qualified doctoral graduates to prepare for careers in medical or health research as independent investigators doing part-time research training during the regular academic year while enrolled in a full-time professional degree program in Alberta. Part-time awards will consist of a stipend only and will be pro-rated to the amount of time the candidate is prepared to commit to research. The full-time fellowship rate will be used as the basis for this calculation.
Requirements: Candidates must hold a PhD and must be engaged in a professional degree program (usually the MD program).

Restrictions: Part-time fellowship awards are tenable only at an Alberta-based university.
Amount of Grant: $5,000-$30,000
Date(s) Application is Due: Mar 1; Oct 1
Contact: Pamela Valentine, Director of Grants and Awards; (780) 423-5727; fax: (780) 429-3509; email: pamela.valentine@ahfmr.ab.ca
Internet: http://www.ahfmr.ab.ca/grants/PT-fellow.php
Sponsor: Alberta Heritage Foundation for Medical Research
10104-103 Avenue, Suite 1500
Edmonton, Alberta T5J 4A7 Canada

AHFMR Part-Time Studentships 420

AHFMR Part-Time Studentships enable students enrolled in a full-time professional degree program (usually the M.D. program) in Alberta to engage in part-time research training during the regular academic year. The award consists of a stipend only. Part-time students may elect to engage in full-time research during the summer; these students should contact AHFMR to request that full-time summer trainee status be implemented.
Requirements: Candidates must normally have been accepted into, or be currently engaged in, a full-time graduate program at an Alberta-based university in a health-related discipline leading to a Master's or doctoral degree. It is expected that the amount of time the candidate will commit to research will not be less than the equivalent of 1 day per week, taken either as 1 full day or 2 half days.
Restrictions: The award is tenable at an Alberta-based university only.
Date(s) Application is Due: Mar 1; Oct 1
Contact: Pamela Valentine, Director of Grants and Awards; (780) 423-5727; fax: (780) 429-3509; email: pamela.valentine@ahfmr.ab.ca
Internet: http://www.ahfmr.ab.ca/grants/PT-student.php
Sponsor: Alberta Heritage Foundation for Medical Research
10104-103 Avenue, Suite 1500
Edmonton, Alberta T5J 4A7 Canada

AHFMR Personnel Funding Grants 421

AHFMR has developed a program to support personnel costs associated with the Canada Research Chairs (CRC) program. In general, support will be given to Heritage personnel who receive CRC awards consistent with AHFMR guidelines concerning competitive personnel funding contributions from other agencies. Though it does not duplicate personnel funding, the Foundation has approved stipend incentives to partner with other personnel awards received. Currently the stipend incentive program provides funding to about 50 AHFMR researchers who have been successful in both AHFMR Personnel Award competitions and the personnel award competitions of other agencies.
Requirements: Researchers recruited to Alberta with a CRC award are eligible to apply for AHFMR Personnel Awards. If they are successful, the latter may be used as a contribution to matching the CFI allocation for CRC awards.
Contact: Pamela Valentine, Director of Grants and Awards; (780) 423-5727; fax: (780) 429-3509; email: pamela.valentine@ahfmr.ab.ca
Internet: http://www.ahfmr.ab.ca/grants/crcinfo.php#1
Sponsor: Alberta Heritage Foundation for Medical Research
10104-103 Avenue, Suite 1500
Edmonton, Alberta T5J 4A7 Canada

AHFMR Polaris Awards 422

The Alberta Heritage Foundation for Medical Research (AHFMR) Polaris Award was established as a means to help Alberta universities recruit outstanding mid-career health researchers of exceptional international calibre to the province. Individuals supported via this program will be known as AHFMR Polaris Investigators. The goal of this Award is to accelerate research activity in key areas that are priorities for Albertans. AHFMR will provide up to $1 million per year per award for up to ten years. This funding is to be at least matched by the institution nominating the AHFMR Polaris Investigator. The institutional support can include funding provided from other partners including government, the private sector and non-profit agencies. It generally cannot include funding originating from standard grants and awards programs of research organizations such as CFI, CIHR, and NSERC. The first AHFMR Polaris Award will be made in 2008. It is anticipated that initially up to three awards will be available.
Requirements: AHFMR Polaris Investigators must meet a number of eligibility criteria, including being: mid-career investigators with outstanding records of accomplishments in health research. Investigators active in any one (or more) of the four pillars of health research (basic, clinical, health services research, population health) are eligible for consideration; leaders who have histories of accomplishment in science and research demonstrated via executive positions in scientific societies, editorship of a prestigious journal, and invitations to speak at international conferences (they would be recognized as candidates for, or would have recently received, a major international recognition award); nominated by the University of Alberta, the University of Calgary, or the University of Lethbridge. Nominations may also be considered for individuals to be based within an Alberta Regional or Provincial Health Authority, but it is anticipated that they will have academic appointments at one of the universities as well; currently based outside Alberta and prepared to relocate to Alberta full-time for the duration of the AHFMR Polaris Award; and prepared to commit a majority of their time to research.
Amount of Grant: $1,000,000
Contact: Pamela Valentine, Director of Grants and Awards; (780) 423-5727; fax: (780) 429-3509; email: pamela.valentine@ahfmr.ab.ca
Internet: http://www.ahfmr.ab.ca/grants/polaris.php
Sponsor: Alberta Heritage Foundation for Medical Research

10104-103 Avenue, Suite 1500
Edmonton, Alberta T5J 4A7 Canada

AHFMR Population Health Investigator (PHI) Awards 423

The Award allows independent investigators in the early stages of their careers in health research to commit the majority of their time (greater than 75%) to research. This award includes a $110,000 annual contribution toward salary and benefits, a $10,000 annual research prize, a $15,000 annual research allowance, and additional funds for renovations, operations, relocation, and equipment. Potential areas of research are: quantitative and population sciences (epidemiology and biostatistics, operations research, decision sciences, and computer sciences); health services research; social and behavioural sciences (sociology, anthropology, psychology, economics, and political science); and the humanities (law, philosophy, history, and theology). In all cases, the relevance of the applicant's proposed research to human health should be apparent.
Requirements: Canadian citizens or permanent residents of Canada who hold an MD or DDS degree, have completed all requirements for clinical specialty recognition, and qualify to hold a full-time position in a clinical department of the sponsoring institution are eligible to apply.
Amount of Grant: $135,000
Date(s) Application is Due: Sep 15
Contact: Pamela Valentine, Director of Grants and Awards; (780) 423-5727; fax: (780) 429-3509; email: pamela.valentine@ahfmr.ab.ca
Internet: http://www.ahfmr.ab.ca/grants/PopHealthInvest.php
Sponsor: Alberta Heritage Foundation for Medical Research
10104-103 Avenue, Suite 1500
Edmonton, Alberta T5J 4A7 Canada

AHFMR Proposals for Special Initiatives 424

AHFMR welcomes proposals for innovative initiatives that accelerate the achievement of its overall objectives. Institutions recruiting internationally-recognized researchers to lead new initiatives should consult AHFMR, if current programs are inappropriate. AHFMR will also consider proposals for pilot projects in patient or population-based research. These projects must have the potential to develop information, and the investigators must have the expertise to attract grants from other agencies. The review process and financial support for special initiatives may be shared by AHFMR and other agencies.
Restrictions: Applications that are normally eligible for regular operating grants from other agencies will NOT be considered under this program.
Contact: Pamela Valentine, Director of Grants and Awards; (780) 423-5727; fax: (780) 429-3509; email: pamela.valentine@ahfmr.ab.ca
Internet: http://www.ahfmr.ab.ca/grants/initiatives.php
Sponsor: Alberta Heritage Foundation for Medical Research
10104-103 Avenue, Suite 1500
Edmonton, Alberta T5J 4A7 Canada

AHFMR Research Prize 425

The AHFMR Research Prize is intended to maintain and improve AHFMR Personnel Awards as a potent vehicle for the recruitment and retention of highly qualified, internationally competitive investigators at Alberta institutions. The Prize will be made to every eligible investigator supported by AHFMR through the AHFMR Personnel Support Programs (Population Health Investigator, Clinical Investigator, Scholar, Senior Scholar, Scientist).
Requirements: To be eligible, the investigator must be currently supported by an unconditional AHFMR investigator award. Investigators receiving AHFMR Terminal support or those on Leaves of Absence are not eligible to receive the Research Prize. The Research Prize is not determined by institutional rank and salary scale. It is provided in recognition of the individual investigator's research accomplishments, and of his/her success in obtaining an AHFMR Independent Investigator award. The Research Prize is a personal prize to the recipient.
Restrictions: Successful applicants for an Independent Establishment Grant are not eligible to receive the AHFMR Research Prize.
Amount of Grant: $10,000-$20,000
Contact: Pamela Valentine, Director of Grants and Awards; (780) 423-5727; fax: (780) 429-3509; email: pamela.valentine@ahfmr.ab.ca
Internet: http://www.ahfmr.ab.ca/grants/resprize.php
Sponsor: Alberta Heritage Foundation for Medical Research
10104-103 Avenue, Suite 1500
Edmonton, Alberta T5J 4A7 Canada

AHFMR Scholar Awards 426

The Scholars grant allows independent researchers in the early stages of their careers in health research to commit the majority of their time (greater than 75%) to research. Scholars are defined as investigators who have recently completed their postdoctoral research training, and are currently seeking their first faculty-level appointments, or who are completing their term as AHFMR Population Heath Investigator or Clinical Investigator awards. This award includes a $110,000 annual contribution toward salary and benefits, a $10,000 to $20,000 annual research prize, a $35,000 annual research allowance, and additional funds for renovations, operations, relocation, and equipment. Potential areas of research are: quantitative and population sciences (epidemiology and biostatistics, operations research, decision sciences, and computer sciences); health services research; social and behavioral sciences (sociology, anthropology, psychology, economics, and political science); and the humanities (law, philosophy, history, and

theology). In all cases, the relevance of the applicant's proposed research to human health should be apparent. All new awards are for seven years.
Requirements: Candidates must possess a M.D., D.D.S., D.V.M., Ph.D. or equivalent qualification. Furthermore, they must be eligible to hold full-time faculty appointments with the sponsoring institutions.
Amount of Grant: $155,000 minimum per year for seven years
Date(s) Application is Due: Sep 15
Contact: Pamela Valentine, Director of Grants and Awards; (780) 423-5727; fax: (780) 429-3509; email: pamela.valentine@ahfmr.ab.ca
Internet: http://www.ahfmr.ab.ca/grants/scholars.php
Sponsor: Alberta Heritage Foundation for Medical Research
10104-103 Avenue, Suite 1500
Edmonton, Alberta T5J 4A7 Canada

AHFMR Scientist Awards   427
The Scientist award is the most senior investigator award offered by AHFMR. Scientists must be established, internationally recognized scientific leaders in their fields of study. Their stature should be demonstrated through: the quality of their publications; the originality and vision of their research proposals; the quality and number of their trainees; their ability to attract and retain research grants; and the leadership qualities they exhibit locally, nationally, and internationally. This award includes a $160,000 annual contribution toward salary and benefits, a $20,000 annual research prize, a $50,000 annual research allowance, and additional funds for renovations, operations, relocation, and equipment. In all cases, the relevance of the applicant's proposed research to human health should be apparent. All new awards are for seven years.
Requirements: Candidates for the award must possess a M.D., D.D.S., D.V.M., Ph.D. or equivalent qualification. Furthermore, they must be eligible to hold full-time faculty appointments with the sponsoring institutions. At time of implementation, candidates will typically have ten to fifteen years experience as independent investigators and are expected to have attained full professor level faculty appointments.
Amount of Grant: $230,000 minimum
Date(s) Application is Due: Sep 15
Contact: Pamela Valentine, Director of Grants and Awards; (780) 423-5727; fax: (780) 429-3509; email: pamela.valentine@ahfmr.ab.ca
Internet: http://www.ahfmr.ab.ca/grants/scientists.php
Sponsor: Alberta Heritage Foundation for Medical Research
10104-103 Avenue, Suite 1500
Edmonton, Alberta T5J 4A7 Canada

AHFMR Senior Investigator Awards   428
Like the Scientist Award, this is the most senior investigator award offered by AHFMR. Applicants must be established, internationally recognized scientific leaders in their fields of study. Their stature should be demonstrated through: the quality of their publications; the originality and vision of their research proposals; the quality and number of their trainees; their ability to attract and retain research grants; and the leadership qualities they exhibit locally, nationally, and internationally. This award includes a $160,000 annual contribution toward salary and benefits, a $40,000 annual research prize, a $20,000 annual research allowance, and additional funds for renovations, operations, relocation, and equipment. In all cases, the relevance of the applicant's proposed research to human health should be apparent. All new awards are for seven years.
Requirements: Candidates for the award must possess a M.D., D.D.S., D.V.M., Ph.D. or equivalent qualification. Furthermore, they must be eligible to hold full-time faculty appointments with the sponsoring institutions. At time of implementation, candidates will typically have ten to fifteen years experience as independent investigators and are expected to have attained full professor level faculty appointments.
Amount of Grant: $220,000 minimum
Contact: Pamela Valentine, Director of Grants and Awards; (780) 423-5727; fax: (780) 429-3509; email: pamela.valentine@ahfmr.ab.ca
Internet: http://www.ahfmr.ab.ca/grants/independent-investigator-revised.php
Sponsor: Alberta Heritage Foundation for Medical Research
10104-103 Avenue, Suite 1500
Edmonton, Alberta T5J 4A7 Canada

AHFMR Senior Medical or Health Scholars Investigator Grants   429
Heritage Senior Medical or Health Scholars must have an excellent track record of independent research over several years. They are more senior than Heritage Scholars and have usually been previously supported via an external salary support award (e.g. as an AHFMR Scholar, CIHR New Investigator, or equivalent). At the time of application, candidates for a Senior Scholar Award will typically have four to six years as an independent investigator and will be appointed at either the Assistant Professor or Associate Professor level. The Senior Scholarship award includes a contribution toward salary and benefits, and may also include a research allowance and funds for renovations, and relocation.
Requirements: Candidates for the Heritage Senior Scholar award must possess a M.D., D.D.S., D.V.M., Ph.D. or equivalent qualification. Furthermore, they must be eligible to hold a full-time appointment with the sponsoring institution. AHFMR expects a time commitment of not less than 75% to health research activities. The department chair and the dean will designate the time expected for teaching, clinical duties, examining, and other activities.
Amount of Grant: $30,000 maximum
Date(s) Application is Due: Sep 15
Contact: Pamela Valentine, Director of Grants and Awards; (780) 423-5727; fax: (780) 429-3509; email: pamela.valentine@ahfmr.ab.ca

Internet: http://www.ahfmr.ab.ca/grants/seniorscholars.php
Sponsor: Alberta Heritage Foundation for Medical Research
10104-103 Avenue, Suite 1500
Edmonton, Alberta T5J 4A7 Canada

AHFMR Senior Scholar Awards   430
The Senior Scholars award allows independent researchers in the later stages of their careers in health research to commit the majority of their time (greater than 75%) to research. Applicants are more senior than Scholars and have usually been previously supported via salary support awards (e.g. as an AHFMR Scholar, CIHR New Investigator, or equivalent). At time of implementation, candidates for a Senior Scholar award will typically have five to ten years experience as independent investigators and are expected to have attained an associate professor level faculty appointment. This award includes a $135,000 annual contribution toward salary and benefits, a $20,000 annual research prize, a $50,000 annual research allowance, and additional funds for renovations, operations, relocation, and equipment. In all cases, the relevance of the applicant's proposed research to human health should be apparent.
Requirements: Candidates must possess a M.D., D.D.S., D.V.M., Ph.D. or equivalent qualification. Furthermore, they must be eligible to hold a full-time faculty appointment with the sponsoring institution.
Amount of Grant: $205,000 minimum
Contact: Pamela Valentine, Director of Grants and Awards; (780) 423-5727; fax: (780) 429-3509; email: pamela.valentine@ahfmr.ab.ca
Internet: http://www.ahfmr.ab.ca/grants/seniorscholars.php
Sponsor: Alberta Heritage Foundation for Medical Research
10104-103 Avenue, Suite 1500
Edmonton, Alberta T5J 4A7 Canada

AHIMA Dissertation Assistance Grants   431
The Dissertation Assistance Award program supports research undertaken as part of an academic program to qualify for a doctorate in areas relevant to health information management (HIM). Each recipient shall be limited to one funded grant per year. Ordinarily, the scope of the proposal shall be such that it can be completed within 18 months from the date of funding. There are no minimum or maximum grant request expectations, but the range of grants historically has been between $5,000 and $10,000. Submissions that address one or more of the AHIMA research priorities will receive priority consideration for funding.
Requirements: To qualify for an award under this program the student (principal investigator) must be enrolled in an accredited doctoral degree program in an area related to HIM (computer science, business management, education, public health, and so forth) and must be an active, associate, or student member of AHIMA. All requirements for the doctoral degree, other than the dissertation, must be completed by the award date.
Samples: Ellen Jacobs, MEd, RHIA—titled In Search of a Message to Promote the Personal Health Record (2007); Melinda Wilkins, MEd, RHIA—titled Adoption of Electronic Health Records: Perceptions of Health Information Managers (2007); Rebecca Reynolds, MHA, RHIA—titled The Intentions of First Year Medical Students to use Electronic Health Records (2007).
Amount of Grant: $5,000-$10,000
Date(s) Application is Due: Oct 31
Contact: Carol Nielsen, Director; (312) 233-1175; email: carol.nielsen@ahimafoundation.org
Internet: http://www.ahimafoundation.org/Scholarships/dissertation.aspx
Sponsor: Foundation of Research and Education (FORE) of the American Health Information Management Association (AHIMA)
233 N. Michigan Avenue, 21st Floor
Chicago, IL 60601

AHIMA Faculty Development Stipends   432
FORE Faculty Development Stipends are intended to encourage excellence in effective teaching and leadership in areas related to e-HIM. Program goals are to: expedite curriculum innovation and development in e-HIM content areas; research and develop new methods of teaching, evaluation, and assessment in the HIM learning process; support faculty in maintaining the highest standards of knowledge and practice in e-HIM; integrate new technologies effectively into the HIM learning process; and strengthen the capacity of HIM academic programs to address emerging issues in design and implementation of the electronic health record at all levels of health care delivery, in support of an electronic health information infrastructure.
Requirements: Application should be for activity planned during the four (4) month period following the expected notification date.
Date(s) Application is Due: Mar 14; Jul 11; Oct 3
Contact: Susan H. Fenton, PhD, RHIA, Director of Research; (312) 233-1532 or (312) 233-1100; fax: (210) 479-1043; email: susan.fenton@ahima.org or fore@ahima.org
Internet: http://www.ahimafoundation.org/Scholarships/Faculty.aspx
Sponsor: Foundation of Research and Education (FORE) of the American Health Information Management Association (AHIMA)
233 N. Michigan Avenue, 21st Floor
Chicago, IL 60601

AHIMA Grant-In-Aid Research Grants   433
The FORE Grant-in-Aid program is directed toward supporting the development of HIM professionals as leaders in defining and validating the unique body of knowledge encompassed by HIM. The results of these studies not only provide information for the HIM professional to apply in meeting current and future challenges, but also support

policy initiatives and the redefinition of the roles of HIM practitioners. Ordinarily, the scope of the proposal shall be such that it can be completed within 18 months from the date of funding. There are no minimum or maximum grant request expectations, but the range of grants historically has been between $15,000 and $40,000. Submissions that address one or more of the AHIMA research priorities will receive priority consideration for funding.

Requirements: The primary or secondary investigator must be an active, associate, or student member of AHIMA. Each recipient shall be limited to one grant per year.
Amount of Grant: $15,000-$40,000 average range
Date(s) Application is Due: Sep 19
Contact: Susan H. Fenton, PhD, RHIA, Director of Research; (312) 233-1532 or (312) 233-1100; fax: (210) 479-1043; email: susan.fenton@ahima.org or fore@ahima.org
Internet: http://www.ahima.org/fore/research/grantinaid.asp
Sponsor: Foundation of Research and Education (FORE) of the American Health Information Management Association (AHIMA)
233 N. Michigan Avenue, 21st Floor
Chicago, IL 60601

Ahmanson and Getty Postdoctoral Fellowships   434
Theme-based, residential fellowships are awarded for periods of two consecutive academic quarters for participation in the interdisciplinary, cross-cultural programs of the Center for 17th- & 18th-Century Studies/William Andrews Clark Memorial Library. The theme for each academic year is announced the preceding fall. An organized research unit of the University of California, the center provides a forum for the discussion of central issues in the study of the 17th and 18th centuries, facilitates research and publication, and encourages the creation of programs that advance understanding of this time period. The William Andrews Clark Memorial Library, which is administered by the center, is known for its collections on 17th- and 18th-century Britain, Oscar Wilde and the 1890s, the history of printing, and certain aspects of Western Americana. Application materials are available upon request.
Requirements: Scholars who have received a Ph.D. in the last six years and are engaged in research pertaining to the announced theme are eligible to apply. Fellows are expected to make a substantive contribution to the Center's workshops and seminars. Awards are for one full academic year in residence at the Clark Library.
Amount of Grant: $35,000 stipend
Date(s) Application is Due: Feb 1
Contact: Fellowship Coordinator, (310) 206-8552; fax: (310) 206-8577; email: c1718cs@humnet.ucla.edu
Internet: http://www.humnet.ucla.edu/humnet/c1718cs/Postd.htm#AhmGet
Sponsor: Center for 17th- & 18th-Century Studies
310 Royce Hall, UCLA
Los Angeles, CA 90095-1404

Ahmed Zewail Award in Ultrafast Science and Technology   435
The Award is intended to recognize outstanding and creative contributions to fundamental discoveries or inventions in ultrafast science & technology in areas of physics, chemistry, biology, or related fields. Areas of interest include physics, chemistry, biology, or related fields without regard to age or nationality. The recipient will have conducted original and insightful research that has had a significant impact on the field of ultrafast science and technology. The award will consist of $5,000 and a certificate.
Requirements: Any individual, except a member of the award committee, may submit one nomination or seconding letter for the award in any given year. Nominating documents consist of a letter of not more than 1000 words containing an evaluation of the nominee's accomplishments and a specific identification of the work to be recognized, a biographical sketch including date of birth, and a list of publications and patents authored by the nominee.
Restrictions: Self-nominations are not accepted.
Samples: Graham R. Fleming—award winner, $5,000 (2008); Robin M. Hochstrasser—award winner, $5,000 (2007).
Amount of Grant: $5,000 plus travel expenses to award meeting
Date(s) Application is Due: Nov 1
Contact: Felicia Dixon, Awards Administrator; (800) 227-5558 or (202) 872-4408; fax: (202) 776-8008; email: f_dixon@acs.org or awards@acs.org
Internet: http://portal.acs.org/portal/acs/corg/content?_nfpb=true&_pageLabel=PP_ARTICLEMAIN&node_id=1319&content_id=CTP_004488&use_sec=true&sec_url_var=region1
Sponsor: American Chemical Society
1155 Sixteenth Street, NW
Washington, DC 20036

Ahn Family Foundation Grants   436
The Ahn Family Foundation, established by Sangwoo and Laura Ahn in 1997, has as its primary purposes to support aging, religion, human and social services, art, culture, education, and the environment. Giving is primarily limited to Connecticut, Massachusetts, and New York, although the Foundation has also supported projects in Maryland and Washington. There are no particular guidelines or deadlines with which to adhere, and applicants should begin by contacting the Foundation in writing. This letter of inquiry should outline the purpose for the request, the mission of the applicant organization, and the amount requested.
Geographic Focus: Connecticut; Massachusetts; New York
Samples: Holy Apostles Soup Kitchen, New York, New York, $20,000 (2006); Saint Lukes LifeWorks, Stamford, Connecticut, $20,000 (2006); All Saints Church, Chevy Chase, Maryland, $5,000 (2006).

Amount of Grant: $1,000-$20,000 range
Contact: Alison D. Ahn, Treasurer; (203) 869-4875
Sponsor: Ahn Family Foundation
901 Hillsboro Mile
Hillsboro Beach, FL 33062-2801

AHNS Alando J. Ballantyne Resident Research Pilot Grant   437
The purpose of this award is to support basic, translational, or clinical research projects in head and neck oncology. Clinical or translational research studies are strongly encouraged and should be specifically related to the prevention, diagnosis, treatment, outcomes, or pathophysiology of head and neck neoplastic disease. Research supported by this award should be specifically directed toward the pathogenesis, pathophysiology, diagnosis, prevention, or treatment of head and neck neoplastic disease, and may be either basic or clinical/translational in approach. While not specifically required, proposals which aim to introduce new knowledge and methodology from other disciplines to research in head and neck disease, or which demonstrate collaborative effort with members of other related disciplines are encouraged. Grants range up to $10,000 maximum, are for one year, and are not renewable. Letters of intent must be received by December 15, and completed applications must be submitted by January 15.
Requirements: This grant is open to resident in U.S. or Canadian training programs. Previous AHNS or AAO-HNS Foundation research grant recipients are eligible to compete for this grant. However, candidates who have successfully obtained funding from a private or federal funding agency for the same research are ineligible. Candidates who have applied for support of the same research from other funding sources, and who are notified of an award from both another agency and from AHNS must choose only one of the awards.
Amount of Grant: $10,000 maximum
Date(s) Application is Due: Dec 15; Jan 15
Contact: Stephanie L. Jones, Assistant Director; (703) 519-1586 or (703) 836-4444; email: sljones@entnet.org
Internet: http://www.headandneckcancer.org/research/grants.php
Sponsor: American Head and Neck Society (AHNS)
11300 W. Olympic Boulevard, Suite 600
Los Angeles, CA 90064

AHNS Pilot Grant   438
The purpose of this award is to support basic, translational, or clinical research projects in head and neck oncology. Clinical or translational research studies are strongly encouraged and should be specifically related to the prevention, diagnosis, treatment, outcomes, or pathophysiology of head and neck neoplastic disease. Research supported by this award should be specifically directed toward the pathogenesis, pathophysiology, diagnosis, prevention, or treatment of head and neck neoplastic disease, and may be either basic or clinical/translational in approach. While not specifically required, proposals which aim to introduce new knowledge and methodology from other disciplines to research in head and neck disease, or which demonstrate collaborative effort with members of other related disciplines are encouraged. Grants range up to $10,000 maximum, are for one year, and are not renewable. Letters of intent must be received by December 15, and completed applications must be submitted by January 15.
Requirements: Candidates for this award should reside in the U.S. or Canada, be medical students, residents, Ph.D.s or faculty members at the rank of associate professor or below.
Amount of Grant: $10,000 maximum
Date(s) Application is Due: Dec 15; Jan 15
Contact: Stephanie L. Jones, Assistant Director; (703) 519-1586 or (703) 836-4444; email: sljones@entnet.org
Internet: http://www.headandneckcancer.org/research/grants.php
Sponsor: American Head and Neck Society (AHNS)
11300 W. Olympic Boulevard, Suite 600
Los Angeles, CA 90064

AHNS/AAO-HNSF Surgeon Scientist Combined Award   439
The Grant is open to surgeons beginning a clinician-scientist career track to support research in the pathogenesis, diagnosis, prevention, or treatment of head and neck neoplastic disease. This grant is only available during odd numbered years. Awards are two year, nonrenewable, with a possible $70,000 maximum total costs ($35,000 per year). One award available per year. Letters of intent must be received by December 15, and completed applications must be submitted by January 15.
Requirements: Applicants must be members or candidate members of the American Academy of Otolaryngology-Head and Neck Surgery and/or the American Head and Neck Society.
Amount of Grant: $70,000 maximum total costs ($35,000 per year)
Date(s) Application is Due: Dec 15; Jan 15
Contact: Stephanie L. Jones, Assistant Director; (703) 519-1586 or (703) 836-4444; email: sljones@entnet.org
Internet: http://www.headandneckcancer.org/research/grants.php
Sponsor: American Head and Neck Society (AHNS)
11300 W. Olympic Boulevard, Suite 600
Los Angeles, CA 90064

AHNS/AAO-HNSF Young Investigator Combined Award   440
The purpose of this award is to support a collaborative AHNS/AAO-HNSF research project by fostering the development of contemporary basic or clinical research skills focused on neoplastic disease of the head and neck among new full-time academic

surgeons. The award is intended as a preliminary step in clinical investigator career development and is expected to facilitate the recipient, preparation of a more comprehensive individualized research plan suitable for submission to the National Institutes of Health or comparable funding agency. This is a two year, non-renewable award, offering $40,000 maximum ($20,000 per year). One award is available annually. Letters of intent must be received by December 15, and completed applications must be submitted by January 15.

Requirements: Applicants must be physicians with demonstrated potential for excellence in research and teaching and serious commitment to an academic research career in head and neck surgery. Applicants must be members or candidate members of the American Academy of Otolaryngology-Head and Neck Surgery and/or the American Head and Neck Society. Priority will be given to fellows or junior faculty who have completed residencies or fellowships within four years of the application receipt date. All candidates must be sponsored by the Chair of his/her Division or Department and by an official representative of the institution which would administer the Award and in whose name the application is formally submitted.

Amount of Grant: $40,000 maximum ($20,000 per year)
Date(s) Application is Due: Dec 15; Jan 15
Contact: Stephanie L. Jones, Assistant Director; (703) 519-1586 or (703) 836-4444; email: sljones@entnet.org
Internet: http://www.headandneckcancer.org/research/grants.php
Sponsor: American Head and Neck Society (AHNS)
11300 W. Olympic Boulevard, Suite 600
Los Angeles, CA 90064

AHRF Eugene L. Derlacki, M.D. Research Grants 441

The Eugene L. Derlacki, M.D. Grant is awarded for excellence in the field of hearing research. This grant provides $25,000 per year for two years, for a total award of $50,000. It is dedicated to hearing research and is intended for more significant research projects requiring funding that extends beyond the normal AHRF grant of $25,000. Researchers are invited to apply for this grant using the Research Grant Application Guidelines. According to the rules of the grant, the recipient will receive the first year funding of $25,000. After the first year, a progress report must be submitted to the Research Committee. If the report is approved, the second year funding of $25,000 will then be awarded.

Samples: Steven H. Green, Ph.D., University of Iowa—for Role of JNK Signaling in the Death of Spiral Ganglion Neurons After Hair Cell Loss (2005-06); Keiko Hirose, M.D., The Cleveland Clinic Foundation, Ohio—for Cellular Repair of the Murine Cochlea After Acoustic Injury (2004).
Amount of Grant: $25,000 per year for two years
Date(s) Application is Due: Aug 3
Contact: Kristen Madhuizen, Development and Communication Associate; (312) 726-9670; fax: (312) 726-9695; email: kristen.madhuizen@american-hearing.org or ahrf@american-hearing.org
Internet: http://www.american-hearing.org/research/derlacki_grant.html
Sponsor: American Hearing Research Foundation (AHRF)
8 South Michigan Avenue, Suite 814
Chicago, IL 60603-4539

AHRF Georgia Birtman Grant 442

The American Hearing Research Foundation (AHRF), together with the Northwestern Memorial Foundation (NMF) give the Georgia Birtman Grant, named in honor of long-time supporter of the AHRF. The one-year $75,000 grant ($25,000 each from the Birtman Fund, the AHRF and Northwestern Memorial Foundation) supports the advancement of research and education in otology and neurotology. The grant is awarded to an exceptional researcher in audiology, otology or neurotology who will work in a lab at Northwestern University. The research topic involves some aspect of the diagnosis, treatment, and rehabilitation of hearing and balance disorders related to the inner ear. The research has the potential to generate clinical care innovations, facilitate translational clinical studies, and develop creative educational programs.

Samples: Timothy C. Hain, M.D., Northwestern University, Feinberg School of Medicine, Chicago, Illinois—for Vestibular Evoked Myogenic Potentials (2007); Claus-Peter Richter, M.D., Ph.D., Northwestern University, Feinberg School of Medicine, Chicago, Illinois—for Electrical Stimulation of Spiral Ganglion Cells (2006).
Amount of Grant: $75,000 maximum
Date(s) Application is Due: Aug 3
Contact: Kristen Madhuizen, Development and Communication Associate; (312) 726-9670; fax: (312) 726-9695; email: kristen.madhuizen@american-hearing.org or ahrf@american-hearing.org
Internet: http://www.american-hearing.org/research/birtman_grant.html
Sponsor: American Hearing Research Foundation (AHRF)
8 South Michigan Avenue, Suite 814
Chicago, IL 60603-4539

AHRF Regular Research Grants 443

The American Hearing Research Foundation funds five to ten $20,000 research grants each year. Research Grants should relate to the hearing or balance functions of the ear. Both basic and clinical studies may be proposed. Priority is given to providing start-up funds for new projects. Applications are reviewed by a Research Committee and awards begin in January. Applications are due no later that noon on August 3 of the previous year.

Amount of Grant: $20,000 maximum
Date(s) Application is Due: Aug 3

Contact: Kristen Madhuizen, Development and Communication Associate; (312) 726-9670; fax: (312) 726-9695; email: kristen.madhuizen@american-hearing.org or ahrf@american-hearing.org
Internet: http://www.american-hearing.org/research/grant_guidelines.html
Sponsor: American Hearing Research Foundation (AHRF)
8 South Michigan Avenue, Suite 814
Chicago, IL 60603-4539

AHRF Wiley H. Harrison Memorial Research Award 444

The purpose of this award is to support clinical research projects in otology or neurotology designed to increase understanding of hearing disorders. Research supported by this award should be specifically directed toward the clinical identification, diagnosis, prevention, or treatment of diseases, disorders, or conditions of the ear. While not specifically required, proposals which aim to introduce new knowledge and methodology from other disciplines to research in otology or neurotology, or which demonstrate collaborative effort with members of other related disciplines are encouraged. This is a one year, non-renewable grant of $25,000 maximum, with only one award available annually. Completed applications must be submitted by August 3.

Requirements: Candidates for this award should be physicians (M.D.) at the resident, fellow, or junior faculty stage, or PhD scientists. Previous AHRF or AAO-HNS Foundation research grant recipients are eligible to compete for this grant.
Amount of Grant: $25,000 maximum
Date(s) Application is Due: Aug 3
Contact: Kristen Madhuizen, Development and Communication Associate; (312) 726-9670; fax: (312) 726-9695; email: kristen.madhuizen@american-hearing.org or ahrf@american-hearing.org
Internet: http://www.american-hearing.org/research/harrisongrant.html
Sponsor: American Hearing Research Foundation (AHRF)
8 South Michigan Avenue, Suite 814
Chicago, IL 60603-4539

AHRQ Assessment of Quality Improvement Strategies in Health Care 445

The projects undertaken as a result of this RFA will analyze the relative utility and costs of various approaches to health care quality improvement. The fundamental long-term goal of this effort is to strengthen the evidence base underlying the choice of strategies to employ when attempting to improve the quality of clinical care. Studies should focus on comparing improvement efforts that target those areas where the greatest improvements in health and functional status can occur, reliable and valid quality measures exist, and a variety of strategies are being employed. Partnerships between academic and other research organizations with existing health care quality improvement efforts through established mechanisms such as Peer Review Organizations (PROs), Quality Improvement Organizations (QIOs), purchaser groups, health plans, and accrediting bodies are required under this RFA. Special preference will be accorded to applications from investigators not recently or currently funded as principal investigators of an AHCPR grant for research on quality improvement strategies. RFA: HS-99-002.

Requirements: Applications may be submitted by public or private nonprofit organizations. For-profit entities may participate as members of consortia or subcontractors if the applicant is nonprofit.
Restrictions: Organizations described in section 501(c)4 of the Internal Revenue Code that engage in lobbying are not eligible.
Amount of Grant: $5000-$2.8 million; $310,000 average
Contact: Mable Lam, Grants Management Officer, (301) 427-1448
Internet: http://www.grants.nih.gov/grants/guide/rfa-files/RFA-HS-99-002.html
Sponsor: Agency for Healthcare Research and Quality
2101 E Jefferson Street
Rockville, MD 20852

AHRQ Centers for Education and Research on Therapeutics Grants 446

CERTs is a three-year program that will support demonstration centers. These centers will evaluate and develop options and methods, and conduct and perform pilot studies. Studies will consist of state-of-the-art clinical, health services, or laboratory research to increase awareness of the benefits, risks and effectiveness of new uses, existing uses, or combined uses of therapeutics. This demonstration program seeks new and more effective ways to develop, translate, and disseminate objective information on therapeutics to health care providers and other decision makers to improve practice. In addition, CERTs may selectively develop protocols and possibly undertake pilot studies on the comparative cost effectiveness and safety of medical products. This will be accomplished with data on appropriate therapeutic usage and outcomes; and the identification and prevention of medical errors and adverse effects. The long-term goal of the program will be to improve the quality of care while reducing costs. RFA: HS-99-004.

Requirements: Applications may be submitted by public or private nonprofit organizations. For-profit organizations may participate as members of consortia or subcontractors if the applicant is nonprofit.
Restrictions: Organizations described in section 501(c)4 of the Internal Revenue Code that engage in lobbying are not eligible.
Amount of Grant: $5000-$2.8 million; $310,000 average
Contact: Mable Lam, Grants Management Officer, (301) 427-1447
Internet: http://www.grants.nih.gov/grants/guide/rfa-files/RFA-HS-99-004.html
Sponsor: Agency for Healthcare Research and Quality
2101 E Jefferson Street, Executive Office Ctr
Rockville, MD 20852

AHRQ Centers of Excellence for Patient Safety Research and Practice   447

The purpose of this program is to support the development of multidisciplinary research teams to build the knowledge base on the scope and impact of medical errors; identify the root causes of threats to patient safety and effective system approaches to prevent the occurrence of errors; study the effectiveness of various interventions to capture information on medical errors; and evaluate the outcomes of promising interventions in a variety of healthcare settings. A letter of intent is requested by January 3; full application is due January 24. Annual deadlines may vary; contact program staff for exact dates.

Requirements: Applications may be submitted by domestic, both public and private, nonprofit organizations; units of state and local governments; and eligible agencies of the federal government.

Restrictions: Foreign institutions are not eligible to apply.

Amount of Grant: $5000-$2.8 million; $310,000 average

Date(s) Application is Due: Jan 3; Jan 24

Contact: Mable Lam, Grants Management Officer, (301) 427-1447

Internet: http://grants.nih.gov/grants/guide/rfa-files/RFA-HS-01-002.html

Sponsor: Agency for Healthcare Research and Quality

2101 E Jefferson Street

Rockville, MD 20852

AHRQ Health Care Access, Quality, and Insurance for Low-Income   448
Children Grants

AHCPR and the David and Lucile Packard Foundation invite applications for cooperative agreements to conduct research that will generate information useful to purchasers and designers of health insurance and health care delivery systems for low-income children. Specifically, answers are sought to the policy question: How do features of U.S. health care insurance and delivery systems improve health care access and quality for low-income children, particularly racial and ethnic minority children or those with special health care needs? Studies will examine how the features of U.S. insurance programs and the organization of the health care delivery systems associated with these programs affect access to services and the quality of care received by low-income children. Studies will fall into one of two categories: those examining the impact on low-income children enrolled in these insurance programs (enrollee impact), and those examining the impact on a low-income community's health care delivery system and all the children that it serves (community impact). Applicants may propose research that falls into one or both of these categories. Care should be taken not to replicate the work of other major health care studies. Application kits are available at most institutional offices of sponsored research and may be obtained from the Division of Extramural Outreach and Information Resources, Office of Extramural Research, National Institutes of Health, 6701 Rockledge Drive, MSC 7910, Bethesda, MD 20892-7910, telephone (301) 435-0714, email: Grantsinfo@nih.gov. RFA: HS-99-005

Requirements: Applications may be submitted by domestic and foreign, public and private nonprofit organizations.

Restrictions: For-profit organizations are not eligible as applicants, but may participate as members of consortia or as subcontractors. Organizations described in section 501(c)4 of the Internal Revenue Code that engage in lobbying are not eligible to receive grant/cooperative agreement awards.

Amount of Grant: $5000-$2.8 million; $310,000 average

Contact: Mable Lam, Grants Management Officer, (301) 427-1447

Internet: http://www.grants.nih.gov/grants/guide/rfa-files/RFA-HS-99-005.html

Sponsor: Agency for Healthcare Research and Quality

2101 E Jefferson Street, Executive Office Ctr

Rockville, MD 20852

AHRQ Independent Scientist Award   449

The award is a special salary-only grant designed to provide protected time for newly independent scientists who currently have non-research obligations such as heavy teaching loads, clinical work, committee assignments, service, and administrative duties that prevent them from having a period of intensive research focus. The award is targeted to persons with doctoral degrees who have completed their research training, have independent peer-reviewed research support, and who need a period of protected research time in order to foster their research career development. New applications are due February 12, June 12, and October 12, while renewal applications are due March 12, July 12, and November 12. Up to $90,000 per year of base salary plus fringe benefits is available. No other research development support funds are provided.

Requirements: The following organizations and institutions are eligible to apply: public/state controlled institutions of higher education; private institutions of higher education; nonprofits with 501(c)3 IRS status; nonprofits without 501(c)3 IRS status; small businesses; for-profit organizations; State governments; U.S. territories or possessions; Indian/Native American tribal governments (Federally recognized and other than Federally recognized); Indian/Native American tribally designated organizations; Hispanic-serving institutions; historically Black colleges and universities (HBCUs); tribally controlled colleges and universities (TCCUs); Alaska Native and Native Hawaiian serving institutions; regional organizations; and faith-based or community based organizations. The candidate must have a doctoral degree and peer-reviewed, independent research support at the time the award is made. The candidate must spend a minimum of 75 percent effort conducting research during the period of the award.

Restrictions: The award is not intended for investigators who already have full time to perform research, or have substantial publication records or considerable research support indicating that they are well established in their fields. Foreign institutions are not eligible to apply.

Amount of Grant: Up to $90,000 annually, plus associated fringe benefits.

Date(s) Application is Due: Feb 12; Mar 12; Jun 12; Jul 12; Oct 12; Nov 12

Contact: Kay Anderson, Ph.D.; (301) 427-1555; fax: (301) 427-1562; email: Kay. Anderson@ahrq.hhs.gov

Internet: http://grants.nih.gov/grants/guide/pa-files/PAR-07-444.html

Sponsor: Agency for Healthcare Research and Quality

540 Gaither Road

Rockville, MD 20850

AHRQ Individual Awards for Postdoctoral Fellows Ruth L. Kirschstein   450
National Research Service Awards (NRSA)

The purpose of the postdoctoral fellowship award is to provide support to promising postdoctoral applicants who have the potential to become productive and successful independent research investigators. The proposed postdoctoral training must offer an opportunity to enhance the applicant's understanding of health services research and must be responsive to AHRQ, mission, which is to improve the quality, safety, efficiency, and effectiveness of health care for all Americans. The research sponsored and conducted by AHRQ develops and presents scientific evidence regarding all aspects of health care. It addresses issues of organization, delivery, financing, utilization, patient and provider behavior, outcomes, effectiveness and cost. It evaluates both clinical services and the system in which these services are provided. These scientific results improve the evidence base to enable better decisions about health care, including such areas as disease prevention, appropriate use of medical technologies, improving diagnosis and treatment utilizing comparative effectiveness research, and reducing racial and ethnic disparities. In addition, AHRQ is interested in the application of health information technology (health IT), as well as reducing medical errors and improving patient safety.

Requirements: The following organizations and institutions are eligible to apply: public or non-profit private institutions, such as a university, college, or a faith-based or community-based organization; units of local or State government; eligible agencies of the Federal government; Indian/Native American tribal governments (Federally recognized); Indian/Native American tribal governments (other than Federally recognized); and Indian/Native American tribally designated organizations.

Restrictions: Awards for fellowship training may only be made to domestic institutions.

Date(s) Application is Due: Apr 8; Aug 8; Dec 8

Contact: Shelley M. Benjamin; (301) 427-1528; fax: (301) 427-1562; email: Shelley. Benjamin@ahrq.hhs.gov

Internet: http://grants.nih.gov/grants/guide/pa-files/PA-09-229.html

Sponsor: Agency for Healthcare Research and Quality (AHRQ)

540 Gaither Road

Rockville, MD 20850

AHRQ Translating Research into Practice Grants   451

AHCPR invites applications to conduct research related to implementing evidence-based tools and information in diverse health care settings among practitioners caring for diverse populations. Applications are sought for studies that apply innovative strategies for implementing evidence-based tools and information and demonstrate improved clinical practice and sustained practitioner behavior change. Evidence-based tools and information include findings from rigorously conducted research and from clinical practice guidelines, algorithms, treatment protocols, practice parameters, quality indicators, and continuous quality improvement initiatives that are developed using a systematic approach to evidence synthesis. AHCPR is especially interested in studies that implement AHCPR-supported, evidence-based tools and information, including Patient Outcome Research Team (P.O.RT) and other research findings; AHCPR-supported clinical practice guideline recommendations; and evidence reports and technology assessments produced by the AHCPR Evidence-based Practice Centers. The goal of this RFA is to improve the translation and use of research findings and evidence-based tools by developing and validating innovative principles, methods, and tools that work in diverse settings, populations, and payment systems. Applicants are encouraged to form public-private partnerships or consortia, such as between academic and other research organizations and health plans and purchasers, to perform this research. Such partnerships may help to more quickly translate research findings into actual practice settings. Applicants may also seek collaborative funding for projects funded under this RFA. Roles of collaborators should be clearly defined in the application. RFA: HS-99-003

Requirements: Applications may be submitted by public or private nonprofit organizations. For-profit organizations may participate as members of consortia or subcontractors if the applicant is nonprofit.

Restrictions: Organizations described in section 501(c)4 of the Internal Revenue Code that engage in lobbying are not eligible.

Amount of Grant: $2 million maximum total for the first year for four to six grants

Contact: Dr. Harold Goldstein, Division of Services and Intervention Research, NIMH (6001 Executive Boulevard, Bethesda, MD 20892), (301) 443-3747; fax: (301) 443-4045; email: goharold@nih.gov; Joan Metcalfe, Grants Management Specialist, (301) 594-1841

Internet: http://www.grants.nih.gov/grants/guide/rfa-files/RFA-HS-99-003.html

Sponsor: Agency for Healthcare Research and Quality

2101 E Jefferson Street, Executive Office Ctr

Rockville, MD 20852

AHRQ Understanding and Eliminating Minority Health Disparities   452

The purpose of this program is to conduct research on racial and ethnic disparities in health that are amenable to improvements in health services. Projects funded under this RFA will build on previous research that has identified disparities in access to, and utilization, quality and outcomes of health care services and the excess burdens of illness and death for African Americans, Hispanic Americans, American Indians,

Alaska Natives, Asian Americans and Pacific Islanders compared to the United States nonminority population. Projects funded by AHCPR will analyze causes and contributing factors for the inequalities that are related to the delivery and practice of health care, and identify and implement strategies to eliminate them. A letter of intent is requested by December 22; full application is due January 21. Annual deadline dates may vary; contact program staff for exact dates. RFA: HS-00-003

Requirements: Applications may be submitted by domestic or foreign, public or private nonprofit organizations, including American Indian/Alaska Native organizations, universities, clinics, units of state, tribal and local governments, and eligible agencies of the federal government. AHCPR, by statute, can make grants only to nonprofit organizations; however, for-profit organizations may participate in grant projects as members of consortia or as subcontractors.

Restrictions: 501(c)4 organizations that engage in lobbying are not eligible.

Amount of Grant: $4.35 million total

Date(s) Application is Due: Jan 21; Dec 22

Contact: Joan Metcalfe, Grants Management Specialist, (301) 594-1841; fax: (301) 594-3210; email: jmetcalf@ahcpr.gov

Internet: http://grants.nih.gov/grants/guide/rfa-files/RFA-HS-00-003.html

Sponsor: Agency for Healthcare Research and Quality
2101 E Jefferson Street
Rockville, MD 20852

AIA Latrobe Prize — 453

Named in honor of one of America's first professional architects, the Latrobe Prize supports path-finding research designed to advance the art and science of architectural practice. The prize winner will receive $100,000 to support a two-year program of research selected by jury review for its promise to advance professional knowledge in architecture.

Requirements: Criteria for the evaluation of proposals include: (a) relevance of the proposed research program to the general goals and objectives of the College of Fellows, the Latrobe Prize program, and the current theme; (b) breadth, depth, and innovativeness of the research program; (c) projected benefits of research outcomes to practice; (d) benefits of research outcomes to professional and public constituencies; (e) qualifications, expertise, prior achievements, maturity, and performance record of the applicant(s); (f) demonstrated capacity to administer a sustained research program, including the formulation and management of the budget; (g) quality and content of supporting documentation; and, (h) quality and content of letters of reference. See the website for detailed listing of required proposal components.

Amount of Grant: $100,000

Date(s) Application is Due: Dec 1

Contact: Pauline Porter, College of Fellows; (202) 626-7521; fax: (202) 626-7527; email: pporter@aia.org

Internet: http://www.aia.org/latrobe_prize/index.html

Sponsor: American Institute of Architects
1735 New York Avenue, NW
Washington, DC 20006-5292

AIAR Andrew W. Mellon Foundation Fellowships — 454

Fellowships are open to those in ancient Near Eastern studies, including the fields of archaeology, anthropology, art history, Bible, epigraphy, historical geography, history, language, literature, philology, and religion or related disciplines from Pre-history, through the early Islamic period. The program is open to Bulgarian, Czech, Estonian, Hungarian, Latvian, Lithuanian, Polish, Romanian, and Slovak scholars.

Requirements: Fellowships are available to Bulgarian, Czech, Estonian, Hungarian, Latvian, Lithuanian, Polish, Romanian, and Slovak scholars. Candidates should not be permanently resident outside the nine countries concerned, and should have obtained a doctorate by the time the fellowship is awarded.

Amount of Grant: $34,500 for three awards ($11,500 each)

Date(s) Application is Due: Apr 2

Contact: Dr. Joan Branham; (401) 865-1789; fax: (401) 865-1036; email: jbranham@providence.edu

Internet: http://www.aiar.org/deadlines.html

Sponsor: W.F. Albright Institute
Providence College
Providence, RI 02918

AIAR Annual Professorships — 455

Open to postdoctoral scholars in Near Eastern archaeology, geography, history, and Biblical studies. Residence at the institute is required. Period of appointment is 10 months. The professorship period should be continuous, without frequent trips outside the country. The professorship carries a stipend plus room and half-board for appointee and spouse. Annual deadline dates may vary; contact program staff for exact dates.

Requirements: Postdoctoral scholars may apply. U.S. citizens are eligible for the entire award. Non-U.S. citizens may apply but are only eligible for nongovernmental funds. Residence at the institute is required.

Amount of Grant: $30,000

Date(s) Application is Due: Oct 1

Contact: Dr. Joan Branham, Department of Art and Art History, (401) 865-1789; fax: (401) 865-1036; email: jbranham@providence.edu

Internet: http://www.aiar.org/deadlines.html

Sponsor: W.F. Albright Institute of Archaeological Research (AIAR) Jerusalem
Providence College
Providence, RI 02918

AIAR Educational and Cultural Affairs Fellowships — 456

Junior Research Fellowships are open to those in ancient Near Eastern studies, including the fields of archaeology, anthropology, art history, Bible, epigraphy, historical geography, history, language, literature, philology and religion or related disciplines from Pre-history, through the early Islamic period. The research period should be continuous, without frequent trips outside the country. Fellowships support 10 months of research. Residence at the Albright is required. Guidelines are available online.

Requirements: Doctoral students and recent PhD recipients who are U.S. citizens are eligible.

Amount of Grant: $16,000

Date(s) Application is Due: Oct 15

Contact: Dr. Joan Branham, (401) 865-1789; fax: (401) 865-1036; email: jbranham@providence.edu

Internet: http://www.aiar.org/deadlines.html

Sponsor: W.F. Albright Institute of Archaeological Research (AIAR) Jerusalem
Providence College
Providence, RI 02918

AIAR Ernest S. Frerichs Fellow and Program Coordinator Grant — 457

The program supports predoctoral students and postdoctoral scholars specializing in near Eastern archaeology, geography, history, and biblical studies. The recipient is expected to assist the Albright's director in planning and implementing the Ernest S. Frerichs Program for Albright Fellows. Residence at the institute for the 10-month research period is required. The research period should be continuous, without frequent trips outside the country. The award covers stipend, room, and one-half board at the institute. Annual deadline dates may vary; contact program staff for exact dates.

Amount of Grant: $19,000 ($10,900 stipend; $8100 room and half-board)

Date(s) Application is Due: Oct 1

Contact: Dr. Joan Branham, Department of Art and Art History, (401) 865-1789; fax: (401) 865-1036; email: jbranham@providence.edu

Internet: http://www.aiar.org/deadlines.html

Sponsor: W.F. Albright Institute of Archaeological Research (AIAR) Jerusalem
Providence College
Providence, RI 02918

AIAR George A. Barton Fellowships — 458

This fellowship is open to seminarians, predoctoral students, and recent PhD recipients specializing in Near Eastern archaeology, geography, history, and biblical studies. Research period is for five months. Awards consist of a stipend plus room and half-board at the institute. The research period should be continuous, without frequent trips outside the country.

Amount of Grant: $7000

Date(s) Application is Due: Oct 1

Contact: Dr. Joan Branham, Department of Art and Art History, (401) 865-1789; fax: (401) 865-1036; email: jbranham@providence.edu

Internet: http://www.aiar.org/deadlines.html

Sponsor: W.F. Albright Institute of Archaeological Research (AIAR) Jerusalem
Providence College
Providence, RI 02918

AIAR Samuel H. Kress Fellowship — 459

The dissertation research fellowship supports students specializing in architecture, art history and archaeology. The research project must have a clear focus on art history or architecture. The 10-month research period should be continuous, without frequent trips outside the country. Guidelines are available online.

Requirements: Applicants must be U.S. citizens, or North American citizens studying at U.S. universities.

Amount of Grant: $18,500

Date(s) Application is Due: Oct 15

Contact: Dr. Joan Branham, Department of Religious Studies, Department of Art and Art History, (401) 865-1789; fax: (401) 865-1036; email: jbranham@providence.edu

Internet: http://www.aiar.org/deadlines.html

Sponsor: W.F. Albright Institute of Archaeological Research (AIAR) Jerusalem
Providence College
Providence, RI 02918

AIAR Samuel H. Kress Traveling Fellowship — 460

The program awards a doctoral dissertation research fellowship for students specializing in architecture, art history, archaeology and classical studies. The fellowship supports travel for 10 months of research, including five months at the Albright, and five months at the American Center of Oriental Research in Amman, the Cyprus American Archaeological Research Institute in Nicosia, or the American School of Classical Studies at Athens. Applicants must demonstrate the necessity of being resident at the Albright and at one of the other three institutions mentioned above in order to complete their research. Guidelines are available online.

Requirements: Applicants must be U.S. citizens or students at U.S. universities.

Amount of Grant: $18,500

Contact: Dr. Joan Branham, Department of Art and Art History, (401) 865-1789; fax: (401) 865-1036; email: jbranham@providence.edu

Internet: http://www.aiar.org/deadlines.html

Sponsor: W.F. Albright Institute of Archaeological Research (AIAR) Jerusalem
Providence College
Providence, RI 02918

AIAR/Council of American Overseas Research Centers Fellowships for Advanced Multi-Country Research 461

The fellowships support scholars pursuing research on broad questions of multicountry significance in the fields of humanities, social sciences, and related natural sciences in countries in the Near and Middle East and South Asia. For information and application, contact CAORC, (202) 842-8636; email: caorc@caorc.si.edu; web: www.caorc.org.

Requirements: Doctoral candidates and established scholars with U.S. citizenship, as individuals or as teams, are eligible.

Amount of Grant: $9000 maximum
Date(s) Application is Due: Oct 1
Contact: 202-633-1599
Internet: http://www.caorc.org/fellowships/multi/index.html, fellowships@caorc.org, 202-633-1599.
Sponsor: W.F. Albright Institute of Archaeological Research (AIAR) Jerusalem
Council of American Overseas Research Centers (CAORC)
P.O. Box 37012, NHB, CE-123, MRC 178, Washington, DC 20013-7012

AICF Mellon Faculty Career Enhancement Fellowships 462

The three-year program offers doctoral fellowships to current faculty of tribal colleges and universities (TCU) who are already in pursuit of a terminal degree and will benefit most if given the time and support to complete degree requirements, unfettered by financial considerations and professional demands. Fellows may use stipends to meet living expenses, tuition, and other expenses that enable dissertation research to be carried out. Additionally, each fellow will be provided an annual travel allowance to aid in travel expenses relating to research. Fellows will be expected to complete their terminal degree within a reasonable time upon receiving the award. Guidelines and application are available online.

Requirements: Applicant must be/will: a current TCU member; considered all-but-dissertation in an accredited terminal degree program (or within one year of attaining an MFA degree); possess a demonstrated commitment to American Indian education and scholarship; agree to continue serving as faculty for at least two years at a TCU after completion of his/her degree; submit an approved dissertation prospectus from her/his institution, which will aid the selection process; work throughout the year with selected senior faculty member, who will serve as mentor; participate in an expense-paid annual fellow/mentor retreat in Arpil of the following year; complete the terminal degree within a reasonable time upon receiving this award; and be able to document that he/she is within one year of completing his/her terminal degree or MFA (all fields of study eligible).

Amount of Grant: $30,000 stipend; $2250 travel allowance
Date(s) Application is Due: May 2
Contact: Kara Anderson, (800) 776-3863 or (303) 426-8900; fax: (303) 426-1200; email: kanderson@collegefund.org
Internet: http://www.raconline.org/funding/funding_details.php?funding_id=519
Sponsor: American Indian College Fund
8333 Greenwood Boulevard
Denver, CO 80221

AIChE Alpha Chi Sigma Award 463

The AIChE Alpha Chi Sigma Award for Chemical Engineering Research are sponsored by The Alpha Chi Sigma Educational Foundation. The award recognizes an individual's outstanding accomplishments in fundamental or applied chemical engineering research. The award recognizes research carried out during the past ten years. This research will normally have been carried out in North America. When appropriate, the winner will be invited to present his or her work at a symposium arranged in his or her honor. When this is not practical, the recipient will be invited to present a paper at an AIChE meeting.

Requirements: The recipient does not have to be a member of AIChE or the Alpha Chi Sigma Fraternity.

Amount of Grant: $5000 and a plaque
Date(s) Application is Due: Feb 15
Contact: Gordon Ellis, Awards Administrator; (646) 495-1348; fax: (646) 495-1503; email: gorde@aiche.org or awards@aiche.org
Internet: http://www.aiche.org/About/Awards/AlphaChiSigmaAward.aspx
Sponsor: American Institute of Chemical Engineers
3 Park Avenue
New York, NY 10016

AIChE Andrew Chase Forest Bioproducts Division Award in Chemical Engineering 464

The AIChE Andrew Chase Forest Bioproducts Division Award in Chemical Engineering is sponsored by Forest Bioproducts Division. The award recognizes an individual's outstanding chemical engineering contribution in the forest products and related industries. The award is presented at a Division event held at the AIChE Annual Meeting

Requirements: Selection criteria include: significant discoveries, research, or development, successfully implemented on a commercial scale; outstanding chemical engineering contributions in the field of design, operation or production management which led to significant technological improvements; distinguished service as a chemical engineering educator, with an emphasis on application of chemical engineering principles to forest products industry technologies; outstanding service to the division; nominees must have presented a paper at one of the division's symposia describing his or her original contributions in the application of chemical engineering in the forest products or related industries.

Amount of Grant: A plaque and a cash award

Date(s) Application is Due: Jun 1
Contact: Gordon Ellis, Awards Administrator; (646) 495-1348; fax: (646) 495-1503; email: gorde@aiche.org or awards@aiche.org
Internet: http://www.aiche.org/About/Awards/Divisions/ForestProductsDivision.aspx
Sponsor: American Institute of Chemical Engineers
3 Park Avenue
New York, NY 10016

AIChE CAST Outstanding Young Researcher Award 465

This Award will be given to an individual under the age of 40 to recognize outstanding contributions to the chemical engineering computing and systems technology literature. The individual must be age 39 or less on December 31 of the award year. An individual age 40 or over will be eligible for this award if, on December 31 of the award year, 12 years or less have elapsed since this individual received the Ph. D. degree.

Amount of Grant: $3,000
Date(s) Application is Due: Apr 15
Contact: Gordon Ellis, Awards Administrator; (646) 495-1348; fax: (646) 495-1503; email: gorde@aiche.org or awards@aiche.org
Internet:http://www.aiche.org/About/Awards/Divisions/CASTOutstanding YoungResearcher.aspx
Sponsor: American Institute of Chemical Engineers
3 Park Avenue
New York, NY 10016

AIChE Charles M.A. Stine Award 466

The Charles M. A. Stine Award is sponsored by E.I. duPont de Nemours & Company. It recognizes an individual's outstanding contribution to the scientific, technological, educational or service areas of materials engineering and science.

Requirements: Nominees must be members of AIChE.

Samples: Sharon Glotzer—for her pioneering simulations of glass-forming liquids and self-assembled nanomaterials, and for her leadership and service to the materials community (2008).

Amount of Grant: $1,500
Date(s) Application is Due: Feb 15
Contact: Gordon Ellis, Awards Administrator; (646) 495-1348; fax: (646) 495-1503; email: gorde@aiche.org or awards@aiche.org
Internet: http://www.aiche.org/About/Awards/Divisions/CharlesMAStineAward.aspx
Sponsor: American Institute of Chemical Engineers
3 Park Avenue
New York, NY 10016

AIChE Clarence (Larry) G. Gerhold Award 467

The AIChE Clarence (Larry) G. Gerhold Award is sponsored by UOP. The award recognizes an individual's outstanding contribution in research, development, or in the application of chemical separations technology. The award is presented at the Separations Division Dinner held during the AIChE Annual Meeting.

Requirements: Selection criteria include: a sustained record (at least 15 years) of contributions to separations technology. Such activity may include demonstrated leadership in research, teaching, or engineering; a unique and valuable contribution to separations technology which has had a major impact on one of the separation technologies. The contribution should represent a turning point in a particular technology; acontinuing and extensive record of service to the Separations Division.

Restrictions: Nominees must be members of AIChE.

Amount of Grant: $3,000, a plaque plus a travel allowance
Date(s) Application is Due: Jun 1
Contact: Gordon Ellis, Awards Administrator; (646) 495-1348; fax: (646) 495-1503; email: gorde@aiche.org or awards@aiche.org
Internet: http://www.aiche.org/About/Awards/Divisions/ClarenceGGerholdAward.aspx
Sponsor: American Institute of Chemical Engineers
3 Park Avenue
New York, NY 10016

AIChE Computational Molecular Science and Engineering Forum Graduate Student Award 468

The award recognizes significant contributions to research in computational molecular science and engineering by graduate students. The award consists of a plaque and a $500 honorarium.

Requirements: The nominee must be a graduate student at the time of the poster presentation, and the faculty nominator must be a member of CoMSEF. Nominations should consist of a nominating letter from the student's research advisor and the curriculum vitae of the nominee.

Amount of Grant: $500
Date(s) Application is Due: Sep 1
Contact: Gordon Ellis, Awards Administrator; (646) 495-1348; fax: (646) 495-1503; email: gorde@aiche.org or awards@aiche.org
Internet: http://www.aiche.org/About/Awards/Divisions/CoMSEFGradAward.aspx
Sponsor: American Institute of Chemical Engineers
3 Park Avenue
New York, NY 10016

AIChE Computing in Chemical Engineering Award 469

The Computing in Chemical Engineering Award is sponsored by The Dow Chemical Company & Mitsubishi Chemical Company. The Computing in Chemical Engineering Award recognizes outstanding contributions in the application of computing and systems technology to chemical engineering. The award consists of a plaque and $3,000.

Requirements: The awardee is invited to deliver an address at the CAST Division dinner at the AIChE Annual Meeting.

Amount of Grant: $3,000

Date(s) Application is Due: Apr 15

Contact: Gordon Ellis, Awards Administrator; (646) 495-1348; fax: (646) 495-1503; email: gorde@aiche.org or awards@aiche.org

Internet: http://www.aiche.org/About/Awards/DivisionsComputingChemical EngineeringAward.aspx

Sponsor: American Institute of Chemical Engineers
3 Park Avenue
New York, NY 10016

AIChE Computing Practice Award 470

The award, which is sponsored by Aspen Technology, Inc., and ExxonMobil Chemical Company, recognizes outstanding contributions in the application of chemical engineering to computing and systems technology.

Amount of Grant: $3000

Date(s) Application is Due: Apr 15

Contact: Gordon Ellis, Awards Administrator; (646) 495-1348; fax: (646) 495-1503; email: gorde@aiche.org or awards@aiche.org

Internet: http://www.aiche.org/awards/awarddtl.asp?AwardID=29

Sponsor: American Institute of Chemical Engineers
3 Park Avenue
New York, NY 10016-5901

AIChE CRE Travel Awards 471

The Catalysis and Reaction Engineering Division will make up to 10 travel awards to assist graduate students with their travel expenses so that they can attend the AIChE Annual meeting and present the results of their research. The awards will consist of $400, plus a ticket to the CRE Division Dinner, where the recipients will be officially recognized.

Requirements: To be eligible, the student must be a member of the AIChE and present a paper at the Annual Meeting in one of the CRE Division sessions.

Amount of Grant: $400. plus price of ticket

Date(s) Application is Due: July 31

Contact: Gordon Ellis, Awards Administrator; (646) 495-1348; fax: (646) 495-1503; email: gorde@aiche.org or awards@aiche.org

Internet: http://www.aiche.org/About/Awards/Divisions/CRETravelAwards.aspx

Sponsor: American Institute of Chemical Engineers
3 Park Avenue
New York, NY 10016

AIChE David Himmelblau Award for Innovations in Computer-Based 472
Chemical Engineering Education

The David Himmelblau Award for Innovations in Computer-Based Chemical Engineering Education is sponsored by CAChE Corporation. The award recognizes an individual or group making new and novel contributions to computer aids for chemical engineering education. Educational innovators working in industry or in a company that develops computer-based educational aids are also eligible for the award. Specifically, the nomination and the citation for the award must refer to a significant contribution to computer-based chemical engineering education within the past decade. The award is presented at the CAST Division dinner at the AIChE Annual Meeting. The award winner will give an invited talk at the AIChE Annual Meeting.

Amount of Grant: $1,000

Date(s) Application is Due: Apr 15

Contact: Gordon Ellis, Awards Administrator; (646) 495-1348; fax: (646) 495-1503; email: gorde@aiche.org or awards@aiche.org

Internet: http://www.aiche.org/About/Awards/DivisionsCASTHimmelblauAward. aspx

Sponsor: American Institute of Chemical Engineers
3 Park Avenue
New York, NY 10016

AIChE Donald Q. Kern Award 473

The AIChE Donald Q. Kern Award is sponsored by Heat Transfer Research, Inc. The award will be presented at the AIChE Spring National Meeting or the Summer Heat Transfer Conference. The award is given in honor of Donald Q. Kern, pioneer in process heat transfer, the Division recognizes an individual's expertise in a given field of heat transfer or energy conversion.

Requirements: Selection criteria include: significant contributions to applied heat transfer or conversion or in the translation of research results into useful technological applications; the recipient is required to prepare a written review of a topic selected by the recipient. The review is usually published in an AIChE publication.

Amount of Grant: A plaque and $1,500.

Date(s) Application is Due: Mar 16

Contact: Gordon Ellis, Awards Administrator; (646) 495-1348; fax: (646) 495-1503; email: gorde@aiche.org or awards@aiche.org

Internet: http://www.aiche.org/About/Awards/Divisions/DonaldQKernAward.aspx

Sponsor: American Institute of Chemical Engineers
3 Park Avenue
New York, NY 10016

AIChE Environmental Division Graduate Student Paper Award 474

The Environmental Division Graduate Student Paper Award recognizes outstanding graduate student contributions to environmental protection through chemical engineering. The award is based on the annual graduate student paper competition.

Requirements: Selection criteria include: the graduate student must be a member of AIChE at the time of submission.; the work presented must be carried out while the graduate student is enrolled at a university with an accredited chemical engineering program.; the graduate student must be the primary author,; the paper must describe original research or design; papers consisting of literature reviews are not permitted.; the paper must be suitable for publication in a refereed journal.; the paper must represent a contribution to environmental protection through the application of chemical engineering. Nomination for the award should include: A single nomination letter detailing the student, strengths and accomplishments and a single paper contributing to environmental science and engineering fundamentals or applications. Papers will be judged on quality, technical content, writing and organization. Nominations may be submitted by any member of AIChE.

Amount of Grant: $500, $300 and $200 for first, second and third place, respectively

Contact: Gordon Ellis, Awards Administrator; (646) 495-1348; fax: (646) 495-1503; email: gorde@aiche.org or awards@aiche.org

Internet: http://www.aiche.org/About/Awards/Divisions/EnvironmentalDivision GraduateStudentPaper.aspx

Sponsor: American Institute of Chemical Engineers
3 Park Avenue
New York, NY 10016

AIChE Environmental Division Undergraduate Student Paper Award 475

The Environmental Division Undergraduate Student Paper Award is awarded to a full-time undergraduate student who prepares the best original paper based on the results of research or an investigation related to the environment.

Requirements: The nominee must be a full-time undergraduate in good standing at a college or university with an accredited program in chemical engineering, and must be a member of an AIChE Student Chapter.; the work must be performed during the student's undergraduate enrollment, and the paper must be submitted prior to or within six months of graduation.; the student must be the sole author of the paper.

Amount of Grant: A plaque and $300, $200, and $100 for first, second and third place.

Contact: Gordon Ellis, Awards Administrator; (646) 495-1348; fax: (646) 495-1503; email: gorde@aiche.org or awards@aiche.org

Internet: http://www.aiche.org/About/Awards/Divisions/EnvironmentalDivision UndergraduateStudentPaper.aspx

Sponsor: American Institute of Chemical Engineers
3 Park Avenue
New York, NY 10016

AIChE Excellence in Process Development Research Award 476

The AIChE Excellence in Process Development Research Award is sponsored by Pfizer. The award recognizes individuals who have made significant technical contributions to the advancement of process development within research, teaching, or regulatory activities. Accomplishments must be disseminated by means of well-documented materials. Emphasis will be placed on accomplishments and advances made within the last ten years although the award can also be given to someone for an outstanding career. The awardee is invited to deliver an address at the Process Development Division Dinner at the AIChE Annual Meeting

Amount of Grant: $1,000

Date(s) Application is Due: Jun 15

Contact: Gordon Ellis, Awards Administrator; (646) 495-1348; fax: (646) 495-1503; email: gorde@aiche.org or awards@aiche.org

Internet: http://www.aiche.org/About/Awards/Divisions/ExcellenceProcess DevelopmentResearchAward.aspx

Sponsor: American Institute of Chemical Engineers
3 Park Avenue
New York, NY 10016

AIChE Food, Pharmaceutical and Bioengineering Division Awards in 477
Chemical Engineering

The AIChE Food, Pharmaceutical and Bioengineering Division Awards in Chemical Engineering is sponsored by Merck. The award recognizes an individual's outstanding chemical engineering contribution in the food, pharmaceutical and/or bioengineering industry. These contributions may have been made in industry, government, or academic areas, or with other organizations. The award is presented at the Food, Pharmaceutical & Bioengineering Division plenary session of the AIChE Annual Meeting.

Amount of Grant: $4000

Date(s) Application is Due: Feb 15

Contact: Gordon Ellis, Awards Administrator; (646) 495-1348; fax: (646) 495-1503; email: gorde@aiche.org or awards@aiche.org

Internet: http://www.aiche.org/About/Awards/Divisions/FoodPharmaceutical BioengineeringDivisionAwardChemicalEngineering.aspx

Sponsor: American Institute of Chemical Engineers
3 Park Avenue
New York, NY 10016

AIChE Food, Pharmaceutical and Bioengineering Division Distinguished Service Award in Chemical Engineering   **478**

The purpose of this award is to recognize a division member for his or her dedication and service to the profession of food engineering, pharmaceutical engineering, and/or bioengineering in general, and to Division 15 in particular. The award consists of a plaque and a monetary portion, the amount of which is determined by the executive committee. The award is to be given to the individual in a public session at the Annual Meeting of AIChE.
Date(s) Application is Due: Apr 15
Contact: Gordon Ellis, Awards Administrator; (646) 495-1348; fax: (646) 495-1503; email: gorde@aiche.org or awards@aiche.org
Internet: http://www.aiche.org/About/Awards/Divisions/FoodPharmaceutical BioengineeringDistinguishedService.aspx
Sponsor: American Institute of Chemical Engineers
3 Park Avenue
New York, NY 10016

AIChE Institute Award for Excellence in Industrial Gases Technology   **479**

The AIChE Institute Award for Excellence in Industrial Gases Technology is sponsored by Praxair, Inc. The award recognizes an individual's sustained excellence in contributing to the advancement of technology in the production, distribution and application of industrial gases. The award is presented at the Honors Ceremony of the AIChE Annual Meeting and consists of a plaque and $3,000, plus a $500 travel allowance.
Requirements: The recipient will have a record of sustained contributions that have advanced the frontier of industrial gases technology. These contributions may be characterized by a sustained record of important fundamental research, innovation, technological development or the novel application of technology, either fostering or leading to important commercial results. The recipient does not have to be a member of AIChE.
Amount of Grant: $3000
Date(s) Application is Due: Feb 15
Contact: Gordon Ellis, Awards Administrator; (646) 495-1348; fax: (646) 495-1503; email: gorde@aiche.org or awards@aiche.org
Internet: http://www.aiche.org/About/Awards/InstituteAwardforExcellencein IndustrialGasesTechnology.aspx
Sponsor: American Institute of Chemical Engineers
3 Park Avenue
New York, NY 10016

AIChE John A. Tallmadge Award for Contributions to Coating Technology   **480**

The AIChE John A. Tallmadge Award for Contributions to Coating Technology is sponsored by Eastman Kodak Company; International Society of Coating Science and Technology, and AIChE. The award recognizes an individual's significant contributions to the understanding or improvement of the technology of the coating of continuous webs. The award is presented bi-annually, in even numbered years.
Requirements: Nominations must follow the nomination form posted on the ISCST website.
Amount of Grant: $1000 and a plaque
Date(s) Application is Due: Jul 1
Contact: Gordon Ellis, Awards Administrator; (646) 495-1348; fax: (646) 495-1503; email: gorde@aiche.org or awards@aiche.org
Internet: http://www.aiche.org/About/Awards/Committee/JohnATallmadgeAward. aspx
Sponsor: American Institute of Chemical Engineers
3 Park Avenue
New York, NY 10016

AIChE Kazutoshi Fujimura Award for Lifetime Achievement   **481**

The Kazutoshi Fujimura Award for Lifetime Achievement is sponsored by Hyperion Catalysis International and Bioveris, Cambridge, Massachusetts, U.S.A. This award recognizes the work of an individual in helping to develop and propagate new technology internationally. The individual should have played a prominent role for many years in identifying, developing and introducing new chemical engineering technologies to the international market place. The award is not intended to honor a single event but instead to recognize the lifetime achievement of an individual in international chemical engineering development. The Fujimura Award is overseen by AIChE, Management Division. A selection committee consisting of representatives of the three international chemical engineering confederations and Dr. Norman Li, the first recipient, plus Management Division representatives and AIChE staff will make the final selection.
Amount of Grant: $10,000
Date(s) Application is Due: Jan 31
Contact: Gordon Ellis, Awards Administrator; (646) 495-1348; fax: (646) 495-1503; email: gorde@aiche.org or awards@aiche.org
Internet: http://www.aiche.org/About/Awards/Divisions/KazutoshiFujimuraAward. aspx
Sponsor: American Institute of Chemical Engineers
3 Park Avenue
New York, NY 10016

AIChE Lawrence K. Cecil Award in Environmental Chemical Engineering   **482**

The Lawrence K. Cecil Award in Environmental Chemical Engineering is sponsored by B.P. America, Inc. The award recognizes an individual's outstanding chemical engineering contribution and achievement in the preservation or improvement of the environment.
Requirements: The recipient must be a member of AIChE, have 15 years of chemical engineering experience in the environmental field, and demonstrate leadership in research, teaching, engineering, or regulatory activities in either the public or private sector.
Amount of Grant: $2,500
Date(s) Application is Due: Feb 28
Contact: Gordon Ellis, Awards Administrator; (646) 495-1348; fax: (646) 495-1503; email: gorde@aiche.org or awards@aiche.org
Internet: http://www.aiche.org/About/Awards/Divisions/LawrenceKCecilAward EnvironmentalChemicalEngineering.aspx
Sponsor: American Institute of Chemical Engineers
3 Park Avenue
New York, NY 10016

AIChE Lectureship Award in Fluidization   **483**

The AIChE Lectureship Award in Fluidization is sponsored by Particulate Solid Research, Inc. The award recognizes an individual's outstanding scientific/technical research contributions with impact in the field of fluidization and fluid-particle flow systems. The award is presented at a Forum sponsored event during the AIChE Annual Meeting.
Requirements: Selection criteria include: an outstanding contribution advancing fluidization or fluid-particle flow systems; the awardee is required to deliver a keynote paper at the Fundamentals of Fluidization and Fluid-Particle Systems session of Area 3b during the AIChE Annual Meeting. The awardee is also required to submit a written manuscript; membership in the Particle Technology Forum or AIChE is not required.
Amount of Grant: $1,000
Date(s) Application is Due: Jun 15
Contact: Gordon Ellis, Awards Administrator; (646) 495-1348; fax: (646) 495-1503; email: gorde@aiche.org or awards@aiche.org
Internet: http://www.aiche.org/LectureshipAwardFluidization.aspx
Sponsor: American Institute of Chemical Engineers
3 Park Avenue
New York, NY 10016

AIChE Nanoscale Science and Engineering Forum Young Investigator Award   **484**

This award recognizes outstanding interdisciplinary research in nanoscience and nanotechnology by engineers or scientists in the early stages of their professional careers (within 10 years of completion of highest degree). The awardee is invited to deliver an address during the NSEF Plenary Lectures at the AIChE Annual Meeting.
Amount of Grant: A plaque, a $750 award, and travel expenses up to $1000.
Date(s) Application is Due: May 31
Contact: Gordon Ellis, Awards Administrator; (646) 495-1348; fax: (646) 495-1503; email: gorde@aiche.org or awards@aiche.org
Internet: http://www.aiche.org/About/Awards/Divisions/NSEFYoungInvestigator. aspx
Sponsor: American Institute of Chemical Engineers
3 Park Avenue
New York, NY 10016

AIChE North American Mixing Forum Award   **485**

The AIChE North American Mixing Forum Award is sponsored by The Procter & Gamble Company. It is Awarded for Excellence and Sustained Contributions to Mixing Research and Practice. The award is presented at the award lecture and a dinner ceremony during the AIChE Annual Meeting.
Amount of Grant: A plaque, $1,000, plus a $500 travel allowance
Date(s) Application is Due: May 1
Contact: Gordon Ellis, Awards Administrator; (646) 495-1348; fax: (646) 495-1503; email: gorde@aiche.org or awards@aiche.org
Internet: http://www.aiche.org/About/Awards/Divisions/NorthAmericanMixing ForumAward.aspx
Sponsor: American Institute of Chemical Engineers
3 Park Avenue
New York, NY 10016

AIChe North American Mixing Forum Start-up Grant   **486**

The AIChE North American Mixing Forum Start-Up Grant is sponsored by North American Mixing Forum. The purpose of the grant is to support a young, non-tenured professor seeking to do research in the field of mixing or mixing related processes.
Date(s) Application is Due: May 1
Contact: Gordon Ellis, Awards Administrator; (646) 495-1348; fax: (646) 495-1503; email: gorde@aiche.org or awards@aiche.org
Internet: http://www.aiche.org/About/Awards/Divisions/NorthAmericanMixing ForumGrant.aspx
Sponsor: American Institute of Chemical Engineers
3 Park Avenue
New York, NY 10016

AIChE North American Mixing Forum Student Award   **487**

The AIChE North American Mixing Forum Student Award is sponsored by Rohm & Haas and AIChE's North American Mixing Forum. The NAMF Student Award was established to encourage, recognize and reward students for quality research in the area of mixing. Any graduate or undergraduate student doing research in the field of Fluid Mixing at an accredited University in North America is eligible. The winner will be notified during the conference and will be presented with a plaque and an award of $500 at the NAMF Award Banquet.

Requirements: An applicant must submit a paper in electronic form, describing his or her work in mixing to the NAMF Student Awards Committee Chair. See website for the contact information.

Amount of Grant: $500

Date(s) Application is Due: Mar 15

Contact: Gordon Ellis, Awards Administrator; (646) 495-1348; fax: (646) 495-1503; email: gorde@aiche.org or awards@aiche.org

Internet: http://www.aiche.org/About/Awards/Divisions/NorthAmericanMixing ForumStudent.aspx

Sponsor: American Institute of Chemical Engineers

3 Park Avenue

New York, NY 10016

AIChE Nuclear Engineering Division Outstanding Student Paper & Presentation Awards
488

The AIChE Nuclear Engineering Division Outstanding Student Paper & Presentation Awards are sponsored by Westinghouse Electric Company. The awards were established to encourage, recognize and reward chemical engineering students for quality research in the nuclear science and technology. Any graduate or undergraduate student doing research in this field is eligible for these awards.

Requirements: All eligible applicants must submit a paper describing their work in nuclear science and technology to the NED Outstanding Student Awards Committee Chair. Finalist(s) are selected and invited to present the paper(s) at the Annual AIChE meeting. The finalist(s) are selected no later than 4 weeks before the conference. NED covers both the travel expenses (up to $700) and the registration fee for the finalist(s). The winner for the Student Paper Award will be selected by the Award Committee Members before the conference. See website for detailed description of the application process.

Amount of Grant: $700 and the registration fee for the finalist(s).

Date(s) Application is Due: Abstract due May 19; Draft Paper due Jun 31; Final Paper due Sep 18

Contact: Gordon Ellis, Awards Administrator; (646) 495-1348; fax: (646) 495-1503; email: gorde@aiche.org or awards@aiche.org

Internet: http://www.aiche.org/About/Awards/Divisions/NuclearEngineeringDivision StudentAward.aspx

Sponsor: American Institute of Chemical Engineers

3 Park Avenue

New York, NY 10016

AIChE Process Development Practice Award
489

The AIChE Process Development Practice Award is sponsored by Zeton Inc. The award recognizes individuals with outstanding contributions in the practice or application of chemical engineering to process development. This award hopes to focus on applicants from industrial settings where documentation of contributions is not as readily provided.

Requirements: Awardees will be selected based on their contributions to the discovery and application of innovative solutions to technological problems, and /or commercialization of new products and processes. The award may be given to someone for an outstanding career.

Restrictions: Candidate must be an AIChE member.

Amount of Grant: $1,000

Date(s) Application is Due: Dec 15

Contact: Gordon Ellis, Awards Administrator; (646) 495-1348; fax: (646) 495-1503; email: gorde@aiche.org or awards@aiche.org

Internet: http://www.aiche.org/About/Awards/Divisions/ProcessDevelopmentPractice Award.aspx

Sponsor: American Institute of Chemical Engineers

3 Park Avenue

New York, NY 10016

AIChE Professional Progress Award for Outstanding Progress in Chemical Engineering
490

This award recognizes outstanding progress in the field of chemical engineering.

Requirements: The awardee will have made a significant contribution to the science of chemical engineering through one of the following means: a theoretical discovery or development of a new principle in the chemical engineering field; development of a new process or product in the chemical engineering field; an invention or development of new equipment in the chemical engineering field; distinguished service rendered to the field or profession of chemical engineering.

Restrictions: The recipient must be less than 45 years of age at the end of the calendar year in which the award is presented.

Amount of Grant: $4000, and $500 travel allowance

Date(s) Application is Due: Feb 15

Contact: Gordon Ellis, Awards Administrator; (646) 495-1348; fax: (646) 495-1503; email: gorde@aiche.org or awards@aiche

Internet: http://www.aiche.org/awards/awarddtl.asp?AwardID=60

Sponsor: American Institute of Chemical Engineers

3 Park Avenue

New York, NY 10016-5901

AIChE Research Excellence In Sustainable Engineering Award
491

The AIChE Research Excellence In Sustainable Engineering Award is annually presented for basic or applied research results relative to the sustainability of products, processes, or the environment. The award is to recognize the one who has made significant technical contributions to the advancement of sustainable engineering in research, teaching, and development activities. Emphasis should be placed on accomplishments and advances made within the last five years, although the award can also be made to someone who has had an outstanding career. The awardee will receive $1,000 and a plaque.

Amount of Grant: $1000

Date(s) Application is Due: Aug 1

Contact: Gordon Ellis, Awards Administrator; (646) 495-1348; fax: (646) 495-1503; email: gorde@aiche.org or awards@aiche.org

Internet: http://www.aiche.org/About/Awards/Divisions/SEFExcellenceAward.aspx

Sponsor: American Institute of Chemical Engineers

3 Park Avenue

New York, NY 10016

AIChE Separations Division Graduate Student Research Award
492

AIChE Separations Division Graduate Student Research Award recognizes six outstanding graduate students in any of the Separations Division's areas: adsorption and ion exchange; crystallization and evaporation; distillation and absorption; extraction; fluid-particle separations; membrane-based separations; bioseparations and general separations. Awards will be made at the next AIChE Annual Meeting.

Requirements: For consideration for the award, the following should be submitted: a single paper contributing to separations fundamentals or applications; a single nomination letter detailing the student's strengths and accomplishments; the student's CV. To be eligible for this year, awards, the nominees must currently be graduate students or have been graduate students since the last AIChE Annual meeting last November. The selection criteria include: the paper must report on research, investigation or design, and must be part of the student's work for a graduate degree; he paper may be co-authored by others, but the student nominee must have been the primary author; the paper should be of a quality acceptable for publication in journals such as the AIChE Journal and Chemical Engineering Science; papers will be judged on technical content, quality of writing and organization; the person nominating a student must be a member of AIChE.

Amount of Grant: $300

Date(s) Application is Due: Jun 1

Contact: Gordon Ellis, Awards Administrator; (646) 495-1348; fax: (646) 495-1503; email: gorde@aiche.org or awards@aiche.org

Internet: http://www.aiche.org/Students/Awards/SeparationsDivisionGraduateStudent ResearchAward.aspx

Sponsor: American Institute of Chemical Engineers

3 Park Avenue

New York, NY 10016

AIChE Thomas Baron Award in Fluid-Particle Systems
493

The AIChE Thomas Baron Award in Fluid-Particle Systems is sponsored by Shell Global Solutions, Inc. The award recognizes an individual's outstanding scientific/technical accomplishment which has made a significant impact in the field of fluid-particle systems or in a related field with potential for cross-fertilization. The award is presented at a Forum sponsored event at the AIChE Annual Meeting.

Requirements: Selection criteria include: an outstanding contribution advancing fluid-particle systems, or a related field; the awardee is invited to deliver a Plenary Lecture at an AIChE Annual Meeting session. The awardee is also required to submit a written manuscript.

Amount of Grant: $1,000

Date(s) Application is Due: Jun 15

Contact: Gordon Ellis, Awards Administrator; (646) 495-1348; fax: (646) 495-1503; email: gorde@aiche.org or awards@aiche.org

Internet: http://www.aiche.org/About/Awards/Divisions/ThomasBaronAwardFluid ParticleSystems.aspx

Sponsor: American Institute of Chemical Engineers

3 Park Avenue

New York, NY 10016

AIChE Transport and Energy Processes Division Award
494

The AIChE Transport and Energy Processes Division Award is sponsored by Fauske & Associates. The award recognizes an individual's outstanding chemical engineering contribution and achievement in heat transfer or energy conversion. The recipient is invited to speak at the AIChE Annual Meeting.

Requirements: Selection criteria include: distinguished service of a chemical engineering nature, in heat transfer or energy conversion as a professional engineer, educator, or AIChE leader; outstanding contributions of a chemical engineering nature, in the field of design, construction, operation, or management of heat transfer or energy conversion facilities or enterprises; research or development of new processes or equipment relating to chemical engineering applications in heat transfer or energy conversion.

Amount of Grant: $500 and a plaque

Date(s) Application is Due:

Contact: Gordon Ellis, Awards Administrator; (646) 495-1348; fax: (646) 495-1503; email: gorde@aiche.org or awards@aiche.org

Internet: http://www.aiche.org/About/Awards/Divisions/TransportEnergyProcesses DivisionAward.aspx

Sponsor: American Institute of Chemical Engineers

3 Park Avenue

New York, NY 10016

AIChE W. David Smith, Jr. Graduate Student Paper Award   **495**
This Award will be given to an individual for outstanding published work in chemical engineering computing and systems technology. The publication shall consist of a single paper based on work done by the individual while pursuing graduate or undergraduate studies.
Amount of Grant: $1,500
Date(s) Application is Due: Apr 15
Contact: Gordon Ellis, Awards Administrator; (646) 495-1348; fax: (646) 495-1503; email: gorde@aiche.org or awards@aiche.org
Internet: http://www.aiche.org/About/Awards/Divisions/WDavidSmithGraduate StudentPaperAward.aspx
Sponsor: American Institute of Chemical Engineers
3 Park Avenue
New York, NY 10016

AIChE Warren K. Lewis Award for Contributions to Chemical   **496**
   Engineering Education
Sponsored by ExxonMobil Research and Engineering Company and administered by AIChE, this award is given to recognize distinguished and continuing contributions to chemical engineering education. Recipients need not be members of the institute.
Amount of Grant: $5000, $500 for travel expenses
Date(s) Application is Due: Feb 15
Contact: Gordon Ellis, Awards Administrator; (646) 495-1348; fax: (646) 495-1503; email: gorde@aiche.org or awards@aiche.org
Internet: http://www.aiche.org/About/Awards/WarrenKLewisAward.aspx
Sponsor: American Institute of Chemical Engineers
3 Park Avenue
New York, NY 10016

AIChE William H. Walker Award for Excellence in Contributions to   **497**
   Chemical Engineering Literature
AIChE William H. Walker Award for Excellence in Contributions to Chemical Engineering Literature is sponsored by John Wiley and Sons. The award is presented to a member of AIChE who has made an outstanding contribution to chemical engineering literature. The contribution may consist of a review, a history of the development of a process, a theoretical contribution, a research report, or other material of interest and importance to the chemical engineering profession.
Requirements: The recipient must be the author or co-author of an outstanding work in chemical engineering.
Amount of Grant: $5000 plus $500 travel allowance
Date(s) Application is Due: Feb 15
Contact: Gordon Ellis, Awards Administrator; (646) 495-1348; fax: (646) 495-1503; email: gorde@aiche.org or awards@aiche.org
Internet: http://www.aiche.org/About/Awards/WilliamHWalkerAward.aspx
Sponsor: American Institute of Chemical Engineers
3 Park Avenue
New York, NY 10016

AICPA Fellowships for Minority Doctoral Students   **498**
The primary objective of the fellowships is to enable more minorities to enter and move ahead in the accounting profession and academe. Recognizing the fact that professors serve as role models, a second objective is to increase the number of CPA role models who can positively influence the career decisions of a college student. These competitive fellowships are available to minority candidates who have been accepted into a doctoral accounting program. Fellowships are awarded once a year to full-time minority accounting scholars who show significant potential to become accounting educators. Renewals will be considered based on satisfactory progress, as a full-time student, toward completion of the degree requirements, up to a total of five years.
Requirements: The applicant must be a minority student (African American, Native American, Pacific Island races, or of Hispanic ethnic origin) who has applied to or has been accepted into an accounting doctoral program; have earned a masters degree and/or completed a minimum of three years full-time experience in the practice of accounting; be attending or planning to attend school on a full-time basis and, once admitted, work consistently and forthrightly to attain the doctoral degree; agree not to work full time in a paid position or accept responsibility for teaching more than one course per semester as a teaching assistant or, dedicate more than one quarter of the time as a research assistant; and be a U.S. citizen.
Amount of Grant: $12,000 maximum per year
Date(s) Application is Due: Apr 1
Contact: Fellowship Administrator, (212) 596-6270; fax: (212) 596-6213
Internet: http://www.aicpa.org/members/div/career/mini/fmds.htm
Sponsor: American Institute of Certified Public Accountants
1211 Avenue of the Americas
New York, NY 10036-8775

AICPA John L. Carey Scholarships in Accounting   **499**
This program is designed to encourage liberal arts undergraduates to consider professional accounting careers by providing scholarships for graduate accounting study to liberal arts undergraduates who plan to work toward a career in professional accounting. The recipients may attend U.S. graduate schools only.
Requirements: Applicants must be liberal arts degree holders of a regionally accredited institution in the United States.
Amount of Grant: $5000 per year, renewable for a second year

Date(s) Application is Due: Apr 1
Contact: Academic and Career Developments, (212) 596-6221; fax: (212) 596-6292; email: educat@aicpa.org
Internet: http://www.aicpa.org/members/div/career/mini/jlcs.htm
Sponsor: American Institute of Certified Public Accountants
1211 Avenue of the Americas
New York, NY 10036

AICPA Minority Accounting Scholarships   **500**
Scholarships are awarded to minority-group students who are U.S. citizens majoring in accounting in graduate or undergraduate courses of study. Students should contact the institute for definitive application instructions.
Amount of Grant: $5000 maximum
Date(s) Application is Due: Jun 1
Contact: Academic and Career Developments, (212) 596-6221; fax: (212) 596-6292; email: info@aicpa.org
Internet: http://www.aicpa.org/members/div/career/mini/smas.htm
Sponsor: American Institute of Certified Public Accountants
1211 Avenue of the Americas
New York, NY 10036-8775

AICR Investigator-Initiated Research Grants   **501**
Grants are awarded to nonprofit institutions for research relevant to understanding the effects of dietary and nutritional practices on the etiology, pathogenesis, and treatment of cancer. Applications for this program should fall within the scope of diet and prevention of cancer and diet and treatment of cancer. New ideas in research that focus on dietary and nutritional means of preventing and treating cancer or improving the quality of life of the cancer patient are encouraged. Funded organizations receive two years of support with possible renewal for two additional years. Preproposal letters of intent are encouraged but are not required. Application forms, instructions, and policy information may be obtained from the institute.
Restrictions: Indirect costs cannot exceed 10 percent of direct cost budget.
Amount of Grant: $75,000 per year (plus 10% for indirect costs)
Date(s) Application is Due: Jul 1; Dec 16
Contact: Research Department, (202) 328-7744 or (800) 843-8114; fax: (202) 328-7226; email: research@aicr.org
Internet: http://www.aicr.org/research/investigator.lasso
Sponsor: American Institute for Cancer Research
1759 R Street NW
Washington, DC 20009

AICR Postdoctoral Award   **502**
These international fellowships are awarded to beginning investigators to stimulate innovative and new research on the prevention, etiology, or treatment of cancer by dietary or nutritional methods. Applications should propose relevant feasibility studies to obtain data in support of a new hypothesis that then could be expanded to increase understanding of the role of dietary and nutritional factors in the etiology, pathogenesis, or treatment of cancer.
Requirements: The principal investigator must have a PhD or MD degree that was awarded no more than three years prior to the date of the application and must hold an academic appointment no higher than assistant professor. The applicant must be sponsored by a professor in whose laboratory the applicant is to perform his or her research.
Amount of Grant: $25,000 plus 10 percent of annual indirect costs
Date(s) Application is Due: Jul 1; Dec 16
Contact: Research Department, (202) 328-7744 or (800) 843-8114; fax: (202) 328-7226; email: research@aicr.org
Internet: http://www.aicr.org/research/post_doc.lasso
Sponsor: American Institute for Cancer Research
1759 R Street NW
Washington, DC 20009

Aid to Scholarly Publications (ASPP)   **503**
The ASPP is designed to assist the publication of works of advanced scholarship which make an important contribution to knowledge, but which are unlikely to be self-supporting. Manuscripts in the various disciplines in the humanities and social sciences are considered for the ASPP process if they meet ASPP eligibility criteria. An award from the ASPP is designed to help offset costs of marketing and promoting a scholarly work. All funds are paid directly to a book's publisher. Also, five subventions per year will be devoted to translations. Successful candidates applying for a translation subvention will also be eligible for a grant of up to $4,000 to help offset the costs of translating the work.
Requirements: With few exceptions, ASPP-funded authors must be Canadian citizens or landed immigrants. In general, only book-length manuscripts (at least 100 typescript pages) written by Canadian citizens or landed immigrants / permanent residents, and to be published in Canada by approved Canadian scholarly presses, are eligible for consideration. Authors wishing to work with overseas presses should consult page 3 of these Guidelines, item 3, "Submitting Party". Authors seeking ASPP support are strongly advised not to sign a contract with a non-Canadian publisher before obtaining ASPP approval of that publisher. The only modern languages in which submissions will be considered are English and French. Classical-language works may also be admissible. Guidelines and registrations forms are available at the sponsor's website.
Restrictions: The ASPP does not award research grants. The ASPP does not provide grants for the publication of: text books; technical reports; original works of poetry,

fiction and drama; scholarly journals and articles; or, conference proceedings. Not all publishers are eligible to work with the ASPP (please see List of Eligible Publishers available at the website).
Samples: A searchable database of authors and titles is available at http://www.fedcan.ca/english/aspp/titles/database/.
Amount of Grant: $8,000 CAN
Contact: Secretariat; (613) 238-6112, ext. 350; fax: (613) 238-6114; email: secaspp@fedcan.ca
Internet: http://www.fedcan.ca/english/aspp/about/general/
Sponsor: Canadian Federation for the Humanities and Social Sciences
151 Slater Street, Suite 415
Ottawa, ON K1P 5H3 Canada

AIEA Honors Scholarships   504
These scholarships are intended to encourage and support self-supporting students who have achieved high academic standing to continue their studies in Japan. Grantees are selected on the basis of the recommendations of their Japanese universities' scholarship committees. Each year approximately 4400 scholarships are offered for the 12-month academic year. Application form is obtained from the university which forwards the completed form to the association. Each university has its own deadline date for receipt of the completed application form.
Requirements: These scholarships are open to non-Japanese students who are under 30 years of age and in their junior or senior year at a Japanese university or college. Application should be made through the school the student is attending in Japan.
Amount of Grant: @Y52,000 per month undergraduate students; @Y70,000 graduate students
Contact: Student Affairs Division, 03 6407-7454; fax: 03 6407-7459
Internet: http://www2.jasso.go.jp/study_j/scholarships_e.html
Sponsor: Association of International Education, Japan
4-5-29 Komaba, Meguro-ku
Tokyo 153-8503 Japan

AIEA Research Grants   505
The association sponsors grants to encourage research projects that address the priorities of its international research agenda. Projects that include statewide, regional, or national frameworks, control groups, or comparative analyses are strongly recommended.
Requirements: Applications are welcome from U.S. and international researchers at higher education institutions or nonprofit educational associations.
Samples: Dr. Carolyn Lavely, U of South Florida (FL)—for her project titled The Development, Piloting, and Validation of a Set of Critical Competencies and Skills for Persons Starting Careers in International Settings, $20,000.
Amount of Grant: $20,000 maximum
Contact: Stephen Dunnett, President, State University of New York at Buffalo, (716) 645-2368; fax: (716) 645-2528; email: aiea@buffalo.edu
Internet: http://wings.buffalo.edu/intled/aiea
Sponsor: Association of International Education
411 Capen Hall, Box 601604
Buffalo, NY 14260-1604

AIHP Sonnedecker Visiting Scholar Grants   506
The Sonnedecker Visiting Scholar Program offers assistance for short-term historical research related to the history of pharmacy, including the history of drugs, at the University of Wisconsin-Madison. The program provides assistance for travel, maintaining temporary residence in Madison, and meeting research expenses associated with utilizing the collection. A brochure is available on request that describes the pharmaco-historical collections, which have been developed in Madison during more than a century by the University of Wisconsin-Madison, the State Historical Society of Wisconsin, and the American Institute of the History of Pharmacy. At least $1,000 becomes available annually to defray part of the expenses of a recipient, for whatever period of residence is appropriate. Grants are made throughout the year on the basis of the merit of previous historical work and on the appropriateness of historical resources on the University of Wisconsin campus to the research proposed.
Requirements: Eligible applicants are: Historians, Pharmacists, and other scholars working in the Pharmacy field.
Amount of Grant: $1,000
Contact: Gregory Higby, Director; (608) 262-5378; email: grants@aihp.org
Internet: http://cms.pharmacy.wisc.edu/aihp/programs/sonnedecker
Sponsor: American Institute of the History of Pharmacy
777 Highland Avenue
Madison, WI 53705-2222

AIHP Thesis Support Grants   507
The Program offers a grant-in-aid totaling $2,000 or more annually to a graduate student to reinforce historical investigations of some aspect of pharmacy, whether ancient or modern, to pay research expenses not normally met by the university granting the degree. Any thesis project devoted to the history of pharmacy, history of drugs, or other humanistic study utilizing a pharmaco-historical approach, is eligible if based in an institution of higher learning of the U.S.A. The maximum grant in this program of the Institute will be $2,500.
Requirements: Any graduate student in good standing at an institution of the United States may apply, regardless of the department through which the Doctor of Philosophy degree will be granted. The graduate student need not be an American citizen; nor does the research topic need to be in the field of American history.

Restrictions: Examples of ineligible expenses would be: living expenses of the applicant at the home university; routine typing to produce research notes or the thesis manuscript; routine illustrations for the manuscript; and publication of research results. Indirect expenses, such as overhead and other institution-related costs, may not be included in a grant application.
Amount of Grant: $2,500 maximum
Date(s) Application is Due: Feb 1
Contact: Gregory Higby, Director; (608) 262-5378; email: grants@aihp.org
Internet: http://cms.pharmacy.wisc.edu/aihp/programs/thesisgrant
Sponsor: American Institute of the History of Pharmacy
777 Highland Avenue
Madison, WI 53705-2222

AIIS Junior Fellowships   508
Fellowships are awarded to graduate students from all academic disciplines whose dissertation research requires study in India. Awards are for up to 11 months. Fellowships for four months or less have significant travel restrictions. Fellowships for six months or more may include limited dependent coverage if funds are available. Junior fellows will have formal affiliation with Indian universities and Indian research supervisors. Awards are announced by the beginning of October; the earliest possible departure date for India by awardees is eleven (11) months following the date of application submission.
Requirements: The fellowships are open to U.S. citizens and noncitizen residents at American colleges and universities who are studying in the humanities, social sciences, or natural sciences. All AIIS-sponsored research projects and programs must receive the approval of the government of India.
Date(s) Application is Due: Jul 1
Contact: Fellowship Coordinator, (773) 702-8638; fax: (773) 702-6636; email: aiis@uchicago.edu
Internet: http://www.indiastudies.org/fellow.htm
Sponsor: American Institute of Indian Studies
1130 E 59th Street, Foster Hall
Chicago, IL 60637

AIIS Performing and Creative Arts Fellowships   509
A limited number of fellowships are available each year to accomplished practitioners of the performing and creative arts of India who demonstrate that studying in India will enhance their skills, develop their capabilities to teach or perform in the United States, enhance American involvement with India's artistic traditions, and strengthen their links with peers in India. These fellowships can be either short-term (up to four months) or long-term (six to nine months).
Requirements: U.S. citizens and noncitizen residents who are engaged in research or teaching at American colleges and universities are eligible. Selection of fellows is made without regard to race, color, national origin, sex, or religion of applicants. All AIIS-sponsored research projects and programs must receive the approval of the government of India.
Restrictions: Graduate students are ineligible.
Date(s) Application is Due: Jul 1
Contact: Fellowship Coordinator, (773) 702-8638; fax: (773) 702-6636; email: aiis@uchicago.edu
Internet: http://www.indiastudies.org/fellow.htm
Sponsor: American Institute of Indian Studies
1130 E 59th Street, Foster Hall
Chicago, IL 60637

AIIS Professional Development Fellowships   510
A few fellowships are awarded to established scholars who have not previously specialized in Indian studies, and to established professionals who have not previously worked or studied in India. Proposals should have a substantial research or project component, and the anticipated results should be clearly defined. Awards are available for six to nine months. While in India, each fellow will be formally affiliated with an Indian university.
Requirements: Eligible to apply are U.S. citizens and resident aliens who are engaged in research or teaching at American colleges or universities. Selection is made without regard to race, color, national origin, sex, or religion of applicants. All AIIS-sponsored research projects and programs must receive the approval of the government of India.
Date(s) Application is Due: Jul 1
Contact: Fellowship Coordinator, (773) 702-8638; fax: (773) 702-6636; email: aiis@uchicago.edu
Internet: http://www.indiastudies.org/fellow.htm
Sponsor: American Institute of Indian Studies
1130 E 59th Street, Foster Hall
Chicago, IL 60637

AIIS Senior Research Fellowships   511
Fellowships are awarded to academic specialists in Indian studies possessing the PhD or its equivalent. The fellowships are designed to enable scholars specializing in South Asian studies to pursue further research in India. Each senior fellow will be formally affiliated with an Indian university for the period of the award, ranging from four to nine months.
Requirements: U.S. citizens who hold the PhD or its equivalent and noncitizen residents who are engaged in research or teaching full time at American colleges or universities are eligible. All AIIS-sponsored research projects and programs must receive the approval of the government of India.

Date(s) Application is Due: Jul 1
Contact: Fellowship Coordinator, (773) 702-8638; fax: (773) 702-6636; email: aiis@uchicago.edu
Internet: http://www.indiastudies.org/fellow.htm
Sponsor: American Institute of Indian Studies
1130 E 59th Street, Foster Hall
Chicago, IL 60637

AIIS Short-Term Senior Fellowships 512
A few fellowships are awarded to academic specialists in Indian studies who have a PhD or the equivalent. Formal affiliation is established for the fellow with an Indian university. Fellowships are offered for up to four months.
Requirements: Eligible to apply are U.S. citizens and permanent residents who are engaged in research or teaching at American colleges or universities. Selection is made without regard to race, color, national origin, sex, or religion of applicants. All AIIS-sponsored research projects and programs must receive the approval of the government of India.
Date(s) Application is Due: Jul 1
Contact: Fellowship Coordinator, (312) 702-8638; fax: (312) 702-6636; email: aiis@uchicago.edu
Internet: http://www.indiastudies.org/fellow.htm
Sponsor: American Institute of Indian Studies
1130 E 59th Street, Foster Hall
Chicago, IL 60637

AINSE International Conference Travel Scholarships 513
Travel scholarships will be awarded to students and post doctoral fellows from AINSE member universities who wish to present their AINSE supported research at an international meeting which is being conducted outside Australia. Successful applicants will be awarded up to $900. More support will be given when you are able to indicate an intention to publish the results being presented in a prestigious journal.
Requirements: Applicants should provide their personal details, name, university, supervisor, etc. as well as the following information: (1) full name of the conference; (2) location of the conference; (3) dates of the conference; (4) are you presenting an oral or poster? If so, provide evidence of the status of your presentation oral/poster, refereed/not refereed, accepted/not accepted; (5) details of (or plans for) publication of the results; note that the committee looks more favorably on applications with an intention to publish in high impact journals; (6) the AINSE Award number(s) which relate to the research being presented; (7) abstract of your presentation; and, (9) your budget including any additional support you have secured for the conference. Applications should be sent by email to ainse@ansto.gov.au.
Amount of Grant: up to 900 AUD
Contact: Dr. Dennis Mather, Executive Secretary; +61 2 9717 3388; fax: +61 02 9717 9268; email: ainse@ansto.gov.au
Internet: http://www.ainse.edu.au/ainse/for_graduate_students/travel_scholarships.html
Sponsor: Australian Institute of Nuclear Science and Engineering (AINSE)
Private Mail Bag 1
Menai, NSW 2234 Australia

AINSE Postgraduate Research Awards 514
Awards are offered by the Australian Institute of Nuclear Science and Engineering (AINSE) for suitably qualified persons wishing to undertake studies in AINSE's field of interest for a higher degree at a member university of AINSE. Applications for PGRAs will also be considered, as in past years, where the research aligns with the broad spectrum of research foci supported by AINSE. A full list of over 100 major and minor facilities facilities is available on request. Research Supplements ($7,500 pa) are the principal form of award and are offered to scholars who are (or will be) in receipt of an Australian Postgraduate Award (APA) or equivalent scholarship. The award also provides $5,500 pa for costs involved in using the facilities and services at Lucas Heights. Certain travel and accommodation costs to enable students to work at Lucas Heights are also provided.
Requirements: Candidates for an AINSE Postgraduate Research Supplement must be nominated by the Australian or New Zealand university where PhD enrollment is held or proposed for tenure commencing before 30 June of the year of application. The university must be a member of AINSE. Guidelines and applications can be found at the website.
Amount of Grant: 7,500 AUD
Date(s) Application is Due: Apr 15
Contact: Dr. Dennis Mather, Scientific Secretary; +61 2 9717 3388; fax: +61 02 9717 9268; email: ainse@ansto.gov.au
Internet: http://www.ainse.edu.au/ainse/for_academic_researchers/ainse_awards.html
Sponsor: Australian Institute of Nuclear Science and Engineering
Private Mail Bag 1
Menai, NSW 2234 Australia

AINSE Postgraduate Research Awards at ISIS 515
Applications for a postgraduate research award will be considered from postgraduate students wishing to do research at ISIS, the world's leading pulsed neutron and muon source situated at the CCLRC Rutherford Appleton Laboratory. The postgraduate award will be modified to provide additional support for travel to and accommodation while conducting experiments at ISIS as follows. This subsidy will be $7,000 per annum; there will be no ANSTO facility costs.
Requirements: For award applications to access ISIS there is no requirement to have an ANSTO co-supervisor. Candidates for an AINSE Postgraduate Research Supplement

must be nominated by the Australian or New Zealand university where PhD enrollment is held or proposed for tenure commencing before 30 June of the year of application. The university must be a member of AINSE.
Amount of Grant: 7,000 AUD
Date(s) Application is Due: Apr 15
Contact: Dr. Dennis Mather, Executive Secretary; +61 2 9717 3388; fax: +61 02 9717 9268; email: ainse@ansto.gov.au
Internet: http://www.ainse.edu.au/ainse/for_graduate_students/post_graduate_awards.html
Sponsor: Australian Institute of Nuclear Science and Engineering (AINSE)
Private Mail Bag 1
Menai, NSW 2234 Australia

AINSE Research Fellowships 516
AINSE Research Fellowship are offered at academic level B for outstanding and enthusiastic scientists with 3 - 8 years of postdoctoral experience to undertake research in areas of science and technology requiring use of the research facilities at the Australian Nuclear Science and Technology Organisation (ANSTO). Fellows will be expected to build a research group in one of the AINSE member universities and be attached to ANSTO for experiments on the new, state of the art facilities on the OPAL research reactor or other facilities at ANSTO.
Requirements: AINSE is highlighting three research directions in calling for applications for research fellowships: (1) neutron scattering; (2) radiopharmaceuticals; (3) and, high resolution climate records using nuclear techniques. Applications will also be considered which fall outside these topics which but still require use of ANSTO facilities and which align with the joint research interests of ANSTO and AINSE member universities.
Samples: 2008: Dr. Duncan McGillivray and Dr. Moeava Tehei.; 2007: Dr. Daniel Riley and Dr. Darren Goossens.
Date(s) Application is Due: May 31
Contact: Dr. Dennis Mather, Executive Secretary; +61 2 9717 3388; fax: +61 02 9717 9268; email: ainse@ansto.gov.au
Internet: http://www.ainse.edu.au/ainse/for_academic_researchers/ainse_awards2.html
Sponsor: Australian Institute of Nuclear Science and Engineering (AINSE)
Private Mail Bag 1
Menai, NSW 2234 Australia

AIR National Summer Data Policy Institute Fellowships 517
The Association for Institutional Research (AIR), with support from the National Center for Educational Statistics (NCES) and the National Science Foundation (NSF), has developed a program, Improving Institutional Research in Postsecondary Educational Institutions. The goals of the program are to provide professional development opportunities to doctoral students, institutional researchers, educators and administrators, and to foster the use of the federal databases for institutional research in postsecondary education. The program has four main components: a National Data Policy Institute held each summer in the Washington, D.C. area to study the national databases of NSF and NCES; dissertation fellowships for doctoral students; research grants for practitioners and faculty; and a focused grant program supported by the National Postsecondary Education Cooperative. Dissemination of research findings is an integral part of this program. Fellowships are available to cover roundtrip transportation from the Fellow's home to Washington, D.C., hotel accommodations, a fixed per diem reimbursement for meals and incidental expenses for the duration of the Institute, and a one-year AIR membership. Acceptance to the Institute is made on the basis of a short proposal.
Requirements: Applicants must have a basic knowledge of statistical methods and be experienced in the use of software packages (i.e., SPSS, SAS). Applicants should also have an interest in using national databases for studies in institutional research. The National Data Policy Institute is open to institutional researchers, faculty, graduate students, and educators affiliated with a U.S. postsecondary institution or governance agency. A limited number of Fellowships are available to similar personnel in non-U.S. institutions. Unaffiliated individuals should contact AIR before preparing a proposal for submission. Prospective Institute Fellows are strongly encouraged to study the databases of NSF and NCES before submitting a proposal. Proposals must be submitted via the online proposal submission tool(http://airweb.org/DataPolicySubmission.html). Guidelines and detailed instructions are available at the website.
Restrictions: Proposals may not be considered for review by AIR for the following reasons: (1) inappropriate for AIR funding; (2) failure to meet AIR requirements for proposal content, format, etc.; or (3) failure to meet announced proposal deadline requirements.
Date(s) Application is Due: Jan
Contact: Bethany Sumrow, Project Manager; (850) 385-4155 x105; fax: (850) 385-5180; email: bsumrow@airweb2.org
Internet: http://www.airweb.org/?page=1084
Sponsor: Association for Institutional Research (AIR)
1435 E. Piedmont Drive, Suite 211
Tallahassee, FL 32308

AIR Research Grants/Dissertation Fellowships 518
The Association for Institutional Research operates a grant program to support research on a wide range of issues of critical importance to U.S. higher education. The funding for these grants comes from the National Center for Education Statistics (NCES), the National Science Foundation (NSF) and the National Postsecondary Education Cooperative (NPEC). The program supports research of national relevance for the improvement of higher education theory and practice. Proposals that use one or more

of the national datasets of NCES and NSF or that address the current NPEC focus topic will be considered. Grant funds may be used to pay for salary support, benefits (if allowable by the institution), graduate assistant salaries, travel to conferences, dissemination of results (printing, etc.), materials and supplies needed for the project, consultant services, and computer software.

Requirements: Doctoral students enrolled at a U.S. postsecondary institution are eligible for dissertation fellowships of up to $20,000 for one year to support dissertation research under the guidance of a faculty dissertation advisor. Fellowships are based on a submitted budget and paid in two installments (May 1st and January 3rd). Faculty and practitioners are eligible for research grants of up to $40,000 for one year of independent research. All grant recipients must be affiliated with a U.S. postsecondary institution or relevant nonprofit higher education organization without regard to citizenship. Unaffiliated individuals may be eligible to submit a proposal, but should contact the AIR Grant Staff before submitting a proposal. Research grants are issued to the applicant, sponsored research unit to be dispersed according to an approved budget. To qualify for funding, the proposal submission must meet one or both of the following criteria: (1) Use data from one or more of the national NCES and/or NSF datasets (Appendix A); and/or (2) Address the current NPEC focus topic (Appendix B). Proposals may only be submitted electronically at http://airweb.org/grants/. The online submission site will be available in late October. The guidelines, online submission, and review process have been completely revised and streamlined, so please review them carefully.

Restrictions: Independent research grants are not available to students. Grant recipients may not purchase computer hardware, but may purchase software necessary for the project.

Samples: Alicia Betsinger, Southwestern University, $21,733 - One Ranking to Rule Them All: Modeling U.S. News & World Report, Predicted Graduation Rate.; John Cheslock, University of Arizona, $39,960 - A Nonparametric Examination of the Prices Low-Income Students Face and the Payment Strategies They Utilize.; Joy Gayles, North Carolina State University, $39,781 - Effects of College on Degree Attainment: Women/Underrepresented Minorities in Sciences at Four-Year Colleges and Universities.; Nicholas Wolfinger, University of Utah, $40,000 - Mid Career and Beyond in Academia

Amount of Grant: up to $40,000
Date(s) Application is Due: Jan 16
Contact: Randy Swing, Ph.D., Executive Director; (850) 385-4155; fax: (850) 385-5180; email: rswing@airweb2.org
Internet: http://www.airweb.org/?page=40
Sponsor: Association for Institutional Research (AIR)
1435 E. Piedmont Drive, Suite 211
Tallahassee, FL 32308

Air Transportation Centers of Excellence Grants 519

The program supports long-term, continuing research in specific areas of aviation-related technology. Responsibilities of centers include, but are not limited to, the conduct of research concerning catastrophic failure of aircraft, airspace and airport planning and design, airport capacity enhancement techniques, human performance in the air transportation environment, aviation safety and security, the supply of trained air transportation personnel including pilots and mechanics, and other aviation issues pertinent to developing and maintaining a safe and efficient air transportation system, and the interpretation, publication, and dissemination of the results of such research. FAA intends to support a center in three-year increments; each center will undergo a reassessment every three years. Contact the office for application deadlines.

Requirements: Applicants are limited to colleges and universities with the financial resources to meet statutory requirements for matching federal funds and maintenance of effort. Locations shall be geographically equitable.

Amount of Grant: $500,000 average per year, per center
Contact: Dr. Patricia Watts, Program Director, FAA Technical Center, (609) 485-5043; fax: (609) 485-9430; email: patricia.watts@faa.gov
Internet: http://www.dot.gov
Sponsor: Department of Transportation
AAR-201, FAA Technical Ctr
Atlantic City International Airport, NJ 08405

AIR/IES-NCES Postdoctoral Policy Fellowships in Washington 520

The program is a project funded by the National Center for Education Statistics (NCES). Postdoctoral Fellow projects will be responsible for undertaking analysis that results in improvements to the quality, comparability, and usefulness of the Integrated Postsecondary Education Data System (IPEDS). Fellows will be introduced to the Washington higher education policy community through meetings, seminars and conversations. Fellowships provide $5,000 monthly for support while the Fellow is in residence in Washington, for a period of from nine to twelve months. An additional $5,000 will be provided for expenses for the life of the fellowship. It is expected that Fellows will be advanced graduate students completing dissertations, postdoctoral students, or institutional research professionals. The award amount is set with the expectation that the Fellowship will provide support for an individual working in residence at NCES for a period of 9 to 12 months. Proposals with other residency formats may be acceptable, and individuals may propose a project of shorter duration.

Requirements: Applications may be submitted at any time. Institutional researchers, graduate students, postdoctoral students, faculty, and others may submit applications. Fellows must be U.S. citizens or permanent residents, and will be employees of AIR during the period of the Fellowship. Before formal submission, the application should be discussed with the AIR staff. Detailed guidelines and application instructions are available at the website. Applications must be submitted electronically, as a Microsoft

Word or PDF email attachment. Individuals unable to submit in this manner should contact the Project Manager for an alternative method.

Restrictions: Financial assistance for Fellowships is subject to certain statutory and other general requirements, including compliance with Circular A-133 Audit requirements and various federal statutes: Title VI of the Civil Rights Act of 1964, Title IX of the Education Amendments of 1972, Section 504 of the Rehabilitation Act of 1973, the Age Discrimination Act of 1975, and other laws and regulations prohibiting discrimination; prohibition of misconduct in science and engineering; Drug-Free Workplace requirements; restrictions on lobbying; patent and copyright requirements; cost-sharing; and the use of U.S.-flag carriers for international travel. Intellectual property developed under the Postdoctoral Policy Fellow is in the public domain.

Amount of Grant: $50,000 - $60,000
Contact: Bethany Sumrow, Project Manager; (850) 385-4155 x105; fax: (850) 385-5180; email: bsumrow@airweb2.org
Internet: http://www.airweb.org/?page=1349
Sponsor: Association for Institutional Research (AIR)
1435 E. Piedmont Drive, Suite 211
Tallahassee, FL 32308

AIR/NCES Fellowships for Graduate Study 521

The Association for Institutional Research (AIR), as part of its contract to improve the quality of institutional research in U.S. postsecondary education systems and hence, the Integrated Postsecondary Education Data System (IPEDS), offers a graduate fellowship program funded by the National Center for Education Statistics (NCES). The fellowships are for graduate study advancing or leading to careers in institutional research, thereby increasing the national level of expertise of institutional research officers and the data that they produce. The program awards individual fellowships for full- and part-time study for up to $10,000 annually for full-time study, $3,000 annually for part-time study. Award funds may be used for tuition, fees and other educational expenses while enrolled in a graduate program.

Requirements: Candidates must demonstrate knowledge and commitment to IR through work experience (current or prior) or previous coursework. Currently employed IR professionals are urged to apply. Fellowships may be used at any regionally accredited U.S. institution. The applicant's proposal must make the case that the course of study is appropriate for preparing for a career in institutional research, by showing that the program addresses the core competencies of institutional research. Three levels of credit-bearing graduate work will be considered eligible for the awards: certificate, masters and doctorate.

Restrictions: There is a three year maximum for the fellowship.

Samples: Wendi Clouse Springs - Ph.D., University of Colorado at Colorado.; Jean Constable - IR Certificate, Texas Lutheran University.; Katie Daniels - Ph.D., Davenport University.; Mary Flagg - IR Certificate, Reading Area Community College.; Mary Jo Harmon - IR Certificate, Owens Community College.; Rebecca Heverly - IR Certificate, San Jose/Evergreen Community College District.; John Hofmann - IR Certificate, Institute of Transpersonal Psychology.; Gina Johnson - Ph.D., Mid Western Higher Education Impact.; Faye Jones - Ph.D., Florida State University.; Jerold Laguilles - Ed.D., University of Massachusetts.; David Leavitt - IR Certificate, Bunker Hill Community College.; Andrew Mauk - Ph.D., Florida State University.; Thomas McGuinness - Ph.D., Tufts University.; Salley Mikel - IR Certificate, University of Illinois at Urbana-Champaign.; Robert Miller - Ph.D., Centenary College.; Marget Moreman - IR Certificate, Piedmont College.; MaryAnn Morgan-Cox - Ph.D., Baylor University.; Kelli Rainey - IR Certificate, Johnson C. Smith University.; Lindsay Renfro - Ph.D. Baylor University.; Jimmy Roberts - Ed.D., Temple College.; Andrea Shaw - Masters, University of Minnesota.; Sean Simone - Ph.D., Maryland Independent College and University Association.; Dale Swartzentruber - IR Certificate, Ohio Wesleyan University.; Marilyn Valencia - IR Certificate, Cuyahoga County, Ohio.; Christopher Ward - Masters, University of Michigan, Ann Arbor.; Alexander Yin - Ph.D., Pennsylvania State University

Amount of Grant: $9,000 - $30,000
Date(s) Application is Due: Mar 14
Contact: Bethany Sumrow, Project Manager; (850) 385-4155 x105; fax: (850) 385-5180; email: bsumrow@airweb2.org
Internet: http://www.airweb.org/?page=1381
Sponsor: Association for Institutional Research (AIR)
1435 E. Piedmont Drive, Suite 211
Tallahassee, FL 32308

AIR/NRC Summer Institute on First-Year Assessment Undergraduate 522
Student Fellowships

The National Conference on First-Year Assessment, formerly the Summer Institute on First-Year Assessment, is designed to maximize opportunities for participant learning and networking and structured to create a cumulative learning experience. The Conference, purpose is to provide a comprehensive introduction to the theory and practice of first-year assessment. Exploring both quantitative and qualitative assessment practices, conference concurrent sessions allow participants to gain a breadth of knowledge while workshops allow an in-depth examination of major issues. The National Resource Center for The First-Year Experience and Students in Transition, in collaboration with the Association for Institutional Research (AIR), is offering up to five fellowships that will cover the registration fee for the National Conference on First-Year Assessment. The fellowships pay the participant, conference registration fee, but no travel or lodging expenses.

Requirements: The Annual Conference on The First-Year Experience Fellowship Program is open to all undergraduate students who are enrolled for the current academic year. All students who are interested in first-year programs, mentoring first-year students, and/or any other aspect of first-year student life and leadership are encouraged to apply. The purpose and aim of this fellowship program is to advance the leadership skills and knowledge base of undergraduate students so that they may in turn use what they learn at the conference to enhance and encourage first-year student development on their respective campuses.

Restrictions: Only one applicant from a single institution is eligible to receive a Fellowship.

Amount of Grant: $270 - $580
Date(s) Application is Due: Oct 24
Contact: Tricia Kennedy, Graduate Assistant for Conference Planning; National Resource Center; (803) 777-3984; email: KENNEDTL@mailbox.sc.edu
Internet: http://www.airweb.org/?page=1189
Sponsor: Association for Institutional Research (AIR)
1435 E. Piedmont Drive, Suite 211
Tallahassee, FL 32308

AJA Bernard and Audre Rapoport Fellowships in American Jewish Studies   523
Fellowships are available to postdoctoral candidates for research and writing at the American Jewish Archives for a one-month period. Annual deadline dates may vary; contact the program office for exact dates.

Requirements: Applicant must provide an up-to-date curriculum vita, a research proposal (not more than five typewritten pages, double-spaced), evidence of published research, and two recommendations from academic colleagues. Typically, Marcus Center fellowships will be awarded to postdoctoral candidates, PhD candidates who are completing dissertations, and senior or independent scholars.

Date(s) Application is Due: Mar 18
Contact: Kevin Proffitt, Program Contact, (513) 221-7444 ext 304; fax: (513) 221-7812; email: kproffitt@huc.edu or aja@huc.edu
Internet: http://www.americanjewisharchives.org/aja/programs/index.html
Sponsor: American Jewish Archives
3101 Clifton Avenue
Cincinnati, OH 45220-2408

AJA Ethel Marcus Memorial Fellowship in American Jewish Studies   524
The Marcus Center, Fellowship Program was founded with the intent of creating a forum where students and scholars of the American Jewish experience could gather together to research, discuss, and study their chosen topics. Under the auspices of this unique program scholars come to Cincinnati to conduct in-depth research at the American Jewish Archives and to take part in the academic community of the Hebrew Union College-Jewish Institute of Religion. The program provides fellows with an opportunity not only to pursue their own research, but also to interact and exchange ideas with research peers as well as with the faculty and students of HUC-JIR.

Requirements: This fellowship is available to ABDs for one month of research or writing at the American Jewish Archives. Applicant must provide an up-to-date curriculum vita, a research proposal, and two faculty recommendations (including one from the dissertation supervisor). Typically, Marcus Center fellowships will be awarded to postdoctoral candidates, PhD candidates who are completing dissertations, and senior or independent scholars.

Date(s) Application is Due: Mar 18
Contact: Kevin Proffitt, Program Contact, The Jacob Rader Marcus Center, (513) 221-7444 ext 304; fax: (513) 221-7812; email: kproffitt@huc.edu or aja@huc.edu
Internet: http://www.americanjewisharchives.org/aja/programs/index.html
Sponsor: American Jewish Archives
3101 Clifton Avenue
Cincinnati, OH 45220-2408

AJA Lowenstein-Wiener Summer Fellowships in American Jewish Studies   525
Fellowships are available to ABDs or postdoctoral candidates for one month of research or writing at the American Jewish Archives during the stipend year. Applicant must provide an up-to-date curriculum vita, a research proposal, and evidence of published research, where possible. ABDs must provide three faculty recommendations (including one from the dissertation supervisor), and postdoctoral candidates must provide two recommendations from academic colleagues.

Requirements: Typically, Marcus Center fellowships will be awarded to postdoctoral candidates, PhD candidates who are completing dissertations, and senior or independent scholars.

Date(s) Application is Due: Mar 18
Contact: Kevin Proffitt, Fellowship Director, (513) 221-7444 ext 304; fax: (513) 221-7812; email: kproffitt@huc.edu
Internet: http://www.americanjewisharchives.org/aja/programs/index.html
Sponsor: American Jewish Archives
3101 Clifton Avenue
Cincinnati, OH 45220-2408

AJA Marguerite R. Jacobs Memorial Postdoctoral Award in American Jewish Studies   526
This award is made to postdoctoral candidates for one month of active research or writing at the American Jewish Archives. Applicant must provide an up-to-date curriculum vita; a research proposal; evidence of published research, where possible; and two recommendations from academic colleagues.

Requirements: Typically, Marcus Center fellowships will be awarded to postdoctoral candidates, PhD candidates who are completing dissertations, and senior or independent scholars.

Amount of Grant: $2000
Date(s) Application is Due: Mar 18
Contact: Kevin Proffitt, Director of the Fellowships Program, (513) 221-7444 ext 304; fax: (513) 221-7812; email: AJA@cn.huc.edu
Internet: http://www.americanjewisharchives.org/aja/programs/index.html
Sponsor: American Jewish Archives
3101 Clifton Avenue
Cincinnati, OH 45220-2408

AJA Rabbi Frederic A. Doppelt Memorial Fellow in American Jewish Studies   527
This award is available to ABDs for one month of research or writing at the American Jewish Archives during the stipend year. Applicant must provide an up-to-date curriculum vita, a research proposal, and two faculty recommendations (including one from the dissertation supervisor).

Requirements: Typically, Marcus Center fellowships will be awarded to postdoctoral candidates, PhD candidates who are completing dissertations, and senior or independent scholars.

Amount of Grant: $1000
Date(s) Application is Due: Mar 18
Contact: Kevin Proffitt, Director of the Fellowship Program, (513) 221-7444 ext 304; fax: (513) 221-7812; email: AJA@cn.huc.edu
Internet: http://www.americanjewisharchives.org/aja/programs/index.html
Sponsor: American Jewish Archives
3101 Clifton Avenue
Cincinnati, OH 45220-2408

AJA Rabbi Levi A. Olan Memorial Fellowship in American Jewish Studies   528
Fellowships are available to postdoctoral candidates for research and writing at the American Jewish Archives for a one-month period. Applicant must provide an up-to-date curriculum vita, a research proposal (not more than five typewritten pages, double-spaced), evidence of published research, and two recommendations from academic colleagues.

Requirements: Typically, Marcus Center fellowships will be awarded to postdoctoral candidates, PhD candidates who are completing dissertations, and senior or independent scholars.

Date(s) Application is Due: Mar 18
Contact: Kevin Proffitt, Director of the Fellowship Program, The Jacob Rader Marcus Center, (513) 221-7444 ext 304; fax: (513) 221-7812; email: kproffitt@huc.edu or aja@huc.edu
Internet: http://www.americanjewisharchives.org/aja/programs/index.html
Sponsor: American Jewish Archives
3101 Clifton Avenue
Cincinnati, OH 45220-2408

AJA Rabbi Theodore S. Levy Tribute Fellowship in American Jewish Studies   529
Fellowships are available to postdoctoral candidates for research and writing at the American Jewish Archives for a one-month period. Applicant must provide an up-to-date curriculum vita, a research proposal (not more than five typewritten pages, double-spaced), evidence of published research, and two recommendations from academic colleagues.

Requirements: Typically, Marcus Center fellowships will be awarded to postdoctoral candidates, PhD candidates who are completing dissertations, and senior or independent scholars.

Date(s) Application is Due: Mar 18
Contact: Kevin Proffitt, Director of the Fellowship Program, The Jacob Rader Marcus Center, (513) 221-7444 ext 304; fax: (513) 221-7812; email: AJA@cn.huc.edu
Internet: http://www.americanjewisharchives.org/aja/programs/index.html
Sponsor: American Jewish Archives
3101 Clifton Avenue
Cincinnati, OH 45220-2408

AJA Starkoff Fellowship in American Jewish Studies   530
This fellowship is available to ABDs for one month of research or writing at the American Jewish Archives. Applicant must provide an up-to-date curriculum vita, a research proposal, and two faculty recommendations.

Requirements: Typically, Marcus Center fellowships will be awarded to postdoctoral candidates, PhD candidates who are completing dissertations, and senior or independent scholars.

Amount of Grant: $1000
Date(s) Application is Due: Mar 18
Contact: Kevin Proffitt, Director of the Fellowship Program, The Jacob Rader Marcus Center, (513) 221-7444 ext 304; fax: (513) 221-7812; email: AJA@cn.huc.edu
Internet: http://www.americanjewisharchives.org/aja/programs/index.html
Sponsor: American Jewish Archives
3101 Clifton Avenue
Cincinnati, OH 45220-2408

AJCC Cancer Staging Grants   531
AJCC seeks proposals to develop and evaluate improved staging algorithms for specific cancer sites and types. Successful proposals will present new approaches and utilize data sets available to the applicant to test and validate current or revised staging algorithms. Applicants may use elements currently used for AJCC/UICC TNM staging, additional or modified anatomic and non-anatomic factors to define prognosis, and incorporate

validated factors that identify response to treatment. Proposals that address and validate the use of non-anatomic factors to improve the utility of staging and define the response to therapy are encouraged. The AJCC will provide funding to a number of groups for a variety of disease sites to promote innovative strategies for enhancing the clinical utility of staging while maintaining compatibility with current staging data collection systems. The listed application deadline is for letters of intent. Guidelines are available online.

Requirements: Eligible institutions include for-profit and nonprofit organizations; public or private institutions, such as universities, colleges, laboratories, and hospitals; units of state and local government; eligible agencies of the federal government; and domestic or foreign institutions/organizations.
Amount of Grant: $150,000 maximum
Date(s) Application is Due: Dec 15
Contact: Valerie Vesich, (312) 202-5420; fax: (312) 202-5009; email: AJCC@facs.org
Internet: http://www.cancerstaging.org/initiatives/rfp.html#
Sponsor: American Joint Committee on Cancer
633 N Saint Clair Street
Chicago, IL 60611

Akonadi Foundation Anti-Racism Grants   532
The foundation's mission is to work with others to eliminate racism, with a particular focus on structural and institutional racism. Grants have supported programmatic approaches including research, policy work, advocacy, litigation, organizing, media, arts, diversity training, education, and other tools in their anti-racism work. The foundation awards general operating grants and project grants. Grants are made to organizations in the San Francisco Bay area and to national organizations with national reach. Letters of interest will be accepted year round. Full proposals should only be submitted upon request.
Geographic Focus: California
Samples: Asian Pacific Environmental Network (Oakland, CA)—to involve Asian-Pacific Islanders in grassroots organizing to fight environmental racism, $20,000.; Ctr for Third World Organizing (Oakland, CA)—for general support, $30,000.; Leadership Excellence (Oakland, CA)—for leadership development for black youths and for critical analysis of racism, $25,000.; Poverty and Race Research Action Council (Washington DC)—to support a civil-rights teaching curriculum, institutes on civil rights, and teacher workshops, $25,000.
Amount of Grant: $10,000-$50,000
Date(s) Application is Due: Feb 18
Contact: Grants Administrator, (510) 663-3867; email: info@akonadi.org
Internet: http://www.akonadi.org/application_guidelines.html
Sponsor: Akonadi Foundation
469 9th Street, Suite 210
Oakland, CA 94607

ALA ACRL Doctoral Dissertation Fellowship   533
This award fosters research in academic librarianship by encouraging and assisting doctoral students in the field with their dissertation research.
Requirements: The recipient of the fellowship must meet the following qualifications: be an active doctoral student enrolled in an accredited degree-granting institution; be engaged in researching a topic related to academic librarianship; have completed all coursework; have had a dissertation proposal accepted by the institution; a recipient of the fellowship may not receive it a second time; applicant need not be an ACRL member
Amount of Grant: $1,500
Date(s) Application is Due: Dec 5
Contact: Megan Griffin, (800) 545-2433 ext. 2514 or (312) 280-2514; email: mgriffin@ala.org
Internet: http://www.ala.org/ala/mgrps/divs/acrl/awards/doctoraldissertation.cfm
Sponsor: ALA ACRL Doctoral Dissertation Fellowship
50 East Huron Street
Chicago, IL 60611

ALA ACRL WESS Coutts Nijhoff International West European Specialist   534
Study Grant
This grant supports research pertaining to Western European studies, librarianship, or the book trade. The grant covers travel to and from Europe, transportation in Europe, and lodging and board for no more than fourteen consecutive days.
Requirements: The nominee must be a librarian employed in a university, college, community college, or research library in the year prior to application for the award.
Restrictions: Funds may not be used for salaries, research-related supplies, publication costs, conference fees, or equipment purchases.
Samples: Dale Askey, Kansas State University—award recipient (2006); Charlene Kellsey, University of Colorado—award recipient (2005); Helene S. Baumann, Duke University—award recipient (2004); Michael Olson, Harvard University—award recipient (2003).
Amount of Grant: $3,000
Date(s) Application is Due: Dec 5
Contact: Megan Griffin, (800) 545-2433 ext. 2514 or (312) 280-2514; email: mgriffin@ala.org
Internet: http://www.ala.org/ala/mgrps/divs/acrl/awards/nijhoffstudy.cfm
Sponsor: American Library Association
50 E Huron Street
Chicago, IL 60611-2795

ALA ALSC Louise Seaman Bechtel Children's Library Collection Fellowship   535
The Louise Seaman Bechtel Fellowship provides a $4,000 grant to a qualified children, librarian to spend a month or more reading and studying at the Baldwin Library of the George A. Smathers Libraries, University of Florida, Gainesville. The Baldwin Library contains a special collection of 85,000 volumes of children, literature published mostly before 1950. The fellowship is endowed in memory of Louise Seaman Bechtel and Ruth M. Baldwin.
Requirements: Applicants must: be personal members of ALSC as well as ALA; currently be working in direct service to children, or retired members who complete their careers in direct service to children, for a minimum of eight years; have a graduate degree from an ALA-accredited program; be willing to write a report about his/her study; the report will be submitted to the ALSC Office, for distribution to the Bechtel Committee and for possible inclusion in Children and Libraries, and to the Director of the Smathers Libraries and the Director of the Center for the Study of Children, Literature and Media at the University of Florida; if selected, retired Fellowship winners would agree to present a minimum of three public programs based on their research project to children in libraries or schools following the completion of their Fellowship period
Restrictions: Organizational members are not eligible.
Amount of Grant: $4000
Date(s) Application is Due: Dec 3
Contact: ALA Awards Staff Liaison, (312) 280-3247; fax: (312) 280-3257; email: alsc@ala.org
Internet: http://www.ala.org/Template.cfm?Section=Grants_and_Fellowships&template=/ContentManagement/ContentDisplay.cfm&ContentID=22345
Sponsor: American Library Association
50 E Huron Street
Chicago, IL 60611

ALA Annual Diversity Research Grant   536
Continuing the American Library Association's commitment to diversity, the ALA Office for Diversity began sponsorship of a Diversity Research Grant program in 2002 to address critical gaps in the knowledge of diversity issues within library and information science. The Diversity Research Grant consists of a one-time $2000 annual award for original research and a $500 travel grant to attend and present at ALA Annual Conference. Three grants are awarded each year.
Requirements: Persons submitting a proposal must be a current member of the American Library Association. You will be required to supply your membership ID number with your proposal. Only proposals demonstrating relevance to the year's designated research topics will be fully considered. See website for complete proposal requirements.
Amount of Grant: $2,500
Date(s) Application is Due: Apr 30
Contact: Tracie D. Hall, Director of the ALA Office for Diversity; (800) 545-2433; email: thall@ala.org.
Internet: http://www.ala.org/Template.cfm?Section=awards&template=/ContentManagement/ContentDisplay.cfm&ContentID=132643
Sponsor: ALA Office for Diversity
50 E Huron Street
Chicago, IL 60611

ALA Carroll Preston Baber Research Grant   537
This grant is given annually to one or more librarians or library educators who will conduct innovative research that could lead to an improvement in services to any specified group(s) of people. The project should have the potential to serve as a model for the library community. Attention to the uses of technology will be given preference, and efforts involving cooperation between libraries, between libraries and other agencies, or between librarians and persons in other disciplines will also receive special consideration.
Requirements: Any ALA member may apply. Preferential consideration will be given to projects that involve a practicing librarian.
Amount of Grant: $3000
Date(s) Application is Due: Dec 10
Contact: Denise M. Davis, Director, (800) 545-2433, ext. 1-4273; fax: (312) 280-4392; email: dmdavis@ala.org
Internet: http://www.ala.org/ala/aboutala/offices/ors/orsawards/baberresearchgrant/babercarroll.cfm
Sponsor: American Library Association
50 E Huron Street
Chicago, IL 60611

ALA Frances Henne YALSA/VOYA Research Grant   538
This grant annually provides $1,000 in seed money for small scale projects which will encourage research that responds to the YALSA Research Agenda.
Requirements: Applicants must be personal members of YALSA, including student members, although the research project may be undertaken by an individual, an institution, or by a group. The proposed research must be a response to the vision, mission, goals, and research agenda of YALSA, and for the general area of library service to young adults. A proposal of 1-2 pages, and one page of biographical information, should be submitted and must include the following: title; objectives; problem statement/questions to be answered; methodology, including how data will be collected and analyzed; significance of the project, in light of previous research; projected time line (project should be completed within a 12 to 18 month period, and a report filed with YALSA within six months of completion of the study.); indicate the way that the money will be used; brief biographical data on researcher on a separate sheet, with contact information including email, as well as the researcher's ALA membership number.

Restrictions: Previous winners are not eligible.
Amount of Grant: $500
Date(s) Application is Due: Dec 1
Contact: ALA Awards Staff Liaison, (312) 545-2433 ext 4387; email: yalsa@ala.org or ngilbert@ala.org
Internet: http://www.ala.org/ala/yalsa/awardsandgrants/franceshenne.htm
Sponsor: American Library Association
50 E Huron Street
Chicago, IL 60611

ALA Information Today Library of the Future Award   539
The purpose of the award, consisting of $1,500 and a 24k gold-framed citation of achievement, is to honor an individual library, library consortium, group of librarians, or support organization for innovative planning for, applications of, or development of patron training programs about information technology in a library setting. Selection criteria should include the benefit to clients served; benefit to the technology information community; impact on library operations; public relations value; and impact on the perception of the library or librarian in the work setting and to the specialized and/or general public. Applications are available on the Web site.
Samples: Pace University Library (Pleasantville, NY)—award recipient (2006); Oberlin's Community Technology Center—award recipient (2005); Border Health Information and Education Network—award recipient (2004); Rowan Public Library (Salisbury, CA)—award recipient.
Amount of Grant: $1,500
Date(s) Application is Due: Dec 1
Contact: Cheryl Malden, Program Officer; (800) 545-2433, ext. 3247; fax: (312) 440-9374; email: awards@ala.org
Internet: http://www.ala.org/Template.cfm?Section=awards&template=/Content Management/ContentDisplay.cfm&ContentID=164985
Sponsor: American Library Association
50 East Huron Street
Chicago, IL 60611-2795

ALA Ingenta Research Awards   540
The Ingenta Research Award is given annually by the Library Research Round Table of the American Library Association to support research projects about acquisition, use, and preservation of digital information. The grant consists of up to $6,000 for research and up to $1,000 for travel to a national or international conference to present the results of the research. Expenditures must directly support research; the award does not cover indirect costs or overhead. Half of the research amount will be paid within one month of the selection of the awardee; the remaining half will be provided approximately six months later upon the receipt of a satisfactory progress report as determined by the Ingenta Award Jury Chair and the ALA staff liaison to the Ingenta Award Jury.
Requirements: Applications are welcome from practicing librarians, faculty and students at schools of library and information science, and independent scholars. Open to non-U.S. researchers. International projects not coordinated through a U.S. institution may be required to have a U.S. account for funds deposits. Please contact the award liason for additional information.
Amount of Grant: $7,000
Date(s) Application is Due: Jan 31
Contact: Denise M. Davis Director, Office for Research and Statistics, (800) 545-2433 ext. 4273; fax: (312) 280-4392; email: dmdavis@ala.org
Internet: http://www.ala.org/ala/aboutala/offices/ors/orsawards/ingentaresearchaward/ingentaresaward.cfm
Sponsor: American Library Association
50 East Huron Street
Chicago, IL 60611

ALA Phyllis Dain Library History Dissertation Award   541
The award, named in honor of a library historian widely known as a supportive advisor and mentor as well as a rigorous scholar and thinker, recognizes outstanding dissertations in the general area of library history. $500 and a certificate are given for a work that embodies original research on a significant topic relating to the history of books, libraries, librarianship or information science. The biennial award is given in the odd numbered years. Dissertations completed and accepted during the preceding two academic years are eligible.
Requirements: Entries are judged on: clear definition of the research questions and/or hypotheses; use of appropriate source materials; depth of research; superior quality of writing; ability to place the subject within its broader historical context; and significance of the conclusions.
Samples: Bonnie Mak—award recipient (2005): for (re)Defining the Page for a Digital World.; Mildred L. Jackson—award recipient (2001): for Do What You Can: Creating an Institution, Ladies' Library Associations in Michigan, 1850-1900.
Amount of Grant: $500
Date(s) Application is Due: Mar 1
Contact: Letitia Earvin Office for Research & Statistics, (800) 545-2433
Internet: http://www.ala.org/Template.cfm?Section=awards&template=/Content Management/ContentDisplay.cfm&ContentID=138732
Sponsor: American Library Association
50 E Huron Street
Chicago, IL 60611

ALA PLA DEMCO New Leaders Travel Grants   542
These grants, sponsored by Demco Inc., are designed to enhance the professional development and improve the expertise of public librarians new to the field by making their attendance at major professional development activities possible. The grants were established to enable PLA members who are new to the profession, and who have not had the opportunity to attend a major PLA continuing education event in the last five years, to attend such an event. Eligible events are the PLA Spring Symposium workshops; PLA National Conferences; and other PLA events, such as preconferences, held in conjunction with ALA Annual Conferences.
Requirements: Applicant must: be a current member of the Public Library Association; have been a practicing librarian (with an MLS from an accredited institution) for five years or less. Applicant cannot be a current: officer or member of the PLA Board of Directors; member of the New Leaders Travel Grant Jury. Applicant, supervisor or supervising authority cannot be a current member of the New Leaders Travel Grant Jury.
Restrictions: See website for complete listing of special restrictions.
Amount of Grant: $1500
Contact: PLA Office, (800) 545-2433 ext. 5027; email: pla@ala.org
Internet: http://www.ala.org/ala/mgrps/divs/pla/plaawards/demconewleadersgrant/index.cfm
Sponsor: American Library Association
50 E Huron Street
Chicago, IL 60611-2795

ALA Samuel Lazerow Fellowship For Research in Collections and   543
Technical Services in Academic and Research Libraries
This award fosters advances in acquisitions or technical services by providing fellowships to librarians for travel or writing in those fields. Proposals will be judged with an emphasis on potential significance to acquisitions or technical services, originality and creativity, clarity and completeness of the proposal, and evidence of an interest in scholarship (previous publication record). Fellowship recipients are required to submit reports of the results of their research to C&RL for possible publication in C&RL News.
Requirements: Brief proposal (five pages or less) should include a description of research, travel, or writing project; project schedule; an estimate of expenses; and an up-to-date curriculum vitae.
Restrictions: Research projects in collection development or the compilation of bibliographies will not be supported.
Samples: Sara Marcus $1,000 - for her research project on the change of terms and terminology over several editions of the Sears List of Subject Headings (2009); Ping Situ and Shuyong Jiang $1,000 - for their research project on vendor-provided records and the experience of a research library in outsourcing cataloging service for its Chinese language materials (2008);
Amount of Grant: $1000
Date(s) Application is Due: Dec 4
Contact: Megan Griffin, (800) 545-2433 ext. 2514 or (312) 280-2514; email: mgriffin@ala.org
Internet: http://www.ala.org/ala/mgrps/divs/acrl/awards/samuellazerow.cfm
Sponsor: American Library Association
50 E Huron Street
Chicago, IL 60611

Alabama Humanities Foundation Mini Grants   544
Designed for flexibility, Mini Grants support a wide variety of projects from workshops to radio programs and reading/discussion series. Up to $2,000 in outright and $2,000 in matching funds is available. Preliminary application can be completed anytime, with final proposals due the first working day of the following months: February, May, August, and November.
Requirements: Applications must be submitted three months prior to start date of the program. Grants are made to nonprofit public or private organizations. Funds are to be used in state only.
Restrictions: Grants are not made to individuals.
Geographic Focus: Alabama
Amount of Grant: $2,000 maximum
Contact: Susan Perry, Grants Director; (205) 558-3980 or (205) 558-3989; fax: (205) 558-3981; email: sperry@ahf.net
Internet: http://www.ahf.net/programs/grantsProgram.html
Sponsor: Alabama Humanities Foundation
1100 Ireland Way, Suite 101
Birmingham, AL 35205

Alabama Humanities Foundation Planning/Consultant Grants   545
Available for planning Major Grants only, these Planning/Consultant Grants provide funds for humanities scholar honoraria and travel, as well as other planning expense. These grants are awarded in support of organizations that do not have extensive grant writing experience or expertise in planning public humanities projects. Application deadlines are the first working day of February, May, August, and November.
Requirements: Grants are made to nonprofit public or private organizations. Funds are to be used in state only.
Restrictions: Grants are not made to individuals. Colleges and universities are ineligible.
Geographic Focus: Alabama
Samples: Hoover Public Library (AL)—to fund an annual three-day conference, Southern Voices: A Celebration of Southern Culture, $5,000.; Institute of Language and Culture

(Clanton, AL)—to plan and script a documentary on the lives of slaves entitled In Their Own Voices—The WPA Ex-Slave Narratives, $3,000.
Amount of Grant: $1,000 maximum
Contact: Susan Perry, Grants Director; (205) 558-3980 or (205) 558-3989; fax: (205) 558-3981; email: sperry@ahf.net
Internet: http://www.ahf.net/programs/grantsProgram.html
Sponsor: Alabama Humanities Foundation
1100 Ireland Way, Suite 101
Birmingham, AL 35205

Alabama Library Association CU.S. Research Grants   546

Sponsored by the Research Committee of the College, University, and Special Libraries (CU.S.) Division, Alabama Library Association, research supported through this program must relate to college, university, or special libraries. The Committee looks for the following: originality, validity, relevance to the profession, and applicant's need for funding support. Grant funds may be used for expenses such as duplicating, mailing, travel, office supplies, consulting services, and computer services. The deadline for submitting a grant application is normally December 1.
Requirements: Competition for the grant is open to CU.S. members employed in Alabama college, university, and special libraries, and to CU.S. members attending an ALA-accredited library school.
Geographic Focus: Alabama
Amount of Grant: $250
Date(s) Application is Due: Dec 1
Contact: Sonja McAbee, Head of Library Services; (256) 782-5757; fax: (256) 782-5872; email: smcabee@jsu.edu
Internet: http://alla.associationsonline.com/membership_grants.cfm
Sponsor: Alabama Library Association
Houston Cole Library, 700 Pelham Road North
Jacksonville, AL 36265

Alaska Conservation Foundation Watchable Wildlife Conservation Grants   547

The Watchable Wildlife Conservation Trust is a donor-advised fund developed in cooperation with the Alaska Department of Fish and Game. The purpose of the Trust is to fund those projects that emphasize non-consumptive use of wildlife, expand wildlife conservation efforts, and broaden public support for conserving Alaska's wildlife resources. Areas of interest include: interpretation (e.g., signage, viewing guides, media programming); education (e.g., curriculum development or publication development and distribution); facilities construction; applied ecosystem research; and projects involving a state wildlife sanctuary. The Trust makes grants once per calendar year. The annual deadline is October 1. Prospective grantees should contact ACF to determine if the project fits within the grant making program and when to submit their proposal. The Trust generally funds projects in the $1,000 to $4,000 range.
Requirements: As a 501(c)3 public foundation, ACF accepts requests for funding from incorporated, tax-exempt organizations, non-incorporated organizations, and individuals.
Restrictions: Ineligible requests include: projects outside the state of Alaska; general operating support; litigation or lobbying; endowments; and scholarships.
Geographic Focus: Alaska
Amount of Grant: $1,000-$4,000 range
Date(s) Application is Due: Oct 1
Contact: Polly Carr, Program Officer; (907) 276-1917; fax: (907) 274-4145; email: pcarr@alaskaconservation.org or grants@akcf.org
Internet: http://alaskaconservation.org/_pages/grants_amp_awards/types_of_grants/watchable_wildlife.php
Sponsor: Alaska Conservation Foundation
441 West Fifth Avenue, Suite 402
Anchorage, AK 99501-2340

Albany Medical Center Prize in Medicine and Biomedical Research   548

The prize recognizes a physician or biomedical scientist (or group of physicians or scientists) who has made extraordinary and sustained leadership contributions to improving healthcare and patient care; or who has successfully pursued innovative biomedical research with demonstrated translational benefits applied to improved patient care. Each year's prize winner will have demonstrated significant outcomes that offer medical value of national or international importance. Prize winner activities will include but not be limited to disease and injury management, clinical research, and basic science investigations of diseases and injuries, leading to new discoveries and improved clinical outcomes. Nomination guidelines are available online.
Requirements: Any physician or scientist or group whose work has led to significant advances in the fields of health care and scientific research with demonstrated translational benefits applied to improved patient care may be nominated. Those honored will be practitioners and/or scientists whose accomplishments and outcomes have been demonstrated in the past quarter century, with preference to demonstrated accomplishments in the past decade.
Samples: Robert J. Lefkowitz, MD, James B. Duke Professor of Medicine and Howard Hughes Medical Institute Investigator at Duke University Medical Center in Durham, NC; Solomon H. Snyder, MD, Distinguished Service Professor in the Department of Neuroscience at Johns Hopkins School of Medicine in Baltimore, MD; and Ronald M. Evans, PhD, Howard Hughes Medical Institute Investigator at The Salk Institute for Biological Studies in La Jolla, CA (2007, joint recipients).
Amount of Grant: $500,000
Date(s) Application is Due: Jan 5

Contact: Fardin Sanai, Secretary, (518) 262-8043; fax: (518) 262-4769; email: AMCprize@mail.amc.edu
Internet: http://www.amc.edu/Academic/AlbanyPrize/prize_criteria.html
Sponsor: Albany Medical Center
Center Building, 628 Madison Avenue - 1st Floor
Albany, NY 12208

Albert and Margaret Alkek Foundation Grants   549

The foundation awards grants to Texas nonprofit organizations to support charitable, religious, scientific (primarily medical), literary, cultural and educational organizations and programs. Preference will be given to research and education-related projects that will pay lasting dividends in terms of new discoveries and improved quality of life. One application per 12-month period will be considered. There are no application deadlines or forms. Applicants should submit a one- to two-page letter of inquiry that includes a brief description of the organization and the project for which funds are being considered, and the amount of funding needed in total as well as the amount being requested. Inquiries should be sent via U.S. postal mail; fax or email applications are not accepted.
Requirements: Texas nonprofit organizations are eligible.
Restrictions: The foundation does not make grants to individuals or loans of any type. The foundation does not make direct scholarships to students. All scholarship programs are administered through educational institutions. The foundation prefers not to fund: organizations that in turn make grants to others; grants intended to influence legislation or to support candidates for political office; fund-raising events such as luncheons, dinners, galas, advertising in programs, or other similar activities; charities operated by service clubs; memorials for individuals; student organizations; or purchase of uniforms, equipment, or trips for school-related organizations or sports teams.
Geographic Focus: Texas
Samples: Baylor College of Medicine (Houston, TX)—for research on cancer, cardiovascular science, diabetes, pharmacogenomics, and other biomedical topics, $31.25 million (2005).
Contact: Grants Administrator, (713) 951-0019; fax: (713) 951-0043; email: info@alkek.org
Internet: http://www.alkek.org/grantguidelines.htm
Sponsor: Albert and Margaret Alkek Foundation
1221 McKinney, Suite 4525
Houston, TX 77010-2023

Albert and Mary Lasker Foundation Awards   550

The major purpose of the awards is to honor the individual(s) who have made significant contributions in basic or clinical research in diseases that are the main cause of death and disability. Three categories of awards are made. The Basic Medical Research Award honors the scientist or scientists who have made fundamental investigations that open new areas of biomedical science. The Clinical Medical Research Award honors the scientist(s) whose contributions, directly or indirectly, have led to the improvement of the clinical management or treatment of patients and to the alleviation or elimination or one of the major medical causes of disability or death. The Special Achievement Award in Medical Science honors a scientist whose contributions to research are of unique magnitude and immeasurable influence on the course of science, health, or medicine, and whose professional career has engendered extreme respect within the biomedical community. Nomination forms are available online.
Samples: Pierre Chambon, Institute of Genetics and Molecular and Cellular Biology (Strasbourg, France)—for the 2004 Lasker Award for Basic Medical Research, $50,000 (2004); Charles Kelman, International Retinal Research Foundation (Birmingham, AL)—for the 2004 Lasker Award for Clinical Medical Research, $50,000 (2004).
Amount of Grant: $50,000 medical research award; $50,000 clinical medical research award; $25,000 special achievement award
Date(s) Application is Due: Feb 1
Contact: David Keegan, (212) 286-0222; fax: (212) 286-0924; email: dkeegan@laskerfoundation.org
Internet: http://www.laskerfoundation.org/awards/awards.html
Sponsor: Albert and Mary Lasker Foundation
110 E 42nd Street, Suite 1300
New York, NY 10017

Albert Ellis Institute Clinical Fellowships   551

This fellowship program is a unique one- to two-year part-time course of study intended to help the therapist develop clinical proficiency in an active psychotherapy approach emphasizing rational emotive behavior psychotherapy and allied cognitive behavioral techniques. Fellows receive highly diversified training in clinical experience, supervision, workshop experience, and research. By special arrangement, persons on sabbatical or from out of town may complete the program in one year.
Requirements: The fellowship is open to those who hold doctorates in psychology or counseling (or near completion of PhD); to MSWs, RNs, and MDs; and to licensed mental health counselors. All candidates must be licensed or license-eligible. Predoctoral interns pursue the same program as postdoctoral fellows and receive the same stipend.
Amount of Grant: $6000 per year for two-year fellowships
Contact: Dr. Kristene Doyle, Program Contact, (212) 535-0822 or (800) 323-4738; fax: (212) 249-3582; email: krisdoyle@albertellis.com or info@albertellis.org
Internet: http://www.rebt.org/titlepages/professionals.asp
Sponsor: Albert Ellis Institute
45 E 65th Street
New York, NY 10021

Albert M. Greenfield Foundation Dissertation Fellowship 552
This fellowship supports dissertation research in residence at the Library Company on any subject relevant to its collections. The term of the fellowship is from September to May, with a stipend of $20,000. The award may be divided between two applicants, each of whom would spend a semester in residence. To apply, an applicant should send five copies each of a brief resume, a two- to four-page description of the proposed research, two letters of reference, and a writing sample of about 25 pages. Deadline for receipt of applications is March 1, 2008.
Amount of Grant: $20,000
Date(s) Application is Due: Mar 1
Contact: James Green, Library Company of Philadelphia; (215) 546-3181; fax: (215) 546-5167; email: jgreen@librarycompany.org
Internet: http://www.librarycompany.org/fellowships/greenfield.htm
Sponsor: Albert M. Greenfield Foundation
1314 Locust Street
Philadelphia, PA 19107

Alberta Heritage Preservation Heritage Awareness Grants 553
Grants support tangible initiatives that promote awareness of Alberta's history and pre-history and have a lasting impact. Contact the foundation for application materials.
Requirements: Applicants must be resident Albertans or have a permanent address in Alberta.
Restrictions: Provincial government departments and their employees are not eligible to apply.
Amount of Grant: $C5000 maximum
Date(s) Application is Due: Feb 1; Sep 1
Contact: Monika McNabb, Coordinator, Heritage Preservation Partnership Program, (780) 431-2305; fax: (780) 427-5598; email: monika.mcnabb@gov.ab.ca
Internet: http://www.cd.gov.ab.ca/preserving/heritage/ahrf/partnership/Heritage Awareness/index.asp
Sponsor: Alberta Historical Resources Foundation
8820-112 Street
Edmonton, AB T6G 2P8 Canada

Alberta Historical Publications Grants 554
Grants are awarded to support the printing or reprinting of historical publications dealing with Alberta's human history or prehistory or the printing of brochures, pamphlets, or other materials of a commemorative or informative nature.
Requirements: Applicants must be resident Albertans or have a permanent address in Alberta.
Restrictions: Provincial government departments and their employees are not eligible to apply. Publisher/printers are not eligible to apply; and personal family histories are ineligible for funding.
Amount of Grant: $C3000 maximum for brochures and pamphlets; $C5000 maximum for books
Date(s) Application is Due: Feb 1; Sep 1
Contact: Program Contact, (780) 431-2305; fax: (780) 427-5598; email: monika. mcnabb@gov.ab.ca
Internet: http://www.cd.gov.ab.ca/preserving/heritage/ahrf/partnership/publications/index.asp
Sponsor: Alberta Historical Resources Foundation
8820 112 Street
Edmonton, AB T6G 2P8 Canada

Alberta Historical Resources Foundation Research Grants 555
Grants support research that will produce new understandings or add to the knowledge base of Alberta's history and pre-history. Projects such as oral histories and historic site inventories may be considered in this category. Contact the foundation for application materials.
Requirements: Applicants must be resident Albertans or have a permanent address in Alberta.
Restrictions: Provincial government departments and their employees are not eligible to apply.
Amount of Grant: $C15,000 maximum
Date(s) Application is Due: Feb 1; Sep 1
Contact: Community Resources Officer, (780) 431-2305; email: monika.mcnabb@gov.ab.ca
Internet: http://www.cd.gov.ab.ca/preserving/heritage/ahrf/partnership/research/index.asp
Sponsor: Alberta Historical Resources Foundation
8820 112 Street
Edmonton, AB T6G 2P8 Canada

Alberta Historical Resources Foundation Roger Soderstrom Scholarship 556
Scholarships are awarded to encourage research that will increase the knowledge of Alberta's history and historic resources. Applicable fields of study include anthropology, archaeology, history, historical geography, architectural history and preservation, and restoration architecture. One scholarship per Alberta university is awarded annually for the duration of one year, renewable for one year. Further information may be obtained through an Alberta university or from the foundation.
Requirements: Applicants must be resident Albertans or have a permanent address in Alberta.

Restrictions: Provincial government departments and their employees are not eligible to apply.
Amount of Grant: $C3000
Date(s) Application is Due: Feb 1; Sep 1
Contact: Heritage Preservation Partnership Program, (780) 431-2305
Internet: http://www.cd.gov.ab.ca/preserving/heritage/ahrf/partnership/rogerscholarship/index.asp
Sponsor: Alberta Historical Resources Foundation
8820-112 Street
Edmonton, AB T6G 2P8 Canada

Alberta Law Foundation Grants 557
The objectives of the foundation are to conduct research into and recommend reform of law and the administration of justice; establish, maintain, and operate law libraries; contribute to the legal education and knowledge of the people of Alberta and provide programs and facilities for those purposes; provide assistance to native people's legal programs, student legal aid programs, and programs of like nature; and contribute to the costs incurred by the Legal Aid Society of Alberta to administer a plan to provide legal aid. To be considered for funding, programs or projects must fall within these objectives. Operating grants and project grants are awarded. The application process begins with a discussing the program or project idea with the executive director.
Restrictions: Grants will not be made to an individual or for the support of a commercial venture. Funds are not available for bursaries, fellowships, sabbatical leave support, endowments, building funds, etc.
Contact: David Aucoin, Executive Director, (403) 264-4701; fax: (403) 294-9238; email: contact@albertalawfoundation.org
Internet: http://www.albertalawfoundation.org/Apply/general.html
Sponsor: Alberta Law Foundation
407 8th Avenue SW, Suite 300
Calgary, AB T2P 1E5 Canada

Alberta Research Council Scholarship 558
One scholarship is offered annually in open competition to a graduate student for research in areas of interest to the Alberta Research Council, such as agriculture, biotechnology, energy, forestry, environment, information technology, and manufacturing. The scholarship is awarded for 12 months duration; a student may apply in open competition for a subsequent year's award.
Requirements: Applicants must be graduate students engaged in thesis research at the master's or doctoral level who are (or will be) registered full-time in a thesis-based graduate program in the Faculty of Graduate Studies at the University of Calgary.
Amount of Grant: $C17,300
Contact: Connie Busch, Faculty of Graduate Studies, (403) 220-5690; fax: (403) 289-7635; email: cbusch@ucalgary.ca
Internet: http://www.grad.ucalgary.ca/funding/internal_scholarships/lev_4/ab_research.htm
Sponsor: University of Calgary
2500 University Dr NW, Earth Sciences Building, Room 720
Calgary, AB T2N 1N4 Canada

Albuquerque Community Foundation Grants 559
The foundation seeks to improve the quality of life in the greater Albuquerque, New Mexico, area by providing support for projects and organizations that serve the community in arts and culture, education, environmental and historic preservation, children and youth, and health and human services. Through its grant program, the foundation supports projects that are innovative, meet the needs of underserved segments of the community, encourage matching funds or additional gifts, promote cooperation among agencies, empower the disadvantaged and disabled, and enhance the effectiveness of local charitable organizations. Types of support include continuing support, general operating support, program development, publication, seed grants, scholarships funds and scholarships to individuals, and technical assistance.
Requirements: IRS 501(c)3 organizations based in Albuquerque, NM, are eligible. Proposals are reviewed on the basis of the following priorities: impact, innovation, leverage, management, and nonduplication.
Restrictions: Grants are generally not made to or for individuals, political or religious purposes, debt retirement, payment of interest or taxes, annual campaigns, endowments, emergency funding, to influence legislation or elections, scholarships, awards, or to private foundations and other grantmaking organizations.
Geographic Focus: New Mexico
Amount of Grant: $10,000 maximum
Date(s) Application is Due: Apr 16; Aug 15
Contact: R. Randall Royster, Executive Director; (505) 883-6240; fax: (505) 883-3629; email: rroyster@albuquerquefoundation.org
Internet: http://www.albuquerquefoundation.org/grants/grant-home.htm
Sponsor: Albuquerque Community Foundation
P.O. Box 36960
Albuquerque, NM 87176-6960

Alces Foundation Grants 560
The Alces Foundation, established in Massachusetts and managed by Welch and Forbes, is focused on the support of multi-purpose art centers, elementary and secondary education, higher education, the environment, performing arts, theater, and interfaith religious issues in Massachusetts. There are no specific application formats or deadlines with which to adhere, and potential applicants should forward a letter of inquiry.

Geographic Focus: Massachusetts

Samples: Lloyd Center for the Environment, Dartmouth, Massachusetts, $12,000—for naturalists, transportation, and supplies (2008); Essex County Greenbelt, Essex, Massachusetts, $10,000—costs for land conservation outreach (2008).

Contact: Jay Emmons, Executive Vice President; (617) 523-1635; fax: (617) 742-6243; email: info@welchforbes.com

Sponsor: Alces Foundation

45 School Street, Old City Hall, 5th Floor

Boston, MA 02108-3204

Alcoa Foundation Grants 561

General priorities of the foundation include safe and healthy children and families—ensuring that children and their families have the tools, the knowledge and the services to remain healthy and safe at home, in the community and in the workplace; conservation and sustainability—educating young leaders on conservation issues, protecting forests, promoting sound public policy research, and understanding the linkages between business and the environment; skills today for tomorrow—providing individuals with critical skills and services to be economically connected, workplace-ready, and productive in a changing economy; business and community partnerships—strengthening the nonprofit sector and developing meaningful partnerships among nonprofits, the private sector, and local government; and global education in science, engineering, technology and business—broadening student participation in areas central to Alcoa to prepare a diverse cross-section of our communities for a global workplace. Types of support include capital grants, building funds, challenge grants, matching gifts, general support, research grants, scholarships, and seed money. Initial contact should be a letter of inquiry.

Requirements: The foundation awards grants to nonprofit public charities in communities where Alcoa has a presence. Local Alcoans work within their communities to evaluate organizations and make recommendations for funding to Alcoa Foundation. Nonprofit organizations that serve localized communities should find the Alcoa facility nearest to them and write a one-page letter describing their mission, nature of request, connection to the areas of excellence and offering contact information. If interested, the Alcoa location contact will notify the requesting organization and invite them to submit more information. Areas of operation include western Pennsylvania; Davenport, Iowa; Evansville, Indiana; Massena, New York; New Jersey; Cleveland, Ohio; Knoxville, Tennessee; and Rockdale, Texas.

Restrictions: The foundation does not make gifts to local projects other than those near Alcoa plant or office locations; endowment funds, deficit reduction, or operating reserves; hospital capital campaign programs unless the hospital presents a comprehensive area analysis that justifies, on a regional rather than an individual institutional basis, the need for the capital improvement; individuals, except for the scholarship program for children of Alcoa employees; tickets and other promotional activities; trips, tours, or student exchange programs; or documentaries and videos.

Amount of Grant: $1,000-$50,000 average

Contact: Meg McDonald, President and Treasurer; (412) 553-2348; fax: (412) 553-4498; email: alcoa.foundation@alcoa.com

Internet: http://www.alcoa.com/global/en/community/foundation.asp

Sponsor: Alcoa Foundation

201 Isabella Street

Pittsburgh, PA 15212-5858

Alcohol Misuse and Alcoholism Research Grants 562

The foundation's research interests include factors influencing transitions in drinking patterns and behavior, effects of moderate use of alcohol on health and well-being, mechanisms underlying the behavioral and biomedical effects of alcohol, biobehavioral/interdisciplinary research on the etiology of alcohol misuse. Applications may be obtained from the Web site or by contacting the office. Grantees are usually notified within two weeks following the advisory council meetings, which are held in April and November.

Requirements: Applications may be submitted by public or private nonprofit organizations such as universities, colleges, hospitals, research institutes and organizations, and governmental research agencies and laboratories in the United States and Canada.

Restrictions: Non-research activities such as education projects, public awareness efforts and treatment or referral services are not eligible for support. The Foundation also does not support the training of pre- and post-doctoral fellows, undergraduates, graduate students, medical students, interns or residents. It does not fund thesis or dissertation research. The Foundation does not encourage applications on treatment of the complications of advanced alcoholism. However, research involving treatment intended to elucidate the pathogenesis of alcohol-related problems will be considered.

Samples: Boston U (MA)—to study alcohol abuse among college students, $38,500.; Bowman School of Medicine (Winston-Salem, NC)—for research on mechanisms in the transition and development of chronic alcohol consumption-induced cardiomyopathy, $39,000.; Cleveland Clinic Foundation (OH)—for research in sub-cellular homeostasis in the heart: effects of chronic ethanol consumption, $40,000.

Amount of Grant: $100,000 maximum over two years

Date(s) Application is Due: Feb 1; Sep 1

Contact: Research Grants Administrator, (410) 821-7066; fax: (410) 821-7065; email: info@abmrf.org

Internet: http://www.abmrf.org/grants.htm

Sponsor: Alcoholic Beverage Medical Research Foundation (ABMRF)

1122 Kenilworth Drive, Suite 407

Baltimore, MD 21204

Alcon Foundation Grants Program 563

The foundation supports organizations in the fields of health care, leadership programs, research, education and community responsibility. Programs that advance the education and skill levels of eye care professionals are given special consideration. General operating grants to organizations and institutions improving education and research in the areas of specialization of Alcon Laboratories—ophthalmology and vision care. Grants are also awarded to community activities that benefit company employees. Applications may be submitted at any time.

Requirements: Grants are not made for building programs.

Restrictions: Non 501(c)3organizations, individuals and scholarship programs, religious, veterans or fraternal organizations, political causes, capital campaigns, matching gifts, trips, tournaments and tours, and endowments are not supported.

Samples: National Sjogrens Syndrome Assoc (Phoenix, AZ)—for general operating support, $2500.

Amount of Grant: $100-$50,000 average

Contact: Mary Dulle, Chair, (817) 293-0450; email: Mary.Dulle@Alconlabs.com

Internet: http://www.alconlabs.com/corporate-responsibility/alcon-foundation.asp

Sponsor: Alcon Foundation

6201 S Freeway

Fort Worth, TX 76134

ALCTS Bowker-Ulrich's Serials Librarianship Award 564

This award, donated by the CSA/Ulrich Company, is given for contributions to serials librarianship in areas of professional association participation, library education, serials literature, research, or development of tools leading to better understanding of the field of serials.

Requirements: Applications or nominations are invited; appropriate forms may be obtained through ALCTS. The award may be divided among two or more individuals who have participated jointly in the achievement for which it is granted.

Restrictions: Employees of the CSA/Ulrich Company are not eligible to receive the award.

Samples: Karen Hunter—award recipient (2006); Dan Tonkery—award recipient, $1500 (2005); Pamela Bluh—award recipient, $1500 (2004); Frieda Rosenberg—award recipient, $1500 (2003).

Amount of Grant: $1500

Date(s) Application is Due: Dec 1

Contact: Mary Page, Chair, CSA/Ulrich's Serials Librarianship Award Committee, (732) 445-5894; fax: (732) 445-5888; email: mspage@rcirutgers.edu

Internet: http://www.ala.org/Template.cfm?Section=awards&template=/ContentManagement/ContentDisplay.cfm&ContentID=48736

Sponsor: Association for Library Collections and Technical Services

50 E. Huron Street

Chicago, IL 60611-2795

Alex C. Walker Educational and Charitable Foundation Grants 565

The foundation awards grants to investigate the causes of economic imbalances; investigate the effect of the monetary system in fostering a sustainable economy; investigate causes tending to destroy or impair the free-market system; explore and develop free-market solutions; and disseminate information on these issues. The foundation funds projects dealing with the research and development of innovative ideas, education of market principles, the application of ideas developed through seed grants and research, and ecological economics with a free-market orientation. Projects with the following qualities are favored: innovation, impact, longevity, collaboration, solution oriented, trackability, objectivity, and sustainability. Grant requests must be submitted on a formal application form through this website. Guidelines are available online.

Requirements: U.S. 501(c)3 tax-exempt organizations are eligible.

Restrictions: Ideological or political activities and general educational programs are not funded. Grants do not support endowment, building funds, or individuals.

Amount of Grant: $5000-$89,500

Date(s) Application is Due: Apr 1; Oct 1

Contact: Grants Administrator, c/o Barret P. Walker, (404) 378-2752

Internet: http://walker-foundation.org/page.aspx?s=5534.0.69.5316

Sponsor: Alex C. Walker Educational and Charitable Foundation

1729 Coventry Pl

Decatur, GA 30030

Alex Stern Family Foundation Grants 566

The foundation awards grants to North Dakota and Minnesota nonprofits in its areas of interest, including arts and culture, child welfare, the elderly, alcohol abuse, community funds, family and social services, education, minorities, hospices, and cancer research. Types of support include general operating support, continuing support, annual campaigns, building construction/renovation, equipment acquisition, emergency funds, program development, scholarship funds, research, and matching funds. Applications are reviewed in June and November.

Requirements: Moorhead, MN, and Fargo, ND, nonprofit organizations are eligible.

Restrictions: Grants are not awarded to individuals or for endowments.

Geographic Focus: Minnesota; North Dakota

Amount of Grant: $1000-$50,000 range

Date(s) Application is Due: Mar 31; Aug 31

Contact: Donald Scott, Executive Director, (701) 237-0170

Sponsor: Alex Stern Family Foundation

609 1/2 1st Avenue N., Suite 205

Fargo, ND 58102

Alexander and Margaret Stewart Trust Grants　　　　**567**

The trust awards grants in the greater Washington, DC, area for cancer treatment, especially equipment used in diagnosis and treatment; and caring for children who are physically ill, mentally ill, or disabled. Grants also support research, education, and prevention of common childhood diseases, including negative societal behavioral patterns that impact children. Proposals for projects aiding the economically deprived receive preference. The trust awards start-up funding. Applications may be submitted at any time; requests received by the listed deadline date are reviewed by the end of the year.

Requirements: Nonprofits in the greater District of Columbia area are eligible.

Restrictions: Requests for endowments, buildings, or capital campaigns are denied.

Geographic Focus: District of Columbia

Samples: Childrens National Medical Center, Washington, DC, $420,000—for Center for Cancer and Blood Disorders (2007); DC Campaign to Prevent Teen Pregnancy, Washington, DC, $75,000—for general support (2007).

Amount of Grant: $50,000-$150,000 average

Date(s) Application is Due: Sep 15

Contact: William J. Bierbower; (202) 785-9892; fax: (202) 785-0918; email: wbierbower@stewart-trust.org

Internet: http://www.stewart-trust.org/guidelines.htm

Sponsor: Alexander and Margaret Stewart Trust

888 17th Street NW, Brawner Building, Suite 610

Washington, DC 20006-3321

Alexander von Humboldt Foundation Fellowships for Japan-Related Research　　**568**

The primary aim of the program, which was established jointly by the Alexander von Humboldt Foundation and the Humboldt Associations in Japan, is to strengthen academic cooperation between Japan and Germany. The Fellowship allows the applicant to undertake a research visit to Japan. Doctoral candidates working at institutes engaged in Japan-related research should submit applications to the Foundation by March 1 and September 1 each year. Sponsorship comprises a single grant of 5,000 EUR.

Requirements: Applicants who are German doctoral candidates with a degree from a German university and are currently conducting doctoral research on a Japan-related subject can apply.

Amount of Grant: 5,000 EUR

Date(s) Application is Due: Mar 1; Sep 1

Contact: Program Contact; (+49) 0228-833-0; fax: (+49) 0228-833-199; email: info@avh.de

Internet: http://www.humboldt-foundation.de/en/programme/stip_sonst/japan.htm

Sponsor: Alexander von Humboldt Foundation

Jean-Paul-Street 12

Bonn D-53173 Germany

Alexander von Humboldt Foundation Feodor-Lynen Research Fellowships　　**569**
for Experienced Researchers

The Fellowship for experienced researchers allows a researcher to carry out a long-term project (6???????18 months) selected in cooperation with an academic host at a research institution in Germany. The fellowship is flexible and can be divided into as many as three stays within three years. The Foundation tries to involve the host in financing the research fellowship. Scientists and scholars from all disciplines may apply to the Foundation directly at any time. The Foundation awards up to 150 Fellowships for postdoctoral researchers and experienced researchers annually.

Requirements: Applicants must furnish proof of work published in recognized academic journals; submit a research plan agreed with the foreign host, including confirmation that research facilities can be made available; and possess a good knowledge of the host country's language or at least a very good knowledge of English, provided they can carry out their research project in that language. The host must be an academic working abroad who has already been sponsored by the Humboldt Foundation.

Contact: Program Contact; (+49) 0228-833-0; fax: (+49) 0228-833-199; email: info@avh.de

Internet: http://www.humboldt-foundation.de/en/programme/stip_deu/flf_e.htm

Sponsor: Alexander von Humboldt Foundation

Jean-Paul-Street 12

Bonn D-53173 Germany

Alexander von Humboldt Foundation Feodor-Lynen Research Fellowships　　**570**
for Postdoctoral Researchers

Fellowships are offered to highly qualified German scientists and scholars for long-term research projects (6-24 months) at foreign institutes. The host must be a foreign scholar formerly sponsored by the foundation (research fellow or research award winner). Scientists and scholars from all disciplines may apply for any target country abroad. The academic host must be an academic working abroad who has already been sponsored by the Humboldt Foundation. Candidates choose their own research projects and their host abroad and prepare their own research plan.

Requirements: Applicants must be less than 38 years old; have successfully completed their doctorate with very good or good ratings; furnish proof of work published in recognized academic journals; submit a research plan agreed with the foreign host, including confirmation that research facilities can be made available; and possess a good knowledge of the host country's language or at least a very good knowledge of English, provided they can carry out their research project in that language.

Contact: Program Contact; (+49) 0228-833-0; fax: (+49) 0228-833-199; email: info@avh.de

Internet: http://www.humboldt-foundation.de/web/2983.html

Sponsor: Alexander von Humboldt Foundation

Jean-Paul-Street 12

Bonn D-53173 Germany

Alexander von Humboldt Foundation Fraunhofer-Bessel Research Awards　　**571**

Scientists and scholars from all non-European countries, who have completed their doctorates less than twelve years ago, who are internationally recognized for their achievements in the field of applied research, and who in future are expected to continue producing cutting-edge achievements which will have a seminal influence on their discipline beyond their immediate field of work, are eligible to be nominated for an Award. Winners are invited to spend a period of up to one year cooperating on a long-term research project with specialist colleagues at one of the Fraunhofer Institutes in Germany. The stay may be divided up into blocks. The Foundation and the Fraunhofer Society for the Advancement of Applied Research jointly grant up to three Awards annually. The award is valued at 45,000 EUR.

Requirements: Nominations may be submitted by the heads of institutes or senior academics at institutes belonging to the Fraunhofer Society. Direct applications are not accepted.

Amount of Grant: 45,000 EUR

Contact: Program Contact; (+49) 0228-833-0; fax: (+49) 0228-833-199; email: info@avh.de

Internet: http://www.humboldt-foundation.de/web/5174.html

Sponsor: Alexander von Humboldt Foundation

Jean-Paul-Street 12

Bonn D-53173 Germany

Alexander von Humboldt Foundation Friedrich Wilhelm Bessel Awards　　**572**

The Foundation awards annually about 25 Friedrich Wilhelm Bessel Research Awards, funded by the Federal Ministry for Education and Research, to internationally renowned scientists and scholars from abroad in recognition of their outstanding accomplishments in research to date and their exceptional promise for the future. Scientists and scholars of all disciplines or research areas, regardless of their nationality, may be nominated for an Award, provided they have received their doctoral degree within the past 12 years. The Foundation particularly encourages the nomination of female scholars. In addition, awardees are invited to conduct an original research project of their own design in close collaboration with an appropriate colleague in Germany over a period of six to twelve months, which may be divided into segments.

Requirements: A candidate must be nominated by distinguished scientists/scholars employed by any university or research institution in Germany. Direct applications are not accepted. Scientists and scholars who are nominated must have attained international reputation through research in their chosen field; they must also show promise that their further research will remain on the cutting edge and have impact beyond their area of specialization.

Amount of Grant: 45,000 EUR

Contact: Program Contact; (+49) 0228-833-0; fax: (+49) 0228-833-199; email: info@avh.de

Internet: http://www.humboldt-foundation.de/en/programme/preise/bessel.htm

Sponsor: Alexander von Humboldt Foundation

Jean-Paul-Street 12

Bonn D-53173 Germany

Alexander von Humboldt Foundation Georg Forster Fellowships for　　**573**
Experienced Researchers

Fellowships are available to scholars from developing countries in humanities and social sciences; political science and economics; projects in the public health sector and in the fields of agriculture, forestry, and geosciences; and interdisciplinary projects relating to environmental and resource protection. Fellowships are designed to promote the transfer of knowledge and methods and to contribute to further development in fellows' home countries. Applications may be submitted for long-term research stays of between six and 18 months. The fellowship is flexible and can be divided up into as many as three stays within three years. The Foundation grants up to 60 Fellowships for postdoctoral researchers and experienced researchers annually.

Requirements: Scholars from developing countries (excluding Turkey, India, and the Peoples' Republic of China) may apply. Applicants must have completed their doctorates less than twelve years ago. Scholars in the humanities or social sciences and physicians must have a good knowledge of German if it is necessary to carry out the project successfully; otherwise a good knowledge of English; scientists and engineers must have a good knowledge of German or English.

Restrictions: Short-term study tours, participation in conferences, or educational visits cannot be funded. Applications for extension of research stays already commenced in Germany cannot be considered.

Amount of Grant: 2,450 EUR

Contact: Program Contact; (+49) 0228-833-0; fax: (+49) 0228-833-199; email: info@avh.de

Internet: http://www.humboldt-foundation.de/web/georg-forster-research-fellowship-experienced.html

Sponsor: Alexander von Humboldt Foundation

Jean-Paul-Street 12

Bonn D-53173 Germany

Sponsor: Alexander von Humboldt Foundation

Jean-Paul-Street 12

Bonn D-53173 Germany

Alexander von Humboldt Foundation Georg Forster Fellowships for 574
Postdoctoral Researchers
Fellowships for postdoctoral researchers are the instrument with which the Foundation enables highly-qualified scientists and scholars who are just embarking on their academic careers and who completed their doctorates less than four years ago1 and who to spend extended periods of research (6-24 months) in Germany. Scientists and scholars from all disciplines from developing and threshold countries (excluding People, Republic of China, India and Turkey) may apply. The research proposal must deal with issues of major relevance to the future development of the candidate's country of origin and, in this context, particularly suited to transferring knowledge and methods to developing and threshold countries. Research projects are carried out in cooperation with academic hosts at research institutions in Germany. Candidates choose their own research projects and their host in Germany and prepare their own research plan.
Amount of Grant: 2,250 EUR
Contact: Program Contact; (+49) 0228-833-0; fax: (+49) 0228-833-199; email: info@avh.de
Internet: http://www.humboldt-foundation.de/web/georg-forster-research-fellowship-postdoc.html
Sponsor: Alexander von Humboldt Foundation
Jean-Paul-Street 12
Bonn D-53173 Germany

Alexander von Humboldt Foundation Helmholtz-Humboldt Research Awards 575
The Award honors internationally renowned scientists and scholars from abroad (up to six per year). These awards are based primarily on a scientist's/scholar's entire academic record. Furthermore, award-winners are invited to carry out research projects of their own choice in the fields of energy, earth and environment, health, key technologies, structure of matter, transportation and space in cooperation with German scientists at one of the 15 member-centers of the Helmholtz Association for periods of between six months and one year (may be divided into segments). At least one further partner from a German university is to be integrated in the research project. The Award amounts to EUR 60,000. An additional amount of EUR 25,000 is made available for the promotion of cooperation in particular with university partners, if the awardee accepts the invitation to undertake research in Germany.
Requirements: The nominee, scholarly qualifications must be internationally recognized and documented. Nominating scientists/scholars must be able to provide sufficient non-financial resources, equipment, and/or access to archives and libraries in order that the nominee may conduct the collaborative research project in Germany.
Amount of Grant: 60,000 EUR
Contact: Program Coordinator; (+49) 0228-833-0; fax: (+49) 0228-833-199; email: info@avh.de
Internet: http://www.humboldt-foundation.de/web/542426.html
Sponsor: Alexander von Humboldt Foundation
Jean-Paul-Street 12
Bonn D-53173 Germany

Alexander von Humboldt Foundation Humboldt Research Awards 576
The Award honors up to one hundred internationally renowned scientists and scholars from abroad, annually. These awards are based primarily on the scientist's/scholar's entire academic record. Scientists and scholars who are nominated for an Award must have contributed fundamental discoveries, new theories, or insights that significantly impact their own research area and also reach beyond their specialized field of research. Nominees should be active researchers from whom one might anticipate the same high level of achievement in the future as is evident in their past work. A candidate must be nominated by distinguished scientists/scholars employed by any university or research institution in Germany. Direct applications are not accepted. The Award amounts to EUR 60,000.
Requirements: Scientists and scholars of all disciplines or research areas, regardless of their nationality, may be nominated. The Foundation particularly encourages the nomination of female scholars.
Amount of Grant: EUR 60,000
Contact: Program Coordinator; (+49) 0228-833-0; fax: (+49) 0228-833-199; email: info@avh.de
Internet: http://www.humboldt-foundation.de/web/6446.html
Sponsor: Alexander von Humboldt Foundation
Jean-Paul-Street 12
Bonn D-53173 Germany

Alexander von Humboldt Foundation Humboldt Research Awards 577
Based on Reciprocity for Top German Researchers
Agreements have been made with partner organizations in various countries (Belgium, Brazil, Canada, Denmark, Finland, France, Hungary, India, Israel, Japan, Korea, The Netherlands, Poland, South Africa, Spain, Sweden, and Taiwan) stating that awards based on reciprocity may be granted each year by these partners and the Humboldt Foundation. These awards are based primarily on the scientist's/scholar's entire academic record. Scientists and scholars who are nominated for an Award must have contributed fundamental discoveries, new theories, or insights that significantly impact their own research area and also reach beyond their specialized field of research. The nomination procedure will be handled and the final decision on each award will be made by the respective partner organization independently.
Requirements: Scientists and scholars of all disciplines or research areas, regardless of their nationality, may be nominated. The Foundation particularly encourages the nomination of female scholars.

Amount of Grant: EUR 60,000
Contact: Program Coordinator; (+49) 0228-833-0; fax: (+49) 0228-833-199; email: info@avh.de
Internet: http://www.humboldt-foundation.de/en/programme/preise/ptg.htm
Sponsor: Alexander von Humboldt Foundation
Jean-Paul-Street 12
Bonn D-53173 Germany

Alexander von Humboldt Foundation Institutional Academic 578
Cooperation Grants
Under the follow-up program of the Foundation, institutional partnerships for long-term research collaboration can be created by combining various sponsorship opportunities. Special emphasis is placed on the exchange of younger scientists. The aim is to give these scientists access to the world-wide network of scholars sponsored by the Foundation and to alert them to the Humboldt Research Fellowship Program. The program includes: reciprocal research visits by applicants and their co-workers at the partner institutes (including travel and living expenses); jointly hosted academic conferences and workshops (including travel and living expenses for participants from the partner institutes, as well as expenses for organizing and realizing the events); and funds for the purchase of scientific equipment and additional material (for the foreign institutes).
Requirements: Young scientists and scholars who are involved in the co-operation should preferably have completed a PhD or should at least be doctoral candidates and must be under 35 years of age.
Restrictions: Funds up to a maximum of 50,000 Euro. Funds for material costs (scientific equipment and additional material) are limited to a maximum of 50% of the total funds requested. The purchase price of scientific equipment may not exceed a total amount of 20,000 Euro.
Amount of Grant: 55,000 EUR maximum
Contact: Program Coordinator; (+49) 0228-833-0; fax: (+49) 0228-833-199; email: info@avh.de
Internet: http://www.humboldt-foundation.de/web/31390.html
Sponsor: Alexander von Humboldt Foundation
Jean-Paul-Street 12
Bonn D-53173 Germany

Alexander von Humboldt Foundation International Climate Protection 579
Fellowships
The Alexander von Humboldt Foundation is granting up to twenty International Climate Protection Fellowships annually funded under the Federal Environment Ministry's (BMU) International Climate Protection Initiative. The fellowships target prospective leaders from non- European threshold and developing countries (see list of countries, on the Foundations website) who are engaged in the field of climate protection and resource conservation in academia, business and administration in their countries. The fellowship will enable the recipients to conduct a research-related project of their own choice with hosts in Germany whom they are free to choose themselves. The Alexander von Humboldt Foundation is running the programme, which is currently scheduled to last until 2012/2013, in cooperation with the Federation of German Industries (BDI), the Centre for International Postgraduate Studies in Environmental Management (CIPSEM) at TU Dresden, the German Academic Exchange Service (DAAD), the Deutsche Bundesstiftung Umwelt (DBU), the Deutsche Gesellschaft fur Technische Zusammenarbeit (GTZ) and the Renewables Academy AG (RENAC).
Requirements: Applicants must provide a clearly visible leadership potential either by experience in a first leadership position or be able to provide appropriate references. They must also have completed their first university degree (Bachelor, or comparable academic degree) with outstanding results and hold a further academic or professional qualification (Master,, PhD, LL.M., MBA etc.) or have extensive professional experience in a leadership role (at least 48 months at the time of application). Furthermore, they are expected to have gained initial practical experience through involvement in projects related to climate protection and resource conservation. In addition to applicants who have been trained in the natural and engineering sciences, candidates who have been engaged in legal, economic and societal issues relating to climate change are encouraged to apply for this programme. Their first degree - comparable to a Bachelor, - may not have been completed more than 12 years prior to the start of the fellowship (1 September 2010).
Amount of Grant: 2,150 - 2,750 EUR per month
Date(s) Application is Due: Jan 15
Contact: Program Coordinator; (+49) 0228-833-0; fax: (+49) 0228-833-199; email: info@avh.de
Internet: http://www.humboldt-foundation.de/web/ICF.html
Sponsor: Alexander von Humboldt Foundation
Jean-Paul-Street 12
Bonn D-53173 Germany

Alexander von Humboldt Foundation JSPS Research Fellowships 580
The Fellowship assists promising, highly-qualified, young foreign researchers, who hold a doctoral degree. This program aims at providing opportunities for such researchers to pursue collaborative research under the leadership of Japanese host researchers in Japanese universities and research institutes, thereby allowing them to advance their own research while promoting scientific advancement in Japan and the counterpart countries through close collaboration in scientific activities. All fields of humanities, social sciences, and natural sciences are eligible. Fellows must arrive in Japan during the period of April 1 through November 30.

Requirements: Each candidate for the Fellowship must: possess either the nationality, permanent resident or citizenship of the nominating country; hold a doctorate degree when the Fellowship goes into effect, which must have been received within six years prior to April 2,; and have arranged in advance a research plan with his/her Japanese host.
Contact: Program Coordinator; (+49) 0228-833-0; fax: (+49) 0228-833-199; email: info@avh.de
Internet: http://www.humboldt-foundation.de/web/3586.html
Sponsor: Alexander von Humboldt Foundation
Jean-Paul-Street 12
Bonn D-53173 Germany

Alexander von Humboldt Foundation Konrad Adenauer Research Award   581
This award, established and administered through the Alexander von Humboldt Foundation, is made annually to one Canadian scholar in the humanities or social sciences. The aim of the award is to promote academic relations between Canada and Germany. The award is given to a highly qualified Canadian scholar whose research work in the humanities or social sciences has brought international recognition and who belongs to a group of leading scholars in the respective area of specialization. Awardees will carry out research work at German research institutes for up to one year. Award winners are included in the Foundation's sponsorship program and will be invited to take part in all functions arranged by the foundation. The Award includes travel for awardees and family members.
Requirements: Candidates are nominated by their universities for this award. Nomination should include a letter from the candidate's institution regarding the candidate's academic qualifications; a brief statement of the candidate's research proposal; three references who can provide information on the candidate's academic qualifications; and the names of German scholars who would provide guidance to the scholar during his/her stay in Germany.
Restrictions: Self-applications will not be accepted. Applications will not be considered if the candidate holds or has held another award from the Humboldt Foundation within the previous three years.
Amount of Grant: EUR 60,000
Date(s) Application is Due: Jan 31
Contact: Program Coordinator; (+49) 0228-833-0; fax: (+49) 0228-833-199; email: info@avh.de
Internet: http://www.humboldt-foundation.de/web/31764.html
Sponsor: Alexander von Humboldt Foundation
Jean-Paul-Street 12
Bonn D-53173 Germany

Alexander von Humboldt Foundation Max Planck Research Award   582
The Award is conferred upon exceptionally highly-qualified German and foreign scientists. The researchers are expected to have already achieved international recognition and to continue to produce outstanding academic results in international collaboration - not least with the assistance of this award. On an annually-alternating basis, the call for nominations addresses areas within the natural and engineering sciences, the life sciences, and the humanities. Recipients of the Award will be granted funding of up to 750,000 Euros over a period of three to a maximum of five years in order to pursue research of their own choosing.
Amount of Grant: 750,000 Euros over three to five years
Date(s) Application is Due: Oct 26
Contact: Program Coordinator; (+49) 0228-833-0; fax: (+49) 0228-833-199; email: leilani.orate@avh.de
Internet: http://www.humboldt-foundation.de/web/6674.html
Sponsor: Alexander von Humboldt Foundation
Jean-Paul-Street 12
Bonn D-53173 Germany

Alexander von Humboldt Foundation NSC Taiwan Fellowships   583
The Fellowship assists promising, highly-qualified, young foreign researchers, who have held a doctoral degree for less than twelve years. This program aims at providing opportunities for such researchers to pursue collaborative research under the leadership of Taiwanese host researchers in Taiwan universities and research institutes, thereby allowing them to advance their own research while promoting scientific advancement in Taiwan and the counterpart countries through close collaboration in scientific activities. All fields of humanities, social sciences, and natural sciences are eligible.
Requirements: Applicants must be a post-doctoral researcher from Germany and only completed a doctorate in the last twelve years
Contact: Program Coordinator; (+49) 0228-833-0; fax: (+49) 0228-833-199; email: info@avh.de
Internet: http://www.humboldt-foundation.de/web/3626.html
Sponsor: Alexander von Humboldt Foundation
Jean-Paul-Street 12
Bonn D-53173 Germany

Alexander von Humboldt Foundation Philipp Franz von Siebold Award   584
for Japanese Researchers
The Award is given annually to a Japanese academic for outstanding services to enhancing mutual understanding of culture and society in Germany and Japan. The award-winner will be invited to spend up to a year researching in Germany. The prize is valued at 50,000 EUR. Apart from the presidents of Japanese universities, the right of nomination lies with former Siebold award winners, the heads of the Goethe Institutes in Japan and of the

German Institute for Japanese Studies, as well as the German Ambassador in Tokyo and the German Consul General in Osaka. Humboldt fellows and award-winners may make nominations via the Vice Chancellor of a Japanese university. A decision on the award is made by a special committee under the chairmanship of the German Ambassador in Japan. Self-nominations are not accepted.
Requirements: The scientist or scholar should be under 50 years of age and have a working knowledge of German.
Amount of Grant: 50,000 EURO
Contact: Program Administrator; 03-3582-5962; fax: 03-3582-5554; email: daad-tokyo@daadjp.com
Internet: http://www.humboldt-foundation.de/web/31771.html
Sponsor: Alexander von Humboldt Foundation
Akasaka 7-5-56, Minato-ku
Tokyo 107-0052 Japan

Alexander von Humboldt Foundation Professorships   585
Academics of all disciplines from abroad, who are internationally recognized as leaders in their field and who are expected to contribute to enhancing Germany's sustained international competitiveness as a research location in consequence of the award, are eligible to be nominated. The Professorship, which is being financed by the Federal Ministry of Education and Research through the International Research Fund for Germany, enables award winners to carry out long-term and ground-breaking research at universities and research institutions in Germany. The prize money, totaling 5 million EUR for academics in experimental disciplines and 3.5 million EUR for researchers in theoretical disciplines, is being made available for a period of five years. Nominations may be made by German universities; non-university research institutions may also submit nominations jointly with a German university.
Requirements: The nominee must be: recognized internationally as an outstandingly qualified academic; be established as an academic abroad; and be nominated by an institution entitled to make nominations.
Contact: Program Contact; (+49) 0228-833-0; fax: (+49) 0228-833-199; email: avh-professur@avh.de
Internet: http://www.humboldt-foundation.de/en/programme/preise/ahp.htm
Sponsor: Alexander von Humboldt Foundation
Jean-Paul-Street 12
Bonn D-53173 Germany

Alexander von Humboldt Foundation Reimar Lust Award   586
The Award recognizes the contribution of outstanding humanities scholars and social scientists from abroad to promoting cultural and academic relations between Germany and their own countries. Nominations for the award may be made on behalf of highly respected academics who, as multipliers in and through the field of academic studies, have made an exceptional contribution to the enduring promotion of bilateral relations between Germany and their own countries. These services include, for example, the transfer of traditions of and approaches to academic studies from Germany to other countries and from other countries to Germany, translation work, or the promotion of institutional cooperation. In order, furthermore, to strengthen and extend bilateral cooperative relations to their partners and other specialist colleagues in Germany on a sustainable basis, the award winners are invited to spend several months in Germany (total duration approximately 6 months to one year, may be divided into segments). The Award is valued at 50,000 Euro, and deadline for submission is October 31.
Requirements: Academics who have already been granted an award by the Alexander von Humboldt Foundation are not eligible to be nominated for the Award.
Amount of Grant: 50,000 EURO
Date(s) Application is Due: Oct 31
Contact: Program Coordinator; (+49) 0228-833-0; fax: (+49) 0228-833-199; email: info@avh.de
Internet: http://www.humboldt-foundation.de/web/6806.html
Sponsor: Alexander von Humboldt Foundation
Jean-Paul-Street 12
Bonn D-53173 Germany

Alexander von Humboldt Foundation Research Fellowships for   587
Experienced Researchers
These Fellowships for experienced researchers are the instrument with which the Foundation enables highly-qualified scientists and scholars from abroad, who completed their doctorates less than twelve years ago to spend extended periods of research (6-18 months; may be divided up into a maximum of three blocks) in Germany. Candidates are expected to have their own, clearly defined research profile. This means they should usually be working at least at the level of Assistant Professor or Junior Research Group Leader or be able to document independent research work over a number of years. Scientists and scholars from all disciplines and countries may apply. The fellowship is worth 2,450 EUR per month. This includes a mobility lump sum and a contribution towards health and liability insurance.
Requirements: Application requirements include: a doctorate or comparable academic degree (Ph.D., C.Sc. or equivalent), completed less than twelve years prior to the date of application; the candidate's own research profile documented by a comprehensive list of academic publications reviewed according to international standards and printed in journals and/or by publishing houses; and a good knowledge of German if it is necessary to carry out the project successfully (otherwise a good knowledge of English; scientists and engineers must have a good knowledge of German or English).
Restrictions: Short-term visits for study and training purposes or for attending conferences are not eligible for sponsorship.

Amount of Grant: 2,450 EUR per month
Contact: Program Coordinator; (+49) 0228-833-0; fax: (+49) 0228-833-199; email: info@avh.de
Internet: http://www.humboldt-foundation.de/web/humboldt-research-fellowship-experienced.html
Sponsor: Alexander von Humboldt Foundation
Jean-Paul-Street 12
Bonn D-53173 Germany

Alexander von Humboldt Foundation Research Fellowships for   588
Postdoctoral Researchers

These Fellowships for postdoctoral researchers are the instrument with which the Foundation enables highly-qualified scientists and scholars from abroad who are just embarking on their academic careers and who completed their doctorates less than four years ago to spend extended periods of research (6-24 months) in Germany. Scientists and scholars from all disciplines and countries may apply. Research projects are carried out in cooperation with academic hosts at research institutions in Germany. Candidates choose their own research projects and their host in Germany and prepare their own research plan. Details of the research project and the time schedule must be agreed upon with the prospective host in advance. Candidates are selected solely on the basis of their academic record. There are no quotas for individual disciplines or countries. The fellowship is worth 2,250 EUR per month. This includes a mobility lump sum and a contribution towards health and liability insurance.
Requirements: Application requirements include: a doctorate or comparable academic degree (Ph.D., C.Sc. or equivalent), completed less than twelve years prior to the date of application; the candidate's own research profile documented by a comprehensive list of academic publications reviewed according to international standards and printed in journals and/or by publishing houses; and a good knowledge of German if it is necessary to carry out the project successfully (otherwise a good knowledge of English; scientists and engineers must have a good knowledge of German or English).
Restrictions: Short-term visits for study and training purposes or for attending conferences are not eligible for sponsorship.
Amount of Grant: 2,250 EUR per month
Contact: Program Coordinator; (+49) 0228-833-0; fax: (+49) 0228-833-199; email: info@avh.de
Internet: http://www.humboldt-foundation.de/web/humboldt-research-fellowship-postdoc.html
Sponsor: Alexander von Humboldt Foundation
Jean-Paul-Street 12
Bonn D-53173 Germany

Alexander von Humboldt Foundation Sofja Kovalevskaja Awards   589

The Award is granted in recognition of the outstanding academic achievements of exceptionally promising junior researchers from abroad and enables them to establish their own groups of junior researchers at research establishments in Germany. Scientists and scholars from abroad, whose research records to date have already qualified them to be recognized as top-level junior researchers and who are expected to continue producing outstanding results as recipients of the Award, are eligible to submit applications. Virtually unaffected by administrative constraints, the award winners are able to concentrate on high-level, innovative research work of their own choice in Germany and thus strengthen the internationalization of research in Germany. The award money allows the winners to finance their own working group at a university or non-university research establishment of their own choice in Germany and also covers their living expenses.
Requirements: The program is open to scientists and scholars from all countries and disciplines who completed their doctorates with distinction less than six years ago and have published work in prestigious international journals or publishing houses. Foreign academics staying in Germany are eligible to apply if they have not been working in an academic capacity in Germany for more than two years at the time of application. German academics are eligible to apply if they have been working in an academic capacity abroad for at least five years at the time of application. The Foundation particularly welcomes applications from qualified, female junior researchers.
Amount of Grant: 1,650,000 Euro
Date(s) Application is Due: Jan 4
Contact: Program Coordinator; (+49) 0228-833-0; fax: (+49) 0228-833-199; email: info@avh.de
Internet: http://www.humboldt-foundation.de/web/7360.html
Sponsor: Alexander von Humboldt Foundation
Jean-Paul-Street 12
Bonn D-53173 Germany

Alexander von Humboldt Foundation TransCoop Program Grants   590

The program provides funds for cooperative research between German, U.S. and/or Canadian scholars in the fields of the humanities, social sciences, law, and economics. The maximum duration of sponsorship is three years. Researchers working at universities or other research institutions in Germany, the U.S.A, and Canada may apply for funding. Scientific questions, including any from engineering or life sciences, may be considered providing the subject of the proposed research is convincingly related to the humanities and social sciences. In the selection procedure, priority is given to new research cooperation. The funds can be granted for a maximum of three years and may be used for: short term research stays at the partner's institute for up to three months per year; travel expenses; conference organization (maximum 20% of total amount); material and equipment; printing costs; and staff costs for research assistants working on the cooperative project (maximum 20% of total amount).

Requirements: Applications should be submitted jointly by at least one German and one U.S. or Canadian scholar. The PhD degree is required.
Amount of Grant: 55,000 Euro maximum
Date(s) Application is Due: Apr 30; Oct 31
Contact: Program Coordinator; (+49) 0228-833-0; fax: (+49) 0228-833-199; email: info@avh.de
Internet: http://www.humboldt-foundation.de/web/8175.html
Sponsor: Alexander von Humboldt Foundation
Jean-Paul-Street 12
Bonn D-53173 Germany

Alf Heggoy Memorial Book Award   591

This award is given annually in recognition for the best book published during the previous year dealing with the French colonial experience from the 16th to the 20th century. Books from any academic discipline will be considered but they must approach the consideration of the French colonial experience from an historical perspective. Applicants or their publishers should submit three copies of books published (date of publication is determined by the copyright page of the book), one to each of the book prize committee members: Prof Sue Peabody, Heggoy Prize Committee Chair, Washington State University VMMC 202D, 14204 NE Salmon Creek Avenue, Vancouver, WA, U.S.A 98686, email: peabody@vancouver.wsu.edu; Prof Peter Moogk, Department of History, University of British Columbia, Buchanan Tower 1121, 1873 East Mall, Vancouver, BC V6T 1Z1, Canada; and Prof Eric Jennings, Department of History, University of Toronto, 100 Saint George Street, Toronto, Ontario M5S 3G3, Canada. Annual deadline dates may vary; contact program staff for exact dates.
Amount of Grant: $350
Date(s) Application is Due: Mar 1
Contact: Bill Shorrock, Chair, Alf Heggoy Book Prize Committee; email: w.shorrock@csuohio.edu
Internet: http://www.frenchcolonial.org/Heggoy.html
Sponsor: French Colonial Historical Society
Department of History, Cleveland State University, Euclid Ave/24th Street
Cleveland, OH 44115

Alfred Bader Award in BioInorganic or BioOrganic Chemistry   592

This award is given annually to recognize outstanding contributions to bioorganic or bioinorganic chemistry. The award will be granted for outstanding research accomplishments without regard to age or nationality. The award is intended to recognize significant accomplishments that are at the interface between biology and organic or inorganic chemistry. Special consideration will be given to applications of the fundamental principles and experimental methodology of chemistry to areas of biological significance. The award consists of $5,000 and a certificate.
Requirements: Any individual, except a member of the awards committee, may submit one nomination or seconding letter for the award in any given year. The nominating documents consist of a letter of not more than 1000 words containing an evaluation of the nominee's accomplishments and a specific identification of the work to be recognized, a biographical sketch including date of birth, and a list of publications and patents authored by the nominee.
Restrictions: Self-nominations are not accepted.
Samples: Kevin M. Smith—award recipient, $5,000 (2009); Lawrence Que, Jr.—award recipient, $5,000 (2008); Eckard Munck—award recipient, $5,000 (2007).
Amount of Grant: $5,000 and reimbursement of travel expenses to award meeting
Date(s) Application is Due: Nov 1
Contact: Felicia Dixon, Awards Administrator; (800) 227-5558 or (202) 872-4408; fax: (202) 776-8008; email: f_dixon@acs.org or awards@acs.org
Internet: http://portal.acs.org/portal/acs/corg/content?_nfpb=true&_pageLabel=PP_ARTICLEMAIN&node_id=1319&content_id=CTP_004490&use_sec=true&sec_url_var=region1
Sponsor: American Chemical Society
1155 Sixteenth Street, NW
Washington, DC 20036

Alfred Burger Award in Medicinal Chemistry   593

This award, supported by GlaxoSmithKline in 1978, is awarded biennially in even-numbered years to recognize outstanding contributions to research in medicinal chemistry. The award is granted without regard to age or nationality, and the recipient presents an award address at the spring meeting of the ACS Division of Medicinal Chemistry. Applications are accepted in odd-numbered years. The award consists of $3,000 and a certificate.
Requirements: Any individual, except a member of the award committee, may submit one nomination or seconding letter for the award in any given year. The nominating documents consist of a letter of not more than 1000 words containing an evaluation of the nominee's accomplishments and a specific identification of the work to be recognized, a biographical sketch including date of birth, and a list of publications and patents authored by the nominee.
Restrictions: Self-nominations are not accepted.
Samples: Magid Abou-Gharbia—award recipient, $3,000 (2008); Joel R. Huff—award recipient, $3,000 (2006); William Greenlee—award recipient, $3,000 (2004).
Amount of Grant: $3,000 and travel to award meeting
Date(s) Application is Due: Nov 1
Contact: Felicia Dixon, Awards Administrator; (800) 227-5558 or (202) 872-4408; fax: (202) 776-8008; email: f_dixon@acs.org or awards@acs.org

Internet: http://portal.acs.org/portal/acs/corg/content?_nfpb=true&_pageLabel=PP_ARTICLEMAIN&node_id=1319&content_id=CTP_004491&use_sec=true&sec_url_var=region1
Sponsor: American Chemical Society
1155 Sixteenth Street, NW
Washington, DC 20036

Alfred D. Bell, Travel Grants 594

The mission of the society is to advance understanding of the historical interaction of people with the forest environment. To this end, an average of six travel grants are awarded annually to researchers wishing to study at the society's library and archives. Research topics should be in the areas of forest or conservation history. Applications may be submitted at any time as grants are awarded year-round.
Requirements: Preference is given to graduate students, but grants also are awarded to postgraduates and professionals pursuing career advancement.
Amount of Grant: $950 maximum
Contact: Grants Administrator, (919) 682-9319; fax: (919) 682-2349
Internet: http://www.lib.duke.edu/forest/Research/bellgnt.html
Sponsor: Forest History Society
701 Vickers Avenue
Durham, NC 27701-3162

Alfred E. Driscoll Publication Prize 595

This prize is awarded in even-numbered years for an outstanding doctoral dissertation on any aspect of New Jersey history. Unsuccessful applicants may resubmit their dissertations for consideration for the prize. One award is given each year. To apply, submit one copy of the dissertation by deadline date. Use the prize nomination form in the guidelines booklet.
Restrictions: An applicant may not apply for a publication grant and the Driscoll Prize for the same work. Dissertations that have already been accepted for publication are ineligible.
Amount of Grant: $1000
Date(s) Application is Due: Jan 2
Contact: Mary Murrin, Director Grants Program, (609) 984-0954; email: mary.murrin@sos.state.nj.us
Internet: http://www.state.nj.us/state/history/grants_t.html
Sponsor: New Jersey Historical Commission
225 W State Street, 4th Fl, P.O. Box 305
Trenton, NJ 08625-0305

Alfred Hodder Fellowship 596

The fellowship was established for the promotion of independent work in the humanities and is awarded to humanists of exceptional talent. The award is designed to offer one year of independent study in residence at Princeton University. The appointment of the Hodder Fellow is made each February by the president of Princeton on the recommendation of the Committee on Humanistic Studies at which time an announcement is sent to all applicants.
Requirements: Preference is given to candidates outside of academia. Applicant must submit a resume, a sample of previous work (10-page maximum, nonreturnable), and a project proposal of two to three pages along with a self-addressed, stamped envelope.
Restrictions: Funds are not to be used to pursue a PhD degree.
Samples: Mary Jo Bang, poet—fellowship recipient, author of Apology for Want.; Lan Samantha Chang, fiction writer—fellowship recipient, author of Hunger: A Novella and Stories.; Sharona Ben-Tov, writer—fellowship recipient for a memoir about her father's work in the secret Israeli Science Corps.; Naomi Iizuka, playwright—fellowship recipient for a new play about what it means to call something genuine.
Amount of Grant: $55,000 approximately for academic year
Date(s) Application is Due: Nov 1
Contact: Hodder Fellowship, Council of the Humanities, (609) 258-4717; fax: (609) 258-2783; email: dsteidl@princeton.edu
Internet: http://www.princeton.edu/~humcounc/hodderfellows.shtml
Sponsor: Princeton University
122 E Pyne
Princeton, NJ 08544

Alfred I. DuPont Foundation Grants 597

Grants are awarded primarily to elderly adults requiring health, economic, or educational assistance. Support is also given for higher education and medical research. All grants to the elderly are made to individuals in the southeastern United States. General operating support and grants to individuals are awarded. An application form is required. Applications are accepted at any time and are dealt with promptly.
Requirements: Nonprofit organizations in the southeastern U.S. are eligible. Preference is given to those in Florida.
Geographic Focus: Florida
Samples: Gulf Coast Community College Fdn (Panama City, FL)—for operating support, $50,000.; Dreams Come True of Jacksonville (Jacksonville, FL)—for operating support, $10,000.
Amount of Grant: $1000-$35,000 average for general operating support; $1000-$2000 average for individuals
Contact: Rosemary Cusimano Wills, Secretary, (904) 232-4123
Sponsor: Alfred I. DuPont Foundation Inc
4600 Touchton Rd E, Building 200, Suite 120
Jacksonville, FL 32246

Alfred P. Sloan Foundation Business Organizations Grants 598

This program (formerly "Role of the Corporation") supports academic research and scholarship aimed at painting a realistic picture of how corporations and other business organizations function, with special emphasis on how the people in them actually behave, how they are motivated, and how they are rewarded. The Foundation has sought to increase understanding of these organizations because of the enormous effect they have on the standard of living and quality of life for most people in the United States and around the world. Grant requests can be made at any time for support of activities related to Foundation program areas and interests. The Foundation is generally limited to supporting tax-exempt organizations. The Foundation has no deadlines or standard forms. The Foundation accepts proposals sent by email. A brief letter of inquiry, rather than a fully developed proposal, is an advisable first step for an applicant, conserving his or her time and allowing for a preliminary response regarding the possibility of support.
Requirements: Concise, well-organized proposals are preferred. In no case should the body of the proposal exceed 20 double-spaced pages.
Restrictions: The Foundation's activities do not normally extend to religion, the creative or performing arts, elementary or secondary education, medical research or health care, the humanities or to activities outside the United States. Grants are not made for endowments or for buildings or equipment.
Contact: Gail M. Pesyna, Program Director; (212) 649-1649; fax: (212) 757-5117; email: pesyna@sloan.org
Internet: http://sloan.org/programs/stndrd_role.shtml
Sponsor: Alfred P. Sloan Foundation
630 Fifth Avenue, Suite 2550
New York, NY 10111

Alfred P. Sloan Foundation Census of Marine Life Research Grants 599

The goal of this project is to advance a major new international observational program to be completed by 2010 to assess and explain the diversity, distribution, and abundance of marine life. An international Scientific Steering Committee and Secretariat based at the Consortium for Oceanographic Research and Education in Washington DC now guide the program. Grant-making occurs in conjunction with the National Ocean Partnership Program. The Foundation seeks opportunities to work with the institutions and media that can build public interest and with maritime industries and environmentalists to assure their meaningful participation. Grant requests can be made at any time for support of activities related to Foundation program areas and interests. The Foundation is generally limited to supporting tax-exempt organizations. The Foundation has no deadlines or standard forms. The Foundation accepts proposals sent by email. A brief letter of inquiry, rather than a fully developed proposal, is an advisable first step for an applicant, conserving his or her time and allowing for a preliminary response regarding the possibility of support.
Requirements: Concise, well-organized proposals are preferred. In no case should the body of the proposal exceed 20 double-spaced pages.
Restrictions: The Foundation's activities do not normally extend to religion, the creative or performing arts, elementary or secondary education, medical research or health care, the humanities or to activities outside the United States. Grants are not made for endowments or for buildings or equipment.
Contact: Jesse H. Ausubel, Program Director; (212) 649-1649; fax: (212) 757-5117; email: ausubel@sloan.org
Internet: http://sloan.org/programs/scitech_supresearch.shtml
Sponsor: Alfred P. Sloan Foundation
630 Fifth Avenue, Suite 2550
New York, NY 10111

Alfred P. Sloan Foundation Civic Program Grants 600

The goal of the Program is to make a contribution to the Foundation's home area, New York City. There are two directions to the Program: to respond to special opportunities in New York City; and to fund high-leverage projects in New York City that are related to other parts of our program. Interested readers should refer to the descriptions of other program areas in the Foundation website. Grant requests can be made at any time for support of activities related to Foundation program areas and interests. The Foundation is generally limited to supporting tax-exempt organizations. The Foundation has no deadlines or standard forms. The Foundation accepts proposals sent by email. A brief letter of inquiry, rather than a fully developed proposal, is an advisable first step for an applicant, conserving his or her time and allowing for a preliminary response regarding the possibility of support.
Requirements: Concise, well-organized proposals are preferred. In no case should the body of the proposal exceed 20 double-spaced pages.
Restrictions: The Foundation's activities do not normally extend to religion, the creative or performing arts, elementary or secondary education, medical research or health care, the humanities or to activities outside the United States. Grants are not made for endowments or for buildings or equipment.
Geographic Focus: New York
Samples: City U of New York system (NY) and George Mason U (NY)—for the September 11 Digital Archives, $700,000 jointly.
Contact: Ted Greenwood, Program Director; (212) 649-1649; fax: (212) 757-5117; email: greenwood@sloan.org
Internet: http://sloan.org/programs/pg_civic.shtml
Sponsor: Alfred P. Sloan Foundation
630 Fifth Avenue, Suite 2550
New York, NY 10111

Alfred P. Sloan Foundation Education and Careers in Science and Technology Grants

601

Programs to strengthen education in science and technology, to increase interest in these fields, and to understand and communicate to others the nature of careers in these fields have long been supported by the Foundation. Increasingly important are opportunities presented by electronic technologies for learning outside the classroom. This program is divided into the following sections: anytime, anyplace learning; information about careers; professional science master's degrees; the science and engineering workforce; increasing Ph.D.s for underrepresented minorities; education for underrepresented groups; retention of students in higher education; and public understanding of science and technology. Grant requests can be made at any time for support of activities related to Foundation program areas and interests. The Foundation is generally limited to supporting tax-exempt organizations. The Foundation has no deadlines or standard forms. The Foundation accepts proposals sent by email. A brief letter of inquiry, rather than a fully developed proposal, is an advisable first step for an applicant, conserving his or her time and allowing for a preliminary response regarding the possibility of support.

Requirements: Concise, well-organized proposals are preferred. In no case should the body of the proposal exceed 20 double-spaced pages.

Restrictions: The Foundation's activities do not normally extend to religion, the creative or performing arts, elementary or secondary education, medical research or health care, the humanities or to activities outside the United States. Grants are not made for endowments or for buildings or equipment.

Samples: American Assoc for the Advancement of Science (Washington, DC)—to help establish a center that will provide consultant services to universities and colleges seeking to increase the participation of U.S. students, particularly women and minorities, in science and engineering careers, $400,000 over three years.; U of Illinois (Springfield, IL)—to develop and offer five new undergraduate-degree programs and one new master's degree program, using asynchronous-learning technology, $500,000.; Standford U (CA)—for an online professional master's degree program in bioinformatics, $120,154.

Contact: Jesse H. Ausubel, Program Director; (212) 649-1649; fax: (212) 757-5117; email: ausubel@sloan.org

Internet: http://sloan.org/programs/pg_education.shtml

Sponsor: Alfred P. Sloan Foundation
630 Fifth Avenue, Suite 2550
New York, NY 10111

Alfred P. Sloan Foundation Higher Education as an Industry Grants

602

The goal of this program is to produce understanding of how institutions of higher education actually work and how the set of institutions functions together as an industry. Grant requests can be made at any time for support of activities related to Foundation program areas and interests. The Foundation is generally limited to supporting tax-exempt organizations. The Foundation has no deadlines or standard forms. The Foundation accepts proposals sent by email. A brief letter of inquiry, rather than a fully developed proposal, is an advisable first step for an applicant, conserving his or her time and allowing for a preliminary response regarding the possibility of support.

Requirements: Concise, well-organized proposals are preferred. In no case should the body of the proposal exceed 20 double-spaced pages.

Restrictions: The Foundation's activities do not normally extend to religion, the creative or performing arts, elementary or secondary education, medical research or health care, the humanities or to activities outside the United States. Grants are not made for endowments or for buildings or equipment.

Contact: Jesse H. Ausubel, Program Director; (212) 649-1649; fax: (212) 757-5117; email: ausubel@sloan.org

Internet: http://sloan.org/programs/stndrd_universi.shtml

Sponsor: Alfred P. Sloan Foundation
630 Fifth Avenue, Suite 2550
New York, NY 10111

Alfred P. Sloan Foundation History of Science and Technology Grants

603

The goal of this program is to preserve the raw material of history by supporting archival projects now centered on Charles Darwin, Thomas A. Edison, and Kurt G'del, and via new projects based on the World Wide Web. Two specific areas of interest: the archival program; and recent history of science and engineering on the web. Grant requests can be made at any time for support of activities related to Foundation program areas and interests. The Foundation is generally limited to supporting tax-exempt organizations. The Foundation has no deadlines or standard forms. The Foundation accepts proposals sent by email. A brief letter of inquiry, rather than a fully developed proposal, is an advisable first step for an applicant, conserving his or her time and allowing for a preliminary response regarding the possibility of support.

Requirements: Concise, well-organized proposals are preferred. In no case should the body of the proposal exceed 20 double-spaced pages.

Restrictions: The Foundation's activities do not normally extend to religion, the creative or performing arts, elementary or secondary education, medical research or health care, the humanities or to activities outside the United States. Grants are not made for endowments or for buildings or equipment.

Contact: Doron Weber, Program Director; (212) 649-1649; fax: (212) 757-5117; email: weber@sloan.org; Jesse H. Ausubel, Program Director; (212) 649-1649; fax: (212) 757-5117; email: ausubel@sloan.org

Internet: http://sloan.org/programs/scitech_historysci.shtml

Sponsor: Alfred P. Sloan Foundation
630 Fifth Avenue, Suite 2550
New York, NY 10111

Alfred P. Sloan Foundation Indoor Environment Research Grants

604

The goal of this program is to understand the human indoor environment at the microbial level. The Foundation believes that this understanding may ultimately help to make this artificial environment more hospitable to human life or resistant to biological attacks. Their plan has three parts. First, the Foundation wants to establish a basic understanding of the indoor microbial environment. Second, it wants to help develop tools for understanding and probing the indoor environment. And lastly, it wants to apply the knowledge and tools to various human indoor environments, such as hospitals. Grant requests can be made at any time for support of activities related to Foundation program areas and interests. The Foundation is generally limited to supporting tax-exempt organizations. The Foundation has no deadlines or standard forms. The Foundation accepts proposals sent by email. A brief letter of inquiry, rather than a fully developed proposal, is an advisable first step for an applicant, conserving his or her time and allowing for a preliminary response regarding the possibility of support.

Requirements: Concise, well-organized proposals are preferred. In no case should the body of the proposal exceed 20 double-spaced pages.

Restrictions: The Foundation's activities do not normally extend to religion, the creative or performing arts, elementary or secondary education, medical research or health care, the humanities or to activities outside the United States. Grants are not made for endowments or for buildings or equipment.

Samples: Marine Biological Laboratory, Woods Hole, MA—to develop a novel combinatorial technology for imaging and distinguishing between members of a microbial community (2007); J. Craig Venter Institute, in collaboration with Lawrence Berkeley National Laboratory—to convene a workshop on Genomic Aerobiology (2006).

Contact: Paula J. Olsiewski, Program Director; (212) 649-1649; fax: (212) 757-5117; email: olsiewski@sloan.org

Internet: http://sloan.org/programs/scitech_supresearch.shtml

Sponsor: Alfred P. Sloan Foundation
630 Fifth Avenue, Suite 2550
New York, NY 10111

Alfred P. Sloan Foundation Industry Studies Fellowships

605

These Fellowships are intended to recognize and support junior faculty members in a wide range of academic disciplines. Awards are made to scholars who show the most outstanding promise of making important contributions to understanding the complex systems of companies, product and labor markets, institutions and their interactions that shape the multifaceted environment of modern industrial enterprises. Fellowships will be awarded to up to five (5) junior faculty members who are conducting such research on a topic important to a specific industry. At its discretion, the Foundation may choose to award more than five fellowships in a given year. Fellowships are awarded for a two-year period, with possible extension for another two years.

Requirements: Candidates must hold a PhD or equivalent in chemistry, physics, mathematics, computer science, economics, neuroscience or computational and evolutionary molecular biology, or in a related interdisciplinary field, and must also be a faculty member at a U.S. or Canadian university. Candidates must be no more than six years from the completion of the most recent PhD or equivalent as of the year of their nomination.

Restrictions: Direct applications are not accepted. Candidates must be nominated by department heads or other senior scholars.

Amount of Grant: $45,000 for two years

Date(s) Application is Due: Oct 15

Contact: Gail M. Pesyna, Program Director; (212) 649-1649; fax: (212) 757-5117; email: pesyna@sloan.org

Internet: http://sloan.org/programs/fellow_announ.shtml

Sponsor: Alfred P. Sloan Foundation
630 Fifth Avenue, Suite 2550
New York, NY 10111

Alfred P. Sloan Foundation Industry Studies Grants

606

The primary missions of this Program are to encourage research cooperation between academics and industry, and to support the integration of observation-based research with appropriate theory and analysis among a growing community of industry studies scholars. Grant requests can be made at any time for support of activities related to Foundation program areas and interests. The Foundation is generally limited to supporting tax-exempt organizations. The Foundation has no deadlines or standard forms. The Foundation accepts proposals sent by email. A brief letter of inquiry, rather than a fully developed proposal, is an advisable first step for an applicant, conserving his or her time and allowing for a preliminary response regarding the possibility of support.

Requirements: Concise, well-organized proposals are preferred. In no case should the body of the proposal exceed 20 double-spaced pages.

Restrictions: The Foundation's activities do not normally extend to religion, the creative or performing arts, elementary or secondary education, medical research or health care, the humanities or to activities outside the United States. Grants are not made for endowments or for buildings or equipment.

Samples: Boston College (Chestnut Hill, MA)—to produce a series of international briefs on labor-market incentives for older workers in industrial nations, $45,000.; Georgetown U (Washington, DC)—to analyze the structural and legal obstacles to a better alignment between the demands of the workplace and the needs of the work force, $45,000.; U of Northern Iowa (Cedar Falls, IA)—to study work-family issues among nonprofessional employees, $44,000.; Wayne State U (Detroit, MI)—for activities of the Trucking Industry Program related to trucking security, $45,000.

Contact: Gail Pesyna, Program Director; (212) 649-1649; fax: (212) 757-5117; email: pesyna@sloan.org
Internet: http://sloan.org/programs/IndustryStudies.shtml
Sponsor: Alfred P. Sloan Foundation
630 Fifth Avenue, Suite 2550
New York, NY 10111

Alfred P. Sloan Foundation Known, Unknown, and Unknowable Grants 607

The goal of this program is the exploration of what is known, unknown, and unknowable in a variety of fields. It is very valuable to know what you do not know and why. Research has been funded on limits to knowledge in a broad spectrum of academic areas. Grants have supported such studies in plant molecular biology and genetics, ecology, computational economics, history of science, and prehistoric linguistics. The Foundation would like to explore limits to knowledge in fields with obvious practical implications, such as health and finance, where it is important for knowledge consumers, such as regulators or investors, to know what can or cannot be known about therapies or market movements. Grant requests can be made at any time for support of activities related to Foundation program areas and interests. The Foundation is generally limited to supporting tax-exempt organizations. The Foundation has no deadlines or standard forms. The Foundation accepts proposals sent by email. A brief letter of inquiry, rather than a fully developed proposal, is an advisable first step for an applicant, conserving his or her time and allowing for a preliminary response regarding the possibility of support.
Requirements: Concise, well-organized proposals are preferred. In no case should the body of the proposal exceed 20 double-spaced pages.
Restrictions: The Foundation's activities do not normally extend to religion, the creative or performing arts, elementary or secondary education, medical research or health care, the humanities or to activities outside the United States. Grants are not made for endowments or for buildings or equipment.
Samples: Columbia University—for a Workshop on the Known, the Unknown, and the Unknowable (2007).
Contact: Jesse H. Ausubel; (212) 649-1649; fax: (212) 757-5117; email: ausubel@sloan.org
Internet: http://sloan.org/programs/scitech_supresearch.shtml
Sponsor: Alfred P. Sloan Foundation
630 Fifth Avenue, Suite 2550
New York, NY 10111

Alfred P. Sloan Foundation Making Municipal Governments More Responsive to Their Citizens Grants 608

The program to make municipal governments more responsive to their citizens has two components: promoting performance measurement and reporting; and enabling direct service requests. Grant requests can be made at any time for support of activities related to Foundation program areas and interests. The Foundation is generally limited to supporting tax-exempt organizations. The Foundation has no deadlines or standard forms. The Foundation accepts proposals sent by email. A brief letter of inquiry, rather than a fully developed proposal, is an advisable first step for an applicant, conserving his or her time and allowing for a preliminary response regarding the possibility of support.
Requirements: Concise, well-organized proposals are preferred. In no case should the body of the proposal exceed 20 double-spaced pages.
Restrictions: The Foundation's activities do not normally extend to religion, the creative or performing arts, elementary or secondary education, medical research or health care, the humanities or to activities outside the United States. Grants are not made for endowments or for buildings or equipment.
Samples: National Center for Civic Innovation, New York, NY—to continue implementing citizen-based municipal government performance measurement and reporting around the country (2006).
Contact: Ted Greenwood, Program Director; (212) 649-1649; fax: (212) 757-5117; email: greenwood@sloan.org
Internet: http://sloan.org/programs/stndrd_performance.shtml
Sponsor: Alfred P. Sloan Foundation
630 Fifth Avenue, Suite 2550
New York, NY 10111

Alfred P. Sloan Foundation Research Fellowships 609

Fellowships stimulate fundamental research by young scholars of outstanding promise at a time in their careers when their creative abilities are especially high and when government or other support is difficult to obtain. Fellows are free to pursue whatever lines of inquiry that are of the most compelling interest to them. Funds are awarded directly to the fellow's institution and may be used by the fellow for such purposes as equipment, technical assistance, professional travel, trainee support, or activities directly related to the fellow's research. The foundation welcomes nominations of all candidates who meet the traditional high standards of this program and strongly encourages the participation of women and members of underrepresented minority groups. Awards are made in the fields of physics, chemistry, neuroscience, economics, pure mathematics, applied mathematics, computer science, and computational and evolutionary molecular biology. Fellowships are awarded for a two-year period, with possible extension for another two years. Each year 118 fellows are selected from the nominations received. Nominations are due September 15 for awards to begin the following September. Nomination forms are available upon request.
Requirements: Candidates must hold a PhD or equivalent in chemistry, physics, mathematics, computer science, economics, neuroscience or computational and evolutionary molecular biology, or in a related interdisciplinary field, and must also be

a faculty member at a U.S. or Canadian university. Candidates must be no more than six years from the completion of the most recent PhD or equivalent as of the year of their nomination.
Restrictions: Direct applications are not accepted. Candidates must be nominated by department heads or other senior scholars.
Amount of Grant: $45,000 for two years
Date(s) Application is Due: Sep 15
Contact: Michael Teitelbaum, Program Director; (212) 649-1649; fax: (212) 757-5117; email: teitelbaum@sloan.org
Internet: http://sloan.org/programs/scitech_fellowships.shtml
Sponsor: Alfred P. Sloan Foundation
630 Fifth Avenue, Suite 2550
New York, NY 10111

Alfred P. Sloan Foundation Selected National Issues Grants 610

The foundation attempts to contribute to studies of major issues of our time in a way appropriate to its expertise and size. A special approach to the study and understanding of broadly recognized problems is a requirement for foundation support. Grant requests can be made at any time for support of activities related to Foundation program areas and interests. The Foundation is generally limited to supporting tax-exempt organizations. The Foundation has no deadlines or standard forms. The Foundation accepts proposals sent by email. A brief letter of inquiry, rather than a fully developed proposal, is an advisable first step for an applicant, conserving his or her time and allowing for a preliminary response regarding the possibility of support.
Requirements: Concise, well-organized proposals are preferred. In no case should the body of the proposal exceed 20 double-spaced pages.
Restrictions: The Foundation's activities do not normally extend to religion, the creative or performing arts, elementary or secondary education, medical research or health care, the humanities or to activities outside the United States. Grants are not made for endowments or for buildings or equipment.
Samples: City U of New York, American Social History Project/Ctr for Media and Learning (New York, NY) and George Mason U, Ctr for History and New Media (Fairfax, VA)—for the September 11 Digital Archives, an electronic repository for emails, digital images, online diaries, and other electronic media generated in response to the September 11 terrorist attacks on the World Trade Center and the Pentagon, $700,000 jointly.; New York State Office of the State Comptroller (Albany, NY)—to help the Office of the State Comptroller introduce citizen-based performance assessments in local governments in New York State, $45,000.; Connecticut Public Expenditure Foundation (Hartfod, CT)—for planning and a pilot study on an interactive Web site that would solicit citizen feedback regarding local government services, $45,000.
Contact: Ted Greenwood, Program Director; (212) 649-1649; fax: (212) 757-5117; email: greenwood@sloan.org
Internet: http://sloan.org/programs/pg_national.shtml
Sponsor: Alfred P. Sloan Foundation
630 Fifth Avenue, Suite 2550
New York, NY 10111

Alfred P. Sloan Foundation Workplace, Work Force & Working Families Grants 611

The Foundation plays a vital role in developing work-family scholarship and supporting effective workplaces that meet the needs of working parents and older workers. The foundation awards grants for research into conflicts between job and home responsibilities. It will support studies on balancing parental responsibilities, as well as improving access to child care and elderly care. Grant requests can be made at any time for support of activities related to Foundation program areas and interests. The Foundation is generally limited to supporting tax-exempt organizations. The Foundation has no deadlines or standard forms. The Foundation accepts proposals sent by email. A brief letter of inquiry, rather than a fully developed proposal, is an advisable first step for an applicant, conserving his or her time and allowing for a preliminary response regarding the possibility of support.
Requirements: Concise, well-organized proposals are preferred. In no case should the body of the proposal exceed 20 double-spaced pages.
Restrictions: The Foundation's activities do not normally extend to religion, the creative or performing arts, elementary or secondary education, medical research or health care, the humanities or to activities outside the United States. Grants are not made for endowments or for buildings or equipment.
Samples: Cornell U (New York)—for the Sloan Employment and Family Careers Institute, $652,903.; ETV Endowment of South Carolina (Spartanburg, SC)—to compile, produce, and broadcast a two-hour public-television documentary on work and family issues, to be accompanied by a concerted outreach campaign, $2.2 million.; Boston College (MA)—for research on families, $399,922.; U of Michigan at Ann Arbor (MI)—for fellowship support to Margaret Wooldridge, and to help the Department of Mechanical Engineering and Applied Mechanics develop ways to deal with work-family issues, $25,000.
Contact: Kathleen Christensen, Program Director; (212) 649-1649; fax: (212) 757-5117; email: christensen@sloan.org
Internet: http://sloan.org/programs/stndrd_dualcareer.shtml
Sponsor: Alfred P. Sloan Foundation
630 Fifth Avenue, Suite 2550
New York, NY 10111

Alfred W. Bressler Prize in Vision Science 612

The Prize recognizes a professional in the field of vision science whose leadership, research and service have resulted in important advancements in the treatment of eye disease or rehabilitation of persons with vision loss. A panel of distinguished vision science professionals will select the winner who receives a prize of $40,000.

Requirements: The application process is open to established professionals in the field of vision science whose contributions have advanced vision care, the treatment of eye disease, or the rehabilitation of persons with visual disabilities or blindness and whose further work is expected to contribute significantly. Candidates from the United States and countries around the world are eligible for the award.

Samples: Roy W. Beck, MD, PHD, University of South Florida, Tampa, FL— for his work in the field of Epidimiology (2009); Jonathan C. Horton, MD, PhD, University of California, San Francisco— for his work in the field of Neuro - Ophthalmology (2008).

Amount of Grant: $40,000
Date(s) Application is Due: Dec 31
Contact: Program Administrator, (212) 769-7801; email: bressler@jgb.org
Internet: http://www.jgb.org/programs_bressler.asp
Sponsor: Jewish Guild for the Blind
15 West 65th Street
New York, NY 10023

Alice Fisher Society Fellowships   613

The Barbara Bates Center for the Study of the History of Nursing offers a fellowship of $2,500 to support two weeks in residence at the Center and ongoing collaboration with Center historians. Selection of Alice Fisher Society scholars will be based on evidence of interest in and aptitude for historical research related to nursing. It is expected that the research and new materials produced by Alice Fisher Society scholars will help ensure the growth of scholarly work focused on the history of nursing. Fisher scholars will participate in Center activities and will present their research at a Center seminar.

Requirements: The scholarships are open to individuals with master's and doctoral level preparation. The application should be sent via email to either Patricia D'Antonio, PhD, RN, FAAN (dantonio@nursing.upenn.edu) or Barbra Mann Wall, PhD, RN (wallbm@nursing.upenn.edu).

Amount of Grant: $2,500
Date(s) Application is Due: Dec 31
Contact: Dr. Karen Buhler-Wilkerson, Director; (215) 898-4725; fax: (215) 573-2168; email: karenwil@nursing.upenn.edu
Internet: http://www.nursing.upenn.edu/history/Pages/Alice_Fisher_Fellowship.aspx
Sponsor: University of Pennsylvania
School of Nursing, 418 Curie Boulevard
Philadelphia, PA 19104-6020

Alice Tweed Tuohy Foundation Grants Program   614

The foundation promotes organizations that promote young people; that provide outstanding opportunities for performance, growth, and creativity; that nurture personal integrity and ambition; and that reward high achievement. The foundation assists organizations offering services to children whose choices might otherwise be unfairly restricted by need; supported are activities both academic and extracurricular that challenge young people while encouraging the growth of responsibility and personal integrity. Organizations dedicated to improving the quality of life by meeting the vital needs of the community are also supported. Types of support include building construction/ renovation, scholarship funds, and matching funds. Financing priority is accorded those organizations with the least in-house capacity to raise capital, assisting these groups to surmount critical monetary obstacles and continue productive service to the community. Proposals may be submitted annually between July 1 and September 15. However, the foundation has announced that a three-year partial moratorium period on awarding grants will commence July 1, 2009.

Requirements: Applications are considered only from Santa Barbara public, tax-exempt organizations. Priority consideration is given to applications from organizations serving young people, education, health and medicine, community affairs, and the arts.

Restrictions: Excluded from consideration are applications for the benefit of specific individuals, organizations outside the Santa Barbara area, organizations in overpopulated nonprofit areas, national campaigns, fund-raising normally carried out by the organization, operating expenses, or budgetary support.

Geographic Focus: California
Samples: Scholarship Foundation of Santa Barbara (Santa Barbara, CA)—for general support, $97,125 (2004); Mental Health Association of Santa Barbara (Santa Barbara, CA)—for capital campaign, $25,000 (2004).
Amount of Grant: $750-$98,000 range
Date(s) Application is Due: Sep 15
Contact: Program Contact; (805) 962-6430; fax: (805) 962-7135; email: atuohyfdn@aol.com
Sponsor: Alice Tweed Tuohy Foundation
P.O. Box 1328, 205 E. Carrillo Street, Room 219
Santa Barbara, CA 93102-1328

ALISE Doctoral Students' Dissertation Competition Awards   615

Awards provide an opportunity for the exchange of research ideas between established researchers and doctoral students who have recently graduated in the field of education for library and information science. Doctoral students who have recently graduated, or who are about to finish their dissertations, are invited to submit papers summarizing their dissertation research in areas dealing with substantive issues in library and information science. Guidelines are available upon request.

Amount of Grant: $400
Date(s) Application is Due: Sep 22

Contact: Deborah York, Executive Director, (865) 425-0155; fax: (865) 481-0390; email: dyork@infointl.com
Internet: http://www.alise.org/awards/doctoralstudents.html
Sponsor: ALISE National Office
P.O. Box 4219, 1009 Commerce Park Drive, Suite 150
Oak Ridge, TN 37839

ALISE Research Grants   616

ALISE accepts proposals for grants to support research broadly related to education for library and information science. Proposals should include an abstract of the project; a problem statement and literature review; project objectives and description; research design, methodology, and analysis techniques; detailed budget; expected benefits and impact from the research; and vita of project investigators.

Requirements: Applicants must be personal members of ALISE.
Restrictions: Grants will not be given to support doctoral dissertations.
Samples: Darlene Weingand and Rebecca Watson-Boone, U of Wisconsin-Madison (WI)—for research entitled Dimensions of Effectiveness of Schools of Library and Information Studies, $2500.
Amount of Grant: $5000
Date(s) Application is Due: Oct 3
Contact: Program Contact, (865) 425-0155; fax: (865) 481-0390; email: contact@alise.org
Internet: http://www.alise.org/awards/researchgrants.html
Sponsor: Association for Library and Information Science Education
P.O. Box 4219, 1009 Commerce Park Drive, Suite 150
Oak Ridge, TN 37839

ALISE Research Paper Competition Awards   617

Research papers concerning any aspect of librarianship or information studies are eligible for this award. Competition is not limited to research regarding education for librarianship and information studies. Any research mode is acceptable. All research papers submitted must represent completed research not previously published, though manuscript may have been submitted and be in the process of publication. Papers generated as a result of a research grant or some other source of funding are eligible. The same author may submit for both the research grant award and the research paper competition, but the same work cannot be submitted for both categories. In case of joint authorship, one honorarium will be awarded. Annual deadline dates may vary; contact program staff for exact dates.

Requirements: Authors must be members of ALISE as of deadline date. Papers prepared by joint investigators are eligible, but at least one author must be a member of the association. Only one research paper per entrant will be considered.

Restrictions: Research papers completed in the pursuit of master's or doctoral studies are not eligible for entry, but research utilizing data gathered by a master's or doctoral student is eligible unless the research report is taken directly from a paper submitted for degree requirements. Papers that are spinoffs of such research are eligible.

Amount of Grant: $500
Date(s) Application is Due: Oct 3
Contact: Program Contact, (865) 425-0155; fax: (865) 481-0390; email: contact@alise.org
Internet: http://www.alise.org/awards/methodology.html
Sponsor: ALISE National Office
P.O. Box 4219, 1009 Commerce Park Drive, Suite 150
Oak Ridge, TN 37839

All Saints Educational Trust Grants   618

The main purposes of the trust is to help increase the number of qualified teachers, improve the skills and qualifications of teachers, encourage research that would assist teachers in their work, and support specifically the teaching of Religious Studies and Home Economics and related areas. There is no deadline date for the preliminary questionnaire form but there is a deadline for the full application form. Annual deadline dates may vary; contact program staff for exact dates. The Trustees normally only give financial assistance when the course of study is undertaken at recognized educational establishments within the United Kingdom.

Requirements: Applicants must be over 18 years of age when they begin their studies.
Restrictions: Applications are not accepted from people outside the U.K. except residents of Commonwealth nations. The trust cannot support general or core funds of any organization, public appeals, school buildings, equipment or supplies, the establishment of courses or departments in universities and colleges, and general Bursary funds of other organizations.
Contact: Dr. K. G. Riglin, Chairmain; 44 (0) 20 7283 4485; fax: 44 (0) 20 7621 9758; email: aset@aset.org.uk
Internet: http://www.aset.org.uk
Sponsor: All Saints Educational Trust
St. Katharine Cree Church, 86 Leadenhall Street
London EC3A 3DH United Kingdom

All Souls College Senior Research Fellowships   619

The college supports senior research fellowships, one in Classical Studies, and one in Literature in the English Language (both subjects broadly conceived). The fellowship program is of comparable academic standing to an Oxford University professorship, and applicants are expected to have a correspondingly distinguished record of achievement in research. The primary duty of a fellow is to pursue a program of advanced study and research in Oxford, approved by the college and at a level acceptable to it. The college will renew the fellowship only on evidence of satisfactory achievement and on presentation

of a satisfactory program of research for the following seven years (or until retirement). Candidates should submit as part of their application a full curriculum vita with a list of published work. Application and guidelines are available online.
Amount of Grant: 55,713 pounds-50,699 pounds, 2640 pounds academic allowance, 4929 housing allowance
Date(s) Application is Due: Sep 12
Contact: Warden's Secretary
Internet: http://www.all-souls.ox.ac.uk
Sponsor: All Souls College
Oxford OX1 4AL United Kingdom

Allen Foundation Educational Nutrition Grants 620

The foundation supports projects that benefit human nutrition in the areas of education, training, and research. Priorities include training programs for children and young adults to improve their health and development; training programs for educators and demonstrators concerned with good nutritional practices; programs for the education and training of mothers during pregnancy and after the birth of their children, so that good nutritional habits can be formed at an early age; and programs that aid in the dissemination of information regarding healthful nutrition practices. Applications are available by fax or mail. There are no application deadlines; proposals are reviewed throughout the year.
Requirements: 501(c)3 tax-exempt organizations nationwide may apply. In certain circumstances, the foundation will consider requests from the following: hospitals or medical clinics; social, religious, fraternal, or community organizations; private foundations; and K-12 public, parochial, or private schools.
Amount of Grant: $5,000-$250,000
Date(s) Application is Due: Dec 31
Contact: Dr. Dale Baum, Secretary; (989) 832-5678 or (979) 695-1132; fax: (989) 832-8842; email: dbaum@allenfoundation.org or Lucille@allenfoundation.org
Internet: http://www.allenfoundation.org/commoninfo/aboutus.asp
Sponsor: Allen Foundation
P.O. Box 1606
Midland, MI 48641-1606

Allen Lane Foundation Grants 621

The foundation operates in the United Kingdom and Ireland in the field of social welfare and awards grants to institutions at the local, regional, and national levels. Areas of interest include projects in Scotland, Northern Ireland, Wales, and regions outside the London area; groups supporting refugees and asylum-seekers; advisory and information services; and the coordination of small groups. The broad areas of work which are priorities for the foundation include the provision of advice, information, and advocacy; community development; employment and training; mediation, conflict resolution, and alternatives to violence; research and education aimed at changing public attitudes or policy; and social welfare.
Requirements: Grants may be made for project costs or revenue costs. The foundation no longer has closing dates.
Restrictions: The foundation only very rarely makes grants to national organizations with an income of more than $500,000 per annum or to local organizations with an income of more than about $150,000. Grants are not made to individuals.
Amount of Grant: L500-L15,000 range
Contact: Gill Aconley, Grants Officer, 01-904-613-223; fax: 01-904-613-133; email: info@allenlane.org.uk
Internet: http://www.allenlane.org.uk
Sponsor: Allen Lane Foundation
90 The Mount
York YO24 1AR United Kingdom

Alliance for Justice Internship Program 622

The program awards year-round, semester, winter-term, and summer internships. The organization's work consists of promoting reform of the legal system to ensure access to the courts. Projects include extensive legal, legislative, and policy research; special projects, such as helping to organize symposia and producing user-friendly legal publications for nonprofit organizations; attending congressional committee meetings on nonprofit advocacy, tax reform, takings, and judicial selection; contributing to the Alliance's newsletter, Justice First; and attending relevant public interest events around Washington, DC. Applications are accepted at any time.
Requirements: Internships are available to law students and graduate and undergraduate students.
Contact: Marissa Brown, (202) 822-6070; fax: (202) 822-6068; email: marissa@afj.org
Internet: http://www.allianceforjustice.org/about_AFJ/jobs/index.html
Sponsor: Alliance for Justice
11 Dupont Circle NW, 2nd Fl
Washington, DC 20036

Allstate Foundation Grants 623

The foundation awards grants to nonprofit organizations in the categories of Safe and Vital Communities, Economic Empowerment, and Tolerance, Inclusion and Diversity. In the area of Tolerance, Inclusion and Diversity, the foundation seeks to support programs that address teaching tolerance to youth, ending hate crimes, and alleviating discrimination. Programs with a focus on Safe and Vital Communities should address catastrophe response, youth anti-violence, neighborhood revitalization, and teen safe driving. Economic Empowerment programs should address financial and economic literacy, insurance education, and empowerment for victims of domestic violence.

Programs should focus on teaching tolerance to youth, alleviating discrimination, and ending hate crimes. There are no application deadlines. Proposals are accepted throughout the year. Contact local branch offices or the foundation. Grant committees usually meet in March, June, September, and December.
Requirements: The Allstate Foundation makes grants to nonprofit, tax-exempt organizations under Section 501(c)3 of the Internal Revenue Code.
Restrictions: The foundation does not support the following: individuals; fundraising events, sponsorships; capital and endowment campaigns; equipment purchase unless part of a community outreach program; athletic events; memorial grants; athletic teams, bands, and choirs; organizations that advocate religious beliefs or restrict participation on the basis of religion; groups or organizations that will re-grant the foundation???s gift to other organizations or individuals; scouting groups; private secondary schools; requests to support travel; grant requests for production of audio, film, or video; multiyear pledge requests; or nondomestic (international) causes.
Samples: Boys and Girls Clubs of Greater Washington (District of Columbia)—for efforts to promote the acceptance of diversity among young people, including working to eliminate gang violence, $50,000 (2005); Junior Achievement (Colorado Springs, CO)—for financial-literacy programs for youths, $1.5 million (2004).
Amount of Grant: $7 million total; $5000-$10,000 typically
Contact: Executive Director; (847) 402-5502; fax: (847) 326-7517; email: allfound@allstate.com
Internet: http://www.allstate.com/foundation
Sponsor: Allstate Foundation
2775 Sanders Road, Suite F4
Northbrook, IL 60062-6127

Allyn Foundation Grants 624

The foundation places emphasis on higher and other education, including adult basic education/literacy; and support for charitable purposes, including youth and social service agencies, hospitals, medical research, and community development in the central New York area. Types of support include capital campaigns, building and equipment, seed money, scholarship funds, and matching funds. There are no application deadlines. The board meets four times annually.
Requirements: Nonprofit organizations in Onondaga and Cayuga Counties, NY, may submit grant applications.
Restrictions: Grants are not awarded for religious purposes, endowment funds, loans, or to individuals.
Geographic Focus: New York
Samples: Literacy Partners, Cayuga Chapter (New York, NY)—program support, $4000.; Alzheimer's Disease and Related Disorders Assoc (New York, NY)—program support, $3500.; United Negro College Fund (New York, NY)—for scholarships, $25,000.
Amount of Grant: $500-$27,000 range
Contact: Margaret O'Connell, Executive Director, (315) 685-5059
Sponsor: Allyn Foundation
P.O. Box 22
Skaneateles, NY 13152

ALO Promoting Higher Education Partnerships for Global Development 625

The program invites applications for the U.S.-Middle East University Partnerships Program. ALO will award up to five grants, with funding over a three-year period, to implement cooperative partnerships between higher education institutions in the United States and Arab universities located in Algeria, Bahrain, Egypt, Jordan, Kuwait, Lebanon, Morocco, Oman, Qatar, Saudi Arabia, Tunisia, the United Arab Emirates, the West Bank and Gaza, and Yemen. ALO seeks applications for higher education partnerships to strengthen Arab universities' programs in one of the following disciplines: business administration and economics; gender studies; government; information and communication technologies; legal studies; and teacher education. Application and guidelines are available online.
Amount of Grant: $200,000 maximum over three years
Date(s) Application is Due: Aug 10
Contact: Tony Wagner, (202) 478-4704; email: wagner@aascu.org
Internet: http://www.aascu.org/alo/RFPs/mepi05/mepi05.htm
Sponsor: Association Liaison Office for University Cooperation in Development
1307 New York Avenue NW, Suite 500
Washington, DC 20005-4701

ALO U.S.-Japan Trilateral Program for Basic Education in Africa 626

ALO, with funding support from U.S.AID, is issuing this call for concept papers for institutional partnerships in basic education involving a U.S. college or university (or a group of colleges and universities), public and private sector partners, Japanese universities, and institutions in Africa. Five countries have been identified as potential locations for program implementation: Ethiopia, Ghana, South Africa, Tanzania, and Uganda. Concept papers should indicate the specific results and impacts that will follow from the proposed activities, including the projected impact of the activity on basic education and basic education capacity in the target country. Evidence that such matters have received rigorous consideration will increase the competitiveness of the concept paper. The Call for Concept Papers, with complete information, is available online.
Requirements: ALO welcomes concept papers from the member institutions of ACE, AACC, AASCU, AAU, NAICU, and NASULGC; and from other regionally accredited, degree-granting, U.S. higher education institutions. U.S. colleges and universities may respond individually or in partnership with other institutions. Minority serving institutions are encouraged to submit papers.
Restrictions: faxed and electronic submissions are not accepted.

Date(s) Application is Due: Mar 31
Contact: Tony Wagner, (202) 478-4700; fax: (202) 478-4715; email: wagner@aascu.org
Internet: http://www.aascu.org/alo/RFPs/RFPMain.htm
Sponsor: Association Liaison Office for University Cooperation in Development
1307 New York Avenue NW, Suite 500
Washington, DC 20005-4701

Alpha Chi Sigma Award for Chemical Engineering Research   **627**
This award, sponsored by the Alpha Chi Sigma fraternity and administered by AIChE, recognizes outstanding recent accomplishments by an individual in fundamental or applied research in the field of chemical engineering. Research awarded must have been carried out during the 10 years preceding the award year. The recipient need not be a member of AIChE or Alpha Chi Sigma.
Requirements: All members of AIChE as well as other interested persons in North America are urged to nominate deserving candidates. All information and supporting documents are to be included with the completed form; nomination forms and instructions may be obtained from the institute.
Samples: Michael Doherty—award recipient, $5000 (2004).
Amount of Grant: $5000
Date(s) Application is Due: Feb 15
Contact: AIChE Awards Programs, (212) 591-7107; fax: (212) 591-8890; email: awards@aiche.org
Internet: http://www.aiche.org/awards/awarddtl.asp?AwardID=55
Sponsor: American Institute of Chemical Engineers
3 Park Avenue
New York, NY 10016-5991

Alpha Kappa Alpha Educational Advancement Foundation Fellowships   **628**
Educational Advancement Foundation fellowships are awarded during even numbered years. These awards help to improve the quality of life of others by funding research and projects with practical applications. Foundation fellowships have been awarded to individuals for a wide range of purposes, from assistance in finding non-violent solutions to human disputes to conducting research on hypertension and the impact of nuclear testing on marine life. The Helen Cromer Cooper/Rosa Parks Fellowship is earmarked to support individuals and projects involved in finding non-violent solutions to human problems. The Dr. Dorri Phipps Fellowship is for students pursuing degrees in medicine or research pertaining to lupus. Applications, along with supporting documents, must be postmarked by January 15 of the award year.
Samples: Regina Randall, Australasian College of Natural Therapies, Big Lake, Alaska (2008).
Date(s) Application is Due: Jan 15
Contact: Andrea Kerr, Program/Scholarship Coordinator; (800) 653-6528 or (773) 947-0026; fax: (773) 947-0277; email: akerr@akaeaf.net or akaeaf@akaeaf.net
Internet: http://www.akaeaf.org/programsandinitiatives/
Sponsor: Alpha Kappa Alpha Educational Advancement Foundation
5656 South Stony Island Avenue
Chicago, IL 60637

Alpha Omega Foundation Grants   **629**
The Foundation was established to advance the dental profession through research, grants, and scholarships. Fields of interest include: dental care, dental education, and medical research. Grants are given to individuals, program development, and higher education institutions. The application, which can be found at the web site, must be typed and submitted by email by September 1 each year. Notification will be completed by the following December.
Amount of Grant: $1,000-$36,000 range
Date(s) Application is Due: Sep 1
Contact: Ben Williamowsky; (877) 368-6326 or (301) 294-2773; fax: (301) 738-6403; email: bawilly@verizon.net or foundation@ao.org
Internet: http://www.ao.org/index.php?option=com_content&task=view&id=80&Itemid=147
Sponsor: Alpha Omega Foundation
50 W. Edmonston Drive, Suite 404
Rockville, MD 20852-1274

Alpha Research Foundation Grants   **630**
The primary purpose of the Foundation is to encourage and support biomedical research for the welfare of the general public. There are no specific deadlines or application formats with which to adhere. Applicants should contact the foundation prior to submission, in order to discuss their proposed project. If approved, applicants should then submit an outline of the proposed research along with a proposed budget.
Geographic Focus: Maryland
Amount of Grant: $2,000 maximum
Contact: c/o Barry Rueben Fierst; (301) 762-8872; fax: (301) 762-8874; email: bfierst@aol.com
Sponsor: Alpha Research Foundation
7118 Glenbrook Road
Bethesda, MD 20814-1238

Alpha Sigma Nu Book Award Competition   **631**
The book awards operate on a three-year cycle. The category for 2007 is "The Professional Studies", and includes the following disciplines: Architecture, Business

and Administration, Communication, Education, Engineering, Foreign Service, Law, and Social Work. Four prizes will be awarded. The category for 2008, will be "The Humanities" with one award each for Theology, Philosophy/Ethics, Literature/Fine Arts, and History. In 2009, the category will be "The Sciences" with one award for Social Sciences, Natural Sciences, Mathematics/Computer Science, and Health Sciences. Information and application forms are available from academic vice presidents and deans of the Jesuit colleges and universities, or from the Association of Jesuit Colleges and Universities.
Requirements: Candidates must be full- or part-time staff or faculty or anyone who has emeritus status at a Jesuit college or university in the United States.
Amount of Grant: $1000
Date(s) Application is Due: Mar 1
Contact: Program Contact, (202) 862-9893; email: agarner@ajcunet.edu
Internet: http://www.marquette.edu/dept/ASN/book_award_competition.html
Sponsor: Alpha Sigma Nu
One Dupont Cir, Suite 405
Washington, DC 20036

ALS Clinical Management Research Grants   **632**
The mission of the association is to find a cure for and improve living with amyotrophic lateral sclerosis (often called Lou Gehrig's disease). Through this program, the association is encouraging new research, the results of which will build an evidence base and demonstrate measurable, positive effects on the clinical management and lives of patients with ALS. Clinical problems relating to the following priority areas for clinical management research include breathing; swallowing/ nutrition and PEG issues/ choking/excess saliva; speech/communication; musculoskeletal symptoms/treatment; mobility/ activities of daily living; and psychosocial/mental health. The association also will consider abstracts on other clinical management topics.
Requirements: Applicants must be faculty members or scientists at reputable scientific institutions.
Amount of Grant: $30,000 maximum
Date(s) Application is Due: Jan 7
Contact: Dr. Mary Lyon, Grants Administrator, (818) 880-9007 ext 217; fax: (818) 880-9006; email: mary@alsa-national.org
Internet: http://www.alsa.org/patient/research.cfm?CFID=102432&CFTOKEN=86382360
Sponsor: Amyotrophic Lateral Sclerosis Association
27001 Agoura Road, Suite 150
Calabasas Hills, CA 9136491301

ALS Research Grants   **633**
Research into the cause, means of prevention and possible cure of ALS is the driving priority of the ALS Association. The ALS Association annually awards $2 million or more in new research grants and currently holds two calls for abstracts each year. After review of the abstracts, selected investigators are asked to submit full applications which are also peer reviewed by ALSA's scientific review committee. Annual deadline dates may vary; contact program staff for exact dates.
Requirements: Applicants must be faculty members or scientists at reputable scientific institutions.
Amount of Grant: $80,000 maximum; $40,000 maximum for starter grants
Contact: Grants Administrator, (888) 949-2577 or (818) 880-9007; email: researchgrants@alsa-national.org
Internet: http://www.alsa.org/research/process.cfm?CFID=102432&CFTOKEN=86382360
Sponsor: Amyotrophic Lateral Sclerosis Association
27001 Agoura Road, Suite 150
Calabasas Hills, CA 91301

ALSAM Foundation Grants   **634**
The foundation awards grants in its areas of interest, including agriculture, Christian agencies and churches, higher education, health care and medical research, human services, minorities, and the economically disadvantaged. Types of support include building construction/renovation, general operating costs, and scholarships. The board meets in January and October. Contact the office for application forms.
Requirements: Higher education institutions, nonprofit organizations, religious organizations, and research institutions are eligible.
Restrictions: No grants to individuals.
Samples: University of Utah, Moran Eye Center (Salt Lake City, UT)—for building fund, $2.5 million (2003); Saint George Catholic Church (Saint George, UT)—for building fund, $200,000 (2003); Harvard University (Cambridge, MA)—for research of Neurofibromatosis and Allied Disorders, $600,000.; Verbum Dei High School (Los Angeles, CA)—for conversion, $2,374,378.
Amount of Grant: $5000-$50,000 average
Contact: Ron Cutshall, Chair, (801) 266-4950
Sponsor: ALSAM Foundation
6190 Moffat Farm Lane
Salt Lake City, UT 84121

ALSC Bound to Stay Bound Books Scholarship   **635**
Four annual scholarships assist individuals who wish to work in the field of library service to children. The scholarships may be used for study toward the MLS or graduate study beyond the MLS degree at an ALA-accredited library school. The scholarships

are sponsored by Bound to Stay Bound Books Inc, and administered by the Association for Library Service to Children.

Requirements: Only online applications will be accepted (there will be no exceptions). Online application forms are available on the Web site.

Samples: Libby Fry (York Haven, PA)—$6000 (2003); Alison Kelly (Los Angeles, CA)—$6000 (2003); Tessa Michaelson (Madison, WI)—$6000 (2003); Diana Ricciardone (Southington, CT)—$6000 (2003).

Amount of Grant: $6500

Date(s) Application is Due: Mar 1

Contact: Linda Mays, Program Officer, (800) 545-2433 ext 1398; fax: (312) 944-7671; email: lmays@ala.org

Internet: http://www.ala.org/ala/alsc/awardsscholarships/alscschol/alscscholarship.htm

Sponsor: American Library Association
50 E Huron Street
Chicago, IL 60611-2795

ALSC Frederic G. Melcher Scholarships 636

Two annual scholarships assist individuals in the field of library service to children. They are awarded to qualified candidates who have been accepted for admission to a graduate library school program accredited by the ALA but who have not yet begun the program. Application and additional information are available from the ALSC office, which administers the scholarships.

Requirements: Only online applications will be accepted. Online applications are available on the Web site.

Samples: Tami Edwards (Ocala, FL)—scholarship recipient (2006); Sharon Kieffer (North Haledon, NJ)—scholarship recipient (2006). Jonathan Hunt (Hughson, CA)—scholarship recipient, $6000 (2005); Wendy Torrence (Mount Gilead, OH)—scholarship recipient, $6000 (2005); Lauren Anduri (Lafayette, CA)—scholarship recipient, $6000 (2004).

Amount of Grant: $6000 each (2)

Date(s) Application is Due: Mar 1

Contact: ALA Awards Staff Liaison, (800) 545-2433 ext 3247; fax: (312) 944-6131; email: alsc@ala.org or lmays@ala.org

Internet: http://www.ala.org/alsc/scholars.html

Sponsor: American Library Association
50 E Huron Street
Chicago, IL 60611-2795

Alternatives Research and Development Foundation Grants 637

The foundation awards grants to research centers, educational institutions, and other nonprofit organizations exploring non-animal testing. Grants will be awarded for research projects that use human, rather than nonhuman, vertebrae tissue; do not use intact, vertebrae animals; and can be completed within one year. Multiyear projects are considered on a case-by-case basis. The foundation often provides partial grants to initiate projects, with continuation funding available at a later date. The foundation will not consider proposals if program staff uses animals acquired from a shelter, or if individuals are employed who use such animals in their personal research programs. Winners are announced on July 15. Telephone inquiries are discouraged.

Requirements: Individuals attending or employed by U.S. universities and research institutions may apply. Applications from non-U.S. institutions or investigators may be considered.

Restrictions: The foundation does not make grants for in vitro projects that use nonhuman animal serum, indirect costs, purchase of personal computers, salary supplements, fringe benefits, travel expenses, or publication costs for the principal investigators. Phone calls to the foundation are discouraged. Guidelines and application are available online.

Samples: Dr. Anthony Hickey, School of Pharmacy, U of North Carolina at Chapel Hill (NC)—for research, Physiologically Relevant Lung Model of Regional Aerosol Deposition and Drug Transport Utilizing Human Respiratory Tract Cells.; Dr. Katherine Ralls, Department of Conservation Biology, Smithsonian Institution (Washington, DC)—for research, Dogs, Scats, and DNA: A Noninvasive Approach for Carnivore Field Studies.; Dr. Carol Reinisch, Marine Biological Laboratory (Woods Hole, MA)—for research, Regulation of p53 Gene Family Expression in Clam Leukemia Cells.; Dr. Bingfang Yan, Department of Biomedical Sciences, U of Rhode Island (Kingston, RI)—for research, Prediction of Human CYP3A Induction by a Receptor-Activator Based Method.

Amount of Grant: $40,000 maximum

Date(s) Application is Due: Apr 30

Contact: Dr. John McArdle, Director, (215) 887-8076; fax: (215) 887-0771; email: grants@ardf-online.org

Internet: http://www.ardf-online.org

Sponsor: Alternatives Research and Development Foundation
801 Old York Road, Suite 316
Jenkintown, PA 19046

Alton E. Bailey Award 638

This award is given annually for research and/or service in the fields of oils, fats, and related disciplines by the North Central Section of AOCS.

Samples: F.D. Gunstone—award recipient, (2005); W.W. Christie—award recipient (2004).

Date(s) Application is Due: Nov 1

Contact: Kathleen Atchley, Awards, (217) 359-2344; fax: (217) 351-8091; email: awards@aocs.org

Internet: http://www.aocs.org/member/awards/awd-bailey.asp

Sponsor: American Oil Chemists' Society
P.O. Box 3489
Champaign, IL 61826-3489

Alton Ochsner Award 639

The Award is presented to one or more clinical or basic science investigators, without regard to age, race, gender, or nationality, for outstanding and exemplary original scientific investigations that relate tobacco consumption and health. This scientific work may be clinical, fundamental, epidemiological or preventive in scope. The prime criterion for award selection is its scientific impact on this major health threat. The $15,000 award is presented at the Annual Convocation of the American College of Chest Physicians.

Requirements: All nominations, whatever the category of scientific inquiry, must be supported by letters and copies of peer-reviewed scientific publications.

Amount of Grant: $15,000

Date(s) Application is Due: Mar 31

Contact: Edward D. Frohlich, Alton Ochsner Distinguished Scientist, (504) 842-3000; fax: (504) 842-3258

Internet: http://www.ochsner.org/homepage.cfm

Sponsor: Ochsner Clinic Foundation
1514 Jefferson Highway
New Orleans, LA 70121

Alvin and Lucy Owsley Foundation Grants 640

The foundation gives grants in to organizations in the state of Texas in various interest areas including (but not limited to) arts, community development, education.

Requirements: There are no specific deadlines, but the board meets in March, June, September and December. Submit your proposal before the board meets. The application process is a letter no more than 2 pages in length with no attachments.

Geographic Focus: Texas

Samples: Dallas Museum Of Art (Dallas, Texas) - $90,500.; Botanical Research Institute Of Texas (Ft Worth, Texas) - $40,000.; B S A Sam Houston Area (Houston, Texas) - $25,000.; Fort Worth Public Library Foundation (Ft Worth, Texas) - $25,000.; Houston Ballet Foundation (Houston, Texas) - $11,500.; Presbyterian Children's Homes (Austin, Texas) - $10,000.; Boys & Girls Harbor (Laport, Texas) - $10,000.; San Jacinto G S Council (Houston, Texas) - $10,000.; First Presbyterian Church (Houston, Texas) - $7,500.; Boys & Girls Country (Hockley, Texas) - $7,500.; De Pelchin Children's Center (Houston, Texas) - $6,000 # National Center For Policy Analysis (Dallas, Texas) - $ 5,000.; Chinquapin School (Highlands, Texas) - $5,000.; Austin College (Sherman, Texas) - $5,000.; Camp For All (Houston, Texas) - $5,000.; Gathering Place (Houston, Texas) - $5,000.; St. John's School (Houston, Texas) - $5,000.; Volunteer Houston (Houston, Texas) - $5,000.; Texas Heart Institute (Houston, Texas) - $5,000

Amount of Grant: $500 - $100,000

Contact: Alvin Owsley, Trustee; (713) 229-1272

Sponsor: Alvin and Lucy Owsley Foundation
65 Brair Hollow Lane
Houston, TX 77027

ALVRE Casselberry Award 641

The Casselberry Award has been established to encourage the advancement of the art and science of Laryngology and Rhinology. The award is given for outstanding manuscripts or accomplishments in Laryngology and Rhinology and consists of a $1,000 award and a certificate from the Association. Competition for this award will be limited to those persons whose abstracts are selected in consideration for inclusion in the Annual Scientific Program.

Amount of Grant: $1,000

Date(s) Application is Due: Nov 30

Contact: Maxine Cunningham, Administrator; (615) 322-6326; fax: (615) 322-9102; email: Maxine@alahns.org or ala-hns@comcast.net

Sponsor: American Laryngological Association / American Laryngological Voice and Research Education Foundation
1215 21st Avenue South, 7302 MCE South
Nashville, TN 37232-8783

ALVRE Grant 642

The purpose of this award is to support basic, translational, or clinical research projects in laryngology, voice, outcomes, and related subjects. Research supported by this award should be specifically directed toward the pathogenesis, pathophysiology, diagnosis, prevention, or treatment of diseases, disorders, or conditions of the larynx and may be either basic or clinical/translational in approach. While not specifically required, proposals which aim to introduce new knowledge and methodology from other disciplines to research in laryngology or neurolaryngology, or which demonstrate collaborative effort with members of other related disciplines are encouraged. Projects must be designed so as to yield useful information within the period of award, but priority will be given to projects that are also innovative with promise to develop into new long-range or expanded research programs capable of attracting funding from other sources. A single, one-year, non-renewable award of $25,000 maximum is available annually. The foundation will consider requests to cover travel expenses up to $1,000 for the principal investigator to present his/her results at the ALA annual meeting.

Requirements: Candidates for this award should be otolaryngologists in the U.S. or Canada who have completed their training at an ACGME accredited program in otolaryngology—head and neck surgery. The prinicipal investigator should be a physician faculty member of a recognized department, division, or section of otolaryngology-head and neck surgery.

Amount of Grant: $25,000 maximum

Contact: Maxine Cunningham, Administrator; (615) 322-6326; fax: (615) 322-9102; email: Maxine@alahns.org or ala-hns@comcast.net

Internet: http://www.entnet.org/EducationAndResearch/coreGrants.cfm

Sponsor: American Laryngological Association / American Laryngological Voice and Research Education Foundation

1215 21st Avenue South, 7302 MCE South

Nashville, TN 37232-8783

ALVRE Seymour R. Cohen Award 643

The Seymour R. Cohen Award was established with a bequest by the Cohen family during Dr. Cohen's presidency in 1989. In order to qualify for the Award, the candidate must perform basic science research in the area of pediatric laryngology and/or pediatric neurolaryngology and be a citizen of the United States or Canada. The material must be approved and accepted for presentation at the Annual Meeting. Should multiple candidates collaborate in the research, all authors must comply with the citizenship requirements. Dr. Cohen wishes this Award only to be used for the funding of research in pediatric laryngology and/or pediatric neurolaryngology. The Award is to be presented bi-annually.

Requirements: The scope of the Award is limited to citizens of the United States of America and Canada.

Contact: Maxine Cunningham, Administrator; (615) 322-6326; fax: (615) 322-9102; email: Maxine@alahns.org or ala-hns@comcast.net

Internet: http://www.alahns.org/i4a/pages/index.cfm?pageid=3332

Sponsor: American Laryngological Association / American Laryngological Voice and Research Education Foundation

1215 21st Avenue South, 7302 MCE South

Nashville, TN 37232-8783

ALZA Corporate Contributions Grants 644

The corporation awards grants to eligible California nonprofit organizations in four areas of interest: education—K-12 after school programs, professional development for teachers, and health care workers, nonprofit educational outreach programs; health and human services—projects that particularly access to healthcare for underserved and underrepresented members of our community; arts & culture—arts programs that have an education component or art related to health and healing are of particular interest; and environment—projects that focus on sustainable solutions, stewardship, & education.

Requirements: Nonprofit organizations in ALZA's California operating areas are eligible.

Restrictions: Grants will not be made to individuals, or to organizations whose activities or policies include sectarian or denominational religious activities, political campaigns, or organizations that discriminate on the basis of religion, race, nationality, sexual preference, or gender.

Geographic Focus: California

Samples: Santa Clara U (CA)—for scholarships and research by undergraduate students pursuing careers in biology, chemistry, medicine, or the pharmaceutical industry, $500,000 over five years.

Amount of Grant: $10,000 maximum

Contact: Ellen Rose, Director, Corporate Communications, (650) 564-5000; fax: (650) 564-7070

Internet: http://www.alza.com/alza/community

Sponsor: ALZA Corporation

P.O. Box 7210, 1900 Charleston Road

Mountain View, CA 94043

Alzheimer's Association Everyday Technologies for Alzheimer Care 645
Research Grants

Everyday Technologies for Alzheimer Care (ETAC) is a cooperative research funding initiative sponsored by the Alzheimer's Association and Intel Corporation. ETAC seeks proposals on personalized diagnostics, preventive tools and interventions for adults coping with the spectrum of cognitive aging and neurodegenerative disease, particularly Alzheimer, disease. ETAC is designed to support exploratory multidisciplinary research that would not typically be funded by national health and science granting foundations. Minor iterations in testing plans or populations will not be considered for funding. Collaboration between social science/medical/public health and computer science/engineering researchers is valued. Mobile computing, high bandwidth sensing, robotics, imaging, face recognition, natural language processing, statistical modeling and a host of other technology advances allow unprecedented opportunities to study disease progression and therapeutic strategies in the context of everyday life. ETAC supports research that integrates such emerging technology capabilities with leading directions in behavioral science and biomedical research. Grants that merely create Internet-based versions of existing services or paper tools will not be considered. Submissions must be original ideas, not continuations of previously funded ETAC projects. Please see links provided below for examples of studies that have been funded by ETAC. Letters of Intent (LOIs) must be received by December 1, with completed applications due by January 8.

Requirements: Researchers with full-time staff or faculty appointments are encouraged to apply.

Restrictions: The Alzheimer's Association will not accept new research grant applications from currently funded Alzheimer's disease investigators who are delinquent in

submitting interim/final scientific or interim/final financial reports on active grants. ETAC applications from post-doctoral candidates will not be accepted.

Samples: Alex Mihailidis, Ph.D., University of Toronto, Toronto, Ontario, Canada— Toward a Pervasive Prompting System: Improving and Expanding the COACH, $196,324 over three years (2008); Arlene Astell, Ph.D., University of St. Andrews, St. Andrews, United Kingdom—Prompting to Support Independence in Dementia, $179,634 over two years (2008).

Amount of Grant: $200,000 maximum

Date(s) Application is Due: Dec 1; Jan 8

Contact: Nico Stanculescu; (312) 335-5747 or (312) 335-5729; fax: (312) 335-4034; email: Nico.Stanculescu@alz.org or grantsapp@alz.org

Internet: http://www.alz.org/professionals_and_researchers_everyday_technologies_ for_alzheimer_care.asp

Sponsor: Alzheimer's Association

225 N. Michigan Avenue, Floor 17, Suite 1700

Chicago, IL 60601-7633

Alzheimer's Association Investigator-Initiated Research Grants 646

The program is structured to provide one to three years of sustained project support for independent, ongoing research. Proposals are solicited for basic, clinical, and social/ behavioral research relevant to degenerative brain diseases such as Alzheimer's disease. Allowable costs for this award include the purchase and care of laboratory animals; small pieces of laboratory equipment and laboratory supplies; and salary for the principal investigator, scientific (including postdoctoral fellows) and technical staff (including laboratory technicians and modest secretarial support). It is required that most of the funds awarded under this program be used for direct research support. Letters of Intent (LOIs) must be received by December 1, with completed applications due by January 8.

Requirements: Public, private, domestic and foreign research laboratories, medical centers and hospitals, and universities are eligible to apply. Investigators from all stages of their research career development are encouraged to apply.

Restrictions: This program does not support travel to scientific and professional meetings, computer hardware or software, or construction or renovation costs.

Samples: Mathieu Lesort, Ph.D., University of Alabama at Birmingham, Birmingham, Alabama—Pathological Interactions in Diabetes and Alzheimer's Disease, $200,000 over three years (2008); David H. Cribbs, Ph.D., University of California, Irvine, Irvine, California—Reducing the Risk of Cerebral Vascular Adverse Events in Alzheimer's Disease, $240,000 over three years (2008).

Amount of Grant: $240,000 total for up to three years; $100,000 maximum per year

Date(s) Application is Due: Dec 1; Jan 8

Contact: Nico Stanculescu; (312) 335-5747 or (312) 335-5729; fax: (312) 335-4034; email: Nico.Stanculescu@alz.org or grantsapp@alz.org

Internet: http://www.alz.org/professionals_and_researchers_investigatorinitiated.asp

Sponsor: Alzheimer's Association

225 N. Michigan Avenue, Floor 17, Suite 1700

Chicago, IL 60601-7633

Alzheimer's Association Mentored New Investigator Research Grants 647
to Promote Diversity

The Mentored New Investigator Research Grant to Promote Diversity (MNIGD) is a three-year award intended to be a research-based and mentoring investment in the process of closing the health disparities gap between diverse and non-diverse investigator populations. The Alzheimer, Association feels strongly that the mentoring and involvement of diverse researchers in independently funded Alzheimer, research is a pressing need. The MNIRGD is intended to enhance the capacity of diverse and non-diverse scientists to conduct basic, clinical and social/behavioral research. Each MNIRGD award is limited to $170,000. A total of $150,000 will be awarded for costs related to the proposed research for up to three years (direct and indirect costs). Letters of Intent (LOIs) must be received by December 1, with completed applications due by January 8.

Requirements: Eligibility for this grant competition is restricted to investigators who have less than 10 years of research experience after receipt of their terminal degree.

Amount of Grant: $170,000 maximum

Date(s) Application is Due: Dec 1; Jan 8

Contact: Nico Stanculescu; (312) 335-5747 or (312) 335-5729; fax: (312) 335-4034; email: Nico.Stanculescu@alz.org or grantsapp@alz.org

Internet: http://www.alz.org/professionals_and_researchers_mentored_new_ investigator_promote_diversity_grant.asp

Sponsor: Alzheimer's Association

225 N. Michigan Avenue, Floor 17, Suite 1700

Chicago, IL 60601-7633

Alzheimer's Association Molecular Imaging Grants 648

The Alzheimer's Association is launching an initiative to stimulate further research and development of new approaches to image molecular changes associated with early neurodegenerative processes in living humans, animal models and cells. The Association's Request for Applications (RFAs) is aimed at supporting new high-risk exploratory approaches to stimulate new directions of research in molecular imaging. The RFA is designed to enable preliminary pilot research or proof-of-principle studies that can provide data for further research support by other funding agencies. The Association anticipates funding up to 3 Molecular Imaging awards. Each award is limited to $400,000 (direct and indirect costs) for two to three years. Letters of Intent (LOIs) must be received by December 1, with completed applications due by January 8.

Requirements: Researchers with full-time staff or faculty appointments are encouraged to apply. Molecular Imaging applications from post-doctoral candidates will not be accepted.
Amount of Grant: $400,000 (direct and indirect costs) for two to three years
Date(s) Application is Due: Dec 1; Jan 8
Contact: Nico Stanculescu; (312) 335-5747 or (312) 335-5729; fax: (312) 335-4034; email: Nico.Stanculescu@alz.org or grantsapp@alz.org
Internet: http://www.alz.org/professionals_and_researchers_molecular_imaging_grant. asp
Sponsor: Alzheimer's Association
225 N. Michigan Avenue, Floor 17, Suite 1700
Chicago, IL 60601-7633

Alzheimer's Association New Investigator Research Grants 649

The purpose of this program is to provide new investigators with funding that will allow them to develop preliminary or pilot data, to test procedures, and develop hypotheses which will then underpin the preparation of research grant applications to NIH, NSF, and other funding agencies and groups, including the Alzheimer's Association. All applications submitted to the program must focus on a question or questions in interventions for Alzheimer's disease to be considered responsive to the program announcement. Thirty awards will be made under this program. Letters of Intent (LOIs) must be received by December 1, with completed applications due by January 8.
Requirements: Public, private, domestic and foreign research laboratories, medical centers and hospitals, and universities are eligible to apply. Eligibility is restricted to investigators who have less than 10 years of research experience, including postdoctoral fellowships or residencies, after receipt of the doctoral degree. Applications from graduate and doctoral students for research projects, which will be used for the thesis or dissertation, will be accepted.
Samples: Valentina Echeverria Moran, Ph.D., Bay Pines Foundation, Bay Pines, Florida— Molecular Mechanisms Underlying the Neuroprotective Actions of Cotinine, $100,000 over two years (2008); Grace Stutzmann, Ph.D., Rosalind Franklin University of Medicine and Science, North Chicago, Illinois—Neuronal Ca2+ Dysregulation as a Pathogenic Factor in Alzheimer's Disease, $100,000 over two years (2008).
Amount of Grant: $100,000 maximum for two years; $60,000 maximum per year
Date(s) Application is Due: Dec 1; Jan 8
Contact: Nico Stanculescu; (312) 335-5747 or (312) 335-5729; fax: (312) 335-4034; email: Nico.Stanculescu@alz.org or grantsapp@alz.org
Internet: http://www.alz.org/professionals_and_researchers_new_investigator_ research_grants.asp
Sponsor: Alzheimer's Association
225 N. Michigan Avenue, Floor 17, Suite 1700
Chicago, IL 60601-7633

Alzheimer's Association New Investigator Grants to Promote Diversity 650

The New Investigator Research Grant to Promote Diversity (NIRGD) in Alzheimer, research is a two-year award to investigators who are currently underrepresented at academic institutions in Alzheimer, or related dementias research. The objective of this award is to increase the number of highly trained investigators from diverse backgrounds whose basic, clinical and social/behavioral research interests are grounded in the advanced methods and experimental approaches needed to solve problems related to Alzheimer, and related dementias in general and in health disparities populations. Each NIRGD award is limited to $100,000 (direct and indirect costs) for up to two years. Requests in any given year may not exceed $60,000 (direct and indirect costs). Letters of Intent (LOIs) must be received by December 1, with completed applications due by January 8.
Requirements: Eligibility to apply for this grant competition is restricted to investigators who have less than 10 years of research experience after receipt of their terminal degree.
Amount of Grant: $60,000 maximum per year; $100,000 for two years
Date(s) Application is Due: Dec 1; Jan 8
Contact: Nico Stanculescu; (312) 335-5747 or (312) 335-5729; fax: (312) 335-4034; email: Nico.Stanculescu@alz.org or grantsapp@alz.org
Internet: http://www.alz.org/professionals_and_researchers_new_investigator_ promote_diversity_grant.asp
Sponsor: Alzheimer's Association
225 N. Michigan Avenue, Floor 17, Suite 1700
Chicago, IL 60601-7633

Alzheimer's Association Senator Mark Hatfield Award 651

The Hatfield Award is aimed at those investigators whose goal is to establish research careers focused on clinical issues in Alzheimer, disease. The Alzheimer's Association recognizes the need to increase the number of scientists from underrepresented groups in the research enterprise. Researchers from these groups are encouraged to apply. The Association anticipates funding one award under this program. The total award is limited to $225,000 (direct and indirect costs) for up to three years. Requests in any given year may not exceed $112,500 (direct and indirect costs). Indirect costs are capped at 10 percent (rent for laboratory/office space is expected to be covered by indirect costs paid to the institution). Letters of Intent (LOIs) must be received by December 1, with completed applications due by January 8.
Requirements: It is required that most of the funds awarded under this program be used for direct research support.
Samples: Ken A. Paller, Ph.D., Northwestern University, Evanston, Illinois—Memory Processing During Sleep in Alzheimer's Disease, $240,000 over three years (2008).

Amount of Grant: $225,000 maximum
Date(s) Application is Due: Dec 1; Jan 8
Contact: Nico Stanculescu; (312) 335-5747 or (312) 335-5729; fax: (312) 335-4034; email: Nico.Stanculescu@alz.org or grantsapp@alz.org
Internet: http://www.alz.org/professionals_and_researchers_mark_hatfield_award.asp
Sponsor: Alzheimer's Association
225 N. Michigan Avenue, Floor 17, Suite 1700
Chicago, IL 60601-7633

Alzheimer's Association Zenith Fellows Awards 652

Five awards will be granted to talented scientists who have already contributed substantially to the advancement of Alzheimer's research and who are likely to continue to make significant contributions for many years. Proposals are solicited for basic, biomedical research only, and research should address fundamental problems related to early detection, etiology, pathogenesis, treatment, and prevention of Alzheimer's disease. Awardees receive a two-year grant, with a provision for possible competitive renewal upon review of research progress. Applications will be evaluated by an expert panel of senior scientists, already well recognized for their own accomplishments in Alzheimer's research. Letters of Intent (LOIs) must be received by December 1, with completed applications due by January 8.
Requirements: Only established independent investigators are eligible as evidenced by (examples are for illustrative purposes only) academic appointment; major, peer-reviewed, external multi-year grant support on which the applicant is the principal investigator (PI); independent laboratory operation; and quality and independence of publication record.
Restrictions: Previous recipients of Zenith Awards, Alzheimer's Disease Center Directors (P50 and P30), Medical and Scientific Advisory Council, and members of the National Board of the Alzheimer's Association are not eligible to apply.
Samples: Mary Jo LaDu, Ph.D., University of Illinois, Chicago, Illinois—The Effect of ApoE Isoform on Intraneuronal ApoE/Abeta42 Interactions, $450,000 over three years (2008); Michael S. Wolfe, Ph.D., Brigham and Women's Hospital, Boston, Massachusetts—Regulation of RNA Splicing in Alzheimer's and Related Dementias, $450,000 over three years (2008).
Amount of Grant: $250,000 maximum for two years; $150,000 maximum for a single year
Date(s) Application is Due: Dec 1; Jan 8
Contact: Nico Stanculescu; (312) 335-5747 or (312) 335-5729; fax: (312) 335-4034; email: Nico.Stanculescu@alz.org or grantsapp@alz.org
Internet: http://www.alz.org/professionals_and_researchers_zenith.asp
Sponsor: Alzheimer's Association
225 N. Michigan Avenue, Floor 17, Suite 1700
Chicago, IL 60601-7633

Alzheimer, Association Conference Grants 653

The Alzheimer, Association has a long history of supporting scientific conferences that advance research on Alzheimer, disease. One of the principal goals of the Association from its inception has been to increase public awareness and to facilitate the exchange of information through the scientific and clinical communities. The support of conferences, workshops and meetings has been a key vehicle in achieving this goal. The objectives for conference support are to: facilitate and speed the exchange of information relevant to Alzheimer, disease research; convene experts to address emerging issues in Alzheimer, disease research; offer opportunities for new investigators and graduate students to participate in scientific meetings; facilitate the creation of networks among investigators in related areas; and increase visibility of the research interests and programs of the Alzheimer's Association.
Requirements: Requests for conference support may be submitted at any time. It is recommended that requests be submitted at least three months before the conference.
Amount of Grant: $10,000 maximum
Contact: Rachel Souris; (312) 335-5807; fax: (312) 335-4034; email: rachel.souris@alz.org
Internet: http://www.alz.org/professionals_and_researchers_conference_grants.asp
Sponsor: Alzheimer, Association
225 N. Michigan Avenue, Floor 17, Suite 1700
Chicago, IL 60601-7633

AMA Foundation Seed Grants for Research 654

The AMA Foundation established the Seed Grant Research Program to encourage medical students, physician residents and fellows to enter the research field. The program provides $2,500-$5,000 grants to help them conduct small basic science, applied, or clinical research projects. These funds will round out new project budgets, rather than sustain current initiatives. One-year grants will be awarded in the following research categories: Cardiovascular/pulmonary diseases; Leukemia; Neoplastic diseases; Secondhand smoke. Grants in Cardiovascular/Pulmonary Diseases, HIV/AIDS, Leukemia, and Neoplastic Diseases will be $2,500. Grants in the Secondhand smoke category will be $5,000.
Requirements: Applicants must be a medical student, physician resident or fellow of an accredited U.S. medical school or institution; they must also be either a U.S. citizen or a permanent resident of the U.S.. Projects must be applicant-conceived, rather than ongoing research of their mentor or Principal Investigator.
Restrictions: Seed grant funds cannot be used for salary or stipend, indirect/administrative costs, to hire a consultant or contractor, and solely for travel expenses. Seed grants will not be awarded to any applicant who has previously received an AMA Foundation seed grant in the research category in which they are applying.

Amount of Grant: $2,500 - $5,000
Date(s) Application is Due: Dec 15
Contact: Seed Grants Program; (312) 464-4200; fax: (312) 464-5973; email: seedgrants@ama-assn.org
Internet: http://www.ama-assn.org/ama/pub/category/7785.html
Sponsor: American Medical Association Foundation
515 N. State Street
Chicago, IL 60654

AMA-MSS Research Poster Award 655
The MSS/RFS Joint Research Poster Symposium is held annually at the MSS Interim Meeting (November). Research is presented in eight categories (biochemistry/cell biology, cancer biology, cardiology/vascular biology, clinical/epidemiological/health care, immunology/microbiology, neurobiology/neuroscience, radiology/imaging, surgery). The overall winner receives a free trip to the MSS Annual Meeting in Chicago.
Date(s) Application is Due: Sep 1
Contact: Rebecca Gierhahn, Director; (800) 262-3211 ext. 4753; email: rebecca.gierhahn@ama-assn.org or mss@ama-assn.org
Internet: http://www.ama-assn.org/ama/pub/about-ama/our-people/member-groups-sections/medical-student-section/opportunities/grants-awards-scholarships.shtml
Sponsor: American Medical Association
515 N. State Street
Chicago, IL 60654

AMA-WPC Joan F. Giambalvo Memorial Scholarships 656
The American Medical Association (AMA) Foundation in association with the AMA Women Physicians Congress (WPC) has established the Joan F. Giambalvo Memorial Scholarship Fund with the goal of advancing the progress of women in the medical profession and strengthening the ability of the AMA to identify and address the needs of women physicians and medical students. Proposals for the Joan F. Giambalvo Memorial Scholarship Fund will be accepted between Nov. 1 and Feb. 15. The AMA-WPC Joan F. Giambalvo Memorial Scholarship seeks innovative research proposals focusing on professional work/practice issues that affect women physicians, including, but not limited to: part-time working strategies; leadership training protocols; gender-based physician practice patterns; physician satisfaction or burnout; retention incentives; practice re-entry issues. Proposals for projects with concurrent/complementary funding and/or plans to use this grant as seed money for larger studies to follow will be received favorably, as will proposals for independent new projects. This award is for a maximum of $10,000. A budget submitted with each applicant, proposal should reflect the anticipated use of funds for the amount requested. Proposals will be evaluated and awardee(s) selected based on a variety of factors, including, but not limited to: the innovation, quality and/or feasibility of the idea; project/research methodology; potential to produce action, change or more comprehensive studies; and career goals of the applicant.
Requirements: Applicants/awardees shall be: female or male; working alone or collaboratively on the specific research project; physicians, medical students, other health professionals, or individuals working or doing graduate work in an applicable profession such as public health, sociology, psychology, etc.
Restrictions: Members of the AMA-WPC Governing Council, AMA Foundation Board of Directors, AMA Board of Trustees and Joan F. Giambalvo Memorial Scholarship Fund selection committee, and AMA and AMA Foundation staff are not eligible for this award.
Amount of Grant: $10,000 maximum
Date(s) Application is Due: Feb 15
Contact: Alice Reed, Scholarship Contact; (312) 464-5523; email: alice.reed@ama-assn.org
Internet: http://www.ama-assn.org/ama/pub/about-ama/our-people/member-groups-sections/women-physicians-congress/about-wpc/joan-f-giambalvo-memorial-scholarship.shtml
Sponsor: American Medical Association
515 N. State Street
Chicago, IL 60654

Amarillo Area/Harrington Foundations Grants 657
The community foundation seeks to improve the quality of life in the 26 northernmost counties of the Texas Panhandle. Grants are awarded to nonprofits in support of arts and culture, education, health care, and social services. Types of support include research, building construction/renovation, equipment acquisition, program development, seed grants, scholarship funds, and matching funds.
Requirements: Nonprofit 501(c)3 organizations located in the northernmost 26 counties of the Texas Panhandle, including Dallam, Sherman, Hansford, Ochiltree, Lipscomb, Hartley, Moore, Hutchinson, Roberts, Hemphill, Oldham, Potter, Carson, Gray, Wheeler, Deaf Smith, Randall, Armstrong, Donley, Collingsworth, Parmer, Castro, Swisher, Briscoe, Hall, and Childress, are eligible.
Restrictions: The foundation does not make grants to or for religious activities; political lobbying or legislative activities; endowments; debt retirement; deficit financing, reduction of operating deficit, etc.; private or parochial schools; national, state, or local fund-raising activities; general operating expenses for United Way agencies; or umbrella funding organizations that would distribute requested funds at their own discretion.
Geographic Focus: Texas
Samples: U of Texas at Austin (TX)—for an international fellowship program for faculty members and graduate students, $30 million.; U of Texas at Austin (TX)—to endow an international fellowship program that will provide salary, research, and travel grants to approximately five faculty members and 16 graduate students annually, $30 million.

Date(s) Application is Due: Jan 5; Jul 6
Contact: Kathy Grant, (806) 376-4521; email: kathie@aaf-hf.org
Internet: http://aaf-hf.org/grants/guidelines.htm
Sponsor: Amarillo Area/Harrington Foundations
801 S Fillmore Street, Suite 700
Amarillo, TX 79101

Amazon Basin Scholarship Program 658
Affiliated with Harvard University, LASPAU administers scholarships for 24 graduate students seeking to complete a master's degree or a one-year graduate certificate program. The purpose of the program is to develop a well-prepared cadre of professionals from the Amazon Basin countries who will return to their home institutions on completion of their studies and share their expertise with colleagues to foster research collaboration between countries.
Requirements: Eligible applicants must be citizens and residents of Brazil, Bolivia, Colombia, Ecuador, Peru, or Venezuela; have a bachelor's degree in any natural science, social studies, or public policy relating to environmental issues; and have field experience in the Amazon region.
Restrictions: Candidates must be nominated by a university, research institution, nongovernmental organization, or government agency in their home country.
Contact: Fay Henderson de Diaz, Director, Program Relations and Management, (617) 495-0511; fax: (617) 495-8990; email: fay_henderson@harvard.edu
Internet: http://www.laspau.harvard.edu/fb-oas.htm
Sponsor: LASPAU: Academic and Professional Programs for the Americas
25 Mount Auburn Street
Cambridge, MA 02138-6095

Amelia Peabody Charitable Fund 659
This fund gives grants for higher education, hospitals and health services, the environment, and culture primarily in Massachusetts and New England. Types of support include capital campaigns, building construction/renovation, equipment acquisition, endowment funds, and research. Initial approach should be a proposal giving pertinent information concerning the structure of the requesting organization, amount requested, purpose of the project, budget, personnel involved, and other foundations with whom proposals have been filed for this project. Application guidelines are available upon request.
Requirements: Nonprofit organizations operating in Massachusetts are eligible to apply.
Restrictions: Awards will not be made to individuals, for films, or for periods of more than one year, nor will grants be awarded to religious or political organizations or organizations maintained by local, state, or federal governments. Salaries will not be supported and very seldom will operating grants be awarded.
Geographic Focus: Massachusetts
Samples: Boston Public Library Foundation (Boston, MA)—program support, $100,000.; YMCA of Cambridge Family (Cambridge, MA)—program support, $100,000.
Amount of Grant: $10,000-$100,000 average
Date(s) Application is Due: Feb 1; Jun 1; Oct 1
Contact: Cheryl Gideon, Program Contact, (617) 451-6178
Sponsor: Amelia Peabody Charitable Fund
10 P.O. Sq, North Suite 995
Boston, MA 02109-4603

America-Norway Heritage Fund Grants 660
The Lutheran Brotherhood Insurance Company in Minneapolis, Minnesota, hands out this award that aims to cover travel expenses as well as an honorarium for a two week stay in Norway. It was established in 1985 by the Lutheran Brotherhood Insurance Society in the hope of giving special grant to Norwegian-Americans who have contributed in a particular way to American culture. The student can be of all ages, but he or she must present works on American culture in Norway during their stay. Also, the stay must be between October and April, and the applicant must be an American of Norwegian descent. Students wanting to apply for this grant must present a proposal to the company; not an application.
Requirements: Americans of Norwegian descent who have made significant contributions to American culture are eligible to apply.
Contact: Lutheran Brotherhood Insurance Company; (612) 340-7037; email: mail@norway.com
Internet: http://www.norseman.no/grants.asp
Sponsor: Norwegian Information Service in the 625 Fourth Avenue South, MS 856
Minneapolis, MN 55415

American Academy of Advertising Doctoral Dissertation Competition 661
The competition exists to promote doctoral research in advertising. Each award is in the range of $1,000 to $2,500. In addition to the standard awards, the Dunn Award is given for outstanding proposals in the area of international advertising. Awards are based on a competitive review of dissertation proposals. Any topic in advertising may be addressed.
Requirements: Only members of the American Academy of Advertising working on their dissertation at the time of proposal submission are eligible for these awards. If funded, you must also maintain membership until you complete your project. Applicants must be currently enrolled in a graduate program. Recipients have three years to complete their dissertation from the time of the award to receive the second half of their award. Applicants must submit a proposal package including the required documents. See the sponsor's website for the guidelines and required documentation. The submission package - including the proposal and letter of endorsement — must reach the Chair of the Research Committee by 5pm of the deadline date.

Restrictions: Submissions based on completed or near completed dissertations are not eligible; submissions should be in the proposal stage.
Samples: Hyunjae Yu, Grady College of Journalism and Mass Communication, The University of Georgia: Food Advertising and Children: Understanding the Role Television Advertising Plays in Conflicts between Parents and Children Regarding Healthy Food Choices.; Courtney Carpenter, The University of Alabama: What is Most Important to Kids? Developmental Differences in Response to Spokescharacter Appearance and Behavior Associated with Nutritional Content of Food Products in Advertisements Targeting Children.; Stevie Watson, Mississippi State University: The Relevance of Skin Tone: Viewers' Responses to Black Models in Advertising.; Juran Kim, The University of Tennessee: Developing an Integrated Model of Interactivity in the Context of Travel Related Websites.
Amount of Grant: $1,000 -$2,500
Date(s) Application is Due: Nov 7
Contact: Dr. Janas Sinclair, School of Journalism and Mass Communication; University of North Carolina at Chapel Hill; (919) 843-5638; email: sinclair@unc.edu
Internet: http://www.aaasite.org/Call_for_-_Doctoral_Dissertation_Competition.html
Sponsor: American Academy of Advertising
24710 Shaker Boulevard
Beachwood, OH 44122

American Academy of Advertising Research Fellowships 662

The American Academy of Advertising Research Fellowship Competition promotes the continued scholarship of professors and advertising professionals who have completed their education and are doing research in advertising. Each award, typically, is in the range of $1,000 to $3,000. Any topic that is appropriate for potential publication in the Journal of Advertising is eligible for the competition.
Requirements: Candidates must comply with the following *Requirements*: (1) Persons submitting proposals must be current members of the American Academy of Advertising. If funded, you must also maintain membership until you complete your project; (2) Winners must grant the Journal of Advertising first right of refusal on any papers resulting from the supported research. Research fellows receive half of the award at the time of selection and half of the award when the first completed paper is submitted to the Journal of Advertising; (3) Winners must complete the awarded research project in three years to receive the second half of the award. If the project is not completed in three years, the second half of the award is automatically forfeited; (4) Winners should acknowledge that the project was funded by an AAA Research Fellowship in all publications resulting from the project. Winners are asked to publicize the award on their campuses, in their communities, etc., attaining as much publicity as possible. See the guidelines available online for the requirements for submission of proposals. Electronic submissions must be received no later than 5pm of the deadline date.
Restrictions: Doctoral students are not eligible to participate in this competition, and joint research with doctoral students is likewise specifically excluded, unless that research will not be part of the student's dissertation.
Amount of Grant: $1,000 - $3,000
Date(s) Application is Due: Aug 29
Contact: Dr. Janas Sinclair, School of Journalism and Mass Communication; University of North Carolina at Chapel Hill; email: sinclair@unc.edu
Internet: http://www.aaasite.org/Research_Fellowship_Competition.html
Sponsor: American Academy of Advertising
24710 Shaker Boulevard
Beachwood, OH 44122

American Academy of Arts and Sciences Hellman Fellowships 663

As part of the Initiative for Science, Engineering, and Technology, the American Academy has established the the Hellman Fellowship in Science and Technology Policy for an early-career professional with training in science or engineering who is interested in transitioning to a career in public policy and administration. While in residence, the Hellman Fellow will work with senior scientists and policy experts on critical national and international policy issues related to science, engineering, and technology. The focus of the work will be on one or more of the ongoing projects under the Academy's Initiative for Science, Engineering, and Technology to which the Hellman Fellow will contribute substantively. The mission of the Initiative is to examine, in broad terms, how the world of science and technology is evolving, how to help the public understand these changes, and how society can better adapt. The Initiative brings together scientists and public policy experts in a neutral setting, outside of the constraints of the political process.
Requirements: Applicants must have a Ph.D. in an area of science or engineering and have some experience or a demonstrated interest in an area related to science and technology policy. Masters degrees may be considered in the fields of engineering and computer science. Strong writing and organizational skills are desired. Candidates must be U.S. citizens, permanent residents, or current employees of an academic or professional organization in the United States.
Date(s) Application is Due: Jan 15
Contact: Patricia McGarry, Program Coordinator; (617) 576-5068; fax: (617) 576-5050; email: pmcgarry@amacad.org or vsp@amacad.org
Internet: http://www.amacad.org/hellman.aspx
Sponsor: American Academy of Arts and Sciences
136 Irving Street
Cambridge, MA 02138-1996

American Academy of Arts and Sciences Visiting Scholars Program 664

The program supports research fellowships at the academy's visiting scholars center. Fellowships will be awarded to individuals who can demonstrate that their work will make a substantial contribution in one or more of the academy's four major research areas: science and global security; social policy and American institutions; humanities and culture; and education. The academy also welcomes proposals that explore the impact of scientific and technological advances over the past two centuries on American institutions, humanities and culture in America, American foreign policy, and global security. Proposals should take into account the academy's emphasis on interdisciplinary work, as well as its interest in broadening public understanding of important intellectual trends and contemporary policy choices. In addition to pursuing individual projects, scholars participate in activities of the academy; collaboration with academy fellows is encouraged. Visiting scholars remain in residence at the academy throughout the academic year (September-May), with some fellowships eligible for a one-year renewal. Faculty remain in residence for one academic year, although one-semester fellowships may be considered. Scholars receive office space, computer services, library privileges, and assistance in locating housing. Health benefits can be arranged. Applicants are responsible for contacting three references and ensuring that their letters are mailed to the academy by the postmark deadline. Faculty can receive up to $60,000 (not to exceed one-half of current salary). Postdocs receive an annual stipend of up to $40,000. The Academy provides office space, computer services, library privileges, and information on locating housing.
Requirements: Preference is given to untenured junior faculty but the program is also open to qualified postdocs. Candidates must be U.S. citizens, permanent residents, or current employees of an academic or professional organization in the United States. The Ph.D., J.D, or equivalent professional training (e.g., public policy) should have been completed within the last 10 years (although exceptional circumstances will be taken into consideration). Graduate student applicants must complete all degree requirements by August 1.
Restrictions: Research trips, interviews, attendance at scholarly meetings or speaking engagements must be limited to no more than twenty days during the fellowship term, and scholars must not accept teaching appointments or other major commitments during their fellowship. Applications must be submitted electronically.
Samples: David Ekbladh, Ph.D., Columbia University-for The Great American Mission: Development and the Creation of an American World Order (2007-08); Lisa Fluet, Assistant Professor of 20th Century British and Anglophone Literature, Boston College—for Modernism, Human Rights, and the Novel, 1921-1961 (2007-08).
Amount of Grant: $40,000 maximum for postdocs; $60,000 maximum for faculty
Date(s) Application is Due: Oct 16
Contact: Patricia McGarry, Program Coordinator; (617) 576-5068; fax: (617) 576-5050; email: pmcgarry@amacad.org or vsp@amacad.org
Internet: http://www.amacad.org/visiting.aspx
Sponsor: American Academy of Arts and Sciences
136 Irving Street
Cambridge, MA 02138-1996

American Academy of Neurology Clinical Research Training Fellowship 665

This fellowship is a two-year award to support research in clinical neurosciences. The award supplies a stipend plus tuition coverage of formal education in clinical research methodology at the institution or elsewhere. Supplementation of the stipend by other grants is allowable.
Requirements: The fellowship is open to non-U.S. neurologists with interest in an academic career in clinical research who have completed residency training less than five years prior to application. Preference will be given to applicants who have completed one or two years subspecialty training in an area of clinical research interest. Only one application per department will be considered. The applicant must supply a letter of nomination from the chair of the institution's department of neurology; a letter of intent to pursue a two-year program indicating specific clinical focus, proposed institution and preceptor, and future goals; a three-page research proposal; a copy of current curriculum vita and three letters of reference; letter from proposed preceptor detailing any ongoing clinical research in the area of interest, manner the applicant will interact in the service and research components of the fellowship, assurance of restricted clinical responsibilities, and assumption of fellow's research costs; and documentation of formal coursework to include quantitative clinical epidemiology, biostatistics, study design, data analysis, and ethics. Ten copies of all materials must be submitted.
Amount of Grant: $50,000 per year for two years plus $7000 tuition
Date(s) Application is Due: Oct 3
Contact: Martin Schaefer, Executive Director, (651) 695-2759; fax: (651) 695-2791; email: foundation@aan.com or mschaefer@aan.com
Internet: http://www.neurofoundation.org/research/research.cfm
Sponsor: American Academy of Neurology
1080 Montreal Avenue
Street Paul, MN 55116

American Academy of Neurology International Scholarship Program 666

The purpose of this scholarship is to enable a young investigator to attend the academy's annual meeting to deliver a platform or poster presentation at the scientific sessions. A stipend, travel and lodging expenses, and program admission fees are provided. Application forms are available on the Web site. Annual deadline dates may vary; contact program staff for exact dates.
Requirements: This scholarship is available to residents of countries other than the United States or Canada who are investigators in the field of neurology. Applications are especially welcome from qualified individuals in underdeveloped countries.

Restrictions: Foreign residents living in the United States or Canada are not eligible.
Amount of Grant: Up to $2500 toward hotel accommodations and a round-trip coach airfare to the American Academy of Neurology Annual Meeting.
Date(s) Application is Due: Dec 1
Contact: Cheryl Alementi, (651) 695-2737; email: calementi@aan.com
Internet: http://www.aan.com/professionals/awards/award/awa_res_scho.cfm
Sponsor: American Academy of Neurology
1080 Montreal Avenue
Saint Paul, MN 55116

American Academy of Nursing Building Academic Geriatric Nursing   **667**
   Capacity (BAGNC) Scholarships
The Foundation's overall goal is to increase the nation's capacity to provide effective and affordable care to its rapidly increasing older population. Specifically, the Foundation seeks to enhance the training of physicians, nurses, social workers and other health professionals who care for older adults, and promote innovations in the integration and delivery of services. The goal of the Scholarship program is to increase academic geriatric nursing capacity in the United States. BAGNC focuses on the development of academic leadership in gerontological nursing through strong mentorship in the components of academic geriatric nursing (research, teaching and community service); leadership development, a national network of scholars and academic geriatric nurses; and exposure to a wide range of experts in gerontology and geriatrics. Scholars in collaboration with their identified mentor will design and implement a tailored professional development plan designed to support development of new competencies and enhanced effectiveness as an academic leader.
Requirements: Registered nurses who are U.S. citizens or U.S. permanent residents and who hold a degree(s) in nursing are eligible. Predoctoral applicants must: be registered nurses; hold degree(s) in nursing; be United States citizens or permanent U.S. residents; plan an academic and research career; and demonstrate potential for long-term contributions to geriatric nursing.
Amount of Grant: $50,000 maximum per annum for two years
Date(s) Application is Due: Jan 15
Contact: Patricia Archbold, Program Director; (202) 777-1172; fax: (202) 777-0107; email: parchbold@aannet.org
Internet: http://www.aannet.org/i4a/pages/Index.cfm?pageID=3295
Sponsor: American Academy of Nursing / John A. Hartford Foundation
888 17th Street, NW Suite 800
Washington, DC 20006

American Academy of Nursing Claire M. Fagin Fellowships   **668**
The Fellowship supports two years of full time advanced research and leadership training for doctorally prepared faculty committed to careers in academic geriatric nursing by providing $120,000 for the 2-year fellowship ($60,000 per annum). Program focuses on the development of academic leadership in gerontological nursing through such activities as: research; focused study; networking among scholars, mentors and colleagues in other fields; demonstration of growth in ability to transform self and organizations by moving outside of traditional modes of success; completion and write-up of a significant research project; and by success in achieving funding from other sources. Selected fellows, in collaboration with their mentor, will design and implement an individual professional development plan that will support them in developing new competencies and enhanced effectiveness as an academic leader and researcher. Award programs must begin between July 1st and September 1st of the award year. The program is committed to advancing well-qualified applicants from under-represented minority groups to improve the nation's ability to provide culturally competent care to its increasingly diverse aging population.
Requirements: Applicants must: be doctorally-prepared registered nurses; hold degree(s) in nursing; be United States citizens or permanent U.S. residents; be doctorally-prepared registered nurses,; have the potential to develop into independent investigators; and demonstrate potential for long-term contributions to geriatric nursing. Applications will be accepted from doctoral students who will complete their doctoral program prior to the award. Faculty members in accredited Schools of Nursing who hold the rank of assistant professor or associate professor may apply for fellowships.
Amount of Grant: $120,000 for the 2-year fellowship ($60,000 per annum)
Date(s) Application is Due: Jan 13
Contact: Patricia Archbold, Program Director; (202) 777-1172; fax: (202) 777-0107; email: parchbold@aannet.org
Internet: http://www.geriatricnursing.org/applications/cmf-fellowship.asp
Sponsor: American Academy of Nursing / John A. Hartford Foundation
888 17th Street, NW Suite 800
Washington, DC 20006

American Academy of Nursing Mayday Fund Grants   **669**
The Foundation's overall goal is to increase the nation's capacity to provide effective and affordable care to its rapidly increasing older population. Specifically, the Foundation seeks to enhance the training of physicians, nurses, social workers and other health professionals who care for older adults, and promote innovations in the integration and delivery of services. This program is aimed at candidates whose research includes the study of pain in the elderly. Award programs must begin between July 1st and September 1st.
Requirements: Predoctoral applicants must: be registered nurses; hold degree(s) in nursing; be United States citizens or permanent U.S. residents; plan an academic and research career; and demonstrate potential for long-term contributions to geriatric nursing.
Amount of Grant: $5,000 maximum

Date(s) Application is Due: Jan 9
Contact: Patricia Archbold, Program Director; (202) 777-1172; fax: (202) 777-0107; email: parchbold@aannet.org
Internet: http://www.geriatricnursing.org/applications/applications.asp
Sponsor: American Academy of Nursing / John A. Hartford Foundation
888 17th Street, NW Suite 800
Washington, DC 20006

American Academy of Religion Collaborative Research Grants   **670**
These grants are intended to stimulate cooperative research among scholars in different institutions, with a focus on a clearly identified research project. They may also be used for interdisciplinary work with scholars outside the field of religion, especially when such work shows promise of continuing beyond the year funded. Grants can provide funds for networking and communication. In addition, grants may be used to support small research conferences. Conference proposals will be considered only if they are designed primarily to advance research. Conferences presenting papers that report on previous research will not be considered. Award notification letters will be sent by the end of October. There are no application forms; guidelines are available online.
Requirements: Applicants must be current AAR members who have been in good standing for the previous three years. In the case of proposals involving scholars from other disciplines, not all participants need to hold AAR membership.
Restrictions: Applicants will not be considered who have received an AAR research grant in the previous five years.
Amount of Grant: $500-$5,000
Date(s) Application is Due: Aug 1
Contact: Jessica Davenport, (404) 727-4707; fax: (404) 727-7959; email: jdavenport@aarweb.org or info@aarweb.org
Internet: http://www.aarweb.org/Programs/Grants/Research/collaborative.asp
Sponsor: American Academy of Religion (AAR)
825 Houston Mill Road, Suite 300
Atlanta, GA 30329-4205

American Academy of Religion Individual Research Grants   **671**
The grants provide support to academy members for important aspects of research, such as travel to archives and libraries, research assistance, field work, and release times. Proposals are judged on the project's contribution to scholarship in religious studies and on the significance of that contribution for advancing the understanding of religion or for advancing discussion between religion and other humanistic and social science disciplines. Scholars in small institutions, departments, or programs without research support and unaffiliated scholars are encouraged to apply.
Requirements: Applicants: must be current AAR members who have been in good standing for the previous three years; will not be considered if they have received an AAR Research Award in the previous five years; must have completed the doctorate.
Restrictions: Funds are not provided for dissertation research, nor for travel to attend the AAR annual meeting.
Samples: Grace G. Burford, Prescott College: B. Horner and the Transmission of Buddhism to the West; Jamsheed K. Choksy, Indiana University, Bloomington: Whither the Zoroastrian Minority Amidst Sectarian Sociopolitics in Contemporary Iran; Jane Naomi Iwamura, University of Southern California: Altared States: A Cultural History of the Japanese American Home Shrine
Amount of Grant: $500-$5,000
Date(s) Application is Due: Aug 1
Contact: Jessica Davenport; (404) 727-4707; fax: (404) 727-7959; email: jdavenport@aarweb.org or info@aarweb.org
Internet: http://www.aarweb.org/Programs/Grants/Research/individual.asp
Sponsor: American Academy of Religion (AAR)
825 Houston Mill Road, Suite 300
Atlanta, GA 30329-4205

American Academy of Underwater Science Student Scholarship   **672**
The AAU.S. awards scholarships to graduate students engaged in, or planning to begin, a research project in which diving will be used as a principal research tool. One scholarship will be awarded to a masters program student and the other to a PhD candidate.
Requirements: Applicant must be a member of AAU.S. and submit a three- to five-page proposal and a statement of support from a faculty member.
Amount of Grant: $2500
Date(s) Application is Due: Jun 30
Contact: Roy Houston, AAU.S. Scholarship Committee Chair, (310) 338-7343; email: rhouston@lmu.edu
Internet: http://www.aaus.org/scholarshipweb.cfm
Sponsor: American Academy of Underwater Sciences
430 Nahant Road
Nahant, MA 01908

American Association of Petroleum Geologists Foundation Grants in Aid   **673**
The purpose of the program is to support graduate students in the earth sciences whose research has application to the search for and development of petroleum and energy-minerals resources and to related environmental geology issues. Eighteen special grants named in recognition of individuals or entities that have made substantial contributions to the foundation, petroleum and energy-mineral sciences, and teaching are awarded each year. Grants are to be applied to expenses directly related to the student's thesis work, such as summer fieldwork, analytical analyses, etc.

Requirements: The program focuses on support of qualified candidates for master's and equivalent degrees. Qualified doctoral candidates with expenses outside the usual scope of funding by other agencies are also encouraged to apply.
Amount of Grant: $500-$2000
Date(s) Application is Due: Jan 31
Contact: Rebecca Griffin, Grants Coordinator, (888) 945-2274 ext 409 or (918) 560-9409; fax: (918) 560-2642; email: rgriffin@aapg.org
Internet: http://www.aapg.org/foundation/gia/index.cfm
Sponsor: American Association of Petroleum Geologists Foundation
P.O. Box 979
Tulsa, OK 74101

American Astonomical Society Chretien International Research Grants   674
AAS administers this award in honor of the late Henri Chretien to further international collaborative projects in observational astronomy with emphasis upon long-term international visits and the development of close working relationships with astronomers in other countries. Awards are open to astronomers throughout the world and may be used to cover any reasonable costs associated with astronomical observational research including travel costs, salary, publication costs, and small pieces of research equipment. If appropriate, the recipient's family is encouraged to accompany him/her. Preference will be given to individuals of high promise who are otherwise unfunded. Innovative technical approaches including the development and use of new optics, new devices, and new techniques will count heavily in the applicant's favor.
Requirements: Astronomers with a Ph.D or equivalent. Applications should A description of the research project (less than three pages in length) including an assessment of its importance to that particular subfield of astronomy and a statement enumerating all the aspects of international collaboration. A statement of the candidate's ability to do the proposed research. Special emphasis should be placed on international collaboration and foreign visits which have been arranged. Include facilities available and observing time allocations, if any. A proposed budget with brief justification for the amount requested. A description of other financial resources available. The candidate's curriculum vitae and bibliography of recent papers. Two letters of reference from astronomers who know the candidate's work. The candidate is responsible for ensuring that these letters reach the Committee by the application deadline. Any special circumstances which might help in the decision process.
Restrictions: These awards will normally not be given to supplement a major research project that is funded elsewhere. There is no institutional overhead included in this award. Graduate students are not eligible.
Amount of Grant: $20,000 maximum for one or more individuals or groups
Date(s) Application is Due: Apr 1
Contact: Grants Manager, (202) 328-2010; fax: (202) 234-2560; email: aas@aas.org
Internet: http://www.aas.org/grants/chretien.html
Sponsor: American Astronomical Society
2000 Florida Avenue NW, Suite 400
Washington, DC 20009-1231

American Astronomical Society Small Research Grants   675
These small research grants, sponsored jointly by the American Astronomical Society and NASA and administered by AAS, are awarded to postdoctoral astronomers to cover costs associated with any type of astronomical research. Eligible costs include page charges, computing costs, equipment, shipping of equipment, and travel to some observatories.
Requirements: Applicants must be astronomers with a PhD or equivalent degree. Astronomers from smaller, less well-endowed institutions will be given priority. Proposals will be accepted from individuals not associated with an institution. Proposals are reviewed at January and June AAS meetings.
Restrictions: Graduate students are ineligible. No salaries or overhead will be paid.
Amount of Grant: $1000-$7000
Date(s) Application is Due: May 6; Dec 2
Contact: Susana Deustua, Grant Administrator, (202) 328-2010; fax: (202) 234-2560; email: deustua@aas.org
Internet: http://www.aas.org/grants/smrg.html
Sponsor: American Astronomical Society
2000 Florida Avenue NW, Suite 400
Washington, DC 20009

American Australian Association Australia To U.S.A Fellowships   676
The fellowships are available for Australian scholars doing advanced research or study in the fields of business, science, technology, medicine or engineering in the United States. In return, the fellows are expected to add to Australia, intellectual capital as well as contribute to the country, overall social and economic well-being upon their return. The Fellowships will support part of the costs of one year of research or study in the United States. Applicants must submit a complete budget. The Association, budget form lists acceptable expenses.
Requirements: Applicants must: be Australian citizens doing research or study at the graduate level. Applicants may already be in the U.S.; conduct a well-defined research project or study in one of the following fields—business, science, technology, medicine or engineering; demonstrate why travel to the United States is important to their research or study; demonstrate how their research or study will benefit Australia and intend to return to work in Australia; show proof of the arrangements made at a United States university or institution where the research or study will be conducted for the academic year beginning in September of that year; are expected to devote full time to their research or study; and be fluent in English.

Amount of Grant: $25,000 maximum
Date(s) Application is Due: Mar 15
Contact: Diane Sinclair, Director, (212) 338-6860; email: diane.sinclair@aaanyc.org
Internet: http://www.americanaustralian.org/Educational/australia-usa.php
Sponsor: American Australian Association
599 Lexington Avenue, 18th Floor
New York, NY 10022

American Australian Association U.S.A to Australia Fellowships   677
Fellowships are available for Americans who will benefit from doing advanced research or study in the fields of life sciences, engineering, medicine, or mining; there will be particular interest in the fields of oceanography/marine sciences and stem cell research. In return, the fellows are expected to add to the United States and Australia, intellectual capital as well as contribute to those countries‚Äô overall social and economic well-being upon their return.
Requirements: Applicants must: be American citizens or permanent residents of the U.S.A doing research or study at the graduate level (applicants may already be in Australia); conduct a well-defined research project or study in one of the following fields: life sciences, engineering, medicine or mining. There will be particular interest in the fields of oceanography/marine sciences and stem cell research; demonstrate how their research or study will benefit the United States and Australia and that they intend to return to work in the United States; demonstrate why travel to Australia is important to their research or study; show proof of the arrangements made at an Australian university or institution where the research or study will be conducted for the next academic year; devote full time to their research or study; and be fluent in English.
Amount of Grant: $20,000 maximum
Date(s) Application is Due: Oct 31
Contact: Diane Sinclair, Director, (212) 338-6860; email: diane.sinclair@aaanyc.org
Internet: http://www.americanaustralian.org/Educational/usa-australia.php
Sponsor: American Australian Association
599 Lexington Avenue, 18th Floor
New York, NY 10022

American Automatic Control Council John R. Ragazzini Award   678
The award is made for outstanding contributions to automatic control education in any form. The awardee normally is a teacher, but there is no formal requirement that nominees be members of a university faculty.
Requirements: Nomination packages should consist of a letter of nomination, three to five reference letters, current resume, current list of publications and patents, any other supporting material, and a nomination form. The nomination letter should clearly identify the primary reason that the nominee should receive the award, as well as four ancillary reasons. All materials should be submitted in a single package.
Date(s) Application is Due: Dec 1
Contact: Professor Pradeep Misra, Electrical Engineering Department, Wright State University, (937) 775-5062; fax: (937) 775-3936; email: pmisra@cs.wright.edu
Internet: http://www.a2c2.org/awards/ragazzini/nomination.php
Sponsor: American Automatic Control Council
3640 Col Glenn Hwy
Dayton, OH 45435

American Chemical Society Award for Team Innovation   679
This award was established to highlight the value and importance of technical teams and teamwork to the chemical and allied industries by recognizing a multidisciplinary team for successfully moving an innovative idea to a product now in commercial use. For each team member, the award consists of $3,000 and a certificate. Up to $1,000 for travel expenses to the meeting at which the award will be presented will be reimbursed to each team member. A certificate will also be provided to the employers of the team.
Requirements: The team shall be multidisciplinary in nature and consist of not fewer than two nor more than five members. The team's work leading to this award must have been carried out primarily in the United States and the technical accomplishments of the team must be documented in the technical literature as a publication(s) or a patent(s). The output of the team must also demonstrate innovation, commercialization of a product or process, be of a special value to society, and of a nature that could only be achieved by professionals working together effectively. The rate of commercialization of the team's output will also be considered in the selection of the recipients of this award.
Restrictions: Application or self-nomination is not acceptable.
Samples: 2008 - Stanley Collins (3M) and Scott R. Culler (3M); 2007 - Michael R. Barbachyn (Pfizer Inc.), Steven J. Brickner (Pfizer Inc.), Douglas K. Hutchinson (Abbott Laboratories) and Peter R. Manninen (Eli Lilly and Company); 2006 - Charles A. Harbert (retired), B. Kenneth Koe (retired), Reinhard Sarges (retired), Albert Weissman (retired) and Willard M. Welch (retired), Pfizer Inc..; 2005 - S. Randall Holmes-Farley (Genzyme Drug Discovery and Development) and W. Harry Mandeville (Peptimmune, Inc. for work done at Genzyme Drug Discovery and Development); 2004 - Steven A. Van Slyke and Ching W. Tang (Eastman Kodak Company).
Amount of Grant: $3000 per team member + travel expenses
Date(s) Application is Due: Nov 1
Contact: Awards Office; (202) 872-4408; fax: (202) 776-8008; email: awards@acs.org
Internet: http://portal.acs.org/portal/acs/corg/content?_nfpb=true&_pageLabel=PP_ARTICLEMAIN&node_id=1319&content_id=CTP_004554&use_sec=true&sec_url_var=region1
Sponsor: American Chemical Society
1155 16th Street NW
Washington, DC 20036-4800

American Chemical Society Chemical Technology Partnership Mini Grants   680

The American Chemical Society (ACS) offers mini grants to help collaboration activities among industry, academia, and the community that are essential to the education and professional development of chemical technicians. The program aims to: (1) Raise community and industry awareness of the changing needs of professional technicians; (2) Highlight opportunities for industry, academia, professional societies, and the community to collaborate on meeting those needs; (3) Increase involvement of current and future technicians in the American Chemical Society.

Requirements: Eligible programs or activities must meet the following conditions: (1) Program/activity must support technician education or career development; (2) Representatives from at least two different sectors of the chemical enterprise (academia, industry, workforce organizations, professional societies, etc.) must be involved; (3) Program/activity must take place during the year of application.

Samples: Spring 2008: Chemical Laboratory Technician Major‚ÄìStudent Employability Project - Bidwell Training Center.; Video Project: Process Technology in the Great Lakes Region - Great Lakes Process Technology Alliance, Inc., Marathon Oil, Dow Corning, Delta College.; Career Development Talks - Mid-Michigan Technician Group, Dow Chemical Company, Dow Corning, Delta College.; BioTech Community Day - Piedmont Community College (Roxboro, NC), Argos Therapeutics, Inc.

Date(s) Application is Due: Feb 20

Contact: Blake J. Aronson; (800) 227-5558 or (202) 872-4600; email: b_aronson@acs.org

Internet: http://portal.acs.org/portal/acs/corg/content?_nfpb=true&_pageLabel=PP_TRANSITIONMAIN&node_id=1772&use_sec=false&sec_url_var=region1

Sponsor: American Chemical Society
1155 Sixteenth Street, NW
Washington, DC 20036

American Chemical Society Claude S. Hudson Awards   681

This award, supported by the National Starch and Chemical Company, is given biennially in odd-numbered years to recognize outstanding contributions to carbohydrate chemistry, whether in education, research, or application. The award is granted without regard to age or nationality. Applications are accepted in even-numbered years. The award consists of $5,000 and a certificate. Up to $1,000 for travel expenses to the meeting at which the award will be presented will be reimbursed.

Requirements: Any individual, except a member of the award committee, may submit one nomination or seconding letter for the award in any given year. Nominating documents consist of a letter of not more than 1000 words containing an evaluation of the nominee's accomplishments and a specific identification of the work to be recognized, a biographical sketch including date of birth, and a list of publications and patents authored by the nominee.

Restrictions: Self-nominations are not accepted.

Samples: Peter H. Seeberger—award recipient, (2009); Pierre Sinay—award recipient, (2007); David Bundle—award recipient, (2005).

Amount of Grant: $5,000 and up to $1,000 for travel expenses to award meeting

Date(s) Application is Due: Nov 1

Contact: Felicia Dixon, Awards Administrator; (202) 872-4408; fax: (202) 776-8008; email: f_dixon@acs.org or awards@acs.org

Internet: http://portal.acs.org/portal/acs/corg/content?_nfpb=true&_pageLabel=PP_ARTICLEMAIN&node_id=1319&content_id=CTP_004501&use_sec=true&sec_url_var=region1

Sponsor: American Chemical Society
1155 Sixteenth Street, NW
Washington, DC 20036

American Chemical Society Corporate Associates Seed Grants   682

The American Chemical Society recognizes activities at the community level and provides grants to support programs that advance the public's understanding of chemistry. Awards are generally $1,000 up to a maximum of $10,000 per calendar year. CA Seed Grants should not be considered as continuous support.

Requirements: Proposals will be considered in the following areas: (a) Education in the field of chemical sciences; (b) Education of the public regarding the chemical industry; (c) Enhancement of professionalism or safety in chemistry. Specific proposal guidelines and application instructions are available at the sponsor's website.

Amount of Grant: $10,000 maximum

Date(s) Application is Due: Feb 15; Jul 1

Contact: Nelufar Mohajeri, Manager, Industry Member Programs; (202) 872-4441, or 1-800-227-5558 ext. 4441; email: n_mohajeri@acs.org

Internet: http://portal.acs.org/portal/acs/corg/content?_nfpb=true&_pageLabel=PP_TRANSITIONMAIN&node_id=1454&use_sec=false&sec_url_var=region1

Sponsor: American Chemical Society
1155 Sixteenth Street, NW
Washington, DC 20036

American Chemical Society GCI Pharmaceutical Roundtable Research Grants   683

The Roundtable is a partnership with the ACS GCI (Green Chemistry Institute) and pharmaceutical corporations united by a shared commitment to integrate the principles of green chemistry and engineering into the business of drug discovery and production. The program was created to support research in the key green chemistry research areas. (Ten chemical transformations and two process related operations were identified and subsequently published - download the list from the sponsor's website). The primary purpose of this grant is to publish research to make information publicly available.

Requirements: Proposals will only be accepted from public and private institutions of higher education. The grant is not limited to institutions in the United States. Proposals must be submitted through the appropriate institutional office for external funding. Proposed research must focus on the key targeted areas identified by the Roundtable. Proposals must clearly describe how the proposed research will provide improvements over existing technology that are consistent with the goals of Green Chemistry including improved atom economy, use of less hazardous reagents, more energy efficient processes, elimination of by-products, reduced waSuite generation, improved environmental, health, and safety and life cycle impacts, catalytic versus stoichiometric methods, and development of inherently safer transformations. Illustrations of the relevance of the proposed technology to synthetic and process related problems within the pharmaceutical industry are encouraged. Complete guidelines can be downloaded at the sponsor's website.

Restrictions: Applicants may have only one research grant with the ACS GCI Pharmaceutical Roundtable at a time. Current research grant holders may not apply for a new grant until their active grant or no-cost time extension officially expires.

Amount of Grant: up to $130,000

Date(s) Application is Due: Sep 9

Contact: Michael Bernstein; (202) 872-4400 or (202) 872-6102; fax: (202) 776-8009; email: m_bernstein@acs.org or gcipr@acs.org

Internet: http://portal.acs.org/portal/acs/corg/content?_nfpb=true&_pageLabel=PP_TRANSITIONMAIN&node_id=1456&use_sec=false&sec_url_var=region1

Sponsor: American Chemical Society
1155 Sixteenth Street, NW
Washington, DC 20036

American Chemical Society GCI PRF Green Chemistry Grants   684

The American Chemical Society (ACS) Green Chemistry Institute (GCI) sponsors a green chemistry grant program with funds allocated from the ACS Petroleum Research Fund (PRF). The activities funded by the program will be used for green chemistry research or education to be conducted in accordance with the PRF Trust Agreement. Grants are made to nonprofit institutions of higher education in the United States and other countries in response to proposals.

Requirements: Projects will support green chemistry activities with the environmental objectives of intrinsic hazard reduction, biosphere protection and process change in areas of pure science connected with the petroleum field. Projects must: (1) Be related to the petroleum field; (2) Advance scientific educational and fundamental research; and/or, (3) Be of strategic importance and broad applicability. Further guidelines and downloadable application can be found at the sponsor's website.

Restrictions: No overhead costs may be charged, which includes secretarial and/or administrative salaries. Funds may not be used to support laboratory technicians, contractors, consultants, or visiting faculty. The principal investigator (PI) is ineligible to submit a grant application to GCI-PRF if the PI has submitted a grant application to PRF or GCI-PRF within the past 12 months. Further, a principal investigator who holds an active PRF, GCI-PRF, or other GCI grant (including a grant on time extension) may not submit an application for a new grant.

Amount of Grant: $10,000 - $60,000

Date(s) Application is Due: Jan 11

Contact: Gayle Peterman, Senior Grants Administrator; (202) 872-6092; fax: (202) 872-6319; email: g_peterman@acs.org

Internet: http://portal.acs.org/portal/acs/corg/content?_nfpb=true&_pageLabel=PP_SUPERARTICLE&node_id=1289&use_sec=false&sec_url_var=region1

Sponsor: American Chemical Society
1155 Sixteenth Street, NW
Washington, DC 20036

American Chemical Society Herman Frasch Foundation Grants   685

The foundation grants are intended for research in the field of agricultural chemistry. Supported research is primarily chemical or biochemical in nature. Because such research is often interdisciplinary, a joint effort by two scientists with expertise in different complementary fields may achieve the desired results; joint projects have been endorsed by the Frasch Committee. Projects which focus on nutrition, agronomy, genetics, or entomology are not usually favored. Proposals must provide opportunities for training students at the graduate, undergraduate, and/or postdoctoral levels. Grants will be made in the amount of $50,000 per year or $250,000 for the maximum five year period. Proposals are considered only once in five years and grants are allocated at five-year intervals.

Requirements: Research must be conducted in the field of Agricultural Chemistry. The results obtained must be of potential practical benefit to the agricultural development of the United States. Grantee must be a non-profit institution incorporated and existing within the United States.

Restrictions: Funds cannot be used for tuition, overhead, administrative expense, large items of equipment, salary expenses or benefits for the principal investigator.

Samples: (2007-2012) Paul Chirik, Cornell University - Fertilizer Synthesis from Atmospheric Nitrogen; Markus W. Ribbe, University of California, Irvine - The Fascinating Metalloclusters in Nitrogen Fixation: A Case Study of the Vanadium Nitrogenase; Justin P. Gallivan, Emory University - Engineering Bacteria that seek and Destroy WaSuite Herbicides; Suzanne L. Tobey, Wake Forest University - Enantioselective Synthesis of Alpha-Amino Phosphonates using an Ene Reaction.

Amount of Grant: $50,000 per year

Contact: (800) 227-5558 (U.S.) or (202) 872-4600 (Worldwide); email: fraschinfo@acs.org

Internet: http://portal.acs.org/portal/acs/corg/content?_nfpb=true&_pageLabel=PP_TRANSITIONMAIN&node_id=1290&use_sec=false&sec_url_var=region1
Sponsor: American Chemical Society
1155 Sixteenth Street, NW
Washington, DC 20036

American Chemical Society Irving Langmuir Award in Chemical Physics   686

This award, supported by the General Electric Global Research, is presented in even-numbered years to recognize and encourage outstanding interdisciplinary research in chemistry and physics. A nominee must have made an outstanding contribution to chemical physics or physical chemistry within the 10 years preceding the year in which the award is made. The award will be granted without restriction, except that the recipient must be a resident of the United States and the monetary prize must be used in the United States or its possessions. Applications are accepted in odd-numbered years.
Requirements: Any individual, except a member of the award committee, may submit one nomination or seconding letter for the award in any given year. Nominating documents consist of a letter of not more than 1000 words containing an evaluation of the nominee's accomplishments and a specific identification of the work to be recognized, a biographical sketch including date of birth, and a list of publications and patents authored by the nominee.
Restrictions: Self-nominations are not accepted.
Samples: Daniel M. Numark—award winner, $10,000 (2008); F. Fleming Crim, Jr.—award winner, $10,000 (2006); Mark Ratner—award winner, $10,000 (2004).
Amount of Grant: $10,000 and allowance for traveling expenses to award meeting
Date(s) Application is Due: Nov 1
Contact: Felicia Dixon, Awards Administrator; (800) 227-5558 or (202) 872-4408; fax: (202) 776-8008; email: f_dixon@acs.org or awards@acs.org
Internet: http://portal.acs.org/portal/acs/corg/content?_nfpb=true&_pageLabel=PP_ARTICLEMAIN&node_id=1319&content_id=CTP_004534&use_sec=true&sec_url_var=region1
Sponsor: American Chemical Society
1155 Sixteenth Street, NW
Washington, DC 20036

American Chemical Society Local Section YCC Starter Grants   687

The Younger Chemists Committee (YCC) funds starter grants to help Local Section Younger Chemists Committees (LSYCCs) sponsor their first big event or launch a new program. Recipients receive a cash award to start a LSYCC; reports and share the program ideas with other LSYCCs; and opportunities to network with scientists in their area while engaging in fun activities with other chemists.
Requirements: Applications are reviewed year round. Creativity is encouraged.
Amount of Grant: $250
Contact: Alvin Collins, Ph.D.; (800) 227-5558, ext. 8724 or (202) 872-8724; fax: (202) 776-8003; email: a_collins@acs.org
Internet: http://portal.acs.org/portal/acs/corg/content?_nfpb=true&_pageLabel=PP_ARTICLEMAIN&node_id=120&content_id=WPCP_006959&use_sec=true&sec_url_var=region1
Sponsor: American Chemical Society
1155 Sixteenth Street, NW
Washington, DC 20036

American Chemical Society Mentoring Program Grants   688

The American Chemical Society Committee on Minority Affairs offers minigrants of up to $500 to local sections to support their implementing programs that encourage the participation of underrepresented minority scientists.
Requirements: To find the local section in your area, go to http://membership.acs.org/y/ycc/lsycc/lsycc.htm.
Amount of Grant: $500
Contact: Brenda Thompson; (800) 227-5558, ext. 4065 or (202) 872-4065; fax: (202) 776-8003; email: b_thompson@acs.org
Internet: http://portal.acs.org/portal/acs/corg/content?_nfpb=true&_pageLabel=PP_ARTICLEMAIN&node_id=120&content_id=WPCP_006959&use_sec=true&sec_url_var=region1
Sponsor: American Chemical Society
1155 Sixteenth Street, NW
Washington, DC 20036

American Chemical Society PRF Doctoral New Investigator Grants   689

The goals of the American Chemical Society Petroleum Research Fund are (1) the support of fundamental research in the petroleum and energy fields, and (2) development of the next generation of engineers and scientists through support of advanced scientific education. The emphasis of the DNI grants is focused on providing start-up funding for scientists and engineers who are within the first three years of their first academic appointment, those who have limited or no preliminary results for a research project they wish to pursue, with the intention of using the preliminary results obtained to seek continuation funding from other agencies. An estimated 90 awards will be given each year. Proposals are considered at three advisory board meetings per year (early fall, mid-winter, and late spring), which are coordinated with the ACS Grants & Awards board meetings.
Requirements: The program is seeking investigator-initiated, original research across the spectrum of our mission. Original research is defined as being different from that performed previously by the PI as part of their graduate or postdoctoral studies. Excluded from consideration are proposals in which the ideas being presented are a mere extension

of research from the PI, graduate or postdoctoral experience. Research projects must be unique. Regularly-appointed faculty members at U. S. academic institutions are eligible to apply. Research or adjunct faculty holding non-tenured, long-term scientific appointments also may be eligible. Current eligibility criteria are: (a) The non-profit institution submitting the DNI proposal must certify that the individual listed as a principal investigator on the cover page qualifies as a principal investigator under the institution's policies. (b) Each principal investigator must be eligible to serve as the formal, official supervisor of the students for whom he/she is seeking support. (c) The terms of appointment of the principal investigator must promise reasonable continuity of service. The appointment should continue at least through the period of funding requested in the proposal. The full application packet and eligibility requirements can be downloaded at the sponsor's website.
Restrictions: Although a PI may send the same proposal to more than one agency, PRF will not support a project having overlap, or partial overlap, with research funded by another agency. New investigators may submit only one research proposal for consideration per meeting and may not hold more than one active PRF research grant at a time. New investigators may have only three (3) DNI grant applications considered in their career. Thereafter, the new investigator may apply to the New Directions (ND) Grant Program. A principal investigator with an active PRF research grant, including a grant on time extension, may not submit an application for a new grant.
Amount of Grant: $100,000 over 2 years
Date(s) Application is Due: Mar 28; Jun 13; Nov 14
Contact: Dr. W. Christopher Hollinsed, Director; (202) 872-6207; fax: (202) 872-6319; email: prfinfo@acs.org
Internet: http://portal.acs.org/portal/acs/corg/content?_nfpb=true&_pageLabel=PP_TRANSITIONMAIN&node_id=1264&use_sec=false&sec_url_var=region1
Sponsor: American Chemical Society
1155 Sixteenth Street, NW
Washington, DC 20036

American Chemical Society PRF New Directions Grants   690

This program replaces the previous PRF Type AC grant program. The emphasis of the New Directions (ND) grants is focused on providing funds for scientists and engineers who have very limited or even no preliminary results for a research project they wish to pursue with the intention of using the PRF-driven preliminary results to seek continuation funding from other agencies. The ND grants are to be used to illustrate proof of concept, that is, feasibility, and, accordingly, are to be viewed as seed money for new research ventures. The complete application packet is available for download at the sponsor's website.
Requirements: The lead principal investigator must demonstrate that this is an entirely new research direction from what the lead principal investigator has done to date. Proposals that are considered to not be a new direction of research will be denied without external review, or may be rejected. Regularly-appointed faculty members at academic institutions in countries where ACS PRF can administer grants are eligible. In addition, certain other long-term scientific appointments may be eligible. Current eligibility criteria are: (a) The non-profit institution submitting the ND proposal must certify that each individual listed as a principal investigator on the cover page qualifies as a principal investigator under the institution's Principal Investigator policies. (b) In view of the long-standing policy of The Petroleum Research Fund to give priority to support of students (undergraduate, graduate, or postdoctoral), each principal investigator (lead and co-PI) must be eligible to serve as the formal, official supervisor of the students for whom he/she is seeking support. (c) The terms of appointment of the principal investigator (and if appropriate the co-PI) must promise reasonable continuity of service through the life of the grant. Proposals are due by 5:00 pm of the published deadline date.
Restrictions: Principal investigators (lead or co-) may submit only one research proposal per meeting. Principal investigators may not hold more than one active PRF research grant at a time. Principal investigators with an active PRF research grant, including a grant on time extension, may not submit an application for a new grant. Principal investigators may have only one ND grant application considered in a 12-month period. Therefore, a principal investigator who has an ND proposal denied may not submit another ND proposal until the Advisory Board meeting one year later.
Amount of Grant: $100,000 over 2 years
Date(s) Application is Due: Mar 28; Jun 13; Aug 1; Nov 14; Jan 30
Contact: Dr. W. Christopher Hollinsed, Director; (202) 872-6207; fax: (202) 872-6319; email: prfinfo@acs.org
Internet: http://portal.acs.org/portal/acs/corg/content?_nfpb=true&_pageLabel=PP_TRANSITIONMAIN&node_id=1262&use_sec=false&sec_url_var=region1
Sponsor: American Chemical Society
1155 Sixteenth Street, NW
Washington, DC 20036

American Chemical Society PRF Scientific Education (Type SE) Grants   691

The program assists with the travel expenses of foreign (defined as non-U.S. and non-Canadian) scientists who participate in major symposia in this country (or in Canada). The symposium topic must be 'fundamental research in the petroleum field' to be eligible for support. The PRF Advisory Board makes recommendations on Type SE proposals during meetings three times per year: September, January, and May. Funding for foreign travel assistance is limited to $1,200 per foreign speaker with a maximum of $3,600 per symposium (three or more visiting speakers). Download the application packet from the sponsor's website.
Requirements: Researchers at colleges and universities are eligible. Type SE proposals for ACS symposia should be written on behalf of an ACS division, rather than the principal investigator, institutional affiliation; proposals for support of non-ACS symposia should

be written on behalf of the institution organizing the symposium. In order to be eligible, researchers must meet the following *Requirements*: (1) The symposium topic must be 'fundamental research in the petroleum field'. (2) Each eligible speaker must be from a foreign (non-U.S. and non-Canadian) non-profit institution, and must give a petroleum-relevant or alternate-energy presentation. (3) The sponsor of the symposium must be a non-profit scientific or educational institution. (4) The meeting or symposium must be accessible and open; symposia whose attendance is by invitation only, or where facilities sharply limit attendance have not been favored for support by the PRF Advisory Board. (5) The results of the meeting or symposium must be publishable in peer-reviewed literature.

Restrictions: The program does not support travel of American scientists to similar meetings in foreign countries, on the rationale that more American scientists derive benefit from having foreign speakers travel to North America, and that support helps American organizations to enhance the value of their programs and, in some instances, to meet their obligations as hosts for international meetings.

Amount of Grant: $1,200 - $3,600

Date(s) Application is Due: Mar 28; Jun 13; Aug 1; Nov 14; Jan 30

Contact: Dr. W. Christopher Hollinsed, Director; (202) 872-6207; fax: (202) 872-6319; email: prfinfo@acs.org

Internet: http://portal.acs.org/portal/acs/corg/content?_nfpb=true&_pageLabel=PP_TRANSITIONMAIN&node_id=1265&use_sec=false&sec_url_var=region1

Sponsor: American Chemical Society
1155 Sixteenth Street, NW
Washington, DC 20036

American Chemical Society PRF Undergraduate Faculty Sabbatical Grants 692

This program is designed to provide year-long, full-time, research sabbaticals for qualified faculty performing fundamental research in the petroleum field during a sabbatical. This is a matching grants program that can provide up to $45,000 in faculty salary and qualified benefits, and up to $5,000 in relocation and research expenses for a faculty member's sabbatical year. The maximum individual award is $50,000. Applications are considered each year at the September PRF Advisory Board meeting.

Requirements: An eligible sabbatical must be at least one academic year (i.e., 9 months or more) and must be carried out in its entirety at a location other than the applicant's home institution. Sabbaticals may be taken anywhere in the world. Academic, national lab, and industrial sabbaticals are eligible, as are sabbaticals that involve more than one institution. Faculty with tenured appointments in departments that do not offer a doctoral degree and faculty at two- and four-year institutions are eligible to apply. Current ACS PRF grant recipients, as well as those with pending PRF proposals are also eligible. Complete eligibility requirements and application packet can be downloaded from the sponsor's website.

Restrictions: Fundamental research is required. ACS PRF cannot support applied research or methods development. Note also that the proposed research need not be directly connected to petroleum, but rather may provide a basis for subsequent research directly connected with the petroleum field. No overhead or indirect costs may be charged to the grant.

Amount of Grant: $50,000 maximum

Contact: Dr. W. Christopher Hollinsed, Director; (202) 872-6207; fax: (202) 872-6319; email: prfinfo@acs.org

Internet: http://portal.acs.org/portal/acs/corg/content?_nfpb=true&_pageLabel=PP_TRANSITIONMAIN&node_id=1266&use_sec=false&sec_url_var=region1

Sponsor: American Chemical Society
1155 Sixteenth Street, NW
Washington, DC 20036

American Chemical Society PRF Undergraduate New Investigator Grants 693

The program provides funds for scientists and engineers who are beginning their independent careers in academia and have limited or no preliminary results for a research project they wish to pursue. The UNI grants are to be used to illustrate proof of principle, i.e., feasibility, and accordingly, are to be viewed as seed money for generating preliminary results that can be used to apply for continuation funding from the PRF (Type UR grants) or other agencies such as the NSF, DOD, etc. Proposals are considered at two Panel meetings per year (January and June). The submission window for the January meeting is from mid-June to late July. The submission window for the June meeting is from mid-December to late January.

Requirements: Eligibility for a UNI grant requires that a PI is in a department without a doctoral program, and that the students receiving stipends for the work to be done are undergraduates (M.S.-level students can also be supported IF one or more undergraduates are also supported from this grant). Accordingly, the research being proposed need not be high risk but it should be of publishable quality. The research opportunities afforded must be of the highest caliber, and provide a compelling educational experience for the student. To be eligible as a principal investigator for a UNI grant, the applicant must be a member of the faculty of a college or university within the United States; be within the first three years of their first academic appointment as a regular faculty member; have completed all requirements for the Ph.D. (however, an application may be submitted before the degree has been awarded); and be appointed at the rank of Assistant Professor or the equivalent. In addition, applicants must meet the following three criteria: (a) The non-profit institution submitting the UNI proposal must certify that the person listed as a principal investigator on the cover page qualifies as a principal investigator under the institution, policies. (b) In view of the long-standing policy of The Petroleum Research Fund to give priority to support of students, the principal investigator must be eligible to serve as the formal, official supervisor of the students for whom he/she is seeking support. (c) The term of appointment of the principal investigator must promise

reasonable continuity of service. The appointment should continue at least through the period of funding requested in the UNI proposal.

Restrictions: A principal investigator may submit only one research proposal per meeting. Only one UNI grant may be held during an investigator, career.

Amount of Grant: $50,000 over 2 years

Contact: Dr. W. Christopher Hollinsed, Director; (202) 872-6207; fax: (202) 872-6319; email: prfinfo@acs.org

Internet: http://portal.acs.org/portal/acs/corg/content?_nfpb=true&_pageLabel=PP_TRANSITIONMAIN&node_id=1796&use_sec=false&sec_url_var=region1

Sponsor: American Chemical Society
1155 Sixteenth Street, NW
Washington, DC 20036

American Chemical Society PRF Undergraduate Research Grants (UR) 694

This program replaces the previous PRF Type B grant program. The emphasis of the UR grants is to provide funding for scientists and engineers with established programs of research at non-doctoral departments. The program supports the research programs of established scientists and engineers at non-doctoral research departments and provides financial support for students at those institutions to become involved in advanced investigative research activities, in preparation for continued study in graduate school or employment. An estimated number of 45 awards will be made each year. Membership in the American Chemical Society is not a requirement or a factor in awarding ACS PRF grants.

Requirements: ACS PRF type UR proposals are considered at two Panel meetings per year (January and June). The submission window for the January meeting is from mid-June to late July. The submission window for the June meeting is from mid-December to late January. To be eligible as a principal investigator for a UR grant, the applicant must hold a primary appointment in a college or university department that does not grant the Ph.D. or any doctoral degree. Research or adjunct faculty holding non-tenured, long-term scientific appointments also may be eligible. Current eligibility criteria are: (a) The non-profit institution submitting the Type B proposal must certify that each individual listed as a principal investigator on the cover page qualifies as a principal investigator under the institution, policies; (b) The principal investigator must be eligible to serve as the formal, official supervisor of students for whom he/she is seeking support; (c) The terms of appointment of each principal investigator must promise reasonable continuity of service. The appointment should continue at least through the period of funding requested in the UR proposal. Download more eligibility requirements at the sponsor's website.

Restrictions: A principal investigator may have only one research proposal active or under consideration per meeting. This limitation extends to those who have co-PI status on an application. A PI must close an active grant, after the full grant period has been completed, before submitting another proposal.

Amount of Grant: $65,000 over 3 years

Contact: Dr. W. Christopher Hollinsed, Director; (202) 872-6207; fax: (202) 872-6319; email: prfinfo@acs.org

Internet: http://portal.acs.org/portal/acs/corg/content?_nfpb=true&_pageLabel=PP_TRANSITIONMAIN&node_id=1263&use_sec=false&sec_url_var=region1

Sponsor: American Chemical Society
1155 Sixteenth Street, NW
Washington, DC 20036

American Chemical Society Travel Grants for Chemists with Disabilities 695

The Committee on Chemists with Disabilities can assist students and postdoctoral researchers with disabilities with Travel Grant funding to defray some or all of the additional travel and lodging expenses to attend scientific meetings. Recipients receive: up to $2,500 to cover additional travel costs often incurred when attending technical meetings/conferences by people with a disability; and, opportunities to build lasting professional relationships through networking.

Amount of Grant: up to $2,500

Contact: Tony Scurry; (800) 227-5558, ext. 6262 or (202) 872-6262; fax: (202) 776-8003; email: w_scurry@acs.org

Internet: http://portal.acs.org/portal/acs/corg/content?_nfpb=true&_pageLabel=PP_ARTICLEMAIN&node_id=120&content_id=CTP_005836&use_sec=true&sec_url_var=region1

Sponsor: American Chemical Society
1155 Sixteenth Street, NW
Washington, DC 20036

American College of Bankruptcy Grants 696

The college awards grants for existing bankruptcy or consumer-debtor programs, programs in development, and research projects to improve the delivery of pro bono bankruptcy services. Grants also support organizations for broad educational purposes. There are no application guidelines. Guidelines and application are available online.

Samples: The Center for Disability and Elder Law (CDEL) (IL)—$5,000 (2006); Legal Aid Society (KY)—$5,000 (2006); Volunteer Legal Services, Judy Sobin, Executive Director (HI)—$1000 (2005); Idaho Volunteer Lawyers Program, Carol Craighill (ID)—$1000 (2005); Alaska Legal Services Corp, Erick Cordero (Anchorage, AK)—$4000 (2004).

Amount of Grant: $1000-$10,000

Date(s) Application is Due: Mar 1

Contact: Shari Bedker, Executive Director; (703) 934-6154; fax: (703) 802-0207; email: sbedker@amercol.org or college@amercol.org

Internet: http://www.amercol.org/acb_foundation.cfm

Sponsor: American College of Bankruptcy
PMB 626A, 11350 Random Hills Road, Suite 800
Fairfax, VA 22030-6044

American College of Rheumatology Fellows-in-Training Travel Scholarships   697

The American College of Rheumatology Fellows-in-Training (FIT) Travel Scholarship is designed to provide funding to assist Fellows- In-Training at U.S. and Canadian programs in rheumatology attend the Annual Meeting. Preference will be given to those applicants who are currently a first or second year adult or first, second or third year pediatric fellows in a training program approved by the Accreditation Council of Graduate Medical Education or the Royal College of Physicians and Surgeons of Canada. Applicants in a third year of an adult or fourth year of a pediatric fellowship will be given consideration after the registration deadline, space permitting on a first come first serve basis.

Requirements: Applicant must be first or second year adult or first, second or third year pediatric fellows in a training program approved by the Accreditation Council of Graduate Medical Education or the Royal College of Physicians and Surgeons of Canada. Applicants must be in an accredited fellowship position at the time of the Annual Meeting.

Restrictions: Individuals with a faculty appointment will not be considered.
Date(s) Application is Due: Jul 24
Contact: Fellowship Coordinator; (404) 633-3777; fax: (404) 633-1870; email: fittravel@rheumatology.org
Internet: http://www.rheumatology.org/fellows/fitscholarship.asp
Sponsor: American College of Rheumatology (ACR)
1800 Century Place, Suite 250
Atlanta, GA 30345-4300

American College of Surgeons and The Triological Society Clinical Scientist Development Awards   698

The Triological Society (TRIO) and the American College of Surgeons (ACS) announce a competitive research career development program that will provide supplemental funding to otolaryngologist-head and neck surgeons who receive NIH Mentored Clinical Scientist Development Awards (K08) or Mentored Patient-Oriented Research Development Awards (K23). The TRIO and ACS are offering these awards as a means to facilitate the research career development of otolaryngologists-head and neck surgeons, with the expectation that awardees will have sufficient pilot data to submit a competitive R01 proposal prior to the conclusion of the K awards. Awards will use the K08 or K23 mechanisms. Planning, direction, and execution of the program will be the responsibility of the candidate and her/his mentor on behalf of the applicant institution. The project period will be for a period of up to five years with a minimum of three years. Awards are not renewable. TRIO and ACS will provide salary and fringe benefits for the NIH K08/K23 recipient. A combined total of $80,000 per year in direct costs will be awarded to the institution of the awardee.

Requirements: Applicants must be MD otolaryngologist-head and neck surgeons and members of the American College of Surgeons. Applicants must meet all requirements as set forth in the K08 and K23 application criteria and must be recipients of a new K08 or K23 award or have an existing K08/K23 award with a minimum of three years remaining in the funding period as of June.

Restrictions: Pre-existing applications and awards are not eligible for consideration.
Geographic Focus: Alabama; Alaska; Arizona; Arkansas; California; Colorado; Connecticut; Delaware; District of Columbia; Florida; Georgia; Guam; Hawaii; Idaho; Illinois; Indiana; Iowa; Kansas; Kentucky; Louisiana; Maine; Maryland; Massachusetts; Michigan; Minnesota; Mississippi; Missouri; Montana; Nebraska; Nevada; New Hampshire; New Jersey; New Mexico; New York; North Carolina; North Dakota; Ohio; Oklahoma; Oregon; Pacific Northwest; Pennsylvania; Puerto Rico; Rhode Island; South Carolina; South Dakota; Tennessee; Texas; U.S. Virgin Islands; Utah; Vermont; Virginia; Washington; West Virginia; Wisconsin; Wyoming
Amount of Grant: $80,000
Date(s) Application is Due: Apr 15
Contact: Kate Early, Administrative Assistant; (312) 202-5281; email: kearly@facs.org
Internet: http://www.facs.org/memberservices/acstriol.html
Sponsor: American College of Surgeons / Triological Society
633 N. Saint Clair Street
Chicago, IL 60611-3211

American College of Surgeons Co-Sponsored NIH Supplement Awards   699

The American College of Surgeons announces a program that will provide supplemental funding to up to five individuals who receive a Mentored Clinical Scientist Development Award (K08/K23). It is directed at surgeon-scientists working in the early stages of their research careers. The award requires co-sponsorship with an approved surgical society of a three, four or five year period of supervised research experience that may integrate didactic studies with laboratory or clinically based research. This award program will offer a means to facilitate the career development of individuals pursuing careers in surgical research by enhancing salary support over and above that offered by the K08/K23 mechanism. The application deadline is June 12, with funding to begin July 1.

Requirements: Awardees must be members in good standing of both the College and a co-sponsoring surgical society. Participating surgical societies include American Association of Plastic Surgeons (AAPS), American Head and Neck Society (AHNS), American Society of Transplant Surgeons (ASTS), American Vascular Association (AVA), Society of Gynecologic Oncologists (SGO), Society of University Surgeons (SU.S.), and Thoracic Surgery Foundation for Research and Education (TSFRE).
Date(s) Application is Due: Jun 12

Contact: Kate Early, Administrative Assistant; (312) 202-5281; email: kearly@facs.org
Internet: http://www.facs.org/memberservices/acs-nih.html
Sponsor: American College of Surgeons
633 North Saint Clair Street
Chicago, IL 60611-3211

American College of Surgeons Faculty Career Development Award for Neurological Surgeons   700

The American College of Surgeons (ACS) and the Neurosurgery Research and Education Foundation of the American Association of Neurological Surgeons (NREF-AANS) are offering a two-year faculty career development award to neurological surgeons. The award is to support the establishment of a new and independent research program in an area of neurological surgery. The award is $40,000 per year for each of two years, to support the research, and is not renewable thereafter.

Requirements: The award is open to surgeons who are members or candidate members in good standing of both the American College of Surgeons (ACS) and the American Association of Neurological Surgeons (AANS); and have completed specialty training within the preceding five years and have received a full-time faculty appointment at a medical school accredited by the Liaison Committee on Medical Education in the United States or by the Committee for Accreditation of Canadian Medical Schools in Canada.

Restrictions: Applicants may not be current recipients of major research grants.
Amount of Grant: $40,000 per year for two years
Date(s) Application is Due: Dec 1
Contact: Michele S. Gregory, Director of Development; (847) 378-0540 or (847) 378-0500; fax: (847) 378-0600; email: msg@aans.org or info@aans.org
Internet: http://www.facs.org/memberservices/acsnrefaans07.pdf
Sponsor: American College of Surgeons
633 N. Saint Clair Street
Chicago, IL 60611-3211

American College of Surgeons Faculty Research Fellowships   701

The American College of Surgeons is offering two-year faculty research fellowships to surgeons entering academic careers in surgery or a surgical specialty. The fellowship is to assist a surgeon in the establishment of a new and independent research program. Applicants should have demonstrated their potential to work as independent investigators. The fellowship award is $40,000 per year for each of two years, to support the research.

Requirements: The fellowship is restricted to surgeons who have recently (usually within three years) completed formal surgical education and entered the field of full-time academic surgery and received a faculty appointment in a department of surgery, or one of the surgical specialties, at an approved medical school in the United States or Canada. Preference will be given to applicants who directly enter academic surgery following residency or fellowships. Approval of the application is required from the administration (dean or fiscal officer) and the head of the department under whom the recipient will be studying during the fellowship.

Amount of Grant: $40,000 per year
Date(s) Application is Due: Nov 2
Contact: Kate Early, Administrative Assistant; (312) 202-5281; email: kearly@facs.org
Internet: http://www.facs.org/memberservices/acsfaculty.html
Sponsor: American College of Surgeons
633 North Saint Clair Street
Chicago, IL 60611-3211

American College of Surgeons Resident Research Scholarships   702

ACS scholarships will be awarded each year to encourage residents to pursue careers in academic surgery. Priority will be given to residents beginning full-time investigative activities, with no clinical responsibilities, for the two-year period of the scholarship. Study outside the United States or Canada is permissible. The scholarship is for the personal use of the recipient and is not to diminish or replace the usual or expected compensation; award is made directly to the scholar and not to the institution. Renewal for the second year is contingent upon acceptable progress and study protocol for the second year.

Requirements: The applicant must have completed two postdoctoral years in an accredited surgical training program in the United States or Canada at the time the scholarship is awarded and shall not complete formal residency training before the end of the scholarship. Approval of the application is required from the administration (dean or fiscal officer) and the head of the department under whom the recipient will be studying.

Restrictions: Only in exceptional circumstances will more than one ACS scholarship be granted in a single year to applicants from the same institution.
Amount of Grant: $30,000 per year
Date(s) Application is Due: Sep 1
Contact: Administrator, Scholarships Division; (800) 621-4111 or (312) 202-5000; fax: (312) 202-5001
Internet: http://www.facs.org/memberservices/acsresident.html
Sponsor: American College of Surgeons
633 N. Saint Clair Street
Chicago, IL 60611-3211

American College of Surgeons Traveling Fellowship to Germany   703

The International Relations Committee of the American College of Surgeons announces the availability of the ACS Traveling Fellowship to Germany. The purpose of this fellowship is to encourage international exchange of surgical scientific information. The ACS Traveling Fellow will visit Germany, and a German Traveling Fellow will visit North America.

Requirements: The scholarship is available to a Fellow of the American College of Surgeons in any of the surgical specialties who meets the following *Requirements*: has a major interest and accomplishment in clinical and basic science related to surgery; holds a current full-time academic appointment in Canada or the United States; is under 45 years of age on the date the application is filed; and is enthusiastic and personable and possesses good communication skills. The Fellow is required to spend a minimum of two weeks in Germany.

Amount of Grant: $6,000 maximum
Date(s) Application is Due: Apr 1
Contact: Kate Early, Administrative Assistant; (312) 202-5281; email: kearly@facs.org
Internet: http://www.facs.org/memberservices/acsgermany.html
Sponsor: American College of Surgeons
633 North Saint Clair Street
Chicago, IL 60611-3211

American College of Surgeons Traveling Fellowship to Japan 704

The International Relations Committee of the American College of Surgeons announces the availability of the ACS Traveling Fellowship to Japan. The purpose of this fellowship is to encourage international exchange of surgical scientific information. The ACS Traveling Fellow will visit Japan, and a Japanese Traveling Fellow will visit North America. The College will provide the sum of $7,500 to the successful applicant, who will also be exempted from registration fees for the annual meeting of the Japan Surgical Society.

Requirements: The scholarship is available to a Fellow of the American College of Surgeons in any of the surgical specialties who meets the following *Requirements*: has a major interest and accomplishment in clinical and basic science related to surgery; holds a current full-time academic appointment in Canada or the United States; is under 45 years of age on the date the application is filed; and is enthusiastic and personable and possesses good communication skills. The Fellow is required to spend a minimum of two weeks in Japan.

Amount of Grant: $7,500 maximum
Date(s) Application is Due: Jun 1
Contact: Kate Early, Administrative Assistant; (312) 202-5281; email: kearly@facs.org
Internet: http://www.facs.org/memberservices/acsjapan.html
Sponsor: American College of Surgeons
633 North Saint Clair Street
Chicago, IL 60611-3211

American College of Surgeons Wound Care Management Award 705

The American College of Surgeons is offering a two-year faculty research award through KCI U.S.A, to a general surgeon engaged in a research project addressing wound care management. The purpose of this fellowship will be to acquire knowledge leading to new clinical applications or projects that will provide the medical community with a better understanding of the use of advanced wound healing therapies. The fellowship award is $85,000 per year, with the possibility of an extension for a second year if satisfactory progress is made. Specifically, translational projects such as new methods to mechanically stimulate wound healing, design of clinically applicable skin substitutes, methods to enhance specific, difficult-to-heal wounds related to general or extremity trauma, various soft-tissue injuries, quality limb salvage, or methods to reduce surgical intervention to heal a variety of wounds, are being sought. This award may be used by the recipient for support of his/her research or academic enrichment in any fashion that the recipient deems maximally supportive of his/her investigations. This may include faculty salary replacement, support for clinical research personnel, supplies directly related to the research activity, consumable equipment, and research related travel costs up to $2,000 per year.

Requirements: The fellowship is open to Fellows or Associate Fellows of the College who have: (1) completed the chief residency year or accredited fellowship training within the preceding 10 years; and (2) received a fulltime faculty appointment in a department of surgery or a surgical specialty at a medical school accredited by the Liaison Committee on Medical Education in the United States or by the Committee for Accreditation of Canadian Medical Schools in Canada. Preference will be given to applicants who directly enter academic surgery following residency or fellowship.

Amount of Grant: $85,000 per year
Date(s) Application is Due: May 1
Contact: Administrator, Scholarships Division; (800) 621-4111 or (312) 202-5000; fax: (312) 202-5001
Internet: http://www.facs.org/memberservices/woundcareposter2008.pdf
Sponsor: American College of Surgeons
633 North Saint Clair Street
Chicago, IL 60611-3211

American Council of the Blind Scholarships 706

The council funds scholarships to assist outstanding blind and visually impaired students to continue their education at the postsecondary level. Education may be contemplated in an academic, technical, vocational, or professional training program. These scholarships are one-time awards with no renewal. Contact office for availability.

Requirements: Applicant must be legally blind in both eyes; must be a U.S. citizen or resident alien; and must be enrolled in or under consideration for admission at the postsecondary level.

Date(s) Application is Due: Mar 1

Contact: Terry Pacheco, (800) 424-8666 or (202) 467-5081 ext 19; fax: (202) 467-5085; email: TerryPach@aol.com or info@acb.org
Internet: http://www.acb.org/magazine/2005/bf022005-5.html
Sponsor: American Council of the Blind
1155 15th Street NW, Suite 1004
Washington, DC 20005

American Council on Germany Journalism Fellowship 707

The fellowship enables recipients to conduct research in Germany on current political, economic, and social issues. Proposed research projects should address topics with current political or policy significance in Germany's relations with the United States or within the European Union. The program seeks to create a better understanding of transatlantic matters among American journalists and to enable fellows to gain new perspectives on such topics. While applicants should demonstrate an interest in German and European issues, no prior experience in Germany or Europe is required. ACG will make travel arrangements and develop research itineraries in consultation with the fellows. If desired, the fellows may consult with the German Information Center in New York to help implement their projects. Following the completion of the trip, fellows will submit a report summarizing their findings. Fellows are also encouraged to publish articles on their findings. The fellowship award covers the costs of pre-approved international and domestic travel and a per diem allowance for 14 to 28 days in Germany. Applications are reviewed on a rolling basis. Guidelines are available online.

Requirements: The fellowship program aims to serve American print or broadcast journalists who are in relatively early stages of their careers, including those with only limited exposure to Europe. Knowledge of the German language is not a prerequisite for the program.

Amount of Grant: $150 per diem allowance
Contact: Emily Gildersleeve, Fellowship Coordinator, (212) 826-3636; email: egildersleeve@acgusa.org
Internet: http://acgusa.org/acg_journalism_fellowships.htm
Sponsor: American Council on Germany
14 E 60th Street, Suite 606
New York, NY 10022

American Dietetic Association Foundation Scholarships 708

As the largest provider of dietetic scholarships, the program awards funds to deserving students at all levels of study. Additional funds are granted in the form of continuing education stipends, research grants, and recognition awards to practicing dietitians and nutritionists. Additional program information is available online. Application deadlines vary, and applications are typically made available four to six months before the deadline. To receive an application for a particular award or for more information contact the office.

Requirements: Applicant must be a U.S. citizen. Scholarships are awarded in the field of dietetics at all levels: undergraduate, graduate, internship, technician, and for preprofessional practice programs.

Amount of Grant: $500-$5000
Contact: Grants Administrator, (800) 877-1600 ext 5400; email: education@eatright.org.
Internet: http://www.eatright.org/Public/7772.cfm
Sponsor: American Dietetic Association Foundation
120 S Riverside Plz, Suite 2000
Chicago, IL 60606

American Education Research Association Grants 709

The American Educational Research Association, with support from the National Science Foundation (NSF), and the National Center for Education Statistics (NCES), award grants to stimulate quantitative policy- and practice-related research using nationally representative NCES and NSF data sets, improve knowledge of the range of data available at the two agencies and how to use them, and increase the number of educational researchers using the data sets. Research topics include school persistence and career entry, policies and practices related to achievement, policies and practices influencing student and parental attitudes, middle school education, educational finance, and the quality of educational institutions. One- and two-year projects will be funded, excluding indirect costs. Minorities are encouraged to apply.

Requirements: Grants are available for faculty and postdoctoral researchers.
Amount of Grant: $20,000 maximum for one-year projects; $35,000 maximum for two-year projects
Date(s) Application is Due: Jan 3; Mar 1; Sep 1
Contact: Jeanie Murdock, (805) 964-5264; email: jmurdock@aera.net
Internet: http://www.aera.net/grantsprogram/res_training/res_grants/RGFly.html
Sponsor: University of California, Santa Barbara
1190 Phelps Hall
Santa Barbara, CA 93106-9490

American Family Insurance Corporate Giving Program Grants 710

The corporation awards grants to nonprofits in its areas of interest, including education and youth, arts and culture, health, and human services.

Requirements: Nonprofit organizations and K-12 and higher education institutions located in the companies area of interest.

Restrictions: Individuals or teams and organizations outside the 18-state operating territory will not be supported.

Amount of Grant: $2 million total
Contact: Community Relations, (608) 249-2111; email: info@amfam.com

Internet: http://www.amfam.com/company/commitment_contribution.asp
Sponsor: American Family Insurance Corporation
6000 American Pkwy
Madison, WI 53783

American Foundation for Aging Research Fellowhips 711

AFAR awards fellowships to undergraduate, graduate and pre-doctoral (MD, PhD, DDS, etc.) students who are actively involved or are planning active involvement in a biomedical age-related research project. Awards are to alleviate the costs of tuition and fees at a United States degree granting institution. If such costs are already covered by another source of funds, other than personal funds, the awards are to be used for research purposes. Awards are made each semester and are renewable once. Preapplications may be submitted at any time; renewal requests should be made at least 30 days before the next semester.
Requirements: Applicants must be undergraduate, graduate and pre-doctoral (MD, PhD, DDS, etc.) students who are enrolled at degree programs at colleges or universitites in the United States.
Restrictions: Sociology, psychology, and health-professional related research such as physical therapy and exercise physiology are not currently being funded.
Samples: Dr. Ralph Snyderman, Duke U Medical Ctr (Durham, NC)—for aging-related research, $1000.
Amount of Grant: $1000 graduate fellowships; $500 undergraduate scholarships
Contact: Executive Secretary, (919) 515-5679; fax: (919) 515-2047; email: afar_office@ncsu.edu
Internet: http://www4.ncsu.edu/unity/users/a/agris/afar/afar.htm
Sponsor: American Foundation for Aging Research
Campus Box 7622, North Carolina State University
Raleigh, NC 27695-7622

American Foundation for Suicide Prevention Grants 712

The research grants are awarded to investigators conducting clinical, biological, or psychosocial research on the problem of suicide, for one- or two-year periods, and for up to three years for research fellowships. Grant applicants compete in five categories: Pilot Grants provide seed money for new projects and are awarded to individual investigators without regard to academic rank or pervious research experience with suicide; Young Investigator Grants are awarded for two years of support to investigators at the level of assistant professor or lower; Standard Research Grants are awarded to individual investigators for a two-year period; Distinguished Researcher Awards are awarded to investigators at the level of associate professor or higher with a proven history of research in the area of suicide (to fund new directions and initiatives in suicidology); Postdoctoral Research Fellowships are awarded for full-time training projects by investigators who have received a PhD degree within the preceding three years and have not had more than three years of fellowship support. Application deadlines are December 15 for Young Investigator Grants, Standard Research Grants, Distinguished Researcher Awards, and Postdoctoral Fellowships; and April 15, August 15, and December 15 for Pilot Grants. Guidelines and application forms are available online.
Amount of Grant: $20,000 maximum Pilot Grants; $70,000 maximum Young Investigator Grants; $60,000 maximum Standard Research Grants; $100,000 maximum Distinguished Researcher Awards
Date(s) Application is Due: Apr 15; Aug 15; Dec 15
Contact: Grants Administrator, (888) 333-2377 or (212) 363-3500; fax: (212) 363-6237; email: bkoestner@afsp.org
Internet: http://www.afsp.org/research/grants.htm
Sponsor: American Foundation for Suicide Prevention
120 Wall Street, 22nd Fl
New York, NY 10005

American Historical Association Albert B. Corey Prize in Canadian-American Relations 713

The prize is awarded biennially in even-numbered years to an American or Canadian resident for the best book on the history of Canadian-American relations or on the history of both countries. The prize is awarded jointly by the Canadian Historical Association and the American Historical Association.
Requirements: Books bearing an imprint of the previous two years are eligible for the prize.
Samples: John J. Bukowczyk, Wayne State University; Nora Faires, Western Michigan University; David R. Smith, University of Michigan at Ann Arbor; and Randy William Widdis, University of Regina—Permeable Border: The Great Lakes Basin as Transnational Region, 1650‚Äì1990 (University of Pittsburgh Press and University of Calgary Press, 2005); Stephen High, Nipissing University—Industrial Sunset: The Making of North America's Rust Belt, 1969-1984 (University of Toronto Press, 2003).
Date(s) Application is Due: Jan 15
Contact: Book Prize Administrator, (202) 544-2422; fax: (202) 544-8307; email: aha@theaha.org
Internet: http://www.historians.org/prizes/index.cfm?PrizeAbbrev=COREY
Sponsor: American Historical Association
400 A Street, SE
Washington, DC 20003

American Historical Association Albert J. Beveridge Award 714

The Award is given annually for the best book in English on the history of the United States, Latin America, or Canada from 1492 to the present. Books that employ new methodological or conceptual tools or that constitute a significant reinterpretation of an important historical problem are given preference in the awarding of this prize. Biographies, monographs, and works of synthesis or interpretation are eligible.
Requirements: U.S. and Canadian citizens and permanent residents are eligible. Books published after May 1 of the previous year and before April 30 of the award year are eligible.
Restrictions: Translations, anthologies, and collections of documents are not eligible.
Samples: Louis S. Warren, University of California at Davis—Buffalo Bill, America: William Cody and the Wild West Show (Knopf, 2005); Melvin Patrick Ely, College of William and Mary—Israel on the Appomattox: A Southern Experiment in Black Freedom from the 1790s through the Civil War (Knopf, 2004); Edward L. Ayers, University of Virginia—In the Presence of Mine Enemies: War in the Heart of America, 1859-1863 (W.W. Norton, 2003).
Date(s) Application is Due: May 15
Contact: Book Prize Administrator, (202) 544-2422; fax: (202) 544-8307; email: aha@theaha.org
Internet: http://www.historians.org/prizes/AWARDED/BeveridgeWinner.htm
Sponsor: American Historical Association
400 A Street, SE
Washington, DC 20003

American Historical Association Albert J. Beveridge Grants for Research in the History of the Western Hemisphere 715

The grants support research in the history of the Western hemisphere (United States, Canada, and Latin America). The funds for this program come from the earnings of the Albert J. Beveridge Memorial Fund. Only members of the Association are eligible. The grants are intended to further research in progress and may be used for travel to a library or archive, for microfilms, photographs, or photocopying-a list of purposes that is meant to be merely illustrative, not exhaustive. Preference will be given to those with specific research needs, such as the completion of a project or completion of a discrete segment thereof. Preference will be given to Ph.D. candidates and junior scholars.
Restrictions: Within a five-year period, no individual is eligible to receive more than a combined total of $1,000 from the Beveridge, Kraus, and Littleton-Griswold grant programs.
Samples: Matthew Rothwell, University of Illinois at Chicago—Transpacific Revolutionaries: The Chinese Cultural Revolution in Latin America (2006); Justin Roberts, Johns Hopkins University—Sunup to Sundown: A Comparative Study of Slave Plantation Labor in the Late Eighteenth and Early Nineteenth Centuries (2006); Sarah Cornell, New York, New York—Americans of Two Souths: A Social and Cultural History of African, African American, White, and Indigenous U.S. Southerners and Mexico, 1810-1920 (2006).
Amount of Grant: $1,000 maximum
Date(s) Application is Due: Feb 15
Contact: Research Grant Administrator, (202) 544-2422; fax: (202) 544-8307; email: aha@theaha.org
Internet: http://www.historians.org/prizes/BeveridgeGrantInfo.htm
Sponsor: American Historical Association
400 A Street, SE
Washington, DC 20003

American Historical Association Award for Scholarly Distinction 716

According to the selection criteria of this Award, recipients must be senior historians of the highest distinction who have spent the bulk of their professional careers in the United States. Generally, they must also be of emeritus rank, if from academic life, or equivalent standing otherwise. Under normal circumstances the award is not intended to go to former presidents of the Association; rather, the intent is to honor persons not otherwise recognized by the profession to an extent commensurate with their contributions.
Samples: David Brion Davis, Yale University (2006); Lloyd Gardner, Rutgers University-New Brunswick (2006); Fritz Stern, Columbia University (2006).
Date(s) Application is Due: Apr 30
Contact: Sharon K. Tune, Assistant Director, (202) 544-2422; fax: (202) 544-8307; email: aha@theaha.org
Internet: http://www.historians.org/prizes/ScholarlyDistinction.htm
Sponsor: American Historical Association
400 A Street, SE
Washington, DC 20003-3889

American Historical Association Bernadotte E. Schmitt Research Grants 717

Modest grants support research in the history of Europe, Africa, and Asia. The grants are intended to further research in progress and may be used for travel to a library or archive; for microfilms, photographs, or photocopying; or for coding and keypunching; and other approved research-related activities.
Requirements: Preference will be given to those with specific research needs, such as the completion of a project or completion of a discrete segment thereof. Preference will be given to Ph.D. candidates and junior scholars.
Restrictions: Only members of the Association are eligible to apply.
Samples: Jacqueline Fewkes, University of Pennsylvania—A Woman, Space in Islam: Examining the History of Women's Mosques in Asian Muslim Communities (2006); Elizabeth Mellyn, Harvard University— Madness, Medicine, and the Law in Florence, 1350-1600 (2006); Holly Hurlburt, Southern Illinois University at Carbondale—Caterina Veneta: Gender and Power in the Venetian Mediterranean Affiliation (2006).
Amount of Grant: $1,000 maximum
Date(s) Application is Due: Feb 15

Contact: Research Grant Administrator, (202) 544-2422; fax: (202) 544-8307; email: aha@theaha.org
Internet: http://www.historians.org/prizes/SchmittGrantInfo.htm
Sponsor: American Historical Association
400 A Street, SE
Washington, DC 20003

American Historical Association Clarence H. Haring Prize　　　**718**

This prize is awarded every five years to a Latin American who, in the opinion of the committee, has published the most outstanding book on Latin American history during the preceding five years ending June 1 of the award year. Books published between May 1, 2006, and April 30, 2011, will be eligible for the prize in 2011. There is no language limitation on works submitted. Authors should obtain guidelines from the association.

Requirements: Books published between May 1, 2006, and April 30, 2011, will be eligible for the prize in 2011.

Samples: Marial Iglesias Utset, University of Havana—Las Met??foras del Cambio en la Vida Cotidiana: Cuba, 1898,Äi1902 (Ediciones UNION, 2003); Hilda Sabato, University of Buenos Aires—La pol????tica en las calles: Entre el voto y la movilizacion; Buenos Aires, 1862,Äi1880 (Editorial Sudamericana, 1998); Joao Jose Reis, Federal University of Bahia, Brazil—A morte e uma festa: Ritos funebres e revolta popular no Brasil do seculo XIX (Companhia das Letras, 1993).

Date(s) Application is Due: May 15
Contact: Book Prize Administrator, (202) 544-2422; fax: (202) 544-8307; email: aha@theaha.org
Internet: http://www.historians.org/prizes/index.cfm?PrizeAbbrev=HARING
Sponsor: American Historical Association
400 A Street, SE
Washington, DC 20003

American Historical Association George L. Mosse Prize　　　**719**

The prize is designated for an outstanding major work of extraordinary scholarly distinction, creativity, and originality in the intellectual and cultural history of Europe since the Renaissance. This prize was established with funds donated by former students, colleagues, and friends of Dr. Mosse. Only books of a high scholarly distinction should be submitted. Research accuracy, originality, and literary merit are important selection factors.

Requirements: Books published after May 1 of the previous year and before April 30 of the award year are eligible.

Samples: Sandra Herbert, University of Maryland Baltimore County—Charles Darwin, Geologist (Cornell University Press, 2005); Jonathan Sheehan, Indiana University—The Enlightenment Bible: Translation, Scholarship, Culture (Princeton University Press, 2005); Siep Stuurman, Erasmus University Rotterdam—Francois Poulain de la Barre and the Invention of Modern Equality (Harvard University Press, 2004).

Date(s) Application is Due: May 15
Contact: Book Prize Administrator, (202) 544-2422; fax: (202) 544-8307; email: aha@theaha.org
Internet: http://www.historians.org/prizes/index.cfm?PrizeAbbrev=MOSSE
Sponsor: American Historical Association
400 A Street, SE
Washington, DC 20003

American Historical Association George Louis Beer Prize　　　**720**

The prize is awarded in recognition of outstanding historical writing in European international history since the year 1895 that is submitted by a scholar who is a United States citizen or permanent resident. The phrase - European international history since the year 1895 - may be understood to mean any study of international history since the year 1895 with a significant European dimension. Research accuracy, originality, and literary merit are important factors.

Requirements: Books published after May 1 of the previous year and before April 30 of the award year are eligible. Only books of a high scholarly historical nature should be submitted.

Samples: Mark Atwood Lawrence, University of Texas at Austin—Assuming the Burden: Europe and the American Commitment to War in Vietnam (University of California Press, 2005); Carole Fink, Ohio State University—Defending the Rights of Others: The Great Powers, the Jews, and International Minority Protection, 1878,Äi1938 (Cambridge University Press, 2004); Kate Brown, University of Maryland, Baltimore County—A Biography of No Place: From Ethnic Borderland to Soviet Heartland (Harvard University Press, 2004).

Date(s) Application is Due: May 15
Contact: Book Prize Administrator, (202) 544-2422; fax: (202) 544-8307; email: aha@theaha.org
Internet: http://www.historians.org/prizes/index.cfm?PrizeAbbrev=BEER
Sponsor: American Historical Association
400 A Street, SE
Washington, DC 20003

American Historical Association Helen and Howard R. Marraro Prizes　　　**721**

The prizes are given annually by the American Historical Association, the American Catholic Historical Association, and the Society for Italian Historical Studies for the best works on any epoch of Italian cultural history or on Italian-American relations written by American and Canadian resident citizens. The associations appoint a joint selection committee. One copy of each work, together with a brief curriculum vita and bibliography of the author, must be submitted to designated committee members. The American Historical Association will send instructions upon request.

Requirements: Books published after May 1 of the previous year and before April 30 of the award year are eligible. Entries must first have been published in English by a historian whose usual residence is North America.

Samples: Frank M. Snowden, Yale University—The Conquest of Malaria: Italy, 1900,Äi1962 (Yale University Press, 2006); Thomas V. Cohen, York University—Love and Death in Renaissance Italy (University of Chicago Press, 2004); Ronald G. Musto, American Council of Learned Societies and Italica Press—Apocalypse in Rome: Cola di Rienzo and the Politics of the New Age (University of California Press, 2003).

Date(s) Application is Due: May 15
Contact: Book Prize Administrator, (202) 544-2422; fax: (202) 544-8307; email: aha@theaha.org
Internet: http://www.historians.org/prizes/index.cfm?PrizeAbbrev=MARRARO
Sponsor: American Historical Association
400 A Street, SE
Washington, DC 20003

American Historical Association Herbert Baxter Adams Prize　　　**722**

The association annually awards this prize for an author's first book in the field of European history. Chronological coverage follows a two-year cycle: in even-numbered years, subject matter includes ancient, medieval, or early modern European history through 1815; in odd-numbered years, the subject matter should deal with 1816 through the 20th century. The submission of an entry may be made by an author, a third party, or a publisher. Annual deadline dates may vary; contact the association or visit the Web site for exact dates.

Requirements: U.S. and Canadian citizens and permanent residents are eligible. Books published after May 1 of the previous year and before April 30 of the award year are eligible.

Restrictions: Textbooks in the strict sense of the word are not eligible, but a work of wide scope that interprets a major period or area qualifies. Pamphlets, anthologies, edited works, and other small-scale efforts do not qualify.

Samples: Stephanie Siegmund, University of Michigan, The Medici State and the Ghetto of Florence: The Construction of an Early Modern Jewish Community (2006); Maureen Healy, Oregon State University, Vienna and the Fall of the Habsburg Empire: Total War and Everyday Life in World War I (2005); Ethan H. Shagan, Northwestern University. Popular Politics and the English Reformation (2004).

Date(s) Application is Due: May 15
Contact: Book Prize Administrator, (202) 544-2422; fax: (202) 544-8307; email: aha@theaha.org
Internet: http://www.historians.org/prizes/index.cfm?PrizeAbbrev=ADAMS
Sponsor: American Historical Association
400 A Street, SE
Washington, DC 20003

American Historical Association Herbert Feis Award　　　**723**

This prize is offered annually to recognize the scholarly interests of historians outside academe and the importance of the work of independent scholars in the United States. The terms of the award define both ¬ïcontribution¬î and ¬ïpublic history¬î broadly. Contributions could include work as the administrator of a public history group or agency (such as a historical society, a historic site, or a community history project) or as the creator or producer of a public history product or products (such as a museum exhibit, radio script, web site, oral history collection, or film). The contribution may be the result of years of effort in the field, but the prize might also recognize a singular contribution of major importance such as a path breaking museum exhibit. Public history is defined as work primarily directed at non-academic, non-school-based audiences. Those audiences could be very broad (e.g., television viewers) or highly specialized (e.g., policymakers). Although the audience should be primarily outside of academia, the recipient of the award could be employed at a university.

Requirements: Works published or issued in-house between May 1 of the previous year and April 30 of the upcoming year are eligible.

Samples: Victoria A. Harden, American University and National Institutes of Health (retired); Mark Landsman, Independent Scholar—Dictatorship and Demand: The Politics of Consumerism in East Germany (Harvard University Press, 2005); Jonathan Martin, Brooklyn New York—Divided Mastery, Slave Hiring in the American South (Harvard University Press, 2004).

Date(s) Application is Due: May 15
Contact: Book Prize Administrator, (202) 544-2422; fax: (202) 544-8307; email: aha@theaha.org
Internet: http://www.historians.org/prizes/index.cfm?PrizeAbbrev=FEIS
Sponsor: American Historical Association
400 A Street, SE
Washington, DC 20003

American Historical Association J. Franklin Jameson Fellowship　　　**724**

This fellowship is offered collaboratively each year by the Library of Congress and the Association to support significant scholarly research by young historians for one semester in the collections of the Library of Congress. The applicant's project in American history must be one for which the general and special collections of the Library of Congress offer unique research support. The fellowship will be awarded for one semester or as much of an academic year (September-May) as the fellow desires to spend in residence at the Library of Congress, but the fellow is required to spend at least three months. Working space

will be provided by the Library of Congress. Fellows are not required to complete their projects during tenure, nor need they necessarily publish the results as a discrete work.
Requirements: Applicants must hold the PhD degree or equivalent, must have received this degree within the last seven years, and must not have published or had accepted for publication a book-length historical work. Letters of application should include a vita, a statement concerning the proposed project and its relationship to the Library of Congress holdings, a tentative schedule for tenure of the fellowship, and the names and addresses of three persons qualified to judge the project and the applicant's fitness to undertake it.
Restrictions: The fellowship will not be awarded to permit completion of a doctoral dissertation.
Samples: Lisa Tetrault—The Memory of a Movement: Woman Suffrage and Reconstruction America, 1865‚Äì1890 (2006-07); Chad A. Goldberg—Social policy innovations and political struggles over citizenship from the late 19th century to the 1990s (2005-06); Christopher Capozzola—Political and cultural history of the relationship between citizenship and obligation in 20th-century American public life (2004-05).
Amount of Grant: $5,000
Date(s) Application is Due: Mar 15
Contact: Arnita A. Jones, Executive Director, (202) 544-2422, ext. 100; fax: (202) 544-8307; email: ajones@theaha.org or aha@theaha.org
Internet: http://www.historians.org/prizes/Jameson_fellowship.htm
Sponsor: American Historical Association
400 A Street, SE
Washington, DC 20003

American Historical Association J. Russell Major Prize   **725**
The Prize will be awarded annually for the best work in English on any aspect of French history. The prize was established in memory of J. Russell Major, the distinguished scholar of French history who died on December 12, 1998 at the age of 77. One copy of each entry must be received by each of the following committee members.
Requirements: Books published after May 1 of the previous year and before April 30 of the award year are eligible.
Samples: Todd Shepard, Temple University—The Invention of Decolonization: The Algerian War and the Remaking of France (Cornell University Press, 2006); Barbara Diefendorf, Boston University—From Penitence to Charity: Pious Women and the Catholic Reformation in Paris (Oxford University Press, 2004); Steven Englund, Paris France—Napoleon: A Political Life (Scribner, 2004).
Date(s) Application is Due: May 15
Contact: Book Prize Administrator, (202) 544-2422; fax: (202) 544-8307; email: aha@theaha.org
Internet: http://www.historians.org/prizes/index.cfm?PrizeAbbrev=MAJOR
Sponsor: American Historical Association
400 A Street, SE
Washington, DC 20003

American Historical Association James A. Rawley Prize in Atlantic History   **726**
This AHA Prize is intended to recognize outstanding historical writing that explores aspects of integration of Atlantic worlds before the twentieth century. The prize was established in accordance with the terms of a gift from James A. Rawley, Carl Adolph Happold Professor of History Emeritus at the University of Nebraska at Lincoln. Only books of high scholarly and literary merit will be considered. Research accuracy and originality also will be important factors in the evaluation of the books. Books published between May 1 of the previous year and April 30 of the current year are eligible.
Samples: Christopher Leslie Brown, Rutgers University-New Brunswick—Moral Capital: Foundations of British Abolitionism (University of North Carolina Press for the Omohundro Institute of Early American History and Culture, 2006); Londa Schiebinger, Stanford University—Plants and Empire: Colonial Bioprospecting in the Atlantic World (Harvard University Press, 2004).
Date(s) Application is Due: May 15
Contact: Book Prize Administrator, (202) 544-2422; fax: (202) 544-8307; email: aha@theaha.org
Internet: http://www.historians.org/prizes/index.cfm?PrizeAbbrev=ATLANTIC
Sponsor: American Historical Association
400 A Street, SE
Washington, DC 20003

American Historical Association James Harvey Robinson Prize   **727**
The Prize is awarded biennially for the teaching aid which has made the most outstanding contribution to the teaching and learning of history in any field for public or educational purposes. Teaching aid encompasses textbooks, source and reference materials, audiovisuals, computer-assisted instruction, and public history or museum materials. Monographs and revisions will not be considered. The winner will receive a one-year membership in the Association. The work should: have the potential to influence history education (this influence could be in the form of a model that would have wide adaptability and/or could affect teachers and students through widely taught courses; demonstrate recent and good historical scholarship; and be well written and attractively presented.
Requirements: Monographs and revisions will not be considered. Books published after May 1 two years prior and before April 30 of the award year are eligible.
Samples: World History Matters (worldhistorymatters.org), produced by the Center for History and New Media, George Mason University (2006); World History Matters (worldhistorymatters.org), produced by the Center for History and New Media, George Mason University (2004).
Date(s) Application is Due: May 15

Contact: Book Prize Administrator, (202) 544-2422; fax: (202) 544-8307; email: aha@theaha.org
Internet: http://www.historians.org/prizes/index.cfm?PrizeAbbrev=ROBINSON
Sponsor: American Historical Association
400 A Street, SE
Washington, DC 20003

American Historical Association James Henry Breasted Prize   **728**
The prize is offered for the best book in English in any field of history prior to 1000 AD and rotates annually among the following geographical areas: Near East and Egypt; Far East and South Asia; Africa, North America, and Latin America; and Europe. Only works of high scholarly and literary merit will be considered. One copy of each entry must be submitted to designated committee members.
Requirements: Books published after May 1 of the previous year and before April 30 of the award year are eligible.
Samples: Chris Wickham, All Souls College, University of Oxford—Framing the Early Middle Ages: Europe and the Mediterranean, 400‚Äì800 (Oxford University Press, 2005); Callie Williamson, Independent Scholar—The Laws of the Roman People: Public Law in the Expansion and Decline of the Roman Republic (University of Michigan Press, 2005); Kurt Raaflaub, Brown University—The Discovery of Freedom in Ancient Greece (University of Chicago Press, 2004).
Date(s) Application is Due: May 15
Contact: Book Prize Administrator, (202) 544-2422; fax: (202) 544-8307; email: aha@theaha.org
Internet: http://www.historians.org/prizes/index.cfm?PrizeAbbrev=BREASTED
Sponsor: American Historical Association
400 A Street, SE
Washington, DC 20003

American Historical Association Joan Kelly Memorial Prize in Women's History   **729**
This prize is awarded annually for the book in women's history and/or feminist theory that best reflects the high intellectual and scholarly ideals exemplified by the life and work of the late Joan Kelly. Books should demonstrate originality of research, creativity of insight, graceful stylistic presentation, analytical skills, and a recognition of the important role of sex and gender in the historical process. The inter-relationship between women and the historical process should be addressed. Authors must obtain application forms and guidelines for submission from the association.
Requirements: Books in any chronological period, any geographical location, or in an area of feminist theory that incorporates a historical perspective are eligible for consideration. Books published after May 1 of the previous year and before April 30 of the award year are eligible.
Samples: Dorothy Ko, Barnard College, Columbia University—Cinderella, Sisters: A Revisionist History of Footbinding (University of California Press, 2005); Afsaneh Najmabadi, Harvard University—Women with Mustaches and Men without Beards: Gender and Sexual Anxieties of Iranian Modernity (University of California Press, 2005); Laura Gowing, King's College London—Common Bodies: Women, Touch and Power in Seventeeth-Century England (Yale University Press, 2003).
Date(s) Application is Due: May 15
Contact: Book Prize Administrator, (202) 544-2422; fax: (202) 544-8307; email: aschulkin@theaha.org or aha@theaha.org
Internet: http://www.historians.org/prizes/index.cfm?PrizeAbbrev=KELLY
Sponsor: American Historical Association
400 A Street, SE
Washington, DC 20003-3889

American Historical Association John E. Fagg Prize   **730**
The Prize is conferred annually for the best publication in the history of Spain, Portugal, or Latin America. The prize will be awarded annually for a period of 10 years beginning in 2001 and ending in 2011. Only works of high scholarly and literary merit will be considered. One copy of each entry must be submitted to designated committee members.
Requirements: Applicant must be a published author. To be eligible, a book must be published after May 1 of the previous year and before April 30 of the current year.
Samples: David J. Weber, Southern Methodist University—Barbaros: Spaniards and Their Savages in the Age of Enlightenment (Yale University Press, 2005); Brian A. Catlos, University of California at Santa Cruz—The Victors and the Vanquished: Christians and Muslims of Catalonia and Aragon, 1050‚Äì1300 (Cambridge University Press, 2004); Aline Helg, University of Geneva—Liberty and Equality in Caribbean Colombia, 1770‚Äì1835 (University of North Carolina Press, 2004).
Date(s) Application is Due: May 15
Contact: Book Prize Administrator, (202) 544-2422; fax: (202) 544-8307; email: aha@theaha.org
Internet: http://www.historians.org/prizes/index.cfm?PrizeAbbrev=FAGG
Sponsor: American Historical Association
400 A Street, SE
Washington, DC 20003

American Historical Association John H. Dunning Prize   **731**
The prize is awarded in odd-numbered years by the association to a young scholar for an outstanding monograph in manuscript or in print on any subject relating to United States history. Research accuracy, originality, and literary merit are important factors. One copy of each entry must be submitted to each committee member.

Requirements: An entry must be the author's first or second book published or completed. Books published after May 1 of the previous year and before April 30 of the deadline year are eligible.
Samples: Jon T. Coleman, University of Notre Dame—Vicious: Wolves and Men in America (Yale University Press, 2004); Michael Willrich, Brandeis University—City of Courts: Socializing Justice in Progressive Era Chicago (Cambridge University Press, 2003); Ernest Freeberg, Colby-Sawyer College—The Education of Laura Bridgman: First Deaf and Blind Person to Learn Language (Cambridge, Mass.: Harvard University Press, 2001).
Date(s) Application is Due: May 15
Contact: Book Prize Administrator, (202) 544-2422; fax: (202) 544-8307; email: aha@theaha.org
Internet: http://www.historians.org/prizes/index.cfm?PrizeAbbrev=DUNNING
Sponsor: American Historical Association
400 A Street, SE
Washington, DC 20003

American Historical Association John K. Fairbank Prize in East Asian History   732
The prize is awarded annually for an outstanding book on the history of China proper, Vietnam, Chinese Central Asia, Mongolia, Manchuria, Korea, or Japan substantially since the year 1800. Only books of high scholarly and literary merit will be considered. One copy of each entry must be submitted to each committee member. Application forms are available from the association.
Requirements: Books published after May 1 of the previous year and before April 30 of the award year are eligible.
Restrictions: Anthologies, edited works, and pamphlets are ineligible for the competition.
Samples: Madeleine Zelin, Columbia University—The Merchants of Zigong: Industrial Entrepreneurship in Early Modern China (Columbia University Press, 2006); Ruth Rogaski, Vanderbilt University—Hygienic Modernity: Meanings of Health and Disease in Treaty-Port China (University of California Press, 2004); Jordan Sand, Georgetown University—House and Home in Modern Japan: Architecture, Domestic Space, and Bourgeois Culture, 1880-1930 (Harvard University Asia Center, 2003).
Date(s) Application is Due: May 15
Contact: Book Prize Administrator, (202) 544-2422; fax: (202) 544-8307; email: aha@theaha.org
Internet: http://www.historians.org/prizes/index.cfm?PrizeAbbrev=FAIRBANK
Sponsor: American Historical Association
400 A Street, SE
Washington, DC 20003

American Historical Association Leo Gershoy Award   733
The award is given annually by the association to the author of the most outstanding work in English on any aspect of the field of 17th- and 18th-century western European history. Only books of high scholarly and literary merit will be considered. One copy of each entry must be submitted to designated committee members.
Requirements: Books published after May 1 of the previous year and before April 30 of the award year are eligible.
Samples: Howard G. Brown, State University of New York at Binghamton—Ending the French Revolution: Violence, Justice and Repression from the Terror to Napoleon (University of Virginia Press, 2006); Pamela H. Smith, Columbia University—The Body of the Artisan: Art and Experience in the Scientific Revolution (University of Chicago Press, 2004); Ronald Schechter, College of William & Mary—Obstinate Hebrews: Representations of Jews in France, 1715-1815 (University of California Press, 2003).
Date(s) Application is Due: May 15
Contact: Book Prize Administrator, (202) 544-2422; fax: (202) 544-8307; email: aha@theaha.org
Internet: http://www.historians.org/prizes/index.cfm?PrizeAbbrev=GERSHOY
Sponsor: American Historical Association
400 A Street, SE
Washington, DC 20003

**American Historical Association Littleton-Griswold Prize in American   734
Law and Society**
This prize is offered annually for the best book in any subject on the history of American law and society. The $1,000 prize is administered by a joint committee of the American Historical Association and the American Society for Legal History. Only books of high scholarly and literary merit will be considered. Recipients are announced at the association's annual meeting.
Requirements: Books published between May 1 of the previous year and April 30 of the award year are eligible.
Samples: Daniel J. Hulsebosch, New York University School of Law—Constituting Empire: New York and the Transformation of Constitutionalism in the Atlantic World, 1664‚Äì1830 (University of North Carolina Press, 2006); Mary Sarah Bilder, Boston College—The Transatlantic Constitution: Colonial Legal Culture and the Empire (Harvard University Press, 2004); Mae M. Ngai, University of Chicago—Impossible Subjects: Illegal Aliens and the Making of Modern America (Princeton University Press, 2003).
Amount of Grant: $1,000
Date(s) Application is Due: May 15
Contact: Book Prize Administrator, (202) 544-2422; fax: (202) 544-8307; email: aha@theaha.org

Internet: http://www.historians.org/prizes/index.cfm?PrizeAbbrev=LITTLETON%2DGRISWOLD
Sponsor: American Historical Association
400 A Street, SE
Washington, DC 20003

American Historical Association Littleton-Griswold Research Grants   735
Grants are given annually for research in U.S. legal history and in the general field of law and society. The funds for this program come from the earnings of the Littleton-Griswold Fund. Only members of the Association are eligible to apply. The grants are intended to further research in progress and may be used for travel to a library or archive, for microfilms, photographs, or photocopying-a list of purposes that is meant to be merely illustrative, not exhaustive.
Requirements: Preference will be given to those with specific research needs, such as the completion of a project or completion of a discrete segment thereof. Preference will be given to Ph.D. candidates and junior scholars.
Restrictions: Specifically excluded from funding are use of research assistants except in cases of technical skills such as data entry, coding of data, and other procedures relating to the preparation of data for machine computation; help with typing except for typing of camera-ready copy for an article, dissertation, or book, which would qualify as a technical expense; and use of funds for partial salary replacement.
Amount of Grant: $1,000 maximum
Date(s) Application is Due: Feb 15
Contact: Research Grant Administrator, (202) 544-2422; fax: (202) 544-8307; email: aha@theaha.org
Internet: http://www.historians.org/prizes/Littleton-GriswaldGrantInfo.htm
Sponsor: American Historical Association
400 A Street, SE
Washington, DC 20003

**American Historical Association Michael Kraus Research Grant in   736
American Colonial History**
This annual grant seeks to recognize the most deserving proposal relating to work in progress on a research project in American colonial history, with particular reference to the inter-cultural aspects of American and European relations. Only members of the Association are eligible. The grants are intended to further research in progress and may be used for travel to a library or archive, for microfilms, photographs, or photocopying—a list of purposes that is meant to be merely illustrative, not exhaustive.
Requirements: Applicants must be members of the association. Preference will be given to those with specific research needs such as the completion of a project or a discrete segment thereof. Preference will be given to Ph.D. candidates and junior scholars.
Restrictions: Specifically excluded from funding are use of research assistants except in cases of technical skills such as keypunching, coding of data, and other procedures relating to the preparation of data for machine computation; help with typing except for typing of camera-ready copy for an article, dissertation, or book, which would qualify as a technical expense; and use of funds for partial salary replacements. Within a five-year period, no individual is eligible to receive more than a combined total of $1,000 from the Beveridge, Kraus, and Littleton-Griswold grant programs.
Amount of Grant: $800 maximum
Date(s) Application is Due: Feb 15
Contact: Research Grant Administrator, (202) 544-2422; fax: (202) 544-8307; email: aha@theaha.org
Internet: http://www.historians.org/prizes/KrausGrantInfo.htm
Sponsor: American Historical Association
400 A Street, SE
Washington, DC 20003

American Historical Association Morris D. Forkosch Prize   737
The prize is awarded annually in recognition of the best book in English in the field of British, British Imperial, or British Commonwealth history. Submission of books relating to the shared common law heritage of the English-speaking world is particularly encouraged. One copy of each entry must be submitted to designated committee members. Prize recipients are announced at the association's annual meeting.
Requirements: Books published after May 1 of the previous year and before April 30 of the award year are eligible.
Samples: Christopher Leslie Brown, Rutgers University-New Brunswick—Moral Capital: Foundations of British Abolitionism (University of North Carolina Press for the Omohundro Institute of Early American History and Culture, 2006); Bernard Porter, University of Newcastle—The Absent-Minded Imperialists: Empire, Society, and Culture in Britain (Oxford University Press, 2004); Robert Bickers, University of Bristol—Empire Made Me: An Englishman Adrift in Shanghai (Columbia University Press, 2004).
Date(s) Application is Due: May 16
Contact: Book Prize Administrator, (202) 544-2422; fax: (202) 544-8307; email: aha@theaha.org
Internet: http://www.historians.org/prizes/index.cfm?PrizeAbbrev=FORKOSCH
Sponsor: American Historical Association
400 A Street, SE
Washington, DC 20003-3889

American Historical Association NASA Research Fellowship in Aerospace History   738

The fellowship, supported by NASA, annually funds one or more research projects for six to nine months. Proposals of advanced research in history related to all aspects of aerospace, from the earliest human interest in flight to the present, are eligible, including cultural and intellectual history, economic history, history of law and public policy, and history of science, engineering, and management. The Fellowship term is for a period of at least six months, but not more than one year. The Fellow will be expected to devote the term entirely to the proposed research project.

Requirements: The fellowship is open to any U.S. citizen who holds a doctoral degree in history or a closely related field, or who is enrolled in and has completed all course work for a doctoral degree-granting program.

Restrictions: Funds may not be used to support tuition or fees. A Fellow may not hold other major fellowships or grants during the fellowship term, except sabbatical and supplemental grants from their own institutions, and small grants from other sources for specific research expenses.

Samples: Victoria Vantoch—Ambassadors of the Air: The Airline Stewardess, Glamour, and Technology in the Cold War, 1945-1969 (2006-07); Alexander Brown—Accidents, Engineering and History at NASA, 1967,Äì2003 (2005-06); Amy Foster—Sex in Space: The First Class of Women Astronauts (2004-05).

Amount of Grant: $20,000

Date(s) Application is Due: Mar 1

Contact: Arnita A. Jones, Executive Director, (202) 544-2422, ext. 100; fax: (202) 544-8307; email: ajones@theaha.org or

Internet: http://www.historians.org/prizes/NASA.htm

Sponsor: American Historical Association

400 A Street, SE

Washington, DC 20003

American Historical Association Paul Birdsall Prize in European Military and Strategic History   739

The prize is offered biennially in even-numbered years for a major work in European military and strategic history since 1870. Preference will be given to the international aspects of military history (military/diplomatic), but the political, economic, and social impact of technological developments, strategic planning, and military events on society will also qualify.

Requirements: Authors must be citizens of the United States or Canada; preference will be given to younger academics, but older scholars and nonacademic candidates will not be excluded. Books published after May 1 of the previous year and before April 30 of the award year are eligible.

Samples: Mark Atwood Lawrence, University of Texas at Austin—Assuming the Burden: Europe and the American Commitment to War in Vietnam (University of California Press, 2005); Robert M. Citino, Eastern Michigan University—Blitzkrieg to Desert Storm: The Evolution of Operational Warfare (University Press of Kansas, 2004); Matthew Connelly, Columbia University—A Diplomatic Revolution: Algeria's Fight for Independence and the Origins of the Post-Cold War Era (Oxford University Press, 2002).

Date(s) Application is Due: May 15

Contact: Book Prize Administrator, (202) 544-2422; fax: (202) 544-8307; email: aha@theaha.org

Internet: http://www.historians.org/prizes/index.cfm?PrizeAbbrev=BIRDSALL

Sponsor: American Historical Association

400 A Street, SE

Washington, DC 20003

American Historical Association Premio Del Rey   740

The prize is awarded biennially for a distinguished book in English in the field of early Spanish history. It was endowed by a gift of Robert I. Burns S.J., from his Llull and Catalonia prizes and covers the medieval period in Spain's history and culture CE 500-1516. The terms of the prize include works on Hispanic history and culture, including the Islamic and Jewish communities of Medieval Spain as well as early New World topics prior to 1516. Only books of a high scholarly historical nature should be submitted. Research accuracy, originality and literary merit are important factors. One copy of each entry must be received by each of the committee members.

Requirements: Books published after May 1 of the previous year and before April 30 of the award year are eligible.

Samples: Brian A. Catlos, University of California at Santa Cruz—The Victors and the Vanquished: Christians and Muslims of Catalonia and Aragon, 1050,Äì1300 (Cambridge University Press, 2004); Jeffrey A. Bowman, Kenyon College—Shifting Landmarks: Property, Proof, and Dispute in Catalonia around the Year 1000 (Cornell University Press, 2004); Adam J. Kosto, Columbia University—Making Agreements in Medieval Catalonia: Power, Order, and the Written Word, 1000,Äì1200 (Cambridge University Press, 2001).

Date(s) Application is Due: May 15

Contact: Book Prize Administrator, (202) 544-2422; fax: (202) 544-8307; email: aha@theaha.org

Internet: http://www.historians.org/prizes/index.cfm?PrizeAbbrev=DEL%2DREY

Sponsor: American Historical Association

400 A Street, SE

Washington, DC 20003

American Historical Association Waldo J. Leland Prize   741

The prize is awarded every five years (next in 2011) by the American Historical Association to an American resident for the most outstanding reference tool in the field of history. The award is honorific. The term reference tool encompasses bibliographies, indexes, encyclopedias, and other scholarly apparatus. Guidelines must be obtained from the association.

Requirements: Books published after May 1 of the previous award year and before April 30 of the current award year are eligible.

Samples: Encyclopedia of Women and Religion in North America, Indiana University Press (2006); American National Biography, 24 vols., edited by John A. Garraty, Columbia University, and Mark C. Carnes, Barnard College, Columbia University (Oxford University Press under the auspices of the American Council of Learned Societies, 1999).

Date(s) Application is Due: May 16

Contact: Book Prize Administrator, (202) 544-2422; fax: (202) 544-8307; email: aha@theaha.org

Internet: http://www.historians.org/prizes/index.cfm?PrizeAbbrev=LELAND

Sponsor: American Historical Association

400 A Street, SE

Washington, DC 20003

American Historical Association Wesley-Logan Prize   742

The prize is jointly sponsored by the association and the Association for the Study of Afro-American Life and History. The prize is offered for a book on some aspect of the history of the dispersion, settlement, adjustment, and/or return of peoples originally from Africa. Eligible for consideration are books in any chronological period and any geographical location. Only books of high scholarly and literary merit will be considered. One copy of each entry must be submitted to designated committee members. Recipients will be announced at the association's annual meeting.

Requirements: Books published between May 1 of the previous year and April 30 of the current year are eligible.

Samples: Kenneth M. Bilby, Virginia Foundation for the Humanities—True-Born Maroons (University Press of Florida, 2005); Melvin Patrick Ely, College of William and Mary—Israel on the Appomattox: A Southern Experiment in Black Freedom from the 1790s through the Civil War (Knopf, 2004); James H. Sweet, University of Wisconsin-Madison—Recreating Africa: Culture, Kinship, and Religion in the African-Portuguese World, 1441-1770 (University of North Carolina Press, 2004).

Date(s) Application is Due: May 15

Contact: Book Prize Administrator, (202) 544-2422; fax: (202) 544-8307; email: aha@theaha.org

Internet: http://www.historians.org/prizes/index.cfm?PrizeAbbrev=WESLEY%2DLOGAN

Sponsor: American Historical Association

400 A Street, SE

Washington, DC 20003

American Historical Association William Gilbert Award   743

The Award recognizes outstanding contributions to the teaching of history through the publication of journal articles. Eligible for consideration in a given year are articles by members of the AHA, published in the United States between June 1 and May 31 of the award year. Journals and individual members may submit nominations on the teaching of history (including scholarship of teaching and learning, methodology and theory of pedagogy) for each biennial cycle of this award. Journals, magazines, and other serials can submit up to two articles for each award cycle. Each nominator is required to provide a brief letter of support (no more than two pages) with the article.

Samples: Mark C. Carnes, Barnard College, Inciting Speech, Change Magazine (March/April 2005); Carl Guarneri, Saint Mary's College of California, Internationalizing the United States Survey Course: American History for a Global Age, The History Teacher 36:1 (November 2002: 37-64); Daniel A. Segal, Pitzer College, Western Civ and the Staging of History in American Higher Education, American Historical Review 105:3 (June 2000).

Date(s) Application is Due: Jul 16

Contact: Program Administrator, (202) 544-2422; fax: (202) 544-8307; email: aha@theaha.org

Internet: http://www.historians.org/prizes/Gilbert.htm

Sponsor: American Historical Association

400 A Street, SE

Washington, DC 20003-3889

American Honda Foundation Grants   744

American Honda Foundation seek out those programs and organizations with a well-defined sense of purpose, demonstrated commitment to making the best use of available resources and a reputation for accomplishing their objectives. Grants are provided in the fields of youth education and science education to the following: educational institutions, K-12; accredited higher education institutions (colleges and universities); community colleges and vocational or trade schools; scholarship and fellowship programs at selected colleges and/or universities or through selected non-profit organizations; other scientific and education-related, non-profit, tax-exempt organizations; gifted student programs; media concerning youth education and/or scientific education; private, non-profit scientific and/or youth education projects; other non-profit, tax-exempt, institutions in the fields of youth education and scientific education; and programs pertaining to academic or curriculum development that emphasize innovative educational methods and techniques.

Requirements: An applicant must be a nonprofit, tax-exempt educational organization.

Restrictions: The foundation does not consider proposals for service clubs, arts and culture, health and welfare issues, research papers, social issues, medical or educational research, trips, attempts to influence legislation, advocacy, annual funds, hospital operating funds, student exchanges, marathons, sponsorships, political activities, conferences, or fundraising events. An organization may submit only one proposal per year.

Amount of Grant: $10,000-$100,000; $40,000-$80,000 average

Date(s) Application is Due: Feb 1, May 1, Aug 1, Nov 1

Contact: Program Director; (310) 781-4090; fax: (310) 781-4270

Internet: http://corporate.honda.com/america/philanthropy.aspx?id=ahf

Sponsor: American Honda Foundation

1919 Torrence Boulevard, P.O. Box 2205

Torrance, CA 90509-2205

American Horticultural Society Internship Program   745

Interns spend from three to nine months at River Farm, a 25-acre estate in Virginia, learning about horticulture and gardening. The program is designed to expose interns to a wide range of horticultural practices, activities, and experiences, thereby allowing them the opportunity to integrate academic theories with practical, real-world experiences. Interns also become involved in society programs and projects such as an annual seed program and catalog, open house events, and lectures and seminars. Applications are accepted at any time, but must be received at least two months before the internship is to begin. Applications are available at the Web site.

Requirements: Interns must be undergraduate or graduate students majoring in horticulture or a related field, or an adult making a career change into horticulture. The application includes a completed application form; resume; the name, address, and phone number of three references; and a copy of applicant's college transcript (an official original is not required).

Date(s) Application is Due: Mar 1; Jul 1; Nov 1

Contact: Tom Underwood, Director and Curator of Gardens and Buildings, (703) 768-5700 ext 112; fax: (703) 765-6032; email: tunderwood@ahs.org

Internet: http://www.ahs.org/river_farm/internships_employment.htm

Sponsor: American Horticultural Society

7931 E Boulevard Drive

Alexandria, VA 22308-1300

American Indian Graduate Center Scholarships   746

Scholarships are available each year to American Indian students meeting the following qualifications: a member of a federally recognized American Indian tribe or Alaska Native group, or possess one-fourth degree federally recognized Indian blood; demonstrate financial need after exhausting available aid at the college financial aid office; and attending an accredited graduate school in the United States. The April deadline is for summer enrollment; the June deadline is for fall enrollment. Applications are available upon request.

Amount of Grant: $250-$10,000

Date(s) Application is Due: Apr 15; Jun 1

Contact: Scholarship Administrator, (800) 628-1920 or (505) 881-4584; fax: (505) 884-0427; email: aigc@aigc.com

Internet: http://www.aigc.com

Sponsor: American Indian Graduate Center

4520 Montgomery Blvd NE, Suite 1B

Albuquerque, NM 87109

American Indian Science and Engineering Scholarships   747

The A.T. Anderson Memorial Scholarship is awarded to Native American/Alaskan Native graduate or undergraduate students pursuing programs in the sciences, engineering, health-related fields, business, natural resources, and math and science secondary education. Awards are made for one academic year. Deadlines, guidelines, and application forms are available on the Web site.

Requirements: Applicants must be society members who are at least one-quarter Native American or recognized as a member of a tribe and full-time students at the undergraduate or graduate level.

Amount of Grant: $5000 per academic year

Date(s) Application is Due: Jun 15

Contact: Scholarship Department, (505) 765-1052 ext 105; fax: (505) 765-5608; email: scholarship@aises.org

Internet: http://www.aises.org/highered/scholarships/index.html

Sponsor: American Indian Science and Engineering Society

P.O. Box 9828

Albuquerque, NM 87119-9828

American Indian Studies Research Institute Fellowship   748

Each year, the Institute offers a fellowship to a worthy graduate student interested in studying American Indian culture, language, or history. The fellowship is designed to attract and retain top-quality graduate students. Usually, though not always, it is awarded to an incoming graduate student. The fellowship carries with it a tuition waiver, university health insurance, and a stipend. There are no duties formally required of the student; however, the fellowship recipient is encouraged to engage in the ongoing scholarly research at the Institute.

Geographic Focus: Indiana

Contact: Director; (812) 855-4086; fax: (812) 855-7529; email: aisri@indiana.edu

Internet: http://www.indiana.edu/~aisri/index.shtml

Sponsor: American Indian Studies Research Institute / Indiana University

422 North Indiana Avenue

Bloomington, IN 47408

American Institute for Economic Research Summer Fellowships   749

These fellowships are awarded to further the development of economic scientists. Fellows participate in an four-week program and are provided with room and board. They attend seminars that primarily focus on scientific procedures of inquiry and on monetary economics. Attention also is given to business cycle analysis and forecasting. Assigned readings provide the basis for written assignments and seminar discussions. Applications may be downloaded from the Web site.

Requirements: Graduating seniors applying to doctoral programs in economics and students already enrolled in such programs are eligible to apply. (Preference is given to graduate students.) U.S. citizens are given first priority; foreign students must be able to speak and write in English with native fluency.

Restrictions: The program is not designed for students wishing to pursue graduate work in a business school program (e.g. MBA.).

Amount of Grant: $250 per week basic stipend, plus room and board

Date(s) Application is Due: Mar 31

Contact: Susan Gillette, Assistant to the President, (413) 528-1216; fax: (413) 528-0103; email: fellowship@aier.org

Internet: http://www.aier.org/summer.html

Sponsor: American Institute for Economic Research

P.O. Box 1000

Great Barrington, MA 01230

American Institute of Bangladesh Studies (AIBS) Postdoctoral Research   750
and Dissertation Research Grants

The institute awards stipends of two to 12 months duration for postdoctoral research and six to 12 months for dissertation research in Bangladesh. The award includes round-trip transportation and a small cash supplement. Applications must be received by the deadline for study beginning in the summer or fall.

Requirements: Applicants must have completed at least one year of graduate study in a recognized Ph.D. granting institution in either the social sciences or the humanities. Grantees must be U.S. citizens or permanent residents of the United States.

Contact: Administrator, (814) 865-0436; fax: (814) 865-8299; email: sxr17@psu.edu

Internet: http://www.aibs.net/aibsfellowship.htm

Sponsor: American Institute of Bangladesh Studies

111 Sowers Street, Suite 501

State College, PA 16801

American Institute of Physics Science-Writing Awards in Physics and Astronomy   751

Four awards are given to professional writers or to scientists to stimulate and recognize distinguished writing that improves public understanding of physics and astronomy. The Journalist Award is for professional writers writing for the general public; the Scientists Award is for the work of a physicist or astronomer writing for the general public as opposed to writing for strictly scientific, technical, or trade publications; the Children's Books Award is for physics and astronomy writing aimed at children up to 15 years of age; and the Broadcast Media Award is for journalists of scripted radio or television programming aimed at the general public. Entry forms are available on the Web site.

Requirements: Articles or books must have been published within the last year. Entries must be in English or English translations.

Samples: Robyn Suriano and Todd Halvorson, Florida Today—Journalist Award for Cassini: Debating the Risks, $3000.; Donald Silver, children's author—for Extinction Is Forever, $3000.

Amount of Grant: $3000

Date(s) Application is Due: Mar 1

Contact: Marc Brodsky, Executive Director, (301) 209-3131; fax: (301) 209-0846; email: brodsky@aip.org

Internet: http://www.aip.org/aip/writing

Sponsor: American Institute of Physics

1 Physics Ellipse

College Park, MD 20740-3843

American Jewish Historical Society Awards and Prizes   752

In cooperation with the Jewish Historical Society of New York, the society grants awards and prizes to assist researchers to carry on projects in the area of American Jewish history, including immigration, economic, social, political, and religious history. For current awards, details, and deadlines, see the AJHS website (above).

Contact: David Solomon, Executive Director, (212) 294-6160; fax (212) 294-6161; email: dsolomon@ajhs.cjh.org

Internet: http://www.ajhs.org/academic/Awards.cfm

Sponsor: American Jewish Historical Society

15 W 16th Street

New York, NY 10011

American Legacy Foundation National Calls for Proposals Grants   753

The American Legacy Foundation provides grants to support innovations in tobacco control. National calls for proposals, such as the Priority Populations Initiative, are issued annually to address a variety of tobacco prevention and control issues. Comprehensive assistance and training are provided to the foundation, grantees as well as to state and local tobacco programs. Application forms, guidelines, and procedures are available online. There are no application deadlines.

Requirements: Funding is available only to state or local political subdivisions and legally constituted tax-exempt 501(c)3 organizations based in the 46 states, the District of Columbia, and five territories (American Samoa, Guam, Northern Mariana Islands, Puerto Rico, and the Virgin Islands) identified in the MSA with tobacco product manufacturers. An Indian reservation, Indian tribe, or tribal organization located within the 46 settling states or a non-governmental entity that serves such a reservation may also apply for funding.

Restrictions: The foundation will not award grants to applicants that are in current receipt of grant monies or in-kind contributions from any tobacco manufacturer, distributor, or other tobacco-related entity.

Contact: Katherine Wilson, Assistant Vice President of Grants; (202) 454-5555; fax: (202) 454-5599; email: grantsinfo@americanlegacy.org

Internet: http://www.americanlegacy.org/64.aspx

Sponsor: American Legacy Foundation

1724 Massachusetts Avenue, NW

Washington, DC 20036

American Legacy Foundation Small Innovative Grants 754

The program supports projects that advance creative, promising solutions based on sound principles of tobacco control to remedy the harm caused by tobacco use in America. The program was created to seed new projects or enable an organization to pilot a new idea or approach. The proposed project must demonstrate an element of creativity, ingenuity or innovation and must distinguish itself from the large number of solid programs proposed to the foundation in each grant round.

Requirements: Funding is available only to state or local political subdivisions and legally constituted tax-exempt 501(c)3 organizations based in the 46 states, the District of Columbia, and five territories (American Samoa, Guam, Northern Mariana Islands, Puerto Rico, and the Virgin Islands) identified in the MSA with tobacco product manufacturers. An Indian reservation, Indian tribe, or tribal organization located within the 46 settling states or a nongovernmental entity that serves such a reservation may also apply for funding. Successful applications submitted under the Small Innovative Grants Program must: address one or both of Legacy, goals; demonstrate innovative or new tobacco prevention or cessation efforts; demonstrate a strong likelihood for a sustainable effort after the grant period; demonstrate that the project may be replicated; address the Healthy People 2010 risk reduction objectives related to tobacco use; and, incorporate the CDC's Best Practices for Comprehensive Tobacco Control Programs as appropriate. The foundation will give special consideration to applications addressing these current areas of interest.

Restrictions: The foundation will not award a grant to any applicant that is in current receipt of any grant monies or in-kind contribution from any tobacco manufacturer, distributor, or other tobacco-related entity. In addition, the foundation expects that a grantee will not accept any grant monies or in-kind contribution from any tobacco manufacturer, distributor, or other tobacco-related entity over the duration of the grant. Additionally, will not consider applications for: Projects focusing on youth prevention, cessation, activism, or education (up to 18 years old); Nicotine replacement therapy (NRT) and pharmaceuticals (as the sole or primary focus of the grant); Conference support (as the sole or primary focus of the grant); Media or marketing campaigns (as the sole or primary focus of the grant); Research projects EXCEPT community-based participatory research, which is allowed; Projects focusing on substances other than tobacco; Replication of an existing program; Expansion of an existing program; and Replacement funds; Grants to individuals, for religious activities, to build endowments, to support operating deficits, to retire debt, for capital purchases for building improvements, for construction, for lobbying, or for real estate purchase or development.

Amount of Grant: $20,000-$100,000 annually

Date(s) Application is Due: Feb 27; Aug 8

Contact: Katherine Wilson, Assistant Vice President of Grants; (202) 454-5555; fax: (202) 454-5599; email: grantsinfo@americanlegacy.org

Internet: http://www.americanlegacy.org/2530.aspx

Sponsor: American Legacy Foundation

1724 Massachusetts Avenue, NW

Washington, DC 20036

American Liver Foundation Innovative Hepatology Seed Grant 755

The goal of this program is to foster development of an imaginative research program in Alpha 1 Antitrypsin Deficiency and/or directly related areas of scientific investigation.

Requirements: To be eligible candidates must be a faculty member in an academic institution in the United States or its possessions. Applicants who hold research awards directed at salary support, such as Veterans Administration, Research Associates, Clinical Investigator, AGA Industry Research Awards, Glaxo Institute Awards, and ALF Liver Scholar Awards are also eligible.

Restrictions: Individuals with an RO1, Merit Review, or NIH FIRST Award are ineligible if their grant overlaps with the present application.

Amount of Grant: $100,000 over two years

Date(s) Application is Due: Oct 1

Contact: Joan Gallagher, Program Director, (973) 256-2550 ext 224 or (800) 465-4837; email: jgallagher@liverfoundation

Internet: http://www.liverfoundation.org/db/grants/114

Sponsor: American Liver Foundation

1425 Pompton Avenue, Suite 3

Cedar Grove, NJ 07009

American Liver Foundation Liver Scholar Awards Program 756

The goal of the program is to permit scientists with liver research training to bridge the gap between completion of research training and attainment of status as independent research scientists. This additional research experience should enable scientists to compete for research grants from national sources, particularly NIH. The program aims to attract well-trained, basic scientists, who hold MD, PhD, or MD/PhD degrees, to a career in liver disease research. These awards are for the purpose of encouraging research into liver physiology and disease research and developing the potential of young, outstanding scientists; therefore, individuals who already are well established in the fields are not considered eligible. The awards will be made to eligible institutions and are for salary only. Each awardee must continue to hold a full-time faculty appointment in the sponsoring institution; devote at least 75 percent of time to their research project; attend the annual meeting of the American Association for the Study of Liver Diseases; acknowledge support by the ALF in any publications resulting from research performed during the tenure of the award; and submit an annual progress report describing results, their significance, and pending or planned applications for research support.

Requirements: Candidates must be sponsored by a nonfederal public or private nonprofit institution engaged in health care and health-related research within the United States and its possessions. Applicants must have had three to four years of relevant postdoctoral experience prior to the beginning date of the award; have obtained their doctorate degree within the last 10 years; and have institutional confirmation of a faculty appointment at the time the award commences and throughout its duration. Applicants who hold or have applied for nonfederal research awards must provide the title and specific aims of the awards with their application.

Restrictions: At the time of application, the applicant cannot hold or have held any of the following awards: NIH R01, R29, K11, K08; Veterans Administration Merit Review; Associate Investigation; Research Associate; Clinical Investigator; or AGA Industry Award.

Amount of Grant: $75,000 per year over three years

Date(s) Application is Due: Oct 1

Contact: Joan Gallagher, (973) 256-2550 ext 224; fax: (973) 256-3214; email: jgallagher@liverfoundation.org

Internet: http://www.liverfoundation.org/db/grants/113

Sponsor: American Liver Foundation

1425 Pompton Avenue, Suite 3

Cedar Grove, NJ 07009

American Liver Foundation Physician Research Development Award 757

The purpose of this award is to support and develop clinical research training of fellows seeking junior faculty appointment (or in their first years as a junior faculty member). Preference is given to translational research in pathobiology and treatment of viral hepatitis toward developing preclinical and clinical applications as well as addressing natural history, prevention, and epidemiology of viral hepatitis.

Amount of Grant: $225,000-$270,000 over three or four years

Date(s) Application is Due: Oct 1

Contact: Joan Gallagher, (973) 256-2550 ext 224; fax: (973) 256-3214; email: jgallagher@liverfoundation.org

Internet: http://www.liverfoundation.org/db-list/grants/0/descend/ID/Validated

Sponsor: American Liver Foundation

1425 Pompton Avenue, Suite 3

Cedar Grove, NJ 07009

American Liver Foundation Postdoctoral Research Fellowships 758

The foundation offers postdoctoral research fellowships to encourage the development of individuals with research potential who require additional research training and experience, specifically in investigational work relating to liver physiology and disease, in preparation for careers of independent research in this field. The goal of the program is to encourage MD, PhD, and MD/PhD postdoctoral fellows to enter an academic career in liver disease research. Fellows are expected to devote at least 75 percent of time to their specific research project, attend the annual meeting of the American Association for the Study of Liver Diseases, acknowledge support by the ALF in any publications resulting from the research performed during the tenure of the award, and submit a progress report describing results obtained and their significance. The application must be written by the applicant.

Requirements: Candidates must be sponsored by a nonfederal public or private nonprofit institute engaged in health care and health-related research within the United States and its possessions. Physician applicants (MD and MD/PhD) must submit documentation that all clinical training has been completed. All applicants must be in the first year of appointment as a postdoctoral research fellow or trainee.

Restrictions: Individuals with more than two years of postdoctoral research training or those already well established in the field of hepatology are ineligible. In addition, the fellowship is designed as a supplement to augment NIH or nonfederal fellowship stipends and will not be awarded to an applicant who has no other source of research salary support.

Amount of Grant: $12,500 for one year

Date(s) Application is Due: Oct 1

Contact: Joan Gallagher, (973) 256-2550 ext 224; fax: (973) 256-3214; email: jgallagher@liverfoundation.org

Internet: http://www.liverfoundation.org/db/grants/131

Sponsor: American Liver Foundation

1425 Pompton Avenue, Suite 3

Cedar Grove, NJ 07009

American Liver Foundation Student Research Fellowship Award 759

The foundation offers awards to MD and PhD students to encourage them to gain exposure to the research laboratory, and possibly consider liver research as a career option. Each fellowship is for a three-month period. The candidate must be in full-time research for a period of 10-12 weeks under the supervision of a preceptor in hepatic physiology or disease.

Requirements: Full-time students of U.S. undergraduate, graduate, or medical school programs are eligible to apply.

Amount of Grant: $2500

Date(s) Application is Due: Oct 1

Contact: Joan Gallagher, (973) 256-2550 ext 224; fax: (973) 256-3214; email: jgallagher@liverfoundation.org

Internet: http://www.liverfoundation.org/db-list/grants/0/descend/ID/Validated

Sponsor: American Liver Foundation
1425 Pompton Avenue, Suite 3
Cedar Grove, NJ 07009

American Lung Association Career Investigator Award 760

The award supports individuals who have demonstrated success in research and who show great promise for a career in investigation. Awards are subject to annual review and may be granted for up to three years. Awardees are expected to devote full time, and in no case less than 75 percent of their time, to research. Awardees may undertake limited administrative, teaching, and clinical responsibilities if they are directly related to the nature of the research supported by this award.

Requirements: U.S. citizens and foreign nationals with appropriate visa immigration status are eligible. At the time of application, an applicant must hold a doctoral degree and faculty appointment at the level of assistant or associate professor and be undertaking a project related to lung disease.

Restrictions: Individuals who have attained the rank of full professor or who have more than eight years of faculty experience at the time of application are ineligible.

Amount of Grant: $60,000 maximum per year

Date(s) Application is Due: Sep 1

Contact: Evita Mendoza, Grants Administrator, (212) 315-8793; email: emendoza@lungusa.org or info@lungusa.org

Internet: http://www.lungusa.org/site/apps/s/content.asp?c=dvLUK9O0E&b=34706&ct=67676

Sponsor: American Lung Association
1740 Broadway
New York, NY 10019

American Lung Association Clinical Patient Care Research Grants 761

The objective of the grant is to support investigators working in patient-focused research. The areas of research may include the disciplines of health services research, behavioral research, epidemiology, pharmacology, and medical ethics. The outcome of these research activities is to develop new insights into clinical, social, behavioral, and biological factors that affect the manifestation of lung health and disease. Grants are subject to annual review and may be granted for up to two years.

Requirements: U.S. citizens and foreign nationals with appropriate visa immigration status are eligible. At the time of application, an applicant must hold a doctoral degree and faculty appointment with an academic institution and have completed two years of postdoctoral research training.

Restrictions: Residents, interns, postdoctoral fellows, graduate students, and established investigators are ineligible.

Amount of Grant: $40,000 maximum per year

Date(s) Application is Due: Sep 1

Contact: Evita Mendoza, Grants Administrator, (212) 315-8793; email: emendoza@lungusa.org or info@lungusa.org

Internet: http://www.lungusa.org/site/apps/s/content.asp?c=dvLUK9O0E&b=34706&ct=67676

Sponsor: American Lung Association
1740 Broadway
New York, NY 10019-4374

American Lung Association Dalsemar Research Grant 762

The grant supports up to two years of research in interstitial lung disease. Grants are subject to annual review and may be granted for up to two years. The second year of support is based on demonstrating satisfactory progress as well as the availability of American Lung Association funding.

Requirements: U.S. citizens and foreign nationals with appropriate immigration visa status are eligible. At the time of application, an applicant must hold a doctoral degree and a faculty appointment with an academic institution and have completed two years of research training.

Restrictions: Residents, interns, postdoctoral fellows, students enrolled in degree-granting programs, and established investigators are not eligible.

Amount of Grant: $40,000 maximum per year

Date(s) Application is Due: Sep 1

Contact: Evita Mendoza, Grants Administrator, (212) 315-8793; email: emendoza@lungusa.org or info@lungusa.org

Internet: http://www.lungusa.org/site/apps/s/content.asp?c=dvLUK9O0E&b=34706&ct=67676

Sponsor: American Lung Association
1740 Broadway
New York, NY 10019

American Lung Association Research Grants 763

The grants provide starter or seed money to new investigators working in areas relevant to the prevention of lung disease and the promotion of lung health. The research supported may be clinical, laboratory, epidemiologic, social, environmental, or any other kind as long as it is relevant to lung biology. Six individuals will be designated Edward Livingston Trudeau Scholars. Awards are for one year and may be renewed for an additional year, depending on the availability of funds.

Requirements: At the time of application, an applicant must hold a doctoral degree, faculty appointment with an academic institution, and have completed two years of research training. U.S. citizens, Canadian citizens, and permanent residents of the United States training in U.S. institutions are eligible.

Restrictions: An American Lung Association grantee cannot simultaneously hold any other national ALA award. An investigator who has previously received a research grant or a Dalsemer Award is not eligible for additional national research grant support. An ALA research grant awardee may not hold an award of established investigatorship or large grant during the tenure of her or his ALA award.

Amount of Grant: $35,000 maximum

Date(s) Application is Due: Sep 1

Contact: Evita Mendoza, Grants Administrator, (212) 315-8793; email: emendoza@lungusa.org or info@lungusa.org

Internet: http://www.lungusa.org/site/apps/s/content.asp?c=dvLUK9O0E&b=34706&ct=67676

Sponsor: American Lung Association
1740 Broadway
New York, NY 10019-4374

American Lung Association Senior Research Training Fellowship 764

The fellowship supports the training of scientific investigators in the fields of adult and pediatric pulmonary medicine and lung biology. Preference is given to applicants who demonstrate a program of training that will enable them to pursue academic careers. Fellows are expected to devote full time or at least 75 percent of their time to research training. Fellowships may be renewed for an additional year.

Requirements: U.S. citizens and foreign nationals with appropriate immigration visa status are eligible. At the time of application, an applicant must hold a doctoral degree or have comparable qualifications. MD applicants must have completed their clinical training, have some research experience, and be entering the third or fourth year of research fellowship training. PhD applicants must not be beyond the third postdoctoral year at the time of application.

Amount of Grant: $32,500 maximum per year

Date(s) Application is Due: Sep 1

Contact: Evita Mendoza, Grants Administrator, (212) 315-8793; email: emendoza@lungusa.org

Internet: http://www.lungusa.org/site/apps/s/content.asp?c=dvLUK9O0E&b=34706&ct=67676

Sponsor: American Lung Association
61 Broadway, 6th Fl
New York, NY 10006

American Mathematical Society Centennial Fellowships 765

The program makes awards annually to outstanding mathematicians to help further their careers in research. Fellowship holders may use their stipend as full support for a year or may combine it with half-time teaching and use it as half support over a two-year period. Fellows are expected to spend some of the fellowship period at another institution that has a stimulating research environment suited to the candidates' research development. Preference will be given to candidates who have not had extensive fellowship support in the past. Applications should include a short research plan describing both an outline of the research to be pursued and a program for using the fellowship, including the institutions at which it will be used and reasons for the choices. Completed applications and references should not be sent to this address but to the address given on the application form. Applications are available on the Web site.

Requirements: A recipient of the fellowship will have held his or her doctoral degree for at least three years and not more than 12 years at the inception of the award (that is, received between September 1, 1994, and September 1, 2003). Applications will be accepted from those currently holding a tenured, tenure track, postdoctoral, or comparable (at the discretion of the selection committee) position at an institution in North America.

Restrictions: Recipients may not hold the fellowship concurrently with other research fellowships (e.g., Sloan Foundation fellowships or NSF postdoctoral fellowships), and they may not use the stipend solely to reduce teaching at the home institution.

Amount of Grant: $64,000 and $3250 (approximately) expense allowance

Date(s) Application is Due: Dec 1

Contact: Executive Director, (401) 455-4107; fax: (401) 331-3842; email: ams@ams.org

Internet: http://www.ams.org/employment/centflyer.html

Sponsor: American Mathematical Society
201 Charles Street
Providence, RI 02940-3842

American Montessori Society Thesis and Dissertation Awards 766

The American Montessori Society offers annual awards for graduate level work that furthers the public understanding of Montessori education. The awards are - Outstanding Doctoral Dissertation: First Place: $1,000, Second Place: $500; Outstanding Master's Thesis: First Place, $750, Second Place: $250.

Requirements: Candidates whose works have been accepted by an accredited university and meet Assessment Criteria (see website) are invited to submit a CD (plus hard copy)

of their dissertation or thesis, for consideration. To ensure anonymity in reading by the committee, all identifying factors should be removed from the copies wherever possible.

Geographic Focus: Alabama; Alaska; Arizona; Arkansas; California; Colorado; Connecticut; Delaware; District of Columbia; Florida; Georgia; Guam; Hawaii; Idaho; Illinois; Indiana; Iowa; Kansas; Kentucky; Louisiana; Maine; Maryland; Massachusetts; Michigan; Minnesota; Mississippi; Missouri; Montana; Nebraska; Nevada; New Hampshire; New Jersey; New Mexico; New York; North Carolina; North Dakota; Ohio; Oklahoma; Oregon; Pennsylvania; Puerto Rico; Rhode Island; South Carolina; South Dakota; Tennessee; Texas; U.S. Virgin Islands; Utah; Vermont; Virginia; Washington; West Virginia; Wisconsin; Wyoming

Samples: 2009: Erin Hennigan, Assessment and Instructional Decision-Making in Montessori Early Childhood Classrooms (Master, thesis, first-place award); 2008: Linda Gatewood Massey, Pilgrims and Guides: A Phenomenological Study of Montessori Teachers Guiding and Being Guided by Children in Public Montessori Schools (Doctoral Dissertation); Michelle K. Yezbick, How Montessori Educators in the U.S. Address Culturally Responsive Education (Master's thesis, first-place award); Alison Stern, Observational Assessment of Literacy Development: The Use of Running Records in the Montessori Classroom (Master's thesis, second-place award).

Amount of Grant: $250 - $1,000
Date(s) Application is Due: Nov 1
Contact: Dr. Phyllis Povell, Committee Chair; (212) 358-1250; fax: (212) 358-1256; email: wwprof@optonline.net
Internet: http://www.amshq.org/society_thesis.htm
Sponsor: American Montessori Society
281 Park Avenue South
New York, NY 10010-6102

American Numismatic Society Doctoral Fellowships 767

The ANS administers a doctoral fellowship program, as well as a museum internship, and other grants. The society will award a doctoral fellowship to a university graduate student in the United States or Canada in the fields of the humanities or the social sciences, who will be writing a dissertation during the academic year on a topic in which the use of numismatic evidence plays a significant part. Contact the Society for details.

Requirements: Candidates must have completed the general examinations or the equivalent for the doctorate and must have attended the society's graduate seminar.
Date(s) Application is Due: Mar 1
Contact: Joanne Isaac, Museum Administrator; (212) 571-4470, ext. 112; fax: (212) 571-4479; email: isaac@numismatics.org or info@amnumsoc.org
Internet: http://www.amnumsoc.org/about/study.html
Sponsor: American Numismatic Society
75 Varick Street, 11th Floor
New York, NY 10013

American Numismatic Society Donald Groves Numismatic Research Grants 768

The fund supports research and publication in the field of early U.S. numismatics involving material dated no later than 1800. Funds are available for travel and other expenses in association with research, as well as for publication costs. Applications in letter form are welcome at any time, setting forth the proposed research project, the method of accomplishing the research, and a budget supporting the funding requested.

Contact: Joanne Isaac, Museum Administrator; (212) 571-4470, ext. 112; fax: (212) 571-4479; email: isaac@numismatics.org or info@amnumsoc.org
Internet: http://www.amnumsoc.org/about/grove.html
Sponsor: American Numismatic Society
75 Varick Street, 11th Floor
New York, NY 10013

American Numismatic Society Eric P. Newman Graduate Seminar Grants 769

The program provides graduate students and university instructors in the United States and Canada with a deeper understanding of the indispensable contributions numismatics makes to other fields of study. The program includes attendance at lectures and research on individual topics at museum headquarters, culminating in oral and written presentations. Applications are due on February 13th.

Requirements: Applicants must have completed at least one year of graduate study at a U.S. or Canadian university in classics, archaeology, history, art history, or related disciplines. Applications will also be accepted from junior college or university instructors with a degree in one of those fields.
Amount of Grant: $4,000
Date(s) Application is Due: Feb 13
Contact: Dr. Peter van Alfen, Seminar Director; (212) 571-4470, ext. 153; fax: (212) 571-4479; email: vanalfen@numismatics.org or info@amnumsoc.org
Internet: http://www.numismatics.org/about/study.html
Sponsor: American Numismatic Society
96 Fulton Street
New York, NY 10038

American Numismatic Society Frances M. Schwartz Fellowship 770

The fellowship supports the work and study of numismatic and museum methodology at the society. The purpose of establishing the fellowship was so that the funds would be used to educate qualified students in museum practice and to train them in numismatics, as well as to provide for curatorial assistance in the Greek, Roman, and Byzantine departments. The stipend will vary with the term of tenure (normally the academic year) but will not exceed $5,000.

Requirements: Candidates must have completed the BA or the equivalent.
Amount of Grant: $5,000 maximum
Date(s) Application is Due: Mar 1
Contact: Joanne Isaac, Museum Administrator; (212) 571-4470, ext. 112; fax: (212) 571-4479; email: isaac@numismatics.org or info@amnumsoc.org
Internet: http://www.numismatics.org/about/schwafell.html
Sponsor: American Numismatic Society
75 Varick Street, 11th Floor
New York, NY 10013

American Nurses Foundation Nursing Research Grants 771

The program is designed to award grants for nursing research conducted by registered nurses who are either beginning nurse researchers or experienced nurse researchers entering a new field of investigation. Priority is based upon the scientific merit of the proposal, with consideration given to the investigator's ability to conduct the study.

Requirements: Principal investigators must be baccalaureate-prepared registered nurses; however, the coinvestigator does not have to be a nurse as long as the proposal is for a nursing research project.
Restrictions: Some ANF grants are restricted as to their use by corporations and organizations that have contributed the funds.
Amount of Grant: $2700
Date(s) Application is Due: May 3
Contact: Leo Schargorodski, Executive Director, (301) 628-5230; email: anf@ana.org or lschargo@ana.org
Internet: http://www.nursingworld.org/anf/nrggrant.htm
Sponsor: American Nurses Foundation
8515 Georgia Avenue, Suite 400 W
Silver Spring, MD 20910

American Oil Chemists' Society Honored Student Award 772

This international award is given annually to a graduate student in fats and lipid chemistry who is currently attending a college or university. An applicant must submit a research paper for presentation at the annual meeting of the society. The awards provide funds equal to travel costs, complementary registration, and hotel accommodations, plus an additional stipend to permit attendance at the AOCS annual meeting. Application forms and guidelines are available upon request.

Date(s) Application is Due: Oct 15
Contact: Membership Department, (217) 359-2344; fax: (217) 351-8091; email: awards@aocs.org
Internet: http://www.aocs.org/member/awards/awd-honstud.asp
Sponsor: American Oil Chemists' Society
P.O. Box 3489
Champaign, IL 61826-3489

American Otological Society Research Fund Fellowhips and Grants 773

Research grants and fellowships are awarded to U.S. and Canadian physicians and doctoral-level investigators in all aspects of otosclerosis, Meniere's disease, and related ear disorders. The one-year grants, which begin July 1, are renewable for a second year and are tenable at U.S. and Canadian institutions. Awards are made to an institution on behalf of a candidate. Appropriate areas of research include diagnosis, management, and pathogenesis of otosclerosis and Meniere's disease as well as underlying processes. The applicant should describe correlations between proposed research with the clinical pathological entities of otosclerosis and Meniere's disease.

Requirements: Physician and nonphysician investigators are eligible for research grants. Fellowships are open to physicians (residents and medical students).
Restrictions: Research grants do not support funding for investigator's salary.
Samples: Timothy Doerr, Wayne State U (Detroit, MI)—for fellowship proposal, Regulatory Mechanisms of Vestibular Blood Flow, $39,621.; Dr. Bradley William Kesser, U of Virginia (Charlottesville, VA)—for fellowship proposal, The Pathophysiology and Pathogenesis of Changes in the Inner Ear During Meningitis, $39,750.
Amount of Grant: $40,000
Date(s) Application is Due: Jan 31
Contact: Dr. Jeffrey Harris, Otolaryngology—Head and Neck Surgery, (619) 543-7896; fax:(619) 543-5521; email: jpharris@ucsd.edu
Internet: http://itsa.ucsf.edu/~ajo/AOS/SchGra.html
Sponsor: American Otological Society
200 W Arbor Drive, 8895
San Diego, CA 92103-8895

American Parkinson's Disease Association Cotzias Fellowships 774

This three-year fellowship is intended to recruit and assist young neurologists in establishing careers in research, teaching, and patient service relevant to the problems, causes, prevention, diagnosis, and treatment of Parkinson's disease and other neurological movement disorders.

Requirements: An applicant must be a U.S. citizen, have an MD degree obtained within 10 years of application, and must have training in a clinical discipline concerned with disorders of the nervous system. Rank of assistant professor is required. The applicant must be sponsored by a nonprofit institution in the United States or its territories.
Samples: Dr. Russell H. Swerdlow, U of Virginia Health System (Charlottesville, VA)—fellowship recipient, for research entitled Genetic Studies in Parkinson's Disease. Dr. Mel Feany, Brigham and Women's Hospital (Boston, MA)—fellowship recipient, for research entitled Genetic Model of Parkinsonian Movement Disorders. Dr. Serge Przedborski, Columbia U (New York, NY)—fellowship recipient, for research entitled

Transplanted Dopamine Cell Survival in Animal Model of Parkinson's Disease. Dr. David Standaert, Massachusetts General Hospital (Boston, MA)—fellowship recipient, for research entitled NMDA Receptors: Regulation of Basal Ganglia Function and Pathogenesis of Parkinsonism.
Amount of Grant: $80,000 per year for three years
Date(s) Application is Due: Mar 1
Contact: Grants Administrator, (718) 981-8001 or (800) 223-2732; fax: (718) 981-4399; email: apda@apdaparkinson.org
Internet: http://www.apdaparkinson.org/user/ViewFellowshipAndGrants.asp
Sponsor: American Parkinson's Disease Association
1250 Hylan Boulevard, Suite 4B
Staten Island, NY 10305-1946

American Parkinson's Disease Association Research Grants   775
Qualified scientists may apply for grants for research in the area of Parkinson's disease. Grants are awarded for one year for direct costs of the research project; indirect costs, salary for the principal investigator, travel expenses, and institutional overhead are not covered. Grants may be renewed for a second year on deadline date, but reapplication must be made. Description of the research proposal should not exceed two pages and should include a statement of how research is related to Parkinson's disease.
Requirements: Scientists affiliated with U.S. institutions are eligible.
Restrictions: Publication costs may not exceed $300, and equipment budget may not exceed $5000.
Samples: Dr. David Albers, Massachusetts General Hospital (Boston)—for research entitled Oxidative Mechanisms of Injury in an In Vitro Model of Parkinson's Disease. Dr. Han-Xiang Deng, Northwestern U (Chicago, IL)—for research entitled Mapping and Cloning of a Novel Gene for Parkinson's Disease. Dr. Alice Weaver Flaherty, Massachusetts General Hospital (Boston)—for research entitled Functional MRI of Movement in Parkinsonian Patients. Dr. Alice Lazzarini, UMDNJ-RWJ Medical School (New Brunswick, NJ)—(renewal) for research entitled Allelic Association Studies of a Tau Polymorphism and Parkinson's Disease. Dr. Efthimia Kokotos Leonardi, Mount Sinai School of Medicine (New York, NY)—for research entitled Regulation of Dopamine Neurons by GDNF and BDNF in GTP Cyclohydrolase Deficiency.
Amount of Grant: $50,000
Date(s) Application is Due: Mar 1
Contact: Grants Administrator, (718) 981-8001 or (800) 223-2732; fax: (718) 981-4399; email: apda@apdaparkinson.org
Internet: http://www.apdaparkinson.org/user/ViewFellowshipAndGrants.asp
Sponsor: American Parkinson's Disease Association
1250 Hylan Boulevard, Suite 4B
Staten Island, NY 10305-1946

American Philosophical Society Daland Fellowships in Clinical Investigation   776
A limited number of fellowships are awarded annually for research in clinical medicine including the fields of internal medicine, neurology, psychiatry, pediatrics, and surgery. Patient-oriented research is emphasized. Essentially 100 percent of the fellow's time will be devoted to research. Teaching or clinical service of a limited amount is permitted. Additional salary may be granted by the institution at which the fellow is located, from another fellowship, or from a similar award during the tenure of the fellowship. The term of the fellowship is one year, with possible renewal for another year. Applications are available on the Web site, or may be obtained by written request, stating when the MD or MD/PhD degree was awarded; include a self-addressed mailing label.
Requirements: These fellowships are designed for qualified persons who have held an MD, or MD/PhD degree for less than eight years. The fellowship is generally intended to be the first post-clinical fellowship; but each case will be decided on its merits. Preference is generally given to candidates who have not more than two years of post-doctoral training and research. Applicants must expect to perform their research at an institution in the United States, under the supervision of a scientific adviser. Candidates are to be nominated by their department chairman, in a letter providing assurance that the nominee will work with the guidance of a scientific adviser of established reputation who has guaranteed adequate space, supplies, etc. for the Fellow. The adviser need not be a member of the department nominating the Fellow, nor need the activities of the Fellow be limited to the nominating department. As a general rule, no more than one fellowship will be awarded to a given institution in the same year of competition.
Restrictions: The society does not provide funds for institutional overhead.
Amount of Grant: $40,000 per year for up to two years
Date(s) Application is Due: Sep 1
Contact: Linda Musumeci, Research Administrator; (215) 440-3429; email: LMusumeci@amphilsoc.org
Internet: http://www.amphilsoc.org/grants/daland.htm
Sponsor: American Philosophical Society
104 South Fifth Street
Philadelphia, PA 19106-3387

American Philosophical Society Franklin Research Grants   777
The society awards small grants to scholars to support the cost of research leading to publication in all areas of knowledge. The program is particularly designed to help meet the costs of travel to libraries and archives for research purposes; the purchase of microfilm, photocopies, or equivalent research materials; the costs associated with fieldwork; or laboratory research expenses. Franklin grants are made for noncommercial research. They are not intended to meet the expenses of attending conferences or the costs of publication. The society is particularly interested in supporting the work of young

scholars who have recently received their PhDs. Applicants who have received Franklin grants may reapply after an interval of two years.
Requirements: Applicants are expected to have a doctorate, or to have published work of doctoral character and quality. American citizens and residents of the United States may use their Franklin awards at home or abroad. Foreign nationals must use their Franklin awards for research in the United States.
Restrictions: Predoctoral graduate students are not eligible. The society does not pay overhead or indirect costs to any institution. Grants will not be made to replace salary during a leave of absence or earnings from summer teaching; pay living expenses while working at home; cover the costs of consultants or research assistants; or purchase permanent equipment such as computers, cameras, tape recorders, or laboratory apparatus.
Amount of Grant: $6,000 maximum
Date(s) Application is Due: Oct 1; Dec 1
Contact: Linda Musumeci, Research Administrator; (215) 440-3429; email: LMusumeci@amphilsoc.org
Internet: http://www.amphilsoc.org/grants/franklin.htm
Sponsor: American Philosophical Society
104 South Fifth Street
Philadelphia, PA 19106-3387

American Philosophical Society John Hope Franklin Dissertation Fellowships   778
The fellowship is designed to support an outstanding African-American graduate student attending any PhD- granting institution in the United States, in any field of knowledge. There is no residential requirement. The stipend for this fellowship is $25,000 for a twelve-month period, plus $5,000 to support the cost of residency in Philadelphia, for a total award of $30,000. The twelve-month period is flexible.
Requirements: Candidates must have completed all course work and examinations preliminary to the doctoral dissertation and be prepared to devote full time for 12 months—with no teaching obligations—to research on their dissertation projects or the writing of their dissertations.
Amount of Grant: $30,000
Date(s) Application is Due: Apr 1
Contact: Linda Musumeci, Research Administrator; (215) 440-3429; email: LMusumeci@amphilsoc.org
Internet: http://www.amphilsoc.org/grants/johnhopefranklin.htm
Sponsor: American Philosophical Society
104 South Fifth Street
Philadelphia, PA 19106-3387

American Philosophical Society Judson Daland Prize   779
The $20,000 Judson Daland Prize recognizes outstanding achievement in patient-oriented research. The prize will be awarded at the meeting of the American Philosophical Society in November of each year. The Society reserves the right to award a shared prize. Nominations and all three letters of support must be received by March 15.
Requirements: Nominees must have done their work in an institution in the United States. Nominees need not, however, be citizens of the United States. Nominees should be no more than 15 years beyond receipt of the M.D. degree.
Amount of Grant: $20,000
Date(s) Application is Due: Mar 15
Contact: Linda Musumeci, Research Administrator; (215) 440-3429; email: LMusumeci@amphilsoc.org
Internet: http://www.amphilsoc.org/prizes/daland.htm
Sponsor: American Philosophical Society
104 South Fifth Street
Philadelphia, PA 19106-3387

American Philosophical Society Lewis and Clark Fund for Exploration and Field Research   780
The program encourages exploratory field studies for the collection of specimens and data and to provide the imaginative stimulus that accompanies direct observation. Applications are invited from disciplines with a large dependence on field studies, such as archeology, anthropology, biology, ecology, geography, geology, linguistics, and paleontology, but grants will not be restricted to these fields. Budgets should be limited to travel and related expenses, including personal field equipment. Application and guidelines are available online.
Requirements: Grants are available to graduate students, postdoctoral students, junior and senior scientists, and social scientists who wish to participate in field studies for their theses or for other purposes. A graduate student applicant should ask his or her academic supervisor or field trip leader to write one of the two letters of recommendation, specifying the role of the student in the field trip and the educational contribution of the trip. The competition is open to U.S. residents wishing to carry out research anywhere in the world. Foreign applicants must either be based at a U.S. institution or plan to carry out their work in the United States.
Restrictions: Undergraduates are not eligible.
Amount of Grant: $5,000 maximum
Date(s) Application is Due: Feb 15
Contact: Linda Musumeci, Research Administrator; (215) 440-3429; email: LMusumeci@amphilsoc.org
Internet: http://www.amphilsoc.org/grants/lewisandclark.htm
Sponsor: American Philosophical Society
104 South Fifth Street
Philadelphia, PA 19106-3387

American Philosophical Society Lewis and Clark Fund for Exploration and Field Research in Astrobiology 781

The Lewis and Clark Fund for Exploration and Field Research in Astrobiology is open to field studies in any area of interest to astrobiology. Astrobiology is the study of the origin, evolution, distribution, and future of life on Earth and in the universe. It encompasses research in, among others, the fields of astronomy, chemistry, evolutionary biology, field and population biology, geology, microbiology, molecular biology, oceanography, paleontology, and planetary science. Astrobiology includes investigations of the geologic and fossil record to understand the conditions of the early Earth when life arose. Its scope also includes research of contemporary locations on Earth that might be similar to early earth and to environments elsewhere in our Solar System (such as on Mars, Europa, and Titan), which may be, or have been in the past, suitable for life. Astrobiology is also about understanding the characteristics of life, which requires investigations into extreme natural environments on Earth and, eventually, elsewhere. Budgets should be limited to travel and related expenses, including personal field equipment. Amounts will depend on travel costs, but will ordinarily be in the range of several hundred dollars up to about $5,000.

Requirements: Grants will be available to graduate students and post-doctoral and junior scientists who wish to participate in field studies for their theses or for other purposes. Eligibility for applicants with doctorates is limited to those five years or fewer beyond their Ph.D. or equivalent degree, although, rarely, exceptions may be made. The competition is open to those affiliated with a U.S.-based institution, who may carry out research anywhere in the world.

Amount of Grant: $5,000 maximum
Date(s) Application is Due: Feb 1
Contact: Linda Musumeci, Research Administrator; (215) 440-3429; email: LMusumeci@amphilsoc.org
Internet: http://www.amphilsoc.org/grants/astrobiology.htm
Sponsor: American Philosophical Society
104 South Fifth Street
Philadelphia, PA 19106-3387

American Philosophical Society Library Resident Research Fellowships 782

Short-term residential fellowships for up to three months encourage research in the library's collections. Outstanding historical collections and subject areas include the papers of Benjamin Franklin; the American Revolution; 18th- and 19th-century natural history; western scientific expeditions and travel including the journals of Lewis and Clark; polar exploration; the papers of Charles Wilson Peale, including family and descendants; history of genetics, eugenics, and evolution; history of biochemistry, physiology, and biophysics; 20th-century medical research; and history of physics. Application procedures are undergoing revision, and will be entirely online. You may submit a message of interest to Libfellows AT amphilsoc DOT org and you will be notified when the new procedures are in place and ready for use. The next application deadline is March 1st.

Requirements: U.S. citizens and foreign nationals who hold the PhD or the equivalent, PhD candidates who have passed their preliminary exams, and independent scholars are eligible to apply. The fellowships are intended to encourage research in the library collections by scholars who reside beyond a 75-mile radius of Philadelphia.

Amount of Grant: $2,000 per month
Date(s) Application is Due: Mar 1
Contact: Library Resident Research Fellowships; (215) 440-3443; fax: (215) 440-3423; email: Libfellows@amphilsoc.org
Internet: http://www.amphilsoc.org/grants/resident.htm
Sponsor: American Philosophical Society
105 South Fifth Street
Philadelphia, PA 19106-3386

American Philosophical Society Phillips Fund Grants for Native American Research 783

The Phillips Fund of the American Philosophical Society provides grants for research in Native American linguistics, ethnohistory, and the history of studies of Native Americans, in the continental United States and Canada. Grants are not made for projects in archaeology, ethnography, psycholinguistics, or for the preparation of pedagogical materials. The committee distinguishes ethnohistory from contemporary ethnography as the study of cultures and culture change through time. The grants are intended for such costs as travel, tapes, films, and consultants' fees but not for the purchase of books or permanent equipment. The average award is about $2,500; grants do not exceed $3,500.

Requirements: Applicants may be graduate students who have passed their qualifying examinations for either the master's or doctorate degrees; postdoctoral applicants are eligible.

Restrictions: Grants are not made for projects in archaeology, ethnography, or psycholinguistics, nor for the preparation of pedagogical materials.

Amount of Grant: $3,500 maximum
Date(s) Application is Due: Mar 1
Contact: Linda Musumeci, Research Administrator; (215) 440-3429; email: LMusumeci@amphilsoc.org
Internet: http://www.amphilsoc.org/grants/phillips.htm
Sponsor: American Philosophical Society
104 South Fifth Street
Philadelphia, PA 19106-3387

American Philosophical Society Sabbatical Fellowship in the Humanities and Social Sciences 784

The society supports fellowships in the humanities and the social sciences for faculty of U.S. universities and four-year colleges. The awards are intended for those, usually in mid-career, who have been granted a sabbatical year/research leave by the university or college, but for whom full financial support from the parent institution is available for only half an academic year. The fellowship helps defray expenses for the second half of an awarded sabbatical year. There is no restriction on where the fellow resides during the fellowship year, but an indication of the appropriateness of the available library resources should be given.

Requirements: Applicants should be based in the United States, but need not be U.S. citizens, and must not have had a financially supported leave during the past three years. The application requires a form for the cover sheet and three letters of support; details are available on the Web site.

Restrictions: The fellowship cannot be used to supplement external awards of similar purposes. Institutions are not eligible to apply.

Amount of Grant: $30,000-$40,000 for second half of awarded sabbatical year
Date(s) Application is Due: Oct 15
Contact: Linda Musumeci, Research Administrator; (215) 440-3429; email: LMusumeci@amphilsoc.org
Internet: http://www.amphilsoc.org/grants/sabbatical.htm
Sponsor: American Philosophical Society
104 South Fifth Street
Philadelphia, PA 19106-3387

American Press Institute Seminar Fellowships 785

The institute conducts seminars on all phases of newspaper operation at announced times and locations. In conjunction with these seminars, the API offers a few fellowships for attendance by faculty members in good standing in journalism departments at four-year colleges and universities. API also offers a few fellowships to newspaper employees; marketing or general management executive, city editor, female reporter, or editor with a daily circulation below 25,000. Periodically, announcements are made through departments of journalism; information also is available from the institute. The listed deadline is for seminars in the following calendar year. Some fellowships will be awarded to journalism faculty members who belong to a minority group. Funding includes tuition, room and board, and a travel subsidy if funds permit.

Requirements: Applicants must be U.S. citizens.
Amount of Grant: $650-$1175
Date(s) Application is Due: Nov 18
Contact: Linda Kepner, Fellowship Coordinator, (703) 620-3611; fax: (703) 620-5814; email: lkepner@americanpressinstitute.org
Internet: http://www.americanpressinstitute.org/fellowships
Sponsor: American Press Institute
11690 Sunrise Valley Drive
Reston, VA 22091

American Psychiatric Association Award for Research in Psychiatry 786

The Award is given in recognition of a single distinguished contribution, a body of work, or a lifetime contribution that has had a major impact on the field and/or altered the practice of psychiatry. The Award is intended to cover the full spectrum of psychiatric research. The Award consists of a $5,000 prize and an honorary plaque to be presented at APA's Annual Meeting in May. The Award also includes an honorary lecture by the awardee.

Requirements: Candidates for the Award must be citizens of the United States or Canada and be nominated by a sponsor. Sponsors must be members of the American Psychiatric Association.

Restrictions: Members of the Award Committee are excluded from submitting nominations.

Amount of Grant: $5,000
Date(s) Application is Due: Aug 28
Contact: Harold Goldstein; (703) 907-8623; email: goharold@psych.org
Internet: http://www.psych.org/research/apire/training_fund/psychaward.cfm
Sponsor: American Psychiatric Association
1000 Wilson Boulevard, Suite 1825
Arlington, VA 22209-3901

American Psychiatric Association Early Career Award 787

APA's early career award recognizes the best nominated paper published during the past year by an early career psychiatrist (defined by the APA as under 40 years of age or within five years of training). Nominations can be either from the individual or a colleague in the field, such as a department chairperson, division chief, or other health services researcher. While the proposed applicant must be an APA member, the nominating individual need not be a member and may be from any discipline. The award will be presented at the Health Services Research Breakfast which is held in conjunction with APA's Institute for Psychiatric Services (IPS) meeting.

Requirements: Nomination letters should succinctly indicate the contributions that are the basis for the nomination and the nature of the relationship of the nominator and nominee. A curriculum vita of the nominee should accompany the letter, along with the nominated paper (in the case of early career award) or one to two papers of greatest significance (for senior scholar award).

Amount of Grant: $1,000
Date(s) Application is Due: Aug 29
Contact: Harold Goldstein; (703) 907-8623; email: goharold@psych.org

Internet: http://psych.org/MainMenu/Research/ResearchTrainingandFunding/APAResearchAwards/EarlyCareerandSeniorScholarHealthServicesResearchAwards.aspx
Sponsor: American Psychiatric Association
1000 Wilson Boulevard, Suite 1825
Arlington, VA 22209-3901

American Psychiatric Association Minority Fellowships Program   **788**
The APA offers psychiatric residents one-year fellowships in the APA organization to increase the knowledge of cultural factors influencing psychiatric diagnosis and treatment, provide opportunities for these residents to participate in the deliberations and decision-making processes, and provide role models for the fellows. Each fellow is appointed to a component of the APA's organizational structure and attends the association's annual meeting as an observer and active participant. Fellows are selected on the basis of their commitment to serving underrepresented populations, demonstrated leadership abilities, and interest in the interrelationship between mental health/illness and transcultural factors.
Requirements: Psychiatric residents who are starting their second year of psychiatric training, or third year if they are in a four-year residency program, are eligible to apply.
Contact: Marilyn King, Fellowship Coordinator, (703) 907-8653; fax: (703) 907-7849; email: mking@psych.org
Internet: http://www.psych.org/edu/other_res/apa_fellowship/cmhs_index.cfm
Sponsor: American Psychiatric Association
1000 Wilson Boulevard, Suite 1825
Arlington, VA 22209-3901

American Psychiatric Association Minority Medical Student Fellowship   **789**
    in HIV Psychiatry
The APA invites ethnic minority medical students who have an interest in psychiatric issues to apply. The program is intended to identify minority medical students who have primary interests in services related to HIV/AIDS and substance abuse and its relationship to the mental health or psychological well being of ethnic minorities.
Requirements: These programs are open to ethnic minority medical students who are U.S. citizens or permanent residents currently enrolled in a U.S. medical school.
Date(s) Application is Due: Mar 31
Contact: Carol Svoboda, (703) 907-8642; email: csvoboda@psych.org; Diane Pennessi, (703) 907-8668; email: dpennessi@psych.org
Internet: http://www.psych.org/edu/other_res/apa_fellowship/cmhs_index.cfm
Sponsor: American Psychiatric Association
1000 Wilson Boulevard, Suite 1825
Arlington, VA 22209-3901

American Psychiatric Association Minority Medical Student Summer   **790**
    Externship in Addiction Psychiatry
The APA invites ethnic minority medical students who have an interest in psychiatric issues to apply. This clinical shadowing program identifies minority medical students who may have a specific interest in services related to substance abuse treatment/prevention and provide a setting where the student can work closely with a mentor who specializes in addiction psychiatry for one month.
Requirements: These programs are open to ethnic minority medical students who are U.S. citizens or permanent residents currently enrolled in a U.S. medical school.
Date(s) Application is Due: Feb 28
Contact: Marilyn King, (703) 907-8653; fax: (703) 907-7849; email: mking@psych.org; Rosa Bracey, (703) 907-8539; email: rbracey@psych.org
Internet: http://www.psych.org/edu/other_res/apa_fellowship/cmhs_index.cfm
Sponsor: American Psychiatric Association
1000 Wilson Boulevard, Suite 1825
Arlington, VA 22209-3901

American Psychiatric Association Minority Medical Student Summer   **791**
    Mentoring Program
The APA invites ethnic minority medical students who have an interest in psychiatric issues to apply. This program is intended to identify ethnic minority medical students who have an interest in psychiatric issues and expose students to a setting where they can work closely with a psychiatrist mentor for one month.
Requirements: These programs are open to ethnic minority medical students who are U.S. citizens or permanent residents currently enrolled in a U.S. medical school.
Date(s) Application is Due: Feb 28
Contact: Marilyn King, (703) 907-8653; fax: (703) 907-7849; email: mking@psych.org; Rosa Bracey, (703) 907-8539; email: rbracey@psych.org
Internet: http://www.psych.org/edu/other_res/apa_fellowship/cmhs_index.cfm
Sponsor: American Psychiatric Association
1000 Wilson Boulevard, Suite 1825
Arlington, VA 22209-3901

American Psychiatric Association Program for Minority Research   **792**
    Training in Psychiatry (PMRTP)
The program is designed to increase the number of underrepresented minority men and women in the field of psychiatric research. Research training offers the opportunity to engage in scientific investigation across the full array of disciplines, from basic neuroscience, genetics, and pharmacology to the cognitive behavioral, and social sciences, clinical psychiatry, and mental health services research. Research exposure can help students and trainees develop sound skills for clinical assessment and treatment planning. The program provides funding for short and long-term training opportunities at three levels: Medical School, Residency and Post-residency. National competitions also enable qualified mini-fellows to attend research-oriented meetings of psychiatric organizations. For medical students and residents, the duration can be two months to one year. For post-residency fellows, the duration is generally two years. A third year of fellowship support may be available if appropriate to a trainee's career development. Support from PMRTP falls into three categories: stipends, travel, and tuition and fees. Stipends are based on the trainee's years of relevant experience and the length of the research training experience.
Requirements: Preference in selection is given to underrepresented minorities such as American Indians, Asian-Americans, Blacks/African-Americans, Hispanics, Pacific Islanders, or other ethnic or racial group members found to be underrepresented in biomedical or behavioral research.
Amount of Grant: $19,968-$51,036
Contact: Ernesto Guerra, (703) 907-7300; fax: (703) 907-1085; email: eguerra@psych.org or apa@psych.org
Internet: http://www.psych.org/research/APIRE/pmrtp5302.cfm
Sponsor: American Psychiatric Association
1000 Wilson Boulevard, Suite 1825
Arlington, VA 22209-3901

American Psychiatric Association Research Colloquium for Junior Investigators   **793**
The purpose of the colloquium is to provide guidance, mentorship and encouragement to young investigators in the early phases of their training. Junior investigators will have an opportunity to obtain feedback about their past, present, and future research interests from mentors who are tops in their field in a small group setting as well as general information about research career development and grantsmanship. Candidates whose research interests are similar to those listed below are also encouraged to apply. An all-day workshop for junior psychiatric investigators will focus on these three areas: childhood disorders including autism, ADHD, conduct, eating, and mood disorders; genomics, epigenetics, and proteomics; and treatment of major psychiatric disorders: from substance abuse to schizophrenia.
Requirements: Psychiatrists who are senior residents, fellows, or junior faculty, and who have an interest and potential in developing research careers in the areas of research listed above. Participants should hold a medical degree or be a member of the APA; or be eligible to become members of the APA.
Restrictions: Those with individual federal research awards are not eligible.
Amount of Grant: $1,000 stipend
Date(s) Application is Due: Nov 16
Contact: Ernesto Guerra; (703) 907-7300; fax: (703) 907-1085; email: eguerra@psych.org or apa@psych.org
Internet: http://psych.org/MainMenu/Research/ResearchTrainingandFunding/ResearchColloquiumforJuniorInvestigators.aspx
Sponsor: American Psychiatric Association
1000 Wilson Boulevard, Suite 1825
Arlington, VA 22209-3901

American Psychiatric Association Senior Scholar Health Services Award   **794**
The Award recognizes singular or sustained research accomplishments by a researcher beyond early career status which have made an important contribution to the field of mental health services research. Nominations can be either from the individual or a colleague in the field, such as a department chairperson, division chief, or other health services researcher. While the proposed applicant must be an APA member, the nominating individual need not be a member and may be from any discipline. The award will be presented at the Health Services Research Breakfast which is held in conjunction with APA's Institute for Psychiatric Services (IPS) meeting.
Requirements: Nomination letters should succinctly indicate the contributions that are the basis for the nomination and the nature of the relationship of the nominator and nominee. A curriculum vita of the nominee should accompany the letter, along with the nominated paper (in the case of early career award) or one to two papers of greatest significance (for senior scholar award).
Amount of Grant: $1,000
Date(s) Application is Due: Aug 29
Contact: Harold Goldstein; (703) 907-8623; email: goharold@psych.org
Internet: http://psych.org/MainMenu/Research/ResearchTrainingandFunding/APAResearchAwards/EarlyCareerandSeniorScholarHealthServicesResearchAwards.aspx
Sponsor: American Psychiatric Association
1000 Wilson Boulevard, Suite 1825
Arlington, VA 22209-3901

American Psychiatric Association Travel Scholarships for Minority   **795**
    Medical Students
The APA invites ethnic minority medical students who have an interest in psychiatric issues to apply. The program supports travel and related costs for approximately 10 minority medical students interested in psychiatric to attend either the APA annual meeting in May or the Institute on Psychiatric Services (IPS) meeting in October. This program is a way for medical students to witness organized psychiatry at work and to learn more about the field. Not only will students attend sessions for experts and trainees alike, but they will be assigned to a mentor who will help them maximize their annual meeting or IPS experience and discuss career plans and resident training programs.
Requirements: These programs are open to ethnic minority medical students who are U.S. citizens or permanent residents currently enrolled in a U.S. medical school.

Date(s) Application is Due: Jan 26
Contact: Marilyn King, (703) 907-8653; fax: (703) 907-7849; email: mking@psych.org;
Rosa Bracey, (703) 907-8539; email: rbracey@psych.org
Internet: http://www.psych.org/edu/other_res/apa_fellowship/cmhs_index.cfm
Sponsor: American Psychiatric Association
1000 Wilson Boulevard, Suite 1825
Arlington, VA 22209-3901

American Psychiatric Association/AACDP Research Mentorship Award　　　796
This award honors an academic psychiatrist who has in a significant traditional or
innovative manner, fostered the pursuit of student research within his/her university
department. The nominee's contribution may be through direct mentorship of individual
students, or by the promotion of novel research-oriented training activities within a
department or residency program. The award consists of an inscribed plaque and a $1,500
honorarium, to be presented at the APA Annual Meeting.
Requirements: Candidates need not be limited to senior, well-established candidates;
innovativeness and dedication will be honored any academic level. Nominees must be
members of APA.
Restrictions: Only one nomination from any department will be considered.
Amount of Grant: $1,500 honorarium
Date(s) Application is Due: Aug 29
Contact: Harold Goldstein; (703) 907-8623; email: goharold@psych.org
Internet: http://psych.org/MainMenu/Research/ResearchTrainingandFunding/
APAResearchAwards/APAAACDPAwardforResearchMentorship.aspx
Sponsor: American Psychiatric Association / American Association of Chairs of
Departments of Psychiatry
1000 Wilson Boulevard, Suite 1825
Arlington, VA 22209-3901

American Psychiatric Association/AstraZeneca Young Minds in　　　797
　　Psychiatry International Awards
The program recognizes and supports promising international young psychiatrists within
five years of completing a psychiatric residency. Four unrestricted career development
awards of $45,000 (U.S.D) will be available. Awards will be made to two promising
physicians from the U.S. with one in Bipolar Disorder research and one on research in
Schizophrenia. An additional two awards will be made to promising physicians from
countries outside the U.S., with one in Bipolar Disorder research and one on research
in Schizophrenia. Three other awards of $30,000 (U.S.D) in either Bipolar Disorder or
Schizophrenia research will specifically focus on applicants from developing countries
whose economies are classified by the World Bank as low or lower middle income.
Requirements: The U.S. applicants must be citizens or permanent residents of the
United States. Applications are evaluated based on evidence of academic promise; how
the proposal will advance the applicant's career; and innovative or original concepts,
approaches, or methods for developing the applicant's career.
Amount of Grant: $45,000
Date(s) Application is Due: Oct 31
Contact: Ernesto Guerra, (703) 907-7300; fax: (703) 907-1085; email: eguerra@psych.
org or apa@psych.org
Internet: http://www.psych.org/research/apire/res_careerdev/youngminds_announce.
cfm
Sponsor: American Psychiatric Association
1000 Wilson Boulevard, Suite 1825
Arlington, VA 22209-3901

American Psychiatric Association/Janssen Resident Psychiatric Scholars　　　798
The program is intended to identify promising PGY-1, PGY-2, and PGY-3 psychiatric
residents with the potential to become leaders in clinical and health services research
in all areas of psychiatric research. Emphasis will be placed on special mentoring and
career enrichment programs both at the APA Annual Meeting and throughout the year.
An individual research mentor will be assigned to oversee the resident's fellowship.
The mentors will be chosen among nationally known leaders in clinical and health
services research. The mentor will advise and encourage the Scholar during the two-year
fellowship. The scholar will receive $2,500 during the second year of the fellowship, to
assist in their research career development (e.g., for use in developing a pilot research
project, obtaining statistical consultation, or visiting potential research training
programs). The program also provides funding for travel to the American Psychiatric
Association (APA) Annual Meeting during both years of the fellowship.
Amount of Grant: $2,500
Date(s) Application is Due: Jan 15
Contact: Ernesto Guerra, (703) 907-7300; fax: (703) 907-1085; email: eguerra@psych.
org or apa@psych.org
Internet: http://www.psych.org/edu/res_fellows/res_training/janssen.cfm
Sponsor: American Psychiatric Association
1000 Wilson Boulevard, Suite 1825
Arlington, VA 22209-3901

American Psychiatric Association/Kempf Fund Award for Research　　　799
　　Development in Psychobiological Psychiatry
This award recognizes a senior researcher who has made a significant contribution
to research on the causes and treatment of schizophrenia as both a researcher and a
mentor. A stipend will support the research career development of a young research
psychiatrist working in a mentor-trainee relationship with the award winner on further
research in this field. Submissions will be judged on the excellence of the nominee's

overall contribution to the body of research in schizophrenia, including submission
of the nominee's most significant paper or book; a description of the nominee's role
as a mentor to younger colleagues in the field; and a description of the qualifications
of the young research psychiatrist, as well as a detailed description of his or her career
development plan.
Amount of Grant: $1,500 to senior researcher; $20,000 to young research psychiatrist
Date(s) Application is Due: Oct 14
Contact: APA/Kempt Fund Award, (202) 682-6316 or (800) 852-1390; email: apa@
psych.org
Internet: http://www.psych.org/research/apire/res_careerdev/kempf.cfm
Sponsor: American Psychiatric Association
1000 Wilson Boulevard, Suite 1825
Arlington, VA 22209-3901

American Psychiatric Association/Lilly Psychiatric Research Fellowship　　　800
This one-year fellowship is awarded to two postgraduate psychiatry trainees specifically
to focus on research and personal scholarship. Minimal time will be devoted to teaching,
patient care, consultation, or other duties. The fellowship is designed for a resident who
demonstrates significant research potential, has not had extensive research training
prior to residency, and is not already an established investigator. Each chairman of a
department of psychiatry is invited to nominate one outstanding eligible resident.
Requirements: Individuals who have received their MD or DO degree and who have
completed residency training in general psychiatry or child psychiatry immediately
prior to the time the fellowship commences are eligible. The individual must also be a
member of APA.
Amount of Grant: $45,000
Date(s) Application is Due: Oct 14
Contact: Darrel A. Regier, (202) 682-6316 or (800) 852-1390; email: apa@psych.org
Internet: http://www.psych.org/research/apire/training_fund/fellow/lilly.cfm
Sponsor: American Psychiatric Association
1000 Wilson Boulevard, Suite 1825
Arlington, VA 22209-3901

American Psychiatric Association/Merck Co. Early Academic Career　　　801
　　Research Award
The Award is intended to help support the research of a junior faculty member with
an interest in sleep disorders or schizophrenia. Two separate awards will be made to
candidates who have completed a psychiatry residency, at least one year of a psychiatry
research fellowship, and are seeking to make a commitment to a research career with the
end goal of successfully transitioning to that of an independent investigator. The Award
is intended to assist in this key transition period by providing one year of funding of
$45,000. Salary support provided by this award will allow junior faculty in departments
of psychiatry at U.S. academic institutions to devote more time to research.
Requirements: Candidates must be citizens or permanent residents of the U.S. and should
also be APA members. Eligible candidates will be trained psychiatrists with an MD,
MD/Ph.D. or DO degree who have completed residency training in general psychiatry
or child psychiatry in the U.S. Candidates also should have completed at least one year of
a psychiatry research fellowship but are not 3 or more years post fellowship completion,
and do not currently hold an academic rank higher than Assistant Professor.
Restrictions: Individuals who have obtained K-awards or are considered an independent
Principal Investigator on R-01 or R-21 research grants from NIH are not eligible.
Amount of Grant: $45,000
Date(s) Application is Due: Oct 14
Contact: Ernesto Guerra, (703) 907-7300; fax: (703) 907-1085; email: eguerra@psych.
org or apa@psych.org
Internet: http://www.psych.org/research/apire/res_careerdev/apamerck.cfm
Sponsor: American Psychiatric Association
1000 Wilson Boulevard, Suite 1825
Arlington, VA 22209-3901

American Psychiatric Association/SmithKline Beecham Young Faculty　　　802
　　Award for Research Development in Biological Psychiatry
This award is designed to support research by a junior faculty member in the biology and
psychopharmacology of mood disorders and/or anxiety disorders. Applicants must hold
a tenure track position as an assistant professor in the psychiatry department at a school
of medicine in the United States.
Requirements: APA members with a MD or DO degree who have completed residency
training in general or child psychiatry are eligible for this award.
Amount of Grant: $45,000
Date(s) Application is Due: Oct 14
Contact: Darrel A. Regier, (202) 682-6316 or (800) 852-1390; email: apa@psych.org
Internet: http://www.psych.org/research/apire/training_fund/beecham.cfm
Sponsor: American Psychiatric Association
1000 Wilson Boulevard, Suite 1825
Arlington, VA 22209-3901

American Psychiatric Association/Wyeth Pharmaceuticals M.D./Ph.D.　　　803
　　Psychiatric Research Fellowships
The Fellowship provides one year of funding and is designed to support two postgraduate psychiatry trainees with research experience, specifically to focus on research
and personal scholarship. Minimal time will be devoted to teaching, patient care,
consultation, or other duties. The protection of time for research should be assured by
the department chairman.

Requirements: Individuals who have received their MD or DO and a PhD degree and who have completed residency training in general psychiatry or child psychiatry immediately prior to the time the fellowship commences are encouraged to apply. The Fellowship is designed for residents who: (a) have demonstrated significant research potential, (b) have had research training (i.e. as part of their work towards their PhD), and (c) are not already an established investigators. These individuals must also be members of the APA.
Amount of Grant: $45,000
Date(s) Application is Due: Oct 14
Contact: Darrel A. Regier, (202) 682-6316 or (800) 852-1390; email: apa@psych.org
Internet: http://www.psych.org/research/apire/training_fund/fellow/wyeth.cfm
Sponsor: American Psychiatric Association
1000 Wilson Boulevard, Suite 1825
Arlington, VA 22209-3901

American Psychological Foundation F. J. McGuigan Young Investigator Prize   804
The American Psychological Foundation awards this biennial prize to recognize the efforts of a young psychological science investigator in the areas of research consistent with those pursued by Frank Joseph McGuigan, PhD. The prize is focused to support research to explicate the concept of the human mind. The approach must be a materialistic one fostering both empirical and theoretical research. Empirical research would primarily be psychophysiological, but physiological and behavioral research may also qualify for support. The recipient will be selected based on the excellence of the full breadth of research conducted and published to date, as well as the promise of research planned for five years. The prize will be awarded to the recipient's institution for the benefit of his or her research. Nomination guidelines are available online.
Requirements: Nominees must have earned a doctoral degree in psychology or a related field, and be nine or fewer years post doctoral degree at the time of the nomination deadline. Nominees must have an affiliation with an accredited college, university, or other research institution.
Restrictions: Faculty salaries and indirect costs (i.e., overhead) may not be requested. Dualistic approaches such as espoused by many contemporary cognitive psychologists do not qualify for support.
Amount of Grant: $25,000
Date(s) Application is Due: Mar 1
Contact: Science Directorate, (202) 336-6000; fax: (202) 336-5953; email: science@apa.org
Internet: http://www.apa.org/science/mcguigan.html
Sponsor: American Psychological Association
750 First Street NE
Washington, DC 20002-4242

American Psychological Foundation Todd E. Husted Memorial Award   805
This award supports dissertation research that indicates the most potential to contribute toward the development and improvement of mental illness services for those with severe and persistent mental illness. Relevant topics include those that foster the development of a more comprehensive, humane, and responsive system of mental health care; develop a protective and humane sequencing of interventions that prevents the deterioration, homelessness, and premature deaths of those with serious mental illness; develop effective methods of improving patient compliance with medication and treatment for those having impaired insight as a result of schizophrenia and bipolar affective disorder; demonstrate practical methods of improved identification, diversion, and treatment of persons with mental illness who, as a result of that illness, enter the criminal justice system; foster methods to improve training and social attitudes of professionals in the criminal justice system (attorneys, public defenders, judges) regarding the role of serious mental illness in the behaviors of mentally ill offenders; and increase access to and utilization of appropriate services and supports for the most treatment-resistant and severely mentally ill persons. Application and guidelines are available online.
Requirements: APA student affiliates enrolled in a psychology graduate program are eligible to apply.
Amount of Grant: $1000
Date(s) Application is Due: Sep 15
Contact: Science Directorate, (202) 336-6000; fax: (202) 336-5953; email: science@apa.org
Internet: http://www.apa.org/science/dissinfo.html
Sponsor: American Psychological Association
750 First Street NE
Washington, DC 20002-4242

American Psychosomatic Society Herbert Weiner Early Career Awards   806
These annual awards honor individuals early in their career who have contributed significantly to the field of psychosomatic medicine and show substantial promise of continued meritorious academic accomplishments in the field. Recipients have the opportunity to present their research findings during the annual meeting of the society.
Requirements: Nominees of any nationality must be fewer than 10 years past their final academic degree and must be members of the society.
Amount of Grant: $1000
Date(s) Application is Due: Nov 1
Contact: Program Contact, (703) 556-9222; fax: (703) 556-8729; email: info@psychosomatic.org
Internet: http://www.psychosomatic.org/awards/index.htm
Sponsor: American Psychosomatic Society
6728 Old McLean Village Drive
McLean, VA 22101

American Rivers Anthony A. Lapham River Conservation Fellowship   807
The Anthony A. Lapham River Conservation Fellowship at American Rivers (AR) provides an excellent professional development opportunity for talented post-graduates pursuing careers as leaders in the field of conservation advocacy. The program will develop the next generation of skilled conservation leaders who can promote practical solutions that benefit natural and human communities. Recent master's degree graduates will focus on an applied research project that will make a tangible contribution to AR's mission. The 12-month Fellowship will be supported by a team of conservation staff and members of our Scientific and Technical Advisory Committee, which includes some of the nation's foremost experts on freshwater and other conservation science and policy.
Amount of Grant: $35,000 plus health and vacation benefits
Date(s) Application is Due: Feb 1
Contact: Serena McClain; (202) 347-7550, ext. 3004; fax: (202) 347-9240; email: smcclain@americanrivers.org or rivergrants@americanrivers.org
Internet: http://www.americanrivers.org/newsroom/press-releases/2007/new-lapham-fellowship.html
Sponsor: American Rivers
1101 14th Street, NW, Suite 1400
Washington, DC 20005

American Schlafhorst Foundation Grants   808
The foundation awards grants to eligible North Carolina nonprofit organizations in its areas of interest, including arts, children and youth, elderly, education, health care, science, and social services delivery. Types of support include building construction/renovation, equipment acquisition, general operating grants, research grants, scholarships, and seed money grants.
Requirements: 501(c)3 tax-exempt organizations serving the greater Charlotte, NC, area are eligible.
Restrictions: Individuals are not eligible.
Geographic Focus: North Carolina
Amount of Grant: $1000-$40,000
Contact: Grants Administrator, (704) 554-0800; email: info@schlafhorst.com
Sponsor: American Schlafhorst Foundation
8801 South Boulevard
Charlotte, NC 28224

American Society for Dermatologic Surgery Cutting Edge Research Grants   809
The goal of the program is to encourage research in areas of specific relevance to dermatologic surgery and/or cutaneous oncology. It is the hope of the ASDS Research Grant Committee to encourage well-conceived basic and clinical research projects to stimulate the transfer of new technologies from the clinical setting or laboratory to dermatologic surgery practice.
Restrictions: Only one application for an ASDS grant award will be considered from an institution or practice.
Amount of Grant: $5000-$15,000
Date(s) Application is Due: Jul 15
Contact: Dermatology Foundation, Medical and Scientific Committee, (847) 956-0900; fax: (847) 956-0999
Internet: http://www.asds-net.org/education/ResearchGrant/education-research.html
Sponsor: American Society for Dermatologic Surgery
5550 Meadowbrook Dr. Suite 120
Rolling Meadows, IL 60008

American Society for Enology and Viticulture Scholarships   810
Numerous scholarships are awarded annually to undergraduate and graduate students in North America pursuing a degree in enology, viticulture, or in a curriculum emphasizing a science basic to the wine and grape industry. Candidates are required to submit transcripts of previous education, information on financial need, and a written statement of intention to pursue a career in the wine or grape industry. Previous recipients are eligible each year in open competition with all other applicants.
Requirements: Applicant must be a resident of a North American country and accepted in a full-time college or university degree program in the required fields at the junior unit level or higher. Undergraduates must have at least a 3.0 GPA; graduate students must have at least a 3.2 GPA.
Date(s) Application is Due: Mar 1
Contact: Executive Director, (530) 753-3142; fax: (530) 753-3318; email: society@asev.org
Internet: http://www.asev.org
Sponsor: American Society for Enology and Viticulture
P.O. Box 1855
Davis, CA 95617-1855

American Society for Surgery of the Hand Outcome Studies Grants   811
Outcome studies grants are awarded to encourage young investigators to initiate projects that will lead to additional funding from other sources. There are no specific eligibility requirements, although the society encourages applications from surgeons and other allied health personnel who have a primary interest in hand surgery. All proposals must make use of a recognized and well-documented outcomes instrument, such as the SF-36, Sickness Impact Profile, Arthritis Impact Measurement Scale, etc., and must address an issue of importance to hand surgery. The society encourages applications from surgeons and other allied health personnel who have a primary interest in hand surgery.

Restrictions: Grants may not be used for the personal compensation of investigators, travel or conferences, publication costs, the payment of hospital care for patients under study, or the purchase of major pieces of equipment.
Samples: Dr. Thomas Trumble, Department of Orthopedic Surgery, U of Washington Medical Ctr (Seattle, WA)—for clinical trial of hand therapy following carpal tunnel surgery, $10,000.; Dr. Jeffrey Katz, Department of Rheumatology, Brigham and Women's Hospital (Boston, MA)—for research on carpal tunnel syndrome and shoulder disorders, $10,000.
Amount of Grant: $10,000 per year
Date(s) Application is Due: Jul 11
Contact: Julie Quinn, (847) 384-9250; fax: (847) 384-1435; email: jquinn@assh.org
Internet: http://www.assh.org/Content/NavigationMenu/Hand_Surgery_Professionals/Outcome_Studies/Translational_and_Clinical_Research.htm
Sponsor: American Society for Surgery of the Hand
6300 N River Road, Suite 600
Rosemont, IL 60018-4256

American Society for Surgery of the Hand Research Grants   812

The society has established grant funding of research projects to encourage young investigators. It is anticipated that the funding will enable investigators to initiate projects that will lead to additional funding from other sources. Submission of proposals intended for NIH or OREF is discouraged. There are no specific guidelines within the framework of hand surgery; in the past, funding has been provided for a range of subjects, including microvascular, tendons, nerves, artificial joints, biochemical, biomechanical, rehabilitation, and other subjects.
Requirements: Awards are made to universities, medical schools, research institutions, or individuals. Grants may be used for salaries of technical and professional assistants, purchase of supplies, or minor pieces of equipment.
Restrictions: Grants may not be used for the personal compensation of investigators, travel or conferences, publication costs, the payment of hospital care for patients under study, or the purchase of major pieces of equipment. No overhead or indirect costs may be imposed by the recipient institution.
Samples: Dr. Steven Moran—for Effects of S-Fluoracil on Flexor Tendon Repairs, $6000.; Dr. Jack Choucka and Dr. Daniel Mass—for Cyclical Failure of Cadaveric Zone II Human Flexor Tendon Repairs, $8000.
Amount of Grant: $16,000 maximum
Date(s) Application is Due: May 2
Contact: Julie Quinn, (847) 384-9250; fax: (847) 384-1435; email: jquinn@assh.org
Internet: http://www.assh.org/Content/NavigationMenu/Hand_Surgery_Professionals/Research/Basic_Science_Research.htm
Sponsor: American Society for Surgery of the Hand
6300 N River Road, Suite 600
Rosemont, IL 60018-4256

American Society of Criminology Gene Carte Student Paper Competition   813

This competition is held to recognize outstanding scholarly work done by students in the field of criminology. To enter the competition, students must submit papers, conceptual and/or empirical, directly related to criminology. Papers must be 7500 words or less, typewritten double-spaced, on standard-size white paper using standard format for the organization of text, citations, and references. Submissions must be accompanied by a letter indicating the author(s) enrollment status and cosigned by the dean, department chair, or program director. Author name(s), department(s), and advisor(s) must appear only on the title page since papers will be evaluated by blind review. Eight copies of the paper must be provided. Papers should be submitted to the office.
Requirements: Any student currently enrolled on a full-time basis in an academic program at either the undergraduate or graduate level may apply.
Samples: David Kirk, U of Chicago (IL)—first place for The Neighborhood Context of Racial and Ethnic Disparities in Arrest, $500 (2005); Callie Harbin Burt, U of Georgia (GA)—second place for A Longitudinal Test of Low-Self Control Theory with an African-American Sample, $300 (2005); Benjamin Steiner, U of Cincinnati (OH)—third place for Assessing Static and Dynamic Influences on Inmate Misconduct Levels Over Time, $200 (2005).
Amount of Grant: $500 first prize plus a travel award to annual meeting; $300 second prize; $200 third prize
Date(s) Application is Due: Apr 15
Contact: Crystal Garcia, (317) 274-7006; fax: (317) 274-7860; Awards Committee, (614) 292-9207; fax: (614) 292-6767
Internet: http://www.asc41.com/cartesp.html
Sponsor: American Society of Criminology
1314 Kinnear Road, Suite 212
Columbus, OH 43212

American Society of Naval Engineers Scholarship Program   814

This scholarship program is intended to encourage college students to enter into the field of naval engineering and to seek advanced education in this field. Programs of study that are supported include naval architecture; marine, mechanical, civil, aeronautical, ocean, electrical, and electronic engineering; and the physical sciences. Naval engineering includes the design, construction, and repair of ships and their installed systems and equipment, as well as research, logistic support, and the management of acquisition and maintenance. Candidates may apply for support for either the last year of a full-time or co-op undergraduate program or for one year of full-time graduate study leading to a designated engineering or physical science degree in an accredited college or university of the student's choice. Application forms are available on the Web site.
Requirements: This scholarship program is open to U.S. citizens who have expressed an interest in a career in naval engineering.
Restrictions: Doctoral candidates or those who have an advanced degree are ineligible for this scholarship. A scholarship leading to an undergraduate degree will not be continued for enrollment in a graduate program; however, a student may apply for a new award for a graduate scholarship.
Amount of Grant: $3500 for graduate students; $2500 for undergraduate students
Date(s) Application is Due: Feb 15
Contact: Dennis Pignotti, Operations Manager, (703) 836-6727; fax: (703) 836-7491; email: dpignotti@navalengineers.org
Internet: http://www.navalengineers.org/Programs/Scholarships/sc_info.htm
Sponsor: American Society of Naval Engineers
1452 Duke Street
Alexandria, VA 22314-3458

American Society on Aging Graduate Student Research Award   815

The award is presented annually to a graduate student for research relevant to aging and applicable to practice. The winner will be expected to attend the joint conference of the ASA and the National Council on the Aging to present the winning paper. The award comprises a certificate, complimentary registration and one-night's lodging for the conference, presentation of research findings at a highlighted session at the conference, and complimentary one-year membership in ASA.
Requirements: Only graduate-level research will be considered. Findings must be from completed research; submission of the conceptual framework alone is not sufficient. Applicants must either be currently enrolled in a graduate degree program or must have completed their studies no more than one year before the time of submission. Applicants must be sponsored by a faculty member. Applicants and faculty sponsors need not be ASA members.
Restrictions: Papers that have been published are not eligible for submission.
Date(s) Application is Due: Oct 1
Contact: Nancy Decia, Coordinator, Education & Training; (415) 974-9610; fax: (415) 974-0300; email: awards@asaging.org
Internet: http://www.asaging.org/asav2/awards/description_grad.cfm?submenu1=grad
Sponsor: American Society on Aging
833 Market Street, Suite 511
San Francisco, CA 94103-1824

American Sociological Association Awards   816

Each year the American Sociological Association honors outstanding scholars and scholarship in the field of Sociology. There are curently nine ASA awards presented each August, including: Distinguished Book Award; Dissertation Award; Excellence in the Reporting of Social Issues; Jessie Bernard Award; Cox-Johnson-Frazier Award; Award for the Public Understanding of Sociology; Distinguished Career Award for the Practice of Sociology; Distinguished Contributions to Teaching Award; and W.E.B. DuBois Career of Distinguished Scholarship Award. he deadline for submission of nominations is January 31st of each year unless noted otherwise in the individual award criteria.
Contact: Margaret Weigers Vitullo, Director; (202) 383-9005, ext. 323; fax: (202) 638-0882; email: spivack@asanet.org or apap@asanet.org
Internet: http://www.asanet.org/cs/awards
Sponsor: American Sociological Association
1430 K Street NW, Suite 600
Washington, DC 20005

American Sociological Association Congressional Fellowships   817

The Fellowship brings a PhD-level sociologist to Washington, DC, to work as a staff member on a congressional committee or in a congressional office. Some congressional agencies may allow fellow placements as well. This intensive six month experience reveals the intricacies of the policy making process to the sociological fellow, and shows the usefulness of sociological data and concepts to policy issues. Each applicant should have a general idea about the area of interest, some experience in client-driven work, good writing skills, and a commitment to the policy process. The stipend for the Fellowship is $20,000 for six months and $30,000 for 11 months.
Amount of Grant: $20,000-$30,000
Date(s) Application is Due: Feb 1
Contact: Margaret Weigers Vitullo, Director; (202) 383-9005, ext. 323; fax: (202) 638-0882; email: spivack@asanet.org or apap@asanet.org
Internet: http://www.asanet.org/cs/root/leftnav/funding/asa_congressional_fellowship
Sponsor: American Sociological Association
1430 K Street NW, Suite 600
Washington, DC 20005

American Sociological Association Fund for the Advancement of the Discipline   818

The American Sociological Association invites submissions for the Fund for the Advancement of the Discipline (FAD) awards. Supported by the American Sociological Association through a matching grant from the National Science Foundation, the goal of this project is to nurture the development of scientific knowledge by funding small, groundbreaking research initiatives and other important scientific research activities such as conferences. FAD awards provide scholars with seed money for innovative research that has the potential for challenging the discipline, stimulating new lines of research, and creating new networks of scientific collaboration. The award is intended to provide opportunities for substantive and methodological breakthroughs, broaden the dissemination of scientific knowledge, and provide leverage for acquisition of additional research funds. The amount of each award shall not exceed $7,000.

Amount of Grant: $7,000 maximum
Date(s) Application is Due: Jun 15; Dec 15
Contact: Margaret Weigers Vitullo, Director; (202) 383-9005, ext. 323; fax: (202) 638-0882; email: apap@asanet.org
Internet: http://www.asanet.org/cs/funding/FAD
Sponsor: American Sociological Association
1430 K Street NW, Suite 600
Washington, DC 20005

American Sociological Association Minority Fellowships 819
Through its Minority Fellowship Program (MFP), the American Sociological Association (ASA) supports the development and training of sociologists of color in mental health and drug abuse research. Funded by a training grant sponsored by the National Institute of Mental Health (NIMH) and co-funded by the National Institute of Drug Abuse (NIDA), MFP seeks to attract talented doctoral students to ensure a diverse and highly trained workforce is available to assume leadership roles in research related to the nation, mental health and drug abuse research agendas.
Requirements: American citizens and permanent visa residents including, but not limited to, persons who are African American, Latino/Hispanic (e.g., Chicano, Cuban, Puerto Rican), Native American, and Asian American (e.g., Chinese, Japanese, Korean), and Pacific Islanders (e.g., Hawaiian, Guamanian, Samoan, Filipino) are eligible. The program is open to students beginning or continuing study in graduate sociology departments. New students must qualify for acceptance at accredited institutions of higher learning and express a commitment to sociological research on mental health. Upon completion of their support, recipients are expected to engage in behavioral research or teaching, or a combination thereof, for a period equal to the period of support beyond 12 months.
Amount of Grant: $20,000-$25,000 stipend
Date(s) Application is Due: Jan 31
Contact: Margaret Weigers Vitullo, Director; (202) 383-9005, ext. 323; fax: (202) 638-0882; email: spivack@asanet.org or apap@asanet.org
Internet: http://www.asanet.org/cs/root/leftnav/funding/minority_fellowship_program
Sponsor: American Sociological Association
1430 K Street NW, Suite 600
Washington, DC 20005

American Sociological Association Sydney S. Spivack Program in Applied 820
Social Research and Social Policy
The program is intended to encourage sociologists to undertake community action projects that bring social science knowledge, methods, and expertise to bear in addressing community-identified issues and concerns. Sociologists are expected to work in relevant community organizations, local public interest groups, or community action projects doing such activities as needs assessments, empirical research relevant to community activities or action planning, the design and implementation of evaluation studies, or analytic review of the social science literature related to a policy issue or problem. Grants cover direct costs associated with the project. Approximately four awards are made annually.
Requirements: Applications are encouraged from sociologists in academic settings, research institutions, private and nonprofit organizations, and government.
Restrictions: Advanced graduate students are eligible, but the funding cannot be used to support doctoral dissertation research.
Amount of Grant: $1,000-$2,500 average
Date(s) Application is Due: Feb 1
Contact: Margaret Weigers Vitullo, Director; (202) 383-9005, ext. 323; fax: (202) 638-0882; email: spivack@asanet.org or apap@asanet.org
Internet: http://www.asanet.org/cs/root/leftnav/funding/community_action_research_initiative
Sponsor: American Sociological Association
1430 K Street NW, Suite 600
Washington, DC 20005

American Sociological Association Teaching Enhancement Fund Grants 821
The goal of this program is to support projects that extend the quality of teaching in the United States and Canada. TEF small grants can support an individual, a program, a department, or a committee of a state/regional association. ASA will award up to two grants, each up to $2,000. Proposals limited to a maximum of five pages should describe the project and the intended audience or beneficiaries, explain how the financial support would be used, describe the expected benefits of the project including systemic impacts, and indicate how the project might have a lasting benefit.
Requirements: An individual, department, program, or committee of a state/regional association may apply. Individuals applying for the award must be members of ASA.
Amount of Grant: $2,000 maximum
Date(s) Application is Due: Feb 2
Contact: Margaret Weigers Vitullo, Director; (202) 383-9005, ext. 323; fax: (202) 638-0882; email: apap@asanet.org
Internet: http://www.asanet.org/cs/root/leftnav/funding/teaching_enhancement_fund
Sponsor: American Sociological Association
1430 K Street NW, Suite 600
Washington, DC 20005

American Studies in Japan Research Grants 822
Through their public school system and media, the Japanese people have achieved a level of general knowledge about and interest in the United States. Nevertheless, the formal or integrated study of U.S. history and civilization, and of the economic, social, and political institutions of the United States, has been a relatively recent development in Japanese universities. Moreover, opportunities for Japanese scholars to track major developments and changes in contemporary U.S. society need to be expanded. The commission, therefore, has committed itself to the long-term development of both institutional and individual expertise in the Japanese academic world in U.S. studies, as well as to curriculum development at the more general undergraduate level. The commission will give high priority to research projects that investigate the study of the United States itself, particularly on how the Japanese acquire basic knowledge about the United States (its politics, society, and economy) through both formal and informal channels, such as classroom instruction and media, and how to increase that knowledge and make it more accurate. The following project areas are of interest: research center development, research projects in and about U.S. studies, faculty development, curriculum development, and conferences and seminars. Projects receiving matching grants from other appropriate U.S. or Japanese sources will be given high priority.
Requirements: Citizens and permanent residents of the United States and Japan are eligible. Grants may be made to individual universities or local organizations. Awards made to individuals under all programs normally will be made through academic, professional, artistic, or other appropriate organizations that will examine, recommend, and, in most instances, select the individuals to be supported financially by the commission.
Restrictions: Japanese institutions and U.S. academic organizations only may apply. This award is granted twice a year.
Samples: U of Tokyo, Ctr for American Studies (Tokyo, Japan)—to provide assistance in promoting center activities, @Y2 million.; Doshisha U, Ctr for American Studies (Japan)—to promote activities at the center, @Y5 million.
Date(s) Application is Due: Mar 1; Aug 1
Contact: Grant Administrator, (202) 418-9800; fax: (202) 418-9802; email: grants@jusfc.gov
Internet: http://www.jusfc.gov/commissn/BrochureJan03.htm#_The_Study_of
Sponsor: Japan-United States Friendship Commission
1201 15th Street NW, Suite 330
Washington, DC 20005

American Translators Association German Prize 823
The prize is awarded biennially in odd-numbered years for works translated from German and published in the United States during the preceding two years. The award includes a monetary prize, certificate of recognition, and travel allowance to attend the annual meeting. Ten pages from the original and two copies of the published translation must be submitted for consideration along with any other documentation.
Amount of Grant: $1000 and $500 maximum toward travel expenses
Date(s) Application is Due: May 15
Contact: Walter Bacak, Honors and Awards Committee, (703) 683-6100 ext. 3006; fax: (703) 683-6122 ext 3066; email: Walter@atanet.org or ata@atanet.org
Internet: http://www.atanet.org/News_Award_German.htm
Sponsor: American Translators Association
225 Reinekers Ln, Suite 590
Alexandria, VA 22314

American Wine Society Educational Foundation Scholarship Program 824
The foundation provides academic scholarships and research grants to graduate students based on academic excellence and genuine interest in pursuing a career in enology, viticulture, or the responsible use and health aspects of wine.
Requirements: Awards are made to North American (U.S., Canada, Mexico, and islands of Caribbean) citizens who have been accepted into a graduate program in enology, viticulture, or a wine-related area. Applicants must complete the undergraduate degree with a 2.8 overall academic average and a 3.0 in their major area of study.
Samples: Jonathan Licker, Cornell U (New York, NY)—for the study Flavors via Brettanomyces, $2000.; Elwyn Gladstone, U of California Davis (CA)—for a study on canopy density, $2000.
Amount of Grant: $2500
Date(s) Application is Due: Mar 31
Contact: Dr. Les Sperling, (610) 758-3845 or 865-2401 (7-9 pm Eastern time preferred); fax: (610) 758-3526; email: lhs0@lehigh.edu or Sperling@AmericanWineSociety.com
Internet: http://www.americanwinesociety.com/web/scholarship_eligibility.htm
Sponsor: American Wine Society Educational Foundation
1134 Prospect Avenue
Bethlehem, PA 18018-4914

American-Italian Cancer Foundation Fellowships 825
The mission of the program is to further the advancement of cancer research by sponsoring the research and training of promising young scientists in Italy and in the United States. A major activity is the award of fellowships to Italian and American scientists so that they might spend a period of research and advanced training in preclinical or clinical research centers in the United States or Italy. The fellowships are intended to augment the training of motivated scientists who have the potential to conduct path-breaking original research upon the return to their respective countries. The one-year fellowships are renewable for an additional year.
Requirements: Individuals in the United States and Italy who have been awarded the degrees of PhD, MD, DSc, or DVM may apply in conjunction with a host institution. Candidates must have received their degree no more than three years prior to the date of application.
Amount of Grant: $25,000 Italian fellowships; $30,000 U.S. fellowships
Date(s) Application is Due: Mar 1

Contact: Franca Gaudio, (212) 628-9090; fax: (212) 517-6089; email: fgaudio@aicfonline.org
Internet: http://www.aicfonline.org
Sponsor: American-Italian Cancer Foundation
112 E 71st Street, Suite 2B
New York, NY 10021

American-Italian Cancer Foundation Prize for Scientific Excellence   826

The prize recognizes a major scientific discovery in basic or translational cancer research advances that emanate from outstanding scientific work that has reached the early stages of clinical trials and shows real promise. The foundation is keenly interested in research that promises to make a difference in the quality of life for cancer patients. Prize winners will have made a significant contribution to understanding cancer, continue to be active in research, and their work holds promise for future outstanding contributions to the field. Each year two awards will be presented to one U.S. and one European scientist. Nomination guidelines are available online.
Requirements: Nominations may be made by any scientist currently or previously affiliated with a cancer research institution or an advocacy organization.
Restrictions: Institutions or organizations are not eligible for this award, and candidates may not nominate themselves.
Amount of Grant: $50,000
Contact: Dr. Esther Dyer, Executive Director, (212) 628-9090; fax: (212) 517-6089; email: aicf@aicfonline.org
Internet: http://www.aicfonline.org
Sponsor: American-Italian Cancer Foundation
112 E 71st Street, Suite 2B
New York, NY 10021

American-Scandinavian Foundation Awards for Advanced Study or   827
Research in the United States

ASF offers a limited number of awards, usually at the graduate level, for Scandinavians wishing to undertake research or studies in the United States. The program also is designed to foster understanding and further cooperation between the nations. Candidates are recommended by cooperating Scandinavian organizations such as the Denmark-America Foundation, the League of Finnish-American Societies, the Icelandic-American Society, the Norway-America Association, and the Sweden-America Foundation.
Requirements: Applicants must be citizens of a Scandinavian country (Denmark, Finland, Iceland, Norway, or Sweden).
Contact: Ellen McKey, Director of Fellowships and Grants; (212) 879-9779; fax: (212) 249-3444; email: grants@amscan.org
Internet: http://www.amscan.org/fellowship.html
Sponsor: American-Scandinavian Foundation
58 Park Avenue
New York, NY 10016

American-Scandinavian Foundation Fellowship for Study in Scandinavia   828

The American-Scandinavian Foundation (ASF) offers fellowships (up to $23,000) to individuals to pursue research or study in one or more Scandinavian country for up to one year. Awards are made in all fields.
Requirements: Applicants must: have a well-defined research or study project that makes a stay in Scandinavia essential; be United States citizens or permanent residents; have completed their undergraduate education by the start of their project in Scandinavia. Team projects are eligible, but each member must apply as an individual, submitting a separate, fully-documented application. The ASF considers it desirable that all candidates have at least some ability in the language of the host country, even if it is not essential for the execution of the research plan.
Restrictions: Application may be made for either an ASF fellowship or a grant, not for both. Funds will not be provided for research assistants, dependent support, loan obligations, publication costs, equipment purchase, institutional overhead charges, study at English-language institutions, beginning studies of any subject matter, conference attendance, foregone salary, or supplementation of substantial sabbatical support.
Amount of Grant: $23,000 max.
Date(s) Application is Due: Nov 2
Contact: Ellen McKey, Director of Fellowships and Grants; (212) 879-9779; fax: (212) 249-3444; email: grants@amscan.org
Internet: http://www.amscan.org/study_scandinavia_details.html
Sponsor: American-Scandinavian Foundation
58 Park Avenue
New York, NY 10016

American-Scandinavian Foundation Grant for Study in Scandinavia   829

ASF encourages advanced study and research in Scandinavian countries. Grants are considered especially suitable for postgraduate scholars, professionals, and candidates in the arts to carry out research or study visits of one to three months. Outstanding proposals from all sources are invited. ASF desires that all candidates have at least some ability in the language of the country in which they plan to study.
Requirements: Applicants must have: a well-defined research or study project that makes a stay in Scandinavia essential; be United States citizens or permanent residents; completed their undergraduate education by the start of their project in Scandinavia. Team projects are eligible, but each member must apply as an individual, submitting a separate, fully documented application. The ASF considers it desirable that all candidates have at least some ability in the language of the host country, even if it is not essential for the execution of the research plan.

Restrictions: Application may be made for either an ASF grant or a fellowship, not for both.
Amount of Grant: $5,000
Date(s) Application is Due: Nov 2
Contact: Ellen McKey, Director of Fellowships and Grants; (212) 879-9779; fax: (212) 249-3444; email: grants@amscan.org
Internet: http://www.amscan.org/study_scandinavia_details.html
Sponsor: American-Scandinavian Foundation
58 Park Avenue
New York, NY 10016

American-Scandinavian Foundation Short-Term Training Program Grants   830

Grants enable U.S. citizens and permanent residents who are at least 20 years of age to live, work, and train in Scandinavia on a temporary basis. Work assignments of eight to 12 weeks or longer (from spring through fall), are available to U.S. students in certain fields, principally engineering, chemistry, and agriculture. The emphasis of the program is the cultural and educational experience rather than financial gain.
Requirements: An applicant should be a full-time student majoring in the field in which training is sought; have at least three years of undergraduate studies completed; and have some previous related work experience. Knowledge of a Scandinavian language is not necessary.
Restrictions: Applications are not accepted in the medically related professions involving patient care.
Contact: Ellen McKey, Director of Fellowships and Grants; (212) 879-9779; fax: (212) 249-3444; email: grants@amscan.org
Internet: http://www.amscan.org/training.html
Sponsor: American-Scandinavian Foundation
58 Park Avenue
New York, NY 10016

Ametek Foundation Grants   831

Corporate contributions are made through the foundation to nonprofit organizations in company-operating areas. The foundation supports programs and projects in the categories of health, education, and social services, with education being the largest area of support. Under the category of health, grants are awarded to hospitals, health care facilities, and for medical research. Support for education is given to colleges and universities and technical schools, as well as scholarship funds. Welfare funding supports philanthropic organizations, including the United Way. Arts groups, museums, and civic groups also receive support. Types of support include general operating support, annual campaigns, building construction/renovation, equipment acquisition, endowment funds, scholarship funds, research, technical assistance, and matching funds. Annual application deadline dates may vary; contact the office for specific dates. The board meets in November and May to consider requests.
Requirements: IRS 501(c)3 organizations are eligible.
Restrictions: Grants are not made to individuals or to political, fraternal, or veterans organizations.
Samples: Rochester City School District (Rochester, NY)—grant recipient, $116,231.; Gnaden Huetten Memorial Hospital (Lehighton, PA)—grant recipient, $1000.; Abilities, Inc of Florida (Clearwater, FL)—grant recipient, $15,000.
Amount of Grant: $1000-$25,000 average
Date(s) Application is Due: Feb 28; Sep 1
Contact: Kathryn Londra, (610) 647-2121; fax: (610) 296-3412
Sponsor: Ametek Foundation
P.O. Box 1764, 37 N Valley Road, Building 4
Paoli, PA 19301-0801

amfAR Fellowships   832

An amfAR fellowship is a grant that encourages the postdoctoral (M.D., Ph.D., or equivalent) investigator with limited experience in the field to advance a career in HIV/AIDS research. Fellowships are awarded for two years and may not be renewed for additional funding. amfAR fellows and mentors must be affiliated with the same nonprofit institution. The applicant's interest in a career in HIV/AIDS will be demonstrated by previous relevant work at the postdoctoral fellow or instructor level and will be carefully evaluated. Each fellowship is funded at a total of up to $125,000: a maximum of $110,000 is allowed for personnel (salary and fringe benefits) and other research-related direct costs. It is expected that a fellow will devote the decided majority of his or her time to the approved fellowship project. Personnel costs supported by the fellowship must represent a minimum of 85% effort and be consistent with institution policy for other institution personnel of similar rank and title, regardless of source(s) of support. An additional $3,636 is provided to support attendance at amfAR-approved professional development activities, for a direct cost maximum of $113,636. The period of performance for fellowships awarded under this RFP will be for two years starting January 1.
Requirements: The fellowship applicant must be mentored by an experienced investigator who: is qualified to oversee the proposed research; has successfully supervised postdoctoral fellows; and is at the associate professor level or higher.
Restrictions: Institutional indirect costs may not exceed 10% of direct costs.
Amount of Grant: $125,000 total
Date(s) Application is Due: Jul 7
Contact: Grant Administrator; (212) 806-1696 or (212) 806-1600; fax: (212) 806-1601; email: grants@amfar.org
Internet: http://www.amfar.org/lab/grants/default.aspx?id=7455&terms=Fellowships
Sponsor: American Foundation for AIDS Research

120 Wall Street, 13th Floor
New York, NY 10005-3908

amfAR Mathilde Krim Fellowships in Basic Biomedical Research   833

amfAR, The Foundation for AIDS Research, is pleased to announce the availability of support for Mathilde Krim Fellowships in Basic Biomedical Research. Each fellowship is funded at a total of up to $125,000 (phase I): a maximum of $110,000 is allowed for personnel (salary and fringe benefits) and other research-related direct costs. It is expected that a Krim fellow will devote the decided majority of his or her time to the approved fellowship project. Personnel costs supported by the fellowship must represent a minimum of 85% effort and be consistent with institution policy for other institution personnel of similar rank and title, regardless of source(s) of support. An additional $3,636 is provided to support attendance at amfAR-approved professional development activities, for a direct cost maximum of $113,636. Contingent upon subsequent application and peer review, phase II funding for an additional consecutive year may be approved to support basic biomedical HIV/AIDS research costs (up to $50,000 total) during the first year of a tenure-track position at any U.S. or international nonprofit research institution. The period of performance for Mathilde Krim Fellowships awarded under this RFP will be from January 1 to December 31 of the following year.
Restrictions: Institutional indirect costs may not exceed 10% of direct costs.
Amount of Grant: $125,000 total
Date(s) Application is Due: Jul 7
Contact: Grant Administrator; (212) 806-1696 or (212) 806-1600; fax: (212) 806-1601; email: grants@amfar.org
Internet: http://www.amfar.org/lab/grants/default.aspx?id=7455&terms=Fellowships
Sponsor: American Foundation for AIDS Research
120 Wall Street, 13th Floor
New York, NY 10005-3908

amfAR Research Grants   834

Grants are awarded to support basic and clinical research projects in biomedical, humanistic, and social sciences research relevant to AIDS. In general, funds are applied to direct costs of salaries and fringe benefits for professional and technical personnel, laboratory supplies and equipment, travel, and the publication of findings. Research grants are awarded for one year without assurance of continued funding. $100,000 in direct costs plus up to 20% for indirect costs is available. Investigators are required to submit a pre-application letter of intent (LOI). Any postdoctoral investigator who is affiliated with a nonprofit institution may submit an LOI.
Requirements: Principal investigators for research grants must be faculty-level researchers affiliated with a nonprofit institution. Research grants are given to nonprofit institutions worldwide to support investigator-led projects approved by the Foundation.
Samples: Columbia U (New York, NY)—to support the work of Susana Valente, for studies on microbicide development and AIDS-related research, $90,000.
Amount of Grant: $100,000
Contact: Grant Administrator; (212) 806-1696 or (212) 806-1600; fax: (212) 806-1601; email: grants@amfar.org
Internet: http://www.amfar.org/rfp/
Sponsor: American Foundation for AIDS Research
120 Wall Street, 13th Floor
New York, NY 10005-3908

Amgen Foundation Grants   835

Amgen seeks to: advance science education, improve quality of care and access for patients, and support resources that create sound communities where Amgen staff members live and work. Requests must be received at least 90 days in advance of the desired contribution date. Guidelines are available online.
Requirements: 501(c)3 tax-exempt organizations located in Amgen communities are eligible. Eligible grantees may include public elementary and secondary schools, as well as public colleges and universities, public libraries and public hospitals.
Restrictions: In general, Amgen does not consider requests for the following: support to individuals, fundraising or sports-related events, corporate sponsorship requests, religious organizations unless the program is secular in nature and benefits a broad range of the community, political organization or lobbying activity, labor unions, fraternal, service or veterans' organizations, private foundations, or organizations that are discriminatory.
Samples: California State U (Camarillo, CA)—for scientific equipment, $950,000 (2005); Institute for Systems Biology (Seattle, WA)—for the endowment and operations, $3 million (2005); International Federation of the Red Cross and Red Crescent Societies (Geneva)—for relief efforts in South Asia and Africa, $1 million (2005).
Contact: Program Contact, (805) 447-4056 or (805) 447-1000; fax: (805) 447-1010
Internet: http://wwwext.amgen.com/citizenship/apply_for_grant.html
Sponsor: Amgen Foundation
1 Amgen Center Drive, MS 38-3-B
Thousand Oaks, CA 91320

AMI Semiconductors Corporate Grants   836

The company makes grants to nonprofit organizations in support of the performing arts, economic development, business education, health cost containment, and social services for senior citizens. Types of support include conferences and seminars, general operating support, matching grants, multiyear grants, professorships, research, and scholarships. There are no application deadlines. Submit a letter of inquiry that includes a description of the organization and program, amount of funds requested, purpose of the request,

recently audited financial statement, and proof of tax-exempt status. Contact office for grant availability.
Restrictions: The company does not support political or lobbying groups.
Amount of Grant: $1000-$2500
Contact: TAMERA DRAKE, (208) 234-6890; fax: (208) 234-6795 email: tamera_Drake@amis.com
Internet: http://www.amis.com/about
Sponsor: AMI Semiconductors
2300 Buckskin Road
Pocatello, ID 83201

AMNH Collection Study Grants   837

These grants provide financial assistance to enable predoctoral and recent postdoctoral investigators to study any of the scientific collections at the American Museum in the departments of anthropology, earth and planetary sciences, entomology, herpetology, ichthyology, invertebrates, mammalogy, ornithology, and vertebrate paleontology. The awards partially support travel and subsistence; visits are arranged through and approved by the appropriate scientific department of the museum and are expected to be four days or longer in duration. Applicants should first contact the staff to discuss the feasibility of the proposed visit prior to requesting the special application form. Applications may be submitted any time during the year.
Requirements: Predoctoral and recent postdoctoral investigators are eligible to apply. Only one collection study grant will be awarded to an individual.
Restrictions: Grants are not available to investigators residing within daily commuting distance of the American Museum.
Amount of Grant: $500-$1500 range
Date(s) Application is Due: May 1; Nov 1
Contact: Grants Administrator, Office of Grants and Fellowships, (212) 769-5467; fax: (212) 769-5495; email: grants@amnh.org
Internet: http://research.amnh.org/grants/grantsprog.html
Sponsor: American Museum of Natural History
Central Park W at 79th Street
New York, NY 10024

AMNH Frank M. Chapman Memorial Fellowships   838

These fellowships are intended for postdoctoral scientists to support one year of ornithological research, both neontological and paleontological, with the possibility of renewal for a second year. These are salaried positions and have a benefits package. Fellows are expected to do the majority of their work at the museum.
Requirements: Graduate students and established ornithologists are eligible to apply for grants. Recent postdoctoral scholars and distinguished ornithologists may apply for fellowships to support a year of research at the museum or one of its field stations.
Amount of Grant: $500-$2000; $1000 average
Date(s) Application is Due: Nov 15
Contact: Grants Administrator, Office of Grants and Fellowships, (212) 769-5467; fax: (212) 769-5495; email: grants@amnh.org
Internet: http://research.amnh.org/ornithology/grants.htm
Sponsor: American Museum of Natural History
Central Park W at 79th Street
New York, NY 10024-5192

AMNH Graduate Student Fellowships   839

The program is an educational partnership with selected universities and is dedicated to the training of PhD candidates in those scientific disciplines practiced at the Museum. The university exercises educational jurisdiction over the program and awards the degree. The museum curator serves as a graduate advisor, co-major professor, or major professor. Joint programs are with Columbia University, providing students opportunities in vertebrate and invertebrate paleontology, astrophysics, earth and planetary sciences, and evolutionary biology; Cornell University in entomology; City University of New York in the Evolutionary Biology Program; Yale University in molecular biology/systematics. Fellowships cover stipend and health insurance, and awards are for one year, renewable annually for up to a maximum of four years.
Requirements: U.S. and non-U.S. citizens are eligible. Students in developing nations are particularly encouraged to apply. Applicants must have bachelors' degrees and be able to fulfill university admission requirements, which include TOEFL and Graduate Record Examinations.
Restrictions: Candidates for the master's degree are ineligible.
Date(s) Application is Due: Nov 30
Contact: Grants Administrator, Office of Grants and Fellowships, (212) 769-5467; fax: (212) 769-5495; email: grants@amnh.org
Internet: http://research.amnh.org/grants/gradprog.html
Sponsor: American Museum of Natural History
Central Park W at 79th Street
New York, NY 10024-5192

AMNH Lerner-Gray Grants for Marine Research   840

The grants provide financial assistance to highly qualified persons starting careers in marine zoology. Support is limited to projects dealing with systematics, evolution, ecology, and field-oriented behavioral studies of marine animals. Research projects need not be carried out at the American Museum. Application forms and guidelines are available from the museum. Awards are announced in May.
Requirements: Eligible to apply are advanced graduate students and postdoctoral researchers at the beginning of their careers.

Amount of Grant: $200-$2000; $1400 average
Date(s) Application is Due: Mar 15
Contact: Grants Administrator, Office of Grants and Fellowships, (212) 769-5467; fax: (212) 769-5495; email: grants@amnh.org
Internet: http://research.amnh.org/grants/grantsprog.html
Sponsor: American Museum of Natural History
Central Park W at 79th Street
New York, NY 10024-5192

AMNH Research and Museum Fellowships 841

Fellowships provide support to recent postdoctoral investigators, established scientists, and other scholars to carry out a specific project within a limited time period. The project must fit into the areas of vertebrate zoology, invertebrate zoology, paleozoology, anthropology, mineral sciences, astronomy, or museum education. The program is designed to advance the training of the participant by having her/him pursue a project in association with museum professionals in a museum setting. Grants provide funds to be used for travel, expendable supplies, and living expenses. Appointment is usually made for a one-year term but may be for a longer or shorter period depending on the source of funds. Candidates are expected to be in residence at the museum or at one of its field stations. Limited relocation, research, and publication support is often available. Interested researchers should obtain a special application form from the Office of Grants and Fellowships. Appointments are effective between July 1 and September 1.
Requirements: Graduate and undergraduate students and scientists who have recently received the doctorate are eligible.
Amount of Grant: $500-$2000
Date(s) Application is Due: Nov 15
Contact: Grants Administrator, Office of Grants and Fellowships, (212) 769-5467; fax: (212) 769-5495; email: grants@amnh.org
Internet: http://research.amnh.org/grants/resprog.html
Sponsor: American Museum of Natural History
Central Park W at 79th Street
New York, NY 10024

AMS Alfred Einstein Award 842

This award is given annually to the author of the article on a musicological subject selected by a committee of scholars to be the most significant published in a periodical during the preceding calendar year. Nominations should include the name of the author, the title of the article, and the name and year of the periodical or other collection in which it was published. Additionally, the award committee will solicit the curriculum vita of each nominee.
Requirements: Eligible are Canadian and U.S. scholars in the early stages of their careers.
Samples: David Rothenberg—for The Marian Symbolism of Spring, ca. 1200,Äica. 1500: Two Case Studies, Journal of the American Musicological Society (2007); Gundula Kreuzer—for Oper im Kirchengewande Verdi's Requiem and the Anxieties of the Young German Empire, Journal of the American Musicological Society (2006).
Amount of Grant: $400
Date(s) Application is Due: May 1
Contact: Robert Judd, Executive Director; (207) 798-4243 or (877) 679-7648; fax: (207) 798-4254; email: rjudd@ams-net.org or ams@sas.upenn.edu
Internet: http://www.ams-net.org/awards/einstein.php
Sponsor: American Musicological Society
6010 College Station
Brunswick ME 04011-8451

AMS Alvin H. Johnson 50 Dissertation Fellowships 843

The society supports research in the various fields of music, and makes available four dissertation-year fellowships each year. Dissertation fellowships support full-time study and completion of the dissertation within the fellowship year. Preliminary application forms are sent to directors of graduate study at all North American doctorate-granting institutions. Application must be submitted online via the AMS website.
Requirements: Anyone is eligible to apply who is registered in good standing for a doctorate at a North American university and has completed all formal degree requirements except the dissertation at the time of full application.
Amount of Grant: $19,000
Date(s) Application is Due: Jan 15
Contact: Georgia Cowart, Committee Chair; (207) 798-4243; fax: (877) 679-7648 or (207) 798-4254; email: a50-recomms@ams-net.org
Internet: http://www.ams-net.org/ams50.php
Sponsor: American Musicological Society
6010 College Station
Brunswick ME 04011-8451

AMS Claude V. Palisca Award 844

The Claude V. Palisca Award honors each year a scholarly edition or translation in the field of musicology published during the previous year in any language and in any country by a scholar who is a member of the AMS or a citizen or permanent resident of Canada or the United States, deemed by a committee of scholars to best exemplify the highest qualities of originality, interpretation, logic and clarity of thought, and communication.Three categories of musicological works are eligible for the Palisca award: translations into English of musicologically significant texts; editions of music; or editions of musicologically significant texts. The winner will receive a monetary prize

and a certificate, and the finalists a certificate, conferred at the Annual Business Meeting of the Society by the chair of the committee.
Requirements: Eligible are Canadian and U.S. scholars in the early stages of their careers.
Samples: Jeffrey Taylor—for Earl Fatha Hines, Selected Piano Solos, 1928-1941 (2007); David Lawton—for Verdi, Macbeth (2006); Ross W. Duffin—for Shakespeare's Songbook (2005).
Date(s) Application is Due: May 1
Contact: Robert Judd, Executive Director; (207) 798-4243 or (877) 679-7648; fax: (207) 798-4254; email: rjudd@ams-net.org or ams@sas.upenn.edu
Internet: http://www.ams-net.org/awards/palisca.php
Sponsor: American Musicological Society
6010 College Station
Brunswick ME 04011-8451

AMS Eileen Southern Travel Fund Grants 845

Beginning with the 1995 Annual Meeting in New York City, the Society and the Committee on Cultural Diversity have hosted minority students at each annual conference to expose them to various aspects of the field. These elements might include introductions to representatives of graduate programs in musicology from across the country, future colleagues and mentors, listening to papers covering a broad spectrum of musicological interests, exposure to the many interest groups within the Society, and opportunities to attend a variety of concerts. The AMS also has funds available to assist minority undergraduates and terminal masters degree candidates (African-Americans, Asian-Americans, Hispanic-Americans, Native-Americans, and other traditionally under-represented groups) with travel and lodging expenses.
Samples: Joy Doan, University of Michigan (2007); Bo kyung (Blenda) Im, University of California, Los Angeles (2007); Luci Mok, University of Toronto (2007).
Contact: Ingrid Monson, Co-Chair; (617) 495-2791; email: imonson@fas.harvard.edu; George Lewis, Co-Chair; (212) 854-5837; email: gl2140@columbia.edu
Internet: http://www.ams-net.org/ccd/ccdtf-background.php
Sponsor: American Musicological Society
6010 College Station
Brunswick ME 04011-8451

AMS Eugene K. Wolf Travel Fund for European Research Grants 846

The Eugene K. Wolf Travel Fund is intended to encourage and assist Ph.D. candidates in all fields of musical scholarship to travel to Europe to carry out the necessary work for their dissertation on a topic in European music. The fund will award one or more travel grants each year to students attending North American universities who have completed all requirements except the dissertation for the Ph.D. in any field of musical scholarship and who need to undertake research in Europe toward the dissertation. The application should be in the form of a statement of up to 1,000 words in length describing the dissertation topic, research plan, projected itinerary, and institutions where research would occur. It should also provide a budget for travel and lodging.
Requirements: Letters of support are required from the dissertation advisor and one other scholar.
Samples: Adeline Mueler, University of California, Berkeley (2008); Amber Youell-Fingleton, Columbia University (2008); Ewelina Boczkowska, UCLA (2007); Kimberly A. Francis, University of North Carolina (2007); Loren M. Ludwig, University of Virginia (2007).
Date(s) Application is Due: Mar 3
Contact: Robert Judd, Executive Director; (207) 798-4243 or (877) 679-7648; fax: (207) 798-4254; email: rjudd@ams-net.org or ams@sas.upenn.edu
Internet: http://www.ams-net.org/awards/wolf.php
Sponsor: American Musicological Society
6010 College Station
Brunswick ME 04011-8451

AMS Graduate Fellowships in the History of Science 847

The goal of the graduate fellowship is to generate a dissertation topic in the history of the atmospheric, or related oceanic or hydrologic sciences, and to foster close working relations between historians and scientists. An effort will be made to place the student into a mentoring relationship with an AMS member at an appropriate institution. The fellowship is sponsored by member donations to the AMS 21st Century campaign. Fellowships cannot be deferred and must be used for the year awarded, but can be used to support research at a location away from the student, institution provided the plan is approved by the student, thesis advisor. The award carries a $15,000 stipend and will support one year of dissertation research.
Requirements: Candidates wishing to apply must be a graduate student in good standing who proposes to complete a dissertation.
Amount of Grant: $15,000
Date(s) Application is Due: Feb 22
Contact: Donna Sampson, Program Manager; (617) 227-2426, ext. 246; fax: (617) 742-8718; email: dfernand@ametsoc.org
Internet: http://www.ametsoc.org/amsstudentinfo/scholfeldocs/
Sponsor: American Meteorological Society
45 Beacon Street
Boston, MA 02108-3693

AMS H. Colin Slim Award 848

The H. Colin Slim Award honors each year a musicological article of exceptional merit, published during the previous year in any language and in any country by a scholar who is past the early stages of her or his career and who is a member of the AMS or a citizen or permanent resident of Canada or the United States. Early stages of the career is typically indicated by time from completion of the Ph.D. degree, or academic appointment at a non-tenured level, or position of the article among the initial items of the author's bibliography. The winner will receive a monetary prize and a certificate, conferred at the Annual Business Meeting and Awards Presentation of the Society by the chair of the committee.

Samples: Anne Walters Robertson—for The Savior, the Woman, and the Head of the Dragon in the Caput Masses and Motet (2007); Ralph Locke—for Beyond the Exotic: How Eastern is Aida? (2006); Jann Pasler—for The Utilitiy of Musical Instruments in the Racial and Colonial Agendas of Late Nineteenth-Century France (2005).
Date(s) Application is Due: May 1
Contact: Robert Judd, Executive Director; (207) 798-4243 or (877) 679-7648; fax: (207) 798-4254; email: rjudd@ams-net.org or ams@sas.upenn.edu
Internet: http://www.ams-net.org/awards/slim.php
Sponsor: American Musicological Society
6010 College Station
Brunswick ME 04011-8451

AMS Harold Powers World Travel Fund Grants 849
The Harold Powers World Travel Fund is intended to encourage and assist Ph.D. candidates, post-docs, and junior faculty in all fields of musical scholarship to travel anywhere in the world to carry out the necessary work for their dissertation or other research. The Fund honors the polymathic scholar and distinguished longtime AMS member whose publications have ranged from music and language to medieval mode to Indian music to Puccini and whose interests are wider still, but always with the communicative aspects of music at their base. The application should be in the form of a statement of up to 1,000 words describing the research topic, a research plan, projected itinerary, and institutions where research would occur. The current maximum award is set at $2,000.
Requirements: Eligible applicants must currently attend or have graduated from a doctoral program in a North American university. If they seek to conduct research for their dissertation, they must have completed all other requirements for the Ph.D. (or the equivalent doctoral degree in any field of music scholarship). If they seek to conduct post-doctoral research, they should have completed the Ph.D. within the past five years.
Samples: Max Katz, University of California, Santa Barbara, $2,000 (2008); Joshua Walden, Columbia University, $2,000 (2007).
Amount of Grant: $2,000 maximum
Date(s) Application is Due: Mar 3
Contact: Robert Judd, Executive Director; (207) 798-4243 or (877) 679-7648; fax: (207) 798-4254; email: rjudd@ams-net.org or ams@sas.upenn.edu
Internet: http://www.ams-net.org/awards/powers.php
Sponsor: American Musicological Society
6010 College Station
Brunswick ME 04011-8451

AMS Howard Mayer Brown Fellowships 850
Intended to increase the presence of minority scholars and teachers in musicology, the fellowship will support one year of graduate work for a member of a group historically underrepresented in the discipline. Applications are encouraged from African Americans, Native Americans, Latinos/Hispanics, and Asian Americans. Nominations may come from a faculty member of the institution at which the student is enrolled, from a member of the AMS at another institution, or directly from the student. The fellowship is awarded in alternating years. Inquiries should be addressed to the chair of the committee. Applications should be made using the online submission form.
Requirements: The fellowship will be awarded to a student who has completed at least one year of academic work at an institution with a graduate program in musicology, and who intends to complete a PhD in the field.
Amount of Grant: $17,000 stipend
Date(s) Application is Due: Jan 15
Contact: Martha Feldman, Committee Chair; (207) 798-4243; fax: (877) 679-7648 or (207) 798-4254; email: rore@uchicago.edu or hmb-apps@ams-net.org
Internet: http://www.ams-net.org/hmb.php
Sponsor: American Musicological Society
6010 College Station
Brunswick ME 04011-8451

AMS Janet Levy Fund for Independent Scholars 851
The purpose of the Janet Levy Fund is to support professional travel and research expenses for independent scholars who are members of the American Musicological Society. Examples of projects supported by the Levy Fund include travel to the Annual Meeting of the AMS, to a conference to read a paper or participate in an official capacity, or to archives or research libraries; research expenses; microfilms; and specialized research materials. Award amounts will range between $500 and $2,000. Proposals are evaluated by a committee of three appointed by the AMS president.
Requirements: The award is open to members of the American Musicological Society who hold the Ph.D. or similar degree (e.g., D.M.A. in Historical Performance), who are not employed as a full-time faculty member in an institution of higher learning, and who are not emeritus faculty.
Restrictions: No individual may receive a Levy Award more than once in a three-year period.
Samples: Peter Poulos—travel funds to deliver one paper each at meetings of the Italian Musicological Society and the Renaissance Society of America (2007); Vera Deak—

travel to Budapest to work on an edition of the complete source catalogue of Bart??k's folksong settings (2007).
Amount of Grant: $500-$2,000
Date(s) Application is Due: Jan 25; Jul 25
Contact: Robert Judd, Executive Director; (207) 798-4243 or (877) 679-7648; fax: (207) 798-4254; email: rjudd@ams-net.org or ams@sas.upenn.edu
Internet: http://www.ams-net.org/awards/levy.php
Sponsor: American Musicological Society
6010 College Station
Brunswick ME 04011-8451

AMS Lewis Lockwood Award 852
The Lewis Lockwood Award honors each year a musicological book of exceptional merit published during the previous year in any language and in any country by a scholar in the early stages of his or her career who is a member of the AMS or a citizen or permanent resident of Canada or the United States. Early stages of the career is typically indicated by time from completion of the Ph.D. degree, or academic appointment at a non-tenured level, or position of the book among the initial items of the author''s bibliography. The winner will receive a monetary prize and a certificate, and the finalists a certificate, conferred at the Annual Business Meeting and Awards Presentation of the Society by the chair of the committee.
Requirements: Eligible are Canadian and U.S. scholars in the early stages of their careers.
Samples: Susan Boynton—for Shaping a Monastic Identity: Liturgy and History at the Imperial Abbey of Farfa, 1000,Äï1125 (2007); Kate van Orden—for Music, Discipline, and Arms in Early Modern France (2006); Marc Perlman—for Unplayed Melodies: Javanese Gamelan and the Genesis of Music Theory (2005).
Date(s) Application is Due: May 1
Contact: Robert Judd, Executive Director; (207) 798-4243 or (877) 679-7648; fax: (207) 798-4254; email: rjudd@ams-net.org or ams@sas.upenn.edu
Internet: http://www.ams-net.org/awards/lockwood.php
Sponsor: American Musicological Society
6010 College Station
Brunswick ME 04011-8451

AMS M. Elizabeth C. Bartlet Fund for Research in France 853
The M. Elizabeth C. Bartlet Fund for Research in France is a memorial to M. Elizabeth C. Bartlet, one of the foremost scholars of French music from the eighteenth and nineteenth centuries. Her research was centered in Paris, where she was a commanding and beloved presence in the libraries and archives. It is to be awarded annually to one or more doctoral students at or graduates of universities in the United States and Canada to conduct doctoral or post-doctoral musicological research in France. The application should be in the form of a statement of up to 1,000 words describing the research topic, a research plan, projected itinerary, and institutions where research would occur. The current maximum award is set at $2,000.
Requirements: Eligible applicants must currently attend or have graduated from a doctoral program in a North American university. If they seek to conduct research for their dissertation, they must have completed all other requirements for the Ph.D. If they seek to conduct post-doctoral research, they should have completed the Ph.D. within the past five years. Preference will be given to applicants whose home institutions do not offer financial support for musicological research.
Samples: Will Gibbons, University of North Carolina, $2,000 (2008); Jennifer Saltzstein, University of Oklahoma, $2,000 (2008); Willa J. Collins, Cornell University. $2,000 (2007).
Amount of Grant: $2,000
Date(s) Application is Due: Mar 3
Contact: Robert Judd, Executive Director; (207) 798-4243 or (877) 679-7648; fax: (207) 798-4254; email: rjudd@ams-net.org or ams@sas.upenn.edu
Internet: http://www.ams-net.org/awards/bartlet.php
Sponsor: American Musicological Society
6010 College Station
Brunswick ME 04011-8451

AMS Noah Greenberg Award 854
The award is given annually for a distinguished contribution to the study and performance of early music, up to the end of the 17th century. The award is intended as a grant-in-aid to stimulate active cooperation between scholars and performers by recognizing and fostering outstanding contributions to historical performing practices. Both scholars and performers may apply, since the award may subsidize the publication costs of articles, monographs or editions, as well as performance, recordings, or other projects. Applicants must submit, in triplicate, a description of the project, a detailed budget, and relevant supporting materials to the committee chair.
Samples: Elisabeth Le Guin—for Audience Performance Practice: A Pilot Project (2007); Christopher Wolverton, Vox Early Music Ensemble, and Honey Meconi, University of Rochester—for Extreme Singing: Very Low Music of the Renaissance (2006).
Amount of Grant: $2000 maximum
Date(s) Application is Due: Aug 15
Contact: Robert Judd, Executive Director; (207) 798-4243 or (877) 679-7648; fax: (207) 798-4254; email: rjudd@ams-net.org or ams@sas.upenn.edu
Internet: http://www.ams-net.org/awards/greenberg.php
Sponsor: American Musicological Society
6010 College Station
Brunswick ME 04011-8451

AMS Otto Kinkeldey Award 855

The Otto Kinkeldey Award will honor each year a musicological book of exceptional merit published during the previous year in any language and in any country by a scholar who is past the early stages of his or her career and who is a member of the AMS or a citizen or permanent resident of Canada or the United States. Early stages of the career is typically indicated by time from completion of the Ph.D. degree, or academic appointment at a non-tenured level, or position of the book among the initial items of the author's bibliography. Nominations, including self-nominations, may be submitted at any time.

Requirements: Canadian and U.S. scholars in early stages of their careers are eligible.

Samples: Philip Gossett—for his work entitled Divas and Scholars: Performing Italian Opera (2007); Richard Taruskin—for his work entitled Stravinsky and the Russian Traditions (2006).

Amount of Grant: $400

Date(s) Application is Due: May 1

Contact: Robert Judd, Executive Director; (207) 798-4243 or (877) 679-7648; fax: (207) 798-4254; email: rjudd@ams-net.org or ams@sas.upenn.edu

Internet: http://www.ams-net.org/awards/kinkeldey.php

Sponsor: American Musicological Society

6010 College Station

Brunswick ME 04011-8451

AMS Paul A. Pisk Prize 856

This prize is awarded each year to a graduate music student for a scholarly paper, to be read at the annual meeting of the society. Any paper for which the abstract has been submitted to the program committee, and the paper accepted for inclusion in the annual meeting, is eligible. Five copies of the complete text of the paper together with supporting materials must be submitted to the Philadelphia AMS office by the deadline.

Requirements: The submission must be accompanied by a statement from the student's academic advisor affirming graduate-student status of the applicant as of the date of the paper's acceptance by the program committee.

Samples: Emily Abrams Ansari, University of Western Ontario/Harvard University (2007); Jesse Rodin, Harvard University (2006); Paul Berry, Yale University (2005).

Amount of Grant: $1,000

Date(s) Application is Due: Oct 1

Contact: Robert Judd, Executive Director; (207) 798-4243 or (877) 679-7648; fax: (207) 798-4254; email: rjudd@ams-net.org or ams@sas.upenn.edu

Internet: http://www.ams-net.org/awards/pisk.php

Sponsor: American Musicological Society

6010 College Station

Brunswick ME 04011-8451

AMS Robert M. Stevenson Award 857

The Robert M. Stevenson Award recognizes outstanding scholarship in Iberian music. The designation Iberian music is here meant to include music composed, performed, created, collected, belonging to, or descended from the musical cultures of Spain, Portugal, and all Latin American areas in which Spanish and Portuguese are spoken. The prize will be awarded annually to a book, monograph, edition, or journal article by a member of the AMS. The publication must be written in English and must have been published during the preceding three calendar years. The award will consist of a monetary prize and a certificate.

Samples: Kenneth Kreitner—for The Church Music of Fifteenth-Century Spain; Walter Aaron Clark—for Enrique Granados: Poet of the Piano; Cristina Magaldi—for Music in Imperial Rio de Janeiro: European Culture in a Tropical Milieu.

Date(s) Application is Due: May 1

Contact: Robert Judd, Executive Director; (207) 798-4243 or (877) 679-7648; fax: (207) 798-4254; email: rjudd@ams-net.org or ams@sas.upenn.edu

Internet: http://www.ams-net.org/awards/stevenson.php

Sponsor: American Musicological Society

6010 College Station

Brunswick ME 04011-8451

AMS Ruth A. Solie Award 858

The Ruth A. Solie Award honors each year a collection of musicological essays of exceptional merit published during the preceding calendar year in any language and in any country and edited by a scholar or scholars who are members of the AMS or citizens or permanent residents of Canada or the United States. Established to honor the editor of Musicology and Difference: Gender and Sexuality in Music Scholarship (University of California Press, 1995), a field-defining and field-changing book, the award acknowledges the value of the individual authors' contributions to the volume while recognizing the central role of the editor(s) in conceiving and shaping the whole. The winning editor(s) will receive a monetary prize and a certificate, conferred at the Annual Business Meeting and Awards Presentation of the Society by the chair of the committee.

Samples: Martha Feldman & Bonnie Gordon—for The Courtesan's Arts: Cross-Cultural Perspectives (2007).

Date(s) Application is Due: May 1

Contact: Robert Judd, Executive Director; (207) 798-4243 or (877) 679-7648; fax: (207) 798-4254; email: rjudd@ams-net.org or ams@sas.upenn.edu

Internet: http://www.ams-net.org/awards/solie.php

Sponsor: American Musicological Society

6010 College Station

Brunswick ME 04011-8451

AMS Undergraduate Scholarships 859

Scholarships are available to full-time students entering their final year of undergraduate study in the fall of the application year. The evaluation of applicants will be based on applicant, performance as an undergraduate student, including academic records and recommendation. Selection will be made by the AMS Executive Committee based on recommendations from the AMS Committee of Judges for Undergraduate Awards.

Requirements: Applicants must: be majoring in the atmospheric or related oceanic or hydrologic science, and/or must show clear intent to make the atmospheric or related sciences their career; be enrolled full time in an accredited U.S. institution; and have a cumulative grade-point average of at least a 3.25 on a scale of 4.0.

Date(s) Application is Due: Feb 22

Contact: Donna Sampson, Program Manager; (617) 227-2426, ext. 246; fax: (617) 742-8718; email: dfernand@ametsoc.org

Internet: http://www.ametsoc.org/amsstudentinfo/scholfeldocs

Sponsor: American Meteorological Society

45 Beacon Street

Boston, MA 02108-3693

AMS/Industry Government Graduate Fellowships 860

The graduate fellowships are sponsored by major high-technology firms and government agencies and are designed to attract promising young scientists to prepare for careers in the meteorological, oceanic, and hydrologic fields. Candidates currently studying chemistry, computer sciences, engineering, environmental sciences, mathematics, and physics who intend to pursue careers in the atmospheric, oceanic, or hydrologic sciences also are encouraged to apply. Awards are based on the applicant's performance as an undergraduate student and his or her qualifications to pursue a career in the atmospheric and related oceanic and hydrologic sciences.

Requirements: Students entering their first year of graduate study who wish to pursue advanced degrees in the atmospheric and related oceanic and hydrologic sciences are eligible.

Amount of Grant: $23,000 nine-month stipend

Date(s) Application is Due: Feb 22

Contact: Donna Sampson, Program Manager; (617) 227-2426, ext. 246; fax: (617) 742-8718; email: dfernand@ametsoc.org

Internet: http://www.ametsoc.org/amsstudentinfo/scholfeldocs

Sponsor: American Meteorological Society

45 Beacon Street

Boston, MA 02108-3693

AMS/Industry Minority Scholarships 861

The American Meteorological Society (AMS) Minority Scholarships help support the college educations of minority students traditionally underrepresented in the sciences, especially Hispanic, Native American, and Black/African American students, who intend to pursue careers in the atmospheric or related oceanic and hydrologic sciences. The two-year scholarships are for $3,000 for a nine-month period in the freshman year and an additional $3,000 for a nine-month period in the sophomore year. Second year funding depends on successful completion of the first academic year.

Requirements: Minority students who will be entering their freshman year of college in the fall of the application yearare eligible to apply.

Amount of Grant: $6,000 over two years

Date(s) Application is Due: Feb 22

Contact: Donna Sampson, Program Manager; (617) 227-2426, ext. 246; fax: (617) 742-8718; email: dfernand@ametsoc.org

Internet: http://www.ametsoc.org/amsstudentinfo/scholfeldocs

Sponsor: American Meteorological Society

45 Beacon Street

Boston, MA 02108-3693

AMSSM Foundation Research Grants 862

The purpose of the AMSSM Foundation Research Grant Awards is to foster original scientific investigations by members of AMSSM. in this program, the Foundation welcomes research grant proposals that investigate issues within the broad discipline of sports medicine, including clinical practice, injury prevention and rehabilitation, basic science, epidemiology, and education. Grant awards are designed to provide partial support of research projects. Grantee institutions are expected to provide all necessary basic facilities and services that normally would be expected to exist in any institution qualified to undertake such research. Overhead or indirect costs will be supported to a maximum of 10% of direct costs. It is anticipated that the requested support will fall into one of three ranges. 1) $500 to $2,500 total costs annually for mailings, supplies, equipment, and technical support of small or pilot projects; 2) $2,500 to $10,000 total costs annually for a project larger in scope than the previous category and may provide partial support of research assistants or study coordinators; and 3) $10,000 to $25,000 total costs annually for a significant project that includes well documented pilot or previous studies and may provide partial salary support for key investigators. The maximum grant award is $25,000.

Requirements: The primary investigator of the grant must be an AMSSM member at the time of grant submission. Resident and fellow AMSSM members may apply as the principal investigator but must have at least one full AMSSM member listed as a co-investigator at the time of application.

Amount of Grant: $25,000 maximum

Date(s) Application is Due: Oct 1

Contact: Grants Administrator; (913) 327-1415; fax: (913) 327-1491; email: office@amssm.org

Internet: http://www.newamssm.org/Research%20Grant%20Information.html
Sponsor: American Medical Society for Sports Medicine
11639 Earnshaw
Overland Park, KS 66210

AMWA Anne C. Carter Student Leadership Award   863

AMWA honors the memory of Anne C. Carter, MD, with an annual award for outstanding student leadership. Nominees must meet the following criteria: national AMWA medical student member; demonstrated exceptional leadership skills through vision, inspiration, innovation, and coordination of local projects that further the mission of AMWA by improving women's health and/or supporting women in medicine; and must be nominated by an AMWA student chapter. Awardees will be honored during AMWA's annual meeting. Nomination guidelines are available online.
Amount of Grant: $1500 ($750 to awardee, $750 to nominating chapter)
Date(s) Application is Due: Oct 31
Contact: Award Administrator, (703) 838-0500; fax: (703) 549-3864; email info@amwa-doc.org
Internet: http://www.amwa-doc.org/index.cfm?objectId=855A0811-D567-0B25-5094E9176D8FFF58
Sponsor: American Medical Women's Association Foundation
801 N Fairfax Street, Suite 400
Alexandria, VA 22314

AMWA Carroll L. Birch Award   864

The annual award, sponsored by the Chicago Branch of AMWA, is presented for the best original research paper written by an AMWA student member. Award criteria include: original research paper written by a national AMWA medical student member; and applicant must attend an accredited U.S. allopathic or osteopathic medical school. The paper may have been previously published. The recipient receives a cash award, a plaque presented at AMWA's annual meeting, and an article noting the award winner will appear in one of the association's publications. The award recipient is strongly encouraged to attend AMWA's annual meeting. Submit the following materials to the office: cover letter stating applicant's medical school, graduation date, and permanent mailing address; an abstract of 250 words or less; four copies of the manuscript; and a letter from the faculty research sponsor.
Amount of Grant: $1000
Date(s) Application is Due: Jun 30
Contact: Award Administrator, (703) 838-0500; fax: (703) 549-3864; email info@amwa-doc.org
Internet: http://www.amwa-doc.org/index.cfm?objectId=E5F39C87-D567-0B25-5377ABBB15148B2E
Sponsor: American Medical Women's Association Foundation
801 N Fairfax Street, Suite 400
Alexandria, VA 22314

AMWA Glasgow-Rubin Essay Award   865

The annual award is presented to a medical student for the best essay about a mentor. Award criteria include: AMWA national student member; attends an accredited U.S. allopathic or osteopathic medical school; original research paper written about a personal relationship with a woman physician mentor; and essay must be approximately 1000 words. Submitted essays should be of publishable quality; typed, double-spaced; and title page to include title, author's name, address and phone number, email address, and name medical school.
Amount of Grant: $1000
Date(s) Application is Due: May 31
Contact: Award Administrator, (703) 838-0500; fax: (703) 549-3864; email: info@amwa-doc.org
Internet: http://www.amwa-doc.org/index.cfm?objectId=288C83B1-D567-0B25-539C27C31D5C35EC
Sponsor: American Medical Women's Association Foundation
801 N Fairfax Street, Suite 400
Alexandria, VA 22314

AMWA Kathryn C. Bemmann New Investigator Grant on Violence   866
Against Women

The grant supports a physician early in his or her career to pursue research on violence against women. The candidate should be a recent graduated physician; have a serious interest in research on violence against women; and have the ability to pursue an independent career in research. Proposals will be judged on scientific quality, research facility, and significance to the field. The grant recipient will be recognized at the annual meeting. Transportation, one-night's hotel accommodation, and complimentary meeting registration will be provided. Guidelines and application are available online.
Requirements: Physicians who have received their medical degree within the last five years are eligible.
Amount of Grant: $3000-$5000
Date(s) Application is Due: Apr 30
Contact: Julie Dogil, (703) 838-0500; fax: (703) 549-3864; email: jdogil@amwa-doc.org
Internet: http://www.amwa-doc.org/index.cfm?objectId=E22EC179-D567-0B25-5D6E0F68BB316ECB
Sponsor: American Medical Women's Association Foundation
801 N Fairfax Street, Suite 400
Alexandria, VA 22314

Anderson Center Residency Program   867

The Anderson Center develops, fosters, and promotes the creation of works by artists of all kinds, and provides leadership and services that help to insure a strong and healthy arts community and a greater recognition of the value of arts in society by providing residencies to artists. The Center offers short-term residencies of two weeks or one month from May-October to artists, writers, scholars, and scientists. Through a grant from the Jerome Foundation of St. Paul, the Center annually devotes the month of August to encourage the work of emerging artists from New York City and Minnesota. Each resident is provided room, board, and workspace for the length of the stay during the residency period. Each resident is asked to serve the community of Red Wing or its surrounding area, by giving a talk, lecture, reading, performance, or classroom activity. Anderson Center residencies begin on the 1st or 16th of each month and end on the 15th or last day of each month for periods of 2-4 weeks.
Restrictions: If you have been a prior resident of the Anderson Center, you must wait two years from the time of your residency to apply again.
Date(s) Application is Due: Feb 1; Mar 1
Contact: Dawn Erickson; (651) 388-2009; fax: (651) 388-5538; email: dawn@andersoncenter.org or info@andersoncenter.com
Internet: http://www.andersoncenter.org/resapp.htm
Sponsor: Anderson Center for Interdisciplinary Studies
163 Tower View Drive, P.O. Box 406
Red Wing, MN 55066

Andrew W. Mellon Curatorial Fellowships   868

These two-year fellowships, renewable for a third year, will provide curatorial training and support scholarly research related to the gallery's collections of European and American art. Fellows will be fully integrated into a specific curatorial department with duties, privileges, and status equivalent to an assistant curator; divide their time between specific research projects and more general curatorial work within the department, including research on the collection and new acquisitions, work on the presentation of the collection, participation in aspects of special exhibition projects, and opportunities to give public lectures and gallery talks; and, in consultation with the supervising curator, develop a concrete project intended to complement their own research interests. Fellowships include stipend, annual travel allowance, and eligibility for medical and term life insurance through the federal government. Guidelines are available online.
Requirements: Consideration will be given to candidates in the fields of European, American, and British paintings, drawings, prints, photographs, and sculpture from the 13th century to the present. Applicants must have completed a PhD before beginning a fellowship and within the last five years.
Amount of Grant: $41,815 annually approximately
Date(s) Application is Due: Jan 14
Contact: Grants Administrator, (202) 842-6257; email: intern@nga.gov
Internet: http://www.nga.gov/education/fellowed.htm
Sponsor: National Gallery of Art
2000B S Club Drive
Landover, MD 20785

Andrew W. Mellon Fellowships in Humanistic Studies   869

The fellowships are issued to attract especially promising students into the humanities and to support students while preparing for teaching and other scholarly careers in humanistic disciplines. Selection procedures include review by regional and national committees. Winners of fellowships may take their awards to graduate schools of their choice in the United States and Canada. Fellowships are not renewable.
Requirements: College seniors or recent graduates with outstanding academic records and plans to enter PhD programs in the humanities are eligible for the fellowships, providing that they are U.S. citizens or permanent residents. PhD programs should begin in September.
Restrictions: Individuals who have been candidates in a previous year are not eligible. Applicants must not be already enrolled in graduate or professional study or hold the MA degree. Ineligible fields of study include archaeology (except within art history), education (or any studies leading to the EdD), fine arts, performing arts, international studies, law, political science, psychology, public policy, science and medicine, sociology, or theology for pastoral ministry (or any DDiv).
Amount of Grant: $17,500 plus tuition and fees for the first year
Date(s) Application is Due: Dec 1
Contact: Robert Weisbuch, Program Director, Mellon Fellowships in Humanistic Studies, (800) 899-9963 ext 127 or (609) 452-7007; fax: (609) 452-0066; email: bobweis@woodrow.org or mellon@woodrow.org
Internet: http://www.woodrow.org/mellon
Sponsor: Woodrow Wilson National Fellowship Foundation
P.O. Box 5329
Princeton, NJ 08543-5329

Andrew W. Mellon Postdoctoral Fellowship in American Cultures   870

The college invites applications for a two-year Mellon Postdoctoral Fellowship in Native American Studies, to begin in the fall. The program seeks scholars who have an interest in teaching in an undergraduate liberal arts college and who would welcome the opportunity to participate in the development of a multidisciplinary concentration in Native American Studies within the college's American Culture Program. The fellow will teach or team-teach one course in the fall semester entitled Introduction to Native American Studies, and will help facilitate a monthly faculty seminar throughout the academic year intended to extend and deepen the interests in Native American Studies of faculty from several disciplines. The fellowship includes salary plus benefits, as well as

support for research, professional travel, and relocation expenses. Application guidelines are available online.
Requirements: Candidates should be relatively new PhDs with training in Native American Studies or any relevant discipline.
Amount of Grant: $44,000 stipend
Date(s) Application is Due: Feb 15
Contact: Ron Sharp, Dean of the Faculty, (845) 437-7485; fax: (845) 437-7204
Internet: http://americanculture.vassar.edu/index.html?posting=39
Sponsor: Vassar College
Box 746, 124 Raymond Avenue
Poughkeepsie, NY 12604-0739

Andrew W. Mellon Postdoctoral Fellowship in Medieval Studies   871
The Mellon Fellow's principal obligation will be to pursue his or her research. Though the fellowship carries no teaching responsibilities, it is expected that the Fellow will take advantage of the opportunity to participate in the intellectual life of the institute and the multidisiciplinary activities that it sponsors for the medievalist community at Notre Dame. The Fellow will be provided an office in the Medieval Institute, full library and computer privileges, and access to the Institute's research tools.
Requirements: Mellon scholars must hold a regular appointment at a U.S. institution and plan to return to their institution following their fellowship year. Applicants must have the PhD in hand as of the application date and must not be more than five years beyond the PhD. The fellow will be expected to live in South Bend, IN.
Amount of Grant: $37,500
Date(s) Application is Due: Jan 15
Contact: Roberta Baranowski, (574) 631-8304; email: Roberta.Baranowski.7@nd.edu
Internet: http://www.nd.edu/~medinst/funding/funding.html#mellon
Sponsor: University of Notre Dame
715 Hesburgh Library
Notre Dame, IN 46556-5629

Andrew W. Mellon Postdoctoral Fellowships at UCLA Humanities Consortium   872
These fellowships are devoted to the theme Nations and Identities. The topic for the current year is The Secularization Thesis. Fellows are required to be in residence and to participate in the consortium's Mellon Seminar and Conference. Fellows also will teach, through relevant departments or programs, one course in their first year and two courses in their second, and will be expected to participate in the intellectual life of these programs. One fellow will be appointed in each of three historical periods: AD 600 to 1600, 1600 to 1800, and 1800 to present.
Requirements: Fellows must have earned their doctorates after September 1, 2001, and must have the doctorate in hand by September 1, 2004.
Amount of Grant: $35,000 stipend
Date(s) Application is Due: Mar 1
Contact: Mark Pokorski, (310) 206-0559; email: mpok@humnet.ucla.edu
Internet: http://www.humnet.ucla.edu/humnet/consortium/fships.html
Sponsor: UCLA Humanities Consortium
310 Royce Hall
Los Angeles, CA 90095-1461

Andrew W. Mellon Postdoctoral Fellowships at Washington University   873
The program encourages interdisciplinary scholarship and teaching across the humanities and social sciences. It supports new and recent PhDs who wish to strengthen their own advanced training and to participate in the university's ongoing interdisciplinary programs and seminars. Fellows receive a two-year appointment with stipends. Fellows will outline a plan for their own continuing research in association with a senior faculty member at the university. Over the course of the two-year appointment, fellows will teach three undergraduate courses in their home discipline and collaborate each spring semester in leading a seminar in theory and methods of interdisciplinary research. There is no application form. Submit a cover letter, a description of the research program (three-page maximum, single spaced), a brief proposal for the seminar and theory and methods, a curriculum vita, and three letters of recommendation.
Amount of Grant: $40,700 minimum
Date(s) Application is Due: Dec 1
Contact: Steven Zwicker, Department of English, (314) 935-5190; email: szwicker@artsi.wustl.edu
Internet: http://www.artsci.wustl.edu/~szwicker/Mellon_Postdoctoral_Program.html
Sponsor: Washington University
Campus Box 1122, 1 Brookings Drive
Saint Louis, MO 63130

Andrew W. Mellon Postdoctoral Fellowships in the Humanities at   874
Stanford University
The fellowship is designed to give the best recent PhD recipients in the humanities a unique opportunity to develop as scholars and teachers. Each year, up to six fellowships are awarded for a two-year term. Fellows teach one course and contribute a second course-equivalent per year in one of Stanford's 15 humanities departments. In addition, fellows participate in the intellectual life of the program by sharing work in progress, meeting regularly as a group and with faculty, and generally contributing to the community of humanists at Stanford. It is expected that fellows are in residence during the term of their appointment. The total number of fellows in the program will typically be between 12 and 15. In addition to the stipend, the fellowship includes additional support for computer assistance, research, and relocation expenses. Application and guidelines are available online.

Requirements: Candidates must have received their PhD degree between January 1, 2002, and June 30, 2005.
Amount of Grant: $50,000 annual stipend plus benefits
Date(s) Application is Due: Nov 28
Contact: Dr. Seth Lerer, (650) 723-3054; fax: (650) 725-0755; email: lerer@stanford.edu
Internet: http://fellows.stanford.edu
Sponsor: Stanford University
Stanford University, 450 Serra Mall, Building 460, Room 201
Stanford, CA 94305-2070

Andrew W. Mellon Predoctoral Fellowship   875
The fellowship is for research in fields other than Western art, to be held partly in residence at the National Gallery of Art, Center for Advanced Study in the Visual Arts, and partly elsewhere in the United States or abroad. The fellow is expected to spend the second year of the fellowship at the center to complete the dissertation. Application may be made only through the chairs of the graduate departments of art history or other appropriate departments, who should act as sponsors for applicants from their respective schools. Departments should limit nominations to one candidate. Fellowships begin September 1 and are not renewable.
Requirements: Applicants must have completed their residence requirements and coursework for the PhD and general or preliminary examinations before the date of application and know two foreign languages related to the topic of the dissertation. Applicants must be either U.S. citizens or enrolled in a university in the United States.
Samples: Gennifer Weisenfeld, Princeton U—for dissertation, Murayama, Mavo, and Modernity: Constructions of the Modern in Taisho Period Avant-Garde Art.; Heghnar Zeitlian, U of California, Los Angeles—for dissertation, The Image of an Ottoman City: Urban Space, Social Structure, and Civic Identity in 17th- and 18th-Century Aleppo.
Amount of Grant: $20,000 per year; $4000 housing subsidy
Date(s) Application is Due: Nov 15
Contact: Grants Administrator, Fellowships Program, Center for Advanced Study in Visual Arts, (202) 842-6482; fax: (202) 789-3026; email: advstudy@nga.gov
Internet: http://www.nga.gov/resources/casvapre.htm
Sponsor: National Gallery of Art
Fourth Street and Constitution Avenue NW
Washington, DC 20565

Andrew W. Mellon Predoctoral Fellowships at the University of Pittsburgh   876
Fellowships are awarded to students of exceptional ability and promise who wish to enroll, or are enrolled, at the University of Pittsburgh in programs leading to the PhD in various fields of the humanities, the natural sciences, and the social sciences. Awards are made on an annual basis but may be renewed; fellows who wish to renew must file new applications. Annual deadline dates may vary; contact the office listed for exact dates.
Requirements: Applicant must submit the formal application form supported by transcripts of all previous academic work, three letters of recommendation from persons able to judge the applicant's qualification for graduate study, a three-page summary describing the applicant's proposed research or study program, and any other materials that would show promise of distinguished achievement.
Restrictions: Individuals holding the fellowships are expected to engage in full-time study during the periods of their fellowships; no additional duties will be required or permitted.
Amount of Grant: $15,000 for 12 months plus tuition
Date(s) Application is Due: Jan 15
Contact: Financial Aid Office, (412) 624-7488
Internet: http://www.fas.pitt.edu/financial.htm#apply
Sponsor: University of Pittsburgh
2604 Cathedral of Learning
Pittsburgh, PA 15260

Andrew W. Mellon/ACLS Dissertation Completion Fellowships   877
ACLS invites applications for the annual competition for the Mellon/ACLS Dissertation Completion Fellowships. These fellowships are to assist graduate students in the humanities and related social sciences in the last year of Ph.D. dissertation writing. This program aims to encourage timely completion of the Ph.D. Applicants must be prepared to complete their dissertations within the period of their fellowship tenure or shortly thereafter. ACLS will award 65 Fellowships in this competition for a one-year term beginning between June and September for the following academic year. The Fellowship tenure may be carried out in residence at the Fellow's home institution, abroad, or at another appropriate site for the research. The total award of up to $33,000 includes a stipend plus additional funds for university fees and research support.
Requirements: Applicants must: be Ph.D. candidates in a humanities or social science department in the United States. Applicants from other departments may be eligible if their project is in the humanities or related social sciences, and their principal dissertation supervisor holds an appointment in a humanities field or related social science field. (Students completing master's degrees are not eligible, even if they are the terminal degree in the field); have all requirements for the Ph.D. except the dissertation completed before beginning fellowship tenure; and be no more than six years in the degree program (awardees can hold this Fellowship no later than their seventh year).
Restrictions: These Fellowships may not be held concurrently with any other major fellowship or grant.
Amount of Grant: $33,000 maximum
Date(s) Application is Due: Nov 12

Contact: Nicole Stahlmann, Director of Fellowship Programs; (212) 697-1505, ext. 134; fax: (212) 949-8058; email: nstahlmann@acls.org
Internet: http://www.acls.org/grants/Default.aspx?id=512
Sponsor: American Council of Learned Societies
633 Third Avenue
New York, NY 10017-6795

Andrew W. Mellon/ACLS Recent Doctoral Recipients Fellowships 878

ACLS invites applications for the annual competition for the Mellon/ACLS Recent Doctoral Recipients Fellowships. This is the second stage of the Andrew W. Mellon Foundation/ACLS Early Career Fellowship Program, which provides support for young scholars. This part of the program provides support for a year following the completion of the doctorate for scholars to advance their research. A grant from The Andrew W. Mellon Foundation supports the program. The Fellowships provide a stipend of $30,000 to allow the Fellow to devote an academic year to research.
Requirements: Applicants for this program must be scholars who have been: awarded Mellon/ACLS Dissertation Completion Fellowships; designated Alternates in the Mellon/ACLS Dissertation Completion Fellowship program; or awarded dissertation completion fellowships in another competitive program of national stature. Applicants must be in the final year of dissertation completion at the time of application, and must be Ph.D. candidates in a humanities or social science department in the United States.
Restrictions: The Mellon/ACLS Recent Doctoral Recipients Fellowships may not be held concurrently with any other major fellowship or grant.
Amount of Grant: $30,000
Date(s) Application is Due: Dec 10
Contact: Nicole Stahlmann, Director of Fellowship Programs; (212) 697-1505, ext. 134; fax: (212) 949-8058; email: nstahlmann@acls.org
Internet: http://www.acls.org/grants/Default.aspx?id=514
Sponsor: American Council of Learned Societies
633 Third Avenue
New York, NY 10017-6795

Anesthesia Clinical Research Starter Grants 879

The grants offer start-up money to institutions on behalf of faculty members who seek to obtain further funding for continuation of their anesthesiology-related projects that focus on scholarly, hypothesis-based clinical studies. Institutional matching funds are required.
Requirements: Faculty members may be at the associate, instructor, or assistant professor level.
Amount of Grant: $35,000 first year; $50,000 second year
Date(s) Application is Due: Feb 15; Aug 15
Contact: Thomas Bruckman, Executive Director, (507) 266-6866; fax: (507) 284-0120; email: grunewald.nathan@mayo.edu
Internet: http://www.faer.org/grants.php
Sponsor: Foundation for Anesthesia Education and Research
200 First Street SW, Charlton Building, Mayo Clinic
Rochester, MN 55905

Angels Baseball Foundation Grants 880

Since its inception in 2004, the Angels Baseball Foundation has awarded many grants to worthy organizations throughout the community, ranging from music programs to at-risk youth shelters. Currently, it focuses on initiatives aimed to create and improve education, health care, arts and sciences, and community related youth programs, in addition to providing children the opportunity to experience the game of baseball. Giving is limited to the greater Los Angeles, California, area. Fields of interest include: the arts, athletics/sports activities, children and youth, cancer detection, cancer research, community and economic development, education (at all levels), health care, and community service programs. Applicants should begin by contacting the Foundation directly.
Requirements: Applicants must serve the residents of the greater Los Angeles area.
Geographic Focus: California
Contact: Anne Bafus, Community Relations; (714) 940-2174 or c; fax: (714) 940-2244
Internet: http://losangeles.angels.mlb.com/ana/community/baseball_foundation.jsp
Sponsor: Angels Baseball Foundation
2000 E. Gene Autry Way
Anaheim, CA 92806-6100

Animal Behavior Society Developing Nations Research Grants 881

Awards provide financial support for scientific studies of animal behavior conducted by current members of the Animal Behavior Society. Applications are invited from student members as well as more established members of the research community. Electronic application forms are available on the Web site. Grants are awarded for research to be conducted within a one-year period from the date of award.
Requirements: Only members of the Animal Behavior Society who are residents of a developing nation and are conducting research at an institution in a developing nation are eligible to apply. The following nations shall not be considered as developing nations: the United States, Canada, Israel, Japan, Australia, New Zealand, Iceland, Norway, Sweden, Finland, Denmark, Germany, the Netherlands, Belgium, Luxembourg, Ireland, Switzerland, Austria, Italy, France, the United Kingdom, Spain, and Portugal.
Amount of Grant: $700 maximum
Date(s) Application is Due: Jan 12
Contact: Stephen Nowicki, Department of Biology, Duke University, (919) 684-6950; email: snowicki@duke.edu
Internet: http://www.animalbehavior.org/ABS/Grants/DNG/dng_info2005.html

Sponsor: Animal Behavior Society
Box 90325
Durham, NC 27708-0325

Animal Behavior Society Student Research Grants 882

The program provides financial support for scientific studies of animal behavior. Grants are ordinarily awarded for research to be conducted within a one-year period from the date of award. An electronic submission form is available on the Web site. Annual deadline dates may vary; contact program staff for exact dates.
Requirements: Society members who are currently enrolled in graduate programs may apply.
Amount of Grant: $1000 maximum
Date(s) Application is Due: Jan 28
Contact: Dr. Stephen Nowicki, (919) 684-6950; email: snowicki@duke.edu or plsch@ou.edu
Internet: http://www.animalbehavior.org/ABS/Grants/SRG/ABSSRG_announcement_2004_2005.html
Sponsor: Animal Behavior Society
700 College Street
Beloit, WI 53511

ANL Faculty Research Leave Program 883

Argonne National Laboratory, one of DOE's major research centers, offers faculty research leave appointments (FRLA) to college and university faculty members. Faculty members spend their sabbatical leave, typically 9-12 months, at the Argonne Laboratory. The purpose of the program is to involve college/university faculty in DOE research programs conducted at Argonne, and stimulate continuing research collaboration between faculty and Argonne scientists. Participation takes the form of individual collaboration with an Argonne staff member in some part of an ongoing project of interest to the faculty participant. The research experience is augmented by seminars, independent study, and other appropriate activities. Faculty research projects are available in the basic physical and life sciences, mathematics, computer science, and in engineering, as well as in a variety of applied research programs. Interaction of faculty with students in the research programs is strongly encouraged. Typically, Argonne reimburses the university for 50 percent of salary and fringe benefits for the academic year (9 months) and full salary and fringe benefits for the summer (three months). In addition, the laboratory may negotiate reimbursement for certain travel, moving, and housing expenses. Applications are accepted at any time. Write for additional information and application forms.
Requirements: An appointee must be a full-time faculty member of an accredited U.S. college or university and must have a commitment to continue in teaching and research as a career. The applicant's objectives for the Argonne tour should be clearly specified, and these objectives must be endorsed by the department head or dean. The expression of interest in participation by an individual faculty member, along with university endorsement of such participation, should be transmitted to the Cross Division Program Leader, Division of Educational Programs. Initial expressions of interest should include a curriculum vita, a publication list, and a brief statement of the research interests of the faculty member.
Contact: Carol Przyzycki, Faculty Research Leave (Sabbatical), Division of Educational Programs, (630) 252-5448; email: cprzyzy@dep.anl.gov
Internet: http://www.dep.anl.gov/crossdiv/frl.htm
Sponsor: Argonne National Laboratory
9700 S Cass Avenue, DEP 223
Argonne, IL 60439-4845

ANL Faculty Research Participation Program 884

Faculty participants in the program spend 10-12 weeks collaborating with an Argonne staff scientist or engineer on an existing project of interest to the faculty member. Argonne research falls into four broad categories: engineering research (including advanced nuclear reactors, batteries, and fuel cells); physical research (materials science, physics, chemistry, mathematics, and computer science); energy and environmental science and technology research (biology, alternate energy systems, and environmental and economic impact assessments); and technology transfer (moving the benefits of Argonne's publicly funded research to the marketplace). Appointments are made principally for the summer; however, similar arrangements can be made during the academic year. Stipends are based on academic-year salary and round-trip travel is provided. Additional information and application forms are available upon request.
Requirements: An appointee must be a U.S. citizen or permanent resident who is a full-time faculty member of an accredited U.S. college or university and must have a commitment to continue in teaching and research as a career. The applicant's objectives for the Argonne tour should be clearly specified, and these objectives must be endorsed by the department head or dean.
Date(s) Application is Due: Jan 13
Contact: Carol Przyzycki , Faculty Research Participation (Short-Term), Division of Educational Programs, (630) 252-5448; email: cprzyzy@dep.anl.gov
Internet: http://www.dep.anl.gov/p_faculty/frp.htm
Sponsor: Argonne National Laboratory
9700 S Cass Avenue, DEP 223
Argonne, IL 60439

ANL Graduate Thesis Parts Research Appointments 885

Thesis Parts appointments support qualified graduate students who wish to visit Argonne for periods from a few days to a few months so that they may utilize special laboratory facilities or capabilities during the course of their thesis research. Research areas include physical and life sciences, mathematics, computer science, engineering, conservation, environment, fission and fusion energy, and other energy technologies. Application is best made through an Argonne staff member or research division appropriate to the proposed activity.

Requirements: Qualified graduate students in U.S. universities who wish to carry out some thesis research at Argonne are eligible to apply.

Contact: Lisa Reed, Lab-Grad Research Appointments, Division of Educational Programs, (630) 252-3366; email: Lreed@dep.anl.gov

Internet: http://www.dep.anl.gov/p_graduate/thesispa.htm

Sponsor: Argonne National Laboratory

9700 S Cass Avenue, DEP 223

Argonne, IL 60439

ANL Laboratory-Graduate Research Appointments 886

Laboratory graduate research appointments are available for qualified graduate students at U.S. universities who wish to carry out their thesis research at Argonne National Laboratory under the cosponsorship of an Argonne staff member and a faculty member. Research may be conducted in the basic physical and life sciences, mathematics, computer science, and engineering, as well as in a variety of applied areas relating to conservation, environment, fission and fusion energy, and other energy technologies. Lab-grad appointments are for a one-year term with annual renewals being contingent upon satisfactory performance by the appointee. Applications may be submitted at any time, and appointments may commence any time during the year. Completed application should be submitted at least one month prior to proposed starting date. Temporary appointments are available to qualified graduate students so they may work with an Argonne staff member and become familiar with his/her work prior to application for a year-long research appointment. The student's faculty sponsor may also receive payment for limited travel expenses.

Requirements: An applicant must currently be a full-time graduate student in a recognized graduate program in a U.S. university and must be a U.S. citizen or permanent resident alien.

Amount of Grant: $5000 maximum tuition payment, stipend, and travel expenses

Contact: Lisa Reed, Lab-Grad Research Appointments, Division of Educational Programs, (630) 252-3366; email: Lreed@dep.anl.gov

Internet: http://www.dep.anl.gov/p_graduate/labgrad.htm

Sponsor: Argonne National Laboratory

9700 S Cass Avenue, DEP 223

Argonne, IL 60439

Ann and Robert H. Lurie Family Foundation Grants 887

Founded in 1986, the foundation initially had set up six categories: medical services and research; child-related medical organizations; basic services including food and shelter; education; the arts; and so-called wild things. When Ann Lurie's husband, Robert, died in 1990, she was left with a fortune worth hundreds of millions and six kids — ages 5 to 15. She has given away well over $100-million since Robert H. Lurie died at age 48, generally spurning requests from nonprofit groups with undistinguished records and concentrating her gifts on some of the world's biggest problems, including hunger, cancer, and inadequate health care. Today, the foundation supports educating and providing resources to children so they are better prepared for the future.

Requirements: There are no specific deadlines or application forms to fill out. An organization requesting funding should have finances in good order, demonstrating that it is making good use of its resources. Note: Ann Lurie has a preference for identifying her own projects, rather than responding to solicitations.

Geographic Focus: Illinois; Michigan

Samples: University of Michigan College of Engineering (Ann Arbor, Michigan) - $2,500,000.; University of Michigan Entrepreneurial Institute (Ann Arbor, Michigan) - $1,000,000.; Africa Infectious Disease Village Clinic (Chicago, Illinois) - $403,429 Ancient Egypt Research Associates (Cambridge, Massachusetts) - $400,000.; Infant Welfare Society of Chicago (Chicago, Illinois) - $154,457.; K-idea Trust (Chicago, Illinois) - $111,160.; Riders For Health II (Rockville, Maryland) - $100,000.; Greater Chicago Food Depository (Chicago, Illinois) - $65,000.; Robert H. Lurie Comprehensive Cancer Center (Chicago, Illinois) - $63,870.; The Synergos Institute, Inc. (New York, New York) - $25,000.; The Chicago Public Library Foundation (Chicago, Illinois) - $25,000.; Marwen Foundation (Chicago, Illinois) - $25,000.; The Smile Train (New York, New York) - $25,000.; Choate Rosemary Hall (Wallingford, Connecticut) - $25,000.; The Chicago Public Library Foundation (Chicago, Illinois) - $25,000.; Children Affected By AIDS Foundation (Chicago, Illinois) - $25,000.; Children's Memorial Foundation (Chicago, Illinois) - $23,750.; Chicago Cultural Center Foundation (Chicago, Illinois) - $10,000

Contact: Ann Lurie, President; (312) 466-3750; fax: (312) 466-3700

Sponsor: Ann and Robert H. Lurie Family Foundation

2 N. Riverside Plaza, Suite 1500

Chicago, IL 60606-2600

Ann Arbor Area Community Foundation Grants 888

The foundation is interested in funding projects that will improve the quality of life for citizens of the Ann Arbor, Michigan, area. Eligible projects generally fall within the categories of education, culture, social service, community development, environmental awareness, health and wellness, and youth and senior citizens. Types of support include: emergency funds, program development, conferences and seminars, publication, seed grants, scholarship funds, research, and matching funds. Higher priority is given to programs that are preventive rather than remedial, increase individual access to community resources, examine and address the underlying causes of local problems,

promote independence and personal achievement, attract volunteer resources and support, strengthen the private nonprofit sector, encourage collaboration with other organizations, and build the capacity of the applying organization. Organizations interested in applying are strongly encouraged to discuss their project with the program director prior to submitting an application.

Requirements: 501(c)3 nonprofit organizations in the Ann Arbor area, which is the area that falls within the boundaries of the Ann Arbor public schools district, are eligible.

Restrictions: The foundation usually does not make grants for construction projects, annual giving campaigns or capital campaigns, normal operating expenses (except for start-up purposes), religious or sectarian purposes, computer hardware equipment, individuals, advocacy or political purposes, multiyear funding, or regranting.

Geographic Focus: Michigan

Samples: Jazzistry, $9,139 for Finding Your Place in the Jazz Story; Community Respite Center, Inc., $14,079 for development of a Lifespan Respite Center for Washtenaw County; Jewish Family Services, $9,000 for a database implementation project; Washtenaw United Way, $25,000 for a 2-1-1 Call Center; Vineyard Church of Ann Arbor, $8,000 or a mentoring ministry for single parent families; University of Michigan-Turner Geriatric Clinic, $3,500 for the Senior Unmet Needs Fund (all 2006).

Date(s) Application is Due: Feb 11; Oct 7

Contact: Phil D'Anieri, Program Director; (734) 663-0401; fax: (734) 663-3514; email: pdanieri@aaacf.org

Internet: http://www.aaacf.org/grants.asp

Sponsor: Ann Arbor Area Community Foundation

301 North Main Street, Suite 300

Ann Arbor, MI 48104

Ann Peppers Foundation Grants 889

The foundation awards grants to eligible California nonprofit organizations in its areas of interest, including arts and cultural programs, disabled, elderly, health care, private education, and social services. Types of support include capital grants, general operating grants on a temporary basis, matching grants, scholarships, and research grants. Grants are initiated by the foundation manager. There are no application deadlines; the board meets quarterly to consider requests.

Requirements: Southern California 501(c)3 nonprofits, colleges, and universities are eligible. Preference is given to requests from Los Angeles County.

Geographic Focus: California

Samples: Boys Republic (Chino Hills, CA)—for kitchen renovation at residential facility , $5,000.; Pepperdine University (Malibu, CA)—for scholarships, $25,000.; Huntington Medical Research Institute (Pasadena, CA)—to purchase imaging technology for preclinical breast cancer, $5,000.; Marymount College (Rancho Palos Verdes, CA)—for scholarships, $10,000.

Amount of Grant: $2000-$50,000 range

Contact: Jack Alexander, Secretary, (626) 449-0793

Sponsor: Ann Peppers Foundation

625 S Fair Oaks Avenue

South Pasadena, CA 91030

Anna C. and Oliver C. Colburn Fellowship 890

One fellowship will be awarded every other year for an academic year of study and research to an applicant contingent upon his or her acceptance as an incoming associate member or student associate member of the American School of Classical Studies at Athens. Candidates for the fellowship must apply concurrently to the American School for associate membership or student associate membership. Application forms are available from the institute.

Requirements: This competition is open to U.S. or Canadian citizens or permanent U.S. residents who are at the predoctoral stage or who have received the PhD degree within the last five years.

Restrictions: Applicants may not be members of the American School during the year of application. The fellowship may be held for a maximum of one year. Other major fellowships may not be held during the requested tenure of the award.

Amount of Grant: $11,000

Date(s) Application is Due: Jan 15

Contact: Elizabeth Gilgan, AIA Programs Administrator, (617) 353-9361; fax: (617) 353-6550; email: egilgan@aia.bu.edu or aia@aia.bu.edu

Internet: http://www.archaeological.org/webinfo.php?page=10007

Sponsor: Archaeological Institute of America

656 Beacon Street

Boston, MA 02215-2010

Anna Louise Hoffman Award for Outstanding Achievement in Graduate 891 Chemistry Research

At the time of nomination, the candidate must be a full-time female graduate student who is a candidate for a graduate degree at an accredited institution. Research presented by the candidate must be original and of one of the main chemical divisions—analytical, biochemical, inorganic, organic, physical, and/or ancillary. The nominee may be, but need not be, a member of Iota Sigma Pi.

Requirements: Nomination must be made by members of the institution's graduate faculty. Dossier must contain an academic history, including all transcripts; candidate's permanent and school addresses; two letters of recommendation; the nomination; a brief description of the candidate's research; and a list of publications and talks or papers presented outside of degree requirements. Description of the research should be no more than 1000 words and prepared by the candidate.

Amount of Grant: $500

Date(s) Application is Due: Feb 15
Contact: Vicki Grassian, (319) 335-1392; fax: (319) 353-1115; email: vicki-grassian@uiowa.edu
Internet: http://www.iotasigmapi.info//studentAwards.htm
Sponsor: Iota Sigma Pi National Honor Society for Women in Chemistry
Euclid Avenue at East 24th Street
Cleveland, OH 44115

Anna-Greta and Holger Crafoords Foundation Grants   892

The foundation operates in Sweden and the grants promote international basic research within the following subject areas: mathematics and astronomy; geosciences; biosciences, with particular emphasis on ecology; and polyarthritis. In addition to the Crafoord Prize, research grants are awarded within the relevant disciplines to individual persons or researchers in Sweden. Award of a research grant within polyarthritis takes place every third year, but a prize within polyarthritis is only awarded when a specific enquiry has demonstrated that such scientific progress has been made that a prize is warranted.
Contact: Grants Administrator, 046-38 58 80; fax: 046-38 58 85; email: crafoord@crafoord.se
Internet: http://www.crafoord.se/index-e.htm
Sponsor: Anna-Greta and Holger Crafoords Foundation
P.O. Box 137, Malmovagen 8
Lund 22100 Sweden

Anne Louise Barrett Fellowship   893

This fellowship is awarded in music, preferably and primarily for study or research in musical theory, composition, or the history of music. The fellowship may be used abroad or in the United States. Awards are usually made to applicants who plan full-time graduate study for the coming year. Preference will be given to applicants who have not previously held one of Wellesley College's awards. Awards will be based on merit and need. One fellowship is awarded annually; application form must be used. Annual deadline dates may vary; contact program officer for exact dates.
Requirements: The fellowship is open to graduating seniors and graduates of Wellesley College to be used for study at institutions other than Wellesley.
Amount of Grant: $15,000 maximum
Contact: Secretary to the Committee on Graduate Fellowships, Center for Work and Service, (781) 283-3525; email: cws-fellowships@wellesley.edu
Internet: http://www.wellesley.edu/CWS/alumnae/wellfs.html
Sponsor: Wellesley College
106 Central Street
Wellesley, MA 02181-8200

Annenberg Foundation Grants   894

The foundation provides support for projects within its grantmaking interests of education and youth development, arts and culture, civic, community and the environment, and health and human services. It encourages the development of more effective ways to share ideas and knowledge. Letters of inquiry may be submitted at all times during the year and there are no deadlines. Please review the grants database for additional types of grants given.
Requirements: 501(c)3 tax-exempt organizations are eligible.
Restrictions: Full proposals are not accepted unless requested by a Foundation representative. The foundation is not presently considering inquiries for: individuals, individual K-12 schools, for-profit organizations, political activities or attempts to influence specific legislation, individual scholarships, projects focused exclusively on research, or programs outside of its grant-making interests.
Samples: Centro Latino De Educacion Popular, Los Angeles, CA, $50,000 - for the Spanish Literacy Campaign to mobilize, train, equip, and support Spanish speakers to teach their non-literate friends and family members to read and write using an online literacy program (2009); American Patrons of the Tate Gallery Foundation, Inc., New York, NY, $50,000 - to support special exhibitions and education programs at the Tate Modern (2009); National Public Radio, Inc., Washington, DC, $500,000 - for the programs of NPR West to broaden the creation and distribution of high-quality public radio programming (2009).
Contact: Leonard Aube, Executive Director; (310) 209-4560; fax: (310) 209-1631; email: info@annenbergfoundation.org
Internet: http://www.annenbergfoundation.org/grants
Sponsor: Annenberg Foundation
2000 Avenue of the Stars, Suite 1000
Los Angeles, CA 90067

ANS Neurotology Fellows Award   895

The Neurotology Fellows Award was established by the American Neurotology Society in 1996 to reward scientific excellence. The award has been generously endowed through royalties received for the textbook Neurotology, Mosby Publishers, D. Brackmann and R. Jackler, editors. One or more awards of $500 each is intended to subsidize travel expenses incurred while giving a scientific presentation at the Annual Spring Meeting of the American Neurotology Society. The material presented need not have been performed during the Fellowship year; it may derive from earlier work performed during the residency or extracurricular research experience. Applicants should submit electronically a cover letter, a supporting letter from the Fellowship director to verify qualifications, and electronic submission of the abstract by October 15.
Requirements: Applicants must be full-time participants in a post-residency Neurotology Fellowship in the United States or Canada. In addition, the Fellow must be both podium presenter and first author of the paper submitted for publication.

Amount of Grant: $500
Date(s) Application is Due: Oct 15
Contact: Shirley Gossard, Administrator; (352) 751-0932 or (217) 414-4868; fax: (352) 751-0696; email: segossard@aol.com
Internet: http://www.americanneurotologysociety.com/funding.html
Sponsor: American Neurotology Society
3096 Riverdale Road
The Villages, FL 32162

ANS Nicholas Torok Vestibular Award   896

The Nicholas Torok Vestibular Award, established under the aegis of the American Neurotology Society instituting a yearly lecture, is to be presented at the Annual Scientific Spring Meeting of the Society. The subject should be an innovative observation, experience or technique in the field of vestibular basic science or clinical science. The candidate shall be selected preferably from the membership of the Society or upon specific merit of any M.D. or Ph.D. A yearly award of $1,500 will be given for the selected manuscript.
Requirements: The manuscript must be presented electronically in the publication format recommended by Otology & Neurotology, the official journal of the American Neurotology Society.
Amount of Grant: $1,500
Date(s) Application is Due: Oct 15
Contact: Shirley Gossard, Administrator; (352) 751-0932 or (217) 414-4868; fax: (352) 751-0696; email: segossard@aol.com
Internet: http://www.americanneurotologysociety.com/funding.html
Sponsor: American Neurotology Society
3096 Riverdale Road
The Villages, FL 32162

ANS Trainee Award   897

The American Neurotology Society offers an award for the best clinical or basic science paper in Neurotology submitted by a Resident or Fellow in training in the field of Otolaryngology-Head and Neck Surgery. A clinical study or basic research study is acceptable. The Society has established a cash award of $1,000 for the winning paper. The author will be expected to attend the Annual ANS Spring Meeting and present the paper personally.
Requirements: Residents from any approved residency program or Fellows in Neurotology in the United States or Canada are eligible to compete. The Resident or Fellow must be the primary author. He or she must provide the main inspiration for the paper and do the literature search, data collection, original and final drafts of the article, the discussion, and the conclusions.
Restrictions: The paper must be an original contribution, not previously published.
Amount of Grant: $1,000
Date(s) Application is Due: Oct 15
Contact: Shirley Gossard, Administrator; (352) 751-0932 or (217) 414-4868; fax: (352) 751-0696; email: segossard@aol.com
Internet: http://www.americanneurotologysociety.com/funding.html
Sponsor: American Neurotology Society
3096 Riverdale Road
The Villages, FL 32162

ANS/AAO-HNSF Herbert Silverstein Otology and Neurotology   898
Research Award

The purpose of this award, which is jointly sponsored by the AAO-HNS Foundation and the American Neurotology Society, is to support a clinical or translational research project focused on diseases, disorders, or conditions of the peripheral or central auditory and/or vestibular system among new full-time academic surgeons. The award is intended as a preliminary step in clinical investigator career development and is expected to facilitate the recipient's preparation of a more comprehensive individualized research plan suitable for submission to the National Institutes of Health or comparable funding agency.
Requirements: Applicants must be physicians in the United States or Canada with demonstrated potential for excellence in research and teaching and serious commitment to an academic research career in otology or neurotology. Priority will be given to fellows or junior faculty, who have completed residencies or fellowships within four years of the application receipt date, although otolaryngology-head and neck surgery residents are eligible. All candidates must be sponsored by the Chair of his/her Division or Department and by an official representative of the institution which would administer the Award and in whose name the application is formally submitted.
Amount of Grant: $25,000 maximum ($12,500 per year)
Contact: Stephanie L. Jones, Assistant Director; (703) 519-1586 or (703) 836-4444; email: sljones@entnet.org
Internet: http://www.americanneurotologysociety.com/funding.html
Sponsor: American Neurotology Society / American Academy of Otolaryngology-Head and Neck Surgery Foundation
One Prince Street
Alexandria, VA 22314-3357

Anthem Blue Cross and Blue Shield School Nurse Grants   899

The mission of the project is to promote cardiovascular health among school-age children. Anthem's goals are to assist school nurses in initiating programs that may encourage a healthy lifestyle, reduce cardiovascular risk factors, promote physical activity, and encourage healthy eating behaviors; and facilitate the development of innovative school

health programs that are based on research. No awards will be made to individuals, but will be awarded to the school. To apply, the school nurse representing the school must submit a completed application form with the appropriate attachments by the listed application deadline. Guidelines, including program contacts in Kentucky and Indiana, and application are available online.
Requirements: Eligible applicants include school nurses who are licensed in the state in which they practice and who represent a public or private school (elementary, junior high, middle, or high school) in Ohio, Kentucky, or Indiana.
Restrictions: Funds cannot be used for the salary or wages of any staff or employee.
Geographic Focus: Indiana; Kentucky; Ohio
Amount of Grant: $500-$1000
Date(s) Application is Due: Feb 1
Contact: Gabrielle Karpowicz, (614) 529-5068; email: gkarpowicz@aol.com
Internet: http://oasn.org/Awards_Grants/awards.htm
Sponsor: Ohio Association of School Nurses
6167 Heritage Point Ct
Hilliard, OH 43026

Anthony R. Abraham Foundation Grants 900
For more than 30 years, the Anthony R. Abraham Foundation mission has been to help non-profit organizations worldwide. The foundations strives to: provide programs and services to help people around the world become self-productive and give back to their communities; ensure that no child is denied medical treatment due to a lack of insurance; provide education that breaks barriers; guarantee that research into the cure of catastrophic diseases continues; help raise the quality of life. It, been a privilege for Foundation to be able to help ease poverty, raise hospitals, build orphanages and further medical research and the Foundation looks forward to doing even more. Some of the organizations helped include: Domestic organizations -St. Jude Children, Research Hospital; Camillus House; Miami Rescue Mission; Habitat for Humanity of Greater Miami; America, Second Harvest; Miami Children, Hospital; Jackson Memorial Hospital; Florida Heart Research Institute; Big Brothers/Big Sisters; Alonzo Mourning Charities; Honey Shine Mentoring Program; Cancer Link; The Miami Lighthouse for the Blind; Overtown Youth Center. International organizations: Brothers of the Good Shepherd; School for the Blind, Lebanon; Rene Moawad Foundation; Haitian Foundation; Children's International Network; Doctors without Borders; Little Sisters of Nazareth; Maronite Order of the Holy Family.
Requirements: Non-profit 501 (c) 3 organization requesting funding should fill out a funds-request form and submit it at the Foundation's website. The Foundation will contact you, if your organization qualifies under the Foundation, guidelines.
Restrictions: Grants are not made to individuals.
Amount of Grant: $100-$50,000 range average
Contact: Anthony R. Abraham, Chairman; (305) 665-2222
Internet: http://www.abrahamfoundation.com/about
Sponsor: Anthony R. Abraham Foundation
1320 S. Dixie Highway, Suite. 241
Coral Gables, FL 33146-2937

Anthropological Center for Training in Global and Environmental 901
Change (ACT) Grad Assistantships
ACT is an interdisciplinary training and research center on the human dimensions of global environment change. ACT's research and training objectives focus on how particular local populations manage resources and how those activities may be monitored using remote sensing technologies and field studies. There is no formal application process for a graduate assistantship at ACT, but if an individual is applying for a graduate program at Indiana University, and feels that their skills and research interests match that of ACT, then an applicant should share his/her interest in a graduate assistantship.
Geographic Focus: Indiana
Contact: Emilio Moran, Director; (812) 855-6181; fax: (812) 855-3000; email: moran@indiana.edu or act@indiana.edu
Internet: http://www.indiana.edu/~act/newstudents.shtml
Sponsor: Anthropological Center for Training in Global and Environmental Change (ACT)
701 East Kirkwood Avenue, Student Building, Room 331
Bloomington, IN 47405

AO North America Grants 902
Support is given to trauma research projects, training in the management and care of teachers, and to visiting professors in the teaching of orthopedic, maxillofacial, spine, or veterinary trauma treatment.
Samples: Orthopaedic Trauma Assoc (Rosemont, IL)—for research, $75,000. Vanderbilt U, School of Medicine (Nashville, TN)—for research, $25,000. Foothills Hospital (Calgary, Canada)—for research, $7255.
Contact: Grants Administrator, (610) 251-9007; fax: (610) 251-9059; email: ellisa@aona.com
Internet: http://www.aona.org/fellow_preceptor.asp
Sponsor: AO North America Inc
P.O. Box 308
Devon, PA 19333-0308

AOA Clinical Investigator Development Award 903
This program recruits and supports osteopathic physicians for research and teaching careers in the medical sciences. Support is given for three years. Contact program staff for availability.

Amount of Grant: $50,000
Contact: Elizabeth Freeman, (800) 621-1773 ext 8006; email: research@osteopathic.org
Internet: http://do-online.osteotech.org/index.cfm?PageID=res_main
Sponsor: American Osteopathic Association
142 E Ontario Street
Chicago, IL 60611-2864

AOA Gutensohn-Denslow Award 904
The award recognizes and rewards a person who has made an outstanding contribution to the osteopathic profession in the areas of research and education. AOA and the National Osteopathic Foundation administer this program, which is funded by Glaxo Wellcome Inc.
Requirements: A nominee must be either an osteopathic physician or hold a doctoral level degree and be on the faculty or staff of an osteopathic institution. Nominations may come from anyone in the osteopathic profession and from colleagues of osteopathic physicians. Individuals nominated previously must be renominated for consideration. Initial nominations must include a current curriculum vita of the nominee and a detailed account of his/her contributions to research.
Amount of Grant: $5000, consisting of $4250 stipend and $750 travel allowance
Date(s) Application is Due: Mar 15
Contact: Elizabeth Freeman, Division of Research Development, (800) 621-1773 ext 8006; fax: (312) 280-3860; email: research@osteopathic.org
Internet: http://www.osteopathic.org/index.cfm?PageID=ost_resoverview
Sponsor: American Osteopathic Association
142 E Ontario Street
Chicago, IL 60611-2864

AOA Osteopathic Research Fellowships 905
These research fellowships are designed to enable the applicant to conduct a basic science or clinical research project that will make a significant contribution to osteopathic medicine. The fellowship also serves as seed funding to encourage an osteopathic physician to contribute to research throughout his/her career in osteopathic medicine.
Requirements: Applicant must be either a postdoctoral osteopathic medical student, possessing an earned DO degree awarded by an AOA-accredited institution and be enrolled in an internship, residency, or research fellowship; or an undergraduate osteopathic medical student enrolled in an AOA-accredited institution.
Amount of Grant: $5000 stipend
Date(s) Application is Due: Mar 15
Contact: Elizabeth Freeman, Division of Research Developoment, (800) 621-1773 ext 8006; fax: (312) 202-8200; email: research@osteopathic.org
Internet: http://www.do-online.osteotech.org/index.cfm?PageID=res_grant
Sponsor: American Osteopathic Association
142 E Ontario Street
Chicago, IL 60611-2864

AOA Osteopathic Research Grants 906
Grants support projects investigating problems of a biological nature that will lead to a better understanding and a more effective application of the principles and concepts of osteopathic medicine. Grants are supported by AOA and the A.T. Still Osteopathic Foundation and Research Institute and administered by the foundation at the same address.
Requirements: Grants are made to qualified AOA-accredited, -affiliated, or -approved osteopathic institutions, not to individuals. Osteopathic physicians having the DO degree who are at non-osteopathic-accredited, -affiliated, or -approved institutions are also eligible to apply. Grantee must operate within an established institutional research program.
Amount of Grant: $25,000-$50,000 average per year
Date(s) Application is Due: Dec 1
Contact: Elizabeth Freeman, Division of Research Development, (800) 621-1773 ext 8006; email: research@osteopathic.org
Internet: http://www.do-online.osteotech.org/index.cfm?PageID=res_grant
Sponsor: American Osteopathic Association
142 E Ontario Street
Chicago, IL 60611-2864

AOA-Burnett Osteopathic Student Research Award 907
AOA offers this annual award in recognition of a student of osteopathic medicine for the most outstanding concept paper pertaining to an osteopathic-oriented research proposal. The award consists of an all-expense paid trip to the association's research conference plus a cash award and submittal of the paper to the Journal of the American Osteopathic Association for possible publication.
Requirements: Eligibility is limited to students enrolled in AOA-accredited colleges of osteopathic medicine.
Amount of Grant: $100 plus all-expense paid trip to research conference
Date(s) Application is Due: Apr 1
Contact: Elizabeth Freeman, Division of Research Development, (800) 621-1773 ext 8006; fax: (312) 202-8200; email: research@osteopathic.org
Internet: http://www.do-online.osteotech.org/index.cfm?PageID=res_grant
Sponsor: American Osteopathic Association
142 E Ontario Street
Chicago, IL 60611-2864

AORN Foundation Scholarships 908

AORN scholarships provide funds for tuition and registration fees to members enrolled in baccalaureate, master's, and doctoral degree programs. Payment is made directly to the educational institution. The scholarships are granted for a one-year period; reapplication for funds is required annually through a new application and all supporting documents. An applicant must submit evidence of acceptance or enrollment in the particular program with specific information in support of accreditation status and a school catalog description of course of study; a personal statement reflecting contributions in nursing practice, research, and/or education; an official AORN scholarship application; official school transcripts; and three personal reference letters.

Requirements: Applicants must be RNs and currently active or associate members of AORN in good standing; membership must have been maintained continuously for 12 months prior to application deadline. Applicants must maintain a cumulative GPA of 3.0.

Restrictions: Members of the national board of directors during their elected terms, the scholarship board, and AORN headquarters personnel are ineligible. It is not possible to combine an AORN scholarship with a federal nurse traineeship or other total funding awards; other nonconflicting one-time only scholarships received by an applicant will not affect the funds of an AORN scholarship.

Amount of Grant: $500 minimum
Date(s) Application is Due: May 1
Contact: Ingrid Bendzsa, (800) 755-2676 ext 328; fax: (303) 755-4219; email: ibendzsa@aorn.org
Internet: http://www.aorn.org/foundation/scholarship.htm
Sponsor: Association of Operating Room Nurses Foundation
2170 S Parker Road, Suite 300
Denver, CO 80231-5711

AOS Research Grants 909

Research grants are awarded for experimental projects and applied and fundamental research pertaining to orchids. Support is given in biological research including taxonomy, genetics, cytogenetics, physiology, and pathology as well as in conservation and education. The society wishes to further the study of orchids in every aspect including their classification, evolution, conservation, propagation, culture, care, and development. As a first step, prospective applicants should submit a brief letter requesting an application form and instructions for requesting a grant. Grants are usually awarded for one year, possibly renewable, but not to exceed a maximum of three years. Grants received by January 1 are reviewed at the Spring Trustees Meeting and those grants received by July 1 are reviewed at the Fall Trustees Meeting

Requirements: Graduate students or other qualified research personnel associated with accredited institutions of higher learning or appropriate research institutes may apply.

Amount of Grant: $500-$12,000
Date(s) Application is Due: Jan 1; Jul 1
Contact: Pamela Giust, (561) 404-2000; fax (561) 404-2045; email: pgiust@aos.org
Internet: http://www.theaos.org/aos/research/research.aspx?sec=2
Sponsor: American Orchid Society
16700 AOS Ln
Delray Beach, FL 33446-4351

APA Culture of Service in the Psychological Sciences Award 910

The Award recognizes departments that demonstrate a commitment to service in the psychological sciences. Departments selected for this award will show a pattern of support for service from faculty at all levels, including a demonstration that service to the discipline is rewarded in faculty tenure and promotion. Successful Departments will also demonstrate that service to the profession is an integral part of training and mentoring. Service to the discipline includes such activities as departmental release time for serving on boards and committees of psychological associations; editing journals; serving on a review panel; or chairing an IRB. Other culture of service activities that a department would encourage include mentoring students and colleagues; advocating for psychological science's best interests with state and federal lawmakers; and promoting the value of psychological science in the public eye. The focus of this award is a department's faculty service to the discipline and not their scholarly achievements. Both Undergraduate and Graduate Departments of Psychology are eligible. Self-nominations are encouraged.

Amount of Grant: $5,000
Contact: Suzanne Wandersman, Director for Governance Affairs; (202) 336-6000; fax: (202) 336-5953; email: swandersman@apa.org
Internet: http://www.apa.org/science/dept_award.html
Sponsor: American Psychological Association
750 First Street, NE
Washington, DC 20002-4242

APA Dissertation Research Awards 911

Dissertation awards are granted to graduate students in psychology programs to help offset the costs associated with dissertation research. The research may be done in any area of psychology. Applicants who have already defended their dissertations are eligible to apply for these funds, as long as they have not yet received doctoral degrees as of the application deadline. APA provides dissertation research awards to approximately 50 students annually. Dissertations must be approved by the students' committees before application. The application deadline falls in mid-September of each year. Awards are announced in late December.

Requirements: APA student affiliates enrolled in a psychology graduate program are eligible to apply.

Restrictions: Departments may not nominate more than three students per year for dissertation awards.

Amount of Grant: $1000, $3000, $5000
Date(s) Application is Due: Sep 15
Contact: APA Science Directorate, (800) 374-2721 or (202) 336-5500; fax: (202) 336-5953; email: science@apa.org
Internet: http://www.apa.org/science/dissinfo.html
Sponsor: American Psychological Association
750 First Street NE
Washington, DC 20002-4242

APA Distinguished Scientific Award for Early Career Contribution to Psychology 912

The Distinguished Scientific Award for Early Career Contribution to Psychology recognizes excellent young psychologists. Nominations of persons who received doctoral degrees during and since 1998 are being sought in these five areas: animal learning and behavior, comparative; cognition/human learning; developmental psychology; health; and psychopathology. These categories should be interpreted broadly and are not meant to be exclusive; all of psychology is of sufficient merit to be considered for awards. The Award consists of a citation and a cash prize, which are presented at the APA Annual Convention.

Samples: Robert D. Gray, PhD (applied research), Department of Applied Psychology, Arizona State University (2007); Patrik O. Vuilleumier, MD (behavioral /cognitive neuroscience), Department of Neurology, University of California, Davis (2007); R. Chris Fraley, PhD (individual differences), Department of Psychology at University of Illinois at Urbana-Champaign (2007); Jorn Diedrichsen, PhD (perception/motor performance), School of Psychology, Adeilad Brigantia University, Wales (2007); Matthew D. Lieberman, PhD (social), University of California, Los Angeles (2007).

Contact: Suzanne Wandersman, Director for Governance Affairs; (202) 336-6000; fax: (202) 336-5953; email: swandersman@apa.org
Internet: http://www.apa.org/science/sciaward.html
Sponsor: American Psychological Association
750 First Street, NE
Washington, DC 20002-4242

APA Distinguished Scientific Award for the Applications of Psychology 913

The Distinguished Scientific Award for the Applications of Psychology honors psychologists who have made distinguished theoretical or empirical advances in psychology leading to the understanding or amelioration of important practical problems. The Award consists of a citation and a cash prize, which are presented at the APA Annual Convention.

Samples: Karl G. Joreskog, PhD, Emeritus Professor, University of Upssala, Sweden (2007),; Peter M. Bentler, PhD, UCLA Psychology Department, Los. Angeles, CA (2007); John P. Campbell, PhD, Professor and Chair of Psychology and Industrial Relations, University of Minnesota, MN (2006).

Contact: Suzanne Wandersman, Director for Governance Affairs; (202) 336-6000; fax: (202) 336-5953; email: swandersman@apa.org
Internet: http://www.apa.org/science/sciaward.html
Sponsor: American Psychological Association
750 First Street, NE
Washington, DC 20002-4242

APA Distinguished Scientific Contribution Award 914

The Distinguished Scientific Contribution Award honors psychologists who have made distinguished theoretical or empirical contributions to basic research in psychology. The Award consists of a citation and cash prize, presented at the APA Annual Convention.

Samples: Marilynn B. Brewer, PhD, Department of Psychology, Ohio State University, Columbus, OH (2007); Jean M. Mandler, PhD, Research Professor, Department of Cognitive Science, University of California, San Diego (2007); Paul Rozin, PhD, Professor of Psychology, University of Pennsylvania (2007).

Contact: Suzanne Wandersman, Director for Governance Affairs; (202) 336-6000; fax: (202) 336-5953; email: swandersman@apa.org
Internet: http://www.apa.org/science/sciaward.html
Sponsor: American Psychological Association
750 First Street, NE
Washington, DC 20002-4242

APA Distinguished Service to Psychological Science Award 915

This Award recognizes individuals who have made outstanding contributions to psychological science through their commitment to a culture of service. Award recipients will receive an honorarium of $1,000. Nominees will have demonstrated their service to the discipline by aiding in association governance; serving on boards, committees and various psychological associations; editing journals; reviewing grant proposals; mentoring students and colleagues; advocating for psychological science's best interests with state and federal lawmakers; and promoting the value of psychological science in the public eye. Nominees may be involved in one service area, many of the areas, or all of the service areas noted above. An individual's service to the discipline and not a person's scholarly achievements are the focus of this award.

Samples: Robert L. Balster, Ph.D., Professor of Pharmacology, University of Michigan Medical School; Nora Newcombe, Professor of Psychology, Temple University; Robert A. Bjork, Chair of the Psychology Department, University of California, Los Angeles; J. Bruce Overmier, Ph.D., Department of Psychology, University of Minnesota.

Amount of Grant: $1,000

Contact: Suzanne Wandersman, Director for Governance Affairs; (202) 336-6000; fax: (202) 336-5953; email: swandersman@apa.org
Internet: http://www.apa.org/science/serv_award.html
Sponsor: American Psychological Association
750 First Street, NE
Washington, DC 20002-4242

APA Planning Fellowships 916
The goals of this program are both to encourage students of certain minority backgrounds to enter the planning profession and to help such students who would otherwise be unable to continue their studies in planning. The exact amount of the awards offered is contingent upon the number and qualifications of the applicants. The program is open to first- and second-year students; first-year students who receive fellowships are eligible to compete for an award the following year as well.
Requirements: Eligible to apply are U.S. or Canadian citizens who are African Americans, Hispanics, or Native Americans currently enrolled or accepted for enrollment in a PAB-accredited graduate planning program. Applicants must document their need for financial assistance.
Amount of Grant: $1000-$5000
Date(s) Application is Due: Apr 30
Contact: Kriss Blank, Leadership Affairs Associate, Scholarships and Fellowships, (312) 786-6722 or (312) 431-9100; fax: (312) 431-9985; email: kblank@planning.org or fellowship@planning.org
Internet: http://www.planning.org/institutions/scholarship.htm
Sponsor: American Planning Association
122 S Michigan Avenue, Suite 1600
Chicago, IL 60603

APA Scientific Conferences Grants 917
The program seeks proposals for research conferences in psychology. The purpose of this program is to promote the exchange of important new contributions and approaches in scientific psychology. Conference formats include festschrifts—conferences organized as tributes to distinguished scholars; stand-alone conferences—usually two days long, involves psychologists as the primary organizers; and add-a-day conference—meetings that occur at the beginning or end of a scientific conference other than APA (e.g., Society for Ingestive Behavior or Psychonomics Society). The conference must also be supported by the host institution with direct funds, in-kind support, or a combination of the two.
Requirements: One of the primary organizers must be a member of APA. Only academic institutions accredited by a regional body may apply. Independent research institutions must provide evidence of affiliation with such an accredited institution. Joint proposals from cooperating institutions are encouraged. Conferences may be held only in the United States, its possessions, or Canada.
Restrictions: APA governance groups, APA divisions, and related entities are not eligible for funding under this program.
Amount of Grant: $500-$20,000
Date(s) Application is Due: Jun 1; Dec 1
Contact: Deborah McCall, Science Directorate, (202) 336-6000; fax: (202) 336-5953; email: dmccall@apa.org
Internet: http://www.apa.org/science/psa/sep07ann.html
Sponsor: American Psychological Association
750 First Street, NE
Washington, DC 20002-4242

APA Young Investigator Grant Program 918
The program provides financial support to teaching, research, and health care delivery projects in general pediatrics. The number of awards is dependent on available funds and the size of grant requests of selected projects. Proposals should address the goal of the association: to encourage, promote, and facilitate improved patient care, teaching, and research in general pediatrics.
Requirements: The principal investigator must be a member of the APA or have submitted an application for membership. Preference will be given to new investigators, including those in training.
Restrictions: Multiple year funding requests will not be considered.
Amount of Grant: $10,000 maximum
Date(s) Application is Due: Oct 25
Contact: Grants Administrator, (703) 556-9222; fax: (703) 556-8729; email: info@academicpeds.org
Sponsor: Ambulatory Pediatric Association
6728 Old McLean Village Drive
McLean, VA 22101

APF Call for Proposals 919
The program is intended to fund public education programs, as well as information and outreach initiatives that promote the early recognition and treatment of mental illness. Up to $750,000 in grant funds are available over the course of three years. Average grants are in the $50,000 range. Grant making under this program has been temporarily suspended while a new Call for Proposals is being developed. Contact the office for further updates.
Requirements: Organizations that have been in existence for at least two years and currently maintain a 501(c)3 charitable status and American Psychiatric Association District Branches and subsidiaries. Organizations need not be mental health programs.
Amount of Grant: $50,000 average

Contact: Barbara Matos, Director; (703) 907-8517; fax: (703) 907-7851; email: bmatos@psych.org
Internet: http://www.psychfoundation.org/GrantAndAwards/Grants/CallforProposals.aspx
Sponsor: American Psychiatric Foundation
1000 Wilson Boulevard, Suite 1825
Arlington, VA 22209-3901

APF James H. Scully Jr., M.D., Educational Fund Grants 920
The fund supports education and research within the APA, its subsidiary organizations, or affiliated organizations, including district branches, state associations, or other psychiatric organizations that are in partnership with the APA. The fund can be used to support activities that improve the quality of care for patients with mental illness, advance the prevention of mental illness, raise awareness of mental illness and its treatments, and increase access to mental health services. Requests for funding should be sent in writing to the foundation and include the following: identification and significance of need; description of program/activity to address that need; expected outcomes and evaluation process; budget and time line; and background of key project personnel. Funding decisions are made by the foundations board of directors.
Restrictions: The fund will not support public policy or lobbying activities.
Contact: Barbara Matos, Director; (703) 907-8517; fax: (703) 907-7851; email: bmatos@psych.org
Internet: http://www.psychfoundation.org/GrantAndAwards/Grants/JamesHScullyEducationalFund.aspx
Sponsor: American Psychiatric Foundation
1000 Wilson Boulevard, Suite 1825
Arlington, VA 22209-3901

APF Jeanne Spurlock Congressional Fellowships 921
The fellowship provides general psychiatry and child psychiatry residents an opportunity to work in a congressional office or committee on federal health policy, particularly policy related to child and minority issues. The recipient will serve a ten-month fellowship in Washington, DC. The fellow will be introduced to the structure and development of federal and congressional health policy procedures, with a focus on mental health issues affecting minorities and under-served populations, including children. Fellows traditionally help develop legislative proposals, track and analyze legislative initiatives, arrange hearings, and brief Members of Congress and their staff, and interact with their constituents. During the fellowship, recipients have opportunities to interact with health policymakers and advocacy/professional groups, including the APA.
Requirements: PGY-II and III general psychiatry residents, child psychiatry residents, or child psychiatrists who will be out of training for less than one year at the time of the fellowship may apply. Applicants must be U.S. citizens or permanent residents. The recipient will be required to submit a written summary of the Fellowship experience at the end of the fellowship and may make recommendations or suggestions for improving the fellowship.
Amount of Grant: $35,000 stipend; $2,500 maximum moving expenses
Date(s) Application is Due: Mar 14
Contact: Marilyn King; (703) 907-8653; email: mking@psych.org
Internet: http://www.psychfoundation.org/GrantAndAwards/AwardsandFellowships/SpurlockFellowship.aspx
Sponsor: American Psychiatric Foundation
1000 Wilson Boulevard, Suite 1825
Arlington, VA 22209-3901

APHA Fellowship in Printing History 922
The fellowship is awarded for research in any area of the history of printing in all its forms, including all the arts and technologies relevant to printing, the book arts, and letter forms. The subject of research has no geographical or chronological limitations, and may be national or regional in scope, biographical, analytical, technical, or bibliographical in nature. Printing history-related study with a recognized printer or book artist may also be supported. The fellowship can be used to pay for travel, living, and other expenses.
Samples: Lance Hidy—to support research on Boston's Society of Printers, $2000 (2005); Dr. Susanna Ashton—for work on William Stanley Braithwaite (2004).
Amount of Grant: $2000 maximum
Date(s) Application is Due: Dec 10
Contact: Stephen Crook, Executive Secretary, email: sgcrook@printinghistory.org
Internet: http://printinghistory.org/htm/fellowship/2005.html
Sponsor: American Printing History Association
P.O. Box 4519, Grand Central Sta
New York, NY 10163-4519

APhA Foundation Incentive Grants 923
The incentive grants provide seed funding to help pharmacists in all practice settings explore new methods and services that enhance their role as pharmaceutical care providers, and to encourage them to share their experiences with other pharmacists. This year, the incentive grants program will continue to expand to include project proposals specifically for pharmacists interested in establishing or supporting an already existing practice model in the area of pain management. Projects should concentrate on new, innovative patient care service that is of significance to pharmacy care settings and that can be evaluated for its relevance. Application and guidelines are available online.
Requirements: Applicants must be members of APhA, currently licensed, and actively engaged in ambulatory pharmacy practice. Recipients are encouraged to attend the APhA annual meeting and submit present findings, if applicable.

Restrictions: Indirect costs will not be funded. Salaries for recipients are not funded; however, consultant fees, fees for reseach/technician services, and expenses for adminsitrative services may be allowable if essential if essential to development/ implementation of the project.
Amount of Grant: $1000
Date(s) Application is Due: Oct 3
Contact: Grants Administrator, email: info@aphafoundation.org
Internet: http://www.aphafoundation.org/IncentiveGrants/incentive.htm
Sponsor: American Pharmacists Association
2215 Constitution Avenue NW
Washington, DC 20037-2985

APHA Youth Development Foundation Grants 924
The Youth Development Foundation is a nonprofit corporation established in 1980 by the American Paint Horse Association (APHA). The YDF provides grants for equine research on topics such as the lethal white syndrome in foals and the genetic basis of Paint Horse coat patterns. YDF also annually presents scholarships of up to $1000 per year to young horsemen and women who are pursuing their educational goals at a college or university. Deadline listed is for scholarships. Guidelines and application are available online.
Requirements: Scholarship applicants must be participants in the American Junior Paint Horse Association, demonstrate potential for continued achievement, and excel academically. Heavy consideration is also given to students with financial need. Contact the foundation for requirements for research grant applicants.
Restrictions: Children of the YDF board or the executive committee board are ineligible.
Amount of Grant: $1000 scholarships
Date(s) Application is Due: Mar 1
Contact: Program Contact, (817) 834-2742; fax: (817) 834-3152; email: askapha@apha.com
Internet: http://www.apha.com/ydf/ydfscholarships.html
Sponsor: American Paint Horse Association
P.O. Box 961023
Fort Worth, TX 76161-0023

Appalachian Regional Commission Asset-Based Development Grants 925
ARC's Asset-Based Development Initiative seeks to help communities identify and leverage local assets to create jobs and build prosperity while preserving the character of their community. Development strategies include: capitalizing on traditional arts, culture, and heritage; leveraging ecological assets for outdoor sports such as fishing, camping, white-water rafting, and rock climbing; adding value to farming through specialized agricultural development, including processing specialty food items, fish farming, and organic farming; getting the most from hardwood forests by maximizing sustainable timber harvesting and value-added processing; encouraging the development of local leadership and civic entrepreneurs; and converting overlooked and underused facilities into industrial parks, business incubators, or educational facilities. The Commission has allocated $3 million annually for regional initiatives.
Requirements: States, and through states, public bodies and private nonprofit organizations are eligible to apply.
Contact: Jim Stokoe, Land-of-Sky Regional Council; (828) 251-6622 or (202) 884-7799; fax: (202) 884-7691; email: jim@landofsky.org or crea@arc.gov
Internet: http://www.arc.gov/index.do?nodeId=17
Sponsor: Appalachian Regional Commission
1666 Connecticut Avenue, NW, Suite 700
Washington, DC 20009-1068

Appalachian Regional Commission Grants 926
ARC awards program grants for projects that further the four goals identified by the Commission in its strategic plan. Interest Areas include infrastructure, demonstration projects, economic development, education, health, and community capacity. Typically, ARC grants are awarded to state and local agencies and governmental entities (such as economic development authorities), local governing boards (such as county councils), and nonprofit organizations (such as schools and organizations that build low-cost housing). Program grants are made to nonprofit organizations. Most program grants originate at the state level. Potential applicants should contact their state ARC program manager to request a pre-application package. The grant cycle is open, and funds become available in October.
Requirements: States, and through states, public bodies and private nonprofit organizations are eligible to apply.
Contact: Jim Stokoe, Land-of-Sky Regional Council; (828) 251-6622 or (202) 884-7799; fax: (202) 884-7691; email: jim@landofsky.org or crea@arc.gov
Internet: http://www.arc.gov/index.do?nodeId=101
Sponsor: Appalachian Regional Commission
1666 Connecticut Avenue, NW, Suite 700
Washington, DC 20009-1068

Applied Biosystems Grants 927
The company is committed to the communities where its employees live and work. Financial contributions as well as in-kind and product donations are awarded to nonprofit organizations. The focus is on education, health and human services, community outreach programs, environment, arts & culture, and civic. Precollege, college, graduate, and other unique programs that use technology to improve science, math, and engineering education are education programs supported. Organizations providing access to quality health care facilities and innovative solutions to chronic problems through disease and therapeutic research are supported. Outreach programs and civic groups that improve communities with innovative science, technology, and education are supported. Programs maintaining a healthy environment, arts and culture, and civic programs are also supported. Requests are accepted at any time.
Requirements: Nonprofit organizations in communities where Applied Biosystems has operations are eligible.
Contact: Applied Biosystems, (800) 327-3002 or (650) 638-5800; fax: (650) 638-5998
Internet: http://www.appliedbiosystems.com/about/community.cfm
Sponsor: Applied Biosystems
850 Lincoln Centre Drive
Foster City, CA 94404

Appraisal Institute Education Trust Scholarships 928
Graduate and undergraduate scholarships are offered to students majoring in programs in land economics, real estate, real estate appraising, or allied fields. The scholarships are awarded on the basis of scholarly qualities and are intended to help finance the cost of college work leading to a degree in one of the aforementioned professional fields. The institute also awards scholarships to minority and women college students pursuing academic degrees in real estate appraisal or related fields. Application forms must be obtained from the institute.
Requirements: Applicants must be sophomores, juniors, seniors, or graduates, as well as U.S. citizens.
Amount of Grant: $2000 undergraduate; $3000 graduate
Date(s) Application is Due: Mar 15
Contact: Olivia Carreon, Project Coordinator, (312) 335-4100; fax: (312) 335-4400; email: ocarreon@appraisalinstitute.org
Internet: http://www.appraisalinstitute.org/education/scolarshp.asp
Sponsor: Appraisal Institute Education Trust
550 W Van Buren Street, Suite 1000
Chicago, IL 60607

APS Max Delbruck Prize in Biological Physics 929
The Max Delbruck Prize in Biological Physics is intended to recognize and encourage outstanding achievement in biological physics research. The prize consists of $10,000, an allowance for travel to attend the meeting at which the prize is awarded, and a certificate citing the contributions made by the recipient or recipients. It is presented biennially in even-numbered years. Nominations are open to scientists of all nationalities regardless of the geographical site at which the work was done. The prize may be awarded to more than one investigator on a shared basis. Nominations are active for three cycles.
Amount of Grant: $10,000
Date(s) Application is Due: Jul 1
Contact: Jacquelyn Beamon-Kiene, Executive Office Administrator; (301)209-3269 or (301) 209-3268; fax: (301) 209-0865; email: beamon@aps.org or honors@aps.org
Internet: http://www.aps.org/programs/honors/prizes/delbruck.cfm
Sponsor: American Physical Society
One Physics Ellipse
College Park, MD 20740-3844

APS Abraham Pais Prize for History of Physics 930
The American Physical Society (APS) through its Forum on the History of Physics (FHP) and the American Institute of Physics (AIP) through its Center for History of Physics (CHP) established the prize in 2005. Major sponsors include Richard Lounsbery Foundation, John & Elizabeth Armstrong and Virginia Trimble. The Prize is intended to recognize outstanding scholarly achievements in the history of physics. Outstanding scholarly achievements are reflected primarily in the publication of one or more scholarly books or a series of scholarly articles on any topic or chronological period in the history of physics. These can include significant editions of works of physicists that display outstanding editorial scholarship; other edited books or collections of previously published articles will be considered but given less weight. Further outstanding professional achievements such as teaching, lecturing, editing of journals, organization of conferences, and productions on the Web or in other electronic media may be considered if they have contributed significantly to fostering excellence in the history of physics. The Award usually is given to a single person but to no more than three individuals. The prize is given annually and will consist of $10,000 and a certificate citing the contributions of the recipient, plus an allowance for travel to an APS meeting to receive the award and deliver an invited lecture on the history of physics.
Amount of Grant: $10,000
Date(s) Application is Due: Jul 1
Contact: Jacquelyn Beamon-Kiene, Executive Office Administrator; (301)209-3269 or (301) 209-3268; fax: (301) 209-0865; email: beamon@aps.org or honors@aps.org
Internet: http://www.aps.org/programs/honors/prizes/pais-guidelines.cfm
Sponsor: American Physical Society
One Physics Ellipse
College Park, MD 20740-3844

APS Andreas Acrivos Dissertation Award in Fluid Dynamics 931
The program provides recognition to exceptional young scientists who have performed original doctoral thesis work of outstanding scientific quality and achievement in the area of fluid dynamics. The annual award consists of a monetary prize, a certificate citing the accomplishments of the recipient, and an allowance for travel to attend the annual meeting of the Division of Fluid Dynamics at which the award will be presented. Each nominee will be considered in not more than two consecutive cycles.

Requirements: Nominations will be accepted for any doctoral student studying at a college or university in the United States or in a education abroad program of a college or university in the United States. The work to be considered must have been accomplished as part of the requirements for a doctoral degree. Nominees must have completed their dissertations during the previous calendar year, or by March 31 of the year of the award, at a university within the United States.
Amount of Grant: $1,000 award; $1,500 maximum travel allowance
Date(s) Application is Due: May 1
Contact: Jacquelyn Beamon-Kiene, Executive Office Administrator; (301)209-3269 or (301) 209-3268; fax: (301) 209-0865; email: beamon@aps.org or honors@aps.org
Internet: http://www.aps.org/programs/honors/dissertation/acrivos.cfm
Sponsor: American Physical Society
One Physics Ellipse
College Park, MD 20740-3844

APS Andrei Sakharov Prize   932

The Sakharov Prize is named in recognition of the courageous and effective work of Andrei Sakharov on behalf of human rights, to the detriment of his own scientific career and despite the loss of his own personal freedom. The Prize is endowed by contributions from friends of Andrei Sakharov. It is intended to recognize outstanding leadership and/or achievements of scientists in upholding human rights. The Prize consists of $10,000 and an allowance for travel to the meeting of the Society at which the Prize will be presented. It is intended that the Prize be awarded every other year, at a general meeting of the American Physical Society.
Samples: Liangying Xu, Chinese Academy of Sciences, $10,000 (2008); Yuri Orlov, Cornell University, $10,000 (2006).
Amount of Grant: $10,000
Date(s) Application is Due: Jul 1
Contact: Jacquelyn Beamon-Kiene, Executive Office Administrator; (301)209-3269 or (301) 209-3268; fax: (301) 209-0865; email: beamon@aps.org or honors@aps.org
Internet: http://www.aps.org/programs/honors/prizes/sakharov.cfm
Sponsor: American Physical Society
One Physics Ellipse
College Park, MD 20740-3844

APS Aneesur Rahman Prize for Computational Physics   933

The annual prize recognizes and encourages outstanding achievement in computational physics research. The award consists of a monetary prize of $5,000, an allowance for travel to the meeting of the society at which the prize is awarded and at which the recipient will deliver the Rahman Lecture, and a certificate citing the contributions made by the recipient. The prize ordinarily is awarded to one person, but a prize may be shared when all recipients have contributed to the same accomplishments. Nominations are active for three years. Nomination guidelines are available online.
Requirements: Nominations are open to scientists of all nationalities regardless of the geographical site at which the work was done.
Samples: Gary S. Grest, Sandia National Laboratories, $5,000 (2008); Daniel Frenkel, FOM Institute for Atomic and Molecular Physics, $5,000 (2007); David Vanderbilt, Rutgers University, $5,000 (2006).
Amount of Grant: $5,000
Date(s) Application is Due: Jul 1
Contact: Jacquelyn Beamon-Kiene, Executive Office Administrator; (301) 209-3269 or (301) 209-3268; fax: (301) 209-0865; email: beamon@aps.org or honors@aps.org
Internet: http://www.aps.org/programs/honors/prizes/rahman.cfm
Sponsor: American Physical Society
One Physics Ellipse
College Park, MD 20740-3844

APS Arthur L. Schawlow Prize in Laser Science   934

The prize recognizes outstanding contributions to basic research that uses lasers to advance knowledge of the fundamental physical properties of materials and their interaction with light. Some examples of relevant areas of research are: nonlinear optics, ultrafast phenomena, laser spectroscopy, squeezed states, quantum optics, multiphoton physics, laser cooling and trapping, physics of lasers, particle acceleration by lasers, and short wavelength lasers. The award consists of a $10,000 prize plus an allowance for travel to the meeting at which the prize is awarded and a certificate citing the contributions made by the recipient. Nominations are active for three years.
Requirements: Nominations are open to candidates who have made outstanding contributions to basic research using lasers.
Amount of Grant: $10,000
Date(s) Application is Due: Jul 1
Contact: Jacquelyn Beamon-Kiene, Executive Office Administrator; (301)209-3269 or (301) 209-3268; fax: (301) 209-0865; email: beamon@aps.org or honors@aps.org
Internet: http://www.aps.org/programs/honors/prizes/schawlow.cfm
Sponsor: American Physical Society
One Physics Ellipse
College Park, MD 20740-3844

APS Dannie Heineman Prize for Mathematical Physics   935

The annual prize recognizes outstanding publications in the field of mathematical physics. The award consists of a $10,000 prize and a certificate citing the contributions made by the recipient, plus travel expenses to attend the meeting at which the prize is bestowed. The prize may be awarded to more than one person on a shared basis when all recipients have contributed to the same accomplishments. Nominations are active for three years. Nomination guidelines are available online.
Requirements: This prize is awarded solely for valuable published contributions made in the field of mathematical physics with no restrictions placed on a candidate's citizenship or country of residence. Publication is defined as either a single paper, a series of papers, a book, or any other communication that can be considered a publication.
Samples: Mitchell Feigenbaum, Rockefeller University—prize recipient, $7,500 (2008); Joseph Polchinski and Juan Maldacena—prize recipients, $7,500 (2007); Sergio Ferrara, Daniel Freedman and Peter van Nieuwenhuizen—prize recipients, $7,500 (2006); Giorgio Parisi—award recipient, $7,500 (2005).
Amount of Grant: $10,000
Date(s) Application is Due: Jul 1
Contact: Jacquelyn Beamon-Kiene, Executive Office Administrator; (301) 209-3269 or (301) 209-3268; fax: (301) 209-0865; email: beamon@aps.org or honors@aps.org
Internet: http://www.aps.org/programs/honors/prizes/heineman.cfm
Sponsor: American Physical Society
One Physics Ellipse
College Park, MD 20740-3844

APS Davisson-Germer Prize in Atomic or Surface Physics   936

The prize recognizes and encourages outstanding work in atomic physics or surface physics. The award consists of a $5,000 prize and a certificate citing the contributions made by the recipient or recipients. This prize will normally be awarded in even-numbered years for outstanding work in atomic physics and odd-numbered years for outstanding work in surface physics. The prize ordinarily is awarded to one person, but may be shared when all recipients have contributed to the same accomplishments. Nominations are active for three cycles. Nomination guidelines are available online.
Requirements: Nominations are open to scientists of all nationalities regardless of the geographical site at which the work was done.
Samples: Horst Schmidt-Bocking, University of Frankfurt—award recipient, $5,000 (2008); Franz Himpsel—award recipient, $5,000 (2007); C. Lewis Cocke—award recipient, $5,000 (2006); Ernst Bauer—award recipient, $5,000 (2005).
Amount of Grant: $5,000
Date(s) Application is Due: Jul 1
Contact: Jacquelyn Beamon-Kiene, Executive Office Administrator; (301)209-3269 or (301) 209-3268; fax: (301) 209-0865; email: beamon@aps.org or honors@aps.org
Internet: http://www.aps.org/programs/honors/prizes/davisson-germer.cfm
Sponsor: American Physical Society
One Physics Ellipse
College Park, MD 20740-3844

APS Dissertation Award in Nuclear Physics   937

The program recognizes a recent PhD in nuclear physics. The annual award consists of a $2,500 stipend and an allowance for travel to the annual Spring Meeting of the Division of Nuclear Physics of the American Physical Society, where the award will be presented. Nomination guidelines are available online.
Requirements: Nominations are open to any person who has received a PhD degree in experimental or theoretical nuclear physics from a North American university within the two-year period preceding September 1 of the current year.
Samples: Deepshikha Choudhury, Ohio University and Nikolai Tolich, Stanford University—prize recipients, $2,500 (2008); Kathryn K.S. Miknaitis and Magdalena Djordjevic——prize recipients, $2,500 (2007).
Amount of Grant: $2,500
Date(s) Application is Due: Jul 1
Contact: Jacquelyn Beamon-Kiene, Executive Office Administrator; (301)209-3269 or (301) 209-3268; fax: (301) 209-0865; email: beamon@aps.org or honors@aps.org
Internet: http://www.aps.org/programs/honors/dissertation/nuclear.cfm
Sponsor: American Physical Society
One Physics Ellipse
College Park, MD 20740-3844

APS Earle K. Plyler Prize for Molecular Spectroscopy   938

The annual prize recognizes and encourages notable contributions to the field of molecular spectroscopy. The prize may be given for experimental or theoretical achievements, for a single dramatic innovation, or for a series of research contributions which, when integrated, amounts to a major contribution to the field of molecular spectroscopy. The award consists of a $10,000 prize, an allowance for travel expenses to attend the meeting at which the prize is to be presented, and a certificate citing the contributions made by the recipient. The prize ordinarily is awarded to one person, but a prize may be shared when all the recipients have contributed to the same accomplishments. Nominations are active for three years, and guidelines are available online.
Requirements: Nominations are open to scientists in North America.
Samples: Steven G. Boxer, Stanford University—prize recipient, $10,000 (2008); Timothy S. Zwier, Purdue University—prize recipient, $10,000 (2007); Mark Johnson—prize recipient, $10,000 (2006); Robert Tycko—prize recipient, $10,000 (2005).
Amount of Grant: $10,000
Date(s) Application is Due: Jul 1
Contact: Jacquelyn Beamon-Kiene, Executive Office Administrator; (301)209-3269 or (301) 209-3268; fax: (301) 209-0865; email: beamon@aps.org or honors@aps.org
Internet: http://www.aps.org/programs/honors/prizes/plyler.cfm
Sponsor: American Physical Society
One Physics Ellipse
College Park, MD 20740-3844

APS Einstein Prize   939

The prize recognizes outstanding accomplishments in the field of gravitational physics. The award consists of a monetary prize and a certificate citing the contributions of the recipient. It also includes an allowance for the recipient to travel to a meeting of the society to receive the award and deliver a lecture. It will be awarded biennially. The award, usually to a single individual, is for outstanding achievement in theory, experiment, or observation in the area of gravitational physics. Nominations will remain active for three years. Nomination guidelines are available online.

Requirements: Scientists worldwide are eligible for nomination.

Restrictions: Members of the Topical Group on Gravitation Executive Committee are not eligible for nomination while sitting on the committee.

Samples: Rainer Weiss, Massachusetts Institute of Technology and Ronald Drever, California Institute of Technology—prize recipients, $10,000 (2007); Bryce DeWitt—prize recipient, $10,000 (2005); Peter G. Bergmann and John Wheeler—prize recipients, $10,000 (2003).

Amount of Grant: $10,000

Date(s) Application is Due: Jul 1

Contact: Jacquelyn Beamon-Kiene, Executive Office Administrator; (301)209-3269 or (301) 209-3268; fax: (301) 209-0865; email: beamon@aps.org or honors@aps.org

Internet: http://www.aps.org/programs/honors/prizes/einstein.cfm

Sponsor: American Physical Society
One Physics Ellipse
College Park, MD 20740-3844

APS Fluid Dynamics Prize   940

The annual prize recognizes and encourages outstanding achievement in fluid dynamics research. The award consists of $10,000, an allowance for travel to the meeting at which the prize is awarded, and a certificate citing the contributions made by the recipient. The prize ordinarily is awarded to one person but may be shared when all the recipients have contributed to the same achievement. Nominations are active for three years. Nomination guidelines are available online.

Samples: Julio M. Ottino, Northwestern University—prize recipient, $10,000 (2008); Guenter Ahlers, University of California, Santa Barbara—prize recipient, $10,000 (2007); Thomas S. Lundgren, University of Minnesota—prize recipient, $10,000 (2006).

Amount of Grant: $10,000

Date(s) Application is Due: Mar 31

Contact: Jacquelyn Beamon-Kiene, Executive Office Administrator; (301)209-3269 or (301) 209-3268; fax: (301) 209-0865; email: beamon@aps.org or honors@aps.org

Internet: http://www.aps.org/programs/honors/prizes/fluid.cfm

Sponsor: American Physical Society
One Physics Ellipse
College Park, MD 20740-3844

APS Frank Isakson Prize for Optical Effects in Solids   941

The prize recognizes outstanding optical research that leads to breakthroughs in the condensed matter sciences. The award consists of a $5,000 prize, an allowance for travel to the meeting of the society at which the prize is being presented, and a certificate citing the contributions made by the recipient. The prize is awarded biennially in even-numbered years as a memorial to Frank Isakson. The prize ordinarily is awarded to one person, but the prize may be shared among recipients when all recipients have contributed to the same accomplishment. Preference will be given to work that has been published within the past 10 years. Nominations are active for three years. Nomination guidelines are available online.

Requirements: Nominations are open to scientists of all nations regardless of the geographical site at which the work was done.

Samples: Joseph Orenstein, University of California, Berkeley and Zeev Valentine Vardeny, University of Utah—prize recipients, $5,000 (2008); Roberto Merlin, University of Michigan—prize recipient, $5,000 (2006); James Wolfe, University of Illinois, Urbana-Champaign—prize recipient, $5,000 (2004).

Amount of Grant: $5,000

Date(s) Application is Due: Jul 1

Contact: Jacquelyn Beamon-Kiene, Executive Office Administrator; (301)209-3269 or (301) 209-3268; fax: (301) 209-0865; email: beamon@aps.org or honors@aps.org

Internet: http://www.aps.org/programs/honors/prizes/isakson.cfm

Sponsor: American Physical Society
One Physics Ellipse
College Park, MD 20740-3844

APS George E. Pake Prize   942

The annual prize recognizes and encourages outstanding work by physicists combining original research accomplishments with leadership in the management of research or development in industry. The award consists of a $5,000 prize, an allowance for travel to the meeting at which the prize is to be awarded, and a certificate recognizing the contribution of the recipient. Nominations are active for three years, and guidelines are available online.

Requirements: This prize will be awarded to one individual for outstanding achievements in physics research combined with major success as a manager of research or development in industry.

Samples: Julia M. Phillips, Sandia National Laboratories—prize recipient, $5,000 (2008); Mark Kryder, Seagate Research—prize recipient, $5,000 (2007); Charles B. Duke, Xerox Innovation Group—prize recipient, $5,000 (2006).

Amount of Grant: $5,000

Date(s) Application is Due: Jul 1

Contact: Jacquelyn Beamon-Kiene, Executive Office Administrator; (301)209-3269 or (301) 209-3268; email: beamon@aps.org or honors@aps.org

Internet: http://www.aps.org/programs/honors/prizes/pake.cfm

Sponsor: American Physical Society
One Physics Ellipse
College Park, MD 20740-3844

APS George E. Valley, Jr., Prize   943

The prize recognizes one individual in the early stages of his or her career for an outstanding scientific contribution to physics that is deemed to have significant potential for a dramatic impact on the field. The award consists of a $20,000 award and a certificate citing the contribution made by the recipient. Nomination documents must include a statement from the nominator or from the candidate's department certifying the date of the candidate's PhD. The prize is awarded biennially. Nomination guidelines are available online.

Requirements: The nominee must have received his/her PhD no earlier than five years before April 1 of the year of the nomination deadline. Work done by a graduate student for his/her thesis is eligible for consideration if it is demonstrated that the student's contributions have been crucial to an important piece of research.

Samples: Irfan Siddiqi, University of California, Berkeley—prize recipient, $20,000 (2006); Ivo Souza, University of California, Berkeley—prize recipient, $20,000 (2004); David Goldhaber-Gordon, Stanford University—prize recipient, $20,000 (2002).

Amount of Grant: $20,000

Date(s) Application is Due: Jul 1

Contact: Jacquelyn Beamon-Kiene, Executive Office Administrator; (301)209-3269 or (301) 209-3268; fax: (301) 209-0865; email: beamon@aps.org or honors@aps.org

Internet: http://www.aps.org/programs/honors/prizes/valley.cfm

Sponsor: American Physical Society
One Physics Ellipse
College Park, MD 20740-3844

APS Hans A. Bethe Prize   944

To the prize recognizes outstanding work in theory, experiment, or observation in the areas of astrophysics, nuclear physics, nuclear astrophysics, or closely related fields. The prize consists of a $7,500 prize and a certificate citing the contributions made by the recipient. It is presented annually. Nominations are active for three years, and guidelines are available online.

Requirements: This award is made annually to one individual for outstanding accomplishments in the areas of astrophysics, nuclear physics, nuclear astrophysics, or closely related fields. It is open to any scientist working in these areas, worldwide.

Samples: Friedrich K. Thielemann, University of Basel—prize recipient, $7,500 (2008); James R. Wilson, Lawrence Livermore National Laboratory—prize recipient, $7,500 (2007); Alastair G.W. Cameron, Harvard University—prize recipient, $7,500 (2006).

Amount of Grant: $7,500

Date(s) Application is Due: Jul 1

Contact: Jacquelyn Beamon-Kiene, Executive Office Administrator; (301)209-3269 or (301) 209-3268; fax: (301) 209-0865; email: beamon@aps.org or honors@aps.org

Internet: http://www.aps.org/programs/honors/prizes/bethe.cfm

Sponsor: American Physical Society
One Physics Ellipse
College Park, MD 20740-3844

APS Herbert P. Broida Prize   945

The prize recognizes and enhances outstanding experimental advancements in the fields of atomic and molecular spectroscopy or chemical physics. The prize consists of a $5,000 prize, an allowance for travel to the award ceremony, and a certificate citing the contributions made by the recipient. Preference will be granted to an individual whose contributions have displayed a high degree of breadth, originality, and creativity. Nominations are active for three years, and guidelines are available online.

Requirements: The prize is awarded to one individual in recognition of an outstanding contribution to the field of atomic and molecular spectroscopy or chemical physics. Emphasis will be given to work done within the five years prior to the awarding of the prize.

Samples: James C. Bergquist, National Institute of Science & Technology—prize recipient, $5,000 (2007); Hanna Reisler, University of Southern California—prize recipient, $5,000 (2005); George W. Flynn, Columbia University—award recipient, $5,000 (2003).

Amount of Grant: $5,000

Date(s) Application is Due: Jul 1

Contact: Jacquelyn Beamon-Kiene, Executive Office Administrator; (301)209-3269 or (301) 209-3268; fax: (301) 209-0865; email: beamon@aps.org or honors@aps.org

Internet: http://www.aps.org/programs/honors/prizes/broida.cfm

Sponsor: American Physical Society
One Physics Ellipse
College Park, MD 20740-3844

APS I.I. Rabi Prize in Atomic, Molecular, and Optical Physics   946

The prize recognizes and encourages outstanding research in atomic, molecular, and optical physics. It will be awarded in odd-numbered years. The award consists of a $7,500 prize and a certificate citing the contributions made by the recipient. An allowance will be provided for travel expenses of the recipient to the society meeting at which the prize is presented. The prize ordinarily is awarded to one person but may be shared when all recipients have contributed to the same accomplishment. Nominations are active for three years, and guidelines are available online.

Requirements: Nominations are open to scientists of all nationalities regardless of the geographical location at which the work was done. Investigators who have held a PhD for 10 years or less are eligible.

Samples: Jun Ye, JILA—prize recipient, $7,500 (2005); Deborah Jin—prize recipient, $7,500 (2005); Mark Kasevich, Stanford University—prize recipient, $7,500 (2003).

Amount of Grant: $7,500

Date(s) Application is Due: Jul 1

Contact: Jacquelyn Beamon-Kiene, Executive Office Administrator; (301)209-3269 or (301) 209-3268; fax: (301) 209-0865; email: beamon@aps.org or honors@aps.org

Internet: http://www.aps.org/programs/honors/prizes/rabi.cfm

Sponsor: American Physical Society

One Physics Ellipse

College Park, MD 20740-3844

APS Irving Langmuir Prize in Chemical Physics   947

The prize recognizes and encourages outstanding interdisciplinary research in chemistry and physics. This biennial prize consists of a $10,000 award and a certificate citing the contributions made by the recipient. In even-numbered years, the American Chemical Society selects the prize recipient and presents the prize. In odd-numbered years, the American Physical Society selects the prize recipient and presents the prize. An allowance is provided for travel expenses of the recipient to the meeting of the society at which the prize is to be bestowed. The prize is made to one person who has made an outstanding contribution in the field of chemical physics or physical chemistry within the 10 years prior to the prize. Nominations are active for three years, and guidelines are available online.

Requirements: The prize is granted without restriction, except that the recipient must be a resident of the United States at the time of selection and the prize funds must be used in the United States or its possessions.

Samples: Gabor Somorjai, University of California, Berkeley—prize recipient, $10,000 (2007); David Chandler—prize recipient, $10,000 (2005); Phaedon Avouris—award recipient, $10,000 (2003).

Amount of Grant: $10,000

Date(s) Application is Due: Jul 1

Contact: Jacquelyn Beamon-Kiene, Executive Office Administrator; (301)209-3269 or (301) 209-3268; fax: (301) 209-0865; email: beamon@aps.org or honors@aps.org

Internet: http://www.aps.org/programs/honors/prizes/langmuir.cfm

Sponsor: American Physical Society

One Physics Ellipse

College Park, MD 20740-3844

APS J. J. Sakurai Prize for Theoretical Particle Physics   948

The annual prize recognizes and encourages outstanding achievement in particle theory. The award consists of a $5,000 prize, an allowance for travel to the meeting of the society at which the prize is to be awarded, and a certificate citing the contributions made by the recipient. The prize may be awarded to more than one person on a shared basis. Nominations are active for three years, and guidelines are available online.

Requirements: Nominations are open to scientists of all nationalities regardless of the geographical site at which the work was done. The prize will normally be awarded for theoretical contributions made at an early stage of the recipients research career.

Samples: Alexei Smirnov, The Abdus Salam ICTP and Stanislav Mikheyev, Russian Academy of Sciences—prize recipients, $5,000 (2008); Stanley Brodsky, Stanford Linear Accelerator Center—prize recipient, $5,000 (2007); Savas Dimopoulos, Stanford University—prize recipient, $5,000 (2006).

Amount of Grant: $5,000

Date(s) Application is Due: May 1

Contact: Jacquelyn Beamon-Kiene, Executive Office Administrator; (301)209-3269 or (301) 209-3268; fax: (301) 209-0865; email: beamon@aps.org or honors@aps.org

Internet: http://www.aps.org/programs/honors/prizes/sakurai.cfm

Sponsor: American Physical Society

One Physics Ellipse

College Park, MD 20740-3844

APS James C. McGroddy Prize for New Materials   949

The annual prize recognizes and encourages outstanding achievement in the science and application of new materials, including the discovery of new classes of materials, the observation of novel phenomena in known materials leading to both fundamentally new applications and scientific insights, as well as theoretical and experimental work contributing significantly to the understanding of such phenomena. The award consists of a $7,500 prize, plus a certificate citing the contribution of the recipient and an allowance for travel to the meeting of the society at which the award is presented. Nominations are active for three years, and guidelines are available online.

Requirements: The prize is open to scientists of all nationalities irrespective of where their work has been carried out.

Samples: Arthur F. Hebard, University of Florida, Jun Akimitsu, Aoyama-Gakuin University and Robert C. Haddon, University of California, Riverside—prize recipients, $5,000 (2008); Arthur J. Epstein, Ohio State University and Joel S. Miller, University of Utah—prize recipients, $5,000 (2007); Hongjie Dai and Alex Zettl—prize recipients, $5,000 (2006); Yoshinori Tokura—prize recipient, $5,000 (2005).

Amount of Grant: $5,000

Date(s) Application is Due: Jul 1

Contact: Jacquelyn Beamon-Kiene, Executive Office Administrator; (301)209-3269 or (301) 209-3268; fax: (301) 209-0865; email: beamon@aps.org or honors@aps.org

Internet: http://www.aps.org/programs/honors/prizes/mcgroddy.cfm

Sponsor: American Physical Society

One Physics Ellipse

College Park, MD 20740-3844

APS James Clerk Maxwell Prize for Plasma Physics   950

The annual prize recognizes outstanding contributions to the field of plasma physics. The prize is made for outstanding contributions to the advancement and diffusion of the knowledge of properties of highly ionized gases of natural or laboratory origin. The award consists of a $5,000 award and a certificate citing the contributions made by the recipient. A travel allowance to attend the meeting at which the prize is to be presented is also provided. The prize ordinarily is awarded to one person, but a prize may be shared when all the recipients have contributed to the same accomplishments. Nominations are active for three years, and guidelines are available online.

Samples: Ronald C. Davidson, Princeton Plasma Physics Laboratory, $5,000 (2008); John Lindl, Lawrence Livermore National Laboratory, $5,000 (2007); Chandrashekhar J. Joshi, University of California, Los Angeles, $5,000 (2006).

Amount of Grant: $5,000

Date(s) Application is Due: Apr 1

Contact: Jacquelyn Beamon-Kiene, Executive Office Administrator; (301)209-3269 or (301) 209-3268; fax: (301) 209-0865; email: beamon@aps.org or honors@aps.org

Internet: http://www.aps.org/programs/honors/prizes/maxwell.cfm

Sponsor: American Physical Society

One Physics Ellipse

College Park, MD 20740-3844

APS Julius Edgar Lilienfeld Prize   951

The annual prize recognizes a most outstanding contribution to physics. The award consists of a $10,000 prize, a certificate citing the contributions made by the recipient, plus expenses for the three lectures by the recipient given at an APS meeting, a research university, and a predominantly undergraduate institution. Nominations are active for three years, and guidelines are available online.

Requirements: The prize is awarded for outstanding contributions to physics by a single individual who also has exceptional skills in lecturing to diverse audiences.

Samples: H. Eugene Stanley, Boston University, $10,000 (2008); Lisa Randall, Harvard University, $10,000 (2007); Mikhail Shifman, University of Minnesota, $10,000 (2006).

Amount of Grant: $10,000

Date(s) Application is Due: Jul 1

Contact: Jacquelyn Beamon-Kiene, Executive Office Administrator; (301)209-3269 or (301) 209-3268; fax: (301) 209-0865; email: beamon@aps.org or honors@aps.org

Internet: http://www.aps.org/programs/honors/prizes/lilienfeld.cfm

Sponsor: American Physical Society

One Physics Ellipse

College Park, MD 20740-3844

APS Lars Onsager Prize   952

The prize recognizes outstanding research in theoretical statistical physics including the quantum fluids. The prize is open to researchers in statistical physics covering a wide range of physical phenomena, as long as the nominee is active at the time of the award. The award consists of a $15,000 prize as well as a certificate citing the contribution made by the recipient. Nominations are active for three years, and guidelines are available online.

Requirements: Nominations are accepted from all scientists of all nations regardless of geographical location.

Samples: Christopher Pethick, NORDITA, Gordon Baym, University of Illinois and Tin-Lun Ho, Ohio State University, $15,000 (2008); A. Brooks Harris, University of Pennsylvania, $15,000 (2007); Rodney J. Baxter, Australian National University, $15,000 (2006).

Amount of Grant: $15,000

Date(s) Application is Due: Jul 1

Contact: Jacquelyn Beamon-Kiene, Executive Office Administrator; (301)209-3269 or (301) 209-3268; fax: (301) 209-0865; email: beamon@aps.org or honors@aps.org

Internet: http://www.aps.org/programs/honors/prizes/onsager.cfm

Sponsor: American Physical Society

One Physics Ellipse

College Park, MD 20740-3844

APS Marshall N. Rosenbluth Outstanding Doctoral Thesis Award   953

The program recognizes exceptional young scientists who have performed original thesis work of outstanding scientific quality and achievement in the area of plasma physics. The award consists of a $2,000 stipend, a certificate to be presented during the award ceremony at the banquet for the Division of Plasma Physics Annual Meeting, and a $500 travel allowance to the meeting. Nomination guidelines are available online.

Requirements: Nominations will be accepted for any doctoral student of a college or university in the United States or for a United States student abroad who has successfully passed his/her final thesis defense within the preceding 24 months of the current nomination deadline. The work to be considered must have been performed as part of the requirements for a doctoral degree.

Amount of Grant: $2,000; $500 maximum travel allowance

Date(s) Application is Due: Apr 1

Contact: Jacquelyn Beamon-Kiene, Executive Office Administrator; (301)209-3269 or (301) 209-3268; fax: (301) 209-0865; email: beamon@aps.org or honors@aps.org
Internet: http://www.aps.org/programs/honors/dissertation/rosenbluth.cfm
Sponsor: American Physical Society
One Physics Ellipse
College Park, MD 20740-3844

APS Mitsuyoshi Tanaka Dissertation Award in Experimental Particle Physics 954
The award provides recognition to exceptional young scientists who have performed original doctoral thesis work of outstanding scientific quality and achievement in the area of experimental particle physics. The annual award consists of a $1,500 award, a certificate citing the accomplishments of the recipient, and an allowance for travel to attend the annual meeting of the Division of Particles and Fields, at which the award will be presented. Each nominee will be considered in not more than two consecutive cycles, and guidelines are available online.
Requirements: Nominations will be accepted for any doctoral student studying at a college or university in the United States or in an education abroad program of a college or university in the United States for dissertation research carried out at a U.S. laboratory. The work to be considered must have been accomplished as part of the requirements for a doctoral degree. Nominees must pass their thesis defense 12 months or less from the nomination deadline, for the selection cycle in which the nomination is to be considered.
Samples: Jedrzej Biesiada, Princeton University, $1,500 (2008); Jean-Francois Arguin, Lawrence Berkeley National Laboratory, $1,500 (2007).
Amount of Grant: $1,500; $1,000 maximum travel allowance
Date(s) Application is Due: Jun 30
Contact: Jacquelyn Beamon-Kiene, Executive Office Administrator; (301)209-3269 or (301) 209-3268; fax: (301) 209-0865; email: beamon@aps.org or honors@aps.org
Internet: http://www.aps.org/programs/honors/dissertation/tanaka.cfm
Sponsor: American Physical Society
One Physics Ellipse
College Park, MD 20740-3844

APS Nicholas Metropolis Award for Outstanding Doctoral Thesis Work in Computational Physics 955
The purpose of the annual award is to recognize doctoral thesis research of outstanding quality and achievement in computational physics and to encourage effective written and oral presentation of research results. The award consists of a $1,500 prize and a certificate to be presented at an awards ceremony at the Division of Computational Physics annual meeting and an additional allowance to travel to the meeting. The recipient will be invited to present his or her work in an appropriate session of the meeting. An individual can be nominated only once; however, an unsuccessful candidate will be carried over for one year. Nomination guidelines are available online.
Requirements: Nominations will be accepted for any doctoral student (present or past) in any country for work performed as part of the requirements for a doctoral degree. Nominees must pass their thesis defense not more than 18 months before the deadline.
Samples: Soon Yong Chang, University of Illinois, Urbana-Champaign, $1,500 (2008); Chengkun Huang, University of California, Los Angeles, $1,500 (2007).
Amount of Grant: $1,500; $500 maximum travel allowance
Date(s) Application is Due: Nov 1
Contact: Jorge Jose, Committee Chair; email: jjose@buffalo.edu; Jacquelyn Beamon-Kiene, Executive Office Administrator; (301)209-3269 or (301) 209-3268; fax: (301) 209-0865; email: beamon@aps.org or honors@aps.org
Internet: http://www.aps.org/programs/honors/dissertation/metropolis.cfm
Sponsor: American Physical Society
One Physics Ellipse
College Park, MD 20740-3844

APS Oliver E. Buckley Condensed Matter Prize 956
The annual prize recognizes and encourages outstanding theoretical or experimental contributions to condensed matter physics. The prize ordinarily is awarded to one person, but may be shared when all the recipients have contributed to the same accomplishments. The award consists of a $10,000 prize and a certificate citing the contributions made by the recipient or recipients. Nominations are active for three years, and guidelines are available online.
Samples: Mildred Dresselhaus, Massachusetts Institute of Technology, $10,000 (2008); Allan H. MacDonald, James P. Eisenstein and Steven M. Girvin—prize recipients, $10,000 (2007); Noel A. Clark and Robert Meyer—prize recipients, $10,000 (2006).
Amount of Grant: $10,000
Date(s) Application is Due: Jul 1
Contact: Jacquelyn Beamon-Kiene, Executive Office Administrator; (301)209-3269 or (301) 209-3268; fax: (301) 209-0865; email: beamon@aps.org or honors@aps.org
Internet: http://www.aps.org/programs/honors/prizes/buckley.cfm
Sponsor: American Physical Society
One Physics Ellipse
College Park, MD 20740-3844

APS Outstanding Doctoral Thesis Research in Atomic, Molecular, or Optical Physics 957
The annual award recognizes doctoral thesis research of outstanding quality and achievement in atomic, molecular, or optical physics and encourages effective written and oral presentation of research results. The award consists of a $2,500 prize and a certificate citing the contribution made by the recipient. All finalists will receive a travel stipend. A student may be a finalist in the competition only once. Eligible non-finalists may only be renominated by submitting an entirely new package, even if it is the same as the original package. Nomination guidelines are available online.
Requirements: Doctoral students at any university in the United States or abroad who have passed their thesis defense for the PhD in the disciplines of atomic, molecular, or optical physics any time during the prior two calendar years are eligible for the award, except for those whose thesis advisors serve on the current selection committee.
Samples: David Moehring, Max Planck Institute for Quantum Optics, $2,500 (2008); Cindy Regal, University of Colorado, Boulder, $2,500 (2007).
Amount of Grant: $2,500 award; $500 travel allowance
Date(s) Application is Due: Dec 7
Contact: Alexander Cronin, Committee Chair; email: cronin@physics.arizona.edu; Jacquelyn Beamon-Kiene, Executive Office Administrator; (301)209-3269 or (301) 209-3268; fax: (301) 209-0865; email: beamon@aps.org or honors@aps.org
Internet: http://www.aps.org/programs/honors/dissertation/amo.cfm
Sponsor: American Physical Society
One Physics Ellipse
College Park, MD 20740-3844

APS Outstanding Doctoral Thesis Research in Beam Physics Award 958
The Award is intended to recognize doctoral thesis research of outstanding quality and achievement in beam physics and engineering. The annual award consists of $2,500 and a certificate to be presented at an awards ceremony at the Division of Physics of Beams Annual Meeting, and an additional allowance of up to $500 for travel to the meeting. The recipient will be invited to present the work as an invited talk in an appropriate session of the meeting.
Requirements: A nomination will be accepted for any doctoral student of a university world-wide for work performed as part of the requirements for a doctoral degree. Nominees must pass their thesis defense not more than 18 months before the nomination deadline.
Samples: Jeroen van Tilborg, Technische Universiteit Eindhoven and Lawrence Berkeley National Laboratory, $2,500 (2008); Evgenya I. Smirnova, Massachusetts Institute of Technology, $2,500 (2007).
Amount of Grant: $2,500 award; $500 travel allowance
Date(s) Application is Due: Sep 30
Contact: Gennady Shvets, Committee Chair; (512) 471-7371; fax: (512) 471-6715; email: gena@physics.utexas.edu; Jacquelyn Beamon-Kiene, Executive Office Administrator; (301)209-3269 or (301) 209-3268; fax: (301) 209-0865; email: beamon@aps.org or honors@aps.org
Internet: http://www.aps.org/programs/honors/dissertation/beamphysics.cfm
Sponsor: American Physical Society
One Physics Ellipse
College Park, MD 20740-3844

APS Polymer Physics Prize 959
The annual prize recognizes outstanding accomplishment and excellence of contributions in polymer physics research. The award consists of a $10,000 prize and a certificate citing the contributions made by the recipient. The prize ordinarily is awarded to one person, but a prize may be shared when all the recipients have contributed to the same accomplishments. Nominations are active for three years, and guidelines are available online.
Requirements: Nominations are open to all scientists of all nations regardless of membership in the society or the geographical location in which the work was carried out.
Samples: Kenneth S. Schweizer, University of Illinois—award recipient, $10,000 (2008); Glenn Fredrickson, University of California, Santa Barbara—award recipient, $10,000 (2007); Ludwik Leibler—award recipient, $10,000 (2006); Thomas Paul Russell—award recipient, $10,000 (2005).
Amount of Grant: $10,000
Date(s) Application is Due: Jul 1
Contact: Jacquelyn Beamon-Kiene, Executive Office Administrator; (301)209-3269 or (301) 209-3268; fax: (301) 209-0865; email: beamon@aps.org or honors@aps.org
Internet: http://www.aps.org/programs/honors/prizes/polymer.cfm
Sponsor: American Physical Society
One Physics Ellipse
College Park, MD 20740-3844

APS Prize for a Faculty Member for Research in an Undergraduate Institution 960
The annual prize honors a physicist whose research in an undergraduate setting has achieved wide recognition and contributed significantly to physics and who has contributed substantially to the professional development of undergraduate physics students. The award consists of a $5,000 stipend to the prize recipient and a separate unrestricted $5,000 grant for the research to the prize recipient's institution. An additional allowance will be provided for travel expenses to the APS meeting at which the prize ceremony will take place, and a certificate citing the contributions by the recipient. Nominations are active for three years, and guidelines are available online.
Requirements: The prize will be given to a physics faculty member at a U.S. undergraduate institution. The recipient will have been recognized as contributing substantially to physics research and providing inspirational guidance and encouragement of undergraduate students participating in this research. Nominations also should include a publication list that highlights student co-authors, including whether these students are high school, undergraduate, or graduate students.

Restrictions: The nominee's department may offer a program leading to a masters degree but must not have a doctoral program in physics.
Samples: Michael R. Brown, Swarthmore College, $5,000 (2008); William K. Wootters Williams College, $5,000 (2007); Q. Charles Su and Rainer Grobe, $5,000 (2006).
Amount of Grant: $5,000 stipend; $5,000 unrestricted grant
Date(s) Application is Due: Jul 1
Contact: Jacquelyn Beamon-Kiene, Executive Office Administrator; (301)209-3269 or (301) 209-3268; fax: (301) 209-0865; email: beamon@aps.org or honors@aps.org
Internet: http://www.aps.org/programs/honors/prizes/faculty-undergraduate.cfm
Sponsor: American Physical Society
One Physics Ellipse
College Park, MD 20740-3844

APS Prize for Industrial Applications of Physics 961

The purpose of the Prize is to recognize excellence in the industrial application of physics, and thereby to publicize the value of physics in industry, to encourage physics research in industry, and to enhance students' awareness of and interest in the role of physics in commercial product development. The recognized contribution may be a product, a process, or a tool enabling practical application of physics. The Prize consists of $10,000 and a certificate citing the contribution for which the Prize was awarded. In addition, an allowance is made for the expenses incurred by the recipient(s) to attend the APS meeting at which the Prize is presented. The APS Prize will be presented in odd-numbered years.
Requirements: Nominees need not be APS members, and there is no restriction with regard to geography or nationality. It is not necessary for the application to have already achieved commercial success, but it should have demonstrated potential for significant impact. Although nominees need not now be working in industry, the contribution for which they are cited must have been made while their principal employment was in industry.
Amount of Grant: $10,000
Date(s) Application is Due: Apr 1
Contact: Jacquelyn Beamon-Kiene, Executive Office Administrator; (301)209-3269 or (301) 209-3268; fax: (301) 209-0865; email: beamon@aps.org or honors@aps.org
Internet: http://www.aps.org/programs/honors/prizes/industrial.cfm
Sponsor: American Physical Society
One Physics Ellipse
College Park, MD 20740-3844

APS Robert R. Wilson Prize for Achievement in Particle Accelerators 962

The annual prize recognizes and encourages outstanding achievement in the physics of particle accelerators. The award consists of a $7,500 prize, an allowance for travel to the meeting at which the prize is awarded, and a certificate citing the contributions made by the recipient. The prize ordinarily is awarded to one person but may be shared when all recipients have contributed to the same accomplishment. Nominations are active for three years, and guidelines are available online.
Requirements: Nominations are open to scientists of all nations regardless of the geographical site at which the work was done. The prize will normally be awarded for contributions made at an early stage of the recipient's career.
Samples: Lyndon R. Evans, CERN—prize recipient, $7,500 (2008); Lee C. Teng, Argonne National Laboratory—prize recipient, $7,500 (2007); Glen Lambertson—prize recipient, $7,500 (2006); Keith Symon—prize recipient, $7,500 (2005).
Amount of Grant: $7,500
Date(s) Application is Due: Jul 1
Contact: Jacquelyn Beamon-Kiene, Executive Office Administrator; (301)209-3269 or (301) 209-3268; fax: (301) 209-0865; email: beamon@aps.org or honors@aps.org
Internet: http://www.aps.org/programs/honors/prizes/wilson.cfm
Sponsor: American Physical Society
One Physics Ellipse
College Park, MD 20740-3844

APS Tom W. Bonner Prize in Nuclear Physics 963

The prize recognizes and encourages outstanding experimental research in nuclear physics, including the development of a method, technique, or device that significantly contributes in a general way to nuclear physics research. The prize consists of $7,500 prize and a certificate citing the contributions made by the recipient. It is presented annually. The prize ordinarily is awarded to one person but may be shared when all the recipients have contributed to the same accomplishment. Nominations are active for three years, and guidelines are available online.
Requirements: Nominations are open to physicists whose work in nuclear physics is primarily experimental or a particularly outstanding piece of theoretical work. There is no time limitations on work described.
Samples: Arthur M. Poskanzer, Lawrence Berkeley National Laboratory—award recipient, $7,500 (2008); Stuart J. Freedman, University of California, Berkeley—award recipient, $7,500 (2007); John C. Hardy and Ian S. Towner—award recipients, $7,500 (2006); Roy James Holt—award recipients, $7,500 (2005).
Amount of Grant: $7,500
Date(s) Application is Due: Jul 1
Contact: Jacquelyn Beamon-Kiene, Executive Office Administrator; (301)209-3269 or (301) 209-3268; fax: (301) 209-0865; email: beamon@aps.org or honors@aps.org
Internet: http://www.aps.org/programs/honors/prizes/bonner.cfm
Sponsor: American Physical Society
One Physics Ellipse
College Park, MD 20740-3844

APS W.K.H. Panofsky Prize in Experimental Particle Physics 964

The annual prize recognizes and encourages outstanding achievements in experimental particle physics. The award consists of a $10,000 prize, an allowance for travel to the meeting at which the prize is to be awarded, and a certificate citing the contributions made by the recipient. The prize ordinarily is awarded to one person, but the prize may be shared when all recipients have contributed to the same accomplishment. Nominations are active for three years, and guidelines are available online.
Requirements: Nominations are open to scientists of all nations regardless of the geographical site at which the work was accomplished. The prize will normally be awarded for contributions made at an early stage of the recipient's career.
Samples: George Cassiday and Pierre Sokolsky, University of Utah—award recipients, $5,000 (2008); Bruce Winstein, Heinrich Wahl and Italo Mannelli—award recipients, $5,000 (2007); William Ford, John Jaros and Nigel Lockyer—award recipients, $5,000 (2006); Piermaria Oddone—award recipient, $5,000 (2005).
Amount of Grant: $10,000
Date(s) Application is Due: May 1
Contact: Jacquelyn Beamon-Kiene, Executive Office Administrator; (301)209-3269 or (301) 209-3268; fax: (301) 209-0865; email: beamon@aps.org or honors@aps.org
Internet: http://www.aps.org/programs/honors/prizes/panofsky.cfm
Sponsor: American Physical Society
One Physics Ellipse
College Park, MD 20740-3844

APS Will Allis Prize for the Study of Ionized Gases 965

The prize was established in 1989 by contributions from American Telephone and Telegraph, General Electric, General Telephone and Electronics, International Business Machines, and Xerox Corporations in recognition of the outstanding contributions of Will Allis to the study of ionized gases. The prize ordinarily is awarded to one person, but the prize may be shared when all recipients have contributed to the same accomplishment. The prize consists of a $5,000 award and a certificate citing the contributions made by the recipient. An allowance will be provided for travel expenses of the recipient to the society's meeting, where the prize is bestowed. The prize is awarded in even-numbered years. Nominations are active for three years, and guidelines are available online.
Requirements: Nominations are open to scientists of all nations regardless of the geographical location at which the work was done.
Amount of Grant: $5,000
Date(s) Application is Due: Jul 1
Contact: Jacquelyn Beamon-Kiene, Executive Office Administrator; (301)209-3269 or (301) 209-3268; fax: (301) 209-0865; email: beamon@aps.org or honors@aps.org
Internet: http://www.aps.org/programs/honors/prizes/allis.cfm
Sponsor: American Physical Society
One Physics Ellipse
College Park, MD 20740-3844

APS/British Academy Fellowship for Research in London 966

In collaboration with the British Academy, the APS offers an exchange post-doctoral fellowship for up to three months of research in the archives and libraries of London. This award includes travel expenses between the United States and United Kingdom paid by the APS and a monthly subsistence of 1,350 British pounds paid by the British Academy. Candidates should use the Franklin application form, specifying that they are asking for the British Academy Fellowship, and apply by October 1; applicants not selected for the British Academy Fellowship will be considered for a Franklin Research Grant.
Requirements: Applicants are expected to have held the doctorate for at least one year. Applications may be made by residents of the United States, by American citizens on the staffs of foreign institutions, and by foreign nationals whose research can only be carried out in the United States.
Restrictions: Grants will not be made for journalistic or other writing for general readership; the preparation of textbooks, casebooks, anthologies, or other materials for classroom use by students; nor for the work of creative and performing artists. Grants are rarely made to persons who have had the doctorate less than a year and never for predoctoral study or research. The committee will seldom approve more than two grants for the same project within any five-year period.
Amount of Grant: 1,350 British pounds paid by the British Academy
Date(s) Application is Due: Oct 1
Contact: Linda Musumeci, Research Administrator; (215) 440-3429; email: LMusumeci@amphilsoc.org
Internet: http://www.amphilsoc.org/grants
Sponsor: American Philosophical Society
104 South Fifth Street
Philadelphia, PA 19106-3387

APS/Phoenix Suns Energy and Environmental Minigrants 967

The program supports innovative, student-based educational projects that enhance learning with the objective of improving student achievement. Projects should have the ability of being replicated or adapted by other schools. Projects addressing energy, the environment, reading, mathematics or character education will be considered. Annual deadline dates may vary; contact program staff for exact dates.
Requirements: The program is open to educators who teach in public schools in the Arizona Public Service territory. Funds may be applied to equipment, materials and field trips that would not normally be provided by the school or school district.
Geographic Focus: Arizona
Amount of Grant: $500
Date(s) Application is Due: Jan 30

Contact: Louise Moskowitz; (602) 250-2291; fax: (602) 250-2113; email: louise.moskowitz@aps.com
Internet: http://www.aps.com/my_community/EducationTools/LearningCenter_1.html
Sponsor: Arizona Public Service
P.O. Box 53999, MS 8010
Phoenix, AZ 85072-3999

APSA Minority Fellows Program 968

The program is designed primarily for minority students applying to enter a doctoral program in political science for the first time. The association has re-focused and increased its efforts to assist minority students in completing their doctorates by concentrating not only on the recruitment of minorities, but also on the retention of these groups within the profession. The program designates six stipend minority fellows each year. Additional applicants who do not receive funds from the association may also be recognized and recommended for admission and financial support to graduate political science programs. Fellows with stipends receive a fellowship that is disbursed in two payments—one at the end of their first graduate year and one at the end of their second—provided that they remain in good academic standing. Awards are based on students' undergraduate course work, GPA, extracurricular activities, GRE scores, and recommendations from faculty. Guidelines are available online.

Requirements: Applicants must: be members of one of the following racial/ethnic minority groups: African Americans, Latinos/as, and Native Americans (federal and state recognized tribes); demonstrate an interest in teaching and potential for research in political science; be a U.S. citizen at time of award; and demonstrate financial need.
Samples: Angela Boyle, Loyola University New Orleans, (LA) (2007); Michael Castro, University of Rochester, (NY) (2007); Karen Ellis, William Paterson University, (NJ).
Amount of Grant: $4,000
Date(s) Application is Due: Oct 2
Contact: Harold Alan Pincus M.D., Director; (212) 543-6213; email: pincush@pi.cpmc.columbia.edu; Jeffrey R. Biggs, Program Director; (202) 483-2512; fax: (202) 483-2657; email: jbiggs@apsanet.org or cfp@apsanet.org
Internet: http://www.apsanet.org/content_3284.cfm
Sponsor: American Political Science Association
1527 New Hampshire Avenue NW
Washington, DC 20036-1206

APSA Political Science Doctoral Dissertation Awards 969

There are eight named prize awards given annually for specific doctoral dissertations. The Almond Award is for the best doctoral dissertation in the field of comparative politics; the Anderson Award, in the general field of state and local politics, federalism, or intergovernmental relations; the Corwin Award, in the field of public law broadly defined to include the judicial process, judicial biography, judicial behavior, courts, law, legal systems, the American constitutional system, civil liberties, or any other substantive area, or any work that deals in a significant fashion with a topic related to or having substantial impact on the American Constitution; the Lasswell Award, in the field of policy studies; the Reid Award, in the field of international relations, law, and politics; the Schattschneider Award, in the general field of American government and politics; the Strauss Award, in political philosophy; and the White Award, within the general field of public administration including broadly related problems of policy formation and administrative theory.

Requirements: Political science departments are invited to nominate not more than one doctoral dissertation per award. The dissertations must have been completed and accepted during the two calendar years preceding the date of the prize.
Samples: Emmanuel J. Teitelbaum, The George Washington University, Gabriel A. Almond Award (2007); Ronald S. Smith, Hanover College, William Anderson Award (2007).
Amount of Grant: $750-$1,000
Date(s) Application is Due: Jan 15
Contact: Doctoral Dissertation Awards; (202) 483-2512; fax: (202) 483-2657; email: awards@apsanet.org
Internet: http://www.apsanet.org/content_4113.cfm
Sponsor: American Political Science Association
1527 New Hampshire Avenue NW
Washington, DC 20036-1206

APSA Small Research Grants 970

The APSA Small Research Grant Program supports research in all fields of political science. The intent of these grants is to support the research of political scientists who are not employed at PhD granting department in the field and to help further the careers of these scholars. Grant recipients in the last five years have published over 12 books, 16 refereed articles, 8 book chapters, many working papers, and a large number of conference presentations. Others are still working on book, article, and conference projects and many recipients report great benefits for their students. Students have co-authored projects, worked as research assistants, and benefited through the professors' classroom use of their research. Many recipients also were able to use the APSA grant as seed money to get additional funding.

Requirements: Applicants must be APSA members at the time of application and at the time funds are dispersed. In addition, the principal investigator and any co-author must be one of the following: a faculty member at a college or university that does not award a PhD in Political Science, Public Administration, Public Policy, International Relations, Government, or Politics and whose primary appointment is in one of these

departments; or a political scientist not affiliated with an academic institution and is either (a) unemployed or (b) working in a research organization such as a think tank.
Restrictions: This grant does not provide support for dissertation research or writing and graduate students are not eligible to apply.
Amount of Grant: $2,500
Date(s) Application is Due: Feb 1
Contact: Bahram Rajaee; (202) 483-2512, ext. 130; fax: (202) 483-2657; email: brajaee@apsanet.org
Internet: http://www.apsanet.org/content_9222.cfm
Sponsor: American Political Science Association
1527 New Hampshire Avenue, NW
Washington, DC 20036-1206

APSA Travel Grants 971

APSA provides travel grants to assist some members in attending the Annual Meeting. U.S. graduate students, international graduate students studying in the U.S., unemployed APSA members, and international scholars are encouraged to apply. Priority will be given to presenters at the Annual Meeting, first-time applicants, or those who have not received a travel grant for three years. Application forms are available upon request.
Requirements: Only applicants who are APSA members and pre-registered for the APSA Annual Meeting are eligible to receive travel grants.
Amount of Grant: $700 maximum
Date(s) Application is Due: Jun 5
Contact: Program Coordinator; (202) 483-2512; fax: (202) 483-2657; email: travelgrants@apsanet.org
Internet: http://www.apsanet.org/content_11228.cfm
Sponsor: American Political Science Association
1527 New Hampshire Avenue, NW
Washington, DC 20036-1206

APSAA Fellowships 972

The mission of the Fellowship Program is to encourage interest and involvement in psychoanalysis among the future leaders, researchers and educators of mental health and academia. Early-career psychiatrists, psychologists, social workers and academics are eligible for to apply for the Fellowship. The Fellowship offers the opportunity to attend the biannual meetings of APSAA, to meet analysts and Fellows from across the country, to have a Mentor, and to present their clinical work or research at the meetings. The application for the Fellowship Program is available online September 15.
Requirements: Psychiatrists, psychologists, social workers, and academics who meet the specific eligibility requirements (available online) are eligible. Applicants may be nominated by their department chairs or program directors if applicable. When not applicable, self-nominations are encouraged. Applicants must be training or working in the United States during the fellowship year. All applicants should have demonstrated leadership ability in their discipline, or have special aptitude in research, teaching, and/or clinical endeavors; and have special interest in psychodynamics, psychoanalysis, or applied psychoanalysis.
Contact: James Guimaraes, Fellowship Coordinator; (212) 752-0450; fax: (212) 593-0571; email: jguimaraes@apsa.org or info@apsa.org
Internet: http://www.apsa.org/FELLOWSHIPPROGRAM/GENERAL INFORMATION/tabid/298/Default.aspx
Sponsor: American Psychoanalytic Association
309 East 49th Street
New York, NY 10007-1601

APSAA Mini-Career Grants 973

The Fund conceives of psychoanalytic research along the broadest lines, including: scholarly and empirical investigative contributions that can advance knowledge of psychoanalytic theory, practice, and links between psychoanalysis and neighboring disciplines such as developmental psychology or neuroscience. Awards of up to $15,000 annually for up to two years are intended to support the beginning career of a psychoanalytic investigator. These mini-career awards are intended, for example, to buy time for a junior faculty member or clinician just starting a practice so that they may consult with other investigators, join an investigative team, or attend year long seminars on research methodology or specific methods relevant to their research. Applicants for these awards must provide a year-long career development plan with letters from appropriate faculty and mentors describing the candidate's career development plan and how the award will facilitate the candidate's goals.
Requirements: Applicants should meet the following criteria: proposal relevant to psychoanalysis; applicant is either a new applicant to the Fund or resubmitting a previously reviewed application; applicant has a masters degree or higher; and human subject protection and consent procedures must be specified for research studies involving ongoing data collection.
Restrictions: Applicants who have received $80,000 total from the Fund are ordinarily not eligible to apply again.
Amount of Grant: $15,000 maximum annually
Date(s) Application is Due: Apr 1; Nov 1
Contact: Dean K. Stein, Executive Director; (212) 752-0450; fax: (212) 593-0571; email: deankstein@apsa.org or info@apsa.org
Internet: http://www.apsa.org/RESEARCH/FUNDFORPSYCHOANALYTIC RESEARCH/GRANTS/tabid/137/Default.aspx
Sponsor: American Psychoanalytic Association
309 East 49th Street
New York, NY 10007-1601

APSAA Research Grants   974

The Fund conceives of psychoanalytic research along the broadest lines, including: scholarly and empirical investigative contributions that can advance knowledge of psychoanalytic theory, practice, and links between psychoanalysis and neighboring disciplines such as developmental psychology or neuroscience. Grants of one or two years duration at a maximum of $20,000 yearly are intended for a specific project building upon psychoanalytic principles or directly investigating the process and/or outcome of psychoanalytically informed treatments. The Board recognizes that this amount of money is not sufficient usually to cover the costs of a large research study, particularly one that may involve a large number of subjects or a longitudinal design. Hence, the Board encourages applicants to consider feasibility of their request, that is, what they can achieve for this amount of money, and also to think about this type of award as supporting the beginning stages of a research project, the gathering of pilot data, or the refinement of design and methods.

Requirements: Applicants should meet the following criteria: proposal relevant to psychoanalysis; applicant is either a new applicant to the Fund or resubmitting a previously reviewed application; applicant has a masters degree or higher; and human subject protection and consent procedures must be specified for research studies involving ongoing data collection.

Restrictions: Applicants who have received $80,000 total from the Fund are ordinarily not eligible to apply again.

Amount of Grant: $20,000 maximum

Date(s) Application is Due: Apr 1; Nov 1

Contact: Dean K. Stein, Executive Director; (212) 752-0450; fax: (212) 593-0571; email: deankstein@apsa.org or info@apsa.org

Internet: http://www.apsa.org/RESEARCH/FUNDFORPSYCHOANALYTIC RESEARCH/GRANTS/tabid/137/Default.aspx

Sponsor: American Psychoanalytic Association

309 East 49th Street

New York, NY 10007-1601

APSAA Small Beginning Scholar Pre-Investigation Grants   975

The Fund conceives of psychoanalytic research along the broadest lines, including: scholarly and empirical investigative contributions that can advance knowledge of psychoanalytic theory, practice, and links between psychoanalysis and neighboring disciplines such as developmental psychology or neuroscience. These small grants of less than $5,000 for one year are intended to permit a beginning scholar to gather pilot data in preparation for the submission of a full grant to the Fund or to another agency. The Fund accepts revisions of a previously submitted application up to four years after the previous submission date. Up to two revised submissions may be submitted within four years of the original submission date.

Requirements: Applicants should meet the following criteria: proposal relevant to psychoanalysis; applicant is either a new applicant to the Fund or resubmitting a previously reviewed application; applicant has a masters degree or higher; and human subject protection and consent procedures must be specified for research studies involving ongoing data collection.

Restrictions: Applicants who have received $80,000 total from the Fund are ordinarily not eligible to apply again.

Amount of Grant: $5,000 maximum for one year

Date(s) Application is Due: Apr 1; Nov 1

Contact: Dean K. Stein, Executive Director; (212) 752-0450; fax: (212) 593-0571; email: deankstein@apsa.org or info@apsa.org

Internet: http://www.apsa.org/RESEARCH/FUNDFORPSYCHOANALYTIC RESEARCH/GRANTS/tabid/137/Default.aspx

Sponsor: American Psychoanalytic Association

309 East 49th Street

New York, NY 10007-1601

APSAA Small Beginning Scholar Visiting and Consulting Grants   976

The Fund conceives of psychoanalytic research along the broadest lines, including: scholarly and empirical investigative contributions that can advance knowledge of psychoanalytic theory, practice, and links between psychoanalysis and neighboring disciplines such as developmental psychology or neuroscience. Small grants of $3,000 are intended to permit a beginning scholar to spend time visiting and consulting with a more experienced, senior investigator who has agreed to help the junior investigator begin their investigative work. Typically these awards are made to permit the beginning scholar the funds to travel and/or to reimburse the senior investigator for travel and consultation. These grants may also be used to support travel to a research training seminar in the U.S. or abroad. The Fund accepts revisions of a previously submitted application up to four years after the previous submission date. Up to two revised submissions may be submitted within four years of the original submission date.

Requirements: Applicants should meet the following criteria: proposal relevant to psychoanalysis; applicant is either a new applicant to the Fund or resubmitting a previously reviewed application; applicant has a masters degree or higher; and human subject protection and consent procedures must be specified for research studies involving ongoing data collection.

Restrictions: Applicants who have received $80,000 total from the Fund are ordinarily not eligible to apply again.

Amount of Grant: $3,000 maximum

Date(s) Application is Due: Apr 1; Nov 1

Contact: Dean K. Stein, Executive Director; (212) 752-0450; fax: (212) 593-0571; email: deankstein@apsa.org or info@apsa.org

Internet: http://www.apsa.org/RESEARCH/FUNDFORPSYCHOANALYTIC RESEARCH/GRANTS/tabid/137/Default.aspx

Sponsor: American Psychoanalytic Association

309 East 49th Street

New York, NY 10007-1601

Aratani Foundation Grants   977

The foundation awards grants to nonprofits in its areas of interest, including education, health care, museums, recreation, and religion. Preference is given to Japanese-American cultural organizations. Types of support include annual campaigns, building construction/renovation, capital campaigns, conferences and seminars, continuing support, curriculum development, endowments, exchange programs, fellowships, general operating support, program development, scholarship funds, and seed grants. Application forms are not required, but application outlines are available.

Requirements: Giving primarily in California.

Restrictions: Grants are not made to individuals.

Geographic Focus: California

Amount of Grant: $100-$150,000 average range

Contact: George Aratani, President

Sponsor: Aratani Foundation

23505 Crenshaw Boulevard, No 230

Hollywood, CA 90505

Arca Foundation Grants   978

The Arca Foundation believes that access to knowledge, vigorous public education and citizen engagement are essential to democracy. However, there exist structures and private interests that serve to limit the transparency of our government, stifle public debate on critical issues, and foster an environment where government is not effectively serving the interests of its citizens. In order to promote greater social equity and justice at home and abroad, the foundation supports organizations and projects that work to advance transparent, accountable, and just policies. They support strategic initiatives that work to directly affect policies by: Developing and advocating for innovative ideas; Promoting transparency and access to information; Fostering greater public debate on critical issues; Educating key stakeholders; Engaging citizens in strategic organizing and advocacy that builds power and drives change.

Requirements: Domestically, in the current sociopolitical climate, the foundation is concerned about the promotion of a more equitable, accountable, and transparent economic recovery. They are considering proposals that work to advance more just policies on this and other critical issues. Internationally, the foundation has a long history of working to promote greater dialogue between the U.S. and Cuba. They are considering proposals that advance policies that further normalized U.S.-Cuban relations, as well as proposals that work to foster more just policies on a range of international issues. The Arca Foundation has two deadlines annually, on March 1 and September 1 of every year, for consideration in June and December respectively. When deadline dates fall on a weekend, the deadline is effective on the next weekday. The foundation will not respond to letters of inquiry, but will accept complete proposals submitted according to the application guidelines on regular grant deadlines.

Restrictions: The foundation does not fund organizations that provide direct social services, scholarship funds or scholarly research, capital projects or endowments, individuals, or government programs. Proposals received via fax or email will not be considered. Late proposals will not be considered, and extensions are not available.

Samples: Catholics in Alliance for the Common Good (Washington, DC) $50,000 to support 'A Campaign for the Common Good' and related civic participation activities.; Consumer Watchdog (Santa Monica, CA) $75,000 to support OilWatchdog.org, a company-by-company comprehensive watchdog of the oil industry.; Medical Education Cooperation with Cuba (Decatur, GA) $30,000 to support continuing and expanding MEDICC's journalism internship program.; Independent Media Institute (San Francisco, CA) $50,000 to support AlterNet and its news and information coverage of strategic issues.;

Amount of Grant: $50,000 average

Date(s) Application is Due: Mar 1; Sep 1

Contact: Emily Casteel, Program Associate; (202) 822-9193; fax: (202) 785-1446; email: grants@arcafoundation.org

Internet: http://www.arcafoundation.org/howtoapply.htm

Sponsor: Arca Foundation

1308 19th Street, NW

Washington, DC 20036

Arcadia Foundation Grants   979

The foundation awards grants to Pennsylvania nonprofit organizations to improve the quality of life. Areas of interest include hospitals and hospital building funds, health agencies and services, nursing, hospices, early childhood, adult and higher education, libraries, child development and welfare agencies, youth organizations, and social service and general welfare agencies, including care of the handicapped, aged, and hungry. Also supported are family services, environment and conservation, wildlife and animal welfare, religious organizations, historical preservation, and music organizations. Types of support include general operating support, continuing support, annual campaigns, capital campaigns, building construction/renovation, equipment acquisition, endowment funds, program development, scholarship funds, and research. Applications are accepted between September 1 and November 1.

Requirements: Eastern Pennsylvania organizations whose addresses have zip codes of 18000-19000 are eligible. Application form not required. The initial approach should be a letter or proposal—not exceeding two pages.

Restrictions: Grants are not awarded to support individuals, deficit financing, land acquisition, fellowships, demonstration projects, publications, or conferences.
Geographic Focus: Pennsylvania
Samples: Thomas Jefferson U (Philadelphia, PA)—for a new medical education building, $1 million (2005).
Date(s) Application is Due: Nov 1
Contact: Marilyn Lee Steinbright, President, fax: (610) 275-8460
Sponsor: Arcadia Foundation
105 E Logan Street
Norristown, PA 19401

Archaeological and Pre-Colonial History in Eastern Africa Research Grants   980

The institute occasionally offers limited support to qualified scholars undertaking field research in eastern African history, later archaeology, or related studies. The Institute intends to give priority to research in relevant fields which focuses on at least one of the following themes: Histories of Environmental Change; Colonial Encounters; Sacred Space in Eastern Africa; Maritime Heritage of the Western Indian Ocean; or Migrations in Cultural and Historical Perspective.
Requirements: Graduate background or comparable experience in the subject and the relevant research methods are required.
Amount of Grant: @L1000 sterling or its equivalent in local currency
Date(s) Application is Due: May 30; Oct 31
Contact: Paul Lane, Director, 254 2 4343190; fax: 254 2 4343365; email: pjlane@africaonline.co.ke
Internet: http://www.britac.ac.uk/institutes/eafrica
Sponsor: British Institute in Eastern Africa
P.O. Box 30710
Nairobi GP.O. 00100Kenya

Archer Daniels Midland Foundation Grants Program   981

The foundation prefers to fund programs that will directly impact the communities where operating units are located. Nearly a third of the funding is directed toward educational institutions, including elementary, secondary, and higher education. The advocacy category covers world affairs and foreign relations groups that deal with international trade; also supported are projects stressing free enterprise and assistance for women. Support of social services goes to minority group development, cultural activities, and hospital and youth agencies. Health service funding supports hospital and disease association programs. Conservation funding supports programs for protecting the environment and beautification as well as conservation. Religion grants support Christian, Jewish, and Roman Catholic organizations, such as churches, colleges and universities, international ministries/missions, Jewish welfare, and the Salvation Army. Applications are accepted at any time; initial contact by letter of inquiry is encouraged.
Requirements: Tax-exempt organizations are eligible. Current United Way recipients and national organizations are also eligible.
Restrictions: Grants are restricted to public, religious, charitable, philanthropic, benevolent, scientific, literary, artistic, educational, or for the prevention of cruelty to children and animals.
Samples: The German Coast Farmers' Market (LA) to add new location to its farmers market—award recipient 2006; The Audrain County 4-H Foundation (MO) to build a replacement for its 45-year-old building and fulfill its commitments to the children and families in the area—award recipient 2006; The Girl Scouts of U.S.A-Kickapoo Council (IL) to construct a lab for girls from local schools and outreach agencies to learn about career opportunities in agriculture.
Amount of Grant: $10,000 per award
Contact: Brian Peterson, Program Contact, (217) 424-5413
Internet: http://www.admworld.com/naen/
Sponsor: Archer Daniels Midland Foundation
P.O. Box 1470
Decatur, IL 62526

Archibald Wayne Dingman Memorial Graduate Scholarship   982

This scholarship is for four months of study related to the petroleum industry, awarded annually to qualified graduates of any recognized university who are registered in or admissible to a program leading to a master's or doctoral degree at the University of Calgary.
Requirements: To be eligible, applicants must be graduate students that are Canadian citizens or permanent residents.
Amount of Grant: $C3000
Date(s) Application is Due: Feb 1
Contact: Connie Busch, Faculty of Graduate Studies, (403) 220-5690; email: cbusch@ucalgary.ca
Internet: http://www.grad.ucalgary.ca/funding/internal_scholarships/lev_4/dingman.htm
Sponsor: University of Calgary
2500 University Dr NW, Earth Sciences Building, Room 720
Calgary, AB T2N 1N4 Canada

Arcus Foundation Religion and Values Program Grants   983

The Arcus Foundation is a private grantmaking foundation that supports organizations around the world working in two areas - lesbian, gay, bisexual, and transgender (LGBT) human rights; and conservation of the world's great apes. In the former area, the Foundation supports organizations that are working to achieve social justice that is inclusive of sexual orientation, gender identity and race. In the latter area, it supports organizations seeking to ensure respect and survival of great apes and their natural habitat. Specifically, the Religion and Values Program has the goal of achieving long-term change in cultural attitudes and religious institutions that currently stigmatize same-gender sexual orientation and gender variance. The program engages with and advances greater understanding and acceptance of sexual orientation and gender variance by mainstream religious institutions and leaders, and helps to create a positive shift in cultural attitudes and values toward sexuality in general and GLBT (gay, lesbian, bisexual and transgender) issues in particular.
Requirements: Nonprofit organizations are eligible.
Restrictions: No grants to are given to individuals, or for religious or political activities, medical research or film/video production.
Amount of Grant: $1,000-$50,000 average
Contact: Lucia Leandro Gimeno, Program Assistant; (212) 488-3000; fax: (212) 488-3010; email: lucialeandro@arcusfoundation.org
Internet: http://www.arcusfoundation.org/pages_2/home.cfm
Sponsor: Arcus Foundation
119 West 24th Street, 9th Floor
New York, NY 10011

Arecibo Observatory Grants   984

The National Astronomy and Ionosphere Center enables research in the areas of astronomy, planetary studies, and space and atmospheric sciences by providing unique capabilities and state-of-the-art instrumentation for data collection and analysis, together with logistical support to users. Use of the Arecibo Observatory (AO) operated by NAIC is available on an equal, competitive basis to all scientists from throughout the world to pursue research in astronomy, planetary studies and, space and atmospheric sciences. Observing time is granted on the basis of the most promising research, as ascertained by peer review of proposals by external referees.
Date(s) Application is Due: Feb 1; Jun 1; Oct 1
Contact: Ellen Howell, Planetary Science; (787) 878-2612; fax: (787) 878-1861; email: ehowell@naic.edu
Internet: http://www.naic.edu/science/proposals_set.htm
Sponsor: Arecibo Observatory
HC03 Box 53995
Arecibo, Puerto Rico 00612

Arie and Ida Crown Memorial Grants   985

The program supports programs that offer opportunities to the disadvantaged, strengthens the bond of families, and improves the quality of people's lives. As a general rule, the Foundation funds organizations that serve the greater Chicago area as well as organizations that serve the broader Jewish community. Most grants are awarded to organizations within the city of Chicago. Organizations are supported in the areas of arts and culture (concentrating on educational and enrichment programs for youth), civic affairs, education, health (stressing access to services, hospice and health promotion), and human service (focusing on programs which offer assistance for children and families).
Requirements: Nonprofit organizations in Chicago and Cook County, IL, may apply for grant support.
Restrictions: Grants are not made to support individuals, conference expenses, film projects, government programs (50 percent government funded), or research projects.
Geographic Focus: Illinois
Amount of Grant: $1000-$200,000 average
Date(s) Application is Due: Jan 31; Jul 31
Contact: Susan Crown, President, (312) 236-6300; fax: (312) 984-1499; email: AICM@crown-Chicago.com
Internet: http://www.crownmemorial.org/
Sponsor: Arie and Ida Crown Memorial
222 N LaSalle Street, Suite 2000
Chicago, IL 60601

ARIT Fellowships   986

Fellowships support research in ancient, medieval, or modern times, in any field of the humanities and social sciences, including prehistory, history, art, archaeology, literature, and linguistics as well as interdisciplinary aspects of cultural history for applicants who have completed their academic training. Postdoctoral and dissertation research fellowships may be held for terms of four months to one year. Grants for tenures up to one year will be considered, but preference will be given to projects of shorter duration. ARIT maintains two research institutes in Turkey. ARIT-Istanbul has a research library focused on Byzantine, Ottoman, and modern studies of Turkey. ARIT-Ankara focuses on art, archaeology, and ancient history in its library and serves Turkish and American archaeologists through its programs. Both institutes have residential facilities for fellows and provide general assistance as well as introductions to colleagues, institutions, and authorities in Turkey. Predoctoral applicants also may qualify for ARIT's Kress fellowship, and postdoctoral scholars also may qualify for ARIT's NEH fellowships. Applicants will be notified of the fellowship committee's decision by January 25.
Requirements: Applicants must be U.S. citizens or a three-year resident of the U.S., and have completed formal academic training affiliated with U.S. or Canadian institutions. Scholars and advanced graduate students engaged in research on ancient, medieval, or modern times in Turkey, in any field of the humanities and social sciences, are eligible. Student applicants must have fulfilled all preliminary requirements for the doctorate except the dissertation. Turkish law requires all foreign scholars, prior to entering the country, to obtain formal permission for any research to be carried out in Turkey.
Amount of Grant: $16,800-$50,400
Date(s) Application is Due: Nov 1

Contact: Nancy Leinwand; (215) 898-3474; fax: (215) 898-0657; email: leinwand@sas.upenn.edu
Internet: http://ccat.sas.upenn.edu/ARIT/FellowshipPrograms.htm
Sponsor: American Research Institute in Turkey
3260 South Street
Philadelphia, PA 19104-6324

ARIT Fellowships for Intensive Advanced Turkish Language Study 987

ARIT offers fellowships for 10 advanced students for participation in the summer program in intensive advanced Turkish language at Bogazici University in Istanbul. This intensive program offers the equivalent of one full academic year of study in Turkish at the college level. ARIT fellowship- supported courses are offered at the advanced level. Classes are conducted in Turkish, with informal and formal styles introduced and reviewed through instruction, language laboratory work, and open conversations with teaching assistants. Students meet with teaching assistants on an informal basis for additional instruction and free conversation. Participants also attend extracurricular activities including films, lectures, and cultural events both on- and off-campus. The fellowships cover round-trip airfare to Istanbul, application and tuition fees, and a maintenance stipend. Full-time students and scholars affiliated at academic institutions are eligible to apply. Guidelines and application are available online.
Requirements: Applicant must be a U.S. citizen, national, or permanent resident; be currently enrolled in an undergraduate or graduate-level academic program, or be faculty; have a minimum B average, if still a student; and perform at the high-intermediate level on a proficiency-based admissions examination.
Date(s) Application is Due: Feb 1
Contact: Erika H. Gilson, Director; email: ehgilson@princeton.edu
Internet: http://ccat.sas.upenn.edu/ARIT/ARITSummerLanguageProgram.htm
Sponsor: American Research Institute inTurkey
110 Jones Hall, Princeton University
Princeton NJ 08544

ARIT Kenan T. Erim Fellowship for Research at Aphrodisias 988

The institute invites applications for a fellowship to support excavation and/or research in art history and archaeology to be carried out at the site of Aphrodisias in Turkey during the summer. Applicants must submit an application, a letter of acceptance from a director of the excavations at Aphrodisias, and two letters of reference by the listed application deadline. Graduate students should supply a copy of their graduate transcript. Guidelines are available online.
Requirements: Scholars or advanced graduate students engaged in excavation at the site of Aphrodisias or research on material from that site are eligible to apply. Fields of study include the history of art and architecture from antiquity to the present, and archaeology. Graduate student applicants must have fulfilled all preliminary requirements for the doctorate except the dissertation by June, and before beginning any ARIT-sponsored research.
Amount of Grant: $2,375 maximum
Date(s) Application is Due: Nov 1
Contact: Nancy Leinwand; (215) 898-3474; fax: (215) 898-0657; email: leinwand@sas.upenn.edu
Internet: http://ccat.sas.upenn.edu/ARIT/FellowshipPrograms.htm
Sponsor: American Research Institute in Turkey
3260 South Street
Philadelphia, PA 19104-6324

ARIT Mellon Foundation Postdoctoral Fellowships for East European Scholars 989

The postdoctoral fellowships are intended to bring scholars from Central and Eastern Europe into the broader research community, specifically into Turkey. Fellowships, offered in all areas of the humanities and social sciences, have a tenure of two to three months for research to be carried out in Turkey. Preference will be given to scholars in the early stages of their careers. Fellows are expected to devote full time to their projects and to participate in the activities of the institute. Turkish law requires all foreign scholars, prior to entering the country, to obtain formal permission for any research to be carried out. Fellowship recipients are personally responsible for obtaining their own research permission and may acquire forms and procedures through the Turkish diplomatic office within each country included in the competition. Replies for permission may take six months or more. Applicants may contact the office for advice about research permission.
Requirements: The fellowships are available to Czech, Hungarian, Polish, Slovak, Bulgarian, and Romanian scholars holding the PhD or its equivalent who are engaged in advanced research in any field of the social sciences or the humanities involving Turkey. Scholars must be permanent residents of one of the included countries.
Restrictions: Applicants may not have held a prior fellowship under this program.
Amount of Grant: $11,500 maximum stipend for two to three months of research
Date(s) Application is Due: Mar 5
Contact: Nancy Leinwand; (215) 898-3474; fax: (215) 898-0657; email: leinwand@sas.upenn.edu
Internet: http://ccat.sas.upenn.edu/ARIT/MellonFellowships.htm
Sponsor: American Research Institute in Turkey
3260 South Street
Philadelphia, PA 19104-6324

ARIT/NEH Fellowships for Research in Turkey 990

ARIT invites applications for two or three postdoctoral fellowships made possible by support from NEH. The fields of study cover all periods in the general range of the humanities, including humanistically oriented aspects of social sciences, prehistory, history, art, archaeology, literature, and linguistics, as well as interdisciplinary aspects of cultural history. The fellowships for research in Turkey can be held for four to 12 months. ARIT maintains two research institutes in Turkey: ARIT-Istanbul, with a research library focused on Byzantine, Ottoman, and modern studies of Turkey; and ARIT-Ankara, which focuses on art, archaeology, and ancient history in its library and serves Turkish and American archaeologists through its programs. Both institutes have residential facilities for fellows and provide general assistance as well as introductions to colleagues, institutions, and authorities in Turkey.
Requirements: Scholars who have completed their formal training and plan to carry out research in Turkey may apply. They may be U.S. citizens or three-year residents.
Amount of Grant: $13,335-$40,000
Date(s) Application is Due: Nov 1
Contact: Nancy Leinwand; (215) 898-3474; fax: (215) 898-0657; email: leinwand@sas.upenn.edu
Internet: http://ccat.sas.upenn.edu/ARIT/NEHFellowships.htm
Sponsor: American Research Institute in Turkey
3260 South Street
Philadelphia, PA 19104-6324

Arizona Cardinals Grants 991

The National Football League franchise supports programs designed to improve the quality of life and enhance opportunities for children, women, and minorities in the state of Arizona. Specific areas of interest include arts and culture, civic affairs, education, health, science, and social services. The foundation is interested in expanding its giving and looks for new charities to fund. First-time applicant organizations generally will receive grants of $5000 or less.
Requirements: Applicants must be exempt under 501(c)3 of the Internal Revenue Service code.
Geographic Focus: Arizona
Samples: AZCenters for Comprehensive Education and Life Skills; Chrysalis Shelter for Victims of Domestic Violence; Child crisis Center-East Valley; Jewish Family and Children's Services; Juvenile Diabetes Research Foundation; United Food Bank.
Amount of Grant: $2000-$5000
Date(s) Application is Due: Aug 1
Contact: Pat Tankersley, (602) 379-0101; fax: (480) 785-7327
Internet: http://www.azcardinals.com/community/charities.php
Sponsor: Arizona Cardinals
P.O. Box 888
Phoenix, AZ 85001-0888

Arizona Commission on the Arts Education Projects Grants 992

The Arizona Commission on the Arts is committed to making the arts fundamental to education, particularly in programs that serve Pre-K-12 students, classroom teachers, teaching artists, arts specialists and administrators in school, after-school and summer/inter-session programs. Our goal is that applicants present a plan that creatively stimulates arts education in their school/community organizations.
Requirements: An applicant organization must be a 501(c)3 Arizona nonprofit organization or school, or a unit of government. An unincorporated Arizona organization may apply through a fiscal agent, providing that the fiscal agent is an Arizona 501(c)3 or a governmental organization. Grant applications must be submitted online.
Restrictions: No more than one Arts Learning project per individual site is funded per year. Support for staff salaries, funding for insurance or supplies are not provided through this grant.
Geographic Focus: Arizona
Date(s) Application is Due: Mar 22
Contact: Alison Marshall, Director, (602) 771-6523; email: amarshall@azarts.gov
Internet: http://www.azarts.gov/index.htm
Sponsor: Arizona Commission on the Arts
417 W Roosevelt
Phoenix, AZ 85003

Arizona Commission on the Arts Folklorist Residencies 993

In this residency program, folklorists guide students and teachers in the process of researching, identifying, and documenting traditional art forms in their community. Students develop a new sense of community as they learn and experience the significance of family, community, and cultural traditions directly from parents and family members, employers and workers, senior citizens, public officials, and other community members. Residency lengths are variable and are built around a combination of services: training for students and teachers in research methodology; fieldwork conducted by the folklorist with participating students and teachers; classroom workshops; lecture-demonstrations on traditional art forms; interviews with local traditional artists; and presentation of traditional artists/art forms in school or community settings. A travel, lodging, and meal subsidy is provided for artists funded for out-of-town residencies in addition to their payment for services.
Requirements: 501(c)3 tax-exempt Arizona nonprofit organizations, schools, or units of government are eligible. Schools must select an individual artist, performing company, folklorists, or interdisciplinary artists from the Arizona Artist Roster.
Geographic Focus: Arizona
Amount of Grant: $3500 minimum: $3000 for folklorist services; $500 for services by local traditional artists
Contact: Robert Booker, Executive Director, (602) 771-6501; fax: (602) 256-0282; email: rbooker@azarts.gov

Internet: http://www.azarts.gov/localarts/index.htm#
Sponsor: Arizona Commission on the Arts
417 W Roosevelt Street
Phoenix, AZ 85003

Arizona Commission on the Arts Planning and Development Grants   **994**
Planning grants aid in the thoughtful development of artistic projects. The grants do not have to result in a final artistic product, but can be used in an early phase of a project. Examples of projects are: rehearsals with a guest choreographer/composer/playwright in the gradual development of a new work that may not necessarily be ready for production during the grant period; research toward a specific project; identification of guest artists for a specific project; bringing guest artists/collaborators/community partners on-site for planning meetings; and identification of curators for the development of an exhibition. Funding may be requested for the project director, guest artist, consultant, curator and/or collaborator fees; travel, lodging, and meals for planning activities; company/ensemble members' pay involved in the development and rehearsal of work involving a guest artist; and prospectus printing and mailing. Applicants who have received planning grants can apply to the arts commission the following year or in subsequent years to complete the project.
Requirements: An applicant organization must be a 501(c)3 Arizona nonprofit organization or school, or a unit of government. An unincorporated Arizona organization may apply through a fiscal agent, providing that the fiscal agent is an Arizona 501(c)3 or a governmental organization.
Restrictions: Funding may not be requested for attendance at booking conferences, company/ensemble rehearsal fees for projects that do not involve guest artists, artistic director fees for the development of new work, or purchase of materials during the development of a work.
Geographic Focus: Arizona
Amount of Grant: $500 maximum
Contact: Jo Kobert, (602) 255-5882; email: jkobert@ArizonaArts.org
Internet: http://www.arizonaarts.org/organizations/planning_development.htm
Sponsor: Arizona Commission on the Arts
417 W Roosevelt Street
Phoenix, AZ 85003

Arizona Public Service Corporate Giving Program Grants   **995**
Grants are awarded to support Arizona nonprofit organizations in the areas of health and human services, community development, arts and culture, education, and environment. The foundation awards support for project grants, capital building funds, research, employee matching gifts, in-kind services, conferences and seminars, and operating support. Applications are accepted on an ongoing basis.
Requirements: Arizona 501(c)3 nonprofits are eligible.
Restrictions: APS Corporate Giving does not fund individual request, charter or private schools, religious, political fraternal, legislative or lobbying efforts to organizations, travel-related or hotel expenses, private or family foundation, private non-profit organizations, salaries and/or debt reduction.
Geographic Focus: Arizona
Samples: Hospice of the Valley (Phoenix, AZ)—for operating support, $10,000.; Arizona Bridge to Independent Living (Phoenix, AZ)—for program support, $3000.; Arizona Wolf Trap—to sponsor a series of field trips for children to visit performing arts centers.
Contact: Cindy Slick, Manager; (602) 250-4707; fax: (602) 250-2113; email: Cindy.Slick@aps.com
Internet: http://www.aps.com/main/community/dev/default.html
Sponsor: Arizona Public Service Corporation
P.O. Box 53999, MS 8010
Phoenix, AZ 85072-3999

Arkema Foundation Science Teachers Program   **996**
This program is an intensive week-long session for elementary and secondary school teachers. Armed with innovative science experiment kits and the guidance of chemical engineers and scientists, teachers learn new and fascinating ways to illustrate scientific concepts. Scientific topics explored include life, earth and physical science and technology. School principals are asked to nominate two teachers in grades three through six, who are then chosen by the corporate committee to participate in the program. Application forms and contact information are available on the Web site.
Requirements: Teachers in the following geographic areas may participate: Alabama (Mobile county); Kentucky (Graves, Livingston, Lyon, Marshall, McCracken, Carroll, Gallatin, Trimble, Jefferson counties); Michigan (Wayne county); Minnesota (Didge, Mower, Steele counties); New Jersey (Burlington, Camden, Gloucester, and Salem counties); New York (Genesee, Livingston counties); Pennsylvania (Berks, Bucks, Chester, Delaware, Montgomery, and Philadelphia counties); Tennessee (Shelby county); Texas (Jasper, Newton, Jefferson, Orange, Harris, and Brazos counties).
Geographic Focus: Alabama; Kentucky; Michigan; Minnesota; New Jersey; New York; Pennsylvania; Tennessee; Texas
Amount of Grant: $500
Contact: Jane Crawford, (215) 419-7614; email: jane.crawford@arkemagroup.com
Internet: http://www.products.arkemagroup.com/index.cfm?pag=190
Sponsor: Arkema Inc. Foundation
2000 Market Street
Philadelphia, PA 19103-3222

ARM Faculty Fellowships   **997**
The fellowships provide opportunities for college and university faculty members in science and engineering to participate in research and collaborative activities at government and industrial research facilities. Sabbatical and travel awards also are available. Fellowship duration varies with the research project needs and facility approval. Award funding may be allocated for a stipend up to the faculty member's certified college/university salary, travel expenses to and from the facility, and a modest relocation allowance. Award funding may also be allocated for research participation by undergraduate or graduate students. Applications may be submitted at any time.
Requirements: The program is open to full-time faculty members at an accredited college or university. Citizenship restrictions may apply for some awards or facilities. Selection is based on applicant's professional qualifications, references, strength of the research proposal, and the expected benefit of the fellowship experience to the applicant, the applicant's home institution, and the host facility. Preference is given to faculty redirecting their research or establishing new collaborations with facility scientists.
Contact: Mariann Bleazard, Faculty Fellowship Program, (801) 273-8911; fax: (801) 277-5632; email: mariann@armanagement.org
Internet: http://www.awu.org/faculty_default.htm
Sponsor: Academia Resource Management
535 E 4500 S, Suite D120
Salt Lake City, UT 84107

ARM Graduate Fellowships   **998**
Graduate student fellowships provide master's and doctoral degree candidates in science and engineering the opportunity to conduct research toward their thesis or dissertation at one of more than 50 cooperating government and industrial facilities. Institutional affiliation and citizenship restrictions may apply for some awards or facilities. The fellowship term ranges from one to 12 months. The award includes a monthly stipend, tuition assistance, and travel allowance. Renewals are competitive and require reapplication annually. Applications may be submitted at any time. To insure optimum consideration submit by February 22 for summer, by March 20 for fall, and by October 20 for spring.
Requirements: The program is open to master's and doctoral degree candidates enrolled in an accredited graduate program. Institutional affiliation and citizenship restrictions may apply for some awards or facilities.
Amount of Grant: $1300 minimum stipend per month
Contact: Mariann Bleazard, Graduate Fellowship Program, (801) 273-8911; fax: (801) 277-5632; email: info@armanagement.org
Internet: http://www.awu.org/appl_forms/descriptions/lg.htm
Sponsor: Academia Resource Management
535 E 4500 S, Suite D120
Salt Lake City, UT 84107

ARM Postgraduate Fellowships   **999**
The fellowships provide opportunities for post-master's and postdoctoral research associates to participate in research at national laboratories and other research facilities. Stipends are established by the host facility, and travel and relocation expenses may be available. The initial fellowship term is up to one year but may be renewed up to two additional years upon mutual agreement by the fellow and the facility and the availability of funds. Applicants must have identified a host scientist and a research project at a cooperating facility before sending their completed application to AWU. Applications may be submitted at any time but Applications may be submitted at any time, however, it is recommended that applications are submitted two to three months prior to the requested start date.
Requirements: Candidates must have been awarded the master's or doctoral degree within four years of applying. A commitment to a professional career in science or engineering research is expected. U.S. citizenship or permanent resident status is required. Exceptions to these requirements are subject to the approval of the host facility.
Contact: Mariann Bleazard, Postgraduate Fellowship Program, (801) 273-8900; fax: (801) 277-5632; email: info@armanagement.org
Internet: http://www.awu.org/appl_forms/descriptions/pd.htm
Sponsor: Academia Resource Management
535 E 4500 S, Suite D120
Salt Lake City, UT 84107

ARM Visiting Scientist Awards   **1000**
AWU administers awards to Visiting Scientists who are industrial associates and collaborators in science, mathematics, engineering, and technology. They provide the opportunity to participate in and contribute to science and technology at sponsoring facilities and are designed to encourage professional development and provide special expertise to the host facility. Stipend is established by the host facility and varies by experience and discipline; usually, it is based on the applicant's current salary. Funds for relocation, supplemental travel, and professional expenses may be available. Applicants must have identified a host scientist and a project at a cooperating facility before sending their application to AWU. Applications may be submitted at any time. However, it is recommended that you submit your application two to three months prior to the requested start date.
Requirements: Professionals with continued commitment to science and engineering. U.S. citizenship or permanent resident status is required. Nonresident aliens require the approval of the host facility and an appropriate visa status.
Contact: Mariann Bleazard, Postgraduate Fellowship Program, (801) 273-8900; fax: (801) 277-5632; email: info@awu.org
Internet: http://www.awu.org/appl_forms/descriptions/vs.htm

Sponsor: Academia Resource Management
535 E 4500 S, Suite D120
Salt Lake City, UT 84107

Armenian Students' Association Scholarships   1001
The association awards undergraduate and graduate scholarships to applicants of
Armenian ancestry. Scholarship applications must be requested (obtain form from office
or the Web site—application fee required) by February 15; applications must be received
March 15. Guidelines and request form are available online.
Requirements: Students must be enrolled full-time in four-year accredited colleges or
universities in the United States and must be of Armenian descent. The student must be
a sophomore or beyond in the current academic year.
Amount of Grant: $500-$1500
Date(s) Application is Due: Feb 15; Mar 15
Contact: Scholarship Administrator, (401) 461-6114; email: headasa@aol.com
Internet: http://www.asainc.org/national/scholarships.shtml
Sponsor: Armenian Students' Association of America
333 Atlantic Avenue
Warwick, RI 02888

Arnold Fund Grants   1002
The primary focus of the fund is to support charitable organizations located in Newton
County, GA. Fields of interest include performing arts, music, higher education, library,
education, biomedicine, and medical research. Grants are also made to institutions of
higher education in which the founders had a special interest. Proposers are requested
to write a brief letter explaining the project for which the funds are being requested. A
formal application form is not required.
Restrictions: Grants are not made to individuals.
Geographic Focus: Georgia
Amount of Grant: $500-$279,000 range
Contact: John Sawyer, Executive Director, (404) 881-7886
Sponsor: Arnold Fund
1201 W Peachtree Street, Suite 4200
Atlanta, GA 30309-3400

ARO Atomic, Molecular, and Optical (AMO) Physics Research Contracts   1003
Research in atomic, molecular, and optical physics will create fundamentally new
capabilities for the army, as well as providing the scientific underpinnings to enhance
existing technologies. Topics of interest include unconventional optics for enhanced
imaging and detection; atom optics and laser cooling and trapping for ultrasensitive
detectors; nonlinear atomic and molecular processes for sensor protection and optical
processing; and the investigation of plasmas for toxic gas destruction, pollution
reduction, materials processing, and propellant ignition. Image science, automatic target
recognition, and sensor fusion issues are also addressed in the AMO program. Proposals
may be submitted at any time.
Requirements: Organizations and individuals interested in submitting research proposals
are strongly encouraged to make preliminary inquiries as to the general need for the type
of research effort contemplated before expending extensive effort in preparing a detailed
research proposal or submitting proprietary information.
Contact: Dr. Richard Hammond, Physics Division, Army Research Office, (919) 549-
4313; email: richard.hammond@us.army.mil
Internet: http://www.arl.army.mil/main/main/default.cfm?Action=29&Page=203#ato
mic_molecular_physics
Sponsor: Army Research Office
P.O. Box 12211
Research Triangle Park, NC 27709-2211

ARO Biomolecular and Cellular Materials and Processes Research Contracts   1004
The program supports Fundamental studies to define structure-function relationships
and biochemical interactions for enzymes, receptors and other macromolecules exhibiting
mechanisms and properties uniquely relevant to synthetic and degradative pathways of
interest to the military, including establishment of the foundations for manipulation
and exploitation of biocatalysis, ribosomal and non-ribosomal biosynthesis to enhance
permissiveness toward elaboration of useful biomolecular structures and cellular systems
designed with "metabolic engineering" in mind. Also, research to provide insight from
nature on novel theoretical principles and mechanisms in sensory and motor function, as
well as on materials with extraordinary properties, from biological sources. Includes not
only initial molecular events, signal transduction pathways and integrated information
processing for the powerful sensing capabilities exhibited in the biological world, but also
self-assembly processes, hierarchical structure formation, and functional characterization
of biomolecular materials such as those with potential biomimetic utility for nanometer
scale fabrication or for energy and information transfer, among other possibilities.
Proposals may be submitted at any time.
Requirements: Organizations and individuals interested in submitting research proposals
are strongly encouraged to make preliminary inquiries as to the general need for the type
of research effort contemplated before expending extensive effort in preparing a detailed
research proposal or submitting proprietary information.
Contact: Dr. Robert Campbell, Biosciences, (919) 549-4230; email: bob.Campbell@
us.army.mil
Internet: http://www.aro.ncren.net
Sponsor: Army Research Office
P.O. Box 12211
Research Triangle Park, NC 27709-2211

ARO Chemical Kinetics Research Contracts   1005
The army's program in ignition and combustion processes associated with energetic
materials, explosives, detonation phenomena, the control of energy release and energy
transfer processes will benefit from increased understanding of fast reactions of energetic
species. The army is especially interested in the investigation of chemical reactions
using time-resolved techniques to observe transient species and infer reaction pathways
and other experiments and calculations that enable modeling of the time dependent
processes of ignition and combustion. Research on controlled transformation of toxic
materials to relatively benign products in chemical reactors is also of interest. Proposals
are accepted at any time.
Requirements: Organizations and individuals interested in submitting research proposals
are strongly encouraged to make preliminary inquiries as to the general need for the type
of research effort contemplated before expending extensive effort in preparing a detailed
research proposal or submitting proprietary information.
Contact: Dr. Robert Shaw, (919) 549-4293; email: Robert.Shaw@us.army.mil
Internet: http://www.aro.ncren.net
Sponsor: Army Research Office
P.O. Box 12211
Research Triangle Park, NC 27709-2211

ARO Computational Mathematics Research Contracts   1006
The computational mathematics program supports army needs in producing faster,
more stable and accurate solutions to relevant problems in the physical, biological and
engineering sciences, and in operations. The areas of interest in numerical methods include
methods for efficient numerical solution of nonlinear partial differential equations (both
time-dependent and steady state), such as adaptive finite difference and finite element
methods, high order methods, gridless methods and methods for computing interfaces,
such as front tracking and level sets. In the field of optimization, specific topics currently
supported include interior-point trust-region approaches to nonlinearly constrained
optimization, limited-memory approaches for large-scale optimization, and convergence
of optimization methods for singular problems. Proposals are accepted at any time.
Requirements: Organizations and individuals interested in submitting research proposals
are strongly encouraged to make preliminary inquiries as to the general need for the type
of research effort contemplated before expending extensive effort in preparing a detailed
research proposal or submitting proprietary information.
Contact: Dr. Stephen Davis, Mathematics and Computer Science Division, (919) 549-
4284; email: stephen.f.davis@arl.army.mil
Internet: http://www.aro.ncren.net/research/index.htm
Sponsor: Army Research Office
P.O. Box 12211
Research Triangle Park, NC 27709-2211

ARO Conference and Symposia Grants   1007
ARO supports conferences and symposia in special areas of science that bring experts
together to discuss recent research or education findings or to expose other researchers
or advanced graduate students to new research and educational techniques. ARO
encourages the convening in the United States of major international conferences and
assemblies of international alliances.
Requirements: Scientific, technical, or professional organizations that qualify for tax
exemption under 501(c)3 may receive conference and symposia grants. Conference
support proposals should be submitted a minimum of six months prior to the date of
the conference.
Restrictions: Cosponsorship of conferences and symposia with industrial concerns is
not eligible. Funds provided by the ARO cannot be used to support participants from
communist countries, nor can they be used for payment to any federal government
employee for support, subsistence, or services in connection with the proposed conference
or symposium.
Contact: ARO Legal Office, (919) 549-4292; email: Mark.rutter@us.army.mil
Internet: http://www.aro.army.mil
Sponsor: Army Research Office
P.O. Box 12211
Research Triangle Park, NC 27709-2211

ARO Degradation, Reactivity, and Protection of Materials Research Contracts   1008
The program in degradation, reactivity, and protection of materials investigates the
metastable behavior of materials prepared by nonequilibrium surface processing
approaches, including high-pressure, shock, plasma, beam, and self-assembling
techniques. The principal objective is to discover new nonequilibrium structural
materials and novel smart materials, which will enhance the reliability of Army systems
in service. This research includes nondestructive characterization of smart and ultra-
hard/superstrong materials and in-situ process monitoring. Principal emphasis is on
surface or interface control during processing of these materials, characterization of
their near-surface transport behavior and surface properties, and modeling or theoretical
predictions of their properties. The synthesis of novel degradation-resistant, ultra-high
strength nanolaminates and other refractory materials is also of interest. Experimental
and theoretical approaches are sought that provide for characterization of surfaces and
interfaces, modification of surface and near-surface regions, and processing and materials
characterization as this relates to the structure, wear, and environmental resistance
of smart materials and ultrahard/superstrong structural materials. Other phenomena
including spectroscopic and chemical analysis of low level hydrogen and hydrocarbons,
where these are important for understanding the properties of such new materials during
processing and in-service life assessment, are also of interest. Proposals may be submitted
at any time.

Requirements: Organizations and individuals interested in submitting research proposals are strongly encouraged to make preliminary inquiries as to the general need for the type of research effort contemplated before expending extensive effort in preparing a detailed research proposal or submitting proprietary information.
Contact: Dr. John Prater, Materials Science Research, (919) 549-4259; email: John.T.Prater@us.army.mil
Internet: http://www.aro.ncren.net/research/index.htm
Sponsor: Army Research Office
P.O. Box 12211
Research Triangle Park, NC 27709-2211

ARO Discrete Mathematics and Computer Science Research Contracts   1009

The interests in discrete mathematics are the development and analysis of solution procedures for discrete problems in computational geometry, computational algebra, logic, network flows, graph theory and combinatorics. Specific areas of emphasis include robust geometric computation, solid modeling, multiresolution methods, parallel and distributed computing and dynamic interactive visualization techniques. Other areas of interest include distributed algorithms for network flows, randomization in computing, computational algebraic geometry techniques for solution of polynomial systems, discrete methods for combinatorial optimization, symbolic methods for differential equations, mixed symbolic-numerical methods for applied problems, parallel symbolic sparse matrix methods, and algorithmic methods in symbolic mathematics arising in, for example, automated reasoning systems, mathematical logic and formal language theory. The interests in theoretical computer science include fundamental issues in parallel computing such as advanced data structures for parallel architectures, parallel algorithms, graph theoretic methods applied to a parallel and distributed computation and models and algorithms for the control of heterogeneous concurrent computing. Also of interest is research on tools for the development of parallel algorithms and expert systems for computation and visualization of solutions to partial differential equations. Exploring fundamental techniques that optimize I/O communication is a research area of great strategic importance. Proposals are accepted at any time.
Requirements: Organizations and individuals interested in submitting research proposals are strongly encouraged to make preliminary inquiries as to the general need for the type of research effort contemplated before expending extensive effort in preparing a detailed research proposal or submitting proprietary information.
Contact: Michael Coyle, Mathematics and Computer Science Division, (919) 549-4256.; email: joseph.michael.Coyle@us.army.mil
Internet: http://www.aro.ncren.net
Sponsor: Army Research Office
P.O. Box 12211
Research Triangle Park, NC 27709-2211

ARO Electrochemistry and Advanced Energy Conversion Research Contracts   1010

Army relies on compact power sources to support many different weapons systems, communications, and other devices. Power sources under development include batteries and fuel cells, microturbines, thermophotovoltaics, alkali metal thermal to electric converters. This program supports fundamental chemical studies of materials and processes that limit the performance of current or enable future power sources. Topics include ionic conduction in electrolytes, electro catalysis, fuel processing (particularly hydrogen), interfacial electron transfer, transport through coatings, surface films and polymer electrolytes, and activation of carbon-hydrogen bonds. Novel electrochemical synthesis, investigations into the effect of microenvironment on chemical reactivity, and quantitative models of electrochemical systems are also encouraged. Proposals are accepted at any time.
Requirements: Organizations and individuals interested in submitting research proposals are strongly encouraged to make preliminary inquiries as to the general need for the type of research effort contemplated before expending extensive effort in preparing a detailed research proposal or submitting proprietary information.
Contact: Dr. Richard Paur, Chemistry Division, (919) 549-4208; email: Richard.Paur@us.army.mil
Internet: http://www.aro.ncren.net
Sponsor: Army Research Office
P.O. Box 12211
Research Triangle Park, NC 27709-2211

ARO Fluid Dynamics Research Contracts   1011

Research in fluid dynamics supports the development of improved or new technology for advanced helicopters, small gas turbine engines, improved airdrop (parachute) systems, maneuverable high-speed missiles and high performance gun-launched projectiles. While basic research studies that address the fundamental flow physics underlying these devices are solicited, innovative research in the specific topical thrust areas of vortex-dominated flows, unsteady aerodynamics, and thermal science of micro-/mesoscale devices are especially encouraged. Proposals may be submitted at any time.
Requirements: Organizations and individuals interested in submitting research proposals are strongly encouraged to make preliminary inquiries as to the general need for the type of research effort contemplated before expending extensive effort in preparing a detailed research proposal or submitting proprietary information.
Contact: Dr. Thomas Doligalski, Engineering Sciences Division, (919) 549-4251; email: Thomas.Doligalski@us.army.mil
Internet: http://www.aro.ncren.net
Sponsor: Army Research Office
P.O. Box 12211
Research Triangle Park, NC 27709-2211

ARO Information Processing and Fusion and Circuits Research Contracts   1012

This program sponsors research the understanding of image background, including target-competitive clutter, and of how image background compounds the difficulty of target detection and recognition. Available models for background and clutter are currently inadequate. Objective measures of clutter and modeling paradigms that enable the quantification of image properties are needed for effective comparison of scenes, evaluation of algorithm performance, validation of synthetic imagery, and strategies for data fusion. Research is needed which addresses: (i) modeling of background and clutter, (ii) definition and assessment of clutter metrics, (iii) the manner in which clutter degrades the discrimination processes, and (iv) interaction between image background and targets. Proposals may be submitted at any time.
Requirements: Organizations and individuals interested in submitting research proposals are strongly encouraged to make preliminary inquiries as to the general need for the type of research effort contemplated before expending extensive effort in preparing a detailed research proposal or submitting proprietary information.
Contact: Dr. William Sander, Electronics Division, (919) 549-4241; email: william.sander@us.army.mil
Internet: http://www.aro.ncren.net
Sponsor: Army Research Office
P.O. Box 12211
Research Triangle Park, NC 27709-2211

ARO Mechanical Behavior of Materials Research Contracts   1013

The program on Mechanical Behavior of Materials addresses the fundamental relationships between the structure of materials and their mechanical properties as influenced by composition, processing, environment, stress state, and loading rate. The objectives of the subfield are to provide structural materials with improved mechanical properties and quantitative models for predicting both the response and the remaining useful life of a material. Major thrusts include the development of new strengthening, plasticity and toughening mechanisms for preventing or retarding fracture; especially at large strains (1000%) and high strain rates (to 10 6/sec). New knowledge is sought concerning fundamental deformation processes in materials including: load transfer, fatigue, creep, transformation toughening, super plasticity, and shear localization. New processing procedures are required for optimizing the mechanical properties of a material and reducing the inherent variations in the mechanical properties of materials. Additional thrusts include fundamental investigations on biomimetic and hierarchical materials to afford improved mechanical behavior and reliability. A fundamental understanding of the role of grain boundaries and interphases in composites and their hierarchical relationship to the overall mechanical behavior of the material system is desired. Proposals may be submitted at any time.
Requirements: Organizations and individuals interested in submitting research proposals are strongly encouraged to make preliminary inquiries as to the general need for the type of research effort contemplated before expending extensive effort in preparing a detailed research proposal or submitting proprietary information.
Contact: Dr. David Stepp, Materials Science Research, (919) 549-4329; email: David.M.Stepp@us.army.mil
Internet: http://www.aro.ncren.net
Sponsor: Army Research Office
P.O. Box 12211
Research Triangle Park, NC 27709-2211

ARO Microbiology and Biodegradation Research Contracts   1014

This program supports basic research on the biochemical and physiological mechanisms underlying the biodegradative processes in normal, extreme, and engineered environments; basic studies on organisms from extreme environments; the properties of materials that make them susceptible or resistant to biological attack; basic concepts for antifungals; mechanisms for the remediation of contaminated sites; analytical microbiology, including microbial signatures; and general microbial mechanisms with relevance to Army problems. Research into microbial communities and how to study organisms that cannot be grown in the lab. Research into methods to enhance the stabilization of military materiel that would include methods to prevent microbial growth is also sought. Proposals may be submitted at any time.
Requirements: Organizations and individuals interested in submitting research proposals are strongly encouraged to make preliminary inquiries as to the general need for the type of research effort contemplated before expending extensive effort in preparing a detailed research proposal or submitting proprietary information.
Contact: Dr. Shirley Tove, Biosciences, (919) 549-4344; email: sherry.tove@us.army.mil
Internet: http://www.aro.army.mil/research/index.htm
Sponsor: Army Research Office
P.O. Box 12211
Research Triangle Park, NC 27709-2211

ARO Minimum Energy Low Power Electronics Research Contracts   1015

The future army communications, information processing, imaging, and other systems must be portable, functional, versatile, and highly reliable. In addition, the systems will require ultra-high speed capability for handling complex voice, data, and video multimedia signal formats. Many of the new systems will find application in wireless communications, infrared imaging, and portable diagnostic and computational equipment. As these systems become more sophisticated, the necessary electronic circuitry becomes more complex, with increased prime power requirements. Currently, and in the near future, prime power for mobile and light weight system utilization is limited. It is necessary, therefore, to develop a new generation of electronic and other systems that can operate under minimum energy constraints with very low direct current (DC) power dissipation. This research topic goes beyond the simple desirability of efficient devices and circuits. It is aimed at new concepts resulting from considering the multilevel problem; materials, devices, circuits, coding, networking, and an overarching system of systems considerations, in order to reduce overall power dissipation in RF systems. Research can be approached from the component level or the overall design level. Proposals may be submitted at any time.

Requirements: Organizations and individuals interested in submitting research proposals are strongly encouraged to make preliminary inquiries as to the general need for the type of research effort contemplated before expending extensive effort in preparing a detailed research proposal or submitting proprietary information.

Contact: Dr. William Clark, Associate Director, Electronics Division, (919) 549-4314; email: William.W.Clark@us.army.mil

Internet: http://www.aro.ncren.net

Sponsor: Army Research Office

P.O. Box 12211

Research Triangle Park, NC 27709-2211

ARO Mobile, Wireless Communications and Networks Research Contracts   1016

The mobile, wireless communications and networks research program is concerned primarily with establishing the fundamental understanding necessary to support the Army's future mobile, wireless tactical battlefield communications needs. The research in this program primarily targets the tactical battlefield at brigade and below. The Army is interested in communication systems operating in frequency bands traditionally occupied by narrowband radios high frequency (HF), very high frequency (VHF), and ultrahigh frequency (UHF) as well as systems operating in frequencies extending into the millimeter wave region. These systems must support broad-based and highly mobile communications and must perform in environments of impressive diversity, from dense foliage to dense urban obstructions, and unintentional and intentional jamming. Future Army tactical communication systems for the digital battlefield will consist of many different types of networks and must be capable of communicating on the move. These systems will be highly mobile creating highly dynamic network topologies (mobile ad-hoc networks) and routing multimedia (voice, data and video) data. Unlike commercial systems, the communications infrastructure must be mobile. In addition to the highly mobile communications, there is interest in algorithms for small, very energy-limited, stationary, unattended ground sensors. Proposals may be submitted at any time.

Requirements: Organizations and individuals interested in submitting research proposals are strongly encouraged to make preliminary inquiries as to the general need for the type of research effort contemplated before expending extensive effort in preparing a detailed research proposal or submitting proprietary information.

Contact: Dr. Robert Ulman, Electronics Division, (919) 549-4330; email: Robert.Ulman@us.army.mil

Internet: http://www.aro.ncren.net

Sponsor: Army Research Office

P.O. Box 12211

Research Triangle Park, NC 27709-2211

ARO Organic Chemistry Research Contracts   1017

Molecules targeted for special attention include pigments and dyes with emission and reflectance characteristics useful for camouflage and controlled signature applications, potentially important monomers, especially for preparation of elastomer barrier materials and polymer electrolytes. New, more efficient, and environmentally benign organic reactions, both stoichiometric and catalytic, are of interest, especially nitration, and oxidative and nucleophilic displacements of phosphorus and sulfur for destruction of toxic organic compounds. Selected mechanistic studies that promise new insights to the pathways of the above reactions are encouraged as are new synthetic pathways with reduced production of waSuite byproducts. Proposals may be submitted at any time.

Requirements: Organizations and individuals interested in submitting research proposals are strongly encouraged to make preliminary inquiries as to the general need for the type of research effort contemplated before expending extensive effort in preparing a detailed research proposal or submitting proprietary information.

Contact: Dr. Stephen Lee, (919) 549-4365; email: Stephen.Lee2@us.army.mil

Internet: http://www.aro.ncren.net

Sponsor: Army Research Office

P.O. Box 12211

Research Triangle Park, NC 27709-2211

ARO Physical Behavior of Materials Research Contracts   1018

The program of Physical Behavior of Materials seeks research directed at providing an improved understanding of the fundamental mechanisms and key materials and processing variables that determine the electronic, magnetic and optical (EMO) properties of materials and affect the reliability of EMO devices. Emphasis is on research that will facilitate the nanostructuring of materials to realize the materials-by-design concept where new and unique materials are constructed on the atomic scale with application-specific properties. This includes research on understanding the underlying thermodynamic and kinetic principles that control the evolution of microstructures; understanding the mechanisms whereby the microstructure affects the physical properties of materials; and developing insight and methodologies for the beneficial utilization and manipulation of defects and microstructure to improve material performance. Major trends in this subfield include: (i) electronic materials - materials for microelectronics and packaging; fabrication and processing of semi-conductors, interconnects and device structures, and the characterization and control of trace impurities, defects and interfaces in semiconductors, (ii) magnetic materials—bulk and thin-film processing of magnetic

materials for electronic and high frequency communications; and fundamental studies on magnetic coercivity and spin dynamics, and (iii) optical materials - materials and processing methods for detectors, lasers, nonlinear optical materials, refractive and diffractive optics, and optical windows and coatings. Research to improve the long-term stability of EMO materials, develop multifunctional or smart EMO materials, and develop low observable materials is also being sought. Other important areas of interest include new approaches for materials processing, new composite formulations, and surface treatments that minimize environmental impacts; and novel composite concepts, including multifunctional and hierarchical materials. Finally, there is general interest in identifying basic research in the area of manufacturing science, which will address fundamental issues related to the reliability and cost (including environmental) associated with the production and long-term operation of Army systems.

Requirements: Organizations and individuals interested in submitting research proposals are strongly encouraged to make preliminary inquiries as to the general need for the type of research effort contemplated before expending extensive effort in preparing a detailed research proposal or submitting proprietary information.

Contact: Dr. John Prater, Materials Science Research, (919) 549-4259; email: John.T.Prater@us.army.mil

Internet: http://www.aro.ncren.net

Sponsor: Army Research Office

P.O. Box 12211

Research Triangle Park, NC 27709-2211

ARO Polymer Chemistry Research Contracts   1019

The Polymer Chemistry Program seeks novel, fundamental polymer research that may lead to new materials that provide the soldier with critical protection and required materiel. Research of interest may be related to chemical and biological agent protective materials, ballistic protection, transparent materials for eye protection and sensors, including laser protection, materials that have tunable optical properties, and lightweight super-strong materials. Research areas of interest that may be relevant include synthesis of polymers, including polymers with novel architectures and compositions, new approaches to synthesizing polymers, organic/inorganic hybrid polymeric materials, creating light-weight polymeric materials with enhanced strength, and the design and synthesis of environmentally benign polymeric materials that may benefit the soldier. Also of potential interest is characterization of structure/property relationships, diffusion and transport, and fiber properties related to polymer chemistry.

Requirements: Organizations and individuals interested in submitting research proposals are strongly encouraged to make preliminary inquiries as to the general need for the type of research effort contemplated before expending extensive effort in preparing a detailed research proposal or submitting proprietary information.

Contact: Dr. Douglas Kiserow, Chemistry Division, (919) 549-4213; email: Douglas.Kiserow@us.army.mil

Internet: http://www.aro.ncren.net

Sponsor: Army Research Office

P.O. Box 12211

Research Triangle Park, NC 27709-2211

ARO Quantum Information Science Research Contracts   1020

Quantum mechanics provides the opportunity to perform highly nonclassical operations that can result in exponential speed ups in computation or ultra-secure transmittal of information. This workpackage seeks to understand, control, and exploit such nonclassical phenomena for revolutionary advances in computation and secure communication. Major areas of interest are fundamental studies, quantum computation, and quantum communication. Proposals may be submitted at any time.

Requirements: Organizations and individuals interested in submitting research proposals are strongly encouraged to make preliminary inquiries as to the general need for the type of research effort contemplated before expending extensive effort in preparing a detailed research proposal or submitting proprietary information.

Contact: Dr. Henry Everitt, Physics Division, (919) 549-4369; email: henry.o.everitt@us.army.mil

Internet: http://www.aro.ncren.net/research/index.htm

Sponsor: Army Research Office

P.O. Box 12211

Research Triangle Park, NC 27709-2211

ARO Short-Term Innovation Research Program (STIR)   1021

The objectives of the STIR program are to fund innovative ideas in basic research. Proposed research may be for the continuation of or the natural outgrowth of experimental or theoretical explorations.

Requirements: Research proposals are sought from educational institutions, nonprofit organizations, and private industry. Prospective offerors of a STIR proposal are encouraged to contact the ARO program manager in their area of interest to ascertain the extent of interest in the specific research project. Research must be completed within six months of award of the agreement.

Restrictions: No capital equipment may be purchased. Travel costs must not exceed $500. Report preparation costs must not exceed $100. The assessment of indirect costs or fee is unallowable.

Amount of Grant: $50,000 maximum

Contact: Dr. Jim Chang, Director, (919) 549-4201; email: jim.chang@us.army.mil

Internet: http://www.aro.army.mil/research

Sponsor: Army Research Office

P.O. Box 12211

Research Triangle Park, NC 27709-2211

ARO Software and Knowledge-Based Systems Research Contracts   1022

The program in software and knowledge-based systems (SKBS) addresses the theoretical bases for the analysis, design, development, and evolution of advanced information-based systems that enable significant improvements in state-of-the-art engineering for software, software-based applications (including modeling and simulation), machine learning, knowledge/acquisition/representation/synthesis, and knowledgebases/databases. Currently, the research supported by the program is in software prototyping, development and evolution, formal methods for software engineering, and knowledgebase/database sciences. Proposals may be submitted at any time.

Requirements: Organizations and individuals interested in submitting research proposals are strongly encouraged to make preliminary inquiries as to the general need for the type of research effort contemplated before expending extensive effort in preparing a detailed research proposal or submitting proprietary information.

Contact: Dr. David Hislop, Mathematics and Computer Science Division, (919) 549-4255; email: David.W.Hislop@us.army.mil

Internet: http://www.aro.ncren.net

Sponsor: Army Research Office

P.O. Box 12211

Research Triangle Park, NC 27709-2211

ARO Solid Mechanics Research Contracts   1023

Solid mechanics provides the link between optimized material properties on the one hand and the desired structural behavior in terms of changes in shapes under a specific set of constraints on the other. Complete understanding of the behavior of structures made of advanced materials and quantitative description of their behavior allow predictive capability so necessary for design methodology. In situations that are described as ballistic in nature, the army faces unique constraints of very high strain rates, large deformations, high pressures, and rapid changes in temperature. Analytical, experimental, and computational techniques are integrated to solve well-formulated problems. Predictive models, validated by well-characterized experiments, are needed to identify controlling parameters at every scale. These models must be implementable in codes for the analysis of the efficiency of the designed hardware in the performance of military missions. Proposals may be submitted at any time.

Requirements: Organizations and individuals interested in submitting research proposals are strongly encouraged to make preliminary inquiries as to the general need for the type of research effort contemplated before expending extensive effort in preparing a detailed research proposal or submitting proprietary information.

Date(s) Application is Due: Oct 15

Contact: Dr. Bruce LaMattina, (919) 549-4379; email: Bruce.LaMattina@us.army.mil

Internet: http://www.aro.army.mil/research/index.htm

Sponsor: Army Research Office

P.O. Box 12211

Research Triangle Park, NC 27709-2211

ARO Solid State Devices Research Contracts   1024

research area emphasizes efforts to establish a new and comprehensive base of knowledge for the electronic, photonic, acoustic and magnetic properties of solid-state materials, structures and devices. Functions such as very intelligent surveillance and target acquisition; command, control, and communications; electronic warfare; and reconnaissance, must be accomplished with the high data rates and real-time capability that are essential for these applications. To support the U.S. Army vision of Objective Force and Future Combat System of Systems (FCSS), these systems will need to operate at much higher speeds and frequencies, have greatly increased functionality, and have much higher levels of integration than present day technology provides. Therefore, fundamental research in the area of Solid State Devices is the corner stone and an essential requirement in the development of these future systems for military defense. Proposals may be submitted at any time.

Requirements: Organizations and individuals interested in submitting research proposals are strongly encouraged to make preliminary inquiries as to the general need for the type of research effort contemplated before expending extensive effort in preparing a detailed research proposal or submitting proprietary information.

Contact: Dr. Dwight Woolard, Electronics Division, (919) 549-4297; email: Dwight.Woolard@us.army.mil

Internet: http://www.aro.ncren.net

Sponsor: Army Research Office

P.O. Box 12211

Research Triangle Park, NC 27709-2211

ARO Stochastic Analysis, Applied Probability, and Statistics Research Contracts   1025

Army R&D programs directed toward system design, development, testing and evaluation problems generate a need for research in the field of stochastic processes, including stochastic differential equations. Under the stochastic analysis and applied probability subarea, special emphasis is placed on research into methods for the analysis of observations from phenomena modeled by such processes and to numerical methods for stochastic partial differential equations. Research areas of importance to the army in probability and its applications include stochastic optimization and approximation, stochastic control, large deviations, simulation methodology, spatial processes, and image analysis. Ideas are needed from Markov random fields, renormalization of the state space, scaling of time, nonlinear stochastic analysis, and infinite-dimensional stochastic differential equations. The techniques required include Brownian flows, infinite-dimensional stochastic processes driven by Poisson noise, and Levy noise. Under the statistical methods subarea, the army has great interest in statistical methods for very large data sets or very small data sets, sampled from nonstandard, poorly understood

distributions. The extraction of more information from small data sets requires improved methods for combining information from disparate sets, as in meta-analysis. Useful statistical models should be based on a thorough understanding of physical processes combined with sound statistical theory. Thus, it is important to integrate statistical procedures with scientific and engineering information about mechanisms as exemplified by a probabilistic methodology that describes the nature of the growth of cracks in different media and the associated statistical analysis. More research is required in several statistical areas including Bayesian methods, Markov random fields, cluster analysis, change point methods, and Markov chain Monte Carlo methods. It is important to bring novel statistical thinking into resource management and optimization in very large communication and logistics networks. Proposals may be submitted at any time.

Requirements: Organizations and individuals interested in submitting research proposals are strongly encouraged to make preliminary inquiries as to the general need for the type of research effort contemplated before expending extensive effort in preparing a detailed research proposal or submitting proprietary information.

Contact: Dr. Mou-Hsiung Harry, (919) 549-4229; email: mouhsiung.chang@us.army.mil

Internet: http://www.aro.ncren.net

Sponsor: Army Research Office

P.O. Box 12211

Research Triangle Park, NC 27709-2211

ARO Structures and Dynamics Research Contracts   1026

A significant challenge facing Army laboratory engineers is the determination of the influence of inertial, thermal, electrical, magnetic, impact, damping, and aerodynamic forces on the dynamic response of adaptive armament systems, ground vehicles, rotorcraft, missiles, projectiles, gears, parachutes, and shelters. Its resolution is of fundamental importance to the design and construction of affordable, reliable, durable, and maintainable Army equipment with acceptable levels of personnel safety and comfort. Consequently, the ARO is supporting basic research in these areas, with emphasis on air vehicle dynamics, including missile and rotorcraft dynamics; the dynamics, non-linear vibrations, structural control, and simulation of land vehicles and weapon systems; and the dynamic response of structural components and systems fabricated from advanced composite materials, with or without embedded actuators and sensors. Submittal of fundamental research proposals on the general topics described above is encouraged, keeping in view the paramount importance of Army relevance. Proposals may be submitted at any time.

Requirements: Organizations and individuals interested in submitting research proposals are strongly encouraged to make preliminary inquiries as to the general need for the type of research effort contemplated before expending extensive effort in preparing a detailed research proposal or submitting proprietary information.

Contact: Dr. Gary Anderson, Engineering Sciences Division, (919) 549-4317; email: Gary.L.Anderson@us.army.mil

Internet: http://www.aro.ncren.net

Sponsor: Army Research Office

P.O. Box 12211

Research Triangle Park, NC 27709-2211

ARO Synthesis and Processing of Materials Research Contracts   1027

The program on synthesis and processing of materials focuses on the use of innovative approaches for processing high-performance structural materials reliably and at lower costs. Emphasis is placed on the design and fabrication of new materials with specific microstructure, constitution, and properties. Research interests include experimental and theoretical modeling studies to understand the influence of fundamental parameters on phase formation, microstructural evolution, and the resulting properties, in order to predict and control materials structures at all scales ranging from atomic dimensions to macroscopic levels. Trends in this subfield include nonequilibrium materials processing (e.g., rapid solidification); powder synthesis and consolidation; novel processing of ceramics, polymers, metals and composites; welding and joining including composite materials; elastomers; fibers and fabrics; and utilization of microstructural, compositional, or other unique signatures which may provide non-destructive in situ feedback process control to enhance product reproducibility and quality. Supercritical fluid, shock-induced chemical processing and other innovative approaches for processing materials are also of interest. Proposals may be submitted at any time.

Requirements: Organizations and individuals interested in submitting research proposals are strongly encouraged to make preliminary inquiries as to the general need for the type of research effort contemplated before expending extensive effort in preparing a detailed research proposal or submitting proprietary information.

Contact: Dr. William Mullins, (919) 549-4286; email: William.Mullins@us.army.mil

Internet: http://www.aro.army.mil/matsc/matsc.htm

Sponsor: Army Research Office

P.O. Box 12211

Research Triangle Park, NC 27709-2211

ARO Terrestrial Sciences Research Contracts   1028

In general, the terrestrial sciences program is concerned with the impact of the earth's surficial environment on army activities. Program interests cover a broad spectrum, ranging from terrain characterization and analysis, military engineering, and mobility considerations under combat conditions, to the management and stewardship of its installations as regards the impact of army activities on the natural environment. Primary emphasis is directed toward understanding the behavior of the land surface and the near-surface environments, understanding the natural processes operating upon and within these domains, and modeling these environments for predictive and simulation purposes.

Special emphasis is given to the need to better understand, model/simulate, and predict those environments/conditions that are most extreme, dynamic, or restrictive to systems performance or military operations. Proposals may be submitted at any time.
Requirements: Organizations and individuals interested in submitting research proposals are strongly encouraged to make preliminary inquiries as to the general need for the type of research effort contemplated before expending extensive effort in preparing a detailed research proposal or submitting proprietary information.
Contact: Dr. Russell Harmon, Environmental Sciences Division, (919) 549-4326; email: Russell.Harmon@us.army.mil
Internet: http://www.aro.ncren.net
Sponsor: Army Research Office
P.O. Box 12211
Research Triangle Park, NC 20779-2211

ARO Theoretical Physics and Nonlinear Phenomena Research Contracts   **1029**
The theoretical physics and nonlinear phenomena program is very closely coupled to experimental science as well as to ARO's programs in mathematics, chemistry, biological chemistry, materials science, and engineering sciences. The program thus encompasses a broad base including research in electron physics, photon physics, and classical and quantum mechanical systems, and statistical physics. It includes first-principles derivations of thermomechanical strengths of alloys for armor and armor penetrators; and electronic band structure calculations of materials for electronic, magnetic, optical, and optoelectronics applications, including those that result from quantum well and multiquantum well structures for signal generation, signal processing, and propagation and detection of signals. Also of interest are many-body theoretic approaches that address the electron correlation problem in extended molecular and condensed matter systems to provide the means to predict reaction kinetics, nonequilibrium dynamics, and application to the "alloy problem. There is interest in quantum optics research to explore the role of coherent states, squeezed states, etc. which may provide new tools for improved information processing and means to control information. Statistical physics interests go beyond thermodynamics, into nonequilibrium structures and their metastability, into information theoretic formulations, and into decision algorithms to connect the underlying physics to real world applications via proper modeling, instrumentation, and data analysis. Subtopics of interest include theoretical condensed matter physics, nonlinear dynamics, and nonequilibrium dynamics. Proposals may be submitted at any time.
Requirements: Organizations and individuals interested in submitting research proposals are strongly encouraged to make preliminary inquiries as to the general need for the type of research effort contemplated before expending extensive effort in preparing a detailed research proposal or submitting proprietary information.
Contact: Dr. Mikael Ciftan, Physics Division, (919) 549-4236; email: Mikael.Ciftan@us.army.mil
Internet: http://www.aro.army.mil
Sponsor: Army Research Office
P.O. Box 12211
Research Triangle Park, NC 27709-2211

ARO Young Investigator Program/Presidential Early Career Awards   **1030**
The objective of the Young Investigator Program (YIP) is to attract to army research outstanding young university faculty members, to support their research and to encourage their teaching and research careers. Outstanding YIP projects may be considered for a Presidential Early Career Award for Scientists and Engineers (PECASE). PECASE awards are the highest honor bestowed by the Army on outstanding scientists and engineers beginning their independent careers. Awards are granted in all research topic areas in the army Research Office. Support for the YIP is limited to three years; PECASE support is limited to five years. Proposals may be submitted at any time.
Requirements: This program is open to U.S. citizens holding tenure-track positions at U.S. universities and colleges who have held their graduate degrees (PhD or equivalent) for fewer than five years at the time of application. ARO strongly encourages informal discussions with the cognizant ARO technical program manager before submission of a formal proposal.
Amount of Grant: $50,000 maximum for three years
Contact: Kurt Preston, (919) 549-4234; email: Kurt.Preston@us.army.mil
Internet: http://www.aro.ncren.net
Sponsor: Army Research Office
P.O. Box 12211
Research Triangle Park, NC 27709-2211

Arronson Foundation Grants   **1031**
The foundation supports nonprofit organizations in the areas of religion, including churches (Baptist, Christian, Jewish, Roman Catholic, and Salvation Army) and religious education, higher education, health care, hospices, Jewish welfare, international ministries/missions, and youth. The foundation awards grants primarily in New York, NY; Philadelphia, PA; and Israel. Types of support include endowment funds, general operating support grants, research, scholarships, and seed money grants. Applicants should submit a brief letter of inquiry and include information on the organization and its work. There are no application deadlines.
Requirements: Nonprofit organizations in Pennsylvania, with emphasis on the Philadelphia area, are eligible to apply.
Geographic Focus: Pennsylvania
Samples: Federation of Allied Jewish Appeal (Philadelphia, PA)—for operating support, $100,000.
Amount of Grant: $200-$25,000 range

Contact: Joseph Kohn, President & Secretary, (215) 238-1700 or (215) 238-1968; email: jkohn@kohnswift.com
Sponsor: Arronson Foundation
1 S Broad Street, Suite 2100
Philadelphia, PA 19107

ARS New Investigator Award   **1032**
The purpose of this award is to support basic, translational, or clinical research projects in rhinology. Proposed projects may be related to any area of rhinology. Proposed project shall be designed in collaboration with a preceptor investigator and approved by the candidate's department chairperson and institution. One award of up to $25,000 over a two-year period will be given annually. Letters of Interest should be received no later than December 15, with full proposals due on January 15.
Requirements: Any member of the American Rhinologic Society who has not received previous, significant outside funding is eligible. The applicant must have, as a mentor, an established researcher who will provide a letter of support stating the extent of involvement in the project and provide a summary of his/her research experience.
Amount of Grant: $25,000 maximum total (direct and indirect) costs
Date(s) Application is Due: Dec 15; Jan 15
Contact: Stephanie L. Jones, Assistant Director; (703) 519-1586 or (703) 836-4444; email: sljones@entnet.org
Internet: http://www.entnet.org/EducationAndResearch/coreGrants.cfm
Sponsor: American Rhinology Society (ARS)
One Prince Street
Alexandria, VA 22314-3357

ARS Resident Research Grants   **1033**
The purpose of this award is to support basic, translational, or clinical research projects in rhinology. Proposed projects may be related to any area of rhinology. Proposed project shall be designed in collaboration with a preceptor investigator and approved by the candidate's department chairperson and institution. Two awards of up to $8,000 over a one-year period will be given annually. Letters of Interest should be received no later than December 15, with full proposals due on January 15.
Requirements: Any resident in training in an approved program in the U.S. or Canada is eligible to apply for the American Rhinologic Society Research Grant. Resident applicants must have as a co-investigator a supervising faculty who is a member in good standing in the American Rhinologic Society (ARS).
Amount of Grant: $8,000 maximum
Date(s) Application is Due: Dec 15; Jan 15
Contact: Stephanie L. Jones, Assistant Director; (703) 519-1586 or (703) 836-4444; email: sljones@entnet.org
Internet: http://www.entnet.org/EducationAndResearch/upload/2009-ARS-Resideint-Research-Grant.pdf
Sponsor: American Rhinology Society (ARS)
One Prince Street
Alexandria, VA 22314-3357

Art Conservation Advanced Training Fellowships   **1034**
The fellowships are awarded to encourage qualified individuals to prepare for careers as conservators and curators in museums of art. Conservation fellowships support one-year internships in advanced fine arts conservation at a museum or conservation research facility. Curatorial fellowships support one-year internships for curatorial training in European art at an American museum. Fellowship stipends support travel, administrative costs, benefits for the fellow, and other direct costs of the fellowship. Deadlines are January 15 for curatorial fellowships and March 1 for conservation fellowships.
Requirements: Individuals who have completed an MA degree in art conservation are eligible for conservation fellowships. Application must be made by the museum or conservation research facility at which the internship will be based. Individuals who have completed a PhD in the history of European art are eligible for curatorial fellowships.
Restrictions: Fellowships will not be awarded to complete a degree program.
Samples: Art Conservation Research Foundation (New York, NY)—for research by Spike Bucklow for an interactive database on the properties and phenomena of painted surfaces, conducted at the Hamilton Kerr Institute, Cambridge, England, $5000.; Harvard U (Cambridge, MA)—for research by David Bull on the painting style of Leonardo da Vinci conducted at the Villa I Tatti, Florence, Italy, $5000.; U of Iowa (Iowa City, IA)—for research by Timothy Barrett on the role of gelatin in the permanence of paper, $5000.
Amount of Grant: $30,000
Date(s) Application is Due: Jan 15; Mar 1
Contact: Lisa Ackerman, Executive Vice President, (212) 861-4993; fax: (212) 628-3146
Internet: http://www.kressfoundation.org/kressorg/advafell.html
Sponsor: Samuel H. Kress Foundation
174 E 80th Street
New York, NY 10021

Arthritis Foundation of Australia Grants   **1035**
The foundation awards grants in the fields of medicine and health, with emphasis on arthritis research and education. Types of awards include grants-in-aid, fellowships, and scholarships. Grants are awarded in Australia, the United Kingdom, and the United States. Annual deadline dates may vary; contact program staff for exact dates.

Requirements: Awards are available to clinical, scientific, and allied health professionals who are Australian citizens or permanent residents. Scholarships, fellowships, and grants may only be held in universities, hospitals, or recognized research institutes.

Amount of Grant: $A10,000-$A15,000 grants-in-aid

Date(s) Application is Due: Jun 30

Contact: Grant Coordinator, 61 (0)2 9552 6085; fax: 61 (0)2 9552 6078; email: info@arthritisaustralia.com.au

Internet: http://arthritisaustralia.com.au/Research%20Grants

Sponsor: Arthritis Foundation of Australia

P.O. Box 121

Sydney 2001 Australia

Arthritis Medical Research Grants 1036

The foundation's primary purpose is to provide financial support to research studies aimed at discovering new knowledge for the prevention, treatment, and cure of arthritis and other rheumatic diseases. Both clinical and basic studies will be considered. The focus of proposals should be on high-incidence diseases. Grants are awarded for one year; projects requiring multiyear funding must reapply each year. Partial funding may be awarded if the grantee has documentation that the remainder of the funding for proposed research has been acquired. Grants support salaries, supplies, and equipment directly related to the proposed studies. Overhead or indirect costs will not be funded.

Requirements: Priority will be given to newer investigators possessing either a MD and/or PhD degree who are associated with qualified institutions.

Samples: Arunan Kaliyaperuman, Northwestern U (Evanston, IL)—for autoimmune diseases research, $38,000.; Morteza Setareh, Scripps Research Institute (La Jolla, CA)—for research on mechanisms in the human body that trigger autoimmune diseases, $38,000.

Amount of Grant: $20,000-$50,000

Date(s) Application is Due: Jan 15

Contact: Helene Belisle, Executive Director, (800) 588-2873; fax: (562) 983-1410; email: anrf@ix.netcom.com

Internet: http://www.curearthritis.org/grant.htm

Sponsor: Arthritis National Research Foundation

200 Oceangate, Suite 400

Long Beach, CA 90802

Arthritis Society Investigator Awards 1037

A limited number of grants are offered to individuals who have demonstrated their ability as independent research scientists with international impact and are pursuing a full-time academic career in medical science. Candidates will only be considered between three and eight years after receiving their first academic appointment. Registration deadline is August 15; full application is due by September 15.

Requirements: Application is made on behalf of candidates by the chair of the department or director of the research institute in which the scholar will be appointed. Application should be submitted on form NF-109A.

Restrictions: Grants will not be made to replace faculty salary support for well-established investigators who have held faculty appointment for longer than five years at the applying institution.

Amount of Grant: $C70,000

Contact: Medical and Scientific Program, (416) 967-3353 ext 311 or 979-7228; fax: (416) 979-8366; email: can@mtsinai.on.ca

Internet: http://www.arthritis.ca/look%20at%20research/awards/investigators/default.asp?s=1

Sponsor: The Arthritis Society (Canada)

393 University Avenue, Suite 1700

Toronto, ON M5G 1E6 Canada

Arthritis Society Multicenter Grants 1038

Multicenter grants are designed to foster collaboration of individuals, working at different geographic locations, on a research project deemed relevant to the rheumatic diseases. Grants are usually made for a period of three years and are intended to cover the costs of the research for which they are provided. Appropriate space and basic facilities at the institutions concerned are prerequisite to an application. Registration deadline is November 15; full application is due by December 15.

Requirements: Completed applications for grants or renewals must be submitted on form NF-107A; obtain detailed instructions from the society.

Samples: R. McKendry and J. Esdaile, U of Ottawa (Ottawa, Canada)—for Multi-RDU Maintenance Gold Discontinuation Study, $C19,800.; Kiem Oen, U of Manitoba (Manitoba, Canada)—for Predictors of Disease Outcome in JRA, $C60,000.

Amount of Grant: $C60,000 average

Date(s) Application is Due: Nov 15; Dec 15

Contact: Medical and Scientific Program, (416) 967-3353 ext 311 or 979-7228; fax: (416) 979-8366; email: can@mtsinai.on.ca

Internet: http://www.arthritis.ca/look%20at%20research/awards/default.asp?s=1

Sponsor: The Arthritis Society (Canada)

393 University Avenue, Suite 1700

Toronto, ON M5G 1E6 Canada

Arthritis Society New Investigator Award 1039

The awards are intended to support newly appointed junior faculty members (within three years of appointment) who are planning to pursue a career in basic or clinical research related to arthritis. The scholarships assist individuals with the PhD or MD degree who have received extensive postdoctoral research training in their chosen area

and who have clearly demonstrated their potential as independent research scientists to establish themselves. Scholarships are not renewable; prior recipients may apply for research scientist grants. Registration deadline is November 15; full application is due by December 15.

Requirements: Application is made on behalf of candidates by the chair of the department or director of the research institute in which the scholar will be appointed. Application should be submitted on form NF-110A.

Restrictions: The awards are not intended to replace faculty salary support for well-established investigators who have been on faculty for a number of years at the applying institution.

Amount of Grant: $C50,000

Date(s) Application is Due: Nov 15; Dec 15

Contact: Medical and Scientific Program, (416) 967-3353 ext 311 or 979-7228; fax: (416) 979-8366; email: can@mtsinai.on.ca

Internet: http://www.arthritis.ca/look%20at%20research/awards/new%20investigators/default.asp?s=1

Sponsor: The Arthritis Society (Canada)

393 University Avenue, Suite 1700

Toronto, ON M5G 1E6 Canada

Arthritis Society Research Grants 1040

Grants in aid of research are designed to promote and support research proposed by investigators holding staff appointments at Canadian universities or other recognized Canadian institutions where the research is deemed relevant to the rheumatic diseases. Grants are usually made for a period of three years and are intended to cover the costs of the research for which they are provided. Appropriate space and basic facilities at the institution concerned are prerequisite to an application. Requests for all funds on grant applications must be carefully and fully justified.

Requirements: Completed applications for grants or renewals must be submitted on form NF-104A; obtain detailed instructions from the society.

Samples: Suzanne Bernier, U of Western Ontario (Canada)—for Cell Signalling in Chondrocytes and Arthritis, $C59,000.; Elizabeth Theriault, U of Toronto (Canada)—for The Pathogenesis of Inflammatory Joint Pain: Spinal Cord Mechanisms, $C49,380.

Amount of Grant: $C60,000 average

Date(s) Application is Due: Dec 15

Contact: Medical and Scientific Program, (416) 967-3353 ext 311 or 979-7228; fax: (416) 979-8366; email: can@mtsinai.on.ca

Internet: http://www.arthritis.ca/look%20at%20research/awards/d9/default.asp?s=1

Sponsor: The Arthritis Society (Canada)

393 University Avenue, Suite 1700

Toronto, ON M5G 1E6 Canada

Arthur and Rochelle Belfer Foundation Grants 1041

The foundation awards grants to nonprofit organizations of the Jewish faith, with a focus on New York. Grants are targeted toward programs supporting the elderly and women; education/higher education; institutions such as seminaries, synagogues, and temples; hospitals; Jewish welfare; and medical centers. Types of support include general support grants and fellowships. There are no application deadlines. Applicants should send a brief letter of inquiry describing the program.

Restrictions: Grants are not made to individuals.

Geographic Focus: New York

Samples: Joan S. Brugge, PhD, chair, Department of Cell Biology, arvard Medical School—for breast cancer studies (2006); Dana-Farber Cancer Institute (Boston, MA)—for general support, $1 million.; Anti-Defamation League of B'nai B'rith (New York, NY)—for operating support, $22,000.; American Friends of Israel Museum (New York, NY)—for operating support, $5000.

Amount of Grant: $1000-$100,000 average

Contact: Robert Belfer, President; 212-508-6020

Sponsor: Arthur and Rochelle Belfer Foundation, Inc

767 Fifth Avenue, 46th Fl

New York, NY 10153-0002

Arthur C. Cope Award 1042

The award is given to recognize outstanding achievement in the field of organic chemistry, the significance of which has become apparent within the five years preceding the year in which the award will be considered. In addition to the award to the recipient, an unrestricted grant in aid for $150,000 for research in organic chemistry, under the direction of the recipient, will be made to any university or nonprofit institution selected by the recipient. The recipient may choose to assign the Arthur C. Cope Fund Grant to an institution for use by others for research or education in organic chemistry.

Requirements: Any individual, except a member of the award committee, may submit one nomination or seconding letter for the award in any given year. The nominating documents consist of a letter of not more than 1000 words containing an evaluation of the nominee's accomplishments and a specific identification of the work to be recognized, a biographical sketch including date of birth, and a list of publications and patents authored by the nominee.

Restrictions: Self-nominations are not accepted.

Samples: James F. Stoddart—award recipient, $25,000 (2008); Jean M. J.Frechet—award recipient, $25,000 (2007); Peter Schultz—award recipient, $25,000 (2006).

Amount of Grant: $25,000 plus travel to award meeting for recipient

Date(s) Application is Due: Nov 1

Contact: Felicia Dixon, Awards Administrator; (800) 227-5558 or (202) 872-4408; fax: (202) 776-8008; email: f_dixon@acs.org or awards@acs.org

Internet: http://portal.acs.org/portal/acs/corg/content?_nfpb=true&_pageLabel=PP_ ARTICLEMAIN&node_id=1319&content_id=CTP_004495&use_sec=true&sec_ url_var=region1
Sponsor: American Chemical Society
1155 Sixteenth Street, NW
Washington, DC 20036

Arthur C. Cope Scholar Awards 1043

These awards are given to recognize and encourage excellence in organic chemistry. Ten (10) Arthur C. Cope Scholars will be named annually: four (4) between the ages of thirty-six and forty-nine, four (4) age fifty or older, and two (2) thirty-five years old and younger all inclusive before April 30, of the year in which the award is presented. The award consists of $5,000, a certificate and an unrestricted research grant in the amount of $40,000 to be assigned by the recipient to any university or nonprofit institution. The recipient is required to deliver a lecture at the annual Arthur C. Cope Symposium; traveling expenses incidental to participation in the symposium will also be paid.
Requirements: Any individual qualified to evaluate the nominee's accomplishments may submit one nomination for the award in any year; nomination should include a letter containing an evaluation of the nominee's contributions to organic chemistry with identification of the work to be recognized including literature and/or patent references; a biographical sketch of the nominee including date of birth and record of employment; and a list of publications and patents authored by the nominee.
Restrictions: Self-nominations are not accepted. An individual may not receive the award a second time. Recipients of the Arthur C. Cope Award are ineligible to be named Arthur C. Cope Scholars.
Samples: Jeffrey W. Bode, Cynthia J. Burrows, Dieter Enders, Tamio Hayashi, Linda C. Hsieh-Wilson, Colin P. Nuckolls, Melanie S. Sanford, Thomas S. Scanlan, Mukund P. Sibi, and Daniel A. Singleton—award recipients, $5,000 (2008).
Amount of Grant: $5,000 plus $40,000 for unrestricted research to a named institution
Date(s) Application is Due: Nov 1
Contact: Felicia Dixon, Awards Administrator; (800) 227-5558 or (202) 872-4408; fax: (202) 776-8008; email: f_dixon@acs.org or awards@acs.org
Internet: http://portal.acs.org/portal/acs/corg/content?_nfpb=true&_pageLabel=PP_ ARTICLEMAIN&node_id=1319&content_id=CTP_004496&use_sec=true&sec_ url_var=region1
Sponsor: American Chemical Society
1155 Sixteenth Street, NW
Washington, DC 20036

Arthur H. Cole Grants in Aid for Research in Economic History 1044

These grants in aid are designed to be supplemental to other grants or income, and support research in economic history regardless of time or geographic area. Applicants should supply seven copies of a presentation of not more than five pages that includes a description of the project, a curriculum vita, and a brief budget for the project. Annual deadline dates may vary; contact program staff for exact dates.
Requirements: Preference is given to recent PhD recipients.
Amount of Grant: $1500 typically
Date(s) Application is Due: Apr 1
Contact: The Committee on Research in Economic History, (785) 864-2847; fax: (785) 864-5270; email: eha@falcon.cc.ukans.edu
Internet: http://www.eh.net/EHA
Sponsor: Economic History Association
213 Summerfield Hall, University of Kansas, Dept of Economics
Lawrence, KS 66045

Arthur J. Ennis Fellowships 1045

The Arthur J. Ennis Fellowships in the Augustine and Culture: The Villanova Seminar Program were established in 1994 by the College of Liberal Arts and Sciences. Each is a one-year fellowship renewable for up to three years. These positions are available as of August 15. Each holder of the fellowship will teach three courses each semester in the College of Arts and Sciences Villanova Center for Liberal Education (VCLE).
Requirements: Candidates for these positions must have a PhD in a Humanities discipline, a strong commitment to undergraduate teaching in the humanities, and an interest in teaching the thought of Augustine and its influence. Candidates with an interest in Hebrew Scriptures and the New Testament are also encouraged to apply.
Geographic Focus: Pennsylvania
Contact: John A. Doody, Robert M. Birmingham Chair; (610) 519-4691 or (610) 519-7325; email: john.doody@villanova.edu
Internet: http://www.villanova.edu/artsci/vcle/fellowships/ennis.htm
Sponsor: Villanova University
St. Augustine Center Liberal Arts Room 104, 800 Lancaster Avenue
Villanova, PA 19085

Arthur L. Williston Medal 1046

This award is given annually for the best paper or thesis submitted in the Williston Award Contest by an undergraduate or junior engineer fostering a spirit of civil service. Candidates must be student members or associate members with an ASME sponsor for the award. In addition to the Williston Medal, second and third prizes are awarded.
Requirements: Applicant must be a student member or an associate member of ASME who has not graduated more than two years prior.
Amount of Grant: $1000 first prize; $500 second prize; $250 third prize
Date(s) Application is Due: Mar 1
Contact: Grants Administrator, (212) 591-7735; email: persaudl@asme.org

Internet: http://www.asme.org/honors/ms71/list.html
Sponsor: American Society of Mechanical Engineers
3 Park Avenue
New York, NY 10016

Arthur S. Tuttle Memorial Scholarships 1047

Scholarships are for tuition assistance to students during their first year of graduate studies in civil engineering. Any student in good standing who is also a national student member of the society may apply; the funds may be applied only for tuition expenses during the first year of formal graduate civil engineering education in an accredited educational institution. Financial need and educational standing will be considered in selecting recipients. The total sum for awards available will be determined annually based on earnings available from the endowment.
Requirements: Application consists of a completed application form; applicant's justification for applying; plans for continuing formal education; official transcripts, recommendation forms, and supporting statements from the student chapter faculty advisor and two other members of the engineering school faculty; and a detailed financial plan outlining both income and expenses anticipated for the award year. Applicants must be a member of the society, in any grade, and must be in good standing at the time of application and award. Membership applications may be submitted with scholarship application.
Amount of Grant: $2000-$5000
Date(s) Application is Due: Feb 9
Contact: Grants Coordinator, (800) 548-2723; fax: (703) 295-6132; email: student@ asce.org
Internet: http://www.asce.org/inside/stud_tuttle.cfm
Sponsor: American Society of Civil Engineers
1801 Alexander Bell Drive
Reston, VA 20191-4400

Arthur W. Adamson Award for Distinguished Service in the Advancement 1048
of Surface Chemistry

This annual award, supported by Occidental Petroleum Corporation, recognizes distinguished services in the advancement of surface chemistry. Activities recognized by the award may include such fields as teaching, writing, research, and administration. The award consists of $5,000, a medallion with a presentation box, and a certificate.
Requirements: Any individual, except a member of the award committee, may submit one nomination or seconding letter for each award in any given year. The letter of nomination should contain an evaluation of the nominee's accomplishments and a specific identification of the work to be recognized, and may not exceed 1000 words in length.
Restrictions: Self-nominations are not accepted.
Samples: Francisco Zaera—award recipient, $5,000 (2008); Charles T. Campbell—award recipient, $5,000 (2007); Steven Bernasek—award recipient, $5,000 (2006).
Amount of Grant: $5,000 and up to $1,000 travel allowance to the award meeting
Date(s) Application is Due: Nov 1
Contact: Felicia Dixon, Awards Administrator; (800) 227-5558 or (202) 872-4408; fax: (202) 776-8008; email: f_dixon@acs.org or awards@acs.org
Internet: http://portal.acs.org/portal/acs/corg/content?_nfpb=true&_pageLabel=PP_ ARTICLEMAIN&node_id=1319&content_id=CTP_004494&use_sec=true&sec_ url_var=region1
Sponsor: American Chemical Society
1155 Sixteenth Street, NW
Washington, DC 20036

Artists's Exploration Fund Travel Grants 1049

The grants enable individual performing artists to pursue opportunities abroad that further their artistic development. Grants support a variety of activities, including development of relationships with artists and arts organizations, research of significant artistic expression, participation in international conferences and seminars, or creation of new work. Eligible expenses include international and in-country travel, food and lodging, and other essential costs. The Explorations Fund will reimburse up to $2,000 to cover the real cost of one international airfare (based on a 14-day advance fare) and per diem, not to exceed $35 per day for up to 10 days. Per diems will be based on the total amount of days traveled.
Requirements: U.S. citizens and permanent residents are eligible for travel to any country outside the United States and its protectorates.
Restrictions: The fund will not support travel costs related to touring. Applications will not be accepted from students, scholars, curators, presenters, administrators, and critics.
Amount of Grant: $1000-$3000
Date(s) Application is Due: Apr 7; Sep 23
Contact: Olga Martins, Program Coordinator, (212) 924-0771 ext 101; fax: (212) 924-0773; email: omartins@artsinternational.org.
Internet: http://www.performingartsregister.com/opportunities.phtml
Sponsor: Arts International
770 Broadway, 2nd Fl
New York, NY 10003

ArvinMeritor Foundation Health Grants 1050

The foundation provides grants primarily in company-operating locations in the areas of education and training, civic and health, youth organizations, and arts and culture. In the area of Health, the foundation supports programs designed to promote health research, treatment, education, and awareness; and reduce the cost of illness. Types of support include general operating budgets, building funds, continuing support, equipment, and projects. The committee foundation meets every six to eight weeks to review grant requests. Contact program staff for current guidelines.

Requirements: Nonprofit 501(c)3 organizations should submit a one- to two-page letter outlining the purpose and needs of the program, its budget, duration, goals, leadership, and amount requested.

Restrictions: Ineligibility applies to individuals; organizations that limit participation or services based on race, gender, religion, color, creed, age, or national origin; projects without ties to a community that is home to an ArvinMeritor facility; organizations that pose any conflict with the goals and mission of ArvinMeritor, its employees, communities, or products; operating expenses for United Way local agencies, except through the foundation's support of annual United Way campaigns; sponsorships of fund raising activities by individuals (i.e., walk-a-thons); requests for loans or debt retirement; religious or sectarian programs for religious purposes; labor, political, or veterans organizations; fraternal, athletic, or social clubs; or seminars, conferences, trips, and tours.

Contact: Jerry Rush, Senior Director, Government and Community Affairs; (248) 435-7907; fax: (248) 245-1031; email: jerry.rush@arvinmeritor.com
Internet: http://www.arvinmeritor.com/community/community.asp
Sponsor: ArvinMeritor Foundation
2135 W. Maple Road
Troy, MI 48084

ASA & NCHS Research Fellowships   1051

The program is designed to bridge the gap between academic scholars and the federal government's health research programs. Fellowship recipients conduct research in Washington, DC, at the National Center for Health Statistics (NCHS), where they may use NCHS data and interact with NCHS staff. The 12-month fellowship may be extended one additional year under special circumstances. The stipend received is commensurate with current faculty salaries plus moving expenses and a travel allowance. Each applicant is required to submit a curriculum vita; the names and addresses of three references who may be contacted; and two copies of a detailed research proposal that includes a one-half page abstract, statement of relevant work already accomplished, statement citing the significance of the expected results to NCHS, and resource requirements.

Requirements: Applicants must be U.S. citizens.
Restrictions: U.S. government employees are ineligible.
Date(s) Application is Due: Nov 30
Contact: Dr. Phillip S. Kott, Chief Research Statistician; (703) 877-8000, ext. 102; fax: (703) 877-8042; email: phillip_kott@nass.usda.gov; Joyce Narine, ASA Grants Manager; (703) 684-1221, ext. 1837; email: joyce@amstat.org
Internet: http://www.amstat.org/careers/fellowshipsgrants.cfm
Sponsor: American Statistical Association
732 North Washington Street
Alexandria, VA 22314-1943

ASA and U.S.DA/NASS Research Fellow and Associate Program   1052

The purpose of this program is to supplement research in statistics with experience at NASS. The program is designed to provide the selected Fellows and Associates with research opportunities or experience in the application of statistical theory in all phases of large scale agricultural survey operations, including design, collection, quality control, forecasting, estimation, and analysis. Participants in the program will engage in research which is of mutual interest to the participant and NASS. Fellows and Associates conduct research in residence at NASS, gaining unique opportunities for use of extensive NASS data and interaction with NASS staff.

Requirements: Applicants for the fellowship program should have a Ph.D. and have established recognized research record and considerable expertise in their area of proposed research. Applicants for Associates are expected to have completed a Ph.D. in an appropriate field or to have made significant progress toward the degree (at least two years of graduate study).

Contact: Dr. Phillip S. Kott, Chief Research Statistician; (703) 877-8000, ext. 102; fax: (703) 877-8042; email: phillip_kott@nass.usda.gov; Joyce Narine, ASA Grants Manager; (703) 684-1221, ext. 1837; email: joyce@amstat.org
Internet: http://www.nass.usda.gov/research/ASA-NASS.html
Sponsor: American Statistical Association / National Agricultural Statistics Service
732 North Washington Street
Alexandria, VA 22314-3415

ASA Deming Lecturer Award   1053

The Deming Lecturer Award was established in 1995 to honor the accomplishments of W. Edwards Deming, recognize the accomplishments of the awardee, and enhance the awareness among the statistical community of the scope and importance of Deming's contributions. The awardee will give the Deming Lecture (an invited paper) at the Joint Statistical Meetings and receive a $1,000 honorarium, an award plaque, and travel expenses.

Requirements: The individual must have either made significant contributions in fields related to those in which Deming devoted his career-including survey sampling, statistics in the transportation industry, quality management, and quality improvement-or has made significant contributions through effective promotion of statistics and statistical thinking in business and industry.
Amount of Grant: $1,000
Date(s) Application is Due: Dec 31

Contact: Ronald Wasserstein, Executive Director; (703) 684-1221, ext. 1859; fax: (703) 684-6456; email: ron@amstat.org; Pamela Craven, Executive Secretary; (703) 684-1221, ext. 1860; fax: (703) 684-6456; email: pamela@amstat.org
Internet: http://www.amstat.org/careers/deminglectureraward.cfm
Sponsor: American Statistical Association
732 North Washington Street
Alexandria, VA 22314-1943

ASA Jeanne E. Griffith Mentoring Award   1054

The Jeanne E. Griffith Mentoring Award has been established to encourage mentoring of junior staff in the Federal statistical system. It is presented annually to a supervisor who is nominated by co-workers and supervisors, and chosen by the Award Selection Committee. The award is co-sponsored by the Interagency Council on Statistical Policy, the Council for Excellence in Government, the Washington Statistical Society, the Social Statistics and Government Statistics Sections of the American Statistical Association, and the Council of Professional Associations on Federal Statistics. The award will consist of a $1,000 honorarium and a citation, which will be presented at a ceremony arranged by the co-sponsors in June.

Amount of Grant: $1,000
Date(s) Application is Due: Mar 27
Contact: Ronald Wasserstein, Executive Director; (703) 684-1221, ext. 1859; fax: (703) 684-6456; email: ron@amstat.org; Pamela Craven, Executive Secretary; (703) 684-1221, ext. 1860; fax: (703) 684-6456; email: pamela@amstat.org
Internet: http://www.amstat.org/sections/sgovt/jegmaann09.htm
Sponsor: American Statistical Association
732 North Washington Street
Alexandria, VA 22314-1943

ASA Outstanding Statistical Application Award   1055

The Outstanding Statistical Application Award was established in 1986 to recognize the authors of papers that demonstrate an outstanding application of statistics in any substantive field. The award consists of an engraved award and $1,000 that is divided evenly among the winners. The award is given annually if an eligible and worthy work is nominated. The Award is bestowed upon a distinguished individual or individuals based on the following criteria: the impact of the statistical application in addressing a significant problem in a substantive field; and the ingenuity and or novelty of the statistical treatment of the problem. Eligible work includes papers, monographs, reports, and other substantive evidence appearing two years prior to the presentation of the award. All nominated work must have been subject to external peer review and, preferably, to formal refereeing. Nominations are due by April 1 each year. The award is presented at the Joint Statistical Meetings in August the same year.

Requirements: Nominations should include a letter that carefully describes the paper's significance, particularly its impact on the substantive field, along with copies of the nominated work.
Amount of Grant: $1,000
Date(s) Application is Due: Apr 1
Contact: Ronald Wasserstein, Executive Director; (703) 684-1221, ext. 1859; fax: (703) 684-6456; email: ron@amstat.org; Pamela Craven, Executive Secretary; (703) 684-1221, ext. 1860; fax: (703) 684-6456; email: pamela@amstat.org
Internet: http://www.amstat.org/careers/outstandingstatisticalapplicationaward.cfm
Sponsor: American Statistical Association
732 North Washington Street
Alexandria, VA 22314-1943

ASA Samuel S. Wilks Award   1056

The Samuel S. Wilks Award, one of the ASA's most prestigious, was established in 1964 to honor the memory and distinguished career of Samuel S. Wilks. The award recognizes outstanding contributions to statistics that carry on in the spirit of his work. The Wilks Memorial Award is bestowed upon a distinguished individual who has made statistical contributions to the advancement of scientific or technical knowledge, ingenious application of existing knowledge, or successful activity in the fostering of cooperative scientific efforts that have been directly involved in matters of national defense or public interest. The Award includes two cash benefits: $500 is provided by Friends of the U.S. Army to honor Wilks' contributions to Army research programs and $1,000 is provided by a fund created by Alexander M. Mood to honor Wilks' teaching skill and life-long development of outstanding statisticians. This second award was established in 2008 at $1,000, but will increase by 3% annually. Nominations are due by April 1 each year. The award is presented at the Joint Statistical Meetings in August the same year.

Amount of Grant: $1,000
Date(s) Application is Due: Apr 1
Contact: Ronald Wasserstein, Executive Director; (703) 684-1221, ext. 1859; fax: (703) 684-6456; email: ron@amstat.org; Pamela Craven, Executive Secretary; (703) 684-1221, ext. 1860; fax: (703) 684-6456; email: pamela@amstat.org
Internet: http://www.amstat.org/careers/samuelwilksaward.cfm
Sponsor: American Statistical Association
732 North Washington Street
Alexandria, VA 22314-1943

ASA W. J. Youden Award in Interlaboratory Testing   1057

The W.J. Youden Award in Interlaboratory Testing was established in 1985 to recognize the authors of publications that make outstanding contributions to the design and/or analysis of interlaboratory tests or describe ingenious approaches to the planning and evaluation of data from such tests. Each recipient receives an engraved award and a cash prize that are presented at the awards ceremony during the Joint Statistical Meetings. The current amount of the cash prize is $1,000, which is divided evenly among the winners. The award is given every year if, in the opinion of the awards committee, an eligible and worthy publication is nominated. Nominations are due by April 15 each year. The award is presented at the Joint Statistical Meetings in August.

Amount of Grant: $1,000
Date(s) Application is Due: Apr 15
Contact: Ronald Wasserstein, Executive Director; (703) 684-1221, ext. 1859; fax: (703) 684-6456; email: ron@amstat.org; Pamela Craven, Executive Secretary; (703) 684-1221, ext. 1860; fax: (703) 684-6456; email: pamela@amstat.org
Internet: http://www.amstat.org/careers/wjyoudenaward.cfm
Sponsor: American Statistical Association
732 North Washington Street
Alexandria, VA 22314-1943

ASA/BJS Statistical Methodological Research Program Grants 1058

The American Statistical Association (ASA) Committee on Law and Justice Statistics announces a small grant program for the analysis of Bureau of Justice Statistics (BJS) and other justice-related data. Award preference is given to research using BJS data that addresses the priority topics listed below. This program is designed to encourage the creative and appropriate use of these data to inform substantive and methodological issues. Dissertation research may qualify for these funds and young investigators are encouraged to apply. Research is to be completed within a two-year period. Awards are typically in the range of $25,000 to $30,000, although exceptional projects as high as $40,000 will be considered.

Amount of Grant: $25,000-$30,000 range; $40,000 maximum
Date(s) Application is Due: May 23
Contact: Dr. Phillip S. Kott, Chief Research Statistician; (703) 877-8000, ext. 102; fax: (703) 877-8042; email: phillip_kott@nass.usda.gov; Joyce Narine, ASA Grants Manager; (703) 684-1221, ext. 1837; email: joyce@amstat.org
Internet: http://www.amstat.org/careers/bjs.cfm
Sponsor: American Statistical Association
732 North Washington Street
Alexandria, VA 22314-3415

ASA/NSF Federal Statistics Fellowships 1059

The ASA/NSF/Federal Statistics Fellowship program is jointly supported by the National Science Foundation, U.S. Census Bureau, and the Bureau of Labor Statistics. The general objective of the program is to foster collaborative and interdisciplinary research efforts that will continue to stimulate the development and advancement of methodology and social science research relevant to issues on which federal statistical agencies seek to provide information. The program accomplishes this by bringing academic researchers to work with statisticians and social scientists in the three federal agencies for up to one year.

Date(s) Application is Due: Mar 1
Contact: Dr. Phillip S. Kott, Chief Research Statistician; (703) 877-8000, ext. 102; fax: (703) 877-8042; email: phillip_kott@nass.usda.gov
Internet: http://www.amstat.org/careers/fellowshipsgrants.cfm
Sponsor: American Statistical Association / National Science Foundation
732 Washington Street
Alexandria, VA 22314

ASA/NSF/BLS Fellowships 1060

ASA, under a grant from the National Science Foundation and with cosponsorship by the Bureau of Labor Statistics (BLS), seeks senior researchers for this program. Fellows conduct research in residence at BLS, gaining unique opportunities for use of BLS data and interaction with bureau staff. Fellows may pursue research in any area related to BLS data or methodology. Possible areas for research include, but are not limited to, areas under the fields of economic and social measurement and research, and statistical methodology and computing. Any proposed project should have the potential of encouraging further significant, broadly based research and should require hands-on access to BLS data and/or have direct application to BLS programs. Salary is commensurate with qualifications; fringe benefits and a travel allowance are included. A program brochure is available.

Requirements: Applicants for fellowships should have recognized research records and considerable expertise in their area of proposed research. Qualified women and members of minority groups are encouraged to apply for fellowships and associateships.
Date(s) Application is Due: Dec 10
Contact: Dr. Phillip S. Kott, Chief Research Statistician; (703) 877-8000, ext. 102; fax: (703) 877-8042; email: phillip_kott@nass.usda.gov; Joyce Narine, ASA Grants Manager; (703) 684-1221, ext. 1837; email: joyce@amstat.org
Internet: http://www.amstat.org/careers/fellowshipsgrants.cfm
Sponsor: American Statistical Association
732 North Washington Street
Alexandria, VA 22314-1943

ASA/NSF/Census Research Fellowships 1061

The ASA/NSF/Census Bureau Research Program helps bridge the gap between government and academic science by bringing researchers closer to the production of data sets relevant to their research. The program allows for senior statisticians, social scientists, computer scientists, geographers, and others to come to the U.S. Census Bureau as research fellows for 6 to 12 months to use bureau data sets and interact with bureau staff.

Requirements: Applicants should have recognized research records and considerable expertise in their areas of proposed research. They must submit detailed research proposals for competitive evaluation.
Date(s) Application is Due: Dec 10
Contact: Dr. Phillip S. Kott, Chief Research Statistician; (703) 877-8000, ext. 102; fax: (703) 877-8042; email: phillip_kott@nass.usda.gov; Joyce Narine, ASA Grants Manager; (703) 684-1221, ext. 1837; email: joyce@amstat.org
Internet: http://www.amstat.org/careers/fellowshipsgrants.cfm
Sponsor: American Statistical Association
732 North Washington Street
Alexandria, VA 22314-3415

ASCE Samuel Fletcher Tapman Scholarships 1062

These scholarships are awarded for undergraduate tuition assistance to ASCE Student Chapter members in good standing who will employ them to continue their formal undergraduate education in a recognized educational institution.

Requirements: Any undergraduate freshman, sophomore, or junior of an ASCE Student Chapter may apply; previous holders of this scholarship may also apply. Application consists of brief application form; statement of applicant's justification for receiving the award; applicant's plans for continuing his/her formal education; official transcript; and applicant appraisals from student chapter faculty advisor and not less than two other faculty members.
Restrictions: Each student chapter may submit only one application from its membership; three applicants in each society zone (total of 12) are selected each year.
Amount of Grant: $2000
Date(s) Application is Due: Feb 9
Contact: Grants Coordinator, (800) 548-2723; fax: (703) 295-6132; email: student@asce.org
Internet: http://www.asce.org/inside/stud_tin-tap.cfm
Sponsor: American Society of Civil Engineers
1801 Alexander Bell Drive
Reston, VA 20191-4400

ASCO Improving Cancer Care Grants 1063

The ASCO Cancer Foundation Improving Cancer Care Grant, funded by Susan G. Komen for the Cure, will provide extramural research funding to address important issues regarding access to healthcare, quality of care, and delivery of care, with general applicability to breast cancer. The goal of this program is to encourage multi-disciplinary research that will have a major impact on cancer care, with general applicability in the breast cancer arena. This grant opportunity is part of the Susan G. Komen for the Cure/ASCO Cancer Foundation Research Initiative. Susan G. Komen for the Cure and The ASCO Cancer Foundation expect to award up to three research grants, each totaling 1.35 million dollars. This money will be distributed as $450,000 total costs per year over three years. Applicants must submit their letter of intent in the online application system by 5:00 PM EDT on September 1st. Only online applications will be accepted. After review by the Research Initiative Grant Review Subcommittee, applicants will be notified about the status of their letter of intent. Only applicants who have submitted an approved letter of intent will be eligible to submit the full proposal. Applicants can expect to be notified on October 1st.

Requirements: Eligible research teams: will focus on implementing and/or evaluating new solutions to existing problems in quality of, access to, and delivery of care, with general applicability to breast cancer; will be led by a single Principal Investigator, who must be an active ASCO member (or have submitted a membership application) with an MD, DO, PhD or equivalent degree. All categories of active ASCO members are eligible; will have a multidisciplinary team of investigators that may include clinicians, nurses, pharmacists, statisticians, epidemiologists, information technologists, and other research experts; will be allowed to obtain expertise not represented in the core team through consultants and/or sub-contracts.
Amount of Grant: 1.35 million dollars total
Date(s) Application is Due: Sep 1; Dec 1
Contact: Nancy R. Daly, Executive Director; (571) 483-1432; email: dalyn@asco.org
Internet: http://www.ascocancerfoundation.org/TACF/Grants/Grant+Opportunities/Improving+Cancer+Care+Grant
Sponsor: American Society of Clinical Oncology
2318 Mill Road, Suite 800
Alexandria, VA 22314

ASCO State Affiliate Grants 1064

The State Affiliate Grant Program enables state/regional affiliates to engage in activities that will benefit their members, ASCO members, other state societies and the oncology community at large. The purpose of the State Affiliate Grant Program is to provide financial support to help state/regional affiliates to implement specific activities, such as conducting an educational meeting, developing an informational brochure or newsletter, creating a state clinical trials network, or taking part in a quality cancer care initiative collaborating with a patient advocacy group on an awareness campaign. For more information about the State Affiliate Grant Program, email stateaffiliates@asco.org. Applications are available online, with a submission deadline of November 17.

Requirements: All 48 Affiliates are eligible to apply for a grant.
Samples: Denali (Alaska) Oncology Group, $10,000 (2009); Georgia Society of Clinical Oncology, $10,000 (2009); Maryland/DC Society of Clinical Oncology, $10,000 (2009); New York State Society of Medical Oncologists & Hematologists, $10,000 (2009); Northern New England Society of Clinical Oncology, $10,000 (2009).
Amount of Grant: $10,000

Contact: Nancy R. Daly, Executive Director; (571) 483-1432; email: dalyn@asco.org
Internet: http://www.ascocancerfoundation.org/TACF/Professional+Education+and+
Support/Cancer+Policy
Sponsor: American Society of Clinical Oncology
2318 Mill Road, Suite 800
Alexandria, VA 22314

ASCO Advanced Clinical Research Award in Colorectal Cancer 1065

The Advanced Clinical Research Award (ACRA) in Colorectal Cancer is designed to fund investigators who are committed to clinical cancer research and who wish to conduct original research not currently funded. The grant totals $450,000 and is paid in 3 annual increments of $150,000 made on or about July 1 of each year of the grant term. The grant funds will be directed to the sponsoring institution and should be used towards salary support, supplies, equipment, travel, etc. necessary for the pursuit of the recipient's research project.
Requirements: Appplicants must meet the following criteria: be a physician (MD, DO, or international equivalent) who is in the fourth to ninth year of a full-time, primary faculty appointment in a clinical department at an academic medical institution at the time of grant submission; have completed productive post doctoral/post fellowship research and demonstrated the ability to undertake independent investigator-initiated clinical research; be an active member of ASCO or have submitted a membership application with the grant application; be able to commit to 75% of full-time effort in research (applies to total research, not just the proposed project) during the award period.
Amount of Grant: $450,000
Date(s) Application is Due: Nov 3
Contact: Nancy R. Daly, Executive Director; (571) 483-1432; email: dalyn@asco.org
Internet: http://www.ascocancerfoundation.org/TACF/Grants/Grant+Opportunities/
Advanced+Clinical+Research+Award+in+Colorectal+Cancer
Sponsor: American Society of Clinical Oncology
2318 Mill Road, Suite 800
Alexandria, VA 22314

ASCO Advanced Clinical Research Awards in Breast Cancer 1066

The ASCO Foundation and Genentech BioOncology invite physicians with full-time faculty appointments to apply for a clinical oncology research award in Breast Cancer. The program funds clinical investigators who are committed to clinical cancer research. Their research must have a direct patient-oriented focus including clinical trials and/or translational research involving human subjects. The Advanced Clinical Research Award is designed to provide funding beyond the Clinical Research Career Development Award (CDA). By continuing to support proven clinical researchers who are post-CDA but at a critical stage in their early career, the foundation hopes to expand the cadre of expert clinical oncology researchers who are developing promising research initiatives.
Requirements: The applicant must meet the following criteria at the time of grant award: be a physician (MD or DO) who is 5-10 years post final sub-specialty training; have a full-time faculty appointment in a clinical department at an academic medical center; have completed productive postdoctoral/post-fellowship research and demonstrated the ability to undertake independent investigator-initiated clinical research; be an active member of the American Society of Clinical Oncology (ASCO); and expect to spend 75% of time during the award period dedicated to research.
Amount of Grant: $450,000
Date(s) Application is Due: Oct 20
Contact: Nancy R. Daly, Executive Director; (571) 483-1432; email: dalyn@asco.org
Internet: http://www.ascocancerfoundation.org/TACF/Grants/Grant+Opportunities/
Advanced+Clinical+Research+Award+in+Breast+Cancer
Sponsor: American Society of Clinical Oncology
2318 Mill Road, Suite 800
Alexandria, VA 22314

ASCO Advanced Clinical Research Awards in Hematologic Malignancies 1067

The ASCO Foundation and Genentech BioOncology invite physicians with full-time faculty appointments to apply for a clinical oncology research award in Hematologic Malignancies. The program funds clinical investigators who are committed to clinical cancer research. Their research must have a direct patient-oriented focus including clinical trials and/or translational research involving human subjects. The Advanced Clinical Research Award is designed to provide funding beyond the Clinical Research Career Development Award (CDA). By continuing to support proven clinical researchers who are post-CDA but at a critical stage in their early career, the foundation hopes to expand the cadre of expert clinical oncology researchers who are developing promising research initiatives.
Requirements: The applicant must meet the following criteria at the time of grant award: be a physician (MD or DO) who is 5-10 years post final sub-specialty training; have a full-time faculty appointment in a clinical department at an academic medical center; have completed productive postdoctoral/post-fellowship research and demonstrated the ability to undertake independent investigator-initiated clinical research; be an active member of the American Society of Clinical Oncology (ASCO); and expect to spend 75% of time during the award period dedicated to research.
Amount of Grant: $450,000 (paid over three years)
Date(s) Application is Due: Aug 15
Contact: Julia McCormack, Executive Director, (703) 299-0150; fax: (703) 299-1044; email: McCormaj@asco.org
Sponsor: American Society of Clinical Oncology
1900 Duke Street, Suite 200
Alexandria, VA 22303

ASCO Advanced Clinical Research Awards in Lung Cancer 1068

The ASCO Foundation and Genentech BioOncology invite physicians with full-time faculty appointments to apply for a clinical oncology research award in lung cancer. The program funds clinical investigators who are committed to clinical cancer research. Their research must have a direct patient-oriented focus including clinical trials and/or translational research involving human subjects. The Advanced Clinical Research Award is designed to provide funding beyond the Clinical Research Career Development Award (CDA). By continuing to support proven clinical researchers who are post-CDA but at a critical stage in their early career, the foundation hopes to expand the cadre of expert clinical oncology researchers who are developing promising research initiatives.
Requirements: The applicant must meet the following criteria at the time of grant award: be a physician (MD or DO) who is 5-10 years post final sub-specialty training; have a full-time faculty appointment in a clinical department at an academic medical center; have completed productive postdoctoral/post-fellowship research and demonstrated the ability to undertake independent investigator-initiated clinical research; be an active member of the American Society of Clinical Oncology (ASCO); and expect to spend 75% of time during the award period dedicated to research.
Samples: Eric Haura, MD, H. Lee Moffitt Cancer Center and Research Institute—Targeting EGFR and SRC Tyrosine Kinase Signaling in Lung Cancer (2007); Pierre P. Massion, MD, Vanderbilt University Medical Center—Tumor-Specific Antibodies for Lung Cancer Diagnosis (2007).
Amount of Grant: $450,000 (paid over three years)
Date(s) Application is Due: Aug 8
Contact: Julia McCormack, Executive Director, (703) 299-0150; fax: (703) 299-1044; email: McCormaj@asco.org
Sponsor: American Society of Clinical Oncology
1900 Duke Street, Suite 200
Alexandria, VA 22303

ASCO Advanced Clinical Research Awards in Sarcoma 1069

The Advanced Clinical Research Award (ACRA) in Sarcoma is designed to fund investigators who are committed to clinical cancer research and who wish to conduct original research not currently funded. The award is intended to support proposals with a patient-oriented focus, including a clinical research study and/or translational research involving human subjects. The grant totals $450,000 and is paid in three annual increments of $150,000 made on or about January 1 of each year of the grant term. Preference will be given to proposals involving therapeutic vaccine or immunotherapeutic approaches for sarcoma.
Requirements: Applicants must meet the following criteria: be a physician (MD, DO, or international equivalent) who is in the fourth to ninth year of a full-time, primary faculty appointment in a clinical department at an academic medical institution at the time of grant submission; have completed productive post doctoral/post fellowship research and demonstrated the ability to undertake independent investigator-initiated clinical research; be an active member of ASCO or have submitted a membership application with the grant application; be able to commit to 75% of full-time effort in research (applies to total research, not just the proposed project) during the award period.
Restrictions: Proposals with a predominant focus on in vitro or animal studies (even if clinically relevant) are not allowed.
Amount of Grant: $450,000
Date(s) Application is Due: Aug 6
Contact: Nancy R. Daly, Executive Director; (571) 483-1432; email: dalyn@asco.org
Internet: http://www.ascocancerfoundation.org/TACF/Grants/Grant+Opportunities/
Advanced+Clinical+Research+Award+in+Sarcoma
Sponsor: American Society of Clinical Oncology
2318 Mill Road, Suite 800
Alexandria, VA 22314

ASCO Career Development Award 1070

The program provides funds to clinical investigators who have received their initial faculty appointment to establish an independent clinical cancer research program. Since many awards are supported with restricted grants from outside organizations, the ASCO is particularly interested in identifying young researchers working in the following sub-specialties and emerging disciplines: breast cancer; cancer survivorship; geriatric oncology; health disparities; kidney cancer; multiple myeloma; ovarian cancer; pancreatic cancer; sarcoma; survivorship; and young adult cancer. However, The ASCO Foundation welcomes application submissions in all oncology sub-specialties.
Requirements: Clinical investigators who have received their initial faculty appointment to establish an independent clinical cancer research program are eligible.
Amount of Grant: $200,000 (paid over three years)
Date(s) Application is Due: Sep 9
Contact: Nancy R. Daly, Executive Director; (571) 483-1432; email: dalyn@asco.org
Internet: http://www.ascocancerfoundation.org/TACF/Grants/Grant+Opportunities/
Career+Development+Award/Career+Development+Award
Sponsor: American Society of Clinical Oncology
2318 Mill Road, Suite 800
Alexandria, VA 22314

ASCO Community Oncology Research Grants 1071

The Community Oncology Research Grant is open to all eligible, community-based practices. The award is intended to support community-based practices, in their efforts to enhance their clinical trials program. Areas of improvement are based on the American Society of Clinical Oncology Statement on Minimum Standards and Exemplary Attributes of Clinical Trial Sites (Journal of Clinical Oncology, May 20 2008). The grant period begins July 1 and concludes June 30, in addition to a six-month follow-up period for evaluation. Progress reports are required at six and twelve months. A final report is due at eighteen months. Eligible practices are encouraged to apply for the award by completing an online application. Grant proposals will be entertained up to $30,000, including no more than $26,000 for grant support, $1,500 for travel to the ASCO Annual Meeting, and $2,500 for overhead costs. The grant funds will be directed to the community practice, not to an individual. A total of three awards will be presented at the ASCO's Annual Meeting in Chicago, IL.

Requirements: CORG eligibility is based on the following criteria, practices must: have at least one member of ASCO involved in the research program; have at least one member of an ASCO State/Regional Affiliate Program involved in the research program; meet basic standards for a quality clinical trials site (based on Good Clinical Practice standards, the accepted international ethical and scientific quality standard for designing, conducting, recording, and reporting trials); be community-based, with high-quality audit reports and investigators in good standing.

Samples: Billings Clinic Cancer Center, Billings, MT (2009); Marion L Shepard Cancer Center, Washington, NC (2009); Nebraska Cancer Specialists, Omaha, NE (2009).

Amount of Grant: $30,000

Contact: Nancy R. Daly, Executive Director; (571) 483-1432; email: dalyn@asco.org

Internet: http://www.ascocancerfoundation.org/TACF/Grants/Grant+Opportunities/Community+Oncology+Research+Grant

Sponsor: American Society of Clinical Oncology

2318 Mill Road, Suite 800

Alexandria, VA 22314

ASCO Comparative Effectiveness Research Professorship in Breast Cancer   1072

The Comparative Effectiveness Research Professorship (CERP) in Breast Cancer is intended to provide support to qualified individuals to investigate comparative effectiveness in breast cancer and to train future generations of researchers in this field. The CERP will provide funding to outstanding researchers who have made and are continuing to make significant contributions that have changed the direction of breast cancer research and who will provide mentorship to junior researchers. The award totals $500,000 and is paid in five annual increments of $100,000 made on or about July 1 of each year of the grant term. The award is not designed to cover the total cost of a research project or the investigator's entire compensation, but can be used at the discretion of the awardee to support their overall research activities. See website for complete budget guidelines.

Requirements: Applications focusing on all aspects of breast cancer including prevention, diagnosis, treatment, delivery of care, and health economics will be accepted. Applicants must: have an MD, DO, PhD or equivalent degree; have the rank of full professor (or equivalent). If current rank is not that of a full professor but meets all other eligibility criteria, a written explanation must be provided for equivalency of current rank; have a full-time faculty appointment at an academic medical center; have made significant contributions that have changed the direction of breast cancer research; be serving as a research mentor to one or more researcher(s) in training, and must be planning to continue to provide leadership in this area throughout the award period; lead a research team in conducting research on comparative effectiveness in breast cancer; be an active member of ASCO or have submitted a membership application with the grant application; commit to spending 75% of time during the award period dedicated to research (applies to total research, not just the proposed project) including leading a team of researchers (in the lab and/or the clinic) and mentoring physician-scientists. Individuals with an endowed professorship from their own institution are eligible to apply for this award. Individuals with administrative responsibilities (such as a Division or Department Chair) must clearly document their time commitment for performing research activities and mentoring activities. The sponsoring institution must be a not-for-profit institution.

Amount of Grant: $500,000

Date(s) Application is Due: Nov 3

Contact: Nancy R. Daly, Executive Director; (571) 483-1432; email: dalyn@asco.org

Internet: http://www.ascocancerfoundation.org/TACF/Grants/Grant+Opportunities/Comparative+Effectiveness+Research+Professorship

Sponsor: American Society of Clinical Oncology

2318 Mill Road, Suite 800

Alexandria, VA 22314

ASCO Geriatrics/Oncology Training Program Development Grant   1073

The grant is designed to aid institutions in developing three-year fellowship training programs in geriatric medicine and medical oncology. The goal of these grants is to produce academicians who will be teachers of geriatric issues in medical oncology and lead research efforts in geriatric oncology. Grants support fellow stipends, faculty development support (to support investigators who would be mentors to geriatric oncology fellows; to develop geriatric oncology curricula; and/or to lead advancements in clinical care); and support for ongoing geriatric oncology research projects. Up to seven grants will be awarded over three years.

Requirements: Applying institutions must have in place the elements necessary for a geriatrics/medical oncology training program.

Samples: Sharon Levine, MD, Boston Medical Center, Boston, MA; Keith M. Sullivan, MD, Duke University Medical Center; Ross C. Donehower, MD, Johns Hopkins University; William J. Gradishar, MD, Northwestern University.

Amount of Grant: $100,000 year one, $75,000 year two, and $50,000 year three; plus $75,000 institutional matching funds

Contact: ASCO Training Division, (703) 299-1076; fax: (703) 299-1044; email: driscolh@asco.org

Internet: http://www.asco.org/portal/site/ASCO/ menuitem.509189bfd2c2bf5ca7ffa807320041a0/ ?vgnextoid=b4b5e30e7208e010VgnVCM100000ed730ad1RCRD

Sponsor: American Society of Clinical Oncology

1900 Duke Street, Suite 200

Alexandria, VA 22314

ASCO International Development and Education Awards (IDEA)   1074

The award provides support for oncologists in developing countries to attend the ASCO annual meeting and spend additional time at a comprehensive cancer center. The grants are designed to provide continuing medical education, assist in career development, and help establish strong relationships with leading ASCO members who serve as scientific mentors to each recipient. The strongest consideration will be given to candidates who: have submitted an abstract for the most recent Annual Meeting; and have less than 10 years of experience in the field of oncology. Guidelines are available online.

Requirements: Applicants must: complete an application, which includes a 250-word essay, letter of recommendation, and submission of the applicant, CV; be able to travel to the United States and have no visa restrictions or anticipate having any difficulty obtaining a U.S. visa; be citizens of a country identified by ASCO as one of greatest need or with limited resources; and either be or meet the requirements of being either an ASCO Active Member, Active-Junior Member, or an Associate Member.

Restrictions: Applicants cannot: have received one year or more of formal training (for instance, attended medical school, participated in internships, fellowships, or residency programs) in the United States, Western Europe, Australia, New Zealand, Canada, or the United Kingdom; be a previous recipient of the IDEA (formally known as ITG) program; or have previously attended more that one (1) ASCO Annual Meeting.

Date(s) Application is Due: Jan 13

Contact: Amanda Woodell, International Affairs Department; (571) 483-1501; email: idea@asco.org

Internet: http://www.ascocancerfoundation.org/TACF/Awards/Award+Opportunities/International+Development+and+Education+Award+%28IDEA%29

Sponsor: American Society of Clinical Oncology

2318 Mill Road, Suite 800

Alexandria, VA 22314

ASCO Long-Term International Fellowships   1075

The ASCO Cancer Foundation Long Term International Fellowship (LIFe) provides young oncologists in developing nations the support and resources needed to advance their training by deepening their relationship with a U.S. or Canadian colleague and his or her institution. Through a one or two year fellowship the recipient will earn valuable training and experience with which they can affect change in cancer care in their home country. Recipients must return to their home institutions after completing their fellowship and are expected to disseminate the knowledge they have gained. The ASCO Cancer Foundation (TACF) will administer fundraising and grant distribution for the fellowships. Funds will be sent from TACF to the host institution in January of the award year. The host institution will then be responsible for financial management of the fellowship. Any funds not spent on the fellowship will be returned to TACF. Recipients must attend the ASCO Annual Meeting that falls during the first year of their award term where they will be honored at The ASCO Cancer Foundation Grants Brunch. Recipients will also be expected to submit their research to ASCO via an abstract or publication submission following their fellowship. Applications are due to ASCO by October 15 for the fellowship to begin anytime in the next calendar year. Applications should be completed and returned jointly by the prospective host and fellow. A decision will be distributed by December 15.

Requirements: Hosts must: be an ASCO member in good standing; be employed at a U.S. or Canadian institution with sufficient infrastructure to provide for a 1 year fellowship; independently identify fellowship applicants; agree to cover any costs in excess of ASCO, contribution; complete the host section of the LIFe Application; complete fellowship evaluations as requested by ASCO. Applicants must: be an ASCO member in good standing; reside in a country identified by the World Bank as income level low, lower middle, and upper middle; have the support of their home institution including a letter of support from their supervisor; be a medical doctor who has finished a fellowship in clinical oncology or the equivalent; have less than 10 years experience in the field of oncology; complete the applicant section of the LIFe application including submission of a CV of up to 3 pages in length; commit to return to country of origin at the completion of their fellowship.

Date(s) Application is Due: Oct 15

Contact: Nancy R. Daly, Executive Director; (571) 483-1432; email: dalyn@asco.org

Internet: http://www.ascocancerfoundation.org/TACF/Awards/Award+Opportunities/Long-term+International+Fellowship+%28LIFe%29

Sponsor: American Society of Clinical Oncology

2318 Mill Road, Suite 800

Alexandria, VA 22314

ASCO Medical Student Rotation Grants   1076

The ASCO Diversity in Oncology Initiative, funded by Susan G. Komen for the Cure, is designed to facilitate the recruitment and retention of individuals from populations underrepresented in medicine to cancer careers, with particular attention to the development of clinical practitioners and investigators. The ASCO Medical Student Rotation, funded by Susan G. Komen for the Cure, provides 8 - 10 week clinical or clinical research oncology rotations for U.S. medical students from populations underrepresented in medicine who are interested in pursuing oncology as a career. Recipients receive a $5,000 stipend for the rotation plus $1,500 for future travel to the

ASCO Annual Meeting. An additional $2,000 will be provided to support the student's mentor. The application period opened October 7 and will close January 7. Awardees will be announced in early April.

Requirements: To qualify, applicants must be enrolled in an MD or DO, U.S. medical school program and be of an underrepresented population as defined by the program eligibility criteria. Applicants self-identify and recipients are drawn from this pool. Candidates must be U.S. citizens, U.S. nationals, or permanent residents. Candidates also must demonstrate an interest in pursuing oncology as a career and have a record of good academic standing.

Amount of Grant: $8,500 total
Date(s) Application is Due: Jan 7
Contact: Nancy R. Daly, Executive Director; (571) 483-1432; email: dalyn@asco.org
Internet: http://www.ascocancerfoundation.org/TACF/Awards/Award+Opportunities/ASCO+Diversity+in+Oncology+Initiative+funded+by+Susan+G.+Komen+for+the+Cure/The+ASCO+Medical+Student+Rotation%2C+funded+by+Susan+G.+Komen+for+the+Cure¬Æ
Sponsor: American Society of Clinical Oncology
2318 Mill Road, Suite 800
Alexandria, VA 22314

ASCO Merit Awards   1077

The Merit Awards are designed to further promote clinical research by young scientists and to provide fellows with an opportunity to present their research and interact with other clinical cancer investigators at ASCO scientific meetings. A select number of Merit Awards are given annually to recognize outstanding abstracts submitted for consideration for presentation at an ASCO scientific meeting. The awards are given to oncology fellows who are first authors on selected abstracts.

Requirements: The first author of an abstract must meet the following requirements to be considered for an Award: be a physician (MD or DO) in an oncology fellowship training program or a doctoral degree candidate (such as PharmD or PhD) enrolled in an approved oncology specialty training program at the time of abstract submission; work in an oncology laboratory or clinical research setting; agree to present the abstract at the ASCO scientific meeting; check the Merit Award consideration box on the online abstract submitter; provide a letter of support from the Training Program Director indicating eligibility for the award; and submit a two-page curriculum vitae.

Contact: Nancy R. Daly, Executive Director; (571) 483-1432; email: dalyn@asco.org
Internet: http://www.ascocancerfoundation.org/TACF/Awards/Award+Opportunities/Merit+Award
Sponsor: American Society of Clinical Oncology
2318 Mill Road, Suite 800
Alexandria, VA 22314

ASCO Translational Research Professorships   1078

The Translational Research Professorship is designed to provide flexible funding to outstanding translational researchers who have made and are continuing to make significant contributions that have changed the direction of cancer research and who provide mentorship to future translational researchers. The award is intended to support qualified individuals who are dedicated to bringing advances in basic sciences into the clinical arena and to mentoring other translational researchers. The award totals $500,000 and is paid in 5 annual increments of $100,000 throughout the award period. Only online applications will be accepted. In order to complete an application, a Letter of Intent must first be submitted by November 20th. Completed applications must be submitted by January 22nd.

Requirements: The applicant must meet the following criteria at the time of grant award (July 1): be a physician (MD, DO or international equivalent) with a full-time faculty appointment in a clinical department at an academic medical center; currently hold the rank of full professor (or equivalent) at an academic medical center. If applicant does not hold the rank of full professor, but meets all other eligibility criteria, he/she must provide written explanation of why his/her current rank at their institution should be considered equivalent; made significant contributions that have changed the direction of cancer research; be serving as a research mentor to one or more translational researcher(s) in training, and must be planning to continue to provide leadership in this area throughout the award period Lead a research team in the conduct of translational research; be a member of the American Society of Clinical Oncology (ASCO); expect to spend 75% of time during the award period dedicated to translational research activities, leading a team of translational researchers (both in the lab and in the clinic) and mentoring physician-scientists. Individuals with an endowed professorship from their own institution are eligible to apply for this award. Individuals with administrative responsibilities (such as a Division or Department Chair) must clearly document their time commitment for performing research activities and mentoring activities. The application must originate jointly from the sponsoring institution and the applicant. The sponsoring institution must be a not-for-profit institution.

Samples: Merrill Egorin, MD, The Hillman Cancer Institute, Pittsburgh, PA— Preclinical and Clinical Studies of PARP Inhibitors as Cancer Therapy (2009).
Amount of Grant: $500,000 over five years
Date(s) Application is Due: Nov 20; Jan 22
Contact: Nancy R. Daly, Executive Director; (571) 483-1432; email: dalyn@asco.org
Internet: http://www.ascocancerfoundation.org/TACF/Grants/Grant+Opportunities/Translational+Research+Professorship
Sponsor: American Society of Clinical Oncology
2318 Mill Road, Suite 800
Alexandria, VA 22314

ASCO Young Investigator Award   1079

This grant provides funding to promising investigators to encourage and promote quality research in clinical oncology. The purpose of this award is to fund physicians during the transition from a fellowship program to a faculty appointment. Priority consideration will be given to proposals that include patient-oriented and, ultimately, clinical research. Since many awards are supported with restricted grants from outside organizations, the ASCO is particularly interested in identifying young researchers working in the following sub-specialties and emerging disciplines: breast cancer; cancer survivorship; geriatric oncology; health disparities; kidney cancer; multiple myeloma; ovarian cancer; pancreatic cancer; sarcoma; survivorship; and young adult cancer. However, The ASCO Foundation welcomes application submissions in all oncology subspecialties.

Requirements: The recipient must be a physician (MD or DO) who, at the time of grant award is in the final year of his/her final sub-specialty training program or in the first year post his/her final sub-specialty training. The sponsoring facility must be an academic medical institution. The primary mentor must be in the candidate, proposed research field, must assume responsibility, and provide guidance for the research. The applicant must either be a member of ASCO or submit a membership application with the grant application. The applicant should spend at least 60 to 75% of his or her time in research during the award period.

Amount of Grant: $50,000
Date(s) Application is Due: Sep 9
Contact: Nancy R. Daly, Executive Director; (571) 483-1432; email: dalyn@asco.org
Internet: http://www.ascocancerfoundation.org/TACF/Grants/Grant+Opportunities/Young+Investigator+Award
Sponsor: American Society of Clinical Oncology
2318 Mill Road, Suite 800
Alexandria, VA 22314

ASCO/UICC International Cancer Technology Transfer Fellowships   1080

The scholarships are intended primarily for researchers in the early stages of their career as well as for experienced clinical oncologists. The ICRETT objective is to facilitate rapid international transfer of cancer research and clinical technology, and it allows fellows to exchange knowledge and enhance skills in basic, clinical, behavioral and epidemiological areas of cancer research, and cancer control and prevention. Through the program fellows also acquire appropriate clinical management, diagnostic and therapeutic expertise for effective application and use in home organizations upon return. The UICC has also developed the ICRETT for teaching faculty which funds experts to teach at host institutions throughout the world for a duration of 1 week to 1 month.

Requirements: The candidate must possess the appropriate professional qualifications and experience according to the specifications of each fellowship described on the website and must currently be engaged in cancer research, clinical oncology practice, or cancer society work. To permit effective communication at the host institute, the candidate must have adequate fluency in a common language. The candidate must also be on the staff payroll of a university, research laboratory or institute, hospital, oncology unit, or voluntary cancer society (or be accredited volunteers of such societies) to where they will return at the end of a fellowship.

Restrictions: Candidates attached to commercial entities, going to a profit organization or have associations with the tobacco industry not eligible.
Amount of Grant: $3,400
Contact: Nancy R. Daly, Executive Director; (571) 483-1432; email: dalyn@asco.org
Internet: http://www.ascocancerfoundation.org/TACF/Awards/Award+Opportunities/ASCO+%26+UICC+ICRETT+Fellowship
Sponsor: American Society of Clinical Oncology
2318 Mill Road, Suite 800
Alexandria, VA 22314

ASCSA Alison M. Frantz Fellowship   1081

The fellowship supports research in postclassical studies at the Gennadius Library in Greece. Fields of study include late antiquity, Byzantine studies, and modern Greek studies. A limited number of research associateships are awarded for well-defined projects (deadlines April 1, September 1, and December 1). Travel grants for graduate students or postdoctoral scholars from North American institutions working on projects in archeological science in Greece also are awarded. Guidelines and application are available online.

Requirements: PhD candidates and recent PhDs from U.S. or Canadian institutions are eligible. Fellows are expected to be in residence at the school for the full term of the fellowship.

Amount of Grant: $10,000 stipend
Date(s) Application is Due: Jan 15
Contact: Program Contact, (609) 683-0800; fax: (609) 924-0578; email: ascsa@ascsa.org
Internet: http://www.ascsa.edu.gr/fellowship/fellowships.htm
Sponsor: American School of Classical Studies at Athens
6-8 Charlton Street
Princeton, NJ 08540-5232

ASCSA Fellowships   1082

Fellowships awarded include the Heinrich Schliemann and John Williams White Fellowships in archaeology; the Thomas Day Seymour Fellowship in history and literature; two Brunilde Ridgway Fellowships in art history; and seven Fellowships unrestricted as to field—the Virginia Grace, the Michael Jameson, the Philip Lockhart, the Lucy Shoe Meritt, the Martin Ostwald, the James and Mary Ottaway Jr., and the James Rignall Wheeler fellowships. The Bert Hodge Hill fellowship is unrestricted, but with a preference for a student in art history. Fellowships are awarded on the basis of examinations and recommendations to students of classical archaeology, ancient history, and classical languages and literature.

Requirements: Applicants must be students in an American or Canadian university and will, preferably, have taken at least one year of graduate work but will not have completed the PhD before the term of fellowship starts.

Amount of Grant: $10,000

Date(s) Application is Due: Jan 15

Contact: Carolyn Snively, Program Officer, (609) 683-0800; fax: (609) 924-0578; email: ascsa@ascsa.org

Internet: http://www.ascsa.edu.gr/fellowship/fellowships.htm#regular

Sponsor: American School of Classical Studies at Athens
6-8 Charlton Street
Princeton, NJ 08540

ASCSA Harry Bikakis Fellowship 1083

The fellowship is available to graduate students attending North American institutions or Greek graduate students whose research subject is ancient Greek law; or Greek graduate students working on a school excavation. The fellowship is awarded periodically, but not more frequently than once a year.

Amount of Grant: $1875

Date(s) Application is Due: Jan 15

Contact: Fellowship Administrator, (609) 683-0800; fax: (609) 924-0578; email: asca@ascsa.org

Internet: http://www.ascsa.edu.gr/Forms/bikakis.pdf

Sponsor: American School of Classical Studies at Athens
6-8 Charlton Street
Princeton, NJ 08648-5232

ASCSA Jacob Hirsch Fellowship 1084

ASCSA offers this annual fellowship in preclassical, classical, or postclassical archaeology to U.S. or Israeli graduate students who are at the dissertation-writing stage, or to recent recipients of the PhD who are completing the dissertation program with dissertation publication. Project requires substantial residence in Greece. The fellowship is nonrenewable. Requests for applications should be forwarded to the Committee on Admissions and Fellowships at the school.

Requirements: Students in the United States or Israel who are PhD candidates writing their dissertations in archaeology or recent PhD's completing a project are eligible.

Amount of Grant: $10,000 stipend plus room and board

Date(s) Application is Due: Jan 15

Contact: Jacob Hirsch Fellowship Program, (609) 683-0800; fax: (609) 924-0578; email: ascsa@ascsa.org

Internet: http://www.ascsa.edu.gr/fellowship/fellowships.htm

Sponsor: American School of Classical Studies at Athens
6-8 Charlton Street
Princeton, NJ 08540

ASCSA NEH Fellowships 1085

The school is the most significant resource on Greece for American scholars in the fields of ancient and postclassical studies in Greek language, literature, history, archeology, philosophy, and art, from pre-Hellenic times to the present. The fellowships support scholars with distinguished research and teaching careers in the humanities. Fellows will be expected to reside primarily in Athens, contribute to and enhance the scholarly dialog, as well as contribute to and expand scholarly horizons at the school. Guidelines and application are available online.

Requirements: Postdoctoral scholars and professionals in relevant fields who are U.S. citizens or foreign nationals who have lived in the United States for the three years immediately preceding the application deadline are eligible. Applicants must have completed their professional training but do not have to hold the PhD.

Amount of Grant: $17,500 five-month tenure; $35,000 10-month tenure

Date(s) Application is Due: Nov 15

Contact: NEH Fellowship, (609) 683-0800; fax: (609) 924-0578; email: asca@ascsa.org

Internet: http://www.ascsa.edu.gr/fellowship/fellowships.htm

Sponsor: American School of Classical Studies at Athens
6-8 Charlton Street
Princeton, NJ 08648-5232

ASCSA Summer Sessions Scholarships 1086

Scholarships are awarded annually for six-week summer sessions of study and research emphasizing the topography and antiquities of Greece. Graduate and advanced undergraduate students and high school and college teachers may apply. Application information is available on the Web site.

Date(s) Application is Due: Jan 15

Contact: Summer Sessions Committee, (609) 683-0800; fax: (609) 924-0578; email: ascsa@ascsa.org

Internet: http://www.ascsa.edu.gr/fellowship/fellowships.htm#SummerSession

Sponsor: American School of Classical Studies at Athens
6-8 Charlton Street
Princeton, NJ 08540

ASCSA Weiner Laboratory Faunal Studies Research Fellowship 1087

This is a one-year fellowship at the Wiener Laboratory for graduate students or postdoctoral scholars working on projects in skeletal, faunal, geoarcheological, or environmental studies in Greece. Applicants must have a well-defined project that can be undertaken within the given time in the laboratory or in collaboration with local research institutions. In addition to the proposed research, the fellow will be expected to develop and curate the lab's comparative collection, contribute to the development of the lab, assist with queries from excavators, offer a lecture on the work undertaken while at the lab, participate in one regular member school trip, and contribute to seminars on aspects of archaeological science as part of the school's annual curriculum. Applications can be downloaded from the Web site. Athens Office address is 54 Souidias Street, GR-106 76 Athens Greece; phone: 30-1-723-6313; fax: 30-1-725-0584; email: info@ascsa.edu.gr; Web site: www.ascsa.edu.gr.

Requirements: The fellowship is open to scholars with PhDs and those working on doctoral dissertations.

Amount of Grant: $15,500-$25,000

Date(s) Application is Due: Jan 15

Contact: Director, Wiener Laboratory, (609) 683-0800; fax: (609) 924-0578; email: ascsa@ascsa.org

Internet: http://www.ascsa.edu.gr/Wiener/fellowship.htm

Sponsor: American School of Classical Studies at Athens
6-8 Charlton Street
Princeton, NJ 08540-5232

ASECS Aubrey Williams Research Travel Fellowship 1088

The fellowship is awarded annually to support documentary research in 18th-century English literature by U.S.-based scholars. The fellowship is restricted to doctoral students working on the dissertation in the field of 18th-century English literature. Applicants must be members of the society at the time of the award and must be residents of North America. Application forms are available on the Web site.

Requirements: The fellowship is restricted to doctoral students at work on a dissertation in the field of eighteenth-century English literature.

Amount of Grant: $1000

Date(s) Application is Due: Jan 1

Contact: Program Director, (336) 727-4694; fax: (336) 727-4697; email: ASECS@wfu.edu

Internet: http://asecs.press.jhu.edu/travelgr.html

Sponsor: American Society for Eighteenth-Century Studies
P.O. Box 7867, Wake Forest University
Winston-Salem, NC 27109

ASECS Cooperative Fellowships 1089

The society participates in eight fellowship programs to promote and sustain the study of the 18th century, funded jointly by the society and some of North America's leading research institutions: the Folger Shakespeare Library and the Folger Institute of Renaissance and Eighteenth Century Studies; the Newberry Library; the William Andrews Clark Memorial Library of the University of California, Los Angeles; McMaster University Library; Yale Center for British Art; the Houghton Library of Harvard University; the Harry Ransom Center of the University of Texas at Austin; and the American Antiquarian Society. Fellowships will generally be limited to one month's support, although individual institutions may make adjustments depending on the quality of applications and the availability of funds. Evaluation and selection will be by each institution in accordance with its established procedures. Applications are available from ASECS or the participating institutions.

Requirements: An applicant must be a member in good standing of ASECS, be a postdoctoral scholar no more than 10 years beyond receipt of the PhD or equivalent degree at the time of application, and be working on a project concerning the 18th century. Contact should be made to the specific library of interest; complete contact information for each library is available on the ASECS Web site.

Contact: ASECS Fellowship Program, (336) 727-4694; fax: (336) 727-4697; email: ASECS@wfu.edu

Internet: http://asecs.press.jhu.edu/aboutus.html

Sponsor: American Society for Eighteenth-Century Studies
P.O. Box 7867, Wake Forest University
Winston-Salem, NC 27109

ASECS Gwin J. and Ruth Kolb Research Travel Fellowship 1090

The fellowship is awarded annually to supplement costs for 18th-century scholars to travel to distant collections in North America and abroad. There are no restrictions based on applicants' age, sex, race, religion, or academic rank or discipline. Advanced doctoral candidates with demonstrable need for specific collections necessary for their dissertations also are encouraged to apply. Application forms are available on the Web site.

Requirements: Applicants must be members of ASECS at the time of application and faculty or independent scholars within the first five years of receipt of their PhD.

Amount of Grant: $500

Date(s) Application is Due: Jan 1

Contact: Program Director, (336) 727-4694; fax: (336) 727-4697; email: ASECS@wfu.edu

Internet: http://asecs.press.jhu.edu/travelgr.html

Sponsor: American Society for Eighteenth-Century Studies
P.O. Box 7867, Wake Forest University
Winston-Salem, NC 27109

ASECS Irish-American Research Travel Fellowship 1091
The award is given annually to support documentary research in Irish repositories, both in the Republic of Ireland and the North, by U.S.-based scholars of Ireland in the period between the Treaty of Limerick (1691) and the Act of Union (1800). In alternate years, the award will go to Irish-based scholars seeking to travel to North America. The fellowship is restricted to documentary scholars whose research centers on primary sources from the 18th century (printed matter, manuscripts, buildings, works of art, or other artifacts), rather than on extant secondary literature. Applications are available on the Web site.
Requirements: Applicants must be members of the society at the time of the award and must be residents of North America. The fellowship is restricted to documentary scholars, whose research centers on primary sources from the 18th century (printed matter, manuscripts, buildings, works of art, or other artifacts), rather than on the secondary literature already extant.
Amount of Grant: $1500
Date(s) Application is Due: Nov 1
Contact: Program Director, (336) 727-4694; fax: (336) 727-4697; email: ASECS@wfu.edu
Internet: http://asecs.press.jhu.edu/travelgr.html
Sponsor: American Society for Eighteenth-Century Studies
P.O. Box 7867, Wake Forest University
Winston-Salem, NC 27109

ASECS James L. Clifford Prize 1092
The prize is awarded annually for an outstanding article appearing in a journal, festschrift, or other serial publication; the article must be an outstanding study of some aspect of 18th-century culture interesting to any 18th-century specialist, regardless of discipline, and no longer than 15,000 words. Nominations must be accompanied by eight copies of the article.
Requirements: The article may be nominated by any member of the society, by its author, or by an editor of the publishing journal. The author must be a member of the society at the time of presentation of the award.
Amount of Grant: $500
Date(s) Application is Due: Jan 1
Contact: Clifford Prize Committee, (336) 727-4694; fax: (336) 727-4697; email: ASECS@wfu.edu
Internet: http://asecs.press.jhu.edu/awards.html
Sponsor: American Society for Eighteenth-Century Studies
P.O. Box 7867, Wake Forest University
Winston-Salem, NC 27109

ASECS Louis Gottschalk Prize 1093
The prize is awarded annually for an outstanding historical or critical study on a subject of 18th-century interest. Books, which may be written in any modern language, may be commentaries, critical studies, biographies, or critical editions.
Requirements: Applicants must be members of the society at the time of the award. Submission must be made by the publisher and received at ASECS by the listed application deadline.
Restrictions: Books that are primarily translations are ineligible.
Amount of Grant: $1000
Date(s) Application is Due: Nov 15
Contact: Gottschalk Prize Committee, (336) 727-4694; fax: (336) 727-4697; email: asecs@wfu.edu
Internet: http://asecs.press.jhu.edu/awards.html
Sponsor: American Society for Eighteenth-Century Studies
P.O. Box 7867, Wake Forest University
Winston-Salem, NC 27109

ASECS Robert R. Palmer Research Travel Fellowship 1094
The travel fellowship is awarded annually to support documentary research related primarily to the history and culture of France. Applicants must be members of the society at the time of application. There are no restrictions based on age, sex, race, religion, or academic rank. Applications are available on the Web site.
Requirements: All members of ASECS are eligible to apply.
Amount of Grant: $500
Date(s) Application is Due: Jan 1
Contact: Program Director, (336) 727-4694; fax: (336) 727-4697; email: ASECS@wfu.edu
Internet: http://asecs.press.jhu.edu/travelgr.html
Sponsor: American Society for Eighteenth-Century Studies
P.O. Box 7867, Wake Forest University
Winston-Salem, NC 27109

ASECS/Clark Fellowships 1095
One-month resident fellowships at the William Andrews Clark Memorial Library are available to postdoctoral scholars with projects in the Restoration or the 18th century. The library, which is administered by the Center for 17th- and 18th-Century Studies, is known for its collections on 17th- and 18th-century Britain, Oscar Wilde and the 1890s, the history of printing, and certain aspects of the American West. The application deadline is for fellowships to be held between July 1 of the current year and June 30 of the next. Applications are available on the Web site.
Requirements: Scholars who have received a PhD in the last six years and are engaged in research pertaining to the announced theme are eligible to apply.

Amount of Grant: $2000
Date(s) Application is Due: Feb 1
Contact: Program Contact, fax: (310) 206-8577
Internet: http://www.humnet.ucla.edu/humnet/c1718cs/applic3.htm
Sponsor: William Andrews Clark Memorial Library
310 Royce Hall, 405 Hilgard Avenue
Los Angeles, CA 90095-1404

ASEE Army Research Laboratory Postdoctoral Fellowships 1096
The program is designed to significantly increase the involvement of creative and highly trained scientists and engineers from academia and industry in scientific and technical areas of interest and relevance to the Army. Scientists and engineers at the Army Research Laboratory (ARL) help shape and execute the Army's program for meeting the challenge of developing technologies that will support Army forces in meeting future operational needs by pursuing scientific research and technological developments in such diverse fields as applied mathematics, atmospheric characterization, simulation and human modeling, digital/optical signal processing, material science and technology, multifunctional technology, combustion processes, propulsion, and flight physics. Before writing a proposal, applicants are advised to communicate directly with an ARL advisor, who can provide more specific information on the current research and available technical facilities and offer scientific support of proposal development. Applications are available on the Web site.
Requirements: Before appointment, participants must present evidence of having received the PhD, ScD, or other earned research doctoral degree recognized in U.S. academic circles as equivalent to the PhD within seven years of the date of application, or must present evidence of having completed all formal academic requirements for one of these degrees.
Contact: Rick Kempinski, (202) 331-3525; email: r.kempinski@asee.org
Internet: http://www.asee.org/arl
Sponsor: American Society for Engineering Education
1818 N Street NW, Suite 600
Washington, DC 20036-2479

ASEE Naval Research Laboratory Postdoctoral Fellowships 1097
The purpose of this program is to increase the involvement of highly trained U.S. scientists and engineers in disciplines to meet the evolving needs of naval technology. To this goal, the ASEE, as the agent for the Naval Research Laboratory, will award approximately 40 postdoctoral appointments per year for research at 15 participating navy laboratories. Appointments are for one year and renewable for a second. At the discretion of the laboratory, third-year appointments may be arranged if warranted. A person who has received a prior postdoctoral appointment at a navy laboratory under any program may not be eligible to participate in this postdoctoral program. Applications may be downloaded from the Web site. Applications are accepted on an ongoing basis.
Requirements: Before appointment, participants must present evidence of having received the PhD, ScD, or other earned research doctoral degree recognized in U.S. academic circles as equivalent to the PhD within seven years of the date of application, or must present evidence of having completed all formal academic requirements for one of these degrees. Participants must be U.S. citizens.
Amount of Grant: $65,000 maximum stipend, depending on qualifications and experience
Contact: Program Manager, (202) 331-3525; fax: (202) 265-8504; email: r.kempinski@asee.org
Internet: http://www.asee.org/resources/fellowships/nrl/about.cfm
Sponsor: American Society for Engineering Education
1818 N Street NW, Suite 600
Washington, DC 20036

ASGE / ConMed Award for Outstanding Manuscript by a Fellow/Resident 1098
The American Society for Gastrointestinal Endoscopy and ConMed Endoscopic Technologies sponsor the annual ASGE/ConMed Outstanding Manuscript Award. The prize consists of $10,000 cash award to the Fellow or Resident in training in a gastroenterology or a surgical gastrointestinal endoscopic program whose manuscript is accepted in for publication in Gastrointestinal Endoscopy. The award is presented during Digestive Disease Week, the largest and most prestigious meeting in the world for the GI professional.
Samples: Tyler Stevens, MD, Case Western Reserve University, $10,000 (2005).
Amount of Grant: $10,000
Contact: Chair, ASGE Research Committee, (630) 573-0600; fax: (630) 573-0691; email: grants@asge.org
Internet: http://www.asge.org/nspages/about/center/factsheet.cfm#conMed
Sponsor: American Society for Gastrointestinal Endoscopy and ConMed Endoscopic Technologies
1520 Kensington Road, Suite 202
Oak Brook, IL 60523

ASGE Don Wilson Award 1099
The Award provides Advanced Fellows or Junior Faculty with the opportunity to train outside of their home country with a premier GI endoscopist or group in order to advance their training. The award assists in underwriting reasonable and customary travel and living expenses for a period of one to three months. The award includes a $7,500 cash stipend prior to the recipient's travel. In addition, a 20% disbursement will be made to the host institution. A total of three awards will be available annually for North

American and International ASGE members. Note: One training must take place in the United States.

Requirements: Applicants must be the equivalent of Junior Faculty or Advanced Fellows and be proficient in the English Language. Applicants must also: be a current ASGE Member; be a Junior Faculty or Advanced Fellow (3rd or 4th year or international equivalent); have permission from their own institution to undertake the travel; and have permission from an ASGE member at the host institution to accept the application for training.

Restrictions: Individuals applying for 1-2 year advanced fellowships do not qualify for this award.

Amount of Grant: $7,500
Date(s) Application is Due: Sep 1
Contact: Chair, ASGE International Committee, (630) 573-0600; fax: (630) 573-0691; email: membership@asge.org
Internet: http://www.asge.org/nspages/research/applications/donWilson.cfm
Sponsor: American Society for Gastrointestinal Endoscopy
1520 Kensington Road, Suite 202
Oak Brook, IL 60523

ASGE Endoscopic Research Awards   1100

These research awards are offered to physicians for projects in basic and clinical endoscopic technology research, outcomes and effectiveness of endoscopy research. The primary objective is to foster research in gastrointestinal endoscopy both within and outside of academic centers. Two categories of grants may be used for up to two years of study: grants of $1-15,000 (category A) and grants of $15,001 - 50,000 (category B). Requests for funding seed projects that will lead to further research as well as larger requests for definitive clinical trials will be considered. Funding requests may also include: personnel expenses for research assistant and/or faculty salary support (percentage of time for study should be specified and appropriately justified); study supplies; and equipment essential for study.

Requirements: Candidate must be: an ASGE member; an MD (or have equivalent degree); and current in a gastroenterology-related and endoscopic practice in academic institutions or private practice in North America.

Restrictions: Funding will not be provided for: salary support for trainees; computer purchases (unless a unique application is proposed); standard equipment and supplies needed for appropriate patient care (for example, sclerotherapy needles); travel to meetings; or indirect costs.

Amount of Grant: $50,000 maximum per year for ywo years
Date(s) Application is Due: Sep 15
Contact: Chair, ASGE Research Committee, (630) 573-0600; fax: (630) 573-0691; email: grants@asge.org
Internet: http://www.asge.org/nspages/research/applications/roeapp.cfm
Sponsor: American Society for Gastrointestinal Endoscopy
1520 Kensington Road, Suite 202
Oak Brook, IL 60523

ASGE Endoscopic Research Career Development Awards   1101

These awards provide the salary and/or research support necessary for the investigator to enhance his/her career development. The award must be used to acquire new skills for furthering a career in endoscopic research. Examples of such skills include advanced training in endoscopic procedures, training in outcomes research relevant to endoscopy, training in use of large databases such as CORI, and new endoscopic research techniques, including animal models.

Requirements: Applicants must be ASGE members and hold full-time faculty positions at North American (U.S., Canada, or Mexico) universities or professional institutions at the time of application. The award is intended for faculty who have demonstrated promise and have some record of accomplishment in research. Candidates must devote at least 30 percent of their effort to research related to gastrointestinal endoscopy during the period of the award.

Restrictions: The award is not available for fellows. Faculty with principle investigator current federal funding or other concurrent career development support still active at the time of this award are not eligible. Recipients of the ASGE Career Development Awards may not be granted an ASGE Research & Outcomes & Effectiveness Award during their two-year award period.

Amount of Grant: $75,000 per year for two years
Date(s) Application is Due: Dec 14
Contact: Chair, ASGE Research Committee, (630) 573-0600; fax: (630) 573-0691; email: grants@asge.org
Internet: http://www.asge.org/nspages/research/applications/careerapp.cfm
Sponsor: American Society for Gastrointestinal Endoscopy
1520 Kensington Road, Suite 202
Oak Brook, IL 60523

ASGE Given Capsule Endoscopy Research Award   1102

These research awards are offered to physicians for projects specific to capsule endoscopy both within and outside of academic centers. The award must be used for projects directly relating to capsule endoscopy. Requests for funding seed projects that will lead to further research as well as requests for definitive clinical trials will be considered. Funding requests may include: personnel (research assistant and/or faculty salary support—percentage of time for study should be specified and appropriately justified); study supplies; and equipment essential for study.

Requirements: Candidate must be an ASGE member, be an MD (or have equivalent degree), and be current in a gastroenterology-related and endoscopic practice in academic institutions or private practice in North America.

Restrictions: Funding will not be provided for: salary support for trainees; computer purchases (unless a unique application is proposed); standard equipment and supplies needed for appropriate patient care (for example, sclerotherapy needles); travel to meetings; publication costs; or indirect costs.

Amount of Grant: $25,000 maximum
Date(s) Application is Due: Dec 7
Contact: Chair, ASGE Research Committee, (630) 573-0600; fax: (630) 573-0691; email: grants@asge.org
Internet: http://www.asge.org/nspages/research/applications/ceapp.cfm
Sponsor: American Society for Gastrointestinal Endoscopy
1520 Kensington Road, Suite 202
Oak Brook, IL 60523

ASHA Advancing Academic-Research Careers (AARC) Award   1103

This award is given to new faculty in higher education to support their academic and research endeavors in the field of communication sciences and disorders (CSD). Up to five awards of $5,000 each will be granted to support activities such as: Improving teaching skills; Mentoring graduate students; Participating in research activities; Preparing a grant application; Preparing a manuscript or publication; Presenting at a professional meeting (e.g., the ASHA Convention).

Requirements: Candidates must possess: A research doctoral degree (PhD, ScD, CScD, or equivalent, e.g., EdD), which must be conferred by August 1; A full-time faculty appointment; Primary responsibilities of teaching and conducting research; Less than six years of teaching/research experience or are below the level of associate professor; A tenure or non-tenure track appointment. A list of required documents and a downloadable application form are available at the website.

Restrictions: Postdoctoral fellows are not eligible. Only students attending a university in the U.S. are eligible to apply for the award.

Amount of Grant: $5,000
Date(s) Application is Due: May 15
Contact: Margaret Rogers, PhD, Chief Staff Officer for Science & Research; (800) 498-2071 or (301) 296-5706; email: mrogers@asha.org
Internet: http://www.asha.org/students/awards.htm#aarc
Sponsor: American Speech-Language-Hearing Association (ASHA)
2200 Research Boulevard, #245
Rockville, MD 20850-3289

ASHA Research Mentoring-Pair Travel Award   1104

This award is designed to foster the professional development of students, clinicians, and emerging scientists who have expressed an interest in research in communication sciences and disorders. Up to 10 travel awards of $1,000 each will be granted to eligible mentor mentee pairs (mentee= $750; mentor= $250) to help defray the cost of attending the annual research symposium. Special consideration will be given to individuals whose research or research interests are (a) relevant to the symposium topic, (b) interdisciplinary in nature, (c) translational in nature, or (d) related to issues of cultural or linguistic diversity.

Requirements: Mentees must be (a) a current student at the bachelor's, master's, AuD, or PhD level, (b) a postdoctoral fellow, (c) a junior level faculty member (less than 6 years teaching/research experience in a tenure-track position or below the level of associate professor), or (d) a clinician. Mentors must be a seasoned investigator in an academic or health care environment. Download the Mentee and Mentory Application Forms at the website.

Amount of Grant: $1,000
Date(s) Application is Due: Jun 22
Contact: Margaret Rogers, PhD, Chief Staff Officer for Science & Research; (800) 498-2071 or (301) 296-5706; fax: (301) 296-8580; email: mrogers@asha.org
Internet: http://www.asha.org/students/awards.htm#2008%20Research%20Mentoring-Pair%20Travel%20Award
Sponsor: American Speech-Language-Hearing Association (ASHA)
2200 Research Boulevard, #245
Rockville, MD 20850-3289

ASHA Student Research Travel Award   1105

The awards are intended to highlight the research activities of budding scientists and encourage careers in science and research. Approximately 40 student research travel awards of $500 each are available for undergraduate, Master's, and doctoral students who are first authors on a paper (all session types are acceptable: technical, poster, one hour seminar, two hour seminar, or short course). Identification of prospective awardees will be based on the recommendation of Convention co-chairs and topic coordinators. Awardees will receive a $500 travel award; have their exceptional paper highlighted in the convention program; and, be recognized at the Researcher-Academic Town Meeting and other designated events.

Requirements: Each award recipient will be required to: register for the Convention at his or her own expense; present his or her paper at the assigned time (as noted in the Convention program booklet); secure a mentor to shadow throughout the convention; and, attend selected research-related sessions during the convention. In order to apply, students must: submit an abstract summarizing their research through the normal Call for Papers process; check the appropriate box (when prompted) indicating that their paper is 'student-authored'; and, check the appropriate box (when prompted) indicating that they desire to have this abstract considered for the Student Research Travel Award.

Restrictions: Papers must be submitted online by the deadline date and must be designated as a student-authored paper.
Amount of Grant: $500
Date(s) Application is Due: Mar 31
Contact: Margaret Rogers, PhD, Chief Staff Officer for Science & Research; (800) 498-2071 or (301) 296-5706; email: mrogers@asha.org
Internet: http://www.asha.org/students/awards.htm#research
Sponsor: American Speech-Language-Hearing Association (ASHA)
2200 Research Boulevard
Rockville, MD 20850-3289

ASHA Students Preparing for Academic & Research Careers (SPARC) Award   1106

The goal of SPARC is to foster students' interest in the pursuit of PhD education and careers in academia in order to fill faculty/researcher vacancies in communication sciences and disorders (CSD). Students will identify a primary faculty mentor and propose a one-year CSD Career Mentoring Plan. As many as ten students will be awarded up to $1,500 each to be used for teaching and research enhancement activities, such as travel to a research, pedagogy conference, or meeting; travel for a visit to an off-campus site that provides learning opportunities in a research lab and/or a college classroom setting; or course registration to support the mentoring plans outlined in the application.
Requirements: Eligible applicants include junior or senior undergraduates, 1st year master's students, and 1st and 2nd year entry-level clinical doctoral (e.g., AuD) students enrolled part-time or full-time in a CSD program in the United States. At the time of the award, eligible recipients' education status for the upcoming academic year must be undergraduate senior, 1st or 2nd year master's, or 1st, 2nd or 3rd year entry-level clinical doctoral (i.e., AuD) in a full-time or part-time CSD program in the United States. Master's degree and entry-level clinical doctoral students must be in a program accredited by the Council for Academic Accreditation in Audiology and Speech-Language Pathology (CAA). Full requirements (essay question, documentation, etc.) are listed at the website, along with the downloadable application form.
Amount of Grant: $5,000
Date(s) Application is Due: May 15
Contact: Arlene A. Pietranton, Executive Director; email: apietranton@asha.org or academicaffairs@asha.org
Internet: http://www.asha.org/students/awards.htm#sparc
Sponsor: American Speech-Language-Hearing Association (ASHA)
2200 Research Boulevard
Rockville, MD 20850-3289

ASHE/Lumina Foundation Dissertation Fellowships   1107

With support of the Lumina Foundation, the association awards fellowships to support dissertation research on the broad topics of financial aid, student retention and success, and adult learners and learning. The fellowships will support up to one year of activity that will be conducted through the students' home universities and can be used to support costs of supplying data, dissemination of project results, travel, tuition, and salary for the fellows. Eight one-year fellowships will be awarded over a three-year period.
Requirements: Doctoral students affiliated with any accredited doctoral program may submit a proposal. Students in doctoral programs outside the United States may apply if their study is about student financial assistance, student retention and success, and/or adult learning and learners in the United States.
Amount of Grant: $14,000
Date(s) Application is Due: Apr 1
Contact: Kristen Renn, PhD; (517) 353-5979; email: ASHE/Lumina Fellowship
Internet: http://www.ashe.ws/?page=224
Sponsor: Association for the Study of Higher Education / Lumina Foundation
428 Erickson Hall
East Lansing, MI 48824-1034

ASHFoundation Clinical Research Grants   1108

These $50,000‚Äì$75,000 grants support research scientists with a research doctorate within the discipline of communication sciences and disorders to support investigations that will advance knowledge of the efficacy of treatment and assessment practices. Project funding is available for mentored treatment research, independent treatment research, or collaborative treatment research as specified in grant guidelines.
Samples: Swathi Kiran, Assistant Professor, University of Texas at Austin—Semantic Feature Analysis in the Treatment of Lexical Retrieval Deficits in Spanish-English and French-English Bilingual Aphasia, $50,000 (2008); Tiffany Johnson, Assistant Professor, University of Kansas Center for Research, Inc.—New Directions in Clinical Applications of Otoacoustic Emissions. $50,000 (2008); Andrea Pittman, Assistant Professor, Arizona State University—Evaluating Advanced Signal Processing in Children with Hearing Loss, $75,000 (2008); Nancy Pearl Solomon, Research Speech Pathologist, Walter Reed Army Medical Center—Treating Dysphagia in Blast-Injured Patients. $50,000 (2008).
Amount of Grant: $50,000 - $75,000 range
Contact: Emily Diaz, Project Assistant; (301) 296-8703; email: ediaz@asha.org
Internet: http://www.ashfoundation.org/Foundation/grants/research_grants.htm
Sponsor: American Speech-Language-Hearing Foundation
2200 Research Boulevard
Rockville, MD 20850-3289

ASHFoundation Graduate Student Scholarships   1109

The foundation invites full-time graduate students to submit applications in competition for one of seven scholarships. Full-time master's or doctoral students in communication

sciences and disorders programs demonstrating outstanding academic achievement are eligible to compete for these $4,000 awards. Supported in part by Psi Iota Xi National Philanthropic Organization and the Marni Reisberg Memorial Fund. All scholarship recipients will be announced and recognized at the annual convention held in November.
Requirements: Applicants must be accepted for graduate study in a communication sciences and disorders program (master's degree candidates must be in an ASHA Educational Standards Board accredited program; this is not mandatory for doctoral degree candidates); be enrolled for full-time study (12 or more credit hours or the full-time standard); submit official university transcripts of academic coursework, credits, and grades; and be recommended by a committee of two or more past or present college faculty and colleagues (at least one supervisor) at the student's current place of employment. Annual deadline dates may vary; contact program staff for exact dates.
Restrictions: Applicants must not have received a prior scholarship from the foundation.
Samples: Tracy Conner, MA candidate, University of Massachusetts-Amherst , Minority Scholarship, Minority Fund (2008); Beverly A. Collisson, PhD Candidate, University of Connecticut—International Scholarship, Kala Singh Memorial Fund (2008); Stephanie Cochran, MA Candidate, Duquesne University—Disability Scholarship, Leslie Londer Fund (2008).
Amount of Grant: $4,000
Contact: Emily Diaz, Project Assistant; (301) 296-8703; email: ediaz@asha.org
Internet: http://www.ashfoundation.org/Foundation/grants/GraduateScholarships.htm
Sponsor: American Speech-Language-Hearing Foundation
2200 Research Boulevard
Rockville, MD 20850-3289

ASHFoundation Graduate Student Scholarships for Minority Students   1110

Racial/ethnic minority students who are U.S. citizens, who are accepted for graduate study in speech-language pathology or audiology, and who demonstrate outstanding academic achievement are eligible to compete for this scholarship. Applicants should submit a formal paper, such as a term paper, in competition for this award. The paper must be either an integrative literature review or an opinion paper on a current professional issue. All scholarship recipients will be announced and recognized at the ASHA convention.
Requirements: Candidates must be U.S. citizens and members of a racial/ethnic minority, including Native American and Alaska Native, Asian and Pacific Islander, African American, and Hispanic. Applicants must be accepted for master's or doctoral study in a speech-language pathology or audiology program for the upcoming academic year. The student must be enrolled for full-time (12 or more credit hours) study.
Restrictions: Applicants may not have received a prior scholarship from the foundation.
Samples: Tracy Conner, MA candidate, University of Massachusetts,ÄìAmherst, $5,000 (2008); Rhona Galera, CScD candidate, University of Pittsburgh, $5,000 (2008); Derek E. Daniels, PhD candidate, Bowling Green State University, $4,000 (2006).
Amount of Grant: $2,000-$4,000
Contact: Emily Diaz, Project Assistant; (301) 296-8703; email: ediaz@asha.org
Internet: http://www.ashfoundation.org/Foundation/grants/GraduateScholarships.htm
Sponsor: American Speech-Language-Hearing Foundation
2200 Research Boulevard
Rockville, MD 20850-3289

ASHFoundation Graduate Student with a Disability Scholarship   1111

Full-time graduate students with a disability who are enrolled in a communication sciences and disorders program and demonstrate outstanding academic achievement are eligible to compete for a $4,000 scholarship. Applicants must demonstrate superior academic achievement and submit transcripts, essay and references with official application. The Award is intended for the blind, hearing impaired, physically handicapped, or learning impaired. Master's, but not doctoral, students must be enrolled in an ASHA Council on Academic Accreditation approved program.
Requirements: Candidates must be U.S. citizens and either blind, hearing impaired, physically handicapped, or learning impaired. Applicants must be accepted for master's study in a speech-language pathology or audiology program for the upcoming academic year. The student must be enrolled for full-time (12 or more credit hours) study.
Samples: Stephanie Cochran, MA Candidate, Duquesne University (2008); Stephanie P. Hirsh, MA candidate, Northwestern University (2006); Alexandra Eng, MS candidate, University of the Pacific (2003).
Amount of Grant: $4,000
Contact: Emily Diaz, Project Assistant; (301) 296-8703; email: ediaz@asha.org
Internet: http://www.ashfoundation.org/Foundation/grants/GraduateScholarships.htm
Sponsor: American Speech-Language-Hearing Foundation
2200 Research Boulevard
Rockville, MD 20850-3289

ASHFoundation New Century Scholars Program Doctoral Scholarships   1112

The Program is a special funding initiative resulting from the ASHFoundation's Dreams and Possibilities Campaign. The Award is for students accepted or currently enrolled in a research doctoral program in communications sciences and disorders. Full-time study will be given priority; part-time status will be considered. Students should be committed to a teacher-investigator career in communication sciences and disorders in the United States.

Requirements: The Award is available to U.S. citizens only. Applicants must be accepted or currently enrolled in a research doctoral program.

Samples: Sophie Ambrose, PhD Candidate, University of Kansas (2008); Megha Bahl, PhD Candidate, University of Arizona (2008); Angela Yarnell Bonino, PhD candidate, University of North Carolina at Chapel Hill (2008); Deanna Britton, PhD Candidate, University of Washington (2008); Jamie L. Desjardins, PhD Candidate, Syracuse University (2008); Kerry Danahy Ebert, PhD Candidate, University of Minnesota, Twin Cities (2008); Aaron M. Johnson, PhD Candidate, University of Wisconsin, Madison (2008); Karen Le, PhD Candidate, University of Connecticut (2008); Jimin Lee, PhD Candidate, University of Wisconsin, Madison (2008); Megan K. McPherson, PhD Candidate, Purdue University (2008); Kimberly M. Meigh, PhD Candidate, University of Pittsburgh (2008).

Amount of Grant: $10,000 minimum

Contact: Emily Diaz, Project Assistant; (301) 296-8703; email: ediaz@asha.org

Internet: http://www.ashfoundation.org/Foundation/grants/GraduateScholarships. htm

Sponsor: American Speech-Language-Hearing Foundation
2200 Research Boulevard
Rockville, MD 20850-3289

ASHFoundation New Century Scholars Research Grant   1113

The Program is a special funding initiative resulting from the ASHFoundation's Dreams and Possibilities Campaign. The Grant is a one-time award intended to advance the knowledge base in communication sciences and disorders. Applicants must have a PhD or equivalent research doctorate within the discipline. Priority will be given to studies that are innovative, groundbreaking, or that meet research needs not yet addressed.

Restrictions: Students enrolled in a degree program or working on dissertation research are not eligible.

Samples: Jungmee Lee, Assistant Professor, University of Arizona—Proposal: Correlation Between Cochlear Tuning and Otoacoustic Emissions: Exploring Scientific and Clinical Implications, $10,000 (2008); Chang Liu, Assistant Professor, University of Texas at Austin—Auditory Processing: Comparing Phonologically Disordered and Typically Developing Children, $10,000 (2008); John McCarthy, Assistant Professor, Ohio University—Improving Auditory Scanning Interfaces in AAC Devices, $10,000 (2008); Valeriy Shafiro, Assistant Professor, Rush University Medical Center—Effects of Environmental Sound Training on the Perception of Environmental Sounds and Speech in Cochlear Implant Patients, $10,000 (2008).

Amount of Grant: $10,000

Contact: Emily Diaz, Project Assistant; (301) 296-8703; email: ediaz@asha.org

Internet: http://www.ashfoundation.org/Foundation/grants/research_grants.htm

Sponsor: American Speech-Language-Hearing Foundation
2200 Research Boulevard
Rockville, MD 20850-3289

ASHFoundation Research Grant for New Investigators   1114

The grants are designed to help further research activities of new investigators and should have particular clinical relevance to speech-language pathology and audiology. The awards are designed to encourage research activities of new scientists who earned their latest degree within the last five years. Proposals, while not limited in topic, are encouraged in the area of treatment research, particularly efficacy and outcome studies. Include abstract, research plan, bibliography, management plan, and budget with application.

Requirements: The individual must have received the master's or doctoral level degree in the last five years; must not have received prior funding for research, with the exception of internal university funding; and the proposal must be for research to be initiated.

Restrictions: Open to individuals with a Master's degree or doctoral degree who are not currently enrolled in a degree program.

Samples: Nina C. Capone, Associate Professor, Seton Hall University—The Effects of Gesture Cues on Object Word Learning by Children with Language Impairments (2008); Gayle L. DeDe Assistant Professor, University of Arizona—On-Line Sentence Comprehension in Aphasia: Is Reading Different than Listening? (2008); Ciara Leydon, Assistant Professor, Brooklyn College—Construction and Characterization of a Novel Model of Vocal Fold Mucosa (2008); Rita R. Patel, Assistant Professor, University of Kentucky—High Speed Digital Analysis of Vocal Fold Vibration in Children (2008); Yasmeen Faroqi Shah, Assistant Professor, University of Maryland, College Park—Retrieval of Action Names in Aphasia: An Investigation of the Embodied Cognition Framework (2008); Yana Yunosova, Assistant Professor, University of Toronto—Visual Feedback Systems in Speech Rehabilitation: Defining Vocal Tract Targets (2008).

Amount of Grant: $5,000

Contact: Emily Diaz, Project Assistant; (301) 296-8703; email: ediaz@asha.org

Internet: http://www.ashfoundation.org/Foundation/grants/research_grants.htm

Sponsor: American Speech-Language-Hearing Foundation
2200 Research Boulevard
Rockville, MD 20850-3289

ASHFoundation Research Grant in Speech Science   1115

ASLH, in conjunction with the Acoustical Society of America, invites new researchers to submit proposals in competition for one grant every other year. The grant is designed to further research activities of new investigators and promulgate the work of the late Dennis Klatt, a noted researcher and scientist in the area of speech communication. Priority will be given to areas reflecting Dr. Klatt's broad interests, such as proposals studying speech perception, synthesis, and acoustics, with an emphasis on an interdisciplinary research approach. The grant can be used to initiate new research or supplement an existing research project. Funds may be requested for a variety of purposes, i.e., equipment, subjects, research assistants, or research-related travel. The grant recipient will be announced at the ASHA convention.

Requirements: Individuals having received a doctoral degree within the last five years and who wish to further research activities in the areas of speech communication are eligible to compete for the grant.

Samples: Mary K. Fagan, Indiana University School of Medicine—Vocalization and Sound Exploration in Hearing, Deaf, and Cochlear-Implanted Infants, $5,000 (2007); Rajka Smiljanic, Northwestern University—Effect of Clear Speech on Production and Perception of Croatian Rhythm, $4,000 (2005); Lori L. Holt, Carnegie Mellon University—An Investigation of the Perceptual and Learning Influences on Phonetic Category Formation, $5,000 (2002).

Amount of Grant: $5,000

Contact: Emily Diaz, Project Assistant; (301) 296-8703; email: ediaz@asha.org

Internet: http://www.ashfoundation.org/Foundation/grants/research_grants.htm

Sponsor: American Speech-Language-Hearing Foundation
2200 Research Boulevard
Rockville, MD 20850-3289

ASHFoundation Student Research Grants in Audiology   1116

Each year the foundation awards one grant for support of research to be initiated in the area of clinical and/or rehabilitative audiology. One $2,000 award to a master's or doctoral student for research to be initiated in the area of audiology. Applicants must submit up to a ten-page research plan, one-page abstract, two-page management plan and budget, and letter of support. All study must be done in the United States.

Requirements: The competition open to master's or doctoral degree students enrolled in, or accepted for, graduate study in a communication sciences and disorders program in the United States. Master's degree candidates must be in a program accredited by the Council on Academic Accreditation in Audiology and Speech-Language Pathology (CAA); this is not mandatory for doctoral degree candidates. Applicants must be enrolled for full-time study for the full academic year.

Samples: Tara D. Reed, University of Texas at Dallas, Mentor: James F. Jerger—Effects of Aging on Interaural Asymmetry in a Competing Speech (2008); Faith M. Parker, Montclair State University, Mentor: Janet Koehnke—The Effects of Age and Reverberation on Cortical Audiotry Processing (2007); TaskYu-Hsiang Wu, University of Iowa, Mentor: Ruth Bentler—Impact of Visual Cues on Microphone Mode Preference (2006).

Amount of Grant: $2,000

Contact: Emily Diaz, Project Assistant; (301) 296-8703; email: ediaz@asha.org

Internet: http://www.ashfoundation.org/Foundation/grants/research_grants.htm

Sponsor: American Speech-Language-Hearing Foundation
2200 Research Boulevard
Rockville, MD 20850-3289

ASHFoundation Student Research Grants in Early Childhood   1117
Language Development

One $2,000 grant per year is awarded to a graduate or postgraduate student to support research in early childhood language development. The applicant must submit up to a ten-page research plan, one-page abstract, two-page management plan and budget, and letter of support. All study must be done in the United States. The recipient and the recipient's mentor will be announced and recognized at the ASHA Convention.

Requirements: The competition is open to graduate or postgraduate students in the area of communication sciences and disorders. Applicants must submit a proposal according to foundation guidelines, to be received by the indicated deadline date for applications.

Samples: Kathryn Wright Brady, University of Missouri, Mentor: Judith C. Goodman—Clues to Meaning: Exploring Potential Effects of Paired, Congruent Cues on Toddlers' Word Learning (2008); Dawn Vogler-Elias, University of Buffalo, Mentor: Geralyn Timler—A Shared Storybook Reading Intervention for Preschoolers with Autism (2008); Jonathan L. Preston, Syracuse University, Mentor: Mary Louise Edwards—Preliminary Investigation of a Weighted Measure of Speech Sound Accuracy (2007).

Amount of Grant: $2,000

Contact: Emily Diaz, Project Assistant; (301) 296-8703; email: ediaz@asha.org

Internet: http://www.ashfoundation.org/Foundation/grants/research_grants.htm

Sponsor: American Speech-Language-Hearing Foundation
2200 Research Boulevard
Rockville, MD 20850-3289

Ashland Corporate Contributions Grants   1118

The foundation awards grants to eligible Kentucky, Ohio, and West Virginia nonprofit organizations in its areas of interest, including education, arts, communities/civic, disaster relief, environment, and health and human services. Types of support include seed money grants, project grants, endowments, matching gifts, and employee volunteers. The foundation does not provide an application. Most giving is centered on programs that best address the needs of Ashland???s employees, stockholders, customers, and other constituencies, and the communities in which they live. Funding requests are not solicited but will be considered. The primary focus is on education, with an emphasis on mentoring, literacy and/or diversity.

Requirements: 501(c)3 tax-exempt organizations in Boyd and Greenup Counties, KY; Lawrence County, OH; and Cabell and Wayne Counties, WV; are eligible. Charitable groups within church or religious organizations are eligible.

Restrictions: The foundation does not support individuals, capital campaigns for building or equipment, endowments, travel, film or video production, tickets, religious or political activities, or goodwill advertising.

Geographic Focus: Kentucky; Ohio; West Virginia
Samples: U of Kentucky (Lexington, KY)—for the William T. Young Library Endowment, $1 million.
Contact: Program Contact, (859) 815-3333; email: community@ashland.com
Internet: http://www.ashland.com/commitments/contributions.asp
Sponsor: Ashland Inc
P.O. Box 391, 50 E Rivercenter Boulevard
Covington, KY 41012-0391

ASHP Research and Education Foundation Research Grants   1119
The ASHP offers grants in three different categories. Medication Safety Research grants give priority to projects that study medication-use design in hospitals and health systems and result in promotion of safety initiatives in medication use. Junior Investigator Research Grants support health services research in medication use that is conducted by pharmacist junior investigators; and strengthen the skills of newer pharmacist investigators while fostering development of mentoring relationships with more experienced senior investigators. The Pharmacy Resident Medication Safety Grant program is intended to support a worthy research project devoted to improving medication safety. Projects that demonstrate the role of the pharmacist in safe medication use, enhance medication-use systems, and can be duplicated in other health systems will be given high priority. Annual deadline dates may vary; contact program staff for exact dates.
Requirements: Eligibility requirements vary by program; contact program staff for eligibility.
Contact: Daniel Cobaugh, Director of Research, (301) 664-8767; fax: (301) 664-8872; email: foundation@ashp.org
Internet: http://www.ashpfoundation.org/Research/fundingOpps.cfm
Sponsor: American Society of Health-System Pharmacists Research and Education Foundation
7272 Wisconsin Avenue
Bethesda, MD 20814

ASHRAE Graduate Student Grant-in-Aid Program   1120
The Grant is intended to encourage the student to continue his or her preparation for service in the HVAC&R industry. The relevance of the research proposed by the candidate is a consideration for awarding the grant. The Grant-in-Aid is made through the university solely for the support of the student in an amount not to exceed U.S. $10,000.00 per school year per student and is not renewable. Typical expenditures include living expenses, tuition, travel to ASHRAE meetings, experimental equipment, and supplies. The basis for scoring heavily emphasizes ASHRAE involvement and is intended to stimulate advisor activity in ASHRAE and associate societies.
Requirements: Qualified graduate engineering students capable of carrying out appropriate and scholarly research are eligible to apply.
Amount of Grant: $10,000 maximum per year
Date(s) Application is Due: Dec 15
Contact: Manager of Research, (404) 636-8400; fax: (404) 321-5478; email: mvaughn@ashrae.org
Internet: http://www.ashrae.org/template/AssetDetail/assetid/23027
Sponsor: American Society of Heating, Refrigerating, and Air Conditioning Engineers
1791 Tullie Cir NE
Atlanta, GA 30329-2305

ASHRAE Willis H. Carrier Graduate Research Fellowship   1121
This fellowship is awarded for graduate research at Purdue University. The fellowship is for one year with monies dispersed by the University to the student. Completed applications must be mailed to the ASHRAE. Application forms are available on the Web site.
Requirements: Applicants should have a cumulative college grade point average of at least 3.0 (where 4.0 is the highest); current full-time enrollment with at least one full year of studies remaining; potential service to the HVAC and/or refrigeration profession; need for financial assistance; leadership ability; and recommendations from instructors.
Amount of Grant: $21,000
Contact: Graduate Administrator, (765) 494-6900; fax: (765) 494-0539; email: PurdueME@ecn.purdue.edu
Internet: http://www.ashrae.org/template/AssetDetail/assetid/23628
Sponsor: American Society of Heating, Refrigerating, and Air Conditioning Engineers
1791 Tullie Cir NE
Atlanta, GA 30329

Asia Foundation Exchange Programs   1122
The foundation seeks to encourage greater understanding between Asians and Americans with the ultimate aim of contributing towards strengthened U.S.-Asia relations. Exchange programs???administered by the San Francisco-based Asian-American Exchange (AAX) unit???help shape the perceptions of the Asians and Americans who take part in academic studies, conferences, special programs, and study tours tailored to particular needs that address U.S.-Asia interests. The foundation also sponsors the following special exchange programs: the William P. Fuller Fellowship in Conflict Resolution; the Chang-Lin Tien Distinguished Fellowship Program; the L.Z. Yuan Fellowship in Media and International Affairs; India Regional Security Studies Fellowship; and the Freeman Program for Broadening American Engagement with Southeast Asia. Guidelines are available online.
Contact: Grants Administrator, (415) 982-4640; fax: (415) 392-8863; email: info@asiafound.org

Internet: http://www.asiafoundation.org
Sponsor: Asia Foundation
465 California Street, 9th Fl
San Francisco, CA 94104

Asian Pacific American Institute for Congressional Studies Fellowships   1123
The fellowships are designed to provide unique opportunities to outstanding graduate students who have a commitment to the Asian Pacific Islander American communities, and who plan to pursue a career in public service. The fellows will spend nine months in Washington, D.C., in a Congressional office, in another branch of the government, or a non-profit organization. One fellow will spend nine months in the Congressional office of the Chair of the Asian Pacific American Caucus (CAPAC). These programs are made possible by the support of the following companies: Anheuser-Busch/Frank Horton Fellowship; Wal-Mart/Governor George R. Ariyoshi Fellowship; Sodexho Health and Policy Fellow; National Association of Realtors Fellow. APAICS will provide a total stipend of $20,000 to cover travel arrangements, housing and personal expenses for nine months. A separate stipend will be provided for basic health insurance coverage. In order to receive the full-stipend, the fellow will be required to participate in mandatory APAICS events, submit a report of activities, and complete an evaluation at the end of the program.
Requirements: Postgraduate and graduate applicants must meet the following *Requirements*: 1) Demonstrated interest in the political process; 2) Demonstrated commitment to public policy issues and Asian Pacific Islander American community affairs; 3) Demonstrated leadership abilities; 4) Relevant work experience; 5) Excellent oral and written communication skills; 6) Must have a graduate or bachelor, degree from an accredited educational institution. Minimum cumulative grade point average of 3.0 (on a 4.0 scale); 7) U.S. citizenship or legal permanent residency by September 1; 8) Submission of all application materials (Part I & II) by mail and email. All materials must be postmarked by the deadline.
Restrictions: faxed documents will not be accepted.
Amount of Grant: $20,000
Date(s) Application is Due: May 31
Contact: Ruby G. Moy, APAICS Interim Executive Director; (202) 296-9200; email: rubyapaics@gmail.com
Internet: http://www.apaics.org/pages/programs/fellowships.php
Sponsor: Asian Pacific American Institute for Congressional Studies (APAICS)
1001 Connecticut Avenue, NW Suite 530
Washington, DC 20036

Asian, Black, Hispanic and Native American United Methodist History Grants   1124
The purpose of the award is to promote excellence in research and writing in the history of four groups — Asian, Black, Hispanic, and Native American — in the United Methodist Church or its antecedents. Grants totaling $3,000 will be available with each grant being at least $1,000 awarded at the discretion of the committee of judges.
Requirements: The candidate must submit an application in English and include biographical information; a description of the project, including its significance, what resources will be used, and how it will be made available to the public; the expected date of completion; and an itemized budget. The candidate must submit three letters of recommendation in English from persons who can provide evidence of his/her capability. A reference copy of the final product will be donated to the General Commission on Archives and History of the United Methodist Church. Submit materials to the office.
Date(s) Application is Due: Dec 31
Contact: General Secretary, General Commission on Archives and History; (973) 408-3189; fax: (973) 408-3909; email: research@gcah.org
Internet: http://www.gcah.org/AwardsGrants/RacialEthnic.htm
Sponsor: United Methodist Church
P.O. Box 127
Madison, NJ 07940

ASID Foundation Dora Brahms Award   1125
One award is made on a biennial basis to educational institutions on behalf of their students in historic preservation and/or restoration studies, to encourage and support the advancement of professional activities in historic preservation and/or restoration. Institutions must illustrate how the award will best assist the student and the historic preservation and design education community.
Amount of Grant: $3,000
Date(s) Application is Due: Apr 30
Contact: Jack Pruitt, Director of Development; (202) 546-3480; fax: (202) 546-3240; email: education@asid.org
Internet: http://www.asid.org/asidfoundation/Foundation+Scholarships+and+Awards.htm
Sponsor: American Society of Interior Designers Foundation
608 Massachusetts Avenue, NE
Washington, DC 20002-6006

ASID Foundation Grants   1126
The ASID Foundation supports endeavors that capture and disseminate knowledge, encourage innovation, and benefit the health, safety and welfare of the public through interior design research, scholarships and education. Critical topics of interest include: sustainable design; technology; design for learning environments; and the needs of an aging population. Application is open to organizations and individuals.
Amount of Grant: $1,000-$5,000 range
Date(s) Application is Due: Mar 1; Sep 1

Contact: Jack Pruitt, Foundation Director; (202) 675-2365 or (202) 546-3480; fax: (202) 546-3240; email: education@asid.org
Internet: http://www.asidfoundation.org/
Sponsor: American Society of Interior Designers Foundation
608 Massachusetts Avenue, NE
Washington, DC 20002-6006

ASID Foundation Irene Winifred Eno Grant   1127

The Grant award in the amount of $5,000 is intended to provide financial assistance to individuals or groups engaged in the creation of an educational program or an interior design research project dedicated to health, safety and welfare. The grant will be awarded on the basis of the project description, break down of potential use of funds and the marketing plan for the use/distribution of the end product of the project.
Requirements: The grant is open to students, educators, interior design practitioners, institutions or other interior design related group.
Amount of Grant: $5,000
Date(s) Application is Due: Apr 30
Contact: Jack Pruitt, Director of Development; (202) 546-3480; fax: (202) 546-3240; email: education@asid.org
Internet: http://www.asid.org/asidfoundation/Foundation+Scholarships+and+Awards.htm
Sponsor: American Society of Interior Designers Foundation
608 Massachusetts Avenue, NE
Washington, DC 20002-6006

ASID Foundation Joel Polsky Academic Achievement Award   1128

This prize is given annually to recognize an outstanding interior design research or thesis project by an undergraduate or graduate student. Entries should address the needs of the public, designers, and students on such topics as educational research, behavioral science, business practice, design process, theory, or other technical subjects. Material will be judged on bibliography/references, breadth of material, comprehensive coverage of topic, graphic presentation and organization, and innovative subject matter. Entries will be returned upon request with a $10 postage and handling fee.
Amount of Grant: $1,000
Date(s) Application is Due: Apr 30
Contact: Jack Pruitt, Director of Development; (202) 546-3480; fax: (202) 546-3240; email: education@asid.org
Internet: http://www.asid.org/asidfoundation/Foundation+Scholarships+and+Awards.htm
Sponsor: American Society of Interior Designers Foundation
608 Massachusetts Avenue, NE
Washington, DC 20002-6006

ASID Foundation Joel Polsky Prize   1129

The Prize is given by the ASID Foundation to recognize outstanding academic contributions to the discipline of interior design through literature or visual communication. Entries should address the needs of the public, designers and students on such topics as educational research, behavioral science, business practice, design process, theory or other technical subjects. Material will be judged on innovative subject matter, comprehensive coverage of topic, organization, graphic presentation, bibliography and references.
Amount of Grant: $1,000
Date(s) Application is Due: Apr 30
Contact: Jack Pruitt, Director of Development; (202) 546-3480; fax: (202) 546-3240; email: education@asid.org
Internet: http://www.asid.org/asidfoundation/Foundation+Scholarships+and+Awards.htm
Sponsor: American Society of Interior Designers Foundation
608 Massachusetts Avenue, NE
Washington, DC 20002-6006

ASID Foundation Mabelle Wilhelmina Boldt Memorial Scholarship   1130

One scholarship will be awarded to a student who is enrolled in or has applied for admission to a graduate-level interior design program at a degree-granting institution. The scholarship will be awarded on the basis of academic/creative accomplishment, as demonstrated by school transcripts and a letter of recommendation. Preference will be given to students with a focus on design research. Annual deadline dates may vary.
Requirements: Applicants must have been practicing designers for a period of at least five years prior to returning to graduate study.
Amount of Grant: $2,000
Date(s) Application is Due: Apr 30
Contact: Jack Pruitt, Director of Development; (202) 546-3480; fax: (202) 546-3240; email: education@asid.org
Internet: http://www.asid.org/asidfoundation/Foundation+Scholarships+and+Awards.htm
Sponsor: American Society of Interior Designers Foundation
608 Massachusetts Avenue, NE
Washington, DC 20002-6006

ASM Abbott Laboratories Award in Clinical and Diagnostic Immunology   1131

The ASM Abbott Laboratories Award in Clinical and Diagnostic Immunology honors a distinguished scientist in clinical or diagnostic immunology for outstanding contributions to those fields.

Requirements: The nominee must demonstrate significant contributions to the understanding of the functioning of the host immune system in human disease, clinical approaches to diseases involving the immune system or development or clinical application of immunodiagnostic procedures.
Restrictions: Self-nominations will not be accepted.
Amount of Grant: $2,000, commemorative piece, and travel to ASM General Meeting
Date(s) Application is Due: Oct 1
Contact: Awards Committee, (202) 942-9226; fax: (202) 942-9353; email: awards@asmusa.org
Internet: http://www.asm.org/Academy/index.asp?bid=2301
Sponsor: American Society for Microbiology
1752 N Street, NW
Washington, DC 20036-2804

ASM BD Award for Research in Clinical Microbiology   1132

The ASM BD Award for Research in Clinical Microbiology honors a a distinguished scientist for research accomplishments that form the foundation for important applications in clinical microbiology.
Requirements: The nominee must be a distinguished clinical microbiologist. All nomination components must be emailed to awards@asmusa.org. Nominations must consist of the following: curriculum vitae; letter of nomination; letters of support.
Restrictions: Self-nominations will not be accepted.
Amount of Grant: A cash prize of $2,000, a commemorative piece, and travel to the ASM General Meeting
Date(s) Application is Due: Oct 1
Contact: Awards Committee, (202) 942-9226; fax: (202) 942-9353; email: awards@asmusa.org
Internet: http://www.asm.org/Academy/index.asp?bid=2304
Sponsor: American Society for Microbiology
1752 N Street, NW
Washington, DC 20036-2804

ASM bioMerieux Sonnenwirth Award for Leadership in Clinical Microbiology   1133

Given in memory of Alexander Sonnenwirth, Ph.D., this award recognizes a distinguished microbiologist for the promotion of innovation in clinical laboratory science, dedication to ASM, and the advancement of clinical microbiology as a profession.
Requirements: Nominations will be considered without updating for three years. email (see contact info.) all nomination components. Nominations must consist of the following: curriculum vitae, including a list of nominee's publications; letter of nomination, describe the nominee's innovation in clinical laboratory science, dedication to ASM, role as a leader in the field of diagnostic microbiology, and the advancement of clinical microbiology as a profession; letters of support, two letters of support must come from persons, other than the nominator, who are familiar with the nominee's qualifications and accomplishments. No more than one of the three letters may be from the nominee's institution or the same institution.
Restrictions: Self-nominations will not be accepted.
Amount of Grant: A cash prize of $2,000, a commemorative piece, and travel to the ASM General Meeting, where the laureate delivers the Sonnenwirth Award Lecture.
Date(s) Application is Due: Oct 1
Contact: Awards Committee, (202) 942-9226; fax: (202) 942-9353; email: awards@asmusa.org
Internet: http://www.asm.org/Academy/index.asp?bid=36368
Sponsor: American Society for Microbiology
1752 N Street, NW
Washington, DC 20036-2804

ASM Eli Lilly and Company Research Award   1134

ASM's oldest and most prestigious prize, it rewards fundamental research of unusual merit in microbiology or immunology by an individual on the threshold of his or her career.
Requirements: The nominee's work is not judged by comparison with the research of more experienced scientists, and special consideration is given to originality and independence of thought. The nominee must not have reached his or her 45th birthday by April 30 of the year the award is given. The nominee must be working in the United States or Canada and be actively engaged in the line of research for which the award is to be made.
Restrictions: Nominations must consist of the following: curriculum vitae, including a list of publications; verification of date of birth. Photocopy or scanned file of driver's license, passport or birth certificate; letter of nomination, describe the nominee's fundamental research of unusual merit in microbiology or immunology; letters of support, two letters of support must come from persons, other than the nominator, who are familiar with the nominee's qualifications and accomplishments. No more than one of the three letters may be from the nominee's institution or the same institution.
Amount of Grant: A cash prize of $5,000, a commemorative piece, and travel to the ASM General Meeting where the laureate delivers the Eli Lilly Award Lecture.
Date(s) Application is Due: Oct 1
Contact: Awards Committee, (202) 942-9226; fax: (202) 942-9353; email: awards@asmusa.org
Internet: http://www.asm.org/academy/index.asp?bid=2281
Sponsor: American Society for Microbiology
1752 N Street, NW
Washington, DC 20036-2804

ASM International Fellowship for Latin America and the Caribbean   1135

Funded by ASM this award offers fellowships to promising young investigators in South and Central America and the Caribbean who are within five years of obtaining, or are in the process of obtaining, their Masters, PhD, or other equivalent academic degree, and who are working in any of the microbiology disciplines, to visit a host scientist in the U.S.. The International Fellowship Program provides funding to meet the costs of a visit to one institution for between 6 weeks to 6 months. The International Fellowship Program provides funding to meet the costs of a visit to one institution for between 6 weeks to 6 months.

Requirements: The application must be made jointly between a host institution and the visiting fellow. The Investigator must be:a member of ASM or any other national microbiological society; actively involved in research in the microbiological sciences; have obtained, or be in the process of obtaining, their masters, PhD, or other equivalent academic degree within the last five years; a national of a resource-limited Africa, Asian or Latin American country submitting the application from an institution in their home region. Investigators from developed nations, as defined by BOTH the UN high HDI group and the World Bank high income economy group, are ineligible. Investigators submitting an application from the U.S. where they are spending a training period are ineligible; sufficiently proficient in the use of the English language. The Host Scientist must be: an ASM member; actively involved in research in the microbiological sciences in the U.S.; actively involved in teaching in the microbiological sciences in the U.S.; interested in sustaining international collaborations Preference will be given to: applicants who can prove three years of membership in ASM or any other national microbiological society, and; applicants who have not previously had the opportunity to travel to a facility in another country. The application should demonstrate to the Review Committee: academic excellence of the applicant - Honors & Awards, CV, and Letters of Recommendation; depth of the applicant's research experience - Statement of prior Research Experience; quality and originality of work proposed during the Fellowship - Potential Collaboration and Proposed Research Plan; relevance of the work proposed to the applicant's locale - Potential Collaboration and Proposed Research Plan; excellence of the host, and the research and training environment

Amount of Grant: $4,000

Date(s) Application is Due: Apr 15; Oct 15

Contact: International Fellowship Program, (202) 942-9368; fax: (202) 942-9328; email: international@asmusa.org

Internet: http://www.asm.org/International/index.asp?bid=2778

Sponsor: American Society for Microbiology

1752 N Street NW

Washington, DC 20036-2804

ASM International Fellowships for Asia and Africa   1136

Funded by ASM this award offers a maximum of one fellowship to a promising young investigator in Asia, as well as to one young investigator in Africa who is within five years of obtaining, or in the process of obtaining, his/her Masters, PhD, or other equivalent academic degree, and who is working in any of the microbiology disciplines, to visit a host scientist in the U.S.. Both International Fellowships will award a maximum grant of $5,500. The award is not intended to provide travel to obtain a degree at the host institution.

Requirements: The application must be made jointly between a host institution and the visiting fellow. The Investigator must be: a member of ASM or any other national microbiological society; actively involved in research in the microbiological sciences; have obtained, or be in the process of obtaining, their masters, PhD, or other equivalent academic degree within the last five years; a national of a resource-limited Africa, Asian or Latin American country submitting the application from an institution in their home region. Investigators from developed nations, as defined by BOTH the UN high HDI group and the World Bank high income economy group, are ineligible. Investigators submitting an application from the U.S. where they are spending a training period are ineligible; sufficiently proficient in the use of the English language. The Host Scientist must be: an ASM member; actively involved in research in the microbiological sciences in the U.S.; actively involved in teaching in the microbiological sciences in the U.S.; interested in sustaining international collaborations Preference will be given to: applicants who can prove three years of membership in ASM or any other national microbiological society, and; applicants who have not previously had the opportunity to travel to a facility in another country.

Amount of Grant: $5,500

Date(s) Application is Due: Apr 15; Oct 15

Contact: International Fellowship Program, (202) 942-9368; fax: (202) 942-9328; email: international@asmusa.org

Internet: http://www.asm.org/International/index.asp?bid=2778

Sponsor: American Society for Microbiology

1752 N Street NW

Washington, DC 20036-2804

ASM Merck Irving S. Sigal Memorial Awards   1137

The ASM Merck Irving S. Sigal Memorial Awards recognize and award excellence in basic research in medical microbiology and infectious diseases. The awards are presented in memory of Irving S. Sigal, who was instrumental in the early discovery of therapies to treat HIV/AIDS.

Requirements: The nominee must be no more than five years beyond completion of postdoctoral research training in microbiology or infectious diseases at the time of the nomination deadline. Nominations must consist of the following: curriculum vitae, including a list of publications; letter of nomination, describe the nominee's excellence in basic research in medical microbiology and infectious diseases; letters of support, two

letters of support must come from persons, other than the nomination, who are familiar with the nominee's qualifications and accomplishments. No more than one of the three letters may be from the nominee's institution or the same institution.

Amount of Grant: A cash prize of $2,500 and a commemorative piece.

Date(s) Application is Due: Oct 1

Contact: Awards Committee, (202) 942-9226; fax: (202) 942-9353; email: awards@asmusa.org

Internet: http://www.asm.org/academy/index.asp?bid=15056

Sponsor: American Society for Microbiology

1752 N Street, NW

Washington, DC 20036-2804

ASM Microbiology Undergraduate Research Fellowship   1138

The goal of the Microbiology Undergraduate Research Fellowship (MURF) program is to increase the number of underrepresented undergraduate students who wish to, and have demonstrated the ability to pursue graduate careers (Ph.D. or M.D./Ph.D.) in microbiology. Students will have the opportunity to conduct full time summer research with an ASM member at a host institution, and present research results at the Annual Biomedical Research Conference for Minority Students and the ASM General Meeting. Students will: agree to participate in an undergraduate summer research program at a U.S. based institution; conduct a research project for a minimum of 10 weeks; work with a faculty mentor who is an ASM member; submit a research abstract to the Annual Biomedical Research Conference for Minority Students (ABRCMS); submit a research abstract to ASM for presentation at the ASM General Meeting. Students decide the institution, research area, and level of activity for the summer. Based on interests, independence, and ability, students will be placed at a host U.S. Institution of the student's choice to conduct basic science research. From a list provided on the application, students interested in conducting research at a host institution will select three institutions where they would like to conduct their summer research. Every effort will be made to place fellows at their first choice.

Requirements: Eligible student candidates for the fellowship must be from groups that have been determined by the applicant's institution to be underrepresented in the microbiological sciences. The ASM encourages institutions to identify individuals that have been historically underrepresented, and remain underrepresented today in the microbiological sciences nationally. These groups include African-Americans, Hispanics, Native Americans, Alaskan Natives, and Pacific Islanders. In addition, applicants must also: be U.S. citizen or permanent U.S. resident; be enrolled as full-time matriculating undergraduate students during the academic year at an accredited U.S. institution; be either freshmen with college level research experience, sophomores, juniors, or seniors who will not graduate before the completion date of the summer program; be members of an underrepresented group in microbiology; have taken introductory courses in biology, chemistry, and preferably microbiology prior to submission of the application; have strong interests in obtaining a Ph.D., or M.D./Ph.D. in the microbiological sciences, and have lab research experience.

Restrictions: Travel funds are contingent upon acceptance of an abstract for the General Meeting.

Amount of Grant: $5,850

Date(s) Application is Due: Feb 1

Contact: Fellowship Committee, (202) 942-9283; fax: (202) 942-9329; email: fellowships-careerinformation@asmusa.org

Internet: http://www.asm.org/Education/index.asp?bid=4322

Sponsor: American Society for Microbiology

1752 N Street, NW

Washington, DC 20036-2804

ASM Millis-Colwell Postgraduate Travel Grant   1139

This new grant, negotiated between the American Society for Microbiology (ASM U.S.A) and the Australian Society for Microbiology (ASM Australia), allows for the exchange of one student member from each society to present an abstract at the annual General Meeting of the other society and to spend a week at nearby research laboratory. The grant aims to create a long lasting bond between ASM U.S.A and ASM Australia and is designed to benefit PhD students in both countries by giving them the opportunity to travel overseas to present their work and experience the best of microbiology in the partner country. ASM U.S.A will submit the successful applicant, abstract for an oral or poster presentation to ASM Australia for their Annual Scientific Meeting & Exhibition. The applicant will also have negotiated an agreement to visit the research laboratory of an Australian scientist in the period either immediately before or immediately after the ASM Australia annual meeting. The award will be conditional upon ASM Australia's acceptance of the submitted abstract.

Requirements: The successful applicant will be: a U.S. citizen or permanent resident of the U.S.; a student member of the American Society for Microbiology; and enrolled in a PhD program at an American University. The award will cover roundtrip economy airfare between the U.S. and Australia, and any necessary internal flights; per diem living allowance for up to 14 days; and, registration fee for the Australian Society for Microbiology, Annual Scientific Meeting & Exhibition.

Amount of Grant: up to $5,000

Date(s) Application is Due: Jan 1

Contact: Education Programs/Resources; (202) 942-9282; fax: (202) 942-9329

Internet: http://www.asm.org/International/index.asp?bid=53729

Sponsor: American Society for Microbiology

1752 N Street NW

Washington, DC 20036

ASM Procter & Gamble Award in Applied and Environmental Microbiology 1140

The ASM Procter & Gamble Award in Applied and Environmental Microbiology recognizes distinguished achievement in research and development in applied (non-clinical) and environmental microbiology.

Requirements: The nominee must demonstrate outstanding accomplishment in research or development in an appropriate field and be actively engaged in research or development at the time that the award is presented. Nominations must consist of the following: curriculum vitae, including a list of publications; letter of nomination, describe the nominee's outstanding accomplishments in research or development in the appropriate field, including a list of the ten most relevant publications to the award; letters of support, two letters of support must come from persons, other than the nominator, who are familiar with the nominee's outstanding accomplishments. No more than one of the three letters may be from the nominee's institution or the same institution.

Restrictions: Self-nominations will not be accepted.

Amount of Grant: A cash prize of $2,000, a commemorative piece,

Date(s) Application is Due: Oct 1

Contact: Awards Committee, (202) 942-9226; fax: (202) 942-9353; email: awards@asmusa.org

Internet: http://www.asm.org/academy/index.asp?bid=2288

Sponsor: American Society for Microbiology

1752 N Street, NW

Washington, DC 20036-2804

ASM Promega Biotechnology Research Award 1141

The ASM Promega Biotechnology Research Award honors outstanding contributions to the application of biotechnology through fundamental microbiological research and development.

Requirements: A nomination can be a single exceptionally significant achievement or the aggregate of a number of exemplary achievements. Nominations must consist of the following: curriculum vitae, including a list of publications; letter of nomination, describe the nominee's outstanding accomplishments in microbiological research and development, including a list of the ten most relevant publications to the award; letters of support, two letters of support must come from persons, other than the nominator, who are familiar with the nominee's accomplishments. No more than one of the three letters may be from the nominee's institution or the same institution.

Restrictions: Neither self-nominations nor co-nominations will be accepted.

Amount of Grant: A cash prize of $5,000, a commemorative piece, and travel to the ASM General Meeting where the laureate delivers the Promega Biotechnology Research Award Lecture.

Date(s) Application is Due: Oct 1

Contact: Awards Committee, (202) 942-9226; fax: (202) 942-9353; email: awards@asmusa.org

Internet: http://www.asm.org/academy/index.asp?bid=2291

Sponsor: American Society for Microbiology

1752 N Street, NW

Washington, DC 20036-2804

ASM Raymond W. Sarber Awards 1142

The ASM Raymond W. Sarber Awards recognize students at the undergraduate and predoctoral levels for research excellence and potential. The awards honor Raymond W. Sarber for his contributions to the growth and advancement of ASM.

Requirements: Nominees must be a student at the undergraduate or predoctoral level, attending an accredited institution in the United States, in an academic program involving microbiology. Two awards are presented at the ASM General Meeting, one to an undergraduate and one to a graduate student. They will be judged separately.

Restrictions: Nominations must consist of the following: curriculum vitae; statement, no longer than two pages, from the student outlining academic achievements, research accomplishments, and future career goals; letter of nomination, describe the nominee's research excellence and potential; letters of support, two letters of support: one from a supervisor or mentor, and the other letter may be from an academic advisor or colleague.

Samples: Self-nominations will not be accepted.

Amount of Grant: A framed certificate and cash prize of $1,500.

Date(s) Application is Due: Oct 1

Contact: Awards Committee, (202) 942-9226; fax: (202) 942-9353; email: awards@asmusa.org

Internet: http://www.asm.org/academy/index.asp?bid=2294

Sponsor: American Society for Microbiology

1752 N Street, NW

Washington, DC 20036-2804

ASM Robert D. Watkins Graduate Research Fellowship 1143

The goal of the fellowship is to increase the number of underrepresented groups completing doctoral degrees in the microbiological sciences. The ASM Robert D. Watkins Graduate Research Fellowship is aimed at highly competitive graduate students who are enrolled in a Ph.D. program and who have completed their graduate course work in the microbiological sciences. The fellowship encourages students to continue and complete their research project in the microbiological sciences. Students will be required to submit an abstract each year to ASM for presentation at the annual ASM General Meeting and attend the ASM Kadner Institute one time during the three-year tenure of the fellowship.

Requirements: Eligible candidates must be from groups that have been determined by the applicant's institution to be underrepresented in the microbiological sciences.

The ASM encourages institutions to identify individuals that have been historically underrepresented, and remain underrepresented today in the microbiological sciences nationally. These groups include African-Americans, Hispanics, Native Americans, Alaskan Natives, and Pacific Islanders. In addition, applicants must: be formally admitted to a doctoral program in the microbiological sciences in an accredited U.S. institution; have successfully completed the first year of the graduate program (first year graduate students cannot apply); have successfully completed all graduate coursework requirements for the doctoral degree by the date of activation of the fellowship; be a student member of ASM; be mentored by an ASM member; be a U.S. citizen or a permanent resident; not have funding or have funding that will expire by the start date of the fellowship(fellowship cannot run concurrently with other national fellowships from NIH, NSF, HHMI, etc.); three letters of recommendations must be submitted with your application. One letter must be from your research advisor/mentor; Official transcripts from all colleges and universities attended.

Amount of Grant: a total stipend of $63,000 ($21,000 a year) for a three year period

Date(s) Application is Due: May 1

Contact: Education Board, (202) 942-9283; fax: (202) 942-9329; email: fellowships-careerinformation@asmusa.org

Internet: http://www.asm.org/Education/index.asp?bid=6278

Sponsor: American Society for Microbiology

1752 N Street, NW

Washington, DC 20036

ASM sanofi-aventis ICAAC Award 1144

ASM's premier award in antimicrobial chemotherapy research; it stimulates research and honors outstanding accomplishment in antimicrobial chemotherapy.

Requirements: The nominee must be actively engaged in research involving development of new agents, investigation of antimicrobial action or resistance to antimicrobial agents, and/or the pharmacology, toxicology or clinical use of those agents. The nominee must not have served on the ICAAC Program Committee within the past two years. Nominations must consist of the following: curriculum vitae, including a list of publications; letter of nomination, describe the nominee's outstanding accomplishment in antimicrobial chemotherapy, including a list of the ten most relevant publications to the award; letters of support, two letters of support should come from two people, other than the nominator, who are familiar with the nominee's accomplishments. No more than one of the three letters may be from the nominee's institution or the same institution.

Restrictions: Self-nominations will not be accepted.

Amount of Grant: A cash prize of $20,000, a commemorative medal, and travel to ICAAC where the laureate delivers the sanofi-aventis ICAAC Award Lecture.

Date(s) Application is Due: Apr 1

Contact: Awards Committee, (202) 942-9226; fax: (202) 942-9353; email: awards@asmusa.org

Internet: http://www.asm.org/academy/index.asp?bid=2268

Sponsor: American Society for Microbiology

1752 N Street, NW

Washington, DC 20036-2804

ASM Siemens Healthcare Diagnostics Young Investigator Award 1145

The ASM Siemens Healthcare Diagnostics Young Investigator Award honors outstanding laboratory research in clinical microbiology or antimicrobial agents and is intended to further the career development of a young clinical scientist and promote awareness of clinical microbiology as a career.

Requirements: The nominee should be conducting outstanding research in clinical microbiology, automation in clinical laboratories, development of novel antimicrobial agents, mechanisms of action of antimicrobial agents or mechanisms of resistance to antimicrobial agents. The nominee must be no more than five years beyond completion of postdoctoral research training in microbiology, infectious diseases or related disciplines at the time of the nomination deadline. Nominations must consist of the following: curriculum vitae, including a list of publications, abstracts, and manuscripts in preparation; letter of nomination, describe the nominee's outstanding clinical research and potential; letters of support, two letters of support must come from persons, other than the nominator, who are familiar with the nominee's qualifications and accomplishments. No more than one of the three letters may be from the nominee's institution or from the same institution.

Restrictions: Self-nominations will not be accepted.

Amount of Grant: A cash prize of $2,000, commemorative piece, and travel to the ASM General Meeting.

Date(s) Application is Due: Oct 1

Contact: Awards Committee, (202) 942-9226; fax: (202) 942-9353; email: awards@asmusa.org

Internet: http://www.asm.org/Academy/index.asp?bid=2307

Sponsor: American Society for Microbiology

1752 N Street, NW

Washington, DC 20036-2804

ASM Undergraduate Research Fellowship 1146

The ASM Undergraduate Research Fellowship (URF) is aimed at highly competitive students who wish to pursue graduate careers (Ph.D. or MD/Ph.D) in microbiology. Students will have the opportunity to conduct full time research at their home institutions with an ASM member and present research results at the ASM General Meeting the following year. Students will: conduct a research project for a minimum of 10 weeks beginning in the summer; work with faculty mentors who are ASM members and who are employed at the students home institutions, and submit a research abstract for presentation at the ASM General Meeting. The Fellowship provides: up to $4000 for student stipend; two-year ASM student membership, and up to $1000 in travel support for students to present the results of the research project at the ASM General Meeting. Travel funds are contingent upon acceptance of an abstract for the General Meeting.

Requirements: Eligible student candidates for the fellowship must: be enrolled as full-time matriculating undergraduate students during the academic year at an accredited U.S. Institution; be involved in a research project; have an ASM member at their home institutions willing to serve as a mentor, and not receive financial support for research (i.e., Council for Undergraduate Research, Minority Access to Research Careers, Sigma Xi) during the fellowship. Applicants who do not meet all eligibility requirements will not be considered.

Restrictions: The program requires a joint application from both the student and a faculty mentor. It is student's responsibility to submit a completed application.

Amount of Grant: $4,000 - $5,000 range

Date(s) Application is Due: Feb 1

Contact: Fellowship Committee, (202) 942-9283; fax: (202) 942-9329; email: fellowships-careerinformation@asmusa.org

Internet: http://www.asm.org/Education/index.asp?bid=4319

Sponsor: American Society for Microbiology
1752 N Street, NW
Washington, DC 20036-2804

ASM William A Hinton Research Training Award 1147

The ASM William A Hinton Research Training Award honors outstanding contributions toward fostering the research training of underrepresented minorities in microbiology. It is given in memory of William A. Hinton, a physician-research scientist and one of the first African-Americans to join the ASM.

Requirements: The nominee must contribute to the research training of undergraduate students, graduate students, postdoctoral fellows or health professional students and efforts leading to the increased participation of underrepresented minorities in microbiology. Nominations must consist of the following: curriculum vitae, including a list of nominee's publications; letter of nomination, describe the nominee's training of undergraduate students, graduate students, postdoctoral fellows, or health professional students. Detail specific examples of excellence in training and mentoring of underrepresented minorities. Describe the duration, extent, and impact of the nominee's training of underrepresented minority students, graduate students, postdoctoral fellows, or health professional students; letters of support, two letters of support must come from persons, other than the nominator, who are familiar with the nominee's outstanding contributions to the training of underrepresented minorities. No more than one of the three letters may be from the nominee's institution or from the same institution.

Amount of Grant: A cash prize of $2,000, a commemorative piece, and travel to the ASM General Meeting.

Date(s) Application is Due: Oct 1

Contact: Awards Committee, (202) 942-9226; fax: (202) 942-9353; email: awards@asmusa.org

Internet: http://www.asm.org/Academy/index.asp?bid=2317

Sponsor: American Society for Microbiology
1752 N Street, NW
Washington, DC 20036-2804

ASM-PAHO Infectious Diseases Epidemiology and Surveillance Fellowships 1148

Funded by the Pan American Health Organization and ASM, this award offers fellowships to young scientists from Bolivia, Colombia, Costa Rica, Dominican Republic, Ecuador, El Salvador, Guatemala, Honduras, Nicaragua, Panama, Paraguay, or Peru, who have obtained their Masters, PhD, or other equivalent academic degree, and who have at least five years of laboratory experience in the area of antimicrobial resistance, to visit a host scientist in the U.S.. The ASM-PAHO Infectious Diseases Epidemiology and Surveillance Fellowship will award a maximum grant of $4,000. The award is not intended to provide travel to obtain a degree at the host institution.

Requirements: The application must be made jointly between a host institution and the visiting fellow. The Investigator must be: a member of ASM or any other national microbiological society; actively involved in research in the microbiological sciences; have obtained, or be in the process of obtaining, their masters, PhD, or other equivalent academic degree within the last five years; a national of a resource-limited Africa, Asian or Latin American country submitting the application from an institution in their home region. Investigators from developed nations, as defined by BOTH the UN high HDI group and the World Bank high income economy group, are ineligible. Investigators submitting an application from the U.S. where they are spending a training period are ineligible; sufficiently proficient in the use of the English language. The Host Scientist must be: an ASM member; actively involved in research in the microbiological sciences in the U.S.; actively involved in teaching in the microbiological sciences in the U.S.; interested in sustaining international collaborations Preference will be given to: applicants who can prove three years of membership in ASM or any other national microbiological society; applicants who have not previously had the opportunity to travel to a facility in another country. The application should demonstrate to the Review Committee: academic excellence of the applicant - Honors & Awards, CV, and Letters of Recommendation; depth of the applicant's research experience - Statement of prior Research Experience; quality and originality of work proposed during the Fellowship - Potential Collaboration and Proposed Research Plan; relevance of the work proposed to the applicant's locale - Potential Collaboration and Proposed Research Plan; excellence of the host, and the research and training environment

Amount of Grant: $4,000

Date(s) Application is Due: Apr 15; Oct 15

Contact: International Fellowship Program, (202) 942-9368; fax: (202) 942-9328; email: international@asmusa.org

Internet: http://www.asm.org/International/index.asp?bid=2778

Sponsor: American Society for Microbiology
1752 N Street NW
Washington, DC 20036-2804

ASM-UNESCO Leadership Grant for International Educators 1149

This new program, sponsored jointly by ASM and UNESCO, has been developed to enable a select group of educators from resource-limited countries to attend the ASM Conference for Undergraduate Educators (ASMCUE) and a pre-conference workshop to provide leaders in education with the resources to build innovative teaching modules that engage students and lead to enduring understanding in microbiology. Successful applicants will receive financial support to cover airfare (economy class), room, board and registration to the pre-conference workshop and ASMCUE. Participants will be reimbursed for airfare via wire transfer upon submission of a confirmed invoice. Room, board and registration will be arranged by ASM.

Requirements: Preference will be given to ASM members. Applicants must be a citizen of, and residing in, a resource-limited country in Latin America, Asia, Africa, and Central and Eastern Europe. Resource-limited countries are classified as upper-middle income and below by the World Bank Development Group. (To find out if your country fits this criteria, visit http://tinyurl.com/7he3u.) Participants must have a minimum of eight years teaching in microbiology/biology at a university level and must also have regular and sustained access to the Internet.

Restrictions: Members of the ASM International Board, International Education Committee, International Membership Committee, International Laboratory Capacity Building Committee, and the ASM International Ambassador Network are not eligible for this program.

Date(s) Application is Due: Dec 1

Contact: Michelle Godinez, Coordinator, Education Programs and Events; (202) 942-9317; fax: (202) 942-9328; email: mgodinez@asmusa.org

Internet: http://www.asm.org/International/index.asp?bid=61306

Sponsor: American Society for Microbiology
1752 N Street NW
Washington, DC 20036

ASM/CCID Fellowships 1150

The program is sponsored by the American Society for Microbiology (ASM) and the Coordinating Center for Infectious Diseases (CCID), formerly known as the National Center for Infectious Diseases (NCID). The goal of ASM/CCID Fellowship is to support the development of new approaches, methodologies and knowledge in infectious disease prevention and control in areas within the public health mission of the CDC. The fellowship allows one to perform research in residence at the CCID which is headquartered at the Centers for Disease Control and Prevention (CDC) in Atlanta, GA. Eligible fields of study include: bacterial and Mycotic Diseases; viral and Rickettsial Infections; nosocomial Infections; HIV/AIDS; vector-borne Infectious Diseases; parasitic Diseases. The Fellowship provides an annual stipend (up to $42,188), health benefits (up to $3000 annually) for a maximum of 2 years, relocation benefits (up to $500), and (up to $2000 annually) for professional development for a maximum of 2 years.

Requirements: The program is intended for individuals who either earned their doctorate degree or completed a primary residency within three years of their proposed start date. Applicants may not have a faculty position or be enrolled in a graduate degree program during the fellowship. Considerations will be given to individuals with more experience if there are compelling reasons. Qualified applicants will receive consideration without regard to race, creed, color, age, sex, or national origin.

Amount of Grant: $41,000 - $42,000 range

Date(s) Application is Due: Jan 15

Contact: Fellowship Committee, (202) 942-9283; fax: (202) 942-9329; email: fellowships-careerinformation@asmusa.org

Internet: http://www.asm.org/Education/index.asp?bid=15497

Sponsor: American Society for Microbiology
1752 N Street NW
Washington, DC 20036-2804

ASME Gas Turbine Award 1151

The award is given annually in recognition of an outstanding individual or multiple-author contribution to the literature of combustion gas turbines or gas turbines thermally combined with nuclear or steam power plants. The paper may be devoted to the design aspects of overall gas turbines or individual components and/or systems such as compressors, combustion systems, turbines, controls and accessories, bearings, regenerators, inlet air filters, silencers, etc. It may cover topics specifically related to gas turbines, such as high-temperature materials or fuel considerations including erosion and corrosion complications. It also can be devoted to application or operational aspects of gas turbines for aircraft propulsion and ground power units, or automotive, electric utility, gas pipeline pumping, locomotive, marine, oil field pumping, petrochemical, space power, steel, and similar uses.

Requirements: Papers published anywhere in the world are eligible; authors are not restricted by nationality, age, profession, or membership in any engineering society or other organizations.

Samples: Frank Hummel—for his paper entitled: Wake-Wake Interaction and its Potential for Clocking in a Transonic High Pressure Turbine, $1000 (2003).

Amount of Grant: $1000

Date(s) Application is Due: Jan 1

Contact: Honors Department, (800) 843-2763 or (212) 591-7722; fax: (212) 591-7674; email: infocentral@asme.org
Internet: http://www.asme.org/honors/ms71/sla/turbine.html
Sponsor: American Society of Mechanical Engineers
3 Park Avenue
New York, NY 10016-5990

ASME Heat Transfer Memorial Award   **1152**
This award is bestowed on individuals who have made outstanding contributions to the field of heat transfer through teaching, research, design, or publications. Each award is based on papers in an area of heat transfer or on a paper dealing with the science or art of heat transfer. Awards are made annually in each of the three following categories: the science of heat transfer, the art of heat transfer, and the general subject of heat transfer.
Requirements: There are no restrictions by age, nationality, or society membership.
Samples: Abdolhossein Haji-Sheikh, Michael Modest, Wei Shyy—award winners, $1000 (2005); Mohammad Faghri, Yildiz Bayazitoglu—award winners, $1000 (2004); Dimos Poulikakos, M. Michael Yovanovich, James R. Welty—award winners, $500 (2003); Massoud Kaviany, Je-Chin Han, Roop L. Mahajan—award winners, $500.
Amount of Grant: $1000
Date(s) Application is Due: Oct 1
Contact: Gilda DiTullio, Manager, (212) 591-7736; fax: (212) 705-7739; email: ditulliog@asme.org
Internet: http://www.asme.org/honors/ms71/saa/heat.html
Sponsor: American Society of Mechanical Engineers
3 Park Avenue
New York, NY 10016

ASME Internal Combustion Engine Award   **1153**
This award is given in recognition of eminent achievement or distinguished contribution over a substantial period of time, which may result from research, innovation, or education in advancing the art of engineering in the field of internal combustion engines or in directing the efforts and accomplishments of those engaged in engineering practice in the design, development, application, and operation of internal combustion engines.
Samples: Karl Springer—award winner, $1000 (2005); Humphrey Niven—award winner, $1000 (2004); Rodica A. Baranescu—award winner, $1000 (2003); Warren E. Snyder—award winner, $1000.; Charles A. Amann—award winner, $1000.
Amount of Grant: $1000
Date(s) Application is Due: Feb 1
Contact: Gilda DiTullio, Manager, (212) 591-7736; fax: (212) 705-7739; email: ditulliog@asme.org
Internet: http://www.asme.org/honors/ms71/saa/ice.html
Sponsor: American Society of Mechanical Engineers
3 Park Avenue
New York, NY 10016

ASME Machine Design Award   **1154**
The award is made to recognize eminent achievement or distinguished service in the field of machine design, which includes application, research, development, or teaching of machine design.
Requirements: Any individual member, group of members, or committee may nominate candidates for this award.
Samples: Bahram Ravani—award winner, $1000 (2005); Sridhar Kota—award winner, $1000 (2004); Richard F. Salant—award winner, $1000 (2003).
Amount of Grant: $1000
Date(s) Application is Due: Feb 1
Contact: Gilda DiTullio, Manager, (212) 591-7736; fax: (212) 705-7739; email: ditulliog@asme.org
Internet: http://www.asme.org/honors/ms71/saa/md.html
Sponsor: American Society of Mechanical Engineers
3 Park Avenue
New York, NY 10016

ASME Medal   **1155**
One medal may be awarded annually to an individual of any age or nationality for a lifetime of service in engineering or related fields. Any individual member, group of members, or committee may nominate candidates. Application forms are available on the Web site.
Samples: Robert E. Uhrig—award winner, $10,000 (2005); Bradford W. Parkinson—award winner, $10,000 (2004); Norman R. Augustine—award winner, $10,000 (2003).
Amount of Grant: $10,000
Date(s) Application is Due: Mar 1
Contact: Gilda DiTullio, Manager, (212) 591-7736; fax: (212) 705-7739; email: ditulliog@asme.org
Internet: http://www.asme.org/honors/ms71/daa/asme.html
Sponsor: American Society of Mechanical Engineers
3 Park Avenue
New York, NY 10016

ASME Melvin R. Green Codes and Standards Medal   **1156**
This award is given annually for outstanding contribution to codification, standardization, and certification for service on ASME, ANSC, or ISO/TAC Committee administered by ASME.

Requirements: The recipient may be from industry, government, education, or private professional practice, and need not be an ASME member.
Samples: James Turner—award winner, $1500 (2005); Edward A. Donoghue—award winner, $1500 (2004); James A. Perry—award winner, $1500 (2003).
Amount of Grant: $1500
Date(s) Application is Due: Jan 1
Contact: Gilda DiTullio, Manager, (212) 591-7736; fax: (212) 705-7739; email: ditulliog@asme.org
Internet: http://www.asme.org/honors/ms71/saa/green.html
Sponsor: American Society of Mechanical Engineers
3 Park Avenue
New York, NY 10016

ASME Performance Test Codes Medal   **1157**
This medal is awarded to an individual (or individuals in exceptional circumstances) who has made an outstanding contribution to the development and promotion of ASME performance test codes including the supplements on instruments and apparatus.
Requirements: Any individual member, group of members, or committee may nominate candidates for this award.
Amount of Grant: $1000
Date(s) Application is Due: Jan 1
Contact: Gilda DiTullio, Manager, (212) 591-7736; fax: (212) 705-7739; email: ditulliog@asme.org
Internet: http://www.asme.org/honors/ms71/saa/ptc.html
Sponsor: American Society of Mechanical Engineers
3 Park Avenue
New York, NY 10016

ASNS Bio-Serv Award in Experimental Animal Nutrition   **1158**
The award is given for meritorious research in nutrition by an investigator who received the doctoral degree in the 10 years preceding the month the award is presented. The work recognized must involve the nutrition of experimental animals used as models. The award and an engraved plaque are made available by Bio-Serv Inc. Nominations should include a letter stating the basis for the nomination, a selected bibliography that supports the nomination, and the reprint or series of reprints on which the nomination is based.
Requirements: Nominations may be made by anyone, including members of the respective nominating committees and nonmembers of the institute. Nominations will be retained for two years. Candidates may be renominated. Candidates need not be members of the institute. However, the awards are usually given to professionally active nutrition scientists.
Restrictions: An individual who has received one AIN award is not eligible to receive another award unless it is for accomplishments not covered by the first award.
Amount of Grant: $1000
Date(s) Application is Due: Sep 1
Contact: Secretariat, (301) 530-7051; fax: (301) 571-1892; email: nnotes@asns.faseb.org
Internet: http://www.asns.org/invitation.html
Sponsor: American Society for Nutritional Sciences
9650 Rockville Pike, Suite L-4500
Bethesda, MD 20814

ASNS Centrum Center for Nutrition Science Award   **1159**
The award is given for recent investigative contributions of contemporary significance to the understanding of human nutrition. The contributions need not be restricted to investigative work with humans as long as they have relevance to human nutrition and health. Preference is usually given to scientists in the Western Hemisphere. The award and an engraved plaque are made available by Wyeth Consumer Healthcare. Nomination should include a letter stating the significance of the work, a selected bibliography that supports the nomination, and a reprint or series of reprints reporting such research.
Requirements: Nominations may be made by anyone, including members of the respective nominating committees and nonmembers of the institute. Nominations will be retained for two years. Candidates may be renominated. Candidates need not be members of the institute. However, the awards are usually given to professionally active nutrition scientists.
Restrictions: An individual who has received one ASNS award is not eligible to receive another award unless it is for accomplishments not covered by the first award.
Amount of Grant: $1500
Date(s) Application is Due: Sep 1
Contact: Program Officer, (301) 530-7050; fax: (301) 634-7892; email: sec@asns.org
Internet: http://www.asns.org/awards.html
Sponsor: American Society for Nutritional Sciences
9650 Rockville Pike, Suite L-4500
Bethesda, MD 20814

ASNS Conrad A. Elvehjem Award for Public Service in Nutrition   **1160**
The award is given for specific and distinguished service to the public through the science of nutrition. Such service could be rendered through governmental, industrial, private, or international institutions, but contributions of an investigative character are not excluded. The award and an engraved plaque are made available by Kraft Foods. Nominations should include a letter stating the basis for the nomination, a selected bibliography indicating the candidate's contributions to public service, and the candidate's curriculum vita.

Requirements: Nominations may be made by anyone, including members of the respective nominating committees and nonmembers of the institute. Nominations will be retained for two years. Candidates may be renominated. Candidates need not be members of the institute. However, the awards are usually given to professionally active nutrition scientists.
Restrictions: An individual who has received one ASNS award is not eligible to receive another award unless it is for accomplishments not covered by the first award.
Amount of Grant: $1500
Date(s) Application is Due: Oct 1
Contact: Secretariat, (301) 530-7050; fax: (301) 637-7892; email: sec@asns.org
Internet: http://www.asns.org/awards.html
Sponsor: American Society for Nutritional Sciences
9650 Rockville Pike, Suite L-4500
Bethesda, MD 20814

ASNS Mead Johnson Nutrition Award 1161
The award is given to an investigator for a single outstanding piece of nutrition research or a series of papers on the same subject completed within 10 years of postgraduate training. The award and an inscribed scroll are made available by Mead Johnson Nutritionals. Nominations should include a letter stating the significance of the work, a selected bibliography that supports the nomination, and a reprint or series of reprints reporting such research.
Requirements: Nominations may be made by anyone, including members of the respective nominating committees and nonmembers of the institute. Nominations will be retained for two years. Candidates may be renominated. Candidates need not be members of the institute. However, the awards are usually given to professionally active nutrition scientists.
Restrictions: An individual who has received one ASNS award is not eligible to receive another award unless it is for accomplishments not covered by the first award.
Amount of Grant: $2500
Date(s) Application is Due: Sep 1
Contact: Program Officer, (301) 530-7050; fax: (301) 634-7892; email: sec@asns.org
Internet: http://www.asns.org/awards.html
Sponsor: American Society for Nutritional Sciences
9650 Rockville Pike, Suite L-4500
Bethesda, MD 20814

ASNS Osborne and Mendel Nutrition Award 1162
The award is given for outstanding recent basic research accomplishments in nutrition. Nominations should include a letter stating the significance of the work, a selected bibliography of all papers relating to the research on which the nomination is based, and a reprint or series of reprints reporting this research.
Requirements: Nominations may be made by anyone, including members of the respective nominating committees and nonmembers of the institute. Nominations will be retained for two years. Candidates may be renominated. Candidates need not be members of the institute. However, the awards are usually given to professionally active nutrition scientists.
Restrictions: An individual who has received one ASNS award is not eligible to receive another award unless it is for accomplishments not covered by the first award.
Amount of Grant: $2500
Date(s) Application is Due: Sep 1
Contact: Program Officer, (301) 530-7050; fax: (301) 634-7892; email: sec@asns.org
Internet: http://www.asns.org/awards.html
Sponsor: American Society for Nutritional Sciences
9650 Rockville Pike, Suite L-4500
Bethesda, MD 20814

ASNS Predoctoral Fellowship Program 1163
This fellowship is designed to support research in human and clinical nutrition. Preference will be given to work of immediate relevance to human nutrition information and human nutrition status. Contact the institute for application materials.
Requirements: Applicants must be enrolled in a graduate program registered in the AIN Directory of Graduate Programs in Nutritional Sciences.
Amount of Grant: $5000
Date(s) Application is Due: Dec 1
Contact: Secretariat, (301) 530-7052; fax: (301) 364-7892; email: sec@asns.org
Internet: http://www.asns.org/winners.html
Sponsor: American Society for Nutritional Sciences
9650 Rockville Pike, Suite L-4500
Bethesda, MD 20814

ASOR Fellowships at Cyprus American Archaeological Research Institute 1164
ASOR offers opportunities for study at CAARI in humanistic disciplines in studies of the Middle East from prehistoric times to the modern era. One Stuart Swiny fellowship is available to support participation in any phase or aspect of a project in Cyprus that has been approved by ASOR's Committee on Archaeological Policy (CAP). A $750 stipend is included to help cover room and board at CAARI. This program is open to scholars of any nationality (application deadline February 1). One Anita Cecil O'Donovan fellowship ($750 maximum) is available for a one- to three-month period to assist in partial payment of essential expenses for an undergraduate or graduate student to conduct research in Cyprus. Residence at CAARI is mandatory (application deadline February 1). A number of Fulbright fellowships are also available. Interested persons should contact the Fulbright program advisors at the Council for International Exchange of Scholars,

3007 Tilden Street NW, Suite 5M, Box GP.O.S, Washington, DC 20008-3009, (202) 686-7877 (application deadline August 1).
Requirements: All eligible persons are encouraged to apply for as many fellowships and professorships as they wish, but a person may hold only one award at a time. Persons who receive awards in one year can reapply for the same or other awards the following year, but new applicants will have priority.
Amount of Grant: $750
Date(s) Application is Due: Feb 1
Contact: Program Contact, (617) 353-6574; fax: (617) 353-6575; email: caari@bu.edu
Internet: http://www.caari.org/fellow.html
Sponsor: American Schools of Oriental Research
656 Beacon Street, 5th Fl
Boston, MA 02215-2010

ASOR Mesopotamian Fellowship 1165
The Mesopotamian Fellowship provides support for one three-to-six month period of research. This fellowship is primarily intended to support field research in ancient Mesopotamian civilization carried out in the Middle East, but other research projects such as museum or archival research, related to ancient Mesopotamian studies may be considered. The Mesopotamian Fellowship is based on a 1 July 2005 to 30 June 2006 fiscal year. A recipient who does not use the fellowship for at least three months must forfeit and return a prorated amount of the stipend. Fellowship time should be continuous, without frequent trips outside the Middle East. It is open to qualified predoctoral students and postdoctoral students and scholars from any country. Prospective applicants are encouraged to consult with the ASOR coordinator of academic programs for inquiries.
Requirements: Fellowship is open to predoctoral students or postdoctoral scholars for research on a project dealing with ancient Mesopotamian civilization. Applicants must become professional members of ASOR and be either affiliated with an institution that is a corporate member of ASOR or have been an individual professional member for more than two years. Priority is given to applicants whose projects are affiliated with ASOR.
Samples: Kathryn Slanski (Trumbull, CT)—for research on Kudurru's inscribed Mesopotamian boundary stones in museums in London, Paris, Berlin, and Istanbul, $5000.
Amount of Grant: $7000
Date(s) Application is Due: Apr 1
Contact: Director, Mesopotamian Fellowship, (617) 353-6570; fax: (617) 353-6575; email: asor@bu.edu
Internet: http://www.asor.org/bagdam.html
Sponsor: American Schools of Oriental Research
656 Beacon Street, 5th Fl
Boston, MA 02215-2010

ASPEN C. Richard Fleming Grant 1166
The Foundation provides annual grant support to nutrition researchers at all stages of their careers. This grant, funded by the donations of ASPEN members, is intended to assist a nutrition investigator by providing preliminary funding for promising new research in the field of nutrition and metabolic support and related areas of clinical nutrition. Applicants still in training positions may apply. Corporate employees are eligible to receive grants if the research project is not part of their normal duties and they have at least a part-time academic appointment or are associated with a professional institute that conducts nutrition research. In making choices among otherwise high-quality applicants, the reviewers will take into account the geographic distribution of grant winners; their ethnic background and gender, should the applicant choose to supply such information; their professional discipline; and other criteria as appropriate to ensure an equitable system. Funds may be used only for technician salary, equipment, supplies, animals, clinical research costs or other expenses directly related to the conduct of the proposed research.
Requirements: Applicants must submit a letter from their supervisor or department head at the institution confirming his/her commitment to the project. If the project involves human subjects, a letter pledging support in recruiting patients from the primary care provider and the institutional review board overseeing human studies is required. For junior faculty and applicants who do not hold faculty positions, three letters of reference are required.
Restrictions: This grant is not intended to support pursuit of an additional degree. Grant funds may not be used for: comparison of commercial products; indirect costs or overhead; costs of patient care; constructing or renovating facilities; furniture or office equipment; secretarial services; honoraria or membership dues; textbooks or periodicals; repair or service contract costs on institutional equipment; entertainment; travel; or salary support for the Principal Investigator.
Amount of Grant: $3,500
Date(s) Application is Due: Sep 8
Contact: Paula Bowen, Research Program Administrator; (301) 587-6315, ext. 132 or (301) 920-9132; fax: (301) 587-2365; email: paulab@aspen.nutr.org
Internet: http://www.nutritioncare.org/wcontent.aspx?id=784
Sponsor: American Society for Parenteral and Enteral Nutrition (ASPEN) Rhoads Research Foundation
8630 Fenton Street, Suite 412
Silver Spring, MD 20910

ASPEN Douglas Wilmore Grant 1167
The Foundation provides annual grant support to nutrition researchers at all stages of their careers. The grant is intended to assist a nutrition investigator by providing preliminary funding for promising new research in the field of nutrition and metabolic support

and related areas of clinical nutrition. Priority consideration is given to applications that involve a multidisciplinary team or that investigate the efficacy of parenteral or enteral nutrition. Applicants still in training positions may apply for the Fleming Grant. Corporate employees are eligible to receive grants if the research project is not part of their normal duties and they have at least a part-time academic appointment or are associated with a professional institute that conducts nutrition research. In making choices among otherwise high-quality applicants, the reviewers will take into account the geographic distribution of grant winners; their ethnic background and gender, should the applicant choose to supply such information; their professional discipline; and other criteria as appropriate to ensure an equitable system. Funds may be used only for technician salary, equipment, supplies, animals, clinical research costs or other expenses directly related to the conduct of the proposed research. The grant can be renewed for a second year of funding, assuming satisfactory progress.

Requirements: Applicants must submit a letter from their supervisor or department head at the institution confirming his/her commitment to the project. If the project involves human subjects, a letter pledging support in recruiting patients from the primary care provider and the institutional review board overseeing human studies is required. For junior faculty and applicants who do not hold faculty positions, three letters of reference are required.

Restrictions: Individuals who have received other research grants in excess of $25,000 are ineligible to apply, as are those who completed training more than 5 years prior to the start of the grant period. This grant is not intended to support pursuit of an additional degree. Grant funds may not be used for: comparison of commercial products; indirect costs or overhead; costs of patient care; constructing or renovating facilities; furniture or office equipment; secretarial services; honoraria or membership dues; textbooks or periodicals; repair or service contract costs on institutional equipment; entertainment; travel; or salary support for the Principal Investigator.

Amount of Grant: $25,000 maximum

Date(s) Application is Due: Sep 8

Contact: Paula Bowen, Research Program Administrator; (301) 587-6315, ext. 132 or (301) 920-9132; fax: (301) 587-2365; email: paulab@aspen.nutr.org

Internet: http://www.nutritioncare.org/wcontent.aspx?id=784

Sponsor: American Society for Parenteral and Enteral Nutrition (ASPEN) Rhoads Research Foundation

8630 Fenton Street, Suite 412

Silver Spring, MD 20910

ASPEN Dudrick Research Scholar Award 1168

A.S.P.E.N. members are invited to nominate individuals to be considered by the Dudrick Award Committee for the Stanley Dudrick Research Scholar Award. A.S.P.E.N. presents this annual award to recognize and support a mid-career investigator who has shown significant achievements in nutrition support, is an A.S.P.E.N. member, and demonstrates exceptional research productivity and the potential to continue to make contributions in the field of nutrition therapy. The award includes $5,000, plus recognition during the Rhoads Lecture and Awards Symposium.

Requirements: The applicant should: be an A.S.P.E.N. member; have an advanced degree (MD, PharmD, MS, PhD, or equivalent); currently be in a training program or have completed formal training in nutritional research/therapy within the past 10-15 years, e.g. mid-career; have demonstrated exceptional research productivity during and/or following formal training.

Amount of Grant: $5,000

Date(s) Application is Due: Oct 15

Contact: Paula Bowen, Research Program Administrator; (301) 587-6315, ext. 132 or (301) 920-9132; fax: (301) 587-2365; email: paulab@aspen.nutr.org

Internet: http://www.nutritioncare.org/wcontent.aspx?id=878

Sponsor: American Society for Parenteral and Enteral Nutrition

8630 Fenton Street, Suite 412

Silver Spring, MD 20910

ASPEN Harry M. Vars Award 1169

The Award is made annually for the best research presentation by an investigator at Nutrition Week. The recipient of the award is selected by the Vars Award Selection Subcommittee based upon a review of manuscripts submitted by qualified candidates. The author of the best paper presents his or her paper at the Premier Paper Session and is given a $1,000 cash prize, a travel grant and a plaque during the Research Awards Ceremony at Nutrition Week. The winning manuscript is published in the Journal of Parenteral and Enteral Nutrition (JPEN). The Award is open to all disciplines.

Samples: Zhi Yong Peng, Natalie J. Serkova, Douglas J. Kominsky, Jaimi L. Brown, and Paul E. Wischmeyer—Glutamine-Mediated Attenuation of Cellular Metabolic Dysfunction and Cell Death After Injury Is Dependent on Heat Shock Factor-1 Expression (2006).

Amount of Grant: $1,000

Contact: Michelle Spangenburg, Program Director for Education; (301) 587-6315, ext. 127; fax:(301) 587-2365; email: michelles@aspen.nutr.org

Internet: http://www.nutritioncare.org/wcontent.aspx?id=878

Sponsor: American Society for Parenteral and Enteral Nutrition

8630 Fenton Street, Suite 412

Silver Spring, MD 20910

Aspen Institute FIELD Financing Products Grants 1170

The Microenterprise Fund for Innovation, Effectiveness, Learning and Dissemination (FIELD) seeks proposals from microenterprise programs that have developed or are developing innovative financing products targeted to the specific capital needs of low-income borrowers. FIELD wishes to support applicants who have taken or will take deliberate, innovative approaches to financing for poor entrepreneurs. Grants are awarded for two years. Each applicant should submit a single letter of intent (three-page maximum, single-spaced) and address the organizational history/capacity, statement strategy, anticipated outcomes, and leveraging potential.

Requirements: Applicants must be 501(c)3 nonprofits with a three-year operating history providing microenterprise services to a significant number of low-income clients. Experience in providing access to financing for low-income entrepreneurs also is required.

Samples: Community Ventures Corp (Lexington, KY)—to create three new loan products that will expand the organization's continuum of loan products for low-income borrowers, $500,000.; ACCION U.S. Network (Cambridge, MA)—to significantly increase the scale of the network's lending activities by reengineering the loan delivery process, $500,000.

Amount of Grant: $100,000 maximum over two years

Date(s) Application is Due: Jan 15

Contact: Kirsten Moy, Program Coordinator, (202) 736-1071; fax: (202) 467-0790; email: fieldweb@aspeninstitute.org

Internet: http://www.fieldus.org

Sponsor: Aspen Institute

1 Dupont Circle NW, Suite 700

Washington, DC 20036

Aspen Institute FIELD Technical Assistance Assessment Grants 1171

The Microenterprise Fund for Innovation, Effectiveness, Learning and Dissemination (FIELD) seeks proposals that will advance the industry's understanding of how to assess the effectiveness of training and technical assistance activities directed to low-income clients. The grants support creative approaches to documentation and analysis of outcomes. Proposed programs also may include funding for training and/or technical assistance services to pools of low-income clients. Grantees are required to raise local matching funds to promote the sustainability of the program at the local level. Each applicant should submit a single letter of intent (three-page maximum, single-spaced) and address the organizational history/capacity, statement strategy, anticipated outcomes, and leveraging potential.

Requirements: Applicants must be 501(c)3 nonprofits with a three-year operating history providing microenterprise services to a significant number of low-income clients.

Samples: ACCION Texas (San Antonio, TX)—to bring in a professional media consulting firm to develop and implement a comprehensive marketing strategy that promotes ACCION Texas services in target markets, double the number of clients (from 571 to 1194), and increase the number of loans five-fold by the end of the grant period, $70,000-$100,000.

Amount of Grant: $100,000 maximum over two years

Date(s) Application is Due: Jan 15

Contact: Program Coordinator, (202) 736-1071; fax: (202) 467-0790; email: fieldus@aspeninst.org

Internet: http://www.fieldus.org

Sponsor: Aspen Institute

1 Dupont Cir NW, Suite 700

Washington, DC 20036

ASPEN Maurice Shils Grant 1172

The Foundation provides annual grant support to nutrition researchers at all stages of their careers. This grant, funded by a donation from Baxter Health Care and Nestle Clinical Nutrition, is intended to assist a nutrition investigator by providing preliminary funding for promising new research in the field of nutrition and metabolic support and related areas of clinical nutrition. Persons applying for the grant must be in a post-training position and commit at least 20% of their time to research. Corporate employees are eligible to receive grants if the research project is not part of their normal duties and they have at least a part-time academic appointment or are associated with a professional institute that conducts nutrition research. In making choices among otherwise high-quality applicants, the reviewers will take into account the geographic distribution of grant winners; their ethnic background and gender, should the applicant choose to supply such information; their professional discipline; and other criteria as appropriate to ensure an equitable system. Funds may be used only for technician salary, equipment, supplies, animals, clinical research costs or other expenses directly related to the conduct of the proposed research. Grants can be renewed for a second year of funding, assuming satisfactory progress.

Requirements: Applicants must submit a letter from their supervisor or department head at the institution confirming his/her commitment to the project. If the project involves human subjects, a letter pledging support in recruiting patients from the primary care provider and the institutional review board overseeing human studies is required. For junior faculty and applicants who do not hold faculty positions, three letters of reference are required.

Restrictions: Individuals who have received other research grants in excess of $25,000 are ineligible to apply, as are those who completed training more than 5 years prior to the start of the grant period. This grant is not intended to support pursuit of an additional degree. Grant funds may not be used for: comparison of commercial products; indirect costs or overhead; costs of patient care; constructing or renovating facilities; furniture or office equipment; secretarial services; honoraria or membership dues; textbooks or periodicals; repair or service contract costs on institutional equipment; entertainment; travel; or salary support for the Principal Investigator.

Amount of Grant: $18,000

Date(s) Application is Due: Sep 8

Contact: Paula Bowen, Research Program Administrator; (301) 587-6315, ext. 132 or (301) 920-9132; fax: (301) 587-2365; email: paulab@aspen.nutr.org
Internet: http://www.nutritioncare.org/wcontent.aspx?id=784
Sponsor: American Society for Parenteral and Enteral Nutrition (ASPEN) Rhoads Research Foundation
8630 Fenton Street, Suite 412
Silver Spring, MD 20910

ASPEN Norman Yoshimura Grant 1173

The Foundation provides annual grant support to nutrition researchers at all stages of their careers. This grant, supported by B. Braun Medical, Inc. funding, is intended to assist a nutrition investigator by providing preliminary funding for promising new research in the field of nutrition and metabolic support and related areas of clinical nutrition. Applicants still in training positions may apply. Corporate employees are eligible to receive grants if the research project is not part of their normal duties and they have at least a part-time academic appointment or are associated with a professional institute that conducts nutrition research. In making choices among otherwise high-quality applicants, the reviewers will take into account the geographic distribution of grant winners; their ethnic background and gender, should the applicant choose to supply such information; their professional discipline; and other criteria as appropriate to ensure an equitable system. Funds may be used only for technician salary, equipment, supplies, animals, clinical research costs or other expenses directly related to the conduct of the proposed research.

Requirements: Applicants must submit a letter from their supervisor or department head at the institution confirming his/her commitment to the project. If the project involves human subjects, a letter pledging support in recruiting patients from the primary care provider and the institutional review board overseeing human studies is required. For junior faculty and applicants who do not hold faculty positions, three letters of reference are required.

Restrictions: This grant is not intended to support pursuit of an additional degree. Grant funds may not be used for: comparison of commercial products; indirect costs or overhead; costs of patient care; constructing or renovating facilities; furniture or office equipment; secretarial services; honoraria or membership dues; textbooks or periodicals; repair or service contract costs on institutional equipment; entertainment; travel; or salary support for the Principal Investigator.
Amount of Grant: $15,000
Date(s) Application is Due: Sep 8
Contact: Paula Bowen, Research Program Administrator; (301) 587-6315, ext. 132 or (301) 920-9132; fax: (301) 587-2365; email: paulab@aspen.nutr.org
Internet: http://www.nutritioncare.org/wcontent.aspx?id=784
Sponsor: American Society for Parenteral and Enteral Nutrition (ASPEN) Rhoads Research Foundation
8630 Fenton Street, Suite 412
Silver Spring, MD 20910

ASPEN Nutrition Research Grants 1174

The Foundation provides annual grant support to nutrition researchers at all stages of their careers. This grant, funded by a donation from the Abbott Foundation, is intended to assist a nutrition investigator by providing preliminary funding for promising new research in the field of nutrition and metabolic support and related areas of clinical nutrition. Persons applying for the grant must be in a post-training position and commit at least 20% of their time to research. Corporate employees are eligible to receive grants if the research project is not part of their normal duties and they have at least a part-time academic appointment or are associated with a professional institute that conducts nutrition research. In making choices among otherwise high-quality applicants, the reviewers will take into account the geographic distribution of grant winners; their ethnic background and gender, should the applicant choose to supply such information; their professional discipline; and other criteria as appropriate to ensure an equitable system. Funds may be used only for technician salary, equipment, supplies, animals, clinical research costs or other expenses directly related to the conduct of the proposed research. Grants can be renewed for a second year of funding, assuming satisfactory progress.

Requirements: Applicants must submit a letter from their supervisor or department head at the institution confirming his/her commitment to the project. If the project involves human subjects, a letter pledging support in recruiting patients from the primary care provider and the institutional review board overseeing human studies is required.

Restrictions: Individuals who have received other research grants in excess of $25,000 are ineligible to apply, as are those who completed training more than 5 years prior to the start of the grant period. This grant is not intended to support pursuit of an additional degree. Grant funds may not be used for: comparison of commercial products; indirect costs or overhead; costs of patient care; constructing or renovating facilities; furniture or office equipment; secretarial services; honoraria or membership dues; textbooks or periodicals; repair or service contract costs on institutional equipment; entertainment; travel; or salary support for the Principal Investigator.
Amount of Grant: $25,000 maximum
Date(s) Application is Due: Sep 8
Contact: Paula Bowen, Research Program Administrator; (301) 587-6315, ext. 132 or (301) 920-9132; fax: (301) 587-2365; email: paulab@aspen.nutr.org
Internet: http://www.nutritioncare.org/wcontent.aspx?id=784
Sponsor: American Society for Parenteral and Enteral Nutrition (ASPEN) Rhoads Research Foundation
8630 Fenton Street, Suite 412
Silver Spring, MD 20910

ASPEN Scientific Abstracts Awards for Papers or Posters 1175

ASPEN will consider all Scientific Abstract submissions for awards. Applicants will be notified if their abstract is a candidate. The first authors of the three highest ranked abstracts from this category (paper or poster) will receive travel grants to attend the Research Workshop, and to present their work at Clinical Nutrition Week. In addition, cash awards in the amounts of $750, $500 and $250 will be offered to the presenters of the three best papers. Awards are presented during the Rhoads Lecture and Research Awards Ceremony at Clinical Nutrition Week.
Amount of Grant: $250-$750 plus travel expenses
Contact: Paula Bowen, Research Program Administrator; (301) 587-6315, ext. 132 or (301) 920-9132; fax: (301) 587-2365; email: paulab@aspen.nutr.org
Internet: http://www.nutritioncare.org/Index.aspx?id=724
Sponsor: American Society for Parenteral and Enteral Nutrition
8630 Fenton Street, Suite 412
Silver Spring, MD 20910

ASPEN Scientific Abstracts Promising Investigator Awards 1176

ASPEN will consider all Scientific Abstract submissions for awards. Applicants will be notified if their abstract is a candidate. These awards are presented to the top three Scientific Abstract authors, who are also in the early stages of their careers. Winners receive $750 in travel funds to offset their travel expenses to and from Clinical Nutrition Week. Vars Award winners are eligible for these awards if they are in the early stages of their careers.
Amount of Grant: $750 in travel funds
Contact: Paula Bowen, Research Program Administrator; (301) 587-6315, ext. 132 or (301) 920-9132; fax: (301) 587-2365; email: paulab@aspen.nutr.org
Internet: http://www.nutritioncare.org/wcontent.aspx?id=878
Sponsor: American Society for Parenteral and Enteral Nutrition
8630 Fenton Street, Suite 412
Silver Spring, MD 20910

ASPET Bernard B. Brodie Award in Drug Metabolism 1177

This biennial award (even-numbered years), established by CIBA-GEIGY Corporation to honor the fundamental contributions of Bernard B. Brodie, is presented to recognize outstanding original research contributions in drug metabolism and disposition, particularly those having a major impact on future research in the field. Only one nominator is necessary although more are acceptable. The award consists of a $2,000 honorarium, a commemorative medal, hotel and economy airfare to the award ceremony at the annual meeting.
Requirements: Nominees must be members of ASPET; nominators need not be members of ASPET.
Restrictions: Supporting research accomplishments must not have been used to win any other major award.
Amount of Grant: $2,000
Date(s) Application is Due: Sep 15
Contact: Christine Carrico, Ph.D., Executive Officer; (301) 634-7060; fax: (301) 634-7061; email: ccarrico@aspet.org or info@aspet.org
Internet: http://www.aspet.org/public/awards/brodie_award.html
Sponsor: American Society for Pharmacology and Experimental Therapeutics
9650 Rockville Pike
Bethesda, MD 20814-3995

ASPET Division for Drug Metabolism Early Career Achievement Award 1178

The ASPET Division for Drug Metabolism Early Career Achievement Award has been established to recognize excellent original research by early career investigators in the area of drug metabolism and disposition. The award is presented biennially in odd-numbered years. The award consists of $1,000, a plaque, and complimentary registration plus travel expenses (to a maximum of $1,000) for the winner to attend the award ceremony at the annual meeting. The awardee will deliver a lecture at the annual meeting describing their relevant research accomplishments. The awardee will be invited to publish a review article on the subject matter of the award lecture in Drug Metabolism and Disposition.
Requirements: Nominees for this award must have a doctoral degree (e.g. Ph.D., M.D., Pharm. D., D.V.M.) and must be within 15 years of having received their final degree, as of December 31 of the year of the award. There are no restrictions on institutional affiliation and a candidate need not be a member of ASPET.
Amount of Grant: $1,000 honorarium; $1,000 maximum for travel
Date(s) Application is Due: Sep 15
Contact: Christine Carrico, Ph.D., Executive Officer; (301) 634-7060; fax: (301) 634-7061; email: ccarrico@aspet.org or info@aspet.org
Internet: http://www.aspet.org/public/divisions/drugmetab/early_achievement_award.htm
Sponsor: American Society for Pharmacology and Experimental Therapeutics
9650 Rockville Pike
Bethesda, MD 20814-3995

ASPET Epilepsy Research Award for Outstanding Contributions to the Pharmacology of Antiepileptic Drugs 1179

The International League Against Epilepsy (ILAE) has sponsored this biennial award of $2,000 and a Certification of Citation to be awarded by the American Society for Pharmacology and Experimental Therapeutics, for the purpose of recognizing and stimulating outstanding research leading to better clinical control of epileptic seizures. This research may include the basic screening and testing of new therapeutic agents, studies on mechanisms of action, metabolic disposition, pharmacokinetics, and clinical pharmacology studies. The recipient will be selected by the Epilepsy Award Committee appointed by the President of ASPET, with representation of ILAE.

Requirements: Nominations for the Award may be submitted by members of any recognized scientific association, domestic or foreign.

Amount of Grant: $2,000

Date(s) Application is Due: Sep 15

Contact: Christine Carrico, Ph.D., Executive Officer; (301) 634-7060; fax: (301) 634-7061; email: ccarrico@aspet.org or info@aspet.org

Internet: http://www.aspet.org/public/awards/epilepsy_award.html

Sponsor: American Society for Pharmacology and Experimental Therapeutics
9650 Rockville Pike
Bethesda, MD 20814-3995

ASPET Goodman and Gilman Award in Drug Receptor Pharmacology   1180

This award is given biennially (even-numbered years) to recognize and stimulate outstanding research in the pharmacology of biological receptors. It is the hope that such research might provide a better understanding of the mechanism of biological processes and potentially provide the basis for the discovery of drugs useful in the treatment of diseases. The award includes a monetary prize and travel support for winner and spouse to attend the award ceremony. Nominations must be received September 15 of the year prior to the year in which the award will be made.

Requirements: Nominations must be made by ASPET members; however, there are no restrictions as to age, sex, nationality, or institutional affiliation. Nominee need not be a member of ASPET.

Amount of Grant: $2,500 honorarium

Date(s) Application is Due: Sep 15

Contact: Christine Carrico, Ph.D., Executive Officer; (301) 634-7060; fax: (301) 634-7061; email: ccarrico@aspet.org or info@aspet.org

Internet: http://www.aspet.org/awards/goodman_and_gilman/

Sponsor: American Society for Pharmacology and Experimental Therapeutics
9650 Rockville Pike
Bethesda, MD 20814-3995

ASPET John J. Abel Award in Pharmacology   1181

This award, sponsored by Eli Lilly and Company and administered by ASPET, is given annually to stimulate original and outstanding fundamental research in pharmacology and experimental therapeutics by young investigators. Candidates may not have passed their 39th birthday as of April 30 of the year of the award, or previously received an award from the sponsor for the same technical accomplishment.

Requirements: Candidates need not be members of the society; however, nominations must be made by ASPET members.

Amount of Grant: $2,500

Date(s) Application is Due: Sep 15

Contact: Christine Carrico, Ph.D., Executive Officer; (301) 634-7060; fax: (301) 634-7061; email: ccarrico@aspet.org or info@aspet.org

Internet: http://www.aspet.org/public/awards/abel_award.html

Sponsor: American Society for Pharmacology and Experimental Therapeutics
9650 Rockville Pike
Bethesda, MD 20814-3995

ASPET P. B. Dews Lifetime Achievement Award for Research in   1182
Behavioral Pharmacology

ASPET, Division of Behavioral Pharmacology sponsors the P. B. Dews Award for Research in Behavioral Pharmacology to recognize outstanding lifetime achievements in research, teaching and professional service in the field of Behavioral Pharmacology and to honor Peter Dews for his seminal contributions to the development of behavioral pharmacology as a discipline. The Award consists of $750, a plaque, and partial travel expenses to the award ceremony at the ASPET annual meeting. The recipient will be invited by the Chair of the Division of Behavioral Pharmacology to deliver a special lecture on this occasion. The lecture will be published subsequently in an appropriate ASPET-sponsored publication.

Requirements: Nominations may be made by members of ASPET or of any relevant scientific society.

Restrictions: There are no restrictions on nominees for this award.

Amount of Grant: $750

Date(s) Application is Due: Sep 15

Contact: Christine Carrico, Ph.D., Executive Officer; (301) 634-7060; fax: (301) 634-7061; email: ccarrico@aspet.org or info@aspet.org

Internet: http://www.aspet.org/public/awards/dews_award.html

Sponsor: American Society for Pharmacology and Experimental Therapeutics
9650 Rockville Pike
Bethesda, MD 20814-3995

ASPET Paul M. Vanhoutte Award   1183

The Paul M. Vanhoutte Award in Vascular Pharmacology was established to honor Dr. Vanhoutte, lifelong scientific contributions to our better understanding and appreciation of the importance of endothelial cells and vascular smooth muscle function in health and disease and for his mentoring of countless prominent endothelial and vascular biologists and pharmacologists. This is a biennial award, consisting an honorarium of $1,000, a custom-designed crystal bowl depicting the named Lectureship, and up to $2,000 travel expenses including registration to the annual spring ASPET meeting.

Requirements: There are no restrictions on institutional affiliation, nationality, or age of the candidate, but the recipient must be a member of the ASPET.

Amount of Grant: $1,000 honorarium; $2,000 maximum for travel

Date(s) Application is Due: Sep 15

Contact: Christine Carrico, Ph.D., Executive Officer; (301) 634-7060; fax: (301) 634-7061; email: ccarrico@aspet.org or info@aspet.org

Internet: http://www.aspet.org/public/divisions/cardiovascular/vanhoutte_award.htm

Sponsor: American Society for Pharmacology and Experimental Therapeutics
9650 Rockville Pike
Bethesda, MD 20814-3995

ASPET Summer Undergraduate Research Fellowships (SURF)   1184

The purpose of the Fellowship program is to introduce undergraduate students to pharmacology research in order to heighten interest in science as a career, with an emphasis on pharmacology graduate training to increase the number of young scientists entering the research discipline of pharmacology. A group of at least five ASPET Regular members from one institution may wish to apply for support of up to $9,000 for an undergraduate fellowship program, to include up to five students stipends. Students are to receive at least a $2,800 stipend for a minimum of ten weeks participation. It is anticipated that the request for funding will include matching funds of at least $5,000 from local resources. Awards are normally made for three (3) years.

Restrictions: No indirect costs will be provided. All funds shall be used for student stipends/wages only.

Amount of Grant: $2,800 per student; $9,000 per group

Date(s) Application is Due: Mar 1; Oct 1

Contact: Christine Carrico, Ph.D., Executive Officer; (301) 634-7060; fax: (301) 634-7061; email: ccarrico@aspet.org or info@aspet.org

Internet: http://www.aspet.org/public/surf/surf.htm

Sponsor: American Society for Pharmacology and Experimental Therapeutics
9650 Rockville Pike
Bethesda, MD 20814-3995

ASPET Torald Sollmann Award in Pharmacology   1185

This award, established by Wyeth-Herst Labs, is given in odd-numbered years to commemorate the pioneer work in the United States of Dr. Torald Sollmann in the fields of pharmacological investigation and education. The award is made to a nominee for significant contributions over many years to the advancement and extension of knowledge in the field of pharmacology. The award consists of an honorarium of $3,500, a medal, hotel and economy airfare for the winner and spouse to the annual meeting. The formal presentation of this biennial award and medal will be made at the annual meeting of ASPET. The recipient will be invited by the President of the Society to deliver a lecture to the membership that may be published in an appropriate ASPET journal.

Requirements: Nominations must be made by ASPET members; however, the nominee need not be a member. There are no restrictions as to age or institutional affiliation.

Amount of Grant: $3,500

Date(s) Application is Due: Sep 15

Contact: Christine Carrico, Ph.D., Executive Officer; (301) 634-7060; fax: (301) 634-7061; email: ccarrico@aspet.org or info@aspet.org

Internet: http://www.aspet.org/public/awards/sollmann_award.html

Sponsor: American Society for Pharmacology and Experimental Therapeutics
9650 Rockville Pike
Bethesda, MD 20814-3995

ASPET-Astellas Awards in Translational Pharmacology   1186

The ASPET-Astellas Awards in Translational Pharmacology are intended to extend fundamental research closer to applications directed towards improving human health. The awards will be given to: recognize those individuals whose research has the potential to lead to the introduction of novel pharmacologic approaches or technologies that may offer significant advances in clinical medicine in the future; and to facilitate that translational process. Three (3) awards of $30,000 each will be made to individuals. The money may be used for supplemental research funding, travel, training, or in any way that furthers the goals of the organization. Nominations should be made electronically.

Requirements: Any ASPET member in good standing may nominate an individual for this award.

Restrictions: Self-nominations will not be accepted.

Amount of Grant: $30,000

Date(s) Application is Due: Sep 15

Contact: Christine Carrico, Ph.D., Executive Officer; (301) 634-7060; fax: (301) 634-7061; email: ccarrico@aspet.org or info@aspet.org

Internet: http://www.aspet.org/public/awards/Astellas_awards.html

Sponsor: American Society for Pharmacology and Experimental Therapeutics
9650 Rockville Pike
Bethesda, MD 20814-3995

ASP.O. Daiichi Innovative Technology Grant   1187

The Daiichi Innovative Technology Grant will fund exploratory or hypothesis-generating projects that are not well-suited to a formal grant application (eg, based on statistical analysis and sample size specification) such as: development of new surgical or diagnostic instruments; survey or quality of life measures; new use of internet technology or computer software such as CDs or DVDs; educational brochures, materials, software for patients or physicians; and other applications of innovative technology for education or research in pediatric otolaryngology. Full patent and copyright control must be retained by the applicant and the applicant's institution. If patented innovations funded by this award generate more than $5,000, the applicant may be required to return funds to the Society. One award of up to $5,000, to be used over a one-year period, will be given annually. Letters of Intent should be received no later than December 15, with full proposals due on January 15.

Requirements: Researchers (MD, PhD, DMD, DO) in disciplines who will conduct research directly relevant to innovative technology in pediatric otolaryngology are eligible to apply. Applications submitted by otolaryngologists or demonstrating collaborations with otolaryngologists are preferred. Participation of an ASP.O. member is not required, but is preferred.

Amount of Grant: $5,000
Date(s) Application is Due: Dec 15; Jan 15
Contact: Stephanie L. Jones, Assistant Director; (703) 519-1586 or (703) 836-4444; email: sljones@entnet.org
Internet: http://www.aspo.us/information.php?info_id=12
Sponsor: American Society for Pediatric Otolaryngology (ASP.O.)
One Prince Street
Alexandria, VA 22314-3357

ASP.O. Fellowships 1188

ASP.O. exists to foster excellence in the care of children with otorhinolaryngologic disorders through education and research and thereby enhance the profession of Pediatric Otolaryngology. Its fellowship program intends to: facilitate the creation and dissemination of knowledge about the care of infants and children with ORL disorders; serve as an advocate for infants and children with ORL disorders, ASP.O. members and others with shared goals; and preserve and promote dedication to excellent and humane care for infants and children with ORL disorders. See the web site for details and all available fellowships.

Contact: Stephanie L. Jones, Assistant Director; (703) 519-1586 or (703) 836-4444; email: sljones@entnet.org
Internet: http://www.aspo.us/fellowships.php
Sponsor: American Society for Pediatric Otolaryngology (ASP.O.)
One Prince Street
Alexandria, VA 22314-3357

ASP.O. Research Grants 1189

The American Society of Pediatric Otolaryngology (ASP.O.) awards funds annually to support innovative research in pediatric otolaryngology. ASP.O. will consider applications from both individuals and institutions. Preference is given to proposed projects that are to be completed within one year, although exceptional proposals that have duration in excess of one year will be considered. Two awards of up to $15,000, to be used over a one-year period, will be given annually. Letters of Intent should be received no later than December 15, with full proposals due on January 15.

Requirements: Researchers (MD, PhD, DMD) in disciplines who will conduct research directly relevant to pediatric otolaryngology are eligible to apply.

Restrictions: No portion of any grant may be used for travel expenses or for principal investigator salaries.

Amount of Grant: $15,000
Date(s) Application is Due: Dec 15; Jan 15
Contact: Stephanie L. Jones, Assistant Director; (703) 535-3747 or (703) 836-4444; email: sljones@entnet.org
Internet: http://www.aspo.us/information.php?info_id=12
Sponsor: American Society for Pediatric Otolaryngology (ASP.O.)
One Prince Street
Alexandria, VA 22314-3357

ASPRS Leica Geosystems Internship 1190

The award is an eight-week internship for graduate students in photogrammetry. The selected intern will work with LH Systems personnel at its facilities in San Diego, Denver, Heerbrugg, or elsewhere. The internship provides the award winner with an opportunity to carry out a small research project of his or her own choice or to work on an existing LH Systems project as part of a team. Application form is available on the Web site. The intern will provide ASPRS with a final report of his or her research accomplishments during the period of the internship.

Requirements: The internship is open to graduate students of photogrammetry and remote sensing who are also members of ASPRS.

Amount of Grant: $2500 stipend plus expenses
Date(s) Application is Due: Dec 1
Contact: ASPRS Awards Director, (301) 493-0290; fax: (301) 493-0208; email: scholarships@asprs.org
Internet: http://www.asprs.org/membership/scholar.html#LH
Sponsor: American Society for Photogrammetry and Remote Sensing
5410 Grosvenor Ln, Suite 210
Bethesda, MD 20814

ASPRS Robert E. Altenhofen Memorial Scholarship 1191

This award, given in memory of Robert E. Altenhofen, who was an outstanding practitioner of photogrammetry and who made notable contributions to the mathematical aspects of the science, is to encourage and commend college students who display exceptional interest and ability in the theoretical aspects of photogrammetry.

Requirements: Application may be made by an undergraduate or graduate student in a recognized college or university in the United States or elsewhere, who is either a student member or active member of the ASPRS. The recipient is obligated to provide a final report to ASPRS of his or her scholastic accomplishments during the period for which the award is granted.

Samples: Eva Paska, Ohio State University—scholarship recipient, $2000 (2005); Benton Anderson, U of Maine (Orono, ME)—scholarship recipient, $1000.
Amount of Grant: $2000
Date(s) Application is Due: Dec 1
Contact: ASPRS Awards Director, (301) 493-0290 ext 101; fax: (301) 493-0208; email: scholarships@asprs.org
Internet: http://www.asprs.org/membership/scholar.html#altenhofen
Sponsor: American Society for Photogrammetry and Remote Sensing
5410 Grosvenor Ln, Suite 210
Bethesda, MD 20814

ASPRS Ta Liang Memorial Award 1192

This award is given annually to one graduate student in remote sensing. The award is to be used for research-related travel. Application forms are available on the Web site.

Requirements: The award is restricted to student members of the society who are currently pursuing graduate-level studies at an accredited college or university. The recipient is obligated to provide ASPRS and Ta Liang's family, a report of his or her accomplishments during the travel for which the award is granted.

Samples: Jingli Yang, U of Maryland (MD)—award recipient, $500.
Amount of Grant: $500
Date(s) Application is Due: Dec 1
Contact: Award Administrator, (301) 493-0290 ext 101; fax: (301) 493-0208; email: scholarships@asprs.org
Internet: http://www.asprs.org/membership/scholar.html#ta%20liang
Sponsor: American Society for Photogrammetry and Remote Sensing
5410 Grosvenor Ln, Suite 210
Bethesda, MD 20814

ASPRS William A. Fischer Memorial Scholarship 1193

This scholarship, created in memory of William A. Fischer, a pioneer in the use of remote sensing from space for the study of the earth, is presented annually to a worthy student from the United States or another country to facilitate graduate studies and career goals judged to address new and innovative uses of remote sensing that relate to the natural, cultural, or agricultural resources of the earth.

Requirements: The award is restricted to society members. It is intended for a student who is currently pursuing graduate-level studies or who plans to enroll in a graduate program in an accredited college or university in the United States or elsewhere. The recipient is obligated to provide ASPRS with a final report of his/her scholastic accomplishments during the period for which the award is granted.

Samples: Timothy Warner, Purdue U (IN)—scholarship recipient, $2000.
Amount of Grant: $2000
Date(s) Application is Due: Dec 1
Contact: ASPRS Awards Director, (301) 493-0290; fax: (301) 493-0208; email: scholarships@asprs.org
Internet: http://www.asprs.org/membership/scholar.html#fischer
Sponsor: American Society for Photogrammetry and Remote Sensing
5410 Grosvenor Ln, Suite 210
Bethesda, MD 20814

ASR Fichter Research Grant 1194

Applications are invited from scholars involved in promising research on women and religion, gender issues, feminist perspectives on religion, and new religious movements. Dissertation research qualifies for funding. Scholars at the beginning of their careers are particularly encouraged to apply. Send four copies of the proposal, budget, and curriculum vita to the address listed.

Requirements: Applicants must be members of the Association for the Sociology of Religion at the time of submission. A membership form is available online.

Amount of Grant: $12,500 total
Date(s) Application is Due: Mar 1
Contact: William Swatos, Jr., Executive Director, (309) 932-2727; fax: (309) 932-2282
Internet: http://www.sociologyofreligion.com/FICHEXLP2004.html
Sponsor: Association for the Sociology of Religion
618 SW 2nd Avenue
Galva, IL 61434-1912

ASR Robert J. McNamara Student Award 1195

The annual competition recognizes an outstanding student paper in the sociology of religion. The paper must not have appeared in print prior to receiving the award. Students who wish their papers considered for the program must submit paper abstracts to the program chair following the guidelines of all standard paper submissions. Submissions should be by one of the following methods: IBM-formatted, virus-free disk, with the text in WordPerfect, Microsoft Word, or plain ASCII; or four paper copies, typed double spaced, and single sided. Submission should be in the form of articles with a maximum length of 40 single-sided pages inclusive of all material. The title page should include an abstract of no more than 200 words. Text should not exceed 12,000 words, i.e., approximately 36 double-spaced pages of 12 point (or 10 cpi) type. Sociology of Religion has the right of first review of award-winning papers.

Requirements: Authors must be currently enrolled students who have not defended the PhD when the paper is submitted. Membership in the Association for the Sociology of Religion is required either at the time of application or previously (membership form available online).

Amount of Grant: $500
Date(s) Application is Due: Jun 1

Contact: Prof. Lutz Kaelber, email: lkaelber@zoo.uvm.edu
Internet: http://www.sociologyofreligion.com/MCNAMARA2004.html
Sponsor: Association for the Sociology of Religion
1291 University of Oregon
Eugene, OR 97403-1291

ASSE Liberty Mutual Safety Research Fellowship   1196
The program supports two four- to six-week summer research fellowships at the Liberty Mutual Research Center for Safety and Health, in Massachusetts. The goals of the program are to encourage research activity in the field of safety; familiarize graduate students and faculty members with current research projects for application in teaching situations; provide a forum for linking safety professionals, industry needs, and quality research programs; and lay the groundwork for graduate students and faculty members to pursue safety/health applied research projects of their choice. Fellows receive a weekly stipend to cover transportation to and from the fellow's home to Hopkinton, MA; room and board; and rental car as needed. Members of the society receive special consideration.
Requirements: Application is open to U.S. citizens or permanent residents who possess the PhD degree or are working toward a master's or PhD degree and are enrolled or teaching at an accredited U.S. college or university in a safety or safety related field. Preference will be given to applicants holding appointments or enrolled in a department with an ABET accredited program.
Amount of Grant: $2000 maximum for the first week; $1000 maximum per week thereafter; $8000 total
Date(s) Application is Due: Mar 1
Contact: Mary Goranson, (847) 699-2929; fax: (847) 296-3769; email: mgoranson@asse.org
Internet: www.asse.org/foundat.htm
Sponsor: American Society of Safety Engineers
1800 E Oakton Street
Des Plaines, IL 60018-2187

ASSE Safety Research Grants   1197
The foundation supports safety and health research projects, particularly in the following areas: effective interventions for prevention of traumatic injuries; effectiveness of incentive programs; developing management commitment to safety/risk control; criteria to measure risk control effectiveness; effective safety programs for temporary workers; proactive incorporation of risk control principles in early design of facilities, processes, and management systems; study cultures within organizations with outstanding safety records; and safety in public places.
Samples: Dr. Daniel Della-Giustina, West Virginia U—for the project How to Cope with Terrorism in Schools, $3000.; N. Kumar Kittusamy, Marshall U—for the project Assessment of Ergonomic Risk Factors among Operators of Heavy Earth Moving Machinery, $3000.
Amount of Grant: $20,000
Date(s) Application is Due: Oct 1
Contact: Mary Goranson, Foundation Manager, (847) 768-3412; fax: (847) 296-3769; email: mgoranson@asse.org
Internet: http://www.asse.org/foundat.htm
Sponsor: American Society of Safety Engineers
1800 E Oakton Street
Des Plaines, IL 60018-2187

ASSE Scholarships   1198
The foundation's mission is to promote the advancement of the safety, health, and environmental professions engaged in protecting people, property, and the environment by providing the necessary resources. The program supports a variety of undergraduate and graduate scholarships. Applications are available September 1. Winners names are posted on the foundation's website on or around April 1.
Amount of Grant: $1000-$5000
Date(s) Application is Due: Dec 1
Contact: Grants Administrator, (847) 699-2929; fax: (847) 768-3434
Internet: http://www.asse.org/foundat.htm
Sponsor: American Society of Safety Engineers
1800 E Oakton Street
Des Plaines, IL 60018-2187

Assisi Foundation Grants   1199
The foundation supports organizations in its areas of interest, including health and human services—promote the health and well-being of the Mid-South community and help the healthcare system respond more effectively to community needs; education and literacy—projects/programs that build organizational capacity of provider agencies, provide professional development to service providers, promote collaboration among provider agencies, and leverage resources (local, state, and federal); social justice/ethics—projects/programs that strengthen ethical values among Mid-South citizens and promote social justice leading to a better understanding of and a more effective response to economic or social threats to the community; and cultural enrichment and the arts—projects/programs that foster an appreciation of the arts in the Greater Memphis community. Religious organizations seeking funding for religious programs also are eligible. Types of support include general grants, mini grants, and capital project grants. Deadlines are set to coordinate with quarterly meetings of the Board of Directors of the Foundation. Information and application are available online.

Requirements: Memphis-area and Shelby, Fayette, and Tipton Counties, TN, Crittenden County, AK, and Desota County, Mississippi nonprofit organizations are eligible.
Restrictions: Grants are not made for individuals, national fundraising drives, projects that address the needs of only one congregation, tickets for benefits, political organizations or candidates for public office, lobbying activities, recurring budget deficits, or tournament fees and/or travel for athletic competitions.
Geographic Focus: Tennessee
Samples: National Ornamental Metal Museum (Memphis, TN)—for a new library, $75,000 (2004); Memphis Urban League (TN)—to provide basic-skills, workplace-behavior, and computer training, $25,000 (2004); Christian Brothers U (Memphis, TN)—for its LaSallian Assoc of New Catholic Educators Program, $90,000 over three years (2004); Saint Jude Children's Research Hospital (Memphis, TN)—for research in cell and gene therapy, $750,000 (2004).
Amount of Grant: $20,000 maximum
Contact: Jan Young, Executive Director, (901) 684-1564; email: joung@assisifoundation.org
Internet: http://www.assisifoundation.org/generalgrants.html
Sponsor: Assisi Foundation
515 Erin Drive
Memphis, TN 38117

Association for Research on Nonprofit Organizations and Voluntary   1200
Action Awards
ARNOVA invites nominations for the current year's awards, to be presented at the annual conference in Los Angeles, CA. The following awards will be presented: Award for Outstanding Book in Nonprofit and Voluntary Action Research—books published in the three calendar years preceding the award that contribute to the advancement of theory, conceptualization, research, or practice; Award for Distinguished Achievement and Leadership in Nonprofit and Voluntary Action—for significant contributions to the field through research and leadership; and the Gabriel G. Rudney Memorial Award for an Outstanding Dissertation in Nonprofit and Voluntary Action Research—for a PhD dissertation completed and/or defended in the three calendar years preceding the award that contributes to the advancement of theory, conceptualization, research, or practice. Guidelines and nominating form are available online.
Amount of Grant: $1000
Date(s) Application is Due: Mar 17
Contact: Katherine Finley, Executive Director, (317) 684-2120; fax: (317) 684-2128; email: kmfinley@arnova.org
Internet: http://www.arnova.org/award_nominations_news.php
Sponsor: Association for Research on Nonprofit Organizations and Voluntary Action
550 W North Street, Suite 301
Indianapolis, IN 46202-3162

Association for the Study of Higher Education Dissertation Fellowships   1201
With support of the Lumina Foundation for Education, the association awards fellowships to support dissertation research on the broad topics of financial aid, student retention and success, and adult learners and learning. The fellowships will support up to one year of activity that will be conducted through the students' home universities and can be used to support costs of supplying data, dissemination of project results, travel, tuition, and salary for the fellows. Eight one-year fellowships will be awarded. Guidelines are available online.
Requirements: Doctoral students affiliated with any accredited doctoral program may submit a proposal. Students in doctoral programs outside the United States may apply if their study is about student financial assistance, student retention and success, and/or adult learning and learners in the United States. Recipients of the fellowship must have completed their course work and any required qualifying examinations and have their dissertation proposal accepted by their institution.
Amount of Grant: $12,500
Date(s) Application is Due: May 31
Contact: Dr. Dennis Brown, Executive Director, (517) 432-8805; fax: (573) 884-2197; email: ashemsu@msu.edu
Internet: http://35.8.168.242/fellowship/aboutfellowship.htm
Sponsor: Michigan State University
424 Erickson Hall
East Lansing, MI 48824

Association for Women in Mathematics Mentoring Travel Grants   1202
The program helps junior women develop long-term working and mentoring relationships with senior mathematicians. The relationship should help the junior mathematician establish her research program and eventually receive tenure. The applicant's research may be conducted in any field funded by the NSF Division of Mathematical Sciences. Up to seven grants will be awarded, each to fund travel, subsistence, and other required costs for an untenured woman to travel to an institute and do research with a senior mentor for one month. Specific guidelines and instructions are available on the Web site.
Requirements: Applicants holding a doctorate or equivalent experience are eligible.
Amount of Grant: $4000
Date(s) Application is Due: Feb 1
Contact: Grants Administrator, (703) 934-0163; email: awm@awm-math.org
Internet: http://www.awm-math.org/travelgrants.html#mentor
Sponsor: University of Maryland
4114 Computer and Space Sciences Bldg
College Park, MD 20742-2461

Association of Teachers of Latin American Studies Grants 1203

Grants are intended to improve the teaching about Latin America in U.S. secondary schools by permitting high school social studies teachers and graduate students to participate in six-week summer seminars in a Latin American country. This program is sponsored by the Department of Education, and the number and amount of grants available each year varies according to program funding. A letter of application should include a self-addressed, stamped envelope for reply.

Requirements: Applicants must be high school teachers, college faculty, supervisors, or curriculum developers. The grants are aimed at teachers from secondary level social studies and Spanish language who can enrich their programs through participation in an overseas experience.

Date(s) Application is Due: May 1
Contact: Dr. Daniel Mugan, President, (718) 428-1237; email: djmugan@aol.com
Sponsor: Association of Teachers of Latin American Studies
P.O. Box 754
Flushing, NY 11362

ASTD Dissertation Award 1204

The award is given to foster and disseminate research in the practice of workplace learning and performance. It is presented annually to the person who has submitted the best dissertation completed during the previous academic year (September 21 through September 20). Illustrative areas of concentration include training and development, performance analysis, career development, organizational development/learning, workplace design, and human resource planning. All materials submitted must be in English and include the following: letter of application from candidate; recommendation from committee chair, on letterhead, with the dissertation completion date; abstract of dissertation (five to 15 pages, double-spaced) including summary of the problem addressed by the study, critique of relevant literature, synopsis of the findings, and implications for practice and research.

Requirements: The candidate must be recommended and sponsored by his or her committee chair. A committee chair may nominate more than one candidate who meets the criteria.

Samples: Anna Rowe (Brooks AFB, TX)—for a comparative evaluation of methods for measuring mental model knowledge, $500.
Amount of Grant: $500
Date(s) Application is Due: Sep 20
Contact: Ray Rivera , (703) 683-8144; email: rrivera@astd.org
Internet: http://www.astd.org/astd/About_ASTD/Awards/dissertation.htm
Sponsor: American Society for Training and Development
P.O. Box 1443, 1640 King Street
Alexandria, VA 22313-2043

ASTD Research Award 1205

The award is given to encourage the publication of research with practical implications for practitioners of workplace learning and performance. It is presented annually to the author(s) of the best article published in a refereed journal during the calendar year. Illustrative areas of concentration include training and development, career development, work design, performance analysis, organizational development/learning, and human resource planning. Applications must include letter of application; proof of publication—either a galley proof or reprint of the article in its entirety or a copy of the article with its letter of acceptance; and supporting documents, including the name and issue of the refereed journal in which the article appeared or will appear and contact information for the editor of the journal (name, address, and telephone number). The award includes a cash prize, designated place on the international conference program to present the research, and a commemorative plaque(s) presented at the awards ceremony.

Requirements: The article must have appeared, or will appear, in print in a refereed journal between January 1 and December 31 of the current year. All materials submitted, including the article, must be in English.

Amount of Grant: $500
Date(s) Application is Due: Oct 31
Contact: Ray Rivera, Research Department, (703) 683-8144.; email: rrivera@astd
Internet: http://www.astd.org/astd/About_ASTD/Awards/research_award.htm
Sponsor: American Society for Training and Development
P.O. Box 1443, 1640 King Street
Alexandria, VA 22313-2043

Astellas Foundation Research Grants 1206

The Foundation was established with the objectives of contributing to the improvement of national health and welfare, and to the progress of therapeutic medicines. In fulfilling the objectives, the Foundation aims to conduct pioneering research in new fields of diseases and metabolisms of medicines, by elucidating the mechanism of diseases and their treatment, in particular the relationship between therapeutic medicines and in-situ metabolisms. Major activities include: subsidizing research on metabolic disorders related to the diagnosis of diseases and therapeutic medicine; publication and editing of periodicals for research achievements; supporting production of educational movies; and other projects to achieve the objectives of the Foundation.

Contact: Grants Manager; +81-3-3244-3397; fax: +81-3-5201-8512
Sponsor: Astellas Foundation
2-3-11 Nihonbashihoncho, Chuo-ku
Tokyo 103-8411 Japan

Astronomical Society of the Pacific Amateur Achievement Award 1207

The award and a plaque are given annually to recognize significant contributions to astronomy or amateur astronomy by those not employed in the field of astronomy in a professional capacity. Nominations may be submitted by any individual or group. All nominations will remain active for three years of competition; additional support letters or nomination updates will be accepted at any time.

Restrictions: Self nominations or nomination by a family member will not be accepted.
Samples: Nik Szymanek (United Kingdom)—for imaging and image processing, $500 (2004); Kyle E. Smalley (Cambridge, Massachusetts)—for near-earth asteroids, $500 (2003); M. Daniel Overbeek (Edenvale, South Africa)—for variable stars.; Donald Parker (Coral Gables, FL)—for planetary imaging.
Amount of Grant: $500
Date(s) Application is Due: Dec 15
Contact: Marilyn Delgado, Program Contact, (415) 337-1100; fax: (415) 337-5205; email: mdelgado@astrosociety.org
Internet: http://www.astrosociety.org/membership/awards/amateur.html
Sponsor: Astronomical Society of the Pacific
390 Ashton Avenue
San Francisco, CA 94112

ASU Graduate Scholar Awards 1208

The awards recognize outstanding students with high credentials such as GRE scores, GPA, publications, and prestigious awards. Students are nominated by the program areas to which they are applying. Three-year award packages include a stipend plus tuition and fees. A recruiting visit to ASU is available to finalists. Members of groups traditionally underrepresented in graduate education are strongly encouraged to apply.

Requirements: Recipients of the award must be U.S. citizens or permanent residents.
Date(s) Application is Due: Nov 18
Contact: Dr. Marjorie Zatz, (480) 965-5906 or (480) 965-3521; email: marjorie.zatz@asu.edu
Internet: http://www.asu.edu/graduate/generalinfo/UGS/index.html
Sponsor: Arizona State University Graduate College
P.O. Box 871003
Tempe, AZ 85287-1003

ASU Institute for Humanities Research Visiting Fellowships 1209

The program is for scholars from other institutions of higher education in the United States and abroad to spend spring semester (January 15 through May 15) in residence at the Institute for Humanities Research (IHR), participating in the intellectual life of the IHR and the university community. The theme is broadly defined as Humanities in Times of Crises to encourage applications from a wide range of intellectual and scholarly approaches and chronological time frames within the humanities. Fellows will be provided a stipend, an office, and support services. The fellowship provides the opportunity to conduct research and write. Visiting fellows will participate in weekly meetings around the theme with the working groups of ASU fellows, and will give public lectures and seminars on their research topics while in residence. Application and guidelines are available online.

Amount of Grant: $20,000 maximum stipend
Date(s) Application is Due: Feb 20
Contact: Rachel Fuchs, (480) 965-3000; email: ihr@asu.edu
Internet: http://www.asu.edu/clas/ihr/faculty/fellows/index.html
Sponsor: Arizona State University
P.O. Box 876505
Tempe, AZ 85287-6505

ASU SIRLS Scholarships and Grants 1210

A variety of grants and scholarships are awarded through the university's School of Information Resources and Library Science. Graduate aid is available to students enrolling in SIRLS based on financial need, individual accomplishments, prior experiences and expertise, academic record, and contributions to the diversity of the SIRLS student community. Application and guidelines are available online.

Requirements: Eligible students must have been accepted for admission or be currently enrolled in a SIRLS graduate degree program. Students receiving financial aid are required to maintain full-time enrollment status (nine credits) and maintain successful progress toward their degree, defined at a minimum as a GPA of at least 3.0

Date(s) Application is Due: Apr 1; Sep 1
Contact: Scholarships Administrator, (520) 621-3565; email: sirls@email.arizona.edu
Internet: http://www.sir.arizona.edu/program/admissions/financial.html
Sponsor: Arizona State University
1515 E First Street
Tucson, AZ 85719

AT&T Bell Labs Graduate Research Program 1211

The Graduate Research Program for Women (GRPW) is designed to increase the number of minorities and women in the fields of science, math, engineering and technology. Financial support is provided to outstanding women students from the beginning of their full-time graduate studies leading to a PhD. The program includes two types of financial awards, fellowships and grants, which may be renewed on a yearly basis for the normal duration of the graduate program, subject to the participant's satisfactory progress toward the doctoral degree. Employment at AT&T Bell Laboratories is offered to fellowship and grant holders for the summer preceding their entry into graduate school as well as for subsequent summers during graduate school. If a fellowship holder chooses to remain on campus for supervised university research or study during the summer, her fellowship support will be continued through the calendar year. Application forms must be obtained from AT&T Bell Laboratories. Applications and all supporting materials are due January 13.

Requirements: Women who are applying to graduate school to eventually obtain a PhD in chemical engineering, chemistry, communications science, computer science/engineering, electrical engineering, information science, materials science, mathematics, mechanical engineering, operations research, physics, or statistics are eligible. Applicants must be U.S. citizens or permanent residents and, by the time the award is made, be admitted to full-time study in a doctoral program approved by AT&T Bell Laboratories.

Amount of Grant: $25,000
Date(s) Application is Due: Jan 13
Contact: Fellowship Program Manager, GRPW, (908) 582-4822; email: coopgraduate@lucent.com
Internet: http://www.lucent.com/social/blgrfp
Sponsor: AT&T Bell Laboratories
P.O. Box 297, 1505 Riverview Road
St. Peter, MN 56082

AT&T Education Grants   1212
The foundation supports initiatives that improve student achievement, teacher preparedness, minority student success and that increase the use of new technologies, from kindergarten to the university. The foundation also supports NCCEP and GEAR UP. Primary support is provided to projects in math, technology, science and reading that provide the skills and knowledge that students need in order to succeed in an ever-expanding global economy. The request for proposals process is invitational for some special grants programs and an open, competitive process for others. Those who wish to submit an unsolicited proposal should send a brief letter of introduction and description of their organization and project to the office. Guidelines and application are available online.

Requirements: The foundation prefers to work with 501(c)3 public charity or government instrumentalities that have clearly stated objectives, long-range planning, active participation of the governing board and strategies that incorporate diversified sources of support. The foundation also considers grants to organizations that qualify as government instrumentalities, including public libraries, police and fire departments and publicly funded social services agencies.

Contact: Program Contact; (800) 591-9663
Internet: http://www.att.com/foundation
Sponsor: AT&T Foundation
32 Avenue of the Americas, 6th Floor
New York, NY 10013

AT&T Industrial Ecology Faculty Fellowships   1213
The foundation awards fellowships to academic researchers in the emerging, multidisciplinary field of industrial ecology. The program is intended to stimulate interdisciplinary research and curriculum development that involve social issues, engineering, the sciences, economics, management, business, law, and public policy issues. The overarching objectives of the program are (1) to produce university faculty and students who can contribute to solving global and regional environmental problems and help shape environmentally and economically efficient strategies that have a firm scientific and engineering basis and (2) to advance the theoretical basis for the field of industrial ecology and its application in service sectors, industrial activity, and regulatory arenas. Application guidelines are available online and applications are solicited electronically.

Requirements: Individuals working independently or in collaboration at U.S. universities in the following fields are eligible: business, economics, engineering, law, management, physical sciences, public policy, and social sciences.

Amount of Grant: $25,000
Date(s) Application is Due: Nov 20
Contact: Michael Blazek, Administrator, (908) 221-4191; email: mblazek@att.com
Internet: http://www.corp.att.com/ehs/ind_ecology/fellow_guidelines.html
Sponsor: AT&T Foundation
1 AT&T Way
Bedminster, NJ 07921

ATA Doctoral Fellowships in Education   1214
Each year, the Alberta Teachers Association awards two fellowships, to Alberta teachers entering the first year of a full-time, campus-based doctoral program in education. The fellowship program is intended to recognize academic excellence and to help defray the financial costs of university study.

Requirements: Each applicant must be a current member and have a permanent Alberta teaching certificate.

Restrictions: A teacher who is not an active ATA member at the time of application must be an associate member and either must have been an associate member for the last three consecutive years, or must have been an active member at some time within the three years preceding the deadline date for application.

Amount of Grant: $15,000 annually each
Date(s) Application is Due: Mar 10
Contact: Corinne Anderson, Administrative Assistant, (800) 232-7208 or (780) 447-9470; email: corrine.anderson@ata.ab.ca
Internet: http://www.teachers.ab.ca/Professional+Development/Grants+and+Scholarships/Fellowships+and+Scholarships.htm
Sponsor: Alberta Teachers Association
11010 142 Street NW
Edmonton, AB T5N 2R1 Canada

ATA Education Research Award   1215
The award is presented annually to a faculty member or sessional lecturer at an Alberta university who has undertaken high-quality research on classroom teaching and learning. Research must be directly related to school and classroom practice; be focused on school teaching and/or learning; be current (either ongoing or completed within the last two years); be related to critical issues; have involved classroom teachers and/or students; be applicable to the Alberta context and must be of practical benefit to teachers in improving their professional practice, and be of high quality in terms of purpose, approach, methodology, originality and clarity.

Requirements: Applicant must be a faculty of education member or sessional lecturer at an Alberta university who has undertaken high quality research on classroom teaching and learning to be eligible.

Amount of Grant: $5000 (Canadian)
Date(s) Application is Due: May 11
Contact: Corinne Anderson, Program Contact, (780) 447-9400 or (800) 232-7208; email: corrine.anderson@ata.ab.ca
Internet: http://www.teachers.ab.ca/Professional+Development/Grants+and+Scholarships
Sponsor: Alberta Teachers Association
11010 142nd Street NW
Edmonton, AB T5N 2R1 Canada

ATA Nadene M. Thomas Graduate Research Bursary   1216
The research bursary is offered annually to current members of the association. The award is based on a research focus on challenges to public education and professional practice, the applicant's contribution to the association, the research proposal, a letter of support from the university advisor, and the applicant's commitment to public education. The recipient is required to submit a copy of the completed research to the association. The award is payable in two equal installments, half upon receipt of the award and half upon receipt of the completed research.

Requirements: An applicant must hold a permanent Alberta teaching certificate and have completed at least five years of successful teaching in Alberta; be either an active or associate ATA member; state intention to continue a career in education in Alberta; and be registered as a graduate student in a specialty in education at an Alberta university.

Restrictions: The applicant must not have received an ATA fellowship.

Amount of Grant: $C5000
Date(s) Application is Due: Mar 10
Contact: Corinne Anderson, Administrative Assistant, (780) 447-9470 or (800) 232-7208; email: corrine.anderson@ata.ab.ca
Internet: http://www.teachers.ab.ca/Professional+Development/Grants+and+Scholarships
Sponsor: Alberta Teachers Association
11010 142nd Street NW
Edmonton, AB T5N 2R1 Canada

Athenaeum of Philadelphia Senior Fellowships   1217
The fellowship trust supports the study, recording, and preservation of early American architecture and building technology and the teaching of conservation skills in American schools of architecture. Research is not subject to geographical restrictions, although preference is given to Delaware Valley topics. Applications are reviewed by a committee of architects, architectural historians, and educators appointed by the board of directors. Applications should be submitted in the form of a single-page letter setting forth a brief statement of the project, with attached budget, schedule for completion, and professional resume. Two letters of reference should be requested by the applicant and submitted directly to the committee. There are no application forms. A clear statement of objectives is necessary, and a final report is expected. Successful applicants may be invited to give a public lecture or participate in a seminar at the Athenaeum sharing the results of the project.

Requirements: Senior fellows must hold a terminal degree and possess a distinguished record of accomplishment.

Restrictions: Grants may not be used for international travel.

Amount of Grant: $5000 maximum
Date(s) Application is Due: Mar 1
Contact: Eileen Magee, Assistant Director for Programs, (215) 925-2688; fax: (215) 925-3755; email: magee@PhilaAthenaeum.org
Internet: http://www.philaathenaeum.org/grants.html
Sponsor: Athenaeum of Philadelphia
219 S Sixth Street
Philadelphia, PA 19106-3794

Athwin Foundation Grants   1218
Areas of interest include arts and humanities, education, human services, the natural environment, and organizational capacity building. General operating grants, capital grants, and special project grants will be awarded. Applications are accepted throughout the year.

Requirements: Tax-exempt organizations in the Twin Cities area of Minnesota and Western Montana are eligible.

Restrictions: Grants do not support individuals, scholarships, fellowships, or loans.

Geographic Focus: Minnesota; Montana
Samples: Claremont McKenna College (CA)—for program support, $30,000.; Minnesota Children's Museum (Saint Paul, MN)—to support the capital campaign, $5000.
Amount of Grant: $1000-$100,000 range
Date(s) Application is Due: Mar 1: Oct 1

Contact: Bruce Bean, Trustee, (952) 915-6165
Sponsor: Athwin Foundation
5200 Wilson Road, Suite 307
Minneapolis, MN 55424

ATLA Bibliography Grant 1219

The grant is awarded annually to one or more bibliographers or indexers to aid in the development of a work that provides access to a significant body of literature within the fields of theological and/or religious studies. The scope, subject matter, length, and format are broadly conceived, with the intent of encouraging bibliographical or indexing work at all levels, especially by persons undertaking their first major project in this area. The applicant(s) should complete and submit the application form (available online) along with the two requisite letters of reference by the specified due date.
Requirements: This grant is open to applicants inside and outside of the association.
Amount of Grant: $1500 maximum
Date(s) Application is Due: Jan 15
Contact: Publications Committee; (888) 665-2852.
Internet: http://www.atla.com/pub_com/grant.html
Sponsor: American Theological Library Association
300 S Wacker Drive, Suite 2100
Chicago, IL 60606-6701

Atlantic Fellowships in Public Policy 1220

The fellowships provide an opportunity for outstanding mid-career professionals to study and gain practical experience in a variety of public policy areas in the United Kingdom, as well as a firsthand introduction to the European Union. Approximately 10 awards are made each year, and fellows spend between six and 10 months in the UK. Brochures and applications are available on the Web site.
Requirements: Fellowships are open to U.S. citizens with at least five years experience in their professions. Candidates should be between their late 20s and early 40s.
Restrictions: Fellowships are not awarded to support basic research or to enable study for an academic degree.
Contact: Scholarships and Exchanges Officer, (202) 588-7844; fax: (202) 588-7918; email: atlantic.fellow@us.britishcouncil.org
Internet: http://www.britishcouncil-usa.org/learning/policy/atlantics.shtml
Sponsor: The British Council
3100 Massachusetts Avenue NW
Washington, DC 20008

ATS Lilly Theological Faculty Fellowships 1221

The Faculty Fellowships, supported by the Lilly Endowment, encourage scholarly research that contributes to theological education, informs the life of the church, develops a greater public voice for theology in society, collaborates with other academic disciplines, and offers new perspectives on Christianity in a pluralistic setting. Faculty Fellowships will support up to six faculty members with awards of $30,000 each during an institutionally approved research leave of at least one full semester.
Requirements: All full-time faculty members at ATS-accredited and candidate schools are eligible to apply. For faculty fellowships applicants must have a research leave of at least one term from all teaching and institutional responsibilities during the award year.
Restrictions: Project proposals for less than six months will not be considered for faculty fellowships. No more than two applicants from a single institution may receive awards in any given competition.
Amount of Grant: $30,000 maximum
Date(s) Application is Due: Jan 5
Contact: William Myers, Program Contact; (412) 788-6505, ext. 252; fax: (412) 788-6510; email: myers@ats.edu
Internet: http://www.ats.edu/LeadershipEducation/pages/LillyTheologicalResearchGrants.aspx
Sponsor: Association of Theological Schools
10 Summit Park Drive
Pittsburgh, PA 15275-1103

ATS Henry Luce III Fellowships 1222

The grants, supported by the Henry Luce Foundation, are awarded to scholars of distinction for research and publication that offer an innovative and substantial contribution to theological scholarship. The foundation supports the research of junior and senior scholars whose projects offer significant and innovative contributions to theological studies; meet high scholarly standards; enhance the theological understanding of people of faith and enrich the experience of church life in North America; and develop ways for scholarship to inform contemporary culture. No more than one fellow will be selected per the following categories: bible and the church, Christianity and contemporary culture, constructive theology, history of Christianity and the church today, ministry and practice of communities of faith, plus topics that do not fall into the above categories. Applicants must specify the category in which they are requesting appointment. Fellows present their research at the annual conference, held in the fall. Annual deadline dates may vary; contact the program officer for exact dates.
Requirements: The program is open to full-time faculty members of ATS-accredited and candidate schools who have an appropriate period of leave from all teaching and institutional responsibilities. Applicants also must be eligible to receive at least one-half salary support from their institutions.
Restrictions: Fellows may not hold any other major fellowship or grant during the tenure of the award without express approval of the association.

Amount of Grant: $75,000 maximum
Date(s) Application is Due: Dec 1
Contact: William Myers, Program Contact; (412) 788-6505, ext. 252; fax: (412) 788-6510; email: myers@ats.edu
Internet: http://www.ats.edu/LeadershipEducation/pages/LuceFellowsInTheology.aspx
Sponsor: Association of Theological Schools in the United States and Canada
10 Summit Park Drive
Pittsburgh, PA 15275-1103

ATS Lilly Theological Collaborative Research Grants 1223

The Theological Collaborative Research Grants, supported by the Lilly Endowment, are designed to enhance the skill and capacity of faculty in ATS schools as theological researchers and scholars. It supports research efforts of faculty, particularly at junior and non-tenured levels in order to nurture the development of their scholarship; seeks to enlarge the pool of faculty actively engaged as theological researchers; works to increase knowledge about grant seeking and the craft of theological research; and nurtures the habit of research as an ongoing aspect of scholarly life. The Program will provide grants of up to $16,000 each to as many as four collaborative research projects.
Requirements: All full-time faculty members at ATS-accredited and candidate schools are eligible to apply.
Restrictions: No more than two applicants from a single institution may receive awards in any given competition.
Amount of Grant: $16,000 maximum
Date(s) Application is Due: Jan 5
Contact: William Myers, Program Contact; (412) 788-6505, ext. 252; fax: (412) 788-6510; email: myers@ats.edu
Internet: http://www.ats.edu/LeadershipEducation/pages/LillyTheologicalResearchGrants.aspx
Sponsor: Association of Theological Schools
10 Summit Park Drive
Pittsburgh, PA 15275-1103

ATS Lilly Theological Research Expense Grants 1224

The Theological Expense Grants, supported by the Lilly Endowment, are designed to enhance the skill and capacity of faculty in ATS schools as theological researchers and scholars. It supports research efforts of faculty, particularly at junior and non-tenured levels in order to nurture the development of their scholarship; seeks to enlarge the pool of faculty actively engaged as theological researchers; works to increase knowledge about grant seeking and the craft of theological research; and nurtures the habit of research as an ongoing aspect of scholarly life. The Program will give up to $5,000 each to as many as six scholars for direct research expenses.
Requirements: All full-time faculty members at ATS-accredited and candidate schools are eligible to apply.
Restrictions: No more than two applicants from a single institution may receive awards in any given competition.
Amount of Grant: $5,000 maximum
Date(s) Application is Due: Jan 5
Contact: William Myers, Program Contact; (412) 788-6505, ext. 252; fax: (412) 788-6510; email: myers@ats.edu
Internet: http://www.ats.edu/LeadershipEducation/pages/LillyTheologicalResearchGrants.aspx
Sponsor: Association of Theological Schools
10 Summit Park Drive
Pittsburgh, PA 15275-1103

ATS Lilly Theological Scholars Grants 1225

The Theological Scholars Grants, supported by the Lilly Endowment, are designed to enhance the skill and capacity of faculty in ATS schools as theological researchers and scholars. It supports research efforts of faculty, particularly at junior and non-tenured levels in order to nurture the development of their scholarship; seeks to enlarge the pool of faculty actively engaged as theological researchers; works to increase knowledge about grant seeking and the craft of theological research; and nurtures the habit of research as an ongoing aspect of scholarly life. This program will provide as many as six grants of up to $12,000 each for research apart from a formal research leave.
Requirements: All full-time faculty members at ATS-accredited and candidate schools are eligible to apply.
Restrictions: No more than two applicants from a single institution may receive awards in any given competition.
Amount of Grant: $12,000 maximum
Date(s) Application is Due: Jan 5
Contact: William Myers, Program Contact; (412) 788-6505, ext. 252; fax: (412) 788-6510; email: myers@ats.edu
Internet: http://www.ats.edu/LeadershipEducation/pages/LillyTheologicalResearchGrants.aspx
Sponsor: Association of Theological Schools
10 Summit Park Drive
Pittsburgh, PA 15275-1103

AUAF Health Services Research Scholar Grants in Urology 1226

The program supports urology residents or trained urologists (MD/DO) within five years of residency who aspire to conduct research in health care services as it relates to urology. The candidate must be committed to a career in academic urology and able to

provide evidence of current interest and/or accomplishments in matters relating to the delivery of health services and health economics and policy. The project should include didactic experience and may include plans for obtaining an advanced degree in public health, health policy, health financing and economics, or other relevant areas. In view of the multidisciplinary nature of health services research and health care policy, the proposal should include an explanation of how researchers from different clinical and nonclinical disciplines will be involved in the research project. The candidate must dedicate a minimum of 80 percent of his/her time to the research project. This experience should prepare the scholar for a subsequent career in academic urology as it relates to health care policy, health financing, and health economics. Applications are available on the Web site. Annual deadline dates may vary; contact program staff for exact dates.
Requirements: An accredited medical education research institution and/or department within such an institution must sponsor the candidate. A preceptor, an established health services researcher in a urology department or a nonurology clinician/researcher in a school of public health/health policy, must have prior experience in health services research with funded grant support and must make a two-year commitment to sponsor the scholar.
Amount of Grant: $44,000 per year
Date(s) Application is Due: Sep 1
Contact: Anthony Caputi, Manager, Research Scholar Program, (800) 828-7866; fax: (410) 468-1808; email: anthony@afud.org
Internet: http://www.afud.org/research/index.asp
Sponsor: American Urologic Association Foundation, Inc.
1128 N Charles Street
Baltimore, MD 21201

AUAF PhD Research Scholars Program   1227
These fellowships are offered to postdoctoral basic scientists with a research interest in urologic or related diseases and dysfunctions. Research projects focus on many different diseases including, but not limited to: prostate cancer, bladder cancer, kidney stones, incontinence, impotence, BPH, prostatitis, urinary tract infections, interstitial cystitis, and pediatric dysfunctions. A commitment to dedicate two years in the program as a full-time researcher is required. The sponsoring urology department is required to provide space, laboratory equipment, and supplies to enable the recipient to perform the proposed research. Continued support is subject to yearly review and evaluation by the scientific and education committee. Application forms are available on the Web site. Annual deadline dates may vary; contact program staff for exact dates.
Requirements: An accredited medical education research institution or department within such an institution must sponsor the candidate by guaranteeing research time commitment and adequate financial support, including responsibility for the adequacy of the scientific environment, laboratory equipment, and supplies to perform the proposed research.
Amount of Grant: $10,000-$60,000
Date(s) Application is Due: Sep 1
Contact: Anthony Caputi, Manager, Research Scholar Program, (410) 468-1803; fax: (410) 468-1808; email: anthony@afud.org
Internet: http://www.afud.org/research/index.asp
Sponsor: American Urologic Association Foundation, Inc.
1128 N Charles Street
Baltimore, MD 21201

AUAF Practicing Urologist Research Awards   1228
To aid and encourage practicing urologists with research ideas to undertake collaborative investigations at a urological research laboratory, the AUAF grants research awards to be matched by the sponsoring institution. Applications are available on the Web site. Annual deadline dates may vary; contact program staff for exact dates.
Requirements: An accredited medical research institution or department must sponsor the candidate by guaranteeing adequate support, including responsibility for the adequacy of the environment, laboratory equipment, and supplies needed to perform the proposed research.
Amount of Grant: $5000 matched by sponsoring institution
Date(s) Application is Due: Sep 1
Contact: Anthony Caputi, Manager, Research Scholar Program, (410) 468-1803; fax: (410) 468-1808; email: anthony@afud.org
Internet: http://www.afud.org/research/index.asp
Sponsor: American Urologic Association Foundation, Inc.
1128 N Charles Street
Baltimore, MD 21201

AUAF Research Scholar Program   1229
The fellowship program supports urology residents in training or post residency (not more than five years) for one year of full-time investigation in laboratory research of urologic diseases. An established basic science academician in either a basic science or clinical department with substantial experience in grant support and mentoring of scientific trainees must make a one-year commitment to sponsor the scholar. One-year awards can be renewed for a second year. Annual deadline dates may vary; contact program staff for exact dates.
Requirements: An accredited medical education research institution or department within such an institution must sponsor the candidate by guaranteeing research time commitment and adequate financial support, including responsibility for the adequacy of the scientific environment, laboratory equipment, and supplies to perform the proposed research.
Amount of Grant: $25,000 for one year

Date(s) Application is Due: Sep 1
Contact: Anthony Caputi, Manager, Research Scholar Program, (410) 468-1803; fax: (410) 468-1808; email: anthony@afud.org
Internet: http://www.afud.org/research/scholars/current.asp
Sponsor: American Urologic Association Foundation, Inc.
1128 N Charles Street
Baltimore, MD 21201

AUAF/NIH Intramural Urology Research Training Grants   1230
This intramural program, developed by NIDDK and NCI in collaboration with AFUD, provides an opportunity for selected individuals to complete a research project under the direction of a senior investigator at NIDDK or NCI. Two positions each are available in NIDDK and NCI for laboratory-based research. Up to two positions are available in the surgery branch of NCI, of which six months are dedicated to clinical training in urologic oncology. The program consists of a two-year fellowship with possible extension for a third year. Applications are available on the Web site. Annual deadline dates may vary; contact program staff for exact dates.
Requirements: Eligibility is open to physicians in urology residency programs, or those who have recently completed a urology residency, who desire intensive exposure to research experience utilizing basic molecular and cellular biologic techniques in projects relevant to urology. Applicants must have less than five years of postdoctoral research experience (clinical residency training is not counted as postdoctoral research).
Amount of Grant: $40,000-$44,000 per year, depending on experience
Date(s) Application is Due: Sep 1
Contact: Kym Liddick, Director, Research Scholar Program, (410) 468-1812; fax: (410) 468-1808; email: kym@afud.org
Internet: http://www.afud.org/research/index.asp
Sponsor: American Urologic Association Foundation, Inc.
1128 N Charles Street
Baltimore, MD 21201

Auburn Foundation Grants   1231
The purpose of the Foundation is to stimulate giving and cooperative leadership among the citizens of Auburn; help improve the lives of all community residents, especially those who are most vulnerable; and enrich the cultural environment and community life. Of special interest are projects that bring together all ages and sections of the town, or that contribute to healthy, active living. Application is available online.
Requirements: Any nonprofit organization that serves residents of Auburn is invited to apply.
Restrictions: Grants will not be awarded to for-profit businesses or expenses already incurred by the applicant.
Geographic Focus: Massachusetts
Samples: Bancroft School (Worcester, MA)—for general support, $45,000.; New England Science Ctr, Rutland House (Worcester, MA)—for repairs, $20,000.; Tower Hill Botanic Garden (Boylston, MA)—for general support, $30,000.
Amount of Grant: $5,000
Date(s) Application is Due: Apr 15
Contact: Lois Smith, Senior Program Officer; (508) 755-0980, ext. 107; email: lsmith@greaterworcester.org
Internet: http://www.greaterworcester.org/grants/Auburn.htm
Sponsor: Auburn Foundation
370 Main Street, Suite 650
Worcester, MA 01608-1738

Auburn University President's Graduate Opportunities Program   1232
The major purpose of the program is to recruit, retain, and support African-American students engaged in graduate study leading to a doctoral degree from Auburn University. Successful applicants will receive a fellowship in addition to a stipend provided by the department, school, or college in which recipients are enrolled. Both the fellowship and the departmental stipend are renewable for up to four years of doctoral study. Application and guidelines are available online.
Requirements: Candidates must hold an earned degree from an accredited institution, meet admission standards of the graduate school and the doctoral program in which they seek to enroll, and be recommended for the award by that academic department or unit.
Amount of Grant: $10,000 stipend plus $5000 minimum from recipients' department, school, or college
Date(s) Application is Due: Feb 1
Contact: Dr. Christine Curtis, Chair; (334) 844-5771; email: curticw@auburn.edu
Internet: http://media.cla.auburn.edu/english/gs/handbook/policies/index.cfm?include=president
Sponsor: Auburn University
2003 RBD Library, 231 Mell Street
Auburn University, AL 36849-5168

AUCC Education Funding Programs   1233
On behalf of governments, foundations, and private sector companies, the association administers Canadian and international scholarships, exchange programs, work placements, and research opportunities. With a focus on international opportunities that enhance other countries' resources and lead to improvements in their living standards, the association administers more than 150 scholarship, fellowship, and internship programs.

Requirements: Canadian citizens and permanent residents of Canada can apply for AUCC-administered programs. Certain programs are also open to foreign professors.
Contact: International and Canadian Programs Division, (613) 563-1236; fax: (613) 563-9745; email: awards@aucc.ca
Internet: http://www.aucc.ca/programs/index_e.html
Sponsor: Association of Universities and Colleges of Canada
350 Albert Street, Suite 600
Ottawa, ON K1R 1B1 Canada

Audio Engineering Society Graduate Scholarships 1234

Graduate students in audio engineering and related fields may apply for the scholarships, which will be awarded on the basis of demonstrated past interest and achievements in the field and on two faculty recommendations. Awards are made annually, in August, and each successful applicant may apply for a one-time renewal after the successful completion of at least one year of graduate studies. Payments for the original award or the renewal will be sent directly to the graduate school on behalf of the student's account. Completed applications including recommendations will be accepted from March 15 to the listed application deadline.
Requirements: Applicants must have acceptance or pending application for graduate studies leading to a master's or higher degree or an internationally recognized equivalent. Applications are accepted from students worldwide.
Amount of Grant: $3000
Date(s) Application is Due: May 15
Contact: Coordinator, (212) 661-8528; fax: (203) 682-0477; email: HQ@aes.org
Internet: http://www.aes.org/education/edu_foundation.html
Sponsor: Audio Engineering Society Educational Foundation
60 E 42nd Street
New York, NY 10165

Audubon Naturalist Society Research Grants 1235

This program provides grants for either conservation-related field work or conservation and education projects. The project should be located outside the United States. One need not be affiliated with an institution or organization to apply. There are no restrictions on nationality. Preference given to start-up projects. Applied research projects are given preference over basic research projects.
Amount of Grant: $500-$2000
Contact: Education Program Coordinator, (301) 652-9188; fax: (301) 951-7179; email: Contact@audubonnaturalist.org
Internet: http://www.audubonnaturalist.org
Sponsor: Audubon Naturalist Society of the Central Atlantic States
8940 Jones Mill Road
Chevy Chase, MD 20815

AUPHA Baxter International Foundation Prize for Health Services Research 1236

The prize acknowledges national or international contributions of health services research, defined as a product of the application of analytic methods to the organization, financing, and/or delivery of health services. An individual's specific contribution or a career-long achievement may be recognized. The prize recognizes a person who has made significant and demonstrable contributions to the health of the public in three primary focus areas: health services management, health policy development, and health care delivery.
Requirements: Nominations must include a letter of nomination, two letters of recommendation, and a current curriculum vita. Nominations should focus on the nominee's contribution in the areas of health services management, health policy development, and health care delivery. Nominations should be sent to and further information can be obtained from the Secretary, HSR Prize Committee, AUPHA, 2000 N. 14th Street, Suite 780, Arlington, VA 22201.
Amount of Grant: $10,000 personal award and $15,000 to a nonprofit institution designated by the recipient, to support his or her work
Date(s) Application is Due: Dec 15
Contact: Lydia Reed, (703) 894-0940; email: lydia.reed@aupha.org
Internet: http://www.e-guana.net/organizations.php3?action=printContentItem&orgid=75&typeID=506&itemID=3744
Sponsor: Association of University Programs in Health Administration
2000 14th Street N, Suite 780
Arlington, VA 22201

Australian Academy of Science Grants 1237

The objectives of the Academy are to promote science through a range of activities. It has defined four major program areas as their focus. The Academy focus is on recognition of outstanding contributions to science, education and public awareness, science policy, and international relations. Guidelines are available online.
Contact: Executive Secretary, 61-2-6201-9400; fax: 61-2-6201-9494; email: eb@science.org.au
Internet: http://www.science.org.au
Sponsor: Australian Academy of Science
G.P.O. Box 783
Canberra, ACT 2601 Australia

Australian Cancer Foundation for Medical Research Grants 1238

The foundation awards grants in Australia in the field of cancer research. This includes, but is not limited to, cell or molecular biology, epidemiology, prevention or drug development. Types of support include capital projects and equipment.
Amount of Grant: $A2 million total
Date(s) Application is Due: May 12

Contact: Lorraine McNuff, (02) 9223 7833; fax: (02) 9223 1800; email: lmcnuff@acrf.com.au
Internet: http://www.acrf.com.au/page/research_grants.html
Sponsor: Australian Cancer Foundation for Medical Research
Strand Arcade, Suite 403
Sydney NSW 2000 Australia

Australian Institute of Aboriginal and Torres Strait Islander Studies Research Grants 1239

Grants cover costs associated with conduct of research for projects concerned with Australian Aboriginal and Torres Strait Island cultures in fields such as health, human biology, social anthropology, linguistics, ethnomusicology, material culture, rock art, archaeology, ethnobotany, psychology, education, and Aboriginal history including oral history. Annual deadline dates may vary; contact program staff for exact dates.
Requirements: Nationals of any country are eligible for assistance.
Samples: Glynn Barratt—for Russian Texts Translation and Annotation project, $9800.; Andree Grau—for Tiwi Dance and the Embodiment of Knowledge project, $11,747.
Amount of Grant: $A45,000 maximum generally
Date(s) Application is Due: Dec 21
Contact: Research Grants Administrator, 61 2 6246 1145; fax: 61 2 6249 7714; email: grants@aiatsis.gov.au
Internet: http://www.aiatsis.gov.au/rsrch/rsrch_grnts/rg_abt.htm
Sponsor: Australian Institute of Aboriginal and Torres Strait Islander Studies
G.P.O. Box 553
Canberra, ACT 2601 Australia

Australian National University Humanities/Research Center Visiting Fellowships 1240

The fellowships are awarded to established academics who wish to do research in the humanities worldwide. Each year a particular theme is announced for the next academic year, and most fellowships are awarded on some variation of that theme. Fellowships are awarded for the academic year and are not usually renewable. Approximately 20 awards are made annually.
Requirements: Applicants must complete a formal application form and send it with curriculum vita and three references.
Restrictions: Candidates currently working for an advanced degree in the humanities are not eligible.
Amount of Grant: $A550 stipend per week and travel allowance
Date(s) Application is Due: Mar 31
Contact: Leena Messina, (02) 6125 4357; fax: (06) 248-0054; email: leena.messina@anu.edu.au
Internet: http://www.anu.edu.au/hrc/grants/index.php
Sponsor: Australian National University
Humanities Research Centre, Old Canberra House Building 73
Canberra, ACT 0200 Australia

Australian Postdoctoral Research Fellowships 1241

The specific purpose of the fellowship is to strengthen Australia's national research and development capability by providing opportunities for postdoctoral level researchers to undertake research of national and international significance and to broaden their research experience.
Requirements: Applicants must have submitted their PhD thesis before commencement of the fellowship. Candidates must obtain Australian citizenship or temporary residency status at the time of commencing the fellowship.
Contact: Dr. Andrew Smith, Australian Postdoctoral Fellowships, 61 2 6284 6629; email: andrew.smith@arc.gov.au
Internet: http://www.arc.gov.au/grant_programs/linkage_australian.htm
Sponsor: Australian Research Council
G.P.O. Box 2702
Canberra ACT 2601 Australia

Australian Postgraduate Research Fellowships 1242

The program supports postgraduate training with stipends available through Linkage'Projects (APAIs), ARC Centres of Excellence (APACs), and Discovery'Projects. International opportunities for postgraduate researchers are also offered through Linkage'International awards, and project support is offered through Discovery'Indigenous Researchers Development. Guidelines are available online.
Requirements: The applicant must have a PhD degree and at least three but no more than eight years of postdoctoral experience at the time of application.
Restrictions: Postgraduate students are not normally eligible to apply for funding: applications seeking support for research training must be submitted by eligible researchers and research teams.
Contact: Program Contact, Research Grant and Training Section, 61-2-6284-6600; fax: 61-2-6284-6601; email: info@arc.gov.au
Internet: http://www.arc.gov.au/info_users/researcher_postgraduate.htm
Sponsor: Australian Research Council
G.P.O. Box 2702
Canberra, ACT 0200 Australia

Australian Research Fellowships/Queen Elizabeth II Fellowships 1243

ARFs and QEIIs provide opportunities for established researchers to undertake research of national and international significance. QEIIs encourage research in Australia by postdoctoral graduates of exceptional promise and proven capacity for original work.

Requirements: Australian Research Fellowships (ARF) and Queen Elizabeth II Fellowships (QEII) are available to researchers with three to eight years of research experience since the award of the PhD or equivalent research doctorate.
Contact: Program Contact, Research Grant and Training Section, 61-2-6284-6600; fax: 61-2-6284-6601; email: info@arc.gov.au
Internet: http://www.arc.gov.au/info_users/researcher_fellowships.htm
Sponsor: Australian Research Council
GP.O. Box 2702
Canberra, ACT 2601 Australia

Australian Spinal Research Foundation Grants   1244

The foundation invites applications for grants to support research concerning chiropractic. Priority will be given to projects that investigate the vertebral subluxation complex and how chiropractic contributes to wellness. In addition, priority will be given to chiropractic-based research projects that cover topics as diverse as those comparing the clinical outcome of various chiropractic techniques, or the effect of the chiropractic adjustment (e.g. sleep patterns, colic or immune function), or an analysis of how ergonomics may benefit posture, whether this be lying, sitting, or standing.
Amount of Grant: $2000
Date(s) Application is Due: May 31
Contact: Grants Administrator, 61 (7) 3808 4098; fax: 61 (7) 3808 8109; email: asrf@spinalresearch.com.au
Internet: http://www.spinalresearch.com.au/Research/research.html
Sponsor: Australian Spinal Research Foundation
P.O.B 1047
Springwood, Queensland QLD 4127 Australia

Austro-American Association of Boston Scholarship   1245

The scholarship is awarded to a student at any level (degree or nondegree) or to any other person demonstrating a capacity for scholarly research, creative work, or cultural mediatorship. The basic requirement is that the project relate to Austrian literature, music, art, history, or to other aspects of the Austrian culture. The grant may be used to travel to Europe or within the United States, to purchase books and other materials, or to facilitate publication. The scholarship also supports dissertation research and faculty/professional development. The award recipient will be announced by May 15.
Requirements: Applicants should be Massachusetts residents and may be asked to appear for an interview. Application should include a curriculum vita, a detailed description of the proposed project, and two letters of recommendation from persons well acquainted with the applicant's background and potential.
Amount of Grant: $1000 maximum
Date(s) Application is Due: Apr 15
Contact: Dr. George Hauser, Chairperson, Scholarship Committee, (617) 332-4055; email: george_hauser@hms.harvard.edu
Internet: http://www.austria-boston.org/scholar.asp
Sponsor: Austro-American Association of Boston
47 Windermere Road
Newton, MA 02166-2521

AVDF Religion Grants   1246

The foundations's principal commitment in the field of religion is to graduate theological education. All grants to institutions of graduate theological education will be to schools or seminaries that are fully accredited by the Association of Theological Schools; primarily produce persons prepared for ordination to pastoral or pulpit ministry; are known for academic excellence; and have solid records of continued alumni/trustee support and institutional financial stability. All denominations are eligible. Proposals from recognized consortia representing several seminaries of this type also will be considered. Similarly, joint proposals from two or more seminaries sharing programs or facilities are also of interest. No preference exists for any particular category or type of project for which grants are requested. Grantees should wait at least four years from the time of an award before reapplying for a grant.
Requirements: Proposals should be signed by the president or dean and should represent the leader's highest priority for reinforcing institutional excellence. The head of a seminary or divinity school should be in office for at least one year prior to submitting a grant request.
Restrictions: Proposals normally will not be considered from institutions in transition.
Samples: Candler School of Theology (Atlanta, GA)—to design a library building, $150,000 (2005); Good Shepherd Mission (Fort Defiance, AZ)—for improvements to the Good Shepherd Church, $75,700 (2005); Northern Baptist Theological Seminary (Lombard, IL)—to improve information-technology systems, $150,000 (2005); Phillips Theological Seminary (Tulsa, OK)—for an integrated library-management system, $128,000 (2005).
Amount of Grant: $100,000—$150,000 typically
Contact: Dr. Jonathan Howe, Executive Director, (904) 359-0670; email: arthurvining@msn.com
Internet: http://jvm.com/davis/PROGRAMS.HTM#religion
Sponsor: Arthur Vining Davis Foundations
225 Water Street, Suite 1510
Jacksonville, FL 32202-5185

AVDF Health Care Grants   1247

This program supports efforts to encourage caring attitudes in the delivery of patient care. Projects should have potential for widespread practical application and should be of interest to other groups. New ideas are encouraged, especially if they facilitate communication between patients (and their families) and doctors, nurses and other caregivers; ameliorate patient anxieties; strengthen trust and cooperation; and foster caring attitudes. Proposals, must have potential for wide application and not merely local improvement of requesting institutions or their communities.
Requirements: Proposals should be submitted by the head of the institution under whose auspices the project would be accomplished.
Samples: Southern Illinois U School of Medicine (IL)—for a program designed to foster caring attitudes toward aging patients, $195,523 (2005); U of Chicago, Pritzker School of Medicine (IL)—to expand the community-based teaching of cross-cultural patient-doctor communication, $199,986 (2005); U of Massachusetts Medical School (Worcester, MA)—for efforts to identify the unique attributes of a caring attitude in health-care settings, $198,314 (2005).
Amount of Grant: $100,000-$200,000
Contact: Dr. Jonathan Howe, Executive Director, (904) 359-0670; email: arthurvining@msn.com
Internet: http://jvm.com/davis/PROGRAMS.HTM#health
Sponsor: Arthur Vining Davis Foundations
225 Water Street, Suite 1510
Jacksonville, FL 32202-5185

AVDF Private Higher Education Grants   1248

The higher education program seeks to strengthen private four-year liberal arts institutions that place strong emphasis on teaching and whose students choose majors in the humanities, science, and math. Support generally will be reserved for schools of broadly acknowledged academic excellence in the liberal arts and sciences and a solid record of financial strength. A few grants are also made to larger teaching and research universities with a national reputation for excellence in graduate and undergraduate education. The foundations also have an interest in helping to improve programs at historically black colleges, those that primarily have a Native American student body, colleges in Appalachia, and similar schools that provide opportunities for traditionally underserved students. Joint proposals from two or more colleges that share a program or facilities are also of interest.
Requirements: Successful applicants in higher education should have a recognized position of leadership in the liberal arts with emphasis on the humanities and science/math majors, rather than career or vocational studies; adequate financial resources; and a record of outstanding support from trustees and alumni/ae.
Samples: College of the Atlantic (Bar Harbor, ME)—to endow a chair in human ecology, $200,000 (2005); Reed College (Portland, OR)—for a quantitative-skills center, $200,000 (2005); Clark University (Worcester, MA)—for construction of Biosciences building and renovation of existing Biophysics building, $200,000 (2005); Hendrix College(Conway, AR)—for an Award for Excellence, $200,000 (2005).
Amount of Grant: $100,000-$200,000
Contact: Dr. Jonathan Howe, Executive Director, (904) 359-0670; email: arthurvining@bellsouth.net
Internet: http://jvm.com/davis/PROGRAMS.HTM#Private
Sponsor: Arthur Vining Davis Foundations
225 Water Street, Suite 1510
Jacksonville, FL 32202-5185

AVDF Secondary Education Grants   1249

The goal of the program is to strengthen teachers of grades nine through 12 and their teaching; examples are teacher professional development and use of new technologies. Collaboration between school districts and higher education institutions is encouraged. Special consideration will be given to projects addressing education issues on a national level.
Requirements: Nonprofit organizations are eligible.
Restrictions: Projects mainly benefiting a local school or district will not be approved, and only in rare situations are applications from individual schools and districts accepted.
Samples: Cast Inc (Wakefield, MA)—to improve reading skills and methods, $150,000 (2005); Marine Biological Laboratory (Woods Hole, MA)—for an endowment for the Semester in Environmental Science, $150,000 (2005); Rural School and Community Trust (Washington, DC)—for a rural teacher-development center, $150,000 (2005); Yale-New Haven Teachers Institute (CT)—for national seminars for high-school teachers, $150,000 (2005).
Amount of Grant: $100,000-$150,000 average
Contact: Dr. Jonathan Howe, Executive Director, (904) 359-0670; email: arthurvining@bellsouth.net
Internet: http://jvm.com/davis/PROGRAMS.HTM#secondary
Sponsor: Arthur Vining Davis Foundations
225 Water Street, Suite 1510
Jacksonville, FL 32202-5185

Aviation Research Grants   1250

The program objectives are to encourage and support innovative and advanced research in the areas of potential benefit to the long-term technical needs of the National Airspace System; areas related to research on the prevention of catastrophic failures; and areas related to research, development, and implementation of technologies and procedures to counteract terrorist acts against civil aviation. The grants may be used to support research and development projects in the following areas: communications, navigation, and surveillance; capacity and air traffic control technology; aviation weather; airport technology; aircraft safety technology; system security technology; human factors and aviation medicine; environment and energy; and operations research.

Requirements: Colleges, universities, and nonprofit institutions are eligible. Profit-making organizations may be eligible in the area of aviation security.
Restrictions: Discretionary funds are not currently available.
Amount of Grant: $25,000-$5 million; $50,000-$150,000 average
Contact: Aviation Research Grants Program, (609) 485-4424; fax: (609) 485-6509
Internet: http://www.tc.faa.gov/logistics/grants
Sponsor: Department of Transportation
800 Independence Avenue SW
Washington, DC 20591

Avon Foundation Breast Care Fund Grants 1251

The fund awards grants for the development of breast cancer education and outreach activities, particularly breast health education and early detection services for under-served women. AFBCF supports programs that: recruit women for both first time screening and annual screening; develop partnerships between community-based outreach providers and local medical providers; work with health care providers to ensure proper clinical follow-up of abnormal screening results; and educate older women (65+ years old) about Medicare coverage of annual screening mammograms and assist them in obtaining the service from providers who accept Medicare.
Requirements: Private, non-government, nonprofit organizations in the United States and Puerto Rico are eligible. Grants are awarded to community-based programs and/or health care agencies that provide medically under-served women aged 40 and older with direct access to breast cancer education, annual clinical screening services, and prompt follow-up care. All programs must utilize the three-part approach to breast cancer early detection including regular screening mammography, clinical breast examination, and monthly breast self-examination.
Samples: Family Services Center of Coffee County, Enterprise, AL (2007); The Witness Project, University of Arkansas for Medical Sciences, Little Rock, AR (2007); Latino Community Development Agency, Oklahoma City, OK (2007).
Amount of Grant: $30,000-$50,000
Date(s) Application is Due: Aug 28
Contact: Coordinating Center; (212) 244-5368; fax: (212) 695-3081; email: admin@ avonbreast care.org
Internet: http://www.avonbreastcare.org/fundinginfo.htm
Sponsor: Avon Foundation
505 Eighth Avenue, Suite 1601
New York, NY 10018-6505

Avon Products Foundation Grants 1252

The foundation's two-fold focus is to support education, community and social services, and arts organizations and programs that provide economic opportunities for women and girls; and to support breast cancer and other women's health organizations and programs. The foundation awards grants in cities and regions with a large concentration of representatives and business operations, with the majority of funds going to U.S.-based institutions. National, international, and New York metropolitan area programs are administered through the foundation's headquarters in New York. For regional funding support, contact the following locations and ask for the corporate contributions department: Atlanta, GA, (770) 271-6100; Morton Grove, IL, (847) 583-5169; Newark, DE, (302) 453-7700; Pasadena, CA, (626) 578-8000; Rye, NY, ((914) 935-2000; Suffern, NY, (845) 369-2000; Caguas, Puerto Rico, (787) 476-6161; and Springdale, OH, (513) 551-2000.
Requirements: Applying organizations must be tax-exempt; national and municipal organizations are eligible. Request the guidelines brochure prior to submitting a formal proposal.
Restrictions: Grants do not support individuals; memberships; lobbying organizations; political activities and organizations; religious, veteran, or fraternal organizations; fundraising events; and journal advertisements.
Samples: World Trade Ctr Memorial Foundation (new York, NY)—to build the World Trade Center Memorial and a museum to remember the September 11, 2001, and February 26, 1993, terrorist attacks, $325,000 (2005); Red Cross and Red Crescent Societies (Geneva, Switzerland) and U.S. Fund for UNICEF (New York, NY)—to assist tsunami-relief efforts, $500,000 jointly (2005); Family Violence Prevention Fund (San Francisco, CA)—to develop an educational program available online and in print on relationship violence, $100,000 (2004); Georgetown U, Lombardi Cancer Ctr (Washington, DC)—to provide breast-health education and care at the Capital Breast Care Ctr, $500,000 (2004).
Amount of Grant: $2.3 million total
Contact: Grants Administrator, (866) 505-2866; email: info@avonfoundation.org
Internet: http://www.avoncompany.com/women/avonfoundation
Sponsor: Avon Products Foundation
1345 Avenue of the Americas
New York, NY 10105

Award for Achievement in Research for the Teaching & Learning of Chemistry 1253

The purpose of the Award is to recognize outstanding contributions to experimental research that have increased our understanding of chemical pedagogy and led to the improved teaching and learning of chemistry. The award will recognize research contributions involving a significant body of work rather than a single project or contribution. The award consists of $5,000 and a certificate. Reasonable travel expenses to the meeting at which the award will be presented will be reimbursed. This is an international award.
Requirements: Any individual, except a member of the award committee, may submit one nomination or seconding letter for the award in any given year. The nominating documents consist of a letter of not more than 1000 words containing an evaluation of the nominee's accomplishments and a specific identification of the work to be recognized, a biographical sketch including date of birth, and a list of publications and patents authored by the nominee.
Restrictions: Self-nominations are not accepted.
Samples: Dorothy L. Gabel, $5,000 (2008); J. Dudley Herron, $5,000 (2007).
Amount of Grant: $5,000
Date(s) Application is Due: Nov 1
Contact: Felicia Dixon, Awards Administrator; (800) 227-5558 or (202) 872-4408; fax: (202) 776-8008; email: f_dixon@acs.org or awards@acs.org
Internet: http://portal.acs.org/portal/acs/corg/content?_nfpb=true&_pageLabel=PP_ARTICLEMAIN&node_id=1319&content_id=CTP_004484&use_sec=true&sec_url_var=region1
Sponsor: American Chemical Society
1155 Sixteenth Street, NW
Washington, DC 20036

Award for Affordable Green Chemistry 1254

The purposes of the Award are: to recognize outstanding scientific discoveries that lay the foundation for environmentally friendly products or manufacturing processes at a cost comparable to or less than that of current technologies, or discoveries that deliver new applications with compelling cost/benefit profiles; and to identify and recognize discovery of new eco-friendly chemistries with the potential to enable products or manufacturing processes that are less expensive than existing alternatives. Nominees may include: individuals; research teams (maximum 3 individuals); representatives from an organization (company, university or national lab); or professors or other academic-based teams whose work has been commercialized. The award will consist of $5,000, and a certificate. Reasonable travel expenses to the meeting at which the award will be presented will be reimbursed.
Requirements: Nominee eligibility is neither contingent on the timeframe in which the technology was developed, nor on demonstrated commercial success. Discoveries must have significant commercial potential, however, to be considered for award. Any individual, except a member of the award committee, may submit one nomination or seconding letter for the award in any given year. The nominating documents consist of a letter of not more than 1000 words containing an evaluation of the nominee's accomplishments and a specific identification of the work to be recognized, a biographical sketch including date of birth, and a list of publications and patents authored by the nominee.
Restrictions: Self-nominations are not accepted.
Amount of Grant: $5,000
Date(s) Application is Due: Nov 1
Contact: Felicia Dixon, Awards Administrator; (800) 227-5558 or (202) 872-4408; fax: (202) 776-8008; email: f_dixon@acs.org or awards@acs.org
Internet: http://portal.acs.org/portal/acs/corg/content?_nfpb=true&_pageLabel=PP_ARTICLEMAIN&node_id=1319&content_id=CTP_004487&use_sec=true&sec_url_var=region1
Sponsor: American Chemical Society
1155 Sixteenth Street, NW
Washington, DC 20036

Award for Computers in Chemical or Pharmaceutical Research 1255

The purpose of the Award is to recognize and encourage the use of computers in the advancement of the chemical and biological sciences. The award consists of $5,000 and a certificate. Up to $1,000 for travel expenses to the meeting at which the award will be presented will be reimbursed.
Requirements: Any individual, except a member of the award committee, may submit one nomination or seconding letter for the award in any given year. The nominating documents consist of a letter of not more than 1000 words containing an evaluation of the nominee's accomplishments and a specific identification of the work to be recognized, a biographical sketch including date of birth, and a list of publications and patents authored by the nominee.
Restrictions: Self-nominations are not accepted.
Samples: James A. McCammon—award recipient, $5,000 (2008); Emily A. Carter—award recipient, $5,000 (2007); Johann Gasteiger—award recipient, $5,000 (2006); Peter Willett—award recipient, $5,000 (2005).
Amount of Grant: $5,000 and travel expenses up to $1,000
Date(s) Application is Due: Nov 1
Contact: Felicia Dixon, Awards Administrator; (800) 227-5558 or (202) 872-4408; fax: (202) 776-8008; email: f_dixon@acs.org or awards@acs.org
Internet: http://portal.acs.org/portal/acs/corg/content?_nfpb=true&_pageLabel=PP_ARTICLEMAIN&node_id=1319&content_id=CTP_004503&use_sec=true&sec_url_var=region1
Sponsor: American Chemical Society
1155 Sixteenth Street, NW
Washington, DC 20036

Award for Creative Advances in Environmental Science and Technology 1256

The purpose of the Award is to encourage creativity in research and technology or methods of analysis to provide a scientific basis for informed environmental control decision-making processes, or to provide practical technologies that will reduce health risk factors. The award consists of $5,000 and a certificate. Reasonable travel expenses to the meeting at which the award will be presented will be reimbursed.

Requirements: Any individual, except a member of the award committee, may submit one nomination or seconding letter for the award in any given year. The nominating documents consist of a letter of not more than 1000 words containing an evaluation of the nominee's accomplishments and a specific identification of the work to be recognized, a biographical sketch including date of birth, and a list of publications and patents authored by the nominee.

Restrictions: The award will be granted without regard to age or nationality. Self-nominations are not accepted.

Samples: Margaret A. Tolbert—award recipient, $5,000 (2009); Susan D. Richardson—award recipient, $5,000 (2008); Richard C. Flagan—award recipient, $5,000 (2007).

Amount of Grant: $5,000 and allowance of up to $1000 for travel expenses to award meeting

Date(s) Application is Due: Nov 1

Contact: Felicia Dixon, Awards Administrator; (800) 227-5558 or (202) 872-4408; fax: (202) 776-8008; email: f_dixon@acs.org or awards@acs.org

Internet: http://portal.acs.org/portal/acs/corg/content?_nfpb=true&_pageLabel=PP_ARTICLEMAIN&node_id=1319&content_id=CTP_004504&use_sec=true&sec_url_var=region1

Sponsor: American Chemical Society
1155 Sixteenth Street, NW
Washington, DC 20036

Award for Creative Invention   1257

The purpose of the Award is to recognize a single inventor for the successful application of research in chemistry and/or chemical engineering that contributes to the material prosperity and happiness of people. The award consists of $5,000 and a certificate. Up to $1,000 for travel expenses to the meeting at which the award will be presented will be reimbursed.

Requirements: An inventor/nominee must be a resident of the United States or Canada. A patent must have been granted for the work to be recognized, with the nominee listed as an inventor, and it will have been developed during the 17 years ending January 1, 2009. A copy of the patent must be submitted with the nominating documents. Any individual, except a member of the award committee, may submit one nomination or seconding letter for the award in any given year. The nominating documents consist of a letter of not more than 1000 words containing an evaluation of the nominee's accomplishments and a specific identification of the work to be recognized, a biographical sketch including date of birth, and a list of publications and patents authored by the nominee.

Restrictions: Self-nominations are not accepted.

Samples: Adam Heller—award recipient, $5,000 (2008); Bruce Ganem—award recipient, $5,000 (2007); Alan Davison—award recipient, $5,000 (2006); Joseph DeSimone—award recipient, $5,000 (2005).

Amount of Grant: $5,000 and up to $1,000 travel expenses

Date(s) Application is Due: Nov 1

Contact: Felicia Dixon, Awards Administrator; (800) 227-5558 or (202) 872-4408; fax: (202) 776-8008; email: f_dixon@acs.org or awards@acs.org

Internet: http://portal.acs.org/portal/acs/corg/content?_nfpb=true&_pageLabel=PP_ARTICLEMAIN&node_id=1319&content_id=CTP_004506&use_sec=true&sec_url_var=region1

Sponsor: American Chemical Society
1155 Sixteenth Street, NW
Washington, DC 20036

Award for Creative Research and Applications of Iodine Chemistry   1258

The purpose of the Award is to support, promote, and motivate global research of iodine chemistry and develop its use and knowledge through applications. Applications may include but are not limited to its uses in medicine, catalysis, food, and photography. The award consists of $10,000 and a certificate. Up to $1,000 for travel expenses to the meeting at which the award will be presented will be reimbursed. The award is presented biennially in odd-numbered years.

Requirements: A nominee must have performed outstanding and creative research related to iodine chemistry or its applications. Any individual, except a member of the award committee, may submit one nomination or seconding letter for the award in any given year. The nominating documents consist of a letter of not more than 1000 words containing an evaluation of the nominee's accomplishments and a specific identification of the work to be recognized, a biographical sketch including date of birth, and a list of publications and patents authored by the nominee.

Restrictions: Self-nominations are not accepted.

Samples: Peter J. Stang—award recipient, $10,000 (2007); Robert M. Moriarty—award recipient, $10,000 (2005).

Amount of Grant: $10,000 and travel expenses up to $1,000

Date(s) Application is Due: Nov 1

Contact: Felicia Dixon, Awards Administrator; (800) 227-5558 or (202) 872-4408; fax: (202) 776-8008; email: f_dixon@acs.org or awards@acs.org

Internet: http://portal.acs.org/portal/acs/corg/content?_nfpb=true&_pageLabel=PP_ARTICLEMAIN&node_id=1319&content_id=CTP_004507&use_sec=true&sec_url_var=region1

Sponsor: American Chemical Society
1155 Sixteenth Street, NW
Washington, DC 20036

Award for Creative Work in Fluorine Chemistry   1259

The purpose of the Award is to recognize outstanding contributions to the advancement of the chemistry of fluorine. The award consists of $5,000 and a certificate. Reasonable

travel expenses to the meeting at which the award will be presented will be reimbursed. The award is presented in odd-numbered years at the Biennial Winter Fluorine Conference. The award is presented in even-numbered years during an ACS national meeting at which the ACS Division of Fluorine Chemistry meets.

Requirements: A nominee must have made an outstanding contribution or contributions to the field of fluorine chemistry. Any individual, except a member of the award committee, may submit one nomination or seconding letter for each award in any given year. The nominating documents consist of a letter of not more than 1000 words containing an evaluation of the nominee's accomplishments and a specific identification of the work to be recognized, a biographical sketch including date of birth, and a list of publications and patents authored by the nominee.

Restrictions: Self-nominations are not accepted.

Samples: Dennis P. Curran—award recipient, $5,000 (2008); Kenji Uneyama—award recipient, $5,000 (2007); Boris Zemva—award recipient, $5,000 (2006); Shlomo Rozen—award recipient, $5,000 (2005).

Amount of Grant: $5,000

Date(s) Application is Due: Nov 1

Contact: Felicia Dixon, Awards Administrator; (800) 227-5558 or (202) 872-4408; fax: (202) 776-8008; email: f_dixon@acs.org or awards@acs.org

Internet: http://portal.acs.org/portal/acs/corg/content?_nfpb=true&_pageLabel=PP_ARTICLEMAIN&node_id=1319&content_id=CTP_004505&use_sec=true&sec_url_var=region1

Sponsor: American Chemical Society
1155 Sixteenth Street, NW
Washington, DC 20036

Award for Creative Work in Synthetic Organic Chemistry   1260

The purpose of the Award is to recognize and encourage creative work in synthetic organic chemistry. The award consists of $5,000 and a certificate. Up to $1,000 for travel expenses to the meeting at which the award will be presented will be reimbursed.

Requirements: A nominee must have accomplished outstanding creative work in synthetic organic chemistry that has been published. Any individual, except a member of the award committee, may submit one nomination or seconding letter for the award in any given year. The nominating documents consist of a letter of not more than 1000 words containing an evaluation of the nominee's accomplishments and a specific identification of the work to be recognized, a biographical sketch including date of birth, and a list of publications and patents authored by the nominee.

Restrictions: Self-nominations are not accepted.

Samples: Masakatsu Shibasaki—award recipient, $5,000 (2008); Steven V. Ley—award recipient, $5,000 (2007); Stephen Buchwald—award recipient, $5,000 (2006); Chi-Huey Wong—award recipient, $5,000 (2005).

Amount of Grant: $5,000 and up to $1,000 for travel

Date(s) Application is Due: Nov 1

Contact: Felicia Dixon, Awards Administrator; (800) 227-5558 or (202) 872-4408; fax: (202) 776-8008; email: f_dixon@acs.org or awards@acs.org

Internet: http://portal.acs.org/portal/acs/corg/content?_nfpb=true&_pageLabel=PP_ARTICLEMAIN&node_id=1319&content_id=CTP_004508&use_sec=true&sec_url_var=region1

Sponsor: American Chemical Society
1155 Sixteenth Street, NW
Washington, DC 20036

Award for Distinguished Service in the Advancement of Inorganic Chemistry   1261

The purpose of the Award is to recognize individuals who advanced inorganic chemistry by significant service in addition to performance of outstanding research. Activities recognized by the award may include such fields as teaching, writing, research, and administration. The award consists of $5,000 and a certificate. Up to $1,000 for travel expenses to the meeting at which the award will be presented will be reimbursed.

Requirements: A nominee must have demonstrated extensive contributions to the advancement of inorganic chemistry. Any individual, except a member of the award committee, may submit one nomination or seconding letter for the award in any given year. The nominating documents consist of a letter of not more than 1000 words containing an evaluation of the nominee's accomplishments and a specific identification of the work to be recognized, a biographical sketch including date of birth, and a list of publications and patents authored by the nominee.

Restrictions: Self-nominations are not accepted.

Samples: Tobin J. Marks—award recipient, $5,000 (2008); Robert J. Angelici—award recipient, $5,000 (2007); Edward Solomon—award recipient, $5,000 (2006); Thomas G. Spiro—award recipient, $5,000 (2005).

Amount of Grant: $5,000 and up to $1,000 for travel expenses to award meeting

Date(s) Application is Due: Nov 1

Contact: Awards Administrator, (202) 872-4408; fax: (202) 872-6317; email: awards@acs.org

Internet: Felicia Dixon, Awards Administrator; (800) 227-5558 or (202) 872-4408; fax: (202) 776-8008; email: f_dixon@acs.org or awards@acs.org

Sponsor: American Chemical Society
1155 Sixteenth Street, NW
Washington, DC 20036

Award for Encouraging Disadvantaged Students into Careers in the   1262
Chemical Sciences

The purpose of the Award is to recognize individuals who have significantly stimulated or fostered the interest of students, especially minority and/or economically disadvantaged students, in chemistry, thereby promoting their professional development as chemists or chemical engineers, and/or increasing their appreciation of chemistry as the central science. Nominees for the award may come from any professional setting: academia, industry, government, or other independent facility. The award is intended to recognize significant accomplishments in the United States by individuals in stimulating students, especially those currently underrepresented in the profession, to elect careers in the chemical sciences and engineering, and in generating a broader appreciation of chemistry as the central science. The award consists of $5,000 and a certificate. A grant of $10,000 will be made to an academic institution, designated by the recipient, to strengthen its activities in meeting the objectives of the award. Up to $1,500 for travel expenses to the meeting at which the award will be presented will be reimbursed.

Requirements: Any individual, except a member of the award committee, may submit one nomination or seconding letter for the award in any given year. Nominating documents consist of a letter of not more than 1000 words containing an evaluation of the nominee's accomplishments and a specific identification of the work to be recognized, a biographical sketch including date of birth, and a list of publications and patents authored by the nominee.

Restrictions: Self-nominations are not accepted.

Samples: Susan V. Olesik—award recipient, $5,000 (2008); Robyn E. Hannigan—award recipient, $5,000 (2007); Susan Fahrenholtz—award winner, $5,000 (2006); Jeannette Brown—award winner, $5000 (2005).

Amount of Grant: $5,000 personal prize; $10,000 will be made to an academic institution; and $1,500 travel allowance to award meeting

Date(s) Application is Due: Nov 1

Contact: Felicia Dixon, Awards Administrator; (800) 227-5558 or (202) 872-4408; fax: (202) 776-8008; email: f_dixon@acs.org or awards@acs.org

Internet: http://portal.acs.org/portal/acs/corg/content?_nfpb=true&_pageLabel=PP_ARTICLEMAIN&node_id=1319&content_id=CTP_004509&use_sec=true&sec_url_var=region1

Sponsor: American Chemical Society
1155 Sixteenth Street, NW
Washington, DC 20036

Award for Encouraging Women into Careers in the Chemical Sciences   1263

The purpose of the Award is to recognize individuals who have significantly stimulated or fostered the interest of women in chemistry, thereby promoting their professional development as chemists or chemical engineers, and/or increasing their appreciation of chemistry as the central science. Nominees for the award may come from any professional setting: academia, industry, government, or other independent facility. The award is intended to recognize significant accomplishments by individuals in stimulating women to elect careers in the chemical sciences and engineering, and in generating a broader appreciation of chemistry as the central science. The award consists of $5,000 and a certificate. A grant of $10,000 will be made to an academic institution, designated by the recipient, to strengthen its activities in meeting the objectives of the award. Up to $1,500 for travel expenses to the meeting at which the award will be presented will be reimbursed.

Requirements: Any individual, except a member of the award committee, may submit one nomination or seconding letter for the award in any given year. Nominating documents consist of a letter of not more than 1000 words containing an evaluation of the nominee's accomplishments and a specific identification of the work to be recognized, a biographical sketch including date of birth, and a list of publications and patents authored by the nominee.

Restrictions: Self-nominations are not accepted.

Samples: Esther M. Conwell—award recipient, $5,000 (2008); Bojan H. Jennings—award recipient, $5,000 (2007); Catherine H. Middlecamp—award winner, $5,000 (2006).

Amount of Grant: $5,000 and $1,500 travel allowance to award meeting; $10,000 to an academic institution

Date(s) Application is Due: Nov 1

Contact: Felicia Dixon, Awards Administrator; (800) 227-5558 or (202) 872-4408; fax: (202) 776-8008; email: f_dixon@acs.org or awards@acs.org

Internet: http://portal.acs.org/portal/acs/corg/content?_nfpb=true&_pageLabel=PP_ARTICLEMAIN&node_id=1319&content_id=CTP_004517&use_sec=true&sec_url_var=region1

Sponsor: American Chemical Society
1155 Sixteenth Street, NW
Washington, DC 20036

Award for Research at an Undergraduate Institution   1264

The purpose of the Award is to recognize the importance of research with undergraduates. The award will honor a chemistry faculty member whose research in an undergraduate setting has achieved wide recognition and contributed significantly to chemistry and to the professional development of undergraduate students. This award recognizes fundamental research that constitutes advances in science as evidenced by refereed publications with undergraduate coauthors in leading scientific research journals, external research grant support, and the subsequent professional development of students who have participated in the research program. The award will be given for significant work over a long period of time rather than for a specific, limited project. The award consists of $5,000 and a certificate. Reasonable travel expenses to the meeting at which the award will be presented will be reimbursed. the program will also provide a grant of $5,000 directly to the recipient's institution.

Requirements: The nominee's department may offer work leading to the master's degree but cannot have a doctoral program. A nominee must be a tenured faculty member of

a predominantly undergraduate institution. Any individual, except a member of the award committee, may submit one nomination or seconding letter for the award in any given year. The nominating documents consist of a letter of not more than 1000 words containing an evaluation of the nominee's accomplishments and a specific identification of the work to be recognized, a biographical sketch including date of birth, and a list of publications and patents authored by the nominee.

Restrictions: Self-nominations are not accepted.

Samples: John T. Gupton—award recipient, $5,000 (2008); Cheryl D. Stevenson—award recipient, $5,000 (2007); Charles Beam—award recipient, $5,000 (2006).

Amount of Grant: $5,000 plus travel expenses to award meeting; $5,000 will be awarded to the recipient's institution

Date(s) Application is Due: Nov 1

Contact: Felicia Dixon, Awards Administrator; (800) 227-5558 or (202) 872-4408; fax: (202) 776-8008; email: f_dixon@acs.org or awards@acs.org

Internet: http://portal.acs.org/portal/acs/corg/content?_nfpb=true&_pageLabel=PP_ARTICLEMAIN&node_id=1319&content_id=CTP_004549&use_sec=true&sec_url_var=region1

Sponsor: American Chemical Society
1155 Sixteenth Street, NW
Washington, DC 20036

Award for Volunteer Service to the American Chemical Society   1265

The purpose of the Award is to recognize the volunteer efforts of individuals who have served the ACS, contributing significantly to the goals and objectives of the Society. The volunteerism to be recognized may comprise a variety of activities, including, but not limited to, the initiation or sponsorship of a singular endeavor or exemplary performance as a member or chair of a national level ACS committee, or as an elected division or local section officer, or outstanding service in a leadership role in regional meeting(s) or a local section. The award consists of $3,000, a certificate, and an inscription of the recipient, name on a plaque displayed at ACS Headquarters in Washington, DC.

Requirements: A nominee will have been a member of the ACS for at least 15 years, and will have made significant contributions to the ACS.

Restrictions: Past and present members of the ACS Board of Directors and staff are ineligible for the award.

Samples: Thomas L. Netzel—award recipient, $3,000 (2008); Morton Z. Hoffman—award recipient, $3,000 (2007); E. Gerald Meyer—award recipient, $3,000 (2006).

Amount of Grant: $3,000

Date(s) Application is Due: Nov 1

Contact: Felicia Dixon, Awards Administrator; (800) 227-5558 or (202) 872-4408; fax: (202) 776-8008; email: f_dixon@acs.org or awards@acs.org

Internet: http://portal.acs.org/portal/acs/corg/content?_nfpb=true&_pageLabel=PP_ARTICLEMAIN&node_id=1319&content_id=CTP_004556&use_sec=true&sec_url_var=region1

Sponsor: American Chemical Society
1155 Sixteenth Street, NW
Washington, DC 20036

Award in Analytical Chemistry   1266

This award, sponsored by Battelle Memorial Institute and administered by the American Chemical Society, is given annually to recognize and encourage outstanding contributions to the science of analytical chemistry, pure or applied, carried out in the United States or Canada. Special consideration will be given to the independence of thought and the originality shown, or to the importance of the work when applied to public welfare, economics, or the needs and desires of humanity. The award consists of $5,000 and a certificate. Reasonable travel expenses to the meeting at which the award will be presented will be reimbursed.

Requirements: A nominee must be a resident of the United States or Canada and must have made an outstanding contribution to analytical chemistry. Any individual, except a member of the award committee, may submit one nomination or seconding letter for the award in any given year. The nominating documents consist of a letter of not more than 1000 words containing an evaluation of the nominee's accomplishments and a specific identification of the work to be recognized, a biographical sketch including date of birth, and a list of publications and patents authored by the nominee.

Restrictions: Self-nominations are not accepted.

Samples: Robert Mark Wightman—award winner, $5,000 (2008); James W. Jorgenson—award winner, $5,000 (2007); Milos Novotny—award winner, $5.000 (2006).

Amount of Grant: $5,000 and travel expenses to award meeting

Date(s) Application is Due: Nov 1

Contact: Felicia Dixon, Awards Administrator; (800) 227-5558 or (202) 872-4408; fax: (202) 776-8008; email: f_dixon@acs.org or awards@acs.org

Internet: http://portal.acs.org/portal/acs/corg/content?_nfpb=true&_pageLabel=PP_ARTICLEMAIN&node_id=1319&content_id=CTP_004492&use_sec=true&sec_url_var=region1

Sponsor: American Chemical Society
1155 Sixteenth Street, NW
Washington, DC 20036

Award in Applied Polymer Science   1267

This award, sponsored by Eastman Chemical Company and administered by the American Chemical Society, is given annually to recognize and encourage outstanding achievements in the science or technology of plastics, coatings, polymer composites, adhesives, and related fields. The recipient will be selected primarily on the basis of scientific contributions made to the specified fields during the 10-year period preceding date of selection. To avoid repeating specific areas of technology, preference will be given whenever recognized by the granting of this award in the two preceding years. The award consists of $5,000 and a certificate. Reasonable travel expenses to the meeting at which the award will be presented will be reimbursed.

Requirements: Any individual, except a member of the award committee, may submit one nomination or seconding letter for the award in any given year. The nominating documents consist of a letter of not more than 1000 words containing an evaluation of the nominee's accomplishments and a specific identification of the work to be recognized, a biographical sketch including date of birth, and a list of publications and patents authored by the nominee.

Restrictions: Self-nominations are not accepted.

Samples: P. Anne Hiltner—award winner, $5,000 (2008); Harry R. Allcock—award winner, $5,000 (2007); Christopher Ober—award winner, $5,000 (2006).

Amount of Grant: $5,000 plus travel expenses to award meeting

Date(s) Application is Due: Nov 1

Contact: Felicia Dixon, Awards Administrator; (800) 227-5558 or (202) 872-4408; fax: (202) 776-8008; email: f_dixon@acs.org or awards@acs.org

Internet: http://portal.acs.org/portal/acs/corg/content?_nfpb=true&_pageLabel=PP_ARTICLEMAIN&node_id=1319&content_id=CTP_004493&use_sec=true&sec_url_var=region1

Sponsor: American Chemical Society
1155 Sixteenth Street, NW
Washington, DC 20036

Award in Chromatography   1268

This award, sponsored by SUPELCO, Inc., and administered by the American Chemical Society, is given annually to recognize outstanding contributions to the field of chromatography. The award consists of $5,000 and a certificate. Reasonable travel expenses to the meeting at which the award will be presented will be reimbursed.

Requirements: A nominee must have made an outstanding contribution to the field with particular consideration given to developments of new methods. Any individual, except a member of the award committee, may submit one nomination or seconding letter for the award in any given year. The nominating documents consist of a letter of not more than 1000 words containing an evaluation of the nominee's accomplishments and a specific identification of the work to be recognized, a biographical sketch including date of birth, and a list of publications and patents authored by the nominee.

Restrictions: Self-nominations are not accepted.

Samples: Frantisek Svec—award winner, $5,000 (2008); J. Michael Ramsey—award winner, $5,000 (2007); John Dorsey—award winner, $5000 (2006).

Amount of Grant: $5,000 and travel expenses to award meeting

Date(s) Application is Due: Nov 1

Contact: Felicia Dixon, Awards Administrator; (800) 227-5558 or (202) 872-4408; fax: (202) 776-8008; email: f_dixon@acs.org or awards@acs.org

Internet: http://portal.acs.org/portal/acs/corg/content?_nfpb=true&_pageLabel=PP_ARTICLEMAIN&node_id=1319&content_id=CTP_004500&use_sec=true&sec_url_var=region1

Sponsor: American Chemical Society
1155 Sixteenth Street, NW
Washington, DC 20036

Award in Colloid or Surface Chemistry   1269

The award, sponsored by the Procter and Gamble Company, is given annually to recognize and encourage outstanding scientific contributions to colloid or surface chemistry in the United States or Canada. Recognition will also be given to originality and independence of thought, and to the technological impact of the nominee's contribution. In odd-numbered years, awards are made for contributions in surface chemistry. In even-numbered years, awards are made for advances in colloid chemistry. The award consists of $5,000 and a certificate.

Requirements: The nominee must be a resident of the United States or Canada and must have made outstanding scientific contributions to colloid or surface chemistry. Any individual, except a member of the award committee, may submit one nomination or seconding letter for the award in any given year. The nominating documents consist of a letter of not more than 1000 words containing an evaluation of the nominee's accomplishments and a specific identification of the work to be recognized, a biographical sketch including date of birth, and a list of publications and patents authored by the nominee.

Restrictions: Self-nominations are not accepted.

Samples: Lee R. White—award winner, $5,000 (2008); William B. Russel—award winner, $5,000 (2007); Alice Gast—award winner, $5,000 (2006).

Amount of Grant: $5,000 and travel to award meeting

Date(s) Application is Due: Nov 1

Contact: Felicia Dixon, Awards Administrator; (800) 227-5558 or (202) 872-4408; fax: (202) 776-8008; email: f_dixon@acs.org or awards@acs.org

Internet: http://portal.acs.org/portal/acs/corg/content?_nfpb=true&_pageLabel=PP_ARTICLEMAIN&node_id=1319&content_id=CTP_004502&use_sec=true&sec_url_var=region1

Sponsor: American Chemical Society
1155 Sixteenth Street, NW
Washington, DC 20036

Award in Industrial Chemistry   1270

The award, sponsored by ACS Division of Business Development and Management, is given annually to recognize outstanding contributions to industrial chemistry resulting in the commercialization of an economically significant new product or process. Any field of chemical, chemical engineering, or biochemical research is appropriate if it is of general interest and reflects the concerns of modern society. Any chemical research, whether industrial, governmental, or academic, is eligible, provided the work was done in North America and yielded significant commercial results for a period of more than one year. The award consists of $5,000 and a certificate.

Requirements: Any individual, except a member of the award committee, may submit one nomination or seconding letter for the award in any given year. The nominating documents consist of a letter of not more than 1000 words containing an evaluation of the nominee's accomplishments and a specific identification of the work to be recognized, a biographical sketch including date of birth, and a list of publications and patents authored by the nominee.

Restrictions: Self-nominations are not accepted.

Samples: Timothy J. Wallington—award winner, $5,000 (2008); Margaret M. Wu—award winner, $5,000 (2007); James Stevens—award winner, $5,000 (2006).

Amount of Grant: $5,000 and travel to award meeting

Date(s) Application is Due: Nov 1

Contact: Felicia Dixon, Awards Administrator; (800) 227-5558 or (202) 872-4408; fax: (202) 776-8008; email: f_dixon@acs.org or awards@acs.org

Internet: http://portal.acs.org/portal/acs/corg/content?_nfpb=true&_pageLabel=PP_ARTICLEMAIN&node_id=1319&content_id=CTP_004531&use_sec=true&sec_url_var=region1

Sponsor: American Chemical Society
1155 Sixteenth Street, NW
Washington, DC 20036

Award in Inorganic Chemistry   1271

This award, sponsored by the Aldrich Chemical Company and administered by the American Chemical Society, is given annually to recognize and encourage fundamental research in the field of inorganic chemistry. A nominee must have accomplished outstanding research in the preparation, properties, reactions, or structure of inorganic substances. Special consideration shall be given to the independence of thought and originality shown. The award consists of $5,000 and a certificate.

Requirements: Any individual, except a member of the award committee, may submit one nomination or seconding letter for the award in any given year. The nominating documents consist of a letter of not more than 1000 words containing an evaluation of the nominee's accomplishments and a specific identification of the work to be recognized, a biographical sketch including date of birth, and a list of publications and patents authored by the nominee.

Restrictions: Self-nominations are not accepted.

Samples: Kenneth N. Raymond—award winner, $5,000 (2008); Sheldon G. Shore—award winner, $5,000 (2007); Karl Wieghardt—award winner, $5,000 (2006).

Amount of Grant: $5,000 and up to $1,000 for travel to award meeting

Date(s) Application is Due: Nov 1

Contact: Felicia Dixon, Awards Administrator; (800) 227-5558 or (202) 872-4408; fax: (202) 776-8008; email: f_dixon@acs.org or awards@acs.org

Internet: http://portal.acs.org/portal/acs/corg/content?_nfpb=true&_pageLabel=PP_ARTICLEMAIN&node_id=1319&content_id=CTP_004532&use_sec=true&sec_url_var=region1

Sponsor: American Chemical Society
1155 Sixteenth Street, NW
Washington, DC 20036

Award in Organometallic Chemistry   1272

The Award, sponsored by the Dow Chemical Company Foundation and administered by the American Chemical Society, is given annually to recognize a recent advancement that is having a major impact on research in organometallic chemistry. A nominee must have shown outstanding research in the preparation, reactions, properties, or structure of organometallic substances. Special consideration will be given to demonstrated creativity and independence of thought. The award consists of $5,000 and a certificate.

Requirements: Preference will be given to U.S. citizens. Any individual, except a member of the award committee, may submit one nomination or seconding letter for the award in any given year. The nominating documents consist of a letter of not more than 1000 words containing an evaluation of the nominee's accomplishments and a specific identification of the work to be recognized, a biographical sketch including date of birth, and a list of publications and patents authored by the nominee.

Restrictions: Self-nominations are not accepted.

Samples: Gerard F.R. Parkin—award winner, $5,000 (2008); David Milstein—award winner, $5,000 (2007); John Hartwig—award winner, $5,000 (2006).

Amount of Grant: $5,000 plus $1,000 toward travel to award meeting

Date(s) Application is Due: Nov 1

Contact: Felicia Dixon, Awards Administrator; (800) 227-5558 or (202) 872-4408; fax: (202) 776-8008; email: f_dixon@acs.org or awards@acs.org

Internet: http://portal.acs.org/portal/acs/corg/content?_nfpb=true&_pageLabel=PP_ARTICLEMAIN&node_id=1319&content_id=CTP_004542&use_sec=true&sec_url_var=region1

Sponsor: American Chemical Society
1155 Sixteenth Street, NW
Washington, DC 20036

Award in Polymer Chemistry   1273

The Award, supported by the ExxonMobil Chemical Company, is intended to recognize outstanding contributions to polymer chemistry. The award consists of $5,000 and a certificate. Up to $1,000 for travel expenses to the meeting at which the award will be presented will be reimbursed.

Requirements: The award will be granted without regard to age or nationality. Any individual, except a member of the award committee, may submit one nomination or seconding letter for the award in any given year. The nominating documents consist of a letter of not more than 1000 words containing an evaluation of the nominee's accomplishments and a specific identification of the work to be recognized, a biographical sketch including date of birth, and a list of publications and patents authored by the nominee.

Restrictions: Self-nominations are not accepted.

Samples: James E. McGrath—award winner, $5,000 (2008); Ludwik Leibler—award winner, $5,000 (2007); Egbert W. Meijer—award winner, $5,000 (2006).

Amount of Grant: $5,000 and travel expenses up to $1,000

Date(s) Application is Due: Nov 1

Contact: Felicia Dixon, Awards Administrator; (800) 227-5558 or (202) 872-4408; fax: (202) 776-8008; email: f_dixon@acs.org or awards@acs.org

Internet: http://portal.acs.org/portal/acs/corg/content?_nfpb=true&_pageLabel=PP_ARTICLEMAIN&node_id=1319&content_id=CTP_004544&use_sec=true&sec_url_var=region1

Sponsor: American Chemical Society
1155 Sixteenth Street, NW
Washington, DC 20036

Award in Pure Chemistry 1274

This award, sponsored by Alpha Chi Sigma Fraternity and the Alpha Chi Sigma Educational Award Foundation, is given annually to recognize and encourage fundamental research in pure chemistry carried out in North America by young men and women. Special consideration is given to independence of thought and the originality shown in the research. The award consists of $5,000 and a certificate. Up to $1,000 for travel expenses to the meeting at which the award will be presented will be reimbursed.

Requirements: A nominee must have been born after April 30, 1973 and must have accomplished research of unusual merit for an individual on the threshold of her or his career. Any individual, except a member of the award committee, may submit one nomination or seconding letter for the award in any given year. The nominating documents consist of a letter of not more than 1000 words containing an evaluation of the nominee's accomplishments and a specific identification of the work to be recognized, a biographical sketch including date of birth, and a list of publications and patents authored by the nominee.

Restrictions: Self-nominations are not accepted.

Samples: Rustem F. Ismagilov—award winner. $5,000 (2008); Xiaowei Zhuang—award winner, $5,000 (2007); David Liu—award winner, $5,000 (2006).

Amount of Grant: $5,000, plus $1,000 toward travel to awards meeting

Date(s) Application is Due: Nov 1

Contact: Felicia Dixon, Awards Administrator; (800) 227-5558 or (202) 872-4408; fax: (202) 776-8008; email: f_dixon@acs.org or awards@acs.org

Internet: http://portal.acs.org/portal/acs/corg/content?_nfpb=true&_pageLabel=PP_ARTICLEMAIN&node_id=1319&content_id=CTP_004546&use_sec=true&sec_url_var=region1

Sponsor: American Chemical Society
1155 Sixteenth Street, NW
Washington, DC 20036

Award in Separations Science and Technology 1275

This award is given annually to recognize outstanding accomplishments in fundamental or applied research directed to separations science and technology. The award shall be granted to an individual without regard to age or nationality. The scope of the award is to be as broad as possible, covering all fields where separations science and technology is practiced, including but not limited to biology, chemistry, engineering, geology, and medicine.

Requirements: Any individual, except a member of the award committee, may submit one nomination or seconding letter for the award in any given year. The nominating documents consist of a letter of not more than 1000 words containing an evaluation of the nominee's accomplishments and a specific identification of the work to be recognized, a biographical sketch including date of birth, and a list of publications and patents authored by the nominee.

Restrictions: Self-nominations are not accepted.

Samples: Abraham M. Lenhoff—award recipient, $5,000 (2009); Allan S. Myerson—award winner, $5,000 (2008); William H. Pirkle—award winner, $5,000 (2004).

Amount of Grant: $5,000 and reimbursement of travel expenses to award meeting

Date(s) Application is Due: Nov 1

Contact: Felicia Dixon, Awards Administrator; (800) 227-5558 or (202) 872-4408; fax: (202) 776-8008; email: f_dixon@acs.org or awards@acs.org

Internet: http://portal.acs.org/portal/acs/corg/content?_nfpb=true&_pageLabel=PP_ARTICLEMAIN&node_id=1319&content_id=CTP_004552&use_sec=true&sec_url_var=region1

Sponsor: American Chemical Society
1155 Sixteenth Street, NW
Washington, DC 20036

Award in the Chemistry of Materials 1276

This award, supported by E.I. du Pont de Nemours and Company, is given annually to recognize and encourage creative work in the chemistry of materials. A nominee must

have made outstanding contributions to the chemistry of materials. Particular emphasis will be placed on research relating to materials of actual or potential technological importance, where a fundamental understanding of the chemistry associated with materials preparation, processing, or use is critical. The award consists of $5,000 and a certificate.

Requirements: Any individual, except a member of the award committee, may submit one nomination or seconding letter for the award in any given year. The nominating documents consist of a letter of not more than 1000 words containing an evaluation of the nominee's accomplishments and a specific identification of the work to be recognized, a biographical sketch including date of birth, and a list of publications and patents authored by the nominee.

Restrictions: Self-nominations are not accepted.

Samples: Thomas E. Mallouk—award winner, $5,000 (2008); Robert S. Langer—award winner, $5,000 (2007); Stephen T. Wilson—award winner, $5000 (2006).

Amount of Grant: $5,000 and reimbursement of travel expenses to award meeting

Date(s) Application is Due: Nov 1

Contact: Felicia Dixon, Awards Administrator; (800) 227-5558 or (202) 872-4408; fax: (202) 776-8008; email: f_dixon@acs.org or awards@acs.org

Internet: http://portal.acs.org/portal/acs/corg/content?_nfpb=true&_pageLabel=PP_ARTICLEMAIN&node_id=1319&content_id=CTP_004499&use_sec=true&sec_url_var=region1

Sponsor: American Chemical Society
1155 Sixteenth Street, NW
Washington, DC 20036

Award in Theoretical Chemistry 1277

The purpose of this annual award, which is sponsored by IBM Corporation, is to recognize innovative research in theoretical chemistry that either advances theoretical methodology or contributes to new discoveries about chemical systems. Emphasis in the selection process will be on work characterized by depth, originality, and scientific significance. The award is given without regard to age or nationality, and consists of $5,000 and a certificate.

Requirements: Any individual, except a member of the award committee, may submit one nomination or seconding letter for each award in any given year. The letter of nomination may not exceed 1000 words and should contain an evaluation of the nominee's accomplishments and a specific identification of the work to be recognized.

Restrictions: Self-nominations are not accepted.

Samples: William A. Goddard—award winner, $5,000 (2008); Rodney J. Bartlett—award winner, $5,000 (2007); Hans C. Andersen—award winner, $5,000 (2006).

Amount of Grant: $5,000 plus travel to award meeting

Date(s) Application is Due: Nov 1

Contact: Felicia Dixon, Awards Administrator; (800) 227-5558 or (202) 872-4408; fax: (202) 776-8008; email: f_dixon@acs.org or awards@acs.org

Internet: http://portal.acs.org/portal/acs/corg/content?_nfpb=true&_pageLabel=PP_ARTICLEMAIN&node_id=1319&content_id=CTP_004555&use_sec=true&sec_url_var=region1

Sponsor: American Chemical Society
1155 Sixteenth Street, NW
Washington, DC 20036

AWHONN Research Grants 1278

The AWHONN small grants program is designed primarily for novice researchers, serving first-time efforts or beginning development of a program of research. The purpose of the funding is to provide seed money, pilot funding, or total funding for small projects with promising contributions to nursing knowledge in clinical practice. The focus of the research is women's health, obstetric, or neonatal nursing phenomena.

Requirements: Applicants must be members of the association. An application for membership may accompany the proposal.

Restrictions: Primary investigators of current federally funded research and of previous AWHONN research are not eligible. Grant funds may not be used for indirect costs, tuition, computer hardware/printers.

Amount of Grant: $2500-$10,000

Date(s) Application is Due: Nov 1

Contact: Daniella McCarthy, Research Program Manager, (202) 261-2434; fax: (202) 728-0575; email: ResearchPrograms@awhonn.org or daniellam@awhonn.org

Internet: http://www.awhonn.org/awhonn/?pg=0-874-2240

Sponsor: Association of Women's Health, Obstetric, and Neonatal Nurses
2000 L Street NW, Suite 740
Washington, DC 20036

AWIS Educational Foundation Predoctoral Awards 1279

Grants provide funds for women graduate students enrolled in a program in the life, physical, or social sciences or in an engineering program leading to a PhD degree. The award can be used for any aspect of education—tuition, books, housing, research, expenses, equipment, and so forth. U.S. citizens may use the money for study in the United States or abroad; non-U.S. citizens must be enrolled in a U.S. institution of higher education. Each year regular awards and named awards are granted; at times, citations of merit also are awarded.

Requirements: Women students enrolled in any life, physical, or social science or in an engineering program leading to a PhD degree are eligible. Candidates should have passed all qualifying exams and expect to complete the dissertation within two years.

Restrictions: Students pursuing a professional doctorate other than the PhD are not eligible to apply.

Amount of Grant: $1000 grant
Date(s) Application is Due: Jan 26
Contact: Dr. Barbara Filner, Foundation President, (202) 326-8940; fax: (202) 326-8960; email: awisedfd@awis.org
Internet: http://www.awis.org/resource/edfoundation.html
Sponsor: Association for Women in Science Educational Foundation
1200 New York Avenue, Suite 650
Washington, DC 20005

AWM Mentoring Travel Grants for Women   1280

The objective of the program is to help junior women develop a long-term working and mentoring relationship with a senior mathematician. This relationship should help the junior mathematician to establish her research program and eventually receive tenure. Grants support travel, subsistence, and other required expenses for an untenured woman mathematician to travel to an institute or a department to do research with a specified individual for one month. Any unexpended funds could be used for further travel to work with the same individual during the following year. Applicants may receive up to three grants through their careers, possibly in successive years. Submit five copies of the cover letter; curriculum vita; research proposal of approximately five pages in length that specifies why the proposed travel is particularly beneficial; supporting letter from the proposed mentor (who must promise to be available at the time of the proposed travel and may be either a man or a woman), together with the curriculum vita of the proposed mentor; approximate budget; and information about other sources of funding available.
Requirements: Applicants must be women holding a doctorate or equivalent experience and with a work address in the United States (or home address in case of unemployed). The applicant's research may be in any field that is funded by the Division of Mathematical Sciences of NSF.
Amount of Grant: $4000 maximum
Date(s) Application is Due: Feb 1
Contact: Association for Women in Mathematics, Travel Grant Selection Committee, (703) 934-0163; email: awm@awm-math.org
Internet: http://awm-math.org/travelgrants.html
Sponsor: University of Maryland
4114 Computer and Space Sciences Bldg
College Park, MD 20742-2461

AWM Travel Grants for Women Researchers   1281

This program is supported by NSF and the Association for Women in Mathematics. Grants provide full or partial support to women researchers for travel to attend research conferences in their fields of specialization. An applicant should send five copies of a description of her current research and how the proposed travel will benefit her research program; curriculum vita; budget for the proposed travel; and information about all other sources of travel funding. In a cover letter, indicate the conference name, dates, and location (city/state/country). There are three award periods per year.
Requirements: The research conference must be in an area supported by the Division of Mathematical Sciences of NSF. Applicants must be women holding a doctorate (or equivalent experience) and with a work address in the United States.
Restrictions: Individuals who were awarded an AWM-NSF travel grant in the past two years, or who have other sources of external funding, such as any type of NSF grant, are ineligible. Partial support from the applicant's institution or from a nongovernmental agency does not make the applicant ineligible.
Amount of Grant: $1500 maximum for domestic travel; $2000 maximum for foreign travel
Date(s) Application is Due: Feb 1; May 1; Oct 1
Contact: Association for Women in Mathematics, (703) 934-0163; email: awm@awm-math.org
Internet: http://awm-math.org/travelgrants.html
Sponsor: University of Maryland
4114 Computer and Space Sciences Bldg
College Park, MD 20742-2461

AWRA Richard A. Herbert Memorial Scholarship   1282

This scholarship is to be used for the enhancement of education in water resources. One scholarship is awarded to a full-time undergraduate student working toward his/her first undergraduate degree and enrolled in a program related to water resources. The second scholarship is awarded to a full-time graduate student enrolled in a program related to water resources. In addition to the scholarships, each winner receives a complimentary membership in AWRA for the year.
Requirements: Full-time graduate and undergraduate students enrolled in a program related to water resources are eligible to apply.
Amount of Grant: $2000 maximum
Date(s) Application is Due: Apr 29
Contact: Harriette Bayse, (540) 687-8390; fax: (540) 687-8395; email: harriette@awra.org
Internet: http://www.awra.org/student/herbert.html
Sponsor: American Water Resources Association
P.O. Box 1626, 4 W Federal Street
Middleburg, VA 20118

AWWA Abel Wolman Doctoral Fellowships   1283

The fellowships are designed to encourage promising students to pursue advanced training and research in the field of water supply and treatment. Applicants are evaluated on the basis of the quality of their academic record, the significance of the proposed research to water supply and treatment, and the applicant's potential to perform high-quality research.
Requirements: Applicants must be citizens of Canada, Mexico, or the United States and anticipate completion of the requirements for the PhD degree within two years of the award.
Amount of Grant: $20,000 maximum annually
Date(s) Application is Due: Jan 15
Contact: Stepanie Wheeler, Scholarship Coordinator, (303) 347-6206; email: swheeler@awwa.org
Internet: http://www.awwa.org/About/scholars
Sponsor: American Water Works Association
6666 W Quincy Avenue
Denver, CO 80235

AWWA Academic Achievement Awards   1284

The awards recognize outstanding academic achievement on a graduate level through the selection of excellent master's theses and doctoral dissertations that relate to the field of potable water. First- and second-place awards are given for master's theses and also for doctoral dissertations. An announcement and form must be obtained from the training programs manager prior to submitting an application.
Requirements: All master's theses and doctoral dissertations that are relevant to the water supply industry are eligible. Applicants must have completed their theses or dissertations within the specified year.
Amount of Grant: $3000 each, doctoral dissertation and master's thesis; $1500 for second place in each category
Date(s) Application is Due: Oct 1
Contact: Linda Moody, Scholarship Coordinator, (303) 347-6201; fax: (303) 347-0804; email: swheeler@awwa.org
Internet: http://www.awwa.org/About/scholarships.cfm
Sponsor: American Water Works Association
6666 W Quincy Avenue
Denver, CO 80235

AWWA Holly A. Cornell Scholarship   1285

The annual scholarship is made to encourage master's-level female and/or minority students to pursue advanced training and careers in the field of water supply and treatment. The recipient is expected to complete all requirements for the MS degree no sooner than December 1 of the year of the award.
Requirements: Female and/or minority (as defined by the Equal Employment Opportunity Commission) U.S. citizens who are currently master's degree students and anticipate completion of the requirements for a master's degree in engineering no sooner than December 1 of the year of the award are eligible to apply.
Amount of Grant: $5000
Date(s) Application is Due: Jan 15
Contact: Linda Moody, Scholarship Coordinator, (303) 347-6201; email: swheeler@awwa.org
Internet: http://www.awwa.org/About/scholarships.cfm
Sponsor: American Water Works Association
6666 W Quincy Avenue
Denver, CO 80235

AWWA Larson Aquatic Research Support (LARS) Scholarships   1286

These scholarships, given in honor of Dr. Thurston E. Larson, are intended to provide encouragement to outstanding graduate students in the fields of water treatment; aquatic, analytic, and environmental chemistry; and corrosion control. Guidelines and application are available online.
Requirements: Applicant must be pursuing a degree (MS or PhD) at an institution of higher learning in the United States, Canada, Puerto Rico, or Mexico. MS candidates must complete the requirements of their degree sometime after August 1 of the year of the award, and PhD candidates must complete degree requirements sometime after December 1 of the year of the award.
Amount of Grant: $5000 for MS candidates; $7000 for PhD candidates
Date(s) Application is Due: Jan 15
Contact: Linda Moody, Scholarship Coordinator, (303) 347-6201; email: swheeler@awwa.org
Internet: http://www.awwa.org/About/scholarships.cfm
Sponsor: American Water Works Association
6666 W Quincy Avenue
Denver, CO 80235

AWWA Thomas R. Camp Scholarship   1287

The annual scholarship is presented to an outstanding graduate student doing applied research in the drinking water field and planning a career in this field. The scholarship is awarded to master's-level students in odd-numbered years and to doctoral-level students in even-numbered years. The recipient should not anticipate completion of his/her degree program before May of the year following the award.
Requirements: Students pursuing a graduate degree at an institution of higher education located in Canada, Guam, Puerto Rico, Mexico, or the United States are eligible.
Amount of Grant: $5000
Date(s) Application is Due: Jan 15
Contact: Linda Moody, Scholarship Coordinator, (303) 347-6201; email: swheeler@awwa.org

Internet: http://www.awwa.org/About/scholarships.cfm
Sponsor: American Water Works Association
6666 W Quincy Avenue
Denver, CO 80235

Axe-Houghton Foundation Grants 1288

The foundation operates in the New York City metropolitan area exclusively for charitable, educational, and scientific purposes to foster and encourage an appreciation of the English language, with major emphasis on the spoken language. Priority is given to projects for the improvement of speech and its uses in the areas of public affairs, education, theater, poetry, debate, and the oral interpretation of literature. A portion of available funds is devoted to speech remediation and to scientific research pertaining to speech. Types of support include program grants and seed money grants.
Requirements: Initial inquiries should be by full proposal and should contain a brief statement of objective, duration of time required to complete this objective, qualifications of personnel responsible for conducting the project, and the general amounts of funds needed. The proposal should also indicate the degree of internal support that would be committed.
Restrictions: Grants are made only to tax-exempt institutions and not to individuals, private foundations or organizations outside of the United States. Grants are not made for scholarships, fellowships, capital, or general support programs.
Geographic Focus: New York
Amount of Grant: $1000-$7500
Date(s) Application is Due: Sep 1
Contact: Claire Brook, President, (212) 909-8304
Internet: http://www.foundationcenter.org/grantmaker/axehoughton/index.html
Sponsor: Axe-Houghton Foundation
919 3rd Avenue, 2nd Fl
New York, NY 10022

BA British Conference Grants 1289

These awards are intended as contributions toward the expenses of conferences held in the United Kingdom. Awards may be used for any aspect of the conference costs, but the awarding committee, in reaching its decisions, will pay particular attention to the overall budget for the conference. Applications should be submitted on the British Conference Grant form, available from university research offices, or from the academy's Research Grants Department. Applications should not be submitted more than one year before the date of the conference. Annual deadline dates may vary; contact program staff for exact dates.
Requirements: Applications may be submitted by British academics or organizations responsible for organizing conferences in Britain.
Restrictions: Applications for support for one-day conferences will not normally be accepted. Applications on behalf of bilateral conferences between two institutions will not be considered unless it is shown that there will be open and widespread national participation.
Amount of Grant: @L500-@L2000
Date(s) Application is Due: Jan 15; Apr 15; Oct 15
Contact: Research Grants Department, 020 7969 5217; fax: 020 7969 5414; email: grants@britac.ac.uk
Internet: http://www.britac.ac.uk/funding/guide/bcg.html
Sponsor: British Academy
10 Carlton House Terrace
London SW1Y 5AH United Kingdom

BA Central and Eastern Europe and the Former Soviet Union Exchange Grants 1290

The primary purpose of the exchange programs is to facilitate scholarly research in the humanities and social sciences with academies in Central and Eastern Europe and the former Soviet Union. The academy will meet the British scholar's travel expenses to the receiving country. The host institution will provide accommodation, a maintenance allowance, and local travel expenses. Applicants should bear in mind that the maintenance allowance may not cover all necessary expenses. The academy will not undertake to supplement the allowance, but applications may be made to other sources of support. No contribution can be made by either side towards the expenses of accompanying partners, although host institutions may be asked to facilitate their visit. Application must be made on the East Europe Exchanges application form, available from the academy.
Requirements: Applicants must be ordinarily resident in the United Kingdom, the Isle of Man, or the Channel Islands. Most awards are made to staff employed in universities and other institutions of higher education, but applicants are not restricted by either academic or employment status.
Restrictions: Awards are not available for the completion of doctoral research projects, nor for the support of courses of study leading to professional qualifications.
Amount of Grant: @L2500
Date(s) Application is Due: Apr 30; Sep 31; Dec 31
Contact: International Relations Department, 020 7969 5220; fax: 020 7969 5414; email: overseas@britac.ac.uk
Internet: http://www.britac.ac.uk/funding/guide/intl/ceefsujp.html
Sponsor: British Academy
10 Carlton House Terrace
London SW1Y 5AH United Kingdom

BA Elie Kedourie Memorial Fund Research Grants 1291

The object of this fund, established by the family of Elie Kedourie, FBA, is to promote the study of Middle Eastern and modern European history, and the history of political thought by recent postdoctoral scholars of any nationality. Applications should be submitted on the standard research grant application form, available from university research offices, or from the academy. Applicants will be notified approximately three months after the closing date.
Restrictions: Funds are not available to support travel to or attendance at conferences, workshops, or seminars, either in the United Kingdom or elsewhere.
Amount of Grant: @L1000 maximum typically
Date(s) Application is Due: Apr 15
Contact: Research Grants Department, 020 7969 5217; fax: 020 7969 5414; email: grants@britac.ac.uk
Internet: http://www.britac.ac.uk/funding/guide/ekmf.html
Sponsor: British Academy
10 Carlton House Terrace
London SW1Y 5AH United Kingdom

BA Elisabeth Barker Fund 1292

The fund was established in memory of Elisabeth Barker (1910-1986), diplomatic correspondent and historian of modern Europe. It is intended to support studies in recent European history, particularly the history of central and eastern Europe. Grants may be made to support individual, collective, or institutional projects (including conferences). Awards to individuals may take the form of grants for private study and research in other European countries. British scholars and institutions may also apply for funds to assist scholars from other European countries to visit the United Kingdom, or to extend a stay arranged under other auspices. Application is made by letter, with appropriate supporting documentation, including a curriculum vita, a detailed research proposal describing the project and the scheme of research, a breakdown of the costs, and details of the plans for publication or dissemination of the research. Two referees, from outside the employing institution of the applicant, should be asked to submit references on the applicant and the project direct to the academy. Applicants will be notified of the outcome approximately three months after the closing date.
Requirements: Applicants must be ordinarily resident in the United Kingdom, the Isle of Man, or the Channel Islands. Most awards are made to staff employed in universities and other institutions of higher education, but applicants are not restricted by either academic or employment status.
Restrictions: PhD candidates are not eligible to apply, whether or not the project is related to the topic of their thesis. Awards are not available for the support of courses of study leading to professional qualifications.
Amount of Grant: @L1000 maximum typically
Date(s) Application is Due: Jan 15; Apr 15; Oct 15
Contact: Research Grants Department, 020 7969 5217; fax: 020 7969 5414; email: grants@britac.ac.uk
Internet: http://www.britac.ac.uk/funding/guide/intl/ebf.html
Sponsor: British Academy
10 Carlton House Terrace
London SW1Y 5AH United Kingdom

BA Exchange Grants 1293

The academy provides opportunities, through exchange agreements, for British scholars to undertake approved programs of research in the humanities and social sciences in certain countries overseas. The programs vary in nature and structure, from formal agreements, mainly with the countries of Central and Eastern Europe, the former Soviet Union and China, to less formal arrangements with individual institutions in particular countries, including France, Israel, Japan, Korea, Sweden, Switzerland, and the United States. Contact the academy or visit the Web site for details.
Requirements: Applicants must be ordinarily resident in the United Kingdom, the Isle of Man, or the Channel Islands. Most awards are made to staff employed in universities and other institutions of higher education, but applicants are not restricted by either academic or employment status.
Restrictions: Awards are not available for the completion of doctoral research projects, nor for the support of courses of study leading to professional qualifications.
Date(s) Application is Due: Jan 15; Apr 15; Oct 15
Contact: International Relations Department, 020 7969 5220; fax: 020 7969 5414; email: overseas@britac.ac.uk
Internet: http://www.britac.ac.uk/funding/guide/research.html
Sponsor: British Academy
10 Carlton House Terrace
London SW1Y 5AH United Kingdom

BA Larger Research Grants 1294

Grants are available for primary research in the humanities and social sciences for which support is not normally available from other funding agencies. Eligible costs include travel and maintenance, research assistance, workshops, consumables, specialist software, and costs of interpreters in the field. Applications for collaborative or individual research projects are equally welcome. Applications from international groups of scholars are welcome, provided there is a UK-based scholar as lead applicant. The maximum grant period is three years. Guidelines are available online.
Restrictions: The following items are not eligible for support: institutional overheads; computer hardware; books and other permanent resources; the preparation of camera-ready copy; any editorial-related task; publication subventions; costs of publication in electronic media; payment to the principal researcher(s) in lieu of salary, or for personal

maintenance at home; replacement teaching costs; travel and maintenance expenses for purposes such as lecture tours, or to write up the results of research; or attendance at conferences or organization of conferences either in the UK or abroad to disseminate the results of research.
Amount of Grant: @L15,000-@L100,000
Date(s) Application is Due: Oct 15
Contact: Research Grants Department, 020-7969 5217; fax: 020-7969 5414; email: grants@britac.ac.uk
Internet: http://www.britac.ac.uk/funding/guide/lrg.html
Sponsor: British Academy
10 Carlton House Terrace
London SW1Y 5AH United Kingdom

BA Neil Ker Memorial Fund Research Grants 1295

The object of this fund, established by the family and friends of Neil Ker, FBA, is to promote the study of Western medieval manuscripts—and in particular, those of British interest—by means of awards in support of all aspects of research, including travel and publication, made to both younger and established scholars of any nationality. Applications should be submitted on the standard research grant application form, available from university research offices, or from the academy. Applicants will be notified approximately three months after the closing date.
Requirements: Younger and established scholars of any nationality are eligible.
Restrictions: The program does not fund travel to or attendance at conferences, workshops, or seminars, either in the United Kingdom or elsewhere.
Amount of Grant: @L2000 maximum typically
Date(s) Application is Due: Apr 15
Contact: Research Grants Department, 020 7969 5217; fax: 020 7969 5414; email: grants@britac.ac.uk
Internet: http://www.britac.ac.uk/funding/guide/nkmf.html
Sponsor: British Academy
10 Carlton House Terrace
London SW1Y 5AH United Kingdom

BA Postdoctoral Fellowships 1296

This fellowship is designed to enable outstanding younger scholars to obtain experience of research and teaching in the university environment, which will strengthen their curriculum vita and improve their prospects of obtaining permanent posts by the end of the fellowship. Awards will be tenable for three years (not renewable) and are offered with a salary starting at spine point six on the University Lecturer's Grade A scale or equivalent scale. Applicants nominate the institution at which they wish to hold their award; that institution will be asked to give its consent before any award made by the academy is confirmed. Fellows will be employees of their host institution and will be subject to the terms and conditions of that institution. Applications should be made on the postdoctoral fellowship application form, available from the academy.
Requirements: These awards are open to scholars in the humanities and social sciences. Eligible applicants must be ordinarily resident in the United Kingdom; must be under the age of 30 at time of award (but special factors may be taken into account and this discretionary criterion set aside); and must have obtained or expect to obtain their doctorate by the award year.
Restrictions: Applicants must not have held an established teaching post in an institution of higher education.
Date(s) Application is Due: Feb 28
Contact: Dr. Ken Emond, Assistant Secretary (Research Appointments), 020 7969 5265; fax: 020 7969 5414; email: posts@britac.ac.uk if
Internet: http://www.britac.ac.uk/funding/guide/pdfells.html
Sponsor: British Academy
10 Carlton House Terrace
London SW1Y 5AH United Kingdom

BA Research Readerships 1297

These awards are aimed at established scholars in United Kingdom universities who are in midcareer, and are designed to allow the award holders to undertake or to complete an approved program of sustained research, while relieved of their normal teaching and administrative commitments. The principal purpose of the awards is to enable the completion of a major piece of research which will not only be an important contribution to knowledge and understanding but also help to enhance the future career and career prospects of the award holder. Grants are made direct to the award holder's employing institution to cover the costs of replacement teaching. Readerships are tenable for two years.
Requirements: Readerships are open to scholars in the humanities and social sciences. Applicants are expected normally to be aged 55 or under at time of application unless they can show that exceptional circumstances have delayed the advancement of their academic career. Awardees will be expected to be able to disseminate the results of their research not only through publications, but also through the teaching of future courses after the end of their awards.
Date(s) Application is Due: Sep 30
Contact: Research Appointments Department, 020 7969 5265; fax: 020 7969 5414; email: posts@britac.ac.uk
Internet: http://www.britac.ac.uk/funding/guide/readfell.html
Sponsor: British Academy
10 Carlton House Terrace
London SW1Y 5AH United Kingdom

BA Senior Research Fellowships 1298

These awards are aimed at established scholars in United Kingdom universities who are in midcareer, and are designed to allow the award holders to undertake or to complete an approved program of sustained research, while relieved of their normal teaching and administrative commitments. The principal purpose of the awards is to enable the completion of a major piece of research that will not only be an important contribution to knowledge and understanding but also help to enhance the future career and career prospects of the award holder. Grants are made direct to the award holder's employing institution to cover the costs of replacement teaching. Fellowships are tenable for one year.
Requirements: Fellowships are open to scholars in the humanities and social sciences. Applicants are expected normally to be aged 55 or under at time of application unless they can show that exceptional circumstances have delayed the advancement of their academic career. Awardees will be expected to be able to disseminate the results of their research not only through publications, but also through the teaching of future courses after the end of their awards.
Date(s) Application is Due: Sep 30
Contact: Research Appointments Department, 020 7969 5265; fax: 020 7969 5414; email: posts@britac.ac.uk
Internet: http://www.britac.ac.uk/funding/guide/readfell.html
Sponsor: British Academy
10 Carlton House Terrace
London SW1Y 5AH United Kingdom

BA Sino-British Fellowship Trust 1299

The academy is able to bid for funds from the Sino-British Fellowship Trust (SBFT) to support individual or cooperative research projects, which may be conducted either in Britain or in China, or in both countries, and must involve person-to-person contacts. Applications should first be made to the academy; if approved, the bids will be submitted to the SBFT. Advice on the nature and structure of applications should be sought from the academy's International Relations Department. The academy will be unable to offer support should the SBFT decline to offer funding.
Amount of Grant: @L10,000 maximum
Date(s) Application is Due: Jan 15; Apr 15; Oct 15
Contact: International Relations Department, 020 7969 5220; fax: 020 7969 5414; email: overseas@britac.ac.uk
Internet: http://www.britac.ac.uk/funding/guide/intl/sbft.html
Sponsor: British Academy
10 Carlton House Terrace
London SW1Y 5AH United Kingdom

BA Sir Ernest Cassel Educational Trust Fund Grants 1300

The academy administers funds on behalf of the Sir Ernest Cassel Educational Trust, to promote research in any field falling within the humanities or social sciences. Awards are offered to support travel costs relating to a research project, and are particularly aimed at recent postdoctoral scholars. Awards for travel costs rarely exceed @L1000; however, an application to this fund for travel costs may be combined with an application for a personal research grant for other costs, up to a total of @L7500. Applications should be submitted on the standard research grant application form, available from university research offices, or from the academy. In cases where an applicant seeks both travel and other costs, a single application form should be used. Applicants should not complete two separate forms when submitting a dual application to the Cassel fund for travel and to the research grants fund for other elements of a project. Applicants will be notified of the outcome approximately three months after the closing date.
Requirements: Applicants must be ordinarily resident in the United Kingdom, the Isle of Man, or the Channel Islands. Most awards are made to staff employed in universities and other institutions of higher education, but applicants are not restricted by either academic or employment status. These awards are aimed particularly at recent postdoctoral scholars.
Amount of Grant: @L1000 maximum; @L7500 maximum for dual applications
Date(s) Application is Due: Apr 15
Contact: Research Grants Department, 020 7969 5217; fax: 020 7969 5414; email: grants@britac.ac.uk
Internet: http://www.britac.ac.uk/funding/guide/cetf.html
Sponsor: British Academy
10 Carlton House Terrace
London SW1Y 5AH United Kingdom

BA Small Research Grants 1301

Small research grants are available to support primary research in the humanities and social sciences. Eligible costs include travel and maintenance, research assistance, workshops, consumables, specialist software, and costs of interpreters in the field. Applications for collaborative or individual projects are equally welcome. Applications from international groups of scholars are welcome, provided there is a UK-based scholar as lead applicant. Grants are tenable for up to 24 months.
Amount of Grant: @L500-@L7500
Date(s) Application is Due: Jan 15; Apr 15; Oct 15
Contact: Research Grants Department, 020-7969 5217; fax: 020 7969 5414; email: grants@britac.ac.uk
Internet: http://www.britac.ac.uk/funding/guide/srg.html
Sponsor: British Academy
10 Carlton House Terrace
London SW1Y 5AH United Kingdom

BA Stein-Arnold Exploration Fund Research Grants   1302

The fund was established according to the terms of the will of Sir Aurel Stein, FBA, to commemorate his friendship with Sir Thomas Arnold, FBA. Eligible projects include research on the antiquities, historical geography, early history, or arts of those parts of Asia that come within the sphere of the ancient civilizations of India, China, and Iran, including Central Asia, or of one or more of these. Special consideration shall be paid, if possible, to research of this character bearing upon the territories comprised in the present Kingdom of Afghanistan, including the region of ancient Bactria and in the northwestern frontier region of India. Research should be so far as possible by means of exploratory work.

Requirements: Applicants should be British or Hungarian subjects.
Amount of Grant: @L2500 maximum
Date(s) Application is Due: Apr 15
Contact: Research Grants Department, 020 7969 5217; fax: 020 7969 5414; email: grants@britac.ac.uk
Internet: http://www.britac.ac.uk/funding/guide/saef.html
Sponsor: British Academy
10 Carlton House Terrace
London SW1Y 5AH United Kingdom

BA Thank-Offering to Britain Fellowship   1303

The fellowship is funded by the Thank-Offering to Britain fund from part of a generous endowment arising from the proceeds of a Thank-You Britain Appeal, initiated by the Association of Jewish Refugees as a mark of gratitude for Britain's provision of a home for Jews persecuted by the Nazi regime. These awards are aimed at established scholars in United Kingdom universities who are in midcareer, and are designed to allow the award holders to undertake or to complete an approved program of sustained research, while relieved of their normal teaching and administrative commitments. The principal purpose of the awards is to enable the completion of a major piece of research which will not only be an important contribution to knowledge and understanding but also help to enhance the future career and career prospects of the award holder. Grants are made direct to the award holder's employing institution to cover the costs of replacement teaching. Appropriate subjects for research will be related to human studies widely interpreted and their bearing on the well-being of the inhabitants of the United Kingdom.

Requirements: Fellowships are open to scholars in the humanities and social sciences. Applicants are expected normally to be aged 55 or under at time of application unless they can show that exceptional circumstances have delayed the advancement of their academic career. Awardees will be expected to be able to disseminate the results of their research not only through publications, but also through the teaching of future courses after the end of their awards.
Date(s) Application is Due: Sep 30
Contact: Research Grants Department, 020 7969 5265; fax: 020 7969 5414; email: posts@britac.ac.uk
Internet: http://www.britac.ac.uk/funding/guide/readfell.html
Sponsor: British Academy
10 Carlton House Terrace
London SW1Y 5AH United Kingdom

BA Visiting Professorships and Fellowships   1304

The program enables distinguished scholars from overseas to be invited to spend a minimum of two weeks in the United Kingdom. The academy grants the title of British Academy Visiting Professor or (for a more junior scholar) British Academy Visiting Fellow and awards a sum of money towards the estimated travel and maintenance costs. All arrangements are undertaken by the visitor's British sponsor. While the delivery of lectures and participation in seminars is not precluded, the main purpose of the visit should be to enable the visitor to pursue research. It is not intended that the fellowships and professorships should be used in conjunction with a nonstipendiary university fellowship. The normal maximum length of visit will be one month, but applications for longer periods will be considered, although it will be expected that the weekly budget for longer visits will be set at a more moderate level.

Requirements: Candidates for nominations must be from the United Kingdom and must be either established scholars of distinction or younger people who show great promise and who would benefit from time to pursue their research.
Amount of Grant: @L700 per week, usually for up to one month; travel expenses to the United Kingdom
Date(s) Application is Due: Dec 15
Contact: International Relations Department, 020 7969 5220; fax: 020 7969 5414; email: overseas@britac.ac.uk
Internet: http://www.britac.ac.uk/funding/guide/intl/visprof.html
Sponsor: British Academy
10 Carlton House Terrace
London SW1Y 5AH United Kingdom

BA Worldwide Congress Grants   1305

Grants are available for the congresses of major subject areas or disciplines, ordinarily occurring every three, four, or five years, and involving an extensive program and large attendance from all over the world. It should clearly be the British turn to host such a congress, and the event should not have been held in the UK for a considerable period. Typical examples would be the International Congress of Historical Sciences, or the Congress of the International Geographical Union. British scholars contemplating the possibility of extending an invitation for a worldwide congress to be held in Britain are invited to contact the academy to discuss possible support. Please note that it is expected

that preliminary proposals should be submitted at least three years in advance of the date of the Congress.

Restrictions: Applications from individual scholars for conferences supported by a block grant will not be accepted.
Amount of Grant: @L15,000 maximum
Date(s) Application is Due: Jan 15; Apr 15; Oct 15
Contact: Research Grants Department, 020 7969 5217; fax: 020 7969 5414; email: grants@britac.ac.uk
Internet: http://www.britac.ac.uk/funding/guide/wcg.html
Sponsor: British Academy
10 Carlton House Terrace
London SW1Y 5AH United Kingdom

Bailey Family Foundation Grants   1306

The foundation's primary mission is to expand the availability and enhance the quality of postsecondary education. The foundation also conducts research directed toward improving the state of higher education. Financial assistance in the form of scholarships is awarded to students based on their academic record, financial need, and level of community involvement. Scholarship programs include educational scholarships to high school seniors and college students.

Requirements: Applicants must possess a minimum cumulative GPA of 2.5, demonstrate financial need, be a graduating senior from a participating high school/college listed on the application, be U.S. residents, and be pursuing their undergraduate degree.
Amount of Grant: $5000
Date(s) Application is Due: Mar 31; Sep 30
Contact: Grants Administrator, (703) 971-6203; fax: (703) 971-6205; email: Bailey@Bailey-Family.org
Internet: http://www.bailey-family.org
Sponsor: Bailey Family Foundation
P.O. Box 803
Newington, VA 22122-0803

Bailey Wildlife Foundation Grants   1307

The foundation awards grants to support research projects in the natural sciences and arts and for wildlife and environmental conservation and protection in the eastern United States.

Restrictions: Individuals are not eligible.
Samples: Wildlife Conservation Society (Bronx, NY)—$3000.; Smithsonian Institution, Conservation and Research Ctr (Front Royal, VA)—$8027.; Southern Environmental Law Ctr (Charlottesville, VA)—$100,000.
Amount of Grant: $1890-$260,000
Contact: H. Whitney Bailey, (978) 901-3471
Sponsor: Bailey Wildlife Foundation
10223 Bushveld Ln
Raleigh, NC 27612-6149

Ball Brothers Foundation Grants   1308

The foundation seeks to build and sustain a high quality of life in Indiana by awarding grants to nonprofit organizations in broad subject areas, including elementary, secondary, higher, and adult basic education and literacy skills; cultural activities; community betterment; and health and human services. Usually Muncie and Delaware Counties receive a higher priority for funding than requests from within East Central Indiana or across the state. Types of support include general operations, annual campaigns, capital campaigns, building construction/renovation, program development, conferences and seminars, professorships, publication, curriculum development, research, fellowships, matching funds, seed grants, and technical assistance. Preference will be given to catalytic grants that will stimulate others to participate in problem solving or in matching fund programs and to innovative approaches for addressing either traditional or emerging community needs. Applications are reviewed by the board of directors in January, May, and September of each calendar year. Proposals are encouraged to be submitted from February to May. Grant seekers may send a preliminary proposal, complete proposal, or ask for a personal visit to discuss a potential grant request.

Requirements: Indiana nonprofit institutions and organizations are eligible.
Restrictions: The Foundation will not support: direct assistance to individuals or scholarships; applications coming from outside of Indiana; booster organizations; on-going salary requests of staff personnel to support an organization; services that the community-at-large should normally underwrite (i.e. roads, bus transportation, etc.); capital building projects; research projects (except for philanthropic studies); or unsolicited proposals (all requests must begin with a preliminary proposal).
Geographic Focus: Indiana
Samples: Muncie Center for the Arts (Muncie, IN)—for general operating support, $15,000.; Ball State U (Muncie, IN)—for the university's museum, $2 million and artwork valued at $8 million.; Minnetrista Cultural Foundation (IN)—for general support, $2.6 million.
Contact: Donna Munchel; (765) 741-5500; fax: (765) 741-5518; email: donna.munchel@ballfdn.org
Internet: http://www.ballfdn.org
Sponsor: Ball Brothers Foundation
222 South Mulberry Street
Muncie, IN 47305

Ball Brothers Foundation Rapid Grants 1309

Ball Brothers Foundation offers funding opportunities for a limited number of Ball Rapid Grants that are designed to provide funding to organizations requiring immediate funding for the following types of needs, but not limited too: continue a project; provide professional development; buy equipment or materials for a project; travel to meet representatives to advance ideas for a current or future project; to formulate a project idea; to carry out a mandated law or event; or for seed money to begin a new project. Generally requests up to $5,000 are considered. Ball Rapid Grants do not fall within the normal granting period. Requests may be submitted at any time between February 1 and November 30; decisions will be made within four business days and awards sent within 7 to 10 business days.

Requirements: Ball Brothers Foundation is restricted by its charter to grants to nonprofit institutions and organizations within Indiana.

Restrictions: The Foundation will not support: direct assistance to individuals or scholarships; applications coming from outside of Indiana; booster organizations; on-going salary requests of staff personnel to support an organization; services that the community-at-large should normally underwrite (i.e. roads, bus transportation, etc.0; capital building projects; research projects (except for philanthropic studies); or unsolicited proposals (all requests must begin with a preliminary proposal).

Geographic Focus: Indiana
Amount of Grant: $5,000 range
Date(s) Application is Due: Nov 30
Contact: Donna Munchel; (765) 741-5500; fax: (765) 741-5518; email: donna.munchel@ballfdn.org
Internet: http://www.ballfdn.org/index/staff/generalinfo.asp
Sponsor: Ball Brothers Foundation
222 South Mulberry Street
Muncie, IN 47305

Baltimore Washington Center Adult Psychoanalysis Fellowships 1310

The Adult Psychoanalysis Fellowship Program is a one-year program designed for individuals who have a significant interest in psychoanalysis as a body of knowledge and as a framework with which to understand and carry out therapeutic efforts. Intended for advanced graduate students and residents as well as recent graduates in Psychiatry, Psychology and Social Work, the program at its core consists of monthly seminars presented by psychoanalysts on clinically-related topics. Fellows are also welcome to attend Forums on Psychoanalysis and other conferences offered by the Institute.

Date(s) Application is Due: Jul 15
Contact: Elizabeth Manne, Executive Director; (301) 470-3635 or (410) 792-8060; fax: (410) 792-4912; email: admin@bwanalysis.org
Internet: http://www.bwanalysis.org/Fellowship%20Program-Adult.html
Sponsor: Baltimore Washington Center for Psychoanalysis
14900 Sweitzer Lane, Suite 102
Laurel, MD 20707

Baltimore Washington Center Child Psychotherapy Fellowships 1311

The Child Psychotherapy Fellowship program is a one-year program specifically designed for child mental health professionals who are interested in deepening their understanding of working with children from a psychoanalytic or psychodynamic perspective. The program is tuition-free. Detailed discussions of actual clinical work with children and reviews of relevant psychoanalytic literature will be used as a framework to enhance understanding and ability to carry out therapeutic work. Fellowship meetings will be held monthly, September through May, on the second Wednesday of every month, at the homes of Baltimore and Washington faculty members.

Requirements: Professionals, including psychiatrists, psychologists, social workers, and counselors currently working with children in psychotherapy should apply. Recent graduates of training programs should also apply.

Date(s) Application is Due: Jul 15
Contact: Elizabeth Manne, Executive Director; (301) 470-3635 or (410) 792-8060; fax: (410) 792-4912; email: admin@bwanalysis.org
Internet: http://www.bwanalysis.org/Fellowship%20Program-Child.html
Sponsor: Baltimore Washington Center for Psychoanalysis
14900 Sweitzer Lane, Suite 102
Laurel, MD 20707

Bancroft Prizes 1312

Two annual prizes are awarded to the authors of distinguished works in either or both of the following categories: American history (including biography) and diplomacy. The awards in any given year are for books first published in the previous year. The word American is interpreted to include all of the Americas—North, Central, and South; however, the award is confined to works originally written in English or where there is a translation published in English.

Samples: Christine Leigh Heyrman, U of Delaware—prize winner for Southern Cross: The Beginnings of the Bible Belt, $4000.; Walter LaFeber, Cornell U—prize winner for The Clash: U.S.-Japanese Relations throughout History, $4000.; Thomas Sugrue, U of Pennsylvania—prize winner for The Origins of the Urban Crisis: Race and Inequality in Postwar Detroit, $4000.

Amount of Grant: $4000
Contact: Bancroft Prizes Program Support, (212) 854-2271; fax: (212) 854-9099
Internet: http://www.columbia.edu/cu/lweb/eguides/amerihist/bancroft.html
Sponsor: Columbia University
517 Butler Library, 535 W 114th Street
New York, NY 10027

Banfi Vintners Foundation Grants 1313

The foundation awards general operating grants to nonprofits in its areas of interest, including higher education, civic and public affairs, arts and humanities, wildlife protection, health (hospitals and disease research/prevention), religion, science, social services, and international. Grants are made nationwide, with preference given to requests from Massachusetts and the New York, NY, area.

Geographic Focus: Massachusetts; New York
Samples: Colgate U (Hamilton, NY)—for operating support, $75,000.; Friends for Long Island's Heritage (Syosset, NY—for operating support, $25,000.; Huntington Hospital Assoc (Huntington, NY)—for operating support, $25,000.; Cornell U (Ithaca, NY)—for operating support, $465,000.

Amount of Grant: $100-$465,000 range
Contact: Philip Calderone, Executive Director, (516) 626-9200
Sponsor: Banfi Vintners Foundation
1111 Cedar Swamp Road
Glen Head, NY 11545

Banfield Charitable Trust Grants 1314

This trust offers grants annually to nonprofit organizations and educational institutions that make life better for Pets and their families. Special considerations will be given to collaborative programs with Pet-related organizations and Banfield team members working together towards a shared goal. At this time, Banfield Charitable Trust is pursuing the following funding priorities: promotion of preventative healthcare for Pets; educating children about veterinary medicine and the Pets they love; programs based on the human-Pet bond and how this relates to longer, healthier lives for Pets and people; and veterinary education programs. The Trust does not generally fund an entire project, but expects to be one of multiple funding sources. Typically, the Trust will not fund more than 50% of the project's entire budget.

Requirements: 501(c)3 tax-exempt organizations and educational institutions are eligible.

Restrictions: Funding will not be considered for the following purposes: Spay and neuter or adoption programs; General operating expenses, deficit reduction, or general administrative overhead expenses; Fundraising campaigns, including special events; Grants to individuals, or to provide support for business enterprise.

Samples: Redland Elementary School (Oregon City, OR)—$210 (2005); Nebraska Poodle Rescue (Omaha, NE)—$6000 (2005); Ohio State U (Columbus, OH)—for the College of Veterinary Medicine, $10,000 (2005).

Date(s) Application is Due: Jun 30; Nov 30
Contact: Grants Administrator, (503) 922-5801 or (866) 802-0566; email: charitabletrust@banfield.net
Internet: http://www.banfieldcharitabletrust.net/guidelines.html
Sponsor: Banfield Charitable Trust
8000 NE Tillamook Sreet, P.O. Box 13998
Portland, OR 97213

Bank of America Charitable Foundation Anchor Institutions Grants 1315

Bank of America's local leaders, working within their communities, look for opportunities to support community stakeholders; opportunities that will further economic and community development, that will do more than put bricks and mortar on your block. When it does find those opportunities, the Foundation invites anchor institutions to apply for a grant. Such institutions might include: arts and cultural institutions such as museums, zoos, and performance halls; colleges, universities, and community colleges that attract and educate the workforce; and hospitals that provide excellent health care.

Samples: Apollo Theater, New York, New York (2007); Children's Hospital, Boston, Massachusetts (2007); Cook Children's Hospital, Fort Worth, Texas (2007); New Jersey Performing Arts Center, Newark, New Jersey (2007).

Contact: Anne Finucane; (800) 218-9946 or (617) 434-9405; email: anne.m.finucane@bankofamerica.com; Andrew Plepler, President; (800) 218-9946 or (800) 432-1000; email: andrew.plepler@bankofamerica.com
Internet: http://www.bankofamerica.com/foundation/index.cfm?template=fd_anchorinstitutions
Sponsor: Bank of America Charitable Foundation
100 North Tryon Street
Charlotte, NC 28202

Bank of Sweden Tercentenary Foundation Grants 1316

The foundation focuses on research aimed at expanding knowledge of the effects of technical, economic, and social changes on Sweden and its people. The foundation awards project grants to individual scientists or groups of scientists. Grants generally support large, long-term projects.

Requirements: Applicants outside of Sweden must work with Swedish researchers or research institutes.

Contact: Dr. Dan Brandstrom, Director, 46-8-506 264 02; fax: 46-(0)8-506-264-30; email: rj@rj.se
Internet: http://www.rj.se/default.asp?ItemID=22770
Sponsor: Bank of Sweden Tercentenary Foundation
P.O. Box 5675
Stockholm 114 86 Sweden

Barbara Thom Postdoctoral Fellowships 1317

The Huntington Library and Art Gallery is an independent research center with holdings in British and American history, literature, art history, the history of science, and photography. The program awards fellowships to support nontenured faculty members

pursuing scholarship in a field appropriate to the Huntington's collections while revising a manuscript for publication. Recipients are expected to be in continuous residence at the Huntington during the nine- to 12-month fellowship tenure and to participate in the intellectual life of the center. There is no application form. Applications consist of cover sheet, description of the project, curriculum vita, and three letters of recommendation.
Requirements: Preference will be given to scholars who received the PhD three years prior to the fellowship.
Amount of Grant: $40,000
Date(s) Application is Due: Dec 15
Contact: Robert Ritchie, Director of Research, (626) 405-2194; fax: (626) 449-5703; email: cpowell@hungtington.org
Internet: http://www.huntington.org/ResearchDiv/Fellowships.html
Sponsor: Huntington Library and Art Gallery
1151 Oxford Road
San Marino, CA 91108

BARD Postdoctoral Fellowships   1318
The United States-Israel Binational Agricultural Research and Development Fund (BARD) was established to promote and support research and development in agriculture for the mutual benefit of both countries. The postdoctoral program was initiated to promote cooperative agricultural research between postdoctoral fellows from one country and senior scientists from the other; to assist young scientists to become professionally established in the scientific community; and to provide BARD with input into new research areas and to enhance scientific competence in these areas. The program emphasizes innovation in agriculture, particularly in the areas of cellular and molecular biology and the use and development of explanatory models in agriculture. Also included in the fellowship is a $5000 allowance for children, and, if necessary, a small allowance not to exceed $3000 to host institution for research-related expenses. The duration of the fellowship is one year. Requests for application kits and information may be obtained from the Maryland address or BARD, P.O. Box 6, Bet Dagan, Israel 50250; phone: 03-968-3230, fax: 03-966-2506, or email: bard@bard-isus.com.
Requirements: Applicants must be U.S. or Israeli citizens. Applicants should have fulfilled the requirements for a PhD degree within the last three years. Candidates must file an application with BARD in English naming only one senior scientist. Application kits may be obtained at either BARD office. Only one application per year for each candidate will be accepted.
Restrictions: Israelis already in the United States and Americans already in Israel (for a period of six months or longer) are not eligible.
Amount of Grant: $32,000 to cover travel and living expenses
Date(s) Application is Due: Jan 15
Contact: BARD, U.S.DA-ARS-OIRP, (301) 504-4584; fax: (301) 504-4619; email: lea@bard-isus.com
Internet: http://www.bard-isus.com
Sponsor: United States-Israel Binational Agricultural Research and Development Fund
5601 Sunnyside Avenue
Beltsville, MD 20705

BARD Research Grants   1319
The United States-Israel Binational Agricultural Research and Development Fund (BARD) was established by the governments of the United States and Israel for the purpose of promoting and supporting cooperative research and development projects in agriculture for the mutual benefit of both countries. Proposals may be unidisciplinary or interdisciplinary and may cover any or all phases of research and development. BARD will consider financing research and development projects, exploratory or basic research studies, and small initial feasibility studies and other closely related activities in the areas of soil and water conservation management and utilization; crop production including new technology; crop protection; animal production including aquaculture; veterinary medicine; crop and animal genetic improvement; recycling of wastes; postharvest sciences covering operations from production through processing; agricultural engineering; agricultural economics; and cellular and molecular biology in agriculture. Investigators of either country may request information or mail proposals to Lynn Gipe, U.S.DA-ARS-BARD at the Maryland address, or BARD Executive Director, Volcani Center, P.O. Box 6, Bet Dagan, Israel 50250, 03-9683230. Guidelines and applications are also available on the Web site.
Requirements: Affiliates of public or private nonprofit research institutions that demonstrate the necessary research and development capabilities are eligible to apply for funding. Scientists who wish to apply for grants should submit their proposals through such legally constituted institutions or agencies. The research proposal should be prepared jointly by at least one U.S. and one Israeli investigator and should describe the areas of anticipated cooperative endeavor between them. BARD will assist scientists who are unable to find collaborators; the BARD office in Israel or its liaison office in the United States should be contacted if such assistance is necessary.
Restrictions: BARD will not consider more than one application from the same investigator in a given year or fund the same investigator in more than one concurrent project.
Amount of Grant: $100,000-$350,000; $300,000 average for three years
Date(s) Application is Due: Sep 1
Contact: Program Contact, BARD, c/o Department of Agriculture, (301) 504-4522; fax: (301) 504-4619; email: bard@bard-isus.com
Internet: http://www.bard-isus.com
Sponsor: United States-Israel Binational Agricultural Research and Development Fund

5601 Sunnyside Avenue
Beltsville, MD 20705

Barker Award   1320
Open to qualified graduate students registered full-time in the Faculty of Graduate Studies at the University of Calgary.
Requirements: Applicants must be studying in the fields of Business Administration with emphasis on Entrepreneurship, New Venture Development and Marketing. The award is tenable during a student's first year of the MBA program.
Amount of Grant: $C1500
Date(s) Application is Due: May 30
Contact: Connie Baines, (403) 220-5690; email: cbaines@ucalgary.ca
Internet: https://pr1web.ucalgary.ca/FGS_SAM/public/InternalSearchResults.aspx
Sponsor: University of Calgary
2500 University Drive, NW, Earth Sciences Building, Room 720
Calgary, AB T2N 1N4 Canada

Barra Foundation Project Grants   1321
Project grants are one-time grants generally for amounts above $10,000. The Foundation considers grants for innovative projects that aid research in advancing the frontiers of human services, arts and culture, health and education. Grants are not made for ongoing or expanding programs where substantial initial support was previously provided from other sources. Three principal criteria are strictly adhered to in judging the merits of a proposed project. They are: innovation; evaluation; and dissemination. Proposals may be submitted at any time of the year. Initial requests for project grants should be in the form of a preliminary concept paper, generally not to exceed two pages, summarizing: the principal focus and objectives of the proposed project; uniqueness of the concept; the overall methodology; estimated timetable; preliminary budget data; and other sources of support for the project.
Restrictions: The foundation does not provide grants for: ongoing operating budgets; staff salaries; budget deficits; endowments; capital campaigns and projects; international programs; environmental and religious organizations; scholarships and fellowships; or audio/video projects, publications, catalogs, and exhibitions.
Geographic Focus: Pennsylvania
Samples: U of Pennsylvania (Philadelphia, PA)—to build a facility to house the Center for Early American Studies at the School of Arts and Sciences, and to endow the building's operating costs, $2.5 million.; Woodmere Art Museum (Philadelphia, PA)—to endow the position of curator of education, $1 million.
Amount of Grant: $10,000-$500,000 range
Contact: William Harral, III, President; (215) 233-5115; fax: (215) 836-1033; email: william.harral@verizon.net
Internet: http://www.barrafoundation.org/grants/index.html
Sponsor: Barra Foundation
8200 Flourtown Avenue, Suite 12
Wyndmoor, PA 19038-7976

Bart Harvey Enterprise Fellowship   1322
The Bart Harvey Enterprise Fellowship is named in honor of Enterprise, former chairman and CEO. The two-year fellowship exposes rising professionals to the spirit and mission of Enterprise's affordable housing and community development work. The Harvey Fellow should provide: substantial and creative support to the CEOs, Chairs and other senior staff; contribution to the progress and success of assigned projects; and demonstration of technical, teamwork, leadership and communications skills.
Samples: My Trinh, UCLA School of Law—helped set up law clinics to protect the rights of low-income tenants, immigrant communities and disadvantaged workers (2008).
Amount of Grant: $40,000 average
Contact: Kathy Holmes, Program Associate; (410) 772-2411; fax: (410) 964-1918; email: kholmes@enterprisecommunity.org
Internet: http://www.enterprisecommunity.org/programs/awards_and_fellowships/fellowships/harvey_fellowship/
Sponsor: Enterprise Community Partners / Enterprise Foundation
10227 Wincopin Circle, Suite 500
Columbia, MD 21044

Barth Syndrome Foundation Research Grants   1323
Barth syndrome is a serious X-linked recessive condition associated with cardiomyopathy, neutropenia, skeletal muscle weakness, exercise intolerance, growth delay, and diverse biochemical abnormalities (including defects in mitochondrial metabolism and phospholipid biosynthesis). Because many clinical and biochemical abnormalities of Barth syndrome remain poorly understood, the program is seeking proposals for research that will advance knowledge on any aspect of the syndrome. The foundation prefers to award seed grants to experienced investigators for testing of initial hypotheses and collection of preliminary data leading to successful long-term funding by NIH and other major granting institutions. The foundation also encourages investigators new to the field of Barth Syndrome research. Send an electronic version of the full application and all the attachments by the listed application deadline. The Foundation anticipates awarding several one- or two-year grants of up to $40,000 each. Funds will be available as soon as the successful grant applicants have been notified.
Requirements: Principal investigators who are affiliated with nonprofit institutions are eligible.
Amount of Grant: $10,000-$40,000
Date(s) Application is Due: Oct 31

Contact: Matthew J. Toth, Ph.D., Science Director; (617) 469-6769; fax: (617) 849-5695; email: mtoth@barthsyndrome.org
Internet: http://www.barthsyndrome.org/english/View.asp?x=1635
Sponsor: Barth Syndrome Foundation
675 VFW Parkway, #372
Chestnut Hill, MA 02467

Batchelor Foundation Grants 1324

The foundation awards grants to Florida nonprofits in its areas of interest, including medical research, childhood diseases, and natural resource conservation. There are no application forms. Initial approach should be a letter that details the grant proposal.
Requirements: Florida area nonprofits are eligible.
Restrictions: Individuals are not eligible.
Geographic Focus: Florida
Samples: Audubon of Florida (Miami, FL)—to expand its scientific work, including efforts to deal with sustainable-development issues affecting wildlife habitats statewide, $3 million partial challenge grant (2004).
Amount of Grant: $5000-$3 million range
Contact: Anne Batchelor-Robjohns, Treasurer, (305) 416-9066
Sponsor: Batchelor Foundation Inc
111 NE 1st Street, Suite 820
Miami, FL 33132

Baxter International Foundation Grants 1325

The foundation supports the development of more accessible and affordable healthcare. The foundation funds initiatives that improve the access, quality and cost-effectiveness of healthcare. Focusing on these priorities, the foundation makes grants in and near communities throughout the world where significant numbers of Baxter employees live and work.
Requirements: Nonprofits in the United States, Europe, Latin America, and Mexico, with some emphasis on Chicago and Puerto Rico, are eligible.
Restrictions: In general, the foundation does not make grants to capital and endowment campaigns; disease-specific organizations; educational institutions (except in instances where a grant would help to achieve other goals, such as increasing the skills and availability of health care providers); hospitals; individuals; fraternal, veterans, or religious organizations; lobbying and political organizations; organizations soliciting contributions for advertising space, tickets to dinners, benefits, social and fund-raising events, sponsorships and promotional materials; organizations seeking travel support for individuals or groups, medical missions or conferences; magazines, professional journals, documentary, film, video or radio productions.
Samples: Brigadas de Amistad (Toluca, Mexico)—for salary support of a psychologist who will provide counseling to ex offenders recovering from substance abuse and to their families, $20,000 (2005); Casa del Peregrino Aguadilla (Puerto Rico)—to establish a mental-health clinic for mentally ill and elderly homeless people in Aguadilla, $41,032 (2005); Sigma Theta Tau (Indianapolis, IN)—for a biannual award that recognizes outstanding research conducted by a nurse, $15,000 (2005); To assist tsunami-relief efforts, $1 million (2005).
Amount of Grant: $2.5 million total
Date(s) Application is Due: Jan 12; Apr 2; Jun 27; Sep 19
Contact: (847) 948-4605; email: fdninfo@baxter.com
Internet: http://www.baxter.com/about_baxter/foundation/index.html
Sponsor: Baxter International Foundation
Baxter International Foundation, One Baxter Parkway - DF2-2E
Deerfield, IL 60015-4625

Bay and Paul Foundations, Inc Grants 1326

General operating and project grants are awarded for support of children's services and precollege educational programs, with an emphasis on technology and enhancement of science, math, and writing curricula; preservation of cultural and natural history collections and collections care training in museums, zoos, libraries, and botanical gardens; advocacy and research programs for preserving biodiversity; and Native American cultural heritage preservation and economic development programs. There are no application forms. Proposals should be directed to the executive director and should include a brief description of the applicant organization and of the program for which funding is requested, including objectives and numbers served; a project budget or financials; other expected sources of support; qualifications of key personnel; and evidence of tax-exempt status. The foundation's directors meet three times a year, usually in January, May, and October, to consider grant proposals. Requests are not accepted via fax. Regular postal service delivery is the preferred method of proposal acceptance.
Requirements: Nonprofits in Connecticut, Massachusetts, Maine, New Hampshire, New Jersey, New York, Rhode Island, and Vermont are eligible.
Restrictions: Grants do not support requests for endowments, building campaigns, building construction or maintenance, sectarian religious programs, books or studies, individual scholarships or fellowships, loans, travel, film, television or video productions, programs consisting primarily of conferences, for annual fund appeals, or to other than publicly recognized charities. First time grants for K-12 arts-in-education programs and K-12 science and math programs are currently geographically restricted to the New York City metropolitan area.
Geographic Focus: Connecticut; Maine; Massachusetts; New Hampshire; New Jersey; New York; Rhode Island; Vermont
Samples: American Assoc for State and Local History (Nashville, TN)—to update the book Starting Right: A Guide to Museum Planning, $13,100 (2003).

Contact: Frederick Bay, Executive Director, (212) 663-1115; fax: (212) 932-0316; info@bayandpaulfoundations.org
Internet: http://www.bayandpaulfoundations.org/areas.html
Sponsor: Bay and Paul Foundations, Inc
17 West 94th Street
New York, NY 10025

Bayer Clinical Scholarship Award 1327

This award is intended to facilitate the development of specific clinical expertise in the field of hemophilia for applicants who have completed medical training and have an interest in pursuing a career as a hemophilia treater/researcher. The award will support a mentored physician in training for two years. Clinical duties will encompass diagnosis, evaluation, and the planning of management strategies for patients with hereditary bleeding disorders. In addition to the clinical experience, the applicant may pursue a research project in the field of hemostasis. Clinical scholarships will provide funding for two years. Up to five new awards will be made each year. Guidelines and application are available online.
Requirements: The applicant should have earned his/her medical degree within the previous eight years. The commitment of the mentor to the grantee and the research project (if applicable) is a vital element of the application, as is the quality of the clinical environment at which the applicant will undertake the scholarship.
Amount of Grant: $70,000 annually for two years
Contact: Program Administrator, email: programadministrator@bayer-hemophilia-awards.com
Internet: http://www.bayer-hemophilia-awards.com/awards.cfm#clinical
Sponsor: Bayer HealthCare Corporation
100 Bayer Road
Pittsburgh, PA 15205-9741

Bayer Foundation Grants 1328

The Bayer Foundation supports programs that enhance the quality of life, provide unique and enriching opportunities that connect diverse groups and ensure preparedness for tomorrow's leaders. The Foundation welcomes proposals from 501(c)3 organizations whose programming matches at least one of the following areas: Civic and Social Service Programs; Education and Workforce Development; Arts and Culture and Health and Human Services.
Requirements: 501(c)3 nonprofit organizations in Bayer operating communities are eligible. Submit proposals to regional offices in California, Connecticut, Georgia, Indiana, Kansas, Massachusetts, New Jersey, New York, North Carolina, Ohio, Pennsylvania, Texas and West Virginia.
Restrictions: The Foundation will not fund: For-profit organizations or those without Internal Revenue Service code 501(c)3 nonprofit, tax-exempt status; Organizations that discriminate on the basis of race, color, creed, gender, sexual orientation or national origin; General operating support for United Way affiliated agencies. Only support for special projects will be considered; Organizations or programs designed to influence legislation or elect candidates to public office; General endowment funds; Deficit reduction or operating reserves; Religious organizations; Charitable dinners, events or sponsorships; Community or event advertising; Individuals; Student trips or exchange programs; Athletic sponsorships or scholarships; Telephone solicitations; Organizations outside of the United States or its territories
Geographic Focus: California; Connecticut; Georgia; Indiana; Kansas; Massachusetts; New Jersey; New York; North Carolina; Ohio; Pennsylvania; Texas; West Virginia
Samples: Robert Morris College (Moon Township, PA)—for the Bayer Ctr for Nonprofit Management, which awards master's degrees in business administration with a focus on nonprofit management and that offers technical and consulting services, educational programs, and research and referral services to nonprofit organizations in Pennsylvania, $500,000.
Contact: Bayer Foundation, (800) 422-9374
Internet: http://www.bayerus.com/about/community/i_foundation.html
Sponsor: Bayer Foundation
100 Bayer Road, Building 4
Pittsburgh, PA 15205-9741

Bayer Hemophilia Early Career Investigator Award 1329

This award will fund salary support and research funds for a junior faculty member who wishes to undertake a mentored basic and/or clinical research project in the bleeding disorders field. The applicant would be expected to dedicate a significant amount of time to the project. Examples of topics for research projects that might be considered for these awards include, but are not limited to: clinical studies; properties and delivery of clotting factor proteins; assays and models; genetics and epidemiology; and molecular aspects and mechanisms of clotting factor inhibitor formation. Awards will provide funding for two years. Up to five new awards will be made each year. Part of the award may fund salary support. The candidate must spend at least 25 percent of his/her time on the project in order to request salary support. Applicants should submit a letter of intent via email describing the proposed project in 500 words or fewer; full proposals are by invitation. Guidelines are available online.
Requirements: The applicant should have an entry-level academic or clinical appointment within his or her institution. This award is open to applicants with a medical degree and/or PhD. Applicants should have earned their terminal degree within the previous 10 years.
Amount of Grant: $100,000 annually
Date(s) Application is Due: Oct 31
Contact: Corporate Communications, (412) 777-2000

Internet: http://www.bayer-hemophilia-awards.com/awards.cfm#early
Sponsor: Bayer HealthCare Corporation
100 Bayer Road
Pittsburgh, PA 15205-9741

Bayer Hemophilia Special Project Award   1330

The award is designed to support a wide range of research projects in the field of hemophilia. Examples of the types of projects that might be considered for these awards include, but are not limited to, those related to: clinical research; basic research; assessment and intervention in psychosocial issues facing patients and their families; assessment of quality of life and other health economic outcomes in patients with bleeding disorders; and the effects of treatment modalities on such outcomes. The award is designed to encourage hypothesis-driven research, where the investigator is attempting to prove or disprove a set of assumptions. It is not designed to support studies such as the collection of epidemiological data. The funding will be awarded to allow the project to run for one or two years. Part of the award may fund salary support. The candidate must spend at least 25 percent of his/her time on the project in order to request salary support. Applicants should submit a letter of intent via email describing the proposed project in 500 words or fewer; full proposals are by invitation.
Requirements: Any individual affiliated with a facility that carries out research in inherited bleeding disorders, or provides care to patients with those disorders, may make a request for these grants. Such facilities may include medical universities, hospitals, treatment centers, blood centers, and other laboratories. All applications must be in English.
Amount of Grant: $200,000 maximum
Date(s) Application is Due: Oct 31
Contact: Corporate Communications, (412) 777-2000
Internet: http://www.bayer-hemophilia-awards.com/awards.cfm
Sponsor: Bayer HealthCare Corporation
100 Bayer Road
Pittsburgh, PA 15205-9741

Baylor Institute for Faith and Learning Visiting Fellows Program   1331

The Institute for Faith and Learning sponsors a visiting fellows program that brings outstanding scholars to the university for one-year appointments. Fellows pursue research in their academic disciplines informed by Christian intellectual traditions. They serve as participants in the full life of the university, assisting the institute in its aim to encourage academic research that integrates faith and learning, and helping to cultivate a context conducive to the pursuit of such scholarship. Fellows may teach one course per term within a cohosting department, help lead institute-sponsored colloquia, and otherwise further the work of the institute. Applicants should submit a curriculum vita, three recommendation letters, a 750-word description of the project to be undertaken while at the institute, and a published paper.
Requirements: Senior fellows typically hold senior faculty rank at their home institutions, possess records as accomplished, nationally recognized scholars, and pursue research programs of the highest caliber integrative of their academic discipline and the Christian faith. Junior fellows typically hold junior but tenure-track faculty rank at their home institutions, possess records showing clear promise commensurate with their time in rank, and also pursue research programs of the highest caliber integrative of the Christian faith and their area of expertise.
Amount of Grant: $50,000 maximum for senior fellows; $40,000 maximum for junior fellows
Date(s) Application is Due: Nov 1
Contact: Dr. Douglas Henry, (254) 710-4805; fax: (254) 710-4813; email: IFL@baylor.edu
Internet: http://www.baylor.edu/ifl
Sponsor: Baylor University
P.O. Box 97270
Waco, TX 76798

BC Center for Retirement Research Dissertation Fellowships   1332

The purpose of the program is to promote the next generation of retirement research scholars and to improve the quality of scholarship in Social Security and income studies. Applicants must demonstrate their dissertation focuses on the Social Security Administration research priorities including Social Security and retirement; macroeconomic analyses of Social Security; wealth and retirement income; program interactions; international research; and demographic research. Doctoral candidates from a wide variety of academic disciplines, including actuarial science, demography, economics, finance, gerontology, political science, public administration, public policy, sociology, social work, and statistics are encouraged to submit a proposal. Awards will be announced in March. Fellowships for successful applicants are renewable for a second-year based on adequate progress. Application and guidelines are available online.
Requirements: Applicants are required to be enrolled in a qualified doctoral program at a U.S. university; have completed all coursework for a PhD by the time funding would start; and have a dissertation advisor and/or committee.
Amount of Grant: $20,000 ($16,000 stipend, $4000 tuition and summer support)
Date(s) Application is Due: Jan 31
Contact: Fellowship Administrator, (617) 552-1762; fax: (617) 552-0191; email: crr@bc.edu
Internet: http://www.bc.edu/centers/crr/dissertation.shtml
Sponsor: Boston College
140 Commonwealth Avenue, Fulton Hall 550
Chestnut Hill, MA 02467-3808

BC Steven H. Sandell Grant Program   1333

The program's purpose is to promote research on retirement issues including: Social Security and retirement; macroeconomic analyses of Social Security; wealth and retirement income; program interactions; international research; and demographic research. Within the six broad areas, four topics of particular interest include: retirement planning/risk management decisions of workers and retirees; measures of retirement well-being; trends in labor force participation and their implications for Social Security and for the well-being of the elderly; Social Security reform.
Requirements: The principal investigator must: have a PhD in a related discipline or a comparable professional certification; and be a nontenured junior scholar or senior scholar working in a new area. All applicants must complete the online submission form.
Amount of Grant: $40,000
Date(s) Application is Due: Mar 31
Contact: Paige Eppenstein, Center for Retirement Research, Sandell Grant Program, (617) 552-1092; email: eppense@bc.edu
Internet: http://www.bc.edu/centers/crr/sandellguidelines.shtml
Sponsor: Boston College
140 Commonwealth Avenue, Fulton Hall 550
Chestnut Hill, MA 02467-3808

BCBS Improving Men's Health Grants   1334

The purpose of this initiative is to encourage research, as well as demonstration and evaluation projects, on improving men's health through the early detection and screening of disease. The foundation will focus on increasing screening and prevention for men in the following five clinical areas: diabetes, hypertension, coronary heart disease, colorectal cancer, and prostate cancer. Proposed projects should focus on one or more of three clinical areas: screen men for specific diseases and/or conditions; refer men with abnormal screens for diagnostic and treatment services in a timely and efficient manner; and follow-up on men for subsequent testing. Funds are available for salary support, supplies and office operations, as well as limited staff travel expenses and consultant fees for Michigan-based researchers working on initiative to improve men's health through the early detection of disease. The listed application deadline is for letters of intent; full proposals are by invitation.
Requirements: Letters of interest are sought from Michigan-based physicians and doctoral-level researchers based at universities, academic medical settings, community hospitals, health systems, and community-based nonprofit organizations.
Geographic Focus: Michigan
Amount of Grant: $25,000-$50,000; $100,000 maximum over two years
Date(s) Application is Due: Jan 15
Contact: Grants Administrator, (313) 225-8706; fax: (313) 225-7730; email: foundation@bcbsm.com
Internet: http://www.bcbsm.com/foundation/grant_programs.shtml
Sponsor: Blue Cross Blue Shield of Michigan Foundation
600 E Lafayette, X520
Detroit, MI 48226-2998

BCBS of Massachusetts Foundation Grants   1335

With a mission to expand access to health care, the foundation awards grants intended to make a significant impact on the health of Massachusetts's low income and uninsured residents. Grant programs include Innovation Fund for the Uninsured; Connecting Consumers with Care; Strengthening the Voice for Access; Pathways to Culturally Competent Health Care; Catalyst Fund; Building Bridges in Children's Mental Health; and Policy Research and Analysis. Grant awards in particular focus areas are made during specific grant cycles. Deadlines and additional information are available online.
Requirements: Massachusetts 501(c)3 nonprofit organizations or government agencies with a mission that includes improving access and removing barriers to healthcare for low-income and uninsured Massachusetts's residents are eligible.
Restrictions: The foundation will not fund individuals; for-profit organizations; capital campaigns, endowments, or building drives; fundraising drives and events; retiring debt or operating deficits; direct political or lobbying activity; projects outside of Massachusetts; or religious organizations for religious purposes (proposals for secular programs of faith-based organizations that meet funding criteria will be considered). Grants are rarely made for curriculum development, conferences, film or video production, or scholarships.
Geographic Focus: Massachusetts
Samples: Beth Israel Deaconess Medical Ctr (Boston, MA)—to help managers select, retain, and promote highly qualified, diverse staff members, $40,000 (2004); Boston U, Goldman School of Dental Medicine (Boston, MA)—to develop policy recommendations designed to increase access to fluoridated water, with a focus on uninsured and low-income people and minorities, $46,716 (2004); John Snow Research and Training Institute (Boston, MA)—to study the extent to which a subsidized health-insurance program for people in the Massachusetts fishing industry can be duplicated in other industries, $50,000 (2004); Massachusetts Department of Mental Health (Boston, MA)—to integrate primary care and mental-health services for underserved Hispanics in the Worcester, MA, area, $50,000 (2004).
Date(s) Application is Due: Jan 20; Sep 7
Contact: Grant Requests, (617) 246-3744; fax: (617) 246-3992; email: grantinfo@bcbsmafoundation.org
Internet: http://www.bcbsmafoundation.org/foundationroot/en_U.S./grants/focusArea.jsp#aInnovation%20Fund%20for%20the%20Uninsured
Sponsor: Blue Cross Blue Shield of Massachusetts Foundation
401 Park Drive
Boston, MA 02215

BCBS of Minnesota Healthy Together Grants   1336

Healthy Together: Creating Community with New Americans is a grantmaking initiative designed to reduce health disparities for immigrants and improve the health and vitality of the entire community. The foundation awards grants to projects that foster exchanges and interactions between newcomers and the receiving community, strengthen the capacity of immigrant-led organizations and their attention to health, and address social adjustment and mental health.

Requirements: Eligible applicants include community- and faith-based organizations; state, county and municipal agencies; tribal governments and agencies; professional associations or collaboratives; and policy and research organizations. Applicants must be located in Minnesota or serve Minnesotans. Eligible applicants include units of government as well as organizations designated as nonprofit under section 501(c)3 of the IRS code.

Geographic Focus: Minnesota

Samples: Episcopal Community Services, Inc., $25,000 to engage Native Americans in the Bemidji community in early childhood development and stable housing issues; Folwell Neighborhood Association, $25,000 to improve school readiness of 500 children; National Center for Healthy Housing, $150,000 to demonstrate green building principles and tenant education in Worthington, Minnesota; Resources for Child Caring, $25,000 to improve home environments of children in the Payne-Phalen neighborhood (2006).

Amount of Grant: $1200-$250,000

Date(s) Application is Due: Feb 1

Contact: Jocelyn Ancheta, program officer, (651) 662-2894; email, Jocelyn_L_Ancheta@bluecrossmn.com.

Internet: http://www.bluecrossmn.com/public/foundation/index.html

Sponsor: Blue Cross Blue Shield of Minnesota Foundation

3535 Blue Cross Road, Rte M459

Eagan, MN 55122

BCBSM Foundation Excellence in Research Awards for Students   1337

The annual Excellence in Research Award for Students acknowledges doctoral candidates or medical students enrolled in Michigan universities who have published research papers that represent significant contributions to health policy or clinical care. Entries will be judged based on the subject matter's potential to make a contribution to health policy and medical care as well as the quality and originality of the research. There will be three awards: 1st place - $1,000; 2nd place - $750; 3rd place - $500.

Requirements: Doctoral candidates or medical students enrolled in Michigan universities are eligible. The nomination must be made by a faculty member at the student's university. If the paper has multiple authors, the student must be the first author.

Restrictions: Basic research papers are not eligible.

Geographic Focus: Michigan

Amount of Grant: $500 - $1,000 range

Date(s) Application is Due: Jan 1

Contact: Ira Strumwasser, Director; (313) 225-8706; fax: (313) 225-7730; email: foundation@bcbsm.com

Internet: http://www.bcbsm.com/foundation/grant.shtml

Sponsor: Blue Cross Blue Shield of Michigan Foundation

600 East Lafayette Boulevard, X520

Detroit, MI 48226

BCBSM Foundation Physician Investigator Research Awards   1338

The Physician Investigator Research Award program provides $10,000 in seed money to physicians who a propose a pilot, feasibility or small research study related to quality, cost or appropriate access to health and medical care for Michigan residents. See BCBSM Foundation website to obtain application form.

Requirements: Applicants must be a physician interested in research, who is licensed and domiciled in the state of Michigan. Applicants may include physicians working in research environments such as medical schools or university-affiliated hospitals, health care systems, or nonprofit agencies.

Restrictions: The program does not support basic science, biomedical research including drug studies, or studies using animals.

Geographic Focus: Michigan

Amount of Grant: $10,000

Contact: Nora Maloy, Senior Program Officer; (313) 225-8706; fax: (313) 225-7730; email: foundation@bcbsm.com

Internet: http://www.bcbsm.com/foundation/grant.shtml

Sponsor: Blue Cross Blue Shield of Michigan Foundation

600 East Lafayette Boulevard, X520

Detroit, MI 48226

BCBSM Foundation Primary and Clinical Prevention Grants   1339

The Blue Cross Blue Shield of Michigan Foundation requests letters of interest from Michigan physicians and members of the research community interested in preventing the three leading causes of death in Michigan: heart disease, cancer and stroke. The purpose of this grant initiative is to increase the use of evidence-based prevention efforts to improve the health of Michigan residents. The BCBSM Foundation has allocated $500,000 to fund several projects for up to two years. Funds will be available for salary support, program costs, supplies, office operations and other costs related to the proposed project. Computer equipment expenses, including PC hardware and software, will not be supported. In addition to the overall quality of the project, the feasibility and appropriateness of the budget is a factor in the funding decisions. The required application cover page and the terms and conditions may be downloaded for the BCBSM web site.

Requirements: Applicants must be doctoral level researchers (including physicians) based at Michigan universities, academic medical settings, community hospitals, health systems and community based nonprofit organizations. The principal investigator must have appropriate research and clinical credentials.

Geographic Focus: Michigan

Date(s) Application is Due: Sep 1

Contact: Nora Maloy, Senior Program Officer; (313) 225-8706; fax: (313) 225-7730; email: nmaloy@bcbsm.com

Internet: http://www.bcbsm.com/foundation/grant.shtml

Sponsor: Blue Cross Blue Shield Michigan Foundation

600 East Lafayette Boulevard, X520

Detroit, MI 48226

BCBSM Foundation Student Award Program   1340

The program offers a one year stipend to fund a wide range of health care projects, including applied research, pilot programs, or demonstration and evaluation projects.

Requirements: All doctoral and medical students enrolled in Michigan universities are eligible. For consideration, the proposed project must focus geographically on the state of Michigan and address the BCBSM Foundation's objectives.

Restrictions: Completed or substantially completed dissertations or research projects are not eligible. Students who previously received this award are not eligible. Blue Cross and Blue Shield of Michigan affiliates and subsidiaries are not eligible. Investigation of pharmaceutical efficacy, basic research or research involving non-human subjects is not eligible. Grants are not intended to support field placements, practica or internships.

Geographic Focus: Michigan

Amount of Grant: $3000

Date(s) Application is Due: Apr 30

Contact: Ira Strumwasser, Director; (313) 225-8706; fax: (313) 225-7730; email: foundation@bcbsm.com

Internet: http://www.bcbsm.com/foundation/

Sponsor: Blue Cross Blue Shield of Michigan Foundation

600 Lafayette East-X520

Detroit, MI 48226

BCBSM Frank J. McDevitt, DO, Excellence in Research Awards   1341

The Frank J. McDevitt Excellence in Research Awards for Health Services, Policy and Clinical Care recognize Michigan-based researchers and physicians who have published research in research journals that contributes to improving health and medical care in Michigan. The BCBSM Foundation will award $10,000 for unrestricted research to physician and doctoral level researchers for research on: health policy; health services; clinical care.

Requirements: Applicants must be from Michigan based researchers with terminal research degrees (e.g. PhD, DrPH) and from Michigan based physicians (MD or DO). In the case of multiple authorship, the nominee must be the first author. All authors will be nominated and will constitute one team nomination. The nomination research must have been published, or accepted for publication in a refereed health or medical care journal, within the previous two years. The financial award will be made to the recipient's 501 (c)(3) non-profit or educational organization.

Restrictions: Only one article per nominee will be accepted for consideration each year.

Geographic Focus: Michigan

Amount of Grant: $10,000

Date(s) Application is Due: Jan 1

Contact: Nora Maloy, Senior Program Officer; (313) 225-8706; fax: (313) 225-7730; email: nmaloy@bcbsm.com

Internet: http://www.bcbsm.com/foundation/grant.shtml

Sponsor: Blue Cross Blue Shield of Michigan Foundation

600 Lafayette E, X520

Detroit, MI 48226

BCBSM Investigator-Initiated Research Program Grants   1342

This program encourages Michigan-based applied research projects designed to improve health care in the state of Michigan. Projects that focus on the quality and cost of health care and appropriate access are considered priority and include: organization and delivery of health care services; evaluation of new methods or approaches to containing health care costs; evaluation of new methods or approaches to providing access to high quality health care; assessment and assurance of quality care; identification and validation of clinical protocols and evidence based practice guidelines. Applications are excepted at any time and are reviewed at the next possible Board of Directors meeting.

Requirements: Applications will be accepted from medical and doctoral-level researchers based in hospitals, university settings, non-profit health care organizations, health systems, medical or nursing schools, schools of public health, or other relevant academic disciplines. A complete application must include the original (unbound) and four copies of the supporting documents.

Restrictions: This program does of support basic science or biomedical research, including drug studies or studies using animals.

Geographic Focus: Michigan

Amount of Grant: $50,000-$75,000 per year

Contact: Administrator, (313) 225-8706; fax: (313) 225-7730; email: foundation@bcbsm.com

Internet: http://www.bcbsm.com/foundation/gp_iip.shtml

Sponsor: Blue Cross Blue Shield of Michigan Foundation

600 Lafayette East, X520

Detroit, MI 48226

BCBSNC Foundation Grants 1343

The foundation's primary objective is to fund clearly defined, innovative grants that further the foundation mission of improving the health and well-being of North Carolinians. The foundation funds programs typically possessing the following characteristics: programs and/or services designed to produce measurable, long-term impact; programs that are sustainable and designed to be ongoing, rather than one-time or sporadic events; programs and/or services that are replicable. The three primary focus areas include health of vulnerable populations, healthy active communities and community impact through nonprofit excellence.

Requirements: Applicants must meet the following criteria to be eligible for funding: organization is located within North Carolina; organization is a 501(c)3 organization or an educational or governmental entity with tax-exempt status, that is not a private foundation or Type III supporting organization; organization must be able to provide its most recent IRS Form 990. Depending on the type of grant and the size of the organization, an audit may be required as part of the submitted proposal.

Restrictions: The foundation will not provide funding for annual campaigns, political campaigns, religious purposes, individuals, endowments, purchase of advertisements, or for the sole purpose of receiving goods or entitlements from a charitable organization.

Geographic Focus: North Carolina

Contact: Grant Review Committee, (919) 765-7347; email: foundation@bcbsnc.com

Internet: http://www.bcbsnc.com/foundation/grants.html?previouslyOver=true&currentlyOver=true

Sponsor: Blue Cross Blue Shield of North Carolina Foundation
P.O. Box 2291
Durham, NC 27702

BCRF Research Grants 1344

The goal of the grant process is a serious peer review of all proposals. Proposals are invited by the Medical Advisory Board, rather than accepted as unsolicited requests. The MAB generally reviews proposals in the late summer. There is no set format, but the BCRF requests a brief narrative (no more than 4-5 pages), lay language summary, annual budget (no more than $225,000 initially, with a maximum of 20% in indirect costs) and budget narrative. Considerable latitude is given to the investigators in terms of the work proposed. Both the Board of Directors and the Medical Advisory Board concur that some of the most important advances in understanding the disease will most likely occur by enabling brilliant minds to pursue some of their most creative theories.

Restrictions: While most of the grants are unrestricted, the Foundation has donors requesting restricted grants for specific purposes. When this occurs, the Medical Advisory Board will meet to determine whether the area of interest merits further exploration. If so, they will identify various research projects meeting the donor, criteria.

Amount of Grant: $225,000 maximum

Contact: Program Administrator, (646) 497-2600; fax: (646) 497-0890; email: bcrf@bcrfcure.org

Internet: http://www.bcrfcure.org/rese_meet_grantapp.html

Sponsor: Breast Cancer Research Foundation
60 East 56th Street, 8th Floor
New York, NY 10022

BCRF-AACR Grants for Translational Breast Cancer Research 1345

The grants will provide direct support for innovative cancer research projects designed to accelerate the discovery, development, and application of new agents to treat breast cancer and/or for pre-clinical research with direct therapeutic intent. Special emphasis will be placed on research that holds promise for leading to individualized therapeutic options for treatment in the near future.

Requirements: Researchers who are affiliated with any institution involved in cancer research, cancer medicine, or cancer-related biomedical science anywhere in the world may apply. There are no geographic, national, or residency status restrictions. Applicants must have acquired a doctoral degree in a related field. If an applicant has obtained an equivalent degree at a foreign institution, information on the nature of the degree must be provided at the time of application.

Restrictions: Employees or subcontractors of a national government or the for-profit private industry are not eligible to serve as the Principal Investigator for the purpose of these grants. However, collaborations with such individuals are encouraged. Neither the members of the Scientific Review Committee nor the members of their individual laboratories are eligible for these grants.

Amount of Grant: $200,000

Date(s) Application is Due: May 20; Jul 20

Contact: Hanna Hopfinger, Program Assistant; (267) 646-0665 or (215) 440-9300; fax: (215) 440-9372; email: hanna.hopfinger@aacr.org or awards@aacr.org

Internet: http://www.aacr.org/home/scientists/research-funding—fellowships/bcrf-aacr-grants.aspx

Sponsor: Breast Cancer Research Foundation and the American Association for Cancer Research
615 Chestnut Street, 17th Floor
Philadelphia, PA 19106-4404

Beatrice Laing Trust Grants 1346

The trust was founded for the relief of poverty and for the advancement of the evangelical faith internationally. Grants are awarded for general charitable purposes at the discretion of the Trustees, with an emphasis on social services and relief, and medical aid and research. Grant recipients typically are charities working in deprived sections of the community in the United Kingdom, to missionary societies and, less frequently, to individuals working in the field of missions in United Kingdom and abroad.

Samples: ActionAid, 10,000 pounds to improve access to education for pastoralist families in Somaliland; OxFam, 10,000 pounds to the "You Are Here" program; Sense International, 15,000 pounds for work with deaf-blind people in Romania.

Contact: Trust Administrator, 020 8238 8890; fax: 020 8238 8897

Sponsor: Beatrice Laing Trust
33 Bunns Ln
London NW7 2DX United Kingdom

Beckman Coulter Community Relations Grants 1347

The corporation funds nonprofit organizations, charities and educational institutions that support and promote progress and interest in the areas of science, research-related health care and science education in our plant and customer communities.

Requirements: Organizations seeking funding should submit a letter of request providing the following information: full legal name of the organization; brief description of the organization, including a mission statement and services provided; amount requested; description of how the funds will be used. Additional attachments should include: history of previous support by Beckman Coulter; description of any involvement by Beckman Coulter employees; statement as to why you consider Beckman Coulter an appropriate donor; list of governing board members.

Restrictions: Grants do not support golf tournaments, galas, or religious or political activities.

Contact: Cathy Doherty, Community Relations, (714) 961-4478

Internet: http://www.beckman.com/hr/ourcompany/oc_communityRelations.asp#

Sponsor: Beckman Coulter Corporation
Community Relations, Beckman Coulter, Inc., 200 S. Kraemer Boulevard
Brea, CA 92821

Beckman Scholars Program Grants 1348

The Program is an invited program for accredited universities and four-year colleges in the United States. It provides scholarships that contribute significantly in advancing the education, research training and personal development of select students in chemistry, biochemistry, and the biological and medical sciences. The sustained, in-depth undergraduate research experiences and comprehensive faculty mentoring are unique in terms of program scope, content and level of scholarship awards. The amount given is $19,300 for two summers and one academic year.

Samples: Christa Jordan Laser, Arizona State University—Photosynthetic Reaction Center Mutagenesis (2007-08); Diana Ahn Tran, Harvey Mudd College—Characterization of Conserved Regulatory Elements in the Telomerase Promoter in Primates (2007-08); Donald Bungum, University of Chicago—Synthesis of 61Ni Compounds for Mossbauer Spectroscopy (2007-08).

Amount of Grant: $19,300

Contact: Program Administrator, (949) 721-2222; fax: (949) 721-2225; email: beckmanscholars@beckman-foundation.com or k.williams@beckman-foundation.com

Internet: http://www.beckman-foundation.com/bsp.html

Sponsor: Arnold and Mabel Beckman Foundation
100 Academy Drive
Irvine, CA 92617

Beckman Young Investigators Program Grants 1349

The foundation makes grants to nonprofit research institutions to promote research in chemistry and the life sciences, broadly interpreted, and particularly to foster the invention of methods, instruments, and materials that will open up new avenues of research in science. The program is intended to provide research support to the most promising young faculty members in the early stages of academic careers in the chemical and life sciences. The program is open to persons with tenure-track appointments in academic and nonprofit institutions that conduct fundamental research in the chemical and life sciences. The program is intended primarily for U.S. institutions. Only proposals of exceptional merit from foreign institutions will receive consideration. Projects are normally funded for a period of two years. When extraordinary circumstances warrant, support may be provided over a one-year or three-year period. Application materials are available on the Web site.

Requirements: To be eligible, an applicant should not have completed more than three full years in his or her tenure-track or other comparable independent research appointment on the anniversary date of initial appointment in the year in which application is to be made. Individuals may not apply more than three times.

Restrictions: Funding will not be considered for general institutional expenses or general fund-raising campaign expenses such as dinners and mass mailings. The foundation does not provide funds for overhead or indirect costs.

Samples: Christopher W. Bielawski, Ph.D., Department of Chemistry and Biochemistry, University of Texas at Austin (2007); Sean F. Brady, Ph.D., Laboratory of Genetically Encoded Small Molecules, Rockefeller University (2007); Joshua J. Coon, Ph.D., Department of Chemistry and Biomolecular Chemistry, University of Wisconsin, Madison (2007).

Amount of Grant: $264,000 average

Date(s) Application is Due: Sep 28

Contact: Program Administrator, (949) 721-2222; fax: (949) 721-2225; email: younginvestigators@beckman-foundation.com or k.williams@beckman-foundation.com

Internet: http://www.beckman-foundation.com/byi_guides.html

Sponsor: Arnold and Mabel Beckman Foundation
100 Academy Drive
Irvine, CA 92617

Becton Dickinson and Company Grants 1350

The company supports national and international activities in keeping with its commitment to advance the quality of medical practice and patient care in four areas: the global healthcare fund; the local initiatives fund; the matching gifts program; and the BD product donation program. Call for guidelines.

Contact: Community Relations, (201) 847-6800

Internet: http://www.bd.com/responsibility/contributions

Sponsor: Becton Dickinson and Company

1 Becton Drive

Franklin Lakes, NJ 07417

Bedding Plants Foundation Horticulture Scholarships 1351

Scholarships are available to graduate students in the following categories: majors in horticulture or related fields with a specific interest in bedding or flowering potted plants; students interested in international horticulture marketing; work/study abroad programs; students who intern or work for public gardens; and students in horticulture with a business emphasis or business with a horticulture emphasis. Application for all scholarships is available from the BPFI office, or students may apply via Fastweb.com.

Amount of Grant: $500-$2000 range

Date(s) Application is Due: May 1

Contact: William Willbrandt, Executive Director, (517) 333-4617; fax: (517) 333-4494; email: first@firstinfloriculture.org

Internet: http://www.firstinfloriculture.org/scholarships.htm

Sponsor: Bedding Plants Foundation

P.O. Box 280

Lansing, MI 48826

Bedding Plants Foundation Research Grants 1352

Grants provide funding for scientific research related to the production of bedding and potted plants. Topics include genetic engineering for diseases and insect resistance, management of greenhouse waste, increasing product shelf life and garden life, evaluation of biocontrol methods and techniques, water recirculation and reuse, water quality/alkalinity control, IPM in the greenhouse, insect problems (aphids, whitefly, thrips, fungus gnats), disease problems (botrytis, damping off, root rot, tomato spotted wilt virus, pythium), energy efficient greenhouse designs, and evaluation of alternative media.

Requirements: Work must be carried out by a nonprofit organization with IRS tax-exempt status.

Restrictions: The foundation does not fund overhead or travel expenses.

Amount of Grant: $100,000 average total

Date(s) Application is Due: Jan 15

Contact: Executive Director, (517) 333-4617; fax: (517) 333-4494; email: bpfi@aol.com

Sponsor: Bedding Plants Foundation

P.O. Box 280

Lansing, MI 48826

Beez Foundation Grants 1353

The foundation's mission is to raise money for pediatric brain cancer research and pediatric patient services. Research and/or support grants are awarded for brain cancer research projects or studies that are designed to advance the treatment, cure, outcome, and etiology of pediatric brain cancer; and any patient services that are focused on making the lives of children with cancer and their families a little easier. The foundation's particular interest is in the initial funding of projects that are likely to generate data that can be used in larger scale projects or studies. Preference is made to pediatric cancer-related requests. Applicants are requested to forward a one-page summary (in Microsoft Word or Adobe Acrobat format), the amount requested, and how the funds will be spent.

Amount of Grant: $40,000 total

Date(s) Application is Due: Nov 20

Contact: Grants Administrator, (732) 563-1144; email: proposals@beezfoundation.org

Internet: http://www.beezfoundation.org/Pages/Grants.html

Sponsor: Beez Foundation

26 H Worlds Fair Drive

Somerset, NJ 08873

Beim Foundation Grants 1354

The foundation awards grants to eligible nonprofit organizations in its areas of interest, including arts and culture, environmental conservation, education, and social services. Types of support include program grants, general operating grants, capital campaigns, building construction/renovation, land acquisition, equipment acquisition, and seed grants. High priority is given to the following types of projects: capital drives and equipment purchases; innovative start-up programs that require a moderate amount of grant money; intergenerational projects that involve community service; cooperative projects that involve several agencies or volunteers; on-going programs that have proven themselves unique and essential; and matching funds drives. Low priority is given to the following types of projects: medical research, debt retirement, national fundraising programs, and requests from public schools and governmental agencies due to the lack of good financial data. Deadlines are February 3 for education, and human services; and July 3 for arts, arts small capital equipment, and environment. Guidelines are available online.

Requirements: 501(c)3 tax-exempt organizations located within Minnesota, as well as the city of Denver, CO; counties of Park and Gallatin in Montana; county of Santa Fe in New Mexico; and county of Cumberland in Maine are eligible.

Restrictions: The foundation does not fund individuals; private foundations; political organizations or campaigns; religious organizations, including schools, except for secular human service activities; memberships, subscriptions, tickets for benefits, conferences, fundraising events, or annual campaigns; organizations that have as a substantial part of their purpose the influencing of legislation; endowment; multi-year commitments; or international efforts.

Geographic Focus: Colorado; Maine; Minnesota; New Mexico

Amount of Grant: $2000-$10,000 average

Date(s) Application is Due: Jan 17; Jul 18

Contact: Grants Administrator, (612) 605-8192; email: contact@beimfoundation.org

Internet: http://www.beimfoundation.org/guide.html

Sponsor: Beim Foundation

3109 W 50th Street, Suite 120

Minneapolis, MN 55410-2102

Beinecke Library Resident Fellowships 1355

The Beinecke Rare Book and Manuscript Library offers short-term fellowships to support visiting scholars pursuing postdoctoral or equivalent research in its collections. The library is Yale's principal repository for literary papers, and for early manuscripts and rare books in the fields of literature, theology, history, and the natural sciences. The fellowships, which pay for travel costs to and from New Haven and a living allowance, are designed to provide access to the library for scholars who live outside the greater New Haven area. Fellowships, normally granted for one month, must be taken up between September and May. Recipients are expected to be in residence during the period of their award and are encouraged to participate in the activities of Yale University. Successful applicants normally explain in extensive and specific detail the relationship of the Beinecke collections to the project and the significance of the project to its larger field of scholarly concern. Application form and guidelines are available online.

Restrictions: Students enrolled in degree programs are ineligible.

Amount of Grant: $3800 for one month plus travel

Date(s) Application is Due: Dec 15

Contact: Program Contact, Beinecke Rare Book and Manuscript Library, (203) 432-2956; fax: (203) 432-4047; email: beinecke.fellowships@yale.edu

Internet: http://www.library.yale.edu/beinecke/brbleduc/brblapplyvisiting.html

Sponsor: Yale University Library

Box 208240, Yale Station

New Haven, CT 06520-8240

Benign Essential Blepharospasm Research Grants 1356

The foundation supports research to find the cause and cure for benign essential blepharospasm/meige and related disorders of the facial musculature. A number of grants are issued annually. Proposals detailing research projects to be considered for support are welcomed.

Samples: Dr. John McCann, U of Utah (Salt Lake City, UT)—for the project Trial of Cervical Sympathetic Blockade for the Treatment of Benign Essential Blepharospasm, $13,000.; Dr. Mark Hallett, NINDS, NIH (Bethesda, MD) and Dr. Robert Daroff, Case Western Reserve U (Cleveland, OH)—for the blepharospasm workshop, $25,763.; Dr. Michael Hutchinson, NYU Medical Ctr (New York, NY)—for the project Neuroimaging of Blepharospasm, $62,325.

Amount of Grant: $150,000 maximum

Date(s) Application is Due: Jul 1

Contact: Office Manager (409) 832-0788; fax: (409) 832-0890; email: bebrf@blepharospasm.org

Internet: http://www.blepharospasm.org/res-prop.html

Sponsor: Benign Essential Blepharospasm Research Foundation

P.O. Box 12468

Beaumont, TX 77726-2468

Berea College Appalachian Music Fellowships 1357

The program is designed to support graduate students, faculty, public school teachers, and/or performers for a period of one to six months in order to conduct research in Berea's collection of noncommercial traditional music and to promote the preservation of and access to that music. Proposals should address the applicant's background and interest in Appalachian music, the reason(s) for seeking an opportunity for concentrated study in Berea's traditional music collections, anticipated contribution to the preservation or promotion of these musical resources (e.g., assistance with sound digitization; website development; writing contextual historical summaries, biographies, musical annotations; public performances, publications), and the length of time needed for the project (one month minimum, six months maximum). Contact information for three references should be included with proposals. Guidelines are available online. The fellowships must be taken up between January and August 2006. Fellows are expected to be in residence during the term of the fellowship.

Restrictions: Berea College employees are ineligible.

Amount of Grant: $5000

Date(s) Application is Due: Dec 5

Contact: Steve Gowler, (859) 985-3272; email: steve_gowler@berea.edu

Internet: http://www.berea.edu/hutchinslibrary/specialcollections/MusicFellowship2006.asp

Sponsor: Berea College

Berea, KY 40404

Berks County Community Foundation Grants 1358
The foundation supports a broad range of community projects in this Pennsylvania county, including the arts and culture, economic development, education, the environment, and health and human services. Types of support include general operating grants, capital campaigns, demonstration grants, seed grants, and program grants. Although applicants do not have to be located in Berks County, they must provide programs and services within the county.
Requirements: Tax exempt or public benefit organizations, individuals, associations and private or public agencies are eligible to apply. The grant must be used for charitable purposes only. Organizations are eligible to apply to more than one grant program in the same year.
Geographic Focus: Pennsylvania
Contact: Richard Mappin, Vice President for Grantmaking; (610) 685-2223; fax: (610) 685-2240; email: info@bccf.org
Internet: http://www.bccf.org/pages/grants.html
Sponsor: Berks County Community Foundation
P.O. Box 212, 501 Washington Street, Suite 801
Reading, PA 19603-0212

Berlex Foundation Reproductive Scientist Development Program Grants 1359
The program supports the postdoctoral career development of young obstetrician-gynecologists who are committed to careers in academic medicine. Grants support two to three years of full-time research under the mentorship of internationally recognized senior scientists. Applicants should contact Dr. Robert Jaffee, University of California, San Francisco, Box 0922, San Francisco, CA 94145-0922, (415) 476-9047.
Amount of Grant: $25,000 annually
Contact: Grants Administrator, (201) 342-4441; email: llisanti@berlex-foundation.org
Internet: http://www.berlex-foundation.org/aw04.asp
Sponsor: Berlex Foundation
433 Hackensack Avenue, 9th Floor
Hackensack, NJ 07601

Berlex Foundation Scholar Award in Basic Science Research 1360
The award provides the opportunity for clinician-investigators who wish to initiate new studies or who are currently conducting promising basic research in the area of reproductive medicine to continue work on their projects for an additional year. One or more awards may be made annually.
Requirements: Candidates must have completed an approved residency or fellowship in the field of obstetrics/gynecology. The applicant should be junior faculty. If not yet faculty, the individual should have reasonable assurance of such a position within a department of obstetrics and gynecology as evidenced by a Letter of Endorsement from the department head.
Restrictions: Candidates who are in an active training program as of January 1 are ineligible.
Samples: Dr. Robert Barbieri—for research, The Role of Hyperinsulinemia in the Pathogenesis of PCO, $40,000.; Dr. Jill Flood—for research, Identification of Factor(s) Required for Embryonic Developmental Competence, $40,000.
Amount of Grant: $40,000 to the scholar; $10,000 for laboratory support
Date(s) Application is Due: Oct 1
Contact: Grants Administrator, (201) 342-4441; email: llisanti@berlex-foundation.org
Internet: http://www.berlex-foundation.org/aw01.asp
Sponsor: Berlex Foundation
433 Hackensack Avenue, 9th Floor
Hackensack, NJ 07628

Berlex Foundation Scholar Award in Clinical Research 1361
The award is intended to support studies that are specifically focused on clinical research in obstetrics and gynecology. Its goal is to promote excellent clinical research by young investigators in the areas of diagnosis, treatment, and prognosis. The foundation plans to give more weight to proposals that incorporate more rigorous methods such as randomized controlled trials. Analytic studies, including cohort and case-control studies, also may be considered. One or more awards may be presented annually. Support is awarded to the institution in which the recipient will conduct the research. The recipient must identify the total amount required to complete the study and designate the specific allocation of funds. It is expected that the candidate devote a minimum of 50 percent of his/her time to the research activity.
Requirements: Candidates must have full-time faculty appointments in schools of medicine or public health. Preference will be given to proposals by academicians who are early in their careers.
Restrictions: Applicants should not currently be in training programs. The foundation is not seeking proposals for descriptive studies. This award is not intended for support of basic science research or animal studies.
Samples: Dr. Debra Guinn—for research, Multicenter, Randomized, Placebo-Controlled Trial of Serial Doses of Antenatal Corticosteroids, $50,000.; Andrea Witlin, DO—for research, The Efficacy of Magnesium Sulfate in Women with Mild Preeclampsia, $50,000.
Amount of Grant: $50,000 maximum awarded to the institution
Date(s) Application is Due: Oct 1
Contact: Svetlana Lisanti, Administrator, (201) 342-4441; email: llisanti@berlex-foundation.org
Internet: http://www.berlex-foundation.org/aw02.asp
Sponsor: Berlex Foundation

433 Hackensack Avenue, 9th Floor
Hackensack, NJ 07628

Bernard and Audre Rapoport Foundation Grants 1362
The mission of the foundation is to support programs that have the broadest possible impact in meeting important human needs and aspirations, with an emphasis on the needs of the least advantaged members of society. Preference is given to five primary areas: education, including early learning through the elementary years, adult education and training initiatives, and programs that enhance the capabilities of teachers and other professionals in public schools; arts and culture, especially activities that encourage the participation and enrich the lives of children and disadvantaged members of the community; healthcare— especially services to women, children, and those who do not have access to conventional medical resources; community-based outreach initiatives such as immunization programs; community building and social services—i.e., programs that build grassroots neighborhood networks, provide job training and job opportunities for the unemployed and underemployed, or provide a comprehensive safety net of social services for the least advantaged citizens; democracy and civic participation—efforts to make government more responsive, encourage citizens to take an active interest and role in political life, encourage intergovernmental cooperation, broaden public awareness of public policy issues and alternatives, build skills necessary for political leadership, and provide opportunities for community service; and Rapoport Service Scholarships at the University of Texas at Austin. There are no deadlines for proposals.
Requirements: Applicant must provide proof of tax exempt status.
Geographic Focus: Texas
Samples: Economic Policy Institute (Washington, DC)—to increase and improve the quality of research and analysis that is disseminated into the mainstream, $50,000 (2002); Central Texas Senior Ministries (Waco, TX)—to help fund 8163 Meals on Wheels (33 meals per day for one year) to Waco???s low-income homebound seniors, $20,000 matching grant (2002); Intermountain Therapy Animals (Salt Lake City, UT)—to support the R.E.A.D. program, whose goal is to improve the literacy skills of children through the assistance of registered therapy teams as literacy mentors, $10,000 (2002); Waco Symphony Assoc (Waco, TX)—to help cover the salary for a director, fees for core singers and accompanist and music, and to subsidize violinist Joshua Bell for the 2003-2004 season, $23,500 (2002).
Contact: Carole Jones, (254) 741-0510; fax: (254) 741-0092; email: carole@rapoportfdn.org
Internet: http://www.rapoportfdn.org/priorities.asp
Sponsor: Bernard and Audre Rapoport Foundation
5400 Bosque Boulevard, Suite 245
Waco, TX 76710

Bernard F. Langer Nuclear Codes and Standards Award 1363
Recognition of an individual(s) who has contributed to the nuclear power plant industry through the development and promotion of ASME nuclear codes and standards or the ASME nuclear certification program is made through the presentation of this award. The award includes a bronze plaque and certificate.
Requirements: Any individual member, group of members, or committee may nominate candidates.
Samples: Richard E. Gimple—award winner, $1000 (2005); Yasuhide Asada—award winner, $1000 (2004); Richard W. Barnes—award winner, $1000, (2003).
Amount of Grant: $1000
Date(s) Application is Due: Feb 1
Contact: Award Administrator, (212) 591-7158; fax: (212) 591-7739; email: soukupd@asme.org
Internet: http://www.asme.org/honors/ms71/saa/langer.html
Sponsor: American Society of Mechanical Engineers
3 Park Avenue
New York, NY 10016

Bernard L. Schwartz Fellows Program 1364
The foundation awards a number of fellowships to help individuals establish themselves as credible new voices in the country's public debate. Regular and senior fellowships are available. Fellowships are awarded for one year, subject to renewal. Fellows choosing to work out of the Washington, DC, office will receive fully equipped offices, health benefits, research and editorial assistance, and the opportunity to formulate their ideas in a well-knit community. Applications are available on the Web site.
Requirements: Regular fellows show promise in writing for student, academic, or popular publications, but have yet to establish themselves as leading commentators. Senior fellows are generally established academics, policy analysts, or journalists wishing to make departures in their work.
Amount of Grant: $25,000-$45,000 fellowships; $75,000 maximum stipend for senior fellows
Contact: Sherle Schwenninger, Program Director, (202) 986-2700; fax: (202) 986-3696; email: schwenninger@newamerica.net
Internet: http://www.newamerica.net/index.cfm?pg=fellows
Sponsor: New America Foundation
1630 Connecticut Avenue NW, 7th Floor
Washington, DC 20009

Bernard Lowy Fund for the Study of Tropical Botany in Latin America 1365
The Department of Plant Biology of Louisiana State University (LSU) and the family of the late mycologist, Bernard Lowy, established the Bernard Lowy Fund for the Study of Tropical Botany in Latin America. The fund is designed to provide much-needed

additional funding for research travel by graduate students and postdoctoral researchers studying diverse aspects of tropical botany in Latin America.

Requirements: Applicant either must be studying or working at a university or research organization in Latin America or be studying, working at, or have received the doctoral degree from LSU at Baton Rouge. Support for applicants from Latin American institutions is for travel and living expenses for research conducted at LSU. These applicants must seek sponsorship from a faculty member in the Plant Biology Department at LSU.

Amount of Grant: $1000 maximum
Date(s) Application is Due: May 1
Contact: Dr. Meredith Blackwell, Lowy Fund Committee, (225) 388-1557; fax: (504) 388-8459; email: mblackwell@lsu.edu
Sponsor: Bernard Lowy Fund
Dept of Plant Biology, Louisiana State University
Baton Rouge, LA 70803-1705

Bernard Van Leer Foundation Grants   1366

The foundation operates in 34 countries in Europe, the Far East, Australia, Central and South America, the United States, and Africa to further the holistic development of young children. Grantmaking focuses on thematic initiatives, which may include partners from countries where the foundation does not otherwise have a grantmaking program. Current themes are: children affected by HIV/AIDS; respect for diversity; and growing up in indigenous societies. The foundation funds projects that promote a holistic approach to early childhood development; the enhancement of parental capacity to support their children's development; a development strategy that is rooted in the local context and is culturally, socially, and economically appropriate; and the building of capacity, local ownership, and working in partnership. Unsolicited proposals are rarely funded. To be considered, submit information about the organization; the project's objectives, strategies, beneficiaries, scope, location, and duration; and the overall budget and the amount required.

Requirements: Grants are made to governmental and nongovernmental organizations located in the list of eligible countries, which include: Brazil, Colombia, the East Caribbean region, Germany, Greece, India, Indonesia, Israel, Kenya, Mexico, Morocco, the Netherlands, Peru, Poland, South Africa, Tanzania, Thailand, Turkey, Uganda, United States of America and Zimbabwe.

Restrictions: The foundation does not fund applications for support to individual children; projects that concentrate solely on one aspect of children's development or learning; projects that specifically focus on children with special needs such as mental and/or physical handicap; proposals for the construction and maintenance of buildings, or the purchase of equipment and materials; isolated requests for scholarships, conferences, media or theatre events; or general support to organisations, or requests to cover recurrent costs or deficits

Contact: Peter Laugharn, Executive Director; 31-70-351-20-40; fax: 31-70-350-23-73; email: proposal.administration@bvleerf.nl
Internet: http://www.bernardvanleer.org/about/Applications_for_funding
Sponsor: Bernard Van Leer Foundation
Eisenhowerlaan 156, P.O. Box 82334
The Hague 2508 EH Netherlands

Bersted Foundation Grants   1367

The foundation awards grants to Illinois nonprofit organizations in response to changing community needs. Areas of interest include health care and mental health, social services, children and youth services, family services, the economically disadvantaged and homeless, community development, and environment. Types of support include scientific research, general operating support, continuing support, building construction/renovation, and technical assistance. The board meets in January, April, July, and October. Submit a letter of intent detailing the request and organization.

Requirements: Illinois tax-exempt organizations in DeKalb, DuPage, Kane, and McHenry Counties are eligible.

Restrictions: Grants do not support religious houses of worship, institutions of higher education, endowment funds, or deficit financing.

Geographic Focus: Illinois
Samples: Aurora Foundation (IL)—for general support, $50,000.; South East Assoc for Special Parks and Recreation (Downers Grove, IL)—for general support, $5000.
Amount of Grant: $5000-$35,000 average
Contact: M.Catherine Ryan, (312) 828-1785
Sponsor: Bersted Foundation
231 S LaSalle Street
Chicago, IL 60697

Beryl Henderson Foundation Grants   1368

The Beryl Henderson Foundation and the Australian Federation of University Women will offer a grant of up to $1500 to assist in completion of a research project being undertaken by a female student or researcher in an Australian university. The grant is made in alternate years to honour the memory of Beryl Henderson (1897-1990), a leading women's rights campaigner in Britain and later in Australia. Applications are invited from women who are enrolled either in Honours IV, or programs at an Australian tertiary institution leading to the award of a degree, or embarked on research work with publication in Australia in view. The purpose of the grant is to assist with practical aspects of research, such as intra- or interstate travel to consult library sources, or to consult with research collaborators. All or part of the grant could also be used for obtaining information (e.g. archival material) from overseas which is not available via the internet. It is not envisaged that the grant would be available for equipment such as

computer hardware. The award of the grant is not limited to particular disciplines, but research focusing on issues related to the rights and welfare of women and children will be viewed favorably. Selection will be based on merit.

Amount of Grant: $1,500 maximum
Date(s) Application is Due: Jul 31
Contact: Dr. Gwen Woodroofe, Fellowships Officer; (02) 6295 6970; email: jocelyn. eskdale@daff.gov.au; Gaynor Reeves, Fellowships Convener; AFUW-SA Trust Fund; (02) 4952 3174; fax: (02) 4952 3070; email: gay.reeves@bigpond.com or gaynor.reeves@gmail.com
Internet: http://www.afuw.org.au/ScholarshipsFS.htm
Sponsor: Beryl Henderson Foundation / Australian Federation of University Women
138 LaPerouse Street
Red Hill ACT, Australia 2603 Australia

Bethesda Foundation Grants   1369

The foundation awards grants to nonprofits in Hornell, NY, in its areas of interest, including education, hospitals, health care, substance abuse services, nutrition, AIDS research, and social services. Types of support include general operating support, equipment acquisition, program development, and scholarship funds. Application forms must be obtained from the office. The board meets in March, June, September, and December.

Requirements: Nonprofit organizations in Hornell, NY, are eligible.
Geographic Focus: New York
Amount of Grant: $1000-$24,000
Contact: Grants Administrator, (513) 745-1616; fax: (513) 745-1623; email: bethesdafoundation@trihealth.com
Internet: http://www.bethesdafoundation.com
Sponsor: Bethesda Foundation
10506 Montgomery Road, Suite 304
Cincinnati, OH 45242

Bettina Bahlsen Memorial Graduate Scholarship   1370

The scholarship competition is open to full-time graduate students registered in or eligible to register in the Department of Biological Sciences at the university in the field of cellular, molecular, microbial, or biochemical biology. Selection will be based on academic excellence. Preference will be accorded to a foreign student and, if possible, to a student entering the first year of graduate studies. The scholarship is awarded annually for one-year duration with possibility of renewal in open competition.

Requirements: To be eligible applicants must be graduate students in the department of biological sciences
Amount of Grant: $C17,500 maximum
Date(s) Application is Due: Feb 1
Contact: Connie Baines, Faculty of Graduate Studies, (403) 220-5690; email: cbaines@ucalgary.ca
Internet: http://www.grad.ucalgary.ca/funding/internal_scholarships/lev_4/bahlsen_bettina.htm
Sponsor: University of Calgary
2500 University Dr NW, Earth Sciences Building, Room 720
Calgary, AB T2N 1N4 Canada

Beverly Willis Architecture Foundation Grants and Fellowships   1371

The foundation offers funding to individuals and institutions to support innovative projects that advance the study and expand the recognition of women in architecture and related professions, and that lead to the dissemination of this knowledge to professional and public audiences alike. Grant funding is divided into the following categories: fellowships for scholarly research, publication, exhibition, or documentary in film or other media; and grants for honoraria to plenary session speakers at professional meetings, conferences, or symposia whose focus matches the mission of BWAF; and travel grants for research trips or professional conferences at which the recipient will be making a presentation related to the purpose of the BWAF. Successful candidates receive their fellowship awards in two equal installments. Applicants are strongly encouraged to send a preliminary inquiry to the director regarding their proposal and its eligibility for funding. Application guidelines are available online. Although not mandatory, applicants are encouraged to send a preliminary inquiry to the Director of BWAF regarding their proposal and its eligibility for funding.

Requirements: Applications should contain, as their cover page, a summary sheet that includes the following: project title; name, address, telephone number, and email address of the applicant; a concise abstract (150 words or less) of the proposed project; a specific amount (in U.S. dollars) sought from the Foundation; a description of the anticipated final product(s) that would result from the proposed project, including plans for distribution of the completed work; names and contact information of the people from who letters of support have been requested. Tax-exempt organizations must include with their grant request a copy of the IRS Determination Letter, which indicates the particular paragraph of the Internal Revenue Code that governs their exempt activities.

Amount of Grant: $10,000 maximum fellowships; $3000 maximum grants; $1500 maximum travel grants
Date(s) Application is Due: Mar 15
Contact: Director, (212) 577-1200; email: director@bwaf.org
Internet: http://www.bwaf.org/grants.html
Sponsor: Beverly Willis Architecture Foundation
2 Columbus Avenue, Suite 3A
New York, NY 10023

BFGoodrich Collegiate Inventors Prize   1372

The national competition is designed to encourage undergraduate and graduate college students active in science, technology, and creative invention, while stimulating their problem-solving abilities. Prizes are awarded to recognize students and their faculty advisors for research and innovative discoveries that can be patented. Winners are honored alongside inductees of the National Inventors Hall of Fame in September.

Requirements: Students enrolled full-time in any U.S. college or university as of the date of entry submission are eligible to participate.

Amount of Grant: $30,000 maximum total

Date(s) Application is Due: Sep 1

Contact: Ray DePuy, (330) 849-6887; email: rdepuy@invent.org

Internet: http://www.invent.org/collegiate/overview.html

Sponsor: BFGoodrich Collegiate Inventors Program

221 S Broadway Street

Akron, OH 43308-1505

BIA Native American Graduate Student Fellowships   1373

Graduate fellowship grants are provided to supplement financial assistance to eligible American Indian/Alaska Native students pursuing a postbaccalaureate degree. OIEP contracts with the American Indian Graduate Center (AIGC). Application requirements and time frames for submitting an application are available with the AIGC by calling (505) 881-4584, writing to 4520 Montgomery Blvd NE, Suite 1-B, Albuquerque, NM 87109, or visiting their web site at www.aigc.com.

Requirements: Graduate fellowship grants are available to individuals who are pursuing a master's or doctorate degree as a full time student at a U.S. accredited graduate school; able to demonstrate financial need; and an enrolled member of a federally recognized American Indian tribe or Alaskan Native group, or possess one- fourth degree federally recognized Indian blood.

Contact: American Indian Higher Education Grant Program, (202) 208-3478; fax: (202) 208-3312

Internet: http://www.aigc.com/fellowship-application/what-is-graduate-education.html

Sponsor: Bureau of Indian Affairs

1849 C Street NW, MS 3512 MIB

Washington, DC 20240

Bibliographical Society of America Fellowships   1374

Short-term (one- or two-month) fellowships are available in support of bibliographical inquiry as well as research in the history of the book trades and in publishing history. Topics may concentrate on books and documents in any field but should focus on the book or manuscript as historical evidence, whether for establishing a text or understanding the history of book production, publication, distribution, collection, or consumption. Stipends support travel, living, and research expenses.

Samples: Jacqueline Goldsby (University of Chicago)—A Book of its Time: James Weldon Johnson's The Autobiography of an Ex-Colored Man, a Critical Edition (2005); Margit Smith (University of San Diego)—The Medieval Girdle Book Documentation Project (2005).

Amount of Grant: $2000 maximum per month

Date(s) Application is Due: Dec 1

Contact: Executive Secretary, (212) 452-2710; fax: (212) 452-2710

Internet: http://www.bibsocamer.org

Sponsor: Bibliographical Society of America

P.O. Box 1537, Lennox Hill Sta

New York, NY 10021

Bilateral Scholarships   1375

Scholarships are offered for three to nine months to foreign students, research workers, and specialists in various fields for studies and research in Finland on a reciprocal basis to nationals of the following countries: Australia, Austria, Belgium, Bulgaria, Canada, China, Cuba, Czech Republic, Denmark, Egypt, France, Germany, Great Britain, Greece, Hungary, Iceland, India, Ireland, Israel, Italy, Japan, Luxembourg, Mexico, Mongolia, the Netherlands, Norway, Poland, Portugal, Republic of Korea, Romania, Slovakia, Spain, Sweden, Switzerland, Turkey, and the United States. Application must be obtained from the diplomatic representatives or directly from CIMO.

Requirements: Applicant must have been awarded a degree at a university or institute of higher education in his/her own country; applicant must also have a working knowledge of Finnish, Swedish, English, or German. Selection of scholars is primarily in the hands of the relevant authorities in the candidate's own country who will make the selection and propose the candidates to the Ministry of Education for final approval.

Restrictions: Scholars may not do full-time paid work during the scholarship period.

Contact: Ulla-Maija Anttila, Exchanges Coordinator, 358-9-7747 7680; helpline: 358-1080-6767; email: cimoinfo@cimo.fi

Internet: http://www.cimo.fi/Resource.phx/cimo/services/scholarships.htx

Sponsor: Center for International Mobility (CIMO)

P.O. Box 343, Hakaniemenkatu 2

Helsinki 00531 Finland

Bill and Melinda Gates Foundation Grand Challenges in Global Health Grants   1376

A grand challenge is a call for a specific scientific or technological innovation that removes a critical barrier to solving an important health problem in the developing world with a high likelihood of global impact and feasibility. Grand challenges address the diseases and health conditions that cause the greatest morbidity and mortality in the developing world, thus accounting for the enormous health disparities between the developing and the developed world, and that receive disproportionately less attention from the scientific and technical community than their consequences demand. The goals are broad, encompassing prevention, detection, diagnosis, treatment, rehabilitation, and surveillance and control of diseases. Challenges include to improve childhood vaccines; create new vaccines; control insects that transmit agents of disease; improve nutrition to promote health; improve drug treatment of infectious diseases; cure latent and chronic infections; and measure disease and health status accurately and economically in developing countries. Letters of intent are due by the listed application deadline; form and guidelines are available online.

Requirements: For-profit, nonprofit, academic, domestic, and foreign institutions may submit letters of intent.

Restrictions: Individuals not affiliated with an institution are ineligible.

Contact: Program Administrator, (301) 402-4968; fax: (301) 480-1661; email: Grants@GrandChallengesGH.org

Internet: http://www.grandchallengesgh.org/challenges.aspx?SecID=258

Sponsor: Foundation for the National Institutes of Health

Natcher (Building 45), Room 3AN-44

Bethesda, MD 20892-6300

Billie Jean King Foundation Grants   1377

The mission of the foundation is to inspire humankind in the pursuit of excellence, ensure equal opportunity, and enhance the quality of life for all individuals-'regardless of gender, race, religion, appearance, or sexual orientation. Areas of interest include health, education, and athletics. Programs eligible for support include public education, training, service provision, research, legal services, advocacy, and legislative reform. Priority will be given to projects focusing on women; multicultural, lesbian, gay, bisexual, and transgendered individuals; minorities; and youth. Innovative, entrepreneurial efforts where new funding is likely to have a broad and lasting impact receive preference. There are no application deadlines. Submit a brief letter of inquiry; full applications are by invitation.

Requirements: 501(c)3 nonprofit organizations are eligible.

Amount of Grant: $1000-$25,000

Contact: Billie Jean King, President, (623) 362-7208

Internet: http://vpr2.admin.arizona.edu/rso/02010411.htm

Sponsor: Billie Jean King Foundation

P.O. Box 10777

Phoenix, AZ 85064-0777

Blackall Machine Tool Award   1378

This award is given annually for the best paper or papers clearly concerned with or related to the design or application of machine tools or dimensional measuring instruments, submitted to ASME for presentation and publication. Papers by multiple authors are eligible.

Requirements: Authors are not restricted by nationality, age, or society membership.

Amount of Grant: $1000

Date(s) Application is Due: Feb 1

Contact: Gilda DiTullio, Manager, (212) 591-7736; fax: (212) 705-7739; email: ditulliog@asme.org

Internet: http://www.asme.org/honors/ms71/sla/bmtg.html

Sponsor: American Society of Mechanical Engineers

3 Park Avenue

New York, NY 10016

Blakemore Freeman Fellowships for Advanced Asian Language Study   1379

The fellowships are awarded for one year of advanced language study in East or Southeast Asia in structured language programs or private tutorial programs where the primary focus is on study of the modern language. Eligible languages include Chinese, Vietnamese, Tibetan, Japanese, Indonesian, Thai, Korean, Khmer, Burmese, and Malaysian. Consideration will be given to other East or Southeast Asian languages on an individual basis. The program also supports short-term refresher grants, which are awarded to college professors, postdoctoral professionals, and indiviudals who have previously completed certain designated advanced language study abroad. Application and guidelines are available online.

Requirements: An applicant must be pursuing an academic, professional, or business career that involves the regular use of a modern East or Southeast Asian language; have a college undergraduate degree; be at or near an advanced level in the language (a minimum of three academic years of regular language study at the college level, or a minimum of one academic year of full-time intensive language study at the college-level, or proof of equivalent competency); be able to devote oneself exclusively to language study during the term of the grant (grants are not made for part-time study or research); and be a U.S. citizen or permanent resident of the United States.

Restrictions: Grants will not be made for the study of classical forms of Chinese, Japanese, Korean, Tibetan, or other languages not in current use in Asia that are studied primarily for academic purposes.

Date(s) Application is Due: Dec 30

Contact: Cathy Scheibner, Administrative Assistant, (206) 583-8778; fax: (206) 359-9778; email: blakemore@perkinscoie.com

Internet: http://www.blakemorefoundation.org/language.htm

Sponsor: Blakemore Foundation

1201 Third Avenue, Suite 4800

Seattle, WA 98101-3266

Bliss Prize Fellowships in Byzantine Studies 1380

This award is intended to provide encouragement, assistance, and training to outstanding college seniors who plan to enter the field of Byzantine studies. The fellowship covers graduate school tuition and living expenses as estimated by the graduate school in which the successful candidate enrolls. In addition, it includes summer travel for the intervening summer to areas that are important for an understanding of Byzantine civilization and culture. Students must be nominated by their advisors by October 15 and must furnish an application stating their future plans by November 1. Official undergraduate transcripts as well as two letters of recommendation and a writing sample must also be received by November 1. Write to the office for application information.

Requirements: The fellowship is normally restricted to candidates currently enrolled in U.S. or Canadian universities or colleges. Applicants must be in their last year of undergraduate education or have a BA, must have completed at least one year of Greek by the end of the senior year, and must be applicants to graduate school in any field or area of Byzantine studies.

Amount of Grant: $5000 maximum for summer travel, plus tuition and living expenses for two years

Date(s) Application is Due: Oct 15; Nov 1

Contact: Alice-Mary Talbot, Director's Office, (202) 339-6940; fax: (202) 339-6419; email: DumbartonOaks@doaks.org or Byzantine@doaks.org

Internet: http://www.doaks.org/Blissprize.html

Sponsor: Dumbarton Oaks

1703 32nd Street NW

Washington, DC 20007

Blowitz-Ridgeway Foundation Early Childhood Development Award 1381

The foundation supports nonprofit organizations and programs in the areas of medicine, psychology, residential care, and education, and for research in medicine, psychology, social science, and education. Types of support include capital grants, endowments, program development grants, research grants, and scholarships. The Foundation prefers prospective grantees whose programs or services benefit persons who have not yet reached their majority and/or are for the care of individuals or elderly persons who lack sufficient resources to provide for themselves. Although grants may be made to organizations outside the state of Illinois, preference will generally be given to applicants from Illinois. Application information is available on the Web site.

Requirements: Applicants must be classified as 501(c)3 by the IRS.

Restrictions: Grants will not be made for religious or political purposes, nor generally for the production or writing of audio-visual materials.

Samples: Catholic Charities, Diocese of Joliet (IL)—for support of its Back to School Fairs in DuPage and Will counties, which provide low-income families with immunizations and services, $10,000 (2003); Children???s Memorial Hospital/Northwestern U, Dr. Emerick (IL)—for a research project which will study the mechanisms of growth failure in children with different forms of liver disease, $63,700 (2003); Cabrini-Green Tutoring Program (IL)—for general operating support of its tutoring programs, $5000 (2003); Science and Arts Academy (IL)—for scholarships of varying amounts to help economically disadvantaged, gifted students attend its school, $10,500 (2003).

Contact: Serena Moy, Administrator, (847) 330-1020; fax: (847) 330-1028; email: megan@blowitzridgeway.org

Internet: http://www.blowitzridgeway.org/information/information.html

Sponsor: Blowitz-Ridgeway Foundation

1701 E. Woodfield Road, Suite 201

Schaumburg, IL 60173

Blowitz-Ridgeway Foundation Grants 1382

The foundation supports nonprofit agencies that provide medical, psychiatric, and psychological care to economically disadvantaged children and adolescents. Program and capital grants are awarded, primarily in Illinois, in support of medical, psychiatric, psychological, and/or residential care; and research programs in medicine, psychology, social science, and education. The foundation supports operating budgets, and applicants may request commitments that extend beyond one year, but requests for annual funding will not be considered. Applications are accepted throughout the year and are reviewed in the order in which they are received. Guidelines and applications are available online.

Requirements: 501(c)3 nonprofit organizations that offer services to people who lack resources to provide for themselves may apply.

Restrictions: Grants will not be awarded to government agencies or to organizations that subsist mainly on third-party funding and have demonstrated no ability or expended little effort to attract private funding. Grants will not be made for religious or political purposes or for the production or writing of audio-visual materials.

Geographic Focus: Illinois

Samples: Elizabeth Ann Seton Program, $5,000 in general operating support of programs for pregnant and parenting mothers; The Enterprising Kitchen, $15,000 for programs improving self-sufficiency and employability of low-income women; Faith in Action of McHenry County, $7,000 to recruit volunteers; Gospel Rescue Mission, $10,000 for children's social and recreational activities; Guardian Angel Community Services, $5,000 for the Groundwork Domestic Violence Program (2006).

Amount of Grant: $5000-30,000

Contact: Serena Moy, Administrator, (847) 330-1020; fax: (847) 330-1028; email: serena@blowitzridgeway.org

Internet: http://www.blowitzridgeway.org/information/information1.html

Sponsor: Blowitz-Ridgeway Foundation

1701 E Woodfield Road, Suite 201

Schaumburg, IL 60173

Blue Shield of California Grants 1383

Consideration for funding will be given exclusively to organizations that pursue activities directly related to the foundation's program goals, including domestic violence prevention through service provision, education, and outreach; research and education regarding medical best practices and health technologies; and direct or indirect provision of medical insurance or health care to those populations that are uninsured or underinsured, and related policy development.

Requirements: The Foundation funds organizations that are non-profit and tax-exempt under 501(c)3 of the Internal Revenue Service Code (IRC) and defined as a public charity under 509(a)1, 2, or 3 (types I, II, or a functionally integrated type III); accredited schools; units of government/public agencies; tribal governments. The foundation will only support projects that meet the following criteria: the mission of the grantee organization is consistent with the goals and mission of the foundation; the grant is used primarily to serve Californians; the grant seeking organization has a reputation for credibility and integrity; the grant seeking organization is pursuing activities directly related to one of the Foundation's three Program Areas: Health Care and Coverage, Health and Technology and Blue Shield Against Violence.

Restrictions: The foundation does not fund award dinners, athletic events, competitions, special events, or tournaments; conferences or seminars; capital construction; television/film/media production; religious organizations for religious purposes; political causes, candidates, organizations or campaigns; capital projects over $50,000; multi-year projects (generally); grants to individuals (with the exception of the regulated Blue Shield of California Employee Scholarship Program); grants to 509(a) 3, type III supporting organizations that are not "functionally integrated".

Geographic Focus: California

Samples: U of California at San Francisco Medical Ctr (CA)—to participate in a program that will use cutting-edge monitoring and control technology designed to prevent hospital-borne infections, $90,000 (2005); 304 California nonprofit organizations (CA)—for projects to prevent domestic violence, increase access to health-care services, and assess medical technologies that can contribute to improved health, $6.7 million (approximately) divided (2004).

Contact: Grants Administrator, (415) 229-5785; fax: (415) 229-6268

Internet: http://blueshieldcafoundation.org/grant-center/index.cfm

Sponsor: Blue Shield of California

50 Beale Street

San Francisco, CA 94105-1808

Blum-Kovler Foundation Grants 1384

The foundation awards general operating grants to eligible Illinois nonprofit organizations in its areas of interest, including child welfare, civic affairs, cultural programs, health care, higher education, hospitals, medical research, and social services. Grants are awarded primarily in the Chicago metropolitan area. There are no application forms or deadlines. Submit a letter of request.

Requirements: Illinois nonprofit organizations are eligible.

Geographic Focus: Illinois

Amount of Grant: $2500-$50,000 average

Contact: Hymen Bregar, Secretary, (312) 664-5050

Sponsor: Blum-Kovler Foundation

875 N. Michigan Avenue, Suite 3400

Chicago, IL 60611

BMI Woody Guthrie Fellowships 1385

The foundation offers short-term fellowships to support scholarly use of the Woody Guthrie Archives Research Collection. The program invites applicants who are pursuing research topics or themes related to Woody Guthrie which explore his creative work and contribution to American music and culture. Disciplines may include, but are not limited to: American musicology; historical musicology; ethnomusicology; cultural studies, social sciences, humanities; and American history. A limited number of fellowships will be selected with a value of up to $2,500 per recipient to help defray travel to New York City and residence expenses for the duration of the fellowship. The length of the fellowship will depend on the applicant's research proposal, but is normally limited from one to six months.

Amount of Grant: $2,500 maximum

Date(s) Application is Due: Dec 1

Contact: Fellowships Administrator; (212) 586-2000; fax: (212) 245-8986; email: fellowship@woodyguthrie.org

Internet: http://bmi.com/foundation/program/woody_guthrie_fellowship_program/

Sponsor: Broadcast Music, Inc. (BMI) Foundation

320 West 57th Street

New York, NY 10019

BMW of North America Charitable Contributions 1386

The corporation funds charitable programs that seek to benefit society. Areas of interest include education, road traffic safety, and environment. Education—at all levels, from the very young to those pursuing advanced degrees; intercultural learning for K-12 students and their teachers; automotive technology, mechanics, and career and repair programs (in high schools, technical schools, and community colleges); research in the areas of safety design, ergonomics, and new materials. Road traffic safety—drivers' education programs geared at teenagers and new drivers; basic maintenance programs for women; consumer education on general road safety issues; and programs to promote the safety of children and young people on the road. Environment—conserve/preserve natural resources, in particular parklands and waterways; research/promote the use of alternative fuels; and provide environmental education for K-12 students. In general, grants are awarded for specific projects rather than for general operating support, although some operating and capital grants are given consideration. Applicants must submit an online application; guidelines are available online. Telephone solicitations are not considered.

Requirements: Organizations that have been approved by the IRS as 501(c)3 charities or 501(c)(9) organizations are eligible.

Restrictions: Grants do not support non-tax-exempt organizations; individuals; religious organizations for religious purposes; political candidates or lobbying organizations; organizations with a limited constituency, such as fraternal, labor, or veterans groups; travel by groups or individuals; national or local chapters of disease-specific organizations; national conferences, sports events, and other one-time, short-term events; sponsorships or advertising; anti-business groups; team sponsorships or athletic scholarships; or organizations outside the United States or its territories.

Contact: Corporate Contributions, (201) 307-4000

Internet: http://www.bmwusa.com/About/philanthropy.htm

Sponsor: BMW of North America, LLC

300 Chestnut Ridge Road

Woodcliff Lake, NJ 07677-7731

Board of Directors Distinguished Service Award for Senior Administrators 1387

The purpose of the Award is to recognize distinguished service to the Society over a period of years. The award is given at irregular intervals at the discretion of the ACS Board of Directors to recognize outstanding service to the Society by a senior staff member over a period of years. It may be given to the widow, widower, child, or children of the person recognized. The award consists of a cash amount to be determined by the ACS Board of Directors and an appropriate certificate or medallion.

Requirements: Any individual, except a member of the award committee, may submit one nomination or seconding letter for the award in any given year. The nominating documents consist of a letter of not more than 1000 words containing an evaluation of the nominee's accomplishments and a specific identification of the work to be recognized, a biographical sketch including date of birth, and a list of publications and patents authored by the nominee.

Restrictions: Self-nominations are not accepted.

Samples: Halley A. Merrell—award recipient (2002).

Date(s) Application is Due: Nov 1

Contact: Felicia Dixon, Awards Administrator; (800) 227-5558 or (202) 872-4408; fax: (202) 776-8008; email: f_dixon@acs.org or awards@acs.org

Internet: http://portal.acs.org/portal/acs/corg/content?_nfpb=true&_pageLabel=PP_ARTICLEMAIN&node_id=1319&content_id=CTP_004497&use_sec=true&sec_url_var=region1

Sponsor: American Chemical Society

1155 Sixteenth Street, NW

Washington, DC 20036

Boat U.S. Foundation Clean Water Grants 1388

Grants are available for projects that encourage boaters to learn to love their waterways. Projects of interest are helping boaters understand and appreciate their local boating habitat, as well as learn hands-on boating strategies that will keep the water and local habitat healthy and accessible for future boaters. Topics can range from petroleum pollution prevention to pumpout education to keeping trash out of our waterways. Guidelines and application are available online.

Requirements: The Foundation is interested in funding small, local, volunteer based nonprofits. The Foundation will fund volunteer boating groups, clubs, and associations, as well as local nonprofit organizations, including local chapters of national organizations.

Restrictions: Government agencies, national or international organizations, for-profit businesses, individuals, and private clubs not open to the general public are not eligible.

Samples: Baltimore County Sailing Ctr (Baltimore, MD)—to conduct an on-the-water research program to collect and identify nonpoint source pollution in the Herring Run watershed.; Junior Sailing Assoc of Long Island Sound (Larchmont, NY)—to distribute prizes to junior sailors who exhibit environmentally sensitive behavior while boating.; Schooner Sound Learning (New Haven, CT)—to produce Seven Sound Tips brochures for anglers, sailors, and power boaters to complement their Dirty Dozen traveling displays of marine debris.; U.S. Coast Guard Auxiliary ??? Flotilla 87 (Lake Anna, VA)—to use signs to educate boaters about trash disposal and pump-out laws.

Amount of Grant: $4000 maximum

Date(s) Application is Due: Feb 1

Contact: Grants Administrator, (703) 823-9550; email: cleanwater@boatus.com

Internet: http://www.boatus.com/cleanwater/grants

Sponsor: Boat U.S. Foundation for Boating Safety and Clean Water

147 Old Solomon's Island Road, Suite 513

Annapolis, MD 21401

Bodenwein Public Benevolent Foundation Grants 1389

The foundation awards grants to Connecticut nonprofit organizations to support social service and health agencies, including AIDS support, mental health services, health associations, and cancer and AIDS research; arts and culture, including performing arts and fine arts; youth and child welfare agencies; community development; education, including early childhood education, adult basic education, and continuing education; libraries and literacy and reading programs; the environment; animal rights and welfare; legal services; children, youth, women, and family services; religion; and minority services. Types of support include capital campaigns, building/renovations, equipment, program/project development, conferences and seminars, publication, seed money,

scholarship funds, research, consulting services, and matching funds. The board meets in January and July of each year to consider requests. Applicants may submit one grant application per calendar year.

Requirements: Nonprofit organizations in Lyme, Old Lyme, East Lyme, Waterford, New London, Montville, Groton, Ledyard, Stonington, and North Stonington, CT, are eligible to apply.

Geographic Focus: Connecticut

Date(s) Application is Due: May 15; Nov 15

Contact: Grants Administrator, c/o Bank of America (800) 841-4000

Sponsor: Bodenwein Public Benevolent Foundation

777 Main Street, CTEH40222B

Hartford, CT 06115

Bodman Foundation Grants Program 1390

The Foundation concentrates its programs in New York City, but also makes some grants in northern New Jersey. Funding is concentrated in six program areas: Arts and Culture, Education, Employment, Health, Public Policy, and Youth and Families. In both Health and Arts and Culture, the Foundation tends to focus its grants on leading institutions, generally where there is a longstanding relationship. In other areas, special consideration is given to competition, self-help, volunteerism, leadership and character development, parental involvement, consumer choice, economic empowerment, independent living, prevention and earlier intervention, independent research, faith-based programs, free markets, social entrepreneurship, advancing the state of the art, measuring participant outcomes and programs results, and strengthening traditional marriages and families.

Requirements: Nonprofit organizations based in New York City and northern New Jersey that are tax-exempt under Section 501(c)3 of the Internal Revenue Code and fall within the program areas of the Foundations are welcome to submit an inquiry or proposal letter. Initial contact should only include: an inquiry or proposal letter briefly summarizing the history of the project, need, research, objectives, time period, key staff, project budget, and evaluation plan emphasizing measurable outcomes and specific program results; latest annual report; current and complete audited financial statements; copy of IRS 501(c)3 tax-exemption letter.

Restrictions: The Foundations generally do not offer the following types of support, nor participate in the following program areas: nonprofit organizations outside of New York and New Jersey; annual appeals, dinner functions, and fundraising events; endowments and capital campaigns; loans and deficit financing; direct grants to individuals (such as scholarships and financial aid); individual day-care and after-school programs; housing; international; films and travel; projects for the elderly; small art, dance, music, and theater groups; independent or public K-12 schools (except charter schools); national health and mental health organizations; government agencies and nonprofit programs and services significantly funded or substantially reimbursed by government.

Geographic Focus: New Jersey; New York

Samples: Catholic Relief Services (Baltimore, MD)—for relief efforts for refugees and other victims of the fighting in Afghanistan, $20,000.; STRIVE/East Harlem Employment Services (New York, NY)—for its Community Partnership, a citywide job training and placement effort by 15 nonprofit groups to assist workers displaced by the terrorist attacks and the economic recession, $75,000 (2001); Manhattan Institute for Policy Research (New York, NY)—for research, articles, panels, and other events on rebuilding New York, as part of a round of grants responding to the September 11 attacks and their aftermath, $100,000.; New York Historical Society (NY)—for an exhibition and public program series in collaboration with the Skyscraper Museum on the conception, design, engineering, building, and destruction of the World Trade Center, $20,000.

Amount of Grant: $10,000-$100,000 average

Contact: Joseph S. Dolan, Executive Director, (212) 644-0322; fax: (212) 759-6510; email: main@achelis-bodman-fnds.org

Internet: http://fdncenter.org/grantmaker/achelis-bodman

Sponsor: Bodman Foundation

767 Third Avenue, 4th Floor

New York, NY 10017-2023

Boeing Company Contributions Program Grants 1391

The Boeing U.S. contributions program welcomes applications in five focus areas: education; health and human services; arts and culture; civic and the environment. Education is a priority for Boeing. The largest single block of charitable contributions goes toward supporting programs and projects related to education. Boeing also believes that the health of a community is measured by the well being of all its citizens. Boeing looks for innovative initiatives that promote the economic well-being of the community and neighborhood revitalization. Boeing invests in programs that promote participation in arts and cultural activities and experiences, programs that increase public understanding of and engagement in the processes and issues that affect communities and programs that protect and conserve the natural environment. Boeing accepts applications for cash grants, in-kind donations, and services.

Requirements: To apply for support you must be a U.S. based IRS 501(c)3 qualified charitable or educational organization or an accredited K-12 educational institution. U.S. grant guidelines and applications are available online.

Restrictions: Grants do not support: an individual person or families; adoption services; political candidates or organizations; religious activities, in whole or in part, for the purpose of further religious doctrine; memorials and endowments; travel expenses; nonprofit and school sponsored walk-a-thons, athletic events and athletic group sponsorships other than Special Olympics; door prizes or raffles; U.S. hospitals and medical research; school-affiliated orchestras, bands, choirs, trips, athletic teams, drama groups, yearbooks and class parties; general operating expenses for programs within

the United States; organizations that do not follow our application procedures; follow-on applications from past grantees that have not met our reporting requirements or satisfactorily completed the terms of past grants; fundraising events, annual funds, galas and other special-event fundraising activities; advertising, t-shirts, giveaways and promotional items; documentary films, books, etc.; debt reduction; dissertations and student research projects; loans, scholarships, fellowships and grants to individuals; for-profit businesses; gifts, honoraria, gratuities; capital improvements to rental properties.
Samples: Mercy Corps (Portland, OR)—to expand its disaster-relief efforts to provide food, water, and other supplies to people in Pakistan affected by the recent earthquake, $1 million (2005); Muntu Dance Theatre of Chicago (IL)—to provide this African-dance company with a permanent performance space, $3 million (2005); Millennium Park (Chicago, IL)—to construct two outdoor art-exhibition spaces, $5 million (2005); Mercy Corps (Portland, OR)—to assist tsunami-relief efforts, $300,000 (2005).
Contact: Corporate Contributions Manager, (312) 544-2000
Internet: http://www.boeing.com/companyoffices/aboutus/community/charitable.htm
Sponsor: Boeing Company Contributions Program
100 North Riverside
Chicago, IL 60606

Bogliasco Foundation Liguria Study Center Residencies 1392
The fellowships are awarded, without regard to nationality, to qualified persons doing advanced creative work or scholarly research in the following disciplines: archaeology, architecture, classics, dance, film or video, history, landscape architecture, literature, music, philosophy, theater, and visual arts. Applicants are expected to demonstrate significant achievement in their disciplines, commensurate with their age and experience. In addition, they must submit descriptions of the projects that they intend to pursue in Bogliasco. An approved project is presumed to lead to the completion of an artistic, literary, or scholarly work, followed by publication, performance, exhibition, or other public presentation. Approximately 50 fellowships are awarded during two semesters each year—September 8 to December 13 (fall-winter), and February 9 to May 16 (winter-spring). Fellowships usually are for one month (31-32 days) or, in some cases, a half semester (48 days). Fellows may be accompanied by spouses (or spouse-equivalent companions) during their stay at the Liguria Study Center (separate application required). Information and application procedures are available online.
Date(s) Application is Due: Jan 15; Apr 15
Contact: Grants Administrator, email: info@bfny.org
Internet: http://www.liguriastudycenter.org/english/home.htm
Sponsor: Bogliasco Foundation
885 Second Avenue, Room 3100
New York, NY 10017

Booth-Bricker Fund Grants 1393
The Foundation makes contributions for the purposes of promoting, developing and fostering religious, charitable, scientific, literary and educational programs. Requests are welcomed for capital needs, special projects and other one-time requirements. Applications should be made by letter. There are no forms or deadlines. Requests should include complete information about the applicant organization, including its history, purpose, finances, current operations, governing board and tax status. A detailed explanation of the proposed use of the funds must be provided.
Requirements: Requests are accepted for the funding of projects within the state of Louisiana. Priority is given to the New Orleans area.
Restrictions: The Foundation generally does not provide sustaining (operations and maintenance) funding. No grants are made to individuals.
Geographic Focus: Louisiana
Samples: YMCA (New Orleans, LA)—program support, $51,700.; Tulane U (New Orleans, LA)—for the Medical Ctr Cerise Chair, $30,000.; Saint Dominic School (New Orleans, LA)—program support, $15,000.
Amount of Grant: $5000-$50,000
Contact: Gray S. Parker, Chairperson, (504) 581-2430
Sponsor: Booth-Bricker Fund
826 Union Street, Suite 300
New Orleans, LA 70112

Bosque Foundation Grants 1394
The foundation awards grants to support higher education and medical research in Texas. Types of support include capital grants and research grants. There are no application forms or deadlines. Applicants should submit a one-page letter of intent that describes the program and request.
Requirements: Texas nonprofit and for-profit organizations are eligible.
Restrictions: Individuals are not eligible.
Geographic Focus: Texas
Amount of Grant: $5000-$20,000 average
Contact: Grants Administrator, (214) 956-6732; fax: (214) 956-6733
Sponsor: Bosque Foundation
5950 Cedar Springs Boulevard, Suite 210
Dallas, TX 75235

Boston Athenaeum Fellowships 1395
The program supports up to seven short-term fellowships. One fellowship is available for research on topics concerning the confederate states and the Civil War, and one fellowship is offered through the American Society for Eighteenth-Century Studies. Grants will support use of the Athenaeum collections for research, publication, curriculum and program development, or other creative projects. Guidelines are available online.

Requirements: Fellowships are open to advanced scholars, graduate students, independent scholars, teaching faculty, and professionals in the humanities, with applications encouraged from teachers and librarians in secondary public, private, and parochial schools.
Amount of Grant: $1500
Date(s) Application is Due: Apr 1
Contact: Stephen Nonack, Head Reference Librarian, (617) 227-0270 ext 250; email: nonack@bostonathenaeum.org
Internet: http://www.bostonathenaeum.org/fellowships.html
Sponsor: Boston Athenaeum
10 1/2 Beacon Street
Boston, MA 02108-3777

Boston Foundation Grants 1396
The Boston Foundation has a particular concern for low income and disenfranchised communities and residents and supports organizations and programs whose work helps advance the Foundation's high priorities in a variety of subject areas: Arts and Culture; Civic Engagement; Community Safety, Economic Development; Education/Out-of-School Time, Health and Human Services; Housing and Community Economic Development; the Nonprofit Sector, Urban Environment and Workforce Development. The Foundation generally makes the following types of grants: Project or program support for community-based efforts that improve the quality of life in the community, test new models, and promote collaborative and innovative ventures; advocacy and public policy research that is linked to specific action; support for planning to enable organizations and residents to assess community needs, respond to new challenges and opportunities, and provide for the inclusion of new populations; organizational support to develop and build the capacity of nonprofit organizations - support that helps organizations keep pace with the changing requirements and demands of their communities and broader environments; small grants awarded on a rolling basis for one-time organizational development needs through the Vision Fund. In addition, on a very limited basis, the Foundation will consider development grants and strategic alliances.
Requirements: Grants are made only to tax-exempt organizations in Massachusetts.
Restrictions: The committee does not consider more than one proposal from the same organization within a 12-month period. Discretionary grants are generally not made to the following applicants: city or state government agencies or departments; individuals; medical research; endowments; equipment; replacement of lost/expired government funding or gap funding to cover the full cost of providing services; scholarships and fellowships; video and film production; construction and renovation projects and capital campaigns; programs with religious content; travel; summer camps and lobbying. Activities that are generally lower priorities for the Foundation are conferences, lectures, one-time events, programs benefiting only a small number of participants or routine service delivery and/or operating expenses.
Geographic Focus: Massachusetts
Samples: Raw Art Works in Lynn, MA, $75,000 - for arts-based youth development programs to young people in Lynn (2009); Boston Private Industry Council, Boston, MA, $200,000 - to administer the Accuplacer assessment to all program participants to determine the need for remedial writing support, and provide college transition coaching to 75 participating students headed to Bunker Hill and Roxbury Community College (2009);
Date(s) Application is Due: Jan 5; Jul 1
Contact: Corey Davis, Grants Manager; (617) 338-1700; fax: (617) 338-1604; email: info@tbf.org
Internet: http://www.tbf.org
Sponsor: Boston Foundation
75 Arlington Street, 10th Floor
Boston, MA 02116

Boston Psychoanalytic Society and Institute Fellowship in Child 1397
Psychoanalytic Psychotherapy
The Child Fellowship Program offers participants an intensive, one year course of study in psychoanalytic approaches to child and adolescent psychotherapy. The curriculum includes both a theoretical course and a clinical course to be taught weekly. The theory courses will introduce the fundamentals of child analytic theories and will review historical as well as modern theoretical approaches. The clinical courses are organized around different developmental stages of childhood. Those who complete the program and wish to continue on in the Child and Adolescent Advanced Training Program will receive credit for their participation and can apply for advanced standing. A printable application form is available for download. Tuition will be $1677 for the academic year.
Requirements: The program is open to mental health clinicians who have completed or are in an advanced phase of their clinical training (residency, graduate internship) and who are interested in further didactic and clinical education in psychoanalytic work with children. Other professionals who work with children and their families such as pediatricians, teachers, and ancillary therapists are also encouraged to apply.
Date(s) Application is Due: May 15
Contact: Elizabeth Jordan, Program in Psychoanalytic Studies; (617) 266-0953; fax: (617) 266-3466; email: office@bostonpsychoanalytic.org
Internet: http://www.bostonpsychoanalytic.org/child_fellowship
Sponsor: Boston Psychoanalytic Society and Institute
15 Commonwealth Avenue
Boston, MA 02116

Boston Psychoanalytic Society and Institute Fellowship in   **1398**
Psychoanalytic Psychotherapy
The Fellowship is a one-year program designed to enhance participants‚Äô grasp of the principles and practice of psychoanalytic psychotherapy. It is also well suited for clinicians considering more extensive training in psychoanalytic psychotherapy or training in psychoanalysis, but who are either uncertain or not yet ready for a multi-year training commitment. The Fellowship offers two seminars weekly, one theoretical and one clinical, led by outstanding teachers from the Boston Psychoanalytic Society and Institute. The theoretical seminar focuses on Fundamental Concepts of Psychoanalytic Psychotherapy for the first term, on Comparative Theories of Psychotherapeutic Technique for the second term and on Selected Topics in Psychoanalytic Psychotherapy for the third term. The clinical seminar consists of discussion of case material presented by instructors and class members, with senior faculty as discussants. The seminars meet on Thursday evenings for 30 weeks starting in September. Each Fellow is matched with a faculty advisor who is available to discuss individual training goals and options. There is also an opportunity for individual supervision at reduced fees with a wide choice of experienced supervisors from the Boston Psychoanalytic Society and Institute. Tuition will be $1,677 for the academic year. The application fee is $25.
Requirements: The Fellowship is designed especially for mental health clinicians who have completed or are in an advanced phase of their requisite clinical training (psychiatric residency, psychology internship, MSW, masters in psychiatric nursing, etc.) and who desire further didactic and clinical education in psychoanalytic psychotherapy.
Amount of Grant: $1,677
Date(s) Application is Due: May 15
Contact: Elizabeth Jordan, Program in Psychoanalytic Studies; (617) 266-0953; fax: (617) 266-3466; email: office@bostonpsychoanalytic.org
Internet: http://www.bostonpsychoanalytic.org/fellowship
Sponsor: Boston Psychoanalytic Society and Institute
15 Commonwealth Avenue
Boston, MA 02116

Bower Award and Prize for Achievement in Science   **1399**
The Award recognizes outstanding achievement in life, physical, and applied sciences; innovation in the sciences; and training of scientists. The award is given without regard for nationality and includes a gold medal and prize of $250,000. The theme for the 2008 Bower Award for Achievement in Science is Robotics. The Institute seeks to honor an individual who has played a seminal role in either the design and construction of robotic systems or the advancement of enabling technologies as related to robotics such as mechanical structure, sensors, and control algorithms.
Requirements: This is an international competition for individuals whose work has had a significant impact on the field of Robotics. In cases of equal scientific merit, the factor of current economic value of the discovery or application will weigh favorably on behalf of the candidate. Candidates for the Award must be living, and the winner must participate in the April Awards Ceremony in Philadelphia.
Restrictions: This award is for an individual rather than for a group.
Samples: Stuart K. Card, Ph.D., Palo Alto Research Center, Palo Alto, CA—Bower Award for Achievement in Science (2007); John Deibel, Chair and CEO, Meade Instruments Corp (Irvine, CA)—Bower Award for Business Leadership for courage and insight and leadership of a commercial venture founded on the premise of making astronomy accessible and affordable to the public.; Sir Martin Rees, Astronomer Royal, Royal Society Professor, Institute of Agronomy, Cambridge U (Cambridge, UK)—Bower Award and Prize for Achievement in Science for contributions to elucidating the nature of quasars, black holes, X- and gamma-ray sources, and many other phenomena in high-energy astrophysics and cosmology.
Amount of Grant: $250,000 cash prize
Date(s) Application is Due: May 31
Contact: Awards Program Director, (215) 448-1329; fax: (215) 448-1364; email: awards@fi.edu
Internet: http://www.fi.edu/tfi/exhibits/bower/07/bscience_nominate.html
Sponsor: The Franklin Institute
222 N. 20th Street
Philadelphia, PA 19103-1194

BP Conservation Programme Future Conservationist Awards   **1400**
The aim of the awards is to develop leadership capacity for biodiversity conservation as a fundamental contribution to sustainable development. The program provides annual grants to passionate people developing innovative projects addressing biodiversity issues of global importance. Projects should address three key areas: development of team capabilities and skills; practical high-priority conservation projects combining research and action; and demonstrate long-term conservation benefits contributing to sustainable development. Application and guidelines are available online.
Requirements: Teams must include only members less than 35 years of age with no more than two years professional conservation and include a minimum of three people.
Restrictions: The program does not fund conference attendance, tuition fees or scholarships, salaries, costly laboratory analyses or gene storage, captive breeding projects, or high school level expeditions. Projects that are specifically for PhD research or master's dissertations will not be supported. Employees from any of the BPCP partner organizations are not eligible to apply.
Amount of Grant: $12,500 maximum
Date(s) Application is Due: Dec 16
Contact: Marianne Dunn, Conservation Program Manager, (44 01223) 277318; fax: (44 01223) 277200; email: bp-conservation-programme@birdlife.org.uk
Internet: http://conservation.bp.com/applications/default.asp

Sponsor: BP Conservation Programme
Wellbrook Ct, Griton Road
Cambridge CB3 0NA United Kingdom

BP Conservation Programme Grants   **1401**
This initiative, which is the result of a collaboration between BirdLife International, Fauna and Flora International, the Wildlife Conservation Society, Conservation International, and BP, aims to support and encourage long-term conservation projects that address global conservation priorities at a local level. The program provides advice, training, and financial awards, primarily targeting university students. Each year, the program awards gold, silver, and bronze awards; follow-up awards; and consolidation awards. Eligible projects must take place in Africa, Asia, the Pacific Islands, the Middle East, Eastern Europe, or Latin America; address a globally recognized conservation priority; involve local people; be approved by the host government; and last for less than one year but show potential for follow-up work.
Requirements: The program is open to teams for all over the world. Teams should include individuals from the host country and members from more than one country, and demonstrate collaboration with local and national conservationists. All applicant projects must address a wildlife conservation priority of global importance (preferably linking with established work-plans—e.g., national biodiversity action plan); have a strong link with the country where the project will take place (local people participating in all parts of project planning and implementation); and have a majority of team members in full- or part-time university education (undergraduate or postgraduate, and of any age).
Restrictions: The program does not offer scholarships, pay for any salaries, or fund projects organized by other organizations.
Amount of Grant: $17,500 Gold Awards; $12,500 Silver Awards; $7500 Bronze Awards; $20,000-$55,000 Follow-up Awards; $75,000 Consolidation Awards
Date(s) Application is Due: Oct 31
Contact: Marianne Dunn, Conservation Program Manager, (44 01223) 277318; fax: (44 01223) 277200; email: bp-conservation-programme@birdlife.org.uk
Internet: http://conservation.bp.com/aboutus/default.asp
Sponsor: BP Conservation Programme
Wellbrook Ct, Griton Road
Cambridge CB3 0NA United Kingdom

BPW Career Advancement Scholarships   **1402**
These scholarships assist mature women who need further education or training to reenter the workforce, enter a new career field, or to improve their chances for advancement. Applicants are strongly encouraged to seek training at the undergraduate or graduate level in computer science, education, science, engineering, paralegal studies, or professional (JD, MD, DDS) degrees. Scholarships are awarded for a one-year period to cover tuition, fees, and school-related expenses such as child care and transportation. Application requests must include a #10 self-addressed double-stamped (first-class) envelope. The listed application deadline date is to request an application.
Requirements: The scholarships are open to women who are U.S. citizens and 25 years of age or older; demonstrate financial need; are officially accepted into a program or course of study in an accredited institution in the United States, Puerto Rico, or the U.S. Virgin Islands; and will graduate within 24 months from date of receipt of funds.
Restrictions: Doctoral studies, study abroad, or correspondence courses are not supported.
Amount of Grant: $750-$1000
Date(s) Application is Due: Apr 15
Contact: Tricia Dwyer-Morgan, (202) 777-8932; fax: (202) 861-0298; email: bpwfoundation@act.org or tdwyermorgan@bpwusa.org
Internet: http://www.bpwusa.org/i4a/pages/index.cfm?pageid=4553
Sponsor: Business and Professional Women's Foundation
1900 M Street, NW, Suite 310
Washington, DC 20036

Brain Trust Translational Research Grants   **1403**
The program, a collaborative effort of Accelerate Brain Cancer Cure, the Alzheimer's Association, the Michael J. Fox Foundation for Parkinson's Research, and the Robert Packard Center for ALS Research at Johns Hopkins University, seeks to identify common challenges, explore possible solutions, and fund development of new therapeutic approaches with application to diverse brain diseases. The program seeks applications focused on technologies that can achieve selective targeting and/or delivery of therapeutic agents to specific regions/cells in the brain, including overcoming the limitations imposed by the blood-brain barrier, which can be applied in the clinical setting in the next 24 months. Funding is for one year, and the funds can only be applied to direct costs of research.
Requirements: For-profit and nonprofit organizations, individuals affiliated with nonprofit and for-profit organizations, and unaffiliated individuals are eligible.
Date(s) Application is Due: Jun 1
Contact: John Reher, (650) 685-2202; email: info@brain-trust.org
Internet: http://www.brain-trust.org/html/rfa.html
Sponsor: Brain Trust
800 Airport Boulevard, Suite 508
Burlingame, CA 94010

Brain Tumor Research Grants   **1404**
Grants support basic scientific research directed at finding a cure for brain tumors. Grants are awarded for up to a two-year period and may be used for start-up projects

or supplementary funding. Research guidelines and application are available on the Web site.
Amount of Grant: $100,000 maximum per year
Date(s) Application is Due: Mar 16
Contact: Carrie Treadwell, Research Manager, (800) 770-8287 ext 10 or (480) 575-8388; email: grants@tbts.org
Internet: http://www.tbts.org/itemDetail.asp?categoryID=300&itemID=16402
Sponsor: The Brain Tumor Society
124 Watertown Street
Watertown, MA 02472-2500

Brainerd Foundation Grants   1405
The foundation is dedicated to protecting the environmental quality of the Pacific Northwest, including Washington, Oregon, Idaho, Montana, Alaska, and British Columbia. Program grants are made in the following areas: endangered ecosystems—conservation biology, conservation assessment, and mining reform and roadless areas; and communications and capacity building—organizational development, and allied voices. Program grants are awarded to cover costs associated with activities such as public education and grassroots outreach, media strategies, litigation, scientific and economic studies, computer networking, and building organizational capacity. Opportunity Fund grants are awarded to organizations for support such as outreach, litigation, applied research, and other unexpected needs. Additional types of support include general operating support, continuing support, equipment acquisition, conferences and seminars, seed money, research, technical assistance, and employee matching gifts. Applications are available online.
Requirements: Nonprofit organizations in the Pacific Northwest are eligible.
Restrictions: The foundation does not favor proposals for school education programs, land acquisition, endowments, capital campaigns, projects sponsored by government agencies, basic research, fellowships, or books or videos that are not part of a broader strategy.
Geographic Focus: Alaska; Idaho; Montana; Oregon; Pacific Northwest; WashingtonCanada
Samples: Cook InletKeeper, $1,500 to protect Alaskan salmon streams and intact ecosystems from a one-billion ton coal strip mining project in the Beluga Coal fields; Georgia Strait Alliance, Nanaimo, BC, $8,500 for a summer outreach program to educate citizens in the lower mainland about untreated sewage; Institutes for Journalism & Natural Resources, Missoula, MT, $30,000 to increase organizational capacity and bolster fundraising and programmatic successes; Sierra Club of British Columbia Foundation $3,000 to launch an outreach campaign to raise public awareness and build leverage to enable a well-funded and legitimate recovery strategy for Mountain Caribou.
Amount of Grant: $250-$25,000 range
Contact: Ann Krumboltz, Executive Director, (206) 448-0676; fax: (206) 448-7222; email: annk@brainerd.org
Internet: http://www.brainerd.org/grants/intro.php
Sponsor: Brainerd Foundation
1601 Second Avenue, Suite 610
Seattle, WA 98101-1541

Bread and Roses Community Fund Grants and Scholarships   1406
The program awards grants to a broad spectrum of organizations and individuals working on social change. Programs include general fund grants—for organizations whose chief aim is to take collective action against a problem affecting the community, to work for social change, emergency/discretionary grants, Lax scholarships (graduate)—available to gay men, and the Phoebus Criminal Justice Initiative.
Requirements: Pennsylvania nonprofit organizations in the Delaware Valley (Philadelphia, Chester, Montgomery, Bucks, and Delaware Counties) and Camden County, NJ, are eligible.
Geographic Focus: New Jersey; Pennsylvania
Amount of Grant: $2000-$10,000
Contact: Grants Administrator, (215) 731-1107; fax: (215) 731-0453; email: info@breadrosesfund.org
Internet: http://www.breadrosesfund.org/grants/grants.html
Sponsor: Bread and Roses Community Fund
1500 Walnut Street, Suite 1305
Philadelphia, PA 19102

Breast Cancer Fund Grants   1407
Traditionally, the fund has awarded three types of grants: innovative research grants support projects investigating cutting-edge scientific approaches to the detection, treatment, and prevention of breast cancer; community model grants (non-research grants) to encourage and support the start-up of truly creative, innovative, and replicable U.S. programs that are developing new methods for addressing the support, education, health care access, and advocacy needs of women affected by breast cancer; and discretionary grants that generally fund smaller education, support, and advocacy initiatives. The fund gives preference to projects that address the needs of an ethnically, economically, and geographically diverse representation of women, including persons who are underserved or have low incomes. Initial application for research and community model grants is by letter of intent. If interested, the fund will invite a full proposal. Applicants for discretionary grants may submit an abbreviated application form available from the fund. Contact program staff for project availability.
Requirements: The fund will consider applications from individuals or organizations, whether public or private, for-profit or nonprofit.

Restrictions: No support is available for the direct cost of medical services, and no funds may be used for any lobbying efforts or political expenditures.
Contact: Jeanne Rizzo, Executive Director: (415) 346-8223; fax: (415) 543-2975; email: ed@breastcancerfund.org or info@breastcancerfund.org
Internet: http://www.breastcancerfund.org
Sponsor: Breast Cancer Fund
1388 Sutter Street, Suite 400
San Francisco, CA 94109-5400

Brico Fund Grants   1408
The mission of the fund is to effect systemic change—to change attitudes, policies and societal patterns. Grants are made to secure full participation in society for women and girls; restore and sustain the earth's natural systems; promote a just and equitable society; and nourish the creative spirit. Types of support include general operating, program, and rarely, capital and endowment grants. Applicants should complete the fund's preliminary application form and a two-page letter of intent, describing the organization's intended project or program.
Requirements: The fund supports organizations with projects and programs within the Greater Milwaukee community. Some funding is done statewide or nationally for programs of broader scope.
Restrictions: Grants do not support conferences and meetings, disease-specific programs, educational institutions, individuals, media projects, medical institutions, religions, or organizations with a focus on animals.
Geographic Focus: Wisconsin
Samples: 1000 Friends (WI)—for the opening of a Milwaukee office to explore transportation and smart growth issues, $30,000 (2003); Fondy Market—to support building of the market, $50,000 (2003); Growing Power—for operating support, $115,000 (2003); Midtown Neighborhood Assoc—to support an urban tree house and education program, $35,000 (2003).
Amount of Grant: $1000-$500,000 range
Date(s) Application is Due: Jan 15; Jul 15
Contact: Melissa Nimke, Grants Administrator, (414) 272-2747; fax: (414) 272-2036; email: mbn@bricofund.org or bricofund@bricofund.org
Internet: http://www.bricofund.org
Sponsor: Brico Fund
205 E Wisconsin Avenue, Suite 200
Milwaukee, WI 53202

Bright Family Foundation Grants   1409
The foundation awards grants to eligible California nonprofit organizations in its areas of interest, including children and youth, education (medical, business, and other), health, medical research, and religion. Types of support include general operating support, building construction/renovation, scholarship funds, and special projects. Grants are awarded for one year, with possible renewal.
Requirements: Stanislaus County, CA, 501(c)3 tax-exempt organizations within 30 miles of Modesto, CA, are eligible.
Geographic Focus: California
Samples: California Youth Soccer (CA)—$200.; Slavic Gospel Assoc (CA)—$7500.; Big Valley Grace Church (CA)—$10,000.; Ctr for Human Services (CA)—$2500.
Amount of Grant: $377,500 total
Date(s) Application is Due: Dec 1
Contact: Calvin Bright, President, (209) 526-8242
Sponsor: Bright Family Foundation
1620 N Carpenter Road, Building B
Modesto, CA 95351

Brinson Foundation Grants   1410
The foundation supports education, public health, and scientific research programs that engage, inform, and inspire committed citizens to confront the challenges that face humanity. Grantmaking priorities are education—awareness and outreach, democracy and citizenship, economically disadvantaged, and libraries and literacy; public health—awareness and outreach, and economically disadvantaged; and scientific research—astrophysics, cosmology, geophysics, medical research (i.e., Alzheimer's disease, cancer, Lou Gehrig's Disease (ALS), and stroke). Types of support include general operating grants and project grants. The foundation does not accept unsolicited grant applications. Grantseekers are asked to review the foundation's mission, vision, beliefs, priorities (accessed from the Who We Are link), and guidelines. If a grantseeker believes the request would match one or more of the foundation's grantmaking priorities, they can make an inquiry by completing the online Grantseeker Information Form. The completed form should be emailed to the office. Further application is by invitation.
Requirements: The foundation will consider inviting grant applications from organizations: whose request matches one or more of the Foundation's grantmaking priorities; located in the United States of America that are exempt from tax under Section 501(c)3 of the Internal Revenue Code and are defined as charitable organizations as described in Section 509(a)(1), (2) or (3) or 170(b)(1)(A); located outside the United States of America provided they produce a written legal opinion stating that they are a charitable equivalency to a qualifying U.S. organization and/or a written affidavit containing sufficient information for the Foundation to make a reasonable judgment that the organization is charitable.
Restrictions: The Foundation will not consider grant inquiries from organizations that: discriminate on the basis of race, gender, religion, ethnicity or sexual orientation. The Foundation will not consider grant inquiries that request funding for: activities that attempt to influence public elections; voter registration; political activity; lobbying

efforts; promotion of a specific religious faith; medical research involving human cloning. The Foundation discourages grant inquiries requesting funds for: capital improvements; endowments; fundraising events.

Contact: Grants Administrator, (312) 799-4500; fax: (312) 799-4310; email: mail@brinsonfoundation.org
Internet: http://www.brinsonfoundation.org/grants/index.html
Sponsor: Brinson Foundation
737 North Michigan Avenue, Suite 1850
Chicago, IL 60611

Bristol-Myers Squibb / Meade Johnson Award for Distinguished Achievement in Nutrition Research   1411
The Award was first presented in 1978. The Award includes a $50,000 cash prize and a silver commemorative medallion. The winners are selected by an independent peer-review selection committee whose members are grant administrators of current Bristol-Myers Squibb/Mead Johnson Unrestricted Nutrition Research Grants. The annual application deadline is in late April.
Samples: Steven Zeisel, Kenan Distinguished Professor of Nutrition and Pediatrics and associate dean for research at the School of Public Health, $50,000 (2006).
Amount of Grant: $50,000
Date(s) Application is Due: Apr 24
Contact: John Damonti, President; (212) 546-4000; fax: (212) 546-9574
Internet: http://www.bms.com/aboutbms/grants/data/nutrit.html
Sponsor: Bristol-Myers Squibb Foundation
345 Park Avenue, Suite 4364
New York, NY 10154-0037

Bristol-Myers Squibb Clinical Outcomes and Research Grants   1412
Bristol-Myers Squibb's mission is to extend and enhance human life. To help achieve that mission, the Company has established programs to support Investigator Sponsored Trials (ISTs). ISTs must be medically appropriate and scientifically valid. While the Company will consider requests for clinical research trials in all clinical and therapeutic areas, it currently gives priority to proposals in the following therapeutic areas: Cardiovascular/Metabolics, Infectious Diseases, Neuroscience, Oncology, Immunology, and Virology. Bristol-Myers Squibb maintains a strict policy of not exercising any influence or control over the design of any investigator initiated clinical research trial supported by BMS.
Requirements: Individuals in the following settings are eligible for support: private practice, hospitals, community health centers, cooperative groups, physician networks, and academic medical centers and universities.
Contact: Amit Duggal; (212) 546-4000; fax: (212) 546-9574; email: amit.duggal@bms.com
Internet: http://www.bms.com/sr/content/data/clinical.html
Sponsor: Bristol-Myers Squibb Company
777 Scudders Mill Road
Plainsboro, NJ 08536

Bristol-Myers Squibb Foundation Global HIV/AIDS Initiative Grants   1413
The intent of the Program is to develop new models in awareness, in medical care, in community development and in prevention and treatment in poor and resource-limited areas of the world, where the need for all such efforts is greatest. Health care infrastructures must be developed and enhanced, stigmatization must be overcome, health care worker capacity must be built and preserved and local people must be empowered to generate and sustain local solutions to this global problem.
Requirements: The Bristol-Myers Squibb Foundation considers requests for support only from tax-exempt organizations that satisfy the requirements of section 501(c)3 of the U.S. Internal Revenue Code.
Restrictions: The Foundation does not award funds to: individuals; political, fraternal, social or veterans' organizations; religious or sectarian organizations unless engaged in a significant project benefiting the entire community; organizations receiving support through United Way or other federated campaigns; endowments; courtesy advertising; or conferences/special events/videos.
Contact: John Damonti, President; (212) 546-4000; fax: (212) 546-9574
Internet: http://www.bms.com/sr/philanthropy/data/globhiv.html
Sponsor: Bristol-Myers Squibb Foundation
345 Park Avenue, Suite 4364
New York, NY 10154-0037

Bristol-Myers Squibb Foundation Health Disparities Grants   1414
One mission of the Foundation is to reduce health disparities by strengthening community-based health care worker capacity, integrating medical care and community-based supportive services, and mobilizing communities to fight disease. To this end, this Program attempts to address health disparities in four strategic disease areas representing major public health burdens and in four highly affected geographies: hepatitis in Asia, HIV/AIDS in Africa, serious mental illness in the U.S., and cancer in Europe. Additional areas of concern include: metabolic diseases, infectious diseases; rheumatoid arthritis; cardiovascular diseases; substance abuse; women's health issues; and overal health care giving.
Requirements: Nonprofit organizations in communities where Bristol-Myers Squibb maintains a facility should submit their requests for company contributions directly to that location. Contact persons are listed at the company website.
Restrictions: The foundation does not support individuals; conferences, special events, or videos; political, fraternal, social, or veterans organizations; religious or sectarian

activities, unless they benefit the entire community; organizations funded through federated campaigns; endowments; or courtesy advertising.
Geographic Focus: Connecticut; Indiana; Massachusetts; New Jersey; New York
Contact: John Damonti, President, Bristol Myers Squibb Foundation, (212) 546-4000; fax: (212) 546-9574
Internet: http://www.bms.com/foundation/reducing_health_disparities/Pages/default.aspx
Sponsor: Bristol-Myers Squibb Foundation
345 Park Avenue, Suite 4364
New York, NY 10154-0037

Bristol-Myers Squibb Foundation Women's Health Grants   1415
Bristol-Myers Squibb Women's Health Program supports projects that enhance women's health with strategies that improve education, prevention, diagnosis, treatment and access to care for women worldwide. Support has been given to projects that test innovative outreach programs, cultivate multi-sectoral partnerships and add new information to the existing body of knowledge to help define and achieve improved health for women around the world. The goal of the program is to generate initiatives that will help enhance women, health through novel interdisciplinary strategies that improve education, prevention, diagnosis, treatment and access to care for women worldwide. Since its inception, significant resources have been invested in programs that educate women about diseases and conditions that particularly threaten them as women.
Requirements: The Bristol-Myers Squibb Foundation considers requests for support only from tax-exempt organizations that satisfy the requirements of section 501(c)3 of the U.S. Internal Revenue Code.
Restrictions: The Foundation does not award funds to: individuals; political, fraternal, social or veterans' organizations; religious or sectarian organizations unless engaged in a significant project benefiting the entire community; organizations receiving support through United Way or other federated campaigns; endowments; courtesy advertising; or conferences/special events/videos.
Contact: John Damonti, President; (212) 546-4000; fax: (212) 546-9574
Internet: http://www.bms.com/Documents/foundation/women2002.pdf
Sponsor: Bristol-Myers Squibb Foundation
345 Park Avenue, Suite 4364
New York, NY 10154-0037

Bristol-Myers Squibb Virology and Oncology M.D. Research Fellowships   1416
The Program provides support for fellows to gain experience in epidemiological and clinical research as it relates to the care of individuals infected with HIV/AIDS and/or HBV. Goals of the Program include: support of studies that will further strengthen the science and knowledge of HIV/AIDS and HBV; development of a foundation for future prospective and retrospective studies; to provide a forum to share research findings; and support of the development of future clinical researchers. Up to 18 fellows will be selected annually, with grant awards up to $20,000 to support research-related expenses for a one-year research period. Applications must be submitted online.
Requirements: Candidates must: be an active Fellow in good standing in an ACGME-accredited Fellows training program; be a senior Fellow not in the first years of the fellowship; desire to enhance knowledge and skill development in the area of HIV/AIDS, HBV, or Oncology research; and identify a faculty member to serve as a mentor.
Restrictions: Total grant amount is inclusive of indirect costs and associated IRB fees and is not permitted for use towards travel to conferences or for durable equipment.
Amount of Grant: $20,000 maximum
Date(s) Application is Due: Feb 26
Contact: John Damonti, President, Bristol Myers Squibb Foundation, (212) 546-4000; fax: (212) 546-9574
Internet: http://www.bms.com/clinical_trials/investigator_sponsored_trials/fellows_research_program/Pages/default.aspx
Sponsor: Bristol-Myers Squibb Company
345 Park Avenue
New York, NY 10154

British and American History, Literature, Art, and History of Science Fellows   1417
The Huntington Library has established a fellowship fund for graduate and postdoctoral scholars in British and American history, history of science, literature, and art. Special awards of three months or less are available for persons writing doctoral dissertations. Award holders are expected to be in continuous residence at the Huntington throughout their tenure. Applications are accepted from October 1 to December 15 of each year for awards within the 12-month period beginning on the following June 1; results are announced by April 1.
Requirements: Scholars must possess the PhD or equivalent degree.
Amount of Grant: $2000 per month for one to five months
Date(s) Application is Due: Dec 15
Contact: Robert Ritchie, Director of Research, (626) 405-2194; fax: (626) 449-5703; email: cpowell@hungtington.org
Internet: http://www.huntington.org/ResearchDiv/Fellowships.html
Sponsor: Huntington Library and Art Gallery
1151 Oxford Road
San Marino, CA 91108

British-American Transnational History Fellowships   1418
Central Michigan University has joined with the University of Strathclyde, Glasgow, Scotland, in establishing an innovative collaborative degree program that offers the master's and Ph.D. degrees through study under joint British-American faculty. The

program provides international academic study and dialogue beyond ordinary student and faculty exchanges by pooling the faculty of two universities, thereby offering graduate students the opportunity of directed research and study in a truly comparative and transnational environment. Students in the joint history degree program benefit from a small program, in terms of the number of students, with a large combined faculty of 32 members who are able to give considerable attention to individual student coursework and research. The curriculum of the joint history program centers on traditional fields of historical concentration but is strongly international in emphasis, requiring a degree of coursework and research that is transnational and comparative. The joint faculty is strongly grounded in social, cultural and economic history.

Requirements: Candidates are required to select two major fields or one major and two minor fields from the following: the United States, the British Isles, Modern Continental Europe, and the Atlantic World. Minor fields are the Ancient Near East and Mediterranean, Medieval Europe, Latin America, East Asia, and India. Topical fields such as women, ethnicity, race, and poverty are possible. See Web site for MA and PhD requirements.

Contact: Annette Davis, Graduate Programs, (989) 774-3374; email: annette.davis@cmich.edu
Internet: http://www.chsbs.cmich.edu/history/grad.htm
Sponsor: Central Michigan University
History Department, Powers Hall 106
Mount Pleasant, MI 48859

Broad Foundation IBD Research Grants 1419

The program seeks to stimulate innovative research that will lead to both the prevention and successful therapy of inflammatory bowel disease (IBD), including Chrohn's disease and ulcerative colitis. The foundation's goal is to fund basic or clinical research projects that are in the early stages of exploration; propose new directions or ideas; are creative, novel, cutting edge, and imaginative; and are not ready for funding by other more traditional granting agencies. All proposals must be based on sound scientific evidence and careful evaluation of current knowledge in IBD research. Requests must be preceded by a brief (one to three pages)letter of interest, which may be submitted at any time. Funding will be granted for one year; continued funding is based on progress reports and the perceived value of the findings.

Requirements: Nonprofit institutions worldwide, such as universities, hospitals, and research institutes, are eligible.

Restrictions: It is anticipated that the foundation will not fund projects after sufficient progress and maturity have made them fundable by other agencies.

Amount of Grant: $100,000 average
Contact: Dr. Daniel Hollander, (310) 954-5091; email: info@broadmedical.org
Internet: http://www.broadmedical.org/funding.htm
Sponsor: Broad Foundation
10900 Wilshire Boulevard, 12th Floor
Los Angeles, CA 90024-6532

Broadcast Media Awards for Television 1420

These awards recognize outstanding reporting and programming on television that deals with reading and literacy, recognizing the value of reading in today's society, or promoting literacy. Entries must be oriented toward the general public rather than reading education professionals and should be informational, critical, or motivational rather than instructional. Entries may include but are not limited to: journalism on reading in schools, the home, or the community, including accounts of research and educational practices, coverage of reading activities, or appraisals of school reading programs; interview programs on reading in schools or the community or on reading education in general; public service programming that informs about reading, seeks to instill a love of reading, and/or promotes literacy; entertainment programming that informs about reading, seeks to instill a love of reading, and/or promotes literacy.

Requirements: Entries must have aired during the previous calendar year beginning January 1. Only one entry may be submitted per producer. Entries must be sent to Nancy E. Frey, San Diego State University - City Heights, Suite 100, 4283 El Cajon Boulevard, San Diego, CA 92105-1289.

Samples: Monty Haas, Laurie Joy Haas, Leigh Anne Neal—award winners (2005); Marie McMorris, Tira Grey—award winners (2004); Leigh Anne Neal—award winner (2003).

Date(s) Application is Due: Jan 8
Contact: Program Contact, (302) 731-1600 ext 293; fax: (302) 731-1057; email: bcady@reading.org
Internet: http://www.reading.org/association/awards/index.html
Sponsor: International Reading Association
P.O. Box 8139, 800 Barksdale Road
Newark, DE 19714-8139

Broadhurst Foundation Grants 1421

The foundation supports the arts and humanities, community development, and medical research. In addition to program support, funds are awarded for building programs and necessary equipment in these areas. Support for scholarship funds for students training for the Christian ministry at institutions selected by the foundation; grants also to educational and religious institutions, and to medical research institutions, especially those related to pediatric diseases.

Requirements: Organizations in Oklahoma are eligible to apply.

Restrictions: No grants to individuals or for scholarship funds.

Geographic Focus: Oklahoma
Amount of Grant: $100-$18,000

Contact: Ann Cassidy Baker, Chair, (918) 584-0661; fax: (918) 584-5831
Sponsor: Broadhurst Foundation
401 S Boston, Suite 100
Tulsa, OK 74103-4002

Brookdale Foundation Leadership in Aging Fellowships 1422

The Leadership in Aging Fellowship provides two years of support to junior academics to focus on a project that will help establish them in an area of aging research. The fellowship is open to a broad range of disciplines including, but not limited to, medical, biological and basic sciences, nursing, social sciences, the arts and humanities. The award is a two-year grant paid to the candidate's sponsoring institution in support of the candidate, research project. The grant amount of up to $125,000 each year is intended to cover 75% of the fellow's time, base salary and fringe benefits. The award could also be used to include the support of a graduate assistant if necessary as long as the total amount does not exceed $125,000.

Requirements: The fellowship is open to all professionals in the field of aging. Candidates must have: (1) leadership potential; (2) an ongoing commitment to a career in aging; (3) a mentor (or mentors); (4) and, a willingness to commit at least 75% of his or her time for career development during each of the two years of the fellowship.

Samples: (2008) Kate de Medeiros, Ph.D. - Johns Hopkins University School of Medicine.; Jennifer Hagerty Lingler, PhD, MA, FNP - University of Pittsburgh Schools of Nursing and Medicine.; Teresita M Hogan MD, FACEP - Department of Emergency Medicine at the University of Illinois Chicago.; Brie A. Williams, MD, MS - Division of Geriatrics at the University of California, San Francisco.

Amount of Grant: up to $125,000 each year
Contact: Cara Kenien, Fellowship Contact; (212) 308-7355; fax: (212) 750-0132; email: cjk@brookdalefoundation.org
Internet: http://www.brookdalefoundation.org/Leadership/Fellows/fellows.html
Sponsor: Brookdale Foundation
950 Third Avenue, 19th Floor
New York, NY 10022

Brookhaven National Laboratory Sambamurti Memorial Lectureships 1423

The lectureship is awarded yearly to a young (under 40) high-energy or heavy-ion experimentalist of outstanding achievement. The lecture, which should describe the work for which the lecturer is being honored, is to be delivered to students working at BNL during the summer.

Amount of Grant: $500
Contact: Laurence Littenberg, Department of Physics, (516) 344-3811; email: littenbe@bnl.gov
Internet: http://www.phy.bnl.gov/edg/sambamurti.html
Sponsor: Brookhaven National Laboratory
P.O. Box 5000
Upton, NY 11973-5000

Brookings Institution Predoctoral Research Fellowships in Foreign Policy 1424

A limited number of resident fellowships are awarded for policy-oriented predoctoral research in government studies. The fellowships are designed for doctoral candidates whose dissertation topics are directly related to public policy issues and thus to the major interests of the institution. Fellowships will be awarded to scholars whose research will benefit from access to the data, opportunities for interviewing, and consultation with senior staff members afforded by the institution and by residence in Washington, DC. Fellows may participate in appropriate staff conferences and seminars of the institution and have access to the research resources available to resident staff members. Outstanding dissertations will be considered for publication by Brookings. Computer facilities are provided. Stipend is payable on a 12-month basis, 11 months of research in residence and one month of vacation.

Requirements: Candidates must be nominated by the graduate departments of their universities. Departments should nominate no more than two persons who should have completed the preliminary examinations for the doctorate no later than February 15. The institution particularly encourages the participation of women and members of minority groups.

Restrictions: Individual applications are not accepted.

Amount of Grant: $20,500 stipend, supplementary assistance not to exceed $750, and reimbursement for research-related travel up to $750

Date(s) Application is Due: Feb 15; Dec 15
Contact: Sarah Yerkes, (202) 797-6043; fax: (202) 797-2481; email: syerkes@brookings.edu
Internet: http://www.brook.edu/admin/fellowships.htm
Sponsor: Brookings Institution
1775 Massachusetts Avenue NW
Washington, DC 20036

Brookings Institution Sakip Sabanci International Research Award 1425

This annual award is designed to promote fresh thinking, new ideas, and original research relevant to Turkish studies conducted in any field of the humanities and social sciences. The competition calls for original, essay-length studies that address the key issues relating to the changes in Turkey's neighborhood and how Turkey might respond to these changes. Studies focusing on any one or several of the following themes are welcome: Turkey's relations with the European Union, Russia, the Balkans, the Caucasus, Middle Eastern neighbors, and Central Asia, as well as Turkey's potential role in transatlantic relations and democratization in the broader Middle East. The topic changes each year.

Topic for the current year is Turkey's New Geopolitical Environment: Policy Challenges and Opportunities for Engagement. Guidelines are available online.

Requirements: All entries must be new and original works, not published previously in any form. Essays must be approximately 5000 to 6000 words. Essays must be submitted in English only by means of email in the form of an attached Word document to the following two addresses simultaneously: sabanciaward@sabanciuniv.edu; acause@brookings.edu.

Amount of Grant: $20,000 first prize; $10,000 second prize; $5000 third prize

Date(s) Application is Due: Jan 31

Contact: Amanda Cause, Center on the United States and Europe, (202) 797-6227; Katie Busch , Brookings Media Relations Officer, (202) 797-6467

Internet: http://www.brookings.edu/comm/news/20050620sabanciaward.htm

Sponsor: Brookings Institution

1775 Massachusetts Avenue NW

Washington, DC 20036

Brucebo Fine Art Summer Scholarship　　　　1426

The annual, three-month summer scholarship supports a Canadian artist to be used during a three month summer term on the island of Gotland, Sweden. The program supports housing, research, travel, and work-study in the fine arts, handcrafts, and related fields. There are no application forms.

Amount of Grant: $C5000 approximately

Contact: Dr. Jan Lundgren, (514) 398-4111; fax: (514) 398-7437; email: lundgren@felix.geog.mcgill.ca

Internet: http://www.canada-scandinavia.ca/CSFGrants.htm

Sponsor: Canadian-Scandinavian Foundation

McGill University, 805 Sherbrooke Street W

Montreal, PQ H3A 2K6 Canada

Brush Foundation Grants　　　　1427

The primary goal of the foundation is to ensure that family planning worldwide becomes acceptable, available, accessible, affordable, effective, and safe. It focuses on projects that protect and enhance people's ability to manage their reproductive health; carry out public policy analysis and public education in the areas related to reproductive behavior and its social implications; and advance the personal knowledge and purposeful behavior of young people with respect to sexuality within a broad social context. Letters on inquiry are requested by January 15 and June 15.

Requirements: Grants are limited to 501(c)3 organizations.

Restrictions: The foundation will not accept unsolicited proposals.

Samples: Zero Population Growth (Washington, DC)—$15,000.; Women's Health Ctr Foundation (Duluth, MN)—$10,000.; San Benito High School (Hollister CA)—$6037.; Feminist Majority Foundation (Arlington, VA)—$5000.

Amount of Grant: $5000-$50,000

Date(s) Application is Due: Jan 15; Jun 15

Contact: Krystal Fletcher, Grants Administrator, (216) 881-5121; fax: (216) 881-1834; email: brushfoundation@hotmail.com

Sponsor: Brush Foundation

3135 Euclid Avenue, Suite 102

Cleveland, OH 44115

Buell Book Fellowship in American Architecture, Urbanism, or Landscape　　　　1428

These fellowships, sponsored by the Temple Hoyne Buell Center for the Study of American Architecture at Columbia University, offer promising writers and scholars working in an area of American architecture, urbanism, or landscape studies the opportunity to turn a completed book-length manuscript or doctoral dissertation into publishable form. The center will provide office space, library usage fees, and eligibility for housing at Columbia. Fellows will also have access to the resources of the Avery Architectural Library. Fellows are expected to be in residence at Columbia for a substantial part of their tenure, to teach a seminar related to their research, and to contribute to the intellectual life of the Buell Center. Note: Program may not be offered for the current year; contact the center for availability.

Requirements: Applicants must submit a manuscript, accompanied by a summary of content, a project history, and proposed plan of development; a professional resume; and two letters of recommendation by the deadline date.

Samples: Inari Abazos and Juan Herreros—for completion of an English edition of their book, On the History and Technology of the American Skyscraper after World War II, $20,000.

Amount of Grant: $20,000 for one academic year

Date(s) Application is Due: May 9

Contact: Director, The Buell Center, (212) 854-8165; fax: (212) 854-2127; email: buellcenter@columbia.edu

Internet: http://www.arch.columbia.edu/Buell

Sponsor: Columbia University

400 Avery Hall, Columbia University

New York, NY 10027

Bullard Fellowship in Forest Research　　　　1429

These fellowships are offered annually to midcareer individuals in biological and physical sciences and social and political studies to promote the advanced study, research, or integration of subjects relating to forested ecosystems. The fellowships are designed to provide an opportunity for candidates to use the facilities at Harvard University and the Harvard Forest to further their scientific and professional growth.

Amount of Grant: $40,000 maximum

Date(s) Application is Due: Feb 1

Contact: Fellowship Administrator, (978) 724-3302; fax: (978) 724-3595; email: hfapps@fas.harvard.edu

Internet: http://harvardforest.fas.harvard.edu/education/bullard.html

Sponsor: Harvard University

P.O. Box 68, 324 N Main Street

Petersham, MA 01366

Bullitt Foundation Grants　　　　1430

The foundation functions to protect and restore the environment of the Pacific Northwest, including Washington, Oregon, Idaho, western Montana, coastal rainforests in Alaska, and British Columbia, Canada. Program priorities include aquatic ecosystems; terrestrial ecosystems; conservation and stewardship in agriculture; energy and climate change; growth management and transportation; toxic and radioactive substances; training, organizational development, and unique opportunities (including education and public outreach). Areas of interest include air pollution, climate change, endangered species, energy conservation, environmental education and justice, human health, transportation, and tribal communities. The foundation supports challenge/matching, general operating, project/program, seed money, demonstration, and development grants, as well as requests for conferences/seminars and technical assistance support. Grants will be awarded for one year with possible renewal.

Requirements: Nonprofit organizations in the Pacific Northwest, including Washington, Oregon, Idaho, western Montana, coastal rainforests in Alaska, and British Columbia, Canada are eligible.

Geographic Focus: Alaska; Idaho; Montana; Oregon; Pacific Northwest; Washington; Canada

Samples: North Seattle Community College Foundation (WA)—to support a model project for watershed educational activities, $10,000.; Washington's National Park Fund (WA)—to support a program for youth from local school districts, $5000.

Date(s) Application is Due: May 1; Nov 1

Contact: Program Officer , (206) 343-0807; fax: (206) 343-0822; email: info@bullitt.org

Internet: http://www.bullitt.org

Sponsor: Bullitt Foundation

1212 Minor Avenue

Seattle, WA 98101-2825

Bunge y Born Foundation Grants　　　　1431

The foundation awards grants in Argentina in the fields of education, social welfare and studies, the arts and humanities, and science and medicine. Types of support include research prizes, research grants, research scholarships, and support to attend conferences.

Contact: Grants Administrator, 011-4318-6600; fax: 011-4318-6610; email: info@fundacionbyb.org

Internet: http://www.fundacionbyb.org.ar/ingles/index.htm

Sponsor: Bunge y Born Foundation

25 de Mayo 565

Buenos Aires 1002 Argentina

Bunting Fellowship Program　　　　1432

The Mary Bunting Institute of Radcliffe College is a multidisciplinary research center for women scholars, scientists, artists, and writers and is one of the major centers for advanced study in the United States. One fellowship will be awarded to a professional woman in the field of infant and child development, who is conducting research within the framework of, or contributing to, psychoanalysis. Applications will be judged on the quality and significance of the proposed project, the applicant's record of accomplishment, and the stage in the applicant's career. Office or studio space, auditing privileges, and access to libraries and most other resources of Radcliffe College and Harvard University are provided. Residence in the Boston area and participation in the institute community are required during the one-year fellowship appointment. Fellows are expected to present their work-in-progress at public colloquia, performances, or exhibitions.

Amount of Grant: $33,000 stipend

Date(s) Application is Due: Oct 1

Contact: Fellowship Coordinator, (617) 496-1324; fax: (617) 495-8136; email: fellowships@radcliffe.edu

Internet: http://www.radcliffe.edu/fellowships/bunting.php

Sponsor: Bunting Institute of Radcliffe College

34 Concord Avenue

Cambridge, MA 02138

Burden Trust Grants　　　　1433

The trust supports international nonprofit organizations in the fields of medical research and hospitals; schools and training institutions; and care of and homes for the elderly, children, and other individuals in need. Preference will be given to organizations affiliated with the Anglican Church.

Date(s) Application is Due: Mar 31

Contact: Patrick O'Conor, 0117 9628611; email: p.oconor@netgates.co.uk

Sponsor: Burden Trust

51 Downs Park W

Bristol BS6 7QL United Kingdom

Bureau of Broadcast Measurement Research Scholarship 1434

This scholarship is awarded to a graduate student for research into the fields of communications or broadcast journalism as they relate to radio and television. Application materials are available upon request.

Requirements: Applicants must be Canadian citizens who are enrolled in a recognized communications or broadcast journalism course at a Canadian university.

Amount of Grant: $C4000

Date(s) Application is Due: Jun 30

Contact: Scholarships Administrator, (613) 233-4035; fax: (613) 233-6961; email: cab@cab-acr.ca

Internet: http://www.cab-acr.ca/english/about/awards/scholarships/bbm.pdf

Sponsor: Canadian Association of Broadcasters

P.O. Box 627, Sta B, 350 Sparks Street

Ottawa, ON K1P 5S2 Canada

Burlington Northern Santa Fe Foundation Grants 1435

The foundation is focused on the communities where the company operates and areas where its railways pass. The foundation supports education, including scholarships for Native Americans and scholarships for children of employees in conjunction with the National Merit Scholarships program; the arts, including museums, performing arts, and libraries; and civic and public affairs. Support goes to the Nature Conservancy and for local fire departments and law enforcement. Types of support include general operating support, continuing support, annual campaigns, capital campaigns, and program development. Health and human services funding concentrates on the United Way. Awards are made for a single year and for continuing support. The company also matches employee funds given to public and private colleges and universities, cultural organizations, and hospitals in the United States. Requests for applications should describe the purpose for the grant. Requests are reviewed every six weeks.

Requirements: Any 501(c)3 organization located in Schaumburg, IL, and communities where the corporation operates, including 28 states and two Canadian provinces, are eligible to apply.

Samples: Texas Christian University, Fort Worth, TX, $600,000 - for Career Services Center and Neeley Schools Next Generation Leadership Program (2007); Hastings Rural Fire District, Glenvil, NE, $20,000 - for replacement of rescue truck and grass firefighting rig (2007); Foss Waterway Seaport, Tacoma, WA, $50,000 - for restoration and development of waterfront cultural center and maritime museum (2007).

Contact: Deanna Dugas, Manager Corporate Contributions; (817) 867-6407; fax: (817) 352-7924; email: Deanna.dugas@bnsf.com

Sponsor: Burlington Northern Santa Fe Foundation

2650 Lou Menk Drive, 2nd Floor, P.O. Box 961057

Fort Worth, TX 76131-2830

Burton G. Bettingen Grants 1436

The fields of activity of the corporation are education at all levels, mental health, crime and abuse victims and public protection programs, religion (Christian, Roman Catholic, and Salvation Army), environment, and welfare. Top funding priority is children and youth. The current focus is on child prostitutes, runaways, and abandoned children. Nonprofits servicing the economically disadvantaged also may apply. The corporation provides broad types of support, including operating, capital, research, challenge/matching grants, and endowments. A letter of inquiry stating the applicant's background, goals and objectives, and the specific need for funding is welcome. Unsolicited submissions are considered but receive low priority.

Requirements: IRS 501(c)3 organizations are eligible. Giving primarily, but not limited to, Southern California.

Restrictions: The corporation does not award grants to individuals; for general fund-raising events, dinners, or mass mailings; or to grantmaking organizations.

Geographic Focus: California

Samples: Childrens Hospital of Los Angeles (Los Angeles, CA)—for capital campaign, $700,000.; Phoenix House (New York, NY)—for general support, $50,000.; Children's Defense Fund (Washington, DC)—for general support, $750,000.

Amount of Grant: $5000-$200,000 average

Contact: Patricia Brown, Executive Director, (323) 938-8478; fax: (323) 938-8479; email: burtonbet@aol.com

Sponsor: Burton G. Bettingen Corporation

134 S Mansfield Avenue

Los Angeles, CA 90036-3019

Bush Foundation Artist Fellowships 1437

The fellowships enable artists to further their work and contribution to their communities. Fellows may decide to take time for solitary work or reflection, engage in collaborative community projects, embark on travel or research, or pursue any other activity that contributes to their lives as artists. Artists may use the fellowship in many ways—to explore new directions, continue work already in progress, or accomplish work not financially feasible otherwise. New to the program, professional development activities have been added including a two-day orientation retreat along with annual workshops combined with artist gatherings. Grant categories rotate on a two-year cycle. Check the website for this year's categories, which could include: Literature (Fiction, Creative Nonfiction, Poetry); Scriptworks (for stage and screen); Film/Video; or, Music Composition. The program also supports visual artists, performance artists, storytellers and traditional folk artists. Fellowships may last from 12 to 24 months.

Requirements: Eligible applicants must be: residents of the funding region: Minnesota, North Dakota or South Dakota, and are U.S. citizens or permanent residents; have lived in the funding region at least 24 months immediately prior to the application deadline; at least 25-years-old on the application deadline.

Restrictions: Artists are not eligible to apply if they are a student enrolled full- or part-time in a degree-granting program after July 1 of the program year; and a director or staff member of the Bush Foundation, or a spouse, parent, child, or grandchild of a director or staff member of the foundation; have already received two BAFs or an Enduring Vision Award.

Geographic Focus: Minnesota; North Dakota; South Dakota

Amount of Grant: $48,000

Date(s) Application is Due: Oct 30

Contact: Julie Gordon Dalgleish, Program Director; (651) 227-0891 or (800) 605-7315; fax: (651) 297-6485; email: jdalgleish@bushfoundation.org

Internet: http://www.bushfellows.org/artist/bush-artist-fellowships

Sponsor: Bush Foundation

332 Minnesota Street, Suite E-900

Saint Paul, MN 55101

Bushrod H. Campbell and Adah F. Hall Charity Fund Grants 1438

Grants are awarded to organizations in the Boston area devoted to basic needs for the elderly, projects relating to medicine and medical research, health care, hospitals, the blind and deaf, and certain discretionary projects. Grants are also awarded countrywide for projects addressing population control. Types of support include capital grants, general operating grants, program grants, and research grants.

Requirements: Tax-exempt organizations located within Boston and neighboring communities and U.S. tax-exempt organizations devoted to population control are eligible.

Restrictions: Grants are not awarded to individuals.

Geographic Focus: Massachusetts

Amount of Grant: $3000-$7500 average

Date(s) Application is Due: Jan 15; Apr 15; Aug 15; Oct 15

Contact: Brenda Taylor, c/o Palmer & Dodge, (617) 239-0556; fax: (617) 227-4420

Sponsor: Bushrod H. Campbell and Adah F. Hall Charity Fund

111 Huntington Avenue at Prudential Ctr

Boston, MA 02199-7613

Business and Society Foundation Grants 1439

The foundation was established to encourage Spanish and foreign companies working in Spain to improve the quality of life in operating communities. Foundation activities are carried out through research, conferences, and training courses. Types of support include prizes and grants to organizations.

Contact: Francisco Abad, Director, 34-91-4358997; fax: 34-91-4353974; email: fundacion@empresaysociedad.org

Internet: http://www.empresaysociedad.org

Sponsor: Business and Society Foundation

Goya 15, 2 derecha

Madrid 28001 Spain

BWF Ad Hoc Grants 1440

To complement the competitive award programs, BWF also makes modest grants on an ad hoc basis to nonprofit organizations conducting activities intended to improve the general environment for science. These noncompetitive grants are for activities closely related to the Foundation's focus areas. The Foundation places special priority on working with non-profit organizations, including government agencies, to leverage financial support for targeted areas of research, and on encouraging other foundations to support biomedical research. Proposals should be brief and include a description of the focus of the activity, the expected outcomes, qualifications of the organization or individuals involved; provide certification of the sponsor's Internal Revenue Service tax-exempt status; and give the total budget for the activity, including any financial support obtained or promised. Proposals are given careful preliminary review, and those deemed appropriate are presented for consideration by BWF's board of directors.

Contact: Nancy S. Sung, Senior Program Officer; (919) 991-5100; fax: (919) 991-5160; email: info@bwfund.org

Internet: http://204.85.36.154/page.php?mode=privateview&pageID=79&navID=110

Sponsor: Burroughs Wellcome Fund

P.O. Box 13901, 21 T. W. Alexander Drive

Research Triangle Park, NC 27709-3901

BWF Career Awards at the Scientific Interface 1441

This initiative supports the early career development of postdoctoral scientists in the physical and computational sciences, whose work addresses biological questions and who are dedicated to pursuing careers in academic research. Research methods may include any combination of experiment, computation, mathematical modeling, statistical analysis, or computer simulation. Areas of interest include: chemistry, computer sciences, engineering, mathematics, and physics. Awards provide a stipend over five years to support up to two years of advanced postdoctoral training and the first three years of a faculty appointment. During both the postdoctoral and the faculty periods, grants must be made to degree-granting institutions in the United States or Canada on behalf of the award recipient. Award recipients must complete at least one year of postdoctoral training during the period of the award, usually at the institution that nominates the candidate.

Requirements: U.S. and Canadian academic institutions, including medical schools, graduate schools, affiliated hospitals, and research institutes are eligible. Candidates must hold a PhD degree in the fields of mathematics, physics, chemistry (physical, theoretical, or computational), computer science, statistics, or engineering. Exceptions will be made only if the applicant can demonstrate significant expertise in one of these areas, evidenced by publications or advanced course work. Candidates must have completed at least six months but not more than 48 months of postdoctoral training at the time of application.

Restrictions: Candidates must not hold or have accepted a faculty appointment as a tenure-track assistant professor at the time of application.

Amount of Grant: $500,000 total over five years

Contact: Dr. Nancy S. Sung, Senior Program Officer; (919) 991-5100; fax: (919) 991-5160; email: info@bwfund.org

Internet: http://204.85.36.154/page.php?mode=privateview&pageID=129&navID=160

Sponsor: Burroughs Wellcome Fund

P.O. Box 13901, 21 T. W. Alexander Drive

Research Triangle Park, NC 27709-3901

BWF Clinical Scientist Awards in Translational Research 1442

The Clinical Scientist Awards in Translational Research program supports established independent physician-scientists who are dedicated to translational research,Äithe two-way transfer between work at the laboratory bench and patient care. The program is intended to help protect award recipients,Äô time to pursue the vital link between basic and clinical research. Importantly, the program aims to identify and reward proven mentors and to increase their capacity to train the next generation of investigators skilled in translational research. In this way, BWF hopes to increase the ranks of experienced physician-scientists critically positioned to bridge the gap between bench and bedside. Selection criteria include the candidate's qualifications, quality of the proposed activities, the ability to conduct innovative research, originality of the proposed research, and the potential to advance research in the scientific fields of study. Applications must be submitted electronically.

Requirements: U.S. or Canadian nonprofit hospitals and educational, scientific, research, or health institutions may submit applications. Candidates must have an MD or MD/PhD degree; hold an appointment or a joint appointment in a sub-specialty of clinical medicine; present evidence of an established research career; and be investigators at the assistant-professor level, associate-professor level, or an equivalent tenure-track position. Institutions may nominate up to four candidates, as long as at least one of the candidates is female, and at least one of the candidates is a member of an underrepresented minority group (Hispanic-American, African-American, Native American).

Samples: Gilbert Chu, Stanford U Medical Ctr (Palo Alto, CA)—to support researchers working to bridge the gap between basic research and patient treatment and care, $750,000 over five years.

Amount of Grant: $750,000 average over five years ($150,000 per year)

Contact: Nancy S. Sung, Ph.D., Senior Program Officer; (919) 991-5100; fax: (919) 991-5160; email: info@bwfund.org

Internet: http://204.85.36.154/page.php?mode=privateview&pageID=177&navID=180

Sponsor: Burroughs Wellcome Fund

P.O. Box 13901, 21 T. W. Alexander Drive

Research Triangle Park, NC 27703-8472

BWF Institutional Program Unifying Population and Laboratory Based Sciences 1443

The program offers five-year institutional training awards of $500,000 per year to bridge the gap between the population and computational sciences and the laboratory-based biological sciences. The award will support the training of researchers between existing concentrations of research strength in population approaches to human health and in basic biological sciences. The goal is to establish training programs by partnering researchers working in schools of medicine and schools (or academic divisions) of public health.

Requirements: Degree-granting institutions in the U.S. or Canada may submit applications.

Restrictions: For-profit companies may not participate in the application, but could be valuable partners in such training programs.

Amount of Grant: $500,000 per year for five years

Contact: Victoria McGovern, Ph.D., Senior Program Officer; (919) 991-5100; fax: (919) 991-5160; email: info@bwfund.org

Internet: http://204.85.36.154/page.php?mode=privateview&pageID=159&navID=164

Sponsor: Burroughs Wellcome Fund

P.O. Box 13901, 21 T. W. Alexander Drive

Research Triangle Park, NC 27709-3901

BWF Investigators in the Pathogenesis of Infectious Disease 1444

These five-year awards provide $500,000 for opportunities for accomplished investigators at the assistant professor level to study pathogenesis, with a focus on the intersection of human and microbial biology. The program is intended to shed light on the overarching issues of how human hosts handle infectious challenge. The awards are intended to give recipients the freedom and flexibility to pursue new avenues of inquiry and higher-risk research projects that hold potential for advancing significantly the biochemical, pharmacological, immunological, and molecular biological understanding of how infectious agents and the human body interact.

Requirements: Candidates must: generally have an M.D., D.V.M., or Ph.D. degree; have an established record of independent research and hold a tenure-track position as an assistant professor or equivalent (at the time of application) at a degree-granting institution; and be nominated by accredited, degree-granting institutions in the United States or Canada.

Restrictions: Applications from non-tenure track investigators at tenure-offering, degree-granting institutions will not be accepted.

Amount of Grant: $500,000 over five years

Contact: Victoria McGovern, Ph.D., Senior Program Officer; (919) 991-5100; fax: (919) 991-5160; email: info@bwfund.org

Internet: http://204.85.36.154/page.php?mode=privateview&pageID=105&navID=135

Sponsor: Burroughs Wellcome Fund

P.O. Box 13901, 21 T. W. Alexander Drive

Research Triangle Park, NC 27709-3901

Bydale Foundation Grants 1445

The foundation emphasizes international understanding, public policy research, environmental quality, cultural programs, the law and civil rights, social services, higher education, and economics. Funding includes support for: conference/seminars, continuing support, general operating support, matching/challenging support, program development, publication, research, and seed money. An application form is not required. Submit an initial approach in the form of a letter or proposal.

Requirements: U.S. 501(c)3 nonprofits are eligible.

Samples: Clean Water Fund (Washington, DC)—$5000.; Greenhouse Crisis Foundation (Washington, DC)—$20,000.; Greenpeace U.S.A (Washington, DC)—$2500.

Date(s) Application is Due: Nov 1

Contact: Milton Solomon, Vice President, (914) 428-3232; fax: (914) 428-1660

Sponsor: Bydale Foundation

11 Martine Avenue

White Plains, NY 10606

Byron W. and Alice L. Lockwood Foundation Grants 1446

The foundation awards grants to Washington nonprofit organizations in its areas of interest, including arts, biomedical research, higher education, hospitals and health care organizations, housing and homelessness, museums, religion and religious welfare programs, and social services. Types of support include capital improvements, continuing support, general operating grants, professorships, projects grants, and research grants. There are no application forms; submit a letter of request.

Requirements: Washington nonprofit organizations are eligible. Grants are awarded primarily in Washington's Seattle and Puget Sound areas.

Samples: Seattle's Union Gospel Mission (Seattle) for capital improvements at the men's shelter, $50,000; Youth Suicide Prevention Plan for Washington State, (2006).

Amount of Grant: $900-$56,000 range

Date(s) Application is Due: Oct 31

Contact: Lee Kraft, Executive Director, (206) 230-8489

Sponsor: Byron W. and Alice L. Lockwood Foundation

P.O. Box 4

Mercer, WA 98040

BYU Religious Studies Center Research Grants 1447

The mission of the BYU Religious Studies Center is to encourage and sponsor serious, faithful, gospel-related scholarship and the ensuing publication of that scholarship. The mission is accomplished by providing grants to assist worthwhile projects; encouraging faculty members to submit their manuscripts for publication; and publishing and marketing high-quality religious books and periodicals of a scholarly nature.

Contact: Richard Neitzel Holzapfel, Managing Director (801) 422-2332; email: holzapfel@byu.edu

Internet: http://religion.byu.edu/rsc_about.php

Sponsor: Brigham Young University

Brigham Young University, 167 HGB

Provo, UT 84602

C. Alma Baker Trust Grants 1448

The trust operates in the fields of agriculture and education and supports students studying for agricultural degrees at Massey University. Types of support include postgraduate scholarships for agricultural and horticultural research, grants for research to further the science of agriculture and horticulture, and travel grants for overseas trips of three to six weeks. Travel grant recipients are expected to contribute to the New Zealand horticultural or agricultural community upon return. Deadlines vary for each program; contact program staff for exact dates.

Amount of Grant: $10,000 per year for Masters; $20,000 per year for PhD

Date(s) Application is Due: Feb 1

Contact: Trust Administrator, 01480 411331; fax: 01480 459012

Internet: http://www.massey.ac.nz

Sponsor: C. Alma Baker Trust

Private Bag, Massey University

Palmerston North New Zealand

C. Wright Mills Award 1449

The award is an annual prize for a book that best exemplifies social science research and scholarship. The books nominated must have been published in the year preceding the year of the award.

Amount of Grant: $500

Date(s) Application is Due: Jan 15

Contact: Michele Koontz, Administrative Officer, (865) 689-1531; fax: (865) 689-1534; email: mkoontz3@utk.edu

Internet: http://www.sssp1.org/index.cfm?contentID=566

Sponsor: C. Wright Mills Award Committee

901 McClung Tower, SSSP, Univ of TN

Knoxville, TN 37996-0490

C.R. Bard Foundation Grants 1450
The foundation's areas of interest are health care; arts, cultural, and community life; and education and research in company-operating areas. In the area of health care, the foundation supports organizations and programs specializing in vascular medicine, urology, and oncology. In the arts/cultural/community category, the foundation considers requests from local civic, recreational, youth, and art and cultural organizations. In the area of education and research, the foundation considers requests from institutions of higher education that offer medical and managerial curricula, as well as organizations whose programs and services benefit the health care industry. Types of support include program development, scholarship funds, and employee matching gifts. Requests for funding must be in writing. Contact the office for application requirements.
Requirements: 501(c)3 organizations in primarily New Jersey, Massachusetts, and California may apply for grant support.
Geographic Focus: California; Massachusetts; New Jersey
Amount of Grant: $1.4 million total
Contact: Program Contact, (908) 277-8182, (800) 367-2273; fax: (908) 277-8240
Internet: http://www.crbard.com/about/com_relations/foundations/index.cfm
Sponsor: C.R. Bard Foundation
730 Central Avenue
Murray Hill, NJ 07974

Cabot Corporation Foundation Grants Program 1451
The objective of the company's corporate giving is to support community outreach objectives, with priority given to science and technology education; safety health and environment (SH&E); and community and civic improvement efforts in the communities where the company has major facilities or operations. Types of support include capital grants, challenge grants, employee matching gifts, fellowships, general support, professorships, project support, research, scholarships, and seed money. The board meets in January, April, July, and October to consider requests. Applications must be received at least 30 days before a board meeting.
Requirements: The foundation supports nonprofit 501(c)3 organizations in Georgia, Illinois, Louisiana, Massachusetts, Pennsylvania, Texas, and West Virginia. Support is available for international organizations that qualify under U.S. tax regulations.
Restrictions: Contributions are not made to individuals; fraternal, political, athletic or veterans organizations; religious institutions (except for projects that are open to individuals without regard to religious preference); capital and endowment campaigns; sponsorships of local groups/individuals to participate in regional, national, or international competitions, conferences or events; advertising, dinner-table sponsorship, or fund-raising events.
Geographic Focus: Georgia; Illinois; Louisiana; Massachusetts; Pennsylvania; Texas; West Virginia
Samples: Gilbertsville Elementary School (PA)—to enhance science curriculum with LEGO products and training materials, $3700.; Pact Malaysia (Malaysia)—to support literacy programs in Malaysia, $5000.; Children's Hands-On Art Museum (MA)—equipment grant, $2500.
Amount of Grant: $1,000-$25,000; $5000 average
Contact: Karen Morrissey, Executive Director, (617) 345-0100; fax: (617) 342-6312; email: Karen_Morrissey@cabot-corp.com
Internet: http://w1.cabot-corp.com/controller.jsp?N=21+3030
Sponsor: Cabot Corporation Foundation
Two Seaport Lane, Suite 1300
Boston, MA 02210

CALA C.C. Seetoo/CALA Conference Travel Scholarship 1452
One scholarship is awarded each year to a student of Chinese nationality or Chinese descent currently enrolled in an ALA-accredited library and information science program in the United States to attend the ALA annual conference. The purpose of the award is to provide its recipient mentoring and networking opportunities and professional development at the annual conference. Selection will be based on academic achievement and leadership potential.
Requirements: Students who are currently enrolled in ALA-accredited master's or doctoral programs of Library and Information Science in degree-granting institutions in North America are eligible to apply. Applicants must be of Chinese nationality or of Chinese descent.
Amount of Grant: $500
Date(s) Application is Due: Apr 15
Contact: Y. Diana Wu, (408) 808-2087; email: dwu@sjsu.edu
Internet: http://www.cala-web.org
Sponsor: Chinese-American Librarians Association
One Washington Sq
San Jose, CA 95192-0028

California and Michigan Nonprofit Research Program Grants 1453
The Nonprofit Sector Research Fund of the institute makes grants to expand understanding of nonprofit activities, including philanthropy. The fund encourages projects that will inform public policy toward nonprofits and invites proposals from scholars and practitioners. Nine priority areas for national and international grantmaking activity include the role of nonprofits in fostering democratic values, their role in society, advocacy issues, philanthropy, governance, public accountability, financing, workforce, and international nonprofit activities. The California and Michigan programs support studies with the potential to inform public policy and benefit nonprofit practices in those states. The listed application deadline is for preproposals; full proposals are by invitation.
Samples: Gayle Lawn-Day and David Carlson, Northern Michigan U (Marquette, MI)—to create a comprehensive database of nonprofit groups on Michigan's upper peninsula, $13,750.; Leobardo Estrada and Marcos Vargas, U of California (Los Angeles, CA)—to assess how advocacy and social services organizations that serve immigrants in California are faring, $39,916.
Amount of Grant: $25,000 maximum
Date(s) Application is Due: Jul 15
Contact: Rachel Mosher-Williams, (202) 736-2501; fax: (202) 293-0525; email: rwilliams@aspeninst.org
Internet: http://www.nonprofitresearch.org/newsletter1531/newsletter.htm
Sponsor: Aspen Institute
One Dupont Circle NW, Suite 700
Washington, DC 20036

California Breast Cancer Research Program Grants 1454
The state-funded breast cancer research effort supports a partnership comprising community members (such as a breast cancer advocacy organization, a community clinic or organization serving women with breast cancer, or a member of a California community affected by breast cancer) and experienced research scientists. Two funding mechanisms are available. The 18-month pilot award supports the initial phase of the project, including feasibility of methods, strengthening collaborations, development of tools and methods, and/or collecting pilot data. The three-year full award is for projects with a fully developed research plan with supporting preliminary data, carried out by a well integrated team of scientific and community members with a previous work relationship. The partnership works together to identify the research question, develop the research plan, carry out the research, interpret the results, and disseminate information to the community. The listed application deadline is for concept papers. Guidelines are available online.
Requirements: A partnership consists of California community members (organization or individual) and experienced research scientists.
Geographic Focus: California
Amount of Grant: $600,000 maximum, full award; $150,000 maximum, pilot award
Date(s) Application is Due: Nov 3
Contact: Natalie Collins, MSW, (510) 987-0646; email: natalie.collins@ucop.edu
Internet: http://www.cbcrp.org/apply/crcCall
Sponsor: University of California
300 Lakeside Drive, 6th Fl
Oakland, CA 94612-3550

California Stories Grants 1455
The fund is one of the components of California Stories, the California Council for the Humanities' statewide initiative that seeks to strengthen California communities through story-based public humanities projects. Through grants awarded quarterly through a competitive process, the fund supports public humanities programs that will bring to light compelling stories from California's diverse communities and provide opportunities for collective reflection and discussion. The current competition invites proposals for projects that document the stories of Gulf Coast residents evacuated to California in the aftermath of Hurricane Katrina and make use of the humanities in exploring those stories. Guidelines and application materials are available online.
Requirements: Projects must strengthen communities through public sharing and discussion of stories; focus on California communities (defined as such on the basis of geography, ethnicity, culture, or shared experience); use grant funds to gather stories from community members, for archival research, or to provide a forum to share and discuss stories; use one or more formats to present and communicate these stories, including photography and interpretive exhibit, radio documentary, digital media, dramatic presentation, interpretive artwork, author/poetry readings, storytelling events, film festivals, and community conferences; include a community discussion component that engages people in discussion of the stories and their relevance to the community; involve at least one humanities expert in the design and implementation of the project; produce photos, transcripts of stories, or audio and video recordings that can be archived and shared with larger audiences through broadcast, publication, or website dissemination; have a total budget of no more than $30,000; and include plans for assessing project outcomes. Culminating events should be free and open to participation by the general public. Project activities should not start before October 1.
Geographic Focus: California
Amount of Grant: $7500; $15,000 maximum Hurricane Katrina evacuee stories
Date(s) Application is Due: Dec 1
Contact: Susana Loza, (415) 391-1474 ext 314; email: info@calhum.org
Internet: http://www.californiastories.org/guidelines/guidelines_ca_story_intro.htm
Sponsor: California Council for the Humanities
312 Sutter Street, Suite 601
San Francisco, CA 94108-4371

California Teachers Association Scholarships 1456
The CTA Scholarship gives support for study in higher education to active members and/or dependent children of active members and/or dependent children of deceased members. Twenty-five scholarships are awarded annually. Applications are available in October, and may be obtained by contacting the CTA Human Rights Department in Burlingame or any CTA Regional Resource Center office.
Requirements: Applicant must be an active member of California Teachers Association or applicant must be the dependent child of an active, retired-life or deceased member of California Teachers Association depending on which program they are applying for.
Geographic Focus: California

Amount of Grant: $2000
Date(s) Application is Due: Feb 15
Contact: Manuel Ayan, (650) 552-5204; fax: (650) 552-5002; email: scholarships@cta.org
Internet: http://www.cta.org/InsideCTA/ScholarshipsWorkshops/ScholarshipsWorkshops.htm
Sponsor: California Teachers Association
P.O. Box 921, 1705 Murchison Drive
Burlingame, CA 94011-0921

California Wellness Foundation Work and Health Program Grants 1457

The foundation focuses its activities on specific priority areas where it has a significant, long-term commitment. Within each priority area, the foundation allocates the majority of its funds toward initiatives. Initiatives are targeted grantmaking programs with distinct objectives and are generally announced through requests for proposals. The foundation awards general grants and project grants. Under general grants, requests for core operating support for organizations that provide direct services to Californians for disease prevention or health promotion are of primary interest. Priority areas under general grants are diversity in the health professions, environmental health, healthy aging, mental health, teenage pregnancy prevention,, violence prevention, women's health, and work and health. Special projects grants are awarded to areas that fall outside the priority areas. Of particular interest are proposals to help California communities respond to cutbacks in federally funded programs. Activities commonly supported under special projects include strengthening traditional safety-net providers, educating consumers about changes in health care systems, advocating for underserved communities in health policy debates, and informing public decision making through policy analysis.

Requirements: Eligible applicants are California 501(c)3 nonprofit organizations, or organizations with a preapproved fiscal sponsor. An organization should first write a succinct letter of interest (one to two pages) that describes the organization, its leadership, the region and population(s) served, and the activities for which funding is needed, including the amount requested.

Geographic Focus: California

Samples: California Institute for Rural Studies (Davis, CA)—to articulate and distribute research findings on the health of farmworkers to policy makers, opinion leaders, the news media, and the public, $125,000 over three years (2005); California State U, Stanislaus Foundation (Modesto, CA)—to provide holistic, prevention-based health-education services to Southeast Asian women and children living in Stanislaus County, $180,000 over three years (2005); California State U, Auxiliary Services (CA)—to strengthen community organizations working to meet the health needs of residents of East Los Angeles and California's West San Gabriel Valley, $135,000 over three years (2005); California Institute for Nursing and Healthcare (Berkeley, CA)—to develop and disseminate a plan for diversifying the nursing work force in California, $120,000 over three years (2005).

Amount of Grant: $5000-$200,000 over one to three years
Contact: Grants Administrator, (818) 702-1900; fax: (818) 593-6614
Internet: http://www.tcwf.org/grants_program/index.htm
Sponsor: California Wellness Foundation
6320 Canoga Avenue, Suite 1700
Woodland Hills, CA 91367-7111

Calvin Institute of Christian Worship Renewal Grants 1458

The grants program is designed to foster well-grounded worship renewal in congregations throughout North America. Made possible through support of the Lilly Endowment, Inc., these grants are intended to serve a grass-roots constituency of those concerned for the vitality of the worship life in their local Christian communities in a variety of denominations. Through its grantmaking, the institute intends to stimulate thoughtful and energetic work that will result in worship services that exhibit renewed creativity, theological integrity, and relevance. The institute solicits proposals for grants to support a wide variety of projects. Grants will be awarded both to individuals and to nonprofit organizations, such as churches, colleges, seminaries, or other church-related organizations. Grants to individuals, however, will be awarded through the church or organization with which they are affiliated.

Requirements: Any individual or team of people with leadership role(s) who are directly related to the worship life of a congregation, school, church-related organization, or other ministry is encouraged to apply. This includes pastors, educators, church staff persons, church musicians, directors of worship, liturgists, artists, architects, and scholars.

Samples: Choristers Guild, Garland, TX, to teach children the structure of worship, develop their skills for leading various aspects of worship, and work with them to prepare worship services through summer worship, arts, and music camps; LaGrave Avenue Christian Reformed Church, Grand Rapids, MI, to study and implement ways in which participation in formal worship can be made more accessible to children and youth; Winnetka Presbyterian Church, Winnetka, IL, to study and apply successful techniques for integrating multiply-challenged children into corporate worship (2005).

Amount of Grant: $5000-$15,000 range
Date(s) Application is Due: Jan 10
Contact: Betty Grit, Grants Administrator, (616) 526-7168; fax: (616) 957-7168; email: worshipgrants@calvin.edu
Internet: http://webapps.calvin.edu/worship/grants/wrgp.php
Sponsor: Calvin Institute of Christian Worship
3201 Burton SE
Grand Rapids, MI 49546

Camargo Foundation Fellowships 1459

The program provides, at no cost, furnished apartments and a reference library in the city of Cassis for scholars who wish to pursue projects in the humanities and social sciences related to French and francophone cultures. The foundation also sponsors creative projects by visual artists, photographers, composers, and writers. The term of residence is one semester (early September to mid-December or mid-January to May 31), precise dates being announced each year. An informational brochure and application form are available upon request.

Requirements: Applicants may include members of university and college faculties, including professors emeriti; independent scholars working on specific projects; secondary school teachers; graduate students completing the dissertation required for their degree; and writers, photographers, visual artists, and composers.

Amount of Grant: $3500 stipend
Date(s) Application is Due: Jan 15
Contact: Program Contact, (651) 238-8805
Internet: http://www.camargofoundation.org/aboutus.asp
Sponsor: Camargo Foundation
400 Sibley Street, Suite 125
Saint Paul, MN 55101-1928

Cambridge Foundation For Peace Prize 1460

The foundations announces a call for papers that explore the role of diasporas as transnational actors in preventing violence, crisis response, or postconflict peacebuilding. The prize will be awarded for the winning proposal, which will contribute to systematic research on diasporas. In particular, the paper should concentrate on identifying and specifying the mechanisms by which diasporas can be engaged in policy planning and implementation. The regional and/or country focus is open, but preference will be give to Southeastern Europe and/or the Eastern Mediterranean. The CFP will distribute the winning paper to its worldwide audience of policymakers, political leaders, academics, and journalists. The winning paper also will be posted as a working paper on CFP-web.

Requirements: The competition is open to public policy professionals and academics (including graduate students).

Amount of Grant: $3000
Date(s) Application is Due: Apr 1
Contact: Grants Administrator, (617) 254-5437; fax: (617) 254-5437; email: schmitz@cfp-web.org
Internet: http://www.cfp-web.org
Sponsor: Cambridge Foundation for Peace
1770 Massachusetts Avenue, Box 303
Cambridge, MA 02140-2808

Camille Dreyfus Teacher-Scholar Awards Program 1461

The Camille Dreyfus Teacher-Scholar Awards Program supports the research and teaching careers of talented young faculty in the chemical sciences. Based on institutional nominations, the program provides discretionary funding to faculty at an early stage in their careers. Criteria for selection include an independent body of scholarship attained within the first five years of their appointment as independent researchers, and a demonstrated commitment to education, signaling the promise of continuing outstanding contributions to both research and teaching. The Camille Dreyfus Teacher-Scholar Awards Program provides an unrestricted research grant of $75,000. Of the total amount, $7,500 is for departmental expenses associated with research and education. Guidelines and required information are available online.

Requirements: The Camille Dreyfus Teacher-Scholar Awards Program is open to academic institutions in the States, Districts, and Territories of the United States of America that grant a bachelor's or higher degree in the chemical sciences, including biochemistry, materials chemistry, and chemical engineering. Nominees must hold a full-time tenure-track academic appointment, and are normally expected to have been appointed no earlier than mid-year 2004. Awardees are from Ph.D. granting departments in which scholarly research is a principal activity. Undergraduate education is an important component of the nominee's activities. Institutions may submit only one Camille Dreyfus nomination annually.

Restrictions: Defrayal of academic-year salary is not permitted.

Amount of Grant: $75,000
Date(s) Application is Due: Feb 11
Contact: Mark Cordillo, Executive Director; (212) 753-1760; fax: (212) 593-2256; email: mcardillo@dreyfus.org or admin@dreyfus.org
Internet: http://www.dreyfus.org/awards/camille_dreyfus_teacher_award.shtml
Sponsor: Camille and Henry Dreyfus Foundation
555 Madison Avenue, 20th Floor
New York, NY 10022-1301

Campaign for the Civic Mission of Schools Grants 1462

The program is a long-term effort to renew and elevate civic education in the schools, based on recommendations in The Civic Mission of Schools report. The report identifies six promising approaches to good civic education: instruction in government, history, law, and democracy; class discussion of current local, national, and international issues and events; community service and service-learning linked to curriculum and class instruction; extracurricular opportunities to get involved in the school and community; participation in school governance; and simulations of democratic processes and procedures. A major component of the campaign focuses on education policy at the state and local level. The program seeks applications from state-level coalitions for grants in two categories: small grants—two-year grants to state coalitions that propose a specific project or activity that promotes policy-related work that is tied to one or more of the recommendations of the report (July 19 deadline); and large grants—two-year grants to state coalitions that have the capacity and will to implement several of the six recommended approaches (June 4 deadline). Guidelines are available online.

Requirements: An applicant should represent a coalition of education and civic engagement stakeholders and other interested parties, including for example: government agencies, legislators, state and local school board members, student leaders, student associations, government relations and strategic communications firms, business alliances, colleges and universities, and philanthropic institutions.

Samples: Pennsylvania Coalition for Representative Democracy (Philadelphia, PA)—to promote high-quality civic education at public schools in Philadelphia, $150,000 (2004).

Amount of Grant: $150,000 over two years, large grants; $20,000 over two years, small grants

Date(s) Application is Due: Jun 4; Jul 19

Contact: Richard Russo, Center for Democracy and Citizenship, Council for Excellence in Government, (202) 530-3260; email: rrusso@excelgov.org

Internet: http://www.civicmissionofschools.org

Sponsor: Carnegie Corporation of New York
437 Madison Avenue
New York, NY 10022

Campbell Foundation HIV/AIDS Research Grants 1463

The foundation awards grants for research projects on ways to prevent and treat HIV, AIDS, and related conditions or illnesses. Proposals focusing on alternative, nontraditional avenues of study receive preference. Most grants support clinical research, although other types of work are considered. After initial screening, requests are reviewed by a peer review board of noted HIV/AIDS physicians and then sent to the board for final decisions. There are no application deadlines, but as all grants are handled chronologically, the foundation suggests submitting grant requests as early as possible.

Requirements: U.S. nonprofit organizations are eligible.

Amount of Grant: $20,000-$80,000 average

Contact: Program Contact, (954) 493-8822; fax: (954) 493-8801; email: campfound@aol.com

Internet: http://members.aol.com/campfound

Sponsor: Campbell Foundation
5975 N Federal Hwy, Suite 126
Fort Lauderdale, FL 33308

CampusRN/AACN Nursing Scholarship Fund 1464

The scholarship program supports students who are seeking a baccalaureate, master's, or doctoral degree in nursing. Special consideration will be given to students enrolled in a master's or doctoral program with the goal of pursuing a nursing faculty career; completing an RN to baccalaureate program (BSN); or enrolled in an accelerated baccalaureate or master's degree nursing program. Applicants must submit an online application and create a complete profile on CampusRN. All applicants must already be enrolled (not just accepted) at an AACN member institution. Applications are accepted only by email.

Requirements: Nursing students enrolled in a baccalaureate, master's or doctoral degree program in nursing are eligible.

Amount of Grant: $2500

Date(s) Application is Due: Jan 1; Mar 1; May 1; Jul 1; Sep 1; Nov 1

Contact: Scholarship Administrator, (202) 463-6930; fax: (202) 785-8320; email: info@campuscareercenter.com

Internet: http://aacn.campusrn.com/scholarships/scholarship_rn.asp

Sponsor: American Association of Colleges of Nursing
One Dupont Circle NW, Suite 530
Washington, DC 20036

CAN Autism Treatment-Related Study Awards 1465

The organization recognizes the urgent need to develop effective therapies to treat both core and domain specific features of autism. Research proposals focused on all aspects of treatment, from basic research models, to clinical trials, to biomarker identification, are acceptable for submission. Due to the current lack of effective treatment and absence of known biomarkers, applications are reviewed as they are received. Funding amount and duration is determined per individual proposal. Indirect costs are limited to 10 percent of the total amount. Letters of intent are required prior to invitation of full proposals.

Amount of Grant: $60,000 per year for one, two, or three years

Date(s) Application is Due: Sep 1

Contact: Therese Finazzo, Grants Officer, (888) 828-8476 or (323) 549-0500 ext 18; fax: (323) 549-0500; email: therese@cureautismnow.org

Internet: http://www.cureautismnow.org/research/funding/3523.jsp

Sponsor: Cure Autism Now
5455 Wilshire Boulevard, Suite 715
Los Angeles, CA 90036

CAN Pilot Research Projects 1466

Pilot research programs will provide funding for investigators at any stage in their career to work on innovative pilot projects. The purpose of these awards is to encourage innovative approaches towards elucidating the causes, prevention, and appropriate therapy of autism. They are intended to provide support for innovative research that may not have reached a stage where funding can be obtained from federal sources. Annual deadline dates may vary; contact program staff for exact dates.

Requirements: Investigators whose focus has been outside the field of autism as well as innovative investigators from within the field are encouraged to apply.

Restrictions: Members of CAN's Board of Directors are not eligible as investigators in CAN supported research. Members of CAN's Scientific Advisory Committee are eligible, provided they comply with foundation policies regarding the avoidance of conflict of interest.

Amount of Grant: $120,000 maximum two-year award

Date(s) Application is Due: Mar 3; Jun 9

Contact: Therese Finazzo, Grants Officer, (888) 828-8476 or (323) 549-0500 ext 18; fax: (323) 549-0500; fax: (323) 549-0547; email: therese@cureautismnow.org

Internet: http://www.cureautismnow.org/research/funding/3522.jsp

Sponsor: Cure Autism Now
5455 Wilshire Boulevard, Suite 715
Los Angeles, CA 90036-4234

CAN Young Investigator Awards 1467

The purpose of these awards is to encourage innovative approaches towards elucidating the causes, prevention, and appropriate therapy of autism. These awards are designed to support outstanding projects that involve innovative approaches and the application of cutting-edge technologies. All proposals must have direct and immediate relevance to autism and related disorders.

Requirements: Awards provide for the work of outstanding candidates to be carried out under the supervision of a mentor who is an established investigator. The investigator need not be directly involved in autism research but must provide an environment in which the CAN fellow can pursue a project with relevance to autism or related disorders.

Restrictions: Applicants should be no more than four years out of a MD or PhD program on the date the fellowship would begin.

Amount of Grant: $40,000 and $1000 travel allowance per year for two years

Date(s) Application is Due: Mar 3; Jun 9

Contact: Therese Finazzo, Grants Officer, (323) 549-0500 ext 18 or (888) 828-8476; fax: (323) 549-0500; email: therese@cureautismnow.org

Internet: http://www.cureautismnow.org/research/funding/3522.jsp

Sponsor: Cure Autism Now
5455 Wilshire Boulevard, Suite 715
Los Angeles, CA 90036-4234

Canada Council for the Arts Grants 1468

The council awards a wide range of grants to professional Canadian artists and arts organizations in dance, interdisciplinary work and performance art, media arts, music, theater, writing, publishing, architecture, and the visual arts. The council also administers the Killam Program of prizes and fellowships to Canadian scholars of exceptional ability engaged in significant research projects.

Requirements: Professional Canadian artists and arts organizations are eligible.

Contact: Lise Rochon, Information Officer, (800) 263-5588 ext 4138 or (613) 566-4414 ext 4138; fax: (613) 566-4390; email: info@canadacouncil.ca

Internet: http://www.canadacouncil.ca/grants/default.asp

Sponsor: Canada Council for the Arts
P.O. Box 1047, 350 Albert Street
Ottawa, ON K1P 5V8 Canada

Canada Council John G. Diefenbaker Award 1469

This annual award enables a distinguished German scholar to spend up to 12 months in Canada, which may include brief periods in the United States. While research must be the primary activity during the tenure of the award, the recipient will also be encouraged to participate in the activities of the host institution and to interact more extensively with the research communities in Canada and the United States by visiting other institutions. Candidates may not apply for this award, but instead must be nominated by a host university department or research institute in Canada.

Requirements: Candidates must be German citizens and must have a sound working knowledge of one of Canada's two official languages. German scholars who have demonstrated outstanding ability, especially through a substantial publication record over several years, are eligible to apply.

Amount of Grant: $C75,000 maximum plus $C20,000 for travel expenses

Date(s) Application is Due: Dec 1

Contact: Nathalie Lauzon, Administrative Assistant, (800) 263-5588 or (613) 566-4414 ext 4083; fax: (613) 566-4430; TTY: (613) 565-5194; email: endowments.prizes@canadacouncil.ca

Internet: http://www.canadacouncil.ca/prizes/john_g_diefenbaker/hn1272230384 93593750.htm

Sponsor: Canada Council for the Arts
P.O. Box 1047, 350 Albert Street
Ottawa, ON K1P 5V8 Canada

Canada Graduate Scholarships and NSERC Postgraduate Scholarships 1470

Canada Graduate Scholarships and NSERC Postgraduate Scholarships provide financial support to high-calibre scholars who are engaged in master's or doctoral programs in the natural sciences or engineering. The Canada Graduate Scholarships will be offered to the top-ranked applicants at each level (master, and doctoral) and the next tier of meritorious applicants will be offered an NSERC Postgraduate Scholarship. This support allows these scholars to fully concentrate on their studies and to seek out the best research mentors in their chosen fields. NSERC encourages interested and qualified Aboriginal students to apply.

Requirements: To be considered eligible for support, applicants must: (1) be a Canadian citizen or a permanent resident of Canada; (2) hold, or expect to hold (at the time you take up the award), a degree in science or engineering from a university whose standing is acceptable to NSERC (if you have a degree in a field other than science or engineering, NSERC may accept your application at its discretion); (3) intend to pursue in the following year full-time graduate studies and research at the master's or doctoral level in an eligible program in one of the areas of the natural sciences and engineering supported by NSERC; and (4) have obtained a first-class average (a grade of 'A-') in each of the last two completed years of study (full-time equivalent). Any exceptions to this requirement must be accompanied by supporting comments that justify the submission.
Restrictions: Scholarships are not available to students entering graduate studies in a qualifying year. Students who, after receiving an NSERC scholarship, change their field of study and research to a field that is not supported by NSERC, cease to be eligible and their awards will be cancelled.
Amount of Grant: $17,300 - $35,000 CAN per year
Date(s) Application is Due: Oct 15
Contact: PGS Program Officer; (613) 995-5521; fax: (613) 996-2589; email: schol@nserc.ca
Internet: http://www.nserc-crsng.gc.ca/Students-Etudiants/PG-CS/BellandPostgrad-BelletSuperieures_eng.asp
Sponsor: Natural Sciences and Engineering Research Council of Canada
350 Albert Street, 14th Floor
Ottawa, ON K1A 1H5 Canada

Canada-U.S. Fulbright International Science and Technology Award 1471
Canada-U.S. Fulbright awards offer outstanding students and scholars in Canada and the United States a unique opportunity to participate in one of the world's premier educational exchange programs. The International Fulbright Science and Technology Award Ph.D. offers support for study in science, technology, or engineering at top U.S. institutions. The award carries a value of, on average, U.S.$180,000 over three years.
Requirements: Each applicant is required to submit an online application, along with all supporting documents.
Amount of Grant: $180,000 over three years
Date(s) Application is Due: May 31
Contact: Jennifer Regan, Senior Program Officer & NNASC Coordinator; (613) 688-5517 or (613) 688-5540; fax: (613) 237-2029; email: jregan@fulbright.ca
Internet: http://www.fulbright.ca/en/award.asp
Sponsor: Foundation for Educational Exchange between Canada and the United States of America
350 Albert Street, Suite 2015
Ottawa, ON K1R 1A4 Canada

Canada-U.S. Fulbright Senior Specialists Program Grants 1472
The Fulbright Senior Specialists Program is designed to provide U.S. and Canadian faculty and professionals with opportunities to collaborate on curriculum and faculty development, institutional planning and a variety of other activities. During the course of their grant, Fulbright Senior Specialists may engage in any of the following activities at their Canadian host institution: conduct needs assessments, surveys, institutional or programmatic research; take part in specialized academic programs and conferences; consult with administrators and instructors of post-secondary institutions on faculty development; present lectures at graduate and undergraduate levels; participate in or lead seminars or workshops at overseas academic institutions; develop and/or assess academic curricula or educational materials; and conduct teacher-training programs at the tertiary level. These are short-term grants of two to six weeks to that are offered on a continuous basis.
Requirements: Eligibility requirements include: Canadian or American citizenship at time of application (permanent resident status or landed immigrant status are not sufficient); a Ph.D. or equivalent professional/terminal degree by December 31 of the application year or equivalent professional experience; and English language proficiency.
Restrictions: Applicants are not eligible for a Canada-U.S. Fulbright Award if he or she: currently holds permanent residency status in the intended host country; is currently residing in or enrolled at a university in the intended host country; has resided abroad for five or more consecutive years in the six year period immediately preceding the date of application; has received a Fulbright Award within the last five years; has had recent substantial experience in the intended host country (recent substantial experience is defined as study, teaching, research or employment for a period of more than an academic year (nine months) during the past five years); or is a Canadian applicant and you are a dual citizen of Canada and the United States.
Contact: Jennifer Regan, Senior Program Officer & NNASC Coordinator; (613) 688-5517 or (613) 688-5540; fax: (613) 237-2029; email: jregan@fulbright.ca
Internet: http://www.fulbright.ca/en/seniorspecialists.asp
Sponsor: Institute of International Education, Council for International Exchange of Scholars
350 Albert Street, Suite 2015
Ottawa, ON K1R 1A4 Canada

Canada-U.S. Fulbright Traditional Scholar Awards 1473
Grants of four to nine months are available to Canadian and American scholars, post-doctoral researchers and experienced professionals to conduct research and/or teach at a host institution in the United States or Canada, respectively. Canada-U.S. Traditional Fulbright awards offer a unique opportunity to explore important contemporary issues relevant to Canada and the United States and the relationship between the two countries. While the competition is officially field open, applications in the following areas are given

preference: comparative public policy; international trade; North American integration; security; communications and culture; ecology and the environment; indigenous issues; law; border issues; public health; Canada-U.S. relations; and Canadian and American studies. The deadline for American scholars is August 1 and the deadline for Canadian scholars is November 16.
Requirements: Eligibility requirements include: Canadian or American citizenship at time of application (permanent resident status or landed immigrant status are not sufficient); a Ph.D. or equivalent professional/terminal degree by December 31 of the application year or equivalent professional experience; and English language proficiency.
Restrictions: Applicants are not eligible for a Canada-U.S. Fulbright Award if he or she: currently holds permanent residency status in the intended host country; is currently residing in or enrolled at a university in the intended host country; has resided abroad for five or more consecutive years in the six year period immediately preceding the date of application; has received a Fulbright Award within the last five years; has had recent substantial experience in the intended host country (recent substantial experience is defined as study, teaching, research or employment for a period of more than an academic year (nine months) during the past five years); is a Canadian applicant and you are a dual citizen of Canada and the United States.
Amount of Grant: $25,000
Date(s) Application is Due: Aug 1; Nov 16
Contact: Jennifer Regan, Senior Program Officer & NNASC Coordinator; (613) 688-5517 or (613) 688-5540; fax: (613) 237-2029; email: jregan@fulbright.ca
Internet: http://www.fulbright.ca/en/seniorscholar.asp
Sponsor: Foundation for Educational Exchange between Canada and the United States of America
350 Albert Street, Suite 2015
Ottawa, ON K1R 1A4 Canada

Canadian Dental Research Foundation Award 1474
This research award is offered biennially to encourage research related to dentistry conducted by graduate or postgraduate students in Canada. The monetary prize will be presented to the student who will have submitted the best report on a research project at the CDA annual meeting and the paper will be presented at the ACFD/CADR biennial meeting.
Requirements: Research papers entered are limited to graduate or postgraduate students who have conducted their work in association with the dental faculty of a Canadian university.
Contact: Richard Munro, Executive Director, (613) 236-4763; fax: (613) 236-3935; email: Information@dcf-fdc.ca
Internet: http://www.dcf-fdc.ca
Sponsor: Dentistry Canada Fund
427 Gilmour Street
Ottawa, ON K2P OR5 Canada

Canadian Foundation for Dietetic Research Awards 1475
Within the broad category of practice-based dietetic research, the foundation prefers to support research that is of direct relevance to the nutritional well-being of populations within Canada. Priority research directions are outcomes of intervention, new roles for dietitians in meeting the health needs of all Canadians, identification of vulnerable groups and their nutritional needs, and determinants of food choice. An active member of Dietitians of Canada must be a member, if not the leader, of the research team. Applications for projects extending beyond one year will not be considered. The researchers must be associated with an institution or agency with charitable status (such as a teaching hospital or university) that will administer the funds awarded. Grants are intended to finance the direct operating costs of dietetic practice-based research projects. Calls for proposals are posted on the Web site.
Requirements: Applications will only be considered where a practicing dietitian who is an active member of Dietitians of Canada and who is delivering direct or indirect client/patient/public care of service is the principal investigator or a co-investigator.
Amount of Grant: $C10,000 maximum for larger projects; $C5000 maximum for smaller projects
Date(s) Application is Due: Sep 30
Contact: Awards Manager, (416) 642-9307; fax: (416) 596-0603; email: cfdr@dietitians.ca
Internet: http://www.dietitians.ca/cfdr/grants.htm
Sponsor: Canadian Foundation for Dietetic Research
480 University Avenue, Suite 604
Toronto, ON M5G 1V2 Canada

Canadian Friends of Hebrew University Research and Study Grants 1476
Grants are awarded for one year of undergraduate or graduate study to Canadian citizens or residents of Canada who are able to fulfill the entrance requirements of the Hebrew University in Jerusalem in their chosen fields of study, which may include law, dentistry, social sciences and economics, humanities, arts, science, agriculture, pharmacy, and library science.
Amount of Grant: $C1000-$C4000
Date(s) Application is Due: May 1
Contact: Dina Wachtel, National Director, (416) 485-8000; fax: (416) 485-8565; email: dwachtel@cfhu.org
Internet: http://www.cfhu.org
Sponsor: Canadian Friends of Hebrew University
3080 Yonge Street, Suite 5024
Toronto, ON M4N 3P4 Canada

Canadian Library Association Dafoe Scholarship   1477

This scholarship is awarded annually, when merited, to a student entering an accredited Canadian library school. Consideration is given to academic achievement, leadership potential, and demonstrated interest in the profession. Support is for one year.

Requirements: Scholarship candidates must be commencing studies for their first professional library degree and must be Canadian citizens or possess landed immigrant status.

Amount of Grant: $C5000

Date(s) Application is Due: May 1

Contact: Brenda Shields, Member Services Associate, (613) 232-9625 ext 318; fax: (613) 563-9895; email: bshields@cla.ca

Internet: http://www.cla.ca/awards/dafoe.htm

Sponsor: Canadian Library Association
328 Frank Street
Ottawa, ON K2P 0X8 Canada

Canadian Library Association Library Research and Development Grants   1478

Grants support theoretical and applied research that advances the fields of library and information science in their broadest context. Grants are meant to cover only the costs associated with the research, not the salary of the principals. Some examples of typical costs that will be considered are travel, computer time, costs of hiring clerical support, or costs of purchasing supplies.

Requirements: All personal members of the CLA, except current members of the R&D Committee, are eligible for the grants.

Amount of Grant: $C1000 maximum

Date(s) Application is Due: Feb 28

Contact: Brenda Shields, Member Services Associate, (613) 232-9625 ext 318; fax: (613) 563-9895; email: bshields@cla.ca

Internet: http://www.cla.ca/awards/grants.htm

Sponsor: Canadian Library Association
328 Frank Street
Ottawa, ON K2P 0X8 Canada

Canadian Liver Foundation Grants   1479

The foundation awards organization grants, fellowships, and studentships to support research into nephrology, hepatic function or disease, and other related disciplines.

Requirements: Canadian citizens and organizations are eligible.

Date(s) Application is Due: Mar 31

Contact: Billie Potkonjak, (800) 563-5483; fax: (416) 491-4952; email: bpotkonjak@liver.ca

Internet: http://www.liver.ca/Research/Grant_Program

Sponsor: Canadian Liver Foundation
2235 Sheppard Avenue E, Suite 1500
Toronto, ON M2J 5B5 Canada

Canadian Nurses Foundation Study Awards   1480

Grants provide funding for nurses undertaking university studies in nursing and health care and for research related to nursing practice and health care. Awards are given for university study at the baccalaureate, master's, and doctoral levels. Priority will be given to nurses studying in programs with a nursing focus. Once recipients have completed their degrees, they must return to nursing and assist in the development of improved nursing and health care in Canada.

Requirements: Applicants must be members of the Canadian Nurses Foundation and must have gained acceptance into a university program of study; applicants must identify the practice area in which they plan to study and have definite career goals.

Amount of Grant: $C3000-$C6000

Contact: Linda Piazza, Executive Director, (613) 237-2159 ext 250; fax: (613) 237-3520; email: lpiazza@cna-aiic.ca orcnf@cnursesfdn.ca

Internet: http://www.canadiannursesfoundation.com/welcome.htm

Sponsor: Canadian Nurses Foundation
50 Driveway
Ottawa, ON K2P 1E2 Canada

Canadian Patient Safety Institute (CPSI) Research Grants   1481

The Canadian Patient Safety Institute (CPSI) announces its annual Research Competition to continue its strategic mandate of increasing the scope and scale of research and evaluation activities in patient safety. The primary goal for CPSI Research Competition is to develop knowledge about patient safety that can be helpful in a variety of settings and circumstances in organizations across Canada. It is the applicant team, responsibility to identify and secure matching funding from other contributing organization(s). Applicant teams may obtain matching funding from universities, foundations, voluntary health charities, provider associations, provincial government departments, regional health authorities, hospitals, research centers or the private sector.

Requirements: A 1:1 funding ratio is the acceptable minimum required for any project over $20,000.

Amount of Grant: $120,000

Date(s) Application is Due: Dec 2

Contact: Laurel Taylor, Director of Operations; (866) 421-6933 or (780) 409-8090; fax: (780) 409-8098; email: ltaylor@cpsi-icsp.ca

Internet: http://www.patientsafetyinstitute.ca/news/2008researchcompetition.html

Sponsor: Canadian Patient Safety Institute (CPSI)
10235 - 101 Street
Edmonton, AB T5J 3G1 Canada

Canadian Political Science Association Parliamentary Internships   1482

The work-study program gives university graduates an opportunity to supplement their theoretical knowledge of Parliament with firsthand experience of the day-to-day work of the members and of Parliament. Following an orientation period, the interns' responsibilities include working with members of the House of Commons on both sides, conducting scholarly research dealing with Parliament, and undertaking study travel. Tenures begin in September for a period of 10 months. It is expected that 10 internships will be awarded each year. Request application forms in writing. Application is due the last friday in January.

Requirements: The program is open to Canadian citizens who are recent graduates of a university and have an interest in Parliament.

Amount of Grant: $C16,500 estimated plus travel subsidies

Contact: Parliamentary Internship Program, (613) 995-0764; fax: (613) 995-5357; email: cartwj@parl.ga.ca

Internet: http://www.pip-psp.org

Sponsor: Canadian Political Science Association
151 Sparks Street, House of Commons, Room 1200
Ottawa, ON K1A 0A6 Canada

Canadian Science Writers Association Awards   1483

The program comprises a handful of awards including the Canadian Forest Service-Ontario Journalism Award for accurate and informed reporting on research and/or its application to the aesthetics, recreation, timber, and wildlife habitat of Ontario's forests. This award is open to writers whose writings have been published in English or French in a news story, feature, or editorial of a daily or weekly magazine or periodical. The Science in Society Journalism Award considers books that deal with aspects of basic or applied science or technology, historical or current, in areas such as health, social or environmental issues, and regulatory trends. Books will be evaluated by their literary excellence and scientific content and accuracy. Applicants need not be members of the association. Entry forms are available on the Web site.

Requirements: Canadian citizens and residents are eligible.

Amount of Grant: $C1000 each Canadian Forest Service and Science in Society awards

Contact: Program Contact, (800) 796-8595; fax: (416) 408-1044; email: awards@sciencewriters.ca

Internet: http://www.sciencewriters.ca/awards/index.html

Sponsor: Canadian Science Writers Association
P.O. Box 75, Station A
Toronto, ON M5W 1A2 Canada

Canadian Studies Conference Grants   1484

This program has as its purpose the support of major conferences addressing important and timely Canadian or Canada-U.S. issues. It is intended to secure a greater understanding of the background, complexity, and ramifications of these issues. The grants are designed to assist an institution in holding a conference and to publish the resultant papers and proceedings in a scholarly fashion. Application materials are available on the Web site.

Requirements: This grant is intended for four-year U.S. colleges and universities, research and policy-planning institutes, or other established research institutions that undertake a major conference on a Canadian or Canada-U.S. issue.

Restrictions: Provisions are not made for released time stipends nor for overhead costs to the institution.

Amount of Grant: $5000-$15,000 per year

Date(s) Application is Due: Jun 15

Contact: Daniel Abele, Academic Relations Officer, (202) 682-7727; email: daniel.abele@international.gc.ca

Internet: http://www.canadianembassy.org/education/grantguide-en.asp#conference

Sponsor: Canadian Embassy
501 Pennsylvania Avenue NW
Washington, DC 20001

Canadian Studies Faculty Enrichment Grants   1485

The Faculty Enrichment Program (Course Development) provides faculty members an opportunity to develop or redevelop a course(s) with substantial Canadian content that will be offered as part of their regular teaching load, or as a special offering to select audiences in continuing and/or distance education. Priority topics include bilateral trade and economics, Canada-U.S. border issues, cultural policy and values, environmental, natural resources, and energy issues, and security cooperation (in alphabetical order). In addition, projects that examine Canadian politics, economics, culture, and society as well as Canada's role in international affairs are welcome. We especially encourage the use of new internet technology to enhance existing courses, including the creation of instructional Web sites, interactive technologies, and distance learning links to Canadian universities. Applications on official forms must be accompanied by an institutional commitment that the department will offer the new course to be taught by the applicant at least two times during the four following years.

Requirements: This program is intended for full-time, tenured or tenure-track faculty members at accredited four-year U.S. colleges and universities. The candidates should be able to demonstrate that they are already teaching, or will be authorized to teach, courses with substantial Canadian content (33 percent or more). Team teaching applications are welcome. Applicants are required to conduct research in Canada for a minimum of three weeks during the award period.

Restrictions: Provisions are not made for released time stipends nor for overhead costs to the institution. This award is not available two consecutive years to the same recipient.

Amount of Grant: $U.S.5000 maximum

Date(s) Application is Due: Oct 31

Contact: Academic Relations Office, (202) 682-7727; email: AcademicRelations@CanadianEmbassy.org
Internet: http://www.canadianembassy.org/education/grantguide-e.asp#faculty
Sponsor: Canadian Embassy
501 Pennsylvania Avenue NW
Washington, DC 20001

Canadian Studies Graduate Student Fellowships   1486

The Graduate Student Fellowship Program promotes research in the social sciences and humanities with a view to contributing to a better knowledge and understanding of Canada and its relationship with the United States and/or other countries of the world. Priority topics include bilateral trade and economics, Canada-U.S. and values, environmental, natural resources, and energy issues, and security cooperation (in alphabetical order). In addition, projects that examine Canadian politics, economics, culture, and society as well as Canada's role in international affairs are welcome. The purpose of the fellowship is to offer graduate students an opportunity to conduct part of their doctoral research in Canada. Annual deadline dates may vary; contact the program office for exact dates.

Requirements: This program is intended for full-time doctoral students at accredited four-year U.S. and Canadian colleges and universities whose dissertations are related in substantial part to the study of Canada, Canada/United States or Canada/North America. Candidates must be citizens or permanent residents of the United States and should have completed all doctoral requirements other than the dissertation when they apply for a grant.

Amount of Grant: $10,000 maximum
Date(s) Application is Due: Oct 31
Contact: Academic Relations Office, (202) 682-7727; email: AcademicRelations@CanadianEmbassy.org
Internet: http://www.canadianembassy.org/education/grantguide-e.asp#graduate
Sponsor: Canadian Embassy
501 Pennsylvania Avenue NW
Washington, DC 20001

Canadian Studies Matching Grants   1487

The Government of Canada, through the Canadian Embassy, Washington, DC, in cooperation with Canadian Consulates General throughout the United States, at its discretion, may provide a matching grant to support Canadian Studies Programs and projects which have been funded by a major foundation or other funding institution. It is usually a partial match of a larger award (e.g., one to four ratio) provided for a specific project, e.g., the creation of a Canada/U.S. trade institute, or the establishment of a permanent Canadian Studies professorship. Applications are accepted at any time.

Requirements: Four-year, fully accredited U.S. colleges and universities and research institutes which demonstrate a serious interest in Canada, Canada/United States or Canada/North America, are eligible. The institution must estimate the longevity of the program or project after Canadian government support is exhausted.

Contact: Academic Relations Office, (202) 682-7727
Internet: http://www.canadianembassy.org/education/grantguide-e.asp#matching
Sponsor: Canadian Embassy
501 Pennsylvania Avenue NW
Washington, DC 20001

Canadian Studies Program Enhancement Grants   1488

This grant is designed to encourage scholarly inquiry and multidisciplinary professional academic activities that contribute to the development and/or expansion of a Canadian Studies Program. We are particularly interested in innovative projects that promote awareness among students and the public about Canadian society, culture, and values as well as Canada-U.S. bilateral relations and Canada's role in international affairs. Linkages with Canadian institutions, such as student and faculty exchanges or joint academic programs, are especially welcome.

Requirements: This grant is intended for four-year U.S. colleges, universities, and research institutions, which undertake professional academic activities to further the development of a Canadian Studies Program at their institution. Long-term program development proposals spanning several years will be considered. The Canadian Government should be approached as a partial funder. Institutions must demonstrate that they are bringing new sources of funds and innovative ideas to the program.

Restrictions: Provisions are not made for released time stipends nor for overhead costs to the institution.
Amount of Grant: $18,000 maximum per year
Date(s) Application is Due: Jun 15
Contact: Dr. Daniel Abele, Academic Relations Office, (202) 682-7717; fax: (202) 682-7791; email: AcademicRelations@CanadianEmbassy.org
Internet: http://www.canadianembassy.org/education/grantguide-en.asp#enhancement
Sponsor: Canadian Embassy
501 Pennsylvania Avenue NW
Washington, DC 20001

Canadian Studies Research Grants   1489

The Research Grant Program promotes research in the social sciences and humanities with a view to contributing to a better knowledge and understanding of Canada and its relationship with the United States and/or other countries of the world. Priority topics include bilateral trade and economics, Canada-U.S. border issues, cultural policy and values, environmental, natural resources, and energy issues, and security cooperation. In addition, projects that examine Canadian politics, economics, culture, and society as well as Canada's role in international affairs are welcome. The purpose of the grant is to assist individual scholars, or a team of scholars, in writing an article-length manuscript of publishable quality and reporting their findings in scholarly publications, with a view to contributing to the development of Canadian Studies in the United States. We welcome efforts to integrate the research findings into the applicant's teaching load. Annual deadline dates may vary; contact the program officer for exact dates.

Requirements: This program is intended for full-time faculty members at accredited four-year U.S. colleges and universities, as well as scholars at American research and policy-planning institutes who undertake significant Canada, Canada-United States or Canada/North America research projects. Recent PhD recipients who are citizens or permanent residents of the United States are also eligible to apply. Applicants are ineligible to receive the same grant in two consecutive years or to receive two individual category Canadian Studies grants in the same grant period.

Restrictions: Provisions are not made for released time stipends nor for overhead costs to the institution.
Amount of Grant: $15,000 maximum
Date(s) Application is Due: Sep 30
Contact: Dr. Daniel Abele, Academic Relations Office, (202) 682-7717; fax: (202) 682-7791; email: daniel.abele@international.gc.ca
Internet: http://www.canadianembassy.org/education/grantguide-en.asp#research
Sponsor: Canadian Embassy
501 Pennsylvania Avenue NW
Washington, DC 20001

Canadian Studies Senior Fellowships   1490

The Senior Fellowship (awarded every other year) provides senior scholars with an opportunity to complete and publish a major study which will significantly benefit the development of Canadian Studies in the United States. A limited number of fellowships are awarded to academics who have a lengthy track record in teaching, researching and publishing on Canada, Canada/United States or Canada/North America; and on the social, cultural, political and economic issues that impact on these relationships. Preference in such instances is to fund a book project after a publisher has indicated an interest.

Requirements: This program is intended for full-time tenured faculty members at accredited four-year U.S. colleges and universities who are fully involved in Canadian Studies. These Canadianists should be in the process of completing research for a book or major monograph. The study must be on a subject of widespread interest to the Canadian Studies community in the United States as well as in Canada. This fellowship is awarded only once to any one recipient and is in recognition of an academic career dedicated to the promotion of Canadian Studies. It is expected that a recipient will be granted a leave of absence or a sabbatical during the award period.

Restrictions: This award is not available two consecutive years to the same recipient. An individual may not receive two individual category Canadian Studies grants during the same grant period.
Amount of Grant: $3000 maximum monthly stipend for up to six months
Date(s) Application is Due: Jun 16
Contact: Dr. Daniel Abele, Academic Relations Office, (202) 682-7717; fax: (202) 682-7791; email: daniel.abele@dfait-maeci.gc.ca
Internet: http://www.dfait-maeci.gc.ca/can-am/washington/studies/grantguide-en.asp
Sponsor: Canadian Embassy
501 Pennsylvania Avenue NW
Washington, DC 20001

Cancer Research Postdoctoral Fellowships for Basic and Physician Scientists   1491

This program grants initial postdoctoral fellowships to support promising young investigators pursuing careers in cancer research. The fund encourages all theoretical and experimental research that is relevant to the study of cancer and the search for cancer causes, mechanisms, therapies, and prevention. Candidates must apply for the fellowships under the guidance of a sponsor—a senior member of the scientific research community. The sponsor should be actively engaged in the planning, execution, and supervision of the proposed research and should encourage the fellow to report the results of the research in scientific journals and meetings. Awards are made to institutions for the support of the fellow under direct supervision of the sponsor.

Requirements: Candidates must have completed one or more of the following degrees or equivalent: MD, PhD, DDS, or DVM. For level one funding, basic and physician scientist applicants must have received their degrees no more than one year prior to the SAC meeting at which their applications are to be considered and DVM applicants no more than three years. For level two funding, physician scientists (MD, DDS, DVM, or the equivalent) must have completed their residencies and clinical training no more than three years prior to the SAC meeting at which their applications are to be considered.

Restrictions: Only one application will be accepted from a given sponsor or fellow per review session; there is no limit, however, on the number of applications from an institution. No more than two Runyon-Winchell Fellows will be funded to work with the same sponsor at any time.

Samples: Dr. Hong Liao, with Dr. Matthew Scott, Stanford U School of Medicine (Stanford, CA)—for The Sonic Hedgehog Signaling Pathway: Mechanisms of Cerebellar Development and Tumorigenesis, $125,500.; Dr. Weei-Chin Lin, with Dr. Joseph Nevins, Duke U Medical Ctr (Durham, NC)—for The Role of E2F1 in Genotoxic Stress, $174,000

Amount of Grant: $41,000 first year, $43,000 second year, $44,000 third year for level 1; $55,000 first year, $56,000 second year, $57,000 third year for level 2
Date(s) Application is Due: Mar 15; Aug 15

Contact: Lorraine Egan, Executive Director, (212) 455-0541; fax: (212) 697-4950; email: lorraine.egan@drcrf.org or fellowship@cancerresearchfund.org
Internet: http://www.drcrf.org/apFellowship.html
Sponsor: Cancer Research Fund of the Damon Runyon-Walter Winchell Foundation
675 Third Avenue, 25th Fl
New York, NY 10017

Cancer Research Society Fellowships and Grants   1492
Grants and fellowships are supported for projects researching the cause and cure of cancer. Recipients are selected annually by the society's medical advisory board on the basis of the scientific value of the research projects submitted.
Requirements: Research projects supported must be carried out in Canada.
Date(s) Application is Due: Feb 15
Contact: Program Director, (514) 861-9227; fax: (514) 861-9220; email: grants@cancerresearchsociety.ca
Internet: http://src-crs.ca/main.php?lang=2&id=305
Sponsor: Cancer Research Society
402-625 Av Du President-Kennedy
Montreal, QC H3A 3S5 Canada

Cancer Research UK Fellowships   1493
The fund operates in the United Kingdom and worldwide to investigate all matters connected with or bearing on the causes, prevention, treatment, and cure of cancer. Principal disciplines include cellular and molecular biology, clinical cancer research, and epidemiology. Types of support include research fellowships, visiting fellowships, clinical research fellowships, and graduate studentships.
Contact: Grants Administrator, 020 7242 0200; fax: 020 7269 3100; email: research_enquiries@cancer.org.uk
Internet: http://science.cancerresearchuk.org/fandm/grantapplications/?version=1
Sponsor: Cancer Research UK
P.O. Box 123, Lincoln's Inn Fields
London WC2A 3PX United Kingdom

Cancer Research UK Grants   1494
The organization supports cancer research throughout the United Kingdom. Grants are awarded to universities, medical schools, and hospitals in support of research projects. Additional types of support include life fellowships, career development awards, clinical and nonclinical scientists fellowships, and capital grants for new research buildings and major items of equipment.
Contact: Grants Administrator, 020 7242 0200; fax: 020 7269 3100; email: research_enquiries@cancer.org.uk
Internet: http://science.cancerresearchuk.org/gapp/?version=3
Sponsor: Cancer Research UK
P.O. Box 123, Lincoln's Inn Fields
London WC2A 3PX United Kingdom

Cancer Society of New Zealand Grants   1495
The society supports cancer research through grants to individuals and organizations in New Zealand. Types of support include research grants and traveling fellowships to enable New Zealand graduates to travel abroad for study.
Date(s) Application is Due: Nov 1
Contact: Grants Administrator, (04) 494 7270; fax: (04) 494 7271; email: admin@cancernz.org.nz
Internet: http://www.cancernz.org.nz
Sponsor: Cancer Society of New Zealand
P.O. Box 10847
Wellington New Zealand

Cancer Treatment Research Foundation Clinical Investigation Grants   1496
The grants support new and pilot feasibility clinical projects in cancer therapy. Grants are intended to stimulate innovative research relevant to cancer therapy, such as new agents, innovative anticancer therapies, biological response modifiers, immunotherapy, and gene therapy; and quality of life, cancer education, and nutritional oncology. Funds may be used for salaries, supplies, and nonreimbursable research-related patient care costs, including extraordinary laboratory and imaging studies. There are no application deadlines. Preliminary grant request forms are available on the foundation's Web site. The foundation will invite a formal application.
Requirements: New and established investigators are encouraged to apply: young investigators working in established research programs relevant to clinical research who are without support from NIH or other cancer research agencies; and established, clinical researchers who wish to embark on innovative studies directly relevant to the foundation's mission with novel, new, or pilot projects distinctly removed from their currently funded research projects.
Amount of Grant: $25,000-$300,000 range
Date(s) Application is Due: Aug 15
Contact: Kristine Nelson, Grants Administrator, (847) 342-7450; email: grants@ctrf.org
Internet: http://www.ctrf.org/grantapp.cfm
Sponsor: Cancer Treatment Research Foundation
1336 Basswood Road
Schaumburg, Illinois 60173

Canon Foundation Grants   1497
The foundation functions to facilitate mutual understanding in all fields of interest and the development of scientific expertise, particularly between Europe and Japan. Types of support include visiting research fellowships and professorships to enable European students to visit Japan and Japanese students to visit Europe. Up to 15 fellowships are awarded each year.
Requirements: Young, highly qualified postgraduate European and Japanese researchers, preferably holding a PhD and not older than 40 years, are eligible.
Amount of Grant: 22,500-27,500 Euro per year
Date(s) Application is Due: Sep 15
Contact: Corrie Siahaya-Van Nierop, Director, 31-20-545-8934; fax: 31-20-712-8934; email: foundation@canon-europa.com
Internet: http://www.canonfoundation.org/pages/main.htm
Sponsor: Canon Foundation in Europe
P.O. Box 2262
Amstelveen 1180 EG Netherlands

Canon National Parks Science Scholars Program Grants   1498
The program is underwritten by Canon U.S.A Inc and awards scholarships to eight doctoral students each year for up to three years of support to conduct research on critical problems facing national parks worldwide. Research projects in the biological sciences, physical sciences, social sciences, cultural sciences, and technology innovation in support of conservation science are eligible. National parks are those protected areas officially recognized and identified as national parks by the national government of a country in the Americas. Scholarship applicants must submit dissertation proposals that address these topics. The program will award eight scholarships based on two separate competitions: four scholars will be selected from students studying at U.S. universities (one winner in each category) and four scholars will be selected from students studying at universities throughout the Americas, but outside the United States (one winner in each category).
Requirements: Proposals can be considered only from currently enrolled PhD students at an accredited university within the Americas. The Americas include Canada, the United States, Mexico, the countries of Central and South America, and the Caribbean. At least some portion of the student's proposed research must take place in—or be significantly and specifically relevant to—one or more national parks in the student's country of citizenship. Students must have completed a majority of their coursework, and be able to complete the proposed dissertation research within three years of receiving an award.
Amount of Grant: $80,000 scholarship; $1000 honorable mention scholarship
Date(s) Application is Due: May 3
Contact: Dr. Gary Machlis, Program Coordinator, (208) 885-7054; email: gmachlis@uidaho.edu
Internet: http://www.nature.nps.gov/canonscholarships
Sponsor: National Park Service
P.O. Box 441133
Moscow, ID 83844-1133

CAORC Andrew W. Mellon East-Central European Fellowship Program   1499
The program of fellowships enables Bulgarian, Czech, Estonian, Hungarian, Latvian, Lithuanian, Polish, Romanian and Slovak scholars in the humanities and allied social sciences to carry out research at institutes of advanced study in other countries. Each cycle will fund short-term residencies for up to three Mellon Research Fellows at each of the 17 designated institutes in Austria, England, France, Germany, Greece, India, Israel, Italy, Jordan, the Netherlands, Norway, Scotland, Spain, Turkey, and Yemen. Each institute will issue its own announcement and will handle all matters concerning application and selection. Guidelines are available online.
Requirements: The fellowships are intended to serve younger scholars who have already obtained a PhD or have equivalent experience and who wish to undertake a specific research project at one of the participating institutes. Bulgarian, Czech, Estonian, Hungarian, Latvian, Lithuanian, Polish, Romanian and Slovak scholars who wish to apply should contact the participating institute directly for an application.
Contact: Grants Administrator, (202) 633-1599; fax: (202) 786-2430; email: fellowships@caorc.org
Internet: http://www.caorc.org/fellowships
Sponsor: Council of American Overseas Research Centers
P.O. Box 37012, NHB Room CE-123, MRC 178
Washington, DC 20113-7012

CAORC Regional Research Fellowships   1500
Fellowships are available to U.S. doctoral and postdoctoral candidates researching in the fields of the humanities, social sciences, or allied natural sciences. The research will have regional significance and must be conducted in more than one country. At least one of the following countries must be included: Scholars must carry out research in at least one of the countries which host overseas research centers: Bangladesh, Cambodia, Cyprus, Egypt, Greece, India, Iran, Israel, Italy, Jordan, Mexico, Morocco, Pakistan, Senegal/West Africa, Sri Lanka, Tunisia, Turkey, West Bank/Gaza Strip and Yemen, as well as in other countries unless subject to official security and/or travel restrictions or warnings. Research in Nepal is possible via the Center for South Asia Libraries; contact CAORC for more information. Fellows are required to obtain their own research permissions in countries that do not host centers. Application forms are available on the Web site.
Requirements: U.S. doctoral candidates and postdoctoral scholars in the humanities, social sciences, or allied natural sciences are eligible. Candidates may apply either as individuals or in teams.
Amount of Grant: $9000 maximum stipend
Date(s) Application is Due: Jan 13

Contact: Director, (202) 633-1599; fax: (202) 786-2430; email: caorc@caorc.org
Internet: http://www.caorc.org/fellowships
Sponsor: Council of American Overseas Research Centers
1100 Jefferson Dr SW, NHB Room CE-123, MRC 178
Washington, DC 20013-7012

CAPAA Mini Grant for Research Related to African Americans in Education   1501
The program awards mini grants to support theoretical and empirical research that examines the current status of, challenges to, and strategies for improving African Americans' access to the University of California system. The College Access Project for African Americans (CAPAA)awards mini grants to support short-term (two years or less) research projects. CAPAA encourages researchers with diverse perspectives to develop ideas and approaches that extend the conventional boundaries of a research question, area, or method. Request for proposals to conduct specific projects are being supported. 1) Survey at the community level and in school districts with students, parents, and teachers designed to elicit perceptions on factors affecting access to and achievement in higher education by African American students. 2) The educational landscape in school districts and institutions of higher education in California, especially the UC and CSU systems, and the extent to which it creates an environment conducive to learning and achievement by African American students. 3) Studies that investigate the role of African American community resources (networks of support, educational enhancement activities/programs, family support) in determining students' choices, references, aspirations, and performance vis-a-vis higher education. Guidelines are available online.
Requirements: Studies must relate to issues of access and equity for African Americans in California's secondary and higher education institutions. Individual as well as collaborative efforts will be considered.
Amount of Grant: $15,000 maximum each
Date(s) Application is Due: Nov 12
Contact: Dr. Ana-Christina Famon, (310) 794-4997; email: acramon@bunche.ucla.edu
Internet: http://www.bunche.ucla.edu/frames/index.html
Sponsor: UCLA Ralph J. Bunche Center
160 Haines Hall, Box 951545
Los Angeles, CA 90095

Cape Branch Foundation Grants   1502
The foundation awards grants to New Jersey nonprofit organizations in its areas of interest, including education and secondary education, natural resource conservation, and museums. Types of support include general operating support, building construction/renovation, land acquisition, scholarship funds, and research grants. There are no application forms or deadlines. A letter should be submitted outlining purpose and amount of request.
Requirements: New Jersey nonprofit organization are eligible.
Geographic Focus: New Jersey
Samples: Edison Wetlands Assoc (Edison, NJ)—for the Raritan River project, $10,000.
Amount of Grant: $1000-$329,268 range
Contact: Dorothy Frank, c/o Danser, Balaam and Frank, (609) 987-0300; fax: (609) 452-1024
Sponsor: Cape Branch Foundation
P.O. Box 86
Oldwick, NJ 08858

Cardiac Arrhythmias Research and Education Foundation Research Grants   1503
The foundation's two primary goals are to raise funds for clinical research of cardiac arrhythmias and to educate the public and the medical community about sudden death. Awards will be granted to promising young investigators working on the genetics, mechanisms, therapy, and prevention of sudden death due to cardiac arrhythmias. Two research grants are awarded each year.
Samples: Johns Hopkins U—to support the research of David Charles Johns on the mechanisms and treatment of cardiac arrhythmias and the prevention of sudden death caused by heart rhythm disorders, $100,000 over two years.
Amount of Grant: $50,000
Contact: Grants Administrator, (800) 404-9500 or (425) 788-1987; fax: (425) 788-1927; email: care@longqt.org
Internet: http://www.longqt.org/fundingmedical.html
Sponsor: Cardiac Arrhythmias Research and Education Foundation
P.O. Box 369, 26425 NE Allen Street, Suite 103
Duvall, WA 98019

Carl and Eloise Pohlad Family Foundation Grants   1504
The mission of the foundation is to improve the lives of economically disadvantaged children and youth and participate in projects that positively impact the quality of life in the Minneapolis/St.Paul area. The foundation awards grants to Minnesota nonprofits in its areas of interest, including arts and culture, economic development, education, environment, health care, housing, and social services. Types of support include general operating support, continuing support, capital campaigns, building construction/renovation, endowments, emergency funds, scholarship funds, and research.
Requirements: Minnesota nonprofits are eligible.
Restrictions: Individuals are not eligible. Capital request are considered only for physical plant improvements or significant technology investments. Capital requests for housing construction, endowment, program start-up or expansion or to establish operating reserves are not considered.
Geographic Focus: Minnesota
Samples: Abbott Northwestern Hospital (Minneapolis, MN)—to construct a cardiology hospital, $3 million.
Amount of Grant: $3.2 million total
Contact: Rose Peterson, Program Manager, (612) 661-3903; fax: (612) 661-3715; email: rpeterson@pohladfamilygiving.org
Internet: http://www.pohladfamilyfoundation.org/pff/pff_default.aspx
Sponsor: Carl and Eloise Pohlad Family Foundation
60 S Sixth Street, Suite 3900
Minneapolis, MN 55402

Carl and Lily Pforzheimer Foundation Grants   1505
The foundation awards grants to U.S. nonprofit organizations in its areas of interest, including higher and secondary education, cultural activities and programs, public affairs, and healthcare. Types of support include capital campaigns, fellowships, internship funds, matching grants, professorships, program development, publication, scholarship funds, and seed grants. There are no application forms or deadlines. The board meets in April, June, October, and December to consider requests.
Restrictions: Grants are not awarded to individuals or for building construction/renovation.
Samples: Bank Street College of Education (New York, NY)—for scholarships and unrestrictd support, $2 million.
Amount of Grant: $10,000-$150,000 average
Contact: Grants Administrator, (212) 223-6500
Sponsor: Carl and Lily Pforzheimer Foundation
950 Third Avenue, 30th Fl
New York, NY 10022

Carl C. Icahn Foundation Grants   1506
The foundation awards grants to New York and New Jersey nonprofits in the areas of education, arts and culture, health care, child welfare, and Jewish temples and organizations. Types of support include general operating support, annual campaigns, building construction/renovation, and matching funds. There are no application deadlines or forms.
Requirements: New York and New Jersey nonprofits are eligible to apply.
Restrictions: No grants are provided to individuals.
Geographic Focus: New Jersey; New York
Samples: Randall's Island Sports Foundation (New York, NY)—for a new track-and-field stadium, $10 million (2004); Mount Sinai Medical Ctr (New York, NY)—for medical research at the Mount Sinai School of Medicine, $25 million (2004)
Amount of Grant: $1000-$500,000 range
Contact: Gail Golden, Secretary-Treasurer, c/o Icahn Associates, (212) 702-4300
Sponsor: Carl C. Icahn Foundation
767 5th Avenue, 47th Fl
New York, NY 10153-0023

Carl Gellert and Celia Berta Gellert Foundation Grants   1507
The foundation funds religious, charitable, scientific, literary or educational purposes restricted in the nine counties of the greater San Francisco Bay Area (Alameda, Contra Costa, Marin, Napa, San Francisco, San Mateo, Santa Clara, Solano and Sonoma). No grants are made to individuals. Types of support include general operations, annual and capital campaigns, building construction/ renovation, equipment acquisition, debt reduction, program/project development, medical research, publication, and scholarships.
Requirements: California 501(c)3 tax-exempt nonprofit organizations that are not private foundations are eligible.
Restrictions: Grants are not awarded to individuals.
Geographic Focus: California
Samples: U of San Francisco, School of Business and Management (CA)—to renovate and add a new wing to the McLaren Center, home of the School of Business and Management, $500,000.
Amount of Grant: $1000-$10,000 typically
Date(s) Application is Due: Aug 15 at noon
Contact: Jack Fitzpatrick, Executive Director, (415) 255-2829
Internet: http://home.earthlink.net/~cgcbg
Sponsor: Carl Gellert and Celia Berta Gellert Foundation
1169 Market Street, Suite 808
San Francisco, CA 94103

Carl J. Herzog Foundation Grants   1508
The foundation awards grants, primarily in Connecticut, for research and general operating support in its areas of interest, including medical research, dermatology research, hospitals, and education. There are no application deadlines or forms.
Requirements: 501(c)3 tax-exempt organizations are eligible.
Geographic Focus: Connecticut
Amount of Grant: $1000-$200,000; $5000-$50,000 average
Contact: David Babson, (203) 629-2424
Sponsor: Carl J. Herzog Foundation
321 Railroad Avenue
Greenwich, CT 06836-0788

Carlos and Marguerite Mason Trust Grants 1509

The trust awards grants in Georgia with a focus on organ transplants. The trust's interests are patient services—needs relating to the transplantation process for the patient and the immediate family members, as well as donor family members and living donors; donation and transplantation—related education (donor education, professional education, and technical education for transplantation providers; programs to improve consent rates; and programs to promote living organ donations); and research—medical research relating to the area of transplantation. Types of support include capital improvements, start-up grants for new projects, and challenge/matching grants. Grants will be awarded for up to three years. Applicants are encouraged to contact the trustee for an appointment prior to submitting a grant request.

Requirements: Georgia nonprofit organizations are eligible.

Restrictions: Requests for general operating or administrative funds are discouraged. The trust generally does not fund endowment, nor does it fund general goodwill advertising; make grants that would replace existing sources of funding; or pay indirect overhead expenses for projects at colleges, universities, governmental units, or other established organizations.

Geographic Focus: Georgia

Samples: MCG Health System (Augusta, GA)—to educate the public about the importance of transplants for people with chronic kidney disease, and to help meet the needs of people in rural Georgia communities who are without transplant services, $1.1 million (2005).

Amount of Grant: $1.5 million maximum

Date(s) Application is Due: Jun 1

Contact: Randy Karesh, c/o Wachovia Bank of Georgia, (404) 332-6677; email: grantinquiriesga@wachovia.com

Internet: http://www.wachovia.com/corp_inst/charitable_services/0,,4269_3300,00. html

Sponsor: Carlos and Marguerite Mason Trust

3414 Peachtree Road, 5th Fl, MC GA8023

Atlanta, GA 30326

Carnegie Corporation of New York Grants 1510

The foundation provides research, study, and support for projects to improve government at all levels, to increase public understanding of social policy issues, to equalize opportunities for minorities and women, and to increase participation in political and civic life. Also supported are projects that promote electoral reform; education reform from early childhood through higher education; early childhood development; and urban school reform. The foundation will also fund research on the increasing availability and success of after-school and extended service programs for children and teenagers, particularly those in urban areas, that promote high academic achievement. Dissemination of best practices in teacher education will also be emphasized. There is no formal procedure for submitting a proposal. To apply under any of the corporation's grantmaking programs, applicants should submit a full proposal that describes the project's aims, duration, methods, amount of financial support required, and key personnel. The board meets four times a year, in October, February, April, and June.

Restrictions: Grants are not made for construction or maintenance of facilities or endowments. The Corporation does not generally make grants to individuals except through the Carnegie Scholars Program, that supports the work of select scholars and experts conducting research in the foundation's fields of interest.

Samples: American Academy of Arts and Sciences, Cambridge, Massachusetts, $500,000—toward an initiative on minimizing the risks of nuclear energy expansion (2009); George Washington University, Washington, DC, $349,900— toward research and writing on the effects of domestic foreign policy debates on U.S. global engagement (2009).

Amount of Grant: $80 million total approximately annually

Contact: Sarina Cipriano, Grants Manager; (212) 371-3200; fax: (212) 754-4073

Internet: http://www.carnegie.org/sub/program/grant.html

Sponsor: Carnegie Corporation of New York

437 Madison Avenue

New York, NY 10022

Carnegie Developmental Biology Research Fellowships 1511

Fellowships are awarded, primarily at the postdoctoral level, for research at the Carnegie Institution Department of Embryology, Baltimore, MD, under the guidance of the department's staff scientists. Topics include the expression of genes through development and growth, chromosome structure and function, and intracellular communication. Facilities include extensive laboratories for biochemistry and molecular biology. Applicants are encouraged to submit materials a year in advance of the proposed starting date. Fellowships are normally awarded for one year with possible renewal for a second year.

Requirements: Applicant must be postdoctoral researcher. Predoctoral candidates must obtain special permission to apply. Application materials should be sent to the director, Carnegie Institution Department of Embryology, 3520 San Martin Drive, Baltimore, MD 21218.

Contact: Grants Administrator, (410) 246-3001; fax: (410) 243-6311; email: carnegiecollaborative@ciwemb.edu

Internet: http://www.ciwemb.edu

Sponsor: Carnegie Institution of Washington

1530 P Street NW

Washington, DC 20005

Carnegie Endowment for International Peace Junior Fellows Program 1512

Each year the endowment offers eight to 10 one-year fellowships to uniquely qualified graduating seniors and individuals who have graduated during the past academic year. They are selected from a pool of nominees from close to 300 colleges (list of participating colleges available online). Junior fellows provide research assistance to associates working on the Carnegie Endowment's projects such as nonproliferation, democracy building, trade, U.S. leadership, China-related issues, and Russian/Eurasian studies. Junior fellows have the opportunity to conduct research for books, co-author journal articles and policy papers, participate in meetings with high-level officials, contribute to congressional testimony, and organize briefings attended by scholars, activists, journalists, and government officials. The endowment's annual nomination deadline is the listed application deadline; colleges generally set an earlier application deadline (contact the college's Career Services/Placement Office to learn more about the college application process). Information to participating universities generally is sent by the second week of October each year. All fellowships begin on August 1.

Requirements: The Carnegie Endowment accepts applications only through participating universities via designated nominating officials. The program relies on participating universities to nominate uniquely qualified students. Students need not be U.S. citizens; however, all applicants must be eligible to work in the United States from August 1 through July 31 following graduation.

Restrictions: Students who have started graduate studies are not eligible for nomination.

Amount of Grant: $2,750 per month plus benefits

Date(s) Application is Due: Jan 15

Contact: Fellowship Administrator; (202) 483-7600; fax: (202) 483-1840; email: jrfellowinfo@CarnegieEndowment.org

Internet: http://www.carnegieendowment.org/about/index.cfm?fa=jrFellows

Sponsor: Carnegie Endowment for International Peace

1779 Massachusetts Avenue NW

Washington, D.C. 20036-2103

Carnegie Plant Biology Fellowships 1513

Fellowships primarily at the postdoctoral level are available for research at the Carnegie Institution Department of Plant Biology in Stanford, CA, under the guidance of the department's staff scientists. Topics range from biochemical and molecular experiments to broad-ranging ecological research. Facilities include greenhouses, controlled-growth chambers, biochemical and molecular biology laboratories, and varied specialized apparatus. The period of appointment starts July 1. Fellowships are normally awarded for one year and may be renewed for up to two additional years.

Requirements: The successful applicant must have completed, or nearly completed, requirements for the PhD prior to assuming the fellowship. Applications should include educational record, professional positions held, and a list of papers published and submitted. Applicants should write a brief essay describing their interests and what they hope to achieve during the tenure of the fellowship. Applicants should arrange for letters of recommendation from three professional scientists. Materials should be sent to the director, Carnegie Institution Department of Plant Biology, 260 Panama Street, Stanford, CA 94305-4101.

Contact: Tina McDowell, Publications Office, (202) 939-1120; fax: (202) 387-8092; email: tmcdowell@pst.carnegiescience.edu

Internet: http://www-ciwdpb.stanford.edu

Sponsor: Carnegie Institution of Washington

1530 P Street NW

Washington, DC 20005

Carnegie Postdoctoral Research Fellowships 1514

Primarily at the postdoctoral level, fellowships support research at the Geophysical Laboratory and the Department of Terrestrial Magnetism, both in Washington, DC, under the guidance of staff scientists. The Geophysical Laboratory is a private research and educational institution emphasizing high pressure/high temperature research, metamorphic, igneous, and experimental petrology, stable isotope geochemistry, crystallography, mineral physics, theoretical and experimental condensed matter physics, materials research, organic geochemistry, and biogeochemistry. Fellowships are normally awarded for one year and may be renewed for one and occasionally two additional years.

Requirements: The applicant must have completed, or nearly completed, requirements for the PhD prior to assuming the fellowship. Applications should include a curriculum vita, description of thesis research, a list of publications, transcript/grade report, a two- to three-page research proposal for the fellowship period, and three letters of recommendation.

Date(s) Application is Due: Dec 31

Contact: Dr. Wesley Huntress Jr., Director, Geophysical Laboratory, (202) 478-8900; fax: (202) 478-8901; email: w.huntress@gl.ciw.edu

Internet: http://www.gl.ciw.edu/employment/postdoc1.php

Sponsor: Carnegie Institution of Washington

5251 Broad Branch Road, NW

Washington, DC 20015

Carnegie Postdoctoral Research Fellowships in Astronomy 1515

Applications are invited for one or more fellowships at the Observatories of the Carnegie Institution for research in theoretical astronomy and/or instrumentation to begin annually in September. Fellows will have access to the observatories' research facilities in Pasadena and on Cerro Las Campanas, Chile. Fellows will be expected to devote a major portion of their time to astronomical research. Fellowships are awarded for one year and may be renewed for up to two additional years. Foreign applicants will be sponsored under the Carnegie Institution's visitor exchange program.

Requirements: The successful applicant must have completed, or nearly completed, requirements for the PhD prior to assuming the fellowship. Persons who have received a PhD degree within the past three years are particularly urged to apply. Applications should include educational record, professional positions held, and a list of papers published and submitted. Applicants also should write a brief essay describing recent astronomical research and indicate what they would hope to accomplish during tenure of the fellowship. Applicants should arrange for separate transmittal of letters of recommendation from three professional scientists. Materials should be sent to The Chair, OCIW Fellowship Committee, The Observatories of the Carnegie Institution, 813 Santa Barbara Street, Pasadena, CA 91101.

Date(s) Application is Due: Dec 1
Contact: Dr. John S. Mulchaey, email: cfellow@ociw.edu
Internet: http://www.ociw.edu/ociw/fellow
Sponsor: Carnegie Institution of Washington
1530 P Street NW
Washington, DC 20005

Carole Fielding Film/Video Production or Research Grants   1516
Founded in 1947 as the University Film Producers Association, the UFVA has developed into an organization of almost 800 professionals and institutions involved in the production and study of film, video, and other media arts. Few individuals have been as dedicated to UFVA and its members as Carole Fielding. She attended every Association conference for more than twenty-five years, contributed to the Association, activities and made a legion of friends in the organization. Her work to advance the Association and its mission increased further when her husband, Ray, became president of the organization in 1967, and she accompanied him on his trips to Europe when he served as Vice President of the International Congress of Schools of Cinema and Television. Her enthusiastic work for UFVA and affection for its members made her one of the organization, most popular people and she is remembered with fondness and appreciation. The UFVA, annual Carole Fielding Scholarship was created to honor her and has continued to fund outstanding student work for over 20 years.

Restrictions: An applicant must be a student (undergraduate/graduate) at the time the application is made. A faculty member who is an active member of the University Film & Video Association or staff at a UFVA member institution must sponsor the applicant
Geographic Focus: Illinois
Samples: Katherine Hacket, UCLA; "Mesmerize Me"(narrative film) 2008   Sushrut Jain, U.S.C; "Andheri" (narrative film) 2008
Date(s) Application is Due: Dec. 15
Contact: Peter J. Bukalski; (866) 647-8382; email: ufvahome@aol.com
Internet: http://www.ufva.org/content.php?type_id=7&article_id=28
Sponsor: University Film and Video Association
Box 1777
Edwardsville, IL 62026

Carolina Minority Postdoctoral Scholars Program   1517
University funds are available each year to award five or more postdoctoral research appointments for periods of up to two years on the UNC-CH campus. Postdoctoral scholars may teach not more than one course per year, and will spend essentially full-time in research. The stipend includes some funds available for research expenses, including travel. Applications may be for study in any discipline represented at Carolina, preference being given to applicants in the areas of the humanities, social sciences, and fine arts where postdoctoral opportunities are seldom available.

Requirements: Minority students who will have completed the doctoral degree within last four years and by July 1st are eligible to apply. Preference will be given to U.S. citizens and permanent residents. This program is funded by the University of North Carolina and places emphasis on underrepresented minorities. The primary criterion for selection is evidence of scholarship potentially competitive for tenure track appointments in research universities.
Amount of Grant: $35,625 per calendar year with some funds available for research expenses, including travel
Date(s) Application is Due: Jan 6
Contact: Holli Wilson, Program Manager, (919) 962-1319; email: hjwilson@unc.edu
Internet: http://research.unc.edu/red/postdoc.html
Sponsor: University of North Carolina at Chapel Hill
CB 4000, South Bldg
Chapel Hill, NC 27599-4000

Carpenter Foundation Grants   1518
The foundation's primary areas of interest include the arts, education, public interest, and human services. The foundation is deeply concerned with the well-being of children and families and their relationship to their neighborhoods and communities. Also of concern is the health of the web of agencies and organizations which serve them. Grants are awarded for general operating support, program development, capital campaigns, equipment acquisition, scholarship funds, seed money, matching funds, and technical support. Deadlines are generally about six weeks before the quarterly board meetings, held in January, March, June, and September.

Requirements: Tax exempt agencies in the Jackson and Josephine Counties of Oregon may submit proposals.
Restrictions: Grants are not made to individuals. The foundation rarely makes grants for historical applications, hospital construction or equipment, group or individual trips, or activities for religious purposes.

Geographic Focus: Oregon
Samples: Community Health Center (Medford, OR)—fir a program to provide discounted medications to low-income/uninsured patients, $15,000.; Science Works Hands-On Museum, (Ashland, OR)—in support of a life science hands-on exhibition for local schools and students, $6,000.; Southern Oregon State College (Ashland, OR)—for faculty development opportunities which improve teaching, $25,000.; Oregon Water Trust, (Portland, OR)—for a community outreach and education project to locally distribute the General Elections Voter's Guide, $12,000.
Amount of Grant: $250-$25,000 range
Contact: Polly Williams, Program Officer, (541) 772-5732; fax: (541) 773-3970; email: carpfdn@internetcds.com
Internet: http://www.carpenter-foundation.org
Sponsor: Carpenter Foundation
711 E Main Street, Suite 10
Medford, OR 97504

Carr P. Collins Award   1519
This award is given annually to honor a nonfiction book that, in the opinion of the judges, is considered the most outstanding of the previous year's output. Annual deadline dates may vary; contact program staff for exact dates.

Requirements: The author must have lived in Texas for two years, or the book must be about a Texas subject. One copy of each entry must be mailed to each of the three judges. Names and addresses of the judges are available from the institute after October 15 each year.
Geographic Focus: Texas
Amount of Grant: $5000
Date(s) Application is Due: Jan 7
Contact: Program Contact, (512) 245-2428; email: MB13@swt.edu.
Internet: http://www.wtamu.edu/til/awards.htm
Sponsor: Texas Institute of Letters
217 Wood Street, Houston House
San Marcos, TX 78666

Carrier Corporation Contributions Grants   1520
Carrier donates approximately $2 million around the world to registered nonprofit organizations. In the United States, Carrier funds only qualified 501(c)3 organizations that meet its eligibility criteria and operate in locations where the company has a significant employee base. Carrier believes in helping people in the communities where they live, work and do business. To better serve those communities and to better align its corporate contributions with mission and values, it focuses giving on the following these areas: environment and sustainability; civic & community; education; arts & culture; health & human services. All U.S. non-profits are required to complete an online grant application. Applications are accepted from March 1 through June 1 of each year, and are reviewed for funding to be paid the following year. Applicants will receive notification in the first quarter of the calendar year in which funding will occur.

Requirements: Carrier funds only qualified 501(c)3 organizations that meet eligibility criteria and operate in locations where it has a significant employee base.
Restrictions: Carrier will not fund: individuals; religious organizations; alumni groups, sororities or fraternities; booster clubs; political groups; any organization determined by Carrier to have a conflict of interest; any organization whose practices are inconsistent with the company's Code of Ethics
Geographic Focus: Alabama; Arizona; Connecticut; Georgia; Illinois; Indiana; Michigan; Nevada; New York; North Carolina; South Carolina; Tennessee; Texas
Date(s) Application is Due: Jun 1
Contact: Rajan Goel, VP; (860) 674-3420; fax: (860) 622-0488
Internet: http://www.corp.carrier.com/vgn-ext-templating/v/index.jsp?vgnextoid=6afa 80757d7e7010VgnVCM100000cb890b80RCRD
Sponsor: Carrier Corporation
One Carrier Place
Farmington, CT 06034-4015

Carter G. Woodson Institute Research Fellowships   1521
The Carter G. Woodson Institute for Afro-American and African Studies located at the University of Virginia sponsors residential pre- and postdoctoral research fellowships. All fellowships are for the completion of research on race, ethnicity, and society in Africa and the Atlantic world (broadly defined as the African diaspora). Preference will be given to applicants whose research is substantially complete. Tenure for the fellowships begins in August. Postdoctoral fellowships are for one year; predoctoral fellowships are for two years. Application guidelines are available on the Web site. Annual deadline dates may vary; contact program staff for exact dates.

Requirements: Applicants for the fellowships must have been awarded their PhD by the time of application or must furnish proof that it will have been received prior to June 30. The competition is open to qualified candidates without restriction as to citizenship or current residence.
Restrictions: Employees of the University of Virginia and former university employees whose termination dates are less than one year prior to the time of application are not eligible.
Amount of Grant: $25,000 postdoctoral fellowships; $15,000 per year predoctoral fellowships
Date(s) Application is Due: Dec 1
Contact: William Jackson, Associate Director for Research, (434) 924-3109; fax: (434) 924-8820; email: woodson@gwis.virginia.edu
Internet: http://www.virginia.edu/woodson/programs/fellowships.html

Sponsor: Carter G. Woodson Institute for Afro-American and African Studies
P.O. Box 400162
Charlottesville, VA 22904-4162

Carthage Foundation Grants   1522

The foundation awards grants to nonprofit organizations for programs that address public questions concerned with national and international issues, such as public policy, research, government, and international affairs. There are no application deadlines; proposals are reviewed at quarterly meetings.
Restrictions: Proposals for the following are usually declined: event sponsorships, endowments, capital campaigns, renovations, or government agencies. Grants are not made to individuals.
Samples: Heritage Foundation (Washington, DC)—$200,000.; Institute for Foreign Policy Analysis (MA)—$100,000.; U of Virginia Law School Foundation (VA)—$79,000.
Amount of Grant: $50,000-$100,000 average
Contact: Michael Gleba, Treasurer, (412) 392-2900
Internet: http://www.scaife.com/carthage.html
Sponsor: Carthage Foundation
301 Grant Street, Suite 3900
Pittsburgh, PA 15219

Castle Rock Foundation Grants   1523

The foundation's grants are awarded to promote a better understanding of the free-enterprise system; preserve principles on which democracy was founded, ensure a limited role for government, and protect individual rights; encourage personal leadership; and uphold traditional American values. Types of support include general operating grants and special project grants.
Requirements: U.S. 501(c)3 organizations are eligible.
Restrictions: Grants requests from human service agencies, museums, organizations primarily supported by tax-derived funding, endowments, scientific or medical research, publications or media projects, churches, debt retirement, special events, or individuals will be denied.
Samples: College Fund/United Negro Fund, Fairfax, VA, $40,000 for scholarships; Congressional Medal of Honor Society, Mt. Pleasant, SC, $20,000 for general operating expense; Foundation for Research on economics and the Environment, $50,000 for general operating expense; Fund for American Studies, Washington, DC, $25,000 for general operating expense (2006).
Amount of Grant: $25,000-$75,000 average
Date(s) Application is Due: Mar 15
Contact: Sally W. Rippey, Executive Director, (303) 388-1683; fax: (303) 388-1684; email: generalinfo@castlerockfdn.org
Internet: http://www.castlerockfoundation.org
Sponsor: Castle Rock Foundation
4100 E Mississippi Avenue, Suite 1850
Denver, CO 80246

Catherine Holmes Wilkins Foundation Grants   1524

The foundation awards grants in Washington's Puget Sound region to nonprofit organizations in its areas of interest. Funding priorities include medical research and education—medical and academic centers conducting research and training in areas such as cancer, heart disease, and mental illness; physically handicapped and mentally ill—community nonprofit agencies providing direct social services to people with physical disabilities or mental illness; services for the needy—community-based programs providing immediate support to the needy, with particular emphasis on services for abused women and children. Preference will be given to project support, rather than ongoing operating expenses. Proposals are accepted throughout the year. Grants are awarded quarterly.
Requirements: 501(c)3 nonprofit organizations that operate within or significantly affect the residents of the greater Seattle region—Tacoma to Everett, Seattle, and the Eastside—are eligible.
Restrictions: Grants are not made for multi-year projects, debt retirement, operational deficits, or to individuals or for scholarships.
Amount of Grant: $3000-$10,000
Contact: c/o Bank of America, Charitable Investment Services
Internet: http://fdncenter.org/grantmaker/wilkins
Sponsor: Catherine Holmes Wilkins Foundation
P.O. Box 24565
Seattle, WA 98124

Catholic University Academic Scholarships, Grants, and Awards   1525

The university offers a wide range of scholarships that recognize and reward outstanding academic performance in high school, exceptional leadership, and service school, church, and the community. Awards include academic scholarships and grants—Archdiocesan/Trustees Scholarships, University Merit Scholarships, and CUA Incentive Grants; high school nominated scholarships—Award of Excellence, and Leadership/Service Scholarship; alumni grant—CU graduates may recommend new first-time, full-time freshmen for the grant, which is renewable for four years; parish scholarship—parishioners of Catholic churches who are admitted for full-time undergraduate study at CUA; scholarships with special eligibility requirements; and school and departmental scholarships. Deadlines and scholarship amounts vary. Guidelines are available online.
Amount of Grant: $1000-$10,000

Contact: Office of University Admissions, (800) 673-2772 or (202) 319-5305; fax: 202-319-6533; email: cua-admissions@cua.edu
Internet: http://admissions.cua.edu/undergrad/finaid/
Sponsor: Catholic University of America
620 Michigan Avenue NE
Washington, DC 20064

Catholic University Life Cycle Institute Research Fellowship   1526

The Life Cycle Institute, a multidisciplinary research center for social sciences at the university, offers facilities for visiting research fellows to work for a period of one semester to one year on research projects that develop wider dimensions of topics to which the institute is devoted. The institute is particularly interested in developing global, transnational, and comparative dimensions of current research interests in youth development, gender and education, social movements and civil society, volunteerism and NGO careers including religious vocations, the demographics of professions, and the social frontiers of the information revolution. The institute currently houses specialists in sociology, anthropology, politics, demography, developmental psychology, and religious studies who are on the CUA faculty or doing project-funded research. Fellows will be provided office space, secretarial and other services (including computer, Internet, and library facilities and on-line databases), and administrative support for funded projects within the university's guidelines. Guidelines are available online.
Requirements: Applicants with PhDs in the social sciences and related disciplines, including history and economics, should send a letter of intent outlining a proposed project.
Amount of Grant: $6000
Date(s) Application is Due: May 1
Contact: Dr. John White, Director, (202) 319-5999; fax: (202) 319-6267; email: white@cua.edu
Internet: http://lifecycle.cua.edu
Sponsor: Catholic University of America
620 Michigan Avenue NE
Washington, DC 20064

CBO Economic Policy Fellowships   1527

The program provides experts in macroeconomics, health economics, financial economics, public finance, and other specialties with a unique opportunity to address complex budgetary and economic policy issues. While in residence, fellows will conduct research based on a submitted proposal, use the agency's data and facilities, and work daily with colleagues at CBO to contribute to analyses and publications. Fellows will also be able to participate in professional development opportunities and will be in an excellent position to draw on the resources of experts at CBO, at other federal agencies, and within the extended community of policy analysts in Washington, DC. Fellowship appointments typically last one year but may vary depending on the project and individual circumstances. Appointment terms are flexible and can be full-time or part-time, and fellows may be employees of CBO or may maintain employment with their permanent employer depending on the availability of such an agreement. Salaries will be commensurate with qualifications and experience. Applicants must submit application materials via email. Guidelines are available online.
Requirements: In addition to expertise in macroeconomics, health economics, financial economics, or public finance, economists or budget policy analysts should have a PhD or equivalent schooling; considerable expertise in their area of proposed research; a commitment to analyze interesting real-world issues in economics and public policy; and a recognized record of research and publications.
Date(s) Application is Due: Mar 1
Contact: Nancy Fahey, (202) 226-2628; email: jobs@cbo.gov
Internet: http://www.cbo.gov/employment/fellowships.shtml
Sponsor: Congressional Budget Office
Second and D Sts SW, Ford House Office Building, 4th Fl
Washington, DC 20515-6925

CCA Visiting Scholars Grants   1528

The Study Centre is an international institute for advanced research at the postdoctoral level on all aspects of architectural thought, promoting a broad range of inquiry spanning the boundaries of the field and related disciplines. The Visiting Scholar Program promotes research of very specialized as well as interdisciplinary kinds, offering a unique study environment and full access to the outstanding resources of the Collection and Library. Each year the program appoints seven to 15 scholars at various stages of their careers, with highly diverse academic and professional accomplishments. Visiting scholars are provided with generous stipends for periods of residency at the CCA ranging from three to eight months. Applicants are encouraged to apply on the basis of an independent research project based on or supported by the extensive holdings of the CCA library and collection. Submissions should be made in either English or French. Application is available online.
Date(s) Application is Due: Nov 1
Contact: 939-7000; fax: (514) 939-7020; email: studium@cca.qc.ca
Internet: http://www.cca.qc.ca/pages/Niveau3.asp?page=application_details,â©=eng
Sponsor: Canadian Centre for Architecture
1920 rue Baile
Memorial, Quebec H3H 256 Canada

CCCF Dora MacIlelan Brown Seminary Scholarships   1529

Scholarships are awarded to Christian men and women from the greater Chattanooga area who are theologically and biblically conservative, and are seeking a Masters of

Divinity or Masters of Theology degree at a theological Seminary approved by the foundation. The scholarship committee's preference is for candidates for the pulpit ministry, although all applications will be considered. Scholarships are awarded based on merit and need. They are granted usually for three or four years depending on the duration of the degree program. Guidelines and application are available online.
Geographic Focus: Tennessee
Date(s) Application is Due: Feb 15
Contact: Henry Henegar, Scholarship Coordinator, (423) 266-5257 ext 101; fax: (423) 265-0949; email: henryhenegar@cccfdn.org
Internet: http://www.cccfdn.org/page18325.cfm
Sponsor: Chattanooga Christian Community Foundation
736 Market Street, Suite 1706
Chattanooga, TN 37402

CCFF Clinic Incentive Grants   1530
The grants are intended to enhance the standard of clinical care available to Canadians with cystic fibrosis by providing funds to initiate a comprehensive program for cystic fibrosis patient care, research, and teaching; or to strengthen an existing program. Clinic incentive grants are awarded for a term of one year, and are renewable on an annual basis. The grants provide an honorarium for the clinic director, as compensation for his or her administrative responsibilities. Requests may be made for salary support for other nonphysician clinic personnel. A component of the grants is allocated to travel.
Requirements: Canadian hospitals and/or medical schools are eligible to apply for grants.
Date(s) Application is Due: Oct 3
Contact: Grants Administrator, (416) 485-9149; fax: (416) 485-0960; email: info@cysticfibrosis.ca
Internet: http://www.ccff.ca/page.asp?id=87
Sponsor: Canadian Cystic Fibrosis Foundation
2221 Yonge Street, Suite 601
Toronto, ON M4S 2B4 Canada

CCFF Fellowships   1531
A limited number of competitive fellowships are offered by the foundation each year for basic or clinical research training in areas of the biomedical or behavioral sciences pertinent to cystic fibrosis. Fellowships are tenable at approved universities, hospitals, and research institutes in Canada. Initial fellowships are awarded for a two-year period. Fellows may apply for a one-year renewal. No one may receive more than three years of support under a CCFF fellowship. Canadian fellowship applicants of exceptional quality requesting funding to study abroad will be considered. Applicants are expected to demonstrate that comparable training is not available in Canada. Equitable consideration is given to applicants from outside of Canada who intend to return to their own country on completion of the fellowship.
Requirements: Individuals who hold MD or PhD degrees are eligible to apply.
Restrictions: Applicants who have already completed six or more years of postgraduate study or training are not eligible.
Date(s) Application is Due: Oct 1
Contact: Grants Administrator, (416) 485-9149; fax: (416) 485-0960; email: info@cysticfibrosis.ca
Internet: http://www.ccff.ca/page.asp?id=87
Sponsor: Canadian Cystic Fibrosis Foundation
2221 Yonge Street, Suite 601
Toronto, ON M4S 2B4 Canada

CCFF Research Grants   1532
Research grants are intended to facilitate the scientific investigation of all aspects of cystic fibrosis. Applications may be submitted by groups of individuals who plan to collaborate. Grants are awarded for up to three years. Investigators are eligible to hold more than one research grant. Research grants may be allocated to personnel (research assistant(s), technicians, and specified other personnel); materials and supplies; equipment not in excess of $10,000; and travel.
Requirements: A principal investigator should hold a recognized, full-time faculty appointment in a relevant discipline at a Canadian university or hospital.
Restrictions: Research grants do not provide support for graduate students, postdoctoral fellows, or summer students; construction costs; institutional overheads for laboratory facilities; or purchase of equipment in excess of $10,000.
Date(s) Application is Due: Oct 1
Contact: Grants Administrator, (416) 485-9149; fax: (416) 485-0960; email: info@cysticfibrosis.ca
Internet: http://www.ccff.ca/page.asp?id=87
Sponsor: Canadian Cystic Fibrosis Foundation
2221 Yonge Street, Suite 601
Toronto, ON M4S 2B4 Canada

CCFF Scholarships   1533
Scholarships provide salary support to gifted investigators in CF research, offering them the opportunity to develop outstanding cystic fibrosis research programs. Scholarships are held in conjunction with a CCFF research grant. Applications are available online.
Requirements: Applications are accepted from candidates who have received their first faculty appointment within the preceding five calendar years.
Date(s) Application is Due: Oct 1
Contact: Grants Administrator, (416) 485-9149; fax: (416) 485-0960; email: info@cysticfibrosis.ca

Internet: http://www.ccff.ca/page.asp?id=87
Sponsor: Canadian Cystic Fibrosis Foundation
2221 Yonge Street, Suite 601
Toronto, ON M4S 2B4 Canada

CCFF Small Conference Grants   1534
Small conference grants are offered to support medical/scientific conferences which focus on subjects of direct relevance to cystic fibrosis. Such grants will be up to a maximum of $C2500 and are intended to supplement other funding sources. The small conference grants will also support inter-clinic exchanges which are used to facilitate the exchange of special expertise between larger, university-based CF clinics and smaller, more remote clinics. Requests of this type will involve fewer individuals, and will not normally exceed $C1000. Applications may be submitted at any time, but the foundation should be consulted in advance with respect to the availability of funds. Grants are on a first-come, first-served basis.
Requirements: CF clinic directors and CCFF-funded investigators are eligible to apply.
Amount of Grant: $C2500 for medical/scientific conferences; $C1000 for inter-clinic exchanges
Contact: Grants Administrator, (416) 485-9149; fax: (416) 485-0960; email: info@cysticfibrosis.ca
Internet: http://www.ccff.ca/page.asp?id=87
Sponsor: Canadian Cystic Fibrosis Foundation
2221 Yonge Street, Suite 601
Toronto, ON M4S 2B4 Canada

CCFF Studentships   1535
A limited number of competitive studentships are offered by the foundation each year to highly qualified graduate students who are registered for a higher degree and who are undertaking full-time research training in areas of the biomedical or behavioral sciences relevant to cystic fibrosis. Students are expected to spend at least 75 percent of their time on the research training described in their application.
Requirements: Awards are tenable only at Canadian universities.
Date(s) Application is Due: Apr 1; Oct 1
Contact: Grants Administrator, (416) 485-9149; fax: (416) 485-0960; email: info@cysticfibrosis.ca
Internet: http://www.ccff.ca/page.asp?id=87
Sponsor: Canadian Cystic Fibrosis Foundation
2221 Yonge Street, Suite 601
Toronto, ON M4S 2B4 Canada

CCFF Summer Studentship   1536
Summer studentships are intended to provide support to students engaged in summer research projects in areas of the biomedical or behavioral sciences relevant to cystic fibrosis, under the direction of CF clinic directors or principal investigators. The research project should be attainable within the three-month term of the award.
Requirements: Full-time students pursuing an undergraduate degree in an appropriate discipline are eligible to receive this award. Under certain circumstances, other special academic situations, may be considered for a summer studentship.
Restrictions: Applications must be made by the proposed supervisor on behalf of a student. Applications directly from students cannot be accepted.
Date(s) Application is Due: Feb 1
Contact: Grants Administrator, (416) 485-9149; fax: (416) 485-0960; email: info@cysticfibrosis.ca
Internet: http://www.ccff.ca/page.asp?id=87
Sponsor: Canadian Cystic Fibrosis Foundation
2221 Yonge Street, Suite 601
Toronto, ON M4S 2B4 Canada

CCFF Transplant Center Incentive Grants   1537
The grants are intended to enhance the quality of care available to cystic fibrosis transplant candidates by providing eligible centers with supplementary funding for support of personnel directly involved in the provision of patient services; travel to the annual North American Cystic Fibrosis conference; and administrative costs associated with providing data to the Canadian Cystic Fibrosis Lung Transplant Registry. Grants are renewable on an annual basis.
Requirements: Canadian lung transplant centers that currently have one or more individuals with cystic fibrosis listed for transplant may apply.
Date(s) Application is Due: Oct 1
Contact: Grants Administrator, (416) 485-9149; fax: (416) 485-0960; email: info@cysticfibrosis.ca
Internet: http://www.ccff.ca/page.asp?id=87
Sponsor: Canadian Cystic Fibrosis Foundation
2221 Yonge Street, Suite 601
Toronto, ON M4S 2B4 Canada

CCFF Visiting Scientist Awards   1538
Awards are made to senior investigators from abroad who are invited to engage in CF research at a Canadian institution; or junior and senior Canadian investigators who wish to work in another laboratory in Canada or abroad. It is intended that this experience as a CCFF visiting scientist will, in some way, benefit the Canadian CF research effort. Applications may be submitted at any time, but the foundation should be consulted in advance with respect to the availability of funds.

Requirements: Senior investigators from abroad or junior and senior Canadian investigators who wish to work in another laboratory in Canada or abroad are eligible to apply.
Contact: Grants Administrator, (416) 485-9149; fax: (416) 485-0960; email: info@cysticfibrosis.ca
Internet: http://www.ccff.ca/page.asp?id=87
Sponsor: Canadian Cystic Fibrosis Foundation
2221 Yonge Street, Suite 601
Toronto, ON M4S 2B4 Canada

CCIS Visiting Research Fellowships   1539
CCIS will offer a limited number of Visiting Research Fellowships at both the predoctoral and postdoctoral levels for the upcoming academic year. These awards are to support advanced research and writing on any aspect of international migration and refugee flows, in any of the social sciences, history, law, and comparative literature. Visiting Research Fellows are provided with shared office space and a computer, as well as full access to all UCSD academic, institutional, and recreational resources. Stipends are $2,250 per month for predoctoral fellows and $3,000 to $4,000 per month for recent postdoctoral fellows (Ph.D. received within the last 6 years), depending on seniority. Stipends for more senior scholars are negotiable.
Restrictions: Due to funding constraints, CCIS is able to award fellowships only to scholars who have a current or former affiliation with a University of California campus (as a graduate student, faculty member, or researcher).
Amount of Grant: $2,250 per month for predoctoral fellows; $3,000-$4,000 per month for recent postdoctoral fellows (PhD received within the last six years)
Date(s) Application is Due: Jan 15
Contact: John D. Skrentny, Director; (858) 822-4447 or (858) 534-0484; fax: (858) 822-4432 or (858) 534-4753; email: jskrentn@ucsd.edu or ccis@ucsd.edu
Internet: http://www.ccis-ucsd.edu/Programs/fellowships.htm
Sponsor: Center for Comparative Immigration Studies (CCIS)
University of California-San Diego
La Jolla, CA 92093-0548

CCT Fellowships for Community Leaders   1540
The program affords outstanding nonprofit and public sector leaders opportunities to expand their knowledge base through a self-directed course of learning, travel, observation, and reflection. The program gives professionals up to 15 months' time off from their day-to-day professional responsibilities. The fellowship includes a stipend covering current salary and benefits, and associated expenses. Once they return to their work, these leaders draw on new experiences to help improve the entire community. Individuals who want to apply must submit a letter of interest and a current resume. The letter and resume may be sent by email or by postal mail. Guidelines are available online.
Requirements: The fellowship is open to applicants who currently are employed full-time in the not-for-profit or public sectors in metropolitan Chicago; have 10 years of professional experience in the sector; and have demonstrated leadership in the field.
Geographic Focus: Illinois
Amount of Grant: $150,000 maximum
Date(s) Application is Due: Jul 8
Contact: Anne Blanton, (312) 616-8000; email: fellowship@cct.org
Internet: http://www.cct.org/grantsseekers/specialprogramsandawards/fcl_programinfo.html
Sponsor: Chicago Community Trust
111 E Wacker Drive, Suite 1400
Chicago, IL 60601

CCWH Catherine Prelinger Award   1541
The award is made to a scholar who has not followed a traditional academic path of uninterrupted and completed secondary, undergraduate, and graduate degrees leading into a tenured faculty position. Although the recipient's degrees do not have to be in history, the recipient's work should clearly be historical in nature. In accordance with the general goals of CCWH, the award is intended to recognize or to enhance the ability of the recipient to contribute significantly to women in history, whether in the profession in the present or in the study of women in the past. Candidates should submit the application package to Carol Gold, Department of History, University of Alaska Fairbanks, P.O. Box 756460, Fairbanks, AK 99775-6460. Telephone: 907-474-6509. Applications may be requested from the same address or on email at ffcg@uaf.edu. Application and guidelines also are available online. faxed or emailed applications will not be considered.
Requirements: Application is open to scholars with a PhD or ABD degree.
Amount of Grant: $20,000
Date(s) Application is Due: Mar 13
Contact: Carol Gold, Chair, Catherine Prelinger Award Committee, (907) 474-6509; email: ffcg@uaf.edu
Internet: http://theccwh.org/preapp.htm
Sponsor: Coordinating Council for Women in History
P.O. Box 9715
Portland, ME 04104-5015

CCWH Ida B. Wells Graduate Student Fellowship   1542
The fellowship is awarded to an ABD woman graduate student working on a historical dissertation, not necessarily in a history department. History graduate students who apply for the award will automatically be considered for the CCWH/Ida B. Wells award. No additional application is necessary. Request applications from Dr. Ann Le Bar,

Department of History, Eastern Washington University, Patterson Hall 200, Cheney, WA 99004; email: alebar@mail.ewu.edu.
Requirements: The applicant must be a woman graduate student in a U.S. institution; must have passed to ABD status by the time of application; may specialize in any field, but must be working on an historical project; may hold this award and others simultaneously; and need not attend the award ceremony to receive the award.
Amount of Grant: $500
Date(s) Application is Due: Sep 1
Contact: Ann Le Bar, Department of History, Eastern Washington University, email: alebar@mail.ewu.edu
Internet: http://theccwh.org
Sponsor: Coordinating Council for Women in History
P.O. Box 9715
Portland, ME 04104-5015

CCWH/Berkshire Conference of Women Historians Graduate Fellowship   1543
The fellowship is awarded to a woman graduate student completing a dissertation in a history department. History graduate students who apply for the CCWH/Berkshire Award will automatically be considered for the CCWH/Ida B. Wells Award. No additional application is necessary. Request applications from Dr. Ann Le Bar, Department of History, Eastern Washington University, Patterson Hall 200, Cheney, WA 99004; email: alebar@mail.ewu.edu. Information and application form also are available online.
Requirements: The applicant must be a woman graduate student historian in a history department in a U.S. institution; must have passed to ABD status by the time of application; may specialize in any field of history; may hold this award and others simultaneously; and need not attend the award ceremony to receive the award.
Amount of Grant: $500
Date(s) Application is Due: Sep 1
Contact: Ann Le Bar, Department of History, Eastern Washington University, email: alebar@mail.ewu.edu
Internet: http://theccwh.org/awards.htm
Sponsor: Coordinating Council for Women in History
P.O. Box 9715
Portland, ME 04104-5015

CDC Evidence-Based Laboratory Medicine: Quality/Performance   1544
Measure Evaluation
The purpose of the program is to evaluate clinical laboratory practice by identifying evidence-based laboratory medicine quality/performance measures associated with the pre- and post-analytic stages of the total testing process, and to identify and address gaps and opportunities for improvement consistent with national health care priorities to improve public health. The primary objectives are to: provide an evidence base which systematically identifies and defines important gaps in laboratory medicine quality related to individuals or populations and/or patient safety health outcomes with demonstrated impacts (i.e., clinical and/or economic); identify, develop, and define laboratory medicine performance/quality measures that can be broadly implemented to evaluate performance associated with patient outcomes; and identify interventions effective in improving performance.
Requirements: Eligible applicants include: public nonprofit organizations; private nonprofit organizations; small, minority, women-owned businesses; universities; colleges; research institutions; hospitals; community-based organizations; faith-based organizations; federally recognized Indian tribal governments; Indian tribes; Indian tribal organizations; state and local governments or their Bona Fide Agents (this includes the District of Columbia, the Commonwealth of Puerto Rico, the Virgin Islands, the Commonwealth of the Northern Mariana Islands, American Samoa, Guam, the Federated States of Micronesia, the Republic of the Marshall Islands, and the Republic of Palau); and political subdivisions of States (in consultation with States).
Restrictions: Recipients may: not use funds for research; not use funds for clinical care; only expend funds for reasonable program purposes, including personnel, travel, supplies, and services, such as contractual. Awardees may not generally use HHS/CDC/ATSDR funding for the purchase of furniture or equipment (any such proposed spending must be identified in the budget). The direct and primary recipient in a cooperative agreement program must perform a substantial role in carrying out project objectives and not merely serve as a conduit for an award to another party or provider who is ineligible. Reimbursement of pre-award costs is not allowed.
Amount of Grant: $100,000-$500,000
Date(s) Application is Due: Jul 17
Contact: Yolanda Sledge, Grants Management Specialist, (770) 488-2787; email: yiso@cdc.gov
Internet: http://www.cdc.gov/od/pgo/funding/FOAs.htm
Sponsor: Centers for Disease Control and Prevention
2920 Brandywine Road, Mail stop: E-14
Atlanta, GA 30341-3717

CDC Antimicrobial Resistance in Hospital-Acquired Infections   1545
Among Intensive Care Unit Patients
CDC offers funds to conduct studies on antimicrobial resistance in hospital-acquired infections among intensive care unit patients. Grants will be awarded to conduct both economic and outcome evaluations on the extent of the problem, as well as on factors that lead to different outcomes.
Requirements: Applications may be submitted by public and private nonprofit organizations and by governments and their agencies.

Amount of Grant: $60,000-$100,000
Contact: Nealean Austin, Centers for Disease Control and Prevention, (770) 488-2700
Internet: http://edocket.access.gpo.gov/2004/04-9808.htm
Sponsor: Centers for Disease Control and Prevention
2920 Brandywine Road, Room 3000
Atlanta, GA 30341

CDC Cancer Prevention and Control Programs   1546
The purpose of the program is to improve and to promote health among at-risk cancer populations and to reduce cancer morbidity and mortality.
Requirements: Assistance will be provided only to the organizations listed. The Healthcare Association of New York to develop an integrated model for the delivery of comprehensive breast cancer services; the Health Choice Network, Miami/Dade County, Florida to administer the Jesse Trice Cancer Prevention Project; the East Tennessee State University, Cancer Prevention Research Center, James H. Quillen College of Medicine to address cancer care in the rural Appalachian region; the University of Rhode Island, Cancer Prevention Research Center to provide interactive interventions to at-risk populations; the Sisters of Charity Health Care System, to ensure that patients have access to early detection of gastrointestinal cancers; and Marin County, California to evaluate high incidence of breast cancer in the San Francisco Bay Area.
Geographic Focus: California; Florida; New York; Rhode Island; Tennessee
Amount of Grant: $4.1million total
Contact: Grants Administrator, (770) 488-4751; email: cancerinfo@cdc.gov
Internet: http://www.cdc.gov/cancer
Sponsor: Centers for Disease Control and Prevention
4770 Buford Hwy NE, MS K-64
Atlanta, GA 30341-3717

CDC Community Responses to Prevent Intimate Partner Violence   1547
The program's goals are to create and enhance community coalitions and responses for addressing intimate partner violence, to establish community programs for the primary prevention of intimate partner violence, and to evaluate the process and impact of using a coordinated community response to reduce this type of abuse. Officials are particularly interested in projects that focus on preventing violence against adolescents and women by people known to the victim. Other priorities are for projects that implement coalition-building and that expand existing coalitions and associated primary prevention activities. Two types of grants will be awarded: to organizations in rural or tribal areas that do not have intimate partner violence prevention coalitions or where those coalitions are in the early stages of development; and to urban or rural areas with established intimate partner violence prevention coalitions that have broad-based representation. An optional letter of intent is requested by March 13; full application is due April 12. Annual deadline dates may vary; contact the program officer to confirm exact dates. Refer to announcement 00042.
Requirements: Applications may be submitted by public and private nonprofit and for-profit organizations and by governments and their agencies.
Amount of Grant: $300,000 maximum
Date(s) Application is Due: Mar 13; Apr 12
Contact: Ted Jones, Program Manager, (770) 488-4810; fax (770) 488-1011; email: FIVPINFO@cdc.gov
Internet: http://www.cdc.gov/ncipc/factsheets/ipvfacts.htm
Sponsor: Centers for Disease Control and Prevention
4770 Buford Hwy NE
Atlanta, GA 30341

CDC Cooperative Agreement for Continuing Enhanced National Surveillance for Prion Diseases in the United States   1548
The purpose of the funding is to continue an active surveillance program similar to that conducted by the National Prion Disease Pathology Surveillance Center since 1997 to monitor the occurrence of potentially emerging human TSEs in the United States. This program addresses the Healthy People 2010 focus area(s) of Immunizations and Infectious Diseases. Eligible applicants should have experience in conducting prion disease surveillance and must be capable of fulfilling the ongoing needs for enhancing such surveillance at the national level.
Requirements: Eligible applicants include: public nonprofit organizations; private nonprofit organizations; small, minority, women-owned businesses; universities; colleges; research institutions; hospitals; community-based organizations; faith-based organizations; federally recognized Indian tribal governments; Indian tribes; Indian tribal organizations; state and local governments or their Bona Fide Agents (this includes the District of Columbia, the Commonwealth of Puerto Rico, the Virgin Islands, the Commonwealth of the Northern Marianna Islands, American Samoa, Guam, the Federated States of Micronesia, the Republic of the Marshall Islands, and the Republic of Palau); and political subdivisions of States (in consultation with States).
Restrictions: Recipients may: not use funds for research; not use funds for clinical care; only expend funds for reasonable program purposes, including personnel, travel, supplies, and services, such as contractual. Awardees may not generally use HHS/CDC/ATSDR funding for the purchase of furniture or equipment (any such proposed spending must be identified in the budget). The direct and primary recipient in a cooperative agreement program must perform a substantial role in carrying out project objectives and not merely serve as a conduit for an award to another party or provider who is ineligible.
Amount of Grant: $2,550,000.00 first year; $12,750,000.00 over five years
Date(s) Application is Due: Jun 30

Contact: Mattie Jackson, Grants Management Specialist, (770) 488-2696; email: MIJ3@cdc.gov
Internet: http://www.cdc.gov/od/pgo/funding/FOAs.htm
Sponsor: Centers for Disease Control and Prevention
2920 Brandywine Road, Mail stop: E-14
Atlanta, GA 30341-3717

CDC Cooperative Agreement for the Development, Operation, and Evaluation of an Entertainment Education Program   1549
The purpose of this program is raise awareness and behavioral change concerning public health issues through the use of the television media as means of accessing the television viewing public by providing: accurate public health information and public health issues, as well as accurate depictions of healthy living at all stages of life, to entertainment industry leadership for possible inclusion in television story lines; assistance with public health outreach, including providing additional information concerning public health topics depicted on television story lines, and assistance with public service announcements and informational short videos to be aired in conjunction with drama presentations; and evaluation of public health television story lines and the effect on the viewing public.
Requirements: Eligible applicants include: public nonprofit organizations; private nonprofit organizations; universities; colleges; research institutions; federally recognized Indian tribal governments; Indian tribes; Indian tribal organizations; state and local governments or their Bona Fide Agents (this includes the District of Columbia, the Commonwealth of Puerto Rico, the Virgin Islands, the Commonwealth of the Northern Marianna Islands, American Samoa, Guam, the Federated States of Micronesia, the Republic of the Marshall Islands, and the Republic of Palau); and political subdivisions of States (in consultation with States).
Date(s) Application is Due: Jun 7
Contact: Sharon Robertson, Grants Management Officer, (770) 488-2748; email: SRobertson1@cdc.gov
Internet: http://www.cdc.gov/od/pgo/funding/FOAs.htm
Sponsor: Centers for Disease Control and Prevention
2920 Brandywine Road, Mail stop: E-14
Atlanta, GA 30341

CDC Core Capacity Program Grants   1550
This program seeks to develop basic cardiovascular health promotions and strategies at the state level. It should include statewide partnerships and program coordination among health agencies aimed at primary and secondary heart disease prevention.
Requirements: State health agencies, the District of Columbia, Puerto Rico, and U.S. territories are eligible to apply.
Restrictions: Funds may not be used to supplant state or local funds, to provide inpatient care or personal health services, nor to support construction or renovation of facilities.
Amount of Grant: $250,000-$500,000 range; $300,000 average
Contact: Nealean Austin, Team Lead, Branch B, Procurement and Grants Office, Centers for Disease Control and Prevention, (770) 488-2754
Internet: http://www.cdc.gov
Sponsor: Centers for Disease Control and Prevention
2920 Brandywine Road
Altanta, GA 30341

CDC Disabilities Prevention Demonstration-Epidemiology Project Grants   1551
The objectives of this program are to provide a national focus for the prevention of disabilities in targeted disability groups: developmental disabilities, injury disabilities from health and spinal cord trauma, selected adult chronic conditions, and secondary disabilities in persons with physical disabilities; to build capacity at the state and community level to coordinate disabilities prevention activities and establish surveillance; to employ epidemiological methods to set priorities and target interventions; and to quantify and conduct programs to prevent secondary conditions in persons with primary disabilities. For demonstration/epidemiology projects, grant support may be used to implement and evaluate specified project activities to identify and quantify preventable secondary disabilities and measure the effectiveness and costs of preventive interventions. Funds may be used to support personnel services, equipment, supplies, travel, and services directly related to project activities. Projects will be supported for two to three years depending on availability of funds. Contact the office for deadline dates.
Requirements: Eligible applicants are public and private nonprofit entities, including disabilities services organizations (such as independent living centers), local health departments, other local governmental agencies including local organizations, units of state agencies, voluntary agencies, universities, colleges, medical facilities, research institutions, and federally recognized Native American tribal governments.
Restrictions: Project funds may not be used to supplant state or local funds available for disabilities prevention, for construction costs, or to lease or purchase facilities or space.
Amount of Grant: $130,000-$460,000 range; $310,000 average
Contact: Donald Betts, Sr., Division of Human Development and Disability, (404) 498-3957; email: dib3@cdc.gov
Internet: http://www.cdc.gov
Sponsor: Centers for Disease Control and Prevention
255 E Paces Ferry Road
Atlanta, GA 30305

CDC Evaluation of the Use of Rapid Testing For Influenza in Outpatient Medical Settings   1552
The purpose of this project is to evaluate how rapid tests for influenza are being implemented and used in clinical practice in outpatient medical settings such as

community clinics, solo and group practice physician offices, and hospital emergency rooms across the United States. This evaluation will include a determination of the scope of rapid influenza test use, the types of tests in use and how they are selected, the personnel performing testing, the extent to which good laboratory practices and testing guidelines are being followed, how results are reported and interpreted, how results are used for patient care and antiviral and antibiotic prescribing practices, and the presence of linkages between these outpatient settings and the public health system. Additionally, this project may identify potential opportunities to provide guidance to assist sites in making decisions on the appropriate use of these tests and ways to enhance the connectivity with the public health system. Connectivity with public health is especially important in light of the possibility of an influenza pandemic. This project will identify and evaluate practices used in outpatient settings related to processes such as specimen collection, testing, reporting and referral for influenza.

Requirements: Eligible applicants include: public nonprofit organizations; private nonprofit organizations; small, minority, women-owned businesses; universities; colleges; research institutions; hospitals; community-based organizations; faith-based organizations; federally recognized Indian tribal governments; Indian tribes; Indian tribal organizations; state and local governments or their Bona Fide Agents (this includes the District of Columbia, the Commonwealth of Puerto Rico, the Virgin Islands, the Commonwealth of the Northern Marianna Islands, American Samoa, Guam, the Federated States of Micronesia, the Republic of the Marshall Islands, and the Republic of Palau); and political subdivisions of States (in consultation with States).

Restrictions: Recipients may: not use funds for research; not use funds for clinical care; only expend funds for reasonable program purposes, including personnel, travel, supplies, and services, such as contractual. Awardees may not generally use HHS/CDC/ATSDR funding for the purchase of furniture or equipment (any such proposed spending must be identified in the budget). The direct and primary recipient in a cooperative agreement program must perform a substantial role in carrying out project objectives and not merely serve as a conduit for an award to another party or provider who is ineligible. Reimbursement of pre-award costs is not allowed. Reimbursement of construction costs is not allowed.

Amount of Grant: $200,000
Date(s) Application is Due: Aug 6
Contact: Yolanda Sledge, Grants Management Specialist, (770) 488-2787; email: yiso@cdc.gov
Internet: http://www.cdc.gov/od/pgo/funding/FOAs.htm
Sponsor: Centers for Disease Control and Prevention
2920 Brandywine Road, Mail stop: E-14
Atlanta, GA 30341-3717

CDC Foundation Applied Epidemiology Fellowship 1553

The fellowship provides medical students with an applied hands-on training experience in epidemiology and public health. Eight competitively selected third- and fourth-year medical students from around the country will spend up to one full year at the Centers for Disease Control and Prevention (CDC) in Atlanta, Georgia. While at CDC, they will participate in an orientation to CDC, applied epidemiology, the national public health system, and the role of physicians in that system. With the guidance of experienced CDC epidemiologists, they will perform epidemiologic analyses and research, design public health interventions, and assist in field investigations. Annual deadline dates may vary; contact program staff for exact dates.

Requirements: U.S. citizens currently enrolled in an allopathic or osteopathic school of medicine in the United States and would be in the third or fourth year of training are eligible.
Date(s) Application is Due: Dec 2
Contact: Karen Torghele, (404) 639-2696: email: ktorghele@cdc.com
Internet: http://cdcfoundation.org/pages.html?page=303
Sponsor: CDC Foundation
1600 Clifton Rd NE, Mailstop D-18
Atlanta, GA 30333

CDC Grants for Violence-Related Injury Prevention Research 1554

Funded projects should identify effective ways of preventing violence-related injuries and expand the use of new and current intervention methods for preventing these types of injuries. In the areas of suicide and assaultive behavior, projects may focus on understanding the factors that affect this behavior, such as the nature of suicide among gay and lesbian people as compared to the general population or how unequal access to criminal justice, health care, and education is related to violent behavior. Family and intimate violence prevention projects may examine the intervention strategies that are most effective in preventing injuries. CDC encourages projects to study the needs of mothers and children in families where intimate partner violence takes place and to use population-based research to quantify injury and disability among women as a result of partner violence. Officials are also interested in research that defines the cost of violent injuries and the cost effectiveness of prevention or intervention methods. Contact the program office for deadline dates.

Requirements: Public and private nonprofit and for-profit agencies may apply for grants. This includes state and local agencies, hospitals, and universities.
Amount of Grant: $300,000 maximum; $250,000 average
Contact: Paul Smutz, Injury Prevention and Control Research Projects, (770) 488-1508
Internet: http://www.cdc.gov/ncipc/res-opps/VIOLENCE_04045.htm
Sponsor: Centers for Disease Control and Prevention
4770 Buford Hwy NE, MS K65
Atlanta, GA 30341-3724

CDC HIV Demonstration, Research, Public and Professional Education Project Grants 1555

Grants are awarded to states, political subdivisions of states, and other public and nonprofit entities for research on the prevention of HIV infection at the community level. Funds may be used to develop, implement and evaluate new intervention, including those targeting people who are infected with HIV. Applicants are encouraged to have research groups participate in the program. Contact the program office for exact deadline dates.

Requirements: States, political subdivisions of states, other public and nonprofit private entities are eligible to apply.
Amount of Grant: $50,000-$1 million; $293,253 average
Contact: Allyn Nakashima, National Center for HIV, STD, and TB Prevention, (404) 639-0900; Cheryl Maddux, Grants Management Branch, (770) 498-1911
Internet: http://www.cdc.gov
Sponsor: Centers for Disease Control and Prevention
2920 Brandywine Road
Atlanta, GA 30341

CDC HIV Prevention among Ethnic Minority Populations 1556

The purpose of this program is to create community-based organizations aimed at HIV prevention among ethnic minority populations.

Requirements: State health departments and public and private institutions of higher education may apply, along with any community-based organization (CBO). Minority-oriented CBO's must have at least 50 percent of their governing board composed of the racial minority that will be served and at least 50 percent of the program's positions be held by persons who reflect the intended racial demographics.
Amount of Grant: $6.45 million total
Date(s) Application is Due: Mar 7
Contact: Grant Administrator, (404) 639-2072
Internet: http://www.cdc.gov/about/funding.htm
Sponsor: Centers for Disease Control and Prevention
1600 Clifton Rd NE
Atlanta, GA 30333

CDC HIV Prevention Projects for Community-Based Organizations 1557

CDC announces the availability of funds to support community-based organizations (CBOs) to develop, implement, and evaluate state-of-the-art, model community-based HIV prevention programs for populations at risk for HIV infection, especially racial/ethnic minority populations at risk. The goals of this program are to reduce the disproportionate impact of the HIV epidemic on racial/ethnic minority populations and other at-risk populations; to improve and expand community-based HIV prevention services by supporting community-based HIV prevention programs that address priorities described in applicable state and local comprehensive HIV prevention plans (that is, the plans developed by the official HIV prevention community planning groups for the jurisdiction in which the CBO is located) or that adequately justify addressing other priorities; enhance CBOs' incorporation of scientific theory and data, and validated program experience into the design, implementation, and evaluation of HIV prevention services; and support collaboration and coordination of HIV prevention efforts among CBOs, community planning groups, other local organizations, local and state health departments, and managed care organizations serving populations at risk for HIV infection.

Requirements: Applicant organization must have current, valid 501(c)3 tax-exempt status; must be located in the community; and have an established record of at least two years of service to the proposed target population. Contact program staff for a complete list of requirements.
Amount of Grant: $225,000 maximum
Contact: Allyn Nakashima, National Center for HIV, STD, and TB Prevention, (404) 639-0900; Chreyl Maddux, Grants Management Officer, (770) 498-1911
Internet: http://www.cdc.gov/about/funding.htm
Sponsor: Centers for Disease Control and Prevention
2920 Brandywine Road
Atlanta, GA 30341

CDC HIV Prevention: Community-Based Organizations 1558

CDC invites comments under a planned request for applications to support community-based organizations (CBO) to develop, implement, and evaluate effective community-based HIV prevention programs for at-risk populations, especially racial and ethnic minority populations.

Requirements: Applicants must be tax-exempt. They may apply as minority organizations that intend to serve predominantly racial or ethnic minority populations at risk or as CBOs that will serve high-risk populations in general. Applicants must also have two years of service with the proposed target population.
Amount of Grant: $17.6 million total
Date(s) Application is Due: Oct 29
Contact: Division of HIV/AIDS Prevention, Intervention, Research and Support, National Center for HIV, STD and TB Prevention, (404) 639-5230; fax: (404) 639-2072; email: hivmail@cdc.gov
Internet: http://www.cdc.gov/od/pgo/funding/grantmain.htm
Sponsor: Centers for Disease Control and Prevention
1600 Clifton Rd NE
Atlanta, GA 30333

CDC Injury Control Research Centers Grants 1559

Grants are intended to support injury control research and demonstrations on priority issues; to integrate aspects of engineering, public health, behavioral sciences, medicine, and other disciplines to prevent and control injuries more effectively; to rigorously apply and evaluate current and new interventions, methods, and strategies that focus on the prevention and control of injuries; to stimulate and support injury control research centers in academic institutions that will develop a comprehensive and integrated approach to injury control research and training; and to bring the knowledge and expertise of injury control research centers to bear on the development of effective public health programs for injury control. Refer to announcement number 01007. Annual deadline dates may vary; contact the office for exact dates.

Requirements: Any nonprofit or for-profit organization in eligible regions may apply. Eligible regions include 34 states as well as the District of Columbia, the Virgin Islands, and Puerto Rico. The 34 states are: Alabama, Alaska, Arizona, Arkansas, California, Delaware, Florida, Georgia, Hawaii, Idaho, Illinois, Indiana, Kentucky, Louisiana, Maryland, Michigan, Minnesota, Mississippi, Nevada, New Jersey, New Mexico, New York, North Carolina, Ohio, Oklahoma, Oregon, Pennsylvania, South Carolina, Tennessee, Texas, Virginia, Washington, West Virginia, and Wisconsin.

Restrictions: Grantees may not award subgrants but may enter into contracts as necessary to achieve the aims of the program.

Geographic Focus: Alabama; Alaska; Arizona; Arkansas; California; Delaware; District of Columbia; Florida; Georgia; Hawaii; Idaho; Illinois; Indiana; Kentucky; Louisiana; Maryland; Michigan; Minnesota; Mississippi; Nevada; New Jersey; New Mexico; New York; North Carolina; Ohio; Oklahoma; Oregon; Pennsylvania; South Carolina; Tennessee; Texas; Virginia; Washington; West Virginia; Wisconsin

Amount of Grant: $905,500 average

Contact: Program Manager, National Center for Injury Prevention and Control, (770) 488-1506; email: OHCINFO@cdc.gov

Internet: http://www.cdc.gov/ncipc/profiles/icrcs/default.htm

Sponsor: Centers for Disease Control and Prevention
4770 Buford Hwy NE, MS K65
Atlanta, GA 30341-3724

CDC Injury Prevention and Control Research and State and Community Based Programs 1560

Research grants support injury control research on priority issues; projects that integrate aspects of engineering, public health, behavioral sciences, medicine, and other disciplines in order to prevent and control injuries more effectively; projects that rigorously apply and evaluate current and new interventions, methods, and strategies that focus on the prevention and control of injuries; projects that stimulate and support Injury Control Research Centers (ICRC) in academic institutions which will develop a comprehensive and integrated approach to injury control research and training; and projects that bring the knowledge and expertise of ICRC's to bear on the development of effective public health programs for injury control. State and Community Program grants are to develop and evaluate new methods or to evaluate existing methods and techniques used in injury surveillance by public health agencies; and to develop, expand, or improve injury control programs to reduce morbidity, mortality, severity, disability, and cost from injuries.

Requirements: Applications may be submitted by all public and private nonprofit and for-profit organizations and by governments and their agencies.

Contact: Program Officer, Injury Prevention and Control Research Projects, (770) 488-1506; fax: (770) 488-1667; email: OHCINFO@cdc.gov

Internet: http://www.cdc.gov/ncipc/profiles/icrcs/default.htm

Sponsor: Centers for Disease Control and Prevention
4770 Buford Hwy NE, MS K65
Atlanta, GA 30341-3724

CDC Intervention Epidemiologic Research Studies of HIV/AIDS 1561

CDC announces the availability of funds for a cooperative agreement program to support intervention epidemiologic research studies of AIDS and HIV infection. These awards will help support researchers in two areas: the development and evaluation of innovative interventions for preventing and reducing the transmission of HIV infection in young and recently initiated injection drug users (IDUs); and the development and evaluation of an intervention study to improve access to antiretroviral therapy in HIV-infected disadvantaged populations.

Requirements: Applications may be submitted by public and private nonprofit organizations and by governments and their agencies.

Amount of Grant: $5000-$270,000 range; $200,265 average

Contact: Allyn Nakashima, Epidemiology Branch, Center for Disease Control and Prevention, (404) 639-0900

Internet: http://www.cdc.gov/epo

Sponsor: Centers for Disease Control and Prevention
1600 Clifton Rd NE, MS-E-45
Atlanta, GA 30333

CDC Minority Health Statistics Dissertation Research Grants 1562

The purpose of the program is to make awards for the conduct of special surveys or studies on the health of racial and ethnic populations or subpopulations; analysis of data on ethnic and racial populations and subpopulations; and research on improving methods for developing statistics on ethnic and racial populations and subpopulations. The deadline date listed is for letters of intent. Annual dates may vary; contact the program office for exact dates.

Requirements: Eligible applicants may be public or private nonprofit institutions that will administer the grant on behalf of the proposed principal investigator. The proposed principal investigator must be a registered doctoral candidate in resident or nonresident status. All requirements for the doctoral degree other than the dissertation must be completed by the time of the award. Students seeking a doctorate in any relevant research discipline are eligible. An applicant institution may be either the degree-granting institution or another nonprofit institution with which the proposed principal investigator is professionally affiliated.

Amount of Grant: $15,000-$30,000; $20,000 average

Contact: Program Contact, Minority Health Statistics Grants, National Center for Health Statistics (3311 Toledo Road, Hyattsville, MD 20782); (301) 458-4000; email: MGP@cdc.gov

Internet: http://www.cdc.gov/nchswww/about/grants/grants.htm

Sponsor: Centers for Disease Control and Prevention
1600 Clifton Road
Atlanta, GA 30333

CDC National Breast and Cervical Cancer Early Detection Program 1563

The purpose of this program is to apply a state, territorial, or tribal public health approach to increase access to and use of screening services. Funded programs will establish a comprehensive breast and cervical cancer early detection screening program that includes breast and cervical cancer screening, tracking, follow-up and case management; public education and outreach; professional education; quality assurance and improvement; surveillance and evaluation; coalitions and partnerships; and management.

Requirements: Assistance will be provided only to the official health departments of states or their bona fide agents, including the District of Columbia, the Commonwealth of Puerto Rico, the Virgin Islands, the Commonwealth of the Northern Mariana Islands, the Republic of Palau, and federally recognized Indian Tribal governments.

Restrictions: Alaska, Arizona, Arkansas, American Samoa, California, Colorado, Connecticut, Florida, Georgia, Illinois, Iowa, Kansas, Louisiana, Maine, Maryland, Massachusetts, Michigan, Minnesota, Missouri, Nebraska, New Jersey, New Mexico, New York, North Carolina, Ohio, Oklahoma, Oregon, Pennsylvania, Rhodes Island, South Carolina, Texas, Utah, Vermont, Washington, West Virginia, Wisconsin, Puerto Rico, and Guam are not eligible for funding. See announcement for complete listing.

Amount of Grant: $145,000-$8.4 million; $2.1 million average

Contact: Carlos Smiley, Grants Management Branch, Procurement and Grants Office, (770) 488-2754

Internet: http://www.cdc.gov/cancer/nbccedp/index.htm

Sponsor: Centers for Disease Control and Prevention
2920 Brandywine Road
Atlanta, GA 30341-4146

CDC National Poison Prevention and Control Program Grants 1564

The purpose of the program is to support an integrated system of poison prevention and control services including: coordination of all poison control centers (PCCs) through development, implementation, and evaluation of standardized public education; and development of a plan to improve national toxicosurveillance and development of a single, nationwide toll-free telephone number and related public service media campaign.

Requirements: Applications may be submitted by public and private nonprofit organizations, governments and their agencies, other public and private nonprofit organizations, state and local governments or their bona fide agents, and federally recognized American Indian tribal governments, American Indian tribes, or American Indian tribal organizations.

Contact: Cheryl Maddux, Grants Management Officer, Centers for Disease Control and Prevention, (770) 488-2645

Internet: http://www.cdc.gov

Sponsor: Centers for Disease Control and Prevention
4770 Buford Hwy NE
Atlanta, GA 30341

CDC National Programs That Build the Capacity of Schools To Prevent Foodborne Illness Through Coordinated School Health Programs 1565

The purpose of this program is to develop a national program to build the capacity of state and local education and health agencies, and others to prevent foodborne illness and other important health problems as part of a coordinated school health program.

Requirements: Assistance will be provided to national organizations that are private health, education, or social service agencies (professional, or voluntary); qualify as a nonprofit 501(c)3 entity; have the capacity and experience to assist their local affiliates; and have affiliate offices or local, state, or regional membership constituencies in a minimum of ten states and territories.

Restrictions: National organizations that are funded currently by CDC/Division of Adolescent and School Health (DASH) under program announcements 99023, 97065, 00026, 00081, 00109, 00719, 98885, 99072, 00079 or 00618 are not eligible for this program announcement.

Amount of Grant: $299,000 average

Contact: Nealean Austin, Grants Management Officer, (770) 488-2754

Internet: http://www.cdc.gov/healthyyouth

Sponsor: Centers for Disease Control and Prevention
2920 Brandywine Road, MS E-18
Atlanta, GA 30341-4146

CDC National Training Information and Exchange Program 1566

The purpose of this program is to make data and information on trauma care in the United States more accessible to a broad spectrum of individuals and organizations, including trauma care professionals and professional associations, trauma centers and other acute care hospitals, trauma care systems, emergency medical services systems, injury researchers and research organizations, public health agencies, health care payers, and the general public.

Requirements: Applications may be submitted by public and private nonprofit organizations and by governments and their agencies.

Samples: (1) Injury Control Research Centers (ICRC) have undertaken a broad range of work. For example, the John Hopkins University ICRC has sponsored Summer Training Institutes for injury control researchers and practitioners. Harvard has been a key planning, training, and program resource for injury control programs in the New England States. Work at the University of North Carolina ICRC has led to the creation of an injury control unit in the North Carolina State Health Department. Harborview ICRC serves as a State and regional resource in trauma and burn care and is a leader of efforts to reduce pedestrian injuries and injuries associated with motorcycles and bicycles. (2) Funded Injury Prevention and Control Projects address priority research concerns encompassing acute care, biomechanics, prevention, epidemiology, and rehabilitation. As examples, researchers are investigating the mechanism of traumatic brain injury due to impact, other investigators are defining risk factors for intimate partner violence and another group is evaluating the effectiveness of trauma systems. (3) Surveillance programs address E-coded hospital discharge data; model surveillance systems to address nonfatal injuries resulting from intentional and unintentional injuries.

Contact: Cheryl Maddux, Grants Management Officer, (770) 488-2645

Internet: http://www.cdc.gov

Sponsor: Centers for Disease Control and Prevention

2920 Brandywine Road, Suite 3000

Atlanta, GA 30341

CDC Public Health Conference Support Grant Program 1567

The purpose of the grants is to provide partial support for specific non-federal conferences in the areas of health promotion and disease prevention information/education programs. Funds may be used for direct cost expenditures, such as salaries, speaker fees, rental of necessary equipment, registration fees, and transportation costs (not to exceed economy-class fare) for non-federal individuals. The program continues through March 8, 2009.

Requirements: Public and private organizations, including colleges and universities, and units of state and local government are eligible to apply. Applicants must provide a portion of the conference cost.

Restrictions: Funds may not be used for indirect costs, equipment purchases, honoraria, entertainment, or personal expenses.

Contact: Sharon Robertson, Grants Management Officer, (770) 488-2748; email: SRobertson1@cdc.gov

Internet: http://www.cdc.gov/od/pgo/funding/FOAs.htm

Sponsor: Centers for Disease Control and Prevention

2920 Brandywine Road, Room 3000

Atlanta, GA 30341

CDC Research, Treatment, and Education Programs on Lyme Disease in 1568
the United States Cooperative Agreements

The objective of the program is to develop and implement practical and effective measures for the primary and secondary prevention of Lyme disease. Funds will be available to develop disease surveillance, conduct ecological and epidemiological studies, develop prevention and control activities, develop better diagnostic tests, and develop and disseminate educational materials and programs. Cooperative agreements are usually awarded for a three-year project period. Initial awards are made for a one-year budget period with continuation awards for up to an additional two years. Contact the office listed for application deadline dates. PA-04-008

Requirements: Public and nonprofit organizations able to provide services to geographical areas where Lyme disease is endemic or found to be newly emerging are eligible to apply. Funding preference will be given to proposals that incorporate evaluation of effectiveness, cost, and acceptability of strategies, or combinations of strategies, for population-based control of tick-borne diseases.

Samples: Examples of some funded projects include: development of materials for education intervention; conducting studies in integrated pest management; conducting studies on the pathogenesis of Lyme neuroborreliosis using the monkey model; implementation of active surveillance systems; and studies to improve serologic diagnostic techniques.

Amount of Grant: $100,000-$700,000; $350,000 average

Contact: Dr. Brian Mahy, National Center for Infectious Diseases, (404) 639-0043; fax: (404) 639-2469; email: bxm1@cdc.gov

Internet: http://www.cdc.gov/od/pgo/funding/Expired/04008.htm

Sponsor: Centers for Disease Control and Prevention

P.O. Box 2087

Fort Collins, CO 80522

CDC Sexually Transmitted Diseases Research, Demonstration, Public 1569
Information, and Education Grants

The grants are meant to develop, improve, apply, and evaluate methods for the prevention and control of syphilis, gonorrhea, and other sexually transmitted diseases through demonstration and applied research. Also, the grants are meant to develop, improve, apply, and evaluate methods and strategies for public information and education about these diseases; to support professional (including appropriate allied health personnel) education, training, and clinical skills improvement activities; and to support particularly deserving public information and education programs that cannot be supported through other grant programs. Grant funds may be used for the costs associated with planning, organizing, and conducting applied research, demonstrations and education programs,

and to reimburse individuals asked to be participants in the applied research. Contact the program office for deadline dates.

Requirements: Any state, political subdivision of a state, or any other public or private nonprofit institution may apply.

Amount of Grant: $125,000-$450,000; $313,809 average

Contact: Sevgi Aral, Director, Division of STD Prevention, (404) 639-8259; Gladys Gissentanna, Grants Management Officer, (770) 498-1912

Internet: http://www.cdc.gov

Sponsor: Centers for Disease Control and Prevention

255 E Paces Ferry Rd NE

Atlanta, GA 30305

CDC Using Genetic Information to Prevent Disease and Improve 1570
Health Research Grants

The purpose of this program is to strengthen science for public health action, collaborate with health care partners for prevention, and promote healthy living at every stage of life. The program will provide funding for conducting population-based research to assess how risk for disease and disability in well-defined populations is influenced by the interaction of human genetic variation with modifiable risk factors, and ensure that genetic tests and services are incorporated in population-based interventions that promote health and prevent disease and disability. Annual deadline dates may vary; contact the program office for exact dates.

Requirements: Applications may be submitted by public and private nonprofit organizations and by governments and their agencies, other public and private nonprofit organizations, state and local governments or their bona fide agents, and federally recognized American Indian tribal governments, American Indian tribes, or American Indian tribal organizations.

Contact: Nealean Austin, Grants Management Branch, Centers for Disease Control and Prevention, (770) 488-2700

Internet: http://www.cdc.gov

Sponsor: Centers for Disease Control and Prevention

4770 Buford Hwy NE

Atlanta, GA 30341

Cen Cal Diving and Aquatic Studies Scholarship 1571

This scholarship is awarded annually to an outstanding individual who is engaging in studies of underwater habitats. Aquatic-related programs in the disciplines of biology, physical sciences, marine education, maritime archaeology, historical and social aspects of marine resources, or the science of diving are relevant for consideration. Application forms are available on the Web site. Annual deadline dates may vary; contact program staff for exact dates.

Requirements: Applicants must be California residents enrolled in an undergraduate or graduate program at a California college or university, be over 18 years of age, maintain a high academic standing with a minimum GPA of 3.0, and have a national/regional diver certification.

Geographic Focus: California

Amount of Grant: $1000

Date(s) Application is Due: Apr 30

Contact: Jim Kaller, (415) 362-9134 ext 12; email: jameskaller@batnet.com

Internet: http://www.cencal.org/requirements.html

Sponsor: Central California Council of Diving Clubs

P.O. Box 779, Cen Cal

Daly City, CA 94017

Center for Advanced Judaic Studies Postdoctoral Fellowships 1572

The program is designed to engage scholars from within Jewish studies, who specialize in Jewish life under Islamic rule, in fruitful conversation with scholars in Arabic, Syriac, Persian, and Ottoman studies. By bringing together experts from these disparate disciplines who deal with aspects of Jewish, Christian, and Muslim life within pre-modern Islamic polities, the program hopes to encourage a broad view of Islamic societies and to foster new approaches to their religious, ethnic, and linguistic diversity. Guidelines are available online. The 2006-2007 program is Jewish, Christian, and Muslim Life under Caliphs and Sultans. The 2007- 2008 program will be Jewish and Other Imperial Cultures in Late Antiquity: Literary, Social, and Material Histories

Requirements: Any individual or group of scholars in the fields of Jewish studies may submit a proposal for an annual theme.

Amount of Grant: $33,000

Date(s) Application is Due: Nov 1

Contact: David B. Ruderman, Director, (215) 238-1290; fax: (215) 238-1540; email: ruderman@sas.upenn.edu

Internet: http://www.cjs.upenn.edu/program/fellowship.htm

Sponsor: Center for Jewish Studies

Center for Advanced Judaic Studies, 420 Walnut Street

Philadelphia, PA 19106

Center for Advanced Study in the Behavioral Sciences Postdoctoral Fellowships 1573

The center offers nine- to 12-month residential postdoctoral fellowships for scientists and scholars of proven accomplishment or exceptional promise. Fellowships have been offered in psychology, sociology, anthropology, political science, history, economics, philosophy, psychiatry-psychoanalysis, linguistics, humanities, law, and education; and certain biomedical, mathematical, and statistical specialties. The selection process is started by nominations being received from well-known behavioral scientists, academic administrators, and former fellows. Nominees are screened by a review panel and the final selections are made by the center's board of trustees. The selection process normally takes several years, with the fellowship roster for any given year usually prepared more than a year in advance. Usually 45 to 50 fellowships are awarded annually. Nominations may be submitted at any time.

Requirements: Qualified scholars in behavioral and biological sciences, humanities, and certain areas of statistics and computer science from any country are eligible. Resident participants usually have completed university training, hold university appointments, are engaged in research and research training, and are in the top 5 to 10 percent of the scholars in their field.

Contact: Robert Scott, Associate Director, (650) 321-2052; fax: (650) 321-1192; email: info@casbs.org

Internet: https://casbs.stanford.edu

Sponsor: Center for Advanced Study in the Behavioral Sciences

75 Alta Road

Stanford, CA 94305

Center for Advanced Study in the Visual Arts Senior Fellowships 1574

Applications will be considered from scholars for study in the history, theory, and criticism of the visual arts (painting, sculpture, architecture, landscape architecture, urbanism, graphics, film, photography, decorative arts, industrial design, etc.) of any geographical area and of any period. Applications are also solicited from scholars in other disciplines whose work examines physical objects or has implications for the analysis and criticism of physical form. In addition, applications are solicited from scholars who are specifically interested in curatorial research related to objects in the painting, sculpture, graphics, and other collections of the gallery. One Paul Mellon Senior Fellowship and four to six Ailsa Mellon Bruce and Samuel H. Kress Senior Fellowships will normally be awarded for an academic year; applications for a single academic term or a quarter are also possible. In exceptional cases, application may be made for a period of two years. The Samuel H. Kress Senior Fellowships are intended to support primarily research related to objects in the collection of the gallery. Senior fellows receive a monthly stipend with additional allowances for research materials, round-trip travel, and local expenses.

Requirements: Applicants must have held the PhD for five years or more or possess a record of professional accomplishment at the time of application. Senior fellowships are awarded without regard to the age or nationality of the applicant. Fellowships will normally be awarded for an academic year, early fall to late spring, but fellowships for a single term or quarter are also possible. Scholars are expected to reside in Washington, DC, throughout their fellowship and participate in the activities of the center.

Restrictions: Fellowships may not be postponed or renewed.

Amount of Grant: $50,000 maximum; $12,000 maximum housing allowance

Date(s) Application is Due: Oct 1

Contact: Fellowships Program, Center for Advanced Study in the Visual Arts, (202) 842-6482; fax: (202) 789-3026; email: advstudy@nga.gov

Internet: http://www.nga.gov/resources/casvasen.htm

Sponsor: National Gallery of Art

2000B S Club Drive

Landover, MD 20785

Center for Advanced Study in the Visual Arts Visiting Senior Fellowships 1575

Applications will be considered from scholars for short-term fellowships (maximum of 60 days) for study in the history, theory, and criticism of the visual arts (painting, sculpture, architecture, landscape architecture, urbanism, graphics, film, photography, decorative arts, industrial design, etc.) of any geographical area and of any period. Applications are also solicited from scholars in other disciplines whose work examines physical objects or has implications for the analysis and criticism of physical form. In addition, applications are solicited from scholars who are specifically interested in curatorial research related to objects in the painting, sculpture, graphics, and other collections of the gallery. The notable resources represented by the collections of the National Gallery of Art, its library, photographic archives, the Library of Congress, and other specialized research libraries and collections in Washington will be available to the visiting senior fellows; lectures, colloquia, and informal discussions will complement the program. Fellows are provided with a monthly stipend that includes round-trip travel and local expenses. Fellows may be eligible for a $1500 per month housing allowance. Application forms must be obtained from the center. This award is granted twice per year.

Requirements: Applicants must have held the PhD for five years or more or possess a record of professional accomplishment at the time of application. Fellowships are awarded without regard to age or nationality. All grants are based on individual need. Scholars are expected to reside in Washington, DC, throughout their fellowship and participate in the activities of the center.

Restrictions: Fellowships may not be postponed or renewed.

Amount of Grant: $6000-$8000

Date(s) Application is Due: Mar 21; Sep 21

Contact: Fellowships Program, Center for Advanced Study in the Visual Arts, (202) 842-6482; fax: (202) 789-3026; email: advstudy@nga.gov

Internet: http://www.nga.gov/resources/casvavissen.htm

Sponsor: National Gallery of Art

Fourth Street and Constitution Avenue NW

Washington, DC 20565

Center for African Studies Postdoctoral Fellowships 1576

Funded by the Rockefeller Foundation, postdoctoral fellowships on the theme of Globalization and Higher Education in Africa are offered by the university's Center for African Studies and the College of Education. This three-year program offers fellows the opportunity to explore and interrogate, through interdisciplinary inquiries, the role of higher education in and on Africa. Applicants for the 10-month fellowships need not be U.S. citizens. Interdisciplinary, comparative, and regional interests are especially welcome. All fellows will be based in Champaign-Urbana, IL, and some support for relocation expenses, housing, and benefits is also available. Complete information is available online.

Requirements: Applications are welcome from the humanities, education, and social sciences fields, and applicants must have a PhD or its equivalent.

Amount of Grant: $37,500-$40,000

Date(s) Application is Due: Jan 5

Contact: Rockefeller Postdoctoral Fellowships, (217) 333-6335; fax: (217) 244-2429; email: swisher@uiuc.edu

Internet: http://www.afrst.uiuc.edu/postdoc04-05.htm

Sponsor: University of Illinois at Urbana-Champaign

910 S Fifth Street, 210 International Studies Bldg

Champaign, IL 61801

Center for Alternatives to Animal Testing Grants 1577

The center seeks to develop innovative non-whole animal methods to evaluate fully commercial and/or therapeutic products to ensure the health and safety of the public. The program funds research that will lead to the refinement, replacement, or reduction of animals in toxicity testing and by disseminating scientifically correct information about these methods and their applications. The center encourages the development of in vitro approaches to toxicity evaluation including, but not limited to, methods using human cells/cell lines and studies in the areas of skin hypersensitivity/toxicity, phototoxicity, target organ toxicity, and structure-activity relationships. Grants support one year of research with possible renewal for another year. Preproposals must meet the application deadline; full proposals are invited.

Requirements: Applicants must have an established mechanism for handling private funding.

Restrictions: Projects focusing on mutagenicity or carcinogenicity are not funded.

Amount of Grant: $25,000 maximum

Contact: Alan Goldberg, Center Director, (410) 223-1692; fax: (410) 223-1603; email: caat@jhsph.edu or goldberg@jhsph.edu

Internet: http://caat.jhsph.edu/programs/grants/grants.htm

Sponsor: Johns Hopkins University

111 Market Pl, Suite 840

Baltimore, MD 21202-6709

Center for Black Music Research Rockefeller Fellowships 1578

Under the auspices of the Rockefeller Foundation, the Center for Black Music Research (CBMR) will offer a series of resident research fellowships under the theme of Researching the Circum-Caribbean. The CBMR will accept applications to pursue work on musico-cultural issues related to Latin-American and West Indian cultures in the circum-Caribbean region. Applications will be considered from scholars in the fields of music and music librarianship, history, Caribbean studies, and American cultural studies. Two fellowships will be offered. Fellows will spend half their time in residence at the AMRI in Saint Thomas, U.S. Virgin Islands, from which they will do field or archival research, and the other half at the CBMR's Chicago site preparing for or writing the results of the work they performed while at the Caribbean site or other sites of field research. The fellows will present work-in-progress at colloquia and will have opportunities to interact with faculty, students, and members of the Chicago and Saint Thomas arts, cultural, and educational communities. Each fellowship includes a stipend ($25,000), an allowance ($4000 maximum) for moving to and from the fellow's home and the CBMR (or the Virgin Islands); an allowance $2000 maximum) for inter-island or intra-Caribbean research travel; and free housing in both locations. Application guidelines are available online.

Requirements: Applicants must have completed work on their terminal degree (PhD degree or its equivalent) at the time of application and should be able to demonstrate their familiarity with the research of or related to the circum-Caribbean region.

Amount of Grant: $37,750 total for stipend, travel allowances, and housing

Date(s) Application is Due: Feb 2

Contact: Linda Hunter, Fellowship Administrator, (312) 344-7559; fax: (312) 344-8029; email: lhunter@cbmr.colum.edu

Internet: http://www.cbmr.org/fellows0405.htm

Sponsor: Columbia College Chicago

600 S Michigan Avenue

Chicago, IL 60605-1996

Center for Chicana/o Studies Rockefeller Fellowships 1579

The center seeks applicants for postdoctoral fellowships from scholars and artists conducting research on Chicana/o culture, and in particular, from scholars who focus on the interaction of hybridity, cultural mobility, and literacy in a transnational context. Projects on the ways Chicana/o culture responds to, as well as maps, the flows and conflicts among forms of cultural literacy are especially welcome, as are studies exploring the ways diverse global processes affect this cultural literacy. Fellows are expected to remain in residence at the center during the regular academic year (September-June), to participate in the ongoing activities of the center, and to live in the immediate area of the university. During the academic year, fellows will be required to present their research through lectures and workshops, to engage with faculty and students at colloquia and conferences, and to draw on the variety of resources available at the university. Fellows will also be expected to participate in a conference at the end of the four-year program.

Requirements: Junior, senior and independent scholars and artists of any nationality who are conducting research on Chicana/o culture and the interplay of hybridity, cultural mobility and literacy are eligible.
Amount of Grant: $35,000 stipend plus $1575 maximum toward moving allowance, benefits, and research assistance funds
Date(s) Application is Due: Feb 1
Contact: Carl Gutierrez-Jones, Center for Chicano Studies, (805) 893-3895; email: carlgj@english.ucsb.edu
Internet: http://research.ucsb.edu/ccs/rockefeller.html
Sponsor: University of California, Santa Barbara
125 Park Sq Crt, 400 Sibley Street
Santa Barbara, CA 93106-6040

Center for Cultural Studies Resident Scholars Program Grants   1580
The center invites applications from postdoctoral scholars in the humanities, social sciences, and arts. Residencies vary from several months to a full academic year. Resident scholars are provided an office and have access to the UC library and other facilities. The center cannot provide a stipend. To apply, send a project statement, curriculum vita, two letters of recommendation, and a cover letter indicating the desired period of residency. Annual deadline dates may vary; contact program staff for exact dates.
Requirements: Applicants should hold a doctorate or the equivalent.
Date(s) Application is Due: Apr 29
Contact: Resident Scholars Program, Center for Cultural Studies, (831) 459-4899; fax: (831) 459-1349; email: cult@ucsc.edu
Internet: http://humwww.ucsc.edu/CultStudies/PROG/RSCHOL/residence.html
Sponsor: Center for Cultural Studies
University of California, Oakes College
Santa Cruz, CA 95064

Center for Defense Information Internships   1581
The center offers an intern program to undergraduate and graduate students and recent graduates who have strong interests in U.S. military issues and related public policy issues. Interns perform a variety of professional support functions but serve mainly as research and outreach assistants. Interns also assist with the production of CDI's weekly television show on military-related current affairs. They work closely with the center staff, usually with considerable responsibility. Intern projects have included such diverse research subjects as U.S. military nuclear wastes, military spending, and U.S. arms transfer policy, as well as media activities. There is no preferred academic background. Fields of study might encompass the social and policy sciences, physical sciences, and preprofessional curricula. Although prior coursework in U.S. military issues and related public policy areas is not required, high academic achievements are important. Experience with video production is a plus. Writing skills are essential. Interns are expected to work on a full-time basis. Internships run from January through May, June through August, and September through December. Application must include personal letter, resume, college transcript, writing samples, and two letters of recommendation.
Requirements: Internships are open to undergraduates, graduates, and graduate students with strong interests in military policy, national security, foreign affairs, and related public policy issues and/or interest in broadcast communications.
Amount of Grant: $1000 monthly stipend
Date(s) Application is Due: Mar 1; Jul 1; Oct 1
Contact: Internship Coordinator, (202) 332-0600; fax: (202) 462-4559; email: internships@cdi.org
Internet: http://www.cdi.org/about/internships.cfm
Sponsor: Center for Defense Information
1779 Massachusetts Avenue NW
Washington, DC 20036-2109

Center for East Asian Studies Fellowships   1582
The center was formed to increase interdisciplinary communication among linguists, historians, art historians, political scientists, anthropologists, and others whose research, teaching, or study focuses on China, Korea, or Japan. Some students accepted into the graduate program in East Asian studies at Stanford receive financial assistance through these fellowships. The center also offers a limited number of student awards to Stanford students for summer research and/or study. The most up-to-date information about fellowships is available in the CEAS newsletter which can be accessed on the Web site. Deadlines vary for individual programs; contact program staff for exact dates.
Requirements: Graduate students at Stanford University specializing in East Asia may apply for fellowships at the time of initial application for admission.
Restrictions: Students attending institutions other than Stanford are ineligible.
Contact: Connie Chin, Program Administrator, (650) 723-3362; fax: (650) 725-3350; email: csquare@stanford.edu
Internet: http://www.stanford.edu/dept/CEAS
Sponsor: Stanford University
Room 51-L, Building 50
Stanford, CA 94305-5013

Center for Education of Women Visiting Scholar Research Grants   1583
The Center for Education of Women (CEW) invites applications from scholars and practitioners interested in being in residence at the university for a period of one to 12 months to pursue research, writing, and publication interests. Applicants in the following general areas are invited to apply: Women in Higher Education, Women's Work, Women of Color in the Academy, Gender and Education, Career Development, Women in Non-Traditional Fields, Leadership, Gender Equity in Education and Employment,

and Gender and Poverty. Other areas of research may also be considered. Scholars will have the opportunity to conduct their research and will be asked to prepare a working paper and/or give a seminar based on this work. CEW will provide office space, access to facilities and programs, modest stipends, and/or research support funds. Applicants should send letter of interest, a one- to two- page outline of proposed project, funding requirements, anticipated dates of stay, and a vita.
Requirements: Applicants should hold an earned PhD, EdD, JD, etc., or have equivalent experience in a relevant field.
Amount of Grant: $7500 maximum per scholar
Date(s) Application is Due: Mar 1; Jun 1
Contact: Beth Sullivan, (734) 998-7225; email: bsulliva@umich.edu
Internet: http://www.umich.edu/~cew/faculty-staff/vs.htm
Sponsor: University of Michigan
330 E Liberty Street
Ann Arbor, MI 48104-2289

Center for Hellenic Studies Fellowships   1584
Fellowships for research in ancient Greek history, philosophy, literature, language, and religion are offered at the center. Letters of recommendation are due separately at the Director's office by November 1. Application forms are available upon request.
Requirements: Applicants must have PhDs or equivalent, be professionally competent in ancient Greek, and be published writers.
Samples: Fellows for the 2006-2007 year include: Felix Budelmann (Germany)—The Open University Edition with commentary of a selection from Greek lyric; Anne Duncan University of Nebraska—Tyranny and Theater in Ancient Greece; Thomas Jenkins, Trinity University—American Classics: Transformations of Antiquity in Postwar America; Polly Low (Great Britain) University of Manchester— Political Speech and Political Writing in Fourth-century BC Athens; Francesca Schironi (Italy) Harvard University— Aristarchus of Samothrace.
Amount of Grant: $26,000 maximum stipend plus $1000 for travel and research assistance
Date(s) Application is Due: Oct 15
Contact: Program Contact, (202) 234-3738; fax: (202) 797-3745; email: chs@fas.harvard.edu
Internet: http://www.chs.harvard.edu/fellowships.sec
Sponsor: Center for Hellenic Studies
3100 Whitehaven Street NW
Washington, DC 20008

Center for Inquiry Science and the Public Visiting Research Fellowships   1585
Each academic year, the center awards visiting research fellowships at the postdoctoral or senior level in the area of science and the public. All relevant disciplines are welcome. Fellowships include an appointment in the appropriate department at State University of New York-Buffalo and a stipend plus accommodations (optional). Candidates should send a research project proposal (1000 to 2000 words), curriculum vita, two references (for recent doctorates), and a writing sample. Guidelines and current research themes are available online.
Amount of Grant: $15,000 per semester
Date(s) Application is Due: Mar 1
Contact: Austin Dacey, Fellowship Committee, fax: (716) 636-1733; email: adacey@centerforinquiry.net
Internet: http://www.scienceandthepublic.org/programs.htm
Sponsor: Center for Inquiry
P.O. Box 741
Amherst, NY 14226

Center for Italian Renaissance Studies Fellowships   1586
The center awards up to 15 fellowships for independent study on any aspect of the Italian Renaissance. Each fellow is given a place to study, use of the Biblioteca and Fototeca Berenson, weekday lunch, and an opportunity to meet scholars from various countries working in related fields. The selection committee looks for scholarly excellence and promise, requires a project of importance that is suitable to the resources of I Tatti and Florence, and assesses the candidate's ability to contribute to collegiality and the intellectual life of the other fellows. Projects do not have to be devoted to Florentine subjects, but it must be possible for the greater part of the project to be accomplished in Florence. Fellows will be in residence for a full academic year. Renewals are rarely granted. Applicants should send completed application form, curriculum vita, project description, and three letters of recommendation to Professor Walter Kaiser, Director, Villa I Tatti, Via di Vincigliata 26, 50135 Florence, Italy (telephone: 39 055 603 251, fax: 39 055 603 383), with duplicates to the Villa I Tatti office in Cambridge (listed). Completed application and supporting materials must be received at each address by the specified deadline date for applications.
Requirements: Fellowships are for scholars of any nationality, normally postdoctoral (or equivalent), and in the earlier stages of their careers. Preference is given to scholars beginning new research projects and to applicants who have not previously had residencies in Florence.
Restrictions: faxed applications and letters of recommendation will not be accepted.
Samples: Luca Mola, Johns Hopkins U—fellowship recipient (history) for study entitled The Market for Innovations in Renaissance Italy.; Philippe Costamagna, Universite Paris IV-Sorbonne—fellowship recipient (art history) for study entitled Florentine Portraits of the XVI Century.; Alessandro Arcangeli, Liceo Classico Scipione Maffei Verona—fellowship recipient (history) for Studies in the Historical Anthropology of Leisure

in Renaissance Italy.; Carlo Falciani, Universita di Firenze—fellowship recipient (art history) for study Il significato della decorazione del Castello di Fontainebleau.
Amount of Grant: $45,000-$50,000 range
Date(s) Application is Due: Oct 15
Contact: Amanda Smith, Villa I Tatti Fellowship Program, (617) 495-8042; fax: (617) 495-8041; email: amanda_smith@harvard.edu
Internet: http://www.itatti.it/fellow_home.html
Sponsor: Harvard University Center for Italian Renaissance Studies
124 Mount Auburn Street
Cambridge, MA 02138-5795

Center for Judaic Studies Fellowships 1587
The center invites applications from scholars engaged in all fields of Judaic studies and from scholars in other fields interested in approaching the general topic from a comparative and interdisciplinary perspective. Outstanding graduate students in the final stages of writing their dissertations also may apply. Fellowship recipients also may receive a contribution toward travel expenses. Annual research topics vary; contact the center for current topic. Application materials are available upon written request.
Samples: Fellows in the 2006-2007 program include: Haggai Ben-Shammai, Hebrew University— Scriptural Exegesis in Medieval Judaism and Islam; Jonathan Decter, Brandeis University— Hebrew Panegyric, Jewish Court Culture, and the Articulation of Legitimacy; Jessica Goldberg, University of Pennsylvania—Merchant trade and trading networks in the medieval Mediterranean; Alfred Ivry, New York University—Shi¸Âôism and Judaism in the Middle Ages.
Amount of Grant: $33,000 maximum for academic year
Date(s) Application is Due: Nov 15
Contact: Administrator, Fellowship Program, Center for Judaic Studies, (215) 238-1290; fax: (215) 238-1540; email: allenshe@sas.upenn.edu
Internet: http://www.cjs.upenn.edu/program/fellowship.htm
Sponsor: University of Pennsylvania
420 Walnut Street
Philadelphia, PA 19104-6906

Center for Khmer Studies Fellowships 1588
The center invites applications from U.S. scholars from all disciplines who wish to conduct their research in Cambodia. The program supports dissertation research fellowships and senior fellowships. Dissertation research fellowships enable doctoral candidates to pursue their dissertation research in Cambodia. Awards are available for periods of up to 11 months. Senior research fellowships enable scholars in all disciplines focusing on Southeast Asia to pursue further research in Cambodia. Short-term awards are available for up to four months. Long-term awards are available for six to nine months. Fellowships for four months or less have some travel restrictions. Application and guidelines are available online.
Requirements: Dissertation research fellowships are available to doctoral candidates at U.S. colleges and universities in all fields of studies. Senior research fellowships are available to scholars who hold the PhD or its equivalent. Non-U.S. citizens are welcome to apply if they are teaching full-time at U.S. colleges and universities.
Date(s) Application is Due: Nov 1
Contact: Dr. Thak Chaloemtiarana, Southeast Asia Program, Cornell University, email: fellowships@khmerstudies.org
Internet: http://www.khmerstudies.org/fellowships/fellows.htm
Sponsor: Center for Khmer Studies
180 Uris Hall
Ithaca, NY 14853

Center for Medieval and Renaissance Studies Assistantships, Grants, and 1589
Fellowships
The center has six available programs. Research assistantships are awarded to four graduate students working in the field of medieval and Renaissance studies at UCLA. During the academic year, recipients work with several faculty members on research and publication projects. The summer fellowship is awarded each year to a scholar (holding a PhD or the equivalent) who wishes to pursue research within the Los Angeles area in the field of medieval and Renaissance studies; the Lynn White fellowship is awarded each year to an outstanding UCLA graduate student in medieval and Renaissance studies who has advanced to PhD candidacy; this one-year fellowship provides the recipient with support for research and related travel in order to complete the dissertation. The Portuguese Fellowship will be awarded annually on a competitive basis to one outstanding UCLA graduate student to support interdisciplinary research in any area pertaining to Portugal during the Middle Ages and Renaissance, that is, between AD 400 and 1650. Fredi Chiappelli travel fellowships provide a travel grant to assist the recipient with research in any area of medieval and Renaissance Italian studies; the fellowship is open to UCLA graduate students and travel is not restricted to Italy. Grants for interdisciplinary research are awarded to interdisciplinary teams of two or more CMRS faculty and/or graduate students to pursue collaborative research. These grants are not intended to replace Senate research grants or instructional resources grants, but may be used in conjunction with research support from other sources. Applicant teams may include faculty and/or graduate students but must represent more than one department. Annual deadline dates may vary; contact program staff for exact dates.
Amount of Grant: $500 summer fellowships; $15,000 for the Lynn White fellowship; $1500 for the Fredi Chiappelli travel fellowships
Contact: Program Contact, (310) 825-1880; email: cmrs@humnet.ucla.edu
Internet: http://www.humnet.ucla.edu/humnet/CMRS/Awards/Awards_default.htm

Sponsor: Center for Medieval and Renaissance Studies - RA Coordinator
10745 Dickson Plz, 302 Royce Hall
Los Angeles, CA 90095-1485

Center for National Policy Internships 1590
One intern per semester or summer is employed for a position dealing with the overall general operations of CNP. The program promotes practical opportunities to advance the public policy process, as experts and decisionmakers debate significant national objectives and determine how government can best serve American interests both at home and abroad. The intern will conduct research, write briefings, prepare press materials, participate in fund raising, and perform general office support work. Applications must be received two months prior to the semester.
Requirements: Applicants must be undergraduate students.
Date(s) Application is Due: Apr 15; Aug 12 Nov 12
Contact: Alex Sunshine, Intern Coordinator, (202) 682-1800; fax: (202) 682-1818; email: asunshine@cnponline.org or thecenter@cnponline.org
Internet: http://cnponline.org/employment.htm
Sponsor: Center for National Policy
1 Massachusetts Avenue NW, Suite 333
Washington, DC 20001

Center for Science in the Public Interest Internships 1591
The center is a national consumer organization that focuses on health and nutrition issues. CSPI offers internships for a small number of qualified students in undergraduate, graduate, law, and medical schools each summer and during the school year. Generally, an internship is for 10 weeks. The specific dates of internship are flexible and depend on the center's needs and the applicant's schedule. Application materials should include a cover letter indicating issues of interest, future plans, available dates; resume; writing sample; two letters of recommendation; and official transcript of courses and grades. Application materials should be submitted as early as possible.
Amount of Grant: $6 per hour undergraduates; $7 per hour graduates
Contact: Human Resources Manager, (202) 332-9110; fax: (202) 265-4954; email: cspi@cspinet.org
Internet: http://www.cspinet.org/about/jobs_internship_2005_2006.html
Sponsor: Center for Science in the Public Interest
1875 Connecticut Avenue NW, Suite 300
Washington, DC 20009-5728

Center for the Critical Analysis of Contemporary Culture Faculty Fellowships 1592
The center, at Rutgers University, invites applications proposing studies of all kinds— empirical, analytical, theoretical, historical, literary, social, etc.—on the theme of the current year's seminar. Investigation of a range of topics connected to the overarching theme is encouraged. Each year the center brings together approximately 15 internal fellows (faculty and graduate students) and two external fellows from many of the established disciplines to study a single defined problem of large significance. The fellows will be expected to present their work in center seminars, to meet regularly with the center's internal fellows, and to teach one course on the subject in the department to which their work most closely relates. Applications must be requested in early December, with applications due in early January.
Amount of Grant: $40,000
Date(s) Application is Due: Jan 7
Contact: External Fellowship Program, Center for the Critical Analysis of Contemporary Culture, (732) 932-8426; fax: (732) 932-8683; email: theccacc@aol.com
Internet: http://www.criticalanalysis.rutgers.edu
Sponsor: Rutgers, The State University of New Jersey
8 Bishop Pl
New Brunswick, NJ 08903

Center for the Evangelical United Brethren Heritage Audrie E. Reber 1593
Memorial Award
The Center for the Evangelical United Brethren Heritage is offering an annual award to an undergraduate, seminary, or graduate student enrolled in an institution of higher education who submits the best paper on some aspect of Evangelical United Brethren Church history, theology or church life. The Evangelical United Brethren heritage includes the history of its predecessor bodies, the Evangelical Church and the Church of the United Brethren in Christ, as well as the wider Pietist heritage from which these bodies derive. The award is in memory of Audrie E. Reber, former missionary, author and active church-woman who was a Founding Member of the Center and a Member of its Advisory Board. The winning paper may be published in Methodist History, and a cash award of $700 will be given.
Requirements: Three clean copies of each paper must be submitted accompanied by a letter of nomination from the instructor under whose supervision it was written. The paper must not be longer than 20 double-spaced typewritten or computer processed pages. It must be properly footnoted and bear evidence of thorough and reliable handling of sources using the annotation standards accepted at the school. A bibliography also must be included. By April 1st each year, all competing papers must be in the hands of the Director of The Center for the Evangelical United Brethren Heritage, United Theological Seminary, 4501 Denlinger Road, Dayton, OH 45426.
Restrictions: The decision of the judges is final. No award will be made if the judges determine that no paper submitted merits the award.
Amount of Grant: $700
Date(s) Application is Due: Apr 1

Contact: General Secretary, General Commission on Archives and History; (973) 408-3189; fax: (973) 408-3909; email: research@gcah.org
Internet: http://www.gcah.org/site/c.ghKJI0PHIoE/b.4619069/k.552C/EUB_Audrie_E_Reber_Award.htm
Sponsor: United Methodist Church
475 Riverside Drive
New York, NY 10115

Center for the Humanities Andrew W. Mellon Postdoctoral Fellowship   1594

The purpose of this fellowship is to provide scholars who have lately completed their PhDs with free time to further their own work in a cross-disciplinary setting, and to associate them with a distinguished faculty. Current themes are The Natural, the Supernatural, and the Unnatural (fall); and Domesticity, Past, Present, and Future (spring). At least one and possibly two fellows will be appointed to the Center for the Humanities for the entire academic year, and each fellow will be awarded a stipend. He or she will teach a one-semester undergraduate course; participate in the collegial life of the Center for the Humanities, which sponsors conferences, lectures, and colloquia; and give one public lecture. The fellow will be provided with an office at the Center for the Humanities, and will be expected, during the term of the fellowship, to work there on weekdays while the university is in session, and also to reside in Middletown. Guidelines are available online.
Requirements: Scholars who have received their PhD degree after June 2003 in any field of inquiry in the humanities or humanistic social sciences'broadly conceived'are invited to apply.
Amount of Grant: $45,000 stipend
Date(s) Application is Due: Nov 10
Contact: Dr. Henry Abelove, Director
Internet: http://www.wesleyan.edu/chum/mellon.html
Sponsor: Wesleyan University
95 Pearl Street
Middletown, CT 06459

Center for the Humanities Mellon Postdoctoral Fellowship   1595

The college's Center for the Humanities offers a two-year Mellon postdoctoral fellowship to candidates who have broad training in the humanities, which might include disciplines such as comparative literature or rhetoric, or interdisciplinary fields such as women's studies. The successful applicant will teach one course per semester and will be expected to participate in the annual fall faculty seminar, directed by the center's distinguished visiting professor. In addition to a salary, which is comparable to that of a new assistant professor, fellowships include benefits and support for research and travel. Fellows will be matched with an experienced faculty mentor.
Requirements: Candidates should have received the PhD within the past three years.
Amount of Grant: $45,000 stipend
Date(s) Application is Due: Nov 11
Contact: Alan Schrift, Director, (641) 269-3161; email: schrift@grinnell.edu
Internet: http://www.grinnell.edu/offices/dean/supfac/oncampusopportunities/mellon/postdoc
Sponsor: Grinnell College
Grinnell, IA 50112-1690

Center for the Study of Books and Media Visiting Research Fellowships   1596

The Center for the Study of Books and Media at the university invites applications for a visiting research fellowship in any field related to the history of books. The fellow is expected to pursue his or her research at Princeton throughout the academic year and is encouraged to participate in the activities of the center. The stipend depends on qualifications.
Requirements: Candidates should have completed a dissertation in book history within the last three years and should propose a project involving research on subjects such as publishing, the book trade, and reading.
Date(s) Application is Due: Mar 1
Contact: John Hintermaier, (609) 258-8295; email: johnmh@princeton.edu; Thierry Rigogne, email: trigogne@princeton.edu
Internet: http://web.princeton.edu/sites/english/csbm/postdoc_fellowship.htm
Sponsor: Princeton University
129 Dickinson Hall
Princeton, NJ 08544

Center for the Study of Philanthropy International Fellowships   1597

The center provides leadership training through applied research and professional mentorships for young scholar-practitioners in the nonprofit sector. The program offers leadership training to scholar-practitioners outside the U.S. as a means to help build Third-Sector capacity in the fellows' home countries. Specific topical areas are chosen each year. Special attention will also be given to diaspora philanthropy. Fellows participate in a three-month seminar on the U.S. and international voluntary sectors. Fellows are expected to produce a 30-50 page paper on their findings to be presented in the seminar. They will learn about the work of key agencies and meet with foundation and nonprofit representatives. They may also have the opportunity to attend selected conferences. Fellowship includes tuition and monthly stipend, housing, and round-trip air travel to and from the U.S..
Requirements: The program is open to practitioners and researchers under the age of 36 who are citizens of countries other than the U.S.. Applicants must have a college or university degree and speak and write English fluently. Preference will be given to candidates with a strong institutional base and demonstrated research skills.

Amount of Grant: $1300 monthly stipend
Date(s) Application is Due: Sep 13
Contact: Dr. Kathleen McCarthy, Director, (212) 817-2010; fax: (212) 817-1572; email: info@philanthropy.org
Internet: http://www.philanthropy.org/programs/intnl_fellows_program.html
Sponsor: City University of New York
365 Fifth Avenue, Room 5401
New York, NY 10016-4309

Center for the Study of Philanthropy Senior International Fellowships   1598

The center provides leadership training through applied research and professional mentorships for young scholar-practitioners in the nonprofit sector. The program offers leadership training to scholar-practitioners outside the United States as a means to help build Third-Sector capacity in the fellows' home countries. Specific topical areas are chosen each year. The current year's theme is community foundations. Fellows participate in a three-month seminar on the U.S. and international voluntary sectors. Fellows attend weekly seminars, learn about the work of key agencies and foundations, meet with nonprofit representatives, and study U.S. and international community foundation models. Fellowship includes tuition and monthly stipend, housing, and round-trip air travel to and from the United States.
Requirements: The program is open to senior-level practitioners and researchers under the age of 36 who are citizens of countries other than the United States. Applicants must have a college or university degee and speak and write English fluently. Preference will be given to candidates with a strong institutional base and demonstrated research skills.
Amount of Grant: $1300 monthly stipend
Date(s) Application is Due: Jun 15; Oct 15
Contact: Senior International Fellows Program, Graduate School and University Center, fax: (212) 817-1572; email: csp@gc.cuny.edu
Internet: http://www.philanthropy.org/programs/intnl_fellows_program.html
Sponsor: City University of New York
365 Fifth Avenue, Room 5116
New York, NY 10016-4309

Center for the Study of Professional Military Ethics Resident Fellowship   1599

The goal of this full-time, in-residence program is to help prepare select military officers, career civil servants, academics, and others to teach, apply, and practice ethics in a variety of professions and institutional settings. Each fellow will be required to complete a research project and to participate in a weekly fellows seminar; be encouraged to participate in other activities and programs of the center; and receive office space, computer facilities, library privileges, and a stipend. Applicants should send a letter describing their interests in ethics in the areas listed; a proposal for a research project to be undertaken during the fellowship; a curriculum vita; and copies of relevant publications. Two letters of reference also should be sent directly to the center. Recipients are announced by the end of April.
Requirements: Ideal candidates from academia will have a background in ethics, international affairs, security studies, and/or public policy, with an interest in pursuing issues of ethics in military affairs, national defense, and international security in their teaching and research.
Amount of Grant: $35,000 maximum
Date(s) Application is Due: Oct 14
Contact: Dr. Albert Pierce, Director, Center for the Study of Professional Military Ethics, (410) 293-6057; email: acpierce@gwmail.usna.edu
Internet: http://www.usna.edu/Ethics/Programs/Residentfellowship/fellows.htm
Sponsor: U.S. Naval Academy
112 Cooper Road
Annapolis, MD 21402

Center for the Study of Religion Fellowships   1600

The Center brings a select number of pre-tenure scholars and recent Ph.D. graduates to Princeton University to study religion and religious history. The fellows, who are appointed by the Dean of the Faculty, devote time to servicing the intellectual life of the Center and the University through mentoring graduate and undergraduate students and participating in one of the Center's weekly interdisciplinary seminar. For 2007-2008, fellows will be appointed in two areas: Christian Thought and Practice, special emphasis on the religious life of American Christians, congregations, or clergy; and Public Theology: special emphasis on contemporary issues bridging theology and the social sciences, such as just war theory, pluralism, or social welfare.
Requirements: To apply, submit CV, 3 letters of recommendation, a 5 page proposal, and published or unpublished paper demonstrating scholarly command of topic of one of the three themes. Applications available on Web site.
Restrictions: Princeton University PhDs are not eligible.
Samples: 2006-2007 Visiting Fellows include: Rebecca L. Davis, who received her Ph.D. in History at Yale University in 2006 and will join the history department at the University of Delaware as an assistant professor, specializing in the histories of sexuality, immigration, and ethnicity in the United States; K. Healon Gaston, who will receive her Ph.D. in History from the University of California at Berkeley in 2006, and whose research focuses on the role of religion in American public life, with particular emphasis on the relationship between theology and democratic theory. She is currently working on a manuscript, entitled America, Judeo-Christian Moment: Religion, Secularism, and the Cold War Redefinition of Democracy, 1938-1973.
Date(s) Application is Due: Jan 5
Contact: Anita Klein, Director, The Center for the Study of Religion, (609) 258-5545; fax: (609) 258-6940

Internet: http://www.princeton.edu/~csrelig/opportunities/undergrad_ops.html
Sponsor: Princeton University
5 Ivy Lane
Princeton, NJ 08544-1013

Center of Excellence (COE) in General Aviation Grants 1601

The FAA has identified a need for a Center of Excellence in general aviation. The center will conduct research, which includes the entire spectrum (i.e., basic research through engineering development, prototyping, and testing) within the scope of general aviation. This scope includes, but is not limited to airport technology; propulsion and structures; aging aircraft; flight safety; fire safety; and training. The FAA intends to provide long-term funding to establish and operate a prestigious partnership with academia, industry, and government. To this end, the FAA encourages offerors to team with organizations that complement their expertise from academia, industry, state/local government, and other governmental agencies. The successful applicant is required to match FAA grant funds with nonfederal funding over the term of the cooperative agreement. Cost sharing also is required for any orders placed under the IDIQ contract.
Requirements: Colleges and universities are eligible to apply.
Restrictions: Individuals are not eligible for a COE designation and do not qualify for grants under this program.
Amount of Grant: $300,000 minimun per year
Contact: Pat Watts, Grants Manager, (609) 485-5043; email: patricia.watts@faa.gov
Internet: http://www.cgar.org/about_program.asp
Sponsor: Department of Transportation
William J Hughes Technical Ctr
Atlantic City International Airport, NJ 08405

Center on Philanthropy and Civil Society's Emerging Leaders International 1602
   Fellows Program

The program provides leadership training through applied research and professional mentorships for young scholar-practitioners in the nonprofit sector. The program is open to scholars and practitioners interested in building Third-Sector capacity in the United States and overseas. This year???s fellows will be selected from abroad and also from communities of color under-represented in the United States grantmaking sector. Fellows are based at the Graduate Center of The City University of New York, where they design and pursue an individualized research project and participate in a seminar with Third-Sector leaders. Specific topical areas are chosen each year.
Requirements: The program is open to practitioners and researchers under the age of 36. Applicants must hold a college or university degree and speak and write English fluently. Preference will be given to candidates with strong ties to a Third-Sector institution and demonstrated research skills.
Amount of Grant: $1300-per month stipend plus tuition
Date(s) Application is Due: Dec 7
Contact: Dr. Kathleen McCarthy, Director , (212) 817-2010; fax: (212) 817-1572; email: csp@gc.cuny.edu
Internet: http://www.philanthropy.org/programs/ifp/application.html
Sponsor: City University of New York
365 Fifth Avenue, Suite 5401
New York, NY 10016-4309

Central European University Fellowships 1603

The university offers full fellowships for master's, doctoral, and doctoral support programs in Budapest, Hungary and Warsaw, Poland. 625 fellowships covering tuition and living expenses are available to individuals from Central and Eastern Europe and the former Soviet Union. A wide variety of other financial aid programs are available to all students on a competitive basis.
Date(s) Application is Due: Jan 15
Contact: Grants Administrator, (36-1) 327-3000 ext 2311; email: finaid@ceu.hu
Internet: http://www.ceu.hu/financial_forms.html
Sponsor: Central European University
Nador u 9
Budapest 1051 Hungary

Century Foundation Grants 1604

The foundation sponsors and supervises research on significant economic, social, and political issues. The foundation's research projects usually are designed to produce analytic, book-length manuscripts containing public policy recommendations. They are aimed at an audience that includes the informed public, the press, policymakers, and the academic community. The foundation commissions individuals (project directors) to carry out the projects, and the book-length manuscripts that result are edited and then published by the foundation's press or placed with commercial publishers or university presses. Proposals are welcome for reports and books. Preliminary proposals should summarize in no more than three pages the central argument of the proposed publication, the main themes that would be developed in support of the argument, and examples of evidence that would be explored in making the project director's case.
Restrictions: The foundation does not award scholarships or support dissertation research, nor does it make grants to individuals or institutions.
Contact: Carl Robichaud, Program Officer, (212) 452-7712; fax: (212)535-9803; email: robichaud@tcf.org
Internet: http://www.tcf.org/about.asp
Sponsor: Century Foundation
41 East 70th Street
New York, NY 10021

CES Predissertation Fellowships 1605

The fellowships enable graduate students in the social science disciplines to pursue two to three months of exploratory research in Europe to determine the viability and to better define the scope of their proposed dissertation projects. The geographic scope for these fellowships includes Western, Central and Eastern Europe, including Russia and Turkey as they relate to Europe. Recipients will test the research design of their dissertation, determine the availability of archival materials, and contact European scholars in the relevant field.
Requirements: Applicants must be doctoral candidates at an American or Canadian university in anthropology (excluding archaeology), economics, history (post-1750), geography, political science, sociology, social psychology, or urban planning; must be citizens or permanent residents of the United States or citizens or landed immigrants of Canada; and should have completed at least two years of full-time graduate study prior to the beginning date of their proposed research.
Restrictions: Funds may be expended only for research in Europe on the project detailed in the applicant's proposal to the council. Students who are advanced in their dissertation research or whose dissertation prospectuses have received formal approval from their academic departments are ineligible, as are students already in Europe at the time of application.
Amount of Grant: $4000 for travel and living expenses
Date(s) Application is Due: Feb 1
Contact: Fellowship Administrator, (212) 854-4172; fax: (212) 854-8808; email: ces@columbia.edu
Internet: http://www.columbia.edu/cu/ces/frames/overall.html
Sponsor: Council for European Studies
420 W 118th Street, 1203 International Affairs Building, Columbia University, MC 3310
New York, NY 10027

CES Travel Subsidies for European Scholars 1606

Under this program, the council subsidizes the travel of European social scientists who are temporarily in the United States and who have been invited to lecture at a university or college that is an institutional member of the council. Invitations are initiated by the council's institutional representative at the member university or by the faculty member responsible for arranging the European scholar's visit. The council should be contacted for the application form once details of the visit are completed. Except in unusual circumstances, the European scholar's visit should be scheduled during the academic year. The visit should include classroom or university-wide lectures and/or participation in graduate seminars. The council reimburses the European scholar's transportation expenses within North America; the balance of travel costs, if any, as well as lodging, meals, and honoraria, are to be provided by the host university. An application form is available on the Web site.
Amount of Grant: $300 maximum per visit
Contact: Executive Director, (212) 854-4172; fax: (212) 854-8808; email: ces@columbia.edu
Internet: http://www.columbia.edu/cu/ces/frames/overall.html
Sponsor: Council for European Studies
420 W 118th Street, 1203 International Affairs Building, MC 3310
New York, NY 10027

Cesar Chavez Dissertation Fellowship for U.S. Latina or Latino Scholars 1607

The fellowship will support a U.S. Latina or Latino scholar for a year-long residency (September 1 through August 31) at Dartmouth College. The fellowship offers an opportunity for scholars who plan a career in college teaching and who have completed all other PhD requirements to finish the dissertation with access to the outstanding library, computing facilities, and Dartmouth faculty members. In addition, the fellow may participate in classroom activities with scholars who are dedicated to undergraduate teaching. Each fellow will be affiliated with a department or program at the college. The fellow will be expected to complete the dissertation during the tenure of the fellowship, and will have the opportunity to participate in teaching, either as a primary instructor or as part of a team. Office space and library privileges are provided.
Requirements: The fellow may be taking the PhD degree in any discipline or area taught in the Dartmouth undergraduate arts and sciences curriculum.
Amount of Grant: $25,000 plus office space, library privileges, and a $2500 research assistance fund
Date(s) Application is Due: Feb 1
Contact: Sandy Spiegel, Director of Grad Recruiting and Diversity, (603) 646-6578; fax: (603) 646-3488; email: Sandra.J.Spiegel@Dartmouth.edu
Internet: http://www.dartmouth.edu/~gradstdy/funding/fellowships/cem.html
Sponsor: Dartmouth College
6062 Wentworth Hall, Room 304
Hanover, NH 03755

CF Clinical Research Trainee Awards 1608

This is a training program for physicians enrolled in a U.S. or Canadian subspecialty training program. Ten awards will be granted for clinical research in thrombosis, asthma, critical care, lung cancer, and women's health. Applicants may apply for only one type of clinical research award, and each program director may support only one applicant for an award in each area. Requests should be written to include goals and objectives that may be accomplished within one year of the starting date (July 1) of the research project. Research funding may be used as the research mentor and program director deem appropriate (e.g., salary support and supplies directly related to the research project).

Requirements: Applicants must be citizens of the United States or Canada or have a J-1 or H-1 visa and be working in a certified U.S. or Canadian institution and enrolled in a subspecialty training program at the time of application; hold an MD degree or its equivalent; be enrolled in a U.S. or Canadian subspecialty training program in allergy/immunology, cardiac electrophysiology, critical care anesthesiology, critical care intensive care, critical care medicine, infectious disease, cardiology, pediatric critical care, pediatric pulmonary disease, pulmonary disease, surgical critical care, or thoracic surgery; be affiliate members of the American College of Chest Physicians (contact Angela Popa-Padilla at 847/ 498-1400 for an application); show promise of significant contribution to his/her specialty; and apply for only one type of award and be the sole applicant from the subspecialty program to apply for that type of award.
Restrictions: Awards are to support clinical research and not basic or bench level research.
Amount of Grant: $10,000
Date(s) Application is Due: Jun 30
Contact: Sue Ciezadlo, Grants Administrator, (847) 498-8363; fax: (847) 498-5460; email: sciezadlo@chestnet.org
Internet: http://www.chestfoundation.org/researchAwards/index.php
Sponsor: Chest Foundation
3300 Dundee Road
Northbrook, IL 60062-2348

CF Roger C. Bone Award for Advances in End-of-Life Care 1609
The award was created to annually recognize members of the American College of Chest Physicians who have demonstrated leadership in end-of-life care. Recipients of this award have shown substantial improvement in the care of patients as documented in the related sections of the application form. Application forms and guidelines are available online.
Requirements: Applicants must be members of the American College of Chest Physicians and have the endorsement of another ACCP member.
Amount of Grant: $10,000
Date(s) Application is Due: Apr 29
Contact: Sue Ciezadlo, Senior Project Coordinator, (847) 498-8363; fax: (847) 498-5460; email: sciezadlo@chestnet.org
Internet: http://www.chestfoundation.org/criticalCare/index.php
Sponsor: Chest Foundation
3300 Dundee Road
Northbrook, IL 60062-2348

CFF Clinical Research Grants 1610
These awards offer support of clinical research projects directly related to cystic fibrosis. Projects may address diagnostic or therapeutic methods related to cystic fibrosis or the pathophysiology of cystic fibrosis. However, studies performed in animal models or cell/tissues removed from patients will not be supported under this award. Applicants must demonstrate access to a sufficient number of cystic fibrosis patients and appropriate controls. A letter of intent to apply must be submitted by June 1 with applications required by December 1.
Amount of Grant: $80,000 maximum per year for single center; $150,000 maximum for multicenter
Date(s) Application is Due: Jun 1; Dec 1
Contact: Office of Grants Management, (800) 344-4823 or (301) 951-4422; fax: (301) 951-6378; email: grants@cff.org
Internet: http://www.cff.org/research/cystic_fibrosis_foundation_grants/research_grants
Sponsor: Cystic Fibrosis Foundation
6931 Arlington Road, 2nd Fl
Bethesda, MD 20814

CFF First- and Second-Year Clinical Fellowships 1611
The foundation offers competitive clinical fellowships for up to three years for physicians interested in cystic fibrosis and other chronic pulmonary and gastrointestinal diseases of children and adolescents/adults. The intent of this award is to encourage specialized training early in a physician's career and to prepare well-qualified candidates for careers in academic medicine. Training must take place in one of the foundation's accredited centers and must provide thorough grounding in diagnostic and therapeutic procedures, comprehensive care, and cystic fibrosis-related research. All first- and second-year programs must commit a significant portion (at least 30 percent over two years) to research training. Fellows funded by other sources for their first year of training may apply to the foundation for subsequent training support. Third-year fellowships are available for additional basic and/or clinical research training. Research fellowships are also available from the foundation to support physicians beyond the fellowship experience; salary will be commensurate with experience.
Requirements: Applicants must be U.S. citizens or have permanent U.S. resident visas, and must have completed pediatric training and be eligible for board certification in pediatrics by the time the fellowship begins. Candidates with prior training in internal medicine may also apply but must have completed at least two years of an approved adult pulmonary or GI fellowship; be jointly sponsored by the departments of medicine and pediatrics; and be willing to commit at least 75 percent of their time to cystic fibrosis and related problems of young adults.
Amount of Grant: $42,000 stipends first year, $43,750 second year
Date(s) Application is Due: Oct 1
Contact: Office of Grants Management, (800) 344-4823 or (301) 951-4422; fax: (301) 951-6378; email: grants@cff.org

Internet: http://www.cff.org/research/cystic_fibrosis_foundation_grants/training_grants
Sponsor: Cystic Fibrosis Foundation
6931 Arlington Road, 2nd Fl
Bethesda, MD 20814

CFF Leroy Matthews Physician-Scientist Awards 1612
This award will provide up to six years of support for outstanding newly trained pediatricians and internists (MDs and MD/PhDs) to complete subspecialty training, develop into independent investigators, and initiate research programs. Institutional and individual grants are available.
Requirements: U.S. citizenship or permanent resident status is required.
Amount of Grant: $42,000 stipend plus $10,000 for research and development for years one through five; up to $70,000 stipend plus $15,000 for research and development for year six
Date(s) Application is Due: Sep 1
Contact: Office of Grants Management, (800) 344-4823 or (301) 951-4422; fax: (301) 951-6378; email: grants@cff.org
Internet: http://www.cff.org/research/cystic_fibrosis_foundation_grants/research_grants
Sponsor: Cystic Fibrosis Foundation
6931 Arlington Road, 2nd Fl
Bethesda, MD 20814

CFF Pilot and Feasibility Awards 1613
These grants are for developing and testing new hypotheses and/or new methods, and to support promising new investigators as they establish themselves in research areas relevant to cystic fibrosis. Proposed work must be hypothesis driven and must reflect innovative approaches to critical questions in cystic fibrosis research. Two years of support may be requested.
Restrictions: The award is not meant to support continuation of programs begun under other granting mechanisms. Electronic submission is required.
Amount of Grant: $40,000 per year maximum
Date(s) Application is Due: Sep 1
Contact: Office of Grants Management, (800) 344-4823 or (301) 951-4422; fax: (301) 951-6378; email: grants@cff.org
Internet: http://www.cff.org/research/cystic_fibrosis_foundation_grants/research_grants
Sponsor: Cystic Fibrosis Foundation
6931 Arlington Road, 2nd Fl
Bethesda, MD 20814

CFF Postdoctoral Research Fellowships 1614
Competitive postdoctoral fellowships in basic or clinical research are made on an annual basis, renewable for a second year. The fellowships are awarded in areas of basic cellular and metabolic research or in other problem areas pertinent to cystic fibrosis and related chronic and recurrent pulmonary and gastrointestinal diseases of childhood. Preference is shown to recent graduates or those just beginning their investigative careers. A third year is offered on a limited basis to highly qualified candidates. Interested individuals are encouraged to contact the foundation for application procedures and/or to discuss the potential relevance of their work to the objectives of the foundation.
Requirements: These awards are offered to MDs, PhDs, and MD/PhDs interested in conducting basic or clinical research related to cystic fibrosis.
Restrictions: Electronic submission only.
Amount of Grant: $33,000 first year, $34,100 second year, $36,300 optional third year
Date(s) Application is Due: Sep 1
Contact: Office of Grants Management, (800) 344-4823 or (301) 951-4422; fax: (301) 951-6378; email: kcurley@cff.org
Internet: http://www.cff.org/research/cystic_fibrosis_foundation_grants/training_grants
Sponsor: Cystic Fibrosis Foundation
6931 Arlington Road, 2nd Fl
Bethesda, MD 20814

CFF Research Grants 1615
Grants are offered in support of high-quality research projects ranging from basic cellular and metabolic mechanisms to therapy of cystic fibrosis and related chronic and recurrent pulmonary and gastrointestinal diseases of childhood. These grants are broadly oriented to the support of projects for developing and initially testing new hypotheses and/or new methods, or those being applied to problems of cystic fibrosis for the first time. The intent of this award is to enable the investigator to collect sufficient data to compete successfully for long-term support from NIH. Interested individuals are encouraged to contact the foundation for guidelines and/or to discuss the potential relevance of their work to the objectives of this program.
Requirements: Applications may be submitted by established investigators or new investigators starting their independent research careers but may not represent support for continuation of a line of research already established by the applicant.
Amount of Grant: $90,000 maximum per year for a two-year period
Date(s) Application is Due: Sep 1
Contact: Office of Grants Management, (800) 344-4823 or (301) 951-4422; fax: (301) 951-6378; email: grants@cff.org
Internet: http://www.cff.org/research/cystic_fibrosis_foundation_grants/research_grants

Sponsor: Cystic Fibrosis Foundation
6931 Arlington Road, 2nd Fl
Bethesda, MD 20814

CFF Student Traineeships   1616

The traineeships are offered to qualified students to introduce them to cystic fibrosis research. Each recipient must work with a faculty sponsor on a research project related to cystic fibrosis. Interested individuals are encouraged to contact the foundation for application procedures and/or to discuss the potential relevance of their work to the objectives of the program. Applications are accepted throughout the year. A maximum of $300 of the award may be used for laboratory expenses with the remainder used as a stipend for the trainee.
Requirements: Applicants must be students in or about to enter a doctoral program (MD, PhD, or MD/PhD); senior-level undergraduates planning to pursue graduate training may also apply.
Amount of Grant: $1500
Contact: Office of Grants Management, (800) 344-4823 or (301) 951-4422; fax: (301) 951-6378; email: grants@cff.org
Internet: http://www.cff.org/research/cystic_fibrosis_foundation_grants/training_grants
Sponsor: Cystic Fibrosis Foundation
6931 Arlington Road, 2nd Fl
Bethesda, MD 20814

CFF Therapeutics Development Program   1617

The purpose of this program is to provide funds to businesses that will develop commercial products to benefit individuals with cystic fibrosis. Structured as a matching grants program, funds will be awarded only if they are matched by the recipient. First, the grantee will examine the scientific potential of new products under component 1. Component 2 studies will involve support for the continuation of component 1 developments and the initiation of patient clinical studies. Component 1 and component 2 awards each will receive a maximum of two years of support. Funds will be provided for research prior to, but not including, phase 3 clinical trials. Applications are accepted on an ongoing basis.
Amount of Grant: $100,000 maximum per year for component I; up to $750,000 per year for component II
Contact: Office of Grants Management, (800) 344-4823 or (301) 951-4422; fax: (301) 951-6378; email: grants@cff.org
Internet: http://www.cff.org/research/cystic_fibrosis_foundation_therapeutics
Sponsor: Cystic Fibrosis Foundation
6931 Arlington Road, 2nd Fl
Bethesda, MD 20814

CFF Third-Year Clinical Fellowships   1618

The foundation offers competitive clinical fellowships for up to three years for physicians interested in cystic fibrosis and other chronic pulmonary and gastrointestinal diseases of children and adolescents/adults to encourage specialized training early in a physician's career and to prepare well-qualified candidates for careers in academic medicine. The third-year fellowship award offers support for additional intense basic and/or clinical research training related to cystic fibrosis.
Requirements: Applicants and sponsors must submit proposals of the research studies to be undertaken and other specialized training that will be offered during this year. Preference will be given to applicants whose training was supported by the foundation.
Restrictions: Recipients who do not enter a career of academic medicine will be subject to payback provisions.
Amount of Grant: $62,600 maximum; $52,600 stipend for fellow and up to $10,000 for research costs
Date(s) Application is Due: Oct 1
Contact: Office of Grants Management, (800) 344-4823 or (301) 951-4422; fax: (301) 951-6378; email: grants@cff.org
Internet: http://www.cff.org/research/cystic_fibrosis_foundation_grants/training_grants
Sponsor: Cystic Fibrosis Foundation
6931 Arlington Road, 2nd Fl
Bethesda, MD 20814

CFF-NIH Funding Grants   1619

The objective of this award is to support excellent cystic fibrosis-related research projects that have been submitted to and approved by NIH but cannot be supported by available NIH funds. Applications must fall within the upper 40th percentile with a priority score of 200 or better. Investigators will be required to resubmit applications to NIH. Interested individuals are encouraged to contact the foundation for application procedures and/or to discuss the potential relevance of their work to the objectives of this program. Applications may be submitted on an ongoing basis.
Amount of Grant: $75,000-$125,000 per year for up to two years
Contact: Office of Grants Management, (800) 344-4823 or (301) 951-4422; fax: (301) 951-6378; email: grants@cff.org
Internet: http://www.cff.org/research
Sponsor: Cystic Fibrosis Foundation
6931 Arlington Road, 2nd Fl
Bethesda, MD 20814

CFPC Family Medicine Resident Leadership Awards   1620

The CFPC Research and Education Foundation honors outstanding educational and research initiatives in the area of family medicine through conferral of Family Medicine Resident Awards. Canadian Research Awards for Family Medicine Residents provide national recognition for original research carried out by postgraduate trainees. The Murray Stalker Memorial Lecture Award is given to promote and recognize scholarly activities of family medicine residents. The Nadine St. Pierre Award recognizes the outstanding achievement of a francophone family practice resident. Award recipients receive return air fare, complimentary registration for the Section of Teachers Annual Workshop, and a stipend for expenses. Deadline dates vary between programs; contact program staff for exact dates.
Amount of Grant: $C250-$C1000
Date(s) Application is Due: Jun 1
Contact: Kiki Ziten, Coordinator, (905) 629-0900 ext 432; fax: (905) 629-0893; email: kziten@cfpc.ca
Internet: http://www.cfpc.ca/English/cfpc/programs/awards%20program/default.asp?s=1
Sponsor: College of Family Physicians of Canada
2630 Skymark Avenue
Mississauga, ON L4W 5A4 Canada

CFPC Family Physician of the Year Award   1621

The annual award recognizes an individual for outstanding contribution to family medicine. Criteria for the award include excellence in family practice over time, community medical service, contribution to the community, and involvement with CFPC. The award is presented at the time of the CFPC Convocation in conjunction with the National Annual Scientific Assembly in Halifax, Nova Scotia. The recipient will present a short addresss at the time of convocation and will be involved as a spokesperson for Canadian family medicine throughout the year following receipt of the award. The recipient will receive return air fare, complimentary registration, and a stipend for expenses. The deadline for submission varies according to each Provincial Chapter. Please contact the Provincial Chapter of your nominee to confirm the deadline date.
Requirements: Each provincial chapter may select a nominee for the award. Names must be submitted to the National Honours and Awards Committee.
Amount of Grant: $C2000
Contact: Kiki Ziten, Coordinator, Honours and Awards Program, (905) 629-0900 ext 432; fax: (905) 629-0893; email: kziten@cfpc.ca
Internet: http://www.cfpc.ca/English/cfpc/programs/awards%20program/members%20can%20apply%20for/default.asp?s=1#Perkinawards
Sponsor: College of Family Physicians of Canada
2630 Skymark Avenue
Mississauga, ON L4W 5A4 Canada

CFPC Fellowships   1622

The program awards fellowships to CFPC members to recognize achievement in the discipline of family medicine. The conferral of fellowships will take into consideration the nominee's exemplary performance as a family physician and exceptional contributions to community activities that enhance health in its broadest sense; outstanding leadership in the academic discipline of family medicine and major administrative responsibilities within a university department of family medicine; and significant elected or appointed executive positions at the national or chapter level of the CFPC or involvement in major committee or other college activities.
Requirements: Nominees must hold certification in family medicine, have a minimum of 10 years continuous membership, and be members in good standing at the time of nomination.
Date(s) Application is Due: Feb 15
Contact: Kiki Ziten, Coordinator, (905) 629-0900 ext 432; fax: (905) 629-0893; email: kziten@cfpc.ca
Internet: http://www.cfpc.ca/English/cfpc/research/section%20of%20researchers/home/default.asp?s=1
Sponsor: College of Family Physicians of Canada
2630 Skymark Avenue
Mississauga, ON L4W 5A4 Canada

CFPC Research and Education Foundation Grants   1623

The program is supported by the CFPC Research and Education Foundation and industry partners and comprises research grants, scholarships, and training programs for educational and research initiatives in the area of family medicine. The Royal Canadian Legion One-Month Traineeship grants are awarded to improve the quality of medical education and the provision of care to elderly patients. The Hollister King grant provides an opportunity for a practicing family physician to attain or enhance special skills in the delivery of rural medical care. The Douglas M. Robb research grant is given to a community-based member who plans to conduct research on a topic relevant to the practice of family medicine. Research/development grants provide seed money to members to initiate or complete research/development projects in family medicine. Practice Enrichment grants enable members in active practice to participate in a course of study under the direction of a Canadian university for a minimum of three months in fields related to family medicine. Traveling Scholarship grants enable members to pursue clinical studies for a minimum of three weeks under the direction of a clinical department at a Canadian university. Graduate Study grants enable members to pursue a course of study at a Canadian university for a period of at least six months on a part-time basis.
Amount of Grant: $C1000-$C5000

Contact: Kiki Ziten, Coordinator, (905) 629-0900 ext 432; fax: (905) 629-0893; email: kziten@cfpc.ca
Internet: http://www.cfpc.ca/English/cfpc/programs/awards%20program/awardshome%20pdf/default.asp?s=1
Sponsor: College of Family Physicians of Canada
2630 Skymark Avenue
Mississauga, ON L4W 5A4 Canada

CFUW Beverley Jackson Fellowship 1624
Founded in 1919, the Canadian Federation of University Women (CFUW) is a voluntary, non-profit, self-funded bilingual organization offering a variety of fellowships to women university graduates. This fellowship is awarded annually to assist a Canadian woman to pursue graduate work at an Ontario university. The applicant must have been accepted into her place of study at time of application. Guidelines and application forms are available from the federation in the spring.
Requirements: Applicant must be a woman over the age of 35 pursuing graduate work in an Ontario university, must hold at least a bachelor's degree or its equivalent from a recognized university, and must be a Canadian citizen or have held landed immigrant status for one year prior to submitting application.
Restrictions: Eligibility for these fellowships and awards is restricted to women.
Samples: Evelyn Rupert, York University—fellowship recipient, $C3000.; Marta Scythes, M.Sc.BMC (Biomedical Communications), University of Toronto, $C2,000 (2007-08).
Amount of Grant: $C2,000
Date(s) Application is Due: Nov 1
Contact: Betty Dunlop, Fellowships Program Manager; (613) 234-8252 or (613) -234-2732; fax: (613) 234-8221; email: cfuwfls@rogers.com
Internet: http://www.cfuw.org/index.php?option=com_content&task=view&id=73&Itemid=159
Sponsor: Canadian Federation of University Women
251 Bank Street, Suite 305
Ottawa, ON K2P 1X3 Canada

CFUW Bourse Georgette LeMoyne Award 1625
Founded in 1919, the Canadian Federation of University Women (CFUW) is a voluntary, non-profit, self-funded bilingual organization offering a variety of grants and fellowships to women university graduates. Grants awarded by the CFUW in this category are for graduate study at a university where one of the languages of administration and instruction is French. Application forms are available from the federation in the spring.
Requirements: At time of application, applicant must hold a bachelor's degree or its equivalent from a recognized university, must be a Canadian citizen or have held landed immigrant status for one year prior to submitting application, and must have been accepted at her place of study. Finally, the candidate must be studying in French.
Restrictions: Eligibility for these fellowships and awards is restricted to women.
Samples: Dr. Cheryl Harris, U of Ottawa (Canada)—award winner, $C6000 (2004); Annie Paquet, Ph.D. Psychology, University of Quebec, Montreal, $C7,000 (2007-08).
Amount of Grant: $C7,000
Date(s) Application is Due: Nov 1
Contact: Betty Dunlop, Fellowships Program Manager; (613) 234-8252 or (613) -234-2732; fax: (613) 234-8221; email: cfuwfls@rogers.com
Internet: http://www.cfuw.org/index.php?option=com_content&task=category&sectionid=7&id=85&Itemid=88,â©=eng
Sponsor: Canadian Federation of University Women
251 Bank Street, Suite 305
Ottawa, ON K2P 1X3 Canada

CFUW Canadian Home Economics Association (CHEA) Fellowship 1626
Founded in 1919, the Canadian Federation of University Women (CFUW) is a voluntary, non-profit, self-funded bilingual organization offering a variety of fellowships to women university graduates. The candidate for this Fellowship must be studying one or more aspects in the field of home economics, at the masters or doctoral level. Home economics is a field of study that promotes quality home and family life and includes human development, family relations, human sexuality, financial and resource management, consumerism, human ecology, food science, human nutrition, clothing and textiles, housing and shelter and aesthetics. The applicant must also be already accepted into or be enrolled in her postgraduate program in Canada.
Requirements: Candidates must be studying full-time at the master's or doctoral level in Canada or abroad.
Restrictions: Eligibility for these fellowships and awards is restricted to women.
Samples: Laura Templeton, Ph.D. Sociology, University of Alberta, $C6,000 (2007-08).
Amount of Grant: $C6,000
Date(s) Application is Due: Nov 1
Contact: Betty Dunlop, Fellowships Program Manager; (613) 234-8252 or (613) -234-2732; fax: (613) 234-8221; email: cfuwfls@rogers.com
Internet: http://www.cfuw.org/index.php?option=com_content&task=view&id=76&Itemid=159
Sponsor: Canadian Federation of University Women
251 Bank Street, Suite 305
Ottawa, ON K2P 1X3 Canada

CFUW Dr. Alice E. Wilson Awards 1627
These awards are intended to assist women to do refresher work in their chosen field, to do specialized study, or to retrain in new techniques applicable to their field. Special consideration is given to candidates returning to studies after at least three years absence. Work may be undertaken at a Canadian institution or elsewhere. For guidelines and application forms, contact the federation in the spring.
Requirements: Applicant must have a bachelor's degree or equivalent from a recognized university, must be a Canadian citizen or have held landed immigrant status for one year prior to submitting application and documentation, and must have been accepted at her place of study.
Restrictions: Eligibility for these fellowships and awards is restricted to women.
Samples: Helen Baulch, Ph.D. Science, Watershed Ecosystems, Trent University, $C6,000 (2007-08); Gloria Desantis, Ph.D. Canadian Plains Studies (Interdisciplinary Studies), University of Regina, $C6,000 (2007-08); Francoise Papillon, D.M.A. (Doctor of Musical Arts) Piano Performance, University of Washington, Seattle, $C6,000 (2007-08).
Amount of Grant: $C6,000
Date(s) Application is Due: Nov 1
Contact: Betty Dunlop, Fellowships Program Manager; (613) 234-8252 or (613) -234-2732; fax: (613) 234-8221; email: cfuwfls@rogers.com
Internet: http://www.cfuw.org/index.php?option=com_content&task=view&id=75&Itemid=159
Sponsor: Canadian Federation of University Women
251 Bank Street, Suite 305
Ottawa, ON K2P 1X3 Canada

CFUW Ecole Polytechnique Commemorative Award 1628
Founded in 1919, the Canadian Federation of University Women (CFUW) is a voluntary, non-profit, self-funded bilingual organization offering a variety of awards and fellowships to women university graduates. This award is made annually for graduate studies in any field, with special consideration given in the study of issues related particularly to women. Application forms are available in August.
Requirements: At the time of application, an applicant must hold at least a bachelor's degree or equivalent from a recognized university, have been accepted into the proposed place of study, and be a Canadian citizen or have held landed immigrant status for at least one year.
Restrictions: Eligibility for this award is restricted to women.
Samples: Keri Flesaker, Ph.D. Counseling Psychology, University of Alberta, $C7,000 (2007-08).
Amount of Grant: $C7,000
Date(s) Application is Due: Nov 1
Contact: Betty Dunlop, Fellowships Program Manager; (613) 234-8252 or (613) -234-2732; fax: (613) 234-8221; email: cfuwfls@rogers.com
Internet: http://www.cfuw.org/index.php?option=com_content&task=view&id=68&Itemid=159
Sponsor: Canadian Federation of University Women
251 Bank Street, Suite 305
Ottawa, ON K2P 1X3 Canada

CFUW Margaret Dale Philp Award 1629
This award, donated by the Kitchener-Waterloo Club of CFUW, is open to any woman who holds, at the time of application, a bachelor's degree or equivalent from a recognized university and who wishes to enter, or continue, a program leading to an advanced degree in the field of humanities or social sciences. Special consideration will be given to candidates who wish to specialize in Canadian history. Guidelines and application forms are available from the federation in August. These grants are not awarded every academic year. Check with the office or Web site for current offerings.
Requirements: Applicant must: reside in Canada; and must be a Canadian citizen or have held landed immigrant status for one year prior to submitting application and documentation, must reside in Canada, and must have been accepted at her place of study.
Restrictions: Eligibility for these fellowships and awards is restricted to women.
Samples: Carolan Wood, Ph.D. Biological and Forensic Anthropology, University of Toronto, $C3,000 (2007-08).
Amount of Grant: $C3,000
Date(s) Application is Due: Nov 1
Contact: Betty Dunlop, Fellowships Program Manager; (613) 234-8252 or (613) -234-2732; fax: (613) 234-8221; email: cfuwfls@rogers.com
Internet: http://www.cfuw.org/index.php?option=com_content&task=view&id=78&Itemid=159
Sponsor: Canadian Federation of University Women
251 Bank Street, Suite 305
Ottawa, ON K2P 1X3 Canada

CFUW Margaret McWilliams Predoctoral Fellowship 1630
This predoctoral fellowship is open to any woman who is a full-time student and has completed at least one full calendar year in doctoral studies at the time of application. An applicant may be studying abroad when applying, and the fellowship may be used in Canada or elsewhere. Contact the federation for guidelines and application forms in the spring.
Requirements: Applicant must be a Canadian citizen or have held landed immigrant status for one year prior to submitting application and documentation and must be enrolled in a master's degree program.

Samples: Nicole Dinaut, University of Toronto, Canada, $C10,000.; Amy Quark, University of Wisconsin-Madison, $C13,000 (2007-08).
Amount of Grant: $13,000
Date(s) Application is Due: Nov 1
Contact: Betty Dunlop, Fellowships Program Manager; (613) 234-8252 or (613) -234-2732; fax: (613) 234-8221; email: cfuwfls@rogers.com
Internet: http://www.cfuw.org/index.php?option=com_content&task=view&id=56&Itemid=159
Sponsor: Canadian Federation of University Women
251 Bank Street, Suite 305
Ottawa, ON K2P 1X3 Canada

CFUW Marion Elder Grant Fellowship 1631
Founded in 1919, the Canadian Federation of University Women (CFUW) is a voluntary, non-profit, self-funded bilingual organization offering a variety of fellowships to women university graduates. At the time of application for this fellowship, the candidate must be enrolled in a full-time course of studies at any level of a second masters or doctoral program. One of her letters of reference must be provided by her graduate supervisor. All else being equal, preference will be given to the holder of an Acadia University degree.
Requirements: Candidates must be studying full-time at the master's or doctoral level in Canada or abroad.
Restrictions: Eligibility for these fellowships and awards is restricted to women.
Amount of Grant: $C11,000
Date(s) Application is Due: Nov 1
Contact: Betty Dunlop, Fellowships Program Manager; (613) 234-8252 or (613) -234-2732; fax: (613) 234-8221; email: cfuwfls@rogers.com
Internet: http://www.cfuw.org/index.php?option=com_content&task=view&id=57&Itemid=159
Sponsor: Canadian Federation of University Women
251 Bank Street, Suite 305
Ottawa, ON K2P 1X3 Canada

CFUW Memorial Fellowship 1632
Founded in 1919, the Canadian Federation of University Women (CFUW) is a voluntary, non-profit, self-funded bilingual organization offering a variety of fellowships to women university graduates. A candidate for the Memorial Fellowship must be enrolled in a masters degree program in science, mathematics, or engineering in the upcoming academic year. She may be studying abroad.
Requirements: Candidates must be studying full-time at the master's or doctoral level in Canada or abroad.
Restrictions: Eligibility for these fellowships and awards is restricted to women.
Samples: Melisa Hamilton, M.Sc. Experimental Medicine, University of British Columbia, $C10,000 (2007-08).
Amount of Grant: $C10,000
Date(s) Application is Due: Nov 1
Contact: Betty Dunlop, Fellowships Program Manager; (613) 234-8252 or (613) -234-2732; fax: (613) 234-8221; email: cfuwfls@rogers.com
Internet: http://www.cfuw.org/index.php?option=com_content&task=view&id=67&Itemid=159
Sponsor: Canadian Federation of University Women
251 Bank Street, Suite 305
Ottawa, ON K2P 1X3 Canada

CFUW Ruth Binnie Fellowship 1633
Founded in 1919, the Canadian Federation of University Women (CFUW) is a voluntary, non-profit, self-funded bilingual organization offering a variety of fellowships to women university graduates. At the time of application, the candidate must be enrolled in masters or doctoral studies with a focus on one or more aspects of the field of home economics. According to the CFUW, home economics is a field of study that promotes quality home and family life and includes human development, family relations, human sexuality, financial and resource management, consumerism, human ecology, food science, human nutrition, clothing and textiles, housing and shelter and aesthetics. She may be studying abroad.
Requirements: Candidates must be studying full-time at the master's or doctoral level in Canada or abroad.
Restrictions: Eligibility for these fellowships and awards is restricted to women.
Samples: Cara Linzmayer, Ph.D., Human Ecology, University of Alberta, $C6,000 (2007-08).
Amount of Grant: $C6,000
Date(s) Application is Due: Nov 1
Contact: Betty Dunlop, Fellowships Program Manager; (613) 234-8252 or (613) -234-2732; fax: (613) 234-8221; email: cfuwfls@rogers.com
Internet: http://www.cfuw.org/index.php?option=com_content&task=view&id=77&Itemid=159
Sponsor: Canadian Federation of University Women
251 Bank Street, Suite 305
Ottawa, ON K2P 1X3 Canada

Chancellor's Minority Postdoctoral Fellowship 1634
Program assists underrepresented minority faculty members in developing their careers as scholars. For those members of underrepresented minorities committed to university teaching and research this fellowship program provides a stipend, close association with faculty at the university and assistance in furthering the fellow's development as a productive scholar. Application materials are available on the Web site.
Requirements: An applicant must demonstrate promise for a tenure-track appointment at a research college or university, must be a U.S. citizen or permanent resident, and must have received a doctorate or appropriate terminal degree within the past four years, or have completed this requirement by June of the fellowship year. Afro-American Studies must be the candidate's primary research focus. Contact Afro-American Studies and Research Program or check Web site for application deadline announcement.
Amount of Grant: $42,000 stipend; $5000 maximum for travel and other research-related expenses
Date(s) Application is Due: Jan 15
Contact: Carla Bloom, Afro-American Studies, (217) 333-7781; fax: (217) 244-4809; email: c-bloom@uiuc.edu
Internet: http://www.aasrp.uiuc.edu/education/postDoc.html
Sponsor: University of Illinois at Urbana-Champaign
1201 W Nevada
Urbana, IL 61801

Changemakers Grants 1635
Changemakers makes grants to the following types of community-based philanthropic organizations: local, regional, national, and international public foundations; alternative community foundations; affinity groups/networking organizations; groups that administer donor-advised funds; donor organizing/service providers; federations/alternative workplace giving programs; groups that recruit and train grassroots development staff; groups that provide research, media, education, or advocacy on philanthropy; and fundraising and grantmaking collaboratives. Types of support include capacity building grants, collaborative initiatives, general support grants, and technical assistance.
Requirements: Community-based philanthropic organizations that are committed to the principles of community-based philanthropy, have a record of success or a potential for increased success, and demonstrate that they have a vision for the future are eligible.
Restrictions: Grants do not support service providers.
Samples: Ctr for Third World Organizing (Oakland, CA)—to expand and update its fundraising and communications training program, $25,000.; eGrants.org (San Francisco, CA)—for general support, $15,000.; Pride Foundation (Seattle, WA)—to strengthen and link lesbian, gay, bisexual, and transgendered individuals and groups throughout Washington State, $20,000.; Boston Women's Fund (Boston, MA)—to provide technical assistance programs to women and girls in the Boston area, $20,000.
Contact: Grants Administrator, (415) 561-2363; fax: (415) 561-2366; email: grants@changemakers.org
Internet: http://www.changemakersfund.org/grantsprograms.htm
Sponsor: Changemakers
1550 Bryant Street, Suite 850
San Francisco, CA 94103

Chapin Hall International Fellowships in Children's Policy Research 1636
The program is designed to increase research and development capacity in the field of child and family policy and to develop leadership for this enterprise. Fellows will work on existing Chapin Hall projects in one of three broadly defined areas of research: developing conceptual bases for and evaluating implementation of community-based supports and services for all children; developing and using data to improve analysis, planning, and management of services for children; and developing and testing new ideas, policies, and programs for children. Some senior fellows may also work on projects of their own design that fit the Chapin Hall research agenda. The two- to 12-month appointments will provide competitive stipends and housing. A form for the fellowship letter of inquiry is available on the Web site.
Requirements: The fellowships are open to individuals from any country with any of a wide range of disciplinary interests and training, including economics, education, history, human development, law, medicine, psychology, public policy, social work, and sociology. All fellows will be expected to conduct work in English.
Contact: Christin Glodek, Fellowship Coordinator, (773) 256-5151; fax: (773) 256-5351; email: internationalprogramcoordinator@chapinhall.org
Internet: http://www.about.chapinhall.org/intprograms/intprograms.html
Sponsor: Chapin Hall Center for Children at the University of Chicago
1313 E 60th Street
Chicago, IL 60637

Charitable Leadership Foundation Education Grants 1637
The foundation supports educational programs and organizations that address long-term goals, including: improvement of early childhood literacy and improvement of children, ability to maintain grade level reading skills through eighth grade; improvement of science, technology, engineering and math (STEM) skills for K-12 students and the increase in the number of children who pursue these fields in college, and; improvement of management, involvement, and engagement in the public school system. Types of support include grants, loans, and technical assistance for projects, programs, and capacity building. Letters of inquiry and formal proposals are reviewed on an on-going basis. Applicants are required to discuss formal proposals with the staff prior to submission. Preference will be given to projects in upstate New York; however, projects will be considered in any geographic area of the United States.
Requirements: 501(c)3 charitable organizations are eligible.
Restrictions: The foundation does not normally fund organizations with substantial amounts of other resources, or arts or cultural organizations, unless the proposed project presents a compelling need or opportunity and substantially matches the funding criteria

described above. Any grants to religious organizations will be restricted to nonsectarian purposes. Grants are not awarded to individuals.
Samples: State U of New York, Empire State College (Saratoga Springs, NY)—to start a Master of Arts in Teaching degree program designed for early retirees and others who are interested in changing careers in order to teach in urban school districts, $1.2 million (2004); U of New York, School of Social Welfare (Albany, NY)—to evaluate the Supported Housing Program, administered by Unity House, $258,000 over three years (2003); Unity House (Troy, NY)—to enhance the Supported Housing Program, which serves chronically homeless adults and families, $1,041,557 over five years (2003).
Amount of Grant: $20,000 minimum
Contact: Rosemary Weaver McKenna, Esq., General Counsel & Senior Program Officer, (518) 877-6701, ext. 309; fax: (518) 877-6260; email: rmckenna@charitableleadership.org
Internet: http://www.charitableleadership.org/education.aspx?ID=13&subid=13
Sponsor: Charitable Leadership Foundation
747 Pierce Road
Clifton Park, NY 12065

Charles A. Eastman Dissertation Fellowship for Native American Scholars 1638

The immediate goal of the fellowship is to increase the number of Native American faculty in U.S. higher education by supporting Native American scholars in completing the final academic requirement, the dissertation. The second goal is to bring to the college more role models for potential Native American graduate students among Dartmouth undergraduates. The one-year residential fellowship provides access to Dartmouth's outstanding library, computing facilities, and faculty. In addition, fellows will participate in classroom activities with scholars who are dedicated to undergraduate teaching. Fellows may be taking the PhD degree in any discipline or area taught in the undergraduate arts and sciences curriculum. Each fellow will be affiliated with a department or program at the college. The dissertation fellowship will generally run from September 1 through August 31. Each fellow will be expected to complete the dissertation during the tenure of the fellowship and will have the opportunity to participate in teaching either as a primary instructor or as part of a team.
Requirements: The college invites applications from U.S. citizens of Native American descent who plan careers in college or university teaching.
Amount of Grant: $25,000 stipend, office space, library privileges, and $2500 research assistance fund
Date(s) Application is Due: Feb 1
Contact: Sandy Spiegel, Director of Grad Recruiting and Diversity, (603) 646-6578; fax: (603) 646-3488; email: Sandra.J.Spiegel@Dartmouth.edu
Internet: http://www.dartmouth.edu/~gradstdy/funding/fellowships/index.html
Sponsor: Dartmouth College
6062 Wentworth, Room 304
Hanover, NH 03755-3526

Charles Abrams Scholarships 1639

This scholarship is awarded annually to a student enrolled in a graduate planning program leading to a master's degree at one of the five schools at which Charles Abrams taught: Division of Urban Planning, Columbia University; Department of City and Regional Planning, Harvard University; Department of Urban Studies and Planning, Massachusetts Institute of Technology; Department for Urban Affairs and Policy Analysis, New School for Social Research; and Department of City and Regional Planning, University of Pennsylvania. The program is designed to aid students who will pursue careers as practicing planners.
Requirements: Applicant must be a U.S. citizen, must be enrolled in a graduate planning program at one of the five eligible schools, must be able to demonstrate a genuine financial need, and must be nominated by his/her school.
Geographic Focus: New York
Amount of Grant: $2000
Date(s) Application is Due: Apr 30
Contact: Kriss Blank, Scholarships and Fellowships, (312) 786-6722; email: kblank@planning.org
Internet: http://www.planning.org/institutions/scholarship.htm
Sponsor: American Planning Association
122 S Michigan Avenue, Suite 1600
Chicago, IL 60603

Charles Edison Fund Grants 1640

The fund's grants are equally divided among medical research projects, science education, and historic preservation. Science kits designed to experiment for classroom have been distributed. Additionally, but with some important and increasing exceptions, institutions and organizations assisted are based principally in the New York-New Jersey metropolitan area. Grant requests should be submitted on the requesting organization's letterhead and be signed by an official on behalf of the governing board. There are no application forms. The request should be detailed, complete and include background information about the organization, a full explanation of the project and its costs, and a financial report, current budget and evidence of tax-exempt status of the requesting organization. The fund meets three times a year, usually in February or March, June and December at which time requests which have been submitted at least three weeks prior to the meeting will be considered. Progress reports and a final accounting of the use of grant funds will be required of all grant recipients.
Requirements: Public and private schools, universities and colleges, and nonprofits are eligible.
Amount of Grant: $2 million total

Contact: Fund Administrator, (973) 648-0500; fax: (973) 648-0400; email: info@charlesedisonfund.org
Internet: http://www.charlesedisonfund.org/thefund.html
Sponsor: Charles Edison Fund
One Riverfront Plz, 4th Fl
Newark, NJ 07102

Charles G. Koch Charitable Foundation Grants 1641

The foundation provides funding for academic and public policy research directed at solving social problems through voluntary action and free enterprise. In the area of research, the foundation primarily funds organizations working with doctorate-level investigators in disciplines such as economics, history, philosophy, political science, and organizational behavior. Types of support include general operating, scholarship funds, conferences and seminars, research, special projects, and seed money. There are no application deadlines. Submit preproposal letters (three-page limit).
Samples: George Mason U (Fairfax, VA)—to recruit faculty members, and for programs in experimental economics, $3 million.
Amount of Grant: $25,000-$300,000 average
Contact: Kelly Young, Vice President, (202) 393-2354; fax: (202) 393-2355; email: email@cgkfoundation.org
Internet: http://www.cgkfoundation.org
Sponsor: Charles G. Koch Charitable Foundation
655 15th Street NW, Suite 445
Washington, DC 20005-2001

Charles H. Farnsworth Trust Grants 1642

Grants are awarded to nonprofits whose programs and activities provide services to seniors in their residences for health care, homemaker services, nutrition, services to elderly persons in collective/supportive housing, and research and communication to better inform individuals and institutions of ways to improve the quality and quantity of housing and support services for the elderly. Types of support include equipment, general operating budgets, renovation projects, seed money, research, and building funds. Applicants should submit a letter of intent (three-page limit).
Requirements: Only nonprofits in Massachusetts are eligible to apply.
Amount of Grant: $80,000 maximum
Contact: Grants Administrator, (617) 451-0049 ext 702; fax: (617) 423-4619; email: research@tmfnet.org
Sponsor: Charles H. Farnsworth Trust
95 Berkeley Street, Suite 201
Boston, MA 02116

Charles H. Hood Foundation Child Health Research Grants 1643

The grants assist in the promotion of child health in New England. Interest areas include pediatrics, pediatric surgery, pediatric medical and surgical subspecialties, and perinatal obstetrics. Emphasis is on medical research contributing to a reduction of the health problems and health needs of large numbers of children. The foundation seeks innovative, start-up projects that are not yet candidates for government or large foundation support. Guidelines are published twice each year.
Requirements: Applicants must have completed postdoctoral training and must not be receiving salary support from training funds, including fellowships, at the time of the grant award. Giving is limited to Connecticut, Massachusetts, Maine, New Hampshire, Rhode Island, and Vermont.
Restrictions: Grants are not given for building funds, endowments, regular budgets, or public fund-raising campaigns. Grants are not awarded to individuals.
Geographic Focus: Connecticut; Maine; Massachusetts; New England; New Hampshire; Rhode Island; Vermont
Samples: Dr. David Ludwig, Beth Israel Hospital (Boston, MA)—for research on the neuroendocrine regulation of obesity, $50,000.; Dr. Samuel Varghese, Saint Francis Hospital and Medical Ctr (Hartford, CT)—for research on regulation of collagenase gene, $49,994.
Amount of Grant: $75,000 average
Contact: Raymond Considine, Executive Director, (617) 695-9439; fax: (617) 423-4619; email: chhoodfdn@tmfnet.org
Sponsor: Charles H. Hood Foundation
95 Berkeley Street, Suite 201
Boston, MA 02116

Charles H. Revson Fellowships 1644

The program awards fellowships to those who have made significant contributions to New York City or to another large metropolitan center and who can be expected to make even greater contributions in the future, after using Columbia University's instructional, research, and other resources for an academic year. Ten awards are given each year within this program. Award includes tuition.
Amount of Grant: $23,000 stipend
Date(s) Application is Due: Feb 1
Contact: Coordinator, Revson Program, (212) 280-4023; fax: (212) 663-7537; email: revson@columbia.edu
Internet: http://www.columbia.edu/cu/revson
Sponsor: Columbia University
420 W 116th Street, Office 1A
New York, NY 10027

Charles H. Revson Foundation Grants   **1645**

The foundation awards grants nationwide in its areas of interest, including urban affairs and public policy, education and higher education, biomedical research policy, and Jewish education and philanthropy. Types of support include capital campaigns, continuing support, fellowships, internship funds, program development, and research. Preference is given to requests serving New York, NY. There are no application forms or deadlines. The board meets in April, June, October, and December.

Restrictions: Grants do not support local or national health appeals or direct service programs, individuals, building construction/renovation, book projects, charity events, travel expenses, or budgetary support.

Samples: U of North Carolina at Chapel Hill, Ctr for Jewish Studies (NC)—for outreach activities on campus and throughout North Carolina, $250,000 (2004).

Contact: Grants Administrator, (212) 935-3340; fax: (212) 688-0633; email: info@revsonfoundation.org

Internet: http://www.revsonfoundation.org/guidelines.htm

Sponsor: Charles H. Revson Foundation

55 E 59th Street, 23rd Fl

New York, NY 10022

Charles Lafitte Foundation Grants   **1646**

The foundation is committed to helping groups and individuals foster lasting improvement on the human condition by providing support to education, children's advocacy, medical research, and the arts. Children's advocacy grants support organizations working to improve the quality of life for children, particularly in relation to child abuse, literacy, foster housing, hunger, and after-school programs. Education grants support innovative programs that work to resolve social service issues, address the needs of students with learning disabilities, provide technology and computer-based education, offer leadership skills education, and support at-risk students. Colleges and universities also receive support for research and conferences. The foundation's medical issues and research grants support healthcare studies, with emphasis on cancer research and treatment, children's health, health education, and promoting healthy living and disease prevention. Art grants support emerging artists and educational art programs.

Requirements: 501(c)3 tax-exempt organizations are eligible.

Samples: American Cancer Society, Eastern Division (North Brunswick, NJ)—for the Pain Initiative, an educational campaign that promotes chronic-pain relief as an integral part of comprehensive patient care, $500,000.

Contact: Jennifer Vertetis, President, email: jennifer@charleslafitte.org

Internet: http://www.charleslafitte.org

Sponsor: Charles Lafitte Foundation

29520 2nd Avenue SW

Federal Way, WA 98023

Charles Lathrop Parsons Award   **1647**

This award is given to recognize outstanding public service by a member of the American Chemical Society. The award is normally given every two years; however, the board of directors may at its discretion reduce the interval to one year if, in its judgment, circumstances in a given year warrant such action. Neither the scientific nor the record of scientific achievement of a member affects his or her eligibility for this award, which is not directed toward recognition of scientific accomplishment or stature. The public service to be recognized may be performed either as a part of or completely outside the regular duties and activities of the nominee's employment.

Requirements: A nominee must be a member of ACS and a citizen of the United States, and must have performed outstanding public service. Any individual, except a member of the award committee, may submit one nomination or seconding letter for the award in any given year. Nominating document consists of a letter of not more than 1000 words containing an evaluation of the nominee's accomplishments and a specific identification of the public service to be recognized.

Restrictions: Current members of the ACS board of directors are ineligible to receive this award.

Samples: S. Allen Heininger—award winner, $3,000 (2007); Marye Anne Fox—award winner, $3,000 (2005); Zafra Lerman—award winner, $3,000 (2003).

Amount of Grant: $3,000

Date(s) Application is Due: Nov 1

Contact: Felicia Dixon, Awards Administrator; (800) 227-5558 or (202) 872-4408; fax: (202) 776-8008; email: f_dixon@acs.org or awards@acs.org

Internet: http://portal.acs.org/portal/acs/corg/content?_nfpb=true&_pageLabel=PP_ARTICLEMAIN&node_id=1319&content_id=CTP_004498&use_sec=true&sec_url_var=region1

Sponsor: American Chemical Society

1155 Sixteenth Street, NW

Washington, DC 20036

Charles Russ Richards Memorial Award   **1648**

This award, jointly sponsored by Pi Tau Sigma Honorary Mechanical Engineering Fraternity and ASME, is given annually to the engineering graduate who has demonstrated outstanding achievement in mechanical engineering 20 years or more after graduation from the regular engineering course of a recognized college or university. Achievement shall be all or in part in any field including industrial, educational, political, research, civic, and artistic. The candidate's achievements will be examined for an application of basic engineering methods or principles. A special nomination form must be acquired from ASME. Award includes certificate and expense supplement.

Requirements: Applicants must be engineering graduates.

Samples: Warren DeVries—award winner, $1000 (2005); Roop Mahajan—award winner, $1000 (2003); Salvatore Torquato—award winner, $1000.

Amount of Grant: $1000

Date(s) Application is Due: Feb 1

Contact: Gilda DiTullio, Manager, (212) 591-7736; fax: (212) 705-7739; email: ditulliog@asme.org

Internet: http://www.asme.org/honors/ms71/gaa/russ.html

Sponsor: American Society of Mechanical Engineers

3 Park Avenue

New York, NY 10016

Charles S. Sydnor Award   **1649**

The prize is given in even-numbered years for a distinguished book in southern history published in odd-numbered years. The application due date is March 1 of the year in which the prize is to be awarded. Books must be submitted by the publishers. Books should be sent directly to committee members listed on the Web site, not the national office.

Amount of Grant: $1000

Date(s) Application is Due: Mar 1

Contact: Dr. John Inscoe , Secretary-Treasurer, (706) 542-8848; fax: (706) 542-2455; email: jinscoe@uga.edu

Internet: http://www.uga.edu/%7Esha/charle~1.htm

Sponsor: Southern Historical Association

Room 111A , LeConte Hall

Athens, GA 30602

Charles Stewart Mott Foundation Anti-Poverty Program   **1650**

This program focuses on improving education, expanding economic opportunity, building organized communities, and special initiatives as pathways out of poverty. The overall goal is to help people vocalize and mobilize around local concerns, grow through participation in educational opportunities, and attain economic self-sufficiency by engaging more fully in the economy. Types of support include challenge/matching grants, conferences and seminars, demonstration grants, general operating grants, program grants, seed money grants, technical assistance, and training grants. The board meets in March, June, September, and December. Organizations should apply at least four months prior to the start date of the project for which they are seeking funds.

Requirements: Nonprofits and K-12 organizations are eligible to apply. A proposal may be submitted by a church-based or similar organization if the project falls clearly within program guidelines and is intended to serve as broad a segment of the population as the program of a comparable nonreligious organization.

Samples: Children's Aid Society (New York, NY)—to expand its Carrera program, a long-term, holistic approach to working with adolescents who are at high risk for teenage pregnancy, $873,782 (2005); Corp for a Skilled Workforce (Ann Arbor, MI)—to strengthen workforce development organizations in Michigan, $250,000 (2005); Pacific Institute for Community Organization (Oakland, CA)—for the Louisiana Interfaith Together project's work to engage residents of low-income communities in improving the educational outcomes of children in Louisiana, $220,000 over two years (2005); Public/Private Ventures (Philadelphia, PA)—to assist and evaluate the Fathers at Work Initiative, which seeks to reduce poverty by increasing employment, earnings, and responsible fatherhood among young, low-income fathers, $787,047 (2005).

Amount of Grant: $139.5 million total

Contact: Office of Proposal Entry, (810) 238-5651; fax: (810) 766-1753; email: info@mott.org

Internet: http://www.mott.org/programs/poverty.asp

Sponsor: Charles Stewart Mott Foundation

503 S Saginaw Street, Suite 1200

Flint, MI 48502-1851

Charles Stewart Mott Foundation Grants   **1651**

The focus of the foundation's grant making is organized in four programs: civil society; environment; Flint, MI; and poverty. Flexibility to investigate new opportunities is maintained through an exploratory and special projects program. The civil society program promotes and supports civil society in the United States; Central/Eastern Europe and South Africa. The environment program supports efforts to achieve a healthy global environment capable of sustaining all forms of life. The Flint program seeks to strengthen the capacity of local institutions, including schools and school districts, in the foundation's home community of Flint, MI, to respond to economic and social needs. The poverty program addresses issues that contribute to improved life outcomes for children, youth, and families in low-income communities. Programs in low-income communities that connect schools and communities through systemic reform, improved teaching, leadership development, networking, technical assistance, and advocacy are also applicable. In all grant making, particular interest will be given to fresh approaches to solving community problems in the defined program areas; approaches that can generate long-term support from other sources and/or can be replicated in other communities; public policy development and research and development activities to further existing programs as well as to explore new fields of interest; and approaches and activities that lead to systemic change. Although proposals may be submitted at any time, applicants are strongly encouraged to submit during the first quarter of the year for which funding is requested. Grant expenditures are determined by September 1 of each year. The review process takes up to four months from the time the proposal is received. Therefore, proposals should be submitted at least four months prior to the start of the proposed grant period. Funding for unsolicited proposals is limited. It is recommended that letters of inquiry be submitted instead of a full proposal.

Requirements: Only 501(c)3 organizations are eligible, including schools and school districts.

Restrictions: Grants are not made to/for individuals; religious activities or programs that serve, or appear to serve, specific religious groups or denominations; or local projects outside the Flint area unless the projects are part of a national demonstration or foundation-planned network of grants that have clear and significant implications for replication in other communities.

Samples: Fulcrum Foundation (Moscow, Russia)—for a small-grants program that promotes ethnic tolerance among Russian youths, $20,000 (2005); U of Virginia (Charlottesville, VA)—to support the writing of a book that will examine philanthropy and democracy in the United States, $25,000 over 32 months (2005); Virginia Aspen Institute (Washington, DC)—to expand MicroMentor, an online mentorship program for low-income entrepreneurs, $25,000 (2005); U of KwaZulu-Natal (South Africa)—for its efforts to deal with the racial, ethnic, and cultural challenges inherent in its recent merger of two universities, one formerly black and one formerly white, $180,000 (2005).

Amount of Grant: $139.5 million total

Contact: Office of Proposal Entry, (810) 238-5651; fax: (810) 766-1753; email: info@mott.org

Internet: http://www.mott.org/about/programs.aspx

Sponsor: Charles Stewart Mott Foundation

503 S Saginaw Street, Suite 1200

Flint, MI 48502-1851

Charles W. Finley Visiting Scholar Education Grant   1652

The American Academy of Periodontology Foundation established Finley Visiting Scholar Education Grants to honor the memory of Dr. Charles W. Finley and all of the great periodontal leaders of the past, present, and future. Grants are awarded to accredited periodontal programs to offset honoraria, travel, promotion, and seminars by speakers in periodontology. As required, the grants are being matched by funds from each department or institution. Though the speakers will spend the majority of their time with the periodontics department, both students and faculty, the grant requires that the speaker present a formal program to the faculty and students of the institution on a subject of interest to the entire dental institution. Applications are available on the Web site.

Requirements: Accredited periodontal programs are eligible to apply.

Amount of Grant: $1500

Date(s) Application is Due: Feb 14

Contact: Sharon Mellor, Executive Director, (800) 282-4867 ext 256; fax: (312) 573-3272; email: sharon@perio.org

Internet: http://www.perio.org/foundation/research.html

Sponsor: American Academy of Periodontology Foundation

737 N Michigan Avenue, Suite 800

Chicago, IL 60611

Charlotte Geyer Foundation Cancer Research Grants   1653

The foundation supports research into the cause, prevention, and treatment of cancer. The purpose of the awards is to provide one year's funding to exceptional proposals to give investigators the opportunity of advancing and improving projects to the point where they are able to successfully compete for an NCI RO1 or other award. Proposals are reviewed three times per year. Application guidelines and procedures are available online.

Requirements: The foundation will review recent proposals—either the original proposal, a revised proposal, or a portion of the original proposal deemed to have special merit—that were submitted to NCI; received a peer-review ranking in the 10 percentiles above the NCI payline; and did not receive funding.

Amount of Grant: $100,000 maximum

Date(s) Application is Due: Feb 1; Jun 1; Oct 1

Contact: Nancy Falletta, Executive Director, (716) 632-6448; fax: (716) 632-6098

Internet: http://www.charlottegeyer.org

Sponsor: Charlotte Geyer Foundation

P.O. Box 1276

Williamsville, NY 14231-1276

Charlotte W. Newcombe Doctoral Dissertation Fellowships   1654

This program provides approximately 28 dissertation awards annually to encourage original and significant study of ethical and religious values in all fields. The selection process and all administration for this continuing fellowship program are handled by the Woodrow Wilson National Fellowship Foundation and is supported by the Charlotte W. Newcombe Foundation. The awards are intended to finance the last full year of dissertation writing. Annual deadline dates may vary; contact program staff for exact dates.

Requirements: An applicant must be a candidate for a PhD or ThD degree in a doctoral program at a graduate school in the United States.

Samples: Recent fellows include: Kimberly Arkin Anthropology, University of Chicago— Racism and the Politics of Identities: Constructing Adolescent Jewishness in Contemporary France; Jeremy Berndt, History, Northwestern University— Closer Than Your Jugular Vein: Islamic Learning, Rural Life, and History in Gimbala, Central Mali; Paul Dilley, Religious Studies, Yale University— The Crisis of Conversion: Monastic Identity Formation and Culture in Late Antiquity; Hussein Fancy, History, Princeton University—Boundary Crossing, Boundary Making: Muslim and Christian Mercenaries in the Western Mediterranean; Emily Russell, English, University of California at Los Angeles—Embodied Citizenship: Disability in the National Imagination; Don Selby,

Anthropology, Johns Hopkins University—The Politics and Morality of Human Rights in Thailand (2006).

Amount of Grant: $18,000

Date(s) Application is Due: Nov 1

Contact: Program Director, (609) 452-7007, ext 131; fax: (609) 452-7828; email: charlotte@woodrow.org

Internet: http://www.woodrow.org/newcombe/

Sponsor: Charlotte W. Newcombe Fellowships/Woodrow Wilson National Fellowship Foundation

P.O. Box 5281

Princeton, NJ 08543-5281

Chateaubriand Exact Sciences, Engineering, and Medicine Research   1655
Scholarships

Individuals currently pursuing the PhD or who have completed it in the last three years may qualify for a scholarship from the French government to conduct research at a French university, school of engineering, or state-funded laboratory. Candidates must contact the host institution to secure a position before applying. Scholarships are available for periods of six to 18 months and carry a stipend, health insurance, and round-trip travel expenses. Application forms are available on the Web site.

Requirements: Applicants must be U.S. citizens and registered in a U.S. university, and must be PhD candidates or recent PhD holders. Applicants must obtain written agreement from the French hosting institution before applying; an online registration form is available on the Web site.

Amount of Grant: Ff8700 minimum per month

Date(s) Application is Due: Jan 15

Contact: Chateaubriand Fellowship Program, Mission pour la Science et la Technologie, Ambassade de France, (202) 944-6246; fax: (202) 944-6244; email: Chateaubriand@amb-wash.fr

Internet: http://www.frenchculture.org/education/support/chateaubriand/

Sponsor: French Embassy Cultural Services

4101 Reservoir Rd NW

Washington, DC 20007-2176

Chatham Valley Foundation Grants   1656

The foundation awards grants to nonprofits in the metropolitan Atlanta, GA, area in the areas of arts (performing arts and cultural programs), museums, education and higher education, health care and health associations, medical research (cancer), crime prevention and law enforcement, Jewish services (nursing homes, religious welfare, and temples), and social services (homeless and economically disadvantaged). Types of support include general operating support, annual campaigns, capital campaigns, building construction/renovation, endowment funds, and program development.

Requirements: Nonprofit organizations in the greater metropolitan Atlanta, GA, area are eligible.

Geographic Focus: Georgia

Samples: Jewish Federation of Greater Atlanta (Atlanta, GA)—for general support, $272,500 (2003); The Temple (Atlanta, GA)—for general support, $250,000 (2003); Camp Sunshine (Decatur, GA)—for general support, $20,000 (2003).

Amount of Grant: $1 million total

Contact: Avery Tucker, c/o Wachovia Charitable Services, (404) 266-3081

Sponsor: Chatham Valley Foundation

3414 Peachtree Road

Atlanta, GA 30326

Chatlos Foundation Grants Program   1657

The foundation gives consideration to requests for program support. Grants to bible colleges total 33 percent of the foundation distribution per year, and 30 percent is allowed to religious causes, 7 percent to liberal arts colleges, 26 percent for medical concerns with emphasis placed on the purchase of equipment, and 4 percent to social concerns. Types of support include operating budgets, emergency funds, equipment, land acquisition, matching funds, scholarship funds, special projects, publications, and renovation projects. Requests for funding must be submitted in writing. Although there is no deadline for receipt of requests, those received later than one month prior to board meetings may be carried forward. The preliminary review committee meets monthly. Organizations may submit proposals six months from date of denial but are limited to one grant award within any 12-month period.

Requirements: Applicants must be U.S. tax-exempt, nonprofit organizations that provide services in the following areas: bible colleges, religious causes, medical concerns, liberal arts colleges, and social concerns. Proposals must include cover letter, specific request, tax-exemption letter, and budget. If proposal is to be considered at board level, additional information will be requested.

Restrictions: The foundation will not accept requests from individual church congregations, individuals, organizations in existence for less than two years as indicated by IRS tax-exempt letter of determination, for education below the college level, for medical research projects, or for support of the arts.

Samples: Nazareth College (Rochester, NY)—for its capital campaign to renovate an academic complex, $15,000 (2004).

Amount of Grant: $10,000 maximum for initial request

Contact: Grants Administrator, (407) 862-5077; email: cj@chatlos.org

Internet: http://www.chatlos.org/AppInfo.htm

Sponsor: Chatlos Foundation

P.O. Box 915048

Longwood, FL 32791-5048

Chautauqua Region Community Foundation Grants 1658

Grants support projects serving communities in Chautauqua County in its areas of interest, including arts and culture, libraries, education, housing and shelters, children and youth, human services and general charitable giving, and government and public administration. Types of support include general operating support, continuing support, building construction/renovation, equipment acquisition, conferences and seminars, publication, seed grants, emergency funds, and undergraduate and graduate scholarships to individuals.

Requirements: Nonprofit organizations may apply for grants in support of projects serving communities in Chautauqua County, excluding the Fredonia/Dunkirk area, which is served by the Northern Chautauqua Community Foundation.

Geographic Focus: New York

Samples: Salvation Army (NY)—to support the purchasing of infant formula and diapers, $500.; James Prendergast Library Assoc (NY)—to support acquisition of books on and for small business, $2500.; Research and Planning for Human Services (NY)—to support human services coordination activities, $4000.

Amount of Grant: $1.6 mil total

Date(s) Application is Due: Jan 31; Mar 1; Nov 1

Contact: June Diethrick, Grants Coordinator, (716) 661-3392; fax: (716) 488-0387; email: jdiethrick@crcfonline.org

Internet: http://www.crcfonline.org/

Sponsor: Chautauqua Region Community Foundation

418 Spring Street

Jamestown, NY 14701

CHCF Grants 1659

The foundation's mission is to expand access to affordable, quality health care for underserved individuals and communities in California and to promote fundamental improvements in the health status of Californians. The foundation has five funding areas: improving care delivery, business of healthcare, healthcare quality, California health policy, and California's uninsured. The one-year, renewable grants will focus on areas where the foundation's resources can initiate meaningful policy recommendations, innovative research, and the development of model programs. Philanthropic activities include foundation-initiated projects, requests for proposals, and unsolicited proposals. emailed letters of inquiry should include, in two to three pages, a brief description of the proposed project, along with an estimated time line and budget. A full proposal may be requested.

Requirements: California 501(c)3 nonprofit organizations are eligible.

Restrictions: The foundation does not generally support the cost of direct clinical care, ongoing general operating expenses, capital campaigns, annual appeals or other fund-raising events, construction, purchase or renovation of facilities, or purchase of equipment.

Geographic Focus: California

Date(s) Application is Due: Oct 14

Contact: Grants Administrator, (510) 238-1040; fax: (510) 238-1388; email: grants@chcf.org

Internet: http://www.chcf.org/grantinfo

Sponsor: California HealthCare Foundation

476 Ninth Street

Oakland, CA 94607

Chest Foundation Eli Lilly and Company Distinguished Scholar in Critical Care Medicine Award 1660

The Eli Lilly and Company Distinguished Scholar in Critical Care Medicine will have a 3-year opportunity to examine issues that relate to critically ill patients. The award is intended to permit the investigation of issues that are not easily supported by traditional funding, such as clinical trials or basic science. Rather, the award would support activities such as the development of public policy, patient education models, or economic analysis of treatment or care delivery in this patient group. A stipend of $50,000 annually over the 3-year term of the program will be provided. A fourth year stipend of $10,000 will also be awarded to the individual who will serve in a mentorship role to the fourth Distinguished Scholar position during his/her inaugural year of service.

Requirements: An applicant must be: an FCCP and hold the degree of MD, DO, MBChB, MBBCh, MBBS, DNSc, PharmD, PhD, or EdD; board-certified in critical care medicine; and a recognized clinician and/or scientist with a specialty in sepsis and/or critical care as evidenced by publications, presentations, and/or peers.

Amount of Grant: $50,000 per year for first three years, $10,000 for fourth year

Date(s) Application is Due: Apr 30

Contact: Sue Ciezadlo, Senior Project Coordinator; (847) 498-8363; fax: (847) 498-5460; email: sciezadlo@chestnet.org

Internet: http://www.chestfoundation.org/foundation/clinical/dsCriticalCare.php

Sponsor: Chest Foundation

3300 Dundee Road

Northbrook, IL 60062-2348

Chest Foundation Geriatric Development Research Awards 1661

The two-year award is intended to provide the impetus required for long-term career development focused on integrating geriatrics into the sub-specialties of internal medicine. The program awards the grant to an academic internist to develop and implement a basic, clinical, or health services research project focused on a geriatric aspect of chest medicine. The funding can support the salary of the award recipient and/or the purchase of supplies, the salaries of technical personnel, and other resources necessary for the completion of the research project. The award also includes a one-time travel grant to attend the meetings of the American Geriatrics Society and the American College of Chest Physicians in the second year of the award.

Requirements: To be eligible for the award, applicants must: have U.S. citizenship or permanent resident status; hold a degree of MD or its equivalent; have completed a sub-specialty internal medicine fellowship leading to certification in his/her sub-specialty by the American Board of Internal Medicine and be within the first three years of his/her faculty appointment; possess a faculty appointment by July 1 (this faculty appointment should be documented in the required letters of recommendation from the applicant's department chair and division director); be a member of the American College of Chest Physicians; commit 75 percent of his/her professional effort to research activities; develop and implement a basic, clinical, or health services research project focused on a geriatric aspect of chest medicine; and generate and implement a career development plan focused on the geriatrics aspects of chest medicine. This plan must include organizing and interacting with a mentorship team made up of a minimum of four members—the applicant's research mentor as the leader of the research team, a sub-specialist in the applicant's field of chest medicine, a geriatrician, and one other member at the applicant's discretion.

Restrictions: The funding cannot be used to acquire administrative or clerical support. Funding is to cover total costs; no indirect cost funds will be provided.

Amount of Grant: $50,000 per year for two years plus $3,000 travel grant

Date(s) Application is Due: Apr 30

Contact: Sue Ciezadlo, Senior Project Coordinator; (847) 498-8363; fax: (847) 498-5460; email: sciezadlo@chestnet.org

Internet: http://www.chestfoundation.org/foundation/clinical/geriatricAward.php

Sponsor: Chest Foundation

3300 Dundee Road

Northbrook, IL 60062-2348

Chest Foundation Grant in Venous Thromboembolism 1662

The Chest Foundation is issuing this Request for Proposal (RFP) that will fund two projects with up to a total grant amount of $180,000 each for a 1-year period. The goal of the grant process is to confer awards to projects that include evidence-based replicable programs and tools to address the current gaps in the management and prophylaxis in venous thromboembolism (VTE). The deadline is February 20.

Requirements: To be considered a candidate for The Chest Foundation Grant in Venous Thromboembolism, an individual must: be a current ACCP member and hold the degree of MD, DO, MBChB, MBBCh, MBBS, DNSc, PharmD, PhD, or EdD; and be a recognized scientist focused on an area of VTE or clinical decision making as evidenced by academic productivity and/or publications and presentations.

Amount of Grant: $180,000 maximum

Date(s) Application is Due: Feb 20

Contact: Sue Ciezadlo, Senior Project Coordinator; (847) 498-8363; fax: (847) 498-5460; email: sciezadlo@chestnet.org

Internet: http://www.chestfoundation.org/foundation/clinical/VT_Grant.php

Sponsor: Chest Foundation

3300 Dundee Road

Northbrook, IL 60062-2348

Chest Foundation/LUNGevity Foundation Clinical Research in Lung Cancer Grants 1663

The goal of this joint clinical research program is, ultimately, to save the lives of those people afflicted with lung cancer by funding innovative research designed to treat and cure lung cancer. A clinical research award of $75,000 will be granted to a single award recipient for clinical research in lung cancer over a period of 2 years. The amount of $37,500 will be granted after July 1. Upon receipt of the Progress Report received before July 15, the remaining $37,500 will be granted.

Requirements: To be considered a candidate for The Chest Foundation/LUNGevity Foundation Clinical Research Award in Lung Cancer, an individual must: be an ACCP member who has completed at least 2 years of pulmonary or oncology fellowship and be within 5 years of fellowship or no more than 5 years in practice; hold a degree of MD, DO, MB,BCh, PharmD, PhD or its equivalent; can be a citizen of the United States, or have a J-1 or H-1 visa, and be working in a certified United States institution or organization, or be a Canadian or international member who is working in a certified institution or organization in his/her country; and be a clinician in lung cancer research.

Amount of Grant: $75,000

Date(s) Application is Due: Apr 30

Contact: Sue Ciezadlo, Senior Project Coordinator; (847) 498-8363; fax: (847) 498-5460; email: sciezadlo@chestnet.org; Beth Ida Stern, Executive Director; email: bstern@lungevity.org

Internet: http://www.chestfoundation.org/foundation/clinical/LUNGevityAward.php

Sponsor: Chest Foundation / LUNGevity Foundation

3300 Dundee Road

Northbrook, IL 60062-2348

Chestnut Hill Charitable Foundation, Inc Grants 1664

The foundation awards grants to nonprofit organizations in the greater Boston, MA, area in support of medical research, children and youth, and arts and education. Specifically, the foundation's interests include visual arts, museums, humanities, arts and culture, early childhood education and development, education at all levels, hospitals and medical research (cancer, heart and lungs, AIDS, and diabetes), social services, and general charitable giving. Types of support include general operating support, annual campaigns, capital campaigns, building construction/renovations, programs and projects, seed grants, curriculum development, research, and matching funds. There are no application forms; submit a letter of request.

Requirements: Nonprofit organizations in the greater Boston, MA, area may submit letters of request.

Restrictions: Grants are not awarded to support individuals or political or religious organizations.

Geographic Focus: Massachusetts

Samples: Leslie Berg, U of Massachusetts Medical Ctr (Worcester, MA)—for the Chestnut Hill Award for Excellence in Medical Research, administered by the Medical Foundation, in Boston, $15,000 (2003).

Amount of Grant: $1000-$15,000 average

Contact: Kay Kilpatrick, Director Corporate Giving, (617) 630-2415

Sponsor: Chestnut Hill Charitable Foundation, Inc

27 Boylston Street

Chestnut Hill, MA 02167-1700

CHF Charles C. Price Fellowship in Polymer History   1665

Through the Beckman Center, CHF offers long-term fellowships for periods of up to 9 months to support scholars in residence. Generally, research projects must be in an area of the chemical and molecular sciences, technologies, and industries, broadly construed in terms of subject and time period. The Price Fellowship in particular is open to scholars pursuing research in the history of the chemical sciences and technologies. Preference is given to applicants with projects on the history of polymers.

Requirements: To be eligible applicants must either have a Ph.D. (or equivalent) or be a doctoral candidate at the dissertation stage.

Samples: Slawomir Lotysz, research fellow and lecturer, University of Zielona Gora, Poland, $20,000 (2007-08); Thomas Faith, Ph.D. candidate, George Washington University, Washington, D.C., $20,000 (2006-07).

Amount of Grant: $20,000; travel allowance also available

Date(s) Application is Due: Feb 15

Contact: Ashley Augustyniak, Fellowship Coordinator; (215) 873-8269; fax: (215) 925-1954; email: aaugustyniak@chemheritage.org or fellowships@chemheritage.org

Internet: http://www.chemheritage.org/research/research-nav1-price.html

Sponsor: Chemical Heritage Foundation

315 Chestnut Street

Philadelphia, PA 19106

CHF Glenn E. and Barbara Hodsdon Ullyot Scholarship   1666

The Ullyot Scholarship sponsors historical research that promotes public understanding of the chemical sciences. Applications are invited from scholars, graduate students, science writers, and journalists. The scholar will spend at least two months in residence at CHF. There is no specific application form to apply for the fellowship. All printed materials and letters of recommendations must be postmarked by the deadline.

Samples: Augustin Cerveaux, Ph.D. candidate, Universite Louis Pasteur, Strasbourg, France, $4,500 (2007); Eric S. Hintz, Ph.D. candidate, University of Pennsylvania, $4,500 (2005).

Amount of Grant: $4,500; travel allowance also available

Date(s) Application is Due: Feb 15

Contact: Ashley Augustyniak, Fellowship Coordinator; (215) 873-8269; fax: (215) 925-1954; email: aaugustyniak@chemheritage.org or fellowships@chemheritage.org

Internet: http://www.chemheritage.org/research/research-nav3-ullyot.html

Sponsor: Chemical Heritage Foundation

315 Chestnut Street

Philadelphia, PA 19106

CHF Long-Term Fellowships   1667

Through the Beckman Center, CHF offers long-term fellowships for periods of up to 9 months to support scholars in residence. Research projects must be in an area of the chemical and molecular sciences, technologies, and industries, broadly construed in terms of subject and time period. Named fellowships within this group include the following: the Robert W. Allington Fellowship—for general research in the history of chemical and molecular sciences, technologies, and industries; the Gordon Cain Fellowship in Technology, Policy, and Entrepreneurship—for historical research on the development of the chemical industries; the Sydney M. Edelstein Fellowship—for general research in the history of chemical & molecular sciences, technologies, and industries; and the John C. Haas Fellowships—for research that will enhance public understanding of the chemical industries in relation to environmental, societal, health, and safety issues. There is no specific application form to apply.

Requirements: To be eligible applicants must either have a Ph.D. (or equivalent) or be a doctoral candidate at the dissertation stage.

Amount of Grant: $43,000; travel allowance also available

Date(s) Application is Due: Feb 15

Contact: Ashley Augustyniak, Fellowship Coordinator; (215) 873-8269; fax: (215) 925-1954; email: aaugustyniak@chemheritage.org or fellowships@chemheritage.org

Internet: http://www.chemheritage.org/research/research-nav1.html

Sponsor: Chemical Heritage Foundation

315 Chestnut Street

Philadelphia, PA 19106

CHF Robert W. Gore Fellowship in Materials Innovation   1668

The Fellowship is open to advanced Ph.D. candidates who possess the research expertise, writing skills, and technical background to conduct historical case studies of materials innovation of the past thirty years and to communicate findings to academic, policy,

and industrial audiences. The Fellow will undertake one such case study with input from members of the Center for Contemporary History and Policy. The result will be a white paper published by CHF; the fellow will be encouraged (but not required) to publish variants of the study as dissertation chapters or scholarly articles. The fellowship is one component of the larger Gore Innovation Case Studies Program, which will culminate in a conference and an edited volume. In addition to their participation in these and other Case Studies Program activities, the fellow will be given ample opportunity to pursue their own research and academic interests while at CHF. There is no specific application form to apply.

Requirements: To be eligible applicants must either have a Ph.D. (or equivalent) or be a doctoral candidate at the dissertation stage.

Samples: Doogab Yi, Ph.D. candidate, Department of History, Princeton University, $35,000 (2007-08).

Amount of Grant: $35,000; plus small research allowance

Date(s) Application is Due: Feb 15

Contact: Ashley Augustyniak, Fellowship Coordinator; (215) 873-8269; fax: (215) 925-1954; email: aaugustyniak@chemheritage.org or fellowships@chemheritage.org

Internet: http://www.chemheritage.org/research/research-nav1-gore.html

Sponsor: Chemical Heritage Foundation

315 Chestnut Street

Philadelphia, PA 19106

CHF Roy G. Neville Short-Term Fellowships   1669

Roy G. Neville Fellowships are open to historians of science, technology, and allied fields, as well as to historians of the book and of print culture, bibliographers, and librarians, who will make use of the Roy G. Neville Historical Chemical Library. The Neville Library contains approximately 6,000 titles dating from the 15th to the 19th centuries and covers all aspects of the history of chemistry and allied fields. Ph.D., Ph.D. candidates, or equivalent preferred but not required.

Amount of Grant: $3,000 per month, up to four months

Date(s) Application is Due: Feb 15

Contact: Ashley Augustyniak, Fellowship Coordinator; (215) 873-8269; fax: (215) 925-1954; email: aaugustyniak@chemheritage.org or fellowships@chemheritage.org

Internet: http://www.chemheritage.org/research/research-nav2.html

Sponsor: Chemical Heritage Foundation

315 Chestnut Street

Philadelphia, PA 19106

CHF Societe de Chimie Industrielle (American Section) Fellowship 1670

The Societe de Chimie Industrielle (American Section) Fellowship is designed to stimulate public understanding of the chemical industries. Applications are encouraged from writers, journalists, educators, and historians of science, technology, or business. Multimedia, popular book projects, and Web-based projects are encouraged. Applicants must specify how the outcomes of their projects will reach a broad audience. The period of fellowship is three months in residence at CHF anytime from September to August.

Samples: David Caudill, Professor, Villanova University School of Law, $15,000 (2007); Jo Ann Caplin, Temple University, $15,000 (2006).

Amount of Grant: $15,000; travel allowance also available

Date(s) Application is Due: Feb 15

Contact: Ashley Augustyniak, Fellowship Coordinator; (215) 873-8269; fax: (215) 925-1954; email: aaugustyniak@chemheritage.org or fellowships@chemheritage.org

Internet: http://www.chemheritage.org/research/research-nav3-chimie.html

Sponsor: Chemical Heritage Foundation

315 Chestnut Street

Philadelphia, PA 19106

CHF Travel Grants   1671

The Beckman Center for the History of Chemistry at CHF offers grants to cover travel and lodging expenses for researchers who wish to use CHF's wealth of resources in the history of the chemical and molecular sciences for short-term research (1 to 4 weeks) at CHF. Travel grant recipients will have access to the collections of the Othmer Library of Chemical History. Researchers are also encouraged to use CHF's oral history collection and its collection of art, artifacts, archives, and images. Philadelphia and the Delaware Valley also boast numerous area resources for scholars. Grants are usually on the order of $750 per week and are intended to help defray the costs of travel and lodging. The application deadline is ongoing.

Requirements: Applicants must reside more than 75 miles from Philadelphia to be eligible. No more than one travel grant per person per fiscal year (1 July to 30 June) will be awarded.

Amount of Grant: $750 per week average

Contact: Ashley Augustyniak, Fellowship Coordinator; (215) 873-8269 or (215) 925-2222; fax: (215) 925-1954; email: aaugustyniak@chemheritage.org or travelgrants@chemheritage.org

Internet: http://www.chemheritage.org/research/research-nav5.html

Sponsor: Chemical Heritage Foundation

315 Chestnut Street

Philadelphia, PA 19106-2702

Chicago Institute for Psychoanalysis Fellowships   1672

This is a program for advanced trainees in psychiatry, social work, and psychology who have significant interest in psychoanalysis as a body of knowledge and as a framework with which to understand and carry out therapeutic work. It is jointly sponsored by the Chicago Psychoanalytic Society and the Institute for Psychoanalysis. A mentor is

assigned to each Fellow, who is available throughout the year-long program to discuss cases, papers, research, etc. In addition, Fellows are guests at Society meetings and conferences held locally. Opportunities are provided for the Fellows to discuss presented papers with their mentors. A monthly conference is held to acquaint Fellows with basic psychoanalytic clinical concepts through readings and discussion. Each Fellow receives a copy of the Annual of Psychoanalysis and other publications.
Date(s) Application is Due: Aug 15
Contact: Christine Susman, Director; (312) 922-7474 ext. 324; email: csusman@ chicagoanalysis.org or admin@chicagoanalysis.org
Internet: http://www.chicagoanalysis.org/fellow.php
Sponsor: Chicago Institute for Psychoanalysis
122 South Michigan Avenue
Chicago, IL 60603

Chicago Reporter Minority Urban Journalism Fellowship 1673
An experienced minority journalist will be selected for this year-long fellowship to work at the Chicago Reporter, an investigative monthly that covers issues of race and poverty in Chicago. Postgraduate coursework will be available through local colleges and universities. The award includes salary plus benefits and educational opportunities. Interested candidates should send a resume and five clips to the office listed.
Requirements: Applicants must have a bachelor's degree, a minimum of three years reporting experience, excellent news judgment, and a strong interest in urban affairs and investigative reporting. Fluency in Spanish a plus.
Contact: Rui Kaneya, Program Contact, (312) 427-4830 ext 3864; fax: (312) 427-6130; email: ruik@chicagoreporter.com
Internet: http://chicagoreporter.com/Gizmos/Navigation/Aboutus/aboutus.htm
Sponsor: Chicago Reporter
332 S Michigan Avenue, Suite 500
Chicago, IL 60603

Chicago Tribune Heartland Prizes for Nonfiction and the Novel 1674
Two annual prizes are awarded to honor a novel and a book of nonfiction written from or about the nation's heartland. Winners are notified in August.
Requirements: Eligible books must be published between August 1 of the current year and July 31 of the next.
Amount of Grant: $7500
Date(s) Application is Due: Jul 31
Contact: Literary Awards, (312) 222-4300; fax: (312) 222-3751; email: ctcommunityrelations@tribune.com
Internet: http://about.chicagotribune.com/community/literaryawards.htm
Sponsor: Chicago Tribune
435 N Michigan Avenue
Chicago, IL 60611-4041

Chicano Dissertation Fellowships 1675
The Chicano Studies Department offers two Chicano dissertation fellowships. Candidates must have been advanced to doctoral candidacy by the beginning of the fellowship term. Duties of fellows include working toward the completion of the dissertation and teaching one undergraduate course in areas of research expertise. In sponsoring these fellowships, the department hopes that it will assist promising scholars to complete their dissertations, prepare for university teaching and research, and achieve increased professional recognition and associations. The usual fellowship duration is nine months. Fellows are required to be in residence during the entire fellowship period.
Requirements: To apply, submit a letter of application describing progress toward PhD including date of advancement to candidacy, a dissertation proposal, a curriculum vita, and a writing sample. Applicants must arrange to have two letters of recommendation sent by March 29.
Amount of Grant: $20,000 for nine months plus benefits
Date(s) Application is Due: Mar 29
Contact: Program Contact, Department of Chicano Studies, (805) 893-5546; fax: (805) 893-4076
Internet: http://www.chicst.ucsb.edu/jobs
Sponsor: University of California, Santa Barbara
1713 S Hall
Santa Barbara, CA 93106

Children's Advocacy Center Grants 1676
The program supports the establishment and expansion of children's advocacy centers, with the goal of helping to prevent child abuse. One-year grants are awarded in the following categories: program support, research, member training, prevention, associate member program development, nonmember training, and recognized chapters. Applicants must designate a steering committee that comprises child-welfare workers, law-enforcement officials, medical doctors, and mental-health professionals. Application and guidelines are available online.
Requirements: 501(c)3 tax-exempt organizations are eligible.
Amount of Grant: $4.989 million total; $5000-$50,000
Date(s) Application is Due: Sep 13
Contact: Grants Administrator, (800) 239-9950 ext 116
Internet: http://www.nca-online.org
Sponsor: National Children's Alliance
516 C Street NE
Washington, DC 20002

Children's Brain Tumor Foundation Research Grants 1677
Science grants are awarded for basic laboratory research on pediatric brain and central nervous system tumors. Priority will be given to research that relates to: priorities I and II of the Research and Scientific Priorities for Pediatric Brain Tumors, as listed in the Report of the Brain Tumor Progress Review Group, which can be found at: http://prg. nci.nih.gov/brain/pediatrics.Submission of a two-page, preapplication form, equivalent to a letter of intent, is required. An original, signed form must be submitted along with five copies. The listed deadline is for preapplication forms; full proposals are by invitation.
Requirements: Funding is currently restricted to principal investigators at institutions within the United States.
Restrictions: The foundation does not award grants to individuals or to private foundations, and does not fund debt reduction, capital improvements, or travel expenses. Overhead expenses are not to exceed 10 percent of the total project cost.
Samples: Princeton U (NJ)—to support the work of Jonathan Eggenschwiler, for research on pediatric brain tumors, $150,000 over two years (2004); U of California at San Francisco (CA)—to support the work of Daphne Haas-Kogan, for research on pediatric brain tumors, $150,000 over two years (2004); U of Chicago (IL)—to support the work of Manuel Utset, for research on pediatric brain tumors, $150,000 over two years (2004).
Amount of Grant: $75,000 per year
Contact: Judy Hurley, Executive Director, (212) 448-9494; fax: (212) 448-1022; email: JHurley@cbtf.org
Internet: http://www.cbtf.org/grant_info.html
Sponsor: Children's Brain Tumor Foundation
274 Madison Avenue, Suite 1301
New York, NY 10016-0701

Children's Leukemia Research Association Research Grants 1678
The association supports research efforts into the causes and cure of leukemia and gives patient aid to families in need while meeting the expenses incurred in leukemia treatment. A medical advisory committee consisting of prominent internationally known and respected hematologists reviews grant proposals submitted for consideration of projects that are not otherwise funded. Grants are available for start-up funding for laboratory or clinical investigations in leukemia. Although the association prefers to fund new investigators, applications from established investigators for new initiatives are also solicited. Renewal for a second year is considered if other funding for promising projects has not been obtained.
Restrictions: Previously funded projects are not usually considered for additional funding.
Amount of Grant: $20,000 maximum per year
Date(s) Application is Due: Jun 30
Contact: Executive Director, (516) 222-1944; fax (516) 222-0457; email: info@ childrensleukemia.org
Internet: http://www.childrensleukemia.org/ResearchPast%20Pres.htm
Sponsor: Children's Leukemia Research Association Inc
585 Stewart Avenue, Suite LL-18
Garden City, NY 11530

Children's Literature Association Research Fellowships 1679
The association awards a number of fellowships and scholarships annually to support research and original scholarship associated with serious literary criticism of children's literature. In honor of the achievement and dedication of Dr. Margaret P. Esmonde, proposals that deal with critical or original work in the areas of fantasy or science fiction for adolescents or children will be awarded the Margaret P. Esmonde memorial scholarship. It is expected that the research undertaken will lead to publication and make a significant contribution to the field of children's literature. The award may be used for transportation, living expenses, materials, and supplies. Grant recipients should be prepared to submit either a progress report or a summary of the completed project to the Scholarship Committee in February of the year following the award.
Requirements: Recipients must be members of the association. Application consists of a detailed description of the research proposal, a vita that includes a bibliography of major publications and scholarly achievements, and three letters of reference. Applications and supporting materials should be written in or translated into English.
Restrictions: Officers of the association may not apply. Recipients of a scholarship are not eligible to reapply until the third year from the date of the first award. The award may not be used for obtaining advanced degrees, for textbook writing, for pedagogical projects, or for researching or writing a thesis or dissertation.
Amount of Grant: $500-$1000
Date(s) Application is Due: Feb 1
Contact: Scholarship Committee, (616) 965-8180; fax: (616) 965-3568
Internet: http://ebbs.english.vt.edu/chla/
Sponsor: Children's Literature Association
P.O. Box 138
Battle Creek, MI 49016-0138

Children's Medical Research Institute Grants 1680
The program, which is designed to promote pediatric research as a career, supports summer scholarships for first- and second-year Oklahoma University College of Medicine students. Each year students are awarded grants to support their pediatric research interests as they assist OU Medical Center Department of Pediatrics physicians. The program also awards small research grants/short-term projects to individual scientists

at the OU Medical Center, who often use the funds as seed money to finance pilot research projects.
Contact: Sana Rettig, (405) 271-8001 option 1, ext 42389; fax: (405) 271-1175; email: sana-rettig@ouhsc.edu
Internet: http://www.cmri.net/scholarships.html
Sponsor: Children's Medical Research Institute
800 Research Pwy, Suite 150
Oklahoma City, OK 73104

Children, Cardiomyopathy Foundation Research Grants   **1681**
The foundation awards funds to support research related to all forms of cardiomyopathy affecting children under the age of 18 years. The goal of CCF's research program is to advance medical knowledge on the disease and develop more accurate diagnostic methods, life-improving therapies and ultimately a cure. The grant program is designed to provide seed funding to investigators for the testing of initial hypotheses and collecting of preliminary data to help secure long-term funding by the National Institutes of Health and other major granting institutions.
Requirements: Proposals will be accepted on an annual basis for innovative basic, clinical or translational research relevant to the cause or treatment of cardiomyopathy in children. Principal investigators must hold a MD, PhD or equivalent degree and reside in the United States or Canada. The investigator must have a faculty appointment at an accredited U.S. or Canadian institution and have the proven ability to pursue independent research as evidenced by original research in peer-reviewed journals. Guidelines and application forms can be downloaded at the website.
Samples: 2007: Tain-Yen Hsia, MD, Medical University of South Carolina - ($50,000) Extracellular Mechanisms in Pediatric Cardiomyopathy; Anne I. Dipchand, MD, Hospital for Sick Children, Toronto, Canada - ($42,642) Outcome of pediatric patients with cardiomyopathy: a multi-centre review of pediatric patients listed for transplant in the Pediatric Heart Transplant Study. 2006: Tracie Miller, MD, University of Miami, Miami, FL - ($45,000) Exercise Intervention in a Pediatric Population with Cardiomyopathy
Amount of Grant: $25,000 - $50,000
Date(s) Application is Due: Oct 5
Contact: Lisa Yue, Executive Director; (866) 808-2873; fax: (201) 227-7016; email: lyue@childrenscardiomyopathy.org info@childrenscardiomyopathy.org
Internet: http://www.childrenscardiomyopathy.org/site/grants.php
Sponsor: Children, Cardiomyopathy Foundation (CCF)
P.O. Box 547
Tenafly, NJ 07670

Chiles Foundation Grants   **1682**
The foundation has a deep concern for and confidence in the future of Oregon and the Pacific Northwest. Although the foundation has made a steady commitment to the improvement of the quality of life for those who live and work in this area, it is not restricted in its grant making to the Pacific Northwest. The foundation has traditionally made grants to certain select institutions of higher education for business schools, scholarships, and athletics; supports basic research in certain select medical institutions; supports religion through divinity schools and religious education; and believes that the arts and cultural activities of a community are important and supports certain select, established institutions. Types of support include building construction/renovation, equipment acquisition, and scholarship funds. Annual deadline dates may vary; contact program staff for exact dates.
Requirements: The preferred initial method of contact is a phone call to the grants administration office to determine whether a prospective proposal is within guidelines; if so, an applicant will be invited to submit a one-page written preliminary proposal. An application form will be sent after approval of the preliminary proposal by the executive committee.
Restrictions: No support for projects involving litigation. Grants are not made to individuals, for deficit financing, mortgage retirement, or projects and conferences already completed.
Samples: Boston U (Boston, MA)—for general operating support, $469,700.; Thomas Edison High School (Portland, OR)—for building construction, $20,000.
Amount of Grant: $1000-$10,000 average
Contact: Grants Administrator, (503) 222-2143; email: cf@uswest.net
Sponsor: Chiles Foundation
111 SW Fifth Avenue, Suite 4050
Portland, OR 97204-3643

Chinese-American Librarians Association Sheila Suen Lai Scholarship   **1683**
The Scholarship is designed to encourage the professional and leadership development in Chinese American librarianship. For more information about the application and the Scholarship, visit the CALA's Web page or contact Y. Diana Wu, Chair of the CALA Scholarship Committee.
Requirements: Students of Chinese nationality or Chinese descent who are enrolled in ALA-accredited master's programs or in doctoral programs in Library and Information Science in degree-granting institutions in North America are eligible. The recipient must be enrolled as a full-time student at the time the scholarship is awarded.
Amount of Grant: $500
Date(s) Application is Due: Apr 15
Contact: Y. Diana Wu, CALA Scholarship Committee, (408) 808-2087; email: dwu@sjsu.edu
Internet: http://www.cala-web.org
Sponsor: Chinese-American Librarians Association

P.O. Box 208240, Sterling Memorial Library
New Haven, CT 06520-8240

Chiron Foundation Community Grants   **1684**
The foundation's focus areas are health and medicine, education, and community. Four imperatives guide health-care giving: accelerating progress toward the prevention and cure or successful management of cancer through research, education, early detection, and public-policy debate; combating infectious disease through prevention-related programs, educational efforts, and therapeutics targeting at-risk populations, with emphasis on the special needs of children and families; ensuring the availability and safety of the blood supply and promoting the highest standards of care for blood donors and recipients; and supporting initiatives in the international medical community to provide vaccines and immunization services to protect at-risk populations, especially children, against the devastation of crippling and lethal diseases. Four imperatives guide education giving: providing training and professional development opportunities to enable classroom teachers to teach science and math more effectively; increasing opportunities for economically disadvantaged students to continue on to postsecondary education; supporting community-based training programs to prepare underrepresented minorities and underprivileged groups for careers in the biosciences; and encouraging and assisting promising scholars to pursue advanced research in the biosciences and medicine. Four imperatives guide community giving: providing early diagnosis, intervention, and support for children with disabilities or other special physical or emotional needs; providing essential social services such as meals and housing to those in greatest need; empowering individuals to achieve self-sufficiency through employment training and job-skill development; and enriching the character and celebrating the distinctives of local communities. Requests are accepted at anytime, and proposals are reviewed quarterly.
Requirements: 501(c)3 tax-exempt organizations in Chiron communities (San Francisco East Bay Area, including Alameda, Contra Costa, and Solano Counties; Seattle; and Philadelphia) are eligible.
Restrictions: In general, Chiron does not support organizations that do not have 501(c)3 tax status; religious, fraternal, service, or veterans' organizations; civic or cultural organizations that do not serve the areas in which Chiron is located; alumni drives and teacher organizations; memorials; municipal and for-profit hospitals; labor unions; city, municipal, or federal government departments; organizations or causes that do not support the company's commitment to non-discrimination and diversity; projects of national scope or from national organizations not related to health care; matching gifts; individuals, including scholarships (other than those awarded as part of the company-sponsored college scholarship program); travel support; fund-raising activities related to individual sponsorship; and fund-raising dinners other than those for health care or medical research organizations aligned with company research and product interests.
Geographic Focus: California; Pennsylvania; Washington
Amount of Grant: $5000-$30,000 average, education and community; $15,000-$75,000 health and medicine
Contact: Community Relations, fax: (510) 601-6952; email: Chiron_foundation@chiron.com
Internet: http://www.chiron.com/foundation/index.html
Sponsor: Chiron Foundation
4560 Horton Street
Emeryville, CA 94608

Christensen Fund Regional Grants   **1685**
The fund (TCF) focuses its grantmaking on maintaining the rich diversity of the world—biological and cultural—over the long run, by focusing on four geographic regions: the greater South West (Southwest United States and Northwest Mexico); Central Asia, Turkey, and Iran; the African Rift Valley (Ethiopia); and Northern Australia and Melanesia. Grants within the regional programs are generally directed to organizations based within those regions or, where appropriate, to internationally based organizations working in support of people and institutions on the ground. In general, grants are of one year or less duration; currently grants up to two years are by invitation only.
Requirements: 501(c)3 nonprofit organizations and non-U.S.A institutions with nonprofit or equivalent status in their country of origin are eligible. Partnerships or associations with U.S.A-based nonprofit organizations are preferred.
Restrictions: The fund does not make grants directly to individuals but rather assists individuals through institutions qualified to receive nonprofit support with which such individuals are affiliated.
Amount of Grant: $200,000 maximum; $5000-$100,000 first-time grants
Date(s) Application is Due: Jan 15; Mar 31; Jun 30; Sep 30
Contact: Grants Administrator, (650) 462-8600 ext 106; email: info@christensenfund.org
Internet: http://www.christensenfund.org/index.html
Sponsor: Christensen Fund
145 Addison Avenue
Palo Alto, CA 94301

Christian Gauss Award   **1686**
This award is offered for books published in the field of literary scholarship or criticism. An edition of a literary work is not eligible unless it contains an introduction amounting to a substantial critical or historical estimate that might have been published independently. Entries must be submitted by the publisher who will have obtained a copy of the conditions of eligibility prior to submission. By April 10, if the submission is eligible, applicants will receive the shipping list for judges, and should ship one book to each judge by April 30.

Requirements: Entries must be published in the United States during the 12-month period between May 1 (of the past year) and the April 30 (of the current year) deadline date. Entries will ordinarily be the work of a single author. Exceptions may be made for books written by small teams of scholars working in close collaboration. Authors must be U.S. citizens or residents.

Restrictions: Works of fiction and unpublished manuscripts are not eligible.

Samples: George Hutchinson—for In Search of Nella Larsen: A Biography of the Color Line (2007); Virginia Jackson—for Dickinson's Misery: A Theory of Lyric Reading (2006); Marjorie Garber—for Shakespeare After All (2005).

Amount of Grant: $2,500 maximum

Date(s) Application is Due: Apr 30

Contact: Sandra Beasley, Awards Coordinator, (202) 265-3808; fax: (202) 986-1601; email: SBeasley@pbk.org or info@pbk.org

Internet: http://staging.pbk.org/AM/Template.cfm?Section=Book_Awards&Template=/CM/HTMLDisplay.cfm&ContentID=2646

Sponsor: Phi Beta Kappa Society

1606 New Hampshire Avenue, NW

Washington, DC 20009

Christine Mirzayan Science and Technology Policy Graduate Fellowships   1687

The Christine Mirzayan Science & Technology Policy Graduate Fellowship Program of the National Academies is designed to engage graduate science, engineering, medical, veterinary, business, public policy, and law students in the analytical process that informs the creation of national policy-making with a science/technology element. As a result, students develop basic skills essential to working in the world of science policy.

Requirements: Applications for the fellowships are invited from graduate students through post-doctoral scholars in any physical, biological, or social science field or any field of engineering, medicine/health, or veterinary medicine as well as business, law, education, and other graduate and professional programs. Graduate students and postdoctoral scholars and those who have completed graduate studies or postdoctoral research within the last 5 years are eligible to apply.

Amount of Grant: $8,000

Date(s) Application is Due: May 1

Contact: Program Director, (202) 334-2455; fax: (202) 334-1667; email: policyfellows@nas.edu

Internet: http://www.nationalacademies.org/internship

Sponsor: National Academies

500 5th Street, NW, Room 508

Washington, DC 20001

Christine O. Gregoire Youth/Young Adult Award for Outstanding   1688
Use of Tobacco Industry Documents

The award recognizes a person 24 years of age or younger who has made a contribution to the health of the public in the recent past through use of tobacco documents. The award also honors innovation in the use and application of tobacco industry documents to improve the public's health and, where applicable, to further the goals of tobacco prevention and control in order to help build a world where young people reject tobacco and anyone can quit. Those nominated should be individuals who have made a notable impact through innovative use of tobacco industry documents as applied to research, policy, or advocacy.

Requirements: At least one of the following two criteria must be met by the nominated individual: Nominees must have made a remarkable research, policy, or advocacy contribution with the use of tobacco industry documents, and/or; Nominees must have employed innovative, creative approaches to the employment of tobacco industry documents that result in an improvement in the health or public awareness of a community or nation. Nominations may be made by individuals such as colleagues, peers, coworkers, instructors/professors, governments, or CBOs that are qualified to make such a nomination because of their familiarity with the nominee's work and contribution. The preferred method of submission is through the Legacy Awards website at www.americanlegacy.org/awards/nominate.asp.

Restrictions: Nominees must not have any affiliation with the tobacco industry. Members of the awards committee may not nominate potential recipients, although they may provide written support as part of a nomination package. Employees and directors of the American Legacy Foundation and their relatives are not eligible.

Samples: Andrew Berndt, 2006 Christine O. Gregoire Youth/Young Adult Award winner, grant manager for the Minnesota Department of Health.

Amount of Grant: $7,500

Date(s) Application is Due: Jul 11

Contact: Jennifer Bramble, (202) 454-5555; fax: (202) 454-5599; email: awards@americanlegacy.org

Internet: http://www.americanlegacy.org/49.htm

Sponsor: American Legacy Foundation

Legacy Awards, 2030 M Street, NW, 6th Floor

Washington, DC 20036

Christopher Columbus Fellowship Foundation Frank Annunzio Award   1689

This program presents one $25,000 award in the Science/Technology field for a cutting edge innovation and one $25,000 award in the Alternative Energy Sources (AES) field. The awards recognize individual Americans who are improving the world through ingenuity and innovation, and provide incentive for continuing research and/or a specific project.

Requirements: U.S. citizens may apply.

Samples: Bruce E. Logan, Ph.D., Kappe Professor of Environmental Engineering, Penn State University, University Park, Pennsylvania (2005); Jennifer West, Ph.D., Isabel C. Cameron Professor of Bioengineering and Professor of Chemical Engineering, Rice University, Houston, Texas (2004); James A. Thomson, V.M.D., Ph.D., Diplomate A.C.V.P., John D. McArthur Professor, Department of Anatomy University of Wisconsin-Madison Medical School and the Wisconsin National Primate Research Center, Madison, Wisconsin (2003).

Amount of Grant: $25,000

Date(s) Application is Due: May 13

Contact: Judith M. Shellenberger, Executive Director, (315) 258-0090; fax: (315) 258-0093; email: judithmscolumbus@cs.com

Internet: http://www.columbusfdn.org/otherprograms/frankannunzio/

Sponsor: Christopher Columbus Fellowship Foundation

110 Genesee Street, Suite 390

Auburn, NY 13021

Christopher Columbus Fellowship Foundation Homeland Security Awards   1690

The Foundation bestows this award upon United States individual citizens or companies that are making a measurable and constructive contribution related to basic and/or advanced research in the area of homeland security which will result in a significant and positive benefit to society. Nominations are categorized in five fields: Biological, Radiological, Nuclear, Chemical and Explosive Attacks; Border and Transportation Security; Cyber Security and Information Sharing; Emergency Response to Natural and Man-Made Disasters; and Other. The winner may be chosen from any of the fields.

Requirements: Nominees must be United States citizens or companies. Nomination materials must consist of an official Nomination form available only online.

Samples: Donna H. Branson, Ph.D., Regents Professor and Director, Institute of Protective Apparel Research and Technology, Oklahoma State University, Stillwater, Oklahoma (2006); Timothy M. Swager, Ph.D., John D. MacArthur Professor and Department Head, Department of Chemistry, Massachusetts Institute of Technology, Cambridge, Massachusetts (2005); Matthew Hanson, Ph.D., LSTAT Team, Integrated Medical Systems, Inc., Signal Hill, California (2005).

Amount of Grant: $25,000

Date(s) Application is Due: May 25

Contact: Judith M. Shellenberger, Executive Director, (315) 258-0090; fax: (315) 258-0093; email: judithmscolumbus@cs.com

Internet: http://www.columbusfdn.org/homelandsecurity/

Sponsor: Christopher Columbus Fellowship Foundation

110 Genesee Street, Suite 390

Auburn, NY 13021

Christopher D. Smithers Foundation Research Grants   1691

Activities of the foundation are concentrated in the field of alcoholism. The foundation supports research on treatment, prevention, and public education to create a public awareness of alcoholism as a treatable disease. Small groups, such as treatment groups, shelters, and halfway houses, may apply for one-year seed grants to help them get started. There are no deadlines; applications are accepted year-round. The foundation requests that organizations first write, call, or fax for an annual report and guidelines before applying.

Restrictions: Grants are not awarded to individuals.

Samples: Illinois Church Action on Alcohol Problems (Springfield, IL)—for the alcohol education program in public and private schools in Illinois, $10,000.; Gratitude House (West Palm Beach, FL)—to help women suffering from alcoholism, $5000.

Contact: Grants Administrator, (516) 676-0067; fax: (516) 676-0323; email: info@smithersfoundation.org

Internet: http://www.smithersfoundation.org

Sponsor: Christopher D. Smithers Foundation

P.O. Box 67, Oyster Bay Road

Mill Neck, NY 11765

Christopher Reeve Paralysis Foundation Individual Research Grants   1692

The mission of the foundation is to support cutting-edge research to develop effective treatments and cures for paralysis caused by spinal cord injury and other central nervous system disorders. International individual grants support research that seeks to promote neuronal growth and survival, encourage the formation of synapses, enhance the production of myelin, and restore conduction capabilities in the acutely and chronically injured spinal cord; evaluate drugs or other interventions that protect against secondary neuronal injury or provide insight into the mechanisms causing such damage; elucidate the biological mechanisms underlying approaches to improve concomitant function; and understand anatomical characteristics of injury in both animals and human spinal cord, documenting neuronal systems that are most vulnerable to injury and resultant losses in function. The program is designed to encourage promising new investigators to undertake research on spinal cord regeneration and recovery; encourage researchers who are well-established in other areas to transfer their efforts to spinal cord research; and enable researchers with novel ideas to test their ideas and develop pilot data for seeking larger awards from NIH and other funding sources.

Requirements: Senior scientists, young investigators, and postdoctoral fellows are eligible to serve as principal investigators.

Samples: 13 researchers nationwide—for research on paralysis resulting from spinal-cord injuries, $1.976 million divided (2003).

Amount of Grant: $75,000 maximum per year, senior scientists and young investigators; $60,000 maximum per year postdoctoral fellowships

Date(s) Application is Due: Jun 15; Dec 15

Contact: Dr. Douglas Landsman, (800) 225-0292; email: dlandsman@crpf.org
Internet: http://www.christopherreeve.org/research/researchmain.cfm
Sponsor: Christopher Reeve Paralysis Foundation
500 Morris Avenue
Springfield, NJ 07081

Christopher Reeve Paralysis Foundation Quality of Life and Health Promotions Grants   1693

The foundation awards grants to organizations nationwide that help improve opportunities, access, and day-to-day quality of life for individuals living with disabilities—primarily paralysis—and their families. This program recognizes the unique and numerous needs of these individuals and the importance of providing services and programs that enable them to participate in all areas of life. Quality of Life grants are given to programs or projects that improve the daily lives of people living with disabilities, particularly spinal cord injuries. Funding is awarded twice yearly to programs that provide assistance through access, advocacy, education, recreation, and technology, among others. Health Promotion Grants seek to remove societal and environmental barriers that limit the abilities of individuals with paralysis to participate in life activities. Participation in these activities improves physical and emotional health and prevents secondary conditions for persons living with paralysis. Priority funding goes to programs that focus on paralysis generally and are not specific to one condition or disease that results in paralysis. Complete guidelines are available online.
Requirements: 501(c)3 tax-exempt organizations are eligible.
Restrictions: No grants will be made to individuals, for benefit tickets, or for courtesy advertising.
Samples: To be distributed among 126 nonprofit groups nationwide—for awards through the Quality of Life program, which seeks to improve the well-being of people who are paralyzed, $779,321 (2004).
Amount of Grant: $5000-$25,000 quality of life grants; $25,000 program awards (health promotion); $5000-$10,000 direct impact awards (health promotion)
Date(s) Application is Due: Mar 1; Sep 1
Contact: Donna Valente, Quality of Life Grants Coordinator, (800) 539-7309 or (973) 467-8270 ext 211; email: qol@crpf.org
Internet: http://www.christopherreeve.org/qlgrants/qlgrantsmain.cfm
Sponsor: Christopher Reeve Paralysis Foundation
500 Morris Avenue
Springfield, NJ 07081

Chrysalis Scholarships   1694

The association offers scholarships each year to aid women in finishing their theses and completing MS or PhD degree programs in geoscience fields. The support can be used for anything necessary to assist the candidate in completing her thesis, such as typing, drafting expenses, field work, or child care. The applicant should write a letter stating her background, career goals and objectives, involvement in both the geosciences and her community, how she will use the money, and explaining the length and nature of the interruption to her education. The applicant should also submit two letters of reference.
Requirements: Applicants must be women whose educations have been interrupted for at least one year, candidates for advanced degrees in geoscience field, and who need the money to complete their theses during the current academic year.
Amount of Grant: $2000
Date(s) Application is Due: Mar 15
Contact: Program Contact; email: chrysalis@awg.org
Internet: http://www.awg.org/AWGFoundation/chrysalis.html
Sponsor: Association for Women Geoscientists
P.O. Box 30645
Lincoln, NE 68503-0645

Churchill Fellowships   1695

The fellowships are awarded each year to Australians to undertake overseas study or to conduct an investigative project needing facilities not available in Australia. The value of an applicant's work to the community and the extent to which it will be enhanced by the applicant's overseas study project are considered, as well as the applicant's merit based on past achievements or on demonstrated ability for future achievements in any field. No prescribed qualifications are required. No specific amount of money is attached to a Churchill Fellowship. The average amount awarded is about $25,000 AUD.
Requirements: Australian citizens over the age of 18 are eligible. Applicants must be able to complete a minimum of four weeks and a maximum of approximately eight weeks overseas travel to complete their research. Application must be made in the State of residence of the applicant, irrespective of the place of employment. Application can be made in one State or Territory only.
Restrictions: Fellowships will not be awarded to enable the applicant to obtain higher academic or formal qualifications. Proposals for professional or academic investigations will be considered where the project would primarily benefit the Australian community.
Amount of Grant: $25,000 AUD
Date(s) Application is Due: Feb 28
Contact: Norman Owens, President; (02) 9683 9900 or 0418 667 208; email: norman. owens@abw.org.au
Internet: http://www.churchilltrust.com.au/index.php
Sponsor: Winston Churchill Memorial Trust (Australia)
GP.O. Box 1536
CANBERRA ACT 2612 Australia

CIA Undergraduate Co-Op - Engineer/Program Management Engineer   1696

The co-op program seeks motivated undergraduate students pursuing degrees in Materials, Chemical, Electrical, Mechanical, Computer Engineering, Telecommunications, or Computer Science to work in the CIA's Directorate of Science and Technology. Students will work in an R&D environment work with experienced Program and Project Managers and will be tasked to help plan and prepare program plans. Program Management Engineers focus on identifying new materials, technologies, and techniques that can be integrated into existing business areas. Students will gain exposure to management of research and development projects in support of materials and technologies, to include: requirements definition, analysis, product design and operational use, fabrication techniques, test procedures, and contract implementation and planning. Candidates must be willing to explore Program Management engineer concepts and procedures. Students selected for this program are often attending academic institutions with established cooperative education programs, though not exclusively. They work on an alternating semester or quarterly basis and are expected to spend a minimum of three semesters or four quarters (this can include a summer tour) on the job prior to graduation. The undergraduate co-op program provides increasingly challenging assignments that are commensurate with the students' academic training and individual ability to assume additional responsibility. To apply for this program, see web page listed and submit resume online.
Requirements: Minimum qualifications: Students should be enrolled in Materials, Chemical, Electrical, Mechanical, Telecommunications, or Computer Engineering Program, or a Computer Science BS degree program at an accredited college or university with a GPA of 3.0 or better. Applications for a summer co-op will be accepted until November 1st. Applications for winter, spring, and fall co-ops should be sent 6-9 months before the desired start date.
Restrictions: To be considered for Agency employment, applicants must generally not have used illegal drugs within the last 12 months. The issue of illegal drug use prior to 12 months ago is carefully evaluated during the medical and security processing.
Date(s) Application is Due: Nov 1
Contact: Central Intelligence Agency Office of Public Affairs; (703) 482-0623; fax: (703) 482-1739
Internet: https://www.cia.gov/careers/jobs/view-all-jobs/college-students-scientists-engineers-technology/undergraduate-co-op-program-engineer-program-management-engineer.html
Sponsor: Central Intelligence Agency
Office of Public Affairs
Washington, D.C. 20505

CIA Undergraduate Internship Program - Analysis   1697

The undergraduate internship program seeks motivated undergraduate students to serve internships in the analytic offices and centers of CIA's Directorate of Intelligence. Undergraduate interns work on teams with our full-time analysts. They research, analyze, write, and brief on international political, military, economic, scientific, technical, and leadership developments. In addition to their analytic responsibilities, undergraduate interns become familiar with the Agency and Intelligence Community by participating in a range of meetings and projects. The program allows participants and the Agency to assess opportunities for a graduate studies internship or permanent employment following the participant's completion of undergraduate school. See website to apply for internship.
Requirements: The CIA is looking for students with a variety of majors, including: international affairs, area studies, economics, geography, physical sciences, or engineering. Students selected for this program must have completed one full year of undergraduate school and be continuing school on a full time basis following this assignment. Interns generally are required to work either a combination of one semester and one summer internship or two 90-day summer internships. A GPA of 3.0 or better is required. All applicants must successfully complete a thorough medical and psychological examination, a polygraph interview, and an extensive background investigation. U.S. citizenship is required. To be considered for Agency employment, applicants must generally not have used illegal drugs within the last 12 months. The issue of illegal drug use prior to 12 months ago is carefully evaluated during the medical and security processing.
Contact: Central Intelligence Agency Office of Public Affairs; (703) 482-0623; fax: (703) 482-1739
Internet: https://www.cia.gov/careers/jobs/view-all-jobs/college-students-analytical-positions/undergraduate-internship-program-analysis.html
Sponsor: Central Intelligence Agency
Office of Public Affairs
Washington, D.C. 20505

CICU James C. Ross Research Fellowship   1698

The nine-month research fellowship focuses on higher education and its relationship to state and national priorities. Preference is given to candidates who have a doctorate or have completed all but their dissertation and have a proven track record of communicating research findings to a general audience. The fellow will work in CICU's Albany office, and is expected to produce a final paper by the first Monday in October of next year. Housing is the responsibility of the fellow. To apply, submit a cover letter, 500-1000 word outline of a proposed course of research, resume, recent publication or unpublished manuscript, and two confidential letters of reference. Annual deadlines may vary; contact program staff for exact dates.
Restrictions: Class work is prohibited during the full-time fellowship, though credit arrangements for fellowship participation are permissible.
Amount of Grant: $20,000 stipend
Date(s) Application is Due: Dec 1

Contact: Sheila Seery, (518) 436-4781; fax: (518) 436-0417; email: sheila@cicu.org
Internet: http://www.cicu.org/CMT/frontpage/04RossApplication.pdf
Sponsor: Commission on Independent Colleges and Universities
P.O. Box 7289, 17 Elk Street
Albany, NY 12224

CIES African Regional Special Program and Country Grants   1699
The program makes up to 12 awards in all academic fields for research in sub-Saharan Africa. Fellows will conduct research in one country for three to nine months or in two or three countries for five to nine months. Grantees are expected to give occasional lectures and seminars in consultation with host universities and U.S. embassies. Lecturing/research awards in specified fields also are available in some countries: contact program staff for a current listing.
Requirements: Applicants must be U.S. citizens with the PhD or equivalent professional/terminal degree at the time of application. Foreign language proficiency in the field of the advertised assignment or proposed lecturing/research activity is required.
Samples: Victoria Bernal, U of California (Irvine, CA)—for research project Women Making History: Exploring Gender, Nationalism, and Culture through the Personal Narratives of Women Fighters at U of Asmara, Eritrea; Craig Stanford, U of Southern California (Los Angeles, CA)—for research project Comparative Ecology of Chimpanzees and Mountain Gorillas at Uganda Wildlife Authority, Uganda.
Contact: Debra Egan, Program Contact, (202) 686-6230; email: degan@cies.iie.org
Internet: http://www.iie.org/cies/download/2006_07AWARDS_CATALOG.pdf
Sponsor: Council for International Exchange of Scholars
3007 Tilden Street NW, Suite 5L
Washington, DC 20008-3009

CIIT Postdoctoral Fellowships   1700
Ten to 15 fellowships are offered annually to support persons holding recently earned PhD, MD, or DVM degrees during further training in genetic toxicology, immunotoxicology, biochemical toxicology, pathology, reproductive toxicology, epidemiology, teratology, and carcinogenesis. Fellowships are tenable at the institute for two to three years. The institute holds 501(c)6 status. Applications are available on the Web site.
Requirements: Applicants who recently were awarded PhD degrees in a discipline related to toxicology are eligible to apply; those holding recently awarded DVM or MD degrees are expected to have substantial research experience.
Restrictions: Traineeships and fellowships are awarded only for research conducted at the CIIT. Awardees may not use their stipends to supplement funds from other awards.
Amount of Grant: $27,000 first year; $28,500 second year
Contact: Rusty Bramlage, Human Resources Manager, (919) 558-1331; fax: (919) 558-1300; email: bramlage@ciit.org
Internet: http://www.ciit.org/careers/openings_postdoc.asp
Sponsor: Chemical Industry Institute of Toxicology
P.O. Box 12137
Research Triangle Park, NC 27709

CIMO Scholarships for Advanced Studies of the Finnish Language   1701
at Finnish Universities
CIMO offers scholarships for advanced studies of the Finnish language at Finnish universities. The scholarships are meant mainly for degree students of Finnish language and literature at universities outside Finland. Preference is given to applicants who are working on their Master's thesis. Scholarships may be applied for one academic semester (3-5 months). The monthly allowance is approximately 725 euros. No additional allowance for housing is paid. The scholarship holder is also responsible for expenses due to international travel to and from Finland.
Requirements: Students apply directly to CIMO for the grant. Application deadlines are in April for the autumn semester and in October for the Spring semester. The applicant must establish contact with the host university department before applying. The application should preferably be made in Finnish.
Amount of Grant: 725 EURO per month
Date(s) Application is Due: Apr 1; Oct 1
Contact: M?'kel?' Tarja, Program Coordinator; +358 (0)207 868 556; email: makela.tarja@cimo.fi
Internet: http://www.studyinfinland.fi/scholarships/ungergraduate_studies/finnish_studies_and_research.html
Sponsor: Center for International Mobility (CIMO)
P.O. Box 343, Hakaniemenkatu 2
Helsinki FI-00531Finland

Cinnabar Foundation Grants   1702
The foundation awards grants to nonprofits to promote environmental and wildlife conservation and protection. Types of support include general operating support, conferences and seminars, research, and scholarships to individuals. There are no application forms.
Requirements: Idaho, Montana, and Wyoming nonprofit organizations may apply.
Geographic Focus: Idaho; Montana; Wyoming
Amount of Grant: $1000-$15,000 range
Date(s) Application is Due: Mar 15
Contact: James Posewitz, c/o Holmes and Turner, (406) 449-2795; fax: (406) 449-9985; email: cinnabar@mt.net
Sponsor: Cinnabar Foundation
P.O. Box 5088
Helena, MT 59604

CINOA Prize   1703
This art history prize is given by the International Confederation of Art Dealers (Confederation Internationale des Negociants en Oeuvres d'Art) to encourage the study of the history of art in the 19 countries where the confederation is represented—Australia, Austria, Belgium, Czech Republic, Denmark, England, France, Germany, Ireland, Italy, the Netherlands, New Zealand, Portugal, San Marino, South Africa, Spain, Sweden, Switzerland, and the United States. An author wishing to be considered in the competition should forward a copy of his/her bound manuscript, a curriculum vita, a brief summary, and a support letter from a professor or other qualified scholar in the field to which the work relates, as well as a letter from an editor/publisher willing to publish the work if it is granted the CINOA Prize. Contact the organization via email for more information.
Requirements: To be eligible, a candidate must be a resident of one of the member nations, and his/her work should preferably be concerned with an aspect of the art or art history of one of the CINOA countries.
Amount of Grant: $10,000 to the publisher and an all-expense paid trip to the next annual CINOA meeting
Date(s) Application is Due: Nov 15
Contact: CINOA Prize, (212) 940-8925; fax: (212) 940-6484; email: secretary@cinoa.org
Internet: http://www.cinoa.org
Sponsor: Art Dealers Association of America
575 Madison Avenue
New York, NY 10022

CIRCLE Civic Education at the High School Level Research Grants   1704
The program seeks research that will help educators and policymakers to improve civic outcomes for U.S. students of high-school age (roughly 14 to 18). Civic outcomes include, but are not limited to knowledge of politics, democracy and civil society; knowledge of social issues; values such as tolerance, trust, patriotism, concern for others' rights and well-being, and efficacy (the belief that one can make a difference); skills and habits of deliberating about public issues and participating in politics and community affairs; volunteering and membership in voluntary and/or nonprofit groups; and intentions to vote or to consider careers in public service (in the government or nonprofit sectors). CIRCLE is interested in research on interventions and reforms that may enhance civic outcomes. These interventions and reforms include, but are not limited to programs of civic education and classes on history, democracy, or law; approaches to the teaching of other disciplines that may have civic benefits; co-curricular activities, including student government and student media; service-learning; games and simulations that involve political or civic issues; student voice or participation in the governance of their schools; the basic structure of high schools (including their size, focus, requirements, climate, admissions criteria, or composition); professional development for teachers, so long as the effects on students can be assessed; and after-school or community-based programs, insofar as these have the potential to reach large numbers of adolescents or to change mainstream education. In most CIRCLE-funded research projects, the outcomes will be civic knowledge, values, skills, or behaviors. However, research that explores whether being civically engaged helps academic outcomes or positive adolescent development also will be considered. The listed application deadline is for letters of inquiry; full proposals are by invitation.
Requirements: CIRCLE welcomes proposals from academics, students (especially PhD candidates at the dissertation stage), independent scholars, practitioners, and research nonprofits and firms. CIRCLE also welcomes proposals from youth of high school age, perhaps working in partnership with adults. Such youth-led research proposals will be evaluated separately and not compared directly to proposals from adults.
Restrictions: CIRCLE funds rigorous research, not advocacy, education, or other forms of practice.
Amount of Grant: $500,000 total; $100,000 maximum
Date(s) Application is Due: Dec 15
Contact: Carrie Donovan, (301) 405-2790; email: cdonovan@umd.edu
Internet: http://www.civicyouth.org/grants/index.htm
Sponsor: University of Maryland
School of Public Policy
College Park, MD 20742

CIU.S. Darcovich Memorial Doctoral Fellowship   1705
The institute invites applications for one doctoral thesis fellowship, nonrenewable. The award is intended to aid a student in completing a thesis on a Ukrainian or Ukrainian Canadian topic in education, history, law, humanities, arts, social sciences, women's studies, and library sciences. The fellowship will be awarded only in the thesis year of an academic program and only for thesis work. Only in exceptional circumstances may an award be held concurrently with other awards.
Requirements: A Canadian citizen or permanent resident may hold the fellowship at any institution of higher learning in Canada or elsewhere. For non-Canadian applicants, preference will be given to students enrolled at the University of Alberta.
Amount of Grant: $C12,000 maximum
Date(s) Application is Due: Mar 1
Contact: Helen Darcovich Memorial Endowment Fund, (780) 492-2973; fax: (780) 492-4967; email: cius@ualberta.ca
Internet: http://www.ualberta.ca/CIU.S./cius-grants.htm
Sponsor: Canadian Institute of Ukrainian Studies
University of Alberta, 450 Athabasca Hall
Edmonton, AB T6G 2E8 Canada

CIU.S. Master's Fellowship 1706

The institute invites applications for one master's thesis fellowship, nonrenewable. The award is intended to aid a student to complete a thesis on a Ukrainian or Ukrainian Canadian topic in education, history, law, humanities, arts, social sciences, women's studies, and library sciences. The fellowship will be awarded only in the thesis year of an academic program and only for thesis work. Only in exceptional circumstances may an award be held concurrently with other awards.

Requirements: A Canadian citizen or permanent resident may hold the fellowship at any institution of higher learning in Canada or elsewhere. For non-Canadian applicants, preference will be given to students enrolled at the University of Alberta.

Samples: Larissa Kotowich, U of Alberta (Edmonton, Alberta)—fellowship recipient in the history and classics, for Catholic/Orthodox Interaction among Ukrainians in Interwar Western Canada.

Amount of Grant: $C10,000 maximum

Date(s) Application is Due: Mar 1

Contact: Marusia and Michael Dorosh Endowment Fund, (780) 492-2973; fax: (780) 492-4967; email: cius@ualberta.ca

Internet: http://www.ualberta.ca/CIU.S./cius-grants.htm

Sponsor: Canadian Institute of Ukrainian Studies

University of Alberta, 450 Athabasca Hall

Edmonton, AB T6G 2E8 Canada

CIU.S. Neporany Research and Teaching Fellowship 1707

A research and training fellowship in Ukrainian studies will be awarded by the Osyp and Josaphat Neporany Educational Fund. The fellowship will be tenable at any university with research facilities at which the fellow's academic Ukrainian studies specialty may be pursued and the fellow enabled to teach a course related to the specialty. The duration of the fellowship will normally be one term, i.e., half of the academic year. This may be extended where the applicant has been successful in receiving supplemental funding from other sources. Institutions and departments who wish to explore the possibility of hosting a fellow also are encouraged to request information.

Requirements: Applicants must hold doctorates, or have equivalent professional achievement, in Ukrainian studies.

Amount of Grant: $C20,000 maximum

Date(s) Application is Due: Mar 1

Contact: Fellowships Coordinator, (780) 492-2973; fax: (780) 492-4967; email: cius@ualberta.ca

Internet: http://www.ualberta.ca/CIU.S./cius-grants.htm

Sponsor: Canadian Institute of Ukrainian Studies

University of Alberta, 450 Athabasca Hall

Edmonton, AB T6G 2E8 Canada

CIU.S. Research Grants 1708

The Michael and Daria Kowalsky Endowment Fund invites applications for research grants in Ukrainian and Ukrainian Canadian studies in history, literature, language, education, social sciences, and library sciences. Application forms and the guide to research applications are available from the office.

Samples: Iaroslav Dashkevych, National Academy of Sciences of Ukraine (Lviv, Ukraine)—for compilation of archival materials of the Shevchenko Scientific Society, held in the Biblioteka Narodowa, Warsaw.

Date(s) Application is Due: Mar 1

Contact: Program Officer, (780) 492-2973; fax: (780) 492-4967; email: cius@ualberta.ca

Internet: http://www.ualberta.ca/CIU.S./cius-grants.htm

Sponsor: Canadian Institute of Ukrainian Studies

University of Alberta, 450 Athabasca Hall

Edmonton, AB T6G 2E8 Canada

Civic Education Project Grants 1709

The Special Project Grants Program will provide support for projects designed to strengthen community, secondary and higher education by introducing innovative content, methods and materials of teaching and research, strengthening academic and scholarly exchange and fostering school and university linkages to the community. The Program will award Special Project Grants to the graduates of Muskie/FSA Graduate Fellowship Program currently residing in the NIS and Baltic States.

Requirements: Applicants must be fluent in English. Teaching experience is preferred. Applicants must have successfully completed a Muskie/FSA Graduate Fellowship Program and reside in one of the following countries: Armenia, Azerbaijan, Belarus, Estonia, Georgia, Kazakhstan, Kyrgyzstan, Latvia, Lithuania, Moldova, Russia, Tajikistan, Turkmenistan, Ukraine, Uzbekistan. Applications from groups led by Muskie/FSA alumni will be accepted.

Restrictions: Members of staff of CEP, OSI, American Councils, other organizations directly involved in the SCOUT or any other Muskie/FSA alumni support program supervision or administration, and individuals holding similar grants or fellowships supporting their academic activity, which overlap in time with the proposed SCOUT grant activity, are not eligible to receive any SCOUT program grants.

Amount of Grant: $1000-$5000

Date(s) Application is Due: Apr 1

Contact: Sara Werth, Program Director, (202) 663-7793; fax: (202) 663-7799; email: swerth2@jhu.edu

Internet: http://www.cep.org.hu/programs/spannouncement.html

Sponsor: Civic Education Project

2360 N Vermont Street

Arlington, VA 22207

CLA Nurses Respiratory Society Fellowships 1710

The objective of the fellowships is to permit nurses to pursue graduate study so that they may be better able to contribute to the Canadian field of respiratory illness and health care. Applicants may register in any master's or doctoral program where knowledge, skill, and expertise may develop toward this objective. The program should show evidence of a plan to foster theory, research, and either a clinical or functional area of specialization. Application forms may be obtained from the association.

Requirements: Applicant must be a Canadian citizen or permanent Canadian resident, be a registered nurse, be enrolled full-time in a graduate program at the master's or doctoral level, and be a member of CNRS.

Amount of Grant: $C3750-$C7500

Date(s) Application is Due: Feb 1

Contact: Grants Administrator, (416) 864-9911; fax: (416) 864-9916; email: orcs@on.lung.ca

Internet: http://www.on.lung.ca/orcs/fellowships.html

Sponsor: Canadian Lung Association

1900 City Park Drive, Suite 508

Gloucester, ON K1J 1A3 Canada

CLA Nurses Respiratory Society Nursing Research Grants 1711

Research grants are offered to those studying respiratory nursing, chronic or acute lung disease, which will result in improved quality of patient care. Research may include clinical investigation of any nursing-related phenomenon; be pertinent to illness assessment, management, or responses; or be aimed at health promotion and prevention issues. Studies may use either quantitative or qualitative methodologies.

Requirements: Applicant must be a Canadian citizen or permanent Canadian resident; a registered nurse; hold an appointment in, or have an affiliation with, a health care agency, educational institution, or other organization in Canada that can administer the funds in an approved manner; and be a member of CNRS.

Amount of Grant: $C3000-$C30,000

Date(s) Application is Due: Nov 1

Contact: Grant Administrator, (613) 747-6776; fax: (613) 747-7430; email: dhogg@cha.ab.ca

Internet: http://www.lung.ca/resp/Fellowships/Fellow_Grants.htm

Sponsor: Canadian Lung Association

1900 City Park Drive, Suite 508

Gloucester, ON K1J 1A3 Canada

CLA Physiotherapy Cardio-Respiratory Society Fellowships 1712

The purpose of the research program is to pursue increased scientific knowledge in the area of cardio-respiratory physiotherapy practice. Fellowships are offered to physiotherapists pursuing postgraduate training, with respiratory research as the major component. The award period is one year.

Requirements: Applicants must be Canadian citizens or permanent Canadian residents, be registered physiotherapists, be enrolled or accepted for full-time study in a graduate program at the master's or doctoral level, and be a member of the Canadian Physiotherapy Cardio-Respiratory Society.

Amount of Grant: $C6000-$C12,000

Date(s) Application is Due: Nov 1

Contact: Administrator, Societies & Special Projects, (613) 569-6411; email: nprt@lung.ca

Internet: http://www.lung.ca/crhp/index.html

Sponsor: Canadian Lung Association

1900 City Park Drive, Suite 508

Gloucester, ON K1J 1A3 Canada

CLA Scholarship for Minority Students in Memory of Edna Yelland 1713

Awards encourage and support ethnic minority group students in the attainment of a graduate library degree in library or information science. Award announcement will be made June 15. Application forms will be sent upon written request.

Requirements: California residents and U.S. citizens and permanent U.S. residents may apply. To be eligible, applicants must be either enrolled or accepted for enrollment in a master's program in an accredited California graduate library school. The applicant must be an ethnic minority of one of the following groups: Native American, African American, Mexican American, Latin/Hispanic origin, Asian American or Pacific Islander, or Filipino. Applicants must also provide evidence of financial need.

Amount of Grant: $2500

Date(s) Application is Due: May 31

Contact: Scholarship Committee Chair, (916) 447-8541; fax: (916) 447-8394; email: info@cla-net.org

Internet: http://cla-net.org/awards/ednayelland.php

Sponsor: California Library Association

717 20th Street, Suite 200

Sacramento, CA 95814

Claneil Foundation Grants 1714

The foundation awards grants to nonprofit organizations in its areas of interest, including community development, education, arts and culture, environment, health, women, domestic abuse, and social services. Types of support include capital campaigns, equipment acquisition, continuing support, building construction/renovation, exchange programs, and research. Deadlines listed are for letters of intent. Application forms are required.

Requirements: Nonprofit organizations are eligible. Preference is given to organizations serving Pennsylvania counties including Bucks, Montgomery, Chester, Delaware, and Philadelphia.

Restrictions: Grants are not awarded to governmental organizations, individuals, or religious organizations.

Geographic Focus: Pennsylvania

Samples: New England Forestry Foundation (Groton, MA)—to purchase the largest forestland easement in U.S. history, which will protect 762,192 acres in Maine from development, $10,000.

Amount of Grant: $3000-$50,000 range

Date(s) Application is Due: Jun 30; Dec 15

Contact: Cathy Weiss, Executive Director, (610) 941-1131; email: cweiss@claneil.com

Internet: http://www.claneil.org

Sponsor: Claneil Foundation Inc
630 W Germantown Pike, Suite 400
Plymouth Meeting, PA 19462

Clapham Prize for Historical Research 1715

The prize is awarded under the auspices of the William Wilberforce Papers Project and will be awarded biannually to the scholar judged to have completed the most outstanding scholarly work related to Wilberforce during the preceding two-year period. All nominees will be contacted for further information.

Amount of Grant: $5000

Date(s) Application is Due: Feb 1

Contact: Debbie Drost, Program Manager, (978) 867-4365; fax: (978) 867-4673; email: drost@gordon.edu

Internet: http://www.gordon.edu/ccs

Sponsor: Gordon College
255 Grapevine Road
Wenham, MA 01984

Clarence E. Heller Charitable Foundation Grants 1716

The charitable foundation supports nonprofit organizations, with priority given to proposals from California, in its areas of interest, including environment and health—to prevent serious risk to human health from toxic substances and other environmental hazards by supporting programs in research, education, and policy development; management of resources—to protect and preserve the earth's limited resources by assisting programs that demonstrate how natural resources can be managed on a sustainable and an ecologically sound basis, and supporting initiatives for sustainable agriculture, and for promoting the long-term viability of communities and regions; music—to encourage the playing, enjoyment, and accessibility of symphonic and chamber music by providing scholarship and program assistance at selected community music organizations and schools, and by helping community-based ensembles of demonstrated quality implement artistic initiatives, diversify and increase audiences, and improve fund-raising capacity; and education—to focus on support for programs that improve the teaching skills of educators and artists in environmental and arts education. Types of support include continuing support, general operating expenses, publications, research, seed money, and special projects. The foundation's board meets typically in March, June, and October. Letters of inquiry are accepted at any time.

Requirements: 501(c)3 tax-exempt organizations are eligible.

Restrictions: Grants are not made to individuals.

Geographic Focus: California

Samples: Environmental Working Group (Oakland, CA)—for a project to research the links between air pollution and respiratory problems in California, $75,000 (2004); California Sustainable Agriculture Working Group (Santa Cruz, CA)—for support of the Sustainable Agriculture Policy Advocacy Project, $30,000 (2004); Young Musicians Program, UC Berkeley (CA)—to provide music instruction for low-income students, $20,000 (2004); San Francisco Education Fund (CA)—for the Literacy Network, a professional development activity to improve reading instruction, $25,000 (2004).

Amount of Grant: $5000-$600,000; $10,000-$50,000 average

Contact: Bruce Hirsch, Executive Director, (415) 989-9839; fax: (415) 989-1909; email: info@cehcf.org

Internet: http://cehcf.org/app_info.html

Sponsor: Clarence E. Heller Charitable Foundation
1 Lombard Street, Suite 305
San Francisco, CA 94111-1130

Clark Foundation Grants 1717

The foundation awards grants in New York, NY, and upstate New York for charitable and educational purposes and in support of health care, youth, cultural, environmental, and community organizations. Types of support include general operating support, continuing support, annual campaigns, capital campaigns, building construction/renovation, equipment acquisition, program development, seed money, and scholarships to Cooperstown County residents.

Requirements: Nonprofit organizations in upstate New York and New York, NY, are eligible.

Geographic Focus: New York

Amount of Grant: $20,000-$200,000 average range

Date(s) Application is Due: Jan 1; Apr 1; Jul 15; Oct 1

Contact: Charles Hamilton, Executive Director, (212) 977-6900

Sponsor: Clark Foundation
1 Rockefeller Plz, 31st Fl
New York, NY 10020

Clark Library Short-Term Resident Fellowships 1718

These short-term fellowships facilitate access to the Department of Special Collections for researchers and scholars residing outside of the Los Angeles area. Fellowships are held April through December for periods of one to three months. The department collects primary resources in the humanities and social sciences. Application requirements include a cover letter, a curriculum vita, a brief outline of the research and specific collections to be used (two-page maximum), dates to be spent in residence, and three letters of recommendation. Information on the library's holdings, application, and guidelines are available on the Web site.

Requirements: Scholars holding PhD or equivalent degrees involved in a project suitable to the collection are eligible to apply.

Samples: Recent fellowship awards include: Chard, Chloe, Newnham College, Cambridge— Laughter and Imaginative Geography, 1700-1830; Chrissochoidis, Ilias, Stanford University—From the London Stage to Westminster Abbey: Cultural Mobility of Handel, Oratorios in Britain, 1732-1784; Christie-Miller, Ian, London University— The Paper Used in London-Printed Books of 1526 at the Clark and Huntington Libraries; Christov, Theodore, UC Los Angeles— Natural Rights, State Sovereignty, and Law of Nations: The Emergence of International Law Theories, 1625-1795; Clark, William, Independent scholar— The Romance of the Scientist: A Modern Hagiography & Demonology

Amount of Grant: $2000 per month

Date(s) Application is Due: Feb 1

Contact: Fellowship Coordinator, (323) 731-8529; fax: (323) 731-8617; email: clarklib@humnet.ucla.edu

Internet: http://www.humnet.ucla.edu/humnet/c1718cs/Postd.htm

Sponsor: William Andrews Clark Memorial Library
2520 Cimarron Street
Los Angeles, CA 90018

Clark-Huntington Joint Bibliographical Fellowship 1719

Sponsored jointly by the Clark Library and the Huntington Library, this two-month fellowship provides support for bibliographical research in early modern British literature and history as well as other areas where the two libraries have common strengths. The Application form is an online submission, and the applicant should forward a description of research plans, specifying any material located at the Clark and the Huntington Libraries (maximum of 1,000 words, double-spaced).

Requirements: Applicants should hold a PhD degree or have appropriate research experience.

Date(s) Application is Due: Feb 1

Contact: Camie Howard-Rock, Fellowship Coordinator; (310) 206-8552; fax: (310) 206-8577; email: camie@humnet.ucla.edu or clarklib@humnet.ucla.edu

Internet: http://www.humnet.ucla.edu/humnet/c1718cs/applic4.htm

Sponsor: UCLA Center for 17th- and 18th-Century Studies
310 Royce Hall, 405 Hilgard Avenue
Los Angeles, CA 90095-1404

Claude Pepper Foundation Grants 1720

The foundation makes grants primarily to support the work of the Claude Pepper Center and the Pepper Institute on Aging and Public Policy, both located at Florida State University, Tallahassee, FL. The foundation makes limited grants to other organizations to continue the work and vision of Claude and Mildred Pepper. The foundation also supports a visiting scholars program and an oratory competition for Florida students. Grants are usually made for a period of one year and except in rare instances no grant will be made for longer than a three-year period of time. Guidelines are available online.

Requirements: 501(c)3 tax-exempt organizations are eligible.

Samples: Florida Council on Aging - To support Special Plenary Sessions at the Annual Aging Network Conferences; National Academy of Social Insurance - To sponsor the Academy's annual conference opening dinner in Washington, D.C., which addressed issues surrounding Social Security reform; Tallahassee Senior Center - To co-sponsor the Claude Pepper Senior Walk/Run, which consists of a competition for walkers over the age of 60, and one for runners, with prizes and refreshments.

Contact: John T. Herndon, Executive Director; 850) 644-9309; fax: (850) 644-9301; email: herndon@claudepepperfoundation.org

Internet: http://www.claudepepperfoundation.org/programs_grants.cfm

Sponsor: Claude Pepper Foundation
636 West Call Street
Tallahassee, FL 32306-1122

Clay Foundation Grants 1721

The primary mission of the foundation is to promote and enhance the quality of life of the citizens of West Virginia. Initially, grants will be confined to organizations or projects within the greater Kanawha Valley that significantly affect its residents. The foundation considers a broad range of organizations, though special interest is given to programs in the field of aging; health care, research, and education; vocational education; and services to disadvantaged youth and their families. Applicants are asked to submit a preliminary letter. If foundation priorities and resources permit consideration of the request, a detailed formal proposal may be requested. There are no deadline dates for the submission of preliminary letters. The board meets four times annually, generally in January, April, July, and October, at which times applications are considered and grants made.

Requirements: IRS 501(c)3 tax-exempt organizations in West Virginia are eligible.

Restrictions: Grants are not made for ongoing normal operations, debt retirement or operational deficits, contributions to endowment or scholarship funds, annual appeals

or other fund-raising events, national fund-raising campaigns, religious organizations for religious purposes, or grants to conduit organizations.
Geographic Focus: West Virginia
Samples: U of Charleston (WV)—to construct the Buckner and Lyell Clay Ctr, $1.15 million.; Charleston Catholic High School (Charleston, WV)—for the science wing, $60,000.; Fund for the Arts (Charleston, WV)—challenge grant, $60,000. West Virginia State Museum—for renovation, $350,000.
Amount of Grant: $5000-$75,000 average
Contact: Charles Avampato, President, (304) 344-8656; fax: (304) 344-3805
Sponsor: Clay Foundation
1426 Kanawha Blvd E
Charleston, WV 25301

Clements Center for Southwest Studies Research Fellowships   1722
The fellowship is open to individuals in any field in the humanities or social sciences doing research on Southwestern America. The fellowships are designed to provide time for senior or junior scholars to bring book-length manuscripts to completion. Fellows are expected to spend the academic year at SMU, teach one course during the two-semester duration of the fellowship, and participate in center activities. Applicants should send two copies of their vita, a description of their research project, and a sample chapter or extract, and arrange to have letters of reference sent from three persons who can assess the significance of the work and the ability of the scholar to carry it out.
Amount of Grant: $37,000 stipend
Date(s) Application is Due: Jan 16
Contact: David Weber, Director, (214) 768-3684; email: swcenter@smu.edu
Internet: http://www.smu.edu/swcenter/announce.htm
Sponsor: William P. Clements Center for Southwest Studies
P.O. Box 750176, Dallas Hall, Room 356
Dallas, TX 75275-0176

Clements-DeGolyer Library Fellowships   1723
The center and the DeGolyer Library offer fellowships to encourage broader and more intensive research on its holdings, which focus largely on the history of the trans-Mississippi west and railroad history.
Requirements: Any serious researcher in any field is invited to apply. In general, the library awards fellowships to those individuals who have demonstrated strong qualifications in scholarship, publication, or teaching.
Restrictions: Fellowships are not awarded to individuals living within the Dallas/Fort Worth metropolitan area.
Amount of Grant: $500 per week
Contact: Russell Martin, Director, (214) 768-1233 or (214) 768-3234; email: swcenter@smu.edu or rlmartin@smu.edu
Internet: http://www.smu.edu/swcenter/clemdeg.htm
Sponsor: William P. Clements Center for Southwest Studies
P.O. Box 750176
Dallas, TX 75275-0176

Cleveland-Cliffs Foundation Grants   1724
Contributions are made to nonprofit organizations to enhance the quality of life of Cleveland-Cliffs Inc employees and in recognition of a corporate responsibility toward educational, health, welfare, civic, and cultural matters within the communities where the company operates. The foundation was formed for the purpose of making contributions to groups organized and operated exclusively for religious, charitable, scientific, literary, or educational purposes and for the prevention of cruelty to children or animals. Types of support include general operating support, annual campaigns, capital campaigns, building/renovation, professorships, scholarship funds, research, and employee matching gifts. The foundation's major emphasis is on supporting education through a matching gift program and direct contributions to educational institutions. Requests for support must be in writing.
Requirements: Nonprofit organizations in the mining communities in which Cleveland-Cliffs Inc operates, including Michigan, Minnesota, and the greater Cleveland area, are eligible.
Geographic Focus: Michigan; Minnesota; Ohio
Samples: Case Western Reserve U (Cleveland, OH)—for operating support, $25,000.; Great Lakes Museum of Science, Environment, and Technology (Cleveland, OH)—for capital support, $15,000.; Cleveland Bicentennial Commission (Cleveland, OH)—for operating support, $20,000.
Amount of Grant: $250-$50,000
Contact: Dana Byrne, Vice President, (216) 694-5700; fax: (216) 694-4880; email: publicrelations@cleveland-cliffs.com
Internet: http://www.cleveland-cliffs.com/general/community/foundation.asp
Sponsor: Cleveland-Cliffs Foundation
1100 Superior Avenue
Cleveland, OH 44114-2589

CLF Fellowships   1725
The fellowships provide support for specialized clinical or experimental training in hepatic function or disease to individuals who have already completed the basic graduate program. The value of the fellowship is based on the qualifications and seniority of the applicant and will include, if applicable, a travel grant equivalent to return economy airfare for the grantee and spouse between the university of origin and the place of training. Guidelines and application forms are available from the office and in medical schools through the office of the Dean of the Faculty of Medicine.

Requirements: Candidates must hold an MD or PhD degree or equivalent. Canadian citizens and landed immigrants who are resident in Canada at the time of application or Canadian citizens who are normally resident in Canada are eligible. Applications must be sponsored by a Canadian Faculty of Medicine or Health Sciences and accompanied by a letter of support from the Dean (or designee).
Amount of Grant: $C30,000 maximum per year
Date(s) Application is Due: Mar 31
Contact: Research Awards Administrator, (800) 563-5483 or (416) 491-3353; fax: (416) 491-4952; email: clf@liver.ca
Internet: http://www.liver.ca/Research/Grant_Program
Sponsor: Canadian Liver Foundation
2235 Sheppard Avenue E, Suite 1500
Toronto, ON M2J 5B5 Canada

CLF Graduate Studentships   1726
Studentships enable academically superior students to undertake full-time studies in Canadian universities in disciplines relevant to liver function, structure, and disease. Awards are for one academic year and may be renewed three times for a total of four years of support. Registrants in an MS program are restricted to a maximum of three years of support. Guidelines are available from the office or at medical schools through the office of the Dean of the Faculty of Medicine.
Requirements: The candidate must be accepted into a full-time university graduate science program in a medically related discipline related to a master's or doctoral degree; hold a record of superior academic performance; and be sponsored by a faculty supervisor with a record of productive medical research and sufficient, competitively acquired research funding to ensure the satisfactory conduct of the student's research during the term of the award.
Amount of Grant: $C16,500 per year
Date(s) Application is Due: Mar 31
Contact: Billie Potkonjak, (800) 563-5483; fax: (416) 491-4952; email: bpotkonjak@liver.ca
Internet: http://www.liver.ca/Research/Grant_Program
Sponsor: Canadian Liver Foundation
2235 Sheppard Avenue E, Suite 1500
Toronto, ON M2J 5B5 Canada

CLF Operating Grants   1727
The grants provide funding to help fully qualified and trained hepatologists and basic scientists begin liver research projects in their own laboratories at Canadian universities. Candidates must be prepared to spend 80 percent of their time in research and to commit the remaining time to active participation in teaching and/or patient care.
Requirements: Candidates must hold an MD or PhD or equivalent and have a proven interest in liver structure, function, or disease. The applicant must have completed a minimum of two and preferably three years of formal research training (post medical specialty training in the case of MDs) and have obtained additional research experience as a clinical investigator or postdoctoral fellow under the supervision of a well-established, competitively funded, senior basic or clinical scientific faculty colleague who holds a faculty level position.
Amount of Grant: $C60,000 maximum per year for two years
Date(s) Application is Due: Mar 31
Contact: Research Awards Administrator, (416) 964-1953; fax: (416) 964-0024; email: bpotkonjak@liver.ca
Internet: http://www.liver.ca/Research/Grant_Program
Sponsor: Canadian Liver Foundation
2235 Sheppard Avenue E, Suite 1500
Toronto, ON M2J 5B5 Canada

Clowes ACS/AAST/NIGMS Jointly Sponsored Mentored Clinical   1728
Scientist Development Award
The American College of Surgeons and American Association for the Surgery of Trauma announce a program that will provide supplemental salary funding of up to $75,000 per year to an individual who has received a Mentored Clinical Scientist Development Award (K08/K23) from the National Institute for General Medical Science (NIGMS). This award is directed at surgeon-scientists working in the early stages of their research careers. The award supports a three-, four-, or five-year period of supervised research experience that may integrate didactic studies with laboratory or clinical research. The award program offers a means to facilitate the career development of individuals pursuing careers in trauma surgery research by enhancing salary support over and above that offered by the K08/K23 mechanism.
Requirements: Awardees must be members in good standing of the College and eligible for membership in AAST.
Restrictions: Pre-existing applications and awards are not eligible for consideration.
Amount of Grant: $75,000 per year maximum
Date(s) Application is Due: Oct 12
Contact: Kate Early, Administrative Assistant; (312) 202-5281; email: kearly@facs.org
Internet: http://www.facs.org/memberservices/acs-aast-nigms.html
Sponsor: American College of Surgeons
633 N. Saint Clair Street
Chicago, IL 60611-3211

CMA Foundation Grants   1729
The foundation seeks to improve the quality of life for central Ohio residents through the promotion of wellness, prevention of disease, and delivery of services and research.

The foundation will review funding requests for matching or challenge grants, multiyear commitments, and capital expenditures on a project-by-project basis. The foundation is the leading local source of funding for health care projects benefiting the local community. Application may be made online.

Requirements: Nonprofit organizations in central Ohio, including Franklin, Delaware, Fairfield, Licking, Madison, Pickaway, and Union Counties, are eligible.

Geographic Focus: Ohio

Samples: Children's Hunger Alliance (Columbus, OH)—to promote school-breakfast programs, $30,000 (2003); Educational Council Foundation (Columbus, OH)—to train middle- and high-school educators to provide information about positive health behaviors, $50,000 maximum (2003); Franklin County Health Dept (Columbus, OH)—to expand its anti-tobacco activities, $30,000 maximum (2003); Mount Carmel Health Systems Foundation (Columbus, OH)—to provide education, information, and self-care skills training to people who have experienced heart failure, $90,000 (2003).

Contact: Contact, (614) 240-7410; email: yourthoughts@goodhealthcolumbus.org

Internet: http://www.cmaf-ohio.org/cmaf/grants.html

Sponsor: Columbus Medical Association Foundation
431 E Broad Street
Columbus, OH 43215

CMAP Physician Advocacy Fellowship 1730

This program supports doctors to develop or enhance their advocacy skills by implementing a project in partnership with an advocacy organization. The fellowship seeks to make advocacy a core professional value for physicians by developing a cadre of advocates with expertise in achieving system or policy-level social change at the local, state, and national level. Projects must be focused within the United States and should identify system or policy level changes as the outcomes of the fellowship work. Application guidelines are available online. Deadline listed is for short proposals; full proposals are by invitation.

Requirements: U.S. physicians at all stages of their careers are eligible; however, the most competitive applicants are practicing physicians. Prospective applicants must secure the commitment of an advocacy organization that is prepared to house, mentor, and support them throughout the fellowship period. A list of advocacy organizations that are interested in participating in the fellowship is available online. Applicants also may apply with organizations other than those listed online.

Samples: Robert Goodman—for efforts to minimize the influence of commercial interests, particularly those of the pharmaceutical industry, in medical education and training, $111,804.

Amount of Grant: $40,000-$80,000 salary support; $2000 travel funds to MAP-sponsored meetings; $5000 maximum overhead support to organizations

Date(s) Application is Due: Oct 7

Contact: Claudia Calhoon, Program Manager, (212) 342-4769; email: cmc100@columbia.edu

Internet: http://www.cmap.columbia.edu/research_fellowship.shtml

Sponsor: Columbia University
630 W 168th Street, P&S Box 11
New York, NY 10032

CMHC External Research Program Grants for Housing Research 1731

The objective of this program is to encourage and enable individuals in the private and nonprofit sectors to carry out independent housing research of high quality. Current housing research priority areas include sustainable development, housing in the national/international economy, housing affordability, people with differing needs, rental markets, housing renovation, and building performance and technical innovation. Competitions are held annually following distribution of application brochures. Successful applicants enter into grant agreements with CMHC, the end products of which are research reports, disseminated free of charge to the public by CMHC.

Requirements: Financial assistance is provided to Canadian citizens or those having permanent resident status in Canada who are independent housing researchers, as well as to those employed in Canadian universities, institutions, private consulting firms, the professions, and the housing industry.

Restrictions: Full-time students at the graduate or undergraduate level are not eligible to apply but may be used as research assistants on the project.

Amount of Grant: $C25,000 maximum

Date(s) Application is Due: Oct 31

Contact: Administrator, CMHC External Research Program, (613) 748-2300 ext. 3061; fax: (613) 748-2402; email: erp@cmhc-schl.gc.ca

Internet: http://www.cmhc-schl.gc.ca/en/prfias/gr/exrepr/index.cfm

Sponsor: Canada Mortgage and Housing Corporation
700 Montreal Road
Ottawa, ON K1A 0P7 Canada

CMS Historically Black Colleges and Universities Health Services Grants 1732

The purpose of the grant program is to support researchers in implementing health services research activities to meet the needs of diverse CMS beneficiary populations. The goals of the grant program are to: 1) encourage HBCU health services researchers to pursue research issues which impact the Medicare, Medicaid, and SCHIP (State Children's Health Insurance Program) programs, 2) assist CMS in implementing its mission focusing on health care quality and improvement for its beneficiaries, 3) assist HBCU researchers by supporting extramural research in health care capacity development activities for the African American communities, 4) increase the pool of HBCU researchers capable of implementing the research, demonstration, and evaluation activities of CMS, and 5) assist in fostering interuniversity communication

and collaboration regarding African American health disparity issues. Funding is available for grants to implement research related to health care delivery and health financing issues affecting African American communities, including issues of access to health care, utilization of health care services, health outcomes, quality of services, cost of care, health and racial disparities, socio-economic differences, cultural barriers, managed care systems, and activities related to health screening, prevention, outreach, and education.

Requirements: To be eligible for grants under this program, an organization must be an HBCU (Historically Black College or University) and meet one of the following three *Requirements*: 1) offer a Ph.D. or Master's Degree Program in one or more of the following disciplines - Allied Health, Gerontology, Health Care Administration, Health Education, Health Management, Nursing, Nutrition, Pharmacology, Public Health, Public Policy, Social Work; or 2) have a School of Medicine; or 3) a member of the National HBCU Network for Health Services and Health Disparities. All proposals should describe research to be conducted with relevance to the CMS Medicare, Medicaid, and SCHIP programs and which area of Healthy People 2010 is served by this project. Applications must be submitted electronically (via grants.gov). CMS will accept a hard copy if the applicant is unable to access grants.gov or is having serious problems in sending the application electronically.

Restrictions: Grant funds may not be used for any of the following: to provide direct services to individuals except as explicitly permitted under the grant solicitation; to match any other Federal funds; to provide services, equipment, or supports that are already the legal responsibility of another party under Federal law.

Amount of Grant: $200,000 - $250,000

Date(s) Application is Due: Jul 2

Contact: Joi Grymes, (410) 786-7251; email: Joi.Grymes@cms.hhs.gov

Internet: http://www.cms.hhs.gov/ResearchDemoGrantsOpt/02_Historically_Black_Colleges_and_Universities.asp#TopOfPage

Sponsor: Centers for Medicare & Medicaid Services
Office of Acquisition and Grants Management, 7500 Security Boulevard, C2-21-15
Baltimore, MD 21244-1850

CMS Research and Demonstration Grants 1733

The general purpose of the Centers for Medicare & Medicaid Services' (CMS) research and demonstration program is to conduct and support projects to develop, test, and implement new health care financing and payment policies and to evaluate the impact of the agency's programs on its beneficiaries, providers, States, and other customers and partners. The scope of the agency's activities embraces all areas of health care: costs, access, quality, service delivery models, and financing and payment approaches.

Requirements: The following themes represent the agency's current priorities in research. Note that all projects must fall into the agency's statutory authorities to operate and improve Medicare, Medicaid, and other CMS programs and activities: (1) Monitoring and Evaluating CMS Programs; (2) Strengthening Medicaid, State Children, Health Insurance Program (SCHIP), and State Programs; (3) Expanding Beneficiaries' Choices and Availability of Managed Care Options; (4) Developing FFS Payment and Service Delivery Systems; (5) Improving Quality of Care and Performance Under CMS Programs; (6) Improving the Health of Our Beneficiary Population; (7) Prescription Drugs; (8) Building Research Capacity. Application packet is available online.

Restrictions: Applicants are expected to contribute towards the project costs. Generally 5 percent of the total project costs is considered acceptable. CMS rarely approves grants or cooperative agreements for research or demonstration projects in which the Federal Government covers 100 percent of the project's costs. The budget may not include costs for construction or remodeling or for project activities that take place before the applicant has received official notification of our approval of the project.

Amount of Grant: $25,000 - $1,000,000; avg. $235,000

Date(s) Application is Due: none

Contact: Grant Officer, Office of Acquisition and Grants Management; (410) 786-5130; email: Jnorris1@cms.hhs.gov

Internet: http://www.cms.hhs.gov/ResearchDemoGrantsOpt/04_Other_CMS-Grant_Opportunities.asp

Sponsor: Centers for Medicare & Medicaid Services
Office of Acquisition and Grants Management, 7500 Security Boulevard, C2-21-15
Baltimore, MD 21244-1850

CMU Arts in Society Fellowship 1734

The program awards postdoctoral fellowships to scholars and artists in any field of humanistic inquiry or artistic endeavor. Two fellows will be appointed to the Center for the Arts in Society and will be awarded a stipend plus a small research grant. Fellows will teach one undergraduate course each semester that contributes to the center's arts-histories curriculum. Fellows also participate in the work of the center, which sponsors conferences, lectures, and colloquia on various topics, and give one public lecture. Candidates should submit a statement of current research interests as they relate to the current year's fellowship theme; a one-page proposal for a one-semester undergraduate course related to at least two disciplines—one in the arts and one in the humanities; a full curriculum vita; and at least three letters of recommendation. Applications should be submitted by postal mail.

Requirements: Scholars and artists who have received their terminal degree within the last four years in any field of humanistic inquiry or artistic endeavor are eligible.

Amount of Grant: $35,000 stipend plus $5000 research grant

Date(s) Application is Due: Feb 3

Contact: Judith Schachter, Director, (412) 268-3239; email: jm1e@andrew.cmu.edu

Internet: http://www.hss.cmu.edu/cas/Content/visiting_fellows.htm

Sponsor: Carnegie Mellon University

Dean's Office, BH 154
Pittsburgh, PA 15213

Coastal Bend Community Foundation Grants   1735

The community foundation awards grants to nonprofit organizations in the Aransas, Bee, Jim Wells, Kleberg, Nueces, Refugio and San Patricio counties of Texas. Grants are usually made to provide seed money for innovative and start-up programs that will generate additional future funding or revenues. Capital projects will also receive favorable consideration. The board of directors approves grant recipients in early November after considering recommendations presented by the Grants Committee. Areas of interest include alcohol and drug abuse, libraries and literacy, arts and culture, higher and other education, adult basic education, child welfare, hospitals, community development, animal welfare, human services, and general charitable giving. Grants are awarded for general operating support, program and project development, equipment, seed money, scholarship funds, and fellowships. Application forms are not required.

Requirements: Nonprofit organizations in Texas in Aransas, Bee, Jim Wells, Kleberg, Nueces, Refugio, and San Patricio counties, may submit grant proposals.
Geographic Focus: Texas
Amount of Grant: $3.4 million total; $250-$50,000
Date(s) Application is Due: Sep 1
Contact: Jim Moloney, Executive Vice President, (361) 882-9745; fax: (361) 882-2865; email: jmoloney@cbcfoundation.org
Internet: http://www.cbcfoundation.org/grant.html
Sponsor: Coastal Bend Community Foundation
600 Building, Suite 1716
Corpus Christi, TX 78473

Coca-Cola Foundation Grants   1736

The foundation has established education as its philanthropic focus and set aside most of its funds to support educational initiatives that address pressing needs. To help prepare youth for life, the foundation gives in three areas: higher education—pipeline programs that connect various levels of education and help students stay in school, scholarships, and minority advancement; classroom teaching and learning—innovative K-12 projects, teacher development, and small projects that deal with classroom activities; and global education—projects that encourage international studies, global understanding, and student exchange. Grants are awarded to both public and private institutions at all levels of education: universities, colleges, and secondary and elementary schools. International educational institutions and health care organizations also receive consideration. Types of support include annual campaigns, donated equipment, employee matching gifts, operating budgets, special projects, capital campaigns, continuing support, fellowships, internships, endowment funds, matching funds, and scholarship funds. Proposals may be submitted at any time.

Requirements: IRS 501(c)3 nonprofits are eligible.
Restrictions: The foundation does not make grants to individuals, religious endeavors, political or fraternal organizations, or organizations without 501(c)3 status.
Amount of Grant: $25,000-$200,000 average
Date(s) Application is Due: Mar 1; Jun 1; Sep 1; Dec 1
Contact: Executive Director, (404) 676-2568; fax: (404) 676-8804
Internet: http://www2.coca-cola.com/citizenship/foundation_guidelines.html
Sponsor: Coca-Cola Foundation
P.O. Box 1734
Atlanta, GA 30301

Cockrell Foundation Grants   1737

The foundation awards grants to Texas nonprofit organizations in its areas of interest, including higher education, cultural programs, social services, youth agencies, religious education and religious welfare, and hospitals. Types of support include annual campaigns, capital campaigns, building construction/renovation, endowment funds, general operating support, program development, fellowships, professorships, scholarship funds, research, and matching grants. There are no application forms or deadline dates. The board meets in the spring and fall each year.

Requirements: Texas 501(c)3 nonprofit organizations in Houston are eligible.
Geographic Focus: Texas
Samples: Barbara Bush (The) Foundation for Family Literacy—$15,000; Child Advocates, Inc.—$12,000; Houston Museum of Natural Science—$20,000; Homes of St. Mark—$10,00.
Amount of Grant: Median award in 2005, $12,000
Contact: M. Nancy Williams, Executive Vice President, (713) 209-7500; email: foundation@cockrell.com
Internet: http://www.cockrell.com/foundation/grant_guidelines.asp
Sponsor: Cockrell Foundation
1000 Main Street, Suite 3250
Houston, TX 77002

Coeta and Donald Barker Foundation Grants   1738

The foundation awards grants to California and Oregon nonprofit organizations in its areas of interest, including arts, children and youth, community development, disabled, environmental conservation, family services, federated giving, health care and health organizations, heart and circulatory research, higher education, hospitals, mental health, and secondary school education. Types of support include building construction/renovation, equipment acquisition, general operating support, program development, and scholarship funds.

Requirements: California and Oregon nonprofit organizations are eligible.

Restrictions: Grants do not support sectarian religious purposes, federal and tax-dependent organizations, individuals, or endowment funds.
Geographic Focus: California; Oregon
Samples: Santa Barbara Zoological Gardens (Santa Barbara, CA), $5000.
Amount of Grant: $100-$20,450 range
Date(s) Application is Due: Mar 1; Aug 1
Contact: Nancy Harris, Executive Administrator, (760) 324-2656; fax: (760) 321-8662
Sponsor: Coeta and Donald Barker Foundation
P.O. Box 936
Rancho Mirage, CA 92270

Coleman Foundation Cancer Care Grants   1739

Since the 1980's, the Foundation has been an active advocate for raising the standards of cancer care in the Midwest region. Cancer care including support services, treatment and research are primary initiatives of the Foundation. The Foundation has funded specific health care programs, community-based and psychosocial support services, cancer research, development of new treatments, and improvement of direct patient services. Additionally, the Foundation has supported both cancer-specific and general health patient programs.

Requirements: Nonprofits across the Midwest are eligible. Preference is given to requests from Illinois and Chicago.
Restrictions: The program does not fund for-profit businesses, individuals, individual scholarships, ad books, tickets, equipment purchases (including computer hardware or software), or advertising.
Geographic Focus: Illinois; Midwest
Samples: Sinai Health System, Illinois, $328,000—towards capital support for Oncology Infusion Center and OB Triage & Ante partum Testing, to improve oncologic and obstetric facilities, enhance direct patient service areas, and streamline delivery of health care (2008); Wellness House , Illinois, $3,000—to support the 2008 Psychosocial Conference for medical and psychosocial oncology professionals to understand the implication of the Cancer Care for the Whole Patient report published by the Institute of Medicine and discuss ways to improve comprehensive care of cancer patients (2008).
Contact: Michael Hennessy, President; (312) 902-7120; fax: (312) 902-7124; email: info@colemanfoundation.org
Internet: http://www.colemanfoundation.org/cancer.html
Sponsor: Coleman Foundation, Inc.
651 West Washington Boulevard, Suite 306
Chicago, IL 60661

College Art Association Professional Development Fellowships   1740

The fellowship provides two years of funding: a grant in the first, and in the second, CAA provides assistance in securing employment or an internship at a museum, university, or art center and subsidizes the position. Contact the office to request an application form. Applications will also be available in most art and art history graduate departments.

Requirements: Artists and art historians from culturally diverse backgrounds who have been underrepresented in the field due to their race, religion, gender, age, national origin, sexual orientation, disability, or history of economic disadvantage who demonstrate distinction in approach, technique, or perspective in their contribution to the discipline of art or art history and will receive the MFA, terminal MA, or PhD degree in the current funding year, who can demonstrate financial need, and who are citizens or permanent residents of the United States are eligible to apply.
Amount of Grant: $5000 for the first year
Date(s) Application is Due: Jan 31
Contact: Stacy Miller, (212) 619-1051 ext 242; fax: (212) 627-2381; email: smiller@collegeart.org or fellowship@collegeart.org
Internet: http://www.collegeart.org/caa/career/fellowship.html
Sponsor: College Art Association
275 Seventh Avenue
New York, NY 10001

College of Human Sciences Scholarships   1741

Human sciences undergraduate and graduate scholarships are available in the College of Human Sciences at Texas Tech University. Programs include human development and family studies, marriage and family therapy, interior design, merchandising, family financial planning, clothing and textiles, restaurant/hotel management, food and nutrition, early childhood, family and consumer sciences education, consumer economics, environmental design, and fashion design. Application deadline is February 1 unless specifically stated otherwise. Check Web site for current details and criteria of available scholarships.

Requirements: Scholarships are awarded only to students who have officially applied to Texas Tech University. Unless the scholarship award is for multiple years, students are required to resubmit scholarship applications during the spring semester to be eligible for renewal of their scholarship.
Amount of Grant: $200-$2000 generally
Date(s) Application is Due: Feb 1
Contact: Beverly Pinson, Coordinator, Human Sciences Scholarship Committee, (806) 742-3031; email: beverly.pinson@ttu.edu
Internet: http://www.hs.ttu.edu/aas/scholarships/default.php
Sponsor: Texas Tech University, College of Human Sciences
Box 41162
Lubbock, TX 79409-1162

College of Saint Rose Dissertation Fellowships   **1742**

The college supports two minority scholars in the completion of dissertations leading to a doctoral degree. In addition to completing the dissertation by the end of the year, fellows will teach one course, usually in their discipline. They will discuss their research and work with Saint Rose faculty and students at selected events throughout the year. After successful completion of the fellowship year, fellows will be invited to apply for full-time faculty positions. To apply, submit a letter of application, curriculum vita, and contact information of three references.

Requirements: Eligibility requirements include: record of outstanding academic achievement; member of historically underrepresented racial or ethnic group, including, but not limited to: African Americans, Alaskan Natives, American Indians or Native Americans, Asian Americans, Latino/as, Chicano/as, and Pacific Islanders; enrollment in a full-time academic program leading to the doctoral degree at the time of application; admission to degree candidacy before the dissertation fellowship is awarded; approval of dissertation proposal by the applicant's committee prior to application; and U.S. citizenship.

Amount of Grant: $20,000 stipend; $5000 research expenses
Date(s) Application is Due: Nov 15
Contact: Dr. Davide Szcerbacki, Provost, email: szcerbd@strose.edu
Internet: http://www.strose.edu
Sponsor: College of Saint Rose
432 Western Avenue
Albany, NY 12203

Colonel Stanley R. McNeil Foundation Grants   **1743**

The foundation awards grants to eligible Illinois nonprofit organizations in its areas of interest, including child welfare, education, health care, and youth. Types of support include building construction/renovation, equipment acquisition, general operating support, matching grants, research, and special projects. There are no application deadlines or forms.

Requirements: Illinois nonprofit organizations serving the Chicago metropolitan area are eligible.
Restrictions: Grants are not made to individuals.
Geographic Focus: Illinois
Samples: ChildServ (Chicago, IL)—to support ChildServ's Lake County Family Service Center, $45,000 (2005).
Amount of Grant: $7000-$250,000 range
Contact: Charles Slamar, Jr., Vice President, Bank of America, (312) 828-8028
Sponsor: Colonel Stanley R. McNeil Foundation
231 S LaSalle Street
Chicago, IL 60697-0246

Colorado Bioscience Discovery Evaluation Grants   **1744**

The program was first created by the Colorado legislature in 2006 with the purpose being to improve and expand the evaluation of new bioscience discoveries at research institutions with the intent of accelerating the development of new products and services ‚Äi essentially a proof of concept program. 2007 legislation modified the original program by providing additional funding and targeting the new funding to SBIR/STTR recipients and biofuels research. Proof-of-Concept projects can receive a maximum grant of $150,000. SBIR/STTR projects can receive a maximum grant of $100,000.

Requirements: Eligible applicants include any institution located and operating in Colorado that is a public or private, nonprofit institution of higher education, a nonprofit teaching hospital, or a private, nonprofit medical and research center. For-profit entities are not eligible for the program. However, an Office of Technology Transfer must be affiliated with one of these research institutions in order for the Office of Technology Transfer to be eligible to apply under the Program. Application forms and guidelines are available for download at the sponsor's website.
Restrictions: For-profit entities are not eligible for the program.
Geographic Focus: Colorado
Date(s) Application is Due: Jun 30
Contact: Sonya Guram or Alice Kotrlik, Business Finance Division; (303) 892-3840; fax: (303) 892-3848; email: sonya.guram@state.co.us or alice.kotrlik@state.co.us
Internet: http://www.colorado.gov/cs/Satellite?c=Page&childpagename=OEDIT%2FOEDITLayout&cid=1167928017742&p=1167928017742&pagename=OEDITWrapper
Sponsor: Colorado Office of Economic Development and International Trade
1625 Broadway, Suite 2700
Denver, CO 80202

Colorado Clean Energy Fund Solar Innovation Grants (SIG)   **1745**

The program was established to support innovative programs that can demonstrate a strategy and implementation plan for breaking down financial, educational, political, and technical barriers to greater penetration of solar electric and solar thermal technologies in the residential and commercial sectors. Approximately three hundred and fifty thousand dollars ($350,000) is available during the current competitive cycle. Individual awards will not exceed $50,000.

Requirements: Funding requests will be considered for the following purposes: (a) Education - Applicants will provide educational resources to targeted sectors with the goal of expanding the quantity and quality of solar installations in Colorado. (b) Integrated Design - Applicants will address opportunities to incorporate solar technology into the commercial and production home sectors in both new and existing infrastructure. (c) Utility Programs - Applicants in this category will work to improve the viability of solar technology as a means to offset energy and power demands at the utility level. GEO is soliciting innovative solutions to promoting solar technology at the utility level. (d) Market Analysis & Policy Recommendations - Applicants will identify policy barriers to the wider implementation of solar technology in Colorado and present solutions in the form of a comprehensive policy report. (e) Financing - Applicants will present innovative financing models that will increase the affordability of integrating solar technology in the residential and commercial sectors. Potential recipients include, but are not limited to: Utilities; Home Builders; Installers / System Integrators; Lenders / Financial Institutions; Non-Profits; Trade Associations. Preference will be given to proposals that demonstrate the ability to generate substantial matching funds, however matching funding is not required should the applicant demonstrate sufficient need.
Restrictions: Grant dollars will not be used to buy down the cost of solar systems. Only electronic applications will be accepted.
Geographic Focus: Colorado
Amount of Grant: up to $50,000
Date(s) Application is Due: Apr 20
Contact: Jeff Lyng, Solar Programs Manager; (303) 866-2264; email: jeff.lyng@state.co.us
Internet: http://www.colorado.gov/energy/resources/SolarInnovationGrantProgramGovernorsEnergyOfficeColorado.asp
Sponsor: Colorado Governor's Energy Office (GEO)
1580 Logan Street, Suite 100
Denver, CO 80203

Columbia Earth Institute Postdoctoral Fellowships   **1746**

The institute seeks applications from innovative postdoctoral candidates interested in pathbreaking disciplinary research as well as multidisciplinary initiatives on sustainable development issues. The program provides young scholars with the opportunity to enhance their foundation in one of the institute's core disciplines—earth sciences, biological sciences, engineering sciences, social sciences, and health sciences—while at the same time acquiring the cross-disciplinary expertise and breadth needed to address critical issues related to reducing poverty, hunger, disease, and environmental degradation, placing special emphasis on the needs of the world's poor. Application and guidelines are available online. The institute prefers applicants to apply online.

Requirements: Candidates are asked to submit a proposal for research, based in one of the core disciplines, that will contribute to the goal of global sustainable development. Applicants also are requested to indicate the general direction of multidisciplinary training and research opportunities that they would like to pursue at the Earth Institute.
Date(s) Application is Due: Dec 1
Contact: Program Coordinator, (212) 854-3893; fax: (212) 854-6309; email: hd6@columbia.edu
Internet: http://www.earthinstitute.columbia.edu/postdoc
Sponsor: Columbia University
2910 Broadway, B-16 Hogan Hall, MC 3277
New York, NY 10025

Columbia Foundation Grants   **1747**

The foundation currently has three program areas: arts and culture, human rights, and sustainable communities and economies. Types of support include general operating support, program development, publication, seed money, and research. Multiyear grants are made on occasion. A complete application includes application, one-page proposal summary, proposal not to exceed five pages, budgets, funding sources, board of directors, certificate of tax-exempt status, and other descriptive materials. The board of directors generally meets twice a year to consider grant applications. Applicants should submit a one- to two-page inquiry letter with cover sheet. Proposals are invited from the foundation. The deadline is May 1 for arts and culture grants; August 1 for human rights grants; and December 1 for sustainable communities and economies grants. Annual deadlines may vary; contact program staff for exact dates.

Requirements: The foundation considers proposals only from organizations certified by the IRS as public charities. Priority will be given to applications from the San Francisco Bay, CA, area. International grants support the arts in London.
Restrictions: The foundation does not customarily provide support for operating budgets of established agencies, recurring expenses for direct services or ongoing administrative costs, individual fellowships or scholarships, or agencies wholly supported by federated campaigns or heavily subsidized by government funds.
Geographic Focus: CaliforniaUnited Kingdom
Samples: Royal National Theatre (London, England)—for the world premiere of The Talking Cure, a new play about Sigmund Freud and Carl Jung by leading UK playwright Christopher Hampton, $50,000 (2003); Gender Public Advocacy Coalition (Washington, DC)—to support its work to end gender-based stereotypes, discrimination and violence at the national level, $25,000.; Ctr for Food Safety (Washington, DC)—for the California Food and Agriculture Initiative, to establish and staff a California office to focus on legal issues, policy initiatives, and a statewide public-education campaign about the opportunities for California to make the transition to sustainable food systems, $40,000 (2003).
Amount of Grant: $25,000-$100,000
Date(s) Application is Due: May 1; Aug 1; Dec 1
Contact: Susan Clark, Executive Director, (415) 561-6880; fax: (415) 561-6883; email: info@columbia.org
Internet: http://www.columbia.org
Sponsor: Columbia Foundation
P.O. Box 29470
San Francisco, CA 94129

Columbia University Center for Comparative Literature and Society   1748
Postdoctoral Fellowship
The main purpose of the center is to rethink comparative literary and cultural studies in their relation to area studies and the historically oriented social sciences. The fellow will be given time and resources to develop his or her scholarship in a broadening and experimental cross-disciplinary and cross-regional context. The fellow will join an intellectually vibrant community of scholars from the humanities, the social sciences, architecture, and law affiliated with the center. Guidelines and application are available online.
Requirements: Applicants must have received the PhD between January 1, 1999, and July 1, 2005, to be eligible.
Amount of Grant: $41,000 stipend, $1000 travel
Date(s) Application is Due: Oct 15
Contact: Director, Center for Comparative Literature and Society, (212) 854-4541; fax: (212) 662-7289; email: ccls@columbia.edu
Internet: http://www.columbia.edu/cu/ccls/academics/postdocs/intro/index.html
Sponsor: Columbia University
2960 Broadway, MC 5700
New York, NY 10027

Columbia University Society of Fellows in the Humanities   1749
The Columbia Society of Fellows, with support from the Andrew W. Mellon Foundation and the Kenan Trust, will appoint a number of postdoctoral fellows in the humanities. The society seeks to enhance the role of the humanities in the university by exploring and clarifying the interrelationship within the humanities, as well as their relationship to the natural and social sciences. The appointment as fellow, considered as equivalent to the rank of lecturer, is for one year with the expectation of renewal for a second year. The award covers costs of independent research and for teaching in the undergraduate program in general education. Normal fringe benefits will be added. Additional funds are available to support research. Contact the director for eligibility information and application forms.
Requirements: Applicants must have received their PhD between January 1, 1999 and July 1, 2005.
Amount of Grant: $50,000 stipend
Date(s) Application is Due: Oct 1
Contact: Judy Huyck, (212) 854-4631; fax: (212) 854-4069; email: sof-fellows@columbia.edu
Internet: http://ci.columbia.edu/w0250/fellowship.html
Sponsor: Columbia University
2960 Broadway, Mail Code 5700
New York, NY 10027

Columbus Foundation Ann Ellis Fund Grants   1750
The Fund is used for eye research, with primary emphasis on sponsoring scientific research on the diagnosis, prevention, and treatment of glaucoma. Once this is accomplished, the fund will assist in the research of other hidden eye problems. For more information, please contact the Community Research and Grant Management Department at the Foundation
Restrictions: Individuals are ineligible. Requests for religious purposes, budget deficits, endowments, conferences, or projects that are normally the responsibility of a public agency are generally not funded.
Date(s) Application is Due: Oct 1
Contact: Dottie Henderson, Executive Assistant for Community Research and Grants Management; (614) 251-4000; fax: (614) 251-4009; email: dhenderson@columbusfoundation.org
Internet: http://www.columbusfoundation.org/gogrants/targeted_needs/specialized_grants.aspx
Sponsor: Columbus Foundation
1234 East Broad Street
Columbus, OH 43205

Commonfund Endowment Management Prize   1751
Commonfund, which manages the world's largest pool of endowment and cash assets for universities, colleges, independent schools, and health care facilities, sponsors this competition for research and scholarship in the advancement of endowment management. The prize is awarded in two installments—upon announcement of the winner and upon acceptance of the final paper. This paper will be published and distributed by Commonfund. The winner is announced in March.
Requirements: Applicants must submit a sample completed paper or working paper written for an academic or other audience, a proposal of 1000 words or less outlining how the applicant would develop a derivative paper written for the endowment management audience, and a curriculum vita.
Amount of Grant: $50,000
Date(s) Application is Due: Dec 31
Contact: John Griswold Jr., Senior Vice President, (203) 563-5185; email: jgriswol@cfund.org
Internet: http://www.commonfund.org/Commonfund/About+Us/pillars_educational_programs_2004.htm
Sponsor: Commonfund
P.O. Box 812, 15 Old Danbury Road
Wilton, CT 06897-7717

Commonweal Foundation Community Assistance Grants   1752
The foundation supports educational programs and projects assisting disadvantaged, at-risk youth. The foundation focuses on secondary and, to a lesser extent, elementary education. The foundation also considers grants for educational research and, to a limited extent, health care. Guidelines are available online.
Requirements: 501(c)3 tax-exempt organizations located in the District of Columbia, Maryland, or Northern Virginia are eligible. Applicant organizations should have an annual budget not exceeding $1 million.
Geographic Focus: District of Columbia; Maryland; Virginia
Amount of Grant: $25,000 maximum
Date(s) Application is Due: Mar 1; Aug 1
Contact: Gloria Dairsow, (240) 450-0000; fax: (240) 450-4115; email: gdairsow@cweal.org
Internet: http://www.cweal.org/cag.htm
Sponsor: Commonweal Foundation
10770 Columbia Pike, Suite 150
Silver Spring, MD 20901

Commonweal Foundation Grants   1753
The foundation awards grants to eligible nonprofit organizations serving disadvantaged youth in the area comprising the corridor between Baltimore, MD, and Washington, DC. It also offers some assistance to elementary education, educational research, and some health care causes. Programs include Partners in Learning (address low literacy skills of economically disadvantaged children), community assistance grants (i.e., after-school tutoring, parenting classes, and shelters for the homeless or abused women), and learning disabilities support program (special education services to economically disadvantaged children). Contact the office for application procedures.
Requirements: 501(c)3 tax-exempt organizations serving disadvantaged youth in the corridor between Baltimore, MD, and Washington, DC, are eligible.
Geographic Focus: District of Columbia; Maryland; Virginia
Amount of Grant: $10,000-$25,000
Date(s) Application is Due: Feb 1; May 1; Aug 1; Nov 1
Contact: Renee Cottman-Reyes, (240) 450-0000; fax: (240) 450-4115; email: gdairsow@cweal.org
Internet: http://www.commonweal-foundation.org/index.html
Sponsor: Commonweal Foundation
10770 Columbia Pke, Suite 150
Silver Spring, MD 20901

Commonwealth Fund Affordable Health Insurance Grants   1754
The Program on Affordable Health Insurance envisions an efficient and equitable health insurance system that makes available to all Americans comprehensive, continuous, and affordable coverage. In support of that vision, the program seeks to: analyze changes in employer-based, private and public insurance coverage for people under age 65, and determine how those changes may affect the affordability and comprehensiveness of coverage, the number of uninsured, the number of under-insured, and churning in and out of coverage; document the consequences of being uninsured, under-insured, and unstably insured, with regard to access to care, health status, personal financial security, and economic productivity; and develop and evaluate federal and state policies to expand and stabilize health insurance, make it more affordable and aligned with incentives to access appropriate and high quality care, and enhance the efficiency with which it is administered.
Requirements: The Commonwealth Fund requests letters of inquiry to initiate the grant application process, and does not wish to review full proposals at this stage. Applicants are encouraged to submit letters of inquiry using our online form.
Restrictions: The Fund makes grants only to tax-exempt organizations and public agencies and does not support: general planning and ongoing activities or existing deficits; endowment or capital costs, including construction, renovation, or equipment; basic biomedical research; conferences, symposia, major media projects, or documentaries, unless they are an outgrowth of one of the Fund's programs; individuals; scholarships; churches or other religious organizations unless the project for which they seek funding is entirely secular in nature; and work for which achievements cannot be measured.
Contact: Sara R. Collins, Ph.D., Assistant Vice-President; (212) 606-3838 or (212) 606-3800; fax: (212) 606-3500; email: src@cmwf.org
Internet: http://www.commonwealthfund.org/Content/Program-Areas/Affordable-Health-Insurance.aspx
Sponsor: Commonwealth Fund
One East 75th Street
New York, NY 10021-2692

Commonwealth Fund Harkness Fellowships in Health Care Policy and Practice 1755
The Harkness Fellowships in Health Care Policy and Practice provide an opportunity for professionals from the Australia, Germany, the Netherlands, New Zealand, Norway, Switzerland, and the United Kingdom to spend four to 12 months in the United States conducting a research study that is relevant to health care policy and practice in both the United States and the fellow's home country, and that is focused on the issues of greatest concern to the fund. Fellowship awards provide up to $107,000, which covers round trip airfare to the United States, a living allowance, funds for project-related travel, research, and conferences, travel to attend the Commonwealth Fund program of fellowship seminars, health insurance, and U.S. and state taxes. A family supplement - including airfare, living allowance, and health insurance - is also provided to Fellows accompanied by a partner and/or children up to age 18.

Requirements: Applicants must be citizens of Australia, Germany, the Netherlands, New Zealand, Norway, Switzerland, and the United Kingdom in their late 20s to early 40s with broad educational backgrounds, not just those in research or academic careers. In order to apply, applicants must be nominated by their institution and submit a formal application, which is available from The Commonwealth Fund in New York City or its representatives in their home country.
Amount of Grant: $107,000 maximum
Date(s) Application is Due: Sep 15
Contact: Robin Osborn, Director; (212) 606-3809 or (212) 606-3800; fax: (212) 606-3875; email: ro@cmwf.org
Internet: http://www.commonwealthfund.org/Fellowships/Minority-Health-Policy-Fellowship.aspx
Sponsor: Commonwealth Fund
One East 75th Street
New York, NY 10021-2692

Commonwealth Fund Health Care Quality Improvement & Efficiency Grants   1756
The goal of the Fund's Program on Health Care Quality Improvement and Efficiency is to improve the quality and efficiency of health care in the United States. To that end, the program supports projects that: promote the development and widespread adoption of health care quality and efficiency measures; assess and enhance the capacity of health care organizations to provide better care more efficiently; and promote the development and adoption of payment and incentive models that encourage providers to improve quality and efficiency.
Requirements: The Commonwealth Fund requests letters of inquiry to initiate the grant application process, and does not wish to review full proposals at this stage. Applicants are encouraged to submit letters of inquiry using our online form.
Restrictions: The Fund makes grants only to tax-exempt organizations and public agencies and does not support: general planning and ongoing activities or existing deficits; endowment or capital costs, including construction, renovation, or equipment; basic biomedical research; conferences, symposia, major media projects, or documentaries, unless they are an outgrowth of one of the Fund's programs; individuals; scholarships; churches or other religious organizations unless the project for which they seek funding is entirely secular in nature; and work for which achievements cannot be measured.
Contact: Anne-Marie J. Audet, M.D.; (212) 606-3800; fax: (212) 606-3500; email: ama@cmwf.org
Internet: http://www.commonwealthfund.org/Content/Program-Areas/Health-Care-Quality-Improvement-and-Efficiency.aspx
Sponsor: Commonwealth Fund
One East 75th Street
New York, NY 10021-2692

Commonwealth Fund Patient-Centered Coordinated Care Program Grants   1757
As defined by the Institute of Medicine, patient-centered care is health care that establishes a partnership among practitioners, patients, and their families (when appropriate), to ensure that decisions respect patients' needs and preferences, and that patients have the education and support they need to make decisions and participate in their own care. In primary care, such care is best provided in a medical home - a primary care practice or health center that ensures patients have enhanced access to their clinicians (for example, through the availability of evening or weekend appointments), coordinates care, and engages in continuous quality improvement. The goal of The Commonwealth Fund's Patient-Centered Coordinated Care Program, established in 2005, is to improve the quality of primary care by making it more patient- and family-centered. The initiative supports projects that: promote the collection of information on patient-centered care and the delivery of care to facilitate public reporting and quality improvement; stimulate adoption of effective practices, models, and tools to make primary care practices patient- and family-centered; and improve policy to encourage patient- and family-centered care in medical homes.
Requirements: The Commonwealth Fund requests letters of inquiry to initiate the grant application process, and does not wish to review full proposals at this stage. Applicants are encouraged to submit letters of inquiry using our online form.
Restrictions: The Fund makes grants only to tax-exempt organizations and public agencies and does not support: general planning and ongoing activities or existing deficits; endowment or capital costs, including construction, renovation, or equipment; basic biomedical research; conferences, symposia, major media projects, or documentaries, unless they are an outgrowth of one of the Fund's programs; individuals; scholarships; churches or other religious organizations unless the project for which they seek funding is entirely secular in nature; and work for which achievements cannot be measured.
Contact: Melinda Abrams; (212) 606-3800; fax: (212) 606-3500; email: mka@cmwf.org
Internet: http://www.commonwealthfund.org/Content/Program-Areas/Patient-Centered-Coordinated-Care.aspx
Sponsor: Commonwealth Fund
One East 75th Street
New York, NY 10021-2692

Commonwealth Fund Payment System Reform Grants   1758
To achieve a high performing health system, the U.S. must curb spending growth and improve the way health care is provided. Payment system reform is critical to accomplishing these objectives. As a nation, we need to align incentives so that health care providers are rewarded for high-value care rather than a high volume of services. Rewarding value over volume will also encourage the development of a more integrated health care delivery system. The Commonwealth Fund, through its Program on Payment System Reform, supports analysis and the development of policy options to accomplish

these goals. Areas of interest include: reforming the existing payment structure to improve the alignment of incentives and to provide a base for more comprehensive payment reform; modeling and analyzing the potential impact of alternative options for payment reform in Medicare and throughout the health system; using payment reform to encourage the development of new models of health care delivery that provide better and more coordinated care; and using comparative effectiveness research to support better decision-making by providers, payers, and patients.
Requirements: The Commonwealth Fund requests letters of inquiry to initiate the grant application process, and does not wish to review full proposals at this stage. Applicants are encouraged to submit letters of inquiry using our online form.
Restrictions: The Fund makes grants only to tax-exempt organizations and public agencies and does not support: general planning and ongoing activities or existing deficits; endowment or capital costs, including construction, renovation, or equipment; basic biomedical research; conferences, symposia, major media projects, or documentaries, unless they are an outgrowth of one of the Fund's programs; individuals; scholarships; churches or other religious organizations unless the project for which they seek funding is entirely secular in nature; and work for which achievements cannot be measured.
Contact: Heather Drake, Program Associate; (212) 606-3800; fax: (212) 606-3500; email: hd@cmwf.org; Stuart Guterman; (202) 292-6735 or (212) 606-3800; fax: (212) 606-3500; email: sxg@cmwf.org
Internet: http://www.commonwealthfund.org/Content/Program-Areas/Payment-System-Reform.aspx
Sponsor: Commonwealth Fund
One East 75th Street
New York, NY 10021-2692

Commonwealth Fund Quality of Care for Frail Elders Grants   1759
The Commonwealth Fund Program on Quality of Care for Frail Elders aims to transform the nation's nursing homes and other long-term care facilities into resident-centered organizations that are good places to live and work, capable of providing the highest-quality care. The projects it supports aim to: identify, evaluate, and spread models of resident-centered care, or care delivered in accordance with the needs and desires of the people who live in nursing homes; equip nursing home operators to lead transformational change; and promote policy options that support resident-centered care.
Requirements: The Commonwealth Fund requests letters of inquiry to initiate the grant application process, and does not wish to review full proposals at this stage. Applicants are encouraged to submit letters of inquiry using our online form.
Restrictions: The Fund makes grants only to tax-exempt organizations and public agencies and does not support: general planning and ongoing activities or existing deficits; endowment or capital costs, including construction, renovation, or equipment; basic biomedical research; conferences, symposia, major media projects, or documentaries, unless they are an outgrowth of one of the Fund's programs; individuals; scholarships; churches or other religious organizations unless the project for which they seek funding is entirely secular in nature; and work for which achievements cannot be measured.
Contact: Mary Jane Koren; (212) 606-3800; fax: (212) 606-3500; email: mjk@cmwf.org
Internet: http://www.commonwealthfund.org/Content/Program-Areas/Quality-of-Care-for-Frail-Elders.aspx
Sponsor: Commonwealth Fund
One East 75th Street
New York, NY 10021-2692

Commonwealth Fund Small Grants   1760
The fund awards small grants for projects in its major program areas, which include: affordable health insurance; international health care policy and practice; improving the quality of health care services; and improving insurance coverage and access to care. Types of support include employee-matching gifts, program development, program evaluation, and research. Preference is given to projects that seek to solve problems, especially those affecting vulnerable groups; that analyze the effects of policies and trends on well-defined health issues; and that develop and test practical solutions. Prospective grantees should submit a letter of inquiry via regular or electronic mail. Program staff will contact applicants if more detailed information is required. Proposals recommended by fund staff are reviewed and voted upon by the board of directors, which meets in July, April, and November. Letters of inquiry should be brief, no more than two pages.
Requirements: The fund makes grants only to tax-exempt organizations and public agencies.
Restrictions: The fund does not support general planning and ongoing activities or existing deficits; endowment or capital costs, including construction, renovation, or equipment; basic biomedical research; conferences, symposia, major media projects or documentaries, unless they are an outgrowth of one of the fund's programs; individuals; scholarships; churches or other religious organizations, unless the project is entirely secular in nature; or work for which achievements cannot be measured.
Contact: Andrea Landes, Director of Grants Management; (212) 606-3800; fax: (212) 606-3508; email: acl@cmwf.org
Internet: http://www.commonwealthfund.org/Grants-And-Programs/Applicant-and-Grantee-Resources/Applying-for-a-Grant.aspx
Sponsor: Commonwealth Fund
One East 75th Street
New York, NY 10021-2608

Commonwealth Fund State High Performance Health Systems Grants   1761
The Commonwealth Fund's State High Performance Health Systems program is designed to help states develop the infrastructure needed to improve health care quality and outcomes - and to draw out and share lessons of national import from the experience

of states that are moving toward comprehensive health care reforms. To enable states to help local health care providers better meet the needs of their patient populations, the program will support approaches that offer providers shared access to the clinical support and practice-management services essential to achieving high performance. Such resources will often be developed though state-initiated, public,Äiprivate partnerships. To inform and support state health care leaders, additional projects will facilitate information-sharing among states working toward health care reform.

Requirements: The Commonwealth Fund requests letters of inquiry to initiate the grant application process, and does not wish to review full proposals at this stage. Applicants are encouraged to submit letters of inquiry using our online form.

Restrictions: The Fund makes grants only to tax-exempt organizations and public agencies and does not support: general planning and ongoing activities or existing deficits; endowment or capital costs, including construction, renovation, or equipment; basic biomedical research; conferences, symposia, major media projects, or documentaries, unless they are an outgrowth of one of the Fund's programs; individuals; scholarships; churches or other religious organizations unless the project for which they seek funding is entirely secular in nature; and work for which achievements cannot be measured.

Contact: Edward L. Schor, Vice President; (212) 606-3800; fax: (212) 606-3500; email: els@cmwf.org
Internet: http://www.commonwealthfund.org/Content/Program-Areas/State-High-Performance-Health-System.aspx
Sponsor: Commonwealth Fund
One East 75th Street
New York, NY 10021-2692

Commonwealth Fund/Harvard University Fellowship in Minority Health Policy 1762

The Commonwealth Fund/Harvard University Fellowship in Minority Health Policy is a one-year, full-time, academic degree-granting program designed to create physician-leaders, particularly minority physician-leaders, who will pursue careers in health policy, public health practice, and academia. It is designed to incorporate the critical skills taught in schools of public health, government, business, and medicine with leadership forums and seminar series conducted by Harvard senior faculty and nationally recognized leaders in minority health and public policy; supervised practicums and shadowing opportunities; and site visits, conferences, and travel. Each fellowship provides: $50,000 stipend; full tuition; health insurance; books; travel; and related program expenses, including financial assistance for a practicum project.

Requirements: Applicants are required to complete applications to both the Commonwealth Fund/Harvard University Fellowship in Minority Health Policy and the Harvard School of Public Health. Applicants must have finished residency and be U.S. citizens.

Amount of Grant: $50,000 stipend
Date(s) Application is Due: Jan 4
Contact: Joan Y. Reede, Director; (617) 432-2922; email: mfdp_cfhuf@hms.harvard.edu
Internet: http://www.commonwealthfund.org/Fellowships/Minority-Health-Policy-Fellowship.aspx
Sponsor: Commonwealth Fund
164 Longwood Avenue, 2nd Floor
Boston, MA 02115-5818

Communities Foundation of Texas Grants 1763

The unrestricted funds of the foundation support programs and projects intended to improve the quality of life for the citizens of the Dallas, TX, area. The funds support education, health and hospitals, social services, youth, and cultural programs. The foundation encourages projects developed in consultation with other agencies and planning groups and that promote coordination, cooperation, and sharing among organizations. Types of support include seed grants, emergency funds, building funds, equipment acquisition, matching funds, technical assistance, research, capital campaigns, and operating budgets. Requests for operating funds generally are not granted.

Requirements: 501(c)3 organizations in the Dallas, TX, area may apply.

Restrictions: Grants are not made to or for individuals, endowments, sectarian religious purposes, political or lobbying efforts, deficit financing, media projects, or for operational expenses of established organizations.

Geographic Focus: Texas

Samples: Dallas Police Dept (TX)—for equipment and efforts to improve safety in the city, which has had the highest crime rate among major U.S. cities for the past years, through the WW Caruth Jr. Foundation Fund, $15 million over three years (2005); Dallas Women's Foundation (TX)—to provide grants to local and national faith-based programs that benefit women and girls, through the Mrs. Ruth Ray Hunt Philanthropic Fund, $4.8 million (2004).

Date(s) Application is Due: Jan 15; Sep 14
Contact: Leslie Parks, (214) 750-4222; fax: (214) 750-4210; email: Rdear@cftexas.org
Internet: http://www.cftexas.org
Sponsor: Communities Foundation of Texas
5500 Caruth Haven Ln
Dallas, TX 75225-8146

Community Campus Partnerships for Health Fellowships 1764

Community-based professionals, academic administrators, and faculty with significant knowledge and expertise in building and sustaining service-learning, community-based participatory research, and community-campus partnerships are encouraged to apply for a fellowship with CCPH. The expected outcome of the program is to advance and support these concepts in the context of health professions education and practice. Fellows will receive a stipend for a one-year fellowship period, with the potential for a second year of funding. Funding provides support for time spent participating in fellow-related

activities. Fellows also receive a certificate of recognition upon completing the fellowship. Application procedures and additional information are available online.

Requirements: CCPH members and nonmembers are eligible.
Amount of Grant: $5000
Contact: Sarena Seifer, Executive Director, (206) 616-4305; fax: (206) 685-6747; email: sarena@u.washington.edu or ccphuw@u.washington.edu
Internet: http://depts.washington.edu/ccph
Sponsor: Community Campus Partnerships for Health
UW Box 354809
Seattle, WA 98195-4809

Community Foundation AIDS Endowment Awards 1765

The foundation administers an annual special grants program called the AIDS Endowment Awards. Grants of $1,000 will be awarded to one or two organizations that demonstrate a commitment to working towards the prevention, education and treatment of AIDS so that they may continue their efforts. Services provided could include education programs for the prevention of HIV/AIDS, medical and social services for those living with HIV/AIDS, housing services for those living with HIV/AIDS and medical research for the treatment and prevention of HIV/AIDS.

Requirements: Selected organizations in the metropolitan Richmond area interested in the field of AIDS education, research or community service and care are encouraged to apply. Organizations interested in applying should submit a brief proposal to foundation which should be no more than two single-spaced pages and include: [1] a description of the organization's mission; and [2] a description of the agency's work in the field of AIDS. Include a list of Board of Governors, a copy of the most recent IRS Form 990, an audited financial statement if available, and an IRS Tax Exempt Letter designating your organization a 501(c)3 non-profit.

Geographic Focus: Virginia
Amount of Grant: $1,000
Date(s) Application is Due: Feb 15
Contact: Susan Hallett, Program Officer, (804) 330-7400; fax: (804) 330-5992; email: shallett@tcfrichmond.org
Internet: http://www.tcfrichmond.org/Page2954.cfm
Sponsor: Community Foundation Serving Richmond and Central Virginia
7501 Boulders View Drive, Suite 110
Richmond, VA 23225

Community Foundation for Greater Buffalo Grants 1766

Grants are made to organizations in and for the benefit of Erie County, NY, residents. The foundation supports charitable, educational, and civic purposes and awards grants for educational institutions, scholarships, family and child health and welfare, and community development. Highest priority is given to projects that support creative and innovative responses to existing or emerging community problems.

Requirements: IRS 509(a) organizations in or benefiting residents of Allegany, Cattaraugus, Chautauqua, Erie, Genesee, Niagara, Orleans, and Wyoming Counties, NY, are eligible.

Restrictions: Grant requests are not considered for endowment funds, religious purposes, projects outside the western New York region, or schools not registered with the state education department.

Geographic Focus: New York

Samples: Greater Buffalo Opera Comp (Buffalo, NY)—to refurbish and renovate a set for a production of Carmen, $10,000.; Bison Fund-Buffalo Inner-City Scholarship Opportunity Network (Buffalo, NY)—for elementary school students, to provide low-income, inner-city parents with the option of choosing a private or parochial school for their children, $10,000.

Amount of Grant: $1000-$25,000 typically
Date(s) Application is Due: Feb 1; Aug 1
Contact: Jean McKeown, Program Officer, (716) 852-2857, ext. 204; fax: (716) 852-2861; email: jeanm@cfgb.org
Internet: http://www.cfgb.org/page17000.cfm
Sponsor: Community Foundation for Greater Buffalo
712 Main Street
Buffalo, NY 14202

Community Foundation for Muskegon County Grants 1767

The community foundation was established to serve the needs of the people of Muskegon County and nearby western Michigan. The foundation's strategic plan for grant making focuses on the prevention of problems rather than the cure; encourages programs that are collaborative, comprehensive, and have the potential to be continuous; encourages leveraging and matching grant opportunities from multiple funders; and supports seed money opportunities for innovative projects. Major grant interest areas include the arts, community development and urban revitalization, education, environment, health/human services, and needs of young children (ages 0-3). Types of support include seed money grants, special projects, matching funds, equipment, scholarship funds, loans, research, publications, conferences and seminars, endowment funds, consulting services, continuing support, emergency funds, internships, professorships, and renovation projects. Although the foundation generally makes one-year grant commitments, it will consider making longer term commitments for new efforts that show strong promise for positive impact. Applications are accepted on specific dates each year; it is recommended that the applicant contact the foundation to discuss their interests and to obtain the next grant application deadline.

Requirements: IRS 501(c)3 organizations and institutions in Muskegon County and nearby western Michigan are eligible.

Restrictions: Support will not be provided for routine operating expenses; capital equipment, computer hardware and software, and motor vehicles; conferences, publications, videos, films, television, or radio programs; endowment campaigns; special fund-raising events; religious programs that serve specific denominations; existing obligations or debts; individual schools or districts; or individuals.
Geographic Focus: Michigan
Samples: Camp Bluebird, $5,000 to support the participation of Muskegon residents at a retreat for cancer patients; Michigan State University—Muskegon County extension Office, $5,000 to support a local match for the Land Use Education Program; West Shore Symphony Orchestra, $3,000 to support year three of the Carnegie LinkUP; First Congregational Church, $3,200 to support the Saturday Morning Breakfast program; Muskegon County Museum, $5,000 for an exhibit of the history of medicine in Muskegon.
Amount of Grant: $250-$50,000 average
Contact: Gina Van Bruggen or Arn Boezaart, Program Officers, (231) 722-4538; fax: (231) 722-4616; email: gvanbruggen@cffmc.org or aboezaart@cffmc.org
Internet: http://www.cffmc.org/grantapply.php
Sponsor: Community Foundation for Muskegon County
425 W Western Avenue, Suite 200
Muskegon, MI 49440

Community Foundation for Southern Arizona Grants   1768

The foundation awards grants in southern Arizona for a broad array of charitable purposes in the areas of arts and humanities, education, the environment, health, social services. Types of support include challenge/matching grants, conferences and seminars, scholarships, equipment acquisition, fellowships, general support, multiyear support, project development, publications, research, seed funding, and technical assistance. Priority is given to proposals that promote collaboration and build on the strengths of individuals and communities. Deadlines are determined on a yearly basis.
Requirements: Nonprofit and grassroots organizations in southern Arizona communities are eligible. Southern Arizona communities include all of Cochise, Santa Cruz, and Pima Counties; and the areas of Yuma, Mariposa, Pinal, Graham, and Greenlee Counties that lie south of the Gila River.
Restrictions: Funds are generally not available for ongoing operating or capital campaigns, debt retirement, endowments, individuals, individual schools, sectarian activities, or underwriting of fund-raising events.
Geographic Focus: Arizona
Samples: Arizona's Children Association (AZ)—for KARE legal services, $7200 (2005); Tucson Clean and Beautiful (Tucson, AZ)—for Adopt-A-Park Program, $2000 (2005); Tucson Children's Museum, Inc (Tucson, AZ)—to cover admission costs for low-income children, $2250 (2004); Tucson Cinema Foundation Inc (Tucson, AZ)—funding for a ten-month program for high school students, $5600 (2004).
Amount of Grant: $1000-$10,000 average range
Contact: Barbara Brown, Executive Vice President, (520) 770-0800 ext 107; fax: (520) 770-1500; email: bbrown@CFSoAZ.org
Internet: http://www.cfsoaz.org/page17480.cfm
Sponsor: Community Foundation for Southern Arizona
2250 E Broadway Boulevard
Tucson, AZ 85719

Community Foundation of Broward Grants   1769

Through its competitive grantmaking process, the foundation supports programs that strengthen families within low socioeconomic communities in Florida's Broward County. Grantmaking priorities include the arts, foster care, out-of-school time, and technology. Proposals also are accepted in specific areas of interest, including animal welfare—protect and educate; cancer and arthritis research—finding a cure; patient care—relief and assistance to individuals with cancer or other terminal diseases; literacy—opening worlds through language, an AIDS/HIV—through issued RFP. Proposals are accepted anytime throughout the year. Application and guidelines are available online.
Requirements: 501(c)3 organizations, public charities as defined by the IRS code, and governmental agencies located in Broward County or directly benefiting the residents of Broward County are eligible.
Restrictions: Grants are not awarded in the following areas: annual fundraising, capital campaign/improvements, deficit financing, endowment efforts, grants to individuals, religious purposes, or routine operating needs.
Geographic Focus: Florida
Samples: Covenant House Florida (Fort Lauderdale, FL)—for child-rearing education and childcare services for at-risk pregnant youths and young parents at its crisis shelter, $10,000 (2003); Eckerd Youth Alternatives (Clearwater, FL)—for after-school counseling and literacy services for youths recently released from residential delinquency-treatment programs, $10,000 (2003); Pines Youth Ctr (Pembroke Pines, FL)—for an extracurricular program for middle-school students that provides tutoring and computer and reading instruction, $12,000 (2003); First Call for Help of Broward (FL)—for the technological support of a crisis and referral hotline, $15,710 (2003).
Amount of Grant: $10,000 average
Contact: Sheri Brown, Vice President of Strategic Community Initiatives, (954) 761-9503 ext 103; fax: (954) 761-7102; email: jbull@cfbroward.org
Internet: http://www.cfbroward.org/new/grant.html
Sponsor: Community Foundation of Broward
1401 E Broward Boulevard, Suite 100
Fort Lauderdale, FL 33301

Community Foundation of Champaign County Grants   1770

The community foundation awards grants from its unrestricted funds to Champaign County, IL, nonprofit organizations to use for charitable purposes. Representative categories include arts and humanities, environmental concerns, education, health, human services, research, urban affairs, church, and youth programs. Types of support include continuing support, annual campaigns, building/renovation, equipment, publication, consulting services, and scholarships.
Requirements: Champaign County, IL, nonprofit organizations are eligible.
Geographic Focus: Illinois
Amount of Grant: $1000-$2000 typically
Date(s) Application is Due: Aug 31
Contact: Joan Dixon, Executive Director, (217) 359-0125; fax: (217) 352-6494; email: cfcc@soltec.net
Sponsor: Community Foundation of Champaign County
404 W Church Street
Champaign, IL 61820

Community Foundation of Greater Tampa Grants   1771

Program interests of the foundation include arts and culture, community enablement, education, environment and animals, health and human services, history, neighborhoods, senior citizens, and youth and families.
Requirements: Florida 501(c)3 organizations in Hillsborough, Pasco and Pinellas counties are eligible.
Restrictions: Political causes/organizations will not be supported. Grants also will not be made to purchase tickets or advertising space in programs or other publications. The foundation generally is not interested in capital campaigns, funding of operating costs, experimental medical research, religious or sectarian purposes, loans, or multiple year funding.
Geographic Focus: Florida
Date(s) Application is Due: Mar 1; Sep 1
Contact: Paula Fraher, Grants Director, (813) 282-1975; fax; (813) 282-3119; email: pfraher@cftampabay.org
Internet: http://www.cftampabay.org/Grants/grants_process.htm
Sponsor: Community Foundation of Greater Tampa
4950 W Kennedy Boulevard, Suite 250
Tampa, FL 33609

Community Foundation of the Fox River Valley   1772

The foundation awards full-time scholarships to graduating high school seniors, undergraduate college students, and graduate school students within its service area, comprising the City of Aurora, Kendall County, and southern Kane County in the State of Illinois. The residency requirement does not apply to graduating high school students whose parent(s) are employed by Nicor Gas; graduating high school students residing within Community High School District 94, West Chicago; and graduating community/ junior college students whose parent(s) are employed by a company that maintains membership with the Valley Industrial Association. The majority of scholarships are based on academic ability and financial need. Some awards are renewable. Students who attend the following high schools must obtain applications from their high school guidance office (deadline date for applications determined by the school): Aurora Central Catholic, Aurora Christian, East Aurora, Hinckley-Big Rock, IMSA Kaneland, Marmion, Oswego, Plano, Rosary Waubonsie Valley, West Aurora, and Yorkville High School. All other students may obtain an application by calling the office. The deadline date for receipt of applications is generally mid-February.
Requirements: Eligibility is restricted to students who will be attending school on a full-time basis and whose permanent residence is within the foundation's service area. The foundation's service area includes the City of Aurora, Kendall County, and southern Kane County in the state of Illinois.
Geographic Focus: Illinois
Amount of Grant: $1000-$3000
Contact: Rhonda Soos, (630) 896-7800; email: rsoos@communityfoundationFRV.org
Internet: http://www.CommunityFoundationFRV.org
Sponsor: Community Foundation of the Fox River Valley
111 W Downer Pl, Suite 312
Aurora, IL 60506-6106

Community Foundation of the Napa Valley Grants   1773

The foundation awards one-year grants to preserve the assets of the community and to fill gaps in funding to Napa Valley nonprofit agencies and organizations. Grants may be renewed. Applicants should send a letter of inquiry outlining the organization, the program, and amount requested. Letters of inquiry should be submitted three months in advance of the project start date.
Requirements: California 501(c)3 tax-exempt organizations serving the Napa Valley are eligible.
Geographic Focus: California
Contact: Patricia Struntz, Executive Director, (707) 254-9565; fax: (707) 254-7955; email: info@cfnv.org
Internet: http://www.cfnv.org/grant/grantguide.html
Sponsor: Community Foundation of the Napa Valley
433 Soscol Avenue, Suite B-151
Napa, CA 94559

Community Foundation of the Virgin Islands Judith A. Towle   1774
Environmental Studies Grants

The Environmental Studies Fund will provide support for internships: sponsored by an institution of higher education, a public sector agency, or an NGO based in the U.S. or British Virgin Islands or in St. Kitts/Nevis; research studies by graduate students: focused on transboundary environmental issues in the U.S. or British Virgin Islands or in St. Kitts/Nevis; workshops, seminars, lectureships: sponsored by an institution of higher education, a public sector agency, or an NGO based in the U.S. or British Virgin Islands or in St. Kitts/Nevis; publications focused on transboundary environmental issues in the U.S. or British Virgin Islands or in St. Kitts/Nevis.

Requirements: Any individual or non-profit organization whose research focuses on the U.S. Virgin Islands, British Virgin Islands, or the Federation of St. Kitts and Nevis may submit a grant application. Preference will be given to graduate students who are residents of the U.S. or British Virgin Islands or of the Federation of St. Kitts/Nevis.

Geographic Focus: U.S. Virgin Islands

Amount of Grant: $3,500 max.

Contact: Dee Baecher-Brown, President; (340) 774-6031; fax: (340) 774-3852; email: dbrown@cfvi.net

Internet: http://www.cfvi.net/subpages/grants.html

Sponsor: The Community Foundation of the Virgin Islands

P.O. Box 11790

St. Thomas, U.S.VI 00801-4790

Community Health Scholars Program Fellowships   **1775**

The Community Health Scholars Program (CHSP) is a postdoctoral fellowship program designed to meet the growing needs of schools of public health and other health professions for faculty with community competency. The 12-month fellowship enables scholars to develop and enhance skills in working with communities and engaging in participatory community-based research at institutions where these skills are present. Applications are welcome from junior faculty as well as those recently completing doctoral-level training so long as the above eligibility requirements are met. Women and other underrepresented groups are encouraged to apply.

Requirements: Applicants must be U.S. citizens located at one of three training sites (Johns Hopkins, University of Michigan, or University of North Carolina) and have completed all formal requirements for professional and postprofessional training (e.g., internships or residencies) at the doctoral level (e.g., PhD, MD, JD, or DrPH) by the date of entry into the program.

Restrictions: Scholars may not hold other training fellowships or be enrolled full time in another degree program during their tenure in this program.

Amount of Grant: $50,000 fellowship award plus $10,000 research fund

Contact: Saundra Bailey, Program Administrator, (734) 647-3065; fax: (734) 936-0927; email: chsp@umich.edu

Internet: http://www.sph.umich.edu/chsp

Sponsor: University of Michigan

109 Observatory Street, M4142

Ann Arbor, MI 48109-2029

Compass Bank Foundation Grants   **1776**

The Foundation, established in 1981, supports organizations involved with arts and culture, higher and law education, health, heart disease, multiple sclerosis, medical research, youth development, human services, and philanthropy. Giving is primarily centered in the states of Alabama, Arizona, Colorado, Florida, New Mexico, and Texas. There are no specific deadlines with which to adhere. Contact the Foundation for further application information and guidelines.

Restrictions: No support is provided for religious organizations, and there are no grants to individuals (except for employee-related scholarships).

Geographic Focus: Alabama; Arizona; Colorado; Florida; New Mexico; Texas

Samples: University of Alabama, Tuscaloosa, AL, $375,000 (2007); United Way, Mile High, Denver, CO, $29,167 (2007); Alabama Symphonic Association, Birmingham, AL, $55,000 (2007); John P. McGovern Museum of Health and Medical Science, Houston, TX, $20,000 (2007);

Contact: Jerry W. Powell, Treasurer; (205) 297--3960

Sponsor: Compass Bank Foundation

P.O. Box 10566, M.C. AL/BI/CH/ACT

Birmingham, AL 35296-0002

Comprehensive Health Education Foundation Grants   **1777**

The foundation awards grants to support programs that address health inequities. The initial grantmaking effort will focus on Clark, Pierce, and Spokane Counties in Washington State. One-year grants of up to $20,000 each will be awarded to culturally appropriate, community-led collaborations to test their best idea on how to make it easier for people who suffer from health inequities to move more and eat healthier. Health inequities are defined as differences in the incidence, prevalence, mortality, and burden of diseases that exist for specific populations in the United States. Low-income individuals and people of color within the United States generally have higher rates of poor health and injury than those who are in higher-income groups and are Caucasian.

Requirements: 501(c)3 tax-exempt organizations located in Clark, Pierce, and Spokane Counties in Washington State and units of government that are nondiscriminatory in policy and practice regarding disabilities, age, sex, sexual orientation, race, ethnic origin, or creed are eligible.

Restrictions: Support will not be provided for building or land acquisitions; equipment or furniture purchases; endowment funds; emergency funds; grants to individuals; fellowships/scholarships; research; debt retirement; fundraising activities; general fund drives; indirect overhead; or CHEF programs or products.

Geographic Focus: Washington

Samples: Vanessa Behan Crisis Nursery (WA)—for parent education classes and support groups, $5000.; Spokane School District #81 (WA)—for a blended curriculum for the fifth grade that includes classroom teachers and fitness and health specialists, $8500.; Market Foundation (WA)—for the Senior Wellness program, which gives impoverished urban seniors the opportunity to learn how to improve their health, $5000.

Amount of Grant: $500-$20,000 typically

Date(s) Application is Due: Jun 15

Contact: Kari L. Lewis, (206) 824-2907 or (800) 323-2433; fax: (206) 824-3072; email: KariL@chef.org

Internet: http://www.chef.org/about/grants.php

Sponsor: Comprehensive Health Education Foundation

22419 Pacific Hwy S

Seattle, WA 98198-5106

Compton Foundation Grants Program   **1778**

The foundation was founded to address community, national, and international concerns in the fields of peace and world order, population, and the environment. Other concerns of the foundation include equal educational opportunity, community welfare and social justice, and culture and the arts. The foundation makes three kinds of grants. Project grants generally are made to national organizations for projects that fall within the primary areas of peace and world order, population, and the environment. These grants may be for regional (Pacific Coastal states), national, or international activities and are usually for projects of limited duration. Project grants are considered by the board two times a year. Discretionary grants are made at the discretion, and usually at the initiation, of individual board members. Most grants in this area are made for community welfare and social justice, and culture and the arts. Renewal grants provide general support to organizations whose activities have been funded by the foundation for many years and whose work continues to be considered particularly effective by the board. Many of the foundation's grants in the area of equal educational opportunity, and some grants in peace and world order and population, are renewal grants.

Requirements: The foundation makes grants only to tax-exempt organizations and institutions.

Restrictions: Grants will not be made to individuals.

Samples: Colorado Environmental Coalition—for Colorado Instream Flow Project, $20,000 (2007); Klamath Forest Alliance—for Klamath Riverkeeper and Karuk Tribe Klamath Dams Campaign, $30,000 (2007); San Francisco Education Fund—for Peer Resources Program, $10,000 (2007).

Amount of Grant: $5000-$50,000 average

Date(s) Application is Due: Mar 7; Sep 7

Contact: Edith T. Eddy, Executive Director; (650) 508-1181; fax: (650) 508-1191; email: info@comptonfoundation.org

Internet: http://www.comptonfoundation.org/application_procedures.html

Sponsor: Compton Foundation

255 Shoreline Drive, Suite 540

Redwood City, CA 94065

Compton Foundation International Fellowships   **1779**

The goal of the Compton Foundation's International Fellowship Program is to contribute to the capacity of developing countries, especially in Central America and Sub-Saharan Africa, as well as Mexico, to improve policies and programs relating to Peace and Security, Population and Reproductive Health and Environment and Sustainable Development. The Foundation strives to accomplish this goal by supporting outstanding graduate students who are committed to careers in the program areas of interest to the Foundation within the developing world. Compton Fellows are chosen by selected university-based programs working in partnership with the Foundation.

Contact: Edith T. Eddy, Executive Director; (650) 508-1181; fax: (650) 508-1191; email: info@comptonfoundation.org

Internet: http://www.comptonfoundation.org/fellows.html

Sponsor: Compton Foundation

255 Shoreline Drive, Suite 540

Redwood City, CA 94065

Compton Foundation Mentor Fellowships   **1780**

The fellowship program focuses on graduating college students from the United States. This program is designed to promote creativity and support the commitment of graduating seniors as they move beyond academic preparation to real-world application and contribution. Each year ten fellows are selected from participating universities and awarded a one-year, fellowship. The stipend is to implement a self-directed project, contributing their talents and energy to real-world situations. At the core of the fellowship is the partnership between a fellow and a mentor, who provides guidance, encouragement, and impetus for continued learning and service.

Amount of Grant: $36,000

Contact: Edith T. Eddy, Executive Director; (650) 508-1181; fax: (650) 508-1191; email: info@comptonfoundation.org

Internet: http://www.comptonfoundation.org/fellowships.html

Sponsor: Compton Foundation

255 Shoreline Drive, Suite 540

Redwood City, CA 94065

Computing in Chemical Engineering Award   **1781**

The award, which is sponsored by Mitsubishi Chemical Company and the Dow Chemical Company, recognizes outstanding contributions in the application of computing and systems technology to chemical engineering. All members as well as other interested persons are urged to nominate deserving candidates. All information and supporting documentation are to be included with completed forms. Nominations will remain active for three years. The awardee is invited to deliver an address at a CAST Division luncheon or dinner at the AIChE annual meeting.

Samples: Ross Taylor—award recipient, $3000 (2004); Liang-Tseng Fan—award recipient, $3000 (2003).
Amount of Grant: $3000
Date(s) Application is Due: Apr 15
Contact: AIChE Awards Programs, (212) 591-7107; fax: (212) 591-8882; email: awards@aiche.org
Internet: http://www.aiche.org/awards/awarddtl.asp?AwardID=28
Sponsor: American Institute of Chemical Engineers
3 Park Avenue
New York, NY 10016-5901

Conference on Latin American History Prize   1782

The prize is awarded annually for a distinguished English-language article on any significant aspect of Latin American history published in the year preceding the award and appearing in journals edited or published in the United States, including Puerto Rico (exclusive of the Hispanic American Historical Review), the Journal of Latin American Studies (Cambridge), or in volumes of collected articles.

Amount of Grant: $500
Contact: CLAH, University of California at Davis, (530) 752-3046; fax: (530) 752-8964; email: clah@ucdavis.edu
Internet: http://www.h-net.msu.edu/~clah/about/index.html
Sponsor: Conference on Latin American History
One Shields Avenue
Davis, CA 95616

Congressional Fellowships for African Americans   1783

This program is a comprehensive, nine-month paid fellowship designed to prepare minority graduate students and professionals for senior-level careers in the public policy arena. The goal of the program is twofold: to offer talented people the opportunity to learn all aspects of policy development by working on congressional committees and to produce scholarly research on critical issues before the U.S. Congress. The program includes an in-depth orientation to Capitol Hill; a nationally recognized lecture series with policy experts including members of Congress; an educational enrichment component that complements practical work experiences; and a professional development component under the guidance of a mentor. Application forms are available on the Web site.

Requirements: To be considered, an applicant must be a full-time graduate student; an individual with five years of professional work experience presently pursuing part-time graduate studies; or a college faculty member with a demonstrated interest in the legislative or public policy process. All applicants must be U.S. citizens who have demonstrated commitment to black political empowerment.

Amount of Grant: $25,000
Date(s) Application is Due: Apr 2
Contact: Program Contact, (800) 784-2577 or (202) 263-2800; fax: (202) 775-0773; email: info@cbcfinc.org
Internet: http://cbcfinc.org/Leadership%20Education/Fellowships/congressional.html
Sponsor: Congressional Black Caucus Foundation
1720 Massachusetts Avenue NW
Washington, DC 20036

Congressional Hispanic Caucus Institute Fellowships   1784

Every year, the fellowship program offers up to 21 promising Latinos from across the country the opportunity to gain hands-on experience at the national level in the public policy area of their choice. Fellows have the opportunity to work in such areas as international affairs, economic development, education policy, housing, or local government. CHCI also aims to develop leaders in the areas of public health administration (Edward Roybal Public Health Fellowship), telecommunications policy (Telecommunications Fellowship), financial services (Financial Services Fellowship), and corporate-public interest (Corporate Fellowship). Fellowships include domestic round-trip transportation to Washington, DC; health insurance; and a monthly stipend to help cover housing and local expenses. Annual deadline dates may vary; contact program staff for exact dates.

Requirements: Applicants must have a BA/BS or graduate degree within one year of the application deadline or be currently enrolled as graduate students. U.S. citizenship is required.
Restrictions: Relatives of CHCI staff or of its board of directors are not eligible.
Amount of Grant: $2061 monthly stipend; $2500 monthly fee to fellows with a graduate degree
Date(s) Application is Due: Mar 1
Contact: Fellowships Administrator, (800) 392-3532 or (202) 543-1771; fax: (202) 546-2143; email: chci@chci.org
Internet: http://www.chci.org/chciyouth/fellowship/fellowshipprogram.htm
Sponsor: Congressional Hispanic Caucus Institute
911 Second Street NE
Washington, DC 20002

Congressional Research Awards Program   1785

The center's primary interest is to fund the study of the leadership in the Congress, both House and Senate. Topics could include external factors shaping the exercise of congressional leadership, institutional conditions affecting it, resources and techniques

used by leaders, and the prospects for change or continuity in the patterns of leadership. The program was developed to support work intended for publication in some form or for application in a teaching or policy-making setting. Types of support include seed grants, travel grants, and dissertation/thesis research support. Grants will normally extend for one year.

Requirements: The competition is open to individuals with a serious interest in studying Congress. Political scientists, historians, biographers, scholars of public administration or American studies, and journalists are among those eligible. The center encourages graduate students to apply.

Restrictions: Grants will not be awarded for purchase of equipment or for subsidizing publication costs. Organizations are not eligible, and the award does not fund undergraduate or pre-PhD study.

Samples: Jessica Gerrity, Indiana U (IN)—for Congressional Behavior and Interest Group Influence: The Case of the Abortion Issue, $2637 (2005); Laura Jensen, U of Massachusetts (MA)—for Congress, the Petitions of the People, and Representation in the Early American Nation, $3500 (2005).

Amount of Grant: $200-$3500
Date(s) Application is Due: Feb 1
Contact: Executive Director, Congressional Research Grants Program, (309) 347-7113; fax: (309) 347-6432; email: fmackaman@dirksencenter.org
Internet: http://www.dirksencenter.org/print_grants_CRAs.htm
Sponsor: Dirksen Congressional Center
2815 Broadway
Pekin, IL 61554-4219

Connecticut Community Foundation Grants   1786

The Connecticut Community Foundation serves the people of Greater Waterbury and Northwest Connecticut by supporting public and nonprofit organizations providing programs and services including those for the arts, human services, health care, environment, youth development and education. The Foundation gives priority to efforts that prevent problems, encourage community solutions or improve the organizational capability and financial stability of nonprofit agencies.

Requirements: Nonprofit organizations in Beacon Falls, Bethlehem, Bridgewater, Cheshire, Goshen, Litchfield, Middlebury, Morris, Naugatuck, New Milford, Oxford, Prospect, Roxbury, Southbury, Thomaston, Warren, Watertown, Washington, Wolcott, and Woodbury, CT, may submit applications.

Restrictions: Grants are not awarded for religious purposes, political activities, deficit financing, continuing support, fund-raising events, annual campaigns, newly established arts organizations, commissioning of new works of art, general operating support, or endowments.

Geographic Focus: Connecticut
Amount of Grant: $8000-$10,000 average
Date(s) Application is Due: Jan 9; Mar 27; Aug 28
Contact: Carol O'Donnell, Associate CEO/Director of Grants; (203) 753-1315; fax: (203) 756-3054; email: info@conncf.org
Internet: http://www.conncf.org/grants/grants.htm
Sponsor: Connecticut Community Foundation
43 Field Street
Waterbury, CT 06702

Connecticut Health Foundation Health Initiative Grants   1787

The foundation awards grants to organizations and institutions that directly respond to its current priority areas and result in improving the health status of Connecticut's underserved and unserved populations. Program priorities include children???s mental health—projects related to children???s mental health, including research, community grants, creating resources for clinical effective practices, and parent advocacy groups; oral health—improving oral health care access, quality, and utilization; and racial and ethnic health disparities—improving the diversity of the health care workforce, and increasing cultural competency in the existing workforce. The foundation awards two major types of grants: strategic and responsive. Application and guidelines are available online.

Requirements: Connecticut state and local units of government, community health centers, health advocacy organizations, community-based organizations, community and cultural groups, schools, and faith-based organizations are eligible. Applicants must have IRS 501(c)3 tax-exempt status or be public entities. Unincorporated organizations may apply through 501(c)3 fiscal agents.

Restrictions: Foundation grants do not support awards to individuals; construction of buildings; capital projects, endowments, or chairs associated with universities, and medical schools; conferences (unless part of a greater project or program); projects that do not benefit Connecticut residents; lobbying or influencing the outcomes of a proposed piece of legislation or election; and indirect cost for discretionary grants.

Geographic Focus: Connecticut
Samples: 1000 Friends of Connecticut, $10,000 to improve public health through the "Smart Growth Education and Communications Campaign"; Asian Family Services, Inc., Hartford $25,000 to integrate health-related policy issues into the community dialogue during the political season in 2006; Asian Family Services, Inc., Hartford $10,000 to merge their services with Community Renewal Team, an anti-poverty multi-service agency that serves families and people throughout the Connecticut River Valley; Association of Yale Alumni in Public Health, New Haven, $30,000 to conduct an assessment of the policy, practices and procedures for recruiting and retaining historically under-represented faculty members at the Yale University School of Public Health (2006).

Amount of Grant: $50,000-$200,000 typically
Date(s) Application is Due: Mar 15; Jun 15; Sep 15; Dec 15

Contact: Onell Jesus Calderas, Grants Administrator, (860) 224-2200; fax: (860) 224-2230; email: onell@cthealth.org
Internet: http://www.cthealth.org/matriarch
Sponsor: Connecticut Health Foundation
74B Vine Street
New Britain, CT 06052

ConocoPhillips Grants Program   1788
ConocoPhillips maintains a philanthropic contributions budget for nonprofit, charitable programs closely tied to its corporate goals and focused primarily in locations of strong business interests. Submit an executive summary outlining the purpose of the program or project, how it will be accomplished, expected results, a budget (noting administrative expenses such as: salaries and fees, program expenses and total income), other sources of financial support and a copy of the IRS tax determination letter that confirms 501(c)3 status. The grant must be used in the United States. Requests for grants in non-U.S. locations should be made directly to the ConocoPhillips international office doing business in that part of the world. Education and youth, civic and arts, employee volunteerism, safety and social services, and the environment are focus areas.
Requirements: Applications are accepted from areas where ConocoPhillips has a strong business presence, e.g., Texas and Oklahoma. All contributions are to be used within the United States.
Restrictions: ConocoPhillips does not award funds to individuals, sectarian or religious organization, promotional sponsorship and advertising (marketing related) or an endowment.
Geographic Focus: Alaska; California; Illinois; Louisiana; Montana; New Jersey; Oklahoma; Pennsylvania; Texas
Samples: Houston Baptist U (Houston, TX)—for the Cultural Arts Ctr capital campaign, $50,000 (2004); U of Texas (Austin, TX)—for scholarships, fellowships, and special programs in business, engineering, law, and the natural sciences, $1 million (2003).
Contact: Program Contact, (281) 293-2685
Internet: http://www.conocophillips.com/about/Contribution+Guidelines/index.htm
Sponsor: ConocoPhillips Corporation
600 N Dairy Ashford, MA 3144
Houston, TX 77079

Conservation, Food, and Health Foundation Grants for Developing Countries   1789
The geographic focus of the foundation is the developing world. Through grants to support research and through targeted grants to help solve specific problems, the foundation helps build capacity within developing countries in three areas of interest: conservation, food, and health. The foundation concentrates its grantmaking on research, technical assistance, and training projects of benefit to the Third World; favors grants for pilot projects and special programs that have a potential for replication; prefers to support projects that employ and/or train personnel from the developing world; and favors research concerning problems of importance to the developing world. Concept papers may be submitted at any time, but must be received in the office by the listed application deadlines for consideration at board meetings; full proposals are by invitation.
Requirements: 501(c)3 tax-exempt organizations and foreign organizations with the equivalent of 501(c)3 status may apply.
Restrictions: The foundation does not normally provide support for buildings or for land purchase; quantity purchases of durable medical equipment; endowments or fundraising activities; famine or emergency relief; films, videos, or web-site production; re-granting through intermediaries; general operating support; or individuals (however, the foundation may support an individual engaged in research on a problem of significance to the developing world where the research is sponsored by an established, nonprofit organization such as an educational institution and conducted in close partnership with a local nongovernmental organization).
Samples: Boston Women's Health Book Collective (MA)—for the translation and adaptation of Our Bodies, Ourselves, a book on women's health, $22,000 (2003); Mano a Mano Medical Resources (Bolivia)—to train 250 health promoters and establish community health promotion programs in 12 new rural medical clinics, $17,000 (2003); Midwives for Midwives and Women's Health International—to support the training and development of midwives and women's health care providers in Antigua Guatemala, $15,000 (2003); Fauna and Flora International—to develop a community-led Siamese crocodile conservation project in the highlands of the Cardamom Mountains of Cambodia, $23,194 (2003).
Amount of Grant: $11,000 average; $25,000 maximum
Date(s) Application is Due: Feb 1; Aug 1
Contact: Prentice Zinn, c/o Grants Management Associates, (617) 426-7172; fax: (617) 426-7087; email: cfh@grantsmanagement.com
Internet: http://www.grantsmanagement.com/cfhguide.html
Sponsor: Conservation, Food, and Health Foundation
77 Summer Street, Suite 800
Boston, MA 02110

Consortium for Graduate Study in Management Minority Fellowships   1790
An 11-university consortium, designed to hasten the entry of minorities into managerial positions in business, is offering fellowships to capable young men and women to allow them the opportunity to pursue a high-quality educational program leading to the MBA. Fellowships are made possible through the support and contributions of American business. Summer opportunities are also provided to each student accepted for the program. The 11 participating universities are University of California-Berkley, Indiana University-Bloomington, New York University, University of Michigan-Ann Arbor, University of North Carolina at Chapel Hill, University of Rochester, University of

Southern California, University of Texas at Austin, University of Virginia, Washington University-Saint Louis, and University of Wisconsin-Madison. Application forms may be downloaded from the Web site. Applications are accepted between December 1 and January 15.
Requirements: Young African Americans, Puerto Ricans, Cubans, Dominicans, Mexican Americans, or Native Americans who are U.S. citizens or nationals who aspire to managerial careers in business are invited to apply. Applications will be received from college seniors and those holding bachelor's degrees from accredited colleges and universities. Degrees may be in the humanities, life, social and physical sciences, and engineering, as well as business or economics. Those already embarked on other careers but wishing to change careers are also encouraged to apply.
Amount of Grant: $2500 per year plus tuition and fees
Date(s) Application is Due: Jan 15; Dec 1
Contact: Jackie Olden, (888) 658-6814 or (314) 877-5540; email: frontdesk@cgsm.org or oldenj@cgsm.org
Internet: http://www.cgsm.org
Sponsor: Consortium for Graduate Study in Management
5585 Pershing, Suite 240
Saint Louis, MO 63112-4621

Consortium of College and University Media Centers Annual Research Grants   1791
Grants support research related to production, selection, distribution, and/or utilization of film/video to better meet the needs of the educational community. The research should be in progress or expected to be completed within 18 months of the submission of the proposal. Selection of an award recipient will be made on the basis of how well the proposed study focuses on needs or opportunities related to the production, selection, cataloging, distribution, and/or utilization of educational media. Grants are made for one year and are not usually renewable. The application must include a one- to two-page description of the study, a proposed budget, and a resume of the investigator. Mail or fax the required materials to Research Awards Committee, CCUMC, 1200 Communications Building, ITC, Iowa State University, Ames, IA 50011-3243.
Requirements: Eligible to apply are undergraduate or graduate student faculty or staff persons or any constituent member of the consortium. Greater consideration will be given to studies with general application to the field.
Amount of Grant: $2000 maximum
Date(s) Application is Due: May 1
Contact: Research Awards Committee, (515) 294-1811; fax: (515) 294-8089; email: ccumc@ccumc.org
Internet: http://www.indiana.edu/~ccumc
Sponsor: Consortium of College and University Media Centers
Iowa State University, 1200 Communications Bldg
Ames, IA 50011-3243

Consumers Energy Foundation   1792
The corporate foundation provides financial support to Michigan nonprofits whose social welfare programs provide solutions to problems faced by individuals and families who are unable to address their own needs without help; Michigan growth and environmental enhancement efforts that protect and enhance Michigan's physical environment; education initiatives that coordinate and develop partnerships for systemic reform in grades K-12 and programs that support schools of higher learning, with special interest in curricula and capital improvements in the study of business, political science, economics, engineering and natural/physical sciences; community and civic endeavors that support programs focusing on Michigan's economic development at the community, regional, and statewide levels, with preference given to volunteer-driven efforts, public/private partnerships and short- or long-term projects with significant evaluation components; culture and the arts programs that increase awareness of the values of artistic and cultural achievements and encourage their growth. The foundation considers requests from qualified organizations to support operating budgets and capital fund programs for construction, refurbishment or purchase of buildings, structures, equipment, or other physical enhancements, and matching grants. The initial application should be in narrative format, not exceeding two pages, and must include confirmation of the applicant organization's 501(c)3 status. The foundation provides support for minority scholarship programs administered by the United Negro College Fund (UNCF) and the Michigan Colleges Foundation (MCF). For each of these programs, scholarships are awarded to women and minority students enrolled in business or science-related curriculums, including math, engineering and computers.
Requirements: Michigan 501(c)3 organizations are eligible. The principal focus includes communities in Michigan's Lower Peninsula where Consumers Energy has a business presence. Projects that benefit the state of Michigan and the communities served by Consumers Energy will be considered from organizations based elsewhere.
Restrictions: Contributions are not made to individuals, including individual scholarships; organizations to which contributions are not tax deductible; organizations that practice discrimination on the basis of sex, age, height, weight, etc.; those whose operating activities are already supported by the United Way; political organizations and campaigns; religious organizations for religious purposes; or labor or veterans organizations, fraternal orders, or social clubs.
Geographic Focus: Michigan
Samples: Spring Arbor College (MI)—for a new library, $40,000.; College Fund/UNCF (MI)—for scholarships for Michigan students, $20,000.
Amount of Grant: $500-$10,000 average
Contact: Carolyn Bloodworth, Secretary/Treasurer, (517) 788-0432; fax: (517) 788-2281; email: foundation@consumersenergy.com
Internet: http://www.consumersenergy.com/ocompany/index.asp?SSID=34&pop=true

Sponsor: Consumers Energy
1 Energy Plz, Room EP8-210
Jackson, MI 49201

Cooley's Anemia Research Fellowships 1793

The grants-in-aid program supports research that will add to current knowledge about Cooley's anemia and related disorders. The ultimate goal is to find the cause and cure or effective treatment of these diseases. This involves a great deal of basic research in hematology, pathology, physiology, biochemistry, physical chemistry, and related sciences. Fellowships are also available and are awarded for one year, with possible renewal for a second year upon reapplication. The deadline for both fellowships and grants in aid of research is mid-March each year; awards are announced by June 1.
Requirements: Grants are made to institutions and researchers of any country. Individual applicant must have a doctoral degree and no more than five years postdoctoral training.
Amount of Grant: $40,000 maximum stipend per year
Date(s) Application is Due: Mar 1
Contact: Chairperson, Medical Advisory Board, (718) 321-2873 or (800) 522-7222; fax: (718) 321-3340; email: info@cooleysanemia.org
Internet: http://www.cooleysanemia.org/sections.php?sec=58
Sponsor: Cooley's Anemia Foundation
129-09 26th Avenue, Suite 203
Flushing, NY 11354

Cooperative Institute for Research in Environmental Sciences Visiting Fellows 1794

With support from the Environmental Research Laboratories of the National Oceanic and Atmospheric Administration, the Cooperative Institute for Research in Environmental Sciences (CIRES) at the University of Colorado at Boulder offers up to six one-year visiting fellowships to scientists with research interests in the areas of advanced observing and modeling systems, climate system variability, geodynamics, planetary metabolism, regional processes, and integrating activities. The program provides opportunities for interactions between CIRES scientists and visiting fellows to pursue common research interests. Selections for this program are based in part on the likelihood of interactions between the visiting fellow and the scientists at CIRES and the degree to which both parties will benefit from the exchange of new ideas. To further this goal, priority is given to candidates with research experience at institutions outside the Boulder scientific community. Salary, benefits, and a research budget are provided.
Requirements: Awards may be made to PhD scientists at all levels; faculty planning sabbatical leave and recent PhD recipients are especially encouraged to apply. The program is open to scientists of all countries.
Date(s) Application is Due: Dec 31
Contact: Karen Dempsey, Program Contact, (303) 492-1168; fax: (303) 492-1149; email: dempsey@cires.colorado.edu
Internet: http://cires.colorado.edu/visfell
Sponsor: Cooperative Institute for Research in Environmental Sciences
Human Resources Dept, 216 UCB
Boulder, CO 80309-0216

Cooperstown Graduate Fellowships 1795

The program, cosponsored by the State University of New York and the association, awards assistantships, fellowships, and scholarships to individuals in the museum field to enhance their professional training. Candidates should apply for admission to the program and for a fellowship at the same time. Write to the program at Cooperstown for a catalog and application.
Requirements: Applicants must be U.S. citizens who are graduates of accredited four-year colleges with a 3.0 average or above in their major field of study for the last two years. In addition, they must have obtained satisfactory scores on the Graduate Record Exam.
Amount of Grant: $2000-$10,000 over four semesters
Date(s) Application is Due: Jan 10
Contact: Director of Admissions, Cooperstown Graduate Program, (607) 547-2586; fax: (607) 547-8926; email: wrightlg@oneonta.edu
Internet: http://cgp.oneonta.edu
Sponsor: New York State Historical Association
P.O. Box 800
Cooperstown, NY 13326

Coors Brewing Corporate Contributions Grants 1796

The company has a firm commitment to giving back to its home-market communities—including Denver, CO; Memphis, TN; and Elkton, VA—and supports grassroots, nonprofit organizations that address community, civic and industry issues. The primary focus is on programs that enhance the quality of life. Vehicles of support include cash grants, Coors products for events, Coors logo items for fund-raisers, volunteer hours by Coors employees or retirees, or used equipment or in-kind services. Preference will be given to groups that focus on issues of national scope. Corporate Contributions usually reviews requests the first Wednesday of every month. A minimum of two months lead time is required before the event or funding need.
Requirements: IRS 501(c)3 nonprofit organizations located in Denver, CO; Memphis, TN; and Elkton, VA; are eligible.
Restrictions: Requests will not be considered for individuals in personal programs; individual scholarships; teams, groups, or races; travel expenses; third-party fund-raisers or sales promotions; political activities; or requests by telephone.
Geographic Focus: Colorado; Tennessee; Virginia

Contact: Buck Boze, Corporate Contributions, (800) 642-6116 or (303) 277-5953; fax: (303) 277-6132
Internet: http://www.coors.com/part_community_funding.asp
Sponsor: Coors Brewing Company
P.O. Box 4030, Dept NH420
Golden, CO 80401

Cord Foundation Grants 1797

The foundation makes grants to a variety of groups in three broad areas: education, social services, and the arts. Grant recipients have included higher education institutions, research groups, youth organizations, community service agencies, religious organizations, and performing and visual arts centers. Types of support include general operating support, building construction/renovation, equipment acquisition, emergency funds, program development, scholarship funds, research, and matching funds. Grants are awarded nationwide, with preference given to requests from northern Nevada. Application forms are not required.
Requirements: Nonprofit organizations are eligible but giving is primarily in the northern NV area.
Restrictions: The foundation does not support general fund-raising events, memorial campaigns, deficit fundings, conferences, dinners, or mass mailings.
Geographic Focus: Nevada
Samples: George Washington U (Washington, DC)—for the medical school and law school, $100,000.; Nevada Self-Help Foundation (Reno, NV)—for administrative support, $30,500.; Saint Johns Military Academy (Delafield, WI)—for operating and enhancement support, $50,000.
Amount of Grant: $10,000-$100,000 average
Date(s) Application is Due: May 15
Contact: William Bradley, Trustee, (775) 323-0373
Sponsor: Cord Foundation
418 Flint Street
Reno, NV 89501

Corning Foundation Community Service Grants 1798

The foundation supports a variety of organizations that serve a broad base of constituents. Included in this category are hospitals and hospices, community foundations, youth and women's centers, YMCAs, local chapters of Girl Scouts and Boy Scouts of America, and selected United Ways.
Requirements: All requests to the foundation for support must be made in writing. Grant seekers are advised to submit a two- to three-page letter of inquiry, signed by the senior administrative officer of the organization.
Restrictions: Grants do not support individuals; political parties, campaigns, or causes; labor or veterans' organizations; religious or fraternal groups; volunteer emergency squads; athletic activities; courtesy advertising; or fundraising events.
Samples: Glove House (Elmira, NY)—to support programs for children and families at risk, $4000.; Corning Meals on Wheels (New York, NY)—general program support, $2000.
Contact: Karen Martin, Associate Director, (607) 974-8746; fax: (607) 974-4756
Internet: http://www.corning.com/inside_corning/foundation.asp
Sponsor: Corning Foundation
MP-BH-07
Corning, NY 14831

Corning Incorporated Foundation Educational Grants 1799

The grants improve education by supporting selected K-12 school districts, community colleges and four-year institutions of higher learning primarily in communities where the sponsor company has operations. Areas of involvement have included community service programs for students, curriculum enrichment, student scholarships, facility improvement, and instructional technology projects for the classroom.
Requirements: Support goes to institutions that are tax-exempt under Section 501 (c)(3) of the Internal Revenue Code and which are public charities as defined in Section 509(a) of the Code. All requests to the foundation for support must be made in writing. Grant seekers are advised to submit a two- to three-page letter of inquiry, signed by the senior administrative officer of the organization.
Restrictions: Grants do not support individuals; political parties, campaigns, or causes; labor or veterans' organizations; religious or fraternal groups; volunteer emergency squads; athletic activities; courtesy advertising; or fundraising events.
Samples: Wilson College (Chambersburg, PA)—for the Learning Resources Ctr, $5000.; Corning City School District (Painted Post, NY)—to support computerization and curriculum enrichment, $33,000.; Junior Achievement of Elmira (Corning, NY)—for general program support, $3000.
Amount of Grant: $1000-$25,000 average
Contact: Karen Martin, Associate Director, (607) 974-8722; fax: (607) 974-4756; email: martinkc@corning.com
Internet: http://www.corning.com/inside_corning/foundation.asp
Sponsor: Corning Incorporated Foundation
MP-LB-02
Corning, NY 14831

Corpus Christi College Research Scholarships 1800

Successful candidates will pursue, as members of college, a course of study in any field leading to a research-based higher degree in the University of Cambridge. Awards are usually made to those studying for the PhD degree. Subject to satisfactory progress, awards may be extended to a maximum of three years.

Requirements: The awards are open to both men and women who have graduated before October 1 of the year of application. Applicants must first apply for university admission to the Board of Graduate Studies at 4 Mill Ln, Cambridge, CB2 1RZ and must have acceptance as a condition of receiving a scholarship.
Restrictions: Those eligible for United Kingdom State Awards may not apply.
Amount of Grant: @L500 maximum
Contact: Tutor for Advanced Students, 44 01223-338038; fax: 44 01223-338057; email: graduate-tutor@corpus.cam.ac.uk
Internet: http://www.admin.cam.ac.uk/univ/gsprospectus/colleges/corpuschristi.html
Sponsor: Corpus Christi College
Trumpington Street
Cambridge, England CB2 1RH United Kingdom

Coughlin-Saunders Foundation Grants Program   1801
The foundation awards grants to nonprofit organizations in its areas of interest, including arts and arts education, higher education, religion, social services, and youth organizations. Types of support include general operating support, capital campaigns, building construction/renovation, equipment acquisition, emergency funds, program development, professorships, and scholarship funds. Proposals are preferred in January or February. Grants are made primarily for projects which benefit Alexandria, Louisiana, and the surrounding area. New Orleans and the surrounding area will be considered.
Requirements: 501(c)3 tax-exempt organizations in central Louisiana may apply.
Restrictions: Grants are not made to individuals or for fund raisers. Endowments will not be funded.
Geographic Focus: Louisiana
Amount of Grant: $100-$40,000 range
Contact: Ed Crump Jr., Grants Administrator, (318) 561-4070; fax: (318) 487-7339; email: csfoundation@kricket.net
Sponsor: Coughlin-Saunders Foundation
2010 Gus Kaplan Drive
Alexandria, LA 71301

Council on Foundations Emerging Philanthropic Leaders Fellowships   1802
The program is designed to help future foundation leaders become more effective within their communities and the larger field of organized philanthropy. Each year, two two-year fellowships are awarded. Each fellow is matched with a mentor who is a recognized leader in philanthropy. The fellows are asked to identify their professional growth goals and organizational development needs that they would like to address as part of their fellowship. The mentors, drawn from the council's membership, are asked to help develop two-year plans for providing information, advice, and referrals. Applications may be sent by mail, fax, or email.
Requirements: The nominee or applicant must have at least two years experience in the philanthropic area, with at least one year in a leadership position. The nominee or applicant's organization must have a focus on increasing and expanding philanthropic programs within communities that are historically underrepresented in institutional philanthropy. All applicants must submit a simple application letter. The chief executive of the organization also must write a brief letter of support or co-sign the applicant's submission. Nominees or applicants must be associated with a Council on Foundations member or member-eligible organization.
Date(s) Application is Due: Dec 31
Contact: Evelyn Gibson; (703) 879-0691 or (703) 879-0600; email: Evelyn.Gibson@cof.org
Internet: http://www.cof.org/files/Documents/Diversity/diversitybrochure.pdf
Sponsor: Council on Foundations
2121 Crystal Drive, Suite 700
Arlington, VA 22202

Council on Legal Education Opportunity Program   1803
The program provides economically and educationally disadvantaged students, many with less than traditional admissions credentials, an opportunity to attend accredited law schools. Students accepted into the program are assigned to a summer institute where immediately prior to commencing law school they are exposed to and prepared for the ensuing law school regimen. Stipend is available for three years while recipient is attending an American Bar Association-approved law school.
Requirements: Applicants must have completed an undergraduate degree program and have taken the LSAT examination.
Restrictions: Persons who are currently enrolled in law school and did not enter through the CLEO program, as well as those individuals who have previously attended law school, are ineligible to apply.
Amount of Grant: $7000 for the first year and approximately $5000 for each of the second and third years
Date(s) Application is Due: Feb 1
Contact: CLEO Admissions Analyst, (202) 662-8630; email: williatm@staff.abanet.org
Internet: http://cleoscholars.com/all_about_cleo/index.htm
Sponsor: Council on Legal Education Opportunity
740 15th Street NW
Washington, DC 20005

Cowles Charitable Trust Grants   1804
The foundation awards grants, primarily in New York, Florida, and on the East Coast, for the arts and culture, including museums and the performing arts; environment; education, including early childhood education, secondary and higher education, medical school

education, adult basic education and literacy, and adult continuing education; hospitals and AIDS programs, including research; social services, including family planning, human services, and federated giving; and community funds, including leadership development, civil rights, and race relations. Types of support include general operating support, capital campaigns, annual campaigns, equipment acquisition, endowment funds, continuing support, seed money, building construction/renovation, matching funds, professorships, and program development. Application forms are required; initial approach should be by letter. The board meets in January, April, July, and October.
Requirements: Nonprofit organizations may apply for grant support. Grants are awarded primarily along the Eastern Seaboard.
Geographic Focus: Florida; New York
Amount of Grant: $1000-$40,000 range
Date(s) Application is Due: Mar 1; Jun 1; Sep 1; Dec 1
Contact: Gardner Cowles III, President, (732) 936-9826
Sponsor: Cowles Charitable Trust
P.O. Box 219
Rumson, NJ 07760

CPB American History and Civics Initiative   1805
The program is a major commitment by the CPB to use its educational mandate, reach, and creative capacity to address critical shortfalls in middle and high school students' knowledge of American history, our political system, and their roles as citizens. This initiative will award grants to forge unique and sustainable partnerships between public television producers and broadcast outlets, the educational community, curriculum developers, the high-tech industry and other appropriate partners, to design, test and create integrated interactive multimedia platforms that improve learning. Grants will be awarded in three phases: research and development, prototype creation, and production and implementation. The current round is for research and development grants. Only research and development grantees may apply for phase 2 and phase 3 grants. Guidelines are available online.
Requirements: Public broadcasting stations are eligible. Any public or private, nonprofit, educational, or commercial entity is eligible to apply for a research and development grant as a managing partner. Any public or private, nonprofit, educational, or commercial entity is eligible to serve as a subordinate key partner in a proposal.
Amount of Grant: $50,000-$250,000
Date(s) Application is Due: Nov 1
Contact: Grants Administrator, (202) 879-9600; email: History.Civics@cpb.org
Internet: http://www.cpb.org/grants/historyandcivics
Sponsor: Corporation for Public Broadcasting
401 Ninth Street NW
Washington, DC 20004-2129

CPC Catherine H. Beattie Fellowship   1806
The fellowship was created to promote the conservation of rare and endangered flora in the United States through the programs of the Center for Plant Conservation. The research grant enables a student in either biology or horticulture to conduct field research on rare plants. Preference is given to students whose projects focus on the endangered flora of the Carolinas and the southeastern United States. Additional information and applications may be obtained by contacting the office.
Requirements: Graduate students in biology, horticulture, or a related field are eligible to apply.
Amount of Grant: $1000-$4000
Date(s) Application is Due: Dec 31
Contact: Conservation Programs Manager; (314) 577-9450; fax: (314) 577-9465; email: cpc@mobot.org
Internet: http://www.centerforplantconservation.org/beattie.html
Sponsor: Center for Plant Conservation
P.O. Box 299, Missouri Botanical Garden
St. Louis, MO 63166-0299

Cranbrook Academy of Art Guest Artist/Scholar Grants   1807
Applications for the position of guest artist/scholar are invited from scholars in the humanities or from writers, poets, musicians, filmmakers, visual artists, etc. who are active in their field, who have manifested a significant body of work, and who can frame their interests in an interdisciplinary manner. The format of each program is flexible. Possible models include a set of five or six public lectures, weekly or biweekly, with follow-up seminars or studio critiques; or a four- to six-week residency on campus with informal performances, readings, and studio visits. A letter of interest; curriculum vita; publications, slides, or other representation of recent work; title, abstract, outline, and format of proposed program; and indication of scheduling requirements comprise application materials.
Contact: Ruth Ann Clark, Assistant Director for Annual Programs, (248) 645-3065; fax: (248) 646-0046; email: RClark@cranbrook.edu
Internet: http://www.cranbrookart.edu
Sponsor: Cranbrook Academy of Art
Box 801, 39221 Woodward Avenue
Bloomfield Hills, MI 48303-0801

Credit Union Foundation of British Columbia Grants   1808
The purpose of the foundation is to administer funds for the encouragement, promotion, and advancement of postsecondary education by means of grants-in-aid assistance to British Columbia resident students in need of financial assistance in pursuing vocational, technical, or academic training within British Columbia. Areas of interest include adult

and continuing education, education institutions, universities, community colleges, and vocational training. Types of support through the program include scholarships, grants, awards, and bursaries. The majority of student grants are considered in the fall. Contact the office for application forms.
Amount of Grant: $300-$2000 range
Contact: email: cufoundation@shaw.ca
Internet: http://www.cufoundation.org
Sponsor: Credit Union Foundation of British Columbia
1441 Creekside Drive
Vancouver, BC V6J 4S7 Canada

CRF Graduate Research Grants 1809
Proposals are welcome from students enrolled in graduate programs in any accredited colleges or universities. Grants will support fully or partially funded projects. Travel expenses may be included if justified by the research. Guidelines are available online.
Amount of Grant: $15,000 maximum
Date(s) Application is Due: Nov 28
Contact: Craft Research Fund, email: info@craftcreativitydesign.org
Internet: http://www.craftcreativitydesign.org/research/grants.php
Sponsor: Center for Craft, Creativity, and Design
P.O. Box 1127
Hendersonville, NC 28793

CRF Project Research Grants 1810
Support is provided for research on American studio craft. Research involving international craft must be in relationship to craft in America. The fund seeks proposals that address the goals of the fund through monographs on individual craft artists or themes, historical research, and cross-disciplinary research involving craft. Applicants requesting support for monographs on individual craft artists should refer to www. aaa.si.edu on how the research will compliment or expand the Smithsonian Laitman Documentation Project for Craft and Decorative Arts in America. Guidelines are available online.
Requirements: Proposals are welcome from academic researchers, independent scholars, doctoral students, and museum curators.
Restrictions: General overhead (indirect administrative expenses) is not eligible for university-based projects. No capital equipment purchases are eligible for support.
Samples: Dr. Sandra Corse (Waynesville, NC)—support for research for a book-length study tentatively titled Craft: Toward and Aesthetic of Useful Objects, $11,800 (2005); Davira Taragin (Racine, WI)—to support research for a catalogue to accompany the traveling exhibition of approximately 60 works of Viola Frey (1933-2004), $15,000 (2005).
Amount of Grant: $10,000 maximum graduate research grants; $15,000 maximum research grants
Date(s) Application is Due: Nov 28
Contact: Craft Research Fund, email: info@craftcreativitydesign.org
Internet: http://www.craftcreativitydesign.org/research/grants.php
Sponsor: Center for Craft, Creativity, and Design
P.O. Box 1127
Hendersonville, NC 28793

CRI Clinical Investigator Award in Cancer Immunology 1811
Investigator awards are available to six qualified scientists at the assistant professor level who are working in the field of clinical cancer immunology. The four-year award may be used at the recipient's discretion for salary, technical assistance, supplies, or capital equipment. CRI has no citizenship restrictions, and research supported by the award may be conducted anywhere in the United States or abroad, excluding for-profit institutions. Applications may be downloaded from the Web site.
Requirements: Candidate must hold a doctoral degree and be a tenure-track assistant professor or equivalent rank at the time of the award.
Amount of Grant: $50,000 a year for four years for each investigator
Date(s) Application is Due: Mar 1
Contact: Lynne Harmer, Grants Administrator, (212) 688-7515; fax: (212) 832-9376; email: grants@cancerresearch.org
Internet: http://www.cancerresearch.org/clinical.html
Sponsor: Cancer Research Institute
681 Fifth Avenue
New York, NY 10022-4209

CRI Postdoctoral Fellowships 1812
Each year the institute awards postdoctoral fellowships to qualified individuals in the formative stages of their career who wish to receive training in cancer immunology with special emphasis on tumor immunology. Work may be carried out in the United States or abroad. The fellowships are awarded for a period of two years, with the possibility of renewal for one additional year. CRI requires all fellowship applicants to complete an electronic application as well as a paper application.
Requirements: Applicant of any nationality must have a doctoral degree and must conduct research under a sponsor who holds a formal appointment at the sponsoring institution.
Amount of Grant: $40,000 first year, $42,000 second year, $44,000 third year; plus $1500 institutional allowance per year
Date(s) Application is Due: Apr 1; Oct 1
Contact: Brian Brewer, Grants Administrator, (212) 688-7515 ext 242 or (800) 992-2623; fax: (212) 832-9376; email: grants@cancerresearch.org

Internet: http://www.cancerresearch.org/postdoc.html
Sponsor: Cancer Research Institute
681 Fifth Avenue
New York, NY 10022-4209

CRI Prostate Cancer Initiative Grants 1813
The program supports Phase I and Phase I/II clinical trials that test novel therapies for advanced prostate cancer, with particular emphasis on hormonal therapy and immunotherapy. CRI does not provide funds for indirect costs. The completed application consists of 15 sets of the following items collated in the order specified: application form; biographical sketches for all key personnel; abstract of proposed project explaining the importance of the proposed research and ramifications; five-page description of research program, including summary of past work and future plans; and IRB-approved protocol. Applications must be typed or printed (using both sides of the paper, if possible). In addition, a self-addressed, stamped postcard is required if applicant would like to receive notification of receipt of application. Annual deadline dates may vary; contact the program office for exact dates.
Amount of Grant: $150,000 over two years
Date(s) Application is Due: Apr 15
Contact: Grants Administrator, (212) 688-7515 or (800) 992-2623; fax: (212) 832-9376; email: info@cancerresearch.org
Internet: http://www.cancerresearch.org/pciinit.html
Sponsor: Cancer Research Institute
681 Fifth Avenue
New York, NY 10022-4209

Crohn's Disease and Colitis Research Fellowships 1814
Three-year fellowships are awarded to encourage the development of individuals with research potential to help them prepare for careers of independent research in the areas of Crohn's disease and ulcerative colitis. Yearly renewal is dependent on the receipt of satisfactory progress reports. Application forms must be obtained from the foundation. Letters of intent are due May 1 and November 1. Submission deadlines for full applications are January 14 and July 1 of each year. Annual deadline dates may vary; contact program staff for exact dates.
Requirements: The applicant must hold an MD, PhD, or equivalent degree and be currently employed by an institution engaged in health care and/or health-related research within the United States. Candidates who have obtained MD degrees must have at least two years of postdoctoral experience, one year of which must be documented research experience relevant to inflammatory bowel disease. Candidates holding PhD degrees must have one year of postdoctoral experience and at least one year of documented research experience relevant to IBD.
Amount of Grant: $58,250 maximum salary per year
Date(s) Application is Due: Jan 14; May 1; Jul 1; Nov 1
Contact: Carol Cox, Director of Research and Scientific Programs, (212) 685-3440 or (800) 932-2423; fax: (212) 779-4098; email: ccox@ccfa.org
Internet: http://www.ccfa.org/research/?LMI=4
Sponsor: Crohn's & Colitis Foundation of America Inc
386 Park Avenue S, 17th Fl
New York, NY 10016-8804

Crohn's Disease and Ulcerative Colitis Senior Research Grants 1815
Grants are awarded to individuals for up to two years for research in the basic biomedical and clinical sciences that will increase the understanding of the cause, mechanisms of symptoms, and treatment of inflammatory bowel disease including Crohn's disease and ulcerative colitis. Support for additional one-year periods will be by competitive renewal. Application forms must be secured from the foundation. Submission deadlines for letters of intent are November 1 and May 1. Submission deadlines for full applications are January 14 and July 1 of each year, unless noted otherwise.
Requirements: Applicants must be established researchers in the field of inflammatory bowel disease.
Amount of Grant: $115,000 maximum
Date(s) Application is Due: Jan 14; May 1; Jul 1; Nov 1
Contact: Carol Cox, Research and Scientific Programs, (212) 685-3440 or (800) 932-2423; fax: (212) 779-4098; email: ccox@ccfa.org
Internet: http://www.ccfa.org/research/?LMI=4
Sponsor: Crohn's & Colitis Foundation of America Inc
386 Park Avenue S, 17th Fl
New York, NY 10016-8804

Croucher Foundation Grants 1816
The foundation promotes education, learning, and research in the areas of natural science, technology, and medicine. Each year, 20 to 25 scholarships and fellowships, tenable in the United Kingdom, Hong Kong, Canada, Australia, or New Zealand, are awarded.
Requirements: The competition is open to permanent residents of Hong Kong.
Contact: Administrator, (852) 2736 6337; fax: (852) 2730 0742; email: cfadmin@croucher.org.hk
Internet: http://www.croucher.org.hk
Sponsor: Croucher Foundation
9 Queen's Rd Central, Suite 501 Hong Kong

Cruise Industry Charitable Foundation Grants 1817
The foundation awards grants to improve the quality of life in U.S. cities and towns where the cruise industry maintains vessel operations, employs a significant number

of individuals, and purchases products and services. Areas of interest include civic and community development, educational assistance and training programs, public health programs, and environmental initiatives. Funding has supported job creation and training programs; efforts to improve access to community services and youth and adult education, particularly for minority and disadvantaged students; literacy and basic life skills education; and mentoring services. Proposals are encouraged for programs serving the needs of at-risk populations. There are no application deadlines. Letters of inquiry are accepted at any time; full proposals are by invitation.

Requirements: U.S. 501(c)3 organizations and state and local government units, such as public schools and child welfare agencies, are eligible.

Restrictions: Requests will not be considered from individuals, fraternal organizations, religious organizations, political organizations, or organizations that conduct lobbying activity.

Samples: Operation Warm—for distribution of new winter coats to at-risk elementary school students in 33 states.; Crisis Center of Tampa Bay—for community family support services and education.; National Marine Sanctuary Fndn—for community outreach and environmental awareness.

Contact: Cynthia Colenda, Executive Director, (703) 522-3160; fax: (703) 522-3161; email: cicf@iccl.org

Internet: http://www.iccl.org/foundation/guidelines.cfm

Sponsor: Cruise Industry Charitable Foundation

2111 Wilson Boulevard, 8th Fl

Arlington, VA 22201

CSRPC Dissertation Fellowships 1818

The goal of the fellowship is to enable an outstanding doctoral student interested in the study of race and ethnicity to devote his or her full energies to the completion of the dissertation. The fellowship carries a stipend; a travel and research budget; and will cover advanced residence tuition, fees, and basic university student health insurance, if needed. The successful applicant will be provided with an office and use of a computer at the center. The fellow will be expected to be in residence during the award year, present his or her work at one of the Reproduction of Race and Racial Ideologies Workshop meetings and to actively participate in the workshop and other activities sponsored by the center.

Requirements: University of Chicago doctoral students who have completed all requirements for the PhD but the dissertation, including formal admission to candidacy, and expect to complete all field work by September 26 are eligible to apply. Any dissertation that has as its central focus issues related to race or racialized groups will be considered. Special consideration will be given to projects that attend to the intersection of race or ethnicity with other identities such as gender, class, sexuality, and nationality.

Restrictions: The fellow may not engage in any remunerative activity, including teaching either on or off campus, while holding the award, and will be ineligible for further internal University funding from any source, including teaching appointments, if the degree is not completed within six months of the end of the CSRPC fellowship tenure.

Amount of Grant: $18,000 stipend; $1000 travel and research budget

Date(s) Application is Due: Mar 4

Contact: Center for the Study of Race, Politics, and Culture, (773) 702-8063; fax: (773) 834-2200; email: csrpc@uchicago.edu

Internet: http://csrpc.uchicago.edu/resources_funding.shtml

Sponsor: University of Chicago

5733 S University Avenue

Chicago, IL 60637

CSRPC Postdoctoral Fellowship for Advanced Scholars 1819

The goal of the fellowship is to support the work of an outstanding advanced scholar whose research focuses on the study of race or ethnicity by allowing the fellow to devote his or her energies to the further development of their research agenda. The fellowship carries a stipend, a travel and research budget, and an allowance for moving expenses. The fellow will be provided with office space, a computer at the center, and full access to University libraries and other facilities. Awardees will be expected to be in full-time residence during the academic year; teach a 10-week course related to race and/or ethnicity (one quarter); give a public lecture; present his or her work at one of the Reproduction of Race and Racial Ideologies Workshop meetings; and actively participate in the workshop and other activities sponsored by the center. Guidelines are available online.

Requirements: Applicants are required to have a PhD; must have been awarded tenure at the time of application; and must be U.S. citizens or permanent residents at the time of application.

Amount of Grant: $55,000 stipend; $5000 travel and research budget; $2500 for moving expenses

Date(s) Application is Due: Feb 11

Contact: Center for the Study of Race, Politics, and Culture, (773) 702-8063; fax: (773) 834-2200; email: csrpc@uchicago.edu

Internet: http://csrpc.uchicago.edu/resources_funding.shtml

Sponsor: University of Chicago

5733 S University Avenue

Chicago, IL 60637

CSRPC Postdoctoral Fellowship for Junior Scholars 1820

The goal of the fellowship is to support the work of an outstanding junior scholar whose research focuses on the study of race or ethnicity by allowing the fellow to devote his or her energies to the further development of their research agenda. The fellowship carries a

stipend, a travel and research budget, and an allowance for moving expenses. The fellow will be provided with office space, a computer at the center, and full access to university libraries and other facilities. Awardees will be expected to be in full-time residence during the academic year; teach a 10-week undergraduate course related to race and/or ethnicity (one quarter); present his or her work at one of the Reproduction of Race and Racial Ideologies Workshop meetings; and actively participate in the workshop and other activities sponsored by the center. Guidelines are available online.

Requirements: Applicants are required to have a PhD; must be non-tenure track scholars or tenure track faculty who do not expect to receive tenure prior to the start of the fellowship year; and must be U.S. citizens or permanent residents at the time of application.

Amount of Grant: $45,000 stipend; $5000 travel and research budget; $2500 maximum for moving expenses

Date(s) Application is Due: Feb 11

Contact: Center for the Study of Race, Politics, and Culture, (773) 702-8063; fax: (773) 834-2200; email: csrpc@uchicago.edu

Internet: http://csrpc.uchicago.edu/resources_funding.shtml

Sponsor: University of Chicago

5733 S University Avenue

Chicago, IL 60637

CTCRI Idea Grants 1821

Idea Grants are designed to encourage unique or original research that has the potential to advance knowledge in tobacco control. Grants will allow investigators with innovative ideas and observations to conduct pilot studies, to perform secondary analysis of data sets or to gather new evidence necessary to determine the viability of research directions or hypotheses. Innovative projects are a priority for the Idea Grant program. Preference will be given to proposals that address research priorities identified by the Canadian Tobacco Control Research Summit. Research employing new or unconventional methodologies is encouraged. Proposed methods must be appropriate to the research question(s). Each successful proposal may receive a one-time grant of up to $50,000. Work is expected to be completed in one year's time.

Requirements: Research proposals must meet the following eligibility criteria: The Principal Applicant (PA) is a Canadian citizen or legal resident (Co-applicants may be citizens or residents of other countries); The research proposal demonstrates basic relevance to tobacco abuse / nicotine addiction / tobacco control and addresses research priorities identified by the Canadian Tobacco Control Research Summit; Applicant(s) has disclosed other sources of funding; Commercial interests have been disclosed or applicant(s) has indicated no commercial interests; Applicant(s) has affirmed lack of support from the tobacco industry; Applicant(s) can demonstrate that he/she works in an environment that adequately supports research through ethical review, administration of funds, provision of space and equipment, etc.; Research plans include gender analysis, or it has been demonstrated that this is not appropriate; Proposal avoids duplication of previous research, unless it can be demonstrated that replication is of value; Research proposal involves pilot testing of research methods, tools, or hypotheses by a research team in order to strengthen proposals prepared for submission to traditional funding sources; Proposal articulates preliminary plans for development of a full research proposal -OR- Research proposal constitutes a novel idea which falls outside the normal scope of traditional research, either in topic or methodology; Proposal demonstrates that adequate funding is not available from other sources for this purpose.

Restrictions: Members of the staff or Board of Directors of the CTCRI, or a staff member of any CTCRI funding partner organization are not eligible to apply.

Samples: Tony George, Centre for Addiction and Mental Health, Toronto, Canada, $49,995 - a nicotinic partial agonist for tobacco dependence treatment in bipolar disorder (2009); Robert Schwartz, University of Toronto, Ontario Tobacco Research Unit, Toronto, Canada, $50,000 - cigarette pack as advertisement: "Beyond light and mild"? (2009).

Amount of Grant: up to $50,000

Date(s) Application is Due: Apr 1; Oct 1

Contact: Carol Bishop, Assistant Director, Research Operations; (416) 934-5640; email: cbishop@cancer.ca

Internet: http://www.ctcri.ca/en/index.php?option=com_content&task=view&id=25&Itemid=44

Sponsor: Canadian Tobacco Control Research Initiative

10 Alcorn Avenue, Suite 200

Toronto, ON M4V 3B1 Canada

CTCRI Knowledge Synthesis Grants 1822

The goal of the Knowledge Synthesis program is to support interdisciplinary teams of researchers and practitioners / decision-makers to conduct collaborative reviews of evidence for particular tobacco control interventions. The maximum amount awarded for a single grant is $120,000. It is anticipated that at least two applications will be supported in the present competition, contingent upon the availability of funds and the quality of proposals received.

Requirements: Teams are requested to develop proposals to carry out reviews addressing ONE of the following topics: (1) Anti-contraband measures - An issue that undermines the effectiveness of a high tobacco tax strategy is contraband activity that includes the smuggling of tobacco products from lower tax jurisdictions, illicit manufacturing and counterfeiting. Though large price increases may provide incentive for some smokers to quit, it will impel others to seek out lower cost sources. A number of smokers are turning to contraband. (2) The effect of tax and price on prevalence and consumption in subpopulations (defined as tobacco users including 15 to 24 year olds, aboriginal people, persons diagnosed with mental health or substance abuse disorders, and lowest categories of income and formal education). (3) Discount cigarettes (defined as lower-priced brands by small manufacturers and cheaper brands by big tobacco companies) - With the price increase of tobacco products, a number of smokers are turning to price discounted cigarettes. Price discounted cigarettes now comprise more than 40% of market share in cigarette sales. Research proposals must meet the following eligibility criteria: The Principal Applicant (PA) is a Canadian citizen or legal resident (Co-applicants may be citizens or residents of other countries); The research proposal demonstrates basic relevance to tobacco abuse / nicotine addiction / tobacco control and addresses research priorities identified by the Canadian Tobacco Control Research Summit; Applicant(s) has disclosed other sources of funding; Commercial interests have been disclosed or applicant(s) has indicated no commercial interests; Applicant(s) has affirmed lack of support from the tobacco industry; Applicant(s) can demonstrate that he/she works in an environment that adequately supports research through ethical review, administration of funds, provision of space and equipment, etc.; Research plans include gender analysis, or it has been demonstrated that this is not appropriate; Proposal avoids duplication of previous research, unless it can be demonstrated that replication is of value; Proposal demonstrates the inclusion of and consultation with end-users (practitioners, policy and decision-makers) throughout the research process; Applicant demonstrates understanding and application of the Better Practices methodology, including definition of the scope of the review.

Restrictions: Members of the staff or Board of Directors of the CTCRI, or a staff member of any CTCRI funding partner organization are not eligible to apply.

Amount of Grant: up to $120,000

Date(s) Application is Due: Apr 1

Contact: Research Grant Programs; (416) 934-5666; fax: (416) 961-4189; email: info@ctcri.ca

Internet: http://ctcri.ca/en/index.php?option=com_content&task=view&id=212&Itemid=72

Sponsor: Canadian Tobacco Control Research Initiative

10 Alcorn Avenue, Suite 200

Toronto, ON M4V 3B1 Canada

CTCRI Policy Research Grants 1823

Policy Research grants are intended to stimulate research that will influence, guide or have a direct impact on policy decisions in tobacco control. Suggested research areas and methodologies include: The full policy process from agenda setting and decision making through to development and implementation, enforcement, evaluation and refinement; All areas in which policy might be used to influence tobacco use including fiscal, legislative, regulatory and educational policies; Public and private policy at all levels of government/locale; Preference will be given to projects that address research priorities identified by the Canadian Tobacco Control Research Summit (2002). Research employing new or unconventional methodologies is encouraged, but proposed methods must be appropriate to the research question(s). Each successful proposal may receive a one-time grant of up to $80,000. The project may be completed over the course of two years.

Requirements: Research proposals must meet the following eligibility criteria: The Principal Applicant (PA) is a Canadian citizen or legal resident (Co-applicants may be citizens or residents of other countries); The research proposal demonstrates basic relevance to tobacco abuse / nicotine addiction / tobacco control and addresses research priorities identified by the Canadian Tobacco Control Research Summit; Applicant(s) has disclosed other sources of funding; Commercial interests have been disclosed or applicant(s) has indicated no commercial interests; Applicant(s) has affirmed lack of support from the tobacco industry; Applicant(s) can demonstrate that he/she works in an environment that adequately supports research through ethical review, administration of funds, provision of space and equipment, etc.; Research plans include gender analysis, or it has been demonstrated that this is not appropriate; Proposal avoids duplication of previous research, unless it can be demonstrated that replication is of value; Research proposal is relevant to public policy issues surrounding tobacco control in Canada. FastTrack proposals only: Applicants must demonstrate that their proposal is dependent upon policy or legislation which is timely and that the project cannot be delayed until the April 1 or October 1 deadlines.

Restrictions: Members of the staff or Board of Directors of the CTCRI, or a staff member of any CTCRI funding partner organization are not eligible to apply.

Amount of Grant: up to $80,000

Date(s) Application is Due: Apr 1; Oct 1

Contact: Research Grant Programs; (416) 934-5666; fax: (416) 961-4189; email: info@ctcri.ca

Internet: http://ctcri.ca/~ctcri/en/index.php?option=content&task=view&id=206&Itemid=87

Sponsor: Canadian Tobacco Control Research Initiative

10 Alcorn Avenue, Suite 200

Toronto, ON M4V 3B1 Canada

CTCRI Research Planning Grants 1824

Research Planning Grants are offered for the purpose of bringing together new, multi-sectoral and interdisciplinary research teams to construct research proposals for submission to traditional open funding competitions. The specific objectives of the program are: To facilitate the development of excellent research proposals which are relevant to the priorities of the CTCRI's research agenda; To support the formation of strong, interdisciplinary teams that can compete for research funding through traditional sources. Each successful proposal may receive a one-time grant of up to $15,000. Work is expected to be completed in one year's time. The number of applications supported each year is contingent upon the availability of funds and the quality of proposals.

Requirements: Preference will be given to proposals that address research priorities identified by the Canadian Tobacco Control Research Summit. Suggested methodologies include: Physically gathering members of a team to develop a research plan; Conducting a literature review or other background work to develop a research project; Hiring a research assistant to help write a well grounded proposal (if applicable). Research proposals must meet the following eligibility criteria: The Principal Applicant (PA) is a Canadian citizen or legal resident (Co-applicants may be citizens or residents of other countries); The research proposal demonstrates basic relevance to tobacco abuse / nicotine addiction / tobacco control and addresses research priorities identified by the Canadian Tobacco Control Research Summit; Applicant(s) has disclosed other sources of funding; Commercial interests have been disclosed or applicant(s) has indicated no commercial interests; Applicant(s) has affirmed lack of support from the tobacco industry; Applicant(s) can demonstrate that he/she works in an environment that adequately supports research through ethical review, administration of funds, provision of space and equipment, etc.; Research plans include gender analysis, or it has been demonstrated that this is not appropriate; Proposal avoids duplication of previous research, unless it can be demonstrated that replication is of value.

Restrictions: Conducting work to pilot surveys or methods does not qualify for support through this program. Members of the staff or Board of Directors of the CTCRI, or a staff member of any CTCRI funding partner organization are not eligible to apply.

Amount of Grant: up to $15,000

Date(s) Application is Due: Feb 28; May 30; Aug 31; Nov 30

Contact: Research Grant Programs; (416) 934-5666; fax: (416) 961-4189; email: info@ctcri.ca

Internet: http://ctcri.ca/~ctcri/en/index.php?option=content&task=view&id=209&Itemid=88

Sponsor: Canadian Tobacco Control Research Initiative

10 Alcorn Avenue, Suite 200

Toronto, ON M4V 3B1 Canada

CTCRI Researcher Travel Grants 1825

Researcher Travel Grants are offered to provide opportunities for graduate and post-doctoral students and individuals affiliated with non-governmental organizations/community groups to attend conferences/meetings related to tobacco abuse and nicotine addiction. Any poster or presentation of research related to tobacco abuse and nicotine addiction is eligible for support. Preference will be given to proposals that address research priorities identified by the Canadian Tobacco Control Research Summit. Research employing new or unconventional methodologies is encouraged, but proposed methods must be appropriate to the research question(s). Grants of up to $3,000 are available to support travel and accommodations. In some cases, the value of the grant may exceed $,3000 (for example, for conferences involving international travel). The number of applications supported annually is contingent upon the availability of funds and the quality of proposals.

Requirements: Research proposals must meet the following eligibility criteria: The Principal Applicant (PA) is a Canadian citizen or legal resident (Co-applicants may be citizens or residents of other countries); The research proposal demonstrates basic relevance to tobacco abuse / nicotine addiction / tobacco control and addresses research priorities identified by the Canadian Tobacco Control Research Summit; Applicant(s) has disclosed other sources of funding; Commercial interests have been disclosed or applicant(s) has indicated no commercial interests; Applicant(s) has affirmed lack of support from the tobacco industry; Applicant(s) can demonstrate that he/she works in an environment that adequately supports research through ethical review, administration of funds, provision of space and equipment, etc.; Research plans include gender analysis, or it has been demonstrated that this is not appropriate; Proposal avoids duplication of previous research, unless it can be demonstrated that replication is of value; The Principal Applicant must be conducting graduate or post-doctoral work and must be affiliated with an accredited academic program. OR The Principal Applicant must be affiliated with a non-governmental organization or community group conducting tobacco control research or evaluation; The Principal Applicant's abstract has been accepted for the upcoming conference for which the funding is requested (funding cannot be applied for retroactively); The proposal demonstrates that any current funding will not adequately support the stated objectives of the proposal; The Principal Applicant has not been awarded another CTCRI travel grant to present at a conference or event in the same calendar year as the current application. This does not include attendance at the special tobacco control conferences such as the World Conference on Tobacco or Health (WCOTH) or the Society for Research on Nicotine and Tobacco (SRNT) conference.

Restrictions: Members of the staff or Board of Directors of the CTCRI, or a staff member of any CTCRI funding partner organization are not eligible to apply.

Amount of Grant: up to $3,000

Date(s) Application is Due: none

Contact: Research Grant Programs; (416) 934-5666; fax: (416) 961-4189; email: info@ctcri.ca

Internet: http://ctcri.ca/~ctcri/en/index.php?option=content&task=view&id=208&Itemid=89

Sponsor: Canadian Tobacco Control Research Initiative

10 Alcorn Avenue, Suite 200

Toronto, ON M4V 3B1 Canada

CTCRI Student Research Grants 1826

The objectives of the program are to support learning opportunities through a one-time grant for students wishing to conduct research projects related to tobacco abuse and nicotine addiction; and to provide opportunities for graduate and post-doctoral students to improve skills in tobacco abuse and nicotine addiction research. The grant program is intended to support a broad range of research topics and areas as long as the proposed research is led by the applicant under professional research supervision. The program is not intended to support the applicant's regular course of study or background work required to prepare a research project, nor is it intended to support the research supervisor's existing project(s). Preference will be given to proposals that address research priorities identified by the Canadian Tobacco Control Research Summit. Each successful proposal may receive a one-time grant of up to $10,000. Work is expected to be completed in one year's time. The number of applications supported each year is contingent upon the availability of funds and the quality of proposals.

Requirements: Graduate and post-doctoral students are eligible for Student Research grant. Research proposals must meet the following eligibility criteria: The Principal Applicant (PA) is a Canadian citizen or legal resident (Co-applicants may be citizens or residents of other countries); The research proposal demonstrates basic relevance to tobacco abuse / nicotine addiction / tobacco control and addresses research priorities identified by the Canadian Tobacco Control Research Summit; Applicant(s) has disclosed other sources of funding; Commercial interests have been disclosed or applicant(s) has indicated no commercial interests; Applicant(s) has affirmed lack of support from the tobacco industry; Applicant(s) can demonstrate that he/she works in an environment that adequately supports research through ethical review, administration of funds, provision of space and equipment, etc.; Research plans include gender analysis, or it has been demonstrated that this is not appropriate; Proposal avoids duplication of previous research, unless it can be demonstrated that replication is of value.; Proposal demonstrates that any current funding will not adequately support the stated objectives of the proposal; The Research Supervisor has provided a letter of support for the applicant's proposal.

Restrictions: Members of the staff or Board of Directors of the CTCRI, or a staff member of any CTCRI funding partner organization are not eligible to apply.

Amount of Grant: up to $10,000

Date(s) Application is Due: Mar 31; Jun 30; Sep 30; Dec 30

Contact: Research Grant Programs; (416) 934-5666; fax: (416) 961-4189; email: info@ctcri.ca

Internet: http://ctcri.ca/~ctcri/en/index.php?option=content&task=view&id=211&Itemid=90

Sponsor: Canadian Tobacco Control Research Initiative

10 Alcorn Avenue, Suite 200

Toronto, ON M4V 3B1 Canada

CTCRI Workshop and Learning Opportunities Grants 1827

Workshop and Learning Opportunity grants are offered to support workshops, meetings and other events to build research capacity and promote collaboration between researchers and end-users. Priority is given to first-time applicants, to regions or populations where development of research is more acutely needed and to events that address research priorities identified by the CTCRI Research Summit: Building capacity for tobacco-related research in the region or population; Providing unique opportunities to bring together individuals from different disciplines and sectors; Supporting the sharing of knowledge, initiation of collaborations or the planning of relevant research activities. Workshops, courses and other educational methods or tools are appropriate for this grant program. Each successful proposal may receive a one-time grant of up to $15,000. Work is expected to be completed in four (4) months time. The number of applications supported annually is contingent upon the availability of funds and the quality of proposals.

Requirements: Proposals to this program must meet the following eligibility criteria: Topic is related to both research and tobacco abuse and/or addiction and intervention; Research could involve identifying research priorities, discussing and/or developing research frameworks and/or disseminating research results; Method involves multiple participants and could be in the form of a workshop or another educational opportunity; Approach is innovative and deals with engaging new topics and methods or involves new populations (e.g., researchers or practitioners new to tobacco control, historically under-represented regions or populations); Proposal demonstrates that any current funding will not adequately support the stated objectives of the proposal. Additionally, the following criteria apply: Research proposals must meet the following eligibility criteria: The Principal Applicant (PA) is a Canadian citizen or legal resident (Co-applicants may be citizens or residents of other countries); The research proposal demonstrates basic relevance to tobacco abuse / nicotine addiction / tobacco control and addresses research priorities identified by the Canadian Tobacco Control Research Summit; Applicant(s) has disclosed other sources of funding; Commercial interests have been disclosed or applicant(s) has indicated no commercial interests; Applicant(s) has affirmed lack of support from the tobacco industry; Applicant(s) can demonstrate that he/she works in an environment that adequately supports research through ethical review, administration of funds, provision of space and equipment, etc.; Research plans include gender analysis, or it has been demonstrated that this is not appropriate; Proposal avoids duplication of previous research, unless it can be demonstrated that replication is of value.

Restrictions: Members of the staff or Board of Directors of the CTCRI, or a staff member of any CTCRI funding partner organization are not eligible to apply.

Amount of Grant: up to $15,000

Date(s) Application is Due: none

Contact: Research Grant Programs; (416) 934-5666; fax: (416) 961-4189; email: info@ctcri.ca

Internet: http://ctcri.ca/~ctcri/en/index.php?option=content&task=view&id=210&Itemid=91

Sponsor: Canadian Tobacco Control Research Initiative

10 Alcorn Avenue, Suite 200

Toronto, ON M4V 3B1 Canada

Cudd Foundation Grants 1828

The foundation awards grants to eligible nonprofit organizations in its areas of interest, including arts, culture, performing arts, children and youth, education, environment, health care, historic preservation, and social services. Types of support include annual campaigns, building construction/renovation, capital campaigns, continuing support, curriculum development, emergency grants, endowments, program development, research, and scholarship funds. The listed application deadline is for letters of intent; full proposals are by invitation.

Requirements: Louisiana and New Mexico nonprofit organizations are eligible.

Geographic Focus: Louisiana; New Mexico

Amount of Grant: $250-$222,088 range

Contact: Amanda Stuermer, (505) 986-8416; fax: (505) 986-8427; email: cuddfdn@aol.com

Sponsor: Cudd Foundation

P.O. Box 2322

Santa Fe, NM 87504

Cushwa Center Research Travel Grants 1829

This program is designed to foster research in the archives and library of the University of Notre Dame. The library collection is particularly rich in the following areas: Catholic newspapers, history of midwestern Catholicism, Catholic literature, and history of Catholicism in the United States. The archives have manuscripts of historical personages, records of 20th-century Catholic organizations, reports of European missionary societies, and other material related to the American Catholic community. Grants to help defray travel and lodging costs are made to scholars of any academic discipline who are engaged in projects which require substantial use of the collections of the library and archives. Submit one copy of the completed application form, current curriculum vita, a 1,000-word description of the project to be undertaken at the center, and a budget estimating travel, lodging, and research expenses. Two letters of recommendation from two people who know your work should be sent directly to the Cushwa Center.

Requirements: The research project must be related to the study of the American Catholicism. Applicants should describe the relevance of the project for American Catholic studies, indicate the specific resources of interest at Notre Dame, and specify any plans for publication of the project.

Amount of Grant: $2000 maximum

Date(s) Application is Due: Dec 31

Contact: Director, Cushwa Center for the Study of American Catholicism, (219) 631-5441; fax: (219) 631-8471; email: cushwa.1@nd.edu

Internet: http://www.nd.edu/~cushwa/grants/

Sponsor: University of Notre Dame

1135 Flanner Hall

Notre Dame, IN 46556-5611

Cystic Fibrosis Research Grants 1830

The trust was founded to find a complete solution for cystic fibrosis and to improve upon current methods of treatment. Research grants are awarded in the United Kingdom and internationally in the fields of science and medicine.

Requirements: Only researchers from recognized academic institutions are eligible.

Contact: Research Administrator, Research Administrator, 020 8290 7906; email: researchgrants@cftrust.org.uk

Internet: http://www.cftrust.org.uk/scope/page/view.go?layout=cftrust&pageid=54

Sponsor: Cystic Fibrosis Trust

11 London Road

Bromley BR1 1BY United Kingdom

D.F. Halton Foundation Grants 1831

The foundation awards grants to nonprofit organizations in Charlotte, NC, and San Miguel County, CO, primarily in the areas of youth, education, social services, and the performing arts. Additional areas of interest include historical preservation, education, vocational education, business school education, substance abuse, cancer research, heart/circulatory diseases and research, human and family services, and community development. Grants are awarded for general operating support, annual campaigns, capital campaigns, and scholarship funds. There are no application deadlines or forms.

Requirements: Nonprofit organizations in North Carolina counties, including Mecklenburg, Union, Cleveland, Cabarrus, Stanly, Lincoln, and Gaston, may submit proposals. Nonprofit organizations in San Miguel county, CO. may also submit proposals.

Restrictions: No grants will be awarded to individuals.

Geographic Focus: Colorado; North Carolina

Contact: Dale Halton, President

Sponsor: D.F. Halton Foundation

P.O. Box 834

Ophir, CO 81426

DAAD Bilateral Programs for Co-Operative Research Grants 1832

These programs aim to intensify academic collaborative research projects by promoting the project-related exchange of academics. These are bilateral programs agreed between the DAAD in Germany and partner organizations in a number of countries. Support is given to persons participating in specific academic collaborative research projects. Particular importance is attached to the continuing education and training and specialization of young academics working within the framework of research collaboration projects. Financial support is only offered to fund specifically person-related additional costs in as far as these arise through the exchange of participating scientists, scholars, graduates and, for some countries, advanced students. Diploma-level candidates can only be included in the support framework if their final dissertation forms part of the project. The German or respectively the foreign side will pay the travel costs and the stay at the partner institute for the participants of their country on the basis of the generally applicable conditions. Payments by the host institute (for example, for accommodation) will be taken into account. Project-related additional costs (for example, computer time, material purchases, documentation, photocopying, printing) cannot be reimbursed by the DAAD, nor is there any element in the grant to cover bench fees or equipment. To date, programs have been implemented with Argentina, Brazil, Bulgaria, Canada, Chile, China, Croatia, Czech Republic, Finland, France, Greece, Hong Kong, Hungary, India, Italy, Mexico, Norway, Poland, Portugal, Slovakia, Spain, Taiwan, Thailand, the United Kingdom and the United States. Country-specific information, dates and deadlines, and application addresses differ from one country to the next.

Requirements: Applications must be accompanied by a specific academic or scientific research project proposal. Academics from Germany and from the respective partner country must be collaborating in this project. Primary financing for the collaborative research project must be provided by other sources.

Restrictions: Proposals based purely on study and training are not funded.

Contact: Scholarship Administrator, (212) 758-3223; fax: (212) 755-5780; email: kim@daad.org or daadny@daad.org

Internet: http://www.daad.de/deutschland/foerderung/stipendiendatenbank/00462.en.html?detailid=10&fachrichtung=15&land=44&status=4&seite=1&daad=1

Sponsor: German Academic Exchange Service (DAAD)
871 United Nations Plaza
New York, NY 10017

DAAD Emigre Memorial German Internship Program (EMGIP) - Bundestag   1833

The Program offers internship opportunities for U.S. and Canadian students in the German parliament, the Bundestag. The internships are two months long in positions matching the student, interest and experience. Interns will be placed with their preferred Fraktion, Ausschuss, with individual members of the Bundestag and their offices. In addition to contributing to the respective offices, interns have the opportunity to study legislative and administrative procedures in the German parliament. The successful applicant will receive compensation of approximately 1,100 Euros per month from the German Bundestag. Subsidized health insurance is available for a monthly fee of about 23 Euros. DAAD can help the interns to obtain housing in Berlin (the average rent for a room is 250 Euros/month) and make contacts with fellow international interns and German students.

Requirements: Applicants for EMGIP - Bundestag should possess outstanding academic records and personal integrity as well as some knowledge of the German legislative process. Participants should be advanced undergraduates or graduate students in fields such as political science, international relations, law, history, economics or German. Students must be able to fully communicate in German. U.S. and Canadian citizens and permanent residents are eligible to apply. International students who are enrolled in a full time course of study in the U.S. or Canada may also apply. Applicants must be younger than 32 at the start of the internship. Applicants who are graduating seniors should be prepared to show acceptance to a graduate school or further affiliation with their college to ensure student status in Germany. Travel expenses are the intern's responsibility.

Restrictions: German nationals are not eligible.

Amount of Grant: Euro 1100 per month

Date(s) Application is Due: Nov 17

Contact: Program Administrator, (212) 758-3223; fax: (212) 755-5780; email: schenkl@daad.org or daadny@daad.org

Internet: http://www.daad.org/page/53287/

Sponsor: German Academic Exchange Service (DAAD)
871 United Nations Plaza
New York, NY 10017

DAAD Faculty Research Visit Grants   1834

This program offers grants for one to three months in all academic disciplines to scholars at U.S. and Canadian institutions of higher education to pursue research at universities, libraries, archives, institutes or laboratories in Germany. Grants are awarded for specific research projects. Stipends consist of a monthly maintenance allowance.

Requirements: Candidate must hold a PhD, have been engaged in teaching or research for at least two years after receipt of the doctorate, be a U.S. citizen or permanent resident affiliated with an American institution for at least five years, and have a previous research record in the proposed field. The period for which the grant is awarded must fall entirely within one calendar year.

Restrictions: Grants cannot be used for travel, attendance at conferences or conventions, editorial meetings, lecture tours, or extended guest-professorships. No extra allowance is given for dependents who might accompany the grantee. Three years must elapse before another application by a former study visit grantee will be considered; applicant may not hold a DAAD grant or a grant from another German or American organization consecutively or concurrently.

Amount of Grant: 1,840 - 2,240 Euro per month

Date(s) Application is Due: Feb 1; Aug 1

Contact: Grants Administrator; (212) 758-3223; fax: (212) 755-5780; email: daadny@daad.org or schenkl@daad.org

Internet: http://www.daad.org/page/68573/

Sponsor: German Academic Exchange Service (DAAD)
871 United Nations Plaza
New York, NY 10017

DAAD German Studies Research Grants   1835

This specialized program offers up to five German Studies Research Grants to highly qualified undergraduate and graduate students who are nominated by their department/program chairs. The grant may be used for short-term research (one to two months) in either North America or Germany. The program is designed to encourage research and promote the study of cultural, political, historical, economic and social aspects of modern and contemporary German affairs from an inter- and multidisciplinary perspective. Research support ranging in value from $1,500 to $2,500 is available to individual scholarship recipients and is intended to offset living and travel costs during the active research phase.

Requirements: Undergraduates with at least junior standing pursuing a German Studies track or minor may be nominated for the grant by their department and/or program chair. Applicants are expected to have completed two years of college German and a minimum of three courses in German Studies (literature, history, politics or other fields) at the time of nomination. Grants are restricted to citizens of the U.S. who are enrolled full time at the university that nominates them. Applicants must be younger than 32 at the start of the grant period.

Restrictions: Support cannot be provided for stays in Germany in the context of study abroad programs.

Amount of Grant: $1,500 to $2,500

Date(s) Application is Due: Nov 1; May 1

Contact: Program Administrator, (212) 758-3223; fax: (212) 755-5780; email: graaff@daad.org or daadny@daad.org

Internet: http://www.daad.org/page/51532/

Sponsor: German Academic Exchange Service (DAAD)
871 United Nations Plaza
New York, NY 10017

DAAD Intensive Language Courses Grants   1836

DAAD offers grants to graduate students at North American universities to attend 8-week intensive language courses at leading institutes in Germany. Extensive descriptions of the institutes, their teaching philosophies, course content, as well as course dates are available on the website of each institute. Scholarships are awarded to students currently enrolled full-time in a graduate or PhD program in all fields of study except English, German, or any other modern language or literature. Applicants will be assessed on the basis of their academic record and statements of projected academic and professional future.

Requirements: As a rule, applicants must be citizens of the U.S. or Canada. Foreign nationals may be eligible if they have been full-time graduate students at a U.S. or Canadian University for at least one academic year at the time of application. Applicants must have completed three semesters of college German or have achieved an equivalent level of language proficiency. Applicants must be younger than 32 at the start of the grant period.

Restrictions: Applicants who have received a DAAD summer language grant in the past three years are not eligible to apply.

Amount of Grant: 2,650 Euro

Date(s) Application is Due: Jan 31

Contact: Program Administrator, (212) 758-3223; fax: (212) 755-5780; email: graaff@daad.org or daadny@daad.org

Internet: http://www.daad.org/page/54773/

Sponsor: German Academic Exchange Service (DAAD)
871 United Nations Plaza
New York, NY 10017

DAAD Learn German in Germany Grants   1837

The Goethe-Institut, through DAAD, offers one to two grants to faculty members in all academic fields except modern languages and literatures to attend intensive language courses at Goethe-Instituts in Germany. Four and eight-week courses are offered year-round. Preference will be given to applicants in the social sciences, the natural sciences, engineering and professional schools, who are in mid-career and are under 46 years of age.

Requirements: Scholars who hold a PhD (or equivalent) and have been working in research or teaching full-time at a United States university or research institution for at least two years after receipt of the doctorate are eligible. Applicants must have a basic knowledge of German and be able to demonstrate a need for acquiring a better proficiency in the German language for their future studies or research.

Restrictions: Applicants must be citizens or permanent residents of the United States. Faculty members who teach in the fields of English, German or any modern languages or literatures are not eligible.

Amount of Grant: 1,700 Euro

Date(s) Application is Due: Jan 31

Contact: Program Administrator, (212) 758-3223; fax: (212) 755-5780; email: graaff@daad.org or daadny@daad.org

Internet: http://www.daad.org/page/47845/

Sponsor: German Academic Exchange Service (DAAD)
871 United Nations Plaza
New York, NY 10017

DAAD Re-invitation Program for Former Scholarship Holders   1838

The Program helps the DAAD maintain contacts with its former one-year scholarship holders and with former scholarship holders who had studied in East Germany (GDR) for at least one year. Former scholarship holders meeting these requirements can apply for re-invitation to Germany to complete a research or work project at a state (public) or state-recognized higher education institution or non-university research institute.

Depending on the applicant's work schedule, the research stay can last between one and three months. Depending on the applicant's academic status, the monthly award will amount to 1,840 Euros for assistant lecturers, assistant professors and young lecturers, and 1,990 Euros for professors. In some rare exceptions, 2,240 Euros may be available. In addition to these payments, the DAAD generally will pay an appropriate flat-rate travel allowance, unless these costs are covered by the home country or by another funding source.

Requirements: Applications for the re-invitation program can only be submitted by former one-year scholarship holders who have been back in their home country for at least three years. The most important selection criterion is a convincing and well-planned research or work project to be completed during the stay in Germany.

Amount of Grant: Euro 1,840 - Euro 2,240 per month

Date(s) Application is Due: Jan 15; Aug 1

Contact: Scholarship Administrator, (212) 758-3223; fax: (212) 755-5780; email: kim@daad.org or daadny@daad.org

Internet: http://www.daad.de/deutschland/foerderung/stipendiendatenbank/00462.en.html?detailid=38&fachrichtung=15&land=44&status=4&seite=1&daad=1

Sponsor: German Academic Exchange Service (DAAD)

871 United Nations Plaza

New York, NY 10017

DAAD Research Grants   1839

Research grants are awarded to highly qualified undergraduate and graduate students who are nominated by their department/program chairs. The grant may be used for short-term research (one to two months) in either North America or Germany. The program is designed to encourage research and promote the study of cultural, political, historical, economic and social aspects of modern and contemporary German affairs from an inter- and multidisciplinary perspective.

Requirements: Master's level graduate students in the humanities and social services earning a certificate or working on a project in German Studies may be nominated for the grant by their department and/or program chair. Applicants are expected to have completed a minimum of three courses in German Studies (literature, history, politics, or other fields) at the time of nomination. Doctoral degree students in the humanities and social science disciplines in the process of preparing their dissertation proposals on modern German topics may be nominated. Students whose dissertation proposals have already been formally accepted are not eligible for nomination. The intent of the program is to provide an opportunity for short-term exploratory research to determine the viability or to delimit the scope of their proposed dissertations. The program is not intended to supplement or substitute for regular dissertation field work abroad which should lag the short-term research stay by at least one semester. All applicants are expected to have completed two years of college level German language studies. Applicants must be younger than 32 at the start of the grant period.

Restrictions: Grants are restricted to citizens of the U.S. who are enrolled full time at the university that nominates them. Support cannot be provided for stays in Germany in the context of study abroad programs.

Amount of Grant: $1,500-$2,500

Date(s) Application is Due: May 1; Nov 1

Contact: Grants Administrator, (212) 758-3223; fax: (212) 755-5780; email: kim@daad.org or daadny@daad.org

Internet: http://www.daad.org/?p=gradresearch

Sponsor: German Academic Exchange Service (DAAD)

871 United Nations Plaza

New York, NY 10017

DAAD Research Grants for Doctoral Candidates and Young Academics and Scientists   1840

The Program provides young foreign academics and scientists with an opportunity to carry out a research project or a course of continuing education and training at a German state (public) or state-recognized higher education institution or non-university research institute. Research grants can be used to carry out; research projects at a German higher education institution for the purpose of gaining a doctorate in the home country; research projects at a German university for the purpose of gaining a doctorate in Germany; or research projects or continuing education and training, but without aiming at a formal degree/qualification. Depending on the project in question and on the applicant's work schedule, grants can be paid generally for between one and ten months, in the case of full doctoral programs in Germany for up to three years, and in exceptions for up to a maximum of four years. Applications for research grants to run for more than six months are decided once a year, and must be submitted by November 15. Applications by musicians, architects and visual artists must be submitted by November 1. Applications for research grants to run for up to six months are decided twice a year, and must be submitted by August 1 or November 15. At the earliest, the grant can begin 6 months after the date of application. Support can only be provided for the completion of a full doctoral program in Germany when special support policy reasons exist. Depending on the award holder's academic level, the program will pay a monthly award of 715 Euros (graduates holding a first degree) or 975 Euros (doctoral candidates). As a rule, the scholarship additionally includes certain payments towards health insurance coverage in Germany. Furthermore, the program generally will pay an appropriate flat-rate travel allowance, unless these costs are covered by the home country or by another funding source.

Requirements: Applications for DAAD research grants are open to excellently-qualified university graduates who hold a Diplom or Master's degree at the time they commence the grant-supported research and, in exceptional cases, graduates holding a Bachelor's degree or already holding a doctorate/PhD (post-docs). It is required that doctoral candidates wishing to take a doctorate/PhD in their home country will already have been admitted to an appropriate course at their home university. Besides previous study achievements, the most important selection criterion is a convincing and well-planned research or continuing education and training project to be completed during the stay in Germany and which has been coordinated and agreed with an academic supervisor at the chosen German host institute. German language skills are generally required, although the required level also depends on the applicant's project and topic, as well as on the available opportunities for learning German in the applicant's home country.

Restrictions: The award of grants is subject to an age limit of 32 years at the time of starting the grant.

Amount of Grant: Euro 715 - Euro 975 per month

Date(s) Application is Due: Aug 1; Nov 1; Nov 15

Contact: Scholarship Administrator, (212) 758-3223; fax: (212) 755-5780; email: kim@daad.org or daadny@daad.org

Internet: http://www.daad.de/deutschland/foerderung/stipendiendatenbank/00462.en.html?detailid=7&fachrichtung=4&land=44&status=3&seite=1&daad=1

Sponsor: German Academic Exchange Service (DAAD)

871 United Nations Plaza

New York, NY 10017

DAAD Research Internships in Science and Engineering (RISE)   1841

DAAD, in cooperation with science organizations in North America and Germany, offers summer internships in Germany for U.S. and Canadian undergraduate students in the fields of biology, chemistry, physics, earth sciences and engineering. RISE fellows work directly with doctoral students in research groups at top German universities and institutions and can expect to gain serious hands-on research experience. RISE placements provide students the opportunity to live and work in an international context, to gain confidence in their practical and theoretical skills, and to improve their (or begin learning!) German. Last but not least, the research internship should be a source of mutual cultural enrichment for both the interns and their hosts. Every RISE intern receives a pro-rated monthly scholarship of approximately 615 Euros for any period of six weeks to three months between June and August. The program also provides health and accident insurance.

Requirements: To apply for a placement an applicant must: be currently enrolled at an United States or Canadian university/college as a full-time student in the field of Biology, Chemistry, Physics, Earth Sciences or Engineering (or a closely related field); be an undergraduate who will have completed at least 2 years of a degree program by the time of the placement; and prove that he/she will be registered as an undergraduate at a university/college for the academic year.

Amount of Grant: Euro 615

Date(s) Application is Due: Feb 1

Contact: Michaela Gottschling; +49 (0)228-882-567; fax: +49 (0)228-882-551; email: knieps@daad.de or rise@daad.de

Internet: http://www.daad.de/rise/en/index.html

Sponsor: German Academic Exchange Service (DAAD)

Referat 315, Kennedyallee 50

Bonn 53175 Germany

DAAD Research Stays for University Academics and Scientists   1842

These grants and scholarships aim to provide foreign academics and scientists working in higher education or at research institutes with an opportunity to carry out a research project at a state (public) or state-recognized higher education institution or non-university research institute in Germany. Depending on the applicant's work schedule, the research stay will last between one and three months. Depending on the applicant's academic status, the monthly award will amount to 1,840 Euros for assistant lecturers, assistant professors and young lecturers, and 1,990 Euros for professors. In some rare exceptions, 2,240 Euros may be available. In addition to these payments, the DAAD generally will pay an appropriate flat-rate travel allowance, unless these costs are covered by the home country or by another funding source. Applications for research grants are decided twice a year. Applications must have been submitted by 1 August or 15 January. At the earliest, grants can begin 4 months after the date of application.

Requirements: Applications for DAAD research stays are open to excellently-qualified academics and scientists who should generally hold a doctorate/PhD. All applicants must be working in higher education or at a research institute in their home country. Besides their previous academic achievements (for example, recent publications), the most important selection criterion is a convincing and well-planned research project to be completed during the stay in Germany. The application must provide proof of a workplace being provided at the host institute.

Restrictions: DAAD support for a research stay can only be awarded once in any three-year period. No travel expenses can be paid from the grant.

Amount of Grant: Euro 1,840 - Euro 2,240 per month

Contact: Scholarship Administrator, (212) 758-3223; fax: (212) 755-5780; email: kim@daad.org or daadny@daad.org

Internet: http://www.daad.de/deutschland/foerderung/stipendiendatenbank/00462.en.html?detailid=39&fachrichtung=15&land=44&status=4&seite=1&daad=1

Sponsor: German Academic Exchange Service (DAAD)

871 United Nations Plaza

New York, NY 10017

DAAD RISE Professional Internships   1843

The Program offers practical, career-building experience in the fields of Biology, Chemistry, Physics, Earth Sciences and Engineering (or closely related subjects). Aimed at Bachelor, Master or PhD students from the U.S. or Canada, the RISE professional internship placements provide students the opportunity to live and work in an international context, to gain confidence in their practical and professional skills, and to improve their German. Further, the internship will be a source of mutual cultural and professional enrichment for both the interns and their host companies. Every intern receives a pro-rated monthly scholarship of approximately 615 Euros for any period of six weeks to three months between June and August. DAAD also provides health and accident insurance.

Requirements: To apply for a placement an applicant must: be currently enrolled at an United States or Canadian university/college as a full-time student in the field of Biology, Chemistry, Physics, Earth Sciences or Engineering (or a closely related field); be an undergraduate who will have completed at least 2 years of a degree program by the time of the placement; and prove that he/she will be registered as an undergraduate at a university/college for the academic year.

Amount of Grant: Euro 615
Date(s) Application is Due: Jan 25
Contact: Martina Ludwig; +49 (0)228) 882-104; fax: +49 (0)228) 882-551; email: rise-pro@daad.de
Internet: http://www.daad.de/rise-pro/en/index.html
Sponsor: German Academic Exchange Service (DAAD)
Referat 315, Kennedyallee 50
Bonn 53175 Germany

DAAD Scholarships for Artists: Study Visits for Academics 1844

The study visits for academics program aims to enable lecturers from the fields of Fine Art, Design, Film, Music and Architecture as well as Drama, Direction, Dance and Choreography to apply for a supported study visit for the purpose of artistic cooperation with a German host institution. Depending on the applicant's work schedule, the study visit will last between one and three months. Depending on the applicant's academic status, the monthly award will amount to 1,840 euros for assistant lecturers, assistant professors and young lecturers, and 1,990 euros for professors, in exceptions, 2,240 euros. In addition to these payments, the DAAD generally will pay an appropriate flat-rate travel allowance, unless these costs are covered by the home country or by another funding source.

Requirements: To qualify for support under the study visits for artists program, the applicant must be lecturing at a higher education institution in the home country. A well-planned and convincing project is an important selection criterion. The application papers must be accompanied by written confirmation of academic supervision or confirmation of agreement having been reached with a German institution of higher education.

Restrictions: DAAD support for study visits by artists can only be awarded once in any three-year period.

Amount of Grant: Euro 1,840 - Euro 2,240 per month
Date(s) Application is Due: Jan 15; Aug 1
Contact: Scholarship Administrator, (212) 758-3223; fax: (212) 755-5780; email: kim@daad.org or daadny@daad.org
Internet: http://www.daad.de/deutschland/foerderung/stipendiendatenbank/00462. en.html?detailid=41&fachrichtung=8&land=44&status=4&seite=1&daad=1
Sponsor: German Academic Exchange Service (DAAD)
871 United Nations Plaza
New York, NY 10017

DAAD Study & Internship Program (SIP) in Germany 1845

A joint program of German Universities of Applied Sciences (UAS7) and DAAD, this program offers students a full academic semester of study abroad at one of the UAS7 universities in Germany followed by a one-semester professional internship experience in a company or research institute in Germany. Scholarships are granted for one semester (6 months). Recipients will be awarded a monthly stipend of about 675 Euro to cover living expenses, including a monthly contribution to cover health and accident liability insurance. In addition, they receive 850 Euro as a contribution to travel expenses. During the internship the company will pay a minimum of 500 Euro/month to help cover the monthly living expenses.

Requirements: Students must: be full-time, in good academic standing, and currently enrolled in an undergraduate degree-granting program at an accredited U.S. or Canadian college or university; be currently enrolled as Sophomores or Juniors in engineering, science, life sciences, business, management, economics, architecture, art, design, journalism, or social work; possess an outstanding academic record and personal integrity, as proven by academic achievement and letters of recommendation; hold U.S. or Canadian citizenship or be a permanent resident thereof (foreign nationals are eligible if they have been full-time students at an accredited U.S. or Canadian university for more than one year at the time of application and intend to return to the U.S. or to Canada after the scholarship period to complete their Bachelor's degree); and meet the age limit of 32 at the start of the grant period.

Amount of Grant: 675 Euro per month, plus 850 Euro travel expense
Contact: Program Administrator, (212) 758-3223; fax: (212) 755-5780; email: daadny@daad.org
Internet: http://www.uas7.org/content/programs__projects/study__internship_ program/index_en.html
Sponsor: German Academic Exchange Service (DAAD)
871 United Nations Plaza
New York, NY 10017

DAAD Study Visits / Study Seminars and Practicals in Germany 1846

The program aims to provide students with subject-related knowledge by arranging appropriate visits, tours and information meetings (Study Visits) or by organizing subject-related seminars and practical courses (e.g. specialist courses, block seminars, workshops) at the invitation of a German university. This university is also responsible for organizing practical courses with universities, companies and, possibly, public institutions (Study Seminars and Practicals). The program goals also include: facilitating meetings with German students, academics and researchers to establish and maintain contacts between German and foreign universities; and giving students a greater understanding of and insight (regional and area studies) into economic, political and cultural life in Germany. Cultural events (e.g. concert tours) can be funded when the focus is on meeting students and university teachers from a relevant academic field and this academic relevance is appropriately documented. Study visits, study seminars or practicals should last no less than 7 days. Funding is available for a maximum of 12 days (including travel days), although the visits themselves may last longer. The DAAD takes out health, accident and public/private liability insurance for each funded group.

Restrictions: Funding cannot be provided for annual repeat visits (by applicants, faculties or departments); each applicant, faculty or department can only be considered for a maximum of one application per year. Funding cannot be provided for required/ obligatory excursions or for measures that have already been completed. The Program regrets that it is unable to pay any international travel costs.

Contact: Katharina Klein; +49 (0)228/882-370; fax: +49 (0)228/882-447; email: k.klein@daad.de
Internet: http://www.daad.de/deutschland/foerderung/stipendiendatenbank/00462. en.html?detailid=15&fachrichtung=15&land=44&status=4&seite=1&daad=1
Sponsor: German Academic Exchange Service (DAAD)
Referat 224, Kennedyallee 50
Bonn D-53175 Germany

DAAD Undergraduate Scholarships 1847

Highly qualified undergraduate students are invited to apply for scholarships funding study, senior thesis research, and/or internships in Germany. The goal of this program is to support study abroad in Germany and at German universities. Preference will be given to students whose projects or programs are based at and organized by a German university. Scholarships are available either as part of an organized study abroad program or as part of an individual, student-designed study abroad semester or year. The scholarship periods must take place during the German academic year. (short term: October 1 through January 31; long term: October 1 through July 31). Recipients receive a monthly stipend and funds to help defray travel and research expenses, as well as health insurance. Guidelines are available online.

Requirements: Eligible students are current sophomores or juniors of all academic fields; are seeking DAAD support for a four- to ten-month period in Germany during the German academic year; possess outstanding academic records and personal integrity, as evidenced by both their grades and letters of recommendation; are able to receive academic credit at their home institutions or ECTS credits for their activity in Germany; are U.S. or Canadian citizens or permanent residents thereof (Foreign nationals are eligible if they have been full-time students at an accredited U.S. or Canadian university for more than one year at the time of application and will return to the U.S. or Canada after the scholarship period to complete their bachelor's degree.); have well-defined study, research, or internship plans for their stay in Germany; submit the DAAD language evaluation form with their application, although German language competency is not mandatory; demonstrate an interest in contemporary German and European affairs and who explain the significance of their project in Germany to their future studies, research, or professional goals; are enrolled, full-time students in an undergraduate degree-granting program at an accredited North American college or university; and are younger than 32 at the start of the grant period.

Amount of Grant: Euro 615 per month
Date(s) Application is Due: Jan 31
Contact: Scholarship Administrator, (212) 758-3223; fax: (212) 755-5780; email: daadny@daad.org
Internet: http://www.daad.org/?p=47220
Sponsor: German Academic Exchange Service (DAAD)
871 United Nations Plaza
New York, NY 10017

DAAD Visiting Professorship Grants 1848

This program serves to strengthen the internationalization of the educational experience for scholars, host institutions and students by welcoming educators from abroad to university campuses in Germany for guest teaching assignments. The recent development of international degree programs and traditional curricula looking to infuse an international aspect provide opportunities for professors from other countries to contribute their expertise in particular subjects and teaching methods. Courses need not be taught in German. As the applicant, the German host institution receives the DAAD funding. Visiting Professors are paid by the host institution according to the salary schemes for German university faculty. Funding is available for a period of one semester to two years. Additional allowances can be made for travel, luggage and health insurance.

Requirements: Highly qualified scholars in all academic disciplines, preferably those who hold a doctoral or other terminal degree and have an affiliation with an institution of higher education, are eligible. Candidates must secure an invitation from a German host institution and teach courses integrated into the regular curriculum.

Restrictions: Funds cannot be made available to replace faculty on sabbatical.
Date(s) Application is Due: Jul 15; Jan 15

Contact: Grants Administrator; (212) 758-3223; fax: (212) 755-5780; email: daadny@daad.org or gastdozentur@daad.org
Internet: http://www.daad.org/page/50132/
Sponsor: German Academic Exchange Service (DAAD)
871 United Nations Plaza
New York, NY 10017

DAAD-AICGS Research Fellowships   **1849**
The program is designed to bring scholars and specialists working on Germany, Europe, and/or transatlantic relations to AICGS for research stays of two months each. Project proposals should address a topic closely related to one or more of the Institute's five research and programming areas: Globalization and the German and American Economies; Germany in Europe; Security and Foreign Policy—Changing Agendas and New Challenges; Culture and Politics; or Transnational Issues and German-American Cooperation. Fellowships include a monthly stipend of up to $4,600, depending on the seniority of the applicant, economy class round-trip airfare and transportation to and from Washington (for a maximum of $770), and office space at the Institute.
Requirements: Applicants must be German or American citizens.
Amount of Grant: $4,600 per month, plus $770 round-trip airfare
Date(s) Application is Due: Apr 30; May 31; Sep 30
Contact: Program Contact, (202) 332-9312; fax: (202) 265-9531; email: kverclas@aicgs.org or info@aicgs.org
Internet: http://www.daad.org/page/98696/
Sponsor: German Academic Exchange Service (DAAD)
871 United Nations Plaza
New York, NY 10017

DAAD-Leo Baeck Institute Fellowships   **1850**
Fellowships are awarded for research in New York or Germany on the social, communal, and intellectual history of German-speaking Jewry. Financial assistance is provided to doctoral students for dissertation research and to young academics for the preparation of a scholarly essay or book. The New York fellowship consists of a stipend of $2,000 paid in two installments of $1,000 each. The fellowship for Germany follows the terms of award for the DAAD Research Grant.
Requirements: Applicants must be U.S. citizens and PhD candidates or recent PhDs (degree awarded within the last two years).
Amount of Grant: $2,000
Date(s) Application is Due: Nov 15
Contact: Program Contact, Leo Baeck Institute; (212) 744-6400; fax: (212) 988-1305; email: lbaeck@lbi.cjh.org
Internet: http://www.daad.org/page/48513/
Sponsor: German Academic Exchange Service (DAAD)
15 West 16th Street
New York, NY 10011

Dade Community Foundation Grants   **1851**
The funding for this program is made available through the Foundation's unrestricted and field of interest funds. This program is designed to honor both the donors interests and address significant community issues such as: education; health; human services; arts and culture; environment; economic development; at-risk youth; abused and neglected children; living with HIV/AIDS; homelessness; social justice; care of animals; heart disease and more.
Requirements: Eligible applicants include nonprofit tax-exempt organizations, as defined by the Internal Revenue Code, which are serving the residents of Miami-Dade County. Preference will be given to or- ganizations based in Miami-Dade County or if located outside the county, are working in partnership with an organization based in Miami-Dade.
Restrictions: The Foundation does not provide grants to individuals, for memberships, fundraising events or memorials. Grants to government agencies are made on a very restricted basis.
Geographic Focus: Florida
Samples: Alliance for Musical Arts Productions, $7,500 - to provide after school and summer camp services to youth from low-income households living in Opa-locka (2007); Chai Lifeline, $6,500 - to support Smile S'more, a unique program designed to bring fun and love to children who are being treated for cancer and other life-threatening illnesses on pediatric specialty units of hospitals (2007); Earth Learning, $10,000 - to create an EarthFest network that will form an umbrella to unite Earth Day celebrations across South Florida under one banner (2007);
Amount of Grant: $7,500
Date(s) Application is Due: Nov 15
Contact: Charisse Grant, Vice President for Programs; (305) 371-2711; fax: (305) 371-5342; email: charisse.grant@dadecommunityfoundation.org
Internet: http://www.dadecommunityfoundation.org/Site/programs/overview.jsp
Sponsor: Dade Community Foundation
200 South Biscayne Boulevard, Suite 505
Miami, FL 33131-5330

DaimlerChrysler Corporation Fund Grants   **1852**
The fund contributes to organizations grouped under the general categories of education, health and human services, civic and community, religion, and culture and the arts. Within these categories, grants are made available for public welfare or for charitable, scientific, educational, environmental, safety, building, and affirmative action purposes. Higher education grants largely support science and engineering education and business

management. A major interest of the corporation is the establishment of national certification standards for elementary and secondary teachers. Another area of concern is the encouragement of early reading skills, and a pilot project has been funded for research in this area. The fund earmarks funds for its future workforce initiatives, which support business and engineering departments, community-based job-skill training, and entry-level work preparation. Types of support include matching gifts, program grants, scholarships, annual campaigns, building construction/renovation, general support, and employee matching gifts. In considering requests, the fund evaluates each applicant organization on its own merits; considered are the programs in which it is engaged, constituencies served, operation procedures, services offered, quality of management, its accountability, finances, and fund-raising practices. Applications are accepted at any time.
Requirements: Eligible for support are nonprofit, tax-exempt educational, health, civic, and cultural organizations primarily in locations where the greatest number of employees of Chrysler and its U.S.-based subsidiaries live and work (Alabama, Delaware, Illinois, Indiana, Michigan, Missouri, New York, Ohio, and Wisconsin). Some support is targeted for national organizations as well.
Restrictions: Grants are not awarded to support endowments, conferences, trips, direct health care delivery, multiyear pledges, capital campaigns, fund-raising activities related to sponsorships, advertising, or debt retirement.
Samples: Charles H. Wright Museum of African American History (Detroit, MI)—$1 million to complete an exhibit area (2005); Second Harvest Food Bank of Central Florida, $25,000 and food via convoy of Dodge Rams, and $50,000 to help residents affected by recent tornadoes (2006).
Contact: Brian Glowiak, Vice President, (248) 512-2502; fax: (248) 512-2503; email: mek@dcx.com
Internet: http://www.fund.daimlerchrysler.com
Sponsor: DaimlerChrysler Corporation Fund
1000 Chrysler Drive
Auburn Hills, MI 48326-2766

Damon Runyon Cancer Foundation Fellowships   **1853**
The foundation encourages all theoretical and experimental research relevant to the study of cancer and the search for cancer causes, mechanisms, therapies, and prevention. Candidates must apply for the fellowships under the guidance of a sponsor—a senior member of the scientific research community. The sponsor should be actively engaged in the planning, execution, and supervision of the proposed research and should encourage the fellow to report the results of the research in scientific journals and meetings. Awards are made to institutions for the support of the fellow under direct supervision of the sponsor.
Requirements: Applicants must have completed one or more of the following degrees or its equivalent: MD, PhD, DDS, and DVM.
Amount of Grant: $41,000 and $55,000 year one (level I and level II); $43,000 and $56,000 year two; $44,000 and $57,000 year three; $2000 in expenses each year
Date(s) Application is Due: Mar 15; Aug 15
Contact: Fellowship Administrator, (212) 455-0520; email: awards@drcrf.org
Internet: http://www.drcrf.org/apFellowship.html
Sponsor: Damon Runyon Cancer Research Foundation
675 Third Avenue, 25th Fl
New York, NY 10017

Dana Clinical Hypotheses Program in Brain and Immuno-Imaging   **1854**
  Research Grants
The program focuses on improving human brain and immune system functioning in health and disease. The program consists of two tracks. Track A is for conventional systems imaging of brain tissues (imaging white or gray matter in the brain). Track B is for the evolving field of cellular and molecular imaging of brain cells, or immune cells, or their interactions (imaging at the level of cells rather than tissues). Applicants undertaking cellular and molecular imaging also can employ systems imaging in their proposals to view the actions of cells within their system. Institutions may submit only one application per track. Both tracks are designed to support pilot-testing of promising but high-risk innovative ideas that have direct clinical application and that, when successful, could be supported on a larger scale by other funders. Although one goal of the program is to make support available to researchers early in their careers, any researcher with a promising new hypothesis is eligible. Guidelines are available online.
Requirements: Each U.S. medical school dean, and the presidents of the few selected biomedical research institutions that have been invited by letter, may nominate a total of two applicants, one for Track A, conventional brain systems imaging research, and one application for Track B, using cellular and molecular imaging techniques alone or in combination with brain tissue (systems) imaging techniques. Each application must be countersigned by the medical school dean or invited biomedical institution's president.
Amount of Grant: $100,000 maximum per year
Date(s) Application is Due: May 27
Contact: Rebecca Husman, email: danainfo@dana.org
Internet: http://www.dana.org/grants/health/proposals/brainimaging.cfm
Sponsor: Charles A. Dana Foundation
745 Fifth Avenue, Suite 900
New York, NY 10151

Dana Foundation Science and Health Grants   **1855**
The Foundation supports research in neuroscience, immunology, and the effects of arts training on cognition. Areas of focus are: Brain and Immuno-imaging; Human

Immunology; Neuroimmunology; Clinical Neuroscience Research; and Arts and Cognition. All other Science and Health Grants are made solely by invitation.

Requirements: The Foundation requires institutions, in many cases, to share the cost of a project or raise matching funds.

Restrictions: The foundation makes no grants directly to individuals; does not support annual operating budgets of organizations, deficit reduction, capital campaigns, or individual sabbaticals; and does not schedule meetings with applicants, other than by specific invitation initiated by the foundation.

Samples: Dartmouth College (Hanover, NH)—to conduct research on how arts education affects learning and the brain, to be carried out by the Dana Arts and Cognition Consortium, a collaboration of six universities, $1.85 million over three years (2004); Cold Spring Harbor Laboratory (NY)—to support the work of Hollis Cline, for research on fragile X syndrome, a genetic condition that causes mental impairment, $100,000 over 18 months (2004).

Contact: Grants Administrator, (212) 223-4040; fax: (212) 317-8721; email: danainfo@dana.org

Internet: http://www.dana.org/grants

Sponsor: Charles A. Dana Foundation

745 Fifth Avenue, Suite 900

New York, NY 10151-0002

Dana Science and Health Research Grants 1856

Science and health grants support brain research in neuroscience and immunology and their interrelationship in human health and disease. Areas of interest include brain and immuno-imaging—anatomical, physiological, or cellular and molecular imaging techniques to pilot-test novel clinical hypotheses of the brain, immune cells, or their interactions; and human immunology—clinical studies that measure human immune system functioning in health and disease, including the measurement of immune system responses to experimental therapeutic trials supported by other sources. Guidelines are available online.

Requirements: RFPs for the brain and immuno-imaging category are sent twice yearly to deans of U.S. medical schools and other invited biomedical research institutions. Individual investigators as well as collaborating investigators are eligible to apply in the human immunology category.

Amount of Grant: $300,000 maximum over three years

Contact: Grants Administrator, (212) 223-4040; fax: (212) 317-8721; email: danainfo@dana.org

Internet: http://www.dana.org/grants/health

Sponsor: Dana Foundation

745 Fifth Avenue, Suite 900

New York, NY 10151-0002

Dance Advance Grants 1857

The program, which supports dance projects in Pennsylvania's five-county region surrounding and including Philadelphia, is designed to cultivate artistic excellence, strengthen creative capacity, and promote professional standards in the area. One-year grants are awarded to individual choreographers and dance artists and dance organizations. Grants are awarded on a project basis (research and development, rehearsal and creation, production and presentation, and capacity-building projects) and not for unrestricted or general operating support. Complete guidelines and application are available online. Letters of intent are due September 21; full applications are due November 9.

Requirements: Grants are awarded to Philadelphia individual dance artists and 501(c)3 dance organizations in Bucks, Chester, Delaware, Montgomery, and Philadelphia counties. Individual applicants who are full-time employees of a university, college, or institution of higher learning may be considered.

Geographic Focus: Pennsylvania

Amount of Grant: $10,000 maximum, individual choreographers and dance artists; $20,000 maximum, organizations with annual budgets up to $200,000

Date(s) Application is Due: Sep 21; Nov 9

Contact: Grants Administrator, (215) 732-9060; fax: (215) 732-9057; email: info@danceadvance.org

Internet: http://www.danceadvance.org/02guidelines/index.html

Sponsor: Dance Advance

1500 Walnut Street, Suite 305

Philadelphia, PA 19102

Daniel Mendelsohn New Investigator Award 1858

The Daniel P. Mendelsohn award is intended to encourage new investigators whose work is compatible with this mission. New investigators in the field of research or education whose work relates to health and well being are eligible for the award. Candidates, who may or may not be members of the FRI organization, must have demonstrated scholarly research potential compatible with the mission of FRI. They must show future promise, evidenced by energy and enthusiasm in research that addresses prevention, treatment or education activities. Candidates for the biennial award are reviewed by the Awards Review Committee of FRI that presents their recommendation to the Board of Directors at their October Annual Meeting. Deadline for recommending candidates is May 1 of even numbered years. The Award is a check for $5000 payable to the recipient for her or his personal or professional use. The award will be presented in the fall of the awarding year.

Amount of Grant: $5,000

Date(s) Application is Due: May 1

Contact: Janet Klein Brown, Chair, (410) 823-5116; fax: (410) 823-5131; email: fri@friendsresearch.org

Internet: http://www.friendsresearch.org/awards_summary.htm

Sponsor: Friends Research Institute

505 Baltimore Avenue, P.O. Box 10676

Baltimore, MD 21285

Danone Institute Belgium Doctoral Scholarship in Nutrition 1859

The doctoral scholarship, given every four years, is allocated to a candidate who has decided on a career in research on nutrition, and to this end, is enrolled in a doctoral program in a Belgian university in either the French-speaking community or in the Flemish-speaking community. All degrees must be obtained with a minimum grade of distinction.

Requirements: Applicants should be either: holders of a degree of doctor of medicine, pharmacologist, chemical and agricultural engineering, or bioengineering; a graduate in nutrition, physical education and physiotherapy, psychology; or be holders of an equivalent recognized degree obtained in a country of the European Union.

Samples: Mirjam Hacquebard, Universit???? Libre de Bruxelles, Laboratoire de Chirurgie Exp????rimentale L. Deloyers—for Affinity of different plasma lipoproteins to acquire alpha-tocopherol: relation to dietary habits. Consequences on endothelial function (2002).

Amount of Grant: 30,000 Euro annually

Contact: Fabienne Trinon; + 32 2 770 63 54; fax: + 32 2 771 98 97; email: institute@danone.com

Internet: http://www.danoneinstitute.be/

Sponsor: Danone Institute Belgium

Rue du Duc - 100 - Hertogstraat

Brussels 1000 Belgium

Danone Institute Belgium Research in Human Nutrition Grants 1860

The grants for human nutrition are designed to help promote research in human nutrition, including applied research on an animal model, in Belgium university-level institutions. Candidates should be members of, or an employee with, a research service, research department, research institute or research body within the institution to which he belongs or is affiliated. Three grant awards are given each year.

Requirements: Candidates must be: less than 40 years of age. He/she must also be a member of the permanent academic or scientific staff at a Belgian university-level institution, a permanent researcher at the Fund for Scientific Research, or a research fellow specifically recommended by an university authority.

Samples: Dr ir Ann Van Loey, Katholieke Universiteit Leuven, Faculty of Applied Bioscience and Engineering, Department of Food and Microbial Technology—for The effect of processing on the anticarcinogenic properties of Brassicaceae (2005); Barbara Soetens, Universiteit Gent, Faculty of Psychology—for Food for thought: thought suppression and attention bias among obese en non-obese youngsters who diet (2004).

Amount of Grant: 12,000 Euro / $16,500 approximately

Contact: Fabienne Trinon; + 32 2 770 63 54; fax: + 32 2 771 98 97; email: institute@danone.com

Internet: http://www.danoneinstitute.be/

Sponsor: Danone Institute Belgium

Rue du Duc - 100 - Hertogstraat

Brussels 1000 Belgium

DAR American Indian Scholarship 1861

The American Indians Committee of the DAR awards this scholarship to Native Americans. This award is intended to help Native American students of any age, any tribe, in any state striving to get an education. All awards are judged based on financial need and academic achievement. This scholarship is applicable to programs of vocational training or college/university at the undergraduate or graduate level. Graduate students are eligible; however, undergraduate students are given preference.

Requirements: Applicants must be Native Americans(proof of American Indian blood is required by letter or proof papers)in financial need, and have a grade point average of 2.75 or higher. There will be no exceptions.

Amount of Grant: $500

Date(s) Application is Due: Apr 1; Oct 1

Contact: Scholarship Administrator, (202) 628-1776

Internet: http://www.dar.org/natsociety/edout_scholar.cfm#amInd

Sponsor: Daughters of the American Revolution

1776 D Street NW

Washington, DC 20006

DARPA Military Research and Technology Development Grants 1862

The program objective is to support and stimulate basic research, applied research, and technology development at educational institutions, nonprofit organizations, and commercial firms, of interest to DOD and the military services. This support may take the form of grants, cooperative agreements, or other transactions. Funds may support symposia and conferences, programs to encourage careers in science and technology, programs assisting laboratory research instrumentation at universities, and programs intended to produce fundamentally different approaches to relevant technologies. The assistance is generally for a three- to five-years period.

Requirements: For grants, eligibility is limited to public and private educational institutions and nonprofit organizations operated for purposes of special interest. For cooperative agreements, eligibility is limited to educational institutions, nonprofit organizations, and commercial firms.

Restrictions: Individuals are ineligible.
Amount of Grant: $100,000-$100 million; average $1.15 million
Contact: Director, Contract Management Office, Defense Advanced Research Projects Agency, (703) 696-2399
Internet: http://www.darpa.mil/index.html
Sponsor: Department of Defense
3701 N Fairfax Drive
Arlington, VA 22203

Dart Center Ochberg Fellowship   1863
The Dart Center provides six or more expense-paid fellowships to midcareer journalists who want to apply knowledge of emotional trauma to improving coverage of violent events. Fellows will attend a two-day seminar on the role emotional trauma plays in coverage of violent events, then will have access to all events and speakers in the annual conference of the International Society for Traumatic Stress Studies (istss.org). Guidelines are available online.
Requirements: Fellowships are open to print and broadcast reporters, photographers, editors, and producers with at least five years of journalism experience.
Date(s) Application is Due: Jul 22
Contact: Fellowship Administrator, (800) 332-0565; fax: (206) 543-9285; email: info@dartcenter.org
Internet: http://www.dartcenter.org/awards_fellowships/index.html
Sponsor: Dart Center for Journalism and Trauma
Box 353740, 102 Communications Bldg
Seattle, WA 98195-3740

Dartmouth College Andrew W. Mellon Postdoctoral Fellowship   1864
The program offers one two-year postdoctoral teaching/research fellowship in the humanities and related social sciences. Though there are no citizenship requirements, preference will be given to individuals likely to make their careers in the United States.
Requirements: Applicants should have received the PhD during the past three years.
Amount of Grant: $46,250 plus benefits; $1500 research allowance; $3500 computer allowance (first year only)
Date(s) Application is Due: Jan 31
Contact: Administrator, Leslie Humanities Center, (603) 646-0896; fax: (603) 646-0998; email: lhc@dartmouth.edu
Internet: http://www.dartmouth.edu/~lhc/mellon.html
Sponsor: Dartmouth College
6240 Gerry Hall, Room 205
Hanover, NH 03755-3526

Datatel Scholars Foundation Scholarship Program   1865
The tax-exempt foundation awards undergraduate and graduate scholarships to eligible students to attend a higher learning institution selected from Datatel's more than 600 college, university, and nonprofit client sites. The program reflects the foundation's long-standing commitment to higher education and to give back to the company's client base by focusing Datatel's corporate charitable giving on scholarship. Applicants apply through the institution, which may nominate up to two students. The scholarships are awarded annually on May 1, in conjunction with Datatel's corporate anniversary.
Requirements: To be eligible to apply for a Datatel scholarship, a student must be planning to attend a Datatel client college or university for the year of the award. Students who transfer to another college or university during the award year only maintain their eligibility if the institution to which they transfer is a Datatel client site. Applicants must also be employed at a Datatel non-education client site and attending a college or university of the applicant's choice for the year of the award.
Amount of Grant: $1000-$2400
Contact: Scholarship Program Administrator, (703) 968-9000 ext 4549; fax: (703) 968-4573; email: scholars@datatel.com
Internet: http://www.datatel.com/objectViewerBlank.cfm?objectID=EA7DA3E2-F913-11D4-AE900002A5070708
Sponsor: Datatel Scholars Foundation
4375 Fair Lakes Ct
Fairfax, VA 22033

Dave Thomas Foundation for Adoption Grants   1866
The foundation supports advocacy and social service organizations working in the field of adoption. The primary goal of the foundation is to raise public awareness about children in the public welfare system awaiting adoption and to form public and private partnerships to make the adoption process easier and more affordable. Types of support include challenge/matching grants, conferences and seminars, demonstration grants, professorships, program development grants, research grants, seed money grants, and technical assistance. National, regional, or statewide projects will be considered. Priority will be given to projects that request seed money; include matching or other material support from other organizations, government agencies, or funders; coordinate service providers; include a measurement component to evaluate the project's success; and are easily replicable. Proposals are accepted throughout the year. Funding decisions are made four times per year. In order to receive serious consideration, proposals must be received by the submission deadline date for each quarter. Guidelines are available online.
Requirements: Eligible applicants are nonprofit 501(c)3 organizations.
Restrictions: The foundation will not consider funding individual adoption expenses; operating budgets or budget deficits; endowments or capital campaigns; adoption searches or reunions; scholarships; special events; institutions that discriminate on the basis of race, creed, gender, national origin, age, disability, or sexual orientation; or

organizations engaged in sectarian religious activities. Unless solicited by the foundation, the following will not be considered: research, educational or promotional videos, publications, television productions, conferences, or public service announcements.
Amount of Grant: $35,000-$50,000
Date(s) Application is Due: Jan 5; Apr 6; Jul 6; Sep 7
Contact: Rita Soronen, Executive Director, (800) 275-3832; fax: (614) 766-3871; email: adoption@wendys.com
Internet: http://www.davethomasfoundationforadoption.org
Sponsor: Dave Thomas Foundation for Adoption
4150 Tuller Road, Suite 204
Dublin, OH 43017

David & Lucile Packard Foundation - Preschool for California's Children   1867
The goal of this program is to secure high-quality preschool opportunities for all three and four-year-olds in the state by funding leadership and constituency-building, technical assistance and systems building, research, and public preschool programs in selected California communities. The Foundation seeks to: expand and strengthen statewide advocacy efforts and engage a diverse cross-section of groups in support of preschool; support further research on topics related to ensuring preschool for California's children; provide ongoing support to local flagship preschool efforts that demonstrate the promise of high-quality preschool when implemented on a large scale.
Requirements: The Foundation accepts grant proposals only for charitable, educational, or scientific purposes, primarily from tax-exempt, charitable organizations. We do not provide funding for projects that benefit specific individuals or that serve religious purposes.
Restrictions: The Foundation does not fund attempts to influence specific legislation or ballot measures. We also do not fund direct service programs (i.e., individual child care centers, preschools, or schools), individual facility construction or renovation, isolated curriculum or professional development efforts, or local programs that do not have a central goal of helping to achieve preschool for every child on a district-, city-, county-, or statewide level.
Geographic Focus: California
Samples: Scotts Valley Unified School District (CA)—for salary support of the district arts coordinator and a music teacher for fourth and fifth grade, $35,000 (2005); American Leadership Forum (San Jose, CA)—for its February planning retreat, $12,000 (2005); Palau Conservation Society (Koror, Palau)—to develop a plan for managing the ecosystem of Palau, a group of islands in the Pacific Ocean, $33,410 (2005); Catholics for Free Choice (Washington, DC)—for general support, $250,000 (2005).
Contact: Children, Families, and Communities Program, (650) 917-7238; fax: 9650) 948-1361; email: cfc@packard.org
Internet: http://www.packard.org/searchGrants.aspx?RootCatID=3&CategoryID=226
Sponsor: David and Lucile Packard Foundation
300 Second Street, Suite 200
Los Altos, CA 94022

David and Lucile Packard Foundation Fellowships   1868
The goal of the program is to provide support for unusually creative researchers early in their careers; faculty members who are well established and well funded are less likely to receive the award. The foundation emphasizes support for innovative individual research that involves the fellows, their students, and junior colleagues, rather than extensions or components of large-scale, ongoing research programs. Guidelines are available online.
Requirements: Every year, the foundation invites the presidents of 50 universities to nominate two professors each from their institutions. Nominations are reviewed by an advisory panel of distinguished scientists and engineers.
Samples: Alexander Badyaev (Tucson, AZ)—fellowship recipient, $625,000 (2005); Elizabeth Chen, (Baltimore, MD)—fellowship recipient, $625,000 (2005).
Amount of Grant: $125,000 per year for five years
Date(s) Application is Due: Mar 15
Contact: Fellowships for Science and Engineering, (650) 917-7275; fax: (650) 948-2957; email: fellows@packard.org
Internet: http://www.packard.org/index.cgi?page=consci-fellow
Sponsor: David and Lucile Packard Foundation
300 Second Street, Suite 200
Los Altos, CA 94022

David Berry Essay on Scottish History Prize   1869
This competition is held every year for a prize to be awarded to the writer of the best essay on a subject, to be selected by the candidate, dealing with Scottish history. The society may require evidence that the person submitting the essay is the author thereof and shall not be liable if any essay is lost or mislaid nor be called upon to return any essay. The essay submitted must be a genuine work of research based on original (manuscript or printed) materials. The essay should be between 6000 and 10,000 words in length, excluding footnotes and appendices, and must be submitted in typescript. The author's name should not appear on the typescript and should be submitted separately.
Restrictions: Previous winners of the prize may not enter.
Amount of Grant: @L250
Contact: Administrative Assistant, 0171-387 7532; fax: 0170-387 7532; email: royalhistsoc@ucl.ac.uk
Internet: http://www.rhs.ac.uk/rhspri.html#david
Sponsor: Royal Historical Society
Gower Street, University College London
London WC1E 6BT United Kingdom

David Bohnett Foundation Grants 1870

The foundation's mission is to improve society through social activism. Focus areas include the promotion of the positive portrayal of lesbians and gay men in the media; voter registration activities; animal language research, animal companions, and eliminating rare animal trade; environmental conservation; the reduction and elimination of the manufacture and sale of handguns in the United States; community-based social services that benefit gays and lesbians; and the development of mass transit and non-fossil fuel transportation. Types of support include general operating support, program specific grants, seed money, and multiyear grants. Applicants should email a letter of inquiry to the program officer.

Requirements: Nonprofit organizations are eligible.

Restrictions: Grants do not support individuals, videos or other film productions, or organizations outside the United States.

Samples: Disability Rights Education and Defense Fund (Berkeley, CA)—for a project to ensure that voting booths in California are accessible to disabled people, $25,000 (2005); Proteus Fund (Amherst, MA)—for the Civil Marriage Collaborative, a group of organizations working to achieve civil marital rights and oppose efforts that limit or deny those rights to lesbian, gay, bisexual, and transgender people, $100,000 (2005); Service Members Legal Defense Network (Washington, DC)—for operating support and for legal fees to support active-duty service members who are gay or lesbian, $35,000 (2005); Bill Foundation (Beverly Hills, CA)—to find permanent, safe homes for abandoned and unwanted dogs as companion animals, $10,000 (2004).

Amount of Grant: $2 million total; $5000-$50,000 typically

Date(s) Application is Due: Jan 27; Jul 28

Contact: Michael Fleming, Executive Director, (310) 277-4611; fax: (310) 203-8111; email: mfpfleming@yahoo.com

Internet: http://www.bohnettfoundation.org/grants/grantapplication.htm

Sponsor: David Bohnett Foundation

2049 Century Park E, Suite 2151

Los Angeles, CA 90067-3123

David E. Finley Predoctoral Fellowship 1871

The fellowship is intended to fund a 36 month period for travel and research in Europe on a well-advanced dissertation in western art and an additional year in residence at the National Gallery of Art. Half the year in residence will be devoted to gallery research projects designed to complement the subject of the dissertation. A primary requirement for the awarding of this fellowship is that the candidate have a real interest in museum work. However, there is no requirement as to the candidate's subsequent choice of a career. The fellowship begins on September 1 and is not renewable. Application must be made through the chair of the graduate department of art history or other appropriate department. Departments should limit nominations to one candidate.

Requirements: Applicants must have completed their residence requirements and coursework for a PhD and general or preliminary examinations before the date of application and know two foreign languages related to the topic of the dissertation. Applicants must be either U.S. citizens or enrolled in a university in the United States.

Samples: Jenny Anger, Brown U—for dissertation, Modernism and the Gendering of Paul Klee.; Marian Feldman, Harvard U—for dissertation, The Role of Luxury Goods in the International Relations of the Eastern Mediterranean during the Late Bronze Age with Specific Reference to the Site of Ras Shamra-Ugarit.

Amount of Grant: $24,000 annually for three years

Date(s) Application is Due: Nov 15

Contact: Fellowships Program, Center for Advanced Study in the Visual Arts, (202) 842-6482; fax: (202) 789-3026; email: advstudy@nga.gov

Internet: http://www.nga.gov/resources/casvapre.htm

Sponsor: National Gallery of Art

Fourth Street and Constitution Avenue NW

Washington, DC 20565

David H. Smith Conservation Research Fellowship 1872

The fellowship program identifies and supports innovative young scientists who are beginning their careers in applied conservation biology. The program will provide two years of postdoctoral support in applied conservation biology. Awards will be made to the individual fellows, who select the academic institution best suited for carrying out the proposed scientific research. Research will focus on one or more of the conservancy's priority conservation sites, or questions germane to these sites. Guidelines may be requested via email.

Requirements: Applicants must either hold a PhD or be awarded a PhD during the academic year in which the application is made.

Amount of Grant: $36,000 annual salary, plus benefits; $30,000 research fund; $8000 travel budget

Date(s) Application is Due: Oct 28

Contact: Smith Conservation Research Fellowship Program, (800) 628-6860; email: info@smithfellows.org

Internet: http://smithfellows.org/proposalguidelines.cfm

Sponsor: The Nature Conservancy

4245 N Fairfax Drive

Arlington, VA 22203

Daviess County Community Foundation Health Grants 1873

The Foundation considers proposals for grants on a yearly cycle, which begins each June. At the start of each cycle, a notice is mailed to nonprofit organizations that have applied for grants in the past, have received grants in the past, or have otherwise requested notification of the start of each cycle. Grants in the area of Health include activities that: improve and promote health outcomes; provide general and rehabilitative health services; offer mental health services; provide crisis intervention programs; strengthen associations or services associated with specific diseases, disorders, and medical disciplines; and support medical research. Applications are due by August 4, and all organizations that have submitted grant proposals will be notified of the outcome of the grants committee's deliberation in writing no later than November 1.

Requirements: The Foundation welcomes proposals from nonprofit organizations that are deemed tax-exempt under sections 501(c)3 and 509(a) of the Internal Revenue Code and from governmental agencies serving the County of Daviess, Indiana. Proposals from nonprofit organizations not classified as a 501(c)3 public charity may be considered provided the project is charitable and supports a community need.

Geographic Focus: Indiana

Samples: Mental Health America of Daviess County—for the Bridges out of Poverty program, $1,780 (2008); Pregnancy Care Cente—for the Essentials for Newborns parenting project, $6,000 (2008).

Amount of Grant: $5,000-$10,000 average

Date(s) Application is Due: Aug 4

Contact: Jeanne Fields, Director; (812) 254-9354; fax: (812) 254-9355; email: jeanne@daviesscommunityfoundation.org

Internet: http://www.daviesscommunityfoundation.org/program-areas

Sponsor: Daviess County Community Foundation

320 E. Main Street, P.O. Box 302

Washington, IN 47501

Davis Family Foundation Grants 1874

The foundation provides grants primarily to Maine-based educational, medical, and cultural/arts charitable organizations in support of a wide variety of worthwhile projects. Eligible educational organizations include: colleges, universities, and other educational institutions. Eligible medical organizations include: hospitals, clinics and medical research organizations (grant requests will also be considered from other similar health organizations for programs designed to increase the effectiveness or decrease the cost of medical care). Eligible cultural and arts organizations include: those agencies whose customary and primary activity is to promote music, theater, drama, history, literature, the arts or other similar cultural activities. Further guidelines are available online.

Requirements: Eligible educational organizations include colleges, universities, and other educational institutions (grants are not made to public elementary and secondary schools, nor to schools whose financial support is derived primarily from a church or other religious organization. Trustees will consider grant requests from other educational organizations whose purpose is to promote systemic change in education or to provide innovative programs whose objectives are to improve education). Medical organizations eligible for support include hospitals, clinics, and medical research organizations. Grant requests will also be considered from other similar health organizations for programs designed to increase the effectiveness or decrease the cost of medical care. Eligible cultural/arts organizations include organizations whose customary and primary activity is to promote music, theater, drama, history, literature, the arts, or other similar cultural activities.

Restrictions: The Foundation does not make grants to individuals, religious programs, fellowships, or in the form of loans. The Foundation does not normally provide support for annual giving campaigns or general operating needs. Grants to endowment campaigns have a low priority.

Geographic Focus: Maine

Samples: Arthur L. Mann Memorial Library, West Paris, ME, $15,000 - expansion and renovation project (2008); Maine Archaeological Society, Inc., Augusta, ME, $8,300 - Maine Archaeology Classroom Initiative (2008); South Bristol Historical Society, South Bristol, ME, $20,000 - schoolhouse restoration (2008).

Date(s) Application is Due: Feb 10; May 10; Aug 10; Nov 10

Contact: Anne Vaillancourt, Executive Director; (207) 781-5504; email: info@davisfoundations.org

Internet: http://www.davisfoundations.org/site/family.asp

Sponsor: Davis Family Foundation

4 Fundy Road

Falmouth, ME 04105

Deafness Research Foundation Research Grants 1875

The grants are intended for research directed to any aspect of the ear, e.g., investigation of function, physiology, biochemistry, genetics, anatomy, or pathology. Grants cover a range of subjects from basic neuroscience research to studies on corrective hearing/speech testing and training. Grants may be renewed for one to two years.

Requirements: Current policy favors awarding grants in support of projects directed by new investigators, so-called seed money support for studies in generally unexplored areas of research. It does not exclude grant support for new research by established researchers. Application should state whether the same project is receiving support from another source and whether an application has been submitted to another source of funding. The principal investigator and grantee institution must notify the DRF when and if this same project receives support from another granting agency.

Amount of Grant: $20,000

Date(s) Application is Due: Dec 1

Contact: Grants Administrator, (800) 829-5934 or (703) 610-9025; email: grants@drf.org

Internet: http://www.drf.org/grants/grants.htm

Sponsor: Deafness Research Foundation

8201 Greensboro Drive, Suite 300

McLean, VA 22102

Dean Witter Foundation Grants 1876

The Foundation supports graduate schools of business and organizations to promote research and higher education in finance. The Foundation makes additional grants, often on a matching basis, to support specific wildlife research and conservation projects in Northern California and seminal opportunities to improve and extend environmental education and to stimulate learning. The Foundation will consider requests for multi-year funding for two or three years into the future to facilitate effective program planning by the institutions supported. Applicants are encouraged to telephone or write the Consultant to determine whether their proposed program falls within the Foundation's areas of interest and grantmaking priorities.

Requirements: The Foundation accepts grant proposals from tax-exempt charitable institutions as defined under Section 501(c)3 of the Internal Revenue Code on a continuing basis. The Foundation does not have a standard application form. Applicants should send one complete proposal to the Consultant with the following elements: cover letter; specific request; personnel information; organizational information; financial information; and addenda.

Restrictions: The Foundation does not: accept funding requests from individuals or award loans, scholarships and grants to specific individuals; assume any obligation to provide continuing support to grantee programs; make grants for annual fundraising events, operating deficits, and capital campaigns; or support sectarian religious activities or sectarian religious facilities.

Geographic Focus: California

Samples: Hoover Institution on War, Revolution and Peace (Stanford, CA)—for support of Dr. Thomas MaCurdy's research on tax reform to examine the role of state income and sales tax, $75,000 (2005); Pacific Research Institute (San Francisco, CA), $25,000 (2005); Action for a Sustainable Earth (Palo Alto, CA)—for the Youth Environmental Education Project, $7500 (2004); Bay Institute of San Francisco, (San Francisco, CA)—for the Ecological Scorecard Project, $20,000 (2004).

Contact: Kenneth Blum, Consultant, (415) 981-2966; fax: (415) 981-5218; admin@deanwitterfoundation.org

Internet: http://www.deanwitterfoundation.org/FundingGuidelines.html

Sponsor: Dean Witter Foundation
57 Post Street, Suite 510
San Francisco, CA 94104

DeBakey Medical Foundation Fellowships and Scholarships 1877

The foundation fosters and encourages medical education and graduate training and promotes medical research in any field of medicine, but it is especially interested in investigations that have direct clinical application, particularly those concerned with cardiovascular disease. Financial aid may be extended to persons in the form of scholarships or fellowships and to institutions in the form of grants or funds for hospital care. Applications are accepted at any time.

Contact: Program Director, (713) 798-8600; fax: (713) 793-1192; email: pa@bcm.tmc.edu

Sponsor: DeBakey Medical Foundation
1 Baylor Plz, Baylor College of Medicine
Houston, TX 77030

DEED Grants 1878

APPA's Demonstration of Energy-Efficient Developments (DEED) program sponsors grants intended for demonstration or early commercialization projects at DEED member utilities that promise either to improve efficiencies or lower costs in the provision of energy services to the consumers of publicly owned electric utilities. DEED grants also are open to all organizations for applied research to early demonstration projects with long-term potential either to improve efficiencies or lower costs in the provision of energy services to the consumers of publicly owned electric utilities; these outside organizations, however, must apply through a DEED member utility. Maximum funding is limited to 25 percent of available DEED funds; amounts greater than $50,000 must be approved by the APPA board after review by the DEED directors. The DEED board of directors meets each year in March and September; therefore, deadlines for receipt of applications are usually in February and August. Call for exact dates and application forms or visit the Web site.

Requirements: Proposals for DEED grants may be submitted only by DEED member utilities that have been DEED members for at least six months prior to grant submission. All proposals shall include a project plan showing tasks, schedule, costs, and decision criteria for termination versus further research.

Amount of Grant: $75,000 maximum; $25,000-$50,000 typically

Date(s) Application is Due: Feb 15; Aug 15

Contact: Michele Ghosh, DEED Program Manager, (202) 467-2960; fax: (202) 467-2992; email: DEED@appanet.org

Internet: http://www.appanet.org/research/index.cfm?ItemNumber=2671&sn.ItemNumber=2415

Sponsor: American Public Power Association
2301 M Street NW
Washington, DC 20037-1484

DEED Student Research Grants and Internships 1879

DEED (Demonstration of Energy-Efficient Developments) student research grants/internships are intended to promote the involvement of students studying in energy-related disciplines in the public power industry, and to provide host utilities with technical assistance. Successful applicants are expected to conduct research on a project approved by the sponsoring utility and submit a final report on the project, describing

activities, cost, bibliography, achievements, problems, results, and recommendations. Additional information is available online.

Requirements: Only graduate or undergraduate students in energy-related disciplines from accredited colleges or universities are eligible. Students must be attending school in a country with at least one DEED member. Applications must be sent from a DEED member utility. Applicants will not be discriminated against on the basis of sex, race, religion, national origin, or citizenship.

Amount of Grant: $4,000

Date(s) Application is Due: Feb 15; Oct 1

Contact: DEED Administrator, (202) 467-2960; email: DEED@APPAnet.org

Internet: http://www.appanet.org/research/index.cfm?ItemNumber=2671&sn.ItemNumber=2415

Sponsor: American Public Power Association
2301 M. Street NW
Washington, DC 200371484

Del E. Webb Foundation Grants 1880

The Foundation applies its resources only for the benefit of the residents of Arizona, California and Nevada for improved and expanded medical services, medical research and education. Within these broad areas of interest, the Foundation draws upon the talent and experience of leaders from many walks of life to select organizations of demonstrated competence which have sound programs that will be able to reach and sustain high levels of performance. In choosing particular projects for support, the Foundation's Board acts on the basis of how the public welfare most effectively may be served. Application information is available online.

Requirements: Grants are confined to the support of nonpartisan, non-profit organizations that are operated in the public interest and have tax-exempt status as charitable organizations granted by the Internal Revenue Service and the states in which the respective organizations are incorporated and/or engage in activities.

Restrictions: Grants are not made to the following: governmental agencies or subdivisions; sectarian or religious organizations whose principle activity is exclusively for the benefit of their own members; organizations soliciting funds in support of projects or programs operated by organizations other than the applicant; expenditures before the recipient incurs and makes such expenditures, and, in the absence of compelling circumstances, the Foundation does not make grants to recipients to liquidate or reduce previously incurred obligations or operating deficits. No grants or loans are awarded or made to individuals for any purpose. Grants are not made for scholarships, student aid or medical assistance. Grants may be made to organizations that provide scholarships, student aid or medical assistance when the organization selects recipients in conformity with accepted standards. The Foundation does not participate in the administration of program(s) that the Foundation funds through the award of the application. The Foundation does not fund or underwrite: galas or gala-like events, testimonial or fund-raising luncheons or dinners, advertising in programs or similar fund-raising activities; organizations that in turn make grants to others; purchase of uniforms, equipment, or trips for school related organizations or amateur sports teams; honoraria for guest speakers or panelists; charities operated by service clubs; or educational seminars.

Geographic Focus: Arizona; California; Nevada

Samples: Grand Canyon National Park Foundation (Flagstaff, AZ)—to renovate the park's education center, $2.5 million.; U of Southern California (Los Angeles, CA)—to establish a laboratory suite in the New Institute for Genetic Medicine, $1 million.

Amount of Grant: $5000-$100,000 average

Date(s) Application is Due: Nov 30; Feb 28; May 31; Aug 31

Contact: Program Contact, (928) 684-7223; fax: (928) 684-5665

Internet: http://www.dewf.com/DEFAULT.shtml

Sponsor: Del E. Webb Foundation
300 E. Willis, Suite C
Prescott, AZ 86301-3110

Deland Fellowships in Health Care and Society 1881

This one-year fellowship provides a training opportunity for graduates interested in pursuing careers in health care management. Appointees will explore the most challenging issues of health care delivery today, including problems of access to quality health care for the poor, uninsured, and underinsured; the changing roles of physicians, nurses, social workers, administrators, and allied health professionals; ethical, legal, and human considerations arising from new biomedical technologies; the impact of for-profit ventures on the work of teaching and community hospitals; and development of integrated delivery systems in the health care marketplace. Stipends are commensurate with the seniority, prior educational experience, and living requirements of the appointees. Fellowship tenure begins in July. For current deadline information, please email Kevin Gahagan at kgahagan@partners.org.

Requirements: It is anticipated that candidates will come from a variety of careers and educational backgrounds including business, law, economics, public policy, and medicine. Candidates are required to have an advanced degree.

Contact: Kevin Gahagan, (617) 732-5500; TTY/TTD: (617) 732-6458; email: kgahagan@partners.org

Internet: http://www.brighamandwomens.org/general/Deland_Fellowship.asp

Sponsor: Brigham and Women's Hospital
75 Francis Street
Boston, MA 02115

Dell Foundation Open Grants 1882

The foundation, the giving arm of Dell Computer, seeks to fund collaborative and innovative solutions to community and children's issues addressing youth (ages newborn to high-school seniors) in Texas, Tennessee, Idaho, and Oregon. Through these efforts, Dell strives to effect change in its local community, while providing lessons and best practices for communities everywhere. Funded areas include arts, education, social services, and health. The foundation supports specific, preventative, and measurable programs with cash, in-kind, and volunteer contributions. Grants are awarded quarterly. An online application form is available on the Web site.

Requirements: Eligibility is open to 501(c)3 nonprofit organizations in Texas—Travis, Williamson County, and McLennan County; Tennessee—Wilson, Davidson County; Idaho—Twin Falls County; and Oregon—Roseburg, Douglas County.

Restrictions: Grants do not support individuals; academic or research projects; civic, religious, or political institutions; school fundraisers; marketing opportunities; or sports events and organizations.

Geographic Focus: Idaho; Oregon; Tennessee; Texas

Samples: Various rescue and relief organizations—to assist organizations providing disaster-relief services related to the September 11 terrorist attacks, $1 million in cash and products.

Amount of Grant: $5000 maximum

Date(s) Application is Due: Jan 15; Apr 15; Jul 15; Oct 15

Contact: Grants Administrator, (512) 338-4400; email: the_dell_foundation@dell.com

Internet: http://www1.us.dell.com/content/topics/global.aspx/corp/foundation/en/open_grants?c=us&l=en&s=corp

Sponsor: Dell Foundation
1 Dell Wy
Round Rock, TX 78682

Della Martin Foundation Grants 1883

The foundation awards grants to organizations located in southern California for research seeking to find the causes of and cures for mental illness. Types of support include challenge/matching grants, endowments, fellowships, professorships, and research grants. There is no application deadline, but the foundation usually makes grants for mental health research at the year's end.

Requirements: Grants are made only to California 501(c)3 organizations.

Geographic Focus: California

Amount of Grant: $263,000

Contact: Laurence Gould Jr., Director, (213) 617-4143; fax: (213) 620-1398

Sponsor: Della Martin Foundation
333 S Hope Street, 48th Fl
Los Angeles, CA 90071

Deloitte Foundation Grants 1884

The foundation, funded by Deloitte and Touche U.S.A LLP, supports accounting, business, and related fields of study within the United States. Its national programs benefit university students and faculty and promote excellence in teaching and research, curriculum innovation, and cooperation among practitioners and the academic community. It funds the Deloitte Doctoral Fellowship Program, Trueblood Seminars for Professors, matching gifts program, and many other higher education initiatives. Applications for the doctoral fellowship program must meet the listed deadline; there are no deadlines or application forms for other requests.

Requirements: Doctoral fellowship applicants must be doctoral candidates pursuing their PhD in accounting.

Restrictions: Grants are not awarded to support general purposes, capital campaigns, special programs, publications, or loans.

Samples: American Red Cross (Washington, DC)—for relief efforts in South Asia and Africa, $1 million (2005); Divided among 10 students—for fellowships for doctoral students in accounting, $250,000 (2004).

Amount of Grant: $5000-$50,000 average

Contact: Janet Butchko, Manager, Academic Development, (203) 761-3474

Internet: http://www.deloitte.com/dtt/section_node/0,2332,sid%253D2257,00.html

Sponsor: Deloitte Foundation
P.O. Box 820, 10 Westport Road
Wilton, CT 06897-0820

Delta Air Lines Foundation Grants 1885

The foundation awards grants to nonprofits in the areas of: Community Enrichment - focusing on involvement and participation in volunteer, civic, and social activities; Health and Wellness - with support to organizations that dedicate their efforts to the health and well-being of communities with a focus on broad-reaching research for cures and education around diseases that affect all walks of life; Youth Development - offering support to organizations focusing on keeping young people interested in math and science and helping them develop leadership skills and positive self esteem; and Arts and Culture to include fine art, theatre, music, or other creative endeavors which enhance a community's quality of life.

Requirements: For proposals which meet the foundation's area of focus, priority will be given to: programs meeting compelling needs in communities where Delta has a presence; proposals that exhibit clear, reasonable goals, and measurable outcomes; distinctive projects where the foundation's involvement will leave a legacy; projects that include collaboration or cooperation with other nonprofit organizations; projects that offer opportunities for Delta employee involvement. The foundation Board of Trustees reviews and approves funding in March, June, September, and November. The deadline for receiving completed proposals is the first day of each of these months.

Restrictions: The foundation will generally not consider: individual applicant's request for support of personal needs; religious activities; political organizations or campaigns; specialized single-issue health organizations; annual or automatic renewal grants; general operating expenses; endowment campaigns; capital campaigns; multi-year commitments; fraternal organizations, professional associations, or membership groups; fundraising events such as benefits; charitable dinners, or sporting events.

Date(s) Application is Due: Mar 1; Jun 1; Sep 1; Nov 1

Contact: Administrator; (404) 715-5487; fax: (404) 715-3267

Internet: http://www.delta.com/about_delta/community_involvement/delta_foundation/

Sponsor: Delta Air Lines
P.O. Box 20706, Dept 979
Atlanta, GA 30320-6001

Delta Dental Master's Thesis Award 1886

The fund invites Master of Science students at the dental schools in Michigan, Ohio, and Indiana to submit proposals for the thesis award program. This program is intended to encourage thesis research that is of direct relevance to the costs or outcomes of dental care. The award provides a stipend to cover costs associated with the conduct of master's thesis research, including materials, supplies, and rental of necessary equipment. A detailed budget with justification for expenditures is required. Partial funding may be approved, dependent on reasonableness of the budget and availability of funds. A copy of the final thesis must be provided to Delta Dental Fund upon its completion. Applications may be submitted at any time. Application and guidelines are available online.

Requirements: Master of Science students at the dental schools in Michigan, Ohio, and Indiana are eligible.

Restrictions: Salaries, wages, indirect costs, and the purchase of equipment will not be covered.

Geographic Focus: Indiana; Michigan; Ohio

Amount of Grant: $3000 maximum

Contact: Dental Master's Thesis Award Program, (517) 347-5333; fax: (517) 347-5320; email: ddfund@ddpmi.com

Internet: http://www.ddpmi.com/ddf/masterThesis.htm

Sponsor: Delta Dental Plan of Michigan
P.O. Box 293
Okemos, MI 48805-0293

Delta Gamma Foundation Fellowships for Graduate Study 1887

Fellowships are offered annually for graduate study in any chosen field. Selection is based on scholarship, potential for achievement, and financial need. Annual deadline dates may vary. Contact the committee for application form.

Requirements: Applicants must be Delta Gammas who will have completed their undergraduate study by June 30 of the year in which the fellowship is granted. Application consists of completed application form, official transcripts, two passport-style photos, statement of graduate school acceptance, autobiographical sketch, self-addressed stamped envelope, and sealed letters of recommendation from two persons familiar with applicant's academic work, one alumna of Delta Gamma, and employer.

Amount of Grant: $2500

Date(s) Application is Due: Apr 1

Contact: Jacquelyn Geving Everson, Scholarship and Fellowship Director, (614) 481-8169; fax: (614) 481-0133; email: ScholarFellowandLoans@deltagamma.org

Internet: http://www.deltagamma.org/scholarships_fellowships_and_loans_2.shtml

Sponsor: Delta Gamma Foundation
P.O. Box Box 21397, 3250 Riverside Drive
Columbus, OH 43221-0397

Delyte and Dorothy Morris Doctoral Fellowships 1888

Graduate students of the highest caliber are invited to compete for fellowships to support doctoral study in any field at SIU at Carbondale. Fellowships will be awarded for up to three years of full-time study. Application materials and more information are available from the graduate school.

Requirements: Applicant must have an undergraduate grade point average of at least 3.25 overall, or a grade point average for the last two years of bachelor's degree work of 3.5 (A=4.0); if prior graduate study has been undertaken, the applicant must have an overall graduate grade point average of at least 3.7; must have a score in the 75th percentile or higher on a standard test such as the GRE, MAT, GMAT, or ATGSB; and may not already be enrolled in a doctoral program. Preference will be given to citizens or permanent residents of the United States, or immigrants to the United States, and to those who have not previously been enrolled in a graduate program or at SIU.

Amount of Grant: $18,000 stipend plus waiver of tuition for up to three years of full-time doctoral study

Date(s) Application is Due: Jan 23

Contact: Assistantship/Fellowship Office, (618) 453-4555; fax: (618) 453-4562; email: gaoffice@siu.edu

Internet: http://www.siu.edu/gradschl/grad_fellowship.htm

Sponsor: Southern Illinois University at Carbondale
Graduate School, Woody Hall B130
Carbondale, IL 62901-4716

DeMatteis Family Foundation Grants 1889

The foundation makes grants in the New York metropolitan area to eligible institutions whose mission involves education, health and human services, medical research, social services, and the arts. Types of support include facilities construction, expansion, renovation; acquisition of capital equipment; scientific/medical research; projects and programs that enable the applicant to expand its mission through new or expanded programs to reach a greater segment of the community served; project-oriented capital campaigns. In general, most grants are made to cover projects that can be accomplished within one year. For construction and longer duration projects, grants may be structured to conform to identified milestones. For major projects, the payment of the grant may be over a number of years.

Requirements: Metropolitan New York 501(c)3 tax-exempt agencies, institutions, and organizations are eligible.

Restrictions: In general, the foundation does not support grants for operating deficits; general operating support; endowments; loans, or financing of any kind; annual appeals, dinner functions, and other special fund raising events; or unrestricted funds.

Geographic Focus: New York

Contact: Grants Administrator, (516) 705-4974

Internet: http://fdncenter.org/grantmaker/dematteis/about.html

Sponsor: DeMatteis Family Foundation

P.O. Box 25

Glen Head, NY 11545

Democracy Fellows Program   1890

The program, coordinated by World Learning Inc and funded primarily by the U.S. Agency for International Development (U.S.AID), seeks applicants for one-year fellowships that promote professional development of the fellow and the advancement of democratic institutions worldwide. Fellows will be placed with U.S.AID missions in transitional or newly emerging democracies or with U.S.AID offices in Washington, DC. Applications are accepted on a rolling basis.

Requirements: U.S. citizens with the JD or master's degree who have expertise in political science, government, law, public administration, human rights, election administration, justice systems, conflict resolution, or other social sciences relevant to the advancement of democratic institutions abroad may apply. Applicants also must have professional-level foreign language proficiency as appropriate. The program is targeted to junior- and mid-level individuals with one to 10 years of experience and interests in international democracy and governance.

Amount of Grant: $79,000 maximum mid-level; $54,000 junior level; $87,000 senior level

Contact: Democracy Fellows Program, (202) 408-5420; fax: (202) 408-5397; email: dfp. info@worldlearning.org

Internet: http://www.worldlearning.org/wlid/cssc/dfp/available.html

Sponsor: World Learning Inc

1015 15th Street, NW, Suite 750

Washington, DC 20005

Dennis Leslie Mahony Prize in Legal Theory   1891

The prize will go to the author or authors of an outstanding published work in the field of jurisprudence that best reflects an approach combining legal theory with sociological inquiry, in the tradition of the jurisprudence of the late Professor Julius Stone. A published work need not necessarily be in the form of a traditional book or journal publication. Other types of publication, including reports or papers, are eligible. The recipient of the prize will receive a cash prize, with the offer of an invitation to participate in the activities of the Faculty of Law at the University of Sydney for a period of up to one semester. He or she may also receive an invitation to deliver the prestigious Julius Stone Address in the year following the award of the prize. Entrants are required to submit an application form and five copies of the work, plus five copies of their curriculum vita. Four copies will be returned following judging, and one will be kept in the archive of the Julius Stone Institute. Guidelines are available online.

Requirements: Entries may be directly submitted by the author(s), or on the nomination of a third party.

Amount of Grant: AU$50,000

Contact: Helen Irving, 61-2-9351- 0232; fax: 61-2-9351-0200; email: heleni@law.usyd. edu.au

Internet: http://www.law.usyd.edu.au/~jurisprudence

Sponsor: University of Sydney

Julius Stone Institute of Jurisprudence Director, Level 12, 173-175 Phillip Street

Sydney 2000 NSW Australia

DENSO North America Foundation Grants   1892

The Foundation is committed to supporting higher education in engineering and business programs. Priority is given to programs that advance automotive engineering and supply-side business practices. Capital funding is available for equipment, lab development, technological advancements and/or installations, building campaigns and expansion projects. Student projects are supported for university-sanctioned student projects and training competitions. Funding is available for tooling and equipment offering a major, sustaining investment.

Requirements: 501(c)3 nonprofit organizations, educational institutions, and universities located throughout North America are eligible. Proposals should contain: details on how funding will advance student development and training; demonstrate principle(s) of innovation and/or training for efficiency gains in the workplace; contain a clearly articulated desired result.

Restrictions: Grants are not awarded for administrative costs, stipends or trips, conferences and travel expenses.

Samples: Univerity of Tennessee, College of Engineering (TN)—for the purchase of a CNC Machining Center, a second equipment investment in the college's manufacturing laboratory, $75,000 (2004); Michigan Technological University, College of Engineering (MI)—supporting the Challenge X Enterprise and the pre-college outreach program, $60,000 (2004).

Amount of Grant: $2000 minimum

Contact: Grants Administrator, (248) 372-8233; fax: (248) 213-2550; email: DENSOfoundation@denso-diam.com

Internet: http://www.densofoundation.org/foundation/foundation.html

Sponsor: DENSO North America Foundation

24777 Denso Drive, MC 4610

Southfield, MI 48086-5047

Dental Teaching and Research Fellowships   1893

Fellowships are awarded to individuals who plan to become dental teachers and/or researchers in dental schools in Canada and who must guarantee to pursue careers in teaching and research at a university for a minimum of two years following fellowship tenure. A formal application form must be filed with official transcripts and letters of support. The award is given for one year with possible renewal for a second year based on satisfactory progress.

Requirements: Candidates must be Canadian citizens or permanent residents who have completed an undergraduate course in dentistry, a dental hygiene program, or a program in science and who are also eligible for admission to a graduate or other advanced education program.

Amount of Grant: $C1800-$C8000

Date(s) Application is Due: Feb 1

Contact: Richard Munro, Executive Director, (613) 236-4763; fax: (613) 236-3935; email: Information@dcf-fdc.ca

Internet: http://www.dcf-fdc.ca/grants.html

Sponsor: Dentistry Canada Fund

427 Gilmour Street

Ottawa, ON K2P OR5 Canada

Denton A. Cooley Foundation Grants   1894

The foundation awards grants in Texas in support of health care, health education, hospitals, and medical research. Types of support include endowments, general operating grants, program grants, and research grants. There are no application forms or deadlines. The board meets quarterly to consider requests.

Restrictions: Grants do not support conferences, loans, individuals, scholarships, fellowships, or publication.

Geographic Focus: Texas

Samples: U of Texas Medical Branch at Galveston (TX)—to establish a professorship in the division of cardiothoracic surgery, $250,000 (2003).

Amount of Grant: $100-$250,000

Contact: Grants Administrator, (713) 799-2700

Sponsor: Denton A. Cooley Foundation

6624 Fannin, Suite 1640

Houston, TX 77030

Department of Black Studies Dissertation Fellowships   1895

One or two dissertation fellowships are available to assist scholars whose research focuses on areas significant to African, Caribbean, and/or African-American studies. Recipients are required to be in residence at UC Santa Barbara during the academic year, to teach one undergraduate course, and to present one public lecture. It is expected that the dissertation will be completed during residence. Applicants should submit curriculum vita, a brief description of the dissertation project, a writing sample (approximately 25 pages), and three letters of reference.

Requirements: Individuals advanced to candidacy at an accredited university are eligible. Candidates from the humanities, social sciences, and interdisciplinary fields are encouraged to apply.

Amount of Grant: $20,000 stipend

Date(s) Application is Due: Mar 1

Contact: Gwenner Miller, Black Studies Department, (805) 893-7624; fax: (805) 893-3597; email: gmiller@blackstudies.ucsb.edu

Internet: http://www.blackstudies.ucsb.edu/student_info/fellowship.html

Sponsor: University of California, Santa Barbara

3631 South Hall, Dept of Black Studies

Santa Barbara, CA 93106-3150

Dept of Ed Adult Education—National Leadership Activities Grants   1896

This program supports research, evaluation, information dissemination, and other activities to help states improve adult education and literacy programs. Types of projects include: professional development; national evaluations and surveys; curriculum development; distance learning; data analysis; professional papers; seminars and colloquia.

Requirements: Eligible applicants: Institutions of Higher Education, Local Education Agencies, Nonprofit Organizations, Other Organizations and/or Agencies, State Education Agencies, Postsecondary education institutions, public or private agencies, or consortia of these institutions, agencies, or organizations are eligible for grants, cooperative agreements, and contracts.

Contact: Dennis Berry, Division of Adult Education and Literacy, (202) 245-7814; fax: (202) 245-7837; email: dennis.berry@ed.gov

Internet: http://www.ed.gov/programs/aenla/index.html

Sponsor: Department of Education

400 Maryland Avenue, S.W., 11131, PCP

Washington, DC 20202-7240

Dept of Ed International Research and Studies Project Grants   **1897**

This program supports surveys, studies, and development of instructional materials to improve and strengthen instruction in modern foreign languages, area studies, and other international fields. In addition to surveys and studies, the program provides funds for the development of foreign language materials designed to improve and strengthen foreign language and area and related studies in the U.S. education system.

Requirements: Public and private agencies, organizations, institutions, and individuals may apply.

Restrictions: Funds awarded may not be used for the training of students and teachers.

Contact: Ed McDermott, International Education Programs Service, (202) 502-7636; fax: (202) 502-7860; email: ed.mcdermott@ed.gov

Internet: http://www.ed.gov/programs/iegpsirs/index.html

Sponsor: Department of Education

1990 K Street, N.W., 6th Floor

Washington, DC 20006-8521

Dept of Ed Jacob K. Javits Gifted and Talented Students Education Program-National Research and Development Center   **1898**

This program conducts research for the purpose of carrying out activities described in Sec. 5464(b) of the statute including research on methods and techniques for identifying and teaching gifted and talented students and for using gifted and talented programs and methods to serve all students. It also conducts program evaluations and surveys. As part of its work, the center collects, analyzes, and develops information about gifted and talented education. Emphasis is given to the identification of and services for students traditionally not included in gifted and talented education, including individuals with limited English proficiency (LEP), individuals with disabilities, and individuals living under economically disadvantaged conditions.

Requirements: Institutions of Higher Education, State Education Agencies and a consortium of IHEs and SEAs may apply.

Contact: Anne Sweet, U.S. Department of Education, IES (202) 219-2043; fax: (202) 219-2030; email: anne.sweet@ed.gov

Internet: http://www.ed.gov/programs/nrdcjavits/index.html

Sponsor: Department of Education

555 New Jersey Avenue, N.W., Suite 615B

Washington, DC 20208-5573

Dept of Ed Rehabilitation Fellowships   **1899**

The purpose of this program is to build research capacity by providing support to highly qualified individuals, including those who are individuals with disabilities, to perform research on the rehabilitation of individuals with disabilities. Fellows may conduct original research in any area relating to rehabilitation services and may address any problem faced by persons with disabilities. Two types of fellowships are available. Merit Fellowships are awarded to individuals who have advanced training or experience in independent study. Distinguished Fellowships are available to individuals who have seven or more years of research experience in related fields.

Requirements: Only individuals are eligible to be recipients of fellowships.

Restrictions: Institutions are not eligible to be recipients of fellowships.

Amount of Grant: $45,000 for Merit fellows; $55,000 for Distinguished fellows

Date(s) Application is Due: Jun 24

Contact: Donna Nangle, (202) 245-7462; email: donna.nangle@ed.gov

Internet: http://www.ed.gov/legislation/FedRegister/announcements/2005-2/042505c.html

Sponsor: Department of Education

600 Independence Avenue SW

Washington, DC 20202

Dept of Ed Safe and Drug-Free Schools and Communities State Grants   **1900**

This program provides support to SEAs for a variety of drug and violence prevention activities focused primarily on school-age youths. Activities may include: developing instructional materials; providing counseling services and professional development programs for school personnel; implementing community service projects and conflict resolution, peer mediation, mentoring and character education programs; establishing safe zones of passage for students to and from school; acquiring and installing metal detectors; and hiring security personnel.

Requirements: State Education Agencies may apply. Local Education Agencies or intermediate education agencies or consortia must apply to the State Education Agency.

Contact: Paul Kesner, Office of Safe and Drug-Free Schools, (202) 205-8134; fax: (202) 260-7767; email: paul.kesner@ed.gov

Internet: http://www.ed.gov/programs/dvpformula/index.html

Sponsor: Department of Education

400 Maryland Avenue, S.W., Room 3E230, FB-6

Washington, DC 20202-6450

Dept of Ed Star Schools Program Grants   **1901**

The purpose of this program is to support distance education projects that: encourage improved instruction in mathematics, science, foreign languages, and other subjects; and serve underserved populations, including disadvantaged, nonreading, and limited English proficient (LEP) populations and individuals with disabilities. Star Schools grants are made to eligible telecommunications partnerships, to enable such partnerships to: develop, construct, acquire, maintain, and operate telecommunications audio and visual facilities and equipment; develop and acquire educational and instructional programming; and obtain technical assistance for the use of such facilities and

instructional programming. Annual deadline dates may vary; contact the program office for exact dates.

Requirements: Eligible applicants include either one of the following that is organized on a statewide or multistate basis: a public agency or corporation established for the purpose of developing and operating telecommunications networks to enhance education opportunities provided by education institutions, teacher training centers, and other entities, except that any such agency or corporation shall represent the interests of elementary schools and secondary schools that are eligible to participate in the program under Title I, Part A, of the Elementary and Secondary Education Act of 1965; or a partnership that will provide telecommunications services and that include three or more of the following entities (a-g), at least one of which must be an agency as described in (a) or (b): (a) a local education agency that serves a significant number of elementary and secondary schools that are eligible for assistance under Title I, Part A, of the ESEA, or elementary and secondary schools operated or funded for Indian children by the Department of the Interior; (b) a state education agency; (c) an adult and family education program; (d) an institution of higher education or a state higher education agency; (e) a teacher-training center or academy that provides teacher preservice and in-service training, and receives federal financial assistance or has been approved by a state agency; (f) a public or private entity with experience and expertise in the planning and operation of a telecommunications network, including entities involved in telecommunications through satellite, cable, telephone, or computer; or a public broadcasting entity with such experience; and (g) a public or private elementary or secondary school.

Contact: Brian Lekander, Office of Innovation and Improvement, (202) 205-5633; fax: (202) 205-5720; email: brian.lekander@ed.gov

Internet: http://www.ed.gov/programs/starschools/index.html

Sponsor: Department of Education

400 Maryland Avenue, S.W., Room 4W226, FB-6

Washington, DC 20202

Dept of Ed Vocational Education National Programs   **1902**

This program supports research, evaluation, information dissemination, and other activities aimed at improving the quality and effectiveness of vocational and technical education. Projects include: research, development, demonstration, dissemination, identification of best methods, capacity building, technical assistance, evaluation, and assessment activities.

Requirements: Institutions of Higher Education, Nonprofit Organizations, Other Organizations and/or Agencies may apply.

Contact: Ricardo Hernandez, U.S. Department of Education, OVAE, (202) 245-7818; fax: (202) 245-7837; email: ricardo.hernandez@ed.gov

Internet: http://www.ed.gov/programs/venp/index.html

Sponsor: Department of Education

400 Maryland Avenue, S.W., 11137, PCP

Washington, DC 20202-7242

Dept of Ed Vocational Rehabilitation Services Demonstration and Training Programs   **1903**

This program provides competitive grants to eligible entities to expand and improve the provision of rehabilitation and other services authorized under the Rehabilitation Act of 1973 , as amended. Funding also is provided to further the purposes and policies of the act. More specifically, the program supports activities that increase the provision, extent, availability, scope, and quality of rehabilitation services under the act. Sec. 303 authorizes support of activities serving individuals with disabilities in an array of project types. These diverse projects may include the effective practices that demonstrate methods of service delivery to individuals with disabilities, as well as activities such as technical assistance, systems change, model demonstration, special studies and evaluations, and dissemination and utilization of findings from successful, previously funded projects. The expansion and improvement of rehabilitation and other services will lead to more employment outcomes for individuals with disabilities.

Requirements: State vocational rehabilitation (VR) agencies, community rehabilitation programs, Indian tribes or tribal organizations, or other public or nonprofit agencies or organizations or, as the Rehabilitation Services Administration (RSA) commissioner determines appropriate, for-profit organizations, may apply.

Contact: Timothy Muzzio, Rehabilitation Services Administration, (202) 245-7458; fax: (202) 245-7591; email: Timothy.Muzzio@ed.gov

Internet: http://www.ed.gov/programs/demotrain/index.html

Sponsor: Department of Education

400 Maryland Avenue, S.W., Room 5052, PCP

Washington, DC 20202-2800

Dept of Ed Women's Educational Equity Program Grants   **1904**

This program promotes education equity for women and girls through competitive grants. The program designates most of its funding for local implementation of gender-equity policies and practices. Research, development, and dissemination activities also may be funded. Projects may be funded for up to four years. Examples of allowable activities include: training for teachers and other school personnel to encourage gender equity in the classroom; evaluating exemplary model programs to advance gender equity; school-to-work programs; guidance and counseling activities to increase opportunities for women in technologically demanding workplaces; and, developing strategies to assist LEAs in evaluating, disseminating, and replicating gender-equity programs.

Requirements: Public agencies; private nonprofit agencies; organizations, including community and faith-based organizations; institutions; student groups; community groups; and individuals developing programs that promote gender equity may apply.

Contact: Beverly Farrar, Women's Educational Equity Program, (202) 205-3145; fax: (202) 205-5630; email: beverly.a.farrar@ed.gov
Internet: http://www.ed.gov/programs/equity/index.html
Sponsor: Department of Education
400 Maryland Avenue, S.W., Room 4W242, FB-6
Washington, DC 20202-5950

Dermatology Clinical Career Development Awards 1905

The purpose of this award is to enhance the academic careers of clinician-scientists in the early stages of their career development. The program is aimed at junior investigators with significant creativity in clinically relevant research. Research must be conducted in the United States. Such studies could include research on diagnostic and/or prognostic indicators, as well as outcome and/or epidemiologic studies. Creativity in developing new treatment modalities, including novel pharmacologic approaches and clinical trials, is encouraged. Studies need not involve direct laboratory investigation. The primary criteria for selection of a successful awardee are the quality of the research proposal, the environment of the candidate, and ability of the proposal to clarify the place of dermatologic practice in a changing health care environment. Award recipients are expected to spend at least 75 percent of their time in clinical research. Awardees are encouraged to seek simultaneous grant support from other agencies to provide for the nonsalary components of the research being performed under the auspices of this award.
Requirements: Applications are accepted from faculty members in a department or division of dermatology who have completed clinical training in a U.S. dermatology residency program. Applicants must demonstrate a strong commitment to skin research and have already had appropriate initial training (a two- to three-year research fellowship or postdoctoral training) in relevant research.
Restrictions: The foundation does not fund awards to be performed as part of the U.S. government research program (with the exception of the Veterans Administration) or awards to private foundations without an academic affiliation to dermatology.
Amount of Grant: $55,000
Date(s) Application is Due: Oct 17
Contact: Sandra Rahn Benz, Executive Director, (847) 328-2256; fax: (847) 328-0509; email: dfgen@dermatologyfoundation.org
Internet: http://www.dermfnd.org
Sponsor: Dermatology Foundation
1560 Sherman Avenue, Suite 870
Evanston, IL 60201

Dermatology Foundation Grants 1906

Limited funds are available for investigators at the early stages of their career development to initiate research projects in dermatology and cutaneous biology. Applications are reviewed by the Medical and Scientific Committee and awarded on a competitive basis for projects not funded from other sources. The research has to be performed under the sponsorship of a division or department of dermatology. The maximum period of funding is for one year. Only one grant application will be considered from a single academic center.
Restrictions: Awards cannot be used for payment of indirect costs.
Amount of Grant: $10,000 average
Date(s) Application is Due: Oct 17
Contact: Sandra Rahn Benz, Executive Director, (847) 328-2256; fax: (847) 328-0509; email: dfgen@dermatologyfoundation.org
Internet: http://dermatologyfoundation.org
Sponsor: Dermatology Foundation
1560 Sherman Avenue, Suite 870
Evanston, IL 60201

Dermatology Postdoctoral Research Fellowships 1907

Financial support is available to postdoctoral fellows with commitments to careers in academic dermatology who desire research training. Applications will be accepted in two categories. Dermatologist Investigator Research Fellowships are designed to support dermatologists who desire research training and have commitments to careers in academic dermatology. Research Fellowships support training in research skills relevant to dermatology and cutaneous biology. Highest priority will be given to new applicants. A limited number of second-year awards may be given. Although it is recognized that fellowships will encompass specific research projects, they are intended primarily for the support of research training. Fellowship recipients must spend at least 75 percent of their total effort in cutaneous research.
Requirements: Applications for Dermatologist Investigator Research Fellowships are accepted from individuals holding the MD, MD-PhD, or DO degree who have completed their clinical training in U.S. dermatology residency programs. Research Fellowship applications are accepted from individuals holding the MD or PhD or equivalent.
Restrictions: In general, investigators at more than four years past their terminal degrees or two years beyond their residencies at the time of the initial application are not eligible. (Consideration will be given to individuals with substantial training in other areas and who are entering into skin research.) Individuals with academic appointments at the level of assistant professor or above are not ordinarily eligible.
Amount of Grant: $25,000 maximum for one year
Date(s) Application is Due: Oct 17
Contact: Sandra Rahn Benz, Executive Director, (847) 328-2256; fax: (847) 328-0509; email: dfgen@dermatologyfoundation.org
Internet: http://dermatologyfoundation.org/rap

Sponsor: Dermatology Foundation
1560 Sherman Avenue, Suite 870
Evanston, IL 60201

Dermatology Research Career Development Awards 1908

Awards are available to assist in the transition from fellowship to established investigator in cancer and other diseases of the skin, hair, and nails. The intent of the program is to advance the research careers of young individuals in dermatology and cutaneous biology, with the emphasis on research benefiting the dermatology community at large. Award recipients are expected to spend at least 75 percent of their time in cutaneous research. Awardees are encouraged to seek simultaneous grant support from other agencies to provide for the nonsalary components of the research being performed under the auspices of this award.
Requirements: Applicants must be faculty members in a department or division of dermatology who demonstrate a strong commitment to skin research and have already had appropriate initial training (a two- to three-year research fellowship or postdoctoral training) in biomedical research.
Restrictions: The foundation does not fund awards to be performed as part of the U.S. government research program (with the exception of the Veterans Administration) or awards to private foundations without an academic affiliation to dermatology.
Amount of Grant: $55,000
Date(s) Application is Due: Oct 17
Contact: Sandra Rahn Benz, Executive Director, (847) 328-2256; fax: (847) 328-0509; email: dfgen@dermatologyfoundation.org
Internet: http://dermatologyfoundation.org/rap
Sponsor: Dermatology Foundation
1560 Sherman Avenue, Suite 870
Evanston, IL 60201

Detroit Lions Charities Grants 1909

The organization supports charitable and community causes in Michigan. Funding interests include child abuse and domestic violence prevention, youth recreation, and spinal cord injury research. Grants have been awarded to support the Never, Never Shake a Baby billboard campaign; Pigskin Geography, a learning initiative for youth; Big Brothers/Big Sisters; Grand Rapids Metropolitan YMCA—for swimming lessons for inner-city youth; and Saint Joseph Mercy Hospital—for a domestic violence education program. Requests are accepted between October 1 and December 31.
Geographic Focus: Michigan
Samples: Dominion Family Services (MI)— to battle child abuse and domestic violence in Michigan, $210,000.
Date(s) Application is Due: Dec 31
Contact: Detroit Lions Community Affairs Information Hotline, (313) 216-4056
Internet: http://detroitlions.com/community/index.cfm?cont_id=49715
Sponsor: Detroit Lions Charities
222 Republic Drive
Allen Park, MI 48101

DHHS AIDS and HIV Epidemiologic Research Studies Grants 1910

The purpose of this program is to encourage studies using the new rapid HIV tests in different settings (Operational Research), specifically focused on African American, Latino, and other racial and ethnic minorities that are underserved and/or disproportionately affected by the HIV epidemic, and conducted by researchers who have experience working with these populations; to learn more about the effects of rapid HIV testing on motivators and barriers to HIV testing at the individual, provider, and system levels; and to foster collaborations between organizations serving minority communities and their respective state and local health departments in the design and implementation of innovative practical strategies using rapid HIV tests to increase knowledge of HIV serostatus and facilitate entry into prevention and care systems. The research should contribute to the health services knowledge base from which empirically based information can be derived by policy makers, both immediately and over the coming decades. Assistance is available for a 12-month budget period within project periods ranging from one to five years. Contact the office for deadline dates.
Requirements: Applications may be submitted by public and private nonprofit organizations; community-based, national, and regional organizations; state and local governments or their bona fide agents or instrumentalities; federally recognized Indian tribal governments; and Indian tribes or organizations.
Amount of Grant: $5000-$270,000; $200,265 average
Contact: JAllyn Nakashima, Epidemiology Branch, Division of HIV/AIDS Prevention/ Surveillance and Epidemiology, (404) 639-0900; Cheryl Maddux, Grants Management, (770) 498-1911
Internet: http://www.dhhs.gov
Sponsor: Department of Health and Human Services
1600 Clifton Rd NE, MS-E-45
Atlanta, GA 30333

DHHS AIDS Education and Training Centers Grants 1911

Grants are awarded for the establishment of AIDS education and training centers. The purpose is to provide education and training to primary care providers and others on the treatment and prevention of AIDS, in collaboration with health professions schools, local hospitals, and health departments; to provide updates of new and timely information about HIV infection to primary and secondary health care providers; and to serve as the support system for area health professionals through the AIDS hotline, clearinghouse, and referral activities. Awards will also support community-based organizations and

community health clinics affiliated with accredited public and private nonprofit entities. Preference will be given to qualified projects that will train or result in the training of health professionals who will provide treatment for minority individuals with HIV, other individuals who are at high risk of contracting HIV, or to minority health professionals and allied health professionals who will provide treatment for persons with HIV. Awards will be made for a three-year project period. Deadlines are available from the HRSA Application Center. Annual deadline dates vary; contact program staff for exact dates.
Requirements: Public and nonprofit entities, schools, and academic health science centers are eligible to apply.
Amount of Grant: $450,000-$5.1 million range; $2.3 million average
Contact: Deborah Willis-Fillinger, HIV Education Branch, (301) 443-6364; fax: (301) 443-9887
Internet: http://hab.hrsa.gov
Sponsor: Department of Health and Human Services
5600 Fishers Ln, Parklawn Bldg
Rockville, MD 20857

DHHS AIDS Project Grants 1912

The grants are intended to fund programs that develop and implement surveillance, epidemiological research, health education, school health, and risk reduction activities of AIDS in states and major cities. The program also gives support for cooperative agreements for AIDS activities. The funding period will be from one to five years, renewable. Applicants should contact the CDC for deadline information.
Requirements: Public and private organizations, both nonprofit and for profit, state and local governments, U.S. territories and possessions, small and minority businesses, and businesses owned by women are encouraged to apply.
Amount of Grant: $45,000-$2 million; $390,174 average
Contact: Allyn Nakashima , Division of HIV/AIDS Prevention, Centers for Disease Control and Prevention, (404) 639-0900; Cheryl Maddux, Grants Management Branch, (770) 498-1911
Internet: http://www.dhhs.gov
Sponsor: Department of Health and Human Services
2920 Brandywine Road
Atlanta, GA 30341

DHHS Chiropractic Demonstration Project Grants 1913

The objective of the program is to make grants and enter into contracts with schools, colleges, and universities of chiropractic for the purposes of carrying out demonstration projects in which chiropractors and physicians collaborate to identify and provide effective treatment for spinal and lower back conditions. Grant funds may be used for personnel, equipment, supplies, domestic travel, consultants and guest lecturers, rental of space, renovations, and other costs directly related to the project as described in the approved application. Application materials are available on the World Wide Web at address: http://www.hrsa.gov/bhpr/grants.html.
Requirements: Public or private nonprofit schools, colleges, and universities of chiropractic medicine are eligible to apply.
Restrictions: Grant funds may not be used for construction of facilities, acquisition of land, foreign travel, or support of students, including fellowships, stipends, tuition, fees, or travel allowances.
Amount of Grant: $381,048 -$389,214; $386,456 average
Contact: Jennifer Hannah, Health Resources and Services Administration, (301) 443-0908
Internet: http://bhpr.hrsa.gov/interdisciplinary/chiro.html
Sponsor: Department of Health and Human Services
5600 Fishers Ln, Parklawn Bldg
Rockville, MD 20857

DHHS Comprehensive Community Mental Health Services Grants 1914
for Children with Serious Emotional Disturbances

This program provides community-based systems of care for children and adolescents with serious emotional disturbances and their families. The program will ensure that services are provided collaboratively across child-serving systems; that each child or adolescent served through the program receives an individualized service plan developed with the participation of the family; that each individualized plan designates a case manager to assist the child and family; and that funding is provided for mental health services required to meet the needs of youngsters in these systems.
Requirements: States; political subdivisions of a state, such as county or local governments; and federally recognized Native American tribal governments are eligible to apply.
Amount of Grant: $200,064-$3.5 million; $1.85 million average
Contact: Gary Blau, Chief, Child Adolescent and Family Branch, (301) 443-1333
Internet: http://mentalhealth.samhsa.gov/publications/allpubs/CA-0013/default.asp
Sponsor: Department of Health and Human Services
5600 Fishers Ln, Parklawn Bldg
Rockville, MD 20857

DHHS Developmental Disabilities University Affiliated Project Grants 1915

Grants assist with the cost of administration and operation of facilities for providing interdisciplinary training for personnel concerned with developmental disabilities, demonstrations of the provision of exemplary services related to the developmentally disabled, demonstration of technical assistance for generic and specialized agencies, dissemination of findings related to the provision of services to researchers and government agencies, and generation of information on the need for further service-related research. Grants may cover salaries for administrators, coordinators, and others needed to operate a training facility, such as clerical and financial personnel, and maintenance and housekeeping personnel; overhead expenses and expenses required to start up new programs; and faculty for training programs that will meet critical personnel shortages and are not eligible for support from other sources. Deadline dates are available from the office.
Requirements: Public and private nonprofit organizations and agencies are eligible.
Amount of Grant: $480,000
Date(s) Application is Due: Sep 1
Contact: Jennifer Johnson, Administration on Developmental Disabilities, Administration for Children and Families, (202) 690-5982
Internet: http://www.dhhs.gov
Sponsor: Department of Health and Human Services
370 L'Enfant Promenade, SW
Washington, DC 20447

DHHS Emergency Medical Services for Children (EMSC) Program 1916

The purpose of this program is to support demonstration projects for the expansion and improvement of emergency medical services for children who need treatment for trauma or critical care. It is expected that maximum distribution of projects among the states will be made and that priority will be given to projects targeted toward populations with special needs, including Native Americans, minorities, and the disabled. Annual deadline dates may vary; contact program staff for exact dates.
Requirements: States and accredited schools of medicine are eligible to apply.
Amount of Grant: $200,000 average
Contact: Dan Kavanaugh, (301) 443-1321; email: dkavanaugh@hrsa.gov
Internet: http://mchb.hrsa.gov/programs/emsc
Sponsor: Department of Health and Human Services
5600 Fishers Lane
Rockville, MD 20857

DHHS Geriatric Training for Physicians and Dentists Grants 1917

These grants are intended to assist in the operation of postdoctoral training preparing current and future faculty for leadership roles in geriatric medicine and dentistry and to provide support, including traineeships and fellowships, for geriatric medicine training projects to train physicians and dentists who plan to teach geriatric medicine or geriatric dentistry. Geriatric training is to be provided through one or both of the following projects: one grant is offered for a one-year retraining program in geriatrics for physicians who are faculty members in departments of internal medicine, family medicine, gynecology, geriatrics, and psychiatry at schools of medicine and osteopathy; and dentists who are faculty members at schools of dentistry or at hospital departments of dentistry. A grant is offered for a two-year internal medicine or family medicine fellowship program with emphasis in geriatric research for physicians who have completed graduate medical education programs in internal medicine, family medicine, psychiatry, neurology, gynecology, geriatrics, or rehabilitation medicine; and dentists who have completed postdoctoral dental education programs. Each project for which a grant is made must be staffed by full-time teaching physicians who have experience or training in geriatric medicine, be staffed by full-time or part-time teaching dentists who have experience or training in geriatric dentistry, and be based in a graduate medical education program in a department of internal medicine or family medicine or a department of geriatrics that has been in existence over a year. Each project must provide participants with exposure to a diversified population of elderly individuals and provide training in geriatrics and exposure to the physical and mental disabilities of elderly individuals through a variety of service rotations, such as geriatric consultation services, acute care services, dental services, geriatric psychiatry units, day and home care programs, rehabilitation services, extended care facilities, geriatric ambulatory care and comprehensive evaluation units, and community care programs for elderly mentally retarded individuals. Applications and deadline date information can be obtained on the Web site or by contacting the office.
Requirements: Grants will be made to accredited public or private nonprofit schools of medicine, schools of osteopathic medicine, teaching hospitals, and graduate medical education programs.
Amount of Grant: $178,721-$632,825; $374,575 average
Contact: Kathleen Bond, Department of Health and Human Services, (301) 443-8681
Internet: http://www.dhhs.gov
Sponsor: Department of Health and Human Services
5600 Fishers Ln, Parklawn Bldg
Rockville, MD 20857

DHHS Great Lakes Human Health Effects Research Program Grants 1918

The objectives of this program are to build upon and amplify the results from past and ongoing research; develop information, databases, and/or research methodology that will provide long-term benefit to the Great Lakes human health research effort; develop directions and methodology for future human health effects research; provide health information to the subjects of the research and their medical professions; and increase the public awareness of the health implications of the toxic pollution problem in the Great Lakes. Grant funds may be used to conduct research on the impact on human health of fish consumption in this region and to extend the knowledge of the effects of contaminants on human reproductive/developmental, behavioral, neurological, and endocrinological adverse health. Contact the office for deadline dates.
Requirements: Eligible applicants are the official public health agencies or their bona fide agents or instrumentalities and political subdivisions thereof, which may include state universities, state colleges, state research institutions, state and local health departments, and federally recognized Native American tribal governments located in the Great

Lakes states (Illinois, Indiana, Michigan, Minnesota, Ohio, Pennsylvania, New York, and Wisconsin).
Geographic Focus: Illinois; Indiana; Michigan; Minnesota; New York; Ohio; Pennsylvania; Wisconsin
Amount of Grant: $200,000 average
Contact: Heraline Hicks, Division of Toxicology, Agency for Toxic Substances and Disease Registry, (404) 489-0717; fax: (404) 498-0094; email: HEH2@cdc.gov; Edna Green, Grants Management Contact, (770) 488-2743; email: ecg4@cdc.gov
Internet: http://www.atsdr.cdc.gov/grtlakes.html
Sponsor: Department of Health and Human Services
1600 Clifton Rd NE
Atlanta, GA 30333

DHHS Health Care Systems Cost and Access Research & Development Grants   1919
Grants are made to support health services research to create new knowledge and better understanding of the process by which health services are made available and how they may be provided more efficiently and effectively. The AHCPR has a broad legislative mandate to support general health services research on problems related to health care cost, quality, and access to health services. Major categories of research issues include consumer decision making, managed care and the health care marketplace, primary care, rural health services, and AIDS. Annual deadline dates may vary; contact program staff for exact dates.
Requirements: Government agencies (federal, state, and local), federally recognized tribal governments, U.S. territories, sponsored organizations, nongovernment organizations, minority groups, specialized groups, public or private institutions of higher education, and other public or nonprofit private agencies, institutions, or organizations are eligible. Research project grants may also be awarded to individuals.
Restrictions: Profit-making organizations are not eligible for grants.
Amount of Grant: $5000-$2.8 million; $310,000 average
Date(s) Application is Due: Feb 1; Jun 1; Oct 1
Contact: Mable Lam, Grants Management Officer, (301) 427-1447
Internet: http://www.dhhs.gov
Sponsor: Department of Health and Human Services
2101 E Jefferson Street, Executive Office Ctr
Rockville, MD 20852

DHHS Health Conference Support for Toxic Substances and Disease Registry   1920
Funds are available through this program for partial support for nonfederal conferences on disease prevention, health promotion, and projects related to hazardous substances. The purpose of this program is to work closely with state, local, and other federal agencies to reduce or eliminate illness, disability, and death resulting from exposure of the public and workers to toxic substances at spill and waSuite disposal sites.
Amount of Grant: $130,000-$300,000; $200,000 average
Contact: Caroline McDonald, Agency for Toxic Substances and Disease Registry, (404) 498-0270; fax: (404) 498-0059; email: COS4@cdc.gov; Mildred Garner, Grants Management Branch, (770) 488-2745; fax: (770) 488-2777
Internet: http://www.dhhs.gov
Sponsor: Department of Health and Human Services
1600 Clifton Rd NE
Atlanta, GA 30333

DHHS Health Professions Preparatory Scholarship Program for Indians   1921
The program makes scholarship grants for a maximum of two years to individuals of Native American or Alaskan Native descent for the purpose of completing compensatory preprofessional education to enable the recipient to qualify for enrollment or re-enrollment in a health professions school. Grants for stipends and books are made directly to the individual applicant; tuition payments are made to the college or university. New applications are usually available in February with an annual submission deadline in April. Annual deadline dates may vary; contact program staff for exact dates.
Requirements: Scholarship awards are made to individuals of Native American or Alaskan Native descent who have successfully completed high school education or high school equivalency and who have been accepted for enrollment in a compensatory, preprofessional general education course or curriculum.
Amount of Grant: $17,500-$26,019; $17,366 average
Date(s) Application is Due: Apr 1
Contact: Harold Jess Brier, Chief, Scholarship Branch, (301) 443-6197; Lois Hodge, Grants Management Officer, (301) 443-0243
Internet: http://www.ihs.gov
Sponsor: Department of Health and Human Services
12300 Twinbrook Pkwy, Twinbrook Metro Plz
Rockville, MD 20852

DHHS Health Professions Scholarship Program   1922
The program makes scholarship grants to Native Americans and other students for the purposes of completing health professional education—nursing; medicine including allopathic, osteopathic, and veterinary medicine; dentistry; x-ray technology; optometry; pharmacy; public health nutrition (graduate); medical social work (graduate); speech pathology/audiology (graduate); podiatry; health care administration; and other allied health professions—to obtain health professionals to serve Native Americans. Not all disciplines participate each year. Upon completion, grantees are required to fulfill an obligated service payback. Maximum length of funding is four years. Grants for stipends and books are made directly to the individual applicant; tuition payments are made to the college or university. New applications are available in February with a submission

deadline in April. Annual deadline dates may vary; contact program staff for exact dates.
Requirements: Priority consideration for scholarship awards is granted to persons of Native American or Alaskan Native descent. Applicants for new awards must be accepted by an accredited U.S. educational institution for a full-time course of study leading to a degree in medicine, osteopathy, dentistry, or other participating health profession that is deemed necessary by the Indian Health Service; be eligible for or hold an appointment as a commissioned officer in the regular or reserve corps of the Public Health Service; or be eligible for civilian service in the Indian Health Service.
Amount of Grant: $24,128-$38,222; $24,694 average
Date(s) Application is Due: Apr 1
Contact: Harold Jess Brier, Chief, Scholarship Branch, (301) 443-6197; Lois Hodge, Grants Management Officer, (301) 443-0243
Internet: http://www.dhhs.gov
Sponsor: Department of Health and Human Services
12300 Twinbrook Pkwy, Twinbrook Metro Plz
Rockville, MD 20852

DHHS Health Promotion and Disease Prevention Research and   1923
Demonstration Centers Grants
Grants are awarded to qualified schools to establish, maintain, and operate academic-based centers for high-quality research and demonstration with respect to health promotion and disease prevention; to establish linkages between ongoing basic research in a wide array of fields and applied research in disease prevention and health promotion; to bring the knowledge and expertise of academic health centers to bear on practical public health problems; to field test and rigorously evaluate more cost-effective methods and strategies for preventing unnecessary illness and promoting good health; and to shorten the time lag between the development of new and proven effective disease prevention and health promotion techniques and their widespread application. The length of assistance will be from one to five years, renewable based on competitive applications and availability of funds. Deadlines will be announced in the Federal Register.
Requirements: Eligible applicants are schools of medicine, schools of osteopathy, and schools of public health.
Restrictions: Grantees may not award subgrants but may enter into consortia agreements or contracts as necessary to achieve the aims of the program.
Amount of Grant: $740,000-$770,000; $755,000 average
Contact: Dr. Eduardo Simoes, Program Director, National Center for Chronic Disease Prevention and Health Promotion, (770) 488-5919; Carlos Smiley, Grants Management Officer, (770) 488-2754
Internet: http://www.cdc.gov/prc
Sponsor: Department of Health and Human Services
4770 Buford Hwy, K-45
Atlanta, GA 30333

DHHS HIV Demonstration Program for Children, Adolescents, and Women   1924
Grants are awarded to support, improve, and expand the system of comprehensive care services for children, youth, women, and families who are infected with or affected by HIV and AIDS and to link comprehensive care systems with clinical research. Projects should promote collaboration between research institutions and primary care medical and social service programs. Funding priority will be given to projects that demonstrate established models of care that are comprehensive and coordinated, sensitive to differing cultural needs, and family and community based. Grants may be made for up to three-year project periods, renewable competitively after the first year. Contact the office for deadline dates.
Requirements: All public and private nonprofit organizations that provide or arrange for primary health care are eligible for grants. This includes state or local health departments, hospitals, community health centers, drug abuse treatment agencies, school-based clinics, tribal health programs, colleges, and hemophilia treatment centers.
Amount of Grant: $230,384-$2.5 million; $766,222 average
Contact: Dr. Jose Rafael Morales, Division of Community Based Programs/Title IV, HIV/AIDS Bureau, (301) 443-9051
Internet: http://www.dhhs.gov
Sponsor: Department of Health and Human Services
5600 Fishers Ln, Parklawn Bldg
Rockville, MD 20857

DHHS Independence Demonstration Program   1925
The objective of the program is to provide for the establishment of demonstration projects designed to determine the social, civic, psychological, and economic effects of providing to individuals and families with limited means an incentive to accumulate assets by saving a portion of their earned income; the extent to which an asset-based policy that promotes saving for postsecondary education, homeownership, and microenterprise development may be used to enable individuals and families with limited means to increase their economic self-sufficiency; and the extent to which an asset-based policy stabilizes and improves families and the community in which the families live. Deadlines are announced in the Federal Register.
Requirements: Nonprofit 501(c)3 tax-exempt organizations and state or local government agencies or tribal governments submitting an application jointly with such a nonprofit organization are eligible to apply.
Amount of Grant: $360,000 average
Contact: James Gatz, Office of Community Services, (202) 401-4626; email: AFIProgram@acf.hhs.gov
Internet: http://www.acf.hhs.gov/grants/open/HHS-2004-ACF-OCS-EI-0027.html

Sponsor: Department of Health and Human Services
370 L'Enfant Promenade SW, Suite 500 W
Washington, DC 20447

DHHS Maternal and Child Health Research and Training Programs Grants   1926

Grants fund special projects that develop new data and improve the professional development of health practitioners who serve women and children. Research grants will fund projects that address minority and disadvantaged populations, health-promoting behaviors, quality outcome measures, and systems integration and reform. Long-term training grants will be awarded to train health professionals at the graduate and postgraduate levels. Training should focus on family-centered, community-based care. Continuing education grants support continuing education for workers in the maternal and child health field. Priority will be given to projects focusing on emergency medical services for children, violence prevention in schools, and core public health. Application kits may be obtained by calling toll-free (877) 477-2123.

Requirements: Training grants may be made to public or nonprofit private institutions of higher learning. Research grants may be made to public or nonprofit institutions of higher learning and public or nonprofit private agencies and organizations engaged in research, maternal and child health, or programs for children with special health care needs.

Amount of Grant: $14,000-$1.5 million; $179,248 average

Contact: Program Contact, Maternal and Child Health Bureau, (301) 443-2170; Lawrence Poole, Director, Division of Grants Management Operations, (301) 443-2385

Internet: http://www.dhhs.gov

Sponsor: Department of Health and Human Services
5600 Fishers Ln, Parklawn Bldg
Rockville, MD 20857

DHHS National Health Service Corps Loan Repayment Awards   1927

To help ensure an adequate supply of trained health professionals for the National Health Service Corps, this program provides for the repayment of educational loans for participants who agree by written contract to serve an applicable period of time in a health-related workforce shortage area or in an Indian health program or facility. Awards provide payments toward participants' qualified government and commercial health professions education loans during each year of practice at a selected NHSC loan repayment service site with a two-year service minimum. Health professions given priority for selection are those determined by the workforce needs of the NHSC and the Indian Health Service. Beneficiaries include primary care physicians; dentists; certified nurse midwives; certified nurse practitioners; physicians assistants; clinical psychologists; clinical social workers; psychiatric nurse specialists; marriage and family therapists; licensed professional counselors; and dental hygienists. Deadline dates are published in the Federal Register.

Requirements: U.S. citizens who are enrolled as full-time students in the final year of study in an accredited health profession education institution or possess a health professions degree and are either enrolled in postgraduate health professions training or are in professional practice are eligible to apply. They must hold an unrestricted health professions license in a state and be eligible for or hold an appointment as a commissioned officer in the regular or reserve corps of the Public Health Service or be eligible for selection for a federal civil service appointment.

Amount of Grant: $25,000 maximum per year plus a tax assistance payment of 39 percent

Date(s) Application is Due: Mar 26

Contact: Chief, Loan Repayment Programs Branch, Division of Health Services Scholarships, (301) 594-4400; Public Information Phone, (800) 435-6464

Internet: http://www.hrsa.gov

Sponsor: Department of Health and Human Services
4350 East-West Hwy
Rockville, MD 20857

DHHS National Health Service Corps Scholarships   1928

To obtain adequate numbers of trained physicians, dentists, and other health-related specialists for the National Health Service Corps, scholarships are given in exchange for service in health personnel shortage areas within the United States. Disciplines have included allopathic and osteopathic medicine, dentistry, nursing (baccalaureate and graduate), public health nutrition (graduate), medical social work (graduate), speech pathology/audiology (graduate), veterinary medicine, optometry, podiatry, and pharmacy. Not all disciplines participate each year. Deadlines for receipt of applications for each academic year are the last Friday in March. A student may not receive more than a total of four years of support. Each award covers one academic year of support (including monthly living stipend, tuition, and fees), as specified by signed contracts, with annual continuation awards if funds are available.

Requirements: Applicants for new awards must be accepted by accredited U.S. educational institutions for full-time courses of study leading to degrees in medicine, osteopathy, or dentistry; be former recipients of the federal scholarship program for first-year students of exceptional financial need; be eligible for or hold appointments as commissioned officers in the regular or reserve corps of PHS or be eligible for civilian service in the NHSC; and submit applications and signed contracts to accept payment of scholarships and to serve for the applicable period of obligated service in health workforce shortage areas.

Restrictions: Each year of scholarship support incurs a year of federal service obligation with the minimum obligation being two years. Failure to fulfill the service obligation incurs financial damages, payable in one year, calculated at three times the scholarship benefits plus interest.

Amount of Grant: $1128 plus tuition, fees and other reasonable costs

Contact: Division of National Health Service Corps, (301) 594-4400 or (800) 221-9393

Internet: http://www.dhhs.gov

Sponsor: Department of Health and Human Services
4350 East-West Hwy
Bethesda, MD 20814

DHHS Native American Health Service Loan Repayment Awards   1929

To help ensure an adequate supply of trained health professionals for IHS facilities, the program provides for the repayment of educational loans for participants who agree by written contract to serve an applicable period of time at a facility IHS has designated as a retention/recruitment priority site or in a designated specialty at a site with an appropriate position. Recipients must agree to serve an applicable period of time in such a site with a minimum period of participation of two years. Contact the program office for deadline dates.

Requirements: Eligible applicants must be individuals who are enrolled as full-time students in the final year of a course of study or program leading to a degree in allopathic or osteopathic medicine, dentistry, or other health profession in a state; are enrolled in an approved graduate training program in allopathic or osteopathic medicine, dentistry, or other health profession; or have a degree in allopathic or osteopathic medicine, dentistry, or other health profession and have completed an approved graduate training program and have a current and valid license to practice such health profession in a state. In addition, applicants must be eligible for appointment as commissioned officers in the regular or reserve corps of Public Health Service (PHS) or be eligible for selection for civilian service in IHS, submit an application to participate in the loan repayment program, and sign and submit it to DHHS at the time of agreeing to accept repayment of educational loans, and serve for the applicable period of service in a retention/recruitment site as determined by DHHS.

Amount of Grant: $3000-$48,000 for a 2-year obligation

Contact: Jackie Sanitago, Chief, Loan Repayment Program, IHS, (301) 443-3396

Internet: http://www.dhhs.gov

Sponsor: Department of Health and Human Services
12300 Twinbrook Pkwy
Rockville, MD 20852

DHHS Native American Programs   1930

To provide financial assistance, training and technical assistance, and research, demonstration and evaluation activities to public and private nonprofit organizations including Indian Tribes, urban Indian centers, Alaska Native villages, Native Hawaiian organizations, rural off-reservation groups, and Native American Pacific Island groups for the development and implementation of social and economic development strategies that promote self-sufficiency. These projects are expected to result in improved social and economic conditions of Native Americans within their communities and to increase the effectiveness of Indian Tribes and Native American organizations in meeting their economic and social goals. Additional competitive areas include (1) Environmental Regulatory Enhancement, designed to assist Tribal and Alaska Village governments in developing environmental programs responsive to tribal needs; and (2) Native Languages Preservation and Enhancement, a program to assist Native American tribes and communities in ensuring the survival and continued vitality of their languages.

Requirements: Public and private nonprofit agencies, including but not limited to, governing bodies of Indian tribes on Federal and State reservations, Alaska Native villages and regional corporations established by the Alaska Native Claims Settlement Act, such public and nonprofit private agencies serving Native Hawaiians, Indian and Alaska Native organizations in urban or rural nonreservation areas, and Native American Pacific Islanders (American Samoan Natives, and indigenous peoples of Guam, the Commonwealth of the Northern Mariana and the Republic of Palau) are eligible to apply.

Amount of Grant: $125,000 average tribal grant; $100,000 average urban grant

Contact: Sheila Cooper, Director of Program Operations, (202) 690-5787

Internet: http://www.acf.hhs.gov/programs/ana/programs/index.html

Sponsor: Department of Health and Human Services
370 L'Enfant Promenade SW, MS HHH 326-F
Washington, DC 20447

DHHS Neurodevelopmental Test Methods Research Grants   1931

The purpose of the program is to determine and validate a battery of neurodevelopmental tests for use in assessing the effects of prenatal or postnatal exposure to developmental toxicants. The battery of tests should be applicable to a wide range of potential neurodevelopmental toxicants found at waSuite sites and in the environment including metals and solvents; be applicable to a wide range of exposure levels found in the environment; and cover a broad range of developmental domains including cognitive function, sensory function, motor function, and complex multi-tasking performance. These research methods will address an agency goal to develop methods and tools for evaluating human health consequences from exposure to toxic substances in the environment.

Requirements: Applications may be submitted by official public health agencies of the states, or their bona fide agents. This includes the District of Columbia, American Samoa, the Commonwealth of Puerto Rico, the Virgin Islands, the Federated States of Micronesia, Guam, the Northern Mariana Islands, the Republic of the Marshall Island, the Republic of Palau, federally recognized Indian tribal governments, public and private nonprofit and for-profit universities, colleges, and research institutions.

Amount of Grant: $130,000-$300,000; $200,000 average

Date(s) Application is Due: Jul 15

Contact: Mildred Garner, Grants Management Officer, (770) 488-2745; fax: (770) 488-2777; email: info@cdc.gov
Internet: http://www.cdc.gov
Sponsor: Department of Health and Human Services
2920 Brandywine Road
Atlanta, GA 30341-4146

DHHS Oral Health Promotion Research Across the Lifespan 1932
The National Institute of Dental and Craniofacial Research has invited proposals for improving the oral health of people of all ages. The research team must include someone with extensive experience in health promotion, behavioral and/or social science research. The health promotion intervention proposed for funding must be based on a previously conducted assessment of the epidemiology, social, behavioral and/or environmental factors related to the disease or condition under study. Research could focus on maternal and child health, adolescent and young adult health, or health of adults with complex diseases. For example, an applicant might propose to study approaches to involving families, social networks, communities, or neighborhoods in behaviors that promote and improve oral health; improving patient-provider communication related to oral preventive measures; or effective ways to train oral health professional students to communicate with diverse patient populations.
Requirements: Any person with the skills, knowledge, and resources necessary to carry out the proposed research as the project director/principal investigator (PD/PI) is invited to work with his/her organization to develop an application for support. Applications must be submitted electronically through Grants.gov (http://www.grants.gov) using the SF424 research and related forms and the SF424 application guide.
Date(s) Application is Due: Mar 5; Jul 5; Nov 5
Contact: Maria Teresa Canto, (301) 594-5497; fax: (301) 480-8322; email: maria.canto@nih.gov
Internet: http://grants.nih.gov/grants/guide/pa-files/PA-07-225.html
Sponsor: National Institute of Dental and Craniofacial Research (NIDCR)
Building 45, Room 4AS43D, 45 Center Drive MSC 6401
Bethesda, MD 20892

DHHS Podiatric Primary Care Residency Training Grants 1933
This program intends to promote the postgraduate education of podiatrists in primary care podiatric practice. Grants are to assist in meeting the costs of the program that cannot be met from other sources. Grants may include support for the program only, residents only, or support for both the program and residents but are not intended to absorb costs of existing or new positions that can be supported from other available funds. Project period is a maximum of 36 months; a noncompeting continuation application is required at the end of each budget period. Contact the office for deadline dates.
Requirements: Public or nonprofit private accredited schools of podiatric medicine and teaching hospitals may apply.
Amount of Grant: $205,291-$286,362; $245,826 average
Contact: Chris McLaughlin, Division of Medicine, (301) 443-1568; email: cmclaughlin@hrsa.gov
Internet: http://www.dhhs.gov
Sponsor: Department of Health and Human Services
5600 Fishers Ln, Parklawn Bldg
Rockville, MD 20857

DHHS Preventive Medicine 1934
The grants promote the postgraduate education of physicians in preventive medicine. Grants are intended to assist in meeting the costs of planning and developing new preventive medicine programs; maintaining or improving existing residency training programs in preventive medicine; and providing financial assistance to trainees enrolled in such programs. Contact the office for deadlines and application materials.
Requirements: Any accredited public or private school of medicine, osteopathy, or public health may apply. To be eligible for a grant, the applicant must demonstrate that it has or will have available, full-time faculty members with training and experience in the fields of preventive medicine.
Restrictions: Grants may not be used for construction or patient services.
Amount of Grant: $101,897-$430,188; $206,768 average
Contact: Dr. Douglas Lloyd, Public Health and Dental Education Branch, (301) 443-0157
Internet: http://www.dhhs.gov
Sponsor: Department of Health and Human Services
5600 Fishers Ln, Parklawn Bldg
Rockville, MD 20857

DHHS Primary Medical Care National Research Service Awards 1935
This program is intended to promote postdoctoral research training programs in primary medical care. NRSAs are made directly to individuals for research training in primary medical care; in addition, grants may be made to institutions to enable them to award NRSAs to individuals selected by them. Each individual who receives an NRSA is obligated, upon termination of the award, to comply with certain service and payback provisions. Recipients agree to engage in primary medical care research and/or teaching for a period equal to the period of the NRSA support in excess of 12 months. Contact the office for deadline dates.
Requirements: Domestic public or private nonprofit organizations may apply for training grants. States or local governments and U.S. territories are eligible. Individual applicants for fellowships must have received doctoral degrees. All persons supported as fellows or trainees must be citizens or noncitizen nationals of the United States or have been lawfully admitted for permanent residence. Eligibility is limited to individuals affiliated with entities having received grants or contracts under Sections 780, 784, or 786 of the Public Health Service Act.
Amount of Grant: $258,100-$538,167; $389,688 average
Contact: Christopher McLaughlin and Anne Patterson, Division of Medicine, (301) 443-6785
Internet: http://www.dhhs.gov
Sponsor: Department of Health and Human Services
5600 Fishers Ln, Parklawn Bldg
Rockville, MD 20857

DHHS Priority Health Conditions Studies Initiative 1936
The primary purpose of this program is to solicit scientific proposals designed to study the occurrence of and risk factors for priority adverse health conditions. The priority health conditions are birth defects and reproductive disorders, cancers, immune function disorders, kidney dysfunction, liver dysfunction, lung and respiratory diseases, and neurotoxic disorders. Each fiscal year, one of the seven priority health conditions will be emphasized; however, applications may address any of the seven. This will improve recipients' ability to address potential public health problems related to exposure to hazardous substances. Contact the office for deadline dates.
Requirements: Eligible applicants are state health departments and the District of Columbia; the Commonwealth of Puerto Rico; the Virgin Islands; Guam; the Federated States of Micronesia; the Republic of the Marshall Islands; the Republic of Palau; the Northern Mariana Islands; American Samoa; the political subdivisions thereof, which may include state universities, state colleges, and state research institutions; and federally recognized Native American tribal governments.
Amount of Grant: $50,000-$500,000; $100,000 average
Contact: Caroline McDonald, Division of Health Studies, Agency for Toxic Substances and Disease Registry, (404) 498-0270; fax: (404) 498-0059; email: COS4@cdc.gov
Internet: http://www.dhhs.gov
Sponsor: Department of Health and Human Services
1600 Clifton Rd NE
Atlanta, GA 30333

DHHS Professional Nurse Traineeships Program 1937
Grants are awarded to eligible institutions to provide financial support through traineeships for registered nurses enrolled in advanced education nursing programs to prepare nurse practitioners, clinical nurse specialists, nurse midwives, nurse anesthetists, nurse administrators, nurse educators, public health nurses and nurses in other specialties determined by the Secretary to require advanced education.
Requirements: Schools of nursing and public health, public or nonprofit private hospitals, and other public or nonprofit entities are eligible for funds. Trainees are selected by participating institutions. A candidate must be a U.S. citizen or have been lawfully admitted to the United States for permanent residence; a graduate of a state-approved school of nursing; licensed as a professional nurse in a state or territory; and able to enroll full-time in a course of study.
Amount of Grant: $1000-$235,000; $45,000 average
Contact: Karen Breeden, Division of Nursing, (301) 443-6333
Internet: http://www.dhhs.gov
Sponsor: Department of Health and Human Services
5600 Fishers Ln, Parklawn Bldg
Rockville, MD 20857

DHHS Promoting Safe and Stable Families Grants 1938
The National Center on Child Abuse and Neglect is interested in research on the impact of community-based family support and family preservation programs on child abuse and neglect. Research should focus on expanding the current knowledge base, build on previous research, and provide insights into new approaches to preventing child maltreatment and preserving families through support and preservation services. The center is particularly interested in projects that address specific populations and outcomes. The populations are families who receive family support services but have had no previous contact with child protective services; families who have been referred to child protective services whose cases were unsubstantiated but were found to need services and were referred to family support programs; families who have been in the system whose child abuse or neglect cases were substantiated, who received family preservation or support services, and whose cases are now closed; and families who have open cases whose children have not been removed and who are receiving family preservation services. Outcomes of interest are case finding, which involves families who were not referred to child protective services, and the impact of family support and/or preservation services on prevention, recidivism, and removal of children. Applicants should plan and design the proposed research in collaboration with state and local CPS and Title IV-B agencies as well as community-based entities providing family support services, such as family resource centers. Contact the office for deadline dates.
Requirements: Agencies of state and local governments, public and private nonprofit agencies, and institutions engaged in child and family welfare activities and research are eligible for these grants.
Amount of Grant: $194,000-$48 million range
Contact: Joseph Bock, Deputy Associate Commissioner, Children's Bureau, (202) 205-8618
Internet: http://www.acf.hhs.gov/programs/cb/programs/fpfs.htm
Sponsor: Department of Health and Human Services
330 C Street, SW
Washington, DC 20447

DHHS Public Health Students Traineeships and Other Graduate Public Health Programs 1939

The programs support traineeships for students in graduate educational programs in schools of public health or other public or nonprofit educational entities that offer graduate programs for training in biostatistics or epidemiology; health administration, health planning, or health policy analysis and planning; environmental or occupational health; dietetics or nutrition; maternal and child health; or approved residency training in preventive medicine or dentistry. Deadlines and application kits are forwarded to all eligible applicants. Annual deadline dates may vary; contact program staff for exact dates.

Requirements: Accredited schools of public health and other public or nonprofit educational entities providing graduate or specialized training in public health may apply. Trainees must be U.S. citizens or noncitizen nationals possessing visas permitting permanent residence in the United States and must be pursuing a graduate degree in an eligible school of public health or be enrolled in an eligible program.

Amount of Grant: $8109-$148,850; $50,150 average

Contact: Cecelia Maryland, Public Health Branch, Division of State, Community and Public Health, (301) 443-1973

Internet: http://www.dhhs.gov

Sponsor: Department of Health and Human Services
5600 Fishers Ln, Parklawn Bldg
Rockville, MD 20857

DHHS Residency Training in Family Practice, General Internal Medicine, or General Pediatrics Grants 1940

Grants are awarded to assist graduate training programs in family medicine, general internal medicine and/or general pediatrics to expand and improve the quality of residency training programs that prepare graduates to enter primary care practice. Residency training programs should emphasize national innovations aimed at primary care residency education across disciplines.

Requirements: An accredited public or private nonprofit school of medicine, a school of osteopathic medicine, and a public or private nonprofit hospital or other entity located in a state is eligible to apply. Each allopathic program must be fully or provisionally approved by the Accreditation Council for Graduate Medical Education. Each osteopathic program must be approved by the American Osteopathic Association.

Restrictions: Grants may not be used for construction, patient services, or student assistance.

Amount of Grant: $27,000-$657,344; $178,803 average

Contact: Dr. P. Preston Reynolds, Division of Medicine and Dentistry of the Bureau of Health Professions, (301) 443-1467

Internet: http://www.dhhs.gov

Sponsor: Department of Health and Human Services
5600 Fishers Ln, Parklawn Bldg
Rockville, MD 20857

DHHS Rural Telemedicine Grants 1941

The objective of this program is to demonstrate and collect information on the feasibility, costs, appropriateness, and acceptability of telemedicine for improving access to health services for rural residents and reducing the isolation of rural practitioners, and to demonstrate how telemedicine can be used as an effective tool for the development of integrated systems of health care for rural areas. Application kits may be obtained by calling toll-free (888)300-4772. Annual deadline dates may vary; contact program staff for exact dates.

Requirements: Eligible applicants are nonprofit public (nonfederal) or private health care providers or a consortium of providers, nonprofit or for-profit, that are members of an existing or proposed telemedicine network.

Restrictions: Not more than 40 percent of grant funds may be expended for equipment. Not more than 20 percent of grant funds may be expended for direct costs. Grant funds may not be used for purchasing and installing telecommunications transmission equipment.

Amount of Grant: $250,000 maximum

Contact: Monica Cowan, Administrative Assistant, Office for the Advancement of Telehealth, (301) 443-1730

Internet: http://www.telehealth.hrsa.gov

Sponsor: Department of Health and Human Services
4350 East-West Hwy
Rockville, MD 20814

DHHS Special Programs for the Aging Training, Research, and Discretionary Projects and Programs Grants 1942

Funds are given to provide adequately trained personnel in the field of aging, improve knowledge of the problems and needs of the elderly, and to demonstrate better ways of improving the quality of life for the elderly. Funds may be used to train persons to work in the field of aging, to increase the availability and accessibility of training and education programs in the field of aging, and to conduct activities for the development of knowledge to improve the circumstances of older people. Deadlines are posted in the Federal Register.

Requirements: Grants may be made to any public or nonprofit private agency, organization, or institution.

Restrictions: Grants are not available to individuals.

Amount of Grant: $250,000 average

Contact: Center for Planning and Policy Development, Administration on Aging, (202) 619-0724

Internet: http://aspe.hhs.gov/SelfGovernance/inventory/Aoa/048.htm

Sponsor: Department of Health and Human Services
330 Independence Avenue SW
Washington, DC 20201

DHHS Tuberculosis Demonstration, Research, Public, and Professional Education Cooperative Agreements 1943

The purpose of the program is to assist states, political subdivisions of states, and other public and nonprofit private entities to conduct research into the prevention and control of tuberculosis, especially research concerning strains of tuberculosis resistant to drugs and research concerning cases of tuberculosis that affect certain populations; conduct demonstration projects for the prevention and control of tuberculosis; provide public information and provide education programs for prevention and control of tuberculosis; and develop education, training, and clinical skills improvement activities in the prevention and control of tuberculosis for health professionals, including allied health personnel. Project periods are for one to five years with 12-month budget periods. Applications will be evaluated on the extent of the tuberculosis problem; the establishment of specific and measurable objectives; and the development of a sound operational plan that will ensure the implementation of each program element. Contact the office for deadline dates.

Requirements: States, political subdivisions of states, and other public and nonprofit private entities are eligible.

Amount of Grant: $150,000-$200,000; $165,000 average

Contact: Dr. Kenneth Castro, National Center for HIV, STD, and TB Prevention, (404) 639-8120; William Ryan, Grants Management Officer, (770) 488-2717

Internet: http://www.dhhs.gov

Sponsor: Department of Health and Human Services
1600 Clifton Rd NE
Atlanta, GA 30333

DHHS Universal Newborn Hearing Screening and Intervention 1944

This program supports the implementation of universal physiologic newborn hearing screening prior to hospital discharge with linkages to medical home, ongoing family-to-family support, diagnostic evaluation by three months of age, and enrollment in a program of early intervention by six months of age for those infants identified with hearing loss. Applicants are expected to notify the Maternal and Child Health Bureau's Division of Services for Children with Special Health Care Needs by November 10; the deadline for receipt of applications is December 8.

Requirements: This program is open to state agencies with the capacity to implement a statewide universal newborn hearing screening and intervention program for all newborn infants in the state.

Amount of Grant: $43,650-$515,382; $161,017. average

Contact: Irene Forsman, Integrated Services Branch, Division of Services for Children with Special Health Needs, (301) 443-2370; email: iforsman@hrsa.gov; Lawrence Poole, Director, Division of Grants Management Operations, (301) 443-0354

Internet: http://www.dhhs.gov

Sponsor: Department of Health and Human Services
5600 Fishers Ln, Parklawn Bldg
Rockville, MD 20857

DHHS Welfare Reform Research, Evaluations, and National Studies Grants 1945

The objectives of the funding are to support research on the benefits, effects, and costs of different welfare reform interventions; and studies such as on the effects of different programs on welfare dependency, illegitimacy, teen pregnancy, employment rates, child well-being, and related areas; and to assist in the development and evaluation of innovative approaches for reducing welfare dependency and increasing the well-being of minority children in welfare families. Grants, cooperative agreements, and contracts are awarded for innovative research, demonstrations, and evaluations that are responsive to the Administration for Children and Families program priorities. Deadlines for grants are announced in the Federal Register.

Requirements: Grants and cooperative agreements may be made to or with governmental entities, colleges, universities, nonprofit, and for-profit organizations (if fee is waived). Contracts may be awarded to nonprofit or for-profit organizations. Grants or cooperative agreements cannot be made directly to individuals.

Amount of Grant: $10,000-$6 million range; $500,000 average

Contact: Karl Koerper, Office of Planning, Research, and Evaluation, Administration for Children and Families, (202) 401-4535; fax: (202) 205-3598; email: KKoerper@acf.dhhf.gov

Internet: http://www.acf.hhs.gov/programs/opre

Sponsor: Department of Health and Human Services
370 L'Enfant Promenade, SW
Washington, DC 20447

DHHS/ASPE Welfare Outcomes Short-Term Policy Research Grants 1946

The purpose of these grants is to support policy relevant research to complement ongoing research and evaluations on the outcomes of welfare reform, and to broaden our understanding of the outcomes of welfare reform. These grants are meant to supplement other leavers grants that ASPE has previously funded and support short-term research and data analysis efforts that are designed to be completed within 12 months. ASPE hopes to support efforts to analyze a variety of information about individuals (adults and children) and their families, including their economic and non-economic well-being and their participation in government programs. ASPE seeks to gain some understanding of the broader issues of the labor market and individual behaviors, such as the differential

effects of the business cycle on subgroups of the eligible population. The identification of important subgroups such as rural residents and individuals with significant barriers to success (e.g., mental illness, domestic violence, substance abuse, illiteracy, people with disabilities) and analyses of outcomes for these groups is also encouraged. Analyses that focus on differential outcomes by race/ethnicity are also encouraged. In addition, while average effects or outcomes are important, it is also important in the context of welfare reform to look at the distribution of the outcomes, to identify and understand the winners and losers, and to understand the reasons why some individuals are winners and others are losers. Thus researchers are encouraged to look beyond averages.
Requirements: Public or private nonprofit organizations may apply.
Amount of Grant: $300,000-$1.2 million; $575,000 average
Contact: Program Contact, Office of the Assistant Secretary for Planning and Evaluation, (202) 690-8794
Internet: http://aspe.os.dhhs.gov
Sponsor: Department of Health and Human Services
200 Independence Avenue SW
Washington, DC 20201

Diaz-Ayala Cuban and Latin American Popular Music Collection Travel Grants   1947
The Diaz-Ayala Cuban and Latin American Popular Music Collection is the most extensive publicly available collection of Cuban music in the United States. The grants will provide graduate students and scholars the opportunity to visit the Diaz-Ayala Music Collection at the FIU Green Library, thereby enhancing its value as a national resource. The program provides a research stipend to offset the costs of a minimum of one-week stay to use the collection. Application and guidelines are available online.
Requirements: Scholars in the humanities and the social sciences whose work will be enhanced by using the resources of the collection are encouraged to apply.
Amount of Grant: $1,500
Date(s) Application is Due: Feb 15
Contact: Alma DeRojas, Coordinator, (305) 348-1991; email: derojasa@fiu.edu
Internet: http://lacc.fiu.edu/centers_institutes/?body=centers_cri_whatsnew&rightbody=centers_cri
Sponsor: Florida International University
Cuban Research Institute, DM 363
Miami, FL 33119

Dibner Institute Postdoctoral Fellows Program   1948
Postdoctoral fellowships are available for advanced research in the history of science and technology. Fellowships run for one year, from September 1 through August 15, and may be extended for a second and final year at the discretion of the institute. Fellows are expected to reside in the Boston area during the fellowship term, to participate in institute activities, and to present their work at appropriate occasions. The institute provides office space, support facilities, full privileges at the libraries of consortium universities as well as Burndy Library, and access to programs, events, and many resources at consortium member institutions, and to the entire spectrum of activities that will take place at the institute.
Requirements: Fellowships are awarded to outstanding young scholars of diverse countries of origin who have obtained the PhD or equivalent within the previous five years.
Date(s) Application is Due: Dec 31
Contact: Trudy Kontoff, Program Coordinator, (617) 253-6989; fax: (617) 253-9858; email: dibner@mit.edu
Internet: http://dibinst.mit.edu/DIBNER/Fellows/FellowsProgram.htm
Sponsor: Dibner Institute for the History of Science and Technology
MIT E56-100, 38 Memorial Drive
Cambridge, MA 02139

Dibner Institute Science Writing Fellowships   1949
This is an opportunity for a senior science writer with a proven track record of writing and/or reporting for a general audience to pursue a substantial project of his or own choosing that bears on history of science or technology. The fellow will pursue his or her project and have the opportunity to interact with the other Dibner fellows and the community of scholars in history of science and technology in the Boston/Cambridge area. He or she will also be able to participate with faculty, students, and fellows in the Graduate Program in Science Writing and the Knight Science Journalism Fellowship program, both at MIT. Fellowships provide a stipend with added expenses negotiable, office space, support facilities, and full privileges at the Burndy Library and at the libraries of the consortium universities. In judging applications, the primary criterion will be the significance of the project. Guidelines are available online.
Requirements: Science writers with a proven track record of writing and/or reporting for a general audience are eligible.
Amount of Grant: $35,000 stipend
Date(s) Application is Due: Dec 31
Contact: Trudy Kontoff, Program Coordinator, (617) 253-6989; fax. (617) 253-9858; email: dibner@mit.edu
Internet: http://dibinst.mit.edu/DIBNER/Fellows/Application/SciWriterFellowship.htm
Sponsor: Dibner Institute for the History of Science and Technology
MIT E56-100, 38 Memorial Drive
Cambridge, MA 02139

Dibner Institute Senior Fellows   1950
The institute is an international center for advanced research in the history of science and technology. Scholars may apply to the program for the fall (beginning September 1), the

spring term (beginning January 1), or both. At the time of application, Term 1 candidates may request an arrival date in August; Term 2 candidates may request an extension into June/July. The institute prefers that senior fellows apply for a two-term, full-year residency if possible. Fellows are expected to reside in the Cambridge/Boston area during the award period, participate in the activities of the Dibner Institute community, and present their current work once during their fellowship appointments.
Requirements: Candidates should have advanced degrees in disciplines relevant to their research and show evidence of substantial scholarly accomplishment and professional experience.
Date(s) Application is Due: Dec 31
Contact: Trudy Kontoff, Program Coordinator, (617) 253-6989; fax: (617) 253-9858; email: dibner@mit.edu
Internet: http://dibinst.mit.edu/DIBNER/Fellows/Application/Introduction.htm
Sponsor: Dibner Institute for the History of Science and Technology
MIT E56-100, 38 Memorial Drive
Cambridge, MA 02139

Dietetic Outcomes Research Study Grants   1951
The program seeks proposals for a prospective, controlled trial study on the effectiveness of nutrition services. The objective is to investigate the cost and clinical effectiveness of nutrition services provided by registered dietitians compared to other health practitioners or health care teams with and without dietitians in an ambulatory setting. The study must focus on one of four disease states: diabetes, hyperlipidemia, congestive heart failure, or obesity with metabolic complications. Key personnel requirements include an experienced research dietitian and ADA members as a principal or co-principal investigator. Specifications are contained in the RFP posted on the Web site.
Amount of Grant: $225,000 maximum
Contact: Elisabeth Puga, (800) 877-1600 ext 4803; email: epuga@eatright.org
Internet: http://www.eatright.org/Public/7713_7808.cfm
Sponsor: American Dietetic Association Foundation
120 S Riverside Plz, Suite 2000
Chicago, IL 60606

Diversifying Higher Education Faculty in Illinois Program (DFI)   1952
The program aims to increase access of underrepresented students to graduate degree programs in Illinois public and private universities and, ultimately, to increase the number of underrepresented faculty and staff in Illinois colleges and universities. Each recipient is awarded an annual stipend and an institutional scholarship that covers tuition and fees. Doctoral recipients may receive awards for up to four years. Those in master's or professional degree programs may receive awards for up to two years.
Requirements: To be considered for an award, an applicant must be an Illinois resident and a United States citizen or permanent resident; a member of an underrepresented group in higher education such as African American, Hispanic, Asian American, or Native American; the recipient of an earned baccalaureate degree; of above-average academic ability as evidenced by admission to a graduate or professional degree program at a participating ICEOP institution; and unable to pursue a graduate or professional degree in the absence of an ICEOP award.
Geographic Focus: Illinois
Amount of Grant: $12,500-$16,000 for full time enrollment
Date(s) Application is Due: Feb 15
Contact: Program Contact, Graduate School, (618) 453-4558; fax: (618) 453-5313; email: fellows@siu.edu
Internet: http://www.dfi.siu.edu
Sponsor: Southern Illinois University at Carbondale
Woody Hall C-224, MC 4723
Carbondale, IL 62901

DOA Agricultural Competitive Research Grants   1953
The grants promote basic research in food, agriculture, and related areas to further the programs of U.S.DA. Funds may be used for costs necessary to conduct research (salaries and wages, scientific equipment, materials and supplies, travel, publication costs, and other allowable direct and indirect costs). Primary responsibility for general supervision of all grant activities rests with the grantee organization; the principal investigator is responsible for the scientific work. Deadlines are announced annually in the Federal Register.
Requirements: State agricultural experiment stations, U.S. colleges/universities, other U.S. research institutions and organizations, federal agencies, private organizations or corporations, and individuals are eligible to submit proposals.
Restrictions: Funds may not be used for purposes other than those specified in the grant.
Amount of Grant: $4000-$5 million; $183,607 average
Contact: Chief Scientist, National Research Initiative Competitive Grants Program, Cooperative State Research, Education, and Extension Service, (202) 401-5022
Internet: http://www.usda.gov
Sponsor: Department of Agriculture
1400 Independence Avenue SW
Washington, DC 20250

DOA Agricultural Special Research Grants   1954
The program supports research to facilitate or expand promising breakthroughs in areas of the food and agricultural sciences of importance to the nation and to facilitate or expand ongoing state-federal food and agricultural research programs. Areas of basic and applied research are generally limited to high-priority problems of a regional or national scope.

Areas currently considered are water quality, integrated pest management, and rangeland research. Application deadlines are announced each fiscal year in the Federal Register and may also be obtained by calling the Office of Extramural Programs below.
Requirements: State agricultural experiment stations, all colleges and universities, other research institutions and organizations, federal agencies, private organizations or corporations, and individuals having a demonstrable capacity to conduct research to facilitate or expand promising breakthroughs in areas of the food and agricultural sciences of importance to the United States are eligible to apply.
Amount of Grant: $56,664-$9.5 million; $46,183 average
Contact: Competitive Programs, (202) 401-5048
Internet: http://www.usda.gov
Sponsor: Department of Agriculture
1400 Independence Avenue SW
Washington, DC 20250

DOA Animal Health and Disease Research Grants   1955
The grants support animal health and disease research to improve the health and productivity of food animals and horses through effective prevention, control, or treatment of disease; to reduce losses from transportation and other hazards; and to protect human health through control of animal diseases transmissible to people. Research can be conducted under the following categories: infectious diseases; internal and external parasites; noninfectious diseases, toxins, poisons, transportation losses, predators, and other hazards; diseases and parasites of wildlife transmissible to food animals and horses; and diseases and parasites of animals transmissible to people. Applications are accepted at any time.
Requirements: Eligibility is restricted to schools and colleges of veterinary medicine and state agricultural experiment stations.
Amount of Grant: $1,586-$408,854; $65,515 average
Contact: Grants Administrator, Cooperative State Research, Education, and Extension Service, (202) 401-4329
Internet: http://www.usda.gov
Sponsor: Department of Agriculture
1400 Independence Avenue SW
Washington, DC 20250

DOA Biological Nitrogen Fixation Research Grants   1956
The grants help build a foundation of basic information concerning nitrogen fixation as it relates to enhancing the process in currently known systems and in providing a base for developing new nitrogen-fixing associations by genetic transfer or other means for crop species not now possessing such capability. Program priorities will be on innovative approaches that may contribute to a thorough understanding of nitrogen cycling encompassing biochemistry, cellular and developmental biology, genetics and genetic manipulation, and other relevant life science disciplines. An understanding of these processes is essential to the development of strategies that maximize nitrogen fixation, minimize inputs of nitrogenous fertilizers, and optimize their utilization in agriculture. Deadlines are announced in the Federal Register, usually November through March.
Requirements: State agricultural experiment stations, U.S. colleges/universities, other U.S. research institutions and organizations, federal agencies, private organizations or corporations, and individuals may submit proposals.
Amount of Grant: $4000-$5 million; $183,607 average
Contact: Chief Scientist, National Research Initiative Competitive Grants Program, (202) 401-5022
Internet: http://www.usda.gov
Sponsor: Department of Agriculture
1400 Independence Avenue SW
Washington, DC 20250

DOA Biological Stress on Plants Research Grants   1957
The grants support research on stresses on plants arising from their interactions with other plants or with other biological agents such as weeds, insects, nematodes, fungi, bacteria, viruses, and mycoplasma-like organisms. The goal is to reduce losses in plant productivity from damage caused by biologically generated stresses. The program seeks to develop an understanding of how stressful interactions are established between plants and other biological agents, how such interactions are influenced by environmental and other factors inherent to the interacting organisms, how the interactions reduce plant productivity and usefulness to man, how plants react to stresses generated by such interactions, and how damage from such interactions may be reduced or eliminated. Deadlines are announced in the Federal Register between December and April.
Requirements: State agricultural experiment stations, U.S. colleges/universities and other research institutions and organizations, federal agencies, private organizations or corporations, and individuals may apply.
Amount of Grant: $4000-$5 million; $183,607 average
Contact: Chief Scientist, National Research Initiative Competitive Grants Program, (202) 401-5022
Internet: http://www.usda.gov
Sponsor: Department of Agriculture
1400 Independence Avenue SW
Washington, DC 20250

DOA Biotechnology Risk Assessment Research Grants   1958
The purpose of this program is to assist agencies in making science-based decisions about genetically modified organisms. Research funded through this program will be relevant to risk assessment and the regulatory process. Although investigators are not required to perform actual risk assessments in the research they propose, they should design studies that will provide information useful to regulators for making science-based decisions in their assessments of genetically-modified organisms. Deadlines are published in the Federal Register.
Requirements: Eligible applicants include any public or private research or educational institution or organization.
Restrictions: Funds may not be used for purposes other than those approved in the grant award documents.
Amount of Grant: $50,220-$223,269; average award $147,583
Contact: Deputy Administrator, Competitive Programs, (202) 401-1761
Internet: http://www.reeusda.gov
Sponsor: Department of Agriculture
1400 Independence Avenue SW
Washington, DC 20250

DOA Cooperative Forestry Research Grants   1959
The grants encourage and assist states in carrying on programs of forestry research at forestry schools and in developing a trained pool of forest scientists capable of conducting needed forestry research. Categories of forestry research supported include reforestation and management of land for the production of crops of timber and other related products of the forest; management of forest and related watershed lands to improve conditions of waterflow and to protect resources against flood and erosion; management of forest and related rangeland for production of forage for domestic livestock and game and improvement of food and habitat for wildlife; management of forest lands for outdoor recreation; protection of forest resources against fire, insects, diseases, and other destructive agents; utilization of wood and other forest products; development of sound policies for the management of forest lands and the harvesting and marketing of forest products; and other studies necessary to obtain the fullest and most effective use of forest resources. Applications are accepted at any time.
Requirements: Applicants must be state or territorial institutions certified as eligible by a state representative designated by the governor.
Amount of Grant: $30,281-$694,074; $322,889 average
Contact: Deputy Administrator, Cooperative State Research Service, (202) 720-4318
Internet: http://www.usda.gov
Sponsor: Department of Agriculture
1400 Independence Avenue SW
Washington, DC 20250

DOA Emerging Markets Program Grants   1960
The purpose of the program is to assist U.S. organizations, public and private, to improve the market access and to develop and promote U.S. agricultural products and/or processes in low- to middle-income countries that offer promise of emerging market opportunities in the near to medium term. This is to be accomplished by providing U.S. technical assistance through projects and activities in those emerging markets consistent with U.S. foreign policy interests. The program funds technical assistance activities to leverage the export and marketing of U.S. agricultural products to emerging markets. The emphasis is on marketing opportunities where there are risks that the private sector would not normally undertake alone, with funding provided on a project-by-project basis. The program is intended to support primarily small- to medium-sized U.S. firms that may need federal assistance in realizing or maintaining access in overseas markets. Application deadline is announced in the Federal Register.
Requirements: Any U.S. agricultural and/or agribusiness organization, university, or state department of agriculture is eligible to participate in the program. Priority will be given to those proposals that include significant support and involvement by private industry. Proposals from research and consulting organizations will be considered if they provide evidence of substantial participation by U.S. industry. U.S. market development cooperators may seek funding to address priority, market-specific issues and to undertake activities not already serviced by or unsuitable for funding under other FAS marketing programs.
Restrictions: No proposal will be considered without the element of cost-sharing.
Amount of Grant: $500,000 maximum
Contact: Emerging Markets Office, Foreign Agricultural Service; (202) 720-4327; fax: (202) 690-4369; email: emo@fas.usda.gov
Internet: http://www.cfda.gov/public/viewprog.asp?progid=1575
Sponsor: Department of Agriculture
6506 S Bldg
Washington, DC 20250

DOA Federal-State Marketing Improvement Program (FSMIP)   1961
The program is a matching fund program designed to assist state departments of agriculture or other appropriate state agencies in conducting studies or developing innovative approaches related to the marketing of agricultural products. Funds can be requested for a wide range of marketing research and marketing service activities, including projects aimed at developing and testing new or more efficient methods of processing, packaging, handling, storing, transporting, and distributing food and other agricultural products; assessing customer response to new or alternative agricultural products or marketing services and evaluating potential opportunities for U.S. producers, processors, and other agribusinesses, in both domestic and international markets; and identifying problems and impediments in existing channels of trade between producers and consumers of agricultural products and devising improved marketing practices, facilities, or systems to address such problems.

Requirements: Only state departments of agriculture or other appropriate state agencies are eligible to apply for funds.
Amount of Grant: $12,000-$74,000; $49,556 average
Contact: Janise Zygmont, Agricultural Marketing Service, (202) 720-2704
Internet: http://www.ams.usda.gov/tmd/fsmip.htm
Sponsor: Department of Agriculture
Dept of Agriculture
Washington, DC 20250

DOA Food and Agricultural Sciences National Needs Graduate Fellowships 1962
This program awards grants to colleges and universities that have superior teaching and research competencies in the food and agricultural sciences to encourage outstanding students to pursue and complete graduate degrees at such institutions in areas of the food and agricultural sciences for which there is a national need for the development of scientific expertise. Therefore, institutions that currently have excellent programs of graduate study and research in these areas dealing with targeted national needs are particularly encouraged to apply. The Cooperative State Research, Education, and Extension Service (CSREES) will provide support on a biennial basis and combine appropriations from two fiscal years into one competition to be held during odd-numbered years. Deadline dates may vary; contact the program office for exact dates.
Requirements: Proposals may be submitted by all U.S. colleges and universities that confer a master's or doctoral degree in at least one area of the food and agricultural sciences targeted for national needs fellowships. Eligibility also applies to research foundations maintained by eligible colleges or universities.
Restrictions: Individuals selected by the institutions may not have been enrolled previously in the program at the same degree level.
Contact: National Program Leader, Higher Education Programs, (202)720-7854
Internet: http://www.reeusda.gov
Sponsor: Department of Agriculture
1400 Independence Avenue SW
Washington, DC 20250

DOA Food and Agricultural Sciences Small Business Innovation Research (SBIR) Grants 1963
Grants are made to stimulate technological innovation in the private sector, strengthen the role of small businesses in meeting federal research and development needs, increase private-sector commercialization of innovations derived from U.S.DA-supported research and development efforts, and foster and encourage minority and disadvantaged participation in technological innovation. Selected areas for research are forests and related resources; plant production and protection; animal production and protection; air, water, and soils; food science and nutrition; rural and community development; aquaculture; and industrial applications. Phase I awards will be made for periods normally not to exceed six months to determine, if possible, the scientific or technical feasibility of ideas in the selected research areas. Phase II awards will be made to firms with approaches that appear sufficiently promising as a result of Phase I studies for a period not to exceed 24 months. Phase III will be nonfederally funded through the exercising of funding commitment to stimulate technological innovation and the national return on investment from research through the pursuit of commercial objectives resulting from the U.S.DA-supported work carried out in Phases I and II. Deadlines are announced in the Federal Register and SBIR program solicitation for each fiscal year.
Requirements: Small businesses that meet the SBIR program specifications may apply.
Amount of Grant: $80,000-$300,000
Contact: Dr. Charles Cleland, National Program Leader, (202) 401-4002; fax: (202) 401-6070; email: ccleland@csrees.usda.gov
Internet: http://www.csrees.usda.gov/fo/fundview.cfm?fonum=1220
Sponsor: Department of Agriculture
1400 Independence Avenue SW
Washington, DC 20250

DOA Forestry Research Grants 1964
The program extends the fundamental research activities of the Forest Service by awarding grants to nonprofit organizations, institutions of higher education, and organizations engaged in renewable resources research. Grants will be used for research in the fields of timber management, watershed management, forest range management, wildlife habitat management, forest recreation, forest fire protection, forest insect and disease protection and control, forest products utilization, forest engineering, forest production economics, forest products marketing, and forest survey. Grants are awarded for one to five years. Contact the regional experiment stations for deadlines.
Requirements: Eligible to apply are state agricultural experiment stations, universities and colleges, state and local governments, U.S. territories, for-profit and nonprofit research institutions and organizations, and international organizations.
Amount of Grant: $2000-$300,000; $35,000 average
Contact: Deputy Chief for Research, Forest Service, (202) 205-1075
Internet: http://www.usda.gov
Sponsor: Department of Agriculture
P.O. Box 96090
Washington, DC 20090

DOA Higher Education Challenge Grants 1965
The objective of the program is to increase institutional capacities to respond to state, regional, national, or international educational needs by strengthening college and university teaching programs in the food and agricultural sciences. Funds may be used only in targeted areas, e.g., curricula design and materials development, faculty

preparation and enhancement for teaching, instruction delivery systems, scientific instrumentation for teaching, student experiential learning, and student recruitment and retention. A dollar-for-dollar match is required from nonfederal sources. Grants are awarded for a one- to three-year period. Deadlines are published in the Federal Register.
U.S.DA-GRANTS-092605-004
Requirements: Any U.S. college or university having a demonstrable capacity to teach the food and agricultural sciences is eligible.
Amount of Grant: $150,000-$400,000
Date(s) Application is Due: Feb 2
Contact: Gregory Smith, National Program Leader, (202) 720 - 2067; fax: (202) 720 - 2030; email: gsmith@csrees.usda.gov
Internet: http://www.csrees.usda.gov/fo/fundview.cfm?fonum=1082
Sponsor: Department of Agriculture
1400 Independence Avenue SW, Stop 2201
Washington, DC 20250-2201

DOA International Agricultural Training Grants 1966
Grants are to assist U.S. colleges and universities in strengthening their capabilities for food, agricultural, and related research training and extension relevant to agricultural development activities in other countries. Projects funded are most often in course development and/or evaluation. In general, cooperative agreements are established for a 12-month period but may be extended with justification. Applications are accepted at any time.
Requirements: U.S. institutions of higher education or nonprofit organizations involved with agricultural development and educational activities are eligible to apply.
Amount of Grant: $10,000-$40,000; average $18,900
Contact: Dr. Frank Fender, FAS/International Cooperation and Development, Food Industries Division, (202) 690-1339
Internet: http://www.fas.usda.gov
Sponsor: Department of Agriculture
2121 K Street NW, 2nd Fl
Washington, DC 20250

DOA International Collaborative Agricultural Research Program Grants 1967
Grant support is provided to U.S. scientists working in cooperation with foreign researchers to implement collaborative research and to maximize the utilization of U.S. agricultural commodities and products in domestic and export markets; to respond quickly to pressing high-priority plant and animal disease or pest problems that have their roots in international origins and may also have trade implications; and to conduct targeted cooperative research with friendly countries having resources or expertise needed to solve urgent U.S. agricultural problems. Projects funded are to be conducted by the United States and cooperating foreign scientists. Collaborating scientists should be identified in the proposal. Generally, cooperative agreements are funded for a 12- to 36-month period. Funding will be provided only to U.S. researchers. Applications are accepted at any time.
Requirements: U.S. institutions of higher education and public/private nonprofit organizations whose primary purpose is scientific research are eligible to apply, including those located in U.S. territories.
Restrictions: The program does not fund the foreign collaborator because foreign institutions are expected to have sufficient interest in collaborating with U.S. scientists to provide for their portion of the research.
Amount of Grant: $10,000-$40,000 average
Contact: Dr. Frank Fender, FASInternational Cooperation and Development, Food Industries Division, (202) 690-1339
Internet: http://www.fas.usda.gov
Sponsor: Department of Agriculture
2121 K Street NW, 2nd Fl
Washington, DC 20250

DOA National Integrated Food Safety Initiative 1968
The purpose of this program is to support competitive projects that address selected priority issues in food safety that are best solved using an integrated approach. Special emphasis is given to research describing multi-functional activities (i.e., research that contains research, education, and extension components). The research component of the National Integrated Food Safety Initiative focuses on applied food safety research. The education component focuses on education and training in a formal classroom setting, which may include elementary, secondary, undergraduate, or graduate education. The extension component addresses education and training outside of the classroom. Where there is no extension program, outreach activities that deliver science-based and informational education to people in a variety of non-formal settings are appropriate. The RFP is usually announced in the Federal Register in February or March of each year, and proposals are due eight weeks following the release of the RFP. Funding opportunity number: U.S.DA-GRANTS-090304-001
Requirements: Faculty at all four year accredited colleges and universities are eligible to apply.
Amount of Grant: $500,000 maximum
Contact: Chris Wozniak , Food Science and Food Safety, (202) 401-6020; fax: (202) 401-6156; email: cwozniak@csrees.usda.gov
Internet: http://www.csrees.usda.gov/fo/fundview.cfm?fonum=1087
Sponsor: Department of Agriculture
1400 Independence Avenue SW
Washington, DC 20250

DOA Photosynthesis Research Grants   1969
The grants support research in aspects of photosynthetic energy conversion including such areas as early events in photon capture by photosynthetic systems and the mechanisms of charge separation, the structure and function of photosynthetic membranes and constituents and associated reactions; photosynthetic carbon assimilation including CO_2 fixation, biochemistry of photosynthetic pathways, photorespiration and aspects of cellular metabolism control of photosynthate partitioning and translocation by hormones and other metabolic factors; factors controlling development and senescence of photosynthetic competence; genetic and cellular manipulation to improve photosynthetic efficiency in plants; and the photosynthetic process in leaves and whole plants including involvement of the stomatal aperture, and water and temperature extremes. Deadlines are announced each fiscal year in the Federal Register, usually falling between November and March.
Requirements: State agricultural experiment stations, U.S. colleges/universities and other research institutions and organizations, federal agencies, private organizations or corporations, and individuals may apply.
Amount of Grant: $4000-$5 million; $183,607 average
Contact: Chief Scientist, National Research Initiative Competitive Grants Program, (202) 401-5022
Internet: http://www.usda.gov
Sponsor: Department of Agriculture
1400 Independence Avenue SW, Ag Box 2241
Washington, DC 20250

DOA Plant and Animal Disease, Pest Control, and Animal Care Grants   1970
The objectives of this program are to protect U.S. agriculture from economically injurious plant and animal diseases and pests, ensure the safety and potency of veterinary biologics, and ensure the humane treatment of animals. Project and training grants are awarded to eligible organizations to carry out these objectives. Initial contact should be a letter outlining the project proposed.
Requirements: Eligible to apply are foreign, state, local, and U.S. territorial government agencies; nonprofit institutions of higher education; and nonprofit associations or organizations requiring federal support to eradicate, control, or assess the status of injurious plant and animal diseases and pests that are a threat to regional or national agriculture and to conduct related demonstration projects.
Contact: Anita Ridley, Budget and Accounting Division, Animal and Plant Health Inspection Service, (301) 734-8792
Internet: http://www.aphis.usda.gov
Sponsor: Department of Agriculture
4700 River Road, Unit 55, Sta 4B80
Riverdale, MD 20737

DOA Rangeland Research Grants   1971
Applications are solicited for projects covering research on rangeland, both open and contained. Project areas may include multiple forage factors; soil and water erosion; climate, wind, and rain; and growth replenishment. Also supported are innovative projects of interest to all those involved in rangeland use. Deadline dates vary; contact the Office of Extramural Programs for exact dates.
Requirements: Land-grant colleges and universities; state agricultural experimental stations; and college, university, and federal laboratories having a demonstrable capacity in rangeland research are eligible to apply.
Amount of Grant: $56,664-$9.5 million; $46,183 average
Contact: Competitive Programs, (202) 401-5048
Internet: http://www.reeusda.gov
Sponsor: Department of Agriculture
1400 Independence Avenue SW, Stop 2245
Washington, DC 20250

DOA Resource Conservation and Development Grants   1972
The grants are intended to encourage and improve the capability of state and local units of government and local nonprofit organizations in rural areas to plan, develop, and carry out programs for resource conservation and development. Assistance is available for the planning and installation of approved measures in RC&D areas for land conservation, water management, community development, and environmental enhancement. Applications are accepted at any time.
Requirements: State and local governments and nonprofit organizations with authority to plan or carry out activities relating to resource use and development in multijurisdictional areas are eligible to apply. The program is also available in Puerto Rico, the Virgin Islands, Guam, and the Mariana Islands.
Contact: Terry D'Addio, National RC&D Program Manager, (202) 720-0557; email: terry.d'addio@usda.gov
Internet: http://www.nrcs.usda.gov/programs/rcd
Sponsor: Department of Agriculture
1400 Independence Avenue SW
Washington, DC 20250

DOA Rural Business Enterprise Grants   1973
Grants are made to finance and facilitate development of small and emerging private business enterprises located in areas outside the boundary of a city or unincorporated areas of 50,000 or more and its immediately adjacent urbanized or urbanizing area. Costs that may be paid from grant funds include the acquisition and development of land and the construction of buildings, plants, equipment, access streets and roads, parking areas, and utility and service extensions; refinancing; fees for professional services; technical assistance and related training for adults; startup operating costs and working capital, providing financial assistance to a third party; and production of television programs to provide information to rural residents and to create, expand, and operate rural distance learning networks. Grants may also be made to establish or fund revolving loan programs.
Requirements: Eligibility is limited to public bodies, private nonprofit corporations, and federally recognized Indian tribal groups. Small and emerging businesses with less than 50 new employees and less than $1 million in gross annual revenues are eligible for assistance.
Amount of Grant: $2000-$500,000; $83,309 average
Contact: Program Contact, Rural Development Specialist, (202) 720-1400
Internet: http://www.rurdev.usda.gov/rbs/busp/rbeg.htm
Sponsor: Department of Agriculture
1400 Independence Avenue SW
Washington, DC 20250

DOA Small Business Innovation Research (SBIR) Grants   1974
Firms with strong scientific research capabilities in the topic areas listed below are encouraged to participate. Objectives of the three-phase program include stimulating technological innovation in the private sector, strengthening the role of small businesses in meeting federal research and development needs, increasing private sector commercialization of innovations derived from U.S.DA-supported research and development efforts, and fostering and encouraging participation of women-owned and socially and economically disadvantaged small business concerns in technological innovation. Research areas include forests and related resources; plant production and protection; animal production and protection; air, water, and soils; food science and nutrition; rural and community development; aquaculture; industrial applications; and marketing and trade. The deadline for Phase I is August 31; for Phase II is February 5.
Amount of Grant: $80,000 maximum for Phase I; $300,000 maximum for Phase II
Date(s) Application is Due: Feb 2; Sep 1
Contact: Dr. Charles Cleland, Director, SBIR Program, Cooperative State Research, Education, and Extension Service, (202) 401-4002; fax: (202) 401-6070; email: ccleland@reeusda.gov
Internet: http://www.csrees.usda.gov/fo/fundview.cfm?fonum=1220
Sponsor: Department of Agriculture
1400 Independence Avenue SW
Washington, DC 20250

DOA Sustainable Agriculture Research and Education Grants   1975
The intent of these grants is to facilitate and increase scientific investigation and education to reduce the use of chemical pesticides, fertilizers, and toxic materials in agricultural production; improve management of on-farm resources to enhance productivity, profitability, and competitiveness; promote crop, livestock, and enterprise diversification and facilitate the conduct of research projects to study agricultural production systems that are located in areas that possess various soil, climatic, and physical characteristics; study farms that have been and continue to be managed using farm production practices that optimize the use of on-farm resources and conservation practices; take advantage of the experience of farmers and ranchers through their direct participation and leadership in projects; transfer practical, reliable, and timely information to farmers and ranchers concerning low-input sustainable practices and systems; and promote a partnership between farmers, nonprofit organizations, agribusiness, and public and private research and extension institutions. Funds may be used for transportation, per diem, salaries, office supplies, printing, and other direct costs for conducting activities approved in cooperative agreements or interagency reimbursable transfers.
Requirements: Land-grant colleges or universities, other universities, state agricultural experiment stations, nonprofit organizations, or federal or state governmental entities that have demonstrated appropriate expertise in agricultural research and technology transfer may apply.
Restrictions: Funds may not be used to pay indirect costs or tuition.
Amount of Grant: $8000-$1.7 million; $855,540 average
Contact: Administrator, Cooperative State Research, Education, and Extension Service, (202) 720-7948
Internet: http://www.reeusda.gov
Sponsor: Department of Agriculture
1400 Independence Avenue SW
Washington, DC 20250

DOA Technical Agricultural Assistance Research Grants   1976
The objective of the program is to identify and apply the most appropriate solutions to international agricultural problems and to increase the capabilities of U.S. educational institutions and nonprofit agencies in agricultural research and technical assistance. Generally, cooperative agreements are funded for a 12- to 24-month period. Applications are accepted at any time.
Requirements: U.S. institutions of higher learning and public/private nonprofit organizations whose primary purpose is scientific research are eligible to apply, including those in U.S. territories.
Amount of Grant: $30,000-$600,000; $160,000 average
Contact: Dr. Howard Anderson, FAS/International Cooperation and Development Office, Development Resources Division, (202) 690-1924
Internet: http://www.fas.usda.gov
Sponsor: Department of Agriculture
2121 K Street NW, 2nd Fl
Washington, DC 20250

DOA Wildlife Services Grants 1977

Grants are given to reduce damage caused by mammals and birds and those mammal and bird species that are reservoirs for zoonotic diseases, except for urban rodent control. Wherever feasible, humane methods will be employed. Recipients are to work closely with state departments of fish and game, agriculture, health, and counties in joining efforts to alleviate wild animal damage. Recipients are expected to conduct surveys and campaigns to reduce wild animal damage, including bird problems at airports; develop methods to control wild animal damage; and provide technical advice and assistance. For direct technical assistance, state fish and game departments should be contacted.
Requirements: State and local governments, federally recognized Indian tribal governments, public/private nonprofit organizations, nonprofit institutions of higher education, and individuals are eligible to apply.
Contact: Anita Ridley, Budget and Accounting Division, Animal and Plant Health Inspection Service, (301) 734-8792
Internet: http://www.aphis.usda.gov/ws
Sponsor: Department of Agriculture
4700 River Road, Unit 55, Sta 4B80
Riverdale, MD 20737

DOC National Technical Assistance: Training, Research, & Evaluation Grants 1978

The purpose of the program is to provide grants and cooperative agreements for technical assistance projects that are useful in creating or retaining jobs and promoting economic growth. Awards are made for national technical assistance projects, local technical assistance projects, and university center projects. Supported activities include feasibility studies; management and operational assistance; demonstration projects; administrative support for local, regional, and national nonprofit economic development organizations; and other forms of technical assistance and support.
Requirements: Eligible applicants are private individuals, firms, colleges, universities, and other institutions; profit and nonprofit organizations are also eligible.
Amount of Grant: $12,000-$209,000 average range
Contact: John McNamee, Director, Research and National Technical Assistance Division, (202) 482-4085
Internet: http://www.commerce.gov/grants.html
Sponsor: Department of Commerce
14th Street and Constitution Avenue NW, Herbert C. Hoover Bldg
Washington, DC 20230

DOC NERRS Graduate Research Fellowships 1979

The National Estuarine Research Reserve System (NERRS) supports research and training opportunities for graduate students in estuarine ecology. The program is intended to fund high-quality research focused on improving coastal zone management while providing students with hands-on training in conducting ecological monitoring. Fellowships are offered for up to three years and require development of training programs with on-site staff at the NERRS sites. Research projects must focus on one of the following issues: non-point source pollution, biodiversity and the effects of invasive species, estuarine ecosystem restoration, and mechanisms for sustaining estuarine resources. Annual deadline dates may vary; contact the program office or visit the Web site for specific dates and application materials.
Requirements: Students admitted to or enrolled in full-time master's or doctoral programs at U.S.-accredited universities are eligible to apply. Students should have completed a majority of their course work at the beginning of their fellowships and have approved thesis research programs. Fellowships must be matched by the applicant by at least 30 percent of the total project cost.
Amount of Grant: $20,000 per year for up to three years
Date(s) Application is Due: Nov 1
Contact: Susan White, Research Coordinator, (301) 713-3155 ext 224; email: susan. white@noaa.gov
Internet: http://nerrs.noaa.gov/Fellowship
Sponsor: Department of Commerce
1305 East-West Hwy, N/ORM5, SSMC4, Sta 10500
Silver Spring, MD 20910

DOC Special American Business Internship Training (SABIT) Grants 1980

The SABIT program awards funding to qualified U.S. companies for training business executives and scientists from the New Independent States of the former Soviet Union. SABIT exposes NIS business managers and scientists to a completely new way of thinking in which demand, consumer satisfaction, and profits drive production. Mid-to senior-level interns visiting the U.S. for internship programs with public or private sector companies will be exposed to an environment that will provide them with practical knowledge for restructuring their enterprises. The program provides firsthand, eye-opening experience to managers and scientists that cannot be duplicated by American managers traveling to their countries. Annual deadline dates may vary; contact the program office for exact dates.
Requirements: Any profit or nonprofit U.S. corporation, association, organization or other public or private entity is eligible.
Restrictions: Agencies or divisions of the federal government are not eligible.
Amount of Grant: $8400-$40,400 range; $18,000 average
Contact: Program Contact, SABIT Program, (202) 482-0073; fax: (202) 482-2443
Internet: http://www.mac.doc.gov/sabit
Sponsor: Department of Commerce
14th Street and Constitution Avenue NW, Herbert C. Hoover Bldg
Washington, DC 20230

DOC Special American Business Internship Training (SABIT): 1981
Sakhalin-Oil and Gas Internship Program

The SABIT program awards funding to qualified U.S. companies for training business executives and scientists from the New Independent States (NIS) of the former Soviet Union. SABIT exposes NIS business managers and scientists to a completely new way of thinking in which demand, consumer satisfaction, and profits drive production. Mid-to senior-level interns visiting the United States for internship programs with public or private sector companies will be exposed to an environment which will provide them with practical knowledge for restructuring their enterprises. The program provides first-hand, eye-opening experience to managers and scientists which cannot be duplicated by American managers traveling to their territories.
Requirements: All for-profit or nonprofit U.S. corporations, associations, organizations, or other public or private entities are eligible to apply. Interns must be from the Sakhalin region in Russia. At the end of the training, interns must return to Sakhalin, Russia.
Restrictions: Agencies or divisions of the federal government are not eligible.
Amount of Grant: $8400-$40,400; $18,000 average
Contact: Tracy Rollins, Director, (202) 482-0073; fax: (202) 482-2443
Internet: http://www.mac.doc.gov/sabit
Sponsor: Department of Commerce
1401 Constitution Avenue NW, 4th Fl, 4100W
Washington, DC 20230

DOC Technology Administration Fellows Grants 1982

The purpose of the ComSci program is to provide a hands-on learning experience for participants and to increase their understanding of: technological innovation as a source of national and international economic growth; the relationship of science and technology to government policies on economics, trade, education, and fiscal matters; the organization of scientific and technological activities in the federal government; and the technical activities and problems that exist in other executive, legislative, and judicial agencies of the government. Two program options are offered: ComSci program with a work assignment— a 10-month intensive and challenging work assignment in an executive or legislative branch office (preferably different from one's home agency) in the Washington metropolitan area; and ComSci program without a work assignment—a nine-month program offering all of the benefits of the program, without the full-time work assignment. This option allows individuals in the Washington metropolitan area to participate on a part-time basis. Participants continue to work at their home agencies throughout the duration of the program while attending ComSci events. For participants in both program options, the program combines intensive education and orientation activities.
Requirements: The program provides federal government employees in a professional or management series with an opportunity to study national and international issues relating to the development, application, and management of science and technology. Candidates are selected through recommendation procedures developed by participating agencies. In general, recommendations are based on candidates' qualifications, accomplishments, and potential for career growth. All agencies of the federal government may participate.
Amount of Grant: $3800 for option with a work assignment, and $3400 for option without a work assignment), and expenses associated with the one-week field trip (estimated to be about $1900)
Date(s) Application is Due: Nov 21
Contact: Cynthia Lynn, (202) 482-6103; fax: (202) 482-4306; email: cynthia.lynn@ technology.gov
Internet: http://www.technology.gov/comsci/Intro.htm
Sponsor: Department of Commerce
14th Street and Constitution Avenue NW, Room 4823
Washington, DC 20230

DOC/ITA Market Development Cooperator Program (MDCP) 1983

The goal of the MDCP is to develop, maintain, and expand foreign markets for nonagricultural goods and services produced in the United States. The intended beneficiaries of the program are U.S. producers of nonagricultural goods or services that seek to export such goods or services. ITA encourages applicants to propose activities that would be most appropriate to the market development needs of their U.S. industry or industries. Examples of activities might include: opening an overseas office or offices to perform a variety of market development services for companies joining a consortium to avail themselves of such services; detailing a private-sector representative to a U.S.&FCS post in accordance with 15 U.S.C 4723(c); commissioning overseas market research, participating in overseas trade exhibitions, and trade missions to promote U.S. exports, and/or hosting reverse trade missions; conducting U.S. product demonstrations abroad; and other eligible activities. The applicant must contribute at least two dollars for each federal dollar provided. Full guidelines are available on the Web site or may be obtained by contacting the office listed.
Requirements: U.S. trade associations; nonprofit industry organizations; state trade departments and their regional associations, including centers for international trade development; and private industry firms or groups of firms in cases where no entity described above represents that industry are eligible.
Amount of Grant: $400,000 maximum; $298,000 average
Date(s) Application is Due: Jul 1
Contact: Brad Hess, Manager, Trade Development, International Trade Administration, (202) 482-2969; fax: (202) 482-5828; email: Brad_Hess@ita.doc.gov
Internet: http://www.ita.doc.gov/mdcp
Sponsor: Department of Commerce
14th Street and Constitution Avenue NW, Herbert C Hoover Bldg
Washington, DC 20230

DOD Breast Cancer Center of Excellence Award 1984

The award mechanism supports multidisciplinary, multi-institutional teams of gifted scientists, clinicians, and consumer advocates in making groundbreaking advances toward the eradication of breast cancer. The Breast Cancer Research Program (BCRP) encourages highly accomplished scientists renowned for their contributions to the proposed areas of research and promising young investigators who can provide fresh insight to work together to accelerate the solution of a central, overarching research problem in a way that could not be accomplished by a single investigator or group. In addition, the BCRP strongly endorses the integration of scientists from nontraditional disciplines such as computer science, mathematics, economics, physics, and other quantitative disciplines. The BCRP also encourages centers of excellence to incorporate study components addressing ethical issues in breast cancer research. Breast cancer consumer/survivor groups must be active participants in all aspects of these awards. The listed application deadline is for preproposals; full proposals are by invitation. Preproposals must be submitted through the CDMRP eReceipt system. Guidelines are available online.

Requirements: All individuals, regardless of ethnicity, nationality, or citizenship status, may apply as long as they are employed by, or affiliated with, an eligible institution. Eligible institutions include for-profit, nonprofit, public, and private organizations. Agencies of local, state, and federal governments are eligible to the extent that proposals do not overlap with their fully funded intramural programs. Federal agencies will be expected to explain how their proposals do not overlap with their intramural programs.

Amount of Grant: $20 million maximum
Date(s) Application is Due: Apr 5
Contact: Commander, U.S. Army Medical Research and Materiel Command, (301) 619-7079; fax: (301) 619-7792; email: cdmrp.pa@det.amedd.army.mil
Internet: http://cdmrp.army.mil/funding/bcrp.htm
Sponsor: Department of Defense
1077 Patchel Street (Building 1077), MCMR-ZB-C (BC05-COE)
Fort Detrick, MD 21702-5024

DOD Clinical Translational Research Grants 1985

The award is designed to sponsor innovative research that will result in substantial improvements over current approaches to breast cancer chemoprevention and therapy by accelerating the progression of recent, highly promising findings in preclinical breast cancer research from the laboratory to the clinic. These awards are intended to support both new and established scientists across a broad spectrum of disciplines. Partnerships between academic institutions and biotechnology companies are encouraged. Proposals are being sought only in the areas of chemoprevention and therapeutics. Successful applicants must initiate a prospective clinical trial and accrue participants for a minimum of one year during the award period. They must also include preliminary data to support the feasibility of their hypotheses and approaches. Funding for these awards may be requested for up to five years. The listed application deadline is for preproposals; full proposals are by invitation. Preproposals must be submitted through the CDMRP eReceipt system. Guidelines are available online.

Requirements: All individuals, regardless of ethnicity, nationality, or citizenship status, may apply as long as they are employed by, or affiliated with, an eligible institution. Eligible institutions include for-profit, nonprofit, public, and private organizations. Agencies of local, state, and federal governments are eligible to the extent that proposals do not overlap with their fully funded intramural programs. Federal agencies will be expected to explain how their proposals do not overlap with their intramural programs.

Restrictions: This award is not intended to support early drug discovery or development, correlative studies, or the study of new combinations of standard breast cancer therapies.

Amount of Grant: $20 million total
Date(s) Application is Due: Apr 5
Contact: Commander, U.S. Army Medical Research and Materiel Command, (301) 619-7079; fax: (301) 619-7792; email: cdmrp.pa@det.amedd.army.mil
Internet: http://cdmrp.army.mil/funding/bcrp.htm
Sponsor: Department of Defense
1077 Patchel Street (Building 1077), MCMR-ZB-C (BC05-CTR)
Fort Detrick, MD 21702-5024

DOD Era of Hope Scholar Award 1986

The Breast Cancer Research Program sponsors this grant to identify individuals with high potential for innovation in breast cancer research early in their careers. Candidates should be exceptionally talented, early-career scientists who have demonstrated that they are the best and brightest in their field(s) through extraordinary creativity, vision, and productivity. They should also exhibit strong potential for leadership in the breast cancer research community that supports a vision for the eradication of breast cancer. These individuals should challenge current dogma and demonstrate an ability to look beyond tradition and convention. Award recipients will be expected to successfully challenge the status quo through creative, high-risk research that may lead ultimately to the eradication of breast cancer. Proposals must focus on the candidate's emerging record of creative and original accomplishments and potential for leadership in the field. Experience in breast cancer is not a requirement; however, the proposal must focus on breast cancer, and the applicant must commit a minimum of 50 percent effort during the award period to breast cancer research. All nominations must be converted into an electronic PDF file for electronic submission. Nominations guidelines are available online.

Requirements: Applicants must be nominated to be considered for this award. Applicants must be independent, nonmentored investigators within six years of their last training position. All individuals, regardless of ethnicity, nationality, or citizenship status, may apply as long as they are employed by, or affiliated with, an eligible institution. Eligible

institutions include for-profit, nonprofit, public, and private organizations. Agencies of local, state, and federal governments are eligible to the extent that proposals do not overlap with their fully funded intramural programs.

Restrictions: Self-nominations will not be accepted. Postdoctoral fellows, clinical fellows (including residents and interns), and other mentored researchers are not eligible.
Amount of Grant: $2.5 million for up to five years
Date(s) Application is Due: Apr 5
Contact: Commander, U.S. Army Medical Research and Materiel Command, (301) 619-7079; fax: (301) 619-7792; email: cdmrp.pa@det.amedd.army.mil
Internet: http://cdmrp.army.mil/bcrp/era/default.htm
Sponsor: Department of Defense
1077 Patchel Street (Building 1077), MCMR-ZB-C (BC05-EHSA)
Fort Detrick, MD 21702-5024

DOD HBCU/MI Partnership Training Award 1987

This program's goal challenges the scientific community to design innovative research that will foster new directions for, address neglected issues in, and bring new investigators to the field of breast cancer research. The program focuses its funding on innovative projects, particularly those involving multidisciplinary and/or multi-institutional collaborations and alliances that have the potential to make a significant impact on breast cancer. Proposals that address the needs of minority, low-income, rural, and other underrepresented and/or medically underserved populations are strongly encouraged. The listed application deadline is for preproposals. Preproposals must be submitted through the CDMRP eReceipt system. Guidelines are available online.

Requirements: Applicants must be HBCU/MI doctoral-level faculty members. All individuals, regardless of ethnicity, nationality, or citizenship status, may apply as long as they are employed by, or affiliated with, an eligible HBCU/MI institution. Eligible institutions are those approved as HBCU/MIs by the Department of Education.
Contact: Commander, U.S. Army Medical Research and Materiel Command, (301) 619-7079; fax: (301) 619-7792; email: cdmrp.pa@det.amedd.army.mil
Internet: http://cdmrp.army.mil/bcrp/default.htm
Sponsor: Department of Defense
1077 Patchel Street (Building 1077), MCMR-ZB-C (BC05-HPT)
Fort Detrick, MD 21702-5024

DOD Idea Award for Breast Cancer Research 1988

The award supports innovative, high-risk/high-reward breast cancer research. Proposals should describe new paradigms and challenge existing dogma in the study of breast cancer. The award also supports efforts to examine existing problems from a new perspective; however, these proposals will face a greater burden to demonstrate innovation. Special consideration will be given to proposals that focus on biological signal integration, and/or systems implications. Proposals must be submitted electronically at https://cdmrp.org/. This website will contain all the information, forms, documents, and links needed to apply.

Requirements: All individuals, regardless of ethnicity, nationality, or citizenship status, may apply as long as they are employed by, or affiliated with, an eligible institution. Eligible institutions include for-profit, nonprofit, public, and private organizations. Agencies of local, state, and federal governments are eligible to the extent that proposals do not overlap with their fully funded intramural programs.

Amount of Grant: $300,000 maximum for up to three years; $625,000 maximum for population-based studies
Date(s) Application is Due: May 24
Contact: Commander, U.S. Army Medical Research and Materiel Command, (301) 619-7079; fax: (301) 619-7792; email: cdmrp.pa@det.amedd.army.mil
Internet: http://cdmrp.army.mil/bcrp/default.htm
Sponsor: Department of Defense
1077 Patchel Street (Building 1077), MCMR-ZB-C (BC05-IDEA)
Fort Detrick, MD 21702-5024

DOD Multidisciplinary Postdoctoral Award for Breast Cancer Research 1989

The award is designed to identify and fund exceptionally talented recent doctoral graduates committed to broadening the scope of their research by adding significant, mentored training in more than one major discipline. Mentors should be identified for each discipline represented in the proposed project. Major disciplines for this award include, but are not limited to, laboratory, clinical, social/behavioral science, and public health research. The Breast Cancer Research Program is especially interested in proposals that integrate training in disciplines currently underrepresented in breast cancer, such as engineering, physics, and mathematics; however, innovative proposals that incorporate any combination of the major disciplines listed above will be considered. Proposals must be submitted electronically at https://cdmrp.org. Guidelines are available online.

Requirements: Eligible applicants should have been in the laboratory or research setting in which the proposed research is to be performed for no longer than two years at the time of submission and should have a total of less than five years of postdoctoral research experience (excluding clinical residency or fellowship training). All individuals, regardless of ethnicity, nationality, or citizenship status, may apply as long as they are employed by, or affiliated with, an eligible institution as defined below. Eligible institutions include for-profit, nonprofit, public, and private organizations. Agencies of local, state, and federal governments are eligible to the extent that proposals do not overlap with their fully funded intramural programs.

Amount of Grant: $375,000 maximum
Date(s) Application is Due: May 24
Contact: Commander, U.S. Army Medical Research and Materiel Command, (301) 619-7079; fax: (301) 619-7792; email: cdmrp.pa@det.amedd.army.mil

Internet: http://cdmrp.army.mil/funding/bcrp.htm
Sponsor: Department of Defense
1077 Patchel Street (Building 1077), MCMR-ZB-C (BC05-MPA)
Fort Detrick, MD 21702-5024

DOD National Defense Science and Engineering Graduate Fellowships　1990
The NDSEG program is a DOD fellowship program sponsored by AFOSR, the Army Research Office, the Office of Naval Research, and the Advanced Research Projects Agency. The DOD selects about 90 fellows per year; the Air Force sponsors about 25 of those fellows. AFOSR has a goal of awarding 10 percent of these fellowships to applicants who are members of an ethnic minority group underrepresented in the advanced levels of the U.S. science and engineering personnel pool (i.e., Native American, African American, Hispanic, Native Alaskan, or Native Pacific Islander). These fellowships are for study and research in areas of interest to the Air Force. Stipends are prorated for fellowship periods of less than 12 months; however, the duration of the fellowship will not be less than nine months. In addition to the stipend, the Air Force pays the student's tuition and fees. Those fellows selected and sponsored by the Air Force will be offered the opportunity to become associated with an Air Force laboratory, but they are not required to spend a summer at an Air Force laboratory.
Requirements: Applicants must be U.S. citizens who have received their baccalaureate degrees. Air Force graduate fellowships are tenable at any U.S. institution of higher education offering a PhD in science or engineering. Fellowships are awarded for study and research in mathematical, physical, biological, ocean, and engineering sciences. Preference is given to applicants in aeronautical and astronautical engineering; biosciences (including toxicology); chemical engineering; chemistry; cognitive, neural, and behavioral sciences; computer science; electrical engineering; mathematics; mechanical engineering; naval architecture and ocean engineering; oceanography; and physics (including optics).
Amount of Grant: $18,000 first year; $19,000 second year; $20,000 third and fourth years
Date(s) Application is Due: Jan 20
Contact: Dr. George Outterson, NDSEG Fellowship Program, (919) 549-8505; fax: (919) 549-8205; email: ndseg@aro-emh1.army.mil
Internet: http://www.acq.osd.mil/ddre/researchtest/opportunities.html
Sponsor: Department of Defense
P.O. Box 13444, 200 Park Drive, Suite 211
Research Triangle Park, NC 27709-3444

DOD National Security Education Program (NSEP) Scholarships　1991
The purpose of this undergraduate scholarship program is to equip Americans with an understanding of less commonly taught languages and cultures; to build a critical base of future leaders both in the marketplace and in government service; to develop a cadre of professionals with more than the traditional knowledge of language and culture; and to enhance institutional capacity and increase the number of faculty who can educate U.S. citizens toward achieving these goals.
Requirements: Any U.S. citizen enrolled in an accredited two or four year public or private U.S. institution of higher education is eligible to apply.
Restrictions: Students enrolled in federal government schools are not eligible.
Amount of Grant: $8000 maximum per term, not to exceed two terms per year
Contact: Dr. Edmond Collier, Department of Defense, Office of the Secretary of Defense, (703) 696-1991; email: collier@nsep.policy.osd.mil
Internet: http://www.defenselink.mil
Sponsor: Department of Defense
1101 Wilson Boulevard
Arlington, VA 22209

DOD Predoctoral Traineeship in Breast Cancer Research　1992
The award prepares promising graduate students with a strong commitment to breast cancer research for successful and competitive careers. This award promotes creative approaches to training breast cancer investigators. Successful trainees will be selected for their talent, potential, and commitment to breast cancer research; the mentor's qualifications and experience in breast cancer research; the strong breast cancer research training program at the trainee's institution; the innovative research environment; and the institution's commitment to training future leaders in breast cancer research. The proposed training must focus on breast cancer research. Historically Black Colleges and Universities/Minority Institutions are highly encouraged to apply. Proposals must be submitted electronically at https://cdmrp.org. Guidelines are available online.
Requirements: Applicants must be graduate students under the guidance of a designated mentor. Individuals enrolled in a PhD or MD/PhD program are encouraged to apply. All individuals, regardless of ethnicity, nationality, or citizenship status, may apply as long as they are employed by, or affiliated with, an eligible institution. Eligible institutions include for-profit, nonprofit, public, and private organizations. Examples include universities, colleges, hospitals, laboratories, and companies. Agencies of local, state, and federal governments are eligible to the extent that proposals do not overlap with their fully funded intramural programs.
Amount of Grant: $30,000 per year for a maximum of three years
Date(s) Application is Due: May 24
Contact: Commander, U.S. Army Medical Research and Materiel Command, (301) 619-7079; fax: (301) 619-7792; email: cdmrp.pa@det.amedd.army.mil
Internet: http://cdmrp.army.mil/funding/archive/05bcrppredoc_pa.pdf
Sponsor: Department of Defense
1077 Patchel Street (Building 1077), MCMR-ZB-C (BC05-PREDOC)
Fort Detrick, MD 21702-5024

DOE Academies Creating Teacher Scientists (ACTS) Grants　1993
This program is based upon research and best practices in teacher professional development and is aligned with the National Standards for Science Education, the National Teaching Standards, and National Board for Professional Teaching Standards. Instruction is offered to middle and high school teachers in four content areas: biotechnology-genetic engineering; fusion-astrophysics; and energy technologies. Each of these programs teaches knowledge and skills derived from science research and provides a context that when shared with students can help them understand how classroom science is linked to the big science. Teachers will have access to a wealth of mentoring talent that will guide and enrich their understanding of the scientific and technological world. Participants and mentors will create supportive relationships that can follow them into their classrooms. The Teacher Researchers (TR) model provides the teacher with four-weeks of hands-on instruction in a content discipline of their choosing in their initial year followed by two, eight-week summer research internships. Teachers in K-12 classrooms and community college faculty of science, technology, engineering, or mathematics are encouraged to apply. Housing allowances, transportation, and stipends are offered to participants at all the laboratories.
Requirements: All applicants must: be teaching full time in a public or private elementary or secondary school and continue to be full time employees of a school system throughout the duration of the program; have a current teaching assignment with at least 3/4 of classroom contact hours in science, mathematics, and/or technology (applies to secondary school teachers and community college instructors only); be at least 21 years old at the time the appointment starts; be U.S. citizens at the time of application; have current health insurance coverage; and commit to a minimum of 3 years involvement.
Amount of Grant: $800 per week
Date(s) Application is Due: Apr 1
Contact: Todd Clark; (202) 586-7174; fax: (202) 586-0019; email: todd.clark@science.doe.gov; Cindy Musick; (202) 586-0987; fax: (202) 586-0019; email: cindy.musick@science.doe.gov
Internet: http://www.scied.science.doe.gov/SciEd/LSTPD/about.htm
Sponsor: U.S. Department of Energy, Office of Science
1000 Independence Avenue SW
Washington, DC 20585

DOE Advanced Detector Research Grants　1994
The purpose of this program is to support the development of the new detector technologies needed to perform future high-energy physics experiments. Such experiments will require higher performance detectors to exploit the higher beam energies and intensities of new or upgraded accelerators. Higher performance detectors are also needed to probe for new physical processes in both accelerator- and nonaccelerator-based-experiments. Proposed detector research should be driven by the anticipated needs of experiments to be built within the foreseeable future, as well as upgrades to current experiments. Interesting technologies would include but not be limited to charged particle track detectors, calorimeters, or particle identification detectors that are less sensitive to radiation, have higher resolution, are lower in cost, or can be read out faster than currently available detectors. Applicants are requested to submit a Letter of Intent (LOI) by November 15 which includes the title of the proposal, the name of the principal investigator(s), the requested funding, and a one-page abstract. Failure to submit a letter of intent will not negatively prejudice a responsive formal application submitted in a timely manner. Full applications are due by December 12.
Requirements: Investigators who are currently involved in experimental high-energy physics are eligible to apply and should do so through a U.S. academic institution.
Amount of Grant: $10,000-$600,000
Date(s) Application is Due: Nov 15; Dec 12
Contact: Dr. Saul Gonzalez, Program Manager; (301) 903-2359; fax: (301) 903-2597; email: Saul.Gonzalez@science.doe.gov
Internet: http://www.science.doe.gov/grants/FAPN07-33.html
Sponsor: U.S. Department of Energy
19901 Germantown Road
Germantown, MD 20874-1290

DOE Advanced Energy Projects/Laboratory Technology Research Grants　1995
The Advanced Energy Projects program funds research to establish the feasibility of novel, energy-related concepts. These concepts are usually derived from recent advances in basic research, but require additional research to establish their feasibility. A common theme for each concept is the initial linkage of new, or previously neglected, research results to a practical energy payoff. The Laboratory Technology Research program conducts high-risk, energy-related research that advances fundamental science and technology toward innovative applications that could significantly impact the United States' energy economy. Scientists at the Office of Science laboratories enter into cost-shared research partnerships with industry to explore energy applications of research advances in areas of mission relevance to both parties. The partners jointly bring technology research to a point where industry or the department's technology development programs can pursue final development or commercialization. Applications are available on the Web site and may be submitted at any time.
Amount of Grant: $10,000-$2 million; $200,000 average
Date(s) Application is Due: May 31
Contact: Martin Rubinstein, Grants and Contracts Division, Office of Science, (301) 903- 5212
Internet: http://www.er.doe.gov/production/grants/guide.html
Sponsor: Department of Energy
19901 Germantown Road
Germantown, MD 20874

DOE Advanced Nuclear Energy Systems Basic Research Grants   1996

Areas of focus in this program include: understanding of nanoscale interactions under extreme conditions; mastering the behavior of actinides and of fission products; solution behavior under extreme conditions of radiation and temperature; and interfacial behavior under extreme environmental conditions. Research funded under this initiative will pursue breakthroughs in scientific understanding that will advance materials design, will improve characterization of materials and processes, will enhance chemical processes under the extreme conditions present in nuclear energy systems, and will extend interdisciplinary theory-modeling-simulation-experimentation methodology to surmount the existing scientific and technical barriers for nuclear energy systems of the future. Pre-applications are required, and are due no later than November 22. A response to the pre-applications encouraging or discouraging formal applications will be communicated to the applicants by January 4, and formal applications are due March 14.

Requirements: Eligible to apply are institutions of higher education, industry (particularly small and disadvantaged businesses), and nonprofit institutions. Unsolicited research proposals are screened by program officials and, if appropriate, are evaluated by peer review. Successful proposers are awarded one-year or multi-year special research contracts or grants with renewals or incremental funding available, provided performance has been satisfactory. A one- to two-year phase-out period is normally granted to those finishing research projects. Equipment items may be purchased from contract funds with the agreement of DOE.

Amount of Grant: $10,000-$2 million; $200,000 average
Date(s) Application is Due: Nov 22; Mar 14
Contact: Dr. Lester Morss; (301) 903-9311; email: lester.morss@science.doe.gov; Dr. Tim Fitzsimmons; (301) 903-9830; email: tim.fitzsimmons@science.doe.gov
Internet: http://www.science.doe.gov/grants/FAPN07-04.html
Sponsor: U.S. Department of Energy, Office of Basic Energy Sciences (BES)
19901 Germantown Road
Germantown, MD 20874-1290

DOE Albert Einstein Distinguished Educator Fellowship Program   1997

The Albert Einstein Distinguished Educator Fellowship Act was signed into law in November 1994. The law gives the Department of Energy responsibility for administering the program of distinguished educator fellowships for elementary and secondary school mathematics and science teachers. Selected teachers spend up to one year in a Congressional Office or a federal agency. Agencies that have participated include: the Department of Energy (DOE), the National Science Foundation (NSF), the National Aeronautics and Space Administration (NASA), the National Institutes of Health (NIH), the Department of Education (ED), National Institute of Standards and Technology (NIST), the White House Office of Science and Technology Policy (OSTP) and the National Oceanic and Atmospheric Administration (NOAA). The Fellows provide their educational expertise, years of experience and personal insights to these offices. Costs for relocation will be reimbursed based on the distance of current location to Washington DC. During the Fellowship, awardees receive funds for travel to conferences, workshops, and other professional development opportunities.

Requirements: To be eligible, applicants must be U.S. citizens at the time of selection; have a minimum of five years full-time classroom teaching experience; be teaching full-time in a public or private elementary or secondary school; have a current teaching assignment with at least 3/4 of classroom contact hours in science, mathematics, and/or technology (applies to secondary school teachers only); and provide a recommendation from a current school administrator and two additional recommendations.

Amount of Grant: $5,000 per month stipend
Date(s) Application is Due: Jan 13
Contact: Todd Clark; (202) 586-7174; fax: (202) 586-0019; email: todd.clark@science.doe.gov; Cindy Musick; (202) 586-0987; fax: (202) 586-0019; email: cindy.musick@science.doe.gov
Internet: http://www.scied.science.doe.gov/scied/Einstein/about.htm
Sponsor: U.S. Department of Energy, Office of Science
1000 Independence Avenue SW
Washington, DC 20585

DOE Atomic, Molecular, and Optical Sciences Research Grants   1998

The AMOS Program supports a balanced portfolio of experiment and theory to study the fundamental properties of atoms, ions and small molecules and the interactions between electrons, photons and ions in collisions with atoms, molecules and surfaces. Research is focused on the most complete quantum mechanical description of these properties and interactions and is intended to provide a basic understanding of physical processes. The AMOS program plays an underpinning role in relation to other programs within BES in Chemical and Materials Sciences, in relation to current and future BES facilities in which matter is probed with photons, electrons or heavy ions, and in relation to applied efforts in plasma science. AMOS also contributes at the most fundamental level to the science-based optimization of current energy sources and the development of new ones. Some topics of current interest: studies of the interactions of intense electromagnetic fields, induced by highly charged ions or lasers, with atoms and molecules; coherent control of quantum mechanical processes; development and application of ultrafast x-ray light sources, both laser-based and at BES synchrotron facilities; and theory and experiment on ultracold collisions and quantum condensates.

Requirements: Eligible to apply are institutions of higher education, industry (particularly small and disadvantaged businesses), and nonprofit institutions. Unsolicited research proposals are screened by program officials and, if appropriate, are evaluated by peer review. Successful proposers are awarded one-year or multi-year special research contracts or grants with renewals or incremental funding available, provided performance has been

satisfactory. A one- to two-year phase-out period is normally granted to those finishing research projects. Equipment items may be purchased from contract funds with the agreement of DOE.

Amount of Grant: $10,000-$2 million; $200,000 average
Contact: Dr. Eric A. Rohlfing, Director; (301) 903-8165; fax: (301) 903-4110; email: Eric.Rohlfing@science.doe.gov
Internet: http://www.science.doe.gov/bes/chm/Programs/programs.html
Sponsor: U.S. Department of Energy, Office of Science
1000 Independence Avenue, SW
Washington, DC 20585-1290

DOE Basic Energy Sciences (BES) Grants   1999

The mission of the program is to foster and support fundamental research to expand the scientific foundations for new and improved energy technologies and for understanding and mitigating the environmental impacts of energy use. Research is sponsored in a variety of disciplines to broaden the energy supply and technological base knowledge. The disciplines represented are biosciences, chemical sciences, geosciences, engineering, and materials sciences. The program also supports exploratory research on advanced energy projects. Research projects are selected for funding on the basis of scientific merit, the possible relevance to meeting BES long-range research goals, and the contribution toward a balanced, responsive research program. The detailed substance of the program is determined by the selection from among many unsolicited proposals of ideas from the university/scientific community. Applications are accepted at any time.

Requirements: Eligible to apply are institutions of higher education, industry (particularly small and disadvantaged businesses), and nonprofit institutions. Unsolicited research proposals are screened by program officials and, if appropriate, are evaluated by peer review. Successful proposers are awarded one-year or multiyear special research contracts or grants with renewals or incremental funding available, provided performance has been satisfactory. A one- to two-year phase-out period is normally granted to those finishing research projects. Equipment items may be purchased from contract funds with the agreement of DOE.

Amount of Grant: $10,000-$2 million; $200,000 average
Contact: Martin R. Rubinstein, Supervisor of Grants and Contracts; (301) 903-4946; fax: (301) 903-4194; email: Martin.Rubinstein@science.doe.gov
Internet: http://www.science.doe.gov/grants/FAPN08-01.html
Sponsor: U.S. Department of Energy, Office of Basic Energy Sciences
19901 Germantown Road
Germantown, MD 20874-1290

DOE Catalysis and Chemical Transformations Research Grants   2000

This Program supports basic research to understand the chemical aspects of catalysis, both heterogeneous and homogeneous; the chemistry of fossil resources; and the chemistry the molecules used to create advanced materials. Catalysts are crucial to energy conservation in creating new, less-energy- demanding routes for the production of basic chemical feedstocks and value-added chemicals. Catalysts are also indispensable for processing and manufacturing fuels that are a primary means of energy storage. Results from a fundamental, molecular-level understanding of the syntheses of advanced catalytic materials have the potential of providing new chemicals or materials that can be fabricated with greater energy efficiency or function as energy-saving media themselves. This activity is the Nation's major supporter of catalysis research, and it is the only activity that treats catalysis as a discipline integrating all aspects of homogeneous and heterogeneous catalysis research.

Requirements: Eligible to apply are institutions of higher education, industry (particularly small and disadvantaged businesses), and nonprofit institutions. Unsolicited research proposals are screened by program officials and, if appropriate, are evaluated by peer review. Successful proposers are awarded one-year or multi-year special research contracts or grants with renewals or incremental funding available, provided performance has been satisfactory. A one- to two-year phase-out period is normally granted to those finishing research projects. Equipment items may be purchased from contract funds with the agreement of DOE.

Amount of Grant: $10,000-$2 million; $200,000 average
Contact: Dr. Eric A. Rohlfing, Director; (301) 903-8165; fax: (301) 903-4110; email: Eric.Rohlfing@science.doe.gov
Internet: http://www.science.doe.gov/bes/chm/Programs/programs.html
Sponsor: U.S. Department of Energy, Office of Science
1000 Independence Avenue, SW
Washington, DC 20585-1290

DOE Chemical Energy and Chemical Engineering Research Grants   2001

This activity supports research on electrochemistry, thermophysical and thermochemical properties, and physical and chemical rate processes. Emphasis is given to improving and/or developing the scientific base for engineering generalizations and their unifying theories. Also included is fundamental research in areas critical to understanding the underlying limitations in the performance of electrochemical energy storage and conversion systems including anode, cathode, and electrolyte systems and their interactions with emphasis on improvements in performance and lifetime. The program covers a broad spectrum of research including fundamental studies of composite electrode structures; failure and degradation of active electrode materials; thin film electrodes, electrolytes, and interfaces; and experimental and theoretical aspects of phase equilibria, especially of mixtures, including supercritical phenomena.

Requirements: Eligible to apply are institutions of higher education, industry (particularly small and disadvantaged businesses), and nonprofit institutions. Unsolicited research proposals are screened by program officials and, if appropriate, are evaluated by peer

review. Successful proposers are awarded one-year or multi-year special research contracts or grants with renewals or incremental funding available, provided performance has been satisfactory. A one- to two-year phase-out period is normally granted to those finishing research projects. Equipment items may be purchased from contract funds with the agreement of DOE.
Amount of Grant: $10,000-$2 million; $200,000 average
Contact: Dr. Eric A. Rohlfing, Director; (301) 903-8165; fax: (301) 903-4110; email: Eric.Rohlfing@science.doe.gov
Internet: http://www.science.doe.gov/bes/chm/Programs/programs.html
Sponsor: U.S. Department of Energy, Office of Science
1000 Independence Avenue, SW
Washington, DC 20585-1290

DOE Chemical Physics Research Grants 2002
The program supports basic research on fundamental molecular processes related to the mission of the Department in such areas as combustion, catalysis, and environmental restoration. It is the Nation's principal supporter of high temperature chemical kinetics and gas phase chemical physics. Specific areas of research emphasis include, but are not limited to: gas phase chemical reaction theory, computational chemistry, experimental dynamics and spectroscopy, thermodynamics of reaction intermediates, chemical kinetics and reaction mechanisms at high temperatures in the gas phase and at surfaces, combustion diagnostics, and chemical dynamics and kinetics at surfaces and with metal and semiconductor clusters.
Requirements: Eligible to apply are institutions of higher education, industry (particularly small and disadvantaged businesses), and nonprofit institutions. Unsolicited research proposals are screened by program officials and, if appropriate, are evaluated by peer review. Successful proposers are awarded one-year or multi-year special research contracts or grants with renewals or incremental funding available, provided performance has been satisfactory. A one- to two-year phase-out period is normally granted to those finishing research projects. Equipment items may be purchased from contract funds with the agreement of DOE.
Amount of Grant: $10,000-$2 million; $200,000 average
Contact: Dr. Eric A. Rohlfing, Director; (301) 903-8165; fax: (301) 903-4110; email: Eric.Rohlfing@science.doe.gov
Internet: http://www.science.doe.gov/bes/chm/Programs/programs.html
Sponsor: U.S. Department of Energy, Office of Science
1000 Independence Avenue, SW
Washington, DC 20585-1290

DOE Chemical Sciences Research Grants 2003
The Chemical Sciences program supports a major portion of the nation's fundamental research in the chemical sciences. The research covers those areas of the chemical sciences that impact the Department's energy and environmental missions. Research supported covers atomic, molecular and optical (AMO) sciences; chemical physics; photo- and radiation chemistry; surface chemistry and heterogeneous catalysis; organometallic chemistry and homogeneous catalysis; analytical and separation science; heavy element chemistry; and aspects of chemical engineering sciences. This research provides a foundation for fundamental understanding of the interactions of atoms, molecules, and ions with photons and electrons; the making and breaking of chemical bonds in gas phase, in solutions, at interfaces, and on surfaces; and understanding the energy transfer processes within and between molecules. Applications are accepted at any time.
Requirements: Eligible to apply are institutions of higher education, industry (particularly small and disadvantaged businesses), and nonprofit institutions. Unsolicited research proposals are screened by program officials and, if appropriate, are evaluated by peer review. Successful proposers are awarded one-year or multi-year special research contracts or grants with renewals or incremental funding available, provided performance has been satisfactory. A one- to two-year phase-out period is normally granted to those finishing research projects. Equipment items may be purchased from contract funds with the agreement of DOE.
Amount of Grant: $10,000-$2 million; $200,000 average
Contact: Dr. Eric A. Rohlfing, Director; (301) 903-8165; fax: (301) 903-4110; email: Eric.Rohlfing@science.doe.gov
Internet: http://www.science.doe.gov/bes/chm/chmhome.html
Sponsor: U.S. Department of Energy, Office of Science
1000 Independence Avenue, SW
Washington, DC 20585-1290

DOE Coal Liquefaction Research Grants 2004
The objectives of the program are to develop advanced technology and an engineering and operations database sufficient for the production of marketable liquid fuels from coal by the year 2015 at a cost competitive with crude oil, between $25 and $30 per barrel. The implementation strategy is based on catalytic staged direct liquefaction, coprocessing of coal/waSuite and petroleum resid, slurry-phase indirect liquefaction technology, and advanced processes based on exploratory research. Areas covered by the program include direct/indirect coal liquefaction processes, coprocessing of coal-petroleum mixture/coal-waSuite mixture, refining and chemicals, and supporting research. Proposals are accepted at any time.
Requirements: States, local governments, universities, governmental entities, consortia, nonprofit institutions, commercial corporations, joint federal/industry corporations, U.S. territories, and individuals are eligible to apply.
Amount of Grant: $50,000-$400,000 range; $154,000 average
Contact: Fred Glaser, Office of Advanced Research, Assistant Secretary of Fossil Energy, (301) 903-2786

Internet: http://www.fe.doe.gov
Sponsor: Department of Energy
19901 Germantown Road, MS FE-3
Germantown, MD 20874

DOE Combustion Systems Research Grants 2005
The program emphasizes the development of technologies which would allow more cost-effective and efficient coal use. The objectives of this program include developing the technology necessary for electric power applications with lower cost, increased efficiency, and emission significantly lower than current NSPS. Applications are accepted at any time.
Requirements: States, local governments, universities, governmental entities, consortia, nonprofit institutions, commercial corporations, joint federal/industry corporations, U.S. territories, and individuals are eligible to apply.
Amount of Grant: $10,000-$25 million range
Contact: Mary Roland, Department of Energy, Fossil Energy Program, (301) 903-3514
Internet: http://www.fe.doe.gov
Sponsor: Department of Energy
19901 Germantown Road, MS FE-3
Germantown, MD 20874

DOE Competitive Financial Assistance for Renewable Energy and Energy 2006
Efficiency Programs
DOE announces a competitive solicitation for information dissemination, public outreach, training, and related technical analysis and technical assistance activities involving renewable energy and energy efficiency. Areas of interest involving renewable energy include wind, photovoltaic, hydrogen, and bioenergy technologies. Energy efficiency areas of interest include energy efficiency in the transportation, buildings, and industrial sectors.
Requirements: Profit organizations, nonprofit institutions and organizations, state and local governments, universities, individuals, Native American organizations, and Alaskan Native corporations are eligible to apply.
Amount of Grant: $6 million total
Contact: James Damm, Contract Specialist, fax: (303) 275-4788; email: gostate@nrel.gov
Internet: http://www.tgci.com/fedrgtxt/01-10503.txt
Sponsor: Department of Energy
1000 Independence Avenue SW
Washington, DC 20585

DOE Computational Science Graduate Fellowships 2007
Sponsored by the DOE Office of Science and Office Defense Programs and administered by the institute, the program supports highly capable students pursuing graduate study at U.S. universities in scientific or engineering disciplines with applications in high-performance computing. Fellows also participate in off-campus research at DOE laboratories. The program offers an annual stipend and payment of tuition and fees for graduate study in scientific and technical disciplines using computational science methods. Appointments are reviewed annually and may be renewed up to a limit of four years.
Requirements: The program is open to U.S. citizens and permanent residents who are in their first or second year of graduate school and working toward a PhD at the time of application. Exceptional senior undergraduates who can meet all of the requirements listed in the application booklet also may apply.
Amount of Grant: $28,000 annual stipend
Date(s) Application is Due: Jan 11
Contact: Graduate Fellowships, (515) 956-3696; email: csgf@krellinst.org
Internet: http://www.krellinst.org/csgf
Sponsor: Krell Institute
1609 Golden Aspen Drive, Suite 101
Ames, IA 50010

DOE Conservation Research and Development Grants 2008
Grants are awarded to programs to conduct a balanced, long-term research effort in the areas of buildings, industry, and transportation. Grants are offered to develop and transfer various energy conservation technologies to the nonfederal sector. The program is willing to suggest potential areas for new research initiatives. Applications are accepted at any time.
Requirements: Profit organizations, private nonprofit organizations, and state and local governments are eligible to apply.
Amount of Grant: $50,000-$500,000 average
Contact: Polly Perando, Office of Building Technology, State, and Community Programs, (202) 586-2300; email: polly.perando@ee.doe.gov
Internet: http://energy.gov/engine/content.do?BT_CODE=ST_SS11
Sponsor: Department of Energy
1000 Independence Avenue SW, Forrestal Bldg
Washington, DC 20585

DOE Control Technology and Coal Preparation Research Grants 2009
The activities of the program emphasize the development of technologies which will allow greater use of coal in an environmentally acceptable manner. These technologies include coal characterization and cleaning, combined flue gas cleanup, processes for

cleanup of hot coal combustion and gasification streams, and management of wastes generated in coal conversion and utilization. Applications are accepted at any time.
Requirements: States, local governments, universities, governmental entities, consortia, nonprofit institutions, commercial corporations, joint federal/industry corporations, U.S. territories, and individuals are eligible to apply.
Amount of Grant: $10,000-$25 million
Contact: Mary Roland, Department of Energy, (301) 903-3514
Internet: http://www.fe.doe.gov
Sponsor: Department of Energy
19901 Germantown Road, MS FE-3
Germantown, MD 20874

DOE Electric Energy Systems Research Grants 2010
The program assists the private sector in the development of advanced technology options for the nation's electric energy networks; methodologies for integrating new technologies into electric utility systems; and analytical procedures to assess and enhance the stability and efficiency of the country's increasingly complex interconnected electric energy system. Supported are innovative research in such areas as high-voltage systems, new materials, and sophisticated control methods. Basic research is supported on increasing the efficiency of electric energy supply and distribution, thus reducing primary energy use per unit of real GNP. Subprograms include reliable electric power delivery, electric field effects, and systems technologies. Proposals are accepted at any time.
Amount of Grant: $50,000-$500,000
Contact: Office of Utility Technologies, (202) 586-4142
Internet: http://www.eere.energy.gov
Sponsor: Department of Energy
1000 Independence Avenue SW, Forrestal Bldg
Washington, DC 20585

DOE Energy Biosciences Research Grants 2011
To provide financial support for fundamental research in the basic sciences and advanced technology concepts, and assessments in fields related to energy. Financial support, in whole or in part, may be provided for such purposes as the salaries, materials and supplies, equipment, travel, publication costs, and services required for conducting research, related activities, and advanced technology projects or assessments. Restrictions on use of funds depend on grant provisions. Applications may be submitted at any time. Specific grant solicitation notices, issued from time to time, usually contain due dates. Applicants are encouraged to contact the office regarding specific grant due dates and other requirements.
Requirements: Colleges and universities, non-profit organizations, for-profit commercial organizations, State and local governments, and unaffiliated individuals.
Amount of Grant: $10,000-$2 million; $200,000 average
Contact: Martin Rubinstein, Grants and Contracts Division, Department of Energy, (301) 903-5212
Internet: http://www.cfda.gov/public/viewprog.asp?progid=875
Sponsor: Department of Energy
19901 Germantown Road
Germantown, MD 20874

DOE Energy Conservation Research and Development Grants 2012
This program supports scientific and technical research in fields where new knowledge can expand the generic technology base underlying energy conservation. This will allow the private sector to develop ways to make more efficient use of available energy and to facilitate use of nonpetroleum energy sources. Programs include basic and applied research in fields of scientific inquiry that can provide a general technology base that the private sector will be able to use in developing advanced conservation applications and actually getting these technology applications into the marketplace. Applications are accepted at any time.
Requirements: Eligible applicants include profit organizations, private nonprofit institutions/organizations, and state and local governments.
Amount of Grant: $50,000-$500,000
Contact: Office of Transportation Technologies, (202) 586-6715; Office of Building Technology, State, and Community Programs, (202) 586-2300; Office of Industrial Technologies, (202) 586-0098
Internet: http://www.eere.energy.gov
Sponsor: Department of Energy
1000 Independence Avenue SW, Forrestal Bldg
Washington, DC 20585

DOE Energy Conversion Technology Research Grants 2013
Under the Energy Conversion and Utilization Technologies Program, the energy conversion technology subprogram's problem areas that require additional research include thermodynamics and fluid mechanics of closed system energy conversion, combustion processes of internal combustion engines, and physical and chemical aspects of alternative fuels for furnaces and boilers. Potential areas for new research include development of a complete understanding of velocities and turbulence levels at various points during engine operation and how they can be controlled to appropriately tailor combustion processes of advanced engine concepts; complete incorporation of technology base information and development of validated multidimensional large-scale computer models of complete engines for use in solving specific design problems; assemblage of an adequate database of physical and chemical properties for select alternative fuels and experimental evaluation of several combustion parameters of interest to designers of furnaces and boilers; and experimental and analytical investigation of thermodynamic

and fluid dynamic aspects of advanced closed-cycle systems and components. Proposals are accepted at any time.
Amount of Grant: $50,000-$500,000
Contact: Office of Transportation Technologies, (202) 586-6715; Office of Building Technology, State, and Community Programs, (202) 586-2300; Office of Industrial Technologies, (202) 586-0098
Internet: http://www.eere.energy.gov
Sponsor: Department of Energy
1000 Independence Avenue SW, Forrestal Bldg
Washington, DC 20585

DOE Energy from Municipal WaSuite Projects Grants 2014
Grants support research to provide a database on liquids and gas production and to provide a basis for environmental and feedstock preparation. Potential areas for new research initiatives include thermochemical studies to provide a generic technology base for systems producing liquid and gaseous fuels; combustion tasks concerned with kinetics, efficiency, and products formation for corrosion and pollution control; pyrolysis tasks concerned with base data development on effects of various feedstock components and of time, temperature, pressure, and so forth; biochemical studies to provide data on biochemistry and life support systems of anaerobes; water and wastewater studies to provide information to improve energy efficiency in treatment systems; mechanical studies to provide generic data on systems design; and improved densification techniques and binders. Proposals are accepted at any time.
Amount of Grant: $50,000-$500,000
Contact: Office of Utility Technologies, (202) 586-4142
Internet: http://www.eere.energy.gov
Sponsor: Department of Energy
1000 Independence Avenue SW, Forrestal Bldg
Washington, DC 20585

DOE Energy Research Analyses Grants 2015
This program supports energy research analyses of the department's basic and applied research activities. Specific objectives include assessments to identify any duplication or gaps in scientific research activities, and impartial and independent evaluations of scientific and technical research efforts. Applications are accepted at any time and are available on the Web site.
Requirements: Profit organizations, nonprofit institutions, intrastate, interstate, local agencies, and universities are eligible for funding. Preapplication coordination is recommended for unsolicited proposals.
Amount of Grant: $10,000-$2 million; $200,000 average
Contact: Martin Rubinstein, Grants and Contracts Division, Office of Science, (301) 903- 5212
Internet: http://www.er.doe.gov/production/grants/guide.html
Sponsor: Department of Energy
1000 Independence Avenue SW, Forrestal Bldg
Washington, DC 20585

DOE Energy-Related Inventions Grants 2016
Grants encourage innovation in developing nonnuclear energy technology by providing assistance in the development of promising energy-related inventions. Assistance provided includes evaluation of energy-related inventions, limited funding assistance where appropriate, advice concerning engineering, marketing, and business planning.
Requirements: There are no restrictions on eligibility. Small businesses, individual inventors, and entrepreneurs are especially invited to participate.
Amount of Grant: $50,000-$500,000
Contact: Lisa Barnett, Department of Energy, (202) 586-2212
Internet: http://www.eere.energy.gov/inventions
Sponsor: Department of Energy
1000 Independence Avenue SW, Forrestal Bldg
Washington, DC 20585

DOE Engineering Research Grants 2017
The objectives of the engineering research program are to extend the body of knowledge underlying current engineering practice in order to open new ways for enhancing energy savings and production, prolonging useful equipment life, and reducing costs while maintaining output and performance quality; and to broaden the technical and conceptual base for solving future engineering problems in the energy technologies. Long-term research topics of current interest include foundations of bioprocessing fuels and energy related wastes; fracture mechanisms; experimental and theoretical studies of multiphase flows; intelligent machines; and diagnostics and control for plasma processing of materials.
Amount of Grant: $10,000-$2 million; $200,000 average
Date(s) Application is Due: May 31
Contact: Martin Rubinstein, Grants and Contracts Division, Office of Science, (301) 903- 5212
Internet: http://www.er.doe.gov/production/grants/guide.html
Sponsor: Department of Energy
19901 Germantown Road
Germantown, MD 20874-1290

DOE Environmental Processes Research Grants 2018
This program addresses global environmental change from increases in atmospheric carbon dioxide and other greenhouse gases. The scope of the global change program

encompasses the carbon cycle, climate modeling and diagnostics, atmospheric sciences and meteorology, ecosystem responses, and impacts on resources. The role of clouds and radiation in climate prediction is a particular emphasis. Applications are available on the Web site and are accepted at any time.
Amount of Grant: $10,000-$2 million; $200,000 average
Contact: Martin Rubinstein, Grants and Contracts Division, Office of Science, (301) 903- 5212
Internet: http://www.er.doe.gov/production/grants/guide.html
Sponsor: Department of Energy
19901 Germantown Road
Germantown, MD 20874

DOE Environmental Remediation Grants   2019
The objectives of the program relate to environmental processes affected by energy production and use. The program develops information on the physical, chemical, and biological processes that cycle and transport energy-related material, particularly contaminates that arose during nuclear weapons production, through the earth's surface and subsurface. Emphasis is put on the development of a strong basis for understanding and implementing the appropriate and efficient use of bioremediation, particularly at the department's sites. Applications are accepted at any time and are available on the Web site.
Amount of Grant: $10,000-$2 million range; $200,000 average
Contact: Martin Rubinstein, Grants and Contracts Division, Office of Science, (301) 903- 5212
Internet: http://www.science.doe.gov/production/grants/grants.html
Sponsor: Department of Energy
19901 Germantown Road
Germantown, MD 20874

DOE Experimental Program to Stimulate Competitive Research   2020
The objective of the EPSCoR program is to enhance the capabilities of EPSCoR states to conduct nationally competitive energy-related research and to develop science and engineering manpower to meet current and future needs in energy-related fields. This program addressees research needs across all of the Department of Energy research interests. Research supported by the EPSCoR program is concerned with the same broad research areas addressed by the Office of Science programs that are described above. States eligible for DOE/EPSCoR support include: Alabama, Alaska, Arkansas, Delaware, Hawaii, Idaho, Kansas, Kentucky, Louisiana, Maine, Mississippi, Montana, Nebraska, Nevada, New Mexico, North Dakota, Oklahoma, South Carolina, South Dakota, Tennessee, Vermont, West Virginia, Wyoming, Commonwealth of Puerto Rico, and the Virgin Islands. Applicants are encouraged, but not required, to submit a preliminary proposal to DOE by April.
Geographic Focus: Alabama
Amount of Grant: $750,000 maximum
Date(s) Application is Due: Sep 21
Contact: Dr. Matesh Varma, Office of Science, (301) 903-3209; fax: 301-903-9513; email: matesh.varma@science.doe.gov
Internet: http://www.science.doe.gov/bes/EPSCoR/OVER1.HTM
Sponsor: Department of Energy
19901 Germantown Road
Germantown, MD 20874-1290

DOE Faculty and Student Teams (FaST) Program Grants   2021
The program provides hands-on research opportunities in DOE national laboratories during the summer, and will support a team comprised of one faculty member and 2 - 3 undergraduate students. The faculty member identifies a mutually beneficial research area amenable to collaboration by the faculty member and the laboratory scientist. Potential areas of collaboration are based upon the Project Descriptions described at the specific DOE Office of Science laboratory. Faculty from colleges and universities with limited research facilities and those institutions serving populations, women, and minorities underrepresented in the fields of science, engineering, and technology are encouraged to apply. The amounts targeted for the student stipends are $4,500 for each student (allocated as ten weekly stipends of $400, and up to $500 for travel), and faculty stipends up to 2/9 academic year salary (up to $12,000) for faculty team members. Both faculty and student team members will receive funding assistance with travel. Student team members will receive funding assistance with housing.
Requirements: Faculty applicants must be United States Citizens or Permanent Resident Aliens. Student team members must: be currently enrolled as an undergraduate student and completed at least one semester of college course work; be 18 years or older at the start of the program; have earned a high school diploma or GED; be United States Citizens or Permanent Resident Aliens; participate in a maximum of two FaST internships; and have coverage under a health insurance plan.
Amount of Grant: $4,500 maximum per student; $12,000 maximum per faculty member
Date(s) Application is Due: Feb 1
Contact: Todd Clark; (202) 586-7174; fax: (202) 586-0019; email: todd.clark@science. doe.gov; Cindy Musick; (202) 586-0987; fax: (202) 586-0019; email: cindy.musick@ science.doe.gov
Internet: http://www.scied.science.doe.gov/scied/PST/about.htm
Sponsor: U.S. Department of Energy, Office of Science
1000 Independence Avenue SW
Washington, DC 20585

DOE Fossil Advanced Research and Technology Development Grants   2022
The Advanced Research and Technology Development (AR&TD) Program is directed toward the scientific and technical areas that underlie the development of all fossil energy coal technologies. The AR&TD coal science program conducts fundamental research into coal combustion and conversion and addresses scientific and engineering problems that are barriers to fossil energy technological goals. Areas covered by current projects include materials research; fluids and solids handling; instrumentation and diagnostics; bioprocessing of coal; component development; structure and reactions of coal; multiphase flow and thermodynamics; fundamental problems in combustion of pulverized coal, synfuels, and coal-liquid mixtures; environmental aspects of coal use and conversion; control technology and coal preparation; coal gasification; and coal liquefaction. Applications are accepted at any time.
Requirements: States, local governments, universities, governmental entities, consortia, nonprofit institutions, commercial corporations, joint federal/industry corporations, U.S. territories, and individuals are eligible to apply.
Amount of Grant: $10,000-$25 million average
Contact: Mary Roland, Department of Energy, Fossil Energy Program, (301) 903-3514
Internet: http://www.fe.doe.gov
Sponsor: Department of Energy
19901 Germantown Road, MS FE-3
Germantown, MD 20874

DOE Fossil Energy Research and Development Grants   2023
The fossil energy program's role is to support long-term research toward providing an improved capability to convert coal and oil shale to liquid and gaseous fuels; increase domestic production of coal, oil, and gas; ensure that current and new facilities that burn coal can do so in an economically feasible and environmentally acceptable manner; and allow more efficient and more economically attractive utilization of fossil energy resources. Applications are accepted at any time.
Requirements: States, local governments, universities, recognizant industrial governments, U.S. territories, and individuals are eligible to apply.
Amount of Grant: $10,000-$25 million
Contact: Mary Roland, Department of Energy, Fossil Energy Program, (301) 903-3514
Internet: http://www.fe.doe.gov
Sponsor: Department of Energy
19901 Germantown Road
Germantown, MD 20874

DOE Fusion Energy Science Division Grants   2024
Basic and applied research is carried out in the following areas: (1) basic plasma science research directed at furthering the understanding of fundamental processes in plasmas; (2) improving the theoretical understanding of fusion plasmas necessary for interpreting results from present experiments and the planning and design of future confinement devices; (3) obtaining the critical data on plasma properties, atomic physics, and new diagnostic techniques for support of confinement experiments; (4) supporting exploratory research into concepts that are alternatives to the tokamak; and (5) carrying out research on issues that support the development of inertial fusion energy, for which target development is carried out by the Department of Energy's defense programs. Application forms are available on the Web site. Applications may be submitted at any time.
Requirements: Unsolicited research proposals should be preceded by an informal preproposal of no more than two or three pages that briefly describes the research project and the proposed level of funding.
Amount of Grant: $10,000-$2 million; $200,000 average
Date(s) Application is Due: May 31
Contact: Martin Rubinstein, Grants and Contracts Division, Office of Science, (301) 903- 5212
Internet: http://www.er.doe.gov/production/grants/guide.html
Sponsor: Department of Energy
19901 Germantown Road
Germantown, MD 20874

DOE Fusion Energy Science Technology Division Grants   2025
The technology division's science-oriented goal is to provide the technologies that are required to successfully design, build, and operate near-term experiments aimed at producing, understanding, and optimizing the fusion energy process. The division's energy-oriented goal is to develop the technologies that will be needed in the long term for an economically and environmentally attractive fusion energy source. These goals are pursued through multi-institutional domestic programs and international collaboration partnerships. Applications are accepted at any time and are available on the Web site.
Requirements: Unsolicited research proposals should be preceded by an informal preproposal of no more than two or three pages that briefly describes the research project and the proposed level of funding.
Amount of Grant: $10,000-$2 million; $200, 000 average
Date(s) Application is Due: May 31
Contact: Martin Rubinstein, Grants and Contracts Division, Office of Science, (301) 903- 5212
Internet: http://www.science.doe.gov/production/grants/grants.html
Sponsor: Department of Energy
19901 Germantown Road
Germantown, MD 20874-1290

DOE Geosciences Research Grants 2026

The Program supports research aimed at developing an understanding of fundamental Earth processes that can be used as a foundation for efficient, effective, and environmentally sound use of energy resources, and provide an improved scientific basis for advanced energy and environmental technologies. The program funds projects that develop a quantitative and predictive understanding of the energy-related aspects of processes within the earth. Emphasis is on the upper levels of the earth's crust and the focus is on geophysics and geochemistry of rock-fluid systems and interactions. Specific topical areas receiving high emphasis include high-resolution geophysical imaging; rock physics; fundamental properties of rocks, minerals, and fluids; scientific drilling; and sedimentary basin systems. The resulting improved understanding and knowledge base are needed to assist efforts in the utilization of the nation's energy resources in an environmentally acceptable fashion. Applications are accepted at any time and are available on the Web site.

Requirements: Eligible to apply are institutions of higher education, industry (particularly small and disadvantaged businesses), and nonprofit institutions. Unsolicited research proposals are screened by program officials and, if appropriate, are evaluated by peer review. Successful proposers are awarded one-year or multi-year special research contracts or grants with renewals or incremental funding available, provided performance has been satisfactory. A one- to two-year phase-out period is normally granted to those finishing research projects. Equipment items may be purchased from contract funds with the agreement of DOE.

Amount of Grant: $10,000-$2 million; $200,000 average

Contact: Dr. Eric A. Rohlfing, Director; (301) 903-8165; fax: (301) 903-4110; email: Eric.Rohlfing@science.doe.gov

Internet: http://www.science.doe.gov/bes/geo/geohome.html

Sponsor: U.S. Department of Energy, Office of Science

19901 Germantown Road

Germantown, MD 20874-1290

DOE Health Effects and Life Sciences Research Grants 2027

The objectives of this program of basic and biological research are to create and apply new technologies and resources in mapping, sequencing, and information management for characterizing the molecular nature of the human genome; to develop and support DOE national user facilities for use in fundamental structural biology; to use model organisms to understand human genome organization, human gene function and control, and the functional relationships between human genes and proteins; to characterize and exploit the genomes and diversity of microbes with potential relevance for energy, bioremediation, or global climate; to understand and characterize the risks to human health from exposures to low levels of radiation and chemicals; to develop novel technologies for high throughput determination of protein structure; and to anticipate and address ethical, legal, and social implications arising from genome research. Applications are accepted at any time and are available on the Web site.

Requirements: Unsolicited research proposals should be preceded by an information preproposal of no more than two or three pages that briefly describes the research project and the proposed level of funding.

Amount of Grant: $10,000-$2 million range; $200,000 average

Contact: Martin Rubinstein, Grants and Contracts Division, (301) 903- 5212

Internet: http://www.science.doe.gov/production/grants/grants.html

Sponsor: Department of Energy

19901 Germantown Road

Germantown, MD 20874

DOE Heavy Element Chemistry Research Grants 2028

This Program supports research in actinide and fission product chemistry. Areas of interest include aqueous and non-aqueous coordination chemistry; solution and solid-state speciation and reactivity; measurement of chemical and physical properties; synthesis of actinide-containing materials; chemical properties of the heaviest actinide and transactinide elements; theoretical methods for the prediction of heavy element electronic and molecular structure and reactivity; and the relationship between the actinides, lanthanides, and transition metals. This activity represents the Nation's only funding for basic research in the chemical and physical principles of actinide and fission product materials. The program is primarily based at the national laboratories because of the special licenses and facilities needed to obtain and safely handle radioactive materials. However, research in heavy element chemistry is supported at universities, and collaborations between university and laboratory programs are encouraged. The training of graduate students and postdoctoral research associates is viewed as an important responsibility of this activity.

Requirements: Eligible to apply are institutions of higher education, industry (particularly small and disadvantaged businesses), and nonprofit institutions. Unsolicited research proposals are screened by program officials and, if appropriate, are evaluated by peer review. Successful proposers are awarded one-year or multi-year special research contracts or grants with renewals or incremental funding available, provided performance has been satisfactory. A one- to two-year phase-out period is normally granted to those finishing research projects. Equipment items may be purchased from contract funds with the agreement of DOE.

Amount of Grant: $10,000-$2 million; $200,000 average

Contact: Dr. Eric A. Rohlfing, Director; (301) 903-8165; fax: (301) 903-4110; email: Eric.Rohlfing@science.doe.gov

Internet: http://www.science.doe.gov/bes/chm/Programs/programs.html

Sponsor: U.S. Department of Energy, Office of Science

1000 Independence Avenue, SW

Washington, DC 20585-1290

DOE High-Energy Physics Research Grants 2029

The primary objectives of this program are to understand the nature of and relationships among fundamental forces of nature and to understand the ultimate structure of matter in terms of the properties and interrelations of its basic constituents. The research falls into three broad categories: experimental research, theoretical research, and technology R&D in support of the high-energy physics program. Applications are accepted at any time and are available on the Web site.

Requirements: Unsolicited proposals from universities for research support are analyzed by the program staff and undergo external peer review on an individual basis. Independently, experimental research proposals are submitted to national accelerator laboratories for review. The final decision to make accelerator facilities available for a particular experiment rests with the laboratory management. Successful research proposals usually receive one-year contracts (project grants) with provisions for review and renewal on an annual basis.

Amount of Grant: $10,000-$2 million; $200,000 average

Date(s) Application is Due: May 31

Contact: Martin Rubinstein, Grants and Contracts Division, Office of Science, (301) 903- 5212

Internet: http://www.science.doe.gov/production/grants/grants.html

Sponsor: Department of Energy

19901 Germantown Road

Germantown, MD 20874-1290

DOE Human Genome Program—Ethical, Legal, and Social Implications 2030

DOE announces its interest in receiving applications in support of the Ethical, Legal, and Social Implications (ELSI) subprogram of the Human Genome Program (HGP). Applications should focus on issues of genetics and the workplace, storage of genetic information and tissue samples, education, or complex or multigenic traits. The HGP is a coordinated, multidisciplinary, directed research effort aimed at obtaining a detailed understanding of the human genome at the molecular level.

Amount of Grant: $10,000-$2 million range; $200,000 average

Contact: Martin Rubinstein, Grants and Contracts Division, Office of Science, (301) 903-5212

Internet: http://www.sc.doe.gov/production/grants/grants.html

Sponsor: Department of Energy

19901 Germantown Road

Germantown, MD 20874

DOE Hydrogen Program 2031

In a supplemental announcement to a broad-based solicitation, DOE is seeking research and development (R&D) proposals that can advance hydrogen production, storage, and utilization technologies. The solicitation contains information that must be used in conjunction with this supplemental announcement when applying for an award. Thus, in order to prepare a complete application, it is mandatory to comply with the requirements of the overall broad-based solicitation document, DE-PS36-00GO10482, which can be found on the Web site.

Date(s) Application is Due: Dec 15

Contact: Steven Chalk, Program Manager, (202) 586-3388; email: steven.chalk@ee.doe.gov

Internet: http://www.eere.energy.gov/hydrogenandfuelcells/financial.html

Sponsor: Department of Energy

1617 Cole Boulevard

Golden, CO 80401

DOE Industrial Materials for the Future Program 2032

This program is a national effort to research, design, develop, engineer, and test new and improved materials to achieve improvements in energy efficiency, emissions and waSuite reduction, productivity, product quality, and global competitiveness.

Requirements: Proposals are solicited from universities, and nonprofit research institutes for research and development leading to new materials and processing methods for eventual use in the Industries of the Future. Universities, and nonrofits are required to form partnerships for technology development and to work with industry to ensure that core activities will ultimately lead to successful applications in industry.

Amount of Grant: $50,000-$500,000

Date(s) Application is Due: May 31

Contact: Polly Perando, Contract Specialist; email: polly.perando@ee.doe.gov

Internet: http://www.tgci.com/fedrgtxt/01-11274.txt

Sponsor: Department of Energy

850 Energy Drive, MS 1221

Idaho Falls, ID 83401-1563

DOE Inventions and Innovation Grant Program (I&I) 2033

The goal of the program is to improve energy efficiency through the promotion of innovative ideas and inventions that have a significant potential energy impact and a potential future commercial market. Additionally, DOE provides awardees with nonfinancial support by assisting them with business development and commercialization planning through a network of national and regional resource providers. This assistance is provided at three levels: Up to $50,000 for technologies in early-stage development, up to $250,000 for technologies approaching the point of prototype, and up to $500,000 for technology demonstrations. Cost-share is strongly encouraged to receive a Category 1 or 2 award, and cost-share is required to receive a Category 3 award. Funding Opportunity Number: DE-PS36-06GO96001

Requirements: U.S. individual inventors, small businesses (profit or nonprofit with less than 500 employees) may apply for category 1 or 2 applications. Universities and nonprofit research institutes may apply for category 1 applications.
Amount of Grant: $50,000-$500,000
Date(s) Application is Due: Oct 11
Contact: Michael Schledorn, Grant Specialist, (303) 275-4993; email: goii@go.doe.gov
Internet: http://www.eere.energy.gov/inventions
Sponsor: Department of Energy—Golden Field Office
1617 Cole Boulevard
Golden, CO 80401

DOE Low Dose Radiation Research Program   2034
Research is specifically sought for pilot projects that involve innovative collaborations between experimentalists and modelers to model the mechanisms of key radiation-induced biological responses and to describe or identify strategies for developing biologically-based risk models that incorporate information on mechanisms of radiation-induced biological responses. Potential applicants should submit a one-page preapplication referencing Program Notice 01-17 by February 1; formal applications are due May 1. Annual deadline dates may change; contact program staff for exact dates.
Amount of Grant: $10,000-$2 million range; $200,000 average
Date(s) Application is Due: Feb 1; May 1
Contact: Martin Rubinstein, Grants and Contracts Division, Office of Science, (301) 903-5212
Internet: http://www.science.doe.gov/grants
Sponsor: Department of Energy
19901 Germantown Road
Germantown, MD 20874

DOE Materials Sciences and Engineering Research Grants   2035
Research support is directed toward understanding materials properties and phenomena of importance to all energy systems. Emphasis is placed on areas where problems are known to exist or are anticipated. Research is supported in metallurgy, ceramics, solid state physics, materials chemistry, and related disciplines where the emphasis is on the science of materials. Applications are accepted at any time and are available on the Web site.
Requirements: Eligible to apply are institutions of higher education, industry (particularly small and disadvantaged businesses), and nonprofit institutions. Unsolicited research proposals are screened by program officials and, if appropriate, are evaluated by peer review. Successful proposers are awarded one-year or multi-year special research contracts or grants with renewals or incremental funding available, provided performance has been satisfactory. A one- to two-year phase-out period is normally granted to those finishing research projects. Equipment items may be purchased from contract funds with the agreement of DOE.
Amount of Grant: $10,000-$2 million; $200,000 average
Contact: Martin R. Rubinstein, Supervisor of Grants and Contracts; (301) 903-4946; fax: (301) 903-4194; email: Martin.Rubinstein@science.doe.gov
Internet: http://www.science.doe.gov/grants/FAPN08-01.html
Sponsor: U.S. Department of Energy, Office of Basic Energy Sciences
19901 Germantown Road
Germantown, MD 20874-1290

DOE Mathematical, Information, and Computational Sciences Grants   2036
This program supports a spectrum of fundamental research in applied mathematical sciences, computer science, and networking from basic through prototype development. Results of these efforts are used to form partnerships with users in scientific disciplines to validate the usefulness of the ideas and to develop them into tools. Testbeds on important applications for DOE are supported by this subprogram. Applications are accepted at any time, and application forms can be found on the Web site.
Amount of Grant: $10,000-$2 million; $200,000 average
Date(s) Application is Due: May 31
Contact: Martin Rubinstein, Grants and Contracts Division, Office of Science, (301) 903- 5212
Internet: http://www.er.doe.gov/production/grants/guide.html
Sponsor: Department of Energy
19901 Germantown Road
Germantown, MD 20874-1290

DOE Medical Applications and Measurement Science Grants   2037
The objectives of this program are to develop technologies for the beneficial applications of radiation and in-vivo radiotracer detection in the study, diagnosis, and treatment of human diseases and disorders; to develop new instrumentation for biological and medical research; and to develop new concepts and techniques for detecting and measuring the hazardous agents of biochemical, physical, and environmental consequences related to energy production. Applications are accepted at any time; application forms are available on the Web site.
Amount of Grant: $10,000-$2 million range; $200,000 average
Contact: Martin R. Rubinstein, Grants and Contracts Division, Office of Science, (301) 903- 5212
Internet: http://www.er.doe.gov/production/grants/guide.html
Sponsor: Department of Energy
19901 Germantown Road
Germantown, MD 20874

DOE Million Solar Roofs (MSR) Program   2038
The purpose of this initiative is to install solar energy systems on one million U.S. buildings. The department hopes to achieve this goal by establishing state and local MSR partnerships involving states and local governments, businesses, and community-based organizations to promote the use of solar roofs. Matching grants are not required, but will be considered in the application process.
Requirements: Existing and new MSR state and local partnerships are eligible to apply.
Restrictions: Partnerships that received funding under this program or the State Energy Program Special Projects Solicitation in the last fiscal year are ineligible.
Amount of Grant: $10,000-$50,000
Contact: Sandra Burton, (215) 656-6983; fax: (215) 656-6981; email: sandra.burton@ee.doe.gov
Internet: http://www.millionsolarroofs.com
Sponsor: Department of Energy
19901 Germantown Road
Germantown, MD 20874

DOE Natural and Accelerated Bioremediation Research Program   2039
DOE announces its interest in receiving applications for research grants in the Natural and Accelerated Bioremediation Research (NABIR) Program. Applications should describe research projects in one of the following categories: research projects that address the scientific aims of individual NABIR elements including biogeochemistry, biotransformation, community dynamics, biomolecular science and engineering, and assessment; or research projects to be performed at a field research center addressing field scale biostimulation of microbiological processes that immobilize metals and/or radionuclides. Interdisciplinary teams should include, at a minimum, experts in the fields of microbiology, geochemistry, and hydrology. Refer to program notice 02-12.
Amount of Grant: $10,000-$2 million range; $200,000 average
Contact: Dr. Anna Palmisano, Environmental Sciences Division, (301) 903-9963; fax: (301) 903-8519; email: anna.palmisano@science.doe.gov
Internet: http://www.sc.doe.gov/production/grants/Fr02-12.html
Sponsor: Department of Energy
19901 Germantown Road
Germantown, MD 20874

DOE Nonproliferation and National Security Research   2040
The purpose of the program is to conduct basic and applied research and development on verification technologies needed for effective treaty negotiations and for international agreements on the control of special nuclear materials, nuclear weapons, and weapons of mass destruction. Financial support, in whole or in part, may be provided for salaries, materials, supplies, equipment, travel, publication costs, services required for conducting research, and for developing advanced detection technologies. Contact DOE for deadlines.
Requirements: Eligible to apply are public and private universities and institutions of higher education with postdoctoral programs.
Amount of Grant: $100,000-$250,000 per year
Contact: Paul Morrison, Office of Nonproliferation Research and Engineering (NA-22), (202) 586-5751
Internet: http://www.nnsa.doe.gov/na-20
Sponsor: Department of Energy
19901 Germantown Road
Germantown, MD 20874

DOE Nuclear Energy Research Initiative (NERI)   2041
NERI is designed to support innovative research, primarily to address the principal technical and scientific obstacles to future use of nuclear power in the United States. NERI is also intended to reinvigorate the vital nuclear scientific and engineering infrastructure within U.S. universities, industry, and DOE national laboratories.
Requirements: U.S. universities or other institutions of higher learning, industry, nonprofit and R&D organizations are eligible for grant or cooperative agreement awards under this program. DOE national laboratories are eligible to participate, but not as the lead organization in the application. All segments of the U.S. private sector (nonfederal) are eligible to apply. Non-citizens employed by U.S. institutions also are eligible.
Amount of Grant: $100,000-$500,000 range; $350,000 average
Contact: Denise Berry, Contract Specialist, (510) 637-1873; fax: (510) 637-2025
Internet: http://neri.ne.doe.gov
Sponsor: Department of Energy
1301 Clay Street, 700N
Oakland, CA 94612

DOE Nuclear Physics Research Grants   2042
The primary objectives of this program are an understanding of the interactions and structures of atomic nuclei and nuclear matter, and an understanding of the fundamental forces of nature as manifested in nuclear matter. Applications are accepted at any time and are available on the Web site.
Requirements: The DOE Nuclear Physics Program supports university-based user groups that plan experiments at the home institution, execute and partially analyze experiments at the national facility, and complete the analyses and publication of results at the home institution. Support of work under Nuclear Theory is almost equally divided between national laboratory-based theorists and university-based theorists. Written proposals for direct funding for DOE are sent to the DOE. Written proposals for access to the accelerator installations should be sent to the director of the facility.
Amount of Grant: $10,000-$2 million; $200,000 average

Date(s) Application is Due: May 31
Contact: Martin Rubinstein, Grants and Contracts Division, Office of Science, (301) 903- 5212
Internet: http://www.science.doe.gov/production/grants/grants.html
Sponsor: Department of Energy
19901 Germantown Road
Germantown, MD 20874-1290

DOE On-Line Temperature Measurement Instrumentation for Gasification Process Control Development Grants   2043

DOE seeks applications for innovative technical approaches to develop an accurate, reliable, robust, and cost-effective real-time temperature-monitoring system capable of measuring temperatures in the high temperature (typically ranging from 2000-2600F) section of pressurized, coal-fired slagging gasifiers. A variety of approaches, including the use of thermocouple or optically based techniques, are acceptable as long as they offer the clear potential to meet the aforementioned objectives. Furthermore, all proposed temperature measurement instrumentation must be suitable for use on large-scale slagging gasification systems and applications must address issues such as laboratory scale-up, potential placement, and method of mounting on actual operating systems. Solicitation DE-PS26-99FT40565.
Amount of Grant: $10,000-$25 million range
Contact: John Augustine, Program Director, (412) 386-4524
Internet: http://www.netl.doe.gov
Sponsor: Department of Energy
19901 Germantown Road
Germantown, MD 20874

DOE Photochemistry and Radiation Research Grants   2044

The program supports fundamental molecular-level research on interactions of radiation with matter in the condensed phase. The photochemistry research effort emphasizes fundamental processes aimed at the capture and chemical conversion of solar energy. The radiation sciences research effort supports fundamental studies on chemical effects produced by absorption of energy from ionizing radiation.
Requirements: Eligible to apply are institutions of higher education, industry (particularly small and disadvantaged businesses), and nonprofit institutions. Unsolicited research proposals are screened by program officials and, if appropriate, are evaluated by peer review. Successful proposers are awarded one-year or multi-year special research contracts or grants with renewals or incremental funding available, provided performance has been satisfactory. A one- to two-year phase-out period is normally granted to those finishing research projects. Equipment items may be purchased from contract funds with the agreement of DOE.
Amount of Grant: $10,000-$2 million; $200,000 average
Contact: Dr. Eric A. Rohlfing, Director; (301) 903-8165; fax: (301) 903-4110; email: Eric.Rohlfing@science.doe.gov
Internet: http://www.science.doe.gov/bes/chm/Programs/programs.html
Sponsor: U.S. Department of Energy, Office of Science
1000 Independence Avenue, SW
Washington, DC 20585-1290

DOE Program for Ecosystem Research   2045

The mission of this program is to improve the scientific basis for predicting or detecting effects of simultaneous changes in climate and atmospheric composition on terrestrial ecosystems and their component organisms and processes. Ecosystem processes and components of importance to humanity are of special concern. Climatic and atmospheric changes of key interest include (but need not be limited to): warming (and changes in diurnal, seasonal, and interannual temperature cycles), changes in precipitation and evapotranspiration (e.g., intensification of the hydrologic cycle), changes in frequency and/or magnitude of extreme weather events and patterns, and rising atmospheric carbon dioxide and ozone concentrations. Potential applicants are strongly encouraged (but not required) to submit a preapplication for programmatic review. The deadline for preapplications is July 2; deadline for receipt of formal applications is August 13. Annual deadline dates may vary; contact program staff for exact dates.
Amount of Grant: $10,000-$2 million; $200,000 average
Date(s) Application is Due: Jul 2; Aug 13
Contact: Martin Rubinstein, Grants and Contracts Division, Office of Science, (301) 903- 5212
Internet: http://www.tgci.com/fedrgtxt/01-12539.txt
Sponsor: Department of Energy
19901 Germantown Road
Germantown, MD 20874-1290

DOE Programs in Fusion Energy Sciences   2046

DOE announces its interest in receiving grant applications for new research in fusion energy sciences. The specific areas of interest are Magnetic Fusion Concept Exploration Experiments; Inertial Fusion Energy Concept Exploration Research; Inertial Fusion Energy Chamber and Target Research; Magnetic Fusion Liquid Wall Experiments; and Fusion Materials Modeling. Annual deadlines may vary; contact program staff for exact dates. Refer to program notice 00-07.
Requirements: Colleges and universities, nonprofit organizations, for-profit commercial organizations, State and local governments, and unaffiliated individuals.
Amount of Grant: $10,000-$2 million; $200,000 average
Contact: Martin Rubinstein, Grants and Contracts Division, Office of Science, (301) 903-5212

Internet: http://www.science.doe.gov/production/grants/grants.html
Sponsor: Department of Energy
19901 Germantown Road
Germantown, MD 20874

DOE Rebuild America Energy-Efficiency Grants   2047

The Rebuild America program aims to increase the energy efficiency of commercial buildings and multifamily housing in the United States. Partnerships must include at least one state or local government member, as well as businesses, educational institutions, and nonprofits. Projects must apply retrofit improvements to a substantial portion of the floor space of targeted buildings, such as schools, offices, libraries, churches, stores, hospitals, and correctional facilities, within five years of the grant award.
Contact: John Meeker, Procurement, (303) 275-4748; fax: (303) 275-4754
Internet: http://www.rebuild.org/index.asp
Sponsor: Department of Energy
1617 Cole Boulevard
Golden, CO 80401

DOE Reliable Electric Power Delivery Research Grants   2048

Under this program of the Electric Energy Systems Division, recent breakthroughs in new materials and control concepts show the potential for significantly improving our ability to maintain electric power continuity during periods of severe emergencies. New research is needed in such areas as electromagnetic pulse protection, basic mechanisms of insulation failure, emergency detection and control, new systems analysis and equipment research, and system dynamic performance. Proposals are accepted at any time.
Amount of Grant: $50,000-$500,000; $200,000 average
Contact: Office of Utility Technologies, (202) 586-4142
Internet: http://www.eere.energy.gov
Sponsor: Department of Energy
1000 Independence Avenue SW, Forrestal Bldg
Washington, DC 20585

DOE Separations and Analysis Research Grants   2049

This Program supports fundamental research covering a broad spectrum of separation concepts, including membrane processes, extraction under both standard and supercritical conditions, adsorption, chromatography, photo-dissociation, and complexation. Also supported is work to improve the sensitivity, reliability, and productivity of analytical determinations and to develop entirely new approaches to analysis. This activity is the Nation's most significant long-term investment in many aspects of separations and analysis, including solvent extraction, ion exchange, and mass spectrometry.
Requirements: Eligible to apply are institutions of higher education, industry (particularly small and disadvantaged businesses), and nonprofit institutions. Unsolicited research proposals are screened by program officials and, if appropriate, are evaluated by peer review. Successful proposers are awarded one-year or multi-year special research contracts or grants with renewals or incremental funding available, provided performance has been satisfactory. A one- to two-year phase-out period is normally granted to those finishing research projects. Equipment items may be purchased from contract funds with the agreement of DOE.
Amount of Grant: $10,000-$2 million; $200,000 average
Contact: Dr. Eric A. Rohlfing, Director; (301) 903-8165; fax: (301) 903-4110; email: Eric.Rohlfing@science.doe.gov
Internet: http://www.science.doe.gov/bes/chm/Programs/programs.html
Sponsor: U.S. Department of Energy, Office of Science
1000 Independence Avenue, SW
Washington, DC 20585-1290

DOE Small Business Innovation Research Commercialization Assistance   2050

The purpose of this program is to accelerate the development, demonstration, and commercialization of products, services, or technology resulting from the research of SBIR Phase II financial assistance award recipients. The major objective of the program is to provide individualized assistance to SBIR Phase II financial assistance recipients that will lead to the successful commercialization of products, services, or technology developed in the SBIR program.
Requirements: Small businesses are eligible to apply.
Amount of Grant: $750,000 maximum
Date(s) Application is Due: Dec 2
Contact: Grants Administrator, (301) 903-1414; email: sbir-sttr@science.doe.gov
Internet: http://www.science.doe.gov/sbir
Sponsor: Department of Energy
9800 S Cass Avenue
Argonne, IL 60439-4899

DOE State Energy Program Special Projects   2051

The goal of this project is to assist states in deploying energy-efficient and renewable energy technologies; to facilitate the acceptance of emerging and underutilized technologies; and to increase the responsiveness of federally funded technology development efforts to private sector energy needs. DOE is providing funding in the areas of Clean Cities/Alternative Fuels, Industrial Technologies, Codes and Standards, Rebuild America, Federal Energy Management Program, Hydrogen Reformer Field Verification, Building America, Wind Energy Case Studies, Biomass Power Projects, and Photovoltaic Projects.

Requirements: All 50 states plus the District of Columbia, the U.S. Virgin Islands, Puerto Rico, Guam, Samoa, and the Commonwealth of the Northern Mariana Islands may apply.
Amount of Grant: $20,000-$1 million range; $17 million total
Contact: John Millhone, Weatherization and Intergovernmental Program, Department of Energy, (202) 586-1510
Internet: http://12.46.245.173/pls/portal30/CATALOG.PROGRAM_TEXT_RPT. show
Sponsor: Department of Energy
1000 Independence Avenue SW
Washington, DC 20585

DOE University Coal Research Grants 2052
The grants are awarded to improve scientific and technical understanding of the chemistry and physics involved in the conversion and utilization of coal, furnish technical support for the ongoing and developing coal conversion process, produce clear fuels in an environmentally acceptable manner, and develop new approaches to the design of future coal conversion and utilization technologies. Proposals should be submitted to the National Energy Technology Laboratory, Attn: MS921-107, Department of Energy, P.O. Box 10940, 626 Cochrans Mill Road, Pittsburgh, PA.
Requirements: U.S. institutions of higher education may apply for these grants.
Amount of Grant: $50,000-$400,000; $154,000 average
Contact: Fred Glaser, Office of Advanced Research, (301) 903-2786
Internet: http://www.netl.doe.gov/coal/Advanced%20Research
Sponsor: Department of Energy
1000 Independence Avenue SW, Forrestal Bldg
Washington, DC 20585

DOE University Nuclear Science and Reactor Support 2053
The purpose of this program is to provide financial support for research design, analysis, and assessments in science and technology in fields related to nuclear energy. Project grant funds may be used in support of nuclear energy-related research and development financial support, in whole or in part, may be provided for salaries, materials, supplies, equipment, travel, publication costs, supporting costs required for technical activities, market analyses, financing plans, and other activities necessary to achieve the objective.
Requirements: Any individual, partnership, corporation, association, joint venture, institution of higher education, or nonprofit organization may apply.
Amount of Grant: $50,000-$300,000 range
Contact: Nancy Hebron-Isreal , Office of Nuclear Energy, Science and Technology, (301) 903-1536
Internet: http://nuclear.gov
Sponsor: Department of Energy
19901 Germantown Road
Germantown, MD 20874

DOE University Reactor Instrumentation (URI) Program 2054
The purpose of this program is to upgrade and improve the U.S. university nuclear research and training reactors and to contribute to strengthening the academic community's nuclear engineering infrastructure.
Requirements: U.S. colleges and universities having a duly licensed, operating nuclear research or training reactor are eligible to apply.
Contact: Suzette Olson, DOE Idaho Operations Office, (208) 526-7385; fax: (208) 526-5548; email: Suzette.Olson@nuclear.energy.gov
Internet: http://www.id.doe.gov/doeid/psd/proc-div.html
Sponsor: Department of Energy, Idaho Operations Office
P.O. Box 1625, MS 3860
Idaho Falls, ID 83415-3860

DOE Vehicle and Engine Research and Development Grants 2055
The grants support research to reduce the transportation sector's vulnerability to petroleum shortages by developing more energy-efficient vehicle propulsion system options and by developing the ability to switch from petroleum to electricity and alternative fuels. Emphasis is placed on advancing heat engine technology through the development of component and materials technologies; supporting long-term, high-risk technology research and development for petroleum savings in the heavy-duty transport sector; advancing electric and hybrid vehicle technology; and developing means of using alternative fuels in vehicles. Identified potential areas for new research initiatives may be requested from one of the program information contacts. Proposals are accepted at any time. For further information on specific programs and prior to submitting formal proposals, contact Advanced Heat Engines, (202) 252-8012; Heavy Duty Transport, (202) 252-8055; Alternative Fuels, (202) 252-8055; and Electric and Hybrid Vehicles, (202) 252-8044.
Amount of Grant: $50,000-$500,000
Contact: Polly Perando, Energy Efficiency and Renewable Energy Program, email: polly. perando@ee.doe.gov
Internet: http://www.eere.energy.gov
Sponsor: Department of Energy
1000 Independence Avenue SW, Forrestal Bldg
Washington, DC 20585

DOI African Elephant Conservation Grants 2056
The program objectives are to provide financial assistance to any organization or individual responsible for African elephant conservation and to any organization or individual with experience in African elephant conservation, for approved projects to support research, conservation, management, and protection of African elephants. Funds may be used for approved conservation projects. Funds are matched by nonfederal funds or in-kind support, which must be equal to or exceed the amount of federal funds provided.
Requirements: Individuals and public or private organizations with experience in African elephant conservation may apply. All private sector project proposals must contain evidence of support by governmental entities of countries where the project is to be conducted.
Amount of Grant: $50,000 average
Contact: Division of International Conservation, (703) 358-1754
Internet: http://international.fws.gov/grants/grants.html
Sponsor: Department of the Interior
4401 N Fairfax Drive, Room 730
Arlington, VA 22203

DOI Biological Resource Division Brucellosis Program 2057
Applications are invited for a research project on the improvements in ballistic delivery systems for brucellosis vaccination of free-ranging elk and bison of the Greater Yellowstone area. The purpose of this project is to develop methods of ballistic delivery that improve the distance, reliability, ease, and/or rapidity of brucella vaccine parenteral delivery.
Requirements: Applications may be submitted by educational institutions, private firms, private foundations, individuals, and agencies of state and local governments.
Amount of Grant: $1000-$582,000; $50,000 average
Contact: Acquisition and Federal Assistance; (703) 648-7478 or (703) 648-7361
Internet: http://www.usgs.gov/contracts
Sponsor: Department of the Interior
P.O. Box 25046, MS 204B
Denver, CO 80225

DOI Earthquake Hazards Reduction Research Grants 2058
Applications are being sought for grants for research projects under the National Earthquake Hazards Reduction Program to mitigate earthquake losses that can occur in many parts of the nation by providing earth science data and assessments essential for warning of imminent damaging earthquakes, land-use planning, engineering design, and emergency preparedness decisions. Specific objectives are identified in the annual program announcement.
Requirements: Colleges, universities, for-profit and nonprofit organizations, and state and local governments may submit applications for support by a named principal investigator.
Amount of Grant: $5887-$1.1 million; $76,474 average
Date(s) Application is Due: May 10
Contact: Michael Blanpied, Earthquake Hazards Program Office, U.S. Geological Survey, National Center, (703) 648-6696; fax: (703) 648-6642; email: gd-erp-coordinator@usgs.gov
Internet: http://aspe.os.dhhs.gov/cfda/p15807.htm
Sponsor: Department of the Interior
12201 Sunrise Valley Drive
Reston, VA 22092

DOI Geological Research and Data Acquisition Grants 2059
Project grants are awarded to support research in any field of study that helps fulfill the Geological Survey's mission, which is to collect, organize, interpret, and publish information about the nation's energy, minerals, water, and land resources; to determine the geologic structure of the United States; and to develop an understanding of earth processes and hydrologic principles. Applications must be for scientific research projects that are within the survey's area of responsibility. Conferences and symposia will be supported only if it is clear that equivalent results cannot be obtained at regular meetings of professional societies. Cost sharing is encouraged.
Requirements: Colleges, universities, profit-making and nonprofit organizations, and state and local governments may make application for support by a named principal investigator.
Restrictions: Office furniture, office equipment, and foreign travel are not normally considered for support.
Amount of Grant: $1000-$582,000; $50,000 average
Contact: Assistant Director for Research, U.S. Geological Survey, National Center, (703) 648-4460
Internet: http://www.usgs.gov/contracts
Sponsor: Department of the Interior
12201 Sunrise Valley Drive
Reston, VA 22092

DOI National Center for Preservation Technology and Training Grants 2060
The objectives of the program are to develop and distribute preservation and conservation skills and technologies for the identification, evaluation, conservation, and interpretation of prehistoric and historic resources; to develop and facilitate training for federal, state, and local resource preservation professionals, cultural resource managers, maintenance personnel, and others working in the preservation field; to apply preservation technology benefits from ongoing research by other agencies and institutions; to facilitate the transfer of preservation technology among federal agencies, state and local governments,

universities, international organizations, and the private sector; and to cooperate with related international organizations including, but not limited to, the International Council on Monuments and Sites, the International Center for the Study of Preservation and Restoration of Cultural Property, and the International Council of Museums.
Requirements: Eligible applicants are universities and colleges; nonprofit organizations, including museums, research laboratories, professional societies, and similar organizations that are directly associated with educational or research activities; federal and nonfederal laboratories; offices, units, and Cooperative Park Study Units of the National Park System; federal, state, local, and tribal preservation offices; and Native Hawaiian organizations. For-profit organizations and individuals are eligible provided their proposal partners with an eligible entity to be the recipient.
Amount of Grant: $7856-$50,000; $36,000 average
Contact: Kirk Cordell, Executive Director, (318) 356-7444 ext 222; fax: (318) 356-9119; email: kirk_cordell@nps.gov
Internet: http://www.ncptt.nps.gov/default.aspx?m=36
Sponsor: Department of the Interior
1849 C Street NW
Washington, DC 20240

DOI National Spatial Data Infrastructure Cooperative Agreements Grants 2061
The objectives of the program are to help the Federal Geographic Data Committee form partnerships with the nonfederal sector that will further the development and implementation of the National Spatial Data Infrastructure. NSDI means the technology, policies, standards, and human resources necessary to acquire, process, store, distribute, and improve utilization of geospatial data. Partnership activities must be directed towards the creation of the National Geospatial Data Clearinghouse for finding and accessing geospatial data; the development and promulgation of the use of standards in data collection, documentation, transfer, and search and query; the development and creation of a National Digital Geospatial Data Framework; and/or the development and implementation of educational outreach programs to increase awareness and understanding of the major NSDI initiatives.
Requirements: Proposals from federal, state and local government agencies, educational institutions, private firms, private foundations, nonprofit organizations and federally acknowledged or state-recognized Native American tribes or groups are invited. Proposals must demonstrate that more than one organization or agency is involved in the project.
Amount of Grant: $20,000-$75,000 range; $29,000 average
Date(s) Application is Due: May 15
Contact: David Painter, FGDC CAP Coordinator, (703) 648-5513; fax: (703) 648-5755; email: dpainter@fgdc.gov
Internet: http://www.fgdc.gov/funding/cap2002.html
Sponsor: Department of the Interior
Mail Stop 590, 12201 Sunrise Valley Drive
Reston, VA 20192

DOI Rhinoceros and Tiger Conservation Grants 2062
The objectives of the program are to increase conservation of rhinoceros and tigers through strengthening habitat/ecosystem management; surveys and monitoring; conservation education; wildlife inspection, law enforcement, and forensic skills; protected area/reserve management; sustainable development in buffer zones surrounding tiger/rhinoceros habitat; management of human behavior and livestock to decrease conflicts with tigers/rhinoceros; and the use of substitute for tiger/rhinoceros products in oriental medicine. Projects which provide training to strengthen capabilities in these areas will receive priority. Funds may be used for approved conservation projects. Projects proposed should have the support of the local government(s) and have matching funds (cash) or in-kind support (salaries, equipment, etc.) provided by the organization receiving the grant or other partners. Project duration must be one year or less. Proposals may be submitted at any time.
Requirements: U.S. public and private nonprofit institutions and organizations are eligible.
Amount of Grant: $50,000 maximum
Contact: Division of International Conservation, (703) 358-1754; fax: (703) 358-2849
Internet: http://www.fws.gov/international/grants/grants.html
Sponsor: Department of the Interior
4401 N Fairfax Drive
Arlington, VA 22203

DOI State Partnership Program Grants 2063
The purpose of the SPP is to provide support through grants and cooperative agreements to states and tribal agencies whose primary focus is on gathering, analyzing, and distributing biological science information needed for natural resource management decision making. This program requires complementary study participation and interaction between state/tribal institutions and science centers or cooperative research units of the U.S.GS, Eastern Region. Eligible institutions may request a preproposal solicitation package. Preproposals must be submitted to U.S.GS by state/tribe institutions only, but must include information on participating U.S.GS Science Center or Cooperative Research Unit. Full proposals will be requested in writing by the U.S.GS from institutions that have submitted preproposals of high merit and who have met all of the preproposal requirements. Guidelines are available online.
Requirements: State, tribal, and/or U.S. territories and possessions that conduct natural resources studies and associated information management are eligible to apply.
Restrictions: No federal or private agencies may apply.
Amount of Grant: $4000-$300,000 range; $50,000 average

Contact: Gary Brewer, State Partnership Program Coordinator, (304) 724-4507; fax: (304) 724-4505; email: gary_brewer@usgs.gov
Internet: http://www.doi.gov
Sponsor: Department of the Interior
1849 C Street NW
Washington, DC 20240

DOI Upper Mississippi River System Long-Term Resource Monitoring Grants 2064
The mission of this program is to provide decision makers with information needed to maintain the Upper Mississippi River System (UMRS) as a sustainable large river ecosystem given its multiple-use character. The long-term goals of this program are to: understand the UMRS ecosystem, monitor trends and effects with respect to selected resources, develop resource management alternatives, manage information, and develop useful products. This is a 15-year program. There are no deadlines.
Requirements: Eligible to apply are states, local governments, intrastate/interstate agencies, sponsored organizations, private nonprofit institutions/organizations.
Geographic Focus: Illinois
Amount of Grant: $392,426 average
Contact: Dr. Leslie Holland-Bartels, Center Director, Upper Midwest Environmental Sciences Center, (608) 781-6221; fax: (608) 783-6066; fax: (608) 783-6066; email: Leslie_Holland-Bartels@usga.gov
Internet: http://www.umesc.usgs.gov/ltrmp.html
Sponsor: Department of the Interior
2630 Fanta Reed Road
La Crosse, WI 54603

DOJ Corrections Technical Assistance Grants 2065
This program provides assistance for upgrading the operation of correctional facilities, programs, and services at State and local levels. Services are available to the entire range of correctional agencies, including probation, parole, institutions, jails, and community programs. The purpose is to encourage and assist federal, state, and local government programs and services, and programs and services of other public and private agencies, institutions, and organizations in their efforts to develop and implement improved corrections programs. It also seeks to assist and serve in a consulting capacity to federal, state, and local courts, departments, and agencies in the development, maintenance, and coordination of programs, facilities, services, training, treatment, and rehabilitation with respect to criminal and juvenile offenders.
Requirements: States, general units of local government, public and private agencies, educational institutions, organizations, and individuals involved in the development, implementation or operation of correctional programs and services are eligible.
Amount of Grant: $1500-$50,000; $7500 average
Date(s) Application is Due: Jul 30
Contact: Grants Administrator, (202) 307- 3106 or (800) 995-6423; fax: (202) 307-3361; TDD: (202) 307- 3156
Internet: http://www.usdoj.gov
Sponsor: Department of Justice
320 First Street NW
Washington, DC 20534

DOJ Grants to Reduce Violent Crimes Against Women on Campus 2066
The purpose of this program is to encourage institutions of higher education to adopt comprehensive, coordinated responses to violence against women, including sexual assault, stalking, dating and domestic violence. Funds are authorized to enhance the apprehension, investigation, and adjudication of persons committing violent crimes against women on campuses; train campus administrators, security personnel, and disciplinary or judicial boards to effectively identify and respond to violent crimes against women on campuses; implement and operate education programs for the prevention of violent crimes against women on campuses; develop, enlarge or strengthen support services for victims; create, disseminate, or otherwise provide assistance and information about victims' options on and off campus to bring disciplinary or other legal action; develop and implement more effective campus policies, protocols, orders, and services specifically devoted to prevent, identify, and respond to violent crimes against women on campuses; develop, install, or expand data collection and communication systems, including computerized systems, linking campus security to the local law enforcement for the purpose of identifying and tracking arrests, protection orders, violations of protection orders, prosecutions, and convictions; develop, enlarge or strengthen victim service programs for the campus and to improve the delivery of victim services on campus; provide capital improvements (including improved lighting and communication facilities but excluding construction of buildings) on campuses to address violent crimes against women; and to support improved coordination among campus administrators, campus security personnel, and local law enforcement to reduce violent crimes against women on campuses.
Requirements: Eligible applicants for this program are institutions of higher education as defined under the Higher Education Amendments of 1998. A consortium of higher education institutions also may apply for these grants provided that each individual consortium member is also eligible to apply.
Amount of Grant: $143,000-$550,000; $360,000 average
Contact: Violence Against Women Office, (202) 307-6026; fax (202) 307-3911; TTY: (202) 307-2277
Internet: http://www.ojp.usdoj.gov/vawo/campus_desc.htm
Sponsor: Department of Justice
810 Seventh Street NW
Washington, DC 20531

DOJ Juvenile Justice and Delinquency Prevention Special Emphasis Grants   2067
Grants support investigative research into and the development and implementation of programs that design, test, and demonstrate effective approaches, techniques, and methods for preventing and controlling juvenile delinquency, such as community-based alternatives to institutional confinement; developing and implementing effective means of diverting juveniles from the traditional juvenile justice and correctional system; developing and supporting programs stressing advocacy activities aimed at improving services to youth impacted by the juvenile justice system; developing model programs to strengthen and maintain the family unit, including self-help programs; developing and implementing special emphasis prevention and treatment programs relating to juveniles who commit serious crimes; developing programs to prevent hate crimes; and developing and implementing further a coordinated, national law-related education program of delinquency prevention. Deadlines are published in program announcements.
Requirements: Public and private nonprofit agencies, organizations, individuals, state and local units of government, and combinations of state and local units are eligible.
Contact: Office of Juvenile Justice and Delinquency Prevention, (202) 307-5914; email: AskDOJ@usdoj.gov
Internet: http://www.usdoj.gov
Sponsor: Department of Justice
950 Pennsylvania Avenue NW
Washington, DC 20530-0001

DOJ National Incident-Based Reporting System   2068
The purpose of this program is to provide funding for states and units of local government to help create a uniform national standard for reporting crime trends that go beyond simple tallies of crimes and arrests, including as much detailed information about the individual crimes as possible. Funds may be used to develop or enhance NIBRS data collection programs; develop or provide NIBRS training programs, including attending conferences; and develop, implement, or license software that supports NIBRS data collection. A 10 percent matching grant is required from applicants.
Requirements: Eligible applicants are states applying at the state level or on behalf of one or more cities or counties. States may apply on behalf of more than one jurisdiction.
Restrictions: Funds may not be used to purchase equipment or to pay salaries or overtime for persons attending NIBRS training sessions.
Contact: Dr. Charles Kindermann, Senior Statistician, (202) 616-3489; Carol Kaplan, Chief, Criminal History Improvement Programs, (202) 307-0759
Internet: http://www.ojp.usdoj.gov/bjs/ibrs.htm
Sponsor: Department of Justice
Bureau of Justice Statistics, Room 2406, 810 7th Street
Washington, DC 20531

DOJ National Institute for Juvenile Justice and Delinquency Prevention Grants   2069
The objective of the grants is to encourage, coordinate, and conduct research and evaluation of juvenile justice and delinquency prevention activities; to provide for public and private agencies, institutions, justice system agencies, a clearinghouse and information center for collecting, disseminating, publishing, and distributing information on juvenile delinquency; to conduct national training programs of juvenile-related issues and provide technical assistance and training to federal, state, and local governments, courts, corrections, law enforcement, probation, public and private agencies, institutions, and individuals in the planning, establishment, funding, operation, or evaluation of juvenile delinquency programs. Deadlines are published in program announcements.
Requirements: Public or private agencies, organizations, and individuals are eligible.
Contact: Office of Juvenile Justice and Delinquency Prevention, (202) 307-5940; email: AskDOJ@usdoj.gov
Internet: http://www.usdoj.gov/10grants/index.html
Sponsor: Department of Justice
950 Pennsylvania Avenue NW
Washington, DC 20530-0001

DOJ National Institute of Corrections Training and Staff Development Grants 2070
The grants provide assistance to devise and conduct in various geographical locations seminars, workshops, and training programs for law enforcement officers, judges and judicial personnel, probation and parole personnel, correctional personnel, welfare workers, and other personnel, including lay ex-offenders and paraprofessionals, connected with the treatment and rehabilitation of criminal and juvenile offenders. In addition, the grants develop technical training teams to aid in the development of seminars, workshops, and training programs within the several states and with the state and local agencies that work with prisoners, parolees, probationers, and other offenders.
Requirements: States, general units of local government, as well as public and private agencies, educational institutions, organizations, and individuals involved in the development, implementation, or operation of correctional programs and services are eligible to apply.
Amount of Grant: $1500-$300,000; $100,000 average
Contact: Staff Development Branch, National Institute of Corrections, (202) 307-3106 or (800) 995-6423; fax: (202) 307-3361; email: AskDOJ@usdoj.gov
Internet: http://www.usdoj.gov/10grants/index.html
Sponsor: Department of Justice
950 Pennsylvania Avenue NW
Washington, DC 20530-0001

DOJ National Institute of Justice Graduate Research Fellowships   2071
The fellowships are intended to improve the quality and quantity of knowledge about crime and the criminal justice system, while helping to increase the number of persons who are qualified to teach in collegiate criminal justice programs; to conduct research related to criminal justice issues; and help such persons perform more effectively within the criminal justice system. A fellowship is funded for up to two years. Applications may be submitted at any time throughout the year. Proposals will be due and subsequently reviewed three times a year, in February, June, and October, with funding decisions made within 60-90 days of the review date.
Requirements: Applicants must be accredited institutions of higher education offering doctoral degree programs. Eligible students are doctoral candidates engaged in dissertation research and writing on a problem relating to law enforcement, crime, or criminal justice. This competitive program provides fellowship stipends, major project costs, and certain university fees.
Amount of Grant: $20,000 stipend
Contact: Sarah Hart, Director, (202) 307-2942; fax: (202) 307-6394
Internet: http://www.ojp.usdoj.gov/nij/funding.htm
Sponsor: Department of Justice
810 Seventh Street, NW
Washington, DC 20531

**DOJ National Institute of Justice Research, Development, and Evaluation   2072
   Project Grants**
The grants encourage and support research and development for further understanding of the causes and control of crime and improvement of the criminal justice system. The funds may be used to conduct research and development pertaining to these objectives, including the development of new or improved approaches, techniques, and systems; carry out programs of research on the causes of crime and means of preventing crime; and evaluate criminal justice programs and procedures. In addition to larger grant requests, the institute invites proposals for smaller funding amounts, including grants of less than $50,000 across all goals and subject areas in criminal justice. Each year, NIJ announces its program plan of research opportunities with specific deadline dates and contacts for each program. Contact the program office for guidelines.
Requirements: Grants and contracts may be made to and with state and local governments, profit and nonprofit organizations, institutions of higher education, and qualified individuals. Territories of the U.S. are also eligible to participate in this program.
Contact: National Institute of Justice, (202) 307-2942; fax: (202) 307-6394
Internet: http://www.ojp.usdoj.gov/fundopps.htm
Sponsor: Department of Justice
810 Seventh Street NW
Washington, DC 20531

DOJ National Institute of Justice Visiting Fellowships   2073
Fellowships provide opportunities for experienced criminal justice practitioners and researchers to pursue projects aimed at improved understanding of crime, delinquency, and criminal justice administration by sponsoring research projects of their own creation and design. Funds may be used to conduct research on crime causation, crime measurements, crime prevention, law enforcement, criminal justice administration, and the effectiveness and efficiency of anti-crime programs. Juvenile delinquency research projects also are eligible for support under this program. Fellows conduct their studies while based at NIJ. Concept papers may be submitted at any time. Applicants should anticipate a decision time frame of six to nine months from concept paper to award.
Requirements: Fellowships are awarded to individuals or to their parent agencies or organizations. Generally, professionals working in the criminal justice field, including university- or college-based academic researchers and upper-level managers in criminal justice agencies are eligible.
Contact: National Institute of Justice, Department of Justice, (202) 307-2942
Internet: http://www.usdoj.gov/nij
Sponsor: Department of Justice
810 Seventh Street NW
Washington, DC 20531

DOJ/NIJ Violence Against Women Research Grants   2074
NIJ's Violence Against Women and Family Violence program provides funds for research and evaluation, including research fellowships. The program's objectives are to estimate the scope of violence against women and family violence, identify their causes and consequences, evaluate promising prevention and intervention programs, disseminate research results to the field, and build partnerships among a wide variety of disciplines to accomplish these objectives. Visit the Web site for current solicitations for these programs.
Amount of Grant: $1.25 million total
Contact: Leora Rosen, Manager, (202) 616-2452; email: Leora.Rosen@usdoj.gov
Internet: http://www.ojp.usdoj.gov/nij/vawprog/funding.html
Sponsor: Department of Justice
810 Seventh Street NW
Washington, DC 20531

Dolores Zohrab Liebmann Fellowships   2075
The fellowships are awarded to graduate degree candidates who have outstanding undergraduate records, have demonstrated a need for financial assistance, are enrolled in selected U.S. college and universities, are U.S. citizens, and have received baccalaureate degrees. Applications are mailed to selected schools in late fall.
Requirements: The dean of the school must nominate qualified candidates.
Date(s) Application is Due: Jan 31
Contact: Fellowships Administrator, c/o JPMorgan Private Bank, Philanthropic Services, (212) 789-5682; email: jones_ed_l@jpmorgan.com

Internet: http://fdncenter.org/grantmaker/liebmann
Sponsor: Dolores Zohrab Liebmann Fund
345 Park Avenue, 4th Fl
New York, NY 10154

Donaghue Investigator Program Grants 2076

The program supports particularly promising medical researchers holding faculty appointments at Connecticut institutions. The funding emphasis is on the investigator and his or her program of research rather than on a specific study. Approximately four to six awards will be made each year.

Requirements: A prototypical recipient will be committed to pursuing his or her career in Connecticut for the foreseeable future; show potential for an outstanding independent research career, and for leadership in his or her field(s) of research; be accepting of the importance of earning support for research by actively explaining to the public his or her field of work and specific results and their value; be enthusiastic about and committed to disciplinary and interinstitutional collaboration in integrative research; be strongly supported by his or her institution; and possess evidence of external recognition of his or her work.

Geographic Focus: Connecticut
Amount of Grant: $100,000 maximum for five years
Contact: Lynne Garner, (860) 521-9011 ext 11; fax: (860) 521-9018; email: garner@donaghue.org
Internet: http://www.donaghue.org/granttype.htm
Sponsor: Patrick and Catherine Weldon Donaghue Medical Research Foundation
18 N Main Street
West Hartford, CT 06107

Donaghue Practical Benefit Initiatives Grants 2077

The foundation initiates research projects showing particular promise for producing practical benefit to human health and works in an interactive process with prospective investigators to develop projects for funding. Organizations with project ideas aimed at productive ways to deliver practical benefit to the public through health research should submit a short summary letter. There is no specific timeline for application or designated grant award amount.

Contact: Director, (860) 521-9011; fax: (860) 521-9018; email: office@donaghue.org
Internet: http://www.donaghue.org
Sponsor: Patrick and Catherine Weldon Donaghue Medical Research Foundation
18 N Main Street
West Hartford, CT 06107-1919

Donald and Sylvia Robinson Family Foundation Grants 2078

The foundation awards grants to U.S. nonprofit organizations in its areas of interest, including animal and wildlife protection, arms control, arts (general, performing arts, and visual arts), environmental conservation and protection, eye diseases and eye research, family planning and human reproductive health, food distribution, international affairs, Israel, Jewish social services, and social service delivery programs. Types of support include annual campaigns, building construction/renovation, capital campaigns, and general operating support. There are no application deadlines or forms.

Requirements: National nonprofit organizations are eligible to apply.
Restrictions: Individuals are ineligible.
Geographic Focus: Pennsylvania
Contact: Donald Robinson, (412) 661-1200; fax: (412) 661-4645
Sponsor: Donald and Sylvia Robinson Family Foundation
6507 Wilkins Avenue
Pittsburgh, PA 15217

Donald P. Eckman Award 2079

This award is given for outstanding accomplishments by a young engineer in the field of automatic control. Nominees must be younger than 35 years on January 1 of the year of the award and have made contributions to the field while residing in the United States. Contributions may be technical or scientific publications, theses, patents, inventions, or combinations or the above in the field of automatic control. The award consists of a certificate and honorarium, which are presented at the annual conference. Nomination forms as well as additional information may be obtained from the AACC secretary.

Requirements: Nomination packages should consist of letter of nomination, three to five letters of reference, current resume, current list of publications and patents, any other supporting material, and nomination form. The nomination letter should clearly identify the primary reason that the nominee should receive the award, as well as up to four ancillary reasons. All materials should be submitted in a single package.

Date(s) Application is Due: Dec 1
Contact: AACC Secretariat, c/o Department of Electrical Engineering, Wright State University, (937) 775-5062; fax: (937) 775-3936; email: pmisra@cs.wright.edu
Internet: http://www.a2c2.org/awards/index.php
Sponsor: American Automatic Control Council
3640 Col Glenn Hwy
Dayton, OH 45435

Donald Q. Kern Award in Heat Transfer or Energy Conversion 2080

This award is given annually by the Heat Transfer and Energy Conversion Division of AIChE for expertise in a given field of heat transfer or energy conversion. Special emphasis is given to recognition of contributions that have had significant practical applications. A plaque accompanies the award. The recipient will be required to prepare a written review covering an area selected by the recipient.

Requirements: Three to five supporting letters are required that discuss specific contributions by the nominee and the reasons for their value.
Amount of Grant: $1500
Date(s) Application is Due: Jun 1
Contact: AIChE Awards Programs, (212) 591-7107; fax: (212) 591-8882; email: awards@aiche.org
Internet: http://www.aiche.org/awards/awarddtl.asp?AwardID=40
Sponsor: American Institute of Chemical Engineers
3 Park Avenue
New York, NY 10016-5901

Doreen Kronick Scholarship 2081

This scholarship is awarded annually to a graduate student in a program that will lead to the recipient's being able to assist persons with learning disabilities. Application forms are available from the association or the faculty of graduate studies at Canadian universities.

Requirements: Any student, 18 or older, enrolled in any graduate program of any university in Canada is eligible.
Restrictions: Scholarship recipients will not be eligible to reapply for three years.
Amount of Grant: $C500
Date(s) Application is Due: May 15
Contact: Kronick Scholarships, (613) 238-5721; fax: (613) 235-5391; email: information@ldac-taac.ca
Internet: http://www.ldac-taac.ca/Scholarships/scholarships-e.asp
Sponsor: Learning Disabilities Association of Canada
323 Chapel Street, Suite 200
Ottawa, ON K1N 7Z2 Canada

Doris Duke Charitable Foundation Clinical Interfaces Award Program 2082

The program seeks to catalyze activity at the interface of clinical and other research disciplines by: supporting the formation of new collaborations and strengthening existing collaborations of outstanding scientists across disciplines; demonstrating successful models for clinical research at the interface of multiple disciplines; and supporting interdisciplinary and inter-institutional endeavors that go beyond the program project mindset. Full grants, awarded over five years, are made to established teams with key investigators from at least three disciplines. Planning grants are awarded to new teams for the development of full proposals over 18 months. The listed application deadline is for pre-proposals; full proposals are by invitation. Though new grants are not being offered continually, potential applicants can sign up to be notified of future competitions.

Requirements: Teams of at least three key investigators whose primary expertise lie in different disciplines are eligible to apply. Key investigators must have advanced degrees (MD, PhD, MD/PhD, or the equivalent), and one of the key investigators must be a clinical researcher. The team leader must work in a U.S. nonprofit institution, such as an academic medical center. The team may include investigators at other institutions in the United States and overseas.

Restrictions: Planning grants will not be awarded in the current competition.
Amount of Grant: $2.25 million maximum over 5 years
Date(s) Application is Due: Nov 2
Contact: Elaine K. Gallin, Program Director; (212) 974-7104; fax: (212) 974-7590; email: egallin@ddcf.org
Internet: http://www.ddcf.org/page.asp?pageId=299
Sponsor: Doris Duke Charitable Foundation
650 Fifth Avenue, 19th Floor
New York, NY 10019

Doris Duke Charitable Foundation Clinical Research Fellowships for Medical Students 2083

The fellowship program is designed to encourage medical students to pursue careers in clinical research by giving exceptional students the opportunity to take a year to experience clinical research first hand. The CRF program is available at the following medical schools: Columbia University College of Physicians and Surgeons; Harvard Medical School; Mount Sinai School of Medicine; University of California at San Francisco School of Medicine; University of Iowa Roy J. and Lucille A. Carver College of Medicine; University of North Carolina at Chapel Hill School of Medicine; University of Pennsylvania School of Medicine; University of Texas Southwestern Medical Center at Dallas; Washington University Medical School; and Yale University School of Medicine. An additional 11-12 fellowships will be available to medical students interested in conducting clinical research in Africa. Each participating medical school provides medical students with a one-year fellowship experience in clinical research that includes both didactic and research components; matches students to outstanding clinical research mentors; and offers fellowships to at least five students per year.

Requirements: Medical students matriculated at any U.S. medical school who have completed two or more years of medical school prior to the start of the fellowship and who have completed some clinical experience are eligible to apply to any of the participating schools.

Amount of Grant: $27,000 plus health insurance
Date(s) Application is Due: Jan 18
Contact: Elaine K. Gallin, Program Director; (212) 974-7104; fax: (212) 974-7590; email: egallin@ddcf.org
Internet: http://www.ddcf.org/page.asp?pageId=292
Sponsor: Doris Duke Charitable Foundation
650 Fifth Avenue, 19th Floor
New York, NY 10019

Doris Duke Charitable Foundation Clinical Research Systems Grants   2084
The Program is founded on the premise that clinical research requires not only human and research capital, facilities, and equipment, but it also requires robust support systems, including appropriate regulatory organizations and processes, common language, analytical tools, databases and information systems, as well as mechanisms to distribute findings and interface with the public. The Program is interested in supporting activities that seek to identify ways to strengthen the regulatory processes and systems for clinical research, such as the protection of human subjects. The foundation is willing to consider any exceptional opportunities that may arise.
Samples: Consortium to Examine Clinical Research Ethics, three grants totaling $930,000 (2000-05).
Contact: Elaine Gallin, Program Director for Medical Research; (212) 974-7104; fax: (212) 974-7590; email: egallin@ddcf.org
Internet: http://www.ddcf.org/page.asp?pageId=302
Sponsor: Doris Duke Charitable Foundation
650 Fifth Avenue, 19th Floor
New York, NY 10019

Doris Duke Charitable Foundation Clinical Scientist Development Award   2085
Grants are awarded to junior physician-scientists to facilitate their transition to independent clinical research careers. The program is designed to help prepare and support new investigators with an MD or MD/PhD as they begin their careers as independent clinical researchers. The program is aimed at conducting clinical research in any disease area. The listed application deadline is for prepoposals, which must be submitted electronically.
Requirements: Applicants must: be a physician-scientist conducting clinical research in any disease area; have received an M.D. or a foreign equivalent from an accredited institution; be working in a U.S. degree-granting institution, but do not have to be a U.S. citizen; have a full-time faculty level position not higher than the Assistant Professor level; and have been appointed to their first full-time faculty level position between January 31, 2002 and January 31 of the grant year. (All full-time post-fellowship Instructor level positions will be considered full-time faculty level appointments).
Restrictions: Funds cannot be used on experiments that utilize animals or primary tissues derived from animals. An award will not be made if, prior to the commencement of the award, the applicant becomes the principal investigator on a federal government, peer-reviewed, research or career development award or any non-government award averaging more than $126,000 per year in direct costs and of a duration of three or more years.
Amount of Grant: $125,000 (direct costs) per year and $10,000 (indirect costs) per year for three years
Date(s) Application is Due: Jan 23
Contact: Elaine K. Gallin, Program Director; (212) 974-7104; fax: (212) 974-7590; email: egallin@ddcf.org
Internet: http://www.ddcf.org/page.asp?pageId=291
Sponsor: Doris Duke Charitable Foundation
650 Fifth Avenue, 19th Floor
New York, NY 10019

Doris Duke Charitable Foundation Clinical Scientist Development   2086
Award (CSDA) Bridge Grants
The purpose of the bridge grant is to help Doris Duke Clinical Scientist Development Award (CSDA) recipients successfully transition to independent National Institutes of Health (NIH) R01 funding. The Medical Research Program created these grants to bridge the time between when a CSDA investigator learns of a very strong but unfunded score on an NIH application to when s/he is able to successfully obtain an R01 grant. The bridge grants provide protected time and support for CSDA recipients to continue their research projects, collect supportive data, keep patients enrolled if applicable, and strengthen their proposals to resubmit to the NIH. The foundation hopes that the bridge grants will help keep promising physician-scientists committed to clinical research careers by providing stable support during their critical transition to independent R01 funding. The foundation will award up to four bridge grants of $135,000 each to be spent only on direct costs over 12 to 18 months. Eligible CSDA recipients are invited to submit applications throughout the year up to the deadline of September 1.
Requirements: This competition is open only to former CSDA grantees or current CSDA grantees in the last year of their award (or in a no-cost extension period).
Amount of Grant: $135,000 maximum
Date(s) Application is Due: Sep 1
Contact: Elaine Gallin, Program Director for Medical Research; (212) 974-7104; fax: (212) 974-7590; email: egallin@ddcf.org
Internet: http://www.ddcf.org/page.asp?pageId=291
Sponsor: Doris Duke Charitable Foundation
650 Fifth Avenue, 19th Floor
New York, NY 10019

Doris Duke Charitable Foundation Distinguished Clinical Scientist Awards   2087
The purpose of the program is to recognize outstanding physician-scientists who are engaged in applying the latest basic science advances to the prevention, diagnosis, treatment, and cure of disease, and to enable the physician-scientist to support and mentor the next generation of physician-scientists conducting translational clinical research. Awards will be granted to physician-scientists conducting translational clinical research in any disease area; this award cycle is not limited to specific disease areas.
Requirements: Grantees must hold an M.D. degree from an accredited institution in the United States (holders of M.D./Ph.D. degrees are also eligible, as are holders of M.D.-equivalent degrees from non-U.S. institutions); hold a full-time university faculty

appointment at the level of Associate Professor or above as of the date of nomination; have been appointed to their first full-time, faculty-level position for no more than 15 years; and have an established translational clinical research program.
Restrictions: Experiments that utilize animals or primary tissues derived from animals are not eligible for support through this award program. An award will not be made if, prior to the commencement of the foundation's award, the applicant becomes the principal investigator of a federal government, peer-reviewed, research or career development award or of any nongovernment award averaging $100,000 or more per year and of three years or more duration.
Amount of Grant: $1.5 million maximum
Date(s) Application is Due: Feb 14
Contact: Elaine Gallin, Program Director for Medical Research; (212) 974-7104; fax: (212) 974-7590; email: egallin@ddcf.org
Internet: http://www.ddcf.org/page.asp?pageId=297
Sponsor: Doris Duke Charitable Foundation
650 Fifth Avenue, 19th Floor
New York, NY 10019

Doris Duke Charitable Foundation Operations Research on AIDS   2088
Care and Treatment in Africa (ORACTA) Grants
The Program seeks to improve the care and treatment of AIDS patients in resource-limited settings, inform antiretroviral therapy (ART) policy and practice, and improve outcomes of the roll-out and scale-up of ART in Africa. ORACTA grants provide two-year grants of up to $100,000 per year to teams of investigators conducting health operations research on AIDS care and treatment in Africa. Though new grants are offered only periodically, applicants should sign up for the Medical Research Program's mailing list.
Restrictions: Indirect costs are not supported by this award.
Amount of Grant: $100,000 per year for two years
Contact: Elaine Gallin, Program Director for Medical Research; (212) 974-7104; fax: (212) 974-7590; email: egallin@ddcf.org
Internet: http://www.ddcf.org/page.asp?pageId=486
Sponsor: Doris Duke Charitable Foundation
650 Fifth Avenue, 19th Floor
New York, NY 10019

Dorothea Haus Ross Foundation Grants   2089
The foundation awards grants to eligible nonprofit organizations that work to relieve suffering among children who are sick, handicapped, injured, disfigured, orphaned, or otherwise vulnerable. Types of support include direct services, medical research, equipment and supplies, and small renovation projects. There are no application deadlines. Grants have supported projects for children in Native American schools and schools that serve populations of children who are disadvantaged or underserved.
Requirements: U.S. Charities may apply if: they have 501(c)3 status; they are listed in the current edition of the Cumulative List of Charities published by the U.S. Department of the Treasury; they are a Catholic organization listed in the current edition of the Catholic Director; or they are listed in the Free Methodist Yearbook, or other Protestant Denomination Directory that has a group ruling for tax exemption from the IRS. Although grants are made internationally, Foundation by-laws prohibit sending money directly to foreign charities.
Restrictions: The Foundation does not fund day-to-day operations, individuals, conferences, day care, or public education. Although the Foundation makes international grants, there are restrictions in some countries for the following reasons: war, widespread violence, or breakdown of law and order; or countries where grants are restricted by the U.S. Government due to a boycott or other reason.
Contact: Wayne Cook, Executive Director, (716) 473-6006; fax: (716) 473-6007; email: rossfoundation@frontiernet.net
Internet: http://www.dhrossfoundation.org/index.html
Sponsor: Dorothea Haus Ross Foundation
1036 Monroe Avenue
Rochester, NY 14620

Dorothea Klumpke-Roberts Award   2090
An award and plaque are given annually for outstanding contributions to public understanding and appreciation of astronomy. The contributions may be in the form of popular books and articles; lectures; radio, TV, or movie productions; or service to public education in astronomy of any other nature. Nominations are welcome from all members of the Society and the public at large. Nominations are made by a special committee of the Astronomical Society of the Pacific; suggestions from the public are welcome.
Amount of Grant: $500
Date(s) Application is Due: Dec 31
Contact: Marilyn Delgado, Awards Committee, (415) 337-1100; fax: (415) 337-5205; email: mdelgado@astrosociety.org
Internet: http://www.astrosociety.org/membership/awards/klumpke.html
Sponsor: Astronomical Society of the Pacific
390 Ashton Avenue
San Francisco, CA 94112

Dorothy Leet Grants   2091
This competition is held in even-numbered years to award grants that assist women graduates from countries with low per capita income, or they may be given to other women graduates who wish to work as experts in these countries or whose research is of value to such countries. Grant is to be used preferably in a country other than that in which the

applicant received her education or habitually resides. Grants may be used for obtaining special training essential to research and survey work; training in new techniques in group research and further study; or carrying out independent research or surveys, including completion of projects well advanced at the time of application. Members of the IFUW should apply through their national IFUW unit (in the United States, the American Association of University Women, 1111 16th Street NW, Washington, DC 20036; info@aauw.org). Deadline for applications is determined by each affiliate, but normally falls between early September and mid-October of the year preceding the competition. Contact national headquarters for the exact deadline date.

Amount of Grant: Sf3000-Sf6000 average

Date(s) Application is Due: Oct 1

Contact: Fellowships Officer, 41-22-731-23-80; fax: 41-22-738-04-40; email: info@ifuw.org

Internet: http://www.ifuw.org/fellowships/international.htm

Sponsor: International Federation of University Women

8 rue de l'Ancien Port

Geneva CH 1201

Switzerland

Dorothy Rider Pool Health Care Grants   2092

The Foundation's intent is to serve as a means to improve the quality of life in the Lehigh Valley community, to build on community strengths and add to its vitality, and to increase the capacity of the community to serve the needs of all its citizens. Within this objective the Foundation's funding program is focused on education, health and welfare, culture and art and community development. Interested applicants should submit a letter of intent of five pages or less.

Requirements: Allentown, PA, nonprofit organizations are eligible.

Restrictions: The Foundation is restricted from providing funds to individuals, legislative or lobbying efforts, political or fraternal organizations or organizations outside the United States and its territories. The Foundation as a policy does not provide operating or capital funds to Sectarian institutions, organizations or programs in which funds will be used primarily for the propagation of religion, hospitals or United Way member agencies. Further, the Foundation does not underwrite charitable or testimonial dinners, fund-raising events or related advertising or the subsidization of books, mailings or articles in professional journals.

Geographic Focus: Pennsylvania

Samples: Lehigh Valley Hospital and Health (Allentown, PA)—for its department of obstetrics and gynecology, $225,000 over three years.; Lehigh Valley Hospital and Health Network (Allentown, PA)—for the department of emergency medicine's sexual assault response team, $150,000 over three years.; Lehigh Valley Hospital and Health Network (Allentown, PA)—for the Physician Leaders in the Lehigh Valley program, $120,000 over three years.

Amount of Grant: $100,000

Date(s) Application is Due: Apr 1; Aug 15

Contact: Ronald Dendas, Program Manager, (610) 770-9346; fax: (610) 770-9361; email: drpool@ptd.net

Internet: http://www.pooltrust.com

Sponsor: Dorothy Rider Pool Health Care Trust

1050 S. Cedar Crest Boulevard, Suite 202

Allentown, PA 18103

Dorr Foundation Grants   2093

Grants are made primarily for programs designed to develop new science curricula from sixth-12th grade. Support is also given to special education projects for youth relating to conservation and the environment if such projects involve the school's curriculum. In addition, some grants are made available to promote research and disseminate information on chemical, metallurgical, and sanitation engineering. Grants are awarded on a national basis, with emphasis in the Northeast. Types of support include equipment, emergency funds, program development, seed money, curriculum development, scholarship funds, and research. Initial contact should be a phone call. There is no deadline. No response can be expected unless there is interest on the part of the trustees. Applications are accepted at any time.

Requirements: 501(c)3 tax-exempt organizations are eligible.

Restrictions: Grants are not made to individuals or for operation budgets, continuing support, annual campaigns, deficit financing, endowment funds, or conferences and seminars.

Samples: Council on the Environment of New York City (NY)—to support chemistry curriculum revisions for High School for Environmental Studies, $12,000.; Sage Colleges (Troy, NY)—to support two scholarships per year for three years, $10,000 per year.; PS 116, New York City (NY)—for microscopes for a new science room, $8000.

Amount of Grant: $1000-$40,000 range

Contact: Barbara McMillan, Chairperson, (212) 433-6438

Sponsor: Dorr Foundation

P.O. Box 328

Eastchester, NY 10709

Dorrance Family Foundation Grants   2094

The foundation awards general support grants in grant-making priority areas, including arts and culture, civic affairs, education, health, religion, science, and social services. Specific interests include arts associations, museums, galleries, music, and theater; botanical gardens and zoos; higher education, precollege private education, and special education; AIDS/HIV, arthritis, kidney diseases, hospices, and hospitals; religious organizations; science museums; and United Way organizations. Capital campaigns

also are supported. Priority is given to requests from Arizona. There are no application deadlines. Contact Program Officer before application.

Geographic Focus: Arizona

Samples: Phoenix Art Museum (Phoenix, AZ)—to design and construct an outdoor sculpture court, and for an endowment for educational activities, $3 million partial challenge grant (2004).

Contact: Carolyn O'Malley, Program Officer, (480) 367-7000

Sponsor: Dorrance Family Foundation

7600 East Doubletree Ranch Road, Suite 300

Scottsdale, AZ 85258

DOS Freedom Support Educational Partnerships with Eurasia   2095

The Office of Global Educational Programs of the Bureau of Educational and Cultural Affairs announces an open competition for the Freedom Support Educational Partnerships Program with Eurasia. The objectives of the program are to support democratic systems and market economies in Armenia, Azerbaijan, Belarus, Georgia, Kazakhstan, Kyrgyzstan, Moldova, Russia, Tajikistan, Ukraine, and Uzbekistan, and to strengthen mutual understanding and cooperation between these countries and the United States. The means of achieving these objectives may include faculty exchange, curriculum development, and outreach to professionals and other members of the communities served by the participating institutions. The purpose of the program is to support the development or revision of courses, curricula, outreach programs and programs of study at participating institutions in ways that strengthen democracy and free markets in Eurasia as well as mutual understanding between the people of the United States and those of Eurasia. Applicants are strongly encouraged to discuss their project ideas during the proposal development process with the relevant program officer, who may be able to provide additional insight into priorities by country as well as background information on what types of projects are most competitive for funding.

Requirements: Accredited, postsecondary educational institutions meeting the provisions described in Internal Revenue Code section 26 U.S.C 501(c)3 may submit proposals to pursue institutional or departmental objectives in partnership with foreign counterpart institutions.

Amount of Grant: $250,000 maximum

Contact: Office of Global Educational Programs, (202) 647-4000

Internet: http://exchanges.state.gov/education/partnership

Sponsor: Department of State

2201 C Street NW

Washington, DC 20520

DOS Partners in Education Program   2096

Partners in Education (PiE), a project of the Office of Global Educational Programs, brings secondary level social science teachers, teacher trainers and administrators from targeted regions in Russia, Ukraine, Kyrgyzstan, Uzbekistan, Armenia, Azerbaijan and Georgia for a six-week, school-based internship program focusing on citizenship education. Host institutions, often university level schools of education, also provide professional development seminars, cultural activities and homestays. U.S. teachers from host communities return the visit to their NIS counterparts. Both participants and U.S. host institutions are selected through an open merit-based competition. Since 1998, the program has brought to the U.S. more than 500 educators from the Eurasia to learn about different approaches to civic education, curriculum development, and teaching methodologies. The program is administered by the American Councils for International Education.

Requirements: Public and private nonprofit organizations meeting the provisions described in IRS regulation 26 CFR 1.501(c) are eligible to apply.

Restrictions: Proposals relating to the teaching of English or English as a foreign language (EFL) are not eligible.

Amount of Grant: $150,000

Contact: Office of Global Educational Programs, (202) 619-5289; fax: (202) 401-1433

Internet: http://exchanges.state.gov/education/partnership/fulbright.htm

Sponsor: Department of State

301 Fourth Street SW

Washington, DC 20547

DOS Study of the United States Institutes   2097

The program seeks grant proposals from U.S. colleges and universities to develop and implement summer institutes. Each institute is a six-week U.S.-based seminar, including up to two weeks of travel, designed for multinational groups of 18 foreign university teaching faculty or 30 secondary school educators. Institute include Study of the United States for Foreign Secondary School Educators; Religious Pluralism in the United States; U.S. Foreign Policy; American Politics and Political Thought; and Contemporary American Literature. The application of the Study of the United States for Foreign Secondary Educators institute is December 13; the January 10 deadline applies to the other institutes. Guidelines are available at the website.

Requirements: U.S. colleges and universities are eligible.

Date(s) Application is Due: Jan 10; Dec 13

Contact: Jennifer Phillips, (202) 453-8537; fax: (202) 453-8533

Internet: http://exchanges.state.gov/education/rfgps/menu.htm

Sponsor: Department of State

301 4th Street, SW, Room 314

Washington, DC 20547

DOS Women's Political, Educational, and Economic Development for Afghanistan Grants 2098

The program supports a series of exchanges and training programs promoting women's political, educational, and economic development in Afghanistan. Eligible U.S. applicants may submit proposals to develop and implement exchanges and training programs involving participants from Afghanistan, including training conducted in Afghanistan. These U.S. organizations should have a current presence in Afghanistan, or experience working in Afghanistan, and work in conjunction with Afghan NGO partners. The program solicits proposals for exchange projects that involve the following priority themes: women's leadership, educational development and literacy for women and girls, women-led small business development, job skills training, and NGO management. Up to five grants may be awarded.

Requirements: U.S. public and private nonprofit organizations meeting the provisions described in IRS Code section 26 U.S.C 501(c)3 may submit proposals.
Amount of Grant: $150,000 maximum
Date(s) Application is Due: Feb 6
Contact: Office of Citizen Exchanges, (202) 619-5320; fax: (202) 619-4350
Internet: http://exchanges.state.gov/education/rfgps
Sponsor: Department of State
301 4th Street SW, ECA/PE/C/NEA-AF, Room 216
Washington, DC 20547

DOT Dwight David Eisenhower Transportation Fellowship Program 2099

The program's goals are to attract people to the field of transportation, to enhance the careers of transportation professionals by encouraging them to seek advanced degrees, and to retain top talent in the transportation community. The fellowship awards are: the Eisenhower Graduate Fellowships to allow students to pursue master's degrees or doctorates in transportation-related fields; the Eisenhower Historically Black Colleges and Universities Fellowships to provide HBCU students with additional opportunities to enter careers in transportation; the Eisenhower Hispanic Serving Institutions Fellowships to provide HSI students with opportunities to enter transportation careers; the Eisenhower Tribal Colleges Initiatives, which provide fellowships for Native American students and faculty at tribal colleges; and the Eisenhower Faculty Fellowships to provide faculty in transportation fields with opportunities to improve their transportation knowledge by attending conferences, courses, seminars, or workshops.

Date(s) Application is Due: Apr 15; Oct 15
Contact: Gwen Sutton, Program Manager, (703) 235-0535; fax: (703) 235-0593; email: transportationedu@fhwa.dot.gov
Internet: http://www.nhi.fhwa.dot.gov/uandg.asp
Sponsor: Federal Highway Administration
4600 N Fairfax Drive, Suite 800
Arlington, VA 22203

DOT Recreational Boating Safety Grants 2100

These grants are used to fund projects on various subjects promoting boating safety on the national level. It targets specific boat market segments and recreational boating safety topics.

Requirements: National nongovernmental, nonprofit, public service organizations are eligible to apply.
Amount of Grant: $3 million total
Contact: Capt. Scott Evans, Office of Boating Safety, U.S. Coast Guard, (202) 267-0950; Vickie Hartberger, Office of Boating Safety, U.S. Coast Guard, (202) 267-0974
Internet: http://www.uscgboating.org/grants/state/rbs.htm
Sponsor: Department of Transportation
2100 Second Street SW
Washington, DC 20593

DOT Transit Planning and Research Grants 2101

The objectives of the program are to foster innovation in public transit systems, through local demonstrations of promising, but risky, new technologies and service or operational concepts to provide information that can be used nationally; to address economic and social issues resulting from human impacts on the environment, and develop risk assessment methodologies, integrated assessments, and other analytical tools for effective policy formulation; to develop practical know-how for solving fundamental industry-wide problems, such as how to accommodate the travel needs of persons with disabilities, how to finance transit infrastructure construction and maintenance, and how to meet requirements of the Clean Air Act; and to support development of information and technical assistance to convey results of research, technology development, and innovative demonstrations for adaptation and local implementation.

Requirements: Eligible applicants include public bodies, nonprofit institutions, state and local agencies, universities, and legally constituted public agencies and operators of public transportation services.
Contact: Associate Administrator for Research, Demonstration, and Innovation, Federal Transit Administration, (202) 366-4052
Internet: http://www.fta.dot.gov
Sponsor: Department of Transportation
400 Seventh Street SW
Washington, DC 20590

DOT Transportation Statistics Research Grants 2102

The purpose of this grant program is to provide financial assistance to eligible organizations to help advance the discipline of transportation statistics. The Bureau of Transportation Statistics (BTS) is an operating administration within the DOT. Its mission is to lead in developing transportation data and information of high quality, and to advance their effective use in public and private transportation decisionmaking. The ultimate goal is to make transportation better—to enhance safety, mobility, economic growth, the human and natural environment, and national security.

Requirements: Public and private nonprofit institutions of higher education are eligible to apply.
Amount of Grant: $25,000-$200,000; $75,000 average
Contact: Mary Hutzler, Acting Director, Bureau of Transportation Statistics, (202) 366-9913; fax: (202) 366-3640; email: answers@bts.gov
Internet: http://www.bts.gov
Sponsor: Department of Transportation
400 Seventh Street SW
Washington, DC 20590

Douglass Institute Fellowships 2103

The Frederick Douglass Institute for African and African American Studies supports a program of predoctoral and postdoctoral fellowships on the prospects for cultural, social, and economic development in Africa and its diaspora. Annual themes vary; contact the associate director of the institute for details of the program.

Requirements: Scholars holding a PhD degree in a field related to the African and African-American experience are eligible for the postdoctoral fellowship. Graduate students of any university who are studying aspects of the African and African-American experience are eligible for the predoctoral fellowship. Students who want to begin their graduate work in African and African-American studies at the university are eligible for the graduate fellowship.
Amount of Grant: $35,000 postdoctoral fellowship stipend, $18,000 predoctoral fellowship stipend
Date(s) Application is Due: Jan 31
Contact: Associate Director for Research Fellowships, Frederick Douglass Institute, (585) 275-7235; fax: (585) 256-2594; email: fdi@troi.cc.rochester.edu
Internet: http://www.rochester.edu/College/AAS/fellowships.php
Sponsor: University of Rochester
RC Box 270440, 302 Morey Hall
Rochester, NY 14627-0440

Dover Foundation Grants 2104

The primary purpose of the foundation is to serve the interests of Cleveland County and other areas of North Carolina. Foundation grants seek to strengthen the spiritual, mental, and moral fiber of the community; give priority to those initiatives and programs that will elevate the life and educational opportunities of deserving recipients; and discover ways to encourage others to come together to address needs for the common welfare of the community. Types of support include general operating support, annual campaigns, capital campaigns, building/renovation, emergency funds, scholarship funds, and research. A formal grant request must be submitted by letter. The board of directors meets four times a year to consider grant awards—January, April, July, and October.

Requirements: Applicants must reside in North Carolina.
Restrictions: The foundation ordinarily does not make grants to organizations whose principal activities are outside the United States; political activities or entities; individuals or their projects; advertising; newsletters, magazines, or books; or trips or tours.
Geographic Focus: North Carolina
Samples: Gardner-Webb U, School of Business (Boiling Springs, NC)—to endow a faculty position at the School of Business, $500,000.; Heineman Medical Research Center of Charlotte (Charlotte, NC)—for support, $50,000.
Amount of Grant: $500-$25,000 average
Contact: Hoyt Bailey, President, (704) 487-8888; fax: (704) 482-6818; email: doverfnd@shelby.net
Sponsor: Dover Foundation
P.O. Box 208
Shelby, NC 28151

Dr. Bob and Jean Smith Foundation Grants 2105

The foundation awards grants to Texas nonprofit organizations in its areas of interest, including arts and culture, higher education, medical education, medical research, and health care. Types of support include annual campaigns, building construction/renovation, capital campaigns, general operating support, matching/challenge grants, and scholarship funds. There are no application deadlines; the board meets quarterly.

Requirements: Texas nonprofit organizations are eligible. Preference is given to Dallas-based organizations.
Restrictions: Individuals are ineligible.
Geographic Focus: Texas
Samples: Southwestern Medical Foundation (Dallas, TX)—to conduct neuromuscular research at the U of Texas Southwestern Medical Ctr at Dallas, $1 million (2004).
Amount of Grant: $100-$500,000 range
Contact: Sally Smith, Grants Administrator, (214) 521-3461
Sponsor: Dr. Bob and Jean Smith Foundation
3811 Turtle Creek Ctr, No 2150 LB 53
Dallas, TX 75219

Dr. Courtney W. Shropshire Scholarship Grant 2106

All qualified applicants will be considered without regard to race, creed, sex, or national origin. The grant is to be used only for tuition and approved purposes that are invoiced from the school. Necessary forms are available annually after September 1 from the

Civitan International Foundation; please include a self-addressed business envelope with postage to cover two-ounce U.S. mailing.
Requirements: Candidates must be a Civitan member (or a Civitan's immediate family member) and must have been a Civitan for at least two years and/or must be or have been a junior Civitan for no less than two years. Candidates must be enrolled in a degree or certificate program at an accredited community college, vocational school, four-year college, or graduate school.
Amount of Grant: $1000 maximum
Date(s) Application is Due: Jan 31
Contact: Shropshire Scholarship Grant, (205) 591-8910; fax: (205) 592-6307; email: civitan@civitan.org
Internet: http://www.civitaninternational.com/templates/cuscivitan/details.asp?id=23844&PID=155457
Sponsor: Civitan International Foundation
P.O. Box 130744
Birmingham, AL 35213-0744

Dr. J. Roberto Villavicencio Foundation Grants   2107
The foundation operates in Rosario and its surrounding area in the fields of education, medicine and health, and science and technology. Types of support include research grants to institutions and individuals, scholarships and fellowships, prizes, conferences, training courses, and publications.
Contact: Dra. Ana Maria Uriarte, Director, fax: 54-341-449 0152; email: info@villavicencio.org.ar
Internet: http://www.villavicencio.org.ar
Sponsor: Dr. J. Roberto Villavicencio Foundation
Alvear 854
Rosario, Santa Fe 2000
Argentina

Dr. Robert H. Goddard Historical Essay Award   2108
A competition is held annually for essays dealing with any significant aspect of the historical development of rocketry and astronautics. The essay may, through research, bring new information to light or may cast a new and different light upon events or individuals influencing rocketry and astronautics in the United States. Essays should not exceed 5000 words and should be fully documented. Complete contest rules are available; send a self-addressed envelope with inquiries.
Requirements: Applicant must be a U.S. citizen.
Restrictions: Previous winners are not eligible.
Amount of Grant: $1000
Date(s) Application is Due: Dec 1
Contact: Goddard Historical Essay Award, (202) 973-8661
Internet: http://www.nscfl.com/awards.html
Sponsor: National Space Club
2025 M Street NW, Suite 800
Washington, DC 20036-4907

Dr. Robert H. Goddard Research Fellowships   2109
These one-year, nonrenewable fellowships are awarded for full-time graduate study at the institute. In addition to the fellowship stipend, the award includes up to 20 credits of tuition, $1,591/month for 12 months, one year fellowship. Interested persons should contact the graduate admissions office for details on enrolling for graduate study and eligibility for the fellowships. Preference given to doctoral applicants.
Requirements: Applicants must be U.S. citizens with acceptance to a WPI graduate-degree program. GRE or GMAT test scores are required.
Amount of Grant: $1,591 annual stipend
Date(s) Application is Due: Feb 15
Contact: Graduate Admissions, (508) 831-5301; fax: (508) 831-5717; email: gse@wpi.edu
Internet: http://www.grad.wpi.edu/Financial/fellowships.html
Sponsor: Worcester Polytechnic Institute
100 Institute Road
Worcester, MA 01609-2280

Dr. Robert H. Goddard Scholarship   2110
The club awards a scholarship for the academic year to stimulate the interest of talented students in the opportunity to advance scientific knowledge through space research and exploration. The award is given at the Goddard Memorial Dinner in the latter part of March. The National Space Club will pay travel and lodging costs so that the winner may attend the dinner. Applicants should apply by letter and provide the data requested. Funds awarded are paid to the winner through his/her university before the next academic year begins. The winner is eligible to compete for a second year if the circumstances and his/her accomplishments warrant it. Deadline usually is the first Friday in January.
Requirements: Applicant must be a U.S. citizen in at least the junior year of an accredited university and have the intention of pursuing undergraduate or graduate studies in science or engineering during the interval of the scholarship. Selection is based on official transcripts, letters of recommendation from faculty, accomplishments demonstrating personal qualities of creativity and leadership, scholastic plans that would lead to future participation in some phase of the aerospace science and technology, and proven past research and participation in space-related science and engineering. Personal need is considered but is not controlling.
Amount of Grant: $10,000

Date(s) Application is Due: Jan 8
Contact: Goddard Scholarship, (202) 973-8661
Internet: http://www.wpi.edu/Academics/FS/entrydetail.php?award_id=139
Sponsor: National Space Club
2000 L Street NW, Suite 710
Washington, DC 20036-4907

Dr. Sydney Segal Research Grants   2111
Research grants are awarded to enable students to pursue full-time higher degree studies at a Canadian institution researching the possible causes, effects, and/or prevention of sudden infant death syndrome. The research grants are awarded for one year in any discipline—medical, psychological, biological, sociological, nursing, or other. Contact the foundation for applications.
Requirements: Applications are invited from suitably qualified graduate students who are undertaking full-time training in research in the health sciences leading to an MSC, PhD, or equivalent degree.
Amount of Grant: $C35,000 maximum
Date(s) Application is Due: Jun 1
Contact: Executive Director, (800) 363-7437 or (416) 488-3260; fax: (416) 488-3864; email: sidsinfo@sidscanada.org
Internet: http://www.sidscanada.org/segal_research_grants.htm
Sponsor: Canadian Foundation for the Study of Infant Deaths
586 Eglinton Avenue E, Suite 308
Toronto, ON M4P 1P2 Canada

Draper Richards Foundation Fellowships   2112
The foundation awards three-year grants to selected social entrepreneurs to start new nonprofit organizations that are national or global in scope and that have broad social impact. Selected projects will demonstrate innovative ways to solve existing social problems. Proposals are accepted for a variety of public service areas, including, but not limited to, education, youth and families, the environment, arts, health, and community and economic development. The foundation offers financial support as well as strategic and organizational assistance. Proposals are accepted at any time; full proposals are by request.
Requirements: Experienced, dedicated social entrepreneurs with a developed idea for a U.S. nonprofit organization are invited to apply.
Restrictions: The fund does not support research, scholarships, or local community-based organizations.
Samples: Room to Read (San Francisco, CA)—to recognize and help expand this program that builds schools, libraries, and computer labs; donates books; and establishes scholarships for girls in Cambodia, India, Nepal, and Vietnam, $300,000 over three years,; Taproot Foundation (San Francisco, CA)—awarded in recognition and support of its model program that uses teams of volunteers from the business community to provide professional services to San Francisco Bay area nonprofit groups, $300,000 over three years.
Amount of Grant: $100,000 annually
Contact: Jenny Shilling Stein, Executive Director, (415) 616-4050; fax: (415) 616-4060; email: proposals@draperrichards.org
Internet: http://www.draperrichards.org/process/index.html
Sponsor: Draper Richards Foundation
50 California Street, Suite 2925
San Francisco, CA 94111

Dreyfus Postdoctoral Program in Environmental Chemistry   2113
The Foundation seeks to further the development of scientific leadership in the field of environmental chemistry with a postdoctoral fellowship program. The award is given to a principal investigator who submits a proposal judged to be exceptional, both in its potential for leading edge contributions to environmental science, and in the arrangements for the education of the Fellow. Successful applicants are expected to recruit excellent young Ph.D. graduates from the fields of physical, organic, inorganic, biological chemistry or associated fields in chemical engineering, and provide them with the highest caliber of research experience and broad education in environmental science. Instructions for application submission are available online.
Requirements: The program is open to all academic and other not-for-profit organizations that have well-established research efforts in environmental science or engineering in the States, Districts, and Territories of the U.S.. These research activities need not be located in traditional departments in the chemical sciences, and collaboration across departments and institutions is encouraged.
Restrictions: No part of the award may be used for institutional administrative purposes.
Amount of Grant: $120,000 over two years
Date(s) Application is Due: Aug 13
Contact: Mark Cordillo, Executive Director; (212) 753-1760; fax: (212) 593-2256; email: mcardillo@dreyfus.org or admin@dreyfus.org
Internet: http://www.dreyfus.org/awards/postdoctoral_program.shtml
Sponsor: Camille and Henry Dreyfus Foundation
555 Madison Avenue, 20th Floor
New York, NY 10022-1301

Dreyfus Prize in the Chemical Sciences   2114
The Camille and Henry Dreyfus Foundation announces the establishment of the Dreyfus Prize in the Chemical Sciences. The prize, to be awarded biennially, will consist of a citation, a medal, and a monetary award of $250,000. The prize will be awarded to an

individual in a selected area of chemistry to recognize exceptional and original research that has advanced the field in a major way.

Requirements: Any person may nominate a candidate for the Dreyfus Prize in the Chemical Sciences. Self-nominations are not accepted.

Restrictions: The prize is restricted to U.S. residents. Nobel laureates are not eligible. Dreyfus Foundation Advisors and reviewers who serve in the year of the selection are not eligible. There is no restriction on the number of nominees who are at a given institution, nor is institutional approval required.

Amount of Grant: $250,000
Date(s) Application is Due: Feb 13
Contact: Mark Cordillo, Executive Director; (212) 753-1760; fax: (212) 593-2256; email: mcardillo@dreyfus.org or admin@dreyfus.org
Internet: http://www.dreyfus.org/awards/prize.shtml
Sponsor: Camille and Henry Dreyfus Foundation
555 Madison Avenue, 20th Floor
New York, NY 10022-1301

Dreyfus Senior Scientist Mentor Program Grants   **2115**
The program enables undergraduates to conduct research in the chemical sciences in a close working relationship with emeritus faculty members. Only one faculty member may apply per application, although more than one application per department or institution is permitted. A letter of support must be sent directly to the Foundation from a colleague, preferably from outside the institution, who is familiar with the applicant's research and teaching and who can speak to the applicant's experience in mentoring and advising undergraduates. Letters of support cannot be accepted via electronic mail or facsimile.

Requirements: Faculty with emeritus status on or before January of the current year, and who maintain active research programs in the chemical sciences may apply for one of a limited number of awards that will allow undergraduates to do research under their guidance. Successful applicants, who are expected to be closely engaged in a mentoring relationship with the students will receive grants annually for two years for undergraduate stipends and modest research support.

Amount of Grant: $10,000 annually for up to two years
Date(s) Application is Due: Nov 12
Contact: Mark Cordillo, Executive Director; (212) 753-1760; fax: (212) 593-2256; email: mcardillo@dreyfus.org or admin@dreyfus.org
Internet: http://www.dreyfus.org/awards/senior_scientist_mentor.shtml
Sponsor: Camille and Henry Dreyfus Foundation
555 Madison Avenue, 20th Floor
New York, NY 10022-3301

Dreyfus Special Grants Program in the Chemical Sciences   **2116**
This program offers support to eligible institutions for projects that propose to advance the science of chemistry in innovative ways. Proposals are invited in any area consistent with the Foundation's broad objective of advancing the chemical sciences. For projects that require sustaining support, note that this program is intended to seed the initial phases, with the expectation that grant recipients will find continuing funding from other sources. Inquiries should be made by the June 4 deadline, and proposals should be received by November 12.

Requirements: Institutions in the States, Districts, and Territories of the U.S. that have a focus in the chemical sciences are eligible and may submit a preliminary inquiry to the Special Grant Program in the Chemical Sciences. Institutions include schools, colleges and universities, as well as other not-for-profit organizations, such as museums and libraries.

Restrictions: Awards are not made directly to individuals, nor, in general, to private foundations.

Samples: Texas Wesleyan U (Fort Worth, TX)—for a software-based program that will enable students to better access chemical-sciences information, $30,000.; Cornell U (NY)—for a project on computational analysis in the chemistry curriculum, $25,000.; California Institute of Technology (CA)—to support summer undergraduate research fellowships, $15,000.; Columbia U (NY)—for a project on computer-assisted problem-solving tools for general chemistry, $25,000.

Date(s) Application is Due: Jun 4; Nov 12
Contact: Mark Cordillo, Executive Director; (212) 753-1760; fax: (212) 593-2256; email: mcardillo@dreyfus.org or admin@dreyfus.org
Internet: http://www.dreyfus.org/awards/postdoctoral_program.shtml
Sponsor: Camille and Henry Dreyfus Foundation
555 Madison Avenue, 20th Floor
New York, NY 10022-3301

Drs. Bruce and Lee Foundation Grants   **2117**
The Foundation's goal is to advance the general welfare and the quality of all life in the Florence, S.C. area by providing economic support to qualified programs and non-profit organizations. The Foundation will support a broad range of charitable purposes including, but not limited to: medical; health; human services; education; arts; religion; civic affairs; and the conservation, preservation and promotion of cultural, historical and environmental resources. There are no application deadlines. Contact the office for application materials and guidelines.

Requirements: Florence, SC, nonprofit organizations are eligible.
Restrictions: The Foundation does not purchase tickets for fundraising events. Grants to individuals will not be considered.
Geographic Focus: South Carolina
Samples: Francis Marion U (Florence, SC)—to construct a facility for the new bachelor's-degree nursing program, $5 million (2004).

Contact: L. Bradley Callicott, Executive Director, (843) 664-2870; email: blfound@bellsouth.net
Sponsor: Drs. Bruce and Lee Foundation
181 East Evans Street, BTC Box 022
Florence, SC 29506

DSO Cognitive Technology Threat Warning System (CT2WS) Grants   **2118**
The objective of the CT2WS program is to drive a breakthrough in soldier-portable visual threat warning devices. Recent developments and discoveries in the disparate technology areas of flat-field, wide-angle optics, large pixel-count digital imagers, cognitive visual processing algorithms, neurally-based target detection signatures and ultra-low power analog-digital hybrid signal processing electronics have led DARPA to believe that focused technology development, system design, and system integration efforts may produce revolutionary capabilities for the warfighter. The final objective of the DARPA CT2WS program is the development of prototype soldier-portable digital imaging threat queuing systems capable of effective detection ranges of 1-10 km against dismounts and vehicles while simultaneously surveying a 120-degree or greater field of view (FOV). DARPA requests proposals for the full scope of development (e.g., an end-to-end system designed by a team of multidisciplinary research organizations, plus an integrator for coordination and implementation support). Proposals addressing only individual component-level technologies may be considered non-compliant to this BAA. This BAA affords proposers the choice of submitting proposals for the award of a Grant, Cooperative Agreement, Procurement Contract, Technology Investment Agreement, Other Transaction for Prototype Agreement, or other such appropriate award instrument.

Requirements: The Government anticipates proposals submitted under this BAA will be Unclassified.

Date(s) Application is Due: Apr 11
Contact: Amy Kruse, Program Manager; (571) 218-4338; fax: (703) 248-1908; email: amy.kruse@darpa.mil or BAA07-25@darpa.mil
Internet: http://www.darpa.mil/baa/BAA07-25.html
Sponsor: Defense Sciences Office within the Defense Advanced Research Projects Agency (DARPA)
3701 North Fairfax Drive
Arlington, VA 22203-1714

DSO Computer Science Study Group (CSSG) Grants   **2119**
This Research Announcement (RA) initiates Phase 1 of the 2008 Computer Science Study Group (CSSG). The objective of the CSSG is to rapidly identify ideas in the field of computer science that will provide revolutionary advances, rather than incremental benefit, to the Department of Defense. Phase 1 (12 months) consists of participation on the CS Study Panel. The CS Study Panel anticipates meeting at least four times throughout the 2008 calendar year, totaling approximately twenty days. Two of the meetings will occur during the academic year and will take place in the Washington, D.C. metropolitan area. Two extended meetings, lasting approximately one week each, will take place during the summer, and will involve travel throughout the United States. For successful proposers, host institutions will receive grants or other assistance instruments for up to $100,000. The funding to the host institution will support participant travel expenses for Panel meetings; it will also support participant salary, to include time spent while participating in the CS Study Panel and time to connect the participants‚Äô computer science research to critical DoD needs identified during their Panel trips. The funding may also support research assistants‚Äô summer salary and costs for administrative support.

Requirements: All responsible sources may submit a proposal that shall be considered by DARPA. Historically Black Colleges and Universities (HBCUs) and Minority Institutions (MIs) are encouraged to submit proposals.

Restrictions: Only U.S. institutions of higher learning are eligible for awards under this RA, and participants must be junior faculty members who are able to receive a U.S. Department of Defense SECRET security clearance.

Date(s) Application is Due: Aug 10
Contact: Dr. Benjamin Mann, Program Manager; (571) 218-4246; fax: (571) 218-4553; email: benjamin.mann@darpa.mil or RA07-43@darpa.mil
Internet: http://www.darpa.mil/baa/RA07-43.html
Sponsor: Defense Sciences Office within the Defense Advanced Research Projects Agency (DARPA)
3701 North Fairfax Drive
Arlington, VA 22203-1714

DSO Controlling Pathogen Evolution Workshop Grants   **2120**
The goal of the workshop is to explore: the potential of computational methods to precisely predict pathogen evolution; and vaccination and/or therapeutic strategies to control direction of pathogen evolution. The workshop will investigate tools for predicting protective epitopes, pathogen evolution and protein structure, and then review strategies to force pathogens into an evolutionary trap that prevents escape through mutation. From there, vaccines and therapies can be developed against the particular pathogen strain. Workshop participants are strongly encouraged to prepare posters describing previous or potential research in the area in order to facilitate discussions and/or formation of well-rounded teams.

Requirements: Attendees should have specific expertise in infectious disease, epitope prediction (computational and/or immunomic), vaccinology, mathematical modeling, protein structure, pathogen evolution, drug design, and/or animal modeling.

Date(s) Application is Due: Mar 16

Contact: Dr. Michael V. Callahan, Program Manager; (571) 218-4596; fax: (571) 218-4553; email: michael.callahan@darpa.mil
Internet: http://www.darpa.mil/baa/SN07-21.html
Sponsor: Defense Sciences Office within the Defense Advanced Research Projects Agency (DARPA)
3701 North Fairfax Drive
Arlington, VA 22203-1714

DSO Defense Sciences Research and Technology Grants   2121

TThe mission of the DSO is to identify and pursue high-risk/high-payoff research initiatives throughout a broad spectrum of the science and engineering disciplines, and to transform these initiatives into important, radically new military capabilities. To carry out this mission, DSO seeks research ideas and areas that might lead to innovations in science and engineering. Therefore, the office is soliciting proposals for advanced research and development in a variety of enabling technical areas. Proposals may be either basic or applied research. However, in all cases, proposers should demonstrate that their proposed effort is aimed at high-risk/high-payoff technologies that have the potential for making, in the 5-10 year time frame, revolutionary rather than incremental improvements to national security, including emerging threats and operational challenges. Multiple awards are anticipated. The amount of resources made available to this BAA will depend on the quality of the proposals received and the availability of funds. While there is no specific requirement for cost and duration of the proposed effort, it is recommended that proposers include a Phase I of 12 to 18 months in length that addresses the most critical issues on the path to success.
Requirements: All responsible sources may submit a proposal that shall be considered by DARPA.
Date(s) Application is Due: Feb 14
Contact: Barbara K. McQuiston, Deputy Director; (703) 526-4759; fax: (703) 248-1916; email: barbara.mcquiston@darpa.mil or BAA07-21@darpa.mil
Internet: http://www.darpa.mil/baa/baa07-21mod1.html
Sponsor: Defense Sciences Office within the Defense Advanced Research Projects Agency (DARPA)
3701 North Fairfax Drive
Arlington, VA 22203-1714

DSO Radiation Biodosimetry (RaBiD) Grants   2122

The DSO is seeking proposals for new technologies for rapid, high-throughput, portable and low-cost biodosimeters to determine radiation dose to individuals after acute radiation exposure. This technology would provide rapid identification of individuals who have been exposed to high-dose radiation in order to accurately assess radiation exposure levels. The Program is a single phase, 15-month program with a goal of revolutionizing radiation exposure detection. The first 12 months consist of scientific research and development during which the proposer will develop a non- or minimally-invasive radiological biodosimeter and demonstrate accurate radiological detection from biological samples into quartiles of doses for humans with a detection time less than 10 minutes. This will include radiological dose/detection curves, as well as multipoint data regarding decay of biological signal as a function of time after radiation exposure. A successful proposal will demonstrate a clear path to these deliverables, including a timeline for technology development, experimentation, and delivery of results.
Requirements: All proposers must provide evidence of a competent team capable of developing a new radiation biodosimetry technology meeting all program objectives and milestones. The Program requires that each proposal include a team with demonstrated (or established) capability. At a minimum, proposers are expected to possess expertise or demonstrate collaboration with professionals in the following areas: radiation biologist/radiation oncologist—with expertise in radiation exposure to animal models; and engineering—with developmental expertise needed to translate biological sampling to radiation readout.
Restrictions: Only unclassified proposals will be accepted in response.
Date(s) Application is Due: Jul 9
Contact: Dr. Mildred Donlon, Program Manager; (703) 696-2289; fax: (703) 741-3896; email: mildred.donlon@darpa.mil
Internet: http://www.darpa.mil/dso/solicitations/baa07-29.htm
Sponsor: Defense Sciences Office within the Defense Advanced Research Projects Agency (DARPA)
3701 North Fairfax Drive
Arlington, VA 22203-1714

DuBois-Mandela-Rodney Fellowship Program   2123

The center invites applications for the fellowship program from scholars working on Africa or the African diaspora. Consideration will be given to all disciplines including - but not limited to - the humanities, social sciences, physical sciences and professional schools. Scholars from or who study the Gullah speaking Sea islands, Cape Verde islands, the Anglophone Caribbean, the Canary Islands, and Madagascar and/or other less studied areas are especially encouraged to apply. Successful candidates can expect to maintain affiliations with CAAS as well as with departments and research institutes that relate to their projects. Each fellow will also be expected to conduct a CAAS work-in-progress seminar during one semester in residence.
Requirements: Candidates must have the PhD in hand and be no more than 5 years beyond the completion of the degree.
Amount of Grant: $45,000 total
Date(s) Application is Due: Nov 30
Contact: V. Robin Grice, Center for African American and African Studies, (734) 647-5361; email: gricer@umich.edu

Internet: http://www.umich.edu/~iinet/caas/fellow&grants/dubois-mandela-intro.htm
Sponsor: University of Michigan
106 W Hall Bldg
Ann Arbor, MI 48109-1092

Duchossois Family Foundation Grants   2124

The foundation focuses its efforts in the Chicago metropolitan area and gives primary consideration to nonprofit organizations that contribute to the community in the area of health. Contributions to health programs are pledged to organizations and agencies whose services include research and education in the primary area of cancer and cancer research. A one-page summary-request letter should include a description of the organization and its specific needs and purposes, the amount requested, and a list of members of the board of directors and their business/professional affiliations.
Requirements: 501(c)3 tax-exempt public charities serving the Chicago metropolitan area are eligible.
Geographic Focus: Illinois
Samples: Northwestern Memorial Hospital, Core Virology Department (Chicago, IL)—for general support, $100,000.
Contact: Iris Krieg, Executive Director, (312) 641-5765
Sponsor: Duchossois Family Foundation
203 N. Wabash, Suite 1800
Chicago, IL 60601

Duke Energy Foundation Grants   2125

The Duke Energy Foundation gives primarily in areas of company operations in Indiana, Kentucky, North Carolina, Ohio and South Carolina. At the core of The Duke Energy Foundation is its commitment to the community, with a focus in three areas: Environment and Energy Efficiency; Economic Development; and Community Vitality.
Requirements: Organizations with a 501(c)3 verification from the IRS or are a part of a governmental entity are eligible. All organizations applying for a grant must have: completed the Online Grant Application; a clear reason for making the contribution that relates to the areas of focus; regular reports on the measurable results of the project.
Restrictions: Foundation funds are not provided for: organizations that discriminate by race, creed, gender, age or national origin; political activities and organizations; grants to individual agencies of the United Way or the Charlotte Arts and Science Council; capital campaigns and endowments, except in extremely rare and specialized situations that relate directly to our areas of expertise in business; individuals; athletics, including individual sports teams and all-star teams; underwriting of films, video and television productions; reducing the cost of utility service; sectarian or religious activities; conferences, trips or tours; fraternal, veteran or labor groups serving only their members; advertising; membership fees or association fees, either personal or corporate; dinners or tables at fund-raisers are rarely considered; family foundations.
Geographic Focus: Indiana; Kentucky; North Carolina; Ohio; South Carolina
Samples: Arts and Science Council of Charlotte-Mecklenburg, Charlotte, NC., $1,250,000 - for capital support (2007); United Way of Cleveland County, Shelby, NC., $12,610 - for operating support (2007); Greater Cincinnati Arts and Education Center, Cincinnati, OH., $550,000 - for capital support (2007);
Amount of Grant: $1,000-$1,250,000 range
Contact: Deanita McCall, Vice President; (704) 382-7200; fax: (704) 382-7600;
Internet: http://www.duke-energy.com/community/foundation.asp
Sponsor: Duke Energy
P.O. Box 35469
Charlotte, NC 28254

Duke Foundation Fellowships   2126

Beginning in the fall semester of 2009, a limited number of graduate fellowships will be available for students in the Ph. D. Program in Public Policy Analysis in the Terry Sanford Institute of Public Policy who intend to focus their dissertation on the theory of foundations, foundation strategy, the measurement of foundation impact, and foundation accountability.
Contact: Bruce Kuniholm, Director; (919) 613-7309; email: bruce.kuniholm@duke.edu
Internet: http://www.pubpol.duke.edu/centers/dfrp/fellowship.php
Sponsor: Duke Foundation, Sanford Institute of Public Policy
Box 90239
Durham NC 27708-0239

Duke University Adult Cardiothoracic Anesthesia and Critical Care   2127
Medicine Fellowships

Nine one year fellowship positions are available each year in the ACGME Accredited Fellowship. Second year research and clinical fellowship opportunities are available; programs are individualized to meet the goals of successful applicants. Generally clinical experience is the focus of the first fellowship year, along with involvement in clinical research. Fellows opting for a two year fellowship continue with one clinical day per week in their second year, in addition to a structured plan involving laboratory and/or clinical research and teaching. Second year clinical specialization in pediatric cardiac anesthesiology or intensive care is also available. The Adult cardiothoracic Anesthesia Fellowship is highly recommended as formal training for specialization in cardiac and/or thoracic anesthesiology. The Fellowship is designed to develop clinical, consultant, and in-depth research expertise. Applications are usually considered starting 18 months prior to the start date
Requirements: Complete application (see Duke University website), in addition to the references required, include a letter from your Chairperson/Program Director and two letters from other faculty members who have worked with you closely.

Contact: Mark S. Smith; (919) 681-5046 or (919) 416-3853; fax: (919)681-8993
Internet: http://medschool.duke.edu/modules/som_rt/index.php?id=1
Sponsor: Duke University Medical Center, Department of Anesthesiology
P.O. Box 3094
Durham, NC 27710

Duke University Ambulatory and Regional Anesthesia Fellowships   2128

The Fellowship in Ambulatory/Regional Anesthesia is designed to provide advanced training in anesthesia in the ambulatory surgery setting. This program will be an extension to the previous training with an emphases on the continued educational component to include research, clinical and teaching responsibilities.
Requirements: The selection process for fellows includes the completion of a Fellowship Application (see website) which is to be substantiated by three letters of recommendation from experts in the field. In addition to the letters of reference the application is to include a current CV and a letter of affirmation stating interest in the program.
Date(s) Application is Due: Aug 15
Contact: Stephen M. Klein, Director; (919) 668-2056; email: klein006@mc.duke.edu; LuAnne Latta, Staff Assistant; (919) 681-9941; email: latta005@mc.duke.edu
Internet: http://anesthesiology.duke.edu/modules/anes_flwshp/index.php?id=1
Sponsor: Duke University Medical Center, Department of Anesthesiology
P.O. Box 3094 #4
Durham, NC 27710

Duke University Clinical Cardiac Electrophysiology Fellowships   2129

The Clinical Cardiac Electrophysiology fellowship training program is a comprehensive two-year program including training in catheter ablation of supraventricular tachycardias, ventricular tachycardias, and complex arrhythmias such as atrial fibrillation, ischemic VT, and arrhythmias associated with congenital heart disease. Comprehensive training is also provided in lead extraction and device therapy including permanent pacing, implantable defibrillators, and bi-ventricular devices. Fellows are provided with significant opportunity for research activity through protected block time and are expected to participate in and complete a research project with a faculty mentor. Applicants must have completed at least three years in an ACGME approved Cardiology fellowship program.
Requirements: Applications for training to begin July 1, will be accepted beginning January 1; the application deadline will be September 1. To apply, please include with the completed CCE (EP) Application form (available on the Duke University website), a current curriculum vitae, personal statement (not to exceed 250 words), and the $40 (check or money/order no cash please) processing fee.
Date(s) Application is Due: Sep 1
Contact: Dr. Patrick Hranitzky, Program Director; (919) 684-2304; fax: (919) 684-4322; email: Patrick.hranitzky@duke.edu
Internet: http://fellowships.medicine.duke.edu/modules/fellows_cardio/index.php?id=8
Sponsor: Duke University School of Medicine
P.O. Box 3878
Durham, NC 27710

Duke University Hyperbaric Center Fellowships   2130

The clinical Hyperbaric fellow works in the Hyperbaric Center and F.G. Hall Environmental Laboratory, both of which are an integral part of Duke University Medical Center. Fellowships are available for 12 months although they may be extended for additional time. During the period of training the Hyperbaric fellow will be trained in the clinical use of hyperbaric oxygen. This includes the emergency treatment of clostridial myonecrosis, necrotizing fasciitis, arterial gas embolism, decompression sickness, carbon monoxide poisoning and certain other less common conditions such as acute vasculitis with ischemia. The fellow will receive training in the appropriate critical care interventions necessary for the treatment of these patients. This will include hemodynamic monitoring, ventilatory support, fluid resuscitation and the interpretation of arterial blood gases during hyperbaric therapy. The fellow will also receive training in the use of hyperbaric oxygen therapy for non-emergency conditions. The fellow will be responsible for all inpatient and outpatient consults, under the direction of the clinical Hyperbaric faculty. During the training period the fellow will also be required to complete a research project. This will be completed under the guidance of one of the faculty members in the Hyperbaric Center. Part of the didactic program includes regular hyperbaric conferences. The fellow will participate in these and normally be required to present at least one topic during his year of training. The fellow will also have the opportunity to attend the annual two-week course in Hyperbaric and Diving Medicine given by the National Oceanic and Atmospheric Administration in Seattle, Washington. In addition to these activities the applicant may participate in numerous other teaching activities within Duke Medical Center, for example those sponsored by the Departments of Anesthesiology, Medicine, Surgery and Emergency Medicine. Clinical participation under the aegis of those departments can be arranged in some cases, depending upon the training of the individual concerned.
Requirements: Applicants must be eligible for license as a trainee in the state of North Carolina. The Duke Hyperbaric Center Fellowship Application can be downloaded at: http://hyperbaric.mc.duke.edu/education.htm
Contact: John J. Freiberger M.D., Program Director; (919) 668-0005; email: freib002@mc.duke.edu; Ms. Tonya Manning, Assistant; (919) 681-1685 or (919) 684-6726
Internet: http://hyperbaric.mc.duke.edu/education.htm
Sponsor: Duke University Medical Center for Hyperbaric Medicine and Environmental Physiology
P.O. Box 3823
Durham, NC 27710

Duke University Interventional Cardiology Fellowships   2131

The Interventional Cardiology Fellowship Training Program at Duke University Medical Center focuses on training leaders in Interventional Cardiology committed to a career in academic cardiology. The length of the fellowship is 12 months. During the fellowship, all fellows will develop both clinical and technical skills in interventional cardiology. All fellows will gain experience in diagnosing, selecting therapies, performing diagnostic and interventional procedures, and judging the effectiveness of treatments for inpatients and outpatients with chronic ischemic heart disease, acute ischemic syndromes, and valvular heart disease. All fellows are expected to participate in research efforts. During the fellowship, each fellow will work with a mentor on a research project which will help foster their long-term goals.
Requirements: Individuals applying for only a 1 year Interventional Cardiology Fellowship position will be required to provide his/her own funding resources for salary support during their training. This funding must be institutional funding (no personal or private funds). A processing fee of $40 (check or money/order made payable to Duke University- no cash please), letter documenting funding support, current curriculum vitae, and personal statement will be required to accompany the completed Interventional Cardiology Application (see Duke University website for application).
Date(s) Application is Due: Feb 1
Contact: Sherolyn Patterson, Fellowship Program Coordinator; (919) 684-2304; fax: (919) 684-4322; email: sherolyn.patterson@duke.edu
Internet: http://fellowships.medicine.duke.edu/modules/fellows_cardio/index.php?id=8
Sponsor: Duke University School of Medicine
P.O. Box 3878
Durham, NC 27710

Duke University Libraries Franklin Collection Travel Grants   2132

The Franklin Collection offers travel grants for research involving the use of the collections of the John Hope Franklin Collection or for research with a focus on African and African-American studies. The maximum award per applicant is $1,000. Applicants are strongly encouraged to speak with the public services librarian before applying for the grant.
Requirements: Any faculty member, graduate or undergraduate student, or independent scholar with a research project directly related to the collections held by the Bingham Center or to women's history and culture is eligible to apply.
Restrictions: All applicants must reside beyond a 50 mile radius of Durham, North Carolina.
Geographic Focus: North Carolina
Amount of Grant: $1,000 maximum
Date(s) Application is Due: Jan 15
Contact: Karen Jean Hunt, Director; (919) 660-5922; fax: (919) 660-5934; email: franklin-collection@duke.edu
Internet: http://library.duke.edu/specialcollections/franklin/grants.html
Sponsor: Duke University Libraries
Box 90185
Durham, NC 27708-0185

Duke University Libraries Hartman Center Travel Grants   2133

The Grants provide up to $750 to support those individuals whose research focus includes the use of any Hartman Center collections and whose research trips may last any length of time. Grant money may be used for travel to the Rare Book, Manuscript, and Special Collections Library, costs of copying pertinent archival resources, and living expenses while pursuing research here. Travel must originate from at least 60 miles away from Duke University. All applicants must submit a 250-500 word essay summarizing their project and detailing why the Hartman Center collections are essential to their research.
Requirements: Any faculty member, graduate or undergraduate student, or independent scholar with a research project directly related to the collections held by the Bingham Center or to women's history and culture is eligible to apply.
Geographic Focus: North Carolina
Amount of Grant: $750 maximum
Date(s) Application is Due: Jan 15
Contact: Jacqueline Reid, Director; (919) 660-5836; fax: (919) 660-5934; email: j.reid@duke.edu
Internet: http://library.duke.edu/specialcollections/hartman/travel-grants/index.html
Sponsor: Duke University Libraries
Box 90185
Durham, NC 27708-0185

Duke University Libraries J. Walter Thompson Research Fellowships   2134

The Fellowships provide $1000 in funds for up to three individuals planning research trips for a minimum of two weeks primarily use the J. Walter Thompson Archives. The JWT archives hold over 5,000 linear feet dating as far back as the 19th century and as recent as the 21st century. The collections document the pioneering role of JWT in the advertising industry and illustrates the changes in the broader industry over the past 100 years through its own growth and diversification. Grant money may be used for travel to the Rare Book, Manuscript, and Special Collections Library, costs of copying pertinent archival resources, and living expenses while pursuing research here. Travel must originate from at least 60 miles away from Duke University. All applicants must submit a 250-500 word essay summarizing their project and detailing why the Hartman Center collections are essential to their research.

Requirements: Any faculty member, graduate or undergraduate student, or independent scholar with a research project directly related to the collections held by the Bingham Center or to women's history and culture is eligible to apply.
Geographic Focus: North Carolina
Amount of Grant: $1,000 maximum
Date(s) Application is Due: Jan 15
Contact: Jacqueline Reid, Director; (919) 660-5836; fax: (919) 660-5934; email: j.reid@duke.edu
Internet: http://library.duke.edu/specialcollections/hartman/travel-grants/index.html
Sponsor: Duke University Libraries
Box 90185
Durham, NC 27708-0185

Duke University Libraries Mary Lily Research Grants 2135

The Grants are for research involving the use of their collections or research into other collections with a focus on women's history and culture. Grant money may be used for travel to the Rare Book, Manuscript, and Special Collections Library, costs of copying pertinent archival resources, and living expenses while pursuing research, including accommodations and meals. The maximum award per applicant is $1,000. Applicants are strongly encouraged to speak with the public services librarian before applying for the grant.
Requirements: Any faculty member, graduate or undergraduate student, or independent scholar with a research project directly related to the collections held by the Bingham Center or to women's history and culture is eligible to apply. All applicants must reside beyond a 50 mile radius of Durham, North Carolina.
Amount of Grant: $1,000 maximum
Date(s) Application is Due: Jan 15
Contact: Kelly Wooten; (919) 660-5967; fax: (919) 660-5934; email: kelly.wooten@duke.edu
Internet: http://library.duke.edu/specialcollections/bingham/grants/index.html
Sponsor: Duke University Libraries
Box 90185
Durham, NC 27708-0185

Duke University Obstetric Anesthesia Fellowships 2136

The Department of Anesthesiology at Duke University Medical Center offers 1 - 2 positions of a one year Fellowship in Obstetric Anesthesia. Emphasis is on clinical research, advanced clinical training and teaching. Duke University is a tertiary center with approximately 3,000 deliveries per year. The division is currently made up of 8 faculty staff with diverse international pedigree.
Contact: Yemi Olufolabi, Fellowship Director; (919) 681-6535 or (919) 668-6266; fax: (919) 668-6265; email: olufo001@mc.duke.edu
Internet: http://anesthesiology.duke.edu/modules/anes_flwshp/index.php?id=1
Sponsor: Duke University Medical Center, Department of Anesthesiology, Division of Women's Anesthesia
P.O. Box 3094
Durham, NC 27710

Duke University Pain Management Fellowships 2137

The Fellowship Program is based in the Duke Health Center at Morreene Road (MRC). Fellows will: cover the Pain Clinic and Durham VAMC, and Acute Pain Service at Duke North; evaluate new patients; perform comprehensive or neurologic studies; formulate diagnosis and plan; become familiar with procedures; learn intervention of the lumbar spine and its application e.g. facet rhizotomy, discography, and intradiscal annuloplasty; implantation therapies; dictation of clinic and procedure notes. Fellows will be asked to participate in research, journal clubs, round table discussions, grand rounds and lectures. There are opportunities for the fellow to become involved in research projects already in progress or to develop an original project. The clinical objectives are to build knowledge, skills and professionalism.
Requirements: Applicants must meet eligibility specified by the ACGME. Acceptance into the program is based on demonstrated academic qualification, interpersonal skills, and evidence of ability to function in a multidisciplinary setting. Fellowship applications may be downloaded from http://anesthesia.duhs.duke.edu/. Submit application, CV and three letters of reference.
Date(s) Application is Due: Jun 30
Contact: Dianne L. Scott, M.D., Fellowship Director; (919) 684-6736;
Internet: http://anesthesiology.duke.edu/modules/anes_flwshp/index.php?id=1
Sponsor: Duke University, Duke Pain Clinic
932 Morreene Road, RM 232
Durham, NC 27705

Duke University Postdoctoral Research Fellowships in Aging 2138

The goal of the postdoctoral research training program is to produce highly skilled research scientists who have the potential for leadership in gerontological research. In the Duke Research Training Program (RTP), much of the training for each fellow is provided by that person's faculty mentor in a research apprenticeship program. A fellow carries out his/her own research as a junior colleague in the mentor's research program or laboratory. In addition to working in their mentor's programs, all fellows attend a weekly interdisciplinary didactic seminar. The stipend amount is determined by years of relevant postdoctoral experience. The program also provides health insurance for the fellow (using the University's insurance plan), and partial travel support when presenting research at a professional meeting.

Requirements: Applicants must be citizens of the United States, or have been lawfully admitted to the United States for permanent residence. Applicants must have a completed doctoral degree when joining the program. All course work must be completed, and if applying as a Ph.D., final orals must be passed, and the dissertation signed before an applicant can begin the program.
Date(s) Application is Due: May 1
Contact: James A. Blumenthal, Ph.D.; (919) 660-7517; fax: (919) 684-8569; email: blume003@mc.duke.edu
Internet: http://www.geri.duke.edu/post_doc/
Sponsor: Duke University School of Medicine
Box 3003
Durham, NC 27710

Dumbarton Oaks Fellowships 2139

Dumbarton Oaks offers fellowships in the three areas of Byzantine studies (including related aspects of late Roman, early Christian, western medieval, Slavic, and Near Eastern studies); Pre-Columbian studies of Mexico, Central America, and Andean South America; and studies in landscape architecture. Fellowships are for scholars who hold or expect to have, at the beginning of the fellowship, a doctorate or who have done comparable advanced work and wish to pursue research on projects of their own. Fellowships are not renewable, but initial application may be made for two successive years (two annual fellowships and one intervening summer fellowship, part of which may be spent away from Dumbarton Oaks). Applications not postmarked by due date will be returned.
Requirements: Fellowships are open to scholars of any nationality holding a PhD or relevant advanced degree and pursuing research on a project of their own or one sponsored by the center.
Restrictions: Applications from former fellows are accepted if five years have elapsed since tenure of previous fellowship.
Amount of Grant: $25,990 plus housing, lunch on weekdays; $2000 (if needed) to assist with the cost of bringing and maintaining dependents; $975 maximum for research expenses; $1300 maximum travel reimbursement
Date(s) Application is Due: Nov 1
Contact: Carol Sellery, Director's Office, (202) 339-6410; fax: (202) 339-6419; email: DumbartonOaks@doaks.org
Internet: http://www.doaks.org/fellowships.html
Sponsor: Dumbarton Oaks
1703 32nd Street NW
Washington, DC 20007

Dumbarton Oaks Junior Fellowships 2140

Junior fellowships are awarded to students who at the time of application have fulfilled all preliminary requirements for a higher degree and plan to work at Dumbarton Oaks on a dissertation or a final project under the direction of a faculty member at their own university. In exceptional cases, applications will be accepted from students before fulfilling preliminary requirements. Junior fellowships are awarded in the three areas of Byzantine studies (including related aspects of late Roman, early Christian, western medieval, Slavic, and Near Eastern studies); Pre-Columbian studies of Mexico, Central America, and Andean South America; and studies in landscape architecture. Applications not postmarked by due date will be returned.
Requirements: Fellowships are open to individuals of any nationality who have passed all preliminary examinations for a higher degree and are writing a dissertation or equivalent. Fellows are expected to have a working knowledge of the languages necessary for their research and be sufficiently advanced to pursue research on their own.
Amount of Grant: $14,635 plus housing, lunch on weekdays, contribution to health insurance; $975 research allowance; $2000 (if needed) to assist with the cost of bringing and maintaining dependents; $1300 maximum travel expense
Date(s) Application is Due: Nov 1
Contact: Carol Sellery, Director's Office, (202) 339-6410; fax: (202) 339-6419; email: DumbartonOaks@doaks.org
Internet: http://www.doaks.org/fellowships.html
Sponsor: Dumbarton Oaks
1703 32nd Street NW
Washington, DC 20007

Dumbarton Oaks Project Grants 2141

Dumbarton Oaks makes grants to assist with scholarly projects in the three areas of Byzantine studies (including related aspects of late Roman, early Christian, western medieval, Slavic, and Near Eastern studies); Pre-Columbian studies of Mexico, Central America, and Andean South America; and studies in landscape architecture. Support is generally for archaeological research or for the recovery, recording, and analysis of materials that would otherwise be lost. Grants may cover modest expenses such as photography, supplies, special services, and sometimes travel, although Dumbarton Oaks does not make travel grants as such. Applicants must contact Dumbarton Oaks by October 1 to determine whether the project is within the institution's purview; applications must be received by November 1.
Restrictions: Project awards are not offered purely for the purpose of travel, nor for work associated with a degree, for library or archive research, for catalogues, or for conservation and restoration per se.
Amount of Grant: $3000-$10,000 typically
Date(s) Application is Due: Oct 1; Nov 1
Contact: Carol Sellery, Director's Office, (202) 339-6410; fax: (202) 339-6419; email: DumbartonOaks@doaks.org

Internet: http://www.doaks.org/project.html
Sponsor: Dumbarton Oaks
1703 32nd Street NW
Washington, DC 20007

Dumbarton Oaks Summer Fellowships 2142

Dumbarton Oaks offers summer fellowships for periods of six to nine weeks in the three areas of Byzantine studies (including related aspects of late Roman, early Christian, western medieval, Slavic, and Near Eastern studies); Pre-Columbian studies of Mexico, Central America, and Andean South America; and studies in landscape architecture. Awards are for scholars (on any level of advancement) who are not incumbent fellows. Applications not postmarked by due date will be returned.
Amount of Grant: $230 per week plus housing and lunch on weekdays; $1300 maximum travel expenses
Date(s) Application is Due: Nov 1
Contact: Carol Sellery, Director's Office, (202) 339-6410; fax: (202) 339-6419; email: DumbartonOaks@doaks.org
Internet: http://www.doaks.org/fellowshipsann.html
Sponsor: Dumbarton Oaks
1703 32nd Street NW
Washington, DC 20007

Duncan L. Gordon Fellowships 2143

Fellowships are available for research in pediatric medicine including nutrition, infectious diseases, mental retardation, environmental health, and clinical pharmacology, tenable in any agreed institution within Canada for one or two years. Candidates should be nominated by the head of the department in which they are employed or in which they will be employed upon completion of the fellowship. Nominating institutions must affirm their intention to employ the fellow upon completion of studies and to continue or establish a program in the field concerned. Award includes a full-time stipend, paid on a quarterly basis, plus a modest transportation allowance to take up the fellowship if required.
Requirements: Fellowships are available to Canadian citizens or landed immigrants at a post-doctoral level and who are of outstanding academic achievement in a pediatric related health care field and who can provide evidence of aptitude for pediatric patient care, teaching, and research.
Amount of Grant: $C 47,500 maximum
Date(s) Application is Due: Oct 1
Contact: Pam Gilliland, Coordinator, National Grants Program, (416) 813-6166 ext 2354; fax: (416) 813-7311; email: national.grants@sickkids.ca
Internet: http://www.sickkidsfoundation.com
Sponsor: Sick Kids Foundation
555 University Avenue
Toronto, ON M5G 1X8 Canada

Dysautonomia Research Grants 2144

Grants are awarded to promote research into control and/or cure for familial dysautonomia with particular emphasis on genetic screening; prenatal and/or carrier tests are necessary. Funds awarded are variable and depend on research proposed.
Requirements: Postdoctoral candidates must be connected with a recognized medical and/or teaching institution. There are no geographic limitations.
Amount of Grant: $100,000
Date(s) Application is Due: Nov 30
Contact: Grants Administrator, (212) 949-6644; fax: (212) 682-7625; email: info@familialdysautonomia.org
Internet: http://www.familialdysautonomia.org/researchgrantprogram.htm
Sponsor: Dysautonomia Foundation
633 Third Avenue, 12th Fl
New York, NY 10017

Dystonia Medical Research Grants and Fellowships 2145

The research grants provide assistance in supporting investigations in specified areas of biomedical research pertinent to dystonia and related disorders. Research grants are being sought to develop important tools for DYT1 dystonia research; these include cell culture models, proteomics, genetic animal models of DYT1 dystonia, including transgenic and knock-in/ knock-out mice, Drosophila, C. elegans, and Zebrafish models. Grants are also being sought to develop new assays suitable for high throughput drug screening. Funding is available up to $40,000 for one-year pilot studies. A limited number of two-year grants are also available for larger studies with funding up to $75,000 per year. Fellowships assist post-doctoral students in establishing careers in dystonia research.
Requirements: Eligible for research grants are all nonprofit organizations or institutions. Eligible for research fellowships are investigators with MD or PhD degrees. For all investigations involving humans, approval by the institution's human subject protection committee is necessary. The regulations pertaining to recombinant DNA research and animal welfare as established by PHS have been adopted by DMRF.
Restrictions: Grant support will not be given for medical care of dystonia patients or for constructions and alterations.
Amount of Grant: $50,000 per year for two years for fellowships; $75,000 per year for two years for research grants and contracts
Date(s) Application is Due: Dec 30
Contact: Dr. Robert McAlister, Executive Director, (312) 755-0198; fax: (312) 803-0138; email: dystonia@dystonia-foundation.org

Internet: http://www.dystonia-foundation.org/research/other.asp?id=1
Sponsor: Dystonia Medical Research Foundation
One E Wacker Drive, Suite 2430
Chicago, IL 60601-2001

E. Bright Wilson Award in Spectroscopy 2146

Supported by Rohm and Haas Company, this award recognizes outstanding accomplishments in fundamental or applied spectroscopy in chemistry. The award shall be granted to an individual without regard to age or nationality. The scope of the award is to cover all fields of spectroscopy in chemistry, both fundamental and applied. The recipient will deliver a lecture at the spring awards symposium of the ACS Division of Physical Chemistry.
Requirements: Any individual, except a member of the award committee, may submit one nomination or seconding letter for the award in any given year. Nominating documents consist of a letter of not more than 1000 words containing an evaluation of the nominee's accomplishments and a specific identification of the work to be recognized, a biographical sketch including date of birth, and a list of publications and patents authored by the nominee.
Restrictions: Self-nominations are not accepted.
Samples: Jack H. Freed—award recipient, $5,000 (2008); Michael D. Fayer—award recipient, $5,000 (2007); Donald Levy—award recipient, $5,000 (2006).
Amount of Grant: $5,000 and up to $1,500 travel expenses
Date(s) Application is Due: Nov 1
Contact: Felicia Dixon, Awards Administrator; (800) 227-5558 or (202) 872-4408; fax: (202) 776-8008; email: f_dixon@acs.org or awards@acs.org
Internet: http://portal.acs.org/portal/acs/corg/content?_nfpb=true&_pageLabel=PP_ARTICLEMAIN&node_id=1319&content_id=CTP_004511&use_sec=true&sec_url_var=region1
Sponsor: American Chemical Society
1155 Sixteenth Street, NW
Washington, DC 20036

E.A. Baker Foundation for the Prevention of Blindness Grants 2147

The foundation awards grants to Canadian hospitals and universities conducting research in ophthalmology and eye diseases and awards fellowships for advanced training for Canadian ophthalmologists. The advanced training may be undertaken outside of Canada, but the ophthalmologist must return to Canada to teach or practice, particularly in remote areas.
Amount of Grant: $C 40,000 average
Date(s) Application is Due: Dec 1
Contact: Barbara Marjeram, Canadian National Institute for the Blind, (416) 486-2500 ext 7586; email: barbara.marjeram@cnib.ca
Internet: http://www.cnib.ca/eng/eabaker/eabf
Sponsor: E.A. Baker Foundation for the Prevention of Blindness
1929 Bayview Avenue
Toronto, ON M4G 3E8 Canada

E.B. Hershberg Award for Important Discoveries in Medicinally Active Substances 2148

This award, supported by Schering-Plough Corporation, is given biennially in odd-numbered years to recognize and encourage outstanding discoveries in the chemistry of medicinally active substances. The discovery for which the award is given should have been made during the past two decades. Nominations are accepted in even-numbered years, and the award consists of $3,000 and a certificate.
Requirements: Any individual, except a member of the award committee, may submit one nomination or seconding letter for the award in any given year. The nominating documents consist of a letter of not more than 1000 words containing an evaluation of the nominee's accomplishments and a specific identification of the work to be recognized, a biographical sketch including date of birth, and a list of publications and patents authored by the nominee.
Restrictions: Self-nominations are not accepted.
Samples: Choung Un Kim—award recipient, $3,000 (2009); John A. Katzenellenbogen—award recipient, $3,000 (2007); Christopher Lipinski—award recipient, $3,000 (2005).
Amount of Grant: $3,000 and reimbursement of up to $1,000 for travel expenses to award meeting
Date(s) Application is Due: Nov 1
Contact: Felicia Dixon, Awards Administrator; (800) 227-5558 or (202) 872-4408; fax: (202) 776-8008; email: f_dixon@acs.org or awards@acs.org
Internet: http://portal.acs.org/portal/acs/corg/content?_nfpb=true&_pageLabel=PP_ARTICLEMAIN&node_id=1319&content_id=CTP_004515&use_sec=true&sec_url_var=region1
Sponsor: American Chemical Society
1155 Sixteenth Street, NW
Washington, DC 20036

E.D. Thomas Fellowship 2149

The foundation supports research into the diagnosis, prevention, and cure of leukemia and related hematological malignancies. The fellowship is awarded for one year, renewable yearly for two additional years upon satisfactory performance. The award supports costs not to exceed 8% and salary not to exceed $35,000 including fringe benefits, with the remainder for supplies and/or equipment. Applications must be requested by October 15.

Requirements: Candidates must hold an MD or PhD degree and have completed at least three years postdoctoral training but must be less than 10 years post their first doctoral degree when the award begins. There are no restrictions based on nationality, but only one application will be considered from each sponsoring institution. Candidates must be committed to the research goals of the foundation, must be able to devote at least 80% of their time to the project and must have a sponsoring institution with the academic environment to provide adequate support for the proposed project.
Amount of Grant: $50,000
Date(s) Application is Due: Nov 2
Contact: Grants Administrator, 34 93 414 55 66; fax: 34 93 201 0588; email: fundacio@fcarreras.es
Internet: https://www.fcarreras.org/eng/seccio.php?opcion=5
Sponsor: Jose Carreras International Leukemia Foundation
Muntaner, 383 2n
Barcelona 08021 Spain

E.J. Sierleja Memorial Fellowship 2150
The fellowship is available annually to a graduate student pursuing advanced studies in the area of transportation, with priority consideration given to, but not limited to, students focusing on rail transportation. The institute supports the advancement of engineering education and research. The program is open to individuals in the United States, Canada, and Mexico.
Requirements: Candidates for these awards must be active institute members, as reflected on the October chapter roster or before. New member applications must be completely processed prior to the end of September in order for students to be eligible. Students must be enrolled full-time in graduate or undergraduate industrial engineering programs for the upcoming year and must have an overall point-hour average of 3.40 on a scale of 4.00; graduate students must have had a 3.40 average as undergraduates to qualify. Students may not apply directly for this scholarship, they must be nominated.
Amount of Grant: $600 for academic year
Date(s) Application is Due: Nov 15
Contact: Bonnie Cameron, Headquarters Operations Administrator, (770) 449-0461 ext 105; email: bcameron@iienet.org
Internet: http://www.iienet.org/public/articles/index.cfm?cat=525
Sponsor: Institute of Industrial Engineers
3577 Parkway Ln, Suite 200
Norcross, GA 30092

E.L. Wiegand Foundation Grants 2151
The foundation provides grants to develop and strengthen programs and projects at Arizona, California, the District of Columbia, Idaho, Nevada, New York, Oregon, Utah, and Washington educational institutions in the academic areas of science, business, fine arts, and law; and medicine and health organizations in the areas of heart, eye, and cancer surgery, treatment, and research. The foundation also considers requests for projects that enrich children, communities, public policy, and art. Grants for education, including funds for computers and scientific equipment, are awarded. The board of trustees meets in February, June, and October to choose recipients, but applications may be submitted at any time. Application guidelines are available upon request.
Requirements: Nonprofit organizations in Arizona, California, District of Columbia, Idaho, Nevada, New York, Oregon, Utah, and Washington State are eligible.
Geographic Focus: Arizona; California; District of Columbia; Idaho; Nevada; New York; Oregon; Utah; Washington
Samples: Roman Catholic Diocese of Reno (NV)—to help build a new dining hall at a planned complex several blocks east of the downtown casino district that will provide services to homeless people, $1 million (2005); Saint Bonaventure U, School of Business (NY)—to develop internships at 10 local nonprofit organizations for business students, $43,000 (2004).
Amount of Grant: $10,000-$200,000 average
Contact: Kristen Avansino, Executive Director, (775) 333-0310; fax: (775) 333-0314
Sponsor: E.L. Wiegand Foundation
165 W Liberty Street, Suite 200
Reno, NV 89501

E.V. Murphree Award in Industrial and Engineering Chemistry 2152
This award, sponsored by the ExxonMobile Research and Engineering Company, is given annually to stimulate fundamental research in industrial and engineering chemistry, the development of chemical engineering principles, and their application to industrial processes. A nominee must have accomplished outstanding research of a theoretical or experimental nature in the fields of industrial chemistry or chemical engineering. The award is granted without regard to age or nationality, and consists of $5,000 and a certificate.
Requirements: Any individual, except a member of the award committee, may submit one nomination or seconding letter for the award in any given year. Nominating documents consist of a letter of not more than 1000 words containing an evaluation of the nominee's accomplishments and a specific identification of the work to be recognized, a biographical sketch including date of birth, and a list of publications and patents authored by the nominee.
Restrictions: Self-nominations are not accepted.
Samples: Georges Belfort—award recipient, $5,000 (2008); Wolfgang F. Holderich—award recipient, $5,000 (2007); Liang-Shih Fan—award recipient, $5,000 (2006).
Amount of Grant: $5,000 and $1,000 maximum for travel to award meeting
Date(s) Application is Due: Nov 1

Contact: Felicia Dixon, Awards Administrator; (800) 227-5558 or (202) 872-4408; fax: (202) 776-8008; email: f_dixon@acs.org or awards@acs.org
Internet: http://portal.acs.org/portal/acs/corg/content?_nfpb=true&_pageLabel=PP_ARTICLEMAIN&node_id=1319&content_id=CTP_004520&use_sec=true&sec_url_var=region1
Sponsor: American Chemical Society
1155 Sixteenth Street, NW
Washington, DC 20036

Earhart Foundation Institutional Grants 2153
The foundation supports research fellowships for individual projects in economics, history, philosophy, international affairs, and political science. Fellowships are awarded upon direct application to faculty members. Grants for research in these areas also are awarded to educational and research organizations. H.B. Earhart Fellowships are awarded for graduate study (recipients must be nominated). Fields of interest include history and archaeology, philosophy/ethics, economics, political science, and international studies. Types of support include professorships, publication, fellowships, research, grants to individuals, and scholarships to individuals. Application forms are not required.
Restrictions: Grants are not awarded for capital, building, endowment funds, operating budgets, continuing support, annual campaigns, seed money, ermegency funds, deficit financing, matching gifts, or loans.
Amount of Grant: $1000-$25,000 average
Contact: Ingrid Gregg, President, (734) 761-8592
Sponsor: Earhart Foundation
2200 Green Road, Suite H
Ann Arbor, MI 48105

Earle B. Barnes Award for Leadership in Chemical Research Management 2154
This award, supported by the Dow Chemical Company, is awarded yearly to recognize outstanding achievements in chemical research management. The award is intended to recognize those individuals who have demonstrated outstanding leadership and creativity in promoting the sciences of chemistry and chemical engineering in research management. Nominees should have demonstrated success in research management by exhibiting the proven ability to manage research projects and people. This leadership and creativity must have been shown by a record of successful research projects and by a strong motivation of the researchers on those projects. Recognition of these accomplishments by peers is essential.
Requirements: Any individual, except a member of the award committee, may submit one nomination or seconding letter for the award in any given year. The nominating documents consist of a letter of not more than 1000 words containing an evaluation of the nominee's accomplishments and a specific identification of the work to be recognized, a biographical sketch including date of birth, and a list of publications and patents authored by the nominee. A nominee must be a citizen of the United States.
Restrictions: Self-nominations are not accepted.
Amount of Grant: $5,000
Date(s) Application is Due: Nov 1
Contact: Felicia Dixon, Awards Administrator; (800) 227-5558 or (202) 872-4408; fax: (202) 776-8008; email: f_dixon@acs.org or awards@acs.org
Internet: http://portal.acs.org/portal/acs/corg/content?_nfpb=true&_pageLabel=PP_ARTICLEMAIN&node_id=1319&content_id=CTP_004512&use_sec=true&sec_url_var=region1
Sponsor: American Chemical Society
1155 Sixteenth Street, NW
Washington, DC 20036

Early American Industries Research Grants 2155
Awards are made to support research conducted by serious students and scholars for the study and better understanding of early American industries in homes, shops, farms, and on the sea; as well as the discovery, identification, classification, and preservation of obsolete tools, implements, and mechanical devices. Three to five grants will be awarded. Application must be made on forms supplied by the Grants in Aid Committee.
Requirements: Grants in aid are available to all qualified applicants in general, but those who have completed the graduate level of their education will be given preference. Successful applicant is required to file a project report on a form supplied by the association that must include a statement of finances as well as a half-page abstract of the grantee's research.
Restrictions: Grants may not be used to pay for salaries in whole or in part.
Amount of Grant: $2000 maximum
Date(s) Application is Due: Mar 15
Contact: Grants in Aid Committee, (302) 652-7297
Internet: http://www.eaiainfo.org/Grants.htm
Sponsor: Early American Industries Association
1324 Shallcross Avenue
Wilmington, DE 19806

Early Southern History and Decorative Arts Summer Institute 2156
The summer institute offers a four-week curriculum in a museum setting with lectures, object and room studies, research projects, field trips, and workshops, all dealing with early (pre-1821) southeastern U.S. material culture. The program is sponsored by the museum and the University of North Carolina at Greensboro. Six hours of graduate credit in history can be earned. The program is designed for persons interested in American art, art history, American history, American studies, and museum studies.

Requirements: Applicant must have graduate student status or be presently employed in a museum-related profession.
Amount of Grant: $150-$350 for partial tuition
Date(s) Application is Due: Apr 20
Contact: Sally Gant, Director of Education, (336) 721-7360; fax: (336) 721-7367; email: sgant@oldsalem.org
Internet: http://www.oldsalem.org/about/mesda.htm
Sponsor: Museum of Early Southern Decorative Arts
P.O. Box 10310, Salem Street
Winston-Salem, NC 27108

Earthwatch Institute Student Fellowships   2157
Earthwatch Student Fellowships offer students with limited exposure to science the opportunity to experience research first hand. Award recipients will: work and learn alongside inspiring professional scientists; discover how science and technology can advance our understanding of the world; interact with talented and committed peers and mentors.
Requirements: High school teachers and counselors are encouraged to nominate one to two students with limited opportunities for enrichment and resources, and no prior experience helping a scientist conduct research. Current juniors are preferred, but sophomores and seniors are welcome to apply. Students with disabilities may be eligible for this program.
Contact: Fellowship Program Officer, (800) 776-0188 ext 118; fax: (978) 461-2332; email: EducationAwards@earthwatch.org
Internet: http://www.earthwatch.org/site/pp.asp?c=dsJSK6PFJnH&b=2221765
Sponsor: Earthwatch Institute
3 Clock Tower Place, P.O. Box 75
Maynard, MA 01754-0075

East Stroudsburg University Graduate Assistantships   2158
Graduate assistantships are available in a variety of academic disciplines including pedagogy, sciences (biology, computer science, and general), social sciences (history and political science), and health sciences and human performance (public health, community health, movement studies, athletic training, cardiac rehabilitation, and exercise sciences). Stipends are based on 10-hour and 20-hour assignments with tuition waiver for the academic year. Minorities and women are strongly encouraged to apply.
Requirements: A minimum 2.5 GPA and 3.0 GPA in applicant's major field are required.
Amount of Grant: $2500-$5000
Date(s) Application is Due: Jun 15
Contact: Graduate Assistantships, Office of the Graduate School, (866) 837-6130 or (570) 422-3536; fax: (570) 422-3506; email: grad@po-box.esu.edu
Internet: http://www.esu.edu/grants/gradcat/newasst.shtml
Sponsor: East Stroudsburg University
200 Prospect Street
East Stroudsburg, PA 18301-2999

East-West Center Graduate Fellowships   2159
The fellowship program is designed to give students the opportunity to undertake graduate studies at the University of Hawaii and interact with staff of the East-West Center on research subjects of interest to the center. Awards to master's candidates are for up to 24 months, to doctoral candidates up to 48 months.
Requirements: Applicant must be a citizen or permanent resident of the United States or an Asian/Pacific Islander on the Exchange Visitor Program and must meet center guidelines and academic requirements for completing his or her degree. Recipients of the fellowship who are single or married with no accompanying dependents below 18 years of age are required to reside in Center dormitories.
Amount of Grant: $17,000 approximately
Date(s) Application is Due: Nov 1
Contact: Awards Services Officer, EWC-UHM Scholarship Office, (808) 944-7111; fax: (808) 944-7730; email: EWCUHM@EastWestCenter.org
Internet: http://www.eastwestcenter.org/semedu-program.asp?program_ID=1&Topic=Student&Area=Education
Sponsor: East-West Center
1601 East-West Road
Honolulu, HI 96848

East-West Center Jefferson Fellowships   2160
Each spring the center invites 12 to 14 midcareer American and Asian journalists in print and broadcasting to apply for the fellowships. The journalists spend four weeks at the center in an intensive program of seminars that examine Asian and American political, economic, energy, environmental, population, cultural, and security issues. The American fellows then embark on a four-week trip through Asia while the Asian fellows travel to the U.S. mainland. All return to Honolulu for a final session of discussion and review. The selection committee gives priority to gatekeepers such as news editors, editorial editors and writers, broadcast producers, and assignment editors and producers. A stipend in Honolulu for five weeks at the center plus travel and a modest per diem for the travel portion of the program are provided; employers are expected to share costs where feasible.
Requirements: The program is open to qualified journalists from the United States and from the Asian/Pacific area, ranging from Japan to Pakistan, to Australia and New Zealand, including the Pacific Island nations.
Date(s) Application is Due: Oct 26

Contact: Ann Hartman, Coordinator, (808) 944-7384; fax: (808) 944-7600; email: seminars@EastWestCenter.org
Internet: http://www.eastwestcenter.org//semedu-program.asp?program_ID=10&Topic=Media%20Program&Area=Seminars
Sponsor: East-West Center
1601 East-West Road
Honolulu, HI 96848

Eastern Europe and the Independent States of the Former Soviet Union   2161
    Research and Training Grants
Grants are awarded to organizations to support national programs of advanced research; graduate training; language training; public dissemination of research data, methods, and findings; contact and collaboration among government and private specialists; and/or firsthand experience of the area by U.S. specialists. Organizations that receive the grants act as intermediaries for the federal funds by conducting their own national, open competitions to make awards at the graduate level and above to individual students, scholars, or other institutions. Grant recipients have up to three years to spend their awards. Application deadline will be published in the Federal Register.
Requirements: Applicants must be nonprofit organizations or institutions of higher education with an established track record in conducting research and training programs on the independent states of the former Soviet Union and countries of Eastern Europe. These organizations must run national programs of advanced research; graduate training; language training; public dissemination of research data, methods, and findings; contact and collaboration among Government and private specialists; and/or firsthand experience of the area by U.S. specialists.
Amount of Grant: $129,000-$1.2 million range
Contact: Kenneth Roberts, Executive Director, (202) 736-4572; fax: (202) 736-4851
Internet: http://www.state.gov
Sponsor: Department of State
2201 C Street NW, Room 2251
Washington, DC 20520

Eastman School of Music Graduate Awards Program   2162
The program offers financial support through the work opportunities that come with the awards. The award includes variable stipends plus approximately one-third to full tuition. The deadline of January 1 is for support of study beginning the following September.
Requirements: Eligibility is restricted to persons who have been accepted for graduate study at the Eastman School of Music of the University of Rochester.
Contact: Office of Grad Studies, (585) 274-1560; email: cmccamman@esm.rochester.edu
Internet: http://www.esm.rochester.edu/financialaid/grad.php
Sponsor: University of Rochester, Eastman School of Music
26 Gibbs Street
Rochester, NY 14604

Eaton Charitable Fund Grants   2163
The Fund is dedicated to supporting programs that improve the quality of life in communities where the company operates. The Fund gives primary consideration to requests for programs located in an Eaton community, recommended by an Eaton manager and where employees demonstrate leadership involvement. Programs selected for funding will have clearly defined objectives, measurable end results, and provide a positive return on the Funds investment. The Fund's primary interests are in support of community improvement, education, and arts and cultural programs. Program, project and capital grants are awarded. Capital grants are made for special purposes that meet specific community needs within the company's funding focus. On occasion, operating grants are awarded. Proposals should be sent to the manager of the Eaton facility located in an Eaton community.
Requirements: Applicant organizations must be 501(c)3 tax exempt charities and be located in communities where the company has operations.
Restrictions: Eaton does not make contributions to: annual operating budgets of United Way agencies or hospitals; medical research; endowment funds; debt retirement; religious organizations unless they are engaged in a significant program benefiting the entire community; fraternal or labor organizations; individuals or individual endeavors; fund raising benefits, sponsorships or other events.
Geographic Focus: Arkansas; Colorado; Florida; Georgia; Kansas; Kentucky; Louisiana; Maine; Maryland; Michigan; Minnesota; New Jersey; New York; North Carolina; Ohio; Pennsylvania; South Carolina; Texas; Wisconsin
Contact: (216) 523-4944; fax: (216) 479-7013
Internet: http://www.eaton.com/EatonCom/OurCompany/AboutUs/Corporate Responsibility/SocialCommitment/CorporateGiving/index.htm
Sponsor: Eaton Charitable Fund
1111 Superior Avenue
Cleveland, OH 44114-2584

EBSCO Community College Learning Resources/Library Achievement Awards   2164
Two annual awards are given to recognize significant achievement in the areas of program development and leadership. Nominees for the program award should demonstrate significant achievement in development of an innovative learning resources/library program. Nominees for the leadership award should demonstrate significant achievement in advocacy of learning resources/library programs or services or leadership in professional organizations that are associated with the mission of community, junior, or technical colleges.
Requirements: Individuals or groups from two-year institutions, as well as the two-year institutions themselves, are eligible to receive the awards.

Amount of Grant: $500 each
Date(s) Application is Due: Dec 1
Contact: Matthew Burrell, Librarian, (850) 769-1551; email: mburrell@gulfcoast.edu
Internet: http://www.ala.org/ala/acrl/acrlawards/ebscoawards.htm
Sponsor: Association of College and Research Libraries
50 E. Huron Street
Chicago, IL 60611

Echoing Green Fellowships   2165
The program awards full-time fellowships to emerging entrepreneurs to create innovative domestic or international public service projects that seek to catalyze positive social change. The proposed project may be in any public service area, including but not limited to, the environment, arts, education, health, youth service and development, civil and human rights, and community and economic development. The fellowship provides a two-year stipend, health care benefits, online connectivity, access to Echoing Green's network of social entrepreneurs, training, and technical assistance.
Requirements: Applicants must be at least 18 years old and commit to leading the project for at least two years. Partnerships of up to two individuals also are eligible.
Restrictions: Faith-based and research projects and lobbying activities are not eligible.
Amount of Grant: $30,000 per year for two years, individual fellowships; $45,000 per year (per project, not per individual) for two years, partnership fellowships
Date(s) Application is Due: Dec 2
Contact: Maritza Martinez, Senior Associate; (212) 400-3958; fax: (212) 689-9010; email: Maritza@echoinggreen.org or apply@echoinggreen.org
Internet: http://www.echoinggreen.org/index.cfm?fuseaction=Page.viewPage&pageId=41
Sponsor: Echoing Green Foundation
494 Eighth Avenue, Second Floor
New York, NY 10001

Edilia and Francois-AuguSuite De Montequin Fellowship in Iberian   2166
and Latin American Architecture
The fellowship is awarded each year in support of travel costs for research on Spanish, Portuguese, or Ibero-American architecture. The fellowship is intended for graduate students and junior scholars, but senior scholars may also apply.
Amount of Grant: $2000
Date(s) Application is Due: Oct 15
Contact: Pauline Saliga, Executive Director, (312) 573-1365; fax: (312) 573-1141; email: psaliga@sah.org or info@sah.org
Internet: http://www.sah.org/index.php?module=ContentExpress&func=display&title=CE&mid=&ceid=59
Sponsor: Society of Architectural Historians
1365 N Astor Street
Chicago, IL 60610-2144

Edmund Niles Huyck Preserve Graduate and Postgraduate Research Grants   2167
Graduate and postgraduate investigators may apply for grants for the support of research utilizing the natural resources of the Biological Station of the Huyck Preserve. Included within the preserve are 2000 acres of natural and reforested woodlands, old fields, Lake Myosotis (100 acres), Lincoln Pond (10 acres), and approximately three miles of permanent and intermittent streams. Housing and laboratory space are provided. Write for instructions on application and additional information.
Requirements: Applicants must be predoctoral and postdoctoral scientists whose research utilizes the natural resources of the Edmund Niles Huyck Preserve.
Amount of Grant: $2500 maximum
Date(s) Application is Due: Feb 1
Contact: Richard Wyman, Executive Director, (518) 797-3440; fax: (518) 797-3440; email: rlwyman@capital.net
Internet: http://www.huyckpreserve.org
Sponsor: Edmund Niles Huyck Preserve
P.O. Box 189
Rensselaerville, NY 12147

Education Through Astronomy and Space Science Initiative   2168
This program was developed to establish a comprehensive approach to providing education and public outreach to enhance the publics understanding of space science. The spirit of the program is to provide start-up funding to explore innovative, creative ways to integrate astronomy and space science into U.S. education and public outreach venues through partnerships between the astronomers/space scientists and education professionals.
Requirements: Professionals in astronomy or space science with an affiliation to an institution within the United States are eligible to apply. The proposal must incorporate a partnership with at least one professional educator as coinvestigator.
Amount of Grant: $20,000 maximum for small projects; $50,000 maximum for large projects
Contact: Grant Administrator, Office of Public Outreach, (410) 338-4798; email: ideas@stsci.edu
Internet: http://oposite.stsci.edu/pubinfo/edugroup/grants.html#top
Sponsor: Space Telescope Science Institute
3700 San Martin Drive
Baltimore, MD 21218

Educational Communications and Technology Awards   2169
The awards given through the ECT Foundation promote scholarship and leadership in the field of educational communications and technology. Awards are made to defray doctoral dissertation research expenses, for the papers reporting studies addressing a question related to educational technology, and for the papers describing research findings that could be used to improve the process of instructional design, development, and evaluation. Contact the awards program office for official guidelines for presenting a nomination or application.
Contact: K.J. Saville, AECT Awards Chair, (906) 227-2413; email: ksaville@nmu.edu or aect@aect.org
Internet: http://www.aect.org/Foundation/Awards/Awards.htm
Sponsor: Association for Educational Communications and Technology
1800 N Stonelake Drive, Suite 2
Bloomington, IN 47408

Educational Ventures International Foundation Grants   2170
The foundation funds nonprofits addressing important educational issues in areas where adequate funding from other sources is unavailable. The foundation's areas of interest include grants for educational technology, which provide grants for hardware, software, and peripherals for K-12 and underserved areas; grants for educational resources, which provide funding for resources to enhance teaching and learning environments in underserved areas; grants to encourage the development of new and innovative uses of technology to enhance teaching and learning; and grants for research, which provide funds to support research in math and technology education. Submit a three- to four-page proposal stating the program's goals and budgetary needs. Full proposals are by invitation.
Requirements: Nonprofit organizations are eligible.
Contact: Grants Administrator, fax: (801) 659-2233
Sponsor: Educational Ventures International Foundation
P.O. Box 5887
Carefree, AZ 85377-5887

Educause Jane N. Ryland Fellowships   2171
This fellowship expands opportunities for information technology professionals to attend Educause events, thus helping to build future leaders. Fellowships recognize a combination of past achievement, personal and institutional commitment, potential benefit, and financial need. Awards are made on an annual basis, depending on an applicant's educational goals and the needs of his or her institution.
Requirements: Recipients are selected annually on a competitive basis. Applicants may serve their institutions in any information technology management area: central IT organizations, academic units, or administrative departments.
Amount of Grant: $500-$3000
Date(s) Application is Due: Nov 1
Contact: Fellowships Administrator, (303) 449-4430; fax: (303) 440-0461; email: ryland-fellowships@educause.edu
Internet: http://www.educause.edu/awards/fellow
Sponsor: Educause
4772 Walnut Street, Suite 206
Boulder, CO 80301

Edward Bangs Kelley and Elza Kelley Foundation Grants   2172
The Foundation's interest is primarily in Barnstable County, Massachusetts. The Foundation has been the leader in improving the health and welfare of the community. Grants are made to a great variety of health, social and human service agencies, as well as to cultural and environmental organizations. Town libraries, theatre and art groups, and musical organizations have been supported. Grants are sometimes utilized as seed money by young organizations.
Requirements: Grant applicants must be tax-exempt organizations which are not private foundations and must be located in Barnstable County. The proposed project/program must have a direct benefit to the inhabitants of Barnstable County. If the tax-exempt organization is located outside of Barnstable, but the proposed project/program will have a direct and substantial benefit to the inhabitants of Barnstable, the grant application may be eligible for funding.
Restrictions: Applicants must be residents of Barnstable County, Massachusetts, or must demonstrate very significant ties to Barnstable County.
Geographic Focus: Massachusetts
Contact: Henry Murphy Jr., Administrative Manager, (508) 775-3117; email: contact@kelleyfoundation.org
Internet: http://www.kelleyfoundation.org
Sponsor: Edward Bangs Kelley and Elza Kelley Foundation
243 South Street, P.O. Drawer M
Hyannis, MA 02601

Edward C. and Ann T. Roberts Foundation Grants   2173
The foundation awards grants to support arts (visual arts and performing arts) and culture, including education and programs that benefit the disadvantaged and children in Hartford, CT. Types of support include capital campaigns, building construction/renovation, and program development. The board meets in March, June, September, and December. There are no application forms.
Requirements: 501(c)3 nonprofit organizations in Hartford, CT, may apply.
Restrictions: Grants are not made to individuals or for scholarship aid.
Geographic Focus: Connecticut

Samples: Hartford Stage (Hartford, CT)—for Hartford StageFest, a festival that features new plays and solo artists who present new performance pieces, $100,000.; U of Hartford, Hartt School and Hartford Art Museum (Hartford, CT)—for the Roberts Ctr for Young Leadership in the Arts, which enables young artists interested in dance, music, theater, and the visual arts to pursue in-depth study in their fields, $100,000 jointly.
Date(s) Application is Due: Feb 1; May 1; Aug 1; Nov 1
Contact: Elizabeth Normen, Executive Director, (860) 233-0228; fax: (860) 236-8098; email: EJNormen@aol.com
Internet: http://www.fdncenter.org/grantmaker/e&aroberts
Sponsor: Edward C. and Ann T. Roberts Foundation
P.O. Box 271588
West Hartford, CT 06127-1588

Edward F. Kook-U.S.ITT Endowment Fund Grants   2174
The goal of this program is to further the objectives of the institute. Only new projects and research with the purpose of advancement of the industry will be considered for annual grants. The program must benefit the industry of theatrical design and technology and clearly seek new knowledge through experimentation, research, or collection of resources; serve a cross-section of the institute; and provide results that are made available to institute members. Projects may involve theoretical research, applied research, statistical survey, or historical research. A formal application form must be obtained from the Web site or directly from the institute.
Requirements: An applicant must be a current member of U.S.ITT to apply; membership is open to anyone.
Amount of Grant: $10,000 for project support grants; $15,000 maximum for fellowships
Date(s) Application is Due: Jan 10
Contact: Grants Administrator, Marketing and Development, (800) 938-7488; fax: (866) 398-7488; email: info@office.usitt.org
Internet: http://www.usitt.org/activities/GrantsFellowshipsIntro.html
Sponsor: United States Institute for Theater Technology Inc
6443 Ridings Road
Syracuse, NY 13206-1111

Edward G. McDowell Traveling Scholarships   2175
The scholarships are awarded annually for a one-year period of travel, research, and study in Paris and Europe; and are open to unmarried members of the league of either sex, who must have pursued a course of study at the league for at least 16 months in a full-time class of which a minimum six months must have been during the current school year. Applications are available in mid-March, and recipients are announced in early May.
Requirements: Candidates must be league members for at least 16 months.
Restrictions: Winners are not eligible for subsequent McDowell competitions.
Amount of Grant: $10,000 approximately
Contact: Gaetano Scognamillo, Scholarship Coordinator, (212) 247-4510; fax: (212) 541-7024
Internet: http://www.theartstudentsleague.org/Navigation/Home/HP-FRAME.html
Sponsor: Art Students League of New York
215 W 57th Street
New York, NY 10019

Edward S. Moore Foundation Grants   2176
The foundation awards grants to nonprofits in Connecticut and New York in its areas of interest, including youth, hospitals, education, cultural programs, museums, and Christian religion. Types of support include operating budgets, continuing support, annual campaigns, seed money, emergency funds, building funds, equipment, land acquisition, endowment funds, matching funds, internships, scholarship funds, special projects, and research. There are no application deadlines. The board meets in January, April, July, and October to consider proposals.
Requirements: Nonprofits in Connecticut and New York may submit proposals.
Restrictions: Grants are not awarded to individuals or for deficit financing, publications, or conferences.
Geographic Focus: Connecticut; New York
Samples: Children's Aid Society (New York, NY)—$50,000.; Metropolitan Opera (New York, NY)—$35,000.; Stamford Hospital Foundation (Stamford, CT)—$430,000.
Amount of Grant: $10,000-$50,000 average
Contact: John W. Cross III, President, (203) 629-4591
Sponsor: Edward S. Moore Foundation
30 Lismore Ln
Greenwich, CT 06831

Edwin F. Church Medal Award   2177
This award is given annually to an individual who has rendered eminent service in increasing the value, importance, and attractiveness of mechanical engineering education. Mechanical engineering is used here in its broadest sense of preparation for any aspect or level of mechanical engineering through any appropriate mechanism including universities, technical institutes, professional societies and private groups, in-house professional development programs of industrial concerns and governmental agencies, or programmed learning and self-instruction systems. The qualifications for this medal are detailed and rigorous and a prospective nominator may secure the nomination form and details from the ASME. The award is not intended to recognize professional educators on that merit alone.
Requirements: The award may be made to one recipient of any age who need not be a member of ASME.

Samples: Vincent Wilczynski—award winner, $2500 (2005); David Lavery—award winner, $2500 (2004.; Devendra Garg—award winner, $2500 (2003).
Amount of Grant: $2500
Date(s) Application is Due: Feb 15
Contact: Gilda DiTullio, Manager, (212) 591-7736; fax: (212) 705-7739; email: ditulliog@asme.org
Internet: http://www.asme.org/honors/ms71/gaa/church.html
Sponsor: American Society of Mechanical Engineers
3 Park Avenue
New York, NY 10016

Edwin O. Reischauer Institute of Japanese Studies Postdoctoral Fellowships   2178
The program awards postdoctoral fellowships to give recipients the opportunity to turn dissertations into publishable manuscripts. Fellowships are for 10 to 12 months, beginning July 1 or September 1, and include a stipend and health insurance coverage for grantee. Fellows receive office space and access to libraries and resources of Harvard University.
Requirements: Applicants should have received the PhD in Japanese studies in any area of the humanities or social sciences since 1999. Those without the degree must complete all requirements by July 1 of the current year.
Amount of Grant: $20,000 stipend
Date(s) Application is Due: Mar 1
Contact: Ruiko Connor, Staff Assistant Administrative Officer, (617) 495-3220; email: rconnor@fas.harvard.edu
Internet: http://www.gsas.harvard.edu/pdfs/reischauer_dcapp.pdf
Sponsor: Harvard University
1737 Cambridge Street
Cambridge, MA 02138

Edwin S. Webster Foundation Grants   2179
The policy of the foundation is to support charitable organizations that are well known to the trustees, with emphasis on special projects and capital programs, or operating income for hospitals, medical research, education, youth agencies, cultural activities, and programs addressing the needs of minorities. Types of support include operating budgets, continuing support, annual campaigns, building funds, equipment, land acquisition, endowment funds, matching funds, scholarship funds, professorships, internships, fellowships, special projects, and research.
Requirements: The Foundation confines its grants primarily to the New England area. Grantees must provide evidence of their tax-exempt status. The AGM common proposal format, available on the Internet at http://agmconnect.org is suitable for submission of proposals but not required. There are no set deadlines, but for consideration at the spring meeting, proposals should arrive by May 1 and by November 1 for consideration at the fall meeting. Proposals received after the trustees meet will be held for consideration at the next meeting.
Restrictions: Grants are not made to organizations outside the United States or to individuals.
Geographic Focus: Massachusetts
Samples: Massachusetts Institute of Technology, Office of Minority Education, Cambridge, MA, $40,000 - Mentor Advocate Partnership (2008); Eastern Virginia Medical School, Norfolk, VA, $25,000 - general operations and interdisciplinary research (2008); Museum of Science, Boston, MA, $50,000 - capital support for the Hayden Planetarium (2008).
Amount of Grant: $15,000-$50,000 average
Date(s) Application is Due: May 1; Nov 1
Contact: Michelle Jenney, Administrator; (617) 391-3087; fax: (617) 426-7080; email: mjenney@gmafoundations.com
Sponsor: Edwin S. Webster Foundation
c/o GMA Foundations, 77 Summer Street, 8th Floor
Boston, MA 02110-1006

EF Behavioral Sciences Student Fellowships   2180
This fellowship stimulates individuals to pursue careers in epilepsy in either research or practice settings. Appropriate fields include sociology, social work, psychology, anthropology, nursing, economics, vocational rehabilitation, counseling, political science, and others relevant to epilepsy research or practice. Annual deadline dates may vary; contact program staff for exact dates.
Requirements: Applicant must be actively enrolled in a degree program at the time the fellowship is to be undertaken. The fellowship must be undertaken during a free period in the student's year, e.g., during the summer. Both graduate and undergraduate students are eligible provided all other eligibility criteria are met. The project must be carried out at an approved facility.
Samples: Cathleen Buchanan, U of Wisconsin, Madison (WI)—for an examination of attention deficit hyperactivity disorder in children with epilepsy.; David Sbarra, U of Virginia (VA)—for an assessment of patterns of self-reliance, social and emotional adaptation, and parent-child interactions and how they affect children with epilepsy.
Amount of Grant: $3000 stipend
Date(s) Application is Due: Mar 1
Contact: Fellowship Programs, Research and Professional Education, (301) 459-3700; fax: (301) 577-2684; email: grants@efa.org
Internet: http://www.epilepsyfoundation.org/research/grants.cfm
Sponsor: Epilepsy Foundation
4351 Garden City Drive, Suite 406
Landover, MD 20785-7223

EF Health Sciences Student Fellowships 2181

Three-month fellowships are awarded to medical and health science students for work on epilepsy study projects. The project may be carried out at the U.S. institution of the student's choice where there are ongoing programs of research, training, or service in epilepsy. Application forms and guidelines are available upon request. Annual deadline dates may vary; contact program staff for exact dates.

Requirements: Predoctoral training students in the Health Sciences may be accepted at any point in their schooling-following acceptance but before beginning the first year, or in the period immediately following their final year. A supervisor or preceptor must accept responsibility for the supervision of the student program and for the submission of a full report to the institute upon completion of the program.

Samples: Rosalind Carter Brown, U of Virginia Medical School (VA)—for an examination of the mechanisms of action of diazepam (Valium) in the treatment of status epilepticus, prolonged seizures.; Daniel Campbell, Pennsylvania State U College of Medicine (PA)—for an analysis of the role played by one type of calcium-related molecule in a genetic model of epilepsy.

Amount of Grant: $3000 stipend
Date(s) Application is Due: Mar 1
Contact: Fellowship Programs, Research and Professional Education, (301) 459-3700; fax: (301) 577-2684; email: grants@efa.org
Internet: http://www.epilepsyfoundation.org/research/grants.cfm
Sponsor: Epilepsy Foundation
4351 Garden City Drive, Suite 406
Landover, MD 20785

EF Postdoctoral Research Training Fellowships 2182

The foundation awards grants and fellowships for cutting-edge research into the causes of epilepsy. Program goals include understanding basic mechanisms, developing new therapeutic approaches, understanding the behavioral and psychosocial aspects of having epilepsy, and encouraging the professional development of scientists and healthcare professionals. The purpose of the fellowship is to support predoctoral students with dissertation research related to epilepsy. Applications from all fields of research pertinent to epilepsy will be considered.

Requirements: Physicians or PhD neuroscientists who desire postdoctoral research experience are eligible for these fellowships, which must be carried out at an approvable facility where there is an ongoing epilepsy research program. Applicants must be licensed to practice medicine in the United States or Canada.

Restrictions: Individuals holding faculty appointments at the level of assistant professor or higher are ineligible.

Samples: Dr. Jon Backstrom, Vanderbilt U School of Medicine (TN)—for an investigation of the anatomical interaction between GABA and seratonin, and its effect on seizures.; Dr. Sarah Caddick, Duke U Medical Ctr (NC)—for an investigation of the function of zinc in certain systems of the brain that exhibit high concentrations of ion zinc and are highly prone to seizures.

Amount of Grant: $40,000 maximum
Date(s) Application is Due: Sep 1
Contact: Research and Professional Education, (301) 459-3700; fax: (301) 577-2684
Internet: http://www.epilepsyfoundation.org/research/grants.html
Sponsor: Epilepsy Foundation
4351 Garden City Drive, Suite 406
Landover, MD 20785

EF Predoctoral Research Training Fellowship 2183

The predoctoral research training fellowship supports predoctoral students with dissertation research related to epilepsy, thus strengthening their interest in establishing epilepsy research as a career direction. Annual deadline dates may vary; contact program staff for exact dates.

Requirements: Graduate students pursuing a PhD degree in neuroscience, physiology, pharmacology, psychology, biochemistry, genetics, nursing, or pharmacy may apply.

Amount of Grant: $20,000 ($19,000 stipend plus $1000 travel allowance)
Date(s) Application is Due: Sep 1
Contact: Research and Professional Education, (301) 459-3700; fax: (301) 577-2684; email: grants@efa.org
Internet: http://www.epilepsyfoundation.org/research/grants.cfm
Sponsor: Epilepsy Foundation
4351 Garden City Drive, Suite 406
Landover, MD 20785

EF Research and Training Fellowships for Clinicians 2184

These fellowships offer qualified individuals the opportunity to develop expertise in epilepsy research through a one-year training experience and involvement in an epilepsy research project. The purpose of the fellowships is to develop academic clinicians to teach patient care of persons with epilepsy and advance knowledge about epilepsy through clinical investigation. Applications from individuals interested in acquiring experience in the conduct of human clinical studies and who plan to study epilepsy as a human disorder are encouraged. Emphasis is placed on individuals who will be trained in clinical research in epilepsy rather than use epilepsy as a tool in their research in other fields. Applications from women, members of minority groups, and people with disabilities are especially encouraged.

Requirements: Individuals with the MD or DO degree who will have completed residency training in neurology, neurosurgery, pediatrics, internal medicine, or psychiatry by the time the fellowship commences and desire additional postdoctoral clinical research

experience are eligible. Applicants planning to work in the United States receive preference.

Restrictions: These are training fellowships and will not be granted to individuals who have received faculty appointments at the level of assistant professor or higher. Individuals may not receive concurrent support for more than one EFA or AES grant or fellowship.

Amount of Grant: $50,000
Date(s) Application is Due: Oct 14
Contact: Research and Professional Education, (301) 459-3700; fax: (301) 577-2684
Internet: http://www.epilepsyfoundation.org/research/grants.cfm
Sponsor: Epilepsy Foundation
4351 Garden City Drive, Suite 406
Landover, MD 20785

EFWA Accounting Research Grants 2185

The foundation funds issues papers by faculty, individuals, or business to research technical accounting issues or other issues furthering women in the accounting profession. The foundation also provides research grants for women faculty members to contribute to the field of accounting and assist faculty in their careers.

Amount of Grant: $3000
Date(s) Application is Due: May 1
Contact: Cynthia Hires, (610) 407-9229; fax: (610) 644-3713; email: info@efwa.org
Internet: http://www.efwa.org/research.htm
Sponsor: Educational Foundation for Women in Accounting
P.O. Box 1925
Southeastern, PA 19399-1925

EFWA Accounting Scholarships 2186

The program provides a one-year academic scholarship for women who are pursuing a PhD in accounting. Applications are posted online in January.

Amount of Grant: $1000-$5000
Date(s) Application is Due: Mar 15
Contact: Cynthia Hires, (610) 407-9229; fax: (610) 644-3713; email: info@efwa.org
Internet: http://www.efwa.org/scholarships.htm
Sponsor: Educational Foundation for Women in Accounting
P.O. Box 1925
Southeastern, PA 19399-1925

Eileen J. Garrett Scholarship 2187

The foundation awards an annual scholarship for research and study in parapsychology. Any student attending an accredited college or university who has already demonstrated an academic interest in parapsychology and plans to pursue the subject as a career may apply. Applications and collateral material may be submitted via post, fax, or as MS Word or Adobe Acrobat attachments. Guidelines are available online.

Amount of Grant: $3000
Date(s) Application is Due: Jul 15
Contact: Garrett Scholarship, (212) 628-1550; fax: (212) 628-1559; email: office@parapsychology.org
Internet: http://www.parapsychology.org/dynamic/040200.html
Sponsor: Parapsychology Foundation
P.O. Box 1562
New York, NY 10021

Eisenhower Exchange International Fellowships 2188

These competitive fellowships are for U.S. citizens to go abroad for a professional exchange program. The program includes travel throughout the selected country/region for research/inquiry purposes. Countries participating have included Argentina, Germany, Hungary, Ireland, Malaysia, Peru, and Taiwan. Fellowships include travel and living expenses and program arrangements/appointment scheduling for a one- to three-month program. Spouse participation is encouraged. Fellowships are awarded in the year preceding departure. Contact the office or visit the Web site for the most current information on participating countries and deadline dates.

Requirements: Midcareer professionals who are U.S. citizens with demonstrated leadership ability and significant contributions to their fields are eligible to apply.

Contact: U.S.A Fellowships, (215) 546-1738; fax: (215) 546-4567; email: ike@eisenhowerfellowships.org
Internet: http://eisenhowerfellowships.org
Sponsor: Eisenhower Exchange Fellowships
256 S 16th Street
Philadelphia, PA 19102

Ekhagastiftelsen Foundation Grants 2189

The foundation operates in Sweden and supports research into agricultural production without the use of poisons and artificial means; and research into the production of natural foods and medical methods based on the capacity of human beings to cure themselves. Types of support include grants to organizations and scholarships and fellowships to individuals.

Contact: Grants Administrator, 46 70 240 81 81; fax: 46-08-21-83-76; email: info@ekhagastiftelsen.se
Internet: http://www.ekhagastiftelsen.se/eng/stiftelsen.html
Sponsor: Ekhagastiftelsen Foundation
Box 34012
Stockholm SE-100 26 Sweden

El Pomar Leadership Development and Community Engagement Postgraduate Fellowship   2190

The program educates young leaders about the role of nonprofit organizations and the foundation within the communities of Colorado. The two-year postgraduate program also focuses on professional development and prepares fellows for positions of leadership in Colorado and the nation. Fellows serve the foundation's operating programs, such as EPYCS, a high school philanthropy initiative. In addition, fellows will be thoroughly educated in the nature of the nonprofit sector and the role of foundations. At a minimum, fellows participate in monthly courses in public speaking, financial and investment management, nonprofit management, fundraising and strategic planning, and an annual Outward Bound course in the central mountain ranges of Colorado.

Requirements: Applicant must be a graduate of a four-year university or college, have a Colorado affiliation (e.g., be a state resident, have attended an in-state university or college, or have immediate family who are residents or past residents), and demonstrate a strong leadership capability and potential.

Geographic Focus: Colorado

Amount of Grant: $27,000 starting salary

Contact: Grants Administrator, (800) 554-7711 or (719) 633-7733; email: fellowship@elpomar.org

Internet: http://www.elpomar.org/page.asp?pageid=0|6|30&id=0|fellowship

Sponsor: El Pomar Foundation
10 Lake Cir
Colorado Springs, CO 80906

Eleanor Naylor Dana Charitable Trust Grants   2191

The trust supports biomedical research and the performing arts. In the biomedical program, grants are awarded to support clinical investigation by established scientists to pursue innovative projects designed to improve medical practice or prevent disease. Grants do not support large-scale field studies of a therapeutic or epidemiological nature. Grants for the performing arts support U.S. programs benefiting the public with long-range potential. Giving primarily is in areas east of the Mississippi River. The foundation is interested in the organization's history and its public service. The board meets in March, June, September, and December.

Requirements: Established scientists working under a qualified U.S. institution's approval and U.S. arts organizations are eligible to apply.

Restrictions: The biomedical program does not make grants for instrumentation alone or solely for conference or individual travel. The arts program will not fund deficits; exhibits, publications, or conclaves; or individuals.

Samples: New York City Opera (NY)—program support, $100,000.; Lincoln Ctr Theater (NY)—program support, $50,000.

Amount of Grant: $5000-$50,000 average

Date(s) Application is Due: Feb 1; May 1; Sep 1; Nov 1

Contact: Trust Administrator, (212) 754-2890; fax: (212) 754-2892

Sponsor: Eleanor Naylor Dana Charitable Trust
c/o The Trustees, P.O. Box. 1803, Murray Hill Sta
New York, NY 10156

Eli Lilly & Company Foundation Grants Program   2192

Discretionary grants are awarded in two categories: nonprofit groups aligned with company interests, and grants in Indianapolis and several other communities with significant employee populations. Company-aligned giving focuses on public policy research, health and human services aligned with major therapeutic interests, and academic relations. Community-aligned giving focuses on culture, K-12 education and youth development, locally aligned healthcare organizations, fencerow neighborhood groups, and diversity. The foundation supports organizations within these categories that have a well-defined sense of purpose, a demonstrated commitment to maximizing available resources, and a reputation for meeting objectives and delivery quality programs and services.

Restrictions: Grants do not support individuals; endowments; debt reduction; religious or sectarian programs for religious purposes; bands or fraternal, labor, athletic, or veterans organizations; political contributions; beauty or talent contests; fundraising activities related to individual sponsorship; conferences or media productions; nonaccredited education groups; or memorials.

Samples: American National Red Cross (Washington, DC)—for relief efforts in South Asia and Africa, $1 million (2005); Oklahoma Medical Research Foundation Foundation (Oklahoma City, OK)—to endow two professorships in biomedical research, $6 million (2004).

Date(s) Application is Due: Jun 30; Dec 31

Contact: Thomas King, President, (317) 276-3177; fax: (317) 277-2025

Internet: http://www.lilly.com

Sponsor: Eli Lilly & Company Foundation
Lilly Corporate Ctr
Indianapolis, IN 46285

Eli Lilly & Company Reintegration Scholarships   2193

The goal of the scholarship is to help people with schizophrenia and related schizophrenia-spectrum disorders to acquire the educational and vocational skills necessary to reintegrate into society, secure jobs, and regain their lives. The program is designed to offer financial assistance for a wide range of educational opportunities in which students work to attain a certificate or degree, including high school equivalency programs; trade or vocational school programs; and associate, bachelor, and graduate degrees. Guidelines and application are available online.

Requirements: Applicants must be diagnosed with schizophrenia, schizophreniform, or schizoaffective disorder; be currently receiving medical treatment for the disease, including medications and psychiatric follow-up; be actively involved in rehabilitative or reintegrative efforts, such as clubhouse membership, part-time work, volunteer efforts, or school enrollment; and complete an application package that includes an application form, essay, transcripts (if applicable), recommendation forms from three references, and school financial requirements.

Contact: Lilly Secretariat, (800) 809-8202; email: lillyscholarships@reintegration.com

Internet: http://www.reintegration.com/resources/scholarships

Sponsor: Eli Lilly & Company
310 Busse Hwy, PMB 327
Park Ridge, IL 60068-3251

Elias J. Corey Award for Outstanding Original Contribution in Organic Synthesis by a Young Investigator   2194

The Award, supported by the Pfizer Endowment Fund, is intended to recognize original and insightful work by a young investigator that has had significant impact on the field of synthetic organic chemistry. The award will be granted to an individual under the age of 40 on November 1 of the Award year who has demonstrated outstanding creativity in the field of synthetic organic chemistry. A nominee must have accomplished research that is of exceptional merit and adds value to the field, such as the contribution of a novel and highly useful reagent, reaction, or catalytic process that solves a synthetic problem of considerable importance. Special consideration will be given to the application of such novel chemistry toward the synthesis of natural products and organic molecules of medicinal importance, and to the discovery of novel reactions or processes with broad utility.

Requirements: Any individual, except a member of the award committee, may submit one nomination or seconding letter for the award in any given year. The nominating documents consist of a letter of not more than 1000 words containing an evaluation of the nominee's accomplishments and a specific identification of the work to be recognized, a biographical sketch including date of birth, and a list of publications and patents authored by the nominee.

Restrictions: Self-nominations are not accepted.

Samples: Brian M. Stoltz—award recipient, $5,000 (2009); F. Dean Toste—award recipient, $5,000 (2008); Michael J. Krische—award recipient, $5,000 (2007); Justin Du Bois—award recipient, $5,000 (2006).

Amount of Grant: $5,000 and travel to award meeting

Date(s) Application is Due: Nov 1

Contact: Felicia Dixon, Awards Administrator; (800) 227-5558 or (202) 872-4408; fax: (202) 776-8008; email: f_dixon@acs.org or awards@acs.org

Internet: http://portal.acs.org/portal/acs/corg/content?_nfpb=true&_pageLabel=PP_ARTICLEMAIN&node_id=1319&content_id=CTP_004516&use_sec=true&sec_url_var=region1

Sponsor: American Chemical Society
1155 Sixteenth Street, NW
Washington, DC 20036

Elisabeth Severance Prentiss Foundation Grants Program   2195

The foundation awards the majority of its support in the field of health and medicine. Scientific and medical research, hospitals, health projects, medical education, and care and support of the elderly are funded. Types of support include building programs, equipment needs, research, operating budgets, continuing support, seed grants, endowment funds, and projects/programs. Decision makers favor specific projects over general operating support. The maximum term of a commitment is five years, with preference for proposals for three years or less. Special consideration will be given to requests from applicants in the Cuyahoga County area. The board meets in June and December.

Requirements: Grants are awarded to promote and improve medical services in the greater Cleveland, OH, area.

Restrictions: Grants are not awarded to individuals for scholarships, fellowships, or grants in aid; or to organizations for fund-raising campaigns, surveys, assessments, studies, or planning activities.

Geographic Focus: Ohio

Samples: University Hospitals of Cleveland (OH)—to create an inpatient psychiatric unit for children and adolescents at Rainbow Babies and Children's Hospital, $5 million (2005).

Amount of Grant: $10,000-$100,000 average

Date(s) Application is Due: May 15; Nov 15

Contact: Michael Galland, c/o National City Bank, (216) 222-2736

Sponsor: Elisabeth Severance Prentiss Foundation
P.O. Box 94651
Cleveland, OH 44104

Elizabeth Glaser Pediatric AIDS Foundation International Leadership Awards   2196

The Award program is designed to invest in trained individuals in resource-poor countries who have the potential to develop programs which will have a direct impact on the pediatric HIV epidemic in their country, but that lack the resources to do so. This program will complement existing training programs by providing support after training is completed. The Award specifically seeks to identify individuals who are likely to have an enduring impact on control of the epidemic. Applicants may apply for support to and or scale-up effective pediatric HIV/AIDS research and/or implementation programs including those addressing epidemiology, prevention of mother-to-child transmission, treatment of infected mothers and infants, continuum of care and scale up of prevention services to include care and treatment, development of national policies or strategies to combat pediatric HIV, assessment of the economic impact of disease and its prevention, or to help train additional in-country and regional scientists and program leaders. Award recipients will be asked to mentor a minimum of three additional people who would benefit from the experience and could help them achieve their goals. These are individuals who would be hired and trained to work on the project proposed by the applicant. The Award provides up to $100,000 in total costs per year for three years for a total of $300,000. Funding in years 2 and 3 is dependent on progress in year one and is not guaranteed. Letters of Intent are due by August 24th, with full applications solicited by August 31st and full applications due on September 28th. Awards are made to institutions on behalf of individuals.

Requirements: Applicants must be from either Africa or India, and performing work in either Africa or India in order to apply. Special interested will be given to applications from the African nations of South Africa and Tanzania, but the foundation will consider applicants from other African countries as well. Applicants must demonstrate commitment to continuing to work in a developing country on HIV/AIDS when the award is complete, which is a requirement of the award. Applicants are expected to be in the process of establishing successful careers and programs. Most applicants will have completed training within the past 5 years. Senior investigators such as Full Professors will usually not be eligible for the award. Applicants must hold a position at a public or private non-profit institution that will allow him or her adequate time and provide appropriate facilities to perform the work proposed.

Restrictions: U.S. citizens or European citizens living in these countries are not eligible to apply. Previous International Leadership Award recipients will not be eligible for this round of funding. Grants can not be made directly to individuals.

Amount of Grant: $100,000 in total costs per year for three years

Date(s) Application is Due: Mar 20

Contact: Chris Hudnall, Senior Program Coordinator; (310) 491-3154 or (310) 314-1459, ext. 154; fax: (310) 314-1469; email: Chris@pedaids.org or research@pedaids.org

Internet: http://www.pedaids.org/GrantsandAwards/Awards/International-Leadership-Award.aspx

Sponsor: Elizabeth Glaser Pediatric AIDS Foundation

11150 Santa Monica Boulevard, Suite 1050

Los Angeles, CA 90025

Elizabeth Glaser Pediatric AIDS Foundation Operations Research Grants   **2197**

The Foundation, in its commitment to improve programs to prevent mother-to-child transmission of HIV (PMTCT), announces the availability of a second and final round of funding for operations research (OR) projects under the new Operations Research for Enhancing PMTCT Program. The RFP will further develop an OR program that is designed to utilize Foundation-supported sites and data to generate studies capable of ultimately improving PMTCT program activities. For the purposes of this EGPAF Request for Proposals (RFP), Operations Research is defined broadly and includes the use of analytical techniques to achieve better health outcomes, assessment of the relative contribution of components of service delivery to achieve desired outcomes, and the measurement of cost-effective systems to achieve desired goals. One year awards will be funded up to a maximum of $150,000 inclusive of 5% (maximum) indirect costs. Two year awards will provide up to $125,000 per year to a total of $250,000 in research support per 2 year award including 5% indirect costs. Letters of Intent (LOIs) are due April 14th, with full applications due at the Foundation office by June 6.

Requirements: Proposals will be accepted from investigators who are working with health care teams that are providing PMTCT services to EGPAF-supported sites around the world. Foundation partners including host-country government and non-governmental organizations and/or combinations of the above are eligible to apply.

Amount of Grant: $150,000-$250,000

Contact: Chris Hudnall, Senior Program Coordinator; (310) 491-3154 or (310) 314-1459, ext. 154; fax: (310) 314-1469; email: Chris@pedaids.org or research@pedaids.org

Internet: http://www.pedaids.org/GrantsandAwards/Awards/Operations-Research-Award.aspx

Sponsor: Elizabeth Glaser Pediatric AIDS Foundation

11150 Santa Monica Boulevard, Suite 1050

Los Angeles, CA 90025

Elizabeth Glaser Pediatric AIDS Foundation Pediatric HIV Vaccine   **2198**
Research Grants

The Foundation, Pediatric HIV Vaccine Research Program seeks to address basic and pre-clinical research relevant to pediatric HIV vaccine design and development, including, but not limited to, the following topics: the study of breast-milk transmission of HIV; pediatric immune responses to HIV; identification of potential vaccines for pediatric populations; and identification of obstacles to the conduct of clinical trials in infants.

Samples: Marylyn Addo, M.D., Ph.D., Massachusetts General Hospital, Charlestown, Massachusetts (2008); Grace Aldrovandi, M.D. Children's Hospital, Los Angeles, U.S.C Los Angeles, California (2008).

Contact: Chris Hudnall, Senior Program Coordinator; (310) 491-3154 or (310) 314-1459, ext. 154; fax: (310) 314-1469; email: Chris@pedaids.org or research@pedaids.org

Internet: http://www.pedaids.org/GrantsandAwards/Awards/Pediatric-HIV-Vaccine-Program.aspx

Sponsor: Elizabeth Glaser Pediatric AIDS Foundation

11150 Santa Monica Boulevard, Suite 1050

Los Angeles, CA 90025

Elizabeth Glaser Pediatric AIDS Foundation Two-Year International   **2199**
Scholar Award

The Foundation is offering a postdoctoral fellowship for clinicians/scientists from developing countries. The program is aimed at health care professionals who have specific training or experience with HIV/AIDS and hold an M.D. or Ph.D. Applicants must demonstrate commitment to continuing to work in a developing country on HIV/AIDS when the award is complete, as the emphasis of this program is on building long-term scientific capacity in developing countries. The subjects for these projects could include, but are not limited to, the following: epidemiology; issues related to implementation of either prevention or therapy programs; transmission of the HIV virus from mother to infant, including breast-milk transmission and prevention; pediatric HIV vaccine development; antiretroviral drug treatment including pharmacology, complications, and resistance; and microbicide research.

Requirements: Applicants must be from a developing country, and performing work in a developing country. The Foundation encourages those who are determined to improve care in the international setting, as well as those interested in basic science, to apply.

Restrictions: United States or European citizens living in developing countries are not eligible to apply.

Samples: Herbert Kiyingi, M.B.Ch.B., Nsambya Hospital/Medbiotech Lab/EGPAF Collaboration—The Effect of Intermittent Presumptive Treatment With Sulfadoxine/Pyrimethamine to Control Malaria, on the Quality of Life and Survival in Ugandan Children Living With HIV/AIDS (2005); Clive Maurice Gray, Ph.D.National Institute of Communicable Diseases, Johannesburg, South Africa—Understanding Mechanisms of T-Cell Immunity in HIV-1 Infected and Exposed Newborns and Children (2004).

Amount of Grant: $57,225

Contact: Chris Hudnall, Senior Program Coordinator; (310) 491-3154 or (310) 314-1459, ext. 154; fax: (310) 314-1469; email: Chris@pedaids.org or research@pedaids.org

Internet: http://www.pedaids.org/GrantsandAwards/Awards/Two-Year-International-Award.aspx

Sponsor: Elizabeth Glaser Pediatric AIDS Foundation

11150 Santa Monica Boulevard, Suite 1050

Los Angeles, CA 90025

Elizabeth Glaser Pediatric AIDS Foundation Two-Year Scholar Awards   **2200**

The Awards are postdoctoral fellowships that provide two years of salary support, plus supplies and/or travel costs toward medical research on pediatric HIV/AIDS. Annual salaries‚Äîbased on years of postdoctoral experience‚Äîrange from $30,000 to $46,000. These awards provide a powerful incentive for a new generation of scientists to take an early professional interest in studying pediatric HIV/AIDS. Scientists are supported by a mentor experienced in HIV/AIDS research. Through this mentoring relationship, a growing number of talented investigators are focusing their efforts on helping to bring an end to HIV/AIDS in children.

Requirements: Scholars must select an experienced sponsor with appropriate qualifications to oversee the proposed research. The sponsor should be a faculty member with an advanced degree (M.D. or Ph.D.) who will oversee the research, provide guidance to the applicant, and can provide evidence of the institution's commitment to the career development of the applicant.

Samples: Melissa Farrow, Ph.D., College of the Holy Cross—Investigation of the Expression of Anti-HIV AP.O.BEC3 Family Members (2005); Vandana Kalia, Ph.D., Emory University School of Medicine—Development of CD8 T Cell Immunological Memory (2005); Surojit Sarkar, Ph.D., Emory University School of Medicine—Immune Reconstitution and Reversal of T Cell Exhaustion During Chronic Infection (2005).

Amount of Grant: $60,000 maximum

Contact: Chris Hudnall, Senior Program Coordinator; (310) 491-3154 or (310) 314-1459, ext. 154; fax: (310) 314-1469; email: Chris@pedaids.org or research@pedaids.org

Internet: http://www.pedaids.org/GrantsandAwards/Awards/Two-YearScholar.aspx

Sponsor: Elizabeth Glaser Pediatric AIDS Foundation

11150 Santa Monica Boulevard, Suite 1050

Los Angeles, CA 90025

Elizabeth M. Irby Foundation Grants   **2201**

The foundation awards grants to Mississippi nonprofits in its areas of interest, including arts and culture, elementary education, secondary education, higher education, social services, and Christian churches and religious organizations. Types of support include general operating support, continuing support, annual campaigns, capital campaigns, building construction/renovation, endowment funds, emergency funds, program development, research, scholarship funds, and matching funds. There are no application forms or deadlines.

Requirements: Mississippi nonprofit organizations may apply.

Restrictions: Individuals are ineligible.

Geographic Focus: Mississippi

Amount of Grant: $250-$500,000 range

Contact: Stuart Irby, President, (601) 989-1811

Sponsor: Elizabeth M. Irby Foundation

P.O. Box 1819

Jackson, MS 39215

Elizabeth McGraw Foundation Grants   **2202**

The foundation awards grants nationally, with a focus on the U.S. Northeast, in its areas of interest, including cancer research, elementary and secondary education, art museums, performing arts (opera and theater), and culture. There are no application deadlines or specific guidelines.

Geographic Focus: Florida; Massachusetts; New York

Samples: Peabody Essex Museum—Luxury and Innovation: Furniture Masterworks by John and Thomas Seymour, (2003). Uiniversity of Puget Sound (Tacoma, WA)—to renovate Collins Memorial Library, $500,000.
Contact: Charles B. Fischer, Jr., Managing Director, (561) 659-8844; fax: (561) 659-8829; email: charles.b.fischer@bd.com
Sponsor: Elizabeth McGraw Foundation
c/o Deutsche Bank, 350 Royal Palm Way
Palm Beach, FL 33480

Elizabeth Ordway Dunn Foundation Grants   2203
The foundation's grants program awards funds for special projects in conservation and environmental concerns that benefit Florida. Some support is also given for historic preservation. Special interests in the environmental field include preservation of biological diversity, protection of coastal and fresh water ecosystems, conservation of land and wildlife resources, comprehensive planning and growth management, environmental education, environmental health, energy conservation and the development of renewable resources, marine conservation, and pollution prevention and toxics use reduction. Special interests in the field of historic preservation include education and appropriate use of architecturally and historically significant properties, particularly those that are located in fragile and threatened natural areas. Types of support include programs/projects and seed grants. Applicants are urged to submit concept papers; full proposals will be invited. Grants are often awarded in partial support of a project.
Requirements: 501(c)3 tax-exempt organizations that are not 509(a) private charities are eligible.
Restrictions: The foundation does not award grants to support individuals or sectarian religious activities or for capital purposes, operating budgets, endowments, or deficit financing.
Amount of Grant: $5000-$30,000 range
Date(s) Application is Due: Mar 15; Sep 15
Contact: Robert Jensen, Managing Director, (305) 667-5521; email: eodf@worldnet.att.net
Sponsor: Elizabeth Ordway Dunn Foundation
P.O. Box 016309
Miami, FL 33101-6309

Ellison Medical Foundation Senior Scholar Awards   2204
The awards enable established investigators working at U.S. institutions to conduct up to four years of research in the basic biological and clinical sciences relevant to understanding aging processes and age-related diseases and disabilities. The award is intended to provide significant support to established investigators in order to allow the development of new, creative research programs by investigators who may not currently be conducting aging research or who may wish to develop new research programs in aging. Acceptable uses for award funds include salary, other personnel, equipment, supplies, resource acquisition, and travel. (Carry-overs in excess of $25,000 must be approved by the Ellison Medical Foundation Scholars Program Office.) Full indirect costs at the NIH negotiated rate will be provided. Ten awards are made annually. Application deadline is for letters of intent.
Requirements: Established investigators working at U.S. institutions are eligible.
Samples: To be divided among 11 established investigators at U.S. institutions—to pursue innovative research on aging that might not otherwise be financed by traditional resources because of its perceived novelty or risk, $6.6 million maximum over four years (2005).
Amount of Grant: $150,000 maximum per year
Date(s) Application is Due: Mar 9
Contact: Dr. Richard Sprott, Executive Director, (301) 657-1830; fax: (301) 657-1828; email: rsprott@ellison-med-fn.org
Internet: http://www.ellison-med-fn.org/emf_applications.jsp
Sponsor: Ellison Medical Foundation
4710 Bethesda Avenue, Suite 204
Bethesda, MD 20814

Ellison Medical Foundation Senior Scholar Awards in Global Infectious   2205
Diseases
The program funds up to four years of basic research on molecular and cellular mechanisms of parasitology and infectious diseases that result from microbial, protozoan, or viral pathogenesis—with special focus on tuberculosis, malaria, and parasitoses. Awards are intended to inspire new directions that may entail substantial risk. Ten awards are made annually. The listed application deadline is for letters of intent; full applications are by invitation.
Requirements: Investigators at least four years beyond their postdoctoral fellowship, employed by U.S. 501(c)3 institutions, or U.S. colleges or universities, are eligible to apply. Whereas the foundation only makes awards to U.S. nonprofit institutions, the program encourages formation of research consortia between U.S. institutions and those in other disease-endemic countries, as through a subcontract mechanism, when such collaborations will benefit the proposed research.
Restrictions: The awards are not intended to supplement ongoing, already funded programs.
Samples: To be divided among 10 established investigators at U.S. institutions—to pursue innovative research on global infectious diseases that might not otherwise be financed by traditional resources because of its perceived novelty or risk, $6 million maximum over four years (2005).
Amount of Grant: $150,000 maximum per year
Date(s) Application is Due: Mar 9

Contact: Dr. Richard Sprott, Executive Director, (301) 657-1830; fax: (301) 657-1828
Internet: http://www.ellison-med-fn.org/emf_applications.jsp
Sponsor: Ellison Medical Foundation
4710 Bethesda Avenue, Suite 204
Bethesda, MD 20814

Ellison Medical Foundation/AFAR Julie Martin Mid-Career Award   2206
in Aging Research
The Ellison Medical Foundation and AFAR developed this program for outstanding mid-career scientists who propose novel directions of high importance to biological gerontology. Proposals in areas where NIH awards or other traditional sources are unlikely because the research is high risk, are particularly encouraged if they have the potential for leading to major new advances in our understanding of basic mechanisms of aging. Projects investigating age-related diseases are also supported, but only if approached from the point of view of how basic aging processes may lead to these outcomes. Projects concerning mechanisms underlying common geriatric functional disorders are also encouraged, as long as these include connections to fundamental problems in the biology of aging. Two four-year awards of $500,000 will be given, at the level of $125,000 per year. In addition, up to 10% ($50,000) may be requested for administrative/indirect costs. Recipients of this award are expected to attend the AFAR Grantee Conference. The purpose of the meeting is to promote scientific and personal exchanges among recent AFAR grantees and experts in aging research.
Requirements: The applicant must be an Associate Professor who achieved tenured status after December 1st. Non-tenured Associate Professors at institutions with tenure (even if tenure is only offered at the Full Professor level) are not eligible. Applicants at institutions that do not offer tenure must demonstrate that their appointment is equivalent to that of an Associate Professor who received tenure status after December 1st. The proposed research must be conducted at any type of not-for-profit setting in the United States. The deadline of receipt of applications and all supporting materials is December 16th at 5:00 p.m. EST.
Restrictions: Projects that deal strictly with clinical problems such as the diagnosis and treatment of disease, health outcomes, or the social context of aging are not eligible. Applicants who are employees in the NIH Intramural program are not eligible.
Amount of Grant: $125,000 per year
Date(s) Application is Due: Dec 16
Contact: Grants Manager; (212) 703-9977 or (888) 582-2327; fax: (212) 997-0330; email: grants@afar.org or info@afar.org
Internet: http://afar.org/Ellison%20Mid-Career.html
Sponsor: American Federation for Aging Research
55 West 39th Street, 16th Floor
New York, NY 10018

Ellison Medical Foundation/AFAR Postdoctoral Fellows in Aging   2207
Research Program
The Ellison Medical Foundation, in partnership with the American Federation for Aging Research (AFAR), created the Ellison/AFAR Postdoctoral Fellows in Aging Research Program to encourage and further the careers of postdoctoral fellows with outstanding promise in the basic biological and biomedical sciences relevant to understanding aging processes and age related diseases and disabilities. The award is intended to provide significant support to permit these postdoctoral fellows to become established in the field of aging. Projects concerned with understanding the basic mechanisms of aging will be considered. Projects investigating age related diseases are also supported, if approached from the point of view of how basic aging processes may lead to these outcomes. Projects concerning mechanisms underlying common geriatric functional disorders are also considered. It is anticipated that up to 15 one year grants will be awarded, ranging from $44,850 for a first year fellow, up to $58,850 for a fellow with more than 7 years of training. Of the award, up to $7,850 may be requested for expenses such as research supplies, equipment, health insurance and travel to scientific meetings. Refer to the Ellison/AFAR instruction sheet and application on the AFAR website for complete application procedures. Recipients of this award are expected to attend the AFAR Grantee Conference. The purpose of the meeting is to promote scientific and personal exchanges among recent AFAR grantees and experts in aging research.
Requirements: The applicant must be a postdoctoral fellow (MD and/or PhD degree) at the start date of the award (July 1). Any former Ellison/AFAR postdoctoral award recipient may apply for this award. All candidates must submit applications endorsed by their institutions. The proposed research must be conducted at any type of not-for-profit setting in the United States.
Restrictions: Employees in the NIH Intramural program are not eligible. Fellows may not hold any concurrent foundation or not-for-profit funding.
Amount of Grant: $44,000 - $58,000 range
Date(s) Application is Due: Dec 16
Contact: Grants Manager; (212) 703-9977 or (888) 582-2327; fax: (212) 997-0330; email: grants@afar.org or info@afar.org
Internet: http://www.afar.org/ellisonpostdoc.html
Sponsor: American Federation for Aging Research
55 West 39th Street, 16th Floor
New York, NY 10018

Ellison Medical Foundation/AFAR Senior Postdoctoral Fellows   2208
Research Program
The program seeks to encourage and further the careers of senior postdoctoral fellows with outstanding promise in the basic biological and biomedical sciences relevant to understanding aging processes and age-related diseases and disabilities. The award is

intended to provide significant support to permit these postdoctoral fellows to become established in the field of aging. Projects concerned with understanding the basic mechanisms of aging will be considered. Projects investigating age-related diseases are also supported, if approached from the point of view of how basic aging processes may lead to these outcomes. Projects concerning mechanisms underlying common geriatric functional disorders are also considered. It is anticipated that three two-year grants will be awarded each year.

Requirements: Postdoctoral fellows, both MDs and PhDs, are eligible. Fellows with at least three and not more than five years of prior postdoctoral training are eligible.

Amount of Grant: $100,000

Date(s) Application is Due: Dec 15

Contact: Grants Manager, (212) 703-9977; fax: (212) 997-0330; email: grants@afar.org

Internet: http://www.afar.org/ellison.html

Sponsor: American Federation for Aging Research

70 W 40th Street, 11th Fl

New York, NY 10018

Elmer O. and Ida Preston Educational Trust Scholarships 2209

The trust awards scholarships to Iowa male college students of the Protestant faith who currently are at the sophomore through graduate level of study. The applicant must be a current student or plan on attending a college and/or university in Iowa.

Requirements: Applications are accepted from male college students (currently at the freshman through graduate level) who are residents of Iowa, for study at a college/university in Iowa.

Geographic Focus: Iowa

Amount of Grant: $500-$700

Contact: Trust Administrator, (515) 243-4191

Sponsor: Elmer O. and Ida Preston Educational Trust

801 Grand Avenue, Suite 3700

Des Moines, IA 50309

ELP Fellowships 2210

The fellowship is an innovative national program designed to build the leadership capacity of the environmental field's most promising emerging professionals from the United States and U.S. territories. Each year, a new group of fellows is chosen to join the ELP community of U.S. environmental professionals from diverse backgrounds, sectors, and areas of expertise. The three-year fellowship offers unique networking opportunities, intensive leadership and skills training, project seed money, support, and time for personal and professional reflection. Fellows join a community of diverse emerging leaders, and have the opportunity to meet established environmental practitioners and experts in the field. Fellows receive a stipend; travel and accommodations for four fellowship retreats; access to funding for leadership-building projects; and national recognition through the program. Application materials are available online.

Requirements: Applicants must be residents of the United States or U.S. territories. Individuals should be relatively new to the environmental field, with approximately three to 10 years of experience.

Amount of Grant: $10,000 maximum

Date(s) Application is Due: Oct 1

Contact: Fellowship Administrator, (413) 268-0035; fax: (413) 268-0036; email: info@elpnet.org

Internet: http://www.elpnet.org/fellowship_application.html

Sponsor: Environmental Leadership Program

P.O. Box 446

Haydenville, MA 01039

Elsa U. Pardee Foundation Grants 2211

The foundation has three main grantmaking focuses: ongoing research grants to established institutions for cancer research, the funding of worthwhile new cancer research projects, and grants to committees providing financial assistance to cancer victims in specified areas. The foundation particularly welcomes innovative, small-scale, short-term projects that may be difficult to fund elsewhere until some interesting results are obtained. Application forms are available on written request and must accompany all project proposals. Applications are accepted at any time.

Requirements: Application must include completed form, description of project, itemized budget, approval from the institution where the project will be done, amount requested, duration of project, and any other supportive material deemed helpful.

Restrictions: Grants are limited to research toward the cure and control of cancer and in general do not provide for building funds, equipment (except that used in a specific project), fellowships, or fund-raising campaigns.

Amount of Grant: $50,000-$200,000 average

Contact: James Kendall, Secretary, (989) 832-3691; email: info@pardeefoundation.org

Internet: http://www.pardeefoundation.org

Sponsor: Elsa U. Pardee Foundation

P.O. Box 2767

Midland, MI 48641-2767

EMBO / EMBL Science Writing Prize 2212

The Science Writing Prize is given annually for an outstanding piece of science writing designed to communicate a topical issue to a non-scientific audience. There is one winner, and the prize may not be awarded twice to the same person. The theme varies from year to year. The selection jury consists of 15 people, both scientists and non-scientists: the EMBO Science and Society Committee members and EMBO staff. The prize of Euro $1,500 is awarded in June at the EMBO Media Workshop in Heidelberg, where the author is invited to talk about his/her motivation for writing the article and for communicating the topic to the public.

Samples: Edwin Rydberg, Istituto di Ricerche di Biologia Molecolare P. Angeletti, Pomezia (Rome), Italy, for Through the illusions, $1,500 Euro (2005).

Amount of Grant: $1,500 Euro

Contact: Hermann Bujard, Executive Director; +49 (if outside Germany) (0) 6221-8891, ext. 101; fax: +49 (if outside Germany) (0) 6221 8891-200; email: embo@embo.org

Internet: http://www.embo.org/awards/writing.html

Sponsor: European Molecular Biology Organization / European Molecular Biology Laboratory

Meyerhofstrasse 1

Heidelberg D-69117 Germany

EMBO Award for Communication in the Life Sciences 2213

The Organization launched this award in 2002 in order to give recognition to the huge efforts that some scientists make to communicate their science to the public while remaining fully active in research. The winners of the EMBO Award are nominated for the EU Descartes Prize for science communication.

Samples: Christian Sardet and Ali Saib, French cell biologists, $5,000 Euro (2007); Armand Marie Leroi, UK scientist, author, and broadcaster, $5,000 Euro (2006).

Amount of Grant: $5,000 Euro

Contact: Hermann Bujard, Executive Director; +49 (if outside Germany) (0) 6221-8891, ext. 101; fax: +49 (if outside Germany) (0) 6221 8891-200; email: embo@embo.org

Internet: http://www.embo.org/awards/communications.html

Sponsor: European Molecular Biology Organization

Meyerhofstrasse 1

Heidelberg D-69117 Germany

EMBO Installation Grants 2214

The aim of this scheme is to strengthen science in participating EMBC member states. At present applications are accepted from Croatia, the Czech Republic, Estonia, Hungary, Portugal, Poland and Turkey. Additional member states have expressed an interest to join in the near future. Grants will help scientists to relocate, set up their labs and rapidly establish a reputation in the European scientific community. The program is entirely funded by the participating EMBC member states and successful applicants receive an annual support of Euro $50,000 for three to five years. Each participating country sets an upper limit to the number of grants available per year. Application forms should be completed by the individual scientists and the receiving institute.

Requirements: Eligible scientists should: have an excellent publication record; have spent at least two consecutive years prior to the application deadline, outside the country in which they are applying to establish their lab; and be negotiating a full-time position at an institute/university in participating member states by the date of application. No age limit applies.

Amount of Grant: $50,000 Euro per year for 3-5 years

Date(s) Application is Due: Apr 15

Contact: Hermann Bujard, Executive Director; +49 (if outside Germany) (0) 6221-8891, ext. 101; fax: +49 (if outside Germany) (0) 6221 8891-200; email: embo@embo.org

Internet: http://www.embo.org/sdig/index.html

Sponsor: European Molecular Biology Organization

Meyerhofstrasse 1

Heidelberg D-69117 Germany

EMBO Long-Term Fellowships 2215

Long-Term Fellowships are awarded for prolonged visits (12 to 24 months) and are intended for advanced training through research. As the aim of the EMBO is to promote international research, it follows that mobility is a crucial element in deciding the eligibility of an application. All applications must involve a laboratory of origin or a receiving institute or applicant's nationality from one of the member states. Stipend rates depend on the country being visited, marital status, etc.

Requirements: Applicants must hold a doctorate degree or equivalent before the start of the fellowship but not necessarily when applying. For clarity, it is deemed that this requirement is fulfilled only when the thesis has been stamped as submitted for final examination by the university. As a minimum, applicants must have at least one first author publication in press or published in an international peer reviewed journal at the time of application.

Restrictions: Applications are only considered from candidates with a maximum of 3 years postdoctoral experience.

Date(s) Application is Due: Feb 15; Aug 15

Contact: Liselott Maidment; +49 (if outside Germany) (0) 6221-8891, ext. 116; fax: +49 (if outside Germany) (0) 6221 8891-200; email: fellowships@embo.org

Internet: http://www.embo.org/fellowships/long_term.html

Sponsor: European Molecular Biology Organization

Meyerhofstrasse 1

Heidelberg D-69117 Germany

EMBO Short-Term Fellowships 2216

Short-term fellowships are established to advance molecular biology research by helping scientists to visit another laboratory with a view to applying a technique not available in the home laboratory. Such fellowships are not awarded for exchanges between two laboratories within the same country, and are intended for joint research work rather than consultations. The fellowships cover travel plus subsistence of the fellow only and not of any dependents. These fellowships are intended for visits of one week to three months duration (for non-European applicants the short-term fellowships are intended for a fixed duration of 90 days). Stipends vary widely from country to country,

Requirements: Applicants must be either: post-doctoral scientists with less than 10 years of professional experience since finishing their Ph.D. degree (or the equivalent); or pre-doctoral scientists who have not yet completed a Ph.D. degree (or the equivalent).

Restrictions: All applications must involve either a laboratory of origin or a receiving institute from one of the EMBC member states.

Contact: Agnes Visser-de Matteis; +49 (if outside Germany) (0) 6221-8891, ext. 115; fax: +49 (if outside Germany) (0) 6221 8891-200; email: fellowships@embo.org

Internet: http://www.embo.org/fellowships/short_term.html

Sponsor: European Molecular Biology Organization

Meyerhofstrasse 1

Heidelberg D-69117 Germany

Emerson Charitable Trust Grants   2217

Established in 1944 in Missouri as the Emerson Electric Manufacturing Company Charitable Trust, the Foundation supports: arts and culture—fine arts and cultural institutions to enrich the diversity, creativity, and liveliness of the community; education—programs designed to promote educational systems at all levels; health and human services—programs designed to help individuals and families in times of need, including sickness, old age, family crisis, and natural disasters; civic affairs—programs designed to protect citizenry; further economic health of the community; and build and maintain assets such as parks and zoos; and youth—programs designed to give young people the opportunity to recognize their potential, confidence, and skills to achieve their dreams. The foundation awards college scholarships to children and step-children of employees of Emerson Electric. No specific application form is required, and initial approach should be the complete proposal,

Samples: Alvin J. Siteman Ctr, Washington U School of Medicine and Barnes-Jewish Hospital (Saint Louis, MO)—for a new cancer-research facility and for cancer research, $6 million challenge grant.

Amount of Grant: $5,000-$300,000 average

Contact: Jo Ann Harmon, Senior Vice President; (314) 553-3722; fax: (314) 553-1605

Internet: http://www.emerson.com/en-us/about_emerson/company_overview/pages/our_approach_to_corporate_philanthropy.aspx

Sponsor: Emerson Charitable Trust

P.O. Box 4100, 8000 W. Florissant Avenue

Saint Louis, MO 63136

EMIERT Coretta Scott King Awards   2218

Two awards are given annually to recognize outstanding inspirational and educational contributions toward the goals of peace and brotherhood. The category for the awards is creative writing and illustration. Books must have been published one year preceding the year of the award. Nominations should be sent to the office.

Requirements: African American authors and illustrators are eligible to apply.

Amount of Grant: $1000 to author; $1000 to illustrator

Date(s) Application is Due: Dec 1

Contact: ALA Awards Staff Liaison, (800) 545-2433; fax: (312) 280-3257; email: olos@ala.org

Internet: http://www.ala.org/ala/emiert/corettascottkingbookawards/corettascott.htm

Sponsor: American Library Association

50 E Huron Street

Chicago, IL 60611

Emil Buehler Perpetual Trust Grants   2219

The Trust accepts inquiries from organizations with requests that promote the field of aviation science and technology. Grants are made in support of specific programs or projects. Grants are occasionally made for specific capital purchases if a clear need is demonstrated. Priority is given to cost-effective programs that offer the broadest benefit to the community. The Board of Trustees meets monthly to make grant decisions. Proposals must be received by the 10th day of the month for consideration at that month's meeting.

Requirements: Grants are only made to organizations qualifying for tax exemption under Section 501(c)3 of the Internal Revenue Code.

Restrictions: The Trust does not provide grants for general operating support; endowments; individuals; political activities; loans; or fund-raising events including dinners, benefits and athletic events.

Samples: Bergen Community College (Paramus, NJ)—to construct a two-domed observatory as part of the new Technology Education Center, $288,875 (2004).

Contact: Deborah Janof, Grants Administrator, (201) 967-8040

Sponsor: Emil Buehler Perpetual Trust

305 Route 17 South

Paramus, NJ 07652

Emma B. Howe Memorial Foundation Grants   2220

The Emma B. Howe Memorial Foundation makes grants through the Minneapolis Foundation. The focus of all grants is to improve: the health and well-being of children, youth and families; opportunities for educational achievement; access to quality affordable housing; and economic vitality throughout the region. Information on the application process is available online.

Requirements: Eligible organizations include 501(c)3 nonprofits; public institutions; and emerging groups organized for nonprofit purposes.

Restrictions: Funds are not available for: individuals; organizations/activities outside of Minnesota; conference registration fees; memberships; direct religious activities;

political organizations or candidates; direct fundraising efforts; telephone solicitations; courtesy advertising; financial deficits.

Geographic Focus: Minnesota

Samples: Kids in Distressed Situations (KIDS)—(New York, NY)—to provide donated products to low-income children and families in the Minneapolis area through Bridging Inc, $75,000 over three years.

Contact: Community Philanthropy, (612) 672-3836; email: grants@mplsfoundation.org

Internet: http://www.mplsfoundation.org/partners/emma.htm

Sponsor: Emma B. Howe Memorial Foundation

80 South Eighth Street

Minneapolis, MN 55402

Endangered Language Fund Grants   2221

The Fund provides grants for language maintenance and linguistic field work. The work most likely to be funded is that which serves both the native community and the field of linguistics. Work which has immediate applicability to one group and more distant application to the other will also be considered. Publishing subventions are a low priority, although they will be considered. Proposals can originate in any country. The language involved must be in danger of disappearing within a generation or two. Endangerment is a continuum, and the location on the continuum is one factor in our funding decisions. Eligible expenses include consultant fees, tapes, films, travel, etc. Grants are normally for a one year period, though extensions may be applied for. The Fund expects grants to be less than $4,000 in size, and to average about $2,000.

Restrictions: Overhead expenses are not allowed.

Samples: Erin Debenport, University of Chicago, Chicago, IL—Community Language Documentation At Sandia Pueblo (2006); Pastor Dawari Braide, Port Harcourt, Nigeria—Printing of Kalabari-English dictionary for Kalabari children (2006); Amrendra Kumar Singh, Jawaharlal Nehru University—Description Of Pasi (2006).

Amount of Grant: $4,000 maximum

Date(s) Application is Due: Apr 21

Contact: Grants Coordinator; (203) 865-6163, ext. 265; fax: (203) 865-8963; email: elf@endangeredlanguagefund.org

Internet: http://www.endangeredlanguagefund.org/request.html

Sponsor: Endangered Language Fund

300 George Street, Suite 900

New Haven, CT 06511

Energy Foundation Grants   2222

The foundation's mission is to assist in the nation's transition to a sustainable energy future by promoting energy efficiency and renewable energy. To this end, grants are awarded in five areas: utility—competitive structures for the electric industry, energy efficiency, clean energy, and fair competition; buildings—energy-efficient structures, better codes and equipment standards; transportation—alternatives to single-occupancy vehicle use, new highways, and increased vehicle efficiency; renewable energy—accelerate the commercialization of renewable energy; and integrated issues—deepen understanding of energy and its relationship to the health of the economy. The foundation's board meets in March (first week), June (last week), and November (first week) to review proposals, which are accepted at any time. Guidelines and forms are available online.

Requirements: The foundation makes grants to nonprofit 501(c)3 charitable organizations.

Restrictions: The foundation does not support individuals; for-profit organizations; local projects, unless they have been designed for replication or have broad-based regional or national implications; candidates for political office; legislation; sectarian or religious purposes; research and development of technology; demonstration projects; endowments; the planning, renovation, maintenance, or purchase of buildings; or the purchase of equipment or the acquisition of land, even if the intent is to save energy.

Amount of Grant: $2000-$850,000 typically; $11.2 million total approximately

Contact: Program Office, (415) 561-6700; fax: (415) 561-6709; email: energyfund@ef.org

Internet: http://www.ef.org/applications.cfm

Sponsor: Energy Foundation

1012 Torney Avenue, #1

San Francisco, CA 94129

Energy Foundation State and Regional Climate Program Grants   2223

The foundation's funding area focuses on state and regional climate policy initiatives. The efforts supported will have reducing greenhouse gas emissions as their main objective. Potential areas of work include state and regional carbon cap-and-trade programs; state and regional greenhouse gas plans and targets; greenhouse gas reporting and reduction initiatives; and financial mechanisms like incentives or carbon taxes. The foundation's geographic focus is the United States, with special emphasis on regional initiatives. There is no fixed format for proposals. The foundation's Board of Directors meets three times a year (the first week of March, the last week of June, and the first week of November). Proposals are accepted on an ongoing basis. There are no specific deadlines; however, proposals should be received 12 weeks in advance of the next board meeting.

Requirements: 501(c)3 nonprofit, charitable organizations are eligible.

Restrictions: The foundation does not support individuals or for-profit organizations; local projects, unless they have been consciously designed for further replication or have broad regional or national implications; candidates for political office; to influence legislation, or to support sectarian or religious purposes; research and development of technology (e.g., funds to develop hybrid automobiles or commercialization of an invention); demonstration projects (e.g., model solar homes); community energy projects;

endowments or debt reduction; general-support grants; annual fundraising campaigns or capital construction; planning, renovation, maintenance, retrofit, or purchase of buildings; the purchase of equipment; or the acquisition of land, even if the intent is to save energy.

Samples: Climate Solutions (Olympia, WA)—to encourage states in the Pacific Northwest to take a leadership role on improving climate-change policies, $50,000 (2003); Energy Research Institute, Ctr for Renewable Energy Development (Beijing, China)—to develop national mandatory market share policies for renewable energy, $70,000 (2003); Natural Resources Defense Council (New York, NY)—to educate policymakers about tax incentives for energy-efficient buildings and equipment, $10,000 (2003); Northeast States of Coordinated Air Use Management (Boston, MA)—to develop a greenhouse-gas registry for the Northeast that is compatible with the California Climate Action Registry, $200,000 (2003).

Contact: Administrator, (415) 561-6700; fax: (415) 561-6709; email: energyfund@ef.org
Internet: http://www.ef.org/programs.cfm
Sponsor: Energy Foundation
1012 Torney Avenue, No 1
San Francisco, CA 94129

Engineering Awards of Excellence 2224

The purpose of this program is to give national recognition to structural engineering excellence and innovation in steel-framed building projects.
Requirements: Projects must be located in the United States, Canada, or Mexico. A significant part of the framing system must be steel wide-flange structural shapes or hollow structural sections. Both new buildings and renovation projects are eligible. Building construction must have been completed between January 1, 2001, and December 31, 2003.
Date(s) Application is Due: Mar 1
Contact: Becky LeDonne, Excellence Awards, (312) 670-5433; fax: (312) 670-5403; email: ledonne@aisc.org
Internet: http://www.aisc.org/Content/NavigationMenu/About_AISC/Competition_and_Awards/Project_Awards/Engineering_Awards_of_Excellence/Engineering_Awards_of_Excellence.htm
Sponsor: American Institute of Steel Construction
One E Wacker Drive, Suite 3100
Chicago, IL 60601-2001

Engineering Information Foundation Grants 2225

The Foundation supports developmental projects, instructional projects, and training programs in engineering education and research that fit their fields of interest. These currently include the availability and use of published information, women in engineering, and information access in developing countries. The Foundation has a specific interest in innovative projects with measurable results, projects that promote significant and lasting change, projects that can be successfully replicated elsewhere and methodologies that are specific, well-defined and cost-effective.
Requirements: U.S. 501(c)3 tax-exempt organizations are eligible.
Restrictions: Funds are not available for: operating expenses; requests solely for equipment; general overhead; campaigns; conferences; scholarships, fellowships, assistantships; support for doctoral candidates; loans.
Samples: U of California (Los Angeles, CA)—for the development of a learning portfolio for accessing engineering information, $28,100.; U of Colorado (Boulder, CO)—for an energy education workshop for girls, $7250.; Stevens Institute of Technology (Hoboken, NJ)—for a seminar and follow-up for middle-school and high-school guidance counselors to educate them about careers in engineering, $18,600.
Amount of Grant: $5000-$25,000
Contact: Program Contact, (212) 579-7596; fax: (212) 579-7517; email: info@eifgrants.org
Internet: http://www.eifgrants.org
Sponsor: Engineering Information Foundation
180 West 80th Street, Suite 207
New York, NY 10024-6301

Ensign-Bickford Foundation Grants 2226

The corporate foundation gives primarily in areas of company operations, with emphasis on the Simsbury and Avon, CT, areas. Grants are awarded to nonprofit organizations for welfare, education, cultural programs, employee-related scholarships, and community development activities. Types of support include continuing support, annual campaigns, building/renovation, equipment acquisition, land acquisition, programs and projects, conferences and seminars, publication, seed money, internships, scholarships, research, and employee matching gifts. Applications are accepted throughout the year.
Requirements: Nonprofit organizations in company operating areas of Connecticut are eligible to apply.
Restrictions: Grants are not awarded to individuals (except for employee-related scholarships) or for operating budgets, endowment funds, emergency funds, deficit financing, or loans.
Geographic Focus: Connecticut
Amount of Grant: $50-$50,000 for organizations; $500-$2500 for individuals
Contact: Jan Delissio, (860) 843-2388
Internet: http://www.ensign-bickfordind.com/community.html
Sponsor: Ensign-Bickford Foundation
P.O. Box 7, 100 Grist Mill Road
Simsbury, CT 06070-0007

Entente Cordiale Scholarships 2227

Scholarships are awarded to 28 outstanding British or French postgraduates to study or undertake research in France. Applicants may study any discipline at any French institution of higher education or research. The scholarship is for one academic year. Expenses covered are tuition and registration fees, books, and arrival and departure costs. A stipend is also included. Deadline is February 10th for French application and March 31 for British application. An application form is available on the Web site.
Requirements: Applicants must be British or French nationals in their final year of undergraduate study, so that when the scholarship is awarded they will go on to postgraduate doctoral or postdoctoral studies or research on the other side of the Channel. Candidates should contact their chosen institutions before applying.
Amount of Grant: @L8000 maximum, paid in three monthly installments
Contact: French Cultural Department, 020 7073 1312; fax: 020 7073 1326; email: entente.cordiale@mail.ambafrance.org.uk
Internet: http://www.francealacarte.org.uk/entente
Sponsor: Entente Cordiale
French Embassy, 23 Cromwell Road
London SW7 2EL United Kingdom

Environmental Excellence Awards 2228

The awards recognize the outstanding efforts of students across the country who are working at the grassroots level to protect and preserve the environment. Eight projects will be selected. Each winning school will receive: all-expenses-paid trip for three students and one chaperon/teacher to one of the SeaWorld or Busch Gardens parks for a special awards event; 100 t-shirts to share with school and community partners; and award trophy for the school and certificates for every student/teacher participant. From the eight projects, one outstanding environmental educator will be recognized. That educator will receive: all-expenses-paid trip for him/herself and one guest to one of the SeaWorld or Busch Gardens parks for a special awards event in April; all-expenses-paid trip to the National Science Teachers Association (NSTA) national conference; and award trophy and certificate. Previous award-winning projects have been in the areas of habitat restoration, school yard beautification, energy and waSuite reduction, environmental education and community outreach, wildlife protection, and natural resource conservation.
Requirements: Public and private elementary, secondary, and home schools in the United States and Canada are eligible to apply. Community-based projects, such as those managed and operated by community service organizations, public recreation centers, 4-H clubs and other public, nonprofit groups working to protect the environment at the grassroots level, also are eligible.
Restrictions: Individual students and previous award-winning schools are ineligible.
Amount of Grant: $10,000; $5000 to educator
Date(s) Application is Due: Nov 30
Contact: Program Officer, (877) 792-4332 or (407) 363-2389; (877) 792-4332 (toll free)
Internet: http://www.seaworld.org/conservation-matters/eea
Sponsor: Anheuser-Busch Adventure Parks
7007 Sea World Drive
Orlando, FL 32821

Environmental Protection Agency Summer Faculty Fellowships 2229

EPA Summer Faculty Fellowships are awarded to faculty with research experience and will include assistant, associate, and full professors. Each award is for one summer. The participant may reapply for and possibly receive awards for up to two additional summers (three total). Participants receive a stipend based on the level of the award and may receive a daily expense allowance, where applicable. Holders of summer fellowships must devote their full-time effort to the research program proposed in their applications and must be in residence at the sponsoring laboratory during the entire period of the summer fellowship. Holders of summer fellowships must devote their full-time effort to the research program proposed in their applications and must be in residence at the sponsoring laboratory during the entire period of the summer fellowship. Guidelines are available online.
Requirements: U.S. citizens and permanent residents with the PhD in science or engineering who are full-time faculty at accredited baccalaureate-granting U.S. institutions are eligible.
Amount of Grant: $1250 assistant professor; $1450 associate professor; $1650 full professor
Date(s) Application is Due: Nov 1
Contact: National Academies Summer Faculty Fellowships, (202) 334-2760; fax: (202) 334-2759c; email: rap@nas.edu
Internet: http://www4.nationalacademies.org/pga/rap.nsf/vwLabInformation/FDF66DEFC1427CD385256B11006A213E?OpenDocument#8
Sponsor: National Research Council
500 Fifth Street NW
Washington, DC 20001

EPA Baseline Assessment of Risks of Exposure to Lead Poisoning of Native American Children 2230

This program supports efforts to identify the risks of lead poisoning among children by conducting a baseline assessment of potential lead exposures. Tribes may then use this data to determine whether there is a need to implement an authorized lead-reduction program.
Requirements: Federally recognized Indian tribes, tribal consortia, and tribal organizations are eligible to apply.

Restrictions: Funds may not be used to purchase property, provide lead hazard reduction activities, conduct renovation or construction activities, or cover case management costs.
Amount of Grant: $75,000 maximum
Contact: Darlene Watford, (202) 566-0516; fax: (202) 566-0469; email: Watford. Darlene@epa.gov
Internet: http://www.epa.gov/lead
Sponsor: Environmental Protection Agency
1200 Pennsylvania Avenue, NW
Washington, DC 20460

EPA Children's Health Protection Grants   2231
The objectives of this program are to catalyze community-based and regional projects and other actions that enhance public outreach and communication; assist families in evaluating risks to children and in making informed consumer choices; build partnerships that increase a community's long-term capacity to advance the protection of children's environmental health and safety; leverage private and public investments to enhance environmental quality by enabling community efforts to continue past EPA's ability to provide assistance to communities; and promote protection of children from environmental threats through lessons learned. There are no deadline dates.
Requirements: Eligible applicants include community groups, public nonprofit institutions/organizations, tribal governments, specialized groups, profit organizations, private nonprofit institutions/organizations, and municipal and local governments. Potential applicants are strongly encouraged to discuss proposed projects with or submit preapplications to program staff prior to the completion of a full proposal.
Amount of Grant: $5000-$250,000 average
Contact: Bettina Fletcher, Office of Children's Health Protection, (202) 564-2188; fax: (202) 564-2733; email: fletcher.bettina@epa.gov
Internet: http://yosemite.epa.gov/ochp/ochpweb.nsf/content/grants.htm
Sponsor: Environmental Protection Agency
1200 Pennsylvania Avenue NW, Mail Code 1107A, Room 2512 Ariel Rios N
Washington, DC 20004

EPA Five-Star Restoration Challenge Grant Program   2232
Projects must include a strong on-the-ground wetland or riparian restoration component, and should also include education, outreach, and community stewardship. Projects involving only education, research, outreach, monitoring, or planning are not eligible for funding. Projects must demonstrate measurable ecological, educational, social, and/or economic benefits resulting from the completion of the project. Projects may be a discrete part of a larger restoration effort but must be ready to complete within a one-year time frame upon receipt of funding. Preference will be given to projects that are part of a larger watershed or community stewardship effort, include specific provisions for long-term management and protection, and demonstrate the value of innovative, collaborative approaches to restoring the nation's waters. The stars in Five-Star are the partners, funders, and/or participants necessary to complete the project, including schools or youth organizations (e.g. state or local youth conservation corps or county job training programs); local or tribal governments (e.g. boards of county commissioners, departments of planning, environment, or parks and recreation); local businesses or corporations; conservation organizations or local citizen groups; state and federal resource management agencies; and foundations or other funders. Projects must therefore involve diverse partnerships of ideally five organizations that contribute funding, land, technical assistance, workforce support, and/or other in-kind services. Annual deadline dates may vary; contact program staff for exact dates.
Requirements: Any public or private entity in a community with wetlands, riverbanks, or shoreline areas in need of help are eligible to apply.
Restrictions: Projects that are part of a mitigation requirement are not eligible for funding.
Amount of Grant: $5000-$20,000; $10,000 average
Date(s) Application is Due: Mar 1
Contact: Tom Kelsch, Wetlands Division, (202) 857-0166; fax: (202) 857-0162; email: kelsch@nfwf.org
Internet: http://www.nfwf.org/programs/5star-rfp.htm
Sponsor: Environmental Protection Agency
1120 Connecticut Avenue NW, Suite 900
Washington, DC 20036

EPA Grants for Research   2233
The EPA research programs focus on reducing uncertainty associated with risk assessment and risks to human health and ecosystems. Areas where risk assessors are most in need of new concepts, methods, and data will be given highest priority. Projects are invited in the following areas: ecological assessment; exposure of children to pesticides; air quality; analytical and monitoring methods; drinking water; environmental fate and treatment of toxic and hazardous wastes; environmental statistics; high-performance computing; and exploratory research. In all areas, EPA is interested in research that recognizes issues relating to environmental justice, the agency's effort to achieve equal protection from environmental and health hazards for all people without regard to race, economic status, or culture. For specific program deadlines and information, contact the program office or visit the Web site.
Requirements: Academic and nonprofit institutions and state and local governments are eligible for EPA research grants in all areas. For-profit firms are eligible only under certain programs and then under restrictive conditions, including the absence of any profit from the project.
Restrictions: Federal agencies and federal employees are not eligible to participate.

Contact: Office of Research and Development, National Center for Environmental Research and Quality Assurance, (202) 260-3837 or (800) 490-9194; fax: (202) 260-2039; email: ord.grants@epamail.epa.gov
Internet: http://www.epa.gov/epahome/program2.htm
Sponsor: Environmental Protection Agency
1200 Pennsylvania Avenue NW, Ariel Rios Bldg
Washington, DC 20460

EPA Interagency Project to Clean Up Open Dumps on Tribal Lands   2234
This project is intended to demonstrate the federal government's ability to work closely with tribal governments to provide comprehensive solid waSuite management funding and technical support for the closure or upgrade of high-priority waSuite disposal sites and the development and strengthening of tribal or multitribal solid waSuite management programs. In determining whether a site is high priority, the Workgroup will generally rely on the Indian Health Service's Report to Congress on open dumps on American Indian lands. For further information contact the regional solid waSuite Indian coordinator in your area (see web site for details).
Requirements: Federally recognized tribes and multitribe 501(c)3 organizations whose membership consists of federally recognized tribes are eligible to apply.
Contact: Carol Jorgensen, Director, American Indian Environmental Office, (202) 564-0303; email: jorgensen.carol@epa.gov
Internet: http://www.epa.gov/epaoswer/non-hw/tribal/finance.htm#dumps
Sponsor: Environmental Protection Agency
401 M Street SW
Washington, DC 20460

EPA Lead Awareness Outreach for Native American Tribes   2235
This program provides financial assistance to tribes to launch outreach efforts that would educate Native American families about the dangers of exposure to lead-based paint hazards among children, distribute educational information and encourage families to have their children screened and homes tested for lead hazards. Activities may include providing medical training, developing culturally specific lead outreach materials, distributing pamphlets and establishing an in-home education program.
Requirements: Federally recognized Indian tribes, tribal consortia, and tribal organizations are eligible to apply.
Restrictions: Funds may not be used to purchase property, provide lead hazard reduction, conduct renovation or construction activities, or cover case management costs.
Amount of Grant: $50,000 maximum per outreach project
Contact: Darlene Watford, Office of Pollution Prevention and Toxics, (202) 566-0516; fax: (202) 566-0469; email: Watford.Darlene@epa.gov
Internet: http://www.epa.gov/lead
Sponsor: Environmental Protection Agency
1200 Pennsylvania Avenue, NW
Washington, DC 20460

EPA National Environmental Education Grant Program   2236
The purpose of this program is to provide training and related support services to education professionals who are or can become leaders in ensuring the quality and long-term sustainability of coordinated and comprehensive environmental education efforts across a state or states.
Requirements: U.S. institutions of higher education or nonprofit institutions or a consortia of such institutions are eligible to apply.
Amount of Grant: $2 million-$3 million range; $15,000 average
Date(s) Application is Due: Nov 15
Contact: Diane Berger or Sheri Jojokian, (202) 260-8619; email: berger.diane@epa.gov or jojokian.sheri@epa.gov
Internet: http://www.epa.gov/enviroed/grants.html
Sponsor: Environmental Protection Agency
401 M Street SW
Washington, DC 20460

EPA National Estuary Program Grants   2237
Project grants and cooperative agreements are awarded to assist qualified organizations and/or persons to develop programs to protect and restore coastal resources in priority estuaries. The program objective is to develop a comprehensive conservation and management plan for each estuary based on established relationships between pollutant loading and impacts on living resources. Actions to be taken under the plan would implement basin-wide reduction of point and non-point sources of pollution adversely impacting the uses of the estuary. Funds are to be used for projects and tasks needed to address priority problems of the estuary as identified within a five-year period and an annual action plan developed through the management conference structure for the priority estuary under study. Each year 12 conferences are formally convened. Funds are available for the priority estuaries at Long Island Sound, Buzzards Bay, Narragansett Bay, New York-New Jersey Harbor, Delaware Bay, Delaware Inland Bays, Sarasota Bay, Galveston Bay, Santa Monica Estuary, Albemarle-Pimlico Sounds, Puget Sound, San Francisco Bay, Casco Bay, Massachusetts Bays, Indian River Lagoon, Tampa Bay, Peconic Bay, San Juan Harbor Bay, Corpus Christi, Tillamook Bay, the Barataria-Terribonne Estuary Complex, New Hampshire Estuaries, Barnegat Bay, Maryland Coastal Bays, Charlotte Harbor, Mobile Bay, Morro Bay, and Columbia Bay. Projects and tasks will include gathering and analyzing existing and historical data, identifying and conducting further research to acquire new and additional data, targeting efforts to increase regional public understanding of the problems and complexities of the estuary and bring public input to the decision-making and management processes, and assessing

existing state and local initiatives in water sediment quality and living resource protection programs.

Requirements: State water pollution control agencies, interstate and intrastate agencies, other public or nonprofit private agencies, research institutions, and individuals within the geographic area of a priority estuary are eligible to apply.

Restrictions: Profit-making organizations are not eligible.

Date(s) Application is Due: Jun 1

Contact: Darrell Brown, Chief, Coastal Management Branch; (202) 566-1256; fax: (202) 566-1366; email: brown.darrell@epa.gov

Internet: http://www.epa.gov/nep

Sponsor: Environmental Protection Agency

1200 Pennsylvania Avenue, NW, Mail Code: 4504T

Washington, DC 20460

EPA Pollution Prevention Information Network 2238

The Pollution Prevention Act provides funds to states to strengthen the efficiency and effectiveness of state technical assistance programs in providing source reduction information to businesses. These funds will be targeted for applicants that are willing to work as part of a collective service providing pollution prevention information to state and local governments' technical assistance providers. The purpose of this request for proposals is to coordinate work among new and existing grantees to minimize duplication of effort in information collection and synthesis, training for the promotion of pollution prevention technologies, and establish information standards that will facilitate information exchange among centers. Grants for this program are competed through a Federal Register notice or through a more limited announcement usually sent to state technical assistance providers every three years. These grants are intended for state entities as described under the Pollution Prevention Act of 1990. Since this program has operated for seven years, there are eight existing P2 information centers. New applicants would need to work with established centers and be willing to collaborate on some cooperative projects the group has in progress. New proposals should address work with the existing "P2Rx centers" (see www.p2rx.org) as well as any new tasks or projects that would fit into this program.

Requirements: Eligible applicants for purposes of funding under this grant program include the 50 states, the District of Columbia, the U.S. Virgin Islands, the Commonwealth of Puerto Rico, any territory or possession of the United States, any agency or instrumentality of a state including state universities, and all federally recognized American Indian tribes.

Restrictions: Local governments, private universities, private nonprofit entities, private businesses, and individuals are not eligible to apply.

Amount of Grant: $120,000 maximum

Date(s) Application is Due: Jul 6

Contact: Jean Waters, Pollution Prevention Resource Exchange, (402) 595-2381; fax: (402) 554-6260; email: jwaters@mail.unomaha.edu

Internet: http://www.epa.gov/p2/grants/ppin/ppin.htm

Sponsor: Environmental Protection Agency

1200 Pennsylvania Avenue NW, Ariel Rios Bldg

Washington, DC 20460

EPA Senior Environmental Employment Program Project Grants 2239

The objective of the program is to use the talents of Americans 55 years of age or older to provide technical assistance to federal, state, and local environmental agencies for projects of pollution research, prevention, abatement, and control. Applications are accepted at any time. Prediscussion with the office is advisable.

Requirements: Private, nonprofit organizations designated by the Secretary of Labor under Title V of the Older Americans Act are eligible to apply.

Amount of Grant: $184,900 average

Contact: Susan Street, Director, SEE Program, (202) 564-0410; fax: (202) 564-0735; email: street.susan@epa.gov

Internet: http://www.epa.gov

Sponsor: Environmental Protection Agency

401 M Street SW

Washington, DC 20460

EPA Solid WaSuite Management Assistance Grants 2240

The objective of this program is to promote use of integrated solid waSuite management systems to solve municipal solid waSuite generation and management problems at the local, regional, and national levels. Funds are available for the allowable direct cost expenditures incident to program performance plus allocatable portions of allowable indirect costs of the institution. Projects can include training, surveys, public education programs, studies, and demonstrations. Grants will normally be funded on a 12-month basis; total approved project period may not exceed three years. Applications are accepted at any time.

Requirements: Eligible to apply are nonprofit entities including public authorities (federal, state, interstate, and local); public agencies and institutions; private agencies, institutions, and individuals; and Indian tribes.

Restrictions: For-profit organizations are not eligible to apply.

Amount of Grant: $100,000 average

Contact: Nick Vizzone, (703) 308-8460; email: vizzone.nick@epa.gov

Internet: http://www.epa.gov/epaoswer/osw

Sponsor: Environmental Protection Agency

401 M Street SW

Washington, DC 20460

EPA Targeted Watersheds Grants 2241

Established in 2003, the Targeted Watersheds Grant Program is designed to encourage successful community-based approaches and management techniques to protect and restore the nation's watersheds. The Targeted Watersheds Grant program is a competitive grant program based on the fundamental principles of environmental improvement: collaboration, new technologies, market incentives, and results-oriented strategies. The Targeted Watersheds Grant Program focuses on multi-faceted plans for protecting and restoring water resources that are developed using partnership efforts of diverse stakeholders. There are two distinct funding avenues: Targeted Watersheds Implementation Grants are focused on individual watershed organizations; and Capacity Building Grants are awarded to leading organizations with a national or regional focus that are able to provide training, technical assistance and education to local watershed groups and to teach the critical skills necessary to improve watershed health. The goal of these grants is to better serve the needs of local watershed groups.

Contact: Erin Collard, EPA Headquarters; (202) 566-2655; fax: (202) 566-1366; email: initiative.watershed@epa.gov

Internet: http://www.epa.gov/twg/twg_basic.html

Sponsor: Environmental Protection Agency

1200 Pennsylvania Avenue

Washington, DC 20460

Epilepsy Postdoctoral and Clinical Research Fellowships 2242

A few fellowships will be awarded to scientists (PhD or MD) wishing to carry out full-time research projects in the field of epilepsy. The period of support may begin at any time between May and October. The grant will be for one year but will be renewable once and exceptionally twice, upon request.

Requirements: Fellowships are available to Canadian citizens or for projects conducted in Canada.

Amount of Grant: $C30,000

Date(s) Application is Due: Jan 15

Contact: Caroline Savoy, (450) 358-9779; fax: (450) 346-1045; email: epilepsy@savoy-foundation.ca

Internet: http://www.savoy-foundation.ca/eng/mission/defaut.htm

Sponsor: Savoy Foundation

230 Foch Street

Street Jean Sur Richelieu, PQ J3B 2B2 Canada

Epilepsy Research Foundation New Therapy Grants 2243

The foundation supports innovative research leading to new insights into a cure for epilepsy. The primary focus of this program is to bring new approaches and therapies to patients through translational research. Other areas of consideration include but are not limited to: meaningful grants to senior-level scientific and clinical investigators on the brink of new discoveries, working at the nation's leading academic and research institutions, and in private industry; seed funding for preliminary work necessary to explore novel approaches; building innovative platform technologies; research to bring new approaches and therapy to children; innovative cutting edge projects that could lead to breakthrough discoveries; support for applying work in other areas with potential for promise in epilepsy; research programs that might not otherwise be funded through traditional sources; projects that encourage collaboration among scientists and industry; and proposals dealing with commercializing academic research projects. Grant funds may be used to support the direct costs of research such as salary support for personnel directly engaged in the project, consumable supplies and services, and travel essential to the conduct of the research. Multi-year awards will be considered. Guidelines are available online.

Requirements: Investigators must hold a relevant advanced degree (MD and/or PhD) and have completed all research training. The program is open to investigators at corporations as well as the academic/university arena. All applications will be judged on a competitive basis, without regard to the academic rank or title of the investigator. Applicants whose research will involve patient care or direct involvement with patients must be licensed to practice their profession at their institution and must obtain Institutional Review Board (IRB) approval of their research. IRB approval must be included with the original application.

Restrictions: Grant funds may not be used for indirect costs or institutional overhead, and the purchase of permanent equipment that exceeds 10 percent of the budget without prior approval.

Amount of Grant: $100,000-$200,000

Date(s) Application is Due: Apr 1; Oct 1

Contact: Research Department, (800) 332-1000 or (301) 459-3700; email: grants@efa.org or EpilepsyCure@aol.com

Internet: http://www.epilepsytdp.org/sec/support_translational

Sponsor: Epilepsy Research Foundation

4351 Garden City Drive

Landover, MD 20785-7223

Episcopal Church Foundation Graduate Fellowships 2244

Fellowships are awarded to encourage doctoral study by recent seminary graduates to qualify themselves for the teaching ministry in seminaries of the Episcopal Church. Fellowships may be held at accredited institutions and are renewable for a second and third year. Applications should be obtained from the dean's office at any of the 11 accredited Episcopal seminaries, Harvard Divinity, or Union Theological Seminary.

Requirements: Applicants should be seniors or graduates from an accredited Episcopal seminary or an Episcopal candidate from Harvard Divinity School or Union Theological Seminary. Applicants must be recommended by the deans of their theological seminaries

in order to apply. If the applicant is a graduate of a non-Episcopal seminary, nomination for a fellowship must be approved by the dean of an accredited Episcopal seminary.
Amount of Grant: $10,000 per year
Date(s) Application is Due: Dec 1
Contact: Elizabeth Rauen Sciaino, Fellows Program Manager, (800) 697-2858 or (212) 716-6246; fax: (212) 297-0142; email: beth@episcopalfoundation.org
Internet: http://www.episcopalfoundation.org/education/doctoral.html
Sponsor: Episcopal Church Foundation
815 Second Avenue, Room 400
New York, NY 10017

Eppendorf & Science Prize for Neurobiology   2245

The Eppendorf & Science Prize for Neurobiology acknowledges the increasingly active and important role of neurobiology in advancing our understanding of the functioning of the brain and the nervous system — a quest that seems destined for dramatic expansion in the coming decades. This international prize, established in 2002, encourages the work of promising young neurobiologists by providing support in the early stages of their careers. It is awarded annually for the most outstanding neurobiological research by a young scientist, as described in a 1,000-word essay based on research performed during the past three years. The winner is awarded money and publication of his or her essay in 'Science'. The essay and those of up to three finalists are also published on 'Science Online'. The award is announced and presented at a ceremony at the annual meeting of the Society for Neuroscience. Eppendorf provides financial support to help enable the grand prize winner to attend the meeting.
Requirements: Entrants must be a neurobiologist with an advanced degree and not older than 35 years. The research described in the entrant's essay must be based on the methods of molecular and cell biology. The entrant must have performed or directed the work described in the essay. The research must have been performed during the previous three years.
Restrictions: Employees of Eppendorf AG, its subsidiaries, Science and AAAS, and their relatives are not eligible for the prize.
Samples: Past Grand Prize Winners: 2005, Pingxi Xu, 'A Drosophila OBP Required for Pheromone Signaling'; 2004, Miriam Goodman, 'Deconstructing C. elegans Sensory Mechanotransduction'; 2003, Michael Ehlers, 'Deconstructing the Synapse by Ubiquitin-Dependent Protein Turnover'
Amount of Grant: $25,000 and publication of his or her essay in 'Science'.
Date(s) Application is Due: Jun 15
Contact: Maryrose Police, email: mpolice@aaas.org
Internet: http://www.sciencemag.org/feature/data/prizes/eppendorf/eppenprize.dtl
Sponsor: American Association for the Advancement of Science
1200 New York Avenue, NW, Room 1049B
Washington, DC 20005

Eppley Foundation Postdoctoral Grants   2246

The foundation supports research in advanced scientific subjects, both in the biological and physical sciences, and is interested in supporting research where federal support is not available or to provide seed money toward larger undertakings. Work must be original and independent. Approximately 12 grants are awarded each year.
Requirements: Individuals who have several years of postdoctoral research experience in the biological and physical sciences are eligible. Grant proposals from foreign countries are considered only when a U.S. institution will administer the grant on their behalf.
Restrictions: Grants are not made to individuals and rarely for support of newly awarded PhDs or MDs. Social sciences, computer sciences, and education programs will not be supported. Under most circumstances, the foundation will not support heart, HIV, cancer, or DNA research.
Amount of Grant: $15,000 average
Contact: Huyler Held, Secretary-Treasurer, (212) 448-1100 or (212) 371-1660; fax: (212) 448-6260; email: bblauner@Mclaughlinstern.com
Sponsor: Eppley Foundation for Research
260 Madison Avenue
New York, NY 10016

Equal Sweetener Foundation Grants   2247

The foundation invests in programs that improve the health and well being of people throughout the world, and supports programs in which Merisant employees are actively involved as contributors or volunteers. The foundation prefers to support projects that fit within its focus area—diabetes research and education. Collaborative projects that achieve stated, measurable objectives receive preference. There are no deadlines for preliminary requests; full proposals are by invitation.
Requirements: 501(c)3 tax-exempt organizations are eligible. The grantseeking organization must be reputable, experienced, and show capable, responsible financial management.
Restrictions: Grants do not support individual aid or personal support; religious activities, politically partisan groups or activities, or fraternal orders; veterans or labor organizations; athletic teams; capital campaigns; endowments; scholarship programs; or projects from start-up organizations. The foundation will not contribute to any organization that discriminates on the basis of race, color, creed, sex, marital status, or physical ability.
Contact: Jo-Ann Digman, President
Internet: http://www.merisant.com/pages/about_merisant/career_center/community_involve.asp
Sponsor: Equal Sweetener Foundation
#1 N Brentwood Boulevard, Suite 510
Clayton, MO 63105

Equipment Leasing and Finance Foundation Research Grants   2248

Grants are available to encourage academics in all fields of scholarship to study topics of interest to the equipment leasing and finance industry. Research topics of interest include: transitioning the business model and avoiding the commoditization of a finance product through the bundling of services; best practice case studies; analysis of publicly held lease financing companies—does the current model work?; changes in accounting rules—how does this affect the lessee and lessor in the financial decision making process?; what will finance companies of the future look like?; study of successful industry consolidates; corporate size—when does efficiency of size become inefficient?; limited purpose bank charters—are they the vehicle of the future in leasing?; FASB-and changes to synthetic leases; specialty finance in banking—what is the future?; and maintaining residual values—the challenges and solutions. General topics of research interest are: strategic management/planning, accounting/financial reporting, and finance; tax and legislative/regulatory; credit/collections; economics; portfolio management; contracts/documentation; technology benchmarking; strategic importance of captives; operations; and best practices. Grants may be used for research studies; booklets, pamphlets and video production; surveys and statistical gathering; and other initiatives that are considered on a case-by-case basis.
Requirements: For-profit and nonprofit organizations and applicants—which may include individuals, universities, foundations, associations, and corporations—are eligible.
Amount of Grant: $5000-$50,000 range; $12,000 average
Date(s) Application is Due: Jan 15; May 15; Sep 1
Contact: Lisa Levine, Executive Director, (703) 516-8363; email: llevine@elamail.com
Internet: http://www.leasefoundation.org/grant
Sponsor: Equipment Leasing and Finance Foundation Grants
4301 N Fairfax Drive, Suite 550
Arlington, VA 22203

Erasmus Institute Fellowships   2249

The institute will admit six postdoctoral fellows to its center on the campus of the University of Notre Dame. The institute invites applications from academics who by their scholarship and/or faith seek to cultivate Catholic intellectual and cultural traditions. Fellows devote their time in residence to concentrated research and writing on individual projects related to the institute's goals. The fellows will also be expected to teach one course during the academic year, likely in the spring term of next year.
Requirements: Postdoctoral applicants must have received the PhD after May 1, 2001; ABD applicants must fulfill the PhD requirements by June 1 in order to be eligible for the fellowship.
Restrictions: Dissertation and salaried senior and junior faculty fellowships are no longer offered.
Amount of Grant: $35,000
Date(s) Application is Due: Jan 27
Contact: Residential Fellowships, (574) 631-9346; fax: (574) 631-3585; email: erasmus@nd.edu
Internet: http://www.nd.edu/~erasmus
Sponsor: University of Notre Dame
1124 Flanner Hall
Notre Dame, IN 46556-5611

Erasmus Prize   2250

This prize is awarded annually to a person or institution that has made an exceptionally important contribution to European culture in the fields of humanities and the arts or social and social-scientific sphere.
Requirements: There are no restrictions as to age, sex, or citizenship of the recipient.
Samples: Dr. F. Amtenbrink—for The Democratic Accountability of Central Banks: A Comparative Study of the European Central Bank.; Dr. J. Bohnemeyer—for The Worm, the Demon, and the Virgin: Medieval Theories of an Extraordinary Generation (c 1100-c 1350).
Amount of Grant: NLG150,000
Contact: Program Contact, 31-0-20-6752753; fax: 31-0-20-6752231; email: spe@erasmusprijs.org
Internet: http://www.erasmusprijs.org/eng/index.htm
Sponsor: Erasmus Prize Foundation (Stichting Praemium Erasmianum)
Jan van Goyenkade 5
Amsterdam 1075 HN Netherlands

EREF Grants   2251

The Foundation awards several grants each year for research or education in topics pertaining to any aspect of solid waSuite management. Applications are accepted three times each year. There is no formal application. Your submission, in English, should include a description of proposed activity, project timeline, budget, and resumes of principal investigators.
Restrictions: Grants will not be provided for capital campaigns; political contributions; religious causes; operating funds; loans; or support of lobbying activities. The Foundation will not pay overhead in excess of 25%.
Samples: URS Dames and Moore (United Kingdom)—to research the regulatory, economic, and technical aspects of greenhouse gas emission reductions in the waSuite industry, $185,000.; Michigan State U (MI)—to design and operate a bioreactor landfill cell and quantify its emissions, $500,000.; North Carolina State U (NC)—to develop a global assessment of recycling policies and markets, $140,625.
Amount of Grant: $10,000-$500,000
Contact: Michael Cagney, President, (703) 299-5139; fax: (703) 299-5145; email: mjcagney@erefdn.org

Internet: http://www.erefdn.org/guide.html
Sponsor: Environmental Research and Education Foundation
901 N. Pitt Street, Suite 270
Alexandria, VA 22314

EREF Scholarships 2252

Scholarships are awarded in memory of Francois Fiessinger, PhD, to support the work of outstanding students interested in excellence in environmental research. Awards are based on academic or professional performance; relevance of one's work to the advancement of environmental science; and potential for success. Applicants should request an application packet or download the online pdf application.
Requirements: Applications will be considered from full-time PhD students who have a clearly demonstrated interest in environmental research. Awards are made without regard to race, religion, national or ethnic origin, citizenship, or disability. Applications will be considered from students outside the United States or studying abroad.
Amount of Grant: $12,000 maximum per year paid monthly, renewable for two additional years for a total of $36,000
Date(s) Application is Due: Aug 1
Contact: Michael Cagney, President, (703) 299-5139; fax: (703) 299-5145; email: mcagney@erefdn.org
Internet: http://www.erefdn.org/scholar.html
Sponsor: Environmental Research and Education Foundation
120 S Fayette Street
Alexandria, VA 22314

Eric Berne Fund for the Future Grants 2253

Grants support research or projects that demonstrate an interest in the evaluation of the effectiveness of various applications of transactional analysis (TA) theory. Preference is given to projects that are expected to generate particular benefits within a region or community, that extend TA to new areas of the world, support development of measurable professional standards of competency, and/or build bridges between TA and other disciplines. The deadline for making nominations for the EBMA is December 1 of the year preceding the award.
Requirements: Grants are open to individuals or nonprofit organizations. If profit-making organizations wish to submit proposals, special royalty or profit-sharing agreements may be required.
Contact: Eric Berne Fund Distribution Committee, (925) 600-8110; fax: (925) 600-8112
Internet: http://www.itaa-net.org/itaa/grants.htm
Sponsor: International Transactional Analysis Association
2186 Rheem Dr #B-1
Pleasanton, CA 94588

Erikson Institute Child Development Doctoral Fellowships 2254

The PhD program, offered jointly with the Department of Psychology at Loyola University Chicago, prepares students for careers in college teaching and research, administration, program design and evaluation, and child advocacy. Named fellowships include the Fellowship in Applied Research in Child Development and the Irving B. Harris Leadership Fellowship in Child Development. The fellowships include tuition plus a research/teaching assistantship or living stipend and are awarded on the basis of academic promise, experience, and research interest.
Date(s) Application is Due: Apr 1
Contact: Dr. Robert Halpern, (312) 893-7142; email: rhalpern@erikson.edu
Internet: http://www.erikson.edu/students.asp?file=applyfinaid&highlight=fellowships
Sponsor: Erikson Institute
420 N Wabash Avenue
Chicago, IL 60611

Ernest Guenther Award in the Chemistry of Natural Products 2255

This award, supported by Givaudan, is given annually to recognize and encourage outstanding achievements in analysis, structure elucidation, and chemical synthesis of natural products. Special consideration will be given to the independence of thought and originality shown. The award consists of $6,000, a medallion, and a certificate.
Requirements: Any individual, except a member of the award committee, may submit one nomination or seconding letter for each award in any given year. Nominating documents consist of a letter of not more than 1000 words containing an evaluation of the nominee's accomplishments and a specific identification of the work to be recognized, a biographical sketch including date of birth, and a list of publications and patents authored by the nominee.
Restrictions: Self-nominations are not accepted.
Samples: David G. Kingston—award recipient, $6,000 (2008); Dale L. Boger—award recipient, $6,000 (2007); William A. Fenical—award recipient, $5,000 (2006).
Amount of Grant: $6,000 and an allowance of $2,500 toward traveling expenses to award meeting
Date(s) Application is Due: Nov 1
Contact: Felicia Dixon, Awards Administrator; (800) 227-5558 or (202) 872-4408; fax: (202) 776-8008; email: f_dixon@acs.org or awards@acs.org
Internet: http://chemistry.org/portal/Chemistry?PID=acsdisplay.html&DOC=awards\guenther.html
Sponsor: American Chemical Society
1155 Sixteenth Street, NW
Washington, DC 20036

Ernest M. Eller Prize in Naval History 2256

The Naval Historical Center, in cooperation with the Naval Historical Foundation, offers this prize to encourage excellence in research and writing on the history of the U.S. Navy. The annual prize goes to the author of the best article on U.S. naval history published in a scholarly journal. Entry deadline is June 1 of the year following publication.
Amount of Grant: $1000
Date(s) Application is Due: Jun 1
Contact: Senior Historian, Washington Naval Yard
Internet: http://www.history.navy.mil/prizes/grants.htm
Sponsor: Naval Historical Center
805 Kidder Breese SE, Washington Navy Yard
Washington, DC 20374-5060

Erskine A. Peters Dissertation-Year Fellowship 2257

This initiative has two overall goals: to enable two outstanding African-American doctoral candidates (at the ABD level) to devote their full energies to the completion of the dissertation; and to provide an opportunity for African-American scholars at the beginning of their academic careers to experience life at a major Catholic research university. The fellowship is tenable for a full academic year beginning in August and concluding in May. Each fellow is provided with office space, use of a personal computer, an official academic home in the department of the fellow's specialization, access to a faculty mentor in the fellow's discipline, and access to all university facilities. The candidate is expected to be in residence at the university and to devote most of her/his time to the completion of dissertation research. In addition, she/he will be expected to present a public lecture dealing with her or his current research at the end of the spring term, to give one lecture in the university's introductory course in African and African-American Studies, and to sponsor a short-term mutual learning event for faculty and students (e.g., brown bag series, roundtable discussion, etc.).
Requirements: The program invites applications from African-American doctoral candidates in the humanities, social sciences, and theological disciplines who have completed all degree requirements with the exception of the dissertation.
Amount of Grant: $25,000 stipend; $2000 research budget
Date(s) Application is Due: Dec 3
Contact: African and African-American Studies Program, (574) 631-5628 or (219) 631-5666; email: astudies@nd.edu
Internet: http://www.nd.edu/~astudies/erskinepeters.htm
Sponsor: University of Notre Dame
O'Shaughnessy Hall
Notre Dame, IN 46556

ESF Latsis Prize 2258

The foundation invites nominations for the European Latsis Prize. Presented each year at the ESF's Annual Assembly, the European Latsis Prize is awarded to an individual scholar or research group in recognition of outstanding and innovative contributions in a selected field of European research. Nomination information and nomination forms are available on the Web site.
Requirements: Nominations may be received from institutions and individuals for an individual scholar, or from institutions generally for research groups.
Restrictions: Self-nominations are not accepted.
Amount of Grant: Sf100,000 (EU62,000)
Contact: Secretary General, 33-0-3-88-76-71-00; fax: 33-0-3-88-37-05-32; email: prize@esf.org
Internet: http://www.esf.org/prize
Sponsor: European Science Foundation
1 quai Lezay Marnesia, BP 90015
Strasbourg Cedex F-67080
France

ESRI Conservation Program Grants 2259

The conservation grants program provides donations and discounts of GIS software, data, books, and training. There are no application deadlines. The program does not grant hardware or cash, but partners with many other groups who do (list available online). General grants and basic grants are available. Grants are awarded to organizations that can demonstrate a strong commitment to conservation; compelling reasons to use GIS in obtaining their objectives; and the organizational capacity to use GIS effectively over a long period of time. All applicants must present a credible case for how GIS and the support from ESRI will advance their cause of conservation and environmental protection, broadly defined. Successful applicants will effectively make their case about how they propose to integrate and analyze disparate spatial data to accomplish their goals. To request application materials, send a blank email message (no subject, no content) to ecpgrant@esri.com, or visit the Web site.
Requirements: Eligible applicants are U.S.-based 501c(3) nonprofit organizations that actively engage the public in resource conservation and environmental protection. This includes, but is not limited to, grassroots conservation and environmental organizations, community action groups, economic development organizations, sustainable development groups, community-based conservation groups, growth management organizations, environmental justice groups, and tribal conservation organizations. International groups must have a U.S.-based tax-exempt organization as their sponsor.
Restrictions: The following types of organizations are not eligible to apply: colleges and universities; public schools or school districts; local, state, and federal government agencies or natural resource conservation districts; association and coalitions organized to fund-raise for or to benefit local, state, and federal government agencies; natural

resource conservation districts; and CTSP's nonprofit sponsors and their associated members.
Contact: Charles Convis, Program Director; (909) 793-2853 ext 2488; fax: (909) 307-3025; email: ecp@esri.com
Internet: http://www.conservationgis.org/aaesrigrants.html
Sponsor: Environmental Systems Research Institute
380 New York Street
Redlands, CA 92373

Ethical Research Grants and Fellowships   2260
Graduate fellowships are available for investigations into the discovery, development, validation, and implementation of alternatives to the use of live animals in research, testing, and teaching. Areas of interest include tissue cultures, bacteria cultures, protozoan studies, gas chromatography, mass spectrometry, radioimmunoassay, mathematical and computer models, quantum pharmacology, mechanical models, and clinical and epidemiological surveys. Awards include stipend plus supplies. Students from all over the world are eligible to apply.
Requirements: Nonprofit educational and research institutions are eligible for grants. Students who are enrolled in master's and PhD programs in the sciences, humanities, psychology, and journalism are eligible for graudate fellowships.
Amount of Grant: $15,000 maximum
Date(s) Application is Due: Mar 15
Contact: Peter O'Donovan, Executive Director, (312) 427-6025; fax: (312) 427-6524; email: ifer@navs.org
Internet: http://www.ifer.org
Sponsor: International Foundation for Ethical Research
53 W Jackson Boulevard, Suite 1552
Chicago, IL 60604

Ethics and Excellence in Journalism Foundation Grants   2261
This Foundation was founded for charitable, scientific and educational purposes, including the improvement of the quality of the practice of journalism among various media. Support includes funding for creative projects and research that promotes excellence in journalism and instills and encourages high ethical standards in journalism. Funding is available to organizations via an application and review process.
Requirements: U.S. nonprofit organizations, including universities and colleges, are eligible.
Restrictions: Individuals are not eligible.
Samples: Putnam City Public Schools Foundation (Oklahoma City, OK)—for photography equipment and year-book programs at local high schools, $25,000 (2005); U of Central Oklahoma (Edmond, OK)—for a 20-station digital-photography computer lab, $113,000 (2005); U of Maryland at College Park (MD)—for a new building at the Philip Merrill College of Journalism, $500,000 (2005); U of Oklahoma (Norman, OK)—to establish the Institute for Research and Training at the Gaylor College of Journalism and Mass Communication, $600,000 (2005).
Date(s) Application is Due: Apr 15; Oct 15
Contact: Nancy Hodgkinson, Senior Program Officer, (405) 604-5388; fax: (405) 604-0297; email: nancy.hodgkinson@journalismfoundation.org
Internet: http://www.journalismfoundation.org
Sponsor: Ethics and Excellence in Journalism Foundation
210 Park Avenue, Suite 3150
Oklahoma City, OK 73102

ETS Postdoctoral Fellowships   2262
Up to three fellows are selected annually to conduct independent research in Princeton, NJ, in one of the following areas: psychology; education; teaching; learning; literacy; statistics; computer science; educational technology; minority issues; and testing issues, including new forms of assessment and alternate forms of assessment for special populations. The program goals are to provide research opportunities to individuals who hold a doctorate in the fields indicated and to increase the number of women and underrepresented minority professionals conducting research in educational measurement and related fields. Prior to applying, candidates need to send a one-page abstract of their research via email. Guidelines are available online.
Requirements: The program is open to any individual who holds a doctorate in a relevant discipline and provides evidence of prior research.
Amount of Grant: $50,000 stipend
Date(s) Application is Due: Feb 1
Contact: Linda DeLauro, (609) 734-1806 or 734-5949; fax: (609) 497-6032; email: fellowships@ets.org
Internet: http://www.ets.org/research/fellowships.html
Sponsor: Educational Testing Service
MS 09R, Rosedale Road
Princeton, NJ 08541-0001

ETS Predoctoral Summer Program   2263
Students chosen for the eight-week program will participate in ongoing independent research with access to a mentor in one of the following areas: education; educational technology; learning; linguistics; literacy; minority issues; new constructs; policy research; psychology; psychometrics; statistics; teaching; and testing issues (including alternate forms of assessment for special populations and new forms of assessment). An explicit goal of the program is to increase the number of women and minority professionals in educational measurement and related fields. Applicants must submit their application materials electronically. Guidelines are available online.

Requirements: Graduate students who are currently enrolled in a doctoral program and have completed a minimum of two years of full-time graduate study in a program emphasizing one of the areas specified are eligible. A letter of recommendation from the major advisor or committee chair and transcripts, undergraduate and graduate, must be included with completed application form.
Amount of Grant: $5000 stipend
Date(s) Application is Due: Feb 1
Contact: Grants Administrator, (609) 734-5543; email: internfellowships@ets.org
Internet: http://www.ets.org/research/fellowships.html
Sponsor: Educational Testing Service
MS 07R, Rosedale Road
Princeton, NJ 08541-0001

ETS Sylvia Taylor Johnson Minority Fellowship in Educational Measurement   2264
The two-year fellowship, renewable after the first year by mutual agreement, is designed to provide talented minority scholars an opportunity to carry out independent research under the mentorship of ETS senior researchers. The stipend will be set in relation to the successful applicant's compensation at the home institution. Applicants may obtain the center's brochure before they apply.
Requirements: Applicants must have received their doctoral degree within the past ten years and must be U.S. citizens or permanent residents.
Date(s) Application is Due: Feb 1
Contact: Linda DeLauro, (609) 734-1806; fax: (609) 734-5410; email: internfellowships@ets.org
Internet: http://www.ets.org/portal/site/ets/
menuitem.c988ba0e5dd572bada20bc47c3921509/
?vgnextoid=1dadaf5e44df4010VgnVCM10000022f95190RCRD&vgnextchannel=4
9f5be3a864f4010VgnVCM10000022f95190RCRD
Sponsor: Educational Testing Service
MS-09R, Rosedale Road
Princeton, NJ 08541-0001

EU Center for Humanistic Inquiry Junior and Postdoctoral Fellowships   2265
The purpose of the program is to stimulate and support humanistic research by providing scholars in early stages of their careers with the necessary time, space, and other resources. Research projects must be humanistic, but fellows may hold the PhD in any discipline. Fellows are expected to offer an upper-level undergraduate course on a subject of their choice during the spring of their fellowship year. The program provides stipends, research budgets, office space, and access to the university libraries. Application forms and further information are available from the office and online.
Requirements: Junior fellows are scholars who, at the beginning of the fellowship year, will be at least three years beyond receipt of the PhD, and normally no more than 10. Postdoctoral fellows, who must have the PhD before the submission of their applications, are awarded to those who have held the PhD for no more than three years before receiving the fellowship.
Amount of Grant: $38,500 stipend
Date(s) Application is Due: Feb 23
Contact: Center for Humanistic Inquiry, (404) 727-6424; fax: 404-727-1669; email: chi@emory.edu
Internet: http://www.chi.emory.edu/Fellowships/index.html
Sponsor: Emory University
1715 N Decatur Road
Atlanta, GA 30307

EU Robert W. Woodruff Library Research Fellowships   2266
Short-term fellowships support scholarly use of the library's research collections in the areas of modern literature and African American studies. The Special Collections Department has extensive holdings related to the Irish literary renaissance and the finest collection outside of Ireland for the study of contemporary Irish poetry. The library also holds the literary archive of the late poet laureate of England, Ted Hughes, and related British literary collections. The department also houses extensive collections focusing on black print culture, the civil rights and post-civil rights movements, communism and the Left, and African American religion and culture. Fellowship support helps defray expenses in traveling to and residing in Atlanta during the fellowship tenure, usually one month.
Amount of Grant: $2000 maximum
Date(s) Application is Due: May 31
Contact: Fellowship Program, fax: (404) 727-0360; email: speccollref@emory.edu
Internet: http://web.library.emory.edu/services/hr/fellows.html
Sponsor: Emory University
Robert W. Woodruff Library, 540 Asbury Cir, Emory University
Atlanta, GA 30322-2870

Eugene and Agnes E. Meyer Foundation Grants   2267
The Foundation awards grants to nonprofit organizations that serve the people and communities in the Washington, D.C. region. Grants are awarded in ten program areas: arts; heritage, and culture; children, youth and families; civic engagement; education; employment and skills training; health and mental health; homelessness and hunger; housing and community development; immigrant communities; and law and justice. Candidates are required to first submit a letter of inquiry. Instructions for completion of LOI are available online.
Requirements: Eligible applicants must be a nonprofit organization with tax-exempt status and be located within and primarily serve the Washington, D.C. region which

is defined as: the District of Columbia; the Maryland counties of Montgomery, Prince George's, Calvert, Charles, and St. Mary's; and the Virginia counties of Arlington, Fairfax, Loudoun, Prince William, and Stafford and cities of Alexandria, Falls Church, Manassas, and Manassas Park in Virginia.
Restrictions: The Foundation generally will not consider requests to fund: individuals, either through scholarships or other forms of financial assistance; scientific or medical research; sectarian purposes (programs that promote religious doctrine); special events or conferences; endowments.
Geographic Focus: District of Columbia; Maryland; Virginia
Samples: Theatre Downtown, Washington Stage Guild (Washington, DC)—for general operating support, $20,000.; Empower Program (Washington, DC)—to support local youth-services programs, $20,000.; Community Family Life Services (Washington, DC)—to revamp and document its financial accounting and reporting systems, $10,000.; Independent Sector (Washington, DC)—for its conference program, $10,000.
Date(s) Application is Due: Feb 9; Jun 8; Oct 1
Contact: Julie Rogers, President, (202) 483-8294; fax: (202) 328-6850; email: jrogers@meyerfdn.org
Internet: http://www.meyerfoundation.org
Sponsor: Eugene and Agnes E. Meyer Foundation
1400 16th Street, N.W., Suite 360
Washington, DC 20036

Eugene M. Lang Foundation Grants   2268
The foundation awards grants in New York and Pennsylvania in its areas of interest, including education (early childhood education and higher education), medical and health programs, arts, health organizations, medical research, minorities, and performing arts. Types of support include annual campaigns, conferences and seminars, continuing support, fellowships, general operating support, internship funds, professorships, program development, scholarship funds, and seed money. The foundation favors social services such as those helping homeless or single mothers. Locally based groups wanting support must involve a Lang family member. There are no application deadlines; initial approach should be by letter.
Requirements: Organizations in New York and Pennsylvania are eligible to apply.
Restrictions: Grants are not made to individuals, or for building funds, equipment and materials, capital or endowment funds, deficit financing, publications, or matching gifts.
Geographic Focus: New York; Pennsylvania
Samples: New York-Presbyterian Hospital (New York, NY)—for a six-year health-sciences education program for selected inner-city youths, designed to boost their academic success and facilitate their entry into health careers, $1.25 million (2003).
Amount of Grant: $500-$50,000 average
Contact: Program Contact, (212) 949-4100
Sponsor: Eugene M. Lang Foundation
535 5th Avenue, Suite 906
New York, NY 10017

Eugene McDermott Foundation Grants   2269
The foundation awards grants to Texas nonprofit organizations in its areas of interest, including children and youth, community development, education (early childhood through higher education), health care/organizations, international human rights, medical research, minorities, and social service delivery programs. Types of support include annual campaigns, building construction/renovation, capital campaigns, equipment/land acquisition, general operating grants, matching/challenge grants, professorships, programs/project support, research grants, scholarship funds, and seed grants. There are no application deadlines or forms. The board meets quarterly to consider requests.
Requirements: Texas nonprofit organizations are eligible.
Geographic Focus: Texas
Samples: Nasher Sculpture Center (Dallas, TX)—for endowment, $500,000 (2004); U of Texas Southwestern Medical Center (Dallas, TX)—to refurbish Medical School, $400,000 (2004); U of Texas Southwestern Medical Ctr (Dallas, TX)—to establish a chair in cardiothoracic anesthesiology, $1 million (2003).
Amount of Grant: $1000-$25,000 average
Contact: Grants Administrator, (214) 521-2924
Sponsor: Eugene McDermott Foundation
3808 Euclid Avenue
Dallas, TX 75205

Eugene S. Pulliam Fellowship for Editorial Writing   2270
The fellowship is awarded to an outstanding editorial writer to help broaden his or her journalistic horizons and knowledge of the world. The annual award can be used to cover the cost of study, research, and/or travel in any field. In some cases, the fellowship results in editorials and other writings, including books. All entries must be in English.
Requirements: A candidate must hold a position as a full-time editorial writer at a news publication located in the United States; have at least three years experience as an editorial writer; demonstrate outstanding writing and analytical abilities; and secure assurances by the editor or publisher that the applicant will be allowed sufficient time to pursue the fellowship without jeopardizing employment. (Fellows do not have to leave their jobs.)
Amount of Grant: $75,000
Date(s) Application is Due: Jul 1
Contact: Bobby Deckard, Awards Coordinator, (317) 927-8000 ext 215; email: bdeckard@spj.org
Internet: http://www.spj.org/fellowships_pulliam.asp
Sponsor: Sigma Delta Chi Foundation
3909 N Meridian Street
Indianapolis, IN 46208-4045

Eurasia Foundation Grantmaking Program   2271
The grantmaking program supports innovative projects in the foundation's three program areas: private enterprise development, public administration and policy, and civil society. The foundation supports projects aimed at strengthening human capital, developing locally sustainable forms of financing, and promoting a favorable legal and regulatory environment. Projects that cross over programmatic areas and geographic boundaries are particularly encouraged. The Washington office supports partnership projects that are developed and implemented jointly between U.S. or other foreign institutions and New Independent States (NIS) organizations. Proposals are accepted at any time. The foundation occasionally sponsors targeted initiatives. Grant seekers should contact the nearest field office (listed on the Web site) for more information. The foundation strongly encourages applicants to submit a two- to three-page letter of inquiry describing the objectives of the program before making a formal application.
Requirements: Grants are made to 501(c)3 nonprofit organizations as well as universities and government agencies. Organizations must be involved in a U.S.-NIS partnership.
Restrictions: Grants are not made to individuals, nor for scholarships.
Samples: Community Credit Union First Far Eastern (Russia)—to improve access to capital for small and medium businesses by encouraging the development of credit unions, $15,087.; Yakutsk Business and Technology Incubator (Sakha Republic, Russia)—to accelerate the development of small enterprises working in the area of innovative technology through the establishment of an informational and consulting center, $25,771.; Zhytomyr Oblast Branch of International Society of Human Rights (Zhytomyr, Ukraine)—To promote the empowerment of citizens and the bottom-up development of rule of law by providing qualified, cost-free legal information and advice to local residents and nongovernmental organizations, $13,470.
Amount of Grant: $20,000 field grants
Contact: Grantmaking Program, (202) 234-7370; fax: (202) 234-7377; email: eurasia@eurasia.org
Internet: http://www.eurasia.org/grant.html
Sponsor: Eurasia Foundation
1350 Connecticut Avenue NW, Suite 1000
Washington, DC 20036

Eureka-Boston Fellowships   2272
The program awards two-year fellowships to CEOs of Massachusetts-based nonprofit organizations serving children, youth, and families in Boston, Cambridge, Chelsea, and Somerville.
Requirements: CEOs of Massachusetts nonprofit organizations serving Boston, Cambridge, Chelsea, and Somerville are eligible.
Geographic Focus: Massachusetts
Contact: Grants Administrator, (617) 859-8218; fax: (617) 507-7738; email: info@eureka-boston.org
Sponsor: Eureka-Boston
32 Rutland Street
Boston, MA 02118

Evan and Susan Bayh Foundation Grants   2273
Established in 2002 by Senator Evan Bayh and his wife Susan, the foundation gives to various education and human services programs mostly in Indiana.
Requirements: There are no deadlines and no specific forms to use. It is suggested that you contact the foundation prior to submitting an application.
Geographic Focus: Indiana
Samples: Indiana University, Matching the Promise Campaign - $50,000.; (2007) American Cancer Society, Atlanta, GA - $2,500.; Deaconess Hospital, Evansville, IN - $5,000.; Fisher House Foundation, Rockville, MD - $5,000.; Munster Medical Research Foundation, Munster, IN - $5,000.; Nature Conservancy, Arlington, VA - $2,500.; St. Albans School of Public Service, Washington, DC - $5,000.; Salvation Army, Indianapolis, IN - $2,500.; United Way of Central Indiana, Indianapolis, IN - $2,500.
Amount of Grant: $2,500 - $5,000
Contact: G. Frederick Glass; (317) 237-0300; email: fred.glass@bakerd.com
Sponsor: Evan and Susan Bayh Foundation
c/o G. Frederick Glass, 300 N. Meridian Street, Suite 2700
Indianapolis, IN 46204-1750

Evan Frankel Foundation Grants   2274
The foundation awards grants in its areas of interest, including higher education in the humanities and the environment. Giving is primarily in Manhattan and Suffolk County, NY and Los Angeles, CA. Submit a letter to request guidelines and deadline dates.
Restrictions: Individuals are not eligible.
Samples: Yale U (New Haven, CT)—to provide scholarships for students in master's degree programs in the School of Forestry and Environmental Studies, $200,000.
Amount of Grant: $500-$100,000 average
Contact: Nancy Wendell, (631) 329-2833; fax: (631) 329-7102; email: frankelfound@hamptons.com
Sponsor: Evan Frankel Foundation
P.O. Box 5072
East Hampton, NY 11937

Evanston Community Foundation Grants   **2275**
Grants awarded by the Foundation: encourage and support new initiatives and innovative approaches to addressing community needs; build the capacity of local nonprofit organizations to fulfill their missions more effectively; encourage collaborative ventures that will strengthen the community; provide initial support of projects that will have impact beyond the scope and timeline of the proposed project; build community partnerships and resources; and strengthen the area's nonprofit community. Grants typically provide initial seed money to launch new projects, capstone dollars for a larger project that will grow over a longer period, and support a one-time activity or initial phase of a new program. The Foundation's area of interests include: arts and culture; basic human needs; community development; education; environment; health; women and girls and youth and families. The Foundation's RFP and application instructions are available online.
Requirements: IRS 501(c)3 organizations serving the Evanston, IL, community are eligible.
Geographic Focus: Illinois
Samples: Evanston Youth Bank (Evanston, IL)—for Northwestern University Undergraduate Leadership Program, $1000 (2003).
Date(s) Application is Due: Feb 26
Contact: Sara Schastok, Executive Director, (847) 492-0990; fax: (847) 492-0904; email: schastok@evcommfdn.org or info@evcommfdn.org
Internet: http://www.evcommfdn.org/grant_making.htm
Sponsor: Evanston Community Foundation
1007 Church Street, Suite 108
Evanston, IL 60201

Evjue Foundation Grants   **2276**
The Foundation contributes each year to worthy educational, cultural and charitable organizations that contribute to the quality of life in Madison and Dane County. The Foundation's interests include: theater, hunger, troubled youth, and universities. The Foundation looks favorably upon projects responding to overall community needs and priorities that do not duplicate existing services. Application and guidelines are available online.
Requirements: Grants are only made to nonprofit organizations.
Restrictions: Grants are not made to individuals. All grants for scholarships are given to educational institutions, which have full responsibility for selecting the individual recipients. Grants are not usually made to establish or add to endowment funds. Grants are not ordinarily made to fund specific medical or scientific research inquiries, nor to support operating expenses and general administrative expenses of organizations.
Geographic Focus: Wisconsin
Samples: U of Wisconsin at Madison (WI)—to endow a professorship at its school of journalism, $1 million (2005).
Date(s) Application is Due: Mar 12
Contact: Arlene Hornung, Executive Director, (608) 252-6401; email: ahornung@madison.com
Internet: http://www.madison.com/tct/evjue/policy/index.php
Sponsor: Evjue Foundation
1901 Fish Hatchery Road
Madison, WI 53713

Ewing Halsell Foundation Grants   **2277**
The foundation awards grants to eligible Texas nonprofit organizations in its areas of interest, including education, environment, health care and health organizations, medical research, social services, and youth services. Types of support include annual campaigns, building construction/renovation, equipment acquisition, land acquisition, publication, research, seed grants, and technical assistance. There are no application deadlines or forms.
Requirements: Texas nonprofit organizations are eligible. Preference is given to requests from southwestern Texas, particularly San Antonio.
Restrictions: The foundation's grants do not support individuals or requests for deficit financing, emergency funds, general endowments, matching gifts, scholarships, fellowships, demonstration projects, general purposes, conferences, or loans.
Geographic Focus: Texas
Samples: San Antonio Botanical Society (TX)—to endow the Lucile Halsell Conservatory, $1 million matching grant (2005).
Contact: Grants Administrator, (210) 223-2640
Sponsor: Ewing Halsell Foundation
711 Navarro Street, Suite 537
San Antonio, TX 78205

Explorers Club Exploration Fund Grants   **2278**
The Explorers Club is an international multidisciplinary professional society dedicated to the advancement of field research and the ideal that it is vital to preserve the instinct to explore. Since its inception in 1904, the Club has served as a meeting point and unifying force for explorers and scientists worldwide. The Exploration Fund of the Explorers Club provides grants in support of exploration and field research. Grants are made primarily to graduate students. Awards grants range from $500 to $1,500.
Requirements: Individuals at the advanced graduate and professional levels are eligible to apply for support. Applicants need not be members of the Explorers Club.
Amount of Grant: $500-$1,500 average
Date(s) Application is Due: Mar 31
Contact: Matthew Williams, Administrative Assistant; (212) 628-8383; fax: (212) 288-4449; email: mwilliams@explorers.org

Internet: http://explorers.org/resources/funding/funding.php
Sponsor: The Explorers Club
46 East 70th Street
New York, NY 10021

Explorers Club Scott Pearlman Field Awards   **2279**
The Scott Pearlman Field Awards provide grants for professional artists, writers, photographers, filmmakers, and journalists in support of scientific expeditions sponsored by The Explorers Club. Grants in amounts up to $1,500 are made to professionals who are recommended by the leader of an expedition. Applications will be judged solely on the merit of the proposal, the competence and ability of the candidate, and the appropriateness of the budget.
Requirements: Samples of the candidates work must be submitted with this application, along with two references from individuals familiar with the candidate's skills, personality, and ability to provide reproduction-quality documentation of the specified expedition.
Amount of Grant: $1,500 maximum
Contact: Matthew Williams, Administrative Assistant; (212) 628-8383; fax: (212) 288-4449; email: mwilliams@explorers.org
Internet: http://www.explorers.org/resources/funding/pearlmanawards/pearlmanawards.php
Sponsor: The Explorers Club
46 East 70th Street
New York, NY 10021

ExxonMobil Education Foundation Grants   **2280**
The program addresses the use of mathematics specialists in grades K-3, the status of mathematics instruction at the college level, and the analysis of policy issues in mathematics education. Types of support include general operating grants, matching gifts, scholarship funds, and program grants. The foundation also funds the Elementary and Secondary School Improvement program, which is intended to help the educational community gain a better understanding of the changing demographics of the school-age population and the implications for changes in the schools; to help educators learn how to restructure schools to make them instructionally more effective with more students, particularly with educationally at-risk and minority students; and to improve teacher education to help teachers take better advantage of changing school structures. The foundation also sponsors a program addressing the reform of undergraduate science, technology, engineering, and mathematics education.
Requirements: The foundation makes grants only to tax-exempt organizations. A two-page letter of inquiry is required.
Restrictions: The foundation rarely contributes to endowments or make grants for construction or remodeling of facilities. Funds are not provided for equipment acquisition. Scholarships and grants to individuals are not awarded.
Contact: Program Contact, (972) 444-1106; fax: (972) 444-1405; email: contributions@exxonmobil.com
Internet: http://www.exxonmobileurope.com/corporate/Citizenship/Corp_Cit_EducationInter.asp
Sponsor: ExxonMobil Education Foundation
5959 Las Colinas Boulevard
Irving, TX 75039-2298

ExxonMobil Grants   **2281**
The corporation favors grants that relate to the conduct of U.S. business in general and the conduct of the petroleum and chemical industries in particular. The company is also concerned about social and economic conditions in corporate areas. Philanthropic giving is concentrated in the areas of the environment, public policy and public research; health; united appeals and civic and community-service organizations; minority- and women-oriented service organizations; arts, museums, and historical associations; and education. There are no application forms; submit a letter of inquiry that includes a brief history of the organization, the operating budget, a list of the board of directors and contributors, and a copy of the tax-exemption letter.
Requirements: Grants are made to tax-exempt organizations within principal company-operating areas.
Restrictions: Grants are not made to individuals for scholarships, fellowships, research, or travel. Organizations that are primarily religious in nature are ineligible.
Amount of Grant: $126 million total
Contact: Public Affairs, Contributions, Exxon Mobil Corporation, (972) 444-1103; fax: (972) 444-1405; email: contributions@exxonmobil.com
Internet: http://www.exxon.mobil.com
Sponsor: ExxonMobil
5959 Las Colinas Boulevard
Irving, TX 75039-2298

Ezra Jack Keats/Kerlan Collection Memorial Fellowship   **2282**
This fellowship provides a talented writer and/or illustrator of children's books with funds from the Children's Literature Research Collections at the University of Minnesota-Twin Cities Campus to use the Kerlan Collection for artistic development. Special consideration will be given to someone who would find it difficult to finance the visit to the Kerlan Collection. The fellow will receive transportation and a per diem allotment.
Requirements: Grants are awarded to individuals based on need. Candidates may be of any nationality.
Amount of Grant: $1,500
Date(s) Application is Due: Jun 1

Contact: Ezra Jack Keats/Kerlan Collection Memorial Fellowship Committee, (612) 624-4576; fax: (612) 625-5525; email: clrc@tc.umn.edu
Internet: http://special.lib.umn.edu/clrc/awards.php
Sponsor: University of Minnesota Children's Literature Research Collections
222 21st Avenue South, 113 Andersen Library
Minneapolis, MN 55455

F. Albert Cotton Award in Synthetic Inorganic Chemistry   2283

The award recognizes outstanding synthetic accomplishment in the field of inorganic chemistry and will be granted without regard to age or nationality. Creativity and imagination will be especially valued. The award consists of $5,000 and a certificate. Reasonable travel expenses to the meeting at which the award will be presented will be reimbursed.
Requirements: Any individual, except a member of the award committee, may submit one nomination or seconding letter for the award in any given year. The nominating documents consist of a letter of not more than 1000 words containing an evaluation of the nominee's accomplishments and a specific identification of the work to be recognized, a biographical sketch including date of birth, and a list of publications and patents authored by the nominee.
Restrictions: Self-nominations are not accepted.
Samples: Kenneth D. Karlin—ward recipient, $5,000 (2009); John D. Corbett—award recipient, $5,000 (2008); Christopher C. Cummins—award recipient, $5,000 (2007); Richard R. Schrock—award recipient, $5,000 (2006).
Amount of Grant: $5,000 and travel to award meeting
Date(s) Application is Due: Nov 1
Contact: Felicia Dixon, Awards Administrator; (800) 227-5558 or (202) 872-4408; fax: (202) 776-8008; email: f_dixon@acs.org or awards@acs.org
Internet: http://portal.acs.org/portal/acs/corg/content?_nfpb=true&_pageLabel=PP_ARTICLEMAIN&node_id=1319&content_id=CTP_004489&use_sec=true&sec_url_var=region1
Sponsor: American Chemical Society
1155 Sixteenth Street, NW
Washington, DC 20036

F.J. O'Neill Charitable Corporation Grants   2284

The corporation awards general operating grants to eligible Ohio nonprofit organizations in its areas of interest, including higher and secondary education, medical research, and Roman Catholic organizations and churches. There are no application deadlines or forms.
Requirements: Ohio nonprofit organizations serving the Cleveland area are eligible.
Geographic Focus: Ohio
Amount of Grant: $25,000-$100,000 average
Contact: Grants Administrator, (216) 464-2121
Sponsor: F.J. O'Neill Charitable Corporation
3550 Lander Road
Cleveland, OH 44124

F.W. Olin Foundation Construction Grant   2285

Each year, the foundation accepts applications for a grant to pay the entire cost of constructing and equipping a new academic building or library on a U.S. independent college or university campus with a full-time undergraduate enrollment of 500 or more; the cost will not be shared with any other donor. Submit the original application to the New York office and one copy to the Minneapolis office c/o William Horn, 1500 Forshay Tower, Minneapolis, MN 55402; (612) 341-2581; fax: (612) 341-3801.
Samples: Franklin W. Olin College of Engineering (Needham, MA)—for the endowment, $29.7 million (2005).
Contact: Dr. Richard Miller, (212) 832-0508; email: richard.miller@olin.edu
Internet: http://www.olin.edu/about_olin/olin_foundation.asp
Sponsor: F.W. Olin Foundation
780 Third Avenue, Suite 3403
New York, NY 10017-7090

FACHE Albert W. Dent Scholarships   2286

These scholarships have been established to provide financial aid and to increase the enrollment of minority students in health care management graduate programs and to encourage students, through structured, formalized study, to obtain positions in middle and upper levels of health care management. Applications are accepted between January 1 and March 31, and scholarship awards are announced in July. Interested persons should request application forms from the director of their graduate program in health care administration or from the foundation.
Requirements: To be eligible, an applicant must be a U.S. or Canadian student associate in good standing in the ACHE, a minority student who has been enrolled for full-time study for the upcoming fall term, which is his/her final year in a health care management graduate program. Financial need also must be demonstrated.
Restrictions: Previous recipients are ineligible.
Amount of Grant: $3500
Date(s) Application is Due: Mar 31
Contact: Membership, (312) 424-2800; fax: (312) 424-0023; email: membership1@ache.org
Internet: http://www.ache.org/Faculty_Students/dent_scholarship.cfm
Sponsor: Foundation of the American College of Healthcare Executives
1 N Franklin Street, Suite 1700
Chicago, IL 60606-3491

Fairbank Center An Wang Postdoctoral Fellowships in Chinese Studies   2287

The fellowships support well-designed research projects in Chinese studies. The award includes a stipend and research fund for 12 months. Priority will be given to candidates working in late imperial and 20th-century fields and candidates who have had no previous postdoctoral fellowships. The center supports projects at any stage from initial research to revision for publication. Application guidelines are available on the Web site.
Requirements: Candidate must have a PhD at the time of application and may not have held the PhD for more than five years.
Amount of Grant: $40,000 stipend
Date(s) Application is Due: Feb 1
Contact: Postdoctoral Fellowship Coordinator, (617) 495-4046; fax: (617)495-9976; email: fairbank@fas.harvard.edu
Internet: http://www.fas.harvard.edu/~fairbank/An_Wang.html
Sponsor: Harvard University
625 Massachusetts Avenue, 2nd Fl
Cambridge, MA 02139

Fairlawn Foundation Grants   2288

The Foundation is dedicated to supporting excellence and innovation in health care delivery, education and research in the Worcester area. The Foundation was established to: provide financial support to organizations, both public and private, that have specific, well-defined proposals to improve and/or expand the practice and delivery of medical and allied health care services in the Worcester area; and to support the education of Worcester area residents in health care related areas.
Requirements: Funding preference is given to institutions involved in higher education, medical or scientific research, hospitals and health care providers, and other organizations capable of making a long-term impact on health care through innovation or research. Applicants must be tax-exempt, nonprofit organizations as recognized by IRS code 501(c)3.
Restrictions: Except in unusual circumstances, the Foundation generally will not approve grants: to any program or individual in excess of $120,000; to programs that require more than five years to complete; for projects supported by other funds in which the Fairlawn Foundation contribution would comprise less than half of total funding; or for pre-existing projects as a minor participant.
Geographic Focus: Massachusetts
Amount of Grant: $120,000 maximum
Date(s) Application is Due: Mar 15; Sep 15
Contact: Lois Smith, Senior Program Officer; (508) 755-0980, ext. 107; email: lsmith@greaterworcester.org
Internet: http://www.greaterworcester.org/grants/Fairlawn.htm
Sponsor: Greater Worcester Community Foundation
370 Main Street, Suite 650
Worcester, MA 01608-1738

Families U.S.A Wellstone Fellowship for Social Justice   2289

The annual, one-year fellowship is designed to foster the advancement of social justice through participation in health care advocacy work that focuses on the unique challenges facing many communities of color. The fellowship, from August through July, will be based in the Families U.S.A office in Washington, DC, and will afford the fellow the opportunity to learn about Medicare, Medicaid, efforts to achieve universal coverage, and other important health policy issues. At the same time, the fellow will learn about conducting health care campaigns through communication and collaboration with a network of state grassroots advocates and organizations. The goals of the program are to address disparities in access to health care; inspire fellows to continue to work for social justice throughout their lives; and increase the number and racial and ethnic diversity of up-and-coming social justice advocates and leaders. Guidelines are available online.
Requirements: Candidates must demonstrate an interest in both health care and grassroots organizing and a commitment to contributing to social justice work following their year of hands-on experience as a fellow. Preference will be given to applicants who have experience with, or demonstrate a keen interest in, working with communities of color. While there is no bias in favor of any specific academic discipline, a college degree is preferred.
Amount of Grant: $35,000 stipend plus health benefits
Date(s) Application is Due: Jan 7
Contact: Melissa Rosenblatt, (202) 628-3030; fax: (202) 347-2417; email: wellstonefellowship@familiesusa.org
Internet: http://www.familiesusa.org/site/PageServer?pagename=Wellstone_Fellowship_About
Sponsor: Families U.S.A
1334 G Street NW
Washington, DC 20005

Fan Fox and Leslie R. Samuels Foundation Grants Program   2290

The foundation's areas of funding are the performing arts and healthcare. Healthcare funding supports patient-based and social service activities that directly help the elderly of New York City. The foundation supports performing arts organizations in the City of New York, principally, but not exclusively, in the borough of Manhattan. The foundation's primary mission is to support major performing arts institutions of national or international eminence. In addition to providing direct support, the foundation also assists presenting entities that have the requisite expertise, knowledge, and artistic judgment to present groups or individuals, new works, varied repertoire, and arts-in-education projects that will be contributions to the aesthetic and intellectual life of New York. Application guidelines are available on the Web site or upon request.

Requirements: The foundation funds organizations in the New York City area only. Only 501(c)3 tax-exempt organizations are invited to apply.

Restrictions: The foundation does not give grants to individuals or for scholarships, and does not support research, film, or video, nor does it fund education or social services. The foundation no longer actively solicits applications for support of arts-in-education programs at the primary and secondary level.

Geographic Focus: New York

Samples: Hebrew Home for the Aged at Riverdale (New York, NY)—for a comprehensive center dedicated to preventing and treating elder abuse, including a shelter for abused senior citizens, $200,000 over two years (2005); Jamaica Hospital Medical Ctr (New York, NY)—to establish an interdisciplinary family-medicine program for predominantly poor, minority, and elderly patients who do not have access to palliative-care services, $150,000 (2005); Metropolitan Jewish Health System (New York, NY)—to establish a palliative-medicine program based on home visits, $201,000 over two years (2005); Calvary Hospital (New York, NY)—for a program to increase access to comprehensive hospice and palliative care for nursing-home residents, $250,000 (2005).

Amount of Grant: $25,000-$250,000 average

Date(s) Application is Due: Mar 1; Jun 1: Sep 1; Dec 1

Contact: Joseph Mitchell, President , (212) 239-3030; fax: (212) 239-3039; email: info@samuels.org

Internet: http://www.samuels.org

Sponsor: Fan Fox and Leslie R. Samuels Foundation

350 Fifth Avenue, Suite 4301

New York, NY 10118

Fannie and John Hertz Foundation Graduate Fellowship Program   2291

The foundation supports graduate students working towards the PhD degree in the applied physical sciences. Fellows must have proposed fields of graduate study in engineering, applied physics, computer science, math, chemistry, or other area in the applied physical sciences; have excellent previous scholastic performance, typically including at least an A- grade average during the last two years of undergraduate work; be a U.S. citizen or a permanent resident of the United States; and be ready to make themselves, their skills, and abilities available for the common defense in the event of national emergency.

Requirements: Applicant must be a U.S. citizen and have a bachelor's degree and propose to complete a program of graduate study leading to an advanced degree, but not in a professional degree program or a joint PhD/professional degree program.

Restrictions: Joint-professional degree programs are not funded.

Amount of Grant: $28,000-$33,000

Date(s) Application is Due: Oct 28

Contact: Graduate Fellowship Program, (510) 373-1642; fax: (510) 373-6329; email: askhertz@aol.com

Internet: http://www.hertzfndn.org/awards.shtml

Sponsor: Fannie and John Hertz Foundation

2456 Research Drive

Livermore, CA 94550-3850

Fannie E. Rippel Foundation Grants   2292

The foundation aids, assists, funds, equips, and provides maintenance for corporations, institutions, associations, organizations, or societies maintained for the relief and care of aged women; provides funds for the building, equipping, and maintenance of hospitals; and provides funds for corporations, institutions, and other organizations existing for treatment of and/or research on heart disease or cancer. The foundation gives emphasis to the equipment and programmatic needs of major teaching medical centers and local rural hospitals, particularly where opportunities exist for leveraging the expertise or capabilities of the medical centers/rural hospitals. Programs should reach underserved rural and urban groups, advocate preventive care, present strategies to change behaviors of the people served, and promote humanistic medicine and mind-body-spirit connections in the healing process. Preference also is given to proposed projects where the benefits can be leveraged through challenge grants.

Requirements: Organizations, associations, institutions, and hospitals in the Northeast are eligible.

Restrictions: Grants are not awarded to individuals.

Geographic Focus: Northeast

Samples: Cold Spring Harbor Laboratory (NY)—for efforts to detect, treat, and prevent ovarian cancer, $350,000 (2005); Wake Forest U, Baptist Medical Ctr (NC)—to support health education among minorities in North Carolina through its Maya Angelou Research Center, $450,000 (2005).

Amount of Grant: $50,000-$300,000 average

Contact: Barbara Vanderkolk Gardner, Trustee, (908) 766-0404; fax: (908) 766- 0527; email: rippel@attglobal.net

Internet: http://fdncenter.org/grantmaker/rippel/index.html

Sponsor: Fannie E. Rippel Foundation

180 Mount Airy Road, Suite 200

Basking Ridge, NJ 07920

Fannie Mae Foundation Grants for Research   2293

The foundation's Innovation, Research, and Community Technology (IRCT) division focuses on housing finance topics, such as affordable mortgage products and services and alternative mortgage finance structures and systems; ways to expand homeownership opportunities for low- and moderate-income and minority households and communities; the production and ongoing maintenance and management of affordable housing; fair housing and fair lending practices; community development; housing demography;

and urban and regional policy. IRCT makes research grants and a limited number of grants to support research conferences. Research grants are awarded to individuals and organizations with the ability to make a significant contribution to the state of knowledge in housing economics and finance, housing and urban policy, housing demography, or community development issues related to the foundation's areas of focus. Grants to organizations also support research conferences disseminating new information in the areas of the foundation's research interests. Guidelines are available online.

Requirements: U.S. 501(c)3 nonprofit organizations are eligible.

Restrictions: Grants are not awarded to fund individuals; private foundations; organizations/projects/programs that do not fit within the foundation's funding priorities; organizations that channel grant funds to third parties; organizations whose dominant purpose is to influence legislation or participate/intervene in political campaigns on behalf of or against any candidate for public office; organizations/projects/programs for which the foundation is asked to serve as the sole funder; organizations that already have an active Fannie Mae Foundation grant; sectarian purposes (i.e., programs that promote religious doctrine or exclude participants on the basis of religion); endowment or capital campaigns; existing program or organizational deficits; or local affiliates of national organizations with which the foundation has a relationship.

Contact: Program Contact, (202) 274-8057; email: grants@fanniemaefoundation.org

Internet: http://www.fanniemaefoundation.org/grants/grants_research.shtml

Sponsor: Fannie Mae Foundation

4000 Wisconsin Avenue NW

Washington, DC 20016

Fannie Mae Foundation Kennedy School of Government Fellowships   2294

The foundation offers up to 35 fellowships annually as part of its partnership with the John F. Kennedy School of Government at Harvard University. The program enhances the management and decision-making skills of senior public and nonprofit officials committed to improving affordable housing opportunities in the United States. The program engages accomplished leaders experienced in managing housing and related community development programs. The curriculum focuses on organizational strategy, political management, policy development, management control and operations, and management of human resources. Fellowship funds cover the cost of the admission deposit, program tuition, and room and board for the session. Applicants must be able to attend the entire three-week session and are responsible for their own transportation and other incidental costs.

Requirements: Admissions decisions are made solely by the Kennedy School of Government. Applicants must satisfy the requirements for admission to be accepted by the Kennedy School to attend one of the state and local program sessions.

Date(s) Application is Due: Apr 1

Contact: Fellowship Administrator, (202) 274-8000; fax: (202) 274-8100

Internet: http://www.fanniemaefoundation.org/grants/kennedy_school.shtml

Sponsor: Fannie Mae Foundation

4000 Wisconsin Avenue NW, N Tower, Suite 1

Washington, DC 20016-2804

Fannie Mae James A. Johnson Fellowships   2295

Each year, the program selects up to six fellows. These fellows design and pursue development plans that can include research, travel, study, self-designed internships, and other activities that enhance their skills and knowledge. The program's chief goals are to recognize and reward individual dedication and contribution to the affordable housing and community development fields; increase leadership and professional development of the fellows; foster opportunities for new solutions to affordable housing and community development challenges; and inform the affordable housing and community development knowledge base and influence long-term strategies in the field. The fellowship provides each fellow with a grant and a stipend for travel and education-related expenses. The nonprofit organization with which each fellow is associated (through paid or volunteer employment) may receive a grant of up to $25,000 for transitional costs related to the temporary absence of the employee or volunteer. Fellows are selected through a two-step nomination process. Each January, the foundation requests nominations from national housing and community development organizations and and national neighborhood funders. Regional review committees then select a number of individuals as finalists.

Requirements: Nominees must meet the following criteria: a minimum of eight years in the affordable housing and community development field; current full-time (paid or volunteer) duties in the field; employment and affiliation with a neighborhood-based nonprofit or community-based nonprofit organization focused on affordable housing and community development; demonstrated leadership abilities; demonstrated ability to think creatively and design and implement innovative solutions to community development and affordable housing issues on a local, regional, or national level and in rural or urban settings; a continuing commitment to the field upon completion of the fellowship; a significant impact from the fellowship on the neighborhood/communities in which the nominee serves; no other venues or resources to pursue the opportunities that the fellowship will provide; and validation that timing is optimal for the nominee to utilize this experience, professionally and personally.

Samples: Lee Beaulace of Rural Opportunities (Rochester, NY), Manny Gauna of Chicanos por la Causa (Phoenix, AZ), Sharon Lee of the Low Income Housing Institute (Seattle, WA), and Jaimie Ross of 1000 Friends of Florida (Tallahassee, FL)—fellowship recipients, $90,000 each (2004).

Amount of Grant: $70,000 grant; $20,000 maximum for travel and education-related expenses

Date(s) Application is Due: Dec 31

Contact: Josie Gross, (202) 274-8031; fax: (202) 274-8100; email: jgross@fanniemaefoundation.org

Internet: http://www.fanniemaefoundation.org/grants/johnson.shtml
Sponsor: Fannie Mae Foundation
4000 Wisconsin Avenue NW, N Tower, Suite 1
Washington, DC 20016-2804

FAR Fund 2296

The Fund focuses its resources in three domains: improving and expanding services and systems for people on the autistic spectrum; preventing violence against youth; and preventing homelessness. The Fund encourages the creation of projects that operate across domains. The Fund also encourages projects that increase the participation and influence of consumers in shaping the service system that affect their lives. Particular interest is in projects that incorporate psychodynamic thinking in their program philosophy and service delivery. Proposal information is available on the website.

Requirements: The application proposal should be brief and describe the following: the organization where the project will be located; the problem; the project, including the project title; the expected impact; a plan for project evaluation, including outcome indicators; plans for the project beyond the grant period; project staff; project expense budget and income budget, and an organization expense and income budget; the percentage of the project budget that would be supported by the Fund; a monitorable work plan; and the name of the tax-exempt organization that would receive the grant.

Geographic Focus: New York
Amount of Grant: $25,000
Date(s) Application is Due: Feb 1; Jun 1; Oct 1
Contact: Kaajal Shah, Program Officer
Internet: http://www.farfund.org/
Sponsor: Fund for Social Change
135 East 15th Street
New York, NY 10003

Farm Foundation Extension Graduate Training Fellowships 2297

The foundation provides a number of extension fellowships for graduate training in extension administration and the social sciences, with emphasis on agricultural economics, rural sociology, psychology, political science, and agricultural geography, to improve the capability of active extension workers to function in managerial and supervisory roles in the extension service. The period of study may be one quarter, one semester, or nine months. The amount of each award will be determined individually on the basis of each applicant's qualifications and needs, ranging from tuition only to tuition and additional expenses including family subsistence. Extension fellowships are granted effective June 1 for the ensuing year.

Requirements: Active federal-state agricultural extension workers in the United States are eligible to apply. Priority is given to administrators and supervisory personnel.
Amount of Grant: $5500 maximum for nine months
Date(s) Application is Due: Mar 1
Contact: Walter Armbruster, President, (630) 571-9393; fax: (630) 571-9580; email: walt@farmfoundation.org
Internet: http://www.farmfoundation.org
Sponsor: Farm Foundation
1211 W 22nd Street, Suite 216
Oak Brook, IL 60523

Farmers Insurance Group of Companies Grants 2298

The corporate community relations program awards grants in the areas of education, public safety, arts and culture, civic improvement, and health and human services. Education giving focuses on literacy programs, mentoring programs, adopt-a-school programs, employee matching grants, and aid-to-education undergraduate scholarships. Public safety awards support tougher laws against drunk driving, drug/alcohol-free graduation night parties, neighborhood crime prevention, highway safety, and earthquake relief. Arts and culture funding supports children's programs and public television. Civic improvement focuses on recognizing exemplary youth, voter registration drives, adopt-a-highway programs, and community paint-a-thons. Health and human services giving supports March of Dimes, United Way, aid for families with cancer, aid to migrant farmworkers, and feeding the hungry. There are no application deadlines. Requests for contributions should be in the form of a letter outlining the purpose of the organization or program. The letter also should include the amount requested, its intended use, and a description of how Farmers' support will be recognized. Additional information should include a budget, annual report, proof of tax-exempt status, and a roster of the board of directors.

Requirements: 501(c)3 tax-exempt organizations are eligible.
Restrictions: Farmers does not make charitable contributions to individuals, political candidates, or religious groups or for sports events, advertising or raffle tickets, construction projects, or international programs.
Samples: John Wayne Institute for Cancer Treatment and Research (Santa Monica, CA)—for its cancer research and treatment programs, $50,000.; Los Angeles Urban League (Los Angeles, CA)—to upgrade the Information Training Ctr, which teaches computer skills to low-income, minority residents, $10,000.
Amount of Grant: $4.3 million total
Contact: Angela Easton, Director of Community Affairs and Sponsorship, (213) 932-3518
Internet: http://www.farmers.com/FarmComm/content/CC010153.jsp
Sponsor: Farmers Insurance Group of Companies
4680 Wilshire Boulevard
Los Angeles, CA 90010

FASSE Demonstration Project Grant 2299

This research award is to engage and support educators in strengthening and advocating social studies, where social studies is defined as the integrated study of the social sciences and humanities to promote civic competence. This annual award includes a commemorative gift, annual conference session for research presentation and publicity.

Requirements: Nomination *Requirements*: no self nominations; cover page; letter of nomination supporting, rationalizing the nomination; vitae or professional resume prepared by nominee; nominated study.
Date(s) Application is Due: May 1
Contact: Prema Parmar, Programs Assistant, (301) 588-1800; email: excellence@ncss.org
Internet: http://www.socialstudies.org/awards/research/exemplary/
Sponsor: Exemplary Research in Social Studies Award
8555 Sixteenth Street, Suite 500
Silver Spring, MD 20910

Father James B. Macelwane Annual Awards 2300

To stimulate interest in meteorology among college students through the encouragement of original student papers concerned with some phase of the atmospheric sciences, these awards, supported by Weather Corporation of America and administered by the American Meteorological Society, are given for three papers submitted that are judged to be the best in the previous 12-month period; also considered may be papers from the previous 12-month period if they did not receive any award during the previous competition. To be considered, the paper must be accompanied by a letter of application from the author and a letter from the department head or other faculty member of the major department confirming that the author was an undergraduate student at the time the paper was written and indicating the elements of the paper that represent original contributions by the student.

Requirements: All registered undergraduates in a college or university in the Americas are eligible to participate.
Restrictions: Participating students must be enrolled as undergraduates at the time the paper is written, and no more than two students from any one institution may enter papers in any one contest.
Amount of Grant: $300 first place, $200 second place, $100 third place
Contact: Donna Fernandez, Fellowship and Scholarship Coordinator (617) 227-2426 ext 246; fax: (617) 742-8718; email: dfernand@ametsoc.org; or Stephanie Armstrong, Director of Development, (617) 227-2426 ext 235; email: dfernand@ametsoc.org
Internet: http://www.ametsoc.org/amsstudentinfo/scholfeldocs
Sponsor: American Meteorological Society
45 Beacon Street
Boston, MA 02108-3693

Faye McBeath Foundation Grants 2301

Grants are made within the state of Wisconsin and principally to support projects or programs having primary focus on the welfare of the residents of the greater Milwaukee community including Milwaukee, Waukesha, Ozaukee and Washington counties. Generally, the Foundation supports programs for children, aging and elders, health, health education, and civic and governmental affairs. Capital grants are limited to projects with community-wide impact that reflect the program interests of the Foundation. There are no published deadlines for proposals. The common application form is available online.

Requirements: 501(c)3 nonprofit organizations in the metropolitan Milwaukee area are eligible. The Foundation requires its grant applicants to use the Milwaukee-area common application form when submitting a full proposal.
Restrictions: The Foundation does not consider or acknowledge general solicitation letters. Basic health sciences research is not funded. The Foundation does not award grants for annual fund drives, scholarships, support of individuals, or provide funds on an emergency basis.
Geographic Focus: Wisconsin
Samples: Penfield Children's Ctr (Milwaukee, WI)—to redesign and upgrade a playground for children with special needs, $15,000 (2005); Leaders Forum (Milwaukee, WI)—for its African American Fund-Development Institute, which trains executives and civic leaders for civic and nonprofit leadership positions, $15,000 over two years (2005); Friends of Boerner Botanical Garden (Milwaukee, WI)—for the Horticultural Therapy Program, which serves elderly people, $15,000 (2005); Latino Health Organization (Milwaukee, WI)—for its Health Awareness Program, a series of health-education presentations for children and families, $20,000 (2005).
Amount of Grant: $12.5 million total; $10,000-$50,000 average
Contact: Sarah Dean, Executive Director, (414) 272-2626; fax: (414) 272-6235; email: info@fayemcbeath.org
Internet: http://www.fayemcbeath.org
Sponsor: Faye McBeath Foundation
1020 N Broadway
Milwaukee, WI 53202

FCAR General Scholarships 2302

Scholarships and fellowships are for master's and doctoral studies and for postdoctorate research or improvement in the arts, to encourage the most gifted students residing in the province of Quebec and to emphasize special fields of study where a need is felt in the province. Salary for postdoctoral fellowships depends on area and quality of research. Visit the Web site for details on special requirements and restrictions. Annual deadline dates may vary; contact FCAR for exact dates.

Requirements: Applicants must be Canadian citizens or permanent residents who have been residents of Quebec for at least one year at the time of application. Candidates must have obtained a bachelor's or master's degree in the past five years for master's and doctoral fellowships, and have a cumulative B+ or equivalent grade. Grantees must reside in Quebec for the duration of the grant period.

Restrictions: Applicants for postdoctoral fellowships must not have been involved in research they intend to continue during the training period. The following will not be funded: supplies, publication, development, offices/laboratories, or elaborate projects.

Amount of Grant: $C15,000 maximum per year for master's scholarships; $C20,000 maximum per year for doctoral scholarships

Date(s) Application is Due: Oct 13

Contact: Philippe-Edwin Belanger, Program Administer, (418) 643-8560 ext 3447 or 3446 or (888) 653 6512; fax: (418) 643-1451; email: boursesm@fqrnt.gouv.qc.ca or bourses@fqrnt.gouv.qc.ca

Internet: http://www.fcar.qc.ca/nateq/bourses/regles/boGenerauxAng.htm

Sponsor: Formation de Chercheurs et L'Aide a la Recherche
140 Grande-Allee Est, Bureau 450
PQ G1R 5M8 Canada

FCAR Quebec-France Fellowships   2303

Fellowships are for master's and doctoral studies and for postdoctorate research or improvement in the arts, to encourage the most gifted students residing in the province of Quebec, and to emphasize special fields of study where a need is felt in the province. Salary for postdoctoral fellowships depends on area and quality of research. Visit the Web site for details on special requirements and restrictions. Five grants are awarded to masters students, and five are awarded to doctoral students to encourage advanced study in France. Annual deadline dates may vary; contact the program office for exact dates.

Requirements: Applicants must be Canadian citizens or permanent residents who have been residents of Quebec for at least one year at the time of application. Candidates must have obtained a bachelor's or master's degree in the past five years for master's and doctoral fellowships, and have a cumulative B+ or equivalent grade. Grantees must reside in Quebec for the duration of the grant period.

Restrictions: Applicants for postdoctoral fellowships must not have been involved in research they intend to continue during the training period. The following will not be funded: supplies, publication, development, offices/laboratories, or elaborate projects.

Date(s) Application is Due: Oct 15

Contact: Philippe-Edwin Belanger, Program Administrator, (418) 643-8560 ext 3447 or 3446 or (888) 653 6512; fax: (418) 643-1451; email: boursesm@fqrnt.gouv.qc.ca or bourses@fqrnt.gouv.qc.ca

Internet: http://www.fcar.qc.ca/nateq/bourses/index.htm

Sponsor: Formation de Chercheurs et L'Aide a la Recherche
140 Grande-Allee Est, Bureau 450
PQ G1R 5M8 Canada

FCD Changing Faces of America's Fellowships   2304

The program focuses on understanding the changing faces of America's children and seeks to support a new generation of scholars conducting research on the development of young immigrant children from birth to age 10. The goals of the program are to stimulate both fundamental and policy-relevant research in this area, and to support young investigators from a variety of behavioral or social sciences or in an allied professional field. Fellowship recipients are expected to produce a book or article(s) suitable for publication and to articulate how their research may potentially inform policies regarding young newcomer children. Approximately three or four fellowships are available for support of individual scholarship by junior faculty. The funds, to be used over a one to three years, cover each fellow's research expenses and salary and are paid directly to the recipient's academic institution. Complete guidelines are available online.

Requirements: Fellowships are available to scholars who have earned their PhD within the last 15 years. Applicants must hold a PhD or its equivalent in one of the behavior or social sciences or in an allied professional field (i.e., public policy, public health, education, social work, nursing). Applicants must hold a position as a full-time, tenure-track faculty member of a U.S. college or university.

Restrictions: The foundation does not consider requests for: capital campaigns, the purchase, construction or renovation of buildings, grants for projects outside the United States, the direct provision of preschool education or child care, or health care; or under the foundation's health focus, research, policy, or direct-service projects concerned with specific illnesses.

Samples: Krista Perreira, U of North Carolina (Chapel Hill, NC)—for fellowship support over three years, $150,000 (2003); Wen-Jui Han, Columbia U (New York, NY)—for fellowship support over three years, $150,000 (2003).

Amount of Grant: $150,000 maximum

Date(s) Application is Due: Nov 1

Contact: Changing Faces of America's Children, (212) 213-8337 ext 203; fax: (212) 213-5897; email: inforequest@ffcd.org

Internet: http://www.fcd-us.org/ourwork/y-index.html

Sponsor: Foundation for Child Development
145 E 32nd Street, 14th Fl
New York, NY 10016

FCD Child Development Grants   2305

The foundation awards grants to support programs for children, particularly the disadvantaged, and promote their well-being through basic and policy-relevant research about the factors that promote optimal development of children and adolescents; policy analysis, advocacy, services, and public education to enhance the discussion and adoption of social policies that support families in their important child-raising responsibilities; and leadership development activities linked to the programmatic focus of the foundation. Grants focus on the integration of research, policy, and advocacy in two areas: the availability of and access to early childhood education programs and health care for children. Most grants support research, but a small number of direct service grants are made for New York City-based projects that advance the foundation's research and policy analysis efforts. There are no application deadlines; submit a brief letter of inquiry. Full proposals are by invitation.

Requirements: Nonprofit organizations are eligible.

Restrictions: The foundation does not consider requests for scholarships or support for individuals, capital campaigns, building purchase or renovation, or equipment purchase. The foundation does not make grants outside the United States.

Samples: Barnard College (New York, NY)—to create an internship program that will encourage and prepare undergraduate students to pursue public-service careers that benefit children and their families, $30,000 (2005); Harvard Education Letter (Cambridge, MA)—to produce and distribute nine articles, including a special edition, on the most effective ways to educate children age three to eight, $120,750 (2005); U of Wisconsin at Madison (WI)—to prepare a working paper that describes and assesses issues related to the structure, financing, and implementation of Chicago Parent-Child Centers, $30,000 (2005); National Education Assoc (Washington, DC)—to survey existing contract language, develop a policy brief on the survey results, and recommend strategies for bargaining units of the NEA to use contract language to support improved educational opportunities for children in pre-kindergarten through third grade, $104,479 (2005).

Contact: Grants Administrator, (212) 213-8337; fax: (212) 213-5897; email: info@fcd-us.org

Internet: http://www.ffcd.org/ourwork/g-how.html

Sponsor: Foundation for Child Development
145 E 32nd Street, 14th Fl
New York, NY 10016

FCER Chiropractic Research Grants   2306

The grants support basic and clinical research and research program development related to chiropractic. Types of support include awards/prizes, challenge/matching grants, conferences and seminars, demonstration grants, fellowships, research grants, residencies, and research contracts. Fellowships are for a maximum of five years.

Amount of Grant: $10,000-$130,000 per year

Date(s) Application is Due: Mar 1; Oct 1

Contact: Dr. Anthony Rosner, Director of Research, (617) 734-3397; fax: (617) 734-0989; email: rosnerfcer@aol.com

Internet: http://www.fcer.org/html/Research/Grants/guidelines.htm#Instructions%20for%20Application

Sponsor: Foundation for Chiropractic Education and Research
1330 Beacon Street, Suite 315
Brookline, MA 02446-3202

FCER Research Grants and Fellowships   2307

The fellowships support research training primarily for graduate chiropractors enrolled in academic programs in basic sciences and clinical health-related areas. Grants are for basic research and related expenses.

Requirements: Priority for funding will be based on merit and the relevance/importance of the proposed research of chiropractic theory and practice, as well as the ability of the investigator and sponsoring institutions to bring the project to successful completion.

Samples: Robert Maust, Research Dimensions—for research on referral rates to chiropractors, $80,502.; Deborah Callahan, Foundation for Chiropractic Education and Research—for research on the treatment of back pain, $9961.

Amount of Grant: $10,000-$120,000 range

Contact: Dr. Anthony Rosner, Director of Research & Education, (617) 734-3397; fax: (617) 734-0989; email: rosnerfcer@aol.com

Internet: http://www.fcer.org/research.htm

Sponsor: Foundation for Chiropractic Education and Research
1330 Beacon Street, Suite 315
Brookline, MA 02446-3202

FDA Orphan Products Clinical Studies Grants   2308

The objective of the grants is to support clinical trials on the safety and effectiveness of products to treat a rare disease or condition. These include conditions for which no current therapy exists or the therapy that does exist is in need of improvement. The FDA provides grants to conduct clinical studies intended to provide data that will either result in or substantially contribute to approval of these medical products. All studies of new drugs and biological products must be conducted under the FDA's investigational new drug procedure. Studies of medical devices must be conducted under the investigational device exemption procedures.

Requirements: Public or private for-profit or nonprofit organizations may apply. Profit organizations must commit to excluding fees or profit in their request for support.

Amount of Grant: $100,000 maximum for new awards; $200,000 maximum for continuation awards

Contact: Rosemary Springer, Chief Grants Management Officer, (301) 827-7182; fax: (301) 827-7101; email: rspringe@oc.fda.gov

Internet: http://www.fda.gov

Sponsor: Food and Drug Administration
5630 Fishers Ln, HFA-520, Room 2129
Rockville, MD 20857

FDA Scientific Conferences Grants 2309

The FDA accepts applications for grants for scientific conferences held in the United States and Canada on topics relating directly to the agency's mission and funding priorities. This initiative is intended to support and encourage state food regulatory agencies to establish (or provide support of existing) regularly scheduled Food Safety Task Force meetings. These meetings should foster communication and cooperation within the state among state and local food safety regulatory agencies and is part of the President's Food Safety Initiative (FSI).

Requirements: Awards can be made to any public or private nonprofit university, college, hospital, laboratory, or other institution, including state and local units of government. Commercial and nonprofit organizations are also eligible.

Amount of Grant: $5000-$3 million; $225,970 average

Contact: Rosemary Springer, Chief, Grants and Agreements Management Branch, Division of Contracts and Assistance Management, (301) 827-7182; fax: (301) 827-7101; email: rspringe@oc.fda.gov

Internet: http://www.fda.gov

Sponsor: Food and Drug Administration
5600 Fishers Ln, HFA-520, Room 2129
Rockville, MD 20857

FDHN Bridging Grants 2310

Awards of $50,000 each are available to assist investigators were not awarded their first R01 or the first renewal of their R01 to continue their gastroenterology related research. Two awards will be made following each of the three annual NIH review cycles and availability of Summary Statements. The primary objective of the award is to provide interim support to AGA Members who have submitted grants to NIH that were approved on the basis of scientific merit, but received priority scores out of the funding range.

Requirements: Applicants must: be AGA members who are applying for their first R01 or their second R01 either as a competitive renewal or a new R01 (this includes investigators going from a K award to their first R01); hold an MD, PhD, or equivalent degree (e.g., MB, ChB, MBBS, DO); hold full-time faculty positions at North American universities or professional institutes at the time of application; and have submitted a proposal to NIH that has undergone the peer review process and was either approved and not funded or received high commendation but not funded.

Restrictions: Applicants cannot receive more than $75,000 of any additional extramural research support. Up to 25% of total amount the award may be used for salary support of the Principal Investigator. Indirect costs, including travel are not allowed.

Amount of Grant: $50,000

Date(s) Application is Due: Feb 14; Jun 14; Oct 14

Contact: Research Awards Manager, (301) 222-4012; fax: (301) 652-3890; email: awards@fdhn.org

Internet: http://www.fdhn.org/wmspage.cfm?parm1=119

Sponsor: Foundation for Digestive Health and Nutrition
4930 Del Ray Avenue
Bethesda, MD 20814

FDHN Centocor International Research Fellowship in 2311
Gastrointestinal Inflammation & Immunology

This award provides $50,000 to enable promising young investigators from outside the United States to spend a year at a U.S. institution engaged full-time in research related to the fundamental processes in gastrointestinal inflammation and immunology. The primary objectives of the award are: to provide the opportunity for young investigators from outside the U.S. to participate in basic research on inflammatory disease processes at prominent institutions in the U.S. (the award provides salary support for fulltime research); to initiate future international collaborative research efforts in inflammatory digestive diseases; and to address an important question in such a way as to provide a meaningful answer.

Requirements: Candidates must hold an MD or equivalent degree (eg, MB. ChB, MBBS). Applicants should currently be on the faculty of an academic institution outside the U.S. Applicants must be AGA members or be eligible for and submit an application for membership.

Restrictions: At the time of application, candidates are not allowed to hold any other similar research grant. These awards may not be renewed. No more than one application will be accepted from any institution in a given year.

Samples: Bo-In Lee, MD, PhD, Cleveland Clinic Foundation Cleveland, OH (2007); Seamus J. Murphy, BSc, PhD, MB, BCh, Mount Sinai School of Medicine New York, NY (2006).

Amount of Grant: $50,000

Date(s) Application is Due: Jan 14

Contact: Research Awards Manager, (301) 222-4012; fax: (301) 652-3890; email: awards@fdhn.org

Internet: http://www.fdhn.org/wmspage.cfm?parm1=105

Sponsor: Foundation for Digestive Health and Nutrition
4930 Del Ray Avenue
Bethesda, MD 20814

FDHN Designated Outcomes Award in Geriatric Gastroenterology 2312

Two awards in the amount of $35,000 for one year are available to support investigator-initiated outcomes research in geriatric gastroenterology. In general, outcomes studies examine clinical outcomes, patient satisfaction, quality of life, economic evaluation, quality of care, functional status, appropriateness of care, conformance of recommended/desirable standards of performance, or change in practice patterns. The objective of this award is to promote research by young investigators in the area of outcomes, broadly defined above, as it relates to geriatric gastroenterology. Please review the AGA Future Trends Committee Report: Effects of Aging of the Population on Gastroenterology Practice, Education and Research to learn more about research topics encouraged for study. Funds may be used for salary support of personnel and technicians only, supplies and/or equipment/services. Women and minorities are strongly encouraged to apply.

Requirements: Investigators must possess an MD, PhD or equivalent and must hold faculty positions at accredited North American academic institutions by the time of the start date of the award. MD applicants: No more than five years should elapse following the completion of your clinical training (GI fellowship or equivalent) and the start date of this award. Applicants must be AGA Members.

Restrictions: The award is intended for junior faculty; therefore, established investigators are not eligible. Candidates may not hold awards on a similar topic from other agencies. Indirect costs are not allowed.

Amount of Grant: $35,000

Date(s) Application is Due: Sep 5

Contact: Research Awards Manager, (301) 222-4012; fax: (301) 652-3890; email: awards@fdhn.org

Internet: http://www.fdhn.org/wmspage.cfm?parm1=234

Sponsor: Foundation for Digestive Health and Nutrition
4930 Del Ray Avenue
Bethesda, MD 20814

FDHN Designated Research Award in Geriatric Gastroenterology 2313

This award provides $75,000 per year for three years (total $225,000) for young investigators working toward independent careers in academic research related to geriatric gastroenterology. The overall objective is to enable young investigators to develop independent and productive research careers, with a focus on research related to geriatric gastroenterology, by ensuring that a major proportion of their time is protected for research. Non-recipient applicants for this award will be considered for the Research Scholar Awards. Applicants should review the AGA Future Trends Committee Report: Effects of Aging of the Population on Gastroenterology Practice, Education and Research to learn more about research topics encouraged for study. The award is not intended for fellows, but for young faculty who have demonstrated unusual promise and have some record of accomplishment in research.

Requirements: Candidates must hold an MD, PhD, or equivalent degree (e.g., MB, ChB, MBBS, DO). Applicants must hold full-time faculty positions at North American universities or professional institutes at the time award begins. Applicants must be Members of the AGA. # MD applicants: no more than five years shall have elapsed following the completion of your clinical training (GI fellowship or its equivalent) and the start date of this award. PhD applicants: no more than five years shall have elapsed from the completion of your postdoctoral training and the start date of this award. Candidates must devote at least 70 percent of their efforts to research related to geriatric gastroenterology.

Restrictions: Candidates should be in the beginning years of their careers; therefore, established investigators are not appropriate candidates.

Amount of Grant: $75,000 per year for three years

Date(s) Application is Due: Sep 15

Contact: Research Awards Manager, (301) 222-4012; fax: (301) 652-3890; email: awards@fdhn.org

Internet: http://www.fdhn.org/wmspage.cfm?parm1=224

Sponsor: Foundation for Digestive Health and Nutrition
4930 Del Ray Avenue
Bethesda, MD 20814

FDHN Designated Research Award in Research Related to Pancreatitis 2314

The overall objective of this Award is to enable young investigators to develop independent and productive research careers, with a focus on pancreatic disease, by ensuring that a major proportion of their time is protected for research. The Award provides $75,000 per year for three years (total $225,000) for young investigators working toward independent careers in academic research related to understanding, improving treatments or curing Pancreatitis.

Requirements: Candidates must hold an MD, PhD, or equivalent degree (e.g., MB, ChB, MBBS, DO). Applicants must hold full-time faculty positions at North American universities or professional institutes at the time award begins. Applicants must also be members of the AGA. The award is not intended for fellows, but for young faculty who have demonstrated unusual promise and have some record of accomplishment in research. Candidates should be in the beginning years of their careers; therefore, established investigators are not appropriate candidates. MD applicants: No more than five years shall have elapsed following the completion of your clinical training (GI fellowship or its equivalent) and the start date of this award. PhD applicants: No more than five years shall have elapsed from the completion of your postdoctoral training and the start date of this award.

Amount of Grant: $75,000 per year for three years

Date(s) Application is Due: Apr 17

Contact: Research Awards Manager, (301) 222-4012; fax: (301) 652-3890; email: awards@fdhn.org

Internet: http://www.fdhn.org/wmspage.cfm?parm1=100

Sponsor: Foundation for Digestive Health and Nutrition
4930 Del Ray Avenue
Bethesda, MD 20814

FDHN Fellow Abstract Prizes   2315

Two awards of $1000 each will be given to fellows who have submitted abstracts chosen to be presented during Digestive Disease Week¬Æ. Awards will be presented at a ceremony during DDW. The primary objective of this award is to stimulate interest in GI research careers through competition and recognition

Requirements: Qualified candidates are MD or PhD postdoctoral fellows who are trainee members the AGA. Women and minority investigators are strongly encouraged to apply. Applicants must: be sponsored by an AGA member; and be the first author of an abstract accepted for presentation at DDW and provide evidence of abstract acceptance. Individuals with faculty appointments are not eligible. Applicants may only submit one abstract for consideration.

Restrictions: Fellows who have been awarded must present the abstract. No substitute presenters are allowed. Abstracts must have been selected for presentation at DDW. A letter of recommendation from the sponsor is required.

Amount of Grant: $1,000

Date(s) Application is Due: Mar 21

Contact: Research Awards Manager, (301) 222-4012; fax: (301) 652-3890; email: awards@fdhn.org

Internet: http://www.fdhn.org/wmspage.cfm?parm1=124

Sponsor: Foundation for Digestive Health and Nutrition

4930 Del Ray Avenue

Bethesda, MD 20814

FDHN Fellowship to Faculty Transition Awards   2316

This award provides $40,000 per year for two years for current trainees in gastroenterology related fields so they may gain additional research training in gastrointestinal, liver function or related diseases. The objective is to prepare physicians for independent research careers in digestive diseases. The award provides salary support for additional full-time research training in basic science to acquire modern laboratory skills. The additional two years of research training provided by this award should broaden and expand the scope of investigative tools available to the recipient, generally in basic disciplines such as cell or molecular biology or immunology.

Requirements: Applicants must be MDs or MD/PhDs currently in a gastroenterology-related fellowship at an accredited North American institution, committed to academic careers. They will have completed two years of research training at the start of this award. Applicants must be AGA Trainee Members and be sponsored by an AGA Member.

Restrictions: Individuals who hold a PhD are ineligible. Although the institution may supplement the award, the applicant may not concurrently hold a similar training award or grant from another organization, such as the NIH, ALF, CCFA, or the Glaxo Institute of Digestive Health.

Samples: John Chang, MD, Fellow in Gastroenterology University of Pennsylvania, Philadelphia, PA (2007); Lee Peng, MD, PhD, Massachusetts University, Boston, MA (2007); Carl B. Rountree, Jr., MD, Children, Hospital, Los Angeles, CA (2006).

Amount of Grant: $40,000 per year for two years

Date(s) Application is Due: Sep 5

Contact: Research Awards Manager, (301) 222-4012; fax: (301) 652-3890; email: awards@fdhn.org

Internet: http://www.fdhn.org/wmspage.cfm?parm1=102

Sponsor: Foundation for Digestive Health and Nutrition

4930 Del Ray Avenue

Bethesda, MD 20814

**FDHN Funderburg Research Scholar Award in Gastric Biology Related   2317
to Cancer**

This grant of $25,000 per year for two years (total $50,000) is awarded to an established investigator working on novel approaches in gastric cancer, including the fields of gastric mucosal regeneration and regulation of cell growth (not as they relate to peptic ulcer disease inflammation (including Helicobacter pylori) as precancerous lesions; genetics of gastric oncogenes in gastric epithelial malignancies; epidemiology of gastric cancer; etiology of malignancies; or clinical research in the diagnosis or treatment of gastric carcinoma. The primary objective of the award is to support an active, established investigator in the field of gastric biology who enhances the fundamental understanding of gastric cancer pathobiology in order to ultimately develop a cure for the disease.

Requirements: Applicants must hold faculty positions at accredited North American institutions and must have established themselves as independent investigators in the field of gastric biology. Women and minority investigators are strongly encouraged to apply. Applicants must be Members of the AGA.

Samples: Xiaolu Yang, PhD, Associate Professor, Department of Cancer Biology, Associate Investigator, Abramson Family Cancer Research Institute, University of Pennsylvania School of Medicine, PA (2007); Steven Itzkowitz, MD, FACP, FACG The Dr. Burrill B. Crohn Professor of Medicine, Director, Gastroenterology Fellowship Program, Associate Director, Division of Gastroenterology, Mount Sinai School of Medicine, New York City, NY (2006); JeanMarie Houghton, PhD, University of Massachusetts Medical School, Worchester, MA (2005).

Amount of Grant: $25,000 per year for two years

Date(s) Application is Due: Sep 5

Contact: Research Awards Manager, (301) 222-4012; fax: (301) 652-3890; email: awards@fdhn.org

Internet: http://www.fdhn.org/wmspage.cfm?parm1=90

Sponsor: Foundation for Digestive Health and Nutrition

4930 Del Ray Avenue

Bethesda, MD 20814

FDHN Graduate Student Awards   2318

Two awards of $20,000 a year for two years are offered to fund graduate students undertaking research in the biology and epidemiology of diseases of the gastrointestinal tract, liver or pancreas. The award includes $18,000 for stipend and $2,000 to be used towards fringe benefits such as medical insurance and travel to a national meeting. The primary objective of the award is to provide a salary stipend for graduate students performing doctoral research related to the gastrointestinal tract, liver or pancreas.

Requirements: Applicants should have completed at least one year and no more than three years of training towards the doctoral degree and have selected and confirmed the laboratory or department in which they will conduct doctoral research. Research is to be conducted at an accredited academic institution within North America and the research advisor must be a Member of the AGA. Applicants should be U.S. citizens, permanent residents or overseas students who have a current visa to pursue education within North America.

Restrictions: Indirect costs, excluding travel, are not allowed. Tuition costs are expected to be covered by the institution or host department and laboratory.

Amount of Grant: $20,000 per year for two years

Date(s) Application is Due: Mar 14

Contact: Research Awards Manager, (301) 222-4012; fax: (301) 652-3890; email: awards@fdhn.org

Internet: http://www.fdhn.org/wmspage.cfm?parm1=154

Sponsor: Foundation for Digestive Health and Nutrition

4930 Del Ray Avenue

Bethesda, MD 20814

FDHN Jon I. Isenberg International Research Scholar Award   2319

This award provides a total of $50,000 for one year of study, $25,000 from the AGA and $25,000 in matching funds from the applicant, national GI society. The award provides for non-U.S. citizen young investigators to spend one year performing GI-related research at an American institution under the tutelage of an AGA member. Four awards will be given annually. The primary objective of this award is to promote international scholarship, increase AGA, involvement within the international GI community and foster international collaboration in training GI investigators.

Requirements: An eligible candidate must be nominated by his or her national GI Society. The national GI society must match the $25,000 contribution from the AGA. Before submitting an application, a candidate must identify a sponsoring American institution and a research preceptor who agrees in writing to supervise his or her training and research. Qualified candidates must possess a doctoral degree, either an MD, or equivalent degree, and/or a PhD degree. In keeping with the intent to support the career development of young investigators, the candidate must be within five (5) years of completing GI training or, if a PhD, within five (5) years of the receipt of the degree, at the time of initiation of the award. A documented parental leave of absence will not be counted towards the five (5) years of eligibility.

Samples: Maria Cristina, Almansa Menchero Hospital Clinico Universitario San Carlos, Madrid, Spain, to the Mayo Clinic School of Medicine, Jacksonville, FL (2006); Koji Nozaki, MD, PhD, Department of Gastrointestinal Surgery, The University of Tokyo, Tokyo, Japan (2006); Revital Kariv, MD, Case Western Reserve University, Beachwood, OH (2005).

Amount of Grant: $50,000

Date(s) Application is Due: Mar 1

Contact: Research Awards Manager, (301) 222-4012; fax: (301) 652-3890; email: awards@fdhn.org

Internet: http://www.fdhn.org/wmspage.cfm?parm1=101

Sponsor: Foundation for Digestive Health and Nutrition

4930 Del Ray Avenue

Bethesda, MD 20814

FDHN June & Donald O. Castell MD, Esophageal Clinical Research Award   2320

One award of $35,000 is made annually to provide research and/or salary support for junior faculty involved in clinical research in esophageal diseases. This award is funded by the June and Donald O. Castell, MD, Gastroenterology Research and Education Trust. The primary objective of the award is to support investigators who have demonstrated high potential to develop independent, productive research careers.

Requirements: Candidates must hold a MD or PhD or equivalent. Applicants must hold a full-time faculty position at a North American universities or professional institute. Applicants must be members of the AGA. The recipient must be at or below the level of assistant professor, and his/her initial appointment to the faculty position must have been within seven (7) years of the time of application. This award is not intended for fellows, but for junior faculty who have demonstrated unusual promise; have some record of accomplishment in research; and have established independent research programs at the time of the award. Candidates must devote at least 50 percent of their efforts to research related to esophageal function or diseases.

Restrictions: If an award recipient receives notification of another award with overlapping scientific objectives, prior to the start date of an AGA or FDHN award, the applicant must choose between the two awards.

Samples: John Pandolfino, MD, Northwestern University, Chicago, IL (2007); Marcelo Vela, MD, Medical University of South Carolina, Charleston, SC (2006); Braden Kuo, MD, Massachusetts General Hospital Boston, MA (2005).

Amount of Grant: $35,000

Date(s) Application is Due: Jan 14

Contact: Research Awards Manager, (301) 222-4012; fax: (301) 652-3890; email: awards@fdhn.org

Internet: http://www.fdhn.org/wmspage.cfm?parm1=104

Sponsor: Foundation for Digestive Health and Nutrition
4930 Del Ray Avenue
Bethesda, MD 20814

FDHN Moti L. & Kamla Rustgi International Travel Awards 2321

This program awards grants to young basic, translational and clinical investigators to support their travel and related expenses to attend Digestive Disease Week¬Æ. Two awards of $500 each will be given to selected individuals residing outside North America. The primary objective of the award is to enable young investigators outside of North American (U.S. or Canada) institutions to attend Digestive Disease Week and encourage international trainees to become more involved in digestive tract, pancreatic and liver disease research.

Requirements: Candidates must: be MD or PhD or MD PhD postdoctoral fellows who are international trainee members of the AGA; be sponsored by an International member of AGA; be 35 years of age or younger at the time of the meeting; be fluent in English; and have a sufficient number of scientific papers (impact factor) published and/or poster presentations.

Amount of Grant: $500
Date(s) Application is Due: Mar 21
Contact: Research Awards Manager, (301) 222-4012; fax: (301) 652-3890; email: awards@fdhn.org
Internet: http://www.fdhn.org/wmspage.cfm?parm1=126
Sponsor: Foundation for Digestive Health and Nutrition
4930 Del Ray Avenue
Bethesda, MD 20814

FDHN Non-Career Research Awards 2322

This award provides approximately $10,000 (per symposia) for travel support for young investigators and selected established investigators to participate in symposia on gastrointestinalrelated topics. The primary objective of the award is to foster interactions and enhance the exchange of information between clinical and basic science investigators, and established and junior investigators working in gastrointestinal research. Eligibility women and minority organizers are strongly encouraged to apply. Travel support may be provided for: junior investigators who are within 5 years of completing their clinical training in GI or from completion of their PhD doctoral thesis and are at or below the rank of assistant professor; and up to two established investigators who are invited speakers. Biosketches of these individuals and a letter addressing their contribution to the scientific content and atmosphere of the meeting should be submitted with the application.

Requirements: Within 60 days of the conclusion of the meeting, a one-page summary of the highlights of the meeting and an accounting of the use of funds must be submitted. The names of participants and speakers in the symposium, addresses, dates of birth and academic ranks of the attending scientists and their itemized expenses, must be sent to the Foundation. Any unexpended funds must be returned.

Restrictions: Indirect costs are not allowed.
Amount of Grant: $10,000 average
Date(s) Application is Due: Feb 1; May 1; Oct 1
Contact: Research Awards Manager, (301) 222-4012; fax: (301) 652-3890; email: awards@fdhn.org
Internet: http://www.fdhn.org/wmspage.cfm?parm1=123
Sponsor: Foundation for Digestive Health and Nutrition
4930 Del Ray Avenue
Bethesda, MD 20814

FDHN Non-Career Research Grants 2323

A research initiative grant of $25,000 for one year is offered to investigators to support pilot research projects in gastroenterology- or hepatology-related areas. The primary objective of the award is to provide non-salary funds for new investigators to help them establish their research careers or to support pilot projects that represent new research directions for established investigators. The intent is to stimulate research in gastroenterology- or hepatology-related areas by permitting investigators to obtain new data that can ultimately provide the basis for subsequent grant applications of more substantial funding and duration. Women and minorities are strongly encouraged to apply.

Requirements: Investigators must possess an MD or PhD degree or equivalent and must hold faculty positions at accredited North American institutions. Candidates may not hold awards on a similar topic from other agencies. Applicants must be AGA Members. Applicants for this award may not simultaneously apply for the AGA/Miles and Shirley Fiterman Foundation Basic Research Award or the AGA June and Donald O. Castell, MD, Esophageal Clinical Research Award.

Restrictions: If an award recipient receives notification of another award with overlapping scientfic objectives, prior to the start date of an AGA or FDHN award, the applicant must choose between the two awards.

Amount of Grant: $25,000
Date(s) Application is Due: Jan 14
Contact: Research Awards Manager, (301) 222-4012; fax: (301) 652-3890; email: awards@fdhn.org
Internet: http://www.fdhn.org/wmspage.cfm?parm1=121
Sponsor: Foundation for Digestive Health and Nutrition
4930 Del Ray Avenue
Bethesda, MD 20814

FDHN Research Scholar Awards 2324

These awards provide salary support for young investigators working in any area of gastrointestinal, liver function, or related diseases. The primary intent of the program is to support physician-investigators who have a high potential to develop independent, productive research careers in gastroenterology and hepatology. Candidates must devote at least 70 percent of their effort to research related to the gastrointestinal tract or liver. There must be a strong commitment from the candidate's division and department to support the candidate by protecting time for research and providing adequate laboratory space and facilities.

Requirements: Applicants must hold full-time faculty positions at North American universities or professional institutes. Nonphysician candidates with a PhD will also be considered. Candidates should be early in their research careers and commonly will have recently completed their fellowship training.

Restrictions: Indirect costs are not allowed. Candidates who have been at the assistant professor level or equivalent for more than five years are not eligible. Nor can applicants hold, or have held, an RO1, R29, K11, K08, or VA research award or any award with similar objectives from nonfederal sources (such as ALF, CCFA, or Glaxo Institute of Digestive Health). However, awards or grants obtained after receipt of this award need not be surrendered.

Samples: Michael Beyak, BSc, MD, Queen's University at Kingston—Ion channels underlying the actions of the satiety hormone CCK on gastrointestinal vagal afferents (2007); Sean Koppe, MD, Northwestern University Feinberg School of Medicine—The Role of Fatty Acids in the Pathogenesis of Non-Alcoholic Steatohepatitis (NASH) (2007).

Amount of Grant: $75,000 per year for three years
Date(s) Application is Due: Sep 5
Contact: Research Awards Manager, (301) 222-4012; fax: (301) 652-3890; email: awards@fdhn.org
Internet: http://www.fdhn.org/wmspage.cfm?parm1=103
Sponsor: Foundation for Digestive Health and Nutrition
4930 Del Ray Avenue
Bethesda, MD 20814

FDHN Student Research Fellowships 2325

This program offers financial support for students to spend a minimum of 10 weeks performing research in digestive diseases or nutrition and is intended to stimulate interest in research careers in these areas. The work may take place at any time during the year. Up to 20 fellowships will be available for full-time research with a preceptor, who must be a faculty member who directs a research project in a gastroenterology-related area at an accredited North American institution. A complete financial statement and scientific progress report are required upon completion of the program. The AGA has recognized the need to attract and encourage minority individuals to enter and pursue gastroenterology research careers. In response to this concern, seven of the student research fellowship awards will be reserved for underrepresented minority students. For the purpose of this award, minorities have been defined as African American, Hispanic, Native American/Alaskan Native, and Pacific Islander.

Requirements: Candidates may be high school, undergraduate, medical, or graduate students (not yet engaged in thesis research) in accredited North American institutions. Women and minority students are strongly encouraged to apply. The preceptor must be a faculty member who directs a research project in a gastroenterology-related area at an accredited North American institution.

Restrictions: Candidates may not hold similar salary support from other agencies: e.g., American Liver Foundation, Crohn's and Colitis Foundation.

Samples: Joanna M. Peloquin, Mayo Medical School (2006); Alexandra Berger, University of California, Los Angeles, CA (2006); Jeremy Thoms Hetzel, University of Chicago, Chicago, IL (2006).

Amount of Grant: $2,000-$3,000
Date(s) Application is Due: Mar 5
Contact: Research Awards Manager, (301) 222-4012; fax: (301) 652-3890; email: awards@fdhn.org
Internet: http://www.fdhn.org/wmspage.cfm?parm1=115
Sponsor: Foundation for Digestive Health and Nutrition
4930 Del Ray Avenue
Bethesda, MD 20814

FDHN TAP Endowed Designated Research Award in Acid-Related Diseases 2326

This award provides $75,000 per year for three years (total $225,000) for young investigators working toward independent careers in acid-related diseases. The overall objective is to enable young investigators to develop independent and productive careers in acid-related research by ensuring that a major proportion of their time is protected for research.

Requirements: Candidates must hold an MD, PhD, or equivalent degree (e.g., MB, ChB, MBBS, DO). Applicants must hold full-time faculty positions at North American universities or professional institutes at the time award begins. Applicants must be Members of the AGA. The award is not intended for fellows, but for young faculty who have demonstrated unusual promise and have some record of accomplishment in research. Candidates should be in the beginning years of their careers; therefore, established investigators are not appropriate candidates. Candidates must devote at least 70 percent of their efforts to research related to geriatric gastroenterolgy.

Restrictions: MD applicants: no more than five years shall have elapsed following the completion of your clinical training (GI fellowship or its equivalent) and the start date of this award. PhD applicants: no more than five years shall have elapsed from the completion of your postdoctoral training and the start date of this award.

Amount of Grant: $75,000 per year for three years
Date(s) Application is Due: Sep 5
Contact: Research Awards Manager, (301) 222-4012; fax: (301) 652-3890; email: awards@fdhn.org
Internet: http://www.fdhn.org/wmspage.cfm?parm1=132
Sponsor: Foundation for Digestive Health and Nutrition
4930 Del Ray Avenue
Bethesda, MD 20814

FDHN Translational Research Awards 2327
One award of $100,000 per year for two years will be made annually to support translational research in gastroenterology and/or hepatology. Translational research will be defined as the process of applying ideas, insights and discovery generated through basic science research to the diagnosis, treatment or prevention of human disease. The primary objective of the award is to enhance interaction between researchers with basic science and clinical backgrounds with the goal of accelerating the pace of discovery that is directly applicable to patient care. The creation of teams including both a Ph.D. and an M.D. researcher is particularly encouraged.
Requirements: This award must be applied for jointly by a team of researchers (typically consisting of two members), including at least one researcher with significant training and experience in a basic science discipline (including areas outside the traditional biomedical sciences, such as physical sciences and engineering) and at least one researcher who is qualified to provide direct clinical care to patients. Junior investigators in either or both categories are particularly encouraged to apply. Applicants must be members of the AGA. Research must be conducted at an accredited North American institution.
Restrictions: Candidates may not hold support for the same project from another agency. If a proposed award recipient receives notification of another award with overlapping scientific objectives prior to the start date of an AGA or FDHN award, the applicant must choose only one of the awards to accept.
Amount of Grant: $100,000 per year for two years
Date(s) Application is Due: Mar 14
Contact: Research Awards Manager, (301) 222-4012; fax: (301) 652-3890; email: awards@fdhn.org
Internet: http://www.fdhn.org/wmspage.cfm?parm1=148
Sponsor: Foundation for Digestive Health and Nutrition
4930 Del Ray Avenue
Bethesda, MD 20814

Fellowships at American Schools of Oriental Research—Amman 2328
ASOR offers seven different award opportunities for study in humanistic disciplines of the Middle East from prehistoric times through the 19th century. Islamic studies are especially encouraged. (1) The Kress Fellowship in the Art and Archaeology of Jordan: One or more three- to six-month fellowships for predoctoral students completing dissertation research in an art historical topic are awarded. (History of art is defined to include art history, archaeology, architectural history, and in some cases classical studies.) (2) CAORC Fellowship: Six or more two- to six-month fellowships for predoctoral students and postdoctoral scholars are awarded. Fields of study include all areas of the humanities and the natural and social sciences. Topics should contribute to scholarship in Near Eastern studies. (3) CAORC Post-Graduate Fellowship: Two or more two- to six-month fellowships for postdoctoral scholars pursuing research or publication projects in the natural and social sciences, humanities, and associated disciplines relating to the Near East are awarded. (4) National Endowment for the Humanities (NEH) Fellowship: One four-month fellowship is awarded for scholars who have a PhD or have completed their professional training. Fields of research include modern and classical languages, linguistics, literature, history, jurisprudence, philosophy, archaeology, comparative religion, ethics, and the history, criticism, and theory of the arts. Social and political scientists are encouraged to apply. (5) Jennifer C. Groot Fellowship: Three awards to help support beginners in archaeological fieldwork, who have been accepted as staff members on archaeological projects with ASOR/CAP affiliation in Jordan, are given. (6) Harrell Family Fellowship: One award to help support a graduate student for participation in an ACOR-approved archaeological research project, which has passed an academic review process, is given. Senior project staff whose expenses are being borne largely by the project are ineligible. (7) Pierre and Patricia Bikai Fellowship: One or more awards for one or two months of residency at ACOR during the period of June 1 through May 31 are given. Visit the Web site for further details.
Requirements: Requirements corresponding to the seven awards listed in the description are as follows: (1) Kress applicants must be PhD candidates and U.S. citizens or foreign nationals who have matriculated at U.S. institutions. (2 & 3) CAORC applicants must be U.S. citizens. (4) NEH applicants must be U.S. citizens or foreign nationals living in the United States three years immediately preceding the application deadline. (5) Groot Fellowships applicants must be undergraduate or graduate students who are U.S. or Canadian citizens. (6) Harrell Fellowship applicants must be enrolled graduate students of any nationality. (7) Bikai Fellowship applicants must be graduate students of any nationality participating in an archaeological project or a research project in Jordan.
Amount of Grant: $1500-$25,000
Date(s) Application is Due: Feb 1
Contact: Dr. Britt Hartenberger, (617) 353-6571; fax: (617) 353-6575; email: acor@bu.edu
Internet: http://www.bu.edu/acor/fellowsh.htm
Sponsor: American Schools of Oriental Research
656 Beacon Street, 5th Fl
Boston, MA 02215

Fellowships for Research in Egypt 2329
Fellowships support graduate and postgraduate research in Egypt for single continuous periods. Areas of research include archaeology, architecture, art, economics, egyptology, history, humanistic social sciences, humanities, Islamic studies, literature, political science, and religious studies. Research can cover all periods from ancient times to the present. Fellowships are granted for not less than three months nor more than 12 months. Award recipients must be physically present in Egypt to receive stipends; dependents' allowances will be granted for up to two dependents who are physically present in Egypt. Up to 15 fellowships are awarded annually. Also available from ARCE is the Kress Predoctoral Fellowship in Egyptian Art and Architecture that awards funding for up to eight months of research in Egypt, and special fellowships for Egyptian graduate students enrolled in American universities at the ABD stage who seek maintenance support for research in Egypt on topics relating to Egyptian development. Each fellow receives one round-trip ticket.
Requirements: Applicants must be doctoral candidates or postdoctoral scholars. Candidates must complete application forms (available from ARCE) and submit three letters of recommendation. Predoctoral students must also submit transcripts.
Restrictions: Awards are generally limited to the fields of humanities, fine arts, and social sciences; admission to candidacy is a prerequisite for student fellowships; fellows must devote full time to their research and may not accept outside teaching assignments or employment without written consent of program officers. Doctoral candidates must be either U.S. or Egyptian citizens. Postdoctoral scholars must be either U.S. citizens or have been teaching at an American or Canadian university for a minimum of three years.
Amount of Grant: $1530 per month for students; $3019 per month for full professors
Date(s) Application is Due: Jan 7
Contact: Center for Arabic Study Abroad, (404) 727-2575; fax: (404) 727-6187; email: casa@emory.edu
Internet: http://www.casa.emory.edu/programs
Sponsor: American Research Center in Egypt
1385 Oxford Road
Atlanta, GA 30322

Fellowships in Infectious Diseases 2330
The program encourages and assists young qualified physicians to become specialists and investigators in the field of infectious diseases. The fellowship will be awarded to individuals who do not or will not have training or research grant support during the period of this grant and who are seeking support for one to three years of postdoctoral fellowship experience. Preference will be given to those applying for the third year of such support. The applicant must be sponsored by a university affiliated medical center. A letter from the chair of the infectious diseases department expressing a willingness to assume responsibility for training the applicant must accompany the application. In addition, each application must be accompanied by a letter from the host laboratory and signed by the laboratory director or appropriate department chair attesting willingness to accept the applicant and provide guidance, research space, and necessary research equipment. One thousand dollars of the award amount may be used for travel and supplies.
Requirements: Applicants must be U.S. physicians who have satisfactorily completed three or more years of postgraduate medical training (internal medicine, surgery, pediatrics, epidemiology, etc.).
Restrictions: The fellowship will not be awarded if the applicant has received or will receive a major fellowship, research grant, or traineeship in excess of the amount of this award from the federal government or another foundation.
Amount of Grant: $25,000 stipend
Date(s) Application is Due: Feb 15
Contact: Senior Executive Director, (301) 656-0003; fax: (301) 907-0878; email: info@nfid.org
Internet: http://www.nfid.org/fellow
Sponsor: National Foundation for Infectious Diseases
4733 Bethesda Avenue, Suite 750
Bethesda, MD 20814-5278

Fermilab Postdoctoral Fellowship Programs 2331
Fermilab, a Department of Energy laboratory, offers postdoctoral fellowships for work in experimental physics. The objective of the programs is to broaden or intensify the research experience of recent PhDs through involvement in experimental or theoretical research. Recent PhDs receive research associate appointments generally for two or more years.
Contact: Dr. Michael Albrow, Head-Experimental Physics Projects Department, Particle Physics Division, (630) 840-3922; email: albrow@fnal.gov
Internet: http://www.fnal.gov
Sponsor: Fermi National Accelerator Laboratory
P.O. Box 500, MS 116
Batavia, IL 60510-0500

Fetzer Institute Research Grants 2332
The Foundation uses the bulk of its income to actively run its own programs or services. The Institute's mission is to foster awareness of the power of love and forgiveness through research, education, and service programs. Current focus includes scientific research on altruistic love, compassionate love, and forgiveness. Individuals and organizations invited by the Institute to participate in the creation and implementation of its programs can receive funding. Requests are posted on the Foundation's website. The Foundation does not accept unsolicited proposals.

Contact: Heidi Ihrig, (269) 375-2000; email: educationrfp@fetzer.org
Internet: http://www.fetzer.org/Programs.aspx?PageID=Programs&NavID=1
Sponsor: Fetzer Institute
9292 W. KL Avenue
Kalamazoo, MI 49009-9398

FIC International Cooperative Biodiversity Groups Grants 2333

Funding for this program has been provided by six components of the National Institutes of Health (NIH), the Biological Sciences Directorate of the National Science Foundation (NSF) and the Foreign Agriculture Service of the U.S.DA. The cooperating NIH components are the Fogarty International Center (FIC), National Cancer Institute (NCI), National Institute of Allergy and Infectious Diseases (NIAID), National Institute of Mental Health (NIMH), National Institute on Drug Abuse (NIDA) and the National Heart, Lung, and Blood Institute (NHLBI). The purpose of these groups will be to address the interdependent issues of biodiversity conservation, sustained economic growth, and human health in terms of drug discovery for cancer, infectious diseases including AIDS, cardiovascular diseases, mental disorders, and diseases of primary concern to developing countries. A group, under a single group leader (principal investigator), is expected to be a consortium of associate programs working together to form a multidisciplinary and/or multi-institutional team from academic, nonprofit, and/or commercial organizations. At least one of the group's associate programs must be located in a developing country. It is strongly advised that applicants contact program staff early in the planning process to discuss prospective applications and to obtain supplemental clarifying information and instructions. RFA-TW-04-004
Requirements: Public and private nonprofit institutions, governments and their agencies, and foreign institutions are eligible. Applicant institutions must be in the United States or in a participating developing country. For-profit institutions may participate as members of the group.
Amount of Grant: $500,000-$600,000 per year
Contact: Dr. Joshua Rosenthal, Director, (301)496-1653; fax: (301) 402-2056; email: Joshua_Rosenthal@nih.gov
Internet: http://www.fic.nih.gov/programs/grants.html#research
Sponsor: Fogarty International Center
31 Center Dr MSC 2220, Building 31, B2C39
Bethesda, MD 20892-2220

FIC International Malaria Research Training Program Award 2334

FIC invites applications to train or expand the capabilities of scientists and health professionals from malaria endemic developing countries to engage in malaria research. Proposals are requested that would create innovative, collaborative malaria research training programs and that would contribute to the long-term goal of harnessing scientific knowledge and skills to enhance efforts to prevent malaria-related morbidity and mortality and to control malaria transmission in endemic developing countries. To develop sustainable research capacity in endemic developing countries, FIC will support malaria research training efforts as part of ongoing, productive malaria research projects in collaboration with internationally recognized research institutions in Africa and other highly endemic regions of the world. Annual deadlines may vary; contact program staff for exact dates. RFA: TW-00-006
Requirements: Applications may be submitted by U.S. nonprofit public and private institutions, such as universities, colleges, hospitals, laboratories, and eligible agencies of the federal government capable of meeting the objectives of the RFA.
Restrictions: Principal Investigators currently supported by other FIC malaria focused research training awards (D43) are not eligible to apply.
Amount of Grant: $45,000 maximum stipend per year
Contact: Grants Administrator, (301) 496-2075; fax: (301) 594-1211; email: ficinfo@nih.gov
Internet: http://www.fic.nih.gov/programs/malaria.html
Sponsor: Fogarty International Center
31 Center Drive, Building 31
Bethesda, MD 20892

FIC International Research and Training in Population and Health Grants 2335

FIC, NICHD, and NIA have developed this program to support international research and training in population-related sciences and to enable NIH grant recipients to extend the geographic base of research and training efforts to developing nations, in support of international population priorities. Types of training for foreign scientists or health professionals may include predoctoral training in research related to population; postdoctoral training in laboratory procedures and research projects and techniques conducted at the host U.S. institutions or in the trainee's home country; and participation in advanced research training conducted by U.S. faculty in the host country and also short-term in-country training for foreign scientists and health professionals in the host country. A letter of intent is requested by December 13; full application is due January 13. Annual deadline dates may vary; contact program staff for exact dates. RFA-TW-05-002)
Requirements: Applicants must be U.S. principal investigators on at least one NIH-sponsored research grant, cooperative agreement, or contract. The grantee institution must be a U.S. nonprofit private or public institution.
Amount of Grant: $139,000-$185,000
Date(s) Application is Due: Jan 13; Dec 13
Contact: Dr. Jeanne McDermott, Division of International Training and Research, (301) 496-1492; fax: (301) 402-0779; email: mcdermoj@mail.nih.gov
Internet: http://grants.nih.gov/grants/guide/rfa-files/RFA-TW-05-002.html
Sponsor: Fogarty International Center

31 Center Dr MSC 2220, Building 31, Room B2C39
Bethesda, MD 20892-2220

FIC International Research Fellowship Program 2336

The program provides opportunities for non-U.S. postdoctoral biomedical or behavioral scientists in the formative stages of their research careers to gain further research experience by working in the laboratories of distinguished U.S. scientists on problems of mutual interest. Nominations are made by the national nominating committee in each participating country. These committees submit applications to the Fogarty International Center. More than 50 countries or regions in the Americas, Africa, Asia, the Far East, Australia, Europe, and New Zealand participate in the program. Fellowships are awarded for a minimum of 12 months and provide stipend, travel, and institutional allowance. Fellowships provide round-trip travel for the fellow only. In addition, the host institution receives a modest allowance to cover such costs as the fellow's health insurance, supplies, equipment, and travel to scientific meetings or laboratories in the United States or Canada.
Requirements: Candidate must hold a doctoral degree or equivalent in a health science field at the time of submission of application to a nominating committee, have demonstrated the ability to engage in independent basic or clinical research, have a sponsor in the United States at a nonprofit institution who has accepted the applicant for research on his/her proposed project, and have assurance from a nonprofit institution in the home country that there is a position for him/her upon return. Applicant must be proficient in spoken and written English.
Contact: Program Officer, (301) 496-1653; fax: (301) 402-0779/2056; email: nugentr@mail.nih.gov
Internet: http://www.fic.nih.gov/programs/grants.html#fellowships
Sponsor: Fogarty International Center
31 Center Drive, Building 31
Bethesda, MD 20892

FIC International Research Scientist Development Award 2337

The award supports basic research for behavioral and clinical scientists who are committed to a career in international health research and would benefit from an additional period of mentored research as part of a strong, established collaboration between a U.S. sponsor and leading scientists at a developing country center of scientific excellence. The specific research training may be new to the candidate or in an area that would demonstrably enhance the candidate's pursuit of a career focusing on international research pertinent to health in developing countries. Applications are encouraged to address one of the global health research priorities defined by the World Health Organization (http://www.who.ch), which include both infectious and chronic/degenerative conditions and mental health disorders. It is expected that following this experience, the candidate will be able to pursue an independent and productive international research career involving ongoing collaboration with developing country scientists to more effectively pursue research relevant to stemming a major global health problem. Annual deadlines may vary; contact program staff for exact dates. PAR-04-058
Requirements: An applicant must be a U.S. citizen or non-citizen national; hold doctoral or medical degree or the equivalent in a health science field; have earned the relevant degree within seven years of the application receipt date (not including clinical training); have demonstrated a commitment and competence in international health research as well as the potential to engage in independent and productive basic biomedical, behavioral, or epidemiological/clinical research in the period following the award; and have a sponsor in an internationally recognized developing country research institution addressing one or more of that country's major health problems. Eligible countries are in regions that include Africa, Asia (except Japan, Singapore, South Korea, and Taiwan), Central and Eastern Europe, Latin America and the Caribbean, the Middle East (except Israel) and the Pacific Ocean Islands (except Australia and New Zealand). Contact the program office or visit the Web site for full eligibility requirements.
Amount of Grant: $70,000 maximum direct costs per year
Date(s) Application is Due: Feb 16
Contact: Dr. Rachel Nugent, Division of International Training and Research, (301) 496-8733; fax: (301) 402-0779; email: nugentra@mail.nih.gov
Internet: http://www.fic.nih.gov/programs/irsda.html
Sponsor: Fogarty International Center
31 Center Drive, Building 31
Bethesda, MD 20892

FIC International Training and Research in Emerging Infectious Diseases Grants 2338

The Fogarty Center and NIAID have developed this program to train laboratory scientists and public health workers in developing countries and the United States in emerging and re-emerging infectious diseases research, control, and prevention strategies and their implementation and evaluation; facilitate international collaboration on emerging infectious diseases research, including the conduct of research in developing countries; enhance domestic infectious diseases research programs and improve the protection of the U.S. population from infectious diseases by early detection and response to epidemics internationally and nationally; and develop international leaders in science and public health. Training will include predoctoral and postdoctoral programs, participation in advanced and/or short-term in-country training conducted by U.S. faculty in the host country; and participation in training of health workers in the diagnosis, patient management, control, and prevention of disease. RFA: TW-96-001
Requirements: Applicant must be a U.S. principal investigator of at least one NIH-sponsored research grant related to infectious diseases.

Contact: Dr. Joel Breman, International Training and Research, (301) 496-0815; email: Joel_Breman@.nih.gov
Internet: http://www.fic.nih.gov/programs/erid.html
Sponsor: Fogarty International Center
31 Center Drive, Building 31
Bethesda, MD 20892

FIC International Training and Research in Environmental and Occupational Health   2339

This program, sponsored by the Fogarty International Center, NIEHS, NIOSH, and CDC, was developed to train foreign health scientists, clinicians, epidemiologists, toxicologists, engineers, industrial hygienists, chemists, and allied health workers from developing countries and emerging democracies in both general environmental health and occupational health. Types of training may include training in epidemiology concepts and methods, environmental monitoring, industrial hygiene, field studies, and other research related to environmental and occupational health that will lead to the MS or PhD degree for individuals with previous field research experience; short-term comprehensive courses in epidemiology, toxicology, chemistry, industrial hygiene, and environmental and safety engineering, with an emphasis on control of occupational injuries and illnesses, for health and safety professionals to be given in the United States; training in laboratory procedures and research techniques related to environmental and occupational health for individuals with the MS or PhD degree to be given in the United States; and postdoctoral research training for foreign scientists who want to expand their abilities in the epidemiology, diagnosis, prevention, and treatment of environmental and occupational disease and injury. Postdoctoral training can take place both in the United States and in foreign countries. RFA: TW-01-002.
Requirements: The grantee institution must be a U.S., nonprofit private or public institution. Although most applications will be from academic institutions, a nonacademic, nonprofit institution may apply.
Contact: Dr. Christopher Schonwalder, Senior Environmental Health Advisor to the Director, (919) 541-4794; fax: (919) 541-2583; email: cs64c@nih.gov
Internet: http://www.fic.nih.gov/programs/environ.html
Sponsor: Fogarty International Center
31 Center Drive, Building 31
Bethesda, MD 20892

FIC Israeli Ministry of Health Postdoctoral Research Fellowships   2340

The program provides postdoctoral fellowships to U.S. health scientists to conduct biomedical research in Israel. The purpose is to enhance the exchange of research experience and information in the biomedical and behavioral sciences with emphasis on heart diseases, aging, cancer, human reproduction, and child development. Short- and long-term fellowships are available for appointments at Hebrew University; Hadassah Medical School; Sackler Faculty of Medicine, Tel Aviv University; Weizmann Institute of Science; Technion-Israel Institute of Technology; and Ben Gurion University of the Negev. Information and applications are available from NIH Fogarty International Center; funding is from the Israeli Ministry of Health.
Requirements: Applicants must be citizens or permanent residents of the United States; have doctorates in one of the clinical, biomedical, or behavioral sciences; and have professional experience in the proposed area.
Restrictions: Support is not provided for brief observational visits, attendance at scientific meetings, or independent study.
Contact: Bruce Butrum, Grants Management Officer, (301) 496-1653
Internet: http://www.fic.nih.gov
Sponsor: Fogarty International Center
31 Center Drive, Building 31
Bethesda, MD 20892

FIC Research Collaboration Award   2341

Project grants facilitate collaborative research efforts between U.S. and foreign scientists that will expand and enhance the NIH-supported research program of the U.S. principal investigator, while at the same time benefiting the scientific interests of the collaborating foreign scientist. These small grants will provide funds for up to three years to purchase supplies, materials, and small equipment items necessary to conduct the collaborative research in the foreign scientist's laboratory. Travel support for both the U.S. principal investigator and the foreign collaborator also is provided. Applications will be reviewed on the basis of scientific and technical merit of the research proposal. Collaborative potential, significance of the proposed research, and the competence of the foreign collaborator are also prime factors. All biomedical and behavioral research topics supported by the NIH are eligible for inclusion under this program. Contact the office for deadline dates. PAR: PA-02-057
Requirements: The grant applicant is the U.S. institution where the U.S. principal investigator is employed. The foreign collaborator must hold a position at a public or nonprofit private institution.
Amount of Grant: $32,000 per year in direct costs for up to three years
Date(s) Application is Due: Jan 21; May 21; Sep 21
Contact: Dr. Kathleen Michels, Program Director, (301) 496-1653; fax: (301) 402-0779; email: FIRCA@nih.gov
Internet: http://www.fic.nih.gov/programs/firca.html
Sponsor: Fogarty International Center
31 Center Drive, Building 31
Bethesda, MD 20892

FIC Swedish Medical Research Council Fellowships   2342

The Swedish Medical Research Council (SMRC) provides a limited number of research fellowships to U.S. health scientists to conduct research in Sweden and to enhance the exchange of research experience and information in the biomedical and behavioral sciences. Activities supported include collaboration in basic or clinical research and the familiarization with or utilization of special techniques and equipment not otherwise available to the applicant. The program is administered by the Fogarty International Center and funded by the SMRC; applications may be obtained from the center.
Requirements: Candidate must be a citizen or permanent resident of the United States; have a doctorate in clinical, behavioral, or biomedical science; have 10 years or less of postdoctoral experience; and have professional experience in the health sciences for at least two of the last four years. A letter of invitation from the host sponsor is required.
Restrictions: Support is not provided for brief observational visits, attendance at scientific meetings, or independent study.
Amount of Grant: $30,000-$45,000 per year; $41,000 average
Contact: Bruce Butrum, Grants Management Officer, (301) 496-1653
Internet: http://www.fic.nih.gov/programs/grants.html#fellowships
Sponsor: Fogarty International Center
31 Center Drive, Building 31
Bethesda, MD 20892

FIC/NIH AIDS International Training and Research Program   2343

The primary goal of this program is to build research capacity. This program is intended to complement ongoing HIV/AIDS and TB research efforts of the NIH and to the extent possible, of other government, nongovernment, and international organizations. Programs are encouraged to develop human resources in those developing and other participating countries that currently are or are likely to be sites of HIV/AIDS-related research on HIV and/or TB vaccines, anti-HIV and TB drugs, and other HIV and TB interventions. A letter of intent is requested by November 21; full application is due December 21.
Requirements: Countries in Central and Eastern Europe and the New Independent States are eligible to participate, as are countries in Africa, the Americas, Asia, and the Pacific Region. The grantee institution must be a U.S. nonprofit private or public institution capable of meeting the objectives in this RFA.
Restrictions: Institutions that currently hold an award are not eligible to submit a second application.
Date(s) Application is Due: Nov 21; Dec 21
Contact: Dr. Jeanne McDermott, Program Officer, (301) 496-1653; fax: (301) 402-0779; email: mcdermoj@mail.nih.gov
Internet: http://www.fic.nih.gov/programs/aitrp/aitrp.html
Sponsor: Fogarty International Center
31 Center Drive, Building 31
Bethesda, MD 20892

FIC/ORMH Minority International Research Training Grants   2344

FIC and ORMH jointly support scientific training programs that offer international research training opportunities to qualified minority undergraduates and graduate/medical students underrepresented in biomedical and behavioral research careers. Applications are invited from innovative programs that provide international research and training to encourage underrepresented minority students to pursue biomedical research careers, broaden minority research efforts and scientific training to encompass international health problems, stimulate novel approaches to studying health problems that disproportionately affect underserved groups in the United States or in developing countries, and assist minority scientists to participate in international collaborative relationships and work effectively in the rapidly evolving global scientific environment. A letter of intent is requested by December 10; full application is due January 14. Annual deadline dates may vary; contact program staff for exact dates. RFA: TW-00-001
Requirements: Applicant institution and its associated consortia institutions must be colleges or universities that offer baccalaureate degrees in fields relevant to biomedical science. Only one application per institution will be accepted. Priority is given to consortia which include historically black colleges and universities, hispanic serving institutions, or tribal colleges and universities. Participant students and faculty must be U.S. citizens or permanent residents.
Amount of Grant: $1.8 million total
Date(s) Application is Due: Jan 14; Dec 10
Contact: Dr. Barbara Sina, Division of International Training and Research, (301) 496-1653; fax: (301) 402-0779; email: barbara_sina@nih.gov
Internet: http://www.fic.nih.gov/programs/mirt.html
Sponsor: Fogarty International Center
31 Center Drive, Building 31
Bethesda, MD 20892

Field Museum Graduate Student Fellowships   2345

A limited number of Field Museum fellowships are available for graduate students engaged in dissertation research associated with the museum. These fellowships provide stipend support.
Requirements: Candidates will be expected to have formal involvement with the museum; a curator should serve on the student's academic committee, and the research should rely heavily upon the collections and facilities of the museum. Students must be in residence in the Chicago area and are expected to spend a significant portion of their research time at the museum. The period of appointment is one year and starts on September 1st. Applicants must contact the sponsoring curator well in advance of the application deadline.

Restrictions: Only electronic applications using the museum's web site will be accepted.
Geographic Focus: Illinois
Date(s) Application is Due: Feb 1
Contact: Scholarship Coordinator; (312) 922-9410
Internet: http://www.fieldmuseum.org/research_collections/scholarships/default.htm#grad_fellow
Sponsor: Field Museum
1400 S. Lake Shore Drive
Chicago, IL 60605-2496

Field Museum High School Internships 2346
A few salaried internships per year are available for high school students or recent graduates to work directly with collections and research staff at the Museum. Internships are 10 weeks long and are usually taken in the summer. Other internships, including those in administrative areas, are also available.
Requirements: Applicants for these internships must contact the museum curator who seems most appropriate, since the internship project and work schedule will be arranged with her/him (see website for list of curators and contact information). The museum curators travel frequently and should be contacted well in advance of the application deadline.
Date(s) Application is Due: Feb 1
Contact: Sam Burns; (312) 665-7279; email: sburns@fieldmuseum.org
Internet: http://www.fieldmuseum.org/research_collections/scholarships/default.htm#ungrad_intern
Sponsor: Field Museum
1400 S. Lake Shore Drive
Chicago, IL 60605-2496

Field Museum Undergraduate Internships 2347
A limited number of internships per year are available for undergraduate students or recent graduates to work directly with collections and research staff at the Museum. Internships are 10 weeks long and are usually taken in the summer. Other internships, including those in administrative areas, are also available.
Requirements: Applicants for these internships must contact the museum curator who seems most appropriate, since the internship project and work schedule will be arranged with her/him (see website for list of curators and contact information). The museum curators travel frequently and should be contacted well in advance of the application deadline.
Date(s) Application is Due: Feb 1
Contact: Sam Burns; (312) 665-7279; email: sburns@fieldmuseum.org
Internet: http://www.fieldmuseum.org/research_collections/scholarships/default.htm#ungrad_intern
Sponsor: Field Museum
1400 S. Lake Shore Drive
Chicago, IL 60605-2496

Field Museum Visiting Scholarships 2348
Providing opportunities for scientists who wish to use the Field Museum's collections, funds are earmarked for travel and for subsistence while visitors are conducting their research. Young professionals and graduate students can be funded for periods of a few days to several weeks. Owing to the limited availability of these funds, awards are typically less than $1,500 per scholar. Short or extended-term visits by distinguished national and international scientists can be funded for periods of several weeks up to one year. These stipends are negotiable.
Requirements: Applications must be endorsed by a Field Museum curator (see website for list of curators responsible for the collection you wish to study). Contact the appropriate curator beforehand to communicate your background, goals and objectives. Applications require statements regarding: (a) purpose and significance of the proposed research, (b) relevance of the Field Museum, collections to the project, (c) collaboration(s) with Field Museum curators, if any, (d) procedures and methods used in the project.
Restrictions: Only electronic applications using the museum's web site will be accepted.
Date(s) Application is Due: May 1; Nov 1
Contact: Scholarship Coordinator; (312) 922-9410
Internet: http://www.fieldmuseum.org/research_collections/scholarships/default.htm#visiting_scholar
Sponsor: Field Museum
1400 S. Lake Shore Drive
Chicago, IL 60605-2496

Field Psychology Research Grants 2349
Grants support promising graduate students exploring the history, contributions, and further development of perceptual (field) psychology in relationship to the research and writings of Arthur W. Combs, PhD. Applications of Dr. Combs's theoretical position to the improvement of educational, community, conflict-resolution, therapeutic efforts, etc. are also encouraged.
Amount of Grant: $500-$1500
Date(s) Application is Due: Jan 31; Oct 5
Contact: Anne Richards, Trustee, (770) 834-8143; email: arichard@westga.edu
Internet: http://www.fieldpsychtrust.org/application.asp
Sponsor: Field Psychology Trust
301 Dixie Street
Carrollton, GA 30117

Fight for Sight Research Grants 2350
Grants support basic or clinical research of high scientific merit into diseases of the eye. The award may cover salaries, equipment, necessary travel, and/or research expenses. Awards are tenable at the Institute of Ophthalmology or other institutes or universities in the United Kingdom for up to three years.
Requirements: Awards are open to all medical and scientific researchers.
Samples: To be distribute among 19 universities—for research into the causes of blindness and vision disorders, through the grants-in-aid program, $223,000 (2004).
Contact: Janice Benson, Grants Manager, (212) 679-6060; fax: (212) 679-4466; email: janice@fightforsight.com
Internet: http://www.fightforsight.com
Sponsor: Fight for Sight
381 Park Avenue S, Suite 809
New York, NY 10016

Fight for Sight Research Grants and Fellowships 2351
The primary purpose of the program is to encourage and finance research in ophthalmology, vision, and related sciences. The goal is to restore and preserve sight through research in detection, prevention, treatment, and curing of visual disorders, as well as diseases leading to impaired sight and partial or total blindness. Funding priority is given to pilot projects for research relating to clinically important eye diseases. Fight for Sight will continue to support younger investigators, promoting the development of scientific skills that will lead to more substantial funding from other sources. Preference will be given to clinical and disease-oriented grants and training fellowships in the areas of age-related macular degeneration, diabetic retinopathy, retinitis pigmentosa, glaucoma, amblyopia, strabismus, AIDS-related eye disorders, cataract, and uveal and corneal inflammation. Award mechanisms include grants in aid, postdoctoral fellowships, and student fellowships.
Requirements: Applicant must have a doctoral degree in medicine, optometry, or basic science and have less than one year of postdoctoral training. Fellowships are available to U.S. and Canadian residents and are not offered for study abroad. Undergraduate, graduate, and medical students may apply for student fellowships.
Restrictions: An applicant may not currently be financed by other research funds.
Amount of Grant: $25,000-$50,000
Date(s) Application is Due: Mar 1
Contact: Program Coordinator, Fight for Sight Research Division, (847) 843-2020; fax: (847) 843-8458; email: info@preventblindness.org
Internet: http://www.preventblindness.org/research
Sponsor: Prevent Blindness America
500 E Remington Road
Schaumburg, IL 60173

Finnish Cultural Foundation Grants 2352
The foundation awards grants to scientists and artists in Finland. Internationally, the foundation awards travel and research grants. Long-term projects, especially relating to Finnish cultural history, are also supported. The Mirjam Helin International Singing Competition is offered every five years (e.g., 2000, 2005, etc.).
Contact: Grants Administrator, 358 9 612 810; fax: 358-9-640474; email: info@skr.fi; Mirjam Helin International Singing Competition, 358 9 612 810; fax: 358 9 640 474; email: info@skr.fi
Internet: http://www.skr.fi/english/welcome.html
Sponsor: Finnish Cultural Foundation
P.O. Box 203, Bulevardi 5A
Helsinki 00121
Finland

Finnish Literature Translation Grants 2353
Each year, FILI awards translation grants for the translation of Finnish, Finland-Swedish, and Saami language literature into other languages; the translation of both fiction and non-fiction from abroad into Finnish; the translation of academic publications and nonfiction into Swedish and the translation of fiction and nonfiction written in Finnish into Swedish. In addition, FILI also awards travel grants to translators from abroad. Application and guidelines are available online.
Date(s) Application is Due: Apr 1; Nov 1
Contact: Tiina Lehtoranta, +358 (0)9 1312 3344; email: tiina.lehtoranta@finlit.fi
Internet: http://dbgw.finlit.fi/fili/eng/apurahat/apurahat-1.html
Sponsor: Finnish Literature Information Center
Mariankatu 7A2
Helsinki 17 00170
Finland

FINRA Investor Education Foundation Grants 2354
The Foundation funds research and/or educational projects that support its mission of providing underserved Americans with the knowledge, skills and tools necessary for financial success throughout life. The Foundation seeks proposals that advance its mission through high-impact investor education or investor protection projects that are efficiently delivered and achieve significant and tangible benefits for the target audience; research that expands the body of knowledge and offers solutions in the fields of investor education and investor protection; and initiatives that lead with a research element and follow with a high-impact investor education or investor protection project based on the results of the research. Of particular interest are projects that better prepare older Americans for handling their finances during retirement; encourage Americans to implement steps to achieve basic retirement security and avoid investing pitfalls; encourage women to take

control of their financial future through investor education; advance practice, policy, and thought in the fields of investor education and investor protection; and encourage investors to check the background of financial professionals prior to doing business with them. There is no set minimum or maximum grant amount. Projects should be completed within 24 months of grant award. Those exceeding 24 months will be considered if justification for the additional time is included in the proposal.

Requirements: 501(c)3 tax-exempt organizations are eligible. Qualified individuals also are eligible.

Restrictions: The foundation will not award grants to individuals; organizations affiliated with a current member of the foundation board of directors or NASD board of governors; securities firms regulated by NASD; organizations affiliated with a securities firm or individual regulated by NASD, such as a foundation established by a securities firm; securities regulators, self-regulatory organizations, or securities industry trade associations; organizations that are termed disqualified persons pursuant to Article III, Section 3(d) of the NASD by-laws; foreign organizations; or entities that discriminate on the basis of age, color, disability, marital status, national origin, race, religion, sex, sexual orientation, or veteran status.

Samples: Partnership for After-School Education (New York, NY)—to develop and implement investor-education programs for low-income, young adults that focus on personal finance and saving for retirement, $150,000 (2005).

Amount of Grant: $73,404-$202,708

Date(s) Application is Due: Feb 8; May 2; Jul 11

Contact: Christine Kieffer; (202) 728-6970; fax: (202) 728-8149; email: christine. kieffer@finra.com; Susan Sarver; (202) 728-6948; susan.sarver@finra.com

Internet: http://www.finrafoundation.org/grants.asp

Sponsor: NASD Investor Education Foundation

1735 K Street NW

Washington, DC 20006-1506

Fischelis Grants for Research in the History of American Pharmacy   2355

The institute supports a wide range of scholarly activities as long as they promise to contribute significantly to historical understanding and are clearly related to the modern practice of pharmacy in the United States. Projects must be connected directly to American pharmacy practice. Preference will be given to 20th-century topics, although studies of earlier periods will be considered. Historical discussions of current practice issues are also eligible. An application received after the deadline date will be evaluated individually if any funds remain unawarded; otherwise it will be considered in the next year's program.

Requirements: Each application is referred for evaluation to a committee of the institute. Principal criteria used in evaluating an application are a candidate's qualifications and record relevant to the intended purpose, relevance to the history of modern American pharmacy, and the adequacy of resources to fulfill the stated purpose.

Amount of Grant: $5,000 maximum

Date(s) Application is Due: Mar 1

Contact: Gregory Higby, Director; (608) 262-5378; email: grants@aihp.org

Internet: http://www.pharmacy.wisc.edu/aihp

Sponsor: American Institute of the History of Pharmacy

777 Highland Avenue

Madison, WI 53705-2222

FishAmerica Foundation Marine and Anadromous Fish Habitat   2356
Restoration Grants

The program requests proposals for local efforts to accomplish meaningful on-the-ground restoration of marine, estuarine and riparian habitats, including salt marshes, seagrass beds, mangrove forests, and freshwater habitats important to anadromous fish species (fish like salmon and striped bass that spawn in freshwater and migrate to the sea). Emphasis is on using a hands-on, grassroots approach to restore fisheries habitat across coastal America and the Great Lakes.

Requirements: Nonprofit organizations such as sporting clubs, civic organizations, conservation groups, and to a lesser extent state agencies, in the United States and Canada are eligible. Projects must result in on-the-ground habitat restoration, clearly demonstrate significant benefits to marine, estuarine or anadromous fisheries resources, particularly sportfish, and must involve community participation through an educational or volunteer component tied to the restoration activities.

Amount of Grant: $5,000 - $50,000

Date(s) Application is Due: Feb 5

Contact: Erica George, Grants Manager; (703) 519-9691, ext. 247; fax: (703) 519-1872; email: egeorge@asafishing.org or fishamerica@asafishing.org

Internet: http://www.fishamerica.org/grants/index.html

Sponsor: FishAmerica Foundation

225 Reinekers Lane, Suite 420

Alexandria, VA 22314

FishAmerica Foundation Research Grants   2357

Overall, the Foundation will fund research projects that further the National Fish Habitat Plan. Specifically, the program funds research projects that have regional or national implication, not local. Grants support research in the following areas: fisheries management, water quality, habitat studies, stock enhancement, economic impact studies related to sport fishing, and tagging.

Requirements: Nonprofit organizations such as sporting clubs, civic organizations, conservation groups, and to a lesser extent state agencies, in the United States and Canada are eligible.

Restrictions: Grants do not support the following projects or purposes: donations of fishing tackle; salaries, administration, overhead, or travel for conservation projects; individuals; local stream monitoring programs; political activities that attempt to influence political campaigns or legislation; access projects such as road construction, boat ramps, or fishing piers; loans, endowments, trips, tours, tickets, advertising, or publications; or permitting and related costs.

Samples: Restore America's Estuaries, $10,000—to create the Restoration Marketplace website (2006); Pennsylvania Fish and Boat Commission, $5,745—to study the effect of rearing temperature on the straying behavior of hatchery-raised trout (2006); Stripers Forever, $7,500—to commission a study on the potential economic effects of ending commercial harvests of striped bass and allocating striped bass to recreational harvests (2006).

Amount of Grant: $15,000 average

Date(s) Application is Due: Jul 31

Contact: Erica George, Grants Manager; (703) 519-9691, ext. 247; fax: (703) 519-1872; email: egeorge@asafishing.org or fishamerica@asafishing.org

Internet: http://www.fishamerica.org/grants/index.html

Sponsor: FishAmerica Foundation

225 Reinekers Lane, Suite 420

Alexandria, VA 22314

Fishman Family Foundation Grants   2358

The Foundation considers grants for: research, education, and cultural development of and for the community; scholarships related to Jewish services, education, social, and community activities; medical and scientific research; providing resources to meet critical needs in Israel; and educational grants and scholarships. Proposals are reviewed in April and October. Application information is available online.

Requirements: 501(c)3 nonprofits are eligible.

Date(s) Application is Due: Mar 31; Sep 30

Contact: Betty Fishman, President, email: info@fishman.org

Internet: http://www.fishman.org/apply.html

Sponsor: Fishman Family Foundation

730 E. Cypress Avenue

Monrovia, CA 91016

FIU Wolfsonian Fellowships   2359

The program supports research on the material culture of the modern world. All proposals addressing humanities themes that can be supported by materials in the Wolfsonian's collection are welcome. The Wolfsonian also encourages research projects in areas related to its curatorial program. Fellowships are intended to support full-time research, generally for a period of three to five weeks. All fellows are expected to reside in the greater Miami area duing the fellowship period and are encouraged to participate in the activities of the Wolfsonian and other divisions of Florida International University. The fellowship includes a stipend for living expanses, round-trip travel, and an allowance for making reproductions. The application deadline is for projects beginning no earlier than July 1 of the next year. Candidates are asked to indicate preferred dates in residence on their application. Guidelines and application are available online.

Requirements: Fellowships are granted on the basis of outstanding professional or academic accomplishment and are limited to those with at least a master's degree. Doctoral students may apply for dissertation research related to the Wolfsonian collections.

Date(s) Application is Due: Dec 31

Contact: Fellowship Coordinator, (305) 535-2613; fax: (305) 531-2133; email: research@thewolf.fiu.edu

Internet: http://www.wolfsonian.fiu.edu/education/research/index.html

Sponsor: Wolfson-Florida International University

1001 Washington Avenue

Miami Beach, FL 33139

Five College Women's Research Center Ford Associateships   2360

The center invites applications from international applicants for two one-semester Ford Associateships (fall and spring), which offer a stipend plus a housing/travel allowance in return for teaching (in English) one undergraduate course in the women's studies program at the University of Massachusetts or the Women and Gender Studies Department at Amherst College. UMass and Amherst seek a total of two experienced researchers and teachers to strengthen their undergraduate curriculum and, in the case of the university, to complement their graduate program. For the fall, the University of Massachusetts seeks a researcher with expertise in the Middle East or Latin America with an emphasis on globalization, transnational studies, or postcolonial studies. Research should focus on gender, race, ethnicity, class, and sexuality within the context of globalization. For fall or spring 2006, Amherst College seeks a researcher who is also an experienced teacher and who works on the Middle East, Latin America, Asia, or Africa with expertise in the field of gender in the media especially in the context of war and civil unrest. Ford Associates need not be studying their own region of origin. Ford Associates are expected to be in residence in the Five College area, working full-time at the center on the project outlined in their application, during the period of appointment. Each Ford Associate is expected to give a public presentation on her work, lead a faculty seminar, and participate actively in the daily life of the center. Application and guidelines are available online.

Requirements: Applications are welcomed worldwide from writers, visual artists, film and video makers, as well as academic scholars. Ford applicants may work on any and all regions of the world and need not be studying their own region of origin.

Amount of Grant: $12,000 stipend plus $3000 housing/travel allowance

Date(s) Application is Due: Feb 13

Contact: Ford Associateships Administrator, Mount Holyoke College, (413) 538-2275; fax: (413) 538-3121; email: fcwsrc@wscenter.hampshire.edu
Internet: http://www.fivecolleges.edu/sites/fcwsrc/about
Sponsor: Five Colleges Inc
50 College Street
South Hadley, MA 01075-6406

Five College Women's Studies Ford Associateships 2361
Open to international researchers in women's studies, this special program is supported by the Ford Foundation. Ford Associates spend a semester in residence at the center pursuing their own research, participating in faculty seminars with women's studies and area studies colleagues from the Five Colleges, and forging connections with the feminist community in Western Massachusetts. Each associate also teaches or co-teaches (in English) an undergraduate women's studies course at one of the Five Colleges (Amherst, Hampshire, Mount Holyoke, and Smith Colleges, and the University of Massachusetts, Amherst) during her semester at the center. International applicants may apply for one of the two special one-semester Ford Associateships for fall or spring, which offer a stipend and a housing/travel allowance in return for teaching (in English) one undergraduate women's studies course at Smith College. Ford applicants' research should focus on how the economics of globalization regulate gender, race, ethnicity, nationality, class, and sexuality in Latin America, the Caribbean, Africa, the Middle East, the former Soviet bloc, or Asia. Two Ford Associate positions are available. For one position preference will be given to those whose work focuses on sexuality in a global context, including sex work, global sex trafficking, health issues, international gay and lesbian activism, and advocacy for sexual minorities. For the second position, preference will be given to those whose work focuses on cultural production and resistance, including political performance, the transformation and use of international media, and new technologies. Application and guidelines are available online.
Requirements: Applications are welcomed from writers, visual artists, film and video makers, as well as academic scholars. Ford applicants may work on any and all regions of the world and need not be studying their own region of origin.
Amount of Grant: $12,000 stipend plus $3000 housing/travel allowance
Date(s) Application is Due: Feb 28
Contact: Five College Women's Studies Research Center, (413) 538-2275; fax: (413) 538-3121; email: fcwsrc@fivecolleges.edu
Internet: http://www.fivecolleges.edu/sites/fcwsrc/about
Sponsor: Mount Holyoke College
50 College Street
South Hadley, MA 01075-6406

Five Colleges Fellowship Program for Minority Scholars 2362
The program provides a year in residence for minority graduate students in the final phase of the doctoral degree. Each fellow will be hosted within a particular department or program at one of the five colleges in the consortium—Amherst, Hampshire, Mount Holyoke, Smith College, and the University of Massachusetts. (At Smith, recipients hold a Mendenhall Fellowship.) The purposes of the program are to enable fellows to complete their dissertations; to encourage their interest in college teaching; and to acquaint them with the schools. Although the primary goal is completion of the dissertation, each fellow is expected to do some formal teaching or its equivalent (no more than a single, one-semester course).
Requirements: Eligible to apply are minority graduate students who have completed all of the requirements for the PhD except the dissertation.
Amount of Grant: $30,000 stipend, housing assistance, office space, library privileges, departmental affiliation
Date(s) Application is Due: Dec 1
Contact: Five College Fellowship Program Committee, (413) 256-8316
Internet: http://www.fivecolleges.edu/academic_programs/academprog_fellowship.html
Sponsor: Five Colleges Inc
97 Spring Street
Amherst, MA 01002

Fleishhacker Foundation Eureka Fellowships 2363
The Foundation offers an unrestricted fellowship program for visual artists. Artists must be nominated by a local nonprofit arts organization to qualify for the Fellowship. Awards of $25,000 per artist are determined by a review panel of non-Bay Area arts experts. There are no particular deadlines with which to adhere, and nominating individuals or organizations should begin by contacting the Foundation.
Requirements: Nominees must reside in the San Francisco Bay area.
Geographic Focus: California
Amount of Grant: $25,000
Contact: Christine Elbel, Executive Director; (415) 561-5350; email: info@fleishhackerfoundation.org
Internet: http://www.fleishhackerfoundation.org/eureka.html
Sponsor: Fleishhacker Foundation
P.O. Box 29918
San Francisco, CA 94129-0918

Fletcher Jones Foundation Grants 2364
The foundation considers grants for charitable, scientific, literary, and educational support, with primary emphasis given to private colleges and universities, particularly in California. Prior to any written submission, it is advisable to call or to meet with the executive director in order to discuss a tentative proposal and to determine the suitability of the intended request. The board of trustees meets on a quarterly basis to review and act upon grant proposals. There are no deadlines; proposals are accepted throughout the year. Guidelines are available online.
Requirements: Applicants must be private, accredited colleges or universities, primarily in California.
Restrictions: Grants are not made to individuals, K-12 educational institutions, for operating funds, deficit financing, conferences, traveling exhibits, or for projects supported by government agencies.
Geographic Focus: California
Samples: Art Ctr College of Design (Pasadena, CA)—for an endowed scholarship fund, $250,000 (2005); Chapman U (Orange, CA)—to endow a professorship at its law school, $500,000 (2004).
Amount of Grant: $9.4 million total
Contact: Christine Sisley, Executive Director, (213) 943-4646; fax: (213) 943-4648
Internet: http://www.fletcherjonesfdn.org/Fletcher_Jones_Application.htm
Sponsor: Fletcher Jones Foundation
523 W Sixth Street, Suite 301
Los Angeles, CA 90014

Flinn Foundation Grants Programs 2365
The foundation's grantmaking programs, limited to Arizona, include enhancing community-based solutions to local health care needs, especially those for children and youth; strengthening medical education and biomedical research programs in Arizona; strengthening Arizona's universities through an undergraduate scholarship program for outstanding Arizona high school students; and enhancing the visibility and long-term artistic mission of Arizona's principal visual and performing arts organizations. Types of support include program development, seed money, scholarship funds, and research grants.
Requirements: Arizona-based institutions or organizations whose programs are operated for the benefit of Arizona institutions and individuals are eligible to apply. Applications are accepted at any time. There is no application form, but a preliminary letter of inquiry or phone call is requested to determine the appropriateness of a full submission.
Restrictions: The foundation rarely provides grants to individuals, building projects (capital campaigns), purchase of equipment, endowment projects, annual fund-raising campaigns, ongoing operating expenses, or deficit needs. Requests to support conferences and workshops, publications, or the production of films and video are considered only when these activities are an integral component of a larger foundation initiative.
Geographic Focus: Arizona
Samples: Northern Arizona U (Flagstaff, AZ)—to expand teacher-preparation programs in the biological sciences, $517,100 over three years.; Homeward Bound (Phoenix, AZ)—to coordinate access to health care services for disadvantaged children, $10,000.; Ballet Arizona (Phoenix, AZ)—for 15 months of support, $200,000.
Amount of Grant: $2500-$150,000 average
Contact: JoAnn Fazio, Grants Coordinator, (602) 744-6800; fax: (602) 744-6815; email: info@flinn.org
Internet: http://www.flinn.org/about/grants.cms
Sponsor: Flinn Foundation
1802 N Central Avenue
Phoenix, AZ 85004-1506

Florence Tan Moeson Fellowships 2366
This fellowship is made possible by the generous donation by Mrs. Florence Tan Moeson, a former cataloger in the Chinese Team of the Regional and Cooperative Cataloging Division for 45 years. The purpose of the fellowship is to provide individuals with the opportunity to pursue research on East, Southeast, and/or South Asia (including the overseas Asian communities), using the unparalleled collections of the Library of Congress in Washington, DC. The grants are for a minimum of five business days of research and are to be used to cover travel to and from Washington, overnight accommodations, and photocopying; or, to cover expenses incurred while traveling to the Library of Congress to engage in scholastic research in the area of Asian studies. Up to 15 awards are available from $300 to $2,500. Awards are announced in December.
Requirements: Graduate students, independent scholars, community college teachers, researchers without regular teaching appointments, and librarians with a need for fellowship support are especially encouraged to apply.
Restrictions: Non-U.S. citizens and permanent residents who do not have a U.S. Social Security number must submit a Form W-7 to the Internal Revenue Service in order to obtain an individual taxpayer identification number (ITIN) prior to commencing their fellowship at the Library, regardless of the taxability of their income under this program or exemption under a treaty with the U.S.
Samples: 2007 - Ellen Huang, University of California, San Diego, China, China: Jingdezhen Porcelain and the Production of Culture in the Nineteenth Century.; Sho Konishi, University of Chicago, Anarchist Democracy in Early 20th-Century Japan.; Sueyoung Park-Primiano, New York University, Identifying Ideologies of Korean Cinema of 1950s and 1960s.; Liladhur Pendse, University of California, Los Angeles, Portuguese Language Print Materials from Goa in Colonial India.; Ronit Ricci, University of Michigan, Ann Arbor, Suluk: Islamic Mystical Poetry in Javanese.; Lily Ann Villaraza, Northern Illinois University, Philippine Protest Theater in the Anti-Martial Law Movement.; Samuel Wright, University of Maryland, College Park, Sanskrit Intellectual History in Early Modern Bengal: the Case of Gadadhara Bhattacharya.; Xiaoqing Zhu, University of Maryland, College Park, Indigenization or Westernization? Chinese Visual Encounters with European Imageries in the Late Ming Period (1580-1620).
Amount of Grant: $300 - $2,500

Date(s) Application is Due: Sep 30
Contact: Anchi Hoh, Ph.D., Co-Chair, Florence Tan Moeson Fellowship Program; (202) 707-5673; fax: (202) 707-1724; email: adia@loc.gov
Internet: http://www.lcasianfriends.org/ftm-fellowship
Sponsor: Library of Congress, Asian Division Friends Society
101 Independence Avenue, SE
Washington, DC 20540-4810

Florida Sea Turtle Grants Program   2367
Funded by a portion of revenues from Florida's Sea Turtle Specialty License Plate, the grants program distributes funds each year to support sea turtle research, conservation, and education programs throughout Florida. Types of grants include competitive grants, funding for ongoing activities, and emergency grants. Application and guidelines are available online.
Requirements: Florida coastal local governments, Florida-based nonprofit organizations, and education and research institutions that actively participate in marine turtle research, conservation, and educational activities within the state of Florida are eligible to apply.
Geographic Focus: Florida
Samples: Hidden Harbor Marine Environmental Project-The Turtle Hospital, Richie Moretti (FL)—for supplies and equipment to support an environmentally controlled facility to isolate ill sea turtles, $13,164 (2004); Longboat Key Turtle Watch, Inc, Gillian Bustard (FL)—to support ongoing community education programs, $1500 (2004); Anna Maria Island Turtle Watch, Suzi Fox (FL)—for printing of educational materials, $4946 (2004); U of Central Florida, Dean Bagley (FL)—for satellite tracking of juvenile green turtles, $32,115 (2004).
Contact: Grants Administrator, (352) 373-6441; fax: (352) 375-2449; email: stgp@helpingseaturtles.org
Internet: http://www.helpingseaturtles.org/stgp.htm
Sponsor: Caribbean Conservation Corporation
4424 N.W. 13th Street, Suite B11
Gainesville, FL 32609

Fluor Foundation Grants   2368
The Fluor Corporation achieves its contribution objectives through the Fluor Foundation and corporate giving. The Foundation's areas of interest are: education; human services; cultural outreach; and public/civic affairs. The Foundation considers requests for operating, program, capital or endowment support. Priority is given to funding organizations with employee volunteer participation. Interested applicants should submit a letter of inquiry. Application information is available online.
Requirements: Funding is considered for community organizations located where Fluor has a permanent office or project site presence.
Restrictions: Funding is not available for: film production/publishing activities; individuals; sports organizations/programs; veterans, fraternal, labor or religious organizations; or lobbying/political organizations or campaigns.
Geographic Focus: California; South Carolina; Texas; Washington
Samples: U of California (CA)—for its math, engineering, science achievement program, $25,000.; South Carolina Foundation of Independent Colleges (SC)—for program support, $5000.
Contact: Margarita Miranda, Senior Community Relations Coordinator, (949) 349-6797; fax: (949) 349-7175; email: community.relations@fluor.com
Internet: http://www.fluor.com/communities/default.asp
Sponsor: Fluor Foundation
One Enterprise Drive
Aliso Viejo, CA 92656-2606

FMC Foundation Grants   2369
The foundation supports education, community improvement, urban affairs, health and human services, and public issues/economic education. Higher education is supported through scholarships and employee matching gifts. Education support tends to be in business, engineering, chemistry, and some minority education programs. Eligible applicants need to contact the FMC in their geograhic area. Each individual FMC location determines how their contributions will be given.
Requirements: Grants are awarded to 501(c)3 organizations in FMC-plant communities. U.S.-based organizations with an international focus are also eligible.
Restrictions: Grants are not made to individuals.
Geographic Focus: Arizona; California; Delaware; Florida; Illinois; Louisiana; Maine; Maryland; Missouri; New Jersey; New York; North Carolina; Pennsylvania; Tennessee; Texas; West Virginia; WyomingCanada
Contact: Program Contact, (215) 299-6000; fax: (215) 299-6140
Internet: http://www.fmc.com
Sponsor: FMC Foundation
1735 Market Street
Philadelphia, PA 19103

FNN Alice P. Kenney Award   2370
This annual prize is awarded to an individual or group that has made a significant contribution to colonial Dutch studies and/or has encouraged understanding of the significance of the Dutch colonial experience in North America by research, teaching, writing, speaking, or in other ways. Reasonable travel expenses will be reimbursed. The winner will be selected by a four-person committee consisting of the director of the New Netherland Project, two members of the Friends of New Netherland, and a representative of the Alice P. Kenney Memorial Trust Fund. The award will be presented each year at

an event in connection with the New Netherland Project. Nominating guidelines are available online.
Requirements: Persons or groups to be considered for this award can be involved in any pursuit of any aspect of Dutch colonial life in North America. Candidates can be nominated by members of the Friends of New Netherland, by historical organizations, or by the general public.
Samples: The Times Union (Albany, NY)—award recipient, $250.; The Horlepiepdansers (Schenectady, NY)—award winner, $250.
Amount of Grant: $500
Date(s) Application is Due: Apr 1
Contact: Award Selection Committee, (518) 474-6067; fax: (518) 473-0472; email: nyslfnn@mail.nysed.gov
Internet: http://www.nnp.org/fnn/kenny.html
Sponsor: Friends of New Netherland
P.O. Box 2536, Empire State Plaza Sta
Albany, NY 12220-0536

Fogarty International Center/Ellison Medical Foundation Overseas   2371
Fellowships in Global Health and Clinical Research
The fellowship is a one-year mentored clinical research training experience for graduate-level U.S. students in the health professions. The program enables individuals to experience mentored research training at top-ranked NIH-funded research centers in developing countries. Fellowship sites include Bangladesh; Botswana; Brazil; Haiti; Chennai, India; China; Vellore, India; Kenya; Mali, West Africa; Lima, Peru; Durban, South Africa; Pretoria, South Africa; Russia; Tanzania; Thailand; Uganda; and Zambia. Each fellowship is for a one-year period. The term will begin with an intensive orientation program on the NIH campus in Bethesda, MD, in July. This will be followed by approximately 10-plus months of intense research training at the foreign site. Application and guidelines are available online.
Requirements: This program is designed for students meeting all of the following qualifications: a strong interest in, and potential for, a career in international health activities and/or clinical research; advanced standing in a U.S. medical (M3) or osteopathic school, or enrollment in a doctoral-level program at a U.S. school of public health, nursing, or dentistry; and support of their home academic institution, including a committed mentor. Applicants must have strong academic records and must be U.S. citizens or permanent U.S. residents. Medical and osteopathic students must have completed their basic science courses and one year of clinical clerkship; public health doctoral students must have completed their coursework and passed their qualifying exams prior to the beginning of the fellowship.
Amount of Grant: $18,000-$20,000 per year (adjusted to cost of living expenses at the foreign site) plus $6000 per year for travel, insurance, and educational materials
Date(s) Application is Due: Dec 28
Contact: Yolanda Thomas, email: fic-fellowship@aamc.org
Internet: http://www.aamc.org/students/medstudents/overseasfellowship/2006fogartyapp.doc
Sponsor: Association of American Medical Colleges
2450 N Street NW
Washington, DC 20037-1127

Folger Institute Consortium Grants   2372
The grants are available to faculty members and graduate students from affiliated universities and are offered to reimburse travel and some lodging expenses in support of participation in institute programs. Awards for released-time may also be made to consortium faculty members.
Date(s) Application is Due: Jan 3; Jun 1; Sep 1
Contact: Virginia Millington, Program Assistant, (202) 675-0333; fax: (202) 544-4623; email: institute@folger.edu
Internet: http://www.folger.edu/template.cfm?cid=1346
Sponsor: Folger Institute
201 E Capitol Street SE
Washington, DC 20003

Folger Institute Grants in Aid   2373
The institute offers an interdisciplinary program of seminars, workshops, symposia, colloquia, and lectures. Faculty and graduate students from affiliated universities may apply for reimbursement of travel and lodging expenses in support of their attendance at any Folger Institute program. Some awards for released time are also made to consortium faculty members. Full-time faculty from any U.S. college or university who attend Folger programs funded by the NEH may apply for NEH at-large grants. A limited number of registration fee waivers are available by application to independent scholars.
Requirements: Grants in aid are awarded for participation in institute programs for advanced graduate students and faculty from institute-affiliated universities.
Amount of Grant: $2000
Date(s) Application is Due: Jan 3; Jun 1; Sep 1
Contact: Virginia Millington, Program Assistant, (202) 675-0333; fax: (202) 544-4623; email: institute@folger.edu
Internet: http://www.folger.edu/template.cfm?cid=1346
Sponsor: Folger Institute
201 E Capitol Street SE
Washington, DC 20003-1094

Folger Long-Term Postdoctoral Research Fellowships   2374

The Folger holds the largest collection outside Britain of books and manuscripts on early English history from the 15th through 18th centuries. The library offers a limited number of residential fellowships for periods of six to nine months. Three National Endowment for the Humanities and two Andrew W. Mellon Foundation fellowships are offered. Fellowships are open to scholars from any country. Applicants should submit seven copies of each of the following: an application form, a 1000-word description of the research project, a curriculum vita, and three letters of reference (under separate cover).

Requirements: Successful candidates will be advanced scholars who have made substantial contributions in their fields and whose research projects are appropriate to the library's collections.

Amount of Grant: $40,000 NEH fellowships; $50,000 and $35,000 Mellon fellowships

Date(s) Application is Due: Nov 1

Contact: Carol Brobeck, Fellowships Coordinator, (202) 675-0348; email: cbrobeck@folger.edu

Internet: http://www.folger.edu/academic/fellows.asp

Sponsor: Folger Shakespeare Library

201 E Capitol Street SE

Washington, DC 20003

Folger Postdoctoral Short-Term Research Fellowships   2375

Applications are welcome in all areas covered by the Folger collection for work on projects that draw significantly from the Folger holdings from the 16th through 18th centuries: English, American, and European literature and drama; English, American, and Continental history; political, economic, and legal history; and history of philosophy, art, music, religion, science and medicine, and exploration. Short-term postdoctoral fellowships are offered for independent research for a term of one to three months during the period of July to June of the next year. Application consists of a completed Folger's application form, four copies of the applicant's curriculum vita, four copies of a 1000-word description of the research project, and three letters of recommendation submitted directly to the fellowship committee.

Requirements: Applicants for NEH fellowships must be U.S. citizens who have completed the PhD or equivalent and have a research project appropriate to the collections at the Folger Library. Applications for the Folger long-term or short-term fellowships are open to scholars from any country. A fellow is expected to be in residence in Washington, DC, during the term of the fellowship and to conduct his/her research at the Folger Library.

Samples: Katherine West Scheil, U of Rhode Island (RI)—fellowship recipient, for Women's Reading Groups of Shakespeare.; Daniel Vitkus, Florida Institute of Technology (FL)-fellowship recipient, for Turning Turk: Marlowe, Shakespeare, and the Representation of Islam.

Amount of Grant: $2000 per month

Date(s) Application is Due: Mar 1

Contact: Carol Brobeck, Fellowships Coordinator, (202) 675-0348; email: cbrobeck@folger.edu

Internet: http://www.folger.edu/academic/fellows.asp

Sponsor: Folger Shakespeare Library

201 E Capitol Street SE

Washington, DC 20003

Food, Pharmaceutical, and Bioengineering Division Award in Chemical Engineering   2376

This award is given to recognize outstanding chemical engineering contributions and achievements by an individual in those industries involved in food, pharmaceutical, and bioengineering activities. The recipient need not be a member of AIChE, and there is no age limit. The individual's contributions may have been made in industrial, governmental, or academic areas, or with other organizations.

Requirements: All members, as well as other interested persons, are urged to nominate deserving candidates.

Amount of Grant: $4000

Date(s) Application is Due: Mar 25

Contact: AIChE Awards Programs, (212) 591-7107; fax: (212) 591-8890; email: awards@aiche.org

Internet: http://www.aiche.org/awards/awarddtl.asp?AwardID=36

Sponsor: American Institute of Chemical Engineers

3 Park Avenue

New York, NY 10016-5910

Ford Foundation International Fellowships   2377

The program supports advanced study by exceptional individuals who will use this education to become leaders in their respective fields, furthering development in their own countries and greater economic and social justice worldwide. Eligible fields of study are any academic discipline or field of study related to the foundation's three grant-making areas: asset building and community development—children, youth and families, sexuality and reproductive health, work-force development, development finance and economic security, environment and development, community development; education, media, and arts and culture—education reform; higher education and scholarship; religion, society and culture; media; and arts and culture; peace and social justice—human rights; international cooperation; governance; and civil society. The international fellowships provide support for up to three years of formal graduate-level study leading to a masters or doctoral degree. Fellows are selected from countries where the foundation maintains active overseas programs. Once selected, fellows may enroll in an appropriate university program anywhere in the world, including their country of residence. The program also enables fellows to undertake short-term language study and training in research and computer skills prior to graduate school enrollment.

Requirements: Applicants must be resident nationals or residents of Asia, Africa, the Middle East, Latin America, and Russia. Successful candidates will demonstrate superior achievement in their undergraduate studies and hold a baccalaureate degree or its equivalent; have substantial experience in community service or development-related activities; possess leadership potential evidenced by their employment and academic experience; propose to pursue a post-baccalaureate degree that will directly enhance their leadership capacity in a practical, policy, academic, or artistic discipline or field corresponding to one or more of the foundation's areas of endeavor; present a plan specifying how they will apply their studies to social problems or issues in their own countries; and commit themselves to working on these issues following the fellowship period.

Restrictions: U.S. nationals are not eligible, although fellows may study in the United States.

Contact: Fellowship Administrator; (212) 984-5558 or (212) 573-5000; fax: (212) 351-3677; email: question@fordifp.org

Internet: http://www.fordifp.net/

Sponsor: Ford Foundation

809 United Nations Plaza, 9th Floor

New York, NY 10017

Ford Foundation Peace and Social Justice Grants   2378

This program area supports two categories: human rights and international cooperation, and governance and civil society. Grants under human rights and international cooperation support projects involving women's rights and civil rights in the United States, international human rights, international law, multilateral conflict prevention, and U.S. foreign policy. Governance and civil society grants support projects concerning civic participation, electoral reform, and innovations in state and local government. Most grants are given to organizations. The foundation also makes grants to individuals, though they are few in number relative to demand and are limited to research, training, and other activities related to the foundation's program interests. Support for graduate fellowships is generally provided through grants to universities and other organizations, which select recipients. Applications are considered throughout the year; a letter of inquiry is the first step of the application process.

Requirements: The letter of inquiry should include the purpose of the project, problems and issues addressed, information about the applicant organization, estimated overall budget, time period for which funds are requested, and qualifications of those engaged in the project.

Restrictions: Support is not normally given for routine operating costs of institutions or for religious activities. Except in rare cases, funding is not available for the construction or maintenance of buildings. The foundation does not award undergraduate scholarships or make grants for purely personal or local needs.

Amount of Grant: $75,000-$250,000 average

Contact: Luis Antonio Ubinas, President; (212) 573-5000; fax: (212) 599-4584; email: office-secretary@fordfound.org

Internet: http://www.fordfound.org/grants

Sponsor: Ford Foundation

320 East 43rd Street

New York, NY 10017

Ford Motor Company Fund Grants Program   2379

The fund supports not-for-profit organizations in three major areas: innovation and education; community development and American legacy; and auto-related safety education. The fund seeks to build partnerships with organizations that have a well-defined sense of purpose, a demonstrated commitment to maximizing available resources, and a reputation for meeting objectives and delivering quality programs and services. Priority is placed on the support and development of organizations that promote diversity and inclusion. Requests for support are accepted and reviewed throughout the year. The fund now implements an online application system. Details are available online.

Restrictions: Ford does not fund: advocacy-directed programs; animal-rights organizations; beauty or talent contests; day-to-day business operations; debt reduction; donation of vehicles; efforts to influence legislation, or the outcome of any elections or any specific election of candidates to public office or to carry on any voter registration drive; endowments; fraternal organizations; general operating support to hospitals and health care institutions; individual sponsorship related to fundraising activities; individuals; labor groups; loans for small businesses; loans to program-related investments; organizations that do not have 501(c)3 status; organizations that unlawfully discriminate in their provision of goods and services based on race, color, religion, gender, gender identity or expression, ethnicity, sexual orientation, national origin, physical challenge, age, or status as a protected veteran; political contributions; private K-12 schools; profit-making enterprises; religious programs or sectarian programs for religious purposes; species-specific organizations; sports teams.

Samples: American Red Cross, Southeastern Michigan Chapter (Detroit, MI)—for disaster-relief efforts in Florida related to Hurricane Charley, $50,000 (2004).

Contact: Ford Fund Coordinator, (888) 313-0102; email: Fordfund@ford.com

Internet: http://www.ford.com/en/goodWorks/fundingAndGrants/fordMotorCompany Fund/default.htm

Sponsor: Ford Motor Company Fund

P.O. Box 1899

Dearborn, MI 48121-1899

Fordham Prizes　　　　　　　　　　　　　　　　　　　　**2380**

The foundation awards two prizes annually. The Thomas B. Fordham Prize for Distinguished Scholarship is awarded to a scholar who has made major contributions to education reform via research, analysis, and successful engagement in the war of ideas. The Thomas B. Fordham Prize for Valor is awarded to a leader who has made major contributions to education reform via noteworthy accomplishments at the national, state, local, and/or school levels. Nominations from individuals or organizations familiar with the work of outstanding education reformers and scholars are welcome. Self-nominations will not be considered. Nomination guidelines are available online.

Requirements: Anyone can be nominated whose work has had a profound impact on education in the United States. Candidates may be nominated either for cumulative lifetime achievement or for extraordinary one-time accomplishments.

Restrictions: Employees and trustees of the foundation are not eligible, nor are members of the prize committee.

Geographic Focus: Ohio

Contact: Jennifer Leischer, Prize Coordinator, (202) 223-5452; fax: (202) 223-9226; email: jleischer@edexcellence.net

Internet: http://www.edexcellence.net/foundation/global/page.cfm?id=305

Sponsor: Thomas B. Fordham Foundation

1701 K Street, N.W., Suite 1000

Washington, DC 20006

Foreign Language and Russian and East European Area Studies Fellowships　　**2381**

Fellowships are awarded in the fall to advance students' knowledge of the Russian or an East European language and culture. Selection of fellows is based on academic record, letters of recommendation, intent of study plan, and priorities of the Department of Education, which gives final approval to awardees.

Requirements: Applicants must be U.S. citizens or permanent residents of the United States and full-time students in an advanced degree program at the University of Pittsburgh as well as in the graduate certificate program of the Center for Russian and East European studies. Students with Russian background must have completed Russian through the second-year level of college to be eligible. Application must be submitted by the dean/chairperson of the school/department in which the student will study.

Amount of Grant: $11,000 stipend plus tuition and fees for two terms

Date(s) Application is Due: Feb 1

Contact: Russian and East European Studies, (412) 648-7407; fax: (412) 648-2199

Internet: http://sli.slavic.pitt.edu/administrative.html#financial

Sponsor: University of Pittsburgh

4G15 Forbes Quadrangle

Pittsburgh, PA 15260

Forest Products Society Wood Award　　　　　　　　　　　　**2382**

The Forest Products Society (FPS), Borden Inc, and Dynea U.S.A Inc, sponsor the annual wood awards. Two awards are made for the best 2000- to 4000-word paper on the forestry industry. Topics may cover any of a wide range of problems from harvesting trees, to development and manufacture of products, to distribution and marketing of such products. Judges will rate the essays on the basis of subject, skill of treatment, conclusions, applicability, and conformance to competition rules. Winners will make a poster presentation at the annual meeting of the FPS. Tentative titles are due by the listed application deadline.

Requirements: The competition is open to all graduate students who are registered for a full quarter or full semester at any university or college the year previous to application.

Amount of Grant: $1000 first place; $500 second place

Date(s) Application is Due: Feb 6

Contact: Vickie Bruce, (608) 231-1361 ext 210; fax: (608) 231-2152; email: vbruce@forestprod.org or info@forestprod.org

Internet: http://www.forestprod.org/awa-wood.html

Sponsor: Forest Products Society

2801 Marshall Ct

Madison, WI 53705-2295

Forrest C. Lattner Foundation Grants　　　　　　　　　　　　**2383**

The Foundation's primary objectives are in six areas of interest: arts and humanities; education; environment; health and social services; historic preservation; and medical research. The Foundation also wishes to encourage the development of innovative model programs. Application guidelines are available online.

Restrictions: The foundation is unable to fulfill grant requests to individuals, or to support programs that are the primary responsibility of the public sector.

Geographic Focus: Florida; Georgia; Kansas; Rhode Island; Texas

Samples: Woodmere Art Museum (Philadelphia, PA)—for programs and exhibitions at the Helen Millard Children's Gallery, $20,000.

Date(s) Application is Due: Mar 1; Sep 1

Contact: Susan Lattner Lloyd, President, (561) 278-3781; fax: (561) 278-3167; email: lattner@bellsouth.net

Internet: http://www.lattnerfoundation.org

Sponsor: Forrest C. Lattner Foundation Inc

777 E. Atlantic Avenue, Suite 317

Delray Beach, FL 33483

Foundation Fighting Blindness Marjorie Carr Adams Women, Career Development Awards　　　　　　　　　　　　　　　　　　**2384**

The Program's primary purpose is to support female clinical research scientists of superior dedication and talent to pursue vigorous research programs to find the cures for retinal degenerative diseases research (RDD) i.e. inherited orphan retinal degenerative diseases and dry age-related macular degeneration. The goal of this program is to facilitate advances in the laboratory and clinical research, to elucidate the mechanisms for the etiology and pathogenesis of RDDs and to develop innovative strategies to prevent, treat and cure these diseases. This award provides support for mentored study and research for clinically trained biomedical research professionals who have the potential to assume leadership roles as clinician-scientists in RDD research.

Requirements: Female clinical scientists possessing an, M.D., D.O., or recognized equivalent foreign degrees and who are in their first, second or third year of a junior faculty appointment are eligible to apply for the Adams WCDA. Applicants do not have to be U.S. citizens.

Amount of Grant: $65,000 annually; $325,000 over five years

Date(s) Application is Due: Nov 14

Contact: Cindy Settar, (410) 568-0150 or (800) 683-5555; email: csettar@Fightblindness.org or grants@Fightblindness.org

Internet: http://www.blindness.org/index.php?option=com_content&view=section&id=9&Itemid=107#career-development

Sponsor: Foundation Fighting Blindness

11435 Cronhill Drive

Owings Mills, MD 21117-2220

Foundation for Endangered Languages Grants　　　　　　　　　**2385**

The Foundation for Endangered Languages exists to support, enable and assist the documentation, protection and promotion of endangered languages. In order to do this, it aims to: raise awareness of endangered languages, both inside and outside the communities where they are spoken, through all channels and media; support the use of endangered languages in all contexts: at home, in education, in the media, and in social, cultural and economic life; monitor linguistic policies and practices, and to seek to influence the appropriate authorities where necessary; support the documentation of endangered languages, by offering financial assistance, training, or facilities for the publication of results; collect together and make available information of use in the preservation of endangered languages; and disseminate information on all of the above activities as widely as possible.

Samples: John Hobson, University of Sydney—Banjalang: Northern Rivers district, NSW, Australia, $1450 (2008); Chun Huang, University of Florida—Siraya (Taiwan), $1000 (2008); Molly Babel, Mono Lake Northern Paiute: Bridgeport, California, U.S.A, $550 (2008).

Amount of Grant: $1,000-$2,000 average

Contact: Nicholas Ostler; email: nostler@chibcha.demon.co.uk

Internet: http://www.ogmios.org/grant.htm

Sponsor: Foundation for Endangered Languages

Batheaston Villa, 172 Bailbrook Lane

Bath, England BA1 7AA England

Foundation for Enhancing Communities Grants　　　　　　　　**2386**

The foundation awards grants to Pennsylvania nonprofit organizations in the greater Harrisburg area. Areas of interest include art/cultural programs, education, health care and health organizations, human services, and community development. Types of support include program development, publication, seed money, and scholarships.

Requirements: 501(c)3 tax-exempt organizations in Pennsylvania's Dauphin, Cumberland, Franklin, and Perry Counties are eligible.

Restrictions: Grants do not support individuals, operating budgets, or capital expenses.

Geographic Focus: Pennsylvania

Samples: Elizabethtown College (PA)—for equipment, the research outreach and management program, and summer research stipends for students in the biology department, through the Whitaker Foundation Regional Program, $102,295 (2003).

Amount of Grant: $10,000 maximum; $2500-$5000 average

Contact: Rosemary Moore, Program Officer, (717) 236-5040; fax (717) 231-4463; email: rosemary@ghf.org

Internet: http://www.ghf.org

Sponsor: Foundation for Enhancing Communities

P.O. Box 678, 200 N 3rd Street

Harrisburg, PA 17108-0678

Foundation for Medical Research Grants　　　　　　　　　　　**2387**

The foundation promotes all forms of scientific medical research, in particular clinical research and that connected with the basic biological sciences directly or indirectly related to medicine. Grants are awarded to individuals and institutions in France and on an international basis. Types of support include subsidies for travel and study abroad and equipment acquisition. Major annual awards include the Rosen Prize for Cancer Research and the Delahautemaison Prize, which is awarded every two years, alternately for research in nephrology and cancer; the annual Grand Prix; and five prizes for molecular biology, endocrinology, immunology, clinical investigation, and neurobiology.

Amount of Grant: $10 million total annually

Contact: Program Contact, 33-1-44-39-75-75; fax: 33-1-44-39-75-99; email: frm@frm.org

Internet: http://www.frm.org

Sponsor: Foundation for Medical Research

54 rue de Varenne

Paris 75007 France

Foundation for Physical Therapy Mary McMillan Doctoral Scholarships   2388

The purpose of this program is to assist physical therapists with outstanding potential for doctoral studies in the first year of graduate studies towards a doctorate.

Requirements: Applicants must be entering post professional full-time or part-time doctoral study during the current academic year. Applicants must possess a license to practice physical therapy in a U.S. jurisdiction or have met all the requirements for licensure in a U.S. jurisdiction, including having received a passing score on the licensure exam.

Samples: Mark Diehl, PT, MPT, Virginia Commonwealth, $5000 (2004); Shauna D Dudley-Javoroski, PT, MPT, U of Iowa (IA)—scholarship recipient, $5000 (2004); Jill Stewart, PT, MPT, MS, NCS, University of Southern California CA)—scholarship recipient, $5000 (2004).

Amount of Grant: $5000

Date(s) Application is Due: Aug 15

Contact: Lucy Dickson, Program Contact, (800) 999-2782 ext 8505; email: foundation@apta.org

Internet: http://www.apta.org/AM/Template.cfm?Section=Awards_and_Scholarships&TEMPLATE=/CM/ContentDisplay.cfm&CONTENTID=23362

Sponsor: Foundation for Physical Therapy

1055 N Fairfax Street, Suite 350

Alexandria, VA 22314

Foundation for Physical Therapy Research Grants   2389

The foundation provides sole support or matching funds to individuals and groups of researchers to pursue scientifically based and clinically related physical therapy research, focused on studies initiated by emerging investigators. Emerging investigators are individuals who have had some previous research experience, may or may not have completed graduate or postgraduate work, or are beginning a new field of research or making strategic changes in research direction within an existing line of research. Proposed studies should add to or refine the body of theoretical, scientific, and clinical knowledge on which physical therapy practice is based, using any of a variety of recognized investigative methods such as experimental, descriptive, or correlational. Annual deadline dates may vary; contact program staff for exact dates.

Requirements: Proposals may be submitted by individuals or groups of investigators independently or through a sponsoring institution or organization with which they are affiliated. Groups must designate one member as the principal investigator responsible for directing the project. This person must be eligible for licensure as a physical therapist. Other members of the group may be physical therapists, physical therapist assistants, or persons from other fields that are relevant to the proposed study.

Amount of Grant: $40,000 maximum per year for up to two years

Date(s) Application is Due: Aug 15

Contact: Lucy Dickson, Scientific Program Administrator, (800) 999-2782 ext 8505; email: foundation@apta.org

Internet: http://www.apta.org/AM/Template.cfm?Section=Awards_and_Scholarships&TEMPLATE=/CM/ContentDisplay.cfm&CONTENTID=23363

Sponsor: Foundation for Physical Therapy

1111 N Fairfax Street

Alexandria, VA 22314-1488

Foundation for Psychocultural Research Doctoral Fellowships   2390

The foundation will competitively award limited dissertation research fellowships aimed at advancing interdisciplinary research projects and scholarship at the intersection of psychology, culture, neuroscience, and psychiatry, with emphasis on psycho-cultural factors as central, not peripheral. The dissertation topic will involve substantial engagement in Trauma and the Interaction of Early Relational Experiences, Social Context, and Developmental Trajectories. The interdisciplinary approach will emphasize the cultural and neuroscientific approaches to the study of PTSD. Grants will be awarded on a one-time basis. An institutional allowance is allotted to the host institution on a monthly basis for one year, to be used at the sponsor's discretion to help pay for the fellow's research supplies, travel to scientific meetings, or health insurance. If a deadline falls on a weekend, applications will be accepted on the following Monday. Application guidelines are available online.

Requirements: Candidates must be enrolled at an accredited U.S. university or institution. Applicant must be a doctoral candidate in anthropology, psychology, neuroscience, or health sciences, and should have interest in completing a dissertation involving interdisciplinary research in culture, neuroscience, and psychiatry.

Amount of Grant: $25,000; $1250 institutional allowance

Date(s) Application is Due: Feb 15

Contact: Director, fax: (310) 454-1417; email: info@thefpr.org

Internet: http://www.thefpr.org/programs_funding/applications/doctoral_app.html

Sponsor: Foundation for Psychocultural Research

P.O. Box 826

Pacific Palisades, CA 90272

Foundation for Psychocultural Research Postdoctoral Fellowships   2391

The foundation funds research aimed at advancing interdisciplinary research projects and scholarship at the intersection of psychology, culture, neuroscience, and psychiatry, with emphasis on psychocultural factors as central, not peripheral. The current topical focus is Interdisciplinary Approach to Post Traumatic Stress and Resilience. The research will integrate cultural and neuroscientific approaches to the study of PTSD, in its design and/or clinical application. Candidates must conduct their proposed research under a sponsor who holds an academic appointment at an accredited U.S. university or institution. The fellowship may be extended up to the maximum of three years, based on the evaluation of the previous year's performance. An institutional allowance is allotted monthly to the host institution, to be used at the sponsor's discretion to help pay for the fellow's research supplies, travel to scientific meetings, or health insurance. If a deadline falls on a weekend, applications will be accepted on Monday. Guidelines are available online.

Requirements: Applicants must have a doctoral or MD degree and should have interest in pursuing a career involving interdisciplinary research in neuroscience, culture, and psychiatry.

Amount of Grant: $40,000 stipend; $1500 per-year for supplies, travel or health insurance

Date(s) Application is Due: Feb 15

Contact: Director, email: foundationpsycul@aol.com

Internet: http://www.thefpr.org/programs_funding/applications/post_doctoral_app.html

Sponsor: Foundation for Psychocultural Research

P.O. Box 826

Pacific Palisades, CA 90272

Foundation for Science and Disability Grant Program for Physically Disabled Students in the Sciences   2392

A grant may be awarded annually to a disabled student who is interested in obtaining a graduate degree in science, mathematics, medicine, engineering, or computer science. Selection is based on financial need, sincerity of purpose, and scholarship and/or research ability. Funds may be used for an assistive device or instrument, or as financial support to work with a professor on an individual research project, or for some other special need. The number of grants awarded each year varies depending on funds available. Application must include a 250-word essay on professional goals and objectives, as well as the specific purpose for which the grant monies will be used, and two letters of recommendation from faculty members, one of whom must be the student's academic research advisor.

Requirements: College seniors who have some physical or sensory disability and who have been accepted to graduate or professional school, as well as students already in graduate or professional schools, are eligible to apply.

Restrictions: Undergraduate students are not eligible.

Amount of Grant: $1000

Date(s) Application is Due: Dec 1

Contact: Dr. Richard Mankin, Co-Chair, Science Student Grant Committee, (352) 374-5774; fax: (352) 374-5804; email: rmankin@nervm.nerdc.ufl.edu

Internet: http://www.as.wvu.edu/~scidis/organize/fsdinfo.html

Sponsor: Foundation for Science and Disability

503 NW 89th Street

Gainesville, FL 32607

Foundation for Seacoast Health Grants   2393

The mission of the foundation is to invest its resources to improve the health and well being of Seacoast residents. The foundation considers a very limited number of new grant initiatives which address one or more of the following prioritized health needs; access to affordable mental health services; access to preventative and restorative dental services; access to affordable child care and after school care; access to affordable primary medical care; and coordination and dissemination of health information related to identified priority needs. The deadline for the submission of the letter of intent is March 1 for Infants, Children and Adolescence and June 1 for Promoting Health and Preventing Disease. Annual deadline dates may vary; contact program staff for exact dates.

Requirements: Nonprofit organizations in cities and towns in the New Hampshire/Maine Seacoast area, including Greenland, New Castle, Newington, North Hampton, Portsmouth, and Rye, NH; and Eliot, Kittery, and York, ME, may submit applications.

Restrictions: Grants do not support ongoing general operating expenses, deficit elimination, political activities, travel, conferences, or lodging.

Geographic Focus: Maine; New Hampshire

Samples: School Administrative Unit #50—for a nutrition education program at Rye Elementary School, $631.; Seacoast Mental Health Ctr—for the New Heights junior staff training, $10,000.

Date(s) Application is Due: Mar 1; Jun 1

Contact: Susan Bunting, President, (603) 422-8200; email: ffsh@communitycampus.org

Internet: http://www.ffsh.org/grants.cfm

Sponsor: Foundation for Seacoast Health

100 Campus Drive, Suite 1

Portsmouth, NH 03801

Foundation for the Advancement of Mesoamerican Studies Grants   2394

The foundation awards grants to support scholarly contributions to the understanding of ancient Mesoamerican cultures (Mexico, Belize, Guatemala, Honduras, and El Salvador). The foundation supports projects in the disciplines of archaeology, art history, epigraphy, linguistics, ethnohistory, ethnography, and sociology. Interdisciplinary projects are encouraged, especially those that combine disciplines in novel and potentially productive ways. The program provides research grants and contingency grants. Contingency grants provide emergency funds for unforeseen situations encountered during initial research, secondary analysis, and/or the final dissemination of findings.

Restrictions: FAMSI does not provide funds for equipment purchases, salaries, stipends, or honoraria.

Amount of Grant: $500-$10,000 research grants

Contact: Sandra Noble, Director; (352) 795-5990; fax: (352) 795-1970; email: sandra@famsi.org

Internet: http://research.famsi.org/grants/
Sponsor: Foundation for the Advancement of Mesoamerican Studies
268 South Suncoast Boulevard
Crystal River, FL 34429-5498

Foundation for the Carolinas Grants   2395

The foundation supports nonprofit organizations in North and South Carolina, especially in the greater Charlotte, NC, area. The foundation awards grants in the following program areas: education, religion, human services, the environment, historic preservation, public and civic affairs, health, the arts, youths, and senior citizens. Building Youth grants help school-age children make the transition from youth to adulthood.
Requirements: The foundation awards grants to nonprofit organizations serving the following areas in North Carolina: Cebarrus, Cleveland, Iredell, Lincoln, Mecklenburg, Richmond, Stanley, and Union Counties and the cities of Lexiington and Salisbury; and Cherokee, Lancaster, and York Counties in South Carolina.
Restrictions: Grants are not awarded for capital campaigns, operating budgets, endowments, publications, equipment, videos, or conferences, nor to individuals.
Geographic Focus: North Carolina; South Carolina
Samples: Family Ctr (Charlotte, NC)—for a 10-week course designed to strengthen family relationships and to increase parents' involvement in their chldren's preschool education, $62,940 (2002); Richmond County Hospice (Rockingham, NC)—to construct a new residential hospice facility and an administrative center, $50,000 (2002); Boys and Girls Clubs (Gaffney, SC)—to purchase a sound system and wireless microphones for the drama program, $2380 (2002).
Contact: Dr. Don Jonas, Senior Vice President of Community Philanthropy, (704) 973-4500 or (800) 973-7244; fax: (704) 973-4599; email: djonas@fftc.org
Internet: http://www.fftc.org/grants
Sponsor: Foundation for the Carolinas
P.O. Box 34769
Charlotte, NC 28234-4769

Foundation for the Future Grants   2396

The program awards grants to support scholars undertaking research that is directly related to a better understanding of the factors affecting the quality of life of the long-term future of humanity. Preliminary Grant Applications must be submitted between January 1 and April 30 of the year in which grant funding is sought. An application form is available on the Web site.
Requirements: Individuals and organizations worldwide may apply.
Amount of Grant: $5000-$25,000
Date(s) Application is Due: Apr 30
Contact: Sesh Velamoor, Deputy Director, Programs, (425) 451-1333; fax: (425) 451-1238; email: seshvelamoor@futurefoundation.org
Internet: http://www.futurefoundation.org/grants/index.html
Sponsor: Foundation for the Future
123-105th Avenue SE
Bellevue, WA 98004-6265

Foundation Global Ethic Grants   2397

The foundation is based on the Declaration toward a Global Ethic endorsed by the Parliament of the World's Religions, which declares a commitment to a culture of nonviolence and respect for life, solidarity and a just economic order, tolerance and a life of truthfulness, and equal rights and partnership between men and women. Fields of interest include arts and humanities, education, international affairs, law and human rights, and social welfare. Types of support include grants to individuals and institutions, scholarships, fellowships, and prizes.
Contact: Stephan Schlensog, Managing Director, 49-7071-640334; fax: 49-7071-610140; email: office@stiftung-weltethos.uni-tuebingen.de
Internet: http://www.weltethos.org/dat_eng/index_e.htm
Sponsor: Foundation Global Ethic
Waldhauser Strasse 23
Tubingen 72076 Germany

Fragile X Syndrome Postdoctoral Research Fellowships Program   2398

The purpose of this program is to promote research aimed at finding a specific treatment for fragile X syndrome, the primary inherited cause of mental retardation. The foundation is particularly interested in preclinical studies of potential pharmacological and genetic treatments for fragile X and studies aimed at understanding the function of the FMR1 gene. The goal is to bring practical treatment into current medical practice as quickly as possible; therefore, preference will be given to research projects that have a clear practical application and the results of which will be shared with other qualified researchers in a timely fashion. Fellowships are awarded for one year and may be renewed for a second year.
Requirements: Tax-exempt 501(c)3 institutions are eligible for funding. Individuals are nominated by applicant institutions for the fellowships and should have training and experience at least equal to the PhD or MD level.
Amount of Grant: $40,000 maximum
Date(s) Application is Due: May 1; Dec 1
Contact: Katie Clapp, (978) 462-1866; fax: (978) 463-9985; email: kclappFRAXA@comcast.net
Internet: http://www.fraxa.org/research_howtoapply.aspx
Sponsor: FRAXA Research Foundation
45 Pleasant Street, 2nd Fl
Newburyport, MA 01950

Frances C. and William P. Smallwood Foundation Grants   2399

The foundation awards grants in support of education, higher education, and social services. Applications will be received from the following organizations; educational institutions, both public and private, primarily at the college or university level, as well as other non-profit organizations involved in American education and is confined to the area within a 100 mile radius of the Dallas/Fort Worth Metroplex, the area within a 100 mile radius of the city of Gainesville, FL, and the area within a 100 mile radius of Minden, Nevada. Potential applicants should request guidelines prior to applying.
Requirements: Contributions are limited to non-profit, tax-exempt organizations which have obtained IRS status under Section 501(c)3 of the IRS code, and, where appropriate, under Section 170(c).
Restrictions: Funding will not be considered for: general endowment funds of an organization; fund-raising events, i.e. tickets, dinner, telethons; corporate memberships or contributions to Chambers of Commerce, taxpayer associations, and other similar bodies; and contributions to political organizations, campaigns and candidates.
Geographic Focus: Florida; Nevada; Texas
Samples: U of North Carolina, School of Information and Library Science (Chapel Hill, NC)—for the Interaction Design Library to support the work of two doctoral students, to sponsor two symposia, and to purchase additional equipment for the library, $50,000.
Date(s) Application is Due: Feb 28
Contact: Rick Piersall, Foundation Contact; Suzanne Stockdale, (775) 782-3678 or Sally Muller, (352) 378-9646
Sponsor: Frances C. and William P. Smallwood Foundation
P.O. Box 2050
Fort Worth, TX 76113

Frances L. and Edwin L. Cummings Memorial Fund Grants   2400

The foundation supports nonprofits in New York City and northern New Jersey in the areas of education, especially programs that serve public school children from disadvantaged backgrounds; social welfare concerns; and campaigns to build endowments through establishment of challenge grants. Other areas of interest include elementary education, secondary education, vocational education, adult basic education and literacy, and higher education; hospitals, medical care, AIDS, and cancer; children, youth, and human services; and community development. Support is provided for endowment funds, seed money, consulting services, matching funds, technical assistance, professorships, and program development. The board meets in June and December. Application forms are not required.
Requirements: Grants are awarded to nonprofit organizations in the metropolitan New York, NY, area, with emphasis on New York City, southern Westchester County, and northern New Jersey.
Restrictions: Grants are not awarded to individuals or for capital building campaigns, general operating support, moving expenses, conferences, surveys, annual fund-raising campaigns, or research conducted by individuals or private institutions.
Geographic Focus: New Jersey; New York
Samples: Little Sisters of the Assumption Family Health Service (New York, NY)—to establish an endowment fund, $100,000.; Neighborhood Initiatives (Bronx, NY)—for staff support to expand a teen center program, $30,000.; Korean Community Service of Metropolitan New York (Woodside, NY)—to support a health educator and outreach worker for an HIV/AIDS prevention and education program, $40,000.
Amount of Grant: $10,000
Contact: Elizabeth Costas, Director, (212) 286-1778; fax: (212) 682-9458
Sponsor: Frances L. and Edwin L. Cummings Memorial Fund
501 Fifth Avenue, Suite 708
New York, NY 10017-6103

Francis Bacon Award in the History and Philosophy of Science & Technology   2401

The institute and the Francis Bacon Foundation support this biennial prize in the history of science, the history of technology, or historically engaged philosophy of science. The prize will be awarded biennially to an outstanding scholar whose work continues to have a substantial impact on any of the three fields. Submit a one-page letter of nomination that addresses the candidate's qualifications and potential; and a copy of the nominee's curriculum vita and contact information.
Amount of Grant: $20,000
Date(s) Application is Due: Dec 1
Contact: Sanja Ilic, Secretary to the Bacon Committee, (626) 395-4065; fax: (626) 405-9841; email: sanja@hss.caltech.edu
Internet: http://www.hss.caltech.edu/humanities/fbaward
Sponsor: California Institute of Technology
1200 E California Boulevard, Baxter 228-77
Pasadena, CA 91125

Francis Butler Simpkins Award   2402

The award is given in odd-numbered years for the best first book by an author or authors in the field of U.S. southern history published during the preceding two years. The application due date is March 1 of the year in which the prize is to be awarded. Books should be forwarded to the committee members listed on the Web site.
Date(s) Application is Due: Mar 1
Contact: Francis Butler Simpkins Award, (706) 542-8848; fax: (706) 542-2455
Internet: http://www.uga.edu/~sha/franci~1.htm
Sponsor: Southern Historical Association
University of Georgia, Department of History
Athens, GA 30602

Francis C. Wood History of Medicine Fellowships 2403

The institute offers fellowships to scholars engaged in projects requiring personal use of the historical collections of the library of the College of Physicians of Philadelphia and/or the collections of the Mutter Museum. The historical collections constitute one of the largest medical history repositories in the United States, documenting the evolution of medicine from the medieval period to the present. Its resources include rare medical books and journals; manuscript case records, papers, and lecture notes of many physicians; and prints, engravings, and photographs of medical subjects. The Mutter Museum is renowned for its collections of medical artifacts, instruments, and pathological and anatomical specimens. Recipients may be requested to present a seminar at the institute and will be required to submit a report on their research. Awards will be announced in March.

Requirements: Letters of application outlining the proposed project (proposal should not exceed five pages); length of residence; historical materials to be used; and a budget with specific information on travel, lodging, and research expenses should be submitted, along with a curriculum vita and two letters of recommendation.

Amount of Grant: $1000 maximum

Date(s) Application is Due: Apr 1

Contact: Margaret Patton, Program Contact, (215) 563-3737 ext 305; email: mpatton@collphyphil.org

Internet: http://www.collphyphil.org/woodfell.htm

Sponsor: College of Physicians of Philadelphia

19 S 22nd Street

Philadelphia, PA 19103

Francis C. Wood Scholar in Residence Program 2404

Short-term grants will be awarded to scholars engaged in projects requiring personal use of the historical collections of the library and/or Mutter Museum. Letters of application outlining the proposed project (proposal should not exceed five pages), necessary length of residence, historical materials to be used and a budget with specific information on travel, lodging, and research expenses should be submitted, along with a curriculum vitae and two letters of recommendation by January 31.

Requirements: The program is restricted to individuals pursuing advanced research. Candidates must have a PhD or an equivalent record of professional experience and scholarly publication.

Restrictions: The award will not be made to degree candidates or to persons seeking support for work leading to a degree.

Amount of Grant: $30,000, office, and research and computer facilities

Contact: Ed Morman, Committee on Awards, (215) 563-3737 ext 265; fax: (215) 561-6477; email: emorman@collphyphil.org

Internet: http://www.collphyphil.org/woodpg1.shtml

Sponsor: College of Physicians of Philadelphia

19 S 22nd Street

Philadelphia, PA 19103

Francis P. Garvan-John M. Olin Medal 2405

This award, sponsored by the the Francis P. Garvan-John M. Olin Medal Endowment and administered by the American Chemical Society, is given annually to recognize distinguished service to chemistry by women chemists who are U.S. citizens. The award consists of $5,000, a medallion with a presentation box, and a certificate. Reasonable travel expenses to the meeting at which the award will be presented will be reimbursed.

Requirements: A nominee must be a citizen of the United States and have performed distinguished service to chemistry. Any individual, except a member of the awards committee, may submit one nomination or seconding letter for the award in any given year. The nominating documents consist of a letter of not more than 1000 words containing an evaluation of the nominee's accomplishments and a specific identification of the work to be recognized, a biographical sketch including date of birth, and a list of publications and patents authored by the nominee.

Restrictions: Self-nominations are not accepted.

Samples: Elizabeth C. Theil—award winner, $5,000 (2008); Laura L. Kiessling—award winner, $5,000 (2007); Lila Gierasch—award winner, $5000 (2006).

Amount of Grant: $5,000 and an allowance of $1,000 toward traveling expenses to award meeting

Date(s) Application is Due: Nov 1

Contact: Felicia Dixon, Awards Administrator; (800) 227-5558 or (202) 872-4408; fax: (202) 776-8008; email: f_dixon@acs.org or awards@acs.org

Internet: http://portal.acs.org/portal/acs/corg/content?_nfpb=true&_pageLabel=PP_ARTICLEMAIN&node_id=1319&content_id=CTP_004521&use_sec=true&sec_url_var=region1

Sponsor: American Chemical Society

1155 Sixteenth Street, NW

Washington, DC 20036

Frank B. Sessa Scholarship for Continuing Education 2406

The scholarship is available to society members who are intending to enroll in programs for continuing education. Application form must be obtained from the society and returned with a resume and explanation of the proposed study or research. Applications for all scholarships and fellowships are accepted beginning November 15. Deadline for receiving all applications is March 15 of every year.

Amount of Grant: $1250

Date(s) Application is Due: Mar 15

Contact: Dr. Jane Robbins, Executive Director, School of Information Studies, (850) 644-3907; fax: (850) 644-9763; email: Beta_Phi_Mu@lis.fsu.edu

Internet: http://www.beta-phi-mu.org/scholarships.html

Sponsor: Beta Phi Mu International Library and Information Science Honor Society

Beta Phi Mu, School of Information Studies,

101 Louis Shores Building, Florida State University

Tallahassee, FL 32306-2048

Frank H. Field and Joe L. Franklin Award for Outstanding Achievement in Mass Spectrometry 2407

This award, sponsored by Bruker Daltonics, is given annually to recognize outstanding achievement in the development or application of mass spectrometry. The award is granted without regard to age, nationality, or the date of the achievement recognized by the award. In odd-numbered years the award will be presented for advances in techniques or fundamental processes in mass spectrometry. Recognition will be given in even-numbered years to development of the applications of mass spectrometry. The award consists of $5,000 and a certificate.

Requirements: Any individual, except a member of the award committee, may submit one nomination or seconding letter for each award in any given year. The nominating documents consist of a letter of not more than 1000 words containing an evaluation of the nominee's accomplishments and a specific identification of the work to be recognized, a biographical sketch including date of birth, and a list of publications and patents authored by the nominee.

Restrictions: Self-nominations are not accepted.

Samples: Catherine C. Fenselau—award winner, $5,000 (2008); Jean H. Futrell—award winner, $5,000 (2007); Richard Caprioli—award winner, $5000 (2006).

Amount of Grant: $5,000, certificate, and an allowance of up to $1,000 for travel expenses to award meeting

Date(s) Application is Due: Nov 1

Contact: Felicia Dixon, Awards Administrator; (800) 227-5558 or (202) 872-4408; fax: (202) 776-8008; email: f_dixon@acs.org or awards@acs.org

Internet: http://portal.acs.org/portal/acs/corg/content?_nfpb=true&_pageLabel=PP_ARTICLEMAIN&node_id=1319&content_id=CTP_004522&use_sec=true&sec_url_var=region1

Sponsor: American Chemical Society

1155 Sixteenth Street, NW

Washington, DC 20036

Frank Knox Memorial Fellowships 2408

The program enables graduates from Canada to spend one year studying at one of the faculties of Harvard University. Fellowships are available only to citizens of Canada. All applications and supporting documents must be submitted in English only.

Requirements: Applicants must be Canadian citizens or permanent residents who have graduated no earlier than the spring of 2003 or who will graduate before September 2005 from an institution which is a member, or affiliated with a member of AUCC.

Restrictions: Applications from students currently studying in the United States will not be considered.

Amount of Grant: $18,500

Date(s) Application is Due: Dec 31

Contact: Canadian Awards Program, (613) 563-1236; fax: (613) 563-9745; email: awards@aucc.ca

Internet: http://www.aucc.ca/programs/scholarships/knox_e.html

Sponsor: Frank Knox Fellowships

350 Albert Street, Suite 600

Ottawa, ON K1R 1B1 Canada

Frank Stanley Beveridge Foundation Grants 2409

The foundation welcomes proposals in the areas of: animal care; arts, culture and humanities; civil rights, social action, advocacy; education; employment/jobs; environmental quality, protection and beautification; food, nutrition and agriculture; health; housing; human services; medical research; mental health; philanthropy; safety; recreation; religion; science; social services; and youth development. The board meets in October and April to consider requests. Multiyear grants are rare. Contact the foundation via the Web site only. No phone or written inquiries will be accepted.

Requirements: Applicants must be 501(c)3 nonprofit organizations or foundations in Massachusetts's Hampden and Hampshire Counties.

Restrictions: The Foundation prefers not to support: awards or prizes; commissioning of new artistic work; conferences/seminars; curriculum development; debt reduction; employee matching gifts; employee-related scholarships; endowment funds; exhibitions; faculty/staff development; fellowship funds; fellowships to individuals; film/video/radio production; foundation administered programs; general operating support; grants to individuals; income development; internship funds; management development; performance/production costs; professorships; program-related investment/loans; publications; scholarships to individuals; student aid; technical assistance.

Geographic Focus: Massachusetts

Amount of Grant: $50,000

Date(s) Application is Due: Feb 1; Aug 1

Contact: Philip Caswell, President; (800) 229-9667; fax: (561) 748-0644; email: administrator@beveridge.org or caswell@beveridge.org

Internet: http://www.beveridge.org/

Sponsor: Frank Stanley Beveridge Foundation

1340 U.S. Highway 1, Suite 102

Jupiter, FL 33469

Franklin Mosher Baldwin Fellowships 2410

The Baldwin fellowship is intended for scholars with citizenship in an African country who seek to obtain an advanced degree or specialized training in an area of study related to human origins research. This award is for a program of approved special training and/or advanced training towards an MA, PhD, or equivalent and is limited to two years of support.

Requirements: Scholars with citizenship in an African country are eligible to apply.

Amount of Grant: $12,000 maximum per year

Date(s) Application is Due: Feb 15

Contact: Grants Officer, (415) 561-4646; fax: (415) 561-4647; email: grants@leakeyfoundation.org

Internet: http://www.leakeyfoundation.org/grants/g3.jsp

Sponsor: L.S.B. Leakey Foundation

P.O. Box 29346, 1002A O'Reilly Avenue

San Francisco, CA 94129-0346

Fraser-Parker Foundation Grants 2411

The foundation awards general operating grants in its areas of interest, including Christian religion organizations, education and higher education, and hospitals. There are no application forms.

Geographic Focus: Georgia

Amount of Grant: $5000-$50,000 average

Date(s) Application is Due: Apr 15; Oct 15

Contact: John Stephenson, Executive Director, (404) 658-9066

Sponsor: Fraser-Parker Foundation

50 Hurt Plz, Suite 850

Atlanta, GA 30303

Frederic Stanley Kipping Award in Silicon Chemistry 2412

This award, supported by Dow Corning Corporation, is given biennially in even-numbered years to recognize distinguished achievement in research in silicon chemistry and to stimulate the creativity of others toward further advancement of this field of chemistry. A nominee must have accomplished distinguished achievement in research in silicon chemistry during the preceding 10 years. The measure of this achievement should focus primarily on the nominee's significant publications in the field of silicon chemistry but may include consideration of contributions to the related field of organometallic chemistry. There are no limits on age or nationality. Applications are accepted in odd-numbered years. The award consists of $5,000 and a certificate.

Requirements: Any individual, except a member of the award committee, may submit one nomination or seconding letter for the award in any given year. Nominating documents consist of a letter of not more than 1000 words containing an evaluation of the nominee's accomplishments and a specific identification of the work to be recognized, a biographical sketch including date of birth, and a list of publications and patents authored by the nominee.

Restrictions: Self-nominations are not accepted.

Samples: T. Don Tilley—award winner, $5,000 (2008); Akira Sekiguchi—award winner, $5,000 (2006); James E. Mark—award winner, $5,000 (2004).

Amount of Grant: $5,000 and allowance for travel expenses to award meeting

Date(s) Application is Due: Nov 1

Contact: Felicia Dixon, Awards Administrator; (800) 227-5558 or (202) 872-4408; fax: (202) 776-8008; email: f_dixon@acs.org or awards@acs.org

Internet: http://portal.acs.org/portal/acs/corg/content?_nfpb=true&_pageLabel=PP_ARTICLEMAIN&node_id=1319&content_id=CTP_004523&use_sec=true&sec_url_var=region1

Sponsor: American Chemical Society

1155 Sixteenth Street, NW

Washington, DC 20036

Frederick Douglass Summer Fellowships 2413

The summer fellowships offer teaching oppotunities to minority graduate students who are pursuing careers in college teaching and are in their final year of a terminal degree program. The fellowships are intended to attract graduate studnts primarily from historically black colleges and universities. Fellows will teach and/or team teach one course during a five-week summer session.

Requirements: Applicants must be U.S. citizens, and must have completed all doctoral work, except the dissertation, by the end of the next academic year.

Amount of Grant: $2381

Date(s) Application is Due: Jan 31

Contact: Fellowships Administrator, (814) 393-2109; email: info@clarion.edu

Internet: http://www.clarion.edu/admin/socequ/fd.shtml

Sponsor: Clarion University

The Frederick Douglass Teaching Scholars, c/o Clarion University, Office of Social Equity, 207 Carrier

Clarion, PA 16214

Frederick Douglass Teaching Scholars 2414

The Pan African Studies program of the university offers teaching opportunities to graduate students entering the final year of terminal degree programs and preparing to be college teachers, and for the purpose of attracting graduate students from historically underrepresented and underserved populations. Application information and form are available on the Web site.

Requirements: Minimum requirements are: master's degree completed or equivalent and currently enrolled in a terminal degree program; academic background in one of the fields taught at the university; and at least three favorable letters of recommendation from faculty or professionals in the student's field.

Date(s) Application is Due: Feb 9

Contact: Dr. Veronica Watson, Coordinator, The Frederick Douglass Institute, (724) 357-2280; email: maat@grove.iup.edu

Internet: http://www.chss.iup.edu/fdi/teach_scholars.htm

Sponsor: Indiana University of Pennsylvania

421 North Walk, 110 Leonard Hall

Indiana, PA 15705

Frederick Gardner Cottrell Foundation Grants 2415

Giving primarily in Woods Hole, Massachusetts, Hunt Valley, Maryland, Brunswick, Maine, and Houston, Texas. The foundation awards grants to educational and scientific organizations to advance research and chair teaching positions. The foundation's areas of interest include eye research, higher education, marine science, and science. There are no application deadline. Submit a written request for application.

Geographic Focus: Maine; Maryland; Massachusetts; Texas

Amount of Grant: $100,000-$300,000 average

Contact: Gary Munsinger, President; (520) 748-4400; fax: (520) 748-0025; email: munsinger@rctech.com

Sponsor: Frederick Gardner Cottrell Foundation

5210 East Williams Circle, Suite 240

Tucson, AZ 85711-4410

Freedom of Information Internships/Kilgore 2416

The foundation accepts applications each year for two freedom of information internships with the Society of Professional Journalists. The 10-week internships enable two students to research and write about freedom of information issues while assisting in the preparation of the annual SPJ Freedom of Information Report. Each intern will work during a mutually agreeable 10-week period in the late spring or summer. A candidate must submit a completed application form, a resume, a two- to three-page essay detailing expectations from experience, up to five samples of writing, a listing of past FOI-related activities, names and phone numbers of three references, and a black-and-white photograph suitable for publication. Applicants' overall GPA and GPA for journalism courses will be considered.

Requirements: Candidates must be senior or graduate journalism students, or law students with journalism backgrounds.

Amount of Grant: $400 per week

Date(s) Application is Due: Mar 1

Contact: Grants Administrator, (317) 927-8000; fax: (317) 920-4789; email: quill@spj.org

Internet: http://www.spj.org/internships_pk.asp

Sponsor: Sigma Delta Chi Foundation

3909 N Meridian Street

Indianapolis, IN 46208-4045

Freedom of the Press Fellowships and Internships 2417

The Reporters Committee hires a journalism fellow each year. The journalism fellow plans, supervises, and edits articles for the website, newsletter, and magazine and works with staff attorneys to ensure that all work is appealing to and useful for journalists, not just lawyers. The Reporters Committee hires two or three legal fellows each year. Fellows will monitor significant developments in media law, assist with responding to legal defense requests from reporters, prepare legal memoranda, and handle other special projects. In addition, fellows will write for the Committee's publications, the quarterly magazine The News Media & The Law, and the bi-weekly newsletter, News Media Update.

Requirements: A journalism fellowship candidate must have at least three years experience as a reporter or editor, and should have a background in free press issues.

Contact: Rebecca Daugherty, Internship Coordinator, (703) 807-2100; fax: (703) 807-2109; email: rfcp@rcfp.org

Internet: http://www.rcfp.org/interns.html

Sponsor: Reporters Committee for Freedom of the Press

1101 Wilson Boulevard, Suite 1100

Arlington, VA 22209

Freeman Fellowships 2418

Grants to aid and encourage young engineers, especially in research work, are made to fund expenses for experiments, observations, and compilations to discover new and accurate data that will be useful in engineering. The fellowships also recognize underwriting fully or in part some of the loss that may be sustained in the publication of meritorious books, papers, or translations pertaining to hydraulic science and art that might, except for some such assistance, remain mostly inaccessible. Traveling scholarships are available to members younger than 45 years of age in recognition of achievement or promise for the purpose of aiding the candidate to visit engineering works in the United States or any other part of the world where there are good prospects of obtaining information useful to engineers. The grant may be used for assisting in the translation or publication in English of papers or books in foreign languages pertaining to hydraulics. The grant is available annually. In a given year there may be more than one recipient, or no recipient if no meritorious applications are received. Stipend is based on funds available annually from the endowment. An application form is available upon request.

Requirements: Application consists of a completed form and a statement in general terms of the purposes for which the funds are expected to be used. Any member of ASCE is eligible to apply.

Amount of Grant: $2000-$5000
Date(s) Application is Due: Feb 9
Contact: Grants Coordinator, (800) 548-2723; fax: (703) 295-6132; email: student@asce.org
Internet: http://www.asce.org/inside/stud_freeman.cfm
Sponsor: American Society of Civil Engineers
1801 Alexander Bell Drive
Reston, VA 20191-4400

Freeman Scholar Award 2419

This program is conducted biennially in even-numbered years; a person of wide experience in fluids engineering is selected as the Freeman scholar. He/she is expected to review a coherent topic in his/her specialty, including a comprehensive statement of the state of the art, and to suggest key future research needs. The results will be presented at the winter meeting and published in the ASME Journal of Fluids Engineering. Deadline date for applications and nominations is February 1 of odd-numbered years.
Requirements: The recipient may be from industry, government, education, or private professional practice, and need not be an ASME member.
Samples: Gary Settles—award winner, $7500 (2004); Efstathios E. Michaelides—award winner, $7500.
Amount of Grant: $7500 and expense supplement
Date(s) Application is Due: Aug 1
Contact: Gulda DiTullio, Manager, (212) 591-7736; fax: (212) 705-7739; email: ditulliog@asme.org
Internet: http://www.asme.org/honors/ms71/sla/freeman.html
Sponsor: American Society of Mechanical Engineers
3 Park Avenue
New York, NY 10016

French-American Studies Fellowships 2420

The Institut Francais de Washington offers two Gilbert Chinard fellowships and an Edouard Morot-Sir fellowship in literature for maintenance (excluding travel) during research in France for a period of at least two months. The applicable fields of French study are art, economics, history, history of science, linguistics, literature, and social sciences. There is no application form; applicants must write two pages maximum describing their research projects and planned trips, and include curriculum vita. A letter from the dissertation director is also required for PhD candidates. Upon return, the awardee will send a brief report to the institute.
Requirements: Applications are invited from U.S. citizens or permanent residents of the United States who are PhD candidates at the final stage of their dissertations or who have held the PhD for no more than six years prior to January 15 of the award year. Candidates should be studying, teaching, or doing postdoctoral research at an American university.
Amount of Grant: $1500
Date(s) Application is Due: Jan 15
Contact: Catherine Maley, Department of Romance Languages, (919) 962-0154; fax: (919) 962-5457; email: cmaley@email.unc.edu
Internet: http://www.unc.edu/depts/institut/institutapp.htm
Sponsor: Institut Francais de Washington
The University of North Carolina, Department of Romance Languages & Literatures, CB# 3170
Chapel Hill, NC 27599-3170

Frese Senior Research Fellowship 2421

The fellowship is awarded for study in the history, theory, and criticism of sculpture, prints and drawings, and the decorative arts of any geographical area and of any period. Applications are also solicited from scholars in other disciplines whose work examines visual material in these media. The fellowship is for full-time research. Fellows are expected to reside in Washington, DC, throughout the academic year and participate in the activities of the center. Lectures, colloquia, and informal discussions complement the fellowship program. The award is limited to one-half of the applicant's salary. Senior Fellows who relocate to Washington are eligible for a housing allowance (up to $12,000.
Requirements: The fellowship is intended for those who have held the PhD for five years or more or who possess an equivalent record of professional accomplishment at the time of application.
Amount of Grant: $50,000 maximum
Date(s) Application is Due: Oct 1
Contact: Fellowships Program, Center for Advanced Study in the Visual Arts, (202) 842-6482; fax: (202) 789-3026; email: advstudy@nga.gov
Internet: http://www.nga.gov/resources/casvasen.htm
Sponsor: National Gallery of Art
Fourth Street and Constitution Avenue NW
Washington, DC 20565

Friedman-Klarreich Family Foundation Grants 2422

The foundation supports innovative projects, programs, and research in educational, social, and cultural areas. Grants fund efforts to improve educational and economic equality for girls and women and to enhance the stability of families.
Requirements: Organizations must have 501(c)3 tax-exempt status to be eligible.
Samples: The Teenage Parent Ctr (Akron, OH)—for program support, $5000.; Family Education Foundation (Saratoga, CA)—for program support, $5000.
Amount of Grant: $5000 maximum

Date(s) Application is Due: Jun 1
Contact: Susan Klarreich, Grants Administrator, fax: (650) 948-3148; email: klarff@worldnet.att.net
Sponsor: Friedman-Klarreich Family Foundation
551 Tyndall Street
Los Altos, CA 94022

Friedrich Ebert Foundation Doctoral Research Fellowships 2423

These fellowships are intended to provide doctoral candidates in the fields of sociology, political science, economics, and history an opportunity to conduct research necessary for their dissertations in Germany. A German university professor competent in the subject of study will assist fellowship holders and provide academic advice during their stay. This assistance also applies in the case of doctoral candidates who intend to spend most of their time engaged in archival, library, or other types of research outside the university. Sociohistorical studies, studies in contemporary history, and studies on current political problems (including comparative studies) will be given special consideration. These fellowships will be granted for stays of between five and 12 months. Guidelines are available from the foundation's U.S. offices in New York City or from the foundation headquarters at Godesberger Allee 149, D-53715 Bonn, Germany; (49) 228 883-0.
Requirements: The applicant must be a qualified PhD candidate at an American university and must have completed all of the prerequisites for a doctorate except for the dissertation, have an approved dissertation proposal and submit a recommendation from his/her academic advisor, indicate a German counterpart who would be available for cooperation and assistance during his/her stay in the Federal Republic, be a U.S. citizen, and provide evidence that his/her knowledge of German is adequate for research purposes.
Amount of Grant: DM1390 monthly maintenance allowance plus airfare between the United States and Berlin; domestic travel expenses; an allowance for luggage costs, health insurance, and books; tuition and fees if applicable; and family allowance
Date(s) Application is Due: Feb 28
Contact: Doctoral Research Fellowships, (212) 687-0208; fax: (212) 687-0261; email: info@fesny.org
Internet: http://www.fesny.org
Sponsor: Friedrich Ebert Foundation
823 United Nations Plz, Suite 711
New York, NY 10017

Friedrich Ebert Foundation Postdoctoral-Young Scholar Fellowships 2424

These fellowships are intended to provide young scholars who have already accumulated a certain amount of experience in teaching and/or research an opportunity to conduct independent research in Germany in the disciplines of political science, sociology, history, and economics. Priority consideration will be given to applicants who wish to carry out studies on politically relevant subjects, particularly those involving a comparative approach. Considerable importance is attached to the establishment of contacts and cooperative relationships with competent German counterparts in the same field. The fellowships will be granted for stays of between five and 12 months. Guidelines are available from the foundation's U.S. offices in New York City or from the foundation headquarters, at Godesberger Allee 149, D-53170 Bonn, Germany; (49) 228 883-0.
Requirements: Applicant must have a PhD or equivalent university degree and at least two years of subsequent experience in research and/or teaching at a university or in a related research institution; be a U.S. citizen; submit copies of relevant academic publications; indicate a German counterpart who would be available for cooperation and assistance during his/her stay in Germany; and have an appropriate knowledge of German.
Amount of Grant: DM1700 monthly maintenance allowance plus airfare between the United States and Germany; domestic travel expenses; an allowance for luggage costs, health insurance, and books; tuition and fees if applicable; and a family allowance
Date(s) Application is Due: Feb 28
Contact: Postdoctoral-Young Scholars Fellowships, (202) 331-1819; fax: (202) 331-1837; email: fesdc@fesdc.org
Internet: http://www.fesdc.org
Sponsor: Friedrich Ebert Foundation
1155 15th Street, NW, Suite 1100
Washington, DC 20005

Friedrich Ebert Foundation Predissertation-Advanced Graduate Fellowships 2425

These fellowships are intended to provide advanced graduates an opportunity for study and research in Germany in the disciplines of political science, sociology, history, and economics; an opportunity to familiarize themselves with the current state of research in their fields in Germany; and to conduct independent research. Applicants should have special interest in contemporary or past German or European affairs and/or German-American relations, particularly comparative studies. These fellowships are designed to enable the holders to engage in advanced studies at a university in Germany under the guidance of a German university professor. The expected outcome is the development of a dissertation proposal or the accomplishment of a specific research issue. The fellowship will be granted for a stay of between five to 12 months. Additional information and application guidelines may be obtained from the foundation's U.S. offices in New York City or from the foundation headquarters at Godesberger Allee 149, D-53170 Bonn, Germany, (49) 228 883-0. Applications must be filed at foundation headquarters.
Requirements: The applicant must be a qualified graduate student intending to pursue a doctoral degree and/or participating in a specific research project, must have successfully completed at least two years of graduate studies at an American university before initiating the intended period of study in Germany, be a U.S. citizen, indicate a German counterpart who would be available for cooperation and assistance during his/her stay in the Federal Republic, and provide proof that he/she has sufficient knowledge of

German. An applicant must also submit a description of his/her study and/or research objective and a recommendation from the graduate advisor at the American university he/she is attending.

Amount of Grant: DM1250 monthly maintenance allowance plus airfare between the United States and Berlin; domestic travel expenses; an allowance for luggage costs, health insurance, and books; and tuition and fees if applicable.

Date(s) Application is Due: Feb 28

Contact: Predissertation-Advanced Graduate Fellowships, (212) 687-0208; fax: (212) 687-0261; email: fesny@fesny.org

Internet: http://www.fesny.org

Sponsor: Friedrich Ebert Foundation

823 United Nations Plaza, Suite 711

New York, NY 10017

Fritz B. Burns Foundation Grants 2426

The foundation supports nonprofit organizations in southern California by awarding grants for education, with an emphasis on buildings, equipment, endowments (except for ordinary operating expenses), student scholarship and loan funds, and faculty fellowships; to/for hospitals, hospital equipment, and medical research; and religious organizations (Christian, Jewish, Latter-day Saints, nondenominational, Presbyterian, Protestant, Roman Catholic, and Salvation Army). Proposals should be concise, containing a brief description of what is planned, with a clear statement of the objective sought; IRS letter certifying tax exemption; financial statements; and a list of officers and directors. No formalized application or proposal format is required. Proposals are considered on a quarterly basis.

Requirements: Nonprofit, tax-exempt organizations in southern California may apply.

Restrictions: Grant requests are not considered from individuals nor from tax-supported entities.

Geographic Focus: California

Samples: Freeman Hospitals Foundation (Inglewood, CA)—to complete a dialysis center, $340,000 over four years.; Mount Saint Mary's College (Los Angeles, CA)—to endow a professorship in education, $1.5 million.

Amount of Grant: $10,000-$250,000 average

Contact: Joseph Rawlinson, President, (818) 840-8802

Sponsor: Fritz B. Burns Foundation

4001 W Alameda Avenue, Suite 203

Burbank, CA 91505-4338

From The Heart Productions Grants 2427

The program's goal is to make films that might otherwise never be made. Grants are made in a variety of categories, including Writers Grant (August 30 deadline); Editing Grant (September 30 deadline); New York Grant (April 30 deadline); LA Film Grant (May 30 deadline); and LA Video Grant (June 30 deadline). The program seeks film and video projects that are unique and benefit society; new projects; and works-in-progress. Length is not a consideration. New York City and Los Angeles grants are not restricted geographically; anyone may apply for either or both grants.

Requirements: This grant is available for shorts and low-budget independents as well as documentary filmmakers. Student filmmakers, independent producers, or independent production companies are all welcome to apply. There is a $38 application fee ($28 for students).

Restrictions: Employee of From the Heart Productions, an employee of any of the sponsors, or a family member of any employee of From the Heart Productions or any of the sponsors are ineligible.

Date(s) Application is Due: Apr 30; May 30; Jun 30; Aug 30; Sep 30

Contact: Carole Dean, email: Caroleedean@att.net

Internet: http://www.fromtheheartproductions.com/index.shtml

Sponsor: From The Heart Productions

1455 Mandalay Beach Road

Oxnard, CA 93035-2845

FTE Dissertation Fellowships 2428

The fellowships support African American PhD or ThD students in their final year of dissertation work in religious or theological studies. The objectives of the program are to increase completion rates of African American students in the field of theological studies; collaborate with graduate programs in identifying and supporting students at this juncture of their studies; and offer doctoral students special opportunities for scholarly exchange and peer contacts, as well as assistance in the enhancement of writing and research skills. The fellowship includes a stipend for living expenses and other educational costs; full tuition grants; and attendance at a summer conference of workshops to assist in the development of research, writing, and teaching skills. Applicants must be able to write full time during the fellowship year. Application materials are available on the Web site.

Requirements: U.S. African American PhD or ThD students at the final writing stage of their graduate work in religious or theological studies are eligible. The applicant's dissertation research proposal and writing plan must have been approved by the dissertation committee.

Amount of Grant: $15,000 stipend

Date(s) Application is Due: Feb 1

Contact: Sharon Watson Fluker, Director, (404) 727-1450; email: sfluker@thefund.org

Internet: http://www.thefund.org/programs/fellowships/dissertation/index.html

Sponsor: Fund for Theological Education

825 Houston Mill Road, Suite 250

Atlanta, GA 30329-4211

FTE Doctoral Fellowships 2429

The fellowships support African American students entering a PhD or ThD program in religious or theological studies. The objectives of the program are to increase the number of qualified African American students in the field of theological studies; establish partnerships with graduate programs and theological schools to strengthen their efforts toward the recruitment and support of these students; provide opportunities for intellectual stimulation and encouragement through conferences, mentoring, and peer relationships; and undertake assessment of how racial/ethnic minority doctoral candidates are best supported in their work and the factors that contribute to successful completion of their programs. The fellowship includes a stipend for living expenses and other educational costs; full tuition grants; and attendance at a summer conference of workshops to assist in the development of research, writing, and teaching skills. Application materials are available on the Web site.

Requirements: African American students entering their first year of an accredited graduate program leading to a PhD or ThD degree in religious or theological studies are eligible. Applicants must be committed to providing leadership within theological education and strongly considering a career in seminary teaching and research.

Amount of Grant: $15,000 maximum

Date(s) Application is Due: Mar 1

Contact: The Rev. Ellen Echols Purdum, Director, (404) 727-1450; email: epurdum@thefund.org

Internet: http://www.thefund.org/programs/fellowships/doctoral/index.html

Sponsor: Fund for Theological Education

825 Houston Mill Road, Suite 250

Atlanta, GA 30329-4211

FTE Expanding Horizons Partnership Grants 2430

The program supports African American students in PhD and ThD programs in religious and theological studies. It links educators, academic institutions, and those pursuing doctoral work for the purpose of promoting quality in the preparation for and practice of scholarly research and teaching in theological seminaries. The program supports doctoral fellows entering their first year of a PhD or ThD program in religious or theological studies; and dissertation support to fellows at the final stage of their doctoral work in religious or theological studies. Dissertation fellowship applications are due February 1; doctoral fellowship applications are due March 1.

Requirements: African American students in PhD and ThD programs in religious and theological studies are eligible.

Amount of Grant: $15,000

Date(s) Application is Due: Feb 1; Mar 1

Contact: The Rev. Ellen Echols Purdum, (404) 727-1450; fax: (404) 727-1490; email: sfluker@thefund.org or fte@thefund.org

Internet: http://www.thefund.org/programs/horizons/index.html

Sponsor: Fund for Theological Education

825 Houston Mill Road, Suite 250

Atlanta, GA 30329-4211

FTE North American Doctoral Fellowships 2431

The fellowships support talented racial or ethnic minority students currently enrolled in a PhD or ThD program in religious or theological studies. Candidates must pursue full-time study with high academic performance; be committed to teaching and scholarship; and demonstrate capacity for leadership in theological education. Preference is given to students nearing the end of their studies. Application materials are available online.

Requirements: U.S. or Canadian PhD or ThD students who are members of racial or ethnic groups traditionally underrepresented in graduate education are eligible.

Restrictions: DMin students are ineligible.

Amount of Grant: $5000 maximum

Date(s) Application is Due: Mar 1

Contact: Sharon Watson Fluker, Director, (404) 727-1450; email: sfluker@thefund.org

Internet: http://www.thefund.org/programs/fellowships/doctoral/northamerican.html

Sponsor: Fund for Theological Education

825 Houston Mill Road, Suite 250

Atlanta, GA 30329-4211

Fulbright Alumni Initiatives Awards Program Grants 2432

The objective of the AIA program is to help translate the individual Fulbright experience into long-term institutional impact. To this end, the program provides small institutional grants to Fulbright alumni to develop innovative projects that will foster institutionally supported linkages and sustainable, mutually beneficial relationships between the Fulbright scholar's home and host institutions. Just about any activity resulting in the creation or fostering of a sustainable institutional relationship that will have an impact on both the Fulbright alum's home institution and the Fulbright host institution abroad and which both institutions are prepared to support in both the long and short terms will be considered. The program has been temporarily suspended, and applicants should check back for updates.

Requirements: Applicants must be U.S. citizens. The program is open to eligible alumni whose grants occurred at any time from the 1998/1999 academic year through the present academic year.

Restrictions: This program is limited to application by a Fulbright alum (whether U.S. or Visiting Scholar) in partnership with his/her host institution colleague. However, other faculty members from both applicants' institutions may be part of the proposal and participate in the project.

Contact: Stacey Bustillos, Program Officer; (202) 686-6252; email: sbustillos@cies.iie.org; Hajra Zahid, Senior Program Associate; (202) 686-6250; email: hzahid@cies.iie.org
Internet: http://www.cies.org/aia/
Sponsor: Institute of International Education, Council for International Exchange of Scholars
3007 Tilden Street, NW, Suite 5L
Washington, DC 20008-3009

Fulbright Beginning Professional Journalism Awards   2433

The program provides five Professional Journalism Awards for beginning professional journalists or recent graduates in journalism or related fields to conduct independent research and serve as interns or guest journalists at newspapers, publishers or television and radio stations in Germany. The grantee will spend the first few months studying and conducting independent research, followed by one or more internships or guest journalist positions at German newspapers, publishers or television and radio stations. All Journalism areas including photo, print, and broadcast journalism as well as electronic media will be considered. Research and internship affiliation will be arranged by the Fulbright Commission in Berlin; grantee preferences will be accommodated as far as possible. Research may be conducted at a German university or other relevant institution. University matriculation in Germany (actual coursework is optional) will be required.
Requirements: Solid proficiency in German commensurate with the project is required.
Amount of Grant: 1300 Euro per month
Contact: Jon Adler, Program Manager; (212) 984-5326; email: jadler@iie.org
Internet: http://us.fulbrightonline.org/program_country.html?id=41#journ
Sponsor: Institute of International Education
809 United Nations Plaza
New York, NY 10017-3580

Fulbright Binational Business Grants in Mexico   2434

This program is designed to enhance the knowledge, expertise, and understanding of business in Mexico for U.S. students in business, law, or engineering. It combines graduate courses (a minimum of 3 and a maximum of 6) in the pertinent academic area (e.g., business, finance, international trade, or comparative law among others) with an internship with a Mexico-based company, firm, or NGO dedicated to international business. Candidates with a BA degree and recent M.B.A., J.D. or master, degree in business administration, finance, economics, international relations, engineering, or accounting, and some work experience are preferred. A Master, degree is recommended but not required. Significant work experience in a business environment is strongly recommended for these candidates and for candidates who do not have a business-related degree.
Requirements: Applicants must: be U.S. citizens at the time of application (permanent residents are not eligible); be in good health (grantees will be required to submit a satisfactory Medical Certificate of Health from a physician); and hold a B.A. degree or the equivalent before the start of the grant (applicants who have not earned a B.A. degree or the equivalent, but who have extensive professional study and/or experience in business, may be considered). Excellent spoken and written Spanish is required at the time of application.
Contact: Jody Dudderar, Program Manager; (212) 984-5565; fax: (212) 984-5325; email: jdudderar@iie.org
Internet: http://us.fulbrightonline.org/program_country.html?id=70#binational
Sponsor: Institute of International Education
809 United Nations Plaza
New York, NY 10017-3580

Fulbright Business Grants in Spain   2435

These awards are co-sponsored by the Instituto de Empresa Foundation in Madrid. Instituto de Empresa is one of Europe's leading business schools and has hosted nearly 28,000 students, representing 71 different nationalities. Awards are for full-time study in the bilingual English/Spanish International MBA program which offers the basic MBA core, two components of elective courses, and a Global Affairs Seminar in which students analyze business problems focusing on five different geographic areas.
Requirements: Applicants must: be U.S. citizens at the time of application (permanent residents are not eligible); be in good health (grantees will be required to submit a satisfactory Medical Certificate of Health from a physician); and hold a B.A. degree or the equivalent before the start of the grant (applicants who have not earned a B.A. degree or the equivalent, but who have extensive professional study and/or experience in business, may be considered). Excellent spoken and written Spanish is required at the time of application.
Contact: Jon Adler, Program Manager; (212) 984-5326; email: jadler@iie.org
Internet: http://us.fulbrightonline.org/program_country.html?id=98#MBA
Sponsor: Institute of International Education
809 United Nations Plaza
New York, NY 10017-3580

Fulbright Critical Language Enhancement Awards   2436

The Award, a component of the National Security Language Initiative, is available to students who have been awarded a Fulbright U.S. Student full grant and who intend to use the eligible languages in their Fulbright project. Students must also demonstrate the equivalent of an academic year of study in that language by the grant start date. Application for an Award is made in conjunction with the Fulbright Program application. Awards will be granted based on recommendations made by the selection committees, and final selections by Fulbright Commissions or U.S. Embassies in the corresponding countries. The purpose of the Award is to cultivate higher levels of language proficiency prior to and during the Fulbright grant period and beyond. Ultimately, awardees will achieve a high level of proficiency in a targeted language and will go on to careers or further study which will incorporate the use of this and/or related languages. The languages available for the Award are: all dialects of Arabic, Azeri, Bengali, Chinese (Mandarin only), Farsi, Gujarati, Hindi, Korean, Marathi, Pashto, Punjabi, Russian, Tajik, Turkish, Urdu, and Uzbek. It is preferred that language training take place in the same country as that of the award.
Requirements: The grant includes pre- and post-testing in the relevant language, a minimum of 20 hours of formal study per week in-country, and a commitment to study beyond the Critical Language Enhancement Award training period. Applicants must: be U.S. citizens at the time of application (permanent residents are not eligible); be in good health (grantees will be required to submit a satisfactory Medical Certificate of Health from a physician); and hold a B.A. degree or the equivalent before the start of the grant (applicants who have not earned a B.A. degree or the equivalent, but who have extensive professional study and/or experience in fields in which they wish to pursue a project, may be considered). In the creative and performing arts area, four years of professional study and/or experience meets the basic eligibility requirement.
Restrictions: Due to scheduling constraints, Fulbright English Teaching Assistants (ETAs) are ineligible for this award. The following persons are ineligible: anyone who has already held a U.S. Department of State-funded Fulbright student grant of any type; anyone who has previously received a Department of Education-funded Doctoral Dissertation Research Abroad (Fulbright Hays grant); employees of the U.S. Department of State, and their immediate families, for a period ending one year following termination of such employment; employees of private and public agencies (excluding educational institutions) under contract to the U.S. Department of State to perform administrative or screening services on behalf of the U.S. Department of State, exchange program, for a period ending one year following the termination of their services for the U.S. Department of State provided such employees have been directly engaged in performing services related to the exchange programs; applicants holding a doctoral degree at the time of application; applicants seeking enrollment in a medical degree program abroad; or applicants currently residing in the countries of Australia, Belgium/Luxembourg, Canada, Chile, Finland, Hungary, Mexico, Netherlands, New Zealand, Sweden, or Switzerland.
Contact: Jermaine Jones, Program Manager; (212) 984-5341; email: jjones@iie.org
Internet: https://us.fulbrightonline.org/thinking_clea.html
Sponsor: Institute of International Education
809 United Nations Plaza
New York, NY 10017-3580

Fulbright Distinguished Chairs Program Grants   2437

The Fulbright Distinguished Chairs Program comprises approximately forty distinguished lecturing, distinguished research and distinguished lecturing/research awards ranging from three to 12 months. Awards in the Fulbright Distinguished Chairs Program are viewed as among the most prestigious appointments in the Fulbright Scholar Program. Candidates should be senior scholars and have a significant publication and teaching record. Applicants should submit hard copies of the Distinguished Chairs Application Form (one page), a letter of interest (about three pages), a curriculum vitae (maximum eight pages) and, if required, a sample syllabus (maximum four pages). Chairs are available in Australia, Austria, Brazil, Canada, Denmark, Finland, France, Germany, Hungary, Ireland, Israel, Italy, Netherlands, Poland, Portugal, Russia, and Sweden. Because an objective of the Fulbright Program is to provide an educational exchange experience for those not previously afforded such an opportunity, preference will usually be given to candidates who have not had substantial recent experience in the country to which they are applying.
Requirements: Applicants must meet all of the following eligibility *Requirements*: U.S. citizenship at the time of application (permanent resident status is not sufficient); possess a Ph.D. or equivalent professional terminal degree at the time of application (for professionals and artists outside academe, recognized professional standing and substantial professional accomplishments); have college or university teaching experience at the level and in the field of the proposed lecturing activity as specified in the award description (for Distinguished Chairs awards, candidates should be senior scholars with a significant publication and teaching record); have foreign language proficiency only if specified in the award description or required for the completion of the proposed project; be of sound physical and mental health; and disclose prior conviction or current indictment for commission of a felony.
Restrictions: Previous Fulbright scholar grantees are eligible to apply only if five years will have elapsed between the ending date of one scholar award and the beginning date of the new scholar award. This rule does not apply if the previous grant was for less than two months. Employees, spouses or dependent children of the United States Department of State or public and private organizations under contract to the United States Department of State are ineligible to apply for a Fulbright grant until one year after the employee's termination.
Date(s) Application is Due: Aug 3
Contact: Maria Bettua; (202) 686-6245; email: mbettua@cies.iie.org; Abby Greenwell; (202) 686-6232; email: agreenwell@cies.iie.org
Internet: http://www.cies.org/Chairs/
Sponsor: Institute of International Education, Council for International Exchange of Scholars
3007 Tilden Street, NW, Suite 5L
Washington, DC 20008-3009

Fulbright English Teaching Assistantships (ETAs) 2438

The Fulbright program has offered opportunities for U.S. students to serve as English teachers and teaching assistants at schools, colleges and universities abroad for many years. Since an applicant may only apply to one country and one program, it is vital to select carefully, based on educational and career goals, academic background and preparation, language proficiency, and geographic interests. ETAs are placed primarily in elementary and secondary schools and knowledge of the host country language at the time of application is not required; however, ETAs in South America will be working with university and adult students and must have proficiency in the host country language. In most cases, ETAs: are placed in schools or universities outside of capital cities; are assigned various activities designed to improve their students' language abilities and knowledge of the United States; are fully integrated into the host community, increasing their own language skills and knowledge of the host country; may pursue individual study/research plans in addition to ETA responsibilities; and have some background or training in education and/or teaching English. Check the guidelines on line for the ever-changing list of host countries.

Requirements: Applicants must: be U.S. citizens at the time of application (permanent residents are not eligible); be in good health (grantees will be required to submit a satisfactory Medical Certificate of Health from a physician); and have sufficient proficiency in the written and spoken language of the host country (where required) to communicate with the people and to carry out the proposed ETA.

Restrictions: The following persons are ineligible: anyone who has already held a U.S. Department of State-funded Fulbright student grant of any type; anyone who has previously received a Department of Education-funded Doctoral Dissertation Research Abroad (Fulbright Hays grant); employees of the U.S. Department of State, and their immediate families, for a period ending one year following termination of such employment; employees of private and public agencies (excluding educational institutions) under contract to the U.S. Department of State to perform administrative or screening services on behalf of the U.S. Department of State, exchange program, for a period ending one year following the termination of their services for the U.S. Department of State provided such employees have been directly engaged in performing services related to the exchange programs; or applicants holding a doctoral degree at the time of application.

Amount of Grant: Varies

Date(s) Application is Due: Oct 20

Contact: Walter Jackson, Program Manager; (212) 984-5327; email: wjackson@iie.org

Internet: http://us.fulbrightonline.org/thinking_teaching.html

Sponsor: Institute of International Education
809 United Nations Plaza
New York, NY 10017-5330

Fulbright European Union Scholar-in-Residence (EU.S.IR) Program Grants 2439

The primary objective of the European Union Scholar-in-Residence (EU.S.IR) program is to strengthen expertise in European Union affairs. Through an arrangement with the EU, grants will be available to bring European scholars specializing in EU affairs to U.S. campuses as resident scholars for one term of the academic year. The scholars will give guest lectures and conduct seminars as appropriate, consult with faculty and students on research, engage in collaborative study and provide outreach to neighboring institutions and the local community. The resident scholars are not expected to teach regular course offerings. Detailed program guidelines, proposal forms, and further information are available upon request; materials are available in May prior to the deadline date.

Date(s) Application is Due: Oct 15

Contact: Alma Ford, Program Officer; (202) 686-6252; fax: (202) 362-3442; email: aford@cies.iie.org

Internet: http://www.cies.org/sir/eusir/

Sponsor: Institute of International Education, Council for International Exchange of Scholars
3007 Tilden Street NW, Suite 5L
Washington, DC 20008-3009

Fulbright German Studies Seminar Grants 2440

The program allows participation in a group seminar on current German society and culture. The program will begin in Berlin and include visits to other cities in Germany. The focus will be on the formation of policies in current issues at the core of modern society such as climate change, food technology, gene technology, stem cell research and the broad scope of education. The seminar will explore how Germany and its European neighbors view the role of science in their societies, examining the factors that result in differing approaches to these issues and the challenges of harmonizing different standards under a uniform EU-policy umbrella. With an eye to the global marketplace and worldwide competition in science, the seminar will further examine the multiple interests which structure relations between national governments, economic corporations, political and supranational bodies as well as research and development institutions. Scholars from U.S. universities, colleges, and community colleges who hold full-time teaching appointments and meet other academic requirements (Ph.D., Ph.D. candidacy or other equivalent degree or qualifications) are eligible. The seminar lasts two weeks, typically during the month of June.

Requirements: Applicants must be U.S. citizens with permanent residence in the United States at the time of application.

Date(s) Application is Due: Nov 1

Contact: Maria Bettua, Assistant Director; (202) 686-6245; email: mbettua@cies.iie.org; Abby Greenwell, (202) 686-6232; email: agreenwell@cies.iie.org

Internet: http://www.iie.org/cies/us_scholars/us_awards/GSS/index.html

Sponsor: Institute of International Education, Council for International Exchange of Scholars
3007 Tilden Street, NW, Suite 5L
Washington, DC 20008-3009

Fulbright International Education Administrators Seminar Program Grants 2441

The IEA seminars are designed to introduce participants to the society, culture and higher education systems of these countries through campus visits, meetings with foreign colleagues and government officials, attendance at cultural events and briefings on education. Participants in the International Education Administrators Program gain a firsthand look into the host country's academic infrastructure and culture. They gain new perspective on the need to internationalize U.S. campuses and insight into how it can be done. The deadline for the German program is February 1, while the deadline for both Japan and Korea is November 1.

Requirements: To be eligible, applicants must: be U.S. citizens; be international education professionals and senior university administrators (e.g., deans, provosts, vice presidents) with significant responsibility for international programs and activities; have an affiliation with an accredited college or university or nonprofit international exchange organization administering postsecondary student or faculty exchange; and have a minimum of three years of work experience in international education. Applicants for the Japan program must be affiliated with a four-year college or university, while Germany and Korea will consider applicants from both two- and four-year institutions.

Restrictions: Employees, spouses or dependent children of the United States Department of State or public and private organizations under contract to the United States Department of State are ineligible to apply for a Fulbright grant until one year after the employee's termination. TEFL administrators are ineligible for these seminar programs, but they are encouraged to review a listing of other Fulbright opportunities.

Date(s) Application is Due: Feb 1; Nov 1

Contact: Maria Bettua, Program Officer for Germany; (202) 686-6245; email: mbettua@cies.iie.org; David B.J. Adams, Senior Program Officer for Japan and Korea, (202) 686-4021; email: dadams@cies.iie.org

Internet: http://www.cies.org/IEA/

Sponsor: Institute of International Education, Council for International Exchange of Scholars
3007 Tilden Street, NW, Suite 5L
Washington, DC 20008-3009

Fulbright mtvU Awards 2442

Up to four awards will be available to pursue projects around an aspect of international contemporary or popular music as a cultural force for expression. Preference will be given to creative projects that are conveyed in a dynamic fashion and are accompanied by a feasible plan. In addition to presenting unique projects on music as a global force for mutual understanding, applicants must submit a documentation and outreach plan describing how they intend to share their activities with their peers during their Fulbright year abroad through mtvU print, broadcast and/or online mediums.

Requirements: Applicants must: be U.S. citizens at the time of application (permanent residents are not eligible); be in good health (grantees will be required to submit a satisfactory Medical Certificate of Health from a physician); have sufficient proficiency in the written and spoken language of the host country (where required) to communicate with the people; and hold a B.A. degree or the equivalent before the start of the grant (applicants who have not earned a B.A. degree or the equivalent, but who have extensive professional study and/or experience in fields in which they wish to pursue a project, may be considered). In the creative and performing arts area, four years of professional study and/or experience meets the basic eligibility requirement. All candidates for the Europe and Eurasia region are required to obtain their own affiliations, generally established with an educational and/or research institution in the host country or countries. Candidates are responsible for securing their own research clearance as required by the host country.

Restrictions: The following persons are ineligible: anyone who has already held a U.S. Department of State-funded Fulbright student grant of any type; anyone who has previously received a Department of Education-funded Doctoral Dissertation Research Abroad (Fulbright Hays grant); employees of the U.S. Department of State, and their immediate families, for a period ending one year following termination of such employment; employees of private and public agencies (excluding educational institutions) under contract to the U.S. Department of State to perform administrative or screening services on behalf of the U.S. Department of State, exchange program, for a period ending one year following the termination of their services for the U.S. Department of State provided such employees have been directly engaged in performing services related to the exchange programs; or applicants currently residing in the countries of Australia, Belgium/Luxembourg, Canada, Chile, Finland, Hungary, Mexico, Netherlands, New Zealand, Sweden, or Switzerland.

Date(s) Application is Due: Mar 1

Contact: Walter Jackson, Program Manager; (212) 984-5327; email: wjackson@iie.org

Internet: https://us.fulbrightonline.org/overview_typesgrants.html

Sponsor: Institute of International Education
809 United Nations Plaza
New York, NY 10017-5330

Fulbright New Century Scholars (NCS) Program Grants 2443

The NCS Program brings 30 top academics and professionals from around the world together each year to collaborate on an issue of global importance. Of the thirty, approximately one-third will be U.S. citizens while the remaining two thirds will be visiting scholars from countries with an operational Fulbright Scholar Program. NCS will provide a platform for scholars from the U.S. and around the world to engage in debate and dialogue based on multidisciplinary research and to develop new global models for understanding the social context within which nations and communities shape their responses to the many challenges of the 21st century. This particular aspect of the New Century Scholars program is a unique feature that distinguishes it from the core Fulbright Scholar Program.

Requirements: Applicants must be conducting current research relevant to the program's theme and objectives, be open to exploring and incorporating comparative, interdisciplinary approaches in their investigations, and interested in developing collaborative activities with other NCS Scholars. U.S. applicants must have U.S. citizenship and be residing permanently in the United States. For academic applicants, a Ph.D. or equivalent terminal degree in a relevant field is required. For applicants in the professional fields, the appropriate terminal degree in a relevant field is required. Non-U.S. applicants must be citizens or permanent residents of and residing in the country from which they are applying at the time of application. All applicants must have fluency in English.

Restrictions: Non-U.S. applicants holding permanent residency green cards, whether or not they reside in the U.S., are not eligible.

Contact: Jonathan Looper, Program Associate; (202) 686-6235; fax: (202) 362-3442; email: jlooper@cies.iie.org

Internet: http://www.cies.org/NCS/

Sponsor: Institute of International Education, Council for International Exchange of Scholars

3007 Tilden Street, NW, Suite 5L
Washington, DC 20008-3009

Fulbright Occasional Lecturer Program (OLP) Grants 2444

The Council for International Exchange of Scholars provides travel awards through the Occasional Lecturer Program (OLP), that enable Fulbright Visiting Scholars to accept guest lecturing invitations at colleges and universities throughout the United States. OLP Travel Awards are granted at three different levels ranging from $250 to $750, which CIES determines based on distances involved.

Amount of Grant: $250-$750

Contact: Jonathan Looper, Program Associate; (202) 686-6235; fax: (202) 362-3442; email: jlooper@cies.iie.org

Internet: http://www.cies.org/sir/olp/

Sponsor: Institute of International Education, Council for International Exchange of Scholars

3007 Tilden Street NW, Suite 5L
Washington, DC 20008-3009

Fulbright Public Policy Initiative Program Grants in Mexico 2445

This program is designed to enhance the knowledge, expertise, and understanding of public policy in Mexico for U.S. students in business, law, or engineering. Pursuant to this interest in public policy, the Fulbright Commission will fund three grants specifically for students conducting fieldwork or research in the areas of public administration or public policy.

Requirements: Applicants must: be U.S. citizens at the time of application (permanent residents are not eligible); be in good health (grantees will be required to submit a satisfactory Medical Certificate of Health from a physician); and hold a B.A. degree or the equivalent before the start of the grant (applicants who have not earned a B.A. degree or the equivalent, but who have extensive professional study and/or experience in business, may be considered). Excellent spoken and written Spanish is required at the time of application.

Contact: Jody Dudderar, Program Manager; (212) 984-5565; fax: (212) 984-5325; email: jdudderar@iie.org

Internet: http://us.fulbrightonline.org/program_country.html?id=70#public

Sponsor: Institute of International Education

809 United Nations Plaza
New York, NY 10017-3580

Fulbright Scholar-in-Residence (SIR) Program Grants 2446

Grants strengthen the institutional commitment of U.S. higher education to international education and scholarship by enabling colleges and universities to invite visiting scholars from abroad in the humanities and social sciences or in professional specializations with strong international focus. A Fulbright Scholar in Residence may teach regular courses from a foreign area perspective, serve as a resource person in interdisciplinary courses, assist in developing new courses, or participate in special seminars. Host institutions are expected to share the scholar's expertise with a wide range of departments and with neighboring institutions, involve him/her in community activities and professional organizations, and provide opportunities for the visitor to pursue personal research interests. Detailed program guidelines, proposal forms, and further information are available upon request; materials are available in May prior to deadline date.

Date(s) Application is Due: Oct 15

Contact: Alma Ford, Program Officer; (202) 686-6252; fax: (202) 362-3442; email: aford@cies.iie.org

Internet: http://www.cies.org/sir/

Sponsor: Institute of International Education, Council for International Exchange of Scholars

3007 Tilden Street NW, Suite 5L
Washington, DC 20008-3009

Fulbright Specialists Program Grants 2447

The Program is designed to provide short-term academic opportunities (two to six weeks) for U.S. faculty and professionals. Shorter grant lengths give specialists greater flexibility to pursue a grant that works best with their current academic or professional commitments. Applications for the Program are accepted on a rolling basis, and peer review of applications is conducted eight times per year. Program goals include: to increase the participation of leading U.S. scholars and professionals in Fulbright academic exchanges; to encourage new activities that go beyond the traditional Fulbright activities of lecturing and research; and to promote increased connections between U.S. and non-U.S. post-secondary academic institutions.

Requirements: Applicants must: be U.S. citizen at the time of application (permanent resident status is not sufficient; if a naturalized citizen, applicant must provide actual date of naturalization); possess a Ph.D. or equivalent professional/terminal degree at the time of application; have a minimum of five years of post-doctoral teaching or professional experience in the field in which you are applying (for professionals and artists outside academe, recognized professional standing and substantial professional accomplishments plus a minimum of five years of professional experience in the field in which you are applying); disclose prior conviction or current indictment for commission of a felony; and be residing in the United States at the time they are approved for a grant and intend to return to their U.S. institution after the grant's completion.

Restrictions: Employees, spouses or dependent children of the United States Department of State or public and private organizations under contract to the United States Department of State exchange programs are ineligible to apply for a Fulbright grant until one year after the employee's termination.

Date(s) Application is Due: Jan 28; Feb 24; Mar 25; Apr 21; May 20; Jul 15; Sep 9; Nov 4; Dec 30

Contact: Ryan Hathaway, Senior Program Coordinator; (202) 686-4026; email: rhathaway@cies.iie.org

Internet: http://www.cies.org/specialists/#top

Sponsor: Institute of International Education, Council for International Exchange of Scholars

3007 Tilden Street, NW, Suite 5L
Washington, DC 20008-3009

Fulbright Traditional Scholar Program in Europe and Eurasia 2448

The traditional Fulbright Scholar Program sends 800 U.S. faculty and professionals abroad each year. Grantees lecture and conduct research in a wide variety of academic and professional fields. Distribution of awards to countries in the region will vary annually according to the caliber of the applicants. Grants are available to: Albania, Andorra, Armenia, Austria, Azerbaijan, Belarus, Belgium, Bosnia and Herzegovina, Bulgaria, Croatia, Cyprus, Czech Republic, Denmark, Estonia, European Union, Finland, France, Georgia, Germany, Greece, Hungary, Iceland, Ireland, Italy, Latvia, Lithuania, Luxembourg, Macedonia, Moldova, Netherlands, Norway, Poland, Portugal, Romania, Russia, Slovakia (Slovak Republic), Slovenia, Spain, Sweden, Switzerland, Turkey, Ukraine, and United Kingdom. Only countries Considered to be in the Eastern Europe or Eurasia regions may be part of a multi-country application (2 to 3 countries). Multi-country proposals are not permitted for Western Europe. Language requirements vary by country, and prior knowledge of the local language may not be required, particularly where language is not commonly taught in the U.S.

Requirements: Applicants must: be U.S. citizens at the time of application (permanent residents are not eligible); be in good health (grantees will be required to submit a satisfactory Medical Certificate of Health from a physician); have sufficient proficiency in the written and spoken language of the host country (where required) to communicate with the people and to carry out the proposed study; and hold a B.A. degree or the equivalent before the start of the grant (applicants who have not earned a B.A. degree or the equivalent, but who have extensive professional study and/or experience in fields in which they wish to pursue a project, may be considered). In the creative and performing arts area, four years of professional study and/or experience meets the basic eligibility requirement. All candidates for the Europe and Eurasia region are required to obtain their own affiliations, generally established with an educational and/or research institution in the host country or countries. Candidates are responsible for securing their own research clearance as required by the host country.

Restrictions: The following persons are ineligible: anyone who has already held a U.S. Department of State-funded Fulbright student grant of any type; anyone who has previously received a Department of Education-funded Doctoral Dissertation Research Abroad (Fulbright Hays grant); employees of the U.S. Department of State, and their immediate families, for a period ending one year following termination of such employment; employees of private and public agencies (excluding educational institutions) under contract to the U.S. Department of State to perform administrative or screening services on behalf of the U.S. Department of State, exchange program, for a period ending one year following the termination of their services for the U.S. Department of State provided such employees have been directly engaged in performing services related to the exchange programs; applicants holding a doctoral degree at the time of application; applicants seeking enrollment in a medical degree program abroad; or applicants currently residing in the countries of Australia, Belgium/Luxembourg, Canada, Chile, Finland, Hungary, Mexico, Netherlands, New Zealand, Sweden, or Switzerland.

Date(s) Application is Due: Aug 1

Contact: Rachel Holskin, Program Manager; (212) 984-5326; email: rholskin@iie.org

Internet: http://us.fulbrightonline.org/program_regions_countries.php?id=3

Sponsor: Institute of International Education

809 United Nations Plaza
New York, NY 10017-5330

Fulbright Traditional Scholar Program in Sub-Saharan Africa   **2449**

The traditional Fulbright Scholar Program sends 800 U.S. faculty and professionals abroad each year. Grantees lecture and conduct research in a wide variety of academic and professional fields. Selection for countries in sub-Saharan Africa, with the exception of South Africa, will be made based on the quality of the applications, rather than per-country quotas. Distribution of awards to countries in the region will vary annually according to the caliber of the applicants. Grants are available to: Benin, Botswana, Burkina Faso, Cameroon, Chad, Eritrea, Ethiopia, Ghana, Guinea, Ivory Coast, Kenya, Madagascar, Malawi, Mali, Mauritius, Mozambique, Namibia, Niger, Nigeria, Senegal, South Africa, Swaziland, Tanzania, Togo, Uganda, Zambia, and Zimbabwe. All countries within the sub-Saharan Africa Region, with the exception of South Africa, may be part of a multi-country application (2 to 3 countries). For unlisted countries, applications may be considered case-by-case, but for dissertation research only.

Requirements: Applicants must: be U.S. citizens at the time of application (permanent residents are not eligible); be in good health (grantees will be required to submit a satisfactory Medical Certificate of Health from a physician); have sufficient proficiency in the written and spoken language of the host country to communicate with the people and to carry out the proposed study; and hold a B.A. degree or the equivalent before the start of the grant (applicants who have not earned a B.A. degree or the equivalent, but who have extensive professional study and/or experience in fields in which they wish to pursue a project, may be considered). In the creative and performing arts area, four years of professional study and/or experience meets the basic eligibility requirement. All candidates for Africa are required to obtain their own affiliations, generally established with an educational and/or research institution in the host country or countries.

Restrictions: Grants are not available to: Burundi, Central African Republic, Guinea-Bissau, Liberia, Somalia, or Sudan. The following persons are ineligible: anyone who has already held a U.S. Department of State-funded Fulbright student grant of any type; anyone who has previously received a Department of Education-funded Doctoral Dissertation Research Abroad (Fulbright Hays grant); employees of the U.S. Department of State, and their immediate families, for a period ending one year following termination of such employment; employees of private and public agencies (excluding educational institutions) under contract to the U.S. Department of State to perform administrative or screening services on behalf of the U.S. Department of State, exchange program, for a period ending one year following the termination of their services for the U.S. Department of State provided such employees have been directly engaged in performing services related to the exchange programs; applicants holding a doctoral degree at the time of application; applicants seeking enrollment in a medical degree program abroad; or applicants currently residing in the countries of Australia, Belgium/Luxembourg, Canada, Chile, Finland, Hungary, Mexico, Netherlands, New Zealand, Sweden, or Switzerland.

Date(s) Application is Due: Aug 1

Contact: Jermaine Jones, Program Manager; (212) 984-5341; email: jjones@iie.org

Internet: http://us.fulbrightonline.org/program_regions_countries.php?id=1

Sponsor: Institute of International Education

809 United Nations Plaza

New York, NY 10017-3580

Fulbright Traditional Scholar Program in the East Asia/Pacific Region   **2450**

The traditional Fulbright Scholar Program sends 800 U.S. faculty and professionals abroad each year. Grantees lecture and conduct research in a wide variety of academic and professional fields. Distribution of awards to countries in the region will vary annually according to the caliber of the applicants. Grants are available to: Australia, Cambodia, China, Hong Kong, Indonesia, Japan, Korea, Laos, Macau, Malaysia, Mongolia, New Zealand, Philippines, Singapore, Taiwan, Thailand, and Vietnam. All countries within the East Asia/Pacific region, with the exception of China, may be part of a multi-country application (2 to 3 countries). For unlisted countries, applications may be considered on a case-by-case basis, but for dissertation research only. Language requirements vary by country, and prior knowledge of the local language may not be required, particularly where language is not commonly taught in the U.S.

Requirements: Applicants must: be U.S. citizens at the time of application (permanent residents are not eligible); be in good health (grantees will be required to submit a satisfactory Medical Certificate of Health from a physician); have sufficient proficiency in the written and spoken language of the host country to communicate with the people and to carry out the proposed study; and hold a B.A. degree or the equivalent before the start of the grant (applicants who have not earned a B.A. degree or the equivalent, but who have extensive professional study and/or experience in fields in which they wish to pursue a project, may be considered). In the creative and performing arts area, four years of professional study and/or experience meets the basic eligibility requirement. All candidates for the East Asia/Pacific region are required to obtain their own affiliations, generally established with an educational and/or research institution in the host country or countries. Candidates are responsible for securing their own research clearance as required by the host country.

Restrictions: Grants are not available to: Brunei, the Cook Islands, East Timor, Fiji, Myanmar, the Pacific Island Nations, Papua New Guinea, or Western Samoa. The following persons are ineligible: anyone who has already held a U.S. Department of State-funded Fulbright student grant of any type; anyone who has previously received a Department of Education-funded Doctoral Dissertation Research Abroad (Fulbright Hays grant); employees of the U.S. Department of State, and their immediate families, for a period ending one year following termination of such employment; employees of private and public agencies (excluding educational institutions) under contract to the U.S. Department of State to perform administrative or screening services on behalf of the U.S. Department of State, exchange program, for a period ending one year following the termination of their services for the U.S. Department of State provided

such employees have been directly engaged in performing services related to the exchange programs; applicants holding a doctoral degree at the time of application; applicants seeking enrollment in a medical degree program abroad; or applicants currently residing in the countries of Australia, Belgium/Luxembourg, Canada, Chile, Finland, Hungary, Mexico, Netherlands, New Zealand, Sweden, or Switzerland.

Date(s) Application is Due: Aug 1

Contact: Jonathan Akeley, Program Manager; (212) 984-5487; email: jakeley@iie.org

Internet: http://us.fulbrightonline.org/program_regions_countries.php?id=2

Sponsor: Institute of International Education

809 United Nations Plaza

New York, NY 10017-3580

Fulbright Traditional Scholar in the Near East and North Africa Region   **2451**

The traditional Fulbright Scholar Program sends 800 U.S. faculty and professionals abroad each year. Grantees lecture and conduct research in a wide variety of academic and professional fields. Selection for countries in the Near East and North Africa region will be made based on the quality of the applications, rather than per-country quotas. Distribution of awards to countries in the region will vary annually according to the caliber of the applicants. Grants are available to: Bahrain, Egypt, India, Jordan, Kuwait, Morocco, Oman, Qatar, Syria, Tunisia, United Arab Emirates, and Yemen. Multi-country applications (2 to 3 countries) are available, except in Egypt, India, Jordan, and Morocco. For unlisted countries, applications may be considered on a case-by-case basis.

Requirements: Applicants must: be U.S. citizens at the time of application (permanent residents are not eligible); be in good health (grantees will be required to submit a satisfactory Medical Certificate of Health from a physician); have sufficient proficiency in the written and spoken language of the host country to communicate with the people and to carry out the proposed study; and hold a B.A. degree or the equivalent before the start of the grant (applicants who have not earned a B.A. degree or the equivalent, but who have extensive professional study and/or experience in fields in which they wish to pursue a project, may be considered). In the creative and performing arts area, four years of professional study and/or experience meets the basic eligibility requirement. Candidates are responsible for securing research clearance, as required. In countries with Fulbright Commissions, assistance may be provided.

Restrictions: Grants are not available to: Algeria, Iran, Iraq, Lebanon, Saudi Arabia, the West Bank, or Gaza. The following persons are ineligible: anyone who has already held a U.S. Department of State-funded Fulbright student grant of any type; anyone who has previously received a Department of Education-funded Doctoral Dissertation Research Abroad (Fulbright Hays grant); employees of the U.S. Department of State, and their immediate families, for a period ending one year following termination of such employment; employees of private and public agencies (excluding educational institutions) under contract to the U.S. Department of State to perform administrative or screening services on behalf of the U.S. Department of State, exchange program, for a period ending one year following the termination of their services for the U.S. Department of State provided such employees have been directly engaged in performing services related to the exchange programs; applicants holding a doctoral degree at the time of application; applicants seeking enrollment in a medical degree program abroad; or applicants currently residing in the countries of Australia, Belgium/Luxembourg, Canada, Chile, Finland, Hungary, Mexico, Netherlands, New Zealand, Sweden, or Switzerland.

Date(s) Application is Due: Aug 1

Contact: Jermaine Jones, Program Manager; (212) 984-5341; email: jjones@iie.org

Internet: http://us.fulbrightonline.org/program_regions_countries.php?id=4

Sponsor: Institute of International Education

809 United Nations Plaza

New York, NY 10017-3580

Fulbright Traditional Scholar Program in the South and Central Asia Region   **2452**

The traditional Fulbright Scholar Program sends 800 U.S. faculty and professionals abroad each year. Grantees lecture and conduct research in a wide variety of academic and professional fields. Selection for countries in the South and Central Asia region will be made based on the quality of the applications, rather than per-country quotas. Distribution of awards to countries in the region will vary annually according to the caliber of the applicants. Grants are available to: Bangladesh, India, Kazakhstan, Kyrgyz Republic, Nepal, Sri Lanka, Tajikistan, and Uzbekistan. Multi-country applications (2 to 3 countries) are not available. Language requirements vary by country. Applicants to India who are recommended for final review will need to submit research visa applications in January or February.

Requirements: Applicants must: be U.S. citizens at the time of application (permanent residents are not eligible); be in good health (grantees will be required to submit a satisfactory Medical Certificate of Health from a physician); have sufficient proficiency in the written and spoken language of the host country to communicate with the people and to carry out the proposed study; and hold a B.A. degree or the equivalent before the start of the grant (applicants who have not earned a B.A. degree or the equivalent, but who have extensive professional study and/or experience in fields in which they wish to pursue a project, may be considered). In the creative and performing arts area, four years of professional study and/or experience meets the basic eligibility requirement. Candidates are responsible for securing research clearance, as required. In countries with Fulbright Commissions, assistance may be provided.

Restrictions: Grants are not available to: Afghanistan, Bhutan, Pakistan, Republic of Maldives, or Turkmenistan. The following persons are ineligible: anyone who has already held a U.S. Department of State-funded Fulbright student grant of any type; anyone who has previously received a Department of Education-funded Doctoral Dissertation

Research Abroad (Fulbright Hays grant); employees of the U.S. Department of State, and their immediate families, for a period ending one year following termination of such employment; employees of private and public agencies (excluding educational institutions) under contract to the U.S. Department of State to perform administrative or screening services on behalf of the U.S. Department of State, exchange program, for a period ending one year following the termination of their services for the U.S. Department of State provided such employees have been directly engaged in performing services related to the exchange programs; applicants holding a doctoral degree at the time of application; applicants seeking enrollment in a medical degree program abroad; or applicants currently residing in the countries of Australia, Belgium/Luxembourg, Canada, Chile, Finland, Hungary, Mexico, Netherlands, New Zealand, Sweden, or Switzerland.

Date(s) Application is Due: Aug 1
Contact: Jonathan Akeley, South Asia Program Manager; (212) 984-5487; email: jakeley@iie.org; Jon Adler, Central Asia Program Manager; (212) 984-5326; email: jadler@iie.org
Internet: http://us.fulbrightonline.org/program_regions_countries.html?id=6
Sponsor: Institute of International Education
809 United Nations Plaza
New York, NY 10017-3580

Fulbright Traditional Scholar Program in the Western Hemisphere   2453

The traditional Fulbright Scholar Program sends 800 U.S. faculty and professionals abroad each year. Grantees lecture and conduct research in a wide variety of academic and professional fields. Selection for countries in Central America and the Caribbean will be made based on the quality of the applications, rather than per-country quotas. Distribution of awards to countries in the region will vary annually according to the caliber of the applicants. Grants are available to: Argentina, Barbados and the Eastern Caribbean, Bolivia, Brazil, Canada, Chile, Colombia, Costa Rica, Dominican Republic, Ecuador, El Salvador, Guatemala, Honduras, Jamaica, Mexico, Nicaragua, Panama, Paraguay, Peru, Trinidad and Tobago, Uruguay, and Venezuela. English Teaching Assistantships are available in Argentina, Brazil, Chile, Uruguay and Venezuela. Language proficiency may be preferred or required. Applicants for English Teaching Assistantships can apply to only one country. Multi-country applications (2 to 3 countries) are available in all other disciplines throughout the Western Hemisphere. For unlisted countries, applications may be considered on a case-by-case basis, but for dissertation research only.

Requirements: Applicants must: be U.S. citizens at the time of application (permanent residents are not eligible); be in good health (grantees will be required to submit a satisfactory Medical Certificate of Health from a physician); have sufficient proficiency in the written and spoken language of the host country to communicate with the people and to carry out the proposed study; and hold a B.A. degree or the equivalent before the start of the grant (applicants who have not earned a B.A. degree or the equivalent, but who have extensive professional study and/or experience in fields in which they wish to pursue a project, may be considered). In the creative and performing arts area, four years of professional study and/or experience meets the basic eligibility requirement. Candidates are responsible for securing research clearance, as required. In countries with Fulbright Commissions, assistance may be provided.

Restrictions: Grants are not available to: Bahamas, Belize, Cuba, French Guiana, Guyana, Haiti, Martinique, or Suriname. The following persons are ineligible: anyone who has already held a U.S. Department of State-funded Fulbright student grant of any type; anyone who has previously received a Department of Education-funded Doctoral Dissertation Research Abroad (Fulbright Hays grant); employees of the U.S. Department of State, and their immediate families, for a period ending one year following termination of such employment; employees of private and public agencies (excluding educational institutions) under contract to the U.S. Department of State to perform administrative or screening services on behalf of the U.S. Department of State, exchange program, for a period ending one year following the termination of their services for the U.S. Department of State provided such employees have been directly engaged in performing services related to the exchange programs; applicants holding a doctoral degree at the time of application; applicants seeking enrollment in a medical degree program abroad; or applicants currently residing in the countries of Australia, Belgium/Luxembourg, Canada, Chile, Finland, Hungary, Mexico, Netherlands, New Zealand, Sweden, or Switzerland.

Date(s) Application is Due: Aug 1
Contact: Jody Dudderar, Program Manager in South America, Mexico and Canada; (212) 984-5565; fax: (212) 984-5325; email: jdudderar@iie.org; Cara Wolinsky, Program Manager in Central America and the Caribbean; (212) 984-5366; email: cwolinsky@iie.org
Internet: http://us.fulbrightonline.org/program_regions_countries.php?id=5
Sponsor: Institute of International Education
809 United Nations Plaza
New York, NY 10017-3580

Fulbright Travel Grants   2454

Travel grants are available only to Germany, Hungary, or Italy. They are available to supplement an award from a non-IIE source that does not provide funds for travel or to supplement a student, own funds for study. Travel grants provide round-trip transportation to the country where the student will pursue studies for an academic year, supplemental health and accident insurance, and the cost of all orientation courses abroad, if applicable.

Requirements: Applicants must be U.S. citizens with at least the bachelor's degree and capable of meeting the eligibility criteria for Fulbright grants.
Contact: Walter Jackson, Program Manager; (212) 984-5327; email: wjackson@iie.org

Internet: https://us.fulbrightonline.org/overview_typesgrants.html
Sponsor: Institute of International Education
809 United Nations Plaza
New York, NY 10017-3580

Fulbright Visiting Specialists Program: Direct Access to the Muslim World   2455

The program aims to promote American understanding of Islamic civilization and the history, politics, and culture of today's Muslim world. The program further aims to strengthen resources for enriching understanding of relevant global issues. The program offers a unique opportunity for U.S. colleges and universities to host specialists from countries with significant Muslim populations for short-term programs of intensive lecturing, public outreach, and consultation. The Fulbright Visiting Specialists will lecture on and off campus in a variety of humanities and social science disciplines, including the arts, with a focus on Islamic society or civilization. The Program encourages inter-religious dialog through on-campus and off-campus activities. The Visiting Specialists will also work together with the U.S. host institutions on projects with lasting benefits such as curriculum consultation, faculty development, and institutional linkage-building.

Requirements: Institutions with few or no programs/courses related to Islam or the Muslim world are encouraged to apply. Institutions that have well-established Islam-related or area studies programs should clearly indicate that their proposed program for the Fulbright Visiting Specialists Program is in a field/subject that is not currently supported at the institution. Community colleges and minority-serving institutions, such as Historically Black Colleges and Universities, Hispanic Serving Institutions, and Tribal Colleges and Universities, are encouraged to apply and will receive preference for an award. At the time of program participation, the Fulbright Visiting Specialist should be a resident of his/her home country.

Restrictions: The program does not support the teaching of languages or research projects. Institutions that have already hosted a Fulbright Visiting Specialist are no longer eligible. While more than one application may be submitted by the same institution, only one award to the same institution will be offered. Persons holding U.S. citizenship or permanent residency in the U.S. are not eligible.

Date(s) Application is Due: Apr 1
Contact: Alma Ford, Program Officer; (202) 686-6252; fax: (202) 362-3442; email: aford@cies.iie.org
Internet: http://www.cies.org/Visiting_Specialists/
Sponsor: Institute of International Education, Council for International Exchange of Scholars
3007 Tilden Street NW, Suite 5L
Washington, DC 20008-3009

Fulbright-Hays Seminars Abroad Bilateral Projects Grants   2456

The program provides short-term study and travel seminars abroad for U.S. educators in the social sciences and humanities for the purpose of improving their understanding and knowledge of the peoples and cultures of other countries. Support is generally made available through inter-agency agreements. There are approximately seven to ten seminars with fourteen to sixteen participants in each seminar annually. Seminars are four to six weeks in duration. All seminars are held in countries outside of Western Europe. Terms of the award include a round-trip economy airfare, room and board, tuition and fees, and program-related travel within the host country.

Requirements: Applicants must: be a citizen of the U.S. or a permanent resident; hold at least a bachelor's degree from an accredited college or university; have at least three years of full-time experience by the time of departure for the seminar; be currently employed full-time in a U.S. school system, institution of higher education, local education agency (LEA), state education agency (SEA), library, or museum; meet the general language requirement; and be physically and psychologically able to participate in all phases of the seminar.

Contact: Gail Holdren, International Education Graduate Programs Service; (202) 502-7691; fax: (202) 502-7859; email: gale.holdren@ed.gov
Internet: http://www.ed.gov/programs/iegpssap/index.html
Sponsor: Institute of International Education
1990 K Street N.W., 6th Floor
Washington, DC 20006-8521

Fulbright/Garcia Robles Grants   2457

The program is especially interested in attracting students pursuing fieldwork or internships in public administration or public policy. Other fields of interest are projects on Mexican culture, society and politics to deepen the understanding of Mexico in the United States. Areas of collaboration go far beyond border issues, which are also of relevance, and can include projects on migration, ecological issues, public health, education, public policy, environmental protection, human rights, and trade policy. The program will also support projects in the arts that further dialogue and contact between U.S. and Mexican arts communities.

Requirements: Applicants must: be U.S. citizens at the time of application (permanent residents are not eligible); be in good health (grantees will be required to submit a satisfactory Medical Certificate of Health from a physician); and hold a B.A. degree or the equivalent before the start of the grant (applicants who have not earned a B.A. degree or the equivalent, but who have extensive professional study and/or experience in business, may be considered). Excellent spoken and written Spanish is required at the time of application.

Restrictions: Fields of study not recommended include medicine and dentistry.
Contact: Jody Dudderar, Program Manager; (212) 984-5565; fax: (212) 984-5325; email: jdudderar@iie.org

Internet: http://us.fulbrightonline.org/program_country.html?id=70#full
Sponsor: Institute of International Education
809 United Nations Plaza
New York, NY 10017-3580

Fund for Folk Culture Artists Support Grants 2458

The Artists Support Program, a program of the Fund for Folk Culture (FFC) underwritten through generous support from the Ford Foundation, The William and Flora Hewlett Foundation and The San Francisco Foundation, offers individual folk and traditional artists in the United States the opportunity to pursue activities that will help them to grow artistically and professionally, to better connect with their cultural communities or to develop new audiences for their work. The program offers grants of up to $5,000 for specific projects or for general expenses necessary to an artist's work.

Requirements: All folk and traditional artists who are at least 18 years of age, and either citizens or permanent residents of the United States. Examples of eligible proposals include (but are not limited to) purchase of materials, supplies or equipment; studio, rehearsal or performance space rental; development of portfolio or marketing materials; CD production; new work development; collaborations with other artists or cultural organizations; international and out-of-state travel for participation in artist gatherings and retreats or conferences; work on collaborative projects; research related to artistic growth; and other creative or professional development activities. Individual folk and traditional artists will be invited to apply through a nomination process. Do not submit a proposal unless you have been invited to do so. Questions about the nomination process or the Artists Support Program can be directed to Betsy Peterson.

Restrictions: In-state apprenticeship activity (frequently supported through state arts agencies or other state folk arts organizations) is not eligible for support.

Samples: 2008: Ahmad Alaadeen, Overland Park, Kansas - $4,844.; Michael Alpert, New York, New York - $4,850.; Daniel Ansotegui, Boise, Idaho - $4,850.; Mensura Berberovic, Erie, Pennsylvania - $4,850.; Kelly Church, Hopkins, Michigan - $4,850.; Mahmoud Conteh, St. Louis, Missouri - $4,850.; Rosa Elena Egipciaco, New York, New York - $4,620.; Linda Goss, Philadelphia, Pennsylvania - $4,845.; Alan Ham, Selma, Alabama - $4,850.; Jerry Jumba, McKees Rocks, Pennsylvania - $4,850.; James Kelly, Miami Springs, Florida - $4,822.; Mythili Kumar, San Jose, California - $5,000.; Anthony LaFromboise, Mandan, North Dakota - $4,850.; Paul Luniw, Terryville, Connecticut - $4,850.; Konstantine Maris, New Port Richey, Florida - $4,850.; Billy McComiskey, Baltimore, Maryland - $4,850.; Jessie McCullum, Milwaukee, Wisconsin - $4,850.; Molly Jean Neptune Parker, Princeton, Maine - $4,850.; Helder Parreira, Berkeley, California - $5,000.; Sani Rifati, Sebastopol, California - $5,000.; Reba Jo Teran, Fort Washakie, Wyoming - $4,850.

Amount of Grant: up to $5,000

Contact: Elizabeth (Betsy) E. Peterson, Ph.D., Executive Director; (512) 472-8757; fax: (512) 472-8758; email: bpeterson@folkculture.org
Internet: http://www.folkculture.org/about/asp.html
Sponsor: Fund for Folk Culture
P.O. Box 303366
Austin, TX 78703-0057

Fund for French-American Academic Partnerships 2459

The program enables French and American researchers, professors, and administrators from all academic disciplines to enter long-term agreements that jointly develop new curricula to be integrated into the programs of their respective graduate schools. In addition to fostering the creation of innovative joint curricula, the fund seeks to give French and American graduate students the opportunity to explore cross-cultural, interdisciplinary approaches in their courses and/or research, thereby training globally minded professionals. The program's long-term goal is to strengthen ties between France and the United States through these structured partnerships. Partnership grants are awarded each year. Guidelines are available online.

Requirements: Only nonprofit institutions may submit applications. U.S. institutions must demonstrate 501(c)3; French organizations must provide proof of equivalent status for eligibility. Projects must be submitted and co-signed jointly by one or several American universities with a matching commitment from one or several French institutions of higher learning, or by one or several French institutions with a matching commitment from one or several American matching institutions.

Restrictions: Grants are not allocated to individuals.

Date(s) Application is Due: Feb 25

Contact: Christian Tual, (202) 944-6414; fax: (202) 944 6268; email: Christian.TUAL@diplomatie.gouv.fr
Internet: http://www.facecouncil.org/word/UniversityFund.doc
Sponsor: French Embassy, Cultural Services
4101 Reservoir Road
Washington, DC 20007

Fund for New Jersey Grants 2460

Grants support nonprofit organizations in New Jersey or organizations that benefit the state, with particular attention given to projects seeking to affect public policy. Although a few grants are provided for local activities, direct services, and general operating support, such proposals are considered usually at the fund's invitation. Grant applicants are asked to submit a single-page proposal cover sheet containing the following information: organization and contact person, summary of request, amount requested, and problem or need addressed by proposed activity. All proposals must be accompanied by a copy of the IRS tax-exemption letter, names and affiliations of the board of directors, and a budget. The board normally meets in March, June, September, and December to consider proposals.

Requirements: 501(c)3 tax-exempt organizations are eligible. Proposals are not accepted via email.

Restrictions: The fund does not accept proposals for support of individuals nor for capital projects such as acquisition, renovation, or equipment. The fund is unable to support day care centers, drug treatment programs, arts programs, health care delivery, or scholarships.

Geographic Focus: New Jersey

Samples: CityWorks (Trenton, NJ)—for its efforts to develop commercial real-estate projects in selected urban neighborhoods, $100,000 (2005); New Jersey Community Capital (Trenton, NJ)—for a campaign intended to attract investments for community- and economic-development projects in low-income communities, $85,000 (2005); Pinelands Preservation Alliance (Southampton, NJ)—for advocacy efforts designed to help protect New Jersey's Pinelands region, $30,000 (2005); New Jersey Immigration Policy Network (Newark, NJ)—to support a coalition of civil-rights, community, immigrant, and labor groups that promote fair policies for recent immigrants, $100,000 (2005).

Amount of Grant: $3.3 million total

Contact: Mark Murphy, Executive Director, (732) 220-8656; fax: (732) 220-8654; email: info@fundfornj.org
Internet: http://www.fundfornj.org/app_guide.html
Sponsor: Fund for New Jersey
94 Church Street, Suite 303
New Brunswick, NJ 08901

Fund for Santa Barbara Grants 2461

The fund gives preference to projects working to address the root causes of social, economic, and environmental problems. Types of support include seed grants to new grassroots projects, general support, or project grants to small organizations, and grants to larger, more established organizations only for specific targeted purposes. Affirmative action considerations are among the criteria used in all funding decisions.

Requirements: Applications are invited from organizations that are working against discrimination based on race, sex/gender, age, religion, economic status, sexual orientation, physical/mental ability, ethnicity, language, or immigration status; struggling for the rights of workers; promoting self-determination in low-income and disenfranchised communities; promoting international peace and organizing locally for a just foreign policy; working on improving the environment, especially organizing a constituency usually without access or input to environmental concerns; and operating in a democratic manner, responsive to and directed by the constituency being served.

Restrictions: Grants do not support projects involved in electoral campaigns on behalf of candidates or parties; private (vs. public) interests; direct labor organizing; projects located outside of Santa Barbara County; projects providing direct services without a social change component; or direct support to individuals, capital ventures, or building improvements.

Geographic Focus: California

Samples: Committee for Social Justice (CA)—to pay for a Spanish-speaking coordinator to extend their reach, and run workshops and clinics, $3750 (2002); Men Against Rape (CA)—for office supplies, training materials, conference attendance, poster printing, and school outreach expenses, $4000 (2002); Clergy United for Equality of Homosexuals (CA)—for seminars and workshops on biblical oppression and its relief, including models for change for pastors and churches, $3750 (2002).

Amount of Grant: $10,000 maximum; $4000 average

Date(s) Application is Due: Mar 12; Sep 9

Contact: Fund Administrator, (805) 962-9164; fax: (805) 965-0217; email: email@fundforsantabarbara.org
Internet: http://www.fundforsantabarbara.org/apply/apply.htm
Sponsor: Fund for Santa Barbara
924 Anacapa Street, Suite 4H
Santa Barbara, CA 93101-2192

Fund for Scientific Research—Flanders Grants 2462

The foundation awards pre- and postdoctoral fellowships and grants for research in Flemish universities and research institutions in all fields of science, including medicine, technology, environmental studies, social science, law, and the humanities. National and international research is supported.

Contact: Secretary General, 32 (2) 512-91-10; fax: 32 (2) 512-58-90; email: post@fwo.be
Internet: http://sun.fwo.be/page1en.php?lang=en
Sponsor: Fund for Scientific Research—Flanders
Egmontstraat 5
Brussels 1000 Belgium

Fund for the Advancement of the Discipline 2463

The American Sociological Association invites submissions for the Fund for the Advancement of the Discipline (FAD) awards. Supported by the American Sociological Association through a matching grant from the National Science Foundation, the goal of this award is to nurture the development of scientific knowledge by funding small, ground breaking research initiatives and other important scientific research activities. FAD awards provide scholars with venture capital for innovative research that has the potential for challenging the discipline, stimulating new lines of research, and creating new networks of scientific collaboration. The award is intended to provide opportunities for substantive and methodological breakthroughs, broaden the dissemination of scientific knowledge, and provide leverage for acquisition of additional research funds.

Requirements: Proposals are reviewed for scientific merit and the importance of the proposed research project. Within this context, specific evaluation criteria include

the following elements: innovativeness and promise of the research idea; originality and significance of research goals; the potential of the study as a building block in the development of future research; appropriateness and significance of the research hypothesis; feasibility and adequacy of project design; plans for analysis and evaluation of data; plans for dissemination of results; appropriateness of requested budget. The awards are limited to individuals with PhD degrees or the equivalent. Preference is given to applicants who have not previously received a FAD award.
Amount of Grant: $7000 maximum
Date(s) Application is Due: Jun 15; Dec 15
Contact: Roberta Spalter-Roth, FAD Awards, (202) 383-9005 ext. 317; email: spalter-roth@asanet.org
Internet: http://asanet.org/page.ww?section=Funding&name=Fund+for+the+Advancement+of+the+Discipline
Sponsor: American Sociological Association
1307 New York Avenue NW, Suite 700
Washington, DC 20005-4107

Furthermore Grants in Publishing   2464
The program is concerned with nonfiction book publishing about the city; natural and historic resources; art, architecture, and design; cultural history; and civil liberties and other public issues of the day. Grants apply to writing, research, editing, design, indexing, photography, illustration, and printing and binding. Work is sought that appeals to an informed general audience; gives evidence of high standards in editing, design, and production; promises a reasonable shelf life; might not otherwise achieve top quality or even come into being; and represents a contribution without which we would be the poorer. Book proposals to which a university press or trade publisher is already committed and for which there is a feasible distribution plan are usually preferred. In geographical reach, the program is drawn, but in no way limited, to New York City and to New York State and its Hudson Valley. Guidelines and application are available online.
Requirements: 501(c)3 tax-exempt organizations are eligible. Trade publishers and public agencies may apply for Furthermore grants in partnership with an eligible nonprofit project sponsor.
Samples: Century Assoc Archives Fdn (New York, NY)—for its archives, $10,000 (2004); White Box (New York, NY)—to edit and design the book It Is: The American Movement of Abstract Art, $10,000 (2004); Natural Resources Defense Council (New York, NY)—for an oral-history project focusing on this environmental organization, $15,000 (2004).
Amount of Grant: $500-$15,000
Date(s) Application is Due: Mar 15; Sep 15
Contact: Grants Administrator, (518) 828-8900; fax: (518) 828-8901; email: info@furthermore.org
Internet: http://www.furthermore.org
Sponsor: J.M. Kaplan Fund
P.O. Box 667
Hudson, NY 12534

Fyssen Foundation Postdoctoral Study Grants   2465
The foundation awards postdoctoral research grants for research on cognitive mechanisms, including thought and reasoning, that underlie animal and human behavior; their biological and cultural bases; and phylogenetic and ontogenetic development. Research should be in ethology and psychology, neurobiology, anthropology/ethnology, and human paleontology. Grants support French scientists for research abroad and foreign researchers conducting research in France. Annual deadline dates may vary; contact program staff for exact dates.
Date(s) Application is Due: Mar 31
Contact: Postdoctoral Study Grants, 331 42975316; fax: 331 42601795; email: secretariat@fondation-fyssen.org
Internet: http://www.fondation-fyssen.org
Sponsor: Fyssen Foundation
194 rue de Rivoli
Paris 75001
France

G. Harold and Leila Y. Mathers Charitable Foundation Grants   2466
The foundation awards grants to support research in the basic life sciences. Requests for general operating support also will be considered. There are no application forms or deadlines. The board meets two or three times each year.
Requirements: U.S. research organizations are eligible.
Samples: Fox Chase Cancer Ctr (Philadelphia, PA)—for genetic research, $660,000 over three years.
Amount of Grant: $10,000-$200,000 average per year
Contact: James Handelman, Executive Director, (914) 242-0465; email: bcheikin@mathersfoundation.org
Internet: http://www.mathersfoundation.org
Sponsor: G. Harold and Leila Y. Mathers Charitable Foundation
118 N Bedford Road, Suite 203
Mount Kisco, NY 10549-2555

G. Unger Vetlesen Foundation Grants   2467
The foundation's mission is to aid and contribute to religious, charitable, scientific, literary, and educational purposes throughout the world. Grants have been concentrated in the fields of oceanography, climate studies, earth sciences, cultural programs emphasizing Norwegian-American relations, environmental programs, libraries, international affairs,

and public policy. A letter of inquiry, 3 pages or less, should be submitted at any time during the year. Proposals are to be submitted only upon invitation. Unsolicited proposals will not be reviewed.
Requirements: 501(c)3 nonprofit organizations are eligible.
Restrictions: The Foundation does not make grants to individuals.
Samples: U of Rhode Island (RI)—for the Graduate School of Oceanography.; American Scandanavian Assoc—$25,000.; Lamont-Doherty Earth Observatory for Climate Ctr—$250,000.
Date(s) Application is Due: Apr 30; Oct 31
Contact: George Rowe Jr, President, (212) 586-0700; fax: (212) 245-1863; email: info@monellvetlesen.org
Internet: http://www.monellvetlesen.org/vetlesen/appguide.htm
Sponsor: G. Unger Vetlesen Foundation
c/o Fulton, Rowe & Hart, One Rockefeller Plaza, Suite 301
New York, NY 10020-2002

G.D. Searle Charitable Trust Grants   2468
Searle awards grants through its charitable trust to tax-exempt organizations for education at all levels, health, social services, arts and humanities, and civic affairs. A major part of the funding is to United Way/Crusade of Mercy and employee matching gift programs, preventive medicine, and medical research. Most projects funded have a duration of two years or less. Applications outlining the project for which funds are needed should be submitted in the first quarter of each year. Address applications to Searle Charitable Trust, 5200 Old Orchard Road, Skokie, IL 60077.
Requirements: Grants are not made to individuals for personal education or to organizations funded by United Way.
Samples: U of Arizona (Tucson, AZ)—to support biotechnology research, $10,000.
Amount of Grant: $500-$30,000 average
Contact: Judith Van der Vort, Administrator, (708) 982-7000
Sponsor: G.D. Searle Charitable Trust
50 S LaSalle Street
Chicago, IL 60675

Gabor A. Somorjai Award for Creative Research in Catalysis   2469
The Award, supported by the Gabor A. Somorjai Endowment Fund, is intended to recognize outstanding theoretical, experimental, or developmental research resulting in the advancement of understanding or application of catalysis. U.S. residents will be given preference. Outstanding contributions to catalysis in any discipline will be considered. The award consists of $5,000 and a certificate.
Requirements: Any individual, except a member of the award committee, may submit one nomination or seconding letter for the award in any given year. Nominating documents consist of a letter of not more than 1000 words containing an evaluation of the nominee's accomplishments and a specific identification of the work to be recognized, a biographical sketch including date of birth, and a list of publications and patents authored by the nominee.
Restrictions: Self-nominations are not accepted.
Samples: Avelino Corma Canos—award winner, $5,000 (2008); Hans-Joachim Freund—award winner, $5,000 (2007); James A. Dumesic—award winner, $5,000 (2006).
Amount of Grant: $5,000 plus $1,000 toward travel to award meeting
Date(s) Application is Due: Nov 1
Contact: Felicia Dixon, Awards Administrator; (800) 227-5558 or (202) 872-4408; fax: (202) 776-8008; email: f_dixon@acs.org or awards@acs.org
Internet: http://portal.acs.org/portal/acs/corg/content?_nfpb=true&_pageLabel=PP_ARTICLEMAIN&node_id=1319&content_id=CTP_004525&use_sec=true&sec_url_var=region1
Sponsor: American Chemical Society
1155 Sixteenth Street, NW
Washington, DC 20036

Gaius Charles Bolin Fellowships for Minority Graduate Students   2470
This fellowship enables a minority graduate student who is working toward the PhD in the humanities or in the natural, social, or behavioral sciences, and who plans to pursue a career in college teaching, to devote the bulk of his or her time during the academic year to the completion of dissertation work. During the year of residence at Williams, the Bolin fellow will be assigned a faculty advisor in the appropriate department and will be expected to teach a one-semester course. Two minority graduate students are awarded fellowships each year. The fellowship includes a stipend and research allowance; housing assistance; and academic support including office space and computer and library privileges.
Requirements: Applicants must be U.S. citizens who have completed all doctoral work except the dissertation by the end of the current academic year.
Amount of Grant: $31,000 for academic year plus $4000 maximum research expense allowance
Date(s) Application is Due: Dec 1
Contact: Thomas Kohut, Dean of Faculty, (413) 597-4351
Internet: http://www.williams.edu
Sponsor: Williams College
P.O. Box 141
Williamstown, MA 01267

Gallaudet University Alumni Association Graduate Fellowship Fund   2471
These fellowships are offered to deaf and hard-of-hearing college graduates who wish to pursue doctoral study at a university for people who hear normally. Recipients must

carry a full-time semester load. The amount awarded varies, depending on the needs of the recipient and the availability of funds. The fellowship is awarded for one year and may be renewed.

Requirements: Applicants must be deaf or hard-of-hearing graduates of accredited colleges and universities who have been accepted for graduate study at colleges or universities for people who hear normally. Preference is given to U.S. applicants who possess a master's degree or equivalent and who are seeking the doctorate.

Date(s) Application is Due: Apr 20

Contact: Peikoff Alumni House; fax: (202) 651-5062; TTY: (202) 651-5081; email: alumni.relations@gallaudet.edu

Internet: http://alumni.gallaudet.edu/article.asp?ID=2646

Sponsor: Gallaudet University Alumni Association

800 Florida Avenue NE

Washington, DC 20002-3695

Galveston Shriners Hospital Postdoctoral Fellowships　　2472

The program awards postdoctoral clinical and research fellowships to individuals interested in trauma, critical care, and burn-related practice or research. Fellowships are for a minimum of one year (clinical) and two years (research). The UTMB Trauma and Burn Services have NIH- supported active investigators with comprehensive laboratories working in the fields of wound healing, metabolism, cardiopulmonary pathophysiology, molecular biology, and immunology. Graduate programs, including master's and PhD degrees, are also available.

Requirements: Applicants who have attained an MD, DO, PhD, or DVM degree are eligible. Applicants for clinical fellowships should be board eligible or board certified in general surgery; however, senior residents from academically oriented university programs will be considered or those who have completed U.S.MLE part III.

Amount of Grant: $30,000-$45,000

Contact: Dr. David Herndon, Shriners Burns Hospital, (409) 770-6731; fax: (409) 770-6919; email: dherndon@utmb.edu

Internet: http://www.shrinershq.org/shc/galveston/postdoc.html

Sponsor: University of Texas Medical Branch-Galveston

815 Market Street

Galveston, TX 77550

Gas Technology Institute Basic Research Program　　2473

The institute supports a broad program of basic research relevant to the U.S. natural gas industry. Research topics include solid oxide fuel cell stacks and new drilling concepts. Contracts may run as long as three years. Research topics may change in content or timing during the year; changes will be posted on the Internet. Only written, fax, and Internet requests for individual proposal requests (RFPs) will be accepted.

Requirements: Potential applicants must contact the office for specific individual requests for proposals.

Contact: GTI Managers, (847) 768-0500; fax: (847) 768-0501; email: vp_rd@gastechnology.org

Internet: http://www.gastechnology.org/webroot/app/xn/xd.aspx?it=enweb&xd=gtihome.xml

Sponsor: Gas Technology Institute

1700 S Mt Prospect Road

Des Plaines, IL 60018-1804

Gates Family Foundation Fellowships　　2474

The John F. Kennedy School of Government at Harvard University and the Gates Family Foundation have partnered since 1980, each year sending up to five individuals to participate in the Senior Executives in State and Local Government Program. The three-week program operates as an interactive classroom, where case studies of real issues are analyzed to find solutions to complex problems, and faculty and participants work together and learn from each other. Gates Fellowships will be awarded to residents of Colorado by an admissions committee at the Kennedy School based upon the recommendations of a nominating panel of distinguished Colorado residents. The program fee is $10,900 which includes tuition, housing, curricular materials, and most meals. The fellowship is $10,000 and covers most program costs. Participants or their employers must fund the balance.

Geographic Focus: Colorado

Date(s) Application is Due: Jan 15; Apr 1; Jul 15; Oct 1

Contact: Karen White Mather, Program Officer/Grants Manager; (303) 722-1881; fax: (303) 316-3038; email: info@gatesfamilyfoundation.org

Internet: http://www.gatesfamilyfoundation.org/www/gates.php?section=gates_fellowship&p=

Sponsor: Gates Family Foundation

3575 Cherry Creek North Drive, Suite 100

Denver, CO 80209-3600

Gates Global HIV/AIDS Vaccine Grants　　2475

The program seeks research proposals to apply new technologies, concepts, and approaches to the design of safe and effective preventive vaccines against HIV/AIDS, in the context of the Scientific Strategic Plan of the Global HIV/AIDS Vaccine Enterprise. The foundation will fund a number of consortia or centers targeting three different priority areas identified in the Scientific Strategic Plan of the Global HIV/AIDS Vaccine Enterprise: Design of Immunogens that Induce Broadly Reactive Neutralizing Antibodies (RFP: GH-HTR-05-01); Design of Immunogens that Induce Persistent High Levels of Cell-mediated Immunity (RFP: GH-HTR-05-02); and Standardization and Development of Laboratory Assays to Comparatively Measure the Immunogenicity

of HIV Vaccine Candidates in Pre-clinical and Clinical Trials (RFP: GH-HTR-05-03). The complete RFPs are available online. The listed application deadline is for letters of intent; full proposals are by invitation.

Requirements: Governmental entities and 501(c)3 tax-exempt organizations are eligible. Foreign governmental agencies and equivalent organization also are eligible.

Restrictions: Grants are not awarded to support individuals or projects that serve an exclusively religious purpose.

Date(s) Application is Due: Apr 1

Contact: Grants Administrator, (206) 709-3140; email: info@gatesfoundation.org

Internet: http://www.gatesfoundation.org/GlobalHealth/Grantseekers/Grantmaking Areas/default.htm

Sponsor: Bill and Melinda Gates Foundation

P.O. Box 23350

Seattle, WA 98102

Gates Millennium Scholars Program　　2476

The foundation funds college scholarships for low-income minority students. Students receiving scholarships may major in any field. Graduate scholarships support postgraduate study up to and including doctoral degrees for work in mathematics, science, engineering, education, and library science.

Requirements: Applicants must be a citizen of the United States; have a minimum GPA of 3.3 on a 4.0 scale; be a high school senior applying to an accredited college or university for the academic year; or a college student planning to continue undergraduate study; or a college senior or college graduate enrolled or about to enroll in graduate school. Graduate students must be enrolled in a degree program in engineering, mathematics, science, education, or library science; and demonstrate significant financial need.

Contact: Gates Millennium Scholars, (877) 690-4677

Internet: http://www.gmsp.org

Sponsor: Bill and Melinda Gates Foundation

P.O. Box 10500

Fairfax, VA 22031-80214

GCA Interchange Fellowship　　2477

The program provides a reciprocal exchange of British and American students in horticulture, landscape architecture, and related fields. The intent of the program is to foster cultural understanding, as well as to promote horticultural studies and the exchange of information in this field. GCA offers an academic year of graduate work for a British student in America. The Martin McLaren Scholarship provides a noncredit work/study program for an American recipient at universities such as Reading, Wye College of University of London, Gloucester College, and the Royal Botanic Gardens, Kew; Royal Horticultural Society Gardens, Wisley; and Royal Botanic Garden, Edinburgh. The program is jointly sponsored with the Institute of Horticulture in the British Isles.

Requirements: Recent college graduates who have earned the bachelor's degree in related fields are eligible. Because of a 26-year-old age limit on student travel vouchers in Great Britain and Europe, it is strongly advised that the applicant be 26 or younger.

Date(s) Application is Due: Nov 15

Contact: Connie Sutton, Scholarship Committee, (212) 753-8287; fax: (212) 753-0134; email: csutton@gcamerica.org

Internet: http://www.gcamerica.org/scholarship/gcainterchange.html

Sponsor: Garden Club of America

14 East 60th Street

New York, NY 10022-1002

GCA Katharine M. Grosscup Scholarships　　2478

Scholarships provide assistance to college juniors, seniors, and graduate students pursuing the study of horticulture, agriculture, and related subjects. Preference is given to students from Ohio, Pennsylvania, West Virginia, Michigan, Kentucky, and Indiana.

Amount of Grant: $3000 maximum per scholarship

Date(s) Application is Due: Feb 1

Contact: Nancy Stevenson, Grosscup Scholarship Committee; fax: (216) 721-2056

Internet: http://www.gcamerica.org/scholarship/grosscup.html

Sponsor: Garden Club of America

11030 East Boulevard

Cleveland, OH 44106

GCA Summer Environmental Studies Scholarships　　2479

Each year the GCA Awards for Summer Environmental Studies provide financial aid toward summer studies doing field work or research in the environmental field. The awards offer students who have demonstrated a keen interest in the betterment of the environment an opportunity for further study in the field of ecology. With these scholarships, young men and women can pursue summer programs beyond the regular course of study to gain additional knowledge and experience. Work may award academic credit but should be in addition to required courses. College students may apply for these awards to pursue study following their freshman, sophomore, or junior year. Guidelines and application are available online.

Amount of Grant: $1500

Date(s) Application is Due: Feb 10

Contact: Connie Sutton, GCA Scholarship Committee, (212) 753-8287; fax: (212) 753-0134; email: csutton@gcamerica.org

Internet: http://www.gcamerica.org/scholarship/summeraward_org.html

Sponsor: Garden Club of America

14 East 60th Street

New York, NY 10022-1002

GCA Tropical Botany Research Awards 2480

The program was established to promote the preservation of tropical forests by enlarging the body of botanists with field experience. The awards enable recipients to pursue independent field research in the tropics. Two awards will be offered annually.
Requirements: PhD candidates are eligible to apply.
Amount of Grant: $5500
Date(s) Application is Due: Dec 31
Contact: Judith Mashinya, GCA Awards in Tropical Botany, World Wildlife Fund, (202) 861-8316; fax: (202) 887-5293; email: judith.mashinya@wwfus.org
Internet: http://www.gcamerica.org/scholarship/tropical.html
Sponsor: Garden Club of America
1250 24th Street NW
Washington, DC 20037

GCSAA Essay Contest 2481

The contest is open to undergraduate and graduate students pursuing degrees in turfgrass science, agronomy, or any field related to golf course management. Essays should be seven to 12 pages and should focus on the relationship between golf courses and the environment.
Requirements: Undergraduate and graduate students pursuing degrees in turfgrass science, agronomy, or any field related to golf course management are eligible to apply. Applicants must be a member of GCSAA.
Amount of Grant: $2000 for first place; $1500 for second place; $1000 for third plac
Date(s) Application is Due: Mar 31
Contact: Scholarship Coordinator, (800) 472-7878 ext 4424; email: ahoward@gcsaa.org
Internet: http://www.gcsaa.org/Students/scholarships/essay/default.asp
Sponsor: Golf Course Superintendents Association of America Foundation
1421 Research Park Drive
Lawrence, KS 66049

GCSAA Watson Fellowships 2482

These awards are available to candidates for master's and doctoral degrees in fields related to golf course management. The goal of this program is to identify tomorrow's leading teachers and researchers.
Requirements: To be eligible applicants must be in the second year of a recognized graduate program in turfgrass science or a related field. Typically, this includes candidates for masters degrees or doctorates in appropriate fields of study. Applicants must also be planning to pursue a career in research, instruction, or extension in a university setting.
Amount of Grant: $5000 minimum
Date(s) Application is Due: Oct 1
Contact: Amanda Howard, Scholarship Coordinator, (800) 472-7878 ext 4424 or (785) 832-3678; email: ahoward@gcsaa.org
Internet: http://www.gcsaa.org/students/scholarships/default.asp
Sponsor: Golf Course Superintendents Association of America Foundation
1421 Research Park Drive
Lawrence, KS 66049

GE & Science Prize for Young Life Scientists 2483

The GE & Science Prize for Young Life Scientists has been established to provide support to scientists at the beginning of their careers, because Science/AAAS and GE Healthcare believe that such support is critical for continued scientific progress. The prize will recognize outstanding graduate students in molecular biology from all regions of the world. This international prize will be awarded for the outstanding thesis in the general area of molecular biology as described in a 1000-word essay. The winning essay will be published in Science; essays of the regional award winners will appear in the online version of Science. For judging purposes, the essays will be grouped according to the geographic location of the degree-granting institution: North America, Europe, Japan, and all other countries. Initial screening of the submissions will be done by regional judges. Essays will be judged on the quality of the research and the entrant's ability to articulate the contribution of the research to the field of molecular biology. The top five essays from each geographic region will be forwarded to a panel of judges. All regional winners will compete for the grand prize of U.S.$25,000. The regional winners who do not receive the grand prize will be awarded U.S.$5,000. Winners will be announced in Science and the prize will be awarded in a location to be announced. The grand prize essay will be published in 'Science', and essays of the regional winners will be published on the online version of 'Science'.
Requirements: Entrants must have been awarded their Ph.D. between 1 January and 31 December 2006. Candidates for M.D./Ph.D. degrees are eligible to compete for the prize in either the year the Ph.D. is awarded or the year the final degree is awarded. The research described in the entrant's thesis must be in the field of molecular biology as described above. The prize will recognize only work that was performed while the entrant was a graduate student. The prize will be awarded without regard to sex, race, or nationality.
Restrictions: Employees of GE, Science and AAAS, and their relatives are not eligible for the prize.
Samples: 2006 Grand Prize Winner - Irene Chen, 'The Emergence of Cells During the Origin of Life'; Regional Winners: North America - Dianne Schwarz, 'Unraveling the Mysteries of Small RNAs'; Europe: Bernhard Loll, 'Photosystem II, a Bioenergetic Nanomachine'; All Other Countries: Ron Milo, 'Simple Building Blocks for Complex Networks'
Amount of Grant: $25,000

Date(s) Application is Due: Jul 15
Contact: Sylvia Kihara, Prize Coordinator, (202) 326-6507; fax: (202) 289-7562; email: skihara@aaas.org
Internet: http://www.sciencemag.org/feature/data/prizes/ge/index.dtl
Sponsor: American Association for the Advancement of Science
1200 New York Avenue, NW, Room 1049B
Washington, DC 20005

GEM MS Engineering Fellowships 2484

The objective of the master's program is to increase the pool of minority MS engineering graduates. Each participant is provided with a summer internship experience for a GEM corporate member and a portable academic fellowship of tuition, fees, and stipend that must be used at a GEM university member. Fellows are supported for three semesters or four quarters.
Requirements: Eligibility includes U.S. citizenship; ethnic identification of Native American, African American, Mexican American, or Puerto Rican; academic classification of junior, senior, or baccalaureate degree recipient in an accredited engineering discipline; and an undergraduate record that indicates the ability to successfully pursue graduate studies in engineering.
Amount of Grant: $10,000 stipend per academic year
Date(s) Application is Due: Nov 1
Contact: Fellowships Program, (219) 631-7771; fax: (219) 287-1486; email: gem@nd.edu
Internet: http://was.nd.edu/gem/gemwebapp/public/gem_01_100.htm
Sponsor: National GEM Center
P.O. Box 537
Notre Dame, IN 46556

GEM PhD Engineering Fellowships 2485

The objective of this program is to offer doctoral fellowships to underrepresented minority students. Fellowships may be used at any participating GEM member university. Tuition, fees, and a yearly stipend are provided. Fellows may be required to accept a research or teaching assistantship. A paid summer internship may be required.
Requirements: Eligibility includes U.S. citizenship and ethnic identification of Native American, African American, Mexican American, or Puerto Rican. Applicants must have or be in the process of attaining a master's degree in engineering and have an academic record that indicates the ability to successfully pursue doctoral studies in engineering.
Amount of Grant: $14,400 per calendar year plus $5500 cost of instruction grant for first year
Date(s) Application is Due: Nov 1
Contact: Fellowships Program, (219) 631-7771; fax: (219) 287-1486; email: gem@nd.edu
Internet: https://ws4.nd.edu/gem/gemwebapp/public/gem_01_100.htm
Sponsor: National GEM Center
P.O. Box 537
Notre Dame, IN 46556

GEM PhD Science Fellowships 2486

The goal of this program is to increase the number of minority students pursuing doctoral degrees in the natural, physical, and life sciences (chemistry, physics, mathematics, and computer science, as well as the earth, biological, and pharmaceutical sciences). Fellowships are portable to any participating GEM member university, and include a stipend, tuition, and fees. The award is renewable through the university for up to four additional years. Fellows may be required to accept a research or teaching assistantship. A paid summer internship is required prior to entering a doctoral program.
Requirements: Eligibility includes U.S. citizenship; ethnic identification of Native American, African American, Hispanic, or Puerto Rican; academic classification of junior, senior, or baccalaureate degree recipient in an accredited science discipline; and an undergraduate record that indicates the ability to successfully pursue doctoral studies in the natural sciences.
Amount of Grant: $14,400 per calendar year plus $5500 cost of instruction grant for first year
Date(s) Application is Due: Nov 1
Contact: Fellowships Program, (219) 631-7771; fax: (219) 287-1486; email: gem@nd.edu
Internet: http://was.nd.edu/gem/gemwebapp/public/gem_01_100.htm#03
Sponsor: National GEM Center
P.O. Box 537
Notre Dame, IN 46556

General L. Kemper Williams Prize in Louisiana History 2487

The Historic New Orleans Collection, in cooperation with the Louisiana Historical Association, awards one prize for the best published work (book or article) dealing with Louisiana history. The award contains an engraved plaque. The nomination form is available on the Web site.
Requirements: Works about all aspects of Louisiana history and culture are eligible, as are works that treat the history of Louisiana in a regional, national, or international context.
Amount of Grant: $1500
Date(s) Application is Due: Jan 15
Contact: Chair, Williams Prize Committee, (504) 523-4662; fax: (504) 598-7108; email: WRC@hnoc.org

Internet: http://www.hnoc.org/index.html
Sponsor: Historic New Orleans Collection
533 Royal Street
New Orleans, LA 70130

General Motors Foundation Grants Support Program 2488

With a strong commitment to diversity in all areas, the targeted areas of focus for the Foundation are: education; health and human services; civic and community; public policy; arts and culture; and environment and energy. Primary consideration is given to requests that meet the following criteria: exhibit a clear purpose and defined need in one of the foundation's areas of focus; recognize innovative approaches in addressing the defined need; demonstrate an efficient organization and detail the organization's ability to follow through on the proposal; and, explain clearly the benefits to the foundation and the plant city communities. Paper applications are no longer accepted. Completion of an online eligibility quiz is the first step in the application process.

Requirements: Nonprofit, tax-exempt organizations and institutions are eligible to apply. Applications must be made online.

Restrictions: The Foundation not not support organizations that discriminate on the basis of race, religion, creed, gender, age, veteran status, physical challenge or national origin. Contributions are generally not provided for: individuals; religious organizations; political parties or candidates; U.S. hospitals and health care institutions (general operating support); capital campaigns; endowment funds; conferences, workshops or seminars not directly related to GM's business interests.

Samples: American National Red Cross (Washington, DC)—for relief efforts in South Asia and Africa, $1 million (2005); Pierre Chambon, College de France (Paris, France) and Ronald Evans, Salk Institute for Biological Studies (La Jolla, CA)—to honor their contributions to the diagnosis, prevention, and treatment of cancer, $250,000 jointly (2003).

Contact: Grant Coordinator, (313) 556-5000
Internet: http://www.gm.com/company/gmability/community/guidelines/index.html
Sponsor: General Motors Foundation
P.O. Box 300, 300 Renaissance Center
Detroit, MI 48265-3000

General Service Foundation International Peace Grants 2489

The foundation awards grants in three program areas: international peace, reproductive health and rights, and resources. International peace grants address the root causes of conflict in Mexico, Central America, and the Caribbean. Reproductive health and rights grants address the issues of rapid population growth and unintended pregnancies in the United States and Latin America. Resources grants support improvement of the use, quality, and management of water in regions west of the Mississippi River. It is recommended that applicants carefully review the foundation's annual report before sending letters of inquiry. Apply online.

Requirements: Grants are made to organizations that are tax-exempt under U.S. laws. The foundation prefers projects and/or programs that give promise of significant contribution and that are new, innovative, demonstrational, and/or research in nature.

Restrictions: In general, contributions are not made to operating budgets, nor to annual campaigns of established organizations. The foundation ordinarily does not contribute to capital expenses (physical plant, equipment, endowment), to individuals, nor for relief.

Amount of Grant: $3.1 million total; $2500-$35,000 average
Date(s) Application is Due: Jan 14; Sep 1
Contact: Program Officer, International Peace, (970) 920-6834; fax: (970) 920-4578; email: peace@generalservice.org
Internet: http://www.generalservice.org
Sponsor: General Service Foundation
557 N Mill Street, Suite 201
Aspen, CO 81611

General Service Foundation Reproductive Health and Rights Grants 2490

The foundation is concerned with the related issues of rapid population growth and unintended pregnancies, particularly the availability of high-quality reproductive health care and access to that care. The foundation funds research for policy analysis, litigation, public education, and advocacy. Grants are made for programs in the United States and Latin America that address the needs of low-income women, women of color, adolescents, and the issue of access. Apply online.

Requirements: Grants are made to organizations that are tax-exempt under U.S. laws. The foundation prefers projects and/or programs that give promise of significant contribution and that are new, innovative, demonstrational, and/or research in nature.

Restrictions: Generally, grants are not made for service delivery, contraceptive development, or university-based research. In addition, contributions are not made to operating budgets or to annual campaigns of established organizations. The foundation ordinarily does not contribute to capital (physical plant, equipment, endowment), to individuals, nor for relief.

Amount of Grant: $3.1 million total; $2500-$35,000 average
Date(s) Application is Due: Feb 1
Contact: Holly Bartling, Program Contact, (202) 232-1005 or (970) 920-6834; fax: (970) 920-4578; email: holly@generalservice.org or rhr@generalservice.org
Internet: http://www.generalservice.org/Reproductive%20Health.htm
Sponsor: General Service Foundation
557 N Mill Street, Suite 201
Aspen, CO 81611

General Service Foundation Western Water Grants 2491

The goal of the program is to preserve, secure, and protect water flows for aquatic and riparian ecosystems in the rivers of the interior West. Within the target states of Idaho, Montana, Wyoming, Colorado, New Mexico, Utah and Arizona, the foundation will support projects with significant promise of basin-wide, state-wide, or regional (multi-state) impact. The foundation will also continue to consider projects addressing water policies and laws affecting the West as a region. While the foundation supports a broad range of activities, including advocacy, litigation, public education and outreach, research, and media, priority will be given to organizations attempting to affect legal and/or policy changes.

Requirements: Grants are made to organizations that are tax-exempt under U.S. laws. The foundation prefers projects and/or programs that give promise of significant contribution and that are new, innovative, demonstrational, and/or research in nature.

Restrictions: In general, contributions are not made to operating budgets or to annual campaigns of established organizations, nor does the foundation ordinarily contribute to capital expenses (physical plant, equipment, endowment), to individuals, or for relief.

Samples: WaterWatch of Oregon (OR)—to support the Streamflow Restoration Program, $10,000 over two years. Apply online.

Amount of Grant: $3.1 million total; $2500-$35, 000 average
Date(s) Application is Due: Jan 14; Sep 1
Contact: Lani Shaw, Executive Director, (970) 920-6834; fax: (970) 920-4578; email: info@generalservice.org
Internet: http://www.generalservice.org
Sponsor: General Service Foundation
557 N Mill Street, Suite 201
Aspen, CO 81611

Genuardi Family Foundation Grants 2492

The Foundation supports direct providers of services in the areas of education, health, human services, and culture. The Foundation gives preference to projects which receive broad-based community support and provides for reasonable costs associated with conducting the proposed project or program. Grant proposals may include the allocation of a reasonable percentage of grant monies towards general support of the recipient organization. On a limited basis, the Foundation will also consider grants in support of capital campaigns for facilities or equipment, and organizational capacity building.

Requirements: The Foundation only considers grant requests from organizations designated as tax exempt under Section 501(c)3 of the Internal Revenue Code. Preference will be given to non-profit organizations based in the Greater Philadelphia Area.

Restrictions: The Foundation's funding will not process grant requests from: individuals; fraternal and/or civic organizations; political candidates or to influence legislation; other foundations; general fund raising and endowment drives; debt reduction; environmental issues or initiatives; annual appeals or letters of solicitation; or public, private or parochial schools that serve the general public.

Geographic Focus: Pennsylvania
Amount of Grant: $5,000-$50,000
Date(s) Application is Due: Aug 1
Contact: Meredith A. Huffman, Executive Director; (610) 834-2030; email: info@genuardifamilyfoundation.org
Internet: http://www.genuardifamilyfoundation.org/genuardi_priorities.html
Sponsor: Genuardi Family Foundation
Blue Bell Executive Campus, 470 Norristown Road, Suite 300
Auburn, NY 13021

Geographic Education Dissertation Award 2493

Awards are made for outstanding doctoral research to encourage research in geography teaching and learning. Applications for the award must be made through a 15-page paper based on the applicant's doctoral dissertation.

Requirements: Scholars eligible for the award must have received the doctoral degree within the previous two years. All applicants are expected to present their research at the annual meeting.

Amount of Grant: $300 first prize, $100 second prize
Date(s) Application is Due: Mar 15
Contact: Executive Director, (256) 782-5293; fax: (256) 782-5336; email: ncge@ncge.org
Internet: http://www.ncge.org/resources/awards/dissertation.cfm
Sponsor: National Council for Geographic Education
Jacksonville State University, 206A Martin Hall
Jacksonville, AL 36265-1602

George A. and Eliza Gardner Howard Humanities Fellowships 2494

These fellowships assist individuals in the middle stages of their professional humanities or social sciences careers. Support is intended to augment paid sabbatical leaves. Preference will be given to individuals who are professionally based in the United States through institutional affiliation or by residence. Awards are granted in a sequence: arts in odd-numbered years and social sciences in even-numbered years; contact the foundation for current subcategories. Nominations must be received by October 17. Details of the nomination procedure are available from the foundation upon request.

Requirements: Candidates must be nominated by a representative of an affiliated college or university, a professional critic, or editor, or the director of a professional society. Grants are available to men and women of any nationality, generally between the ages of 25 and 45, who are below the rank of full professor.

Restrictions: Fellowships are not available for work leading to an academic degree or for private study.

Amount of Grant: $25,000 stipend
Date(s) Application is Due: Oct 17
Contact: Henry Majewski, Director, (401) 863-2640; fax: (401) 863-7341; email: howard_foundation@brown.edu
Internet: http://www.stg.brown.edu/projects/gradschool/howard
Sponsor: George A. and Eliza Gardner Howard Foundation
42 Charlesfield Street
Providence, RI 02912

George A. Olah Award in Hydrocarbon or Petroleum Chemistry   2495

The award is conferred to recognize, encourage, and stimulate outstanding research achievements in hydrocarbon or petroleum chemistry. Special consideration will be given to independence of thought and originality. The award consists of $5,000 and a certificate. Reasonable travel expenses to the meeting at which the award will be presented will be reimbursed.
Requirements: A nominee must be a resident of the United States or Canada and have accomplished outstanding research in the chemistry of hydrocarbons or petroleum and its products. Any individual, except a member of the award committee, may submit one nomination or seconding letter for the award in any given year. The nominating documents consist of a letter of not more than 1000 words, containing an evaluation of the nominee's accomplishments and a specific identification of the work to be recognized, a biographical sketch including date of birth, and a list of publications and patents authored by the nominee.
Samples: Israel E. Wachs—award winner, $5,000 (2008.; Bruce E. Koel—award winner, $5,000 (2007); G.K. Surya Prakash—award winner, $5,000 (2006).
Amount of Grant: $5,000 award and $1,500 travel expenses to meeting
Date(s) Application is Due: Nov 1
Contact: Felicia Dixon, Awards Administrator; (800) 227-5558 or (202) 872-4408; fax: (202) 776-8008; email: f_dixon@acs.org or awards@acs.org
Internet: http://portal.acs.org/portal/acs/corg/content?_nfpb=true&_pageLabel=PP_ARTICLEMAIN&node_id=1319&content_id=CTP_004526&use_sec=true&sec_url_var=region1
Sponsor: American Chemical Society
1155 Sixteenth Street, NW
Washington, DC 20036

George B. Storer Foundation Grants   2496

The foundation awards grants, primarily in Florida, in its areas of interest, including higher education, social services—particularly for the blind, youth organizations, conservation, hospitals, and cultural programs. Types of support include building construction/renovation, capital campaigns, general operating support, matching/challenge grants, and research grants. There are no application forms. Applications should be submitted between October 15 and the listed deadline.
Samples: U of Miami (FL)—for marine-science and business-education programs, $205,000 (2005).
Date(s) Application is Due: Nov 15
Contact: Grants Administrator, c/o Thomas McDonald, CPA
Sponsor: George B. Storer Foundation
P.O. Box 1040
Tavernier, FL 33070

George C. Pimentel Award in Chemical Education   2497

This award, supported by the Dow Chemical Company, is given annually to recognize outstanding contributions to chemical education. A nominee must have made outstanding contributions to chemical education considered in its broadest meaning, including the training of professional chemists; the dissemination of reliable information about chemistry to prospective chemists, to members of the profession, to students in other fields, and to the general public; and the integration of chemistry into our educational system. The activities recognized by the award may lie in the fields of teaching (at any level), organization and administration, influential writing, educational research, the methodology of instruction, establishment of standards of instruction, and public enlightenment. Preference shall be given to U.S. citizens. The Award includes $5,000, a certificate and travel expenses to award meeting.
Requirements: Any individual, except a member of the award committee, may submit one nomination or seconding letter for the award in any given year. The nominating documents consist of a letter of not more than 1000 words containing an evaluation of the nominee's accomplishments and a specific identification of the work to be recognized, a biographical sketch including date of birth, and a list of publications and patents authored by the nominee.
Restrictions: Self-nominations are not accepted.
Samples: Richard N. Zare—award winner, $5,000 (2008); A. Truman Schwartz—award winner, $5,000 (2007); F. Albert Cotton—award winner, $5,000 (2006).
Amount of Grant: $5,000
Date(s) Application is Due: Nov 1
Contact: Felicia Dixon, Awards Administrator; (800) 227-5558 or (202) 872-4408; fax: (202) 776-8008; email: f_dixon@acs.org or awards@acs.org
Internet: http://portal.acs.org/portal/acs/corg/content?_nfpb=true&_pageLabel=PP_ARTICLEMAIN&node_id=1319&content_id=CTP_004528&use_sec=true&sec_url_var=region1
Sponsor: American Chemical Society
1155 Sixteenth Street, NW
Washington, DC 20036

George Foundation Grants   2498

The Foundation prioritizes its funding opportunities across the following areas: basic needs; health and wellness; education; early childhood - 3rd grade; economic success; scholarship programs; security and safety; community enhancement; and preservation of regional heritage. To support family stability by creating an environment in which families can support one another and children can be nurtured, the Foundation strongly favors programs that: target service delivery to populations with less access to resources, with priority given to programming for children; recognize that providing basic human needs is the first step to assisting the most impoverished; seek to understand the needs of children and promote their well-being; understand today's families are complex and may include kin-ship guardians, non-custodial fathers, foster families, and adoptive families; support preventative rather than intervening efforts by identifying the root cause of community issues preventing the growth of self-reliant and productive families; and understand the cultural diversity and values of the families residing in Fort Bend County.
Requirements: Nonprofit organizations in Fort Bend County, Texas, may submit grant proposals.
Restrictions: The Foundation does not fund: grants to organizations that do not have a current 501(c)3 determination letter; churches or other organized religious bodies; grants to another organization that distributes money to recipients of its own selection, i.e., a regranting organization; regional, national or international programs; grants for research or studies; grants for travel, conferences, conventions, group meetings, or seminars; the purchase of event tickets, tables, ads or sponsorships; support to fairs and festivals; religious or private schools; request for funds to develop films, videos, books or other media projects; direct mail campaigns; loans of any kind; grants to individuals; grants to fraternal organizations; political interests of any kind; and institutions that discriminate on the basis of race, creed, gender, national origin, age, disability or sexual orientation in policy or in practice.
Geographic Focus: Texas
Samples: Associated Catholic Charities of the Diocese of Galveston-Houston (TX)—to assist in providing two full-time bilingual counselors to serve Fort Bend families in crisis, 140,000 three-year grant.; Montrose Clinic (TX)—to assist in funding operating expenses of the HIV Primary Care Clinic to serve Fort Bend residents living with HIV/AIDS, $25,000.; Stehlin Foundation for Cancer Research (TX)—to support research projects involving the Camptothecin family of drugs, $50,000 matching grant.; Wharton County Junior College (TX)—to provide scholarships for 60 Fort Bend County students during the 2003-04 academic school year, $50,000.
Date(s) Application is Due: Jan 15; Apr 15; Jul 15; Oct 15
Contact: Dee Koch, Grant Officer, (281) 342-6109; fax: (281) 341-7635; email: dkoch@thegeorgefoundation.org
Internet: http://www.thegeorgefoundation.org
Sponsor: George Foundation
310 Morton Street, PMB Suite C
Richmond, TX 77469

George Frederick Jewett Foundation Grants Program   2499

The foundation is concerned primarily with people and values. The grants program focuses on the future and on stimulating and supporting activities and projects of established, voluntary, nonprofit organizations that are of importance to human welfare. Grants are made in the fields of arts and humanities, conservation and preservation, education, health care and medical services, population, religion, and social welfare. The foundation may support research on and studies of important problems of public concern solely for the purpose of aiding in the gathering and presenting of facts that may assist the public to better understand such problems and to arrive at realistic and effective solutions to them. From time to time, support may be given to the scholarship, fellowship, and research programs of established institutions when sufficient evidence is available to establish clearly that the applicant organization is awarding such grants in accordance with the regulations established by the IRS. Grants are awarded to support activities in progress, research into potential projects, building and equipment, general operations, program development, seed funding, research, technical assistance, and matching funds. Inquiries for clarification of the foundation's policy and program emphasis are encouraged.
Requirements: Preference is given to public charities or nonprivate foundations. The foundation confines its grants largely to requests from eastern Washington and the San Francisco Bay area.
Restrictions: Grants do not support advertising; advocacy, athletic, international, religious, political, or veterans organizations; or individuals.
Geographic Focus: California; Washington
Samples: American U of Beirut (New York, NY)—general support, $15,000.; Committee to Restore the Opera House (San Francisco, CA)—for a three-year commitment for restoration of the San Francisco War Memorial Opera House, $120,000.
Amount of Grant: $5000-$50,000
Contact: Ann Gralnek, (415) 421-1351; fax: (415) 421-0721; email: ADGjewettf@aol.com or tfbjewettf@aol.com
Sponsor: George Frederick Jewett Foundation
235 Montgomery Street, Suite 612
San Francisco, CA 94104

George Gund Foundation Grants   2500

The Foundation's guidelines reflect their long-standing interests in: arts; economic development and community revitalization; education; the environment; and human services. The Foundation supports opportunities that cross program boundaries and that integrate elements of the Foundation's interests. Although the Foundation's focus is centered in Greater Cleveland, a portion of their grantmaking will continue to support state and national policy making that bolsters their work.

Requirements: The foundation makes grants only to nonprofit, tax-exempt organizations or to qualified governmental units or agencies. The Foundation currently does not accept faxed proposals or proposals submitted electronically.

Samples: Parkworks Inc (Cleveland, OH)—for green-space projects in downtown Cleveland and its neighborhoods, $100,000 (2005); Lesbian-Gay Community Service Ctr of Greater Cleveland (OH)—for its Safe Schools Are for Everyone program, $25,000 (2005); Foundation Fighting Blindness (Owings Mills, MD)—for research on retinal degenerative disease, $2 million (2005); American Civil Liberties Union Foundation (New York, NY)—for the Reproductive Freedom Project, $35,000 (2005).

Date(s) Application is Due: Mar 30; Jun 30; Sep 30; Dec 30

Contact: David Abbott, Executive Director, (216) 241-3114; fax: (216) 241-6560; email: info@gundfdn.org

Internet: http://www.gundfdn.org

Sponsor: George Gund Foundation

45 Prospect Avenue West, 1845 Guildhall Building

Cleveland, OH 44115

George H. Hitchings New Investigator Award in Health Research 2501

Provides awards to provide flexible support for young researchers in their second year of graduate school. Students planning to follow a career path in academic teaching and research that is closely tied to the health of North Carolinians and the strength of North Carolina Science are particularly encouraged to apply.

Requirements: Applications/nominations must be submitted by the institution on behalf of the student(s). Nomination applications are accepted from Duke, NCCU, NCSU, and UNC-Chapel Hill.

Restrictions: Only one nomination/application per graduate program/department within the university will be accepted.

Geographic Focus: North Carolina

Amount of Grant: $10,000

Date(s) Application is Due: Mar 5

Contact: Libby Long, Scholarships and Special Projects Coordinator: (919) 474-8370 ext. 134; fax: (919) 941-9208; email: libby@trialglecf.org

Internet: http://www.trianglecf.org/DatasetRecord.cfm?recordID=10001417&returnURL=%2Fpage10000237%2Ecfm&returntoname=View%20all%20Scholarships&sidepageid=10000237&thetitle=George%20H.%20Hitchings%20New%20Investigator%20Award%20in%20Health%20Research&Ds_PagepropId=1

Sponsor: Triangle Community Foundation

324 Blackwell Street, Suite 1220

Durham, NC 27701

George Kress Foundation Grants 2502

The foundation awards grants to eligible Wisconsin nonprofit organizations in its areas of interest, including arts and culture, children and families, Christian churches and organizations, health care/health organizations/hospitals, historical preservation, human and social services, and sports and recreation. Types of support include building construction and renovation, capital campaigns, emergency services, professorships, program development, research grants, and scholarship funds. There are no application deadlines or forms.

Requirements: Wisconsin nonprofit organizations are eligible. Preference is given to requests benefiting Green Bay and Madison

Geographic Focus: Wisconsin

Samples: Saint Mary's Medical Ctr (Green Bay, WI)—to construct an addition that will house an expanded emergency-services department and improved facilities for surgery, $1 million (2005).

Contact: John Kress, Secretary, (920) 433-3109

Sponsor: George Kress Foundation

P.O. Box 408

Neenah, WI 54957-0408

George M. Brooker Collegiate Scholarship for Minorities 2503

The scholarship supports graduate and undergraduate minority students entering careers in real estate, and specifically real estate management, upon graduation. Applicants for the one-time award must have declared a major in real estate or a related field; must have completed two courses in real estate or indicate intent to complete them; must have at least a 3.0 grade point average within the major; and must submit recommendation letters, themed essay, and official transcript.

Requirements: To be eligible, the applicant must be a member of a minority group and a U.S. citizen.

Amount of Grant: $2500 for graduates; $1000 for undergraduates

Date(s) Application is Due: Mar 31

Contact: Kimberly Holmes, Foundation Administrator, (312) 329-6008; fax: (312) 410-7908; email: kholmes@irem.org or foundatn@irem.org

Internet: http://www.irem.org/sec1ins.cfm?sec=iremfoundation&con=scholarships.cfm&par=#4

Sponsor: Institute of Real Estate Management Foundation

430 N Michigan Avenue

Chicago, IL 60611-8775

George S. and Dolores Dore Eccles Foundation Grants 2504

The foundation awards grants to eligible Utah organizations in its areas of interest, including arts, children and youth, economics, higher education, hospitals, medical

research, performing arts, visual arts, and social services. Types of support include building construction/renovation, capital campaigns, equipment acquisition, general operating grants, matching/challenge grants, professorships, program development, research grants, and scholarship funds. A request for application is available online.

Requirements: Giving primarily in Utah.

Restrictions: Funding requests will not be considered from the following types of organizations: those that have not received a tax exemption letter establishing 501(c)3 status from the Internal Revenue Service, unless they are a unit of government, in which case such a letter is not required; other private foundations; those of a political nature that attempt to influence legislation and/or candidacy of persons for elected public office; conduit organizations, unified funds, or those that use funds to make grants to support other organizations; those that do not have fiscal responsibility for the proposed project. Funds will also not be considered for: contingencies, deficits, or debt reduction; general endowment funds; direct aid to individuals; conferences, seminars, or medical research; requests which do not fall within the Foundation's specified areas of interest.

Geographic Focus: Utah

Contact: Director, (801) 246-5340; fax: (801) 350-3510; email: gseg@gseccles.org

Internet: http://www.gsecclesfoundation.org

Sponsor: George S. and Dolores Dore Eccles Foundation

79 South Main Street, 12th Floor

Salt Lake City, UT 84111

George W. Codrington Charitable Foundation Grants 2505

The foundation's support is limited to public charitable or educational projects within Cuyahoga County, OH, and immediately adjacent areas. Areas of interest include museums, performing arts, arts and cultural programs, higher education, other education, hospitals (general), and children and youth services. Types of support include continuing support, general operating support, annual campaigns, capital campaigns, equipment, program development, and research. The application should state fully but briefly the amount requested, need for the grant, area served by the applicant, brief history of the applicant organization, description of applicant's contribution to the area, listing of the applicant's officers and trustees, and evidence of 501(c)3 status. Grants generally are made on an annual basis. Meetings of the supervisory board are usually held in March, June, September, November and December, at which time pending applications are considered.

Requirements: IRS 501(c)3 organizations serving residents of Cuyahoga County, OH, and immediately adjacent areas are eligible.

Restrictions: Grants are not made to individuals.

Geographic Focus: Ohio

Samples: Cleveland Ctr for Economic Education (Cleveland, OH)—for project support, $5000.; Educational Television Assoc of Metropolitan Cleveland (Cleveland, OH)—for project support, $25,000.; Ohio Chamber Orchestra (Cleveland, OH)—for support of the arts, $2000.

Amount of Grant: $500-$25,000 average

Contact: Raymond Sawyer, Chair, Supervisory Board, (216) 566-5837; fax: (216) 566-5800

Sponsor: George W. Codrington Charitable Foundation

127 Public Sq, 39th Fl

Cleveland, OH 44114-1216

George Westinghouse Medals 2506

Medals are bestowed annually for eminent achievement or distinguished service in the power field of mechanical engineering. Considering power in the broad sense, the basis of the awards shall include contributions of utilization, application, design, development, research, and the organization of such activities in the power field. Any individual member, group of members, or committee may nominate candidates for the awards. A certificate accompanies the award.

Requirements: Candidates are not restricted by profession nor by membership in any engineering society or organization. Silver medals are awarded to candidates under 45 years of age.

Amount of Grant: $1500 and vermeil medal; $1000 and silver medal

Date(s) Application is Due: Sep 1

Contact: Gilda DiTullio, Manager, (212) 591-7736; fax: (212) 705-7739; email: ditulliog@asme.org

Internet: http://www.asme.org/honors/ms71/saa/westinghouse.html

Sponsor: American Society of Mechanical Engineers

3 Park Avenue

New York, NY 10016-5990

Georgetown University Chemistry Department Graduate Research Fellowships 2507

Fellowships enable qualified students to pursue graduate work leading to PhD degrees while gaining valuable experience assisting in undergraduate courses during at least one of the years of graduate study. The graduate program provides a firm foundation in all areas of chemistry as well as opportunities for research under 17 faculty members on a considerable variety of topics.

Requirements: Applicant must have a BS degree in chemistry or its equivalent.

Amount of Grant: $21,400

Contact: Chairperson, (202) 687-6073; fax: (202) 687-6209; email: chemad@georgetown.edu

Internet: http://www.georgetown.edu/departments/chemistry/graduate.htm

Sponsor: Georgetown University

Department of Chemistry, Box 571227

Washington, DC 20057-2222

Georgia Institute of Technology Postdoctoral Fellowship 2508

The 12-month postdoctoral fellowship starts in mid-August. Duties include individual/group counseling, outreach, crisis intervention, consultation, and testing/evaluation. The fellowship provides a stipend plus benefits. Applicants should send resumes to the office.

Requirements: Individuals with the PhD in psychology and experience in counseling/clinical psychology are eligible.

Amount of Grant: $22,000

Date(s) Application is Due: Mar 9

Contact: Dr. Irene Dalton, (404) 894-2575; fax: (404) 894-1804

Internet: http://www.ohr.gatech.edu

Sponsor: Georgia Institute of Technology

500 Tech Pkwy

Atlanta, GA 30332-0286

Georgia Power Foundation Grants 2509

Giving is focused on issues that directly affect customers, employees, business and shareholders. These include: improving the quality of education by partnering with organizations to assist students with personal development, mentoring and career exploration; protecting the environment by promoting programs to improve air and water quality, preserve natural resources and protect endangered species; preventing cancer; and promoting diversity. The Foundation gives strong preference to Georgia-based organizations and programs that seek to improve the quality of life for the state's residents. Applicants may apply online or with a written proposal.

Requirements: The foundation makes grants to tax-exempt organizations that seek to improve the quality of life for Georgia's residents.

Restrictions: The Foundation does not provide grants to individuals, private elementary or secondary schools, and religious organizations, nor political campaigns or causes. The Foundation does not provide multi-year funding commitments.

Geographic Focus: Georgia

Amount of Grant: $10,000

Date(s) Application is Due: Feb 15; May 15; Aug 15; Nov 15

Contact: Grants Administrator, (404) 506-6784; email: gpfoundation@southernco.com

Internet: http://www.georgiapower.com/community/apply.asp

Sponsor: Georgia Power Foundation

241 Ralph McGill Boulevard, N.E., Bin 10131

Atlanta, GA 30308-3374

Gerald R. Ford Foundation Research Travel Grants 2510

Grants are available for research that makes significant use of the Gerald R. Ford Library's archival collections, which focus on U.S. domestic affairs, foreign relations, economic policies, and politics in the 1970s. The library can provide free database searches and other information to help scholars determine whether the collections are useful to their research. The grants defray travel, living, and photocopy expenses incurred while conducting research at the library. Applications are available on the Web site.

Amount of Grant: $20,000 total; $2000 each

Date(s) Application is Due: Mar 15; Sep 15

Contact: Grants Coordinator, (734) 205-0555; fax: (734) 205-0571; email: ford.library@nara.gov

Internet: http://www.ford.utexas.edu/library/hpgrants.htm

Sponsor: Gerald R. Ford Foundation

1000 Beal Avenue, Gerald R. Ford Library

Ann Arbor, MI 48109

Geraldine R. Dodge Foundation Frontiers for Veterinary Medicine Fellowships 2511

Fellowships enable motivated and creative veterinary students to pursue either summer projects or research undertakings of their own design centered on endeavors they are passionate about and that hold promise for advancing the humane treatment of animals. The program gives veterinary students the opportunity to step outside the traditional bounds of veterinary education to explore and bring new, creative problem-solving perspectives to pressing, animal-related issues. Proposals may be for work that is project-based (such as developing a humane education program) or research-based. Projects or research must be completed by September of the upcoming year. A wide range of fellowship topics and issues are open for consideration. They include companion animals, wildlife, livestock, laboratory animals, zoo animals, pet overpopulation, the human-animal bond, animal-assisted therapies, conservation medicine, shelter medicine, animal ethics, and humane and/or veterinary education. Preference will be given to projects that occur in the United States. International work, however, will be considered if the proposal is exceptionally compelling. To be considered, an applicant must submit a fellowship proposal, including an abstract and a budget, and have the support of a sponsoring mentor. Applications and mentor letters via standard mail must be postmarked no later than the listed application deadline. Applications via email must be received by the listed deadline.

Requirements: The applicant must be currently enrolled as a full-time veterinary student at a U.S. or Canadian school or college of veterinary medicine accredited by the American Veterinary Medical Association.

Amount of Grant: $7000 maximum

Date(s) Application is Due: Dec 16

Contact: Lisa Bregman, (973) 540-8443 ext 131; fax: (973) 540-1211; email: lbregman@grdodge.org

Internet: http://www.grdodge.org/frontiers_guidelines.htm

Sponsor: Geraldine R. Dodge Foundation

P.O. Box 1239, 163 Madison Avenue

Morristown, NJ 07962-1239

Geraldine R. Dodge Foundation Teacher Fellowships 2512

The program awards fellowships to full-time New Jersey teachers in public and public charter schools in Camden County, NJ. The program seeks to support life-expanding, rejuvenating projects that may even fulfill a lifetime dream in addition to proposals that will expand intellectual capacity and appreciation of diversity and culture. Consideration will be given to teams of one or more teachers with one common proposal. The deadline is January 17, and submissions should be made online.

Requirements: K-12 New Jersey teachers in Camden County are eligible.

Restrictions: The Fellowship program does not fund: training that your school district would normally be expected to fund such as block scheduling or differentiated instruction; projects whose funds are directly used for classroom supplies, classroom activities, or classroom field trips; funds for equipment purchase or rental (such as a video camera, computer, etc.); funds used for college or graduate level classes leading to dissertation work or degree credit; individual project proposals from administrators; or retroactive financial support for projects, workshops, or travel that has already taken place.

Geographic Focus: New Jersey

Samples: Bonnie Durante and Marilyn Larke, Camden County Technical School, Pennsauken, NJ—travel to Cape Town to help organize a leadership connection between the students at Camden County Technical School and students at a school in South Africa (2007); Amanda Cherico, LEAP Academy University Charter School, Camden, NJ—travel to Ecuador to better understand the culture and to study the Galapagos Islands, Amazon Rainforest and Andes Mountains (2007); Katherine Kaplan, Carusi Middle School, Cherry Hill—to explore the journey on the path of the Silk Road (2007).

Amount of Grant: $2,000-$7,500 for individuals; $5,000-$10,000 for teams

Date(s) Application is Due: Jan 17

Contact: Shannon Joerchel, Teacher Fellowship Coordinator; (973) 540-8442, ext. 106; fax: (973) 540-1211; email: tfellows@grdodge.org

Internet: http://www.grdodge.org/initiatives/teacherfellowship/index.htm

Sponsor: Geraldine R. Dodge Foundation

163 Madison Avenue, P.O. Box 1239

Morristown, NJ 07962-1239

Gerber Foundation Grants 2513

The foundation awards grants to support national programs that have a significant impact on issues facing infants and young children. Areas of interest include pediatric health—promoting health and preventing disease, including projects geared toward research or interventions that will reduce the incidence of serious chronic illnesses (e.g., diabetes, heart disease, obesity, or cancer) or improve cognitive, social, and emotional aspects of development; pediatric nutrition—assuring adequate nutrition for infants and young children through projects of research or interventions; and the effects of environmental hazards. The listed deadlines are for letters of inquiry. Guidelines and application are available online.

Requirements: 501(c)3 nonprofit organizations are eligible. Priority is given to projects that improve infant and young children nutrition, care, and development from the first year before birth to three years of age.

Restrictions: The foundation does not make grants or loans to individuals. Outside the West Michigan area, the foundation does not support capital campaigns, operating support, national child welfare programs, international based programs, or food/baby products giveaway programs.

Samples: Children's Hospital of Michigan (Detroit, MI)—to evaluate a pain-assessment technique for use in infants, $59,798 (2005); Cincinnati Children's Hospital (OH)—to evaluate a potential treatment for necrotizing enterocolitis, $247,331 (2005); U of Minnesota-Twin Cities (Minneapolis, MN)—for a study evaluating oxygen therapy in neonates and oxidative stress, $223,288 (2005); Van Andel Research Institute (Grand Rapids, MI)—for a study involving protein profiling in congenital abnormalities of infants, $743,665 (2005).

Amount of Grant: $1 million maximum; $20,000-$500,000 typically

Date(s) Application is Due: Jun 1; Dec 1

Contact: Program Contact; (231) 924-3175; fax: (231) 924-7906; email: cobits@ncresa.org

Internet: http://www.gerberfoundation.org

Sponsor: Gerber Foundation

4747 W 48th Street, Suite 153

Fremont, MI 49412-8119

Gerd Muehsam Award 2514

Papers should be submitted by graduate students for consideration for the Gerd Muehsam Award for the most outstanding research paper on art or visual resources librarianship. Contact the society for complete guidelines.

Amount of Grant: $500

Date(s) Application is Due: Nov 21

Contact: Paul Gabbard, Award Committee Chair, (212) 854-6745; fax: (212) 854-8904; email: gabbard@columbia.edu

Internet: http://www.arlisna.org/about/awards/muehsam_info.html

Sponsor: Art Libraries Society of North America

1172 Amsterdam Avenue, MC 0301

New York, NY 10027

Gerda Henkel Foundation Grants 2515

The Foundation funds activities concentrating on German and foreign academia in the fields of historical humanities. In particular, research projects in the following fields are supported: history, prehistory and early history; archaeology; art history; history of Islam; and legal history. Funds are provided for: specific temporary research projects through grants for personnel, travel and material expenses; research and doctoral scholarships for German and foreign scholars; innovative academic conferences; publications of particularly successful projects supported by the Foundation.

Restrictions: Support is not provided for: permanent positions; academic institutions through endowments; festschrifts or dictionaries; annual meetings and academic anniversaries; publications which are not related to projects already being supported by the Foundation; excursions; exhibition projects which are not related to projects already being supported by the Foundation.

Date(s) Application is Due: Jun 29

Contact: Grants Adminstrator, 49-(0)211-359853; fax: 49-(0)211-357137; email: info@gerda-henkel-stiftung.de

Internet: http://www.gerda-henkel-stiftung.de/02_foerderung/en_foerderung.htm

Sponsor: Gerda Henkel Foundation

Malkastenstrasse 15

Dusseldorf 40211 Germany

German National Scholarship Foundation Program 2516

The foundation awards grants, scholarships, and fellowships to German university students in the fields of education, science and medicine, arts and humanities, and law and other professions. Scholars may participate in summer schools arranged by the foundation, be offered special research grants, and receive tuition at German and many foreign universities.

Requirements: Students must be recommended by professors, academic advisers for doctoral students, headmasters, or school principals. Only the most academically well-qualified students will be considered.

Restrictions: Students cannot directly apply for a scholarship.

Amount of Grant: DM960.00 maximum to cover living expenses; DM100.00 monthly book allowance

Contact: Dr. Gerhard Teufel, 49-228-820960; fax: 49-228-82096103; email: info@studienstiftung.de

Internet: http://www.studienstiftung.org

Sponsor: German National Scholarship Foundation

MirbachstraBe 7

Bonn 53173 Germany

Germanistic Society of America Fellowships 2517

Usually up to six fellowships are awarded annually to enable prospective specialists to study for an academic year in Germany, primarily in art history; economics and banking; German language and literature; history; international law; philosophy; political science; and public affairs. Candidates selected for these awards, administered in the United States by the IIE, will be considered for Fulbright travel grants.

Requirements: A master's degree is desirable, but candidates must have a bachelor's degree and be U.S. citizens by November 1 of the year preceding the award.

Amount of Grant: $12,000

Contact: U.S. Student Programs Division, (212) 984-5330; fax: (212) 984-5325; email: info@iie.org

Internet: http://www.iie.org/Content/NavigationMenu/Fulbright_Demo_Site/U_S__Student_Program/Fulbright_Grant_Opportunities/Germany.htm

Sponsor: Institute of International Education

809 United Nations Plz

New York, NY 10017

Gerrish/FTE Technology Education Graduate Fellowship 2518

The fellowship is available to technology teachers at the K-12 grade level. Candidates must be accepted into a graduate degree program in technology education and beginning or continuing full-time graduate work. All applicants must be members of the International Technology Education Association. Additional information is available upon request.

Amount of Grant: $5000

Date(s) Application is Due: Dec 1

Contact: Kendall Starkweather, Executive Director, (703) 860-2100; fax: (703) 860-0353; email: iteaordr@iris.org

Internet: http://www.doe.mass.edu/tgfa/tech.html#2

Sponsor: Foundation for Technology Education

1914 Association Drive, Suite 201

Reston, VA 22091

Gertrude B. Elion Mentored Medical Student Research Awards 2519

An annual award granted to support women medical students interested in pursuing health-related research projects.

Requirements: Students must have the support of a faculty mentor and must conduct their research at one of the four medical schools in North Carolina. Applications are accepted from Duke University Medical Center, East Carolina University School of Medicine, UNC-Chapel Hill School of Medicine or Wake Forest University Baptist Medical Center only.

Restrictions: Candidates must be women, enrolled as full-time students, and will have completed at least one year of medical school prior to the start of the award. They must conduct their research at the applying institution. Candidates must be citizens or permanent residents of the United States or Canada at the time of application.

Documentation of permanent residency status must be provided with the application. Persons who have applied for permanent residency but have not received their government documentation by the time of application are not eligible.

Geographic Focus: North Carolina

Amount of Grant: $10,000

Date(s) Application is Due: Mar 5

Contact: Libby Long, Scholarships and Special Projects Coordinator; (919) 474-8370 ext. 134; fax: (919) 941-9208; email: libby@trianglecf.org

Internet: http://www.trianglecf.org/page10000237.cfm

Sponsor: Triangle Community Foundation

324 Blackwell Street, Suite 1220

Durham, NC 27701

Getty Graduate Internships 2520

Full-time internships are available to graduate students who intend to pursue careers in art museums and related fields of the visual arts, humanities, and sciences. Internships are offered by several programs of the Getty Trust, including curatorial—prepare exhibitions, as well as engage in research and writing projects; conservation—inspection, care, and treatment of objects; education and research—museum education, professional education of conservators, provenance research, or public programs; information management—museum registration, Web services, institutional archives, or library special collections; and grantmaking—select and award grants in the areas of research, conservation, and education. Each internship is full time (five day, 38-hour week. Most internships are for eight months; conservation internships last 12. The grant includes health benefits and an educational travel allowance. Housing is not provided.

Requirements: Graduate internships at the Getty are open to applicants of all nationalities. Students must either be currently enrolled in a graduate program leading to an advanced degree in a field relevant to the internship(s) for which they are applying or have completed a relevant graduate degree in 2002 or later.

Amount of Grant: $17,300 for eight months, $25,000 for 12 months; plus health benefits and $1200 educational travel allowance

Date(s) Application is Due: Dec 15

Contact: Internships Administrator, (310) 440-7320; fax: (310) 440-7703; email: gradinterns@getty.edu

Internet: http://www.getty.edu/grants/education/grad_interns.html

Sponsor: J. Paul Getty Museum

1200 Getty Center Drive, Suite 800

Los Angeles, CA 90049-1685

Getty Research Grants for Institutions - Archival Grants 2521

These grants support the basic arrangement and description of important archival collections of art-historical material in order to make them more accessible to scholars. Most projects result in the creation of an electronic finding aid. These grants do not support digitization of already catalogued archival materials, nor do they support archival research or transcription projects. Applicants should submit a preliminary letter of inquiry.

Requirements: Nonprofit institutions that own significant archival collections (including both documentary and photographic archives) relevant to the study of art history are eligible to apply for support. The collections must be open for use by interested scholars.

Restrictions: Requests for support to preserve archival collections are not eligible; however, certain preservation costs may be considered as a part of a larger arrangement and description project.

Contact: Research Grants for Institutions, (310) 440-7320; fax: (310) 440-7703

Internet: http://www.getty.edu/grants/research/institutions/archival.html

Sponsor: J. Paul Getty Trust

1200 Getty Center Drive, Suite 800

Los Angeles, CA 90049-1685

Getty Research Grants for Institutions - Art History Research Grants 2522

Grants in this category support projects that advance the scholarly mission of Art History Research Centers. Projects that enhance scholarly access to library and archival resources are particularly encouraged. Applicants should submit a preliminary letter of inquiry to determine eligibility.

Requirements: Grants are reserved for independent advanced research centers for the history of art that have significant library holdings and residential fellowship programs, and are open to scholars worldwide.

Contact: Art History Research Center, (310) 440-7320; fax: (310) 440-7703

Internet: http://www.getty.edu/grants/research/institutions/art_historical.html

Sponsor: J. Paul Getty Trust

1200 Getty Center Drive, Suite 800

Los Angeles, CA 90049-1685

Getty Research Grants for Institutions - Cataloguing of Museum Collections Grants 2523

These grants support full scholarly research and documentation of one or more distinct groups of objects in a museum's permanent collection by experienced scholars. Support is also available for the publication costs of catalogues. Applicants should submit a preliminary letter of inquiry to determine eligibility.

Requirements: Museums and other nonprofit organizations that own and house art collections and are open to the public on a regular basis are eligible to apply.

Restrictions: These grants do not support electronic collection catalogues, special exhibition catalogues, registrars' inventories, checklists, or summary catalogues.

Contact: Research Grants for Institutions, (310) 440-7320; fax: (310) 440-7703
Internet: http://www.getty.edu/grants/research/institutions/cataloguing.html
Sponsor: J. Paul Getty Trust
1200 Getty Center Drive, Suite 800
Los Angeles, CA 90049-1685

Getty Research Grants for Institutions - Critical Reference Resources Grants 2524
These grants support the preparation and publication of scholarly databases, reference works, innovative electronic resources, and other research tools that provide critical resource materials for art-historical scholarship. Priority is given to projects of international importance for which resources are otherwise limited. Applicants should submit a preliminary letter of inquiry to determine eligibility.
Requirements: Nonprofit institutions are eligible to apply. Grants are intended to support a crucial phase or portion of a larger project, not to provide long-term, ongoing support.
Contact: Research Grants for Institutions, (310) 440-7320; fax: (310) 440-7703
Internet: http://www.getty.edu/grants/research/institutions/critical_reference.html
Sponsor: J. Paul Getty Trust
1200 Getty Center Drive, Suite 800
Los Angeles, CA 90049-1685

Getty Research Institute Library Research Grants 2525
These grants provide short-term support to scholars at all levels to pursue independent projects that will benefit from research in the collections housed in the Getty library. Projects need not relate to the theme of the scholar year. Grantees receive partial support for costs related to research, travel, and living expenses for periods ranging rom several days to a maximum of three months.
Requirements: Scholars at any level who demonstrate a compelling need to use materials housed in the research library and whose place of residence is more than 80 miles from the center are eligible to apply.
Amount of Grant: $500-$2500
Date(s) Application is Due: Nov 1
Contact: Library Research Grants, (310) 440-7374; fax: (310) 440-7703; email: researchgrants@getty.edu
Internet: http://www.getty.edu/grants/research/scholars/library_research.html
Sponsor: J. Paul Getty Trust
1200 Getty Center Drive, Suite 800
Los Angeles, CA 90049-1685

Getty Research Institute Nonresidential Grants - Collaborative Research 2526
These grants provide opportunities for teams of scholars to pursue interpretive research projects that offer new explanations of art and its history. Collaborations that foster a cross-fertilization of ideas and methodologies are particularly encouraged. Grants also fund the research in preparation for scholarly exhibitions. Grants provide support for projects during a determined time frame. Grant periods vary according to the needs of the individual projects, but are generally available for research periods of one to two years. Although team members may alternate their period of leave to work on the project, the proposed plan for the project's completion must include a portion of time dedicated to joint study; such periods of joint study may include travel. Applications are welcome from scholars of all nationalities. Potential applicants are strongly encouraged to send a brief letter of inquiry. Grant amounts may vary. Application information is available online.
Requirements: These grants are intended to support established scholars who have attained distinction in their fields. Teams may consist of two or more art historians, or of one or more art historians together with one or more scholars from other disciplines. Teams for exhibition projects should include scholars from both museums and universities.
Restrictions: Individual scholars may not apply as a member of more than one team. Applications for the development of basic research tools, such as computer databases or art-historical reference works, are not eligible in this grant category.
Date(s) Application is Due: Nov 1
Contact: Collaborative Research Grants, (310) 440-7374; fax: (310) 440-7703; email: researchgrants@getty.edu
Internet: http://www.getty.edu/grants/research/scholars/research_collaborative.html
Sponsor: J. Paul Getty Trust
1200 Getty Center Drive, Suite 800
Los Angeles, CA 90049-1685

Getty Research Institute Nonresidential Grants - Curatorial Research Fellowships 2527
This fellowship supports the professional scholarly development of curators by providing them with time off from regular museum duties to undertake short-term research or study projects that advance the understanding of art and its history. Applicants may apply for a research period of one to three consecutive months. Fellowships provide stipends toward salary replacement and travel expenses. Eligible projects may include research for an exhibition or publication, travel to visit sites or collections, revision of a dissertation for publication, an intensive period of reading or writing, or other projects that support professional scholarly development. Application information is available online.
Requirements: Fellowships are reserved for full-time curators of all nationalities who have a minimum of three years' professional experience and are employed at museums with art collections. Applicants must obtain authorization from their museums for the proposed fellowship term.
Restrictions: Funding is not available to cover institutional overhead costs or for carrying out the planning requirements of an exhibition apart from research. Fellowships are not intended to fund research for the completion of an academic degree.
Amount of Grant: $3,500-$5,000

Date(s) Application is Due: Nov 1
Contact: Curatorial Research Fellowships, (310) 440-7374; fax: (310) 440-7703; email: researchgrants@getty.edu
Internet: http://www.getty.edu/grants/research/scholars/research_curatorial.html
Sponsor: J. Paul Getty Trust
1200 Getty Center Drive, Suite 800
Los Angeles, CA 90049-1685

Getty Research Institute Nonresidential Grants - Postdoctoral Fellowships 2528
These fellowships provide support for outstanding scholars in the early stages of their careers, allowing them the flexibility to travel and study wherever necessary to undertake their work. Grants support interpretive research projects that make a substantial and original contribution to the understanding of art and its history. Fellowships provide a stipend for a twelve-month period. Application information is available online.
Requirements: Scholars of all nationalities whose doctoral degrees in art history (or the equivalent in countries outside the United States) have been or will be officially conferred between 1/1/01 and 1/1/07 are eligible to apply. Candidates who hold doctoral degrees in fields outside art history are eligible to apply if they can demonstrate that their work promises to make a substantial and original contribution to the understanding of art and its history.
Restrictions: During the tenure of the fellowship, fellows may not accept other awards or grants, nor may they undertake any form of employment. These awards may not be deferred, and they are not renewable.
Amount of Grant: $40,000
Date(s) Application is Due: Nov 1
Contact: Postdoctoral Fellowships, (310) 440-7374; fax: (310) 440-7703; email: researchgrants@getty.edu
Internet: http://www.getty.edu/grants/research/scholars/research_postdoc.html
Sponsor: J. Paul Getty Trust
1200 Getty Center Drive, Suite 800
Los Angeles, CA 90049-1685

Getty Research Institute Predoctoral & Postdoctoral Residential Fellowships 2529
Pre- and postdoctoral fellowships provide support for emerging scholars to complete work on projects related to the Getty Research Institute's annual theme. The theme for the current program is Duration. Recipients are in residence at the institute, where they pursue research to complete their dissertations or to expand them for publication. Fellows make use of the Getty collections, join in a weekly meeting devoted to the annual theme, and participate in the intellectual life of the Getty Center. Pre- and postdoctoral fellows are in residence for the entire academic year. Both fellowships provide an office at the institute, airfare to Los Angeles, an apartment in the Getty scholar housing complex, and health benefits.
Requirements: Applications are welcome from scholars of all nationalities who are working in the arts, humanities, or social sciences. Predoctoral fellowship applicants must have advanced to candidacy and expect to complete their dissertations during the fellowship period. Postdoctoral fellowship applicants must have received their degree no earlier than 1998.
Amount of Grant: $18,000 predoctoral fellowships; $22,000 postdoctoral fellowships
Date(s) Application is Due: Nov 1
Contact: Pre- and Postdoctoral Fellowships, (310) 310) 440-7374; fax: (310) 440-7703; email: researchgrants@getty.edu
Internet: http://www.getty.edu/grants/research/scholars/pre_post_fellows.html
Sponsor: J. Paul Getty Trust
1200 Getty Center, Suite 800
Los Angeles, CA 90049-1688

Getty Scholars and Visiting Scholars Grants 2530
These grants provide a unique research experience. Recipients are in residence at the Getty Research Institute or the Getty Villa in Malibu where they are able to pursue their own projects free from academic obligations, make use of Getty collections, and participate in the intellectual life of the Getty. Scholars are in residence for the entire academic year. A salary replacement stipend is awarded. The grant also includes an office, research assistance, airfare to and from Los Angeles, an apartment and health benefits. Application information is available online.
Requirements: These grants are for established scholars, artists, or writers who have attained distinction in their fields. Applications are welcome from researchers of all nationalities who are working in the arts, humanities, or social sciences.
Date(s) Application is Due: Nov 1
Contact: Getty Residential Scholar and Visiting Scholar Grants, (310) 440-7374; fax: (310) 440-7703; email: researchgrants@getty.edu
Internet: http://www.getty.edu/grants/research/scholars/research_grischolars.html
Sponsor: J. Paul Getty Trust
1200 Getty Center Drive, Suite 800
Los Angeles, CA 90049-1685

Giant Food Corporation Scholarships 2531
The corporation awards scholarships for pharmacy education leading to a bachelor of science or doctorate degree. Scholarships include the Izzy Cohen Memorial Scholarship and the Esther Peterson Memorial Education Scholarship. Information regarding these programs is posted online in February.
Amount of Grant: $1000 Esther Peterson Memorial Education Scholarship; $4000 Izzy Cohen Memorial Scholarship
Contact: Barry Scher, Vice President, (301) 341-4710; fax: (301) 618-4972

Internet: http://www.giantfood.com/corporate/company_charitable.htm
Sponsor: Giant Food Corporation
6300 Sheriff Road, Dept. 599
Landover, MD 20785

Gilbert F. White Postdoctoral Fellowships 2532

The fellowships support professionals who wish to devote a year to scholarly work on social science or policy problems in the areas of natural resources, energy, or the environment. Two fellowships per year are awarded for a minimum of nine and a maximum of 12 months. The amount of the stipend is commensurate with experience. Fellows will reside in one of three RFF units: the energy and natural resources division, the quality of the environment division, or the center for risk management. Applications via fax are not accepted.
Requirements: The award is open to individuals in any discipline who will have completed their doctoral requirements by the beginning of the academic year for which the fellowship is requested. Faculty members who will be on sabbatical leave during the fellowship are also encouraged to apply.
Restrictions: Individuals holding positions in government as well as at academic institutions are eligible.
Amount of Grant: $1000 for moving and living expenses plus stipend based on current salary
Date(s) Application is Due: Feb 28
Contact: Fellowship Program, (202) 328-5155; fax: (202) 939-3460; email: white-award@rff.org
Internet: http://www.rff.org/rff/About/Fellowships_and_Internships/White/Gilbert-F-White-Postdoctoral-Fellowship-Program.cfm
Sponsor: Resources for the Future
1616 P Street NW
Washington, DC 20036

Gilbreth Memorial Fellowships 2533

These fellowships are awarded to recognize graduate students pursuing advanced degrees in industrial engineering in accredited industrial engineering programs at colleges and universities in the United States, Canada, or Mexico. Consideration is given to proven scholastic ability, character, leadership qualities, potential service to industrial engineering, and the need for financial assistance. Nominations are solicited from department heads at the end of August for candidates for the following academic year. Nomination packets are then sent to the nominees.
Requirements: Institute members who are enrolled full time in graduate industrial engineering programs and have had a 3.4 to 4.0 GPA as undergraduates are eligible. Nominations are accepted only from department heads of accredited industrial engineering programs.
Samples: Janell Cernik, Florida State U (Tallahassee, FL)—fellowship recipient, $2500.; David Fergus, U of Colorado (Boulder, CO)—fellowship recipient, $2500.; Dale Masel, Pennsylvania State U (U Park, PA)—fellowship recipient, $2500.
Amount of Grant: $4000 maximum
Date(s) Application is Due: Nov 15
Contact: Fellowship Coordinator, (800) 494-0460 or (770) 449-0460; fax: (770) 441-3295; email: info@iienet.org
Internet: http://www.iienet.org/public/articles/index.cfm?cat=525
Sponsor: Institute of Industrial Engineers
3577 Parkway Ln Suite 200
Norcross, GA 30092

Gilder Lehrman Institute of American History Scholarly Fellowships 2534

The fellowships support work in one of five archives: the Gilder Lehrman Collection, on deposit at the New-York Historical Society; the Library of the New-York Historical Society; the Columbia University Rare Book and Manuscript Library; the New York Public Library - Humanities and Social Sciences Library; and the Schomburg Center for Research in Black Culture (NYPL). To apply, candidates must submit a cover sheet with name, mailing address, email address, telephone and fax numbers; current rank, department, and institution; title of project; duration and amount of proposed fellowship; and names of recommenders. They must also submit a curriculum vitae, a 2-3-page project proposal that lists the specific holdings in the collection they intend to use (please specify only one archive in the application), two letters of recommendation, and a schedule and proposed budget of expenses during the tenure of the fellowship.
Requirements: Fellowships range in duration from a week to two months and research must be completed within a year of notification of the award.
Restrictions: Fellowships are not available for scholars who live within commuting distance of New York City.
Amount of Grant: $1500-$4000 per month
Date(s) Application is Due: May 6; Dec 1
Contact: Fellowship Program, (646) 366-9666; fax: (646) 366-9669
Internet: http://www.gilderlehrman.org/historians/scholar4.html
Sponsor: Gilder Lehrman Institute of American History
19 W 44th Street, Suite 500
New York, NY 10036

Gilroy and Lillian P. Roberts Charitable Foundation Grants 2535

The foundation awards grants to eligible nonprofit organizations in its areas of interest, including arts and fine arts, health care and health organizations, higher education, Jewish temples and organizations, and social services. Types of support include annual campaigns, capital campaigns, continuing support, fellowships, general operating grants,

professorships, program development, and scholarships. Most grants are awarded in Montgomery and Delaware Counties, PA. There are no application deadlines or forms.
Restrictions: Individuals are not eligible.
Geographic Focus: Pennsylvania
Samples: Temple University (Philadelphia, PA)—for Merves Professorship in Accounting, $155,000 (2003).
Amount of Grant: $408,750 total
Contact: Stanley Merves, Treasurer, (610) 668-1998
Sponsor: Gilroy and Lillian P. Roberts Charitable Foundation
101 W Elm Street, Suite 500
Conshohocken, PA 19428

Gina Finzi Memorial Student Summer Fellowship 2536

Fellowships are awarded to encourage clinical, basic or psychosocial research related to the causes, treatments, prevention, or cure of lupus. Awards are made in May to the top 10 applicants for support during the summer months. Application materials are available in December of each year. There are no renewals; applicant must submit a new proposal.
Requirements: Junior investigators, defined as assistant professor and below rank if in academic medicine, are eligible to apply.
Amount of Grant: $2000
Date(s) Application is Due: Mar 15
Contact: Grants Coordinator, (202) 349-1155; fax: (202) 349-1156; email: Finzifellowship@lupus.org
Internet: http://www.lupus.org/research/studentannounc.html
Sponsor: Lupus Foundation of America
2000 L Street NW, Suite 710
Washington, DC 20036-4916

Girl's Best Friend Foundation Grants 2537

The foundation supports and promotes programs by and for girls and young women in Illinois (ages eight through 21). The foundation is dedicated to effecting change at the grassroots level by funding community-based organizations statewide. Types of grants made include general operating support, project specific support, planning or start-up support, technical assistance, and collaborative action research (research that lays the groundwork for policy changes conducted by a partnership between two nonprofits or a nonprofit contracting with a researcher or academic institution). When multiyear funding is requested, the foundation will consider making two-year grants if the requesting organization has received at least two previous grants from GBF, is seeking two years of funding for the same purpose, and has clearly demonstrated the need for a two-year grant. Applications are due the first Monday in August.
Requirements: Letters of intent will be accepted from 501(c)3 nonprofits for projects that serve or have a direct impact on girls living in the Chicago metropolitan area including Cook, DuPage, Kane, Lake, McHenry, and Will counties.
Restrictions: Organizations with budgets that exceed $650,000 may apply only for technical assistance and/or collaborative action research grants. The foundation generally does not fund individuals, capital campaigns, debt reduction, scholarships, or government or religious organizations.
Geographic Focus: Illinois
Samples: Chicago Health Connections (IL)—for school-based support services for pregnant teenagers and teenage parents, $11,000 (2004); Chicago Women in Trades (IL)—to develop the leadership and advocacy skills of young women enrolled in vocational classes, $16,500 (2004); Family Matters (Chicago, IL)—for a program that focuses on community building, entrepreneurial skills, and academics, $16,500 (2004); Girls in the Game (Chicago, IL)—for sports and leadership programs for girls, $49,400 (2004).
Amount of Grant: $5000-$20,000 typically
Contact: Robin Dixon, Senior Program Officer, (312) 266-2842; fax: (312) 266-2972; email: robin@girlsbestfriend.org or contact@girlsbestfriend.org
Internet: http://www.girlsbestfriend.org/apply/index.html
Sponsor: Girl's Best Friend Foundation
900 N Franklin, Suite 210
Chicago, IL 60610

Gladys Krieble Delmas Foundation Grants for Venetian Research 2538

Predoctoral and postdoctoral grants will be awarded for travel to and residence in Venice and Veneto for research on Venice and the former Venetian empire and for study of contemporary Venetian society and culture. Disciplines of the humanities and social sciences are eligible areas of study, including, but not limited to, archaeology, architecture, art, bibliography, economics, history, history of science, law, literature, music, political science, religion, and theater. Prospective applicants should write to the foundation for application forms.
Requirements: Applicants must be U.S. citizens or permanent residents, have some experience in advanced research, and, if graduate students, have completed all doctoral requirements except for the dissertation at the time of application.
Amount of Grant: $19,900 maximum for academic year
Date(s) Application is Due: Dec 15
Contact: Program Contact, (212) 687-0011; fax: (212) 687-8877; email: info@delmas.org
Internet: http://www.delmas.org/guidelines/v_ir_a.html
Sponsor: Gladys Krieble Delmas Foundation
521 Fifth Avenue, Suite 1612
New York, NY 10175-1699

Glaucoma Foundation Grants   2539

The foundation funds research initiatives to determine the causes of glaucoma, to improve methods of treatment, and to develop cures for the various kinds of glaucoma. Two areas of particular focus are: optic nerve rescue and restoration—research into new approaches designed to protect the optic nerve against glaucomatous damage, to restore vision lost to glaucoma and eventually reverse blindness by restoring or regenerating the function of the optic nerve cells, and to explore the feasibility of achieving transplantation of optic nerve cells; and molecular genetics—research into the genetic causes of the various forms of glaucoma, particularly the identification of the responsible genes, with the long-term goal of finding ways to reverse these genetic defects. Grants are awarded for a one-year period and are renewable. Guidelines and application are available online.

Requirements: Applicants must have a full time faculty position or the equivalent.

Restrictions: The foundation does not provide funds for investigator salaries, travel, overhead, or other indirect costs.

Amount of Grant: $40,000 - $50,000 range

Date(s) Application is Due: Mar 1; Sep 1

Contact: Kira A. Zmuda, Director of Research Programs: (212) 651-2509; email: kzmuda@glaucomafoundation.org

Internet: http://www.glaucomafoundation.org/Grant_Application.htm

Sponsor: Glaucoma Foundation Grants

80 Maiden Lane, Suite 700

New York, NY 10038

Glaucoma Research Pilot Project Grants   2540

These grants provide funds to encourage innovative and pilot research, as well as to aid ongoing studies that seek to protect and restore the optic nerve, accurately monitor glaucoma's progression, find the genes responsible for the glaucoma, understand the intraocular pressure system and develop better treatments, and determine the risk factors for glaucoma damage.

Requirements: Applicants must have a graduate degree.

Restrictions: Funding is not granted for equipment purchases, overhead commercial applications, or indirect costs.

Amount of Grant: $40,000

Date(s) Application is Due: Aug 1

Contact: Jennifer Rulon; (800) 826-6693 or (415) 986-3162; fax: (415) 986-3763; email: research@glaucoma.org

Internet: http://www.glaucoma.org/research/index.php

Sponsor: Glaucoma Research Foundation

251 Post Street, Suite 600

San Francisco, CA 94108

Glenn T. Seaborg Award for Nuclear Chemistry   2541

The purpose of the Award is to recognize and encourage research in nuclear and radiochemistry or their applications. A nominee must have made outstanding contributions to nuclear or radiochemistry or to their applications. The award will be granted without regard to age or nationality. The award consists of $3,000 and a certificate. Reasonable travel expenses to the meeting at which the award will be presented will be reimbursed.

Requirements: Any individual, except a member of the award committee, may submit one nomination or seconding letter for the award in any given year. Nominating documents consist of a letter of not more than 1000 words containing an evaluation of the nominee's accomplishments and a specific identification of the work to be recognized, a biographical sketch including date of birth, and a list of publications and patents authored by the nominee.

Restrictions: Self-nominations are not accepted.

Samples: Romualdo T. deSouza—award winner, $3,000 (2008); Norbert G. Trautmann—award winner, $3,000 (2007); Steven W. Yates—award winner, $3,000 (2006).

Amount of Grant: $3,000 plus travel expenses

Date(s) Application is Due: Nov 1

Contact: Felicia Dixon, Awards Administrator; (800) 227-5558 or (202) 872-4408; fax: (202) 776-8008; email: f_dixon@acs.org or awards@acs.org

Internet: http://portal.acs.org/portal/acs/corg/content?_nfpb=true&_pageLabel=PP_ARTICLEMAIN&node_id=1319&content_id=CTP_004529&use_sec=true&sec_url_var=region1

Sponsor: American Chemical Society

1155 Sixteenth Street, NW

Washington, DC 20036

Glenn/AFAR Breakthroughs in Gerontology Awards   2542

The goal of the program is to provide timely support to a small number of pilot research programs that may be of high risk but offer significant promise of yielding transforming discoveries in the fundamental biology of aging. The hope is that one or more of the funded research projects will lead to major new insights into the molecular factors that coordinate aging in multiple cells and tissues, and the ways in which the aging process is differentially timed in long-lived species. Projects that focus on genetic controls of aging and longevity, on delay of aging by pharmacological agents or dietary means, or that elucidate the mechanisms by which alterations in hormones, anti-oxidant defenses, or repair processes promote longevity are all well within the intended scope of this competition. Projects that focus instead on specific diseases or on assessment of health care strategies will receive much lower priority, unless the research plan makes clear and direct connections to fundamental issues in the biology of aging. Studies of invertebrates, mice, human clinical materials, or cell lines are all potentially eligible for funding. Although preliminary data are always helpful for evaluating the feasibility

of the experiments proposed, the emphasis in review will be on creativity and the likelihood that the findings will open new vistas and approaches to aging research that might merit intensive follow-up studies. Applications from individuals not previously engaged in aging research are particularly encouraged, as long as the research proposals show high promise for leading to important new discoveries in biological gerontology. The proposed research must be conducted at any type of nonprofit setting in the United States. Guidelines are available online.

Requirements: Applicants must at the time they submit their proposal be full-time faculty members at the rank of assistant professor or higher. A strong record of independent publication beyond the postdoctoral level is a requirement.

Restrictions: Applicants who are employees in the NIH Intramural program are not eligible.

Amount of Grant: $200,000

Date(s) Application is Due: Dec 16

Contact: Grants Manager; (212) 703-9977 or (888) 582-2327; fax: (212) 997-0330; email: grants@afar.org or info@afar.org

Internet: http://afar.org/grants.html

Sponsor: American Federation for Aging Research

55 West 39th Street, 16th Floor

New York, NY 10018

Glenn/AFAR Research Grants for Postdoctoral Fellows   2543

AFAR funds research projects concerned with understanding the basic mechanisms of aging. Projects investigating age-related diseases are also supported, especially if approached from the point of view of how basic aging processes may lead to these outcomes. Projects concerning mechanisms underlying common geriatric functional disorders are also encouraged. Examples of promising areas of research include aging and immune function; genetic control of longevity; neurobiology and neuropathology of aging; invertebrate or vertebrate animal models; cardiovascular aging; aging and cellular stress resistance; metabolic and endocrine changes; age-related changes in cell proliferation; caloric restriction and aging; DNA repair and control of gene expression; biology of the menopause; aging and apoptosis; and biodemographic analysis of aging. Funding begins July 1. Guidelines and application are available online.

Requirements: Postdoctoral fellows (MDs and PhDs) who will have had at least two and not more than five years of prior postdoctoral training at the time of the award are eligible.

Restrictions: Funds may not be requested for overhead or indirect costs. Fellows may not hold any concurrent foundation or not-for-profit funding.

Amount of Grant: $60,000

Date(s) Application is Due: Dec 15

Contact: Grants Manager, (212) 703-9977; fax: (212) 997-0330; email: amfedaging@aol.com

Internet: http://www.afar.org/grants.html

Sponsor: American Federation for Aging Research

70 W 40th Street

New York, NY 10018

Global Center for Dispute Resolution Research Grants   2544

The center is dedicated to research that provides insight, information, and guidance on the growth and practice of alternative dispute resolution (ADR) methods throughout the world, particularly in the area of cross-border commerce. The grants program funds research papers. Areas of interest include pattern and trend analysis of the collected data, such as global trends in ADR by industry, by country, by process type, and by forces in the marketplace that lead to caseload fluctuations in different sectors; and designed experiment research that will examine specific aspects of dispute resolution processes, such as process effectiveness, cost savings, time savings, quality of result, party satisfaction, and optimization of results. Research findings and their application will inform policy development in the field and increase process knowledge of practitioners.

Samples: International Law Institute (Washington, DC)—for research to see which economic, legal, and social factors correlate with successful arbitration programs in countries worldwide, $50,000 (2003).

Contact: Grants Administrator, (212) 716-3900; fax: (212) 716-5913; email: woodw@adr.org

Internet: http://www.globalcenteradr.org/english/research.jsp

Sponsor: Global Center for Dispute Resolution Research

335 Madison Avenue, 10th Fl

New York, NY 10017

Global Change Education Program   2545

This program provides research opportunities in areas related to global change such as atmospheric sciences, ecology, global carbon cycles, climatology, and terrestrial processes. Types of support include graduate research environmental fellowships (GREF) and summer undergraduate research experiences (SURE).

Requirements: College juniors, seniors, and graduate students are eligible to apply.

Amount of Grant: $475 weekly, plus travel for undergratuages; $1500 monthly stipend, plus tuition and fees for graduates

Contact: Dr. Milton Constantin, Program Manager, (865) 576-7009; email: constanm@orau.gov; Dr. Jeffrey Gaffney, Argonne National Laboratory, (630) 252-5178; email: constanm@orau.gov

Internet: http://see.orau.org/ProgramDescription.aspx?Program=10125

Sponsor: Oak Ridge Institute for Science and Education

P.O. Box 117, 120 Badger Avenue, MS 36

Oak Ridge, TN 37831-0117

GMFU.S. Grants   2546

The fund's grantmaking promotes the study of international and domestic policies, supports comparative research and debate on key issues, and assists policy and opinion leaders' understanding of these issues. Grants are allocated in the areas of economics, environment, foreign policy, and immigration and integration. On occasion, grants are awarded to fund projects or institutions that are worthy of support but that do not fall within one of the fund's established program areas. Initial inquiries should be submitted in a letter or brief proposal, written in English, and should outline the project's: purpose and importance; relation to the fund's program areas; budget; other potential funding sources; qualifications of the applicant; and plans for dissemination and follow-up.

Restrictions: The fund generally does not offer support for projects chiefly concerned with countries outside Europe and the United States; building and operating funds; education below the doctoral level; scientific research; or the arts.

Contact: Administrator, (202) 745-3950; fax: (202) 265-1662; email: info@gmfus.org
Internet: http://www.gmfus.org/grants/index.cfm
Sponsor: German Marshall Fund of the 1744 R Street NW
Washington, DC 20009

GMFU.S. Transatlantic Environmental Fellowships   2547

The fund aims to promote collaboration between Americans and Europeans on efforts that increase understanding of policy differences between the United States and Europe on transportation, agriculture, and air quality issues. Projects should involve people on both sides of the Atlantic, develop leadership, and transfer experience and innovations between practitioners and policymakers.

Requirements: Nonprofits in the United States and Western Europe are eligible.

Restrictions: The fund generally does not offer support for building and operating funds, graduate or undergraduate education, research, or the arts.

Amount of Grant: $9000–$40,000
Contact: Lea Rosenbohm, Director of Programs, (202) 745-6663; email: lrosenbohm@gmfus.org
Internet: http://www.gmfus.org/fellowships/index.cfm
Sponsor: German Marshall Fund of the 1744 R Street, NW
Washington, DC 20009

GMFU.S. Transatlantic Policy Research Fellowships   2548

The aim of this program is to support research-driven conferences of scholars and policymakers working on transatlantic policy issues that will be selected and change on a yearly basis. GMF will award six grants for transatlantic policy conferences in the academic year. Conference proposals should include scholars from the United States and Europe, and preference will be given to those proposals with a comparative and interdisciplinary approach. Proposals will be evaluated on their intellectual merits, transatlantic cooperation, engagement with the policy community, and potential policy impact through publications or other means. This year, GMF will consider proposals in three areas: democracy promotion, immigration, and economic competitiveness policy.

Requirements: American and European university-based scholars from any discipline may apply.

Amount of Grant: $25,000 maximum
Date(s) Application is Due: Mar 1
Contact: Program Officer, (202) 745-3950; fax: (202) 265-1662; email: epope@gmfus.org or info@gmfus.org
Internet: http://www.gmfus.org/fellowships/research.cfm
Sponsor: German Marshall Fund of the 1744 R Street NW
Washington, DC 20009

Goddess Scholars Grants   2549

The purpose of these grants is to develop young researchers at the entry faculty level (Instructor or Assistant Professor) who will focus a research career on studying the unique aspects of stroke in women. Such research may be basic or clinical in orientation, including areas such as mechanisms of ischemic brain injury, neuroprotective treatments, primary or secondary stroke prevention, epidemiological studies, educational programs for patients or health care providers, rehabilitation, and outcome research. The grant provides funding for two years, with potential non-competitive renewal for a third year should satisfactory progress be demonstrated. These grants are intended to provide bridging funds for individuals who are finishing or have completed residency or postdoctoral training and require additional support before their subsequent establishment as independent investigators.

Requirements: Goddess Scholar grants are targeted at individuals who: hold an M.D., Ph.D., or equivalent degree; are at the fellow, instructor, or assistant professor rank; and are within 5 years of completion of residency or post-doctoral fellowship at the time of grant activation. Individuals who are currently enrolled in a fellowship program may apply if there is assurance of a faculty position at the time of the grant's activation.

Amount of Grant: $65,000 direct salary plus research costs (e.g. research or laboratory assistant, materials, small equipment)
Contact: Erin Tower, Program/Operations Assistant; (212) 713-6789; fax: (212) 288-2160; email: erin@thegoddessfund.org
Internet: http://www.thegoddessfund.org/grants.html
Sponsor: Hazel K. Goddess Fund for Stroke Research in Women
785 Park Avenue
New York, NY 10021-3552

Goethe-Institut Grants   2550

The institute was founded to promote the German language abroad and for the fostering of international cultural relations. It operates internationally in the fields of science and medicine, arts and humanities, and international relations. Types of support include grants to institutions and individuals and scholarships. The institute is particularly concerned with the teaching and promotion of the German language abroad and provides professional assistance to foreign teachers of German and students of German philology, and for the development and improvement of teaching methods and materials.

Contact: Dr. Hilmar Hoffman, President, 49-89-159210; fax: 49-89-15921450; email: zentralverwaltung@goethe.de
Internet: http://www.goethe.de
Sponsor: Goethe-Institut
Helene-Wever-Allee 1
Munich 80637 Germany

Goldhirsh Foundation Brain Tumor Research Grants   2551

The foundation is interested in providing strategic investment in both pediatric and adult brain tumor research to accelerate progress toward more effective treatment for malignant diffuse glioma tumors. Responses are sought from investigators working in the continuum between basic research and clinical application, integrating and translating knowledge in various disciplines into meaningful progress for patients. Examples of funding areas include but are not limited to oncogenomics and proteomics, genetically engineered models, the discovery and testing of small molecule therapies, unusual drug delivery systems, or improved brain imaging techniques. The foundation also encourages submission of research projects at the interface of developmental biology and cancer along the stem cell to glial axis. Eligibility is not limited to those investigators currently working in brain tumor research. Investigators from other fields are encouraged to apply with proposals relevant to malignant diffuse glioma tumors. Projects must be relevant to malignant diffuse gliomas, i.e., diffuse astrocytomas, oligodendrogliomas and oligoastrocytomas, including glioblastoma. The sponsoring institution accepts responsibility for the scientific, administrative, and financial management of the overall project including subcontracts. The candidate may collaborate with investigators from other institutions and these institutions may include for-profit companies. Letters of intent must be submitted online by the listed application deadline; full proposals are by invitation. Guidelines are available online.

Requirements: Applications are accepted from candidates working in the United States or Canada who meet the following eligibility *Requirements*: have an MD and/or PhD degree(s) or equivalent degree; hold at least the position of Assistant Professor or equivalent at a U.S.-based 501(c)3 or equivalent nonprofit Canadian institution.

Restrictions: Projects in benign/WHO grade I gliomas, nonglial neuroepithelial tumors, meningeal tumors, nerve sheath tumors, CNS lymphomas, germ-cell tumors, and tumors of the sella region are beyond the scope of the grant program focus. Projects that characterize genes and gene products of normal cellular development are not eligible, nor are epidemiological studies.

Samples: Robert Darnell, M.D., Ph.D., Rockefeller University, $600,000 (2006); Arnab Chakravarti, M.D., Massachusetts General Hospital, $100,000 (2006); Lynda Chin, M.D., Dana-Farber Cancer Institute, $100,000 (2006).

Amount of Grant: $600,000 three-year and $100,000 one-year pilot studies
Date(s) Application is Due: Jan 10
Contact: Dr. Sally E. McNagny, (617) 279-2254; fax: (617) 423-4619; email: smcnagny@goldhirshfoundation.org
Internet: http://www.goldhirshfoundation.org/application_information.htm
Sponsor: Goldhirsh Foundation
95 Berkeley Street, Suite 201
Boston, MA 02116

Goldman Philanthropic Partnerships Program   2552

Since its founding, the Program has solicited, validated and helped donors co-fund innovative research with the potential to quickly find cures for life-altering diseases, and supported these donors using for-profit business tools to insure that the research and philanthropic goals are met through proper project and resource management. The Program provides funding research for cures for a range of catastrophic diseases by exploring groundbreaking research in all disciplines of medicine, including those areas of research outside of conventional medicine such as alternative medicine. The Program usually co-funds a venture with at least one organizational, institutional, or corporate partner. The Partnerships maintains a strong interest in diseases of children and young adults, although it will fund research into any patient population where it believes there is a high likelihood of return on disease prevention, treatment or cure.

Contact: Dr. Bruce E. Bloom, President; (312) 601-8856 or (312) 780-3440; fax: (312) 780-3459; email: bruce@goldmanpartnerships.org
Internet: http://www.goldmanpartnerships.org/aboutus.html
Sponsor: Goldman Philanthropic Partnerships
70 West Madison Street, Suite 1500
Chicago, IL 60602

Goldsmith Awards Program   2553

The program, funded by the Goldsmith-Greenfield Foundation, is sponsored by the Joan Shorenstein Center on the Press, Politics, and Public Policy at Harvard University's John F. Kennedy School of Government. The program annually awards the Goldsmith Prize for Investigative Reporting and two Goldsmith Book Prizes. The Goldsmith Prize is given annually to the journalist(s) whose investigative reporting in a story or series of related stories best promotes more effective and ethical conduct of government, the making of public policy, or the practice of politics. The Goldsmith Book Prizes are given annually to the author(s) of the best book that aims at improving the quality of government or politics through an examination of the press and government or the intersection of press and politics in the formation of public policy. Both prizes have a submission deadline of

December 31, or a pstmark deadline of January 2. The program also grants Goldsmith Research Awards on an ongoing basis to scholars, graduate students, and journalists in the field of press/politics. Applications are available on the Web site.
Requirements: Publication must have occurred within 12 months (January 1 through December 31) preceding the submission deadline, and all submissions must be in English.
Restrictions: Edited volumes will not be accepted for the book prize. Books will not be accepted as submissions for the investigative reporting prize.
Amount of Grant: $25,000 for investigative reporting prize, $2,000 to finalists; $2500 for book prize
Date(s) Application is Due: Jan 2
Contact: Alison Kommer; (617) 495-1329; fax: (617) 495-8696; email: alison_kommer@harvard.edu
Internet: http://www.hks.harvard.edu/presspol/prizes_lectures/goldsmith_awards/
Sponsor: Harvard University
79 JFK Street
Cambridge, MA 02138

Gordon Allport Intergroup Relations Prize 2554
This prize is awarded annually to the best paper or article submitted on intergroup relations concerning race, sex, age, and socioeconomic status. The manuscript need not have been published. Entries should be submitted in quadruplicate.
Requirements: Papers submitted for consideration must have been written during the current year on intergroup relations. Competitors need not be society members. Graduate students are particularly urged to compete.
Amount of Grant: $1000
Date(s) Application is Due: Nov 1
Contact: Intergroup Relations Prize, email: awards@spssi.org
Internet: http://www.spssi.org/Allport_flyer.html
Sponsor: Society for the Psychological Study of Social Issues
1901 Pennsylvania Avenue NW, Suite 901
Washington, DC 20006-3405

Gottlieb Daimler and Karl Benz Foundation Grants 2555
Grants are awarded to individuals for doctoral research, to German nationals wishing to conduct research abroad, and to foreign young scientists carrying out research at German institutions. In addition, several grants support the scientific cooperation with the Kaliningrad State University (open for applications), with young economists from East Europe, with Peking University, with Vietnam, and with South Africa. Types of support include awards, general operating support, basic research, dissertation research support, fellowships, international exchange programs and grants, seed grants, and conferences and workshops.
Date(s) Application is Due: Mar 1; Oct 1
Contact: Dr. Jorg Klein, Managing Director, 49-6203-1092-0; fax: 49-6203-1092-5; email: klein@daimler-benz-stiftung.de
Internet: http://www.daimler-benz-stiftung.de/home/fellowship/en/start.html
Sponsor: Gottlieb Daimler and Karl Benz Foundation
Dr Larl-Benz-Platz 2
Ladenburg 68526 Germany

GPP Charles E. Culpeper Foundation Biomedical Pilot Projects Initiative Grants 2556
The initiative is designed to encourage the investigation of novel ideas in cures for disease, particularly in the areas of molecular genetics, bio-engineering, and molecular pharmacology. Research into complimentary and alternative medicine will also be considered. Grants will be made on a one-time basis with the possibility for renewal for a second year upon reapplication. No more than 8 percent of the grant may be allocated to cover indirect costs. The purpose of these grants is to explore new and even untested hypotheses, thus substantial preliminary information is not required. Proposals are accepted throughout the year and each successful
Requirements: Nonprofit health care organizations, accredited medical schools, and universities in the United States are eligible.
Samples: Dr. Natalia Broude, Boston U (MA)—to support a study that will develop an innovative test for SARS, which would hook a molecule that glows in the dark to a protein that would attach a specific spot on the SARS gene, $25,000; Ji-Xin Cheng, PhD, Purdue University (West Lafayette, IN)—to support a project that will develop a new highly sensitive microscope for the imaging of molecules using vibrational imaging, $25,000; Dr. Gregory Crawford, Brown U (Providence, RI)—to create a device that will take a photo of the lining of the lower eyelid using a small camera, resulting in accuracy similar to the current lab test when the physician draws blood from the patient, $25,000 .
Amount of Grant: $25,000 maximum
Date(s) Application is Due: Nov 16
Contact: Grants Administrator, (847) 948-5512; fax: (847) 948-5516
Internet: http://www.goldmanpartnerships.org/Culpeper/CulpeperBiomedicalPilot Initiativegrants.htm
Sponsor: Goldman Philanthropic Partnerships
155 N Pfingsten Road, Suite 109
Deerfield, IL 60015

Graduate Institute Hodson Trust Teacher Fellowships 2557
The Hodson Trust Teacher Fellowship Program is funded through the generosity of the Hodson Trust, and offers a grant of 70% of cost during the summer for eligible full-time primary or secondary school teachers. This Fellowship is open to students from the states of Maryland, Virginia, Washington D.C., Delaware, Pennsylvania, New Jersey, Connecticut, and New York. Applicants must complete the online form and attach a letter from the principal of their school verifying their full time employment as a teacher in the academic year prior to enrollment to the institute.
Requirements: Full-time public or private primary and secondary school teachers with a minimum of one year of teaching experience are eligible for this fellowship.
Geographic Focus: Connecticut; Delaware; District of Columbia; Maryland; New Jersey; New York; Pennsylvania; Virginia
Contact: Janet Huang; (410) 626-2502; fax: (410) 626-2885; email: janet.huang@sjca.edu
Internet: http://www.stjohnscollege.edu/GI/finaid.shtml
Sponsor: Saint John's College
P.O. Box 2800
Annapolis, MD 21404-2800

Graduate Institute of International Studies Scholarships 2558
A number of scholarships are awarded each year by the institute for doctoral instruction in international law and politics, international economics, political science, international institutions, and international development. Limited financial assistance is obtained and maintained on the basis of both the academic performance and the financial status of the applicants. The Gallatin Fellowship is awarded competitively to an advanced American student to permit him/her to engage in research work at the institute for one year, and to an advanced doctoral student from the institute to enable him/her to study and do research for 10 months at an American university of his/her choosing. In addition, the Paul Guggenheim Foundation awards a prize every two years to a work in international law that is distinguished by its exceptional quality and that is the first outstanding work of its author. Obtain instructions for scholarships and this award from the secretariat. The Swiss Confederation also grants scholarships to foreign graduates; candidates for such scholarships should apply to the Swiss diplomatic representatives in their countries. For U.S. citizens, information on other scholarships and fellowships is available from the Institute of International Education, 809 United Nations Plaza, New York, NY 10017.
Requirements: Generally, these scholarships are reserved for advanced students, especially those working toward doctorates, who have been enrolled at the institute for at least one year.
Amount of Grant: Sf10,000 per year
Date(s) Application is Due: Jan 10; Mar 1
Contact: Jean-Claude Frachebourg, Secretaire General, 41 22 908 59 59; fax: 41 22 733 30 49; email: info@hei.unige.ch
Internet: http://heiwww.unige.ch/sections/ec/admissions/tuition.html
Sponsor: Graduate Institute of International Studies
Avenue de la Paix 11A
Geneva 1202
Switzerland

Graduate Writing Fellowships 2559
Fellowships are given to talented writers of fiction or poetry intending to make writing their vocation. Five departmental fellowships are awarded annually for the master's program in creative writing: the Raymond Carver, Hayden Carruth, Loring Williams (Academy of American Poets), Delmore Schwartz, and the Jeremy Lake Memorial prizes in poetry and in fiction, the Raymond Carver, the Peter Neagoe, and the Stephen Crane prizes in fiction. These fellowships include an academic-year award stipend and a full-tuition scholarship for 24 credits for the academic year (12 credit hours per semester). Normally, these are awarded to new students and are not renewable.
Requirements: Applicant must have a bachelor's degree, preferably in English.
Amount of Grant: $9800-$12,300 plus remitted tuition for all fellowships
Date(s) Application is Due: Jan 10
Contact: Christopher Kennedy, Director of Creative Writing Program, (315) 443-3755; fax: (315) 443-5390; email: ckennedy@syr.edu
Internet: http://www-hl.syr.edu/depts/english/cwp/cwsupport.htm#10
Sponsor: Syracuse University
English Department, 401 Hall of Languages
Syracuse, NY 13244-1170

Graham Foundation Carter Manny Award 2560
The foundation supports advanced-level architectural projects and other areas of the arts. This award is made for doctoral dissertation research at international schools of architecture.
Requirements: Candidates must be nominated by their departments.
Amount of Grant: $15,000 maximum
Date(s) Application is Due: Mar 15
Contact: Richard Solomon, Director, (312) 787-4071; email: rsolomon@grahamfoundation.org
Internet: http://www.grahamfoundation.org/grants/carter-desc.asp
Sponsor: Graham Foundation
4 W Burton Pl
Chicago, IL 60610

Graham Foundation Grants 2561
Grants are offered to individuals and institutions in support of activities focused on architecture and the built environment; and generally for activities that lead to the public dissemination of ideas through publication, exhibition, or educational programming. The Foundation has supported a variety of endeavors, including research by scholars; grants to

architectural schools for special projects, enrichment programs, or new curricula; grants to museums, schools, and libraries for exhibitions, catalogues, and, in rare cases, for acquisitions; and support for publications, usually to help make an important publication better or more affordable. When large sums are necessary, the foundation is willing to consider a seed grant to explore possibilities, or a challenge grant when feasibility already seems clear and help is needed to corroborate worthiness of purpose and stimulate broad support. Occasionally multiyear grants are offered, though they normally will not extend beyond three consecutive years.

Requirements: Both individuals and organizations may apply. For individuals, consideration is given to those demonstrating mature creative talents who have specific project objectives.

Restrictions: There is no direct scholarship aid. The foundation does not award grants to cover costs of endowments, annual operating expenses, construction, or architectural fees in support of construction or renovation projects.

Amount of Grant: $10,000 average

Date(s) Application is Due: Jan 15; Jul 15

Contact: Richard Solomon, Director, (312) 787-4071; email: rsolomon@ grahamfoundation.org or info@grahamfoundation.org

Internet: http://www.grahamfoundation.org/grants/gen-desc.asp

Sponsor: Graham Foundation

4 W Burton Pl

Chicago, IL 60610

Grammy Foundation Grants 2562

The foundation awards grants in two program areas. Research Projects grants are made to organizations and individuals to support efforts that advance the research and/or broad- reaching implementations of original research projects related to the impact of music study on early childhood development, the human development, and the medical and occupational well-being of music professionals. Archiving and Preservation Projects grants are made to organizations and individuals to support efforts that advance the archiving and preservation of the music and recorded sound heritage of the Americas. The foundation funds preservation of original, preexisting media and source material; preservation projects that follow the recommended methodology; projects of historical, artistic, cultural, and or/national significance; and archiving projects including the rescue, organization of, and access to preexisting media and materials. Application and guidelines are available online.

Restrictions: The foundation does not fund Recording Academy chapters, trustees, governors, officers, or staff; organizations that discriminate on the basis of race, sex, religion, national origin, disability, or age; projects promoting advocacy issues; a single organization or individual for more than three consecutive years; organizations or individuals not based in the Americas; purchase of collections; recording projects, demo tapes, or performance events; proposals for commercial purposes (i.e., CD reissue or textbook/ A/V package); purchase of repairs of equipment; purchase or repairs of musical instruments; maintenance or upgrading of computer systems; competitions or any expense associated with competitions; work toward academic degrees; music education or in-residence programs; documentaries; endowments and fundraising; buildings and facilities; marketing, publicity, design costs; or projects where copyright status is unknown.

Samples: The Institute for Music & Brain Science, Boston, MA, $20,000 - this project will test whether music decreases behavioral, neurophysiological and endocrinological pain and stress caused by medically-necessary procedures. In addition, the hypothesis that humans innately prefer consonant over dissonant music will be tested (2009); Elliott Leib, San Diego, CA, $5,000 - a plan will be developed to digitally preserve material from the Trade Roots Reggae Collection (2009); Bob Moog Memorial Foundation for Electronic Music, Asheville, N.C., $15,000 - musical and historical content relative to the unique legacy of synthesizer pioneer Dr. Robert Moog will be cleaned, restored, re-housed and transferred to digital format for accessibility and long-term storage (2009).

Amount of Grant: $10,000-$40,000

Date(s) Application is Due: Oct 1

Contact: Kristin Murphy, (310) 392-3777 ext. 8662; fax: (310) 392-2188; email: grants@ grammy.com

Internet: http://www.grammy.com/GRAMMY_Foundation/Grants/

Sponsor: Grammy Foundation

3030 Olympic Boulevard

Santa Monica, CA 90405

Grass Foundation Marine Biological Laboratory Fellowships 2563

The foundation awards fellowships to encourage independent research for scientific discovery by investigators early in their careers and to increase research opportunities for persons trained for careers in neurobiological investigation. Neurobiological approaches supported include neurophysiology, membrane biophysics, integrative neurobiology and neuroethology, neuroanatomy, neuropharmacology, systems neuroscience, cellular and developmental neurobiology, and computational approaches to neural systems. Fellowships provide funds to support an investigator, his/her spouse or domestic partner, and dependent children for one summer at the Marine Biological Laboratory in Woods Hole, MA. Laboratory research space, equipment, housing and board, and travel expenses to and from the MBL are covered. Guidelines and application are available online.

Requirements: Applicants may be researchers in the late stages of predoctoral training or postdoctoral researchers. Priority is given to applicants who have demonstrated a commitment to a research career, who have no more than five years of postdoctoral research and who have no prior research experience at the Marine Biological Laboratory. All U.S. citizens, permanent residents, and foreign applicants with or able to obtain J-1

Visas for the duration of the fellowship are eligible to participate in this program. The foundation must have a copy of the J-1 Visa prior to participation in the program.

Restrictions: Applicants should not attempt to combine a fellowship at the Marine Biological Laboratory with writing a dissertation (PhD thesis).

Date(s) Application is Due: Dec 15

Contact: Fellowship Administrator, (781) 843-0219; fax: (781) 843-0474; email: grassfdn@aol.com

Internet: http://www.grassfoundation.org/fellowships/grass_fellowship/index.html

Sponsor: Grass Foundation

400 Franklin Street, Suite 302

Braintree, MA 02184

Grawemeyer Education Awards 2564

This award is given to the author(s) of the theory, policy proposal, technological advance, program initiative, or research study published in the recent past that is judged to have the most potential for educational improvement. The program intent is to stimulate worldwide the dissemination, public scrutiny, and implementation of ideas that have potential to contribute to improvement in educational practice and attainment. The award is intended not only to recognize the individual(s) responsible, but also to draw critical attention to the award-winning ideas, proposals, or achievements. Annual deadline dates may vary; contact program staff for exact dates.

Requirements: The university invites nominations from professional educators, educational institutions and organizations, and editors and publishers of journals and books worldwide.

Restrictions: Self-nominations will not be considered.

Samples: William G. Bowen, Andrew W. Mellon Foundation (NY) and Derek Bok, former president of Harvard U (Cambridge, MA)—for their book The Shape of the River: Long-Term Consequences of Considering Race in College and University Admissions.

Amount of Grant: $200,000 payable in five annual installments of $40,000

Date(s) Application is Due: Feb 1

Contact: Dr. Allan E. Dittmer, Executive Director, (502) 852-0791; fax: (502) 852-1497; email: allan@louisville.edu

Internet: http://www.louisville.edu/ur/onpi/grawemeyer/education/index.html

Sponsor: University of Louisville

School of Education, University of Louisville

Louisville, KY 40292

Great Lakes Fishery Trust Grants 2565

The mission of the trust is to provide funding to nonprofit organizations, educational institutions, and government agencies to enhance, protect, and rehabilitate Great Lakes fishery resources. Preproposals are sought in the following themes: research and related management activities consistent with the goals in the Lake Michigan Fish-Community Objectives; innovative research intended to lead to effective means to prevent the introduction of invasive species in the Great Lakes and limit the range and/or abundance of invasive species already present; research that models critical ecosystem components essential to sustainable fish population management; and research and analysis that documents the contribution (economic, environmental, and/or social) of various elements of the fishery. The trust will accept preproposals only submitted thought eh e-Grant application system.

Requirements: Preproposals are encouraged from educational, governmental, tribal, and 501(c)3 tax-exempt organizations.

Geographic Focus: Midwest

Date(s) Application is Due: Jan 25

Contact: Trust Administrator, (517) 371-7468; fax: (517) 484-6549; email: glft@glft. org

Internet: http://www.glft.org/geninfo.htm

Sponsor: Great Lakes Fishery Trust

600 W Saint Joseph, Suite 10

Lansing, MI 48933-2265

Great Lakes Protection Fund Grants 2566

The Great Lakes Protection Fund welcomes preproposals for projects that enhance the health of the Great Lakes ecosystem. Applicants should propose projects that will return the greatest ecosystem benefits. Current funding interests include preventing biological pollution, restoring natural flow regimes; and using market mechanisms for environmental improvement. Additional projects are sought to add to, and expand fund-supported work in these areas. The fund also welcomes projects that are designed to identify and explore other master variables that if acted upon, will result in tangible improvements to the health of the Great Lakes ecosystem. Proposals may be submitted at any time. The board makes grant decisions at its March, June, September, and December meetings.

Requirements: Nonprofit organizations (including environmental organizations, trade associations, and universities), for-profit businesses, government agencies, and individuals are eligible for fund support.

Restrictions: The fund does not give to general operating funds, environmental education, and groups from Indiana.

Samples: CAMP Inc (Cleveland, OH)—to reduce the use and emission of chlorinated solvents from parts cleaning and degreasing operations, $218,000.; Ohio State U (Columbus, OH)—to evaluate the potential for zebra mussels to transmit human pathogens through food chains in open lakes and near the lake shore, $17,100.

Amount of Grant: $25,000-$420,000 typically

Contact: Program Contact, (847) 425-8150; fax: (847) 424-9832; email: preproposals@ glpf.org

Internet: http://www.glpf.org/interest/index.html
Sponsor: Great Lakes Protection Fund
1560 Sherman Avenue, Suite 880
Evanston, IL 60201

Greater Worcester Community Foundation Jeppson Memorial Fund for   **2567**
Brookfield Grants
The Fund provides money to civic and community projects that help improve the lives of residents and enrich the cultural environment. The Fund provides support for: cultural or artistic performances; public seminars; festivals or exhibitions; services that help frail or vulnerable citizens, or that contribute to public health and safety; opportunities for educational enrichment; youth involvement in recreation, sports and the arts; and projects that foster community awareness and connections among different groups. Application information is available online.
Requirements: Any nonprofit or civic organization that serves the residents of Brookfield may apply.
Restrictions: Grant funds may not be used for expenses already incurred by the applicant. Fund awards are not intended to replace municipal funds.
Geographic Focus: Massachusetts
Date(s) Application is Due: Jul 15
Contact: Pamela B. Kane; (508) 755-0980; email: pkane@greaterworcester.org
Internet: http://www.greaterworcester.org/grants/Jeppson.htm
Sponsor: Greater Worcester Community Foundation
370 Main Street, Suite 650
Worcester, MA 01608-1738

Green Foundation Grants   **2568**
The Foundation's resources are focused on the four specific areas of arts, education, medical/scientific research and special projects. Preferential attention will be given to institutions exhibiting the following positive factors: a history of achievement, good management, and a stable financial condition; significant programs with the promise of making a measurable impact; programs that are self-sustaining and will not necessitate continued dependence on the Foundation. Interested applicants must complete the Foundation's online eligibility questionnaire. Information is available online.
Requirements: 501(c)3 nonprofits (as per the IRS Service Code of 1986) are eligible. Most grantmaking is limited to institutions that serve the Los Angeles community; however the Foundation will consider requests beyond this geographic boundary for those institutions with the potential to impact communities statewide or nationally.
Restrictions: The Foundation does not provide funds for: those with net assets or fund balances of less than $100,000; multi-year commitments; annual meetings, conferences, and/or seminars; religious programs; capital campaigns; direct mail campaigns; conduit institutions, unified funds, fiscal agents, or institutions using grant funds from donors to support other institutions or individuals; private foundations; individuals.
Contact: Kylie Schwerdtfeger, (626) 584-1285; fax: (626) 577-9400; email: kylies@ligf.org
Internet: http://ligf.org/index.html
Sponsor: Green Foundation
201 South Lake Avenue, Suite 605
Pasadena, CA 91101

Greenburg-May Foundation Grants   **2569**
Grants are almost entirely awarded for medical research, primarily in the fields of cancer, heart, and neurological research. In addition, some support is given to hospitals, Jewish welfare funds, temple support, and the aging. Types of support include general operating support, continuing support, annual campaigns, endowment funds, emergency funds, program development, internships, scholarship funds, research, and consulting services. Organizations in Florida and New York City receive the largest part of the funds. Applications may be submitted at any time.
Requirements: Nonprofit organizations in southern Florida and New York are eligible.
Restrictions: Grants are not given to individuals or for endowment funds, special projects, publications, or conferences, and generally not for scholarships and fellowships.
Geographic Focus: Florida; New York
Samples: Miami Jewish Home and Hospital for the Aged (Miami, FL)—$75,200.; Greater Miami Opera (Miami, FL)—$4,220.
Amount of Grant: $295,282 total
Contact: Isabel May, President, (305) 864-8639
Sponsor: Greenburg-May Foundation
P.O. Box 54-5816
Miami Beach, FL 33154

Greenspun Family Foundation Grants   **2570**
The foundation awards grants to eligible Nevada organizations in its areas of interest, including cancer research, children's services, higher education, the physically and mentally disabled, and religious welfare programs that provide for the homeless. There are no application forms or deadlines. Submit a letter of request.
Requirements: Nevada nonprofit organizations are eligible. Preference is given to requests from the Las Vegas area.
Geographic Focus: Nevada
Contact: Grants Administrator
Sponsor: Greenspun Family Foundation
901 N Green Valley Pkwy, Suite 210
Henderson, NV 89074

Greenwall Foundation Bioethics Grants   **2571**
The Foundation provides funding for physicians, lawyers, philosophers, theologians and other professionals to address micro and macro issues in bioethics, providing guidance for those engaged in decision-making at the bedside as well as those responsible for shaping institutional and public policy. The Foundation is especially interested in the work of junior investigators and pilot projects that may lead to NIH support, and it is prepared to address issues regarded by some as sensitive or potentially controversial.
Restrictions: The Foundation is not normally interested in proposals to support equipment purchase, facility construction or renovation, or general operating expenses, and will not normally consider grants to private foundations, endowment funds, or individual applicants.
Amount of Grant: $5000-$50,000 average
Date(s) Application is Due: Feb 1; Aug 1
Contact: Sam Teigen, Grants Administrator; (212) 679-7266; fax: (212) 679-7269; email: steigen@greenwall.org
Internet: http://www.greenwall.org/guidebio.htm
Sponsor: Greenwall Foundation
420 Lexington Avenue, Suite 2500
New York, NY 10170

Greenwall Foundation Faculty Scholars Program in Bioethics   **2572**
The Program is a career development program that enables junior faculty members to carry out original research on policy and moral dilemmas at the intersection of ethics and the life sciences. To maximize scholarly development, three years of support are provided, requiring a 50% time commitment (carefully monitored) in each of the three years. Each year the Scholars participate in a learner-centered educational program that includes required attendance at two workshops (spring and fall). Continued involvement in this activity, after each Scholar's support period is over, provides Scholar-alumni/ae with ongoing professional development and engages them in the training of successor cohorts. Preliminary applications are due on December 3. Approximately 15 to 20 applicants will be invited to submit full proposals due in February.
Requirements: Applicants must be junior faculty members holding at least a 60% appointment at a university or non-profit research institute in the U.S. Priority will be given to applicants who are below the rank of Associate Professor, who have not received a comparable career development award, and whose work will have an impact on public policy or clinical practice. Within this group, priority will be given to applicants whose research addresses innovative and emerging topics.
Samples: Jeffrey Peppercorn, MD, MPH, Assistant Professor of Medicine in the Division of Hematology/Oncology, University of North Carolina (2007); David E Winickoff, JD, MA, Assistant Professor of Bioethics and Society, University of California, Berkeley (2007).
Date(s) Application is Due: Dec 3
Contact: Sam Teigen, Grants Administrator; (415) 476-6241 or (212) 679-7266; fax: (212) 679-7269; email: steigen@greenwall.org or greenwall@medicine.ucsf.edu
Internet: http://www.greenwallfsp.org/
Sponsor: Greenwall Foundation
420 Lexington Avenue, Suite 2500
New York, NY 10170

Greenwall Foundation Fellowship Program in Bioethics and Health Policy   **2573**
The Fellowship program and its faculty cover much of the waterfront of issues and methodologies in bioethics, from ethics and advances in biomedical science, to ethics and public health and health policy, to ethics and clinical care. The defining focus of the Program, however, is bioethics and public policy. The Program is intended for people who are early in their careers. Greenwall applicants are subdivided by discipline (medicine, philosophy and law) and reviewed by a selection committee composed of program faculty with relevant disciplinary backgrounds. Typically, only one Fellow per year is accepted from each discipline.
Amount of Grant: $111,650 over two years
Date(s) Application is Due: Dec 1
Contact: Gail Geller; (410) 955-7894; email: ggeller@jhmi.edu; Maria Merritt; (410) 614-6335; email: mmerritt@jhsph.edu
Internet: http://www.greenwall.org/guideaffil.htm
Sponsor: Greenwall Foundation
420 Lexington Avenue, Suite 2500
New York, NY 10170

Gruber Foundation International Prizes   **2574**
The Peter and Patricia Gruber Foundation honors and encourages educational excellence, social justice and scientific achievements that better the human condition. The Peter and Patricia Gruber Foundation is a private, United States-based philanthropic organization established in 1993 under the 501(c)3 section of U.S. Corporate Law. It is funded entirely by Peter and Patricia Gruber, who serve as its Chairman and President, respectively. The Foundation is headquartered in St. Thomas, United States Virgin Islands, and maintains a small branch office in New York City. A major focus of the Foundation's philanthropy is its International Prize Program, created to recognize excellence in science and humanities by highlighting five fields that create a better world: Cosmology, Genetics, Neuroscience, Justice, and Women's Rights. Each Prize is given annually, and carries an unrestricted cash award of $500,000 and a gold medal. Independent advisory boards comprised of experts from each field review and select a prize recipient at annual selection meetings held for each discipline. Only fully completed nominations are considered, and the Chair of each Selection Advisory Board screens the nominations, presenting to the Board only those of sufficient caliber. The Selection Advisory Board alone is responsible for

the choice of recipient, making its decision from among officially nominated candidates following discussion and deliberation. The Peter and Patricia Gruber Foundation awards its international prizes to recipients only after following a process designed to ensure complete independence and integrity in selecting, as worthy of international recognition, scientific and human rights achievements at the highest levels.

Requirements: Only candidates who have been officially nominated on a fully completed nomination form will be considered for a Peter and Patricia Gruber Foundation International Prize. Additional guidelines and nomination forms are available on line.

Samples: Gruber Cosmology Prize was won by Astronomers Wendy Freedman, Robert Kennicutt, and Jeremy Mould, they led the Hubble Space Telescope Key Project on the Extragalactic Distance Scale, a painstaking ten-year-long effort that resolved the decades-long dispute about the value of the Hubble constant, one of the most important measurements in astronomy. The project, findings have not only effectively determined the age of the universe (about 14 billion years), but have also enabled scientists to more accurately investigate other profound questions about the birth, evolution, and composition of the universe (2009); Gruber Justice Prize was shared by the European Roma Rights Centre and Bryan Stevenson, the executive director of Equal Rights Initiative. Through recourse to litigation, consultation, and other approaches available within the justice and political systems, Bryan Stevenson and the European Roma Rights Centre have courageously helped members of oppressed groups gain access to equal justice (2009);

Amount of Grant: $500,000 plus a gold medal
Date(s) Application is Due: Dec 15
Contact: Bernetia Akin, Foundation Contact; (340) 775-8035 or (340) 775-8014; fax (340) 775 8040; email: info@gruberprizes.org
Internet: http://www.petergruberfoundation.org/GruberPrizes/GruberPrizes.php
Sponsor: The Peter and Patricia Gruber Foundation
6000 Estate Charlotte Amalie, Suite 4
St. Thomas, U.S.VI 00802

GSA Research Grants   2575
The society offers research grants to provide partial support of master's and doctoral thesis research in earth science for graduate students at universities in the United States, Canada, Mexico, and Central America. Application forms are available on the Web site. faxed or email applications will not be accepted.

Requirements: Eligibility is restricted to GSA members. Applicant must be enrolled in a graduate program in the United States, Canada, Mexico, or Central America and must outline the proposed research. Application must be on current GSA forms, which are available from the geology departments of most colleges and universities in the United States or upon request from the Research Grants Administrator of the society. Completed forms must be accompanied by two faculty evaluations of the applicant. Applicants must use forms for the current grant year.

Restrictions: The program does not offer scholarships to be used for tuition, housing, etc., nor are students in seeking acceptance into graduate school given assistance.

Amount of Grant: $1533 average
Date(s) Application is Due: Feb 1
Contact: Program Officer, (303) 357-1028; email: awards@geosociety.org
Internet: http://www.geosociety.org/profdev/grants/index.htm
Sponsor: Geological Society of America
P.O. Box 9140
Boulder, CO 80301-9140

Guggenheim Foundation Charitable Contribution Grants   2576
The Foundation concentrates its efforts in the criminal justice field. Accordingly, requests for grants should only be presented if there is criminal justice content. Grants are awarded as seed money for organizations to use to develop successful programs that continue beyond the life of the grant. Application information is available online.

Requirements: Connecticut, New Jersey, and New York organizations are eligible.
Restrictions: Grants are not awarded to individuals or for scholarships.
Geographic Focus: Connecticut; New Jersey; New York
Date(s) Application is Due: Apr 15
Contact: President, (212) 755-3199
Internet: http://www.dfguggenheimfoundation.org/apply.html
Sponsor: Daniel and Florence Guggenheim Foundation
950 Third Avenue, 30th Floor
New York, NY 10022

GUITS Grants for the Publication of Scholarly Books and Journals   2577
Grants for the Publication of Scholarly Books and Journals are intended to cover part of the publication costs of scholarly books and journals in the field of Turkish Studies to be published in the U.S. Most grants are disbursed as reimbursements and require the presentation of documentation before payment is made. Grant announcements are sent out in mid-May.

Requirements: To be eligible for a grant, applicants should be: faculty in any field of the social sciences and/or humanities; and U.S. citizens or permanent residents at the time of the application.

Date(s) Application is Due: Mar 7
Contact: David C. Cuthell, Director; (202) 687-0295; fax: (202) 687-3780; email: dcc@turkishstudies.org
Internet: http://www.turkishstudies.org/grantinfo.html
Sponsor: Georgetown University Institute of Turkish Studies
Georgetown University, Intercultural Center—Box 571033-Room 305R
Washington, DC 20057-1033

GUITS Grants for Undergraduate Study   2578
The Institute of Turkish Studies is a non-profit, private educational institution devoted solely to the advancement of training, research, and teaching in the field of Ottoman and modern Turkish Studies. Grants for Undergraduate Study are for students who wish to participate in study abroad programs in Turkey. These awards will be made to institutions, which will be responsible for their distribution to qualified applicants. ITS welcomes applications from institutions seeking to either expand or develop overseas programs in Turkey. The maximum award is $10,000.

Requirements: To be eligible for a grant, applicants should be: faculty in any field of the social sciences and/or humanities; and U.S. citizens or permanent residents at the time of the application.

Amount of Grant: $1,000-$3,000
Date(s) Application is Due: Mar 7
Contact: David C. Cuthell, Director; (202) 687-0295; fax: (202) 687-3780; email: dcc@turkishstudies.org
Internet: http://www.turkishstudies.org/grantinfo.html
Sponsor: Georgetown University Institute of Turkish Studies
Georgetown University, Intercultural Center—Box 571033-Room 305R
Washington, DC 20057-1033

GUITS Matching Seed-Money Grants for New Positions in Turkish Studies   2579
Matching Seed-Money Grants for New Positions in Turkish Studies are intended for institutions. ITS offers matching seed money grants to universities for the establishment of these new positions in any discipline in the field of Turkish Studies. Universities are expected to make commitment to establish a tenure-track position for this position. The maximum award is .50 FTE of a junior faculty appointment.

Requirements: To be eligible for a grant, applicants should be: faculty in any field of the social sciences and/or humanities; and U.S. citizens or permanent residents at the time of the application.

Date(s) Application is Due: Mar 7
Contact: David C. Cuthell, Director; (202) 687-0295; fax: (202) 687-3780; email: dcc@turkishstudies.org
Internet: http://www.turkishstudies.org/grantinfo.html
Sponsor: Georgetown University Institute of Turkish Studies
Georgetown University, Intercultural Center—Box 571033-Room 305R
Washington, DC 20057-1033

GUITS Post-Doctoral Summer Travel-Research Grants   2580
The Institute of Turkish Studies is a non-profit, private educational institution devoted solely to the advancement of training, research, and teaching in the field of Ottoman and modern Turkish Studies. The grant is intended to provide partial support for travel and research to Turkey for those who hold a Ph.D. in social sciences or humanities discipline. The time period for the grant varies with each individual application. Normally, the recipients of the grants are expected to spend a minimum of four weeks in Turkey. Maximum award is round-trip airfare to Turkey. Application information is available online.

Requirements: Eligible recipients must be U.S. citizens or U.S. permanent residents and currently live/work in the U.S.

Date(s) Application is Due: Mar 7
Contact: David C. Cuthell, Director; (202) 687-0295; fax: (202) 687-3780; email: dcc@turkishstudies.org
Internet: http://www.turkishstudies.org/grantinfo.html
Sponsor: Georgetown University Institute of Turkish Studies
Georgetown University, Intercultural Center—Box 571033-Room 305R
Washington, DC 20057-1033

GUITS Research Grants in Comparative Studies of Modern Turkey   2581
The Institute of Turkish Studies is a non-profit, private educational institution devoted solely to the advancement of training, research, and teaching in the field of Ottoman and modern Turkish Studies. A significant portion of the project should be devoted comparatively to one or more states or political entities in Europe, Latin America, the Middle East, and Asia in addition to the Turkish Republic. The grants are primarily, but not exclusively, in the field of Political Science. The grant is for the academic year, and permission to postpone the tenure of the grant beyond this period will normally not be given. Application information is available online.

Requirements: Grants are awarded to graduate students at the dissertation research stage and for post-doctoral scholars in the U.S. who study aspects of the Republic of Turkey (post 1922) in a comparative context. Applicants must be U.S. citizens or permanent residents in the U.S. and affiliated with a university at the U.S. at the time of the application.

Amount of Grant: $10,000
Date(s) Application is Due: Mar 7
Contact: David C. Cuthell, Director; (202) 687-0295; fax: (202) 687-3780; email: dcc@turkishstudies.org
Internet: http://www.turkishstudies.org/grantinfo.html
Sponsor: Georgetown University Institute of Turkish Studies
Georgetown University, Intercultural Center—Box 571033-Room 305R
Washington, DC 20057-1033

GUITS Sabbatical Research Grants   2582
The Institute of Turkish Studies is a non-profit, private educational institution devoted solely to the advancement of training, research, and teaching in the field of Ottoman and modern Turkish Studies. These grants are for the academic year and permission

to postpone the tenure of the grant beyond this period will normally not be given. Application information is available online.

Requirements: To be eligible for a grant, applicants should be: faculty in any field of the social sciences and/or humanities; and U.S. citizens or permanent residents at the time of the application. The grants will be awarded only to applicants who certify that they are taking sabbatical leave and will be conducting research in the field of Turkish Studies.

Amount of Grant: $25,000

Date(s) Application is Due: Mar 7

Contact: David C. Cuthell, Director; (202) 687-0295; fax: (202) 687-3780; email: dcc@turkishstudies.org

Internet: http://www.turkishstudies.org/grantinfo.html

Sponsor: Georgetown University Institute of Turkish Studies

Georgetown University, Intercultural Center—Box 571033-Room 305R

Washington, DC 20057-1033

GUITS Summer Research Grants in Turkey for Graduate Students   2583

The Institute of Turkish Studies is a non-profit, private educational institution devoted solely to the advancement of training, research, and teaching in the field of Ottoman and modern Turkish studies. These grants are awarded for summer travel to Turkey to carry out research projects. The time period for grants varies with each individual application. Normally, the recipients of the Summer Research Grants are expected to spend a minimum of two months in Turkey. Application information is available online.

Requirements: Eligible recipients must be: graduate students in any field of the social sciences and/or humanities in the United States; U.S. citizens or permanent residents at the time of the application; and currently not engaged in dissertation writing.

Amount of Grant: $1,000–$2,000

Date(s) Application is Due: Mar 7

Contact: David C. Cuthell, Director; (202) 687-0295; fax: (202) 687-3780; email: dcc@turkishstudies.org

Internet: http://www.turkishstudies.org/grantinfo.html

Sponsor: Georgetown University Institute of Turkish Studies

Georgetown University, Intercultural Center—Box 571033-Room 305R

Washington, DC 20057-1033

Gustavus and Louise Pfeiffer Research Foundation Grants   2584

The Foundation makes grants to U.S. tax-exempt institutions for projects or programs for the advancement of medicine and pharmacy, including scientific research; postgraduate scholarship and fellowship assistance; and studies in nutrition, blindness, deafness and other physical disabilities. The letter of inquiry should not exceed two pages and should state the nature and purpose of the project, the field of interest of the foundation to which it relates, the amount requested, and a general description of the items to be funded; full proposals are by invitation. Applications are considered by the foundation's directors at two meetings a year, one in spring and one in fall. The listed application deadlines are for letters of inquiry; full applications are by invitation.

Requirements: U.S. tax-exempt organizations are eligible.

Restrictions: Grants will not be made for delivery of health care services; that benefit a limited geographical area; for endowment, general institutional support, building programs, chairs, or the like; where a major part of the funds are for equipment, exhibits, conferences, seminars, lectures, workshops, sabbatical leave, or the like; for indirect costs; for general programs of research rather than specific research projects of limited duration; to other foundations or fund raising organizations for the purpose of their making grants to others; that are more appropriate for support by other sources, such as pharmaceutical companies for commercial applications of existing products; that involve animal experimentation (other than insects); or to institutions for projects not in the United States.

Samples: Pfeiffer U (Misenheimer, NC)—to create the Pfeiffer Research Fellows Program in the School of Natural and Health Sciences, and for the university's capital campaign, $1.4 million.

Amount of Grant: $75,000 maximum

Date(s) Application is Due: Jan 8; Jul 25

Contact: Matt Herold Jr., Secretary, (973) 983-0480; fax: (973) 586-3456

Internet: http://fdncenter.org/grantmaker/pfeiffer

Sponsor: Gustavus and Louise Pfeiffer Research Foundation

89 Diamond Spring Road, CN 3050

Danville, NJ 07834

Gustus L. Larson Memorial Award   2585

This award, sponsored jointly by the Pi Tau Sigma Honorary Mechanical Engineering Fraternity and ASME, is given annually to the engineering graduate who has demonstrated outstanding achievement in mechanical engineering within 10 to 20 years after graduating from a regular engineering course of a recognized college or university. Achievement shall be all or in part in any field including industrial, educational, political, research, civic, and artistic. The candidate's achievements will be examined for an application of basic engineering methods or principles. A special nominating form must be obtained from ASME.

Amount of Grant: $1000

Date(s) Application is Due: Feb 1

Contact: Gilda DiTullio, Manager, (212) 591-7736; fax: (212) 705-7739; email: ditulliog@asme.org

Internet: http://www.asme.org/honors/ms71/gaa/larson.html

Sponsor: American Society of Mechanical Engineers

3 Park Avenue

New York, NY 10016

H.A. and Mary K. Chapman Charitable Trust Grants   2586

Grants are provided to support nonprofit organizations involved in a wide variety of charitable purposes, including education, medical research, health and human services, arts and culture, civic and community, and nature and wildlife. Priority is given to: funding projects or programs that will serve a documented need and have a significant impact on the community; financial support of the organization management and board of directors, individuals of the general community affected by the program or project, and other private foundations; sound financial planning for maintenance and continuation of the project or program after the initial funding is complete; projects or programs that potentially benefit a significant segment of people; applicants with committed volunteers and staff leadership. Application information is available online.

Requirements: 501(c)3 nonprofits in Oklahoma, are eligible. Priority will normally be given to Tulsa and surrounding counties and eastern Oklahoma. Depending on the availability of funds, Arkansas and Texas may also be eligible for funding for Colleges and Universities.

Restrictions: Grants are not made for the following purposes: to be passed-through to another organization that is not a 501(c)3 public charity or a governmental entity; to attempt to influence legislation or the outcome of any election, or to political campaigns. Grant requests for the following purposes are not favored: endowments; deficit financing and debt retirement; projects or programs for which the Chapman Trusts would be the sole source of financial support; travel, conferences, conventions, group meetings, or seminars, camp programs and other seasonal activities; religious programs of religious organizations; project or program planning; start-up ventures are not excluded, but organizations with a proven strategy and results are preferred; purposes normally funded by taxation or governmental agencies; requests made less than nine months from the declination of a previous request by an applicant, or within nine months of the last payment made on a grant made to an applicant; requests for more than one project.

Geographic Focus: Arkansas; Oklahoma; Texas

Samples: Oklahoma Methodist Manor (Tulsa, OK)—$100,000.; Bishop Kelly High School (Tulsa, OK)—$75,000.; Oklahoma Baptist Homes for Children (Tulsa, OK)—$50,000.

Contact: Jerry Dickman or Donne Pitman, (918) 496-7882; fax (918) 496-7887; jerry@chapmantrusts.com

Internet: http://www.chapmantrusts.org/grants_programs.html

Sponsor: H.A. and Mary K. Chapman Charitable Trust

6100 South Yale, Suite 1816

Tulsa, OK 74136

H.B. Earhart Fellowships   2587

These fellowships are awarded to move talented individuals through graduate study in the fields of international studies, economics, political science, and history in optimum time to embark upon careers in college or university teaching or in research. Awards are made to graduate students nominated by faculty sponsors whose participation is invited annually. Sponsors also monitor performance. Direct applications from candidates or from noninvited sponsors are not accepted.

Amount of Grant: $1000–$25,000 average

Contact: Ingrid Ann Gregg, President, (734) 761-8592

Sponsor: Earhart Foundation

2200 Green Road, Suite H

Ann Arbor, MI 48105

H.R. Lissner Award   2588

This award has been established in honor of H.R. Lissner for his pioneering contributions in biomechanical research. The award is bestowed annually for outstanding accomplishments in the area of bioengineering in the form of: significant research contributions; development of new methods of measuring; design of new equipment and instrumentation; educational impact in the training of bioengineers; or service to the bioengineering community and/or the ASME Bioengineering Division.

Samples: Steven Goldstein—award recipient, $1000 (2005); John Tarbell—award recipient, $1000 (2004).

Amount of Grant: $1000 honorarium and expense supplement

Date(s) Application is Due: Jan 1

Contact: Honors and Awards Department, (212) 705-7735; fax: (212) 705-7739

Internet: http://www.asme.org/honors/ms71/list.html

Sponsor: American Society of Mechanical Engineers

345 E 47th Street

New York, NY 10017-2392

H.W. Wilson Foundation Grants   2589

The foundation awards grants nationwide in its areas of interest, including libraries, library schools, and library science; higher education; and cultural outreach. Types of support include program grants, scholarships, and research grants. There are no application forms or deadlines. The board meets in January, March, May, August, and October.

Restrictions: Grants do not support building construction/renovation, endowment funds, or operating expenses.

Samples: Reed College (Portland, OR)—to develop electronic resources at the Eric V. Hauser Library, $150,000 over three years (2003).

Contact: William Stanton, President, (718) 588-8400

Sponsor: H.W. Wilson Foundation

950 University Avenue

Bronx, NY 10452

Hadassah Foundation Grants 2590

The foundation's mission is to improve the status, health, and well-being of women and girls; bring their contributions, issues, and needs from the margins to the center of Jewish concern; and encourage and facilitate their active participation in decision-making and leadership in all spheres of life. The foundation funds projects that empower Israeli women by providing them with the tools they need to become economically self-sufficient. The foundation also supports training for girls so that they will develop into self-confident adults and leaders. Letters of intent must meet the listed application deadline.

Requirements: 501(c)3 nonprofit organizations and organizations who have amutah or mossad tziburi status in Israel are eligible.

Samples: Jewish Arab Economic Development (Herzliya Pituach, Israel)—to provide business-skills training and to encourage Jewish and Palestinian Israeli women to work together on business activities, $30,000 (2005); Economic Empowerment for Women (Haifa, Israel)—to teach women marketing skills to help them promote their businesses and reach a broader customer base, $50,000 (2005); FEGS (Syosset, NY)—to implement a program for Jewish girls and young women designed to prevent eating disorders, $37,000 (2005); Tides Ctr (Oakland, CA)—for a youth-led program that seeks to prevent dating violence among Jewish teenagers, $40,000 (2005).

Amount of Grant: $100,000 maximum; $30,000-$40,000 typically

Date(s) Application is Due: Jul 6

Contact: Grants Administrator, (212) 451-6245; fax: (212) 303-8282; email: Hadassahfoundation@hadassah.org

Internet: http://www.hadassahfoundation.org

Sponsor: Hadassah Foundation

50 W 58th Street

New York, NY 10019-2505

Hagley Museum and Library Grants in Aid 2591

These grants support short-term (two to eight weeks) research in the imprint, manuscript, pictorial, and artifact collections of the Hagley Museum and Library, which contain materials on significant businesses and individuals of the Middle Atlantic region and the nation as well as published material with strong research value in the areas of business, economic, technological, and social history.

Requirements: Degree candidates, advanced scholars, and independent scholars may apply.

Amount of Grant: $1400 maximum per month

Date(s) Application is Due: Mar 31; Jun 30; Oct 31

Contact: Carol Lockman, Center for the History of Business, Technology, and Society, (302) 658-2400 ext 243; fax: (302) 655-3188; email: clockman@hagley.org

Internet: http://www.hagley.lib.de.us/grants.html

Sponsor: Hagley Museum and Library

P.O. Box 3630

Wilmington, DE 19807-0630

Hagley Museum and Library Henry Belin DuPont Fellowships 2592

Fellowships enable individual out-of-state scholars to pursue advanced research and study in the library, archival, and artifact collections of the Hagley Museum and Library. Preference will be given to those whose travel costs to Hagley will be higher. Fellows must devote full time to their studies and may not accept teaching assignments or undertake any other major activities during the tenure of their fellowships. Tenure must be continuous and last from two to six months. At the end of their tenure, fellows must submit final reports on their activities and accomplishments. Fellows are expected to participate in seminars, as well as attend noontime colloquia, lectures, concerts, exhibits, and other public programs offered during their tenure.

Requirements: The following information must be submitted for consideration: fellowship application cover sheet; cover letter, noting applicant's social security number and notice of any sabbatical or other funding to be received during tenure at Hagley; copy of current resume or vita, not to exceed five pages in length; description of the proposed project, not more than 15 pages in length, providing methodology, plan of work, Hagley collections to be used, the relevance of Hagley's fields of interest to the project, and the intended product; and two letters of recommendation from persons familiar with the project.

Restrictions: Degree candidates are ineligible.

Amount of Grant: $1400 maximum

Date(s) Application is Due: Mar 31; Jun 30; Oct 31

Contact: Dr. Philip Scranton, Center for the History of Business, Technology, and Society, (302) 658-2400 ext 208; fax: (302) 658-3188; email: pscranton@hagley.org

Internet: http://www.hagley.lib.de.us/grants.html

Sponsor: Hagley Museum and Library

P.O. Box 3630

Wilmington, DE 19807

Hagley/Winterthur Research Fellowships in Arts and Industries 2593

This is a cooperative program of short-term research fellowships for scholars interested in the historical and cultural relationships between economic life and the arts including design, architecture, crafts, and the fine arts. Fellowships are awarded for periods of from one to six months. Fellows receive stipends; make use of the rich research collections of the Winterthur Museum, Garden, and Library and the Hagley Museum and Library; and participate in seminars that meet at both sponsoring institutions. Applicants doing dissertation research are welcome to apply. Guidelines are available upon request from the Research Fellowship Program, Winterthur Museum, Garden, and Library, Winterthur, DE 19735; (302) 888-4649; or from the Hagley Museum.

Amount of Grant: $1400 stipend per month

Date(s) Application is Due: Dec 1

Contact: Dr. Philip Scranton, Center for the History of Business, Technology, and Society, (302) 658-2400 ext 208; fax: (302) 655-3188; email: pscranton@hagley.org

Internet: http://www.hagley.lib.de.us/grants.html

Sponsor: Hagley Museum and Library

P.O. Box 3630

Wilmington, DE 19807

Hague Academy of International Law Doctoral Scholarships 2594

The academy offers a limited number of residential scholarships for doctoral candidates whose theses in areas of international law are in advanced stages of preparation. Scholarships are given for periods of two months beginning July 1. Time is to be spent at the academy to facilitate the completion of the thesis.

Requirements: The doctoral scholarships are intended for candidates from developing countries who reside in their home countries and who do not have access to scientific sources. The upper age limit of applicants is 45. Application must be accompanied by a recommendation from the professor under whose direction the thesis is being written, giving the title of the thesis. Each candidate must show proof that she/he is at postgraduate level and that his/her thesis is necessary to obtain the doctoral degree. The thesis may be concerned with either private or public international law.

Amount of Grant: 2250 Euros; 910 Euros maximum travel

Date(s) Application is Due: Mar 1

Contact: M. Croese, Head of the Secretariat, (070) 302 42 42

Internet: http://www.hagueacademy.nl/eng-home.html

Sponsor: Hague Academy of International Law

Peace Palace, Carnegieplein 2

The Hague 2517 KJ Netherlands

Hague Academy of International Law Summer Scholarships 2595

The Hague Academy of International Law awards scholarships for one of two summer sessions of study in public and private international law taught by selected authorities. Sessions are for three weeks of intensive study in French or in English; diplomas are available if arrangements for taking the tests are made prior to acceptance of the application deadline of March 1 of each year. Forms may be obtained by writing directly to the academy or to the heads of diplomatic missions and other authorities and scientific institutions connected with international law.

Requirements: Application must be submitted by the applicant and must be accompanied by a recommendation from a professor of international law, a curriculum vita, and a statement of evidence that the applicant considers to be of value in support of his/her candidacy. Applicants must be under 40 years of age.

Restrictions: No more than two scholarships will be awarded to applicants from any one country during the same year; an applicant may obtain only one scholarship.

Amount of Grant: Dfl900

Date(s) Application is Due: Mar 1

Contact: Secretariat, 070-302-42-42

Internet: http://www.hagueacademy.nl/eng-home.html

Sponsor: Hague Academy of International Law

Peace Palace, Carnegieplein 2

The Hague 2517 KJ Netherlands

Halliburton Foundation Grants 2596

The foundation supports education at all levels and charitable organizations in the following ways: matching U.S.-based employee donations on a two-for-one basis up to $20,000 annually per employee for accredited junior colleges, colleges, and universities; matching U.S.-based employee donations to accredited elementary and secondary schools on a two-for-one basis up to $500 annually per employee; making direct donations to U.S.-based elementary and secondary schools and colleges and universities; and recognizing and supporting active U.S.-based employee volunteerism with direct donations through the Halliburton Volunteer Incentive Program. The corporate giving program makes donations to tax-exempt nonprofit organizations dedicated to education, health/welfare, civic issues, and arts and culture. The board meets quarterly. There are no application deadlines or forms.

Geographic Focus: Southwest; Texas

Amount of Grant: $250-$20,000 average

Contact: Margaret Carriere, (713) 676-3717; email: fhoufoundation@halliburton.com

Internet: http://www.halliburton.com/about/community.jsp

Sponsor: Halliburton Foundation

4100 Clinton Drive, Building 1, 7th Fl

Houston, TX 77020

Harold Alfond Foundation Grants 2597

The foundation awards grants to eligible organizations in its areas of interest, including secondary education, higher education, medical research, health care, and general charities. Grants support: education—public and private colleges and universities, as well as private secondary schools, to fund athletically oriented capital projects and scholarship endowments; medical research—individual research projects are considered when sponsored by a recognized medical research center; health—community support for capital campaigns and endowment funds; and general charities—community organizations that support youth, the arts, persons with disabilities, underprivileged, substance abuse rehabilitation, and annual fund drives of national organizations focusing on the above areas. There are no application deadlines or forms.

Restrictions: Grants are not made to individuals.

Geographic Focus: Florida; Maine

Samples: Saint Joseph's College of Maine (Standish, ME)—to construct an academic building, $1 million.
Amount of Grant: $250-$15,000 range
Contact: Gregory Powell, c/o Dexter Enterprises
Sponsor: Harold Alfond Foundation
Two Monument Square
Portland, ME 04101

Harold Amos Medical Faculty Development Program   2598

This program offers four-year, postdoctoral research fellowships to minority physicians who have demonstrated superior academic and clinical skills and who are committed to careers in academic medicine. The program seeks to increase the number of minority faculty physician scientists who can achieve senior rank in academic medicine and who will encourage and foster the development of succeeding classes of minority physicians. Applications are accepted from African American, Mexican American, Native American, and mainland Puerto Rican physicians. (Mainland Puerto Ricans are those who have completed at least their postsecondary education on the mainland.) Fellows will pursue biomedical research, clinical investigation, or health services research. Participants will conduct research in association with senior faculty members at academic centers.
Requirements: Eligible to apply are minority physicians who are U.S. citizens at the time of application, have excelled in their postsecondary education within the United States, are now completing or will have completed formal clinical training, and are committed to academic careers.
Samples: Baylor College of Medicine and U of Utah's School of Medicine (UT)—for awards through the Minority Medical Faculty Development Program, $365,400 each.; U of Medicine and Dentistry of New Jersey (NJ)—for programs to help minorities and disadvantaged people pursue careers in healthcare, $200,953.; U of North Carolina at Chapel Hill (NC)—for faculty development programs for minority medical faculty members, $365,400.; New York U (NY)—for faculty development programs for minority medical faculty members, $365,400.
Amount of Grant: $65,000 stipend per year plus $26,350 for research support
Date(s) Application is Due: Mar 23
Contact: National Program Office, (301) 565-4080; email: amfdp@starpower.net
Internet: http://www.amfdp.org/applicat.htm
Sponsor: Robert Wood Johnson Foundation
P.O. Box 2316, Rt 1 and College Rd E
Princeton, NJ 08543-2316

Harold Howe II Youth Policy Fellowship   2599

The annual fellowship is awarded to a promising young scholar or practitioner to support a self-designed project on significant issues in youth policy, practice, research, or program evaluation, focusing particularly on disadvantaged youth. It may support dissertation research in any of the fields related to youth policy. It is expected that in the 12-month period of the fellowship, a report of high quality will be produced. The resulting paper may be the basis for an article in a peer-reviewed journal and may be the basis for one or more AYPF forums to be held on Capitol Hill. The fellow will join the AYPF staff, accruing leave and working a regular 40-hour week. Proposal guidelines are available online.
Requirements: Candidates must have received a MA degree or higher within the five years preceding commencement of the fellowship or completed coursework toward a PhD within the five years preceding commencement of the fellowship. Candidates must possess excellent writing skills and sufficient computer competency to carry out research and writing related to the project.
Amount of Grant: $25,000
Contact: Samuel Halperin, Fellowship Administrator, (202) 775-9731; fax: (202) 775-9733; email: aypf@aypf.org
Internet: http://www.aypf.org/publications/harold_howe.htm
Sponsor: American Youth Policy Forum
1836 Jefferson Pl NW
Washington, DC 20036

Harold Lancour Scholarship for Foreign Study   2600

This scholarship is awarded to a librarian or graduate library science student for foreign study and research in the library sciences. Application form must be obtained from the society and returned with resume and brief letter describing relevance of the proposed foreign study to work or schooling.
Amount of Grant: $1500
Date(s) Application is Due: Mar 15
Contact: Dr. Jane Robbins, Executive Director, (850) 644-3907; fax: (850) 644-9763; email: Beta_Phi_Mu@lis.fsu.edu
Internet: http://www.beta-phi-mu.org/scholarships.html#lancour
Sponsor: Beta Phi Mu International Library and Information Science Honor Society
101 Louis Shores Building, Florida State University
Tallahassee, FL 32306-2048

Harold Simmons Foundation Grants   2601

The foundation awards grants to Texas nonprofit organizations, with emphasis on social services, religion, health, the arts, and youth. Grants also support community programs and projects, child development, and adult basic education/literacy programs. The foundation also supports international development and relief efforts in Third World countries. Grants are awarded for general operating support, annual campaigns, capital campaigns, building construction/renovation, continuing support, seed money, and program development. Application forms are not required, and there are no deadlines.

Requirements: Dallas, TX, nonprofits are eligible.
Restrictions: Grants are not awarded to individuals or for endowment funds or loans.
Geographic Focus: Texas
Samples: Children's Medical Ctr Dallas (TX)—to establish a hospital within the medical center devoted to the needs of child cardiac patients, $5 million (2005); U of Texas Southwestern Medical Ctr (Dallas)—to enhance the Simmons Comprehensive Cancer Center, including supporting the work of its newly recruited director, $15.4 million (2005).
Contact: Lisa Simmons Epstein, President, (972) 233-2134
Sponsor: Harold Simmons Foundation
5430 LBJ Freeway, Suite 1700
Dallas, TX 75240-2697

Harold Whitworth Pierce Charitable Trust Grants   2602

The trust awards grants to Boston-area nonprofit organizations in its areas of interest, including green and public spaces—projects that support community gardens, parks, and other natural areas, and projects that enhance space for recreation; capital projects—projects that reduce the operating costs for an institution and projects that improve and/or restore the physical heritage of Boston; and research—environmental research and/or for making the results of such research available for public policy. Occasionally other areas are supported. Types of support include program development, seed money, capital projects, and operating grants on occasion. The first step in the application process is a phone call to determine whether the project is eligible. The listed application deadlines are for concept letters (two-page maximum). Full proposals are by invitation.
Requirements: 501(c)3 tax-exempt organizations in the Boston, MA, area are eligible.
Restrictions: Grants do not support scholarships, individuals, fund-raising events or training, films, videos, travel, or advocacy.
Geographic Focus: Massachusetts
Amount of Grant: $1000-$100,000
Date(s) Application is Due: Mar 1; Sep 15
Contact: Elizabeth Nichols, c/o Nichols and Pratt, (617) 523-8368; fax: (617) 523-8949; email: piercetrust@nichols-pratt.com
Sponsor: Harold Whitworth Pierce Charitable Trust
50 Congress Street, Suite 832
Boston, MA 02109

Harriet and Leon Pomerance Fellowship   2603

One fellowship will be awarded annually for the following academic year to enable an individual to work on a scholarly project relating to Aegean Bronze Age archaeology. Preference will be given to a candidate whose project requires travel to the Mediterranean. Official application materials are available upon request from the institute in the spring of each year.
Requirements: Applicants must be residents of the United States or Canada.
Restrictions: Previous Harriet Pomerance fellows are ineligible.
Amount of Grant: $4000
Date(s) Application is Due: Nov 1
Contact: Elizabeth Gilgan, AIA Program Administrator, (617) 353-8705; fax: (617) 353-6550; email: aia@aia.bu.edu or egilgan@aia.bu.edu
Internet: http://www.archaeological.org/webinfo.php?page=10008
Sponsor: Archaeological Institute of America
656 Beacon Street
Boston, MA 02215-2006

Harriet Hale Woolley Scholarships   2604

The Fondation des Etats-Unis annually offers four or five Woolley Scholarships for graduate study in art and music in Paris; specifically, grants are for persons studying painting, printmaking and sculpture and for instrumentalists. Grantees live at the Fondation des Etats-Unis and are expected to participate in the cultural and social activities. They should have a keen interest not only in furthering their own careers, but in contributing to international friendship by showing an active interest in meeting persons of other nationalities. A flyer documenting all materials needed to support an application is available upon request. Preference is given to mature students who have already done graduate study.
Requirements: Candidates must be single American citizens; have proficiency in the French language; must be between 21 and 29 years of age; and have a bachelor's degree in arts, fine arts, or music or show evidence of equivalent training. Applicants' projects must include enrollment at recognized schools or instruction with private teachers. Letters of acceptance must accompany application. Although recipients must remain in Paris for the full academic year, they are free to travel during the Christmas and Easter vacation periods.
Restrictions: Grants do not cover transportation costs to and from Europe. Grants are not given for research in art history, or musicology, nor for students of dance or the theater.
Amount of Grant: $8500 stipend paid in five installments in francs from October 1 through June 29 for room, board, transportation in Paris, and miscellaneous expenses
Date(s) Application is Due: Mar 1
Contact: Director, (+33) 1 53 80 68 87; fax: (+33) 1 53 80 68 99; email: elizabeth.askren-brie@feusa.org
Internet: http://www.feusa.org/index.php?page=harriet
Sponsor: Fondation des Etats-Unis
15 blvd Jourdan
Paris, Cedex 14 75690 France

Harriman Institute Postdoctoral Fellowships   2605

The goals of the institute are the preparation of graduate students for scholarly and professional careers in Russian, Soviet, and post-Soviet studies; the promotion of advanced research on Russia, the CIS, and the post-Soviet reality; and the dissemination of information, analyses, and opinions derived from institute-sponsored research and activities. Postdoctoral fellowships allow scholars to spend a period of time at the institute without any obligations other than to revise their dissertations for publication as a book. Fellows are welcome to participate in the intellectual life of the institute. All fellows will have access to the university's libraries, the institute's reading room, archives, lectures, and discussions.

Requirements: Applicants must have completed dissertations within the preceding five years. To apply, a scholar must write the assistant director a letter outlining how she or he will spend time revising a dissertation for publication. Besides forwarding a curriculum vita and substantial portions of the dissertation, applicants should ask three scholars who are familiar with their work to forward letters of evaluation and recommendation.
Amount of Grant: $10,000 maximum per semester
Date(s) Application is Due: Jan 2
Contact: Barbara Singleton, Administrative Aide, Postdoctoral Fellowships, Harriman Institute, (212) 854-6219 or (212) 854-4623; email: harriman@columbia.edu
Internet: http://sipa.columbia.edu/REGIONAL/HI/post-doc.html
Sponsor: Columbia University
420 W 118th Street
New York, NY 10027

Harris and Eliza Kempner Fund Grants   2606

The foundation provides grants primarily in the Galveston, TX, area to qualifying organizations in the broad areas of the arts, historic preservation, community development, education, health, and human services. The foundation gives preference to requests for seed money, operating funds, small capital needs, and special projects partnering with other funding sources. Application information is available online.
Requirements: Grants are made primarily to Texas residents.
Restrictions: Funding is not available for: fund-raising benefits; direct mail solicitations; grants to individuals; and grants to non-U.S.A based organizations.
Geographic Focus: Texas
Samples: U of Texas Medical Branch (Galveston, TX)—for its fundraising campaign to support programs to improve access to care; healtcare education; and research on infectious diseases, vaccines, and chronic diseases, $1 million (2003).
Date(s) Application is Due: Mar 15; Oct 15
Contact: Grants Manager, (409) 762-1603; fax: (409) 762-5435; email: information@kempnerfund.org
Internet: http://www.kempnerfund.org/app/programs.html
Sponsor: Harris and Eliza Kempner Fund
2201 Market Street, Suite 601
Galveston, TX 77550-1529

Harris Graduate School of Public Policy Studies Research Development Grants   2607

The Harris School supports this program for social science scholars interested in food assistance research. Awards will be made to scholars who propose research including, but not limited to interactions between food assistance programs and other welfare programs with respect to participation, administration, budget exposure, and the role of food assistance as a personal and fiscal stabilizer; the effects of the macroeconomic environment on the need for food assistance, level of participation, and food assistance program costs; and the well-being of current and former food assistance recipients. Other topics related to welfare reform and macroeconomic interactions with food assistance will be considered. This program is designed to encourage experienced researchers in other areas to start projects in the area of food assistance; research on food assistance using innovative approaches and research methods; smaller, start-up projects with the potential to make a significant contribution to food assistance research; and younger and junior scholars to develop research agendas in the area of food assistance. Funding may include compensation for the principal investigator's time; research assistance; travel; and purchase of data, computers, or other research related items. Application guidelines are available online.
Requirements: Applicants must hold a PhD degree.
Amount of Grant: $40,000 maximum; $20,000-$25,000 typically
Date(s) Application is Due: May 1
Contact: U.S.DA Research Development Grants Program, (773) 702-2028; email: spopa@uchicago.edu.
Internet: http://harrisschool.uchicago.edu/Research/funding.asp
Sponsor: University of Chicago
1155 E 60th Street
Chicago, IL 60637

Harry A. and Margaret D. Towsley Foundation Grants   2608

The foundation awards grants to Michigan organizations in its areas of interest, including environment, medical and preschool education, social services, continuing education, and research in the health sciences. Types of support include annual campaigns, building construction and renovation, capital campaigns, continuing support, employee matching gifts, endowments, general operating support, matching/challenge support, professorships, program development, research, and seed grants. There are no application forms; submit a letter of inquiry between January and the listed application deadline.
Restrictions: Grants are not awarded to individuals or for travel, scholarships, fellowships, conferences, books, publications, films, tapes, audio-visual or communication media, or loans.

Geographic Focus: Michigan
Samples: U of Michigan at Ann Arbor, School of Music ((MI)—to construct a theater and performing-arts center, $1.5 million (2005); Nature Conservancy, Michigan Chapter (Lansing, MI)—for the Northern Great Lakes Forest Project, which will protect 271,000 acres across eight counties in Michigan's Upper Peninsula and link more than 2.5 million acres of protected federal, state, and natural areas, $2 million (2005).
Date(s) Application is Due: Mar 31
Contact: Lynn White, President, (989) 837-1100; fax: (989) 837-3240
Sponsor: Harry A. and Margaret D. Towsley Foundation
P.O. Box 349, 140 Ashman Street
Midland, MI 48640

Harry and Laura Jacques Scholarship   2609

This endowed award is given annually to a Canadian graduate student in a field of study related to the petroleum industry. Period of time covered by the award is four months. Both academic standing and financial need will be considered in making the selection of a recipient.
Requirements: The award is open to qualified graduates of any recognized university who are registered in or admissible to a program leading to a master's or doctoral degree at the University of Calgary.
Amount of Grant: $C3500
Date(s) Application is Due: Feb 1
Contact: Connie Busch, Faculty of Graduate Studies, (403) 220-5690; email: cbusch@ucalgary.ca
Internet: http://www.ucalgary.ca/pubs/calendar/current/what/Awards/03GradAwards.htm#66
Sponsor: University of Calgary
2500 University Dr NW, Earth Sciences Building, Room 720
Calgary, AB T2N 1N4 Canada

Harry Frank Guggenheim Foundation Dissertation Fellowships   2610

Ten or more dissertation fellowships are awarded each year to individuals who will complete their dissertations within the award year. These fellowships are designed to help doctoral candidates finish writing their dissertations rather than to support dissertation research. Awards are made only for projects clearly relevant to human dominance, aggression, and violence, but are not necessarily restricted to studies of humans. The foundation will consider proposals from any discipline that will further the foundation's intellectual and practical objectives including, but not limited to, anthropology, biology, history, psychology, sociology, and political science.
Requirements: Applicants may be citizens of any country and studying at colleges or universities in any country.
Amount of Grant: $15,000
Date(s) Application is Due: Feb 1
Contact: Dissertation Fellowships, (212) 644-4907; fax: (212) 644-5110
Internet: http://www.hfg.org/df/guidelines.htm
Sponsor: Harry Frank Guggenheim Foundation
527 Madison Avenue
New York, NY 10022-4304

Harry Frank Guggenheim Foundation Research Grants   2611

The foundation invites proposals for natural and social science research that may increase understanding of the causes, manifestations, and control of human violence and aggression. Grants are awarded for one-year support of postdoctoral research. Issues of particular interest are violence and aggression in children, the relationship between drugs and violence, violence in families, techniques for controlling aggression and violence, and political and religious violence. Types of support include salaries, employee benefits, research assistants, supplies, equipment, fieldwork, and technical assistance. Funded programs average two years in duration.
Requirements: The grants are intended to fund individual investigators, not institutional programs.
Restrictions: Funds are not for studies directed toward requirements for the doctoral degree, institutional overhead, nor for conference support.
Amount of Grant: $15,000-$30,000 for one or two years
Date(s) Application is Due: Aug 1
Contact: Program Officer, (646) 428-0971; fax: (646) 428-0981; email: hfgacf@aol.com
Internet: http://www.hfg.org/rg/guidelines.htm
Sponsor: Harry Frank Guggenheim Foundation
25 W 53rd Street
New York, NY 10019-5401

Harry Ransom Humanities Research Center Research Fellowships   2612

The center offers a number of fellowships on an international basis to scholars for the purpose of conducting research based on the center's collections. The center is the principal rare book and manuscript library of the University of Texas at Austin and noted for its collections of British, American, and French literary materials, with major strengths in photography, music, film, and theater arts. Most fellowships are for one month, but a limited number of fellowships may be awarded for periods of residency up to four months to scholars with longer-term projects. Recipients are expected to be in continuous residence for the duration of their awards. The program also awards domestic and foreign travel stipends to scholars with research projects that require less than one month's research at the center (stipends may not be combined with other Ransom Center fellowships); and dissertation fellowships, which are designated for graduate students who are working on doctoral dissertations. Guidelines are available online.

Requirements: Applicants must have received the PhD and be engaged in postdoctoral or equivalent research based on the center's collections. A two-page proposal, one-page curriculum vita, and two letters of recommendation are required for application.

Restrictions: Individuals who have received the HRHRC Research Fellowship are ineligible to reapply for one year. Fellowships do not support work on the dissertation.

Samples: Brooke Kroeger (New York, NY)—for a biography of Fannie Hurst, $5000.; S.P. Rossenbaum (Toronto, Canada)—for a literary history of Georgian Bloomsbury, $1500.; Sally Cline (Cambridge, England)—for a biography of Radclyffe Hall.

Amount of Grant: $3000 per month; $1000 domestic and $1500 foreign travel stipends; $1200 dissertation fellowships

Date(s) Application is Due: Feb 1

Contact: Grants and Research, (512) 471-8944; fax: (512) 471-9646; email: fellows@hrc.utexas.edu

Internet: http://www.hrc.utexas.edu/about/fellowships/application

Sponsor: University of Texas at Austin
P.O. Box 7219
Austin, TX 78713-7219

Harry S. Truman Fellowship in National Security Science and Engineering   2613

The fellowship enables recipients to pursue independent research of their own choosing that supports the national security mission of Sandia National Laboratories. The appointee is expected to foster creativity and to stimulate exploration of forefront science and technology and high-risk, potentially high-value research and development. candidates are expected to have solved a major scientific or engineering problem in their thesis work or will have provided a new approach or insight to a major problem, as evidenced by a recognized impact in their field. The fellowship carries a stipend and a competitive benefits package that includes health, vision, and dental insurance; savings and income plan (401K); and paid holidays and vacation. Fellows may work at either of the laboratories' principal locations, in Albuquerque, NM, and Livermore, CA. The fellowship is a three-year appointment normally commencing on October 1, although exceptions may be made to accommodate special circumstances.

Requirements: Eligibility criteria include: U.S. citizenship, the ability to obtain a DOE Q clearance; research in areas of interest to national security; the candidate must have been awarded a PhD within the past three years at the time of application or will have completed all PhD requirements by commencement of appointment; and, candidates seeking their first national laboratory appointment.

Amount of Grant: $94,200

Date(s) Application is Due: Dec 5

Contact: Yolanda Moreno, (505) 284-2106

Internet: http://www.sandia.gov/employment/special-prog/truman/index.html

Sponsor: Sandia National Laboratories
P.O. Box 5800, MS-1351
Albuquerque, NM 87185-1351

Harry S. Truman Library Institute Book Awards   2614

One prize is awarded biennially in even-numbered years for the best book written during the previous two years, dealing primarily and substantially with some aspect of U.S. history between April 12, 1945, and January 20, 1953, or with the public career of Harry S. Truman. The institute will accept applications after October 1 for funding after January 1. Information concerning the current status of the program may be obtained by writing to the secretary at the address shown.

Requirements: Three copies of each book entered must be submitted to the secretary of the institute.

Amount of Grant: $1000

Contact: Grants Administrator, (816) 833-0425; fax: (816) 833-2715; email: lisa.sullivan@nara.gov

Internet: http://www.trumanlibrary.org/grants.htm

Sponsor: Harry S. Truman Library Institute
500 W U.S. Hwy 24
Independence, MO 64050-1798

Harry S. Truman Library Institute Dissertation-Year Fellowships   2615

To support the writing of doctoral dissertations on the public career of Harry S. Truman and on the history of the Truman administration, the institute is offering two one-year fellowships to individuals who have completed their dissertation research and are ready to begin writing. Persons receiving dissertation-year fellowships will not be required to come to the Truman Library but will be expected to furnish the library with a copy of their dissertation (or any publications resulting therefrom). Persons interested in applying may obtain application forms from the committee on research and education.

Requirements: Applicants should have substantially completed their research and be prepared to devote full time to writing their dissertation. Preference will be given to projects based on extensive research at the Truman Library.

Amount of Grant: $16,000

Date(s) Application is Due: Feb 1

Contact: Grants Administrator, (816) 833-0425; fax: (816) 833-2715; email: lisa.sullivan@nara.gov

Internet: http://www.trumanlibrary.org/grants.htm#diss

Sponsor: Harry S. Truman Library Institute
500 W U.S. Hwy 24
Independence, MO 64050-1798

Harry S. Truman Library Institute Research Grants   2616

The institute offers research grants to graduate students and postdoctoral scholars to enable them to spend one to three weeks conducting research at the library. Applicants must be working on a project pertaining to the public career of Harry S. Truman or to some facet of the history of the Truman administration. Awards are meant to cover round-trip air transportation between the applicant's home and Independence, MO, and a modest sum to cover living expenses while working at the library. Major factors in the awards decision will be the amount of pertinent material available at the Truman Library and the extent to which the particular topic has already been worked on by other scholars. Applications must be submitted on forms available from the office.

Restrictions: Individuals may receive no more than two grants in this category in any one five-year period.

Amount of Grant: $2500 maximum

Date(s) Application is Due: Apr 1; Oct 1

Contact: Grants Administrator, (816) 833-0425; fax: (816) 833-2715; email: lisa.sullivan@nara.gov

Internet: http://www.trumanlibrary.org/grants.htm#ress

Sponsor: Harry S. Truman Library Institute
500 W U.S. Hwy 24
Independence, MO 64050-1798

Harry S. Truman Library Institute Scholar's Awards   2617

In even-numbered years, the institute will make one award, either to an established scholar or to a scholar about to embark on his or her career, who is engaged in a study of either the public career of Harry S. Truman or some aspect of the history of the Truman administration or of the United States during that administration. The scholar's work must be based on major utilization of the resources of the Truman Library and must be designed to result in the publication of a book-length manuscript. One copy of such book (and/or any other publications resulting from work done under this award) shall be deposited by the author with the Harry S. Truman Library. Proposals should be submitted by December 15 of the odd-numbered year preceding the year of the award. The amount of the award shall be based on a proposed budget submitted by the applicant but will not exceed one-half of the applicant's academic year salary (or a comparable figure in the case of persons not affiliated with an institution of higher learning).

Requirements: The application process is in two stages. Applicants should submit an informal proposal, indicating what work has already been done toward completion of the project, what remains to be done, and the specific resources in the Truman Library that will be utilized. Applicants will be advised of the outcome of this preliminary screening no later than February 1. Candidates selected to continue in the second phase of the process will receive forms (including a budget form) to be submitted by February 15. Applicants in this second stage will be notified of the committee's decision no later than April 15.

Amount of Grant: $30,000 maximum

Date(s) Application is Due: Dec 15

Contact: Grants Administrator, (816) 833-0425; fax: (816) 833-2715; email: lisa.sullivan@nara.gov

Internet: http://www.trumanlibrary.org/grants.htm#scholars

Sponsor: Harry S. Truman Library Institute
500 W U.S. Hwy 24
Independence, MO 64050-1798

Harry Shwachman Clinical Investigator Award   2618

This three-year award provides the opportunity for clinically trained physicians to develop into independent biomedical research investigators who have active involvement in cystic fibrosis-related areas. It is also intended to facilitate the transition from postdoctoral training to a career in academic medicine. Applications are due the first Wednesday in September.

Requirements: U.S. citizenship or permanent resident status is required.

Amount of Grant: $70,000 maximum per year plus $15,000 for supplies

Date(s) Application is Due: Sep 1

Contact: Office of Grants Management, (301) 951-4422; fax: (301) 951-6378; email: grants@cff.org

Internet: http://www.cff.org/research/cystic_fibrosis_foundation_grants/research_grants

Sponsor: Cystic Fibrosis Foundation
6931 Arlington Road, 2nd Fl
Bethesda, MD 20814

Hartford Aging and Health Program Awards   2619

The foundation's two principal programs are Health Care Cost and Quality and Aging and Health. The Health Care Cost and Quality program supports the community health management initiative, health care quality measures, and reducing inappropriate health care services. The Aging and Health program supports strengthening physicians' knowledge of geriatrics, reducing medication problems of the elderly, and demonstrating integrated financing and service delivery for comprehensive geriatric services. Types of support include operating budgets, continuing support, projects/programs, research, publications, and conferences and seminars. The foundation also welcomes inquiries regarding projects that may not fit these specific interests but would further its broad goal of improving health care in America. Types of support include general operations, continuing support, projects/programs, research, publications, and conferences and seminars. Applications are accepted at any time and are reviewed four times each year when the board meets.

Requirements: U.S. health, education, and social service organizations may apply.

Restrictions: Requests will be denied for general research or for projects lasting more than three years.

Samples: New York Academy of Medicine (NY)—to help start rotational field-training components, with a focus on working with older adults, in programs that offer master's degrees in social work, $5.1 million over four years (2005); U of California (San Diego, CA) and U of Pittsburgh (PA)—to create new centers of excellence in geriatric medicine in order to increase the number of physician faculty members in the field of geriatrics, $450,000 each over three years (2005); American Federation for Aging Research (New York, NY)—for the Medical Student Summer Research Training in Aging Program, $1.9 million over five years (2005).

Contact: Corinne Rieder, Executive Director, (212) 832-7788; fax: (212) 593-4913; email: mail@jhartfound.com

Internet: http://www.jhartfound.org

Sponsor: John A. Hartford Foundation

55 E 59th Street, 16th Fl

New York, NY 10022-1178

Hartford Geriatrics Health Outcomes Research Scholars Program   2620

This program supports physician-scientists committed to improving the health care of older adults during the critical transition from junior faculty to independent researcher. Outcomes research is defined as the study of functional status, impairments, perceptions, social opportunities, and health services utilization that are influenced by disease, injury, treatment, or health policy, including etiology, diagnosis, and intervention. Research must be focused on older adults and may address clinical strategies and effectiveness, innovative outcomes measures, and quality of life. Scholars must have a research sponsor who should have a strong background in training and outcomes research, as well as expertise and interest in the outcomes research that is being proposed.

Requirements: The applicant must be a U.S. citizen or permanent resident; be sponsored at a U.S. allopathic or osteopathic institution by a full-time faculty member; have an MD or DO degree; hold a full-time faculty appointment at the level of assistant professor for no longer than two years at the time the grant becomes effective; demonstrate the motivation and ability to devote at least 75 percent of his or her time to conduct outcomes research related to improving the care of older adults with the goal of devoting his/her research career to this area; be a geriatrician or geriatric psychiatrist who has completed all of the requirements to be eligible to sit for a Certificate of Added Qualifications by the time the award commences (July 1), or a general internist (or family physician, neurologist, or subspecialist of internal medicine) who is pursuing a career devoted to aging research and who has completed all the requirements to be eligible to sit for board certification in his/her discipline; and have at least two years of training with one year of clinical experience at an accredited U.S. institution, and one year of research training at a U.S. academic institution.

Amount of Grant: $75,000 per year for two years

Date(s) Application is Due: Jan 15

Contact: Funding Administrator, (800) 247-8779 or (212) 308-1414; fax: (212) 832-8646; email: sreinthaler@americangeriatrics.org

Internet: http://www.healthinaging.org/research

Sponsor: American Geriatrics Society Foundation for Health in Aging

350 Fifth Avenue, Suite 801

New York, NY 10118

Harvard Postdoctoral Fellowships on Education   2621

The graduate school invites applications from scholars for a two-year residential postdoctoral fellowship. Applications are encouraged from a broad range of scholarly traditions and backgrounds whose research links theory, practice, and policy in education. Special consideration will be given to applicants committed to careers in university research and teaching related to the improvement of education, particularly K-12 education. Fellows will teach one course during each of the two academic years and will give an annual lecture to the graduate school community about their research. Fellows receive an annual stipend plus a small housing supplement, are eligible for a travel supplement for participation in professional conferences, and receive additional support for course development or research. Applicants should send a letter of application, including a proposal for a two-year plan of research and teaching that identifies a member of the HGSE faculty who might serve as a mentor and provides a rationale for the selection of the mentor. Contact the office for additional information.

Requirements: Applicants must be within two years of receiving the PhD or EdD. Their degree must be completed by July and cannot have been awarded prior to May 2003.

Amount of Grant: $40,000 annual stipend; $1000 possible travel supplement; $3000 maximum for course development or research support

Date(s) Application is Due: Feb 11

Contact: Kenje Ogata, (617) 496-2805; email: kenje_ogata@gse.harvard.edu

Internet: http://www.gse.harvard.edu/~finaid/fellowship.html

Sponsor: Harvard University

122 Longfellow Hall

Cambridge, MA 02138

Harvard Study of New Scholars Postdoctoral Fellowship   2622

The program will evaluate the satisfaction of junior faculty at the colleges and universities where they are currently employed. Researchers will survey all full-time tenure-track faculty members at 12 pilot sites to learn how they view specific institutional policies, structural and cultural barriers, work climate, ability to balance professional and personal lives, current job satisfaction, and estimated likelihood of success in achieving tenure or contract renewal. The purposes of the program are to enable the academy to attract the best and brightest scholars and teachers; and to increase the recruitment, retention,

success, and satisfaction of all new scholars, with special attention to women and faculty of color. Applications are invited for one- or two-year postdoctoral fellowships. Candidates should send a letter of application including: evidence of a scholarly interest in higher education; two examples of academic writing; three references; and a three- to five-page statement that describes the fellow's intended research.

Requirements: Applicants must have a doctoral degree awarded between May 2001 and June 2003 and a demonstrable, research-based interest in faculty work life or American higher education.

Amount of Grant: $40,000 annual stipend plus $500 maximum travel supplement

Date(s) Application is Due: Mar 15

Contact: Graduate School of Education, email: newscholars@gse.harvard.edu

Internet: http://www.gse.harvard.edu/~newscholars

Sponsor: Harvard University

6 Appian Wy, Gutman Library

Cambridge, MA 02138

Harvard University Administrative Fellowships   2623

The program seeks to attract candidates, especially those from underrepresented ethnic minority groups and those committed to addressing the underrepresentation of ethnic minorities in university administration, to administrative careers in higher education. The fellowship offers a 12-month management experience complemented by a professional development program. Fellows from business, government, education, and the professions will be selected. Each participant receives a salary from Harvard University and is placed in a full-time work assignment in Harvard schools or central administration departments for the duration of the program. All costs for the required educational seminar will be covered by Harvard University. Participants must secure and cover the cost of their housing. Guidelines and application are available online.

Requirements: Applicants must have a bachelors degree and at least five years of work experience. The program seeks applicants with degree and backgrounds in the fields of athletics/sports management, information technology, alumni affairs and development, finance, human resources, and library science and management.

Date(s) Application is Due: Feb 10

Contact: Teresa Malonzo, Program Coordinator, (617) 495-8919; email: teresa_malonzo@harvard.edu

Internet: http://www.oap.harvard.edu/afp

Sponsor: Harvard University

1350 Massachusetts Avenue

Cambridge, MA 02138

Harvard University Faculty Fellowships in Ethics   2624

The Center for Ethics and the Professions encourages teaching and research about ethical issues in public life and the professions. Applications are welcomed from scholars and teachers who wish to develop their ability to address questions of moral choice in such areas as business, education, government, law, and medicine. Application forms are information are available online.

Requirements: Applicants should hold a doctorate in philosophy, political theory, theology, or related disciplines; or a professional degree in business, education, public policy, law, or medicine. Preference will be given to applicants at an early stage of their careers, normally no more than 10 years from their terminal degree.

Amount of Grant: $40,000 maximum stipend

Date(s) Application is Due: Nov 1

Contact: Center for Ethics and the Professions, (617) 495-1336; fax: (617) 496-6104; email: ethics@harvard.edu

Internet: http://ethics.harvard.edu/fellowships.php

Sponsor: Harvard University

79 John F. Kennedy Street

Cambridge, MA 02138

Harvard University Graduate School of Education Fellowships and Awards   2625

Harvard's graduate school of education (HGSE) offers several merit-based fellowships and awards. Presidential Fellowships, awarded to entering doctoral students, provide full tuition and fees for four years of study and an annual stipend for the first three years of study. Recipients must be full-time and are determined by the admissions committee based on the strength of the individual's application for admission. Entering Awards are awarded to first-year doctoral students who are either highly promising future professors or school leaders. These awards are for one year only and provide full tuition and fees plus a stipend. Roy E. Larsen Awards are awarded to first-year doctoral students who exhibit superior achievement and outstanding research potential. These awards are for one year only and provide full tuition and fees plus a stipend. Conant Fellowships support the professional growth of outstanding teachers and administrators in the Boston and Cambridge public schools. To be eligible, applicants must be teachers or administrators with professional status, under contract with either the Boston or the Cambridge public schools, and admitted to a degree program at HGSE. These fellowships cover the cost of full tuition are pro-rated for part-time study. Awards for doctoral Conant Fellowships carry the possibility of a one-year renewal.

Amount of Grant: $7000 entering award; $7000 stipend Roy E. Larsen Award

Date(s) Application is Due: Feb 4

Contact: Fellowship Office, (617) 496-2805; fax: (617) 496-0840; email: kenje_ogata@harvard.edu

Internet: http://www.gse.harvard.edu/~finaid/fellowship.html#merit

Sponsor: Harvard University

108 Longfellow Hall, 13 Appian Way

Cambridge, MA 02138

Harvard Women and Public Policy Postdoctoral Fellowship 2626

The program will award a postdoctoral residential fellowship to an outstanding scholar in security affairs from the United States and elsewhere working to promote basic research in the broad area of international security with a particular focus on gender, prevention, peace building, and sustainable security. The fellow will hold his/her fellowship jointly with the International Security Program of the Belfer Center for Science and International Affairs. The fellow will be expected to participate actively in the weekly ISP seminar and in the intellectual life of the Women and Public Policy Program; take an active organizational role in the monthly Boston Consortium meetings on Gender Security and Human Rights and devote some portion of his/her time to collaborative endeavors; and complete a book, monograph, scholarly article, or other significant publication during the period of residence. Residential fellowships also provide for individual health insurance, institutional support, and some research-related travel stipends.

Requirements: The fellow must hold a PhD by September.

Amount of Grant: $30,000 stipend

Date(s) Application is Due: Jan 31

Contact: Theresa Lund, John F. Kennedy School of Government , (617) 496-6609; fax: (617) 496-6154; email: theresa_lund@harvard.edu

Internet: http://www.ksg.harvard.edu/wappp/research/fellowship.html

Sponsor: Harvard University

79 John F. Kennedy Street

Cambridge, MA 02138

Harvey Prizes 2627

Two prizes are awarded annually to outstanding personalities whose achievements in one of the following fields have served as a source of inspiration to many others: science and technology, human health, and outstanding contribution to peace in the Mideast. Recipients are invited to Israel as guests of the Technion and to spend time at the institute teaching their subjects. Those interested in nominating candidates should contact the society for deadline date and guidelines. Inquiries may also be directed to Ruth Eckstein in Israel at 011-972-4-8292863, email: andree@tx.technionac.il.

Requirements: Nominations may be made by members of the board of governors of Technion and by request from the prize committee. Persons making nominations are requested not to inform the proposed candidates of the nomination.

Restrictions: Candidates may not nominate themselves.

Amount of Grant: $35,000 each

Date(s) Application is Due: Oct 1

Contact: Miriam Golzman, Coordinator, Office of Vice President for Development and Harvey Prize, Haifa 32000, Israel; New York office: (212) 407-6300; fax: (212) 753-2925; email: info@ats.org

Internet: http://www.admin.technion.ac.il/harvey/Philosophy.html

Sponsor: American Technion Society—Israel Institute of Technology

55 E 59th Street

New York, NY 10022

Hastings Center International Biomedical Ethics Research Program for International Scholars 2628

This program was inaugurated in response to the growing interest in comparative bioethics and to the number of foreign scholars and medical professionals interested in the work of the center. The center carries out an active research program on timely and crucial subjects, working in a variety of fields—law, medicine, science, philosophy, and religion, among others. The center is presently studying ethical issues related to abortion, neonatal care, surrogate motherhood, organ transplantation, occupational health, health professional-patient relationships, allocation of resources, animal welfare, aging, AIDS, care of the dying and termination of treatment, chronic illness, artificial reproduction, genetic screening, rehabilitation medicine, and justice in health care delivery. International scholars stay at the center for four to six weeks, dividing their time between their own research and the ongoing activities at the center. There is a modest stipend for foreign citizens who require financial assistance. Applications should be submitted four to six months in advance of proposed stay. Applications are accepted at any time.

Requirements: The program is designed for advanced scholars and practitioners who have made or will make a significant contribution to bioethics in their own countries.

Amount of Grant: $500-$2000

Contact: Director of Education, (845) 424-4040 ext 203; fax: (845) 424-4545; email: visitors@thehastingscenter.org

Internet: http://www.thehastingscenter.org/visitors/visitorsint.asp

Sponsor: Hastings Center

21 Malcolm Gordon Road

Garrison, NY 10524-5555

Hastings Center International Visiting Scholars Program 2629

This program responds to the growing interest in comparative bioethics and to the number of foreign scholars and medical professionals interested in the center's work—examining critical ethical issues in medicine, the life sciences, and the professions. The program is designed for advanced scholars and practitioners who have made or will make significant contributions to bioethics in their own countries. International scholars usually stay two weeks to six weeks, dividing their time between their own research and ongoing activities at the center. There is a modest stipend for foreign citizens who require financial assistance. Applications are due four to six months in advance of proposed stay.

Amount of Grant: $500-$1500

Contact: Daniel Callahan, Director, International Programs, (845) 424-4040 ext 222; fax: (845) 424-4545; email: callahand@thehastingscenter.org

Internet: http://www.thehastingscenter.org/visitors/visitorsint.asp

Sponsor: Hastings Center

Rte 9D, 21 Malcolm Gordon Road

Garrison, NY 10524-5555

Hawaii Community Foundation Health Education and Research Grants 2630

The program provides support from three distinct funds: Tobacco Prevention & Control Trust Fund; the Leahi Fund Research Fund; and the Medical Research Program. Funding will be provided for medical education and research in the fields of: cancer; heart disease; lung disease and research. Priority is given to projects that: demonstrate a foreseeable benefit to the people of Hawaii; support new investigators in Hawaii; and support collaborative efforts. Proposal submission information is available online.

Requirements: To be eligible for consideration an organization must be a 501(c)3 organization or a unit of government. The Principal Investigator must be based in Hawaii and conducting the research in Hawaii.

Geographic Focus: Hawaii

Amount of Grant: $25,000-$50,000

Date(s) Application is Due: Feb 27; Jul 17; Aug 14; Sep 10

Contact: Christel Wuerfel, Philanthropic Services Assistant; (808) 537-6333 or (888) 731-3863; email: cwuerfel@hcf-hawaii.org or jsmith@hcf-hawaii.org

Internet: http://www.hawaiicommunityfoundation.org/index.php?id=71&categoryID=23

Sponsor: Hawaii Community Foundation

1164 Bishop Street, Suite 800

Honolulu, HI 96813

Haystack Mountain School of Crafts Summer Scholarships 2631

Scholarships are awarded for research and study at the school in blacksmithing (as related to craftwork), graphics, ceramics, weaving, jewelry, glass, or wood. Candidates must be at least 18 years of age, of any nationality, and capable of doing professional work. Six sessions are held each summer lasting for two to three weeks each. Obtain guidelines before applying. Tuition plus partial room and board are provided.

Date(s) Application is Due: Mar 25

Contact: Director, Scholarship Program, (207) 348-2306; fax: (207) 348-2307; email: haystack@haystack-mtn.org

Internet: http://www.haystack-mtn.org/workshops_summer_financial_assistance.php

Sponsor: Haystack Mountain School of Crafts

P.O. Box 518

Deer Isle, ME 04627-0518

Hazel Hatcher International Home Economics Award 2632

The international award will be granted biennially (even-numbered years) to assist the applicant in returning to the United States for the purpose of updating professional information through attending a conference, seminar, or national professional meeting.

Requirements: Applicant must hold a degree in home economics or one of its specialized areas from a U.S. college/university and have had at least five years of professional service outside of the United States since receiving the degree.

Amount of Grant: $1500

Date(s) Application is Due: Feb 1

Contact: Fellowships Administrator, National Office, (304) 368-0612; email: rickards@mountain.net

Internet: http://www.ianr.unl.edu/phiu/webdoc3.htm

Sponsor: Phi Upsilon Omicron

P.O. Box 329

Fairmont, WV 26555

HCR Manor Care Foundation Grants 2633

Grants are awarded to communities nationwide where Manor Care Health Services has operations. The foundation has two grantmaking interests: organizations involved in research of diseases and disorders affecting the elderly and community programs that offer support and outreach services to seniors. The foundation concentrates on research organizations that treat and look for cures for illnesses such as Alzheimer's disease and stroke. Grants also are made to community organizations that offer support and information services to seniors. Initial contact should be made through a letter of inquiry (five pages or less). Letters are reviewed on an ongoing basis. If the foundation is interested, a formal grant proposal will be invited, with application deadlines as indicated.

Requirements: Nonprofits must have 501(c)3 status, be located in a state where Manor Care Health Services operates, and have at least one other source of support for the project.

Restrictions: The foundation does not support fundraising events, advertising, individuals, for-profit organizations, building or capital campaigns, political purposes or campaigns, endowment funds, multiple year grants, overhead fees, or general purpose/operating grants.

Geographic Focus: Arizona; California; Colorado; Connecticut; Delaware; Florida; Georgia; Illinois; Indiana; Iowa; Kansas; Kentucky; Maryland; Michigan; Minnesota; Missouri; Nevada; New Jersey; New Mexico; North Carolina; North Dakota; Ohio; Oklahoma; Pennsylvania; South Carolina; South Dakota; Tennessee; Texas; Utah; Virginia; Washington; West Virginia; Wisconsin

Samples: Family Service of Champaign County (IL)—for the Senior Resource Center, which helps senior citizens maintain their independence, $10,000 (2005); Myers Research Institute (Beachwood, OH)—to develop and adopt a computer-based activities system

for long-term care residents suffering from dementia who reside in a skilled-nursing unit and two assisted-living facilities, $30,000 (2005); Coda Alliance (San Jose, CA)—to conduct six Respecting Choices workshops, which provide training and education about end-of-life care, $26,000 (2005); National Hospice Foundation (Alexandria, VA)—for an effort to increase access to hospice and palliative care among underserved populations, $56,192 (2005).
Amount of Grant: $8300-$35,000 average
Date(s) Application is Due: Apr 1; Oct 1
Contact: Executive Director, (419) 252-5989; fax: (419) 252-5521; email: foundation@hcr-manorcare.com
Internet: http://www.hcr-manorcare.org
Sponsor: HCR Manor Care Foundation
P.O. Box 10086, 333 N Summit Street
Toledo, OH 43699-0086

Health Care Financing Research, Demonstrations, and Evaluations Grants   2634
Project grants (including cooperative agreements) are awarded to support analyses, experiments, demonstrations, and pilot projects in efforts to resolve major health care financing issues or to develop innovative methods for the administration of Medicare and Medicaid. HCFA has identified priority areas for the current fiscal year, including access and quality of care, coordinated care systems, provider payment, health care systems reform and financing, program evaluation and analysis, service delivery systems, and subacute and long-term care. Awardees are required to share in the cost of projects; normally, the minimum cost-sharing requirement is 5 percent of total project costs. Projects are generally funded on a 12-month basis, with support beyond the first year contingent upon acceptable evidence of satisfactory progress, continuing program relevance, and availability of funds. Applications may be submitted at any time.
Requirements: Grants or cooperative agreements may be made to private or public and nonprofit or for-profit agencies/organizations.
Amount of Grant: $25,000-$1 million; $235,000 average
Contact: Judy Norris, (410)786-5130; email: jnorris1@cms.hhs.gov
Internet: http://www.cms.hhs.gov/contracts
Sponsor: Health Care Financing Administration
7500 Security Boulevard
Baltimore, MD 21244-1850

Health Foundation of Southern Florida Responsive Grants   2635
The foundation awards Responsive Grants through two grant cycles per year. With exceptions, the foundation focuses on providing one to three year grants that do not exceed $300,000 annually. The majority of grants are funded in the $50,000 to $150,000 range over one or two years. Funding is provided in four categories: Project Planning; Health Services; Organizational Capacity Building and Health System/Health Policy Development.
Requirements: Though the foundation welcomes proposal applications anytime, the applications are reviewed on a semi-annual basis. Applicant organizations must be tax-exempt nonprofit under section 501(c)3 of the Internal Revenue Code or a local or state governmental agency. The project must serve exclusively the residents of Broward, Miami-Dade and/or Monroe counties. Initially, a preliminary proposal is required (see the sponsor's website for specific details). If approved, the sponsor will then invite a full proposal. Download the sponsor's grant guide from the website.
Restrictions: The foundation does not fund: Biomedical research or other research that will not impact local residents within the immediate future (1-3 years) or that does not have a direct application to implementing a community-driven health intervention; Capital campaigns of over $1 million (versus grants toward specific health-related equipment or the 'build out' of a specific health-focused space); Secondary and tertiary services (versus preventive and primary medical, oral and behavioral health care services); Health promotion and/or health care with a high per capita cost (this figure will vary depending upon the type of intervention, but over $1,000 per person/year cost may be a rule of thumb); Service expansion or new projects without viable sustainability (unable to reach sustainability without the foundation's resources within a four-year period).
Geographic Focus: Florida
Amount of Grant: $300,000 maximum
Date(s) Application is Due: Mar 13; Apr 24
Contact: Eliane Morales; (305) 374-7200; fax: (305) 374-7003; email: emorales@hfsf.org
Internet: http://www.hfsf.org/ORIGHTML/responsive.html
Sponsor: Health Foundation of Southern Florida
2 South Biscayne Boulevard
Miami, FL 33131

Health Management Scholarships and Grants for Minorities   2636
This financial assistance program provides scholarships and grants to ethnic minority students for graduate study in health care administration/management. Annual tuition scholarships and grants are made to outstanding students in health services management and related programs. Grants are provided to cover the costs of books, tutorial assistance, and other nontuition educational aids. Students are eligible to receive a combination of awards during the same time period.
Requirements: To be eligible, applicants must be members of an ethnic minority group; must be U.S. citizens or have permanent resident status (birth certificate, green card, or current passport); they must be college juniors, seniors or graduate students; they must be health care administration/management majors; they must have a strong academic background (2.5 GPA on a 4.0 scale); and they must have strong extracurricular and community service activities.

Amount of Grant: $2500 scholarship, $500 grant
Contact: Program Contact, (800) 233-0996 or (312) 422-2680; fax: (312) 422-4566
Internet: http://www.diversityconnection.org/diversityconnection_app/career-center/internships/Standard-Internship-page.jsp?fll=S4
Sponsor: Institute for Diversity in Health Management
1 N Franklin, 30th Fl
Chicago, IL 60606

Healthcare Foundation of New Jersey Grants   2637
The Foundation's funding priorities are the vulnerable populations of the greater Newark New Jersey community; the emergent health needs of serving at-risk individuals and families in the MetroWest Jewish community; and clinical research/medical education initiatives that significantly and directly impact these populations. The Foundation seeks grant proposals that promise innovation and change or a significant enhancement of services. Proposals are accepted on a rolling basis throughout the calendar year. Proposal submission instructions are available online.
Requirements: Nonprofit organizations in the Greater Newark area are eligible to apply. The Foundation strongly suggests that application documents be submitted electronically.
Restrictions: Grants are made only to private nonprofit organizations that have tax-exempt status under Section 501(c)3 of the Internal Revenue Code and that are not private foundations. The Foundation does not make grants to individuals or government agencies. The Foundation does not typically fund the following: organizations outside of Essex, Morris or Union County, New Jersey; programs not related to health care; direct support of an individual's healthcare needs; fundraising events or endowment campaigns; advertising campaigns; lobbying; or scholarships.
Geographic Focus: New Jersey
Contact: Program Contact, (973) 921-1210; fax: (973) 921-1274; email: info@hfnj.org
Internet: http://www.hfnj.org
Sponsor: Healthcare Foundation of New Jersey
60 East Willow Street, 2nd Floor
Millburn, NJ 07041

Heart Rhythm Society Fellowships   2638
Fellowships and research grants are available for projects pertaining to cardiac pacing and/or cardiac electrophysiology. All awards are made for one year, renewable in competition. Also available are traveling fellowships, for a one- to three-month duration, to enable a researcher to observe research at other institutions or attend seminars and conferences. One traveling Canadian fellowship is offered; the applicant must be a Canadian citizen or must train at a site in Canada. Annual deadline dates may vary; contact program staff for exact dates.
Requirements: Fellowships are open to applicants who have a doctor of medicine, philosophy, osteopathy, science, or equivalent training or experience in the field of biomedical engineering. Applicants must be citizens or permanent residents of a North American country and include cardiac pacing or cardiac electrophysiology as a central part of his/her long-range career goals.
Amount of Grant: $40,000 for fellowships, $6000 for traveling fellowships, $6000 for traveling Canadian fellowship
Date(s) Application is Due: Sep 15
Contact: Stephanie Mascetta, (508) 647-0100 ext 3046; fax: (508) 647-0124; email: fellowship@HRSonline.org
Internet: http://www.hrsonline.org/society_in_action/fellowships/default.asp
Sponsor: Heart Rhythm Society
Six Strathmore Road
Natick, MA 01760-2499

Hebrew University Lady Davis Trust Fellowships   2639
The trust offers fellowships to visiting professors, postdoctoral researchers, and doctoral students at the Hebrew University of Jerusalem and at the Technion, Israel Institute of Technology in Haifa. Distinguished scholars spend from three months to one year at one of these institutions developing research networks, teaching and enjoying the world-class academic environment.
Requirements: Scholars of any age, in any field, and from any region may apply.
Contact: Coordinator of Graduate Programs, Office of Academic Affairs, (800) 404-8622 or(212) 472-2288; fax: (212) 517-4548; email: admission@roth.mscc.huji.ac.il
Internet: http://www.huji.ac.il/unew/main.html
Sponsor: Hebrew University of Jerusalem
Givat Ram, Jerusalem 91904
Israel

Hedco Foundation Grants   2640
The foundation awards grants domestically and internationally to nonprofit organizations in its areas of interest, including education at all levels; health care, including hospitals and health organizations; and organizations serving children and youth. Types of support include building construction/renovation and equipment acquisition. Past grants have supported biomedical research, but research grants generally are not awarded. There are no application deadlines or forms. An audited financial statement and IRS tax-exemption letter must accompany all proposals.
Requirements: 501(c)3 nonprofit organizations are eligible.
Restrictions: Grants do not support individuals or requests for general operating support, publications, special projects, or loans.
Geographic Focus: California

Samples: U of Oregon, College of Education (Eugene, OR)—to construct an education complex, $10 million (2004).
Amount of Grant: $5000-$75,000 average
Contact: Grants Administrator, (925) 242-0257
Sponsor: Hedco Foundation
P.O. Box 1980
San Ramon, CA 94583

HEI Grants   2641
The institute awards research grants for scientific information about the health effects of motor vehicle emissions on the environment and on human health.
Amount of Grant: $70,000-$300,000
Contact: Daniel Greenbaum, President, (617) 886-9330 ext 311; fax: (617) 886-9335; email: dgreenbaum@healtheffects.org
Internet: http://www.healtheffects.org/funding.htm
Sponsor: Health Effects Institute
120 Second Avenue
Boston, MA 02129-4533

Heineman Foundation for Research, Education, Charitable and Scientific   2642
   Purposes
The purpose of the Heineman Foundation is to provide seed money to start-up projects and new projects within existing organizations for a maximum of three to five years. Preference will be given to organizations that we have not previously funded. The average range of our donations is $20,000.00 to $50,000.00, per annum. An organization must have 501(c)3 status and upload copies of corresponding IRS documents to the online application form in order for the application to be considered. The Foundation's general areas of interest are the following (in no particular order): programs that enable economically challenged women to enter and remain in the workplace; on site day care centers for women in the workplace; job training programs for women; language and leadership skills for women; environmental research that will help prevent, reduce and/or eliminate water degradation; music as education and preserver of culture; research into prevention of and treatment for childhood illnesses; programs that enable youth to think, create and communicate effectively; and programs that support and promote high achievement in music, science, and literature. Applications/proposals must be submitted online no later than September 1st.
Amount of Grant: $20,000-$50,000 range
Date(s) Application is Due: Sep 1
Contact: Simon Rose, President; c/o Brown Brothers Harriman Trust; (212) 493-8000; email: info@heinemanfoundation.org
Internet: http://www.heinemanfoundation.org/guidelines
Sponsor: Heineman Foundation
140 Broadway
New York, NY 10005-1108

Heinz Latin American Social and Public Policy Fellowships   2643
The Center for Latin American Studies awards the fellowships, made possible from the Howard Heinz Endowment, Alcoa Foundation, and the Mine Safety Appliances Company Charitable Foundation, to assist scholars interested in various dimensions of social and public policy to pursue advanced graduate studies in a department or professional school of the University of Pittsburgh. Fellowship support is for a maximum of two years. Applicants must be accepted into one of the following departments or schools to be eligible: anthropology, economics, history, linguistics, political science, religious studies, sociology, business, education, law, public health, public and international affairs, and social work. Recipients are appointed for fall and spring terms and in addition to pursuing their studies as full-time students, work on collaborative projects with Latin American social and public policy specialists. Applications are due in late January.
Requirements: Individuals of all nationalities who are enrolled as full-time graduate students at the University of Pittsburgh and are pursuing master's or PhD degrees in some aspect of social and/or public policy are eligible.
Amount of Grant: $1325 per month stipend, full tuition for fall and spring terms, medical insurance, and $400 book allowance per year
Contact: Luis Van Fossen Bravo, Center for Latin American Studies, (412) 648-7393; fax: (412) 648-2199; email: bravo@ucis.pitt.edu
Internet: http://www.ucis.pitt.edu/clas/students/fellowships_scholarships.html
Sponsor: University of Pittsburgh
University of Pittsburgh, 4E Wesley W Posvar Hall
Pittsburgh, PA 15260

Heiser Leprosy and Tuberculosis Research Postdoctoral Fellowships   2644
The program fosters research on leprosy and tuberculosis by supporting the research training of a limited number of young biomedical scientists interested in the field. The program is designed to encourage national and international cooperation and research interchange in the study of the diseases. A maximum of two years of support will be provided to successful applicants at stipend levels adjusted according to U.S. government guidelines for geographical location at which the training is to take place. There are no application forms. Application may be made either by the individual seeking training or by the head of a laboratory engaged in research related to leprosy. In the latter case, the head of the laboratory must name the prospective fellow in the application.
Requirements: Biomedical scientists with either the PhD or MD degree are eligible for these fellowships. Applicants should be at the beginning or in an early phase of their

postdoctoral training and interested in obtaining research training in a basic science field directly related to the study of leprosy. There are no citizenship requirements.
Amount of Grant: $35,000 maximum stipend per year; $2000 training allowance; $1500 maximum for health insurance
Date(s) Application is Due: Mar 1
Contact: Len McNally, Program Director, (212) 686-0010 ext 556; email: lm@nyct-cfi.org
Sponsor: Heiser Program for Research in Leprosy and Tuberculosis
c/o The New York Community Trust , 909 Third Avenue
New York, NY 10022

Heiser Leprosy Research Grants   2645
The grants provide support to laboratories involved in leprosy research training. Proposals of high scientific caliber related to leprosy will be considered. Seed grants may be requested for new projects or facilities that show promise of support from other sources within one year. Leprosy postdoctoral research fellowships and tuberculosis postdoctoral research fellowships also are awarded. There are no application forms. Applicants should submit proposals and budgets along with curriculum vita and relevant bibliographies of participating scientists.
Requirements: Senior investigators who are experienced in leprosy and tuberculosis research and associated with laboratories providing training opportunities in these areas are eligible.
Restrictions: Grants do not support clinical trials, salaries of personnel, or more than 10 percent of institutional overhead.
Amount of Grant: $35,000 maximum
Date(s) Application is Due: Mar 1
Contact: Len McNally, Program Director, (212) 686-0010 ext 556; email: lm@nyct-cfi.org
Sponsor: Heiser Program for Research in Leprosy and Tuberculosis
c/o The New York Community Trust, 909 Third Avenue
New York, NY 10022

Helen Bader Foundation Grants   2646
The foundation supports innovative pilot projects and programs that are research based, have strong evaluation components, and involve community collaboration. The foundation's primary focus areas are Alzheimer's disease and dementia, early childhood development in Israel, education, economic development, and Jewish life and learning. Modest grants are awarded to support neighborhood renewal and youth development in Milwaukee's Sankofa area. Grants will be awarded primarily in the Delaware River Valley and in particular Milwaukee, WI, with the exception of the Alzheimer's disease and dementia program, which accepts applications from organizations nationwide. International grants are made for projects conducted in Israel. The foundation often funds multiple-year projects but rarely for a period of more than three years. Preliminary proposals must be received by the listed application deadline dates. If the foundation is interested in exploring the grant request further, a member of the program staff will contact the applicant to arrange a meeting or site visit. Applications are available on the Web site.
Requirements: Grants are awarded for projects consistent with one or more of the Helen Bader Foundation's program areas. Grants are given only to U.S. organizations which are tax exempt under Section 501(c)3 of the Internal Revenue Code or to government entities. Grants will only be approved for foreign entities which meet specific charitable status requirements.
Restrictions: The Foundation does not provide direct support for individuals, such as individual scholarships.
Geographic Focus: Wisconsin
Samples: SHILUV-Integration of Children with Special Needs (Jerusalem, Israel)—for capital improvements to create a new integrated day care and kindergarten program that will include children with Downs Syndrome, $30,000 (2005); Local Initiatives Support Corp (WI)—to host its annual Milwaukee Awards for Neighborhood Development Innovation, $3000 (2005); Milwaukee Christian Ctr (WI)—for its Challenge Program, which offers educational support, life-skills development, and other programming for south side youth, $10,000 (2005); Campus Kitchens Project (WI)—for efforts at Marquette U to provide food to Milwaukeeans in need and offer foodservice skills training to underemployed adults, $6000 three-year grant (2005).
Amount of Grant: $10,000-$100,000 average
Date(s) Application is Due: Jun 25; Feb 25
Contact: Program Contact, (414) 224-6464; fax: (414) 224-1441
Internet: http://www.hbf.org
Sponsor: Helen Bader Foundation
233 North Water Street, 4th Floor
Milwaukee, WI 53202

Helen Brach Foundation Grants   2647
The foundation operates to prevent cruelty to animals or to children; for religious, charitable, scientific, literary, and education purposes; and for public safety testing through support of Midwest 501(c)3 tax-exempt organizations carrying out programs and activities in these areas. Brach provides grants nationally and has wide-ranging interests. For example, it supports homeless and women's emergency shelters, teen pregnancy prevention programs, parenting education, summer school for disadvantaged children, job training for welfare mothers, orphanages, and scholarships for economically disadvantaged students. Types of support include annual campaigns, building construction/renovation, equipment, general operating support, publications, research, and special projects. The foundation ordinarily does not make multiyear grants or

commitments. Applicants are required to complete in full a brief application form, which may be obtained from the office. The board of directors gives final consideration to all applications received in a given year at the board's meeting, which is usually held in March.

Requirements: Although 501(c)3 nonprofits from across the nation are eligible, giving is primarily made in California, Massachusetts, Ohio, Pennsylvania, and South Carolina.

Restrictions: Grants are not made to individuals or to organizations outside the United States. Typically grants are not made in excess of 10 percent of a group's operating budget, which automatically excludes start-up grants.

Samples: Zoological Society of Florida (Miami, FL)—to purchase dinosaur and bird artifacts to be used in an educational program called Are Birds Really Dinosaurs?, $20,000.

Amount of Grant: $10,000-$25,000 average
Date(s) Application is Due: Dec 31
Contact: John P. Hagnell, Associate Director; (312) 372-4417; fax: (312) 372-0290
Sponsor: Helen V. Brach Foundation
55 W. Wacker Drive, Suite 701
Chicago, IL 60601-1609

Helen Carr Minority Fellowships 2648

These fellowships provide support to African American engineering students or faculty at Hampton University, Morgan State University, Howard University, North Carolina A&T State University, Prairie View A&M University, Tennessee State University, Tuskegee University, and Southern University who have not yet completed their doctorates and wish to do so. The fellowships are awarded for one year and may be renewed. The amount awarded is dependent on the educational requirements and financial needs of the recipients.

Requirements: Eligible to apply are African American students or engineering faculty who are interested in earning doctoral degrees and returning to one of the historically black engineering colleges to teach.

Amount of Grant: $10,000 maximum per year
Date(s) Application is Due: May 1
Contact: Historically Black Engineering Colleges Committee, (202) 331-3525 or 331-3500; fax: (202) 265-8504; email: projects@asee.org
Internet: http://www.asee.org/fellowships/hbecc.cfm
Sponsor: American Society for Engineering Education
1818 N Street NW, Suite 600
Washington, DC 20036

Helen Hay Whitney Foundation Postdoctoral Research Fellowships 2649

The program provides three-year fellowships for beginning postdoctoral training in basic biomedical research. The program is intended for individuals planning careers in biological or medical research to receive financial support of sufficient duration to help further their professional careers. It is expected that fellowship training will be obtained in an academic setting. The foundation will not consider applicants who plan tenure of the fellowship in the laboratory in which they have already received extensive predoctoral or postdoctoral training; the aim of the fellowship is to broaden postdoctoral training and experience. Applications are available online.

Requirements: Residents of the United States, Canada, and Mexico who are in the final stages of obtaining the MD, PhD, or equivalent degree and are seeking beginning postdoctoral training in basic biomedical research are eligible.

Restrictions: Applications from established scientists or advanced fellows will not be considered. Applicants who have already had one year of postdoctoral laboratory training by the listed deadline date will not be considered.

Samples: Dr. Sean Burgess, Harvard U, Department of Biochemistry and Molecular Biology. Dr. David Wassarman, National Institutes of Health, Department of Cell Biology and Metabolism.

Amount of Grant: $41,000 for first year, $42,500 for second year, and $44,000 for the third year, plus $2500 to the fellow's laboratory to help defray research expenses

Date(s) Application is Due: Jul 15
Contact: Grants Administrator, (845) 639-6799; fax: (845) 639-6798 or (646) 304-7133; email: hhwf@earthlink.net
Internet: http://www.hhwf.org/HTMLSrc/ResearchFellowships.html
Sponsor: Helen Hay Whitney Foundation
20 Squadron Boulevard, Suite 630
New York, NY 10956

Helen Kellogg Institute for International Studies Residential Fellowships 2650

The institute seeks fellows of high scholarly accomplishment and promise, at both senior and junior levels, whose work and presence will contribute creatively to its major research themes. The institute's research reflects commitments to democracy, development, and social justice, and focuses on five major themes: Democratization and the Quality of Democracy, Paths to Development, Religion and the Catholic Church, Social Movements and Organized Civil Society, and Public Policies for Social Justice. Stipends vary according to seniority; fellows from abroad may receive one direct round-trip economy airfare from the institute. Applications and further information are available on the Web site.

Requirements: Applications are welcome from candidates from any country holding the PhD or equivalent degree in any discipline of the social sciences or history. Candidates will be evaluated individually but joint projects will be considered. All fellows from abroad are requested to apply for travel grants to aid with transportation expenses in the United States.

Samples: Nora Segura, Universidad Nacional de Colombia (Bogata)—for Work, Family, and Gender in Displaced Populations in Colombia.

Date(s) Application is Due: Nov 1
Contact: Sharon Schierling, Program Coordinator, (219) 631-6580; fax: (219) 631-6717; email: Sharon.K.Schierling.1@nd.edu
Internet: http://www.nd.edu/~kellogg/guest.html
Sponsor: University of Notre Dame
216 Hesburgh Ctr
Notre Dame, IN 46556-5677

Helen M. Robinson Dissertation Research Award 2651

This award may be made annually to support doctoral students who are IRA members at the early stages of their dissertation research. The award will be made only in those years when the committee concludes that there are applications of very high quality.

Requirements: Doctoral student applicant should submit an abstract, not to exceed 1000 words, of an approved dissertation proposal; a brief statement, not to exceed 500 words, indicating how the grant will be used to support the research; a copy of the full proposal with a cover sheet, including signatures of approval by a supervisory committee; and a letter from the student's major advisor that includes a brief rationale as to why the grant should be awarded to this applicant.

Samples: Sheila Benson, University of Iowa (Iowa City, IA)—for Engaging students in literacy: A multiliteracies orientation to literacy instruction, $1500 (2005); Nancy Charron University of Massachusetts (Lowell, MA)—for Talk at the terminal: The effect of an Internet pen pal program on fourth graders' written language development and attitudes toward writing, $1500 (2004).

Amount of Grant: $1500
Date(s) Application is Due: Jan 15
Contact: Program Contact, (302) 731-1600 ext 423; fax: (302) 731-1057; email: research@reading.org
Internet: http://www.reading.org/association/awards/research_robinson.html
Sponsor: International Reading Association
P.O. Box 8139, 800 Barksdale Road
Newark, DE 19714-8139

Helen M. Woodruff AIA and American Academy in Rome Archaeology Fellowship 2652

A pre- or postdoctoral fellowship for study of archaeology and classical studies has been established by the institute at the American Academy in Rome. This fellowship, combined with other funds from the American Academy in Rome, will support a Rome Prize Fellowship, which will be open to citizens or permanent residents of the United States. For application forms, write to the American Academy in Rome, 7 E 60th Street, New York, NY 10022; (212) 751-7200; or visit the Web site (http://www.aarome.org).

Restrictions: Current officers and governing board members of the Institute at the American Academy in Rome are not eligible.

Contact: Program Contact, (617) 353-9361; fax: (617) 353-6550; email: aia@aia.bu.edu
Internet: http://www.archaeological.org/webinfo.php?page=10008
Sponsor: Archaeological Institute of America
656 Beacon Street
Boston, MA 02215-2010

Helen McWilliam Memorial Scholarship 2653

This scholarship is open to full-time graduate students at the University of Calgary interested in pursuing careers in school psychology. Recipients will be selected on the basis of academic excellence, skill in the handling of children referred for testing or guidance, and evidence of interest and ability in the area of school psychology. The award will be based on grades and observations by faculty members of academic and practical work. Annual deadline dates may vary; contact program staff for exact dates.

Requirements: Applicants must be full-time graduate students enrolled in the school psychology program of the Department of Educational Psychology at the master's or doctoral level. Applications should be made at the end of the first academic year.

Amount of Grant: $C2000
Contact: Program Contact, Faculty of Graduate Studies, (403) 220-5690; email: cbusch@acs.ucalgary.ca or ucawards@ucalgary.ca
Internet: http://www.grad.ucalgary.ca/funding/internal_scholarships/lev_4/mcwilliam.htm
Sponsor: University of Calgary
2500 University Dr NW, Earth Sciences Bldg
Calgary, AB T2N 1N4 Canada

Hendrick Foundation for Children Grants 2654

The foundation supports programs designed to provide health, medical, social welfare, and educational services to benefit children with illness, disease, injury, pain, disability, incapacity, or other disadvantages; and improve quality of life for children with life-threatening or chronic injuries, illness, and disabilities. It is also contemplated that the Foundation will honor grants to other charitable and educational organizations sponsoring programs which are community oriented and focus on improving the quality of life of children with life-threatening or chronic injuries, illnesses and disabilities. All requests shall be given careful and thorough consideration by the Hendrick Foundation for Children Grant Committee. Grant awards will be determined on the merits of the organization, its objectives and mission. All grant requests should be submitted in writing on the organization's letterhead

Amount of Grant: $3,500 to $250,000 range
Contact: Charles V. Ricks, President; (704) 568-5550

Internet: http://www.thehendrickfoundation.org/grants.html
Sponsor: Hendrick Foundation for Children
P.O. Box 240070
Charlotte, NC 28224-0070

Henry Dreyfus Teacher-Scholar Awards Program 2655

The program is intended to support and encourage young scholars who demonstrate excellence in both research and teaching. The award is based on accomplishments in scholarly research with undergraduates, as well as a compelling commitment to teaching with respect to primarily undergraduate institutions. Institutions may make only one nomination annually. The Program provides an unrestricted research grant of $60,000.
Requirements: Institutions in U.S. states, districts, and territories that grant a bachelor's or higher degree in chemistry, chemical engineering, or biochemistry may submit nominations. Nominees must hold a full-time tenure-track academic appointment, be between the fourth and twelfth years of their independent academic careers, and engage in research and teaching primarily with undergraduates.
Restrictions: Awards are not made directly to individuals.
Amount of Grant: $60,000
Date(s) Application is Due: Aug 6
Contact: Mark Cordillo, Executive Director; (212) 753-1760; fax: (212) 593-2256; email: mcardillo@dreyfus.org or admin@dreyfus.org
Internet: http://www.dreyfus.org/awards/henry_dryfus_teacher_award.shtml
Sponsor: Camille and Henry Dreyfus Foundation
555 Madison Avenue, 20th Floor
New York, NY 10022-1301

Henry Hess Award 2656

The award is given annually for an original technical paper submitted to ASME for presentation and publication during the calendar year prior to the year of award by a student or associate member who was not yet 31 years of age at the time the paper was submitted to ASME. Joint authorship is permissible provided all authors meet the requirements. The paper shall be specifically recommended for the award by a review committee or a qualified individual.
Requirements: Author must not be older than 31 at the time of submission and must be an ASME student member or associate member to be eligible to apply.
Amount of Grant: $1000 and expense supplement
Date(s) Application is Due: Mar 1
Contact: Gilda DiTullio, Manager, (212) 591-7736; fax: (212) 705-7739; email: ditulliog@asme.org
Internet: http://www.asme.org/honors/ms71/gla/hess.html
Sponsor: American Society of Mechanical Engineers
3 Park Avenue
New York, NY 10016

Henry J. Kaiser Family Foundation Grants 2657

The foundation concentrates giving in the following areas: U.S. government's role in health, health of low-income and minority groups with major emphasis on HIV/AIDS policy, reproductive health policy, and health system innovation and reform in California. The foundation also operates a major program to improve health and health care and promote social justice in South Africa. Grants are awarded for one to three years and support a range of activities, including policy analysis, applied research to define and measure public health problems, demonstration and pilot projects, and communications activities that help sharpen health care debates and improve quality of health information. Prospective applicants should submit a preliminary letter (two to three pages in length) that briefly describes the proposed project, along with an estimate of the total budget and the amount requested from the foundation. There are no deadlines and no application forms. Inquiries for projects in South Africa may be addressed to Dr. Michael Sinclair, Senior Vice President, The Henry J. Kaiser Family Foundation, 1450 G Street NW, Suite 250, Washington, DC 20005, (202) 347-5270, fax: (202) 347-5274.
Requirements: Grants in response to unsolicited proposals are made only to governmental agencies and to private organizations with IRS 501(c)3 tax-exempt status.
Restrictions: The foundation does not award grants to individuals. Support is not given to ongoing general operating expenses, indirect costs, capital campaigns, annual appeals or other fundraising events, construction, purchase or renovation of facilities, or equipment purchases.
Amount of Grant: $1000-$500,000 typically
Contact: Renee Wells, Grants Manager, (415) 854-9400; fax: (415) 854-4800; email: rwells@kff.org
Internet: http://www.kff.org
Sponsor: Henry J. Kaiser Family Foundation
2400 Sand Hill Road
Menlo Park, CA 94025

Henry J. Kaiser Family Foundation Media Fellowships 2658

The purpose of the fellowship is to help journalists and commentators do the best possible job of keeping the public informed about health issues at this critical time in the evolution of the health care system. The goal of the fellowship is to provide a highly flexible range of opportunities tailored to each fellow's interests; help journalists improve the quality of the work they do; provide time to research specific topics; and deepen participants' commitment to becoming specialists in health reporting. Fellows may want to participate in the fellowship program on a part-time basis and continue writing or reporting during their fellowship. Fellows are encouraged to do reporting based on their fellowship research and to experiment with different media or forms of writing. Most fellows work from home, or base themselves at a local academic or research institution. There is no obligation to pursue academic coursework, although that is an option open to all fellows. There is no application form. Guidelines are available online.
Requirements: Fellowships will be awarded to U.S. journalists interested in U.S. health policy issues.
Amount of Grant: $50,000 maximum for nine-month fellowship
Date(s) Application is Due: Mar 16
Contact: Penny Duckham, Executive Director, (650) 234-9220; fax: (650) 854-4800; email: pduckham@kff.org
Internet: http://www.kff.org/about/mediafellowships.cfm
Sponsor: Henry J. Kaiser Family Foundation
2400 Sand Hill Road
Menlo Park, CA 94025

Henry L. Guenther Foundation Grants 2659

The foundation awards grants to California nonprofit organizations in its areas of interest, including medical services, creating opportunities for youth, and community services. Grants support hospitals, medical research, and social services. The board meets in January and July. Contact the office for application procedures.
Requirements: California nonprofit organizations are eligible. Grants are awarded primarily in Lost Angeles.
Restrictions: Grants do not support government agencies, religious organizations for religious purposes, individuals, or debt-reduction requests.
Geographic Focus: California
Samples: Downey Community Hospital Health Services Corp (CA)—to build a multipurpose medical campus for seniors, $1 million.; White Memorial Medical Ctr Charitable Foundation (Los Angeles, CA)—to purchase imaging equipment, $500,000.; Saint Jude Medical Ctr (Fullerton, CA)—to update the cardiac catheterization laboratory, $600,000.
Amount of Grant: $10,000-$100,000 average
Date(s) Application is Due: May 31; Nov 30
Contact: W.D. Milliken, Secretary, (310) 785-0658
Sponsor: Henry L. Guenther Foundation
2029 Century Park D, Suite 4392
Los Angeles, CA 90067

Henry Luce Foundation Special Projects in Higher Education Grants 2660

This award is a responsive grant made to colleges and universities for special projects which fall outside the context of the Foundation's funding categories. Higher education is an underlying theme for most of the Foundation's programs, through support for research, training, and the creation of new resources. The current primary focus of the program has reflected the Foundation's commitment to strengthen America's capacity for international understanding. No special forms are required for the application process. An initial letter of inquiry should be addressed to the appropriate program contact. Requests are accepted at any time. Awards are determined by the board of directors, which meets three times a year.
Restrictions: The Foundation does not support health care or medical projects and does not fund development assistance work overseas. It does not normally assist journalism, media and film projects or the performing arts.
Samples: Columbia U (New York, NY)—to establish a professorship in modern Tibetan studies; to help develop the Ctr for Integrated Tibetan Studies, which will focus on both classical and modern Tibet; and to promote exchanges with Tibetan scholars, $3 million (2004).
Contact: Special Projects in Higher Education, (212) 489-7700; fax: (212) 581-9541; email: hlf1@hluce.org
Internet: http://www.hluce.org/4heddefm.html
Sponsor: Henry Luce Foundation
111 West 50th Street, Room 4601
New York, NY 10020

Henry Luce Foundation Theology Program Grants 2661

The foundation's theology program emphasizes theological education and reflects a commitment to support the role of religion in daily life. Six areas of funding illustrate the breadth of the program: seminary education, leadership, publications, religion and the arts, ecumenical programs, and special initiatives. The program supports innovative research and publication by full-time seminary faculty. Requests may be submitted at any time. The board meets three times each year to consider requests.
Restrictions: The foundation does not support health care or medical projects and does not fund development assistance work overseas. It does not normally assist journalism, media and film projects, or the performing arts.
Contact: Theology Program, (212) 489-7700; fax: (212) 581-9541; email: hlf@hluce.org
Internet: http://www.hluce.org/3theofm.html
Sponsor: Henry Luce Foundation
111 W 50th Street, Suite 4601
New York, NY 10020

Henry Luce Foundation/ACLS Dissertation Fellowships in American Art 2662

The fellowships, which are made possible by funding from the Henry Luce Foundation, support graduate students in any stage of Ph.D. dissertation research or writing. To be eligible, a student must be a Ph.D. candidate in a department of art history, and the candidate's dissertation must be focused on a topic in the history of the visual arts of the United States. Students whose degrees will be granted by another department may be eligible if their principal dissertation advisor is in a department of art history and their dissertation topic is object-oriented. The fellowships are for a one-year nonrenewable term beginning in the summer.

Requirements: Applicants must have completed all requirements for the Ph.D. except the dissertation before beginning the one-year tenure.

Restrictions: Fellowship funds may not be used to defray tuition costs. Students who received grants of $5,000 or more from the Henry Luce Program previously administered through university art departments are not eligible, nor are students preparing theses for the master of fine arts degree.

Amount of Grant: $25,000
Date(s) Application is Due: Nov 12
Contact: Nicole Stahlmann, Director of Fellowship Programs; (212) 697-1505, ext. 134; fax: (212) 949-8058; email: nstahlmann@acls.org
Internet: http://www.acls.org/grants/Default.aspx?id=516
Sponsor: American Council of Learned Societies
633 Third Avenue
New York, NY 10017-6795

Henry Luce Foundation/ACLS Grants to Individuals in East and   **2663**
    Southeast Asian Archaeology and Early History
Research fellowships and training grants will be awarded for study of the peoples and cultures of early East and Southeast Asia. For the purposes of this program, East and Southeast Asia refers to Brunei, Burma/Myanmar, Cambodia, China, East Timor, Hong Kong, Indonesia, Japan, Korea, Laos, Macau, Malaysia, Mongolia, Philippines, Singapore, Taiwan, Thailand, and Vietnam. Proposals may cover prehistoric or historical periods, but must focus on research or training that involves excavations and/or excavated materials. Comparative projects and those that build scholarly networks are especially encouraged. Application information is available online.

Requirements: Applicants must be residents of the United States or Canada and have their primary professional affiliation at an institution in this region. Applicants must hold an advanced degree in a relevant discipline such as archaeology, anthropology, art history, literature, or history, must be nearing completion of such a degree, or must possess commensurate experience in scholarship, archives, museums, or field sites. Linguistic ability sufficient to conduct the work proposed is required.

Restrictions: Applications will not be accepted by fax or email.

Date(s) Application is Due: Nov 14
Contact: Nicole Stahlmann, Director of Fellowship Programs; (212) 697-1505, ext. 134; fax: (212) 949-8058; email: nstahlmann@acls.org
Internet: http://www.acls.org/grants/Default.aspx?id=518
Sponsor: American Council of Learned Societies
633 Third Avenue
New York, NY 10017-6795

Henry Luce Fund in American Art Grants   **2664**
The program consists of three areas: the Luce Fund in American Art, which places emphasis on scholarly exhibitions and publications; responsive grants, generally for institutional enhancement; and dissertation fellowships, aimed at the next generation of art historians specializing in this field. Grants are made three times a year. Letters of request may be submitted at any time but should be received by March 1 for the annual summer review of approximately 20 proposals. It is helpful if applicant speaks with appropriate staff members in advance of sending a letter or proposal.

Requirements: American museums may apply by invitation only.

Samples: Corcoran Gallery of Art (Washington, DC)—for a scholarly catalog of the pre-1945 American paintings and sculpture collection, $150,000 three-year grant (2003); Lontar Foundation (Jakarta, Indonesia)—for an Indonesian literature translation and publication program, $180,000 three-year grant (2003); HealthCare Chaplaincy (New York, NY)—to expand the Chaplaincy's pastoral research department and library, $150,000 three-year grant (2003).

Date(s) Application is Due: Mar 1; Jun 15
Contact: Helene Redell, Program Director, (212) 489-7700; fax: (212) 581-9541; email: hlf@hluce.org
Internet: http://www.hluce.org/4lfaafm.html
Sponsor: Henry Luce Foundation
111 W 50th Street, Room 4601
New York, NY 10020

Henry Luce Scholars Program   **2665**
This program is one aspect of the foundation's continuing efforts to improve American understanding of Asia. Each year, 15 young Americans of outstanding promise are selected by the foundation and sent to East and Southeast Asia to undertake internships in their chosen fields under the guidance of leading Asian professionals. A distinguishing feature of the program is that it is directed toward those who are not, and do not plan to become, specialists in Asian affairs. The object is to develop among a broad cross-section of future American leaders an understanding of the nature of Asian societies and cultures. At the program's core are individual internships and job placements arranged for the scholars on the basis of their individual career interests, experience, and training. Though it is hoped that the work of the scholars will be useful to their Asian hosts, a scholar's most significant contribution may simply be an honest attempt to assimilate the wealth of Asian experience and in the process to develop a much broader world perspective. The program relies on a network of 65 colleges and universities for its nominations; individual applications submitted directly to the foundation cannot be considered. Applications must reach the foundation from the nominating institutions by the first Monday of December. A guidelines brochure is available upon request.

Requirements: Candidates must be American citizens who have received at least a bachelors degree and are no more than 29 years old on September 1 of the year they enter the program. Institutions may submit two nominations.

Restrictions: Individual applications submitted directly to the foundation cannot be considered. Those who already have significant experience in Asia or Asian studies are not eligible to apply.

Samples: Asia Foundation (San Francisco, CA)—to administer the Asian portion of the scholars program, $1.4 million.

Contact: Luce Scholars Program, (212) 489-7700; fax: (212) 581-9541; email: hlf@hluce.org
Internet: http://www.hluce.org/3scholfm.html
Sponsor: Henry Luce Foundation
111 W 50th Street, Room 3710
New York, NY 10020

Henry P. Kendall Foundation Grants   **2666**
The foundation focuses almost entirely on the environment and securing its physical, biological, and aesthetic wealth for future generations. Geographic priorities for grants are New England and the Maritime Provinces of Eastern Canada and the Pacific Northwest, including Western Canada and Alaska. Current program themes include Gulf of Maine ecosystem, Northeastern landscape, Northeast climate change initiative, North country institution capacity-building, Yellowstone to Yukon conservation, watershed innovations in North America, public lands management, and special project initiatives. The foundation provides funding for general operating needs and for specific programs and initiatives. Activities funded include advocacy, public education, policy research and analysis, on-the-ground resource management experiments, and institutional development. Grants are normally made for one or two years. Unsolicited proposals and inquiries will not be reviewed.

Restrictions: Grants do not support endowments or capital fund campaigns, land acquisition, television and film projects, fellowships, basic scientific research, building construction or maintenance, equipment, debt reduction, or conferences unrelated to current foundation institutional grants. Nor does the foundation normally fund waSuite clean-ups, toxics or air/water pollution prevention or pollution monitoring initiatives, individual land trusts, or species-specific preservation efforts.

Geographic Focus: AlaskaCanada

Samples: Alaska Conservation Foundation (Anchorage, AK)—to create a strategic transition fund dedicated to strengthening and expanding efforts to build a popular majority for conservation values—terrestrial and marine—throughout Alaska over the next two decades, $100,000.; Earth Day Network (Seattle, WA)—to support preparatory organizing efforts in New England for Earth Day 2000, $35,000.; Montana Wilderness Assoc (Helena, MT)—for general support to advance the protection of public wildlands and naturally functioning ecosystems in the Montana portion of the Yellowstone to Yukon landscape, $40,000.

Amount of Grant: $20,000-$50,000 typically
Date(s) Application is Due: Feb 1; May 1; Oct 1
Contact: Jennifer Patrick, Program Officer, (617) 951-2525; fax: (617) 443-1977
Internet: http://www.kendall.org/grants/types.html
Sponsor: Henry P. Kendall Foundation
176 Federal Street
Boston, MA 02110

Henry R. Worthington Medal   **2667**
This award is given annually for eminent achievement in the field of pumping machinery. Such achievement may be, for example, in the areas of research, development, design, innovation, management, education, or literature. Any individual member, group of members, or committee may nominate candidates for this award.

Amount of Grant: $5000; $1000
Date(s) Application is Due: Feb 1
Contact: Gilda DiTullio, Manager, (212) 591-7736; fax: (212) 705-7739; email: ditulliog@asme.org
Internet: http://www.asme.org/honors/ms71/saa/worthington.html
Sponsor: American Society of Mechanical Engineers
3 Park Avenue
New York, NY 10016

Henry W. Bull Foundation Grants   **2668**
The foundation awards grants nationwide in its areas of interest, including higher education, the disabled, healthcare, Christian religion, music, and the arts. Types of support include annual campaigns, building construction/renovation, equipment acquisition, general operating support, continuing support, and research. There is no formal grant application form. Submit a simple, concise statement of needs and objectives with pertinent supportive data.

Requirements: 501(c)3 nonprofits are eligible.

Restrictions: Foreign grants are not considered.

Amount of Grant: $500-$20,000 range
Date(s) Application is Due: Apr 1; Sep 1
Contact: Janice Gibson, Vice-President, Santa Barbara Bank & Trust, (805) 899-8405; fax: (805) 884-1404
Sponsor: Henry W. Bull Foundation
P.O. Box 2340
Santa Barbara, CA 93120-2340

HERA Foundation Ovarian Cancer OSB1 Seed Grant 2669

The Foundation is a non-profit organization whose mission is to prevent ovarian cancer and improve prognosis through innovative scientific research and education. The HERA Foundation will fund two basic, translational and/or clinical research projects: one grant of $25,000 is for projects that include, but are not limited to, those that study high grade ovarian carcinoma etiology and pathogenesis, and those that explore novel approaches to early ovarian cancer detection and targeted therapy; a second grant of $20,000 is earmarked for low grade/micropapillary serous carcinoma/low malignant potential ovarian tumor research for research leading to a better understanding and/or more effective treatment. The goal is to encourage outside the box thinking and solutions and provide seed money for such projects. Chemotherapy has been in use for many years and to date we have not improved overall survival by 1%. New approaches to understanding and treating this disease are needed and are what is of interest to the Foundation.
Requirements: Applicants must hold either a Ph.D. and/or an M.D. degree, and be either a post-doctoral research fellow or clinical fellow. A previous research background in ovarian cancer is not essential, as the HERA Foundation seeks to encourage new young investigators to enter the field. However, candidates should demonstrate the relevance of their previous research background to the proposed work, and indicate a strong long-term commitment to research in the field of ovarian cancer.
Samples: Dr. Brant Wang, Johns Hopkins University (2006); Dr. Hiroyuki Yoshida, Anderson Cancer Center (2006); Dr. Martina Bazzaro, Johns Hopkins University.
Amount of Grant: $20,000-$25,000
Date(s) Application is Due: Jan 15
Contact: Sheridia Daniels, (970) 948-7360; email: sdaniel6@jhmi.edu
Internet: http://www.herafoundation.org/research.cfm
Sponsor: HERA Foundation
P.O. Box 664
Carbondale, CO 81623

Herb Society of America Research and Education Grants 2670

Grants are meant to further the knowledge and use of herbs and contribute the results of the study and research to the records of horticulture, science, literature, history, art, or economics. The research could involve such varied topics as history, medicinal research and usage from a historical perspective, fragrance, flavor, horticulture, landscaping, economic importance, chemurgy, taxonomy, and political and social influence. Grants are made for periods up to one year.
Requirements: Applicants should be graduate students with proposed programs of scientific or academic investigation of herbal plants, or persons capable of demonstrating in-depth knowledge of herbs.
Restrictions: The grant is not given for financial aid to individuals, rather for specific research on herbal projects. The grant does not provide for indirect costs or personal travel for the benefit of any individual.
Samples: For research on traditional dye plants of the Huichol Indians of Mexico. For research on molecular biological analyses of herb action.
Amount of Grant: $5000 maximum per year
Date(s) Application is Due: Jan 31
Contact: Grants Administrator, (440) 256-0514; fax: (440) 256-0541; email: herbs@herbsociety.org
Internet: http://www.herbsociety.org/research.php
Sponsor: Herb Society of America
9019 Kirtland Chardon Road
Kirtland, OH 44094

Herbert C. Brown Award for Creative Research in Synthetic Methods 2671

This award, sponsored by Purdue Borane Research Fund and the Herbert C. Brown Award Endowment, is to recognize and encourage outstanding and creative contributions to research in synthetic methods. A nominee must have accomplished outstanding and creative research that involved the discovery and development of novel and useful methods for chemical synthesis. The synthetic method that is the basis for the award may be related to any area of chemistry. The award consists of $5,000, a medallion with a presentation box, and a certificate.
Requirements: Any individual, except a member of the award committee, may submit one nomination or seconding letter for the award in any given year. The nominating documents consist of a letter of not more than 1000 words containing an evaluation of the nominee's accomplishments and a specific identification of the work to be recognized, a biographical sketch including date of birth, and a list of publications and patents authored by the nominee.
Restrictions: Self-nominations are not accepted.
Samples: Eric N. Jacobsen—award winner, $5,000 (2008); David A. Evans—award winner, $5,000 (2007); Richard Heck—award winner, $5,000 (2006).
Amount of Grant: $5,000 and travel to award meeting
Date(s) Application is Due: Nov 1
Contact: Felicia Dixon, Awards Administrator; (800) 227-5558 or (202) 872-4408; fax: (202) 776-8008; email: f_dixon@acs.org or awards@acs.org
Internet: http://portal.acs.org/portal/acs/corg/content?_nfpb=true&_pageLabel=PP_ARTICLEMAIN&node_id=1319&content_id=CTP_004530&use_sec=true&sec_url_var=region1
Sponsor: American Chemical Society
1155 Sixteenth Street, NW
Washington, DC 20036

Herbert Eugene Bolton Memorial Prize 2672

The prize is awarded for the best book in English on any significant aspect of Latin American history that is published during the imprint year previous to the year of the award. Sound scholarship, grace of style, and importance of the scholarly contribution are among the criteria for the award. Translations, anthologies of selections by several authors, reprints or reeditions of works published previously, and works not primarily historiographical in aim or content are not normally considered for an award. An honorable mention award may be made for an additional distinguished work.
Requirements: To be considered for the prize, a book must bear the imprint of the year prior to the year for which the award is made.
Amount of Grant: $1000
Contact: CLAH, (530) 752-3046; fax: (530) 752-8964; email: clah@ucdavis.edu
Internet: http://www.h-net.org/~clah/about/index.html
Sponsor: Conference on Latin American History
University of California at Davis, One Shields Avenue
Davis, CA 95616

Herbert H. and Grace A. Dow Foundation Grants 2673

The Foundation has charter goals to improve the educational, religious, economic and cultural lives of Michigan's people. Priority is given to organizations that: have clearly stated objectives, strong and purposeful management and are publicly accountable; have needs which are in areas not normally funded by governmental or public financing; are not hesitant to explore, initiate, volunteer, or execute original ideas or concepts; are willing to collaborate with other persons or organizations to give synergy to a common objective or goal; have purposes which tend to advance private enterprise and the preservation of a free, open and self-resourceful society. Application information is available online.
Requirements: Only organizations in Michigan are eligible to apply.
Restrictions: The Foundation does not make grants directly to individuals. It cannot legally support: organizations to which contributions are not tax deductible, according to Internal Revenue Service regulations; organizations that practice discrimination by race, sex, creed, age or national origin; political organizations or organizations whose purposes are to influence legislation.
Geographic Focus: Michigan
Samples: Grand Valley State U, Dorothy A. Johnson Ctr for Philanthropy and Nonprofit Leadership (Grand Rapids, MI)—for the Growing the Greater Good Endowment Campaign, with a focus on identifying and disseminating best practices of nonprofit organizations, $1 million (2005); Nature Conservancy, Michigan Chapter (Lansing, MI)—for the Northern Great Lakes Forest Project, which will protect 271,000 acres across eight counties in Michigan's Upper Peninsula and link more than 2.5 million acres of protected federal, state, and natural ares. $4.5 million (2005); Albion College (MI)—to renovate and expand its science complex, $5 million (2004).
Contact: Margaret Ann Riecker, President, (989) 631-3699; fax: (989) 631-0675; email: info@hhdowfoundation.org
Internet: http://www.hhdowfdn.org
Sponsor: Herbert H. and Grace A. Dow Foundation
1018 West Main Street
Midland, MI 48640-4292

Herbert Hoover Travel Grant Competition 2674

Travel grants are awarded to encourage scholarly use of the Herbert Hoover Presidential Library in West Branch, IA. The program is intended to promote the study of subjects of interest and concern to Herbert Hoover, Lou Henry Hoover, their associates, and other public figures as reflected in the library's collections. It is strongly suggested that the applicant contact the archival staff at (319) 643-5301 to determine whether library holdings are pertinent to the applicant's research. Three separately mailed references must be submitted.
Requirements: The program is open to graduate and postdoctoral scholars and to qualified nonacademic researchers on a competitive basis.
Amount of Grant: $500-$1500
Date(s) Application is Due: Mar 1
Contact: Patricia Hand, Manager of Programs, (800) 828-0475 or (319) 643-5327; fax: (319) 643-2391; email: pathand@hooverassociation.org or info@hooverassoc.org
Internet: http://www.hooverassociation.org/grants.html
Sponsor: Herbert Hoover Presidential Library Association
P.O. Box 696
West Branch, IA 52358

Hereditary Disease Foundation John J. Wasmuth Postdoctoral Fellowships 2675

Support is offered for research projects that contribute to identifying and understanding the basic defect of Huntington's disease. Areas of interest include trinucleotide expansions, animal models, gene therapy, neurobiology and development of the basal ganglia, cell survival and death, and intercellular signaling in striatal neurons. Awards may be renewed for a third year. To receive an application, submit a letter of intent. Letter of intent forms may be submitted electronically via the Web site.
Amount of Grants: $37,000-$56,000 stipend, plus $8,000 for supplies
Date(s) Application is Due: Feb 15; Jun 15; Oct 15
Contact: Dr. Carl D. Johnson; (212) 928-2121; fax: (212) 928-2172; email: carljohnson@hdfoundation.org
Internet: http://www.hdfoundation.org/funding/postdoct.php
Sponsor: Hereditary Disease Foundation
3960 Broadway, 6th Floor
New York, NY 10032

Hereditary Disease Foundation Lieberman Award 2676

The Hereditary Disease Foundation announces a special Lieberman Award, to catalyze innovative proposals leading to the treatment and cure of Huntington's disease. A Lieberman Award can be funded for two years for up to $75,000 per year. Areas of interest include trinucleotide expansions, animal models, gene therapy, neurobiology and development of the basal ganglia, cell survival and death, and intercellular signaling in striatal neurons.

Amount of Grant: $75,000 per year for up to 2 years
Contact: Dr. Carl D. Johnson; (212) 928-2121; fax: (212) 928-2172; email: carljohnson@hdfoundation.org
Internet: http://www.hdfoundation.org/funding/lieberman.php
Sponsor: Hereditary Disease Foundation
3960 Broadway, 6th Floor
New York, NY 10032

Hereditary Disease Foundation Research Grants 2677

The focus of the Hereditary Disease Foundation is on Huntington's disease. Support will be for research projects that contribute to identifying and understanding the basic defect in Huntington's disease. Grants are usually for one year, with the possibility of renewal for up to three years. Areas of interest include trinucleotide expansions, animal models, gene therapy, neurobiology and development of the basal ganglia, cell survival and death, and intercellular signaling in striatal neurons. In addition to its regular grants, the foundation also offers the Lieberman Award for innovative proposals accelerating the discovery of a treatment and cure of Huntington's disease. Grants that receive funding are considered seed money. If the project shows promise, it is hoped that other institutions will fund it thereafter. To obtain an application, submit a one-page letter of intent. Letter of intent forms may be submitted electronically via the Web site.

Samples: Dr. Andrew Dwork, Columbia U—for research entitled Ultrastructural Immunohistochemistry of Cortical Biopsies in HD.; Dr. Nansheng Chen, U of British Columbia—for research entitled Regulation of NMDA Receptor Function by Mutant Huntingtin in Neostriatal Neurons.; Dr. Chris Huang, Massachusetts General Hospital—for research entitled Anti-Huntingtin Aggregation Agents as Potential Therapeutics for Huntingtons Disease.; Dr. Lucius Passani, Massachusetts General Hospital—for research entitled Evaluation of HYPA/FBP-11, HYPB and HYPC as Candidates for Involvement in HD Pathogenesis.

Amount of Grant: $50,000 maximum
Date(s) Application is Due: Feb 15; Jun 15; Oct 15
Contact: Dr. Carl D. Johnson; (212) 928-2121; fax: (212) 928-2172; email: carljohnson@hdfoundation.org
Internet: http://www.hdfoundation.org/funding/grants.php
Sponsor: Hereditary Disease Foundation
3960 Broadway, 6th Floor
New York, NY 10032

Herman Kahn Fellowship 2678

One fellowship is given annually in the area of national security studies, domestic or international economics, education policy, or the study of political institutions to support PhD candidates who have completed their coursework and have only their dissertations remaining. Studies will be carried out in Indianapolis, IN, for one academic year. The fellow's time will be equally divided between dissertation work and policy research assigned by the institute. Application requirements are available upon request. Selected candidates will be interviewed in Indianapolis.

Requirements: Eligible to apply are graduate students who have completed all PhD coursework within the last five years and who can provide evidence that they have been formally admitted to candidacy for the PhD. Consideration will be given only to applicants with outstanding academic records who are strongly recommended by faculty members and have demonstrated interest in and capability for policy-oriented research.

Amount of Grant: $18,000 plus some travel expenses
Date(s) Application is Due: Apr 1
Contact: Director of Research and Programs, (202) 974-2400; fax: (202) 974-2410; email: rsvp@hudson.org
Internet: http://www.hudson.org/learn/index.cfm?fuseaction=mission_statement
Sponsor: Hudson Institute
1015 15th Street NW, 6th Fl
Washington, DC 20005

Heublein Foundation Inc Grants 2679

The foundation offers grant support in the areas of preschool through higher education, including a scholarship program for dependents of employees, health and social care, culture and art, problems of the inner cities, and education to combat alcohol abuse. A large percentage of grants within all categories is targeted at disadvantaged populations. Heublein also contributes directly to the alcoholism prevention programs of several organizations and to research by noted universities. Application deadline for scholarships is December 1.

Restrictions: Grants may not be made to individuals, religious groups, political organizations, or for endowments. The foundation principally supports programs in communities where the company has major facilities: greater Hartford, CT, area; Allen Park, MI; and Madera, Menlo Park, and Napa Valley, CA.

Geographic Focus: Connecticut
Samples: Real Art Ways (Hartford, CT)—continuing support, $25,000. United Way of the National Capital Area (Washington, DC)—for the Combined Health Appeal project, $12,500.

Date(s) Application is Due: Dec 1
Contact: Treasurer, (860) 231-5000
Sponsor: Heublein Foundation Inc
P.O. Box 778
Farmington, CT 06142-0778

HHMI Biomedical Research Grants for International Scientists in 2680
Canada and Latin America

HHMI's International Research Scholars Program provides five-year grants to support promising basic biomedical research scientists working in Canada and Latin America. Eligible epidemiology research includes that directed toward an understanding of disease distribution in populations or of associations that may suggest causality or preventive strategies.

Requirements: The supported research must be conducted in one of the following eight eligible countries: Argentina, Brazil, Canada, Chile, Mexico, Peru, Uruguay, or Venezuela. Applicants must hold a full-time appointment or have a pending full-time appointment at a nonprofit scientific research organization in any one of the eight eligible countries.

Restrictions: Clinical trials and research on health education, health care delivery, or health services are not eligible fields.

Amount of Grant: $50,000-$100,000
Date(s) Application is Due: Sep 14
Contact: Jill G. Conley, Ph.D., Program Director; (301) 215-8873; fax: (301) 215-8888; email: canlatam@hhmi.org
Internet: http://www.hhmi.org/grants/individuals/canlatam.html
Sponsor: Howard Hughes Medical Institute
4000 Jones Bridge Road
Chevy Chase, MD 20815-6789

HHMI Biomedical Research Grants for International Scientists in 2681
the Baltics, Central and Eastern Europe, Russia, and Ukraine

The Program will award five-year grants to support promising basic biomedical research scientists working in the Baltics, Central and Eastern Europe, Russia, and Ukraine. Eligible epidemiology research includes that directed toward an understanding of disease distribution in populations or of associations that may suggest causality or preventive strategies. Allowable expense categories include salaries and stipends for award recipients, graduate students, postdoctoral fellows, and technicians, equipment and supplies, travel, and publication costs. Up to 10 percent of the grant may be used to pay for indirect costs (including administrative costs) of the scientist's institution.

Requirements: The supported research must be conducted in one of the following eligible countries: Bulgaria, Croatia, Czech Republic, Estonia, Hungary, Latvia, Lithuania, Poland, Romania, Russia, Slovak Republic, Slovenia, or Ukraine.

Restrictions: Researchers may not be citizens or permanent residents of the United States. Grant funds may not be used for laboratory renovations.

Amount of Grant: $50,000-$100,000 per year up to five years
Date(s) Application is Due: Nov 17
Contact: Jill G. Conley, Ph.D., Program Director; (301) 215-8873; fax: (301) 215-8888; email: bceeru@hhmi.org
Internet: http://www.hhmi.org/grants/individuals/bceeru.html
Sponsor: Howard Hughes Medical Institute
4000 Jones Bridge Road
Chevy Chase, MD 20815-6789

HHMI Biomedical Research Grants for International Scientists: 2682
Infectious Diseases and Parasitology

The Program provides five-year grants to support promising basic biomedical research scientists working in the fields of infectious diseases and parasitology. The Institute seeks to build the strongest foundation of scientific knowledge and capability by creating a global network of scientists who make important contributions outside the U.S. research community. It support these leaders in their home countries to facilitate the application and dissemination of their knowledge and to strengthen research environments and educational opportunities.

Requirements: Applicants must hold a full-time appointment or have a pending appointment at a nonprofit scientific research organization in any country other than the United States and the United Kingdom. Researchers may not be citizens or permanent residents of the United States.

Restrictions: Only scientists working in the research areas specified in the program announcement are eligible to apply for grant support. The Program does not consider unsolicited grant proposals or award grants for clinical trials and research on health education, health-care delivery, or health services.

Amount of Grant: $50,000-$100,000 per year up to five years
Date(s) Application is Due: Sep 15
Contact: Jill G. Conley, Ph.D., Program Director; (301) 215-8873; fax: (301) 215-8888; email: parasite@hhmi.org
Internet: http://www.hhmi.org/grants/individuals/idap.html
Sponsor: Howard Hughes Medical Institute
4000 Jones Bridge Road
Chevy Chase, MD 20815-6789

HHMI Gilliam Fellowships for Advanced Study 2683

The Fellowships provide full support for up to five years of study toward a Ph.D. for outstanding students who are from groups underrepresented in the sciences or from disadvantaged backgrounds. Gilliam fellows attend the university of their choice and

work alongside distinguished scientists. Chosen for their academic excellence and scientific potential, they will become the leaders of a new generation of biomedical researchers. The Fellowship is available up to 5 years at $44,000 per year.
Requirements: Must be: past participants in the HHMI Exceptional Research Opportunities Program (EXROP); graduating seniors or recent baccalaureate recipients (within two years) who have completed no more than one year of Ph.D. training. Former EXROP students pursuing an M.D./Ph.D. degree may apply, but the fellowship supports only the Ph.D. portion of the studies. There is no citizenship requirement.
Amount of Grant: $44,000
Date(s) Application is Due: Oct 6
Contact: Maryrose Franko, Senior Program Office; (800) 424-9924 or (301) 215-8880; fax: (301) 215-8888; email: gilliam@hhmi.org
Internet: http://www.hhmi.org/grants/individuals/gilliam.html
Sponsor: Howard Hughes Medical Institute
4000 Jones Bridge Road
Chevy Chase, MD 20815-6789

HHMI Grants and Fellowships Programs   2684
The institute administers a grants program that focuses on improving science education from preschool through postdoctoral training, enhancing the science literacy of the general public, and supporting the research of biomedical scientists in selected foreign countries. Currently, grants are administered through five programs: an international program that supports biomedical scientists outside the United States and provides funding for selected courses and workshops; a precollege science-education program that offers grants for educational activities to science museums, botanical and zoological gardens, etc.; a biomedical-research program that supports medical schools and research organizations; a graduate science-education program that provides fellowships for graduate students, medical students, and physicians, and supports special courses; and a biological-sciences education program that awards grants to selected undergraduate institutions.
Samples: United Way of New York City (NY)—for the September 11th Fund, established to support emergency-assistance to organizations and other nonprofit health and human-service groups providing disaster-relief services to victims of the terrorist attacks in New York, Pennsylvania, and Washington, $50,000.
Contact: Jill G. Conley, Ph.D., Program Director; (301) 215-8873; fax: (301) 215-8888; email: grantswww@hhmi.org
Internet: http://www.hhmi.org/grants
Sponsor: Howard Hughes Medical Institute
4000 Jones Bridge Road
Chevy Chase, MD 20815-6789

HHMI Initiative for Colleges Grants   2685
HHMI awards four-year grants to colleges (baccalaureate and master's institutions) to support science education at the undergraduate, K-12, and community college levels. Grants support a range of activities that engage students in research and broaden their access to science; create courses that convey the excitement of contemporary science; develop new, current, and future faculty members; and establish or expand outreach programs that extend to elementary, high school, and community college faculty and students. The registration deadline is May 16, with final proposals due by October 17.
Requirements: Only invited institutions are eligible to apply. Invitations are based on an institution's classification by the Carnegie Foundation for the Advancement of Teaching as a baccalaureate and master's institution and on the institution's record of postbaccalaureate student achievement in the sciences.
Amount of Grant: $800,000-$1.6 million
Date(s) Application is Due: May 16; Oct 17
Contact: Grants Program Officer; (301) 215-8870 or (301) 215-8500; fax: (301) 215-8888; email: grantugr@hhmi.org
Internet: http://www.hhmi.org/grants/institutions/colleges.html
Sponsor: Howard Hughes Medical Institute
4000 Jones Bridge Road
Chevy Chase, MD 20815-6789

HHMI Initiative for Universities Grants   2686
HHMI awards four-year grants to research and doctoral universities to support science education at the undergraduate and K-12 levels. The program supports initiatives that strengthen education in the biological sciences and other scientific disciplines as they relate to biology. Program objectives include: to strengthen research and teaching in undergraduate education and to support student involvement in faculty research; to prepare undergraduates for graduate studies and for careers in biomedical research, medicine, and science education; to bring fresh perspectives to established scientific disciplines, and develop novel programs and courses in emerging areas; to foster linkages between universities and elementary and secondary schools; and to develop approaches that encourage teamwork among faculty, postdoctoral fellows, graduate students, and undergraduates. The registration deadline is May 12, with final proposals due by October 18.
Requirements: Only invited institutions are eligible to apply. Invitations are based on an institution's classification by the Carnegie Foundation for the Advancement of Teaching as a baccalaureate and master's institution and on the institution's record of postbaccalaureate student achievement in the sciences.
Amount of Grant: $1.2-$2.2 million
Date(s) Application is Due: May 14; Oct 1
Contact: Grants Program Officer; (301) 215-8870 or (301) 215-8500; fax: (301) 215-8888; email: ugradcomp@hhmi.org

Internet: http://www.hhmi.org/grants/institutions/universities.html
Sponsor: Howard Hughes Medical Institute
4000 Jones Bridge Road
Chevy Chase, MD 20815-6789

HHMI International Research Scholars Program   2687
This five year grant program supports non-U.S. biomedical scientists, scientific meetings for grant-supported scientists, and other international educational activities. Grants are awarded to promising biomedical scientists who have made significant contributions to fundamental research. Application information is available online.
Requirements: To be considered, individuals must hold a full-time academic or research appointment at a university, medical school, or other nonprofit scientific institution. They are not permitted to have major administrative responsibilities.
Contact: Jill G. Conley, Ph.D., Program Director; (301) 215-8873; fax: (301) 215-8888
Internet: http://www.hhmi.org/grants/for_grantees/
Sponsor: Howard Hughes Medical Institute
4000 Jones Bridge Road
Chevy Chase, MD 20815/6789

HHMI Med into Grad Initiative Grants   2688
HHMI will award grants to institutions to improve the understanding of medicine and pathobiology by scientists conducting biomedical research. Grants will be used to modify existing graduate training or initiate new programs to develop a cadre of Ph.D. researchers who understand pathophysiology and are committed to working at the interface of the basic sciences and clinical medicine. Applications will be accepted from any university in the United States that offers Ph.D.-level training in an appropriate science or engineering discipline. It is expected that as much as $10 million will be awarded in this competition. Awards will range from $400,000 to $1 million, and the grant funds will be allocated over four consecutive years.
Requirements: U.S. institutions that grant PhD degrees in the appropriate science and engineering disciplines are eligible to apply.
Amount of Grant: $400,000-$1 million
Date(s) Application is Due: Apr 27
Contact: Maryrose Franko, Senior Program Officer; (800) 448-4882, ext. 8880; fax: (301) 215-8888; email: medintograd@hhmi.org
Internet: http://www.hhmi.org/grants/institutions/medintograd.html
Sponsor: Howard Hughes Medical Institute
4000 Jones Bridge Road
Chevy Chase, MD 20815-6789

HHMI Physician-Scientist Early Career Award   2689
Through this competitive grant initiative, HHMI awards five-year grants to selected alumni of the HHMI-NIH Research Scholars Program and the HHMI Research Training Fellowships for Medical Students Program to support these individuals as they begin careers as independent physician-scientists. The award provides $375,000 over a five-year period for direct research costs.
Requirements: Only alumni of the HHMI-NIH Research Scholars Program and the HHMI Research Training Fellowship for Medical Students Program who have received an M.D., M.D./Ph.D., D.D.S, or equivalent degree are eligible to apply.
Restrictions: The funds may not be used for the salary of the awardee or institutional indirect costs.
Amount of Grant: $375,000 over a five-year period
Contact: Office of Grants and Special Programs; (800) 448-4882, ext. 8889 or (800) 424-9924; fax: (301) 215-8888; email: earlycareer@hhmi.org
Internet: http://www.hhmi.org/grants/individuals/earlycareer.html
Sponsor: Howard Hughes Medical Institute
4000 Jones Bridge Road
Chevy Chase, MD 20815-6789

HHMI Precollege Outreach Initiative for Biomedical Research   2690
Institutions Grants
HHMI will award five-year grants to support pre-K to 12th-grade science education. Funding will be provided to medical schools, dental schools, veterinary schools, public health schools, hospitals, academic health centers, and independent research institutions to enable these science-rich organizations to work with school systems, museums, and other partners skilled in delivering science content to students. Letters of Intent to submit a proposal are due by June 13, and proposal submissions are due by September 12.
Requirements: HHMI does not make grants directly to individuals, school systems, or to pre-K to 12th-grade schools under this initiative.
Amount of Grant: $200,000-$500,000
Date(s) Application is Due: Jun 13; Sep 12
Contact: Grants Program Officer; (301) 215-8870 or (301) 215-8500; fax: (301) 215-8888; email: grantprc@hhmi.org
Internet: http://www.hhmi.org/grants/institutions/biomedical.html
Sponsor: Howard Hughes Medical Institute
4000 Jones Bridge Road
Chevy Chase, MD 20815-6789

HHMI Research Training Fellowships   2691
Medical fellowships are intended to strengthen and expand the nation's pool of medically trained researchers. These fellowships provide funds to help meet fellows' research-related expenses and education costs during the research training period. Fellowships are

awarded annually to provide support for one year of full-time training in fundamental biomedical, clinical, translational, or applied research. The fellowship includes a stipend, a research allowance to meet research-related expenses, and a fellow's allowance to be used on behalf of the fellow for health care, tuition and fees, and other research-related expenses. Application information is available online.

Requirements: Fellowships are awarded to students enrolled in medical and dental programs in the U.S.

Date(s) Application is Due: Jan 11

Contact: Program Officer; (301) 215-8500 or (800) 424-9924; fax: (301) 215-8888; email: medfellows@hhmi.org

Internet: http://www.hhmi.org/grants/individuals/medfellows.html

Sponsor: Howard Hughes Medical Institute

4000 Jones Bridge Road

Chevy Chase, MD 20815-6789

HHMI Undergraduate Science Education Grants for Professors   2692

The Program awards four-year grants to research-active undergraduate faculty. The grants are intended to empower leading scientists at doctoral and research universities to work more closely with undergraduates at their home institutions and provide other institutions with innovative models for transmitting the excitement and values of scientific research to undergraduate education.

Requirements: Institutions must be invited by HHMI to nominate faculty members to compete for the HHMI professors awards. Invitations are based on the institutions' classifications by the Carnegie Foundation for the Advancement of Teaching as doctoral and research universities and the institutions' records of postbaccalaureate student achievement in the sciences. Invitees are listed in the program announcement.

Contact: Office of Grants and Special Programs; (800) 424-9924 or (301) 215-8873; fax: (301) 215-8888; email: grantugr@hhmi.org

Internet: http://www.hhmi.org/grants/individuals/professors.html

Sponsor: Howard Hughes Medical Institute

4000 Jones Bridge Road

Chevy Chase, MD 20815-6789

HHMI-NIBIB Interfaces Initiative Grants   2693

HHMI and the National Institute of Biomedical Imaging and Bioengineering have formed a partnership to support biomedical research institutions in developing graduate-level research training programs in emerging interdisciplinary fields. The primary goal of this initiative is to train a cadre of PhD scientists who possess the knowledge and skills to conduct interdisciplinary research at the interface between the biomedical sciences and the physical science, computational, engineering, or mathematical disciplines. These fields may include, but are not limited to, chemistry, imaging, science, materials science, nanotechnology, and physics. Another goal is to reduce barriers to interdisciplinary graduate science education. Proposals should reflect the unique educational and scientific capabilities and strengths of the applicant institution(s) as well as address the specific goals of the initiative. The initiative consists of two phases: Phase I supports the establishment of new interdisciplinary training programs; Phase II will sustain the training programs through their critical years. Registration letters of intent to apply are due by January 20, with the proposal submission deadline June 15.

Requirements: All U.S. institutions that grant PhD degrees in appropriate science or engineering disciplines are eligible to apply. Collaborative programs between two or more institutions are acceptable.

Amount of Grant: $1 million maximum for three years

Date(s) Application is Due: Jan 20; Jun 15

Contact: Maryrose Franko, Senior Program Officer; (800) 448-4882, ext. 8880; fax: (301) 215-8888; email: interdisc@hhmi.org

Internet: http://www.hhmi.org/grants/institutions/nibib.html

Sponsor: Howard Hughes Medical Institute

4000 Jones Bridge Road

Chevy Chase, MD 20815-6789

HHMI-NIH Cloister Research Scholars Program   2694

The goal of this program, a joint venture of HHMI and NIH, is to expand the pool of medically trained researchers by encouraging medical students to pursue research careers. The program provides support for an intensive nine months to one year of full-time research, with a possible one-year extension, at NIH in Bethesda, MD. Most students enter after their second year of medical school. The Cloister, a residential facility for scholars, is available on the NIH campus. The research project is selected upon arrival at NIH, after a round of laboratory visits. Application forms may be downloaded from the Web site or requested from the program officer.

Requirements: Applicant must be in good standing at a medical or dental school in the U.S. or Puerto Rico and must receive permission from the school to participate in the program. The school does not have to be a major academic medical center or research-oriented school.

Restrictions: Enrollees in an M.D./Ph.D. or D.D.S./Ph.D. program are not eligible.

Amount of Grant: $17,800 annual salary

Date(s) Application is Due: Jan 10

Contact: Office of Grants and Special Programs; (800) 424-9924 or (301) 215-8873; fax: (301) 215-8888; email: research_scholars@hhmi.org

Internet: http://www.hhmi.org/research/cloister/program.html

Sponsor: Howard Hughes Medical Institute

4000 Jones Bridge Road

Chevy Chase, MD 20815-6789

HHMI/EMBO Start-up Grants for Central Europe   2695

The program encourages young scientists to obtain independent faculty positions in select countries that have ongoing HHMI and EMBO science programs. Applications are encouraged from scientists to set up their first independent laboratories in a Central European EMBC member state (Croatia, Czech Republic, Estonia, Hungary, Poland, and Slovenia). Application forms should be completed by the individual scientist and the receiving institute. It is expected that the institute make the applicant an offer that goes beyond that of a position and laboratory space. HHMI will contribute $50,000, while the participating member state must provide the remaining $25,000. Application and guidelines are available online.

Requirements: Eligible scientists should have an excellent publication record; have a maximum of 10 years postdoctoral experience; be negotiating a position at an institute/ university in Central Europe by the date of the application; and be located outside Central Europe at the time of application.

Amount of Grant: $75,000 per year for three years

Date(s) Application is Due: Aug 1

Contact: Grants Administrator, (301) 215-8500

Internet: http://www.embo.org/projects/yip/embo_hhmi_startup_grants.html

Sponsor: Howard Hughes Medical Institute

4000 Jones Bridge Road

Chevy Chase, MD 20815-6789

HHS Children's Hospitals Graduate Medical Education Payment Program   2696

The program provides funds to freestanding children's hospitals to support the training of pediatric and other residents in graduate medical education (GME) programs. This program compensates for the disparity in the level of federal funding for teaching hospitals for pediatrics versus other types of hospitals.

Requirements: To be eligible, a children's hospital must meet all of the following criteria: participates in an approved graduate medical education (GME) program; has a Medicare provider agreement; is excluded from the Medicare inpatient prospective payment system (PPS) under section 1886(d)(1)(B)(iii) of the Social Security Act, and its accompanying regulations; and ss a freestanding hospital.

Amount of Grant: $40 million total

Date(s) Application is Due: May 1; Aug 1

Contact: Kara Castellow, (301) 443-1058; email: kcastellow@hrsa.gov

Internet: http://bhpr.hrsa.gov/childrenshospitalgme/default.htm

Sponsor: Department of Health and Human Services

5600 Fishers Ln, Parklawn Building, Room 9A-05

Rockville, MD 20857

HIAS Scholarship Awards   2697

The annual competition offers scholarships to deserving students in the United States and Israel. The awarding of scholarships is based on academic excellence, financial need, and involvement within the Jewish community. Guidelines and application are available online.

Amount of Grant: $1500 average award

Date(s) Application is Due: Mar 15; Jun 30

Contact: Scholarship Committee, (212) 613-1358; email: scholarship@hias.org

Internet: http://www.hias.org/Scholarships/apply.html

Sponsor: HIAS Scholarship Program

333 Seventh Avenue, 16th Fl

New York, NY 10001-5004

Higher Education Graduate Scholarships, Fellowships, and Grants   2698

The program in higher education at New York University offers graduate assistantships and fellowships to graduate students. Degree programs in higher education include the MA, EdD (community college leadership focus), and PhD. Information and applications are available on the Web site.

Requirements: These opportunities are open specifically to students enrolled in the master's and doctoral degree programs in higher education at New York University.

Amount of Grant: $1000-$15,000

Contact: Office of Graduate Admissions, School of Education, (212) 998-5030; email: ed.gradadmissions@nyu.edu

Internet: http://www.nyu.edu/financial.aid

Sponsor: New York University

82 Washington Sq E, 2nd Fl

New York, NY 10003-6680

Hilda and Preston Davis Foundation Grants   2699

The Foundation provides funds to charitable organizations whose programs advance the development of all areas of the lives of children and young adults. The Foundation places special emphasis on, and channels most of its financial resources toward, those organizations whose attention is concentrated on eating disorders and education for the underprivileged. Grants may range from $10,000 up to $100,000. In unique circumstances, the Foundation does consider a more significant grant for a program having a major impact in one or more of its areas of interest.

Requirements: 501(c)3 tax-exempt organizations are eligible.

Restrictions: The foundation generally will not provide grants to: non-501(c)3 tax-exempt organizations; individuals; general fundraising drives; endowments; government agencies; or organizations that subsist mainly on third-party funding and have demonstrated no ability or expended little effort to attract private funding.

Amount of Grant: $10,000-$100,000

Contact: Grants Administrator; (203) 629-8552; fax: (203) 547-6112; email: davis@fsllc.net or info@fsllc.net
Internet: http://www.hpdavis.org/application.htm
Sponsor: Hilda and Preston Davis Foundation
640 West Putnam Avenue, 3rd Floor
Greenwich, CT 06830

Hill Crest Foundation Grants 2700
The foundation awards grants to Alabama nonprofits in its areas of interest, including health, the elderly, arts and culture, community development, higher education, and youth. Types of support include general operating support, capital campaigns, building construction/renovation, equipment acquisition, endowment funds, program development, professorships, publication, seed money, scholarship funds, research, technical assistance, and matching funds. There are no application forms or deadlines.
Requirements: Alabama nonprofit organizations may apply.
Geographic Focus: Alabama
Amount of Grant: $5000-$50,000 average
Contact: Charles Terry, Senior Chair, (205) 425-5800
Sponsor: Hill Crest Foundation
P.O. Box 530507
Mountain Brook, AL 35253

Hirtzel Memorial Foundation Grants 2701
The foundation awards grants to eligible organizations in New York and Pennsylvania in its areas of interest, including medical research and health care; community and neighborhood improvement; higher education; and social and human services. Types of support include building construction and renovation, capital campaigns, scholarships, equipment acquisition, general operating support, research, and scholarships. There are no application deadlines; contact the office for appropriate forms.
Requirements: New York nonprofit organizations in Ripley, Chautauqua County, and Pennsylvania nonprofit organizations in North East, Erie County, are eligible.
Geographic Focus: New York; Pennsylvania
Samples: Mercyhurst College (Erie, PA)—to establish the Institute on Aging and Geriatric Health, $1 million.
Amount of Grant: $750-$250,000 range
Contact: Laurie Moritz, Grants Administrator, c/o Mellon Financial Corporation, (412) 234-0023
Sponsor: Orris C. Hirtzel and Beatrice Dewey Hirtzel Memorial Foundation
P.O. Box 185
Pittsburgh, PA 15230

Hispanic Scholarships 2702
Scholarships are intended for undergraduate and graduate students of Hispanic American background. Successful candidates are chosen on the basis of academic achievement, personal strengths, leadership, and financial need. For more information and application requirements, write to the selection committee at the Hispanic Scholarship Fund.
Requirements: Applicants must be U.S. citizens or legal permanent residents who come from Mexican American, Puerto Rican, Cuban, Caribbean, Central American, or South American parentage and who attend college in one of the 50 states or Puerto Rico on a full-time basis, must have completed a minimum of 12 units of college work prior to submission of an application, and must have a GPA of 2.7 on a 4.0 scale. Community college applicants must be enrolled in majors transferable to a four-year institution offering a baccalaureate degree.
Restrictions: HSF is unable to make awards to fully employed persons or practicing professionals.
Amount of Grant: $2000-$10,000
Contact: Sara Martinez Tucker, President, (877) 473-4636 or (415) 445-9930; fax: (415) 808-2302; email: info@hsf.net
Internet: http://www.hsf.net/scholarship/Special.htm#nhfa
Sponsor: Hispanic Scholarship Fund
55 Second Street, Suite 1500
San Francisco, CA 94105

Historical Analysis Senior and Postdoctoral Research Fellowships 2703
The center provides a setting to discuss issues of broad contemporary relevance in historical perspective. Organizing its annual activities around major themes of inquiry or research projects, the center each year welcomes two to three visiting senior and two postdoctoral fellows chosen through an open, international competition, along with a dozen faculty and graduate fellows from within Rutgers University. Senior fellowship applicants are urged to combine the fellowship with their home university's leave programs or other external grants to be in residence at the center for the full academic year. Formal teaching responsibilities are not expected, but past fellows have welcomed opportunities to work with students on projects of mutual interest. Postdoctoral fellows receive 10-month stipend and benefits. They teach one undergraduate seminar each semester. Both senior and postdoctoral fellows are provided with a modest research fund. The center also awards about a dozen fellowships to Rutgers faculty and graduate students from various disciplines who receive modest research grants and office space if available.
Requirements: Applications are welcome from all disciplines and regional specializations. Senior fellows are defined as persons of any academic rank, including independent scholars. Postdoctoral fellows must hold PhDs no more than four years old.
Amount of Grant: $27,500 senior fellowships; $35,000 teaching salary postdoctoral fellowships
Date(s) Application is Due: Dec 15

Contact: Program Administrator, (732) 932-8701; fax: (732) 932-8708; email: rcha@rci.rutgers.edu
Internet: http://rcha.rutgers.edu/fellowshipinfo.php
Sponsor: Rutgers Center for Historical Analysis
88 College Avenue
New Brunswick, NJ 08901

Historical Studies Research Fellowships 2704
Research fellowships are offered for one or two semesters, from September to January and from February to June, to enable scholars in historical studies to pursue postdoctoral research projects at the center and to contribute actively to the seminar on Friday mornings during the term. Proposals that deal with the material, ethical, and symbolic dimensions of the role of animals in human culture are welcomed. Fellows are expected to live in Princeton to take an active part in intellectual interchange with other members of the seminar. Funds are limited; candidates are strongly urged to apply to other grant-giving institutions as well as the center, if they wish to remain in residence for a full year. Travel funds for fellow, spouse, and children are provided.
Requirements: Fellowships are designed for highly recommended younger scholars, as well as senior scholars with established reputations, either from the United States or abroad; candidates must have finished dissertations and have full-time paid positions to which they can return.
Samples: Yemi Akinseye-George, U of Ibadan, Nigeria—for study on the topic Corruption and the Evolution of Democracy in Africa: A Socio-Legal Perspective.; Virginie Coulloudon, Harvard U—for study on the topic Struggle against Corruption in Russia: A Traditional Pattern of Behavior.; Gabriella Etmektsoglou, Institute for Human Sciences, Vienna—for study on the topic Political Justice and Attempted Transitions to Democracy in Greece, 1943-1967.; Robert Gregg, The Richard Stockton College of New Jersey—for study on the topic Uneasy Streets: Police, Corruption and Imperial Progressives in Bombay, London, and New York City.
Amount of Grant: $72,000 maximum per academic year or $36,000 per semester plus maximum of $900 per semester research expenses
Date(s) Application is Due: Dec 1
Contact: Gyan Prakash, Director, (609) 258-4997; fax: (609) 258-5326
Internet: http://dav.princeton.edu
Sponsor: Shelby Cullom Davis Center for Historical Studies
Princeton University, 129 Dickinson Hall
Princeton, NJ 08544

History of Art Graduate Fellowships at the University of Michigan 2705
The Department of the History of Art and the Museum of Art at the University of Michigan offer a variety of financial aid to students who have completed the first year of graduate work. The sources of aid include departmental, graduate school, and museum funds, and grants in aid from external agencies. For first-year students, Regents fellowships provide support for three years, and CIC and University Merit fellowships provide support for four years for those students who intend to continue through the doctorate. Other graduate fellowships and financial assistance include Rackham Graduate School fellowships; Charles L. Freer and Area Center fellowships in non-Western Art (China, South and Southeast Asia, Near East, Japan); Rackham Nontraditional fellowships; teaching, research, and grading assistantships; as well as assistance with photographic and library collections. Some travel stipends are available for seminar and individual research projects, including a fund for students of medieval art. Applications can be obtained from the departmental graduate secretary. In the case of both new applications and renewal applications, the student must see that letters of recommendation are sent directly to the chairperson of the department. Applicants are urged to be thorough and candid in their appraisal of their resources and needs.
Requirements: Those who qualify for national and international grants must exhaust this possibility before applying for departmental funds.
Amount of Grant: $1000-$1500 monthly
Date(s) Application is Due: Jan 1
Contact: Student Services Assistant, Art History Department, (734) 764-5400; fax: (313) 647-4121; email: histartadmiss@umich.edu
Internet: http://www.lsa.umich.edu/histart/graduate/financial
Sponsor: University of Michigan
519 S State Street
Ann Arbor, MI 48109-1357

HKU Foundation Postgraduate Studentships 2706
The purpose of the studentships is to provide selected students with financial assistance in the form of a scholarship to study full-time for the degrees of MPhil or PhD during the prescribed periods of study. A postgraduate studentship holder may be required to undertake, under supervision, duties such as assistance with research; assistance with scheduled laboratory, studio, and fieldwork classes and with tutorials; assistance with the preparation of materials for scheduled classes; assistance with marking practical notebooks and answers from exercise classes; and assistance with invigilation of university degree examinations. Guidelines are available online.
Amount of Grant: HK$13,000-HK$15,100 per month
Date(s) Application is Due: Jan 31
Contact: Graduate School, (852) 2857-3470; fax: (852) 2857-3543; email: gradsch@hkucc.hku.hk
Internet: http://www.hku.hk/rss/index4.htm
Sponsor: University of Hong Kong
Graduate House, Room P403
Pokfulam Road, Hong Kong

Hoffmann-La Roche Foundation Grants 2707

The corporate contributions program supports nonprofits in company operating areas. Funds may be requested in two categories: health programs that focus on health promotion and health education and educational projects that encourage math and science literacy and promote secondary school teacher enrichment. Seed grants are awarded for new activities for which Hoffman-La Roche may spark other funders to pitch-in operating support.

Requirements: Nonprofits in company-operating communities are eligible to apply, with some preference given to groups near the New Jersey headquarters.

Restrictions: Funds will not be awarded in support of endowments; scholarships; facility expansion or construction; purchase of equipment; capital expenditures; good will advertising; and political, veterans, or labor organizations. Telephone, facsimile, and email requests will not be honored.

Amount of Grant: $2500-$50,000

Contact: Vivian Beetle, Administrative Director, (973) 562-2055; fax: (973) 562-2999

Internet: http://www.rocheusa.com/about/responsibility.html

Sponsor: Hoffmann-La Roche Foundation

340 Kingsland Street

Nutley, NJ 07110-1199

Holberg International Memorial Prize 2708

Bergen-born Ludvig Holberg held the chairs of Metaphysics and Logic, Latin Rhetoric and History at the University of Copenhagen, and played an important part in bringing the Enlightenment to the Nordic countries. Through his interdisciplinary and internationally-oriented efforts, Holberg sought to modernize academic subjects and teaching methods. The Ludvig Holberg Memorial Fund was established in 2003. The main purpose of the Fund is to increase the profile of the arts and humanities, the social sciences, law and theology. The Board of the Ludvig Holberg Memorial Fund awards prizes under three headings every year: the Holberg International Memorial Prize, the Nils Klim Prize and the Holberg Prize School Project. The The Holberg International Memorial Prize is equal to NOK 4.5 million (EUR555,000 / U.S.D750,000).

Samples: Ronald Dworkin, Prize Recipient (2007).

Amount of Grant: $750,000

Contact: Kare Rommetveit, Head of Secretariat; +47 55 58 20 00 or +47 55 58 69 92; email: Kare.Rommetveit@bfs.uib.no

Internet: http://www.holbergprisen.no/en_minnefondet.html

Sponsor: University of Bergen

Post Box 7800, Thormohlensgate 51 (2. floor)

Bergen, Norway N-5020 Norway

Holberg Prize School Project 2709

Bergen-born Ludvig Holberg held the chairs of Metaphysics and Logic, Latin Rhetoric and History at the University of Copenhagen, and played an important part in bringing the Enlightenment to the Nordic countries. Through his interdisciplinary and internationally-oriented efforts, Holberg sought to modernize academic subjects and teaching methods. An important objective of the School Project is to promote greater understanding, interest and enthusiasm for the arts and humanities, the social sciences, law and theology among children and young people. Every year, students at upper secondary schools in Norway are invited to carry out a research project in cooperation with established researchers. Three of the projects are awarded prizes (of NOK 15,000, 10,000 and 5,000 respectively), and the winning students are invited to attend the award ceremony for the Holberg International Memorial Prize. A teacher's grant is also awarded as part of the Project.

Amount of Grant: NOK 15,000, 10,000 and 5,000 respectively

Contact: Siri Breistein, Project Coordinator; +47 55 58 90 45; cell: + 47 93 49 98 69; email: siri.breistein@holbergprisen.no

Internet: http://www.holbergprisen.no/skole/en_skoleprosjekt.html

Sponsor: University of Bergen

Post Box 7800, Thormohlensgate 51 (2. floor)

Bergen, Norway N-5020 Norway

Holley Medal 2710

The medal is bestowed on an individual who by some great and unique act(s) of an engineering nature has accomplished a great and timely public benefit. In judging the merits of any candidate for this award, no limitations shall arise out of the nominee's formal degree of education, membership in any society or organizations, or the circumstances of employment or official position. Attention shall be concentrated on the brilliance of the art—not the individual. The achievement should be of such public importance as to be worthy of the gratitude of the nation and to call forth admiration of engineers. More than one individual may be recognized for a single achievement provided that each individual made equal or comparable contribution.

Requirements: Any individual member, group of members, or committee may nominate candidates for this award.

Samples: James Walker—award recipient, $1000 (2005).

Amount of Grant: $1000

Date(s) Application is Due: Mar 1

Contact: Gilda DiTullio, Manager, (212) 591-7736; fax: (212) 705-7739; email: ditullio@asme.org

Internet: http://www.asme.org/honors/ms71/gaa/holley.html

Sponsor: American Society of Mechanical Engineers

3 Park Avenue

New York, NY 10016

Home Depot Foundation Grants 2711

The foundation's funding initiatives include affordable housing—organizations that work to create and rehabilitate affordable housing, such as Habitat for Humanity affiliates, Rebuilding Together affiliates, community development corporations, and neighborhood housing initiatives; and environment—organizations that direct efforts toward protecting the environment, with a focus on forestry, sustainability and green design, cleanup and recycling, and lead poisoning prevention, such as The Nature Conservancy, Institute for Sustainable Development, Keep America Beautiful, and World Wildlife Fund. The foundation gives first priority to organizations that have been invited to apply for a grant. Unsolicited requests that match foundation eligibility requirements also receive consideration.

Requirements: 501(c)3 tax-exempt public charities in the United States and charitable organizations in Canada are eligible. The foundation only funds programs that meet its eligibility test, which is available online.

Restrictions: The foundation does not support organizations that are not 501(c)3 or Revenue Canada designated charities; scholarships or other direct support to individuals; religious, fraternal, political, labor, athletic, or social groups; capital campaigns, endowments, or endowed chairs; institutional overhead/indirect costs; capital investments; equipment purchases not part of a larger program request; projects for political or religious purposes; special events, such as conferences, sports competitions, or art exhibits; courtesy or journal advertising campaigns; film, music, television, video, or media production projects or broadcast underwriting; activities of organizations serving primarily their own membership; or United Way chapters (the company provides support to United Way).

Samples: Friends of Watersheds/Essex Region Conservation Authority (Canada)—for a program that engages volunteers in tree-planting projects, $15,000 (2004); Rocky Mountain Youth Corps (Steamboat Springs, CO)—for a program that provides youths with service opportunities and educational experiences through conservation projects on public lands, $25,000 (2004); Community Services Council (Salt Lake City, UT)—for a program that provides home repair and maintenance assistance to senior citizens, $15,000 (2004); National Housing Conference (Washington, DC)—to enhance its infrastructure and research on three national housing issues, $25,000 (2004).

Amount of Grant: $10,000-$50,000

Date(s) Application is Due: Jan 15; Apr 15; Jul 15; Oct 15

Contact: Community Affairs, (866) 593-7019 or (770) 384-3889; fax: (866) 593-7027; email: hd_foundation@homedepot.com

Internet: http://www.homedepotfoundation.org

Sponsor: Home Depot Foundation

2455 Paces Ferry Road

Atlanta, GA 30339-4024

HomeBanc Foundation Grants 2712

The Foundation has three main goals: provide support to cancer related causes through funding education, advocacy, and research; support the American dream of homeownership for those who might not reach it on their own; and provide college funding to students who strive to achieve their goals through education beyond high school. Scholarship forms are available online. Contact the Foundation to obtain application information for grants.

Requirements: Nonprofit organizations providing services in Georgia and Florida are eligible.

Geographic Focus: Florida; Georgia

Contact: Amanda Albertelli, (404) 459-7618; email: AAlbertelli@HomeBanc.com

Internet: http://www.homebancfoundation.org/grant_information.aspx

Sponsor: HomeBanc Foundation

2002 Summit Boulevard, Suite 100

Atlanta, GA 30319-1497

Horace Mann ALPLM Fellowships 2713

Horace Mann and the Abraham Lincoln Presidential Library and Museum (ALPLM) offer 50 teachers a fellowship to study the life and legacy of the country's 16th president. Winning fellows will attend a five-day institute in June and July, at the presidential library and museum in Springfield, IL. Teachers submit their resumes and answer two essay questions (250 word max). Within the first essay, teachers briefly describe the subject area(s) and grade level(s) they teach, and how they anticipate this experience will benefit them and their students. In the second essay, teachers show why it is essential to understand Abraham Lincoln today. Teachers must submit their application online. Guidelines are available online.

Requirements: The fellowship is open to any full-time teacher, teaching kindergarten through 12th grade, in the United States.

Amount of Grant: $1000

Date(s) Application is Due: Mar 4

Contact: Fellowships Administrator, (800) 999-1030

Internet: http://www.horacemann.com/educator-resources/abraham-lincoln-fellowship.html

Sponsor: Horace Mann Insurance Company

1 Horace Mann Plaza

Springfield, IL 62715-0001

Horowitz Foundation for Social Policy Grants 2714

The foundation supports and advances research and understanding in the major fields of the social sciences, specifically in the fields of psychology, anthropology, sociology, economics, urban affairs, area studies, political science, and other disciplines. Grants provide direct assistance to individual scholars worldwide who require small grants to further their research with emphasis on policy-oriented studies. Preference will be given to projects that deal with contemporary issues in the social sciences or issues of policy relevance, as well as to scholars in the initial stages of work. Five to 10 grants are awarded each year. Guidelines and application are available online.

Amount of Grant: $3000-$5000 range
Date(s) Application is Due: Dec 31
Contact: Mary Curtis, Vice Chairman and Secretary; email: applications@horowitz-foundation.org
Internet: http://www.horowitz-foundation.org
Sponsor: Horowitz Foundation for Social Policy
P.O. Box 7
Rocky Hill, NJ 08553-0007

Horton Research Grants   2715
One or more grants are awarded annually in support of a research project in hydrology and water resources, with the objective of fostering graduate student research leading to the completion of doctoral dissertations. Appropriate topics may be in hydrology including its physical, chemical, or biological aspects, or in water resources policy sciences including economic systems analysis, sociology, and law. A proposal must be signed by the student and by the faculty supervisor; a formal application form must accompany the proposal.
Requirements: Applicant must be a PhD candidate.
Amount of Grant: $10,000 plus travel allowance to attend the awards luncheon
Date(s) Application is Due: Apr 15
Contact: Horton Research Grants, (800) 966-2481 or (202) 462-6900; email: service@agu.org
Internet: http://www.agu.org/inside/honors.html#SH
Sponsor: American Geophysical Union
2000 Florida Avenue NW
Washington, DC 20009

Hospital for Sick Children Foundation External Grants   2716
Grants are awarded for research in child health (up to age 18). Application forms should be obtained from the foundation. Listed application deadlines are for receipt of letters of intent. Annual deadline dates may vary; contact program staff for exact dates.
Requirements: Individuals affiliated with Canadian registered charities (including hospitals and universities) are eligible to apply.
Restrictions: The foundation does not support projects related to support of children outside Canada; established, ongoing research projects; computer hardware; building construction; operating expenses; or general endowments.
Amount of Grant: $C65,000 maximum per year for two years
Date(s) Application is Due: Mar 1; Sep 1
Contact: Grants Officer, (416) 813-6166 ext 2354; fax: (416) 813-5024; email: national.grants@sickkids.ca
Internet: www.sickkids.ca/foundation/grants.asp
Sponsor: Hospital for Sick Children Foundation
555 University Avenue
Toronto, ON M5G 1X8 Canada

Houghton Library Visiting Fellowships   2717
Short-term fellowships are available to scholars pursuing research in literature, history, philosophy, religion, history of science, theater, dance, and printing and graphic arts. Fellows also have access to collections in Widener as well as throughout the world's largest university library.
Amount of Grant: $2750 stipend
Date(s) Application is Due: Jan 13
Contact: Dennis Marnon, Administrative Officer, (617) 495-2441; email: dmarnon@fas.harvard.edu
Internet: http://hcl.harvard.edu/libraries/houghton/public_programs.html#fellowships
Sponsor: Harvard University
Houghton Library
Cambridge, MA 02138

Household International Corporate Giving Program Grants   2718
Grants made in support of Household's corporate social responsibility fall into three main focus areas: education, housing, and community enrichment. Special attention is given to requests from nonprofits that address issues related to financial and credit education, economic development, housing, youth development, and education, particularly for low-income and minority populations. Types of support include annual campaigns, capital campaigns, challenge/matching grants, employee-related scholarships, scholarship funds, equipment, fellowships, general operating support, continuing support, research grants, and technical assistance. Grants are made at three levels: companywide grants, local-facility grants, and the Help for Communities program administered through local branch offices. Each level has its own deadline. Obtain guidelines, which contain a list of contacts for the company's U.S. and foreign subsidiaries.
Requirements: Nonprofit organizations in corporate operating areas are eligible.
Restrictions: The corporation generally does not support the following: nonprofit organizations that do not have 501(c)3 tax deductible status; fraternal, veteran, labor, athletic, or religious organizations serving a limited constituency; for-profit student aid or scholarship programs, aside from those already established by Household or its business groups; political, lobbying, or voter registration programs, or those supporting the candidacy of a particular individual; funds to support travel for groups or individuals;

organizations that might in any way pose a conflict with corporate values, products, customers, or employees; unsolicited requests from secondary, elementary, and post-secondary schools; single disease research initiatives outside of corporate sponsored programs; unsolicited proposals from colleges and universities; advertising; or individuals.
Samples: Beacon Light Community Housing Development Organization (Hampton, VA)—for money-management classes for low- and middle-income people, $109,000 (2003); Catholic Charities Community Development Corp (Saint Petersburg, FL)—to provide homeownership and financial education to low- and moderate-income adults, and for a program to teach youths about budgeting and investing, $215,000 (2003); Vecinos Unidos (Dallas, TX)—for educational courses for low- and middle-income Latinos buying homes, $60,950 (2003); FAIM Economic Development Corp (Dallas, TX)—for a money-management and credit-education program for members of African-American churches and other Dallas residents, $114,000 over two years (2003).
Amount of Grant: $8 million total
Date(s) Application is Due: Dec 15
Contact: Contributions Administrator, (847) 564-6010; fax: (847) 564-7094; email: communityrelations@household.com
Internet: http://www.hsbcusa.com/corporateresponsibility/contributions_grants
Sponsor: Household International Corporate Giving Program
2700 Sanders Road
Prospect Heights, IL 60070

Houston Young Lawyers Foundation Grants   2719
Grants are awarded to support, promote, and encourage charitable organizations and activities in Harris County, TX, including activities for the furtherance of justice and legal education through the sponsorship and encouragement of legal research, publications, and institutes and forums; the institution and maintenance of legal-aid facilities for the indigent or working poor and the acceptance of aid grants from the governmental and private sources; the support of youths through athletic and educational forums; and any other activity that will promote and aid an improved system for the administration of justice and the improvement of the Bench and Bar in Harris County, TX, and elsewhere. Priority is given to organizations and programs serving the people of the Houston metropolitan area.
Requirements: Charitable organizations and projects that primarily benefit the greater Houston metropolitan community are eligible.
Amount of Grant: $500-$2500
Date(s) Application is Due: Jan 7
Contact: Vivica Simmons, Grants Administrator, (713) 546-5637; email: vsimmons@shb.com
Internet: http://www.hyla.org/hyla/Grant_Guidelines_EN.asp?SnID=381093844
Sponsor: Houston Young Lawyers Foundation
P.O. Box 61208
Houston, TX 77208-1208

Howard Francis Cline Memorial Prize   2720
The prize is awarded biennially to the book or article in English, German, or a Romance language judged to make the most significant contribution to the history of Indians in Latin America, referring to any time before the immediate present. Items appearing in the two calendar years just preceding may be considered for a given year's award.
Amount of Grant: $500
Contact: CLAH, (530) 752-3046; fax: (530) 752-8964; email: clah@ucdavis.edu
Internet: http://www.h-net.org/~clah/about/index.html
Sponsor: Conference on Latin American History
University of California at Davis, One Shields Avenue
Davis, CA 95616

Howard Gilman Foundation Grants   2721
The foundation is dedicated to the preservation of natural and cultural resources, with a focus on environmental, especially animal, conservation; the preservation and advancement of artistic and cultural endeavors; and medical research, especially in the fields of HIV/AIDS, cardiology, and sports medicine. The arts and culture component of the program is focused primarily in the New York metropolitan area. Types of support include general operating grants, program grants, research grants, and seed money grants. Staff prefer to respond to a brief letter of inquiry outlining the project's aim and general budgetary requirements before asking potential grantees to prepare full proposals.
Requirements: 501(c)3 nonprofit organizations are eligible.
Restrictions: Religious and political agencies are ineligible.
Geographic Focus: New York
Samples: Brooklyn Academy of Music (Brooklyn, NY)—for endowment and operating support, $5 million.
Amount of Grant: $1000-$100,000 average
Contact: Harry Brown, Program Associate, (212) 307-1073; fax: (212) 262-4108; email: hbrown@gilman.com
Sponsor: Howard Gilman Foundation
111 W 50th Street, 40th Fl
New York, NY 10020

HRB Clinical Research Training Fellowships   2722
These fellowships are designed to enable medical and dental graduates at any stage in their career to gain specialized clinical research training in a biomedical field in Ireland. Awards are available for up to two years. Fellowships are tenable only in institutions

approved of by the board, such as teaching hospitals, universities and research institutes in Ireland. Applications are available on the Web site.

Requirements: Applicants should be graduates in medicine or dentistry from post-registration up to and including senior registrar or equivalent academic level. Applicants should be EU citizens or be graduates from outside the EU with permanent Irish resident status.

Amount of Grant: IR@L46,909-IR@L56,066
Date(s) Application is Due: Sep 16
Contact: Patricia Cranley, Research Grants Officer, (353) 1 6761176 ext 120; fax: (353) 1 6611856; email: pcranley@hrb.ie
Internet: http://www.hrb.ie/display_content.php?page_id=95
Sponsor: Health Research Board
73 Lower Baggot Street
Dublin 2 Ireland

HRB Health Services Research Fellowships   2723

Fellowships provide an opportunity to graduates with appropriate experience to undertake health services research in Ireland. Research may be interdisciplinary in nature and may involve clinical, epidemiological, public health, statistics, health economics, social science, operational, and management disciplines. Awards are available for up to three years. Application forms are available on the Web site.

Requirements: To be eligible, candidates must hold a primary degree in a discipline relevant to health services research; have acquired appropriate postgraduate experience in the field of health services research; have support from an approved academic department or center; have obtained the prior approval of a head of department for the research study being proposed; be European Union citizens or be graduates from outside the European Union with permanent Irish resident status.

Amount of Grant: IR@L41,902; IR@L43,319; IR@L44,727
Date(s) Application is Due: Oct 14
Contact: Patricia Cranley, Research Grants Officer, (353) 1 6761176 ext 120; fax: (353) 1 6611856; email: pcranley@hrb.ie
Internet: http://www.hrb.ie/display_content.php?page_id=96
Sponsor: Health Research Board
73 Lower Baggot Street
Dublin 2 Ireland

HRB Postdoctoral Research Fellowships   2724

Fellowships enable researchers with a PhD, MD, or equivalent research experience to develop research careers at an advanced level in medicine, dentistry, health-related biological sciences, or epidemiology in Ireland. Awards are available for up to three years. Applications are available on the Web site.

Requirements: Applicants must have a PhD or MD in a related biomedical science field or have submitted a thesis for such at the time of application, or have equivalent research experience. Applicants should not have more than five years postdoctoral experience at the time of application and be European Union citizens or be graduates from outside the European Union with permanent Irish resident status. Applications should be made by the prospective fellow with the support of the head of an appropriate sponsoring laboratory in the Republic of Ireland.

Amount of Grant: IR@L41,902; IR@L43,319; IR@L44,727
Date(s) Application is Due: Oct 7
Contact: Patricia Cranley, Research Grants Officer, (353) 1 6761176 ext 120; fax: (01) 6611856; email: pcranley@hrb.ie
Internet: http://www.hrb.ie/display_content.php?page_id=97
Sponsor: Health Research Board
73 Lower Baggot Street
Dublin 2 Ireland

HRDC Employability and Social Partnerships Grants   2725

The program supports initiatives that develop and promote nationally significant research and identify and develop models of best practices or of new service delivery initiatives designed to meet the social development and employability needs of populations who are or may be at risk. Eligible activities include applied research—identification of new and critical social/employment issues that affect the participation of targeted groups and individuals in the community and in the labor market, and the quality of child care; development—testing and evaluation of models or best practices that will enhance the quality of child care and the social development and employability of populations who are most at risk; capacity building—support national nongovernmental organizations whose mandates reflect those of HRDC in serving their target populations and enhancing civic participation through partnerships among nongovernmental organizations and the public and private sectors; and data development and dissemination—of information on social development, employability of populations most at risk, and child care.

Amount of Grant: $C12.5 million total budget
Contact: Employability and Social Partnerships, (819) 997-1452; fax: (819) 997-1359; email: epb-dgpe@hrdc-drhc.gc.ca
Internet: http://www.hrsdc.gc.ca/en/gateways/nav/top_nav/program/gc.shtml
Sponsor: Human Resources Development Canada
140 Promenade du Portage, Phase 1V
Hull, PQ K1A OJ9 Canada

HRSA Basic Nurse Education and Practice Grants   2726

This program is intended to help schools of nursing and other institutions improve the quality and availability of nursing education through projects for specified purposes such as expanding enrollment in professional nursing programs, improving access to primary

health care in noninstitutional settings, providing continuing education for nurses in medically underserved communities, and providing long-term care fellowships for certain paraprofessionals. Grant funds may be used for salaries of personnel specifically employed for the project, consultant fees, supplies and equipment necessary to conduct the project, and essential travel expenses and other expenses related to the project. Indirect administrative costs incurred as a result of the project are allowed but limited to 8 percent of direct costs. Applications are reviewed two times a year by the Advisory Council on Nurses Education. Assistance is available up to five years. Contact the office or visit the Web site for deadline dates.

Requirements: Public and nonprofit private schools of nursing and other public or nonprofit private entities may apply.
Amount of Grant: $250,000 average
Contact: Grants Administrator, Division of Nursing, Bureau of Health Professions, (301) 443-5688; fax: (301) 443-8586
Internet: http://bhpr.hrsa.gov/nursing
Sponsor: Health Resources and Services Administration
5600 Fishers Lane, Room 9-35
Rockville, MD 20857

HRSA Eliminating Disparities in Perinatal Health Grants   2727

The purpose of this program is to enhance a community's capacity to address significant disparities in several perinatal health indicators, with an emphasis on infant mortality rates.

Requirements: Public and private nonprofit community-based organizations are eligible to apply.
Restrictions: Entities that are currently receiving a Healthy Start implementation grant are not eligible to apply.
Amount of Grant: $100,000-$2.3 million; $782,971 average
Contact: Director, Division of Perinatal Systems and Women's Health, (301) 443-0543
Internet: http://www.hrsa.gov
Sponsor: Health Resources and Services Administration
5600 Fishers Ln, Parklawn Bldg
Rockville, MD 20857

HRSA Geriatric Education Centers   2728

The purpose of this program is to develop geriatric education centers to support development of collaborative arrangements involving several health professions schools and healthcare facilities. Geriatric centers facilitate the training of health professional faculty, students, and practitioners in the diagnosis, treatment, and prevention of disease, disability, and other health problems of the aged. Projects must improve training of health professionals; develop and disseminate curricula relating to treatment; support training and retraining of faculty and continuing education of health professionals who provide geriatric care; and provide students with clinical training in geriatrics in nursing homes, hospitals, ambulatory care centers, and senior centers. HRSA-05-077

Requirements: Accredited health professions schools, physicians assistants training programs, allied health schools, and nursing schools are eligible to apply.
Amount of Grant: $200,000 average
Date(s) Application is Due: Mar 1
Contact: Donna Nash, Grants Managment Specialist, (301) 443-6960; fax: (301) 443-6343; email: dnash@hrsa.gov
Internet: http://www.hrsa.gov/grants/preview/guidanceprofessions/hrsa05077.htm
Sponsor: Health Resources and Services Administration
5600 Fishers Ln, Room 11A-05
Rockville, MD 20857

HRSA Health Professions Scholarships for Disadvantaged Students   2729

The program makes funds available for grants to schools of medicine; osteopathic medicine; dentistry; optometry; podiatric medicine; veterinary medicine; nursing (associate, diploma, baccalaureate, and graduate degree); public health; chiropractic; allied health (baccalaureate or graduate degree programs of dental hygiene, medical laboratory technology, occupational therapy, physical therapy, radiologic technology, speech pathology, audiology, and registered dietitians); graduate programs in behavioral and mental health practice (includes clinical psychology, clinical social work, professional counseling, or marriage and family therapy); and programs providing training of physician assistants for the purpose of assisting such schools in providing scholarships to individuals from disadvantaged backgrounds who are enrolled (or accepted for enrollment) as full-time students in the schools. Notice of awards is issued to the schools, which, in turn, select and issue the awards to the eligible students. Consultation and technical assistance is provided to institutions wishing to apply. Annual deadline dates may vary; contact the office to confirm exact dates.

Requirements: Accredited public or private nonprofit schools of medicine, nursing, osteopathic medicine, dentistry, pharmacy, podiatric medicine, optometry, veterinary medicine, public or allied health, or schools offering graduate programs in clinical psychology are eligible.
Amount of Grant: $727-$650,000; $102,468 average
Contact: Andrea Stampone, Bureau of Health Professions, (301) 443-4776
Internet: http://bhpr.hrsa.gov/dsa/sds.htm
Sponsor: Health Resources and Services Administration
5600 Fishers Ln, Parklawn Bldg
Rockville, MD 20857

HRSA Nurse Anesthetist Traineeships Program   2730

Grants are awarded to eligible institutions to provide financial support through traineeships for licensed registered nurses enrolled as full-time students beyond the twelfth month of study in a master's nurse anesthesia program. HRSA-05-062

Requirements: Eligible applicants are schools of nursing, academic health centers, and other public or private nonprofit entities which provide registered nurses with full-time anesthetist master's education and have evidence of earned accreditation status from the American Association of Nurse Anesthetists (AANA) Council on Accreditation of Nurse Anesthesia Education Programs.

Amount of Grant: $7100 average

Date(s) Application is Due: Nov 18

Contact: Karen Breeden, Division of Nursing, Bureau of Health Professions, (301) 443-6333; fax: (301) 443-8586; email: kbreeden@hrsa.gov

Internet: http://www.hrsa.gov/grants/preview/professions.htm#hrsa05062

Sponsor: Health Resources and Services Administration

5600 Fishers Ln, Parklawn Bldg

Rockville, MD 20857

HRSA Nursing Education Opportunities for the Disadvantaged Project Grants   2731

This program is intended to provide financial assistance to eligible schools of nursing and other applicants to meet the costs of special projects to increase nursing education opportunities for individuals from disadvantaged backgrounds. Funds may be used for salaries of personnel specifically employed for the project; consultant fees; supplies and equipment necessary to conduct the project; essential travel expenses; and other expenses related to the project. Indirect administrative costs are allowed, limited to 8 percent of direct costs. Grants may be made for up to three-year project periods and may be renewed competitively for an additional two years.

Requirements: Public and nonprofit private schools of nursing and other public or nonprofit private entities are eligible.

Amount of Grant: $58,800-$518,600; $255,100 average

Contact: Dr. Janice Young, Division of Nursing, Bureau of Health Professions, (301) 443-6193

Internet: http://bhpr.hrsa.gov/dsa/sds.htm

Sponsor: Health Resources and Services Administration

5600 Fishers Ln, Parklawn Bldg

Rockville, MD 20857

HRSA Physician Assistant Training   2732

Grants are awarded for projects for the training of physician assistants and for the training of individuals who will teach in programs to provide such training. A project supported must meet the following definition of a training program for physician assistants as defined under Section 799B of the Public Health Service Act: has as its objective the education of individuals who will, upon completion of their studies in the program, be qualified to provide primary care under the supervision of a physician; extends for at least one academic year and consists of supervised clinical practice and at least four months (in the aggregate) of classroom instruction directed toward preparing students to deliver health care; has an enrollment of not less than eight students; and trains students in primary care, disease prevention, health promotion, geriatric medicine, and home health care. The program assists schools to meet the costs of projects to plan, develop, and operate or maintain programs for the training of physician assistants and for the training of individuals who will teach in programs to provide such training. Programs must develop and use methods designed to encourage graduates of the program to work in health professional shortage areas. Programs also must develop and use methods for placing graduates in positions for which they have been trained.

Requirements: Public or nonprofit private hospitals, schools of medicine or osteopathic medicine, or public or private nonprofit entities are eligible to apply. Eligible physician assistant programs are those that are either accredited by the American Medical Association's Committee on Allied Health Education and Accreditation (AMA-CAHEA) or its successor organization, the Commission on Accreditation of Allied Health Education Programs (CAAHEP).

Amount of Grant: $13,487-$257,297; $120,076 average

Contact: Ellie Grant, (301) 443-5404; email: egrant@hrsa.gov

Internet: http://www.hrsa.gov

Sponsor: Health Resources and Services Administration

5600 Fishers Ln, Parklawn Bldg

Rockville, MD 20857

HSBC in the Community (U.S.A) Grants   2733

The foundation awards grants in its areas of interest, including education and the environment. Education grants focus on K-12 public schools and postsecondary institutions as well as adult education, such as public library programs and welfare-to-work programs. K-12 education may target students, teachers, and/or parents. Higher education grants support scholarship programs, primarily for disadvantaged students. Grants will be awarded to schools, colleges, and universities as well as to other institutions. In the area of environment, the foundation will award grants to nonprofit organizations that have programs targeting conservation, sustainable development, or environmental education. Programs that promote good environmental practices, and programs that increase environmental public awareness will be considered. Proposals also are welcome for programs that strive to prevent potential environmental degradation, those that focus on scientific research on environmental issues, and activities promoting energy conservation, recycling, preservation of green spaces, waSuite reduction, and ecological concerns. Proposals may be submitted throughout the calendar year. Grant requests will be reviewed on a rolling basis.

Requirements: 501(c)3 public charities, public schools and school districts, or other government agencies are eligible.

Restrictions: The foundation will not make grants to non-education or non-environment organizations; individuals, except for scholarships through a third party; or political organizations or for political purposes.

Samples: Foundation for Independent Higher Education (Washington, DC)—to provide scholarships for first-generation college students enrolled at 81 private college and university campuses in California, Florida, Illinois, and New York State, $300,000 over three years (2005),

Contact: Group Public Affairs, (212) 525-8239; TTY/TTD: (800) 898-5999

Internet: http://www.us.hsbc.com/inside/community/foundation/print.html

Sponsor: HSBC in the Community (U.S.A)

452 Fifth Avenue

New York, NY 10018

Hubbard Scholarship   2734

One scholarship is awarded each year for study at an American Library Association-accredited school for up to two years. The purpose of this scholarship is to recruit excellent librarians for Georgia and provide financial assistance toward completing a degree in library science. The scholar is required to work in a library or in a library-related capacity in Georgia for one year following completion of the program or refund the amount awarded plus interest within two years. Applicants should submit the following materials: official application form, proof of acceptance in an accredited library school, three letters of reference sent directly from the reference, and official transcripts of all academic work sent directly from each institution of higher education. Other considerations being equal, residents of Georgia will be given preference.

Requirements: U.S. citizens accepted for admission to a master's program who intend to complete their course of study within two years at an ALA-accredited library school are eligible to apply.

Amount of Grant: $3000

Date(s) Application is Due: May 1

Contact: Elizabeth Bagley, Scholarship Committee Chair, (404) 651-2172; fax: (404) 651-2476; email: ebagley@gsu.edu

Internet: http://www.library.gsu.edu/gla/committees/scholarship

Sponsor: Georgia Library Association

P.O. Box 3967, 100 Decatur Street SE

Atlanta, GA 30303-3967

HUD Community Outreach Partnership Centers Grants   2735

The program provides funds to Institutes of Higher Learning (IHLs) to establish and operate Community Outreach Partnership Centers. These centers carry out research and administer outreach programs in the surrounding communities that focus on addressing the problems of urban areas. Centers should conduct not only outreach but technical assistance, applied research, and empowerment activities for the local neighborhood and community-based organizations. Centers will be required to address at least three urban problems in their outreach programs; these can include issues relating to housing, economic development, neighborhood revitalization, infrastructure, health care, job training, education, crime prevention, planning, and community organization. Outreach activities may include technical assistance, counseling, workshops, and assistance to community and neighborhood organizations.

Requirements: Institutions of higher education with demonstrated ability to carry out eligible activities are eligible. A 50 percent match of the total cost of research activities is required, as well as a 25 percent match of the total cost of establishing and operating outreach activities.

Amount of Grant: $200,000-$400,000 range

Contact: Kinnard Wright, Office of University Partnerships, Policy Development and Research, (202) 708-3061 ext 7495; email: Kinnard_D._Wright@hud.gov

Internet: http://www.hud.gov/grants

Sponsor: Department of Housing and Urban Development

451 Seventh Street SW

Washington, DC 20410

HUD Doctoral Dissertation Research Grants   2736

The program awards grants to doctoral candidates completing their dissertations on housing and urban development issues. HUD encourages research based on relationships among universities, communities, distressed public housing developments, public schools, municipal agencies, and community-based organizations. Applications are encouraged from students of any discipline that can provide policy-relevant insight on HUD's mission and program commitments to reduce homelessness, revitalize severely distressed public housing, increase housing production, reduce racial barriers to residential mobility, and reinvigorate economically distressed communities. Support from the university such as tuition waivers, office space, or computer time is required. HUD sends program announcements to relevant PhD- granting departments of accredited universities and posts the announcements on the Internet.

Requirements: Any currently enrolled student accepted into candidacy in an accredited doctoral program who has a fully developed and approved dissertation proposal that addresses the program's purpose may apply. Applicants must provide proof that they will have satisfactorily completed all written and oral PhD requirements except the dissertation by September 1 of the current year and that the dissertation can be completed within two years.

Amount of Grant: $15,000 maximum

Contact: University Partnerships Clearinghouse, (800) 245-2691

Internet: http://www.hud.gov/progdesc/ddrg.cfm

Sponsor: Department of Housing and Urban Development
451 Seventh Street SW
Washington, DC 20410

HUD General Research and Technology Activity Grants 2737

Grants are awarded to organizations to carry out research and demonstration projects of high priority and preselected by HUD to improve the operations of their programs. Research relating to low-income housing and housing assistance; public finance and urban economic development; national housing needs; evaluation of existing housing and community development programs; and improving the management and planning of state and local governments are areas eligible for award of grants or cooperative agreements. Specific deadline dates are shown in requests for proposals.
Requirements: State and local governments, academic institutions, and public and/or private profit and nonprofit organizations that have the authority and capacity to carry out projects are eligible to compete for these grants.
Amount of Grant: $13,000-$1.4 million range
Contact: Assistant Secretary for Policy Development and Research, (202) 708-1796
Internet: http://www.hud.gov
Sponsor: Department of Housing and Urban Development
451 Seventh Street SW
Washington, DC 20410

HUD Urban Scholars Postdoctoral Fellowships 2738

The program is intended to expand urban research and ensure a continued pipeline of new researchers who will be vitally important to both research and public policy. The fellowships enable recent doctoral recipients to make a positive impact on the scholarship relating to housing and urban development. Applicants must have an academic appointment that will extend beyond the 15-month duration of the fellowship. The institution of higher education where the fellow is employed must designate a faculty adviser with relevant technical and subject expertise to monitor progress on the research project; provide office space and computer access; and waive indirect costs above the 8 percent allowed by the fellowship. Applicants must identify a mentor at their institution or elsewhere who is a respected scholar in the area of the research topic. (The mentor and faculty adviser can be, but do not need to be, the same person.) Fellows will also have a mentor at HUD who conducts research on a related topic and who will keep the fellow advised of policy developments.
Requirements: Applicants must have earned a PhD degree no earlier than January 1, 1997, in a field that relates to the research topic eligible for support.
Restrictions: A fellow may not hold another comparable major fellowship concurrently.
Amount of Grant: $55,000 maximum
Contact: Fellowship Program-HUD, (202) 334-2872; fax: (202) 334-3419; email: infofell@nas.edu
Internet: http://www7.nationalacademies.org/fellowships
Sponsor: Department of Housing and Urban Development
2101 Constitution Avenue
Washington, DC 20418

Hudson River Graduate Fellowships 2739

The foundation will award up to six full-time research graduate fellowships to doctoral- and master's-level students conducting research on the Hudson River system. The student's home university will be expected to be the primary source of support for materials and expenses required to do the thesis research. In special cases, applicants can apply for a 15-month fellowship to extend the proposed project through an additional summer. (Applications will not be accepted for extensions of existing fellowships.) The award is conditional on a full tuition waiver or reimbursement by the university.
Requirements: Applicants must be enrolled in an accredited doctoral or master's program, must have a thesis advisor and advisory committee, and must have a thesis research plan approved by the student's institution or department.
Samples: Larry Boles, Virginia Institute of Marine Science (VA)—for research on the potential for biological control of the zebra mussel in the Hudson River Estuary.; Steven Chillrud, Columbia U (NY)—for research on metal chronologies in the NY-NJ Metropolitan Hudson Drainage Basin.; Geoffrey Gratton, Buffalo State College (NY)—for research on estimating the population size and predatory demand of striped bass in the Hudson River.
Amount of Grant: $15,000 plus $1000 for supplies for doctoral fellowships; $11,000 plus $1000 for supplies for master's fellowships
Date(s) Application is Due: Apr 11
Contact: Grants Administrator, (212) 483-7667; fax: (212) 924-8325; email: info@hudsonriver.org
Internet: http://www.hudsonriver.org/graduate_fellow.htm
Sponsor: Hudson River Foundation
17 Battery Pl, Suite 915
New York, NY 10004

Hudson River Research Grants 2740

The foundation supports basic and applied research in the natural and social sciences and educational programs concerning all aspects of the Hudson River ecosystem, with an emphasis on studies that bear on potential human uses of the system. The geographical area of primary interest is the estuarine portion of the river. General areas of interest for scientific research include resource species, dynamics of Hudson River trophic webs, toxic substances, and hydrodynamics and sediment transport. The program also awards travel grants and expedited grants for the study of emergency situations affecting the Hudson River. Proposals must be preceded by a preproposal consisting of a maximum

three-page, single-spaced project description including an abstract and an estimated budget. Preproposals must be received by the listed application deadline date.
Requirements: Proposals are welcomed from individual researchers, researchers at colleges and universities, nonprofit/academic institutions, profit-making institutions, and governmental agencies. The foundation prefers that individual researchers seek some institutional affiliation for the purpose of carrying out the proposed research.
Samples: Bonnie Brown, Virginia Commonwealth U (VA)—for research on American shad in NJ coastal intercept fisheries, $63,102 over two years.; Frank Bohlen, U of Connecticut (CT)—for research on dredging effects on the particulate matter flux from the Hudson River Estuary, $85,340 over two years.
Amount of Grant: $50,000-$90,000 generally
Contact: Dr. Dennis Suszkowski, Science Director, (212) 483-7667; fax: (212) 924-8325; email: info@hudsonriver.org
Internet: http://www.hudsonriver.org/toc.htm
Sponsor: Hudson River Foundation
17 Battery Pl, Suite 915
New York, NY 10004

Huebner Foundation Fellowships 2741

The purpose of the fellowships is to increase the number of qualified professors specializing in insurance and risk management by providing financial assistance to graduate students aspiring to academic careers at the college or university level. Postdoctoral fellowships also are available. A successful candidate for a Foundation grant must certify that it is his or her intention to (1) follow a career as a full-time college or university faculty member with a teaching or research specialization in insurance economics and risk management; (2) specialize in insurance economics and risk management, and in a related discipline such as finance, for a graduate degree; and (3) not engage in any outside work for pay or profit without the consent of the Executive Director of the Foundation during the period of the fellowship.
Requirements: Applicant must be a citizen of the United States or Canada and have a bachelor degree from an accredited college or university. Candidates for Foundation grants must apply on a special form available from the Foundation. Candidates for Foundation grants must apply separately and directly for admission to the Wharton Doctoral Program. They must take the Graduate Management Examination Test (GMAT) or the Graduate Record Examination (GRE) administered by the Educational Testing Service in Princeton, New Jersey. A copy of the GMAT or GRE results must be sent by ETS to the Wharton Doctoral Program office, located at 1150 Steinberg Hall-Dietrich Hall at the University of Pennsylvania.
Amount of Grant: $15,000 annually for four years
Contact: Executive Director, (215) 898-9631; fax: (215) 573-2218
Internet: http://www.huebnergeneva.org
Sponsor: S.S. Huebner Foundation
3733 Spruce Street, Wharton School, 430 Vance Hall
Philadelphia, PA 19104-6301

Hugh Atkinson Memorial Award 2742

This award recognizes outstanding achievement, including risk taking, by academic librarians that have contributed significantly to improvements in the area of library automation, library management, and/or library development or research. Nominations include a letter that indicates: the name, address, and phone number of both the nominator and the nominee; a narrative supporting the nomination; and a current vita.
Requirements: The nominee must be a librarian employed in a university, college, or community college library in the year prior to application for the award and must have a minimum of five years of professional experience in an academic library. Individuals may nominate themselves or be nominated by others.
Samples: Nancy M. Cline—award recipient (2006); Paul M. Gherman—award recipient (2005); Jerome Yavarkovsky—award recipient (2004); Wendy Pradt Lougee—award recipient (2003); Harold W. Billings—award recipient (2002); Larry Frye—award recipient (2001); Kenneth Frazier—award recipient, $2000.; Susan Nutter—award recipient, $2000.
Amount of Grant: $2000
Date(s) Application is Due: Dec 1
Contact: Carlen Ruschoff, Chair (301) 314-0409, email: ruschoff@umd.edu; Megan Bielefeld (800) 545-2433 ext.2514, email mbielefeld@ala.org
Internet: http://www.ala.org/ala/alcts/divisiongroups/alctsdivawards/atkinsonaward/hughcatkinson.htm
Sponsor: Association of College and Research Libraries
50 E Huron Street
Chicago, IL 60611-2795

Huie-Dellmon Trust Grants 2743

The trust awards grants to Louisiana nonprofits in its areas of interest, including hospitals, higher and secondary education, libraries, and Protestant churches and organizations. Types of support include general operating support, capital campaigns, building construction/renovation, equipment acquisition, program development, scholarship funds, research, and matching funds. There are no application forms or deadlines.
Requirements: Central Louisiana nonprofit organizations are eligible.
Geographic Focus: Louisiana
Samples: Boy Scouts of America (Alexandria, LA)—for child abuse programs, $35,210.; First United Methodist Church (LA)—for operating support, $10,000.; Southern Forest Heritage Museum (LA)—for operating support, $25,942.
Amount of Grant: $250-$100,000 range
Contact: Richard Crowell Jr., Trustee, (318) 748-8141

Sponsor: Huie-Dellmon Trust
P.O. Box 330
Alexandria, LA 71309-0330

Huisking Foundation Grants 2744

The foundation awards grants in its areas of interest, including Catholic higher and secondary education, church support and social services, hospitals, and religion associations. Types of support include general operating support, building construction/renovation, research grants, endowment funds, scholarship funds, special projects, and continuing support. The board meets in April and November; letters of intent are due in February and August.

Samples: KITSAP Humane Society (OR)—for operating support, $1000.; Georgetown U, Alumni Fund, (Washington, DC)—for the scholarship fund, $6000.
Amount of Grant: $200-$60,000 range
Contact: Frank Huisking, Treasurer, (203) 426-8618
Sponsor: Huisking Foundation
291 Peddlers Road
Guilford, CT 06437

Human Growth Foundation Small Grants 2745

The foundation awards grants for investigation of human growth and its disorders, with special consideration given for the chondrodystrophies. Special consideration will be given to new investigators and ideas new to the field. Postdoctoral research dealing with all aspects of normal and abnormal human growth and development including, but not limited to, the areas of biological, psychological, educational, and diet and nutrition will be considered. Up to three grants will be awarded. A synopsis of the research results is required to be submitted the following year. An NIH-type biographical sketch and two-page letter of intent should be sent by May 15; full applications are due by September 1.

Restrictions: No portion of the grant can be used for administrative overhead.
Amount of Grant: $7500-$10,000
Date(s) Application is Due: May 15; Sep 1
Contact: Program Contact, (800) 451-6434 or (516) 671-4041; fax: (516) 671-4055; email: hgf1@hgfound.org
Internet: http://www.hgfound.org/smallgrants.html
Sponsor: Human Growth Foundation
997 Glen Cove Avenue
Glen Head, NY 11545

Human Rights Fellows Program 2746

The aim of this two-year program is to support the further development of a civil liberties network of lawyers in the countries of Eastern Europe and the former Soviet Union. Three lawyers from Central and Eastern Europe and the former Soviet Union will be selected to participate in the program. Fellows reside in the United States for one year and audit courses related to human rights at the American University Washington College of Law during the fall semester and spend the spring and summer semesters in internships with civil liberties organizations or legal services agencies in the United States. Fellows return to their home countries after the first year, where they spend at least one year working with NGOs on human rights advocacy on a nonprofit basis—providing legal services, litigating test cases, training and educating in human rights, etc. Criteria for selection includes the applicant's experience, the potential of the applicant to contribute to the protection and promotion of human rights in Eastern Europe and the former Soviet Union, and the suitability of the applicant's proposed role in the nominating NGO. Preference will be given to applicants under 35 years of age.

Requirements: Applicants from Central and Eastern Europe and the former Soviet Union with a law degree, a strong commitment to human rights, eligibility for legal practice in home country, and English proficiency may apply.
Amount of Grant: $1500 monthly stipend for up to 12 months, round-trip airfare to the United States, $500 textbook allowance, and medical insurance for one year while in the *Contact*: Eszter Filippinyi, Program Coordinator or Zaza Namoradze, Director, 36 1 327-3102; fax: 36 1 327-3103; email: justiceinitiative@sorosny.org
Internet: http://www.justiceinitiative.org/activities/lcd/fellows/index
Sponsor: Open Society Justice Initiative
Oktober 6 u 12
Budapest 1051 Hungary

Humanities Institute Visiting Scholars Fellowships 2747

The institute offers residential fellowships each academic year to one or two visiting scholars to pursue advanced work in the humanities. Projects may contribute to scholarly knowledge or to the general public's understanding of the humanities. Recipients might eventually produce scholarly articles, a monograph on a specialized subject, a book on a broad topic, an archaeological site report, a translation, an edition, or other scholarly tools. Fellowships support projects that can be completed during the tenure of an award or those that are part of a long-term endeavor. Applicants may be faculty or staff members of colleges or universities, or scholars and independent writers. Guidelines are available online.

Requirements: Applicants should have held the PhD for five years or more or possess a record of professional accomplishment.
Restrictions: Fellowships do not support projects to study teaching methods or theories. Neither do they support surveys of courses and programs or the preparation of institutional curricula.
Amount of Grant: $40,000 stipend
Date(s) Application is Due: Jan 15

Contact: UCHI, (860) 486-9057 or 486-9058; fax: (860) 486-9136; email: UCHI@uconn.edu
Internet: http://www.humanities.uconn.edu
Sponsor: University of Connecticut
215 Glenrook Road, U-4234
Storrs, CT 06269-4234

Humanities Summer Research Fellowships at the University of Oregon 2748

The University of Oregon Humanities Center invites applications for summer research fellowships. This is a residential research program supporting high-quality research in the humanities; arts; and interpretive fields of the social sciences, natural sciences, and professional studies. Research fellows are provided offices with computers and support services, use of the Knight Library at the university, and a small stipend for temporary relocation expenses. Fellowships are available for six to eight weeks from mid-June to mid-September. Fellows must pursue research projects full time during the fellowship period and may simultaneously have other support. Write for application form.

Restrictions: Fellowships are not awarded for dissertation research or writing.
Amount of Grant: $1400 maximum
Date(s) Application is Due: Dec 1
Contact: Julia Heydon, Associate Director, Oregon Humanities Center, (541) 346-1001; email: jheydon@uoregon.edu
Internet: http://darkwing.uoregon.edu/~humanctr/fellowship/index.html
Sponsor: University of Oregon Humanities Center
5211 Oregon Humanities Ctr
Eugene, OR 97403-5211

Hunter's Hope Foundation Grants 2749

Hunter's Hope funds projects directly related to Krabbe disease or other leukodystrophies. Eligible studies may investigate new treatment approaches, or study basic mechanisms related to these diseases. Areas of relevant research include, among others, gene therapy, myelin development, neurodegeneration and transplantation, developmental neurobiology, and molecular biology. Applications are accepted in three project categories: postdoctoral fellowships provide stipend support for individuals who are within five years of receiving their terminal degree (MD, DVM, PhD, or equivalent degree). Fellows will be working with senior investigators within their sponsoring institution. Fellowships are awarded for two years. Pilot studies are one-year projects designed to test new concepts or ideas and can serve as a source of preliminary data for future studies. Major research grants provide support to senior investigators conducting major research studies. Projects may be funded for one to three years. Applications are reviewed twice annually.

Amount of Grant: $100,000 maximum major research grants; $40,000-$60,000 postdoctoral fellowships; $30,000 maximum pilot studies
Date(s) Application is Due: Apr 15; Sep 15
Contact: Executive Director, (877) 984-4673 or (716) 667-1200; fax: (716) 667-1212; email: info@huntershope.org
Internet: http://www.huntershope.org/research/default.asp
Sponsor: Hunter's Hope Foundation
P.O. Box 643
Orchard Park, NY 14127

Huntington Library Babara Thom Postdoctoral Fellowships 2750

The Huntington is an independent research center with holdings in British and American history, literature, art history, and the history of science and medicine. This Fellowship is designed to support non-tenured faculty members who are revising a manuscript for publication. The amount of each award is $45,000 for a nine- to twelve-month tenure. Applications must be postmarked by December 15.

Requirements: Applicants must: be pursuing scholarship in a field appropriate to the Huntington's collections; have received the PhD between 2005 and 2007; and be non-tenured faculty.
Amount of Grant: $45,000
Date(s) Application is Due: Dec 15
Contact: Robert C. Ritchie, Director of Research; (626) 405-2194; fax: (626) 449-5703; email: cpowell@huntington.org
Internet: http://www.huntington.org/huntingtonlibrary.aspx?id=566
Sponsor: Huntington Library and Art Gallery
1151 Oxford Road
San Marino, CA 91108

Huntington Library Dibner History of Science Program Fellowships 2751

The Huntington is an independent research center with holdings in British and American history, literature, art history, and the history of science and medicine. The Dibner program offers long- and short-term fellowships which are designed to further study in the Burndy Library and the other history of science and technology resources at The Huntington. Long-term awards are equal to $45,000 for a nine- to twelve-month tenure. Short-term awards are $2,500 per month for a one- to five-month period. Applications must be postmarked by December 15.

Requirements: Applicants for long-term awards must: have a PhD or equivalent; and have received the PhD by June of the application year. Applicants for short-term awards must: have a PhD or equivalent; or be a doctoral candidate at the dissertation stage.
Amount of Grant: $45,000 maximum
Date(s) Application is Due: Dec 15
Contact: Robert C. Ritchie, Director of Research; (626) 405-2194; fax: (626) 449-5703; email: cpowell@huntington.org
Internet: http://www.huntington.org/huntingtonlibrary.aspx?id=566

Sponsor: Huntington Library and Art Gallery
1151 Oxford Road
San Marino, CA 91108

Huntington Library Mellon Fellowships   2752
The Huntington is an independent research center with holdings in British and American history, literature, art history, and the history of science and medicine. This Fellowship is designed to support applicants pursuing scholarship in a field appropriate to the Huntington's collections. The amount of each award is $45,000 for a nine- to twelve-month tenure. Applications must be postmarked by December 15.
Requirements: Applicants must: be pursuing scholarship in a field appropriate to the Huntington's collections; and have received the PhD in June of the application year.
Amount of Grant: $45,000
Date(s) Application is Due: Dec 15
Contact: Robert C. Ritchie, Director of Research; (626) 405-2194; fax: (626) 449-5703; email: cpowell@huntington.org
Internet: http://www.huntington.org/huntingtonlibrary.aspx?id=566
Sponsor: Huntington Library and Art Gallery
1151 Oxford Road
San Marino, CA 91108

Huntington Library NEH Humanities Fellowships   2753
The Huntington Library and Art Gallery, an independent research center with holdings in British and American history, literature, art history, and the history of science, will award a minimum of three fellowships funded by NEH. These fellowships are designed to permit scholars to spend from four to twelve months at the Huntington. Fellowships may be combined with sabbaticals or other stipendiary support. Preference will be given to scholars who have not held major fellowships during the three years preceding the year for which the center is awarding the fellowships. Fellows will be able to pursue their own research and writing while participating in the intellectual life of the Huntington.
Requirements: Applicants must be established scholars at the postdoctoral level or its equivalent, must be U.S. citizens or foreign nationals who have resided in the United States for at least three years, and must be pursuing scholarship in a field appropriate to the Huntington's collections.
Amount of Grant: $50,000 maximum
Date(s) Application is Due: Dec 15
Contact: Robert C. Ritchie, Director of Research; (626) 405-2194; fax: (626) 449-5703; email: cpowell@huntington.org
Internet: http://www.huntington.org/huntingtonlibrary.aspx?id=566
Sponsor: Huntington Library and Art Gallery
1151 Oxford Road
San Marino, CA 91108

Huntington Library Short-Term Fellowships   2754
The Huntington is an independent research center with holdings in British and American history, literature, art history, and the history of science and medicine. Within this category, The Huntington awards a number of specialized fellowships, including: Francis Bacon Foundation Fellowships in Renaissance England; Haynes Foundation Fellowships in Los Angeles History; Reese Fellowship in American Bibliography and the History of the Book in the Americas; Trent R. Dames Civil Engineering History Fellowship; and Christopher Isherwood Foundation Fellowships. The amount of each award is $2,500 per month for a one- to five-month tenure. Applications must be postmarked by December 15.
Requirements: Applicant scholars must have the PhD or equivalent degree or be doctoral candidates at the dissertation stage.
Amount of Grant: $$2,500-$12,500
Date(s) Application is Due: Dec 15
Contact: Robert C. Ritchie, Director of Research; (626) 405-2194; fax: (626) 449-5703; email: cpowell@huntington.org
Internet: http://www.huntington.org/huntingtonlibrary.aspx?id=566
Sponsor: Huntington Library and Art Gallery
1151 Oxford Road
San Marino, CA 91108

Huntington Library-British Academy Fellowships for Study in Great Britain   2755
The Huntington is an independent research center with holdings in British and American history, literature, art history, and the history of science and medicine. In cooperation with the British Academy, The Huntington offers a limited number of exchange fellowships in any of the fields in which the Huntington collections are strong. It is a one-month tenure, and applications should be postmarked by December 15.
Requirements: Applicant scholars must have the PhD or equivalent degree.
Date(s) Application is Due: Dec 15
Contact: Robert C. Ritchie, Director of Research; (626) 405-2194; fax: (626) 449-5703; email: cpowell@huntington.org
Internet: http://www.huntington.org/huntingtonlibrary.aspx?id=566
Sponsor: Huntington Library and Art Gallery
1151 Oxford Road
San Marino, CA 91108

Huntington's Disease Society of America Research Fellowships   2756
Fellowships are designed to assist promising young postdoctoral investigators in the early stages of their careers who are engaged in research at the basic and clinical levels relating to the cause and treatment of Huntington's disease. The amount of the award

will depend upon the policy of the sponsoring accredited medical school or university and the training and experience of the applicant. Fellowships are awarded for one year with the possibility of renewal. Application forms and requested information on curriculum vita, publications, research proposal, sponsorship, and other sources of funding are provided by the society.
Requirements: All applicants must have an MD or PhD degree or the equivalent and work must in some way be related to Huntington's disease. There is no citizenship requirement.
Samples: Dr. Tsugn Peng, Emory U (Atlanta, GA)—for research on impaired mitochortinal function, $30,000.; Dr. Nansheng Chen, U of British Columbia (Vancouver, BC)—for research on the role of glutamate receptor-HD protein interactions, $30,000.
Amount of Grant: $80,000 maximum over two years
Contact: Robert Graze, Grants Administrator; (212) 242-1968 or (800) 345-4372, ext. 227; email: rgraze@hdsa.org
Internet: http://www.hdsa.org/research/grant-applications.html
Sponsor: Huntington's Disease Society of America
505 Eighth Avenue, Suite 902
New York, NY 10018

Huntington's Disease Society of America Research Grants   2757
The Society awards research grants for one year for support of basic or clinical research related to Huntington's disease. Grant awards are provided as seed monies for new or innovative research projects in the hope that they will develop sufficiently to attract funding from other sources. Application forms are provided by the society and generally follow the format of NIH grant applications.
Requirements: Applicants may be doctorate or graduate students or equivalent professionals. There is no citizenship requirement.
Samples: Dr. Robert Ferante, Bedford VA Medical Ctr (Bedford, MA)—for research on mechanisms of neurodegeneration in Huntington's Disease, $35,000.; Dr. James Bibb, Rockefeller U (New York, NY)—for research on phosphohuntington, $35,000.
Amount of Grant: $100,000 maximum over two years
Contact: Robert Graze, Grants Administrator; (212) 242-1968 or (800) 345-4372, ext. 227; email: rgraze@hdsa.org
Internet: http://www.hdsa.org/research/grant-applications.html
Sponsor: Huntington's Disease Society of America
505 Eighth Avenue, Suite 902
New York, NY 10018

Hutchinson Community Foundation Grants   2758
While the Foundation provides support for a broad range of causes, particular interest is in the following program areas: projects that focus on early childhood development, youth development, and school to work; projects that promote developmental assets and thriving behavior in children and youth; men and fathers; and family economic success. Grant proposal applications must be requested by email. Instructions are available online.
Requirements: Nonprofit organizations in Kansas are eligible to apply. Special purpose units of government can apply for support of innovative projects located in Reno County.
Restrictions: Grant proposals from individuals or non-qualifying organizations will not be considered.
Geographic Focus: Kansas
Date(s) Application is Due: Aug 13
Contact: Audrey Abbott Patterson, President, (620) 663-5293; fax: (620) 663-9277; email: aubrey@hutchcf.org
Internet: http://www.hutchcf.org
Sponsor: Hutchinson Community Foundation
First National Bank Building, One North Main, Suite 501
Hutchinson, KS 67504

Hyde and Watson Foundation Grants Program   2759
The foundation's primary concern is supporting capital projects that increase the quality, capacity, or efficiency of a grantee's programs or services, such as purchase or relocation of facilities, facilities improvements, capital equipment, instructive materials development, or certain medical research areas. Broad fields include health, education, religion, social services, arts, and humanities.
Requirements: Primary geographic areas served include the New York City metropolitan region and primarily Essex, Union, and Morris Counties in New Jersey.
Restrictions: Requests for endowment or operating support or from fiscal agents generally will be denied. Grants are not made to individuals.
Geographic Focus: New Jersey; New York
Samples: Ctr for Molecular Medicine and Immunology (Newark, NJ)—for building renovation and cancer research equipment. $200,000.; Kent Place School (Summit, NJ)—to alter and modernize the dining hall, $50,000.
Amount of Grant: $5000-$25,000 average
Date(s) Application is Due: Feb 15; Sep 15
Contact: Hunter Corbin, President, (973) 966-6024; fax: (973) 966-6404; email: Hcorbin@HydeandWatson.org
Internet: http://fdncenter.org/grantmaker/hydeandwatson
Sponsor: Hyde and Watson Foundation
437 Southern Boulevard
Chatham Township, NJ 07928

i2 Foundation Grants 2760

The foundation promotes advancements in education, technology, environmental practices, medicine, and economic opportunity through programs improving the quality of life and creating a better society. Although the foundation's main priority is the development and education of youth, proposals related to a wide array of issues are accepted, including illiteracy, youth violence, early childhood development, poverty, improved healthcare, and scientific research. All future grant cycles are currently postponed and new proposals are not being accepted.

Requirements: Nonprofit organizations closely aligned with the foundation's goals are eligible.

Restrictions: Individuals, religious institutions, political organizations, and government entities are ineligible.

Samples: Ctr for Nonprofit Management—to launch the center's Nonprofit Technology program in Texas.; CARE—for its Horn of Africa Emergency Relief Program in Kenya.; Orphan Foundation of America—to fund 15 to 20 postsecondary education scholarships for orphans and foster youth in the United States.; Project Hope—to help expand the maternal and child health programs in Haiti, Dominican Republic, Guatemala, Nicaragua, Peru, Malawi, and Uzbekistan.

Amount of Grant: $5000-$50,000

Contact: Grants Administrator, (469) 357-4200; fax: (469) 357-7777; email: Melis_Jones@i2.com

Internet: http://www.i2.com/company/i2foundation/index.cfm

Sponsor: i2 Foundation

One i2 Pl, 11701 Luna Road

Dallas, TX 75234

IAFF Burn Foundation Research Grants 2761

The IAFF Burn Foundation Research Grant Program continues to lead to new knowledge and innovative approaches to the prevention and treatment of the physical and psychological problems that impair the quality of life for a burn patient. Research grant proposals are accepted annually and awards are distributed during the American Burn Association Annual Meeting. Topics of interest include by are not limited to pain management, physical and psychological rehabilitation, and wound healing and scarring, especially in children. Clinical projects are recommended. The foundation's Board of Medical Advisors oversees the selection process, reviewing each of the submissions. While all of candidates are deserving, only seven to ten grants are chosen and submitted to the IAFF Burn Foundation for consideration. Typically, five are funded.

Requirements: The foundation welcomes research grant applications from members of the American Burn Association. The areas of funded research fall into three categories: (1) Quality of Life - This area of research is the most common, focusing on issues such as pain management, nutrition, and physical and psychological rehabilitation. The major focus in this area is the care of burned children. (2) Burn Prevention - Research in this area is especially important in shaping prevention programs. (3) Basic Science - Research grants funded in this category surround nutrition, wound healing and scar formation. No animal studies are funded. The required application and detailed guidelines can be found at the sponsor's website.

Restrictions: No animal studies are funded. The grant does not include indirect costs.

Geographic Focus: Alabama; Alaska; Arizona; Arkansas; California; Colorado; Connecticut; Delaware; District of Columbia; Florida; Georgia; Hawaii; Idaho; Illinois; Indiana; Iowa; Kansas; Kentucky; Louisiana; Maine; Maryland; Massachusetts; Michigan; Minnesota; Mississippi; Missouri; Montana; Nebraska; Nevada; New Hampshire; New Jersey; New Mexico; New York; North Carolina; North Dakota; Ohio; Oklahoma; Oregon; Pennsylvania; Rhode Island; South Carolina; South Dakota; Tennessee; Texas; Utah; Vermont; Virginia; Washington; West Virginia; Wisconsin; WyomingCanada

Amount of Grant: $25,000

Contact: Patrick Morrison, Director, IAFF Burn Foundation; (202) 824-8620; fax: (202) 637-0839; email: burnfoundation@iaff.org

Internet: http://burn.iaff.org/grants.shtml

Sponsor: International Association of Fire Fighters (IAFF)

1750 New York Avenue, NW

Washington, DC 20006

Ian Axford (New Zealand) Fellowships in Public Policy 2762

The Ian Axford (New Zealand) Fellowships in Public Policy are for outstanding mid-career professionals to research, travel and gain practical experience in public policy in New Zealand for seven months. The program has three goals: to reinforce New Zealand-United States links by enabling Americans of high intellectual ability and leadership potential to come to New Zealand to gain experience and build contacts in the field of public policy development; to help improve the practice of public policy in New Zealand and the United States by the cross-fertilization of ideas and experience in the two countries; and to build a network of public policy experts on both sides of the Pacific, and encourage ongoing policy exchange between New Zealand and the United States. Fellows will be based at a relevant government agency in Wellington and carefully paired with a leading policy expert who will serve as their mentor. In addition, they will be expected to spend a substantial part of their Fellowship in contact with other relevant organizations to gain practical experience in their field in New Zealand. These contacts may include an extended or part-time placement.

Requirements: Applicants must be: a U.S. citizen with at least five years of experience in your profession; a mid-career professional active in any part of the public, business or non-profit sector; a potential leader and opinion-former in your chosen field; and have an interest in learning from your experience in New Zealand and capable of putting to effective use in the U.S., any policy lessons learned.

Restrictions: Fellowships are not awarded to support basic research or to enable study for an academic degree.

Date(s) Application is Due: Mar 1

Contact: Martin Boswell; +64 4 494 1507 or +64 4 472 2065; fax: +64 4 499 5364; email: martin@fulbright.org.nz

Internet: http://www.fulbright.org.nz/awards/am-ian-axford.html

Sponsor: Fullbright New Zealand

Level 8, 120 Featherston Street, P.O. Box 3465

Wellington 6140 New Zealand

Ian N. McKinnon Memorial Fellowship 2763

This four-month fellowship is open to qualified graduates of any recognized university who are registered in or admissible to a full-time program leading to a master's or doctoral degree at the University of Calgary. Fields of study supported are all areas relevant to the effective development and utilization of energy resources with special emphasis on economics, engineering, and geology.

Requirements: Candidates must be Canadian citizens or landed immigrants at the time of taking up the award.

Amount of Grant: $C3500

Date(s) Application is Due: Apr 15

Contact: Faculty of Graduate Studies, (403) 220-5417; email: graduate@ucalgary.ca

Internet: http://www.grad.ucalgary.ca/funding/internal_scholarships/lev_4/mckinnon.htm

Sponsor: University of Calgary

2500 University Dr NW, Earth Sciences Bldg

Calgary, AB T2N 1N4 Canada

IARC Expertise Transfer Fellowship 2764

The Fellowship is intended to enable an established investigator to spend normally from six to twelve months in an appropriate host institute in a low- to medium-resource country in order to transfer knowledge and expertise in a research area relevant for the host country and related to epidemiology, biostatistics, environmental chemical carcinogenesis, cancer etiology and prevention, infection and cancer, molecular cell biology, molecular genetics, molecular pathology and mechanisms of carcinogenesis. Applications should include a proposed collaborative research project, specifying the link to IARC's on-going activities and a letter of support from the host lab giving details of feasibility and anticipated benefit to the receiving institute. Priority will be given to projects directly linked to IARC's on-going research program, involving at least one contact at IARC. There will be an annual remuneration of up to $70,000, which will take into account the on-going salary of the Fellow. This amount may include limited support for the project. The cost of travel will also be met.

Requirements: Applicants should be established cancer researchers actively engaged in the field with appropriate scientific or medical qualifications and an excellent publications' record. They must also belong to the staff of a university or a research institution.

Amount of Grant: $70,000 maximum

Date(s) Application is Due: Nov 30

Contact: Dr. Paolo Boffetta; +33-(0)472-73-84-48; fax: +33 (0)472-73-80-80; email: fel@iarc.fr or vsa@iarc.fr

Internet: http://www.iarc.fr/en/education-training/expertisetransfer.php

Sponsor: International Agency for Research on Cancer

150 cours Albert-Thomas

Lyon Cedex 08 69372 France

IARC Postdoctoral Fellowships for Training in Cancer Research 2765

Applications for training fellowships are invited from junior scientists from low- or medium-resource countries wishing to complete their training in those aspects of cancer research related to coordinate and conduct both epidemiological and laboratory research into the causes of cancer. Disciplines covered include: epidemiology, biostatistics, environmental chemical carcinogenesis, cancer etiology and prevention, infection and cancer, molecular cell biology, molecular genetics, molecular pathology and mechanisms of carcinogenesis, with emphasis on interdisciplinary projects. The fellowship is for a period of one year, with the possibility of an extension for a second year subject to satisfactory appraisal.

Requirements: Candidates should have spent less than five years abroad (including doctoral studies) and have finished their doctoral degree within five years of the closing date for application or be in the final phase of completing their doctoral degree (M.D. or Ph.D.). They must provide evidence of their ability to return to their home country and keep working in cancer research. The working languages at IARC are English and French. Candidates must be proficient in English at a level sufficient for scientific communication.

Restrictions: Candidates already working as a postdoctoral fellow at the Agency at the time of application or who have had any contractual relationship with IARC during the 6 months preceding the application deadline or who have already spent more than one year at IARC cannot be considered.

Amount of Grant: 32,000 Euros

Date(s) Application is Due: Nov 30

Contact: Dr. Paolo Boffetta; +33-(0)472-73-84-48; fax: +33-(0)472-73-80-80; email: fel@iarc.fr

Internet: http://www.iarc.fr/en/education-training/postdoc.php

Sponsor: International Agency for Research on Cancer

150 cours Albert-Thomas

Lyon Cedex 08 69372 France

IARC Visiting Scientist Award for Senior Scientists 2766
The IARC is offering this Award for a qualified and experienced investigator with recent publications in international peer-reviewed scientific journals who wishes to spend from six to twelve months at the IARC working on a collaborative project in a research area related to the Agency's programs: epidemiology, biostatistics, environmental chemical carcinogenesis, cancer etiology and prevention, infection and cancer, molecular cell biology, molecular genetics, molecular pathology and mechanisms of carcinogenesis. There will be an annual remuneration of up to $80,000, which will take into account the on-going salary of the visiting scientist plus the cost of travel.
Requirements: Applicants must belong to the staff of a university or a research institution and should provide written assurance of a post to return to at the end of the period of award.
Amount of Grant: $80,000 maximum
Date(s) Application is Due: Nov 30
Contact: Dr. Paolo Boffetta; +33-(0)472-73-84-48; fax: +33-(0)472-73-80-80; email: vsa@iarc.fr
Internet: http://www.iarc.fr/en/education-training/vsa.php
Sponsor: International Agency for Research on Cancer
150 cours Albert-Thomas
Lyon Cedex 08 69372 France

IAS Mellon Fellowships for Assistant Professors 2767
The School of Historical Studies at the institute, with the support of the Andrew Mellon Foundation, has established a program of one-year memberships for assistant professors at universities and colleges in the United States and Canada to support promising young scholars who have embarked on professional careers. While at the institute they will be expected to engage exclusively in scholarly research and writing. Appointments will be for one full year (July 1 through June 30 with the option of staying through the second summer until August 15) and will carry all the privileges of membership at the Institute for Advanced Study. The stipend will match the combined salary and benefits at the member's home institution at the time of application. Application and instructions are available online.
Requirements: Assistant professors in areas represented in the School of Historical Studies (Greek and Roman civilization, the history of Europe, the Islamic world, East Asian studies, the history of art, and modern international relations) may apply, provided at the time of their arrival they will have served at least two and not more than four years as assistant professors in institutions of higher learning in the United States or Canada and provided they can return to their institution.
Date(s) Application is Due: Nov 15
Contact: Administrative Officer, email: mzelazny@ias.edu
Internet: http://www.hs.ias.edu/mellon.htm
Sponsor: Institute for Advanced Study
Einstein Drive
Princeton, NJ 08540

IATA Research Grants 2768
In order to enhance the National Athletic Trainer's Association's (NATA) Research and Education Foundation, the Illinois Athletic Trainers' Association has provided limited funding for athletic training related research in the state of Illinois. Grant monies from the IATA are awarded for projects concerning research in the domains relevant to athletic training within the state of Illinois. Grant applications will be reviewed and monies will be awarded during the Spring and Fall IATA BOD meetings.
Requirements: Applicants must be Certified/Licensed, members in good standing not only with the NATA but also the IATA. Applications must be submitted 3 months prior to the beginning of the research study, and the research must be relevant to the profession of athletic training within the state of Illinois. All applications will be analyzed bi-annually and as stated above must be submitted prior to the beginning of the project. Please note that there is limited funding for these awards, and you may be turned down due to lack of funding.
Restrictions: There will be no awards granted for recovery of cost involved with research that has already been initiated.
Geographic Focus: Illinois
Contact: Contact your Regional Representative
Internet: http://www.illinoisathletictrainers.org/Researchgrts.htm
Sponsor: Illinois Athletic Trainers‚Äô Association (IATA)
Wheaton North High School, 701 Thomas Road
Wheaton, IL 60187

IBM Center for the Business of Government Grants 2769
Through grants for research and forums, the organization stimulates research and facilitates discussion on new approaches to improving the effectiveness of government at the federal, state, local, and international levels. Individuals may apply for grants to produce a 30-40 page research report in one of the areas presented in the Research Announcement. The manuscript must be submitted no later than six months after the start of the project. Interested individuals should submit a three-page description of the proposed research and a resume, including a list of publications. Application and guidelines are available online.
Requirements: Individuals working in universities, nonprofit organizations, and journalism are eligible.
Samples: To be distributed among 16 scholars—for research on how technology can improve the effectiveness of federal and state government, $240,000.
Amount of Grant: $20,000 for each research paper
Date(s) Application is Due: Mar 1; Oct 1

Contact: Jonathan D. Breul, Executive Director; (202) 515-4382; fax: (202) 515-4375; email: jonathan.d.breul@us.ibm.com
Internet: http://www.businessofgovernment.org/main/apply/index.asp
Sponsor: IBM Center for the Business of Government
1301 K Street NW, 4th Floor, West Tower
Washington, DC 20005

ICC Faculty Fellowships 2770
Based on Ernest Boyer, concept that community engagement provides an excellent means by which the three components of faculty professional development of teaching, research, and professional service can support and strengthen each other, Indiana Campus Compact awards Faculty Fellowships as a model for faculty professional development through the scholarship of engagement. Up to 7 fellowships are competitively awarded in each academic year, in addition to one Senior Fellow. The fellowship amount is $3,750. Institutions must match at least $1,250. The fellowship fiscal period is twelve months, from June 1 to May 31.
Requirements: Faculty from all disciplines and professional schools at Indiana Campus Compact (ICC) member campuses are eligible to apply. Former Fellows are eligible to apply; however, preference is given to new applicants. Fellows are expected to model the integration of community engagement into all three aspects of their professional development in the following ways: (1) Teaching - Each Fellow will teach a service-learning course within the program period; (2) Research - Each Fellow will design a research, scholarly, or artistic project relating to the service performed with a community organization or to the practice of service-learning, to be initiated during the program period; (3) Professional Service - Each Fellow will provide direct service to a nonprofit organization during the program year to assist in addressing a particular community issue that is related to his/her academic discipline or expertise as an educator. Note: ALL proposals must be properly routed through the fiscal routing system at an applicant, institution prior to electronic submission to Indiana Campus Compact. However, if the institution is an Indiana University campus, routing through Research & Sponsored Programs or Sponsored Research Services is not required.
Geographic Focus: Indiana
Amount of Grant: $3,750
Date(s) Application is Due: Mar 27
Contact: J.R. Jamison, Associate Director; (317) 274-6500; fax: (317) 274-6510; email: jrjamiso@iupui.edu
Internet: http://www.indianacampuscompact.org/Default.aspx?tabid=86
Sponsor: Indiana Campus Compact (ICC)
620 Union Drive, Suite 203
Indianapolis, IN 46202

ICC Scholarship of Engagement Faculty Grants 2771
Eighteen faculty per year will be selected to choose one of three options related to the Scholarship of Engagement: develop or revise a service-learning course to be taught by the recipient, provide scholarly research on service-engagement, or provide a professional service to a local community agency. The selected faculty members will also encourage the development of service-engagement infrastructure on their respective campuses through trainings and workshops of other faculty. Award amounts are $2,250. Institutions are required to cash match at least $750.
Requirements: Adjunct or full-time faculty from all disciplines and professional schools at Indiana Campus Compact (ICC) member campuses are eligible to apply. An adjunct or full-time faculty member applicant must propose a project that focuses on one of three areas: (1) Teaching - Design a new course or revise an existing course to include a service-learning component; (2) Research - Develop and initiate a research project related to his/her discipline that addresses a community issue and advances the field of engagement through service-learning. Preference will be given to projects that are participatory action research; (3) Professional Service - Provide direct service to a nonprofit organization to assist in addressing a particular community issue. This professional service must be related to the faculty member, academic discipline or his/her expertise as an educator. Note: ALL proposals must be properly routed through the fiscal routing system at an applicant, institution prior to electronic submission to Indiana Campus Compact. However, if the institution is an Indiana University campus, routing through Research & Sponsored Programs or Sponsored Research Services is not required.
Restrictions: Grants are awarded on a reimbursement basis. Grant funds cannot be used for indirect costs, items of durable equipment costing more than $100 (unless approved in writing by ICC), payment for participants to do service, fundraising activities, entertainment costs, tickets to recreational events, ball games, zoos, etc., clothing such as T-shirts or hats, or cash incentives.
Geographic Focus: Indiana
Amount of Grant: $2,250
Date(s) Application is Due: Nov 7; Mar 28
Contact: Lindsay Doucette, Program Director for Constituent Development; (317) 274-6500; email: lgdoucet@iupui.edu
Internet: http://www.indianacampuscompact.org/Default.aspx?tabid=86
Sponsor: Indiana Campus Compact (ICC)
620 Union Drive, Suite 203
Indianapolis, IN 46202

ICCP Outstanding Young Investigator Award 2772
The annual award recognizes the recent contributions of a promising young investigator and, in so doing, encourages and facilitates his/her lifelong commitment to the field of spinal cord repair. The award is intended to encourage young scientists to devote their careers to spinal cord research; promote and enrich the international research field in

repair of the damaged spinal cord; and stimulate collaborative research in the field. The award supports inter-laboratory travel and expenses to complete a collaborative new study or a clearly defined project to learn a specific technique. The goal of the award is to benefit the grantee and the field for a spinal cord study or learning experience. Application and guidelines are available online.

Requirements: To be eligible, a candidate must be a postdoctoral fellow who possesses a doctoral degree at the time of proposal submission and who is in his/her first to seventh year of postdoctoral work. The winning nominee must be a postdoctoral fellow during the 12-month award period, and the study must be completed within one year.

Amount of Grant: $10,000
Date(s) Application is Due: Sep 1
Contact: Dr. John Steeves, fax: (604) 822-2924; email: steeves@icord.or
Internet: http://www.campaignforcure.org/Young%20Investigator%20Award.htm
Sponsor: International Campaign for the Cure for Spinal Cord Paralysis
ICORD at UBC, 2469-6270 University Boulevard
Vancouver, BC V6T 1Z4 Canada

ICCS Graduate Students Scholarships 2773

The scholarships enable successful candidates to spend four to six weeks at a Canadian university or research site other than their own doing research related to their thesis or dissertation in the field of Canadian studies. Scholarships are not intended to initiate a thesis or dissertation, but rather to provide access to crucial scholarly information and resources in Canada in support of a thesis/dissertation that is close to or at the point of being written.

Requirements: Each applicant must be nominated by his/her national Canadian studies association The student must obtain in writing the support of a faculty member at a Canadian university who has agreed to act as the student's academic sponsor during the tenure of his/her award.

Amount of Grant: $C5000 maximum
Date(s) Application is Due: Dec 31
Contact: Grants Administrator, (613) 789-7834 ext 242; fax: (613) 789-7830; email: csppec@iccs-ciec.ca
Internet: http://www.iccs-ciec.ca/pages/4_ICCSprogs/a_gradstu.html
Sponsor: International Council for Canadian Studies
75 Albert, Suite 908
Ottawa, ON K1P 5E7 Canada

ICCS Publishing Fund 2774

The fund assists with the publication and distribution in Canada of scholarly monographs on Canada written by foreign Canadianists who are members of a Canadian Studies Association or associate member belonging to the International Council for Canadian Studies. This fund assists foreign Canadianists by granting financial aid to a recognized scholarly press once the work is published. The fund may also grant financial assistance for the translation from English or French into a third language and from a third language into French or English. The funds allocated will apply to the printing or translating costs of a manuscript/book and not to the costs of research, typing, reproduction or any other stage in the manuscript preparation process. Applications must be presented to the ICCS in either of Canada's official languages. Guidelines are available online.

Requirements: Only scholarly manuscripts on Canada written by foreign Canadianists are eligible. Manuscripts must be previously unpublished. Manuscripts consisting of previously published texts, such as excerpts of books or articles, are eligible if a third of the texts have not previously been published. The author must be a member in good standing of an ICCS member association or associate member. A letter of support from a Canadian Studies Association or associate member belonging to the ICCS must accompany the application.

Restrictions: Collections of poetry, novels, plays, magazine articles or conference proceedings are not eligible.

Date(s) Application is Due: Oct 31
Contact: Guy LeClair, (613) 789-7834 ext 228; fax: (613) 789-7830; email: gleclair@cs-ciec.ca
Internet: http://www.iccs-ciec.ca/pages/4_ICCSprogs/f_publishfund.html
Sponsor: International Council for Canadian Studies
75 Albert Street, Suite 908
Ottawa, Ontario K1P 5E7 Canada

ICF Korean Literature Translation Fellowship 2775

The program was established to provide prospective translators with financial assistance for studying at Korean universities or at their home institutions to develop their skills as translators of Korean Literature. The important function is to cultivate translators of Korean literature, with the long-term goal of introducing Korean literature abroad. Fellowship term is one year, renewable. Each applicant should submit a recommendation letter from a professor of Korean literature, English literature, or Creative writing in any university department.

Requirements: Applications are invited from candidates whose English is native and who want to specialize in translating Korean literature into English.

Amount of Grant: $5000 quarterly
Date(s) Application is Due: Sep 30
Contact: Foundation Administrator, 82+2 2280-7233; fax: 82+2 2269-4310; email: yds@icfkorea.net
Internet: http://exam.ybmsisa.com/icf/icfe02_1.asp
Sponsor: International Communication Foundation
#56-15 Jongno 2 ga
Jongno-Gu Seoul 110-122 Korea

ICJ World Affairs Journalism Fellowships 2776

The fellowship is aimed at news managers, editors, commentary writers, and other gatekeepers of news—basically those who decide which stories appear in the paper—from U.S. community-based dailies. Twelve experienced media professionals will be selected to travel overseas for one to three weeks on assignment. While overseas, fellows will explore issues of local importance and submit articles to their home newspapers for publication. Up to 12 fellows will be selected each year. The program covers all domestic and international travel costs and provides support in setting up contacts while overseas. Guidelines and application are available online.

Date(s) Application is Due: Mar 1
Contact: World Affairs Journalism Fellowships, (202) 737-3700; fax: (202) 737-0530; email: wajf@icfj.org
Internet: http://www.icfj.org/worldaffairs.html
Sponsor: International Center for Journalists
1616 H Street NW, 3rd Fl
Washington, DC 20006

Ida Smedley MacLean International Fellowship 2777

The fellowship shall be awarded to encourage advanced scholarship by enabling university women to undertake original research in some country other than that in which the holder has received her education or habitually resides. Fellowships are intended to cover at least eight months of work and should be taken up within a reasonable period from the date of the award. Members of the IFUW should apply through their national affiliates (in the United States, the American Association of University Women, 1111 16th Street NW, Washington, DC 20036). The competition is held in even-numbered years. The deadline for receipt of applications is determined by each affiliate but normally falls between early September and mid-October of the year preceding the competition. Contact the national headquarters for the exact deadline date.

Requirements: An applicant must be a woman graduate who is a member of IFUW and must be well started on the research program to which the application relates.

Restrictions: A fellowship will not be awarded for the first year of a PhD program.

Amount of Grant: Sf8000
Date(s) Application is Due: Oct 1
Contact: Fellowships Officer, 41-22-731-23-80; fax: 41-22-738-04-40; email: info@ifuw.org
Internet: http://www.ifuw.org/fellowships/international.htm
Sponsor: International Federation of University Women
8 rue de l'Ancien Port
Geneva CH 1201 Switzerland

Idaho Humanities Council Fellowships 2778

The goals of this program are to stimulate scholarship in the humanities, to provide support for individuals who need to devote time and money to specific research projects, and to share the results of research with academic and public communities. The council invites research fellowship proposals from Idaho scholars or out-of-state scholars undertaking work on a subject of special relevance to Idaho. Scholar should have a minimum of a master's degree in the humanities, though individuals will be considered by their previous record of research. Up to four fellowships are awarded each year.

Requirements: Eligible applicants include Idaho scholars who are residents working in any area of the humanities and nonstate scholars undertaking work on a subject of special relevance to Idaho. Also, any scholar pursuing work on a special initiative put forth by the council in that year is eligible.

Geographic Focus: Idaho
Amount of Grant: $3500
Date(s) Application is Due: Sep 15
Contact: Program Contact, (208) 345-5346; fax: (208) 345-5347; email: jennifer@idahohumanities.org or rick@idahohumanities.org
Internet: http://www2.state.id.us/ihc/grantfrm.htm
Sponsor: Idaho Humanities Council
217 W State Street
Boise, ID 83702

IDEM Section 319(h) Nonpoint Source Program Grants 2779

This program funds projects that will work on a watershed level to reduce nonpoint source pollution in Indiana, lakes, rivers, and streams. Nonpoint source pollution does not come from a pipe. It results when water (rain or snowmelt) moves across land, such as city streets, agricultural fields and residential backyards, and picks up dirt, fertilizers, pesticides, animal wastes, road salt, motor oil and other pollutants. Nonpoint source pollution is also caused by wind, which like rain, can pick up soil particles and deposit them in lakes and streams. These pollutants have harmful effects on drinking water supplies, recreation, fisheries and wildlife. Nonpoint source pollution is the leading cause of water quality problems in Indiana and is responsible for many of the impairments identified on the 303(d) List of Impaired Waterbodies. Funding for selected projects will be provided by Clean Water Act Section 319 grant funds and match is provided by grant recipients and partners.

Requirements: To be considered for funding, the project sponsor (the entity responsible for the project and its overall success), must be one of the following: 1) Municipality; 2) County Government; 3) State Government; 4) Federal Government; 5) College/University; 6) Nonprofit 501(c)3. The program provides funding and technical assistance to groups that work on the watershed level with citizens to develop locally-based solutions to nonpoint source pollution. Specific ways to address nonpoint source water pollution include education/outreach on watershed management, information gathering activities such as conducting watershed inventories and water quality assessments for the purpose of developing comprehensive watershed management plans and implementing those plans, including implementation of best management practices that directly reduce sources of nonpoint source pollution. IDEM provides sixty percent (60%) of an approved project, total cost with Section 319 funds. A grant recipient must provide the remaining forty percent (40%) of the total project cost as match. Match may be in-kind services or cash. Match cannot come from any federal funding sources. Guidelines and application forms are available for download at the website. Do not use forms from previous years - they will not be accepted by IDEM.

Restrictions: The following is a list of activities that cannot be funded with Section 319 funds and cannot be counted as matching funds for a Section 319 grant: Permit fees; Food for meetings or other events; Purchase of agricultural equipment, or other large pieces of equipment (equipment modifications and leasing are allowable); Purchase of land or land easements (these activities can be counted as matching funds in some cases); Any project which is directed at water quantity rather than water quality, such as dredging, drainage, or flood control; Any practices, equipment, or supplies used to fulfill the requirements of any federal permit (NPDES permit, Section 401 Water Quality Certification, permits from the U.S. Army Corps of Engineers, as examples) or to comply with IDEM's Confined Feeding Operation rule or permit requirements, or to meet enforcement requirements; Wetland mitigation sites; Incentive payments or yield losses; Nonpoint source best management practices not sanctioned by IDEM or not sanctioned by a partner agency of IDEM; Practices not installed in accordance with standards and specifications developed by NRCS, IDNR or other recognized standards; Office furniture; Sales tax. Additionally, the following will not be funded by this grant program: Septic system pump outs, repairs, rehabilitations, or demonstrations of alternative septic systems; Projects whose sole purpose is data collection, research, demonstration of best management practices, or education/outreach. However, these activities may be incorporated as elements into a proposal that meets one of the three priorities.

Geographic Focus: Indiana
Date(s) Application is Due: Sep 1
Contact: Andrew Pelloso, Section Chief, (317) 233-2481; Doug Campbell, Solicitation Coordinator, (317) 233-8491; Laura Bieberich, Team Leader, 317-233-1863
Internet: http://www.in.gov/idem/resources/grants_loans/319h/index.html
Sponsor: Indiana Department of Environmental Management (IDEM)
Indiana Government Center North, 100 N. Senate Avenue
Indianapolis, IN 46204

IDNR C2000 Ecosystem Project Grants   2780
The purpose of the Program is to integrate the interests and participation of local communities and private, public and corporate landowners to enhance and protect watersheds through ecosystem-based management. The program is made up of partnerships, which are coalitions of local stakeholders (private landowners, businesses, scientists, environmental organizations, recreational enthusiasts, and policy makers). Application deadlines are February 28 for online applications and January 31 for hard-copy applications. Applicants are encouraged to pre-submit project applications to the appropriate Local Partnership Councils (LPCs) for their review and comments before completing their formal submissions to IDNR. Guidelines are available online
Geographic Focus: Illinois
Date(s) Application is Due: Jan 31; Feb 28
Contact: Grants Administrator, (217) 782-6232; email: dnr.grants@illinois.gov
Internet: http://www.dnr.state.il.us/orep/c2000/ecosystem/fy06/index.htm
Sponsor: Illinois Department of Natural Resources
One Natural Resources Way
Springfield, IL 62702-1271

IEEE Electrical History Fellowship   2781
This fellowship supports one year of full-time graduate study in the history of electrical engineering and technology at an engineering school of recognized standing in the United States or Canada.
Requirements: Individuals doing graduate or postgraduate work in the history of electrical technology are eligible.
Amount of Grant: $15,000
Date(s) Application is Due: Feb 1
Contact: Director, Center for the History of Electrical Engineering, (732) 932-1066; fax: (732) 932-1193; email: history@ieee.org
Internet: http://www.ieee.org/portal/index.jsp?pageID=corp_level1&path=membership/students/awards&file=sc_ehistory.xml&xsl=generic.xsl
Sponsor: Institute of Electrical and Electronics Engineers
39 Union Street
New Brunswick, NJ 08901-8538

IERF Sepmeyer Research Grant Program   2782
The Foundation's mission is to conduct research and disseminate information on world educational systems and to facilitate the integration of individuals educated outside the United States into the U.S. educational environment and work force. The Foundation conducts and supports comprehensive, quality research on world education systems, shares research findings with the international community, and provides research-based credentials, evaluations, and related services. The purpose of the research grant is to aid the Foundation in carrying out its mission. There is no application form. Applicants should submit a research project summary. Application information is available online.
Requirements: Research grants are available to all persons, irrespective of citizenship and country of residence.
Amount of Grant: $5000 maximum
Date(s) Application is Due: Feb 1

Contact: Susan Bedil, Executive Director, (310) 258-9451; fax: (310) 342-7086; email: grants@ierf.org
Internet: http://www.ierf.org/grants.asp
Sponsor: International Education Research Foundation
P.O. Box 3665
Culver City, CA 90231-3665

IES Scholarships   2783
The institute offers financial aid to students studying for a semester or full year at one of its centers abroad. IES centers are located in Durham, London, Freiburg, Berlin, Dijon, Dublin, Madrid, Milan, Nantes, Paris, Salamanca, and Vienna. Asian programs are located in Adelaide, Beijing, Nagoya, and Tokyo. South American programs are located in La Plata, Argentina. Most financial aid is awarded on the basis of need and scholarship and is applied to the comprehensive fee for the appropriate program. Competitive merit awards are also made to students for all programs. Scholarships are available in the areas of international business, urban issues, science, foreign languages, fine and performing arts, international relations, cross-cultural and comparative studies, leadership, and community involvement; special scholarships are available for study in Beijing, Dijon, Nantes, and Tokyo.
Requirements: Candidates must be accepted to an IES/IAS study abroad program.
Amount of Grant: $1 million total per year
Contact: Scholarships, (800) 995-2300 or (312) 944-1750; fax: (312) 944-1448; email: info@IESabroad.org
Internet: http://www.iesabroad.org/info/finaid.html
Sponsor: Institute for the International Education of Students (IES)
33 N LaSalle Street, 15th Floor
Chicago, IL 60602

IFP Minnesota McKnight Filmmaking Fellowships   2784
IFP Minnesota in partnership with the McKnight Foundation awards $25,000 fellowships to two Minnesota filmmakers annually. These fellowships recognize Minnesota artists for talent in working with film/video as demonstrated by two examples of completed, original works in any of the genres and formats of narrative, documentary, experimental, or animation in feature or short lengths. The program is intended to support mid-career artists by providing financial assistance, professional encouragement, and industry recognition. Work examples and application material are judged by a panel of industry professionals from outside the state of Minnesota. Judges look for consistent artistic excellence and merit, clarity and uniqueness of vision, professional quality in the technical aspects of production, and demonstrable, sustained growth in the artist's career. Additionally, judges assess the ability of the artist to present their application in an articulate and professional manner.
Geographic Focus: Minnesota
Amount of Grant: $25,000
Date(s) Application is Due: Mar 3
Contact: Lu Lippold; (651) 644-1912, ext.106; fax: (651) 644-5708; email: llippold@ifpmn.org
Internet: http://www.ifpmn.org/mcknight.html
Sponsor: Independent Feature Project Minnesota (IFP MN) / McKnight Foundation
2446 University Avenue West, Suite 100
St. Paul, MN 55114

IFP Minnesota McKnight Screenwriters Fellowships   2785
IFP Minnesota in partnership with the McKnight Foundation awards fellowships to two Minnesota screenwriters annually. These fellowships recognize Minnesota artists for talent in writing for the screen as demonstrated by one completed feature-length screenplay. The program is intended to support mid-career writers by providing financial assistance, professional encouragement, and industry recognition. Screenplays and application material are judged by two panels of industry professionals from outside the state of Minnesota. In addition to the monetary award, the winners receive entrance into the IFP Market in New York, as well as a live reading on stage at the Jungle Theater in Minneapolis with professional actors. Screenplays and applicants are judged on originality of ideas and uniqueness of writer's voice, skillful story structure, talent with creating dialog, and talent with character development.
Restrictions: The following are ineligible: organizations and companies; full-time and part-time students in any film-, video- or screenwriting-related degree program; screenplays associated with a degree-granting program; and employees of IFP/MSP or the McKnight Foundation. Fellows cannot attend school full time during the fellowship year. The screenplay may not be an adaptation of any previously written performed or filmed work by another artist; or have been submitted to the fellowship more than three times previously, under any title.
Geographic Focus: Minnesota
Amount of Grant: $25,000
Date(s) Application is Due: Feb 4
Contact: Lu Lippold; (651) 644-1912, ext.106; fax: (651) 644-5708; email: llippold@ifpmn.org
Internet: http://www.ifpmn.org/mcknight.html
Sponsor: Independent Feature Project Minnesota (IFP MN) / McKnight Foundation
2446 University Avenue West, Suite 100
St. Paul, MN 55114

IFP New York Anthony Radziwill Documentary Fund Grants   2786
The fund provides grants to emerging and established documentary filmmakers in the form of development funds (seed money) for specific new projects. Administered by IFP

New York, the fund seeks to provide funding for independent nonfiction filmmakers at the earliest stage of new work, traditionally a difficult point at which to secure funding. Grants will be awarded to documentary projects that are seeking funds needed for research, treatment and script development, initial interviews and shooting, the production of trailers/clips for further funding needs, etc. Typically a grant will be seed money, though in some cases it could be a supplement to another initial small grant.

Requirements: Grants are given to individuals with creative and financial control on the project. The grantee must be a U.S. legal resident who is 18 years of age or older. Grants are for nonfiction projects intended to be: feature length (over 50 minutes); for general audiences; and primarily for wide distribution via theatrical, television, and festival markets.

Restrictions: The following are not eligible: international projects for which neither the producer nor director is a legal U.S. resident; projects by students in undergraduate or graduate degree programs, or for fulfillment of any course requirement by nonmatriculated students; projects intended primarily for the educational or industrial markets or for promotional purposes; projects for which film or video is not the primary artistic component (work designed to accompany a larger theater piece or multimedia installation); individuals and immediate family members who are employees of any IFP chapter, or currently serving as evaluators, panelists, or on the fund advisory committee; and grantees who have received funding from the fund for any grant cycle within the previous two-year period.

Date(s) Application is Due: Mar 1; Sep 1
Contact: Fund Administrator; (212) 465-8200, ext. 830; fax: (212) 465-8525; email: info@ifp.org
Internet: http://www.ifp.org/common/page.php?ref=grants
Sponsor: Independent Feature Project New York (IFP NY)
104 West 29th Street, 12th Floor
New York, NY 10001

IFT Foundation Grants 2787

The foundation awards funding in science communications and career guidance. Science communications delivers the scientific perspective on food issues to media, policy makers, allied societies, and members. Priority topics include food safety, biotechnology, diet and health, food labeling, food and the environment, and support for research in food science and technology. Career guidance includes educational videos on food science and food science careers. Funds available for distribution by the foundation vary yearly. It is recommended that applicants contact the foundation prior to preparing a proposal. The intent of the foundation is to serve as a catalyst to establish programs rather than a permanent funding source for programs.

Contact: Tekla Syers, Director of Foundation Development, (312) 782-8424; email: tasyers@ift.org
Internet: http://www.ift.org/cms/?pid=1000260
Sponsor: Institute of Food Technologists Foundation
525 W Van Buren Street, Suite 1000
Chicago, IL 60601

IFUW Dr. A. Vibert Douglas International Fellowship 2788

This is a triennial fellowship funded by Canadian Federation of University Women (CFUW) and administered by International Federation of University Women (IFUW). This competition is open to women graduates for research for a period of eight months abroad. Members of the IFUW should apply through their national affiliate (in the United States, the American Association of University Women, 1111 16th Street NW, Washington, DC 20036). The deadline for receipt of applications is determined by each affiliate but normally falls between early September and mid-October of the year preceding the competition; contact national headquarters for exact date.

Requirements: An applicant must be a woman graduate and must be well started on the research program to which the application relates.
Restrictions: A fellowship will not be awarded for the first year of a PhD program.
Amount of Grant: $C12,000
Contact: Betty Dunlop, Fellowships Program Manager; (613) 234-8252 or (613) -234-2732; fax: (613) 234-8221; email: cfuwfls@rogers.com
Internet: http://www.cfuw.org/index.php?option=com_content&task=view&id=499&Itemid=159
Sponsor: International Federation of University Women
8 rue de l'Ancien Port
Geneva CH 1201
Switzerland

IFUW Winifred Cullis Grants 2789

This competition is held in even-numbered years and is open to women graduates who are members of the International Federation of University Women (IFUW). Grants are awarded for research and/or training in a country other than the one in which the candidate has received her education or habitually resides. The grants are to assist women graduates to obtain specialized training essential to their research; to train in new techniques in the humanities, social sciences, and natural sciences; or to carry out independent research including completion of a piece of well-advanced research. Members of the IFUW should apply through the IFUW affiliate in their country (in the United States, the American Association of University Women, IFUW Liaison Executive Office, 1111 16th Street NW, Washington, DC 20036). Application deadlines for members applying through on of IFUWs 71 national affiliates vary by country, but normally fall between early September and early October of the year preceding the competition; contact national headquarters for exact dates. Application deadline for independent members and international individual members is November 1.

Amount of Grant: Sf5000
Date(s) Application is Due: Oct 1; Nov 1
Contact: Anamaria Vere, Program Officer; 41-22-731-23-80; fax: 41-22-738-04-40; email: info@ifuw.org
Internet: http://www.ifuw.org/i_fell.htm
Sponsor: International Federation of University Women
10 rue du Lac
Geneva, Switzerland CH 1207
Switzerland

Ignacio Martin-Baro Fund for Mental Health and Human Rights Grants 2790

The organization supports children's programs in health and mental health and is a resource for advocacy, education, prevention, service, and action research. The fund has a strong preference for projects that are developing innovative and progressive ways to deal with the mental health consequences of violence and political repression. Applicants should understand that the foundation does not view mental health and human rights as issues that can be addressed separately, but are seeking to fund activities that explore the links between them. Groups with budgets under $25,000 receive priority consideration. The listed application deadline applies to letters of interest; full proposals are by invitation.

Amount of Grant: $7000 maximum
Date(s) Application is Due: Sep 1
Contact: Grants Administrator, (617) 469-7454; fax: (617) 469-3379; email: apply@martinbarofund.org
Internet: http://www.martinbarofund.org/contact/applying.htm
Sponsor: Ignacio Mart'n-Bar' Fund for Mental Health and Human Rights
P.O. Box 2122
Jamaica Plain, MA 02130

IHS Filmmaking and Writing Scholarships 2791

Scholarships up to $10,000 in tuition and stipend will be awarded to support students who: are pursuing a Master of Fine Arts (MFA) degree in filmmaking, fiction writing, or playwriting; have a demonstrated interest in classical liberal ideas and their application in contemporary society; and demonstrate the desire, motivation, and creative ability to succeed in their chosen profession. In recognition of the important role that films and novels play in the world of ideas, IHS provides support to promising young filmmakers and writers who share an appreciation for the potential and promise of a free society.

Amount of Grant: $10,000 maximum
Date(s) Application is Due: Jan 14
Contact: Scholarship Administrator, (800) 697-8799 or (703) 993-4880; fax: (703) 993-4890
Internet: http://www.theihs.org/subcategory.php/15.html
Sponsor: Institute for Humane Studies
3301 N Fairfax Drive, Suite 440
Arlington, VA 22201

IHS Humane Studies Fellowships 2792

The institute awards fellowships to support the studies of excellent students who have demonstrated an interest in the classical liberal tradition of individual liberties and free markets. The program is open to graduate students and advanced undergraduates pursuing degrees in the social sciences, humanities, jurisprudence, journalism, and related fields. The disciplines of law and business also are eligible if students intend intellectual careers. Application forms are available on the Web site. Annual deadline dates may vary; contact the institute for exact dates.

Requirements: Graduate and advanced undergraduate students enrolled at accredited colleges or universities in the United States or abroad who are intending academic or other intellectual careers, including public policy and journalism, are eligible.
Amount of Grant: $12,000 maximum
Contact: Humane Studies Fellowships, (703) 993-4880 or (800) 697-8799; fax: (703) 993-4890; email: ihs@gmu.edu
Internet: http://www.theihs.org/subcategory.php/2.html
Sponsor: Institute for Humane Studies
3301 N Fairfax Drive, Suite 440
Arlington, VA 22201-4432

IHS Native American and Alaskan Native Health Research Centers 2793

The purpose of this program is to establish centers that would encourage research on diseases that affect Native American populations and other health issues of importance to tribes and tribal organizations. Funds may be used to reduce health disparities that affect Native Americans and Alaskan Natives, train scientists and health professionals, help encouraging partnerships among health organizations, and for health research. A letter of intent is requested by October 1; full application is due by December 12. Annual deadline dates may vary; contact program staff for exact dates. RFA: GM-00-007

Requirements: Federally recognized tribes, tribal consortia, and nonprofit Indian health organizations are eligible to apply.
Amount of Grant: $27,000-$300,000; $115,000 average
Date(s) Application is Due: Oct 1; Dec 12
Contact: Lois Hodge, Grants Management Officer, (301) 443-5204
Internet: http://www.ihs.gov
Sponsor: Indian Health Service
12300 Twinbrook Plz
Rockville, MD 29852

IIE Cintas Fellowships 2794

Administered for Cintas Foundation by IIE, these fellowships are intended to acknowledge demonstrated creative accomplishments and to encourage the professional development of talented creative artists in the fields of visual arts, including film installation, art, painting, photography, sculpture, and other plastic arts. Each year fellowships are offered in different artistic disciplines. Fellows are free to pursue their arts activities either in the United States or in other countries approved by Cintas Foundation.

Requirements: Eligibility is limited to professionals in the arts who are of Cuban citizenship or lineage (at least one parent a Cuban) living outside of Cuba and who have completed their academic and technical training.

Restrictions: Individuals wishing to pursue academic programs are not eligible for awards nor are performing artists as opposed to creative artists. The fellowships, under ordinary circumstances, are not awarded more than twice to the same person.

Amount of Grant: $10,000

Date(s) Application is Due: Mar 10

Contact: U.S. Student Programs Division, (212) 984-5330; fax: (212) 984-5325; email: cintas@iie.org

Internet: http://www.iie.org/Template.cfm?&Template=/programs/fulbright/cintas/default.htm

Sponsor: Institute of International Education
809 United Nations Plz
New York, NY 10017

IIE Professional Development Fellowships 2795

Support is provided for young American researchers in professional, policy, and public administration-related fields who want to develop specialized knowledge of East Central Europe, the Baltic States, and the New Independent States (NIS). Candidates may apply for grants to Albania, Armenia, Azerbaijan, Belarus, Bulgaria, Croatia, Estonia, Georgia, Kazakhstan, Kyrgyzstan, Latvia, Lithuania, Macedonia, Moldova, Romania, Russia, Slovak Republic,Tajikistan, Turkmenistan, Ukraine, or Uzbekistan. The program is intended to support young specialists in the fields of business and economics, law, journalism, public administration, and international relations. Grants are available for periods of three to seven months. No provisions will be made for fellows' dependents. Round-trip international airfare, monthly living stipend, insurance, and book and travel allowance are provided. Annual deadline dates may vary; contact program staff for exact dates.

Requirements: Applicants must be U.S. citizens and be at least in the second (or terminal) year of a graduate or professional degree program, have graduated within five years from a graduate or professional degree program, or have language ability sufficient to carry out the proposed project by the time of departure from the United States.

Restrictions: Fellows may not receive any other Title VIII support during the fellowship tenure.

Date(s) Application is Due: Mar 15

Contact: U.S. Student Programs Division, (212) 984-5326; fax: (212) 984-5325; email: pdfnis@iie.org

Internet: http://www.iie.org/pgms/pdfnis

Sponsor: Institute of International Education
809 United Nations Plaza
New York, NY 10017

IIPP Graduate Fellowship 2796

The program supports graduate education toward a masters degree in international affairs or a related field. Fellows apply to and enroll in the nation's leading graduate programs, most members of the Association of Professional Schools of International Affairs (APSIA). Participating graduate schools provide monetary support over the course of a two-year program. IIPP, in turn, provides each of its fellows with a matching grant to be applied in accordance with an agreement between IIPP and the institution the fellow attends.

Requirements: To date, IIPP fellows have enrolled at the following institutions: American University, Columbia University, Georgetown University, George Washington University, Harvard University, Johns Hopkins University, Princeton University, Syracuse University, Tufts University, University of Denver, University of Maryland, University of Michigan, University of Washington, and Yale University. Non-APSIA schools include Cornell University, Pepperdine University, New York University, and the Bush School at Texas A&M.

Amount of Grant: $15,000 matching grants ($30,000 total)

Date(s) Application is Due: Mar 15

Contact: Fellowship Administrator, (800) 530-6232 or (703) 205-7623; fax: (703) 205-7645

Internet: http://161.58.87.106/content/program.cfm#grad

Sponsor: Institute for International Public Policy
2750 Prosperity Avenue, Suite 600
Fairfax, VA 22031

Ildaura Murillo-Rohde Scholarships 2797

The association awards these scholarships for tuition to outstanding Hispanic associate, diploma, baccalaureate, or graduate students. Recipients will be selected on the basis of potential to graduate, current academic standing in an accredited school of nursing (minimum 3.0 GPA), potential for contribution to the profession, and financial need. The amount awarded each year depends on the availability of funds.

Requirements: An applicant must be currently enrolled in an accredited school of nursing and be a U.S. citizen or legal resident of the United States, Puerto Rico, or the U.S. territories. Applicant must be a member of the association.

Date(s) Application is Due: Apr 30

Contact: Scholarships, (202) 387-2477; fax: (202) 483-7183; email: info@thehispanicnurses.org

Internet: http://www.thehispanicnurses.org

Sponsor: National Association of Hispanic Nurses
1501 16th Street NW
Washington, DC 20036

Illinois Schoolyard Habitat Action Grants 2798

The program is a means of funding for teachers and students who are interested in creating or enhancing schoolyard habitat areas. Examples include: trail development; vegetation planting; designing, establishing and maintaining a schoolyard prairie plot; butterfly garden; watering station; designing and building a bird feeder or feeding station. Guidelines are available online.

Requirements: Schools, nature centers, and youth groups may apply. Although not required, it is preferred that an educator trained in either Project WILD, Learning Tree or WET be involved with the project (contact the Illinois Department of Natural Resources, Educational Services Section at (217) 524-4126) or visit http://dnr.state.il.us.

Restrictions: Funding cannot be used for consultant fees, bird seed, fuel, equipment (shovels, rakes, trowels), labor, books, web site development, or land acquisition.

Geographic Focus: Illinois

Amount of Grant: $300 average; $600 maximum

Date(s) Application is Due: Oct 15

Contact: Grants Administrator, (217) 782-6232; email: dnr.grants@illinois.gov

Internet: http://dnr.state.il.us/lands/education/CLASSRM/Grants/Brochure.htm

Sponsor: Illinois Department of Natural Resources
One Natural Resources Way
Springfield, IL 62702-1271

Illinois Space Grant Consortium (ISGC) Graduate Fellowships 2799

These graduate fellowships support outstanding graduate students pursuing aerospace, space science, Earth system science and other interdisciplinary space-related science, engineering or mathematics fields. A total of up to $80,000 is available for fellowship awards each academic year. Individual awards will range from $5,000 to $10,000. Deadline for submission is February 2, with awards being announced by February 22.

Requirements: Applicants must have: U.S. citizenship; full-time current enrollment at an ISGC institution; a minimum 3.0 to 4.0 GPA; a good academic record; and research or design experience in a field related to space or aerospace.

Geographic Focus: Illinois

Amount of Grant: $5,000 - $10,000

Date(s) Application is Due: Feb 2

Contact: Diane Jeffers, Associate Director; (217) 244-8048; fax: (217) 244-0720; email: dejeffer@uiuc.edu

Internet: http://www.ae.uiuc.edu/ISGC/funding/grad-fellowship.html

Sponsor: Illinois Space Grant Consortium (ISGC)
306 Talbot Laboratory, 104 South Wright Street
Urbana, IL 61801-2935

Illinois Space Grant Consortium (ISGC) Individual Undergraduate 2800
Research Project Grants

These grants are designed to facilitate undergraduate participation in research in aerospace engineering, space or Earth system science at ISGC institutions. Awards of up to $4,000 will be allocated. The objective of this grant is to encourage undergraduate participation in research. The majority of requested funds must be used for undergraduate student stipends. A small amount may be used for salary, supplies, and travel support. Proposals will be reviewed by a ISGC Review Panel and selected based on adherence to the described criteria, scientific merit, and equitable distribution of resources across the Consortium. Deadline for submission is February 2, with awards being announced by February 22.

Requirements: The undergraduate student must be a U.S. citizen or permanent resident, and be identified in the proposal. The grants are available to faculty members or research scientists at all ISGC institutions. Grants must be matched (at least one-to-one) with funds from non-federal sources. The source of matching funds must be identified and confirmed by a letter from the appropriate authority. Waived indirect costs and faculty effort can qualify as matching funds. Women, under-represented minorities and persons with disabilities are encouraged to apply.

Restrictions: Funds cannot be used for the purchase of equipment.

Geographic Focus: Illinois

Amount of Grant: $4,000 maximum

Date(s) Application is Due: Feb 2

Contact: Diane Jeffers, Associate Director; (217) 244-8048; fax: (217) 244-0720; email: dejeffer@uiuc.edu

Internet: http://www.ae.uiuc.edu/ISGC/funding/ind-ug.html

Sponsor: Illinois Space Grant Consortium (ISGC)
306 Talbot Laboratory, 104 South Wright Street
Urbana, IL 61801-2935

Illinois Space Grant Consortium (ISGC) Multiple Undergraduate 2801
Research Program Grants

These grants are designed to fund programs of undergraduate research in aerospace engineering, space or Earth system science at ISGC institutions. The objective of this grant is to encourage undergraduate participation in research. Proposals can include requests for salary, supplies, and travel support. The program should involve multiple students working with multiple research advisors. The grants are available to faculty members or research scientists at all ISGC institutions. A total of $140,000 has been allocated to these grants.

Requirements: The undergraduate students must be U.S. citizens or permanent residents. Grants must be matched (at least one-to-one) with funds from non-federal sources. The source of matching funds must be identified and confirmed by a letter from the appropriate authority. Waived indirect costs and faculty effort can qualify as matching funds. Women, under-represented minorities and persons with disabilities are encouraged to apply. Participation in this program is limited to faculty (research and professorial) at ISGC affiliate and lead institutions.
Restrictions: ISGC funds cannot be used for the purchase of equipment.
Geographic Focus: Illinois
Date(s) Application is Due: Feb 2
Contact: Diane Jeffers, Associate Director; (217) 244-8048; fax: (217) 244-0720; email: dejeffer@uiuc.edu
Internet: http://www.ae.uiuc.edu/ISGC/funding/multi-undergrad.html
Sponsor: Illinois Space Grant Consortium (ISGC)
306 Talbot Laboratory, 104 South Wright Street
Urbana, IL 61801-2935

Illinois Space Grant Consortium (ISGC) Research Seed Grants   **2802**
These seed grants are designed to facilitate the development of research expertise in aerospace engineering/space science at ISGC institutions and allow the award recipients to obtain the preliminary results needed to support larger proposals to other federal or non-federal funding agencies. The objective of this grant is to support new areas of research. Selected proposals may reapply for a second year of funding. Proposals can include requests for salary, supplies, and travel support. Waived indirect costs and faculty effort can qualify as matching funds. The grants are available to faculty members or research scientists at all ISGC institutions. A total of $40,000 has been allocated to these grants for the period March 1 to February 29. The funding level of each seed grant is up to $10,000. Proposals will be reviewed by a ISGC Review Panel and selected based on adherence to the described criteria, scientific merit, and equitable distribution of resources across the Consortium. Deadline for submission is February 2, with awards being announced by February 22.
Requirements: Seed grant must support research in an area of interest to NASA (aerospace engineering, space or Earth system science). Grants must be matched (at least one-to-one) with funds from non-federal sources. The source of matching funds must be identified and confirmed by a letter from the appropriate authority. Women, under-represented minorities and persons with disabilities are encouraged to apply. Participation in this program is limited to faculty (research and professorial) at ISGC affiliate and lead institutions.
Restrictions: Funds cannot be used for the purchase of equipment.
Geographic Focus: Illinois
Amount of Grant: $10,000 maximum
Date(s) Application is Due: Feb 2
Contact: Diane Jeffers, Associate Director; (217) 244-8048; fax: (217) 244-0720; email: dejeffer@uiuc.edu
Internet: http://www.ae.uiuc.edu/ISGC/funding/seedgrant.html
Sponsor: Illinois Space Grant Consortium (ISGC)
306 Talbot Laboratory, 104 South Wright Street
Urbana, IL 61801-2935

Illinois Space Grant Consortium (ISGC) Undergraduate Scholarships   **2803**
Scholarships support outstanding undergraduate students pursuing aerospace, space science, Earth system science and other interdisciplinary space-related science, engineering or mathematics fields. A total of up to $70,000 is available for scholarship awards each academic year. Individual awards of approximately $2,500 per scholarship will be made. Award winners will be chosen based on student qualifications, scientific and educational relevance of the proposed activity, faculty member's expressed commitment to the project, and the application. Deadline for submission is February 2, with awards being announced by February 22.
Requirements: Applicants must have: U.S. citizenship; full-time current enrollment at an ISGC institution; a minimum 2.5 to 4.0 GPA; a good academic standing; and participation in an extra-curricular activity (i.e. Design/Build/Fly project, outreach, research, etc.). Women, under-represented minorities, and persons with disabilities are strongly encouraged to apply.
Geographic Focus: Illinois
Amount of Grant: $2,500
Date(s) Application is Due: Feb 2
Contact: Diane Jeffers, Associate Director; (217) 244-8048; fax: (217) 244-0720; email: dejeffer@uiuc.edu
Internet: http://www.ae.uiuc.edu/ISGC/funding/undergrad-scholarship.html
Sponsor: Illinois Space Grant Consortium (ISGC)
306 Talbot Laboratory, 104 South Wright Street
Urbana, IL 61801-2935

IMLS 21st Century Museum Professionals Grants   **2804**
The 21st Century Museum Professionals program supports a range of activities, including professional training in all areas of museum operations and leadership development. This program provides the museum community with support for a variety of training and personnel development activities for museum staff members across all types of museums, as well as the collection and dissemination of information to museum professionals and

the public. Project design could include direct dissemination of information through workshops, seminars, and courses, or indirect communication through publications and Web sites. Projects should benefit multiple institutions or diverse constituencies.
Requirements: Museums that fulfill the Eligibility Criteria for Museums may apply. Private not-for-profit museum services organizations or associations that engage in activities designed to advance the well-being of museums and the museum profession also may apply. In addition, institutions of higher education, including public and not-for-profit universities, are eligible. Please see Program Guidelines for specific eligibility criteria.
Amount of Grant: $15,000 - $500,000
Date(s) Application is Due: Mar 15
Contact: Christopher J. Reich, Senior Program Officer, 202-653-4685; email: creich@imls.gov
Internet: http://www.imls.gov/applicants/grants/21centuryMuseums.shtm
Sponsor: Institute of Museum and Library Services
1800 M Street NW, 9th Floor
Washington, DC 20036-5802

IMLS Conservation Project Support Grants   **2805**
The grants provide matching funds to help museums identify conservation needs and priorities and perform activities to ensure the safekeeping of their collections. Projects may assess four types of collections: nonliving, systematics/natural history, living plants, and living animals. Grants are available for different types of conservation projects, such as environmental improvements, research, surveys and long-range plans, training, and treatment. Projects may not exceed two years. IMLS will pay up to one-half the cost of the project. A museum's matching funds must come from cash contributions, earned income, in-kind contributions, or nonfederal funds. This program also funds Exceptional Projects. These are projects with broad-reaching effects that benefit multiple institutions. The November 1 deadline apples to programs for museums in counties of Alabama, Florida, Louisiana, and Mississippi that have been declared disaster areas by the federal government. Application and guidelines are available online.
Requirements: Museums located in one of the 50 states, the Commonwealth of Puerto Rico, American Samoa, the Virgin Islands, the Northern Mariana Islands, the Trust Territory of the Pacific Islands, Guam, or the District of Columbia may apply. Applicant organizations must have provided museum services, including exhibiting objects to the general public, on a regular basis for at least two years prior to application. A public or private nonprofit organization such as a municipality, college, or university that is responsible for the operation of a museum may, if necessary, apply on behalf of the museum.
Restrictions: A museum operated by a department or agency of the federal government is not eligible to apply. Funds may not be used for the acquisition of objects, contributions to endowment funds, costs of social activities, major renovations, new construction, or preaward costs.
Samples: U of Wisconsin (Madison, WI)—for conservation project, $24,000.
Amount of Grant: $250,000 maximum
Date(s) Application is Due: Oct 1
Contact: Steven Shwartzman, Senior Program Officer, (202) 653-4641; fax: (202) 606-8591; email: sshwartzman@imls.gov
Internet: http://www.imls.gov/applicants/grants/conservProject.shtm
Sponsor: Institute of Museum and Library Services
1800 M Street NW, 9th Floor
Washington, DC 20036

IMLS Grants to State Library Administrative Agencies   **2806**
Through the program, the Institute of Museum and Library Services provides funds to State Library Administrative Agencies (SLAAs) using a population-based formula. State libraries may use the appropriation to support statewide initiatives and services. They also may distribute the funds through subgrant competitions or cooperative agreements to public, academic, research, school, and special libraries in their state.
Amount of Grant: Based on state population
Date(s) Application is Due: Apr 1
Contact: George V. Smith, Associate Deputy Director, (202) 653-4650; email: stateprograms@imls.gov
Internet: http://www.imls.gov/programs/programs.shtm
Sponsor: Institute of Museum and Library Services
1800 M Street NW, 9th Floor
Washington, DC 20036-5802

IMLS Laura Bush 21st Century Librarian Program Grants   **2807**
The program supports efforts to recruit and educate the next generation of librarians and the faculty who will prepare them for careers in library science. It also supports grants for research related to library education and library staffing needs, curriculum development, and continuing education and training. Program priorities are masters-level programs, doctoral programs, preprofessional programs, research, programs to build institutional capacity, and continuing education and training. The grant period is up to three years, except doctoral programs, which may be up to four years. Applications, guidelines, and examples of successful proposals are available online.
Requirements: Beginning with the FY 2007 grant cycle, all applicants for the Laura Bush 21st Century Librarian program are required to file their applications, including all attachments, online through Grants.gov. All types of libraries, except federal and for-profit libraries, may apply. Eligible libraries include public, school, academic, special, private (not-for-profit), archives, library agencies, library consortia, and library associations. In addition, research libraries that give the public access to services and

materials suitable for scholarly research not otherwise available to the public and that are not part of a university or college are eligible. Institutions of higher education, including public and not-for-profit universities and colleges, also are eligible. Graduate schools of library and information science may apply as part of an institution of higher education.
Amount of Grant: $50,000-$1 million
Date(s) Application is Due: Dec 15
Contact: Stephanie Clark, Senior Program Officer, (202) 653-4662; email: sclark@imls. gov; Karmen Bisher, Program Specialist, (202) 653-4664; email: kbisher@imls.gov
Internet: http://www.imls.gov/applicants/grants/21centuryLibrarian.shtm
Sponsor: Institute of Museum and Library Services
1800 M Street NW, 9th Floor
Washington, DC 20036-5802

IMLS National Leadership Grants (NLG)   2808
National Leadership Grants enable libraries and museums to help people gain the knowledge, skills, attitudes, behaviors, and resources that enhance their engagement in community, work, family, and society. Projects should enable libraries and museums to address current problems in creative ways, develop and test innovative solutions, and expand the boundaries within which cultural heritage institutions operate. The results of these projects will help equip tomorrow, libraries and museums to better meet the needs of a Nation of Learners. Successful proposals will show evidence that they will have national impact and generate results,Äïnew tools, research, models, services, practices, or alliances,Äïthat can be widely adapted or replicated to extend the benefit of federal support. Proposals will reflect an understanding of current issues and needs, showing the potential for far-reaching impact throughout the museum and/or library community. Projects will provide creative solutions to issues of national importance and provide leadership for other organizations.
Requirements: All types of libraries, except federal and for-profit libraries, may apply. Eligible libraries include public, school, academic, special, private (notfor- profit), archives, library agencies, library consortia, and library associations. Research libraries and archives that give the public access to services and materials suitable for scholarly research not otherwise available and that are not part of a university or college also are eligible. Digital libraries that make library materials publicly available and provide services including selection, organization, description, reference, and preservation under the supervision of at least one permanent professional staff librarian are eligible to apply. Institutions of higher education, including public and not-for-profit universities and colleges, also are eligible. An academic unit, such as a graduate school of library and information science, may apply as part of an institution of higher education. Library applicants may apply individually or as partners. Additionally, all types of museums, large and small, are eligible for funding. Eligible museums include aquariums, arboreta and botanical gardens, art museums, youth museums, general museums, historic houses and sites, history museums, nature centers, natural history and anthropology museums, planetariums, science and technology centers, specialized museums, and zoological parks. Private nonprofit museum services organizations or associations that engage in activities designed to advance the well-being of museums and the museum profession also may apply. In addition, institutions of higher education, including public and nonprofit universities, are eligible.
Amount of Grant: $25,000-$1 million
Date(s) Application is Due: Mar 1
Contact: Martha Crawley, Senior Program Officer (libraries), (202) 653-4667; email: mcrawley@imls.gov; Dan Lukash, Senior Program Officer (museums), (202) 606-4644; fax: (202) 606-8591; email: dlukash@imls.gov
Internet: http://www.imls.gov/applicants/grants/nationalLeadership.shtm
Sponsor: Institute of Museum and Library Services
1800 M Street NW, 9th Floor
Washington, DC 20036-5802

Implant Dentistry Research and Education Foundation Grants   2809
The objective of the grant is to encourage new investigators by providing seed or start-up funding for promising research projects. Grants support basic research and education relating to dental implants. Areas include, but are not limited to, biology, biochemistry, biomechanics, biomaterials, and molecular biology. Grants are awarded after thorough review by an independent scientific advisory board.
Requirements: The principal or coprincipal investigator must hold an appropriate terminal degree. A dental resident or PhD candidate may apply if the appropriate department or chair provides a letter confirming he/she is on staff the year a project is funded.
Amount of Grant: $20,000-$25,000
Date(s) Application is Due: Jan 1
Contact: Craig Johnson, Executive Director, IDREF, (800) 442-0525 or (973) 783-6300; fax: (973) 783-1175; email: icoi@dentalimplants.com
Internet: http://www.dentalimplants.com
Sponsor: Implant Dentistry Research and Education Foundation
248 Lorraine Avenue, #3
Upper Montclair, NJ 07043-1454

Inamori Foundation Kyoto Prize   2810
The Kyoto Prize is awarded annually to persons who have made significant contributions in the three categories of Advanced Technology, Basic Sciences, and Arts and Philosophy. Through this Prize, the and groups worldwide Foundation seeks not only to recognize outstanding achievements but also to promote academic and cultural development and to contribute to mutual international understanding.
Samples: U.S. immunologist and geneticist Dr. Leonard A. Herzenberg, of Stanford, California (2006); Japanese statistical mathematician Dr. Hirotugu Akaike, of Tokyo

(2006); and Japanese designer Issey Miyake, an artist whose innovative creations transcend time, culture and social status, Tokyo (2006).
Amount of Grant: 50 million yen
Contact: Program Director, (075) 353-7272; fax: (075) 353-7270
Internet: http://www.inamori-f.or.jp/e_fd_out_out.html
Sponsor: Inamori Foundation
620 Suiginya-cho
Shimogyo-ku, Koyoto 600-8411 Japan

Inamori Foundation Research Grants   2811
Through the Inamori Grants and other research funding, the Foundation supports a wide variety of activities by young researchers in Japan in the fields of natural sciences, humanities, and social sciences. It aims to cultivate and nurture human resourses who will contribute to the future of our global society, to promote academic and cultural development, and to contribute to international understanding. In the founder's own words, "It is my sincere hope that the activities of the Foundation contribute to the progress and development of humankind and thereby allow me to repay, in some way, my indebtedness to all those in my local community, in my country, and throughout the world who have helped make me and my company what they are today."
Requirements: Must be a resident researcher living in Japan.
Contact: Research Program Director, (075) 353-7272; fax: (075) 353-7270
Internet: http://www.inamori-f.or.jp/e_fd_out_out.html
Sponsor: Inamori Foundation
620 Suiginya-cho
Shimogyo-ku, Koyoto 600-8411 Japan

Independent Institute Olive W. Garvey Fellowship   2812
The program awards fellowships to outstanding college students around the world through a competitive essay contest on the meaning and significance of economic and personal liberty. The program aims to encourage critical thinking and educational excellence by college students in examining the nature and relevance of human liberty. Fellowships are awarded in two categories: students up to age 35, and junior faculty members not yet tenured. The institute publishes the winning essays on its Web site and seeks to have them published elsewhere in major magazines and journals. Complete guidelines, an entry form, and additional information are available online.
Requirements: College or university students who are part-time, full-time, undergraduate, graduate, and from all nations and academic disciplines are eligible. (Student must be enrolled on date of essay submission.) Students must be no older than 35 years. Junior faculty members must hold a position of assistant professor or higher, not yet tenured, and may not be older than 35 years.
Restrictions: Previous fellowship recipients, judges, and current or past institute employees are ineligible.
Samples: David Mitchell, Graduate Student, Dept of Economics, George Mason U—for Property Rights, Interests Groups, and the New Economy, $2500.; Craig Smith, Graduate Student, Dept of Politics, U of Glasgow—for Does the New Economy Require a Free Economy?, $1500.
Amount of Grant: $2500 first prize, $1500 second prize, $1000 third prize to students; $10,000 first prize, $5000 second prize, $1500 third prize to junior faculty
Date(s) Application is Due: May 1
Contact: Carl Close, (510) 632-1366; fax: (510) 568-6040; email: CClose@independent. org
Internet: http://www.independent.org/students/garvey
Sponsor: Independent Institute
100 Swan Way
Oakland, CA 94621-1428

India Studies Language Training Fellowships   2813
The institute, with funding from the Canadian International Development Agency, will award a number of language training fellowships for the summer term to students in Canadian universities. The fellowships are tenable at North American institutions in cities other than the one in which the student studied in the previous term and that offer training in the Indian language of choice.
Requirements: Applicants must be Canadian citizens or permanent residents enrolled as full-time students in a Canadian university working toward a degree in the social sciences or the humanities (including management and law) with a substantial India studies or development studies component.
Samples: Martin Adam, U of Calgary (Canada)—to study Sanskrit.; Richard Coutinho, U of Toronto (Canada)—to study Hindi.
Amount of Grant: $C2500
Date(s) Application is Due: Jun 30
Contact: Program Officer, (403) 220-7467; fax: (403) 289-0100; email: sici@ucalgary. ca
Internet: http://www.ucalgary.ca/~sici/2004shastri/english/indiastudies.htm
Sponsor: Shastri Indo-Canadian Institute
2500 University Dr NW
Calgary, AB T2N 1N4 Canada

Indian Council of Social Science Research Grants   2814
The council funds research projects of scientists in the social sciences and sponsors projects on its own initiatives (e.g., research on the northeastern region of India, women's studies, and entrepreneurship). Types of support include research fellowships, contingency grants, grants for study and publications, and conferences and seminars. The council collaborates in research and exchange programs with a number of countries

and evaluates research proposals submitted by foreign nationals intending to undertake research in India.

Contact: Professor Andre Beteille, Chair, (0) 26179679; email: chairman@icssr.org
Internet: http://www.icssr.org/rp_main.htm
Sponsor: Indian Council of Social Science Research
35 Ferozeshah Road
New Delhi 110 001 India

Indiana 21st Century Research and Technology Fund Awards 2815

The Indiana 21st Century Research and Technology Fund of the Indiana Economic Development Corporation (IEDC) is open to proposals from all public and private entities for technology-based commercialization activities encompassing science/technology creation, innovation, and transfer intended to have commercial impacts. The fund intends to increase the numbers, and rates of development, of new and expanding technology-based companies by funding promising opportunities that, in some cases, the financial markets might find too risky. The Fund makes awards in two broad categories: Science and Technology Commercialization and Centers of Excellence. In addition, the Fund provides cost-share on behalf of Federal proposals submitted by Indiana-based entities. Generally awards are made in multiples of $50,000 up to $2,000,000. Support for awards in excess of $2,000,000 will be rare.

Requirements: The IEDC defines a technology-based company as one that is involved in transferring advanced technology into products, developing technologies with the near-term intention of creating products, or using new or advanced technologies in its design, development, and/or manufacturing of products. The Fund emphasizes the creation of academic-sector - commercial-sector partnerships. In making awards, the Fund expects significant leverage from the partners involved in the projects. Important: before applying, contact Fund staff (email preferred) to discuss your interest in submitting a proposal and to discuss your technology and commercialization goals. While not a review criterion, the fund encourages the inclusion of interns from any academic institution, or participating commercial sector partner, in order to increase project-related involvement of students at all levels.

Restrictions: Only direct costs will be supported. Institutions will not be provided indirect (overhead) cost support. Entities with previous Fund awards that are not current with regard to financial or technical reporting requirements will be disqualified from making new submissions to the Fund. Resubmissions of previously declined proposals will be considered only if substantive changes have been made to the proposal. Fund staff will determine whether to review resubmissions.

Geographic Focus: Indiana
Samples: 2K Corporation: $400,000 - Commercialization of a Neutron Based Explosives Detection Device (Car Bomb Detection). Arxan Technologies, Inc.: $1,944,096 - Protecting Critical IP in the 21st Century: Advancing Anti-Tamper Technologies. BioVitesse Inc.: $1,300,000 - Rapid Detection and Identification of Live Bacteria. CIS LLC: $1,075,000 - Commercialization of Satellite Radio for Cell Phone Applications.
Amount of Grant: $500,000 - $2,000,000
Contact: Linda Peterson-Roe, (317) 234-4652; email: lpeterson-roe@iedc.in.gov
Internet: http://www.21fund.org/
Sponsor: Indiana Economic Development Corporation
One North Capitol Avenue, Suite 900
Indianapolis, IN 46204

Indiana Library Federation Graduate Scholarships 2816

The purpose of the program is to give financial assistance upon availability of funding, in the form of scholarship grants, for library science education to candidates who give sufficient evidence of possessing exceptional talent for librarianship and who show economic need. Scholarship may be used at any ALA-accredited graduate library education program for the master's degree in library science (MLS).

Requirements: A recipient must be a legal resident of Indiana; must accept employment in an Indiana library within three months after completing library education; and must continue to work in an Indiana school or a public or university/college library for at least one year.
Geographic Focus: Indiana
Amount of Grant: $1000 maximum
Date(s) Application is Due: Jan 2
Contact: Jason Gilbert, Program Manager, (317) 257-2040; fax: (317) 257-1389; email: jgilbert@ilfonline.org
Internet: http://www.ilfonline.org/Scholarship.htm
Sponsor: Indiana Library Federation
941 E 86th Street, Suite 260
Indianapolis, IN 46220

Indiana SBIR/STTR Commercialization Enhancement Program (ISCEP) 2817

ISCEP provides funds to enhance commercialization activities of Indiana-based SBIR/STTR (Small Business Innovation Research/Small Business Technology Transfer) awardees. Specific ISCEP goals include: Support of thoughtfully structured commercialization plans of SBIR/STTR Phase II awardees; Accelerate and enhance commercial impacts of SBIR/STTR technologies, and; Establish and enhance successful technology-based businesses in Indiana. Proposals will be accepted from a small business that has received a federal Phase II SBIR or STTR and at least 50% of the federal program dollars must be expended prior to submission of a proposal in response to the RFP. Review considers both the technology development stage and the related business plan. The final stage of review involves a presentation to the IEDC. Between two and four awards of up to $350,000 may be made per funding cycle. The awards will be made with performance periods of up to 2 years and are contingent on the availability of funds.

Requirements: Eligible applicants must have a principal place of business in Indiana and the benefits from commercialization must accrue to an Indiana small business. Applicants must be a small business that has received a federal Phase II SBIR or STTR and at least 50% of the federal program dollars must be expended prior to submission of a proposal in response to the RFP, because an essential metric used in judging suitability for an award will involve assessing current progress toward achievement of Phase II objectives. The technology being commercialized must be directly related to the technology funded under the SBIR or STTR Phase II award. There is no limit as to the number of proposals that a single company can submit. However, only one award per company will be made per RFP cycle. The RFP, proposal tips and application are available at the website.
Restrictions: Awardees must maintain a principal place of business in Indiana for a term of 10 years beginning on the "Effective Date" of the agreement. Otherwise forfeiture and repayment the award plus interest will be required. Successful applicants must adhere to specific reporting requirements, including: quarterly progress reports, annual reports, annual site visits and a final report.
Geographic Focus: Indiana
Amount of Grant: $350,000 maximum
Date(s) Application is Due: Jul 20
Contact: Brooke Pyne, Director, SBIR/STTR Program; office: (812) 384-3078; mobile: (812) 381-0350; fax: (812) 384-3487; email: bpyne@iedc.in.gov
Internet: http://www.in.gov/iedc/sbir/index.html
Sponsor: Indiana Economic Development Corporation
SBIR/STTR Program Office, 32 E. Main Street
Bloomington, IN 47424

Indiana Space Grant Consortium Discovery Program Crosscutting Research-Team Grants 2818

The mission of the Indiana Space Grant Consortium is to promote science, technology, engineering and mathematics (STEM) workforce development, formal and informal education, and research by the dissemination of NASA related activities, content, and opportunities to the residents of the State of Indiana. The Vision for Space Exploration lays out a plan of action to support this mission. As NASA moves forward to develop the science and technology to fully implement the Vision, the Discovery Program for Research offers a means for INSGC members to address that broad mission. These funding opportunities seek to expand the realm of knowledge through discovery of new ideas and new techniques. These specific awards are intended to provide funds for developing new synergies for research collaboration, or a significant inter-institutional/inter-organizational funding initiative meeting INSGC and NASA priorities. Strong student participation in STEM workforce development type activities is encouraged. Amount of up to $25,000 are available for multiple institutions, and up to $15,000 for multiple departments / units at a single institution. The recommended match is 1.5 to 1 for all-academic partnerships, and 2 to 1 if at least one partner is from industry.

Requirements: Faculty and students receiving NASA/INSGC funds for direct support (salary or travel) must be U.S. citizens. At least two participating organizations must be named with cost share amounts coming from each of the participants.
Geographic Focus: Indiana
Amount of Grant: $25,000 maximum
Date(s) Application is Due: Feb 13
Contact: Dr. Barrett Caldwell, Director; (765) 494-5873; fax: (765) 496-3449; email: bscaldwell@purdue.edu or director@insgc.org
Internet: http://www.insgc.org/?q=node/108
Sponsor: Indiana Space Grant Consortium (INSGC)
203 S. Martin Jischke Drive, Gerald D. and Edna E. Mann Hall, Room 160
West Lafayette, IN 47907-1971

Indiana Space Grant Consortium Discovery Program Research Grants 2819

The mission of the Indiana Space Grant Consortium is to promote science, technology, engineering and mathematics (STEM) workforce development, formal and informal education, and research by the dissemination of NASA related activities, content, and opportunities to the residents of the State of Indiana. The Vision for Space Exploration lays out a plan of action to support this mission. As NASA moves forward to develop the science and technology to fully implement the Vision, the Discovery Program for Research offers a means for INSGC members to address that broad mission. These funding opportunities seek to expand the realm of knowledge through discovery of new ideas and new techniques. These awards are intended to support early career faculty (less than 6 years experience) members as they establish their research agenda on the national level, or for faculty with established careers attempting to develop new lines for research prior to requesting external funding. For these research grants, the amount available for early career faculty and research initiation is up to $10,000; faculty salary is limited to $3,000, with a match of 1:1 required.

Requirements: Faculty and students receiving NASA/INSGC funds for direct support (salary or travel) must be U.S. Citizens. Significant opportunities for student research experiences are required, and must be described prominently in the proposal.
Geographic Focus: Indiana
Amount of Grant: $10,000 maximum
Date(s) Application is Due: Feb 13
Contact: Dr. Barrett Caldwell, Director; (765) 494-5873; fax: (765) 496-3449; email: bscaldwell@purdue.edu or director@insgc.org
Internet: http://www.insgc.org/?q=node/108
Sponsor: Indiana Space Grant Consortium (INSGC)
203 S. Martin Jischke Drive, Gerald D. and Edna E. Mann Hall, Room 160
West Lafayette, IN 47907-1971

Indiana Space Grant Consortium Discovery Program Student   **2820**
Hands-On Research Grants
The mission of the Indiana Space Grant Consortium is to promote science, technology, engineering and mathematics (STEM) workforce development, formal and informal education, and research by the dissemination of NASA related activities, content, and opportunities to the residents of the State of Indiana. The Vision for Space Exploration lays out a plan of action to support this mission. As NASA moves forward to develop the science and technology to fully implement the Vision, the Discovery Program for Research offers a means for INSGC members to address that broad mission. These funding opportunities seek to expand the realm of knowledge through discovery of new ideas and new techniques. These types of project support are geared toward workforce development, with opportunities supporting faculty and student research by providing assistance with funded activities. Up to $3,500 per semester is available, including summer semester.
Requirements: A minimum 1 to 1 match is required. A higher match or leverage is recommended, particularly if there is support from ongoing sponsored research or contract projects. All students receiving NASA/INSGC funds for direct support (salary or travel) must be U.S. Citizens. Student activities must be research driven, and must focus on actual research rather than clerical or janitorial support. A student must be the project planner and active researcher.
Geographic Focus: Indiana
Amount of Grant: $3,500 maximum
Date(s) Application is Due: Feb 13
Contact: Dr. Barrett Caldwell, Director; (765) 494-5873; fax: (765) 496-3449; email: bscaldwell@purdue.edu or director@insgc.org
Internet: http://www.insgc.org/?q=node/108
Sponsor: Indiana Space Grant Consortium (INSGC)
203 S. Martin Jischke Drive, Gerald D. and Edna E. Mann Hall, Room 160
West Lafayette, IN 47907-1971

Indiana Space Grant Consortium Fellowships   **2821**
The mission of the Indiana Space Grant Consortium is to promote science, technology, engineering and mathematics (STEM) workforce development, formal and informal education, and research by the dissemination of NASA related activities, content, and opportunities to the residents of the State of Indiana. Available fellowships include: doctoral fellowships in Science Education of $15,000; doctoral fellowships in science and engineering disciplines of $15,000; and MA/MS fellowships of $7,500. Funds for fellowships may be used to offset costs for project expenses or student stipends. It is the applicant's responsibility to work with the institution to make the necessary arrangement for the type of funds being sought. INSGC Fellowships do not include tuition or fee remission. If the student receives a fee remission or any other form or support from their institution as part of their fellowship award, that is considered matching funds. Doctoral and masters level students may receive an award for up to two consecutive years, but must reapply separately for each year of fellowship support.
Requirements: To be eligible for a fellowship, an applicant must be: a U.S. citizen; enrolled in good standing as an active full-time undergraduate student at an INSGC affiliate institution throughout the scholarship period; and major in a science, technology, engineering, or mathematics (STEM) discipline or STEM education.
Restrictions: Indirect costs may not be charged to fellowships or scholarships. Administrative costs associated with fellowship and scholarship programs may be provided as matching funds by the institution.
Geographic Focus: Indiana
Amount of Grant: $7,500-$15,000 annually
Date(s) Application is Due: Feb 13
Contact: Dr. Barrett Caldwell, Director; (765) 494-5873; fax: (765) 496-3449; email: bscaldwell@purdue.edu or director@insgc.org
Internet: http://www.insgc.org/?q=node/109
Sponsor: Indiana Space Grant Consortium (INSGC)
203 S. Martin Jischke Drive, Gerald D. and Edna E. Mann Hall, Room 160
West Lafayette, IN 47907-1971

Indiana WaSuite Tire Fund (WTF) Program Grants   **2822**
The Indiana WaSuite Tire Fund (WTF) Program was created to help Indiana businesses undertake research, development, and/or commercial manufacturing projects that develop markets for use of scrap tires. Funds are made available through grants of up to $100,000 with a 50% match requirement are available for innovative research and development projects. Key elements include project partnerships and technology transfer. Provisions will be made for confidentiality of proprietary information.
Requirements: Eligible applicants are limited to Indiana businesses and not-for-profits. Interested applicants must contact OPPTA (Office of Pollution Prevention and Technical Assistance) to receive a grant application. Activities that may be funded through the program include the following: Civil engineering applications using shredded tires; Asphalt-rubber applications using ground rubber; Development of advanced technology and processes for production of products from scrap tires and; Tire-derived-fuel.
Restrictions: Projects must take place within the state of Indiana.
Geographic Focus: Indiana
Amount of Grant: $100,000 maximum
Contact: Call 800-988-7901 to find the Regional Grant Representative in your area.; email: recycling@idem.in.gov
Internet: http://www.in.gov/recycle/funding/wtf.html
Sponsor: Indiana Department of Environmental Management (IDEM)
100 North Senate Avenue, MC 64-01
Indianapolis, IN 46204-2251

Infinity Foundation Grants   **2823**
The foundation awards grants in a broad array of areas including charitable, scientific, religious, educational, and holistic healing activities by organizations and public agencies that work to better the lives of people. Grants also support individuals' research and development of educational materials to improve the authenticity of the portrayal of Indic traditions in the educational system. Proposed projects could result in one or more of the following: books, curriculum development, articles, conferences, CD-Roms, digital slide shows, Internet presentations, and audio/video materials. Topics covered may include philosophy, history, religion, science, art, and sociology, as they pertain to the educational curricula on Indic traditions. Proposals should be submitted by email.
Requirements: Grantee may be a scholar, teacher, visionary, or spiritual leader whose work in the designated topics would be enhanced by a foundation grant.
Contact: Rajiv Malhotra, (609) 683-0548; fax: (609) 683-0478; email: rm.infinity@gmail.com
Internet: http://www.infinityfoundation.com/callforgrantproposals.htm
Sponsor: Infinity Foundation
66 Witherspoon Street, Suite 400
Princeton, NJ 08542

Informix for Innovation Software Grants   **2824**
The program objective is to provide education institutions around the world and nonprofit institutions in the United States free access to the world's leading database technology. The Informix for Innovation Educational Grant Program offers educators the opportunity to use the latest technology as a teaching tool in the classroom and as an engine for academic research. Software is provided to encourage breakthroughs in technology and the development of next-generation data management applications. The Informix for Innovation Philanthropic Grant Program awards grants to U.S. nonprofits to help the organizations deliver their messages and information to further their cause. Projects should directly contribute to the nonprofit organization's vision and be visible to the public. Most software grants are for the development and enhancement of Web sites. The grant includes software, support, and training.
Requirements: The project or organization must be located in a community where Informix has a business presence. The project must be highly visible in the organization and the community and must have a projected time span, specific goals, and measurable results. Nonprofit organizations must be located in the United States.
Amount of Grant: $30,000 average
Contact: Amy Fenstermaker, Grant Programs Manager, (510) 628-3932; fax: (510) 628-3951; email: grants@informix.com
Internet: http://www-306.ibm.com/software/data/informix
Sponsor: Informix Software Inc
300 Lakeside Drive, Suite 2700
Oakland, CA 94612

Innovations Deserving Exploratory Analysis (IDEA) Program Grants   **2825**
The Transportation Research Board of the National Research Council funds projects testing innovative highway and intelligent transportation technologies and products. IDEA projects should examine the feasibility of innovative concepts or review new applications of advanced technologies or products from other industries that could be converted to transportation use. Requests for support may be submitted at any time. Applications are reviewed after March 31 and September 30 each year. Program announcements are posted on the Web site as are contacts for each area.
Requirements: U.S. and international investigators, businesses, universities, and research institutions may apply.
Amount of Grant: $25,000-$250,000
Date(s) Application is Due: Mar 1; Sep 1
Contact: Debbie Irvin, IDEA Administrative Assistant, (202) 334-3310; email: dirvin@nas.edu or ideaprogram@nas.edu
Internet: http://www4.trb.org/trb/dive.nsf/web/idea_programs
Sponsor: National Research Council
500 Fifth Street NW, Keck Ctr, WS 401
Washington, DC 20001

Innovative Technology Development Award Program   **2826**
The center annually funds biotechnology research projects that promise development of commercializable technologies. In this way, the center aims to further establish and strengthen ties between academic scientists and industry. Grants are for one year, with a continuation seed grant offered for the second and third year if necessary. Proposals are evaluated based on the quality of the science, the commercial potential, and how the project might impact New York state economy. Applications are posted on the Web site or are available by fax or mail.
Requirements: Faculty at any New York state research institution may apply.
Geographic Focus: New York
Samples: Roswell Park Cancer Institute—for DNA damage kits, in collaboration with Zeptogen Corp, $78,000.; Dept of Pathology, SUNY (Brooklyn, NY)—for rational design and testing of new anticancer agents, $32,000.
Amount of Grant: $40,000-$80,000
Contact: Dr. Anil Dhundale, Director of Scientific Affairs, (516) 632-8521; fax: (516) 632-8577; email: anil.dhundale@sunysb.edu
Internet: http://www.biotech.sunysb.edu/techDev/ITD/index.html
Sponsor: Center for Biotechnology
Psychology A Building, 3rd Fl, SUNY Stony Brook
Stony Brook, NY 11794-5208

Institute Award for Excellence in Industrial Gases Technology 2827

Sponsored by Praxair Inc and administered by AIChE, the award is given to recognize sustained excellence in the contribution to the advancement of technology for the production, distribution, and application of industrial gases. The recipient of the award shall have a record of sustained contribution recognized for its impact on advancing the frontier of industrial gases technology. The contribution can be characterized by a sustained record of important fundamental research, innovation, technological development, or the novel application of technology, either fostering or leading to important commercial results. Recipients need not be members of the institute.

Requirements: All members of AIChE as well as other interested persons are urged to nominate deserving candidates. All information and supporting documents are to be included with the completed form; nomination forms and instructions may be obtained from the institute.

Amount of Grant: $3000 plus $500 travel allowance

Date(s) Application is Due: Feb 15

Contact: AIChE Awards Programs, (212) 591-7107; fax: (212) 591-8890; email: awards@aiche.org

Internet: http://www.aiche.org/awards/awarddtl.asp?AwardID=59

Sponsor: American Institute of Chemical Engineers

3 Park Avenue

New York, NY 10016-5901

Institute for Advanced Studies in Culture Postdoctoral Fellowships 2828

The Institute supports research addressing the complex and dynamic relationship between religion and democratic ideals, institutions, and practices. Residential and nonresidential fellowships are available. Application materials and information are available online.

Requirements: Applicants may be from any discipline, but must be working on a project concerned with religion and public life, and must have defended their dissertation by May 1.

Amount of Grant: $35,000

Date(s) Application is Due: Jan 15

Contact: James Davison Hunter, Executive Director, (434) 924-7705; email: iasc@virginia.edu

Internet: http://www.virginia.edu/iasc/programs.html

Sponsor: University of Virginia

Institute for Advanced Studies in Culture, P.O. Box 400816

Charlottesville, VA 22904-4816

Institute for Ecumenical and Cultural Research Resident Scholars Program 2829

The institute is a postdoctoral residential research center for study, reflection, and writing, located on the campus of Saint John's Abbey and University. The institute encourages constructive and creative thought in theology and religious studies and, more generally, in the humanities, natural sciences, and social sciences as they relate to Christian tradition, including the interplay of Christianity and culture. Applicants may apply for a semester (September-December or January-May) or an entire academic year (September-May). Furnished apartments are provided for resident scholars and their families. Application materials are available upon request.

Requirements: The resident scholars program is normally for persons who already have academic doctoral degrees. The Admissions Committee occasionally considers persons with other qualifications. Applicants without a doctoral degree should seek advice from the institute's executive director prior to applying.

Amount of Grant: $3500 per year

Contact: Program Contact, (320) 363-3366; fax: (320) 363-3313; email: iecr@iecr.org

Internet: http://www.iecr.org/resident.htm

Sponsor: Institute for Ecumenical and Cultural Research

P.O. Box 2000

Collegeville, MN 56321-2000

Institute for European History Research Fellowships 2830

The institute annually awards 20 fellowships to young historians from Europe and abroad for research on the history of occidental religion and the history of Europe from the 16th to 20th centuries. Fellows live and work at the institute; their rooms serve both as work and living quarters. An open-stack library provides fellows with over 195,000 volumes pertaining to their specialized fields. A separate reading room for periodical literature offers a recent selection of more than 550 journals and newspapers.

Requirements: All fellows must have completed their undergraduate education. Candidates must have a thorough command of German and be either at the advanced stages of their dissertations or already in possession of their doctorates.

Contact: Dr. Heinz Duchhardt, European History Awards, +49 6131 39 93 60; fax: +49 6131 23 79 88; email: ieg2@inst-euro-history.uni-mainz.de

Internet: http://www.ieg-mainz.de/02-Zielsetzung/Index.htm

Sponsor: Institute for European History (Institut fur Europaische Geschichte)

Alte Universitatstrasse 19

Mainz 55116 Germany

Institute for Palestine Studies, Publishing, and Research Organization Grants 2831

The institute operates in the United States, Lebanon, France, and Jerusalem and awards research grants on all aspects of the Palestine problem, including the history and development of Palestine, the Palestinian problem and the Arab-Israeli conflict, and possible ways of arriving at a peaceful resolution.

Contact: Linda Butler, Acting Director, (202) 342-3990; fax: (202) 342-3927; email: ipsdc@palestine-studies.org

Internet: http://www.palestine-studies.org/final/en/intro/research.php

Sponsor: Institute for Palestine Studies, Publishing, and Research Organization

3501 M Street NW

Washington, DC 20007

Institute for Research in the Humanities Postdoctoral Fellowships 2832

The institute invites applications for postdoctoral Friedrich Solmsen Fellowships, tenable at the institute in the upcoming academic year in literary and historical studies with a European focus, antiquity through the 17th century.

Requirements: Candidates must have doctoral degrees by the time of application; expectation of the degree by the beginning of the fellowship is insufficient.

Amount of Grant: $30,000

Date(s) Application is Due: Oct 15

Contact: Loretta Freiling, Executive Secretary, Institute for Research in the Humanities, (608) 262-3855; fax: (608) 265-4173; email: freiling@facstaff.wisc.edu

Internet: http://www.wisc.edu/irh/research.html

Sponsor: University of Wisconsin—Madison

1401 Observatory Drive

Madison, WI 53706

Institute of American Cultures Graduate and Predoctoral Fellowships 2833

The UCLA Center for Asian American Studies, American Indian Studies Center, Center for African American Studies, and the Chicano Studies Research Center, in conjunction with the Institute of American Cultures, offer a limited number of predoctoral fellowships to support scholarly work in relevant ethnic studies. The awards are for one year. Reapplication for a second year of support is permitted. The deadline for African American and Chicana/o studies is January 13 and the deadline for American Indian and Asian American studies is March 30. On occasion, centers have chosen to divide the graduate fellowship between two or among three students.

Requirements: The fellow must be a full-time UCLA graduate student and make a contribution to the ongoing activities of the center. Application for the fellowship in African American Studies is open only to doctoral students who will have advanced to candidacy by the beginning of the fellowship year.

Amount of Grant: $17,000 per year stipend plus registration fees

Date(s) Application is Due: Jan 13; Mar 30

Contact: Program Contact, (310) 825-1233; email: IACcoordinator@gdnet.ucla.edu

Internet: http://www.gdnet.ucla.edu/iacweb/preweber.htm

Sponsor: UCLA Institute of American Cultures

1237 Murphy Hall, Box 951419

Los Angeles, CA 90095-1419

Institute of American Cultures Postdoctoral and Visiting Scholar Fellowships 2834

The UCLA Center for Asian American Studies, American Indian Studies Center, Center for African American Studies, and the Chicano Studies Research Center, in conjunction with the Institute of American Cultures, offers a limited number of postdoctoral fellowships. Currently, the IAC, in cooperation with UCLA's four Ethnic Studies Research Centers and the UCLA Library's Center for Oral History Research offers four postdoctoral fellowships (one per Center) that focus on conducting and theorizing oral history research in African American, American Indian, Asian American, and Chicano communities. Awards will depend on experience for those who have recently received a PhD. Senior scholars will receive a supplement to their sabbatical salaries to support work in ethnic studies. Application and guidelines are available online.

Requirements: Applicants must be citizens or permanent residents of the United States and hold a PhD from an accredited college/university (or, in the case of the arts, a terminal degree) in the appropriate field at the time of appointment. Fellows must teach or do research in the programs of the center.

Restrictions: UCLA faculty, staff, and currently enrolled students are not eligible.

Amount of Grant: $32,000-$34,000 per year plus health benefits and up to $4000 in research support

Date(s) Application is Due: Jan 13

Contact: IAC Coordinator, (310) 825-1233; email: iaccoordinator@gdnet.ucla.edu

Internet: http://www.gdnet.ucla.edu/iacweb/pstweber.htm

Sponsor: UCLA Institute of American Cultures

Box 951419, 1237 Murphy Hall

Los Angeles, CA 90095-1419

Institute of Current World Affairs Fellowships 2835

The purpose of the Institute of Current World Affairs, also known as the Crane-Rogers Foundation, is to identify promising young (under 36) internationalists with the drive, self-discipline, and ability to spend a minimum of two years studying and writing about areas and issues outside the United States. Full support is provided for the fellows and their immediate families, requiring that they write monthly reports or newsletters to the executive director. Applications are accepted at any time. It is suggested that applicant write for a brochure explaining the fellowship program, application procedure, and current areas of particular interest. Listed deadlines are for letters of interest; full application is by invitation. Annual deadlines vary; contact program staff for exact dates.

Date(s) Application is Due: Feb 28; Aug 1

Contact: Peter Martin, Executive Director, (603) 643-5548; fax: (603) 643-9599; email: icwa@valley.net

Internet: http://www.icwa.org/#anchor110081

Sponsor: Institute of Current World Affairs

4 W Wheelock Street

Hanover, NH 03755

Institute of Food Technologists Graduate Fellowships and Undergraduate Scholarship Program   **2836**

The IFT-administered scholarships/fellowships program is available each year to outstanding undergraduate and graduate students. These awards are given by food industry companies, organizations, divisions, and sections of the IFT, and the IFT to assist young scientists in obtaining the education and training necessary to fill positions in industry, government, and education. Graduate and junior/senior applications are due to department heads February 1; freshman applications, February 15; and sophomore applications, March 1. Instructions and applications are available on the Web site, or from the fax-on-demand system: (800) 234-0270.

Requirements: Applicants for undergraduate scholarships must be enrolled or planning to enroll in an IFT-approved food science/technology program. Graduate fellowship applicants can be enrolled in any college or university that is conducting fundamental studies to further food science/technology.

Restrictions: Research in such disciplines as genetics, horticulture, nutrition, microbiology, biochemistry, engineering, chemistry, etc., is not eligible unless it is directly related to food science/technology.

Amount of Grant: $1000-$5000 graduate fellowships; $1000-$2250 undergraduate scholarships

Date(s) Application is Due: Feb 1; Feb 15; Mar 1

Contact: Patti Pagliuco, Fellowships Administrator, (312) 782-8424; email: pgpagliuco@ift.org

Internet: http://www.ift.org/cms/?pid=1000438

Sponsor: Institute of Food Technologists

221 N LaSalle Street

Chicago, IL 60601

Institute of Global Conflict and Cooperation Research, Conference, and Teaching Grants   **2837**

Institute of Global Conflict and Cooperation (IGCC) research and teaching grants are awarded to University of California (UC) faculty to stimulate independent research and education projects on international issues of contemporary importance. Grants support research and teaching on the nine UC campuses on the causes of international conflict and opportunities to promote international cooperation. IGCC research, research conference, and teaching grants are designed to stimulate independent research and education projects on international issues of contemporary importance. Topics of interest are posted on the Web site, as are application procedures.

Requirements: Applications are accepted from UC faculty in all disciplines. Collaborative work is encouraged; however, the principal investigator must have an integral role in the project. Collaborative projects between UC faculty and the scientific and the technical staff of the national laboratories are eligible. Research projects may be either theoretical or policy analytical.

Amount of Grant: $35,000 maximum

Date(s) Application is Due: Mar 7

Contact: Kim Newin, IGCC Campus Programs Representative, (858) 534-8602; email: ktnewin@ucsd.edu

Internet: http://www-igcc.ucsd.edu/cprograms/fac_grant_menu.php

Sponsor: University of California, San Diego

9500 Gilman Drive, Dept 0518

La Jolla, CA 92093-0518

Institute of Human Values in Health Care Interdisciplinary Fellowships   **2838**

The program is designed to provide non-clinical professionals who work in health related fields (e.g., law, government, economics, sociology, philosophy, journalism, religion, and history) with first-hand exposure to and involvement with intensive inpatient and ambulatory care of the critically ill within a major academic medical center. The fellowship runs from January through May.

Requirements: Professionals who are active and established in their fields and can devote full time to the program for five months may submit fellowship applications. Postdoctoral candidates also will be considered.

Amount of Grant: $7500

Date(s) Application is Due: Sep 15

Contact: Dr. Robert Sade, Director, (843) 792-5278; fax: (843) 792-8286; email: sader@musc.edu

Internet: http://www.values.musc.edu/news/if_announcement.htm

Sponsor: Medical University of South Carolina

P.O. Box 250612, 96 Jonathan Lucas Street, Suite 409

Charleston, SC 29425

Institute of Mental Hygiene Grants   **2839**

The program invests in children through its grantmaking programs, active involvement with grantees, and leadership in improving mental health programs and policies. Grants support organizations that are tax-exempt and provide services in New Orleans. The foundation supports the following programs: early childhood mental health grants, children's mental health grants, early childhood mini-grants, and technical assistance grants. Applications and guidelines, including deadline dates, are available online.

Requirements: Louisiana nonprofit organizations are eligible.

Restrictions: Capital projects, general fund drives, or grants to individuals will not be considered.

Geographic Focus: Louisiana

Samples: Safehouse Youth Services, Lifestyle Counseling Program (New Orleans, LA)—to work with African American youth ages nine - 21 who are enticed by or at risk of becoming involved in the drug trade, $50,000 (2003); Juvenile Justice Project of Louisiana (New Orleans, LA)—to visit model programs and to create a new model for effective legal representation and treatment for troubled youths that includes mobilizing their families and communities,$14,250.; Louisiana State U Health Sciences Ctr, Dept of Psychiatry (New Orleans, LA)—to expand on-site, comprehensive mental-health consultation and treatment services to several early childhood centers, $194,000 over two years.

Contact: Nancy Freeman, Executive Director, (504) 566-1852; fax: (504) 566-1853; email: nfreeman@imhno.org or imh@imhno.org

Internet: http://www.imhno.org/imh_guidelines.html

Sponsor: Institute of Mental Hygiene

1055 Street Charles Avenue, Suite 350

New Orleans, LA 70130

Institute of Paper Science and Technology Fellowships   **2840**

Fellowships are awarded in the interdisciplinary program in chemistry, chemical engineering, mechanical engineering, physics, biology, and mathematics leading to the MS and PhD degrees. Alumni have obtained responsible positions in research, development, production, and management in the pulp and paper industry and in chemical and related industries, with a small but increasing number in advanced education.

Requirements: Applicants generally have BA or BS degrees in chemistry, paper science, chemical engineering, or mechanical engineering, but well-qualified individuals with degrees in other physical sciences and engineering areas will be considered. Citizens or permanent legal residents of the United States, Canada, or Mexico enrolled at the institute are eligible.

Amount of Grant: $16,000 per year plus tuition waiver, master's level; $18,000 per year plus tuition waiver, doctoral level

Contact: Project Manager, (800) 558-6611; email: energy.challenge@ipst.gatech.edu

Internet: http://www.ipst.gatech.edu/energy_challenge/website/challenger/ipst.html

Sponsor: Institute of Paper Science and Technology

500 10th Street NW

Atlanta, GA 30318

Institute of Russian and East European Studies Postgraduate Studentships   **2841**

The studentships, open to applicants worldwide, will pay university fees for a one-year course of full-time study for the postgraduate degree of M.Phil in Russian and East European Studies. The M.Phil consists of three courses and a dissertation, plus a language exam. Language tuition is provided. Program may be temporarily suspended; contact the institute for current availability.

Requirements: Applicants should have an undergraduate degree or an equivalent qualification in a social science subject. Candidates must commence their study at the institute in Glasgow.

Contact: Richard Berry, Director ICEES, Postgraduate Studentships, 44 141 330 5585; fax: 44 141 330 5594; email: R.R.Berry@socsci.gla.ac.uk

Internet: http://www.gla.ac.uk/departments/dcees

Sponsor: University of Glasgow

Hetherington Building, Bute Gardens

Glasgow, Scotland G12 8RS United Kingdom

Institute of Textile Technology Graduate Fellowships   **2842**

Up to 10 students will each be awarded a tuition scholarship and a nonservice fellowship. Each ITT fellowship provides the student with a stipend for the nine months of the academic year. The Academic Committee will decide which students receive these financial awards. Students also may apply for Federal aid in the form of student loans. Fellowships require no service, such as assistance with laboratory work or teaching, and are continued for the second year, provided the student's performance during the first year is satisfactory. All students are expected to maintain a minimum average of B in the course work.

Requirements: Fellowships are intended for U.S. citizens holding bachelor's degrees and interested in careers in the textile industry.

Restrictions: If at the end of the first year the required average is not attained, the fellowship cannot be renewed for the second year.

Amount of Grant: $12,000 stipend plus tuition and fees per academic year

Date(s) Application is Due: Mar 15

Contact: Program Officer, (919) 513-7583; fax: (919) 882-9410

Internet: http://www.itt.edu/Education/FinancialAid.asp

Sponsor: Institute of Textile Technology

2401 Research Drive, Box 8301

Raleigh, NC 27695-8301

Institute of Turkish Studies Dissertation Writing Grants   **2843**

The institute is devoted solely to the advancement of training, research, and teaching in the field of Ottoman and modern Turkish studies. These grants are intended for advanced students who have finished the research stage of their dissertation, and they may not be used for dissertation research. The dissertation writing grants will be awarded only to applicants who certify that they will not be involved in teaching beyond the half-time level. Application and guidelines are available online.

Requirements: Graduate students in any field of the social sciences and/or humanities, who are U.S. citizens or permanent residents at the time of application; currently enrolled in a PhD degree program in the United States; and expecting to complete all PhD requirements except their dissertation by March are eligible.

Amount of Grant: $5000-$10,000 stipend

Contact: David Cuthell, (202) 687-0295; fax: (202) 687-3780; email: dcc@turkishstudies.org

Internet: http://www.turkishstudies.org/grand_app.html
Sponsor: Institute of Turkish Studies
Intercultural Center, Box 571033-Room 305R
Washington, DC 20057-1033

Institute of Turkish Studies Research Grants in Comparative Studies of Modern Turkey   2844

The program supports and encourages the development of research, scholarship, and learning in the field of Turkish Studies in the United States. This grant is for graduate students at the dissertation research stage and for post-doctoral scholars in the U.S. who study aspects of the Republic of Turkey (post-1922) in a comparative context. A significant portion of the project should be devoted comparatively to one or more states or political entities in Europe, Latin America, the Middle East, and Asia in addition to the Turkish Republic. These grants are primarily, but not exclusively, in the field of Political Science. Applications and further information are available online.
Requirements: Applicants must be U.S. citizens or permanent residents in the U.S. and affiliated with a university in the U.S at the time of the application.
Amount of Grant: $10,000
Contact: David Cuthell, (202) 687-0295; fax: (202) 687-3780; email: dcc@turkishstudies.org
Internet: http://www.turkishstudies.org/grand_app.html
Sponsor: Institute of Turkish Studies
Intercultural Center, Box 571033-Room 305R
Washington, DC 20057-1033

Institute of Turkish Studies Summer Language Study Grants in Turkey   2845

These grants are for summer travel to Turkey for language study in preparation for graduate research. The time period for these grants varies with each individual application. Normally, recipients are expected to spend a minimum of two months in Turkey at an established Ottoman or Turkish language training facility. Application guidelines are available online.
Requirements: Applicants must be graduate students in any field of the social sciences and/or humanities; U.S. citizens or permanent residents at the time of the application; and currently enrolled in a university in the United States.
Amount of Grant: $1000-$2000
Date(s) Application is Due: Mar 11
Contact: David Cuthell, (202) 687-0295; fax: (202) 687-3780; email: dcc@turkishstudies.org
Internet: http://www.turkishstudies.org/grantsprogram.html
Sponsor: Institute of Turkish Studies
Box 571033, Intercultural Ctr
Washington, DC 20057-1033

Institute of Turkish Studies Summer Research Grants in Turkey   2846

These grants are for summer travel to Turkey to carry out projects. The time period for these grants varies with each individual application. Normally, grant recipients are expected to spend a minimum of two months in Turkey. Application and guidelines are available online.
Requirements: Applicants must be graduate students in any field of the social sciences and/or humanities in the United States; U.S. citizens or permanent residents at the time of the application; and currently not engaged in dissertation writing.
Amount of Grant: $1000-$2000
Contact: David Cuthell, (202) 687-0295; fax: (202) 687-3780; email: dcc@turkishstudies.org
Internet: http://www.turkishstudies.org/grand_app.html
Sponsor: Institute of Turkish Studies
Intercultural Center, Box 571033-Room 305R
Washington, DC 20057-1033

Intel Bachelor's and Master's Degree Internships   2847

Several Intel manufacturing facilities, in cooperation with participating universities, offer paid internships for students enrolled full-time in specific undergraduate and graduate-level electronics and technology programs. These internships offer hands-on technical education and experience and valuable networking opportunities. Undergraduate and graduate level internships are available at all our major U.S. sites. One of the great advantages of internships is the flexible scheduling, designed to work around the typical school schedule. Interns receive: competitive salaries; vacation and holiday time; access to Intel University classes for professional and personal development; networking with Intel managers and executives; relocation assistance; and earned credit toward sabbaticals. Application details are available through the career centers of participating colleges.
Requirements: In order to be eligible applicants must: be enrolled in bachelor's, master's, or doctoral programs in engineering, science, and business-related fields with a 3.0 or higher GPA on a 4.0 scale.
Contact: Internship Coordinator; (408) 765-8080; fax: (408) 765-9904
Internet: http://www.intel.com/jobs/usa/students/internships/degree.htm
Sponsor: Intel Corporation
2200 Mission College Boulevard
Santa Clara, CA 95054-1549

Intel Foundation PhD Fellowships   2848

The Intel Foundation PhD Fellowship Program awards fellowships to PhD candidates doing work in fields related to Intel's business and research interests. These fellowships, only available at select U.S. universities, include a cash award (tuition/fees/stipend), an Intel PC, introduction to an Intel Mentor, and the opportunity to conduct research or an internship at Intel. Approximately 35 fellowships are awarded annually. In addition, Intel Foundation collaborates with The National Consortium for Graduate Degrees for Minorities in Engineering and Science (GEM) to help support graduate education for underrepresented minorities at the master's and PhD levels. Contact your school's engineering or computer science department to receive additional information.
Date(s) Application is Due: Nov 15
Contact: Fellowship Coordinator; (408) 765-8080; fax: (408) 765-9904
Internet: http://www.intel.com/jobs/usa/students/internships/fellowship.htm
Sponsor: Intel Corporation Foundation
2200 Mission College Boulevard
Santa Clara, CA 95054-1549

Intel Technician Internships   2849

Several Intel manufacturing facilities, in cooperation with participating colleges, offer paid internships for students enrolled full-time in specific two-year electronics and technology programs. These internships offer hands-on technical education and experience and valuable networking opportunities. There are a limited number of internships available, so early application is encouraged. One of the great advantages of technician internships is the flexible scheduling, designed to work around your school schedule. The Compressed Work Week offers a three or four day week that gives an Intern time for all other interests and activities. Interns receive: competitive salaries; vacation and holiday time; access to Intel University classes for professional and personal development; networking with Intel managers and executives; relocation assistance; and earned credit toward sabbaticals. Application details are available through the career centers of participating colleges.
Requirements: In order to be eligible for Intel's technical college internship, applicants must: have declared in a specific full-time degree program such as semiconductor manufacturing, electronics technology or automation, etc. at a participating college; maintain a 3.0 or higher GPA on a 4.0 scale; be able to work a minimum number of hours as set by their Intel manager; and be enrolled in classes and making significant progress toward a degree.
Contact: Internship Coordinator; (408) 765-8080; fax: (408) 765-9904
Internet: http://www.intel.com/jobs/usa/students/internships/technician.htm
Sponsor: Intel Corporation
2200 Mission College Boulevard
Santa Clara, CA 95054-1549

Inter-University Board for Chinese Language Studies   2850

This year-long program provides graduate and undergraduate students with intensive oral-aural language instruction and furthers reading ability in general materials and in disciplinary or professional fields. Acceptance is granted to students of all countries affiliated with recognized academic institutions and to unaffiliated students who provide evidence of serious intention to pursue advanced work after attending the program or use Chinese in their profession. Full tuition may be provided, and fellowships are available.
Requirements: Admission is open to all qualified students from the United States. Applicants from other English-speaking countries are also eligible for admission, as are others who speak and read English fluently. Applicants must demonstrate proficiency in Chinese.
Contact: Karen Cheong, Program Administrator, (510) 642-3873; fax: (510) 643-7062; email: iub@socrates.berkeley.edu
Internet: http://ieas.berkeley.edu/iup
Sponsor: Inter-University Board for Chinese Language Studies
2223 Fulton Street, Suite 2318
Berkeley, CA 94720-2318

International Affairs Fellowship Program   2851

The program is designed to bridge the gap between theory and practice in international relations and to encourage the better use of scholarly or reflective wisdom in decisions on international problems. The fellowships also provide career experiences for young foreign policy professionals. Interested qualified individuals should write to the council in the spring of the year and briefly describe the proposed project and methods to carry it out. Nominations, including self-nominations, must meet the listed deadline. Annual deadline dates may vary; contact program staff for exact dates.
Requirements: The council awards a number of fellowships to U.S. citizens between the ages of 27 and 35 with demonstrated intellectual ability and promise who come from academic, government, business, and/or professional communities. While a PhD or its equivalent is not a firm requirement, successful candidates generally hold advanced degrees and possess a solid record of work experience.
Restrictions: The program does not support research toward a postgraduate degree. The program will not support research as traditionally understood, or the gathering of new information and the development of generalizations concerning it, if the results will be of primary interest only to scholars or theoreticians.
Amount of Grant: $40,000 maximum per year
Date(s) Application is Due: Oct 24
Contact: Aysha Ghadiali; (212) 434-9489; fax (212) 434-9801; email: Fellowships@cfr.org
Internet: http://www.cfr.org/about/fellowships/iaf.html
Sponsor: Council on Foreign Relations
58 E 68th Street
New York, NY 10021

International Cancer Technology Transfer Fellowships 2852

The fellowships are funded by a group of cancer institutes, leagues, societies, associations, foundations, and governmental agencies in North America, Europe, and Australia. The aims of the awards are threefold: to facilitate rapid international transfer of cancer research and clinical technology; exchange knowledge and enhance skills in basic, clinical, behavioral, and epidemiological areas of cancer research and in cancer control and prevention; and acquire up-to-date clinical management, diagnostic, and therapeutic expertise. Applications are particularly encouraged for projects in the fields of cancer control and prevention, epidemiology and cancer registration, public education, and behavioral sciences. The short-term fellowships permit successful candidates to spend up to three months at a suitable host institute abroad. They are particularly aimed at investigators and clinicians working in places where such teaching is not yet available and where the necessary facilities exist to apply and disseminate the new skills upon return. Applications may be submitted at any time, and funding decisions are generally communicated to candidates within 60 days of receipt of a complete application.

Requirements: Qualified cancer investigators should be at the early stages in their careers; clinicians should be well established in their oncology practice. Experts from any country who have been invited to teach specialized skills abroad also may apply.

Amount of Grant: $3000 average

Contact: International Cancer Research Program, (4122) 809 18 40; fax: (41-22) 809 18 10; email: fellows@uicc.org

Internet: http://fellows.uicc.org/fell4icr.shtml

Sponsor: International Union against Cancer

3, rue de Conseil-General

Geneva 1205

Switzerland

International College of Surgeons Grants and Scholarships 2853

The ICS makes grants and scholarships available to surgeons who wish to enhance their surgical skills through postgraduate training. Applicants are responsible for making their own arrangements for the program of study; must clearly indicate the place of study, with whom they will be studying, and acceptance by the institution and surgeon for the course of study as defined; and must indicate what it is they expect to achieve upon completion of the study program (e.g. learn a particular procedure, etc.). Guidelines are available online. No formal application form is required.

Requirements: Applicants must have graduated from an accredited medical school, completed their residency, and be licensed to practice surgery in their home country (documentation of licensure must be provided). Applicants need not be ICS fellows.

Contact: Max Downham, Executive Director, (312) 787-6274; fax: (312) 787-9289; email: max@icsglobal.org

Internet: http://www.icsglobal.org/RandS/index.asp

Sponsor: International College of Surgeons

1516 N Lake Shore Drive

Chicago, IL 60610-1694

International Crane Foundation Internships 2854

This is a formal stipend program with internships being offered in the fields of aviculture and restoration. In the aviculture program, interns will receive intensive hands-on training in the care and management of endangered cranes, including husbandry, handling techniques, behavior, stimulating reproduction, incubation, chick rearing, artificial insemination, health care, and genetic management. The associate also manages the volunteer chick parent program. Each intern is responsible for developing and completing a research project with assistance from the staff. Internships usually run from March through May, June through August, September through November, and December through February. Application deadlines are for spring, summer, fall, and winter internships. Under the restoration program, interns will be involved in the ecosystems restoration program, which directs the restoration of prairie, savanna, and wetland communities on the ICF property. Interns will develop important ecological, hydrological, botanical, and horticultural skills. Depending on the season, interns may be involved in prescribed burns, field studies of wild sandhill care populations; seed collecting, planting, and site preparation for community restorations; and interaction with foreign scientists. The associate also will manage the seed collection program. Internships usually run for six months from March through August and from June through November. Application deadlines are January 1 and April 1 for the spring-summer and summer-fall internships, respectively. The program runs for nine months from March through November.

Requirements: Applications may be made by recent college sophomores through college graduates in biology, zoology, or botany who have a willingness to work at manual tasks and can work in a self-directed manner. Individuals seeking graduate training are especially encouraged.

Amount of Grant: $350 per month plus housing for interns

Date(s) Application is Due: Jan 1; Apr 1; Jul 1; Oct 1

Contact: Korie Klink, (608) 356-9462 ext 127; fax: (608) 356-9465; email: korie@savingcranes.org

Internet: http://www.savingcranes.org/about/get_involved/intern.cfm

Sponsor: International Crane Foundation

P.O. Box 447

Baraboo, WI 53913-0447

International Doctoral Scholarships in Jewish Studies 2855

Scholarships assist in training future Jewish scholars for careers in Jewish scholarship and research and help religious, educational, and other Jewish communal workers obtain advanced training for leadership positions in the Jewish community. Scholarships are awarded for one academic year, renewable up to a maximum of four years. References are required on the background and scholastic ability of the applicant. Final action is taken by the foundation in July with applicants being advised in August. Applicant must submit a written request for an application form.

Requirements: Any graduate student specializing in a Jewish field who is officially enrolled or registered in a doctoral program at a recognized university is eligible to apply. Preference is given to students at the dissertation level.

Amount of Grant: $2000-$7500

Date(s) Application is Due: Oct 31

Contact: Dr. Jerry Hochbaum, Executive Vice President, (212) 425-6606; fax: (212) 425-6602; email: office@mfjc.org

Internet: http://www.mfjc.org

Sponsor: Memorial Foundation for Jewish Culture

50 Broadway, 34th Fl

New York, NY 10004

International ePhilanthropy Awards 2856

These awards honor those individuals, organizations, and companies working in the ePhilanthropy field. Nominees will have demonstrated extraordinary talent, creativity, and insight in drawing the public's attention to the important use of the Internet for philanthropic purposes and have created services or strategies that support this effort. Award categories include: Best ePhilanthropy Research Project; Best Online Donations/Fundraising (revenue generating) Campaign; Best Event Registration and/or Membership Campaign; and Best Community Building/ Volunteerism and/or Activism (non-revenue) Campaign. It is acceptable to nominate individuals, companies, and organizations in multiple categories.

Requirements: Individuals, organizations, and companies working in the ePhilanthropy field who have demonstrated extraordinary talent, creativity and insight in drawing the public's attention to the important use of the Internet for philanthropic purposes and/or have created services or strategies that support this effort are encouraged to apply.

Restrictions: Employees of the ePhilanthropy Foundation, members of the International ePhilanthropy Awards Judging panel, and their immediate family (spouse, siblings, parents, and children) are not eligible for consideration.

Amount of Grant: $500 to charities of awardees' choice

Date(s) Application is Due: Aug 1

Contact: Awards Administrator, (877) 536-1245 ext 1; fax: (202) 478-0910; email: Awards@ephilanthropy.org

Internet: http://www.ephilanthropy.org/awards

Sponsor: ePhilanthropy Foundation

1101 15th Street NW, Suite 200

Washington, DC 20005

International Essential Tremor Foundation Grants 2857

The foundation awards grants for clinical or basic research specific to essential tremor. Data to be derived should lead to grant proposals to governmental or other sources of funds. Two grants are available in the current year. Under exceptional circumstances, funding may be allowed for a second year as a competing renewal. No more than one award can be given to an individual investigator per year. Funding begins in June or July.

Requirements: Grants are awarded to MDs and/or PhDs.

Amount of Grant: $20,000 per year

Contact: Catherine Rice, (888) 387-3667 or (913) 341-3880; fax: (913) 341-1296; email: crice@essentialtremor.org

Internet: http://www.essentialtremor.org/research/index.php

Sponsor: International Essential Tremor Foundation

P.O. Box 14005

Lenexa, KS 66285-4005

International Federation of University Women Research Fellowships 2858

This competition is open to women graduates for research for a period of at least eight months, preferably in a country other than the one in which the candidate has received her education or habitually resides. In addition, the Study and Action Program Fellowship is awarded for research related to the IFUW Study and Action Program—Woman's Role in Changing Society. Members of the IFUW should apply through their national affiliates (in the United States, the American Association of University Women, 1111 16th Street NW, Washington, DC 20036). Competitions are held in even-numbered years. The deadline for receipt of applications is determined by each affiliate but normally falls between early September and November of the year preceding the competition.

Requirements: An applicant must be a woman graduate who is an IFUW member and must be started on the research program to which the application relates.

Restrictions: A fellowship will not be awarded for the first year of a PhD program.

Amount of Grant: Sf3000-Sf6000 approximately

Date(s) Application is Due: Nov 1

Contact: Fellowship Office, (022) 731 23 80; fax: (022) 738 04 40; email: info@ifuw.org

Internet: http://www.ifuw.org/fellowships/index.htm

Sponsor: International Federation of University Women

8 rue de l'Ancien Port

Geneva CH 1201

Switzerland

International Fellowships in Jewish Studies 2859

Fellowships assist qualified individuals to carry out independent scholarly, literary, or art projects in a field of Jewish specialization that will make a significant contribution to the understanding, preservation, enhancement, or transmission of Jewish culture. Fellowships are awarded for one academic year, renewable for one more year. An application form is available upon written request.

Requirements: Any recognized and/or qualified scholar, researcher, writer, or artist who possesses knowledge and experience to formulate and implement a project in a field of Jewish specialization may apply.

Amount of Grant: $7500 maximum per year

Date(s) Application is Due: Oct 31

Contact: Dr. Jerry Hochbaum, Executive Vice President, (212) 425-6606; fax: (212) 425-6602; email: office@mfjc.org

Internet: http://www.mfjc.org/index.htm

Sponsor: Memorial Foundation for Jewish Culture

50 Broadway, 34th Fl

New York, NY 10004

International Fellowships in Medical Education 2860

The program provides opportunities for faculty from schools of medicine outside the United States to study aspects of medical education in the United States that have the potential to improve medical education in their home country institutions. Mentoring will be provided by preceptors in U.S. basic and clinical science departments, medical education departments, and health system institutions. Eligible study areas include educational methodology, curriculum design, evaluation systems, medical school governance, development of basic and clinical science departments, and the design and operation of health care and public health system programs linked to medical education. Fellows may concurrently pursue collaborative research interests with U.S. basic science faculty or observe some clinical activity with clinical faculty. Fellowship allowance generally includes a monthly stipend, round-trip economy-class air fare for the fellow, and travel to one scientific meeting in the United States. Programs may range from six months to one year. The interactive, web-based application is now available online; paper applications are no longer accepted.

Requirements: Candidates must reside and work in their home countries at the time of application; have a graduate or professional degree in medicine or in a basic medical science; have not less than three years of work experience in their chosen field in the home country; hold an academic appointment as a faculty member in a school of medicine; have the ability to communicate effectively in English; have the endorsement of a home country medical school or organization; and have a position to return to in the home country medical school or organization.

Restrictions: Fellowships are not provided for programs in basic or clinical research; degree-granting educational programs that require acceptance to an institution and tuition payments; tuition grants for short-term courses; specialty training in residency programs, or solely for training in clinical procedures.

Amount of Grant: $2400 monthly stipend

Contact: Grants Administrator, (215) 386-5900; fax: (215) 386-9196; email: info@ecfmg.org

Internet: http://www.ecfmg.org/about.html

Sponsor: Educational Commission for Foreign Medical Graduates

3624 Market Street

Philadelphia, PA 19104-2685

International Human Rights Funders Grants 2861

Human rights grantmaking supports a wide range of efforts to ensure that all people have the opportunity to enjoy a genuinely human existence. As affirmed in the Universal Declaration of Human Rights (UDHR) and the international and domestic laws to which it gave rise, the people of the world are endowed with inalienable civil, cultural, economic, political and social rights. What gives human rights work such power is that all governments, including the United States, have a recognized obligation to respect, protect and fulfill these rights, both domestically and internationally. Increasingly, non-state actors - corporations and individuals - are also being held accountable. Funders contribute to the advancement of human rights around the world through their grantmaking. Grants varying in type and amount support a broad range of approaches, including: public education to inform people about their human rights and how to exercise them; documenting, reporting and fact-finding to expose human rights violations; litigation to uphold human rights and hold abusers accountable; policy advocacy to ensure that states and non-state actors conform to human rights standards; research and scholarship to define the content of rights; networking and coalition building to further the effectiveness of a global human rights movement; and capacity building for organizations engaged in the above work, locally and internationally.

Requirements: The applicant must be an organization that works to defend or promote human rights; and is based in India, Bangladesh, Pakistan, and be working, in whole or in part, on human rights issues in the country in which the organization is based, or is based in Algeria, Tunisia, or Morocco, and be working, in whole or in part, on human rights issues in the country in which the organization is based.

Restrictions: The fund does not support stand-alone conferences, individuals, businesses, scholarships, fundraising events, university-based research, government agencies, or activities directly or indirectly intended to support candidates for political office.

Contact: Michael Hirschhorn, Director; (202) 609-2631; fax: (202) 609-2633; email: MichaelH@ihrfg.org or info@IHRFG.org

Internet: http://www.hrfunders.org/hrfunding/index.html

Sponsor: International Human Rights Funders Group

c/o Wellspring Advisors, 1410 Broadway, 23rd Floor

New York, NY 10018

International League of Antiquarian Booksellers Bibliographical Prize 2862

Every four years (next in 2006) the ILAB awards a prize to the author of the best work, published or unpublished, of learned bibliography or of research into the history of books or of typography, and books of general interest on the subject. Two copies of each work to be considered must be deposited at the office of the secretary at the very latest 16 months before date of award. The award period covers the four years prior to the year of deadline date.

Requirements: Entries must be submitted in a language that is universally used. A work already published is eligible only if publication occurred within the four years immediately preceding the closing date for submission or if it has an imprint bearing a date within those four years.

Restrictions: Entries in the form of a specialized catalog of one or more books destined for sale are not eligible, nor are periodicals or public library catalogs.

Amount of Grant: $10,000

Date(s) Application is Due: Dec 31

Contact: Raymond Kilgarriff, Prize Secretary, +44 (0)1424 426146; email: rmkilgarriff@btinternet.com

Internet: http://www.ilab-lila.com/english/prize.htm

Sponsor: International League of Antiquarian Booksellers

Hauptstrasse 19 A

Bad Honnef D-53604 Germany

International Lelio Basso Foundation for the Rights and Liberation of 2863 Peoples Grants

The foundation awards research grants internationally to individuals in the fields of environment, development studies, and human rights issues in developing countries. The work of the foundation concentrates on the environment and development, industrial hazards and human rights, refugee rights, and the rights of children and young people.

Contact: Elmar Altvater, President, 39-6-68801468; fax: 39-6-6877774; email: filb@iol.it

Internet: http://www.grisnet.it/filb/filbeng.html

Sponsor: International Lelio Basso Foundation for the Rights and Liberation of Peoples

Via della Dogana Vecchia 5

Rome 00186 Italy

International Max Planck Research School Mathematics in the 2864 Sciences Fellowships

The research school offers PhD fellowships for qualified students with an excellent background in mathematics or related fields. It will lead interested graduate students towards research problems in the physical and life sciences. This involves a broad range of mathematical fields, including geometry, partial differential equations and functional analysis, stochastics, and discrete mathematics. Specific subject areas are: a) Partial Differential Equations and Material Sciences; b) Numerical Analysis and Scientific Computing; c) Riemannian and Symplectic Geometry and Hamiltonian Systems; d) Quantum Field Theory, Particle Physics; e) Algebraic Geometry, String Theory; f) Geometric and Functional Analytic Methods in Mathematical Physics; g) Stochastic Processes, Many Particle Systems; h) Complex Systems in Evolutionary Processes and Neurobiology. The fellowships will include a generous stipend, and there will be no tuition fees. Teaching will be in English. In addition, German language courses will be provided free of charge.

Requirements: There is no application form to fill out. However, interested applications should submit the following in paper form: CV; transcript of examination results; a copy of the master thesis (if applicable); and two letters of recommendation by academic teachers, to reach the school in sealed envelopes or to be sent directly.

Restrictions: For students who did not write a master thesis there are qualifying grants available as an entrance level to the research school.

Date(s) Application is Due: Mar 25; Jun 25; Sep 25; Dec 25

Contact: Dr. Hayk Mikayelyan, Scientific Coordinator; +49 341 9959 50; fax: +49 341 9959 658; email: imprsmis.mpg.de

Internet: http://www.imprs-mis.mpg.de/positions.html

Sponsor: Max Planck Institute

Max-Planck-Institut fur Mathematik in den Naturwissenschaften; IMPRS Mathematics in the Sciences; InselstraBe 22

Leipzig 04103 Germany

International Oncology Nursing Fellowships 2865

The fellowships are funded by the Oncology Nursing Society and provide an opportunity for qualified nurses to augment their professional knowledge and experience through short-term observerships at a renowned comprehensive cancer center in Australia, North America, or the United Kingdom. A project description must provide full details of the particular cancer nursing skills that the candidate wishes to observe. It should elaborate on the treatment and nursing facilities at the home institute and how the newly acquired knowledge and techniques would be used and disseminated to other nursing staff upon return. Projects need the approval of the candidate's home and host supervisors. The maximum fellowship duration is three months; stipend support is for one month only. The program also contributes to the least expensive international two-way airfare or other appropriate form of transport. Financial support is not provided for dependents.

Requirements: English-speaking registered nurses who are actively engaged in the management of cancer patients in their home institutes and who come from developing or Eastern European countries are eligible, as are established oncology nurses from any country who wish to disseminate their skills in these regions.

Amount of Grant: $2800 average

Date(s) Application is Due: Nov 1
Contact: International Cancer Research Program, (41-22) 809 18 40; fax: (41-22) 809 18 10; email: fellows@uicc.org
Internet: http://fellows.uicc.org/fell5ion.shtml
Sponsor: International Union against Cancer
3, rue de Conseil-General
Geneva 1205 Switzerland

International Peace Research Institute, Oslo Grants　　2866
PRIO offers scholarships for graduate students, mainly at Norwegian universities. In addition to a student stipend, the scholarship includes supervision by PRIO's researchers and office space at the institute's building in Oslo. Residence in Oslo during the scholarship period is essential. Guidelines are available online. The scholarship lasts for an agreed period of time, up to one year. If the thesis is finalized within the agreed period, PRIO may offer an extension of the stipend period to secure academic publishing of the results, e.g. by writing an article for an international journal.
Amount of Grant: NOK 3500 per month
Date(s) Application is Due: Oct 1
Contact: Stein Tonnesson, Director, 47-22-54-77-31; fax: 47-22-54-77-01; email: stein@prio.no or info@prio.no
Internet: http://www.prio.no/page/preview/preview/9346/21744.html
Sponsor: International Peace Research Institute, Oslo
Fuglehauggta 11
Oslo 0260 Norway

International Spinal Research Trust Grants　　2867
The trust awards grants worldwide for research into the treatment of paralysis caused by spinal cord injury. Grants also are available for the purchase of relevant equipment.
Contact: John Cavanagh, Head of Research, 01483 898786; fax: 01483 898763; email: research@spinal-research.org or john@spinal-research.org
Internet: http://www.spinal-research.org/display_section.asp?section=researchers
Sponsor: International Spinal Research Trust
Station Road, Bramley
Guildford GU5 0AZ
United Kingdgom

International Studies Graduate Fellowships　　2868
The Graduate Programs in International Studies department at the university offers MA and PhD fellowships to train graduate students and promote scholarship on transnational issues. The program focuses on traditional problems of foreign and security policy, international business and economics, as well as post-Cold-War priorities such as gender, human rights, and migration.
Contact: Graduate Programs in International Studies, (757) 683-5700: fax: (757) 683-5701; email: isgpd@odu.edu
Internet: http://al.odu.edu/gpis
Sponsor: Old Dominion University
Batten Arts and Letters Building, Room 620
Norfolk, VA 23529-0086

InterUniversity Fellowship Program in Jewish Studies　　2869
The program awards fellowships to graduate students in Jewish Studies who wish to spend a year working in the Department of Jewish Studies at one of Israel's universities. Fellows can take courses, do research in Israel's Judaic libraries, and have personal contact with Israeli scholars in the field. In addition, fellows attend an ongoing seminar program coordinated by an on-site mentor. Fellows are expected to spend the full academic year in Israel. Fellowships are renewable upon consideration.
Requirements: Applicants can be earning their degrees in any discipline, but a primary focus must be placed on some aspect of Jewish life. Fellows must normally have enough fluency in Hebrew to enable them to participate in graduate courses and seminars taught in Hebrew. All nationalities accepted.
Amount of Grant: $4500 for doctoral students; $2500 for master's students
Contact: Fellowship Administrator, (718) 951-5146; fax: (718) 951-4639
Internet: http://www.nyu.edu/pages/gsas/files/interuni
Sponsor: Brooklyn College
Room 3612, James Hall
Brooklyn, NY 11210

IPA George Washington Williams Fellowship　　2870
The fellowship funds stories written by journalists of color about issues such as the environment, global trade policy, healthcare, race, and education. Fellows receive access to some research support, consultants, advanced professional training, and a large network of journalists working in the public interest sector. In addition, program staff work closely with fellows to publish their stories in major publications. Individuals may apply for financial and institutional support to write a single story, or they may seek an investigative or depth reporting fellowship of between three and 12 months to research a specific social issue. The fellowship will pay national commercial rates for individual stories or $1500 per month plus expenses for depth reporting fellowships.
Requirements: Any journalist of color with at least three years of solid professional reporting and writing experience may apply. Individuals with backgrounds in investigative or enterprise reporting are preferred. Previous reporting or other experience in the chosen subject area is desirable. The fellowship is open only to U.S. citizens or to foreign journalists who have established relationships with U.S. publications.

Restrictions: College journalism or internship experience do not qualify as professional experience.
Date(s) Application is Due: May 30; Nov 15
Contact: Fellowship Administrator, (415) 643-4401 ext 116 or ext 117; email: gww@indypress.org
Internet: http://www.indypress.org/programs/gwwfellow.html
Sponsor: Independent Press Association
2729 Mission Street, Suite 201
San Francisco, CA 94110-3131

Ipatieff Prize　　2871
This award is given every three years to recognize outstanding chemical experimental work in the field of catalysis or high pressure, carried out by individuals of any nationality and not over 40 years of age (by date of presentation). If experimental investigations in these fields shall have been abandoned to such a degree that no outstanding results have been achieved, then the award may be given for highly meritorious work in a closely allied field of chemistry. Special weight shall be given to the independence of thought and the originality shown. The award may be made for investigations carried out in any country and without consideration of the nationality of the recipient; however, preference will be given to American chemists.
Requirements: Any individual, except a member of the award committee, may submit one nomination or seconding letter for the award in the award year. Nominating documents consist of a letter of not more than 1000 words containing an evaluation of the nominee's accomplishments and a specific identification of the work to be recognized, a biographical sketch including date of birth, and a list of publications and patents authored by the nominee.
Restrictions: Self-nominations are not accepted.
Samples: Raul Lobo—award winner, $5,000 (2004); Joan Brennecke—award winner, $5,000 (2001).
Amount of Grant: $5,000, depending on income from the trust fund, and travel expenses to award meeting
Date(s) Application is Due: Nov 1
Contact: Felicia Dixon, Awards Administrator; (800) 227-5558 or (202) 872-4408; fax: (202) 776-8008; email: f_dixon@acs.org or awards@acs.org
Internet: http://portal.acs.org/portal/acs/corg/content?_nfpb=true&_pageLabel=PP_ARTICLEMAIN&node_id=1319&content_id=CTP_004533&use_sec=true&sec_url_var=region1
Sponsor: American Chemical Society
1155 Sixteenth Street, NW
Washington, DC 20036

IRA Elva Knight Research Grants　　2872
The grant supports reading and literacy research. Research is defined as that which addresses significant questions for the disciplines of literacy research and practice. Projects should be completed within two years. Studies may be carried out using any research method or approach so long as the focus of the project is on research in reading or literacy. Activities such as developing new programs or instructional materials are not eligible for funding except to the extent that these activities are necessary procedures for the conduct of the research. Each year it is expected that at least one grant will be awarded to a researcher outside the United States and Canada as well as one grant being awarded to a teacher-initiated research project. The applicants will be required to complete and submit their applications online using a web-based grant management system from the IRA website.
Requirements: To be eligible, all applicants must be members of the International Reading Association. Guidelines and application forms are available from headquarters.
Samples: Douglas Fisher, San Diego, California (2008); Diane Lapp, La Mesa, California (2008); Kathleen Clark, Fox Point, Wisconsin (2008); Yonas Asfaha, Tilburg, Netherlands (2008).
Amount of Grant: $10,000 maximum
Date(s) Application is Due: Jan 15
Contact: Program Contact, Division of Research; (302) 731-1600, ext. 423; fax: (302) 731-1057; email: research@reading.org
Internet: http://www.reading.org/association/awards/research_knight.html
Sponsor: International Reading Association
800 Barksdale Road, P.O. Box 8139
Newark, DE 19714-8139

IRA Jeanne S. Chall Research Fellowship　　2873
The fellowship was established to encourage and support reading research by promising scholars. The special emphasis of the fellowship is to support research efforts in the following areas: beginning reading (theory, research, and practice that improves the effectiveness of learning to read); readability (methods of predicting the difficulty of texts); reading difficulty (diagnosis, treatment, and prevention); stages of reading development; the relation of vocabulary to reading; and diagnosing and teaching adults with limited reading ability.
Requirements: The fellowship is open to IRA members.
Samples: Tiffany Hogan, U of Kansas (Lawrence, KS)—fellowship recipient, $6000 (2005); Julie Rosenthal, U of New York (NY)—felloship recipient, $6000 (2004).
Amount of Grant: $6000
Date(s) Application is Due: Jan 15
Contact: Program Contact, (301) 731-1600 ext 423; fax: (301) 731-1057; email: research@reading.org
Internet: http://www.reading.org/association/awards/research_chall.html

Sponsor: International Reading Association
P.O. Box 8139, 800 Barksdale Road
Newark, DE 19714-8139

IRA Nila Banton Smith Research Dissemination Support Grants 2874

The grant is intended to assist any IRA member to spend from two to 10 months working on a research dissemination activity. The grant is funded from the Nila Banton Smith Endowment.
Samples: Laurie Henry, U of Connecticut (CT)—for Investigation of literacy skills and strategies used while searching for information on the *Internet*: A comprehensive review and synthesis of research, $5000 (2005); Michelle Commeyras and Helen Inyega, U of Georgia (Athens, GA)—for an integrative review of research on teaching reading in Kenyan primary schools, $5000 (2004).
Amount of Grant: $5000 maximum
Date(s) Application is Due: Jan 15
Contact: Program Contact, Division of Research, (302) 731-1600 ext 423; fax: (302) 731-1057; email: research@reading.org
Internet: http://www.reading.org/association/awards/research_smith_grant.html
Sponsor: International Reading Association
P.O. Box 8139, 800 Barksdale Road
Newark, DE 19714-8139

IRA Outstanding Dissertation of the Year Award 2875

The competition is intended for those doctoral students who have focused their research in the reading field or who have conducted related research having implications for reading. Studies using any research approach (ethnographic, experimental, historical, survey, etc.) are encouraged. Each study will be assessed in the light of this approach, the scholarly qualification of its report, and its significant contributions to knowledge within the reading field. Application forms and guidelines are available from the association.
Requirements: Dissertations must have been completed between May 15 and May 14 of the following year.
Samples: Molly Fuller Collins, Boston U (Boston, MA)—award recipient, $1000 (2005); Kristen Ritchey, U of Maryland—award recipient, $1000 (2004).
Amount of Grant: $1000
Date(s) Application is Due: Oct 1
Contact: Division of Research, (302) 731-1600 ext 423; fax: (302) 731-1057; email: research@reading.org
Internet: http://www.reading.org/association/awards/research_outstanding.html
Sponsor: International Reading Association
P.O. Box 8139, 800 Barksdale Road
Newark, DE 19714-8139

IRA Outstanding Teacher Educator in Reading Award 2876

This award honors an outstanding college or university teacher of reading methods or reading-related courses. Nominee must be an IRA member, affiliated with a college or a university, and engaged in teacher preparation in reading at the undergraduate and/or graduate level. Entry forms and current guidelines are available from the Executive Office, International Reading Association, P.O. Box 8139, Newark, DE 19714-8139.
Samples: Barbara M. Taylor—award recipient, $1000 (2005); Ronald L. Cramer—award recipient, $1000 (2004).
Amount of Grant: $1000
Date(s) Application is Due: Oct 15
Contact: Program Contact, (302) 731-1600 ext 221; fax: (302) 731-1057; email: exec@reading.org
Internet: http://www.reading.org/association/awards/teachers_outstanding.html
Sponsor: International Reading Association
P.O. Box 8139, 800 Barksdale Road
Newark, DE 19714-8139

IRA Print Media Award 2877

The award recognizes outstanding reporting in newspapers, magazines, and wire services. Entries may include in-depth studies of reading instruction, discussion of research, or ongoing coverage of reading programs in the community and must have appeared during the previous calendar year. Entry forms and guidelines are available from the IRA.
Requirements: The competition for the award is limited to professional journalists.
Samples: Molly McCarthy—award recipient (2005); Jolande K. Gumz—award recipient (2004).
Date(s) Application is Due: Jan 7
Contact: Program Contact, (302) 731-1600 ext 293; fax: (302) 731-1057; email: pubinfo@reading.org
Internet: http://www.reading.org/association/awards/media_print.html
Sponsor: International Reading Association
P.O. Box 8139, 800 Barksdale Road
Newark, DE 19714-8139

Ireland Family Foundation Grants 2878

Established in 2000, the Ireland Family Foundation givings primarily in North Carolina with some funding available in California. The Foundations supports non-profit organizations involved with autism research, human services and education.
Requirements: The initial approach should be in the form of a letter, followed up with a proposal containing : a copy of IRS Determination Letter; a detailed description of project and amount of funding requested.
Restrictions: No grants to individuals.

Geographic Focus: California; North Carolina
Samples: ARC of Orange County, Chapel Hill, NC., $35,000 (2005);
Amount of Grant: $10,000-$60,000 range
Contact: Lori Ireland, President; (919) 932-3556;
Sponsor: Ireland Family Foundation
1434 Arboretum Drive
Chapel Hill, NC 27517-9161

Irene and Daisy MacGregor Memorial Scholarship 2879

The scholarship is awarded to a student who has been accepted into an accredited school of medicine to pursue an MD degree. This scholarship is also available to students who have been accepted into or who are pursuing an approved course of study in the field of psychiatric nursing, graduate level, at accredited medical schools, colleges, or universities. There is a preference to females, if equally qualified.
Requirements: Applicants must be U.S. citizens and must attend an accredited U.S. college or university. All applicants must obtain letters of sponsorship from local DAR chapters.
Amount of Grant: $20,000 maximum; $5000 per year
Date(s) Application is Due: Apr 15
Contact: Office of the Committees/Scholarships, (202) 628-1776; fax: (202) 879-3252
Internet: http://www.dar.org/natsociety/edout_scholar.cfm
Sponsor: National Society of the Daughters of the American Revolution
1776 D Street NW
Washington, DC 20006-5303

IREX ECA Alumni Small Grants Program 2880

The Program is a professional and community development program open to alumni of the Eurasian Undergraduate Program (UGRAD) formerly known as FSAU and Edmund S. Muskie Graduate Fellowship Program. Alumni must complete their small grant activities within Eurasia. Currently, Eurasia is defined as Armenia, Azerbaijan, Belarus, Georgia, Kazakhstan, Kyrgyz Republic, Moldova, Russia, Tajikistan, Turkmenistan, Ukraine and Uzbekistan. Possible project ideas include: initiating a public or community service program; launching a pilot program at an NGO or academic institution; organizing training programs or conferences for professional colleagues and/or other alumni; funding travel of U.S. colleagues to eligible Eurasian countries for conferences, training programs and professional collaboration; development and publication of curricula, textbooks, or related reference or educational materials; publication of public information pamphlets or brochures; and conducting any other projects judged by the selection committee and/or ECA to be in the interest of program goals.
Samples: Tural Ahmad, Azerbaijan—initiated an information campaign on the educational challenges facing orphans in Azerbaijan (2006); Kakhor Aminov, Tajikistan—developed training courses in management, finance, marketing and business planning for local small businesses (2006); Menana Gelashvili, Georgia—joint project that created two Scout Clubs in Rustavi (2006).
Amount of Grant: $3,000 maximum
Date(s) Application is Due: Jun 1
Contact: Program Contact, (202) 628-8188; fax: (202) 628-8189; email: asgp@irex.org
Internet: http://www.irex.org/programs/asgp/index.asp
Sponsor: International Research & Exchanges Board
2121 K Street, NW, Suite 700
Washington, DC 20037

IREX Edmund S. Muskie Graduate Fellowship Program (Muskie) 2881

The program provides opportunities for graduate students and professionals from Armenia, Azerbaijan, Belarus, Georgia, Kazakhstan, Kyrgyzstan, Moldova, Russia, Tajikistan, Turkmenistan, Ukraine and Uzbekistan for one-year non-degree, one-year degree or two-year degree study in the United States. Eligible fields of study are: business administration, economics, education, environmental management, international affairs, journalism and mass communication, law, library and information science, public administration, public health, and public policy.
Date(s) Application is Due: Jan 31
Contact: Grants Administrator; (202) 628-8188; fax: (202) 628-8189; email: muskie@irex.org
Internet: http://www.irex.org/programs/muskie/index.asp
Sponsor: International Research & Exchanges Board
2121 K Street, NW, Suite 700
Washington, DC 20037

IREX Eurasian Undergraduate Exchange Program (UGRAD) Grants 2882

Established by the U.S. Congress in 1992 to foster democratization and economic development in Eurasia, the program was originally called the FREEDOM Support Act Undergraduate Program (FSAU). Key objectives are to: foster democratization and economic development; and promote cultural understanding. Project activities include: encouraging youth leadership; promoting community service in the United States and Eurasia; and supporting practical experience for professional development. IREX opened its first field office in Moscow, Russia and since has expanded its field presence to Armenia, Azerbaijan, Belarus, Georgia, Kazakhstan, Kyrgyzstan, Moldova, Turkmenistan, Tajikistan, Turkey, Ukraine, and Uzbekistan. Programs in Eurasia span academic exchanges, educational advising, alumni programming, independent media assistance and development, Internet training and access, professional training, NGO development, and partnership building.
Requirements: Mid-level university faculty, researchers, scholars, and undergraduate students from the United States are eligible.

Date(s) Application is Due: Nov 6; Jan 31
Contact: Program Officers; (202) 628-8188; fax: (202) 628-8189; email: ugrad@irex.org
Internet: http://www.irex.org/programs/ugrad/index.asp
Sponsor: International Research & Exchanges Board
2121 K Street, NW, Suite 700
Washington, DC 20037

IREX Individual Advanced Research Opportunities (IARO) Grants   2883
The IARO program seeks to attract, select, and support in-depth field research by U.S. students, scholars and experts in policy-relevant subject areas related to Southeast Europe and Eurasia, as well as to disseminate knowledge about these regions to a wide network of constituents in the United States and abroad. The IARO Program provides fellows with the means and support necessary to conduct incountry research on contemporary political, economic, historical, or cultural developments relevant to U.S. foreign policy*. The IARO Program plays a vital role in supporting the emergence of a dedicated and knowledgeable cadre of U.S. scholars and experts who can enrich the U.S. understanding of developments in Southeast Europe and Eurasia. In addition to conducting their research, IARO fellows are asked to make themselves available as temporary consultants or experts to schools, local NGOs or the U.S. Embassy in their host country. Fellows with grants of 4 months or less are asked to perform 10 hours of service during their grant period, and fellows with grants of 5 or more months are asked perform 20 hours. Grants will be awarded in EACH of the four categories: Master, Student—must be enrolled in a Master, program during the grant period; Predoctoral Student—must be enrolled in a PhD program during the grant period; Professional—must have one of the following degrees (MA, MS, MFA, MBA, MPA, MLIS, MPH, JD, MD) and must not currently be enrolled as a student; Postdoctoral Scholar—must hold a PhD by the application deadline.
Requirements: Applicants must be U.S. citizens who have a command of the host country language sufficient for the completion of their research. Normally, applicants are required to have a full-time affiliation with a college or university and to be faculty members or doctoral candidates who will have completed all requirements for the PhD except the dissertation by the time of participation.
Date(s) Application is Due: Nov 15
Contact: Program Officers; (202) 628-8188; fax: (202) 628-8189; email: iaro@irex.org
Internet: http://www.irex.org/programs/iaro
Sponsor: International Research & Exchanges Board
2121 K Street, NW, Suite 700
Washington, DC 20037

IREX International Leadership in Education Program (ILEP) Grants   2884
The Program brings outstanding secondary teachers from the Near East, South Asia and Southeast Asia to the United States to further develop expertise in their subject areas, enhance their teaching skills, and increase their knowledge about the United States. ILEP is a program of the Bureau of Educational and Cultural Affairs of the U.S. Department of State. Key issues and objectives are to: contribute to improving the quality of secondary education in participating countries; strengthen the ability of women and under-served populations to play a part in national development; and develop professional and personal relationships between American and international teachers. Project activities include: teacher training; providing leadership programs for women and under-served populations; and forming international partnerships for professional development and education,
Requirements: Applicant must be secondary teachers from the Near East, South Asia and Southeast Asia
Samples: Yousef Abd-Allah, English Teacher, Sohag Military Secondary School, Sohag, Egypt, U.S. Host University: Kent State University (2006); Marwa Abdelmageed, English Teacher, Nefertari International Schools, Cairo, Egypt, U.S. Host University: Georgia Southern University (2006); Sameh Abdel-Seed, English Teacher, Dewina Secondary School, Abu-Tig, Egypt, U.S. Host University: Georgia Southern University (2006).
Contact: Program Officer, (202) 628-8188; fax: (202) 628-8189; email: iep@irex.org
Internet: http://www.irex.org/programs/iep/index.asp
Sponsor: International Research & Exchanges Board
2121 K Street, NW, Suite 700
Washington, DC 20037

IREX MENA Media Emerging Leaders Fellowships   2885
The Middle East and North Africa (MENA) program is open to media managers and supervisors based in the Middle East and North Africa with the desire to enhance their media management skills and increase their exposure to alternate media business practices. Fellows will attend sessions on leadership skills and media management trends at Northwestern University's Media Management Center followed by a fellowship placement with a U.S. media outlet. The key objective is to support independent media in the Middle East and North Africa.
Requirements: Applicants must have at least ten years experience in either broadcast or print outlets in the MENA region and be proficient in English.
Contact: Program Officer; (202) 628-8188; fax: (202) 628-8189; email: mena@irex.org
Internet: http://www.irex.org/programs/MENAmedia/index.asp
Sponsor: International Research & Exchanges Board
2121 K Street, NW, Suite 700
Washington, DC 20037

IREX Policy-Connect Collaborative Research Grants   2886
The program provides fellowships to U.S. scholars and professionals for overseas research on contemporary political, economic, historical, or cultural developments relevant to U.S. foreign policy. Fellowships support collaborative teams of two or three U.S. scholars and professionals for up to 12 months. The program seeks to attract, select, and support advanced research by U.S. experts in policy-relevant subject areas related to Southeast Europe and Eurasia, facilitate collaboration among and between U.S. and international scholars, and disseminate knowledge about Europe and Eurasia to a wide network of constituents in the United States and abroad. Upon completion of the project, scholars will be requested to present their research findings at a Policy Forum at the U.S. Department of State and to write a short policy paper.
Requirements: The principal investigator must be a U.S. citizen or permanent resident. A PhD or equivalent terminal degree is required. Collaborative research programs involving international colleagues are strongly encouraged.
Samples: Collaborators are Mark Basin, Senior Associate, State University of New York, and Paula Pickering, Assistant Professor, College of William and Mary—research in Bosnia-Herzegovina, Croatia, Kosovo, and Macedonia (2007-08); Collaborators are Keith Brown, Associate Professor, Watson Institute, Brown University, and Paul Nuti, Director, External, International and Government Relations, American Anthropological Association—research in Macedonia (2007-08).
Amount of Grant: $30,000 maximum
Date(s) Application is Due: Apr 1
Contact: Program Officers; (202) 628-8188; fax: (202) 628-8189; email: iaro@irex.org
Internet: http://www.irex.org/programs/policy-connect/index.asp
Sponsor: International Research & Exchange Board
2121 K Street, NW, Suite 700
Washington, DC 20037

IREX Regional Policy Symposium Grants   2887
The Program provides U.S. students, scholars, and professionals with a forum to examine and discuss current policy research on the countries of Eurasia and Central and East Europe from multi-disciplinary and multi-regional approaches. The research ultimately results in the development and dissemination of policy recommendations to academic and policy communities. The program has three primary goals: to enable U.S. junior and senior scholars to work together in analyzing complex issues affecting the countries of Eurasia and Central and East Europe from multi-disciplinary and multi-regional approaches; to encourage the cross-fertilization of ideas and networking opportunities among scholars with similar regional interests; and to provide policymaking communities with knowledge of current research on evolving regions and valuable policy conclusions drawn from intensive interaction among scholars.
Samples: The Former Soviet Republics of Central Asia and the Contemporary Silk Road, Alexandria, VA (2007); EU and NATO Member States and their Eastern Borderlands, Chesapeake Beach, MD (2006); Asia Regional Policy Symposium, Chesapeake Beach, MD (2005).
Date(s) Application is Due: Dec 15
Contact: Program Officers; (202) 628-8188; fax: (202) 628-8189; email: iaro@irex.org
Internet: http://www.irex.org/programs/symp/index.asp
Sponsor: International Research & Exchange Board
2121 K Street, NW, Suite 700
Washington, DC 20037

IREX Short-Term Travel Grants   2888
Grants provide support for scholarly projects focusing on Central and Eastern Europe, Eurasia, and Mongolia in the humanities and social sciences disciplines. Projects should demonstrate academic merit and relevance for the American academic community studying these regions, as well as a positive impact on public, cultural, and historical knowledge of these regions through dissemination of research results. Funding is available for individual scholarly research visits to archives, libraries, museums, etc., or for conducting interviews; presentations at scholarly conferences focused on and located in Central and Eastern Europe or Eurasia; and collaborative projects such as joint publications or comparative surveys. The award provides; travel from the United States to the host country; a stipend to cover in-country costs for meals, lodging, and local transportation; and miscellaneous research expenses directly related to the project incidental expenses, such as conference registration fees, visa fees, and research expenses.
Requirements: Applicants must be U.S. citizens or permanent residents (green card holder) of the U.S. for three consecutive years prior to application; submit a research proposal on a topic in one of the academic disciplines listed; and hold a PhD or other terminal degree.
Restrictions: Applicants cannot be a current IREX employee or consultant or their immediate family members (spouses, parents, children, and siblings). Projects must be in the humanities and social science disciplines only. Travel may not exceed a total of 60 days. Individuals may apply for only one trip and one project per application deadline. IREX-funded project activity must be completed within one year of the application deadline.
Samples: Agron Alibali, Visiting Fellow, Center for Democracy and Development, John W. McCormack Graduate School of Policy Studies, Albania—Constitutional Drafting through Foreign Assistance: Striking the Right Balance, The Case of Albania (2007-08); Bethany Barratt, Assistant Professor of Political Science, Department of Political Science and Public Administration, Roosevelt Univeristy, Kyrgyzstan—Regional Security, Aid Allocation and Human Rights (2007-08).
Amount of Grant: $5,000 maximum
Date(s) Application is Due: Feb 1

Contact: Program Officer, (202) 628-8188; fax: (202) 628-8189; email: stg@irex.org
Internet: http://www.irex.org/programs/stg/index.asp
Sponsor: International Research & Exchanges Board
2121 K Street, NW, Suite 700
Washington, DC 20037

IREX University Administration Support Program (UASP)   2889

The Program benefits university administrators at select state universities in Russia by providing access to pertinent university management skills and models of administration. The Program is further designed to provide training to university administrators in the region while providing more concentrated funding to those university administrators who demonstrate they have the initiative, skills, and institutional support necessary to turn their institutions into models of administrative excellence. To this end, IREX implements a two-tiered program. First, fellowships in University Administration are offered that provide funding for senior level administrators to spend approximately 10 weeks at host universities in the United States. Second, after completion of their fellowships in the United States, the university administrators apply for small grants to jump-start reform projects at their home universities.
Samples: Boris Gladkikh, Professor, Computer Science Department, Tomsk State University, Tomsk, Russia—Administrative Reforms Within the Computer Science Department (2005-07); Elena Ryabenkaya, Director, International Relations Department, Southern Federal University, Rostov oblast, Russia—Taking Southern Federal University to the World (2005-07); Mikhail Pevzner, Vice Rector, International Relations Office, Novgorod State University, Novgorod the Great, Russia—Creation of an Advisory Center for International Education (2005-07).
Contact: Program Officer; (202) 628-8188; fax: (202) 628-8189; email: uasp@irex.org
Internet: http://www.irex.org/programs/uasp/index.asp
Sponsor: International Research & Exchanges Board
2121 K Street, NW, Suite 700
Washington, DC 20037

IREX U.S. Embassy Policy Specialist (EPS) Program Grants   2890

The Program (EPS) was established to support U.S. embassies and consulates overseas by providing policy specialists-in-residence. While serving at the embassy or consulate the specialists also conduct their own research. EPS fellows have served missions in: Baku, Azerbaijan; Bishkek, Kyrgyzstan; Dushanbe, Tajikistan; Ekaterinburg, Russia; and Vladivostok, Russia. EPS fellowships are one to two months. Grant length and dates are determined in consultation with the specific embassy or consulate. Examples of embassy service can include conducting field research, writing policy papers, consulting embassy staff, and assisting with grant panels.
Samples: Rebecca Chamberlain-Creanga, PhD Candidate and Postgraduate Teaching Assistant, Department of Anthropology, London School of Economics and Political Science (2007-08); William Clark, Associate Professor, Department of Political Science, Louisiana State University (2007-08); Robin Haarr, Associate Professor, College of Justice and Safty, Criminal Justice and Police Studies, Eastern Kentucky University (2007-08).
Contact: Program Officer; (202) 628-8188; fax: (202) 628-8189; email: eps@irex.org
Internet: http://www.irex.org/programs/eps/index.asp
Sponsor: International Research & Exchange Board
2121 K Street, NW, Suite 700
Washington, DC 20037

IRP Fellowships in International Journalism   2891

The fellowships aim to strengthen the U.S. public's understanding of key international topics by helping to educate early- and mid-career U.S. journalists by providing them with access to leading international experts in the United States, and offering them opportunities to do reporting projects overseas. The program enables U.S. journalists to study international issues in Washington, DC, at The Paul H. Nitze School of Advanced International Studies (SAIS) of The Johns Hopkins University before traveling abroad. During the program, journalists have access to some of the world's leading specialists in international issues at SAIS and other institutions in the nation's capital. As part of their four-month program, fellows travel for five weeks to the country or region of their choice. While overseas, journalists work on an important global story, which they discuss with other fellows on their return to Washington. Application and guidelines are available online.
Requirements: Any U.S. journalist with at least three years of professional journalistic experience is eligible to apply. The program is open to journalists from newspapers, magazines, wire services, radio, television, and on-line news organizations. Freelancers are also invited to apply.
Amount of Grant: $2000 per month during the Washington stay
Date(s) Application is Due: Apr 1; Oct 1
Contact: International Reporting Project , (202) 663-7761; fax: (202) 663-7762; email: irp@jhu.edu
Internet: http://www.pewfellowships.org/fellows/program/program1.htm
Sponsor: Johns Hopkins University
1619 Massachusetts Avenue, NW
Washington, DC 20036

Irvine Fellowships   2892

The program is designed to attract faculty from historically underrepresented groups to teach in a liberal arts college setting. Of particular interest are the following areas: Asian studies, biological sciences, chemistry (analytical/environmental), child development, Chinese language/culture, economics (microeconomics), education (secondary/action

research), English (19th century British/postcolonial), earth science, international relations, mathematics education, religious studies (South Asia and/or Islam), and theater arts (acting/directing). The fellowships are designed to allow scholarly time for writing, while gaining teaching experience. Fellows will teach three courses during the year. Fellows also are expected to work with other faculty to expand curricula that are inclusive of race, class, and gender; to present the substance of one's dissertation in a public lecture; or to serve as mentors to a diverse student population. Renewal for an additional year is possible, and the position may be converted to tenure-track in future years.
Requirements: Applications should be either new PhDs or in the writing stage of the dissertation before beginning the position.
Amount of Grant: $28,500 salary
Contact: Susan Gotsch, (562) 907-4204; email: sgotsch@whittier.edu
Internet: http://www.whittier.edu
Sponsor: Whittier College
P.O. Box 634, 13406 E Philadelphia Street
Whittier, CA 90608

Irvine Health Foundation Grants   2893

The foundation supports IRS nonprofit organizations in support of projects and programs designed to meet the health care needs of Orange County, Ca, residents. Preference is given to activities addressing current unmet health care needs. Programs for youth, the elderly, and other groups also are supported. Types of support include challenge/matching grants, program grants, research grants, and seed money grants. There are no application deadlines; the board reviews requests ongoing.
Requirements: Orange County, CA, nonprofit organizations are eligible.
Restrictions: Grants are not awarded to individuals or for religious purposes, sporting activities, events, or endowments.
Geographic Focus: California
Samples: Laguna Beach Community Clinic (Laguna Beach, CA)—for expanded staffing, including a full-time receptionist and a part-time nurse and a clinical social worker, $37,000.; U of California at Irvine, Biomedical Research Ctr (Irvine, CA)—for construction costs associated with the new Cancer and Genetics facility, $300,000.; Catholic Charities of Orange County (Santa Ana, CA)—for health and social services delivery to poor residents of southern Orange County, CA, $60,000.
Contact: Executive Director, (949) 253-2959; fax: (949) 253-2962; email: info@ihf.org
Internet: http://www.ihf.org/grant/index.asp
Sponsor: Irvine Health Foundation
18301 Von Karman Avenue, Suite 440
Irvine, CA 92612

Irving and Rose Crown Scholarships and Fellowships in American History   2894

These fellowships are awarded for a program of graduate study and research leading to the PhD degree in American History at Brandeis University. The primary goal of the program is to train historical scholars, teachers, and researchers. Full tuition is provided. Fellowships are normally renewable for four years; additional research grants are available for fellows.
Requirements: Applicants must have a BA, MA, or professional training, preferably with a strong background in history, American studies, or related fields. Applicants must be accepted into the PhD program in American History at Brandeis University.
Amount of Grant: $27,345 plus a cash stipend of $16,000
Contact: Program Contact, Graduate Program in American History, (781) 736-2270; fax: (781) 736-2273; email: historia@brandeis.edu
Internet: http://www.brandeis.edu/departments/history/grad/amst-crown.html
Sponsor: Brandeis University - History Department
Dept of History, MS 036
Waltham, MA 02454-9110

Irvington Institute Immunology Postdoctoral Fellowships   2895

The institute will award up to 12 postdoctoral fellowships in immunology for three consecutive years of work in a U.S. laboratory or hospital for the award period beginning January 1 or July 1. Fellowships support basic medical research in the prevention and treatment of AIDS, cancer, diabetes, allergies, lupus, rheumatoid arthritis, and other immune system diseases. These fellowships are available to students from foreign countries if they have the proper visas for their stay.
Requirements: Medical doctors, PhDs, and MD/PhDs are encouraged to apply. Candidates with not more than three years' laboratory experience will receive priority. The institute does not have a citizenship requirement; however, applicants should indicate their citizenship and/or current visa status.
Restrictions: A laboratory may qualify for only one fellow per year.
Samples: Dr. Michael G. Brown, Jewish Hospital of Saint Louis, Washington U School of Medicine (Saint Louis, MO)—for research on the natural killer cell gene complex.; Dr. Shigemi Kinoshita, Department of Molecular Pharmacology, Stanford U School of Medicine (Stanford, CA)—for research on the molecular secrets of long-term survivors of HIV infection.
Amount of Grant: $45,000 per year
Date(s) Application is Due: Jul 12
Contact: Fellowship Coordinator, (212) 576-1005; fax: 576-1006; email: irving1@ix.netcom.com
Internet: http://www.irvingtoninstitute.org/fellowships.html
Sponsor: Irvington Institute for Immunological Research
245 Fifth Avenue, Room 2101
New York, NY 10016

ISA Hyland R. Johns Grant Program 2896

This program provides funds to researchers for projects of interest and benefit to the arboricultural industry. Research must focus on the biology, management, and care of trees, and their relation to environmental, social, and economic benefits. The society encourages research partnerships to enhance industry involvement, increase interdisciplinary interaction, and provide broader funding potential. Write or fax the society for a list of priority areas. Phone calls will not be accepted.

Restrictions: Grant funds cannot be used to pay overhead expenses.
Amount of Grant: $7500-$25,000
Date(s) Application is Due: May 1
Contact: ISA Research Trust; (888) 472-8733 or (217) 355-9411; fax: (217) 355-9516; email: isa@isa-arbor.com
Internet: http://www.treefund.org/GrantSumm.asp#Johns
Sponsor: International Society of Arboriculture
P.O. Box 3129, 1400 W Anthony Drive
Champaign, IL 61826

ISA John Z. Duling Grants 2897

The goal of the program is to provide seed money or partial support for research and technology transfer projects addressing topics that have the potential of benefiting the everyday work of arborists. Proposals in the following priority areas are more likely to be funded: root and soil management; planting and establishment; plant health care; and risk assessment and worker safety. Application information is available online.

Requirements: Proposals must be submitted on the application form or an exact duplicate. No faxed or reduced copies of the original will be accepted. Applications sent electronically will not be accepted.
Restrictions: Funds cannot be used for overhead expenses or student tuition and fees.
Amount of Grant: $7500 maximum
Contact: Executive Director, TREE Fund, (630) 221-8127; fax: (630) 690-0702; email: treefund@treefund.org
Internet: http://www.treefund.org/grants/Grants.aspx
Sponsor: International Society of Arboriculture
711 East Roosevelt Road
Wheaton, IL 60187

Island Foundation Grants 2898

Grants are awarded in the New England states for the support of arts and culture, civic and public affairs, community development, elementary and secondary education, environmental advocacy as it relates to alternatives in wastewater technology, and social services. Types of support include general operating budgets, matching funds, special projects, and research. Interested persons should request a copy of the foundation's annual report and guidelines before submitting their request.

Requirements: Nonprofit organizations in Maine, Massachusetts, and Rhode Island are eligible for grant support.
Restrictions: Grants do not support individuals, international organizations, religious organizations, special events, benefit dinners, or political campaigns.
Geographic Focus: Maine; Massachusetts; Rhode Island
Samples: Woods Hole Oceanographic Institution (Woods Hole, MA)—to support research on solar aquatic systems, $231,000.; Coalition for Buzzards Bay (Buzzards Bay, MA)—for the protection of Buzzards Bay, $45,000.
Amount of Grant: $1.7 million total
Contact: Julie Early, Executive Director, (508) 748-2809; fax: (508) 748-0991; email: islandfdn@earthlink.net
Sponsor: Island Foundation
589 Mill Street
Marion, MA 02738

ISOA/Elan Research Program 2899

The objective of this program, which targets academia and the biotechnology industry, is to facilitate the discovery, development, and clinical evaluation of effective therapies for Alzheimer's disease. The goal is to catalyze and accelerate the development of innovative and effective treatments by funding the development of lead compounds through preclinical in-vitro and in-vivo evaluation, including pharmacology, toxicology, pharmacokinetics, formulation chemistry, and the conduct of preclinical proof-of-concept studies. One-year grants will be awarded. Proposals covering two years may be submitted, although the second year of funding will be contingent on progress of the completed work, the investigator's second-year work plan, and the numbers and quality of competing proposals. Application and guidelines are available online.

Requirements: The principal investigator must hold an MD or PhD degree.
Amount of Grant: $130,000 approximately
Date(s) Application is Due: Oct 14
Contact: Wendy Ramos, Grant and Database Manager, (212) 901-8005; fax: (212) 935-2408; email: wramos@aging-institute.org
Internet: http://www.aging-institute.org
Sponsor: Institute for the Study of Aging
1414 Avenue of the Americas, Suite 1502
New York, NY 10153

ISUD Jacobsen Prizes 2900

ISUD is a group of scholars and philosophers from many parts of the world who have joined together to discuss and debate leading questions concerning the theory and practice of global community. The Jacobsen Prizes recognize continuation of Jens A.B. Jacobsen's quest to look beyond the individual ego and seek universal meaning that can be

shared by all. Submitted papers are eligible for the awards, which will be made during the Congress. All papers presented at the Congress wil be published in the ISUD bi-annual proceedings. Papers must be submitted by the listed deadline. Detailed information is contained on the Web site.

Requirements: Participation in the Congresses, which take place every two years (odd-numbered years), is open to anyone with a philosophical bent and with the love of wisdom in his or her heart.
Samples: Elizabeth Sperry from William Jewell College, Philosophy, and Sally Holt from William Jewell College, Religion, for their paper on Power and Fundamentalist Religion: A Philosophical and Sociological Approach. Victor J. Krebs from Universidad Simon Bolivar Caracas, Venezuela, for paper on Dionysus??? Revenge: Some Thoughts On Cultural Hybris, Suffering and Terror.
Amount of Grant: $1500
Date(s) Application is Due: Nov 15
Contact: Dr. Daniel Shannon, Department of Philosophy, email: deshan@depauw.edu
Internet: http://www.isud.org/papers/jacobsen.shtml
Sponsor: International Society for Universal Dialogue
DePauw University, 212 Asbury Hall
Greencastle, IN 46135,

Italian Academy for Advanced Studies in America Fellowships 2901

The program accepts applications at the postdoctoral and faculty level for fellowships in areas relating to the study of cultural identity, cultural transmission, and cultural memory, particularly—but not exclusively—with regard to Italy. Applications dealing with the scientific, sociological, and technological aspects of culture and memory are also encouraged. Special consideration will be given to projects in the neurosciences relevant to the academy's ongoing project in art and the neurosciences.

Requirements: Complete applications include a cover letter, curriculum vita, a two- to five-page project statement, and two letters of reference.
Date(s) Application is Due: Dec 15
Contact: Elisabetta Assi, Assistant Director, (212) 854-2306; email: ea2146@columbia.edu
Internet: http://www.italianacademy.columbia.edu/fellowships/fellowships.html
Sponsor: Columbia University
1161 Amsterdam Avenue, 6th Fl
New York, NY 10027

Italian History Prizes 2902

To encourage excellence in Italian history, the society in conjunction with the American Historical Association offers the Helen and Howard R. Marraro Prize for the best published book on Italian history and one prize for the best unpublished study (article- or dissertation-length). A double-spaced, typed manuscript in three copies must be submitted for the manuscript award, including name and address of author, brief vita, a short history of her/his interest in Italian history, and the date of completion of the manuscript. For the Marraro Prize, three copies of the book or article must be submitted.

Requirements: Applicant must be a resident of the United States or Canada and be beginning a career in Italian history or, for the Marraro Prize, have published a work on Italian history.
Amount of Grant: $200-$500
Date(s) Application is Due: May 15
Contact: Alan Reinerman, Executive Secretary, Department of History, Boston College, (617) 552-3814; email: alan.reinerman@bc.edu
Internet: http://faculty.valenciacc.edu/ckillinger/sihs/AnnualAwards.htm
Sponsor: Society of Italian Historical Studies
Boston College, Department of History
Chestnut Hill, MA 02167

ITEA Maley/FTE Scholarship 2903

The Foundation for Technology Education established the Maley/FTE Technology Teacher Scholarship in honor of Dr. Donald Maley. Its purpose is to support teachers in their preparation to increase the positive outcomes of technology education. Criteria include: (1) evidence of teaching success, (2) plans for action research, (3) recommendations, (4) plans for professional development, and (5) the applicant's need. The scholarship is provided directly to the teacher and will be presented during the ITEA Annual Conference.

Requirements: Applicant must be a member of the International Technology Education Association. (Membership may be enclosed with scholarship application). Applicant must be a technology teacher at any grade level who is beginning or continuing graduate study. See the sponsor's website for the list of required items for the application package.
Amount of Grant: $1,000
Date(s) Application is Due: Dec 1
Contact: Kendall N. Starkweather, Ph.D., DTE, Executive Director; (703) 860-2100; fax: (703) 860-0353; email: itea@iteaconnect.org
Internet: http://www.iteaconnect.org/Awards/scholarshipmaley.htm
Sponsor: International Technology Education Association (ITEA)
1914 Association Drive, Suite 201
Reston, VA 20191-1539

Ittleson Foundation AIDS Grants 2904

In regards to AIDS, the foundation is particularly interested in new model, pilot, and demonstration efforts: addressing the needs of underserved at-risk populations and especially those programs recognizing the overlap between such programs; responding to the challenges facing community-based AIDS service organizations and those organizations addressing systemic change; providing meaningful school-based sex education; making treatment information accessible, available and easily understandable to those in need of it; or, addressing the psycho-social needs of those infected and affected by AIDS, especially adolescents.

Requirements: Tax-exempt organizations may apply.

Restrictions: The foundation generally does not provide funds for capital building projects, endowments, grants to individuals, scholarships or internships (except as part of a program), direct service programs (especially outside New York City), projects that are local in focus and unlikely to be replicated, continuing or general support, projects and organizations that are international in scope or purpose, or biomedical research.

Samples: AIDS Alliance for Children Youth and Family, Washington, DC, $5,000—to transform the National Consumer Leadership Corps Training Program into one that can be replicated by local AIDS organizations around the nation (2008); Cesar E. Chavez Institute, New York, NY, $40,000—one-time grant to develop new family interventions and a new family-related model of care to reduce risk for HIV and mental health problems in lesbian, gay, bisexual and transgender (LGBT) youth (2008).

Date(s) Application is Due: Sep 1

Contact: Anthony C. Wood, Executive Director; (212) 794-2008; fax: (212) 794-0351

Internet: http://www.ittlesonfoundation.org/aids.html

Sponsor: Ittleson Foundation

15 E. 67th Street, 5th Floor

New York, NY 10021

Ittleson Foundation Environment Grants 2905

The foundation supports innovative pilot, model and demonstration projects that will help move individuals, communities, and organizations from environmental awareness to environmental activism by changing attitudes and behaviors. This program seeks to encourage and nurture environmental action through: supporting the present generation of environmental activists, whether professionals or volunteers through education, training and other activities; educating and engaging the next generation of environmentalists with a special interest in supporting the training of those who are teaching that generation; strengthening the infrastructure of the environmental movement with a particular focus on efforts at the grassroots and statewide levels; activating new constituencies, particularly those focused on environmental equity issues.

Requirements: Tax-exempt organizations may apply.

Restrictions: The foundation generally does not provide funds for capital building projects, endowments, grants to individuals, scholarships or internships (except as part of a program), direct service programs (especially outside New York City), projects that are local in focus and unlikely to be replicated, continuing or general support, projects and organizations that are international in scope or purpose, or biomedical research.

Samples: Antioch University New England, Keene, NH, $45,000—to advance the emerging field of Conservation Psychology (2007); Cornell University Cooperative Extension, New York, NY, $40,000—one-time grant for the "Participatory Development of an Urban Forestry Community Engagement Model" involving the public in the planning planting and stewardship of urban forestry efforts (2008).

Date(s) Application is Due: Sep 1

Contact: Anthony C. Wood, Executive Director; (212) 794-2008; fax: (212) 794-0351

Internet: http://www.ittlesonfoundation.org/enviro.html

Sponsor: Ittleson Foundation

15 E. 67th Street, 5th Floor

New York, NY 10021

Ittleson Foundation Mental Health Grants 2906

For this program, the foundation is interested in innovative, pilot, model and demonstration projects that are: fighting the stigma associated with mental illness and working to change the public's negative perception of people who have mental illness; utilizing new knowledge and current technological advances to improve programs and services for people who have mental illness; bringing the full benefits of this new knowledge and technology to those who presently do not have access to them; or, advancing preventative mental health efforts, especially those targeted to youth and adolescents, with a special focus on strategies that involve parents, teachers, and others in close contact with these populations.

Requirements: Tax-exempt organizations may apply.

Restrictions: The foundation generally does not provide funds for capital building projects, endowments, grants to individuals, scholarships or internships (except as part of a program), direct service programs (especially outside New York City), projects that are local in focus and unlikely to be replicated, continuing or general support, projects and organizations that are international in scope or purpose, or biomedical research.

Samples: Active Minds, Inc., Washington, DC, $10,000—to dramatically expand the network of chapters on college campuses (2008); Partnership with Children, New York, NY, $50,000—one-time grant to develop the Center for Capacity Building to enable the dissemination of the successful counseling and prevention care program for inner-city boys and girls, "Open Heart-Open Mind" to over 100 schools by 2012 (2008); Horticultural Society of New York, New York, NY, $30,000—to launch their new Nonprofit Partnership for Horticultural Therapy to formalize and expand their ability to help a wide range of organizations working with the mentally and physically ill, formerly homeless, HIV+, victims of substance abuse, the elderly, at-risk juveniles and those re-entering society from incarceration, use horticultural therapy programs to address the needs of these marginalized populations (2008).

Date(s) Application is Due: Sep 1

Contact: Anthony C. Wood, Executive Director; (212) 794-2008; fax: (212) 794-0351

Internet: http://www.ittlesonfoundation.org/mental.html

Sponsor: Ittleson Foundation

15 E. 67th Street, 5th Floor

New York, NY 10021

Ittleson Predoctoral Fellowship 2907

The fellowship is awarded for research in fields other than Western art to be held partly in residence at the National Gallery of Art, Center for Advanced Study in the Visual Arts, and partly elsewhere in the United States or abroad. The Ittleson fellow is expected to spend the second year of the fellowship at the center to complete the dissertation. Application must be made through the chair of the graduate department of art history or other appropriate department. Departments should limit nomination to one candidate. Fellowships begin September 1 and are not renewable.

Requirements: Applicants must have completed their residence requirements and coursework for the PhD and general or preliminary examinations before the date of application and know two foreign languages related to the topic of the dissertation. Applicants must be either U.S. citizens or enrolled in a university in the United States.

Samples: Trian Nguyen, U of California, Berkeley—for dissertation, Ninh-phuc Monastery: An Examination of Seventeenth-Century Buddhist Sculpture in Northern Vietnam.

Amount of Grant: $24,000 annually for two years

Date(s) Application is Due: Nov 15

Contact: Fellowships Program, Center for Advanced Study in Visual Arts, (202) 842-6482; fax: (202) 842-6733; email: advstudy@nga.gov

Internet: http://www.nga.gov/resources/casvapre.htm

Sponsor: National Gallery of Art

Fourth Street and Constitution Avenue NW

Washington, DC 20565

ITVS Research and Development/Commissioning Funding 2908

TVS accepts proposals on an ongoing basis for production funding for projects that do not fit within the parameters of its standing initiatives (Open Call, LINCS and DDF), including limited series. ITVS also accepts proposals on an ongoing basis for projects in all genres in need of research and development funding. Examples of funding requests include travel, research, script development or development of a fundraising reel.

Contact: Jonathan Archer, (415) 356-8383 ext 284; fax: (415) 356-8391; email: jonathan_archer@itvs.org

Internet: http://www.itvs.org/producers/funding.html

Sponsor: Independent Television Service

651 Brannan Street, Suite 410

San Francisco, CA 94107

IUCP Archives Program Fellowship 2909

Grants support research in any discipline at Indiana University's Ruth Lilly Philanthropy Archives. To apply, submit a resume or vita and cover letter including the following information: name, contact information (including telephone and email address); title and brief description of the project; description of how the Philanthropy Archival collection will be utilized in the research project; dates and duration of the project; and amount requested and budget summary. Allowable expenses include round-trip travel, temporary lodging, food, photocopying, and other research expenses. Applications are accepted on an ongoing basis. Contact staff prior to proposal submission to ensure that the required research documents are available in the library.

Requirements: Individual scholars and researchers are eligible.

Amount of Grant: $4000 maximum

Contact: Kathy Steinberg, (317) 684-8957; fax: (317) 684-8900; email: ksteinbe@iupui.edu

Internet: http://www.philanthropy.iupui.edu/research.html

Sponsor: Indiana University Center on Philanthropy

550 W North Street, Suite 301

Indianapolis, IN 46202-3162

IUCP Dissertation Fellowship 2910

One-year grants are awarded for doctoral research relating to issues facing philanthropy and nonprofit organizations. Participants are encouraged to obtain matching funds from their home institutions.

Requirements: Applicants from Indiana institutions must be conducting dissertation research relevant to the center's priority research areas. Applicants from universities outside of Indiana may be asked to contact the Aspen Institute's Nonprofit Research Fund.

Amount of Grant: $4000 maximum

Contact: Kathy Steinberg, (317) 684-8957; email: ksteinbe@iupui.edu

Internet: http://www.philanthropy.iupui.edu

Sponsor: Indiana University Center on Philanthropy

550 W North Street, Suite 301

Indianapolis, IN 46202-3272

IUCP Hearst Minority Fellowships 2911

The fellowship is designed to provide members of minority groups the opportunity to engage in reflective thought and study as well as practical application of philanthropy. The fellowship is a nonrenewable 10-month appointment from August 15 to June 15. The fellow will devote a portion of time to formal study and a portion to the practice of philanthropy. The Center on Philanthropy will provide fellows a stipend and full tuition. The student is expected to pay the remaining tuition, housing, and living expenses, as well as other costs (books, insurance, fees, etc.). Application forms for the MA in philanthropic studies and the MPA program can be requested from the center. These applications are used to determine the fellowship recipient.

Requirements: The fellowship is open to members of groups traditionally underrepresented in organized philanthropy who have bachelor's degrees or equivalent from another country in any academic field and are either recent graduates, scholars, active volunteers, or nonprofit practitioners. To be considered, a candidate must apply and be admitted to either the MA in philanthropic studies or the MPA in nonprofit management.

Amount of Grant: $9000 for living and housing expenses.
Date(s) Application is Due: Feb 1
Contact: Hearst Minority Fellowships, (800) 854-1612; fax: (317) 684-8900; email: maphil@iupui.edu
Internet: http://www.philanthropy.iupui.edu/hearst.html
Sponsor: Indiana University Center on Philanthropy
550 W North Street, Suite 301
Indianapolis, IN 46202-3162

IUCP Indiana Research Fund 2912

Grants are awarded for research relating to issues facing philanthropy and nonprofit organizations. Priority will be given to well-designed projects consistent with the center's strategic emphasis on research and practice that may be translated into potential solutions for problems facing philanthropy, fund raising, and the nonprofit sector. Proposals involving collaboration among nonprofit practitioners, academic researchers, independent scholars, and policy analysts are welcome. Annual deadlines dates may vary; contact program staff for exact dates.

Requirements: Grants will be awarded to faculty members affiliated with any of the branch campuses of Indiana University.
Amount of Grant: $50,000 maximum
Contact: Maggie Bowden, (317) 278-8984; email: mtbowden@iupui.edu
Internet: http://www.philanthropy.iupui.edu/giving_opportunities.html
Sponsor: Indiana University Center on Philanthropy
550 W North Street, Suite 301
Indianapolis, IN 46202-3162

IUVSTA Prize for Science 2913

The Prize is given for experimental and/or theoretical research in vacuum science, technique or its applications. The purpose is to recognize and encourage outstanding internationally-acclaimed research in the fields of interest to the IUVSTA. The award will normally be given to an individual; however, in exceptional cases involving team research, multiple awards may be given. The Prize consists of a cash award, a struck medal and a certificate setting forth the reasons for the award. The Prize will be conferred at intervals of not less than three years. Reasonable travel expenses of the recipient to the meeting at which the Prize is presented shall be agreed upon and reimbursed.

Requirements: The nominee must have accomplished outstanding experimental and/or theoretical research in vacuum science, technique or its applications within the ten years preceding the year in which the award is made. Special consideration will be given to nominees currently engaged in an active career of research. Neither nominees nor nominators are required to be members of IUVSTA member societies.
Restrictions: Current officers and members of the Executive Council and Standing Committees, as well as Division Officers of IUVSTA, are not eligible.
Date(s) Application is Due: Apr 2
Contact: Dr. Masatoshi Ono; fax: +81 44 366 6331; email: ono.m@funai-atri.co.jp
Internet: http://www.iuvsta.org/prizeinfo.html
Sponsor: International Union for Vacuum Science, Technique, and Applications
1-12 Minami-Wataridi-Cho
Kawasaki-shi, Kanagawa 210-0855 Japan

IUVSTA Prize for Technology 2914

The purpose of the Award is to recognize and encourage outstanding internationally-acclaimed achievements in technology and instrumentation in the fields of interest to the IUVSTA. The award will normally be given to an individual; however, in exceptional cases involving team research, multiple awards may be given. The Prize consists of a cash award, a struck medal and a certificate setting forth the reasons for the award. The Prize will be conferred at intervals of not less than three years. Reasonable travel expenses of the recipient to the meeting at which the Prize is presented shall be agreed upon and reimbursed.

Requirements: The nominee must have accomplished outstanding results in the area of technology and instrumentation within the ten years preceding the year in which the award is made. Special consideration will be given to nominees currently actively involved in this work. Neither nominees nor nominators are required to be members of IUVSTA member societies.
Restrictions: Current officers and members of the Executive Council and Standing Committees, as well as Division Officers of IUVSTA, are not eligible.
Samples: Dr. Martin P. Seah, National Physical Laboratory, United Kingdom (2004); Wolf-Dieter Munz, Professor of Surface Engineering, Materials Research Institute of Sheffield Hallam University (2001).
Date(s) Application is Due: Apr 2
Contact: Dr. Masatoshi Ono; fax: +81 44 366 6331; email: ono.m@funai-atri.co.jp
Internet: http://www.iuvsta.org/prizeinfo.html
Sponsor: International Union for Vacuum Science, Technique, and Applications
1-12 Minami-Wataridi-Cho
Kawasaki-shi, Kanagawa 210-0855 Japan

IUVSTA Welch Foundation Scholarship 2915

The scholarship is offered for a one-year period, starting in September, to a promising scholar who wishes to contribute to the study of vacuum science techniques or their application in any field. Because of the international nature of the scholarship, strong preference is given to applicants who propose study in a foreign lab in which they have not yet studied. Payment of the scholarship is in three installments totalling $15,000; one of $7500 at the beginning, another of $7000 after six months, and a third of $500 upon delivery of a final report after completion of work. The scholarship holder is encouraged to seek funds in addition to the scholarship, but should obtain authorization of the chairperson of the Welch Committee of the IUVSTA before accepting any additional funds.

Requirements: Candidates should have at least a bachelor's degree; a doctoral degree is preferred. Application should include a curriculum vita; a photocopy of, or attestation of, all diplomas; the name and address of laboratory chosen and a letter indicating that this laboratory's facilities will be available to the applicant; a 200-word abstract describing the proposed research; a declaration that the candidate will not violate any laws of his/her own country during the tenure of the scholarship, as well as a declaration that the candidate will not violate any laws or engage in any political activity in the country where he/she intends working; and two recommendations from present or past professors or research directors. Additionally, candidates must produce satisfactory evidence of reasonable fluency either in the language of the country where he/she will work during the tenure of the scholarship or in English.
Restrictions: If the candidate cannot begin work as scheduled, he or she may begin within three months after September 1. In the case of a delay of more than three months, another candidate may be chosen. The laboratory where the candidate wishes to work must approve any delay in the commencement of work.
Samples: Oren Tal, Tel Aviv, Israel, to study in The Netherlands (2006); Johanna Rosen, Germany, to study in Australia (2005); Maksim Kireitseu, Belarus, to study in the United States (2004).
Amount of Grant: $15,000
Date(s) Application is Due: Apr 15
Contact: Dr. Frank R. Shepherd, Nortel Networks, 3 Grierson Lane, Ottawa ON K2W 1A6 Canada, (613) 763-3285; fax: (613) 763-2404; email: frank_shepherd@avs.org
Internet: http://www.iuvsta.org/welchann.html
Sponsor: International Union for Vacuum Science, Technique, and Applications
30 avenue de la Renaissance
Brussels B-1000 Belgium

IWMF Public Health Fellowship 2916

The goals of the fellowship are to offer women newspaper editors and radio producers the opportunity to enhance their ability to cover public health issues, build their journalism skills, and learn first-hand about standards and practices in the U.S. media. The six-month fellowship includes four months in the United States and two months in the fellows' home countries, where the fellows will implement a public health reporting project. Each applicant must clearly identify and describe a public health reporting project that she will focus on when she returns to her home country for that segment of the fellowship. Those demonstrating original approaches to covering HIV/AIDS will be given top consideration. The fellowship includes application costs for a U.S. visa; round-trip economy airfare from the fellows' home countries to Washington, DC; round-trip economy airfare from Washington, DC, to the fellowship site and health insurance; and a fixed stipend to cover lodging, meals, and ground transportation during the U.S. segment of the fellowship. For the home country segment, a fixed stipend will be provided to support the fellow and cover all costs associated with implementing her project. Application and guidelines are available online.

Requirements: The fellowship is open to women journalists who are working in one of the following countries: Asia (China and Indonesia), West Africa (Benin, Burkina Faso, Cameroon, Cape Verde, Chad, Gambia, Ghana, Guinea, Guinea Bissau, Ivory Coast, Liberia, Mali, Mauritania, Niger, Nigeria, Senegal, Sierra Leone, and Togo), and Southern Africa (Angola, Botswana, Lesotho, Malawi, Mozambique, Namibia, Swaziland, Zambia, and Zimbabwe). A successful applicant will be dedicated to a career in journalism and have five or more years of full-time experience in the profession, with at least three of those years having been focused on public health journalism. She must be currently working as a newspaper editor or radio producer and show a strong commitment to sharing knowledge and skills with colleagues upon returning home. The fellow must also have excellent written and spoken English skills.
Date(s) Application is Due: Apr 15
Contact: Fellowship Administrator, (202) 496-1992; fax: (202) 496-1977; email: fellowship@iwmf.org
Internet: http://iwmf.org/programs/8702
Sponsor: International Women's Media Foundation
1625 K Street NW, Suite 1275
Washington, DC 20006

Izaak Walton Killam Memorial Postdoctoral Fellowships at Dalhousie University 2917

The postdoctoral fellowships are awarded annually at Dalhousie University to recently graduated scholars of superior academic research ability in any discipline. Awards are tenable for two years and include travel costs, a research grant, and a conference travel allowance. Guidelines and forms are available online.

Requirements: Applicants must have recently completed a PhD (January 1 of last year) at a recognized university and have no current affiliation with Dalhousie University.
Restrictions: Because these fellowships are intended to attract new scholars to Dalhousie, scholars already at Dalhousie and DalTech are not eligible to apply, including Dalhousie

and DalTech PhDs, Dalhousie, DalTech or King's employees, and researchers using Dalhousie, DalTech, or King's facilities.
Amount of Grant: $C40,000 plus travel allowance, $C3000 research allowance, and $C1000 conference travel grant
Date(s) Application is Due: Dec 15
Contact: Office of the Dean, Faculty of Graduate Studies, fax: (902) 494-8797; email: graduate.studies@dal.ca
Internet: http://www.dalgrad.dal.ca/kpdf
Sponsor: Dalhousie University
6299 South Street
Halifax, NS B3H 4H6 Canada

Izaak Walton Killam Memorial Postdoctoral Fellowships at the University of British Columbia 2918

These awards are offered to candidates who have shown superior ability in research and who possess doctorates or who are deemed to have similar qualifications. The candidate may pursue any field of research other than the arts. The basis of the award will be special distinction of intellect, with due regard for sound character and personal qualities. Subject to review at the end of the first year, Killam Memorial Postdoctoral Fellowships are awarded for two years. Fellowships are tenable at the University of British Columbia.
Requirements: Applicants must have obtained the PhD no more than two academic years prior to the anticipated commencement date of the fellowship. The doctorate must be from a university other than The University of British Columbia. Application is open to citizens of any country.
Amount of Grant: $C44,000 plus $6000 travel allowance
Date(s) Application is Due: Nov 15
Contact: Killam Secretary, Faculty of Graduate Studies, (604) 822-2848; fax: (604) 822-5802; email: graduate.awards@ubc.ca
Internet: http://www.grad.ubc.ca/awards/index.asp?menu=015,000,000,000
Sponsor: University of British Columbia
180-6371 Crescent Road
Vancouver, BC V6T 1Z1 Canada

Izaak Walton Killam Memorial Predoctoral Fellowships at the University of British Columbia 2919

These fellowships are awarded to highest ranked doctoral candidates in the university's graduate fellowships competitions. Approximately 25 awards are made each year. Applications are available on the Web site.
Requirements: Applicants must be graduate students with a first-class standing in the last two years of full-time study. Students from any country and from all fields are eligible.
Amount of Grant: $C25,000 plus $C1500 travel allowance
Contact: Killam Secretary, Faculty of Graduate Studies, (604) 822-8501 or 822-0976; fax: (604) 822-5802; email: graduate.awards@ubc.ca
Internet: http://www.grad.ubc.ca/awards/index.asp?menu=003,000,000,000
Sponsor: University of British Columbia
180-6371 Crescent Road
Vancouver, BC V6T 1Z2 Canada

Izaak Walton Killam Memorial Scholarships at Dalhousie University 2920

These awards are offered in support of studies leading toward either a master's or doctoral degree. Supported by The Killam Trusts, the scholarships include stipends and funds to assist with transportation costs to Halifax. Scholars may perform instruction or demonstration duties at the discretion of the department, for which additional remuneration is given. Master's students may hold a Killam Scholarship for 12 months and PhD students for up to 36 months.
Requirements: Eligibility is based on a first-class undergraduate degree in the field of study the student wishes to pursue. Candidates are not required to submit application forms but should apply to the Registrar for admission to graduate studies no later than the listed application deadline date. Killam scholars are selected on the basis of nominations made by departments. It is expected that nominees will also have applied for funding from relevant national or international agencies. Canadian students are eligible for nomination for the Killam Scholarships only if they have applied for the relevant national scholarship (NSERC, SSHRC, MRC, etc.).
Amount of Grant: $C19,000 for the master's level; $C23,000 for the PhD level
Date(s) Application is Due: Apr 1; Aug 31; Dec 31
Contact: Margaret Wood, Killam Administrator; fax: (902) 494-2772; fax: (902) 494-8797; email: margaret.wood@dal.ca
Internet: http://www.dalgrad.dal.ca/funding
Sponsor: Dalhousie University
Henry Hicks Academic Administration Building, Room 314
Halifax, NS B3H 4H6 Canada

Izaak Walton Killam Memorial Scholarships at the University of Calgary 2921

Approximately nine awards are made annually to the top ranking students in the Open competition who are eligible to hold them. The scholarship has the highest monetary value of any of the university's awards ($20,100 per annum) and includes a research allowance. The awards are for two years, the second year being subject to satisfactory progress.
Requirements: These scholarships are open for competition among PhD students who apply to the University of Calgary's Open Scholarship competition.
Amount of Grant: $C20,100 per annum, plus $3000 research allowance
Date(s) Application is Due: Feb 1

Contact: Connie Busch, Faculty of Graduate Studies, (403) 220-5690; email: cbusch@ucalgary.ca
Internet: http://www.ucalgary.ca/pubs/calendar/current/what/Awards/03GradAwards.htm#71
Sponsor: University of Calgary
2500 University Dr NW, Earth Sciences Building, Room 720
Calgary, AB T2N 1N4 Canada

Izumi Foundation Grants 2922

The mission of the foundation is to alleviate human suffering through improved health care, especially the poorest and the most vulnerable members of society. The foundation's current geographic focus area is a select number of countries in Africa and Central and South America. Program interests include disease, disabilities, and inadequate health care related to poverty. Within its expressed goals and designated geographic areas, the foundation seeks to support projects and programs that distribute medicines and vaccines, and provide direct services and care; provide interventions that encourage the most effective use of limited, local health care resources; address the underlying causes of disease and persistent health care problems; build health care capacity on the community level; provide cost-effective strategies for prevention; and advance health care partnerships and collaborations. The foundation does not accept unsolicited proposals. Potential applicants are asked to submit a letter of inquiry that does not exceed three pages in length. Letters of inquiry may be submitted any time during the year.
Requirements: Organizations within the United States must have 501(c)3 tax-exempt status to be eligible to apply. Agencies and organizations outside of the United States will also be required to furnish documentation verifying their nonprofit status.
Restrictions: The foundation does not fund medical research or other research-related activities; income generating, endowments, or fundraising activities; ongoing general operating expenses, indirect costs, or existing deficits; capital costs—the acquisition of land, buildings, or vehicles; direct support to individuals, scholarships, school fees, or clothing; or religious activities, media campaigns, or lobbying of any kind.
Amount of Grant: $100,000 maximum
Contact: Catherine Bryant, Program Officer, (617) 303-0345; fax: (617) 303-0339; email: cbryant@izumi.org
Internet: http://www.izumi.org
Sponsor: Izumi Foundation
One Financial Ctr, 28th Fl
Boston, MA 02111

J. Hall Taylor Medal 2923

This medal is presented for distinguished service or eminent achievement in the field of codes and standards pertaining to the broad fields of piping and pressure vessels that are sponsored or undertaken by ASME. The scope shall include contributions to technical advancement and administration.
Samples: Blaine Roberts—award winner, $1000 (2005); J. Robert Sims—award winner, $1000 (2004); Michael Gold—award winner, $1000 (2003).
Amount of Grant: $1000
Date(s) Application is Due: Nov 15
Contact: Gilda DiTullio, Manager, (212) 591-7736; fax: (212) 705-7739; email: ditulliog@asme.org or awards@asme.org
Internet: http://www.asme.org/honors/ms71/saa/taylor.html
Sponsor: American Society of Mechanical Engineers
3 Park Avenue
New York, NY 10016

J. Waldo Smith Hydraulic Fellowship 2924

The fellowship is restricted to research in the field of experimental hydraulics as distinguished from research concentrating on purely theoretical hydraulics. Emphasis is to be placed on practical experiments designed and executed for the purpose of advancing knowledge with respect to the laws of hydraulic flow, rather than to the type of research that proceeds on the theory of mathematical analysis based on assumptions of unknown validity. The purpose of the research is to test the assumptions that are currently made, and also to develop a better understanding of fluid flow. The fellowship is offered every third year (next in 2006) and runs for one full academic year. Administration is in part through the institution that invites cooperation through its engineering faculty. Application packets are available through the student chapter advisor or from ASCE.
Requirements: Fellowships are awarded to graduate students, who are members of the society, who give promise of best fulfilling the ideals of the fellowship. There is a brief application form. Each application shall include a statement in general terms of the purposes for which the funds are expected to be used.
Amount of Grant: $4000 per year, plus maximum of $1000 research equipment allowance
Date(s) Application is Due: Feb 9
Contact: Grants Coordinator, (703) 295-6342; fax: (703) 295-6343; email: student@asce.org
Internet: http://www.asce.org/inside/stud_scholar.cfm
Sponsor: American Society of Civil Engineers
1801 Alexander Bell Drive
Reston, VA 20191-4400

J.B. Reynolds Foundation Grants 2925

The foundation makes grants in the local Kansas City, MO, area, up to a 150-mile radius of the city. Grants are awarded for building and equipment, community development, medical research, social welfare, and the arts and humanities. Some support is given to

colleges and universities. Additional types of support include general operating support, continuing support, annual campaigns, endowment funds, publications, and research. The board meets in April and December of each year to consider letters of requests, and only invited proposals are reviewed. All grants are awarded in December.
Requirements: 501(c)3 organizations in the Kansas City, MO, area may submit applications.
Geographic Focus: Missouri
Amount of Grant: $5,000-$50,000 range
Contact: Richard L. Finn, Secretary-Treasurer; (816) 753-7000; fax: (816) 753-1354
Sponsor: J.B. Reynolds Foundation
P.O. Box 219139
Kansas City, MO 64141-6139

J.H. Stewart Reid Memorial Fellowship for Doctoral Studies   2926
This fellowship is awarded annually for doctoral study in a Canadian university in any field of study. Fellows must reapply for continuation.
Requirements: Applicants must be Canadian citizens or have held permanent resident (landed immigrant) status for one year prior to application; must hold registration in a doctoral program at a Canadian university; must have completed comprehensive exams and have doctoral thesis proposal accepted by the listed deadline date of the year applying; must have a first-class academic record; and must not hold scholarships that exceed in total $25,000 inclusive of the J. H. Stewart Reid Memorial Fellowship Trust.
Samples: David Nielsen, Queen's U—award recipient, $5000 (2005).
Amount of Grant: $C5000 annually
Date(s) Application is Due: Apr 30
Contact: Johanne Smith, Awards Officer, (613) 820-2270; fax: (613) 820-7244; email: stewartreid@caut.ca
Internet: http://stewartreid.caut.ca/English/policy.htm
Sponsor: Canadian Association of University Teachers
2675 Queensview Drive
Ottawa, ON K2B 8K2 Canada

J.I. Staley Prize   2927
This prize is awarded periodically to a living author for an outstanding book in the field of anthropology. The award is intended to acknowledge those innovative works that have gone beyond traditional frontiers in anthropology and given new dimensions to our understanding of the human species. By recognizing these publications and their authors, the prize helps to stimulate research and writing that steps outside the dominant schools of thought and crosses arbitrary boundaries between disciplines. To nominate a book, send a letter of nomination and solicit two additional scholars to do the same; a complete nomination requires three letters. Each letter of nomination should be a maximum of three pages, single-spaced, and should describe the book's innovative thinking; its use of two or more academic disciplines or of two or more subdisciplines of anthropology; its contribution to our understanding of humankind; and its impact within anthropology. Nomination forms are available on the Web site.
Requirements: To be considered for the prize, a book must have been published at least two years but not more than eight years before the year of nomination. (Once a book has been nominated, it will stay in the pool of nominees until it passes the 10-year age limit.) Coauthored books may be nominated, but edited volumes may not. The nomination must clearly be a single book, even if it builds on prior work by the author or others.
Restrictions: Authors and publishers may not nominate their own work.
Samples: Patrick Kirch and Marshall Sahlins—for Anahulu: The Anthropology of History in the Kingdom of Hawaii, $7500.; Margaret Lock—for Encounters with Aging: Mythologies of Menopause in Japan and North America, $7500.
Amount of Grant: $10,000
Date(s) Application is Due: Oct 1
Contact: Staley Prize Coordinator, (505) 954-7201; fax: (505) 989-9809; email: staley@sarsf.org
Internet: http://www.sarweb.org/staley/staley.htm
Sponsor: School of American Research
P.O. Box 2188
Santa Fe, NM 87504-2188

J.S. Mill Fellowships in Economics for Exceptional Students   2928
The fellowship assists students who are enrolled in the graduate program in economics at Washington University. Application materials may be obtained from the department.
Requirements: Applicants must be admitted to the graduate program in the department of economics at Washington University.
Amount of Grant: $16,000 maximum stipend plus full tuition
Date(s) Application is Due: Jan 15
Contact: Karen Rensing, Secretary, Department of Economics, (314) 935-5670; fax: (314) 935-4156; email: karenr@wuecona.wustl.edu or gradsec@wueconc.wustl.edu
Internet: http://economics.wustl.edu/graduate/brochure/brochure.pdf
Sponsor: Washington University
Campus Box 1208, 205 Eliot Hall
Saint Louis, MO 63130-4899

J.W. and Ida M. Jameson Foundation Grants   2929
The foundation supports higher and theological education, hospitals, medical research, cultural programs, and Protestant and Catholic religion. Types of support include research and general operating budgets. Applicants should submit a proposal.
Requirements: California nonprofits may submit applications for grant support.
Geographic Focus: California

Amount of Grant: $5000-$50,000 range
Date(s) Application is Due: Feb 1
Contact: Les Hugn, President, (626) 355-6973
Sponsor: J.W. and Ida M. Jameson Foundation
P.O. Box 397
Sierra Madre, CA 91025

J.W. Kieckhefer Foundation Grants   2930
The foundation awards grants to support medical research; hospice; health; family planning; disabled; social services; higher, medical, and other education; youth; child welfare; conservation; community funds; public policy; and cultural programs. The grants are awarded for operating budgets, continuing support, annual campaigns, emergency funds, building funds, equipment, land acquisition, endowment funds, matching funds, research, publications, conferences and seminars, and special projects. Applications should be submitted between May and November. The board meets in November and December to consider requests.
Restrictions: Grants are not awarded to individuals or for seed money, deficit financing, scholarships, or demonstration projects.
Amount of Grant: $350-$100,000 range
Contact: John Kieckhefer or Eugene Polk, Trustees, (928) 445-4010
Sponsor: J.W. Kieckhefer Foundation
P.O. Box 1151, 116 E Gurley Street
Prescott, AZ 86302

Jackson Foundation Grants   2931
The Jackson Foundation was established October 1960 pursuant to the last will of Maria C. Jackson. Its purpose is to respond to the requests deemed appropriate to promote the welfare of the public of the City of Portland or the State of Oregon, or both. The Foundation considers projects located outside the Portland metropolitan area only if the project is of statewide appeal, rather than of local concern.
Requirements: Grants are awarded to nonprofit 501(c)3 tax-exempt agencies located within the state of Oregon.
Restrictions: Funding is not available for individuals or private businesses. Grants are generally not made to a K-12 school. Do not contact the Foundation by phone to check the status of an application.
Geographic Focus: Oregon
Samples: Emanuel Medical Center Foundation (Portland, OR)—for general support, $10,000.; Portland Community College Foundation (OR)—for program support, $5000.
Amount of Grant: $1000-$25,000 average
Date(s) Application is Due: Mar 31; Jun 30; Sep 30; Dec 31
Contact: Robert H. Depew, c/o U.S. Bank; (503) 275-4414
Internet: http://www.thejacksonfoundation.com
Sponsor: Jackson Foundation
P.O. Box 3168
Portland, OR 97208

Jackson Laboratory Postdoctoral Research Training Program   2932
Fellowships are intended for postdoctoral associateships at Jackson Laboratory allowing the fellows to participate in research into gene regulation, developmental genetics, hematology and immunology, and other fields of mammalian biology, with the purpose of the recipient receiving research training under the guidance of experienced researchers. The fellowships are awarded for one year, renewable for up to two years. Applications are accepted at any time.
Requirements: Recent recipients of doctoral degrees in medicine, veterinary medicine, and the biological sciences are eligible to apply. Applicants may be of any nationality.
Contact: Suzanne Serreze, Manager, Sponsored Training Programs, (207) 288-6420; fax: (207) 288-6079; email: sbs@jax.org
Internet: http://www.jax.org/education/postdoc.html
Sponsor: Jackson Laboratory
600 Main Street
Bar Harbor, ME 04609-1500

Jacob and Charlotte Lehrman Foundation Grants   2933
The foundation makes grants to establish scholarships and fellowships at institutions of learning and to foster research in medicine and science. Grants are also given for Jewish welfare funds, care of the aged and sick, the establishment of trade schools, the fostering of religious observance, recreation, and aid to refugees. Additional types of support include general operating grants and matching funds. The Foundation Board generally makes grants once a year, typically in October. The deadline for RFPs as well as proposals submitted from current grant recipients is April 1.
Requirements: Grants are made primarily to organizations in the greater metropolitan Washington, DC, area.
Restrictions: Grants are not made to individuals.
Geographic Focus: District of Columbia
Amount of Grant: $1000-$10,000 range
Date(s) Application is Due: Apr 1
Contact: Robert Lehrman, Vice President, (202) 338-8400; fax: (202) 338-8405; email: info@lehrmanfoundation.org
Internet: http://www.lehrmanfoundation.org
Sponsor: Jacob and Charlotte Lehrman Foundation
1027 33rd Street NW, 2nd Fl
Washington, DC 20007

Jacob and Hilda Blaustein Foundation Grants 2934

The Foundation promotes social justice and human rights through its five program areas: Jewish life; strengthening Israeli democracy; health and mental health; educational opportunity; and human rights. Support is provided to organizations in the United States and abroad. The Foundation supports organizations that promote systemic change; involve constituents in planning and decision-making; encourage volunteer and professional development; and engage in ongoing program evaluation. Application information is available online.

Requirements: Nonprofit organizations are eligible to apply.

Restrictions: Support is not provided for: grants or scholarships to individuals; unsolicited proposals for academic, scientific, or medical research; direct mail, annual giving, membership campaigns, fundraising and commemorative events. The Foundation rarely makes capital grants unless there is a prior relationship with the applicant organization.

Geographic Focus: Maryland

Samples: Shefa Fund (Wyndmoor, PA)—to leverage investments from Jewish institutions and individuals for use by community development financial institutions, $200,000.; Johns Hopkins Medical Institutions (Baltimore, MD)—to help construct a cancer research building at the Johns Hopkins Oncology Ctr, $10 million.; Sinai Hospital (Baltimore, MD)—$100,000.; Institute for Christian-Jewish Studies (Baltimore, MD)—$80,000.

Amount of Grant: $5,000-$25,000

Contact: Betsy Ringel, Executive Director, (410) 347-7103; fax: (410) 347-7210; email: info@blaufund.org

Internet: http://www.blaufund.org/foundations/jacobandhilda_f.html

Sponsor: Jacob and Hilda Blaustein Foundation

10 East Baltimore Street, Suite 1111

Baltimore, MD 21202

Jacob and Valeria Langeloth Foundation Grants 2935

The Foundation's grantmaking program is centered on the concepts of health and well-being. The Foundation's purpose is to promote and support effective and creative programs, practices and policies related to healing from illness, accident, physical, social or emotional trauma and to extend the availability of programs that promote healing to underserved populations. The Foundation has established priority funding status for proposals that address caregiving, and correctional health care. Application information is available online.

Requirements: The Foundation welcomes proposals from 501(c)3 organizations that promote physical and emotional healing, especially to underserved populations, such as: community-based organizations, health care providers and research institutions.

Restrictions: The Foundation will not consider proposals for annual or capital campaigns, for building or renovation projects, for budgetary relief, or preventive medicine. The Foundation does not support projects that focus on children or end-of-life issues. Neither will the Foundation make grants to individuals or for sectarian or religious purposes, or for political activities or lobbying. The Foundation will not fund any organization that discriminates on the basis of age, gender, national origin, race, or sexual preference.

Geographic Focus: New York

Samples: National Bone Marrow Transplant Link (Southfield, MI)—to help people who are waiting for or recovering from organ transplants and their families make informed decisions and cope emotionally, $294,734 (2004); New York Legal Assistance Group (New York, NY)—to expand a curriculum designed to help physicians identify and respond to legal issues that affect low-income people with chronic or serious illnesses, $358,352 (2003).

Date(s) Application is Due: Feb 1; Aug 1

Contact: Scott Moyer, (212) 687-1133; fax: (212) 681-2628; email: smoyer@langeloth.org

Internet: http://www.langeloth.org/apply.php

Sponsor: Jacob and Valeria Langeloth Foundation

521 Fifth Avenue

New York, NY 10175-1699

Jacobs Research Funds Small Grants Program 2936

Grants are given for anthropological research on living Native Americans on cultural, social, psychological, aesthetic, and linguistic problems. While the fund has a Pacific Northwest focus, research projects within Canada, Mexico, the rest of the continental United States, and Alaska are eligible. The fund emphasizes field research rather than the analysis of previously collected materials. Grant funds may cover fees for consultants, supplies (notebooks, audio- and videotapes, etc.), transportation to and from the field, and lodging in the field. Grants are made for a maximum of one year and are renewed only on the basis of a new application. Application forms are available upon request.

Requirements: Grants are made to individuals. Applicants need not have completed degrees. Formal institutional affiliation is not required; however, in the absence of formal research training or institutional affiliation, cooperation with or supervision by an experienced research scholar is strongly recommended. Applications from Native Americans are encouraged.

Restrictions: Projects in archaeology, physical anthropology, applied anthropology, and applied linguistics are not included in the program. Research on archive materials or museum collections is not ordinarily considered for funding. Funds are not provided for the salary of the researcher, major equipment, or living expenses.

Samples: Andrej A. Kibrik—for Studies in Upper Kuskokwim Athabaskan Grammar, $1200.; Erik D. Gooding—for Songs of the Canadian Nakota, $1200.

Amount of Grant: $1200

Date(s) Application is Due: Feb 15

Contact: Jacobs Research Funds Administrator, (360) 676-6981; fax: (360) 738-7409; email: museuminfo@cob.org

Internet: http://www.whatcommuseum.org/pages/info/info.htm

Sponsor: Whatcom Museum of History and Art

121 Prospect Street

Bellingham, WA 98225

Jake Duerksen Memorial Scholarship 2937

This scholarship is awarded annually to cover four months of graduate study based on the recipient's academic excellence. Students interested in the award should first apply to the Department of Biological Sciences.

Requirements: The scholarship is open to full-time graduate students who have been enrolled in the Department of Biological Sciences for a minimum of one year.

Amount of Grant: $C2000 maximum

Contact: Connie Busch, Faculty of Graduate Studies, (403) 220-5690; email: cbusch@ucalgary.ca

Internet: http://www.ucalgary.ca/pubs/calendar/current/What/Awards/03GradAwards.htm

Sponsor: University of Calgary

2500 University Dr NW, Earth Sciences Building, Room 720

Calgary, AB T2N 1N4 Canada

James Alexander Robertson Memorial Prize 2938

This prize is awarded annually for an article appearing during the year preceding the award in one of the four consecutive issues (beginning with the August issue) of the Hispanic American Historical Review. The article selected for the award is to be one that, in the judgment of the prize committee, makes an outstanding contribution to Latin American historical literature. An honorable mention award (with no cash stipend) may be made for an additional distinguished article deemed worthy of the same by the prize committee.

Amount of Grant: $500

Contact: CLAH, (530) 752-3046; fax: (530) 752-8964; email: clah@ucdavis.edu

Internet: http://www.h-net.org/~clah/about/index.html

Sponsor: Conference on Latin American History

University of California at Davis, One Shields Avenue

Davis, CA 95616

James and Sylvia Thayer Short -Term Research Fellowships 2939

The James and Sylvia Thayer Short-Term Research Fellowships support the use of special collections materials by visiting scholars and UCLA graduate students. Collections that are administered by the newly integrated UCLA Library Special Collections and available for Thayer Fellowship-supported research include materials in the humanities and social sciences; medicine; life and physical sciences; visual and performing arts; and UCLA history. Research residencies may last up to three months between March 1 and December 17. Recipients receive stipends ranging from $500 to $2500. (Awards vary yearly). Those receiving fellowships are expected to provide a report on the results of their research that can be mounted on the UCLA Library Web site. Applications are due December 11 and should include: cover letter; curriculum vitae; outline of research and special collections to be used (two pages maximum); brief budget for travel, living, and research expenses; dates to be spent in residence; two letters of recommendation from faculty or other scholars familiar with the research project.

Requirements: United States citizens and permanent residents with the legal right to work in the U.S. who are engaged in graduate-level, post-doctoral, or independent research are invited to apply.

Amount of Grant: $500-$2500 per month

Date(s) Application is Due: Dec 11

Contact: Eunice MacGill, Administrative Assistant; (310) 794-4408; email: emacgill@library.ucla.edu or lib_thayer@library.ucla.edu

Internet: http://www.library.ucla.edu/special/thayer.cfm

Sponsor: UCLA Library

A1713 Charles E. Young Research Library, P.O. Box 951575

Los Angeles, CA 90095-1575

James Bradford Ames Fellowship 2940

The fellowship was established to stimulate and support research of African American life and history on Nantucket Island including, but not limited to, the study of individual families, social life, occupations, the institution of slavery, and the interaction between African American and Cape Verdean communities. The studies may include those aspects that place Nantucket in a broader regional, national, and global perspective. The emphasis of the program is research, but the fund also may sponsor scholarly lectures, particularly as may allow for the presentation of research that has been funded in whole or in part by the Ames Fund.

Requirements: ABD graduate students, full-time faculty, and professional staff in colleges and universities are invited to submit applications.

Amount of Grant: $500-$2500

Date(s) Application is Due: Apr 15

Contact: Dr. Robert Johnson Jr., Department of African Studies, (617) 287-6794 or 287-6790; email: Robert.Johnson@umb.edu

Internet: http://www.umb.edu/academic_programs/departments/africana_studies/scholarship/index.html

Sponsor: University of Massachusetts Boston

100 Morrissey Boulevard

Boston, MA 02125

James Flack Norris Award in Physical Organic Chemistry　　　**2941**
This award, sponsored by The Northeastern Section of ACS, is given annually to encourage and reward outstanding contributions to physical organic chemistry. The award consists of $5,000 and a certificate. Reasonable travel expenses to the meeting at which the award will be presented will be reimbursed.
Requirements: Any individual, except a member of the award committee, may submit one nomination or seconding letter for the award in any given year. Nominating documents consist of a letter of not more than 1000 words containing an evaluation of the nominee's accomplishments and a specific identification of the work to be recognized, a biographical sketch including date of birth, and a list of publications and patents authored by the nominee.
Restrictions: Self-nominations are not accepted.
Samples: Dennis A. Dougherty—award recipient, $5,000 (2008); Ben L. Feringa—award recipient, $5,000 (2007); Michael Wasielewski—award recipient, $5,000 (2006).
Amount of Grant: $5,000
Date(s) Application is Due: Nov 1
Contact: Felicia Dixon, Awards Administrator; (800) 227-5558 or (202) 872-4408; fax: (202) 776-8008; email: f_dixon@acs.org or awards@acs.org
Internet: http://portal.acs.org/portal/acs/corg/content?_nfpb=true&_pageLabel=PP_ARTICLEMAIN&node_id=1319&content_id=CTP_004536&use_sec=true&sec_url_var=region1
Sponsor: American Chemical Society
1155 Sixteenth Street, NW
Washington, DC 20036

James H. Cummings Foundation Grants　　　**2942**
The Foundation, gives exclusively for charitable purposes in advancing medical science, research, and education in selected cities in the U.S. and Canada, and for charitable work among underprivileged boys and girls, and aged and infirm persons in designated areas. Priority is given to medical proposals. The funding is limited to: Toronto, Ontario; Canada; Hendersonville, North Carolina and; Buffalo, New York. Grants are available, in the following types of support: building/renovation; capital campaigns; equipment; land acquisition; matching/challenge support; research and seed money.
Requirements: The giving program is limited to the vicinity of the cities of Buffalo, NY; Hendersonville, NC; and Toronto, ON, Canada.
Restrictions: No support for national health organizations. Grants are not awarded to individuals, nor for: loans, annual campaigns, program support, endowment funds, operating budgets, emergency funds, deficit financing, scholarships, fellowships, publications, conferences, or continuing support.
Geographic Focus: New York; North CarolinaCanada
Samples: Child and Family Services Conners Children's Ctr (Buffalo, NY)—to renovate this center that provides residential care for severely disturbed boys between the ages of six and 14, $100,000.; YMCA of Henderson and Henderson County (NC)—program support, $62,500.; Margaret R. Pardee Hospital (NC)—capital support, $100,000.
Amount of Grant: $2000-$250,000 range
Contact: William McFarland, Executive Director; (716) 874-0040; fax: (716) 874-0040; email: cummings.foundation@verizon.net
Sponsor: James H. Cummings Foundation
1807 Elmwood Avenue, Room 112
Buffalo, NY 14207

James Harrison Steedman Memorial Fellowships in Architecture　　　**2943**
This biennial fellowship permits well-qualified architectural graduates to benefit from nine months of travel and study of architecture in foreign countries. A design competition, to be completed during January after deadline date, determines the winner; fellowship will begin within nine months of receiving the award. Specific details about the award and application forms may be obtained from the university. Registration forms and a $50 application fee are due December 10; competition entries and research proposal are due January 24. Deadline dates may vary; contact the program office for exact dates. Registration forms are available on the Web site.
Requirements: The fellowships are open to architects who have worked for at least one year in an architect's office for a period of up to eight years after they have received their professional degrees from accredited schools. There are no citizenship requirements.
Amount of Grant: $30,000
Date(s) Application is Due: Mar 12
Contact: Steedman Governing Committee, (314) 935-6293; fax: (314) 935-7656; email: shannonp@architecture.wustl.edu
Internet: http://www.arch.wustl.edu/news_sc.lasso
Sponsor: Washington University
Campus Box 1079, One Brookings Drive
Saint Louis, MO 63130-4899

James Harry Potter Gold Medal　　　**2944**
The medal is awarded in recognition of eminent achievement or distinguished service in the appreciation of the science of thermodynamics in mechanical engineering. The basis of the award shall include contributions involving the teaching, appreciation, or utilization of thermodynamic principles in research, development, and design in mechanical engineering. Any individual member, group of members, or committee may nominate candidates for consideration of this award. Award includes vermeil medal and certificate.
Samples: Amir Faghri—award recipient, $2000 (2005); Van P. Carey—award recipient, $2000 (2004); Aswhani K. Gupta—award recipient, $2000 (2003).
Amount of Grant: $2000

Date(s) Application is Due: Feb 1
Contact: Gilda DiTullio, Manager, (212) 591-7736; fax: (212) 705-7739; email: awards@asme.org
Internet: http://www.asme.org/honors/ms71/saa/potter.html
Sponsor: American Society of Mechanical Engineers
3 Park Avenue
New York, NY 10016

James M. Collins Foundation Grants　　　**2945**
The foundation awards grants to Texas nonprofit organizations in its areas of interest, including: arts, economic development, health organizations, association, higher education, human services, museums, Salvation Army, secondary school/education. Types of support include research grants, program support, and social services. There are no application deadlines or forms.
Requirements: Texas nonprofit organizations are eligible.
Restrictions: Individuals are not eligible.
Geographic Focus: Texas
Samples: Smithsonian Institution, Washington, DC, $50,475 (2004); Wellesley College, Wellesley, MA, $200,250 (2004); Dallas Museum of Art, Dallas, TX, $17,950 (2004).
Contact: Dorothy Dann Collins Torbert, President; (214) 691-2032;
Sponsor: James M. Collins Foundation
8115 Preston Road, Suite 680
Dallas, TX 75225

James Marshall Public Policy Fellowship　　　**2946**
This fellowship is a two-year appointment representing SPSSI in approved policy and advocacy activities. The position is with the public policy office of the American Psychological Association (APA) in Washington, DC. The fellow will participate in supervised activities including using psychological research to analyze specific social policies and develop policy advocacy. In addition, the fellow will serve in various capacities including, but not limited to, serving on SPSSI committees and task forces; monitoring APA boards, committees, and/or task forces relevant to the fellow's main policy area; reporting to and attending SPSSI's semiannual council meetings; writing a public policy column for the SPSSI newsletter; and meeting regularly with the fellow oversight committee. Candidate is required to submit a detailed vita; a 1000-word biographical statement of past experience and interest in policy activities and/or social issues, career goals, and interest in the fellowship and its objectives; a 600-word briefing statement using social science data to inform legislators about a specific social issue; and three letters of reference. All application materials must be submitted in duplicate.
Requirements: Applicant must hold a PhD or PsyD and either be a member of SPSSI and APA or be eligible to become a member. The selected fellow must join SPSSI before the appointment date. Candidates must demonstrate interest or involvement in the application of social science to social issues and policy and be interested in and knowledgeable about at least one current social issue such as homelessness, violence, adolescent pregnancy, child abuse, etc.
Amount of Grant: $42,600 for first year, plus health and vacation benefits
Date(s) Application is Due: Mar 1
Contact: Dr. Shari Miles, (202) 675-6956; fax: (734) 662-5607; email: smiles@spssi.org
Internet: http://www.spssi.org/jms.html
Sponsor: Society for the Psychological Study of Social Issues
P.O. Box 1248
Ann Arbor, MI 48103

James Marston Fitch Charitable Trust Midcareer Grants　　　**2947**
The trust awards research grants to midcareer professionals who have an academic background, professional experience, and an established identity in one or more of the following fields: historic preservation, architecture, landscape architecture, urban design, environmental planning, architectural history, and the decorative arts. Proposals will be considered for the research and/or execution of the preservation-related projects in any of these fields. The foundation endeavors to establish new links between the academic and professional contingents primarily in the field of American historic preservation and to strengthen the connections between theory and practice in conservation technology.
Requirements: Midcareer professionals with established backgrounds in historic preservation or related fields are eligible for grants. Candidate should have an advanced or professional degree and 10 years of professional experience in historic preservation or in related fields, such as architecture, landscape architecture, urban design, environmental planning, archaeology, architectural history, or the decorative arts.
Samples: Marilyn Kaplan—to study the impact of new building codes on historic buildings (2004).
Amount of Grant: $10,000-$25,000
Date(s) Application is Due: Sep 8
Contact: Margaret Evans, Executive Director, (212) 252-6809; fax: (212) 471-9987; email: info@fitchfoundation.org
Internet: http://www.fitchfoundation.org
Sponsor: James Marston Fitch Charitable Trust
232 E 11th Street
New York, NY 10003

James McKeen Cattell Fund Fellowships　　　**2948**
The nonrenewable fellowships advance the science of psychology and its useful applications by enabling faculty members to take a full year sabbatical leave for independent study and research. Annual deadline dates may vary; contact program staff for exact dates.

Requirements: Eligibility is limited to residents of North America. The fellowships are intended for tenured faculty of a psychology department who are eligible for sabbatical leave. Matching institutions' support required.
Amount of Grant: $32,000 maximum
Date(s) Application is Due: Dec 1
Contact: Dr. Christina Williams, Program Contact; email: williams@psych.duke.edu
Internet: http://www.cattell.duke.edu
Sponsor: James McKeen Cattell Fund
9 Flowers Drive, Box 90086, Duke University, Department of Psychological and Brain Sciences
Durham, NC 27708-0086

James N. Landis Medal   2949
The Landis Medal is given for outstanding personal performance related to designing, constructing, or managing the operation of major steam-powered electric stations using nuclear or fossil fuels, coupled with personal leadership in humanitarian pursuits, which may include committee activity, ASME section leadership, or the broad nontechnical professional activity of the nominee's engineering society. The award is presented preferably to a member of ASME, and includes stipend, bronze medal, and certificate.
Samples: Ashwani K. Gupta and Toshiaki Hasegawa—medal winners, $7500 (2004); William D. Magwood, IV—medal winners, $7500 (2003).
Amount of Grant: $7500 and expense supplement
Date(s) Application is Due: Feb 1
Contact: Gilda DiTullio, Manager, (212) 591-7736; fax: (212) 705-7739; email: awards@asme.org
Internet: http://www.asme.org/honors/ms71/saa/landis.html
Sponsor: American Society of Mechanical Engineers
3 Park Avenue
New York, NY 10016

James R. Dougherty Jr. Foundation Grants   2950
The James R. Dougherty, Jr. Foundation, established in 1950, gives to organizations primarily in Texas. Areas of interest include: family services, domestic violence, human services and women. Support is offered in the form of: annual campaigns, building/renovation, capital campaigns, continuing support, curriculum development, endowments, equipment, general/operating support, income development, management development/capacity building, matching/challenge support, program development, program evaluation, research, scholarship funds, seed money and, technical assistance grants. The board meets twice a year, in the Spring and Fall. No application form is required, contact the Foundation, before submitting a proposal.
Requirements: Nonprofit organizations in Texas are eligible to apply.
Geographic Focus: Texas
Date(s) Application is Due: Mar 1; Sep 1
Contact: Daren Wilder, Grants Administrator, (512) 358-3560
Sponsor: James R. Dougherty Jr. Foundation
P.O. Box 640
Beeville, TX 78104-0640

James R. Scobie Memorial Award for Preliminary Doctoral Research 2951
The award permits a short, exploratory research trip abroad to determine the feasibility of a PhD dissertation topic dealing with some facet of Latin American history. One or more travel grants will be awarded each year; the grant must be used during the summer following the award unless prior approval is received.
Requirements: Applicant must submit a comprehensive research prospectus, including a preliminary bibliography; three letters of recommendation, one of which should attest to the language competence (Spanish or Portuguese) of the applicant; and a current curriculum vita.
Restrictions: The award cannot be combined with a research grant for an extended stay.
Amount of Grant: $1000 maximum
Date(s) Application is Due: Apr 1
Contact: CLAH, (530) 752-3046; fax: (530) 752-8964; email: clah@ucdavis.edu
Internet: http://www.h-net.org/~clah/about/index.html
Sponsor: Conference on Latin American History
University of California at Davis, One Shields Avenue
Davis, CA 95616

James S. McDonnell Foundation Brain Cancer Research Collaborative 2952
   Activity Awards
The Foundation offers Collaborative Activity Awards to initiate interdisciplinary discussions on problems or issues, to help launch interdisciplinary research networks, or to fund communities of researchers/practitioners dedicated to developing new methods, tools, and applications of basic research to applied problems. All proposals submitted to the foundation must clearly link the experimental models and questions to human disease. Proposals primarily intending to characterize basic mechanisms of growth and development that may plausibly but are not yet known to be contributory to human brain cancer are not encouraged. Proposals testing molecules as possible treatment interventions should consider including tests designed to uncover unintended biological effects of such molecules that would disqualify future clinical usefulness. The Foundation is particularly interested in supporting novel research that will generate new knowledge leading to increased rates of survival and improve functional recovery for individuals with brain cancer.
Restrictions: The Foundation does not fund: undergraduate tuition, stipends, scholarships, fellowships, research or travel expenses, or other educational expenses; graduate or postdoctoral stipends, scholarships, fellowships, research or travel expenses, or other educational expenses (exceptions include requests that qualify as allowable budget items as part of specific Research Award applications); expenses tied to projects whose explicit goal is the publication of a book or other bound volume (although publication of a book or special issue of a journal may be one goal or outcome of work funded by the Foundation); expenses tied to the establishment or day-to-day running of a journal or small press; scientific meetings or workshops other than those JSMF puts together or which are affiliated with Collaborative Activity Awards; ongoing operational support for university-based centers, programs, or institutes; professional society meetings or specific sessions at such meetings; charitable functions, museum exhibitions, or similar causes or events; or charitable donations to individuals and organizations.
Contact: Cheryl Washington, Grants Manager; (314) 721-1532; fax: (314) 721-7421; email: washington@jmsf.org or info@jmsf.org
Internet: http://www.jsmf.org/programs/bc/index.htm
Sponsor: James S. McDonnell Foundation
1034 S. Brentwood Boulevard, Suite 1850
Saint Louis, MO 63117

James S. McDonnell Foundation Complex Systems Collaborative Activity 2953
   Awards
The Foundation offers Collaborative Activity Awards to initiate interdisciplinary discussions on problems or issues, to help launch interdisciplinary research networks, or to fund communities of researchers/practitioners dedicated to developing new methods, tools, and applications of basic research to applied problems. The Program supports scholarship and research directed toward the development of theoretical and mathematical tools that can be applied to the study of complex, nonlinear systems. It is anticipated that research funded in this program will address issues in fields such as biology, biodiversity, climate, demography, epidemiology, technological change, economic Complex Systems development, governance, or computation. While the program's emphasis is on the development and application of theoretical models used in these research fields and not on particular fields per se, JSMF is particularly interested in projects attempting to apply complex systems approaches to real world problems. Proposals attempting to apply tools and models to problems where such approaches are not yet considered usual or mainstream (for example, differentiating normal physiology from disease) are encouraged.
Restrictions: The Foundation does not fund: undergraduate tuition, stipends, scholarships, fellowships, research or travel expenses, or other educational expenses; graduate or postdoctoral stipends, scholarships, fellowships, research or travel expenses, or other educational expenses (exceptions include requests that qualify as allowable budget items as part of specific Research Award applications); expenses tied to projects whose explicit goal is the publication of a book or other bound volume (although publication of a book or special issue of a journal may be one goal or outcome of work funded by the Foundation); expenses tied to the establishment or day-to-day running of a journal or small press; scientific meetings or workshops other than those JSMF puts together or which are affiliated with Collaborative Activity Awards; ongoing operational support for university-based centers, programs, or institutes; professional society meetings or specific sessions at such meetings; charitable functions, museum exhibitions, or similar causes or events; or charitable donations to individuals and organizations.
Contact: Cheryl Washington, Grants Manager; (314) 721-1532; fax: (314) 721-7421; email: washington@jmsf.org or info@jmsf.org
Internet: http://www.jsmf.org/programs/cs/index.htm
Sponsor: James S. McDonnell Foundation
1034 S. Brentwood Boulevard, Suite 1850
Saint Louis, MO 63117

James S. McDonnell Foundation Research Grants   2954
The Awards are designed to support research projects with a high probability of generating new knowledge and insights. Projects submitted for funding consideration should be at an early, even preliminary stage of development that intend to break new ground or to challenge commonly-held assumptions. Projects submitted should be sufficiently novel, cross-disciplinary, or heterodox so that they have a strong likelihood of influencing the development of new ways of thinking about important problems. Awards provide adequate, flexible funding over a sufficient time period to allow investigators to pursue and develop innovative directions to their research programs. Funds can be expended over a minimum of 3 years or a maximum of 6 years. Smaller amounts of money expended over shorter amounts of time may be requested to help investigators pursue pilot projects or test the feasibility of an experimental approach. The applicant can apply the grant funds towards any research-based expense, including travel, equipment, and supplies. Funds can be used to support collaborative projects. A percentage of the funds can also be used to support small workshops organized by the applicant where the goal of the workshop is to gather expertise in support of the research objective.
Requirements: Applications for grants are considered only from 501(c)3 organizations which are not private foundations. Applications from foreign organizations will be considered only if they provide written legal opinion that they qualify as tax-exempt under these sections of the U.S. IRS code or if a qualifying American organization is authorized to receive funds for them under this code. Individuals also are eligible.
Amount of Grant: $450,000 maximum
Contact: Cheryl Washington, Grants Manager; (314) 721-1532; fax: (314) 721-7421; email: washington@jmsf.org or info@jmsf.org
Internet: http://www.jsmf.org/apply/research/index.htm
Sponsor: James S. McDonnell Foundation
1034 S. Brentwood Boulevard, Suite 1850
Saint Louis, MO 63117

James S. McDonnell Foundation Scholar Awards   2955

The Awards program derives from and is consistent with the foundation's commitment to supporting high quality research and scholarship leading to the generation of new knowledge and its responsible application. Currently, the Awards are only available in the Brain, Mind & Behavior program area, and provide largely unrestricted funding over a sufficient time period to allow investigators to pursue and develop new directions to their research programs.

Requirements: Applications must be sponsored by a nonprofit institution as defined by Section 501(c)3 of the United States Internal Revenue Tax Code. Eligible Scholar Award nominees must have completed all doctoral, postdoctoral, or fellowship training and hold an independent research position. Nominees in the earlier career stages are encouraged.

Restrictions: Researchers with current grant support from JSMF are not eligible to apply for a Scholar Award until all prior JSMF grant funds have been fully expended and a final grant report has been received and approved by the foundation office.

Amount of Grant: $600,000 to be expended over 6 years

Contact: Cheryl Washington, Grants Manager; (314) 721-1532; fax: (314) 721-7421; email: washington@jmsf.org or info@jmsf.org

Internet: http://www.jsmf.org/apply/scholar/index.htm

Sponsor: James S. McDonnell Foundation

1034 S. Brentwood Boulevard, Suite 1850

Saint Louis, MO 63117

James S. McDonnell Foundation Understanding Human Cognition Awards   2956

The long-standing interest of the James S. McDonnell Foundation on human mind/brain is reflected in changes made to the 2008 RFA, including the explicit emphasis on understanding human cognition. The program is intended to help investigators pursue experiments designed to answer well-articulated questions. JSMF Scholar Awards support research studying how neural systems are linked to and support cognitive functions and how cognitive systems are related to an organism's (preferably human) observable behavior. Studies with model organisms should justify why such models were selected and how data obtained from models advances our understanding of human cognition. Proposals proposing to use functional imaging to identify the "neural correlates" of cognitive or behavioral tasks (for example, mapping the parts of the brain that "light up" when different groups of subjects play chess, solve physics problems, or choose apples over oranges) are not funded through this program. In general, JSMF and its expert advisors have taken an unfavorable view of projects attempting too wide a leap in a single bound. Functional imaging studies using poorly characterized tasks as proxies for complex behavioral issues involving empathy, moral judgments, or social decision-making are generally not appropriate responses to this call for proposals. In past competitions, proposals structured along such lines were eliminated from funding consideration early in the review process.

Requirements: Aspects of proposals appropriate to the JSMF UHC program would include, but are not limited to: characterizing the cognitive operations involved in performing a task; studying how the brain identifies, extracts and uses relevant information from complicated environments; examining how manipulations and/or perturbations at one spatial or temporal scale are meaningful at finer or coarser levels of organization (e.g. does a synaptic change account for a change in network function and vice versa?); re-examining "common wisdom" assumptions (such as the existence of critical periods in human learning); evaluating the usefulness of methodologies or improving the usefulness of methodologies commonly used in mind/brain research; applying approaches and knowledge from cognitive psychology or cognitive science to important problems in education, training, or rehabilitation; taking a comparative, evolutionary approach to characterizing the uniqueness of the human brain and of human cognition.

Restrictions: The Foundation does not fund: undergraduate tuition, stipends, scholarships, fellowships, research or travel expenses, or other educational expenses; graduate or postdoctoral stipends, scholarships, fellowships, research or travel expenses, or other educational expenses (exceptions include requests that qualify as allowable budget items as part of specific Research Award applications); expenses tied to projects whose explicit goal is the publication of a book or other bound volume (although publication of a book or special issue of a journal may be one goal or outcome of work funded by the Foundation); expenses tied to the establishment or day-to-day running of a journal or small press; scientific meetings or workshops other than those JSMF puts together or which are affiliated with Collaborative Activity Awards; ongoing operational support for university-based centers, programs, or institutes; professional society meetings or specific sessions at such meetings; charitable functions, museum exhibitions, or similar causes or events; or charitable donations to individuals and organizations.

Contact: Cheryl Washington, Grants Manager; (314) 721-1532; fax: (314) 721-7421; email: washington@jmsf.org or info@jmsf.org

Internet: http://www.jsmf.org/programs/uhc/index.htm

Sponsor: James S. McDonnell Foundation

1034 S. Brentwood Boulevard, Suite 1850

Saint Louis, MO 63117

James T. Grady-James H. Stack Award for Interpreting Chemistry for the Public 2957

This Award is given annually to recognize, encourage, and stimulate outstanding reporting directly to the public, which materially increases the knowledge and understanding of chemistry, chemical engineering, and related fields. This information may have been disseminated through the press, radio, television, films, the lecture platform, or books or pamphlets for the lay public. The award consists of $3,000, a medallion with a presentation box, and a certificate.

Requirements: Any individual, except a member of the award committee, may submit one nomination or seconding letter for the award in any given year. The nominating

documents consist of a letter of not more than 1000 words containing an evaluation of the nominee's accomplishments and a specific identification of the work to be recognized, a biographical sketch including date of birth, and a list of publications authored by the nominee.

Restrictions: Application or self-nomination is not acceptable.

Samples: Harold McGee—award winner, $3,000 (2008); Stuart F. Brown—award winner, $3,000 (2007); Philip Ball—award winner, $3,000 (2006).

Amount of Grant: $3,000 and reimbursement of traveling expenses to award meeting

Date(s) Application is Due: Nov 1

Contact: Felicia Dixon, Awards Administrator; (800) 227-5558 or (202) 872-4408; fax: (202) 776-8008; email: f_dixon@acs.org or awards@acs.org

Internet: http://portal.acs.org/portal/acs/corg/content?_nfpb=true&_pageLabel=PP_ARTICLEMAIN&node_id=1319&content_id=CTP_004537&use_sec=true&sec_url_var=region1

Sponsor: American Chemical Society

1155 Sixteenth Street, NW

Washington, DC 20036

Jane Beattie Memorial Scholarship   2958

The purpose of the fund is to provide scholarships to subsidize travel to the U.S. for purposes of scholarly activity by a foreign scholar in the area of judgment and decision research, broadly defined. Attendance at the annual SJDM meetings is one example of an activity that would be appropriate for support, but by no means the only one. In most years, the Fund awards one or two scholarships in amounts of $400 - $700 each.

Requirements: Applicants should be scholars living and working in a country other than the U.S. who will use the award to help pay for travel to the U.S. for scholarly activities associated with research in judgment and decision making. It is anticipated that most awards will be granted to faculty or graduate students at colleges and universities, especially recent and soon-to-be Ph.D.,, but others will also be considered.

Restrictions: U.S. Scholars are not eligible to apply.

Amount of Grant: $400 - $700

Date(s) Application is Due: Jul 26

Contact: Joshua Klayman; University of Chicago Graduate School of Business; fax: (773) 834-9134; email: joshk@uchicago.edu

Internet: http://www.sjdm.org/content/beattie-award

Sponsor: Society for Judgment and Decision Making

c/o Bud Fennema, College of Business, Florida State University, P.O. Box 3061110

Tallahassee, FL 32306-1110

Jane Coffin Childs Memorial Fund for Medical Research Fellowships   2959

The fellowships are for furthering of research into the causes, origins, and treatment of cancer. The major effort of the fund, at this time, is directed toward the support of a fellowship program for the training of young men and women at the postdoctoral level for independent research in the same field. Stipend, child dependency allowance, travel for fellow and family, and an institutional allowance are provided. Fellowships are awarded for two to three years.

Requirements: Applicants should not have more than one year of postdoctoral experience. They must either hold the MD degree or PhD degree in the field in which they propose to study. Applicants may be citizens of any country, but awards for foreign nationals will be made only for study in the United States. American citizens may hold fellowships either in the United States or in a foreign country.

Amount of Grant: $41,000 first year, $42,000 second year, $44,000 third year; plus $1500 research allowance per year

Date(s) Application is Due: Feb 1

Contact: Kim Roberts, Administrative Director, (203) 785-4612; fax: (203) 785-3301; email: info@jccfund.org

Internet: http://www.jccfund.org

Sponsor: Jane Coffin Childs Memorial Fund for Medical Research

P.O. Box 208000, 333 Cedar Street

New Haven, CT 06520

Japan Doctoral Fellowships   2960

Fellowships are awarded to provide doctoral candidates in the social sciences and humanities with the opportunity to conduct dissertation research in Japan. Fellowships are for a duration of four to 12 months. Travel expenses for the fellow's family are not included. Requests for application forms must state the theme of the project to be conducted in Japan and the applicant's present position and citizenship. Letters of reference must be sent directly to the New York office.

Requirements: Applicants must have completed all requirements for the PhD except the dissertation at the time the fellowship begins and have a functional knowledge of Japanese. Well-qualified candidates in political science, law, economics, business, and journalism are encouraged to apply. Candidates should be citizens or permanent residents of the United States; U.S. citizens residing abroad are eligible and should apply through the foundation's overseas office or Japanese diplomatic mission in the country where they reside. Those residing in the United States who are not citizens or permanent residents should contact the Japan Foundation headquarters in Tokyo directly.

Restrictions: Fellowships are not granted for improvement of proficiency in Japanese and may not be held concurrently with another major grant. Individuals in the fields of the physical, natural, medical, or engineering sciences are not eligible to apply.

Date(s) Application is Due: Nov 1

Contact: Doctoral Fellowships: (212) 489-0299; fax: (212) 489-0409; email: info@jfny.org

Internet: http://www.jfny.org/jfny/arts.html#doc

Sponsor: Japan Foundation in New York
152 W 57th Street, 39th Floor
New York, NY 10019

Japan Foundation Center for Global Partnership (CGP) Grants   2961

Program supports Japan-U.S. collaborative policy-oriented projects in the target areas listed below. In addition, CGP recognizes the value of including a multinational dimension within projects, and therefore gives priority to those that incorporate Asia into its agenda. Prospective grantees must submit letters of inquiry for CGP consideration, 2-3 pages minimum, no later than March 1. Completed applications invited by CGP must be received by May 1.

Requirements: Proposals from U.S. organizations with 501(c)3 status only will be accepted by CGP New York. Appropriate Japan-based institutional collaboration must be demonstrated and secured throughout the course of the project.

Samples: Green Map System (New York, NY)—to create the first volume of a bilingual digital atlas of maps and corresponding narratives, $33,000 (2004); Japan Society (New York, NY)—to make teaching materials about Japan available on the Internet, $40,516 (2004); Laurasian Institution (Tokyo, Japan)—to bring a group of 240 students from U.S. junior high schools and high schools to Tokyo and Kyoto for study tours, $73,700 (2004); Pacific Forum CSIS (Honolulu, HI)—for efforts to strengthen cooperation among China, Japan, and the United States, $50,000 (2004).

Date(s) Application is Due: Mar 1; May 1

Contact: Tomoki Akazawa, Program Director; (212) 489-1255; fax: (212) 489-1344; email: tomoki_akazawa@cgp.org

Internet: http://www.cgp.org/index.php?option=section&id=3

Sponsor: Japan Foundation Center for Global Partnership
152 West 57th Street, 17th Floor
New York, NY 10019

Japan Foundation Center for Global Partnership Institutional Grants   2962

With emphasis on academic/intellectual excellence and national and international public consequence, CGP's institutional grant program supports policy-oriented projects conducted by an array of institutions from the United States and Japan, including non-profit organizations, universities, and think tanks as well as policymakers, scholars, and educators, with a view to creating new networks and providing opportunities for advancing research and candid discussion of these issues. Within the mandate of enhancing Japan-U.S. collaboration, CGP recognizes the value of including a multinational dimension within projects, and gives priority to those that incorporate Asia into its agenda. Applicants must submit letters of inquiry for consideration, 2-3 pages minimum, no later than March 3. Full proposals must be received by May 1.

Requirements: Applicants can include: non-profit organizations; universities; think tanks; policymakers; scholars; and educators.

Date(s) Application is Due: Mar 3; May 1

Contact: Yukiko Ono, Program Director; (212) 489-1255; fax: (212) 489-1344; email: yukiko_ono@cgp.org

Internet: http://www.jpf.go.jp/cgp/e/grant/index.html

Sponsor: Japan Foundation Center for Global Partnership
152 West 57th Street, 17th Floor
New York, NY 10019-3101

Japan Foundation Center for Global Partnership NP.O. Fellowships   2963

The Fellowship supports practical training in the United States for professionals committed to working on social and global issues through the activities of a Japanese nonprofit organization. The program seeks to build long-lasting relationships between fellows and their counterparts in the United States, fostering information exchange between the nonprofit sectors in both countries and global perspectives on the role of civil society in addressing pressing social issues. It is particularly focused on nurturing new nonprofit leaders who have the ability to initiate meaningful change within Japan and lead future collaborative efforts between the United States and Japan. Fellows pursue self-directed training agendas from the home base of a U.S. nonprofit organization (host organization) that is well-aligned with the interests and objectives of the fellowship. Training agendas enable the fellow to develop management skills for nonprofit organizations and knowledge around a selected theme, such as program development and implementation, development (fund raising), staff/volunteer/board management, and communications. The fellow's specialty field provides a context for the development of skills and knowledge. In the past, fellows have come from fields such as volunteer promotion, at-risk youth, domestic violence, social entrepreneurship, and capacity-building for nonprofits.

Requirements: All fellows are required to participate in orientations in Japan. Fellows have a minimum of three years of experience at Japanese nonprofits. Ability to communicate in English and capacity to pro-actively carry out fellowship activities in a foreign culture are also considered.

Samples: Yoko Ayukawa (Hiroko Matsubara), Founder and Executive Director, Tools to Expand Networking Communication by AIDS INDEX, Tokyo, Japan (2006-07); Kanae Doi, Executive Committee Member, Japan Lawyer's Network for Refugees, Tokyo, Japan (2006-07); Eri Ishikawa, Senior Researcher, Japan Association for Refugees, Tokyo, Japan (2006-07).

Contact: Yukiko Ono; (212) 489-1255; fax: (212) 489-1344; email: yukiko_ono@cgp.org

Internet: http://www.cgp.org/index.php?option=article&task=default&articleid=156&id=14

Sponsor: Japan Foundation Center for Global Partnership
152 West 57th Street, 17th Floor
New York, NY 10019-3101

Japan Foundation Center for Global Partnership Security Studies Fellowships   2964

The Fellowship program, administered by the Research Institute for Peace and Security (RIPS) and operated in collaboration with RIPS and CGP, aims to cultivate excellent young researchers with a global viewpoint in the field of security in Japan. The program endeavors to rear experts who will contribute to security studies, domestically and internationally, by providing the young researchers interested in this area with a research scholarship and by holding research seminars to provide support for specialized research. At the same time, the program aims to heighten academic interest relating to international security, arms control, strategic studies, Japanese security issues and other such fields which were late to develop in Japan's spheres of research and education.

Contact: Yoshinobu Yamamoto, Program Director; +81-(0)3-5562-3542; fax: +81-(0)3-5562-3504

Internet: http://www.jpf.go.jp/cgp/e/fellow/ssfp/index.html

Sponsor: Japan Foundation Center for Global Partnership
Ark Mori Building, 20 Floor, 1-12-32, Akasaka
Minato-ku, Tokyo 107-6021 Japan

Japan Heart Foundation Grants   2965

The foundation awards grants in Japan to individuals, organizations, and research institutes for research projects in the area of heart and blood vessel diseases.

Requirements: Research institutes for research in the area of heart and blood vessel diseases, individuals, and organizations in Japan are eligible to apply.

Contact: Grants Administrator, 81-3-3201-0810; fax: 81-3-3213-3920; email: info@jhf.or.jp

Internet: http://www.jhf.or.jp

Sponsor: Japan Heart Foundation
835-A, New Kokusai Bldg,3-4-1, Marunouchi
Chiyoda-ku, Tokyo 100-0005 Japan

Japan Research Fellowships   2966

The fellowships give scholars, researchers, and professionals the opportunity to conduct research in Japan. All projects related substantially to Japan in the humanities and social sciences, including comparative research, are eligible. Research fellowships are awarded for periods of two to 12 months. A monthly stipend with a dependents' allowance, round-trip airfare, traveler's insurance, a departure allowance and a settling-in allowance, a research/cultural activities allowance, and enrollment fees are provided.

Requirements: Scholars applying for research fellowships should hold academic positions in research institutions and have substantial experience in research, teaching, and writing in their respective fields of study.

Restrictions: Projects focused on improving Japanese-language proficiency or on training or self-improvement in the performing arts, fine arts, crafts, the martial arts, or sports, or in the fields of the physical, medical, or engineering sciences will not be considered.

Samples: Shoji Azuma (University of Utah)—Conversational Analysis of Counseling in Japanese.; James Brandon (University of Hawaii)—Translation of Representative Kabuki Plays.

Date(s) Application is Due: Nov 1

Contact: Research Fellowships, (212) 489-0299; fax: (212) 489-0409; email: info@jfny.org

Internet: http://www.jfny.org/jfny/arts.html#res

Sponsor: Japan Foundation in New York
152 W 57th Street, 39th Floor
New York, NY 10019

Japan Study Program Grants   2967

The program comprises two subprograms. The Publication Assistance program provides financial assistance to publishers for the publication of books on or relating to Japan in the humanities, social sciences, and fine arts, in languages other than Japanese. The Translation Assistance program provides financial assistance to publishers for the translation of Japanese works of high quality on or relating to Japan in the humanities, social sciences, and fine arts. Contact the office for specific requirements of each subprogram.

Requirements: Only publishers in the following states are eligible to apply: Alaska, Arizona, California, Colorado, Hawaii, Idaho, Montana, Nevada, New Mexico, Oregon, Utah, Washington, and Wyoming.

Restrictions: Applications from authors, translators, or other individuals will not be accepted. Works in the natural sciences are not eligible.

Geographic Focus: Alaska; Arizona; California; Colorado; Hawaii; Idaho; Montana; Nevada; New Mexico; Oregon; Utah; Washington; Wyoming

Date(s) Application is Due: Dec 1

Contact: Mamiko Nakai, (213) 621-2267 ext 110; fax: (213) 621-2590; email: mamiko_nakai@jflalc.org

Internet: http://www.jflalc.org/?act=tpt&id=229

Sponsor: Japan Foundation and Language Center in Los Angeles
333 S Grand Avenue, Suite 2250
Los Angeles, CA 90071

Japan-United States Friendship Commission Grants   2968

The program promotes scholarly, cultural, and public affairs activities between Japan and the United States. Requests for support will be considered under six project areas: Japanese studies (for Americans), Study of the United States in Japan, the arts, policy-oriented research, public affairs/education, and infrastructure building. These areas have been chosen on the basis of the commission's considered judgment that both highly trained area specialists and a much broader public understanding and cultural involvement are essential to a stable foundation for the Japanese-American relationship in the years ahead. The commission provides support exclusively for institutional development; it does not provide support for individual training, research, or travel.

Requirements: Programs supported are open to citizens and permanent residents of the United States or Japan without regard to race, creed, sex, disability, or national origin. Grants may be made to individual universities or local organizations but primarily in consideration of their contribution to national resources for understanding of the other country. Grants to individuals, under all programs, normally will be made through academic, professional, artistic, or other appropriate organizations that will examine, recommend, and in most instances, select the individuals to be supported financially by the commission. The commission, as a general rule, will work with nonprofit organizations in carrying out its programs.

Restrictions: Projects not considered for support include language and area studies K-12, undergraduate and teenage exchanges, individual travel grants, university chairs or endowment funds; research, teaching, publications, and translations in mathematics, medicine, and the natural sciences; building construction, design, or maintenance costs; symphony or other strictly musical groups, solo performing artists, or amateur and university groups; or support for American museums for regular staff, acquisition of objects, or cataloging of existing collections. This award is granted twice a year.

Samples: Doshisha U, Ctr for American Studies (Tokyo, Japan)—for publication of a library catalog, employment of full-time librarian, and support of graduate students, $44,000.;Mississippi State U (MS)—for the research project entitled Japan-Russia-United States Study Group on Russian Nuclear Dumping in the Sea of Japan, Sea of Okhotsk, and the North Pacific Ocean, $46,590.;Pittsburgh Filmmakers (PA)—for postproduction costs of a film entitled Frank Lloyd Wright and Japanese Art, $20,000.

Date(s) Application is Due: Mar 1; Aug 1

Contact: Eric Gangloff, Executive Director, (202) 418-9800; fax: (202) 418-9802; email: grants@jusfc.gov

Internet: http://www.jusfc.gov/commissn/BrochureJan03.htm#_Japanese_Studies_in

Sponsor: Japan-United States Friendship Commission

1120 Vermont Avenue NW, Suite 800

Washington, DC 20005

Japanese Language Program Grants 2969

The foundation offers a variety of grant programs and services to facilitate Japanese language teaching. Programs include: salary assistance for full-time Japanese language teachers; Japanese language research/conference/seminar grants; Japanese language teaching materials donation program; training programs for teachers of the Japanese language at the Japanese Language Institute, Urawa, Japan; assistance program for the development of Japanese language teaching resources; Japanese language education fellowship program; Japanese speech contest support program; and the Japanese language program for researchers and postgraduate students at the Japanese Language Institute, Kansai, Japan. Applications to each of the above programs are due December 1. Annual deadline dates may vary; contact program staff for exact dates.

Requirements: Any educational, cultural, or public affairs nonprofit organization is eligible to apply.

Restrictions: Programs planned for high school students are not applicable.

Samples: American Council on the Teaching of Foreign Languages (Yonkers, NY)—for the project Japanese OPI Trainer and Tester Recalibration, $8000.

Date(s) Application is Due: Dec 1

Contact: Grants Administrator, (213) 621-2267; fax: (213) 621-2590; email: jflalc@jflalc.org

Internet: http://www.jflalc.org

Sponsor: Japan Foundation and Language Center in Los Angeles

333 S Grand Avenue, Suite 2250

Los Angeles, CA 90071

Japanese Studies in the United States Research Grants 2970

The goal of this program is to educate a broader stratum of American leadership with respect to Japan. The following project areas have been established: faculty research, library support, language training, and general education. Under its library support category, the commission also supports projects that help organize acquisitions of research materials on a national scale and help expand access to research materials in both printed and electronic format. In its support for language training, the commission will fund institutional activities that promise the greatest national or regional return. The commission will also consider collaborative research and general education projects in Japanese studies on a case-by-case basis. Potential applicants should consult with commission staff before submitting applications.

Requirements: Citizens and permanent residents of the United States and Japan are eligible. Grants may be made to individual universities or local organizations. Awards made to individuals under all programs normally will be made through academic, professional, artistic, or other appropriate organizations that will examine, recommend, and, in most instances, select the individuals to be supported financially by the commission. This award is granted twice a year.

Samples: Columbia U, Graduate School of Journalism (New York, NY)—for training two future journalists for reporting in Japan, $5800 and @Y6.93 million.; Massachusetts Institute of Technology, MIT-Japan Program (MA)—for support of the training, research, and public affairs program on Japanese technology and management, $22,200.

Date(s) Application is Due: Mar 1; Aug 1

Contact: Executive Director, (202) 418-9800; fax: (202) 418-9802; email: grants@jusfc.com

Internet: http://www.jusfc.gov/commissn/BrochureJan03.htm#_Japanese_Studies_in

Sponsor: Japan-United States Friendship Commission

1110 Vermont Avenue NW, Suite 800

Washington, DC 20005

Japanese Studies Speakers and Panels Grants 2971

The purpose of this program is to encourage scholarly study of Japan by disciplinary specialists such as political scientists, economists, geographers, musicologists, linguists, historians, and scientists, by providing financial support to organizers of panels at annual conventions of national scholarly organizations to bring Japan experts of any nationality and Japanese scholars to participate in those panels. Grants of up to $1500 are available to organizers of national conventions of a scholarly discipline to bring eminent speakers to address the convention on a Japanese topic. The person may be an academic figure, a public figure, a distinguished performer in the arts, or any person of distinction. The grant may cover domestic and international travel costs, two days' board and room, an honorarium ($500 maximum), and organizing costs. Additional grants for up to $1000 are available to cover travel within North America for up to four participants, per diem expenses limited to two nights' lodging, and administrative costs.

Requirements: Applications for all programs must include the applicant's curriculum vita. Preference will be given to applications that come from the professional associations where Japanese perspectives historically have been neglected.

Restrictions: Any airfare tickets purchased with funds from these grants must be secured in the United States, from American air carriers; grants may not be used to reimburse any expenses incurred in currencies other than the U.S. dollar.

Amount of Grant: $2500 maximum for any one panel; the daily expenses of lodging and/or food to be reimbursed per person will not exceed $100, and administrative costs are limited to $100.

Date(s) Application is Due: Feb 1; Oct 1

Contact: NEAC Grants, (313) 665-2490; email: postmaster@aasianst.org

Internet: http://www.aasianst.org/grants/grants.htm#NEAC-JAPAN

Sponsor: Northeast Asia Council/Association for Asian Studies

1021 E. Huron Street

Ann Arbor, MI 48104

Japanese Studies United States Research Travel Grants 2972

These short-term grants, sponsored jointly by the Northeast Asia Council and the Association for Asian Studies, are open to those engaged in scholarly research on Japan who wish to use museum, library, or other archival materials located in the United States. A portion of the grant may be used for research materials, assistance, and reasonable subsistence costs. An application form is required.

Requirements: Applicant must be a U.S. citizen or permanent resident; grants are primarily for postdoctoral research, but PhD candidates are also eligible.

Restrictions: Grants may not be used for overseas travel.

Amount of Grant: $1500 maximum including $100 per diem

Date(s) Application is Due: Feb 1; Oct 1

Contact: NEAC Grants, (734) 665-2490; fax: (734) 665-3801; email: postmaster@aasianst.org

Internet: http://www.aasianst.org/grants/grants.htm#NEAC-JAPAN

Sponsor: Northeast Asia Council/Association for Asian Studies

1021 E Huron Street

Ann Arbor, MI 48104

JDF Career Development Awards 2973

The program awards grants to applicants who have demonstrated superior scholarship and show the greatest promise for future achievement in diabetes research, including either clinically relevant research or basic research. The research plan should be suitable for a five-year project, and awardees must spend at least 75 percent of time and effort on the research project during the period of the award. Research may be conducted at foreign and domestic, for-profit and nonprofit, and public and private organizations such as universities, colleges, hospitals, laboratories, units of state and local governments, and eligible agencies of the federal government. Letters of reference are due two weeks before the application deadline, which are July 15 and January 16.

Requirements: Candidates must have the doctoral degree or the equivalent from an accredited institution, with three to seven years of total professional postdoctoral clinical and/or research experience by the projected start of the award period. Candidate's institution must be a university, medical school, or comparable institution with strong, well-established research and training programs in the chosen area. Research may be conducted at foreign and domestic, for-profit and nonprofit, public and private laboratories; units of state and local governments; and eligible agencies of the federal government.

Restrictions: Individuals holding academic positions of associate professor or professor at the time of the award are not eligible.

Amount of Grant: $150,000 per year

Date(s) Application is Due: Jan 16; Jul 15

Contact: Grant Administrator, (212) 479-7565; fax: (212) 785-9595; email: fellowships@jdrf.org

Internet: http://www.jdrf.org/index.cfm?page_id=103291

Sponsor: Juvenile Diabetes Foundation International

120 Wall Street, 19th Fl

New York, NY 10005-4001

Jeanne Humphrey Block Dissertation Award 2974

The grant assists dissertation research on the psychological development of women and girls for a doctoral student who is a woman. Proposals should focus on the sex and gender differences or some other developmental issue of particular concern to women or girls. Priority will be given to projects that draw on or contribute to the resources of the Murray Research Center. Research concerned with the life experiences of racially and ethnically diverse populations within the United States is encouraged.

Requirements: Any female student currently enrolled in a doctoral program in a relevant field is eligible to apply. Dissertation proposals must be approved by an advisor or committee before the grant application is submitted.

Amount of Grant: $5000 maximum

Date(s) Application is Due: Apr 1

Contact: Grants Program Administrator, Jeanne Humphrey Block Dissertation Awards Program, (617) 495-8140; fax: (617) 496-3993; email: mrc@radcliffe.com

Internet: http://www.murray.harvard.edu/mra/index.jsp

Sponsor: Radcliffe College Henry A. Murray Research Center
10 Garden Street
Cambridge, MA 02138

Jefferson Science Fellows Program 2975

This program establishes a new model for engaging the American academic science, technology, and engineering communities in the formulation and implementation of U.S. foreign policy. Each fellow will spend one year at the U.S. Department of State for an on-site assignment in Washington, DC, that may also involve extended stays at U.S. foreign embassies and/or missions. All JSF assignments will be designed in consultation with regional and/or functional bureaus within the U.S. Department of State. Following the fellowship year, the fellow will return to his/her academic career, but will remain available to the U.S. government as an experienced consultant for short-term projects over the following five years. Application and guidelines are available online.

Requirements: Nominees and applicants must be U.S. citizens and must hold a tenured faculty position at a U.S. degree granting academic institution of higher learning. Nominations will be accepted from U.S. academic institutions that have signed a JSF/ MOU with the U.S. Department of State.

Amount of Grant: $50,000 stipend

Date(s) Application is Due: Dec 1

Contact: Fellowships Office, (202) 334-2872; fax: (202) 334-3419; email: jsf@nas.edu

Internet: http://www7.nationalacademies.org/fellowships/Jefferson_Science_Fellows. html

Sponsor: National Academies
500 Fifth Street NW, GR 322A
Washington, DC 20001

Jeffress Research Grants 2976

The trust supports research in chemical, medical, or other scientific fields through grants to nonprofit educational, charitable, scientific, literary, and research institutions in the Commonwealth of Virginia. Types of support include assistantships, fellowships, research grants, and program grants. Grants are awarded for one year, with possible renewal for two additional years. The deadline for receipt of applications for consideration at the May meeting is March 1; for the November meeting, September 1.

Requirements: Tax-exempt institutions are eligible.

Geographic Focus: Virginia

Amount of Grant: $30,000 maximum first year; $10,000 maximum renewal per year for two years

Date(s) Application is Due: Mar 1; Sep 1

Contact: Dr. Richard Brandt, Advisor, Bank of America , Jeffress Memorial Trust, (804) 788-3698; fax: (804) 788-2700

Internet: http://www.wm.edu/grants/OPPS/jeffress.htm

Sponsor: Jeffress Memorial Trust
P.O. Box 26688
Richmond, VA 23261-6688

Jeffrey Campbell Graduate Fellow in Anthropology 2977

The university's Department of Anthropology invites applications for a fellowship in biological (physical) anthropology or paleoanthropology for next year's academic year. The successful candidate will be a specialist and qualified to teach in the area of human origins. The Campbell Fellow will receive a stipend per academic year with the possibility of additional funds for travel to conferences and professional meetings. The fellow will also be afforded office space and a personal computer. It is expected that fellows will reside at the university for one academic year and to teach one course per semester in an area of the candidate's research interest (in consultation with the department). Fellows will also present a research-based paper in the fellows' lecture series each year. Send a detailed letter of application, curriculum vita, three letters of reference, and a short (one to two pages) statement of pedagogy to the office. Guidelines are available online.

Requirements: Applicants must have completed coursework and preliminary examinations for the PhD in anthropology. Candidates should be members of racial or ethnic groups historically underrepresented in higher education in the United States.

Restrictions: Geographic area is open, except for South Asia.

Geographic Focus: New York

Amount of Grant: $28,500 stipend

Contact: Dr. Alice Pomponio, Department of Anthropology, (315) 229-5797; email: apomponio@stlawu.edu

Internet: http://web.stlawu.edu/resources/positions/faculty/fy%202005-2006/ anthrojcgf0805.htm

Sponsor: Saint Lawrence University
Department of Anthropology
Canton, NY 13617

JEHT Foundation Community Justice Grants 2978

JEHT stands for the core values that underlie the foundation's mission: Justice, Equality, Human dignity and Tolerance. The foundation's programs work to transform U.S. criminal justice policies and practices; expand the role of international justice and the rule of law both at home and abroad; and democratize the electoral process in this country. The foundation also will consider limited support for issues or opportunities that fall within the broader scope of its programs but are not reflected in the primary program interests; issues that cut across one or more of the interest areas; internal capacity building for organizations that are already receiving funding from the foundation; external expertise for organizations that are already receiving funding from the foundation- (e.g., professional communications, technology, fundraising, or political consultants); and convenings within its primary interest areas that broaden knowledge in the field, promote cooperation and information-sharing, and link related fields to one another. There are no restrictions on the type of support as long as the organization's work is consistent with the foundation's program interests and it is permissible under applicable U.S. charities law. Grant amounts are determined based on the scope of the project, the size of the applicant's budget, the likelihood that other support can be raised, and the foundation's available financial resources in a given year. Letters of inquiry are accepted throughout the year; full proposals are by invitation. Information and guidelines are available online.

Requirements: 501(c)3 are eligible. Organizations that can meet a 501(c)3 financial equivalency test will be considered.

Contact: Grants Administrator, (212) 965-0400; fax: (212) 966-9606; email: info@ jehtfoundation.org

Internet: http://www.jehtfoundation.org/application/applic1_overview.html

Sponsor: JEHT Foundation
120 Wooster Street, 2nd Fl
New York, NY 10012

Jenifer Altman Foundation Grants 2979

The foundation is a small foundation working for a socially just and ecologically sustainable future, innovative research and demonstration projects in mind-body health, and improved child care prospects for at-risk children. Small grants are made in two areas: initiatives concerning the impact of endocrine disrupting chemicals on human health and the preservation of the environment and initiatives supporting the global citizens' movement for a just and sustainable future in relation to international governmental decision-making forums. The foundation awards a few grants for innovative research and demonstration programs in mind-body health, especially those focused on cancer. Local initiatives that contribute to the quality of life in Bolinas, CA, are also of interest. Applicants should send a concept letter to determine if their projects align with the foundations areas of interest.

Requirements: Creative and innovative nonprofits are eligible to apply for one-time grants.

Restrictions: Grants are not made to individuals. The foundation rarely funds programs for the conservation of specific species or habitats and rarely funds research at universities on environmental issues. Grants are not made for civic programs, projects in other countries, or projects that are focused entirely on one city or region in the United States.

Samples: International Society for Ecological Economics (Solomons, MD)—for general support, $108,000.; Earth Action (Fairbanks, AK)—to support citizens groups worldwide to address concerns about climate change, forests, and biodiversity, $5000.; Ashoka: Innovators for the Public (Washington, DC)—to support sending fellows to work with children and to leverage social change and sustainable development in 16 countries, $15,000.

Amount of Grant: $500-$50,000 range; $1500-$20,000 average

Contact: Ashley Iwanaga, Grants Administrator; (415) 561-2182; fax: (415) 561-6480; email: info@jaf.org

Internet: http://www.jaf.org/apply/index.html

Sponsor: Jenifer Altman Foundation
P.O. Box 29209
San Francisco, CA 94129

Jessie Smith Noyes Foundation Grants 2980

The foundation is committed to preventing damage to the natural systems upon which all life depends and to strengthening individuals and institutions committed to protecting natural systems and ensuring a sustainable society. The foundation makes grants primarily in the interrelated areas of environment and reproductive rights. The program components include toxics, sustainable agriculture (both with emphasis on southern and Rocky Mountain states), and U.S. reproductive rights. In addition, a few grants are made in four areas of special concern: sustainable communities, U.S. environmental justice, strengthening the U.S. nonprofit sector, and environmental issues in the metropolitan New York region. Letters of inquiry are received at any time; proposals will be requested from the foundation after review.

Requirements: 501(c)3 tax-exempt organizations are eligible.

Restrictions: Normally, the foundation will not consider requests for direct service, endowments, loans or scholarships to individuals, capital construction funds, conferences, media events, production of media and TV programming, or general fund-raising drives. General research projects are not funded per se.

Samples: Northwest Women's Law Ctr (Seattle, WA)—for its contributions to the field of reproductive-health care and rights in Alaska, Idaho, Montana, Oregon, and Washington State, $100,000 (2005).

Contact: Millie Buchanan, Program Officer, (212) 684-6577; fax: (212) 689-6549; email: noyes@noyes.org

Internet: http://www.noyes.org

Sponsor: Jessie Smith Noyes Foundation
6 E 39th Street, 12th Fl
New York, NY 10016-0112

Jewish Education and Scholarships Program 2981

The federation provides scholarships and special grants for all types of educational initiatives. The federation's dollars to congregations affiliated with its educational partners enable them to expand their programs, augmenting congregations' limited resources. In addition, federation financial incentive and scholarship programs, in partnership with congregations and families, enable parents to send their children to summer camps, as well as to participate in Israel Experience programs. Guidelines are available online.
Geographic Focus: Illinois
Contact: Education and Scholarships Program, (312) 346-6700
Internet: http://juf.org/jewish_identity/ji_ed_fed.asp
Sponsor: Jewish United Fund/Jewish Federation of Metropolitan Chicago
1 S Franklin Street
Chicago, IL 60606-4594

Jewish Federation of Metropolitan Chicago Scholarships 2982

Scholarship funds are available for Jewish college students. The scholarships support full-time students, predominantly those legally domiciled in the metropolitan Chicago area, who are identified as having promise for significant contributions in their chosen careers. Assistance is available for those in the helping professions, i.e., medicine, education, Jewish communal service, the rabbinate, social service, law, and communications at the University of Illinois (Champaign-Urbana).
Requirements: Applicants must be Jewish; meet one or both of the following criteria—born and raised in either Cook County, Chicago metropolitan area or Northwest Indiana, or prior to starting professional education, worked full-time for at least 12 months in Cook County or Chicago metropolitan area; intend to remain in the Chicago metropolitan area after completing school; and be entering a professional postgraduate education program on a full-time basis (school acceptance is not required at time of scholarship application). Undergraduate applicants must be entering the junior or senior year of a professional education program that will allow them to pursue employment in their chosen helping professions at a bachelor's degree level.
Geographic Focus: Illinois
Samples: To be distributed among 15 Jewish day schools in the Chicago area—for scholarships, through the federation's Jewish Day School Guaranty Trust, $72,900.
Amount of Grant: $500,000 total annually
Date(s) Application is Due: Feb 15
Contact: Program Administrator, (312) 673-3457; email: jvsscholarship@jvschicago.org
Internet: http://www.jvschicago.org/scholarship
Sponsor: Jewish Vocational Service
216 W. Jackson Boulevard, Suite 700
Chicago, IL 60606

Jewish Foundation for Education of Women Scholarships 2983

Scholarships are awarded to women pursuing undergraduate or graduate education. Requests for applications must be made in writing and specify whether aid is being sought for undergraduate or graduate studies. Financial assistance is offered every year depending on progress shown.
Requirements: Eligible to apply are women who are permanent residents of the five boroughs of New York City, although schools attended may be located anywhere.
Restrictions: No awards will be made for the pursuit of JD or MBA degrees. Scholarships will not be available for the upcoming school year.
Geographic Focus: New York
Samples: Teachers College, Columbia U (New York, NY)—for scholarships for master's-degree candidates in mathematics education and science education who will teach in the New York City public schools, $135,000 over three years.; Juilliard School (New York, NY)—to provide scholarships to arts students, $250,000 over four years.; Jewish Foundation for Education of Women (New York, NY)—to provide scholarships to emigres from the former Soviet Union who are studying dentistry, medicine, nursing, and pharmacy at New York-area universities, $720,000 over three years.
Amount of Grant: $5000 maximum
Date(s) Application is Due: Jun 1
Contact: Marge Goldwater, Executive Director, (212) 288-3931; fax: (212) 288-5798; email: fdnscholar@aol.com
Internet: http://www.jfew.org
Sponsor: Jewish Foundation for Education of Women
135 E 64 Street
New York, NY 10021

Jewish Foundation for Education of Women Scholarships for Emigres in the Health Professions 2984

Scholarships are awarded to women emigres from the former Soviet Union for graduate and undergraduate education in medicine, dentistry, nursing, or pharmacy. Approximately 48 scholarships are awarded. Applications are available in late October.
Requirements: Eligible to apply are women who are emigres from the former Soviet Union, live within a 50-mile radius of New York City, and demonstrate financial need.
Restrictions: Pre-medical students are not eligible. Applications submitted directly from students are not accepted.
Geographic Focus: New York

Samples: Jewish Foundation for Education of Women (New York, NY)—to administer scholarships for emigres from the New Independent States who are enrolled in occupational therapy, physical therapy, or physician's assistant programs, $300,000 over three years.
Amount of Grant: $5000
Date(s) Application is Due: Jun 1
Contact: Marge Goldwater, Executive Director, (212) 288-3931; fax: (212) 288-5798; email: FdnScholar@aol.com
Internet: http://www.jfew.org/programs.html#HP
Sponsor: Jewish Foundation for Education of Women
135 East 64th Street
New York, NY 10021

Jewish Institute of Religion Graduate Fellowships and Scholarships 2985

The Jewish Institute of Religion at Hebrew Union College awards graduate fellowships and tuition scholarships for MA and PhD studies in Bible, Semitics, History, Rabbinics, and Modern Jewish studies. Also, joint programs are offered with the University of Cincinnati in modern Hebrew, literature, and sociology.
Contact: Graduate Study Admissions, (513) 221-1875, fax: (513) 221-0321
Internet: http://www.huc.edu/admissions/aid.shtml
Sponsor: Hebrew Union College
3101 Clifton Avenue
Cincinnati, OH 45220-2488

Jewish Institutional Grants 2986

The Memorial Foundation for Jewish Culture awards grants to universities and institutions for Jewish scholarship in such areas as the bible and semitics, Jewish history, Jewish philosophy and thought, Talmud, rabbinics and Jewish law, and general Jewish research; for Jewish educational and cultural programs; and for research and publication on the Holocaust. The amount awarded varies depending on the budget of the project; awards are limited to no more than 25 percent of the total cost excluding overhead. Application form must be requested in writing. Grants are made every two years, in even-numbered years. Applications are due February 28 in odd-numbered years.
Restrictions: Grants are made only for team or collaborative projects of limited duration. Grants are not made to individuals.
Contact: Dr. Jerry Hochbaum, Executive Vice President, (212) 425-6606; fax: (212) 425-6602; email: office@mfjc.org
Internet: http://www.mfjc.org
Sponsor: Memorial Foundation for Jewish Culture
50 Broadway, 34th Fl
New York, NY 10004

Jewish International Community Service Scholarships 2987

Scholarships are awarded annually for professional training for careers in Jewish education, Jewish social service, the rabbinate, shehita, and milah. The recipients must commit themselves after their training to serve two to three years in a Jewishly deprived Diaspora community outside the United States, Canada, and Israel where such professional personnel are urgently needed. Recipient should also be knowledgeable in the language and culture of that country or be prepared to learn it. Scholarships are for one year, renewable up to a total of four years. Application forms are available upon written request.
Requirements: Scholarships are open to any individual, regardless of country of origin, who is presently receiving or plans to undertake training in his/her chosen field in a recognized yeshiva, teacher training seminary, school of social work, university, or other educational institution. Application forms must be requested in writing.
Amount of Grant: $1000-$3000 depending on the country in which the recipient is trained
Date(s) Application is Due: Nov 30
Contact: Dr. Jerry Hochbaum, Executive Vice President, (212) 425-6606; fax: (212) 425-6602; email: office@mfjc.org
Internet: http://www.mfjc.org
Sponsor: Memorial Foundation for Jewish Culture
50 Broadway, 34th Fl
New York, NY 10004

Jewish Post-Rabbinic Scholarships 2988

These scholarships are awarded to newly ordained rabbis for advanced training leading to careers as judges on rabbinical courts, heads of institutions of higher learning, or other advanced religious leadership positions.
Requirements: Eligible to apply are recently ordained rabbis enrolled in full-time graduate study in a rabbinical seminary, yeshiva, or other institution of higher Jewish learning. Application forms will be sent on written request from the individual (not the school).
Amount of Grant: $2000-$6000
Contact: Dr. Jerry Hochbaum, Executive Vice President, (212) 425-6606; fax: (212) 425-6602; email: office@mfjc.org
Internet: http://www.mishpacha.com/mfjc.shtml
Sponsor: Memorial Foundation for Jewish Culture
50 Broadway, 34th Fl
New York, NY 10004

Jewish Social Entrepreneurs Fellowships 2989

The fellowship supports women and men who are developing projects that address current social issues in the Jewish community and the United States. Projects can be an

idea in an advanced stage of planning, a young independent nonprofit organization, or an original transformative program within an existing organization. Selected programs can focus on a wide variety of social arenas, including education, social justice, the arts, technology, the environment, and spirituality. The fellowship provides seed capital, entrepreneurial training, mentorships, technical assistance, and Jewish learning. Each fellow is encouraged and supported in raising additional funds for his or her project. Guidelines and application forms are available online.
Requirements: Jewish innovators between the ages of 21 and 35 are eligible.
Amount of Grant: $30,000 per year; $5000 maximum self-directed learning grant
Contact: Fellowship Administrator, (415) 777-4500; fax: (415) 777-4045; email: fellowship@joshuaventure.org
Internet: http://www.joshuaventure.org
Sponsor: Joshua Venture
28 Second Street, Suite 500
San Francisco, CA 94105

Jewish Women's Foundation of New York Grants 2990
The foundation provides support for unmet social, economic, and health needs of Jewish females in the New York metropolitan area and beyond. Strategic grants support cutting-edge projects that address Jewish education, training, and culture—projects related to education, training of educators, leadership development, ethics, spirituality, arts, sports, etc.; mental and physical health—projects involving genetics, self-esteem, violence prevention, issues concerning the end of life, and other areas that relate to women's health as a quality of life; and economic empowerment—projects that give women more independence, knowledge, competency, and responsibility for their economic situations. Grant requests may address innovative programs, services, and/or research. The projects primary target group must be Jewish females. The foundation expects to award three to four grants. The listed deadline is for concept letters; full proposals are by invitation.
Requirements: 501(c)3 tax-exempt organizations in the New York metropolitan area (the five boroughs of New York City, Long Island and/or Westchester) are eligible.
Restrictions: The foundation will not provide grants for scholarships, equipment, capital campaigns, or ongoing support for existing programs.
Geographic Focus: New York
Samples: Jewish Orthodox Feminist Alliance (New York, NY)—to create a gender-sensitive curriculum for use in Jewish Orthodox day schools, $20,000 (2004); Jewish Outreach Institute (New York, NY)—to develop a program for women who are new to Judaism, $20,000 (2004); Sharsheret (New York, NY)—for a symposium on breast-cancer genetics and Jewish women, $10,000 (2004); Women's Electoral Power (Tel Aviv, Israel)—for workshops for women seeking public office in Israel, $10,000 (2004).
Amount of Grant: $25,000 maximum
Date(s) Application is Due: Sep 30
Contact: Sherri Greenbach, Executive Director, (212) 836-1478; fax: (212) 836-1831
Internet: http://www.jewishwomenny.org
Sponsor: Jewish Women's Foundation of New York
130 E 59th Street, Room 563
New York, NY 10022

JFEW Scholarships for Emigres Training for Careers in Jewish Education 2991
Scholarships are awarded to women emigres from the New Independent States for graduate education in the fields of rabbinical, cantorial, Jewish education, or Jewish studies. Approximately seven scholarships are awarded.
Requirements: Eligible to apply are women who are emigres from the New Independent States who live within a 50-mile radius of New York City and demonstrate financial need.
Geographic Focus: New York
Amount of Grant: $5000 annually or $ 10,000-$20,000 one time grants
Contact: Marge Goldwater, Executive Director, (212) 288-3931; fax: (212) 288-5798; email: FdnScholar@aol.com
Internet: http://www.jfew.org/programs.html
Sponsor: Jewish Foundation for Education of Women
135 E 64th Street
New York, NY 10021

JFK Library Foundation Abba P. Schwartz Research Fellowship 2992
The fellowship is intended to support a scholar in the preparation of a substantial work in the areas of immigration, naturalization, or refugee policy.
Requirements: Scholars and students are invited to apply for support of their research and use of the archival, manuscript, and audiovisual holdings of the library.
Amount of Grant: $3100
Date(s) Application is Due: Mar 15
Contact: Grant and Fellowship Coordinator, (617) 514-1600; fax: (617) 514-1652; email: kennedy.library@nara.gov
Internet: http://www.jfklibrary.org/schwartz.htm
Sponsor: John F. Kennedy Library Foundation
Columbia Point
Boston, MA 02125-3313

JFK Library Foundation Arthur M. Schlesinger Jr. Research Fellowship 2993
The fellowship is intended to support scholars in the preparation of substantial works on the foreign policy of the Kennedy years, especially with regard to the western hemisphere, or on Kennedy domestic policy, especially with regard to racial justice and to the conservation of natural resources.

Requirements: Proposals are invited from all sources, but preference will be given to those from applicants specializing in the areas indicated. Preference is also given to projects not supported by large grants from other institutions.
Amount of Grant: $5000
Date(s) Application is Due: Aug 15
Contact: Grant and Fellowship Coordinator, (617) 514-1631; fax: (617) 514-1652; email: kennedy.library@nara.gov
Internet: http://www.jfklibrary.org/schles.htm
Sponsor: John F. Kennedy Library Foundation
Columbia Point
Boston, MA 02125-3313

JFK Library Foundation Hemingway Research Grants 2994
Funds are awarded to help defray living, travel, and related costs incurred while doing research in the Hemingway Collection at the Kennedy Library. Applications are evaluated on the basis of expected utilization of the Hemingway Collection, the degree to which projects address research needs in Hemingway or related studies, and the qualifications of applicants. Preference is given to dissertation research by PhD candidates in newly opened or relatively unused portions of the collection. Applicants are advised to consult the catalog of the Ernest Hemingway Collection at the John F. Kennedy Library and to contact a member of the Kennedy Library staff for more information before applying.
Amount of Grant: $200-$1000
Date(s) Application is Due: Nov 1
Contact: Grant and Fellowship Coordinator, (617) 514-1600; fax: (617) 514-1652; email: kennedy.library@nara.gov
Internet: http://www.jfklibrary.org/ehgrants.htm
Sponsor: John F. Kennedy Library Foundation
Columbia Point
Boston, MA 02125-3313

JFK Library Foundation Marjorie Kovler Research Fellowship 2995
The fellowship is intended to support a scholar in the preparation of a substantial work in the area of foreign intelligence and the presidency, or a related topic. Application forms are available on the Web site.
Requirements: Scholars and students are invited to apply for support of their research and use of the archival, manuscript, and audiovisual holdings of the library.
Amount of Grant: $2500 maximum
Date(s) Application is Due: Mar 15
Contact: Grant and Fellowship Coordinator, (617) 514-1631; fax: (617) 514-1652; email: kennedy.library@nara.gov
Internet: http://www.jfklibrary.org/kovler.htm
Sponsor: John F. Kennedy Library Foundation
Columbia Point
Boston, MA 02125-3313

JFK Library Foundation Research Grants 2996
The John F. Kennedy Library Foundation administers and funds programs on behalf of the Kennedy Library and Museum. Scholars and students are invited to apply for support of their research and use of the archival, manuscript, and audiovisual holdings of the library. Preference is given to dissertation research by PhD candidates working in newly opened or relatively unused collections, and to the work of recent PhD recipients who are expanding or revising their dissertations for publication, but all proposals are welcome and will receive careful consideration.
Amount of Grant: $500-$2500
Date(s) Application is Due: Mar 15; Aug 15
Contact: Grant and Fellowship Coordinator, (617) 514-1600; fax: (617) 514-1652; email: library@kennedy.nara.gov
Internet: http://www.jfklibrary.org/krg.htm
Sponsor: John F. Kennedy Library Foundation
Columbia Point
Boston, MA 02125-3313

JFK Library Foundation Theodore C. Sorensen Research Fellowship 2997
The fellowship is intended to support a scholar in the preparation of a substantial work in the areas of domestic policy, political journalism, polling, or press relations.
Requirements: Scholars and students are invited to apply for support of their research and use of the archival, manuscript, and audiovisual holdings of the library.
Amount of Grant: $3600
Date(s) Application is Due: Mar 15
Contact: Grant and Fellowship Coordinator, (617) 514-1600; fax: (617) 514-1652; email: kennedy.library@nara.gov
Internet: http://www.jfklibrary.org/sorensen.htm
Sponsor: John F. Kennedy Library Foundation
Columbia Point
Boston, MA 02125-3313

JHI International Fellows in Philanthropy Program 2998
The program offers advanced study, research, and training to eight participants annually who are involved in studying or managing private nonprofit or philanthropic organizations outside the United States, or who are working as liaisons for nongovernmental organizations in the public or commercial sectors. Fellowships can be for either an academic year or a semester and are available at both the junior and senior levels. A high degree of English-language fluency is required.

Requirements: Individuals studying nonprofit organizations or managing a nonprofit group outside the United States are eligible. Candidates are expected to have attained a university undergraduate degree and to be capable of carrying out independent research and inquiry.
Date(s) Application is Due: Feb 25
Contact: Lester Salamon, Program Director, (410) 516-5389; fax: (410) 516-8233; email: npmgt@jhu.edu
Internet: http://www.jhu.edu/~philfellow
Sponsor: Johns Hopkins Institute for Policy Studies
3400 N Charles Street, Wyman Building, 5th Fl
Baltimore, MD 21218

JHMI Patrick C. Walsh Prostate Cancer Research Fund 2999
Funding is available to support multidisciplinary research in prostate cancer through the Patrick C. Walsh Prostate Cancer Research Fund. Awards of $50,000 to $100,000 for up to 2 years are available to fund career development and developmental research programs (pilot projects). This fund will attract and support the best and brightest scientists throughout Johns Hopkins to join in an effort to defeat the number one cancer in men. 1) Awards will be for pilot projects, focused on prostate cancer, that develop a new research direction, explore an innovative idea, test an unconventional, but potentially important, new hypothesis, or ascertain the feasibility of a new research approach. These grants will be reviewed by a scientific advisory board which will award funding to projects that hold the most promise.
Requirements: All faculty of the Johns Hopkins University may apply, whether or not they are members of the James Buchanan Brady Urological Institute or of the Sidney Kimmel Comprehensive Cancer Center. However, an investigator may hold only one pilot project award at a time.
Amount of Grant: $50,000-$100,000
Date(s) Application is Due: Jan 8
Contact: Angela Sciuto, Sponsored Projects Officer, (410) 614-0257; email: asciuto2@jhmi.edu
Internet: http://prostatecancerprogram.onc.jhmi.edu/
Sponsor: Johns Hopkins Medical Institute
733 N. Broadway, Suite 117
Baltimore, MD 21205

JILA Visiting Fellowships and Postdoctoral Research Associateships 3000
One-year postdoctoral research associateships are intended to provide advanced research training in the years immediately after the PhD degree. These research associateships are generally renewable for a second year. Visiting fellowships are awarded for independent research and study, usually to those with extensive research experience. For those in doubt as to which type of appointment is more suitable, the institute advises application for a research associateship. If all materials for the completed application are received by November 1, the applicant may request to be considered for both the visiting fellowship program and the postdoctoral research associateship program. Transportation to Boulder may be paid for the appointee and his/her spouse and minor children, and an allowance for shipment of personal effects and an allowance for appropriate professional travel within the United States during the period of the appointment are usually made available.
Date(s) Application is Due: Nov 1
Contact: Secretary, Visiting Scientists Program, (303) 492-7796; fax: (303) 492-5235; email: jilavf@jila.colorado.edu
Internet: http://jilawww.colorado.edu/www/programs/pd.html
Sponsor: Joint Institute for Laboratory Astrophysics (JILA)
University of Colorado, 440 UCB
Boulder, CO 80309-0440

JM Foundation Grants 3001
The foundation awards grants to eligible nonprofit organizations in its areas of interest, including education and research that fosters market-based policy solutions; developing state and national organizations that promote free enterprise, entrepreneurship, and private initiative; and identifying and educating young leaders. Types of support include internships, matching/challenge grants, program grants, publication, research grants, seed grants, and technical assistance. The foundation's board of directors meets bi-annually, usually in May and October. There are no formal proposal deadlines. Inquiries and proposals are processed on an ongoing basis.
Requirements: Public charities, including 501(c)3, 509(a)1, and 170(b)1(a)(vi) nonprofit organizations, that shares the foundation's priority interests is invited to submit a proposal.
Restrictions: Grants do not support individuals, the arts, government agencies, public schools, and international agencies; or requests for operating expenses, annual fundraising campaigns, capital campaigns, equipment, endowment funds, and loans.
Samples: Acton Institute, Grand Rapids, Michigan, $20,000—in support of the Effective Compassion Initiative (2007); Center of the American Experiment, Minneapolis, Minnesota, $40,000—for Intellectual Takeout (2007); National Taxpayers Union Foundation, Alexandria, Virginia, $25,000—toward the National Taxpayers Conference (2007).
Amount of Grant: $5,000-$100,000 range
Contact: Carl Helstrom, Executive Director; (212) 687-7735; fax: (212) 697-5495
Internet: http://foundationcenter.org/grantmaker/jm/guide_jm.html
Sponsor: JM Foundation
654 Madison Avenue, Suite 1605
New York, NY 10065

JMO American Institutions Grants 3002
Grants promote understanding of the moral, cultural, and institutional foundations of free government. Under this program the foundation supports studies of and research on the American Constitution, the operation of American political institutions and the moral and cultural principles underlying these institutions, economics, foreign policy, and security studies. Financial support is awarded for research, institutional support, fellowships, professorships, lectures, conferences/seminars, books, scholarly journals and journals of opinion, and sometimes television and radio programs.
Requirements: A 501(c)3 nonprofit may apply by submitting a concise written proposal describing the project and grant request and offering background on the organization and its current sources of funding.
Restrictions: Grants are not awarded to support endowments or capital campaigns.
Samples: U of Virginia, Department of Government (VA)—for graduate fellowships, $75,000 per year for three years.; Harvard U (MA)—continued support of the John M. Olin Institute for Strategic Studies, $400,000.
Amount of Grant: $5000-$600,000
Contact: William Voegeli, Program Officer, (212) 661-2670; fax: (212) 661-5917; email: wvoegeli@jmof.org
Internet: http://www.jmof.org/grant_programs.html
Sponsor: John M. Olin Foundation
330 Madison Avenue, 22nd Floor
New York, NY 10017

JMO Law and Legal System Grants 3003
Grants are given to deepen the understanding of the American judicial system and to preserve the rule of law as the bedrock of American constitutional government. The foundation supports public interest law and studies related to the judicial system, jurisprudence, and the relationship between law and economics. Financial support is awarded for research, institutional support, fellowships, professorships, lectures, conferences/seminars, books, scholarly journals and journals of opinion, and sometimes for television and radio programs.
Requirements: A 501(c)3 nonprofit may apply by submitting a concise written proposal describing the project and grant request and offering background on the organization and its current sources of funding.
Samples: Harvard Law School (Cambridge, MA)—for research, a fellowship program, and a speaker series at the Center for Law, Economics, and Business, $10 million.
Contact: Caroline Hemphill, Program Officer, (212) 661-2670; fax: (212) 661-5917; email: chemphill@jmof.org or inquiry@jmof.org
Internet: http://www.jmof.org/grant_programs.html
Sponsor: John M. Olin Foundation
330 Madison Avenue, 22nd Floor
New York, NY 10017

JMO Public Policy Research Grants 3004
Grants support research on the formulation, implementation, and evaluation of public policy in the social and economic fields. Grants are made in such areas as regulatory policy, tax policy, fiscal policy, monetary policy, and welfare policy. The larger grants are made to graduate schools at universities in these areas. Financial support is awarded for research, institutional support, fellowships, professorships, lectures, conferences/seminars, books, scholarly journals and journals of opinion, and sometimes for radio and television programs.
Requirements: A 501(c)3 nonprofit may apply by submitting a concise written proposal describing the project, grant request, and background on the organization.
Restrictions: Grants do not support endowments, building campaigns, annual giving programs, or individuals.
Samples: George Mason U School of Law (Fairfax, VA)—for institutes in law and economics for federal judges, $200,000.; Pacific Legal Foundation (Sacramento, CA)—for litigation and public education on freedom and equality before the law, $60,000.
Amount of Grant: $20,000-$300,000
Contact: Caroline Hemphill, Program Officer, (212) 661-2670; fax: (212) 661-5917; email: chemphill@jmof.org or inquiry@jmof.org
Internet: http://www.jmof.org/grant_programs.html
Sponsor: John M. Olin Foundation
330 Madison Avenue, 22nd Floor
New York, NY 10017

JMO Strategic and International Studies Grants 3005
Grants support projects that address the relationship between American institutions and the international context in which they operate. Such projects include studies of national security affairs, strategic issues, American foreign policy, and the international economy. Financial support is awarded for research, institutional support, fellowships, professorships, lectures, conferences/seminars, books, scholarly journals and journals of opinion, and sometimes for television and radio programs.
Requirements: A 501(c)3 nonprofit may apply by submitting a concise written proposal describing the project and grant request and offering background on the organization and its current sources of funding.
Amount of Grant: $10,000-$300,000
Contact: Caroline Hemphill, Program Officer, (212) 661-2670; fax: (212) 661-5917; email: chemphill@jmof.org or inquiry@jmof.org
Internet: http://www.jmof.org/grant_programs.html
Sponsor: John M. Olin Foundation
330 Madison Avenue, 22nd Floor
New York, NY 10017

Joan B. Kroc Institute for International Peace Studies Rockefeller   3006
Foundation Visiting Fellowships

The fellowships support research that explores the complex role of religion in contemporary conflicts, ranging from the legitimation or sacralization of violence, to participation in conflict mediation and reconciliation, to the advocacy and practice of nonviolent resistance as a religious imperative. Research focuses on the relationship between religious ethics, human rights, and attitudes of tolerance and intolerance toward the other; religious roles in conflict resolution, including conflict within and between religious traditions; and the contributions of religious actors to postconflict reconciliation, justice, and peacebuilding. The program seeks to include research by scholars and practitioners with expertise in Hindu, Muslim, Jewish, Buddhist, Sikh, or Christian traditions and movements. Projects that consider program themes from the perspective of a public intellectual are particularly encouraged. Guidelines are available online.

Requirements: Senior and junior scholars in the humanities and social sciences, as well as religious leaders and peacebuilding practitioners, of any nationality are eligible.

Amount of Grant: $35,000 minimum per year

Date(s) Application is Due: Nov 15

Contact: Rashied Omar, Program Coordinator, (574) 631-7740; fax: (574) 631-6973; email: omar.1@nd.edu

Internet: http://www.nd.edu/~krocinst/visiting_fellows/rockvf0607.html

Sponsor: University of Notre Dame

P.O. Box 639

Notre Dame, IN 46556-0639

Joan B. Kroc Institute for International Peace Studies Visiting Fellows   3007

The institute brings together outstanding scholars to conduct peace-related research, broadly defined. Themes include the peacemaking role of international norms, policies, and institutions; approaches to the study and resolution of violence; and the quest for social, economic, and environmental justice. Fellows conduct research and write in their areas of interest, for a semester or a year, while interacting with institute faculty and staff, and with Kroc faculty fellows in the broader Notre Dame community. Fellows are provided an office, library access, communications links, document retrieval services, and housing. Guidelines are available online.

Requirements: Applications at the predoctoral, postdoctoral, and senior scholar level will be considered.

Amount of Grant: $20,000 maximum per semester, junior fellows; $25,000 maximum per semester, senior fellows

Date(s) Application is Due: Nov 1

Contact: Dr. Martha Merritt, Associate Director, (574) 631-7695; email: mmerritt@nd.edu

Internet: http://www.nd.edu/~krocinst/visiting_fellows/visfell0607.html

Sponsor: University of Notre Dame

P.O. Box 639, 100 Hesburgh Center for International Studies

Notre Dame, IN 46556

Joan's Legacy Lung Cancer Research Grants   3008

Joan's Legacy invites grant applications for institutional research that studies lung cancer. The foundation is particularly interested in the genetic basis and biology of bronchoalveolar carcinoma as well as novel therapeutic approaches for the treatment of this disease. This funding is intended primarily as seed money for promising new work. It is not for supporting research where funding has either lapsed or has been previously disapproved. Preference will be given to applications where indirect costs are minimal or nonexistent. Grants are awarded for one or two years. Funding for promising research may be extended further at the discretion of the medical committee. Application and guidelines are available online.

Amount of Grant: $50,000 per year

Date(s) Application is Due: Jul 1

Contact: Grants Administrator, (212) 627-5500; fax: (917) 661-7811

Internet: http://www.joanslegacy.org/grant_app.html

Sponsor: Joan Scarangello Foundation to Conquer Lung Cancer

27 Union Square W, Suite 205

New York, NY 10003

Joe W. and Dorothy Dorsett Brown Foundation Grants   3009

The foundation awards grants to nonprofit organizations in Louisiana and the Gulf Coast of Mississippi. Areas of interest include medical research; housing for the homeless; support for organizations who care for the sick, hungry or helpless; religious and educational institutions; and organizations and groups concerned with improving the local community. Types of support include operating budgets, research, and student aid. The foundation also supports Service Learning, a learn-by-doing approach to the curriculum. Students receive practical, hands-on experience in the subject matter studied by meeting identified community needs through active participation. The listed deadline date is for Service Learning grants.

Requirements: Louisiana and Mississippi nonprofit organizations, with a focus on South Louisiana, the New Orleans area, and the Mississippi Gulf Coast, are eligible. Service Learning grant applications are available yearly to sixth through 12th grades in the following parishes: Orleans, Jefferson, Plaquemines, Saint Bernard, Saint Charles, Tangipahoa, Saint James, Saint John, Saint Tammany, and Washington.

Geographic Focus: Louisiana; Mississippi

Amount of Grant: $5000-$25,000 average

Date(s) Application is Due: Sep 30

Contact: Beth Buscher, (504) 834-3433; email: BethBuscher@thebrownfoundation.org

Internet: http://www.thebrownfoundation.org

Sponsor: Joe W. and Dorothy Dorsett Brown Foundation

320 Hammond Hwy, Suite 500

Metairie, LA 70005

Joel Henry Hildebrand Award in the Theoretical and Experimental   3010
Chemistry of Liquids

This award, supported by the ExxonMobil Research and Engineering Company, is given annually to recognize distinguished contributions to the understanding of the chemistry and physics of liquids. The award shall be granted without regard to age or nationality. The award consists of $5,000 and a certificate.

Requirements: Any individual, except a member of the award committee, may submit one nomination or seconding letter for the award in any given year. Nominating documents consist of a letter of not more than 1000 words containing an evaluation of the nominee's accomplishments and a specific identification of the work to be recognized, a biographical sketch including date of birth, and a list of publications and patents authored by the nominee.

Restrictions: Self-nominations are not accepted.

Samples: Pablo G. Debenedetti—award winner, $5,000 (2008); Keith E. Gubbins—award winner, $5,000 (2007); Kenneth Eisenthal—award winner, $5,000 (2006).

Amount of Grant: $5,000, certificate, and an allowance of up to $1,000 for travel expenses to award meeting

Date(s) Application is Due: Nov 1

Contact: Felicia Dixon, Awards Administrator; (800) 227-5558 or (202) 872-4408; fax: (202) 776-8008; email: f_dixon@acs.org or awards@acs.org

Internet: http://portal.acs.org/portal/acs/corg/content?_nfpb=true&_pageLabel=PP_ARTICLEMAIN&node_id=1319&content_id=CTP_004539&use_sec=true&sec_url_var=region1

Sponsor: American Chemical Society

1155 Sixteenth Street, NW

Washington, DC 20036

Joel L. Fleishman Civil Society Fellowships   3011

The Fellowship provides a select group of leaders from domestic non-profit organizations, international non-governmental organizations, foundations, government, socially responsible businesses, and other civil society groups in the United States and internationally with the opportunity to come in residence at the Sanford Institute for a four-week mini-sabbatical. While at Duke, fellows will perform research and work with institute faculty and other Duke affiliates on issues relating to the development of civil society. Applicants will be selected based upon their proposed research project and how they intend to utilize Duke's resources to benefit their professional work in civil society. The fellowship provides housing and program expenses; a stipend; and access to the university's library, research centers, and recreational facilities. Download the application online.

Requirements: Employees of nonprofit and nongovernmental groups worldwide are eligible.

Restrictions: Individuals employed at government agencies and academic institutions are not eligible. Full-time academics are not eligible to apply.

Samples: Frehiwot Alebachew, Saves Lives Ethiopia (SaLE), Ethiopia (2006); Anthony Ehrenreich, Congress of South African Trade Unions (COSATU), South Africa (2006); Liza Lim, Institute of Social Order, Philippines (2006).

Amount of Grant: $6,000 stipend

Date(s) Application is Due: May 1

Contact: Melynn Glusman, Program Director, (919) 613-7432; email: melynn.glusman@duke.edu

Internet: http://www.pubpol.duke.edu/centers/civil/

Sponsor: Duke University, Terry Sanford Institute of Public Policy

P.O. Box 90239

Durham, NC 27708

John B. Bene Fellowships in Social Forestry   3012

This fellowship covers graduate studies that focus on the relationship of forest resources to the social, economic, and environmental welfare of people, especially those in developing countries. The award must be used, at least in part, to help fund field research in a developing country. The fellowships are awarded for one year and are renewable for up to an additional two years. Applications can be downloaded from Web site.

Requirements: Applicants must be Canadian citizens or permanent residents of Canada enrolled in a Canadian university at the master's or doctoral level and have an academic background that combines forestry or agroforestry with social sciences. Research proposal must be approved by the research/thesis committee. Applicants must provide evidence of affiliation with an institution or organization in the region in which the research will take place, and have completed course work and passed comprehensive examinations by the time of award tenure.

Amount of Grant: $C15,000

Date(s) Application is Due: Mar 1

Contact: Center Training and Awards Program, International Development Research Center, (613) 236-6163 ext 2098; fax: (613) 563-0815; email: cta@idrc.ca

Internet: http://web.idrc.ca/en/ev-23378-201-1-DO_TOPIC.html

Sponsor: International Development Research Center

P.O. Box 8500

Ottawa, ON K1G 3H9 Canada

John B. Hurford Humanities Center Mellon Postdoctoral Fellowships   3013

The John B. Hurford Humanities Center invites applications for a two-year Mellon postdoctoral fellowship from scholars who consider music as a site of humanistic inquiry. Possible areas of research include philosophical, religious, literary, anthropological, and political discourses on music; histories and theories of musical performance; semiotic, psychoanalytic, and historiographic approaches to music; music and ideology; the function of music in the formation of ethnic, racial, gendered, and social identities; and relations between music and other media. Fields of specialization might include musicology, and ethnomusicology, literary studies, cultural studies, anthropology, history, religious studies, philosophy, performance studies. The successful candidate will be expected to teach one course each semester and participate in a year-long faculty seminar. Submit a letter of application, curriculum vita, two course proposals, and a writing sample of not more than 25 pages, and arrange for three letters of recommendation to be sent to the office.

Requirements: Candidates who received the PhD within the last five years as of the listed application deadline date are eligible.

Date(s) Application is Due: Jan 31

Contact: Emily Carey Cronin, (610) 896-1336; email: ecronin@haverford.edu

Internet: http://www.haverford.edu/HHC/mellon.htm

Sponsor: Haverford College

Haverford, PA 19041

John Ben Snow Memorial Trust Grants   3014

The mission of the Foundation is to make grants within specific focus areas to enhance the quality of life in Central and Northern New York State. Historically, the Foundation has made grants in the following program areas: arts and culture, community development, education, environment, historic preservation, and journalism. The Foundation responds to the ever-changing needs of various segments of the population, especially to the needs of young people and people who are disadvantaged either physically or economically. It is the Foundation's general policy to give preference to proposals seeking funds for new or enhanced programs, one-time, short-term grants to sustain a program until funding is stabilized, matching grants used to encourage the participation of other donors, and last dollars towards a capital campaign. There are no minimums or maximum grant amounts; however, most grants range from $5,000 to $15,000.

Requirements: Giving is primarily focused in Maryland, Nevada, and central New York.

Restrictions: The Foundation will not accept proposals from individuals or for-profit organizations. Additionally, the Foundation does not encourage proposals from religious organizations or proposals for endowments, contingency funding, or debt reduction.

Amount of Grant: $5,000-$25,000 average

Date(s) Application is Due: Apr 1

Contact: Jonathan L. Snow; (315) 471-5256; fax: (315) 471-5256

Internet: http://www.johnbensnow.com/jbsmt

Sponsor: John Ben Snow Memorial Trust

50 Presidential Plaza, Suite 106

Syracuse, NY 13202

John Carter Brown Library Long-Term Research Fellowships   3015

These fellowships, supported in part by NEH and the Andrew E. Mellon Foundation, are for five to nine months. The term for these fellowships will typically begin between June 1 and July 15 or between January 15 and March 15. Awards are based on the applicant's scholarly qualifications, the merits of the project, and the appropriateness of the inquiry to the holdings of the library, which is an outstanding collection of primary materials relating to virtually all aspects of the discovery, exploration, settlement, and development of the New World, ranging from the late 15th century to about 1830. Recipients are expected to be in continuous residence at the library and to participate in the intellectual life at Brown University. Application forms may be obtained from the program officer.

Requirements: Applicants must be U.S. citizens or have been residents in the United States for the three years immediately preceding the term of the fellowship.

Restrictions: Graduate students are not eligible.

Amount of Grant: $4000 stipend per month

Date(s) Application is Due: Jan 10

Contact: Director, (401) 863-2725; fax: (401) 863-3477; email: JCBL_Fellowships@Brown.edu

Internet: http://www.brown.edu/Facilities/John_Carter_Brown_Library/pages/right_resfellow2.html

Sponsor: John Carter Brown Library

Box 1894

Providence, RI 02912

John Carter Brown Library Short-Term Research Fellowships   3016

The library, an independently funded and administered center for advanced research in the humanities at Brown University, offers approximately 15 to 20 fellowships each year, extending from two to four months. These fellowships are open to foreign nationals as well as Americans, and to scholars engaged in pre- or postdoctoral, or independent research related to the resources of the library. These consist of printed materials (both European and American) related to the discovery, exploration, settlement, and development of North and South America before 1830. Preference may be given to applicants able to take up the fellowship during the course of the academic year, September through May. Application forms may be obtained from the program officer. The library also offers several long-term fellowships of five to nine months, with NEH support and specific eligibility requirements.

Requirements: Americans and foreign nationals who are engaged in pre- or postdoctoral or independent research are eligible. Graduate students must have passed their preliminary or general examinations at the time of application.

Restrictions: Recipients of all fellowships are expected to be in continuous residence at the library and to participate in the intellectual life of Brown University.

Amount of Grant: $1800 per month

Date(s) Application is Due: Jan 10

Contact: Director, (401) 863-2725; fax: (401) 863-3477; email: JCBL_Fellowships@Brown.edu

Internet: http://www.brown.edu/Facilities/John_Carter_Brown_Library/pages/right_resfellow2.html

Sponsor: John Carter Brown Library

Box 1894

Providence, RI 02912

John D. and Catherine T. MacArthur Foundation Conservation and   3017
Sustainable Development Grants

The foundation is concerned about the deteriorating condition of the biosphere that supports life on earth. This program area is dedicated to conserving biodiversity, to enhancing knowledge of how to use natural resources sustainably over the long term, and to promoting sustainable economic growth with social equity. The area's exclusive focus is on dealing with the problems of endangered tropical ecosystems, primarily moist forests and coral reefs. The program has three main goals: to diminish threats to biological diversity substantially and measurably in two hotspots and selected portions of the major wilderness area in each of the three tropical regions (Africa, Latin America-Insular Caribbean, Asia-Pacific) within 10 years; to test at least four specific conservation tools or methods within five years to help conservation practitioners to learn from the experience of others and to develop operational theories; and to increase the specific skills and knowledge of institutions and individuals in priority countries that will enable them to respond more effectively to threats to biodiversity.

Restrictions: Proposals dealing with pollution, climate change, ozone depletion, and urban sprawl will not be funded.

Samples: ANGAP (Antananarivo, Madagascar)—to develop training courses in the management of protected marine areas, $325,000 over three years (2005); BirdLife International (Cambridge, England)—to work with Aisty, a local nongovernmental organization, to help conserve the wetlands of southwestern Madagascar, $275,000 over three years (2005); Centre Ecologique di Libanona (Fort Dauphin, Madagascar)—to develop the university-level Environmental Management Training Program to meet the resource needs of conservation-development organizations and programs working in southern Madagascar, $50,000 over three years (2005); Princeton U (NJ) and Wildlife Conservation Society (New York, NY)—to develop a national information system that will provide conservation techniques and up-to-date data on biological diversity in Madagascar, $300,000 jointly over three years (2005).

Amount of Grant: $50,000-$2 million

Date(s) Application is Due: Jan 15; Mar 15; Aug 15

Contact: Michael Wright, Director, (312) 726-8000; fax: (312) 920-6258; TDD: (312) 920-6285; email: 4answers@macfound.org

Internet: http://www.macfound.org/programs/gss/csd/about_csd/index.htm

Sponsor: John D. and Catherine T. MacArthur Foundation

140 S Dearborn Street, Suite 1100

Chicago, IL 60603

John D. and Catherine T. MacArthur Foundation Global Challenges Grants   3018

The global challenges area focuses on emerging opportunities and threats in a period of rapid globalization—the development of complex political, social, and economic interconnections that result from the increased capacity for people, goods, capital, and information to move freely across national borders. Grants are made in the areas of human rights and economic governance. Grants are made throughout the year. Guidelines are available on the Web site.

Requirements: Grants are provided to nonprofit organizations, public agencies, universities, research institutes, and networks of such institutions. Only rarely are grants awarded for individual research projects, preference being given to multifaceted programs.

Restrictions: The foundation will not provide funding for publications and conferences, political activities, attempts to influence legislation, foundation development campaigns, or for programs that are the responsibility of governments.

Samples: Ahmadu Bello U, Ctr for Islamic Legal Studies (Nigeria)—for efforts to develop a standardized Islamic criminal-procedure code that would result in transparent procedures in sharia courts, designed to ensure that human rights are not violated, $110,000 (2005); CARE (Atlanta, GA)—for disaster-relief and reconstruction efforts related to the tsunamis in South Asia, and for relief efforts in Sudan's Darfur region and in the Democratic Republic of Congo, $1.5 million (2005); Brookings Institution, Governmental Studies Program (Washington, DC)—for research on intellectual-property rights and on poverty and labor migration in Central Asia and the Caucasus, $225,000 over three years (2004); Institute for International Economics (Washington, DC)—for policy analysis to assess the North American Free Trade Agreement, $250,000 (2004).

Amount of Grant: $50,000-$2 million

Contact: Office of Grants Management, Research, and Information, (312) 726-8000; fax: (312) 920-6258; TDD: (312) 920-6285; email: 4answers@macfound.org

Internet: http://www.macfound.org/programs/gss/GC.htm

Sponsor: John D. and Catherine T. MacArthur Foundation

140 S Dearborn Street, Suite 1100

Chicago, IL 60603

John D. and Catherine T. MacArthur Foundation Population and   3019
  Reproductive Health Program Grants

The program focuses on two areas: reducing maternal mortality and morbidity and advancing the sexual and reproductive health and rights of young people. Under maternal health, the foundation supports civil society organizations that work for changes needed in health, education, and legal policies and help to create the political will needed for their implementation. Special attention is given to initiatives that focus on increased budgets for maternal health, decreased deaths from abortion, and improved access to quality emergency obstetric care. Under reproductive health and rights, the Foundation supports initiatives that increase the availability of gender-sensitive sexuality education, confidential health services, and programs for young men, and that decrease the incidence of sexual violence. In addition, the foundation supports new ways of linking adolescent sexuality and reproductive health programs with other issues that are fundamental to the present and future well-being of the younger generation. Geographic priorities are India, Mexico, and Nigeria. There are no application deadlines.

Samples: Ctr for Reproductive Rights (New York, NY)—to develop legal strategies related to reproductive health in India, Mexico, and Nigeria, $150,000; K'inal Antsetik (Mexico City, Mexico)—for efforts to reduce maternal mortality and morbidity, $150,000; Socio Legal Information Ctr (Mumbai, India)—for efforts to discourage early marriage in India, $100,000 over two years (2004); Youth for Action (Hyderabad, India)—to develop a model for community-based financing of maternal-health services, $100,000 over three years.

Amount of Grant: $50,000-$2 million typically

Date(s) Application is Due: Feb 1

Contact: Office of Grants Management, Research, and Information, (312) 726-8000; fax: (312) 920-6258; email: poooulation@macfdn.org

Internet: http://www.macfound.org/programs/gss/PRH.htm

Sponsor: John D. and Catherine T. MacArthur Foundation

140 S Dearborn Street, Suite 1100

Chicago, IL 60603

John D. and Catherine T. MacArthur Foundation Program on Peace and   3020
  International Cooperation

The program addresses issues in four interrelated areas: U.S. foreign policy and national priorities; arms control, disarmament, and demobilization; international governance and civil society; and sustainable democracy. Grants to organizations and institutions fund public education projects, policy studies, and publications and media productions for specialized and general audiences. Grants also support the global exchange of ideas and contribute to the development of an international civil society by bringing together people with differing national, institutional, professional, and cultural perspectives on peace and security to participate in projects that address issues of common concern. Individuals may receive grants through three separate programs.

Requirements: The program provides support to nonprofit organizations, research institutes, and universities, as well as to individuals.

Samples: Landau Network-Centro Volta, Centro di Cultura Scientifica Alessandro Volta (Como, Italy)—for a project on nuclear-weapons policy in South Asia, $150,000 over three years; American Councils for International Education (Washington, DC)—to help establish the European Humanities U-International in Vilnius, Lithuania, after it was closed in August 2004 by the Belarus government, $900,000; Federation of American Scientists (Washington, DC)—for policy research designed to provide scientific expertise to policy makers on issues of critical national importance, $2.5 million.

Contact: Office of Grants Management, Research, and Information, (312) 726-8000; fax: (312) 920-6258; email: 4answers@macfdn.org

Internet: http://www.macfound.org/programs/gss/overview.htm

Sponsor: John D. and Catherine T. MacArthur Foundation

140 S Dearborn Street, Suite 1100

Chicago, IL 60603

John Dana Archbold Fellowships   3021

The fellowship program supports educational exchange between the United States and Norway. It is administered by the Nansen Fund in cooperation with the Norway-American Association in Oslo and the University of Oslo. Each fellowship provides for a year of graduate or postdoctoral study and research in Norway. The University of Oslo International Summer School offers Norwegian language courses and orientation for six weeks before the regular academic year begins. University of Oslo waives tuition costs; attendance is required. The maintenance stipend is sufficient to meet ordinary expenses of a single person. Travel allowance is provided to cover overseas airfare to and from Oslo. Application must be complete, including references and transcripts. Successful candidates will be notified before March 15. Fellowships are for one year starting in even-numbered years only.

Requirements: Eligibility is limited to individuals who are citizens and residents of the United States, 20 to 35 years old, in good health, and of good character. Applicants are judged competitively, with emphasis on proven ability in their field, seriousness of purpose, and a record of social adaptability. A high degree of proficiency in the Norwegian language is needed only in such fields as Norwegian literature and law; otherwise it is not necessary particularly for persons doing independent research. Undergraduate applicants must have a BA or BS degree before departure for Norway.

Amount of Grant: $5000

Date(s) Application is Due: Jan 15

Contact: Anne Brith Berge, (713) 526-6222; fax: (713) 521-9473; email: nacc@net1.net

Sponsor: Nansen Fund

77 Saddlebrook Ln

Houston, TX 77024

John David Marshall Library Studies Scholarship   3022

This scholarship is awarded annually to a graduate student at the master's level specializing in academic librarianship. The award will be made based on academic achievement and financial need.

Requirements: Applicant must be admitted to the FSU graduate school.

Amount of Grant: $500-$5000

Date(s) Application is Due: Apr 1

Contact: Delores Bryant, Program Assistant, (850) 644-5772; fax: (850) 644-9763; email: bryant@lis.fsu.edu

Internet: http://www.lis.fsu.edu/Prospects/Grads/ssd86_SIS_aid_scholarships.cfm

Sponsor: Florida State University

School of Information Studies, 101-B Louis Shores Bldg

Tallahassee, FL 32306-2100

John Deere Foundation Grants   3023

The foundation invests in programs in education, health/human services, community improvement, arts and culture. Types of support include annual campaigns, building construction/renovation, continuing support, fellowships, general operating support, scholarship funds, and seed money grants. Foundation interest also includes support for Third World development through U.S.-based nonprofits with international building funds, research grants, general operating purposes, and continuing support. There are no application deadlines.

Requirements: Nonprofit organizations in communities with major John Deere operating units, and employee presence are eligible. Eligible U.S. locations are: Augusta, GA; Quad City Region, IL; Des Moines, IA, Dubuque, IA, Iowa Quad Cities, IA, Ottumwa, IA, Waterloo, IA, Coffeyville, KS, Lenexa, KS, Thibodaux, LA, Springfield, MO, Cary, NC, Fuquay-Varina, NC, Fargo, ND, Greeneville, TN, Madison, WI, Horicon, WI. Exceptions include: accredited colleges and universities; organizations focused on international development initiatives related to John Deere Solutions for World Hunger initiative. Because John Deere dealerships are owned and operated independently, their communities are not included in this geographic scope. Also eligible to apply are organizations and institutions of national or international scope that reflect the foundation's concerns.

Restrictions: Funds are not available for the following organizations or purposes: individual initiatives, including scholarships; sports teams, racing teams, athletic endeavors or scholarships designated for athletes; faith-based organizations for sectarian purposes; political candidates, campaigns or organizations; private clubs, fraternities or sororities; other foundations for purposes of building endowment; tax-supported entities.

Geographic Focus: Georgia; Illinois; Iowa; Kansas; Louisiana; Missouri; North Carolina; Tennessee; WisconsinBrazil; Canada; Eastern Europe

Contact: Amy Nimmer Director; (309) 765-8000;

Internet: http://www.deere.com/en_U.S./globalcitizenship/socialinvestment/index.html

Sponsor: John Deere Foundation

1 John Deere Place

Moline, IL 61265-8098

John Dewey Doctoral Fellowships   3024

Applications are accepted from doctoral students interested in pursuing careers in higher education. An application form is available online. Annual deadline dates may vary; contact program staff for exact dates.

Date(s) Application is Due: Jan 15

Contact: Dr. Robert Ronau, (502) 852-0593; email: bob@louisville.edu

Internet: http://www.louisville.edu/edu/fainfo.htm#dewey

Sponsor: University of Louisville

Room 124

Louisville, KY 40292

John G. Duncan Trust Grants   3025

The trust awards grants in the following areas of interest: arts; education; health care; human services; religion; and support: building/renovation; capital campaigns; emergency funds; equipment; program development; seed money; and research. Contact the foundation, for the application guidelines and application form.

Requirements: IRS 501(c)3 organizations in Colorado are eligible.

Restrictions: No support for other grantmaking organizations, individuals, endowments, or for general operating expenses.

Geographic Focus: Colorado

Samples: Gospel Shelters for Women, Colorado Springs, CO, $3,250 (2006); Boulder Valley Womens Health Center, Boulder, CO, $2,000 (2006); Commerce City Community Health Services, Commerce City, CO, $5,000 (2006).

Amount of Grant: $1,000-$10,000 range

Date(s) Application is Due: Jan 31; Apr 30; Jul 31; Oct 31

Contact: Jason Craig, c/o Wells Fargo Bank, N.A.; (720) 947-6820

Sponsor: John G. Duncan Trust

1740 Broadway, MAC C7300-483

Denver, CO 80274-0001

John Glenn Scholars in Service-Learning Grants   3026

The program recognizes scholars whose scholarship efforts contribute to advancing the understanding of or adoption of service-learning, with specific emphasis on K-12 education. The John Glenn Institute will aggressively seek contributions from scholars whose work is addressing gaps in service-learning research as well as in insulated pockets not currently part of the service-learning mainstream. Areas of emphasis will change from year to year and will reflect emergent issues and priorities.

Amount of Grant: $1000

Contact: John Glenn Institute for Public Service and Public Policy, (612) 292-4545; fax: (612) 292-4868; email: JGISSL@osu.edu

Internet: http://www.glenninstitute.org/glenn/scholars_index.asp

Sponsor: Ohio State University

1947 College Road, 400 Stillman Hall

Columbus, OH 43210

John Grenzebach Awards for Outstanding Research in Philanthropy for Education 3027

The purpose of this award, cosponsored by the Council for Advancement and Support of Education (CASE) and the American Association of Fund-Raising Counsel Trust for Philanthropy, is to encourage and disseminate research on philanthropy for education. The sponsors presents awards are in two program areas: Outstanding Doctoral Dissertation Award and the Outstanding Published Scholarship Award. Each program carries cash awards and brings the award recipients international recognition. The deadline for the awards is the end of January each year. In addition to a stipend for the author, travel and lodging are provided for the author and the sponsoring faculty member to attend the CASE annual assembly for award presentation.

Amount of Grant: $2000 stipend for author; travel and lodging for author and sponsoring faculty member

Date(s) Application is Due: Jan 31

Contact: Freddie Cross, (202) 478-5570 or (202) 328-2273; email: cross@case.org

Internet: http://www.case.org/Content/AwardsScholarships/Display.cfm?CONTENTITEMID=3033

Sponsor: Council for Advancement and Support of Education

1307 New York Avenue, Suite 1000

Washington, DC 20005-4701

John Labatt Limited Scholarship 3028

This scholarship, endowed through a gift from John Labatt Limited with a matching grant provided annually from the Alberta Advanced Education Endowment Fund, is available to graduate students engaged in full-time study in business, management, or related areas. Duration of the award is four months. The deadline is February 1.

Requirements: The scholarship is open to qualified graduate students who are registered in or admissible to an MBA program at the University of Calgary. Awardee must be engaged in full-time study during tenure of the award.

Amount of Grant: $C3000

Date(s) Application is Due: Feb 1

Contact: Connie Busch, Faculty of Graduate Studies, (403) 220-5690; email: cbusch@ucalgary.ca

Internet: http://www.ucalgary.ca/pubs/calendar/current/What/Awards/03GradAwards.htm

Sponsor: University of Calgary

2500 University Dr NW, Earth Sciences Building, Room 720

Calgary, AB T2N 1N4 Canada

John M. Olin Fellowships in National Security 3029

Each year the John M. Olin Institute for Strategic Studies awards up to 10 pre- and postdoctoral fellowships to younger scholars. These fellowships are designed to promote basic research in the broad area of security and strategic affairs. Of particular interest is research into the causes and conduct of war, military strategy and history, defense policy and institutions, and the ways in which the United States and other societies can provide for their security in a dangerous world. Talented scholars in security affairs from the United States and elsewhere are eligible to apply. In awarding fellowships, preference is given to: graduate students who have made progress on their dissertations and are likely to complete them during their fellowship and recent PhD recipients.

Requirements: Scholars in security affairs from the United States and elsewhere are eligible to apply.

Amount of Grant: $35,000 maximum postdoctoral fellowship; $20,000 maximum predoctoral fellowship

Date(s) Application is Due: Jan 14

Contact: Ann Townes, Program Coordinator, (617) 496-5495; fax: (617) 384-5653; email: atownes@wcfia.harvard.edu

Internet: http://www.wcfia.harvard.edu/olin/fellowshipinformation/index.htm

Sponsor: Harvard University Weatherhead Center for International Affairs

1737 Cambridge Street, Coolidge Hall

Cambridge, MA 02138

John Merck Scholars Awards 3030

Scholars are chosen from the ranks of the most promising assistant professors currently working, or planning to work, in neurobiological and cognitive sciences relating to the biology of mental disability and developmental disabilities, including developmental studies of cognition, perception, language, reading, learning and motor performance. The foundation will accept one application for a neurobiologist and one for a cognitive scientist from major universities and other research centers. Scholars will receive $75,000 per year for a four-year period, subject to an annual review of research progress. The sponsoring institution is responsible for fiscal management.

Requirements: The foundation will fund the most promising young researchers whose work illuminates neurodevelopmental disorders from the perspectives of (i) synapse formation and synaptic plasticity; (ii) learning and memory, and synaptic plasticity; (iii)

perception, cognition and behavior; (iv) neurogenesis and pattern formation; and (v) genetics and early development. Also encouraged are proposals that (i) investigate the possible role of environmental chemicals in the origins of developmental disabilities, or that (ii) aim to distinguish subgroups within accepted diagnostic categories through the use of sophisticated behavioral and neuroimaging tests of perception, cognition, and emotions based on concepts from modern cognitive neuroscience. In all cases, they seek proposals from young scientists conducting research that is of the highest quality and that has the greatest chance of increasing understanding of neurodevelopmental disorders. Applicants must have the following: Academic rank in a university or medical school, or equivalent standing in a research institute or medical center; A record of research in areas relating to the foundation's interest in the underlying causes of developmental disabilities; Not more than four years of experience in an independent faculty position; Evidence of a commitment to a career in neuroscience or cognitive science. Request application forms by mail, email or phone.

Restrictions: Applicants may not: hold tenured positions or their equivalents; apply in more than two rounds of competition; apply for continued postdoctoral support. Holding other fellowships concurrently with the John Merck Scholars Award is discouraged. Prior approval by the foundation for an overlapping fellowship is required and will be given only in unusual circumstances.

Amount of Grant: $75,000 per year for a four-year period

Date(s) Application is Due: Sep 15

Contact: Jason Bentsman, Staff Assistant; (617) 556-4120; fax: (617) 556-4130; email: j bentsman@jmfund.org

Internet: http://www.jmfund.org/jm_scholars_program.html

Sponsor: John Merck Fund

2 Oliver Street, 8th Floor

Boston, MA 02108

John O. Galloway Memorial Scholarship in Geology and Geophysics 3031

One scholarship is awarded annually for one year of graduate study in the Department of Geology and Geophysics at the University of Calgary. Students interested in the award should apply first to the department.

Requirements: Candidates must be graduate students enrolled in the master's program in the Department of Geology and Geophysics who have completed a minimum of two academic terms of full-time graduate work prior to the commencement of the award.

Amount of Grant: $C5000 per year

Date(s) Application is Due: Feb 1

Contact: Connie Baines, Faculty of Graduate Studies, (403) 220-5690; email: cbaines@ucalgary.ca

Internet: http://www.grad.ucalgary.ca/funding/internal_scholarships/lev_4/galloway.htm

Sponsor: University of Calgary

2500 University Dr NW, Earth Sciences Building, Room 720

Calgary, AB T2N 1N4 Canada

John P. McGovern Foundation Grants 3032

The John P. McGovern Foundation, established in 1961, supports the charitable interests of the donor to support the activities of established nonprofit organizations, which are of importance to human welfare with special focus on children and family health education and promotion, treatment and disease prevention. The Foundation gives primarily in Texas with an emphasis on the Houston area, but also provides funding in the Southwest region as well. The types of support offered are: building/renovation; conferences/seminars; continuing support; curriculum development; emergency funds; endowments; general/operating support; matching/challenge support; professorships; publication; research; and scholarship funds.

Requirements: Non-profits in Texas are eligible to apply.

Geographic Focus: Texas

Samples: University of Texas Health Science Center, Houston, TX. $1,500,000 - for John P. McGovern, MD, Center for Health, Humanities, and Human Spirit/Certificate Program Endowment Fund (2007); Community Family Centers, Houston, TX. $200,000- for John P. McGovern Community Sports and Recreation Building (2007); Salvation Army of Houston, Houston, TX. $150,000 - for general operating fund (2007).

Amount of Grant: $10,000 - $100,000 range

Date(s) Application is Due: Aug 31

Contact: Kathrine G. McGovern, President; (713) 661-4808; fax: (713) 661-3031

Sponsor: John P. McGovern Foundation

2211 Norfolk Street, Suite 900

Houston, TX 77098-4044

John R. Oishei Foundation Grants 3033

The foundation's primary mission is to support medical research and care and education, as well as cultural and social needs existing in the Buffalo Niagara region. The foundation favors creative programs that attempt to advance from the status quo, are strategically sound, and are strongly focused on excellence. Programs that provide opportunities for foundation support to be leveraged into greater support from other sources will be especially favored. The foundation generally does not fund operating expenses, though occasional exceptions may be made for organizations that address basic human needs. Requests for capital funds will not be considered unless they are an integral part of an otherwise eligible proposal. Contact the foundation to discuss eligibility before submitting an application. Grant applications are accepted throughout the year.

Requirements: It is the general policy of the foundation to confine its support to activities located in the Buffalo, NY, metropolitan region.

Restrictions: Grants do not support endowments; capital requests (buildings or equipment); deficit funding or loans; individual scholarships or fellowships (except within specific foundation programs); travel, conferences, seminars, or workshops; fundraising events; or 509(a) private foundations.
Geographic Focus: New York
Amount of Grant: $13 million total; $100,000-$200,000 average
Contact: Blythe T. Merrill Senior Program Officer; (716) 856-9490, ext. 3; fax: (716) 856-9493; email: btmerrill@oisheifdt.org
Internet: http://www.oisheifdt.org/Home/Fund/WhatWeFundOverview
Sponsor: John R. Oishei Foundation
1 HSBC Center, Suite 3650
Buffalo, NY 14203-2805

John Randolph Haynes and Dora Haynes Foundation Grants   3034

The foundation supports study and research in the social sciences with specific application to the Los Angeles area. Support has been given for projects concerned with education, transportation, local government, elections, public safety, demographics, public personal services, and natural resources. The foundation also provides, in the social sciences, funds to selected colleges and universities in the greater Los Angeles area for undergraduate scholarships, graduate fellowships, and fellowships for faculty members. The recipient institutions select the recipients and make the awards; applications for scholarships and graduate fellowships should be made to the institutions and not to the foundation.
Requirements: Grants are made only to colleges and universities or other qualified nonprofit institutions in the greater Los Angeles area. Organizations wishing to apply should first write for guidelines.
Restrictions: Grants are not made to individuals or for equipment purchases or institutional overhead.
Geographic Focus: California
Samples: California State U, Oviatt Library (Northridge, CA)—to preserve and make accessible a portion of the historical records from the library's United Way of Los Angeles Collection, $21,500 (2003).
Contact: Diane Cornwell, Administrative Director, (213) 623-9151; fax: (213) 623-3951; email: info@haynesfoundation.org
Internet: http://www.haynesfoundation.org
Sponsor: John Randolph Haynes and Dora Haynes Foundation
888 W Sixth Street, Suite 1150
Los Angeles, CA 90017-2737

John S. Dunn Research Foundation Grants and Chairs   3035

Established in 1985, the foundation funding area is limited to the Texas, providing support to health and medical-related organizations, especially hospitals; support also for cancer, other medical research, and freestanding clinics.
Requirements: Grants are made to nonprofit organizations within Texas having 501(c)3 tax-exempt status.
Restrictions: No grants to individuals, or for multi-year or seed money grants.
Geographic Focus: Texas
Samples: Gulf Coast Consortia, Houston, TX, $1,180,000 - to support drug discovery program (2007); University of Texas M.D. Anderson Cancer Center, Houston, TX, $650,000 - to fund radiological sciences and metabolic lab (2007); Memorial Hermann Healthcare System, Houston, TX, $250,000 - to support life flight services to patients without insurance (2007).
Amount of Grant: $10,000-$100,000 average
Contact: Dr. Lloyd Gregory, Executive Vice President; (713) 626-0368; fax: (713) 626-3866
Sponsor: John S. Dunn Research Foundation
3355 West Alabama street, Suite 720
Houston, TX 77098-1718

John S. Knight Fellowships   3036

The fellowships are offered each year to 12 U.S. journalists who have demonstrated uncommon excellence in their work and who have the potential of reaching the top ranks in their specialties. Fellowships are awarded for one academic year. Completed U.S. applications must be postmarked no later than February 1; international applications must be postmarked no later than December 15. Contact the office for an application.
Requirements: Professional journalists with at least seven full years of experience are eligible.
Amount of Grant: $55,000 stipend plus book allowance
Date(s) Application is Due: Dec 15; Feb 1
Contact: John S. Knight Fellowships, (650) 723-4937; fax: (650) 725-6154; email: knight-info@lists.stanford.edu
Internet: http://knight.stanford.edu
Sponsor: Stanford University
Building 120, Room 424
Stanford, CA 94305-2050

John Simon Guggenheim Memorial Foundation Fellowships   3037

The foundation offers fellowships for advanced professionals in all fields (natural sciences, social sciences, humanities, creative arts) except the performing arts. There are two competitions: one for citizens and permanent residents of the United States and Canada (deadline October 1); and the other for citizens and permanent residents of Latin America and the Caribbean (deadline December 1). Appointments are ordinarily made for one year and in no instance for a period shorter than six consecutive months. Fellows are chosen by the selection committee on the basis of unusually distinguished

achievement in the past and exceptional promise for future accomplishment. Interested persons should write the foundation to request application forms.
Requirements: Fellowships are awarded to individuals who have demonstrated exceptional capacity for productive scholarship or exceptional creative ability in the arts.
Restrictions: Guggenheim Fellowships are not scholarships, and they are not available to those who seek to complete their training either as undergraduate, graduate, or part-time students. They are also not intended for immediate postgraduate work.
Samples: Divided among 36 artists, scholars, and scientists from Latin America and the Caribbean—for fellowships, $1.118 million (2005). Fellowships for 186 U.S. and Canadian artists, scholars, and scientists, $7,112 million (2005).
Amount of Grant: $37,362 average
Date(s) Application is Due: Oct 1; Dec 1
Contact: Associate Secretary, (212) 687-4470; fax: (212) 697-3248; email: fellowships@gf.org
Internet: http://www.gf.org
Sponsor: John Simon Guggenheim Memorial Foundation
90 Park Avenue
New York, NY 10016

John Stauffer Charitable Trust Grants   3038

The trust awards grants to California higher education institutions and hospitals. Types of support include building construction/renovation, challenge/matching funds, endowments, equipment acquisition, fellowships, professorships, and scholarship funds. There are no application forms or deadlines.
Requirements: California nonprofits are eligible.
Geographic Focus: California
Samples: California Institute of Technology, Pasadena, CA, $500,000 - for construction of research lab in new chemistry building (2008); Loma Linda University, Loma Linda, CA, $250,000 - for classroom for emergency personnel (2008); Occidental College, Los Angeles, CA, $159,320 - for endowment for summer research (2008).
Amount of Grant: $100,000-$500,000 average
Contact: H. Jess Senecal, Treasurer: (626) 793-9400; fax: (626) 793-5900; email: jess@lagerlof.com
Sponsor: John Stauffer Charitable Trust
301 N Lake Avenue, 10th Floor
Pasadena, CA 91101-4108

John Templeton Foundation Templeton Prize   3039

The annual prize is awarded by an international, multifaith panel of judges to a living person of any religious tradition who has made a unique contribution to progress in research or discoveries about spiritual realities. It is awarded for originality in advancing ideas and institutions that deepen an understanding of spiritual realities. Nominated contributions should be innovative, primarily spiritual rather than primarily humanitarian, and result in an appreciable acceleration in spiritual discoveries. The prize seeks to focus attention on the wide variety of endeavors toward discoveries of spiritual realities research. The program does not seek a unity of denominations nor a unity of world religions; but rather it seeks to encourage understanding of the benefits of diversity.
Requirements: Nominations may be submitted by anyone. Leaders of theological and religious institutions and those engaged in innovative and creative research are especially invited to submit nominations.
Amount of Grant: 700,000 pounds sterling (approximately $1 million)
Date(s) Application is Due: Jul 1
Contact: Templeton Prize, (602) 589-2541; fax: (602) 589-2897; email: info@templetonprize.org
Internet: http://www.templetonprize.org/nom_form_info.html
Sponsor: Canyon Institute for Advanced Studies
3300 W Camelback Road
Phoenix, AZ 85017

John W. Alden Trust Grants   3040

The trust awards grants to Massachusetts nonprofit organizations in its areas of interest, including education and therapy for children who are blind, disabled, retarded, or mentally or physically ill. Grants will normally be made for specific projects rather than general operating purposes. There are no application deadlines. Grant applications should be submitted on-line at www.cybergrants.com/alden. Detailed instructions on the web page will guide you through the process. All questions should be directed to the grants coordinator.
Requirements: Eastern Massachusetts 501(c)3 tax-exempt organizations are eligible.
Restrictions: Grants are not made to individuals.
Geographic Focus: Massachusetts
Samples: Hospice of the North Shore, Inc., $6,000 - Music Therapy Program (2007); - University of Massachusetts Foundation Inc., $12,000 - NERCVE Assistive Technology Center (2007); Franciscan Hospital for Children, $13,000 - aquatic aerobic conditioning program (2007).
Amount of Grant: $15,000 maximum
Date(s) Application is Due: Jan 15; Apr 15; Jul 15; Oct 15
Contact: Susan Monahan, Grants Coordinator; (617) 951-1108; fax: (617) 542-7437; email: smonahan@rackemann.com
Internet: http://www.cybergrants.com/alden
Sponsor: John W. Alden Trust
160 Federal Street, 13th Floor
Boston, MA 02110-1700

Johnson & Johnson Corporate Contributions Grants   3041

Johnson & Johnson and its many operating companies support community-based programs that improve health and well-being. The Company works with community-based partners that have the greatest insight into the needs of local populations and the strategies that stand the greatest chances of success. Giving focuses on: saving and improving the lives of women and children; building on the skills of people who serve community health needs, primarily through education; and preventing diseases and reducing stigma and disability in underserved communities where we have a high potential for impact.

Requirements: Grants are awarded to nonprofit and tax-exempt local, national, and international organizations and institutions.

Restrictions: Grants are not awarded to individuals, for deficit funding, capital or endowment funds, demonstration projects, or publications.

Contact: Michael J. Bzdak, Director of Corporate Contributions; (732) 524-3061; fax: (732) 524-3300; Shaun Mickus; (732) 524-2086; email: smickus@corus.jnj.com

Internet: http://www.jnj.com/connect/caring/corporate-giving/

Sponsor: Johnson & Johnson

1 Johnson & Johnson Plaza

New Brunswick, NJ 08933-0001

Johnson F. Hammond Memorial Scholarship   3042

The annual scholarship will be awarded to a medical student of high moral character and outstanding academic achievement with a demonstrated interest and involvement in medical journalism. The scholarship recipient will be selected from among nominees proposed by the deans of AMA approved medical schools, with financial need not being a consideration. Each school may propose one student.

Amount of Grant: $3000

Date(s) Application is Due: Feb 13

Contact: Dina Lindenberg, Program Officer, (312) 464-4193; fax: (312) 464-5973; email: dina.lindenberg@ama-assn.org or amafoundation@ama-assn.org

Internet: http://www.ama-assn.org/ama/pub/category/14772.html

Sponsor: American Medical Association Foundation

515 N State Street

Chicago, IL 60610

Joint Center for Poverty Research Grants   3043

The center seeks research proposals on the causes and consequences of poverty. Research priorities include the changing labor market, substance abuse and teen pregnancy, concentrated urban poverty, and the effects of recent policy changes on immigrants. Grants also focus on the disabled, the well-being of children, and urban populations.

Requirements: Preference is given to researchers without tenures but who have full-time academic employment.

Restrictions: Applicants are not required to have a PhD.

Samples: Julia Henly, University of Chicago, School of Social Service Administration (Chicago, IL)—for the Study on Work-Child Care Fit (2003).

Amount of Grant: $30,000 maximum

Date(s) Application is Due: May 1

Contact: Joint Center for Poverty Research, (847) 491-3395; fax: (847) 491-9916; email: ipr@northwestern.edu

Internet: http://www.jcpr.org/researchfunding.html

Sponsor: University of Chicago

2040 Sheridan Road

Evanston, IL 60208

Joint Center for Poverty Research Visiting Scholars Grants   3044

The Northwestern University/University of Chicago Joint Center for Poverty Research sponsors this program of visiting scholars in residence. Each of the two campuses will host one visiting scholar. Applicants must be actively involved in research directly relevant to poverty in the United States. Preference will be given to scholars who can spend an entire academic year in residence and who can particularly benefit from interaction with the affiliated faculty in one or both of the two universities involved. Candidates are invited to express a preference for one of the two universities. The award includes a salary for the academic year, plus benefits and some research support. There are no teaching requirements. Applicants with partial funding from other sources are welcome.

Requirements: The program is intended for untenured or junior researchers who have completed their PhDs.

Restrictions: PhD graduates of Northwestern University or of University of Chicago are not eligible.

Amount of Grant: $50,000 maximum salary

Contact: Joint Center for Poverty Research, (847) 491-3395;' fax: (847) 491-9916; email: ipr@northwestern.edu

Internet: http://www.jcpr.org/researchfunding.html

Sponsor: Northwestern University

2046 Sheridan Road

Evantson, IL 60208-4105

Jose Marti Scholarship Challenge Grants   3045

The Jose Mart' Scholarship Challenge Grant Fund is a need-based merit scholarship that provides financial assistance to eligible Hispanic students who will attend Florida public or private institutions. Applicants for undergraduate study must apply during their senior year of high school. Graduate students may apply, but priority for the scholarships is given to graduating high school seniors. The number of new awards made each year

is subject to contributions from private sources and funds appropriated by the Florida Legislature.

Requirements: Applicant must be a U.S. citizen or eligible noncitizen; meet residency requirements for receipt of Florida student aid; have been born in, or have a natural parent who was born in, Mexico, Spain, South America, Central America, or the Caribbean; have a 3.0 cumulative grade point average; apply during the applicant's senior year in high school or for graduate study; and demonstrate financial need as determined by a nationally recognized need-analysis service. To receive an award, an eligible applicant must be enrolled as a full-time undergraduate or graduate student at an eligible Florida postsecondary school. Matching funds are required.

Geographic Focus: Florida

Amount of Grant: $2000 per academic year

Date(s) Application is Due: May 15

Contact: Office of Student Financial Assistance, (850) 487-0049 or (888) 827-2004; fax: (850) 488-5966

Internet: http://www.firn.edu/doe/bin00065/jmfactsheet.htm

Sponsor: Florida Department of Education

1940 N Monroe Street, Suite 70

Tallahassee, FL 32303-4759

Joseph C. Wey Fellowship in Medieval Studies   3046

The Basilian Fathers of Toronto fund the fellowship for study at the Pontifical Institute of Mediaeval Studies during the forthcoming academic year. As the Basilians have for many years been engaged in teaching and pastoral activity in Mexico and Columbia, a special fellowship named for the former Superior General, Joseph C. Wey, CSB is intended to extend their mission to graduate study in some aspect of the Middle Ages. Medievalists in Latin America are therefore encouraged to apply for this new fellowship, as are scholars from outside of North America whose intentions are to return to their original country in pursuance of their teaching careers. The fellow will be given an office and full access to the faculty and the library of the institute, and will also have access to the libraries and faculty of the University of Toronto. The holder must make a firm commitment to return to his or her country of origin to pursue teaching there.

Requirements: Any medievalist, including those presently holding academic positions, may apply. All candidates must have at least one postsecondary qualification and show a strong commitment to teaching. A doctorate is highly desirable, as is entry into the institute's programme for the Licence. Candidates will be expected to come prepared with the languages needed for their research. English, in which all candidates should be proficient, will be the language of instruction and discussion.

Amount of Grant: $35,000

Date(s) Application is Due: May 1

Contact: Jeanette Jardine, Secretary to the President, (416) 926-7142; fax: (416) 926-7292; email: jeanettejardine@utoronto.ca

Internet: http://www.pims.ca

Sponsor: Pontifical Institute of Mediaeval Studies

59 Queen's Park Crescent E

Toronto, Ontario M5S 2C4 Canada

Joseph Klingenstein Fellows Program   3047

The Klingenstein Center for Independent School Education at the university will make eight to 12 fellowships for K-12 independent school educators who are looking to strengthen their careers as teachers, deans, or department or division heads. Fellows study emerging educational issues and problems; work on leadership skills; and explore issues in innovative teaching, leadership, and social and educational policy. Fellows receive tuition, stipends, and housing allowances to participate in a year-long academic program or a one 15-week semester. Applications are available online and from the office.

Requirements: Eligible applicants are K-12 teachers or administrators who have at least five years experience in a private, independent school with a nondiscriminatory admissions policy. Fellows must return to their school following the fellowship period.

Date(s) Application is Due: Jan 15

Contact: Sally Mixsell, Project Director, (212) 678-3156; fax: (212) 678-3254; email: mixsell@tc.columbia.edu

Internet: http://www.klingenstein.org

Sponsor: Columbia University Teachers College

204 Main Hall, Box 125, 525 W 120th Street

New York, NY 10027-6696

Joseph L. Fisher Dissertation Awards   3048

Resources for the Future will award several fellowships for the coming year for support of doctoral dissertation research in economics on issues related to the environment, natural resources, or energy. It is expected that fellowship recipients will engage in no employment during the period of fellowship tenure and that RFF will be notified immediately of financial assistance made available from any other source for support of doctoral work. Application materials and additional information may be obtained from the office. Applications sent by fax or email will not be accepted.

Requirements: This fellowship is intended to support graduate students in the final year of their dissertation research. Candidates must have completed the preliminary examinations for the doctorate not later than February 1. Awards will be announced in May. Women and members of minority groups are especially encouraged to apply.

Amount of Grant: $12,000 for academic year

Date(s) Application is Due: Feb 28

Contact: Coordinator for Academic Programs, (202) 328-5155; fax: (202) 328-5155; email: fisher-award@rff.org

Internet: http://www.rff.org/rff/About/Fellowships_and_Internships/Fisher/Joseph-L-Fisher-Doctoral-Dissertation-Fellowships.cfm
Sponsor: Resources for the Future
1616 P Street NW
Washington, DC 20036-1400

Joseph R. Dunlap Memorial Fellowship 3049

This program supports scholarly, creative, and translation projects about William Morris and his designs, writings, and other work. Up to $1000 per year is granted to individuals (there can be multiple, partial awards) for research and other expenses, including travel to conferences. Projects may deal with any subject—biographical, literary, historical, social, artistic, political, typographical—relating to Morris, and may be scholarly or creative in nature. Applications are encouraged from younger members of the society and from beginning researchers. Applicant is asked to submit a resume and one-page proposal. Two letters of recommendation should be sent under separate cover.
Requirements: U.S. citizens or permanent residents are eligible.
Amount of Grant: $1000 maximum
Date(s) Application is Due: Dec 15
Contact: Program Contact, (804) 788-2963; email: us@morrissociety.org
Internet: http://www.morrissociety.org/fellowships.html
Sponsor: William Morris Society in the P.O. Box 53263
Washington, DC 20009

Josiah Macy Jr. Foundation Grants 3050

The foundation awards grants to improve the education of physicians and other health care professionals. Grant making focuses on improving medical and health professionals' education in the context of the changing health care system, increasing the number of minority physicians, boosting teamwork between health care professionals, and using educational strategies to increase care for under-served populations. Current initiatives cover the fields of pediatric residency, inter-professional curriculum, improved clinical training, increasing diversity among health care professions, Macy Scholars program, women's health in medical curricula, women in medicine, and funding for a New York post-baccalaureate program. Applications may be submitted at any time. A preliminary letter of inquiry is useful in assisting the foundation staff in determining whether submission of a full proposal is appropriate.
Requirements: Grants are made to colleges, universities, and other 501(c)3 nonprofit organizations for activities within established program areas. Grant proposals, addressed to the president, should include the name of the sponsoring agency or institution; description of the project; names and qualifications of the person who will be responsible for the project; expected cost and duration of the project including an itemized budget; documents substantiating the tax-exempt status of the sponsoring institution; and a letter of endorsement from the sponsoring institution.
Restrictions: Grants are not awarded to individuals or for general undesignated support or for construction/renovation projects.
Amount of Grant: $5 million per year total
Contact: George E. Thibault, President; (212) 486-2424; fax: (212) 644-0765; email: jmacyinfo@josiahmacy
Internet: http://www.josiahmacyfoundation.org/index.php?section=grant_guidelines
Sponsor: Josiah Macy Jr. Foundation
44 East 64th Street
New York, NY 10021

Josiah W. and Bessie H. Kline Foundation Grants 3051

The foundation was established for charitable, scientific, literary, and educational purposes. The objectives for its grantmaking activities are to aid blind or incapacitated persons or crippled children. Grants are made to Pennsylvania colleges and universities and to hospitals, to institutions for crippled children, or to any other benevolent or charitable institution. Generally, grants are made to such institutions that are located in south-central Pennsylvania. Grants also are made for scientific and medical research. Types of support include continuing support, annual campaigns, capital campaigns, equipment acquisition, building/renovation, curriculum development, scholarship funds, matching funds, research, and emergency funds. The foundation does not have an application form. Instead, a letter of request must be submitted with the following information: a description of the need and purpose of the applicant organization, a project budget, amount requested from the foundation and the dates of the need, and a copy of the 501(c)3 tax-exemption letter. Requests may be submitted at any time. The board generally meets in March and November to consider requests.
Requirements: 501(c)3 organizations serving south-central Pennsylvania communities are eligible.
Restrictions: The foundation does not make loans and does not make grants to individuals or to normal operational phases of established programs, endowments, campaigns or national organizations, or religious programs.
Geographic Focus: Pennsylvania
Samples: Susquehanna University, Selinsgrove, PA, $125,000 (2007); Central Pennsylvania Blood Bank, Hummelstown, PA, $38,000 (2007); Gettysburg Foundation, Gettysburg, PA, $50,000 (2007).
Amount of Grant: $500-$150,000 range
Contact: John Obrock, Grant Administrator; (717) 561-0820 or (717) 561-4373; fax: (717) 561-0826
Sponsor: Josiah W. and Bessie H. Kline Foundation
515 S 29th Street
Harrisburg, PA 17104

Journalism Foundation of Metropolitan Saint Louis Scholarships 3052

Each year a number of scholarships are offered through the foundation for study leading to a journalism degree or graduate study in the field. Selection of scholarship winners will be based on aptitude, interest in a journalism or communications career, academic standing, and financial need. Winners will be honored at the annual reception, held in the spring of each year. Application forms are available from college English departments, schools of journalism/mass communications, campus newspaper editors, financial aid officers, or the foundation.
Requirements: Applicants must be residents of metropolitan Saint Louis (City of Saint Louis; Franklin, Jefferson, Saint Charles, Lincoln, Warren, and Saint Louis Counties in Missouri; Bond, Clinton, Jersey, Madison, Monroe, and Saint Clair Counties in Illinois). Students must enroll in classes in journalism, mass communications, public relations, advertising, or writing. Undergraduate students must have completed the freshman year; graduate students must carry at least six hours in the listed curricula.
Geographic Focus: Illinois; Missouri
Amount of Grant: $1000-$3000 approximately
Date(s) Application is Due: Mar 25
Contact: Joseph Kenny, c/o Saint Louis Review, (314) 792-7508
Sponsor: Journalism Foundation of Metropolitan Saint Louis
20 Archbishop May Drive
Saint Louis, MO 63119-5004

Jovid Foundation Grants 3053

The foundation partners with nonprofits in Washington, DC, to help low-income residents of the district learn how to become self-sufficient. Particularly interest is given to funding neighborhood-based efforts that provide programs and services to DC adults. The foundation has a modest budget for the support of the arts. Mini grants are awarded to help increase capacity for small groups through board training, staff development, membership dues, conferences, or Web site development. Potential applicants should initiate contact with the foundation by writing a one- to two-page letter describing the proposed project. Submit a two-page letter of inquiry; full proposals are by request.
Requirements: Washington, DC, nonprofits are eligible.
Geographic Focus: District of Columbia
Samples: Academy of Hope (Washington, DC)—to support the External Diploma Program for low-income DC residents, $10,000 (2004).
Amount of Grant: $2500-$15,000 average; $1000 maximum mini grants
Date(s) Application is Due: Jan 10; Apr 4; Jul 11; Oct 3
Contact: Bob Wittig, Executive Director, (202) 686-2616; fax: (202) 686-2621; email: jovidfoundation@yahoo.com
Internet: http://fdncenter.org/grantmaker/jovid
Sponsor: Jovid Foundation
5335 Wisconsin Avenue NW, Suite 440
Washington, DC 20015-2003

Joy Family Foundation Grants 3054

The foundation awards grants to nonprofit organizations and individuals in the city of Buffalo, NY, and the counties of Erie and Niagara, NY, to improve the quality of life. Grants are awarded in support of human and family services, education and literacy, AIDS research, substance abuse, alcoholism, the aging, the economically disadvantaged, hospitals, women and children, and Roman Catholic churches and organizations. Grants will be awarded for programs and projects, general operating support, capital campaigns, building/renovation, technical assistance, and consulting services. The board meets quarterly.
Requirements: New York nonprofit organizations are eligible.
Geographic Focus: New York
Samples: Burchfield-Penny Art Center (Buffalo, NY)—for capital campaign, $25,000.; Christ the King Seminary (East Aurora, NY)—grant recipient, $10,000.; Computers for Children (Buffalo, NY)—for computer refurbishment, $10,000.
Amount of Grant: $100-$100,000 range
Contact: Marsha Sullivan, Executive Director, (716) 633-6600; fax: (716) 633-0600; email: info@joyfamilyfoundation.org
Internet: http://www.joyfamilyfoundation.org
Sponsor: Joy Family Foundation
5436 Main Street, Suite 1
Williamsville, NY 14221

Joyce Foundation Grants 3055

The foundation's program areas are education, employment, environment, gun violence, money and politics, and culture. Grantmaking is focused on initiatives that promise to have an impact on the Great Lakes region, specifically the states of Illinois, Indiana, Michigan, Minnesota, Ohio, and Wisconsin. A limited number of environment grants are made to organizations in Canada. Education grantmaking focuses on Chicago, Cleveland, and Milwaukee. Culture grants are primarily focused on the Chicago metropolitan area. The foundation is committed to improving public policy through its grant program. Public policy grants provide general operating support or fund educational advocacy such as nonpartisan research, technical assistance, or examinations of broad social issues. Grant applicants are encouraged to describe the nature of advocacy activities in their grant applications and reports, so the foundation can ensure that it is in compliance with federal tax laws. Guidelines are available online.
Requirements: Applications should be made only by organizations in the Great Lakes region that can demonstrate their tax-exempt status and that can supply complete financial information and permit the results of a grant to be audited.

Restrictions: The foundation generally does not support capital projects, endowment campaigns, religious activities, direct services, or scholarships.

Geographic Focus: Illinois; Indiana; Iowa; Michigan; Minnesota; Ohio; Wisconsin; and Canada

Samples: Environmental Defense (New York, NY)—to implement federal farm-conservation programs in Midwestern states in an effort to protect Great Lakes water resources, $500,000 over two years (2005); Chicago Metropolis 2020 (Chicago, IL)—for efforts to revise the Illinois criminal code, $150,000 over two years (2005); U of Michigan at Ann Arbor, Great Lakes Radio Consortium (MI)—to cover water-related stories in the region, $200,000 over two years (2005); National-Louis U, Ctr for Early Childhood Leadership (IL)—to educate policy makers and others about the importance of highly qualified directors of early-childhood programs in achieving universal preschool, $200,000 over two years (2005).

Amount of Grant: $20,000-$100,000 average

Date(s) Application is Due: Apr 15; Aug 16; Dec 12

Contact: Veronica Salter, Grants Manager, (312) 782-2464; fax: (312) 782-4160; email: info@joycefdn.org

Internet: http://www.joycefdn.org/seekingagrant/seekingmain-fs.html

Sponsor: Joyce Foundation
70 W Madison, Suite 2750
Chicago, IL 60602

JSPS Bilateral Exchange Programs with Foreign Academic Institutions   3056

JSPS administers programs for scientific cooperation and exchange through memoranda of understanding with 75 foreign academic institutions in 42 countries. Cooperation primarily involves exchange visits by senior and junior scientists. The purpose of the program is to develop scientific understanding between Japan and the cooperating countries. Types of support include research, fellowships, and international exchange programs. Awards cover travel expenses, maintenance, and other necessary expenses. Contact the program office for more information regarding the JSPS Bilateral Exchange Programs with Foreign Academic Institutions and the JSPS Cooperation Programs with Southeast Asian Countries.

Requirements: Eligibility depends on the memoranda of understanding between the cooperating countries and the JSPS.

Contact: Program Administrator, Research Cooperation Division, Asian Program Division, 81-03- 3263-2365; fax: 81-03-3263-1673

Internet: http://www.jsps.go.jp/english/e-bilat/main.html

Sponsor: Japan Society for the Promotion of Science
Nihon Gakujutsu Shinko-kai, 5-3-1, Kojimachi, Chiyoda-ku
Tokyo 102-8471 Japan

JSPS Fellowships for Foreign Scientists   3057

JSPS offers these fellowships to facilitate cooperative research between visiting foreign scientists and hosting Japanese scientists. There are two types of fellowships: short-term, granted to eminent scientists, such as Nobel laureates, and other outstanding senior scientists to visit Japan; and long-term, granted to foreign researchers who wish to conduct research in cooperation with, or under the supervision of, scientists at Japanese universities or research institutions. Awards for short-term fellowships include round-trip airfare, a living allowance of @Y18,000 per day, and an additional allowance for travel within Japan. The long-term fellowship includes round-trip airfare, a monthly stipend of @Y240,000-@Y300,000, and additional allowances for housing and travel within Japan.

Requirements: Applications must be submitted by the inviting scientist through the head of the institution concerned. Candidates eligible for short-term fellowships include eminent or senior scientists. Candidates for long-term fellowships are those who want to conduct joint research with Japanese scientists.

Contact: Fellowship Programs, (81-03) 263-1721; fax: (81-03) 222-1986 or 237-8238

Internet: http://www.jsps.go.jp/english/e-fellow/fellow.html

Sponsor: Japan Society for the Promotion of Science
6 Ichibancho, Chiyoda-ku
Tokyo 102-8471 Japan

JSPS International Joint Research Projects and Scientific Seminars   3058

JSPS provides support for Japanese scientists to undertake cooperative research projects with foreign scientists. This program funds joint international ventures in scientific research with Japanese scientists in a wide variety of disciplines.

Requirements: Japanese scientists engaged in joint research projects and seminars with foreign scientists are eligible to apply.

Contact: Program Administrator, Research Cooperation Division, Asian Program Division, 81-03-3263-1724; fax: 81-03-3263-1673

Internet: http://www.jsps.go.jp/english/site_map.html

Sponsor: Japan Society for the Promotion of Science
Sumitomo-Ichibancho Building, 6 Ichibancho, Chiyoda-ku
Tokyo 102-8471 Japan

JSPS Postdoctoral Fellowship for Foreign Researchers   3059

To facilitate cooperative research between young foreign researchers and hosting Japanese researchers, fellowships are awarded to enable postdoctoral researchers to undertake research at Japanese universities and institutions. The program also supports short-term fellowships of 15 days to 11 months. The fellowship includes round-trip airfare, a monthly stipend of @Y270,000, and an additional settling-in allowance. In addition, JSPS may provide a special research grant up to @Y1.5 million, depending on need of applicants.

Requirements: Fellowships are available to citizens of countries that have diplomatic relations with Japan. Candidates must hold the doctorate degree when the fellowship goes into effect; the degree must have been received within six years prior to April of the fiscal year in which the fellowship commences. The candidate must have arranged in advance a research plan with his/her Japanese host researcher, who must be employed at a Japanese university/research institution.

Amount of Grant: $5000-$13,000

Contact: Program Administrator, Foreign Fellowship Division, 81-3-3263-1721; fax: 81-3-3263-1854

Internet: http://www.jsps.go.jp/english/e-fellow/postdoctoral.html#short

Sponsor: Japan Society for the Promotion of Science
Nihon Gakujutsu Shinko-kai, 6-26-3, Kioi-cho, Chiyoda-ku
Tokyo 102-0094 Japan

Judith and Jean Pape Adams Charitable Foundation ALS Grants   3060

The Foundation was established in 2004 as a private foundation and is involved in making distributions to charitable organizations on an annual basis. It encompasses two areas of support: organizations and agencies predominantly in Tulsa County, Oklahoma, and national Amyotrophic Lateral Sclerosis (ALS) research. For the latter, the Foundation encourages research that is collaborative, innovative and aggressive. These initiatives may fill gaps not currently being investigated and may involve novel approaches. The research must be of high scientific merit in finding a cure for ALS. Proposals may be longer than two pages. A lay summary is also required. Applications are reviewed by an Advisory Committee comprised of scientific researchers, neurologists and others who have had close relationships with friends and family members afflicted with ALS. The Committee makes funding recommendations to the Trustees.

Requirements: Public agencies and organizations that are classified as a charitable organization described in Section 501(c)3 of the Internal Revenue Code and as a public charity under Section 509(a) of the Internal Revenue Code may apply.

Restrictions: Support groups, education and medical equipment are not supported.

Date(s) Application is Due: Aug 15

Contact: Marcia Y. Manhart, Executive Director; (830) 997-7347; fax: (830) 997-9888; email: mmanhart@jjpafoundation.com

Internet: http://www.jjpafoundation.com/guidelines.html

Sponsor: Judith and Jean Pape Adams Charitable Foundation
105 South Cherry Street
Fredericksburg, TX 78624

Judith Graham Pool Postdoctoral Research Fellowship in Hemophilia   3061

A limited number of fellowships are awarded through professional and graduate schools or research institutions. The Judith Graham Pool (JGP) program is designed to support research studies of high scientific merit and relevance to hemophilia. Permissible research topics include clinical or basic research on the biochemical, genetic, hematologic, orthopedic, psychiatric, or dental aspects of the hemophilias or von Willebrand disease. Other acceptable topics include rehabilitation, therapeutic modalities, psychosocial issues, or AIDS/HIV as they pertain to hemophilias or von Willebrand disease. Awards are made for a one- or two-year period, with the second-year renewal contingent upon demonstrated progress in the fellow's research project. The fellowships will be awarded for the period between July 1 of one year and June 30 of the next. Applications are available on the Web site.

Requirements: Applicants must be citizens or permanent residents of the United States who have completed doctoral training and must enter the JGP fellowship program from a doctoral, postdoctoral, internship, or residency training program.

Restrictions: Established investigators or faculty members are not eligible.

Amount of Grant: $42,000 per year

Date(s) Application is Due: Dec 1

Contact: Dr. Rita Barsky, Assistant Director of Research, (800) 424-2634 ext 3730 or (212) 328-3730; fax: (212) 328-3788; email: rbarsky@hemophilia.org

Internet: http://www.hemophilia.org/research/jgp.htm

Sponsor: National Hemophilia Foundation
116 W 32nd Street, 11th Fl
New York, NY 10001

Julius Schulich Award for Entrepreneurship   3062

One scholarship is awarded per year for entrepreneurship in the MBA program. Application should include a one page statement of career objectives and the role that the proposed program of study in entrepreneurship will have in reaching those objectives plus two letters of reference evaluating his/her probability of success. Funded by Julius Schulich Foundation.

Requirements: Any full-time student in the MBA program at the University of Calgary is eligible to apply.

Amount of Grant: $C6000 per year

Date(s) Application is Due: May 1

Contact: Louise Hamilton, Executive Secretary, (403) 220-5690 ext 3792; email: hamiltol@ucalgary.ca

Internet: http://www.grad.ucalgary.ca/funding/internal_scholarships/lev_4/schulich.htm

Sponsor: University of Calgary
2500 University Dr NW
Calgary, AB T2N 1N4 Canada

Justin Winsor Essay Prize 3063

This award is given yearly to an author of an outstanding essay embodying original historical research on a significant subject of library history. Application and/or nomination forms are available from the Library History Round Table Staff Liaison at ALA.

Restrictions: Manuscripts submitted should not be previously published, previously submitted for publication, or under consideration for publication or another award. fax and email are not acceptable.

Amount of Grant: $500

Date(s) Application is Due: Mar 1

Contact: ALA Awards Staff Liaison, (312) 280-3247; fax: (312) 280-3257; email: awards@ala.org

Internet: http://www.ala.org/ala/ors/orsawards/winsorjustin/winsorjustin.htm

Sponsor: American Library Association
50 E Huron Street
Chicago, IL 60611

Juvenile Justice Clearing House Grants 3064

The Juvenile Justice Clearing House, of the Department of Justice, awards grants under its Juvenile Justice and Delinquency Prevention Program. Areas of interest include delinquency prevention research, juvenile gangs, serious violent juvenile offenders, and juvenile justice statistics. Types of support include conferences and seminars, development grants, fellowships, internships, project grants, residencies, visiting scholars grants, and research grants. There are no matching requirements. Awardees generally receive from one to three years of support.

Requirements: Public and private organizations, including institutions of higher education, and individuals may submit applications for grant support.

Amount of Grant: $5.63 million total

Contact: Director, (800) 851'3420; fax: (301) 519'5600; email: askjj@ncjrs.org

Internet: http://ojjdp.ncjrs.org/funding/FundingList.asp

Sponsor: Juvenile Justice Clearing House
P.O. Box 6000
Rockville, MD 20849-6000

Subject Index

NOTE: Numbers refer to entry numbers

FIC Israeli Ministry of Health Postdoctoral Research Fellowships, 2340
Glenn/AFAR Breakthroughs in Gerontology Awards, 2542
Glenn/AFAR Research Grants for Postdoctoral Fellows, 2543
Greater Worcester Community Foundation Jeppson Memorial Fund for Brookfield Grants, 2567
Greenburg-May Foundation Grants, 2569
Hartford Aging and Health Program Awards, 2619
Hartford Geriatrics Health Outcomes Research Scholars Program, 2620
Hastings Center International Biomedical Ethics Research Program for International Scholars, 2628
Hawaii Community Foundation Health Education and Research Grants, 2630
HCR Manor Care Foundation Grants, 2633
HRSA Geriatric Education Centers, 2728
James H. Cummings Foundation Grants, 2942
Joy Family Foundation Grants, 3054
Kavli Foundation Research Grants, 3072
King Foundation Grants, 3088
Larry L. Hillblom Foundation Grants, 3135
Legler Benbough Foundation Grants, 3158
Leonard and Helen R. Stulman Charitable Foundation Grants, 3165
Margaret L. Wendt Foundation Grants, 3263
Maurice J. Masserini Charitable Trust Grants, 3303
McLean Contributionship Grants, 3320
Merck Institute of Aging and Health Grants, 3347
Mount Zion Health Fund Grants, 3454
NCI Cancer Communications System Research Grants, 3582
NCI Cancer Control Research Grants, 3584
NCI Diagnostic Imaging and Guided Therapy in Prostate Cancer Award, 3598
NCI Prostate Cancer Aging, Race, and Ethnicity Research Grants, 3622
NHLBI Independent Scientist Award, 3767
NIA Aging Research Dissertation Awards to Increase Diversity, 3829
NIA Aging, Oxidative Stress and Cell Death Grants, 3830
NIA AIDS and Aging: Behavioral Sciences Prevention Research Grants, 3831
NIA Alzheimer's Disease Core Centers Grants, 3832
NIA Archiving and Development of Socialbehavioral Datasets in Aging Related Studies Grants, 3836
NIA Awards to Support Research on the Biology of Aging in Invertebrates Grants, 3837
NIA Bioenergetics, Fatigability, and Activity Limitations in Aging, 3839
NIA Diversity in Medication Use and Outcomes in Aging Populations Grants, 3840
NIA Effects of Gene-Social Environment Interplay on Health and Behavior in Later Life Grants, 3841
NIA Harmonization of Longitudinal Cross-National Surveys of Aging Grants, 3844
NIA Health Behaviors and Aging Grants, 3845
NIA Healthy Aging through Behavioral Economic Analyses of Situations Grants, 3846
NIA Higher-Order Cognitive Functioning and Aging Grants, 3847
NIA Human Biospecimen Resources for Aging Research Grants, 3848
NIA Independent Scientist Award, 3849
NIA Malnutrition in Older Persons Research Grants, 3850
NIA Mechanisms, Measurement, and Management of Pain in Aging: from Molecular to Clinical, 3852
NIA Medical Management of Older Patients with HIV/AIDS Grants, 3853
NIA Mentored Research Scientist Development Award, 3855
NIA Minority Dissertation Research Grants in Aging, 3856
NIA Multicenter Study on Exceptional Survival in Families: The Long Life Family Study Grants, 3857
NIA Nathan Shock Centers of Excellence in Basic Biology of Aging Grants, 3858

NIA NRSAs--Individual Postdoctoral Fellowships, 3859
NIA NRSAs--Institutional Training Awards, 3860
NIA Paul Beeson Career Development Awards in Aging, 3861
NIA Pilot Research Grant Program, 3862
NIA Postdoctoral Training in Research on Aging in Canada, 3863
NIA Promoting Careers in Aging and Health Disparities Research Grants, 3864
NIA Prostate Cancer Aging, Race, and Ethnicity Research Grants, 3865
NIA Renal Function and Chronic Kidney Disease in Aging Grants, 3866
NIA Role of Apolipoprotein E, Lipoprotein Receptors and CNS Lipid Homeostasis in Brain Aging and Alzheimer's Disease, 3868
NIA Role of Nuclear Receptors in Tissue and Organismal Aging Grants, 3869
NIA Significance of Sarcopenia in Old Age Grants, 3870
NIA Sociobehavorial Data Analysis and Archiving in Aging, 3871
NIA Support of Scientific Meetings as Cooperative Agreements, 3872
NIA Thyroid in Aging Grants, 3874
NIA Transdisciplinary Research on Fatigue and Fatigability in Aging Grants, 3875
NIA Translational Research at the Aging/Cancer Interface (TRACI) Grants, 3876
NIA Vulnerable Dendrites and Synapses in Aging and Alzheimer's Disease Grants, 3877
NIA/NIEHS Biomedical Epidemiologic and Intervention Studies Planning Grants, 3878
NIAAA Hepatitis C Infection and Alcoholic Liver Disease, 3880
NIAMS Significance of Sarcopenia in Old Age Grants, 3919
NIEHS Structure-Function Relationships of Environmentally Relevant Genetic Variants Research Grants, 4049
NIH Human Models and Markers of Skeletal Aging Grants, 4118
NIH Midcareer Investigator Award in Mouse Pathobiology Research Grants, 4134
NIH Pathological Gambling: Basic, Clinical, and Services Research Grants, 4146
NIH Research on Sleep and Sleep Disorders, 4157
NIH Roybal Centers for Translational Research on Aging, 4161
NIH Xenobiotics and Cell Death/Injury in Neurodegenerative Disease Research Grants, 4182
NIMH AIDS and Aging: Behavioral Sciences Prevention Research Grants, 4189
NIMH Behavior Genetics in Adulthood and Old Age Research Grants, 4191
NINR Aging Women and Breast Cancer Grants, 4231
NINR AIDS and Aging: Behavioral Sciences Prevention Research Grants, 4232
NSF Biomedical Engineering and Engineering Healthcare Grants, 4501
Orentreich Foundation for the Advancement of Science Grants, 4837
Postdoctoral Research Fellowships in Gerontological Public Health, 4998
RAND Corporation Fellowships in Population Studies and the Study of Aging, 5045
Rhode Island Foundation Grants, 5084
Rosalinde and Arthur Gilbert Foundation New Investigator Awards in Alzheimer's Disease, 5129
STTI Environment of Elder Care Nursing Research Grants, 5473
United Cerebral Palsy Research Grants, 5606
Walter Nicolai Prize in Biomedical Gerontology, 5710
William B. Dietrich Foundation Grants, 5769

Agribusiness
DOA Sustainable Agriculture Research and Education Grants, 1975

Agricultural Collectives/Cooperatives
National Council of Farmer Cooperatives Graduate Awards, 3523

Agricultural Commodities
ADEC Agricultural Telecommunications Grants, 311

Agricultural Economics
BARD Research Grants, 1319
Farm Foundation Extension Graduate Training Fellowships, 2297
National Council of Farmer Cooperatives Graduate Awards, 3523

Agricultural Engineering
BARD Research Grants, 1319

Agricultural Extension
Farm Foundation Extension Graduate Training Fellowships, 2297

Agricultural Finance
National Council of Farmer Cooperatives Graduate Awards, 3523

Agricultural Geography
Farm Foundation Extension Graduate Training Fellowships, 2297

Agricultural Management
DOA Sustainable Agriculture Research and Education Grants, 1975
Farm Foundation Extension Graduate Training Fellowships, 2297

Agricultural Planning/Policy
Clarence E. Heller Charitable Foundation Grants, 1716
DOA Biotechnology Risk Assessment Research Grants, 1958
DOA Emerging Markets Program Grants, 1960
GMFUS Transatlantic Environmental Fellowship Program, 2547
Great Lakes Protection Fund Grants, 2566
The World Food Prize, 5528
Thomas Jefferson Agricultural Institute Food and Society Policy Fellowships, 5539

Agriculture
Alberta Research Council Scholarship, 558
ALSAM Foundation Grants, 634
American Chemical Society Herman Frasch Foundation Grants, 685
American-Scandinavian Foundation Short-Term Training Program Grants, 830
ASA and USDA/NASS Research Fellow and Associate Program, 1052
ASPRS William A. Fischer Memorial Scholarship, 1193
BARD Postdoctoral Fellowships, 1318
BARD Research Grants, 1319
C. Alma Baker Trust Grants, 1448
Canada Graduate Scholarships (CGS) and NSERC Postgraduate Scholarships (PGS), 1470
Canada-U.S. Fulbright International Science and Technology Award, 1471
Canada-U.S. Fulbright Senior Specialists Program Grants, 1472
Canada-U.S. Fulbright Traditional Scholar Awards, 1473
Canadian Friends of Hebrew University Research and Study Grants, 1476
DOA Agricultural Competitive Research Grants, 1953
DOA Agricultural Special Research Grants, 1954
DOA Biological Nitrogen Fixation Research Grants, 1956
DOA Biological Stress on Plants Research Grants, 1957
DOA Emerging Markets Program Grants, 1960

ORISE Postgraduate Research Participation at the Centers for Disease Control and Prevention, 4862
ORISE Student Research Participation at the Centers for Disease Control and Prevention, 4873
OSI Harm Reduction and Drug Use Grants, 4878
Parker Foundation Grants, 4907
Pfizer Foundation Grants, 4946
Pittsburgh Foundation Medical Research Grants, 4987
Ralph M. Parsons Foundation Grants, 5043
Rhode Island Foundation Grants, 5084
Rich Foundation Grants, 5085
Robert Mapplethorpe Foundation Grants, 5103
Shelley and Donald Rubin Foundation Grants, 5247
Smith Family New Investigator Awards, 5280
SSRC HIV/AIDS and Public Health Policy Research in Africa Fellowships, 5438
Streisand Foundation Grants, 5466
United Hospital Fund of New York Grants Program, 5607
Victor E. Speas Foundation Grants, 5671
W.W. Smith Charitable Trust Grants, 5698
W.W. Smith Charitable Trust Grants for Medical Research in Cancer, AIDS, and Heart Disease, 5699
William B. Dietrich Foundation Grants, 5769

AIDS Counseling
Ann and Robert H. Lurie Family Foundation Grants, 887
Community Foundation AIDS Endowment Awards, 1765
NIA Medical Management of Older Patients with HIV/AIDS Grants, 3853
OSI Harm Reduction and Drug Use Grants, 4878

AIDS Education
Abbott Fund Science and Medical Innovation Grants, 216
amfAR Fellowships, 832
amfAR Mathilde Krim Fellowships in Basic Biomedical Research, 833
amfAR Research Grants, 834
Community Foundation AIDS Endowment Awards, 1765
DHHS AIDS Education and Training Centers Grants, 1911
Fairlawn Foundation Grants, 2288
Frances L. and Edwin L. Cummings Memorial Fund Grants, 2400
Ittleson Foundation AIDS Grants, 2904
McCarthy Family Foundation Grants, 3312
NIH Centers for AIDS Research (CFARs), 4091
OSI Harm Reduction and Drug Use Grants, 4878

AIDS Prevention
Abbott Fund Global AIDS Care Grants, 215
Abbott Fund Science and Medical Innovation Grants, 216
CDC Intervention Epidemiologic Research Studies of HIV/AIDS, 1561
Community Foundation AIDS Endowment Awards, 1765
DHHS AIDS Education and Training Centers Grants, 1911
Elizabeth Glaser Pediatric AIDS Foundation Operations Research Grants, 2197
Elizabeth Glaser Pediatric AIDS Foundation Two-Year Scholar Awards, 2200
Frances L. and Edwin L. Cummings Memorial Fund Grants, 2400
Gates Global HIV/AIDS Vaccine Grants, 2475
Irvington Institute Immunology Postdoctoral Fellowships, 2895
Ittleson Foundation AIDS Grants, 2904
NIA AIDS and Aging: Behavioral Sciences Prevention Research Grants, 3831
NIAID HIV Vaccine Research and Design--Program Project Grants, 3897
NIAID Integrated Preclinical/Clinical AIDS Vaccine Development Grants, 3899

NIAID Statistical Methods in HIV and AIDS Research Grants, 3906
NIDDK Transmission of Human Immunodeficiency Virus (HIV) In Semen, 4035
NIH Centers for AIDS Research (CFARs), 4091
NIH Multidisciplinary Approach for Research on Oral Complications of HIV Infection, 4138
NIMH AIDS and Aging: Behavioral Sciences Prevention Research Grants, 4189
NIMH HIV/AIDS, Severe Mental Illness, and Homelessness Research Grants, 4201
NINR AIDS and Aging: Behavioral Sciences Prevention Research Grants, 4232
OSI Harm Reduction and Drug Use Grants, 4878
SSRC HIV/AIDS and Public Health Policy Research in Africa Fellowships, 5438
Sunny and Abe Rosenberg Foundation Grants, 5487

Air Pollution
Bullitt Foundation Grants, 1430
Carrier Corporation Contributions Grants, 1520
DOA Food and Agricultural Sciences Small Business Innovation Research (SBIR) Grants, 1963
DOA Small Business Innovation Research (SBIR) Grants, 1974
DOT Transit Planning and Research Grants, 2101
Environmental Protection Agency Summer Faculty Fellowships, 2229
EPA Grants for Research, 2233
GMFUS Transatlantic Environmental Fellowship Program, 2547
NOAA Climate Dynamics and Experimental Prediction (CDEP) Program Grants, 4329
Orchard Foundation Grants, 4827
Takemi Program in International Health Research and Advanced Training Fellowships, 5512
Young Scientists Summer Program, 5828

Airplanes
ONR Aerospace Science Research Grants, 4748
ONR Naval Materials Division Grants, 4795

Alaska
Brainerd Foundation Grants, 1405
Jacobs Research Funds Small Grants Program, 2936
William and Flora Hewlett Foundation Environmental Grants, 5767

Alaskan Natives
AHRQ Understanding and Eliminating Minority Health Disparities, 452
American Indian Science and Engineering Scholarships, 747
ASM Robert D. Watkins Graduate Research Fellowship, 1143
DHHS Health Professions Preparatory Scholarship Program for Indians, 1921
DHHS Health Professions Scholarship Program, 1922
DHHS Native American Programs, 1930
DOD National Defense Science and Engineering Graduate Fellowship Program, 1990
FDHN Student Research Fellowships, 2325
IHS Native American and Alaskan Native Health Research Centers, 2793
LITA/LSSI Minority Scholarship in Library and Information Technology, 3204
Morris K. Udall Native American Congressional Internship Program, 3450
NAGPRA Museum Documentation Grants, 3489
NAGPRA Museum Repatriation Grants, 3490
NCI Cancer Control Research Grants, 3584
NHLBI Minority Institutional Research Training Program Grants, 3783
NIMH Career Opportunities in Research Education and Training Honors High School Research Education Grant, 4192
NIMH Career Opportunities in Research (COR) Education and Training Honors Undergraduate Research Training Grant, 4193

NRC-Ford Foundation Dissertation Fellowships for Minorities, 4432
NRC-Ford Foundation Predoctoral Fellowships for Minorities, 4434
NSF Tribal Colleges and Universities Program, 4668
Sequoyah Fellowship Program for Native American and Alaska Native Graduate Students, 5235

Albania
ACLS Dissertation Fellowships in East European Studies, 266
ACLS Early-Career Postdoctoral Fellowships in East European Studies, 267
ACLS East European East Heritage Speakers Research Grant, 268
ACLS East European Studies Program Travel Grants, 270
IREX Individual Advanced Research Opportunities (IARO) Grants, 2883

Alcohol Education
Christopher D. Smithers Foundation Research Grants, 1691
Coors Brewing Corporate Contributions Grants, 1796
Fairlawn Foundation Grants, 2288
Heublein Foundation Inc Grants, 2679

Alcohol/Alcoholism
AAA Foundation for Traffic Safety Grants, 15
Adaptec Foundation Grants, 309
Alcohol Misuse and Alcoholism Research Grants, 562
Alex Stern Family Foundation Grants, 566
amfAR Fellowships, 832
amfAR Mathilde Krim Fellowships in Basic Biomedical Research, 833
amfAR Research Grants, 834
Berks County Community Foundation Grants, 1358
Bethesda Foundation Grants, 1369
Bodman Foundation Grants Program, 1390
Carpenter Foundation Grants, 1518
Christopher D. Smithers Foundation Research Grants, 1691
Community Foundation for Southern Arizona Grants, 1768
Community Foundation of Broward Grants, 1769
ConocoPhillips Grants Program, 1788
Coors Brewing Corporate Contributions Grants, 1796
D.F. Halton Foundation Grants, 1831
DHHS/ASPE Welfare Outcomes Short-Term Policy Research Grants, 1946
Farmers Insurance Group of Companies Grants, 2298
Jackson Foundation Grants, 2931
Johnson & Johnson Corporate Contributions Grants, 3041
Joint Center for Poverty Research Grants, 3043
Joy Family Foundation Grants, 3054
Lesbian Health Fund Grants, 3166
Little River Foundation Grants, 3205
Margaret L. Wendt Foundation Grants, 3263
Musgrave Foundation Grants, 3472
Nationwide Foundation Grants, 3558
NIAAA Alcohol Research Centers Grants, 3879
NIAAA Hepatitis C Infection and Alcoholic Liver Disease, 3880
NIAAA Independent Scientist Award, 3881
NIAAA Mechanisms of Alcohol and Drug-Induced Pancreatitis Grants, 3882
NIAAA Mechanisms of Alcohol-Induced Hepatic Fibrosis Research Grants, 3883
NIAAA Peptide Regulation of Alcohol Intake, 3885
NIAAA Secondary Analysis of Existing Health Services Data Sets Grants, 3887
NIAAA Strategies for the Prevention of DUI Recidivism Research Grants, 3888
NIH Exploratory/Development Grants for Mental Health Intervention Research, 4109
NIH NRSAs for Individual Predoctoral Fellowships, 4144
NIH Research on Sleep and Sleep Disorders, 4157
NIJ Graduate Research Fellowship Program, 4187

DOA Plant and Animal Disease, Pest Control, and Animal Care Grants, 1970
Donald and Sylvia Robinson Family Foundation Grants, 2078
Frank Stanley Beveridge Foundation Grants, 2409
Hastings Center International Biomedical Ethics Research Program for International Scholars, 2628
Helen Brach Foundation Grants, 2647
Little River Foundation Grants, 3205
NIH Technologies to Improve the Utility of Animal Models Research Grants, 4178
Park Foundation Grants, 4905
Rhode Island Foundation Grants, 5084
Vancouver Foundation Grants and Community Initiatives Program, 5662
William and Charlotte Parks Foundation for Animal Welfare Grants, 5765

Animal Science
ACVIM Foundation Clinical Investigation Grants, 300
APHA Youth Development Foundation Grants, 924
BARD Research Grants, 1319
DOA Cooperative Forestry Research Grants, 1959
DOA Food and Agricultural Sciences Small Business Innovation Research (SBIR) Grants, 1963
DOA Plant and Animal Disease, Pest Control, and Animal Care Grants, 1970
IMLS Conservation Project Support Grants, 2805
MFA Established Investigator Grants, 3358
MFA Fellowship Training Grant, 3359
MFA First Award Grants, 3360
National Turkey Federation Research Award, 3551
NFWF National Whale Conservation Fund Grants, 3723
NSF Animal Developmental Mechanisms Grants, 4463
NSF Cell Biology Research Grants, 4507
NSF Hydrologic Sciences Research Grants, 4582
Robert J. Kleberg Jr. and Helen C. Kleberg Foundation Grants, 5100

Animal Welfare
Annenberg Foundation Grants, 894
SeaWorld and Busch Gardens Conservation Fund Grants, 5229
William and Charlotte Parks Foundation for Animal Welfare Grants, 5765

Animals as Pets
Banfield Charitable Trust Grants, 1314
Laura J. Niles Foundation Grants, 3138

Anorexia Nervosa
NIMH Early Identification and Treatment of Mental Disorders in Children and Adolescents Grants, 4200

Antarctica
NSF Antarctic Aeronomy and Astrophysics Grants, 4464
NSF Antarctic Earth Sciences Program Grants, 4466
NSF Antarctic Glaciology Program Grants, 4467
NSF Antarctic Organisms and Ecosystems Grants, 4470
NSF Integrative Graduate Education and Research Traineeship (IGERT), 4587
NSF Polar Programs Division Research Grants, 4633
Tinker Foundation Institutional Grants, 5552

Antennae
AFOSR Electromagnetics Research Grants, 345
ARO Mobile, Wireless Communications and Networks Research Contracts, 1016
ONR Antennas, Propagation and Electromagnetic Interactions Grants, 4751

Anthropology
AAA Minority Dissertation Fellowship Program, 16
ACLS Charles A. Ryskamp Research Fellowships, 262

ACLS Frederick Burkhardt Residential Fellowships for Recently Tenured Scholars, 272
AHFMR Population Health Investigator (PHI) Awards, 423
Alberta Historical Resources Foundation Roger Soderstrom Scholarship, 556
Alfred P. Sloan Foundation Known, Unknown, and Unknowable Research Grants, 607
American Philosophical Society Judson Daland Prize, 779
American Philosophical Society Lewis and Clark Fund for Exploration and Field Research, 780
American Philosophical Society Phillips Fund Grants for Native American Research, 783
AMNH Collection Study Grants, 837
AMNH Research and Museum Fellowships, 841
Anthropological Center for Training in Global and Environmental Change Assistantships, 901
Australian Institute of Aboriginal and Torres Strait Islander Studies--Research Grants, 1239
Canada-U.S. Fulbright International Science and Technology Award, 1471
Canada-U.S. Fulbright Senior Specialists Program Grants, 1472
Canada-U.S. Fulbright Traditional Scholar Awards, 1473
Center for Advanced Study in the Behavioral Sciences Postdoctoral Fellowships, 1573
Center for East Asian Studies Fellowships, 1582
Center for Italian Renaissance Studies Fellowships, 1586
CES Predissertation Fellowships, 1605
Douglass Institute Fellowships, 2103
EF Behavioral Sciences Student Fellowships, 2180
Field Museum Visiting Scholarships, 2348
Foundation for Psychocultural Research Doctoral Fellowships, 2390
Fulbright Alumni Initiatives Awards Grants, 2432
Fulbright Distinguished Chairs Grants, 2437
Fulbright European Union Scholar-in-Residence (EUSIR) Program Grants, 2439
Fulbright New Century Scholars Grants, 2443
Fulbright Occasional Lecturer Grants, 2444
Fulbright Specialists Program Grants, 2447
Fulbright Traditional Scholar Program in Europe and Eurasia, 2448
Fulbright Traditional Scholar Program in Sub-Saharan Africa, 2449
Fulbright Traditional Scholar Program in the East Asia/Pacific Region, 2450
Fulbright Traditional Scholar Program in the Near East and North Africa Region, 2451
Fulbright Traditional Scholar Program in the South and Central Asia Region, 2452
Fulbright Traditional Scholar Program in the Western Hemisphere, 2453
Fulbright Visiting Specialists Program: Direct Access to the Muslim World, 2455
Fyssen Foundation Postdoctoral Study Grants, 2465
Harry Frank Guggenheim Foundation Dissertation Fellowships, 2610
Heinz Latin American Social and Public Policy Fellowships, 2643
Horowitz Foundation for Social Policy Grants, 2714
IMLS 21st Century Museum Professionals Grants, 2804
IMLS National Leadership Grants (NLG), 2808
J.I. Staley Prize, 2927
Jacobs Research Funds Small Grants Program, 2936
Jeffrey Campbell Graduate Fellow in Anthropology, 2977
Katrin H. Lamon Resident Scholar Program for Native Americans, 3071
Leakey Foundation Franklin Mosher Baldwin Fellowships, 3150
Leakey Foundation General Grants, 3151
Macalester College Andrew W. Mellon Postdoctoral Fellowship, 3249
Minority Humanities and Social Sciences Dissertation Scholar in Residence Grants, 3406

MPI for Social Anthropology Department II Postdoctoral Grants, 3456
MPI for Social Anthropology Max Planck Fellow Research Group Law, Organizations, Science, and Technology (LOST), 3457
National Geographic Society Expeditions Council (EC) Grants, 3531
National Geographic Society Field Research and Exploration Grants, 3532
NCI Small Grants Program for Behavioral Research in Cancer Control, 3626
North Carolina Arts Council Arts Folklife Internship, 4411
NSF Archaeology and Archeaometry Research Grants, 4474
NSF Arctic Social Sciences Grants, 4484
NSF SBE Doctoral Dissertation Research Improvement Grants, 4647
Pfizer Foundation Health Literacy Grants, 4947
PHF Penn Faculty Research Fellowships, 4950
PHF Regional Mellon Faculty Research and Cultural Fellowships, 4951
PHF Undergraduate Humanities Forum Research Fellowships, 4952
Richard Carley Hunt Memorial Postdoctoral Fellowships, 5088
S.E.VEN Fund WHY Prize, 5186
School of American Research Advanced Seminars, 5221
School of American Research Resident Scholar Fellowships, 5222
Sequoyah Fellowship Program for Native American and Alaska Native Graduate Students, 5235
Smithsonian Senior Fellowships, 5295
SSRC Mellon Mays Predoctoral Research Grants, 5442
USF Richard F. Pride Research Fellowship, 5646
Visiting Scholar in Archaeology Program, 5687
Wenner-Gren Foundation African Fellowships, 5728
Wenner-Gren Foundation Conference and Workshop Grants, 5729
Wenner-Gren Foundation Dissertation Fieldwork Grants, 5730
Wenner-Gren Foundation Historical Archives Grants, 5731
Wenner-Gren Foundation Hunt Postdoctoral Fellowships, 5732
Wenner-Gren Foundation International Collaborative Research Grants, 5733
Wenner-Gren Foundation Post-PhD Research Grants, 5734
Wenner-Gren Foundation Short-Term Fellowships, 5735
Wenner-Gren Foundation Wadsworth International Fellowships, 5736

Anthropology, Cultural
Anthropological Center for Training in Global and Environmental Change (ACT) Grad Assistantships, 901
Australian Institute of Aboriginal and Torres Strait Islander Studies--Research Grants, 1239
Field Museum Visiting Scholarships, 2348
Foundation for the Advancement of Mesoamerican Studies Grants, 2394
Leakey Foundation Paleoanthropology Award, 3153
Macalester College Andrew W. Mellon Postdoctoral Fellowship, 3249
MPI for Social Anthropology Department II Postdoctoral Grants, 3456
NSF Cultural Anthropology Research Grants, 4545
PHF Penn Faculty Research Fellowships, 4950
PHF Regional Mellon Faculty Research and Cultural Fellowships, 4951
PHF Undergraduate Humanities Forum Research Fellowships, 4952
School of American Research Advanced Seminars, 5221
Wenner-Gren Foundation Dissertation Fieldwork Grants, 5730
Woodrow Wilson Practicum Grants, 5798

Lindbergh Grants Program, 3198
Linn-Henley Charitable Trust Grants, 3200
NOAA Collaborative Science, Technology, and
Applied Research (CSTAR) Program Grants, 4338
Ray Foundation Mental Health and Substance Abuse
Prevention Grants, 5051

Aviation History
Marine Corps Historical Center Research Grants,
3274
NASM Guggenheim Fellowships, 3510
NASM Ramsey Fellowship in Aviation History, 3511
NASM Verville Fellowships, 3512

Azerbaijan
DOS Freedom Support Educational Partnerships
with Eurasia, 2095
Fulbright Critical Language Enhancement Awards,
2436
IREX Individual Advanced Research Opportunities
(IARO) Grants, 2883
IREX Policy-Connect Collaborative Research
Grants, 2886
SSRC Eurasian Program Postdoctoral Research
Fellowships, 5437
USIA NIS Educational Advising Centers, 5651

Back Pain
FCER Research Grants and Fellowships, 2307
NIAMS Low-Back Pain and Common Spinal
Disorders Research Grants, 3913
NICHD Low-Back Pain and Common Spinal
Disorders Research Grants, 3928
NINDS Low-Back Pain and Common Spinal
Disorders Research Grants, 4221

Bacteriology
Ethical Research Grants and Fellowships, 2260
NIAID Microbiology and Infectious Diseases
Research Grants and Contracts, 3901
NIAID Preparedness against Illegitimate Use of
Bacterial Pathogens, 3903
NSF Physiology and Ethology Research Grants, 4628
ORISE Faculty Research Participation at the Centers
for Disease Control and Prevention, 4844
ORISE Postgraduate Research Participation at the
Centers for Disease Control and Prevention, 4862
ORISE Student Research Participation at the Centers
for Disease Control and Prevention, 4873

Baking
AACC International Milling & Baking Division's
John C. Halverson Memorial Lectureship, 75

Bangladesh
American Institute of Bangladesh Studies Postdoctoral
Research and Dissertation Research Grants, 750
CAORC Regional Research Fellowships, 1500
Fulbright Traditional Scholar Program in the South
and Central Asia Region, 2452

Banking
Equipment Leasing and Finance Foundation Research
Grants, 2248
Erasmus Prize, 2250
Germanistic Society of America Fellowships, 2517

Bankruptcy
American College of Bankruptcy Grants, 696

Baptist Church
Booth-Bricker Fund Grants, 1393
Southern Baptist Oral History Grants, 5347
Southern Baptist Study Grants, 5348

Basic Research
Auburn Foundation Grants, 1231
DEED Student Research Grants and Internships, 1879
Dept of Ed Vocational Education National Programs,
1902

Drs. Bruce and Lee Foundation Grants, 2117
FASSE Demonstration Project Grant, 2299

Basic Skills Education
Allyn Foundation Grants, 624
Auburn Foundation Grants, 1231
Ball Brothers Foundation Grants, 1308
Bodenwein Public Benevolent Foundation Grants,
1389
Booth-Bricker Fund Grants, 1393
Boston Foundation Grants, 1396
Carl and Lily Pforzheimer Foundation Grants, 1505
Cowles Charitable Trust Grants, 1804
Cruise Industry Charitable Foundation Grants, 1817
Dept of Ed Adult Education--National Leadership
Activities Grants, 1896
Evjue Foundation Grants, 2276
Frances L. and Edwin L. Cummings Memorial Fund
Grants, 2400
George Foundation Grants, 2498
Harold Simmons Foundation Grants, 2601
Household International Corporate Giving Program
Grants, 2718
Little River Foundation Grants, 3205
Lubrizol Foundation Grants Program, 3229
McCune Foundation Grants, 3314
Norcliffe Foundation Grants, 4406
Oppenstein Brothers Foundation Grants, 4825
Richard and Rhoda Goldman Fund Grants, 5086
Robert Bowne Foundation Grants, 5097
San Antonio Area Foundation Grants, 5200
Sterling-Turner Charitable Foundation Grants, 5463
Strake Foundation Grants, 5465
Talbert and Leota Abrams Foundation, 5513

Behavioral Medicine
American Lung Association Dalsemer Research
Grant, 762
DHHS Great Lakes Human Health Effects Research
Program Grants, 1918
International Cancer Technology Transfer
Fellowships, 2852
James S. McDonnell Foundation Research Grants,
2954
James S. McDonnell Foundation Scholar Awards,
2955
NCCAM Predoctoral Research Training in
Complimentary and Alternative Medicine, 3572
NCI Established Investigator Award in Cancer
Prevention, Control, Behavioral, and Population
Sciences, 3605
NCI Stages of Breast Development: Normal to
Metastatic Disease Grants, 3628
NCIC Canadian Cancer Society Research Grants,
3634
NCRR Institutional Animal Resources Developing
and Improving Grants, 3641
NEI Ruth L. Kirschstein National Research Service
Award Short-Term Institutional Research Training
Grants, 3685
NIA AIDS and Aging: Behavioral Sciences
Prevention Research Grants, 3831
NIAID HIV Prevention Trial Units Grants, 3895
NICHD Central Nervous System Effects of HIV
Infection: Neurobiological, Neurovirological, and
Neurobehavioral Studies Grants, 3924
NIDA Behavioral Science Track Awards for Rapid
Transition, 3940
NIDA Minority Institutions Drug Abuse Research
Development Program (MIDARP) Grants, 3951
NIDA National Drug Abuse Treatment Clinical
Trials Network, 3952
NIDCD Neurobiology and Genetics of Autism
Grants, 3969
NIDCR HIV/AIDS Associated Oral Viral Infections,
3978
NIDDK Race/Ethnic Disparities in the Incidence of
Diabetes Complications, 4022
NIEHS Biological Response to Environmental
Health Hazards Research Grants, 4039

NIH Alzheimer's Disease Clinical Trial Planning
Grant, 4080
NIH Genetics of Neurobehavioral Disorders in
Existing Samples Grants, 4114
NIH NRSAs for Individual Predoctoral Fellowships,
4144
NIH Pathological Gambling: Basic, Clinical, and
Services Research Grants, 4146
NIH Research on Sleep and Sleep Disorders, 4157
NIH Ruth L. Kirschstein National Research Service
Awards for Individual Postdoctoral Fellows, 4163
NIMH AIDS and Aging: Behavioral Sciences
Prevention Research Grants, 4189
NIMH Behavior Genetics in Adulthood and Old Age
Research Grants, 4191
NIMH Central Nervous System Effects of HIV
Infection: Neurobiological, Neurovirological, and
Neurobehavioral Studies Grants, 4195
NIMH Minority Research Infrastructure Support
Program Grants, 4207
NIMH Quantitative Approaches to the Analysis of
Complex Biological Systems, 4211
NINDS Central Nervous System Effects of HIV
Infection: Neurobiological, Neurovirological, and
Neurobehavioral Studies Grants, 4219
NINR Aging Women and Breast Cancer Grants,
4231
NINR AIDS and Aging: Behavioral Sciences
Prevention Research Grants, 4232
NINR Community-Based Prevention Intervention
Research in Environmental Health Sciences
Grants, 4237

Behavioral Sciences
ACS Research Scholar Grants in Psychosocial and
Behavioral and Cancer Control Research, 292
Alzheimer's Association Investigator-Initiated
Research Grants, 646
American Psychological Foundation F. J. McGuigan
Young Investigator Prize, 804
ASID Foundation Joel Polsky Academic Achievement
Award, 1128
CCFF Fellowships, 1531
CCFF Studentships, 1535
CCFF Summer Studentship, 1536
CDC Injury Control Research Centers Grants, 1559
Center for Advanced Study in the Behavioral Sciences
Postdoctoral Fellowships, 1573
DOD National Defense Science and Engineering
Graduate Fellowship Program, 1990
Duke University Postdoctoral Research Fellowships
in Aging, 2138
EF Behavioral Sciences Student Fellowships, 2180
FIC International Research Fellowships, 2336
FIC International Research Scientist Development
Award, 2337
FIC Israeli Ministry of Health Postdoctoral Research
Fellowships, 2340
FIC Swedish Medical Research Council Fellowships,
2342
FIC/ORMH Minority International Research
Training Grants, 2344
Fyssen Foundation Postdoctoral Study Grants, 2465
James S. McDonnell Foundation Scholar Awards,
2955
James S. McDonnell Foundation Understanding
Human Cognition Awards, 2956
Leakey Foundation Franklin Mosher Baldwin
Fellowships, 3150
Marine Corps Historical Center Dissertation
Fellowship, 3272
National Academy of Education Spencer Postdoctoral
Fellowships, 3515
NCI Cancer Education and Career Development
Program, 3586
NCI Established Investigator Award in Cancer
Prevention, Control, Behavioral, and Population
Sciences, 3605
NCI Exploratory Grants for Behavioral Research in
Cancer Control, 3606

Biology

NIH Biology, Development, and Progression of Malignant Prostate Disease Research Grants, 4085
NSF Emerging Frontiers in Research and Innovation (EFRI) Grants, 4559
NSF Physics of Living Systems (PoLS) Grants, 4626
NSF Systematic Biology and Biodiversity Inventories Grants, 4662

Biomarkers
NIDDK Development of Disease Biomarkers Grants, 3992

Biomass Fuels
Colorado Bioscience Discovery Evaluation Grants, 1744
DOE State Energy Program Special Projects, 2051
NSF Biotechnology, Biochemical, and Biomass Engineering (BBBE) Grants, 4504
US CRDF CGP ETF 2009 Energy Research Competition For Joint U.S.-Estonian Research Teams, 5630

Biomaterials
ARO Biomolecular and Cellular Materials and Processes Research Contracts, 1004
Implant Dentistry Research and Education Foundation Grants, 2809
NSF Condensed Matter Physics Research Grants, 4538

Biomechanics
American Society for Surgery of the Hand Research Grants, 812
CDC Injury Prevention and Control Research and State and Community Based Programs, 1560
H.R. Lissner Award, 2588
Implant Dentistry Research and Education Foundation Grants, 2809
Musculoskeletal Rehabilitation Fellowships, 3471
NIOSH Occupational Safety and Health Research and Demonstration Grants, 4251
NSF Physiology and Ethology Research Grants, 4628

Biomedical Education
AAAS Award for Public Understanding of Science and Technology, 23
AHFMR Full-Time M.D./Ph.D. Studentships, 407
AHFMR Infrastructure Funding Grants, 413
AHFMR Institutional Support Program Grants, 414
AHFMR Part-Time Studentships, 420
AHFMR Personnel Funding Grants, 421
AHFMR Scholar Awards, 426
AHFMR Scientist Awards, 427
AHFMR Senior Investigator Awards, 428
AHFMR Senior Scholar Awards, 430
Bristol-Myers Squibb Clinical Outcomes and Research Grants, 1412
HHMI Biomedical Research Grants for International Scientists in Canada and Latin America, 2680
HHMI Biomedical Research Grants for International Scientists in the Baltics, Central and Eastern Europe, Russia, and Ukraine, 2681
HHMI Biomedical Research Grants for International Scientists: Infectious Diseases and Parasitology, 2682
HHMI Gilliam Fellowships for Advanced Study, 2683
HHMI Initiative for Colleges Grants, 2685
HHMI Initiative for Universities Grants, 2686
HHMI Physician-Scientist Early Career Award, 2689
HHMI Undergraduate Science Education Grants for Professors, 2692
Max Kade Foundation Grants, 3307
NCI Ruth L. Kirschstein National Research Service Award (NRSA) Institutional Research Training Grants, 3624
NIGMS Minority Access to Research Careers Ancillary Training Activities Grants, 4063
NIH Summer Internship Program in Biomedical Research, 4175

NIH Undergraduate Scholarships, 4180
NIMH Career Opportunities in Research (COR) Education and Training Honors High School Research Education Grant, 4192
NYCT Biomedical Research Grants, 4701

Biomedical Engineering
Heart Rhythm Society Fellowships, 2638
NCI In Vivo Cellular and Molecular Imaging Centers Planning Grants (Pre-ICMICs), 3611
NHLBI Individual Postdoctoral National Research Service Awards, 3768
NHLBI Intramural Research Training Awards, 3771
NIGMS Fellowships in Quantitative Biology, 4054
NSF Biomedical Engineering (BME) Grants, 4500
NSF Biomedical Engineering and Engineering Healthcare Grants, 4501
NSF Emerging Frontiers in Research and Innovation (EFRI) Grants, 4559
ONR Biological & Biomedical Division Grants, 4755
Whitaker International Fellows and Scholars Program, 5744

Biomedical Research
25th Anniversary Foundation Grants, 5
A.O. Smith Community Grants, 14
AAAS Award for Public Understanding of Science and Technology, 23
AACR Award for Outstanding Achievement in Cancer Research, 97
AACR Team Science Award, 116
AAMC Award for Distinguished Research, 169
Achelis and Bodman Foundations Grants, 256
Adaptec Foundation Grants, 309
AFAR CART Fund Grants, 329
Affymetrix Corporate Contributions Grants, 339
AGS Foundation T. Franklin Williams Scholars Award, 381
AHFMR Clinical Fellowships, 396
AHFMR Clinical Investigator Awards, 397
AHFMR Clinical Investigator Awards, 397
AHFMR Community Support Program Grants, 398
AHFMR Dr. Lionel E. McLeod Health Research Scholarship, 399
AHFMR Establishment and Independent Establishment Grants, 400
AHFMR Fast-Track Studentships and Fellowships, 401
AHFMR Forefront Internship Program Grants, 403
AHFMR Full-Time Fellowships, 406
AHFMR Full-Time M.D./Ph.D. Studentships, 407
AHFMR Full-Time Studentships, 408
AHFMR Health Research Career Renewal Awards, 409
AHFMR Health Research Full-Time Studentships, 410
AHFMR Health Research Fund Grants, 411
AHFMR Heritage Summer Studentships, 412
AHFMR Infrastructure Funding Grants, 413
AHFMR Institutional Support Program Grants, 414
AHFMR Interdisciplinary Team Grants, 415
AHFMR Major Equipment Grants, 416
AHFMR Medical or Health Scholars Investigator Grants, 417
AHFMR Medical or Health Scientist Awards, 418
AHFMR Part-Time Fellowships, 419
AHFMR Part-Time Studentships, 420
AHFMR Personnel Funding Grants, 421
AHFMR Polaris Awards, 422
AHFMR Proposals for Special Initiatives, 424
AHFMR Research Prize, 425
AHFMR Scholar Awards, 426
AHFMR Scientist Awards, 427
AHFMR Senior Investigator Awards, 428
AHFMR Senior Medical or Health Scholars Investigator Grants, 429
AHFMR Senior Scholar Awards, 430
AHRQ Individual Awards for Postdoctoral Fellows Ruth L. Kirschstein National Research Service Awards (NRSA), 450

AHRQ Translating Research into Practice Grants, 451
Albany Medical Center Prize in Medicine and Biomedical Research, 548
Albert and Mary Lasker Foundation Awards, 550
Alcoa Foundation Grants, 561
Alfred I. DuPont Foundation Grants, 597
Allyn Foundation Grants, 624
ALSAM Foundation Grants, 634
American Philosophical Society Library Resident Research Fellowships, 782
Ametek Foundation Grants, 831
Amgen Foundation Grants, 835
AMWA Carroll L. Birch Award, 864
Annenberg Foundation Grants, 894
Anthony R. Abraham Foundation Grants, 900
Arizona Cardinals Grants, 991
ASCO Improving Cancer Care Grants, 1063
ASCO State Affiliate Grants, 1064
ASCO Advanced Clinical Research Award in Colorectal Cancer, 1065
ASCO Advanced Clinical Research Awards in Breast Cancer, 1066
ASCO Advanced Clinical Research Awards in Hematologic Malignancies, 1067
ASCO Advanced Clinical Research Awards in Sarcoma, 1069
ASCO Career Development Award, 1070
ASCO Community Oncology Research Grants, 1071
ASCO Comparative Effectiveness Research Professorship (CERP) in Breast Cancer, 1072
ASCO Long-Term International Fellowships, 1075
ASCO Medical Student Rotation Grants, 1076
ASCO Translational Research Professorships, 1078
ASCO/UICC International Cancer Technology Transfer (ICRETT) Fellowships, 1080
ASM Abbott Laboratories Award in Clinical and Diagnostic Immunology, 1131
Australian Academy of Science Grants, 1237
Aviation Research Grants, 1250
Award in Separations Science and Technology, 1275
Barth Syndrome Foundation Research Grants, 1323
Batchelor Foundation Grants, 1324
BCBSNC Foundation Grants, 1343
Beckman Coulter Community Relations Grants, 1347
Beim Foundation Grants, 1354
Blue Shield of California Grants, 1383
Blum-Kovler Foundation Grants, 1384
Bodman Foundation Grants Program, 1390
Booth-Bricker Fund Grants, 1393
Bosque Foundation Grants, 1394
Bright Family Foundation Grants, 1409
Brinson Foundation Grants, 1410
Bristol-Myers Squibb Clinical Outcomes and Research Grants, 1412
Bristol-Myers Squibb Foundation Global HIV/AIDS Initiative Grants, 1413
Bristol-Myers Squibb Foundation Health Disparities Grants, 1414
Bristol-Myers Squibb Virology and Oncology M.D. Research Fellowships, 1416
Broadhurst Foundation Grants, 1421
Brookdale Foundation Leadership in Aging Fellowships, 1422
Bunge y Born Foundation Grants, 1431
Burden Trust Grants, 1433
Bushrod H. Campbell and Adah F. Hall Charity Fund Grants, 1438
BWF Clinical Scientist Awards in Translational Research, 1442
BWF Institutional Program Unifying Population and Laboratory Based Sciences, 1443
Byron W. and Alice L. Lockwood Foundation Grants, 1446
Carl C. Icahn Foundation Grants, 1506
Carl J. Herzog Foundation Grants, 1508
Catherine Holmes Wilkins Foundation Grants, 1524
CCFF Fellowships, 1531
CCFF Studentships, 1535
CCFF Summer Studentship, 1536

Cancer Prevention

Carbohydrates

Carbon Dioxide

Cardiology

Cardiomyocyte Mitochondria

NHLBI Research on the Role of Cardiomyocyte Mitochondria in Heart Disease: An Integrated Approach, 3800

Cardiovascular Diseases

Abbott Fund Science and Medical Innovation Grants, 216

Advancing Newborn Medicine Fellowships, 316

AHA Beginning Grants-in-Aid, 385

AHA New York State Affiliate Postdoctoral Fellowships, 387

AHA New York State Affiliate Scientist Development Grants, 388

AHA Postdoctoral Fellowship, 389

AHA Predoctoral Fellowship, 390

Alcohol Misuse and Alcoholism Research Grants, 562

AMA Foundation Seed Grants for Research, 654

Anthem Blue Cross and Blue Shield School Nurse Grants, 899

Bristol-Myers Squibb Clinical Outcomes and Research Grants, 1412

C.R. Bard Foundation Grants, 1450

Cardiac Arrhythmias Research and Education Foundation Research Grants, 1503

CDC Core Capacity Program Grants, 1550

Chest Foundation/LUNGevity Foundation Clinical Research in Lung Cancer Grants, 1663

Chestnut Hill Charitable Foundation, Inc Grants, 1664

Children's Cardiomyopathy Foundation Research Grants, 1681

Compass Bank Foundation Grants, 1776

D.F. Halton Foundation Grants, 1831

DeBakey Medical Foundation Fellowships and Scholarships, 1877

Dietetic Outcomes Research Study Grants, 1951

Doris Duke Charitable Foundation Clinical Scientist Development Award, 2085

Duke University Interventional Cardiology Fellowships, 2131

E.L. Wiegand Foundation Grants, 2151

Emma B. Howe Memorial Foundation Grants, 2220

Fannie E. Rippel Foundation Grants, 2292

FIC International Cooperative Biodiversity Groups Grants, 2333

FIC Israeli Ministry of Health Postdoctoral Research Fellowships, 2340

Gerber Foundation Grants, 2513

Greenburg-May Foundation Grants, 2569

Hawaii Community Foundation Health Education and Research Grants, 2630

Howard Gilman Foundation Grants, 2721

Jacob and Valeria Langeloth Foundation Grants, 2935

Japan Heart Foundation Grants, 2965

Lesbian Health Fund Grants, 3166

Missouri Foundation for Health Grants, 3408

National Emergency Medicine Association Grants, 3526

National Heart Foundation of New Zealand Grants, 3537

Nell Warren Elkin and William Simpson Elkin Foundation, 3691

NHLBI Ancillary Studies in Clinical Trials, 3745

NHLBI Cardiac Development Consortium Grants, 3751

NHLBI Cardiac Translational Research Implementation Program Grants, 3752

NHLBI Career Transition Award, 3754

NHLBI Elucidating Nature's Solutions to Heart, Lung, and Blood Diseases and Sleep Disorder Processes Research Grants, 3761

NHLBI Improving Heart Failure Disease Management Grants, 3766

NHLBI Mentored Career Award For Faculty At Minority Institutions, 3774

NHLBI Minority Institution Research Scientist Development Award, 3782

NHLBI Minority Institutional Research Training Program Grants, 3783

NHLBI National Research Service Award Programs in Cardiovascular Epidemiology and Biostatistics, 3785

NHLBI Research on the Role of Cardiomyocyte Mitochondria in Heart Disease: An Integrated Approach, 3800

NHLBI Short-Term Research Education Program to Increase Diversity in Health-Related Research, 3812

NHLBI Stress Management Interventions to Reduce Risk of Coronary Artery Disease, 3813

NIEHS Biological Response to Environmental Health Hazards Research Grants, 4039

NIH Mechanisms Underlying Individual Variations in Drug Responses, 4129

NIH Research on Sleep and Sleep Disorders, 4157

NYAM Glorney-Raisbeck Fellowships, 4696

NYAM Glorney-Raisbeck Medical Student Grants, 4697

RACGP Cardiovascular Research Grants in General Practice, 5024

Rich Foundation Grants, 5085

SMH Heart Failure Research Grants, 5279

Smith Family New Investigator Awards, 5280

W.W. Smith Charitable Trust Grants, 5698

W.W. Smith Charitable Trust Grants for Medical Research in Cancer, AIDS, and Heart Disease, 5699

Cardiovascular System

AHA Established Investigator Grants, 386

AHA New York State Affiliate Scientist Development Grants, 388

CDC Core Capacity Program Grants, 1550

Chest Foundation/LUNGevity Foundation Clinical Research in Lung Cancer Grants, 1663

Children's Cardiomyopathy Foundation Research Grants, 1681

Duke University Interventional Cardiology Fellowships, 2131

NHLBI Cardiac Translational Research Implementation Program Grants, 3752

NYAM Glorney-Raisbeck Medical Student Grants, 4697

Simone and Cino del Duca Grand Prix Awards, 5260

Career Education and Planning

AAUW Career Development Grants, 209

Adaptec Foundation Grants, 309

AFAR Paul Beeson Career Development Awards in Aging Research for the Island of Ireland, 332

AHNS/AAO-HNSF Surgeon Scientist Combined Award, 439

AHNS/AAO-HNSF Young Investigator Combined Award, 440

Alfred P. Sloan Foundation Education and Careers in Science and Technology Grants, 601

Alfred P. Sloan Foundation Workplace, Work Force and Working Families Grants, 611

ASTD Dissertation Award, 1204

ASTD Research Award, 1205

Award for Encouraging Disadvantaged Students into Careers in the Chemical Sciences, 1262

Center for Education of Women Visiting Scholar Research Grants, 1583

Chiron Foundation Community Grants, 1684

Corning Incorporated Foundation Educational Grants, 1799

DaimlerChrysler Corporation Fund Grants, 1852

DARPA Military Research and Technology Development Grants, 1862

Dept of Ed Vocational Rehabilitation Services Demonstration and Training Programs, 1903

Dept of Ed Women's Educational Equity Program Grants, 1904

Engineering Information Foundation Grants, 2225

Ford Motor Company Fund Grants Program, 2379

Girl's Best Friend Foundation Grants, 2537

Hartford Aging and Health Program Awards, 2619

IFT Foundation Grants, 2787

Laura J. Niles Foundation Grants, 3138

McCarthy Family Foundation Grants, 3312

MMA Lifchez/Stronach Curatorial Internship, 3437

Morris K. Udall Native American Congressional Internship Program, 3450

NCI Cancer Education and Career Development Program, 3586

NCI Centers of Excellence in Cancer Communications Research, 3592

NCI Established Investigator Award in Cancer Prevention, Control, Behavioral, and Population Sciences, 3605

NCI Mentored Career Development Award to Promote Diversity, 3614

NCI/NIDA Transdisciplinary Tobacco Use Research Centers Grants, 3632

NEI Scholars Program, 3686

NHGRI Individual Mentored Research Scientist Awards in Genomic Research and Analysis, 3741

NHGRI Mentored Research Scientist Development Award, 3742

NIA Mentored Research Scientist Development Award, 3855

NIEHS Mentored Research Scientist Development Award, 4047

NIGMS Institutional Research and Academic Career Development Award, 4058

NIH Transitional Career Development Award in Women's Health Research, 4179

NIMH Career Opportunities in Research (COR) Education and Training Honors High School Research Education Grant, 4192

NIMH Career Opportunities in Research (COR) Education and Training Honors Undergraduate Research Training Grant, 4193

NINR Mentored Research Scientist Development Award, 4242

NSF Activities in Science, Engineering, and Mathematics for Persons with Disabilities, 4452

NSF Faculty Early Career Development (CAREER) Grants, 4565

NSF Presidential Early Career Awards for Scientists and Engineers (PECASE) Grants, 4639

United Airlines Foundation Education Grants, 5605

Walter S. Johnson Foundation Grants, 5711

Caribbean Islands

Center for Black Music Research Rockefeller Fellowships, 1578

DuBois-Mandela-Rodney Fellowship Program, 2123

Kluge Center Kislak Fellowships in American Studies, 3096

Kluge Center Kislak Short-Term Fellowships in American Studies, 3097

OIEAHC Jamestown Prize, 4733

Schomburg Center for Research in Black Culture Scholars in Residence Grants, 5220

Caries

DHHS Oral Health Promotion Research Across the Lifespan, 1932

NIH Ruth L. Kirschstein National Research Service Awards for Individual Postdoctoral Fellows, 4163

Cartography

Kluge Center Kislak Fellowships in American Studies, 3096

Kluge Center Kislak Short-Term Fellowships in American Studies, 3097

NL Short-Term Fellowships in the History of Cartography, 4279

Cataloging and Classification

Getty Research Grants for Institutions - Archival Grants, 2521

Getty Research Grants for Institutions - Art History Research Grants, 2522

Getty Research Grants for Institutions - Cataloguing of Museum Collections Grants, 2523

NEH Preservation Assistance Grants, 3677

Chemistry Education

Chemistry, Analytical

Computer Science Education

Computer Simulation/Modeling

Computer Software

Cultural Diversity

Cultural Heritage Preservation

Weyerhaeuser Company Foundation Contributions
Program, 5743
Whiting Foundation Grants, 5747
WHOI George P. Panteleyev Award, 5752
Willary Foundation Grants, 5762
William G. McGowan Charitable Fund Grants, 5771
William K. Warren Foundation Grants, 5772
William N. and Myriam Pennington Foundation
Grants, 5773
Williams Companies Foundation Grants, 5779
Women's Foundation of California Grants, 5789
Woodward Fund Grants, 5799
World Bank Graduate Scholarship Program, 5801
Wright Center Mini-Fellows, 5806
Xerox Foundation Grants, 5814

Education and Work
Xerox Foundation Grants, 5814

Education Reform
Bodman Foundation Grants Program, 1390
Boeing Company Contributions Program Grants,
1391
Carnegie Corporation of New York Grants, 1510
Charitable Leadership Foundation Education Grants,
1637
Charles Stewart Mott Foundation Grants, 1651
Consumers Energy Foundation, 1792
ETS Sylvia Taylor Johnson Minority Fellowship in
Educational Measurement, 2264
ExxonMobil Education Foundation Grants, 2280
ExxonMobil Grants, 2281
Fordham Prizes, 2380
Grawemeyer Education Awards, 2564
IREX University Administration Support Program
(UASP), 2889
Joyce Foundation Grants, 3055
M.D. Anderson Foundation Grants, 3244
NFIE Foundation Learning and Leadership Grants,
3711
NSF Interagency Education Research Initiative
(IERI) Program Grants, 4588
Phil Hardin Foundation Grants, 4959
Spencer Foundation Major Research Grants, 5354
Walter S. Johnson Foundation Grants, 5711
William and Flora Hewlett Foundation Education
Grants, 5766
William Caspar Graustein Memorial Fund Grants,
5770

Educational Administration
AASA Educational Administration Scholarships, 201
AASL Distinguished School Administrator Award,
204
ACE Fellows Program, 251
AERA Research Grants, 321
CFPC Fellowships, 1622
Harvard University Administrative Fellowships, 2623
Harvard University Graduate School of Education
Fellowships and Awards, 2625
IREX University Administration Support Program
(UASP), 2889
Joseph Klingenstein Fellows Program, 3047
Lumina Foundation for Education Grants, 3234
Meadows Foundation Grants, 3328
NSU Fischler School of Education and Human
ServicesDoctoral Fellowships, 4677
Pennsylvania State University Education Graduate
Assistantships, 4933
Robert R. McCormick Tribune Foundation Grants
Programs, 5105
RSC Sir John William Dawson Medal, 5152
Wallace Foundation Grants, 5706
William and Flora Hewlett Foundation Education
Grants, 5766

Educational Evaluation/Assessment
A.O. Smith Community Grants, 14
AIR National Summer Data Policy Institute
Fellowships, 517

AIR/IES-NCES Postdoctoral Policy Fellowships in
Washington, 520
ETS Postdoctoral Fellowships, 2262
ETS Sylvia Taylor Johnson Minority Fellowship in
Educational Measurement, 2264
Lilly Endowment Community Alliances to Promote
Education (CAPE), 3194
NSF Centers for Learning and Teaching (CLT)
Grants, 4509
RCPSC Royal College Fellowship for Studies in
Medical Education, 5057
Stuart Foundation Grants, 5481
UPS Foundation Grants, 5627

Educational Finance
AERA Research Grants, 321
Allstate Foundation Grants, 623
American Education Research Association Grants,
709
John Grenzebach Awards for Outstanding Research
in Philanthropy for Education, 3027

Educational Humanities
National Academy of Education Spencer Postdoctoral
Fellowships, 3515

Educational Instruction Programs
ACLS Collaborative Research Awards, 264
ACLS Fellowships, 271
American Historical Association William Gilbert
Award, 743
American Montessori Society Thesis and Dissertation
Awards, 766
Andrew W. Mellon/ACLS Dissertation Completion
Fellowships, 877
ASM International Fellowship for Latin America and
the Caribbean, 1135
ASM International Fellowships for Asia and Africa,
1136
Bayer Clinical Scholarship Award, 1327
BCBSM Foundation Student Award Program, 1340
Center for Khmer Studies Fellowships, 1588
Irvine Fellowships, 2892
ITEA Maley/FTE Scholarship, 2903
Lawrence Fellows Program, 3142
Love Thy Neighbor Course Competition for
Secondary School Faculty, 3216
Posen Foundation Grants for the Study of Secular
Judaism, 4997
Roothbert Fund Scholarships, 5128
San Diego Women's Foundation Grants, 5201
TERC Research Fellowships, 5522
Ursula Thrush Peace Seed Grants, 5628

Educational Planning/Policy
Achelis and Bodman Foundations Grants, 256
AERA Dissertation Grants, 320
AERA Research Grants, 321
AIR/IES-NCES Postdoctoral Policy Fellowships in
Washington, 520
American Education Research Association Grants,
709
Campaign for the Civic Mission of Schools Grants,
1462
Center for Education of Women Visiting Scholar
Research Grants, 1583
Charles G. Koch Charitable Foundation Grants, 1641
Congressional Hispanic Caucus Institute Fellowships,
1784
DOE Albert Einstein Distinguished Educator
Fellowship Program, 1997
ETS Postdoctoral Fellowships, 2262
ETS Predoctoral Summer Program, 2263
ExxonMobil Education Foundation Grants, 2280
Grawemeyer Education Awards, 2564
Harvard Postdoctoral Fellowships on Education, 2621
Herman Kahn Fellowship, 2678
IREX Regional Policy Symposium Grants, 2887
IREX University Administration Support Program
(UASP), 2889

IREX US Embassy Policy Specialist (EPS) Program
Grants, 2890
Japan Foundation Center for Global Partnership
(CGP) Institutional Grants, 2962
Joseph Klingenstein Fellows Program, 3047
Lilly Endowment Community Alliances to Promote
Education (CAPE), 3194
Lumina Foundation for Education Grants, 3234
MetLife Foundation Community College Excellence
Award, 3354
Milken Family Foundation Grants, 3391
National Assessment of Educational Progress Visiting
Scholar Program, 3516
NSF Interagency Education Research Initiative
(IERI) Program Grants, 4588
Pennsylvania State University Education Graduate
Assistantships, 4933
Phil Hardin Foundation Grants, 4959
SRC Educational Science Postdoctoral Fellowships,
5379
Stuart Foundation Grants, 5481
Wallace Foundation Grants, 5706
William and Flora Hewlett Foundation Education
Grants, 5766
William Caspar Graustein Memorial Grants, 5770
William T. Grant Foundation Distinguished Fellows
Awards, 5775

Educational Programs
Allstate Foundation Grants, 623
Blowitz-Ridgeway Foundation Early Childhood
Development Research Award, 1381
Third Millennium Foundation Fellowships, 5532
V. Kann Rasmussen Foundation Grants, 5657
Washington Women's Foundation Grants, 5713

Educational Psychology
ETS Predoctoral Summer Program, 2263
Helen McWilliam Memorial Scholarship, 2653
Ittleson Foundation Mental Health Grants, 2906
RCPSC Royal College Fellowship for Studies in
Medical Education, 5057

Educational Technology
Arizona Commission on the Arts Education Projects
Grants, 992
AT&T Education Grants, 1212
AVDF Secondary Education Grants, 1249
Bay and Paul Foundations, Inc Grants, 1326
Corning Incorporated Foundation Educational
Grants, 1799
Dept of Ed Star Schools Program Grants, 1901
Educational Communications and Technology
Awards, 2169
Educational Ventures International Foundation
Grants, 2170
ETS Postdoctoral Fellowships, 2262
Fletcher Jones Foundation Grants, 2364
Grawemeyer Education Awards, 2564
Green Foundation Grants, 2568
Hilda and Preston Davis Foundation Grants, 2699
Informix for Innovation Software Grants, 2824
ITEA Maley/FTE Scholarship, 2903
NSF CISE Educational Innovation Grants, 4532
RCPSC Royal College Fellowship for Studies in
Medical Education, 5057
Sun Microsystems Academic Equipment Grants, 5486
ThinkQuest Scholarship Competition, 5531

Educational Testing/Measurement
ACT Summer Internship Program, 295
AERA Research Grants, 321
ETS Postdoctoral Fellowships, 2262
ETS Predoctoral Summer Program, 2263
National Assessment of Educational Progress Visiting
Scholar Program, 3516
NBME Stemmler Medical Education Research Fund
Grants, 3566
SRC Educational Science Language Checking
Grants, 5376

Health Care Access

Jacob and Hilda Blaustein Foundation Grants, 2934
Jacob and Valeria Langeloth Foundation Grants, 2935
James H. Cummings Foundation Grants, 2942
Joe W. and Dorothy Dorsett Brown Foundation
Grants, 3009
John S. Dunn Research Foundation Grants and
Chairs, 3035
Josiah W. and Bessie H. Kline Foundation Grants,
3051
Joy Family Foundation Grants, 3054
Legler Benbough Foundation Grants, 3158
Leo Goodwin Foundation Grants, 3163
Lillian S. Wells Foundation Grants, 3192
Linn-Henley Charitable Trust Grants, 3200
Little River Foundation Grants, 3205
Longwood Foundation Grants, 3210
Louis Calder Foundation Grants, 3212
Lubbock Area Foundation, Inc., 3228
Lubrizol Foundation Grants Program, 3229
Lucile Horton Howe and Mitchell B. Howe
Foundation Grants, 3230
M.D. Anderson Foundation Grants, 3244
Mabel Y. Hughes Charitable Trust Grants, 3248
Margaret L. Wendt Foundation Grants, 3263
Mary S. and David C. Corbin Foundation Grants,
3289
Matilda R. Wilson Fund Grants, 3298
Max and Victoria Dreyfus Foundation Grants, 3304
Max Goldenberg Foundation Grants, 3306
McGregor Fund Grants, 3317
McLean Contributionship Grants, 3320
Meadows Foundation Grants, 3328
Mervin Bovaird Foundation Grants, 3350
Molson Companies Donations Fund Grants, 3443
National Patient Safety Foundation Grants, 3544
Norcliffe Foundation Grants, 4406
North Carolina Community Development Grants, 4419
Olin Corporation Charitable Trust Grants, 4739
Oppenstein Brothers Foundation Grants, 4825
Pasadena Foundation Average Grants, 4911
Peacock Foundation Grants, 4927
Pott Foundation Grants, 4999
R.C. Baker Foundation Grants, 5019
Rayonier Foundation Grants, 5052
Rich Foundation Grants, 5085
Richard and Susan Smith Family Foundation Grants,
5087
Robert E. and Evelyn McKee Foundation Grants,
5098
Rockwell International Corporate Trust Grants, 5118
Russell Berrie Foundation Grants, 5157
Salem Foundation Grants, 5191
Sensient Technologies Foundation Grants, 5234
Shell Oil Products US Corporate Contributions, 5246
Sid W. Richardson Foundation Grants, 5254
Starr Foundation Grants, 5454
Sterling-Turner Charitable Foundation Grants, 5463
Strake Foundation Grants, 5465
Temple-Inland Foundation Grants, 5519
Texas Instruments Foundation Grants, 5525
Thomas and Dorothy Leavey Foundation Grants, 5535
Thompson Charitable Foundation Grants, 5540
United Hospital Fund of New York Grants Program,
5607
USG Foundation Grants, 5647
Vernon K. Krieble Foundation Grants, 5665
W. C. Griffith Foundation Grants, 5691
W.C. Griffith Foundation Grants, 5694
Willard and Pat Walker Charitable Foundation
Grants, 5761
Winn-Dixie Stores Foundation Grants, 5781

Hotel and Restaurant Management
College of Human Sciences Scholarships, 1741
Coors Brewing Corporate Contributions Grants, 1796

Housing
Adaptec Foundation Grants, 309
Ann Arbor Area Community Foundation Grants, 888
Ann Peppers Foundation Grants, 889

Arizona Cardinals Grants, 991
Batchelor Foundation Grants, 1324
Boston Foundation Grants, 1396
Byron W. and Alice L. Lockwood Foundation Grants,
1446
Carl and Eloise Pohlad Family Foundation Grants,
1504
Carpenter Foundation Grants, 1518
Chautauqua Region Community Foundation Grants,
1658
Christopher D. Smithers Foundation Research
Grants, 1691
CMHC External Research Program Grants for
Housing Research, 1731
Community Foundation AIDS Endowment Awards,
1765
Congressional Hispanic Caucus Institute Fellowships,
1784
Eugene and Agnes E. Meyer Foundation Grants,
2267
Evanston Community Foundation Grants, 2275
Fannie Mae Foundation Grants for Research, 2293
Fannie Mae Foundation Kennedy School of
Government Fellowships, 2294
Fannie Mae James A. Johnson Fellowships, 2295
Frank Stanley Beveridge Foundation Grants, 2409
George Gund Foundation Grants, 2500
Home Depot Foundation Grants, 2711
HomeBanc Foundation Grants, 2712
Household International Corporate Giving Program
Grants, 2718
HUD Community Outreach Partnership Centers
Grants, 2735
HUD Doctoral Dissertation Research Grants, 2736
HUD General Research and Technology Activity
Grants, 2737
Hutchinson Community Foundation Grants, 2758
Jackson Foundation Grants, 2931
Joe W. and Dorothy Dorsett Brown Foundation
Grants, 3009
John Randolph Haynes and Dora Haynes Foundation
Grants, 3034
Lilly Endowment Grants Program, 3195
Luther I. Replogle Foundation Grants, 3237
Mary S. and David C. Corbin Foundation Grants,
3289
McGregor Fund Grants, 3317
Mizuho USA Foundation Grants, 3410
Musgrave Foundation Grants, 3472
NEA Public Works Grants, 3654
NIST Small Business Innovation Research (SBIR)
Grants, 4258
Northwestern Mutual Foundation Grants, 4423
NSF Partnership for Advancing Technologies in
Housing Grants, 4616
Olin Corporation Charitable Trust Grants, 4739
Paul G. Allen Family Foundation Grants, 4915
Peacock Foundation Grants, 4927
Poverty and Race Research Action Council Grants,
5001
Reuter Foundation Grants, 5082
Rhode Island Foundation Grants, 5084
Schramm Foundation Grants, 5224
Sid W. Richardson Foundation Grants, 5254
Starr Foundation Grants, 5454
Surdna Foundation Grants, 5492
Tessie Agan Award Competition, 5524
Thompson Charitable Foundation Grants, 5540
Van Ameringen Foundation Grants, 5660
Vancouver Foundation Grants and Community
Initiatives Program, 5662
Walter S. Johnson Foundation Grants, 5711

Housing Law
HUD Urban Scholars Postdoctoral Fellowships, 2738

Human Connectome
NIH Human Connectome Project Grants, 4117

Human Development
Chapin Hall International Fellowships in Children's
Policy Research, 1636
College of Human Sciences Scholarships, 1741
DHHS Great Lakes Human Health Effects Research
Program Grants, 1918
Grammy Foundation Grants, 2562
Human Growth Foundation Small Grants, 2745
NICHD NRSAs--Institutional, 3931
NICHD Small Grants Program, 3936
NIDA Neurobiological and Behavioral Research on
Nicotine and Tobacco Products, 3953
NIH Endocrine Disruptors: Epidemiologic
Approaches Grants, 4102
NIH Genetic Susceptibility and Variability of Human
Malformations Research Grants, 4113
NIH Research on Sleep and Sleep Disorders, 4157
NSF Behavioral Neurosciences Research Grants, 4491
Spencer Foundation Education Dissertation
Fellowships, 5353
Wellesley Centers for Women Postdoctoral
Fellowships, 5727

Human Factors in Engineering
Aviation Research Grants, 1250
NLM Research Project Grants, 4295
NSF Thermal Transport Processes Research Grants,
4666
ORISE Nuclear Regulatory Commission Historically
Black Colleges and Universities Student Research
Participation, 4855
SME Research Initiation Grants, 5278

Human Genome
AACR Team Science Award, 116
DOE Health Effects and Life Sciences Research
Grants, 2027
DOE Human Genome Program--Ethical, Legal, and
Social Implications, 2030
Lilly Endowment Grants Program, 3195
NHGRI Courses in Genomic Analysis and
Interpretation and Ethical, Legal, and Social
Implications (ELSI) Research Grants, 3737
NHGRI Curriculum Development Award in
Genomic Research and Analysis, 3738
NHGRI Genome Scholar Development and Faculty
Transition Award, 3739
NHGRI Genomic Mapping, Sequencing, and
Analysis Pilot Project Grants, 3740
NHGRI Individual Mentored Research Scientist
Awards in Genomic Research and Analysis, 3741
NHGRI Mentored Research Scientist Development
Award, 3742
NHGRI Technologies for Genomic Mapping,
Sequencing, and Analysis Research Grants, 3743
NHLBI Integrating Lung Genetics and Genomics in
Human Populations Grants, 3770
NIGMS Pilot Projects for the Protein Structure
Initiative (Structural Genomics), 4068
NIH Genetic Architecture of Complex Phenotypes,
4112
NIH Human Connectome Project Grants, 4117
NIH Individual Postdoctoral Fellowships in
Genomics and Related ELSI Topics, 4122
NIH Protein Structure Initiative (Structural
Genomics) Research Grants, 4148
NIH Protein Structure Initiative (Structural
Genomics)--SBIR/STTR GRANTS, 4149
NIH Studies of Ethical, Legal, and Social
Implications of Research into Human Genetic
Variation Grants, 4174
W.M. Keck Foundation Grants Program, 5697

Human Learning and Memory
AASL/Highsmith Research Grant, 205
ASTD Research Award, 1205
ATA Education Research Award, 1215
Center for Judaic Studies Fellowships, 1587
Coca-Cola Foundation Grants, 1736

Humanities Education

CIA Undergraduate Co-Op Program - Engineer/
Program Management Engineer, 1696
CIA Undergraduate Internship Program - Analysis,
1697
Community Foundation of the Virgin Islands Judith
A. Towle Environmental Studies Grants, 1774
Field Museum High School Internships, 2346
Field Museum Undergraduate Internships, 2347
Robert W. Deutsch Foundation Grants, 5108
Skaggs Foundation Grants, 5269
Smithsonian Native American Internships, 5290
SRC Educational Science Doctoral Program Grants,
5371
SRC Educational Science Junior Research Position
Grants, 5375
SRC Educational Science Network Grants, 5378
WWF Russell E. Train EFN Professional
Development Grants, 5808

Interstellar Studies
NSF Stellar Astronomy and Astrophysics (SAA)
Research Grants, 4660

Intervention Programs
ADC Telecommunications STEM Education Grants,
310
BFGoodrich Collegiate Inventors Prize, 1372
Bodman Foundation Grants Program, 1390
CDC Injury Control Research Centers Grants, 1559
DHHS Universal Newborn Hearing Screening and
Intervention, 1944
DOE Energy-Related Inventions Grants, 2016
DOE Inventions and Innovation Grant Program
(I&I), 2033
Klingenstein Third Generation Foundation Grants in
Depression, 3093
Lemelson-MIT Lifetime Achievement Award, 3161
Lemelson-MIT Prize, 3162
NIAAA Strategies for the Prevention of DUI
Recidivism Research Grants, 3888
NIH Biobehavioral Research for Effective Sleep, 4083
NIH National Occupational Research Agenda
Grants, 4140
NIH New Directions in Pain Research Grants, 4142
NIH Pathological Gambling: Basic, Clinical, and
Services Research Grants, 4146
NIMH HIV/AIDS, Severe Mental Illness, and
Homelessness Research Grants, 4201
NIMH Pilot Effectiveness Trials for Mental
Disorders Grants, 4210
NINR Community-Based Prevention Intervention
Research in Environmental Health Sciences
Grants, 4237
NINR Health Risk Reduction: Community-Based
Strategies, 4239
NSF Partnerships for Innovation (PFI) Grants, 4617
William T. Grant Foundation Research Grants, 5776

Investments and Securities
MSI Marketing Strategy Meets Wall Street Research
Competition, 3461

Ionized Gases
APS James Clerk Maxwell Prize for Plasma Physics,
950
APS Will Allis Prize for the Study of Ionized Gases,
965

Iran
IREX Individual Advanced Research Opportunities
(IARO) Grants, 2883

Ireland
CINOA Prize, 1703
Visiting Fellowship in Irish Studies at National
University Galway, 5685
Islam
USIP International Peace Projects Solicited Grants,
5652

Islamic Studies
Ahmanson and Getty Postdoctoral Fellowships, 434
AIAR Andrew W. Mellon Foundation Fellowships,
454
AIAR Educational and Cultural Affairs Fellowships,
456
Center for Advanced Judaic Studies Postdoctoral
Fellowships, 1572
Fellowships at American Schools of Oriental
Research--Amman, 2328
Fellowships for Research in Egypt, 2329
King Faisal Foundation Grants, 3087
Library of Congress Rockefeller Islamic Studies
Fellowships, 3190
Malone Fellowship in Arab and Islamic Studies, 3255
NEH/AIAR Postdoctoral Fellowships, 3680

Isotope/Radiation Technology
AINSE Postgraduate Research Awards, 514
DSO Radiation Biodosimetry (RaBiD) Grants, 2122

Israel
BA Exchange Grants, 1293
BARD Postdoctoral Fellowships, 1318
BARD Research Grants, 1319
Canadian Friends of Hebrew University Research and
Study Grants, 1476
CAORC Regional Research Fellowships, 1500
Donald and Sylvia Robinson Family Foundation
Grants, 2078
FIC Israeli Ministry of Health Postdoctoral Research
Fellowships, 2340
Fishman Family Foundation Grants, 2358
Hadassah Foundation Grants, 2590
Harvey Prizes, 2627
Helen Bader Foundation Grants, 2646
Institute for Palestine Studies, Publishing, and
Research Organization Grants, 2831
Jacob and Hilda Blaustein Foundation Grants, 2934
Jewish Education and Scholarships Program, 2981
Lady Davis Fellowships, 3123
Lucius N. Littauer Foundation Grants, 3231
Millstone Foundation Grants, 3397
Minerva Fellowships, 3400
Minerva Short-Term Research Grants, 3401
Russell Berrie Foundation Grants, 5157
Shalem Graduate Fellowships, 5238
Stella and Charles Guttman Foundation Grants, 5460

Italian Art
CINOA Prize, 1703
Gladys Krieble Delmas Foundation Grants for
Venetian Research, 2538
Paul Mellon Centre Rome Fellowship, 4920
Roberto Longhi Foundation Fellowships, 5109

Italian History
American Historical Association Helen and Howard
R. Marraro Prizes, 721

Italian Language/Literature
Gladys Krieble Delmas Foundation Grants for
Venetian Research, 2538
MLA Aldo and Jeanne Scaglione Prize for Italian
Studies, 3416
NIAF Scholarships, 3890
Robert L. Baker Summer Research Fellowship for
Second Language Acquisition in an Environment
of Immersion, 5101
Rotary Foundation Ambassadorial Scholarships, 5133

Italian Studies
AAR Postdoctoral Fellowships, 187
AAR Predoctoral Fellowships, 188
Center for Italian Renaissance Studies Fellowships,
1586
NIAF Scholarships, 3890

Italy
AAR Classical Society of the American Academy in
Rome Summer Scholarships, 186
AAR Postdoctoral Fellowships, 187
AAR Predoctoral Fellowships, 188
AAR Rome Prize Fellowships, 189
Alexander von Humboldt Foundation Fellowships for
Japan-Related Research, 568
American Historical Association Helen and Howard
R. Marraro Prizes, 721
CAORC Regional Research Fellowships, 1500
CINOA Prize, 1703
Dumbarton Oaks Junior Fellowships, 2140
Dumbarton Oaks Project Grants, 2141
Dumbarton Oaks Summer Fellowships, 2142
Fulbright Travel Grants, 2454
Gladys Krieble Delmas Foundation Grants for
Venetian Research, 2538
Italian History Prizes, 2902
Olivia James Traveling Fellowship, 4741
Roberto Longhi Foundation Fellowships, 5109

Japan
AIEA Honors Scholarships, 504
Alexander von Humboldt Foundation JSPS Research
Fellowships, 580
Alexander von Humboldt Foundation Philipp Franz
von Siebold Award for Japanese Researchers, 584
American College of Surgeons Traveling Fellowship
to Japan, 704
American Historical Association John K. Fairbank
Prize in East Asian History, 732
American Studies in Japan Research Grants, 822
BA Exchange Grants, 1293
Center for East Asian Studies Fellowships, 1582
Edwin O. Reischauer Institute of Japanese Studies
Postdoctoral Fellowships, 2178
Fulbright International Education Administrators
(IEA) Seminar Program Grants, 2441
Japan Doctoral Fellowships, 2960
Japan Foundation Center for Global Partnership
(CGP) Grants, 2961
Japan Foundation Center for Global Partnership
(CGP) NPO Fellowships, 2963
Japan Research Fellowships, 2966
Japan Study Program Grants, 2967
Japan-United States Friendship Commission Grants,
2968
Japanese Language Program Grants, 2969
Japanese Studies in the United States Research
Grants, 2970
Japanese Studies Speakers and Panels Grants, 2971
Japanese Studies United States Research Travel
Grants, 2972
JSPS Bilateral Exchange Programs with Foreign
Academic Institutions, 3056
JSPS Fellowships for Foreign Scientists, 3057
JSPS International Joint Research Projects and
Scientific Seminars, 3058
Kobe College Corporation Graduate Fellowships,
3104
OAH Japanese Residencies Program Grants, 4717
Saison Foundation Grants, 5190
Sasakawa Fellowships, 5207
Short-Term Travel to Japan Grants, 5250
SSRC Japan Society for the Promotion of Science
(JSPS) Fellowships, 5440
SSRC Japan Studies Dissertation Workshop Grants,
5441
Twentieth-Century Japan Research Awards, 5569

Japanese Art
ACC Japan-US Arts Program Fellowships, 235
Fulbright International Education Administrators
(IEA) Seminar Program Grants, 2441
Japan-United States Friendship Commission Grants,
2968

Kodak Community Relations Grants, 3105
MSI Alden G. Clayton Dissertation Proposal
Competition, 3460
MSI Marketing Strategy Meets Wall Street Research
Competition, 3461
MSI Research Support Standard Grants, 3462
National Dairy Promotion and Research Board
Scholarships, 3524
NCI Review and Analysis of Tobacco Industry
Documents Grants, 3623
NCI/NIDA Transdisciplinary Tobacco Use Research
Centers Grants, 3632
Ontario Arts Council Integrated Arts Grants, 4823
RA Kleinhans Fellowship for Research in Tropical
Nontimber Forest Products, 5022
Robert Sterling Clark Foundation Grants, 5107

Marketing Education
Kosciuszko Foundation Tuition Scholarships, 3111
MSI Alden G. Clayton Dissertation Proposal
Competition, 3460

Marketing Research
ACT Summer Internship Program, 295
American Academy of Advertising Doctoral
Dissertation Competition, 661
DOA Federal-State Marketing Improvement Program
(FSMIP), 1961
DOC/ITA Market Development Cooperator
Program (MDCP), 1983
Duke University Libraries Hartman Center Travel
Grants, 2133
Duke University Libraries J. Walter Thompson
Research Fellowships, 2134
MSI Alden G. Clayton Dissertation Proposal
Competition, 3460
MSI Marketing Strategy Meets Wall Street Research
Competition, 3461
MSI Research Support Standard Grants, 3462

Mass Communication
ACIE Host University Edmund S. Muskie/Freedom
Support Act Graduate Fellowships, 258
IREX Edmund S. Muskie Graduate Fellowship
Program (Muskie), 2881
Journalism Foundation of Metropolitan Saint Louis
Scholarships, 3052
Kappa Tau Alpha Frank Luther Mott KTA Research/
Book Award, 3068
SSRC Collaborative Grants in Media and
Communications, 5432
USIA Bosnia and Herzegovina Undergraduate
Development Program Grants, 5650

Mass Media
AAAS Science Journalism Awards, 36
Abe Schecter Graduate Scholarship in Electronic
News, 218
Brainerd Foundation Grants, 1405
Center for Defense Information Internships, 1581
East-West Center Jefferson Fellowships, 2160
Five College Women's Studies Ford Associateships,
2361
Freedom of the Press Fellowships and Internships,
2417
IFT Foundation Grants, 2787
IREX MENA Media Emerging Leaders Fellowships,
2885
John D. and Catherine T. MacArthur Foundation
Program on Peace and International Cooperation,
3020
Lumina Foundation for Education Grants, 3234
Michele Clark Fellowship, 3376
MLA First Book Prize, 3421
MLA James Russell Lowell Prize, 3425
Paul Robeson Fund for Independent Media Grants,
4923
Ploughshares Fund Grants, 4994
Rockefeller Foundation Humanities Fellowships, 5115
RTNDF Capitol Hill News Internships, 5155

SSRC Collaborative Grants in Media and
Communications, 5432
USHMM Summer Research Workshops for Scholars,
5648
USHMM Visiting Scholars Fellowships, 5649
Vada and Barney Oldfield Fellowship for National
Security Reporting, 5658
World Press Institute Fellowships, 5802

Mass Spectrometry
Ethical Research Grants and Fellowships, 2260
Frank H. Field and Joe L. Franklin Award for
Outstanding Achievement in Mass Spectrometry,
2407

Mass Transfer
NIST Fire Research Grants, 4253

Massachusetts
MHS Long Term Research Fellowships, 3367
MHS Short Term Research Fellowships, 3369

Massage Therapy
Massage Therapy Foundation Practitioner Case
Report Contest, 3293
Massage Therapy Foundation Research Grants, 3294
Massage Therapy Foundation Student Case Report
Contest, 3295

Material Sciences
APS Arthur L. Schawlow Prize in Laser Science, 934

Materials Acquisition (Books, Tapes, etc.)
Ball Brothers Foundation Rapid Grants, 1309
Fund for Folk Culture Artists Support Grants, 2458
Vancouver Foundation Grants and Community
Initiatives Program, 5662

Materials Engineering
AFOSR Structural Mechanics Research Grants, 360
AIChE Charles M.A. Stine Award, 466
ARO Synthesis and Processing of Materials Research
Contracts, 1027
AT&T Industrial Ecology Faculty Fellowships, 1213
Nanyang Technological University Lee Kuan Yew
Postdoctoral Fellowship, 3494
NSF Materials World Network: Cooperative Activity
in Materials Research between US Investigators
and their Counterparts Abroad (MWN), 4598
US CRDF CGP-Climate Change and Energy
Competition For Joint U.S.-Ukrainian Research
Teams, 5631

Materials Sciences
AFOSR Metallic Materials Research Grants, 346
AFOSR Molecular Dynamics Research Grants, 347
AFOSR Organic Matrix Composites Research
Grants, 349
AFOSR Physical Mathematics and Applied Analysis
Research Grants, 350
AFOSR Polymer Chemistry Research Grants, 351
AFOSR Quantum Electronic Solids Research Grants,
352
AFOSR Semiconductor Materials Research Grants,
353
AFOSR Space Electronics Research Grants, 357
AFOSR Theoretical Chemistry Research Grants, 361
AINSE Postgraduate Research Awards, 514
American Chemical Society PRF Undergraduate
Research Grants (UR), 694
ANL Faculty Research Participation Program, 884
APS James C. McGroddy Prize for New Materials, 949
ARO Degradation, Reactivity, and Protection of
Materials Research Contracts, 1008
ARO Electrochemistry and Advanced Energy
Conversion Research Contracts, 1010
ARO Mechanical Behavior of Materials Research
Contracts, 1013
ARO Physical Behavior of Materials Research
Contracts, 1018

ARO Polymer Chemistry Research Contracts, 1019
ARO Solid State Devices Research Contracts, 1024
ARO Structures and Dynamics Research Contracts,
1026
ARO Synthesis and Processing of Materials Research
Contracts, 1027
ARO Theoretical Physics and Nonlinear Phenomena
Research Contracts, 1029
ASEE Army Research Laboratory Postdoctoral
Fellowships, 1096
AT&T Bell Labs Graduate Research Program, 1211
Canada Graduate Scholarships (CGS) and NSERC
Postgraduate Scholarships (PGS), 1470
Carnegie Postdoctoral Research Fellowships, 1514
CHF Glenn E. and Barbara Hodsdon Ullyot
Scholarship, 1666
CHF Robert W. Gore Fellowship in Materials
Innovation, 1668
CHF Societe de Chimie Industrielle (American
Section) Fellowship, 1670
CHF Travel Grants, 1671
DOE Advanced Nuclear Energy Systems Basic
Research Grants, 1996
DOE Basic Energy Sciences (BES) Grants, 1999
DOE Engineering Research Grants, 2017
DOE Fossil Advanced Research and Technology
Development Grants, 2022
DOE Materials Sciences and Engineering Research
Grants, 2035
Environmental Protection Agency Summer Faculty
Fellowships, 2229
HHMI-NIBIB Interfaces Initiative Grants, 2693
Materials Research Society Medalist Awards, 3297
Millipore/Charles P. Schaufus Parenteral-Processing
Technology Research Grant, 3396
NPSC Graduate Fellowships in the Physical Sciences,
4430
NSF CISE Increasing the Participation and
Advancement of Women in Academic Science and
Engineering Careers (ADVANCE) Grants, 4518
NSF CISE Accelerating Discovery in Science and
Engineering Through Petascale Simulations and
Analysis (PetaApps) Grants, 4519
NSF CISE Cross-Cutting Program Grants, 4530
NSF Major Research Instrumentation Program
(MRI) Grants, 4595
NSF Materials Research Science and Engineering
Centers (MRSEC), 4597
NSF Materials World Network: Cooperative Activity
in Materials Research between US Investigators
and their Counterparts Abroad (MWN), 4598
NSF Polymers Research Grants, 4634
NSF-NIST Interaction in Chemistry, Materials
Research, Molecular Biosciences, Bioengineering,
and Chemical Engineering, 4674
ONR Atomic and Molecular Physics Grants, 4753
ONR Electronics Division Research Grants, 4773
ONR Human Performance Division Research Grants,
4779
ONR Naval Materials Division Grants, 4795
ORISE Nuclear Regulatory Commission Historically
Black Colleges and Universities Student Research
Participation, 4855
Paired Fellowships for Research in Conservation and
the History of Art and Archaeology, 4900
SRC Natural and Engineering Sciences Grants, 5396
UNCFSP NASA Harriett G. Jenkins Predoctoral
Fellowships, 5602
W.M. Keck Foundation Grants Program, 5697

Materials Testing
ARO Degradation, Reactivity, and Protection of
Materials Research Contracts, 1008
AT&T Industrial Ecology Faculty Fellowships, 1213

Materials, Metals/Alloys
AFOSR Metallic Materials Research Grants, 346
AFOSR Molecular Dynamics Research Grants, 347
AFOSR Quantum Electronic Solids Research Grants,
352

Swiss National Science Foundation Fellowships for
Advanced Researchers, 5503
Swiss National Science Foundation Fellowships for
Prospective Researchers, 5504
Swiss National Science Foundation Marie Heim-
Vogtlin (MHV) Grants, 5505

Medieval History
American Historical Association Herbert Baxter
Adams Prize, 722
Andrew W. Mellon Postdoctoral Fellowship in
Medieval Studies, 871
BA Neil Ker Memorial Fund Research Grants, 1295
Center for Medieval and Renaissance Studies
Assistantships, Grants, and Fellowships, 1589
Dumbarton Oaks Junior Fellowships, 2140
Dumbarton Oaks Project Grants, 2141
Dumbarton Oaks Summer Fellowships, 2142
Erasmus Prize, 2250
Joseph C. Wey Fellowship in Medieval Studies, 3046
Medieval Academy of America John Nicholas Brown
Prize, 3334
Van Courtlandt Elliott Prize, 5661

Mediterranean Area
Harriet and Leon Pomerance Fellowship, 2603

Membrane Structure/Function
NIGMS Molecular Pharmacology of Anesthetic
Action Research Grants, 4065
NIH Structural Biology of Membrane Proteins
Grants, 4173
NSF Cell Biology Research Grants, 4507

Men
BCBS Improving Men's Health Grants, 1334
Lilly Endowment Grants Program, 3195
NIAAA Hepatitis C Infection and Alcoholic Liver
Disease, 3880
NIH New Directions in Pain Research Grants, 4142

Menopause
AFAR Research Grants, 333
J.I. Staley Prize, 2927

Mental Disorders
AACAP Rieger Service Program Award for
Excellence, 58
AACAP Sidney Berman Award for the School-Based
Study and Intervention for Learning Disorders and
Mental Illness, 61
Alexander and Margaret Stewart Trust Grants, 567
American Psychiatric Association/SmithKline
Beecham Young Faculty Award for Research
Development in Biological Psychiatry, 802
American Psychological Foundation Todd E. Husted
Memorial Award, 805
APF Call for Proposals, 919
APF James H. Scully Jr., M.D., Educational Fund
Grants, 920
APF Jeanne Spurlock Congressional Fellowships, 921
Auburn Foundation Grants, 1231
Catherine Holmes Wilkins Foundation Grants, 1524
Community Foundation for Southern Arizona
Grants, 1768
Della Martin Foundation Grants, 1883
DHHS/ASPE Welfare Outcomes Short-Term Policy
Research Grants, 1946
Eli Lilly & Company Reintegration Scholarships,
2193
Fairlawn Foundation Grants, 2288
FIC International Cooperative Biodiversity Groups
Grants, 2333
FIC International Research Scientist Development
Award, 2337
James H. Cummings Foundation Grants, 2942
John W. Alden Trust Grants, 3040
Leonard and Helen R. Stulman Charitable
Foundation Grants, 3165
Margoes Foundation Grants, 3266

NARSAD Distinguished Investigator Award, 3499
NARSAD Independent Investigator Award, 3500
NARSAD Young Investigator Award, 3501
NFX Patient-Oriented Research Grants, 3732
NFX Summer Research Fellowships, 3733
NIEHS Small Business Innovation Research (SBIR)
Program Grants, 4048
NIH Exploratory/Development Grants for Mental
Health Intervention Research, 4109
NIH Genetics of Neurobehavioral Disorders in
Existing Samples Grants, 4114
NIH Mechanisms Underlying Individual Variations
in Drug Responses, 4129
NIMH Collaborative R01s for Clinical Studies of
Mental Disorders, 4196
NIMH Early Identification and Treatment of Mental
Disorders in Children and Adolescents Grants,
4200
NIMH HIV/AIDS, Severe Mental Illness, and
Homelessness Research Grants, 4201
NIMH Interventions Infrastructure Program
Research Grants, 4204
NIMH National Research Service Awards (NRSA)--
Individual, 4208
NIMH Pilot Effectiveness Trials for Mental
Disorders Grants, 4210
NIMH Silvio O. Conte Centers for Neuroscience
Research Grants, 4215
Obsessive-Compulsive Foundation Research Awards,
4729
Oppenstein Brothers Foundation Grants, 4825
Rosalynn Carter Institute John and Betty Pope
Fellowships, 5130
Social Security Research and Demonstration Grants,
5305
UPS Foundation Grants, 5627
Viola W. Bernard Foundation Grants, 5674

Mental Health
AACAP Sidney Berman Award for the School-Based
Study and Intervention for Learning Disorders and
Mental Illness, 61
Abbott Fund Science and Medical Innovation Grants,
216
American Psychiatric Association Early Career
Award, 787
American Psychiatric Association Senior Scholar
Health Services Research Award, 794
American Psychiatric Association/AACDP Research
Mentorship Award, 796
American Psychological Foundation Todd E. Husted
Memorial Award, 805
American Sociological Association Minority
Fellowships, 819
APF Call for Proposals, 919
APF James H. Scully Jr., M.D., Educational Fund
Grants, 920
APF Jeanne Spurlock Congressional Fellowships, 921
BCBS of Massachusetts Foundation Grants, 1335
Bersted Foundation Grants, 1367
Blowitz-Ridgeway Foundation Early Childhood
Development Research Award, 1381
Bodenwein Public Benevolent Foundation Grants,
1389
Burton G. Bettingen Grants, 1436
Community Foundation of Broward Grants, 1769
Comprehensive Health Education Foundation Grants,
1777
Connecticut Community Foundation Grants, 1786
Connecticut Health Foundation Health Initiative
Grants, 1787
DHHS Comprehensive Community Mental Health
Services Grants for Children with Serious
Emotional Disturbances, 1914
Dover Foundation Grants, 2104
EF Behavioral Sciences Student Fellowships, 2180
Eli Lilly & Company Reintegration Scholarships,
2193
Eugene and Agnes E. Meyer Foundation Grants,
2267

Evjue Foundation Grants, 2276
Fairlawn Foundation Grants, 2288
FAR Fund, 2296
Faye McBeath Foundation Grants, 2301
Foundation for Seacoast Health Grants, 2393
Frank Stanley Beveridge Foundation Grants, 2409
Girl's Best Friend Foundation Grants, 2537
Hadassah Foundation Grants, 2590
Hawaii Community Foundation Health Education
and Research Grants, 2630
Ignacio Martin-Baro Fund for Mental Health and
Human Rights Grants, 2790
Institute of Mental Hygiene Grants, 2839
Ittleson Foundation Mental Health Grants, 2906
Jacob and Hilda Blaustein Foundation Grants, 2934
Jewish Women's Foundation of New York Grants,
2990
Klingenstein Third Generation Foundation Grants in
Depression, 3093
Leland Fikes Foundation Grants Program, 3159
Leo Goodwin Foundation Grants, 3163
Leonard and Helen R. Stulman Charitable
Foundation Grants, 3165
Lesbian Health Fund Grants, 3166
M.J. Murdock Charitable Trust Grants, 3245
Margaret L. Wendt Foundation Grants, 3263
Margaret T. Morris Foundation Grants, 3264
Massage Therapy Research Grants, 3296
Mattel Children's Foundation Grants, 3300
Mattel International Grants Program, 3301
McCarthy Family Foundation Grants, 3312
Medtronic Foundation Health Care Grants, 3337
Nationwide Foundation Grants, 3558
NICHD Socioeconomic Status and Health Across the
Life Course, 3937
Nick Traina Foundation Grants, 3939
NIH Exploratory/Development Grants for Mental
Health Intervention Research, 4109
NIH NRSAs for Individual Predoctoral Fellowships,
4144
NIH Research on Sleep and Sleep Disorders, 4157
NIMH Career Opportunities in Research (COR)
Education and Training Honors High School
Research Education Grant, 4192
NIMH Career Opportunities in Research (COR)
Education and Training Honors Undergraduate
Research Training Grant, 4193
NIMH Core Support Program for Mental Health/
AIDS Research Grants, 4197
NIMH Informed Consent in Research Grants, 4202
NIMH Interdisciplinary Behavioral Science Centers
for Mental Health, 4203
NIMH Mental Health Dissemination Research
Grants, 4206
NIMH Minority Research Infrastructure Support
Program Grants, 4207
NIMH National Research Service Awards (NRSA)--
Individual, 4208
NIMH New Minority Faculty Scientist Development
Award, 4209
NIMH Risk Factors for Psychopathology Using
Existing Data Sets Grants, 4212
NIMH Short Courses on Mathematical and
Statistical Tools for the Study of Complex
Phenotypes and Complex Systems, 4214
NIMH Silvio O. Conte Centers for Neuroscience
Research Grants, 4215
NIMH Supplements for Underrepresented
Minorities, 4216
NIMH Therapeutic Effectiveness Protocol
Development Program Grants, 4217
NIMH/NIDA Children's Services Grants, 4218
North Dakota Community Foundation (NDCF)
Grants, 4420
Pasadena Foundation Average Grants, 4911
Ray Foundation Mental Health and Substance Abuse
Prevention Grants, 5051
Rayonier Foundation Grants, 5052
Robert E. and Evelyn McKee Foundation Grants,
5098

Ross Foundation Grants, 5132
Sensient Technologies Foundation Grants, 5234
Sidgmore Family Foundation Grants, 5255
Simone and Cino del Duca Grand Prix Awards, 5260
Society of Neuroscience Minority Fellowships, 5344
Stackpole-Hall Foundation Grants, 5448
Staglin Family Music Festival NARSAD
   Schizophrenia Research Award, 5449
Stella and Charles Guttman Foundation Grants, 5460
Sunny and Abe Rosenberg Foundation Grants, 5487
Van Ameringen Foundation Grants, 5660
VanLobenSels/RembeRock Foundation Grants, 5664
Viola W. Bernard Foundation Grants, 5674
Whiting Foundation Grants, 5747
William T. Grant Foundation Research Grants, 5776
Winn-Dixie Stores Foundation Grants, 5781

Mental Retardation
Duncan L. Gordon Fellowships, 2143
Fragile X Syndrome Postdoctoral Research
   Fellowships Program, 2398
George Foundation Grants, 2498
John W. Alden Trust Grants, 3040
Mental Retardation Distinguished Research Award,
   3345
NFX Patient-Oriented Research Grants, 3732
NFX Summer Research Fellowships, 3733
NICHD NRSAs--Individual, 3930
NICHD NRSAs--Institutional, 3931
Roeher Institute Research Grants, 5120
UCT Mentally Handicapped Citizens Teachers
   Scholarships, 5586

Mentoring Programs
AAAS Mentor Award, 26
AACN Mentorship Grant, 91
AAJA Fellowships, 167
ACE Fellows Program, 251
American Psychiatric Association Minority Medical
   Student Summer Mentoring Program, 791
American Psychiatric Association/Kempf
   Fund Award for Research Development in
   Psychobiological Psychiatry, 799
AMWA Glasgow-Rubin Essay Award, 865
ASA Jeanne E. Griffith Mentoring Award, 1054
ASCO Comparative Effectiveness Research
   Professorship (CERP) in Breast Cancer, 1072
ASCO International Development and Education
   Awards (IDEA), 1074
ASCO Translational Research Professorships, 1078
ASHA Advancing Academic-Research Careers
   (AARC) Award, 1103
ASHA Research Mentoring-Pair Travel Award, 1104
ASHA Students Preparing for Academic & Research
   Careers (SPARC) Award, 1106
Association for Women in Mathematics Mentoring
   Travel Grants, 1202
AWM Mentoring Travel Grants for Women, 1280
Berlex Foundation Reproductive Scientist
   Development Program Grants, 1359
Boeing Company Contributions Program Grants,
   1391
Center for the Study of Philanthropy Senior
   International Fellowships, 1598
CMA Foundation Grants, 1729
Coca-Cola Foundation Grants, 1736
Congressional Fellowships for African Americans,
   1783
Corning Foundation Community Service Grants,
   1798
Cruise Industry Charitable Foundation Grants, 1817
DOD Multidisciplinary Postdoctoral Award for
   Breast Cancer Research, 1989
Dreyfus Senior Scientist Mentor Program Grants,
   2115
Farmers Insurance Group of Companies Grants, 2298
Fogarty International Center/Ellison Medical
   Foundation Overseas Fellowships in Global Health
   and Clinical Research, 2371
FTE Doctoral Fellowships, 2429

Kansas Health Foundation Grants, 3066
Laura J. Niles Foundation Grants, 3138
MAA-Tensor Foundation Women and Mathematics
   Grants, 3246
Marpat Foundation Grants, 3279
Mericos Foundation Grants, 3348
NAGC Masters and Specialists Award, 3488
NCI Mentored Career Development Award to
   Promote Diversity, 3614
NCI Paul Calabresi Award for Clinical Oncology,
   3619
NEI Mentored Clinical Scientist Development
   Program Award (K12), 3683
NHGRI Individual Mentored Research Scientist
   Awards in Genomic Research and Analysis, 3741
NHLBI Mentored Career Award for Faculty at
   Minority Institutions, 3775
NHLBI Mentored Career Development Award to
   Promote Faculty Diversity/Re-Entry in Biomedical
   Research, 3776
NHLBI Mentored Quantitative Research Career
   Development Awards, 3779
NIA Mentored Clinical Scientist Research Career
   Development Awards, 3854
NIDA International Studies on Drug Abuse and
   HIV/AIDS, 3947
NIDDK Mentored Research Scientist Development
   Award, 4008
NIGMS Initiative for Minority Students: Bridges to
   the Doctoral Degree, 4057
NIH Mentored Clinical Scientist Development
   Award, 4131
NSF Presidential Awards for Excellence in Science,
   Math, and Engineering Mentoring, 4638
NSF Undergraduate Research and Mentoring in the
   Biological Sciences (URM), 4669
OSI-Baltimore Community Fellowships, 4882
Society for Imaging Informatics in Medicine (SIIM)
   Emeritus Mentor Grants, 5307
Society of Neuroscience Minority Fellowships, 5344
United Airlines Foundation Education Grants, 5605
USC Center for Religion and Civic Culture
   Undergraduate Awards, 5641
Vancouver Foundation Grants and Community
   Initiatives Program, 5662
VanLobenSels/RembeRock Foundation Grants, 5664
Virginia Tech Dissertation Year/Postdoctoral
   Fellowship for Academic Diversity, 5682
Wellesley Centers for Women Postdoctoral
   Fellowships, 5727

Merchandising
College of Human Sciences Scholarships, 1741
National Turkey Federation Research Award, 3551

Mesoamerica
Kluge Center Kislak Fellowships in American
   Studies, 3096
Kluge Center Kislak Short-Term Fellowships in
   American Studies, 3097

Metabolic Diseases
Astellas Foundation Research Grants, 1206
Bristol-Myers Squibb Clinical Outcomes and
   Research Grants, 1412
CFF Postdoctoral Research Fellowships, 1614
CFF Research Grants, 1615
Leland McSpadden Memorial Fund for Medical and
   Scientific Research Grants, 3160
MDA Neuromuscular Disease Research Grants, 3322
NIDDK Collaborative Interdisciplinary Team
   Science in Diabetes, Endocrinology and Metabolic
   Diseases Grants, 3990
NIDDK Development of Disease Biomarkers Grants,
   3992
NIDDK Diabetes, Endocrinology, and Metabolic
   Diseases NRSAs--Individual, 3995
NIDDK Pilot and Feasibility Clinical Research
   Grants in Diabetes, Endocrine and Metabolic
   Diseases, 4017

NIDDK Seeding Collaborative Interdisciplinary
   Team Science in Diabetes, Endocrinology and
   Metabolic Diseases Grants, 4027
NIDDK Translational Research for the Prevention
   and Control of Diabetes and Obesity, 4034

Metabolism
AFAR Research Grants, 333
ARO Biomolecular and Cellular Materials and
   Processes Research Contracts, 1004
Astellas Foundation Research Grants, 1206
Beckman Young Investigators Program Grants, 1349
CFF Postdoctoral Research Fellowships, 1614
CFF Research Grants, 1615
Dietetic Outcomes Research Study Grants, 1951
Helen Hay Whitney Foundation Postdoctoral
   Research Fellowships, 2649
NCI Application of Metabolomics for Translational
   and Biological Research Grants, 3579
NCI Diagnostic Imaging and Guided Therapy in
   Prostate Cancer Award, 3598
NIAMS Exploratory Developmental Grants for
   Multidisciplinary Clinical Studies of Sarcopenia,
   3911
NIDDK Collaborative Interdisciplinary Team
   Science in Diabetes, Endocrinology and Metabolic
   Diseases Grants, 3990
NIDDK Diet Composition and Energy Balance
   Grants, 3996
NIDDK Race/Ethnic Disparities in the Incidence of
   Diabetes Complications, 4022
NIDDK Small Grants for Underrepresented Minority
   Scientists in Diabetes and Digestive and Kidney
   Diseases, 4031
NIDDK Type 2 Diabetes in the Pediatric Population
   Research Grants, 4036
NIEHS Applied Toxicological Research and Testing
   Grants, 4038
NSF Biomolecular Systems Cluster Grants, 4502
NSF Chemical, Biochemical, and Biotechnology
   Systems Research Grants, 4514
NSF Interagency Opportunities in Metabolic
   Engineering Research Grants, 4589

Metallurgy
Canada Graduate Scholarships (CGS) and NSERC
   Postgraduate Scholarships (PGS), 1470
DOE Materials Sciences and Engineering Research
   Grants, 2035
Dorr Foundation Grants, 2093

Meteorology
AFOSR Physical Mathematics and Applied Analysis
   Research Grants, 350
AGI Minority Participation Program, 379
AMS Graduate Fellowships in the History of Science,
   847
AMS Undergraduate Scholarships, 859
AMS/Industry Government Graduate Fellowships,
   860
AMS/Industry Minority Scholarships, 861
Aviation Research Grants, 1250
Canada Graduate Scholarships (CGS) and NSERC
   Postgraduate Scholarships (PGS), 1470
DOE Environmental Processes Research Grants,
   2018
Father James B. Macelwane Annual Awards, 2300
NCAR Graduate Research Fellowships, 3569
NOAA Collaborative Science, Technology, and
   Applied Research (CSTAR) Program Grants, 4338
NOAA Cooperative Institute for Applied
   Meteorological Studies (CIAMS) Grants, 4342
NOAA Cooperative Institute for Tropical
   Meteorology (CITM) Research Grants, 4343
NSF Arctic Natural Science Grants in Atmospheric
   Sciences, 4475
NSF Magnetospheric Physics Research Grants, 4594
NSF National Space Weather Program, 4604
NSF Physical and Dynamic Meteorology (PDM)
   Grants, 4620

NSF Solar Terrestrial Research Grants, 4656
ONR Marine Meteorology and Atmospheric Effects Research Grants, 4788
ONR Physical Oceanography Research Grants, 4802
ONR Radio Frequency (RF) Sensor Processing Research Grants, 4803
ONR Radio Frequency Solid State Devices Applied Research Grants, 4804
UNCFSP NASA Harriett G. Jenkins Predoctoral Fellowships, 5602

Methodist Church
Asian, Black, Hispanic and Native American United Methodist History Research Grants, 1124
Barra Foundation Project Grants, 1321
Center for the Evangelical United Brethren Heritage Audrie E. Reber Memorial Award, 1593
The World is My Parish United Methodist History Research Grant, 5529
United Methodist John Harrison Ness Memorial Award, 5608
Women in United Methodist History Research Grant, 5787
Women in United Methodist History Writing Award, 5788

Mexico
Dumbarton Oaks Junior Fellowships, 2140
Dumbarton Oaks Project Grants, 2141
Dumbarton Oaks Summer Fellowships, 2142
Foundation for the Advancement of Mesoamerican Studies Grants, 2394
Fulbright Binational Business Grants in Mexico, 2434
Fulbright Public Policy Initiative Program Grants in Mexico, 2445
Fulbright Traditional Scholar Program in the Western Hemisphere, 2453
Fulbright/Garcia Robles Grants, 2457
Jacobs Research Funds Small Grants Program, 2936
John D. and Catherine T. MacArthur Foundation Population and Reproductive Health Program Grants, 3019

Microbial Degradation
NSF Starter Grants for Postdoctoral Fellows in Microbial Biology, 4658

Microbiology
Abbott-ASM Lifetime Achievement Award, 217
ARO Microbiology and Biodegradation Research Contracts, 1014
ASM BD Award for Research in Clinical Microbiology, 1132
ASM bioMerieux Sonnenwirth Award for Leadership in Clinical Microbiology, 1133
ASM Eli Lilly and Company Research Award, 1134
ASM International Fellowship for Latin America and the Caribbean, 1135
ASM International Fellowships for Asia and Africa, 1136
ASM Merck Irving S. Sigal Memorial Awards, 1137
ASM Microbiology Undergraduate Research Fellowship, 1138
ASM Millis-Colwell Postgraduate Travel Grant, 1139
ASM Procter & Gamble Award in Applied and Environmental Microbiology, 1140
ASM Promega Biotechnology Research Award, 1141
ASM Raymond W. Sarber Awards, 1142
ASM Robert D. Watkins Graduate Research Fellowship, 1143
ASM sanofi-aventis ICAAC Award, 1144
ASM Siemens Healthcare Diagnostics Young Investigator Award, 1145
ASM Undergraduate Research Fellowship, 1146
ASM William A Hinton Research Training Award, 1147
ASM-PAHO Infectious Diseases Epidemiology and Surveillance Fellowships, 1148
ASM-UNESCO Leadership Grant for International Educators, 1149

ASM/CCID Fellowships, 1150
Bettina Bahlsen Memorial Graduate Scholarship, 1370
CDC Antimicrobial Resistance in Hospital-Acquired Infections Among Intensive Care Unit Patients, 1545
DOE Energy Biosciences Research Grants, 2011
DOE Health Effects and Life Sciences Research Grants, 2027
DOE Natural and Accelerated Bioremediation Research Program, 2039
Fletcher Jones Foundation Grants, 2364
LRF New Investigator Leukemia Research Grants, 3219
NFID Colin L. Powell Minority Postdoctoral Fellowship in Tropical Disease Research, 3706
NFID New Investigator Matching Grants, 3708
NIAID Microbiology and Infectious Diseases Research Grants and Contracts, 3901
NIDCR HIV/AIDS Associated Oral Viral Infections, 3978
NIDCR Oral Microbiology/Immunology of Type 1 Diabetes, 3980
NIGMS/NIDCR Evolution of Infectious Diseases Research Grants, 4077
NIH Ruth L. Kirschstein National Research Service Awards for Individual Postdoctoral Fellows, 4163
NSF Animal Developmental Mechanisms Grants, 4463
NSF Cell Biology Research Grants, 4507
NSF Chemical, Biochemical, and Biotechnology Systems Research Grants, 4514
NSF Evolution of Developmental Systems Grants, 4563
NSF Genes and Genome Systems Cluster Grants, 4569
NSF Microbial Observatories (MO) and Microbial Interactions and Processes (MIP), 4600
NSF Plant, Fungal and Microbial Developmental Systems Grants, 4631
NYAM Louis L. Seaman Medical Student Research Grants, 4698
Sequoyah Fellowship Program for Native American and Alaska Native Graduate Students, 5235
Waksman Foundation for Microbiology Grants, 5703

Microbiome
NHLBI Microbiome of the Lung and Respiratory Tract in HIV-Infected Individuals and HIV-Uninfected Controls, 3780

Microelectronics
AFOSR Turbulence and Rotating Flows Research Grants, 362
ARO Electrochemistry and Advanced Energy Conversion Research Contracts, 1010
ARO Solid State Devices Research Contracts, 1024
ONR Microelectronics Applied Research Grants, 4792
SME Manufacturing Education Plan Grants, 5277
SRC Education Alliance Undergraduate Research Assistants Program, 5368
SRC IC Design Challenge, 5386
SRC Simon Karecki Fellowship, 5398

Microenterprises
Aspen Institute FIELD Financing Products Grants, 1170
Aspen Institute FIELD Technical Assistance Assessment Grants, 1171

Microfilming and Microforms
New Jersey Historical Commission Grants, 3698

Microscopy
Beckman Young Investigators Program Grants, 1349
Microscopy Society of America Student Scholarships, 3378

Microwave Theory
AFOSR Electromagnetics Research Grants, 345
AFOSR Quantum Electronic Solids Research Grants, 352
ARO Mobile, Wireless Communications and Networks Research Contracts, 1016
ONR Radio Frequency (RF) Sensor Processing Research Grants, 4803
ONR Radio Frequency Solid State Devices Applied Research Grants, 4804

Middle East
AIAR Annual Professorships, 455
AIAR/Council of American Overseas Research Centers Fellowships for Advanced Multi-Country Research, 461
ASOR Fellowships at Cyprus American Archaeological Research Institute, 1164
BA Elie Kedourie Memorial Fund Research Grants, 1291
Center for Advanced Judaic Studies Postdoctoral Fellowships, 1572
Dumbarton Oaks Junior Fellowships, 2140
Dumbarton Oaks Project Grants, 2141
Fellowships at American Schools of Oriental Research--Amman, 2328
Harvey Prizes, 2627
IREX MENA Media Emerging Leaders Fellowships, 2885

Middle School Education
American Education Research Association Grants, 709
Carnegie Corporation of New York Grants, 1510
CPB American History and Civics Initiative, 1805
DaimlerChrysler Corporation Fund Grants, 1852
Eaton Charitable Fund Grants, 2163
MAA-Tensor Foundation Women and Mathematics Grants, 3246
NSF Engineering Research Centers (ERC) Grants, 4560
NSF Presidential Awards for Excellence in Mathematics and Science Teaching, 4637
Stuart Foundation Grants, 5481
United Airlines Foundation Education Grants, 5605

Midwifery
DHHS National Health Service Corps Loan Repayment Awards, 1927

Migration
AED New Voices Fellowship Program, 317
Farmers Insurance Group of Companies Grants, 2298
International Studies Graduate Fellowships, 2868
Luther I. Replogle Foundation Grants, 3237
NICHD Population Research Grants, 3932

Migratory Animals and Birds
BP Conservation Programme Grants, 1401
NEBC Graduate Student Research Awards, 3656

Military History
American Historical Association Paul Birdsall Prize in European Military and Strategic History, 739
American Historical Association Wesley-Logan Prize, 742
John M. Olin Fellowships in National Security, 3029
Marine Corps Historical Center Beeler-Raider Fellowship, 3271
Marine Corps Historical Center Dissertation Fellowship, 3272
Marine Corps Historical Center Master's Thesis Fellowships, 3273
Marine Corps Historical Center Research Grants, 3274
OAH Avery O. Craven Award, 4709
OAH Richard W. Leopold Prize, 4723
Smith Richardson Foundation Junior Faculty Research Grants, 5283

NMF General Electric Medical Scholars Fellowships, 4301

NRC-Ford Foundation Diversity Fellowships, 4433

NSF BIO Minority Postdoctoral Research Fellowships, 4492

NSF Geosciences Research Grants, 4574

NSF Materials World Network: Cooperative Activity in Materials Research between US Investigators and their Counterparts Abroad (MWN), 4598

NSF Postdoctoral Research Fellowships in Biology (PRFB), 4636

NSF SBE Doctoral Dissertation Research Improvement Grants, 4647

NSF Undergraduate Research and Mentoring in the Biological Sciences (URM), 4669

OAH Huggins-Quarles Award, 4715

Oppenstein Brothers Foundation Grants, 4825

Poverty, the Underclass, and Public Policy Research and Training Program Postdoctoral Fellowships, 5002

RAND Corporation Fellowships in Population Studies and the Study of Aging, 5045

Rayonier Foundation Grants, 5052

Rhode Island Foundation Grants, 5084

Richard and Susan Smith Family Foundation Grants, 5087

Rockefeller Foundation Humanities Fellowships, 5115

Rockefeller Foundation Scholarly Residencies, 5117

RTNDF Capitol Hill News Internships, 5155

RWJF Finding Answers: Disparities Research for Change Grants, 5166

RWJF Harold Amos Medical Faculty Development Scholars, 5168

RWJF New Connections Initiative Grants, 5177

Sensient Technologies Foundation Grants, 5234

Smithsonian Minority Students Internships, 5288

Teachers College Minority Postdoctoral Fellowships, 5516

Thomas J. Watson Foundation Fellowships, 5538

UNCFSP NAI-MIRS Summer Sabbatical Awards, 5599

UNCFSP NASA Science and Technology Institute Summer Scholars Program, 5600

Union Carbide Foundation Grants, 5604

Volkswagen of America Corporate Contributions Grants, 5689

Vulcan Materials Company Foundation Grants, 5690

W. C. Griffith Foundation Grants, 5691

W.C. Griffith Foundation Grants, 5694

Westinghouse Charitable Giving Grants, 5742

WILD Foundation Grants, 5756

Women's Foundation of California Grants, 5789

Minority Education

3M Foundation Grants, 8

AAA Minority Dissertation Fellowship Program, 16

AAAS Mentor Award, 26

AACAP Jeanne Spurlock Minority Medical Student Clinical Fellowship in Child and Adolescent Psychiatry, 48

AAG Dissertation Research Grants, 157

AAUW American Dissertation Fellowships, 206

AAUW American Postdoctoral Research Leave Fellowships, 207

AAUW American Summer/Short-Term Research Publication Grants, 208

AAUW Selected Professions Fellowships--Master's and First Professional Awards, 213

ACHE Foster G. McGaw Student Scholarship, 255

ADA Minority Dental Student Scholarships, 306

AGI Minority Participation Program, 379

AICPA Fellowships for Minority Doctoral Students, 498

AICPA Minority Accounting Scholarships, 500

Akonadi Foundation Anti-Racism Grants, 532

Alcoa Foundation Grants, 561

American Indian Graduate Center Scholarships, 746

American Philosophical Society John Hope Franklin Dissertation Fellowships, 778

American Psychiatric Association Minority Fellowships Program, 788

APA Planning Fellowships, 916

APF Jeanne Spurlock Congressional Fellowships, 921

Appraisal Institute Education Trust Scholarships, 928

APSA Minority Fellows Program, 968

ASHFoundation Graduate Student Scholarships, 1109

ASHFoundation Graduate Student Scholarships for Minority Students, 1110

ASM Robert D. Watkins Graduate Research Fellowship, 1143

Auburn University President's Graduate Opportunities Program, 1232

AVDF Private Higher Education Grants, 1248

Award for Encouraging Disadvantaged Students into Careers in the Chemical Sciences, 1262

AWWA Holly A. Cornell Scholarship, 1285

Carolina Minority Postdoctoral Scholars Program, 1517

Chancellor's Minority Postdoctoral Fellowship, 1634

Chicago Reporter Minority Urban Journalism Fellowship, 1673

Chiron Foundation Community Grants, 1684

CLA Scholarship for Minority Students in Memory of Edna Yelland, 1713

CMS Historically Black Colleges and Universities (HBCU) Health Services Research Grants, 1732

Coca-Cola Foundation Grants, 1736

College of Saint Rose Dissertation Fellowships, 1742

Consortium for Graduate Study in Management Minority Fellowships, 1790

Consumers Energy Foundation, 1792

Cruise Industry Charitable Foundation Grants, 1817

DaimlerChrysler Corporation Fund Grants, 1852

Dept of Ed Star Schools Program Grants, 1901

Dept of Ed Women's Educational Equity Program Grants, 1904

Diversifying Higher Education Faculty in Illinois Program (DFI), 1952

DOD National Defense Science and Engineering Graduate Fellowship Program, 1990

ExxonMobil Education Foundation Grants, 2280

FACHE Albert W. Dent Scholarships, 2286

FIC/ORMH Minority International Research Training Grants, 2344

Five Colleges Fellowship Program for Minority Scholars, 2362

FMC Foundation Grants, 2369

FTE Expanding Horizons Partnership Grants, 2430

FTE North American Doctoral Fellowships, 2431

Gaius Charles Bolin Fellowships for Minority Graduate Students, 2470

Gates Millennium Scholars Program, 2476

GEM MS Engineering Fellowships, 2484

GEM PhD Engineering Fellowships, 2485

GEM PhD Science Fellowships, 2486

George M. Brooker Collegiate Scholarship for Minorities, 2503

Health Management Scholarships and Grants for Minorities, 2636

Helen Carr Minority Fellowships, 2648

HRSA Health Professions Scholarships for Disadvantaged Students, 2729

HRSA Nursing Education Opportunities for the Disadvantaged Project Grants, 2731

IPA George Washington Williams Fellowship, 2870

Jeffrey Campbell Graduate Fellow in Anthropology, 2977

Johnson & Johnson Corporate Contributions Grants, 3041

Joseph L. Fisher Dissertation Awards, 3048

Josiah Macy Jr. Foundation Grants, 3050

KFF Barbara Jordan Health Policy Scholars Program, 3085

Kodak Community Relations Grants, 3105

KPMG Foundation Grants, 3112

KPMG Foundation Minority Accounting and Information Systems Doctoral Scholarships, 3113

Lilly Endowment Grants Program, 3195

LITA/LSSI Minority Scholarship in Library and Information Technology, 3204

Lloyd G. Balfour Foundation Grants, 3207

Margoes Foundation Grants, 3266

Marilyn Yarbrough Dissertation/Teaching Fellowship, 3269

Marine Biological Laboratory Summer Research Fellowships, 3270

Martin Luther King Jr. Graduate Memorial Scholarships, 3284

Michele Clark Fellowship, 3376

Miller Brewing Corporate Contributions Program Grants, 3393

Millipore Foundation Grants, 3395

Minority Doctoral Fellowships in Science and Engineering, 3404

Minority Humanities and Social Sciences Dissertation Scholar in Residence Grants, 3406

MLA Graduate Scholarships for Minority Students, 3424

Multicultural Advertising Intern Program, 3467

NACC David Stevenson Fellowships, 3481

NACC William Diaz Fellowships, 3482

National Assessment of Educational Progress Visiting Scholar Program, 3516

National Black MBA Association Annual Scholarships, 3519

National Urban/Rural Fellows Program, 3552

Nationwide Foundation Grants, 3558

NEH Faculty Research Awards for Historically Black, Tribal Colleges, Universities, and Institutions with High Hispanic Enrollment, 3670

NEH Institutional Grants to Historically Black Minority-Serving and Tribal Colleges and Universities, 3675

NHLBI Mentored Career Award For Faculty At Minority Institutions, 3774

NHLBI Minority Institution Research Scientist Development Award, 3782

NHLBI Minority Institutional Research Training Program Grants, 3783

NHLBI Short-Term Research Education Program to Increase Diversity in Health-Related Research, 3812

NIA Minority Dissertation Research Grants in Aging, 3856

NIDCD Minority Dissertation Research Grants in Human Communication, 3967

NIGMS Initiative for Minority Students: Bridges to the Baccalaureate Degree, 4056

NIGMS Initiative for Minority Students: Bridges to the Doctoral Degree, 4057

NIGMS Institutional Research and Academic Career Development Award, 4058

NIGMS MARC Undergraduate Student Training in Academic Research (U-STAR) Program, 4061

NIGMS Minority Access to Research Careers (MARC) Ancillary Training Activities Grants, 4063

NIGMS Minority Access to Research Careers (MARC) Program, 4064

NIH Health Promotion Among Racial and Ethnic Minority Males, 4115

NIH Predoctoral Fellowship Awards for Minority Students, 4147

NIMH Career Opportunities in Research (COR) Education and Training Honors High School Research Education Grant, 4192

NIMH Career Opportunities in Research (COR) Education and Training Honors Undergraduate Research Training Grant, 4193

NMF General Electric Medical Scholars Fellowships, 4301

NMF National Medical Association Awards for Medical Journalism, 4302

NMF National Medical Association Emerging Scholar Awards, 4303

NMF National Medical Association Patti LaBelle Award, 4304

NMF Need-Based Scholarships, 4305

Nitrogen Fixation
DOA Biological Nitrogen Fixation Research Grants, 1956
NSERC John C. Polanyi Awards, 4443

Nonfiction
AAAS Newcomb Cleveland Prize, 27
AAG John Brinkerhoff Jackson Prize, 161
AAS Historic Research Fellowships, 195
Aid to Scholarly Publications (ASPP), 503
Alfred E. Driscoll Publication Prize, 595
Alfred Hodder Fellowship, 596
American Historical Association Clarence H. Haring Prize, 718
American Historical Association George Louis Beer Prize, 720
American Historical Association Helen and Howard R. Marraro Prizes, 721
American Historical Association Herbert Baxter Adams Prize, 722
American Historical Association Herbert Feis Award, 723
American Historical Association Joan Kelly Memorial Prize in Women's History, 729
American Historical Association John H. Dunning Prize, 731
American Historical Association John K. Fairbank Prize in East Asian History, 732
American Historical Association Littleton-Griswold Prize in American Law and Society, 734
American Historical Association Wesley-Logan Prize, 742
American Institute of Physics Science-Writing Awards in Physics and Astronomy, 751
AMS Alfred Einstein Award, 842
AMS Claude V. Palisca Award, 844
AMS H. Colin Slim Award, 848
AMS Lewis Lockwood Award, 852
AMS Robert M. Stevenson Award, 857
AMS Ruth A. Solie Award, 858
ASECS James L. Clifford Prize, 1092
ASECS Louis Gottschalk Prize, 1093
ASNS Osborne and Mendel Nutrition Award, 1162
Bancroft Prizes, 1312
Bernard L. Schwartz Fellows Program, 1364
Blackall Machine Tool Award, 1378
Canadian Science Writers Association Awards, 1483
Canadian Studies Senior Fellowships, 1490
Carr P. Collins Award, 1519
Charles S. Sydnor Award, 1649
Chicago Tribune Heartland Prizes for Nonfiction and the Novel, 1674
Christian Gauss Award, 1686
Conference on Latin American History Prize, 1782
David Berry Essay on Scottish History Prize, 1869
Forest Products Society Wood Award, 2382
Francis Butler Simpkins Award, 2402
General L. Kemper Williams Prize in Louisiana History, 2487
Harry S. Truman Library Institute Book Awards, 2614
Herbert Eugene Bolton Memorial Prize, 2672
Howard Francis Cline Memorial Prize, 2720
International League of Antiquarian Booksellers Bibliographical Prize, 2862
Italian History Prizes, 2902
Louisiana State Arts Council Artist Fellowships, 3214
Medieval Academy of America John Nicholas Brown Prize, 3334
Millard Meiss Publication Fund Grants, 3392
MLA Aldo and Jeanne Scaglione Prize for Comparative Literary Studies, 3413
MLA Aldo and Jeanne Scaglione Prize for French and Francophone Studies, 3414
MLA Aldo and Jeanne Scaglione Prize for Germanic Languages and Literatures, 3415
MLA Aldo and Jeanne Scaglione Prize for Italian Studies, 3416
MLA Aldo and Jeanne Scaglione Prize for Studies in Slavic Languages and Literatures, 3417

MLA First Book Prize, 3421
MLA James Russell Lowell Prize, 3425
MLA Kenneth W. Mildenberger Prize, 3427
MLA Mina P. Shaughnessy Prize, 3428
MLA Prize for Independent Scholars, 3430
NEA Literature Fellowships, 3653
NFJC Gantz-Zahler Grant, 3712
NLM Grants for Scholarly Works in Biomedicine and Health, 4289
OAH James A. Rawley Prize, 4716
OAH Lerner-Scott Dissertation Prize, 4718
OAH Ray A. Billington Prize, 4722
OIEAHC Jamestown Prize, 4733
Orgone Biophysical Research Lab Lou Hochberg Awards, 4839
Phi Beta Kappa Award in Science, 4953
Ralph Waldo Emerson Award, 5044
Richard P. McCormick Prize for Scholarly Publication, 5091
RSC Bancroft Award, 5145
Society for the History of Technology Samuel Eleazar and Rose Tartakow Levinson Prize, 5323
SSHRC Aid to Scholarly Publications (ASPP), 5403
STC Research Grants, 5457
Stuart L. Bernath Book Prize, 5482
Stuart L. Bernath Scholarly Article Prize, 5484
Trevithick Book Prize Award, 5563
Van Courtlandt Elliott Prize, 5661
W. David Smith, Jr. Graduate Student Paper Award, 5692
W. Frank Blair Award, 5693
Watson Davis Prize, 5715
WAWH Sierra Award, 5719
Worcester Reed Warner Medal Award, 5800

Nonprofit Organizations
Alfred P. Sloan Foundation Civic Program Grants, 600
Association for Research on Nonprofit Organizations and Voluntary Action Awards, 1200
Bayer Foundation Grants, 1328
California and Michigan Nonprofit Research Program Grants, 1453
Center on Philanthropy and Civil Society's Emerging Leaders International Fellows Program, 1602
Community Foundation of Broward Grants, 1769
DOA Sustainable Agriculture Research and Education Grants, 1975
Eureka-Boston Fellowships, 2272
Fund for New Jersey Grants, 2460
Helen Bader Foundation Grants, 2646
IDEM Section 319(h) Nonpoint Source Program Grants, 2779
Jessie Smith Noyes Foundation Grants, 2980
Jovid Foundation Grants, 3053
M.J. Murdock Charitable Trust Grants, 3245
NACC David Stevenson Fellowships, 3481
NACC William Diaz Fellowships, 3482
NEFE Fellowships, 3662
Peter F. Drucker Award for Nonprofit Innovation, 4938
Rhode Island Foundation Grants, 5084
Rockefeller Archive Center Research Grants, 5112
Rockefeller Brothers Fund Fellowship in Nonprofit Law, 5114
Surdna Foundation Grants, 5492
Virginia A. Hodgkinson Research Prize, 5675
W.K. Kellogg Nonprofit Research Fellowship, 5695
WWIC/National Service Agency Civil Society Nonprofit Scholars Grants, 5809

North America
John Carter Brown Library Long-Term Research Fellowships, 3015
OAS Regular Training Program Fellowships, 4727

Norway
America-Norway Heritage Fund Grants, 660
John Dana Archbold Fellowships, 3021
NAA Norwegian Marshall Fund Grants, 3476

SASS Travel Grants, 5208
University of Oslo International Summer School (ISS) Scholarships, 5623

Norwegian Americans
America-Norway Heritage Fund Grants, 660

Norwegian Language/Literature
MLA Aldo and Jeanne Scaglione Prize for Germanic Languages and Literatures, 3415
University of Oslo International Summer School (ISS) Scholarships, 5623

Nuclear Astrophysics
APS Hans A. Bethe Prize, 944

Nuclear Chemistry
ORISE University Radioactive Ion Beam (UNIRIB) Consortium, 4875

Nuclear Engineering
AINSE Postgraduate Research Awards, 514
DOE Nuclear Energy Research Initiative (NERI), 2041
DOE University Reactor Instrumentation (URI) Program, 2054

Nuclear Fuels, Waste Disposal
ORISE Civilian Radioactive Waste Management Fellowship Program, 4840

Nuclear Materials Monitoring
ORISE Nuclear Regulatory Commission Historically Black Colleges and Universities Student Research Participation, 4855

Nuclear Medicine
AINSE Research Fellowships, 516
SNM Pilot Research Grants, 5298
SNM Pilot Research Grants in Nuclear Medicine/ Molecular Imaging, 5299
SNM Student Fellowship Awards, 5301
SNM Tetalman Memorial Young Investigator Awards, 5302
SNM/ Covidien Seed Grant in Molecular Imaging/ Nuclear Medicine Research, 5303

Nuclear Physics
APS Hans A. Bethe Prize, 944
APS Tom W. Bonner Prize in Nuclear Physics, 963

Nuclear Power
ANL Faculty Research Participation Program, 884
ANL Graduate Thesis Parts Research Appointments, 885
ANL Laboratory-Graduate Research Appointments, 886
Bernard F. Langer Nuclear Codes and Standards Award, 1363
DOE Advanced Nuclear Energy Systems Basic Research Grants, 1996
DOE Nuclear Energy Research Initiative (NERI), 2041
DOE University Reactor Instrumentation (URI) Program, 2054
James N. Landis Medal, 2949

Nuclear Reactors
DOE University Reactor Instrumentation (URI) Program, 2054

Nuclear Receptors
NIA Role of Nuclear Receptors in Tissue and Organismal Aging Grants, 3869

Nuclear Safety
Ploughshares Fund Grants, 4994

Nuclear Science Education
Ploughshares Fund Grants, 4994

Physical Sciences Education

Physics

Research Resources (Health/Safety/Medical)

Respiratory Diseases

Respiratory System

Restoration and Preservation

Restoration and Preservation, Art Works/Artifacts

Restoration and Preservation, Manuscripts/Books/Music Scores

School Dental Programs

School Food Programs

School Health Programs

School-to-Work Transition

Science

Security

Sedimentology

Seismology

Semantics

Semiconductors

Seminars

Senegal

Senile Dementia

Senior Citizen Programs and Services

Sensing

Sensing Devices and Transducers

Sensors

Sensory System

NIA Research Grants on Chemical Senses: Neurotransmitters and Neuromodulators, 3867
NIDA Drug Addiction Neuroscience Research Grants, 3945
NIDCD Research Grants on Chemical Senses: Neurotransmitters and Neuromodulators, 3973
NIDCD Small Grants Program, 3974
NSF Behavioral Neurosciences Research Grants, 4491
ONR Target Tracking and Sensor Fusion Grants, 4815
Whitehall Foundation Neurobiology Research Grants, 5746

Separation Process
AFOSR Unsteady Aerodynamics and Hypersonics Research Grants, 363
AIChE Clarence (Larry) G. Gerhold Award, 467
DOE Atomic, Molecular, and Optical Sciences Research Grants, 1998
DOE Catalysis and Chemical Transformations Research Grants, 2000
DOE Chemical Energy and Chemical Engineering Research Grants, 2001
DOE Chemical Physics Research Grants, 2002
DOE Chemical Sciences Research Grants, 2003
DOE Heavy Element Chemistry Research Grants, 2028
DOE Photochemistry and Radiation Research Grants, 2044
DOE Separations and Analysis Research Grants, 2049
NSF Chemical, Biochemical, and Biotechnology Systems Research Grants, 4514
NSF Interfacial Processes and Thermodynamics Research Grants, 4590

Serbia and Montenegro
IREX Individual Advanced Research Opportunities (IARO) Grants, 2883

Serology
Orentreich Foundation for the Advancement of Science Grants, 4837

Service Delivery Programs
Administration on Aging Alzheimer's Disease Demonstration Grants to States, 315
Bernard and Audre Rapoport Foundation Grants, 1362
CMS Research and Demonstration Grants, 1733
Community Foundation of Broward Grants, 1769
Consumers Energy Foundation, 1792
Eugene McDermott Foundation Grants, 2269
Health Care Financing Research, Demonstrations, and Evaluations Grants, 2634
Health Foundation of Southern Florida Responsive Grants, 2635
Henry L. Guenther Foundation Grants, 2659
HRDC Employability and Social Partnerships Grants, 2725
ICC Scholarship of Engagement Faculty Grants, 2771
Ittleson Foundation AIDS Grants, 2904
JM Foundation Grants, 3001
Luther I. Replogle Foundation Grants, 3237
Lutheran Foundation Grants, 3239
Max Goldenberg Foundation Grants, 3306
MAZON Grants: A Jewish Response to Hunger, 3311
NHLBI National Research Service Awards for Senior Fellowships, 3786
NIMH HIV/AIDS, Severe Mental Illness, and Homelessness Research Grants, 4201
Richard and Susan Smith Family Foundation Grants, 5087
Robert E. and Evelyn McKee Foundation Grants, 5098
William and Flora Hewlett Foundation Education Grants, 5766

Set Theory
NSF Topology Research Grants, 4667

Sex Education
Brush Foundation Grants Program, 1427
Ittleson Foundation AIDS Grants, 2904
John D. and Catherine T. MacArthur Foundation Population and Reproductive Health Program Grants, 3019
NIH Demographic Research on Sexual Behaviors Related to HIV, 4097

Sex Roles
CIES African Regional Special Program and Country Grants, 1699
Ford Foundation Peace and Social Justice Grants, 2378
Gordon Allport Intergroup Relations Prize, 2554
International Studies Graduate Fellowships, 2868
Jeanne Humphrey Block Dissertation Award, 2974
John D. and Catherine T. MacArthur Foundation Population and Reproductive Health Program Grants, 3019
NCSS Exemplary Reserach in Social Studies Award, 3646
NCSS Jean Dresden Grambs Distinguished Career Research in Social Studies Award, 3647
NCSS Larry Metcalf Exemplary Dissertation Award, 3648
NCSS/FASSE/CUFA Inquiry Grant, 3649
NIDA Neurobiological and Behavioral Research on Nicotine and Tobacco Products, 3953
NIH Pathological Gambling: Basic, Clinical, and Services Research Grants, 4146
Women's Studies Dissertation Fellowships, 5791

Sexism
Dept of Ed Women's Educational Equity Program Grants, 1904
Sioux Falls Area Community Foundation Grants, 5264
SPSSI Grants in Aid Program, 5365
Vanguard Public Foundation Grant Funds, 5663

Sexual Abuse
Burton G. Bettingen Grants, 1436
Community Foundation for Southern Arizona Grants, 1768
DOJ/NIJ Violence Against Women Research Grants, 2074
Ford Foundation Peace and Social Justice Grants, 2378
WHO Sexual Violence Research Initiative, 5749

Sexual Behavior
amfAR Fellowships, 832
amfAR Mathilde Krim Fellowships in Basic Biomedical Research, 833
amfAR Research Grants, 834
Brush Foundation Grants Program, 1427
NIH Demographic Research on Sexual Behaviors Related to HIV, 4097
NIMH HIV/AIDS, Severe Mental Illness, and Homelessness Research Grants, 4201
SSSS Student Research Grants, 5446

Sexuality
Five College Women's Studies Ford Associateships, 2361
Mellon Postdoctoral Fellowships in the Humanities at Cornell University, 3340
Overbrook Foundation Grants, 4888

Sexually Transmitted Diseases
ASM/CCID Fellowships, 1150
CDC HIV Demonstration, Research, Public and Professional Education Project Grants, 1555
CDC Sexually Transmitted Diseases Research, Demonstration, Public Information, and Education Grants, 1569
Ittleson Foundation AIDS Grants, 2904
NFID New Investigator Matching Grants, 3708
NIAID HIV Prevention Trial Units Grants, 3895

NICHD Population Research Grants, 3932
NIH Demographic Research on Sexual Behaviors Related to HIV, 4097
ORISE Faculty Research Participation at the Centers for Disease Control and Prevention, 4844
ORISE Postgraduate Research Participation at the Centers for Disease Control and Prevention, 4862
ORISE Student Research Participation at the Centers for Disease Control and Prevention, 4873
Sexually Transmitted Diseases Postdoctoral Fellowships, 5236

Shelters
Blowitz-Ridgeway Foundation Grants, 1382
Chautauqua Region Community Foundation Grants, 1658
Christopher D. Smithers Foundation Research Grants, 1691
Helen Brach Foundation Grants, 2647
Household International Corporate Giving Program Grants, 2718
Laura B. Vogler Foundation Grants, 3137
Luther I. Replogle Foundation Grants, 3237
Marpat Foundation Grants, 3279
Melville Charitable Trust Grants, 3343
Mervin Bovaird Foundation Grants, 3350
Reuter Foundation Grants, 5082
Thompson Charitable Foundation Grants, 5540
Vancouver Foundation Grants and Community Initiatives Program, 5662
W.W. Smith Charitable Trust Grants, 5698

Sicily
Olivia James Traveling Fellowship, 4741

Sickle Cell Disease
NHLBI Exploratory Studies in the Neurobiology of Pain in Sickle Cell Disease, 3762

Signal Processing
AFOSR Quantum Electronic Solids Research Grants, 352
AFOSR Semiconductor Materials Research Grants, 353
AFOSR Signals Communication and Surveillance Research Grants, 355
ARO Information Processing and Fusion and Circuits Research Contracts, 1012
ARO Mobile, Wireless Communications and Networks Research Contracts, 1016
ARO Theoretical Physics and Nonlinear Phenomena Research Contracts, 1029
ASEE Army Research Laboratory Postdoctoral Fellowships, 1096
ONR Ocean Acoustics Research Grants, 4799
ONR Signal and Image Processing Grants, 4808

Single-Parent Families
Eugene M. Lang Foundation Grants, 2268

Skin Diseases
Center for Alternatives to Animal Testing Grants, 1577
Dermatology Foundation Grants, 1906
Dermatology Postdoctoral Research Fellowships, 1907
Dermatology Research Career Development Awards, 1908
Gina Finzi Memorial Student Summer Fellowship, 2536
Lupus Foundation of America Research Grants, 3235
National Psoriasis Foundation Research Grants, 3546
NIAMS Clinical Trial Planning Grants, 3908
NIAMS High Risk Arthritis and Musculoskeletal and Skin Diseases Research, 3912
NIAMS Skin Diseases Research Core Centers Grants, 3920
Skin Cancer Foundation Research Grants, 5270
Society for Investigative Dermatology Career Development Award, 5312

NASA Undergraduate Student Researchers Project (USRP) Grants, 3508
NASM Verville Fellowships, 3512
NIH Biomedical Research on the International Space Station, 4086
NSF National Space Weather Program, 4604
NSF University Radio Observatories Program (AST-URO), 4670
ONR Radio Frequency (RF) Sensor Processing Research Grants, 4803
ONR Radio Frequency Solid State Devices Applied Research Grants, 4804
Society for the History of Technology NASA Fellowship, 5322
Zonta International Foundation Amelia Earhart Fellowships, 5830

Spain
American Historical Association John E. Fagg Prize, 730
American Historical Association Premio Del Rey, 740
CINOA Prize, 1703
Edilia and Francois-Auguste De Montequin Fellowship in Iberian and Latin American Architecture, 2166
Fulbright Business Grants in Spain, 2435
Tinker Foundation Field Research Grants, 5551
Tinker Foundation Institutional Grants, 5552

Spanish Language/Literature
Association of Teachers of Latin American Studies Grants, 1203
Robert L. Baker Summer Research Fellowship for Second Language Acquisition in an Environment of Immersion, 5101
Rotary Foundation Ambassadorial Scholarships, 5133
Tinker Foundation Field Research Grants, 5551

Special Education
Becton Dickinson and Company Grants, 1350
Doreen Kronick Scholarship, 2081
Dorr Foundation Grants, 2093
Dorrance Family Foundation Grants, 2094
Rural Student Scholarships, 5156
Schlumberger Foundation Grants, 5217
Skaggs Foundation Grants, 5269
UCT Mentally Handicapped Citizens Teachers Scholarships, 5586
Wisconsin American Legion Auxiliary Scholarships, 5785

Spectroscopy
AFOSR Atomic and Molecular Physics Research Grants, 342
APS Arthur L. Schawlow Prize in Laser Science, 934
APS Earle K. Plyler Prize for Molecular Spectroscopy, 938
APS Herbert P. Broida Prize, 945
ARO Degradation, Reactivity, and Protection of Materials Research Contracts, 1008
E. Bright Wilson Award in Spectroscopy, 2146
EF Postdoctoral Research Training Fellowships, 2182
Nakanishi Prize, 3491
NSF Physical Chemistry (PCHEM) Grants, 4622
ONR Atomic and Molecular Physics Grants, 4753

Speech
Axe-Houghton Foundation Grants, 1288
Claude Pepper Foundation Grants, 1720
NIDCD Clinical Trial Planning Grant, 3960
NIDCD Exploratory/Developmental Research Grants, 3963
NIDCD Minority Dissertation Research Grants in Human Communication, 3967
NSF Linguistics Research Grants, 4591

Speech Pathology
ASHA Student Research Travel Award, 1105
ASHFoundation Graduate Student Scholarships, 1109

ASHFoundation Graduate Student Scholarships for Minority Students, 1110
ASHFoundation Graduate Student with a Disability Scholarship, 1111
ASHFoundation New Century Scholars Program Doctoral Scholarships, 1112
ASHFoundation Research Grant for New Investigators, 1114
ASHFoundation Research Grant in Speech Science, 1115
Deafness Research Foundation Research Grants, 1875
Leadership in Education in Neurodevelopmental Disabilities Postdoctoral Fellowships, 3148
Mabel H. Flory Charitable Trust Grants, 3247
NIDCD Small Grants Program, 3974
Regents Professional Opportunity Scholarships, 5068

Speech Therapy
Axe-Houghton Foundation Grants, 1288
Mabel H. Flory Charitable Trust Grants, 3247

Speech/Communication Education
Axe-Houghton Foundation Grants, 1288
Mabel H. Flory Charitable Trust Grants, 3247

Spinal Cord Injury
AO North America Grants, 902
Australian Spinal Research Foundation Grants, 1244
CDC Disabilities Prevention Demonstration-Epidemiology Project Grants, 1551
Christopher Reeve Paralysis Foundation Individual Research Grants, 1692
Christopher Reeve Paralysis Foundation Quality of Life and Health Promotions Grants, 1693
Detroit Lions Charities Grants, 1909
ICCP Outstanding Young Investigator Award, 2772
International Spinal Research Trust Grants, 2867
NIA Symptom Management for Chronic Neurological Conditions Research Grants, 3873
NICHD Symptom Management for Chronic Neurological Conditions Research Grants, 3938
NIDCD Symptom Management for Chronic Neurological Conditions Research Grants, 3975
NINDS Symptom Management for Chronic Neurological Conditions Research Grants, 4229
NINR Symptom Management for Chronic Neurological Conditions Research Grants, 4249
NIOSH Occupational Safety and Health Research and Demonstration Grants, 4251

Spirituality
Templeton Advanced Research Grants, 5520

Sports
AMSSM Foundation Research Grants, 862
Angels Baseball Foundation Grants, 880
Billie Jean King Foundation Grants, 1377
Carl and Eloise Pohlad Family Foundation Grants, 1504
Carl C. Icahn Foundation Grants, 1506
Chiles Foundation Grants, 1682
Detroit Lions Charities Grants, 1909
East Stroudsburg University Graduate Assistantships, 2158
FishAmerica Foundation Marine and Anadromous Fish Habitat Restoration Grants, 2356
FishAmerica Foundation Research Grants, 2357
GCSAA Essay Contest, 2481
George Kress Foundation Grants, 2502
Gerald R. Ford Foundation Research Travel Grants, 2510
Greater Worcester Community Foundation Jeppson Memorial Fund for Brookfield Grants, 2567
Helen Bader Foundation Grants, 2646
IATA Research Grants, 2768
Jewish Women's Foundation of New York Grants, 2990
McCombs Foundation Grants, 3313
NCAA Postgraduate Scholarships, 3568

NIAMS High Risk Arthritis and Musculoskeletal and Skin Diseases Research, 3912
NIAMS Research on Musculoskeletal Fitness and Sports Medicine Grants, 3918
NICHD Research on Musculoskeletal Fitness and Sports Medicine Grants, 3934
NINR Research on Musculoskeletal Fitness and Sports Medicine Grants, 4247
Paul Stock Foundation Grants, 4924
QLT Corporate Sponsorships, 5015
Ray Foundation Mental Health and Substance Abuse Prevention Grants, 5051
Sidgmore Family Foundation Grants, 5255
SSHRC Sport Participation Research Grants, 5425
xyz.net Corporate Giving Grants, 5816

Sports Medicine
AMSSM Foundation Research Grants, 862
Angels Baseball Foundation Grants, 880
Howard Gilman Foundation Grants, 2721
IATA Research Grants, 2768
Musculoskeletal Rehabilitation Fellowships, 3471
SSHRC Sport Participation Research Grants, 5425

Sri Lanka
CAORC Regional Research Fellowships, 1500
Fulbright Traditional Scholar Program in the South and Central Asia Region, 2452

Standards and Standardization
Bernard F. Langer Nuclear Codes and Standards Award, 1363
DOE State Energy Program Special Projects, 2051
NIST Precision Measurement Grants, 4257

Starch
AACC International Alsberg-French-Schoch Memorial Lectureship Awards, 66

State Government
APSA Political Science Doctoral Dissertation Awards, 969
CDC Core Capacity Program Grants, 1550
DOJ National Incident-Based Reporting System, 2068
DOJ National Institute for Juvenile Justice and Delinquency Prevention Grants, 2069
Ford Foundation Peace and Social Justice Grants, 2378
HUD General Research and Technology Activity Grants, 2737
IBM Center for the Business of Government Grants, 2769
Lynde and Harry Bradley Foundation Grants, 3242
NIEHS Applied Toxicological Research and Testing Grants, 4038
OAH Richard W. Leopold Prize, 4723
Robert Sterling Clark Foundation Grants, 5107
Senate Legislative Fellows Program, 5232
SPSSI Applied Social Issues Internships, 5364
Westar Energy Foundation Grants, 5739

Statistical Mechanics
Alfred P. Sloan Foundation Known, Unknown, and Unknowable Research Grants, 607

Statistics
AAUW Selected Professions Fellowships--Master's and First Professional Awards, 213
ACT Summer Internship Program, 295
AFOSR Signals Communication and Surveillance Research Grants, 355
Alfred P. Sloan Foundation Selected National Issues Grants, 610
American Education Research Association Grants, 709
ARO Stochastic Analysis, Applied Probability, and Statistics Research Contracts, 1025
ARO Theoretical Physics and Nonlinear Phenomena Research Contracts, 1029
ASA and USDA/NASS Research Fellow and Associate Program, 1052

DOA Animal Health and Disease Research Grants, 1955

International Crane Foundation Internships, 2854

Jackson Laboratory Postdoctoral Research Training Program, 2932

Max Bell Foundation Grants, 3305

MFA Established Investigator Grants, 3358

MFA Fellowship Training Grant, 3359

MFA First Award Grants, 3360

MHRC Postdoctoral Research Fellowships, 3366

NIGMS/NIDCR Evolution of Infectious Diseases Research Grants, 4077

NIH Midcareer Investigator Award in Mouse Pathobiology Research Grants, 4134

Regents Professional Opportunity Scholarships, 5068

Robert J. Kleberg Jr. and Helen C. Kleberg Foundation Grants, 5100

William and Charlotte Parks Foundation for Animal Welfare Grants, 5765

Winn Feline Foundation Grants, 5780

Veterinary Medicine Education

Banfield Charitable Trust Grants, 1314

Christine Mirzayan Science and Technology Policy Graduate Fellowship Program, 1687

Geraldine R. Dodge Foundation Frontiers for Veterinary Medicine Fellowships, 2511

HRSA Health Professions Scholarships for Disadvantaged Students, 2729

Laura J. Niles Foundation Grants, 3138

NHLBI Short-Term Research Education Program to Increase Diversity in Health-Related Research, 3812

PAHO/WHO Fellowships, 4899

UNC Minority Presence Doctoral Study Grants, 5595

WICHE Professional Student Exchange Program (PSEP), 5754

Winn Feline Foundation Grants, 5780

Video Production

Bogliasco Foundation Liguria Study Center Residencies, 1392

Carole Fielding Film/Video Production or Research Grants, 1516

Center for Defense Information Internships, 1581

Louisiana State Arts Council Artist Fellowships, 3214

National Black Programming Program Development Grants, 3520

NEA Arts on Radio and Television, 3651

North Carolina Arts Council Folklife Public Program Grants, 4412

OAH Erik Barnouw Award, 4713

Paul Robeson Fund for Independent Media Grants, 4923

School of Visual Arts National Scholarship, 5223

Videos

ACC Residency Program in Asia, 240

Bogliasco Foundation Liguria Study Center Residencies, 1392

Carole Fielding Film/Video Production or Research Grants, 1516

Consortium of College and University Media Centers Annual Research Grants, 1791

OAH Erik Barnouw Award, 4713

Vietnam

American Historical Association John K. Fairbank Prize in East Asian History, 732

University of Massachusetts Rockefeller Foundation Humanities Fellowships, 5622

Vietnamese Language

University of Massachusetts Rockefeller Foundation Humanities Fellowships, 5622

Violence in Schools

DHHS Maternal and Child Health Research and Training Programs Grants, 1926

FAR Fund, 2296

Violent Behavior

Allen Lane Foundation Grants, 621

Allstate Foundation Grants, 623

AMWA Kathryn C. Bemmann New Investigator Grant on Violence against Women, 866

California Wellness Foundation Work and Health Program Grants, 1457

CDC Community Responses to Prevent Intimate Partner Violence, 1547

CDC Grants for Violence-Related Injury Prevention Research, 1554

CDC Injury Prevention and Control Research and State and Community Based Programs, 1560

Dart Center Ochberg Fellowship, 1863

David Bohnett Foundation Grants, 1870

DOJ Grants to Reduce Violent Crimes Against Women on Campus, 2066

DOJ/NIJ Violence Against Women Research Grants, 2074

FAR Fund, 2296

Foundation Global Ethic Grants, 2397

Harry Frank Guggenheim Foundation Dissertation Fellowships, 2610

Harry Frank Guggenheim Foundation Research Grants, 2611

Ignacio Martin-Baro Fund for Mental Health and Human Rights Grants, 2790

James Marshall Public Policy Fellowship, 2946

Jewish Women's Foundation of New York Grants, 2990

Joyce Foundation Grants, 3055

Meadows Foundation Grants, 3328

NIAAA Research Grants on Violence against Women and within the Family, 3886

NIJ Graduate Research Fellowship Program, 4187

Overbrook Foundation Grants, 4888

Richard and Rhoda Goldman Fund Grants, 5086

VFH and Institute on Violence and Survival Resident Fellowships, 5667

WHO Sexual Violence Research Initiative, 5749

William Caspar Graustein Memorial Fund Grants, 5770

Women's Foundation of California Grants, 5789

Violent Crime

Allstate Foundation Grants, 623

FAR Fund, 2296

Viral Studies

CDC Evaluation of the Use of Rapid Testing For Influenza in Outpatient Medical Settings, 1552

NIAID Microbiology and Infectious Diseases Research Grants and Contracts, 3901

NSF Starter Grants for Postdoctoral Fellows in Microbial Biology, 4658

Virginia

Virginia Historical Society Research Fellowships, 5681

Virology

American Liver Foundation Physician Research Development Award, 757

ASM/CCID Fellowships, 1150

Bristol-Myers Squibb Clinical Outcomes and Research Grants, 1412

Bristol-Myers Squibb Foundation Global HIV/AIDS Initiative Grants, 1413

Bristol-Myers Squibb Virology and Oncology M.D. Research Fellowships, 1416

LRF New Investigator Leukemia Research Grants, 3219

NIAAA Hepatitis C Infection and Alcoholic Liver Disease, 3880

NICHD Central Nervous System Effects of HIV Infection: Neurobiological, Neurovirological, and Neurobehavioral Studies Grants, 3924

NIDCR HIV/AIDS Associated Oral Viral Infections, 3978

NIH Impact of HAART on HIV/CNS Disease, 4120

NIH Multidisciplinary Approach for Research on Oral Complications of HIV Infection, 4138

NIH Small Business Innovation Research (SBIR): Animal Models of HCV Infection Research Grants, 4171

NINDS Central Nervous System Effects of HIV Infection: Neurobiological, Neurovirological, and Neurobehavioral Studies Grants, 4219

NSF Genes and Genome Systems Cluster Grants, 4569

ORISE Faculty Research Participation at the Centers for Disease Control and Prevention, 4844

ORISE Postgraduate Research Participation at the Centers for Disease Control and Prevention, 4862

Vision

1st Touch Foundation Grants, 4

AFOSR Sensory Systems Research Grants, 354

Alcon Foundation Grants Program, 563

Alfred W. Bressler Prize in Vision Science, 612

Fight for Sight Research Grants and Fellowships, 2351

Foundation Fighting Blindness Marjorie Carr Adams Women's Career Development Awards, 2384

Hawaii Community Foundation Health Education and Research Grants, 2630

NEI Mentored Clinical Scientist Development Program Award (K12), 3683

NEI Scholars Program, 3686

NEI Translational Research Program On Therapy For Visual Disorders, 3687

NEI Vision Research Core Grants, 3688

NYCT Blindness and Visual Disabilities Grants, 4702

OneSight Research Foundation Block Grants, 4744

Sigma Xi Grants in Aid of Research, 5258

Visual Arts

ACC John D. Rockefeller III Award, 236

ACC Starr Foundation Fellowship Program, 241

Amgen Foundation Grants, 835

Andrew W. Mellon Predoctoral Fellowship, 875

ASID Foundation Joel Polsky Prize, 1129

Bogliasco Foundation Liguria Study Center Residencies, 1392

Booth-Bricker Fund Grants, 1393

Bush Foundation Artist Fellowships, 1437

Camargo Foundation Fellowships, 1459

Canada Council for the Arts Grants, 1468

Center for Advanced Study in the Visual Arts Senior Fellowships, 1574

Center for Advanced Study in the Visual Arts Visiting Senior Fellowships, 1575

Chestnut Hill Charitable Foundation, Inc Grants, 1664

Christensen Fund Regional Grants, 1685

Cord Foundation Grants, 1797

Cranbrook Academy of Art Guest Artist/Scholar Grants, 1807

DAAD Research Grants for Doctoral Candidates and Young Academics and Scientists, 1840

DAAD Research Stays for University Academics and Scientists, 1842

DAAD Scholarships for Artists: Study Visits for Academics, 1844

Delta Air Lines Foundation Grants, 1885

Donald and Sylvia Robinson Family Foundation Grants, 2078

Edward C. and Ann T. Roberts Foundation Grants, 2173

EMIERT Coretta Scott King Awards, 2218

Fleishhacker Foundation Eureka Fellowships, 2363

Flinn Foundation Grants Programs, 2365

Frese Senior Research Fellowship, 2421

George S. and Dolores Dore Eccles Foundation Grants, 2504

Georgia Power Foundation Grants, 2509

Getty Graduate Internships, 2520

Harriet Hale Woolley Scholarships, 2604

Henry Luce Foundation/ACLS Dissertation Fellowships in American Art, 2662

End Index by subjects

Program Type Index

NOTE: Numbers refer to entry numbers

Building Construction and/or Renovation

Citizenship Instruction

Community Development

Consulting/Visiting Personnel

Cultural Outreach

Dissertation/Thesis Research Support

Emergency Programs

**Exhibitions, Collections, Performances, Video/
Film Production**

Faculty/Professional Development

Job Training/Adult Vocational Programs

Religious Programs

Scholarships

Seed Grants

Symposiums, Conferences, Workshops, Seminars

Geographic Index

Note: This index lists grants for which applicants must be residents of or located in a specific geographic area. Numbers refer to entry numbers.

Delaware

Missouri

Ohio

Wisconsin

LaVergne, TN USA
24 March 2010
176946LV00001B/2/P